The Geology of North America
Volume G-3

The Cordilleran Orogen: Conterminous U.S.

Edited by

B. C. Burchfiel
Department of Earth, Atmospheric, and Planetary Sciences
Massachusetts Institute of Technology
Cambridge, Massachusetts 02139

P. W. Lipman
U.S. Geological Survey
Mail Stop 910
345 Middlefield Road
Menlo Park, California 94025

M. L. Zoback
U.S. Geological Survey
Mail Stop 977
345 Middlefield Road
Menlo Park, California 94025

1992

Acknowledgment

Publication of this volume, one of the synthesis volumes of *The Decade of North American Geology Project* series, has been made possible by members and friends of the Geological Society of America, corporations, and government agencies through contributions to the Decade of North American Geology fund of the Geological Society of America Foundation.

Following is a list of individuals, corporations, and government agencies giving and/or pledging more than $50,000 in support of the DNAG Project:

Amoco Production Company
ARCO Exploration Company
Chevron Corporation
Cities Service Oil and Gas Company
Conoco, Inc.
Diamond Shamrock Exploration Corporation
Exxon Production Research Company
Getty Oil Company
Gulf Oil Exploration and Production Company
Paul V. Hoovler
Kennecott Minerals Company
Kerr McGee Corporation
Marathon Oil Company
Maxus Energy Corporation
McMoRan Oil and Gas Company
Mobil Oil Corporation
Occidental Petroleum Corporation
Pennzoil Exploration and Production Company
Phillips Petroleum Company
Shell Oil Company
Caswell Silver
Standard Oil Production Company
Oryx Energy Company (formerly Sun Exploration and Production Company)
Superior Oil Company
Tenneco Oil Company
Texaco, Inc.
Union Oil Company of California
Union Pacific Corporation and its operating companies:
Union Pacific Resources Company
Union Pacific Railroad Company
Union Pacific Realty Company
U.S. Department of Energy

Published by The Geological Society of America, Inc.
3300 Penrose Place, P.O. Box 9140, Boulder, Colorado 80301

Printed in U.S.A.

Cover Photo: Looking north along the Keystone thrust fault in the northeastern Spring Mountains, southern Nevada. Well-bedded gray and white Paleozoic rocks have been thrust eastward over red and white early Jurassic Aztec Sandstone that has been folded into a footwall syncline. Photo by B. C. Burchfiel.

10 9 8 7 6 5 4 3 2 1

Library of Congress Cataloging-in-Publication Data
The Cordilleran orogen : conterminous U.S. / edited by B.C. Burchfiel, P.W. Lipman, M.L. Zoback.
p. cm. — (The Geology of North America : v. G-3)
Includes bibliographical references and index.
ISBN 0-8137-5217-5
1. Orogeny—West (U.S.) 2. Geology—West (U.S.) I. Burchfiel, B. C., 1934– . II. Lipman, Peter W. III. Zoback, Mary Lou C. IV. Series.
QE71.G48 1986 vol. G-3
[QE621.5.U6]
557.1 s—dc20
[551.8'2'0978] 92-6030
CIP

Preface

The Geology of North America series has been prepared to mark the Centennial of The Geological Society of America. It represents the cooperative efforts of more than 2,000 individuals from academia, state and federal agencies of many countries, and industry to prepare syntheses that are as current and authoritative as possible about the geology of the North American continent and adjacent oceanic regions.

This series is part of the Decade of North American Geology (DNAG) Project, which also includes six wall maps at a scale of 1:5,000,000 that summarize the geology, magnetic and gravity anomaly patterns, regional stress fields, thermal aspects, and seismicity of North America and its surroundings. Together, the synthesis volumes and maps are the first coordinated effort to integrate all available knowledge about the geology and geophysics of a crustal plate on a regional scale.

The products of the DNAG Project present the state of knowledge of the geology and geophysics of North America through the 1980s, and they point the way toward work to be done in the decades ahead.

A. R. Palmer
General Editor for the volumes
published by The Geological
Society of America

J. O. Wheeler
General Editor for the volumes
published by the Geological
Survey of Canada

Foreword

This volume is an appropriate contribution to the Decade of North American Geology; it was conceived and organized in the early part of 1980, and work on the volume was spread over the entire decade culminating with the completion of the volume in 1992. Based on the outline written for the Perspectives in Regional Geological Synthesis (DNAG Special Publication 1), this volume follows rather closely our original concept. We planned two parts to the volume: a first part that presented the basic geological data for the Cordillera consisting of time-slice chapters and a second part consisting of topical synthesis chapters that dealt with time-independent processes of Cordilleran evolution. The final volume retains that organization. During the 1980s a few of the original chapters were dropped, others added, and the editorship had two changes in mid-stream.

It was hoped that the individual chapters would be well cross-referenced; however, as completion of chapters was spread over five years, this was not possible. Presentations in the chapters vary, a result of giving chapter contributors as much freedom as possible. Differences of interpretation are common throughout the chapters, which we view as a healthy airing of fundamental unresolved Cordilleran issues. These issues are mostly current, although a few have been discussed for several decades, such as the dynamics of the Sevier foreland fold-and-thrust belt or the emplacement of the mid-Paleozoic Antler allochthon. Mainly, the set of unresolved problems is new because of the remarkable advances in understanding Cordilleran evolution. An interested reader need only go back to the years just prior to the development of plate tectonics to see that rarely was the Cordilleran orogen treated as a single dynamic system; geology east and west of the Sierra Nevada was often considered independently. As in most orogenic belts, the plate tectonic revolution profoundly affected Cordilleran geological concepts, and while much basic geology remains to be done, the trend is toward more process-oriented research. We anticipate the next decade of Cordilleran research to be even more exciting than the last, and we hope that this volume can be important in at least pointing toward some new research directions.

The editors wish to express our deep appreciation to all the contributors to this volume. It has taken a major effort to complete the chapters where considerable time was taken away from personal research efforts. We hope that those involved in producing this volume will feel their efforts were worthwhile. We certainly do.

B. C. Burchfiel
P. W. Lipman
M. L. Zoback
February 1992

The Geology of North America
Vol. G-3, The Cordilleran Orogen: Conterminous U.S.
The Geological Society of America, 1992

Chapter 1

Introduction

B. C. Burchfiel
Department of Earth, Atmospheric, and Planetary Sciences, Massachusetts Institute of Technology, Cambridge, Massachusetts 02139
P. W. Lipman
U.S. Geological Survey, Mail Stop 910, 345 Middlefield Road, Menlo Park, California 94025
M. L. Zoback
U.S. Geological Survey, Mail Stop 977, 345 Middlefield Road, Menlo Park, California 94025

INTRODUCTION

The Cordilleran orogen lies on the western part of the North American continent and rims the northeastern Pacific Ocean Basin. That part of the orogen covered in this volume extends between the Mexican and Canadian borders, with some consideration of the geology on both sides of the border, and from the offshore continental borderlands of the Pacific eastward as far as the Black Hills of South Dakota and the mountains of west Texas. It ranges from 800 to 1,600 km wide and is physiographically complex, consisting of high mountains, intervening lowlands, and plateaus that rise from a more gentle continental interior (Fig. 1). The physiography of the Cordillera largely reflects the younger underlying structure, and the present orogenic belt has limits coincident with its physiography.

The continental margin, along which the Cordilleran orogen developed, originated about 600 Ma, and since that time the orogen has evolved at the boundary between oceanic and continental realms, making it one of the longest lived orogenic belts in the world. Its evolution is not complete because the active deformation within the Cordillera is a reflection of ocean-continent plate interaction and the present complex plate geometry is characteristic of past plate boundaries. Precambrian tectonostratigraphic elements of the Northern American craton as old as Archean have been incorporated into the orogen, but elements of Middle Proterozoic and younger ages have had the most important influence on Cordilleran development. From the inception of the Cordilleran margin in latest Precambrian time, the orogen has been the locus of plate-tectonic activity that includes development a passive margin of latest Precambrian and early Paleozoic age; a collage of accreted fragments incorporated intermittently into the orogen from middle Paleozoic time into Cenozoic time; the igneous and tectonic record of successive plate-convergent events, probably beginning with Antler deformation in middle Paleozoic time and culminating in a fully developed Andean margin late in Mesozoic time; eastward Laramide migration of igneous and tectonic features related to rapid plate convergence in latest Cretaceous to early Cenozoic time; and gradual termination of subduction-related activity and inception of extensional and transform tectonic regimes later in Cenozoic time.

IMPORTANCE OF THE CORDILLERAN OROGEN

Geological study of the Cordilleran region began only about 150 years ago. Scientific exploration of the Cordilleran followed closely on the heels of the opening of the west and discovery of gold in California in 1849. Geologists, or more precisely, geological engineers accompanied the railroad surveys of the 1850s. They made many of the first geological observations as exploring parties searched for the best routes for the railroad lines from the Mississippi River to the west coast of the continent. They were followed, after the Civil War, by federally funded geological and geographical surveys of Colorado, Wyoming, and surrounding regions under the direction of F. V. Hayden. At about the same time, Clarence King organized the federally funded geological exploration of the 40th parallel, which was to survey the geology and potential resources on either side of the proposed rail line that followed generally the 40th parallel from Wyoming to California. A third survey was organized and lead by John Wesley Powell into the Rocky Mountains and Colorado Plateau. Powell is perhaps best known for leading the first exploration of the Colorado River in 1869. He was one of the early voices for conservation of both human and material resources of the west. The studies by Powell and his coworkers developed several important geological concepts. Powell himself contributed major ideas about the importance of stream erosion in molding the magnificent topography of the Colorado Plateau. G. K. Gilbert developed the concepts that the Great Basin was the result of block faulting, the

Burchfiel, B. C., Lipman, P. W., and Zoback, M. L., 1992, Introduction, *in* Burchfiel, B. C., Lipman, P. W., and Zoback, M. L., eds., The Cordilleran Orogen: Conterminous U.S.: Boulder, Colorado, Geological Society of America, The Geology of North America, v. G-3.

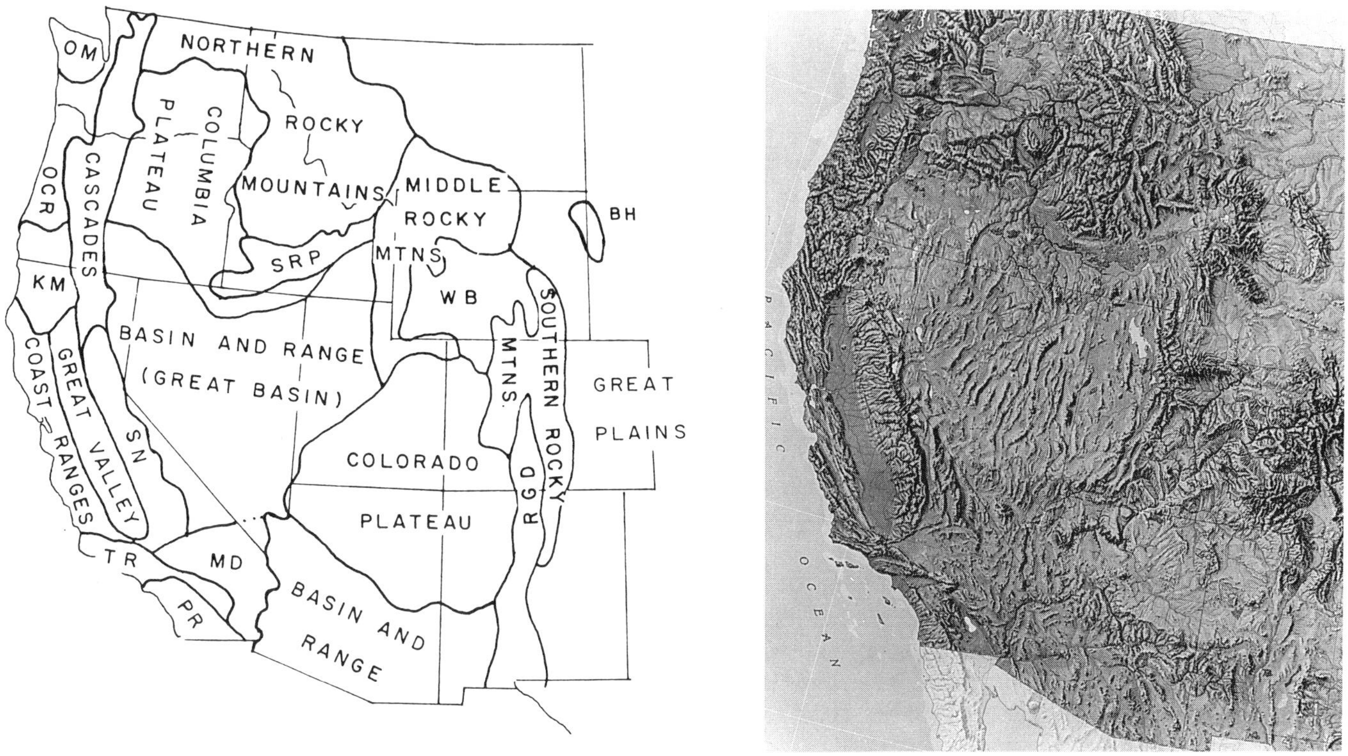

Figure 1. Physiographic regions of the western United States. OM, Olympic Mountains; OCR, Oregon Coast Ranges; KM, Klamath Mountains; TR, Transverse Ranges; SN, Sierra Nevada; MD, Mojave Desert; PR, Peninsular Ranges; SRP, Snake River Plain; WB, Wasatch Basin; RGD, Rio Grande Rift; BH, Black Hills.

Great Salt Lake is a remnant of the extensive older Lake Bonneville, and the Henry Mountains were formed by laccolithic intrusions. This latter work is not only remarkable for the wonderful writing, but presents an almost modern quantitative approach in explaining the developing of laccoliths. The various surveys produced considerable duplication, which led, after considerable intrigue and discussion, in 1879 to their consolidation into the U.S. Geological Survey, first directed by King and one year later by Powell who held the position for 14 years.

Following this colorful and fruitful early period of Cordilleran scientific investigation, the flow of people to the west was accompanied by a steady increase in scientific contributions until today the volume of yearly earth science literature focused on the Cordillera is probably greater than perhaps any orogenic belt in the world. The decade from 1980 to 1990 saw explosive research and publication in two major topics of Cordilleran tectonism: accretionary and extensional tectonics. The Cordillera has also been the focus of related topics, such as the use of paleomagnetic studies to interpret accretionary evolution and the importance of strike-slip displacements in the development of the orogen: topics about which considerable controversy persists today. The importance of accretionary and extensional tectonism was introduced in the 1960s by Warren Hamilton (Hamilton and Myers, 1966; Hamilton, 1969), and saw further but gradual development with seminal papers on accretionary tectonics (e.g., Monger and Ross, 1971; Jones and others, 1972; Speed, 1978; Davis and others, 1978) and on extension (e.g., Armstrong, 1972; Anderson, 1971; Proffett, 1977; Davis and Coney, 1979). But the 1980s became the decade when world-wide attention was riveted on western Cordillera accretionary and extensional tectonics, an appropriate beginning for the second century of Cordilleran tectonic research (for a brief history of Cordilleran tectonic research see Monger and Davis, 1982). Somewhat ironically the intensive field studies spawned by these landmark papers belied the arguments of some earth scientists in the 1970s that little was to be learned from additional fieldwork in the "already mapped" United States Cordillera. Cordilleran studies since 1980 of both accretionary and extensional tectonics, in addition to related fields, not only revolutionized our knowledge of Cordilleran evolution, but they serve as role models for the recognition of similar tectonic phenomena throughout the world. If Alpine tectonics set the pace for a global tectonics marathon in the first quarter of the twentieth century, Cordilleran studies have surely been the front runner in the last quarter.

Similarly, during the last few decades the Cordillera has increasingly become the focus for intensive volcanologic and petrologic studies of continental magmatism, utilizing new and improved field, petrologic, isotopic tracer, geochronologic, and paleomagnetic techniques. One major focus has been on magmatism in relation to tectonic setting; development of plate-tectonic

models has provided a fruitful framework within which to interpret regional variations in magma types, and conversely to use magmatic associations to infer past plate-tectonic configurations. The relations between the complex Cenozoic Cordilleran magmatic record and evolving boundaries between the American and various Pacific plates have been an especially rich source of debate and publication (Atwater, 1970; Christiansen and Lipman, 1972). Another relatively recent breakthrough has been recognition of the pyroclastic nature of large-scale silicic volcanism, and the associated widespread ash-flow sheets with caldera subsidence at sources that mark the upper crustal emplacement of batholith-scale magma chambers (Smith, 1960, 1979; Lipman, 1984). Study of compositional zonations in Cordilleran ash-flow sheets has provided high-resolution control on chemical gradients in silicic magma chambers (Lipman and others, 1966; Hildreth, 1981). Studies of Cordilleran basalts have provided key evidence for multiple sources of mafic continental margins, including upwelling asthenosphere, old lithospheric mantle forming a lithospheric keel beneath old cratonic crust, and variable contamination/assimilation of crustal materials (Leeman, 1970; Dungan and others, 1986; Perry and others, 1987; Fitton and others, 1988). Basaltic maars and diatremes in the Cordilleran region provide natural drill-cutting samples of the continental lithosphere, brought to the surface as xenoliths within rapidly rising alkalic basalt magma (Hearn, 1968; Menzies and others, 1987; Wilshire and others, 1988). Structurally complex ultramafic bodies in the western Cordillera, long recognized as tectonically emplaced fragments of mantle material, have been providing key evidence for accretionary processes and tectonic setting (Coleman, 1977; Moores, 1982; Saleeby, 1982). Cordilleran magmatism has also been the target of much study related to direct and indirect societal application, including volcanic hazard studies such as the one that anticipated the devastating 1980 eruption of Mount St. Helens (Crandell and Mullineaux, 1978), magmas as thermal engines and direct metal sources for ore deposits (Norton, 1982; Christiansen and others, 1983), magmatic systems as geothermal resources (Hearn and others, 1981), and as potential hosts for toxic waste repositories (Roseboom, 1983).

Geologic, tectonic, and volcanological studies of the United States Cordillera in the past few decades have provided the framework for interpreting data on the geophysical character of the crust and lithosphere of this region. An extensive program of seismic reflection profiling throughout the western Cordillera has made significant contributions in delineating the geometry of broad-scale upper crustal deformational structures including: the ~30° dip of the bounding thrust of the Laramide Wind River uplift (Smithson and others, 1979; Brewer and others, 1980; Lynn and others, 1983) indicating lateral compression rather than vertical tectonism, the <20° dip throughout the upper crust of a major late Tertiary (Sevier Desert) detachment fault (McDonald, 1976; Allmendinger and others, 1983; Smith and Bruhn, 1984), and the three-dimensional nature of metamorphic core complexes and large-scale detachment faults (including broad scale antiforms and synforms both parallel and perpendicular to transport direction; McCarthy and others, 1991). The seismic-reflection data have also provided a view of a highly reflective, "layered" lower crust throughout the extensional portions of the western Cordillera (Klemperer and others, 1986; Allmendinger and others, 1987; Hague and others, 1987; Potter and others, 1987; Serpa and de Voogd, 1987; Goodwin and McCarthy, 1990); the source of this layering, be it either metamorphic, magmatic, or deformational in origin, remains a topic of intense interest (e.g., Goodwin and Thompson, 1988). The concept of a young, highly mobile Moho in extensional areas was suggested by reflection profiles crossing the Basin and Range, which showed a remarkably uniform depth to Moho across the province despite large lateral variations in both reconstructed post-compressional crustal thickness and subsequent amounts of extension (Klemperer and others, 1986). These data are now being used to constrain models of a "fluid" lower crust (Gans, 1987; Block and Royden, 1990; Wernicke, 1990, and this volume).

Considerable progress has also been made in understanding the interrelationship between crustal structure, gravity, heat flow, and deformation and its influence on regional topography (e.g., Lachenbruch and Sass, 1978; Lachenbruch and Morgan, 1990). As shown in Figure 2a, much of the western United States has a crustal thickness between 30 and 40 km. A crustal thickness of 30 to 35 km is commonly associated with regions at sea level, however, as also shown in Figure 2b, much of this region has average elevations of >1,000 m above sea level. Much of this high topography region is also characterized by above-normal heat flow (Fig. 2c). In addition, at long wavelengths the Bouguer gravity is nearly a mirror image of topography, implying isostatic compensation (Kane and Godson, 1989). Because the crust is generally so thin, much of the isostatic support of the region must come from buoyancy due to a warm upper mantle and/or lithospheric thinning. Magmatic addition is required for balance of mass lost laterally by extension and to explain the high heat flow of the extended areas; simple steady-state conduction models require massive melting of the lower crust, which is not observed seismically (Lachenbruch and Sass, 1978). Ultimately, the elevation of an extending region is a balance between extensional subsidence resulting from crustal thinning and thermal buoyancy derived from the warm upper mantle, and magmatic additions to the lithosphere of material of near-crustal density (Lachenbruch and Morgan, 1990; Thompson and McCarthy, 1990).

Refinement of plate reconstructions for the Pacific Basin (Engebretson and others, 1985; Stock and Molnar, 1988) have permitted correlations between major orogenic and magmatic episodes and changes in both rate and velocity of relative plate motions (e.g., Engebretson and others, 1984). Present-day plate motions explain the high rate of active seismicity along the San Andreas fault zone; however, the lack of Benioff-zone seismicity related to subduction beneath Oregon and Washington remains an enigma and fuels the debate as to the likelihood of a major subduction earthquake beneath the Pacific Northwest. A broad zone of seismicity characterizes most of the United States Cordillera, and nearly all the deformation in regions over 1,000 m in

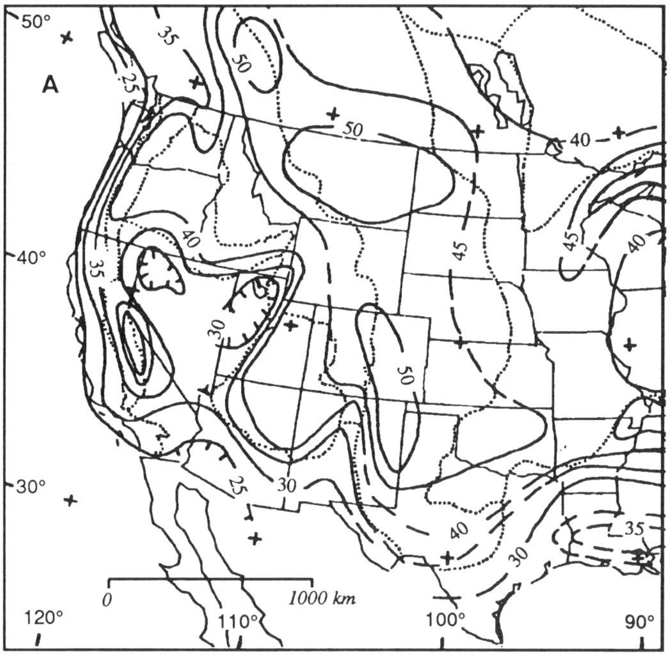

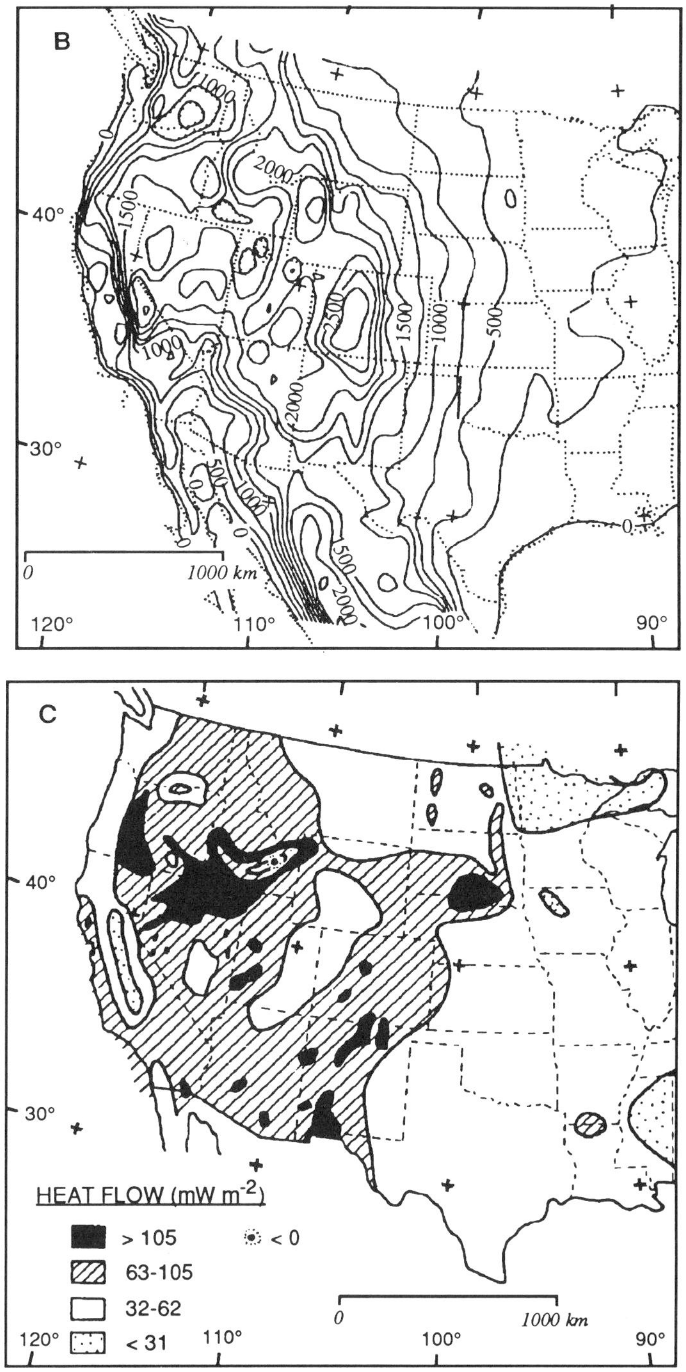

Figure 2. Geophysical setting of the western United States Cordillera. A, crustal thickness in kilometers measured from the surface to the inferred Moho discontinuity, contour interval = 5 km (modified from Braile and others, 1989); B, regional topographic elevation based on 1° × 1° averages, contour interval = 250 m (modified from Braile and others, 1989); C, generalized heat flow (modified from Morgan and Gosnold, 1989).

elevation is due to normal faulting characterized by dominantly west-northwest extension (Zoback and Zoback, 1989). A 90° rotation of least principal stress direction on the Colorado Plateau and southern Great Plains may be related to buoyancy forces derived from abrupt lateral variations in thickness of the upper mantle lithospheric lid (Thompson and Zoback, 1979; Zoback and Zoback, 1989).

Paleomagnetic investigations have indicated the importance of large-scale rotations about a vertical axis; perhaps the best-documented examples are the 70° rotation of the Oregon Coast Range since Eocene time (Simpson and Cox, 1977), and the roughly 90° Miocene and younger rotation of much of southern California west of the San Andreas including the borderland region (Luyendyk and others, 1985). These large-scale examples of tectonic rotations occur along evolving plate boundaries; however, evidence of smaller-scale rotations related to a combination of strike-slip and normal faulting have also been suggested (e.g., Ron and others, 1986).

Heat flow, seismicity, and in-situ stress data have been reconciled with geologic observations of fault-normal compression along a broad zone (125 km wide) adjacent to the San Andreas fault in terms of a low average shear strength for this portion of the North American–Pacific plate boundary (Mount and Suppe, 1987; Zoback and others, 1987). The 70 to 85° angle between the strike of the San Andreas fault and the inferred in-situ stress field is interpreted to be the result of a very broad-scale stress rotation, required so that the San Andreas fault lies close to a principal stress direction, hence minimizing the shear stress on that plane (Zoback and others, 1987). The implied low shear strength for the San Andreas fault is consistent with the lack of a frictionally generated heat flow anomaly, and the near fault-normal compression explains the occurrence of Pliocene and younger folds and thrusts subparallel to the San Andreas system (Page, 1981).

ORGANIZATION OF THE VOLUME

This volume is divided into two sections: time-slice chapters followed by topical synthesis chapters. Within the time-slice chapters the basic data for the Cordillera are presented. The time-slice organization was chosen because ultimately, it is the grouping of events by their contemporaneity that permits an understanding of the kinematic and dynamic systems responsible for these events. This is best exemplified when plate-tectonic interpretations are applied to the geologic evolution of a region; a great variety of geological events and features can be related across diverse geologic terranes and can be grouped into broad plate-tectonic kinematic and dynamic systems. The evolution of the region can then be traced by the time-space relations of these systems.

The time-slice format permits examination of temporally related paleogeographic and geologic features. All areas and tectonostratigraphic units containing rocks and events of a particular time slice are discussed. The time-slice analysis forces us to examine the relationship between diverse geologic elements so that they can be grouped into meaningful genetic systems. Furthermore, it brings into focus those terranes that during that time-slice may have been far removed from their present location. The boundaries of the time units selected are elastic, because certain events and geologic features are time transgressive.

Although not treated in a separate chapter, the Precambrian geology of the Cordillera is summarized in Burchfiel and others (this volume); a more detailed discussion is presented by Link and others (1992). The topical synthesis chapters examine specific categories of geologic features and phenomena from an evolutionary and processual perspective through time. They include summaries, discussions, and interpretations of time-independent features of the Cordillera, such as the evolution of magmatic arcs and metamorphic belts; tectonic processes of convergent, transform, and extensional regimes; relations of sedimentation to tectonism; development of ophiolite belts; and metallogenesis.

Both sets of chapters present diverse opinions about the geology of the Cordillera. By the very nature of the debate on issues, the differences evident in the chapters point to significant unresolved problems and important areas of future research.

How does the reader attack such a massive and detailed presentation of the long and complex evolution of the Cordillera? One suggestion is to read Chapter 8 on tectonic synthesis first, for a general, albeit somewhat detailed overview. This chapter is accompanied by a tectonostratigraphic map (Plate 1A, B) keyed to the text, and a set of one-page evolutionary diagrams so that the reader unfamiliar with the Cordillera can grasp the location and evolution of its major elements. The time-slice chapters, followed by the topical synthesis chapters and a final review of the tectonic synthesis (Chapter 8) should provide a reasonable understanding of the United States Cordillera and its unresolved problems.

Although the geologic, tectonic, and magmatic evolution of the Cordillera are covered in detail in this volume, the modern geophysical setting of this region and its ties to the tectonic evolution have received considerably less detailed treatment. This is due in part to a concurrent effort, the recent publication of an entire volume on that topic, "Geophysical Framework of the Continental United States," (Pakiser and Mooney, 1989). This volume provides an excellent synthesis and interpretation of available geophysical data for the western United States Cordillera (see Pakiser, 1989, for an overall summary) and useful comparisons between the active western Cordillera and the tectonically more stable eastern United States. The volume includes a series of papers focused directly on crustal structure and composition of the crust and upper mantle in the western United States (Couch and Riddihough, 1989; Iyer and Hitchcock, 1989; Mooney and Weaver, 1989; Prodehl and Lipman, 1989; Smith and others, 1989; Thompson and others, 1989), as well as regional interpretations of gravity and magnetic data (Kane and Godson, 1989; Jachens and others, 1989) and geoelectric data (Keller, 1989). Because of that concurrent effort, these topics are not covered in this volume.

REFERENCES CITED

Allmendinger, R. W., and 7 others, 1983, Cenozoic and Mesozoic structure of the eastern Basin and Range Province, Utah, from COCORP seismic-reflection data: Geology, v. 11, p. 532–536.

Allmendinger, R. W., and 7 others, 1987, Overview of the COCORP 40 N Transect, western United States; The fabric of an orogenic belt: Geological Society of America Bulletin, v. 98, p. 308–319.

Anderson, R. E., 1971, Thin-skin distension in Tertiary rocks of southwestern Nevada: Geological Society of America Bulletin, v. 82, p. 43–58.

Armstrong, R. L., 1972, Low angle (denudation) faults, hinterland of the Sevier orogenic belt, eastern Nevada and western Utah: Geological Society of America Bulletin, v. 83, p. 1729–1754.

Atwater, T., 1970, Implications of plate tectonics for the Cenozoic tectonic evolution of western North America: Geological Society of America Bulletin, v. 81, p. 3513–3538.

Block, L., and Royden, L. H., 1990, Core complex geometries and regional scale flow in the lower crust: Tectonics, v. 9, p. 557–567.

Brewer, J. A., Smithson, S. B., Oliver, J. E., Kaufman, S., and Brown, L. D., 1980, The Laramide orogeny; Evidence from COCORP deep crustal seismic profiles in the Wind River Mountains, Wyoming: Tectonophysics, v. 62, p. 165–189.

Braile, L. W., Hinze, W. J., von Frese, R.R.B., and Keller, G. R., 1989, Seismic properties of the crust and uppermost mantle of the conterminous United States and adjacent Canada, *in* Pakiser, L. C., and Mooney, W., Geophysical framework of the continental United States: Geological Society of America Memoir 172, p. 655–680.

Christiansen, E. H., Burt, D. M., Sheridan, M. F., and Wilson, R. T., 1983, The petrogenesis of topaz rhyolites from the western United States: Contributions to Mineralogy and Petrology, v. 85, p. 16–32.

Christiansen, R. L., and Lipman, P. W., 1972, Cenozoic volcanism and plate-tectonic evolution of the western United States, II, Late Cenozoic: Royal Society of London, Philosophical Transactions Series A, v. 271, p. 249–284.

Coleman, R. G., 1977, Ophiolites, ancient oceanic lithosphere: New York, Springer-Verlag, 240 p.

Couch, R. W., and Riddihough, R. P., 1989, The crustal structure of the western continental margin of North America: Geological Society of America Memoir 172, p. 103–128.

Crandell, D. R., and Mullineaux, D. R., 1978, Potential hazards from future eruptions of Mount St. Helens volcano, Washington: U.S. Geological Survey Bulletin 1383–C, 26 p.

Davis, G. A., Monger, J.W.H., and Burchfiel, B. C., 1978, Mesozoic construction of the Cordilleran "collage," central British Columbia to central California, *in* Howell, D. G., and McDougall, K. A., eds., Mesozoic Paleogeography of the Western United States, Pacific Coast Paleogeography Symposium 2: Los Angeles, California, Pacific Section, Society of Economic Paleontologists and Mineralogists, p. 1–32.

Davis, G. H., and Coney, P. J., 1979, Geologic development of the Cordilleran metamorphic core complexes: Geology, v. 7, p. 120–124.

Dungan, M. A., and 5 others, 1986, Open-system evolution of the Taos Plateau volcanic field, Northern New Mexico; 1. The petrology and geochemistry of the Servilleta Basalt: Journal of Geophysical Research, v. 91, p. 5999–6028.

Engebretson, D. C., Cox, A., Thompson, G. A., 1984, Correlation of plate motions with continental tectonics; Laramide to Basin-Range: Tectonics, v. 3, p. 115–119.

Engebretson, D. C., Cox, A., and Gordon, R. G., 1985, Relative motions between oceanic and continental plates in the Pacific Basin: Geological Society of America Special Paper 206, 59 p.

Fitton, J. G., James, D., Kempton, P. D., Ormerod, D. S., and Leeman, W. P., 1988, The role of lithospheric mantle in the generation of late Cenozoic basic magmas in the western United States: Journal of Petrology, Special Lithosphere Issue, p. 331–349.

Gans, P. B., 1987, An open-system, two-layer crustal stretching model for the eastern Great Basin: Tectonics, v. 6, p. 1–12.

Goodwin, E. B., and McCarthy, J., 1990, Composition of the lower crust in west-central Arizona from three-component seismic data: Journal of Geophysical Research, v. 95, p. 20097–20109.

Goodwin, E. B., and Thompson, G. A., 1988, The seismically reflective crust beneath highly extended terranes; Evidence for its origin in extension: Geological Society of America Bulletin, v. 100, p. 1616–1626.

Hague, T. A., and 10 others, 1987, Crustal structure of western Nevada from COCORP deep seismic-reflection data: Geological Society of America Bulletin, v. 98, p. 320–329.

Hearn, B. C., Jr., 1968, Diatremes with kimberlitic affinities in north-central Montana: Science, v. 159, p. 622–625.

Hearn, B. C., Jr., Donnelly-Nolan, J. M., and Goff, F. E., 1981, The Clear Lake volcanics; tectonic setting and magma sources: U.S. Geological Survey Professional Paper 1141, p. 25–46.

Hildreth, W., 1981, Gradients in silicic magma chambers; Implications for lithospheric magmatism: Journal of Geophysical Research, v. 86, p. 10153–10192.

Hamilton, W. B., 1969, Mesozoic California and the underflow of Pacific mantle: Geological Society of America Bulletin, v. 80, p. 2409–2430.

Hamilton, W., and Myers, W. B., 1966, Cenozoic tectonics of the western United States: Reviews of Geophysics, v. 4, p. 509–549.

Iyer, H. M., and Hitchcock, T., 1989, Upper-mantle velocity structure in the continental U.S. and Canada: Geological Society of America Memoir 172, p. 681–710.

Jachens, R. C., Simpson, R. W., Blakely, R. J., and Saltus, R. W., 1989, Isostatic residual gravity and crustal geology of the United States: Geological Society of America Memoir 172, p. 405–424.

Jones, D. L., Irwin, W. P., and Ovenshine, A. T., 1972, Southeastern Alaska; A displaced continental fragment?: United States Geological Survey Professional Paper 800-B, p. B211–B217.

Kane, M. F., and Godson, R. H., 1989, A crust/mantle structural framework of the conterminous United States based on gravity and magnetic trends: Geological Society of America Memoir 172, p. 383–404.

Keller, G. V., 1989, Electrical structure of the crust and upper mantle beneath the United States; Part 2, Survey of data and interpretations: Geological Society of America Memoir 172, p. 425–446.

Klemperer, S. L., Hauge, T. A., Hauser, E. C., Oliver, J. E., and Potter, C. K., 1986, The moho in the northern Basin and Range Province, Nevada, along the COCORP 40°N seismic-reflection transect: Geological Society of America Bulletin, v. 97, p. 603–618.

Lachenbruch, A. H., and Sass, J. H., 1978, Models of an extending lithosphere and heat flow in the Basin and Range Province: Geological Society of America Memoir 152, p. 209–250.

Lachenbruch, A. H., and Morgan, P., 1990, Continental extension, magmatism and elevation; Formal relations and rules of thumb: Tectonophysics, v. 174, p. 39–62.

Leeman, W. P., 1970, The isotopic composition of strontium in late-Cenozoic basalts from the Basin-Range province, western United States: Geochimica et Cosmochimica Acta, v. 34, p. 857–872.

Link and 12 others, 1992, Middle and Late Proterozoic stratified rocks of the western U.S. Cordillera, Colorado Plateau, and Basin and Range Province, *in* Read, J. C., Jr., and 6 others, eds., Precambrian; Conterminous U.S.: Boulder, Colorado, Geological Society of America, The Geology of North America, Volume C-2.

Lipman, P. W., 1984, The roots of ash-flow calderas in western North America; Windows into the tops of granitic batholiths: Journal of Geophysical Research, v. 89, p. 8801–8841.

Lipman, P. W., Christiansen, R. L., and O'Connor, J. J., 1966, A compositionally zoned ash-flow sheet in southwestern Nevada: U.S. Geological Survey Professional Paper 524-F, 47 p.

Luyendyk, B. P., Kamerling, M. J., Terres, R. .R., and Hornafius, J. S., 1985, Simple shear of southern California during Neogene time suggested by paleomagnetic declinations: Journal of Geophysical Research, v. 90, p. 12454–12466.

Lynn, H. B., Quam, S., and Thompson, G. A., 1983, Depth migration and interpretation of the COCORP Wind River, Wyoming, seismic reflection data: Geology, v. 11, p. 462–469.

McCarthy, J., Larkin, S. P., Guis, G. S., Simpson, R. W., and Howard, K. A., 1991, Anatomy of a metamorphic core complex; Seismic refraction/wide-angle reflection profiling in southeastern California and western Arizona: Journal of Geophysical Research, v. 96, p. 12259–12291.

McDonald, R. E., 1976, Tertiary tectonics and sedimentary rocks along the transition; Basin and Range Province to plateau and thrust belt province: Denver, Colorado, Rocky Mountain Association of Geologists Symposium, p. 281–327.

Menzies, M. A., and 7 others, 1987, A record of subduction process and within-plate volcanism in lithospheric xenoliths of the southwestern USA, *in* Nixon, P. W., ed., Mantle Xenoliths: London, John Wiley & Sons, Ltd., p. 59–74.

Monger, J.W.H., and Davis, G. A., 1982, Evolving concepts of the tectonics of the North American Cordillera, *in* Levinton, A. E., Rodda, P. U., Yochelson, E. L., and Aldrich, M. L., eds., Frontiers of geological explorations of the western North Americ: American Association for the Advancement of Science, Pacific Division, p. 215–248.

Monger, J.W.H., and Ross, C. A., 1971, Distribution of fusulinaceans in the western Canadian Cordillera: Canadian Journal of Earth Science, v. 8, p. 259–278.

Mooney, W. D., and Weaver, C. S., 1989, Regional crustal structure and tectonics of the Pacific Coastal States; California, Oregon, and Washington: Geological Society of America Memoir 172, p. 129–162.

Moores, E. M., 1982, Origin and emplacement of ophiolites: Reviews of Geophysical and Space Physics, v. 20, p. 735–760.

Morgan, P., and Gosnold, W. D., 1989, Heat flow and thermal regimes in the continental United States, *in* Pakiser, L. C., and Mooney, W., Geophysical framework of the continental United States: Geological Society of America Memoir 172, p. 493–522.

Mount, V. S., and Suppe, J., 1987, State of stress near the San Andreas fault; Implications for wrench tectonics: Geology, v. 15, p. 1143–1146.

Norton, D. L., 1982, Fluid and heat transport phenomena typical of copper-bearing pluton environments, *in* Titley, S. R., ed., Advances in geology of the porphyry copper deposits: Tucson, University of Arizona Press, p. 59–72.

Page, B., 1981, The southern Coast Ranges, *in* Ernst, W. G., ed., The geotectonic development of California: Englewood Cliffs, New Jersey, Prentice-Hall, p. 329–417.

Pakiser, L. C., 1989, Geophysics of the Intermontane system: Geological Society of America Memoir 172, p. 235–248.

Pakiser, L. C., and Mooney, W. D., 1989, eds., Geophysical framework of the

continental United States: Geological Society of America Memoir 172, 826 p.

Perry, F. V., Baldridge, W. S., and DePaolo, D. J., 1987, Role of asthenosphere and lithosphere in the genesis of late Cenozoic basaltic rocks from the Rio Grande rift and adjacent regions of southwestern United States: Journal of Geophysical Research, v. 92, p. 9193–9214.

Prodehl, C., and Lipman, P. W., 1989, Crustal structure of the Rocky Mountain region: Geological Society of America Memoir 172, p. 249–284.

Proffett, J. M., Jr., 1977, Cenozoic geology of the Yerington District, Nevada, and implications for the nature and origin of Basin and Range faulting: Geological Society of America Bulletin, v. 88, p. 247–266.

Potter, C., Allmendinger, R. W., Hauser, E. C., Oliver, J. E., 1987, COCORP deep seismic reflection traverses of the U.S. Cordillera: Geophysical Journal of the Royal Astronomical Society, v. 89, p. 99–104.

Ron, H., Aydin, A., and Nur, A., 1986, Strike-slip faulting and block rotation in the Lake Mead fault system: Geology, v. 14, p. 1020–1023.

Roseboom, E. H., Jr., 1983, Disposal of high-level nuclear waste above the water table in arid regions: U.S. Geological Survey Circular 903, 21 p.

Saleeby, J., 1982, Polygenetic ophiolite belt of the California Sierra Nevada; Geochronological and tectonostratigraphic development: Journal of Geophysical Research, v. 87, p. 1803–1824.

Serpa, L., and de Voogd, B., 1987, Deep seismic reflection evidence for the role of extension in the evolution of the continental crust: Geophysical Journal of the Royal Astronomical Society, v. 89, p. 55–60.

Simpson, R. W., and Cox, A., 1977, Paleomagnetic evidence for tectonic rotation of the Oregon Coast Range: Geology, v. 5, p. 585–589.

Smith, R. B., and Bruhn, R. L., 1984, Intraplate extensional tectonics of the eastern Basin-Range; Inferences on structural style from seismic reflection data, regional tectonics, and thermal-mechanical models of brittle/ductile deformation: Journal of Geophysical Research, v. 89, p. 5733–5762.

Smith, R. B., Nagy, W. C., Julander, K. A., Viveiros, J. J., Barker, C. A., and Gants, D. G., 1989, Geophysical and tectonic framework of the eastern Basin and Range–Colorado Plateau–Rocky Mountain transition: Geological Society of America Memoir 172, p. 205–234.

Smith, R. L., 1960, Ash flows: Geological Society of America Bulletin, v. 71, p. 795–842.

—— , 1979, Ash-flow magmatism, *in* Chapin, C. E., and Elston, W. E., eds., Ash-flow tuffs: Geological Society of America Special Paper 180, p. 5–27.

Smithson, S. B., Brewer, J. A., Kaufman, S., Oliver, J. E., and Hurich, C. A., 1979, Structure of the Laramide Wind River uplift, Wyoming, from COCORP deep reflection data and gravity data: Journal of Geophysical Research, v. 84, p. 5955–5972.

Speed, R. C., 1978, Paleogeographic and plate tectonic evolution of the early Mesozoic marine province of the western Great Basin, *in* Howell, D. G., and McDougall, K. A., eds., Mesozoic paleogeography of the western United States: Society of Economic Paleontologists Mineralogists, Pacific Section, Pacific Coast Paleogeography Symposium 2, Los Angeles, p. 253–270.

Stock, J. M., and Molnar, P., 1988, Uncertainties and implications of the Late Cretaceous and Tertiary position of North America relative to Farallon, Kula, and Pacific Plates: Tectonics, v. 7, p. 1339–1384.

Thompson, G. A., and McCarthy, J., 1990, A gravity constraint on the origin of highly extended terranes: Tectonophysics, v. 174, p. 197–206.

Thompson, G. A., and Zoback, M. L., 1979, Regional geophysics of the Colorado Plateau: Tectonophysics, v. 61, p. 149–181.

Thompson, G. A., and 7 others, 1989, Geophysics of the western Basin and Range Province: Geological Society of America Memoir 172, p. 177–204.

Wernicke, B., 1990, The fluid crustal layer and its implications for continental dynamics, *in* Salisbury, M., ed., Exposed cross sections of the continental crust, NATO Advanced Studies Institute: Dordrecht, Holland, Kluwer Academic Publishers, D. Riedel, p. 509–544.

Wilshire, H. G., and 6 others, 1988, Mafic and ultramafic xenoliths from volcanic rocks of the western United States: U.S. Geological Survey Professional Paper 1443, 179 p.

Zoback, M. D., and 12 others, 1987, New evidence on the state of stress of the San Andreas fault system: Science, v. 238, p. 1105–1111.

Zoback, M. L., and Zoback, M. D., 1989, Tectonic stress field of the conterminous United States: Geological Society of America Memoir 172, p. 523–539.

Manuscript Accepted by the Society August 2, 1991

Printed in U.S.A.

The Geology of North America
Vol. G-3, The Cordilleran Orogen: Conterminous U.S.
The Geological Society of America, 1992

Chapter 2

Latest Precambrian to latest Devonian time; Development of a continental margin

Forrest G. Poole
U.S. Geological Survey, Box 25046, Denver Federal Center, Denver, Colorado 80225
John H. Stewart
U.S. Geological Survey, 345 Middlefield Road, Menlo Park, California 94025
Allison R. Palmer
Geological Society of America, 3300 Penrose Place, P.O. Box 9140, Boulder, Colorado 80301
Charles A. Sandberg
U.S. Geological Survey, Box 25046, Denver Federal Center, Denver, Colorado 80225
Raul J. Madrid
Conquistador Gold Ltd., 3210 Marthiam Avenue, Reno, Nevada 89509
Reuben J. Ross, Jr.
Department of Geology and Geological Engineering, Colorado School of Mines, Golden, Colorado 80401
Lehi F. Hintze
Department of Geology, Brigham Young University, Provo, Utah 84602
M. Meghan Miller
Jet Propulsion Laboratory, California Institute of Technology, Pasadena, California 91109
Chester T. Wrucke
U.S. Geological Survey, 345 Middlefield Road, Menlo Park, California 94025

INTRODUCTION

Forrest G. Poole

Nearly continuous successions of late Proterozoic through Upper Devonian rocks are widespread in the western United States between the structural fronts of the Sevier and Sonoma orogens (Plates 2-1 to 2-6, 3-1 to 3-4). West of the Sonoma orogenic front in Washington, Oregon, northern and southern California, western Idaho, and western Nevada, rocks of this age are limited to many small areas of exposures, most of which are shown on Plate 3-5. East of the Sevier orogenic front, some large areas on the cratonic platform contain only partial stratigraphic records because of either nondeposition or erosion. The western limit of mapped carbonate-shelf rocks can be obtained from only a few tectonic windows because those rocks disappear beneath thrust plates of western facies rocks of the same age or of younger and older rocks that moved eastward as much as 200 km or more during the Antler and later orogenies. Figure 1 shows Devonian and older paleotectonic features, major post-Devonian faults, and locations of the 12 generalized stratigraphic columns shown in Figure 2.

Correlation, dating, and nomenclature of some late Proterozoic through Upper Devonian rocks are shown in Figure 2 and on Plate 3-6. These utilize unpublished fossil and stratigraphic data of the authors or are from published literature and unpublished theses. Series and stage names are those used on recent published time scales (Harland and others, 1990; Ross and others, 1982; Palmer, 1983; Hintze, 1985). The North American series and stage names of the Ordovician advocated by Ross and others (1982) are used in preference to the customary British subdivisions for continental-shelf and cratonic-platform rocks. Hiatuses shown in Figure 2 are based on both faunal and physical evidence; however, in areas where hiatuses have not been firmly established, they are inferred from regional evidence. The name "dolostone" is used in Figure 2 and throughout this chapter for rocks composed predominantly of the mineral dolomite, although in the literature it has been common practice to use "dolomite" rather than "dolostone" in formal names of dolomitic formations and members.

The two terms "miogeosyncline" and "miogeocline" have been applied to Paleozoic continental-shelf deposits along the Cordilleran continental margin. Form and structure of the shelf are recorded by thickness trends of the deposits, and they provide

Poole, F. G., Stewart, J. H., Palmer, A. R., Sandberg, C. A., Madrid, R. J., Ross, R. J., Jr., Hintze, L. F., Miller, M. M., and Wrucke, C. T., 1992, Latest Precambrian to latest Devonian time; Development of a continental margin, *in* Burchfiel, B. C., Lipman, P. W., and Zoback, M. L., eds., The Cordilleran Orogen: Conterminous U.S.: Boulder, Colorado, Geological Society of America, The Geology of North America, v. G-3.

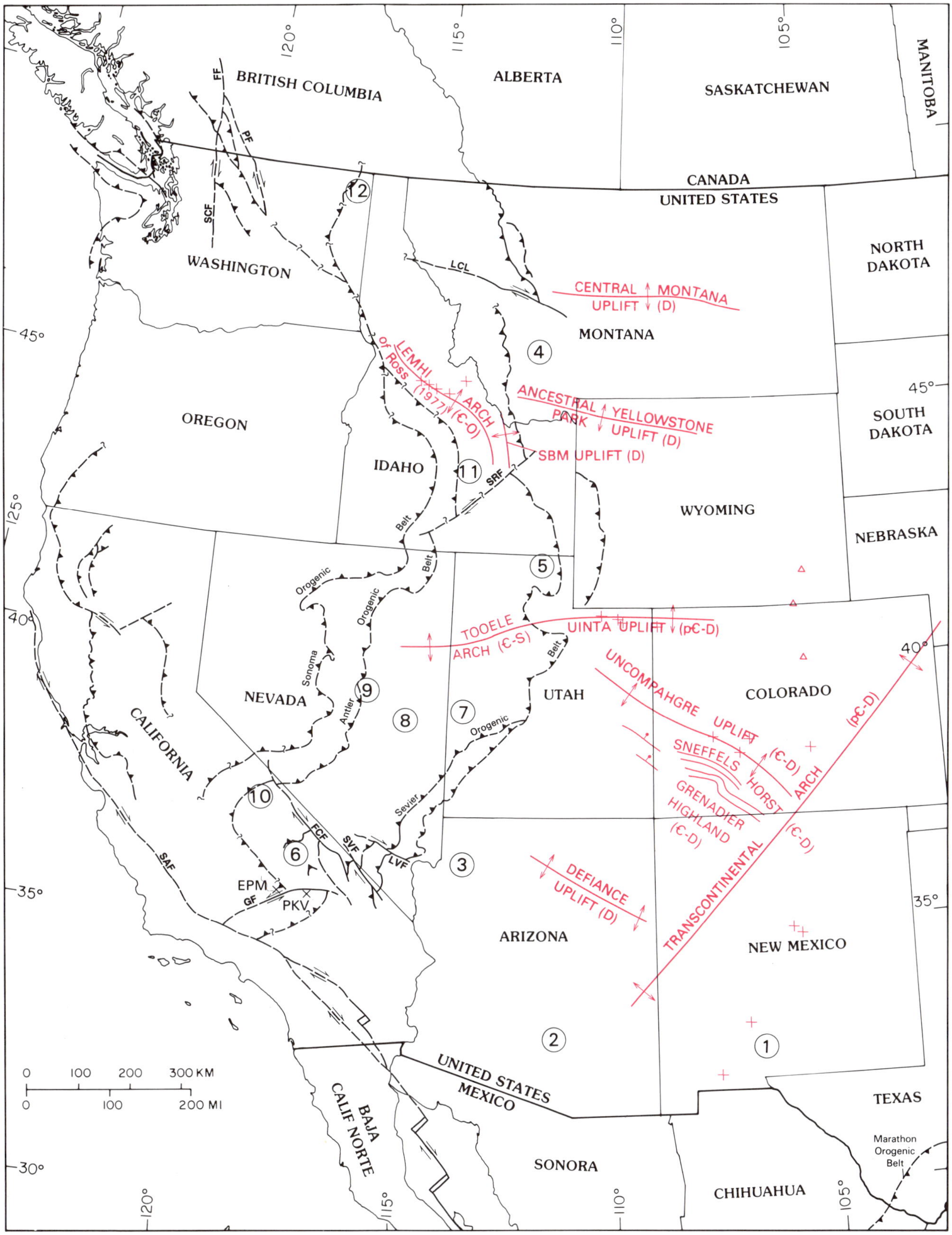

Figure 1. Map showing Devonian and older paleotectonic features, major post-Devonian faults, and locations of stratigraphic columns shown in Figure 2. Explanation on facing page.

EXPLANATION

Younger structural features (black)—Shown by capital-and-lower case lettering; movements occurred in post-Devonian time

Thrust fault—Dashed where inferred; sawteeth on upper plate. Generally depicts leading edge of fault system or major fault within system that significantly displaces Devonian and older rocks

Strike-slip fault—Dashed where inferred; arrows show latest relative horizontal displacement of Devonian and older rocks

Older structural features (red)—Shown by all-capital lettering; movements occurred in pre-Mississippian time and locally influenced Devonian and older sedimentation; maximum development of feature given in parentheses

Normal or reverse fault—Bar and ball on downthrown side

Axis of anticline or uplift

Cambrian and (or) Ordovician alkalic and mafic intrusions

Lower or Middle Devonian kimberlitic diatremes

(7) **Location of stratigraphic column**—Number refers to column on correlation chart (Fig. 2)

SRF **Abbreviations for structural features**—EPM, El Paso Mountains; FF, Fraser fault; FCF, Furnace Creek fault; GF, Garlock fault; LCL, Lewis and Clark line; LVF, Las Vegas Valley fault; PF, Pasayten fault; PKV, Pilot Knob Valley; SAF, San Andreas fault; SBM, Southern Beaverhead Mountains uplift; SCF, Straight Creek fault; SRF, Snake River fault; SVF, Stewart Valley fault

a basis for applying the appropriate morphologic term. Kay (1951) described most miogeosynclines as consisting of asymmetrical lensoid bodies of sediment deposited on the continental shelf. Dietz and Holden (1966) defined miogeocline as a seaward-thickening wedge of sediment deposited on the continental shelf; however, sediments generally are thinner on the continental slope and ocean floor. Moreover, the so-called thickening wedge actually thins rather than thickens seaward from the axis of the Middle Cambrian to Late Devonian continental shelf in the western U.S. In the western U.S., latest Proterozoic and Early Cambrian sedimentation resulted in a miogeoclinal prism of deposits during the initial development of the continental margin (Plate 2-2), whereas Middle Cambrian to Late Devonian sedimentation and subsidence of the shelf resulted in a two-sided miogeosynclinal lensoid-shaped sequence of rocks (Plates 2-3 to 2-5, 3-1 to 3-3). The change from miogeocline to miogeosyncline occurred in the early Middle Cambrian.

In this chapter, the formal terms "Formation" and "Group" are applied to sequences of rocks that possess internal lithic homogeneity or distinctive lithic features that can be delineated at the scale of mapping practiced in the region; the informal term "assemblage" is applied to one or more formations or groups that possess distinctive lithic features but, because of intense folding, faulting, and metamorphism, cannot be separately delineated at the scale of mapping practiced in the region; and the informal terms "terrane" and "belt" are applied to a composite of assemblages of a specific geographic area having major tectonic boundaries and paleogeographic characteristics unlike neighboring areas.

Our 10 combined thickness and lithofacies maps (Plates 2-1 to 2-6, 3-1 to 3-4) are presented side-by-side on two separate plates in the pocket of this volume for ease in comparing the regional changes in sedimentational and structural patterns from the late Proterozoic through the Late Devonian. Plates 2-1 to 2-6 display maps of late Proterozoic through Middle Ordovician rocks, and Plates 3-1 to 3-4 display maps of Upper Ordovician through Upper Devonian rocks. Plate 3-5 shows the areal distribution of allochthonous rocks of the Antler orogen and younger accreted terranes. Plates 3-6 and 3-7 show selected stratigraphic columns for some of the most complete sections of pre-Mississippian rocks recognized in the Roberts Mountains allochthon of Nevada and in younger accreted terranes in northern California.

The thickness and lithofacies maps are based on surface sections measured by the authors and on data from well logs, measured sections, and maps contained in numerous published reports and unpublished theses. In areas of mild to moderate metamorphism, original rock types or protoliths generally have been interpreted from their metamorphic equivalents. Data points are shown on the thickness and lithofacies maps. In areas of abundant data, it was necessary to arbitrarily select 1 point to represent from 2 to 5 data points in order to avoid congestion on the maps. Palinspastic reconstruction of the mapped intervals was not attempted in this synthesis. Space limitations prohibit listing every report or map from which data were obtained, but many of our sources are in the bibliographies of major regional syntheses (Palmer, 1971; Poole and others, 1977; Ross, 1977; Sandberg and Poole, 1977; Stewart and Suczek, 1977; Sandberg and others, 1982; Johnson and others, 1988; Sandberg and others, 1988).

Several roof pendants in the east-central Sierra Nevada (Fig. 3) contain metasedimentary rocks of early and middle Paleozoic age that generally can be linked lithologically, structurally, and chronologically. Although additional stratigraphic and sedimentologic data and fossil ages are needed for many of the pendants to firmly establish affinities and correlations with Cordilleran shelf-margin rocks, we have provisionally integrated data for these pendants in our time-slice intervals (Plates 2 and 3). The pendants include the Saddlebag Lake, Log Cabin mine, Gull Lake, Casa Diablo, Mount Morrison, Pine Creek, Bishop Creek, Big Pine Creek, and Dinkey Creek (Moore and Foster, 1980; Nokleberg and Kistler, 1980; see Nokleberg, 1983, for detailed summary). Ordovician and Silurian(?) graptolites occur in argillites in the Mount Morrison and Bishop Creek pendants, Ordovician graptolites in argillites in the Log Cabin mine pendant, and *Skolithos* and *Cruziana* trace fossils in quartzites of possible Early Cambrian age in the Big Pine Creek pendant. The unfossiliferous roof pendants can be linked by lithic similarities to rocks in the pendants containing fossils, as well as to certain structural episodes that affected these rocks (Brook, 1977; Russell and Nokleberg, 1977). The dominant protoliths of these severely deformed rocks include shale, chert, calcareous sandstone, and minor limestone and dolostone, all of which are interpreted to represent continental-slope and -rise deposits in the Saddlebag Lake, Log

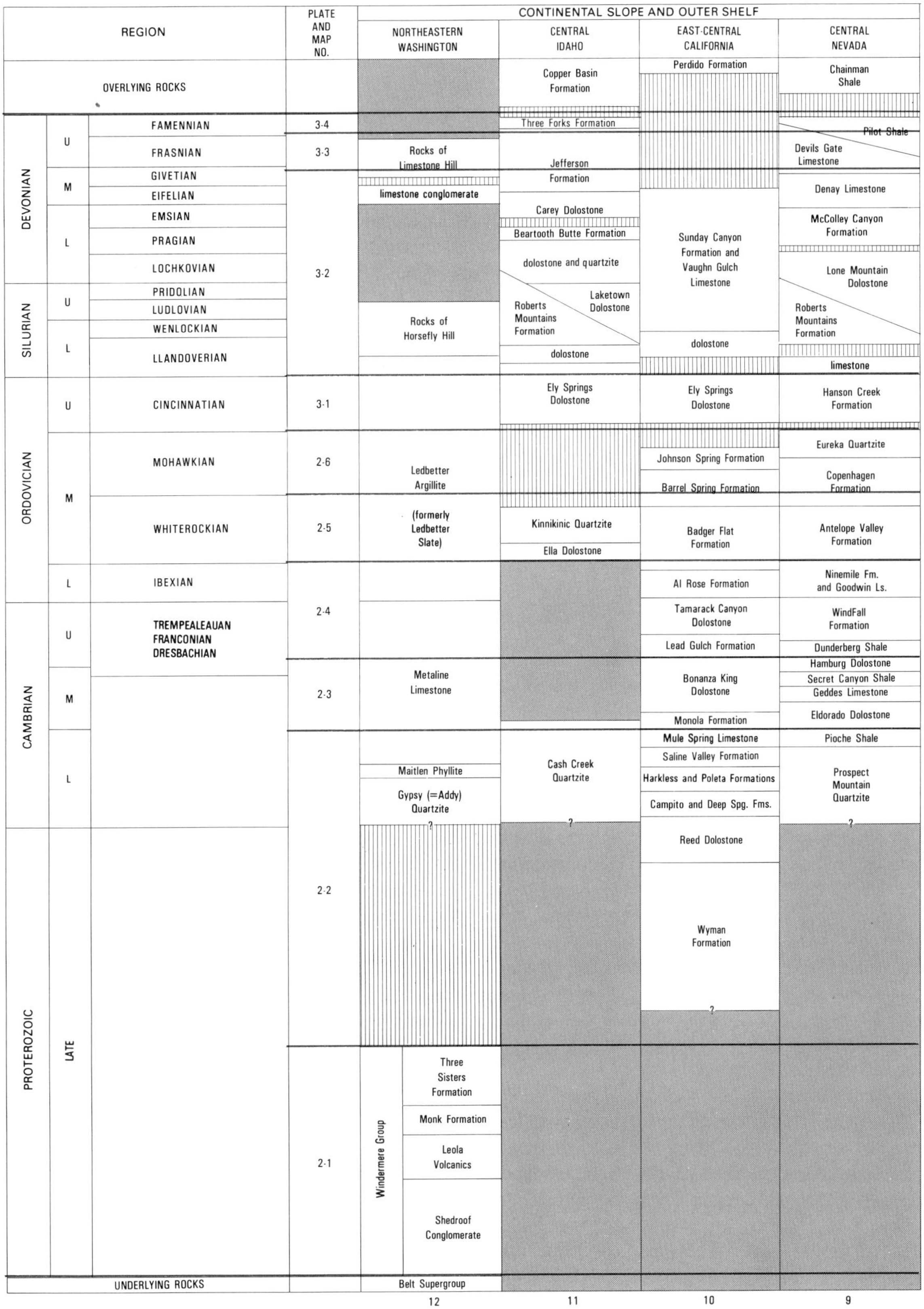

Figure 2. Correlation chart for some representative continental-shelf and cratonic-platform rocks of the North American margin.

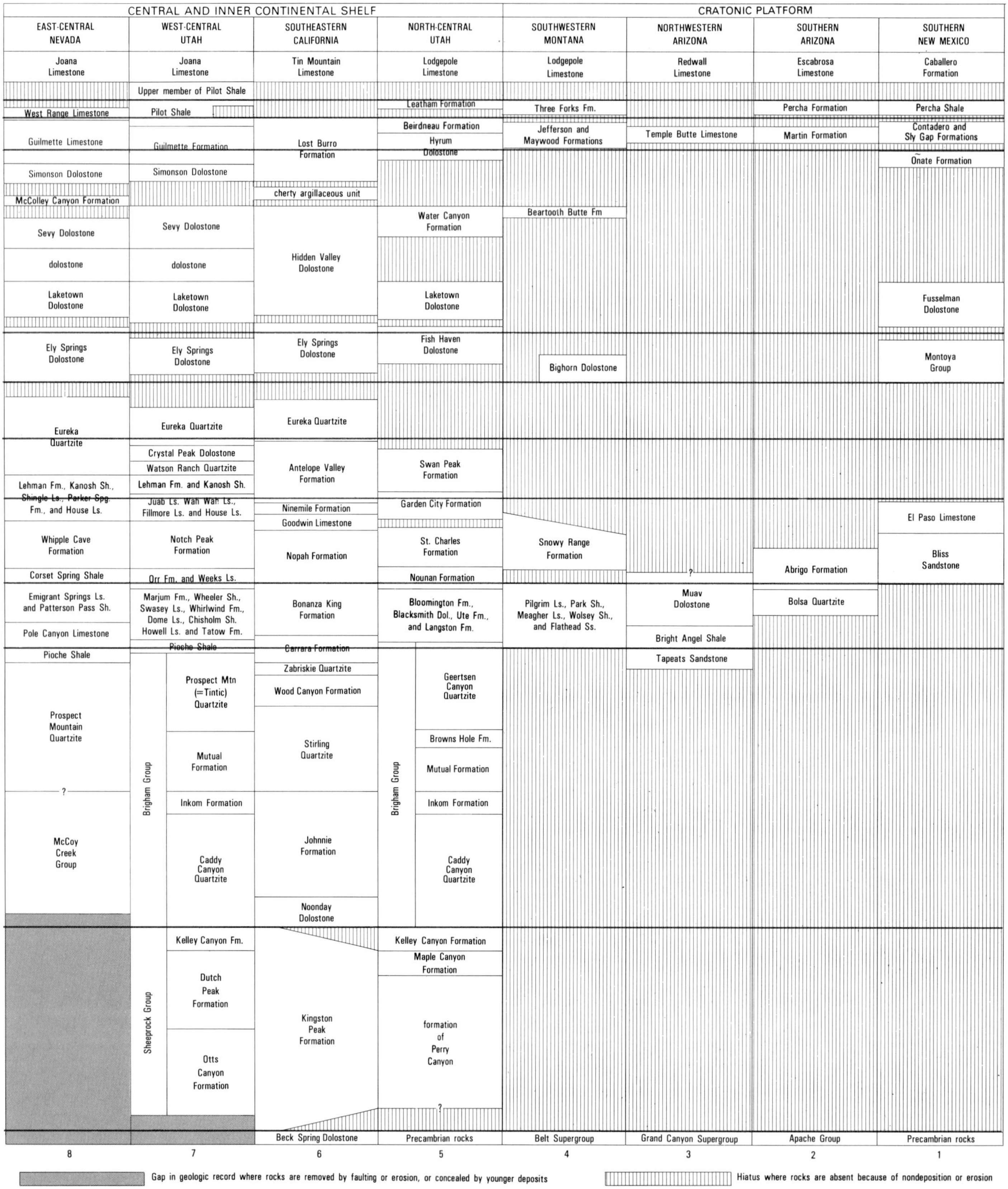
CENTRAL AND INNER CONTINENTAL SHELF
CRATONIC PLATFORM
EAST-CENTRAL NEVADA
Joana Limestone
West Range Limestone
Guilmette Limestone
Simonson Dolostone
McColley Canyon Formation
Sevy Dolostone
dolostone
Laketown Dolostone
Ely Springs Dolostone
Eureka Quartzite
Lehman Fm., Kanosh Sh., Shingle Ls., Parker Spg. Fm., and House Ls.
Whipple Cave Formation
Corset Spring Shale
Emigrant Springs Ls. and Patterson Pass Sh.
Pole Canyon Limestone
Pioche Shale
Prospect Mountain Quartzite
?
McCoy Creek Group
8
WEST-CENTRAL UTAH
Joana Limestone
Upper member of Pilot Shale
Pilot Shale
Guilmette Formation
Simonson Dolostone
Sevy Dolostone
dolostone
Laketown Dolostone
Ely Springs Dolostone
Eureka Quartzite
Crystal Peak Dolostone
Watson Ranch Quartzite
Lehman Fm. and Kanosh Sh.
Juab Ls. Wah Wah Ls., Fillmore Ls. and House Ls.
Notch Peak Formation
Orr Fm. and Weeks Ls.
Marjum Fm., Wheeler Sh., Swasey Ls., Whirlwind Fm., Dome Ls., Chisholm Sh. Howell Ls. and Tatow Fm.
Pioche Shale
Brigham Group
Prospect Mtn (=Tintic) Quartzite
Mutual Formation
Inkom Formation
Caddy Canyon Quartzite
Sheeprock Group
Kelley Canyon Fm.
Dutch Peak Formation
Otts Canyon Formation
7
SOUTHEASTERN CALIFORNIA
Tin Mountain Limestone
Lost Burro Formation
cherty argillaceous unit
Hidden Valley Dolostone
Ely Springs Dolostone
Eureka Quartzite
Antelope Valley Formation
Ninemile Formation
Goodwin Limestone
Nopah Formation
Bonanza King Formation
Carrara Formation
Zabriskie Quartzite
Wood Canyon Formation
Stirling Quartzite
Johnnie Formation
Noonday Dolostone
Kingston Peak Formation
Beck Spring Dolostone
6
NORTH-CENTRAL UTAH
Lodgepole Limestone
Leatham Formation
Beirdneau Formation
Hyrum Dolostone
Water Canyon Formation
Laketown Dolostone
Fish Haven Dolostone
Swan Peak Formation
Garden City Formation
St. Charles Formation
Nounan Formation
Bloomington Fm., Blacksmith Dol., Ute Fm., and Langston Fm.
Brigham Group
Geertsen Canyon Quartzite
Browns Hole Fm.
Mutual Formation
Inkom Formation
Caddy Canyon Quartzite
Kelley Canyon Formation
Maple Canyon Formation
formation of Perry Canyon
?
Precambrian rocks
5
SOUTHWESTERN MONTANA
Lodgepole Limestone
Three Forks Fm.
Jefferson and Maywood Formations
Beartooth Butte Fm
Bighorn Dolostone
Snowy Range Formation
Pilgrim Ls., Park Sh., Meagher Ls., Wolsey Sh., and Flathead Ss.
Belt Supergroup
4
NORTHWESTERN ARIZONA
Redwall Limestone
Temple Butte Limestone
?
Muav Dolostone
Bright Angel Shale
Tapeats Sandstone
Grand Canyon Supergroup
3
SOUTHERN ARIZONA
Escabrosa Limestone
Percha Formation
Martin Formation
Abrigo Formation
Bolsa Quartzite
Apache Group
2
SOUTHERN NEW MEXICO
Caballero Formation
Percha Shale
Contadero and Sly Gap Formations
Oñate Formation
Fusselman Dolostone
Montoya Group
El Paso Limestone
Bliss Sandstone
Precambrian rocks
1
Gap in geologic record where rocks are removed by faulting or erosion, or concealed by younger deposits
Hiatus where rocks are absent because of nondeposition or erosion

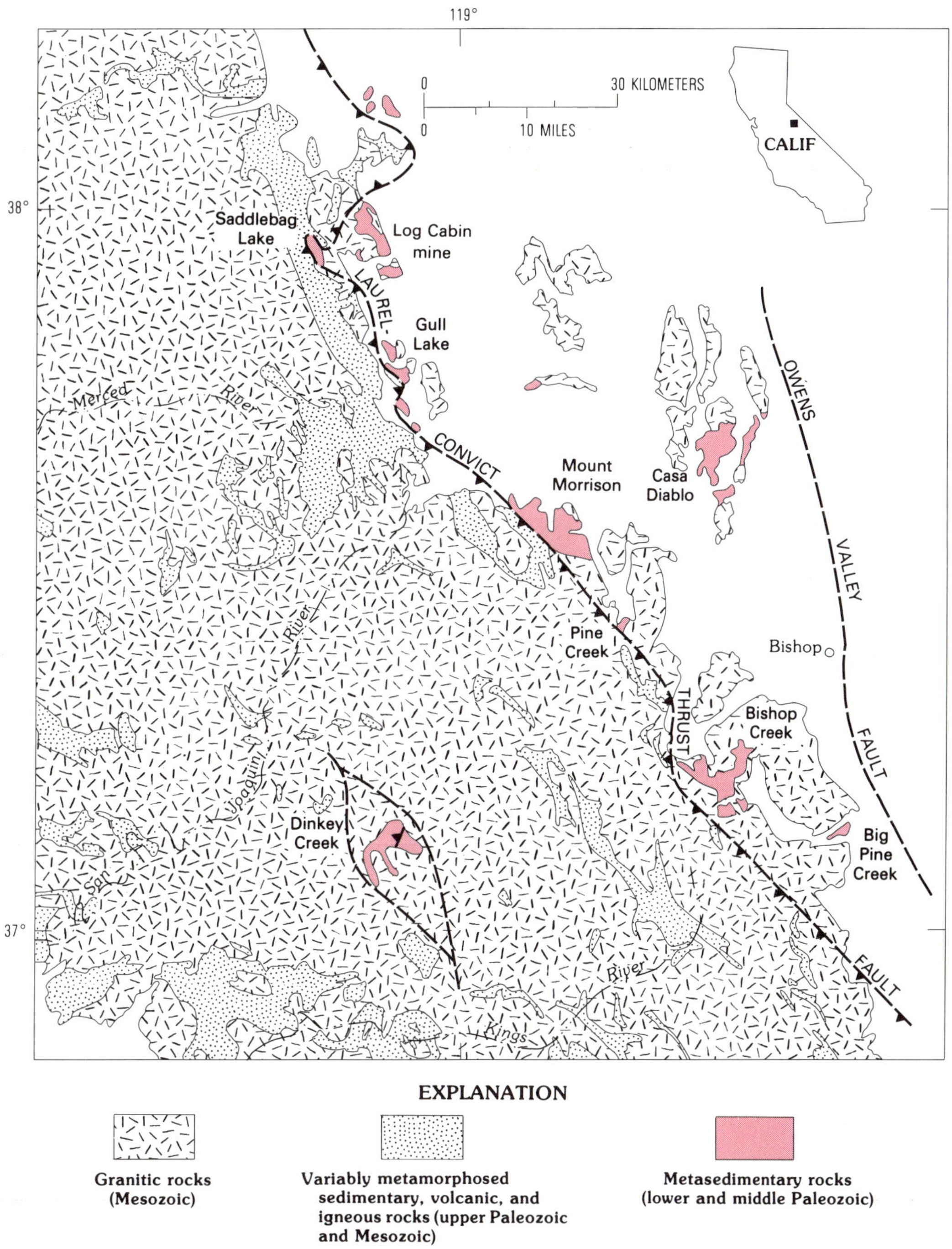

Figure 3. Distribution of roof pendants in the east-central Sierra Nevada that contain lower and middle Paleozoic rocks. Modified from Nokleberg (1983).

Cabin mine, Gull Lake, and Mount Morrison pendants; continental-slope and outer continental-shelf deposits in the Bishop Creek pendant; and continental-shelf deposits in the Big Pine Creek pendant (Nokleberg, 1983). Some pendants contain rocks similar to those in the Antler orogen of Nevada, but metamorphic and structural complexities preclude assignment to the Roberts Mountains allochthon at this time. Most of the metamorphosed, lower and middle Paleozoic rocks in the roof pendants in the east-central Sierra Nevada can be correlated with nonmetamorphosed and only slightly metamorphosed strata east of Owens Valley in California and Nevada (R. W. Kistler, USGS, personal communication, 1985).

REGIONAL STRUCTURAL SETTING

Forrest G. Poole

The late Proterozoic to Late Devonian geologic evolution and paleogeography of the western U.S. is related to continental-plate and oceanic-plate movements that resulted in numerous episodes of uplift and erosion and of subsidence and sedimentation. Important older (pre-Mississippian) and younger (post-Devonian) structural features are shown in Figure 1.

OLDER STRUCTURAL FEATURES

Several paleotectonic features (Fig. 1, shown in red) were in existence during late Proterozoic through Late Devonian time. Important uplifts and arches include the Uinta uplift–Tooele arch in northern Utah, of Precambrian to Devonian age (Hintze, 1988); the Lemhi or Salmon River arch in north-central Idaho, of Precambrian to Ordovician age (Ross, 1977; Stewart and Suczek, 1977; Ruppel, 1985); the Sneffels and Grenadier horsts and Uncompahgre uplift and associated faults in southwestern Colorado and southeastern Utah, of Cambrian and Devonian age (Baars, 1966); and the Central Montana uplift in central Montana, the ancestral Yellowstone Park uplift in northwestern Wyoming and southwestern Montana, the Southern Beaverhead Mountains (SBM) uplift in east-central Idaho, and the Defiance uplift in northeastern Arizona, all of Devonian age (Poole and others, 1967; Sandberg and Mapel, 1967; Sandberg and others, 1975). Cambrian and/or Ordovician alkalic and mafic intrusions (Fig. 1; Plates 2-2 to 2-4, 3-1) in Colorado, Idaho, New Mexico, and Utah occur along extensional fractures cutting the continental crust (Olson and others, 1977; Loring and Armstrong, 1980; Evans and Zartman, 1988; Hansen and others, 1982; Ritzma, 1983; Evans, 1984). Lower or Middle Devonian kimberlitic diatremes (Fig. 1; Plates 2-4 to 2-6) in north-central Colorado and southeastern Wyoming contain Cambrian to Silurian carbonate-rock xenoliths (Chronic and others, 1969).

One of the most conspicuous of the older structural features is the ancestral Uinta uplift–Tooele arch. This was an enduring west-trending prong of the Transcontinental basement arch that extended from Colorado through northern Utah to eastern Nevada (Fig. 1). The trend and antiquity of the Uinta uplift–Tooele arch probably are related to an Archean or Proterozoic discontinuity in the crust that formed along the southern margin of the Archean Wyoming craton. The uplift-arch influenced sedimentation on the continental shelf from Middle Cambrian to Late Devonian time (Plates 2-3 to 2-6, 3-1 to 3-3), and divided the pre–Late Devonian shelf of the Great Basin region into two depositional basins or troughs—the Ibex basin in southwestern Utah and southeastern Nevada and the Tristate (new name) basin in northwestern Utah, northeastern Nevada, and southern Idaho.

YOUNGER STRUCTURAL FEATURES

Among the younger structural features (Fig. 1, shown in black), the middle Paleozoic Roberts Mountains thrust system, related to the Antler orogeny, is the oldest. It displaced Devonian and older continental-rise and outer-shelf rocks of the continental margin toward the continental interior. In Nevada, Devonian and older rocks were moved eastward as much as 200 km on the upper plate of the low-angle Roberts Mountains thrust (Fig. 1). These allochthonous rocks now abut, or rest in thrust contact on, autochthonous or parautochthonous Devonian and older continental-slope and outer-shelf strata exposed in structural windows (Plate 3-6). Because the thrusting probably ended in Early Mississippian time, these structural features are discussed in detail in the following chapter; however, the stratigraphy of the Devonian and older rocks in the allochthon are discussed in this chapter under our section on "Rocks of the Antler orogen" because these rocks were deposited along the western margin of early Paleozoic North America and are part of the continental-margin depositional framework.

The Early Triassic Golconda thrust, related to the Sonoma orogen in Nevada and its possible extensions into California and Idaho, is the oldest regional thrust system that displaced Paleozoic eugeosynclinal rocks of the accreted terranes. Many younger Mesozoic thrust and strike-slip faults west of the Sonoma orogenic belt in California, Oregon, and Washington significantly displaced Paleozoic eugeosynclinal rocks of accreted terranes (Fig. 1). The faults shown in the Pacific states on Figure 1 and on Plates 2 and 3 are after Hamilton (1978).

The Sevier thrust system of Jurassic to Cretaceous age (Fig. 1) can be traced from southeastern California northward into Canada. The mapped and inferred traces of the leading edge or major thrust of this system, shown on Plates 2-1 to 2-6 and 3-1 to 3-5, are based on structural continuity along the thrust belt and on the late Proterozoic to Upper Devonian stratigraphic discontinuity across the leading edge or inferred major thrust. Late Proterozoic and Paleozoic continental-shelf rocks were moved eastward 50 to 100 km in Mesozoic and early Cenozoic time on low-angle thrusts of the Sevier thrust system and now abut, or rest in thrust contact upon, Paleozoic strata that occupy the central part of the area discussed in this chapter.

Several mapped and inferred right-lateral strike-slip fault zones, active in Mesozoic and Cenozoic time, have brought together upper Precambrian to middle Paleozoic rocks of different thicknesses and lithofacies (Plates 2-1 to 2-6, 3-1 to 3-5). Maximum right-lateral displacements of 100 km on the Furnace Creek fault (FCF), 25 km on the Stewart Valley fault (SVF), 65 km on the Las Vegas fault (LVF), and 50 km on the Lewis and Clark line (LCL) (Fig. 1) are demonstrated by bending and offsets in thickness and lithofacies trends of upper Precambrian to middle Paleozoic rocks. Maximum left-lateral displacement of 70 km on the Garlock fault (GF) is demonstrated by the offset of lithofacies and a dike swarm (Smith, 1962) between the El Paso Mountains

(EPM) on the north block and Pilot Knob Valley (PKV) on the south block (Fig. 1). Evidence for the inferred Snake River fault (SRF) in south-central Idaho (Fig. 1) can be demonstrated by offset of the Sevier thrust system across the Snake River Plain, and evidence for both left- and right-lateral movements of that fault can be demonstrated by significantly different amounts of offset of the thickness trends of Devonian and Mississippian rocks (Poole and Sandberg, 1977). Stratigraphic and structural data indicate major left-lateral movement on this fault of about 100 km in earliest Mississippian time, an additional 100 km of left-lateral movement in late Paleozoic time, and right-lateral movement of about 50 km in Cenozoic time, resulting in a net left-lateral offset of 150 km. Note, however, that the arrows on this fault (Plates 2-1 to 2-6, 3-1 to 3-5) show only the direction of latest movement and not the net offset. The discontinuity between depositional-basin axes and thickness trends (Plates 2-6, 3-1 to 3-3) suggests a counterclockwise relative rotation of about 60° north of the Snake River fault (Poole and Sandberg, 1977). Elsewhere in the region, disorientation of thickness and lithofacies trends across major strike-slip fault zones, such as the Furnace Creek–Death Valley fault system in southeastern California, indicates possible significant rotation. Thickness and facies trends of upper Precambrian to middle Paleozoic rocks commonly are curved or arcuate, especially near major strike-slip fault zones (Fig. 1).

The present-day easternmost distribution of eugeosynclinal rocks of the Roberts Mountains allochthon in Nevada and adjacent states (Fig. 1, Plate 3-5) is, in part, the result of reshuffling during Permian and Early Triassic compression, Jurassic and Cretaceous thrusting and strike-slip faulting, and early Tertiary transtensional faulting.

LATE PROTEROZOIC AND LOWER CAMBRIAN ROCKS

John H. Stewart

UNDERLYING ROCKS

Late Proterozoic to Devonian strata deposited along and inland from the North American continental margin in the western U.S. rest unconformably on structurally complex crystalline basement rocks consisting of gneiss, schist, quartzite, marble, greenstone, and granite ranging in age from about 2,500 to about 1,410 Ma, and on younger supracrustal sedimentary and volcanic rocks ranging in age from about 1,450 to 900 or 800 Ma (Fig. 2).

The younger supracrustal rocks, the Belt Supergroup (middle Proterozoic) and related rocks, are composed of argillite, shale, siltstone, and quartzite, locally abundant dolostone, limestone, conglomerate, and thin mafic lavas and gabbroic or diabase sills (1,150 to 1,100 Ma). The Belt Supergroup and Uinta Mountain Group (middle Proterozoic) are 20 and 7 km thick, respectively, and define deep epicratonic troughs. The middle and late Proterozoic Grand Canyon Supergroup (4 km thick) and the Precambrian Crystal Spring and Beck Spring Formations may also have been deposited in troughs, whereas the middle Proterozoic Apache Group and Troy Quartzite, judged from their thinness and wide distribution, may be shelf deposits. The deep epicratonic troughs containing the middle Proterozoic Belt Supergroup and related rocks contrast strongly with the laterally continuous miogeocline or miogeosyncline along the continental margin in late Proterozoic to Devonian time. Locally, between the earlier Precambrian rocks and the terrigenous detrital sequence at the base of the continuous continental-shelf sequences, significant diamictite and volcanic sequences have been found.

DIAMICTITE AND VOLCANIC SEQUENCE

Late Proterozoic and Lower Cambrian rocks in the western United States are divided into a diamictite and volcanic sequence and an overlying terrigenous detrital sequence (Stewart and Suczek, 1977). Each of these sequences is lithologically and tectonically distinct and is described separately.

The diamictite and volcanic sequence contains a variety of rock types that commonly vary abruptly in facies and thickness. The most distinctive rock is diamictite consisting of well-rounded to angular clasts, locally as large as 1 m, set in a clay, silt, and fine-sand matrix. The clasts are of both extrabasinal and intrabasinal origin. The diamictite occurs as massive beds, graded beds, or laminated siltstone with isolated clasts. At some locations, striated and faceted clasts, dropstones with "splash-up" borders, isolated sand and gravel clots, and a striated pavement, all features indicative of a glacial origin, have been noted (Blick, 1979, 1981; Ojakangas and Matsch, 1980; Crittenden and others, 1983). The diamictite is interpreted to have formed in a variety of ways, including deposition as terrestrial or subaqueous lodgement till or flowtill, as debris dropped from a floating ice sheet, and as proximal turbidites in relatively deep water.

Diamictite forms a relatively small to moderately large part of the diamictite and volcanic sequence. It occurs interstratified with phyllite or argillite, graded sandstone, quartzite, pebble to cobble conglomerate, carbonate rocks, and mafic volcanic rocks. The sedimentary rocks associated with the diamictite have been interpreted to include relatively deep as well as shallow marine deposits. The mafic volcanic rocks associated with the diamictite are altered, and their original composition is difficult to determine, although they have generally been considered to be tholeiitic basalts (Stewart, 1972; Miller and Clark, 1975; Yates, 1976; Hammond, 1983; Devlin and others, 1985; Harper and Link, 1985). Pillow basalts are recognized locally.

The diamictite and volcanic sequence occurs discontinuously in western North America (Fig. 2; Plate 2-1). It is widely exposed in northeastern Washington and adjacent parts of British Columbia (Aalto, 1971, 1981; Reesor, 1973; Glover and Price, 1976; Miller, 1983) where it forms the Windermere Group (Supergroup of Canada) and includes such units as the Shedroof Conglomerate, Leola Volcanics, Monk Formation, and Three

Sisters Formation in Washington; in southern Idaho (Link, 1981, 1983) and northwestern Utah (Blick, 1979, 1981; Christie-Blick, 1982, 1983; Crittenden and others, 1983) it forms the Pocatello Formation of southern Idaho and the formation of Perry Canyon, the Mineral Fork Formation, the Sheeprock Group, and the Horse Canyon Formation of northwestern Utah; and in the Death Valley region of eastern California (Wright and others, 1974; Miller and others, 1981; Miller, 1985; Walker and others, 1986) it forms the Kingston Peak Formation.

In some areas, the diamictite and volcanic sequence was deposited in locally deep troughs that were formed in part by syndepositional faulting. In the Death Valley area of eastern California, the Kingston Peak Formation, which varies in thickness from a few meters to more than 3 km over a horizontal distance of less than 4 km, contains abundant coarse detritus and locally intact slabs of older rock as much as 1.5 km in strike length (Troxel and others, 1977, 1987). The abrupt changes in thickness and the abundance of coarse detritus suggest deposition in a fault-bounded trough or rift valley (Wright and others, 1974; Stewart and Suczek, 1977). Wright and others (1974) ascribed emplacement of the gigantic slabs of older rocks to sedimentary processes, whereas Walker and others (1986) suggested tectonic emplacement during transtensional events associated with syndepositional strike-slip faulting. In southern Idaho, Link (1983) described abrupt local facies changes and an abundance of clasts derived from exposed and uplifted parts of the diamictite and volcanic sequence, and he suggested that syndepositional tectonism is responsible for these relations. In northeastern Washington, F. K. Miller (1983) proposed a deep basin bounded by a high-angle fault on one side to account for abrupt changes in facies and thickness in the Windermere Group. Deposition of the diamictite and volcanic sequence in fault-bounded basins has also been demonstrated in other parts of western North America (Christie-Blick and others, 1980; Eisbacher, 1981).

The age of the diamictite and volcanic sequence is poorly constrained. Miller and others (1973) indicated K-Ar isotopic ages of from 918 to 827 Ma for whole rocks and mineral separates from greenstone in the sequence in northeastern Washington. More recent work suggests that the K-Ar system may be disturbed and that these dates are too old (Devlin and others, 1985); a Sm-Nd whole-rock mineral age of about 760 Ma has been obtained for those rocks (Devlin and others, 1988). In northwest Canada, diabase dikes and sills that pre-date diamictite deposits presumably correlative with the diamictite deposits in the western U.S. are approximately 770 m.y. old on the basis of Rb-Sr isochron dates (Armstrong and others, 1982). In British Columbia, Canada, gneissic granite that unconformably underlies diamictite-bearing rocks of the Windermere Supergroup dates from about 730 Ma on the basis of U-Pb analysis of zircons (Parrish and Armstrong, 1983; Evenchick and others, 1984). Paleontological dates based on identifications of microscopic planktonic algae (acritarchs) from interbeds within the diamictite indicate a Vendian age (650 to 570 Ma) in Utah and southwestern Alberta (Moorman, 1974; Knoll and others, 1981).

TERRIGENOUS DETRITAL SEQUENCE

The terrigenous detrital sequence (late Proterozoic and Lower Cambrian) at the base of the Paleozoic miogeosynclinal assemblage consists dominantly of quartzite and interstratified argillite or phyllite. The quartzite characteristically occurs as thick cliff-forming units that dominate the total landscape. In some regions the sequence rests unconformably on the diamictite and volcanic sequence, in other regions it appears to be gradational with this sequence, and in still other areas it rests on sedimentary, igneous, or metamorphic rocks older than the diamictite and volcanic sequence. In much of the western U.S., the terrigenous detrital sequence is conformably overlain by interstratified Cambrian shale, siltstone, phyllite, and limestone that grade upward into thick units of limestone and dolostone.

The terrigenous detrital sequence is widely distributed in the western U.S. (Fig. 2; Plate 2-2). It occurs in northeastern Washington (Miller and Clark, 1975; F. K. Miller, 1983), central and southern Idaho (Hobbs and others, 1968; Crittenden and others, 1971; Oriel and Armstrong, 1971; Ruppel and others, 1975; Lindsey, 1982), western Utah (Crittenden and others, 1971; Blick, 1979; Christie-Blick, 1982), Nevada (Stewart, 1970, 1974), eastern California (Stewart, 1970), and southern California (Cameron, 1982). The sequence probably was originally continuous along the western margin of North America (Fig. 4), although rocks of the sequence have not been recognized in southern Washington and northern Idaho, perhaps because they were removed by rifting sometime in the Phanerozoic.

The quartzite in the terrigenous detrital sequence consists for the most part of light gray, pinkish gray, or pale red, fine- to medium-grained, quartz-rich rock that is either cross stratified, irregularly or wavy laminated, or evenly parallel laminated. Granule to pebble conglomerate with clasts of quartz and quartzite, and locally jasper, occurs locally. Very fine to fine-grained, parallel-laminated quartzite is also common. The quartzite occurs interstratified with reddish brown or greenish gray laminated siltstone, argillite, or phyllite. The siltstone, argillite, or phyllite occurs either as thin laminae or irregular wavy laminae or lenses interstratified with the quartzite or as thick units, some several hundreds of meters thick, between major quartzite units. The terrigenous detrital sequence also contains varied amounts of limestone and dolostone. In places, such rocks are sparse; in other places, carbonate rock, particularly dolostone, occurs in units as thick as about 600 m.

Rocks of the terrigenous detrital sequence are interpreted to have been deposited in shallow-marine environments, probably mainly on a tidal-dominated shelf (Stewart, 1970; Barnes and Klein, 1975; Klein, 1975; Benmore, 1978; Diehl, 1979) and in braided fluvial systems (Diehl, 1979; Mount, 1982; Christie-Blick, 1984; Schneck and McCollum, 1985). Shallow-marine tidal-dominated shelf deposits are indicated by current directions that locally show a bimodal and bipolar character, suggesting the ebb and flow of tidal currents (Fig. 4). Some units may be subtidal sand bodies developed during major storms (Mount, 1982).

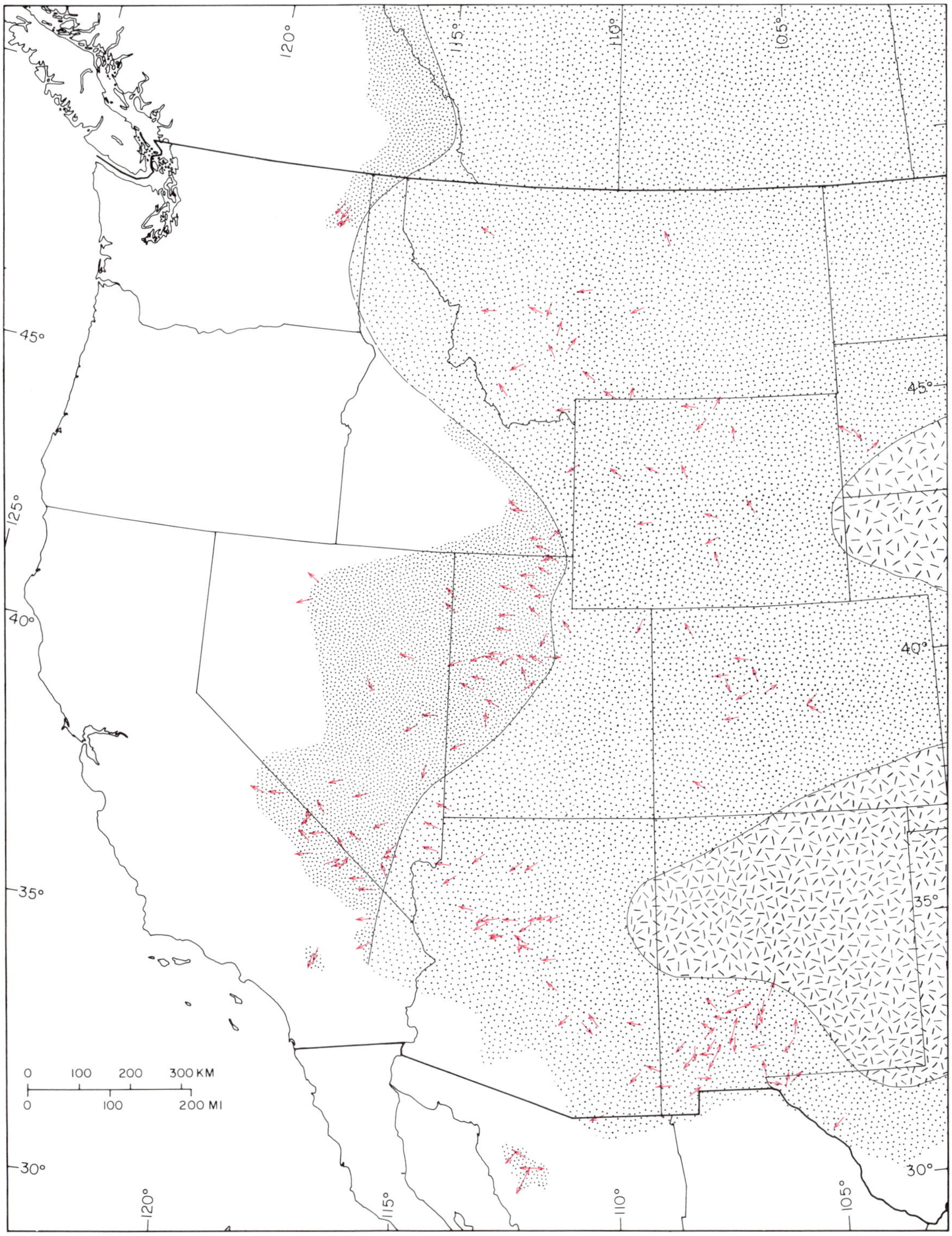

EXPLANATION

Current direction

Bimodal-bipolar current directions

Late Proterozoic and Cambrian rocks absent

Cambrian cratonal rocks, partly restored. Late Proterozoic rocks probably not present

Miogeoclinal Late Proterozoic and Cambrian rocks

Figure 4. Late Proterozoic and Early Cambrian paleogeography. Paleocurrent data mainly from McKee (1940), Wilson (1962), Seeland (1968, 1969), Reynolds (1968), Stewart (1970), Hereford (1977), Thompson and Potter (1981), and Stewart and others (1984).

Peritidal oolite lime sands occur locally (Moore, 1976). Braided-stream deposits are indicated by the presence of shallow channels containing conglomerate, by features that suggest transverse and longitudinal sand bars and overbank fine-grained deposits, and by the widespread distribution of coarse-grained sandstone and conglomeratic sandstone.

Stratigraphic units in the terrigenous detrital sequence are generally lithologically similar and persistent along sedimentary strike, generally in a north-south direction, and they change facies and thickness across sedimentary strike, generally in an east-west direction. The persistance of some of the units is truly remarkable. For example, the Zabriskie Quartzite (Lower Cambrian) is recognized throughout the southern Great Basin region of Nevada and California, in the Victorville–San Bernardino Mountains region of southern California, and in the Caborca region of Sonora, Mexico, where it is called the Proveedora Quartzite (Stewart and others, 1984). Across sedimentary strike, units commonly become more silty and carbonate rich to the west. This relation is particularly evident in the southern Great Basin where units such as the Zabriskie Quartzite change facies from quartzite in the east to interstratified quartzite and siltstone, and finally to almost entirely siltstone (Harkless Formation) in the west. Shallow-marine carbonate units, such as the Reed Dolostone of eastern California and western Nevada, are thickest to the west and lens out eastward within largely terrigenous units.

The terrigenous detrital sequence ranges in thickness from a zero edge in eastern areas to more than 6,000 m locally in western areas (Plate 2-2).

The upper part of the terrigenous detrital sequence contains sparse, but locally abundant, Early Cambrian biotas consisting of trilobites, brachiopods, mollusks and small shelly fossils, sponges, archeocyathids, stromatolites, algae, algal-like microorganisms, and trace fossils (Cloud and Nelson, 1966; Stewart and others, 1984).

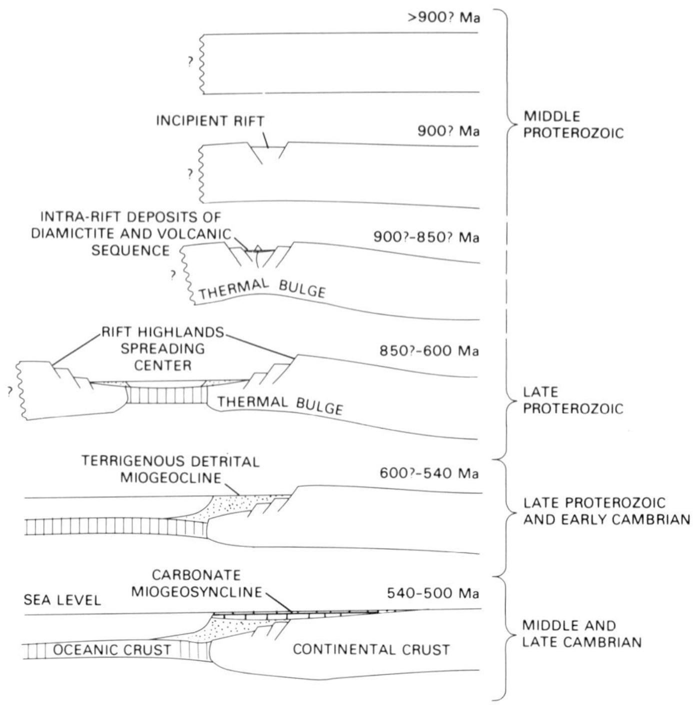

Figure 5. Tectonic model of late Proterozoic and Cambrian development of the western U.S. Modified from Stewart and Suczek (1977).

TECTONIC MODELS

The terrigenous detrital sequence in the western U.S. forms a well-defined, westward-thickening miogeoclinal sedimentary wedge (Fig. 2; Plate 2-2). The extent and shape of this sedimentary wedge differs significantly from the depositional patterns of both earlier Precambrian rocks, such as the Belt Supergroup and Uinta Mountain Group, that accumulated in deep epicratonic troughs and the diamictite and volcanic sequence that appears to have accumulated in fault-bounded basins. This change in depositional patterns is thought to represent a major change in the tectonic setting of western North America related to rifting that created a new, or at least reshaped, continental margin along western North America.

The diamictite and volcanic sequence may represent deposits that accumulated in rift-valley basins that formed prior to the main stage of rifting (Fig. 5). Such an interpretation is supported by the discontinuous geographic extent of the diamictite and volcanic sequence in western North America, by the abundance of presumably rift-related volcanic rocks (tholeiitic? basalt) in the sequence, and by the indication of significant faulting during deposition of the sequence. In British Columbia, the term "East Kootenay orogeny" has been used to describe tectonic events prior to deposition of the diamictite and volcanic sequence (Windermere Supergroup) in Canada. Recent work, however, indicates that the term "East Kootenay orogeny" is better restricted to a 1,350- to 1,300-Ma event that apparently terminated deposition of the Belt Supergroup and correlative Purcell Supergroup and involved folding, regional metamorphism, and granitic intrusion (McMechan and Price, 1982). The term "Goat River orogeny" was used by McMechan and Price (1982) in British Columbia, and "Hayhook orogeny" (Young and others, 1979) in the northern Canadian Cordillera for the event that occurred during deposition of the diamictite and volcanic sequence and involved uplift, block faulting, and low-grade regional metamorphism. The Goat River orogeny is considered to be related to the rift-valley stage of late Proterozoic rifting. In the western U.S., faulting synchronous with deposition of the diamictite and volcanic sequence has been described in northeastern Washington (Miller, 1983), southern Idaho (Link, 1983), and eastern California (Wright and others, 1974; Troxel and others, 1977; Walker and others, 1986). In eastern California, Walker and others (1986) have proposed tectonic events including folding, development of cleavage, and later low-angle normal faults during the

time when the diamictite and volcanic sequence was being deposited. Although their proposals have been challenged by Troxel and others (1987), Walker and others (1986) related these events to transtensional deformation associated with strike-slip faulting.

Timing of this initial stage of rifting is poorly constrained, but, on the basis of dates mentioned earlier, the diamictite and volcanic sequence and the rift-valley stage of rifting are probably younger than 730 Ma.

The main stage of late Proterozoic rifting is interpreted to have started shortly before deposition of the terrigenous detrital sequence, which is considered to be the initial part of the miogeoclinal-miogeosynclinal succession related to a newly formed continental margin (Fig. 5). Analysis of subsidence rates (Stewart and Suczek, 1977; Armin and Mayer, 1983; Bond and others, 1983; Christie-Blick, 1984; Bond and Kominz, 1984; Bond and others, 1985), after removal of isostatic effects of sediment and water, suggests that subsidence was related to initial stretching of the crust and to thermal contraction. Subsidence curves suggest that the main stage of rifting started at about 600 Ma. This age is compatible with the 600-Ma age (J. B. Saleeby, California Institute of Technology, written communication, 1987) of a plagiogranite in ophiolitic rocks in the northern Sierra Nevada of California and with the 565- to 570-Ma age (Wallin and others, 1988) of tonalite from the Trinity ultramafic sheet in the Klamath Mountains of northern California. The ophiolitic rocks may have formed at an oceanic spreading ridge related to the main stage of rifting, although the original position of these rocks relative to Cordilleran miogeoclinal rocks is uncertain.

The apparent long span of time between presumed initial rifting and the main stage may be due to a prolonged episode of tectonic activity (Fig. 5). Rifting may have taken place in two distinct breakup events (Bond and Kominz, 1984), one during deposition of the diamictite and volcanic sequence and one in the Early Cambrian during deposition of the terrigenous detrital sequence, or the breakup events may have been protracted and episodic (Devlin and Bond, 1988). The initial rifting event, as proposed by Walker and others (1986), could be related to transtensional tectonics associated with strike-slip faulting and thus may not be strictly part of the rifting events that formed the Cordilleran continental margin.

The source terrane of the terrigenous detrital sequence may have been the initial rift bulge (uplifted area) that was created by thermal expansion during continental rifting (Fig. 5). Such uplift is well known along the Afro-Arabian rift system (Baker and others, 1972; Kinsman, 1975; Lowell and others, 1975). Cessation of deposition of terrigenous detritus in the miogeocline may be related to the destruction of the bulge by thermal contraction due to cooling, and by erosion. After the bulge had been lowered to a critical level, the Early Cambrian ocean transgressed eastward into cratonic areas. As a consequence, coarse detrital material was trapped in more easterly coastal areas on the craton, and the Middle Cambrian miogeosynclinal-shelf area was an area of carbonate deposition relatively free of terrigenous detritus.

MIDDLE CAMBRIAN TO LOWER ORDOVICIAN ROCKS

Allison R. Palmer and Lehi F. Hintze

Rocks of Middle Cambrian to Early Ordovician age are widespread throughout the U.S. Cordilleran region east of the Sonoma orogenic belt; they record the initial development of a continental-margin carbonate platform (Plates 2-3 and 2-4). Regional summaries of the Cambrian or Ordovician parts of this interval have been made by Palmer (1971), Lochman-Balk (1971, 1972), Hayes (1975), Ross (1976, 1977), and Stewart and Suczek (1977). Data points for analyses in this section of the chapter are shown on Plates 2-3 and 2-4. These points are sufficiently numerous and scattered to provide reasonably good controls on the major patterns of facies distribution and on the regional isopachs and to permit definition of the southwestern part of early Paleozoic North America. At many of the locations, complete or almost complete sections of lower Paleozoic rocks document the shifting of depositional environments across those locations through a time span of perhaps 50 m.y. The facies patterns on Plates 2-3 and 2-4 are representative of specific times within the isopached interval and were chosen as typical examples. Details of some of the facies boundaries would shift if other specific time intervals had been chosen.

The stratigraphic and tectonic history of the region can be described in terms of two stratigraphic packages, one of latest Early Cambrian through early Late Cambrian age and the other of medial Late Cambrian to Early Ordovician age, and three tectonic belts that more or less parallel the regional north-south structural grain.

A regressive event, believed to represent a eustatic sea-level fall in mid-to-late *Bonnia-Olenellus* Zone (late Early Cambrian) time (Palmer, 1981a), marks the termination of the dominantly siliciclastic phase of early Paleozoic sedimentation described by Stewart in the previous section of this chapter. The sedimentary record of this phase is confined to the region west of the Sevier orogenic front. Subsequent sea-level rise, probably coupled with continued continental-margin subsidence related to lithospheric cooling (Bond and Kominz, 1984), brought Paleozoic marine sedimentation eastward onto the cratonic platform for the first time. By early Late Cambrian time and continuing with one brief regional regressive event (see below) into Middle Ordovician time, the present region of the U.S. Cordillera was almost completely submerged.

The sedimentary record of this transgression is dominated by the development of a widespread and persistent carbonate platform. Although the carbonate platform straddles the boundary or hinge line between the cratonic platform and the continental shelf or miogeosyncline, this important tectonic boundary is not reflected by the paleogeography (Plates 2-3 and 2-4). It is reflected in a general way, however, by the thickness trends.

In early Late Cambrian time, the Cordilleran marine record was punctuated by a regressive event, perhaps also the result of a eustatic sea-level fall (Palmer, 1981a). This event is reflected by various sedimentary features in almost all Cambrian sections of the Cordilleran region, and it serves as the boundary between the two stratigraphic packages distinguished here. A similar regressive event reflected by the Middle Ordovician Eureka and Swan Peak Quartzites (Fig. 2) marks the end of the time interval discussed in the following section.

Regionally, the Antler and Sevier orogenic belts serve as approximate boundaries of areas of contrasting lower Paleozoic sequences or stratigraphic packages. Sediments of the late Precambrian to mid *Bonnia-Olenellus* Zone, predominantly siliciclastic sequence described earlier by Stewart, were not deposited east of the Sevier orogenic belt. The Dresbachian/Franconian disconformity that marks the boundary between the two stratigraphic packages distinguished here can be identified in all sections east of the Sevier orogenic belt that have appropriate paleontologic control, and many sections are truncated by later Ordovician or by Devonian erosional events such that the record of the younger package, and sometimes even the older package, is incomplete (Plates 2-3 and 2-4).

Most lower Paleozoic sections lack any major regional sedimentary discontinuities between the Antler and Sevier orogenic belts. These sections are dominated by shallow-water continental-shelf carbonate facies in the eastern part and by thin-bedded, deeper-water carbonate and fine siliciclastic facies in the western part. The seaward margin of the carbonate shelf was fairly sharply defined and sinuous (Plate 2-3 and 2-4) and retained its general regional position throughout most of the early Paleozoic. In contrast, the inner margin of the carbonate shelf was much more mobile. It moved gradually eastward through time but frequently retreated westward more or less synchronously from southern Canada to northern Mexico. During regressions the predominantly carbonate succession was punctuated by widespread shaly or silty marker units that appear abruptly above a particular carbonate unit and grade gradually upward into an overlying carbonate unit. The result is siliciclastic-carbonate couples of various dimensions. The larger couples, commonly a few hundred meters thick, constitute Grand Cycles (Aitken, 1966, 1978). The synchroneity of many of the Grand Cycle tops in the U.S. Cordillera and adjacent parts of Canada and Mexico (Palmer, 1981b), considered with the absence of the Grand Cycles on other margins of Cambrian North America, suggests that many of the tops may reflect differential subsidence of the southwestern quadrant of the North American margin rather than eustatic events (Palmer and Halley, 1979). However, eustatic events may be recorded by abrupt shifts of the outer-shelf facies over the carbonate-platform facies at several times during the Cambrian and Early Ordovician. The tectonic and/or eustatic controls on these facies shifts need further study.

West of the Antler orogenic belt, most sections are dominated by thin-bedded, deeper-marine carbonate and fine siliciclastic facies. Most rocks in this region are moderately to strongly deformed and sparsely fossiliferous so that thicknesses and detailed correlations with sections in the eastern regions are difficult to determine.

Analyses of depositional environments have been made for parts of the carbonate-platform facies (Kepper, 1972, 1976; Ross, 1972; Brady and Rowell, 1976; Koepnick, 1976; Lohmann, 1976; Halley *in* Palmer and Halley, 1979) and for sedimentary sequences interpreted to represent continental-slope, continental-rise, or deep-sea fan deposits (Ketner, 1969; Cook and Taylor, 1977; Rowell and others, 1979). These analyses demonstrate the presence of intertidal to supratidal sedimentation in the shelf interior, islands and significant algal buildups along the oceanward margin of the carbonate shelf, and possible remnants of the depositional margin of the Cambrian continental shelf.

The synthesis of isopach and facies information reveals a major unresolved paleogeographic problem west of the Sevier orogenic front and north of the Great Basin in Idaho. Crustal shortening along the Sevier orogenic front is well defined and seems to increase northward as illustrated by the data on Plate 2-3. This tectonic shortening moved outer-shelf facies close to the Sevier orogenic front in southeastern Idaho. A major paleogeographic anomaly between northern and southeastern Idaho possibly can be explained by dislocation and sinistral rotation across the Snake River strike-slip fault zone. Available evidence indicates no northward continuation of the Tristate basin (Northern Utah basin of Hintze, 1988); instead, Middle Cambrian arkosic sandstones in the Bloomington Formation in the Pocatello area and central Idaho are contemporaneous with basinal or outer carbonate-shelf rocks of southeastern Idaho. Upper Cambrian arkosic sandstones of the Worm Creek Quartzite Member of the St. Charles Formation thicken toward the northwest through Pocatello. Furthermore, the Upper Cambrian Harmony Formation, a coarse arkosic turbiditic sandstone containing interbedded limestone and redeposited North American Late Cambrian trilobites, is present in north-central Nevada (Plate 3-6). The source of the arkosic sediment is believed to be a persistent Cambrian highland, the Lemhi or Salmon River arch (Armstrong, 1975; Stewart and Suczek, 1977), within the miogeosyncline in central and eastern Idaho (Palmer, 1971; Rowell and others, 1979). No data are available to explain the nature of the paleogeographic connection between the north-trending carbonate-shelf/outer-shelf edge of southeastern Idaho and the south-trending carbonate-shelf/outer-shelf edge in northeastern Washington and adjacent parts of northern Idaho and British Columbia (Plate 2-3). Resolution of the paleogeographic anomalies indicated by the Lemhi arch and by the lack of continuity of the carbonate-shelf margin between southern and northern Idaho has an important bearing on the tectonic history of the northern sector of the U.S. Cordillera.

The isopach data suggest that the Middle Cambrian to Early Ordovician Cordilleran continental margin had a series of marginal basins analogous to those that developed along the Atlantic continental margin of North America in Mesozoic and Cenozoic time. In the southern Great Basin region, isopachs of Upper

Cambrian to Lower Ordovician rocks clearly define a shelf basin, the Ibex basin (Hintze, 1988), in the general geographic position of the House Range embayment (Robison, 1960) of Middle and early Late Cambrian age (Plates 2-3 and 2-4). Interpretation of the isopachs of the older interval is ambiguous in the southern Nevada and southeastern California continental margin; however, an interpretation that is consistent with the presence of a shelf basin, based on the isopachs of the younger interval, shows an "arch," analogous to the Tooele arch in the northern Great Basin, projecting westward across Las Vegas, Nevada. A southward projection of the House Range embayment separates the thinner deposits over this "arch" from similar thin deposits in the Inyo Mountains of east-central California. Westward thinning of the southern deposits toward the Inyo Mountains, which is recorded on the Middle Cambrian through early Late Cambrian isopach map (Plate 2-3), suggests that more control might show the area of the House Range embayment to be a discrete shelf basin. North of the Tooele arch in northwestern Utah, shelf basins are indicated by isopachs of both of the stratigraphic packages discussed here (Plates 2-3 and 2-4), and an embayment similar to the House Range embayment is indicated for late Middle to early Late Cambrian time (Plate 2-3).

Contemporaneous tectonic activity is consistent with the record of an evolving continental margin. In the central sector, between the Antler and Sevier orogenic belts, where most Cambrian sections preserve a record of nearly continuous marine sedimentation, areas of thick sediment accumulation were areas of greater subsidence. Stratigraphic analyses for two areas along the southern margin of the House Range embayment suggest that this may have been a fault-controlled margin (Kepper, 1981; Rees, 1986). East of the Sevier orogenic belt, subsurface studies by Baars (1966) document an area of normal faulting beneath the Paradox basin at the margin of the Middle Cambrian sea.

Evidence for volcanic activity is limited to ash-fall tuffs in uppermost Lower Cambrian and lower Middle Cambrian rocks of the cratonic-platform succession in and just west of the Grand Canyon and in uppermost Cambrian miogeosynclinal carbonate rocks of the House Range in western Utah (F. G. Poole, personal communication, 1984). Two ash-fall tuffs in the western Grand Canyon have been dated by the fission-track method (C. W. Naeser, USGS, personal communication, 1984), as 565 ± 32 Ma (latest Early Cambrian) and 535 ± 30 Ma (early Middle Cambrian). Alkalic plutons and alkalic and mafic dikes in Colorado, Idaho, New Mexico, and Utah have been dated as Cambrian and/or Ordovician (Olson and others, 1977; Loring and Armstrong, 1980; Hansen and others, 1982; Ritzma, 1983; Evans, 1984; Evans and Clemons, 1988; Evans and Zartman, 1988), and some of these intrusive events may represent parts of an igneous system that produced the ash-fall tuffs.

Middle Cambrian trilobites of open-shelf biofacies have been discovered in clasts in Pennsylvanian conglomerates derived from a nearby southern source in the Marathon region of western Texas (Palmer and others, 1984), and in miogeosynclinal shallow-subtidal dark limestones near Arivechi in eastern Sonora, about 200 km southeast of the miogeosynclinal Cambrian exposures near Caborca in northwestern Sonora (E. Almazán-Vazquez, Instituto de Geología, personal communication, 1984; F. G. Poole, personal communication, 1985). These new localities have helped define the southern and southwestern margins of Cambrian North America.

MIDDLE ORDOVICIAN—WHITEROCKIAN AND MOHAWKIAN ROCKS

Reuben J. Ross, Jr., Lehi F. Hintze, and Forrest G. Poole

Middle Ordovician Whiterockian and Mohawkian rocks are widespread on the continental shelf that bordered the Ordovician North American craton (Fig. 2). The upward-shallowing limestone and subordinate limy mudstone of the Lower Ordovician Ibexian Series were buried by a procession of subtidal to intertidal and supratidal carbonate facies prograding westerly and northwesterly. These facies, in turn, were buried by late Whiterockian and early Mohawkian quartz sands prograding to the southwest and west (Ross and others, 1989). The thickness and lithofacies maps (Plates 2-5 and 2-6) show the present distribution that has been modified by post-depositional erosion, faulting, and cover by younger deposits. The Tooele arch in Utah and Nevada separated the Ibex and Tristate basins on the continental shelf. Figure 6 shows idealized stratigraphic relations of the principal Whiterockian and Mohawkian units: the Juab Limestone, Kanosh Shale, Lehman Formation, Watson Ranch Quartzite, Crystal Peak Dolostone, and Eureka Quartzite in Utah; the Antelope Valley Limestone and Eureka Quartzite in southern Nevada, and facies equivalents of these units in southeastern California. From the north and northeast, the lowest quartz sandstones of the Kinnikinic Quartzite of central Idaho and of the Swan Peak Formation of southeastern Idaho and adjacent Utah intertongue with coeval carbonate rocks over which they prograde southwestward.

The Whiterockian Series in the Basin and Range province is composed predominantly of upward-shallowing, subtidal to supratidal carbonate rocks. The Whiterockian carbonate shelf of western Utah, Nevada, and southern California developed during a single offlap-onlap cycle, 12 m.y. in duration, while most of the North American continent was exposed to subaerial weathering (Ross and others, 1989). The shallow-bank to tidal-flat deposits in aggregate are as thick as 300 m. Enclosed within them in west-central Utah and east-central Nevada in the Ibex basin is an extensive shelf-basin deposit of graptolitic black shale (Kanosh Shale) interbedded with thin limestone beds rich in oncolites (*Girvanella*), algal biscuits (*Receptaculites*), and trilobites, brachiopods, and ostracodes, all of which occur in the limestone-bank deposits to the south and west, and a great variety of calathid algae and sponges. The limestone banks compose the middle member of the Antelope Valley Limestone in southern and central Nevada (Plate 2-5). The lowest Whiterockian units

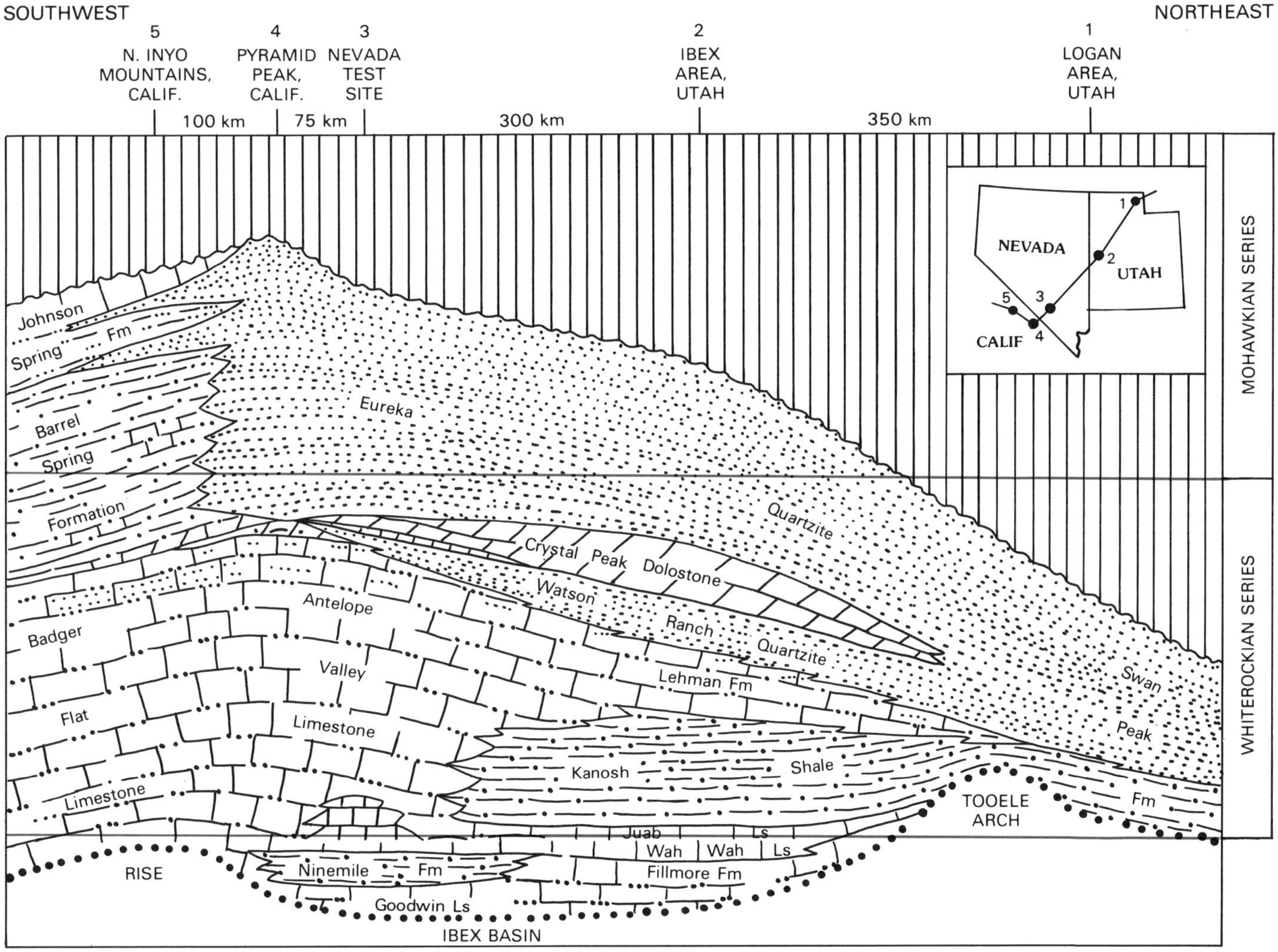

Figure 6. Middle Ordovician facies relations on the continental shelf in the Great Basin. Chronostratigraphic framework from Ross and others (1982).

are thin-bedded limestones of the uppermost Wah Wah Limestone in central Utah, the upper cherty member of the Garden City Formation of northern Utah, and the lower member of the Antelope Valley Limestone in much of Nevada (Figs. 2 and 6).

The lowest Whiterockian (Zone L of Ross, 1951, and Hintze, 1952) Wah Wah beds in the Ibex district of west-central Utah include a fine-grained sandstone containing runzel marks (swash marks) that indicate deposition near sea level. Slightly higher, in the Juab, are abundant *Girvanella*-bearing oncolites. Correlative thin beds to the west are darker in color, more tabular, and interbedded with an increasing amount of black shale.

Several different Whiterockian facies are represented in different thrust sequences in the Toquima Range (Plate 2-5, TR) in central Nevada (McKee, 1976). In the Mill Canyon thrust sequence, exposed in Ikes Canyon, soft-sediment deformation in the lower Whiterockian beds suggests deposition on a slope. In the next more westerly June Canyon thrust sequence in Ikes Canyon, the coeval beds are green shales of the Ninemile Formation that prograded westward from the type area where they are of Ibexian age.

Although little more than reconnaissance has been done, the distribution of low, stacked, and laterally overlapping mounds composed largely of calathid algae and/or sponges in the northern and southern Egan Ranges (Plate 2-5, ER) suggests that they may have formed the first barriers to the ocean, behind which the graptolitic black muds of the Kanosh Shale were deposited.

While the muds formed in western Utah and easternmost Nevada, thick algal banks of the middle part of the Antelope Valley Limestone grew in southern and central Nevada and prograded westward to the Callaghan Ranch area in the northern Toiyabe Range (Plate 2-5, NTR) and then to the Mill Canyon area in the Toquima Range (TR). In the more westerly June Canyon sequence exposed in Ikes Canyon in the Toquima Range, the oncolitic bank limestone is only 1 m thick; all the older beds of the formation form a thin-bedded, tabular limestone slope facies. At this location the underlying green shales of the

Ninemile Formation are of the same age as the lower beds of the Antelope Valley Limestone in more easterly ranges (Ross, 1977; Ross and others, 1982).

In seeming contradiction to these outer continental-shelf and continental-slope deposits, possible subaerial exposure of upper Whiterockian inner carbonate-shelf rocks in the Death Valley region of southern California and adjacent Nevada resulted in dolomitization and local development of a deeply penetrating karst system. Solution cavities contain abundant quartz sand probably derived from the overlying Eureka Quartzite; however, this karst system may have formed much later through subsurface solution during Cenozoic extension.

While the cratonic platform in the Midcontinent region was being eroded on the well-known unconformity beneath the Middle Ordovician St. Peter Sandstone, the Great Basin region was the site of shallow-marine carbonate deposition. More than 300 m of these carbonate rocks accumulated just below sea level. Whether sea level dropped enough to expose the Midcontinent cratonic platform or whether the platform was uplifted, the Cordilleran continental shelf must have been subsiding at a greater rate to accommodate these Whiterockian deposits. The resulting Midcontinent unconformity probably represents only a minor fall in sea level rather than the major retreat that has been so vigorously argued by many stratigraphers.

The thickness and facies map of the upper Middle Ordovician Mohawkian Series (Plate 2-6) shows the extent of quartz sandstones and associated siltstones, mudstones, and limestones that prograded southwestward from the craton during carbonate deposition in the resubmerging Midcontinent region. The Tooele arch in Utah and Nevada was a prominent structural feature during Mohawkian sedimentation and created two shelf basins, the Tristate basin on the north and the Ibex basin on the south (Plate 2-6). Except for the Harding Sandstone, which was deposited on the cratonic platform in Colorado and Wyoming-Montana, the record of Mohawkian sedimentation in the U.S. Cordillera is confined to continental-shelf, continental-slope, and continental-rise settings. Continental-slope sandstones interbedded with graptolitic shales occur in northeastern Washington and in central Nevada as well as in continental-slope and continental-rise facies in the Roberts Mountains allochthon (Plate 3-6).

The spread of remarkably pure quartz sands into the continental-shelf region from the cratonic platform began in central and southeastern Idaho and northern Utah in Whiterockian time. These sandstones were deposited coevally with Kanosh Shale to the south and with thick algal-oncolite banks of the middle part of the Antelope Valley Limestone to the southwest (Fig. 6 and Plate 2-6). The northern areas were buried by sands of the Kinnikinic and Swan Peak Quartzites by medial Whiterockian time. Sands of the Watson Ranch Quartzite (a tongue of the Swan Peak) and the main body of the Eureka Quartzite encroached on and terminated carbonate production southward and westward (Fig. 6). In the south, sand reached the Nevada Test Site and the Death Valley region by the close of Whiterockian time. The blanket of sand did not terminate carbonate deposition of the upper part of the Copenhagen Formation in central Nevada (Fig. 2, column 9) until after medial Mohawkian time. By then, a few westerly tongues of sand reached the northern Inyo Mountains of California and were preserved within the Johnson Spring Formation (Fig. 6) at the outer edge of the continental shelf.

The major source of quartz sand in Mohawkian rocks was older quartzose rocks exposed on the cratonic platform east of the continental shelf, although some geologists have postulated that a major point source of quartz sand was the Lemhi or Salmon River arch in east-central Idaho (Ross, 1977; Suczek, 1977; Ruppel, 1985). Evans and Lund (1981) have questioned the validity of the postulated arch in Idaho. The widespread quartz sandstone package extends from southern Nevada and adjacent California northward as far as the Alberta Rockies in Canada. This distribution requires either many point sources on the inner shelf or a broad source terrain on the cratonic platform of North America.

A regional hiatus followed Mohawkian sand deposition and preceded Cincinnatian carbonate deposition (Figs. 2 and 6). As the Late Ordovician sea encroached over most of North America, there was no longer a supply of sand to prevent the buildup of carbonate deposits that had dominated Cincinnatian time throughout the western U.S.

UPPER ORDOVICIAN—CINCINNATIAN ROCKS

Forrest G. Poole and Charles A. Sandberg

The most complete record of Upper Ordovician rocks in the western United States is on the early Paleozoic continental shelf that extended from southeastern California northward to central Idaho (Fig. 2; Plate 3-1). Locally, Cincinnatian rocks are thin or missing on the outer continental shelf in Nevada. Strongly deformed shaly siliceous rocks containing planktonic and nektonic fossils of Cincinnatian age occur in the Roberts Mountains and younger allochthons in the western U.S. We interpret these rocks to represent continental-rise sediments that were deposited oceanward of the continental shelf and slope and subsequently were thrust eastward onto the outer continental shelf as part of the Roberts Mountains allochthon during the Devonian and Mississippian Antler orogeny.

During Late Ordovician time (about 20 m.y. duration), the continent was submerged to a greater extent than in any previous Paleozoic time. Vast areas were covered by very shallow water, which drowned the source areas of quartz sand that characterizes the underlying Mohawkian rocks. Widespread occurrences of quartz sand in dolostone at the base of the Cincinnatian interval on the cratonic platform and continental shelf indicate significant reworking of the underlying Mohawkian sands by the advancing seas. Regional stratigraphic evidence indicates that a major unconformity separates the Cincinnatian and Mohawkian intervals.

Late Ordovician brachiopods and corals in carbonate-rock inclusions of the Ferris and Sloan diatremes near the Colorado-Wyoming state line indicate the presence of normal marine conditions at these locations (Chronic and others, 1969) and imply widespread Cincinnatian seas in the northern Front Range area of Colorado and across much of the cratonic platform of western North America. Both the distribution and thickness of Cincinnatian rocks on the cratonic platform were greatly affected by post-depositional episodes of epeirogenic uplift and erosion during the Paleozoic and Mesozoic. In most areas, the zero-thickness line represents the post-Ordovician erosional edge because no shoreline facies are recognized in the vicinity of the zero line.

The thickness and lithofacies map for the Cordilleran and northern Rocky Mountains regions (Plate 3-1), which is based both on published and unpublished data of ours and published data of others, is integrated herein for the first time. The central Colorado area is after Sweet (1954, 1961) and Ross and Tweto (1980), and the area of southern New Mexico and adjacent states is after Hayes (1975).

Carbonate-platform rocks of this interval attain a maximum thickness of more than 300 m along the axis of the Tristate basin in southern Idaho and adjacent Nevada and more than 200 m along the axis of the Ibex basin in southern Nevada and adjacent California (Plate 3-1). Thinning of Cincinnatian continental-shelf rocks toward the western margin of the carbonate shelf indicates their deposition in an axial basin on the shelf.

South of the Uinta uplift, Cincinnatian shelf rocks wedge out eastward near the Wasatch hinge line, which also marks the approximate continental shelf-cratonic platform boundary. This wedge edge of Upper Ordovician rocks is generally concealed beneath the Cretaceous Sevier allochthon, which overrides the hinge line in most of Nevada and Utah (Plate 3-1). Northward, the eastern limit of Cincinnatian rocks reflects onlap of sediments on the northwestern and southeastern sides of the northeast-trending Transcontinental arch and subsequent erosion of the shallow-marine deposits from the arch during eustatic regression in Late Silurian to mid–Early Devonian time. Cratonic-platform carbonate rocks of the Cincinnatian interval attain a maximum preserved thickness of 150 m in the northeast-trending seaway that connected the continental shelf in southern Idaho with the intracratonic Williston basin in eastern Montana; 100 m in the seaway that crossed the Transcontinental arch in central Colorado; and 150 m in the north-trending embayment in southern New Mexico and adjacent states (Plate 3-1).

Cincinnatian rocks are absent from the Uinta uplift and are thin on the Tooele arch, which continued to separate major depositional centers on the continental shelf (Plate 3-1). Although rocks of the Cincinnatian interval are now mostly dolostone, they probably were originally deposited as lime mud. The dolostones grade westward into thin-bedded limestones and shales of the outer continental shelf and continental slope, as exposed in structural windows in central and northeastern Nevada and northeastern Washington (Plate 3-1).

The large area of Bighorn and Red River Dolostones north of the Transcontinental arch in the northern Rocky Mountains and the remnant of Fremont Dolostone on the Transcontinental arch in central Colorado are preserved beneath Devonian rocks and indicate widespread erosion related to warping of the continental crust prior to Late Devonian time. Most of the large area of Montoya Dolostone east of the south end of the Transcontinental arch in southern New Mexico and adjacent states is preserved beneath Silurian rocks. The Montoya Dolostone probably represents a northern embayment onto the cratonic platform from the continental shelf in north-central Mexico.

LOWER SILURIAN TO MIDDLE DEVONIAN—LLANDOVERIAN TO GIVETIAN ROCKS

Forrest G. Poole and Charles A. Sandberg

Lower Silurian through Middle Devonian rocks are combined into a single map interval because of their almost uniform dolomitic character and almost continuous depositional record on the continental shelf from southeastern California northward to central Idaho (Plate 3-2). This interval, about 65 m.y. in duration, includes Silurian rocks of Llandoverian, Wenlockian, Ludlovian, and Pridolian age and Devonian rocks of Lochkovian, Pragian, Emsian, Eifelian, and all but latest Givetian age (Fig. 2). The thickness and lithofacies map combines five rock complexes that previously were mapped separately by Poole and others (1977, Figs. 3-7). Information for the map area in southern New Mexico and adjacent Texas is after Hayes (1975, Fig. 42).

Continental-shelf rocks of this interval are well represented in the western U.S. except in northern Idaho and adjacent Montana, where Late Silurian and younger episodes of uplift and erosion have removed most of the depositional record. Elsewhere, the distribution of cratonic-platform rocks east of the Cordilleran miogeosyncline is the result of several episodes of deposition and erosion related to eustatic cycles (Johnson and others, 1985). The episodes of erosion on and adjacent to the northeast-trending Transcontinental arch greatly reduced the original extent of Lower Silurian to Lower Devonian (Lochkovian) rocks (Plate 3-2). The large area in southern New Mexico and adjacent Texas that contains as much as 300 m of Fusselman Dolostone is probably a remnant of a north-trending embayment onto the cratonic platform from the continental shelf in north-central Mexico. The irregular distribution and varied thickness of the Fusselman, like that of underlying Upper Ordovician rocks, is due to post-depositional episodes of uplift and erosion. The zero-thickness line is an erosional wedge edge to the west and north in New Mexico. Locally, thin patches of dolomitic siltstone and silty dolostone assigned to the Middle Devonian Oñate Formation rest unconformably on the Fusselman, but in most places Upper Devonian shales unconformably overlie the Fusselman (Kottlowski, 1963). Local karst development on the Fusselman sur-

face indicates Late Silurian to Middle Devonian erosion on the cratonic platform in New Mexico. Occurrence of the pentamerid brachiopod *Virgiana* in the Fusselman Dolostone suggests that the formation may be mostly Early Silurian (Llandoverian-Wenlockian) in age. Southward (oceanward) the Fusselman Dolostone thickens and may become conformable with overlying Devonian strata on the continental shelf where sedimentation was more continuous. Carbonate-rock inclusions containing *Virgiana* occur in Lower or Middle Devonian kimberlitic diatremes near the Colorado-Wyoming state line (Plate 3-2), indicating shallow-marine limestone deposition on the cratonic platform in this area in Early Silurian time (Chronic and others, 1969). Many isolated Lower Devonian nonmarine deposits on the cratonic platform of Wyoming and Montana are preserved beneath Upper Devonian rocks as fillings in channels, caverns, crevices, and sinkholes developed on the underlying surface of Cambrian to Silurian rocks (Plate 3-2).

Sub–Upper Devonian carbonate-platform rocks of this interval attain a maximum thickness of more than 1,100 m along the axis of the Ibex basin in southeastern Nevada and western Utah, more than 800 m along the axis of the Tristate basin in northeastern Nevada and northwestern Utah, and more than 1,000 m in an unnamed shelf basin in central Idaho (Plate 3-2). In southern Nevada, thinning of Lower Silurian to Middle Devonian continental-shelf rocks toward the western margin of the carbonate shelf suggests they were deposited in an axial basin on the shelf produced by a submarine rise at the shelf edge. The shelf basins probably formed because of isostatic sinking that accompanied sediment loading. Lower Silurian to Middle Devonian shelf rocks wedge out eastward near the Wasatch hinge line in Utah. This erosional wedge edge is generally concealed beneath the Jurassic and Cretaceous Sevier allochthon, which overrides the hinge line in southeastern Idaho and part of Utah (Plate 3-2). The eastern limit of Lower Silurian to Middle Devonian rocks reflects onlap of the western and southern sides of the northeast-trending Transcontinental arch and subsequent erosion of the shallow-marine deposits from the arch during eustatic regression in Late Silurian to mid-Early Devonian time.

Although carbonate-platform rocks of the Lower Silurian to Middle Devonian interval are mostly dolostone, they probably were originally limestone; lime-mud deposition probably dominated this interval, but the muds were replaced by secondary dolomite. Some primary or early diagenetic dolomite may occur in areas of restricted circulation and hypersaline carbonate mudflats, such as the Lower Devonian Sevy Dolostone and Water Canyon Formation on the continental shelf (Fig. 2); these dolostones are laminated and very fine grained and contain some well-rounded and frosted floating quartz sand grains of probable windblown origin. Deeper-water basinal shales probably represent marine transgression, and fossiliferous carbonate rocks represent higher-energy settings close to wave base. Diverse fossil assemblages in the carbonate rocks indicate a shallow-water origin. In the northern Rocky Mountains, outer cratonic-platform Devonian rocks consist of marginal-marine siltstone, sandstone, and minor dolostone and conglomerate. Location of the Early Devonian shoreline in central Idaho is shown on Plate 3-2.

Three major quartz sandstone zones of early Llandoverian, late Lochkovian to Pragian, and late Emsian to early Eifelian age have been widely recognized in the Lower Silurian through Middle Devonian dolostone sequence on the continental shelf from southeastern California northward to northern Utah (Poole and others, 1977, Fig. 2). Most of the quartz sand probably was derived from erosion of quartzose rocks exposed on the Transcontinental arch during major phases of eustatic regression. The related unconformities increase in magnitude shoreward on the platform, decrease in magnitude downslope, and virtually disappear in the deeper parts of shelf basins. The east-trending Tristate shelf basin in southern Idaho and the north-trending unnamed shelf basin in central Idaho may have been continuous during most of Silurian time. As a result of later dislocation and sinistral rotation of the basin axis across the Snake River strike-slip fault (Poole and others, 1977), the basinal axis is now discontinuous and has different trends north and south of this fault.

LOWER UPPER DEVONIAN—FRASNIAN AND LOWER FAMENNIAN ROCKS

Charles A. Sandberg and Forrest G. Poole

The wide distribution of rocks of this interval reflects the maximum transgression by Devonian seas onto the North American craton (Plate 3-3). This transgression, which began in the Givetian as the Taghanic onlap, occurred as part of a 17-m.y. major eustatic cycle (Sandberg and others, 1982) that in western North America was partly coincident with early phases of the Antler orogenic uplift. The thickness and lithofacies map (Plate 3-3) has been updated and areally extended from earlier versions (Sandberg and Poole, 1977, Figs. 4A and B; Poole and Sandberg, 1977, Figs. 8 and 9; Sandberg and others, 1982, Figs. 8 and 9). Carbonate-shelf rocks of this interval attain a maximum thickness of more than 1,000 m along the axis of a northeast-trending shelf basin or miogeosyncline that is continuous across the former position of the Tooele arch. The margins of this basin are outlined by the 200-m isopachs in Nevada, Utah, and Idaho. This shelf basin later was dislocated and rotated sinistrally across the Snake River strike-slip fault zone (Poole and Sandberg, 1977), and its axis is now discontinuous and has different trends north and south of this fault. The eastern limit of Frasnian to lower Famennian rocks reflects onlap of the western and southern sides of the northeast-trending Transcontinental arch and subsequent erosion of the resulting thin nearshore deposits during the regressive phase of the same megacycle later in the middle and late Famennian. An early effect of the Antler orogeny was the formation within the carbonate shelf of the Pilot basin, shown by the hatchet shape of combined lithofacies 5 and 7 (Plate 3-3) in eastern Nevada and western Utah. Other early effects of the Antler orogeny were epeirogenic uplifts on the craton that produced areas of thinning,

such as the Southern Beaverhead Mountains uplift (Sandberg and others, 1975) and Central Montana uplift (Sandberg and Mapel, 1967). The Pilot basin initially occupied only the western part of the "head" of the "hatchet" but eventually extended eastward into the "handle" along the axis of the northeast-trending shelf basin.

Rocks of the Frasnian and lower Famennian interval are more calcareous (Plate 3-3, lithofacies 6 and 7) on the outer carbonate shelf and become progressively more dolomitic (lithofacies 2 through 5 and 8) from the inner shelf toward the former shoreline. The predominantly dolomitic facies is sandier (lithofacies 4) along the axis of the shelf basin, where thick sand bodies are present and locally overwhelm organic buildups and biostromes, and approaching the nearshore erosional limit. Highly disseminated (floating) sand grains in many carbonate rocks of the dolomitic facies are probably windblown. Sandstone and conglomerate (lithofacies 9) mark the area of the Stansbury Formation (Upper Devonian) at the west end of the ancestral Uinta uplift in Utah (Plate 3-3). These clastic rocks were shed not westward from the Uinta uplift but eastward from the ancestral Stansbury uplift, as determined by the direction of fining from boulder conglomerate to very fine sandstone. The local, intense Stansbury uplift (Plate 3-3, SU) may have been the faulted eastern edge of a thrust plate or of a structural cuesta because no detritus was shed to the west. Similar coarse clastic rocks that unconformably overlie an outlier of Pilot Shale (lithofacies 5) in the Newfoundland Range (Plate 3-3, NR), just to the west, were proved by conodonts to be no older than latest Pennsylvanian, although they contain many reworked Late Devonian conodonts.

The location of the anhydritic Potlatch basin in northwestern Montana is depicted by lithofacies 1 (Plate 3-3). Sabkha-related anhydrite precipitation in this area began midway in Frasnian time and gradually spread in all directions (lithofacies 2 and 3) with deposition of the Logan Gulch Member of the Three Forks Formation and equivalents. The special feature that produced the restricted (sabkha and evaporite basin) conditions in this part of the carbonate shelf was the presence on the west of a carbonate bank related to the Palliser Formation of Alberta. This bank is represented by the "False Birdbear" unit of Idaho (Johnson and others, 1985, Fig. 4) and by an outlier of Palliser Formation, just south of the Canadian border at Roosville, Montana (Plate 3-3; Sandberg, 1983), and 45 km to the southeast in the Whitefish Range.

East of the south end of the Transcontinental arch (Plate 3-3), in southwestern New Mexico, the predominantly sandy dolostone nearshore deposits (lithofacies 4) become more shaly (lithofacies 10). Farther east and probably more offshore, in southeastern New Mexico, shale and novaculite (lithofacies 11) predominate. Novaculite may have been formed from oceanic circulation between converging continental plates; this origin is similar to that described by Gutschick and Sandberg (1983) for lower Carboniferous novaculites farther south and east.

UPPER UPPER DEVONIAN—MIDDLE AND UPPER FAMENNIAN ROCKS

Charles A. Sandberg and Forrest G. Poole

The narrow distribution of rocks of this interval reflects their deposition during a general regression that took place in the last 5 m.y. of a 17-m.y. major eustatic cycle (Fig. 2; Sandberg and others, 1982). Middle and upper Famennian rocks have a maximum thickness of about 85 m in western Utah, northern Arizona, and central Idaho (Plate 3-4). The separation of these rocks into nine discrete areas resulted from their erosion during a 3-m.y. episode of continental stability between eustatic cycles in earliest Mississippian time.

The thickness and lithofacies map (Plate 3-4) is a modification of an earlier map (Poole and others, 1977, Fig. 10) that combined three rock complexes previously mapped separately (Sandberg and Poole, 1977, Figs. 10, 12, and 16; Sandberg and others, 1982, Figs. 10, 11, and 12; Sandberg and Dreesen, 1984, Fig. 3). The two older rock complexes were deposited during brief transgressions that interrupted the general regression within the *trachytera* (formerly *velifer*) and Early *expansa* (formerly Upper *styriacus*) conodont Zones (Ziegler and Sandberg, 1990). Definitions of the new Famennian zones and a comparison with the old zones were given by Ziegler and Sandberg (1984). The third complex includes siltstone, sandstone, and other rocks that were deposited during the final Devonian regression within the *Siphonodella praesulcata* Zone; rocks of this complex are shown on Plate 3-4 by lithofacies 3 in a large area mainly in western Montana and a small area in northern Utah and southern Idaho. The previously published third map also shows the rocks of the second transgression that extended into the Middle *expansa* (formerly Lower *costatus*) Zone.

The lower tongue of the Cottonwood Canyon Member (Upper Devonian and Lower Mississippian) of the Madison Limestone is depicted by the NNE–trending area (Plate 3-4, lithofacies 1 and 4) in western Wyoming. This tongue was deposited in an elongate seaway that entered the area from the south during the second minor Famennian transgression. The sandy dolostone and carbonaceous mudstone and siltstone of this tongue were isolated from rocks of the same age in western Utah by erosion resulting from renewed movement of the intervening ancestral Uinta uplift.

An area of sandstone, quartzite, and sandy dolostone mapped in south-central Wyoming has been considered to be of Cambrian age by some workers. New evidence shows that these rocks are at least in part equivalent to the Parting Formation (Upper Devonian) mapped with the Dyer Dolostone (Upper Devonian) in northern Colorado (Sando and Sandberg, 1987; Plate 3-4).

The north-trending elongate areas of thickening portrayed by the 25-, 50-, and 75-m isopachs in western Montana, western

Utah, and northern Arizona and in the large east-trending area in southeastern Arizona and southwestern New Mexico have a common origin. They depict remnant deposits of an epicontinental sea that briefly transgressed the margins of the North American craton during the *trachytera* and Early *expansa* Zones. The broad areas of sandy carbonate-platform rocks (Plate 3-4, lithofacies 4) in Utah and Colorado were deposited during the second minor Famennian transgression, which continued from the Early to Middle *expansa* Zone. Any thin deposits that might have connected these two areas were removed during later rejuvenation of the intervening ancestral Uncompahgre uplift. This northern carbonate platform has a southern erosional limit that is controlled by the ancestral Defiance uplift. South of the area of Percha Shale and equivalents (lithofacies 2 and 3) in southwestern New Mexico and southeastern Arizona is the northern edge of a southern carbonate platform (lithofacies 4) of the same age as the Percha and of the northern carbonate platform. These rocks in the upper part of the Martin Formation in southernmost Arizona are remarkably similar in lithology and faunal content to those in the upper member of the Pinyon Peak Limestone (Upper Devonian) in western Utah.

Westward-thickening, lower-slope, radiolarian-rich, carbonaceous shale and siltstone, deposited during the Early *expansa* Zone, are known from Nevada, Utah, Idaho, and Montana. In eastern Nevada and western Utah, two areas of lithofacies 1 (Plate 3-4) depict the basal black shale of the Leatham Member of the Pilot Shale. The same facies occurs at the base of the Leatham Formation (lithofacies 3) in northern Utah and southern Idaho and at the base of the Sappington Member of the Three Forks Formation (lithofacies 1) along the Montana-Idaho state line. The distribution of time-equivalent continental-rise rocks is less well known. These allochthonous western-facies rocks have been mapped only locally, but six conodont-dated occurrences in central Nevada and central Idaho are named and located on Plate 3-4. The late Famennian (Early *expansa* Zone) age of these rocks is highly significant because no younger Devonian rocks are known from the upper plate of the Roberts Mountains thrust (Sandberg and others, 1982). This lack of younger Devonian marine rocks dates the emergence of the Antler orogenic belt on the continental margin. Locally in Nevada, overlapping Lower Mississippian-barite siliceous facies date emplacement of the Roberts Mountains allochthon (Poole and Sandberg, 1992).

OVERLYING ROCKS

Following the Famennian regression, a 3-m.y.-long episode of continental stability took place and most of the western North American craton was emergent (Fig. 2). During this episode, Devonian and older rocks were deeply eroded except in the area of an elongate epicontinental seaway that trends northeastward across east-central Nevada, northern Utah, and western Wyoming, as mapped by Sandberg and others (1982, Fig. 13). Lowermost Mississippian rocks that overlie Devonian rocks in the area of this seaway are, from south to north, silty argillaceous limestone and calcareous siltstone of the upper member of the Pilot Shale; basinal limestone within the Fitchville Formation; and siltstone, sandstone, and silty sandy carbonate rocks of the upper tongue of the Cottonwood Canyon Member of the Lodgepole and Madison Limestones (Poole and Sandberg, 1991, Fig. 7). Overlying rocks in several areas of the western U.S. are shown in Figure 2.

Outside the area of the earliest Mississippian remnant seaway, the Mississippian rocks that unconformably overlie Devonian rocks are generally the basal transgressive deposits of a major eustatic cycle that began during the Lower *crenulata* Zone (Sandberg and others, 1982). Lower Mississippian carbonate rocks deposited during the lower part of this cycle grade upward from nearshore to slope, ramp, or bank deposits within the Kinderhookian to widespread carbonate-platform and carbonate-bank deposits in the Osagean. The lowermost rocks include the Kinderhookian parts of the Caballero Formation and Lake Valley Limestone in New Mexico; the Escabrosa and Redwall Limestones in Arizona; the Monte Cristo and Joana Limestones in eastern Nevada; the Joana and Lodgepole Limestones and Fitchville Formation in western Utah; the Redwall and Leadville Limestones in southern Utah and Colorado; the Madison Limestone in northeastern Utah and Wyoming; and the Lodgepole Limestone in southeastern Idaho and western Montana. Farther west, in the area of the Antler foreland trough, the basal Mississippian units are the Eleana, Antelope Range, Dale Canyon, and Tripon Pass Formations in Nevada and the McGowan Creek and Copper Basin Formations in east-central Idaho. Most of these units are western-derived clastic flysch deposits; however, the Tripon Pass Limestone in Nevada and northwestern Utah and the Drummond Mine Limestone Member of the Copper Basin Formation, the equivalent of which forms the base of the McGowan Creek in some areas, are eastern-derived calcareous debris flows and turbidites deposited in the flysch trough (Poole and Sandberg, 1991). Exceptionally, in western Utah, the Pilot Shale at the top of the Devonian is unconformably overlain by the mid-Mississippian (Osagean-Meramecian) Delle Phosphatic Member of the Chainman Shale at Granite Mountain at the northern end of the Confusion Range (Plate 3-4, CR) and by uppermost Pennsylvanian and Permian conglomerate at the base of the Pequop Formation in the Newfoundland Range (Plate 3-4, NR).

ROCKS OF THE ANTLER OROGEN—THE ROBERTS MOUNTAINS ALLOCHTHON

Raul J. Madrid, Forrest G. Poole, and Chester T. Wrucke

The Roberts Mountains allochthon (RMA) consists principally of siliceous rocks deposited in a relatively deep-water marine environment along the continental margin of North America. Present distribution of allochthonous rocks in the Antler orogen is shown in green on Plate 3-5.

Rocks of the RMA consist of complexly deformed, rhythmically bedded radiolarian chert, graptolitic shale, quartzarenite,

feldspathic sandstone, siltstone, minor mafic lava flows and dikes, bedded tuff, limestone, and bedded barite (Plate 3-6). They are, in part, representative of continental-rise, continental-slope, and ocean-basin deposits. Sedimentary processes that produced these strata included both slow hemipelagic accumulation of chert and mudstone and rapid deposition of clastic sediments as turbidites (Stanley and others, 1977; Suczek, 1977; Miller and Larue, 1983; Madrid, 1987). Sequences of chert, argillite, and minor limestone are most abundant in the eastern and southern exposures and apparently grade into the more quartzite- and greenstone-rich rocks of the central and northern parts of the allochthon. Sedimentary barite beds occur in Cambrian to Devonian chert and argillite units, the most regionally extensive of which occur in Cincinnatian (Upper Ordovician) and Famennian (Upper Devonian) strata (Poole, 1988). Locally disseminated and layered iron, copper, lead, and zinc sulfide minerals occur in the black shale and chert facies as well as in some mafic lava flow sequences of the allochthon. Sedimentologic patterns, geochemistry of greenstones, and petrology of sandstones suggest that these rocks were depositionally connected both to autochthonous rocks of the Paleozoic miogeosyncline in Nevada and adjacent states and to an emergent arkosic and subarkosic source area west of the basin of deposition. The relation of the RMA to the accreted terranes farther west remains uncertain.

On the basis of stratigraphic, sedimentologic, and geochemical data and detailed geological mapping, strata of the RMA in Nevada are interpreted to have been deposited in a rift basin or marginal ocean basin floored by attenuated continental crust adjacent to the North American continental margin. The stratigraphy records episodes of extension or rifting that affected the continental margin for a time span of 135 to 180 m.y. prior to the Antler orogenic event. Rifting probably produced episodes of alkalic volcanism that were followed by sedimentation in anaerobic or dysaerobic environments and finally by siliciclastic sedimentation. The Upper Cambrian Harmony Formation in Nevada, which was exposed probably in an uplifted block or microcontinent along the western margin of this rift basin, may have been the major source of feldspar-bearing detritus in post-Cambrian rocks. The more quartzose, feldspar-poor detritus was probably derived from the miogeosynclinal margin. The RMA consists, in general, of an older-over-younger stack of thrust nappes, although details of the stack are locally complicated by map-scale fold thrusts, recumbent fold nappes, and later structural dismemberment. The thrust nappes are transported eastward over coeval continental-shelf strata for distances of 75 to 200 km during the Late Devonian and Early Mississippian. In many areas, Devonian and Mississippian deformation is overprinted by Jurassic to early Tertiary deformation.

STRUCTURE

During the Late Devonian and Early Mississippian Antler orogeny, lower and middle Paleozoic siliceous deep-water oceanic strata, including greenstones and some ultramafic rocks, were thrust eastward for a distance of at least 75 km, perhaps as much as 200 km, over coeval shallow-water rocks of the continental shelf or miogeosyncline (Roberts and others, 1958; Stewart and Poole, 1974; Smith and Ketner, 1977). The structurally complex rocks of the RMA were emplaced along the nearly flat-lying Roberts Mountains thrust in a broad area of the continental margin in Nevada and adjacent parts of Idaho and California (Poole and others, 1977, Fig. 1; Carr and others, 1981; Dover, 1980; Poole and others, 1981, 1982). Antler orogenic deposits of Mississippian age onlap and occur along the eastern margin of the allochthon (Plate 3-5) in Nevada and in both southeastern California (El Paso Mountains, EPM) and central Idaho (Pioneer Mountains, PM). In north-central Nevada, where the Antler orogeny was originally defined (Roberts, 1964), rocks of the allochthon and the autochthon are well exposed. The Roberts Mountains thrust that separates them is best exposed in the Cortez Mountains, in the main pit at the Carlin gold mine in the Tuscarora Mountains, and in the Roberts Mountains. Most thrust contacts are not unequivocal, but those exposures of the thrust surface and the regional distribution of windows in the allochthon that expose autochthonous rocks indicate the regional extent of the thrust. The thrust has also been penetrated by several drill holes, notably along the eastern flank of the northern Shoshone Range (Wrucke, 1974).

The maximum thickness of the structural stack in north-central Nevada in Middle Pennsylvanian time appears to have been on the order of 5 km (Fig. 7). The geometry of the allochthon in Middle Pennsylvanian time seems to have been an eastward-tapering wedge (as a consequence of erosion), but its western part remains poorly known.

Early studies (Roberts and others, 1958) suggested that western, transitional, and eastern assemblages, corresponding to contiguous deep-water siliceous, intermediate, and shallow-water carbonate-shelf strata, were telescoped along the Roberts Mountains thrust. Recent detailed geological mapping corroborates this relation. In addition, the detailed mapping indicates that the allochthon generally consists of an older-over-younger tectonic stack of thrust nappes of regional extent, although structural details of the stack are locally complicated by map-scale thrust nappes, fold thrusts, and recumbent fold nappes (Fig. 7). Later structural deformation of the allochthon, including Mesozoic thrusting and folding and Cenozoic high-angle normal faulting, have also contributed to the structural complexity characteristically displayed by rocks in the allochthon.

In northern Nevada, the highest exposed thrust nappe of the allochthon consists of Upper Cambrian Harmony Formation, and the lowest thrust nappes generally involve Devonian Slaven Chert (Fig. 7). Ordovician rocks are predominant within thrust nappes in the upper and middle parts of the allochthon. The Middle Cambrian Shwin Formation forms one of the lowest thrust nappes. Its low structural position is a consequence of its paleogeographic position, 75 km west of its present location in the allochthon, at the time of thrusting (Madrid, 1987). The Shwin Formation represents deposition on a previously downfaulted

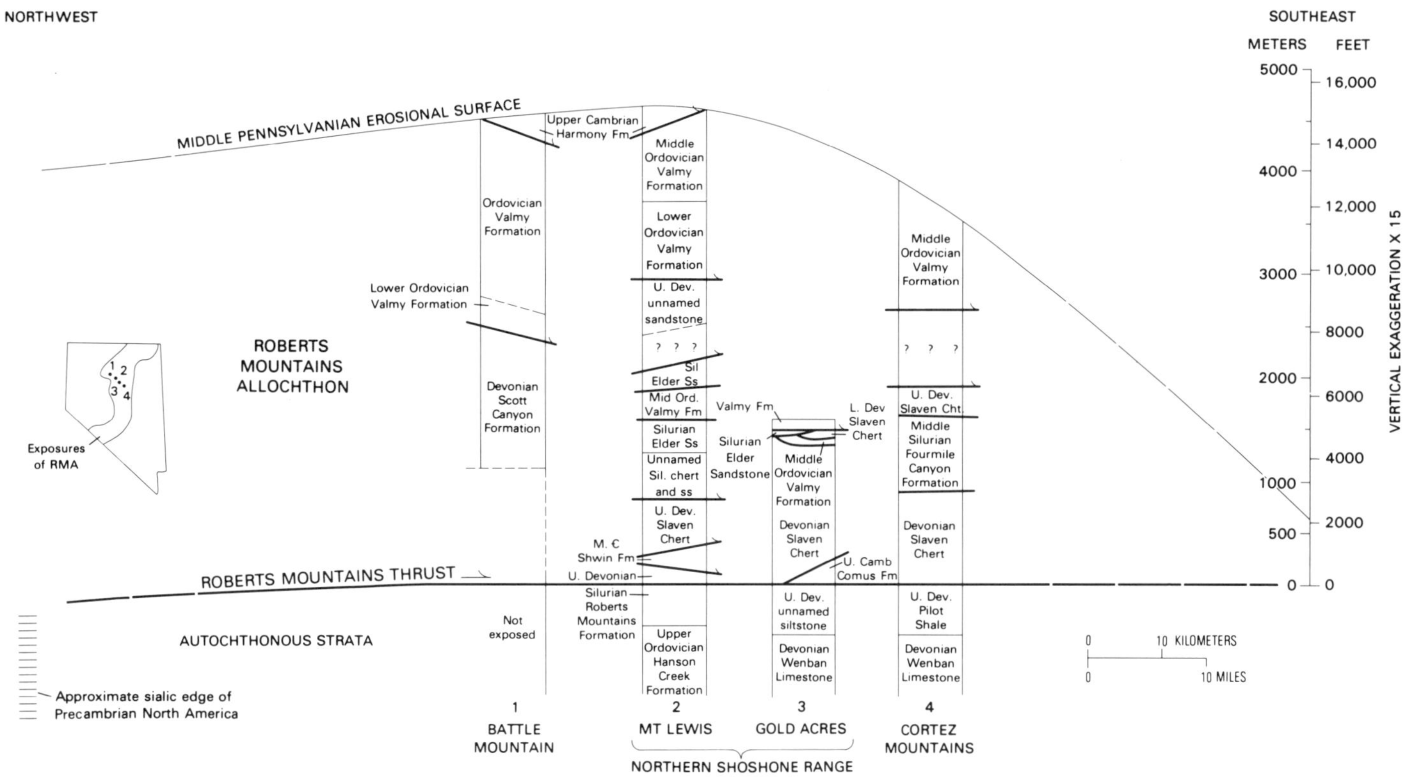

Figure 7. Pre-Middle Pennsylvanian stacking order and thickness of the Roberts Mountains allochthon. The convex upper surface of the allochthon is based on measurements of the angle between the subparallel stratigraphic sequences and dips of thrusts and fold nappes in the allochthon and the horizontal datum of the Middle Pennsylvanian rocks of the Overlap assemblage (Silberling and Roberts, 1962). The approximate edge of North America is based on isotopic work of Farmer and DePaolo (1983) and Kistler (1990). Sections of the allochthon from Battle Mountain, the Mt. Lewis area in the northern Shoshone Range, and the Cortez Mountains are from Madrid (1987, unpublished data). Section from the Gold Acres area from Wrucke (unpublished data).

portion of the continental slope that was thrust over the nearby continental margin during emplacement of the allochthon. These relations, considered together with structural transport indicators (Wrucke and Theodore, 1970; Evans and Theodore, 1978; Oldow, 1983; Madrid, 1987), support a minimum easterly transport distance of 75 km for the entire allochthon. The lateral extent of the highest thrust nappe (the Harmony Formation) of the allochthon is approximately 200 km, as inferred from the present distribution of the Harmony Formation, the distribution of lower-plate autochthonous rocks exposed in windows in the allochthon, and angular Harmony-derived arkosic-rock clasts in the lower part of the Mississippian Chainman Shale to the east. The transport directions and distances (75 to 200 km) confirm that the highest thrust nappes originally were emplaced from areas west of the former and present locations of the lower thrust nappes.

Alpine-type serpentinites occur discontinuously in the RMA. These foliated and serpentinized ultramafic rocks form small, elongate, semiconcordant bodies that are tectonically interleaved with strongly deformed Paleozoic oceanic and continental-rise sedimentary and mafic volcanic rocks. The serpentinites were inferred by Poole and Desborough (1973) to be fragments of Paleozoic and older upper-mantle olivine-rich rocks that may have been detached and intercalated with stratified ocean-basin deposits at the eastern Pacific margin along active subduction or obduction zones during Paleozoic lithospheric plate convergence. Most of the serpentinite bodies were probably emplaced during the Antler orogeny. Mid-Paleozoic K-Ar and Rb-Sr isotopic ages have been determined from a biotite vein within a serpentinite body in central Nevada (Poole and Desborough, unpublished data). Serpentinite detritus in some Pennsylvanian, Permian, and Triassic marine conglomerates and grits in central Nevada (Poole, unpublished data) indicates erosion of exposed serpentinite bodies in late Paleozoic and early Mesozoic time.

STRATIGRAPHY

Sedimentary rocks

The RMA has a distinct geologic record expressed by its stratigraphy and structure and by foreland-basin deposits east of

the allochthon. In Nevada, most strata of the RMA range in age from Ordovician through very late but not latest Devonian (Plate 3-6). The youngest Devonian rocks have well-established late Famennian conodont ages. The Middle and Upper Cambrian Shwin Formation occurs in the allochthon in northern Nevada (Wrucke and Silberman, 1975; Madrid, 1987). In central Nevada, a sequence of argillite, siltite, and minor quartzite and limestone of Early Cambrian age may occur in the RMA (Poole, unpublished data). Locally in central Nevada, Lower Mississippian baritic siliceous rocks dated by conodonts overlap the RMA (Poole and Sandberg, 1992). Further study of other Mississippian rocks, the ages of which are based on radiolarian fossils (B. L. Murchey *in* Coles and Snyder, 1985; Coles, 1986), is necessary to clarify their relation to the RMA. Stratigraphic columns considered representative of the allochthon in Nevada are shown on Plate 3-6, which also shows the locations of the stratigraphic sections in the allochthon. The columns were constructed using available fossil ages of mostly coherent stratigraphic sections found in thrust slices in the RMA. Sections 4 through 8 represent a transect that lies diagonally across the allochthon. Formational names are shown on the left side of each column, and diagnostic fossils are shown on the right side. A central Nevada reference section of coeval autochthonous strata of the carbonate shelf is given for comparison.

Attempts to establish a comprehensive stratigraphy for allochthonous strata have been hindered by lithologic similarity of rocks of different ages, paucity of fossils, and structural complexities in the numerous thrust slices of the allochthon. Some workers (e.g., Schweickert and Snyder, 1981) believe that the RMA developed as an accretionary prism characterized by structural disruption accompanied by stratigraphic complexity. Most studies have utilized limited faunal data in conjunction with gross lithologic characteristics to define stratigraphic units. Faunal ages typically indicate repetition of units by structural disruption (Roberts and others, 1958; Stanley and others, 1977; Nilsen and Stewart, 1980; Stewart, 1980; Madrid, 1987). Consequently, the original stratigraphic sequence is no longer intact, and thicknesses of individual units are not known with certainty, although estimates of several well-exposed, mostly intact stratigraphic sequences have been reported by many workers. Despite these difficulties, a persistent stratigraphic sequence can be traced from northwest to southeast across the allochthon in north-central Nevada (Plate 3-6). Feldspathic and micaceous sandstones characterize the Upper Cambrian Harmony Formation. Considerable lithologic continuity exists in such stratigraphic units as the Vinini and Valmy Formations (Ordovician), Elder Sandstone (Silurian), and Slaven Chert (Devonian). Quartzose and feldspar-rich sandstone and siltstone-sandstone units, chert-argillite sequences, and bedded barite also are regionally persistent. Mafic lava flows and dikes are most abundant and persistent in the central exposures of the allochthon, where they occur in similar positions within several sections (Plate 3-6, columns 2 through 5) of Early and Late Ordovician and Late Devonian age. In southwestern Nevada, strata in the southern exposures of the allochthon, represented by the Miller Mountain–Candelaria Hills section, are very similar to sequences in the Roberts Mountains in east-central Nevada. The sequences in east-central Nevada contain more shale and limestone and less quartzite and greenstone (Vinini Formation) than those to the west, which have more abundant sandstone and greenstone (Valmy Formation).

Petrographic studies of sandstones, together with evidence from detailed mapping and fossil studies, suggest that little reworking of intrabasin sediments took place, except perhaps in the Silurian, during deposition of the stratigraphic units of the allochthon. Madrid (1987) found that Ordovician, Silurian, and Devonian sandstones spatially associated with both mapped and inferred synsedimentary normal faults locally are rich in reworked, intrabasin lithic detritus; however, most sandstones of the allochthon are characterized by only trace amounts of intrabasin detritus. Dubé (1987) observed that chert conglomerates in Devonian rocks most commonly occur in association with bedded barite deposits, but chert fragments in other siliciclastic strata of the allochthon are sparse to absent. Moreover, despite abundant greenstone and radiolarian chert in the RMA in north-central Nevada, these rock types are rare as framework grains in the sandstones of the allochthon; thus, widespread reworking of the sedimentary and volcanic sequences of the allochthon is unlikely. Many sandstones of the allochthon contain first-cycle arkosic and subarkosic sandstone grains that were not derived from reworking of intrabasin sediments (Madrid, 1987). Girty and others (1985) reported numerous chert grains in a portion of the Silurian Elder Sandstone, which suggests that it may have resulted, in part, from reworking of intrabasin lithologies. Most second-cycle grains in the RMA probably were derived, however, from erosion of immature first-cycle feldspar-bearing sandstones of the Harmony Formation on the west. On the other hand, sandstones with greater than 95 percent quartz and quartzose sandstone grains probably came primarily from the continental margin on the east (Madrid, 1987).

Distinctive cobbles, pebbles, and grains of immature first-cycle feldspathic and micaceous sandstone and grit occur in Ordovician and Devonian second-cycle siliciclastic rocks and in volcanogenic debris flows within major mafic volcanic sequences of the allochthon (Madrid, 1987). The presence of these large clasts is indicated by circles to the right of columns 5 and 7 on Plate 3-6. Clast compositions (Madrid, 1987) are modally indistinguishable from sandstones of the Harmony Formation (Suczek, 1977), and the Harmony or similar rocks may also have been a source for rocks of these ages. This evidence provides a sedimentological connection between the Harmony Formation and younger rocks of the RMA.

Comparison of detrital modes of Precambrian rocks of the miogeoclinal assemblage (overlapping fields A and B on Fig. 8A) with those of sandstones of the RMA (overlapping fields C and D on Fig. 8A) suggest that the Precambrian units were not sources for the feldspar-rich sandstones of the RMA. Similarly, Ordovician to Devonian sandstone units of the miogeosynclinal assemblage do not contain sufficient feldspar to have been source rocks.

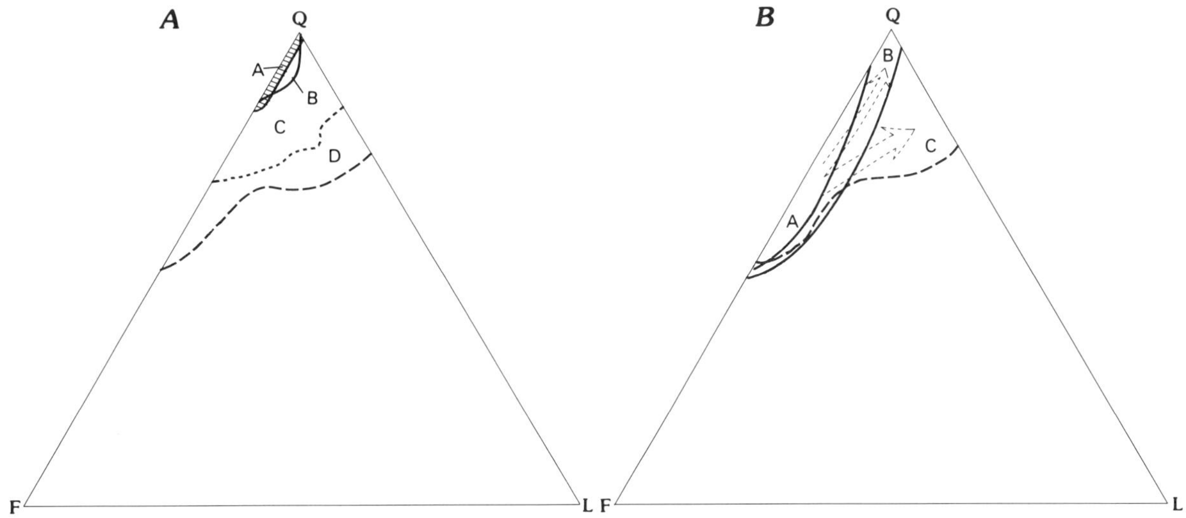

Figure 8. Quartz (Q)–feldspar (F)–unstable lithic fragments (L) ternary diagrams showing detrital modal compositions of the Roberts Mountains allochthon (RMA) as compared to other sandstones discussed in text. A. Detrital modal data of Lobo and Osborne (1976) for Precambrian and Cambrian terrigenous strata of the miogeoclinal rocks: Field A, Stirling Quartzite; Field B, Johnnie Formation; Fields C and D, compositional overlapping fields of the type Elder Sandstone (C) (Silurian; Girty and others, 1985) and the post-Upper Cambrian strata of the RMA (D) (Madrid, 1987). B. Detrital modal compositions of the Upper Cambrian Harmony Formation as compared with data acquired from the RMA (Girty and others, 1985; Madrid, 1987). Data for the Harmony is from (A) Suczek (1977) and (B) T. G. Theodore (USGS, written communication, 1985); C, data for Lower Ordovician to Upper Devonian strata of the RMA. Dashed arrows indicate possible change in detrital modal composition as a function of derivation from the Harmony Formation.

Because Cambrian units were not exposed along the continental margin from Ordovician to Devonian time, feldspar-rich units such as the Upper Cambrian Worm Creek Quartzite Member of the St. Charles Limestone in Utah and southeastern Idaho and the Ordovician or older Clayton Mine Quartzite in central Idaho probably were not sources for younger sandstones of the RMA. Comparison of the detrital modes of the immature, continentally derived sandstones of the Harmony Formation (overlapping fields A and B on Fig. 8B) with those of the combined Lower Ordovician to Upper Devonian sandstones of the RMA (overlapping field C on Fig. 8B) suggest that the first-cycle Harmony was a possible source for the Lower Ordovician to Upper Devonian sandstones of the RMA.

Conodont data from north-central Nevada indicate that Lower Ordovician to lower Upper Devonian strata of the RMA were not extensively reworked and were deposited as unconformity-bound, widespread, discrete stratigraphic sequences that are compatible with continental slope-rise environments and argue against a long-lived, far-traveled accretionary wedge. J. W. Huddle (USGS, written communication, 1972) and A. G. Harris (USGS, written communication, 1981) found that samples of the Elder Sandstone, collected by Wrucke in the eastern Shoshone Range, contain fragments of Ordovician conodonts, an indication that some reworking of older sediments occurred during Silurian sedimentation. Ordovician, Silurian, and Devonian conodont faunas identified by A. G. Harris and J. E. Repetski (USGS, written communication, 1983–85) from allochthonous rocks in the western and central Shoshone Range generally lack evidence of reworking and commonly define very narrow fossil-age ranges (Madrid, 1987). On the other hand, C. A. Sandberg (personal communication, 1988) found that latest Devonian (late Famennian) limestones in the Independence Mountains, Nevada, contain faunas that are dominated by reworked Early Ordovician conodonts. This reworking supports the conclusion (Sandberg and others, 1988) that the continental margin became emergent at this time during the initial phase of Antler orogeny.

Igneous rocks

Recent stratigraphic, petrographic, and trace-element geochemical studies indicate that most Middle Cambrian through Late Devonian greenstones in the RMA represent original sequences of predominantly alkalic basalts interlayered with sparse tholeiites (Madrid, 1987). This conclusion differs from earlier descriptions of greenstones in the allochthon as being tholeiitic ocean-floor basalts (Rodgers and others, 1974; Wrucke and others, 1978). Magmatic affinity, trace-element geochemistry, and paleogeography of the lavas indicate an intraplate tectonic setting in continental crust (Madrid, 1987). The lava flows and associated minor devitrified hyaloclastites, which occur in stratigraphic sequence with siliciclastic rocks, commonly are interbedded with volcanogenic basaltic debris flows that are spatially associated with mapped syndepositional high-angle normal faults (Fig. 9) and were extruded from faults and fractures in the sea

floor. Dubé (1987) reported in association with barite deposits the occurrence of lamprophyric dikes, which he suggested formed in an extensional environment. Alkalic basalt dikes present in the slope and basin facies of the Shwin Formation in the northern Shoshone Range (Madrid, 1987) and layers of tuff in the Comus and Vinini Formations (Poole and G. A. Desborough, unpublished data) indicate volcanic activity in the region during the Late Cambrian and Ordovician.

The volcanogenic debris flows typically contain older clasts of continent-derived (based on detrital modal analyses) pre-Ordovician sandstone of the allochthon, a characteristic that suggests spatial and temporal association with normal faulting related to rifting and attenuation of continental crust near the continental edge. The clasts consist of first-cycle arkosic, subarkosic, and micaceous sandstone. Clast composition suggests a sedimentologic connection and a similar source terrain for the clasts in both the volcanogenic sequences and most siliciclastic rocks in the RMA.

EVENT STRATIGRAPHY

Extensional faulting in rocks of the allochthon continued from Cambrian until either latest Devonian time when the Antler orogeny began or Early Mississippian time when the RMA was emplaced over the continental margin. Available geologic evidence suggests no significant compressional event prior to the Late Devonian.

Tectonic and sedimentary events determined by detailed mapping, stratigraphic analysis, and geochemical studies of rocks in the allochthon in north-central Nevada are summarized in Figure 9. The major sedimentary events inferred from miogeosynclinal strata are shown for comparison. Mapped and inferred early Paleozoic syndepositional faults and their time of formation relative to periods of volcanism as recorded in strata of the allochthon (Fig. 9) suggest that faulting and volcanism were simultaneous. Extrusion of submarine alkalic basalt began in the Middle Cambrian and continued until the Late Devonian, during which

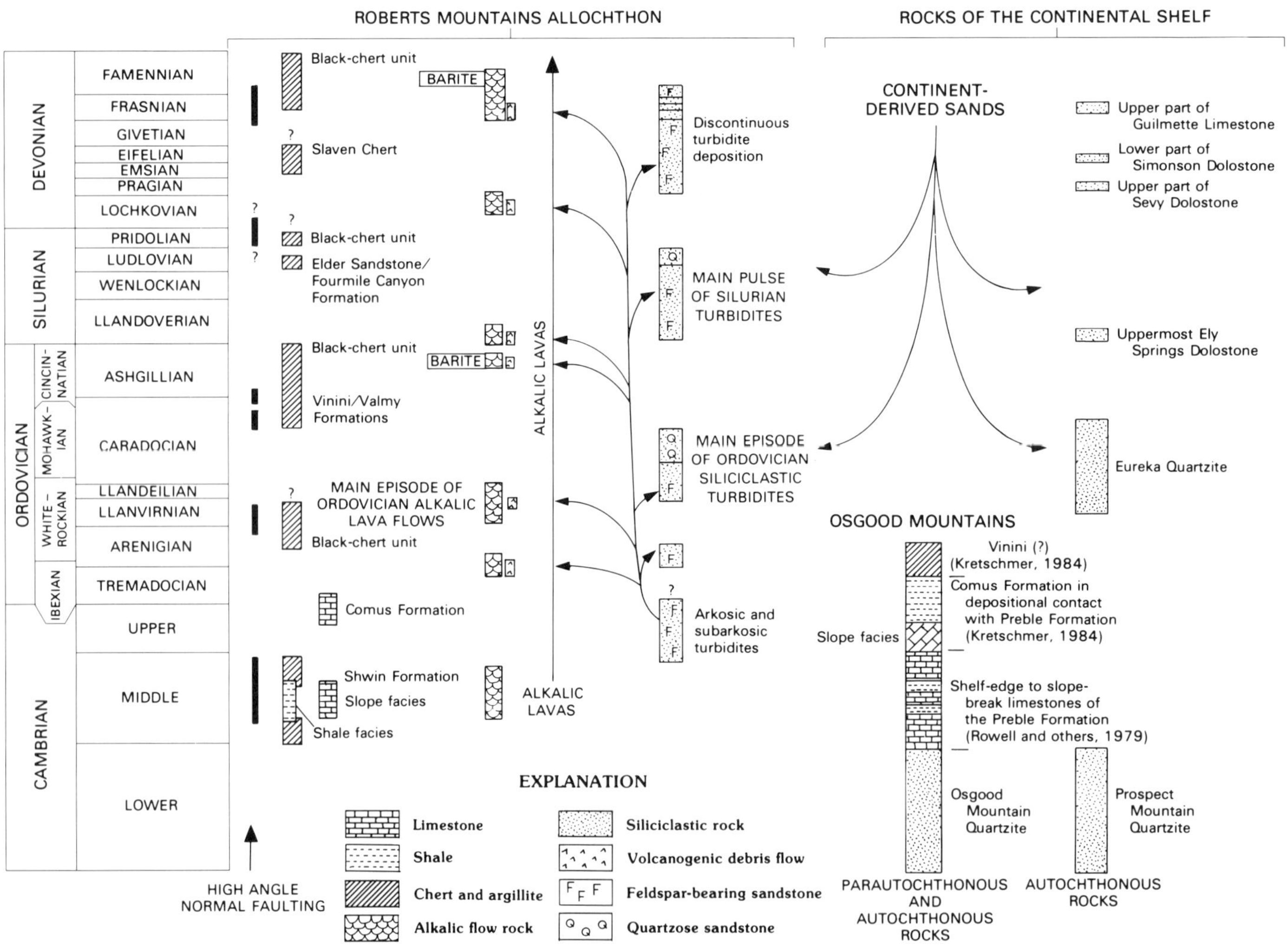

Figure 9. Comparison of sedimentary and tectonic events of the Roberts Mountains allochthon and those of the continental margin and miogeocline–miogeosyncline. Time ranges of rock units and faults are based on stratigraphic relations and on the occurrence of fossils shown on Plate 3-6.

time volcanism and faulting occurred at about 30- to 35-m.y. intervals (Madrid, 1987). Volcanic rocks more highly differentiated than the alkalic basalts have not been observed in the allochthon. Euxinic sedimentation, represented by black chert and shale units, seems to have followed main episodes of faulting and preceded deposition of siliciclastic sequences. The major barite deposits in Nevada are generally coeval with major episodes of volcanism. In north-central Nevada, major barite deposits of both Late Ordovician and Late Devonian age are locally interbedded with basalt flows (Plate 3-6). Some stratiform barites show evidence of reworking of strata along syndepositional faults (Madrid, 1987; Dubé, 1987), and some were deposited at or near sea-floor hydrothermal vents (Poole, 1988).

CRATONIC CONNECTION

Coney and others (1980) and Silberling and others (1987) have depicted the RMA as a lithotectonic terrane, a view totally contradictory to that expressed by the authors of this chapter. Their interpretation suggests that the allochthon consists of rocks not demonstrably related to the North American continental margin. Stratigraphic studies demonstrate, however, that rocks of the RMA are depositionally connected to rocks of the miogeosyncline. Principal lines of evidence for a cratonic connection are as follows. (1) In the Osgood Mountains of north-central Nevada, a depositionally continuous miogeosynclinal section of Cambrian age is conformably overlain by siliceous rocks of the Ordovician Vinini or Comus Formation, a unit found elsewhere in the allochthon in Nevada (Kretschmer, 1984; Madrid and E. L. Kretschmer, unpublished data; Wrucke, unpublished data). (2) In Esmeralda and Mineral Counties, Nevada, the Ordovician Palmetto Formation, which in some areas is allochthonous, appears to be conformable with the underlying miogeosynclinal Middle and Upper Cambrian Emigrant Formation (Albers and Stewart, 1972). (3) Additional stratigraphic evidence on Plate 3-6 and Figure 9 indicates that siliciclastic sedimentary patterns of the strata in the allochthon closely follow those in the autochthon. For example, great volumes of Middle Ordovician quartz sand in the Vinini and Valmy Formations of the allochthon are correlative with the Eureka Quartzite, a regionally extensive unit in the autochthon (also see Miller and Larue, 1983). (4) Similarly, regionally persistent Devonian sandstones or quartzites in the allochthon may be partly time equivalent to sandstones found in formations of the autochthon (Fig. 9). (5) Silurian feldspathic sandstone and siltstone beds, which are persistent in the allochthon, have no direct equivalents in the autochthon, although the Silurian part of the Roberts Mountains Formation does contain some feldspar silt (Mullens, 1980).

Feldspathic sandstones of the Upper Cambrian Harmony Formation have been interpreted by some geologists to be derived from a continental point source (Lemhi or Salmon River arch) in north-central Idaho (Palmer, 1971; Stewart and Suczek, 1977; Suczek, 1977). If sand in the Harmony was derived from the continent, this would provide a sixth line of evidence for a cratonic connection. Older Cambrian rocks also suggest a linkage. The Middle Cambrian Shwin Formation, present as a major thrust nappe in the allochthon, has been correlated with the Preble Formation (Madrid, 1987). The Preble intertongues with the Osgood Mountain Quartzite, a correlative of the Prospect Mountain Quartzite of the miogeosynclinal assemblage. Similarly, Upper Cambrian to Lower Ordovician rocks mapped as part of the RMA along the eastern flank of the northern Shoshone Range are considered correlatives of the Comus Formation in the Osgood Mountains 75 km to the west (Wrucke, unpublished data). The Comus is believed to be directly related to rocks of the continental margin, suggesting another stratigraphic connection. Similar lithologies in the Comus Formation and the allochthonous Ordovician Vinini Formation indicate a possible depositional connection (Stewart, 1980; Poole, unpublished data). Devonian rocks also suggest a linkage. For example, C. A. Sandberg has found a conodont-bearing unit in the Adobe Range, Nevada (Plate 3-6, column 2, top), that is an exact temporal and lithologic equivalent of, and was probably depositionally continuous with, the lower black concretionary shale of the Leatham Formation of Utah and its equivalents elsewhere on the craton (Sandberg and others, 1988). In addition, Poole (unpublished data) considers a conodont- and *Angustidontus*-bearing unit in the northern Roberts Mountains (Plate 3-6, column 8, top) and a *Platyclymenia*- and *Angustidontus*-bearing unit in the northern Piñon Range (locality about 80 km to the north-northeast of former locality) to be lithologic equivalents of the tan-weathering platy siltstones of the lower part of the Pilot Shale of eastern Nevada. These units, which are believed to be in the eastern part of the RMA, suggest that the Pilot basin of eastern Nevada had a westward connection with the open ocean at times during the Late Devonian (see Sandberg and others, 1988, maps 5 and 6).

ROCKS RELATED TO THE ANTLER OROGEN IN OTHER AREAS

Forrest G. Poole

ALLOCHTHONS IN CENTRAL IDAHO

Chert-argillite sequences correlative with those in the Roberts Mountains allochthon of Nevada crop out discontinuously in a belt 75 km long and as wide as 20 km that trends north in Blaine and Custer Counties in central Idaho and extends from just south of the town of Bellevue northward to the East Fork of the Salmon River (Plate 3-5). The southern part of the belt is composed of the Devonian Milligen Formation (Sandberg and others, 1975), and the central and northern parts of the belt are composed of the Ordovician and Silurian Phi Kappa and Wenlockian (Silurian) or younger Trail Creek Formations and an unnamed Silurian and Devonian unit (Dover and others, 1980). Dover and others (1980) described a conformable sequence of dark gray graptolitic shale and argillite and sparse interbedded quartzite and

micritic limestone (Phi Kappa Formation); overlying dark- and light-banded, thinly interbedded siliceous argillite, siltite, and very fine-grained quartzite (Trail Creek Formation); and calcareous siltstone and minor silty limestone (unnamed Silurian-Devonian unit). The postulated depositional contact with the overlying Devonian Milligen Formation has not been observed in the belt due to vagaries of structure and exposure. The Milligen Formation was described by Sandberg and others (1975) as consisting mainly of interbedded, dark gray, siliceous argillite and fine-grained quartzite interbedded with some siltstone, siltite, and micritic silty limestone and minor interbedded dolostone, chert, and granule conglomerate; disseminated and layered base-metal sulfide minerals are present. Conodont faunas date the lower member of the Milligen as Early and Middle Devonian and the upper member as Late Devonian (Sandberg and others, 1975).

The Ordovician to Devonian eugeosynclinal rocks in the central Idaho belt occur in a series of thrust-bounded slices within a broad zone of imbricate Mesozoic thrust faults that includes both structural repetitions of the Ordovician to Devonian rocks as well as other tectonically juxtaposed lithostratigraphic units of diverse ages (Dover and others, 1980). On the west side of the Pioneer window in the Pioneer Mountains (Plate 3-5, PM), Dover and others (1976) mapped a major thrust that separates contrasting Ordovician facies. Here continental-rise argillaceous rocks of the Phi Kappa Formation override the East Fork Formation (Middle and Upper? Ordovician) consisting of metamorphosed equivalents of the outer continental-shelf Middle Ordovician Ella Dolostone and Kinnikinic Quartzite and the continental-slope Middle and Upper Ordovician Saturday Mountain Formation. Although a similar juxtaposed facies contrast is not exposed elsewhere in the belt, a sole thrust is inferred to underlie the Ordovician to Devonian continental-rise rocks in central Idaho. We interpret this thrust to be a northward extension from Nevada of the Roberts Mountains thrust system. We believe that the Ordovician to Devonian continental-rise rocks were intensely folded and faulted during Antler orogenic deformation and were emergent during most of Mississippian time when they formed part of the Antler highlands, which shed flysch and molasse sediments eastward into the Antler foreland basin to form the Copper Basin Formation. The Wood River Formation of Middle and Late Pennsylvanian and Early Permian age was then deposited over a subdued topography on the Ordovician to Devonian allochthonous sequence. Eastward thrusting of the sequence of Phi Kappa, Trail Creek, unnamed unit, Milligen, Copper Basin, and Wood River Formations during the Mesozoic destroyed most of the original depositional contacts between the Ordovician to Devonian formations and the Copper Basin and Wood River Formations (Sandberg and others, 1975; Dover and others, 1980).

ALLOCHTHONS IN SOUTHERN CALIFORNIA

Allochthonous Paleozoic marine rocks of the Garlock assemblage in the El Paso Mountains (EPM) and Pilot Knob Valley (PKV), and equivalent rocks at Goldstone (G) and Lane Mountain (LM) in southern California (Plate 3-5), consist of a tectonically shuffled and diverse assemblage of chert, argillite, siltite, limy sandstone, quartzite, conglomerate, greenstone, and silty to sandy limestone and dolostone (Carr and others, 1981, 1982, 1984; Poole and others, 1981, 1982). The Garlock assemblage is, in part, correlative with rocks of the Antler orogen, foreland basin, and overlap assemblages of Nevada. Most of the Ordovician and Devonian rocks of the Garlock assemblage resemble the Ordovician Vinini Formation and Devonian Slaven Chert in central Nevada and the unnamed Devonian sandy limestone of Stanley and others (1977) at Miller Mountain (Plate 3-5, MM) in southwestern Nevada. In the El Paso Mountains, much of the assemblage contains Ordovician rocks representing two different deep-water marine facies—a probable continental-rise facies in the west and a slightly shallower-water facies that may represent upper continental rise or lower slope to the east. The juxtaposition of these two Ordovician facies was probably the result of pre-Carboniferous deformation because a flysch- and molasse-like sequence of conglomerate, quartzite, siltite, and argillite, similar to Antler foreland-basin deposits of Nevada, overlies Ordovician argillite of the western Ordovician facies on an apparently faulted unconformity in the central part of the range. Juxtaposition of lower and middle Paleozoic continental-rise and continental-slope rocks with continental-shelf and cratonic-platform rocks in the central Mojave Desert may also be an Antler-related event. The entire Garlock assemblage was intensely deformed, metamorphosed, and intruded by granitoid plutons during latest Permian time (Carr and others, 1982).

ALLOCHTHON REMNANTS IN SOUTHERN SIERRA NEVADA

Several roof pendants in the southeastern Sierra Nevada contain metasedimentary and metavolcanic rocks similar to Paleozoic rocks in the Antler orogen. The Indian Wells (IW), Kennedy (K), and Bald Mountain (BM) pendants (Plate 3-5) are composed of metasedimentary rocks similar to Ordovician to Mississippian facies in the nearby El Paso Mountains and in southwestern Nevada (M. D. Carr, USGS, and G. C. Dunne, California State University, personal communication, 1985; Taylor, 1984; Dunne and Suczek, 1991). Because of the apparent affiliation of the metasedimentary rocks in the pendants to the Garlock assemblage, they are here included with rocks of the Antler orogen. The Kennedy and Bald Mountain pendants in southeastern Tulare County contain stratiform barite and iron sulfide minerals in addition to quartz-mica schists, quartzites, phyllites, and minor beds of marble, slate, hornfels and metavolcanic rocks (P. E. Maciulaitis, geologic consultant, personal communication, 1982; Taylor, 1984; G. C. Dunne, personal communication, 1987). Although stratiform barites have not been recognized in the Garlock assemblage of the El Paso Mountains, Ordovician and Devonian stratiform barites occur in similar facies of west-central Nevada (Poole, 1988).

DEVONIAN AND OLDER ROCKS IN ACCRETED TERRANES

Raul J. Madrid, Forrest G. Poole, and M. Meghan Miller

Unlike relatively well-constrained paleogeographic relations of rocks within the Roberts Mountains allochthon (RMA), ties between the North American continent and lithotectonic terranes outboard of the RMA are little understood. These accreted terranes, shown in red on Plate 3-5, are fault bounded and have distinct geologic histories, as revealed by their stratigraphic, igneous, and metamorphic histories. Although the terranes indicated on Plate 3-5 contain rocks that range in age from Proterozoic to Devonian, most of the terranes were metamorphosed and/or finally accreted to the North American continent during late Paleozoic and Mesozoic time. Some of our terrane boundaries indicated on Plate 3-5 are similar to those of Silberling and others (1987). Silberling and others (1987) emphasized that (1) disparities within and between the various terranes and adjacent parts of the North American continent are so great that the original sites of the terranes and their displacement paths through time are difficult or impossible to determine; (2) some terranes of the western U.S. Cordillera may have been displaced great distances before their accretion to the North American continent, whereas others formed in continental-margin settings; and (3) displacement of terranes by strike-slip faulting along or within the continental margin has been especially important in the northern Cordillera, and original accretionary relations have been significantly obscured by subsequent strike-slip fault displacements.

We discuss 14 allochthonous assemblages in the western U.S. Cordillera that include Upper Devonian arc sequences and/or older Ordovician through Devonian oceanic sequences (Plate 3-5). These terranes or belts include the Coyote Mountains terrane (CY) in southern California; Kings River terrane (KR), Northern Sierra Nevada terrane (NSI), eastern Klamath Mountains terranes of the Redding area (RE), Trinity terrane (TR), Yreka terrane (YR), and Central Metamorphic terrane (CM) in northern California; Grindstone terrane (GS) in central Oregon; Roaring Creek terrane (RC) in central Washington; San Juan terrane (SJ), Chilliwack terrane (CK), and Yellow Aster Complex (YA) in northwestern Washington; Kootenay terrane (KO) in northeastern Washington; and Thompson Creek terrane (TC) in central Idaho. Recent detailed studies have greatly improved our knowledge of the ages, stratigraphy, and structure of rocks within these terranes and indicate that sedimentation patterns abruptly changed prior to Middle Devonian time. Complicated structural and metamorphic histories, however, make it difficult to assess paleogeographic positions of the terranes relative to the continental margin of the western U.S. Cordillera. Several sedimentological models linking these terranes are permissible, and some workers (e.g., Miller, 1987) have suggested that many of the arc terranes shared a common convergent plate-margin origin at least during late Paleozoic time. Until paleomagnetics, rock-age constraints, and stratigraphic relations within the arc terranes are better known, however, some original paleogeographic positions remain elusive.

We describe stratigraphic and structural relations and biostratigraphic and radiometric age constraints for only the relatively well-understood terranes within the northern Sierra Nevada and the Klamath Mountains (Plate 3-7) because of the fragmentary data base represented by most lower Paleozoic allochthonous terranes. Other lower Paleozoic terranes record continental-margin, basinal, or oceanic settings; their original paleogeographic relations are little understood. Although only the better known Klamath Mountains and Northern Sierra Nevada terranes are described in detail, the lithologies, structural history, age constraints, and tectonic affinity of each allochthonous terrane shown on Plate 3-5 are discussed, beginning with the Coyote Mountains terrane in southern California and proceeding northward to terranes of northern Washington and southern British Columbia.

COYOTE MOUNTAINS TERRANE

Allochthonous metasedimentary rocks having a minimum structural thickness of 1,400 m in the Coyote Mountains in western Imperial County, southern California (Plate 3-5, CY), have yielded Early Ordovician conodonts (Dockum and Miller, 1982; Gastil and Miller, 1983; Miller and Dockum, 1983). This structurally complex, fault-bounded assemblage is mostly composed of marble and minor amounts of quartz-muscovite schist and metaquartzite. Coarsely crystalline marble southeast of Carrizo Mountain contains a conodont fauna of middle Ibexian age (middle and late Early Ordovician). Miller and Dockum (1983) correlated this Lower Ordovician carbonate unit with the lower part of the Lower and Middle Ordovician Pogonip Group in Nevada and Utah. Similar carbonate-rich strata of probable early Paleozoic age in the Santa Rosa Mountains west of the Salton trough (Miller and Dockum, 1983) may be a northerly extension of the Coyote Mountains terrane. Because of uncertainties of stratigraphic affinity and original geographic position of the Coyote Mountains assemblage, we cannot with confidence physically correlate it with continental-shelf and continental-slope deposits of the western margin of Paleozoic North America; instead, we consider it to be part of an accreted terrane. The assemblage was metamorphosed by Mesozoic intrusions of the Peninsular Ranges batholith (Miller and Dockum, 1983).

NORTHERN SIERRA NEVADA TERRANE

Shoo Fly Complex

In the northern Sierra Nevada (Plate 3-5, NSI; Plate 3-7), complexly deformed Cambrian to Middle Devonian oceanic rocks of the Shoo Fly Complex are unconformably overlain by an Upper Devonian to upper Paleozoic arc sequence (Durrell and Procter, 1948; D'Allura and others, 1977; Varga and Moores, 1981) to the east. The Shoo Fly consists of four allochthons emplaced and exposed in pre–Late Devonian time and it com-

prises a complex eastward-dipping structural stack (Plate 3-7) consisting of, from top to bottom and from east to west, the Sierra City mélange, Culbertson Lake allochthon, Duncan Peak allochthon, and Lang-Halsted sequence (Schweickert and others, 1984). The Shoo Fly Complex trends in a northerly to northwesterly direction within a narrow band approximately 500 km in length (Plate 3-5) along the western flank of the Sierra Nevada batholith and east of the Melones fault zone and the Calaveras–Shoo Fly thrust in the northern and central Sierra Nevada. The Shoo Fly is bounded on the west by Jurassic and older volcanic rocks of the Sierra Nevada foothills and on the east by upper Paleozoic, Triassic, and Jurassic volcanic rocks of the high Sierra Nevada.

The Shoo Fly Complex consists primarily of deep-water siliciclastic rocks including quartz- and feldspar-bearing sandstone, siltstone, mudstone, phosphatic and radiolarian chert, minor greenstone, ultramafic rocks, bedded tuff, and volcanogenic sedimentary rocks; a limestone and sedimentary mélange that was deposited in an oceanic environment is locally present. These rocks have been metamorphosed to greenschist grade, although higher metamorphic grades are present in the southern exposures where protoliths have been correlated with the less metamorphosed rocks in the north (Merguerian, 1985; Sharp, 1988). Later episodes of metamorphism, especially during the Mesozoic, have obscured earlier metamorphic and deformational events. The internal stratigraphy of the Shoo Fly is not well known (Plate 3-7) due to the absence of fossil ages, except in the Sierra City mélange, and the complex and multiple episodes of deformation to which the units have been subjected. Despite this fact, additional data acquired from the Shoo Fly Complex establish it as one of the best understood accreted terranes of the western U.S. Cordillera. Several workers have argued that many of the rocks in the Shoo Fly were deposited in continental slope-rise environments along the western margin of Paleozoic North America (D'Allura and others, 1977; Schweickert and Snyder, 1981; Varga, 1982; Hannah and Moores, 1986), although no definitive stratigraphic studies have been made.

Sierra City mélange. The Sierra City mélange at the top of the structural stack (Plate 3-7, column A) is the most heterogeneous allochthon. It consists of "broken formation" and includes blocks of chert; fossil-bearing limestone; dolostone; lithic-, quartz-, and feldspar-rich sandstone; volcanic rocks; and gabbro and small pieces of serpentinized ultramafic rocks in a matrix of sheared shale and sandstone (Schweickert, 1981; Bond and Schweickert, 1981; Girty, 1983; Girty and Schweickert, 1984). Detrital modes of an arkosic sandstone block within the Sierra City mélange, which yielded a U-Pb intercept age of 506 ± 22 Ma, require a provenance of volcanic and plutonic rocks (Girty and Wardlaw, 1984). Because such an Upper Cambrian to Lower Ordovician provenance terrain is not known to have been exposed in California or Nevada during Middle Ordovician to Middle Devonian time, Girty and Wardlaw (1984) suggested that the source terrain of the volcanic-derived sandstones was rifted away. Quartzose sandstones thought to have been derived from the underlying Culbertson Lake allochthon are present as tectonic blocks in the Sierra City mélange (Girty and Wardlaw, 1984) and suggest a sedimentary link between strata of the two allochthons.

Beds in a dolostone and limestone block in the Sierra City mélange contain oolites and stromatolites indicative of shallow-water deposition. The presence of soft-sediment deformation in interbedded sandstone and shale below the lens suggests that it was emplaced as a slide block while the sediments were still soft (Bond and DeVay, 1980). Although the Sierra City mélange has been interpreted as part of an accretionary prism (D'Allura and others, 1977; Bond and Schweickert, 1981) emplaced over the other allochthons of the Shoo Fly, D'Allura and others (1977) suggested that the mélange is of sedimentary origin.

Fossil and zircon dates indicate that rocks of the Sierra City mélange range from Ordovician to possibly Silurian in age (Varga and Moores, 1981), although a Silurian age is not well defined by diagnostic fossils. Ashgillian (Late Ordovician) megafossils are present in the Montgomery Limestone, which occurs either as exotic tectonic blocks (D'Allura and others, 1977) or as submarine slide blocks (Bond and DeVay, 1980) in the Sierra City mélange. The Montgomery Limestone has also yielded conodonts that indicate an Ordovician age and display North American affinities (A. G. Harris, *in* Hannah and Moores, 1986). Zircon from a submarine tuff in the Sierra City mélange yielded a U-Pb age of 423 +5/−5 Ma (Early Silurian) for the upper part of the Shoo Fly Complex (Saleeby and others, 1989). This age suggests that the Early Silurian part of the Shoo Fly Complex was characterized by volcanism. Zircon obtained from a plagiogranite block from a structural slice of serpentinite-matrix mélange within the Shoo Fly Complex below the depositionally overlying Sierra Buttes volcanic sequence yielded a U-Pb intercept age of 600 ± 10 Ma (J. B. Saleeby and R. A. Schweickert, unpublished data). Saleeby (this volume) suggests that the plagiogranite block represents part of an ophiolite assemblage that presumably floored the Shoo Fly Complex. Its age suggests that this oldest recognized ophiolite underlying the Shoo Fly Complex has an age compatible with continental rifting of ~600 Ma, as suggested by Bond and Kominz (1984).

Culbertson Lake allochthon. The Culbertson Lake allochthon (Plate 3-7, column B) has been studied in detail by Girty (1983), who defined two stratigraphic units: the Bullpen Lake and Bowman Lake sequences. The Bullpen Lake sequence, the lowest exposed unit, is 810 m thick and consists primarily of pillow basalt, massive basalt flows, and basaltic breccia overlain to the east by oolitic and pisolitic limestone, argillite, and spiculitic and radiolarian chert. The depositionally overlying Bowman Lake sequence is 815 m thick and consists of rhythmically bedded radiolarian and spiculitic chert, argillite, and two informal sandstone units—the formation of Poison Canyon and overlying formation of Red Hill (Girty and Schweickert, 1979, 1984; Girty, 1983).

Detrital modes of sandstones and U-Pb isotopic ages of detrital zircons in the Shoo Fly Complex differ considerably,

suggesting compositionally and isotopically heterogeneous source terrains. Sandstones of the Culbertson Lake allochthon display detrital modes that indicate derivation from plutonic or metamorphic rocks of a continental-crust provenance (Bond and DeVay, 1980; Girty and Wardlaw, 1984). These sandstones are similar to the Ordovician and/or Silurian Antelope Mountain Quartzite in the Klamath Mountains and the Upper Cambrian Harmony Formation in the RMA of central Nevada (Bond and DeVay, 1980). Bond and DeVay concluded that the sandstones of the Shoo Fly Complex are not similar to the Ordovician Valmy Formation, but Madrid (1987) concluded that many sandstones in the Valmy and other formations of the RMA are similar to those in the Shoo Fly. The intercept age of detrital zircon in a sample of continent-derived sandstone from the Culbertson Lake allochthon is ~2.09 Ga (Girty and Wardlaw, 1985), similar to the ~2.2-Ga age of zircons from the feldspathic Elder Sandstone in the RMA (Girty and others, 1985). These ages, together with the detrital modal data, support the conclusion that the sandstones of the Shoo Fly and RMA are similar.

Sparsely fossiliferous oceanic rocks of the Shoo Fly Complex in the Culbertson Lake allochthon were deposited in deep-water environments including deep-sea channels (Bond and DeVay, 1980).

U-Pb isotopic data on zircons from igneous bodies that intrude the Culbertson Lake allochthon indicate ages of 385 to 364 Ma for the Bowman Lake batholith (Hanson and others, 1988) and 378 ± 5 Ma for the Wolf Creek stock (Saleeby and others, 1987). These intrusive rocks collectively cut all the allochthons of the Shoo Fly Complex. The Culbertson Lake allochthon appears to have undergone a major deformational event prior to the intrusion of these Middle Devonian plutons (Merguerian, 1985; Schweickert and others, 1984).

Duncan Peak allochthon. The Duncan Peak allochthon (Plate 3-7, column C), structurally above the Culbertson Lake allochthon, consists of "broken formation" at its base overlain by a poorly understood sequence of chert, argillite, and minor sandstone units, which have not yielded fossil ages.

Lang-Halsted sequence. The Lang-Halsted sequence (Plate 3-7, column D) in the central Sierra Nevada has been studied by Merguerian (1985). These rocks are highly metamorphosed but are interpreted to represent an assemblage of feldspathic sandstone, subordinate mudstone, chert, limestone, and rare volcanic rocks (Merguerian, 1985). These unfossiliferous rocks are gradationally interbedded in places and locally faulted, and stratigraphic sequences have not yet been determined. Siliceous rocks in the Lang-Halsted sequence in northern exposures of the Shoo Fly Complex are similar to siliceous rocks in the central Sierra Nevada (Schweickert, 1977; Merguerian, 1985).

Isotopic dating of zircon, sphene, and amphibole from gneissic granitoids of the Lang-Halsted sequence indicate that their protoliths intruded the Shoo Fly Complex at 440 ± 15, 330 ± 10, and 275 ± 10 Ma (Sharp and others, 1982). Revision of the older ages to 370 Ma for the Confluence granitic orthogneiss (Sharp, 1988) and 363 Ma for the syenitic orthogneisses of upper Rose Creek (Brueckner *in* Merguerian and Schweickert, 1987) suggests that the intrusions occurred in the Late Devonian.

Upper Devonian arc sequence

A complex Upper Devonian and upper Paleozoic sequence of magmatic-arc pyroclastic rocks and associated epiclastic turbiditic and hemipelagic rocks unconformably overlies the Shoo Fly Complex (Plate 3-7). The basal part of this sequence, the Upper Devonian Grizzly Formation, is a discontinuous succession of sandstone, conglomerate, and sedimentary breccia interbedded with deep-water radiolarian chert and pelitic rocks. The cherts have yielded Late Devonian (Famennian) radiolarians (B. L. Murchey, USGS, oral communication, 1987). The Grizzly fills local paleobathymetric irregularities along an angular unconformity between the overlying pyroclastic sequence and the underlying Shoo Fly (McMath, 1958; Hanson, 1983; Hanson and Schweickert, 1986). In some places, it consists predominantly of laterally continuous, laminated to thinly bedded chert and argillite, and locally abundant phosphatic nodules. Sandstones of the Grizzly are quartzofeldspathic, rarely chert and lithic rich, and display features suggestive of turbidity-current origin. The Grizzly conglomerates contain well-rounded clasts of distinctive white to gray chert, sandstone, and limestone from the underlying Shoo Fly Complex (Schweickert and others, 1984).

Depositionally above the Grizzly Formation is the Sierra Buttes Formation, a stratigraphically complex assemblage of submarine pyroclastic and epiclastic deposits and radiolarian cherts (Hanson and Schweickert, 1986). The Sierra Buttes at Dugan Pond has yielded ammonoids that are dated as Late Devonian and equated to the middle Famennian *trachytera* conodont Zone (Sandberg and others, 1988, Fig. 16). The lower part of the formation consists of shallow plugs, dikes, and sills of rhyodacite; at Sierra Buttes, these intrusive rocks in the Sierra Buttes Formation are suggestive of a vent complex (Schweickert, 1981).

The upper part of the formation consists predominantly of pyroclastic rocks that grade upward into the Elwell Formation. Deposition of the overlying Elwell, Taylor, and Peale Formations spanned Late Devonian to Early Mississippian time (e.g., Hannah and Moores, 1986). The contact of the Peale with overlying rocks represents a major hiatus that separates predominantly Upper Devonian and Lower Mississippian rocks from rocks of Permian arc magmatism.

KINGS RIVER OPHIOLITE

Rocks of the Kings River ophiolite (Plate 3-5, KR) in the southern Sierra Nevada have yielded a Sm-Nd isochron age of 485 ± 21 Ma, and a plagiogranite has yielded a U-Pb upper intercept age of 430 +200/−60 Ma that lies on the 483 Sm-Nd isochron (Shaw and others, 1987). The samples that yielded the 483-Ma isochron have isotopic ratios compatible with a mid-ocean ridge origin. Saleeby and others (1989) argued that the Kings River ophiolite, the Feather River peridotite belt of Day

and others (1985), and the Trinity ultramafic sheet (Jacobsen and others, 1984) collectively represent a 1,000-km-long belt of early Paleozoic oceanic lithosphere greatly depleted in certain trace elements. They suggested that the Feather River peridotite belt and the Shoo Fly Complex were a composite basement for the Sierra Buttes volcanic arc and that Devonian or older metamorphism in the Feather River peridotite belt may record tectonic juxtaposition of this depleted oceanic lithosphere below the Shoo Fly.

KLAMATH MOUNTAINS REGION

The lower and middle Paleozoic rocks of the Klamath Mountains, like those of the Northern Sierra Nevada terrane, record a complex stratigraphy of varied metamorphic fabrics and repetition by folding and faulting. Despite these complications, recent detailed studies have determined the ages of sequences and the lithic types found within the sequences (Plate 3-7). Lower Paleozoic rocks occur in the two easternmost and structurally highest, arcuate lithologic belts of the Klamath Mountains—the Central Metamorphic belt and the easterly and structurally overlying Eastern Klamath belt (Irwin, 1977; Cashman, 1980). The Eastern Klamath belt contains three terranes: (1) the structurally lowest and westerly Trinity ultramafic sheet, (2) lower Paleozoic sedimentary and volcanic rocks of the Yreka-Callahan area, and (3) Devonian to Jurassic volcanic-arc-related strata in the vicinity of Redding. Rocks of the lithologic belts have been variably metamorphosed from subgreenschist to amphibolite facies.

There are some permissive sedimentologic ties between rocks of the Yreka assemblage and the Redding assemblage, underlying Trinity ultramafic sheet, and possibly the Central Metamorphic belt. Environments of deposition recorded by lower Paleozoic sedimentary rocks of the Klamath Mountains are compatible with continental-rise and continental-slope settings, although in many cases other settings are likely. For instance, volcanic and volcaniclastic rocks compose major parts of the sequences, especially in the lower Middle Ordovician and the Lower to Middle Devonian. Provenance terrains for siliciclastic rocks include either plutonic-metamorphic or predominantly volcanic terrains. Other clastic rocks record intraformational sources. Rocks younger than Middle Devonian have not been recognized in the numerous lithotectonic sequences; pre–Middle Devonian intrusive and deformational events have been described by several workers. Similarities in detrital modes and gross lithic types suggest a sedimentary link between the Klamath Mountains and the Northern Sierra Nevada terranes (Plate 3-7).

Yreka-Callahan area

The Yreka-Callahan area (Plate 3-7, columns E and F) contains a diverse assemblage of lithic types and stratigraphic sequences separated by thrust faults (Potter and others, 1977). The sequences have been subdivided into four major lithotectonic units. The structurally lowest unit includes a basal keratophyre depositionally overlain by Ordovician and Silurian clastic rocks and limestone of the Horseshoe Gulch and Lover's Leap sequences. The keratophyre is probably arc related, whereas overlying clastic rocks range from relatively deep-basin to shallow-shelf deposits, as suggested by fossil evidence including Middle Ordovician graptolites and Late Silurian corals and brachiopods (Potter and others, 1977). In places, Silurian or younger conglomerates commonly contain clasts of trondhjemite, hornblende-biotite quartz diorite, keratophyre, and red and pink limestone containing Middle Ordovician conodonts. The trondhjemite and diorite clasts are believed to have been derived from the Trinity ultramafic sheet (Mattinson and Hopson, 1972; Potter and others, 1977).

Shale, calcareous siltstone, sandstone, and minor bedded chert and limestone of the Moffett Creek Formation are faulted against the Horseshoe Gulch and Lover's Leap sequences (Plate 3-7, columns H and I). According to Potter and others (1977), the Moffett Creek locally consists of true tectonic mélange, but elsewhere it consists of mélange that can be ascribed to soft-sediment deformation. In the few places where original stratigraphic succession is preserved, siltstone and fine sandstone are interbedded and exhibit rare fluting and graded bedding suggesting that part of the Moffett Creek Formation formed as part of a submarine fan. A limestone clast from siliceous mudstone of the Moffett Creek contains abundant branching corals and a Silurian brachiopod. The Moffett Creek is almost everywhere in fault contact with adjacent units; however, at one locality it is depositionally overlain by the Gazelle Formation. There, the age of the Gazelle has been determined to be probably Late Silurian based on brachiopods. Thus, the Moffett Creek Formation is Silurian.

In contrast to the Moffett Creek Formation, the Gazelle Formation (Plate 3-7, column F), although locally complexly faulted, is a relatively coherent package of hemipelagic, pelagic, and turbiditic chert, shale, siltstone, sandstone, and coarser clastic rocks, and minor volcanogenic clastic rocks, volcanic breccia, porphyritic albite-clinopyroxene keratophyre, and massive mafic pillow lavas. The Early Silurian to Middle Devonian age of the Gazelle is well established on the basis of graptolites, conodonts, corals, and brachiopods (Potter and others, 1977). The keratophyre and spilitic pillow lavas occur in the Lower Devonian part of the section. Of three members of the Gazelle, the middle member is the least structurally disrupted and contains well-bedded sequences of siltstone and shale. The lower and upper members are structurally disrupted and display abrupt sedimentary facies changes attributable to mass transport by debris flows and gravity slides. These suggest slope, slope-break, and slope-channel inner-fan environments for much of the Gazelle (Potter and others, 1977).

Coarse clastic rocks of the Gazelle Formation are partly intraformational, but some beds contain clasts derived from volcanic and minor plutonic sources (Condie and Snansieng, 1971; Potter and others, 1977). A permissible source for the plutonic clasts is the Trinity ultramafic sheet. The source for possible quartz keratophyre clasts may have been the Horseshoe Gulch unit described above. The structurally highest lithotectonic unit of

Potter and others (1977) is the Yreka assemblage (Plate 3-7, column E), which previously was included in the Duzel Formation of Wells and others (1959). This assemblage occurs above the Mallethead thrust and consists of the Duzel Phyllite, Sissel Gulch Graywacke, informal Schulmeyer Gulch sequence, Antelope Mountain Quartzite, and the previously discussed Moffett Creek Formation. The ages and stratigraphic relations of these units are not known with certainty.

Overall, the siltstone and sandstone of the Yreka assemblage contain detritus possibly derived from a plutonic and metamorphic provenance (Potter and others, 1977) that was deficient in mafic rocks. The Ordovician and/or Silurian Antelope Mountain Quartzite was derived from a continental-crust provenance and is quartz rich and feldspar poor, similar to parts of the Shoo Fly Complex of the northern Sierra Nevada (Bond and DeVay, 1980). Some sandstones from the Moffett Creek Formation and the Sissel Gulch Graywacke apparently were derived from sedimentary and silicic and mafic volcanic sources. The Sissel Gulch Graywacke does not contain white mica and microcline as do other units of the Yreka assemblage, and it may have had a different source.

Redding area of Eastern Klamath belt

In the Redding area (Plate 3-5, RE; Plate 3-7, column J), a Devonian to Jurassic pyroclastic, siliciclastic, and volcanic sequence overlies the Trinity ultramafic sheet (column K). The Lower to Middle Devonian part of the sequence consists of the Copley Greenstone, which interfingers with and is overlain by rhyolitic flows and tuffs of the Balaklala Rhyolite (Plate 3-7). The Balaklala Rhyolite is gradationally overlain by shales and limestones of the Middle Devonian Kennett Formation. The Copley and Balaklala volcanic units have been metamorphosed to greenschist facies and locally to zeolite and amphibolite facies (Barker and others, 1979). The amphibolite metamorphism may be related to intrusion of the Shasta-Bally pluton, but the greenschist and zeolite facies metamorphism may be related to burial (Kinkel and others, 1956; Barker and others, 1979).

The Copley Greenstone consists of massive flows and pyroclastic rocks overlain by pillow lavas in a sequence approximately 1,800 m thick (Kinkel and others, 1956; Lapierre and others, 1985). Some of the lower part of the Copley was apparently deposited under subaerial conditions in an island-arc environment (Lapierre and others, 1985).

The Balaklala Rhyolite is subdivided into three units (Kinkel and others, 1956) on the basis of phenocryst content. Pyroclastic rocks of the lower unit containing clasts of nonporphyritic rhyolite grade upward to porphyritic rhyolite flows and bedded tuffs of the middle unit. The middle unit is overlain by coarse phenocrystic rhyolite flows of the upper unit, some of which have ignimbritic textures. The highest part of the upper unit consists of bedded pyroclastic rocks that grade upward into the Kennett Formation (Lapierre and others, 1985). Submarine slumping and syndepositional faulting are evidenced in the lower unit, and mainly submarine environments are inferred for the middle unit; part of the upper unit, however, reflects subaerial environments. One of the tuff units in the Balaklala Rhyolite near Redding yielded a Middle Devonian (Eifelian) fish plate that has been used to date the sequence (Boucot and others, 1974). Albers and others (1981) reported a concordant U-Pb zircon isotopic age of 400 Ma and a K-Ar hornblende age of 392 ± 3 Ma for the Mule Mountain trondhjemitic stock (Plate 3-7, column J), which is believed to be cogenetic with the Balaklala Rhyolite. The zircon age is Early Devonian, however, and about 15 m.y. older than the fossil fish age based on the DNAG time scale. Above the Balaklala Rhyolite is the Kennett Formation, which consists of interbedded volcanic and epiclastic rocks. These are overlain by shallow-water limestone that contains late Eifelian coral debris (Kinkel and others, 1956; Watkins and Flory, 1986). Above the Kennett is an unconformity separating these predominantly Middle Devonian rocks from the overlying predominantly Mississippian rocks of the Bragdon Formation (column J).

Trinity ultramafic sheet

The east-dipping Trinity ultramafic sheet occurs in an area 50 km wide and 75 km long (Plate 3-5, TR; Plate 3-7, column K) in the east-central Klamath Mountains. It consists of ultramafic and related rocks that lie structurally between the Yreka-Callahan assemblage to the north and underlying rocks of the Central Metamorphic belt to the west (Quick, 1981). Most of the geochemical data for the Trinity ultramafic sheet suggest that it has tholeiitic affinities; possibly it was diapirically emplaced within the upper crust in the vicinity of an island-arc or back-arc basin (Quick, 1981). This ultramafic sheet has been described by Lindsley-Griffin (1977) as an ophiolite; however, Quick (1981) argued that gabbros described by Lindsley-Griffin (1977) as stratigraphically overlying the ultramafic sheet appear to intrude it. Jacobsen and others (1984) suggested that values of initial ϵ_{Nd} = +6.7 and +7.3 indicate that gabbros of the Trinity sheet are not geochemically related to the ultramafic rocks, which yielded a value of initial ϵ_{Nd} = +10.4 ±0.4. These data support Quick's (1981) conclusion that the gabbros are unrelated to the Trinity ultramafic sheet.

Isotopic dating of crystallization of these rocks is conflicting (column K) but indicates a general Middle Ordovician to earliest Silurian age. Recently obtained Cambrian zircon ages confirm that the Trinity is a polygenetic body of mafic and ultramafic assemblages (Wallin and others, 1988). A minimum age for structural emplacement of the Trinity sheet and overlying rocks is inferred to be 374 to 371 Ma or early Late Devonian, assuming metamorphic Rb/Sr isotopic ages for samples from underlying rocks of the Central Metamorphic belt (Lanphere and others, 1968) represent structural emplacement ages.

Central Metamorphic belt

The Central Metamorphic belt consists of greenschist- to amphibolite-facies marble, quartz-mica schist, and amphibolite of the Grouse Ridge Formation and hornblende schist of the Salmon Formation (Plate 3-7, column L; Davis and others, 1965;

Davis, 1968; Irwin, 1977; Cashman, 1980). Protoliths of the Central Metamorphic belt include gabbro, pyroclastic rocks, alkalic basaltic flow rocks, and minor calcareous sedimentary rocks (Cashman, 1980). Minimum isotopic ages of 399 to 350 Ma (Lanphere and others, 1968; Hotz, 1977; Cashman, 1980) were obtained for metamorphism of the rocks that form the Central Metamorphic belt. These ages also date emplacement of the Trinity ultramafic sheet during initiation of subduction along a long-lived convergent plate margin (Peacock, 1985, 1988).

SAN JUAN ISLANDS TERRANE

Exposed on the San Juan Islands in northwestern Washington (Plate 3-5, SJ) are igneous rocks of the Turtleback Complex and associated volcanogenic rocks of the group of East Sound, an informal name used by Brandon and others (1988). The Turtleback Complex on the San Juan Islands consists of a varied assemblage of gabbro, quartz diorite, and amphibolite. Quartz diorite, which intrudes amphibolitic rocks, has been isotopically dated as 437 ± 10 Ma and 409 ± 25 Ma, or Early and Late Silurian (Mattinson, 1972; Whetten and others, 1978). The amphibolitic rocks may be much older, possibly Precambrian, because a pegmatite dike that cuts the gabbroic rocks yielded a K-Ar hornblende isotopic age of 554 ± 16 Ma, or Early Cambrian (determined by G. B. Dalrymple, *in* Whetten and others, 1978).

Above the Turtleback Complex are volcanogenic rocks of the Devonian group of East Sound (Brandon and others, 1988), which consists of sandstone, turbiditic sandstone, graywacke, argillite, volcanic rocks, limestone, and minor bedded chert and conglomerate (Danner, 1977). The group has a minimum thickness of approximately 450 m. The age of the group is Early to Late Devonian based on a Late Devonian brachiopod and other fossils of Early and Middle Devonian age (Brandon and others, 1988). Depositional settings for the sequence have not yet been assessed, but the lithic types suggest an oceanic environment in which submarine-fan sediments were deposited.

CHILLIWACK AND GRINDSTONE TERRANES

The little-known Chilliwack and Grindstone terranes are composed of Devonian and upper Paleozoic rocks that are in fault contact with older and younger rocks. The Chilliwack terrane is in Washington and southern Canada; the Grindstone terrane crops out in east-central Oregon (Plate 3-5).

The Devonian to Permian Chilliwack Group (Plate 3-5, CK) consists primarily of reefoid limestone associated with volcanic rocks, shale, argillite, graywacke and minor chert. The reef limestone may be as young as early Late Devonian (Frasnian); collectively, the abundant corals, brachiopods, protozoans, and ostracodes suggest that much of the Devonian is represented (Danner, 1977).

The Grindstone terrane in central Oregon (Plate 3-5, GS) was included in the Grindstone-Twelvemile mélange by Dickinson and Thayer (1978). The Grindstone terrane is a sedimentary breccia composed of discrete blocks of middle and upper Paleozoic rocks set in a matrix of mostly Triassic volcaniclastic siltstone, sandstone, and conglomerate (Blome and Nestell, 1988). Some Devonian limestone blocks contain brachiopods, conodonts, and corals. The Triassic turbiditic and debris-flow rocks are unconformably overlain by fossiliferous Lower and Middle Jurassic basin-slope rocks (Blome and Nestell, 1988).

YELLOW ASTER COMPLEX

In the northern Cascades of Washington (Plate 3-5, YA), the Yellow Aster Complex consists of predominantly amphibolite-facies rocks and associated, less mafic quartz diorite, diorite, and pegmatite gneiss. Pyroxene gneiss yields an isotopic age of 1,452 ± 20 Ma (middle Proterozoic), and quartz diorite gneiss and pegmatite gneiss range from 427 to 411 Ma (Silurian; Mattinson, 1972). Some of the rocks have been considered to be Precambrian continental basement (Misch, 1966, 1988), but no consensus exists as to whether the protoliths were sedimentary. No sedimentary or igneous link has been demonstrated between the Yellow Aster Complex and younger rocks of the region. These Proterozoic and Silurian rocks may belong to a mini-continent or terrane that reached its present position relative to cratonic North America in Cretaceous time (Hamilton, 1978).

ROARING CREEK TERRANE

The Roaring Creek terrane in central Washington (Plate 3-5, RC) was defined by Silberling and others (1987) as a composite metamorphic terrane consisting of the Swakane and Mad River terranes of Tabor and others (1987) and the Chelan Mountains terrane of Tabor and others (1980). The Swakane terrane consists of the Swakane Biotite Gneiss, which is a relatively uniform granoblastic gneiss with planar foliation, and sparse, thin layers of hornblende schist, schistose amphibolite, and marble. U-Th-Pb isotopic dates on zircon from these rocks allow various interpretations but seem to indicate a Precambrian protolith age (Mattinson, 1972). The Mad River terrane structurally overlies the Swakane terrane along a low-angle fault and consists of complexly and isoclinally folded hornblende schist, schistose amphibolite, micaceous quartz schist, micaceous quartzite, biotite gneiss, and sparse metaconglomerate, calc-silicate schist, and marble, all of which appear to be intruded by orthogneiss containing zircon of probable Paleozoic or Precambrian age (Mattinson, 1972). The Chelan Mountains terrane is dominated by the Chelan Complex, consisting of migmatite derived mainly from Triassic tonalite (Hopson and Mattinson, 1971) and, to a lesser extent, from Permian or older metasedimentary and metavolcanic rocks (Tabor and others, 1987). The Roaring Creek terrane of Silberling and others (1987) was metamorphosed in Late Cretaceous and earliest Tertiary time and intruded by synkinematic to late kinematic tonalite to granodiorite plutons (Tabor and others, 1987).

KOOTENAY TERRANE

In Stevens County, northeastern Washington, (Plate 3-5, KO), multiply deformed and variably metamorphosed Devonian

and probable older and younger Paleozoic rocks of the Flagstaff Mountain, Pend d'Oreille, Glass Mountain, and correlative sequences of Yates (1964, 1971, 1976) are part of the Kootenay terrane of Silberling and others (1987). Because of the stratigraphic and structural complexities and probable equivalence of these sequences, they are referred to collectively in this report as the Flagstaff Mountain assemblage. The Flagstaff Mountain assemblage consists of argillite, phyllite, slate, and subordinate quartzite, siltite, chert, limestone, marble, tuff, greenstone, conglomerate, bedded barite, and layers of base-metal sulfide minerals (Poole, unpublished data). The conglomerate locally contains cobbles and boulders of granite and metagreenstone. Webster and Beka (1980) reported Middle and Late Devonian conodonts from three different limestone beds in the Flagstaff Mountain assemblage west of the Columbia River near Northport, Washington. These conodont-bearing limestone beds are within the "black shale belt" of Beka (1980).

The Covada assemblage (Covada Group of Campbell and Raup, 1964) in southeastern Ferry County and southwestern Stevens County (Plate 3-5) contains Ordovician and probable older and younger Paleozoic rocks, and is included in the Kootenay terrane. The Covada assemblage consists mainly of variably metamorphosed and complexly intercalated argillite, slate, siltstone, siltite, graywacke, quartzite, limestone, marble, chert, metachert, and lava flows (Campbell and Raup, 1964; Fox and Rinehart, 1974). Along the Columbia River between Kettle Falls and Hunters, Washington, graywacke of the Covada assemblage contains angular to subrounded grains of quartz, sodic plagioclase, potassium feldspar, mica, orthoquartzite, and sparse granitic rock (Fox and Rinehart, 1974). A granitic source terrain seems evident for these graywacke beds.

Although both low- and high-angle faults occur along the east margin of the Kootenay terrane, a well-defined high-angle fault or suture striking about N20 to 30°E parallels the Columbia River and marks the east boundary of the terrane from Fruitland, in the Hunters quadrangle, northward to Northport, in the Northport quadrangle. This fault or suture (Snook, 1982) approximates the western limit of North American continental crust based on an initial $^{87}Sr/^{86}Sr$ ratio of 0.706 (R. J. Fleck, USGS, personal communication, 1985). Although some allochthonous Paleozoic siliceous-assemblage rocks of unknown affinity occur east of Northport, most of the Precambrian to Devonian rocks east of the Kootenay terrane in northeastern Washington represent continental-slope and outer-shelf deposits of the western margin of North America (Fig. 2) and are included on our time-slice maps.

In northeastern Washington, rocks of the Kootenay terrane were metamorphosed and intruded by plutons during Mesozoic and early Cenozoic time (Fox and others, 1977).

THOMPSON CREEK TERRANE

Allochthonous lower and middle Paleozoic oceanic rocks of the Salmon River assemblage (Hobbs, 1990) in the Thompson Creek–Slate Creek area of central Idaho (Plate 3-5, TC) have uncertain affiliation with the Roberts Mountains allochthon and are here provisionally assigned to the Thompson Creek terrane. Rocks of the Thompson Creek terrane are strongly folded, faulted, and metamorphosed. The Salmon River assemblage consists of turbiditic siltite, quartzite, quartzwacke, and limestone interbedded with hemipelagic argillite and siltite (Poole, unpublished data). Many of the dark gray argillite and siltite units are metalliferous; bedded barite and bedded base-metal sulfide minerals occur locally. Samples from limestone turbidite beds, collected by Poole, W. E. Hall, G. A. Desborough, G. N. Green, and S. W. Hobbs, yielded conodonts that range in age from Late Cambrian through Late Devonian (J. E. Repetski and A. G. Harris, USGS, written communication, 1984).

Before the early and middle Paleozoic conodont dating was available, the Salmon River assemblage was considered to be part of the Antler foreland basin sequence. First, it was correlated with the Mississippian Copper Basin Formation by Hobbs and others (1975) and then renamed the Salmon River sequence by Nilsen (1977). The initial age assignment to the Mississippian was based on fossils contained in two limestone blocks collected by S. W. Hobbs near the Salmon River in Deer Spring Gulch and in Mill Creek valley. Subsequent attempts failed to recover additional fossiliferous blocks or to find comparable bedrock lithologies in the Salmon River assemblage. Thus, the fossiliferous blocks seem to be unrelated to the Salmon River assemblage in the Thompson Creek–Slate Creek area and are believed to be exotic blocks derived by erosion from higher structural units containing Mississippian rocks (Poole, unpublished data).

Nilsen (1977) recognized outer-fan and basin-plain facies in the Salmon River assemblage and reported east-directed paleocurrents for turbidities in the assemblage. These sediment transport directions seem incompatible with a continental-rise setting adjacent to the continental shelf of Paleozoic North America, and they suggest that the Salmon River assemblage is part of an accreted lithotectonic terrane rather than part of the Roberts Mountains allochthon.

A major low-angle fault on the east flank of Saturday Mountain in the southwest corner of the Clayton quadrangle (Hobbs and others, 1975; Hobbs, 1990) separates the easterly transported Salmon River assemblage of the Thompson Creek terrane from the underlying Ordovician Saturday Mountain Formation. The type Saturday Mountain Formation and associated stratigraphic sequences east of the Thompson Creek terrane in central Idaho represent continental-slope and outer continental-shelf deposits of the western margin of Paleozoic North America (Fig. 2) and are included on the time-slice maps.

SUMMARY OF ACCRETED TERRANES

Early and middle Paleozoic terranes of the western U.S. Cordillera constitute a diverse assemblage of lithotectonic sequences, which in recent years, have yielded much detailed information about ages, stratigraphy, and sedimentary environments. The most closely allied terranes are those of the Northern Sierra Nevada and the Klamath Mountains. Similarities in ages,

stratigraphic sequences, deformation, metamorphism, compositions of approximately coeval sandstones, and the nature of overlying upper Paleozoic sequences suggest that these terranes may have once constituted a continuous belt as suggested by Davis (1968). The youngest pre-Mississippian age of deformed strata in these terranes is Middle Devonian. These deformed strata are overlain unconformably by a Devonian to early Mesozoic magmatic arc sequence. Probable alkalic lavas and associated limestones, and turbidite sequences occur at similar positions in the poorly dated sections of both the Klamath Mountains and Northern Sierra Nevada terranes. Lower and middle Paleozoic rocks of the Moffett Creek and Gazelle Formations described as slope deposits in the Klamath Mountains may also be related to similar deposits in the northern Sierra Nevada. Pre–Middle Devonian deformation and metamorphism have been documented for the Sierran and Klamath rocks. The Kings River ophiolite and the Trinity ultramafic sheet have similar ϵ_{Nd} values of about +10 and have yielded compatible Ordovician isotopic ages. These ultramafic bodies, together with the late Paleozoic Feather River peridotite belt, constitute a discontinuous belt of ultramafic rocks within the Klamath Mountains and the Sierra Nevada that has a present-day length of approximately 1,000 km.

Other terranes of the western U.S. Cordillera generally exhibit pre–Middle Devonian deformation and metamorphism, but ages and stratigraphic details are less well known than those in the northern Sierra Nevada and the Klamath Mountains.

The ages of deformation and change in sedimentation patterns of the accreted lithotectonic terranes are substantially older than those recorded by rocks of the continental shelf and the Roberts Mountains allochthon. The rocks of the Northern Sierra Nevada and the Klamath Mountains terranes may have been outboard of the strata composing the Roberts Mountains allochthon and hence subjected to earlier deformation, different depositional patterns, and more severe metamorphism. A possible link between these accreted terranes and the Roberts Mountains allochthon is the similarity in lithology and isotopic ages of zircon (around 2.2 Ga) from feldspathic sandstones of the Culbertson Lake allochthon in the northern Sierra Nevada and the feldspathic sandstones of the Silurian Elder Sandstone in the Roberts Mountains allochthon in north-central Nevada. These data, together with the similarities in the detrital modes of sandstones in the Klamath Mountains, the Shoo Fly Complex, and the Roberts Mountains allochthon in Nevada, are permissive for establishing a tie between these terranes, the Roberts Mountains allochthon, and the North American continental margin.

SEDIMENTARY MINERAL DEPOSITS

Forrest G. Poole

A variety of syngenetic and diagenetic mineral deposits occurs in marine sedimentary rocks of late Proterozoic to Devonian age in the western United States. Distribution and type of sedimentary deposits contemporaneous with this time interval are shown in Figure 10. Most of the deposits are facies controlled and occur in specific paleogeographic and paleotectonic settings. Rocks of this time interval are also hosts for younger epigenetic ore deposits.

METALLIFEROUS SHALES

Many sapropel-rich (>10 wt % organic carbon) anoxic black mudstone and siltstone units in the Roberts Mountains allochthon and in younger accreted terranes contain anomalously high concentrations of several metals, including vanadium, chromium, molybdenum, nickel, silver, zinc, and copper. Some metals reside in the kerogen, whereas others reside in sulfide and oxide minerals. Anomalous metals in the Ordovician Vinini Formation and Devonian Woodruff Formation of Nevada have been documented (Desborough and Poole, 1983; Poole and Desborough, 1980, 1981), and preliminary studies of other Paleozoic marine black-shale units in the western U.S. reveal anomalous metal concentrations. For example, the Salmon River assemblage in central Idaho contains anomalous silver, copper, molybdenum, vanadium, and zinc (Fig. 10; Fisher and May, 1983). Although all deposits known to date are economically marginal to submarginal, they constitute a future resource.

BASE-METAL SULFIDES

Stratiform metal sulfide deposits of pre-Carboniferous age occur in the mid-Paleozoic Roberts Mountains allochthon and in several accreted terranes of the western U.S. (Fig. 10). Metal sulfides were deposited as laminae and beds interlayered with mudstone, siltstone, and chert, and locally with tuff and lava. The sulfide intervals are as thick as several meters.

Sources of iron, copper, lead, and zinc were probably sub-sea-floor volcanic and sedimentary rocks that were leached by hot, chloride-bearing, hydrogen-sulfide-rich solutions in magma-driven circulating seawater within convective systems that vented on the sea floor as "black smokers." The vents may have been localized by active synsedimentary faults and deep-seated crustal fractures. A possible syngenetic and diagenetic, exhalative-type iron and copper sulfide deposit in the Ordovician Valmy Formation near Mountain City, Nevada, was considered to be epigenetic in origin by Coats and Stephens (1968). The sulfides may be remobilized from stratiform deposits. Local concentrations of silver-bearing iron, lead, and zinc sulfides have been reported by Ketner (1983) in black chert at the base of the Silurian in several places in Nevada. The Cambrian(?) to Devonian Flagstaff Mountain assemblage in northeastern Washington and the Devonian Milligen Formation and Cambrian to Devonian Salmon River assemblage in central Idaho contain significant stratiform base-metal sulfide deposits (Fig. 10) that may be exhalative in origin. Devonian island-arc volcanic rocks in the Redding area in the southeastern Klamath Mountains (Fig. 10) contain significant disseminated, banded, and massive sulfide deposits consisting of iron, copper, zinc, and minor lead sulfide minerals and some gold and silver. The Klamath deposits are associated with the Middle Devonian Balaklala Rhyolite in the East and West Shasta

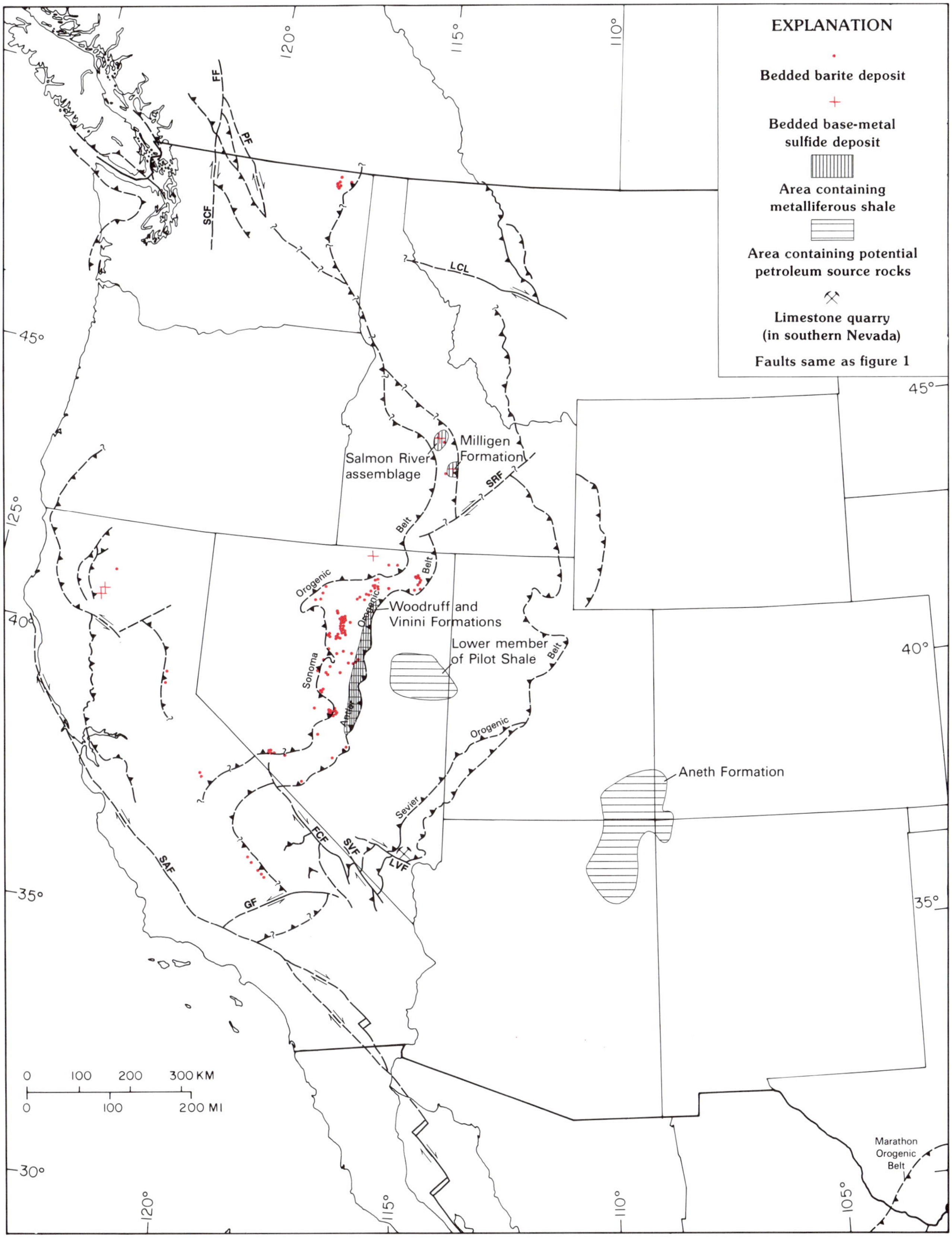

Figure 10. Map showing locations and types of sedimentary mineral deposits in Devonian and older rocks.

copper-zinc districts (Albers and Robertson, 1961; Kinkel and others, 1956). Some of the iron sulfide deposits were mined for the production of sulfuric acid.

BARITE

Stratiform barite deposits of pre-Carboniferous age occur in the mid-Paleozoic Roberts Mountains allochthon and in several younger accreted terranes of the western U.S. Cordillera (Fig. 10). Bedded barite occurs in eugeosynclinal siliceous facies of Cambrian to Devonian age (Poole, 1988). Barite forms light to dark gray beds as thick as 15 m interbedded with organic-rich mudstone, siltstone, and chert, minor limestone or dolostone, sandstone, tuff, and lava. Barite beds consist mainly of laminated finely crystalline rock, layers of rosettes, and some debris-flow conglomerate and turbidite sandstone.

Stratiform barite occurs in the Roberts Mountains allochthon in Elko, Humboldt, Eureka, Lander, Nye, Mineral, and Esmeralda Counties in Nevada; in southeastern Tulare County in southern California; in the Flagstaff Mountain assemblage in northern Stevens County in northeastern Washington; in the Salmon River assemblage in Custer County in central Idaho; in the Redding area of the Eastern Klamath Mountains in western Shasta County in northern California; and in the Northern Sierra Nevada terrane in Plumas and Nevada Counties in northern California (Fig. 10). The Upper Devonian part of the Slaven Chert in central and north-central Nevada contains the largest commercial bedded deposits in the western U.S. (Plate 3-6; Papke, 1984). The Ordovician Valmy, Vinini, and Comus Formations in Nevada also contain commercial bedded deposits.

The source of barium was probably comparable to that for the base-metal sulfides except that during the exhalative process, free barium combined with seawater sulfate to form "white smokers." Sulfur isotopes indicate a seawater sulfate origin for the sulfur (Rye and others, 1978). Although some barites seem to be exhalative "white-smoker" deposits close to sea-floor hydrothermal vents, most barites are hosted by organic-rich dark-gray mudstone, siltstone, and chert, indicating anoxic sea-floor settings possibly some distance from a source vent (Poole, 1988). In vent areas where bottom waters were anoxic, the exhalative warm, dense, and saline brines could flow by gravity for long distances with only minimal mixing with seawater. Subsequent ventilation or oxygenation of the anoxic basins would cause mixing of the brine solution and seawater, resulting in barite precipitation.

Interbedded barite and metal sulfides indicate that there is an overlap of chemical environments in which sulfates (oxidizing) and sulfides (reducing) are precipitated. Similar gray bedded barites are peripheral to many bedded sulfide deposits worldwide and in places are a major component of the massive-sulfide system.

PETROLEUM

Discussions of the geologic framework for each of the 11 western states and a review of the geologic characteristics for petroleum potential of wilderness lands within each of the states were reported as separate chapters in USGS Circular 902 edited by B. M. Miller (1983). Individual chapters on petroleum geology and resource potential of the 11 western states include: Arizona by R. T. Ryder; California by E. W. Scott; Colorado by C. W. Spencer; Idaho by C. A. Sandberg; Montana by W. J. Perry, Jr., D. D. Rice, and E. K. Maughan; Nevada by C. A. Sandberg; New Mexico by R. T. Ryder; Oregon by T. D. Fouch; Utah by C. M. Molenaar and C. A. Sandberg; Washington by T. D. Fouch; and Wyoming by C. W. Spencer and R. B. Powers.

Distribution of major Devonian and older potential petroleum source rocks is shown in Figure 10; more detailed information for each of these areas is presented by Meissner and others (1984). Many Devonian and older shales may be petroleum source rocks in areas where they are immature and marginally mature. Oil and gas generation and migration into Devonian and older rock reservoirs could have occurred much later. Potential petroleum source rocks in the western U.S. are in areas of low thermal history. Organic-rich shales having a low thermal history occur in the Ordovician Vinini Formation and the Devonian Woodruff Formation in the eastern part of the Roberts Mountains allochthon in central Nevada (Poole and others, 1983; Poole and Claypool, 1984). Some of the shales are very organic rich and qualify as oil shales, but no reservoir oils sampled to date are known to contain Vinini or Woodruff oil.

The Upper Devonian lower member of the Pilot Shale in a relatively small shelf basin in east-central Nevada and west-central Utah contains some organic-rich black shale units (Fig. 10), but no reservoir oils are known to contain Pilot oil. The Upper Devonian Aneth Formation in the Four Corners area is considered a possible source for oil in the overlying McCracken Sandstone Member of the Elbert Formation (Fig. 10; Meissner and others, 1984).

Some of the organic-rich black shale units deeply buried within the Roberts Mountains allochthon may have reached maturity during late Paleozoic and early Mesozoic time. In areas of moderate to high thermal history, hydrocarbons were generated and probably migrated into Paleozoic reservoirs in contemporaneous stratigraphic and structural traps; however, most of the oil and gas accumulations were later destroyed or remigrated during Mesozoic and Tertiary faulting. Poole and Claypool (1984) presented maturation data on some Devonian and older shales in Nevada.

SILICA

Naturally pure silica in quartz sandstone and quartzite beds of the miogeoclinal Cambrian Zabriskie Quartzite (Stewart, 1970) and miogeosynclinal Ordovician Eureka Quartzite and equivalent quartzose rocks (Ketner, 1976) makes these available sources of high-grade silica for the manufacture of glass, metallurgical processes, and other industrial uses. These quartzite units crop out sporadically in the western U.S. from southeastern California through eastern Nevada and western Utah to central Idaho (Figs. 2 and 6; Plates 2-2 and 2-6). Equivalent Cambrian and

Ordovician quartz sandstone and quartzite beds also crop out sporadically in Sonora in northwestern Mexico (Plates 2-2 and 2-6).

LIME AND DOLOMITE

Calcite-rich limestone beds, from which high-purity lime (CaO) is derived, and dolostone beds of high-purity dolomite are widely distributed in Middle Cambrian through Devonian miogeosynclinal sequences. A good source for high-purity lime may be the widespread thick limestone-bank deposits in the upper part of the Devonian Guilmette Limestone in northeastern Nevada (Sandberg and others, 1988). The principal use of high-purity lime from these rocks has been for portland cement, metallurgical processes, and other industrial and agricultural uses. Most limestone produced in the U.S. is used in construction, chiefly as aggregate, fill, and road metal. The principal use for high-purity dolostone has been for refractory dolomite where stable lime is produced by calcining dolomite at high temperatures. The product is used chiefly for lining open-hearth steel furnaces.

GEOLOGIC HISTORY—OVERVIEW OF CONTINENTAL-MARGIN DEVELOPMENT

Forrest G. Poole

Late Proterozoic and Cambrian rocks of the western U.S. are mostly shallow-water deposits that formed on a broad shelf along the western margin of the North American craton. Relatively deep-water deposits, some of which are allochthonous, crop out in a few areas near the western margin of the shelf. The late Proterozoic and Cambrian shelf deposits consist of three main lithogenetic sequences, in ascending order: (1) a discontinuous sequence, locally as thick as 3,000 m, of diamictite, mudstone, sandstone, conglomerate, and mafic volcanic rocks of late Proterozoic age; (2) a westward-thickening wedge, locally more than 6,000 m thick, of late Proterozoic and Lower Cambrian terrigenous detrital rocks of shallow-subtidal and intertidal origin; and (3) a Middle and Upper Cambrian sequence of predominantly shallow-subtidal carbonate rocks generally about 2,000 m thick. The rate of accumulation of strata was greatest in late Precambrian and Early Cambrian time. Accumulation and subsidence rates may be related to thermal contraction that followed an uplift related to heating at a spreading center (Stewart and Suczek, 1977; Armin and Mayer, 1983; Bond and others, 1983; Bond and Kominz, 1984). This sequence of events may be related to fragmentation and reshaping of the western margin of the U.S. by rifting in the late Proterozoic (Stewart, 1972). The diamictite and volcanic sequence may have been deposited in rift valleys that formed during the early stage of the rifting; the terrigenous detrital sequence may be related to erosion of a thermally expanded area near the rift; and the carbonate sequence may be related to widespread shallow-water shelf sedimentation that occurred after destruction of the thermally uplifted area by cooling and thermal contraction and by erosion.

The Cordilleran continental margin apparently began development following the late Proterozoic rifting (Stewart, 1972), and by Middle Cambrian time the basic tectonic framework of the western U.S. that controlled Middle Cambrian to Devonian deposition included, from east to west, a cratonic platform, a continental shelf, a continental slope, a continental rise, a marginal ocean basin or inner-arc basin, and possibly an island arc. Because of their tectonic instability, the marginal ocean basin and continental rise contain only fragmentary records of late Proterozoic to Devonian deposits. Preserved ocean-basin and continental-rise deposits are characteristically thin-bedded chert, mudstone, siltstone, limestone, volcanic rocks, and some sandstone and conglomerate, all indicative of neritic, bathyal, and abyssal environments. The continental shelf contains a neary complete record of as much as 6,000 m of Middle Cambrian to Upper Devonian deposits that are characterized by thick units of limestone and dolostone and some units of sandstone and siltstone. Terrigenous clastic sediments were derived mainly from the cratonic platform. Several regionally persistent quartz-sand-bearing carbonate units and quartzose sandstone units are recognized. The continental-shelf sediments were deposited in shallow-subtidal, intertidal, and some supratidal environments. Depositional trends varied considerably on the continental shelf owing to sea-level changes, provenance of sediment, and paleotectonic features. Persistent arches that extended westward across the shelf from the Transcontinental-basement arch influenced lithofacies and thickness trends from the Middle Cambrian to the Upper Devonian (Plates 2 and 3). A transitional zone, perhaps 100 km wide, separated the continental-shelf and continental-rise facies. This zone, which is characterized by as much as 4,000 m of thin-bedded limestone, mudstone, siltstone, and sandstone, including many sediment gravity-flow deposits, apparently represents a gently but irregularly inclined continental slope at moderate water depths. The cratonic platform contains a fragmentary record of Middle Cambrian to Middle Devonian deposits because of widespread vertical tectonic fluctuation and erosion. The Upper Devonian contains mostly intertidal and supratidal rocks, generally less than 300 m thick, characterized by limestone and dolostone and local beds of sandstone and siltstone.

Widespread crustal movements that probably were related to interactions between oceanic and continental plates along the northeast Pacific margin produced several regional unconformities within the Middle Cambrian to Upper Devonian interval. Accelerated sea-floor spreading west of the North American continental margin in Middle to Late Devonian time apparently initiated the Antler orogeny and resultant Mississippian flysch and molasse deposition in the foreland basin east of the orogenic highland. Marginal ocean basin or inner-arc basin, continental-rise, and continental-slope rocks of Cambrian to Devonian age were strongly deformed and thrust eastward over autochthonous continental-slope and continental-shelf rocks in Late Devonian

and Early Mississippian time. These rocks of the Antler orogen are represented in the Roberts Mountains allochthon in Nevada and adjacent states. Overprinting of Devonian and Mississippian deformation by Permian to early Tertiary deformation accounts for the present distribution of eugeosynclinal rocks of the Roberts Mountains allochthon. Many lithotectonic accreted terranes in the western U.S. contain Devonian and older rocks that are not paleogeographically compatible with coeval rocks of the North American continental margin. Most of these terranes were emplaced and sutured to western North America in Mesozoic and Cenozoic time.

REFERENCES CITED

Aalto, K. R., 1971, Glacial marine sedimentation and stratigraphy of the Toby Conglomerate (upper Proterozoic) southeastern British Columbia, northwestern Idaho, and northeastern Washington: Canadian Journal of Earth Sciences, v. 8, no. 7, p. 753–787.

—— , 1981, The late Precambrian Toby Formation of British Columbia, Idaho, and Washington, *in* Hambrey, M. J., and Harland, W. B., eds., Earth's pre-Pleistocene glacial record: Cambridge, United Kingdom, Cambridge University Press, p. 731–735.

Aitken, J. D., 1966, Middle Cambrian to Middle Ordovician cyclic sedimentation, southern Rocky Mountains of Alberta: Bulletin of Canadian Petroleum Geology, v. 14, no. 4, p. 405–441.

—— , 1978, Revised models for depositional Grand Cycles, Cambrian of the southern Rocky Mountains, Canada: Bulletin of Canadian Petroleum Geology, v. 26, no. 4, p. 515–542.

Albers, J. P., and Robertson, J. F., 1961, Geology and ore deposits of East Shasta copper-zinc district, Shasta County, California: U.S. Geological Survey Professional Paper 338, 107 p.

Albers, J. P., and Stewart, J. H., 1972, Geology and mineral deposits of Esmeralda County, Nevada: Nevada Bureau of Mines and Geology Bulletin 78, 80 p.

Albers, J. P., Kistler, R. W., and Kwak, L., 1981, The Mule Mountain stock; An early Middle Devonian pluton in northern California: Isochron/West, no. 31, p. 17.

Armin, R. A., and Mayer, L., 1983, Subsidence analysis of the Cordilleran miogeocline; Implications for timing of late Proterozoic rifting and amount of extension: Geology, v. 11, no. 12, p. 702–705.

Armstrong, R. L., 1975, Precambrian (1,500 m.y. old) rocks of central Idaho; The Salmon River Arch and its role in Cordilleran sedimentation and tectonics: American Journal of Science, v. 275-A, p. 437–467.

Armstrong, R. L., Eisbacher, G. H., and Evans, P. D., 1982, Age and stratigraphic-tectonic significance of Proterozoic diabase sheets, Mackenzie Mountains, northwestern Canada: Canadian Journal of Earth Sciences, v. 19, no. 2, p. 316–323.

Baars, D. L., 1966, Pre-Pennsylvanian paleotectonics; Key to basin evolution and petroleum occurrences in Paradox Basin, Utah and Colorado: American Association of Petroleum Geologists Bulletin, v. 50, no. 10, p. 2082–2111.

Baker, B. H., Mohr, P. A., and Williams, L.A.J., 1972, Geology of the eastern rift system of Africa: Geological Society of America Special Paper 136, 67 p.

Barker, F., Millard, H. T., Jr., and Knight, R. J., 1979, Reconnaissance geochemistry of Devonian island-arc volcanic and intrusive rocks, west Shasta district, California, *in* Trondhjemites, dacites, and related rocks: Developments in Petrology, v. 6, p. 531–545.

Barnes, J. J., and Klein, G. deV., 1975, Tidal deposits in the Zabriskie Quartzite (Cambrian), eastern California and western Nevada, *in* Ginsburg, R. N., ed., Tidal deposits; A casebook of Recent examples and fossil counterparts: New York, Springer-Verlag, p. 163–169.

Beka, F. T., 1980, Upper Paleozoic metasedimentary and metavolcanic rocks and associated mineral deposits between Glasgo Lakes and Little Sheep Creek, Stevens County, Washington [Ph.D. thesis]: Pullman, Washington State University, 187 p.

Benmore, W. C., 1978, Stratigraphy, sedimentology, and paleoecology of the late Paleophytic or earliest Phanerozoic Johnnie Formation, eastern California and southwestern Nevada [Ph.D. thesis]: Santa Barbara, University of California, 263 p.

Berry, W.B.N., 1976, Aspects of correlation of North American shelly and graptolitic faunas, *in* Bassett, M. G., ed., The Ordovician System: Proceedings Palaeontological Association Symposium, September 1974, Birmingham England: University of Wales Press, National Museum of Wales, Cardiff, p. 153–169.

Blick, N. H., 1979, Stratigraphic, structural, and paleogeographic interpretation of upper Proterozoic glaciogenic rocks in the Sevier orogenic belt, northwestern Utah [Ph.D. thesis]: Santa Barbara, University of California, 708 p.

—— , 1981, Late Precambrian glaciation in Utah, *in* Hambrey, M. J., and Harland, W. B., eds., Earth's pre-Pleistocene glacial record: Cambridge, United Kingdom, Cambridge University Press, p. 740–744.

Blome, C. D., and Nestell, M. K., 1988, Paleontologic and tectonic significance of the Grindstone terrane, east-central Oregon: Geological Society of America Abstracts with Programs, v. 20, p. 144.

Bond, G. C., and DeVay, J. C., 1980, Pre-Upper Devonian quartzose sandstones in the Shoo Fly Formation, northern California; Petrology, provenance, and implications for regional tectonics: Journal of Geology, v. 88, no. 3, p. 285–308.

Bond, G. C., and Kominz, M. A., 1984, Construction of tectonic subsidence curves for the early Paleozoic miogeocline, southern Canadian Rocky Mountains; Implications for subsidence mechanisms, age of breakup, and crustal thinning: Geological Society of America Bulletin, v. 95, p. 155–173.

Bond, G. C., and Schweickert, R. A., 1981, Significance of pre-Devonian mélange in the Shoo Fly Complex, northern Sierra Nevada, California: Geological Society of America Abstracts with Programs, v. 13, p. 46.

Bond, G. C., Kominz, M. A., and Devlin, W. J., 1983, Thermal subsidence and eustasy in the lower Palaeozoic miogeocline of western North America: Nature, v. 306, no. 5945, p. 775–779.

Bond, G. C., Christie-Blick, N., Kominz, M. A., and Devlin, W. J., 1985, An Early Cambrian rift to post-rift transition in the Cordillera of western North America: Nature, v. 316, no. 6022, p. 742–745.

Boucot, A. J., Dunkle, D. H., Potter, A. W., Savage, N. M., and Rohr, D. M., 1974, Middle Devonian orogeny in western North America; A fish and other fossils: Journal of Geology, v. 82, no. 6, p. 691–708.

Brady, M. J., and Rowell, A. J., 1976, Upper Cambrian subtidal blanket carbonate of the miogeocline, eastern Great Basin, *in* Robison, R. A., and Rowell, A. J., eds., Paleontology and depositional environments; Cambrian of western North America: Provo, Utah, Brigham Young University Geology Studies, v. 23, part 2, p. 153–163.

Brandon, M. T., Cowan, D. S., and Vance, J. A., 1988, The Late Cretaceous San Juan thrust system, San Juan Islands, Washington: Geological Society of America Special Paper 221, 81 p.

Brook, C. A., 1977, Stratigraphy and structure of the Saddlebag Lake roof pendant, Sierra Nevada, California: Geological Society of America Bulletin, v. 88, p. 321–344.

Cameron, C. S., 1982, Stratigraphy and significance of the upper Precambrian Big Bear Group, *in* Cooper, J. D., Troxel, B. W., and Wright, L. A., eds., Geology of selected areas in the San Bernardino Mountains, western Mojave Desert, and southern Great Basin, California: Shoshone, California, Death Valley Publishing Company, Geological Society of America Cordilleran Section, Anaheim, California, April 1982, Guidebook and Volume, p. 5–20.

Campbell, A. B., and Raup, O. B., 1964, Preliminary geologic map of the Hunters Quadrangle, Stevens and Ferry Counties, Washington: U.S. Geological Survey Mineral Investigations Field Studies Map MF-276, scale 1:48,000.

Carr, M. D., Poole, F. G., Harris, A. G., and Christiansen, R. L., 1981, Western

facies Paleozoic rocks in the Mojave Desert, California, *in* Howard, K. A., Carr, M. D., and Miller, D. M., eds., Tectonic framework of the Mojave and Sonoran Deserts, California and Arizona; Abstracts from a conference held by the U.S. Geological Survey in Menlo Park, California, November 4–6, 1980: U.S. Geological Survey Open-File Report 81–503, p. 15–17.

Carr, M. D., Poole, F. G., and Christiansen, R. L., 1982, Geologic framework of Paleozoic eugeosynclinal and transitional assemblage rocks in the north-central Mojave Desert, southern California: Geological Society of America Abstracts with Programs, v. 14, p. 154.

——, 1984, Pre-Cenozoic geology of the El Paso Mountains, southwestern Great Basin, California; A summary (field trip 7), *in* Lintz, J., Jr., ed., Western geological excursions; Guidebook of field trips for the 1984 Annual Meeting of the Geological Society of America at Reno: Reno, University of Nevada, Mackay School of Mines, Department of Geological Sciences, v. 4, p. 84–93.

Cashman, S. M., 1980, Devonian metamorphic event in the northeastern Klamath Mountains, California: Geological Society of America Bulletin, part 1, v. 91, p. 453–459.

Christie-Blick, N., 1982, Upper Proterozoic and Lower Cambrian rocks of the Sheeprock Mountains, Utah; Regional correlation and significance: Geological Society of America Bulletin, v. 93, p. 735–750.

——, 1983, Glacial-marine and subglacial sedimentation, upper Proterozoic Mineral Fork Formation, Utah, *in* Molnia, B. F., ed., Glacial-marine sedimentation: New York, Plenum Press, p. 703–776.

——, 1984, Latest Proterozoic and Early Cambrian rifting in California, Nevada, and Utah; Support for Early Cambrian continental separation: Geological Society of America Abstracts with Programs, v. 16, p. 471.

Christie-Blick, N., Link, P. K., Miller, J.M.G., Young, G. M., and Crowell, J. C., 1980, Regional geologic events inferred from upper Proterozoic rocks of the North American Cordillera: Geological Society of America Abstracts with Programs, v. 12, p. 402.

Chronic, J., McCallum, M. E., Ferris, C. S., Jr., and Eggler, D. H., 1969, Lower Paleozoic rocks in diatremes, southern Wyoming and northern Colorado: Geological Society of America Bulletin, v. 80, p. 149–156.

Churkin, M., Jr., and Kay, M., 1967, Graptolite-bearing Ordovician siliceous and volcanic rocks, northern Independence Range, Nevada: Geological Society of America Bulletin, v. 78, p. 651–668.

Cloud, P. E., Jr., and Nelson, C. A., 1966, Phanerozoic–Cryptozoic and related transitions; New evidence: Science, v. 154, no. 3750, p. 766–770.

Coats, R. R., and Stephens, E. C., 1968, Mountain City Copper Mine, Elko County, Nevada, *in* Ridge, J. D., ed., Ore deposits of the United States, 1933–1967; The Graton–Sales Volume: American Institute of Mining Metallurgical, and Petroleum Engineers, v. 2, p. 1074–1101.

Coles, K. S., 1986, Deposition and deformation of a mid-Paleozoic siliceous sequence, central Toquima Range, Nevada: Geological Society of America Abstracts with Programs, v. 18, p. 96.

Coles, K. S., and Snyder, W. S., 1985, Significance of lower and middle Paleozoic phosphatic chert in the Toquima Range, central Nevada: Geology, v. 13, p. 573–576.

Condie, K. C., and Snansieng, S., 1971, Petrology and geochemistry of the Duzel (Ordovician) and Gazelle (Silurian) Formations, northern California: Journal of Sedimentary Petrology, v. 41, no. 3, p. 741–751.

Coney, P. J., Jones, D. L., and Monger, J.W.H., 1980, Cordilleran suspect terranes: Nature, v. 288, no. 5789, p. 329–333.

Cook, H. E., and Taylor, M. E., 1977, Comparison of continental slope and shelf environments in the Upper Cambrian and lowest Ordovician of Nevada, *in* Cook, H. E., and Enos, P., eds., Deep-water carbonate environments: Society of Economic Paleontologists and Mineralogists Special Publication 25, p. 51–81.

Crittenden, M. D., Jr., Schaeffer, F. E., Trimble, D. E., and Woodward, L. A., 1971, Nomenclature and correlation of some upper Precambrian and basal Cambrian sequences in western Utah and southeastern Idaho: Geological Society of America Bulletin, v. 82, p. 581–602.

Crittenden, M. D., Jr., Christie-Blick, N., and Link, P. K., 1983, Evidence for two pulses of glaciation during the late Proterozoic in northern Utah and southeastern Idaho: Geological Society of America Bulletin, v. 94, p. 437–450.

D'Allura, J. A., Moores, E. M., and Robinson, L., 1977, Paleozoic rocks of the northern Sierra Nevada; Their structural and paleogeographic implications, *in* Stewart, J. H., Stevens, C. H., and Fritsche, A. E., eds., Paleozoic paleogeography of the western United States, Pacific Coast Paleogeography Symposium 1: Society of Economic Paleontologists and Mineralogists, p. 395–408.

Danner, W. R., 1977, Paleozoic rocks of northwest Washington and adjacent parts of British Columbia, *in* Stewart, J. H., Stevens, C. H., and Fritsche, A. E., eds., Paleozoic paleogeography of the western United States, Pacific Coast Paleogeography Symposium 1: Society of Economic Paleontologists and Mineralogists, p. 481–502.

Davis, G. A., 1968, Westward thrust faulting in the south-central Klamath Mountains, California: Geological Society of America Bulletin, v. 79, p. 911–934.

Davis, G. A., Holdaway, M. J., Lipman, P. W., and Romey, W. D., 1965, Structure, metamorphism, and plutonism in the south-central Klamath Mountains, California: Geological Society of America Bulletin, v. 76, p. 933–965.

Day, H. W., Moores, E. M., and Tuminas, A. C., 1985, Structure and tectonics of the northern Sierra Nevada: Geological Society of America Bulletin, v. 96, p. 436–450.

Desborough, G. A., and Poole, F. G., 1983, Metal concentrations in some marine black shales of the United States, *in* Shanks, W. C., III, ed., Unconventional mineral deposits, Cameron Volume: New York, Society of Mining Engineers of the American Institute of Mining, Metallurgical, and Petroleum Engineers, p. 99–110.

Devlin, W. J., and Bond, G. C., 1988, The initiation of the early Paleozoic Cordilleran miogeocline; Evidence from the uppermost Proterozoic–Lower Cambrian Hamill Group of southeastern British Columbia: Canadian Journal of Earth Sciences, v. 25, p. 1–19.

Devlin, W. J., Bond, G. C., and Brueckner, H. K., 1985, An assessment of the age and tectonic setting of volcanics near the base of the Windermere Supergroup in northeastern Washington; Implications for latest Proterozoic–earliest Cambrian continental separation: Canadian Journal of Earth Sciences, v. 22, no. 6, p. 829–837.

Devlin, W. J., Brueckner, H. K., and Bond, G. C., 1988, New isotopic data and a preliminary age for volcanics near the base of the Windermere Supergroup northeastern Washington, U.S.A.: Canadian Journal of Earth Sciences, v. 25, p. 1906–1911.

Dickinson, W. R., and Thayer, T. P., 1978, Paleogeographic and paleotectonic implications of Mesozoic stratigraphy and structure in the John Day inlier of central Oregon, *in* Howell, D. G., and McDougall, K. A., eds., Mesozoic paleogeography of the western United States, Pacific Coast Paleogeography Symposium 2: Society of Economic Paleontologists and Mineralogists, p. 147–161.

Diehl, P. E., 1979, The stratigraphy, depositional environments, and quantitative petrography of the Precambrian–Cambrian Wood Canyon Formation, Death Valley [Ph.D. thesis]: University Park, Pennsylvania State University, 430 p.

Dietz, R. S., and Holden, J. C., 1966, Miogeoclines (miogeosynclines) in space and time: Journal of Geology, v. 74, p. 566–583.

Dockum, M. S., and Miller, R. H., 1982, Ordovician conodonts from the greenschist facies carbonates, western Imperial County, California: Geological Society of America Abstracts with Programs, v. 14, p. 160–161.

Dover, J. H., 1980, Status of the Antler orogeny in central Idaho; Clarifications and constraints from the Pioneer Mountains, *in* Fouch, T. D., and Magathan, E. R., eds., Paleozoic paleogeography of the west-central United States, Rocky Mountain Paleogeography Symposium 1: Society of Economic Paleontologists and Mineralogists, p. 371–386.

Dover, J. H., and 5 others, 1976, Geologic map of the Pioneer Mountains region, Blaine and Custer Counties, Idaho: U.S. Geological Survey Open-File Report 76–75, map scale 1:62,500.

Dover, J. H., Berry, W.B.N., and Ross, R. J., Jr., 1980, Ordovician and Silurian

Phi Kappa and Trail Creek Formations, Pioneer Mountains, central Idaho; Stratigraphic and structural revisions, and new data on graptolite faunas: U.S. Geological Survey Professional Paper 1090, 54 p.

Dubé, T. E., 1987, Setting and origin of exhalative bedded barite and associated rocks of the Roberts Mountains allochthon in north-central Nevada [M.S. thesis]: Seattle, University of Washington, 76 p.

Durrell, C., and Procter, P. D., 1948, Iron ore deposits near Lake Hawley and Spencer Lakes, Sierra County, California: California Division of Mines and Geology Bulletin 129L, p. 167–192.

Eisbacher, G. H., 1981, Sedimentary tectonics and glacial record in the Windermere Supergroup, Mackenzie Mountains, northwestern Canada: Geological Survey of Canada Paper 80–27, 40 p.

Evans, J. G., and Theodore, T. G., 1978, Deformation of the Roberts Mountains allochthon in north-central Nevada: U.S. Geological Survey Professional Paper 1060, 18 p.

Evans, K. V., 1984, Ordovician plutonism in east-central Idaho; Variation on a Canadian theme: Geological Society of America Abstracts with Programs, v. 16, p. 504.

Evans, K. V., and Clemons, R. E., 1988, Cambrian–Ordovician (500 Ma) alkalic plutonism in southwestern New Mexico; U-Th-Pb isotopic data from the Florida Mountains: American Journal of Science, v. 288, p. 735–755.

Evans, K. V., and Lund, K., 1981, The Salmon River "Arch"?: Geological Society of America Abstracts with Programs, v. 13, p. 448.

Evans, K. V., and Zartman, R. E., 1988, Early Paleozoic alkalic plutonism in east-central Idaho: Geological Society of America Bulletin, v. 100, p. 1981–1987.

Evenchick, C. A., Parrish, R. P., and Gabrielse, H., 1984, Precambrian gneiss and late Proterozoic sedimentation in north-central British Columbia: Geology, v. 12, p. 233–237.

Farmer, G. L., and DePaolo, D. J., 1983, Origin of Mesozoic and Tertiary granite in the western United States and implications for pre-Mesozoic crustal structure; 1, Nd and Sr isotopic studies in the geocline of the northern Great Basin: Journal of Geophysical Research, v. 88, no. B4, p. 3379–3401.

Fisher, F. S., and May, G. D., 1983, Geochemical characteristics of the metalliferous Salmon River sequence, central Idaho: U.S. Geological Survey Open-File Report 83–670, 30 p., map scale 1:24,000.

Fox, K. F., Jr., and Rinehart, C. D., 1974, Comparison of Anarchist and Covada Groups of north-central Washington, *in* Geological Survey Research 1974: U.S. Geological Survey Professional Paper 900, p. 50–51.

Fox, K. F., Jr., Rinehart, C. D., and Engels, J. C., 1977, Plutonism and orogeny in north-central Washington; Timing and regional context: U.S. Geological Survey Professional Paper 989, 27 p.

Gastil, R. G., and Miller, R. H., 1983, Pre-batholithic terranes of southern and peninsular California, U.S.A. and Mexico; Status report, *in* Stevens, C. H., ed., Pre-Jurassic rocks in western North American suspect terranes: Society of Economic Paleontologists and Mineralogists, Pacific Section, p. 49–61.

Gilluly, J., and Gates, O., 1965, Tectonic and igneous geology of the northern Shoshone Range, Nevada, *with sections on* Gravity in Crescent Valley, by D. Plouff, and Economic geology, by K. B. Ketner: U.S. Geological Survey Professional Paper 465, 153 p.

Gilluly, J., and Masursky, H., 1965, Geology of the Cortez Quadrangle, Nevada: U.S. Geological Survey Bulletin 1175, 117 p.

Girty, G. H., 1983, The Culbertson Lake allochthon, a newly identified structural unit within the Shoo Fly Complex; Sedimentologic, stratigraphic, and structural evidence for extension of the Antler orogenic belt to the northern Sierra Nevada, California [Ph.D. thesis]: New York, Columbia University, 184 p.

Girty, G. H., and Schweickert, R. A., 1979, Preliminary results of a detailed study of the "lower" Shoo Fly, Bowman Lake, northern Sierra Nevada, California: Geological Society of America Abstracts with Programs, v. 11, p. 79.

——, 1984, The Culbertson Lake allochthon, a newly identified structure within the Shoo Fly Complex, California; Evidence for four phases of deformation and extension of the Antler orogeny to the northern Sierra Nevada: Modern Geology, v. 8, no. 3, p. 181–198.

Girty, G. H., and Wardlaw, M. S., 1984, Was the Alexander terrane a source of feldspathic sandstones in the Shoo Fly Complex, Sierra Nevada, California?: Geology, v. 12, p. 339–342.

——, 1985, Petrology and provenance of pre-Late Devonian sandstones, Shoo Fly Complex, northern Sierra Nevada, California: Geological Society of America Bulletin, v. 96, p. 516–521.

Girty, G. H., Reiland, D. N., and Wardlaw, M. S., 1985, Provenance of the Silurian Elder Sandstone, north-central Nevada: Geological Society of America Bulletin, v. 96, p. 925–930.

Glover, J. K., and Price, R. A., 1976, Stratigraphy and structure of the Windermere Supergroup, southern Kootenay arc, British Columbia: Geological Survey of Canada Paper 76–1B, p. 21–23.

Gutschick, R. C., and Sandberg, C. A., 1983, Mississippian continental margins of the conterminous United States, *in* Stanley, D. J., ed., The shelfbreak; Critical interface on continental margins: Society of Economic Paleontologists and Mineralogists Special Publication 33, p. 79–96.

Hamilton, W. B., 1978, Mesozoic tectonics of the western United States, *in* Howell, D. G., and McDougall, K. A., eds., Mesozoic paleogeography of the western United States, Pacific Coast Paleogeography Symposium 2: Society of Economic Paleontologists and Mineralogists, p. 33–70.

Hammond, J.L.G., 1983, Late Precambrian diabase intrusions in the southern Death Valley region, California; Their petrology, geochemistry, and tectonic significance [Ph.D. thesis]: Los Angeles, University of Southern California, 281 p.

Hannah, J. L., and Moores, E. M., 1986, Age relationships and depositional environments of Paleozoic strata, northern Sierra Nevada, California: Geological Society of America Bulletin, v. 97, p. 787–797.

Hansen, W. R., Carrara, P. E., and Rowley, P. D., 1982, Geologic map of the Canyon of Lodore North Quadrangle, Moffat County, Colorado: U.S. Geological Survey Geologic Quadrangle Map GQ–1568, scale 1:24,000.

Hanson, R. E., 1983, Volcanism, plutonism, and sedimentation in a Late Devonian submarine island-arc setting, northern Sierra Nevada, California [Ph.D. thesis]: New York, Columbia University, 354 p.

Hanson, R. E., and Schweickert, R. A., 1986, Stratigraphy of mid-Paleozoic island-arc rocks in part of the northern Sierra Nevada, Sierra and Nevada Counties, California: Geological Society of America Bulletin, v. 97, p. 986–998.

Hanson, R. E., Saleeby, J. B., and Schweickert, R. A., 1988, Composite Devonian island-arc batholith in the northern Sierra Nevada, California: Geological Society of America Bulletin, v. 100, p. 446–457.

Harland, W. B., and 5 others, 1990, A geologic time scale 1989: Cambridge, England, Cambridge University Press, 263 p.

Harper, G. D., and Link, P. K., 1985, Geochemistry of rift related metabasalts associated with late Proterozoic diamictites, northern Utah and southeast Idaho: Geological Society of America Abstracts with Programs, v. 17, p. 223.

Hayes, P. T., 1975, Cambrian and Ordovician rocks of southern Arizona and New Mexico and westernmost Texas: U.S. Geological Survey Professional Paper 873, 98 p.

Hereford, R., 1977, Deposition of the Tapeats Sandstone (Cambrian) in central Arizona: Geological Society of America Bulletin, v. 88, p. 199–211.

Hintze, L. F., 1952, Lower Ordovician trilobites from western Utah and eastern Nevada: Utah Geological and Mineralogical Survey Bulletin 48, 249 p.

——, 1985, Correlation of stratigraphic units of North America (COSUNA) project; Great Basin region correlation chart: American Association of Petroleum Geologists, one sheet.

——, 1988, Geologic history of Utah: Provo, Utah, Brigham Young University Geology Studies, Special Publication 7, 202 p.

Hobbs, S. W., 1990, Geologic map of the Bayhorse area, central Custer County, Idaho: U.S. Geological Survey Miscellaneous Geologic Investigations Map I–1882, scale 1:62,500 (in press).

Hobbs, S. W., Hays, W. H., and Ross, R. J., Jr., 1968, The Kinnikinic Quartzite of central Idaho; Redefinition and subdivision: U.S. Geological Survey Bulletin 1254-J, 22 p.

Hobbs, S. W., Hays, W. H., and McIntyre, D. H., 1975, Geologic map of the

Clayton Quadrangle, Custer County, Idaho: U.S. Geological Survey Open-File Report 75–76, scale 1:62,500.

Hopson, C. A., and Mattinson, J. M., 1971, Metamorphism and plutonism, Lake Chelan region, northern Cascades, Washington, *in* Metamorphism in the Canadian Cordillera: Geological Association of Canada Cordilleran Section Programme and Abstracts, p. 13.

Hotz, P. E., 1977, Geology of the Yreka Quadrangle, Siskiyou County, California: U.S. Geological Survey Bulletin 1436, 72 p.

Irwin, W. P., 1977, Review of Paleozoic rocks of the Klamath Mountains, *in* Stewart, J. H., Stevens, C. H., and Fritsche, A. E., eds., Paleozoic paleogeography of the western United States, Pacific Coast Paleogeography Symposium 1: Society of Economic Paleontologists and Mineralogists, p. 441–454.

Jacobsen, S. B., Quick, J. E., and Wasserburg, G. J., 1984, A Nd and Sr isotopic study of the Trinity peridotite; Implications for mantle evolution: Earth and Planetary Science Letters, v. 68, p. 361–378.

Johnson, J. G., Klapper, G., and Sandberg, C. A., 1985, Devonian eustatic fluctuations in Euramerica: Geological Society of America Bulletin, v. 96, p. 567–587.

Johnson, J. G., Sandberg, C. A., and Poole, F. G., 1988, Early and Middle Devonian paleogeography of western United States, *in* McMillan, N. J., Embry, A. F., and Glass, D. J., eds., Devonian of the World: Calgary, Canadian Society of Petroleum Geologists Memoir 14, v. 1, p. 161–182.

Kay, M., 1951, North American geosynclines: Geological Society of America Memoir 48, 143 p.

Kepper, J. C., 1972, Paleoenvironmental patterns in Middle to lower Upper Cambrian interval in eastern Great Basin: American Association of Petroleum Geologists Bulletin, v. 56, no. 3, p. 503–527.

——, 1976, Stratigraphic relationships and depositional facies in a portion of the Middle Cambrian of the Basin and Range Province, *in* Robison, R. A., and Rowell, A. J., eds., Paleontology and depositional environments; Cambrian of western North America: Provo, Utah, Brigham Young University Geology Studies, v. 23, part 2, p. 75–91.

——, 1981, Sedimentology of a Middle Cambrian outer shelf margin with evidence for syndepositional faulting, eastern California and western Nevada: Journal of Sedimentary Petrology, v. 51, no. 3, p. 807–821.

Ketner, K. B., 1969, Ordovician bedded chert, argillite, and shale of the Cordilleran eugeosyncline in Nevada and Idaho: U.S. Geological Survey Professional Paper 650-B, p. B23–B34.

——, 1976, Map showing high-purity quartzite in California, Nevada, Utah, Idaho, and Montana: U.S. Geological Survey Miscellaneous Field Studies Map MF-821, scale 1:1,000,000.

——, 1983, Strata-bound, silver-bearing iron, lead, and zinc sulfide deposits in Silurian and Ordovician rocks of allochthonous terranes, Nevada and northern Mexico: U.S. Geological Survey Open-File Report 83–792, 6 p.

Kinkel, A. R., Jr., Hall, W. E., and Albers, J. P., 1956, Geology and base-metal deposits of West Shasta copper-zinc district, Shasta County, California: U.S. Geological Survey Professional Paper 285, 156 p.

Kinsman, D.J.J., 1975, Rift valley basins and sedimentary history of trailing continental margins, *in* Fischer, A. G., and Judson, S., eds., Petroleum and global tectonics: Princeton, New Jersey, Princeton University Press, p. 83–127.

Kistler, R. W., 1990, Two different lithosphere types in the Sierra Nevada, California, *in* Anderson, J. L., ed., The nature and origin of Cordilleran magmatism: Geological Society of America Memoir 174, p. 271–282.

Klein, G. deV., 1975, Paleotidal range sequences, middle member, Wood Canyon Formation (late Precambrian), eastern California and western Nevada, *in* Ginsburg, R. N., ed., Tidal deposits; A casebook of Recent examples and fossil counterparts: New York, Springer-Verlag, p. 171–177.

Knoll, A. H., Blick, N., and Awramik, S. M., 1981, Stratigraphic and ecologic implications of late Precambrian microfossils from Utah: American Journal of Science, v. 281, no. 3, p. 247–263.

Koepnick, R. B., 1976, Depositional history of the upper Dresbachian–lower Franconian (Upper Cambrian) Pterocephaliid Biomere from west central Utah, *in* Robison, R. A., and Rowell, A. J., eds., Paleontology and depositional environments; Cambrian of western North America: Provo, Utah, Brigham Young University Geology Studies, v. 23, part 2, p. 123–138.

Kottlowski, F. E., 1963, Paleozoic and Mesozoic strata of southwestern and south-central New Mexico: New Mexico Bureau of Mines and Mineral Resources Bulletin 79, 100 p.

Kretschmer, E. L., 1984, Geology of the Pinson and Preble gold deposits, Humboldt County, Nevada: Arizona Geological Society Digest, v. 15, p. 59–66.

Lanphere, M. A., Irwin, W. P., and Hotz, P. E., 1968, Isotopic age of the Nevadan orogeny and older plutonic and metamorphic events in the Klamath Mountains, California: Geological Society of America Bulletin, v. 79, p. 1027–1052.

Lapierre, H., Albarede, F., Albers, J., Cabanis, B., and Coulon, C., 1985, Early Devonian volcanism in the eastern Klamath Mountains, California; Evidence for an immature island arc: Canadian Journal of Earth Sciences, v. 22, p. 214–227.

Lindsey, K. A., 1982, The upper Proterozoic and Lower Cambrian "Brigham Group," Oneida Narrows, southeastern Idaho: Northwest Geology, v. 11, p. 13–21.

Lindsley-Griffin, N., 1977, Paleogeographic implications of ophiolites; The Ordovician Trinity complex, Klamath Mountains, California, *in* Stewart, J. H., Stevens, C. H., and Fritsche, A. E., eds., Paleozoic paleogeography of the western United States, Pacific Coast Paleogeography Symposium 1: Society of Economic Paleontologists and Mineralogists, p. 409–420.

Link, P. K., 1981, Upper Proterozoic diamictites in southeastern Idaho, U.S.A., *in* Hambrey, M. J., and Harland, W. B., eds., Earth's pre-Pleistocene glacial record: Cambridge, United Kingdom, Cambridge University Press, p. 736–739.

——, 1983, Glacial and tectonically influenced sedimentation in the upper Proterozoic Pocatello Formation, southeastern Idaho, *in* Miller, D. M., Todd, V. R., and Howard, K. A., eds., Tectonic and stratigraphic studies in the eastern Great Basin: Geological Society of America Memoir 157, p. 165–181.

Lobo, C. F., and Osborne, R. H., 1976, Petrology of late Precambrian–Cambrian quartzose sandstones in the eastern Mojave Desert, southeastern California: Journal of Sedimentary Petrology, v. 46, no. 4, p. 829–846.

Lochman-Balk, C., 1971, The Cambrian of the craton of the United States, *in* Holland, C. H., ed., Cambrian of the New World: New York, Wiley-Interscience, p. 79–167.

——, 1972, Cambrian System, *in* Mallory, W. W., ed., Geologic atlas of the Rocky Mountain region: Denver, Colorado, Rocky Mountain Association of Geologists, p. 60–75.

Lohmann, K. C., 1976, Lower Dresbachian (Upper Cambrian) platform to deep-shelf transition in eastern Nevada and western Utah; An evaluation through lithologic cycle correlation, *in* Robison, R. A., and Rowell, A. J., eds., Paleontology and depositional environments; Cambrian of western North America: Provo, Utah, Brigham Young University Geology Studies, v. 23, part 2, p. 111–122.

Loring, A. K., and Armstrong, D. G., 1980, Cambrian–Ordovician syenites of New Mexico, part of a regional alkalic intrusive episode: Geology, v. 8, p. 344–348.

Lowell, J. D., Genik, G. J., Nelson, T. H., and Tucker, P. M., 1975, Petroleum and plate tectonics of the southern Red Sea, *in* Fischer, A. G., and Judson, S., eds., Petroleum and global tectonics: Princeton, New Jersey, Princeton University Press, p. 129–153.

Madrid, R. J., 1987, Stratigraphy of the Roberts Mountains allochthon in north-central Nevada [Ph.D. thesis]: Stanford, California, Stanford University, 341 p.

Mattinson, J. M., 1972, Ages of zircons from the northern Cascade Mountains, Washington: Geological Society of America Bulletin, v. 83, p. 3769–3784.

Mattinson, J. M., and Hopson, C. A., 1972, Paleozoic ophiolite complexes in Washington and northern California: Carnegie Institution of Washington, Yearbook 71, p. 578–583.

McKee, E. D., 1940, Three types of cross-lamination in Paleozoic rocks of north-

ern Arizona: American Journal of Science, v. 238, no. 11, p. 811–824.

McKee, E. H., 1976, Geology of the northern part of the Toquima Range, Lander, Eureka, and Nye Counties, Nevada: U.S. Geological Survey Professional Paper 931, 49 p., geologic map scale 1:62,500.

McMath, V. E., 1958, The geology of the Taylorsville area, Plumas County, California [Ph.D. thesis]: Los Angeles, University of California, 199 p.

McMechan, M. E., and Price, R. A., 1982, Superimposed low-grade metamorphism in the Mount Fisher area, southeastern British Columbia; Implications for the East Kootenay orogeny: Canadian Journal of Earth Sciences, v. 19, no. 3, p. 476–489.

Meissner, F. F., Woodward, J., and Clayton, J. L., 1984, Stratigraphic relationships and distribution of source rocks in the Greater Rocky Mountain region, *in* Woodward, J., Meissner, F. F., and Clayton, J. L., eds., Hydrocarbon source rocks of the Greater Rocky Mountain region: Denver, Colorado, Rocky Mountain Association of Geologists, p. 1–34.

Merguerian, C. M., 1985, Stratigraphy, structural geology, and tectonic implication of the Shoo Fly Complex and the Calaveras–Shoo Fly thrust, central Sierra Nevada, California [Ph.D. thesis]: New York, Columbia University, 255 p.

Merguerian, C. M., and Schweickert, R. A., 1987, Paleozoic gneissic granitoids in the Shoo Fly Complex, central Sierra Nevada, California: Geological Society of America Bulletin, v. 99, p. 699–717.

Miller, B. M., ed., 1983, Petroleum potential of wilderness lands in the western United States: U.S. Geological Survey Circular 902-A-P, 197 p.

Miller, E. L., and Larue, D. K., 1983, Ordovician quartzite in the Roberts Mountains allochthon, Nevada; Deep sea fan deposits derived from cratonal North America, *in* Stevens, C. H., ed., Pre-Jurassic rocks in western North American suspect terranes: Pacific Section, Society of Economic Paleontologists and Mineralogists Special Publication, p. 91–102.

Miller, F. K., 1983, Preliminary geologic map of the Gypsy Peak area, Pend Oreille County, Washington, and Bonner and Boundary Counties, Idaho: U.S. Geological Survey Open-File Report 83–601, 14 p., map scale 1:48,000.

Miller, F. K., and Clark, L. D., 1975, Geology of the Chewelah–Loon Lake area, Stevens and Spokane Counties, Washington: U.S. Geological Survey Professional Paper 806, 74 p.

Miller, F. K., McKee, E. H., and Yates, R. G., 1973, Age and correlation of the Windermere Group in northeastern Washington: Geological Society of America Bulletin, v. 84, p. 3723–3730.

Miller, J.M.G., 1985, Glacial and syntectonic sedimentation; The upper Proterozoic Kingston Peak Formation, southern Panamint Range, eastern California: Geological Society of America Bulletin, v. 96, p. 1537–1553.

Miller, J.M.G., Wright, L. A., and Troxel, B. W., 1981, The Late Precambrian Kingston Peak Formation, Death Valley region, California, *in* Hambrey, M. J., and Harland, W. B., eds., Earth's pre-Pleistocene glacial record: Cambridge, United Kingdom, Cambridge University Press, p. 745–748.

Miller, M. M., 1987, Dispersed remnants of a northeast Pacific fringing arc; Upper Paleozoic terranes of Permian McCloud faunal affinity, western U.S.: Tectonics, v. 6, no. 6, p. 807–830.

Miller, R. H., and Dockum, M. S., 1983, Ordovician conodonts from metamorphosed carbonates of the Salton Trough, California: Geology, v. 11, p. 410–412.

Minnick, E. P., 1975, Stratigraphy and structure of the Ordovician Vinini Formation, Tyrone Gap area, Eureka County, Nevada [M.S. thesis]: Athens, Ohio University, 55 p.

Misch, P., 1966, Tectonic evolution of the northern Cascades of Washington State; A west-Cordilleran case history, *in* Gunning, H. C., ed., A symposium on the tectonic history and mineral deposits of the western Cordillera in British Columbia and neighbouring parts of the United States: Canadian Institute of Mining and Metallurgy Special Volume 8, p. 101–148.

——, 1988, Tectonic and metamorphic evolution of the North Cascades; An overview, *in* Ernst, W. G., ed., Metamorphism and crustal evolution of the western United States (Rubey volume 7): Englewood Cliffs, New Jersey, Prentice-Hall, p. 179–195.

Moore, J. N., 1976, Depositional environments of the Lower Cambrian Poleta Formation and its stratigraphic equivalents, California and Nevada, *in* Robison, R. A., and Rowell, A. J., eds., Paleontology and depositional environments; Cambrian of western North America: Provo, Utah, Brigham Young University Geology Studies, v. 23, part 2, p. 23–38.

Moore, J. N., and Foster, C. T., Jr., 1980, Lower Paleozoic metasedimentary rocks in the east-central Sierra Nevada, California; Correlation with Great Basin formations: Geological Society of America Bulletin, part 1, v. 91, p. 37–43.

Moorman, M., 1974, Microbiota of the late Proterozoic Hector Formation, southwestern Alberta, Canada: Journal of Paleontology, v. 48, no. 3, p. 524–539.

Mount, J. F., 1982, Earliest Cambrian paleoenvironments of the southern Great Basin, *in* Cooper, J. D., Troxel, B. W., and Wright, L. A., eds., Geology of selected areas in the San Bernardino Mountains, western Mojave Desert, and southern Great Basin, California: Shoshone, California, Death Valley Publishing Company, Geological Society of America Cordilleran Section, Anaheim, California, April 1982, Guidebook and Volume, p. 187–193.

Mullens, T. E., 1980, Stratigraphy, petrology, and some fossil data of the Roberts Mountains Formation, north-central Nevada: U.S. Geological Survey Professional Paper 1063, 67 p.

Murphy, M. A., Power, J. D., and Johnson, J. G., 1984, Evidence for Late Devonian movement within the Roberts Mountains allochthon, Roberts Mountains, Nevada: Geology, v. 12, p. 20–23.

Nilsen, T. H., 1977, Paleogeography of Mississippian turbidites in south-central Idaho, *in* Stewart, J. H., Stevens, C. H., and Fritsche, A. E., eds., Paleozoic paleogeography of the western United States, Pacific Coast Paleogeography Symposium 1: Society of Economic Paleontologists and Mineralogists, p. 275–299.

Nilsen, T. H., and Stewart, J. H., 1980, Penrose conference report; The Antler orogeny; Mid-Paleozoic tectonism in western North America: Geology, v. 8, p. 298–302.

Nokleberg, W. J., 1983, Wallrocks of the central Sierra Nevada Batholith, California; A collage of accreted tectono-stratigraphic terranes: U.S. Geological Survey Professional Paper 1255, 28 p.

Nokleberg, W. J., and Kistler, R. W., 1980, Paleozoic and Mesozoic deformations in the central Sierra Nevada, California: U.S. Geological Survey Professional Paper 1145, 24 p.

Nolan, T. B., Merriam, C. W., and Williams, J. S., 1956, The stratigraphic section in the vicinity of Eureka, Nevada: U.S. Geological Survey Professional Paper 276, 77 p.

Ojakangas, R. W., and Matsch, C. L., 1980, Upper Precambrian (Eocambrian) Mineral Fork Tillite of Utah; A continental glacial and glaciomarine sequence: Geological Society of America Bulletin, part 1, v. 91, p. 495–501.

Oldow, J. S., 1983, Spatial variability in the structure of the Roberts Mountains allochthon, western Nevada: Geological Society of America Bulletin, v. 95, p. 174–185.

Olson, J. C., Marvin, R. F., Parker, R. L., and Mehnert, H. H., 1977, Age and tectonic setting of lower Paleozoic alkalic and mafic rocks, carbonatites, and thorium veins in south-central Colorado: U.S. Geological Survey Journal of Research, v. 5, no. 6, p. 673–687.

Oriel, S. S., and Armstrong, F. C., 1971, Uppermost Precambrian and lowest Cambrian rocks in southeastern Idaho: U.S. Geological Survey Professional Paper 394, 52 p.

Palmer, A. R., 1971, The Cambrian of the Great Basin and adjacent areas, western United States, *in* Holland, C. H., ed., Cambrian of the New World: New York, Wiley-Interscience, p. 1–78.

——, 1981a, Subdivision of the Sauk sequence, *in* Taylor, M. E., ed., Short papers for the Second International Symposium on the Cambrian System, 1981: U.S. Geological Survey Open-File Report 81–743, p. 160–162.

——, 1981b, On the correlatability of Grand Cycle tops, *in* Taylor, M. E., ed., Short papers for the Second International Symposium on the Cambrian System, 1981: U.S. Geological Survey Open-File Report 81–743, p. 156–159.

——, 1983, compiler, The decade of North American geology 1983 geologic time scale: Geology, v. 11, p. 503–504.

Palmer, A. R., and Halley, R. B., 1979, Physical stratigraphy and trilobite biostratigraphy of the Carrara Formation (Lower and Middle Cambrian) in the southern Great Basin: U.S. Geological Survey Professional Paper 1047, 131 p.

Palmer, A. R., DeMis, W. D., Muehlberger, W. R., and Robison, R. A., 1984, Geological implications of Middle Cambrian boulders from the Haymond Formation (Pennsylvanian) in the Marathon basin, west Texas: Geology, v. 12, p. 91–94.

Papke, K. G., 1984, Barite in Nevada: Nevada Bureau of Mines and Geology Bulletin 98, 125 p.

Parrish, R. R., and Armstrong, R. L., 1983, U-Pb zircon age and tectonic significance of gneisses in structural culminations of the Omineca crystalline belt, British Columbia: Geological Society of America Abstracts with Programs, v. 15, p. 324.

Peacock, S. M., 1985, Inverted metamorphic gradient and fluid evolution of the Central Metamorphic belt, Klamath Mountains province, northern California: Geological Society of America Abstracts with Programs, v. 17, p. 400.

——, 1988, Inverted metamorphic gradients in the westernmost Cordillera, *in* Ernst, W. G., ed., Metamorphism and crustal evolution of the western United States (Rubey Volume 7): Englewood Cliffs, New Jersey, Prentice-Hall, p. 953–975.

Poole, F. G., 1988, Stratiform barite in Paleozoic rocks of the western United States, *in* Zachrisson, E., ed., Proceedings of the 7th Quadrennial IAGOD Symposium, Luleå, Sweden: Stuttgart, E. Schweizerbart'sche Verlagsbuchhandlung (Nägele u. Obermiller), p. 309–319.

Poole, F. G., and Desborough, G. A., 1973, Alpine-type serpentinites in Nevada and their tectonic significance: Geological Society of America Abstracts with Programs, v. 5, p. 90–91.

——, 1980, Oil and metals in Ordovician and Devonian kerogenous marine strata of central Nevada, *in* American Association of Petroleum Geologists Abstracts for National Meeting, Denver, Colorado: American Association of Petroleum Geologists Bulletin, v. 64, no. 5, p. 767.

——, 1981, Geologic factors controlling occurrence of low-temperature metal deposits in Paleozoic marine shales in Nevada: Geological Society of America Abstracts with Programs, v. 13, p. 530.

Poole, F. G., and Claypool, G. E., 1984, Petroleum source-rock potential and crude-oil correlation in the Great Basin, *in* Woodward, J., Meissner, F. F., and Clayton, J. L., eds., Hydrocarbon source rocks of the Greater Rocky Mountain region: Denver, Colorado, Rocky Mountain Association of Geologists, p. 179–229.

Poole, F. G., and Sandberg, C. A., 1977, Mississippian paleogeography and tectonics of the western United States, *in* Stewart, J. H., Stevens, C. H., and Fritsche, A. E., eds., Paleozoic paleogeography of the western United States, Pacific Coast Paleogeography Symposium 1: Society of Economic Paleontologists and Mineralogists, p. 67–85.

Poole, F. G., and 7 others, 1967, Devonian of the southwestern United States, *in* Oswald, D. H., ed., International Symposium on the Devonian System: Calgary, Alberta Society of Petroleum Geologists, v. 1, p. 879–912.

Poole, F. G., Sandberg, C. A., and Boucot, A. J., 1977, Silurian and Devonian paleogeography of the western United States, *in* Stewart, J. H., Stevens, C. H., and Fritsche, A. E., eds., Paleozoic paleogeography of the western United States, Pacific Coast Paleogeography Symposium 1: Society of Economic Paleontologists and Mineralogists, p. 39–55.

Poole, F. G., Christiansen, R. L., Carr, M. D., and Ross, R. J., Jr., 1981, Lower and middle Paleozoic eugeosynclinal rocks in the El Paso Mountains, northwestern Mojave Desert, California, *in* Howard, K. A., Carr, M. D., and Miller, D. M., eds., Tectonic framework of the Mojave and Sonoran Deserts, California and Arizona; Abstracts from a conference held by the U.S. Geological Survey in Menlo Park, California, November 4–6, 1980: U.S. Geological Survey Open-File Report 81-503, p. 85, 86.

Poole, F. G., Carr, M. D., and Christiansen, R. L., 1982, Paleozoic geology of the El Paso Mountains, Kern County, California: Geological Society of America Abstracts with Programs, v. 14, p. 225.

Poole, F. G., Claypool, G. E., and Fouch, T. D., 1983, Major episodes of petroleum generation in part of the northern Great Basin, *in* The role of heat in the development of energy and mineral resources in the northern Basin and Range Province: Geothermal Resources Council Special Report 13, p. 207–213.

Potter, A. W., Hotz, P. E., and Rohr, D. M., 1977, Stratigraphy and inferred tectonic framework of lower Paleozoic rocks in the eastern Klamath Mountains, northern California, *in* Stewart, J. H., Stevens, C. H., and Fritsche, A. E., eds., Paleozoic paleogeography of the western United States, Pacific Coast Paleogeography Symposium 1: Society of Economic Paleontologists and Mineralogists, p. 421–440.

Quick, J. E., 1981, Petrology and petrogenesis of the Trinity peridotite, an upper mantle diapir in the eastern Klamath Mountains, northern California: Journal of Geophysical Research, v. 86, no. B12, p. 11837–11863.

Rees, M. N., 1986, A fault-controlled trough through a carbonate platform; The Middle Cambrian House Range embayment: Geological Society of America Bulletin, v. 97, p. 1054–1069.

Reesor, J. E., 1973, Geology of the Lardeau map-area, east-half, British Columbia: Geological Survey of Canada Memoir 369, 129 p.

Reynolds, R. L., 1968, Paleocurrents, petrology, and source area of the Addy Quartzite (Cambrian), northeastern Washington [B.A. thesis]: Princeton, New Jersey, Princeton University, 48 p.

Ritzma, H. R., 1983, Igneous dikes of the eastern Uinta Mountains, Utah and Colorado: Utah Geological and Mineral Survey Special Studies 56, 23 p.

Riva, J., 1970, Thrusted Paleozoic rocks in the northern and central HD Range, northeastern Nevada: Geological Society of America Bulletin, v. 81, p. 2689–2716.

Roberts, R. J., 1964, Stratigraphy and structures of the Antler Peak Quadrangle, Humboldt and Lander Counties, Nevada: U.S. Geological Survey Professional Paper 459-A, 93 p.

Roberts, R. J., Hotz, P. E., Gilluly, J., and Ferguson, H. G., 1958, Paleozoic rocks of north-central Nevada: American Association of Petroleum Geologists Bulletin, v. 42, no. 12, p. 2813–2857.

Robison, R. A., 1960, Lower and Middle Cambrian stratigraphy of the eastern Great Basin, *in* Boettcher, J. W., and Sloan, W. W., Jr., eds., Guidebook to the geology of east-central Nevada, 1960: Intermountain Association of Petroleum Geologists and Eastern Nevada Geological Society, 11th Annual Field Conference: p. 43–52.

Rodgers, J.J.W., and 7 others, 1974, Paleozoic and lower Mesozoic volcanism and continental growth in the western United States: Geological Society of America Bulletin, v. 85, p. 1913–1924.

Ross, R. J., Jr., 1951, Stratigraphy of the Garden City Formation in northeastern Utah, and its trilobite faunas: New Haven, Connecticut, Yale University, Peabody Museum of Natural History Bulletin 6, 155 p.

——, 1972, Fossils from the Ordovician bioherm at Meiklejohn Peak, Nevada: U.S. Geological Survey Professional Paper 685, 47 p.

——, 1976, Ordovician sedimentation in the western United States, *in* Bassett, M. G., ed., The Ordovician System; Proceedings Palaeontological Association Symposium, September 1974, Birmingham (England): Cardiff, University of Wales Press, National Museum of Wales, p. 73–105.

——, 1977, Ordovician paleogeography of the western United States, *in* Stewart, J. H., Stevens, C. H., and Fritsche, A. E., eds., Paleozoic paleogeography of the western United States, Pacific Coast Paleogeography Symposium 1: Society of Economic Paleontologists and Mineralogists, p. 19–38.

Ross, R. J., Jr., and Tweto, O., 1980, Lower Paleozoic sediments and tectonics in Colorado, *in* Kent, H. C., and Porter, K. W., eds., Colorado geology: Denver, Colorado, Rocky Mountain Association of Geologists, p. 47–56.

Ross, R. J., Jr., and 27 others, 1982, The Ordovician System in the United States: International Union of Geological Sciences Publication 12, 73 p.

Ross, R. J., Jr., James, N. P., Hintze, L. F., and Poole, F. G., 1989, Architecture and evolution of a Whiterockian (early Middle Ordovician) carbonate plat-

form, Basin Ranges of western U.S.A., *in* Crevello, P. D., Wilson, J. L., Sarg, J. F., and Read, J. F., eds., Controls on carbonate platform and basin development: Society of Economic Paleontologists and Mineralogists Special Publication 44, p. 167–185.

Rowell, A. J., Rees, M. N., and Suczek, C. A., 1979, Margin of the North American continent in Nevada during Late Cambrian time: American Journal of Science, v. 279, no. 1, p. 1–18.

Ruppel, E. T., 1985, Lemhi Arch, a late Proterozoic and early Paleozoic landmass, central Idaho; *in* American Association of Petroleum Geologists Rocky Mountain Section Meeting Abstracts; American Association of Petroleum Geologists Bulletin, v. 69, no. 5, p. 864.

Ruppel, E. T., Ross, R. J., Jr., and Schleicher, D. L., 1975, Precambrian Z and Lower Ordovician rocks in east-central Idaho: U.S. Geological Survey Professional Paper 889-B, p. 25–34.

Russell, S. R., and Nokleberg, W. J.,1977, Superimposition and timing of deformations in the Mount Morrison roof pendant in the central Sierra Nevada, California: Geological Society of America Bulletin, v. 88, p. 335–345.

Rye, R. O., Shawe, D. R., and Poole, F. G., 1978, Stable isotope studies of bedded barite at East Northumberland Canyon in Toquima Range, central Nevada: U.S. Geological Survey Journal of Research, v. 6, no. 2, p. 221–229.

Saleeby, J. B., Hannah, J. L., and Varga, R. J., 1987, Isotopic age constraints on middle Paleozoic deformation in the northern Sierra Nevada, California: Geology, v. 15, p. 757–760.

Saleeby, J. B., Shaw, H. F., Niemeyer, S., Moores, E. M., and Edelman, S. H., 1989, U/Pb, Sm/Nd, and Rb/Sr geochronological and isotopic study of northern Sierra Nevada ophiolitic assemblages, California: Contributions to Mineralogy and Petrology, v. 102, p. 205–220.

Sandberg, C. A., 1983, Petroleum potential of wilderness lands in Idaho, *in* Miller, B. M., ed., Petroleum potential of wilderness lands in the western United States: U.S. Geological Survey Circular 902-F, p. F1–F6.

Sandberg, C. A., and Dreesen, R., 1984, Late Devonian icriodontid biofacies models and alternate shallow-water conodont zonation, *in* Clark, D. L., ed., Conodont biofacies and provincialism: Geological Society of America Special Paper 196, p. 143–178.

Sandberg, C. A., and Mapel, W. J., 1967, Devonian of the Northern Rocky Mountains and Plains, *in* Oswald, D. H., ed., International Symposium on the Devonian System: Calgary, Alberta Society of Petroleum Geologists, v. 1, p. 843–877.

Sandberg, C. A., and Poole, F. G., 1977, Conodont biostratigraphy and depositional complexes of Upper Devonian cratonic-platform and continental-shelf rocks in the western United States, *in* Murphy, M. A., Berry, W.B.N., and Sandberg, C. A., eds., Western North America; Devonian: Riverside, California University, Campus Museum Contribution 4, p. 144–182.

Sandberg, C. A., Hall, W. E., Batchelder, J. N., and Axelsen, C., 1975, Stratigraphy, conodont dating, and paleotectonic interpretations of the type Milligen Formation (Devonian), Wood River area, Idaho: U.S. Geological Survey Journal of Research, v. 3, no. 6, p. 707–720.

Sandberg, C. A., Gutschick, R. C., Johnson, J. G., Poole, F. G., and Sando, W. J., 1982, Middle Devonian to Late Mississippian geologic history of the Overthrust belt region, western United States, *in* Powers, R. B., ed., Geologic studies of the Cordilleran thrust belt: Denver, Colorado, Rocky Mountain Association of Geologists, v. 2, p. 691–719.

Sandberg, C. A., Poole, F. G., and Johnson, J. G., 1988, Upper Devonian of western United States, *in* McMillan, N. J., Embry, A. F., and Glass, D. J., eds., Devonian of the World: Calgary, Canadian Society of Petroleum Geologists Memoir 14, v. 1, p. 183–220.

Sando, W. J., and Sandberg, C. A., 1987, New interpretations of Paleozoic stratigraphy and history in the northern Laramie Range and vicinity, southeast Wyoming: U.S. Geological Survey Professional Paper 1450, 39 p.

Schneck, W. M., and McCollum, L. B., 1985, Upper Proterozoic and lower Phanerozoic fluvial deposits in the central Great Basin: Geological Society of America Abstracts with Programs, v. 17, p. 264.

Schweicket, R. A., 1977, Major pre-Jurassic thrust fault between the Calaveras and Shoo Fly Complexes, Sierra Nevada, California: Geological Society of America Abstracts with Programs, v. 9, p. 497.

——, 1981, Tectonic evolution of the Sierra Nevada Range, *in* Ernst, W. G., ed., The geotectonic development of California (Rubey Volume 1): Englewood Cliffs, New Jersey, Prentice-Hall, p. 87–131.

Schweickert, R. A., and Snyder, W. S., 1981, Paleozoic plate tectonics of the Sierra Nevada and adjacent regions, *in* Ernst, W. G., ed., The geotectonic development of California (Rubey Volume 1): Englewood Cliffs, New Jersey, Prentice-Hall, p. 182–201.

Schweickert, R. A., Harwood, D. S., Girty, G. H., and Hanson, R. E., 1984, Tectonic development of the northern Sierra terrane; An accreted late Paleozoic island arc and its basement (field trip 5), *in* Lintz, J., Jr., ed., Western geological excursions; Guidebook of field trips for the 1984 Annual Meeting of the Geological Society of America at Reno: Reno, University of Nevada, Mackay School of Mines, Department of Geological Sciences, v. 4, p. 1–65.

Seeland, D. A., 1968, Paleocurrents of the late Precambrian to Early Ordovician (basal Sauk) transgressive clastics of the western and northern United States [Ph.D. thesis]: Salt Lake City, Utah University, 276 p.

——, 1969, Marine current directions in upper Precambrian and Cambrian rocks of the southwestern United States, *in* Baars, D. L., ed., Guidebook to the geology and natural history of the Grand Canyon region: 5th Four Corners Geological Society Field Conference, p. 123–126.

Sharp, W. D., 1988, Pre-Cretaceous crustal evolution in the Sierra Nevada region, California, *in* Ernst, W. G., ed., Metamorphism and crustal evolution of the western United States (Rubey Volume 7): Englewood Cliffs, New Jersey, Prentice-Hall, p. 823–864.

Sharp, W. D., and 6 others, 1982, Age and tectonic significance of Paleozoic orthogneisses of the Sierra Nevada foothills metamorphic belt, California: Geological Society of America Abstracts with Programs, v. 14, p. 233.

Shaw, H. F., Chen, J. H., Saleeby, J. B., and Wasserburg, G. J., 1987, Nd-Sr-Pb systematics and age of the Kings River ophiolite, California; Implications for depleted mantle evolution: Contributions to Mineralogy and Petrology, v. 96, no. 3, p. 281–290.

Silberling, N. J., and Roberts, R. J., 1962, Pre-Tertiary stratigraphy and structure of northwestern Nevada: Geological Society of America Special Paper 72, 58 p.

Silberling, N. J., Jones, D. L., Blake, M. C., Jr., and Howell, D. G., 1987, Lithotectonic terranes of the western conterminous United States, Part C, *in* Silberling, N. J., and Jones, D. L., eds., Lithotectonic terrane maps of the North American Cordillera: U.S. Geological Survey Miscellaneous Field Studies Map MF-1874–C, scale 1:2,500,000.

Smith, G. I., 1962, Large lateral displacement on Garlock fault, California, as measured from offset dike swarm: American Association of Petroleum Geologists Bulletin, v. 46, no. 1, p. 85–104.

Smith, J. F., Jr., and Ketner, K. B., 1977, Tectonic events since early Paleozoic in the Carlin–Piñon Range area, Nevada: U.S. Geological Survey Professional Paper 867-C, p. C1–C18.

Snook, J. R., 1982, The Kootenay Arc in N.E. Washington: Geological Society of America Abstracts with Programs, v. 14, p. 235.

Stanley, K. O., Chamberlain, C. K., and Stewart, J. H., 1977, Depositional setting of some eugeosynclinal Ordovician rocks and structurally interleaved Devonian rocks in the Cordilleran mobile belt, Nevada, *in* Stewart, J. H., Stevens, C. H., and Fritsche, A. E., eds., Paleozoic paleogeography of the western United States, Pacific Coast Paleogeography Symposium 1: Society of Economic Paleontologists and Mineralogists, p. 259–274.

Stewart, J. H., 1970, Upper Precambrian and Lower Cambrian strata in the southern Great Basin, California and Nevada: U.S. Geological Survey Professional Paper 620, 206 p.

——, 1972, Initial deposits in the Cordilleran geosyncline; Evidence of a late Precambrian (<850 m.y.) continental separation: Geological Society of America Bulletin, v. 83, p. 1345–1360.

——, 1974, Correlation of uppermost Precambrian and Lower Cambrian strata from southern to east-central Nevada: U.S. Geological Survey Journal of Research, v. 2, no. 5, p. 609–618.
——, 1980, Geology of Nevada, a discussion to accompany the geologic map of Nevada: Nevada Bureau of Mines and Geology Special Publication 4, 132 p.
Stewart, J. H., and Poole, F. G., 1974, Lower Paleozoic and uppermost Precambrian Cordilleran miogeocline, Great Basin, western United States, *in* Dickinson, W. R., ed., Tectonics and sedimentation: Society of Economic Paleontologists and Mineralogists Special Publication 22, p. 28–57.
Stewart, J. H., and Suczek, C. A., 1977, Cambrian and latest Precambrian paleogeography and tectonics in the western United States, *in* Stewart, J. H., Stevens, C. H., and Fritsche, A. E., eds., Paleozoic paleogeography of the western United States, Pacific Coast Paleogeography Symposium 1: Society of Economic Paleontologists and Mineralogists, p. 1–17.
Stewart, J. H., McMenamin, M.A.S., and Morales-Ramirez, J. M., 1984, Upper Proterozoic and Cambrian rocks in the Caborca region, Sonora, Mexico; Physical stratigraphy, biostratigraphy, paleocurrent studies, and regional relations: U.S. Geological Survey Professional Paper 1309, 36 p.
Suczek, C. A., 1977, Sedimentology and petrology of the Cambrian Harmony Formation of north-central Nevada: Geological Society of America Abstracts with Programs, v. 9, p. 510.
Sweet, W. C., 1954, Harding and Fremont Formations, Colorado: American Association of Petroleum Geologists Bulletin, v. 38, no. 2, p. 284–305.
——, 1961, Middle and Upper Ordovician rocks, central Colorado, *in* Berg, R. R., and Rold, J. W., eds., Symposium on lower and middle Paleozoic rocks of Colorado: 12th Rocky Mountain Association of Geologists Field Conference, p. 17–24.
Tabor, R. W., and 7 others, 1980, Preliminary geologic map of the Chelan 1:100,000 Quadrangle, Washington: U.S. Geological Survey Open-File Report 80–841, 46 p.
Tabor, R. W., Zartman, R. E., and Frizzell, V. A., Jr., 1987, Possible tectonostratigraphic terranes in the North Cascades crystalline core, Washington, *in* Schuster, J. E., ed., Selected papers on the geology of Washington: Washington Division of Geology and Earth Resources Bulletin 77, p. 107–127.
Taylor, G. C., 1984, Mineral resource potential of the Rockhouse Basin Wilderness Study Area, Kern and Tulare Counties, California: California Geology, v. 37, no. 12, p. 263–272.
Thompson, S., III, and Potter, P. E., 1981, Paleocurrents of Bliss Sandstone (Cambrian–Ordovician), southwestern New Mexico and western Texas: New Mexico Bureau of Mines and Mineral Resources Annual Report, v. 1979–80, p. 36–51.
Troxel, B. W., Wright, L. A., and Williams, E. G., 1977, Late Precambrian history derived from the Kingston Peak Formation, Death Valley region, California: Geological Society of America Abstracts with Programs, v. 9, p. 517.
Troxel, B. W., McMackin, M. A., and Calzia, J. P., 1987, Comment on "Late Precambrian tectonism in the Kingston Range, southern California": Geology, v. 15, p. 274–275.
Varga, R. J., 1982, Implications of Paleozoic phosphorites in the northern Sierra Nevada Range: Nature, v. 297, no. 5863, p. 217–220.
Varga, R. J., and Moores, E. M., 1981, Age, origin, and significance of an unconformity that predates island-arc volcanism in the northern Sierra Nevada: Geology, v. 9, p. 512–518.
Walker, J. D., Klepacki, D. W., and Burchfiel, B. C., 1986, Late Precambrian tectonism in the Kingston Range, southern California: Geology, v. 14, p. 15–18.
Wallin, E. T., Mattinson, J. M., and Potter, A. W., 1988, Early Paleozoic magmatic events in the eastern Klamath Mountains, northern California: Geology, v. 16, p. 144–148.
Watkins, R., and Flory, R. A., 1986, Island arc sedimentation in the Middle Devonian Kennett Formation, eastern Klamath Mountains, California: Journal of Geology, v. 94, no. 5, p. 753–761.
Webster, G. D., and Beka, F. T., 1980, A Devonian age for parts of the Black Shale belt northwest of Northport, Washington: Geological Society of America Abstracts with Programs, v. 12, p. 159.
Wells, F. G., Walker, G. W., and Merriam, C. W., 1959, Upper Ordovician(?) and Upper Silurian formations of the northern Klamath Mountains, California: Geological Society of America Bulletin, v. 70, p. 645–650.
Whetten, J. T., Jones, D. L., Cowan, D. S., and Zartman, R. E., 1978, Ages of Mesozoic terranes in the San Juan Islands, Washington, *in* Howell, D. G., and McDougall, K. A., eds., Mesozoic paleogeography of the western United States, Pacific Coast Paleogeography Symposium 2: Society of Economic Paleontologists and Mineralogists, p. 117–132.
Wilson, W. F., 1962, Sedimentary petrology and sedimentary structures of the Cambrian Hickory Sandstone Member, central Texas [M.S. thesis]: Austin, University of Texas, 282 p.
Wright, L. A., Troxel, B. W., Williams, E. G., Roberts, M. T., and Diehl, P. E., 1974, Precambrian sedimentary environments of the Death Valley region, eastern California, *in* Death Valley region, California and Nevada, Field Trip 1: Shoshone, California, Death Valley Publishing Company, Geological Society of America Cordilleran Section Annual Meeting Guidebook, Las Vegas, 1974, p. 27–36.
Wrucke, C. T., 1974, Geologic map of the Gold Acres–Tenabo area, Shoshone Range, Lander County, Nevada: U.S. Geological Survey Miscellaneous Field Studies Map MF–647, scale 1:15,840.
Wrucke, C. T., and Silberman, M. L., 1975, Cauldron subsidence of Oligocene age at Mount Lewis, northern Shoshone Range, Nevada: U.S. Geological Survey Professional Paper 876, 20 p.
Wrucke, C. T., and Theodore, T. G., 1970, Direction of movement of the Roberts Mountains thrust determined from folds, northern Shoshone Range and Battle Mountain, Nevada: Geological Society of America Abstracts with Programs, v. 2, p. 356.
Wrucke, C. T., Churkin, M., and Heropoulos, C., 1978, Deep sea origin of Ordovician pillow basalt and associated sedimentary rocks, northern Nevada: Geological Society of America Bulletin, v. 89, p. 1272–1280.
Yates, R. G., 1964, Geologic map and sections of the Deep Creek area, Stevens and Pend Oreille Counties, Washington: U.S. Geological Survey Miscellaneous Geologic Investigations Map I–412, scale 1:31,680.
——, 1971, Geologic map of the Northport Quadrangle, Washington: U.S. Geological Survey Miscellaneous Geologic Investigations Map I–603, scale 1:31,680.
——, 1976, Geology of the Deep Creek area, Washington, and its regional significance: U.S. Geological Survey Open-File Report 76–537, 435 p.
Young, G. M., Jefferson, C. W., Delaney, G. D., and Yeo, G. M., 1979, Middle and late Proterozoic evolution of the northern Canadian Cordillera and Shield: Geology, v. 7, p. 125–128.
Ziegler, W., and Sandberg, C. A., 1984, *Palmatolepis*-based revision of upper part of standard Late Devonian conodont zonation, *in* Clark, D. L., ed., Conodont biofacies and provincialism: Geological Society of America Special Paper 196, p. 179–194.

Manuscript Accepted by the Society February 28, 1990

ACKNOWLEDGMENTS

Interpretations of regional structures in western Montana have benefited from our discussions with Mitchell Reynolds and William Perry, Jr. We are grateful to Craig Bannister, who ably handled the many chores related to initial preparation of the manuscript, and to Margaret Clemensen, who expertly utilized the word processor for all versions of the manuscript. Patricia Poole, Gary Poole, and Ezekiel Rivera contributed many tedious hours to coloring and proofing the maps on Plates 2 and 3. We are also grateful to Dennis Welp, who expertly processed the illustrations, and to Joseph and Alice Springfield, who patiently drafted them. The manuscript was improved by careful and beneficial reviews by Clark Burchfiel, Jess Johnson, Jason Saleeby, Daniel Shawe, Norman Silberling, Paul Stone, and Michael Taylor.

NOTES ADDED IN PROOF

The manuscript for this chapter was compiled by individual authors in 1985 and assembled in early 1986. Since then, several important papers have been published that provide additional information on various aspects of the geology of this time slice. In particular, Pacific section SEPM volumes 1 and 2 on *Paleozoic paleogeography of the western United States-II* (Cooper and Stevens, 1991) contain many articles presenting information that was previously unavailable. Also, the Canadian Society of Petroleum Geologists Memoir 14 entitled *Devonian of the world* (McMillan and others, 1988) contains a wealth of new data on the Devonian of the western United States.

Critical new information and reference citations have been integrated into the galley text where rephrasing and additions did not change the length of the containing paragraph. Where modifications were not possible, we have added brief notes and important new references herein. The two color plates (Plates 3 and 4) were printed in 1991, so no physical corrections could be made. However, some important notes pertaining to Plate 3 (Figures 1, 2, 3, 4, and 6) are listed herein.

Text

Cambrian rocks. Any references to the Cambrian age of the Harmony Formation in the Osgood Mountains are incorrect. The unit mapped as Harmony Formation is comprised of debris flows and stratigraphically overlies the late Mississippian Gough's Canyon Formation. Interbedded silty limestone turbidites in its lower part contain Pennsylvanian conodonts, and a dolomite near the top of the unit contains a lower Permian conodont fauna. The limestones with Cambrian trilobites that are found in this formation and that provided the evidence of its "age" are olistoliths (McCollum and McCollum, 1991). Locally, clasts recognized as Harmony Formation are in rocks as old as Meramecian.

Lower Silurian to Middle Devonian–Llandoverian to Givetian Rocks. In text Figure 2b, columns 7 and 8, the Upper Silurian and Lower Devonian unit labeled "dolostone" (between the Laketown and Sevy Dolostones) is not recognized in the type areas of these formations. We consider this unit to be transitional between the originally defined Laketown and Sevy because it is composed of dolostone beds similar to both. Most workers have lumped this intermediate "dolostone" with the Sevy although, in most sections in western Utah and eastern Nevada, the lower part contains medium- and coarsely crystalline dolostone beds similar to those in the Laketown. The dolostone unit occurs in the axial part of the subsiding Paleozoic continental shelf, where sedimentation was continuous between the latest Silurian and earliest Devonian. The Laketown correlates with Units A to C, the "dolostone" correlates with Units D and E, and the Sevy correlates with Unit F of the Dolomite of Spotted Range (Poole and others, 1977, Fig. 2b, col. 32).

Upper Upper Devonian–Middle and Upper Famennian rocks. The general Famennian regression has been attributed to Southern Hemisphere glaciation and the four transgressive interruptions have been attributed to interglacial stages by Sandberg and others (1988). The Late Devonian standard conodont zonation was revised and many of the zones were renamed by Ziegler and Sandberg (1990). The updated zonation was integrated in the chapter text during proofing of the galleys.

Upper Upper Devonian–Middle and Upper Fammennian rocks. Overlying rocks. The nomenclature and paleogeography of the overlying Mississippian rocks throughout the western United States have been revised and reinterpreted by Poole and Sandberg (1991). New maps therein supersede the distribution of Mississippian rocks shown in E. L. Miller and others (this volume, Plate 4B, 4C), particularly in Montana, Wyoming, and Colorado.

Rocks of the Antler Orogen–The Roberts Mountains allochthon. Stratigraphy: sedimentary rocks. Mississippian rocks have not been positively identified within the Roberts Mountains allochthon (RMA). New stratigraphic data from central Nevada show that Lower Mississippian barite and siliceous facies locally overlap the RMA and are not part of the allochthon as previously believed. For more information on the relations between the RMA and overlapping Lower Mississippian rocks in central Nevada, see Poole and Sandberg (1992).

Rocks related to the Antler Orogen in other areas. Allochthon remnants in southern Sierra Nevada. Several roof pendants in the southeastern Sierra Nevada have been described in more detail by Dunne and Suczek (1991).

Plate 3

Figures 1, 2, and 4. The "Transcontinental arch" was inadvertently omitted from Figures 1, 2, and 4. The position of the arch is similar to that shown on Plate 3-3.

Figure 2. On Plate 3-2, the distribution of Lower Silurian rocks was arbitrarily not shown in most of eastern Montana and adjacent states. The presence of *Virgiana* in a well core in western South Dakota, only a few kilometers east of the Montana state line, indicates that Silurian rocks were once continuous across eastern Wyoming between the Interlake Formation and the area of kimberlitic diatremes in southeastern Wyoming and adjacent Colorado, where the same brachiopod occurs.

Figures 3 and 4. Plate 3-3 shows rocks deposited mainly from the Early *falsiovalis* Zone, but locally from the older, Middle Devonian *disparilis* Zone, through the Late *marginifera* Zone, and Plate 3-4 shows rocks deposited from the *trachytera* through *praesulcata* Zones (Ziegler and Sandberg, 1990). These maps were updated to incorporate revised distributions from Sandberg, Poole, and Johnson (1988), which presented paleobiogeographic maps for nine Late Devonian time slices.

Figure 6, column 2. For "Adobe Hills" read "Adobe Range."

ADDITIONAL REFERENCES

Cooper, J. D., and Stevens, C. H., eds., 1991, Paleozoic paleogeography of the western United States-II: Los Angeles, California, Pacific section, Society of Economic Paleontologists and Mineralogists, book 67, 2 vols., 872 p.

Dunne, G. C., and Suczek, C. A., 1991, Early Paleozoic eugeoclinal strata in the Kern Plateau pendants, southern Sierra Nevada, California, *in* Cooper, J. D., and Stevens, C. H., eds., Paleozoic paleogeography of the western United States-II, v. 2: Los Angeles, California, Pacific section, Society of Economic Paleontologists and Mineralogists, book 67, p. 677–692.

Johnson, J. G., and Sandberg, C. A., 1988, Devonian eustatic events in the Western United States and their biostratigraphic responses, *in* McMillan, N. J., Embry, A. F., and Glass, D. J., eds., Devonian of the world: Calgary, Canadian Society of Petroleum Geologists Memoir 14, v. 3, p. 171–178.

McCollum, L. B., and McCollum, M., 1991, Paleozoic rocks of the Osgood Mountains, Nevada, *in* Raines, G. L., Lisle, R. E., Schafer, R. W., and Wilkinson, W. H., Geology and ore deposits of the Great Basin: Reno, Geological Society of Nevada, p. 735–738.

McMillan, N. J., Embry, A. F., and Glass, D. J., eds., 1988, Devonian of the world: Calgary, Canadian Society of Petroleum Geologists Memoir 14, 3 vols.

Poole, F. G., and Sandberg, C. A., 1991, Mississippian paleogeography and conodont biostratigraphy of the western United States, *in* Cooper, J. D., and Stevens, C. H., eds., Paleozoic paleogeography of the western United States-II, v. 1: Los Angeles, California, Pacific section, Society of Economic Paleontologists and Mineralogists, book 67, p. 107–136.

Poole, F. G., and Sandberg, C. A., 1992, Roberts Mountains allochthon and overlapping Lower Mississippian syntectonic foredeep deposits in the southern Hot Creek Range, central Nevada: Geological Society of America Abstracts with Programs, v. 24, Rocky Mountain section [in press].

Ross, R. J., Jr., and Ethington, R. L., 1991, Stratotype of Ordovician Whiterock Series, *with* an appendix on graptolite correlation of the topmost Ibexian by C. E. Mitchell: Palaios, v. 6, p. 156–173.

Ross, R. J., Jr., James, N. P., Hintze, L. F., and Ketner, K. B., 1991, Early Middle Ordovician (Whiterock) paleogeography of Basin Ranges, *in* Cooper, J. D., and Stevens, C. H., eds., Paleozoic paleogeography of the western United States-II, v. 1: Los Angeles, California, Pacific section, Society of Economic Paleontologists and Mineralogists, book 67, p. 39–50.

Ross, R. J., Jr., and Ethington, R. L., 1992, North American Whiterock Series suited for global correlation, *in* Webby, B. D., and Laurie, J. R., eds., Global perspectives on Ordovician geology: Proceedings volume of the 6th International Symposium on the Ordovician System [in press].

Sandberg, C. A., Ziegler, W., Dreesen, R., and Butler, J. L., 1988, Late Frasnian mass extinction; conodont event stratigraphy, global changes, and possible causes, *in* Ziegler, W., ed., 1st International Senckenberg Conference and 5th European Conodont Symposium (ECOS V), Contribution 1: Courier Forschungsinstitut Senckenberg, v. 102, p. 263–307.

Ziegler, W., and Sandberg, C. A., 1990, The Late Devonian standard conodont zonation: Courier Forschungsinstitut Senckenberg, v. 121, 115 p.

Printed in U.S.A.

The Geology of North America
Vol. G-3, The Cordilleran Orogen: Conterminous U.S.
The Geological Society of America, 1992

Chapter 3

Late Paleozoic paleogeographic and tectonic evolution of the western U.S. Cordillera

Elizabeth L. Miller and M. Meghan Miller*
Department of Geology, Stanford University, Stanford, California 94305
Calvin H. Stevens
Department of Geology, San Jose State University, San Jose, California 95192
James E. Wright* and Raul Madrid*
Department of Geology, Stanford University, Stanford, California 94305

INTRODUCTION

Upper Paleozoic rocks of the Cordillera form several contrasting belts that differ in their depositional and tectonic settings (Fig. 1). Cratonic and bordering continental shelf strata constitute the autochthonous portion of the Paleozoic Cordillera and provide an excellent, if only partial, sedimentologic record of late Paleozoic events affecting the western edge of the North American continent. Paleozoic rocks to the west are allochthonous with respect to the Paleozoic shelf and represent the preserved remnants of deep ocean basins and volcanic island arcs. Their less fossiliferous nature and greater degree of deformation make interpretation of their depositional setting and tectonic history more speculative. Paleozoic autochthonous rocks of the western U.S. Cordillera and adjacent regions were deposited in two distinct belts or provinces: the craton and cratonic basins province, and the miogeoclinal shelf province (Fig. 1). Paleozoic allochthonous rocks constitute four additional belts: the lower Paleozoic slope, rise, and ocean basin sequences of the Roberts Mountains allochthon; the late Paleozoic rocks of the Golconda allochthon; Paleozoic volcanic arc sequences exposed primarily in the western Sierra Nevada and the Klamath Mountains; and upper Paleozoic oceanic sequences tectonically included in accretionary subduction complexes. These six belts of rocks have general counterparts in the Canadian and Alaskan Cordillera to the north (Fig. 2) although the exact details of their depositional setting and subsequent tectonic history may differ along strike of the orogen.

*Present addresses: M. M. Miller, Jet Propulsion Laboratory, California Institute of Technology, 4800 Oak Grove Drive, Pasadena, California 91109; Wright, Department of Geology and Geophysics, Rice University, Box 1892, Houston, Texas 77001; Madrid, Conquistador Gold Ltd., 3210 Marthiam Ave., Reno, Nevada 89509.

Craton and cratonic basins province

The western portion of the North American craton was the site of accumulation of thin sequences of marine and nonmarine strata during the Paleozoic. The rise of the ancestral Rocky Mountains in Pennsylvanian time with the attendant formation of fault-bounded sedimentary basins flanking these uplifts was the most important event to affect patterns of sedimentation on the craton. Intracratonal tectonism has been attributed to continental collision along the southern edge of North America in Pennsylvanian time, as represented now by the Ouachita-Marathon orogenic belt (Kluth and Coney, 1981). The modern Rocky Mountains formed during the early Tertiary Laramide orogeny, when thrust faults reactivated many of the older structures. This younger event obscures the geometry of the older faults, whose existence and geometry are inferred largely from the Paleozoic stratigraphic record.

Miogeoclinal shelf province

The shelf province or miogeocline is underlain by a westward-thickening wedge of shallow-marine sedimentary rocks deposited on the slowly subsiding continental margin of western North America, believed to have originated by rifting in latest Precambrian to Early Cambrian time (Stewart, 1972; Bond and Kominz, 1984). The shelf succession is locally as young as Early Jurassic and attains thicknesses in excess of 10 to 15 km. Equivalent rock units occur along the entire length of the northern Cordillera from Mexico to Alaska (Fig. 2). The transition between thin Paleozoic cratonal sections to the east and westward-thickening units of the Paleozoic continental shelf to the west provided control on the locus of Jurassic to Eocene deformation in the foreland fold and thrust belt (Fig. 1). Shortening across this zone in the Mesozoic to early Tertiary is estimated at about 100

Miller, E. L., Miller, M. M., Stevens, C. H., Wright, J. E., and Madrid, R., 1992, Late Paleozoic paleogeographic and tectonic evolution of the western U.S. Cordillera, *in* Burchfiel, B. C., Lipman, P. W., and Zoback, M. L., eds., The Cordilleran Orogen: Conterminous U.S.: Boulder, Colorado, Geological Society of America, The Geology of North America, v. G-3.

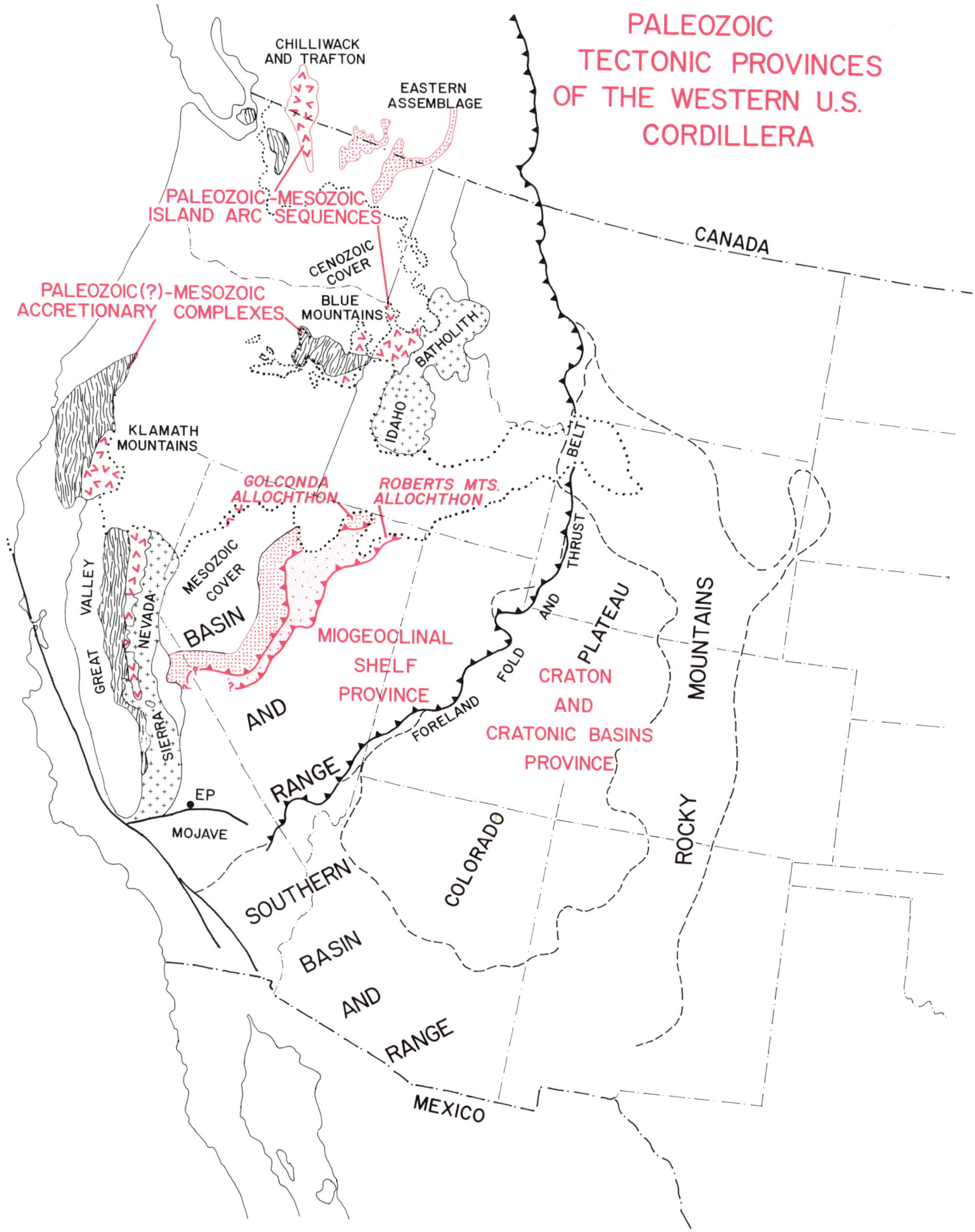

Figure 1. Paleozoic tectonic provinces of the western U.S. Cordillera, shown in red. Important younger features, including Mesozoic-Cenozoic intrusive rocks (crosses) and Cenozoic–Mesozoic cover (white bounded by dots), are shown in black or gray.

to 140 km in the Idaho-Utah-Wyoming portion of the belt (Royce and others, 1975). The Paleozoic depositional hingeline is also roughly coincident with the present-day physiographic boundary between the uplifted Colorado Plateau and the extended and topographically lower Basin and Range province (Fig. 1). Extension across the Basin and Range province, beginning in Eocene-Oligocene time and continuing to the present, has produced as much as 250 to 300 km of dilation in the northern part of the province (Hamilton and Meyers, 1966) and at least 140 km in the southern part of the province (Wernicke and others, 1982). Extension is distributed heterogeneously and is locally as high as several hundred percent (Gans and Miller, 1983) to as low as 15 to 30 percent (Stewart, 1980a). An estimate of the maximum width of the Paleozoic continental shelf across Nevada and Utah prior to Basin and Range extension and Mesozoic thrust faulting is about 300 to 400 km, as compared to its present width of about 500 km. Upper Paleozoic shelf stratigraphic sequences are fairly well represented across the Basin and Range province, but the reconstruction of shelf-basin geometry and structures based on facies and isopachs of units can only be approximate without rigorous palinspastic restoration of younger structures.

Roberts Mountains allochthon

During the earliest Mississippian Antler orogeny, lower Paleozoic chert-shale sequences and associated mafic volcanic rocks of the Roberts Mountains allochthon were complexly deformed

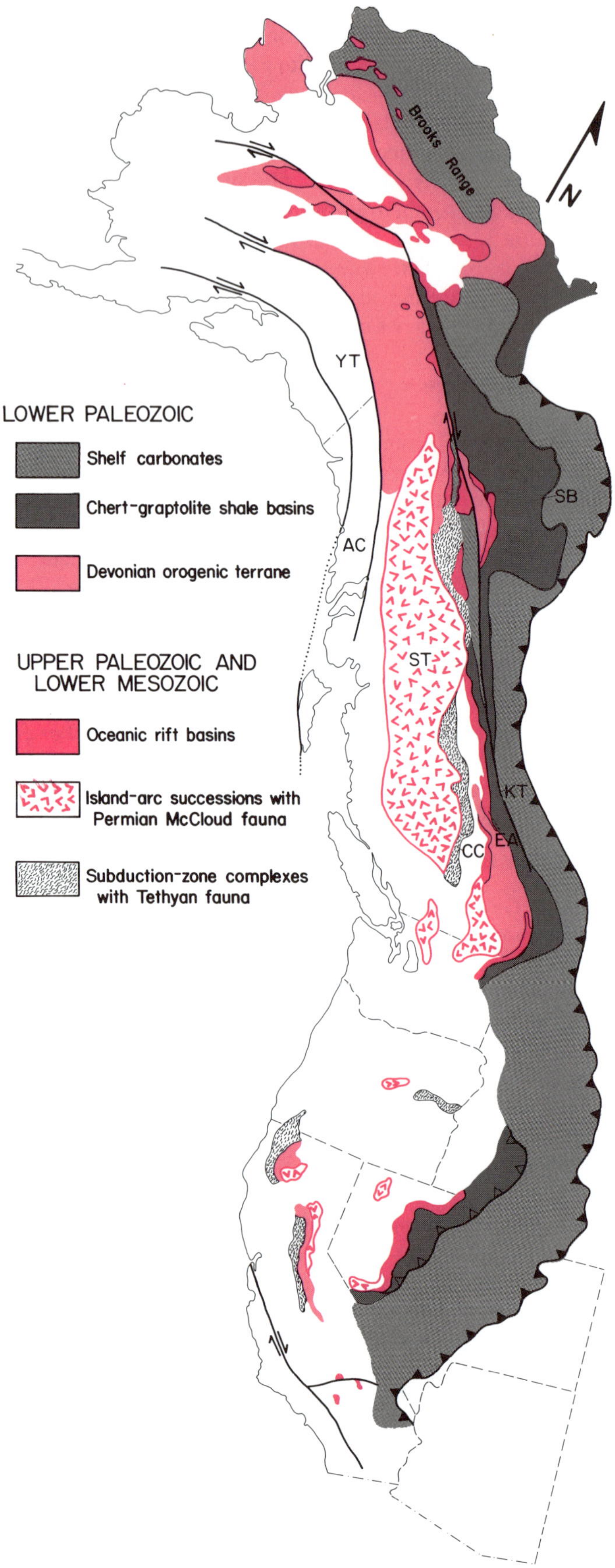

Figure 2. Correlative Paleozoic lithologic and tectonic elements of the North American Cordillera, modified from Jones and others (1987), Silberling and others (1987) and Monger and Berg (1987); base map from King (1969). Paleozoic shelf (light gray), lower Paleozoic shale out (dotted line), and lower Paleozoic shale basins (SB) in Canadian Cordillera from Tipper and others (1981) and in Alaska from Churkin (1973). Devonian orogenic terrane (pink) includes basement rocks of upper Paleozoic arc sequences in the northern Sierra Nevada and eastern Klamath Mountains; the Barkerville terrane and Kootenay terrane of Monger and Berg (1984) and/or Selkirk terrane of Struik (1986) in southern British Columbia (KT); the Yukon-Tanana terrane (YT) in Canada and Alaska; and the Hammond subterrane (Brooks Range) and correlative rocks in Alaska that have undergone tectonism and/or intrusion by plutons in the Devonian and Mississippian. Upper Paleozoic and lower Mesozoic oceanic rift basin sequences (solid red) include the Havallah sequence of the western U.S.; the Eastern assemblage or Slide Mountain terrane of Tipper and others (1981), Monger and Berg (1984), and Struik and Orchard (1985) in the Canadian Cordillera (EA); and the Angayucham and correlative terranes of Jones and others (1987) in Alaska. Upper Paleozoic to lower Mesozoic island-arc successions containing Permian McCloud-type fauna (red v) from Miller (1987) and includes only basement rocks of the dominantly Mesozoic Stikine terrane of Canada (ST). Paleozoic(?) and Mesozoic subduction zone complexes (black squiggles), characterized by Tethyan-type fauna (or mixed Tethyan-McCloud), from Miller (1987) and correlative Cache Creek group in Canada (CC) from Jones and others (1987), Silberling and others (1987), and Monger and Berg (1987). AX, Alexander terrane. For discussion, see text.

and thrust eastward over the edge of the continental shelf. Strata of the Roberts Mountains allochthon represent continental slope, rise, and basinal settings inferred to have been deposited west of the coeval, early Paleozoic continental shelf. Deformed rocks of the allochthon are overlain unconformably by upper Paleozoic shallow-marine sequences. Rocks of similar age and lithology to those in the Roberts Mountains allochthon occur in northern Washington and southern British Columbia, but may have been deformed at a subsequent time. Farther north in the Canadian Cordillera, coeval chert-shale and volcanic sequences, similar in many respects to rocks in the Roberts Mountains allochthon, were deposited in a series of outer shelf basins (Fig. 2).

Golconda allochthon

Deep-marine sedimentary rocks and associated volcanic rocks of late Paleozoic age, the Havallah sequence, were emplaced onto the outer shelf above the Golconda thrust during the Late Permian to Early Triassic Sonoma orogenic event. The Golconda allochthon structurally overlies the Roberts Mountains allochthon and its upper Paleozoic sedimentary cover. The Golconda allochthon is in turn overlain by Mesozoic sequences to the west (Fig. 1). Rocks of similar age and lithology to those in the Golconda allochthon are known to occur in northern Washington and in the Canadian Cordillera. The Canadian equivalents form part of an extensive belt of rocks, termed the "Eastern assemblage" of Monger (1977), interpreted by Struik and Orchard (1985) to be the remains of a series of late Paleozoic oceanic rift basins (Fig. 2).

Upper Paleozoic volcanic arc sequences

Upper Paleozoic rocks exposed discontinuously along the western portion of the Cordillera are volcanic island-arc suites (Fig. 1). Exposures in the northern Sierra Nevada and Klamath Mountains are the most extensive, and these stratified sequences record sporadic but persistent island-arc volcanism during late Paleozoic time. Scattered exposures of coeval island-arc sequences are present in westernmost Nevada; the Blue Mountains, Oregon; and the northern Cascades, Washington (Fig. 1). Similar successions occur in the Canadian Cordillera and form the basement for the largely Mesozoic-age Stikine terrane (Fig. 2).

Upper Paleozoic oceanic sequences

West of the island-arc sequence lies a discontinuous belt of structurally disrupted Mesozoic and Paleozoic oceanic rocks (Fig. 1). Based on structural style and the occurrence of blueschist-facies metamorphic rocks, this belt has been interpreted as a subduction zone complex similar to, and probably once continuous with, the Cache Creek terrane of the Canadian Cordillera (Monger, 1977; Davis and others, 1978) (Fig. 2). Although upper Paleozoic rocks are important tectonic elements of this belt, the formation of the subduction zone complex itself may have occurred largely in early Mesozoic time.

METHODS AND DATA PRESENTATION

We have divided the late Paleozoic history of the western U.S. Cordillera into six time intervals: Early Mississippian, Late Mississippian, Pennsylvanian, Early Permian, Late Permian, and latest Permian–Early Triassic. The time intervals used were chosen because they represented natural or convenient breaks in the geologic history of the Paleozoic continental shelf and craton, and best emphasize the nature of events occurring during a particular time interval. The data for these time intervals are displayed on the paleogeographic maps of Plate 4. In addition, we illustrate data for latest Devonian time along the westernmost portion of the continental margin (Plate 4A) and present data from the craton on two diagrams for the Pennsylvanian (Plate 4D and E) to better chronicle the rapid rise of the ancestral Rocky Mountains during this time span. References used to compile various portions of these figures are listed in Appendix I.

Paleogeographic data for the shelf and craton (lithologies, facies, isopachs, and active structural features) are displayed on present-day geographic base maps that do not account for subsequent Mesozoic shortening and Cenozoic extension. The location of major younger structures, large regions of post-Paleozoic cover, and plutonic rocks are indicated on the base maps to better portray some of the reasons for the scant preservation of Paleozoic rocks. Although the actual exposures of rocks of a particular age on the shelf and craton are scattered, units are sufficiently widespread to allow correlation between exposures for great distances (Plate 4I). For this reason the actual distribution of outcrops of Paleozoic rocks on the shelf is not shown.

The Golconda allochthon (Fig. 1), emplaced in latest Permian to Early Triassic time, contains rocks that span the entire upper Paleozoic. A growing body of data suggests that most of the strata within the allochthon accumulated in a system of offshore basins. Their history thus provides important insight into the Paleozoic paleogeography of the continental margin. Since the allochthonous rocks are complexly deformed and presently overlie shelf sequences along the regionally extensive Golconda thrust, we have cut and exploded our base maps along a line corresponding to the westernmost exposures of Paleozoic shelf strata in order to show data for this offshore region in a schematic fashion (Plate 4). The time-equivalent and remarkably similar sequences found in southern British Columbia (Eastern Assemblage of Monger, 1977), northern Washington (Mount Roberts Formation of Yates, 1976), and somewhat similar rocks in the El Paso Mountains, southern California (Carr and others, 1984) (Fig. 1), provide additional data for this offshore region. It is important to note that rifting, strike-slip, and thrust faulting during the formation and subsequent closure of this basin may have introduced an unknown amount of longitudinal displacement of these rocks with respect to the adjacent shelf.

The exposures of Paleozoic sequences deposited in island-arc and oceanic settings (Fig. 1) are shown in their present geographic positions on the western part of the exploded base maps of Plate 4. Their separation from the continent serves to empha-

size that their position with respect to the shelf during the Paleozoic is not known with certainty. The upper Paleozoic island-arc sequences, however, appear to have been constructed on similar basement terranes, display similar volcanic and depositional histories through time, and share a Permian "McCloud" faunal assemblage, suggesting greater continuity prior to disruption by later tectonism (Miller, 1987). We have generalized the depositional settings and lithologies of the island-arc sequences on our paleogeographic maps beyond their actual areas of outcrop, which are quite limited (e.g., Plate 4A-H).

The distribution of the westernmost belt of rocks, the Cache Creek–affinity subduction zone assemblages (represented by rocks in the northern Sierra Nevada; Klamath Mountains; Blue Mountains, Oregon; and northern Washington), appear only on our last map for the Late Permian–Triassic time interval, when this complex may have been at least partially assembled. Devonian to Permian sequences representing carbonate reef, platform, and pelagic drape deposits and associated volcanic rocks of oceanic island affinity occur as tectonic fragments in this belt, and although we have not shown individual locations on our maps, the best studied of these are described in the appropriate time-slice intervals. Based on their faunal assemblages, most workers suggest that these sequences originated in a more distal position within the paleo-Pacific Ocean than did island-arc successions deposited within the McCloud biogeographic realm (Stevens, 1985; Ross and Ross, 1983).

Within each of the time intervals, we describe rock sequences and structures from east to west, proceeding from a firm data base to one that is, at best, incomplete. The discussion and analysis of data in a time-slice framework forces us to consider rock units of a particular age together, regardless of their depositional setting. This approach emphasizes the great variety of paleodepositional and tectonic settings represented by late Paleozoic rocks of the Cordillera, but at the same time allows one to compare the temporal evolution of these sequences, providing fresh insight to their possible paleogeographic relations.

EARLY MISSISSIPPIAN

Introduction

The Antler orogeny was the first deformational event to significantly alter patterns of shelf sedimentation in the western U.S. during the Paleozoic. The Antler orogeny sensu strictu (Nilsen and Stewart, 1980) involved deformation of the outer edge of the continental shelf during the eastward thrust emplacement of basinal-facies rocks along the Roberts Mountains thrust. Deformation postdates deposition of latest Devonian strata within the Roberts Mountains allochthon and emplacement is synchronous with deposition of clastic strata shed from the allochthon and dated as Early Mississippian. Within the basement complex of the upper Paleozoic island-arc sequences exposed in the western Cordillera, somewhat older, Devonian-age events have been ascribed to the Antler orogeny, but their relation to the Antler orogeny sensu strictu are interpretive. In order to include these events in our coverage of the Antler orogeny, we briefly summarize the pre-Mississippian history of rocks depicted in Plate 4A, while data for the entire western U.S. for the Early Mississippian are shown on Plate 4B. Based on a review of this data, we present an updated tectonic model for Antler-age orogenesis at the end of this section.

The craton

Much of Colorado, New Mexico, and Arizona was emergent in earliest Mississippian time, exposing Precambrian crystalline basement to erosion (Sando, 1976; Chronic, 1979; Armstrong and others, 1980). Sedimentation along the Ouachita continental margin in the extreme southeastern portion of the map area was characterized by a condensed, deep-water sequence of rocks (Armstrong and others, 1980). In Montana, the east-trending Williston Basin was a relatively well-defined feature at this time (Smith and Gilmour, 1979). Craig and others (1979) also recognized thicker accumulations of sedimentary rocks in the areas of the Pedregosa and Delaware Basin.

The continental shelf

During the Early Mississippian, shallow-water carbonates were deposited across the western portion of the craton and the eastern portion of the subsiding continental shelf (Plate 4B and I). Farther west on the shelf, carbonates interfinger with siliciclastic sequences, derived from the uplift and erosion of the Roberts Mountains allochthon and deposited in the foredeep basin related to this thrust system. Deformed lower Paleozoic rocks of the Roberts Mountains allochthon overlie Devonian and older shelf-facies strata along the Roberts Mountains thrust immediately west of the thickest exposed sections of foredeep basin deposits.

Antler foredeep

The Antler foredeep extends from southern Nevada to central Idaho (Plate 4B). To the north of central Idaho, no allochthon exists and a similar clastic wedge is absent on the shelf proper. Still farther north in the Canadian Cordillera, mid-Paleozoic conglomeratic units deposited in a basinal setting are probably related to transcurrent tectonics rather than to thrust faulting (Eisbacher, 1983; Gordey and others, 1987). In southeastern California, a thick clastic section, coeval with the foredeep deposits to the north, is present but was apparently derived from the erosion of shelf-facies rocks rather than the Roberts Mountains allochthon (Stevens and Ridley, 1974). It is thus unclear if these strata represent the southern continuation of the Antler foredeep basin.

Foredeep deposits near the eastern toe of the allochthon in Nevada include the Chainman Shale and Diamond Peak Formations and are as much as 3.5 km thick (Poole and Sandberg, 1977 (Plate 4B, I). Sedimentation in the basin is thought to have

evolved in two distinct phases (Harbaugh and Dickinson, 1981); an Early Mississippian phase of submarine slope and submarine fan deposition, when fine-grained clastic rocks accumulated in steadily deepening waters west of the carbonate shelf; and a Late Mississippian phase of eastward-prograding delta-slope and delta platform sedimentation, when dominantly coarse-grained strata accumulated in gradually shoaling waters.

Using conodont biostratigraphy, Sandberg and others (1983) related the timing of Early Mississippian events on the shelf to the emplacement of the Roberts Mountains allochthon. They discussed this timing in terms of millions of years from the Devonian-Carboniferous boundary:

+0 to 3 m.y. (early Kinderhookian): A period of low sea-level stand, no Antler highlands developed yet. Coastal plains covered most of the shelf and marine sedimentation was ephemeral.

+3 to 4.5 m.y. (mid-Kinderhookian): First evidence for the presence of the Antler highlands and the formation of a narrow marine basin at the leading edge of the Roberts Mountains thrust.

+6 m.y. (late Kinderhookian): Peak of a marine transgression. Only the newly formed Antler highlands were exposed and were the source of clastic debris for the foreland basin. The Roberts Mountains allochthon at this time continued to move eastward across syntectonic deposits (Johnson and Pendergast, 1981). To the east, the Joana carbonate bank was developed (Joana Limestone, Plate 4I) and was the source of limestone turbidites shed westward into the foredeep basin.

+9 m.y. (early Osage): Coarse, western-derived turbidites overwhelmed the eastern-derived limestone turbidites (Chainman Shale and Diamond Peak Formations, Plate 4I). The foredeep basin continued to subside in response to both sedimentary loading and tectonic loading due to thrusting (Speed and Sleep, 1982; Johnson and Pendergast, 1981). As the shelf was tectonically loaded, the foredeep basin widened, and the water depth over the Joana carbonate bank increased.

20*12 m.y. (mid-Osage):* At this time, the second Early Mississippian marine transgression occurred, and the Antler highlands rose to maximum relief, shedding coarse detritus into the foredeep basin.

+24 m.y. (mid-Meramec): During this time the nature of the sediments accumulating in the foredeep changed from deep-water turbidites deposited in a tectonically subsiding basin to shallower water and even continental sedimentation of Harbaugh and Dickinson's (1981) progradational phase.

These data suggest that thrust faulting, resulting in the emplacement of the Roberts Mountains allochthon, began in mid-Kinderhookian time and continued through the Osage. In Meramecian time, the tectonic loading due to thrust faulting ceased, and the foredeep basin filled and shoaled while the Antler highlands continued to be a major source of clastic detritus. Several workers have suggested alternatively that thrust faulting began earlier based on the development of the Pilot Shale basin on the shelf in the Late Devonian (Plate 4I), interpreted to have formed by incipient thrust-loading of the outer shelf (e.g., Johnson and Pendergast, 1981; Speed and Sleep, 1982). As discussed below, and as indicated on Plate 4A, sedimentary sequences in the Roberts Mountains allochthon suggest that the latest Devonian may have been time of extension or rifting along the continental margin. This would imply that sedimentation was still taking place in the most proximal parts of the Antler basin, raising questions about the proposed Late Devonian age of emplacement of the allochthon.

Roberts Mountains allochthon

The Roberts Mountains allochthon consists of complexly deformed Upper Cambrian to uppermost Devonian chert, argillite, sandstone, and greenstone. In Nevada, the easternmost exposures of the allochthon generally coincide with the thickest sections, or depositional axis, of the foredeep (Stewart, 1980b), suggesting that the present eastern limit of allochthonous rocks (Plate 4B) approximates what was once the leading edge of the allochthon during the Early Mississippian. Its present-day structural overlap with autochthonous shelf strata of Devonian and older age is about 140 km (Stewart, 1980b); prior to Basin and Range extension it may have been as little as half of that. In Idaho, the actual Roberts Mountains thrust has not been identified due to the complexity of superimposed Mesozoic thrust faulting, metamorphism, and Cenozoic faulting, but the presence of both basinal-facies lower Paleozoic rocks and a well-developed foredeep succession suggests it extended at least this far north (Skipp and Hall, 1980; Dover, 1980).

The southernmost exposures of rocks similar to those in the Roberts Mountains allochthon occur in the El Paso Mountains in the Mojave Desert (Plate 4B and Fig. 3). Those farther south in Mexico are described in the previous chapter of this volume. In the El Paso Mountains, part of the sequence exposed has been interpreted as transitional between deep-water and shelf-edge sequences (Carr and others, 1984). It is unclear whether these strata are part of the Roberts Mountains allochthon sensu strictu, because Mississippian synorogenic clastic rocks overlie Devonian and older deep-water-facies rocks along an apparent concordant contact in one locality (Carr and others, 1984). Elsewhere the contact is interpreted as an unconformity. Lower Paleozoic rocks in the El Paso Mountains presently lie in an unusual position with respect to trends and isopachs of adjacent shelf strata, suggesting that they may have been translated southward by left-lateral strike-slip faults (Burchfiel and Davis, 1981b) prior to the Late Permian (Miller and Sutter, 1982; Stone and Stevens, 1988a; Walker, 1988). Alternatively, they may not be displaced but represent an Ordovician or older reentrant or basin developed along the edge of the continental shelf.

Stratigraphy. Despite the lithologic monotony, paucity of fossils, poor exposure, and structural complexity of the Roberts Mountains allochthon, a number of age-equivalent and lithologically similar units appear to define a systematic stratigraphy in the allochthon (Fig. 3). Although these units were given formational status by previous workers, they are probably too structurally

disrupted to merit this designation. However, it is useful to continue to use these formation names for the sake of convenience and continuity.

The oldest unit described from the Roberts Mountains allochthon is the Paradise Valley Chert in the Hot Springs Range, reported to underlie the Harmony Formation and to have yielded Cambrian fossils (Hotz and Willden, 1964; Rowell and others, 1979). Subsequent visits to this locality by Madrid (unpublished data) have failed to relocate fossil occurrences and have demonstrated that the contact between the Paradise Valley Chert and the overlying Harmony Formation is a low-angle fault.

The most distinctive strata in the Roberts Mountains belong to the Harmony Formation, a Cambrian(?) turbidite sequence consisting of immature arkosic sandstone, shale, and grit. Palmer (1971), Suzcek (1977), and Stewart and Suzcek (1977) suggest that the source for the Harmony deep-sea-fan sedimentary rocks is represented by Precambrian crystalline rocks of the Salmon River arch, a region of intermittent uplift and erosion on the Idaho shelf during early Paleozoic time.

Ordovician rocks in the Roberts Mountains allochthon consist of basaltic greenstone and chert-argillite sequences interbedded with orthoquartzites (some partly feldspathic) and lesser limestone turbidites of the Valmy and Vinini Formations. The influx of well-rounded, well-sorted, pure quartz sand into deeper-water basin(s) was coeval with the deposition of the Eureka Quartzite on the continental shelf (Churkin, 1974; Miller and Larue, 1983) (Fig. 3) and helps establish a stratigraphic tie between allochthonous Ordovician deep-water facies and autochthonous shelf strata. The composition of more feldspathic sandstones in the Valmy and the occurrence of rounded cobbles of the older Harmony Formation in coarser-grained Ordovician deposits in the Shoshone Range and in the Cortez Mountains (Fig. 3) led Madrid (1987) to conclude that the Harmony deep-sea-fan deposits were faulted, uplifted, and eroded during the Ordovician. The Harmony is now in fault contact everywhere above Ordovician strata of the allochthon, but the presence of Harmony debris in Ordovician rocks suggests paleogeographic continuity between the two rock packages prior to thrust faulting and assembly of the Roberts Mountains allochthon.

The Silurian Elder Sandstone and Four Mile Canyon Formation are units that contain distinctive feldspathic sandstone (Fig. 3). Coarser units, dated as Late Silurian–Early Devonian, also contains clasts of the older Harmony Formation (Madrid, 1987). Based on U-Pb ages of detrital zircons, Girty and others (1985) have demonstrated a 2.2-b.y.-old provenance for feldspathic sands in the Elder, supporting an ultimate cratonic source for these probable recycled sediments.

The youngest rocks in the Roberts Mountains allochthon are Late Devonian in age and consist of basaltic greenstone and associated radiolarian chert and shale of the Slaven Chert. Strata equivalent to the Slaven Chert in the Toquima Range, central Nevada, have yielded fossils as young as very earliest Mississippian (Coles and Snyder, 1985). Here the Slaven Chert contains interbedded phosphatic chert, suggesting deposition in a zone of upwelling proximal to the continental margin (Coles and Snyder, 1985). The age of this succession pre-dates thrust emplacement onto the margin by only a few million years.

Basaltic lavas intercalated with Cambrian, Ordovician, Silurian, and Devonian strata of the allochthon can be shown, at least locally, to be in stratigraphic sequence with well-dated sedimentary rocks (Madrid, 1987). The alkalic affinity of these lavas, regardless of age, indicate an intraplate tectonic setting for sedimentary and volcanic rocks of the allochthon (Madrid, 1987). The presence of stratigraphically intercalated, continentally-derived sediments suggest that sedimentary, volcanic, and pelagic rocks of the Roberts Mountains allochthon accumulated in basins that formed as a result of recurrent extensional or transtensional tectonism in an intraplate setting along the continental margin (Madrid, 1987). Coeval shelf sequences in the western U.S. appear to provide some additional evidence in support of this interpretation, as detailed below.

Lower Paleozoic shelf and transitional sequences exposed in central Nevada and northeastern Washington often indicate abrupt transitions from shelf carbonate deposition during the Cambrian, to basinal chert-shale deposition during the Ordovician, as described by Yates (1976) for northeastern Washington and by Stewart (1980b) in west-central Nevada. Stewart (1980) has also pointed out that some of the rocks mapped as Roberts Mountains allochthon near the southern limit of the allochthon are actually autochthonous or parautochthonous in the sense that they are in stratigraphic continuity with older shelf deposits. A similar rapid transition from shallow-water Cambrian to relatively deep-water Ordovician rocks is recorded in east-central California, where tilting of the outer continental shelf was likely caused by extensional faulting (Stevens, 1986). Local normal faulting and basin formation is documented even on the inner part of the continental shelf in Utah and Nevada during the Middle to Late Cambrian (Rees, 1986, House Range embayment), and Late Cambrian to Ordovician alkalic plutonism accompanied extensional faulting and basin formation on the shelf in east-central Idaho (Evans and Zartman, 1988).

The abrupt transition from shallow- to deeper-water deposition recorded in lower Paleozoic shelf sequences suggests that Cambro-Ordovician subsidence occurred rapidly and locally along the continental margin well *after* its presumed initial rifting in the latest Precambrian or Early Cambrian (Stewart, 1972; Bond and Komintz, 1984). Although the concept of recurrent early Paleozoic extensional tectonism along the continental margin is not widely held by western U.S. geologists (but see Churkin, 1974), it is the accepted view for the Canadian and Alaskan portion of the Cordillera, where the transition from shelf carbonate sequences to basinal shale, chert, and volcanic rocks, though complex, is better documented in many places. In Canada, strata that are coeval and lithologically similar to those in the Roberts Mountains allochthon escaped an Antler-age deformational event, and their relation to shelf/platform sequences is clearcut (Fig. 2). Stratigraphic sequences indicate tectonic instability and recurrent extension throughout the lower Paleozoic (e.g., Abbott

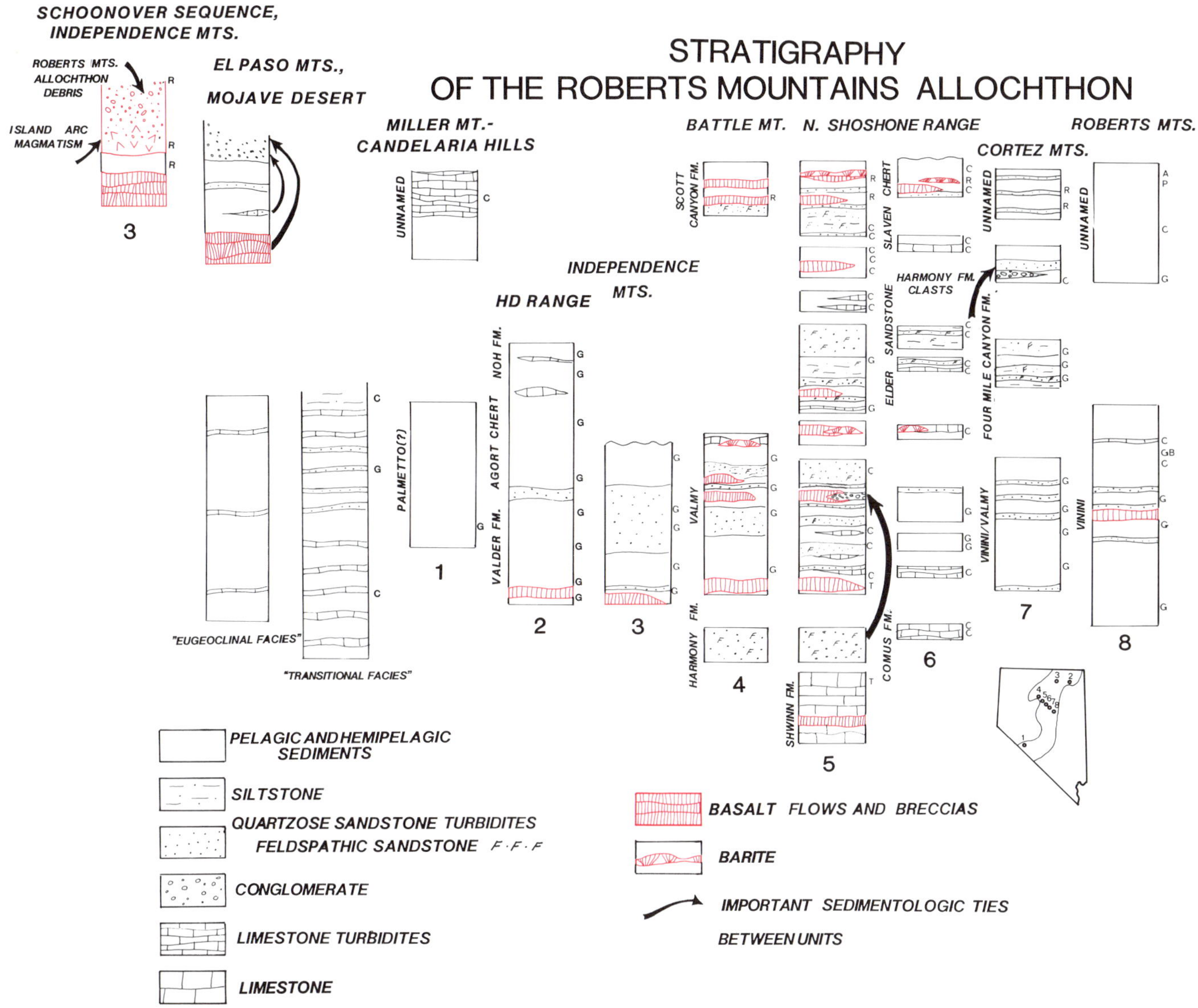

Figure 3. Stratigraphic/lithologic units of the Roberts Mountains allochthon and comparison with stratigraphy of coeval shelf sequences. Letters on sides of columns indicate age control based on fossils: A, *Angustidontus*; B, brachiopods; C, conodonts; G, graptolites; P, *Palmatolepis*; R, radiolarians; T, trilobites. Sources of data: 1. Stanley and others (1977), Stewart (1979), and Stewart (1984); 2. Riva (1970); 3. Churkin and Kay (1967); 4. Roberts (1964) and Madrid (1987); 5. Gilluly and Gates (1965), and Madrid (1987); 6. Gilluly and Gates (1965) and Madrid (1987); 7. Gilluly and Mazursky (1965) and Madrid (1987); 8. Murphy and others (1984) and Mannick (1975); 10. El Paso Mountains (Carr and others, 1984) (for location see Fig. 1) and Schoonover sequence (location 3 above) from Miller and others (1984).

and others, 1986; Turner and others, 1989). The remarkable similarity between the stratigraphy of the Roberts Mountains allochthon and coeval shelf-basin sequences in the Canadian Cordillera has been pointed out by Turner and others (1989), and it seems reasonable to conclude that the rocks of the Roberts Mountains allochthon were also deposited in tectonically unstable shelf basins developed as a result of recurrent extension along the outer margin, as originally proposed by Churkin (1974).

Structure. Map- to outcrop-scale folds and thrust faults characterize the internal structure of the Roberts Mountains allochthon. Deformation has dismembered stratigraphic sequences at all scales of observation, but strain is heterogeneous, and portions of the allochthon preserve sections of intact stratigraphy (Madrid, 1987). Orientation of mesoscopic folds and thrust faults were used to establish the eastward-directed kinematic movement directions shown in Plate 4B. Penetrative cleavage and metamor-

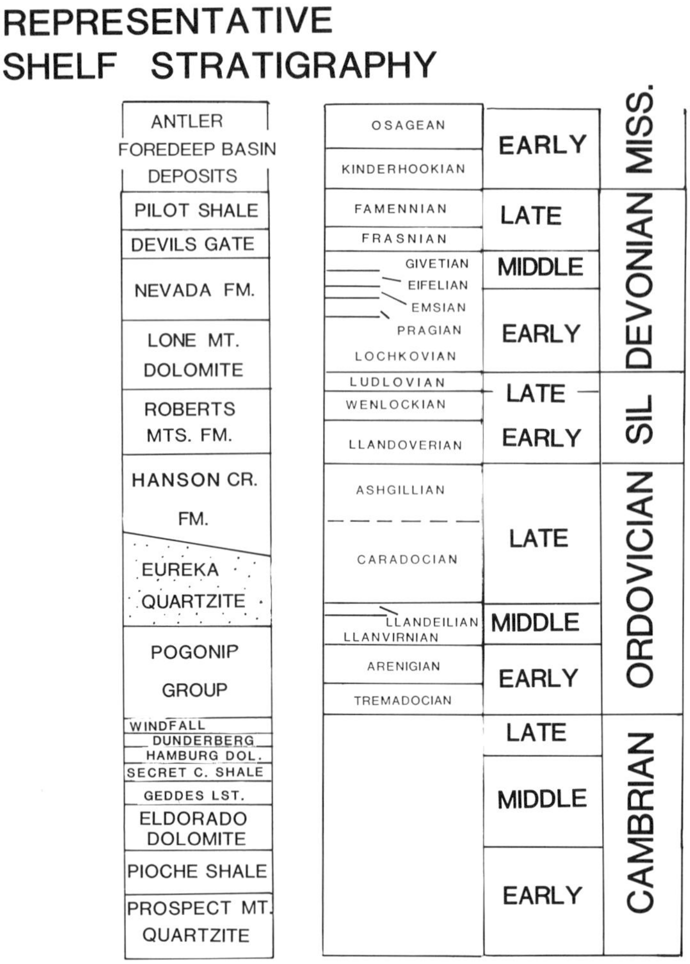

phic recrystallization are generally absent, indicating deformation under fairly low-temperature conditions. Penetrative cleavage of probable Antler age has, however, been reported in the southerly reaches of the allochthon by Oldow (1984), in rocks in the eastern Sierran roof pendants by Schweickert and others (1984), and in some of the westerly exposures of the Harmony Formation near Battle Mountain by Madrid (1987) (Plate 4B). The present maximum structural thickness of the allochthon, measured in east-central Nevada, is about 5 km. Prior to erosion, it was likely somewhat greater than this.

The lack of a straightforward stratigraphy has made it more difficult to delineate map-scale structures in the allochthon. In north-central Nevada, where the entire width of the allochthon has been studied in some detail, the structurally highest and westernmost exposures consist of the Harmony Formation, which is in thrust contact everywhere with Ordovician strata below (Madrid, 1987). In general, Devonian rocks occupy structurally lower and more easterly positions within the nappe pile and occur structurally beneath Ordovician strata (Madrid, 1987). Palinspastic restoration indicates that the structurally highest Harmony Formation has undergone the greatest amount of eastward translation with respect to the continental margin, while Devonian volcanic and sedimentary rocks (the Slaven Chert), present in the structurally lowest positions, were deposited in closer proximity to the continental margin.

The basal Roberts Mountains thrust has been described as a narrow fault zone along which deformed allochthonous rocks were emplaced above nearly undeformed autochthonous rocks. Although this may be true along the eastern or leading edge of the allochthon, farther west, greater structural complexity characterized the basal thrust. In these structurally deeper exposures, thrust-bound slivers of parautochthonous strata are common beneath or along the basal thrust. The fault slivers, as well as more westerly located exposures of autochthonous or parautochthonous strata, show evidence for polyphase folding and locally possess a slaty cleavage developed under greenschist-grade conditions (Madrid, 1987; Miller and others, 1984; Ehman, 1985). These fabrics are difficult to distinguish from younger Mesozoic fabrics that overprint the Roberts Mountains allochthon in its western reaches, but in places such as the Osgood Mountains, Edna Mountain, and Independence and Bull Run Mountains (Plate 4B), penetrative fabrics clearly pre-date deposition of overlying middle Pennsylvanian and Permian strata, and are assumed to be Antler-age fabrics (Hotz and Willden, 1964; Madrid, 1987; Ehman, 1985; Miller and others, 1984).

Conflicting evidence exists for the timing of inception of deformation within the allochthon itself. Madrid (1987) presented evidence that deformation entirely post-dated deposition of the Upper Devonian (Famennian) Slaven Chert. Murphy and others (1984) described fild relations near the eastern toe of the allochthon that suggest thrust faulting began *prior to* deposition of Upper Devonian (Frasnian) sedimentary rocks. More work is needed to resolve this important problem. The widespread preservation of pelagic and hemipelagic sedimentary rocks of Late Devonian age in the allochthon suggests that the younger age of onset of thrusting is likely correct, and alkalic basalts within this youngest sequence indicate an extensional or transtensional environment for the basin in Late Devonian time, immediately prior to thrusting. However, age constraints for the inception of deformation within the structurally higher portions of the allochthon are much poorer, and a multi-stage history of deformation is possible.

In summary, three important points emerge from the data discussed above that provide a somewhat different perspective on the Roberts Mountains allochthon and the mid-Paleozoic Antler orogeny:

1. The Roberts Mountains allochthon probably does not constitute a simple continental slope-and-rise sequence deposited along a passive continental margin formed by rifting in the latest Precambrian or Cambrian, or a truly oceanic volcanic and pelagic sequence, as has been suggested by previous workers. Instead, the rocks that now form the allochthon most likely represent the deposits of a series of extensional basins developed intermittently along the edge of the continental margin in early

Paleozoic time. These basins were at least in part underlain by rifted continental crust (Madrid, 1987; Turner and others, 1989).

2. Pelagic and hemipelagic sediments and alkalic basalt of the Slaven Chert constitute the youngest and structurally lowest units of the Roberts Mountains allochthon, implying that an extensional or transtensional tectonic setting characterized the region immediately offshore the continental margin in the Late Devonian. The Slaven Chert and associated volcanic rocks are coeval with the development of the Pilot Shale basin on the shelf, suggesting an extensional origin for that basin as well, rather than a compressional one as has been previously suggested (e.g., Speed and Sleep, 1982; Johnson and Pendergast, 1981; Sandberg and others, 1983).

3. Earlier descriptions of the Roberts Mountains allochthon have emphasized its complete lack of metamorphism (e.g., Nilsen and Stewart, 1980), but deformation appears to become more penetrative and is associated with lower greenschist-facies metamorphism westward, with structural depth, and perhaps southward in the allochthon. This increase in metamorphic grade may indicate that deformation and thrust faulting occurred in a tectonic setting characterized by a relatively elevated, rather than depressed geotherm, but no detailed work has been done to address this question.

Offshore Havallah basin

In Plate 4B and in Figure 3 we show Devonian-Mississippian strata of the Golconda allochthon in the Independence Mountains, Nevada, and sequences in the El Paso Mountains, California, as representative portions of a series of basins that lay closely adjacent to the Antler orogenic belt, but which escaped Antler-age penetrative deformation. In these two localities, Antler deformation is recorded indirectly by the arrival of Antler-derived clastic material, deposited in apparent continuity or in possible disconformity above Devonian strata.

A comparison of the stratigraphy and timing of Devonian and Mississippian events in the Havallah basin, as documented in the Independence Mountains, Nevada, with events occurring on the continental shelf (Fig. 4), helps illustrate possible stratigraphic ties between the shelf and this basin. The oldest rocks in the Golconda allochthon in the Independence Mountains are basaltic pillow lavas overlain by Upper Devonian (Fammenian) radiolarian chert sequences, which are coeval with the youngest basaltic lavas and sediments of the Roberts Mountains allochthon. Volcanic arc–derived tuff and volcaniclastic deposits containing greenschist-facies metamorphic debris were suddenly shed into this portion of the Havallah basin in Early Mississippian (Kinderhookian) time, synchronous with the first evidence for Antler-age thrusting on the continental shelf. Arc-derived sedimentary rocks are interbedded with, and grade upward into, coarse-grained turbidite sequences and boulder-bearing pebbly mudstones derived strictly from the erosion of rocks similar to those in present exposures of the Roberts Mountains allochthon (Miller and others, 1984; Whiteford, 1990). Within the limits of biostratigraphic resolution, the deposition of these younger turbidites was synchronous with the Early Mississippian stage of deposition in the Antler foredeep basin. These stratigraphic data provide an indirect paleogeographic tie between active arc magmatism and the Antler orogenic belt on the shelf, and require modification of the traditional view that the Antler orogeny was not accompanied by magmatism or metamorphism (e.g., Nilsen and Stewart, 1980).

Volcanic island-arc sequences

Introduction. Devonian and younger volcanic-arc sequences are exposed in the northern Sierra Nevada and the Klamath Mountains of California (Plate 4A, B; Fig. 5). These two areas of exposure, though presently separated by younger cover, are inferred to have once formed a more continuous belt (Davis, 1969). This semi-coherent Paleozoic arc succession unconformably overlies lower Paleozoic rocks that have a complex history. Devonian and older events that affect these basement rocks have been ascribed to the Antler orogeny and include deformation, metamorphism, plutonism, uplift, and erosion (Burchfiel and Davis, 1975; Varga and Moores, 1981; Schweickert and Snyder, 1981). The Quinn River crossing region of northwestern Nevada, the Grindstone terrane of central Oregon, and the Chilliwack Group of Washington and British Columbia contain upper Paleozoic arc sequences that are similar to those exposed in the Klamath Mountains and northern Sierra Nevada, but in these places, older basement rocks are not exposed (Fig. 5). Arc-related volcanism and sedimentation in all of these areas continued into the early Mesozoic, and in general, these arc-related sequences share many similarities, including similar McCloud-type Permian faunal assemblages (Miller, 1987). Below we briefly describe Devonian orogenic events in basement rocks of the arc and discuss evidence for the age of initiation of deposition and magmatism in the overlying Devonian and younger volcanic arc succession.

Northern Sierra Nevada. In the northern Sierra Nevada, a pre–Late Devonian (pre-Famennian) unconformity bevels across four fault-bounded packages of rocks in the lower Paleozoic Shoo Fly Complex (Plate 4A; Fig. 5). These structural units, from east to west and downward from the unconformity, include the Sierra City melange (a sedimentary matrix tectonic melange), Culbertson Lake allochthon (consisting of a younger siliciclastic sequence and an older chert, limestone, and basalt sequence), Duncan Peak allochthon (radiolarian and spiculitic chert with rare sandstone), and Lang sequence (phyllite, quartzose sandstone, chert, and minor marble) (Schweickert and others, 1984; Harwood, 1988) (Fig. 5). Some sandstone blocks within the Sierra City melange have a volcano-plutonic provenance and yield detrital zircon with Cambrian ages (506 ± 22 Ma, U-Pb zircon, Girty and Wardlaw, 1984). Tuff from the melange matrix is Silurian (423 +5/−15 Ma, U-Pb zircon, Saleeby and others, 1987). Late Ordovician fossils have been reported from limestone mélange blocks (D'Allura and others, 1977). Detrital zircon from turbidite sandstones of the Culbertson Lake allochthon yields older provenance ages of 2.09

Ga (Girty and Wardlaw, 1985), and together with the petrography of the enclosing sandstones (Bond and Devay, 1980), indicates a Precambrian crystalline source for some of the clastic rocks in the Shoo Fly Complex.

Folding and local development of slaty cleavage accompanied the thrust-imbrication of the Shoo Fly Complex. Structural depth of exposure and metamorphic grade increase southward (Girty and Schweickert, 1984; Sharp, 1988). Deformation and imbrication of the complex post-dates the age of the youngest rocks involved (Late Silurian) and pre-dates intrusion of Middle and Late Devonian intrusive bodies (Merguerian and Schweickert, 1987; 378 +5/−10 Ma, U-Pb, Saleeby and others, 1987; 364 to 385 Ma, U-Pb, Hanson and others, 1988) (Plate 4A; Fig. 5). In the central Sierra Nevada, rocks of the Shoo Fly Complex are cut by syn-metamorphic peraluminous orthogneisses that yield radiometric data broadly suggestive of a Devonian crystallization age (Sharp, 1988; Merguerian and Schweickert, 1987; Saleeby and others, 1989). Interpretation of vergence and direction of transport along the presently east-dipping thrust faults within the Shoo Fly Complex is ambiguous, but this deformation is believed to have taken place along a convergent plate boundary (Girty and Schweickert, 1984).

Uplift and erosional planation of the Shoo Fly Complex was succeeded by Late Devonian and Early Mississippian subsidence

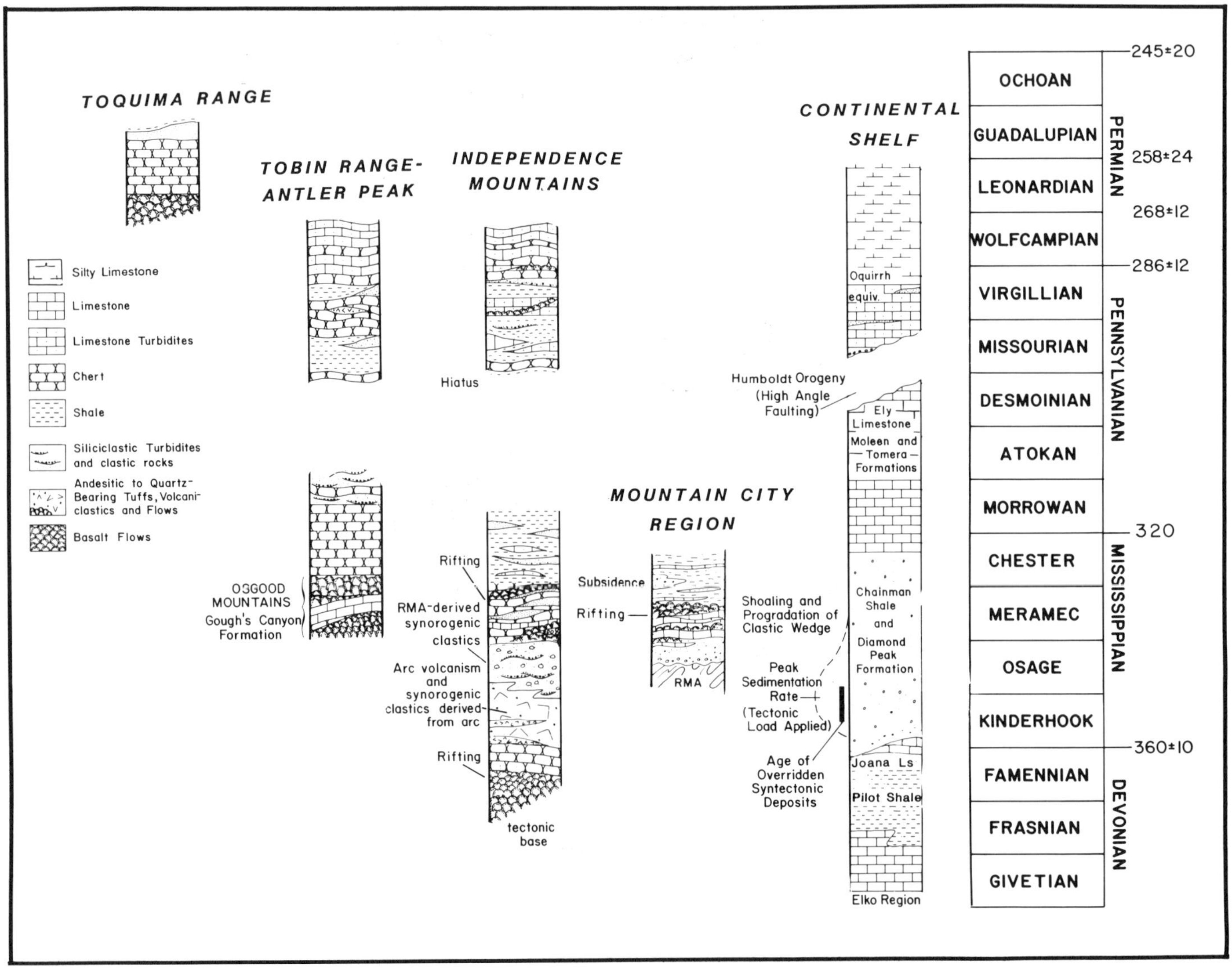

Figure 4. Reconstructed stratigraphy of the Havallah sequence of the Golconda allochthon, Nevada, and comparison with stratigraphy of coeval shelf sequences. Sources of data are Miller and others (1984) for the Independence Mountains; Stewart and others (1977, 1986), Murchey (1982), and Miller and others (1982) for the Tobin Range and Antler Peak areas; Laule and others (1981) for the Toquima Range; Hotz and Willden (1964) for the Osgood Mountains; Little (1987) for the Mountain City region; and Dott (1955), Ketner (1977), Poole and Sandberg (1977), Johnson and Pendergast (1981), Gutschick and others (1980), and Harbaugh and Dickinson (1981) for the shelf stratigraphy.

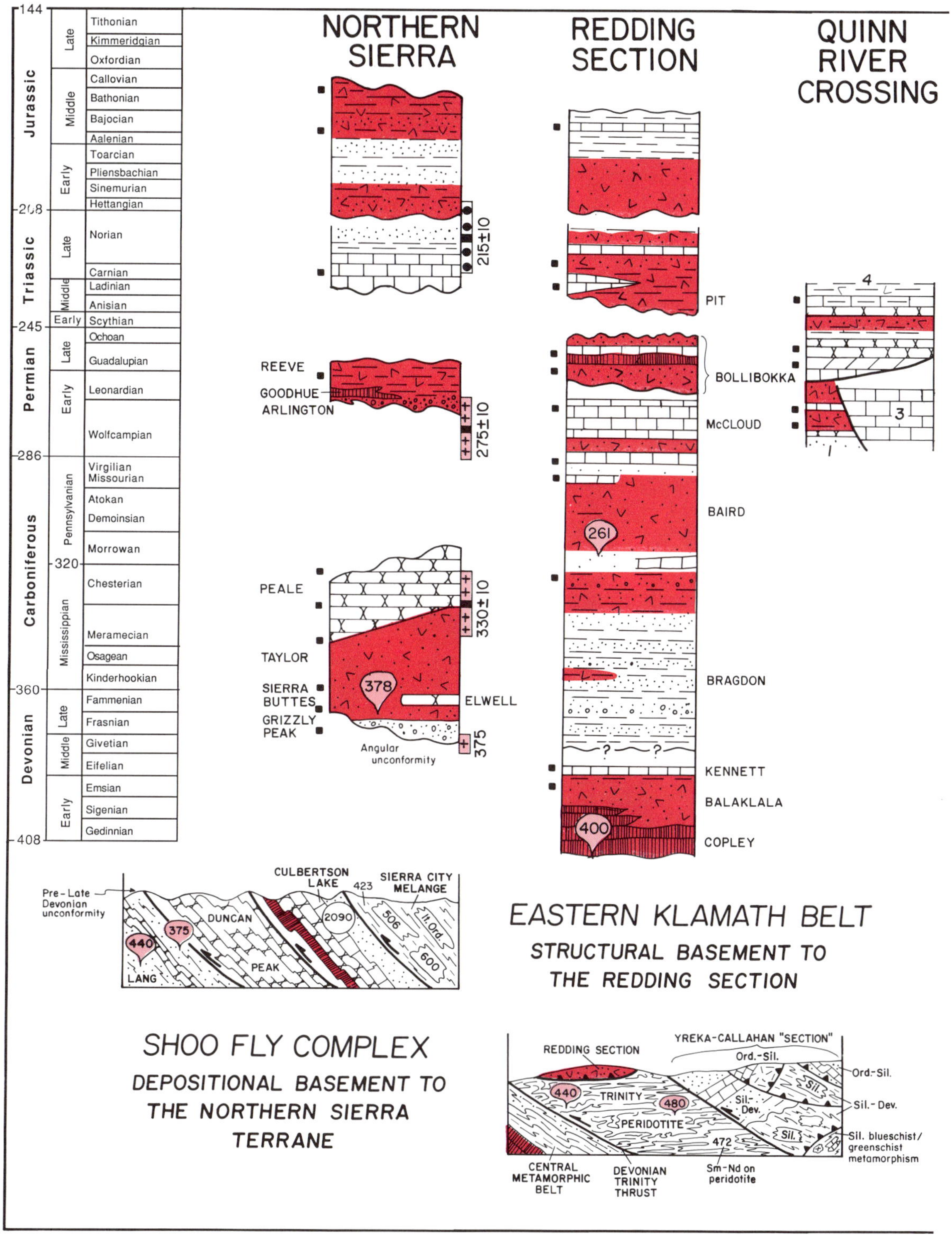

Figure 5. Tectonostratigraphic columns for upper Paleozoic rocks of volcanic island-arc affinity. Compiled from Albers and Robertson (1961), Albers and others (1981), Armstrong and others (1983), Blome and others (1983), Blome and Nestell (1988), Boucot and others (1974), Brown and others (1981), Cashman (1980), Clark and others (1962), Coogan (1960), Curtis (1983), Danner (1977), Davis and others (1965), Davis (1968), Durrell and D'Allura (1977), Fraticelli and others (1985), Girty and Wardlaw (1985), Hanson and others (1988), Harwood (1983, 1988), Holdsworth and Jones

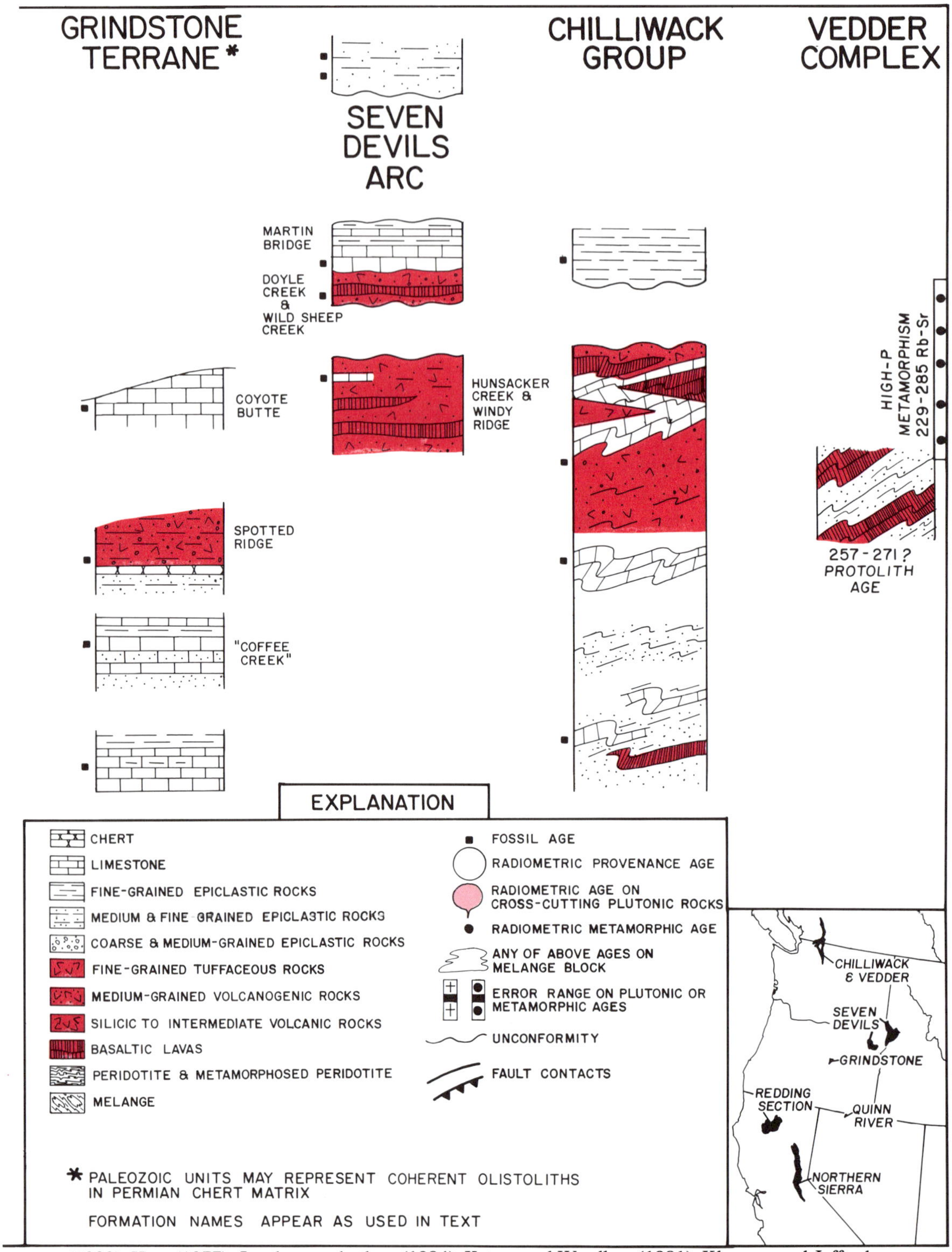

(1980), Hotz (1977), Jacobsen and others (1984), Ketner and Wardlaw (1981), Kleweno and Jeffords (1961), Lanphere and others (1968), Mattinson and Hopson (1972), McMath (1966), Merriam and Berthiaume (1943), Miller (1987, 1989), Monger (1977), Potter and others (1977), Quick (1981), Russell (1981), Sanborn (1960), Savage (1976), Schweickert and others (1984), Sharp, 1988, Silberling and Jones (1982), Skinner and Wilde (1965), Stevens and others (1987), Vallier (1977), and Wardlaw and others (1982).

and transgression synchronous with island-arc volcanism. The Devonian Bowman Lake batholith, which intrudes the northern exposures of the Shoo Fly Complex, is in part cogenetic with unconformably overlying silicic volcanic rocks of the Sierra Buttes Formation dated by fossils as Famennian (Hanson, 1983; Hanson and Schweickert, 1986). The Sierra Buttes Formation contains an increasing percentage of andesite upsection and is locally overlain by mid-Famennian (Anderson and others, 1974; written communication by T. B. Anderson in Hanson and Schweickert, 1986) ammonoid-bearing meta-chert once assigned to the Elwell Formation (Durrell and D'Allura, 1977), but is now included with other phosphatic chert interbeds in the Sierra Buttes Formation (Harwood, 1983, 1988) (Fig. 5). Elsewhere the Sierra Buttes Formation is directly overlain by andesitic turbidites, andesite tuff-breccia, and lapilli tuff of the Devonian-Mississippian Taylor Formation. Early Mississippian brachiopods in the overlying tuffaceous lower Peale Formation (McMath, 1966) indicate that volcanism persisted to this time, although sparse chert-granule conglomerate in the lower Peale marks the introduction of epiclastic debris into this region (Harwood, 1983; Hannah and Moores, 1986).

Klamath Mountains. The Klamath Mountains consist of an imbricate stack of terranes bounded by east-dipping thrust faults of largely Mesozoic age. Paleozoic rocks occur in three broadly defined structural belts. The structurally highest eastern Klamath belt and the more westerly central metamorphic belt contain most of the exposures of Paleozoic rocks (Plate 4A; Fig. 5). Permian limestone is present farther west within the western Paleozoic and Triassic belt of Irwin (1972) (not shown on Plate 4A) as blocks in melange. The two eastern belts are separated by the Devonian Trinity thrust fault (Peacock, 1987). The eastern Klamath belt can be further subdivided into three parts (Davis, 1969; Irwin, 1977; Silberling and Jones, 1987) (Fig. 5). The structurally lowest and most westerly of these is the Trinity ultramafic complex. The Trinity complex is structurally overlain to the north by lower Paleozoic sedimentary and metasedimentary rocks of the Yreka-Callahan area. To the southeast, the Trinity complex is structurally overlain by a Devonian through Jurassic island-arc sequence, the Redding section (Fig. 5).

The Trinity peridotite is Ordovician in age, and preserves evidence for a distinctive, early-depleted mantle reservoir (Jacobsen and others, 1984). Cambrian ages reported from associated plagiogranite come from rocks thrust over the Trinity complex (Wallin and others, 1988) and do not bear directly on the age of the complex. Cross-cutting gabbro and mafic lavas are isotopically and geochemically distinct from the peridotite, implying a polygenetic and protracted origin for the mafic and ultramafic rocks (see summary in Saleeby, this volume).

Rocks of the Yreka-Callahan "section" that structurally overlie the Trinity complex (Potter and others, 1977) were deformed in an event that is inferred to be pre–Late Devonian in age, based on K-Ar dates of 390 and 399 Ma from the central metamorphic belt to the west (Davis, 1968; Irwin, 1977; Lanphere and others, 1968; Lindsley-Griffin and Griffin, 1983) (Plate 4A). Rocks of the Yreka-Callahan area are similar to the Shoo Fly Complex in that both are varied assemblages of metasedimentary rocks, in part derived from continental source terranes, with lesser island-arc volcanic and basinal sequences represented (see Saleeby, this volume). Both regions contain evidence for penetrative deformation, imbricate thrusting and formation of melange during early Late Devonian or earlier times; a coeval Devonian unconformity with overlying arc sequence (the Redding section) has been suggested (Boucot and others, 1974; Charvet and others, 1989), but its presence is debated (Watkins, 1986).

The Lower and Middle Devonian Copley Greenstone contains the oldest rocks in the Redding section (Fig. 5; Plate 4A) (Kinkel and others, 1956). The Copley has been subdivided into two members: a lower unit of massive flows and pyroclastic debris, and an upper unit of pillow lavas, which include high-Mg andesites (LaPierre and others, 1985a). Sandstone containing granitic debris and shaley tuffs is locally interbedded with the pillow lavas. Geochemical data suggests that the Copley lavas are typical of primitive, oceanic island-arc magmas extruded in an extensional tectonic setting (LaPierre and others, 1985a).

In its southern exposures, the Copley Greenstone interfingers with and is overlain by the keratophyric "Balaklala Rhyolite," consisting of tuffs and flows which host Kuroko type massive sulfide deposits (Kinkel and others, 1956; Albers and Robertson, 1961; LaPierre and others, 1985b; Albers and Bain, 1985). A Middle Devonian (Eifelian) fish plate dates the submarine tuffs (Boucot and others, 1974). The trondhjemitic Mule Mountain stock, which intrudes and is inferred to be cogenetic with the Balaklala, yields a concordant 400-Ma U-Pb date from a single zircon fraction (Fig. 5; Albers and others, 1981).

The Middle Devonian (Eifelian) Kennett Formation (Savage, 1976) conformably overlies the Balaklala Rhyolite and consists mostly of tuffaceous shale, volcanogenic breccia, and basinal limestone (Watkins and Flory, 1986). An angular unconformity is reported from above the Kennett (Albers and Robertson, 1961; Charvet and others, 1989). The amount of time represented by this proposed unconformity is unknown, however. In contrast, Watkins (1986) suggests that the Bragdon Formation is conformable with the underlying Kennett Formation and merely drapes across the preexisting, subsided, vent-proximal, basinal-facies topography (Watkins and Flory, 1986).

The Bragdon Formation is a turbidite sequence dominated by reworked sedimentary, metasedimentary, and intermediate-composition volcanic debris, with locally common felsic to mafic tuffs and lavas. Sedimentary clasts in the turbidites include chert, argillite, and quartzose to feldspathic sandstone; metasedimentary clasts include greenschist-grade equivalents of these lithologies. Volcanic debris is characterized by volcanic quartz, plagioclase, and rarer lithics (Miller and Cui, 1989). The sedimentary and metasedimentary detritus marks the influx of sediment from a source other than the arc itself, and is compositionally typical of an orogenic provenance (cf. Dickinson and Suczek, 1979). The most likely source rocks are pre-Devonian basement rocks of the

arc itself, such as the Shoo Fly Complex or Yreka terrane (Watkins, 1986; Miller and Cui, 1989). The Bragdon Formation contains Mississippian-Pennsylvanian fauna near its top and is believed to be mostly Mississippian in age (Fig. 5; Watkins, 1973).

In summary, mid-Paleozoic rocks of the Redding section of the eastern Klamath belt indicate a history of extension-related, primitive arc volcanism in the Middle Devonian. By the Mississippian, orogenic sediments were deposited with possible angular discordance upon these rocks, recording the uplift and erosion of the sedimentary and low-grade metasedimentary terrane that constituted the basement of this arc.

Central Oregon and northwestern Washington. The Grindstone terrane of eastern Oregon and the Chilliwack Group of northwestern Washington (Plate 4A, B; Fig. 5) contain Devonian and Mississippian rocks, but very little is known about their stratigraphy and structural setting. The Grindstone terrane was formerly included in the central melange terrane of the Blue Mountains by Dickinson and Thayer (1978). Although it is internally disrupted, it lacks the serpentinite matrix that characterizes rocks of the central melange terrane (see also Wardlaw and others, 1982) and thus warrants a separate description. It has been most recently described as an olistostrome made up of Paleozoic olistoliths in a disrupted Lower Permian to Lower Triassic sedimentary matrix (Blome and Nestell, 1988; Morris and Wardlaw, 1986; Blome and others, 1986). The oldest rocks consist of Middle Devonian limestone intercalated with chert-argillite and chert-bearing sandstones inferred to be turbidites (Kleweno and Jeffords, 1961; Silberling and Jones, 1987). Lower Mississippian strata, originally assigned to the Coffee Creek Formation, include argillite, sandy limestone, calcareous sandstone, and conglomerate of unknown depositional setting and provenance, which have yielded early Carboniferous brachiopods and corals (Fig. 5; Merriam and Berthiaume, 1943). Formation status is questioned because the only exposure of these rocks is now believed by some to be an olistolith (Blome and Nestell, 1988).

The Chilliwack Group or Chilliwack terrane of northwestern Washington, internally disrupted and metamorphosed under blueschist-facies conditions in late Mesozoic time, is reported to have a coherent upper Paleozoic stratigraphy (Monger, 1977; Brown and others, 1981; Armstrong and others, 1983). The oldest rocks recognized in the Chilliwack terrane are Middle to Upper Devonian reefal and deep-water limestone (probably turbidites) associated with shale, graywacke, argillite, minor chert, and a thick section of volcanic rocks (Fig. 5) (Danner, 1977). The reefal limestone contains a wide variety of shallow-water fauna of probable Frasnian age (Danner, 1977). The limestone is overlain by fine-grained unfossiliferous siliciclastic turbidites, which are in turn overlain by Pennsylvanian strata (Fig. 5) (Monger, 1977; Danner, 1977).

The Roche Harbor terrane, included in the westward Tethyan or Cache Creek–affinity belt, contains Frasnian reefoid and stromatoporoid bank limestones with associated fine and coarse clastic sedimentary rocks and volcanic rocks (Danner, 1977). A single chert locality yielded Mississippian radiolaria (Whetten and others, 1978). Since the final deformation and emplacement of these rocks did not occur until Mesozoic time (Whetten and others, 1978), their location is shown only on the Permo-Triassic map (Plate 4H) where they are labeled as the Tafton Group.

Interpretation of Devonian–Early Mississippian orogenesis

Lower to mid-Paleozoic rocks in the western U.S. Cordillera form two distinct belts of contrasting styles and ages of deformation (Plate 4A, B): the Early Mississippian Antler thrust belt on the shelf, and a more westerly, discontinuously exposed belt of rocks that was subjected to deformation, metamorphism, and plutonism prior to or during the early Late Devonian. This second belt forms the basement for a thick and extensive island-arc succession that spans the entire upper Paleozoic.

Many authors have discussed the plate tectonic framework of Devonian-Mississippian events in the western U.S. Cordillera (e.g., Burchfiel and Davis, 1972, 1975; Churkin, 1974; Silberling, 1973; Dickinson, 1977; Speed, 1977a; Schweickert and Snyder, 1981; Miller and others, 1984) and have generally concluded that the Antler orogeny resulted from the collision of an island-arc system with the western edge of the continent. Opinions differ as to the nature of the arc, the polarity of subduction, and when and where subduction was initiated (Fig. 6).

Dickinson (1977), Speed (1977a), and Schweickert and Snyder (1981) ascribed the Antler orogeny to the collision of an east-facing arc with a long-lived passive continental margin (Fig. 6A, B). They viewed the Roberts Mountains allochthon as the scraped-up remnants of an ocean basin of unknown width that was imbricated with continental slope-and-rise sediments as the arc approached the continent. Schweickert and Snyder (1981) suggested that strata in the Roberts Mountains allochthon were once co-extensive with lower Paleozoic units now represented by the Shoo Fly Complex and correlative rocks in the northern Sierra Nevada and Klamath Mountains. More westerly rocks were first imbricated by east-vergent thrusting (during east-dipping subduction), and imbrication continued as the arc moved eastward toward the continent, forming the Roberts Mountains allochthon (Fig. 7.6 of Schweickert and Snyder, 1981) (Fig. 6). They further considered the possibility that the Alexander terrane in southeastern Alaska (Jones and others, 1972) (AX in Fig. 2) formed part of this arc system; subsequently it was displaced northward by strike-slip faulting.

Burchfiel and Davis (1972, 1975) emphasized the similarity in timing of events in both the Sierra Nevada–Klamath Mountains and Antler orogenic belt. Based on the nature of metamorphism and inferred west-vergence of structures in the western belt, they suggested that the Antler orogeny represented the closure of a back-arc basin behind a west-facing arc system (Fig. 6C). This interpretation assumes that the Klamath-Sierra arc developed and existed offshore for some time prior to the Antler orogeny. Building on the models proposed by these various authors, an updated interpretive model for the plate tectonic setting

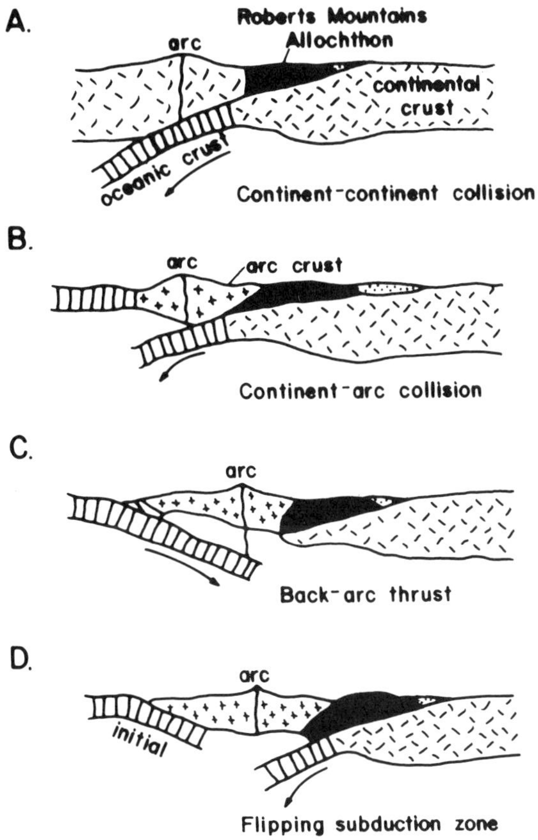

Figure 6. Plate tectonic models proposed for the Antler orogeny after Nilsen and Stewart (1980), all of which are compatible with observed deformation resulting in the emplacement of the Roberts Mountains allochthon. Whether back-arc basin closure results in subduction-zone polarity flip depends on the original (and uncertain) size of the basin.

of Paleozoic events leading to the Antler orogeny (Fig. 7) is discussed below.

The earlier history of tectonic instability, extensional faulting, and basaltic volcanism inferred from the stratigraphy of the Roberts Mountains allochthon is important to any plate tectonic model for the Antler orogeny (Fig. 7A). This history may represent the continentward effects of back-arc rifting or transtension following the establishment of a subducting margin in the Late Cambrian to Early Ordovician (e.g., Churkin, 1974). Alternatively, the western North American margin may have been a long-lived transform boundary between oceanic and continental plates and thus characterized by transtensional tectonics throughout the early Paleozoic (e.g., Eisbacher, 1983). This second possibility shares similarities to the northeastern coast of North America during the Mesozoic, where continued reactivation of fracture zones and normal faults accompanied by alkalic magmatism occurred in a continental-shelf and shelf-basin setting (Jansa and Pe-Piper; Pe-Piper and Jansa, 1987). Similar events along the Cordilleran margin, however, spanned a much greater length of time. The shelf stratigraphic record generated by these two alternative plate tectonic settings is likely to be indistinguishable, but evidence for Cambrian to Devonian arc magmatism, as represented by rocks in the western Cordillera, suggests that the first alternative may be more likely.

Following Schweickert and Snyder (1981), similar North American source terranes are shown for the quartzose and feldspathic sediments of the Harmony Formation in the Roberts Mountains allochthon and for the Shoo Fly and correlative rocks in the northern Sierra and Klamath Mountains (Fig. 7B). The problem of deriving these sediments from a shelf region mostly blanketed by carbonate rocks (Stewart and Suczek, 1977) is alleviated by documentation of normal faulting and uplift along the outer continental margin during the Cambro-Ordovician (e.g., Evans and Zartman, 1988) or within the offshore basin region itself (e.g., Madrid, 1987) (Fig. 7B).

Island-arc volcanic and plutonic sources are represented by some of the lower Paleozoic units of the northern Sierra and Klamath Mountains, as is arc magmatism (Fig. 7B). These rocks are now structurally interleaved with continentally derived sediments. Girty and Wardlaw (1984) have dated zircons in arc-derived sediments that occur as blocks within Shoo Fly melange as 506 m.y. old. In the Klamath Mountains, arc-derived graywacke is present in the Sissel Gulch Graywacke and Siluro-Devonian Gazelle Formation (Condie and Snansieng, 1971; Potter and others, 1977). These units have also yielded Ordovician-Silurian zircon ages (Wallin, 1989). Geochronologic work on the Trinity ultramafic complex indicates that the intrusive rocks that compose the complex range in age from Ordovician to Silurian (Wallin and others, 1988).

The inception of deformation in the northern Sierra-Klamath Mountains is older than the emplacement of the Roberts Mountains allochthon (Girty and Schweickert, 1984) (Fig. 7C). In the Klamath Mountains, Silurian or older greenschist-facies metamorphism affected structurally disrupted sequences that enclose blueschist blocks (Hotz, 1977). Greenschist- to amphibolite-facies metamorphism also affected rocks immediately beneath the eastern Klamath belt and yield K-Ar mica ages of 399 to 371 Ma, indicating a minimum Middle Devonian age for this deformation. The suggested unconformity at the base of the Bragdon Formation may also indicate tectonism of this age (Boucot and others, 1974; Charvet and others, 1989). Plutonic rocks that post-date deformation in the northern Sierra are 378 Ma and younger (Plate 4A, B).

The vergence, or direction of thrust faulting, is a critical piece of information for understanding the plate tectonic setting of Devonian and older subduction-related deformation, metamorphism, and magmatism in the Klamath Mountains and northern Sierra Nevada. Devonian structures are reported as west-vergent in the Klamath Mountains (Burchfiel and Davis, 1975). Specifically, the fault that separates the central metamorphic belt from the Trinity complex dips east and is inferred to represent a "frozen" subduction zone (Peacock, 1987). Structures are depicted as east-vergent in the Shoo Fly Complex of the northern Sierra Nevada after rotation of the Devonian unconformity to horizontal (Schweickert and Snyder, 1981). Until

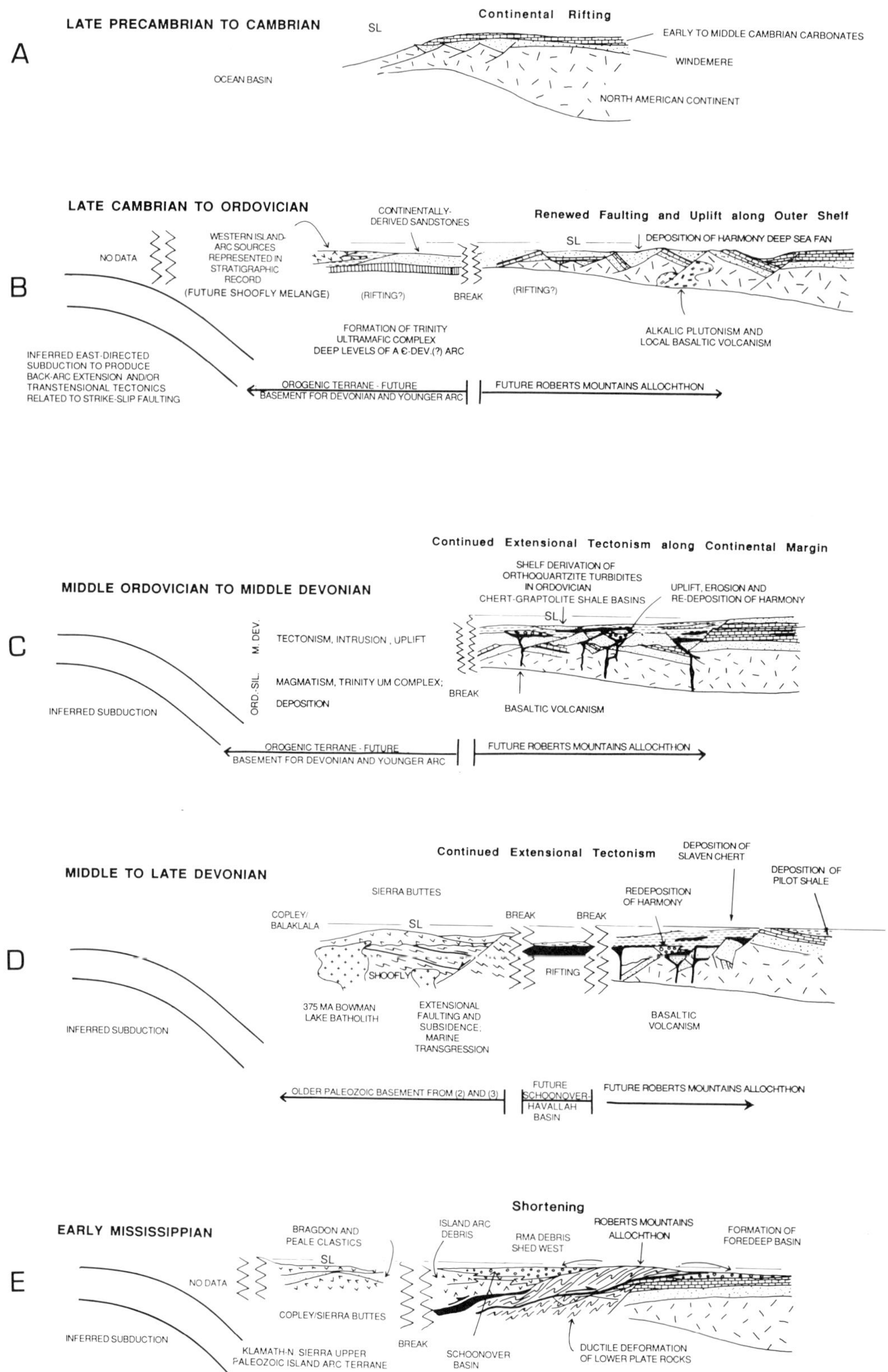

Figure 7. Interpretive model for early to mid-Paleozoic tectonic evolution of the western U.S. continental margin. "Breaks" designate lack of information about distances and/or the possibility of strike-slip motion in and out of the plane of the section. For explanation, see discussion in text.

more is understood about the structural relations between rock units, the polarity of this arc in the northern Sierra Nevada region is unclear from existing geologic data. Based on data from the Klamath Mountains, we indicate east-dipping subduction in Figure 7.

The episodes of deformation, uplift, and erosion documented by this pre-Devonian package of rocks do not appear to be recorded in sequences that compose the Roberts Mountains allochthon, nor are volcanic arc sources represented. We show a "break" in Figure 7C, representing intervening basins and/or strike-slip displacement between coeval assemblages in these two depositional settings.

The Lardeau Group in southern British Columbia (Plate 4A) contains units somewhat similar to lower Paleozoic strata in the northern Sierra and Klamath Mountains, but which are more clearly part of the North American continental margin. Cambro-Ordovician to Devonian(?), pelagic and quartzose to feldspathic, clastic sedimentary rocks and tholeiitic volcanic flows in the Lardeau Group were deformed and metamorphosed under greenschist-facies conditions and unconformably overlain by Mississippian strata of the Milford Group (Wheeler and Gabrielse, 1972; Okulitch and others, 1975; Read, 1975; Bond and Komintz, 1984). Okulitch and others (1975) have dated Middle to Late Devonian calcalkaline granitic plutons that intrude correlative strata in the region of the Shushwap metamorphic complex. The overlying Milford Group contains clasts of granitic material, confirming that Devonian magmatism accompanied deformation of the Lardeau Group. Unlike lower Paleozoic rocks in the Klamath Mountains and northern Sierra Nevada, whose depositional basement is not known, strata of the Lardeau Group are reported to depositionally overlie Cambrian shelf-facies units (Wheeler and Gabrielse, 1972; Okulitch and others, 1975). Thus, Devonian tectonism and calcalkaline magmatism may have involved the actual edge of the continental shelf in British Columbia, suggesting deformation occurred above, or continentward of, an east-dipping subduction zone. Alternatively, the differences in stratigraphy, and the fact that Devonian deformation in the Lardeau Group is not traceable eastward into clearly autochthonous strata of the Cordillera, have been cited in favor of the allochthoneity of this terrane (Monger and others, 1982; Monger, 1984).

Regardless of their present allochthonous nature, the stratigraphic and sedimentologic similarities of the Lardeau Group to lower Paleozoic rocks in the Klamaths and northern Sierra Nevada, together with their similar histories of metamorphism and plutonism, suggest that all of these rocks may have formed part of an extensive and more continuous orogenic terrane in the Devonian (Fig. 2). The cratonal source regions represented by sedimentary rocks in this terrane indicate deposition proximal to North America. The composition and isotopic characteristics of plutonic rocks indicate the presence of evolved continental crust at depth beneath at least parts of this terrane (Rubin and others, 1990). Other regions in the northern Cordillera that may have formed part of this orogenic belt include the Yukon-Tanana terrane in northern Canada and southern Alaska (Aleinikoff and others, 1986) and portions of northern Alaska, including the Brooks Range and Seward Peninsula (Fig. 2). In the southern Brooks Range, metasedimentary rocks intruded by Devonian plutons form the depositional basement for Mississippian and younger North American shelf-margin sequences.

In the northern Sierra Nevada, transgressive sequences as old as Frasnian(?)-Famennian (Late Devonian) bevel across a part of this older orogenic terrane, represented here by the Shoo Fly complex (Fig. 7D). This well-documented overlap in the northern Sierra Nevada pre-dates the emplacement of the Roberts Mountains allochthon by at least 15 m.y. (Girty and Schweickert, 1984) (Fig. 7C, D) but is coeval with basaltic volcanism and pelagic and hemipelagic deposition in the Antler basin, the older part of the Havallah basin, and formation of the Pilot Shale basin on the continental shelf (Fig. 7D; Plate 4A). These data are compatible with the proposal of a Cordilleran-wide rifting event in the Late Devonian (e.g., Turner and others, 1989).

During the Early Mississippian Antler orogeny, arc volcanism and intra-arc sedimentation occurred in the northern Sierra Nevada, represented by the Taylor and the basal part of the Peale Formation (Harwood, 1988), and in the Klamath Mountains by intercalated tuffs in the Bragdon Formation (Fig. 7E). Sedimentary rocks rich in chert and argillite debris are present in the Peale and Bragdon Formations and indicate that uplift of sedimentary, volcanic, and metasedimentary source terranes occurred in or adjacent to the arc terrane at the time of the Antler orogeny on the shelf (Fig. 7E). The Schoonover sequence, now part of the Golconda allochthon, possibly occupied an intermediate position with respect to the arc and Antler orogenic belt (Fig. 7E) or lay in uncertain position somewhere along strike of the Antler orogenic belt. Its stratigraphy provides strong evidence that a western volcanic and metasedimentary source terrane lay adjacent to the continental margin during the Antler orogeny (Fig. 7E).

In conclusion, the geologic data discussed above are compatible with the interpretation that a magmatic arc, constructed upon older deformed rocks and perhaps in part on continental basement, lay offshore of the western U.S. continental margin during Devonian-Mississippian time. Shortening of the basin(s) that lay between this island-arc and the continental margin constitutes the Antler orogeny and may have been facilitated or localized by lithospheric heating associated with rifting immediately prior to the orogeny. The available data do not rigorously constrain the size of the basinal region between the arc and the continent in mid-Paleozoic time, nor the polarity of subduction through time (Fig. 6). The data, however, are compatible with the proposed protracted history of eastward subduction beneath the continental margin depicted in Figure 7.

LATE MISSISSIPPIAN

Introduction

During the Late Mississippian, the Antler orogenic belt and foreland basin remained important paleogeographic elements on

the continental shelf (Plate 4C). Although coarse debris continued to be shed from the Antler belt in the Late Mississippian, stratigraphic and sedimentologic data suggest that thrust-faulting had ceased. Subsidence along the western part of the Antler highlands is recorded by transgressive sequences of limestone and shale. Associated basaltic volcanism may indicate that subsidence was related to extension or rifting. The cessation of thrust-faulting coincided with the inferred onset of extension along the continental margin. In the Klamath Mountains, Lower(?) to Upper(?) Mississippian clastic rocks indicate source region uplift and are succeeded by volcanic rocks of the Upper Mississippian and Pennsylvanian Baird Formation. In the northern Sierra Nevada an important lull in volcanism is recorded by pelagic deposits that span the Late Mississippian.

The craton

Isopachous maps show accumulation of Upper Mississippian sediment in two east-west–trending depositional centers, the Williston Basin and the Uinta Basin. These basins were flanked by two east-west–trending uplifts, the Montana Arch and the Oquirrh-Uinta uplift (Rose, 1976) (Plate 4C). According to Sando (1976), the Williston Basin underwent uplift and was no longer a site of deposition by latest Chesterian time. At about the same time, the Montana Arch farther south ceased to be a positive feature.

Limestone and shallow-water clastic rocks began to accumulate in the Uinta Basin in northern Utah and northwestern Colorado in pre–latest Chesterian time (Rose, 1976), followed by the deposition of the thick Manning Canyon Shale during the latest Mississippian and Early Pennsylvanian (Morrowan) (Plate 4I). The Uinta Basin is considered by Rose (1976) to represent foundering related to the inception of the younger Oquirrh Basin.

Two positive areas in New Mexico and eastern Arizona, the Pedernal uplift and the Defiance-Zuni uplift, were apparently never covered by the Late Mississippian seas and instead were low-relief areas of exposure of Precambrian rocks (Armstrong and others, 1980). Crystalline rocks presumably remained exposed in northern Colorado as well (Sando, 1976; Gutschick and others, 1980). The Ouachita continental margin remained a major region of subsidence and sedimentation in the southern U.S. during the Late Mississippian. The Pedregosa basin (Armstrong and others, 1980) may have had its inception at this time, although its boundaries are poorly defined. Craig and others (1979) showed the Delaware basin to be a well-defined feature at this time.

The continental shelf

The Antler foreland basin remained the most conspicuous depositional feature on the continental shelf during the Late Mississippian (Plate 4C). During this time span, clastic rocks shed eastward from the Antler highlands interfingered with carbonate rocks on the shelf to the east (Plate 4I). In the uppermost Mississippian to lowest Pennsylvanian, shelf carbonates prograded westward across the foreland basin (Ely Limestone and equivalents, Plate 4I). The edge of the Late Mississippian carbonate shelf, or eastern boundary of the Antler foredeep basin, can be traced into eastern California. Here, however, coarse clastic debris was derived from the eastern shelf region, rather than the Antler belt (Plate 4C) (Stevens and others, 1979).

A major regression began in Meramec time (Sandberg and others, 1983). The carbonate platform that had covered the western part of the North American craton was transformed into a broad karst plain, and the starved-basin sedimentation in the eastern part of the Antler foreland basin was overwhelmed by floods of fine clastic sediment from both east and west (Sandberg and others, 1983). According to these authors, the tectonic load related to the thrust emplacement of the Roberts Mountains allochthon (Speed and Sleep, 1982) was removed at this time. Sediments shed from the highlands were thus post-orogenic "molasse" (Sandberg and others, 1983), and represent progradational delta-slope and delta-platform sedimentation in gradually shoaling waters (Harbaugh and Dickinson, 1981).

The Antler highlands

Sparse exposures of rocks of early Late to Late Mississippian (Meramec and Chester) age, consisting of conglomerate, sandstone, limestone, and basalt, unconformably overlie the Roberts Mountains allochthon near its western limits of exposure (Plate 4C). These Mississippian-age "overlap" sequences are not as well represented or studied as the Pennsylvanian and Permian strata referred to formally as the Overlap Assemblage (Roberts and others, 1958), which are more widespread in the region of the Antler highlands.

In southwestern Nevada, Meramec-age carbonate strata are reported to overlie deformed lower Paleozoic rocks at several localities near Candelaria (Plate 4C) (Poole and Sandberg, 1977; Speed, 1977c; Stewart, 1980b; Oldow, 1984; Ferguson and others, 1952; Stanley and others, 1977; Sando and Bamber, 1985), providing a good upper age limit for penetrative deformation of deep-water facies lower Paleozoic strata (Oldow, 1984). However, if these deep-water-facies rocks are actually deposited in sequence with Cambrian shelf strata (Stewart, 1980b), then the unconformity documents penetrative shortening of continental-shelf strata as well.

In the Independence Mountains, northern Nevada (Plate 4C), Chesterian conglomerate, limestone, and siltstone unconformably overlie deformed Ordovocian rocks of the Roberts Mountains allochthon and are derived from the subaerial erosion of these rocks (Miller and others, 1984). Near Mountain City in northernmost Nevada, the Mississippian overlap assemblage, from the base upward, consists of conglomerate derived from the Roberts Mountains allochthon (Grossman Formation), Meramec-age limestone (Banner Formation) interbedded with, overlain by, and intruded by basaltic lavas of the Nelson Formation (Coats, 1969; Little, 1987) (Fig. 4), in turn overlain by a

Mississippian-age clastic sequence similar to the Chainman Shale of the Antler foredeep basin. Little (1987) has suggested that the basaltic volcanism and accompanying transgression indicate rifting and subsidence of the Antler highlands. Following Little (1987), we show the marine basin represented by the Chainman Shale connecting to offshore basins across the Antler highlands region in northern Nevada by Late Mississippian time (Plate 4C).

A sequence of rocks similar to that in the Mountain City region is exposed in the Osgood Mountains (Plate 4C). Here, basaltic lavas are interbedded with Meramec-age limestone (Hotz and Willden, 1964; Sando and Bamber, 1985). Stewart (1980b) considered the Osgood Mountains sequence to be autochthonous and deposited on the Antler highlands because of its similarity to the Mountain City sequence. However, the base of the section in the Osgood Mountains is not exposed, and unlike the autochthonous Mountain City section, radiolarian chert is characteristic of the section, suggesting it may be more analogous to coeval sections present in the northern part of the Golconda allochthon, which also contain abundant basalt flows of identical age (discussion below and Fig. 4). Mafic magmatism in the Late Mississippian may be associated with extension or rifting that caused the foundering and subsidence of the Antler orogenic belt (Little, 1987). The coincidence of this poorly documented event in the western reaches of the Antler highlands and adjacent Havallah basin, and the cessation of thrusting in the Antler indicates that crustal shortening leading to the emplacement of the Roberts Mountains allochthon may have abruptly terminated with the onset of extension in the Meramec.

In southern British Columbia, the Mississippian Milford Group (Plate 4C) unconformably overlies deformed, metamorphosed, and intruded lower Paleozoic continental-margin sequences (Read, 1975; Read and Wheeler, 1976; Okulitch and others, 1975). The Milford Group consists of locally fossiliferous Middle to Upper Mississippian limestone overlain by sandstone, phyllite, chert, and conglomerate, in turn overlain by Mississippian(?) andesitic lavas and tuffs (Okulitch and others, 1975; Klepacki, 1983). The Milford Group is similar to the Early Mississippian overlap sequence in Nevada in that it records subsidence and transgression of a region involved in mid-Paleozoic deformation.

Havallah basin

Except for the Lower Mississippian strata in the Schoonover sequence discussed above, the oldest rocks in the Golconda allochthon elsewhere in Nevada appear to be no older than Late Mississippian in age and consist of radiolarian chert sequences that overlie pillow basalts (e.g., Tobin Range and Antler Peak; Fig. 4 and Plate 4C) (Stewart and others, 1986; Murchey, 1982). In the Independence Mountains, Nevada, well-dated Meramec-age basaltic lavas occur in the Golconda allochthon (Miller and others, 1984). In the structurally lowest thrust plate of the allochthon, these basalts interfinger with lenses of boulder conglomerate containing large, rounded cobbles of sedimentary rocks like those in the Roberts Mountains allochthon. The flows are overlain by, and interbedded with, Mississippian-age fossiliferous limestone, which is in turn overlain by chert, shale, and quartzose- and chert-grit turbidites that also appear to be derived from the erosion of rocks in the Roberts Mountains allochthon (Fagan, 1962; Whiteford, 1990; Miller and others, 1984). This succession is quite similar to the autochthonous Meramec-age sequence in the Mountain City area and to the questionably autochthonous or allochthonous Gough's Canyon Formation in the Osgood Mountains (Fig. 4).

At about the same time, a pronounced change in the volume and composition of siliciclastic detritus supplied to the Havallah basin occurs. Detritus become more quartzose, and volcanic debris is conspicuously absent (Whiteford, 1990), suggesting the removal of the Early Mississippian volcanic arc terrane represented in stratigraphically underlying units (Whiteford, 1990; Miller and others, 1984).

Rock units equivalent to the Havallah sequence, such as the Eastern Assemblage of Monger (1977) in the Canadian Cordillera (Fig. 2), contain basalt and radiolarian chert sequences of Mississippian age (Struik and Orchard, 1985), indicating that rifting and formation of deep-water basins may have characterized much of the length of the Cordilleran margin (Plate 4C).

Volcanic arc and oceanic sequences

Northern Sierra Nevada. In the northern Sierra Nevada, volcanogenic deposits of the lower part of the Peale Formation and sparse interstratified clastic rocks derived from a sedimentary-metasedimentary provenance accumulated in the Early Mississippian. The Late Mississippian is characterized by a pelagic section consisting of radiolarian chert and shale, which marks an end to active arc volcanism (Plate 4C; Fig. 5) (Harwood, 1983; Schweickert and others, 1984; Hannah and Moores, 1986). The upper pelagic member of the Peale is dated by Late Mississippian brachiopods in the north (Harwood, 1983, 1988). The hiatus in volcanism thus persisted into at least the middle Pennsylvanian. This stratigraphic transition is similar to the one documented farther east in the Havallah basin (see above; Harwood and Murchey, 1990).

Klamath Mountains. Thick successions of strata containing sand-rich turbidite facies typify the Upper Devonian(?) and Mississippian Bragdon Formation in the eastern Klamath Mountains (Fig. 5; Plate 4C). These clastic rocks record the mixing of sediment from differing source areas in a submarine-fan setting within a rapidly subsiding basin (Miller and Cui, 1989). Sedimentary, metasedimentary, volcaniclastic, and pyroclastic detritus indicate uplift and erosion of volcanic, pelagic, and clastic successions accompanied by volcanism. The Bragdon Formation contains Late Mississippian fossils near its top and is believed to be mostly Mississippian in age (Watkins, 1973). Based on its conformable relation with the underlying Kennett Formation, it may be in part Late Devonian as well (Watkins, 1986). It interfingers with the overlying volcanogenic Baird Formation, which is Late Mississippian at its base (Fig. 5).

Northwestern Washington. In northwestern Washington, in the Chilliwack Group, no Mississippian rocks are recognized and Lower Pennsylvanian rocks overlie Devonian limestone (Monger, 1977; Danner, 1977; Plate 4A, D).

The only confirmed occurrences of Mississippian rocks in the belt of younger accretion-subduction complexes of the western U.S. are radiolarian-bearing cherts in the Roche Harbor terrane on San Juan Island, northwestern Washington (Plate 4H) (Whetten and others, 1978).

PENNSYLVANIAN

Introduction

During the Pennsylvanian, the rise of the Ancestral Rocky Mountains drastically changed sedimentation patterns across the cratonic region of the Cordillera. Two paleotectonic maps, one for the Atokan, and one for the Virgillian, are shown for the cratonal region in the Pennsylvanian (Plate 4D, E). These maps are based primarily on the compilation of McKee and others (1975), and show the rapid development of the Ancestral Rocky Mountains and their flanking basins during this time span. Further information about the development of these basins can be found in Baars and others (1988) and Frenzel and others (1988). Farther west, faulting caused the formation of basins on the continental shelf and was responsible for renewed uplift of the Antler orogenic belt. Modification of the shelf-edge-to-basin transition by faulting occurred in southern California. Strata deposited in the offshore Havallah basin also record tectonism at this time by the development of unconformities and renewed clastic input. Island-arc sequences record a varied history that is marked by uplift, erosion, and deposition of shallow-water limestone and volcaniclastic sequences coeval with arc volcanism, although pelagic sedimentation characterized the northern Sierra terrane. This uplift was most likely related to contemporary tectonism and/or volcanism.

The craton

Orogenic events leading to the rise of the Ancestral Rocky Mountains began in the Early or Middle Pennsylvanian (Bachman, 1975). Uplifts at this time included the Frontrange uplift, which in its southern reaches includes the Apishapa and Sierra Grande uplifts, the Pathfinder uplift, whose most rapid rise occurred in the Atokan (Mallory, 1975), and the Uncompahgre uplift (Plate 4D). The eastern end of the Oquirrh-Uinta uplift in southern Wyoming, northeastern Utah, and northwestern Colorado was undergoing a large amount of uplift and formed part of the Ancestral Rocky Mountains. The Central Colorado trough and the Paradox basin had also developed by Middle Pennsylvanian time and began to receive coarse conglomerates in Atokan time (Mallory, 1975; Casey, 1980). Deposition of evaporites began in the Eagle basin (northern part of Central Colorado trough) and in the Paradox basin in Desmoinesian time (Chronic, 1979). In west Texas and adjacent New Mexico, the Delaware basin and Central basin platform were well differentiated as the Ouachita continental margin was deformed in what was to become the Marathon fold and thrust belt.

The Williston basin and Montana arch, important features of the Wyoming shelf region during the Late Mississippian, were no longer discernible during Middle Pennsylvanian time. The east-trending Uinta basin, however, continued to develop and apparently extended westward onto the continental shelf forming the Ferguson basin. The Zuni-Defiance and Pedernal uplifts, also present in the Late Mississippian, continued to be active sources of clastic debris in New Mexico and Arizona. The Pedregosa basin apparently became much better defined, and a prong of the Zuni-Defiance uplift, called the Burro uplift, developed. Lane and DeKeyser (1980) indicate its presence as early as Early Mississippian time, separating the Pedregosa basin from the newly formed Orogrande basin. All of western Arizona evidently was exposed in Atokan times and formed a very broad north-trending uplift called the Piute positive (Plates 4D, E).

Most Late Pennsylvanian paleotectonic features of the Ancestral Rocky Mountain region were inherited from those developed in the Middle Pennsylvanian (Plate 4E). The Frontrange and Uncompahgre uplifts were still active during Late Pennsylvanian time (Mallory, 1975; Jordan and Douglass, 1980), and the southern parts of these uplifts may have been most active during this time (Bachman, 1975). According to Welsh and Bissell (1979), Precambrian basement rocks of the Uncompahgre uplift underwent about 6,000 m of uplift during this time. Mallory (1975) indicated that 2,000 m of sediment were deposited in the Paradox basin on the southwest side of the Uncompahgre uplift and almost 4,000 m of sediment were deposited in the narrow Eagle basin (central part of the Colorado trough) on the northeast side of the uplift. The Central Colorado trough was still active, especially the Rowe-Mora basin (McKee, 1975) situated at the very south end of the trough. Evidently the Pathfinder uplift in Wyoming ceased to be active and was largely covered by sediment by Late Pennsylvanian time. At this same time the Wyoming shelf region was fully emergent for the first time, although there had been positive areas in the region prior to this time.

The Zuni-Defiance uplift continued to be an active source of sediment, and the Pedernal uplift probably reached its maximum development during Late Pennsylvanian time. The Pedregosa and Orogrande basins continued to subside and received sediment from the intervening Burro uplift. The Piute positive was in part depressed in Late Pennsylvanian time, receiving some sediment, but it was still in part recognizable as an active source region. The main structural features of west Texas apparently had not changed significantly from Atokan time (early Middle Pennsylvanian).

The exact nature of the faulting that formed the Ancestral Rocky Mountains is unclear. Jordan and Douglass (1980) suggested that this system of uplifts and basins was formed by movement along high-angle faults with some strike-slip move-

ment. As summarized by Kluth (1986) and Kluth and Coney (1981), the uplift-bounding faults appear to have moved in large part as steep reverse faults. The latter authors proposed that these structures formed as a cratonward effect of the collision of South America and Africa with North America in the Pennsylvanian along the Ouachita-Marathon margin. This deformation reactivated older Precambrian zones of weakness far within the continental interior.

The continental shelf

Tectonism along the Cordilleran continental shelf occurred broadly at the same time as the uplift of the Ancestral Rocky Mountains on the craton, causing renewed uplift and erosion of the old Antler highlands and the formation of shelf basins.

The Oquirrh basin developed in Pennsylvanian time (Plate 4D and I) and was the site of deposition of more than 300 m (Mallory, 1975) of shallow-water sedimentary rocks during Atokan time (Jordan and Douglass, 1980). The Oquirrh basin may have cut across older east-west–trending structures, but there are still questions concerning the exact extent and geometry of the basin. The western end of the Oquirrh-Uinta uplift in northeastern Nevada was apparently uplifted and eroded at this time (Plate 4D). Jordan and Douglass (1980) showed that no sediment was deposited along this feature in western Utah in the early Late Pennsylvanian, and that shallow-water or subaerial sediments were deposited here in the late Late Pennsylvanian. These observations suggest that uplift extended across the Oquirrh basin. A several-kilometer-thick section of deep-water sediments began to accumulate across much of the Oquirrh basin area in the Late Pennsylvanian, but there is still some question as to whether or not the Oquirrh basin was a single, north-trending basin. The east-trending Ferguson basin, which almost surely continued into the Uinta basin, lay south of the Oquirrh-Uinta uplift. The broad Deep Creek–Tintic uplift in eastern Nevada and western Utah lay farther south.

Deposition occurred without significant breaks from Mississippian to Pennsylvanian time along most of the continental shelf in the region of the Antler foreland basin. Near Carlin, in northern Nevada, however, Middle to Late Pennsylvanian faulting or folding steeply tilted the older foredeep basin deposits, which were subsequently eroded and unconformably overlain by Upper Pennsylvanian (Missourian) strata (Dott, 1955) (Plate 4D; Fig. 4). This event was termed the "Humboldt Orogeny" by Ketner (1977), who ascribed it to regional compression. In northernmost Nevada, transgressive Lower Permian sequences overlie rocks cut by high-angle faults that post-date emplacement of the Roberts Mountains allochthon. Although the history of movement along these faults is not clear, they likely had a normal component of displacement as they appear to displace rocks of the structurally higher Roberts Mountains allochthon downward with respect to underlying autochthonous strata. The deposition of thick transgressive sequences of Permian rocks across both footwall and hanging-wall rocks in the region of northern Nevada suggests that faulting was followed by subsidence (Little, 1987) (Plate 4D), indicating regional extension as a possible cause of faulting. Similar relations have been documented in other parts of the Antler highlands (Plate 4D), and Smith (1989) has documented Pennsylvanian(?)-Permian(?) normal faulting in the Toiyabe Range, central Nevada.

Faulting modified the continental margin in southern California in the Pennsylvanian and Permian (Stone and Stevens, 1988a; Stevens and Stone, 1988). The west-southwest–trending Mississippian shelf-to-basin transition (Stevens, 1986) apparently was changed to a southeast-trending shelf-basin transition by Middle Pennsylvanian time (Plate 4D). To the northwest and southwest of the shelf break, limestone turbidites accumulated in a basinal environment.

In summary, scattered exposures of Pennsylvanian rocks along the continental shelf-edge record rapid facies changes and indicate modification of older continental-margin paleogeography by active faulting during this time span. In detail, the age constraints on faulting vary from place to place, and the nature of faulting, whether strike-slip, normal, or reverse, is largely unknown. However, the transgressive nature of many of the post-faulting sequences suggests extension followed by subsidence.

Figure 8 summarizes the trends and distribution of structures related to the Ancestral Rocky Mountains in relation to the structural trends of the Cordilleran continental margin. To the west of the Ancestral Rocky Mountains, Cordilleran-margin structural trends are predominantly north-northeast, reflecting the original configuration of the rifted continental margin. Facies and isopachs for lower Paleozoic rocks follow this trend, as does the trace of the Roberts Mountains thrust and its foreland basin. Facies and isopachs for upper Paleozoic rocks in the continental-shelf region similarly trend north-northeast. Several major east-west–trending structures that extended into the cratonic interior affected depositional patterns on the eastern edge of the continental shelf (see Fig. 11). Some of these structures, such as the Uinta basin and Uinta arch, may have been inherited from late Precambrian time. It appears that northwest-trending Ancestral Rocky Mountain structures may bend westward into these older east-west trends. Little is known about the geometry of faulting and basin formation farther west along the edge of the continental margin. Presumably, extensional or transtensional faulting followed north to northwest trends and paralleled the Permian continental margin, as suggested by the geometry of basins that formed in southern California during the Pennsylvanian and Permian (Stone, 1984; Stevens, 1986). Below, we discuss evidence for subsidence and extension in offshore basins and possibly in some portions of the Pennsylvanian to Permian island-arc terrane. When viewed collectively (Fig. 8), it appears that the nature and orientation of structures across the western part of the continent may reflect changing tectonic boundary conditions or far field stresses, from north-south compression along the southern boundary of the continent to east-west extension along the western boundary. The western, extending margin of the continent may have acted as a tectonic "free-face," analogous to the

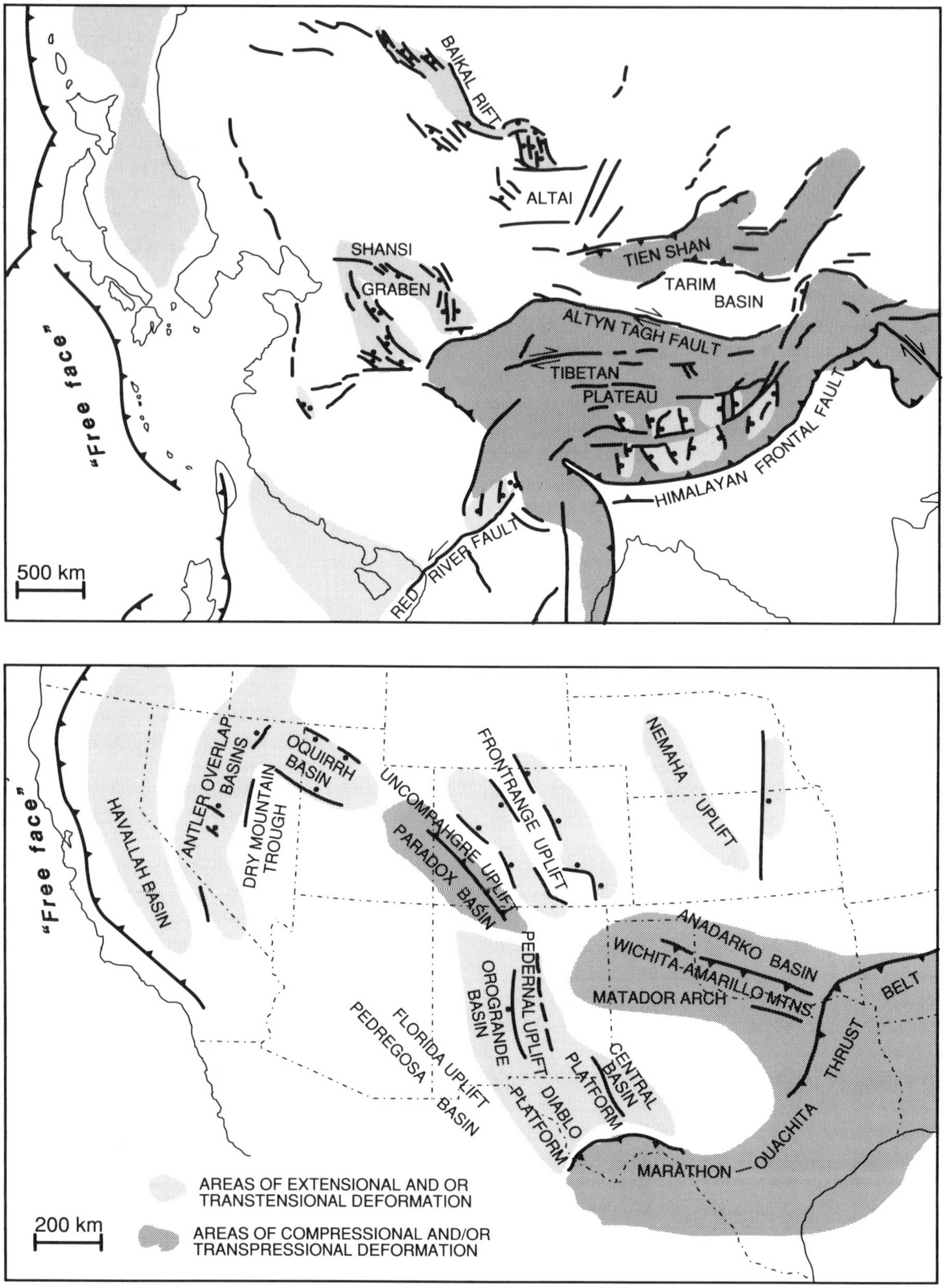

Figure 8. Trends of late Paleozoic age structures of the Ancestral Rocky Mountains and Cordilleran shelf region in relation to both the Ouachita and Cordilleran continental margin tectonics and comparison to the present-day tectonics of Asia (from Smith and Miller, 1990).

eastern margin of southeast Asia today (e.g., Molnar and Tapponier, 1975) (Fig. 8).

Offshore Havallah basin

Beginning in latest Mississippian time, and continuing through Pennsylvanian and Permian time, limestone turbidites were deposited in the offshore Havallah basin (Miller and others, 1984; Stewart and others, 1977, 1986; Fig. 4; Plate 4D). The limestone turbidites are interbedded with both radiolarian chert-argillite sequences as well as with coarse clastic turbidites derived from the continued erosion of rock types similar to those in the Roberts Mountains allochthon. Transported fusulinids in the Permian rocks are species found in coeval shelf sequences (Stevens, 1987). The mix of clastic rocks of Roberts Mountains allochthon provenance and calcareous material in these sequences reflects the development of carbonate shelf conditions along at least parts of the Antler highlands as represented by strata of the overlap assemblage (Plate 4D).

In the Independence Mountains, northeastern Nevada, disconformities and/or gaps in the stratigraphic record, together with renewed input of coarse siliciclastic turbidites suggest tectonic instability in the offshore basin region as well (Fig. 4; Plate 4D). Basaltic pillow lavas may be present in Pennsylvanian sequences of the Golconda allochthon, but their ages and stratigraphic association are not as well documented as those in Mississippian and Permian sections.

Pennsylvanian rocks in the eastern Sierra Nevada (Plate 4D) probably consist primarily of metamorphosed limestone (Rinehart and Ross, 1964; D. M. Willahan, personal communication, 1989) whose protoliths prior to deformation and metamorphism may have been similar to Pennsylvanian rocks in the Inyo Mountains.

In northeastern Washington, Pennsylvanian and possibly Pennsylvanian rocks of the Mount Roberts Formation and correlative sequences are poorly dated but contain andesitic volcanic rocks, conglomerates, graywackes, and tuffs, in addition to black argillite and chert pebble conglomerate. As described by Yates (1976), the conglomerates are clearly of volcanic arc provenance.

In southern British Columbia, Upper Mississippian and Lower Pennsylvanian rocks of the Keen Creek sequence that are part of the Milford Group contain tholeiitic pillow lavas and some limestone (Klepacki, 1984; J. Wheeler, personal communication). Although in fault contact with underlying rocks of the Lardeau Group and overlying limestone of the Milford Group, Klepacki (1984) reasons that these rocks are only parautochthonous because they contain clasts similar to rocks present in the Lardeau and Milford Group. Pennsylvanian strata in the eastern assemblage of Monger (1977) (the Slide Mountain terrane) are dominantly chert-argillite; Late Pennsylvanian strata, if present at all, are confined to a narrow time interval (Struik and Orchard, 1985).

Volcanic island-arc sequences

Northern Sierra Nevada. In the northern Sierra Nevada, radiolarites and shales in the upper member of the Peale Formation indicate waning volcanism and starved-basin sedimentation during Late Mississippian to Middle Pennsylvanian time (Harwood, 1988). Rocks of Late Pennsylvanian and early Early Permian age are missing in the rock record, indicating nondeposition or Early Permian erosion (Harwood, 1988). As such, this section bears many similarities to coeval pelagic sequences in the Golconda allochthon as pointed out by Harwood and Murchey (1989). The Peale Formation is unconformably overlain by the volcanic upper Lower Permian Reeve Formation (Harwood, 1983) or by the intervening or laterally equivalent Goodhue Formation (Harwood, 1988).

To the west of the northern Sierra arc sequence lie ophiolitic rocks of the western Sierra Nevada belt. Radiometric ages clustering around 300 Ma (mostly U-Pb zircon, also Nd/Sm; Saleeby and Sharp, 1980; Saleeby, 1981; Shaw and others, 1987; Saleeby and others, 1989) come from plagiogranite blocks in ophiolite melange, felsic screens within mafic dike complexes, and structurally concordant layers in mafic tectonic blocks or ultramafic schist (Saleeby, 1982). These rocks are spatially associated with or locally intruded across older mafic-ultramafic complexes that likely represent the ophiolitic substrate of the late Paleozoic volcanic arc in the northern Sierra Nevada (Saleeby and others, 1989). The dike intrusion and melange formation reflect development of a transform within the basement of the volcanic arc during the Pennsylvanian and persisting into the Mesozoic (Saleeby, this volume).

Eastern Klamath Mountains. In the eastern Klamath Mountains, sediment gravity flows of the mostly epiclastic Bragdon Formation are gradationally and conformably overlain by shallow-marine volcanogenic debris flows, water-lain tuffs, and lavas of the Baird Formation (Watkins, 1985; Miller, 1986, 1989). The transition to shallow-water deposition and attendant increase in volcanism reflect localized uplift within the island arc. Fossils in the Baird Formation range from Late Mississippian to Late Pennsylvanian–Early Permian(?) in age, although local unconformities are recognized within the formation (Watkins, 1973, 1985).

Central Oregon and northwestern Washington. In the Grindstone terrane of eastern Oregon the Spotted Ridge Formation (Fig. 5; Plate 4D) contains clastic rocks with terrigenous plant debris dated as Pennsylvanian. These clastic rocks contain both volcanic and plutonic debris and are inferred to have been deposited in nonmarine to shallow-marine (estuarine to deltaic) conditions (Merriam and Berthiaume, 1943). Although the Spotted Ridge Formation was believed to disconformably overlie the Coffee Creek Formation, these rocks are now thought by some to be olistoliths within a disrupted terrane; thus their age

relations and formation status have been questioned (Blome and Nestell, 1988; Fig. 5).

Probable Mississippian-age rocks in the Chilliwack Group are succeeded by fine-grained clastic rocks that are conformably and gradationally overlain by argillaceous limestone containing Early Pennsylvanian foraminifera, corals, and crinoid debris (Monger, 1977) (Fig. 7). Volcaniclastic sandstone, argillite, and conglomerate disconformably overlying the limestone contain both shallow-marine fauna and terrestrial organic debris of Late Pennsylvanian or Early Permian age; thus, shallow-marine to nonmarine conditions characterized the Late Pennsylvanian in this region.

EARLY PERMIAN

Introduction

Fault-bound basins continued to form on the outer shelf during the Early Permian, indicating continued tectonic instability and an extensional or transtensional tectonic regime along the western U.S. continental margin. Early Permian transgressive sequences were deposited upon the inner subsiding shelf and in shelf basins, some of which extended far east onto the craton. Deposition and minor uplift persisted in the region of the Ancestral Rocky Mountains. The offshore Havallah basin and coeval island-arc sequences deposited farther west record a distinct two-part history, which we emphasize by dividing the Permian into two time slices, one representing the Early Permian (Wolfcampian to Leonardian) and the other late Early to Late Permian (Leonardian to Guadalupian). There is little earliest Permian (Wolfcampian) arc-related volcanism recorded in the island-arc terranes, which underwent differential uplift and locally experienced shallow-marine conditions and subaerial exposure during this time interval. By the late Early Permian (Leonardian), arc volcanism characterized by basaltic and andesitic magmatism was active. A concomitant influx of arc-derived clastic debris is recorded in the Havallah basin and was apparently synchronous with the inception of thrust faulting.

The craton

In the region of the Ancestral Rocky Mountains, Early Permian paleotectonic features are mostly remnants from paleogeographic patterns in existence during Late Pennsylvanian time (Plate 4F). The Zuni-Defiance, Pedernal, and Burro uplifts were much reduced. A thick evaporite sequence was deposited in the Holbrook basin, which developed on the western edge of the Zuni-Defiance uplift. The Orogrande and Pedregosa basins were still subsiding and accumulating marine sediment. The axis of the Piute positive is well-defined on the isopachous map of Blakey (1980) and the Frontrange and Uncompahgre uplifts continued to shed sediment into the surrounding basins during Early Permian time. The Central Colorado trough is shown as discontinuous by Peterson (1980) but continuous by McKee and others (1967). The Paradox basin was still the site of greater deposition than surrounding areas. The Wyoming shelf expanded and became continuous with the Alberta shelf. The Delaware basin and the remnants of the Ouachita continental margin in west Texas subsided dramatically in Early Permian time, and very thick sequences of deep-water sedimentary rocks were deposited. The formation of these sedimentary basins occurred in response to the collision of Gondwanaland with the southern margin of North America in the Early Permian. Collision began to the east along the Appalachian portion of the margin and progressed westward to the Ouachita and Marthon regions, finally closing the foredeep or foreland basin in the Late Permian (Graham and others, 1975).

The continental shelf

The Ferguson-Uinta basin that lay south of the Oquirrh-Uinta uplift and north of the Deep Creek–Tintic uplift continued to be a prominent locus of deposition in Early Permian time. The western end of the Oquirrh-Uinta uplift shed coarse sediment southward during the Wolfcampian, but by the Leonardian (late Early Permian) it had been drowned (Stevens, 1979). Thick sequences of deep-water sediment accumulated in the Oquirrh basin in early and middle Wolfcampian time (Jordan and Douglass, 1980). Subaerial sediments were deposited in an east-trending region crossing the Oquirrh basin in the general area of the Oquirrh-Uinta uplift (Plate 4F). Before the end of the Wolfcampian, the Oquirrh basin had been filled and shelf depths prevailed (Jordan and Douglass, 1980).

Across the region of the Antler orogenic belt, Lower Permian strata are, in general, part of a transgressive marine sequence that in places is as old as Pennsylvanian (Plate 4F). Local uplift and subsidence are evident based on intra-Permian disconformities and suggest that high-angle normal(?) faulting that began in the Pennsylvanian continued into the Permian. Farther south along the margin, in east-central California, thick turbidite sequences define a system of southeastward-trending margin-parallel basins that extended southward through the northern Mojave Desert (Stone, 1984; Stone and Stevens, 1988a). Essentially contemporaneous basin formation and local thrust faulting have been interpreted by Stevens and Stone (1988) to be the result of left-lateral wrench faulting along the continental margin. To the northwest of the Inyo Mountains, upper Paleozoic strata are present in the Mount Morrison sequence in the eastern Sierra Nevada. These rocks may represent a distal equivalent to the east-central California sequences.

In the western Mojave Desert, the youngest shelf deposits are Wolfcampian and correlate with the widespread Pennsylvanian to Lower Permian Bird Spring Group (Plate 4I). Metamorphosed, unfossiliferous sequences of upper Paleozoic (?) siltstone, shale, limestone turbidites, and sandstone described in the western Mojave Desert are not directly correlative with shelf units and may indicate that Permian turbidite basins similar to those in the Inyo Mountains formed across this part of the shelf as well (Miller and Cameron, 1982; Walker, 1988).

Conglomerate, calcareous sandstone, and sandy limestone of Permian and locally Pennsylvanian age unconformably overlie early Paleozoic slope and basinal strata in the El Paso Mountains and grade upward into basinal sequences with limestone turbidites that range in age from Pennsylvanian to early Late Permian (Leonardian to Guadalupian) (Plate 4I; Carr and others, 1984). Rapid facies changes in these rocks are interpreted as evidence that deposition occurred in structurally controlled troughs and the deepening-upward sequences probably indicate subsidence due to extensional faulting (Carr and others, 1984). Upper Paleozoic strata in the El Paso Mountains are lithologically similar to autochthonous sequences in the Inyo Mountains and surrounding region and contain a similar fusulinid fauna, suggesting they were deposited in a position at least marginal to the continental shelf (Carr and others, 1984; Stone, 1984).

Offshore Havallah basin

Permian sedimentary successions of the Golconda allochthon are dominated by limestone turbidite sequences but also contain coarse siliciclastic horizons and basaltic lavas that may indicate a similar history of faulting and subsidence in the offshore region during the Early Permian (Fig. 4; Plate 4F). In the Antler Peak area, well-dated Lower Permian radiolarian chert sequences are interbedded with quartzose clastic rocks derived from the continental shelf as well as with conglomerate lenses that contain cobbles of volcanic, sedimentary, and metasedimentary rocks (Miller and others, 1982, 1984). As there is no known source for the volcanic and metasedimentary debris on the continental shelf, it was probably derived from the west, or perhaps from uplifted fault blocks within the basin itself. Thick sequences of basaltic composition pillow lava occur in some Permian sections in the Golconda allochthon (Plate F). Intercalated and overlying chert sequences in the Hot Springs Range date the basalts as Early Permian (Sosson, 1984). In the Toquima Range, basalts are dated as late Early Permian (Leonardian) to possibly Late Permian (Guadalupian) (Laule and others, 1981). Well to the east, along the leading edge of the Golconda allochthon, rare greenstones occur in a sequence of phosphatic limestone and sandstone of probable Permian age in the Monitor Range (Kleinhampl and Ziony, 1972). Coarse sandy horizons contain Early Permian fusulinids that are typical North American forms, thus are presumably derived from local carbonate environments along the western edge of the continental shelf.

Previously discussed, poorly dated rocks of the Mount Roberts Formation and correlative units in northeastern Washington (Plate 4F) contain andesite, tuff, argillite, and fossiliferous limestone of either Early or Late Permian age. Rocks of the Eastern assemblage of British Columbia (Slide Mountain terrane) are better dated and contain abundant Lower to Upper(?) Permian mafic volcanic rocks and intrusive bodies, chert, and minor ultramafic rocks that accumulated in a series of late Paleozoic rift basins (Monger, 1977; Struik and Orchard, 1985).

Island-arc and oceanic sequences

Introduction. As noted in the introduction to this chapter, island-arc and oceanic sequence rocks of the western Cordillera can be broadly separated into two belts, based on whether Permian strata have "McCloud" or "Tethyan" faunal affinity, although in some cases, this affinity is not known (Figs. 1 and 9). The eastern of these belts includes terranes composed of relatively coherent stratigraphic sequences, deposited in a volcanic arc setting, and discussed at some length already in this chapter (see also Miller, 1987). The typical McCloud fauna is best developed in the McCloud Limestone of the eastern Klamath Mountains (Skinner and Wilde, 1965; Stevens, 1985; Wilson, 1982, 1984, 1985). Most of the fusulinids and corals of the McCloud Limestone differ from typical North American and Tethyan forms and probably represent transitional faunas (Stevens and Rycerski, 1983; Wilson, 1982, 1985). Rare Tethyan forms are, however, locally present in this eastern belt (Stevens and others, 1987). The second belt includes terranes containing tectonically disrupted oceanic and arc assemblage rocks that contain olistostromal or tectonic limestone blocks of Permian age that are predominantly of Tethyan faunal affinity. These rocks are similar in many respects to the Cache Creek assemblage of the Canadian Cordillera (Fig. 9) and have been interpreted as the preserved remnants of a subduction zone complex or complexes formed largely in early Mesozoic time (e.g., Davis and others, 1978), discussed later in the Permo-Triassic section of this chapter.

The development of contrasting faunal assemblages in Permian time and their preservation in the Cordillera are the results of several factors. During the Permian, the equatorial paleo-Pacific Ocean contained several tropical, shallow-water biogeographic provinces (Ross and Ross, 1983; Stevens, 1985), including the North American, McCloud, and Tethyan provinces (Fig. 10). The isolation of the widespread North American miogeoclinal province of the eastern paleo-Pacific Ocean from the Tethyan province in the western paleo-Pacific is attributed to three principal barriers: the Pangean supercontinent, cold polar waters, and the deep and wide paleo-Pacific Ocean (Fig. 10). The restricted verbeekinid fusulinid and waagenophyllid coral families, which have been considered diagnostic of the Tethyan faunal realm (Monger and Ross, 1971; Stevens, 1985), occur almost exclusively within the subduction-zone complexes, but important exceptions occur in the island-arc sequence of the eastern Klamath Mountains (Stevens and others, 1987) and in the Quesnell terrane of southern British Columbia (Nelson and Nelson, 1985).

Monger and Ross (1971) first suggested that the present distribution of Permian rocks in the Cordillera resulted from tectonic transport related to large-scale plate motions. Since that time, the mobility of many of the terranes that make up the western Cordillera has been better established, based on paleomagnetic, stratigraphic, and structural data (see reviews in Saleeby, 1983; Monger, 1984; Ross and Ross, 1981, 1983; Yancey, 1979). Stevens (1983) and Stevens and Rycerski (1983) have

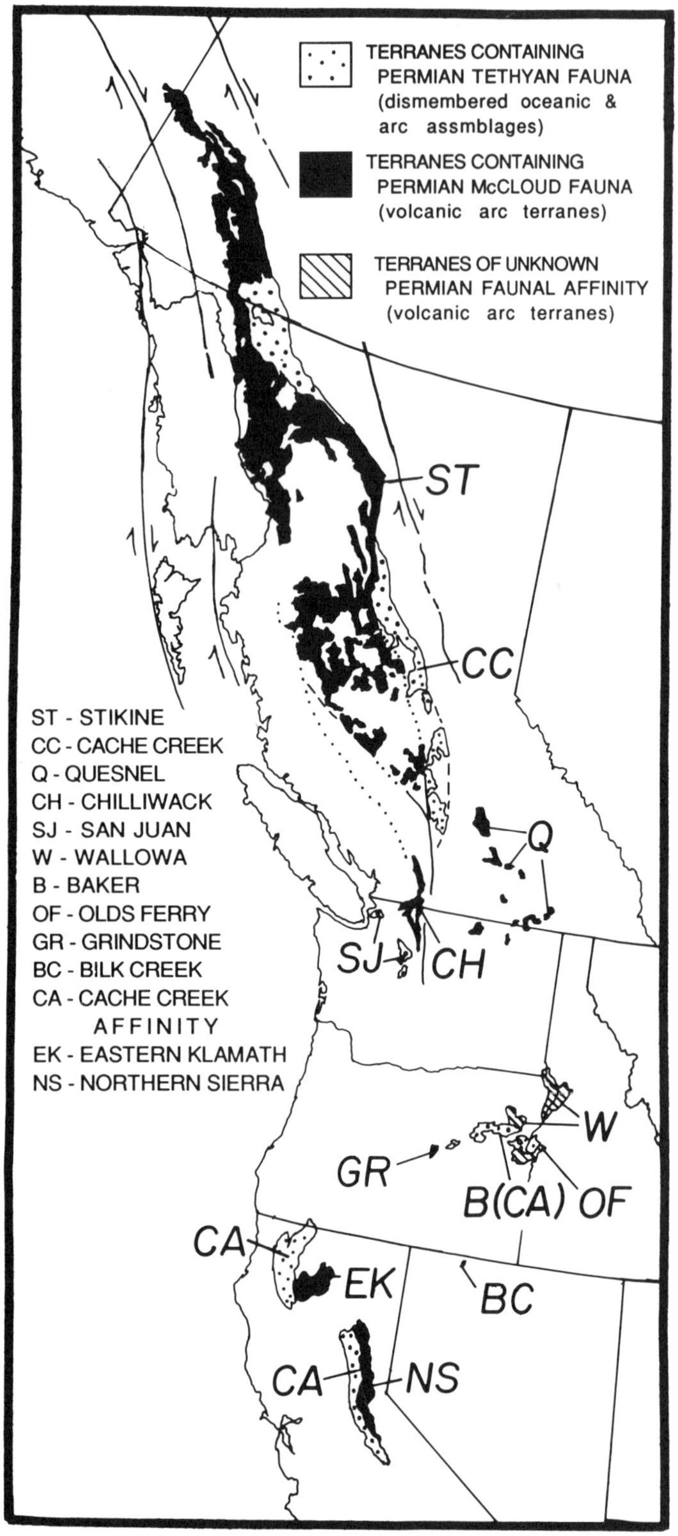

Figure 9. Distribution of Permian Tethyan and McCloud fauna in the North American Cordillera from Miller (1987).

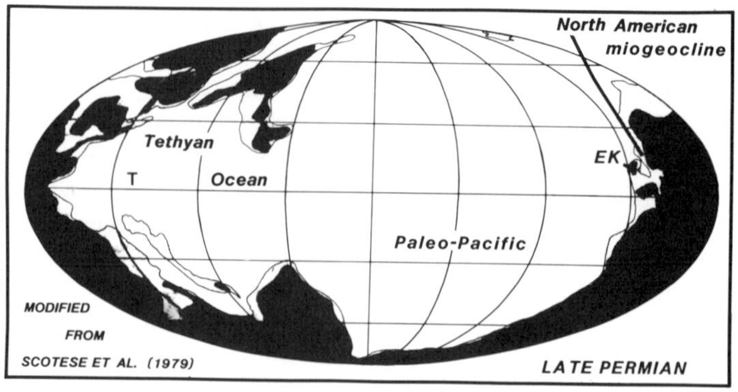

Figure 10. Late Permian distribution of land masses (black), shallow seas (shaded), and open ocean (white). Modified from Scotese and others (1979). Present-day location of the eastern Klamath Mountains shown as EK. Longitude is shown at 30-degree intervals.

since integrated these concepts into the evaluation of the original geographic distribution of the Tethyan and McCloud biogeographic provinces. Although it is agreed that rocks bearing McCloud fauna originated closer to the North American shelf than did those bearing Tethyan fauna, the exact geographic distances and paleogeography implied by these fauna are uncertain and subject to debate (e.g., Saleeby, 1983; Miller and Wright, 1987; Stevens and others, 1990).

When compared to modern coral diversity, the differences observed between McCloud and North American faunas are great, suggesting large geographic separations between the two provinces (Yancey and others, 1988). To these workers, the differences imply that the eastern Klamath terrane lay more than 5,000 km away from western North America during the Early Permian. In contrast, geologic relations suggest that the McCloud belt terranes had an episodic relation to a continental mass that was most likely North America (e.g., Miller and Saleeby, 1991; Harwood and Murchey, 1989).

Similar controversy exists concerning the absolute distance that originally separated the Tethyan and McCloud provinces in the paleo-Pacific Ocean. Based on similarities to faunas preserved in the Tethyan region (also mostly in allochthonous rocks), "Tethyan" faunas in the North American Cordillera are interpreted to have been transported across great distances of the paleo-Pacific Ocean in post-Permian time (e.g., Stevens, 1985), an interpretation compatible with their occurrence mostly in subduction-zone complex settings. The study of these limestone bodies suggests that they were deposited in atolls and oceanic plateau settings and therefore could have originated far east of what is considered the Tethyan Ocean proper (Fig. 10) (e.g., Saleeby, 1981; Miller and Wright, 1987). Newton (1988) evaluated these various hypotheses by using data on the distributions of Cretaceous and Cenozoic tropical biotas. She concluded that the modern occurrence of many Indo-West Pacific faunas in the eastern Pacific might be used to substantiate the existence of a Permian pan-

tropic Tethyan fauna, and that the use of faunal differences to limit longitudinal tectonic displacement is premature. Yancey and others (1988) disagreed with this assessment.

Northern Sierra Nevada. Volcanogenic Permian rocks overlie the Peale Formation along a major disconformity and contain a sporadic, thin, Lower Permian basal epiclastic unit (Fig. 5). In northern exposures, locally thick epiclastic deposits of the Leonardian Arlington Formation rest above the Peale and contain debris derived from the basement Shoo Fly Complex (Harwood, 1988). In central exposures, the upper Peale is conformably overlain by intermediate to mafic volcanic and volcanogenic rocks of the areally restricted Goodhue Formation (Durrell and D'Allura, 1977; Harwood, 1988), which is as young as late Early Permian (Leonardian) (Fig. 5) (Harwood, 1988). Harwood (1988) has suggested that the basaltic magmatism represented in the Goodhue Formation, together with uplift and erosion implied by the unconformity beneath this formation, indicate normal faulting and extension in Late Pennsylvanian(?) and Early Permian time. The Goodhue Formation is succeeded by mid-Permian shallow-water and subaerial andesitic breccia, tuff, and lavas of the Reeve Formation (Durrell and D'Allura, 1977; D'Allura and others, 1977). In the northern part of the terrane, the Reeve Formation has yielded late Early Permian (early Leonardian) fusulinids (Stevens, 1987). It overlies the Peale Formation across a locally angular unconformity that spans Middle Pennsylvanian through early Early Permian time (Schweickert, 1981; Schweickert and others, 1984; Harwood, 1983). Arc-related volcanism was active in the late Early Permian.

Lower Permian rocks of appropriate lithology to preserve fusulinids or other shallow-water fossils are rare in the northern Sierra. Sparsely fossiliferous tuff, volcaniclastic conglomerate, and chert-pebble conglomerate within the Reeve Formation have yielded a few fusulinids like those of the Lower Permian McCloud Limestone (Stevens, 1987), apparently establishing the Permian faunal affinity of this terrane (Fig. 9). Widespread occurrence of McCloud-type fusulinids from limestone lenses west of the Shoo Fly Complex has only recently been recognized (Jayko and others, 1987; Standlee and Nestell, 1985; Harwood and others, 1988). However, these rocks are in fault contact with, and lie structurally below, the Shoo Fly Complex and thus do not bear directly on the biogeographic affinity of Permian rocks of the Taylorsville sequence in the northern Sierra Nevada. These limestone bodies with McCloud fauna appear to represent a previously unrecognized belt of rocks that occurs between the Shoo Fly Complex and the Feather River Peridotite (Harwood and others, 1988; Jayko, 1988); they are very similar to Permian and Triassic (and possibly older) strata within the Redding section of the eastern Klamath terrane, specifically the McCloud Limestone, Pit Formation, and possibly the Baird Formation.

Eastern Klamath Mountains. In the eastern Klamath Mountains, the Pennsylvanian Baird Formation locally interfingers with the overlying Lower Permian (Wolfcampian to lower Leonardian) McCloud Limestone, although in places the contact is disconformable (Watkins, 1985) (Fig. 5). The McCloud Limestone is a fossiliferous, shallow-water platformal limestone. Development of the McCloud carbonate platform during the Early Permian occurred in two stages. A carbonate bank blanketed the southern and central parts of the presently exposed belt in the early Wolfcampian, and a second platformal sequence overlapped it and extended farther north during the late Wolfcampian to early Leonardian (Watkins, 1985). In both stages of platform development, shallower-water facies are present to the south, and distal, steepened ramps developed to the north (Watkins, 1985; Miller, 1989).

The McCloud Limestone contains abundant fusulinids, corals, and brachiopods (Skinner and Wilde, 1965; Watkins, 1973, 1985; Wilson, 1982, 1984, 1985). It ranges from Wolfcampian to early Leonardian in age, and its distinctive fusulinids differ from those of the Tethyan and North American Permian provinces (see discussion above), although some similarities to fusulinids of the southwestern California miogeoclinal shelf have been noted (Stevens, 1985). The corals contained in the McCloud Limestone were considered typical of those found in North American shelf sequences (Wilson, 1982, 1984) but have been shown to be quite different by Stevens and Rycerski (1983).

Northwestern Nevada. In northwestern Nevada, at Quinn River Crossing (Plate 4F; Fig. 5), Ketner and Wardlaw (1981) have divided small exposures of unmetamorphosed Permian and Triassic rocks into four structural blocks, two of which overlap in age. The structurally lowest block (block 1, Fig. 5) includes mafic to intermediate volcaniclastic and volcanic rocks, phosphatic chert and quartz- and chert-rich sandstone, and lenses of limestones with early Leonardian corals and late Wolfcampian to early Leonardian fusulinids, indicating shallow-water deposition. At its northern end, block 1 is structurally overlain by possibly deep-marine clastic rocks composed mostly of chert and quartzose detritus with lesser quartz-mica schist and chalcedony detritus. These clastic rocks are younger than middle Devonian (Jones, 1990) but are otherwise undated and consequently omitted from Figure 5. In its central portion, block 1 is structurally overlain by cherty, fossiliferous biocalcarenite of block 3 (Fig. 5), which has yielded brachiopods, fusulinids, and conodonts that range from Wolfcampian to late Leonardian in age and show strong affinities to the McCloud fauna (Ketner and Wardlaw, 1981; Skinner and Wilde, 1965, 1966). Upper Permian and Middle Triassic silicified limestone, dolomite, and radiolarites depositionally (Jones, 1990) overlie block 1 at its southern extent (block 4, Fig. 5).

Although original stratigraphic contacts are not preserved, Lower Permian rocks at Quinn River Crossing (Fig. 9) were deposited in a shallow-marine setting near centers of mafic to intermediate volcanism.

In the Jackson Mountains, the upper McGill Canyon succession (Plate 4F) contains limestone clasts with late Paleozoic fusulinids and brachiopods together with clasts of Devonian and Mississippian radiolarian chert (Russell, 1981). A late Leonardian brachiopod from the volcanogenic matrix of a debris flow may date deposition or may have been reworked and provide

only a maximum age for these deposits (Russell, 1981). The upper McGill Canyon succession consists of pelitic rocks, chert, calcarenite, quartzose, and volcanogenic turbidites in which quartzose detritus becomes more abundant upsection.

Central Oregon. The Lower Permian Coyote Butte Limestone was thought to overlie Pennsylvanian(?) strata assigned to the Spotted Ridge Formation along an unconformity that locally shows angular discordance (Grindstone terrane; Merriam and Berthiaume, 1943). Now those rocks, together with blocks of Permian chert, are thought to be olistostromes in a Triassic argillaceous matrix (Blome and Nestell, 1988; Fig. 9). The limestone has been subdivided into three lithologic units: (1) a basal massive, crinoidal packstone; (2) a medium- to thick-bedded, finer-grained packstone; and (3) an upper, thin- to medium-bedded, argillaceous wackestone (Wardlaw and others, 1982). Shallow-water carbonate bank deposition in the Early Permian was punctuated by influx of volcanic debris and detrital chert (Fig. 5; Merriam and Berthiaume, 1943; Wardlaw and others, 1982). Although different fossil groups give slightly different ages, the Coyote Butte Limestone probably ranges from late Wolfcampian to late Leonardian in age. Fusulinids assigned to the late Wolfcampian occur with brachiopods and conodonts considered to range from early to late Leonardian (Wardlaw and others, 1982). The fusulinids are similar to those of the McCloud Limestone (Skinner and Wilde, 1965; Wardlaw and others, 1982), and these units show remarkable petrologic and biostratigraphic similarity to Lower Permian limestone at Quinn River Crossing (Wardlaw and others, 1982).

Northwestern Washington. Lower Permian micritic and calcarenitic limestone gradationally and conformably overlies Pennsylvanian rocks of the Chilliwack Group (Plate 4F, Fig. 9). The limestone contains fusulinids, brachiopods, and other shallow-water fossils. The fusulinids are of McCloud faunal affinity (Monger and Berg, 1984; Skinner and Wilde, 1966). This limestone is overlain by, and may interfinger laterally with, basalt flows and intermediate to siliceous tuffs. No direct ages on the volcanic rocks exist. The Lower Permian limestone is disconformably overlain by Upper Triassic argillite and siltstone (Monger, 1977).

Permian arc sequences in the western U.S. Cordillera share many similarities, including the distinctive McCloud fauna, and may thus represent fragments of a once-more-continuous late Paleozoic island arc (Miller, 1987). In all of the terranes of the McCloud belt, except for the northern Sierra Nevada where little record remains, the Early Permian is a period of waning volcanism and uplift, allowing development and growth of shallow-water carbonate banks with only intermittent volcanism. In the northern Sierra Nevada, mafic volcanic centers were locally active. The similarities in the rock stratigraphic record of these scattered exposures suggest that the McCloud belt island-arc system (Plate 4F) underwent some fundamental tectonic changes at this time. Faulting, basaltic volcanism, and formation of carbonate platforms suggest that extension or rifting may have affected the late Paleozoic island arc in Early Permian times.

LATE PERMIAN

Introduction

By the Late Permian, most of the basins that formed during the development of the Ancestral Rockies were filled with sediment, and structural highs were subdued (Plate 4G). No signs of tectonic instability are present on the craton. Marine deposition characterized the continental-shelf region, and basins accumulated cherty and phosphatic sediments in Middle to Late Permian time. Faulting and erosion apparently still occurred in the Antler highlands area, and Guadalupian-age rocks in many places rest unconformably on older strata. Magmatism and an important episode of tectonism began offshore of the continental margin and migrated toward the continent in Late Permian time. Voluminous basaltic to dacitic volcanism is recorded in arc sequences after a period of previous quiescence. At the same time, the Havallah basin began to receive abundant arc-derived detritus prior to its closure in Late Permian time.

The craton

The Alberta Shelf region was a prominent positive region during the Late Permian. The Frontrange and Uncompahgre uplifts were still source regions for some coarse-grained sediment (McKee and others, 1967), but most other older structural highs had become subdued by Late Permian time, and bathymetric basins largely filled with sediment (Hallgarth, 1967; Peterson, 1980). The Zuni-Defiance and Piute uplifts were buried with sediment and, by Late Permian time, were unrecognizable. The Pedregosa and Orogrande basins in southeastern Arizona and southern New Mexico continued to be separated by the Burro uplift and to be clearly defined structural elements. The Delaware basin continued to subside rapidly, and before the end of Permian time, vast amounts of salt were deposited not only in the Delaware basin itself but also across the Central basin platform and other areas to the east. By this time the Ouachita continental margin had been deformed, telescoped by thrust faulting, and was a region of uplift and erosion.

The continental shelf

The Oquirrh-Uinta uplift still existed as an east-trending area of reduced sediment accumulation (Wardlaw and others, 1979) or exposure (Wardlaw, 1980) during middle to Late Permian time. Most of the middle to Upper Permian chert and phosphate accumulated north of this long-lasting feature (Wardlaw, 1980). The Uinta and Ferguson basins were still negative features. Isopachous maps of middle to Upper Permian rocks show that the axis of the Ferguson basin appears to have migrated southward during this time span (Wardlaw, 1980). During the Late Permian, the Oquirrh basin was no longer a recognizable feature.

In the Independence Mountains (part of the old Antler highlands), east-west–trending faults cut a transgressive sequence of

Lower and mid-Permian rocks. Upper Permian turbidites were synchronous with or succeeded movement on these faults (Miller and others, 1981, 1984) (Plate 4 G). Along the east-central California portion of the margin, pre–Upper Permian rocks were uplifted, folded, and eroded. The older, northwest-trending turbidite basins were destroyed in this deformational event, but locally, thin sequences of Guadalupian rocks accumulated (Stone, 1984; Stone and Stevens, 1988b).

The Havallah basin

In latest Permian time, andesitic arc-derived debris entered the Havallah basin, signaling the encroachment or establishment of a volcanic arc system west of the continental shelf (Plate 4G). This is well documented by the presence of widespread Permian volcaniclastic strata in the Golconda allochthon in exposures in the Shoshone, New Pass, and Toiyabe Ranges (MacMillan, 1972; Tomlinson and Wright, 1986; Babaie, 1987) (Plate 4H). In the Toiyabe Range, coarse Permian volcanogenic debris flows contain cobbles and boulders of plutonic rocks dated at 267 Ma (Early Permian, U-Pb zircon; Tomlinson and Wright, 1986), indicating exposure of deeper-levels of the arc in the source terrane. In addition, the 2.2-Ga inherited zircon component found in these granitic boulders indicates that these plutons intruded Precambrian crust or sedimentary sequences ultimately derived from a cratonic source.

Farther south in the El Paso Mountains, Pennsylvanian-Permian, shelf-derived limestone turbidite sequences grade upward into upper Leonardian volcanogenic turbidites and tuff, which grade upward with apparent conformity into a thick (>1,500 m structural thickness) sequence of porphyritic andesite flows and flow breccias (Carr and others, 1984). Reworked tuffs at the base of the andesite are marine, but ash flows higher in the unit may be subaerial (Carr and others, 1984). The age of these volcanic rocks is likely Guadalupian and is constrained as post–latest Leonardian and pre–250 Ma, the age of a cross-cutting granite (Carr and others, 1984). These stratigraphic relations in the El Paso Mountains are important because they provide a good tie between the Permian continental shelf and Permian volcanic arc sequences discussed in the section below.

In the Mina-Candelaria region, a structurally thin section of sedimentary rocks is structurally overlain by volcanogenic sedimentary and andesitic rocks of the Mina and Black Dyke Formations, dated respectively as 256 ± 6 and 258 ± 3 Ma (K-Ar hornblende; Speed, 1977b) (Plate 4G). Volcanic and volcanogenic rocks are considered representative of the arc that collided with the continental margin during closure of the Havallah basin (Speed, 1977a), but are in all respects similar to Permian volcanogenic strata interbedded with sequences in the Golconda allochthon farther north (Tomlinson and Wardlaw, 1988).

In the northernmost exposures of the Golconda allochthon in the Independence Mountains, Nevada, the youngest strata in the Havallah sequence are only Early Permian in age, indicating that basinal deposition must have ceased shortly after this time. We infer from this indirect evidence that deformation and thrusting likely brought an end to deposition and thus were ongoing by the beginning of the Late Permian (Plate 4G).

Volcanic arc and oceanic sequences

Northern Sierra Nevada. In the northern Sierra Nevada, the late Early Permian to early Late Permian (Leonardian to Guadalupian) volcanic succession comprises the laterally equivalent Arlington, Goodhue, and Reeve Formations (Harwood, 1988) (Fig. 5). Locally thick deposits of the Arlington Formation contain material derived from the erosion of underlying Shoo Fly Complex lithologies, suggesting faulting or differential uplift within the northern Sierra Nevada terrane in the early Early Permian. Differential uplift together with contemporaneous basaltic volcanism (the Goodhue Formation) are thought to reflect extensional tectonism (Harwood, 1988). Age constraints on this sequence are provided by early Leonardian fusulinids that occur as solitary fossils within a volcaniclastic horizon of the Reeve in the northern part of the terrane (Stevens, 1987) and Guadalupian fossils from elsewhere in the terrane (Harwood, 1988). Elsewhere in the northern Sierra Nevada, the chert member of the Peale Formation is directly overlain by intermediate to mafic volcanogenic rocks of the Lower Permian (Leonardian) Goodhue Formation, or by shallow-marine to subaerial andesitic tuff, breccia, and lavas of the Reeve Formation (Harwood, 1988). These rocks are all Leonardian to Guadalupian in age, and collectively record an important episode of magmatism during mid-Permian time.

Eastern Klamath Mountains. In the eastern Klamath Mountains, part of the upper Lower Permian (middle and upper Leonardian) rock record is missing in most places across a disconformity that is capped by the Guadalupian Nosoni Formation, which, together with the overlying Dekkas Formation, makes up the volcanogenic Upper Permian Bollibokka Group (Plate 4G; Fig. 5) (Coogan, 1960). Locally, however, Leonardian rocks are thought to be present (C. H. Stevens, in Miller, 1986).

Vent-proximal basaltic to andesitic lavas dominate the southern exposures of the Nosoni and the younger Dekkas (Albers and Robertson, 1961). More distal facies, such as tuffaceous debris flows and other volcanogenic sediment gravity flows, predominate in northern exposures and are inferred to have been deposited in a more basinal setting, although lavas are also present in the section. The section is believed to span most of Leonardian time. In the southern region of its exposure, the Dekkas is overlain by keratophyric tuffs and flows of the Bully Hill "Rhyolite" and is intruded by the Pit River stock, which is inferred to be cogenetic with the Bully Hill (Albers and Robertson, 1961). The Pit River stock has yielded a Late Permian radiometric age of 261 Ma (Pb-Pb from discordant zircon, Fraticelli and others, 1985). The Bollibokka Group is now known to be entirely Permian in age because the uppermost Dekkas contains Late Permian (Guadalupian) corals and fusulinids and early Late Permian (Wordian) brachiopods (T. Dutro, personal communication, 1988) (Fig. 5).

As described by Stevens and others (1987) and Miller and Wright (1987), a Tethyan waagenophyllid coral occurs in the Dekkas Formation. Fusulinids in beds stratigraphically in the same part of the section, however, do not belong to the Tethyan family Verbeekinidae, but are instead more typical of the McCloud faunal association.

In summary, the Upper Permian rock record of the eastern Klamath Mountains indicates that Leonardian uplift and tilting were succeeded by voluminous island-arc volcanism during the early Guadalupian.

Northwestern Nevada. In northwestern Nevada, at Quinn River Crossing (Fig. 5; Plate 4G), Lower Permian rocks are discordantly overlain by silicified limestone, dolomite, and bedded radiolarian and tuffaceous chert that are Late Permian (Guadalupian) in age (Ketner and Wardlaw, 1981). The basal contact may be structural (Ketner and Wardlaw, 1981; A. E. Jones, unpublished data) or an angular unconformity (A. E. Jones, unpublished data). These Upper Permian rocks are paraconformably overlain by Triassic rocks, and the uppermost Permian and perhaps part of the lower Triassic are apparently missing (B. R. Wardlaw, personal communication, 1989).

Central Oregon and northwestern Washington. In central Oregon, no Upper Permian rocks are recognized in stratigraphic continuity with the Lower Permian Coyote Butte Limestone, which is unconformably overlain by Triassic conglomerate. Murchey and others (1983) and Silberling and Jones (1987) have reported that Upper Permian and Lower Triassic chert exists in the Grindstone terrane, but the relation of these rocks to the upper Paleozoic Coyotte Butte Formation is unclear.

Upper Permian and Triassic volcanic rocks are present to the north of the Grindstone terrane in the Seven Devils arc (Plate 4G). The lower part of the section, the undated Windy Ridge Formation and the overlying Hunsaker Creek Formation, is composed of conglomerate, volcanic sandstone, keratophyric water-laid tuff, mafic breccia, and spilitized flows (Fig. 5) (Vallier, 1977) and is much like the Guadalupian Bollibokka Group in the eastern Klamath Mountains. The Hunsaker Creek Formation contains brachiopods that may be as old as late Early Permian (Leonardian), but which are probably Late Permian (Guadalupian) in age (Vallier, 1977). These rocks are overlain by volcanic and volcaniclastic rocks of Triassic age. The Sparta ophiolite has been interpreted as the deep-seated basement for the Permian and Triassic succession of the Seven Devils arc (Phelps and Ave L'Allemant, 1980). The various terranes within the Blue Mountains are thought by some to record the evolution of a single convergent plate margin (e.g., Pessagno and Blome, 1986).

No Upper Permian strata have been dated in the Chilliwack terrane, but Lower Permian limestone is overlain by undated volcanic rocks. These are disconformably overlain by Upper Triassic argillite and siltstone (Fig. 5) (Monger, 1977).

In summary, within the island-arc terranes, the Late Permian was characterized by voluminous andesitic and basaltic magmatism that persisted from the late Early Permian. Development of thick, basinal pyroclastic aprons; reworking of shallow-water faunas into a deeper setting; and in the upper parts of some sections, deposition of shallow-water carbonate rocks and spiculitic chert, are all locally characteristic of the mid-Permian volcanic succession. Most sections are dated by Late Permian or Guadalupian fossils. Where better age resolution exists, volcanism appears to have been restricted to the Wordian (early to mid-Guadalupian or earliest Late Permian) time. Much of the uppermost Permian is locally missing, although in places (e.g., Quinn River Crossing, most of the northern Sierra Nevada, and eastern Klamath Mountains terranes) no angular discordance is present.

PERMO-TRIASSIC

Introduction

Many fundamental changes took place along the outer edge of the continental shelf in latest Permian time (Plate 4H). At this time, sedimentary rocks of the Havallah basin were thrust onto the shelf along the Golconda thrust during the Sonoma orogeny or Sonoman event (Gabrielse and others, 1983). Autochthonous shelf strata in southern California document a more complex sequence of Permo-Triassic events that includes polyphase deformation, greenschist- to amphibolite-facies metamorphism, and intrusion in a continental or Andean arc setting (Plate 4H). Although these events brought an end to marine sedimentation in many places and had an important effect on the evolving paleogeography of the margin, crustal shortening was short-lived. By late Early to Middle Triassic, marine sediments began to accumulate in regions of renewed subsidence (Saleeby and others, this volume).

Some portions of the McCloud belt island-arc record ongoing volcanism from Late Permian through Early Triassic time, but more commonly, rocks that span the Permo-Triassic transition are missing along an unconformity. Typically, Upper Triassic platform limestone or, in places, shale (Chilliwack Group) overlies Upper Permian or older volcanic rocks along an angular unconformity.

Younger Triassic to Jurassic marine deposits in western Nevada, together with Cenozoic cover, obscure the exact structural relations between Permian to Triassic island-arc sequences in the northern Sierra Nevada, Klamath Mountains, Black Rock Desert region and the deformed rocks of the Golconda allochthon. However, the influx and nature of arc-derived detritus into the Havallah basin suggest that the McCloud island arc formed the western paleogeographic margin of the Havallah basin in the Permian.

Scattered exposures of a once more continuous subduction zone complex lie to the west of exposures of island-arc sequences (Plate 4H). Blueschist minerals in this belt generally have minimum metamorphic ages (K-Ar) only as old as Late Triassic, but locally are as old as Early Permian (Rb/Sr). Arguments presented below suggest that deformation within this subduction-zone complex was likely ongoing in Permo-Triassic time,

synchronous with magmatism in the McCloud belt and deformation farther east in the Havallah basin, providing important constraints on the polarity of subduction along the North American plate margin during this time span.

In this final section, we discuss only the three main belts of rocks involved in Permo-Triassic orogenesis: the Golconda allochthon and the outer edge of the continental shelf, the McCloud belt island arc, and the subduction-zone complex. Discussion of events farther east on the shelf and craton beginning in the Early Triassic will be treated in detail in the subsequent chapter (Saleeby and others, this volume).

The Golconda allochthon

The Golconda allochthon extends from the Snake River Plain in northern Nevada to the eastern Sierra Nevada, where it has been recently recognized by Schweickert and Lahren (1987; Plate 4H). Its continuation southward from the eastern Sierra Nevada is unclear. Sequences in the western and northern Mojave (e.g., El Paso Mountains) are somewhat similar to those in the Golconda allochthon, but their style of deformation is different from that which characterizes the Golconda allochthon.

Rocks in the allochthon range from latest Devonian to Late Permian in age and have been discussed in previous sections and summarized in Figure 4. In its northern part, the Golconda allochthon consists mostly of sedimentary rocks and lesser greenstone. In the Mina-Candelaria region, a thin selvage of sedimentary rocks constituting the Golconda allochthon is overthrust by Permian volcanic and volcanogenic sedimentary rocks of the Mina and Black Dyke Formations (Speed, 1977b). Speed (1977a) attributed the great change in the width of the allochthon southward to subduction or underthrusting of the allochthon beneath an overriding island-arc edifice (represented by the Mina and Black Dyke Formations). Alternatively, the decreasing width of the allochthon southward, together with increased evidence for interfingering of Permian shelf and arc-derived detritus in this direction may suggest that the present map geometry of the Golconda allochthon is a function of a basin that originally narrowed southward.

Tight folding and tectonic dismemberment characterize the structure of the allochthon at the outcrop scale, and closely spaced thrust faults characterize the map-scale structure of the Golconda allochthon (MacMillan, 1972; Turner, 1982; Miller and others, 1982, 1984; Snyder and Brueckner, 1983; Babaie, 1984; Brueckner and Snyder, 1985). An eastward direction of tectonic transport derived from mesoscopic structural data is remarkably consistent over entire ranges (Plate 4H). In most of the Golconda allochthon, deformation occurred under fairly low-temperature conditions. Radiolarians in chert are generally well preserved, and conodonts have low color-alteration indices. Although tight to isoclinal mesoscopic folds are abundant in the more thinly bedded units of the allochthon, interbedded shales do not, in general, possess even a weak slaty cleavage. In the central to southern part of the allochthon, however, Babaie (1987) and Speed (1984) reported that the formation of a slaty cleavage accompanied east-vergent folding in the allochthon, indicating higher temperatures during deformation.

The minimum amount of displacement along the basal Golconda thrust, measured between the easternmost exposures of the upper plate and the westernmost exposures of lower plate rocks, is about 50 km, but this figure has been increased somewhat by Basin and Range extension. If one includes the exposures in the Monitor Range (Plate 4H), the distance is about 72 km (Stewart, 1980b). The amount of displacement represented by the imbricate thrust faults within the allochthon itself is unknown. Where detailed mapping and biostratigraphic studies have been carried out, it has been shown that closely spaced thrust faults repeat relatively thin (several hundreds of meters) pelagic and clastic sequences many times (Stewart and others, 1977, 1986; Miller and others, 1984). However, the three-dimensional control on the map-scale geometry of thrust faults is generally lacking, as are distinctive units and facies changes that would enable one to restore the basinal succession with any accuracy.

In the northern exposures of the allochthon, the youngest strata involved in folding and thrusting are Leonardian in age (Miller and others, 1984). The conspicuous absence of younger units likely indicates that deformation of this part of the basin had begun prior to Late Permian time. The youngest overridden autochthonous deposits are Late Permian in age, indicating that actual emplacement of the allochthon occurred in latest Permian or younger time. In the central part of the allochthon, youngest strata are generally late Leonardian to Guadalupian (Miller and others, 1982; Murchey, 1982; Murchey and Harwood, 1986; Stewart and others, 1986), and in the southern reaches of the allochthon, they are generally Guadalupian or younger in age (Speed, 1977b). Early early Triassic conodonts have been reported from the Golconda allochthon in the Candelaria Hills (Speed, 1984) and from exposures mapped as part of the Golconda allochthon by Babaie (1984) and Poole and Wardlaw (1978). The southern progression in the age of youngest units involved in deformation may reflect diachronous deformation that was younger southward in the allochthon, but the data also suggest the possibility that deformation at any latitude may have occurred throughout the interval of time spanning mid–Late Permian to earliest Triassic.

On structural grounds, Brueckner and Snyder (1985) and Snyder and Brueckner (1983) suggest alternatively that the history of deformation in the allochthon was a more protracted one that took place *throughout* the entire late Paleozoic. They noted that strata of all ages in the Golconda allochthon are similarly deformed and appear to have been soft sediments (i.e., had not undergone complete diagenesis to quartz chert) at the time of their deformation. They then reasoned that deformation must have been continuous or intermittent throughout the time span of deposition of strata in the allochthon, because diagenesis of siliceous sediments occurs relatively quickly (within a few tens of millions of years) but deposition of the sequence spanned more than a hundred million years. Their interpretation is not supported by detailed mapping of the Golconda allochthon in the

Independence Mountains and Tobin Range, Nevada, where deformation can be shown to entirely post-date deposition of stratified rock sequences that range from Late Devonian to Early Permian in age (Tomlinson and others, 1987; Miller and others, 1984). Whether or not this is the case everywhere in the Golconda allochthon should be resolved when detailed structural studies such as those of Brueckner and Snyder (1985) are coupled with more complete stratigraphic and biostratigraphic data.

Deformed rocks of the Golconda allochthon as young as late Early Permian are unconformably overlain by the dominantly volcanic Triassic Koipato Group in the Tobin, Sonoma, East, and Humboldt ranges of north-central Nevada (Plate 4H), which is succeeded by the shelf-facies Star Peak Group. The upper part of the Koipato Group has been dated as late Early Triassic (Spathian) (Silberling, 1973). Because no regionally extensive foredeep basin was created east of the Golconda allochthon at this latitude, and no Koipato-equivalent rocks have been documented to the east of the allochthon, the presumed Permo-Triassic age of emplacement of the Golconda allochthon has been questioned. Ketner (1984) proposed that the Golconda allochthon and unconformably overlying Mesozoic strata may have been carried eastward along the basal thrust of the Golconda allochthon at a younger time. Several lines of reasoning, however, argue in favor of a Permo-Triassic age of emplacement and against a significantly younger, Mesozoic age of emplacement.

1. Triassic to Jurassic marine sequences were deposited both east of and depositionally above the Golconda allochthon; if the Golconda allochthon had indeed been emplaced at a subsequent time, Triassic and Jurassic strata would be present locally beneath the Golconda thrust, but they are not.

2. Although Triassic facies are not easily matched from east to west across the Golconda allochthon, Triassic basinal rocks exposed to the west of the Golconda allochthon had source areas on the craton in the Colorado Plateau region (Lupe and Silberling, 1985). The authors reasoned that the allochthon at least must have been close by, if not in its present position, by late Triassic time.

Evidence for an Early Triassic age of emplacement of the Golconda allochthon comes from the Mina-Candelaria region (Plate 4H) (Speed, 1977). There, autochthonous Permian strata of the overlap sequence are conformably overlain by fossiliferous, lowest Triassic, hemipelagic sedimentary rocks that, in turn, are overlain by Lower Triassic submarine-fan deposits derived from the encroaching allochthon. This sequence is then overthrust by the allochthon (Pickhandle allochthon and structurally higher Mina Formation), which contains Upper Permian volcanogenic and volcanic material that in part supplied the sediments deposited on the autochthon (Speed, 1977b).

Schweickert and Lahren (1987) have proposed that the Golconda thrust extends into the eastern Sierra Nevada, California. In the Saddlebag Lake pendant, metasedimentary and metaigneous rocks are correlated to the Havallah sequence on the basis of lithology. The rocks in question are in structural contact above fossiliferous Permian strata and rocks correlated to the Triassic Candelaria Formation described by Speed (1984). The presumed Havallah-equivalent rocks consist of harzburgitic ultramafic tectonite and metagabbro, pillow lava, chert, and argillite.

East of the Sierra Nevada, a middle Permian angular unconformity (Stone and Stevens, 1988b) that could be equivalent to an Early Permian phase of the Sonoma orogeny is present, but Upper Permian rocks are overlain paraconformably by Lower Triassic strata, suggesting that there may be little deformation here that is contemporaneous with the Permo-Triassic emplacement of the Golconda allochthon farther north (Plate 4H).

In the El Paso Mountains, north of the Garlock fault in the Mojave Desert (Plate 4H), mid-Permian (Leonardian-Guadalupian) sequences document the inception of arc volcanism adjacent to the continental margin (Carr and others, 1984), similar to relations inferred from the southern part of the Golconda allochthon. The deformational history of the El Paso Mountains, however, differs from that of the Golconda allochthon in that at least final deformation occurred under upper-greenschist-grade conditions during intrusion of syn- to post-kinematic plutons dated at 252 to 249 Ma (^{40}Ar-^{39}Ar hornblende; U-Pb zircon) (Carr and others, 1984; Cox and Morton, 1980). Folds and thrust faults in the El Paso Mountains and correlative rocks in the northern Mojave are north-northwest trending and are vergent to the west, as compared to characteristically east-vergent structures in the Golconda allochthon (Christiansen, 1961; Miller and Sutter, 1982).

Miogeoclinal strata in the central Mojave Desert were affected by a similar deformational event (Plate 4H). Greenschist- to amphibolite-facies metamorphism, polyphase deformation, and thrust-faulting pre-dated or were synchronous with plutons dated at 241 Ma (U-Pb zircon; Wright, unpublished data) (Miller, 1981; Cameron, 1981; Miller and Cameron, 1982). Deformation occurred in a continental-arc setting characterized by high heat flow (Cameron, 1981). The deformed and intruded rocks are unconformably overlain by the Triassic Fairview Valley Formation from which Spathian conodonts have been reported (E. L. Miller, 1978, 1981; Walker, 1985; Walker and others, 1984).

Considering our inability to compare radiometric and paleontologic ages close to the Permo-Triassic boundary (e.g., Harland and others, 1982), the present data permit the interpretation that compressive deformation related to the Sonoma orogeny may have occurred at about the same time along the length of the continental margin, and that deformation may have spanned the interval from mid-Permian to earliest Triassic time.

No known correlatives to the Golconda thrust have been mapped in Idaho or northeast Washington. However, in southern British Columbia, rocks of the Eastern assemblage of Monger (1977) (Fig. 2) were thrust eastward along the Stubbs fault in Permo-Triassic time, and the fault is intruded by a Late Permian pluton (Klepacki, 1984). The allochthonous rocks (Kaslo) are overlain unconformably by Upper Triassic turbidites of the Slocan Group (Klepacki, 1984, 1985) (Plate 4H).

Island-arc sequences

In part of the northern Sierra Nevada, Permian volcanic rocks of the Reeve Formation are overlain by Upper Triassic limestone across a locally angular unconformity (Harwood, 1983, 1988). Elsewhere, Permian volcanic rocks are overlain disconformably by Lower Jurassic rocks (Schweickert, 1981).

In the eastern Klamath Mountains, the lower part of the Pit Formation may be as old as Early Triassic in places, but the oldest confirmed ages from the base of the Formation are late Middle Triassic (Albers and Robertson, 1961; summary in Miller and Harwood, 1990). Interfingering mafic and intermediate lava flows and waterlaid pyroclastic rocks record ongoing but waning volcanism as shallow-water fauna and volcanogenic facies give way up section to more basinal, turbidite-dominated deposition of clastic and radiolarian-bearing hemipelagic sedimentary rocks (Curtis, 1983). No single continuous section has been documented across the Permo-Triassic boundary in the eastern Klamath Mountains. Until recently, however, many workers inferred from general concordance of units and representation of major epochs that deposition was continuous during this time interval (e.g., Albers and Robertson, 1961; Silberling and Jones, 1982; Miller, 1987; Renne and Scott, 1988). Improved resolution of biostratigraphic ages now suggests that much of the Late Permian (Late Guadalupian and younger) and some of the Triassic (perhaps all of the Early Triassic and the early Middle Triassic) are missing (Miller and Harwood, 1990).

At Quinn River Crossing in the Black Rock Desert region of northwest Nevada, Upper Permian limestone and chert are apparently concordantly overlain by tuffaceous chert, shale, argillite, and sandstone, which yield late Middle Triassic pelecypods and ammonites (Ketner and Wardlaw, 1981). The upper Upper Permian, Lower Triassic, and lower Middle Triassic may be represented by the several meters of unfossiliferous intervening rocks. Alternatively, and more likely, a disconformity is concealed within the section, and rocks of those ages are missing.

In the Grindstone terrane of eastern Oregon, the "Coyote Butte Limestone," a block within an olistostromal deposit, is unconformably overlain by Triassic(?), chert-rich conglomerate that includes clasts of underlying lithologies, Permian and Triassic chert and argillite form the matrix of the olistostrome (Blome and Nestell, 1988). Together, these relations suggest that formation of the olistostrome deposit ended during Triassic time.

Upper Permian volcanic rocks of the Seven Devils Arc are unconformably overlain by volcanic and volcaniclastic rocks of the Wild Sheep Creek and Doyle Formations of Ladinian to Carnian age (late Middle to early Late Triassic). Upper Triassic (Carnian to Norian) platformal limestone of the Martin Bridge Formation overlies these rocks (Vallier, 1977; Follo and Seiver, 1985). In the Chilliwack Group of Washington and British Columbia, Upper Permian volcanogenic rocks are disconformably overlain by Upper Triassic argillite and siltstone (Monger, 1977).

Cache Creek–affinity subduction-zone complex

In British Columbia, the Cache Creek terrane contains a mixture of rocks, including serpentinized peridotite, tholeiitic and alkalic basalts, chert, argillite, and limestone, that are generally interpreted as the remnants of ocean floor, atoll, or oceanic platform sedimentary assemblages deposited on oceanic crust. These oceanic sedimentary sequences range in age from Paleozoic to Late Triassic (Monger, 1984). One of the most distinctive elements of the Cache Creek terrane is the occurrence of conspicuous limestone blocks ranging in age from Carboniferous to Late Triassic that are commonly associated with alkalic pillow lavas. This association of rock types led Monger (1975, 1977) to interpret the limestones as reef deposits that accumulated upon intraplate paleo-Pacific ocean-floor seamounts. The Permian limestone blocks commonly contain a Tethyan verbeekinid fusulinid fauna that most workers have interpreted to represent warm, shallow-water equatorial forms distinct from age-equivalent North American forms and transitional McCloud fauna (Stevens, 1985; Ross and Rsos, 1981, 1983; Danner, 1977). The lithology of the Cache Creek terrane, together with a style of deformation marked by severe structural disruption typical of melange and broken formation, and the occurrence of high-pressure, low-temperature, blueschist-facies prograde metamorphic mineral assemblages in some areas, has led to the interpretation that the Cache Creek terrane represents an accretionary complex formed in the hanging wall of a subduction zone (e.g., Davis and others, 1978; Monger, 1984; Hamilton, 1978). Allochthonous and potentially far-travelled fragments of paleo-Pacific Ocean lithosphere and its hemipelagic cover, along with seamounts and their reefal carbonate deposits, were tectonically accreted and assembled within the accretionary subduction complex represented by the Cache Creek terrane. A summary of available geochronologic data by Davis and others (1978) consists principally of K-Ar dates on white mica, and indicates minimum or cooling ages of about 220 to 225 Ma for high-pressure metamorphism. These data indicate that subduction was underway and conditions for high-P/low-T metamorphism were well established prior to uplift at 220 to 225 Ma.

Extending from southern British Columbia into Washington, Idaho, Oregon, and California are several terranes that have lithologic, structural, and paleontological affinities to the Cache Creek terrane (Plate 4H), including the Vedder Complex (British Columbia and Washington); the central mélange terrane (Idaho and Oregon); the eastern Hayfork, North Fork–Salmon River, Yreka–Fort Jones, and Stuart Fork terranes (Klamath Mountains, California); and the Calaveras Complex (Sierran foothills, California). Following Saleeby (1983), we refer to these as terranes of Cache Creek affinity.

The Vedder Complex is exposed as small tectonic slivers within the North Cascades of Washington and in southern British Columbia on Vedder Mountain. The complex consists of pelitic

and quartzose sedimentary rocks (metachert?) and mafic igneous rocks, metamorphosed to mica schists and amphibolites under high-P/low-T conditions (Armstrong and others, 1983). These rocks are interpreted to represent oceanic assemblages that were metamorphosed within a subduction-zone environment during late Paleozoic time (Armstrong and others, 1983). Rocks from the Vedder complex have yielded metamorphic ages that range from Early Permian to Late Triassic (Rb/Sr 229 to 285 Ma; K-Ar 219 to 279 Ma), with the oldest white mica ages being 242 Ma (Fig. 5).

Within the Blue Mountains region of eastern Oregon and western Idaho, a belt of melange (central melange terrane of Dickinson and Thayer, 1978; dismembered oceanic crust terrane of Brooks and Vallier, 1978) separates the upper Paleozoic and lower Mesozoic Seven Devils arc terrane from the early Mesozoic Huntington arc terrane. The central melange terrane contains elements of Paleozoic oceanic lithosphere such as the Canyon Mountain complex, Mesozoic and Paleozoic(?) hemipelagic sedimentary rocks and volcanic rocks that may belong to island-arc magma series, and limestone blocks and pods that range in age from Devonian to Middle Triassic (Brooks and Vallier, 1978; Dickinson and Thayer, 1978). Permian limestone blocks or clasts in the central melange terrane belong to both the Tethyan and McCloud faunal provinces (Nestell, 1983). The melange locally includes blueschist-facies metamorphic rocks that have K-Ar cooling ages of about 225 Ma (Hotz and others, 1977). Although this terrane has had a complicated history that is not fully understood, the main stage of melange development preceded the deposition of Upper Triassic (Carnian and Norian) sedimentary roks (Dickinson and Thayer, 1978), and blueschist-facies metamorphism indicates that plate convergence was underway prior to 225 Ma in this general area.

In the Klamath Mountain province of northern California, Wright (1982) interpreted the eastern Hayfork, Stuart Fork, Yreka–Fort Jones, and North Fork terranes as related elements of a late Paleozoic and early Mesozoic accretionary subduction complex. Rocks originally included within the North Fork terrane have been subsequently divided into the structurally lower Salmon River terrane composed principally of Permian oceanic lithosphere (Ando and others, 1983), and the North Fork terrane, which now refers to structurally overlying metasedimentary strata (e.g., Mortimer, 1985). Collectively, these terranes include Permian oceanic crust and its overlying hemipelagic cover (Salmon River–North Fork terranes), chert-argillite and mafic volcanic sequences that have been metamorphosed to the blueschist facies (Stuart Fork–Yreka–Fort Jones terranes), and a melange containing disrupted hemipelagic and epiclastic rocks with associated blocks of shallow-water limestone, metachert, and glaucophane schist (eastern Hayfork terrane). Dated chert sequences in the melange range in age from Permian to Late Triassic (Irwin and others, 1978; Irwin and Wardlaw, 1983); the limestone blocks are Late Permian in age (Luken, 1985). Several of the limestone blocks contain a Permian Tethyan fauna (Nestell and others, 1981; Luken, 1985) and have been interpreted as shallow-water reef buildups on seamounts (e.g., Wright, 1982; Mortimer, 1985; Luken, 1985; Stevens and others, 1987). K-Ar cooling ages on blueschist-facies white micas of the Stuart Fork–Yreka–Fort Jones terrane cluster around 220 Ma (Hotz and others, 1977). As many workers have pointed out, the parallels with the Cache Creek terrane of British Columbia are striking, and have led to the inference that the Klamath terranes developed within a similar convergent-margin setting.

In the Sierra Nevada foothills region, the Calaveras Complex may represent a southern extension of this subduction-related belt. Confusion in the literature concerning the distribution, age, and tectonic significance of "Calaveras"-type rocks led Schweickert and others (1977) and Schweickert (1981) to restrict the name "Calaveras Complex" to metamorphic rocks that occur east of the Melones fault zone between 37° and 39°N. Thus, he excluded rocks west of the Melones fault (see Schweickert, 1981, Fig. 5-1) and rocks that are known to be Mesozoic in age from the Calaveras Complex. However, lithologic ages are poorly constrained. Burchfiel and Davis (1981) and Saleeby (1981, 1983) retain a broader definition of Calaveras Complex, and have assigned a Cache Creek affinity to these rocks (Saleeby, 1983). Plate 4H shows the distribution of rocks fitting the broader definition. In this context, the Calaveras Complex contains upper Paleozoic and lower Mesozoic rocks that consist of disrupted sedimentary (chert and argillite) sequences, exotic blocks of limestone with Permian Tethyan faunal affinity (Saleeby, 1983), and Paleozoic olistostromal deposits that contain clasts of McCloud faunal affinity (Watkins and others, 1987). Based on faunal assemblages present, it is realized that some of these elements may have formed in areas far removed from one another during the Permian. Ultramafic and mafic complexes representing fragments of oceanic lithosphere are also present, and range in age from Carboniferous to Early Jurassic (Saleeby, 1981, this volume; Saleeby and others, this volume). Blueschist-facies metamorphic rocks yield disturbed K-Ar ages that are Middle Jurassic, but initial cooling of these rocks may be older (Schweickert and others, 1980).

Collectively, components of this belt of rocks, which stretches from British Columbia to California, share several characteristics. They are composed of Mesozoic and locally Paleozoic chert argillite melange and broken formation, in places contain demonstrably far-travelled oceanic components, such as exotic limestone blocks that contain a Permian Tethyan fauna, and have in places undergone a high-P/low-T subduction-related metamorphism. Although most of the metamorphic ages obtained from blueschists of this belt are Late Triassic (K-Ar), ages as old as Early Permian have also been reported (Rb/Sr; Armstrong and others, 1983). The current age data suggest that this subduction complex was well established prior to Late Triassic time and continued its development at a convergent plate margin into the Late Triassic following the Sonoma orogeny. The location of this belt west of the McCloud belt island-arc assemblage rocks and olistostromal deposits with blocks of McCloud limestone suggest a genetic and paleogeographic link between subduction-zone

complexes of Cache Creek affinity and Late Permian to Triassic island-arc volcanism in the McCloud belt (Burchfiel and Davis, 1972; Wright, 1982; Miller and Wright, 1987; Watkins and others, 1987).

Plate tectonic setting of the Permo-Triassic Sonoma orogeny

Most discussions of the Paleozoic history of the western U.S. Cordillera have pointed out the similarities between the Antler orogeny, with the emplacement of the Roberts Mountains allochthon, and the Sonoma orogeny, which led to the emplacement of the Golconda allochthon. With accumulating data, the similarities between these two events seem even more marked. Thus, the most common questions concerning the nature of Permo-Triassic deformation along the continental margin are similar to those posed by older Antler-age events: (1) Was the Havallah sequence deposited in a back-arc basin? (2) Did deformation of the Havallah sequence occur in a back-arc setting with respect to an oceanward, west-facing island-arc system? (3) If so, was there a flip in subduction polarity during the Sonoma orogeny? (4) Or was deformation caused by collision of a perhaps farther traveled east-facing island-arc system with the edge of the continent? Burchfiel and Davis (1972, 1975) and Silberling (1973) first proposed that the deformation of the Golconda allochthon resulted from the closure of a back-arc basin, with limited subduction of underlying oceanic crust. Speed (1979) proposed that the Golconda allochthon represented an accretionary prism of scraped-up paleo-Pacific ocean-floor sediments developed in front of an east-facing arc, "Sonomia," that collided with the continent. The paired Cache Creek and McCloud island-arc sequences have generally been cited in favor of the back-arc basin closure hypothesis. The east-vergent folds and thrust faults in the Golconda allochthon and their similarity to the internal structure of modern subduction-zone complexes have often been cited in favor of the second hypothesis, although it represents the expected structural style of deformation in both cases.

In addressing the first question above, stratigraphic data suggest that the Havallah basin began to form by rifting in the early Late Mississippian (Meramec). Based on relations in the Independence Mountains, Nevada, it appears that this event may have removed an Antler-age island-arc system from proximity to the margin (Whiteford, 1990; Miller and others, 1984). Basaltic volcanism and possibly rifting may have occurred intermittently in the history of the Havallah basin, as evidenced by lavas interstratified with chert and argillite of Mississippian, Pennsylvanian, and Permian age. The basaltic magmatism and possibly intermittent history of extension recorded by the Havallah sequence are compatible with formation of the basin in a back-arc setting, but the data do not place constraints on the width or size of the basin.

The stratigraphic successions preserved in thrust-bound packages in the Golconda allochthon are more helpful in establishing the paleogeographic setting and dimensions of the Havallah basin. Where detailed studies have been carried out, it appears that basaltic lavas overlain by pelagic sections are commonly followed with little break in time by clastic sequences derived from the North American shelf. True pelagic sections that span significant portions of the history of the basin are lacking. These data would suggest that the Havallah sequence represents the continentward side of a back-arc basin, but again, the data do not constrain the width of the basin. Island-arc volcanic sources are only represented in Lower Mississippian and Upper Permian strata of the Havallah sequence, and for these two time intervals, a strong case may be made for a two-sided back-arc basin of limited width. During the rest of the late Paleozoic, there are no such constraints on the seaward paleogeography of the basin.

Based on the ages of youngest sedimentary rocks involved in thrust faulting (regardless of structural position in the allochthon), we reason that closure of the Havallah basin began in the mid-Permian to the north, and was perhaps somewhat younger southward in the basin. The inception of deformation within the basin would then be coeval with the major pulse in island-arc magmatism represented by Upper Permian volcanic rocks of the McCloud belt, whose presence adjacent to the Havallah basin is recorded by the nature of debris shed into the basin at this time (Plate 4G, H) (Tomlinson and Wright, 1986; Tomlinson and Wardlaw, 1988).

Within island-arc sequences of the McCloud faunal belt, the Permo-Triassic is marked by a locally angular unconformity indicating that regional uplift and erosion post-dated the Late Permian pulse in volcanism. Such a period of uplift and erosion in the arc might be predicted in plate tectonic models for the emplacement of the Golconda allochthon that involve limited subduction of back-arc basin floor and collision of an arc with the continental margin.

The Cache Creek subduction-related complex of deformed oceanic rocks lies west of upper Paleozoic arc sequences in the Klamath Mountains and Northern Sierra Nevada (Plate 4H) and, as discussed above, contains fragmented portions of oceanic crust, oceanic pelagic sections, and seamount-flanking reef limestone with a Tethyan fauna. These fragmented sections span the same time interval as do sedimentary rocks of the Havallah basin, yet are clearly farther traveled or were deposited in a more distal location with respect to North America. As such, they seem to be better candidates for late Paleozoic paleo-Pacific Ocean floor than are the successions in the Golconda allochthon, and indeed, no Cache Creek belt–like rocks are exposed east of Permian island-arc sequences. Based on these relations, it seems more reasonable to interpret the Cache Creek belt, not the Havallah sequence, as the subduction zone complex for the McCloud belt island-arc sequences. The Havallah basin thus occupied a back-arc setting with respect to this arc during Permian time, compatible with the observation that dual source terranes (arc and continent) are represented by Havallah sequence sedimentary rocks of this age. Farther south along the continental margin, deformation, metamorphism and intrusion of plutons into shelf-facies rocks occurred in Late Permian to earliest Triassic time within an Andean arc setting (Cameron, 1981; Miller and Cameron, 1982; Walker, 1988), dictating the presence of an

east-dipping subduction zone beneath the continental margin at this latitude (Plate 4H).

The proposed paleogeography for the Permo-Triassic (Plate 4H) is similar to the subsequent, better-understood, Triassic–Early Jurassic paleogeography of the Cordilleran margin. During the Triassic–Early Jurassic, eastward subduction occurred beneath the margin, but basinal conditions prevailed in the arc and back-arc region to the north, while deposition along the southern portion of the margin was largely nonmarine to shallow marine (see Saleeby and others, this volume).

SUMMARY

Several contrasting depositional and tectonic settings are represented by upper Paleozoic rocks of the western U.S. Cordillera. Strata deposited on the stable craton and bordering continental shelf constitute the autochthonous portion of the Paleozoic Cordillera and provide an excellent, but only partial, sedimentologic record of the many events that have affected the continental margin during this time. To the west of the Paleozoic continental shelf lay a region of poorly understood paleogeography consisting of offshore basins and volcanic island arcs, perhaps in part underlain by rifted, attenuated continental crust. Basin development by extension and rifting, together with island-arc magmatism, occurred intermittently throughout the Paleozoic. Deformation related to crustal shortening affected offshore basins and the edge of the continental shelf twice during the Paleozoic, in the earliest Mississippian and latest Permian–earliest Triassic. These episodes of shortening resulted in the thrust emplacement of basinal rocks onto the shelf as the Roberts Mountains and Golconda allochthons, respectively. Upper Paleozoic rocks in the westernmost part of the U.S. Cordillera form two contrasting but parallel belts. The eastern belt consists of relatively coherent, continuous sequences of Devonian to Jurassic volcanic and sedimentary strata deposited in an island-arc setting above a basement complex of Devonian and older arc and orogenic terranes. The western of the two belts contains allochthonous masses of Paleozoic limestone, chert, and mafic volcanic rocks inferred to have been accreted in a subduction-zone setting. Blueschist-facies metamorphic minerals yield Triassic minimum ages for subduction, but east-dipping subduction is inferred to have been established at least by Late Permian time, the age of voluminous calcalkaline magmatism in the adjacent arc. These two belts of rocks are represented today by scattered exposures in the northern Sierra Nevada, Klamath Mountains, northwestern Nevada, Oregon, Washington, and British Columbia. Based on their geologic history, tectonic setting, and the faunal affinities of late Paleozoic fossils, both the island-arc sequences and accretionary complexes are believed to have once constituted portions of two more continuous and extensive paleogeographic and structural belts: an active late Paleozoic island-arc system developed within the McCloud biogeographic province, and a subduction-zone complex (or complexes) representing the southerly continuation of the Cache Creek belt of the Canadian Cordillera.

The time-slice approach used in this chapter presents some challenging problems when discussing the evolution of these different depositional and tectonic provinces, particularly since younger tectonism and cover obscure the original paleogeographic relations between several of these belts. We have approached this problem by comparing and contrasting the details of the history of autochthonous North America with the stratigraphic and tectonic history of allochthonous rocks in a time-slice framework. Based on stratigraphic and biostratigraphic studies, it is concluded that the Roberts Mountains and Golconda allochthons (presently in fault contact with continental margin strata) have strong sedimentologic and paleogeographic ties to autochthonous sequences. These rocks are considered an integral part of the continental margin and provide added insight into its overall evolution in late Paleozoic time. In addition, possible paleogeographic ties exist between these units and Paleozoic island-arc sequences generally considered to be more far-traveled with respect to continental margin strata.

From our analysis, we conclude that the rocks of the Roberts Mountains allochthon represent the deposits of a series of lower Paleozoic extensional or transtensional rift basins formed along the outer continental shelf of western North America. The age of formation of these basins may be in most part younger than the presumed initial time of rifting of the continental margin in the late Precambrian or Cambrian (Stewart and Suczek, 1977; Bond and Kominz, 1984). Basaltic volcanism, perhaps associated with rifting or extension, occurred intermittently throughout the history of these basin(s). Two important periods of mafic volcanism, one during the Ordovician and one in the Late Devonian, are best recorded. Minor- and trace-element geochemistry indicate magmatism is alkalic and occurred in an intra-plate setting. The sedimentary rocks deposited in these basin(s) consist of pelagic and hemipelagic deposits interbedded with clastic rocks derived from the continental shelf and/or from uplifted fault blocks within the basin itself. Less deformed analogs to the early Paleozoic basin(s) represented by these rocks exist in the Canadian Cordillera and provide important insight to the depositional setting of strata in the Roberts Mountains allochthon (Fig. 1) (Turner and others, 1989). The formation of lower Paleozoic shelf basins apparently occurred synchronously along the entire length of the Cordillera, suggesting a history more complex than the simple single-stage rifting generally proposed based on studies of the miogeoclinal shelf succession (e.g., Bond and Komintz, 1984). Two alternative tectonic settings are considered for the margin during the early Paleozoic. Episodic extension may reflect a long-lived transtensional transform margin or may reflect intermittent spreading or transtensional tectonics in the back-arc region of a subducting plate boundary.

Evidence in favor of a subducting or convergent margin history for the early Paleozoic Cordillera includes plutonic rocks that intrude the basement terrane of the younger, more coherent and extensive, Devonian to Jurassic island-arc system. These are now known to span the entire interval from the Cambrian to the Devonian, and many yield isotopic evidence for assimilation of

even older crust. Exposures of this basement terrane occur in the central metamorphic belt of the Klamath Mountains and within the Shoo Fly Complex in the northern Sierra Nevada. The complex deformational, metamorphic, and intrusive history of these rocks occurred within an arc and subduction-zone setting prior to Late Devonian time. Ordovician and younger rifting and/or strike-slip faulting, as documented by strata of the Roberts Mountains allochthon, may have separated this active-margin orogenic terrane from close proximity to the shelf. Subsidence of the Devonian and older arc basement occurred during the Late Devonian. At this same time, an episode of mafic magmatism occurred in the Antler basin adjacent to the continental shelf. The Pilot Shale basin was established on the shelf during this same time span. All are thought to be the result of a proposed Cordillera-wide rifting event in the Late Devonian (Frasnian-Famennian).

The contents of the Antler basin were shortly thereafter deformed and thrust onto the edge of the continental shelf in earliest Mississippian time (Kinderhookian). Basin collapse by thrust faulting may have been facilitated by lithospheric heating during the prior episode of rifting. The emplacement of the Roberts Mountains allochthon is well dated by the deposits of the Antler foredeep, which indicate that thrusting lasted for about 10 to 20 m.y. at the most. Volcanism and uplift in the McCloud belt island arc at this time are matched by synchronous deposition of arc-derived debris into a remnant arc or back-arc basin, which was to become wider and eventually become the late Paleozoic Havallah basin. These relations are best explained by back-arc thrusting during a period of increased calc-alkaline magmatism in an arc that lay in close proximity offshore the continental margin. During emplacement of the Roberts Mountains allochthon, overridden continental-margin strata underwent polyphase deformation under greenschist- and higher-grade conditions.

Shortening ended abruptly in early Late Mississippian time with the onset of extension. As tectonic loading by thrusting ceased, strata in the Antler foredeep basin accumulated in gradually shoaling waters. The western reaches of the Antler highlands subsided and were overlapped by basalt flows, limestone, pelagic strata, and clastic turbidite sequences. Basaltic volcanism and extension occurred synchronously in the offshore Havallah basin, as suggested by numerous sections that contain well-dated lower Upper Mississippian pillow lavas and radiolarian chert. The Havallah basin and its northern counterparts in the Canadian Cordillera (the Eastern assemblage) formed as a series of rift basins by continued episodic extension into the late Paleozoic (Struik and Orchard, 1985 (Fig. 2). Sedimentary rocks of clear arc provenance are not present in Upper Mississippian and Pennsylvanian strata of the Havallah basin. If island-arc terranes were present along the western margin of the basin during this time interval, they were likely physically separated by rift basins or bathymetric barriers from those basins that developed more closely adjacent to the continental margin. Within the island-arc sequences of the westernmost Cordillera, continuing but reduced volcanism and the initiation of epiclastic deposition is represented by the Mississippian Bragdon Formation in the eastern Klamath Mountains. Deposition of radiolarian chert of the Peale Formation in the northern Sierra Nevada marks a major break in volcanism during early Carboniferous time, although minor epiclastic debris of provenance similar to that of the Bragdon Formation was locally deposited.

During Pennsylvanian time, the development of the Ancestral Rocky Mountains dramatically changed sedimentation patterns across the cratonic region of the western U.S. High-relief, northwest-trending, fault-bounded highlands began to rise in the Early or Middle Pennsylvanian. Fault-bounded basins contain as much as 6 km of sedimentary fill, and uplift-bounding faults are believed to have moved as steep reverse faults with or without a component of strike-slip motion. Kluth and Coney (1981) proposed that these uplifts formed in response to compressional tectonism along the Ouachita-Marathon margin, and represent an intracratonic response to the collision of North America with South America–Africa. Apparently, faults related to the rise of the Ancestral Rocky Mountains reactivated Precambrian zones of weakness. The northwest-trending Rocky Mountain uplifts become more east-west trending toward the edge of the Cordilleran continental margin. This is believed to be due in part to the reactivation of older east-west–trending structures, such as the Uinta axis, which developed during the late Precambrian rifting of the Cordilleran margin. Simultaneous with the uplift of the Ancestral Rockies, a broad zone of complex faulting and basin formation extended across the entire western portion of the North American shelf, including the older Antler orogenic belt. Faulting is suggested to have formed in response to crustal extension along the Cordilleran margin. The complex transition to extensional faulting reflects tectonic activity along the western margin, which may have acted as a "free-face" tectonic boundary at this time—similar to the Southeast Asian margin today in concert with continental collision with India along the southern boundary of Eurasia (e.g., Molnar and Tapponier, 1975). Although the data are sparse, Pennsylvanian and Permian sequences deposited in the offshore Havallah basin also seem to reflect tectonic instability and possible extension. Limestone turbidites flooded offshore basins, reflecting the subsidence and development of carbonate-shelf conditions on the eroded remnants of the Roberts Mountains allochthon.

Pennsylvanian-Permian island-arc sequences of the McCloud biogeographic belt indicate that shallow-marine to deltaic deposition predominated with local erosion of older parts of the island-arc edifice; however, basinal conditions persisted locally until the middle Pennsylvanian. Active arc volcanism is represented by the Baird Formation in the eastern Klamath Mountains and Spotted Ridge Formation in east-central Oregon.

During Permian time, transgressive sequences were deposited across the continental-shelf region, and during mid- to Late Permian time, phosphatic sediments accumulated in subsiding shelf basins. In the region of the Ancestral Rockies, structural relief was quite subdued. The offshore Havallah basin and the island-arc sequences record a distinct two-part Permian history.

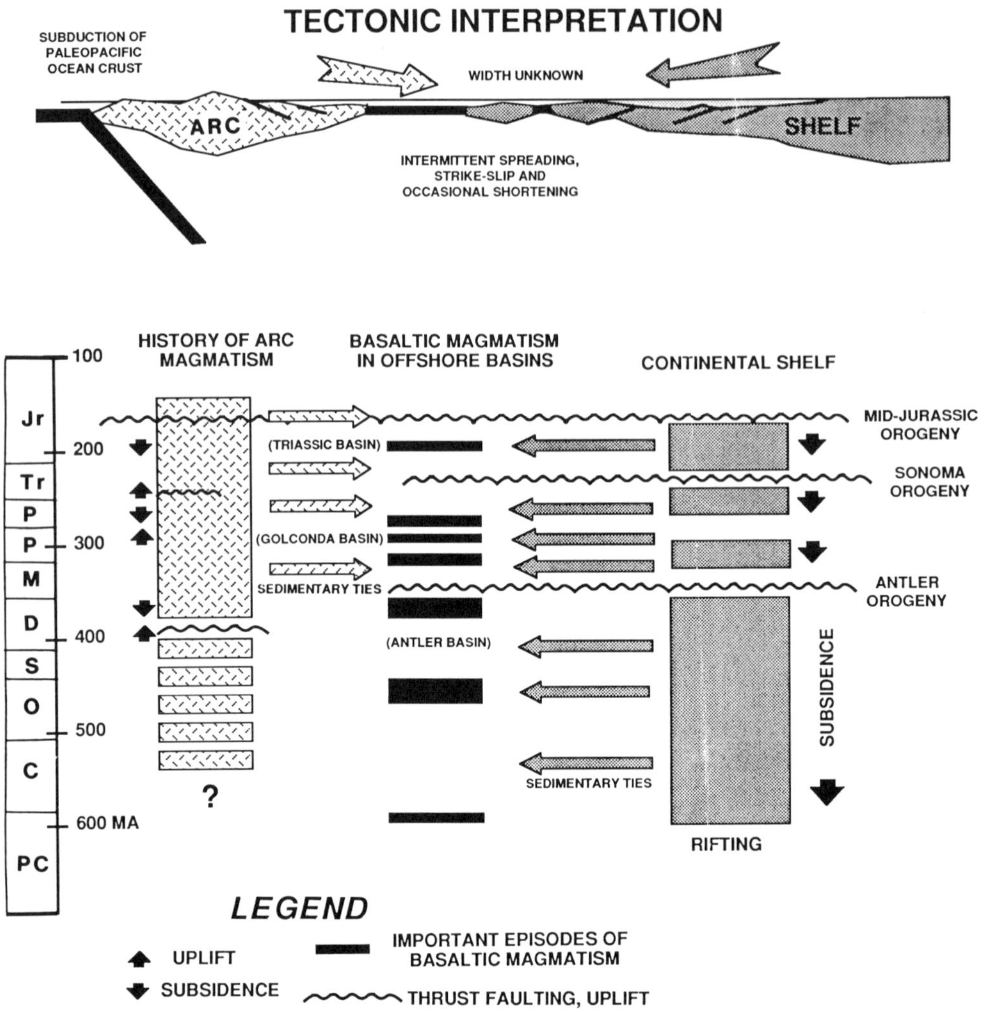

Figure 11. Summary diagram of the timing and nature of tectonism and magmatism in island-arc and offshore basin sequences of the western U.S. Cordillera throughout Paleozoic time and their comparison to events recorded by autochthonous continental margin sequences. The preferred interpretation of these relations is schematically shown. For discussion see text.

In the Havallah basin, basaltic volcanism, subsidence, and sedimentation occurred in the Early Permian. In the arc terrane, waning volcanism and local uplift and subsidence allowed development and growth of shallow-water carbonate banks. These relations are compatible with reduced rates of convergence and possibly significant extension or rifting along the continental margin during the Early Permian. In contrast, mid-Permian time was characterized by increased mafic to silicic volcanism in the arc region and by the influx of abundant arc-derived debris into the Havallah basin. By Late Permian times, the McCloud island arc and Havallah basin constituted parallel paleogeographic elements and, locally, Permian arc-derived debris interfingered with continental shelf-derived sediments. Blueschist-bearing subduction-zone complexes belonging to the Cache Creek belt lie west of the McCloud island arc and contain allochthonous masses of Pennsylvanian and Permian limestone with Tethyan or mixed Tethyan and McCloud faunal assemblages. Based on their faunal affinities, the Tethyan allochthonous masses are thought to be considerably farther traveled with respect to the continental margin than are coeval sequences deposited in the Havallah basin or within the McCloud island arc. The included fragments of late Paleozoic paleo-Pacific floor within these complexes attest to the fact that a vast ocean basin lay to the west of the McCloud island arc.

By Late Permian time, the McCloud island arc moved closer to the continental margin by shortening or subducting the depositional basement of the Havallah sequence, resulting in imbricate thrust faulting of the sedimentary fill of the basin. Shortening of the basinal succession probably began diachronously, possibly in the late Early to mid-Permian in the northern part of the basin, and in the Late Permian in the southern part of the basin. Deformation apparently continued into earliest Triassic time when the allochthon was thrust eastward onto the shelf. Coeval deformation, metamorphism, and intrusion of shelf-facies strata occurred in an Andean arc setting above an east-dipping subduction zone

along the southern part of the margin, as recorded by sequences in the Mojave Desert, California. Permo-Triassic events brought fundamental changes to the paleogeography of the margin. Final destruction of extensive marginal basins resulted in a general configuration of rock units and paleogeography that remained approximately the same throughout the early Mesozoic (Saleeby and others, this volume).

CONCLUSIONS

Some important conclusions regarding the paleogeographic and tectonic setting of the western North American plate margin in the Paleozoic emerge from our compilation. Figure 11 highlights the salient aspects of the tectonic history of the shelf, offshore basins, and island-arc sequences of the western U.S. Cordillera upon which these conclusions are based.

Extensional tectonism, with the formation and modification of shelf basins and offshore regions, seems to have characterized most of the late Paleozoic as well as the early Paleozoic history of the autochthonous and parautochthonous part of the western U.S. continental margin, an impressive time interval spanning about 350 m.y.

Compressional tectonic events along the shelf edge were restricted to two short intervals of about 10 to 20 m.y. duration in the earliest Mississippian and in the Late Permian to earliest Triassic.

Island-arc sequences, although undoubtedly displaced somewhat with respect to their site of origin, are concluded to have developed offshore and/or along strike of the North American plate margin, not at some distant site in the paleo-Pacific Ocean. This reasoning is based on the fact that the arc successions are at least in part built on continentally derived sedimentary rocks and/or plutonic rocks intruded through older, possibly in part Precambrian(?) crust. A long-lived but intermittent history of eastward subduction along the western North American plate boundary in Paleozoic time best explains the history of semicontinuous magmatism and dominantly extensional tectonism punctuated by shorter intervals of shortening. The presence of offshore island arcs is also necessary to account for the occasional westerly derived arc detritus in offshore basin sequences.

The overall nature and timing of Paleozoic events recorded along the western U.S. portion of the margin is similar to the general sequence of events documented along strike of the margin in the Canadian and Alaskan portions of the Cordillera (Fig. 2). This overall similarity suggests that these events may indirectly reflect the nature of larger-scale plate interactions along the western margin of North America. Changing plate motions and rates of convergence along this long-lived paleo-Pacific plate margin were likely responsible for the extensional and compressional events recorded in the overriding North American plate and its offshore systems of Paleozoic basins.

APPENDIX 1. DATA SOURCES FOR PALEOGEOGRAPHIC MAPS (PLATE 4)

Devonian map (Plate 4A)

Volcanic arc and oceanic sequences: distribution of lithologies and ages of rocks in the northern Sierra Nevada from Schweickert (1981), Schweickert and others (1984), Sharp (1988); U-Pb radiometric ages on gneisses intruding the Shoo Fly Complex from Saleeby and others (1987), Hanson and others (1988), Sharp and others (1990), and Merguerian and Schweickert (1987); inset diagram from Saleeby and others (1987). Klamath Mountains data from Albers and Robertson (1961), Cashman (1980), Potter and others (1977), Lanphere (personal communication in Hotz, 1977) for K-Ar ages in northern Klamath Mountains; Quick (1981), Jacobsen and others (1984), Mattinson and Hopson (1972), Lindsley-Griffin (1977), Lindsley-Griffin and Griffin (1983) for Trinity ultramafic body. Rb-Sr data from Lanphere and others (1968) and concordant single zircon fraction from Mule Mt. Stock from Albers and Robertson (1961). Grindstone terrane from Kleweno and Jeffords (1961). Chilliwack Group from Danner (1977). Distribution of Lardeau Group in British Columbia from Tipper and others (1981). Ages of plutons in British Columbia from Okulitch and others (1975), where 372 is upper intercept of chord defined by four zircon fractions with concordia.

Early Mississippian map (Plate 4B)

The craton: compiled from Rose (1976), Sando (1976), Poole and Sandberg (1977), Craig and others (1979), Armstrong and others (1980), Gutschick and others (1980), Smith and Gilmore (1979), and Chronic (1979). *The shelf and antler foredeep:* data in Nevada from Poole and Sandberg (1977), Stewart (1980b); Idaho from Skipp and Hall (1980) and Dover (1980); distribution of lower Paleozoic basinal facies in the Mojave Desert from Carr and others (1984) and Miller and Sutter (1982), in southwestern Nevada from Stewart (1980b), in eastern Sierran roof pendants from Rinehart and Ross (1964), and Nokleberg (1983), and in northeastern Washington from Yates (1976). *The Roberts Mountains allochthon:* distribution of rocks in Nevada from Stewart (1980b); in Idaho from Dover (1980); directions of tectonic transport from unpublished compilation by Madrid; penetrative fabrics from Oldow (1984) in west-central Nevada, Schweickert and others (1984) in eastern Sierra roof pendants, and Madrid and Miller (unpublished data) in central Nevada and northern Nevada. *Offshore Havallah basin:* data from Miller and others (1984). *Volcanic arc and oceanic sequences:* Distribution of rocks in the northern Sierra Nevada from Harwood (1983, 1988), Schweickert and others (1984), D'Allura and others (1977), Durell and D'Allura (1977); eastern Klamath belt from Albers and others (1981), Miller (1989), Miller and Cui (1989); Grindstone terrane from Merriam and Berthiaume (1943).

Late Mississippian map (Plate 4C)

The craton: data compiled from Rose (1976), Sando (1976), Poole and Sandberg (1977), Craig and others (1979), Armstrong and others (1980), Gutschick and others (1980). *The shelf:* data for Antler foredeep from Poole and Sandberg (1977); location of Meramec-age overlap sequences from Poole and Sandberg (1977), Stewart (1980b), Miller and others (1984), Little (1987); Mississippian(?) coral, crinoid and brachiopod-bearing limestone in the eastern Sierran roof pendants from Nokleberg (1983) and Brook and others (1979). *Havallah basin:* data compiled from Miller and others (1984), Stewart and others (1986), Hotz and Willden (1964); for Eastern assemblage (Slide Mountain terrane) in British Columbia from Struik and Orchard (1985). *British Columbia and northeastern Washington:* distribution of "Milford Group" from Tipper and others (1981). *Island-arc and oceanic sequences:* distribution of rocks in the northern Sierra Nevada from Harwood (1983, 1988), Schweickert and others (1984), D'Allura and others (1977), Durrell and D'Allura (1977); eastern Klamath belt from Albers and Robertson (1961), Miller (1989), Miller and Cui (1989), and Watkins (1973); Grindstone terrane from Merriam and Berthiaume (1943).

Pennsylvanian maps (Plate 4D, E)

The craton: data compilation based primarily on maps of McKee and others (1975) modified in Arizona by data from Blakely (1980). *The continental shelf:* distribution and facies of Pennsylvanian rocks from Stewart (1980b) and Rich (1977). Stratigraphy of Pennsylvanian–Permian rocks in the Antler highland region from Stewart (1980b), Dott (1955), Little (1983), Hotz and Willden (1964), Saller (1980), and Madrid (unpublished data) for the Osgood Mountains; Erickson and Marsh (1974), Saller (1980), Madrid (unpublished data), and Roberts (1964) for the Edna Mountain–Battle Mountain region; Muffler (1964), Roberts and others (1967), and Madrid (unpublished data) for the Cortez Range; McKee (1976) for the Toquima Range; Stone and Stevens (1984, 1988a, b) and Stone (1984) for southern California. Skipp and Hall (1980) for Idaho. *The Havallah basin:* compilation of data from Miller and others (1984) and Stewart and others (1986) for Nevada; Carr and others (1980, 1984) for southern California; Rineheart and Ross (1964), Nokleberg (1983) and Tobisch and Fiske (1982) for eastern Sierra Nevada roof pendants; northeastern Washington (Mt. Roberts Formation and correlative rocks) from Yates (1976); distribution of upper Paleozoic to Triassic rocks of the Eastern assemblage (Milford and Kaslo Groups) from Tipper and others (1981). *Volcanic arc and oceanic sequences:* distribution of rocks and lithologies in the northern Sierra Nevada from Harwood (1983, 1988), Schweickert and others (1984). D'Allura and others (1977), and Durrell and D'Allura (1977); eastern Klamath Mountains from Albers and Robertson (1961), Miller (1989), and Watkins (1973); Grindstone terrane from Merriam and Berthiaume (1943) and Buddenhagen (1967); Chilliwack Group from Monger (1977) and Danner (1977).

Permian maps (Plate 4F and G)

The craton: compilation based primarily on the maps of Peterson (1980), modified in Arizona and southern Utah by data from Blakely (1980) and in southern Idaho on the basis of work of Morgan (1980). *The continental shelf:* compilation based primarily on the maps of Peterson (1980) modified in eastern Nevada, California, and western Utah on the basis of the work of Stevens (1977, 1979, 1982), Stone (1984), Stone and Stevens (1984), and Jordan and Douglass (1980). Location of the Antler highlands and overlap assemblage rocks in Nevada from Stewart (1980b); in Idaho from Skipp and Hall (1980). Unconformities in Antler belt from Miller (unpublished data) in the Independence Mountains, northern Nevada; Saller and Dickinson (1982) in the Battle Mountain region; Speed and others (1977) in the Toiyabe Range; Speed and others (1977) in the Mina–Candelaria region; Stone (1984) and Stone and Stevens (1984, 1988a, b) for the Wolfcampian basins in the Inyo Mountains region. *Offshore Havallah basin:* Data compiled from Miller and others (1984); Permian basaltic pillow lavas from Laule and others (1981; Toquima Range, central Nevada) and Sosson (1984; Hot Springs Range, northwestern Nevada). Lower Permian volcanic, metamorphic, and plutonic debris in allochthon from Babaie (1984, 1987) and Tomlinson and Wright (1986) in Toiyabe Range; Miller (unpublished data) in the Antler Peak area. Late Permian arc volcanic rocks from Speed (1977b) in the Mina-Candelaria region; Carr and others (1984) in the El Paso Mountains; rock units and lithologies in the eastern Sierra Nevada roof pendants from Rineheart and Ross (1964), Nokleberg (1983), and Tobisch and Fiske (1982); data for Eastern assemblage in southern British Columbia from Monger (1977), Struik and Orchard (1985), and Klepacki (1984). *Volcanic arc and oceanic sequences:* distribution of rocks and lithologies in the northern Sierra Nevada from Harwood (1983), Schweickert and others (1984), D'Allura and others (1977), Durrell and D'Allura (1977); U-Pb ages in western Sierra Nevada from Sharp and others (1990) and Merguerian and Schweickert (1987). Eastern Klamath belt from Albers and Robertson (1961), Fraticelli and others (1985), Miller (1989), Watkins (1973); northwestern Nevada from Russell (1981) and Ketner and Wardlaw (1981); Grindstone terrane from Merriam and Berthiaume (1943), Buddenhagen (1967), and Wardlaw and others (1982). Central Mélange terrane and Seven Devils terrane of Oregon from Dickinson and Thayer (1978), Vallier and others (1980). Chilliwack Group and Vedder Complex (for location see Fig. 5) from Monger (1977), Danner (1977), Brown and others (1981), and Armstrong and others (1983).

Permo-Triassic map (Plate 4H)

The craton (Guadalupian): compilation based primarily on the maps of Peterson (1980) with modifications based on the work of McKee and others (1967); northern Utah, northeastern Nevada, and southern Idaho modified on the basis of work by Wardlaw and others (1979) and Wardlaw (1980). *The continental shelf:* data compiled from Stone (1984) and Stone and Stevens (1984) for the Inyo Mountain region. Exposures of the Golconda allochthon and Koipato sequence from Stewart (1980b). Thrusting directions in the Golconda allochthon from Miller and others (1984) in the Independence Mountains; Miller and others (1982) in the Antler Peak area; Turner (1982) in the Tobin Range; MacMillan (1972) in the New Pass Range; Babaie (1984) in the Toiyabe Range; Speed (1977b) in the Mina area. Data from eastern Sierra Nevada roof pendants from Rinehart and Ross (1964), Tobisch and Fiske (1982), Nokleberg (1983), and Schweickert and Lahren (1987). Data from the Mojave Desert from Carr and others (1984) in the El Paso Mountains; Miller and Cameron (1982) and Miller and Sutter (1982) in the San Bernardino–Victorville region. Data from Canada and northeastern Washington from the Tectonic Assemblage map of the Canadian Cordillera (Tipper and others, 1981) and from Klepacki (1984). *Volcanic arc and oceanic sequences:* distribution of rocks and lithologies in the northern Sierra from Harwood (1983), Schweickert and others (1984), D'Allura and others (1977), and Durrell and D'Allura (1977); Calaveras Complex from Schweickert (1981); eastern Klamath belt from Albers and Robertson (1961), Miller (1989), and Watkins (1973); northwestern Nevada from Russell (1981) and Ketner and Wardlaw (1981); Grindstone terrane from Merriam and Berthiaume (1943), Buddenhagen (1967), and Wardlaw and others (1982); Central Mélange terrane and Seven Devils of Oregon from Dickinson and Thayer (1978), Vallier and others (1977), Brooks and Vallier (1978), Walker (1983), and Ave Lallemant and others (1980); Chilliwack Group and Vedder Complex from Monger (1977), Danner (1977), Brown and others (1981), Armstrong and others (1983), and Saleeby (1983).

REFERENCES CITED

Abbott, J. G., Gordey, S. P., and Tempelman-Kluit, D. J., 1986, Setting of sedex, sediment hosted Pb-Zn deposits in Yukon and northeastern British Columbia, *in* Morin, J., ed., Mineral deposits of northern Cordillera: Canadian Institute of Mining and Metallurgy Special Volume 37, p. 1–18.

Albers, J. P., and Bain, J.H.C., 1985, Regional setting and new information on some critical geological features of the West Shasta district, California: Economic Geology, v. 80, p. 2072–2091.

Albers, J. P., and Robertson, J. F., 1961, Geology and ore deposits of East Shasta copper-zinc district, Shasta County, California: U.S. Geological Survey Professional 338, 107 p.

Albers, J. P., Kistler, R. W., and Kwak, L., 1981, The Mule Mountain stock; An early Middle Devonian pluton in northern California: Isochron/West, no. 31, p. 17.

Aleinikoff, J. N., Dusel-Bacon, C., and Foster, H., 1986, Geochronology of augen gneiss and related rocks, Yukon–Tanana terrane, east-central Alaska: Geological Society of America Bulletin, v. 97, p. 626–637.

Anderson, T. B., Woodard, G. D., Strathouse, M., and Twitchell, M. K., 1974, Geology of a Late Devonian fossil locality in the Sierra Buttes Formation, Dugan Pond, Sierra City Quadrangle, California: Geological Society of America Abstracts with Programs, v. 6, p. 139.

Ando, C. J., Irwin, W. P., Jones, D. L., and Saleeby, J. B., 1983, The ophiolitic North Fork terrane in the Salmon River region, central Klamath Mountains, California: Geological Society of America Bulletin, v. 94, p. 236–252.

Armstrong, A. K., Mamet, B. L., and Repetski, J. E., 1980, The Mississippian System of New Mexico and southern Arizona, *in* Fouch, T. D., and Magathan, E. R., eds., Paleozoic paleogeography of the west-central United States: Denver, Colorado, Rocky Mountain Section Society of Economic Paleontologists and Mineralogists Rocky Mountain Paleogeography Symposium 1, p. 82–99.

Armstrong, R. L., Harakal, J. E., Brown, E. H., Bernardi, M. L., and Rady, P. M., 1983, Late Paleozoic high-pressure metamorphic rocks in northwestern Washington and southwestern British Columbia; The Vedder Complex: Geological Society of America Bulletin, v. 94, p. 451–458.

Ave Lallement, H. G., Phelps, D. W., and Sutter, J. F., 1980, $^{40}Ar/^{39}Ar$ ages of some pre-Tertiary plutonic and metamorphic rocks of eastern Oregon and their geologic relationships: Geology, v. 8, p. 371–374.

Baars, D. L., and 15 others, 1988, Basins of the Rocky Mountain region, *in* Sloss, L. L., ed., Sedimentary cover–North American craton: U.S.: Boulder, Colorado, Geological Society of America, The Geology of North America, Volume D-2, p. 109–220.

Babaie, H., 1984, Structural and tectonic history of the Golconda allochthon, southern Toiyabe Range, Nevada [Ph.D. thesis]: Evanston, Illinois, Northwestern University, 236 p.

—— , 1987, Paleogeographic and tectonic implications of the Golconda allochthon, southern Toiyabe Range, Nevada: Geological Society of America Bulletin, v. 99, p. 231–243.

Bachman, G. O., 1975, New Mexico, *in* McKee, E. D., and Crosby, E. J., eds., Paleotectonic investigations of the Pennsylvanian System in the United States: U.S. Geological Survey Professional Paper 853, part 1, p. 233–243.

Blakely, R. C., 1980, Pennsylvanian and Early Permian paleogeography, southern Colorado Plateau and vicinity, *in* Fouch, T. D., and Magathan, E. R., eds., Paleozoic paleogeography of the west-central United States: Denver, Colorado, Rocky Mountain Section Society of Economic Paleontologists and Mineralogists Rocky Mountain Paleogeography Symposium 1, p. 239–257.

Blome, C. D., and Nestell, M. K., 1988, Paleontological and tectonic significance of the Grindstone terrane, east-central Oregon: Geological Society of America Abstracts with Programs, v. 20, p. 144.

Blome, C. D., Jones, D. L., and Murchey, B. L., 1983, Paleogeographic implications of radiolarian-rich rocks from eastern Oregon: Geological Society of America Abstracts with Programs, v. 15, p. 371.

Blome, C. D., Jones, D. L., Murchey, B. L., and Liniecki, M., 1986, Geologic implications of radiolarian-bearing Paleozoic and Mesozoic rocks from the Blue Mountains province, eastern Oregon: U.S. Geological Survey Professional Paper 1435, p. 79–101.

Bond, G. C., and Devay, J. C., 1980, Pre-Upper Devonian quartzose sandstones in the Shoo Fly Formation, northern California; Petrology, provenance, and implications for regional tectonics: Journal of Geology, v. 88, p. 285–308.

Bond, G. C., and Kominz, M. A., 1984, Construction of tectonic subsidence curves for the early Paleozoic miogeocline, southern Canadian Rocky Mountains; Implications for subsidence mechanisms, age of breakup, and crustal thinning: Geological Society of America Bulletin, v. 95, p. 155–173.

Boucot, A. J., Dunkle, D. H., Potter, A., Savage, N. M., and Rohr, D., 1974, Middle Devonian orogeny in western North America?; A fish and other fossils: Journal of Geology, v. 82, p. 691–708.

Brook, C. A., Gordon, M., Jr., Mackey, M. J., and Chetelat, G. F., 1979, Fossiliferous upper Paleozoic rocks and their structural setting in the Ritter Range and Saddlebag Lake roof pendants, central Sierra Nevada, California: Geological Society of America Abstracts with Programs, v. 11, p. 71.

Brooks, H. C., and Vallier, T. L., 1978, Mesozoic rocks and tectonic evolution of eastern Oregon and western Idaho, *in* Howell, D. G., and McDougall, K. A., eds., Mesozoic paleogeography of the western United States: Los Angeles, California, Pacific Section Society of Economic Paleontologists and Mineralogists Pacific Coast Paleogeography Symposium 2, p. 133–146.

Brown, E. H., and 6 others, 1981, Metamorphic facies and tectonics in part of the Cascade Range and Puget Lowland of northwestern Washington: Geological Society of America Bulletin, v. 92, p. 170–178.

Brueckner, H. K., and Snyder, W. S., 1985, Structure of the Havallah sequence, Golconda allochthon, Nevada; Evidence for prolonged evolution in an accretionary prism: Geological Society of America Bulletin, v. 96, p. 1113–1130.

Buddenhagen, H. J., 1967, Structure and orogenic history of the southwestern part of the John Day uplift, Oregon: Ore Bin, v. 29, p. 129–138.

Burchfiel, B. C., and G. A., 1972, Structural framework and evolution of the southern part of the Cordilleran orogen, western United States: American Journal of Science, v. 272, p. 97–118.

—— , 1975, Nature and controls of cordilleran orogenesis, western United States; Extensions of an earlier synthesis: American Journal of Science, v. 275-A, p. 363–396.

—— , 1981a, Triassic and Jurassic tectonic evolution of the Klamath Mountains–Sierra Nevada geologic terrane, *in* Ernst, W. G., ed., The geotectonic development of California; Volume Rubey Volume 1: Englewood Cliffs, New Jersey, Prentice-Hall, p. 50–70.

—— , 1981b, Mojave desert and environs, *in* Ernst, W. G., ed., The geotectonic development of California; Rubey Volume 1: Englewood Cliffs, New Jersey, Prentice-Hall, p. 217–255.

Cameron, C. S., 1981, Geology of the Sugarloaf and Delamar Mountains areas, San Bernardino Mountains, California [Ph.D. thesis]: Cambridge, Massachusetts Institute of Technology, 398 p.

Carr, M. D., Poole, F. G., Harris, A. G., and Christiansen, R. L., 1980, Western facies Paleozoic rocks in the Mojave Desert, California, *in* Howard, K. A., and others, eds., Tectonic framework of the Mojave and Sonoran Deserts, California: U.S. Geological Survey Open-File Report 81–503, p. 15–17.

Carr, M. D., Poole, F. G., and Christiansen, R. L., 1984, Pre-Cenozoic geology of the El Paso Mountains, southwestern Great Basin, California; A summary, *in* Lintz, J. Jr., ed., Western geological excursions: Geological Society of America 1984 Annual Meeting Field Trip Guide, Volume 4: Reno, University of Nevada Mackay School of Mines, p. 84–93.

Casey, J. M., 1980, Depositional systems and paleogeographic evolution of the late Paleozoic Taos trough, northern New Mexico, *in* Fouch, T. D., and Magathan, E. R., eds., Paleozoic paleogeography of the west-central United States: Denver, Colorado, Rocky Mountain Section Society of Economic Paleontologists and Mineralogists Rocky Mountain Paleogeography Symposium 1, p. 181–196.

Cashman, S. M., 1980, Devonian metamorphic event in the northeastern Klamath Mountains, California: Geological Society of America Bulletin, v. 91, p. 453–459.

Cecile, M. P., 1982, The lower Paleozoic Misty Creek embayment, Selwyn Basin, Yukon and Northwest Territories: Geological Survey of Canada Bulletin 335, 78 p.

Charvet, J., Lapierre, H., and Campos, C., 1989, Les effets de la phase Antler (Devonien superieur-Carbonifere inferieur) dans les Klamath orientales (Californie, U.S.A.); Implications geodynamiques: Comptes Rendus, l'Academie des Sciences, Paris, Serie II, p. 1629–1635.

Christiansen, R. L., 1961, Structure, metamorphism, and plutonism in the El Paso Mountains, Mojave Desert, California [Ph.D. thesis]: Stanford, California, Stanford University, 180 p.

Chronic, J., 1979, The Mississippian and Pennsylvanian (Carboniferous) Systems in the United States; Colorado: U.S. Geological Survey Professional Paper 1110-M-DD, p. V1–V25.

Churkin, M., Jr., 1973 Paleozoic and Precambrian rocks of Alaska and their role in their structural evolution: U.S. Geological Survey Professional Paper 740, 64 p.

—— , 1974, Paleozoic marginal ocean basin volcanic arc systems in the Cordilleran foldbelt, *in* Dott, R. H., and Shaver, R. H., eds., Modern and ancient geosynclinal sedimentation: Society of Economic Paleontologists and Mineralogists Special Publication 19, p. 174–192

Churkin, M., Jr., and Kay, M., 1967, Graptolite-bearing Ordovician siliceous and volcanic rocks, northern Independence Range, Nevada: Geological Society of America Bulletin, v. 78, p. 651–668.

Clark, L D., Imlay, R. W., McMath, V. E., and Silberling, N. J., 1962, Angular unconformity between Mesozoic and Paleozoic rocks in the northern Sierra Nevada, California: U.S. Geological Survey Professional paper 450-B, p. B15–B19.

Coats, R. R., 1969, Upper Paleozoic formations in the Mountain City area, Elko County, Nevada: U.S. Geological Survey Bulletin 1274-A, p. A22–A27.

Coles, K. S., and Snyder, W. S., 1985, Significance of lower and middle Paleozoic phosphatic chert in the Toquima Range, central Nevada: Geology, v. 13, p. 573–576.

Condie, K. C., and Snansieng, S., 1971, Petrology and geochemistry of the Duzel (Ordovician) and Gazelle (Silurian) Formations, northern California: Journal of Sedimentary Petrology, v. 41, p. 741–751.

Coogan, A. H., 1960, Stratigraphy and paleontology of the Permian Nosoni and Dekkas Formations (Bollibokka Group): University of California Publications in Geological Sciences, v. 36, p. 243–316.

Cox, B. F., and Morton, J. L., 1980, Late Permian plutonism, El Paso Mountains, California: Geological Society of America Abstracts with Programs, v. 12, p. 103.

Craig, L. C., and others, 1979, Paleotectonic investigations of the Mississippian System in the United States: U.S. Geological Survey Professional Paper 1010, 559 p.

Curtis, D. R., 1983, Stratigraphy and origin of the Pit Formation, northern California [M.S. thesis]: Berkeley, University of California, 56 p.

D'Allura, J. A., Moores, E. M., and Robinson, L., 1977, Paleozoic rocks of the northern Sierra Nevada; Their structural and paleogeographic implications, *in* Stewart, J. H., Stevens, C. H., and Fritsche, A. E., eds., Paleozoic paleogeography of the western United States: Los Angeles, California, Pacific Section, Society of Economic Paleontologists and Mineralogists Pacific Coast Paleogeography Symposium 1, p. 395–408.

Danner, W. R., 1977, Paleozoic rocks of northwest Washington and adjacent parts of British Columbia, *in* Stewart, J. H., Stevens, C. H., and Fritsche, A. E., eds., Paleozoic paleogeography of the western United States: Los Angeles, California, Pacific Section, Society of Economic Paleontologists and Mineralogists Pacific Coast Paleogeography Symposium 1, p. 481–502.

Davis, G. A., 1968, Westward thrust faulting in the south-central Klamath Mountains, California: Geological Society of America Bulletin, v. 79, p. 911–934.

—— , 1969, Tectonic correlations, Klamath Mountains and western Sierra Nevada, California: Geological Society of America Bulletin, v. 80, p. 1095–1108.

Davis, G. A., Holdaway, M. J., Lipman, P. W., and Romey, W. D., 1965, Structure, metamorphism, and plutonism in the south-central Klamath Mountains, California: Geological Society of America Bulletin, v. 76, p. 933–966.

Davis, G. A., Monger, J.W.H., and Burchfiel, B. C., 1978, Mesozoic construction of the Cordilleran "collage," central British Columbia to central California, *in* Howell, D. G., and McDougall, K. A., eds., Mesozoic paleogeography of the western United States, Los Angeles, California, Pacific Section Society of Economic Paleontologists and Mineralogists Pacific Coast Paleogeography Symposium 2, p. 1–32.

Dickinson, W. R., 1977, Paleozoic plate tectonics and the evolution of the Cordilleran continental margin, *in* Stewart, J. H., Stevens, C. H., and Fritsche, A. E., eds., Paleozoic paleogeography of the western United States: Los Angeles, California, Pacific Section, Society of Economic Paleontologists and Mineralogists Pacific Coast Paleogeography Symposium 1, p. 137–156.

Dickinson, W. R., and Suczek, C. A., 1979, Plate tectonics and sandstone compositions: American Association of Petroleum Geologists Bulletin, v. 63, p. 2164–2182.

Dickinson, W. R., and Thayer, T. P., 1978, Paleogeographic and paleotectonic implications of Mesozoic stratigraphy and structure in the John Day inlier of central Oregon, *in* Howell, D. G., and McDougall, K. A., eds., Mesozoic paleogeography of the western United States: Los Angeles, California, Pacific Section Society of Economic Paleontologists and Mineralogists Pacific Coast Paleogeography Symposium 2, p. 147–162.

Dott, R. H., Jr., 1955, Pennsylvanian stratigraphy of Elko and northern Diamond Ranges, northeastern Nevada: American Association of Petroleum Geologists Bulletin, v. 39, no. 11, p. 2211–2305.

Dover, J. H., 1980, Status of the Antler Orogeny in central Idaho; Clarifications and constraints from the Pioneer Mountains, *in* Fouch, T. D., and Magathan, E. R., eds., Paleozoic paleogeography of the west-central United States: Denver, Colorado, Rocky Mountain Section Society of Economic Paleontologists and Mineralogists Rocky Mountain Paleogeography Symposium 1, p. 371–386.

Durrell, C., and D'Allura, J. A., 1977, Upper Paleozoic section in eastern Plumas and Sierra Counties, Sierra Nevada, California: Geological Society of America Bulletin, v. 88, p. 844–852.

Ehman, K. D., 1985, Paleozoic stratigraphy and tectonics of the Bull Run Mountains, Elko County, northern Nevada [Ph.D. thesis]: Davis, University of California, 174 p.

Eisbacher, G. H., 1983, Devonian–Mississippian sinistral transcurent faulting along the cratonic margin of western North America; A hypothesis: Geology, v. 11, p. 7–10.

Erickson, R. L., and Marsh, S. P., 1974, Paleozoic tectonics in the Edna Mountain Quadrangle, Nevada: U.S. Geological Survey Journal of Research, v. 2, p. 331–337.

Evans, K. V., and Zartman, R. E., 1988, Early Paleozoic alkalic plutonism in east-central Idaho: Geological Society of America Bulletin, v. 100, p. 1981–1987.

Fagan, J. J., 1962, Carboniferous cherts, turbidites, and volcanic rocks in northern Independence Range, Nevada: Geological Society of America Bulletin, v. 73, p. 595–612.

Ferguson, H. G., Roberts, R. J., and Muller, S. W., 1952, Geologic map of the Golconda Quadrangle, Nevada: U.S. Geological Survey Geological Quadrangle Map GQ–11, scale 1:125,000.

Follo, M. F., and Siever, R., 1985, Carbonate platform margin facies on an evolving suspect terrane; Wallowa Mountains, Oregon: Geological Society of America Abstracts with Programs, v. 17, p. 584–585.

Fraticelli, L. A., Albers, J. P., and Zartman, R. E., 1985, The Permian Pit Rock stock of the McCloud plutonic belt, eastern Klamath terrane, northern California: Isochron/West, no. 44, p. 6–8.

Frenzel, H. N., and 13 others, 1988, The Permian Basin region, *in* Sloss, L. L., ed., Sedimentary cover–North American craton: U.S.: Boulder, Colorado, Geological Society of America, The Geology of North America, Volume D-2, p. 261–306.

Gabrielse, H., 1985, Major dextral transcurrent displacements along the Northern Rocky Mountain trench and related lineaments in north-central British Columbia: Geological Society of America Bulletin, v. 96, p. 1–14.

Gabrielse, H., Snyder, W. S., and Stewart, J. H., 1983, Sonoma Orogeny and Permian to Triassic tectonism in western North America: Geology, v. 11, p. 484–486.

Gans, P. B., and Miller, E. L., 1983, Style of mid-Tertiary extension in east-central Nevada, *in* Geological Society of America Rocky Mountain and Cordilleran Sections Meeting Guidebook, Part 1: Utah Geological and Mining Special Studies, p. 107–160.

Gilluly, J., and Gates, O., 1965, Tectonic and igneous geology of the northern Shoshone Range, Nevada, with sections on Gravity in Crescent Valley by Donald Plouff and Economic Geology by K. B. Ketner: U.S. Geological Survey Professional Paper 465, 153 p.

Gilluly, J., and Mazursky, H., 1965, Geology of the Cortez Quadrangle, Nevada: U.S. Geological Survey Bulletin 1175, 117 p.

Girty, G. H., and Schweickert, R. A., 1984, The Culbertson Lake allochthon, a newly identified structure within the Shoo Fly Complex, California; Evidence for four phases of deformation and extension of the Antler Orogeny to the northern Sierra Nevada: Modern Geology, v. 8, p. 181–198.

Girty, G. H., and Wardlaw, M. S., 1984, Was the Alexander terrane a source of feldspathic sandstones in the Shoo Fly Complex, Sierra Nevada, California?: Geology, v. 12, p. 339–342.

—— , 1985, Petrology and provenenance of pre-Late Devonian sandstones, Shoo Fly Complex, northern Sierra Nevada, California: Geological Society of America Bulletin, v. 96, p. 512–521.

Girty, G. H., Reiland, D. N., and Wardlaw, M. S., 1985, Provenance of Silurian Elder Sandstone, north-central Nevada: Geological Society of America Bulletin, v. 96, p. 925–930.

Gordey, S. P., Abbott, J. G., Templeman-Kluit, D. J., and Gabrielse, H., 1987, "Antler" clastics in the Canadian Cordillera: Geology, v. 15, p. 103–107.

Graham, S. A., Dickinson, W. R., and Ingersoll, R. V., 1975, Himalayan–Bengal model for flysch dispersal in Appalachian–Ouachita system: Geological Society of America Bulletin, v. 86, p. 273–286.

Gutschick, R. C., Sandberg, C. A., and Sando, W. J., 1980, Mississippian shelf margin and carbonate platform from Montana to Nevada, *in* Fouch, T. D., and Magathan, E R., eds., Paleozoic paleogeography of the west-central United States: Denver, Colorado, Rocky Mountain Section Society of Economic Paleontologists and Mineralogists Rocky Mountain Paleogeography Symposium 1, p. 111–128.

Hallgarth, W. E., 1967, Western Colorado, southern Utah, and northwestern New Mexico, *in* McKee, E. D., and Oriel, S. S., eds., Paleotectonic maps of the Permian System: U.S. Geological Survey Miscellaneous Geological Investigations Map I–450, p. 175–197.

Hamilton, W., 1978, Mesozoic tectonics of the western United States, *in* Howell, D. G., and McDougall, K. A., eds., Mesozoic paleogeography of the western United States: Los Angeles, California, Pacific Section Society of Economic Paleontologists and Mineralogists Pacific Coast Paleogeography Symposium 2, p. 33–70.

Hamilton, W., and Myers, W. B., 1966, Cenozoic tectonics of the western United States: Reviews of Geophysics, v. 4, p. 509–549.

Hannah, J. L., and Moores, E. M., 1986, Age relationships and despositional environments of Paleozoic strata, northern Sierra Nevada, California: Geological Society of America Bulletin, v. 97, p. 787–797.

Hanson, R. E., 1983, Volcanism, plutonism and sedimentation in a Late Devonian submarine island-arc setting, northern Sierra Nevada, California [Ph.D. thesis]: New York, Columbia University, 345 p.

Hanson, R. E., and Schweickert, R. A., 1986, Stratigraphy of mid-Paleozoic island-arc rocks in part of the northern Sierra Nevada, Sierra and Nevada Counties, California: Geological Society of America Bulletin, v. 97, p. 986–998.

Hanson, R. E., Saleeby, J. B., and Schweickert, R. A., 1988, Composite Devonian island-arc batholith in the northern Sierra Nevada, California: Geological Society of America Bulletin, v. 100, p. 446–457.

Harbaugh, D. W., and Dickinson, W. R., 1981, Depositional facies of Mississippian clastics, Antler foreland basin, central Diamond Mountains, Nevada: Journal of Sedimentary Petrology, v. 51, p. 1223–1234.

Harwood, D. S., 1983, Stratigraphy of upper Paleozoic volcanic rocks and regional unconformities in part of the northern Sierra terrane, California: Geological Society of America Bulletin, v. 94, p. 413–422.

——, 1988, Tectonism and metamorphism in the northern Sierra terrane, northern California, *in* Ernst, W. G., ed., Tectonic and metamorphic evolution of the U.S. Cordillera; Rubey Volume 71: Englewood Cliffs, New Jersey, Prentice-Hall, p. 764–788.

Harwood, D. S., and Murchey, B., 1990, Biostratigraphic, tectonic, and paleogeographic ties between upper Paleozoic volcanic and basinal rocks in the northern Sierra terrane, California and the Havallah sequence, Nevada, *in* Harwood, D. S. and Miller, M. M., eds., Paleozoic and early Mesozoic paleogeographic relations; Sierra Nevada, Klamath Mountains, and related terranes: Geological Society of America Special Paper 255, 422 p.

Harwood, D. S., Jayko, A. S., Harris, A. G., Silberling, N. J., and Stevens, C. H., 1988, Permian–Triassic rocks slivered between the Shoo Fly Complex and Feather River peridotite belt, northern Sierra Nevada, California: Geological Society of America Abstracts with Programs, v. 20, p. 167–168.

Hintze, L. H., 1985, Correlation of stratigraphic units of North America (COSUNA) Project, Great Basin Region: American Association of Petroleum Geologists CSD 625, 1 plate.

Holdsworth, B. K., and Jones, D. L., 1980, Preliminary radiolarian zonation for Late Devonian through Permian time: Geology, v. 8, p. 281–285.

Hotz, P. E., 1977, Geology of the Yreka Quadrangle, Siskiyou County, California, U.S. Geological Survey Bulletin 1436, 72 p.

Hotz, P. E., and Willden, R., 1964, Geology and mineral deposits of the Osgood Mountains Quadrangle, Humboldt County, Nevada: U.S. Geological Survey Professional Paper 431, 128 p.

Hotz, P. E., Lanphere, M. A., and Swanson, D. A., 1977, Triassic blueschist from northern California and north-central Oregon: Geology, v. 5, p. 659–663.

Irwin, W. P., 1972, Terranes of the western Paleozoic and Triassic belt in the southern Klamath Mountains, California: U.S. Geological Survey Professional Paper 800-C, p. C103–C111.

——, 1977, Review of Paleozoic rocks of the Klamath Mountains, *in* Stewart, J. H., Stevens, C. H., and Fritsche, A. E., eds., Paleozoic paleogeography of the western United States: Los Angeles, California, Pacific Section, Society of Economic Paleontologists and Mineralogists Pacific Coast Paleogeography Symposium, p. 441–454.

Irwin, W. P., and Wardlaw, B. R., 1983, Conodonts of the western Paleozoic and Triassic belt, Klamath Mountains, California and Oregon: Journal of Paleontology, v. 57, p. 1030–1039.

Irwin, W. P., Jones, D. L., and Klaplan, T. A., 1978, Radiolarians from pre-Nevadan rocks of the Klamath Mountains, California and Oregon, *in* Howell, D. G., and McDougall, K. A., eds., Mesozoic paleogeography of the western United States: Los Angeles, California, Pacific Section Society of Economic Paleontologists and Mineralogists Pacific Coast Paleogeography Symposium 2, p. 303–310.

Jacobsen, S. B., Quick, J. E., and Wasserburg, G. J., 1984, A Nd and Sr isotopic study of the Trinity peridotite; Implications for mantle evolution: Earth and Planetary Science Letters, v. 68, p. 361–378.

Jansa, L. F., and Pe-Piper, G., 1985, Early Cretaceous volcanism on the northeastern American margin and its implications for plate tectonics: Geological Society of America Bulletin v. 96, p. 92–107.

Jayko, A. S., 1988, Paleozoic and Mesozoic rocks of the Almanor 15′ Quadrangle, Plumas County, California: U.S. Geological Survey Open-File Report 88–757, 12 p.

Jayko, A. S., Murchey, B. L., Wardlaw, B. R., and Stevens, C. H., 1987, Revised interpretation of the pre-Cenozoic geology of the Butt Lake area, northern Sierra Nevada, California: Geological Society of America Abstracts with Programs, v. 19, p. 391.

Johnson, J. G., and Pendergast, A., 1981, Timing and mode of emplacement of the Roberts Mountain allochthon Antler Orogeny: Geological Society of America Bulletin, v. 92, part 1, p. 648–658.

Jones, A. E., 1990, Geologic and tectonic significance of terranes near Quinn River Crossing, Nevada, *in* Harwood, D. S. and Miller, M. M., eds., Paleozoic and early Mesozoic paleogeographic relations; Sierra Nevada, Klamath Mountains, and related terranes: Geological Society of America Special Paper 255, p. 239–254.

Jones, D. L., Irwin, W. P., and Ovenshine, A. T., 1972, Southeastern Alaska; A displaced continental fragment?: U.S. Geological Survey Professional Paper 800-B, p. B211–B217.

Jones, D. L., Silberling, N. J., Coney, P. J., and Plafker, G., 1987, Lithotectonic terrane map of Alaska: U.S. Geological Survey Miscellaneous Field Studies Map MF-1874A, scale 1:2,500,000.

Jordan, T. E., and Douglass, R. C., 1980, Paleogeography and structural development of the Late Pennsylvanian to Early Permian Oquirrh Basin, northwestern Utah, *in* Fouch, T. D., and Magathan, E. R., eds., Paleozoic paleogeography of the west-central United States: Denver, Colorado, Rocky Mountain Section Society of Economic Paleontologists and Mineralogists Rocky Mountain Paleogeography Symposium 1, p. 217–238.

Ketner, K. B., 1977, Late Paleozoic orogeny and sedimentation, southern California, Nevada, Idaho, and Montana, *in* Stewart, J. H., Stevens, C. H., and Fritsche, A. E., eds., Paleozoic paleogeography of the western United States: Los Angeles, California, Pacific Section, Society of Economic Paleontologists and Mineralogists Pacific Coast Paleogeography Symposium 1, p. 363–369.

——, 1984, Recent studies indicate that major structures in northeastern Nevada and the Golconda thrust in north-central Nevada are of Jurassic or Cretaceous age: Geology, v. 12, p. 483–486.

Ketner, K. B., and Wardlaw, B R., 1981, Permian and Triassic rocks near Quinn River Crossing, Humboldt County, Nevada: Geology, v. 9, p. 123–126.

King, P. B., 1969, Tectonic map of North America: U.S. Geological Survey, scale 1:5,000,000.

Kinkel, A. R., Jr., Hall, W. E., and Albers, J. P., 1956, Geology and base-metal deposits of West Shasta copper-zinc district, Shasta County, California, U.S. Geological Survey Professional Paper 285, 156 p.

Kleinhampl, F. J., and Ziony, J. I., 1972, Possible outlier of the Golconda thrust plate in the Monitor Range (Nevada): U.S. Geological Survey Professional Paper 800-A, p. A42.

Klepacki, D. W., 1983, Stratigraphic and structural relations of the Milford, Kaslo, and Slocan groups, Roseberry Quadrangle, Lardeau map area, British Columbia: Current Research, Part A, Geological Survey of Canada Paper 83–1A, p. 229–233.

—— , 1984, Accretionary tectonics in southern British Columbia; Implications for the Cordillera: Geological Society of America Abstracts with Programs, v. 16, p 560.

—— , 1985, Stratigraphy and structural geology of the Goat Range area, southeastern British Columbia [Ph.D. thesis]: Cambridge, Massachusetts Institute of Technology, 252 p.

Kleweno, W. P., Jr., and Jeffords, R. M., 1961, Devonian rocks in the Suplee area of central Oregon: Ore Bin, v. 23, p. 50.

Kluth, C. F., 1986, Plate tectonics of the Ancestral Rocky Mountains, *in* Peterson, J. A., ed., Paleotectonics and sedimentation in the Rocky Mountain region, United States: American Association of Petroleum Geologists Memoir 41, p. 353–369.

Kluth, C. F., and Coney, P. J., 1981, Plate tectonics of the Ancestral Rocky Mountains: Geology, v. 9, p. 10–15.

Lane, H. R., and DeKeyser, T. L., 1980, Paleogeography of the late Early Mississippian (Tournasian) in the central and southwestern United States, *in* Fouch, T. D., and Magathan, E. R., eds., Paleozoic paleogeography of the west-central United States: Denver, Colorado, Rocky Mountain Section Society of Economic Paleontologists and Mineralogists Rocky Mountain Paleogeography Symposium 1, p. 149–162.

Lanphere, M. A., Irwin, W. P., and Hotz, P. E., 1968, Isotopic age of the Nevadan Orogeny and older plutonic and metamorphic events in the Klamath Mountains, California: Geological Society of America Bulletin, v. 79, p. 1027–1052.

Lapierre, H., Albarede, F., Albers, J., Coulon, C., and Cabanis, B., 1985a, Early Devonian volcanism in the eastern Klamath Mountains, California: Evidence for an immature island arc: Canadian Journal of Earth Sciences, v. 22, p. 214–227.

Lapierre, H., Brouxel, M., Albarede, F., Coulon, C., and Cabanis, B., 1985b, The Paleozoic and Mesozoic volcanism in the eastern Klamath Mountains (N. California, USA); Part 1, The early Devonian intra-oceanic island-arc series: Terra Cognita–the Earth, v. 5, p 320.

Laule, S. W., Snyder, W., and Ormiston, A. R., 1981, Willow Canyon Formation, Nevada; An extension of the Golconda allochthon: Geological Society of America Abstracts with Programs, v. 13, p. 66.

Lindsley-Griffin, N., 1977, Paleogeographic implications of ophiolites; The Ordovician Trinity Complex, Klamath Mountains, California, *in* Stewart, J. H., Stevens, C. H., and Fritsche, A. E., eds., Paleozoic paleogeography of the western United States: Los Angeles, California, Pacific Section, Society of Economic Paleontologists and Mineralogists Pacific Coast Paleogeography Symposium 1, p. 409–420.

Lindsley-Griffin, N., and Griffin, J. R., 1983, The Trinity terrane; An early Paleozoic microplate assemblage, *in* Stevens, C. H., ed., Pre-Jurassic rocks in western North American suspect terranes: Los Angeles, California, The Pacific Section, Society of Economic Paleontologists and Mineralogists, p. 63–75.

Little, T. A., 1983, Structure and metamorphism of upper Paleozoic rocks in the Mountain City Quadrangle, Elko County, Nevada [M.S. thesis]: Stanford, California, Stanford University, 392 p.

—— , 1987, Stratigraphy and structure of metamorphosed upper Paleozoic rocks near Mountain City, Nevada: Geological Society of America Bulletin, v. 98, p. 1–17.

Luken, M. D., 1985, Petrography and origin of limestones in the western Paleozoic and Triassic belt, Klamath Mountains [M.S. thesis]: San Jose, California, San Jose State University, 131 p.

Lupe, R., and Silberling, N. J., 1985, Genetic relationship between lower Mesozoic continental strata of the Colorado Plateau and marine strta of the western Great Basin; Significance for accretionary history of Cordilleran lithotectonic terranes, *in* Howell, D. G., ed., Tectonostratigraphic terranes of the Circum-Pacific region: Circum-Pacific Council for Energy and Mineral Resources, Earth Science Series 1 p. 263–271.

MacMillan, J. R., 1972, Late Paleozoic and Mesozoic tectonic events in western-central Nevada [Ph.D. thesis]: Evanston, Illinois, Northwestern University, 217 p.

Madrid, R., 1987, Stratigraphy of the Roberts Mountains allochthon in north-central Nevada [Ph.D. thesis]: Stanford, California, Stanford University, 336 p.

Mallory, W. W., 1975, Middle and Southern Rocky Mountains, northern Colorado Plateau, and eastern Great Basin region, *in* McKee, E. D., and Crosby, E. J., eds., Paleotectonic investigations of the Pennsylvanian System in the United States, Part 1: U.S. Geological Survey Professional Paper 853, part 1, p. 265–278.

Mannick, E. P., 1975, Stratigraphy and structure of the Ordovician Vinini Formation, Tyrone Gap area, Eureka County, Nevada [M.S. thesis]: Athens, Ohio, Ohio University, 55 p.

Mattinson, J. M., and Hopson, C. A., 1972, Paleozoic rocks from ophiolitic complexes in Washington and northern California: American Journal of Science, v. 53, p. 543.

McKee, E. D., 1975, Interpretation of Pennsylvanian history, *in* McKee, E. D., and Crosby, E. J., eds., Paleotectonic investigations of the Pennsylvanian System in the United States, Part 2: U.S. Geological Survey Professional Paper 853, part 2, p. 1–21.

McKee, E. D., and others, 1967, Paleotectonic maps of the Permian System: U.S. Geological Survey Miscellaneous Geological Investigations Map I–450, 164 p.

McKee, E. D., and 9 others, 1975, Paleotectonic investigations of the Pennsylvanian System in the United States, Parts 1, 2, 3: U.S. Geological Survey Professional Paper 853, part 1, 349 p., part 2, 192 p., part 3, 17 plates.

McKee, E. H., 1976, Geology of the northern part of the Toquima Range, Lander, Eureka and Nye Counties, Nevada: U.S. Geological Survey Professional Paper 931, 49 p.

McMath, V. E., 1966, Geology of the Taylorsville area, northern Sierra Nevada, *in* Bailey, E. H., ed., Geology of northern California: California Division of Mines and Geology Bulletin 190, p. 173–183.

Merguerian, C. and Schweickert, C. A., 1987, Paleozoic gneissic granitoids in the Shoo Fly Complex, central Sierra Nevada, California: Geological Society of America Bulletin, v. 99, p. 699–717.

Merriam, C. W., and Berthiaume, S. A., 1943, Late Paleozoic formations of central Oregon: Geological Society of America Bulletin, v. 54, p. 145–172.

Miller, E. L., 1978, The Fairview Valley Formation; A Mesozoic intraorogenic deposit in the southwestern Mojave Desert, *in* Howell, D. G., and McDougall, K. A., eds., Mesozoic paleogeography of the western United States: Los Angeles, California, Pacific Section Society of Economic Paleontologists and Mineralogists Pacific Coast Paleogeography Symposium 2, p. 163–187.

—— , 1981, Geology of the Victorville region, California: Geological Society of America Bulletin, v. 92, p. 554–608.

Miller, E. L., and Cameron, C. S., 1982, Late Precambrian to Late Cretaceous evolution of the southwestern Mojave Desert, California, *in* Cooper, J. D., ed., Field guidebook, Geological Society of America Cordilleran Section: Shoshone, California, Death Valley Publishing Co., p. 21–34.

Miller, E. L., and Larue, D. K., 1983, Ordovician quartzite in the Roberts Mountains allochthon, Nevada; Deep sea fan deposits derived from cratonal North America, *in* Stevens, C. H., ed., Pre-Jurassic rocks in western North American suspect terranes: Los Angeles, California, Pacific Section, Society of Economic Paleontologists and Mineralogists, p. 91–102.

Miller, E. L., and Sutter, J., 1982, Structural geology and $^{40}Ar/^{39}Ar$

geochronology of the Goldstone-Lane/Mountain area, Mojave Desert, California: Geological Society of America Bulletin, v. 93, p. 1191–1207.

Miller, E. L., and 5 others, 1981, Thrust emplacement of the Schoonover sequence, northern Independence Mountains, Nevada: Geological Society of America Bulletin, v. 92, p. 730–737.

Miller, E. L., and 5 others, 1982, Structural fabric of the Paleozoic Golconda allochthon, Antler Peak Quadrangle, Nevada: Progressive deformation of an oceanic sedimentary assemblage: Journal of Geophysical Research Special Issue on Accretionary Tectonics, v. 87, p. 3795–3804.

Miller, E. L., Holdsworth, B. K., Whiteford, W. B., and Rogers, D., 1984, Stratigraphy and structure of the Schoonover sequence, northern Nevada; Implications for Paleozoic plate margin tectonics: Geological Society of America Bulletin, v. 95, p. 1063–1076.

Miller, M. M., 1986, Tectonic evolution of late Paleozoic island arc sequences in the western U.S. Cordillera; with detailed studies from the eastern Klamath Mountains, northern California [Ph.D. thesis]: Stanford, California, Stanford University, 201 p.

—— , 1987, Dispersed remnants of a northeast Pacific fringing arc–Upper Paleozoic island arcterranes of Permian McCloud faunal affinity, western U.S. Tectonics, v. 6, p. 807–830.

—— , 1989, Intra-arc sedimentation and tectonism; Late Paleozoic evolution of the eastern Klamath terrane, California: Geological Society of America Bulletin, v. 101, p. 170–187.

Miller, M. M., and Cui, B., 1989, Submarine fan characteristics and dual sediment provenance, lower Carboniferous Bragdon Formation, eastern Klamath terrane, California: Canadian Journal of Earth Sciences, v. 26, p. 927–940.

Miller, M. M., and Harwood, D. S., 1990, Paleogeographic setting of upper Paleozoic rocks in the northern Sierra and eastern Klamath terranes, northern California, *in* Harwood, D. S. and Miller, M. M., eds., Paleozoic and early Mesozoic paleogeographic relations; Sierra Nevada, Klamath Mountains, and related terranes: Geological Society of America Special Paper 255, p. 175–192.

Miller, M. M., and Saleeby, J. B., 1991, Continental detrital zircon in Carboniferous ensimatic arc rocks, Bragdon Formation, eastern Klamath terrane, northern California: Geological Society of America Bulletin, v. 103, p. 268–276.

Miller, M. M., and Wright, J. E., 1987, Paleogeographic implications of Permian Tethyan corals from the Klamath Mountains, California: Geology, v. 15, p. 266–269.

Molnar, P., and Tapponnier, P., 1975, Cenozoic tectonics of Asia; Effects of a continental collision: Science v. 189, p. 419–426.

Monger, J.W.H., 1970, Hope map-area, west half British Columbia: Geological Survey of Canada Paper 69–47, 75 p.

—— , 1975, Correlation of eugeosynclinal tectonostratigraphic belts in the North American Cordillera: Geoscience Canada, v. 2, p. 4–9.

—— , 1977, Upper Paleozoic rocks of the western Canadian Cordillera and their bearing on Cordilleran evolution: Canadian Journal of Earth Sciences, v. 14, p. 1832–1859.

—— , 1984, Cordilleran tectonics; A Canadian perspective: Bulletin de la Societie Geologique de France, v. 7 t. XXVI no. 2, p. 255–278.

Monger, J.W.H., and Berg, H. C., 1984, Lithotectonic terrane map of western Canada and southeastern Alaska, *in* Silberling, N. J., and Jones, D. L., eds., Lithotectonic terrane maps of the North American Cordillera: U.S. Geological Survey Open-File Report 84–523, p. B1–B7.

—— , 1987, Lithotectonic terrane map of Alaska: U.S. Geological Survey Miscellaneous Field Studies Map MF-1874B, scale 1:2,500,000.

Monger, J.W.H., and Ross, C. A., 1971, Distribution of fusulinaceans in the western Canadian Cordillera: Canadian Journal of Earth Sciences, v. 8, p. 259–278.

Monger, J.W.H., Price, R. A., and Templeman-Kluit, D. J., 1982, Tectonic accretion and the origin of the two major plutonic welts in the Canadian Cordillera: Geology, v. 10, p. 70–75.

Morgan, W. A., 1980, Euxinic Early Permian sedimentation in the Cassia Basin of southern Idaho, *in* Fouch, T D., and Magathan, E. R., eds., Paleozoic paleogeography of the west-central United States: Denver, Colorado, Rocky Mountain Section Society of Economic Paleontologists and Mineralogists Rocky Mountain Paleogeography Symposium 1, p. 305–326.

Morris, E. M., and Wardlaw, B. R., 1986, Conodont ages for limestones of eastern Oregon and their implications of pre-Tertiary mélange terranes, *in* Vallier, T. L., and Brooks, H. C., eds., Geology of the Blue Mountains region of Oregon, Idaho, and Washington; Geologic implications of Paleozoic and Mesozoic paleontology and biostratigraphy, Blue Mountains province, Oregon and Idaho: U.S. Geological Survey Professional Paper 1435, p. 59–63.

Mortimer, N., 1985, Structural and metamorphic aspects of Middle Jurassic terrane juxtaposition, northeastern Klamath Mountains, California, *in* Howell, D. G., ed., Tectonostratigraphic terranes of the Circum-Pacific region: Houston, Texas, Circum-Pacific Council for Energy and Mineral Resources, p. 201–214.

Muffler, L.J.P., 1964, Geology of the Frenchie Creek Quadrangle, north-central Nevada: U.S. Geological Survey Bulletin 1179, 99 p.

Murchey, B., 1982, Chert facies in the Havallah sequence near Battle Mountain, Nevada: Geological Society of America Abstracts with Programs, v. 14, p. 219.

Murchey, B., and Harwood, D. S., 1986, Correlative chert sequences from the northern Sierra Nevada and Havallah sequences, Nevada: Geological Society of America Abstracts with Programs, v. 18, p. 162.

Murchey, B. L., Jones, D. L., and Blome, C. D., 1983, Comparison of Permian and Lower Mesozoic radiolarian chert in western accreted terranes: Geological Society of America Abstracts with Programs, v. 15, p. 371.

Murphy, M. A., Power, J. D., and Johnson, J. G., 1984, Evidence for Late Devonian movement within the Roberts Mountains allochthon, Roberts Mountains, Nevada: Geology, v. 12, p. 20–23.

Nelson, S. F., and Nelson, E. R., 1985, Allochthonous Permian micro and macrofauna, Kamloops area, British Columbia: Canadian Journal of Earth Sciences, v. 22, p. 442–451.

Nestell, M. K., 1983, Permian foraminiferal faunas of central and eastern Oregon: Geological Society of America Abstracts with Programs, v. 15, p. 371.

Nestell, M. K., Irwin, W. P., and Albers, J. P., 1981, Late Permian (early Djulfian) Tethyan foraminifera from the southern Klamath Mountains, California: Geological Society of America Abstracts with Programs, v. 13, p. 519.

Newton, C. R., 1988, Significance of "Tethyan" fossils in the American Cordillera: Science, v. 242, p. 365–391.

Nilsen, T. J., and Stewart, J. H., 1980, The Antler Orogeny; Mid-Paleozoic tectonism in western North America: Geology, v. 8, p. 298–302.

Nokleberg, W. J., 1983, Wallrocks of the central Sierra Nevada batholith, California; A collage of accreted tectono-stratigraphic terranes: U.S. Geological Survey Professional Paper 1255, p. 1–28.

Okulitch, A. W., Wanless, R. K., and Loveridge, W. D., 1975, Devonian plutonism in south-central British Columbia: Canadian Journal of Earth Sciences, v. 12, p 1760–1769.

Oldow, J. S., 1984, Spatial variability in the structure of the Roberts Mountains allochthon, western Nevada, Geological Society of America Bulletin, v. 95, p. 174–185.

Palmer, A. R., 1971, Cambrian of the Great Basin and adjacent areas, western United States, *in* Holland, C. H., ed., Cambrian of the New World: New York, John Wiley Interscience, p. 1–78.

Peacock, S. M., 1987, Serpentinization and infiltration metasomatism in the Trinity peridotite, Klamath province, northern California; Implications for subduction zones: Contributions to Mineralogy and Petrology, v. 95, p. 55–70.

Pe-Piper, G., and Jansa, L. F., 1987, Geochemistry of Middle Jurassic and Early Cretaceous igneous rocks on the eastern North American margin: Geological Society of America Bulletin, v. 99, p. 803–813.

Pessagno, E. A., Jr., and Blome, C. D., 1986, Tectonogenesis of Mesozoic rocks in the Blue Mountain province of eastern Oregon and western Idaho, *in* Vallier, T. L. and Brooks, H. C., eds., Geology of the Blue Mountains region of Oregon, Idaho, and Washington; Geologic implications of Paleozoic and

Mesozoic paleontology and biostratigraphy, Blue Mountains province, Oregon and Idaho: U.S. Geological Survey Professional Paper 1435, p. 65–78.

Peterson, J. A., 1980, Permian paleogeography and sedimentary provinces, west-central United States, *in* Fouch, T. D., and Magathan, E. R., eds., Paleozoic paleogeography of the west-central United States: Denver, Colorado, Rocky Mountain Section Society of Economic Paleontologists and Mineralogists Rocky Mountain Paleogeography Symposium 1, p. 271–292.

Phelps, D., and Ave Lallement, H. G., 1980, The Sparta ophiolitic complex, northeast Oregon; A plutonic equivalent to low K_2O island-arc volcanism: American Journal of Science, v. 280-A, p. 345–358.

Poole, F. G., and Sandberg, C. A., 1977, Mississippian paleogeography and tectonics of the western United States, *in* Stewart, J. H., Stevens, C. H., and Fritsche, A. E., eds., Paleozoic paleogeography of the western United States: Los Angeles, California, Pacific Section, Society of Economic Paleontologists and Mineralogists Pacific Coast Paleogeography Symposium 1, p. 67–85.

Poole, F. G., and Wardlaw, B. R., 1978, Candelaria (Triassic) and Diablo (Permian) Formations in southern Toquima Range, central Nevada, *in* Howell, D. G., and McDougall, K. A., eds., Mesozoic paleogeography of the western United States: Pacific Section, Society of Economic Paleontologists and Mineralogists Pacific Coast Paleogeography Symposium 2, p. 271–276.

Potter, A. W., Hotz, P. E., and Rohr, D. M., 1977, Stratigraphy and inferred tectonic framework of lower Paleozoic rocks in the eastern Klamath Mountains, northern California, *in* Stewart, J. H., Stevens, C. H., and Fritsche, A. E., eds., Paleozoic paleogeography of the western United States: Los Angeles, California, Pacific Section, Society of Economic Paleontologists and Mineralogists Pacific Coast Paleogeography Symposium 1, p. 421–440.

Quick, J. E., 1981, Petrology and petrogenesis of the Trinity peridotite, an upper mantle diapir in the eastern Klamath Mountains, northern California: Journal of Geophysical Research, v. 86, p. 118376–11863.

Read, P. B., 1975, Lardeau Group, Lardeau map-area, west half, British Columbia: Geological Survey of Canada Paper 75–1, part A, p. 7–8.

Read, P. B., and Wheeler, J. O., 1976, Geology of Lardeau west half map-area: Geological Survey of Canada Open File Report 381.

Rees, M. N., 1986, A fault-controlled trough through a carbonate platform; The Middle Cambrian House Range embayment: Geological Society of America Bulletin, v. 97, p. 1122–1132.

Renne, P. R., and Scott, G. R., 1988, Structural chronology, oroclinal deformation, and tectonic evolution of the southeastern Klamath Mountains, California: Tectonics, v. 7, p. 1223–1242.

Rich, M., 1971, Middle Pennsylvanian rocks of eastern Great Basin: American Association of Petroleum Geologists Bulletin, v. 55, p. 432–453.

—— , Pennsylvanian paleogeographic patterns in the western United States, *in* Stewart, J. H., Stevens, C. H., and Fritsche, A. E., eds., Paleozoic paleogeography of the western United States: Los Angeles, California, Pacific Section, Society of Economic Paleontologists and Mineralogists Pacific Coast Paleogeography Symposium, p. 87–111.

Rinehart, C. D., and Ross, D. C., 1964, Geology and mineral deposits of the Mount Morrison Quadrangle, Sierra Nevada, California: U.S. Geological Survey Professional Paper 385, p. 106.

Riva, J., 1970, Thrusted Paleozoic rocks in the northern and central HD Range, northeastern Nevada: Geological Society of America Bulletin, v. 81, p. 2689–2716.

Roberts, R. J., 1964, Stratigraphy and structure of the Antler Peak Quadrangle, Humboldt and Lander Counties, Nevada: U.S. Geological Survey Professional Paper 459A, p. 93.

Roberts, R. J., Hotz, P. E., Gilluly, J., and Ferguson, H. G., 1958, Paleozoic rocks of north-central Nevada: American Association of Petroleum Geologists Bulletin, v. 42, p. 2813–2857.

Roberts, R. J., Montgomery, K. M., and Lehner, R. E., 1967, Geology and mineral resources of Eureka County, Nevada: Nevada Bureau of Mines and Geology Bulletin 64, 152 p.

Rose, P. R., 1976, Mississippian carbonate shelf margins, western United States: U.S. Geological Survey Journal of Research, v. 4, no. 4, p. 449–466.

Ross, C. A., and Ross, J.R.P., 1981, Biogeographical influences on late Paleozoic faunal distributions, *in* Larwood, G. P., and Nielsen, C., eds., Recent and fossil bryozoa: Fredensborg, Denmark, Olsen and Olsen, p. 199–212.

—— , 1983, Late Paleozoic accreted terranes of western North America, *in* Stevens, C. H., ed., Pre-Jurassic rocks in western North American suspect terranes: Los Angeles, California, Pacific Section, Society of Economic Paleontologists and Mineralogists, p. 7–22.

Rowell, A. J., Rees, M. N., and Suczek, C. A., 1979, Margin of the North American continent in Nevada during Late Cambrian time: American Journal of Science, v. 279, p. 1–18.

Royce, F., Jr., Warner, M. A., and Reese, D. L., 1975, Thrust belt structural geometry and related stratigraphic problems, Wyoming–Idaho–northern Utah, *in* Bolyard, D. W., ed., Symposium on deep drilling frontiers in the central Rocky Mountains: Rocky Mountain Association of Geologists, p. 41–54.

Rubin, C. M., Miller, M. M., and Smith, G., 1990, Tectonic development of Cordilleran mid-Paleozoic volcano-plutonic complexes; Evidence for convergent margin tectonism, *in* Harwood, D. S. and Miller, M. M., eds., Paleozoic and early Mesozoic paleogeographic relations of the Sierra Nevada, Klamath Mountains, and related terranes: Geological Society of America Special Paper 255, p. 1–16.

Russell, B. J., 1981, Pre-Tertiary paleogeography and tectonic history of the Jackson Mountains, northwestern Nevada [Ph.D. thesis]: Evanston, Illinois, Northwestern University, 205 p.

Saleeby, J. B., 1981, Ocean floor accretion and volcano-plutonic arc evolution of the Mesozoic Sierra Nevada, *in* Ernst, W. G., ed., The geotectonic development of California; Rubey Volume 1: Englewood Cliffs, New Jersey, Prentice-Hall, p. 132–181.

—— , 1982, Polygenetic ophiolite belt of the California Sierra Nevada; Geochronologic and tectonostratigraphic development: Journal of Geophysical Research v. 87, p. 1803–1824.

—— , 1983, Accretionary tectonics of the North American Cordillera: Annual Review of Earth and Planetary Science, v. 15, p. 45–73.

Saleeby, J. B., and Sharp, W. D., 1980, Chronology of the structural and petrologic development of the southwest Sierra Nevada foothills, California, 2: Geological Society of America Bulletin v. 91, p. 1416–1435.

Saleeby, J. B., Hannah, J. L., and Varga, R. J., 1987, Isotopic age constraints on middle Paleozoic deformation in the northern Sierra Nevada, California: Geology, v. 15, p. 757–766.

Saleeby, J. B., Shaw, M. F., Niemeyer, S., and Moores, E. M., 1989, U/Pb, Sm/Nd, and Rb/Sr geochronological and isotopic study of northern Sierra Nevada ophiolitic assemblages, California: Contributions to Mineralogy and Petrology, v. 102, p. 205–220.

Saller, A. H., 1980, Depositional setting of post-Antler Pennsylvanian strata in north-central Nevada: [M.S. thesis]: Stanford, California, Stanford University, 118 p.

Saller, A. H., and Dickinson, W. R., 1982, Alluvial to marine facies transition in the Antler overlap sequence, Pennsylvanian and Permian of north-central Nevada: Journal of Sedimentary Petrology, v. 52, p. 925–940.

Sanborn, A. F., 1960, Geology and paleontology of the southwest quarter of the Big Bend Quadrangle Shasta County, California: California Division of Mines and Geology Special Report 63, 26 p.

Sandberg, C. A., Gutschick, R. C., Johnson, J. G., Poole, F. G., and Sando, W. J., 1983, Middle Devonian to Late Mississippian geologic history of the Overthrust belt region, western United States, v. 2, *in* Powers, R. B., ed., Geologic studies of the Cordilleran thrust belt: Denver, Colorado, Rocky Mountain Association of Geologists, p. 691–719.

Sando, W. J., 1976, Mississippian history of the Northern Rocky Mountains region: U.S. Geological Survey Journal of Research, v. 4, no. 3, p. 317–338.

Sando, W. J., and Bamber, E. W., 1985, Coral zonation of the Mississippian System in the Western Interior province of North America: U.S. Geological Survey Professional Paper 1334, 61 p.

Savage, N. M., 1976, Middle Devonian species of *Polygnathus* from the Kennett

Formation, northern California: Journal of Paleontology, v. 50, p. 374–379.

Schweickert, R. A., 1981, Tectonic evolution of the Sierra Nevada range, *in* Ernst, W. G., ed., The geotectonic development of California; Rubey Volume 1: Englewood Cliffs, New Jersey, Prentice-Hall, p. 87–131.

Schweickert, R. A., and Lahren, M. M., 1987, Continuation of Antler and Sonoma orogenic belts to the eastern Sierra Nevada, California, and Late Triassic thrusting in a compressional arc: Geology, v. 15, p. 270–273.

Schweickert, R. A., and Snyder, W. S., 1981, Paleozoic plate tectonics of the Sierra Nevada and adjacent regions, *in* Ernst, W. G., ed., The geotectonic evolution of California; Rubey Volume 1: New Jersey, Prentice-Hall, p. 182–202.

Schweickert, R. A., Saleeby, J. B., Tobisch, O. T., and Wright, W. W., 1977, Paleotectonic and paleogeographic significance of the Calaveras Complex, western Sierra Nevada, California, *in* Stewart, J. H., Stevens, C. H., and Fritsche, A. E., eds., Paleozoic paleogeography of the western United States: Los Angeles, California, Pacific Section, Society of Economic Paleontologists and Mineralogists Pacific Coast Paleogeography Symposium 1, p. 381–394.

Schweickert, R. A., Armstrong, R. L., and Harakal, J. E., 1980, Lawsonite blueschist in the northern Sierra Nevada, California: Geology, v. 8, p. 27–31.

Schweickert, R. A., Harwood, D. S., Girty, G. A., and Hanson, R. E., 1984, Tectonic development of the Northern Sierra terrane; An accreted late Paleozoic island arc and its basement, *in* Lintz, J., Jr., ed., Western geological excursions; Geological Society of America 1984 Annual Meeting Field Trip Guide Volume 4: Reno, University of Nevada Mackay School of Mines, p. 1–65.

Scotese, C. R., Bambach, R. K., Barton, C., Van der Voo, R., and Ziegler, A. M., 1979, Paleozoic base maps: Journal of Geology, v. 87, p. 217–277.

Sharp, W. D., 1988, Pre-Cretaceous crustal evolution in the Sierra Nevada region, California, *in* Ernst, W. G., ed., Metamorphism and crustal evolution of the western United States; Volume Rubey Volume 7: Englewood Cliffs, New Jersey, Prentice-Hall Publishing Company, p. 823–864.

Shaw, H. F., Chen, J. H., Saleeby, J. B. and Wasserburg, G. J., 1987, Nd-Sr-Pb systemics and age of the Kings River ophiolite, California: Contributions to Mineralogy and Petrology v. 96, p. 281–290.

Silberling, N. J., 1973, Geologic events during Permian–Triassic time along the Pacific margin of the United States, *in* Logan, A., and Hill, L. V., eds., The Permian and Triassic Systems and their mutual boundary: Calgary, Canada, Canadian Society of Petroleum Geologists Memoir 2, p. 345–362.

Silberling, N. J., and Jones, D. L., 1982, Tectonic significance of Permian–Triassic strata in northwestern Nevada and northern California: Geological Society of America Abstracts with Programs, v. 14, p. 234.

——, eds., 1987, Lithotectonic terrane map of the western coterminous United States: U.S. Geological Survey Miscellaneous Field Studies Map MF-1874C.

Silberling, N. J., Jones, D. L., Blake, M. C., Jr., and Howell, D. G., 1987, Lithotectonic terrane map of the western coterminous United States: U.S. Geological Survey Miscellaneous Field Studies Map MF-1974C, scale 1:2,500,000.

Skinner, J. W., and Wilde, G. L., 1965, Permian biostratigraphy and fusulinid faunas of the Lake Shasta area, northern California: University of Kansas Paleontological Contributions, Protozoa 6, p. 1–98.

——, 1966, Permian fusulinids from Pacific Northwest and Alaska: University of Kansas Paleontology Contributions, Paper 4, p. 1–16.

Skipp, B., and Hall, W. E., 1980, Upper Paleozoic paleotectonics and paleogeography of the west-central United States: Denver, Colorado, Rocky Mountain Section Society of Economic Paleontologists and Mineralogists Rocky Mountain Paleogeography Symposium 1, p. 387–422.

Smith, D. L., 1989, Structural geology and tectonic development of the Toiyake Range, Louder and Nye Counties, Nevada [Ph.D. thesis]: Stanford, California, Stanford University, 169 p.

Smith, D. L., and Gilmour, E. H., 1979, The Mississippian and Pennsylvanian (Carboniferous Systems) in the United States; Montana: U.S. Geological Survey Profession Paper 1110-M-DD, p. X1-X32.

Smith, D. L., and Miller, E. L., 1990, Late Paleozoic extension in the Great Basin, western United States: Geology, v. 18, p. 712–715.

Snyder, W. S., and Brueckner, H. K., 1983, Tectonic evolution of the Golconda allochthon, Nevada; Problems and perspectives, *in* Stevens, C. H., ed., Pre-Jurassic rocks in western North America suspect terranes: Society of Economic Paleontologists and Mineralogists Pacific Section, p. 103–123.

Sosson, M., 1984, Contribution a l'etude geologique des Cordilleres Nord-Americaines. Une coupe de part et d'autre du chevauchement de Golconda dans le nord-ouest du Nevada, Etats-Unis [Ph.D. thesis]: Memoires des Sciences de la Terre, Academie de Paris, Universite Pierre et Marie Curie, 189 p.

Speed, R. C., 1977a, Island-arc and other paleogeographic terranes of late Paleozoic age in the western Great Basin, *in* Stewart, J. H., Stevens, C. H., and Fritsche, A. E., eds., Paleozoic paleogeography of the western United States: Los Angeles, California, Pacific Section, Society of Economic Paleontologists and Mineralogists Pacific Coast Paleogeography Symposium 1, p. 349–362.

——, 1977b, Excelsior Formation, west-central Nevada; Stratigraphic appraisal, new divisions, and paleogeographic interpretations, *in* Stewart, J. H., Stevens, C H., and Fritsche, A. E., eds., Paleozoic paleogeography of the western United States: Los Angeles, California, Pacific Section, Society of Economic Paleontologists and Mineralogists Pacific Coast Paleogeography Symposium, p. 325–336.

——, 1977c, An appraisal of the Pablo Formation of presumed late Paleozoic age, central Nevada, *in* Stewart, J. H., Stevens, C. H., and Fritsche, A. E., eds., Paleozoic paleogeography of the western United States: Los Angeles, California, Pacific Section, Society of Economic Paleontologists and Mineralogists Pacific Coast Paleogeography Symposium 1, p. 315–324.

——, 1979, Collided Paleozoic microplate in the western United States: Journal of Geology, v. 87, p. 279–292.

——, 1984, Paleozoic and Mesozoic continental margin collision zone features, Minato Candelaria, Nevada, Traverse, *in* Lintz, J., Jr., ed., Western geological excursions, v. 4; Geological Society of America 1984 Annual Meeting, Reno, Nevada: Reno, Nevada, Mackay School of Mines, p. 66–80.

Speed, R. C., and Sleep, N. H., 1982, Antler orogeny and foreland basin; A model: Geological Society of America Bulletin, v. 93, p. 815–828.

Speed, R. C., and MacMillan, J. R., Poole, F. G., and Kleinhampl, F. J., 1977, Diablo Formation, central western Nevada; composite of deep and shallow water upper Paleozoic rocks, *in* Stewart, J. H., Stevens, C. H., and Fritsche, A. E., eds., Paleozoic paleogeography of the western United States: Los Angeles, California, Pacific Section, Society of Economic Paleontologists and Mineralogists Pacific Coast Paleogeography Symposium 1, p. 301–314.

Standee, L. A., and Nestell, M. K., 1985, Age and tectonic significance of terranes adjacent to the Melones fault zone, N. Sierra Nevada, California: Geological Society of America Abstracts with Programs, v. 17, p. 410.

Stanley, K. O., Chamberlain, C. K., and Stewart, J. H., 1977, Depositional setting of some eugeosynclinal Ordovician rocks and structurally interleaved Devonian rocks in the Cordillera mobile belt, Nevada, *in* Stewart, J. H., Stevens, C. H., and Fritsche, A. E., eds., Paleozoic paleogeography of the western United States: Los Angeles, California, Pacific Section, Society of Economic Paleontologists and Mineralogists Pacific Coast Paleogeography Symposium 1, p. 259–274.

Stevens, C. H., 1977, Permian depositional provinces and tectonics, western United States, *in* Stewart, J. H., Stevens, C. H., and Fritsche, A. E., eds., Paleozoic paleogeography of the western United States: Pacific Section, Society of Economic Paleontologists and Mineralogists Pacific Coast Paleogeography Symposium 1, p. 113–136.

——, 1979, Lower Permian of the central Cordilleran miogeosyncline: Geological Society of America Bulletin, v. 90, pt. 2, p. 381–455.

——, 1982, The Early Permian thysanophyllum coral belt; Another clue to Permian plate-tectonic reconstructions: Geological Society of America Bulletin, v. 93, p. 798–803.

——, 1983, Corals from a dismembered late Paleozoic paleo-Pacific plateau: Geology, v. 11, p. 603–606.

——, 1985, Reconstruction of Permian paleogeography based upon distribution

of Tethyan faunal elements, *in* Dutro, J. T., and Pfefferkorn, H. W., eds., Compte Rendu/Neuvieme Congres International de Stratigraphie et de Geologie du Carbonifere, Washington and Champaign-Urbana, 5: Palaeontology, Palaeoecology, Palaeogeography, May 17–26, 1979: Carbondale, Southern Illinois University Press, p. 383–393.

——, 1986, Evolution of the Ordovician through Middle Pennsylvanian carbonate shelf in east-central California: Geological Society of America Bulletin, v. 97, p. 11–25.

——, 1987, Affinities of Permian fusulinid faunas in the Golconda allochthon and northern Sierra Nevada: Geological Society of America Abstracts with Programs, v. 19, p. 455.

Stevens, C. H., and Ridley, A. P., 1974, Mid-Paleozoic off-shelf deposits in southeastern California; Evidence for proximity of the Antler orogenic belt?: Geological Society of America Bulletin, v. 85, p. 27–32.

Stevens, C. H., and Rycerski, B. A., 1983, Permian colonial rugose corals in the western Americas; Aids in positioning of suspect terranes, *in* Stevens, C. H., ed., Pre-Jurassic rocks in western North American suspect terranes: Los Angeles, Pacific Section, Society of Economic Paleontologists and Mineralogists, p. 23–36.

Stevens, C. H., and Stone, P., 1988, Early Permian thrust faults in east-central California: Geological Society of America Bulletin, v. 100, p. 552–562.

Stevens, C. H., Dunne, G. C., and Randall, R. G., 1979, Carboniferous stratigraphy of part of eastern California; The Mississippian and Pennsylvanian (Carboniferous) Systems in the United States; California, Oregon, and Washington: U.S. Geological Survey Professional Paper 1110-CC, p. CC10–21.

Stevens, C. H., Miller, M. M., and Nestell, M., 1987, A new Permian waagenophyllid coral from the Klamath Mountains, California: Journal of Paleontology, v. 61 p. 690–699.

Stewart, J. H., 1972, Initial deposits in the Cordilleran geosyncline; Evidence of a Late Precambrian (850 m.y.) continental separation: Geological Society of America Bulletin, v. 83, p. 1345–1360.

Stewart, J. H., 1979, Geological map of the Miller Mountain and Columbus Quadrangle, Nevada: U.S. Geological Survey Open-File Report 79–1145, 25 p., scale 1:24,000.

——, 1980a, Regional tilt patterns of late Cenozoic basin-range fault blocks, western United States: Geological Society of America Bulletin, v. 91, part 1, p. 460–464.

——, 1980b, Geology of Nevada: Nevada Bureau of Mines and Geology Special Publication 4, p. 136.

——, 1984, Siliceous Ordovicia strata and structurally interleaved Devonia calcerous gravity flow deposits in the Miller Mountain area, western Nevada, in Carr, M. D., ed., Field trip guide to the Paleozoic stratigraphy and tectonics of the southwestern Great Basin, California and Nevada (Field Trip 7): Geological Society of America Field Trip for 1984 Annual Meeting, Reno, Nevada, p. 132–139.

Stewart, J. H., and Suczek, C. A., 1977, Cambrian and latest Precambrian tectonics in the western United States, *in* Stewart, J. H., Stevens, C. H., and Fritsch, A. E., eds., Paleozoic paleogeography of the western United States: Society of Economic Paleontologists and Mineralogists Pacific Section, Pacific Coast Paleogeography Symposium 1, p. 1–18.

Stewart, J. H., MacMillan, J. R., Nichols, K. M., and Stevens, C. H., 1977, Deep-water upper Paleozoic rocks in north-central Nevada; A study of the type area of the Havallah Formation, *in* Stewart, J. H., Stevens, C. H., and Fritsche, A. E., eds., Paleozoic paleogeography of the western United States: Los Angeles, California, Pacific Section, Society of Economic Paleontologists and Mineralogists Pacific Coast Paleogeography Symposium 1, p. 337–347.

Stewart, J. H., Murchey, B., Jones, D. L., and Wardlaw, B. R., 1986, Paleontologic evidence for complex tectonic interlayering of Mississippian to Permian deep water rocks of the Golconda alochthon in Tobin Range; North-central Nevada: Geological Society of America Bulletin, v. 97, p. 1122–1132.

Stone, P., 1984, Stratigraphy, depositional history, and paleogeographic significance of Pennsylvanian and Permian rocks in the Owens Valley–Death Valley region, California [Ph.D. thesis]: Stanford, California, Stanford University, 399 p.

Stone, P., and Stevens, C. H., 1984, Stratigraphy and depositional history of Pennsylvanian and Permian rocks in the Owens Valley, Death Valley region, eastern California, *in* Lintz, J., Jr., ed., Western geological excursions, v. 4; Geological Society of America 1984 Annual Meeting Field Trip Guidebook: Reno, Nevada, Mackay School of Mines, p. 94–118.

Stone, P., and Stevens, C. H., 1988a, Pennsylvanian and Early Permian paleogeography of east-central California; Implications for the shape of the continental margin and the timing of continental truncation: Geology v. 16, p. 330–333.

——, 1988b, An angular unconformity in the Permian section of east-central California: Geological Society of America Bulletin, v. 100, p. 547–551.

Struik, L. C., 1986, Imbricated terranes of the Cariboo gold belt with correlation and implications for tectonics in southeastern British Columbia: Canadian Journal of Earth Science, v. 23, p. 1047–1061.

Struik, L. C., and Orchard, M. J., 1985, Late Paleozoic conodonts from ribbon chert delineate imbricate thrusts within the Antler Formation of the Slide Mountain terrane, central British Columbia: Geology, v. 13, p. 794–798.

Suczek, C. A., 1977, Sedimentology and petrology of the Cambrian Harmony Formation of north-central Nevada: Geological Society of America Abstracts with Programs, v. 9, p. 510.

Tipper, H. W., Woodsworth, G. J., and Gabrielse, H., 1981, Tectonic assemblage map of the Canadian Cordillera: Geological Survey of Canada, scale 1:200,000.

Tobisch, O. T., and Fiske, R. S., 1982, Repeated parallel deformations in parts of the eastern Sierra Nevada, California, and its implications for dating structural events: Journal of Structural Geology, v. 4, p. 177–195.

Tomlinson, A. J., and Wardlaw, B. R., 1988, Permian-Triassic Sonoma Orogeny, western U.S., and late Cenozoic orogenesis in Japan; Examples of back-arc closure: Geological Society of America Abstracts with Programs, v. 20, p. A109.

Tomlinson, A. J., and Wright, J. E., 1986, The Klamath–Sierran Arc: A possible region for Permian volcaniclastic and conglomeratic units within the Golconda allochthon: Geological Society of America Abstracts with Programs, v. 18, p. 193.

Tomlinson, A. J., Miller, E. L., Holdsworth, B. K., and Whiteford, W. B., 1987, Discussion *on* Structure of the Havallah sequence, Golconda allochthon, Nevada; Evidence for prolonged evolution in an accretionary prism: Geological Society of America Bulletin, v. 98 p. 615–616.

Turner, R.J.W., 1982, The geology of the east-central Tobin Range, Nevada [M.S. thesis]: Stanford, California, Stanford University, 113 p.

Turner, R.J.W., Madrid, R.J., and Miller, E.L., 1989, Roberts Mountains allochthon; Stratigraphic comparison with lower Paleozoic outer continental margin strata of the northern Canadian Cordillera: Geology, 17, p. 341–344.

Vallier, T. L., 1977, The Permian and Triassic Seven Devils Group, western Idaho and northeastern Oregon: U.S. Geological Survey Bulletin 1437, p. 1–58.

Vallier, T. L., Brooks, H. C., and Thayer, T. P., 1977, Paleozoic rocks of eastern Oregon and Idaho, *in* Stewart, J. H., Stevens, C. H., and Fritsche, A. E., eds., Paleozoic paleogeography of the western United States: Pacific Section, Society of Economic Paleontologists and Mineralogists Pacific Coast Paleogeography Symposium 1, p. 421–440.

Varga, R. J., and Moores, E. M., 1981, Age and significance of an unconformity that pre-dates island-arc volcanism in the northern Sierra Nevada: Geology, v. 9, p. 512–518.

Walker, J. D., 1985, Permo–Triassic paleogeography and tectonics of the southwestern United States [Ph.D. thesis]: Cambridge, Massachusetts Institute of Technology, 224 p.

——, 1988, Permian and Triassic rocks of the Mojave Desert and their implications for the timing and mechanisms of continental truncation: Tectonics, v. 7, p. 685–709.

Walker, J. D., Burchfiel, B. C., and Wardlaw, B. R., 1984, Early Triassic overlap sequence in the Mojave Desert; Its implications for Permo–Triassic tectonics

and paleogeography: Geological Society of America Abstracts with Programs, v. 16, p. 685.

Walker, N. W., 1983, Pre-Tertiary tectonic evolution of northeastern Oregon and west-central Idaho; Constraints based on U-Pb ages of zircons: Geological Society of America Abstracts with Programs, v. 15, p. 371.

Wallin, E. T., 1989, Provenance of lower Paleozoic sandstones in the eastern Klamath Mountains and the Roberts Mountains allochthon, California and Nevada [Ph.D. thesis]: Lawrence, University of Kansas, 152 p.

Wallin, E. T., Mattinson, J. M., and Potter, A. W., 1988, Early Paleozoic magmatic events in the eastern Klamath Mountains, northern California: Geology, v. 16, p. 144–148.

Wardlaw, B. R., 1980, Middle-Late Permian paleogeography of Idaho, Montana, Nevada, Utah, and Wyoming, *in* Fouch, T. D., and Magathan, E. R., eds., Paleozoic paleogeography of the west-central United States: Denver, Colorado, Rocky Mountain Section Society of Economic Paleontologists and Mineralogists, Rocky Mountain Paleogeography Symposium 1, p. 353–361.

Wardlaw, B. R., Collinson, J. W., and Ketner, K. B., 1979, Regional relations of middle Permian rocks in Idaho, Nevada, Utah, *in* Newman, G. W., and Goode, H. D., eds., 1979 Basin and Range Symposium: Rocky Mountain Association of Geologists and Utah Geological Association, p. 277–283.

Wardlaw, B. R., Nestell, M. K., and Dutro, J. T., Jr., 1982, Biostratigraphy and structural setting of the Permian Coyote Butte Formation of central Oregon: Geology, v. 10, p. 13–16.

Watkins, R., 1973, Carboniferous faunal associations and stratigraphy, Shasta County, northern California: American Association of Petroleum Geologists Bulletin, v. 57, p. 1743–1764.

——, 1985, Volcaniclastic and carbonate sedimentation in late Paleozoic island-arc deposits, eastern Klamath Mountains, California: Geology, v. 13, p. 709–713.

——, 1986, Late Devonian to Early Carboniferous turbidite facies and basinal development of the eastern Klamath Mountains, California: Sedimentary Geology, v. 49, p. 51–71.

Watkins, R., and Flory, R. A., 1986, Island arc sedimentation in the Middle Devonian Kennett Formation, eastern Klamath Mountains, California: Journal of Geology, v. 94, p. 753–761.

Watkins, R., Reinheimer, C. E., Wallace, J. W., and Nestell, M. K., 1987, Paleogeographic significance of a Permian megamictite in the Central belt of the northern Sierra Nevada: Geological Society of America Bulletin, v. 99, p. 771–778.

Welsh, J. E., and Bissell, H. J., 1979, The Mississippian and Pennsylvanian (Carboniferous) Systems in the United States; Utah: U.S. Geological Survey Professional Paper 1110–YY, p. Y1–Y35.

Wernicke, B., Spencer, J. E., Burchfiel, B. C., and Guth, P. L., 1982, Magnitude of crustal extension in the southern Great Basin: Geology, v. 10, p. 645–647.

Wheeler, J. O., and Gabrielse, H., 1972, The Cordilleran structural province; Variations in tectonic styles in Canada: Geological Association of Canada Special Paper 11, p. 1–81.

Whetten, J. T., Jones, D. L., Cowan, D. S., and Zartman, R. E., 1978, Ages of Mesozoic terranes in the San Juan Islands, Washington, *in* Howell, D. G., and McDougall, K. A., eds., Mesozoic paleogeography of the western United States: Los Angeles, California, Pacific Section Society of Economic Paleontologists and Mineralogists, Pacific Coast Paleogeography Symposium 2, p. 117–132.

Whiteford, W. B., 1990, Paleogeographic setting of the Schoonover sequence, Nevada, and implications for the late Paleozoic margin of western North America, *in* Harwood, D. S., and Miller, M. M., eds., Paleozoic and early Mesozoic paleogeographic relations of the Sierra Nevada, Klamath Mountains, and related terranes: Geological Society of America Special Paper 255, p. 115–136.

Wilson, E. C., 1982, Wolfcampian rugose and tabulate corals (Coelenterata: Anthozoa) from the Lower Permian McCloud Limestone of northern California: Natural History Museum of Los Angeles County Contributions in Science, 90 p.

——, 1984, Stratigraphic range extensions and coral province affiliations of stony corals in the Lower Permian McCloud Limestone of California: Palaeontographica Americana 54, p. 437.

——, 1985, Rugose corals (Coelenterata, Anthozoa) from the Lower Permian McCloud Limestone at Tombstone Mountain, northern California: Natural History Museum of Los Angeles County Contributions in Science, 11 p.

Wright, J. E., 1982, Permo–Triassic accretionary subduction complex, southwestern Klamath Mountains, northern California: Journal of Geophysical Research, v. 87, p. 3805–3818.

Yancey, T. E., 1979, Permian positions of the Northern Hemisphere continents as determined from marine biotic provinces, *in* Gray, J., and Boucot, A. J., eds., Historical biogeography, plate tectonics, and the changing environment; Proceedings of the 37th Annual Biology Colloquium and Selected Papers: Corvalis, Oregon State University Press, p. 239–248.

Yancey, T. E., Stevens, C. H., and Hanger, R. A., 1988, Geographic placement of the eastern Klamath terrane in the Pacific Ocean basin during the Early Permian: Geological Society of American Abstracts with Programs, v. 20, p. A308.

Yates, R. G., 1976, Geology of the Deep Creek area, Washington, and its regional significance: U.S. Geological Survey Open-File Report 76–537, p. 435.

Manuscript Accepted by the Society March 20, 1990

ACKNOWLEDGMENTS

A compilation of this scope would not have been possible without the input of innumerable colleagues working on the Paleozoic of the Cordillera. We gratefully thank all of those who have shared their data and ideas with us but assume full responsibility for any errors in compilation and for the interpretations presented herein. In terms of our respective contributions to the chapter, Meghan Miller was responsible for the compilation and discussion of data from the allochthonous island-arc assemblage rocks; Cal Stevens for the compilation and discussion of the craton and continental shelf region, and insight into the "Tethyan" realm of the paleo-Pacific Ocean; Raul Madrid for the section on the stratigraphy, paleogeography, and structural history of the Roberts Mountains allochthon; Jim Wright for the discussion of the subduction-zone complex of Cache Creek affinity and its relevance to Permo-Triassic arc magmatism and tectonism; and Elizabeth Miller for the sections on the Antler orogenic belt, foredeep basin, the Havallah sequence, and the interpretations of the Antler and Sonoma orogenies. Need we add that not all of us agree with the interpretations presented here. We gratefully acknowledge the input and corrections by our first round of reviewers, A. Tomlinson, R. C. Speed, P. Palmer, B. C. Burchfiel, and an especially helpful review by R. Schweickert. Additional final reviews by S. Wyld and J. Saleeby are also much appreciated. Drafting of most of the figures was done by Jeanette Smith, whose energy and time were critical factors for the final assembly of Plate 4. D. L. Smith performed miracles with the reference list on the Mac, for which we are grateful.

Printed in U.S.A.

The Geology of North America
Vol. G-3, The Cordilleran Orogen: Conterminous U.S.
The Geological Society of America, 1992

Chapter 4

*Early Mesozoic tectonic evolution of the western U.S. Cordillera**

Jason B. Saleeby
Division of Geological and Planetary Sciences, California Institute of Technology, Pasadena, California 91125
Cathy Busby-Spera
Department of Geological Sciences, University of California, Santa Barbara, California 93106

With contributions by:
J. S. Oldow
Department of Geology and Geophysics, Rice University, Houston, Texas 77251
G. C. Dunne
Department of Geological Sciences, California State University, Northridge, California 91330
J. E. Wright
Department of Geology and Geophysics, Rice University, Houston, Texas 77251
D. S. Cowan
Department of Geological Sciences, University of Washington, Seattle, Washington 98195
N. W. Walker
Department of Geological Sciences, University of Texas, Austin, Texas 78712
R. W. Allmendinger
Department of Geological Sciences, Cornell University, Ithaca, New York 14853

INTRODUCTION

The early Mesozoic evolution of the U.S. Cordillera differs greatly from its previous history of mainly miogeoclinal sedimentation with outboard marginal-basin–island-arc mobile zones. The Early Mississippian and Permian-Triassic thrust emplacement of eugeoclinal strata across the miogeocline signaled the initial propagation of subduction-related tectonism onto the sialic edge. Following these events, the sialic edge and the resulting accreted terranes became an active continental margin. The active margin history records not only eastward subduction of oceanic crust beneath North America, but also the formation, migration, and accretion of marginal basin and fringing island-arc systems along the continental margin. At the close of the Jurassic, the fringing arc–marginal basin system collapsed, resulting in a more direct interaction of major Pacific basin plates with hte Cordilleran margin. Such interactions are manifested by Andean and San Andreas types of marginal regimes which characterized the Cretaceous and Cenozoic. In this chapter we will discuss the tectonic evolution of the U.S. Cordilleran margin during the early phases of its active margin history (Triassic through Jurassic).

*The reference list for this chapter is on the microfiche included with this volume: Hard copies may be obtained free of charge by writing the senior author.

A number of fundamental problems are highlighted throughout our discussions. These are summarized in the following questions: (1) What constituted the edge of the North American plate through time versus accreted vestiges of smaller marginal basin–fringing island-arc plates? (2) What was the role of collisional tectonics in the accretion of the ensimatic assemblages, and can the structural consequences of collision be distinguished from the consequences of strongly coupled subduction? (3) The time span under consideration included the initiation of a nearly Cordillera-wide foreland deformation belt. Was this a response to subduction, collision, continental-arc development, or some combination of these? (4) There is abundant evidence of major strike-slip faulting and rifting episodes having operated in series and in parallel with convergent tectonics. How severely did these episodes disrupt the record of convergence and continental-margin growth, and how severely have episodes of contractile deformation masked early episodes of strike-slip faulting and extension? (5) From tectonic maps of North America (King, 1966; Muehlberger and others, in preparation) it is clear that significantly more material was accreted to the northern Cordillera than in the southwest. Much of the northern Cordillera was accreted during or shortly after the time span considered here, and paleomagnetic as well as paleobiogeographic data suggests that much of this material was transported northward over

Saleeby, J. B., and Busby-Spera, C., 1992, Early Mesozoic tectonic evolution of the western U.S. Cordillera, *in* Burchfiel, B. C., Lipman, P. W., and Zoback, M. L., eds., The Cordilleran Orogen: Conterminous U.S.: Boulder, Colorado, Geological Society of America, The Geology of North America, v. G-3.

substantial distances. What were the possible relations between the northern Cordilleran accretionary terranes and the southwest Cordilleran margin? In considering the problems outlined above, our prime purpose is not only to offer our preferred solutions and interpretations, but also to point out some of the major contrasting models.

General subdivisions of the early Mesozoic margin

The early Mesozoic continental margin is subdivided on the basis of both preexisting and superposed tectonic elements. The preexisting elements consist of the craton edge and miogeocline (Poole and others, this volume), and marginal-basin–volcanic-arc terranes accreted to the miogeocline during the Early Mississippian Antler orogeny and the Permian-Triassic Sonoma orogeny (Miller and others, this volume). The accreted terranes constitute the main elements of a generally west-facing Paleozoic arc and basinal system referred to as the McCloud arc (Miller, 1987). This is named after a faunally-distinct carbonate reef interval (McCloud limestone) within the eastern Klamath Mountains arc sequence. Together, the accreted fragments of the McCloud arc and back-arc basin, and the adjacent miogeocline and craton edge, formed the basement framework for native rock sequences of the early Mesozoic margin. The known western limit of such sequences and their varied basements lies in the eastern Klamath Mountains, northern and axial Sierra Nevada, and Mojave Desert (Fig. 1). Lower Mesozoic rocks between these areas and the craton span the regions of a continental arc and back-arc basin. The craton edge lies farther to the east, inboard of the foreland fold and thrust belt (Fig. 1). A well-recognized but enigmatic boundary cuts obliquely across the preexisting crustal elements in northern and central California. The boundary has been termed the Foothills suture (Saleeby, 1981). Rocks west of the suture consist primarily of highly disrupted Paleozoic to early Mesozoic ophiolites interleaved with early Mesozoic volcanic-arc and basinal terranes. This zone of disrupted terranes continues into the central and western Klamath Mountains and is collectively called the western Sierra–Klamath belt (Figs. 1 and 2).

Early Mesozoic rocks of the western Sierra–Klamath belt and adjacent Great Valley constitute the fore-arc for the early Mesozoic continental arc, as well as the Cretaceous Sierra Nevada batholith. Fore-arc basement rocks, where preserved, consist of Paleozoic and early Mesozoic ophiolite (Saleeby, this volume). The early Mesozoic sequences of the western Sierra–Klamath belt and the Great Valley are structurally allochthonous, but are considered here not to be exotic to the Cordilleran margin. A widely recognized alternative model that considers these as exotic to North America was presented in Schweickert and Cowan (1975).

Mesozoic tectonic relations between the northern Washington, eastern Oregon, and the California-Nevada regions are obscured by the cover of Cenozoic lava fields. Cretaceous and younger tectonic events have further obscured relations between lower Mesozoic rocks of northern Washington and eastern Oregon and the adjacent craton. The tectonic history of the western Sierra–Klamath belt and Great Valley parallels that of the adjacent continental arc and back-arc basin. The tectonic histories of the Blue Mountains region of eastern Oregon and the northern Washington region, in contrast, are not readily related to nearby continental tectonics to the east. In eastern Oregon and northeastern Washington, the terranes appear to be northern equivalents of the McCloud arc, but facies links to the adjacent miogeocline, such as those observed in lower Mesozoic rocks in the southwestern United States, are lacking. Additionally, the northwestern Washington terranes include exotic elements that are lacking to the south, but that characterize large parts of the Pacific Northwest Cordillera.

Recognition of the plate edge

In general, the southwest Cordilleran plate edge for early Mesozoic time can be approximated through time by a composite belt of continental-arc rocks. Fore-arc basin and subduction complexes cannot be equally delineated through time, however. Geologists tend to look at thick terrigenous sedimentary prisms, like the Cretaceous Great Valley sequence of California, as being representative of fore-arc regimes. Likewise, imbricated wedges of metamorphosed terrigenous and pelagic sedimentary rocks with interleaved oceanic igneous rocks, and high P/T metamorphic rocks, like the Franciscan complex of California, are often sought as the definitive products of subduction. Such fore-arc–subduction assemblages are locally represented in lower Mesozoic rocks of the western Sierra–Klamath belt and in the Blue Mountains of eastern Oregon. But in general, and particularly along the Sierra-Klamath belt, the areal extent and petrogenetic histories of these assemblages do not match the spatial and temporal configurations of the continental arc.

Recent studies in modern fore-arc regimes of the western Pacific indicate that the Great Valley–Franciscan model of fore-arc–subduction complexes has only limited application. First, strongly versus weakly coupled subduction zones are both widely recognized (Kanamori, 1986), and those that are weakly coupled are characterized by fore-arc as well as back-arc extension. Fore-arc and inner trench walls under extension can undergo removal of material by subduction, rifting, and strike-slip faulting, instead of the growth of an accretionary prism. Extensional fore-arcs are also known to be magmatically active; the eruptions of island-arc–type tholeiites, boninite-series rocks, and even lavas of calc-alkaline and shoshinite affinity are known (Bloomer, 1981; Natland, 1981; Kobayashi, 1983; Sharaskin and others, 1983a, 1983b; Jakes and Miyake, 1984; Ishii, 1985; Ogawa and others, 1985; Crawford and others, 1986; Saleeby, this volume). Addi-

Figure 1. Generalized tectonic map of the U.S. Cordillera showing major divisions of the continental margin, locations of columnar sections given in Figures 3 through 7, and palinspastic restoration in the construction of the Figure 2 base. Restoration steps shown in red. Data sources given in text.

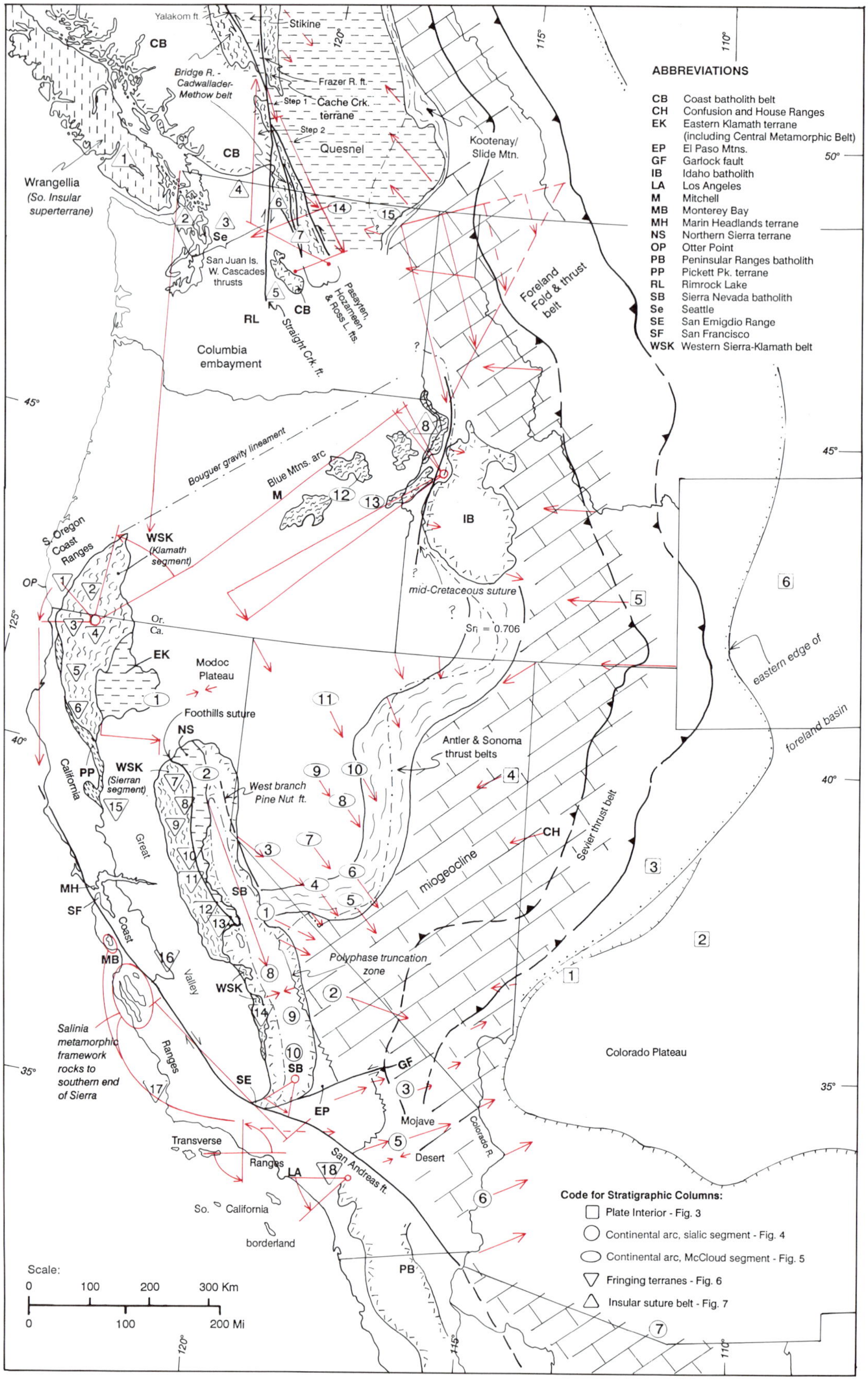
ABBREVIATIONS
CB Coast batholith belt
CH Confusion and House Ranges
EK Eastern Klamath terrane (including Central Metamorphic Belt)
EP El Paso Mtns.
GF Garlock fault
IB Idaho batholith
LA Los Angeles
M Mitchell
MB Monterey Bay
MH Marin Headlands terrane
NS Northern Sierra terrane
OP Otter Point
PB Peninsular Ranges batholith
PP Pickett Pk. terrane
RL Rimrock Lake
SB Sierra Nevada batholith
Se Seattle
SE San Emigdio Range
SF San Francisco
WSK Western Sierra-Klamath belt
Code for Stratigraphic Columns:
Plate Interior - Fig. 3
Continental arc, sialic segment - Fig. 4
Continental arc, McCloud segment - Fig. 5
Fringing terranes - Fig. 6
Insular suture belt - Fig. 7
Yalakom ft.
Stikine
Bridge R. - Cadwallader-Methow belt
Frazer R. ft.
Step 1
Cache Crk. terrane
Step 2
Quesnel
Kootenay/ Slide Mtn.
Wrangellia (So. Insular superterrane)
San Juan Is. W. Cascades thrusts
Pasayten, Hozameen & Ross L. fts.
Straight Crk. ft.
Foreland Fold & thrust belt
Columbia embayment
Bouguer gravity lineament
Blue Mtns. arc
S. Oregon Coast Ranges
WSK (Klamath segment)
mid-Cretaceous suture
Sri = 0.706
Or.
Ca.
Modoc Plateau
Foothills suture
West branch Pine Nut ft.
Antler & Sonoma thrust belts
WSK (Sierran segment)
California
Great Valley
Coast Ranges
miogeocline
Sevier thrust belt
eastern edge of foreland basin
Polyphase truncation zone
Salinia metamorphic framework rocks to southern end of Sierra
Colorado Plateau
Mojave Desert
Colorado R.
Transverse Ranges
San Andreas ft.
So. California borderland
Scale:
0 100 200 300 Km
0 100 200 Mi
45°
40°
35°
50°
125°
120°
115°
110°

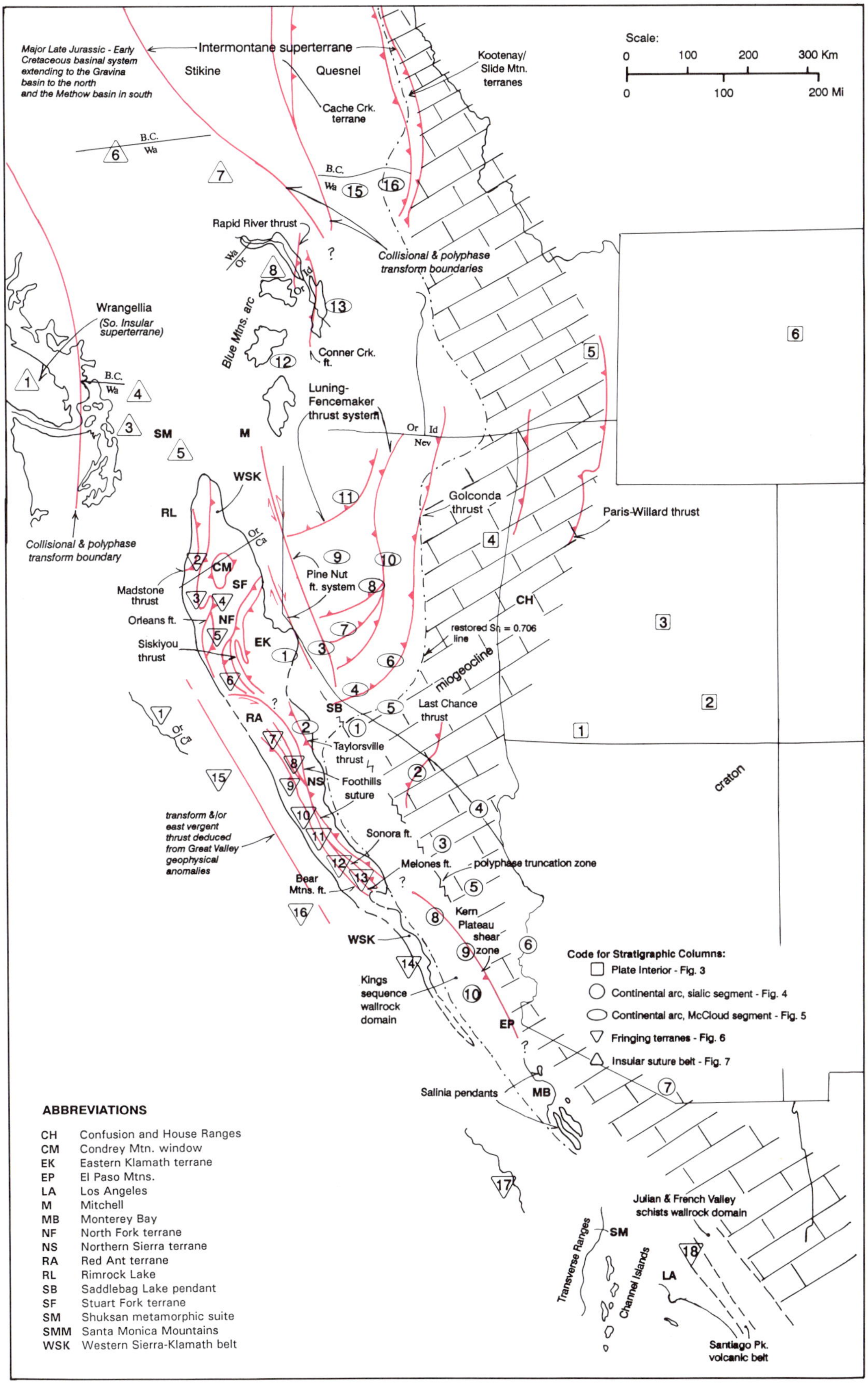
Major Late Jurassic - Early Cretaceous basinal system extending to the Gravina basin to the north and the Methow basin in south
Intermontane superterrane
Stikine
Quesnel
Kootenay/ Slide Mtn. terranes
Scale:
0 100 200 300 Km
0 100 200 Mi
Cache Crk. terrane
B.C.
Wa
Rapid River thrust
Collisional & polyphase transform boundaries
Blue Mtns. arc
Wrangellia
(So. Insular superterrane)
Conner Crk. ft.
Luning-Fencemaker thrust system
Or
Id
Nev
SM
M
WSK
RL
Golconda thrust
Paris-Willard thrust
Collisional & polyphase transform boundary
Madstone thrust
CM
SF
Pine Nut ft. system
Orleans ft.
NF
CH
Siskiyou thrust
EK
restored Sri = 0.706 line
miogeocline
SB
Last Chance thrust
RA
Taylorsville thrust
Foothills suture
NS
craton
transform &/or east vergent thrust deduced from Great Valley geophysical anomalies
Sonora ft.
Melones ft.
polyphase truncation zone
Bear Mtns. ft.
Kern Plateau shear zone
Kings sequence wallrock domain
EP
Code for Stratigraphic Columns:
Plate Interior - Fig. 3
Continental arc, sialic segment - Fig. 4
Continental arc, McCloud segment - Fig. 5
Fringing terranes - Fig. 6
Insular suture belt - Fig. 7
Salinia pendants
MB
ABBREVIATIONS
CH Confusion and House Ranges
CM Condrey Mtn. window
EK Eastern Klamath terrane
EP El Paso Mtns.
LA Los Angeles
M Mitchell
MB Monterey Bay
NF North Fork terrane
NS Northern Sierra terrane
RA Red Ant terrane
RL Rimrock Lake
SB Saddlebag Lake pendant
SF Stuart Fork terrane
SM Shuksan metamorphic suite
SMM Santa Monica Mountains
WSK Western Sierra-Klamath belt
Julian & French Valley schists wallrock domain
Transverse Ranges
Channel Islands
Santiago Pk. volcanic belt

tionally, extensional fore-arcs may undergo copious serpentinite diapirism (Bloomer, 1983; Bloomer and Hawkins, 1983; Ocean Drilling Program, 1989) which may transport high *P/T* metamorphic blocks up and out of subduction-zone metamorphic enviroments (Saleeby, this volume). Large domains of fore-arc crust may also decouple as semirigid microplates bounded by suprasubduction-zone transtensional, transpressional, and/or normal-plate junctures (Curray and others, 1979; Kimura, 1986). Fore-arc microplates may also serve as transient basement for volcanic-arc eruptions and intrusions during migration. The range of tectonic and petrogenetic phenomena known to occur in fore-arc regimes provides an important interpretive basis for this synthesis.

Cordillera-wide considerations

Throughout the early Mesozoic, the southwest Cordilleran plate edge was profoundly influenced by the transport and accretion of several major terranes that together constitute much of the Cordilleran orogen between southern Alaska and central California. Most of these terranes lie north of the U.S.–Canadian border, but geologic, paleomagnetic, and paleobiogeographic data suggest substantial northward transport relative to North America for a number of the terranes (Monger, 1984). We must, therefore, consider the broad stratigraphic affinities and tectonic histories of these terranes and the possible influences they may have had on events in the southwest.

The southern termination of the Pacific Northwest terrane assembly is shown in Figure 1. The first major grouping consists of the Quesnel-Kootenay and Slide Mountain terranes. Quesnel consists mainly of middle Paleozoic to Jurassic-Triassic volcanic-arc strata. The upper Paleozoic strata contain the distinct faunal assemblage that characterizes the McCloud arc sequence of the eastern Klamath Mountains. Along the inner edge of Quesnel lie the Kootenay and Slide Mountain terranes, which consist mainly of lower Paleozoic and Mississippian continental-rise and marginal-basin strata, respectively; these are locally overlain unconformably by Middle(?) and Upper Triassic back-arc–basin strata related to the younger volcanic rocks of Quesnel (Monger, 1984). The Kootenay terrane has affinities to the Roberts Mountains allochthon of the Antler belt, and the Slide Mountain terrane has affinities to the Golconda allochthon of the Sonoma belt (Gehrels and Smith, 1987; Miller and others, this volume; Saleeby, this volume). The Middle(?)-Upper Triassic back-arc basinal strata also resemble early Mesozoic marine strata of northwest Nevada which lap across the Antler-Sonoma belts.

The outer edge of Quesnel is bounded by a highly disrupted oceanic terrane (Davis and others, 1978; Monger, 1984). The Cache Creek terrane occurs discontinuously for ~1,300 km and consists of Mississippian to local Upper Triassic, and rare Lower to Middle(?) Jurassic, chert, argillite, basalt, alpine-type ultramafites, extensive carbonate rocks, and local blueschist with Middle to Late Triassic metamorphic ages. The thicker carbonate sequences are late Paleozoic in age and contain a distinct shallow-water "Tethyan" fauna that formed on ocean islands and seamounts in the equatorial proto-Pacific (Monger and Ross, 1971; Saleeby, 1981, 1983, this volume; Ross and Ross, 1983; Miller and Wright, 1987). Fragments of such carbonates occur in association with Cache Creek–affinity rocks of northern Washington, the Blue Mountains, and along the western Sierra–Klamath belt. The Cache Creek–affinity terranes of the southwest United States lie along the outer edge of McCloud arc rocks. In contrast, the type Cache Creek terrane is structurally surrounded by major belts of McCloud arc rocks, including Quesnel on the east and Stikine on the west. Jurassic-Triassic arc rocks of Stikine and Quesnel are partly correlative (Davis and others, 1978; Monger, 1984). Structural and possible depositional basement for part of Stikine includes lower Paleozoic continental-rise strata of the Yukon-Tanana terrane (Rubin and others, 1990a). The Stikine, Cache Creek, Quesnel, and Kootenay–Slide Mountain terranes together were accreted to North America by Middle-Late Jurassic time. This major accretionary complex is referred to as the Intermontane superterrane. Plutonism and metamorphism of the Omineca crystalline belt was superposed across the Intermontane superterrane in Middle Jurassic to Early Cretaceous time (Monger and others, 1982). The initial phases of Intermontane superterrane accretion, however, may have begun during the Sonoma orogeny (Saleeby, this volume). The later Jurassic phases of accretion are considered in this synthesis to have been driven by the collision of another major superterrane (Insular) to the west (Saleeby and Rubin, 1990).

The Insular superterrane consists of Wrangellia, the Alexander terrane, and the Peninsular terrane, which extends for ~2,500 km between Vancouver Island (Fig. 1) and the Alaska Peninsula (Saleeby, 1983). Wrangellia is characterized by Paleozoic island-arc strata overlain by thick basaltic flows and carbonate rocks of Late Triassic age, and Jurassic and Cretaceous island-arc and basinal strata (Berg and others, 1972; Jones and others, 1977; Saleeby, 1983; Monger, 1984). Paleomagnetic and paleobiogeographic data indicate that the Insular superterrane migrated from subtropical latitudes to northern temperate latitudes in the eastern Pacific between Late Triassic and Late Jurassic time (Hillhouse, 1977; Yole and Irving, 1980; Hillhouse and Gromme, 1984; Taylor and others, 1984; Tipper, 1984, 1985). A possible fragment of Wrangellia occurs as the Wallowa terrane in the northern Blue Mountains outboard of Cache Creek–affinity and McCloud arc rocks (Davis and others, 1978; Wernicke and Klepacki, 1988). The suture between the Insular and Intermontane superterranes is extensively invaded by the Coast batholith belt of primarily Late Cretaceous to Eocene age (Armstrong, 1988).

A narrow belt of smaller terranes lies between the Insular

Figure 2. Palinspastic base map derived from restoration steps shown in Figure 1. Locations of columnar sections are shown in restored positions. Major faults of early Mesozoic age are also shown (in red) in restored positions. Palinspastic base is used in Plate 5 tectonic reconstruction. Data sources are given in text.

superterrane and Coast batholith in southern British Columbia and northern Washington. In southern British Columbia, this belt consists of the Bridge River and Cadwallader terranes and the overlapping Methow basinal sequence. The Bridge River terrane consists of Permian to Middle Jurassic chert, argillite, basalt, Alpine-type ultramafite, and local limestone (Potter, 1983). It has strong similarities to the Cache Creek terrane to the east, but has yet to yield the diagnostic late Paleozoic "Tethyan" fauna. The Cadwallader terrane is characterized by Upper Triassic oceanic-arc lavas and arc-derived sediments which are unconformably(?) overlain by northern deposits of the Methow basin (Rushmore, 1987). The Methow basin intervened between the Insular and Intermontane superterranes from Late Jurassic through much of Early Cretaceous time within a probable extensional basin that formed along the Jurassic suture. This basin collapsed in the mid-Cretaceous when the Bridge River and Cadwallader terranes, the overlying basinal strata, and the Insular superterrane were imbricated with the Intermontane superterrane and the rest of North America (Rubin and others, 1990b). This polyphase accretionary zone is referred to below as the Insular suture belt.

Despite the great discrepancy in relative volumes of accretionary terranes between the Pacific Northwest and the southwestern United States, major accretionary sutures of both regions are characterized by postaccretion batholithic belts. Southern analogues to the Coast and Omineca belts include the Idaho batholith, which stitched terranes of the Blue Mountains to the continent starting in the Late Cretaceous, and the Sierra Nevada batholith, which stitched the western Sierra–Klamath belt to the continent starting in the Early Cretaceous. In close conjunction with the major stitching episodes, Late Cretaceous dextral-slip faults displaced fore-arc and western zone–arc rocks of the California and Blue Mountain regions (Saleeby, 1991). Similar dextral faulting affected the major accretionary terranes of the Pacific Northwest, but there they developed in conjunction with Insular suture-belt tectonics and also persisted into early Tertiary time.

Time-slice paleotectonic-paleogeographic maps

The early Mesozoic tectonic evolution of the southwest U.S. Cordillera for six time intervals is described. These include: (1) Early to Middle Triassic; (2) Late Triassic; (3) Early Jurassic; (4) Middle Jurassic to the base of the Callovian; (5) Callovian-Oxfordian; and (6) Kimmeridgian-Tithonian. The first interval includes the final phases of the Sonoma orogeny and was covered in depth by Miller and others (this volume) and also summarized in Saleeby (this volume). Thus, our treatment of this interval will be brief and will emphasize stratigraphic and tectonic relations which relate most directly to post-Sonoman tectonics and paleogeography. Stratigraphic, structural, and tectonic relations within the Late Triassic through Jurassic intervals constitute the focus of this synthesis. Paleotectonic-paleogeographic maps are presented for each time interval in Plate 5. The maps summarize the main tectonic and petrogenetic events that characterized each time interval. Key stratigraphic sequences and relations are given in Figures 3 through 7, the explanation of symbols is given in Figure 3. The plate interior columns (Fig. 3) resemble conventional stratigraphic columns, whereas columns from the continental arc, fringing terranes, and Insular suture belt are, in many cases, reconstructed from disrupted sequences (Figs. 4–7). Where age data are sparse, the stratigraphic locations of key faunal or isotopic ages are shown along the columns. Correlation between relative and absolute time scales is after Palmer (1983). Within the text, the relative time-scale age is used for isotopic ages where rigorous error analyses are not provided, or where they yield large uncertainties.

The symbols of Plate 5 primarily portray first-order tectonic, depositional, and petrogenetic phenomena. The colors represent major tectonic domains; the lithologic patterns denote the formation of dominant rock types in areas where there are at least reasonable constraints that such rocks formed. Structural symbols pertain to active structures for the time interval considered. Regional metamorphic overprint symbols likewise denote active conditions that are known to have existed at depth. The time-slice maps are constructed on a palinspastic base for the Early Cretaceous.

Palinspastic base. The time intervals covered in the Plate 5 reconstructions predate the Cretaceous Period, during which first-order changes occurred throughout the Cordillera (Oldow and others, 1989; Cowan and Bruhn, this volume). In addition to widespread batholith emplacement, the Cretaceous was characterized by major longitudinal strike-slip displacements. Early Cretaceous displacements, like those of earlier Mesozoic time, are difficult to quantify in sense, time duration, and magnitude. Mid- to Late Cretaceous as well as Cenozoic displacements were clearly dominated by dextral senses and profoundly influenced the distribution of lower Mesozoic rocks of the continental arc and more outboard terranes. Such displacements are accounted for below in the construction of a palinspastic base for Plate 5.

The elements of the Plate 5 palinspastic reconstruction are shown in red in Figure 1 and the resulting base with the restored locations of the key stratigraphic columns and geographic markers are shown in Figure 2. The Colorado Plateau is used as the reference domain. Many of the palinspastic elements have substantial uncertainty; however, we have chosen a sequence of steps that we believe to be fairly accurate, and where various data sets or ideas are in conflict we have resorted to compromises.

The most straightforward step in the reconstruction concerns the San Andreas fault system, which entails a ~300 km dextral slip on the main trace (Huffman, 1972; James and others, 1986). The San Andreas restoration is done in conjunction with the restoration of ~60 km of sinistral slip on the Garlock fault (Davis and Burchfiel, 1973), and 90° clockwise rotation of the western Transverse Ranges and related dextral shear in the Los Angeles–Peninsular Ranges region (Luyendyk and others, 1980). The Garlock fault restoration is, in turn, done in conjunction with the restoration of Cenozoic extension across the Death Valley region and clockwise rotation of the southernmost Sierra Nevada (Kanter and McWilliams, 1982; Plescia and Calderone, 1986).

The Sierran Foothills belt is extended ~100 km southward, in conjunction with the southern Sierra rotation, on the basis of the presence of Jurassic and possibly older metamorphic and plutonic rocks that are typical of the Foothills belt in basement cores of the southernmost Great Valley (Ross, 1979) and in the nearby San Emigdio Range (Reitz, 1983; James and others, 1986). In this restoration, Salinia is returned to the southern end of the Sierra Nevada; this is most consistent with batholithic age and geochemical patterns, and pre-batholithic framework geology (Ross, 1976, 1977, 1978; Saleeby and others, 1978; Mattinson and James, 1985; Silver and Mattinson, 1986; James and Mattinson, 1988).

Cenozoic extension across the Basin and Range province is considered from several perspectives. In the region—underlain by the Paleozoic miogeocline and the Antler and Sonoma thrust belts from southernmost Idaho to east-central California—the restoration is taken from Levy and Christi-Blick (1989), which also accounts for Cretaceous and early Cenozoic shortening along the Sevier thrust belt. Restoration of the central and northwest Nevada region entails less than 10% local extension, consistent with the apparent lack of large displacement normal faults. In the Yerrington district, an additional ~50 km of west-northwest extension has been restored (after the work of Proffett [1977] and Proffett and Dilles [1984]). A limited amount of clockwise rotation of the Blue Mountains related to Basin and Range extension (modified after Hamilton, 1978) is restored in conjunction with the northwest Nevada restoration. Restoration of Cenozoic extension in the Mojave Desert region is based on Walker and others (1990b). In the region between the Colorado River and the southwest Colorado plateau, ~75 km of west-southwest–east-northeast extension is arbitrarily restored, in consideration of the local, well-determined work of Reynolds and Spencer (1985).

The western metamorphic belt and adjacent batholithic rocks of the Sierra Nevada are suggested to have been decoupled from the back-arc region in mid- to Late Cretaceous time by regional dextral transtension along the axis of the range during the most voluminous phases of silicic batholith emplacement (Saleeby, 1981, 1991). The tangential displacement pattern is considered a response of the thermally-softened plate edge to north-directed oblique subduction. Restoration of the western Sierra is done in two steps. About 30 km of transverse extension is restored for the Cretaceous Sierra Nevada batholith, on the basis of evidence spreading within the southern silicic domain of the batholith (Tobisch and others, 1986; Saleeby, 1990a). The batholithic extensional restoration is projected southward into the Mojave region, where synbatholithic low-angle normal faults are at least locally recognized (Walker and Wardlaw, 1989), and northward beneath the Modoc Plateau, where the batholithic belt has been geophysically imaged (Zucca and others, 1986). Restoration of the tangential displacement is based on the proposition that the sharp promontory in the initial $^{87}Sr/^{86}Sr = 0.706$ isopleth along the Sierra Nevada batholith mimics the dextral sense of deformation of the North America sialic edge (Fig. 1; Kistler and Peterman, 1978; Saleeby and others, 1986). The 0.706 isopleth for Mesozoic plutons trends southwestward through Nevada adjacent to the outer edge of the miogeocline, whereas to the west it forms the promontory along a distinctly different north-northwest trend. Restoration of the promontory is justified by the apparent presence of offset miogeoclinal strata within roof pendants of the central Sierra Nevada (Lahren and Schweickert, 1989). The isotopic patterns suggest ~250 km of dextral shear along the Sierra Nevada to account for the promontory, whereas Lahren and Schweickert (1989) suggested ~500 km of dextral offset to match the central Sierra miogeoclinal pendants to similar rocks in the western Mojave region. Considering the uncertainties in both analyses, we adopt a ~300 km displacement in our restoration. This displacement takes into account (1) a more realistic analysis of the initial spatial uncertainties for structural-stratigraphic markers presumably offset in the central Sierra pendants; (2) a more southeasterly limit for the bounds of the offset zone, which is more consistent with interpendant stratigraphic ties; and (3) the known ages and distribution of longitudinal shear zones within the axial and eastern Sierra Nevada (Saleeby and others, 1978; Dunne and others, 1991).

The locus of displacement along the Sierra Nevada is thought to have been complex, and to have involved a family of strike-slip faults as well as obliquely distended batholithic welts and calderas. Discrete mid- to Late Cretaceous age right-slip faults have been mapped and dated in the southern Sierra (Saleeby and Busby-Spera, 1986; Busby-Spera and Saleeby, 1987, 1990), and possibly in the central Sierra (Girty, 1985; Saleeby and others, 1990). A major displacement zone in the east-central Sierra apparently lies adjacent to the eastern margin of the initial Sr promontory, which satisfies Lahren and Schweickert's (1989) proposed offset. Uncertainty is posed here, however, by the assertion that dextral displacement occurred along this zone during Middle Jurassic time (Kistler, 1990). This uncertainty, as discussed below, is compounded by reversals in tangential displacement patterns along the plate edge. The east-central Sierra displacement zone is projected southward into the Owens Valley–northern Mojave region as a polyphase continental truncation zone. The truncation zone is considered to approximate the eastern limit of a series of both sinistral and dextral displacements within the wall rocks and batholithic complexes of the Sierra Nevada. The Cretaceous displacements of the truncation zone are thought to have followed the batholithic belt northward through the Modoc plateau. At these northern latitudes, the east boundary of the Cretaceous displacement zone could coincide with the western branch of the Pine Nut fault system (Oldow and others, 1989).

The Klamath Mountains are believed to have been partially coupled to the western Sierra Nevada during the mid- to Late Cretaceous dextral displacement episode. Antithetic decoupling along an east-west sinistral shear zone between the two ranges is restored on the basis of offset Early Cretaceous facies patterns in the Great Valley sequence (Jones and Irwin, 1971).

The Blue Mountains are restored in conjunction with the western Sierra and Klamath Mountains restoration. Several key

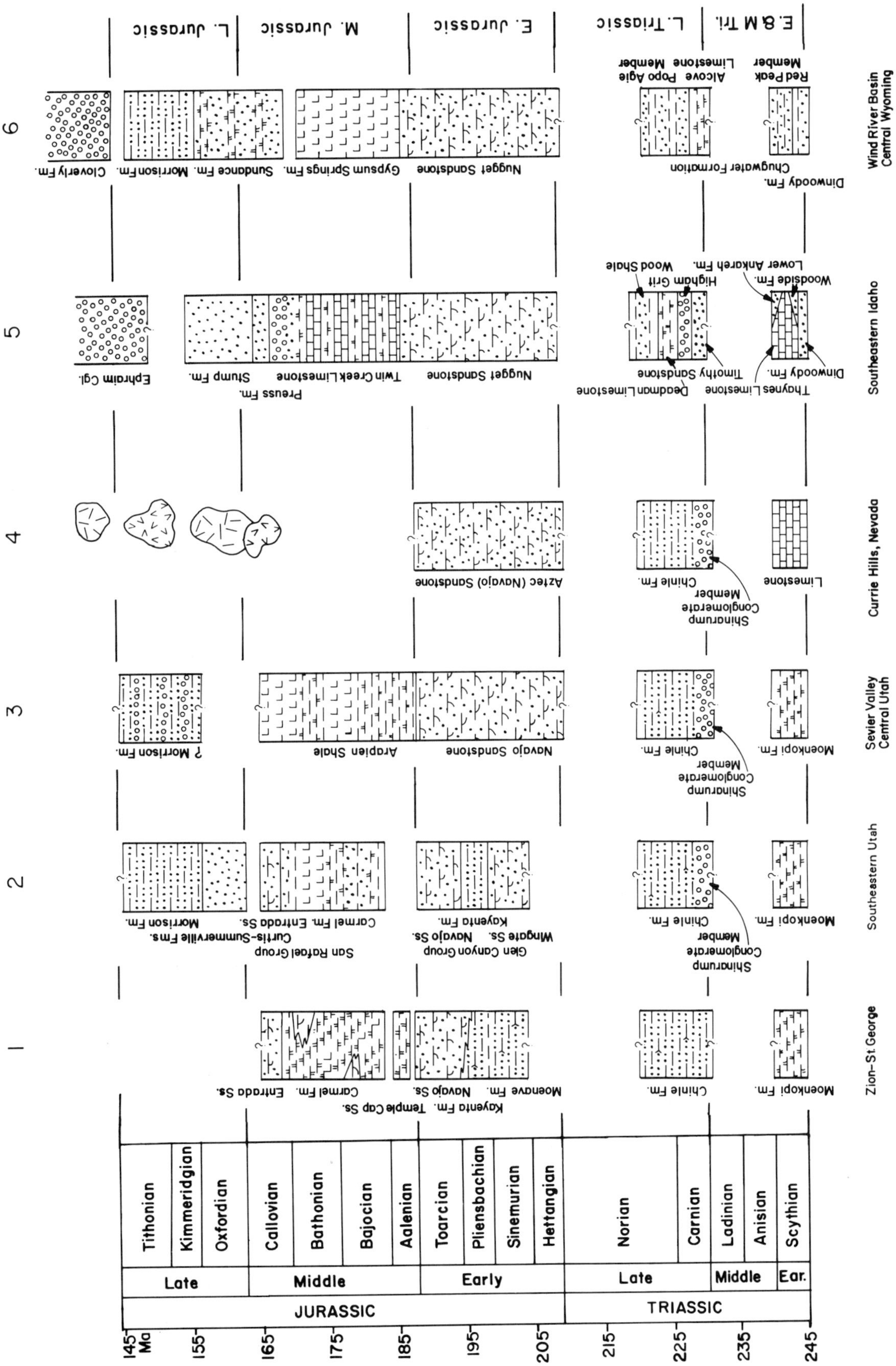

Figure 3. Stratigraphic columns for key locations within plate interior. Locations are shown in Figures 1 and 2. Data sources given in text. Lithologic symbols used in Figures 3 through 7 also shown in explanation.

EXPLANATION OF SYMBOLS FOR FIGURES 3,4,5,6, and 7

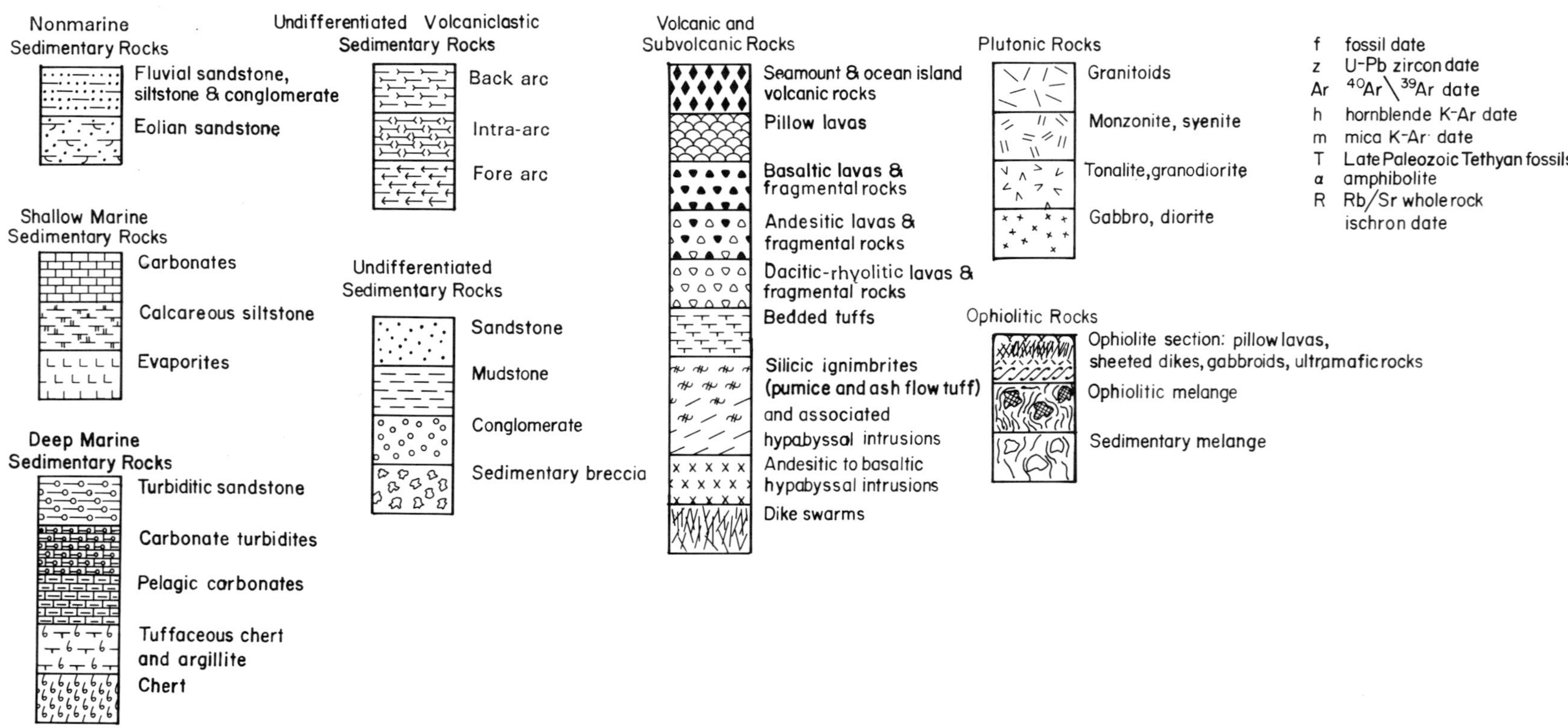

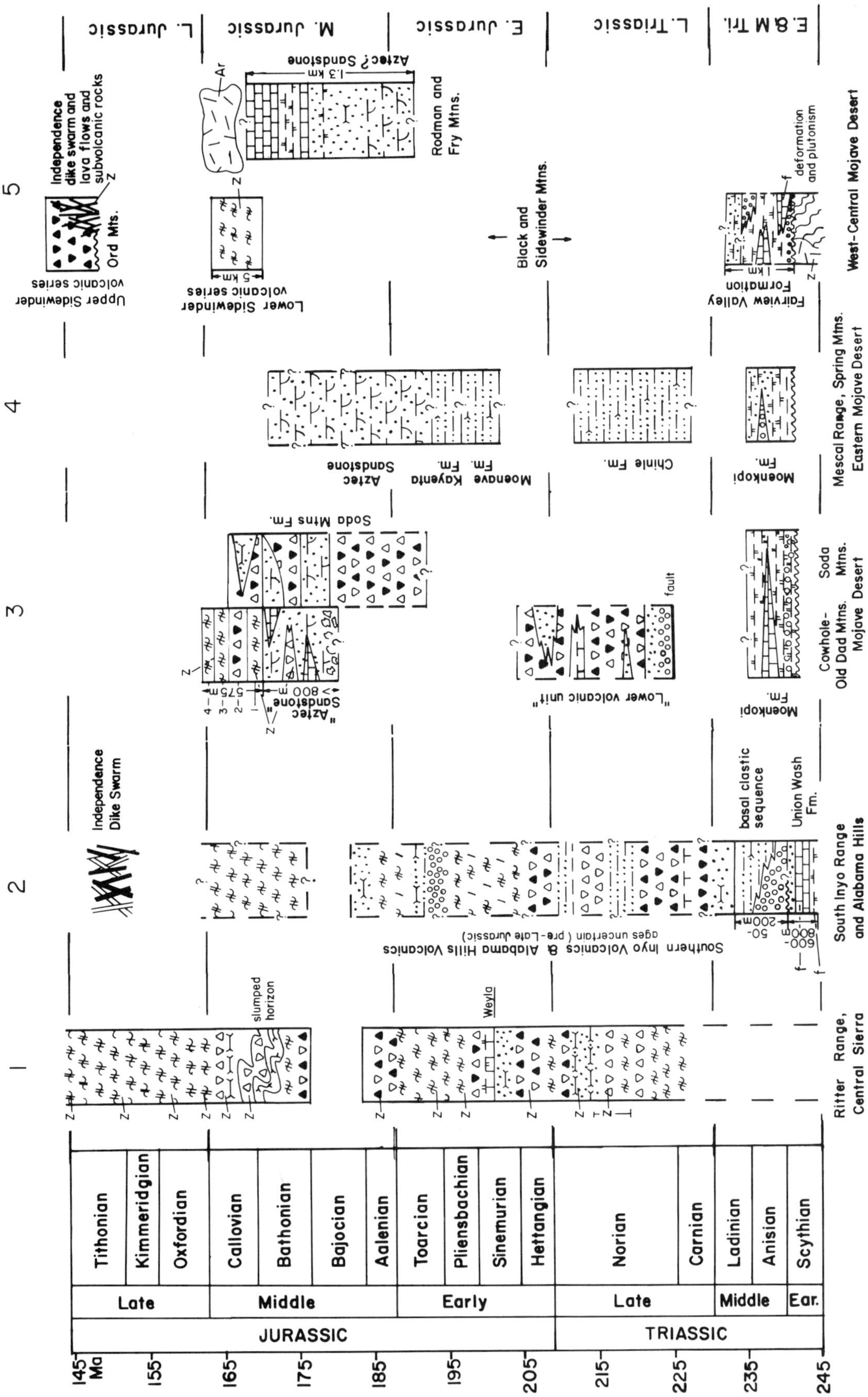

Figure 4. Lithostratigraphic columns for key locations along the sialic segment of the continental arc. Locations are shown in Figures 1 and 2. Data sources given in text. See Figure 3 for explanation of lithologic symbols.

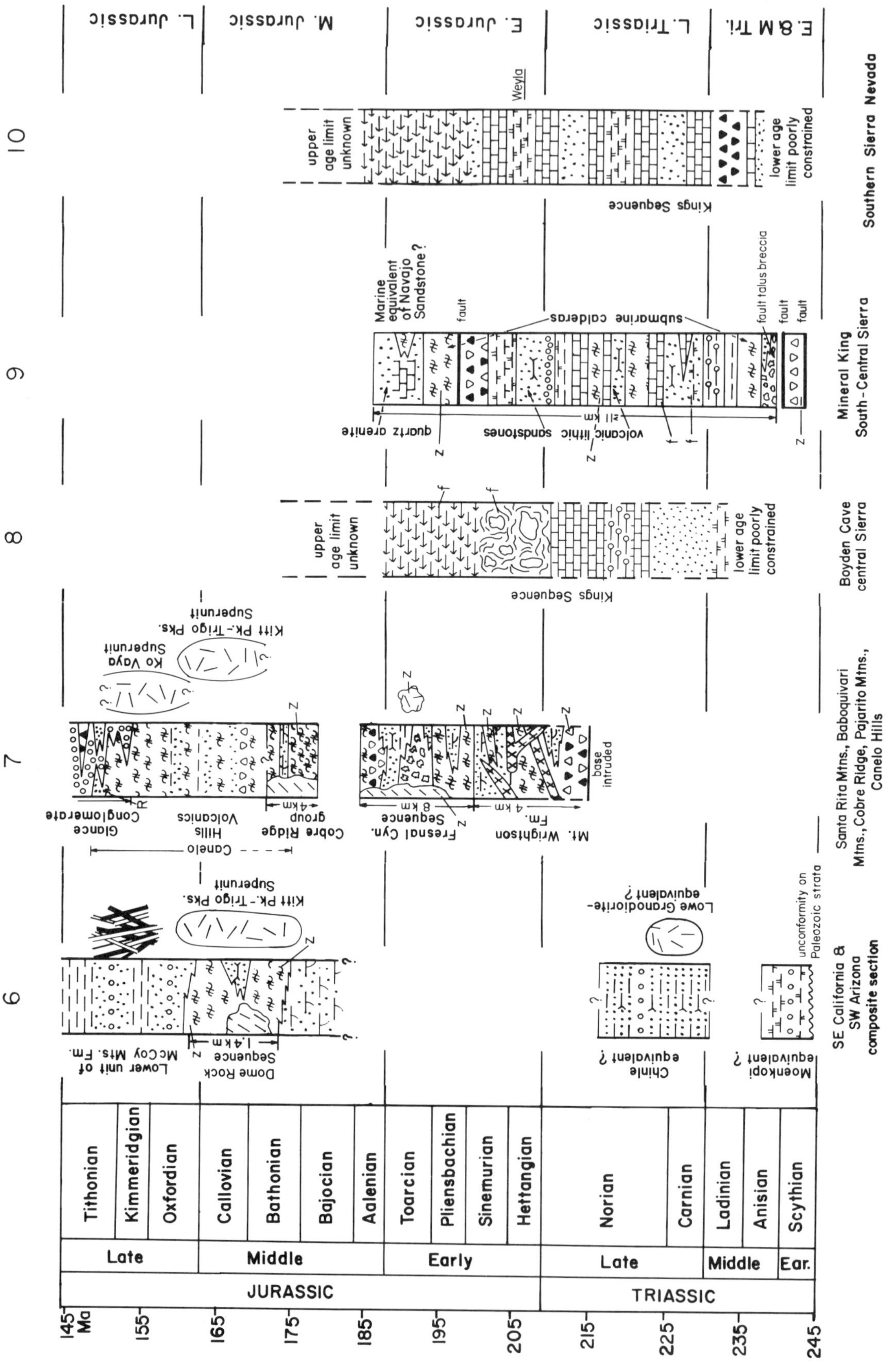
JURASSIC
TRIASSIC
Late
Middle
Early
Ear.
Tithonian
Kimmeridgian
Oxfordian
Callovian
Bathonian
Bajocian
Aalenian
Toarcian
Pliensbachian
Sinemurian
Hettangian
Norian
Carnian
Ladinian
Anisian
Scythian
145 Ma
155
165
175
185
195
205
215
225
235
245
6
7
8
9
10
L. Jurassic
M. Jurassic
E. Jurassic
L. Triassic
E. & M. Tri.
Lower unit of McCoy Mts. Fm.
Dome Rock Sequence
Kitt Pk.-Trigo Pks. Superunit
Chinle equivalent ?
Lowe Granodiorite-equivalent ?
Moenkopi equivalent ?
unconformity on Paleozoic strata
Glance Conglomerate
Canelo Hills Volcanics
Cobre Ridge group
Ko Vaya Superunit
Fresnal Cyn. Sequence
Mt. Wrightson Fm.
base intruded
upper age limit unknown
Kings Sequence
lower age limit poorly constrained
quartz arenite
volcanic lithic sandstones
submarine calderas
Marine equivalent of Navajo Sandstone ?
fault
fault talus breccia
Weyla
SE California & SW Arizona composite section
Santa Rita Mtns., Baboquivari Mtns., Cobre Ridge, Pajarito Mtns., Canelo Hills
Boyden Cave central Sierra
Mineral King South-Central Sierra
Southern Sierra Nevada

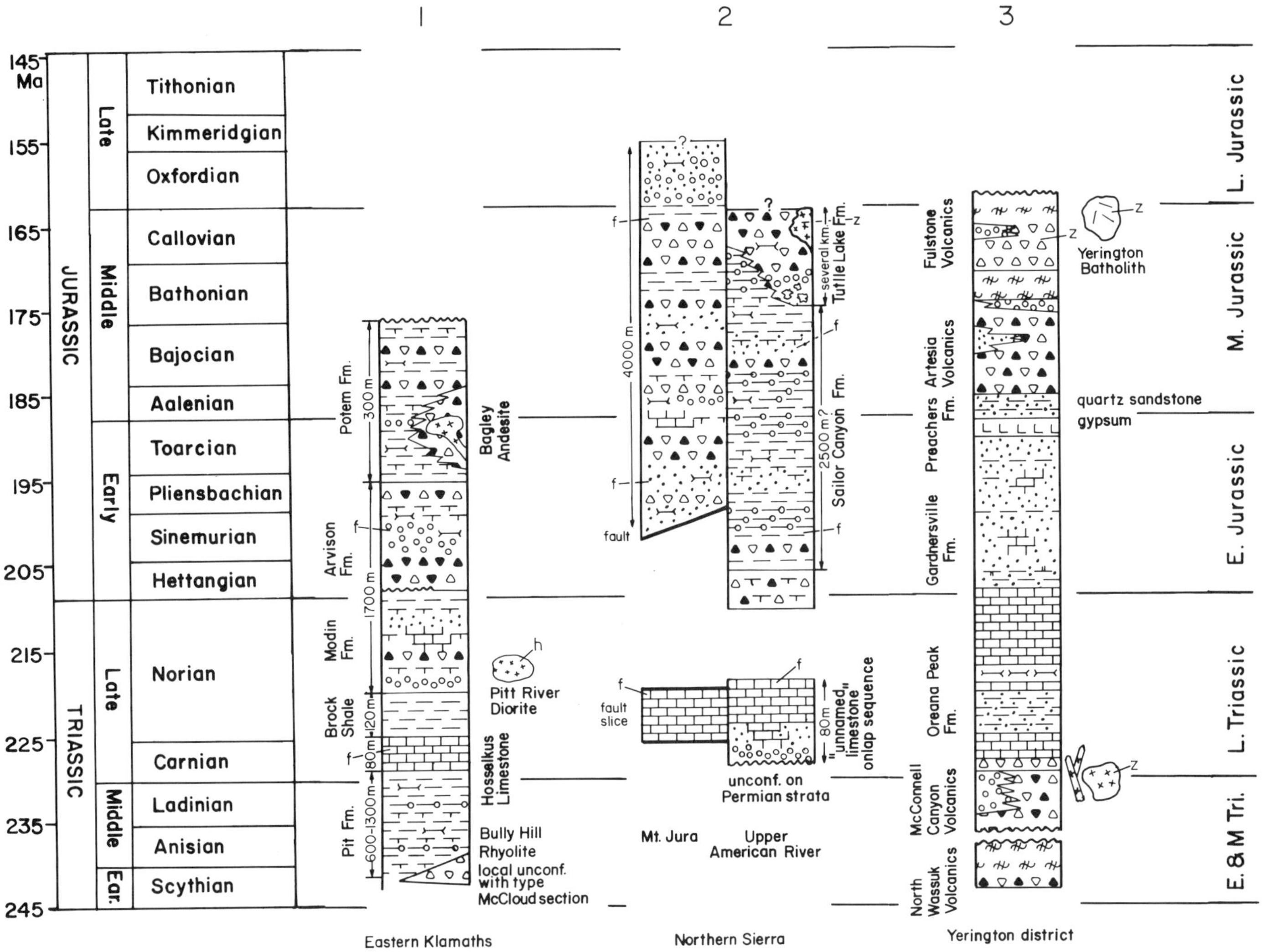

Figure 5 (on this and following three pages). Lithostratigraphic columns for key locations along the McCloud segment of the continental arc. Locations are shown in Figures 1 and 2. Data sources given in text. See Figure 3 for explanation of lithologic symbols.

factors are taken into account: (1) paleomagnetic data in conjunction with regional structural relations suggest that the Blue Mountains underwent ~60° clockwise rotation primarily in mid- to Late Cretaceous time (Wilson and Cox, 1980; Hillhouse and Vallier, 1982; Oldow and others, 1984); (2) the existence of a strong Bouguer gravity lineament between the northern fringes of both the Blue Mountains and Klamath Mountains suggests basement continuity between the two (Riddihough and others, 1986); (3) paleomagnetic data as well as regional structural patterns throughout the Klamath Mountains suggest relatively incoherent clockwise rotations in the northern part of the range and little or no rotation in the central and southern parts of the range (Renne and Scott, 1988); (4) during the mid-Cretaceous, the Blue Mountains arc complex was tectonically emplaced against and intruded by the dextral transpressive western boundary zone of the Idaho batholith (Lund and Snee, 1988). The transpressive boundary zone juxtaposes the very different wall-rock assemblages of the ensimatic arc complex with the tectonically truncated and intruded miogeocline. These factors lead to a restoration whereby a common pole of rotation for the Blue Mountains and northern Klamaths is placed in the central Klamaths and, in conjunction with the ~50° rotation, the Blue Mountains are displaced a moderate distance southward trailing the western Sierra Nevada and Klamath Mountains. The 50° rotation is superposed over the 10° rotation of the Blue Mountains, restored above with Basin and Range extension.

The rotational restoration of the northern Klamath Mountains carries with it a southward restoration of the southern Oregon Coast Range ophiolite and its tectonically underlying blueschist metamorphic terrane. The restoration is poorly constrained, but in general follows the logic that McLaughlin and others (1988) use for California Coast Range ophiolite fragments

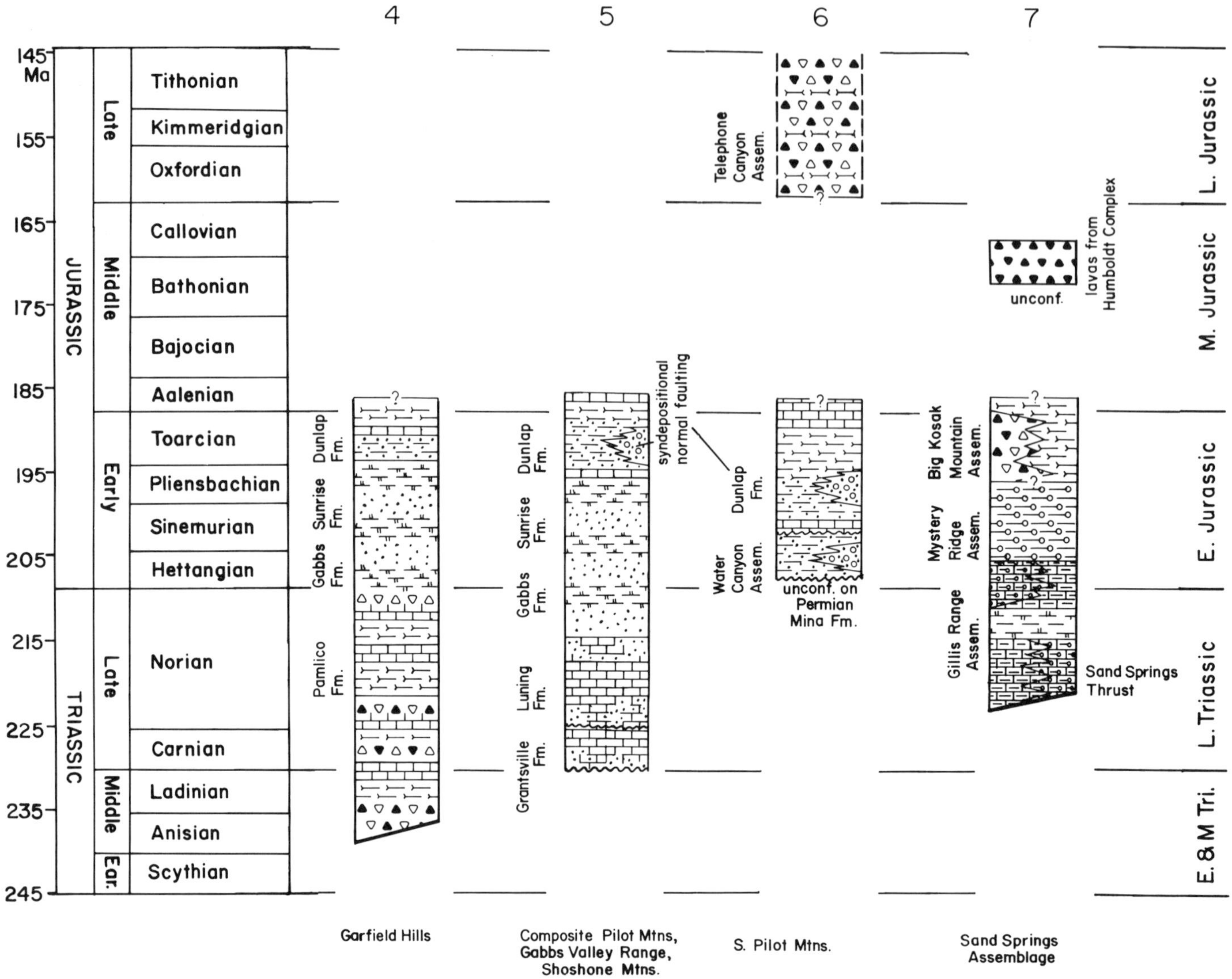

dispersed by dextral faulting throughout the Franciscan complex. The southern Oregon restoration is constrained here by not surpassing the more intact exposures of the Pickett Peak parent terrane for the lower plate metamorphic rocks (Fig. 1), and by an outer basin setting, suggested for the ophiolite by its thin sedimentary cover (Blake and others, 1985; Saleeby, this volume).

Restoration of the Blue Mountains and northern Klamaths lead directly to the consideration of the early Mesozoic positions of terranes which underlie the northern Washington and southern British Columbia region. Large northward displacements of these terranes during Late Cretaceous to possibly Eocene time is indicated by geological and paleomagnetic data. Some authors have chosen to weigh very strongly all paleomagnetic data available, and have accordingly hypothesized displacement from latitudes well to the south of the area in Figure 1 (Umhoefer, 1987; Oldow and others, 1989). Such an analysis does not critically consider the overall analytical and geological uncertainties of the various paleomagnetic data sets, nor does it consider the possibility of regional tilting along the Coast batholithic belt (Butler and others, 1989); this presents a number of geological paradoxes. In our reconstruction, we consider primarily geological observations which results in a restoration that partially closes the Columbia Embayment and returns the northern Washington-southern British Columbia terranes southward as far as southern Oregon.

The northern Washington restoration follows a northward increase in the amount of foreland shortening restoration to ~150 km at the Canadian border (Price and Mountjoy, 1970). In conjunction with this, a nominal 30 km of EW extension is restored along the long dimension of Idaho which approximates the chain of metamorphic core complexes, as well as the Idaho batholith. In northern Idaho, the extensional and shortening restorations are performed in one step. Restoration of tangential displacements of

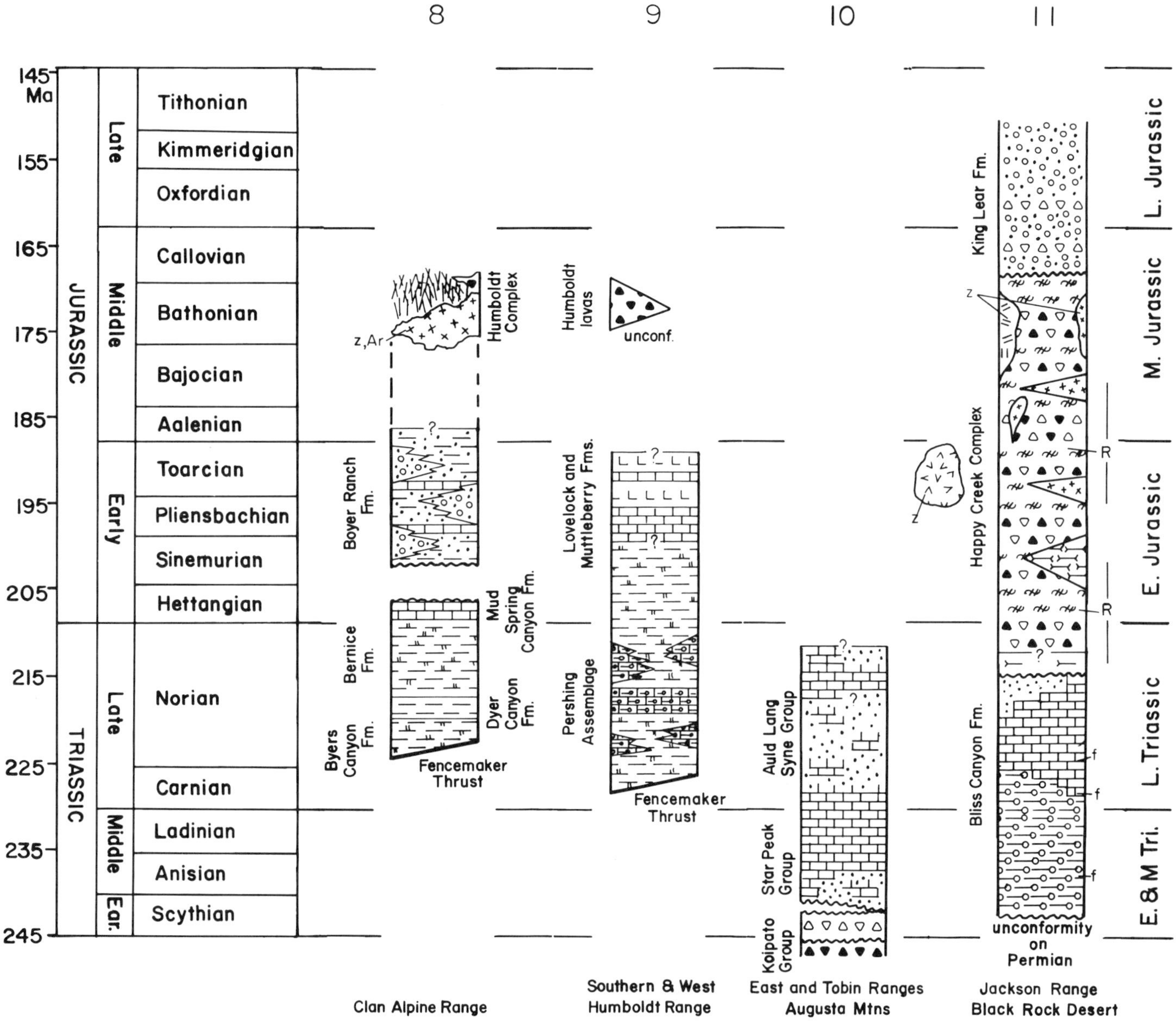

the northern Washington-southern British Columbia terranes is based on the proposed dextral offset of at least 750 km along the Tintina-northern Rocky Mountain trench fault system of northern British Columbia and the Yukon (Gabrielse, 1985). It is not clear how this offset pattern extends through southern British Columbia and northern Washington, and some assumptions are used in conjunction with the available constraints. As discussed in Gabrielse (1985), ~300 km of the northern Rocky Mountain trench offset can be routed into the Frazier River-Straight Creek system. Unfortunately, no more than ~90 km of dextral offset can be carried through the Straight Creek branch (Cowan and Bruhn, this volume); the discrepancy is accounted for below. Using Gabrielse's (1985) minimum figure of 750 km, this leaves at least 450 km of displacement to be accounted for in areas to the east of the Frazier River-Straight Creek system. This is distributed through dextral-sense oblique extension in and around the Shuswap metamorphic complex (Price, 1979), and by distributed shear along the eastern boundary of Quesnel, in basement exposures to the east and within (beneath) the foreland fold and thrust belt (Price, 1981; Gabrielse, 1985; Struik, 1988). This restoration places the Quesnel, Kootenay, and Slide Mountain terranes against the truncated miogeocline in west-central Idaho, but also results in an apparent overlap between the terranes west of the Pasasyten fault and the Blue Mountains. This overlap is considered an artifact of Insular suture-belt structure.

The Insular suture belt consists of small terranes between the

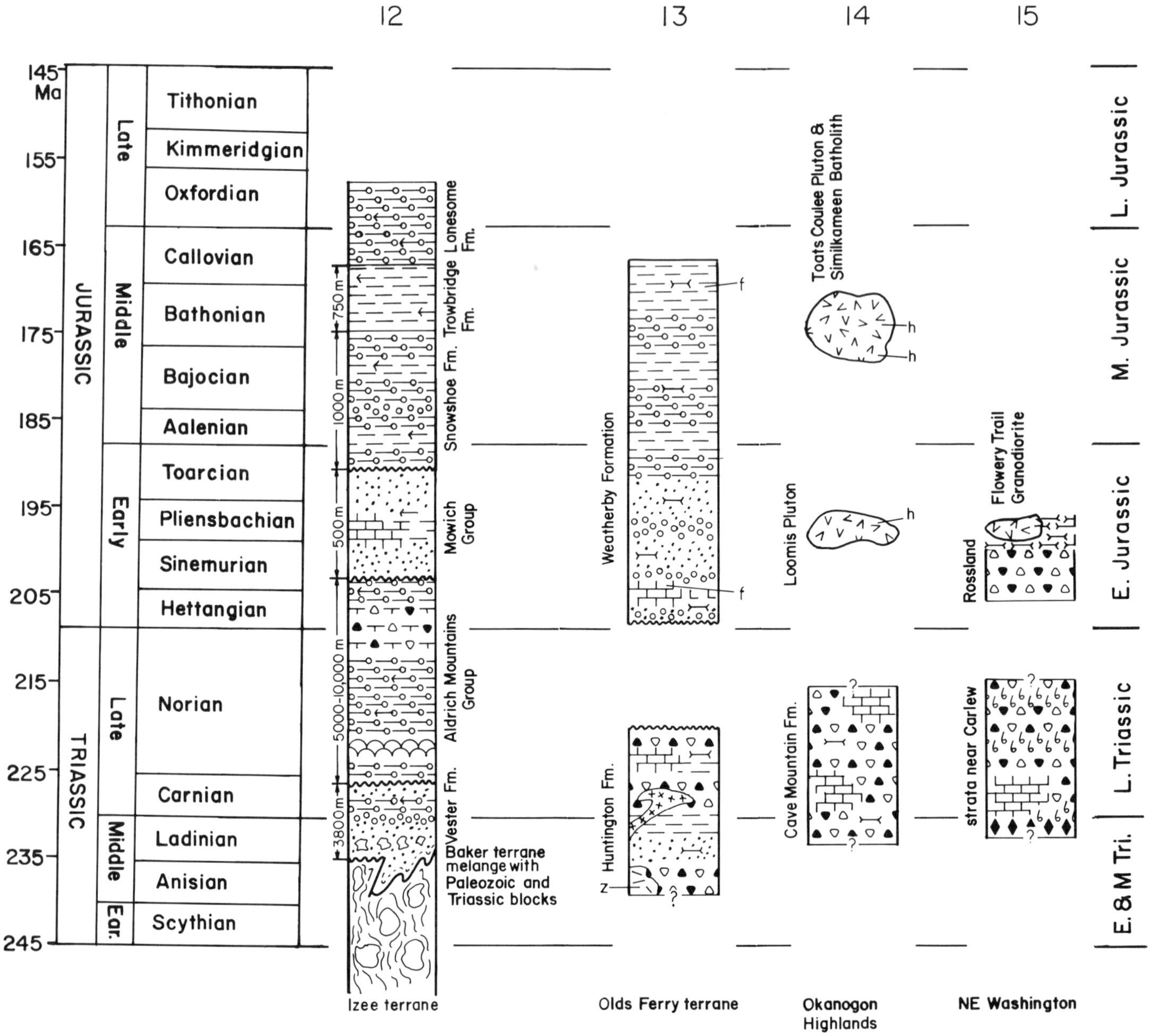

Intermontane and the Insular superterranes, and is characterized by a major mid-Cretaceous west-vergent thrust system that occurs discontinuously for ~1,500 km between southeast Alaska and northern Washington (Rubin and others, 1990b). In the northern Washington–southern British Columbia region, the Yalakom, Straight Creek, Ross Lake, Hozameen, and Pasayten faults lie within the belt, and strike-slip displacements along some or all of these faults accompanied and/or postdated major crustal shortening. This network of major high-angle faults obscures the amount of suture-belt shortening. Three distinct fold-thrust systems are recognized near the southern end of the belt. The eastern Cascades system lies between the Straight Creek and Pasayten faults (Fig. 1) and exhibits a minimum of 50 km of shortening (McGroder, 1989b). The western Cascades system lies west of the Straight Creek fault, and consists of such disparate thrust sheets that shortening reconstructions are not feasible. One major thrust, the Shuksan, exhibits a minimum of 60 km of structural overlap (Misch, 1988). The San Juan Islands system also consists of such an array of disparate thrust sheets, and shortening reconstructions are thus not feasible (Brandon and others, 1988).

A major episode of strike-slip faulting along the margins of Stikine is restored in conjunction with the restoration of Insular suture-belt shortening. Rocks of the Hozomeen Mountain–Ross Lake and Methow Valley regions (6 and 7 in Fig. 7) appear to be part of a west-facing lower Mesozoic fore-arc of the Intermontane superterrane, specifically Stikine. These rocks are restored to the

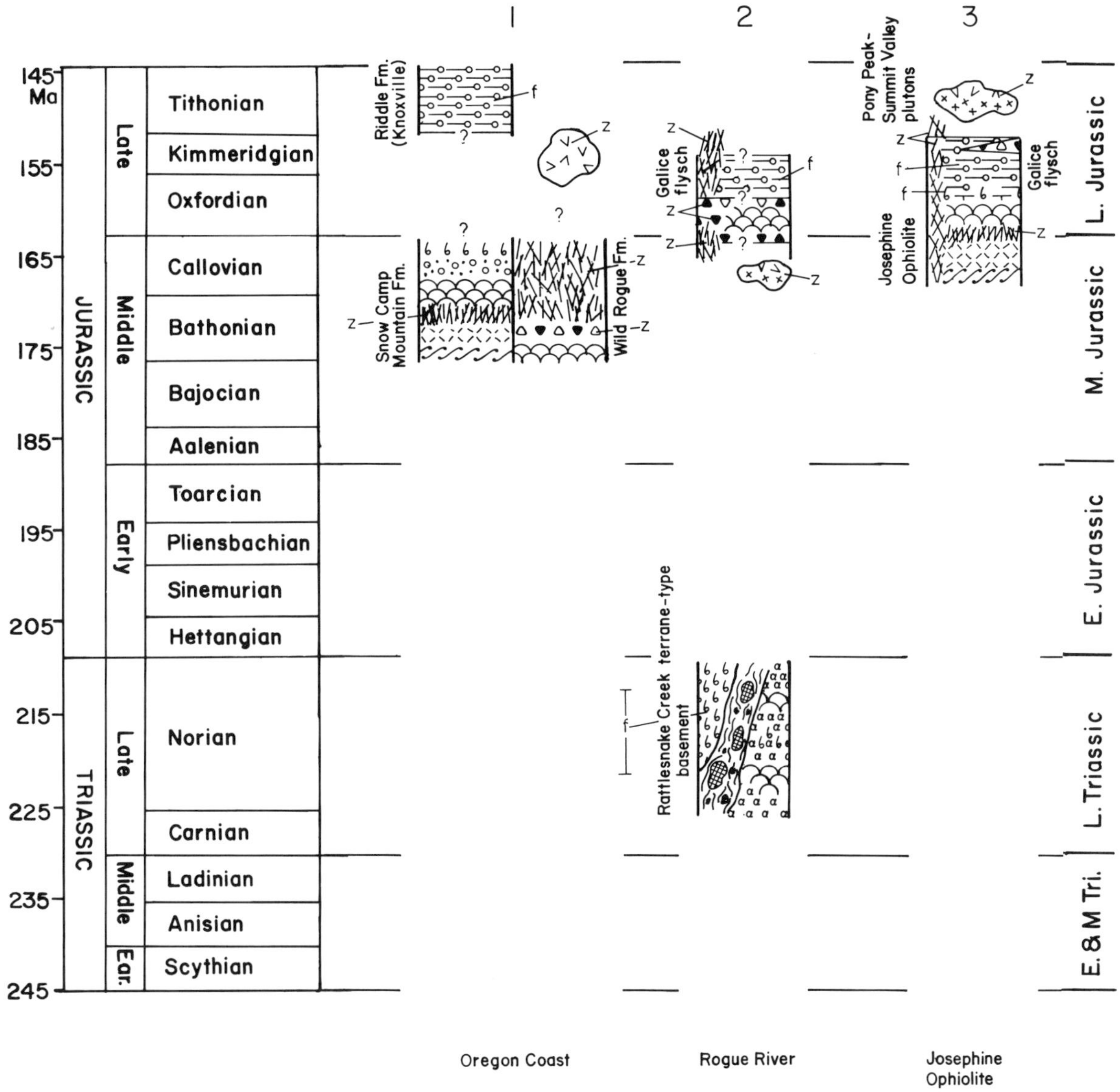

Figure 6 (on this and following three pages). Lithostratigraphic columns for key fringing terranes. Locations are shown in Figures 1 and 2. Data sources given in text. See Figure 3 for explanation of lithologic symbols.

western margin of Stikine in two steps. First, only ~50 km of dextral slip is restored on the Straight Creek–Frazier River system, which roughly aligns the Yalakom and Pasayten faults. The additional 250 km of dextral displacement on the Frazier River system is restored in conjunction with an equal sinistral restoration on the Yalakom-Pasayten system. The sense of the Yalakom-Pasayten restoration is based on the structural studies of M. G. Miller (1988), Greig (1989), and Hurlow (1989), and is consistent with mid-Cretaceous sinistral oblique thrust patterns observed along the suture belt in southern southeast Alaska (Saleeby and Rubin, 1990). The coupled sinistral and dextral displacement patterns along the western and eastern margins of Stikine, respectively, are attributed to the northward escape of Stikine during east-west contraction of the suture belt (modified after Wernicke and Klepacki, 1988). Monger and others (1985) indicated a minimum transverse structural overlap of ~100 km within the Bridge River–Cadwallader-Methow belt related to suture-belt contraction. Restoration of this is added to the minimum of 50 km shortening deduced to the south by McGroder (1989b), primarily within the upper stratigraphic levels of the belt. As discussed below, the fore-arc setting of this belt (basin) changed to an intra-arc setting following the initial accretion of the Insular superterrane during Middle Jurassic time (Saleeby and Rubin, 1990).

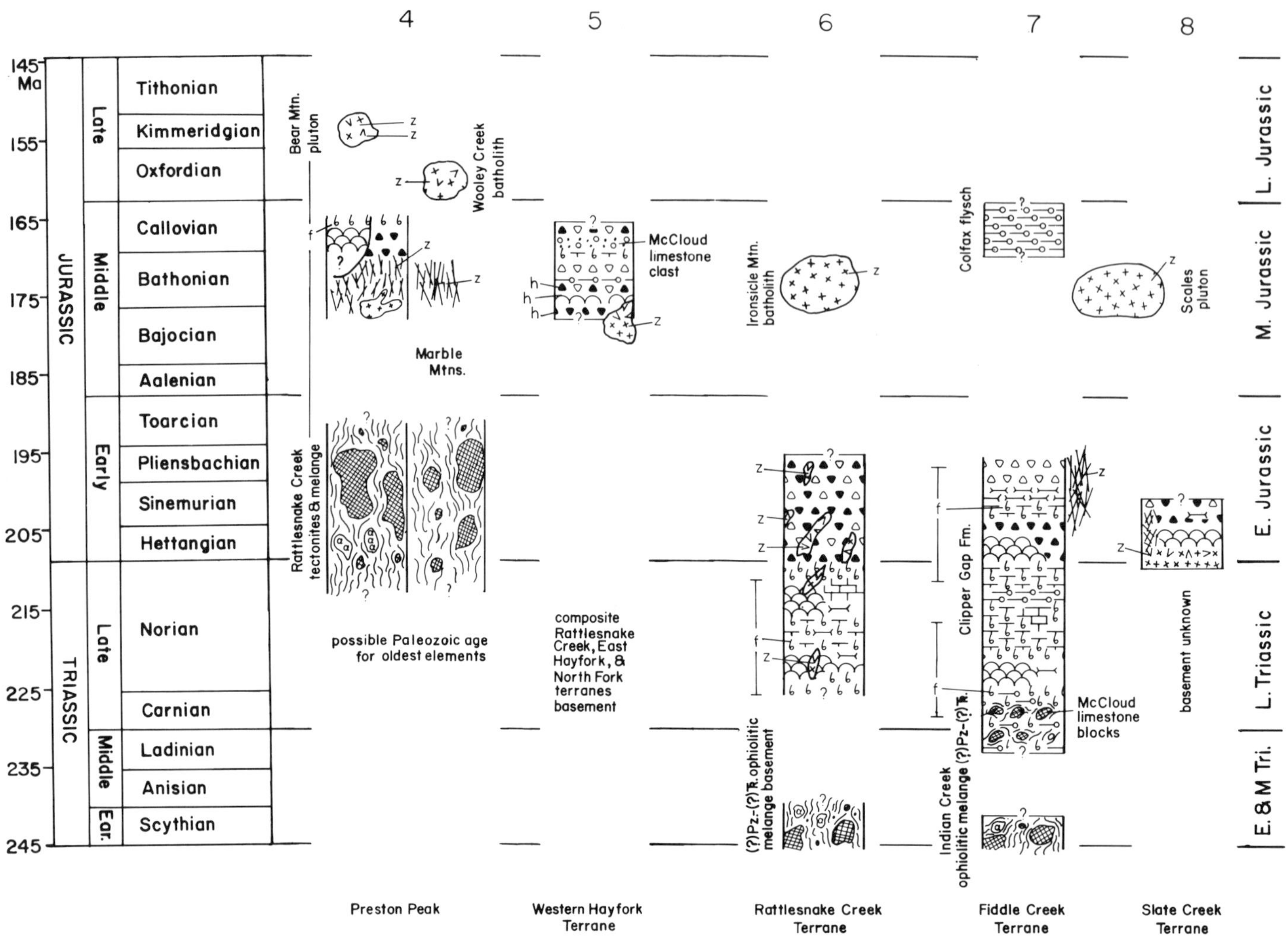

Structural studies in the west Cascades thrust system suggest a strong component of dextral-sense tangential displacement during nappe transport. This is exhibited both in thrust fabrics and in associated high-angle faults (Brown, 1988; Smith, 1988). A number of thrust sheets from both the San Juan Islands and west Cascades have probable source terranes in the northern California Coast Ranges, which is consistent with their dextral-sense of dispersion. The Shuksan metamorphic suite and fragments of Late Jurassic suprasubduction-zone ophiolite may be correlated with the more-intact Pickett Peak metamorphic terrane of the Franciscan complex and the California Coast Range ophiolite, respectively (Brown and Blake, 1987; Garver, 1988). In this synthesis, the protoliths for the metamorphic terrane(s), as well as the ophiolitic fragments, are thought to be remnants of an extensive Late Jurassic inter-arc and extensional arc–fore-arc basin system that extended from at least southwest Oregon to the southern California borderland. A number of other thrust sheets from the San Juan Islands and west Cascades were derived from McCloud arc and Cache Creek–affinity parent terranes; either of which may have been tapped over considerable distances to the north or south along the plate edge.

Restoration of the northwesternmost Washington region is poorly determined, and we have used internal consistency as our guide. The ~750 km of northward displacement restored within and along the eastern margin of the Intermontane superterrane is taken as the total post-Jurassic northward displacement of the southern Insular superterrane (Wrangellia of Vancouver Island). The San Juan–west Cascades thrust complexes are restored with Wrangellia. Thrust displacements throughout these complexes are not restored because there are no data on magnitude, and because first-order tangential displacements along thrust surfaces are as likely as transverse displacements (Brown, 1988; Smith, 1988). The above restoration is generally consistent with the observation that the northwest Washington region contains the southernmost exposures of the Insular superterrane and the Methow basin, as well as the northernmost exposures of the regional belt of Late Jurassic suprasubduction-zone ophiolites and associated Late Jurassic–Early Cretaceous blueschist terranes. The restoration places the San Juan Island–west Cascade ophiolite and blueschist fragments in proximity to the northern termination of more-intact correlative rocks. The Late Jurassic regional ophiolitic basin is suggested to have continued, at least

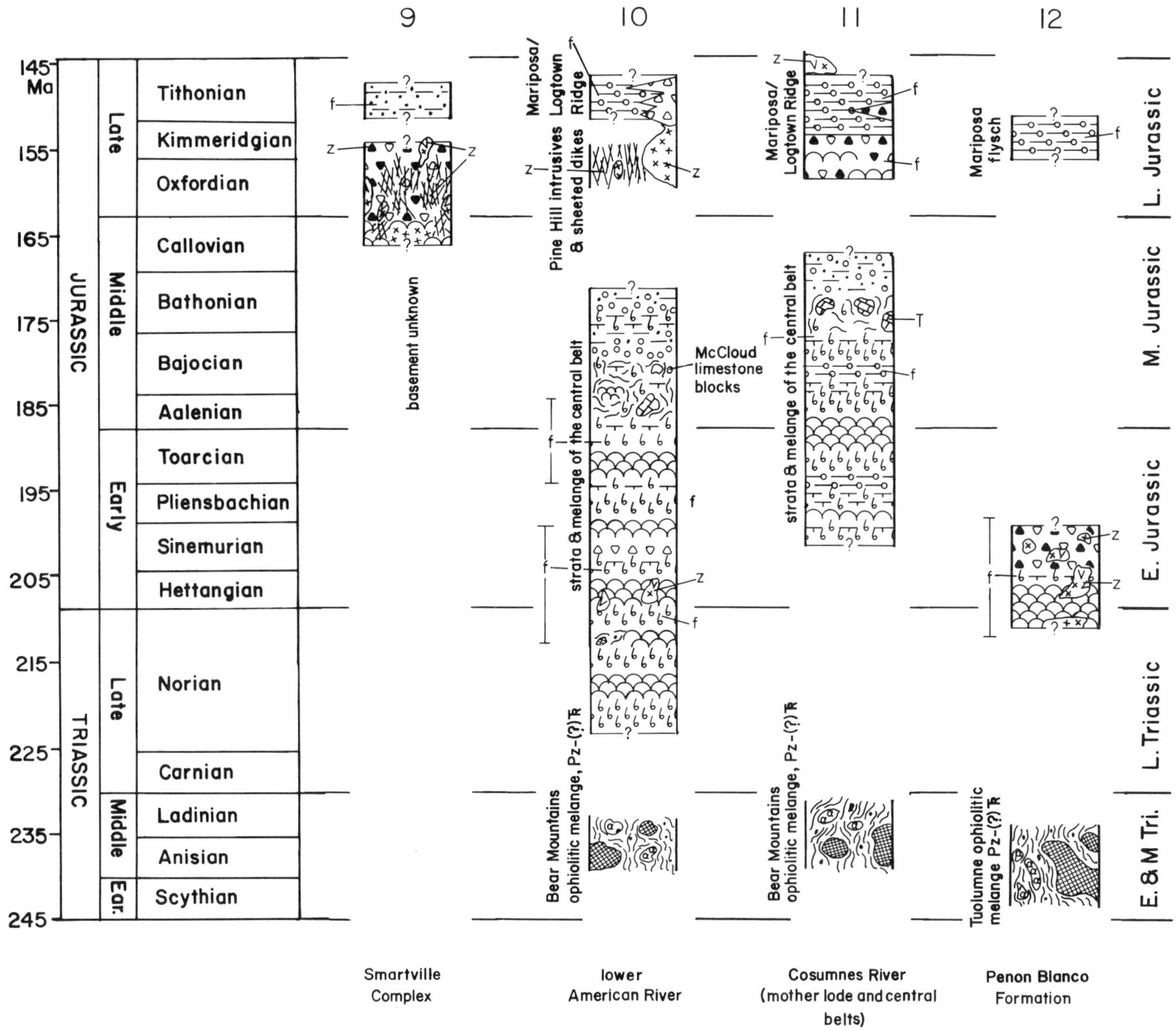

bathymetrically, northward as the Methow basin and its equivalents to the north as far as southern Alaska (Gehrels and Saleeby, 1985).

The palinspastic base of Figure 2 provides a reference for the map relations depicted in Plate 5. Major structures that were active during the Triassic and Jurassic are shown in their Early Cretaceous positions as red in Figure 2. Triassic and Jurassic displacements are much more difficult to quantify than many of the younger displacements. For this reason, a sequence of quantitative palinspastic restorations were not attempted for Plate 5 sequence. Rather, conservative estimates of petrogenetic and depositional localities are propagated through the Plate 5 sequence where structural, stratigraphic, and paleomagnetic data provide some kinematic constraints. These restorations are cumulative in reverse order of the text and Plate 5 sequence, and thus their logic must be found by skimming the text and Plate 5 in reverse.

Structural trends discussed in the text are based on their orientations within the Figure 2 palinspastic base. Plate 5 also displays a series of cross-sectional insets. These are intended to represent general tectonostratigraphic relations and are not to be construed as precise structure sections. The petrogenetically active levels of the inset sections are color coded with the maps; preexisting rocks and structures are not colored.

An important consideration in the Plate 5 reconstructions is the absolute motion and related paleolatitudinal track of the North American plate. Estimates for these are given for Jurassic time based on the North American apparent polar wander path model of May and Butler (1986) and May and others (1989), using their "corrected plateau" paleomagnetic Euler pole data.

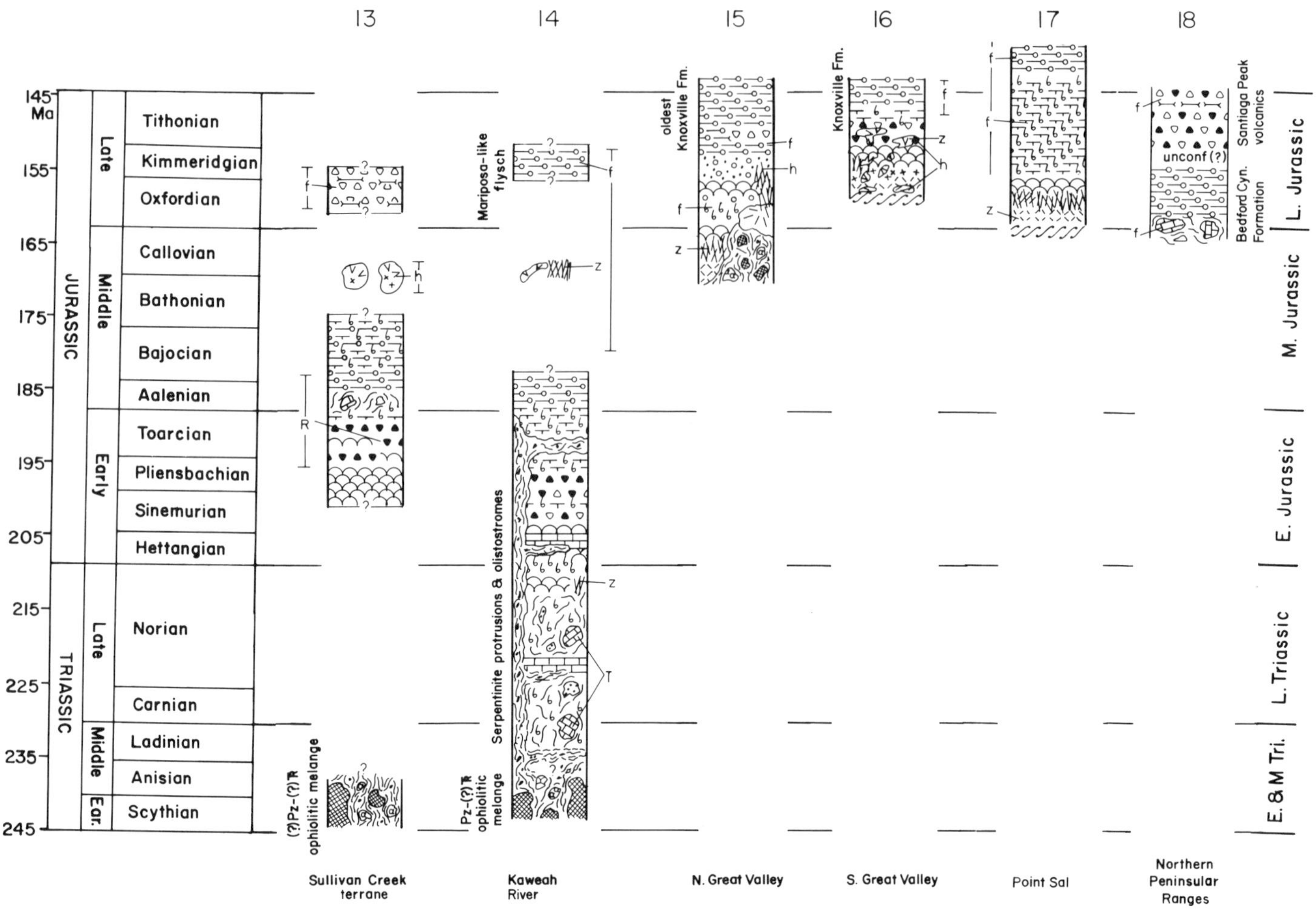

The Triassic azimuth of North America's absolute motion is approximated from Gordon and others (1984). Plate-motion reconstructions for the Pacific basin, based on hot-spot tracks (Engebretson and others, 1985), are not applicable for early Mesozoic time.

EARLY AND MIDDLE TRIASSIC (245–230 Ma)

The major tectonic and paleogeographic features of the Early and Middle Triassic are shown in Plate 5A. Early Triassic tectonic activity was marked by the accretion of a major Paleozoic island-arc terrane in the northwest Nevada–northern California region (Sonomia of Speed, 1979; McCloud belt of Miller and others, this volume). The effects of the Sonoma orogeny were apparently present as far south as the Mojave region (Miller and Cameron, 1982; Snow and Lux, 1990; Miller and others, this volume), and vestiges of it are locally recognized in the Saddlebag Lake pendant of the east-central Sierra Nevada (Schweickert and Lahren, 1987). In the Pacific Northwest, the Sonoma orogeny may be represented by Permian-Triassic east-verging thrusts in the Slide Mountain terrane (Monger, 1977), and regionally extensive unconformable relations and deformation-metamorphic contrasts between Permian and lower Mesozoic strata of both Stikine and Quesnel (Monger and others, 1990).

The Sonoma orogeny in Nevada is defined by the east-directed thrusting of Paleozoic basinal strata (Golconda allochthon) across upper Paleozoic shelf strata which lie above the Antler foredeep and thrust belt (Candelaria inset, Plate 5A). The Sonoma and Antler thrust belts approximately follow the $^{87}Sr/^{86}Sr$ = 0.706 isopleth defined by Mesozoic pluton data (Kistler and Peterman, 1973, 1978). In the Nevada region, the isopleth corresponds to the tapered edge of sialic basement for the Paleozoic miogeocline, and thus the loci of the thrust belts were apparently controlled by the sialic edge. Sonoma-age thrusting is recorded within the miogeocline of the east-central California region by the Last Chance thrust system, where deep-water facies miogeoclinal strata were transported ~50 km southeastward over shallow-water facies (Last Chance inset Plate 5A) (Wernicke and others, 1988; Snow and Lux, 1990). In the Mojave Desert and southeastern California, Lower Triassic conglomerates were shed from western highlands created by Permian-Triassic igneous and deformational events that were presumably Sonoman (Southwest Mojave inset, Plate 5A). Additional Sonoman uplifts are hypoth-

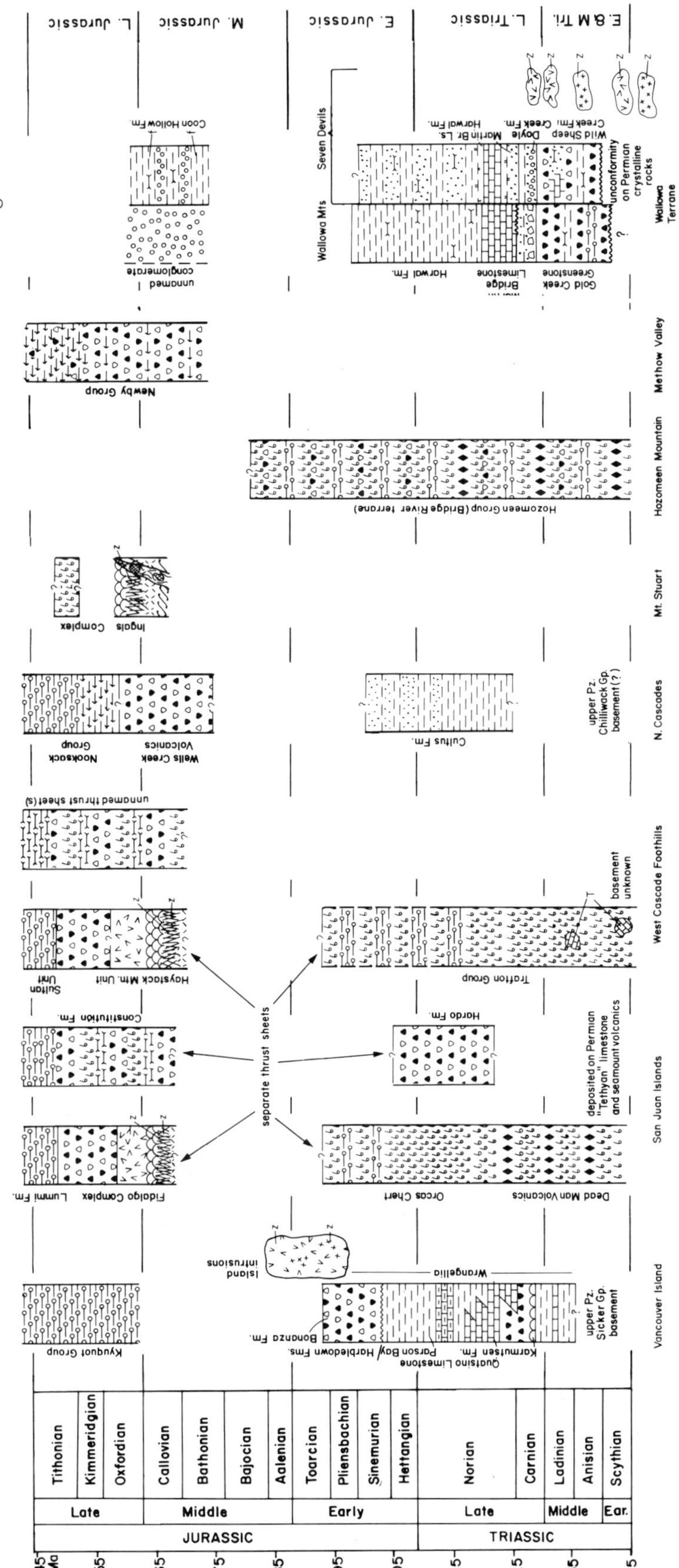

Figure 7. Lithostratigraphic columns for terranes of Insular suture belt. Locations are shown in Figures 1 and 2. Data sources given in text. See Figure 3 for explanation of lithologic symbols.

esized for the central and northern Sierra region and along the Golconda thrust on the basis of stratigraphic breaks (2 in Fig. 5) and the presence of synorogenic deposits (Collinson and Hosenmueller, 1978).

Early to Middle Triassic igneous rocks in the southeastern Sierra Nevada and Mojave regions may be the first arc-magmatic products of a north-northwest–south-southeast–trending subduction zone that dipped eastward beneath the continental margin following, or within the later phases of, the Sonoma orogeny. The subduction zone and magmatic arc appear to have cut obliquely across the Sonoma belt and the trend of the Paleozoic miogeocline.

In the plate interior, Early Triassic shelf depositional patterns continued to follow Paleozoic miogeoclinal trends, and remnants of the ancestral Rockies persisted as a low-relief source area. The last widespread episode of marine sedimentation in the miogeocline was terminated in Early Triassic time, presumably in response to the Sonoma orogeny and the ensuing active-margin tectonic regime.

Major plate kinematic patterns are poorly known for this time interval. The absolute motion of North America is presumed to have been along a westerly trajectory (after Gordon and others, 1984). Transform truncation patterns discussed below and Golconda and Last Chance thrust kinematics (Speed, 1979; Snow and Lux, 1990) suggest a sinistral tangential component along the north-trending plate edge in California for Permian-Triassic time. Sonoman tectonics in the southwestern United States is accordingly depicted on Plate 5A as a regional sinistral transpressive system.

Basement framework of the continental arc

"Continental arc" is used here to denote arc and adjacent fore-arc and back-arc rocks that are known, or appear, to have formed above an east-dipping subduction zone along the edge of western North America. The early Mesozoic continental arc consists of two segments differentiated by basement assemblages: a southern segment, which was constructed primarily across and along the edge of sialic North America and the overlying miogeocline, and a northern segment, constructed across transitional and ensimatic rocks accreted to the North American plate mainly during the Sonoma orogeny. The former is referred to as the sialic segment, the latter the McCloud segment (after Miller, 1987). Representative stratigraphic columns of the sialic segment are in Figure 4, and columns for the McCloud segment are in Figure 5.

The McCloud basement constitutes the main pre-Mesozoic substrate in northwest Nevada, the eastern Klamath Mountains and northern Sierra Nevada, much of the Blue Mountains, and throughout Quesnel and Stikine. This basement is characterized by mid to late Paleozoic island-arc and related rocks, and includes Permian fringing carbonate-reef sequences that contain the distinctive McCloud fauna (Speed, 1979; Miller, 1987; Miller and others, this volume). The known or inferred position of the Golconda thrust approximates its inner boundary in the southwest. We assume here that the Golconda allochthon represents the accretionary prism (Speed, 1979) and/or a back-arc–basin complex (Miller and others, this volume) of the late Paleozoic McCloud arc. The type McCloud fauna occurs in the eastern Klamath Mountains where the Lower Permian McCloud limestone caps an arc volcanic sequence as old as Devonian (Miller, 1987). The McCloud limestone is distinct for its fusilinids and corals, which differ from those of the coeval miogeocline as well as from "Tethyan" oceanic forms (Nestell, 1983). The McCloud fauna represent a transitional form that occupied an eastern Pacific fringing island-arc system (Miller, 1987; Miller and others, this volume).

The McCloud basement was accreted to the North American plate in California and Nevada during the Sonoma orogeny. As discussed below and by Saleeby (this volume), McCloud basement is also considered to have undergone initial phases of accretion in the Pacific Northwest during the Sonoma orogeny, a view that is divergent from most Cordilleran tectonic syntheses. Shortly after accretion in northwest Nevada, a broad marine basin and shelf province formed across the McCloud basement and lapped onto the miogeocline (Speed, 1978). The onset of this sedimentation pattern is poorly dated, but strata as old as Middle Triassic are locally recognized (11 in Fig. 5). The marine province has been considered a successor basin to the Sonoma orogeny, with basinal facies centered over the thermally contracted arc and shelfal facies located atop the subsided thrust belt to the east (Speed, 1978). Alternatively, the basin may have been inherited from the late Paleozoic arc and back-arc bathymetry. The southern margin of the early Mesozoic basinal terrane grades into the sialic segment of the continental arc. To the west in the eastern Klamath Mountains Lower Triassic silicic volcanic rocks interfinger northward with Lower to Middle Triassic basinal deposits (Bully Hill and Pit Formations, respectively; 1 in Fig. 5). Lower and Middle Triassic strata are missing in the northern Sierra (2 in Fig. 5). Early Mesozoic facies relations developed across the McCloud basement are broadly represented in columns 1–11 in Figure 5. The western bounds on the southern McCloud basement are defined by the Sierra Foothills suture and its possible continuation into the Klamath Mountains. West of the suture, pre-Mesozoic basement rocks consist of disrupted Paleozoic ophiolite, and late Paleozoic "Tethyan" limestone blocks are widely dispersed.

The McCloud basement assemblage is shown to continue northward through part of the Blue Mountains and into Quesnel and ultimately Stikine (Plate 5A). Lower Mesozoic rocks of the Blue Mountains occur as several inliers separated by Cenozoic lava flows (Figs. 1 and 2). Vallier and Brooks (1986) interpreted the inliers as the arc, fore-arc, and subduction complex of one system, the Blue Mountains arc. Wernicke and Klepacki (1988), in contrast, interpreted the northern two inliers (Wallowa terrane; 8 in Fig. 7) as a collided fragment of Wrangellia, and the southeasterly inlier (Olds Ferry terrane; 13 in Fig. 5) of McCloud arc affinity, as it is defined here. The Wallowa terrane can alternatively be interpreted as a fragment of the Stikine, or the arc from

which the Cadwallader terrane was derived. Stikine consists of the largest expanse of McCloud limestone and regionally-intact, overlapping lower Mesozoic arc rocks throughout the Cordillera (Davis and others, 1978; Monger and others, 1990). Inclusion of a major belt of Cache Creek rocks between Stikine and Quesnel has been cited as evidence for major Late Triassic to Middle Jurassic collapse of a Tethyan ocean basin culminated by the collision of Stikine with the Cordilleran margin. An alternative view is developed in Saleeby (this volume) and below, wherein Stikine's McCloud basement was initially accreted to the Cordilleran plate edge during the Sonoma orogeny, and thereafter functioned as a loosely coupled basement fragment within the continental arc and/or fore-arc until the Middle Jurassic collapse of the orogen.

Plate edge

Tectonic accretion and truncation. The spatial relations between the contrasting basement regimes for the early Mesozoic continental arc indicate that the late Paleozoic to Early Triassic(?) tectonics involved tectonic truncation and removal processes, in addition to the accretion of the McCloud basement. Tectonic truncation of the southwest Cordillera in late Paleozoic to Early Triassic time has been called on to explain the termination of Paleozoic stratigraphic and structural trends against the continental arc (Hamilton and Meyers, 1966; Burchfiel and Davis, 1972; Saleeby, 1975; this volume; Davis and others, 1978), as well as the change in the orientation of the sialic edge from Nevada into California (Kistler and Peterman, 1973, 1978). Truncation appears to have begun in Pennsylvanian time, when a northwest-trending sinistral strike-slip zone formed across the Antler belt and adjacent miogeocline along east-central and southern California (Walker, 1988). Rocks of the Antler belt were displaced southward toward the Mojave region during Pennsylvanian-Permian time, and sedimentary rocks of this age in the Inyo to Mojave region record an unstable continental borderland reflecting the transform regime (Stone and Stevens, 1984; Walker, 1988; Walker and Wardlaw, 1989). The polyphase truncation zone shown in Figure 1 was first established during the transform event. Displaced rocks of the Mojave region and the adjacent autochthonous rocks were together deformed and intruded in Late Permian time, and overlapped with Triassic marine rocks that Walker (1988) correlated with autochthonous cratonic strata to the east. The transform zone may have extended through the Mojave region and into Sonora, Mexico, where miogeoclinal strata are suspected to have been displaced southward as well (Silver and Anderson, 1974; Saleeby, 1975; Anderson and Schmidt, 1983). Outer miogeocline and Paleozoic deep basinal strata preserved along the eastern wall of the Peninsular Ranges batholith (Gastil and Miller, 1984; Todd and others, 1988) may also have been displaced southward during the transform event.

Late Paleozoic transform tectonics are also recorded along the western Sierra Nevada, where Paleozoic rocks of the McCloud arc are truncated and juxtaposed on the west by a Carboniferous to Permian transform ophiolite belt (Saleeby, 1990b, this volume). The western Sierra truncation zone parallels the Antler belt–miogeocline truncation zone to the east. The western Sierra zone corresponds to the Foothills suture and its probable continuation into the Klamath Mountains. As discussed below, the suture has been remobilized by early Mesozoic thrusting. The transform ophiolite belt consists of ophiolitic melange and tectonite that extend for ~500 km along the suture. Fragments of lower Paleozoic ensimatic basement rocks with strong affinities to basement rocks beneath the type McCloud arc sequence in the eastern Klamath Mountains are embedded within the ophiolitic melange along much of its length. The age relations along the Foothills ophiolite belt are consistent with leaky-transform faulting and/or a system of spreading-ridge–transform junctures truncating the southwest McCloud arc, contemporaneous with the truncation of the Antler belt and miogeocline to the south and east.

The evolution of the transform ophiolite belt and its tectonic significance are detailed in Saleeby (this volume). The conclusions and interpretations are in Plate 5A, where the Foothills suture is shown to have originated as a transform truncation zone parallel to the more eastern truncation zone delineated by Walker (1988). The Foothills truncation zone served as the accretion locus for the transform ophiolite belt and fragments of "Tethyan" oceanic assemblages, which became fore-arc basement for the early Mesozoic continental arc. Accretion of the ophiolite belt occurred when subduction began along the transform zone during, or at the end of, the Sonoma orogeny. This change in relative plate motion led to the development of the continental arc.

Subduction complexes. Complex rock assemblages that are presumed to have formed, or been accreted to, the plate edge along an east-dipping subduction zone in Triassic time are widespread between British Columbia and the southern end of the Sierra Foothills metamorphic belt. The most common rock association is a chaotic mixture of chert and argillite, the various blocks commonly consisting of metabasite, carbonate rock, and subordinate blueschist. The carbonate rocks commonly contain late Paleozoic "Tethyan" fauna (Douglass, 1967; Monger and Ross, 1971; Danner, 1977; Davis and others, 1979; Saleeby, 1979), although some contain McCloud fauna in the Blue Mountains and Sierra Foothills (Wardlaw and others, 1982; Standlee and Nestell, 1985; Watkins and others, 1987). K/Ar ages on the blueschist blocks and larger thrust slices range from Permian to Late Triassic, with a clustering of ages in the Late Triassic and isolated instances of resetting to Middle Jurassic ages (data reviewed in Davis and others, 1978; Brandon and others, 1988; Saleeby and others, 1989a). The presumed subduction complexes do not exhibit the coherent thrust-belt structure that might be expected in an accretionary-prism setting; rather, they are largely melanges with polystage developmental histories, some clearly involving voluminous olistostromes (Cox and Pratt, 1973; Sckweickert and others, 1977).

The various segments of the subduction-complex belt include the Cache Creek terrane within the Intermontane superter-

rane, the Baker terrane of the Blue Mountains, and the eastern Hayfork–North Fork–Stuart Fork terranes, Calaveras complex, and Red Ant terrane of the western Sierra–Klamath belt. Additional fragments of the regional belt, including the Trafton, Deadman Bay, and Garrison terranes, lie within the San Juan–west Cascades thrust system (Danner, 1977; Brandon and others, 1988). Together, all of these terranes are referred to below as Cache Creek–affinity terranes.

Accretion of the Cache Creek–affinity terranes may have occurred in a Triassic west-facing continent-edge subduction zone unrelated to, and in part superposed across, the Sonoma orogenic belt (Davis and others, 1978). Alternatively, accretion may have been more complex and in part related to the Sonoma orogeny. The later phases of a Sonoma-related accretion history are depicted in Plate 5A (after Saleeby, this volume). In this interpretation, the Sonoma orogeny was a Cordilleran-wide event that involved, in addition to the more widely recognized deformations in the southwest, the initial accretion of equatorial ocean-island complexes containing "Tethyan" fauna–bearing carbonate reefs, and a large eastern Pacific remnant arc complex or fore-arc microplate (Stikine), which was capped by extensive McCloud fauna–bearing carbonate reefs. Accretion occurred along a west-facing subduction zone roughly delineated by the Cache Creek–affinity terranes. Throughout most of its length, the Sonoma orogen consisted of submarine fold-thrust belts that bounded residual basins. These basins survived within Triassic to early Middle Jurassic back-arc, arc, and fore-arc environments of the west-facing continental arc. The blueschist metamorphic ages that are widespread along the Cache Creek–affinity terranes represent denudation(s) of subduction-zone metamorphic environment(s) by collisional uplifts, and/or by postcollisional extension, particularly during Late Triassic time (Saleeby, this volume). Accretion of the Cache Creek ocean-island complex and Stikine terrane occurred primarily in northerly latitudes. In the southwestern United States, where Sonoman tectonics followed the transform truncation zone, late Paleozoic transform ophiolite was accreted as part of the hanging wall of the west-facing subduction zone along the Foothills suture. The Calaveras complex is at least partly depositional on this ophiolitic basement complex, and may represent a huge olistostrome tongue(s) within the ophiolite belt, as is typical of transform axial troughs (Saleeby, 1984, 1990b). The contrast between the tectonics of the Sierra foothills and other Cache Creek–affinity terranes to the north is also marked by the southern termination of the Triassic blueschist localities in the northernmost Sierra and the occurrence of higher-temperature amphibolite facies blocks of similar age to the south (Saleeby, 1990b).

Permian-Triassic deformation of the miogeocline edge. Late Paleozoic–early Mesozoic tectonic truncation along the central and southern California regions resulted in the superposing of plate-edge phenomena across the Paleozoic miogeocline. Carbonate platform strata typical of the miogeocline continued to accumulate in the southwest Mojave region until the Early Permian (Miller and Cameron, 1982), although tectonically unstable borderland facies are at least locally present in the northern Mojave (Walker and Wardlaw, 1989). The Upper Permian miogeoclinal section thins and the top of the Permian section becomes older from the Colorado Plateau west to the Mojave region, probably due to broad erosion near the Permian-Triassic boundary (Burchfiel and Davis, 1977; Dunne, 1977). Shallow-marine sedimentation in the Inyo Mountains continued through Permian and into Early Triassic time (2 in Fig. 4; Stone and others, 1987), but Permian strata along much of the plate edge were exposed to erosion at or prior to the beginning of the Triassic.

Evidence for deformation and metamorphism of the truncated miogeocline during the Permian-Triassic is greatest in the western and central Mojave region. Sequences as young as Mississippian were multiply deformed, metamorphosed, and intruded by ~240 Ma plutons (Southwest Mojave inset, Plate 5A) (Miller, 1981; Miller and Cameron, 1982; Karish and others, 1987). Deformation included polyphase folding and thrusting that involved crystalline basement at intermediate temperatures and low pressures within an intra-arc setting (Miller and Cameron, 1982). To the south in the San Bernardino Mountains, Permian-Triassic thrusting of miogeoclinal strata is also documented (Miller and Cameron, 1982). The deformational and magmatic trends in the Mojave region are parallel to Mesozoic trends, crosscutting older paleogeographic elements, and these trends were developed during or after the displacement of Antler belt rocks into the region (Miller and Cameron, 1982).

Deformation trends diverged from the newly established plate edge in regions to the north where the truncation locus approached the Antler and Sonoman belts. Here, northeast-trending, southeast-verging folds and thrust faults of the Last Chance system developed in ranges to the east of the Sierra Nevada and followed the tectonic grain of the Paleozoic miogeocline and the Antler-Sonoman belts (Dunne and others, 1978; Burchfiel and Davis, 1981; Wernicke and others, 1988; Snow and Lux, 1990). The Last Chance thrust system involves strata as young as Permian; structures that are superposed across the system are cut by plutons as old as Middle Triassic, and are apparently overlapped by Lower and Middle Triassic strata (Last Chance inset, Plate 5A). The southeast vergence of the Last Chance system is consistent with the transport patterns of the Sonoma orogeny and the transform truncation event; this suggests a sinistral tangential component of relative plate motion along the north-trending plate edge at the beginning of the Triassic.

Poorly dated deformations in areas to the north and south of the Last Chance thrust system are also interpreted here as having formed during Early Triassic time. These include north-trending folds in the Benton Range (Renne and Turrin, 1987), a pre-Early Jurassic–post Middle Permian thrust fault in the Cowhole Mountains (Novitsky-Evans and Burchfiel, 1973), and east-vergent folds and thrusts cut by a 200 Ma pluton in the Clark Mountains (Burchfiel and Davis, 1972).

Lower Triassic conglomeratic overlap sequence. During the Early to Middle Triassic, marine and nonmarine clastic

rocks, evaporites, and carbonates of the Moenkopi Formation and correlative strata were deposited over a broad area extending from the plate interior to the plate-edge region (Figs. 3 [1, 2, and 3] and 4 [3, 4, 5, and 6]). The ancestral Rockies provided a source for quartzo-feldspathic detritus, but westward-derived conglomerates in Moenkopi-equivalent strata along the plate edge probably had their source in the uplifted, volcanically active orogenic highlands created by Permian-Triassic deformation (Walker, 1985, 1988; Walker and others, 1983). Conglomerates in Moenkopi or equivalent strata occur in the Inyo Mountains region (Last Chance inset, Plate 5; 2 in Fig. 4), the Clark Mountains of the eastern Mojave Desert (4 in Fig. 4; Walker and others, 1983), and Soda Mountains of the east-central Mojave (3 in Fig. 4; Walker, 1985), the Fairview Valley Formation of the southwest Mojave Desert (6 in Fig. 4; Miller and Cameron, 1982), and the Bond Buyer Formation of the El Paso Mountains (Walker, 1988). Conglomerates in possible Moenkopi-equivalent strata also occur in the southeast California–southwest Arizona region (6 in Fig. 4; Miller, 1981; Hamilton, 1982; Le Veque, 1982; Richard, 1983; Reynolds and others, 1986). In this region, the Lower Triassic strata are on cratonic strata.

The importance of the Lower Triassic overlap sequence is that it was deposited across a varied substrate (Walker, 1985, 1988) that consisted of: (1) deformed rocks and Permian-Triassic plutons in the southwest Mojave region (Plate 5A, inset); (2) allochthonous Antler belt rocks that were folded and intruded by syn to post-kinematic plutons of Late Permian age in the El Paso Mountains area (Carr and others, 1984); (3) relatively nondisturbed miogeoclinal sequences in the eastern Mojave and southern Nevada regions; and (4) nondisturbed cratonic strata in the southeast California–western Arizona region. The Lower Triassic strata thus postdate deformation and emplacement of the allochthonous terrane(s) in the Mojave region. In contrast, the Lower Triassic Candelaria Formation of west-central Nevada is considered to be syndeformational and to have been deformed along the base of the Golconda thrust following its overriding of Antler belt rocks (Plate 5A, inset). Lower Triassic marls and limestones of the Candelaria Formation are overlain by upper-Lower Triassic sandstones and conglomerates derived from the advancing Golconda allochthon (Speed, 1977). The final stages of Sonoman deformation along the contiguous thrust belt corresponded in time with post-"Sonoman" overlap sedimentation in the southwest. Such late-stage Sonoman thrusting may also have overlapped in time with movement along the Last Chance thrust system, or may have postdated it (Last Chance inset, Plate 5A).

Clastic sedimentation related to Sonoman uplift and erosion along the truncated miogeocline is also suggested along the trace of the newly established early Mesozoic fore-arc of the southern Sierra–Mojave region. Thick shallow-water and turbiditic sandstone and quartzite-pebble conglomerate of the Kings sequence lie generally beneath presumed and locally known Upper Triassic shallow-water limestone (Saleeby and others, 1978; Girty, 1985; Saleeby and Busby-Spera, 1986). These clastic units, as well as similar rocks of Salinia metamorphic pendants (Ross, 1977, 1978), are interpreted in Plate 5A as recycled miogeoclinal sediments and underlying sialic basement detritus shed copiously into the adjacent fore-arc shelf and basin.

Continental arc. Lower to Middle Triassic arc rocks are scattered in clusters along the plate edge from the Blue Mountains to the Mojave region. Intermediate to mafic submarine-arc volcanism began by Ladinian time in the McCloud segment of the arc, as recorded in the Olds Ferry terrane of the Blue Mountains (13 in Fig. 5). Basement rocks of the Olds Ferry terrane are unknown, but are presumed to be of McCloud affinity, based on the location of the terrane along the inner edge of the Baker terrane, which contains limestone blocks of both Tethyan and McCloud affinity.

The Wallowa terrane of the Blue Mountains records Permian-Triassic through Triassic arc plutonism and volcanism. The volcanic rocks lie unconformably above Pennsylvanian metaplutonic basement rocks (Balcer, 1980; Walker, 1983, 1991). The metaplutonic rocks are atypical of the McCloud basement, and the Upper Triassic stratigraphic record of the Wallowa terrane differs significantly from that of the Olds Ferry terrane and much of Quesnel and Stikine. Such differences, in conjunction with its outboard position relative to the Baker terrane, lead us to consider the Wallowa terrane as exotic relative to other terranes of the Blue Mountains, but not necessarily equivalent to Wrangellia (Wernicke and Klepacki, 1988). The options for a parental terrane are atypical facies of Stikine lower Mesozoic strata and Paleozoic basement rocks, Wrangellia, or possibly a segment of the arc from which the Cadwallader terrane was derived. Below, we consider this problem in the context of both Stikine terrane migration and the evolution of the Insular suture belt.

Middle Triassic arc volcanism is not as well recorded in other areas along the McCloud segment of the arc as it is in the Blue Mountains. The volcanism that formed Quesnel-related arc sequences of the Okanogon Highlands (14 in Fig. 5) and northeast Washington (15 in Fig. 5) could have begun as early as the Ladinian, but no direct age data exist. Ladinian-age arc volcanic rocks and generally fine-grained marine strata occur locally at the base of the Nicola sequence in Quesnel, and similar strata possibly as old as Early Triassic occur at the base of the Takla-Stuhini arc sequences of Stikine (Monger and others, 1990). Chert, argillite, graywacke, and tuffaceous strata of the younger assemblages of the Cache Creek terrane locally contain Early and Middle Triassic fossils (Monger and others, 1990). Such strata are depicted in Plate 5A as having formed in basinal settings above the west-facing subduction zone of the continental arc. In most localities, the younger assemblages are unconformably above or are distinct fault slices from the late Paleozoic "Tethyan" ocean-floor and ocean-island assemblages. The regionally extensive stratigraphic breaks at or near the Paleozoic-Mesozoic boundary throughout the Intermontane superterrane are interpreted here as an expression of the Sonoma orogeny, as they are in the southwest.

In central Nevada, silicic volcanic rocks of the Lower Triassic Koipato Group lie unconformably above the Havallah se-

quence (10 in Fig. 5). The upper part of the Koipato has been dated as late Early Triassic (Silberling, 1973). The unconformity at the base of the Koipato marks the Sonoma orogeny. Middle Triassic shelfal strata of the northwest Nevada marine province lap conformably across the Koipato Group.

In the southeastern Klamath Mountains (1 in Fig. 5), the lower Triassic Bully Hill rhyolite apparently conformably overlies Permian strata of the type McCloud arc. To the north, the rhyolite unit interfingers with the Lower to Middle Triassic Pit Formation, which records basinal sedimentation and intermittent arc volcanism (Miller, 1987). In the northern part of the eastern Klamaths, marine-basin conditions are also recorded by the Middle Triassic Brock shale. The Pit Formation exhibits local angular discordance and a deformation contrast with the underlying Permian strata that probably reflects the Sonoma orogeny (Miller, 1988).

Silicic volcanism began in Early to Middle Triassic time along the central California–Nevada border region, as in regions to the north. Volcanism is recorded in the Yerington region, where the basement rocks are unknown (3 in Fig. 5; Dilles and Wright, 1988). In the eastern Sierra Nevada, volcanic and volcaniclastic rocks rest unconformably upon Paleozoic strata in the Ritter Range pendant (1 in Fig. 4; Huber and Rhinehart, 1965; Fiske and Tobisch, 1978; Kistler, 1966), the Saddlebag Lake pendant (Chesterman, 1975; Brook, 1977; Schweickert and Lahren, 1987,1989), and the Mt. Morrison pendant (Rinehart and Ross, 1964; Russell and Nokleberg, 1977). Meager age data on these rocks indicate that they may be, in part, Early Triassic in age (Kistler and Swanson, 1981). In the Inyo Mountains, Lower(?) or Middle(?) Triassic alluvial-fan deposits overlie a Paleozoic to Lower Triassic marine section (2 in Fig. 4; Osborne and Dunne, 1982). Coarse debris in basal alluvial-fan deposits were derived from the Paleozoic to Lower Triassic substrate; these give way upsection to volcanic debris flows derived from the west (Osborne and Dunne, 1982).

Silicic igneous activity started in the southern California region in Early to Middle Triassic time. Lower Triassic granitic plutons are scattered in the southeastern Sierra Nevada and El Paso Mountains region (Walker, 1988; Kistler, 1990; Dunne and others, 1991). Granitic and alkaline plutons are sparse but widespread in the Mojave region (Beckerman and others, 1982; Karish and others, 1987) and have apparent Middle Triassic ages (C. F. Miller, 1978). A mylonitized metadacite that occurs as a fault-bounded sliver in the Mineral King roof pendant of the south-central Sierra is of Early to early-Middle Triassic age (9 in Fig. 4; Busby-Spera, 1983). These scattered occurrences of largely calc-alkaline igneous rocks record the inception of continental-arc magmatism across displaced Antler belt rocks and more southerly miogeoclinal strata synchronous with regional deformation and metamorphism.

Plate interior

The Early Triassic rocks represent the last widespread episode of marine sedimentation in the Cordilleran miogeocline. Depositional patterns remained fairly constant between late Paleozoic and Early Triassic time (Collinson and Hasenmaueller, 1978; Carr and Paull, 1983). Miogeoclinal carbonates and fine-grained terrigenous rocks of Early Triassic age record four transgressive episodes, and thin eastward toward the craton where they intertongue with redbeds. A widespread *Meekoceras*-bearing limestone unit that occurs in the Moenkopi and Thaynes Formations appears to thin westward, along a north-trending belt in eastern Nevada, and is locally replaced by chert-pebble conglomerate (Collinson and Hasenmueller, 1978), presumably derived from the Golconda allochthon (Plate 5A).

Lower Triassic facies patterns closely mimic those of the Permian, even though a disconformity occurs between them (Carr and Paull, 1983). In Late Permian time, a broad carbonate shelf covered much of Nevada, Utah, and western Wyoming, while relatively deep-water mudstones, cherts, and phosphorites accumulated in the Phosphoria basin centered in southeast Idaho. Basinal facies of the Lower Triassic are also present in southeast Idaho. The southern boundary between basinal and shelf facies, located in northeast Nevada, may be related to vertical tectonics on a basement flaw (Carr and Paull, 1983) that originated as a Precambrian aulacogen (Kistler and others, 1981), and was reactivated in Permian time (Oquirrh-Uinta uplift) and again during the Sevier-Laramide orogeny (Wells fault).

Lower Triassic miogeoclinal facies were divided by Carr and Paull (1983) into: (1) basinal facies—calcareous mudstone and siltstone lacking abundant shelly benthic fauna (6 in Fig. 3); depositional environments were commonly dysaerobic, but ranged from aerobic to anaerobic; (2) outer shelf facies—bioclastic carbonates with less than 10% clastic rocks interpreted to have accumulated in water depths of less than 50 m (4 and 5 in Fig. 3); (3) inner-shelf facies—fine-grained sandstones, coarse-grained siltstone, silty limestone, and dolomite with relatively low faunal diversity representing shallow subtidal to intertidal environments (3 and 4 in Fig. 3); and (4) red-bed sandstone, siltstone, and mudstone representing very shallow marine to continental deposits, including fluvial, deltaic, lacustrine, and shallow tidal flat (1, 2, and 8 in Fig. 3).

Termination of marine sedimentation in the miogeocline may have been related to the Sonoma orogeny accretion and the onset of active-margin tectonism (Carr and Paull, 1983). Relations between the miogeoclinal rocks and those of the Sonoma belt are obscure, however, because outcrops of appropriate age are sparse within a north-south–trending belt 150 km wide separating east and central Nevada (Collinson and Hasenmueller, 1978). There is no evidence of thinning or shoaling of Lower Triassic facies against highlands to the west; this led Carr and Paull (1983) to speculate that marine deposition originally extended west into central Nevada. By Middle Triassic time, however, most of the plate interior was emergent and undergoing erosion (1–7 in Fig. 3; Blakey and Middleton, 1983). Middle Triassic continental deposition may have occurred in parts of Wyoming and Colorado (Red Draw Member of Jelm Formation and Crow Mountain Member of Chugwater Formation; Tohill

and Picard, 1966), but there are no recognized Middle Triassic strata in the western part of the plate interior (McKee and others, 1959; Pipiringos and O'Sullivan, 1978; MacLachlan, 1972). A eustatic drop in sea level may have contributed to subaerial exposure of the region during the later part of the Middle Triassic.

LATE TRIASSIC (230–208 Ma)

The major tectonic and paleogeographic features for Late Triassic time are shown in Plate 5B. The continental arc was well established, having cut obliquely across the McCloud basement and its boundary with the Paleozoic miogeocline. Much of the McCloud basement was submerged beneath a marine basinal province, primarily submarine-arc volcanism being recorded in the Intermontane superterrane, the Blue Mountains, and the eastern Klamaths. Arc plutonic rocks of Late Triassic age are present along the east-central Sierra Nevada, in the White-Inyo Mountains, and in the Mojave region (Evernden and Kistler, 1970; Miller, 1978; Stern and others, 1981; Chen and Moore, 1982). Several lines of evidence, when considered together, suggest that the Late Triassic to Early Jurassic arc of California and Arizona occupied an extensional or transtensional graben system (Busby-Spera, 1988) (Plates 5-B and 5-C). These include: (1) extreme thickening of Upper Triassic to Lower Jurassic continental and shallow-marine successions along the arc, relative to the adjacent craton, requiring high subsidence rates for their accumulation and preservation; (2) dispersal of quartz arenites across the width and along the length of the arc, suggesting that it acted as a funnel and a trap for craton-derived detritus; (3) abundance of high-level intrusions and thick silicic ignimbrite sequences; (4) presence of alkalic volcanic rocks locally interstratified with calc-alkalic volcanic rocks, similar to modern rifted arcs like the Colima graben, Mexico; (5) stratigraphic and structural evidence for syndepositional normal faults, including localized thick sequences of polylithologic breccia that suggest faulting contemporaneous with sedimentation; and (6) occurrences of limestone and calc-silicate rocks interstratified with continental-arc deposits, suggesting depressions occupied by marine embayments and/or large lakes. The regional extensional environment recorded along the sialic segment of the arc may also have extended along the McCloud segment, as recorded by broad basinal conditions and widespread alkaline-affinity volcanic rocks.

To the west of the continental arc, the vestiges of subduction complexes are preserved along the western Sierran–Klamath belt, in the Blue Mountains, and perhaps within or along the western edge of the Intermontane superterrane. A number of the fringing terranes that are imbricated with the subduction complexes of the western Sierra–Klamath belt record mafic submarine volcanism, marking the initiation of a Late Triassic–Early Jurassic extensional fore-arc regime. Fore-arc extension may have led to the initial denudation of the high P/T metamorphic rocks and related setting of their isotopic ages (Saleeby, this volume).

In the plate interior, the Chinle Formation records alluvial-plain and shallow lacustrine sedimentation on the eastern and southern edge of the back-arc basin. Additional basin sediments that may have been deposited in western Utah and Nevada have been eroded away (Blakey and Gubitosa, 1983). Remnants of the ancestral Rockies persisted as a low-relief source area, and volcanic detritus within the southern Chinle indicates volcanism in a source area to the southwest.

Plate-kinematic patterns for Late Triassic time are in general poorly determined, although a major change in the absolute motion of North America is clearly indicated near the Triassic-Jurassic boundary by the J1 cusp in the apparent polar wander (APW) path (May and Butler, 1986). The J1 cusp resulted in a change from generally westward Triassic motion to north-northeast–directed motion in Early and Middle Jurassic time. Uncertainties in the timing of the J1 cusp permit it to have occurred as early as ~220 Ma, which could correspond to the initiation of broad suprasubduction-zone extension throughout the Cordillera. The broad extensional regime probably reflects weakly coupled subduction which may have arisen from the retreat component in North America's motion relative to the plate edge during the J1 to J2 APW track. An additional factor may have been the subduction of old, cold Panthalassa lithosphere inherited from the Pangea regime (Busby-Spera and others, 1991b).

Plate edge

The North American plate edge is best defined by the extensive belt of continental-arc rocks. Fore-arc basin strata possibly extending back to the Late Triassic occur in the Kings sequence of the southern Sierra Nevada (Saleeby and others, 1978) and are well preserved in the Izee terrane of the Blue Mountains (Dickinson and Thayer, 1978). The Cache Creek–affinity terranes presumably record west-facing subduction in the Late Triassic (Davis and others, 1978), although much of the Late Triassic record may be alternatively interpreted as representing an extensional fore-arc.

Subduction complexes. We think that the Cache Creek terrane resided within the McCloud basement framework of the continental arc by Late Triassic time, and therefore it is discussed with the McCloud segment of the arc. The Bridge River terrane is interpreted as a subduction accretion assemblage formed along the western margin of the Intermontane superterrane (Plate 5B). The position of the superterrane relative to the Blue Mountains is poorly known for this time interval, and we suggest that its southern termination was very close to the northern end of the Blue Mountains. From the Blue Mountains southward, the Cache Creek–affinity terranes faced open ocean within a west-facing subduction trench–fore-arc setting.

Blue Mountains—A belt of highly disrupted oceanic rocks extends along the axis of the Blue Mountains arc and is thought to be a subduction complex (Davis and others,1978; Dickinson and Thayer, 1978; Coward, 1982). These rocks consist of the Baker and Grindstone terranes, part of the Izee terrane, and the Mitchell inlier. We refer to all of these rocks as the Baker terrane. We consider the history of the Baker terrane to have been much more

complex than accretion of open ocean-floor assemblages in a west-facing subduction zone. Accretion of the Baker terrane involved the emplacement, and perhaps in situ generation, of arc-related ophiolitic crust, in addition to thrust imbrication of basinal sediments and high P/T metamorphism. The mixed assemblage of Baker terrane rocks was progressively deformed in a subduction complex–fore-arc setting, and at least partly overlapped by the thick fore-arc basin sequence of the Izee terrane (Izee-Baker inset, Plate 5B) (Dickinson and Thayer, 1978). The main belt of Baker terrane rocks currently separates the Wallowa and Olds Ferry terranes.

Principal elements of the Baker terrane consist of: (1) Permian to Upper Triassic and possibly Lower Jurassic chert-argillite melange and broken formation (Elkhorn Ridge argillite) that encases blocks of undated greenstone and Lower Triassic metaplutonic rock, Devonian and Pennsylvanian limestone, and Permian "Tethyan" limestone (Nestell, 1983; Coward, 1982; Blome and others, 1986; Pessagno and Blome, 1990; Walker, 1991); and (2) serpentinite-matrix melange with primarily ophiolitic blocks and slabs, the larger of which (Canyon Mountain complex) exhibits remnants of an internal stratigraphy (Thayer, 1977). The Canyon Mountain complex consists of a low K_2O arc magmatic series (Gerlach and others, 1981) that developed in Early Permian to perhaps Carboniferous time (Walker, 1991). Smaller ophiolitic igneous blocks within the melange range from Early Permian to Middle Triassic in age (Walker, 1991). The Sparta igneous complex occurs as a $\sim$70 km^2 slab of "ophiolitic" rocks along the northwest margin of the Baker terrane. It also constitutes a low K_2O arc volcanic-plutonic series that developed in numerous igneous pulses between Late Permian and Late Triassic time (Phelps and Avé Lallemant, 1980; Walker, 1991). Both the Canyon Mountain and Sparta complexes could represent fragments of oceanic arc, inter-arc basin, or extensional fore-arc complexes (Saleeby, this volume).

The melange assemblage of the Baker terrane continues southward as basement beneath Upper Triassic fore-arc basin strata of the northern Izee terrane (Dickinson and Thayer, 1978), and disrupted rocks of the Grindstone terrane lie unconformably beneath the southern end of the Izee strata. The Grindstone terrane consists of chert-argillite and arc volcanic arenite-matrix melange with limestone and keratophyric blocks (Dickinson and Thayer, 1978). Mixing of the matrix and blocks is attributed primarily to Permian through Early Triassic olistostrome activity (Blome and others, 1986). Limestone blocks are of Devonian, Mississippian, and Permian age; the Permian blocks contain McCloud fauna (Wardlaw and others, 1982). Mafic blueschist of Late(?) Triassic age is exposed west of the Gridstone-Izee area in a small inlier near Mitchell (Hotz and others, 1977).

The relations outlined above indicate that disrupted rocks of the Baker terrane evolved by polystage processes between Permian, Triassic, and possibly Early Jurassic time, and that large-scale olistostromal activity, serpentinite diapirism, possibly in situ ophiolitic magmatism, as well as tectonic imbrication, were involved. Furthermore, late Paleozoic and Triassic oceanic-arc rocks, McCloud arc, and "Tethyan" abyssal crust and seamount basement regimes were all incorporated into the disrupted terrane. Tectonic and sedimentary mixing began in Permian time and continued through Triassic time. Thus, formation and deformation of the Baker terrane occurred during and after the Sonoma orogeny.

Klamath Mountains—The eastern Hayfork, North Fork, and Stuart Fork terranes of the central Klamath Mountains are interpreted as related elements of a Permian-Triassic subduction complex that developed west of Permian-Triassic arc-related strata of the eastern Klamaths (Wright, 1982). The inclusion of abundant Upper Triassic chert suggests that subduction accretion extended into Late Triassic time. As discussed below, however, the west-vergent thrust faults that currently bound the various terranes of the subduction complex are Middle Jurassic in age (Plate 5D, Klamath inset) and represent a distinctly different phase of convergent tectonism. Furthermore, radiolaria locally range into the Early and Middle (?) Jurassic, which permits final assembly of the complex as late as Middle Jurassic time. The present structural sequence in the Klamaths consists of, in ascending order, the eastern Hayfork, North Fork, and Stuart Fork terranes, and then the central metamorphic belt, which is part of the structural basement of the McCloud arc. To the west, the Rattlesnake Creek terrane lies at the structural base of the sequence, although its pre–Upper Triassic ophiolitic melange may also have been initially accreted as part of the complex. As discussed below, the Permian-Triassic subduction complex is interpreted to have undergone a phase of Late Triassic–Early Jurassic extensional denudation prior to Middle Jurassic thrusting (Klamath inset, Plate 5B).

The eastern Hayfork terrane is composed principally of a chert, argillite, and sandstone-matrix melange and broken formation (Wright, 1982). Scattered throughout the terrane are various blocks, some of which were clearly deposited and/or deformed and metamorphosed in an environment that is markedly distinct from the enclosing matrix. Sandstone blocks and matrix materials contain abundant continental detritus. The blocks also include metachert, mica schist, quartzofeldspathic schist, greenstone, marble, and rare glaucophane schist (Wright, 1982; Fahan, 1982). Well-dated cherts from the melange range from Permian to Late Triassic in age (Irwin and others, 1978), and shallow-water limestone blocks locally associated with alkalic basalt range from Permian to Late Triassic in age. Several of the limestone blocks contain Permian "Tethyan" fauna (Irwin and others, 1978; Irwin and Wardlaw, 1983; Miller and Wright, 1987).

The North Fork terrane is lithologically similar to the eastern Hayfork terrane (Wright, 1982; Mortimer, 1985) in that it contains: (1) radiolarian chert that range from Permian to Early or Middle(?) Jurassic in age (Irwin and others, 1977, 1978; Ando and others, 1983); and (2) limestone masses associated with alkalic basalt that have yielded Permian "Tethyan" fauna (Elliott and Bostwick, 1973; Mortimer, 1985). The North Fork terrane, however, for the most part lacks the chaotic structure of the eastern Hayfork terrane, with the important exception of the

chert-argillite olistostrome mass described by Cox and Pratt (1973). The Salmon River terrane constitutes the late Paleozoic ophiolitic basement of the North Fork terrane and consists of basaltic pillow lava, diabase, gabbro, and harzburgite tectonite (Ando and others, 1983). An additional sequence of enriched Mg-rich mafic lavas and underlying continent-derived metaclastic rocks of the Sawyers Bar area is thought to be the northward continuation of the North Fork terrane (Ernst, 1987). These lavas, along with the Salmon River ophiolitic rocks and the alkalic lavas associated with the "Tethyan" limestone masses, record highly diverse oceanic basement assemblages within the subduction complex.

The Stuart Fork terrane consists primarily of metachert, glaucophane, lawsonite, and jadeite-bearing metavolcanic rocks, along with minor marble, metagabbro, serpentinite, and eclogite (Hotz, 1977; Borns, 1980, 1984; Goodge, 1985, 1990). Late Triassic metamorphic ages (Hotz and others, 1977) may record peak metamorphism or, alternatively, transport out of the metamorphic environment. Metabasalts of the Stuart Fork terrane are slightly enriched tholeiites that could have formed in an abyssal, back-arc basin or primitive arc environment.

Sierra Nevada— High *P/T* tectonites of Mitchell inlier and Stuart Fork affinity may extend into the northern Sierra as the Red Ant terrane, which consists of fault-bounded lenses of glaucophane schist derived from chert and mafic metavolcanic and metaclastic rocks (Schweickert and others, 1980; Edelman and others, 1989). The Red Ant terrane is present along a major tectonic break defined by the Paleozoic Feather River ophiolite belt. Most geochronometric systems along this belt have been disturbed by Triassic and Jurassic tectonism and Early Cretaceous plutonism (Saleeby and others, 1989a). Ages reported for the Red Ant terrane (Schweickert and others, 1980) strongly reflect these disturbances, and thus the actual metamorphic age of the terrane is poorly constrained. The Red Ant terrane is locally observed to have been thrust eastward beneath the Feather River belt (Edelman and others, 1988). This may reflect a subduction-related transport pattern, but is more likely a reflection of Middle Jurassic thrusting analogous to that observed in the Klamath Mountains.

A belt of highly deformed Calaveras chert, phyllite, and mafic metavolcanic rock occurs as a fault sliver along the western margin of the Red Ant terrane (Hietanen, 1981a; 1981b; Edelman and others, 1989). This Calaveras assemblage contains lenses of McCloud limestone (Standlee and Nestill, 1985), demonstrating accumulation adjacent to the late Paleozoic McCloud arc. This Calaveras assemblage is nearly continuous with the Calaveras complex of the central Sierra, although high-angle faults intervene (Edelman and Sharp, 1989).

The Calaveras complex of the central Sierra is defined as a structural sequence (from west to east) of basalt-andesite pillowed and pyroclastic rocks, chert and volcanic-rich wackes, phyllitic rocks derived primarily from argillite and chert, a metabasaltic seamount and depositionally(?) overlying marble belt, and chert-argillite melange (Schweickert and others, 1977). The chert-argillite melange and metabasalt-marble belt are probably members of a Triassic to Early Jurassic subduction complex (Saleeby and others, 1986). The chert-argillite is largely chaotic, having been mixed in a soft-sediment state (Schweickert and others, 1977). The more western tracts of the chert-argillite are more orderly and contain depositional beds and lenses of Lower Triassic limestone (Bateman and others, 1985). The Permian-Carboniferous marble blocks within the chert-argillite can be traced to a thick source sequence that rests on and is locally interlayered with basaltic metavolcanic rock. The Calaveras complex was thrust eastward beneath the Shoo Fly complex (Schweickert and Bogen, 1983) between Late Triassic and early-Middle Jurassic time (Sharp, 1988; Sharp and Herzig, 1991). Metamorphism was at high greenschist to amphibolite facies. To the west and northwest in the Sierran Foothills, Permian-Carboniferous ophiolitic rocks with minor chert and limestone were also imbricated along the east-dipping subduction zone and underwent amphibolite facies metamorphism under intermediate *P/T* conditions (Behrman, 1978; Sharp and Leighton, 1987; Sharp, 1988). Subsequently, these disrupted ophiolitic rocks formed basement for the younger, more western phyllitic units of Calaveras-affinity rocks, which locally contain late Paleozoic "Tethyan" limestone blocks.

The boundary between the Calaveras complex and the Shoo Fly complex and its northward continuation as the Feather River ophiolite belt marks the Foothills suture (after Saleeby, 1981). The continuation of the suture northward into the Klamaths may coincide with the boundary between the Permian-Triassic subduction complexes discussed above, and mid- to lower Paleozoic metamorphic and oceanic basement rocks associated with the McCloud arc. As discussed below, however, the suture relations in the northern Sierra and Klamath mountains are complicated by Late Jurassic sinistral strike-slip faulting, in addition to Middle Jurassic thrusting. In the southern Sierra the suture places upper Paleozoic transform ophiolite against early Mesozoic shelfal and fore-arc basin strata (Kings sequence) that were built along the truncated miogeocline and offset Antler belt rocks (Saleeby and others, 1978). Calaveras rocks in this region consist of chert-argillite olistostromal tongues that were apparently channeled down a transform-controlled trough within the ophiolite belt (Saleeby, 1979). The olistostromes contain limestone blocks with Permian "Tethyan" fauna. Quartz-rich olistoliths and turbidites of the western Kings sequence are mixed with Calaveras rocks of this region. Sedimentary facies relations along this segment of the Foothills suture are summarized for the Early Jurassic in the Kaweah inset of Plate 5C, but are probably applicable for the Late Triassic as well. Lower Mesozoic strata are not known to lap across the suture in the central and northern Sierra, although Calaveras-affinity strata along the western Foothills in the central Sierra contain quartz-arenite mixtures that resemble the western facies of the Kings sequence. In addition, the broad geochemical patterns of the southern Sierra batholith discussed below suggest that the original belt of Kings sequence rocks extended into the central Foothills belt.

Fore-arc basin strata. The development of a major fore-arc basin sequence during Late Triassic time is well recorded in the Blue Mountains (12 in Fig. 5; Dickinson and Thayer, 1978). The Izee fore-arc basin terrane was built unconformably across the disrupted Baker terrane (Izee-Baker inset, Plate 5B). The basinal strata consist of a composite ~15-km-thick sequence of volcanogenic turbidites and subordinate lava flows and tuffs. The oldest stratigraphic unit is the Carnian Vester Formation, which contains abundant coarse channel fill and olistostrome deposits. The Aldrich Mountains Group lies in pronounced angular unconformity above the Vester Formation. The unconformity may represent the onset of west-northwest–trending, north-verging contractile deformation which progressively deformed the fore-arc basin in late-Late Triassic through Early Jurassic time. These deformation trends, which were oblique to the plate edge, may mark the initiation of a strong dextral tangential component of relative motion that was prevalent during the Early(?) and Middle Jurassic.

The unconformity between the Vester Formation and Aldrich Mountains Group extends across to Baker terrane basement, where it is recorded by folding and thrusting of the subduction complex and lower fore-arc basin fill. The base of the Aldrich Mountains Group is Norian in age and consists of olistostromes derived from Baker terrane and Vester strata, along with local mafic pillowed flows which are overlain by slope and channel facies turbidite sequences. The Aldrich Mountain Group extends into the Jurassic and exceeds 10 km in thickness. Mafic lava flows within the lower Aldrich Mountains Group could correspond to pulses of fore-arc extensional volcanism, as recorded in the fringing terranes of the western Sierra–Klamath belt. Younger phases of "ophiolitic" magmatism in the Sparta complex of the Baker terrane could likewise represent in situ fore-arc extensional igneous activity.

The Upper Triassic to Lower Jurassic Kings sequence of the southern Sierra Nevada represents primarily shelfal and fore-arc basin strata that developed along the truncated edge of the miogeocline, and/or along offset Antler belt rocks. The basinal facies lapped westward across juxtaposed upper Paleozoic transform ophiolite. Fore-arc sedimentation is best documented for Early Jurassic time, but in several areas it apparently began by Late Triassic. In the Kaweah River area of the southern Foothills (14 in Fig. 6), quartz-rich olistoliths and turbidites as well as local silicic volcaniclastic debris were intermixed with Calaveras-affinity chert-argillite during, or shortly after, the initiation of Late Triassic–Early Jurassic fore-arc volcanism (Saleeby, 1979, and this volume; Saleeby and Sharp, 1980). To the east in the Boyden Cave pendant (8 in Fig. 4) and in a belt of pendants in the Sequoia Park area, thick shallow-water quartzose and quartzo-feldspathic sandstone and carbonate sequences are interbedded with, and depositionally overlain by, basinal turbidites and olistostromes (Moore and Marks, 1972; Saleeby and others, 1978, 1990; Girty, 1985). Likewise, the Kings sequence strata in the southern Sierra (10 in Fig. 4) contain Late Triassic–Early Jurassic fossils at the top of a shelfal quartzite-carbonate sequence that grades abruptly upward into a thick turbidite sequence of probable fore-arc basin strata (Saleeby and Busby-Spera, 1986). Adjacent pendants exhibit similar shallow-water strata, but different upper-level turbidite facies, including monotonous, faintly laminated, fine-grained turbidites and thick-bedded coarse channel-fill sequences with abundant quartz and silicic volcanic detritus. The southern Sierra Kings sequence also contains a number of basaltic flows and sills lying generally beneath the turbidite sequences. The mafic lava flows may represent an early phase of arc activity and/or extensional magmatism in the Late (or earlier) Triassic shelf environment.

Pendants of the Kings sequence lie within a petrochemically distinct domain of the Sierra Nevada batholith. Isotopic, mineralogic, and phase chemical data all characterize the Kings sequence domain of the batholith as having been variably contaminated with craton-derived detritus to the extent that plutons with initial $^{87}Sr/^{86}Sr$ ratios greater than 0.706 attained such rations by a mixture of mantle-derived magma with the melted detritus (Ague and Brimhall, 1988; Saleeby and others, 1987; Kistler, 1990; Chen and Tilton, 1990). These relations suggest that the Kings sequence was originally thick, linear, basin fill floored primarily by oceanic lithosphere that was juxtaposed against the miogeocline. The Kings sequence domain of the batholith is mapped to the central western Foothills, where it ends as a major intrusive border against the Calaveras complex and structurally adjacent Sullivan Creek terrane (13 in Fig. 6). The western phyllitic Calaveras rocks, which lie in the transition between Calaveras chert-argillite and Sullivan Creek volcanic rocks, contain abundant continental and volcanic detritus, as well as reworked Calaveras material (Sharp, 1988; Sharp and Herzig, 1991). The phyllitic rocks are between Middle Triassic and early-Middle Jurassic in age (Bateman and others, 1985; Sharp, 1988; Sharp and others, 1991), and are interpreted on Plates 5-B and C as the northward continuation of the Kings sequence fore-arc basin. The general facies pattern exhibited in the Kings sequence and its hypothetical extension to the north is prevalence of finer and more basinal facies to the west and north, and shallow shelfal facies and more proximal arc facies such as the Mineral King pendant (9 in Fig. 4) to the south and east. The western fringe of the Kings sequence basin was, at least in part, bounded by a volcanic-tectonic ridge elevated by serpentinite diapirism and fore-arc volcanism, as recorded in the fringing terranes.

Metasedimentary roof pendants of Salinia and the axial zone of the northern Peninsular Ranges batholith have overall compositions and meager age data comparable to those of the Kings sequence. Thick marble layers, impure quartzite, and quartzo-feldspathic metasandstones of western and northern Salinia (Ross, 1977, 1978) are similar to the lower stratigraphic levels of the Kings sequence and, like upper stratigraphic levels, the schist of Sierra de Salinas is a monotonous metaturbidite sequence with subordinate quartzite, marble, and amphibolite (Ross, 1976, 1978). The Salinia metasedimentary rocks are determined to be pre-middle Cretaceous in age but, like much of the Kings sequence, they are very different in character from miogeoclinal

strata (also see Ross, 1977, 1978). Late Triassic and Triassic fossils have been reported from the Julian schist and French Valley schist of the northern Peninsular Ranges (Todd and others, 1988). The Julian schist consists of turbiditic and olistostromal pelite, quartzite, marble, calc-silicate rock, amphibolite, and local arc-type metavolcanic rocks. The French Valley schist consists of similar rocks, but also contains slivers of metaserpentinite, suggesting an ophiolitic basement. The Julian and French Valley schist units appear to be mixed into a longitudinal gradation zone with high-grade miogeoclinal strata (Todd and others, 1988). To the east of the gradation zone, miogeoclinal rocks constitute the metamorphic framework for the batholith. Thus, the Julian and French Valley schists lie in a tectonic position analogous to that of the Kings sequence.

In Plate 5B, the Salinia and northern axial Peninsular ranges metasedimentary rocks are interpreted as the southern continuation of the shelfal and fore-arc basin belt that is represented by the Kings sequence in the southern Sierra. As in the southern Sierra region, the fore-arc basin is interpreted as lying along the general trace of the previously truncated miogeocline, and as having at least partly formed above disrupted oceanic crust. The quartzofeldspathic sand components represent unroofing of sialic crystalline basement rocks probably situated along the truncation zone, and possibly uplifted in part during Sonoman tectonism.

Fringing terranes. The fringing terranes consist primarily of structurally complex ensimatic assemblages that lie in the fore-arc region of the continental arc. The terranes constitute much of the western Sierra–Klamath belt, the basement for the Cretaceous Great Valley fore-arc basin, and ophiolite fragments of the California and southern Oregon Coast ranges. In the western Sierra–Klamath belt, the fringing terranes are intimately associated with the early Mesozoic subduction complexes and the upper Paleozoic transform ophiolite belt. These terranes are characterized by early Mesozoic submarine mafic volcanism and basinal sedimentation widely superimposed over older disrupted oceanic crust and upper mantle (Saleeby, 1990b, this volume). The fringing terranes also include relatively complete and intact late-Middle Jurassic ophiolite sequences which formed in a suprasubduction-zone environment.

There are two contrasting plate tectonic models for the fringing terranes. (1) They represent vestiges of an exotic island-arc system that had opposite polarity to, and collided with, the Cordilleran continental arc during Late Jurassic time (Schweickert and Cowan, 1975). (2) They are the remnants of extensional fore-arc and inter-arc basin terranes with short-lived fringing arc volcanic chains which formed and together migrated along the frontal edge of the continental arc (Saleeby, 1981, 1990b, this volume). The petrogenetic associations of the fringing terranes are more typical of extensional fore-arcs (Saleeby, 1990, this volume), and a number of sedimentary provenance and stitching pluton relations suggest that the terranes evolved as part of the North American plate. These terranes are thus interpreted as being primarily parautochthonous.

Two distinct phases of submarine volcanism are recognized in the fringing terranes: Late Triassic–Early Jurassic and late-Middle Jurassic. Both of these episodes correspond in time with phases of regional extension along the plate edge. This extension along the continental arc resulted in the formation of major graben complexes that ponded thick silicic ignimbrite sequences and terrestrial to marine sedimentary sequences (Busby-Spera, 1988). Regional extension along the fore-arc disrupted preexisting subduction complexes and created pseudo-ophiolitic fore-arc crust with polygenetic igneous histories. Serpentinite diapirism was also widespread in the fore-arc extensional environment, derivation being from older hydrated oceanic mantle of the fore-arc basement wedge. The Late Triassic–Early Jurassic fringing terranes of the western Sierra–Klamath belt record well the interplay between diffuse extension, submarine mafic volcanism, and serpentinite diapirism along a semicontinuous belt of ~900 km length (Saleeby, 1990b, this volume). Some of the key locations are outlined below.

Western Klamath Mountains—Internal complexities that typify many of the fringing terranes of the western Sierra–Klamath belt are exhibited in the Rattlesnake Creek terrane (Irwin, 1972; Wright, 1982) and in higher-grade equivalent rocks of the Marble Mountains terrane of the western Klamaths (Rawson and Peterson, 1982; Donato and others, 1982). The Rattlesnake Creek terrane of the southwest Klamaths consists of a multistage serpentinite melange that is depositionally overlain by Upper Triassic–Lower Jurassic metavolcanic and metasedimentary rocks (6 in Fig. 6; Wright and Wyld, 1985). The serpentinite melange represents mid-ocean ridge (MORB)-like oceanic crust and upper mantle that was disrupted prior to the Late Triassic and subsequently remobilized in several episodes. The cover sequence consists of Upper Triassic arc tholeiitic pillow lavas and breccias, interbedded with terrigenous epiclastic strata, chert, argillite, and minor limestone; overlying units are calc-alkaline to possibly shoshonitic basaltic breccias with hemipelagic and terrigenous epiclastic layers (Wright and Wyld, 1985). The melange basement and lower levels of the cover sequence are intruded by uppermost Triassic to Lower Jurassic mafic to plagiogranite stocks and dikes that probably fed the volcanic cover sequence.

The serpentinite melange basement of the Rattlesnake Creek terrane continues for more than 100 km northward to the Marble Mountains region (Donato and others, 1982; Gray and Wright, 1984; Groman, 1985; Donato, 1987), and an additional large fault slice is present in the western Klamaths of southern Oregon (2 in Fig. 6; Roure and DeWever, 1983; Saleeby, 1990b, this volume). Much of the Upper Triassic strata in these areas have been obliterated or obscured by Middle and Late Jurassic thrusting, plutonism, and metamorphism, although the melange retains many of the characteristics observed beneath the Upper Triassic volcanic cover in the south.

The Rattlesnake Creek terrane was imbricated beneath the central Klamath Mountains Permian-Triassic subduction complex in Middle Jurassic time (Wright and Fahan, 1988). The protolith of the serpentinite melange basement may have been accreted as part of the subduction complex. The Klamath inset of

Plate 5B depicts the tectonic setting of the Rattlesnake terrane and its relations to the subduction complex at the end of the Late Triassic (Saleeby, this volume). The subduction complex is shown undergoing extensional denudation, the Rattlesnake Creek melange basement diapirically rising as a fore-arc ridge. A carapace of submarine lavas formed across this ridge. Diffuse volcanism also affected exhumed accretionary prism complexes, represented by the younger mafic lava sequences of the eastern Hayfork and North Fork terranes. Late Triassic setting of isotopic ages within high-*P*/*T* metamorphic rocks of the Stuart Fork terrane may have occurred during extensional denudation.

Western Sierra Nevada—A major belt of Late Triassic–Early Jurassic rocks that closely resembles the Rattlesnake Creek terrane extends along the western margin of the Calaveras complex in the northern Sierra Nevada (Hietanen, 1973, 1981a, 1981b). Edelman and others (1989) have grouped these rocks into a quasistratigraphic sequence named the Fiddle Creek complex (7 in Fig. 6). The base of the complex consists of a serpentinite matrix melange with blocks and slabs of harzburgite tectonite, and metamorphosed gabbro, chert, and basalt, and also includes sedimentary serpentinite, ophicalcite, and serpentinite-argillite polymict breccia. The age of the ophiolitic melange is probably late Paleozoic (Saleeby and others, 1989a; Saleeby, 1990b). A submarine volcanic sequence of olivine phyric pillow basalt and keratophyric tuff unconformably overlies the melange, and lenses of flaser gabbro and boninitic dikes, some of which are in sheeted sets, intrude across the melange structure (Dilek and others, 1990). The volcanic rocks are undated, but the intrusions yield Late Triassic to Early Jurassic ages (Saleeby and others, 1989a). An overlying sequence of chert, siliceous argillite, and volcaniclastic sandstone contains Middle to Late Triassic and Late Triassic to Early Jurassic radiolaria (Hietanen, 1981a, 1981b).

Upper Triassic to Lower Jurassic chert and mafic lava sequences constitute a major element in the central melange belt of the Sierran foothills. The Late Triassic history is best documented in the lower American River area (10 in Fig. 6), where a disrupted sequence of pillow lava and breccia, aquagene tuff, tuffaceous chert and ribbon chert, along with dioritic hypabyssal rocks, was built on Paleozoic ophiolitic melange (Behrman, 1978; Saleeby, 1982; Wright and Saleeby, 1983). The radiolarian cherts have Late Triassic and Early Jurassic ages, and the dioritic hypabyssal rocks yield latest Triassic–earliest Jurassic ages. Chert and siliceous argillite along with local volcanic wackes of the lower American River sequence strongly resemble coeval strata of the Fiddle Creek complex. Ophiolitic basement rocks of the two assemblages are also similar, and the two assemblages can be mapped nearly continuously. Fore-arc volcanism may have begun in the Late Triassic at additional key localities along the central belt (8, 11, and 12 in Fig. 6) but age data clearly indicate activity in Early Jurassic time.

Late Triassic to Early Jurassic boninitic pillow lava, phyric breccia, ribbon and tuffaceous chert, and basaltic to dioritic dikes overlie and intrude upper Paleozoic ophiolitic melange in the lower Kaweah River area (14 in Fig. 6) (Saleeby, 1979, 1982, 1990b). The submarine environment was complex; widespread serpentinite diapirism resulted in the deposition of ophiolitic debris flows and ophicalcites within the lava-chert sequence. Tuffaceous cherts of this sequence grade into siliceous argillite and turbidites of the western Kings sequence fore-arc basin strata. Upper Triassic to Lower Jurassic rocks of the Kaweah melange are in many ways similar to those of the lower American River, Fiddle Creek, and Rattlesnake Creek terranes, but have a greater proportion of continent and sialic arc-derived turbidites as western facies of the Kings sequence are traversed. Such similarities are also present in the Sullivan Creek terrane, traversing eastward into phyllitic rocks of the western Calaveras complex.

We consider the structural geometry depicted in the Klamath inset of Plate 5B to be broadly applicable for the Late Triassic–Early Jurassic setting of the western Sierra fringing terranes. The outer fore-arc ridge is analogous to ophiolitic and younger bonititic to arc tholeiitic volcanic ridges that bound large extensional basins and trench slopes of the modern Bonin-Mariana system (Saleeby, this volume). Pelagic and generally fine terrigenous sedimentation of Late Triassic–Early Jurassic age within the Sierran-Klamath fringing terranes and adjacent "subduction complexes" are thought to have formed across analogous basins and ridges during volcanic quiescence. The thicker terrigenous basin fill present to the south along the western Sierra–Klamath belt reflects greater proximity to the North American continent and its sialic arc.

McCloud segment of the continental arc. Late Triassic igneous activity along the McCloud segment of the continental arc is well recorded from central California to British Columbia. Deposition of volcanic and volcaniclastic rocks occurred primarily under marine conditions with only local subaerial conditions. Deep-marine basins had back-arc, intra-arc and fore-arc settings. The Intermontane superterrane in particular appears to have constituted a complex array of basins and volcanic-island chains that narrowed southward into the Blue Mountains, northern California, and Nevada region, where a simpler fore-arc basin–arc–back-arc basin evolved. Accumulation and preservation of thick Upper Triassic arc sequences in the northern California–Nevada region probably occurred in an extensional setting (Wyld, 1989).

Intermontane superterrane—All major elements of the Intermontane superterrane appear to have been linked to one another, and above the west-facing continent edge subduction zone, during Late Triassic time (Monger, 1984; Monger and others, 1990). Late Triassic island-arc volcanism is well recorded in both Quesnel and Stikine by the Nicola, Takla, Stuhini, and Lewes River groups. Together, these groups constituted a Late Triassic volcanic-island chain ~1000 km in length. Lava flows and pyroclastic rocks were primarily mafic to intermediate in composition, and included abundant calc-alkaline as well as alkaline-series magmas. The volcanic rocks are commonly interbedded with marine volcaniclastic, argillaceous, and carbonate rocks. Plutonic rocks of the Late Triassic arc are also widely exposed.

Arc rocks interpreted as southernmost exposures of the Nicola Group of Quesnel occur in northern Washington State. These include Late Triassic metamorphosed plutonic rocks of mainly quartz-dioritic composition from the Marblemount area and the Holden (U.S.G.S.) quadrangle (Misch, 1966, 1977; Cater and Crowder, 1967; Mattinson, 1972), which are interpreted as plutonic members of the Nicola arc. The stratified arc sequences of northern Washington that appear to be linked to Quesnel, and thus to be of Nicola arc affinity, occur in the Okanogon highlands and the northeast Washington region (14 ad 15 in Fig. 5). In the Okanogan highlands, thick carbonates and intermediate-composition volcanic and volcaniclastic rocks of the Cave Mountain Formation are apparently of Late Triassic age (Rinehart and Fox, 1976). In the Curlew quadrangle of northeast Washington, an unnamed greenstone unit with interbedded chert, mudstone, sandstone, and conglomerate also contains beds of Upper Triassic limestone (Parker and Calkins, 1964). These strata may represent a more basinal facies deposited off the back edge of the arc.

The Upper Triassic arc rocks of the Intermontane superterrane exhibit a number of stratigraphic relations that are critical for regional tectonic analysis: (1) the Takla and Nicola groups together form an overlap sequence between Quesnel and Stikine (Davis and others, 1978); (2) Nicola arc rocks have facies relations with Slocan back-arc basin strata which, in turn, lie depositionally above the Kootney–Slide Mountain terrane (Monger, 1984); (3) Nicola volcanic rocks occur as blocks within Upper Triassic chaotic deposits of the Cache Creek terrane (Monger, 1984; Monger and others, 1990); and (4) sedimentary rocks of the Lewes River Group of northern Stikine appear to be equivalent to the Kutcho Formation, which constitutes the uppermost stratigraphic unit of the northern Cache Creek terrane (Monger, 1984; Monger and others, 1990). Similar volcaniclastic and argillaceous strata occur elsewhere in the Cache Creek terrane with Upper Triassic chert or limestone, indicating that the Cache Creek basin was within or adjacent to the arc system (Cordey and others, 1987; Monger and others, 1990). These relations taken together indicate that the Intermontane superterrane constituted a complex network of back-arc and intra-arc basins, and one or more volcanic-island chains. The accreted "Tethyan" oceanic crust and islands of the Cache Creek terrane resided within the largely McCloud basement framework of the arc-basin system.

Plates 5-A through 5-C depict the early Mesozoic Cache Creek basin as an extensional intra-arc basin which was, in part, inherited from large slabs of previously accreted "Tethyan" oceanic lithosphere, and which also broke the northern McCloud segment of the continental arc into multiple volcanic-island chains and basins. The extensional regime is also widely represented in the Nicola and parts of the Takla groups by alkaline-series volcanic rocks (Preto, 1977; Monger and others, 1990). The Cache Creek basin is interpreted to have continued southward toward the extensional fore-arc system represented in the fringing and spatially associated Cache Creek affinity terranes of the western United States. This interpretation poses the possibility that during early Mesozoic time Stikine migrated as a fore-arc microplate (Saleeby, this volume). The Wallowa terrane of the Blue Mountains may have originally been juxtaposed outboard of Baker terrane rocks as a dislodged fragment of Stikine shed during its migration.

Blue Mountains—The evolutions of the Wallowa and Olds Ferry terranes diverged markedly in the Late Triassic (13 in Fig. 5, 8 in Fig. 7). The Wallowa terrane shows a Carnian lull in arc volcanism and the deposition of the lower Norian Martin Bridge Limestone (Nolf, 1966; Vallier, 1977; Brooks and Vallier, 1978). Fossils indicate a shallow-water environment of deposition, and sedimentary structures and facies relations record a platform-type accumulation with local reefs (Vallier, 1977). By upper Norian time, arc-derived detritus of the Hurwall Formation was deposited conformably above the Martin Bridge Limestone. The Olds Ferry terrane differs by the absence of the Martin Bridge Limestone and the continuation of arc volcanism, indicated by the presence of late-middle Norian ammonites within the volcanic section (Brooks and Vallier, 1978). Arc volcanism appears to have terminated by late Norian time.

Eastern Klamaths–northern Sierra—The earlier Triassic marine conditions recorded by the Pit Formation persisted with the deposition of the Carnian Hosselkus Limestone and the Middle to Upper Triassic Brock Shale in the eastern Klamath Mountains (1 in Fig. 5). A major pulse of arc volcanism is represented by the conformably overlying Upper Triassic Modin Formation, which consists of submarine andesitic pyroclastic deposits, tuffs and lava flows with interbedded argillite, volcanic conglomerate, and limestone (Sanborn, 1960). Upper Triassic arc rocks are apparently absent in the northern Sierra (2 in Fig. 5), where Carnian(?)-Norian limestone and associated shale, sandstone, and conglomerate lie unconformably on Permian arc strata (Crickmay, 1933; Imlay, 1961; Clark and others, 1962; McMath, 1966; Harwood, 1983, 1988). The unconformity is believed to represent the Sonoma orogeny (Harwood, 1988), the Upper Triassic strata representing western nonvolcanic exposures of the marine overlap province.

Northwestern Nevada—Early Mesozoic marine overlap relations are widely preserved in northwestern Nevada (Speed, 1978); detailed facies relations, however, are difficult to reconstruct due to Late Jurassic and Cretaceous thrusting. This thrusting involved substantial southward transport, so some of the northwesterly basinal strata may have originated along the back edge of the Blue Mountains arc in proximity to the southern limit of Slocan back-arc basin exposures. Shallow-marine carbonates are present in the eastern part of the Nevada marine province along the basin margin, and interfinger with arc-derived volcanogenic sedimentary rocks and andesitic lavas in the south and west parts of the basin (3–10 in Fig. 5). The northwest part of the marine province is predominantly a starved basin succession (11 in Fig. 5; Russell, 1981, 1984; Maher, 1989). The intervening regions probably represent deep basinal accumulations with turbidites derived both from continental sources to the east and the volcanic-carbonate regions to the west and south. The early history of the basin is characterized by rapid subsidence in the

central region and the westerly onlap of carbonate bank and continentally derived siliciclastic rocks over volcanic strata to the west (Speed, 1978; Oldow, 1984).

Latest Triassic arc volcanism is well recorded in the northern region of the marine basinal province. Norian basaltic and andesitic lavas and volcaniclastic rocks accumulated in a rapidly subsiding basin under marine conditions in the Pine Forest Range (Wyld, 1989), and subaerial andesitic lavas of late Norian age mark basin filling or uplift in the nearby Jackson Range (11 in Fig. 5; Maher, 1989).

Arc volcanic rocks are well preserved along the southwest margin of the marine basin in the Yerington district (3 in Fig. 5), where andesitic to rhyolitic calc-alkaline volcanic rocks were intruded by comagmatic plutons which have yielded 230 ± 2 Ma ages (Dilles and Wright, 1988). A depositional base for this arc sequence is not recognized, and thus igneous activity could have begun during earlier Triassic time. Upper Carnian and Norian strata at Yerington include thick shallow-marine carbonate, argillite, and siltstone sequences interpreted as the southwest facies of the marine province (Speed, 1978; Oldow, 1984).

To the south of the Yerington district in the Saddlebag Lake pendant of the east-central Sierra, an Upper Triassic rhyolitic caldera complex formed in a probable shallow-marine environment; the complex is unconformably above the western extent of the Antler and Sonoma thrust belts (Schweickert and Lahren, 1987, 1989). The caldera complex formed during a transition from regional compression across the continental arc, marked by Early to early-Late Triassic folding and thrusting, to widespread Late Triassic–Early Jurassic extension and ignimbrite activity. The caldera complex was involved in east-vergent thrusting shortly after its formation, as shown by a crosscutting Upper Triassic pluton (Schweickert and Lahren, 1987, 1989).

Sialic segment of the continental arc. The sialic segment of the continental arc developed across miogeoclinal, displaced Antler belt rocks and cratonal settings inherited from the truncated Paleozoic margin. Late Triassic–Early Jurassic intra-arc extension in the miogeoclinal and displaced Antler belt settings (central and southern Sierra Nevada) resulted in deposition of very thick shallow-marine and thinner deep-marine successions. Coeval intra-arc extension to the south in the cratonal setting (Mojave-Sonoran deserts of California and Arizona) resulted in accumulation and preservation of thick terrestrial volcanic and sedimentary sequences.

Sierra Nevada—Lower Mesozoic metavolcanic rocks of the Saddlebag Lake pendant form a semicontinuous southeast-trending belt that appears to continue into the thick silicic pyroclastic sequence of the Ritter Range pendant (1 in Fig. 4). The Ritter Range pendant is located at the northwest corner of the Paleozoic miogeocline basement; to the north, the miogeocline is tectonically overlain by the allochthons of the Antler and Sonoma belts, and to the west it is truncated. Major ignimbrite activity is particularly well recorded in this area for the later part of Late Triassic and the early part of Early Jurassic time, implying the existence of a family of calderas. Silicic volcanism occurred in an environment undergoing rapid subsidence, resulting in a composite stratigraphic thickness of ~10 km of shallow-marine ignimbrites (Fiske and Tobisch, 1978). The Upper Triassic–Lower Jurassic Ritter Range silicic pyroclastic sequence occurs again between 50 and 100 km to the southeast, at the base of the Goddard pendant (Saleeby and others, 1990). To the east and southeast, volcanism of the continental arc is presumed to be represented by calc-alkaline plutons which extend discontinuously along the eastern flank of the mainly Cretaceous Sierra Nevada batholith (Evernden and Kistler, 1970; Stern and others, 1981; Chen and Moore, 1982). The arc plutonic belt continues southward into the Mojave, southeastern California, and southwest Arizona regions.

In the southern Sierra Nevada, shelfal strata at Mineral King (8 in Fig. 4) yield Carnian and Norian fossils, and an intercalated rhyolite ash flow is dated at 217 ± 2 Ma (Christensen, 1959; Busby-Spera, 1983). Upper Triassic and Lower Jurassic strata of the Mineral King pendant record periods of volcanic activity characterized by large-volume caldera-forming eruptions of rhyolite ignimbrite (Busby-Spera, 1984b), construction of small-volume andesite stratocones (Busby-Spera, 1986), and growth of submarine fans and debris aprons (Busby-Spera, 1985). These periods alternated with periods of local volcanic and tectonic quiescence, characterized by deposition of fine-grained sediment and ash, or progradational shelf sequences. Water depths fluctuated from 10 to >500 m, and subsidence rates were high (~180 m/m.y.; Busby-Spera, 1984a). During periods of local volcanic quiescence, subsidence was even and continuous, whereas the parts of the section that contain proximal volcanic facies (such as calderas) record extreme fluctuations in water depths.

Intra-arc facies patterns in the southern Sierra are complex due to rapid lateral variation. The Mineral King pendant preserves the proximal volcanic facies of the Upper Triassic–Lower Jurassic Kings sequence, whereas pendants to the north, west, and south contain shallow-water quartzite-carbonate, and basinal turbiditic and olistostromal sequences with subordinate water-laid tuffs and tuffaceous quartz-rich sandstones (Saleeby and others, 1978; Saleeby and Busby-Spera, 1986). Silicic volcanic components of the Upper Triassic–Lower Jurassic quartz-rich sedimentary rocks consist primarily of reworked quartz and feldspar phenocrysts, local grit-size fragments of vitrophyre and local thin water-laid tuff beds. The unusual mixture of volcanic arc and mature continental detritus and interstratified shallow-water carbonates is consistent with accumulation of the Kings sequence along the truncated edge of the miogeocline, with fore-arc basin facies lapping westward onto juxtaposed ophiolitic basement (Saleeby and others, 1978).

Eastern California—Predominately subaerial andesitic volcanic rocks of possible Late Triassic age lie upon Lower Triassic strata in the Inyo Mountains (4 in Fig. 4), the New York Mountains (Burchfiel and Davis, 1977), the Old Dad and Cowhole mountains (3 in Fig. 4; Dunne, 1977), the Soda Mountains (Marzolf, 1982), Butte Valley (Walker and others, 1983), and in the southern Owens Valley area (Osborne and Dunne, 1982). Sedi-

mentary rocks interstratified with these arc volcanic rocks are predominantly fluvial and/or lacustrine (Osborne and Dunne, 1982).

Southeastern California and Arizona—The continental arc in this region was built on the margin of cratonic North America. Upper Triassic volcanic rocks are not widespread, although in southern Arizona (7 in Fig. 4) andesitic to dacitic lavas, ignimbrites, and hypabyssal intrusions of the Mount Wrightson Formation began to accumulate in a subsiding intra-arc graben (Riggs and Busby-Spera, 1991; discussed further below). Granitic plutons of probable Late Triassic age are widespread between California and southern Arizona, and record emplacement and synplutonic deformation at deep-crustal levels, and then rapid uplift prior to Early or Middle Jurassic volcanism (Haxel and others, 1985b; Tosdal, 1986; Barth and others, 1988; Reynolds and others, 1989). We interpret this uplift event as unroofing within the Late Triassic–Early Jurassic extensional arc. Volcanoes related to the unroofed plutons may have been the source of volcanic detritus in the Chinle Formation (discussed below).

Late Triassic plutonic rocks in southeast California–southwest Arizona are correlated with the Mount Lowe plutonic complex (Tosdal, 1986), which was partly displaced to the San Gabriel Mountains along the late Tertiary San Andreas fault system. In southeastern California–southwestern Arizona, these plutonic rocks lie in the upper plate of the Cretaceous Mule Mountains thrust, along with upper Paleozoic and lower Mesozoic rocks of cratonal North American affinities. In the San Gabriel Mountains, the Mount Lowe complex intrudes Proterozoic host rock that cannot be correlated with native North American rocks. Correlation of the Upper Triassic plutonic rocks from these two regions implies that the Proterozoic basement was accreted to North America by Late Triassic time, possibly during the late Paleozoic transform truncation of the southwest sialic edge and/or the Sonoma orogeny.

Plate interior

The Upper Triassic Chinle Formation of the Colorado Plateau constitutes the main record of the Late Triassic plate interior. It consists of alluvial-plain and shallow lacustrine sedimentary rocks. Fluvial sediments prograded in pulses to the north and west (Blakey and Gubitosa, 1983). The base of the Chinle Formation fills paleovalleys cut, with up to 80 m of relief, into the Lower Triassic Moenkopi Formation; some paleovalleys are filled with the Shinarump conglomerate (2, 3, and 4 in Fig. 3) and others are filled with mudstone. The depositional evolution of the Chinle Formation was described by Blakey and Gubitosa (1983) as six phases: (1) Shinarump sheet, consisting of sandstones and conglomerates deposited in a laterally shifting braided stream system; (2) widespread flood-basin swamps and meandering streams, associated with regional subsidence and influx of ash; (3) similar to phase 1, but less areally extensive; (4) same as phase 2; (5) lake mudstones and limestones; and (6) dry alluvial plain, with eolian dunes of the Wingate sandstone encroaching from the north and accompanying a drying trend (2 in Fig. 3).

During the Late Triassic, for the first time in the Phanerozoic, voluminous volcanic detritus was deposited in the inland areas of the continent (Stewart and others, 1986). The lower (upper Carnian) Chinle is bentonitic, whereas the upper Chinle (lower Norian) is not (Stewart, 1969; Stewart and others, 1972). Claystones in the lower Chinle Formation consist of volcanic detritus, including pumice, volcanic lithic fragments, and crystals of sanidine, biotite, and plagioclase, in a devitrified ashy matrix (Stewart and others, 1986). These particles probably represent far-traveled wind-blown ash erupted from the magmatic arc to the west (McKee and others, 1959; Stewart and others, 1972; Blakey and Gubitosa, 1983). The source of pebbles and cobbles of volcanic rock in conglomerates of the lower Chinle Formation is more difficult to determine. Although these volcanic clasts constitute only a small percentage of some of the conglomeratic layers, they are locally abundant. The volcanic clasts include tuffs that are too soft to be far-traveled, and some give Late Triassic isotopic ages for the volcanic source region (Stewart and others, 1986; Pierce and others, 1985). The part of the Chinle Formation with volcanic debris thickens and coarsens toward a southern source region, inferred from paleocurrent data, and the volcanic debris disappears toward the north in southern Nevada and Utah (Blakely and Middleton, 1983). Anderson and others (1984) noted that no volcanic source rocks (or intrusive equivalents) of appropriate age are present in southern Arizona, suggesting that such rocks have been displaced by rifting or strike-slip faulting. As discussed above, however, arc plutonic rocks of probable Late Triassic age occur in southeastern California–southwestern Arizona (Tosdal, 1986). In addition, the Mount Wrightson Formation in southern Arizona may have included Chinle-age rocks before the base of the section was intruded (7 in Fig. 4), and correlative arc volcanic rocks in Sonora, Mexico, are as yet undated.

In addition to the Chinle Formation on the Colorado Plateau, two other areas of Late Triassic continental sedimentation have been recognized in the plate interior. In southeastern Idaho, Upper Triassic strata resemble those of the Colorado Plateau; however, the Higham grit (6 in Fig. 3), which is the same age as the Shinarump conglomerate, appears to have had a source area to the west or northwest, from a cryptic orogenic belt rather than the craton (Kummel, 1954; McKee and others, 1959; Hintze, 1973). The second area of Upper Triassic(?) continental sedimentary rocks is in the Currie Hills of Nevada (4 in Fig. 3), the only known locality of such strata in eastern Nevada north of lat 38°N. Although these rocks are undated, they overlie dated Lower Triassic marine strata and are so similar to the Upper Triassic Colorado plateau section that the same nomenclature is used (Clark, 1957; Stewart, 1980).

The great sand sheet of the western interior of the United States, the Navajo, Nugget, and Aztec sandstones, was originally believed to be Triassic and Jurassic in age; however, palynomorphs from the Moenave Formation (4 in Fig. 4 and 1 in Fig. 3) at the base of the sequence have recently been shown to be Early Jurassic and, thus, the sand sheet is now thought to be Jurassic in

age (Peterson and Pipiringos, 1979; Padian, 1989). Because the underlying Chinle Formation is considered to be Carnian and earliest Norian in age (Stewart and others, 1972), the latest Triassic is now represented by an unconformity (Pipiringos and O'Sullivan, 1978). During late-Late Triassic and early-Early Jurassic time, the Chinle Formation in southern Nevada and adjacent Arizona and Utah was substantially eroded (Blakey and Middleton, 1983). This may have been a manifestation of an important structural event, because regional paleoslopes changed from northwest to south and southwest by Early Jurassic time (Harshbarger and others, 1957). In the Spring Mountains of southern Nevada, volcanic clasts are locally abundant in the conglomerate developed on this erosional surface (Wilson and Stewart, 1967); these may have been derived from the continental arc to the west.

EARLY JURASSIC (208–187 Ma)

The major tectonic and paleogeographic features of the Early Jurassic are shown in Plate 5C. Volcanism, intra-arc sedimentation, and plutonism are well recorded along the entire continental arc, with the possible exception of the Mojave region. The position of the North American plate edge is shown to have shifted westward as a result of subduction accretion, and fore-arc extension, sedimentation, and volcanism in Late Triassic time. Regional extension across the arc and fore-arc continued, or was intensified, from the Late Triassic, as shown by the persistence of the graben complex along the arc and the continued eruption of fore-arc volcanic sequences. There is also direct evidence of back-arc extension in northwest Nevada. Marine basinal sedimentation and arc construction persisted widely across the McCloud basement, and arc construction along the truncated miogeocline was characterized by the ponding of thick silicic ignimbrite sequences. A major craton-derived eolian quartz-sand sheet (Navajo-Nugget-Aztec) was deposited in the plate interior and back-arc regions. Entrapment of the craton-derived sands occurred along much of the length of the arc graben depression and in back-arc grabens in Nevada.

There are a number of constraints on major plate kinematic patterns for Early Jurassic time. The absolute motion of North America changed from a westerly trajectory to ~45 km/m.y. north-northeast during latest Triassic to Early Jurassic time (J1 APW Cusp of May and Butler, 1986). The California region began moving towards northern subtropical latitudes. Structural relations along the Intermontane superterrane and the Blue Mountains fore-arc and back-arc (Dickinson and Thayer, 1978; Ellis, 1986; Hansen, 1988, 1989) suggest the initiation of a significant dextral tangential component of relative motion along the plate edge. Widespread extensional tectonism along the continental arc, back-arc, and fore-arc suggest weak coupling along the subduction zone at least in the southwest, and may reflect the retreat component of North American motion relative to the plate edge. Additionally, the Insular superterrane may have migrated northward along the southwest Cordilleran plate edge as a fore-arc microplate during the latter half of the Early Jurassic.

Plate edge

As in the Late Triassic, the Early Jurassic North American plate edge is best defined by the continental arc. Extensional tectonism that was widespread along much of the arc is known to have propagated into the back-arc region during the Early Jurassic (Oldow and Bartel, 1987). The fore-arc region in the southwest remained magmatically active, suggesting the persistence of extensional tectonism there as well. There is no record of high P/T metamorphism for Early Jurassic time and, in general, rocks that could be considered accretionary-prism complexes are absent. The lack of accretionary-prism growth, coupled with fore-arc igneous activity, suggests tectonic erosion by subduction, rifting, or strike-slip faulting along the inner trench wall.

Fore-arc sedimentation, volcanism, and tectonics. *Blue Mountains*—The clearest record of fore-arc basin sedimentation and deformation for the Early Jurassic is within the Izee terrane (12 in Fig. 5). The youngest elements of the Baker terrane melange basement may contain Lower Jurassic radiolarian chert and argillite that were imbricated into the older chaotic rock assemblages. Sedimentation and synchronous deformation continued in the fore-arc basin sequence of the Aldrich Mountains group until the end of the Sinemurian. Axial surfaces of folds in the basin trended west-northwest with north vergence (Dickinson and Thayer, 1978), suggesting a dextral tangential component of relative motion along the plate edge. Following Aldrich Mountains sedimentation and deformation, shallower-water conditions developed, as recorded by the unconformably overlying sands and carbonates of the Mowich Group.

The relations between fore-arc compressional deformation in the Blue Mountains and fore-arc extension in the western Sierra–Klamath belt are poorly understood, as are the kinematics of spreading in the Sierran-Klamath fore-arc. Upper Triassic–Lower Jurassic flaser gneisses and associated serpentinite melange zones in the Fiddle Creek complex (7 in Fig. 6) suggest at least local leaky transform-fault tectonics within the extensional fore-arc (Saleeby, this volume). Perhaps transtensional rifting along the Sierran-Klamath segment of the fore-arc was transferred into transpressional shortening along the Blue Mountains segment (Plate 5C). Alternatively, Stikine may have migrated as a fore-arc microplate with an inner transpressional boundary along the Blue Mountains region.

Western Sierra–Klamath belt—The possibility of transtensional spreading within the fore-arc greatly complicates its geometric analysis, both in terms of the relative positions of the fore-arc volcanic-sedimentary sequences to one another and the disruptions that were superposed over preexisting subduction complexes. We speculate that the Rattlesnake Creek terrane, in addition to the Fiddle Creek complex, was subjected to transform tectonism within the extensional fore-arc during the Late Triassic–Early Jurassic, and the former may have been reactivated in part during Callovian-Oxfordian time as a fore-arc transform as well.

Regardless of the uncertainty posed by possible transform

displacements, the cross-sectional Klamath inset of Plate 5B points out a number of features that were superposed over the central Klamath Triassic subduction complex. The North Fork terrane records tuffaceous chert sedimentation and submarine mafic volcanism of Early Jurassic age atop older disrupted parts of the complex (Ando and others, 1983; Blome and Irwin, 1983). The environment of deposition is depicted as an extensional basin. The Rattlesnake Creek terrane is distinct, its basement having been constructed from copious serpentinite diapirism within the fore-arc wedge.

Submarine mafic and subordinate intermediate volcanism and siliceous and argillaceous sedimentation continued in the Sierran fore-arc into Early Jurassic time (7, 10, 11, and 14 in Fig. 6). In addition, submarine lava shields and thick pyroclastic aprons with subvolcanic wehrlitic to dioritic plutons formed as the Slate Creek complex (8 in Fig. 6) and the Penon Blanco sequence (12 in Fig. 6). The Penon Blanco pyroclastic rocks were deposited conformably on a thin veneer of tuffaceous radiolarian chert that has yielded Jurassic-Triassic radiolaria (Morgan, 1976; Saleeby, 1982; Bogen, 1985). The chert lies conformably on several hundred meters of aphyric arc tholeiitic basaltic to andesitic pillow lava, which was in turn deposited on Paleozoic ophiolitic melange basement (Saleeby, 1982). The Penon Blanco pyroclastic rocks consist of up to ~5 km of basaltic and basaltic-andesite tuff breccia, pillow breccia, and tuff turbidite of boninitic to arc tholeiitic composition. The Penon Blanco Formation is dated as 200 to 196 Ma, based on ages of dioritic hypabyssal rocks which, along with wehrlitic intrusive rocks, cut through the pyroclastic rocks; the hypabyssal rocks are also entrained as cognate xenoliths within tuff breccia (Saleeby, 1982). Possible correlative rocks to the Penon Blanco sequence occur in the Sullivan Creek terrane (12 in Fig. 6) to the east across the Melones fault zone (Sharp, 1983, 1988; Bateman and others, 1985; Herzig and others, 1985; Saleeby and others, 1986; Sharp and Herzig, 1991). The Lower Jurassic submarine lava flows and pyroclastic rocks of the Sullivan Creek terrane are overlain by Lower to Middle Jurassic argillite, chert, quartz wackes, olistostromes, and subordinate volcanic rocks (Sharp, 1988; Sharp and others, 1991). A similar stratigraphic sequence is above the northern end of the Penon Blanco volcanic belt and is laterally equivalent to strata of the central belt melange (11 in Fig. 6; Duffield and Sharp, 1975; Behrman and Parkison, 1978; Saleeby, 1982; Schweichert and Bogen, 1983; Sharp, 1988). These relations record the broad basinal setting in which the lava and pyroclastic sequences formed, as well as the resumption of basinal sedimentation following the Late Triassic–Early Jurassic volcanic pulses.

The Early Jurassic history in the Kaweah melange in the southwest Sierra (14 in Fig. 6) is recorded by a complex assemblage of tuffaceous and siliceous argillites and cherts, olistostromes, and boninitic lava flows and breccias that stratigraphically overlie the Upper(?) Triassic pillow lava–chert carapace that was built over ophiolitic melange. Quartz arenite and wacke, and silicic volcaniclastic turbidites of the western Kings sequence are intermixed with, and depositionally overlie, the olistostromes and fore-arc volcanic rocks. The Kings sequence thus appears to lap from the truncated edge of the Paleozoic miogeocline and offset Antler belt rocks onto juxtaposed transform ophiolite (Saleeby and others, 1978). The relations between the Upper(?) Triassic–Lower Jurassic strata of the Kaweah melange and the western Kings sequence, as well as the apparent basement relations are shown in the Kaweah inset of Plate 5C. Tectonic relations depicted in the inset consider the submarine basaltic pillow-lava flows and breccias of the Kaweah section to be a fore-arc igneous suite, and the turbidite-olistostrome facies of the Kings sequence to be a fore-arc basin that overlapped the truncation suture.

Fore-arc basin sedimentation is hypothesized (Plate 5C) as having continued southward from the western Sierra in areas now represented by the schist of Sierra de Salinas of Salinia, and the Julian and French Valley schists of the northern Peninsular Ranges. No direct age data exists, however, and this interpretation is based on the broad correlation suggested above between these schist bodies and the Kings sequence.

McCloud segment of the continental arc. Early Jurassic arc volcanism is variably recorded in the McCloud segment of the continental arc, although changes in volcanic and depositional patterns were widespread. Most notable changes occurred within the Intermontane superterrane both at the Triassic-Jurassic boundary and during middle-Early Jurassic time.

Intermontane superterrane—The apparent axis of arc volcanism changed from the north-northwest trend of the Takla-Nicola arc to the northwest trend of the Hazelton arc within southern Stikine and Quesnel at the Triassic-Jurassic boundary (Davis and others, 1978). Lower Jurassic arc rocks that lie above the southernmost Nicola arc are referred to as the Rossland Group. The Late Triassic and Early Jurassic arc axes appear to merge near the Rossland exposures. The cause of the apparent change in trend of this part of the continental arc is unclear, but it may be related to the initial phases of collapse of the Intermontane superterrane, which led to the final closure of the Cache Creek and Slocan basins.

Part of the southern Cache Creek basin appears to have survived into Early Jurassic time (Cordey and others, 1987), although the position of the respective Lower to Middle(?) Jurassic strata at the extreme southwestern tip of the terrane permit these strata to have been tectonically introduced into the terrane at a later time from similar rocks of the Bridge River terrane to the west. Davis and others (1978) noted strong similarities between Bridge River and, by analogy, the younger southwest Cache Creek terrane rocks and rocks which we interpret as the extensional fore arc assemblage of the western Klamath Mountains. As in the Late Triassic time interval, we show the Bridge River and youngest Cache Creek basin in broad continuity with the extensional fore-arc of the western Sierra–Klamath belt (Plate 5B and 5C). The northern part of the basin may have closed as transtensional tectonics in the south were transferred into transpressional tectonics to the north.

Broad changes also occurred in volcanic patterns within the

Intermontane superterrane in middle-Early Jurassic time (Monger and others, 1990). Volcanism of the Hazelton-Rossland arc is well recorded for Hettangian-Sinemurian time within Stikine and Quesnel. Volcanism began to wane considerably in early Pliensbachian time, when a broad, shale-rich sequence overlapped much of the superterrane. Thereafter, arc volcanism occurred only sporadically within the superterrane. Late Pliensbachian ammonite paleobiogeographic data from the broad marine deposits record subsequent ~500 km scale northward transports of Quesnel relative to the craton, and Stikine relative to Quesnel (Taylor and others, 1984).

Northern Washington and the Blue Mountains—Arc volcanism is not as widely recorded in these areas for the Early Jurassic as it is for the Late Triassic; nevertheless, the aggregate volcanic, plutonic, and sedimentary histories indicate the persistence of arc activity into the Early Jurassic. In northeastern Washington (15 in Fig. 5), the Rossland Group comprises basaltic and andesitic flows, breccias, and tuffs, and interbedded and overlying clastic sedimentary rocks which, in southernmost British Columbia, are Sinemurian to Toarcian in age. Overall lithology and geochemistry indicate that the Rossland accumulated in an island-arc setting under basinal conditions (Beddoe-Stephens and Lambert, 1981). Most of the group is exposed in British Columbia, where it represents one of the last phases of arc activity in Quesnel (Monger and others, 1982). Quartz dioritic to granodioritic plutons are present in northeastern Washington and in the Okanogan highlands (14 in Fig. 5). Available age data suggest that these are Early Jurassic in age and are part of the Quesnel arc (Rinehart and Fox, 1972; Miller and Clark, 1975). The Flowery Trail pluton in northeastern Washington may represent the plutonic roots of the Rossland Group volcanics.

Proximal facies of arc volcanism are rare or lacking in the Lower Jurassic of the Blue Mountains. In the Olds Ferry terrane (13 in Fig. 5), there is a hiatus near the Triassic-Jurassic boundary where the deposition of the volcanic Huntington Formation ceased; then, after a time, the mainly arc-derived Weatherly Formation began to accumulate in the Sinemurian(?), possibly in a fore-arc setting (Brooks, 1979). The Weatherly Formation consists of conglomerate containing volcanic and plutonic clasts, limestone, sandstone, and minor gypsum and anhydride near its base and deeper-water turbidite sequences of volcanic sandstone and shale higher in the section (Brooks and Vallier, 1978; Brooks, 1979). The Weatherly Formation may represent the southern continuation of the broad shale overlap that is widespread in the Intermontane superterrane to the north. Within the Wallowa terrane (8 in Fig. 7), arc-derived strata of the Hurwall Formation continued to accumulate under marine conditions without a break across the Triassic-Jurassic boundary. Deposition of Hurwall Formation continued into Pleinsbachian time.

Eastern Klamaths–northern Sierra—Marine arc volcanism continued from Late Triassic through Early Jurassic time in the eastern Klamath Mountains (1 in Fig. 5) with the deposition of predominately andesitic pyroclastic and flow rocks and lesser conglomerate and sandstone of the Arvison Formation (Sanborn, 1960; Irwin, 1966). Marine to subaerial arc volcanism and sedimentation are recorded in the northern Sierra (2 in Fig. 3). Here, the Mount Jura section is in tectonic contact with McCloud basement rocks, but the upper American River section is depositionally above the basement. In the upper American River section, andesitic lapilli tuffs overlie Norian limestone and are overlain by marine pelite, feldspathic sandstone, turbidites, and tuff of the Lower to Middle Jurassic Sailor Canyon Formation (Harwood, 1988). The Lower and Middle Jurassic Mount Jura section contains more proximal volcanic rocks, including andesite and dacite lava flows and tuff breccias, as well as volcanic lithic sandstones, conglomerates, and mudstones with marine and nonmarine fossils (McMath, 1966).

Rocks of the upper American River section occur to the south in the Mount Tallac roof pendant southwest of Lake Tahoe (Fisher, 1989, 1990), and in an inlier surrounded by Cenozoic volcanic rocks in Blackwood Canyon west of Lake Tahoe (D. Harwood, 1988, personal commun.). Lower Jurassic quartz arenite occurs at both localities, in gradational contact with Pleinsbachian ammonite-bearing black slate and limestone, which also occurs in the type area of the Sailor Canyon Formation (2 in Fig. 5). The presence of volcanic rocks in the sections at Mount Tallac and the upper American River supports the interpretation that the quartz arenites are intra-arc deposits. These Lower Jurassic quartz arenites provide an important paleogeographic link between the McCloud and sialic segments of the arc (Busby-Spera and others, 1990).

Northwest Nevada—Marine conditions began to diminish in areal extent during the Early Jurassic. Basin filling is recorded in the northern part of the marine province in the Black Rock Desert region (11 in Fig. 5), where a thick sequence of predominately intermediate flows and tuffs of the Happy Creek Formation accumulated under shallow-water to mainly subaerial conditions (Russel, 1981, 1984; Maher and Saleeby, 1986; Maher, 1989). The Happy Creek volcanics have shoshonitic petrochemical characteristics that may reflect an arc to back-arc extensional regime. The southeastern margin of the marine province also became emergent in Early to Middle Jurassic time, in conjunction with extensional tectonism. Here, north-northwest–striking normal growth faults with throws locally in excess of 1.5 km formed half grabens where craton-derived quartz sands of the Dunlap Formation, and possibly the Boyer Ranch Formation, were deposited (Stanley and others, 1971; Oldow and Bartel, 1987). The poorly dated Boyer Ranch Formation is present primarily in the eastern part of the basin. In places it lies depositionally above Norian strata (Speed and Jones, 1969), but it extends well into the Middle Jurassic at other localities. The central part of the marine basin contains distal volcanogenic turbidites, local debris flows, and volcaniclastic lenses in the south (7 in Fig. 5), and calcareous shales and siltstones in the north (9 in Fig. 5). The stratigraphic record throughout much of the central and eastern part of the basin ends at the close of the Early Jurassic.

Central California–Nevada border region—Along the south-

west margin of the marine province in the Yerington district (3 in Fig. 5), Lower Jurassic shallow-marine clastic, carbonate, and evaporitic strata overlie the volcanic-rich Upper Triassic sequence. Quartz arenite of the Preachers Formation occurs in this section above a Toarcian fossil locality, and is similar in age to quartz arenite of the northern Sierra Nevada. The great thickness of Lower Jurassic marine strata at Yerington (~1,800 m; Diles and Wright, 1988) requires rapid subsidence, possibly due to extension. In the east-central Sierra, voluminous silicic volcanism persisted from Late Triassic into Early Jurassic time within the Saddlebag Lake pendant. As is more clearly recorded in the Ritter Range pendant to the south, the largest volume of material appears to have been erupted near the Late Triassic–Early Jurassic boundary probably in a series of large calderas.

Sialic segment of the continental arc. East-central Sierra Nevada—The Upper Triassic–Lower Jurassic strata of the Ritter Range pendant (1 in Fig. 5) and equivalent rocks to the north in the Saddlebag Lake pendant, and to the south in the Mt. Goddard pendant, record a tremendous outpouring and ponding of silicic pyroclastic material. High rates of subsidence are recorded by the accumulation of a 10-km-thick section of mainly shallow-marine ignimbrite in the Ritter Range, and the sparseness of interbedded sedimentary material suggests copious volcanism without significant lulls (Fiske and Tobisch, 1978).

Inyo Mountains–Owens Valley—Subaerial and shallow-water silicic to andesitic volcanism may have continued through Early Jurassic time in this region without influxes of quartz sand, although age controls are poor. Arc plutonic rocks are well represented by a clustering of ages in the earlier part of the Early Jurassic as well as in the Late Triassic (Stern and others, 1981; Chen and Moore, 1982). A number of the more eastern plutons have alkaline characteristics (C. Miller, 1978).

Southern Sierra Nevada—Upper Triassic to Lower Jurassic strata of the Mineral King pendant are dominated by marine silicic to intermediate-composition volcanic rocks and clastic rocks of volcanic provenance described under the Late Triassic time interval (9 in Fig. 4). In the uppermost part of the section, however, metacalcareous quartz arenite overlies a 190 ± 3 Ma ash-flow tuff (Busby-Spera, 1983), and may be a marine equivalent of the Navajo dune field (Busby-Spera, 1984a). Similar calcareous quartz arenites are widespread in the volcanic-poor pendants of the Kings sequence (8 and 10 in Fig. 4). These quartz arenites occur as interbeds within shallow-water carbonate sequences and as turbidites and olistostromal blocks in fore-arc basin strata.

Eastern Mojave Desert—Hewett (1931) first applied the name "Aztec Sandstone" to eolian quartz arenite that overlies Triassic strata in the Spring Mountains of southern Nevada, and correlated it with the Navajo Sandstone. He later extended this correlation to the Mescal Range (Hewett, 1956). The Aztec Sandstone in the Mescal Range (4 in Fig. 4) lies above volcanogenic and epiclastic redbeds that are variably assigned to the Chinle Formation, lower Aztec Formation, or the Kayenta-Moenave equivalent (Hewett, 1956; Marzolf, 1982; Reynolds, 1983). The Aztec sandstone in the Mescal Range has the only dinosaur tracks known in California (Reynolds, 1983). Marzolf (1982, 1983) proposed that the Aztec sandstone, which is relatively thin in the Mescal Range, was eroded at this locality prior to the deposition of the overlying Delfonte volcanics, because the change in paleocurrent directions that occurs halfway through the Navajo section at Zion is not observed here.

The name "Aztec Sandstone" was applied to eolian quartz arenites in the Cowhole Mountains (Novitsky-Evans, 1978) and these were correlated with the Lower Jurassic Navajo Sandstone (Marzolf, 1983, 1990). The Aztec sandstone in the Cowhole Mountains is dominantly eolian in origin, although admixing of quartz sand with shards, phenocrysts, pumice, and volcanic-rock fragments occurred in interdune ephemeral streams and on pumice plains. Isotopic age data on interstratified volcanic rocks indicate that at least the upper half of the "Aztec Sandstone" is equivalent in age to the Carmel Formation of the Middle Jurassic San Rafael Group. The age of the lower half of the "Aztec Sandstone" is not known, and could extend to the Early Jurassic. The Aztec sandstone unconformably overlies a low-angle fault involving Paleozoic strata that was interpreted by Novitsky-Evans (1978) as a thrust, although kinematic constraints are poor. Angular boulders of the Lower Permian Bird Spring Formation, as well as clasts of volcanic rock, quartzite, and jasper, occur as lenses near the base of the Aztec sandstone. Novitsky-Evans (1978) speculated that these breccias were shed from the active thrust fault. The fault, which places younger rocks over older rocks, may alternatively be interpreted as a normal fault related to intra-arc extension.

Southeastern California and southwestern Arizona—Lower Mesozoic strata of this region record a transition from Paleozoic and Triassic(?) cratonal sedimentation to Jurassic arc volcanism, although the details of their age and correlation remain poorly known (6 in Fig. 4; Tosdal and others, 1989). Quartz arenites with large-scale eolian cross-bedding overlie possible Moenkopi-Chinle equivalents and may be equivalent to the Lower Jurassic Navajo Sandstone (Stone and Kelley, 1988; Tosdal and others, 1989). The ~170 Ma Dome Rock volcanic sequence lies conformably above the quartz arenites (6 in Fig. 4), however, so the arenites may be partly or entirely Middle Jurassic in age and equivalent to the San Rafael Group.

Southern Arizona–northern Sonora—Ignimbrites, lava flows, and hypabyssal rocks of the Mount Wrightson Formation continued to accumulate in the regional graben system from latest Triassic to Early Jurassic time (7 in Fig. 4; Riggs and others, 1986; Asmeron and others, 1988; Busby-Spera, 1988). The Mount Wrightson Formation in the Santa Rita Mountains records the evolution of a multivent volcanic complex from an early effusive stage, to a late explosive stage, followed by waning volcanism (Riggs and Busby-Spera, 1989). Rapid intra-arc subsidence resulted in burial of active vents by ignimbrites and reworked tuffs. The Mount Wrightson Formation is bounded to the northeast by the northwest-trending Sawmill Canyon discontinuity. The discontinuity underwent up to ~1 km of normal fault

separation between Permian and Middle Jurassic time (Titley, 1976), and may represent the northeast wall of the graben system in southern Arizona (Busby-Spera, 1989).

Quartz arenites interstratified with volcanic rocks throughout southern Arizona have all been correlated with the Lower Jurassic Navajo Sandstone by Bilodeau and Keith (1986). This interpretation is supported in the Fresnel Canyon sequence (7 in Fig. 4) by a ~190 Ma age on metavolcanic rocks interstratified with quartz arenites in the Baboquivari Mountains and Sil Nakya Hills (Wright and others, 1981). Age data from the Mount Wrightson Formation of the Santa Rita Mountains, however, show that some of these quartz arenites are older (205–210 Ma), perhaps equivalent in age to the Wingate sandstone (Riggs and others, 1986; Asmerom and others, 1988). Additionally, isotopic ages from the Cobre Ridge tuff along the Arizona-Sonora border show that some eolian quartz arenites are 175 Ma, perhaps age-correlative with the Carmel Formation (Riggs and Busby-Spera, 1990; Busby-Spera and others, 1991b). Deposition of craton-derived eolian sand spanned 30 to 40 m.y. within the arc graben depression of southern Arizona.

Several features of the Fresnal Canyon sequence in the Baboquivari Mountains (7 in Fig. 4) suggest that it accumulated in an extensional arc setting (Haxel and others, 1985b; Tosdal and others, 1989). Age data from the bottom and top of the section, and from a hypabyssal intrusion that cuts most of the section, indicate that the 8-km-thick section of continental volcanic and sedimentary rocks accumulated in less than 10 m.y. (Wright and others, 1981). High rates of subsidence are required to preserve such a thick continental succession. A 2.0-km-thick unit of angular to subangular polymict breccia within the succession is interpreted as fault-talus breccia reworked by debris flows (Busby-Spera, 1988). Minor alkalic mafic to silicic volcanic rocks interstratified with calc-alkalic ignimbrites also provide evidence for extensional processes within the arc (Haxel and others, 1985b).

Plate interior

The stratigraphic record east of the continental arc indicates that flooding of the western interior occurred four times in the Jurassic Period, contemporaneous with global highstands in sea level. These were: (1) end of the Toarcian (latest Early Jurassic); (2) end of the Bathonian; (3) end of the Callovian; and (4) Kimmeridgian-Tithonian boundary (Late Jurassic) (Brenner, 1983). The Toarcian event in part corresponded with deposition of the Navajo-Aztec sand sheet, which resulted in a regional coastal dune system and marine reworking of the eolian sands.

In the Early Jurassic, the Moenave and Kayenta formations accumulated in predominantly fluvial and lacustrine environments across southern Nevada and adjacent Arizona and Utah (1 in Fig. 3, 4 in Fig. 4). Relief was low, and gentle paleoslopes were inclined to the south and southwest over much of the region (Blakey and Middleton, 1983). In the late Early Jurassic, a south-closing epeiric sea spread southward into central Nevada. Carbonate, mud and siliciclastic detritus derived from the craton and western areas of tectonic activity accumulated in shallow-marine environments in the western part of the broad basin. Along the east and south margins of the basin, the Navajo-Aztec eolian sand sea spread south and southwest, interfingering with progressively younger parts of the underlying Kayenta Formation (Blakey and Middleton, 1983). This sand sheet was apparently derived primarily from the north (Stanley and others, 1971; Kocurek and Dott, 1983). Equivalent strata of southeastern Idaho and central Wyoming (6, 7 in Fig. 3) consist of the Nugget Formation. The westernmost exposure of eolian sand overlying the plate interior is in the Currie Hills of east-central Nevada. This isolated exposure is an important link between the well-documented continuous sand sheet of the plate interior and possible equivalent sand lenses within continental arc to the west.

The stratigraphic relations discussed above indicate that the Navajo-Nugget-Aztec sand field covered parts of Idaho, Wyoming, Utah, Colorado, Nevada, Arizona, and California. The erg may have been as large as 10^6 km^2. Assuming a conservative average thickness of 100 m, we calculate that 1.5×10^5 km^3 of sand was deposited. Stanley and others (1971) estimated a larger volume of 1.6×10^6 km^3 for the erg. Upper Mississippian, Pennsylvanian, Permian, and Triassic strata are missing from northern Montana and parts of Alberta, Saskatchewan, and Manitoba; as much as 2,000 m of section may have been eroded from this area before Middle Jurassic time. Approximately 50% of this section is clastic in the age-equivalent strata in Montana. Evidence for erosion of this section includes lateral continuation of facies to the erosional edge, and erosional remnants beyond the zero isopach line. If 1,000 m of clastic material was eroded from ~640,000 km^2 of Montana and southern Canada, 6.4×10^5 km^3 of material could have been made available for incorporation into the Navajo erg. With the conservative assumption that at least one-third of the clastic material was quartz, these strata could provide ample source for the erg volume we estimate; however, it may be insufficient for the volume estimates of Stanley and others (1971).

MIDDLE JURASSIC (187–163 Ma)

The major tectonic and paleogeographic features for Middle Jurassic time are shown in Plates 5-D and 5-E. Plate 5D covers Middle Jurassic time to the base of the Callovian, and 5-E covers Callovian to Oxfordian (early Late Jurassic) time. The Callovian-Oxfordian map is separate from the Middle and Late Jurassic maps due to the rapid change that the plate edge underwent during this time. These changes were marked by a widespread rifting event that was interspersed between major thrusting and tectonic accretion events. The first of these accretionary events is well recognized in the Klamath Mountains and is Middle Jurassic in age (Davis and others, 1978; Harper and Wright, 1984; Wright and Fahan, 1988); it has been named the Siskiyou event by Coleman and others (1988). It began in the Klamath Mountains by ~170 Ma (Bathonian time) and was completed by ~161 Ma (Callovian-Oxfordian boundary). It is also recognized along

the western and southern Sierra Nevada. Age relations in the Sierra permit a possible earlier initiation in deformation and metamorphism than in the Klamaths. Structural and stratigraphic relations in the Sierran region arc and back-arc suggest that Siskiyou age contractile deformation worked in conjunction with regional extensional tectonism and arc magmatism. The sialic segment of the arc appears to have remained primarily extensional during Middle Jurassic time, although in its more northerly regions, contractile deformations may have worked in conjunction with regional extension. In the southerly regions of the sialic arc, from the eastern Mojave to Sonora desert regions, extensional and/or transtensional tectonism may have been influenced by, and may have begun to work in conjunction with, Atlantic margin tectonics and the opening of the Gulf of Mexico.

The Middle Jurassic records the onset of regional thrusting and metamorphism along much of the back edge of the continental arc. General transport patterns are eastward, away from the arc axis, although significant southward transport is recorded in northwest Nevada for Late Jurassic time. This pattern in vergence is complimented by the generally west-verging thrusts of the western Sierra–Klamath belt and early thrusts within the Intermontane superterrane, and thus the Middle Jurassic marks the beginning of the regional two-sided tectonic geometry of Cordillera noted by Burchfiel and Davis (1968). During Middle Jurassic time, the back-arc and parts of the arc were also cut by longitudinal dextral strike-slip faults and local related transpressive deformation zones. Magnitudes of displacement along the dextral faults are difficult to determine due to Late Jurassic–Early Cretaceous sinistral faulting and renewed mid- to Late Cretaceous dextral faulting.

The causes of Middle Jurassic regional thrusting appear to have varied significantly from north to south along the Cordillera. In the north, thrusting was related to the collapse of the Intermontane superterrane (Davis and others, 1978; Monger, 1984), which was probably in response to the initial collision of the Insular superterrane (Saleeby and Rubin, 1990). The causes for Siskiyou thrusting in the Klamath Mountains and western Sierra Nevada are unclear; it is conceivable that the Siskiyou zone corresponded to the southern end of Intermontane-Insular collisional system, collided and/or inner collapsed fragments having since escaped northward. Alternatively, Siskiyou thrusting could reflect a period of strongly coupled subduction, although this is inconsistent with widespread extensional tectonism recorded in the continental arc from the northern Sierra region southward, and with the retreat sense in North American motion relative to the plate edge. Furthermore, Middle Jurassic age contractile deformations along the Sierra Nevada back-arc were different from thrust systems to the north. The Sierran back-arc fold-thrust zone(s) show small overall strain magnitudes which, in conjunction with associated regional extensional tectonics and arc magmatism, suggest that they were dynamically linked to voluminous mid- to upper-crust magmatism within an overall extending environment (Eastern Sierra inset, Plate 5D).

The small magnitude of contractile strain along the southern back-arc region is also reflected by the survival of the back-arc epeiric sea and broad plate interior coastal plain throughout Middle Jurassic time. The epeiric sea–coastal-plain environment hosted the last traces of the Navajo-Aztec sand sheets that were widespread during Early Jurassic time. The northern regions of the back-arc to plate-interior regime may have begun to slowly evolve into a foreland deformational-depositional system in Middle to Late Jurassic time with the subsequent rise of an orogenic highland and related foreland basin.

A number of constraints exist for major plate kinematic patterns during Middle Jurassic time. The absolute motion of North America continued at ~45 km/m.y. north-northeast, resulting in the migration of California into northern subtropical latitudes (May and Butler, 1986). A number of deformation features along the arc and back-arc region indicate a significant dextral tangential component of relative motion along the plate edge. Furthermore, a broad base of paleobiogeographic data (Taylor and others, 1984; Blome, 1987) from outer fringing terranes of the western Sierra–Klamath belt, and from marine strata of both the Intermontane and Insular superterranes, indicate a regional northward transport pattern in Middle and Late Jurassic time. The Insular superterrane may have migrated northward along the southwest Cordilleran plate edge as a fore-arc microplate during Middle Jurassic time. Alternatively, it may have been embedded in an oceanic plate, and collided with the plate edge during Middle Jurassic time (Saleeby, this volume).

Plate edge

The position of the North American plate edge for much of Middle Jurassic time cannot be identified in detail due to the paucity of subduction complexes and fore-arc assemblages of appropriate age. The axis of the continental arc from the central Sierra to Klamath Mountains region appears to have shifted westward relative to its Early Jurassic position in conjunction with the incorporation of Triassic to Early Jurassic subduction complexes and fore-arc terranes into the arc basement (Wright, 1982; Harper and Wright, 1984; Saleeby and others, 1986; Donato, 1987). In the central and western Klamath Mountains, structural position and/or crosscutting intrusive relations place the Rattlesnake Creek, eastern Hayfork, North Fork, and Stuart Fork terranes within the basement of the Middle Jurassic continental arc. In the central and southern parts of the western Sierra, the Calaveras complex and the Kaweah melange were incorporated into the continental arc basement as well. The central melange belt, Fiddle Creek, and Slate Creek complexes, and the Penon Blanco Formation and Sullivan Creek terrane, also constituted basement for the Middle Jurassic arc and/or fore-arc. In the later part of Middle Jurassic time, an elongate ocean-basin system formed outboard of the continental arc, represented by the Josephine and Coast Range ophiolites. The accretion of large tracts of this ocean-basin system to the plate edge resulted in the formation of a residual-type fore-arc basin (Dickinson and Seely, 1979) that subsided through the Paleogene and persists as the modern Great Valley.

In addition to the widespread occurrence of Middle Jurassic arc rocks built across older subduction complexes and fore-arc terranes, Middle Jurassic arc activity persisted across the McCloud basement and the truncated miogeocline edge. The overall trend of the arc from the central Klamaths to southeast California is ~N40°W. This is currently disrupted in central California by the Cretaceous Sierra Nevada batholith, which trends ~N25°W (Fig. 1).

Structural complications within the continental arc, the paucity of late Early to early Late Jurassic subduction complexes and the late Middle Jurassic creation of the Josephine-Coast Range ophiolite ocean basin system are all considered to be intimately related to the Siskiyou deformation and metamorphic event.

The Siskiyou event. Within the western Sierra-Klamath belt the phenomena of Middle Jurassic arc magmatism and crustal deformation were closely related in time and space. This is particularly well displayed in the Klamath Mountains where a major west-vergent thrust system imbricated the McCloud arc with lower Mesozoic subduction complex and fore-arc terranes, as well as superposed Middle Jurassic arc volcanic and plutonic rocks (Davis and others, 1978; Harper and Wright, 1984; Mortimer, 1985; Wright and Fahan, 1988). The Klamath inset of Plate 5-E (after Davis and others, 1978; Wright and Fahan, 1988) depicts the structural stacking sequence which is considered here to have evolved directly from the distended state of the Late Triassic-Early Jurassic fore-arc (Klamath inset, Plate 5-C). Coleman and others (1988) offered the term "Siskiyou" for this major thrusting event and described the related regional metamorphism as a low P/T facies series. The Siskiyou thrust is one of the main structures formed during the event. It places McCloud arc structural basement above the previously exhumed Triassic subduction complex. The Siskiyou metamorphic conditions are consistent with the thrust system having developed within the active arc, with synkinematic metamorphism promoted by broad heating within the magmatic arc. The complex temporal and spatial relations between plutonism and thrusting are depicted in the Plate 5-E inset. Wright and Fahan (1988) have demonstrated that thrusting had commenced by ~170 Ma and was probably completed by ~161 Ma.

There is a possibility that Siskiyou thrusting was nearly continuous with Late Jurassic Nevadan thrusting. In the Condrey Mountain region of the central Klamath Mountains, deep structural levels of both Siskiyou and Nevadan structural sequences are exposed (Klein, 1977; Hill, 1985; Coleman and others, 1988; Helper and others, 1989; Harper and others, 1991). Siskiyou event thrusting between the Rattlesnake Creek and western Hayfork terranes is constrained here to be pre-162 Ma, the age of a crosscutting batholith (Barnes and others, 1986). The Condrey Mountain schist terrane was subcreted beneath this structural sequence, including the batholith, starting at ~160 Ma and continuing through Nevadan time. The Condrey Mountain protolith assemblage strongly resembles the 10 to 15 m.y. younger (upper Callovian-lower Kimmeridgian) Josephine ophiolite and overlying Galice flysch sequence which was thrust eastward beneath the Rattlesnake Creek terrane and, by inference the Condrey Mountain terrane, during the Late Jurassic Nevada orogeny. The initial phase of Condrey Mountain terrane thrusting is thus immediately post-Siskiyou and pre-Nevadan, and in structural succession with both Siskiyou and Nevadan thrust sheets. Below, a plate kinematic model is offered, wherein Siskiyou and Nevadan thrusting are linked kinematically to mid-Bathonian to early Oxfordian ophiolitic basin spreading and contemporaneous subduction.

Geophysical data (Zucca and others, 1986) indicate that the west-vergent Siskiyou thrust system roots deeply beneath the eastern Klamaths. Structural studies in the lower Mesozoic arc sequence of the eastern Klamaths indicate a reversal to east-vergence in Siskiyou age structures (Renne and Scott, 1988). Additionally, the long and relatively complete arc-related stratigraphic history of the eastern Klamaths (1 in Fig. 5) ended at the onset of Siskiyou deformation.

Deep eastward rooting of a west-vergent probable Siskiyou-related thrust system is suggested by seismic reflection data in the northern Sierra Nevada which show a prominent east-dipping reflector beneath the northern Sierra terrane (Nelson and others, 1986). The main reflector projects towards, but does not surface, in the western Foothills. It appears to be truncated by a system of high-angle Nevadan faults which exhibit reverse and sinistral displacements. The Nevadan faults also crosscut a low-angle tectonic contact between the Slate Creek (8 in Fig. 6) and underlying Fiddle Creek complexes (7 in Fig. 6, Edelman and others, 1989). The low-angle contact is cut by a 168 Ma arc-related intrusive complex (Saleeby and others, 1989a), and may represent Siskiyou deformation (Edelman and Sharp, 1989); alternatively, the contact could be a low-angle normal fault that formed in conjunction with the fore-arc construction of the Slate Creek complex. Further south along the Fiddle Creek complex lies as an unconformable infold of Callovian-age volcanic-poor turbidites (Imlay, 1961), referred to as the Colfax sequence. The turbidites record Siskiyou uplift of oceanic and arc-basement terranes of the western Sierra-Klamath belt. The basin in which the turbidites accumulated may have resulted from late Callovian extensional tectonism discussed below. The Colfax sequence is similar to the more-extensive upper Oxfordian–lower Kimmeridgian Mariposa Formation farther south of the Sierran Foothills.

Siskiyou deformation in the central Foothills east of the Melones fault is represented by: (1) a family of ~176 Ma crossfolds and related steeply dipping axial surface schistosites which were superposed across the Foothills suture (Sharp, 1988); (2) pre–166 Ma southwest-verging movement along the Sonora thrust (Sharp, 1988); and (3) a related family of steeply plunging asymmetric folds and intersecting foliation surfaces, all of which together indicate dextral transpression (Saleeby and others, 1986). Sonora thrusting resulted in the partial collapse of the Late Triassic(?) to early Middle Jurassic fore-arc basin represented by the Sullivan Creek terrane (13 in Fig. 6) and phyllitic rocks that grade eastward into the Calaveras complex. Deformation in this region was broadly synchronous with arc magmatism, as in the

Klamaths. To the south in the Kings River area, a 170 Ma arc plutonic complex was deformed by dextral transpression during its magmatic emplacement into Paleozoic ophiolitic wall rocks (Saleeby, 1978; Saleeby and Sharp, 1980).

Siskiyou deformation in the central Foothills west of the Melones fault may be represented by modest to locally pronounced deformation and metamorphism of lower Mesozoic fore-arc assemblages that lie beneath Callovian and Oxfordian volcanic and turbiditic sequences and by a number of metamorphic ages on mafic schist blocks of ~175 Ma (Duffield and Sharp, 1975; Sharp, 1988; Sharp and Herzig, 1991). Siskiyou deformation is difficult to resolve in this region because of a pronounced phase of extensional tectonism in the late Callovian–early Oxfordian, and then superposed Nevadan folding and cleavage formation during Kimmeridgian to Early Cretaceous time (Saleeby, 1982, this volume; Schweickert and Bogen, 1983; Schweickert and others, 1984; Tobisch and others, 1989).

Kings sequence rocks may have undergone a major phase of deformation and northward translation during Middle Jurassic time. Pre-Cretaceous east-west–trending folds and related cleavages (Saleeby and Busby-Spera, 1986) may represent Middle(?) Jurassic dextral transpression within the fore-arc. The eastern margin of the southern Kings sequence is also bounded by a major Middle Jurassic ductile shear zone that apparently juxtaposed Kings sequence rocks against offset Antler belt rocks of the southeast Sierra Nevada and El Paso Mountains region (Dunne, 1989; Kistler, 1990; Dunne and others, 1991). The Kern Plateau shear zone was active at ~180 Ma and records basement-involved west-over-east movement along part of the boundary between Kings sequence and offset Antler rocks. This is suggested in Plate 5D to represent transpressive overthrusting of Kings sequence fore-arc basin rocks as an early (southern) manifestation of Siskiyou tectonics. Siskiyou deformation is suggested here to have migrated northward along the Sierran-Klamath belt from ~180 Ma in the south to between 177 and 170 Ma in the western Foothills belt, and to between 170 and 161 Ma in the Klamath Mountains.

The Insular suture belt and the Siskiyou event. The Insular superterrane is interpreted as having collided with the Cordilleran margin during mid-Cretaceous time (Davis and others, 1978). Along much of the present length of the Insular superterrane, the pre-collisional North American margin consisted of the Intermontane superterrane, which underwent a phase of major orogenic collapse during Middle Jurassic time (Monger, 1984). Recent studies in southeast Alaska have shown that regionally diagnostic rocks of both the Insular and Intermontane superterranes were thrust imbricated with one another before, and were in a phase of late- to post-orogenic extension during, Late Jurassic time (Saleeby and Rubin, 1990). We therefore interpret the initial collision of the Insular superterrane as a Middle Jurassic event, and speculate that this collision resulted in the collapse of the Intermontane superterrane (Plate 5D). The present suture belt between the two superterranes is quite linear in regional map view, due to a combination of superposed mid-Cretaceous to early Tertiary strike-slip faulting, intraplate thrusting, and batholith emplacement (Monger, 1984; Rubin and others, 1990b). The complex array of small and large terranes within the suture belt and composing the inner edge of the Insular superterrane, however, suggests that the colliding superterrane may have been irregular in shape. Furthermore, deformational and synorogenic depositional patterns for the Intermontane superterrane (discussed below) suggest that the collision was diachronous.

In Plate 5D and 5E, we interpret the initial collision of the Insular superterrane, collapse of the Intermontane superterrane, and the Siskiyou deformation event of the Sierra Nevada to Klamath Mountains region as one broadly related collisional episode. Paleomagnetic constraints on the early Mesozoic latitudinal tract of the Insular superterrane are highly equivocal at 1,000 km scales (reviewed in May and Butler, 1986), but paleobiogeographic data strongly suggest ~1,000 km northward displacement relative to the Intermontane superterrane between Pleinsbachian and Callovian time (Taylor and others, 1984; Tipper, 1984, 1985). Accordingly, we suggest that some part of the Insular superterrane grazed the Sierra Nevada to Klamath Mountains region, perhaps progressing diachronously northward, during early Middle to mid-Middle Jurassic time, resulting in Siskiyou deformation. It is unknown whether the impact was directly against western Sierra–Klamath fore-arc terranes, or whether the southern end of Stikine intervened as the remnants of a fore-arc microplate. It is also unknown whether the Insular superterrane was itself migrating as a fore-arc microplate prior to and during collision, or alternatively was attached to the oceanic plate that was descending beneath the west-facing Cordilleran plate edge. Alternative models for ocean basin evolution along the inner edge of the Insular superterrane are presented in Saleeby (this volume). We speculate here, however, that the collision of the Insular superterrane was within a weakly coupled subduction system with the development of dispersed fold-thrust belts bounding multiple basins, many of which survived the collisional event. This is consistent with Jurassic deformational and depositional patterns throughout the Insular and Intermontane superterranes that are discussed below, with the sense of retreat motion of North America relative to the plate edge, and by the broadly extensional setting of the continental arc south of the collision zone.

Fore-arc tectonics. Fore-arc basin sequences of Middle Jurassic age are not widely exposed along the southwest Cordillera. As in the earlier time intervals, the most continuous fore-arc sedimentation record is in the Izee terrane of the Blue Mountains (12 in Fig. 5). During Middle Jurassic time, volcaniclastic turbidites of the Snowshoe Formation continued to accumulate until the early Callovian (Dickinson and Thayer, 1978). A local unconformity in the Callovian marks the beginning of renewed mainly volcaniclastic turbidite deposition of the Townbridge Formation, which grades abruptly upward into mixed lithic turbidites of the Lonesome Formation. The Lonesome Formation marks the influx of orogenic sediments from the northeast (Dick-

inson and Thayer, 1978). Deformation at the end of the Callovian marks the end of basinal conditions in both the Izee and Olds Ferry terranes.

The rapidly changing conditions during Callovian time in the Blue Mountains may be a reflection of Siskiyou (Insular) tectonism migrating northward from the western Sierra–Klamath belt to the Blue Mountains and Quesnel. Collapse of Quesnel and its back-arc basin may have provided a source area for the synorogenic deposits of the Lonesome Formation. Juxtapositioning of the Wallowa terrane against the Baker, Izee, and Old Ferry terranes may have occurred at this time as well. The Wallowa terrane may have been derived from the collapsing and northward migrating Stikine terrane, or alternatively from the colliding Insular superterrane.

The grazing collision and/or fore-arc passage of the Insular superterrane along the northern California region and its continued migration to the northwest are considered closely related to the generation and emplacement of a vast track of fore-arc basement that underlies parts of southwest Oregon and California along the western Klamath Mountains, Great Valley, Coast Ranges, and southern California borderland. The main exposures of this oceanic basement consist of the Josephine and Coast Range ophiolites. Petrochemical and stratigraphic data (reviewed in Saleeby, this volume) show a clear suprasubduction-zone origin for these ophiolites (Bailey and others, 1970, Bailey and Blake, 1974; Evarts, 1977; Chen and Shaw, 1982; Saleeby and others, 1982; Sharp and Evarts, 1982; Shervais and Kimbrough, 1985; Harper, 1988; Natland, 1991). Stratigraphic columns from five ophiolite sections scattered along a ~1,000 km belt display the spectrum of primary features typical of the ophiolite belt (1, 3, 15, 16, and 17 in Fig. 6). Fragments of the regional ophiolite belt are also dispersed into the Insular suture belt (2, 5 in Fig. 7).

The Josephine ophiolite (3 in Fig. 6) represents a complete spreading-ridge sequence with depleted upper mantle rocks and mafic crustal rocks containing both tholeiitic and boninitic lavas (Harper, 1980, 1984, 1988). After rapid northward transport in the latest Callovian and Oxfordian, the Josephine ophiolite was partly subducted eastward beneath the Rattlesnake Creek terrane and partly obducted northwestward over the Rogue fringing arc (Harper and others, 1989, 1991). The southern Oregon Coast Range ophiolite (1 in Fig. 6) developed by multiple igneous pulses in a rifted-arc environment, as shown by screens of intermediate pyroclastic rock within its sheeted dikes and an abundance of post-dike small trondhjemitic to dioritic plutons (Saleeby and others, 1984; Blake and others, 1985). The northern Great Valley ophiolite at Paskenta (15 in Fig. 6) represents a MORB-like basin-floor sequence which was uplifted and deformed along a major north-south–trending submarine fault zone (Bailey and others, 1970; Hopson and others, 1981; Shervais and Kimbrough, 1985; Blake and others, 1987a, 1987b) which may represent a transform fault (Saleeby, this volume). The southern Coast Range ophiolite in the Diablo Range (16 in Fig. 6) is a rifted arc assemblage with local tholeiitic sheeted dikes containing screens of calc-alkaline plutonic rocks, pillowed flows consisting of tholeiitic and boninitic series, and an upper calc-alkaline submarine pyroclastic sequence (Evarts, 1977; Sharp and Evarts, 1982; Williams, 1983; Natland, 1991). In the southernmost Coast Ranges at Point Sal (17 in Fig. 6), a deep basin-floor igneous sequence consisting of basaltic andesite, boninite, and altered tholeiite (Hopson and others, 1981; Natland, 1991) is preserved with the pillow lavas. Much or all of the Coast Range ophiolite lies tectonically above the Franciscan complex, a relation established during Cretaceous time.

The Coast Range ophiolite, as exposed along the western Great Valley, appears to extend for a considerable distance eastward in the subsurface as the Great Valley basement (Cady, 1975; Suppe, 1979; Saleeby and others, 1986; Blake and others, 1989). As discussed below, widespread diking, extensional tectonism, and submarine volcanism in the western Sierra Foothills corresponds closely in time with the generation of the Coast Range ophiolite. The Foothills belt is considered here as a fore-arc rift edge to the Great Valley basin. Additional rocks which may be related to the Coast Range ophiolite underlie the Santa Barbara Channel Islands, parts of the southern California borderland, and perhaps the western Transverse Ranges (Hopson and others, 1981; Sorenson, 1985). Geochemical data on these rocks also show a strong suprasubduction-zone setting for their petrogenesis (Sorenson, 1985).

In addition to general tectonic position and ophiolite stratigraphy, the Josephine and Coast Range ophiolites share a number of other common features. Ages on sheeted-dike complexes are tightly clustered between 165 and 162 Ma (Callovian), although some local complexes are as old as ~170 Ma and a number of the complexes were magmatically active in Late Jurassic time (Hopson and others, 1981; Williams, 1983; Saleeby, 1984, 1987, 1990b; Saleeby and others, 1984; Harper and others, 1991). We refer to the ages of these ophiolites collectively as Late Jurassic. Their dike complexes represent their main phases of rifting and sea-floor spreading; ages of a major system of crosscutting sheeted dikes in the Sierran Foothills also cluster at ~162 Ma (Saleeby, 1982; Saleeby and others, 1989a). Post-rifting arc activity as young as early Tithonian is variably recorded by submarine pyroclastic aprons, tephra layers within chert sequences, and crosscutting dikes and small stocks throughout the entire ophiolite belt. Synthesis of structural relations within the Josephine and Coast Range ophiolites suggest a dominant north-northwest–south-southeast component to spreading directions and fracture zone orientations (Saleeby, this volume). Additionally, there is strong evidence for very rapid northward transport immediately following sea-floor spreading (Hopson and others, 1986; Harper and others, 1989; Pessagno and Blome, 1990). Upper Jurassic flysch derived from orogenic highlands laps across all of the ophiolites, as well as the Sierran Foothills submarine volcanic sequences. In general, the base of the flysch increases in age and thickness to the north and east, suggesting greater proximity to the orogenic source.

Synthesis of structural, stratigraphic, and petrotectonic data on the Late Jurassic ophiolites, in conjunction with the Siskiyou

tectonics outlined above, leads to the plate kinematic interpretation of Plate 5E. The ophiolite belt is shown forming along an obliquely spreading basin nucleated along the Cordilleran fore-arc above an east-dipping subduction zone. Local spreading commenced by 172 Ma (mid-Bathonian), but was most vigorous in late Callovian–earliest Oxfordian time. North-directed spreading and plate transport is kinematically linked to the Insular superterrane displacement pattern. The overall plate geometry is similar to that of the modern Andaman Sea, the Insular superterrane being analogous to the Burma microplate (Curray and others, 1979). Sea-floor spreading and transport accreted a major fore-arc basement terrane to the Cordilleran plate edge. Accretion of this basin floor was roughly synchronous with the initial Insular collisional tectonics to the north. The Late Jurassic ophiolite belt projects northward into the southern Insular suture belt, reflecting the creation of juvenile quasioceanic lithosphere in the wake of the Insular superterrane northward migration. In the Oregon-California region, late-stage pyroclastic aprons developed along the outer edge of the ophiolite basin (2 and 16 in Fig. 6); and represent initial phases in fringing arc development. Linear geophysical anomalies along the axial Great Valley may delineate major basin-floor transforms that controlled later ophiolite obduction (Saleeby, this volume). Growth of the ophiolitic basin and its related northward transport ended in Kimmeridgian time with the onset of the Nevadan orogeny.

One of the striking features of the Middle Jurassic of California is an absence of subduction complex rocks. This may have resulted from multiple cycles in fore-arc transtension and transpression, or perhaps Insular superterrane oblique collision and/or fore-arc migration. There are, however, fragments of late-Middle to Late(?) Jurassic high P/T oceanic metamorphic rocks reworked as blocks within Cretaceous melanges of the Franciscan complex (Coleman and Lanphere, 1971; Suppe and Armstrong, 1972; Suppe and Foland, 1978; Ross and Sharp, 1988; Mattinson, 1981, 1986). The isotopic age data suggest that peak metamorphic conditions occurred or were terminated in the source environments at ~162 Ma, the same maximum observed for igneous ages of the Late Jurassic ophiolites. The high P/T metamorphic blocks may represent Siskiyou-age subduction-zone rocks that were initially transported out of the metamorphic environment by fore-arc transtension during Late Jurassic ophiolite formation, or during Insular superterrane accretion. Alternatively, they may represent rapid cycles in subduction and then denudation within or along the edge of the actively spreading ophiolitic basin. The latter interpretation is more consistent with metamorphic paragenetic studies that indicate that the protolith oceanic lithosphere was hot when it was subducted, and then it retrogressed through the high P/T field prior to rapid uplift (Blake and others, 1988; Moore and Blake, 1989). Accordingly, we suggest that part of the Late Jurassic ophiolitic basin was subducted beneath itself and/or adjacent terranes immediately following its formation. Remnants of some of the oldest domains of the basin floor are preserved as subcreted mafic schist around the outer edges of the Condrey Mountain window in the central Klamaths. As noted above, subcretion and metamorphism occurred at the end of what is generally accepted as Siskiyou time (Harper and others, 1991). Underthrusting of the basin floor may have followed in the wake of the northward migration of the Insular superterranes past the western Klamaths.

A vast expanse of oceanic crust was generated by the Late Jurassic suprasubduction-zone spreading event. This is clearly demonstrated by the dispersal of ophiolite fragments for ~2,000 km along the plate edge between the southern Insular suture belt and the southern California borderland. An even greater initial expanse along the plate edge is suggested by the protolith association of a major subcreted schist terrane, which currently underlies much of the southern California region. The Rand-Pelona-Orocopia schist terrane represents a major oceanic basinal assemblage which was thrust beneath Cretaceous batholithic and sialic host rocks during latest Cretaceous time (Haxel and Dillon, 1978; Ehlig, 1981; Jacobson and others, 1988). The overall assemblage strongly resembles the Late Jurassic ophiolitic crust and overlying volcaniclastic and synorogenic flysch sequences that are so widespread to the north. Meager age data indicate at least local sedimentation and igneous activity at 163 ± 2 Ma (Mukasa and others, 1984). Additional subcreted schist bodies that may also represent remnants of the Late Jurassic basin system include the Condrey Mountain terrane of the central Klamaths, the Shuksan suite of the southern Insular suture belt, and the Pickett Peak terrane of the northern California–southern Oregon Coast Ranges (Brown and Blake, 1987). Protolith age constraints of ~170 Ma on the Condrey Mountain terrane are very close to the 172 Ma age of high-grade ophiolitic rocks which lie tectonically above the schist terrane (Hill, 1985; Helper and others, 1989; Harper and others, 1991). Meager protolith age data of ~163 Ma on both the Pickett Peak and Shuksan schist terrane are similar to the widespread Callovian-Oxfordian ophiolite ages (Brown and Blake, 1987). As noted above, subcretion of the Condrey Mountain terrane started in late Siskiyou to early Nevadan time, whereas timing of subcretion of the Pickett Peak and Shuksan terranes was late Nevadan and younger.

Continental arc. The continental arc underwent major changes in Middle Jurassic time, both in terms of comparisons with its Late Triassic–Early Jurassic history and in terms of events during Callovian-Oxfordian time. In general, there was a marked decrease in volcanism throughout much of the McCloud segment of the arc that probably reflects Siskiyou and Insular suture-belt tectonics. This may also correspond with the cessation of marine basinal conditions within the McCloud segment back-arc region. Along the sialic segment of the arc, there is an apparent lull in arc activity between late-Early and early-Middle Jurassic time, followed by a major pulse of activity that continued into the early-Late Jurassic. The apparent lull in arc activity apparently predates the Siskiyou event to the north, and corresponds in time with rapid northward displacement of the Insular superterrane (Taylor and others, 1984; Tipper, 1984, 1985). One possible explanation for these broad relations is that the Insular superterrane was migrating northward along the

southwest Cordilleran plate edge as a fore-arc microplate during late-Early to early-Middle Jurassic time, and the axis of continental arc magmatism shifted outboard to the upper-Lower to lower-Middle Jurassic volcanic and plutonic arc rocks of Wrangellia (1 in Fig. 7) (Armstrong, 1988; Plafker and others, 1989; Monger and others, 1990). In this model, the continued northward displacement of the Insular superterrane became transpressional into the Siskiyou regime and its ultimate suturing to the north. In its wake, the transtensional regime represented by the Late Jurassic ophiolite belt also promoted renewed extension and major ignimbrite activity along the sialic segment of the arc. Presence of a rift arm of the Gulf of Mexico was an additional control on the renewed extension and ignimbrite activity in the Mojave-Sonora desert region of the active sialic arc (Dickinson and others, 1987; Busby-Spera and others, 1991a, 1991b).

Intermontane superterrane—The magmatic and depositional patterns of the Intermontane superterrane changed dramatically during Middle Jurassic time (Monger, 1984; Monger and others, 1990), presumably as a result of the collision of the Insular superterrane (Saleeby and Rubin, 1990). The regional linear chains of arc volcanism that characterized much of early Mesozoic time were replaced by synorogenic basins and arches, fold-thrust belts, and the synkinematic Omineca plutonic-metamorphic belt.

During Bathonian to Callovian time, rocks of the northern Cache Creek terrane were thrust westward over Stikine, and synorogenic deposits of the Bowser basin began to accumulate in a foredeep environment (Eisbacher, 1981; Monger, 1984; Yorath and others, 1990). Transverse arches bounded the northern and southern margins of the basin; the Nechako basin formed above southern Stikine south of the Skeena arch, and received synorogenic deposits as well (Tipper, 1984; Yorath and others, 1990). To the east, the Quesnel, Slide Mountain, and Kootney terranes were imbricated eastward, and then cut by ~170 Ma synkinematic plutons of the Omineca crystalline belt (Monger, 1984).

The stratigraphic record of arc volcanism ended in the northern Washington exposures of Quesnel by Middle Jurassic time (14, 15 in Fig. 5). Middle Jurassic plutonism is recorded in the Okanagon highlands by the dioritic to quartz-monzonite Toats Coulee pluton (Fox and others, 1977) and the granodioritic to monzonitic Similkameen batholith (Rinehart and Fox, 1972). These plutons represent the southern end of the contiguous Omineca crystalline belt.

Blue Mountains—Arc volcanic rocks of Middle Jurassic age are rare or absent in the Blue Mountains. Arc-derived turbidites of the Weatherby Formation continued to accumulate in the Olds Ferry terrane from Early Jurassic until Callovian time, indicating arc activity somewhere in the general region (13 in Fig. 5). Termination of Weatherby deposition roughly corresponds with the Late Jurassic onset of deformation that was widespread throughout the Blue Mountains region. Within the Wallowa terrane, turbiditic and conglomeratic strata of the Coon Hollow Formation and unnamed units were deposited during the Callovian after a major hiatus (8 in Fig. 7). These synorogenic deposits may be analogous to Callovian strata of the Bowser and Nechako basins, which formed above Stikine, and/or synorogenic deposits that are widespread in terranes of the Insular suture belt.

Klamath Mountains—Marine arc volcanism persisted in the eastern Klamaths (1 in Fig. 5) through Bajocian time (Sanborn, 1960; Irwin, 1966). Middle Jurassic arc activity is represented by the Bagley andesite and overlying flows and volcaniclastic rocks of the Potem Formation. The apparent termination of arc activity near the end of the Bajocian (~175 Ma) corresponds to the onset of western Klamath arc activity in the western Hayfork terrane and possibly to Siskiyou thrusting. Two lines of evidence suggest spatial proximity between eastern and western Klamath arc activity: (1) Paleozoic fossils from several limestone cobbles from the western Hayfork arc sequence contain Permian McCloud fauna; and (2) Middle Jurassic arc-related plutons that cut the western Hayfork terrane basement also cut the eastern Klamath structural basement (Harper and Wright, 1984; Wright and Fahan, 1988).

Middle Jurassic arc rocks of the western Klamath Mountains are represented primarily by volcanic rocks of the western Hayfork terrane (5 in Fig. 6) and their contemporaneous mafic-ultramafic intrusive complexes (Harper and Wright, 1984; Wright and Fahan, 1988). The western Hayfork terrane consists of an ~6 km structural thickness of Bathonian to lower Callovian basaltic to basaltic-andesite volcaniclastic rocks with minor pillow lava and breccia. The lower unit is composed of submarine monolithologic breccias and crystal-lithic tuffs with minor flows, tuffaceous and siliceous argillite, and radiolarian chert (Fahan and Wright, 1983; Wright and Fahan, 1988). The upper unit is a mixed assemblage of volcaniclastic, continent-derived epiclastic, and hemipelagic rocks. Volcaniclastic rocks indistinguishable from those of the lower unit are abundant, but siliceous argillite, pebble conglomerate, arenites, and radiolarian chert are also common in this unit.

Basement rocks of the western Hayfork terrane cannot be directly observed due to thrust detachment; however, spatial and age relations, as well as petrologic data, suggest a close tie between widespread Middle Jurassic intrusive complexes and the western Hayfork volcanic rocks (Snoke and others, 1982; Harper and Wright, 1984; Wright and Fahan, 1988). The intrusive complexes are present in the western Hayfork terrane as well as most other pre–Late Jurassic terranes of the Klamath mountains, including the structural basement for the McCloud arc, the Stuart Fork terrane, the North Fork terrane, the eastern Hayfork terrane, and the Rattlesnake Creek terrane (Snoke and others, 1982; Harper and Wright, 1984; Saleeby, 1984; Wright and Fahan, 1988). They contain partial to complete suites composed of: (1) olivine- and clinopyroxene-rich ultramafic cumulates; (2) two-pyroxene–bearing gabbroic to dioritic and monzodioritic rocks; and (3) intermediate to felsic rocks of similar bulk composition containing hornblende but lacking orthopyroxene.

The western Hayfork arc, along with its basement amalgam, has been thought of as a remnant arc complex that ceased its activity during the late Callovian–early Oxfordian opening of the

Josephine ophiolite basin to the west (Snoke, 1977; Harper, 1980; Saleeby and others, 1982; Harper and Wright, 1984). Spatial relations suggest that the basin grew along the fore-arc edge of the western Hayfork arc. The tangential transport pattern of the Josephine ophiolite, however, complicates the interpretation of its primary spatial relations, although there are a number of relations to suggest a Middle Jurassic phase of extension along the frontal edge of the Hayfork arc. These include: (1) the construction of a Bathonian and possibly older arc tholeiitic and pelagic sediment carapace unconformably across Rattlesnake Creek terrane ophiolitic melange in the Preston Peak area (4 in Fig. 6; Snoke, 1977; Snoke and others, 1977; Saleeby and others, 1982; Harper and others, 1991); (2) the formation of a similar age (~172 Ma) sheeted-dike complex which lies along the structural base of the Rattlesnake Creek terrane in the Condrey Mountain region; it may belong to the Rattlesnake Creek terrane or may have been subcreted to it with the Condrey Mountain terrane (Hill, 1985; Harper and others, 1991); and (3) the presence of at least local boninitic lavas within rocks of the Marble Mountains terrane that are interpreted to be metamorphosed western Hayfork terrane (Donato, 1987). The age and structural-metamorphic relations indicate that this phase of frontal and/or fore-arc extension was pre-Siskiyou in age, but conceivably within the upper age limit of the regional Late Jurassic ophiolites. In the plate kinematic model of Plates 5-D and E, the ~170 Ma phase of extension is considered the earliest phase in the growth of the Late Jurassic ophiolitic basin. This ophiolite generation in the suprasubduction-zone environment, along with the tangential basin growth and displacement patterns, makes it quite possible that major extension began prior to, lasted during, and continued after Siskiyou contractile deformation.

Northern and western Sierra Nevada—Submarine volcanic arc activity continued in the northern Sierra region probably throughout Middle Jurassic time (2 in Fig. 5). In the upper American River sequence, arc-derived turbidites of the Sailor Canyon Formation accumulated into the Bajocian (Imlay, 1968), and these are overlain by pre–upper Callovian mainly andesitic breccias (Harwood, 1988; Saleeby and others, 1989a) referred to as the Tuttle Lake Formation (Fisher, 1990). The Tuttle Lake Formation is more than 3 km thick and consists of coarse conglomerate, breccia, lava flow, and hypabyssal rock, in addition to the thick volcanic breccias and tuffs. It is interpreted as an extensional trough fill (Fisher, 1990), indicating that extensional arc activity persisted in the northern Sierra well into Middle Jurassic time. Similar marine andesitic pyroclastic rocks dominate the Mt. Jura section, but here fossiliferous argillaceous interbeds demonstrate volcanic activity into Callovian time (Imlay, 1961; McMath, 1966). Arc activity ceased in the northern Sierra during Callovian time with the deposition of synorogenic conglomerates, which may have continued to accumulate into the Late Jurassic (2 in Fig. 5).

The distinctive suite of Middle Jurassic peridotitic to dioritic intrusive complexes that are widespread in the Klamath Mountains is present in the northern and western Sierra Nevada as well (Snoke and others, 1982). Some of these complexes are closely related to generally coeval volcanic sequences and may be direct plutonic roots. The 164 Ma Emigrant Gap complex intrudes the Tuttle Lake Formation of the upper American River sequence (2 in Fig. 5) and is probably very close in age to the upper andesite rocks (Saleeby and others, 1989a). In the lower American River area (10 in Fig. 6), the Pine Hill intrusive complex is coeval with, and intrudes the basement of, the Callovian Logtown Ridge Formation (Clark, 1964; Springer, 1980; Saleeby, 1982). Similar close age and spatial relations between plutonic and volcanic rocks occur in the Sullivan Creek terrane to the south (13 in Fig. 6; Sharp, 1988; Sharp and Herzig, 1991).

Callovian to early Oxfordian arc activity within an extensional tectonic regime is recorded along the western Sierra Foothills (7, 8, 9, 10, 11, 13, 15 in Fig. 6). The clearest record of this activity is in the Logtown Ridge Formation and the Smartville complex of the northern Foothills. The Logtown Ridge Formation consists of a ~2 km thickness of arc tholeiitic to calc-alkaline basalt and basaltic-andesite submarine tuff, tuff breccia, and pillowed flow and breccia (Clark, 1964; Duffield and Sharp, 1975; Behrman and Parkison, 1978). Its upper levels are intercalated with upper Oxfordian–lower Kimmeridgian flysch of the Mariposa Formation. Together, these units constitute the main stratigraphic sequence of the Mother Lode belt. The Logtown Ridge Formation lies in faulted unconformity above central belt melange. The unconformity may represent Siskiyou, as well as earlier, deformations of the central belt melange, although the unconformity surface is complicated by probable low-angle normal faulting related to Callovian-Oxfordian extension. Initial movement along the Bear Mountains fault zone may have been low-angle normal as a result of this extension. At most localities, the fault zone places older above younger, and the fault zone branches and merges northward with the 164 Ma Folsom sheeted-dike complex (Saleeby, 1982). The sheeted-dike complex invaded its wall rocks preferentially along the north-northwest–trending melange structure of the basement for the Logtown Ridge Formation, and appears to continue northward into sheeted dikes of the Smartville complex (Saleeby, 1982).

Callovian to earliest Oxfordian arc volcanism and extension are well recorded in the Smartville complex (9 in Fig. 6). The Smartville complex consists of basal tholeiitic basaltic pillow lavas overlain by calc-alkaline basalt to basaltic andesite submarine pyroclastic rocks, both of which are cut and underlain by a complex subvolcanic intrusive series (Xenophontos and Bond, 1978; Menzies and others, 1980; Beard and Day, 1987). The entire volcanic sequence is broadly constrained as Middle Jurassic, but an upper-level conformable dacite flow is dated at 164 ± 1 Ma (Saleeby and others, 1989a). The Smartville intrusive series consists of cumulate and static textured peridotitic, dioritic, and trondhjemitic intrusions and a regionally extensive sheeted-dike complex. The dike complex formed at 162 ± 1 Ma and represents a major extensional phase within the complex (Saleeby and others, 1989a). The volcanic section of the Smartville complex appears to have been folded into a large anticline prior to

sheeted-dike emplacement. This may represent Siskiyou deformation.

Extensional arc activity along the Sierran Foothills is considered to have been closely related to the Late Jurassic ophiolite basin to the west. The Foothills volcanic and sheeted-dike complexes closely resemble those of the ophiolite belt, except the Foothills rocks were built on an older Paleozoic ophiolitic melange–lower Mesozoic fore-arc igneous-sedimentary basement complex which had been previously disrupted by transform, extensional, and compressional deformations. In Plate 5E, the Foothills belt is depicted as a fore-arc rift edge that faced the ophiolitic basin. Tangential spreading and plate transport patterns within the basin suggest that the rift edge may have been a boundary transform. The longitudinal orientation of sheeted dikes within the Foothills, however, was controlled by the preexisting structural fabric within the host rocks.

Northwest Nevada—Subaerial to shallow-water volcanism of the Happy Creek complex continued in the Black Rock Desert region through Bathonian time (11 in Fig. 5). During the Bathonian, a roughly east-west–trending sinistral normal-fault system controlled the volcanic and depositional patterns (Maher, 1989). Major volcanism ceased in the Callovian when a system of west-vergent high-angle reverse faults disrupted the volcanic environment, and molasse of the King Lear Formation was shed off of westward advancing fault scarps. The Bathonian fault system may represent part of a back-arc transtensional system represented to the south by the Pine Nut fault, the Humboldt lopolith, and a graben system. The superposed Callovian reverse-fault system may record the northward migration of Siskiyou deformation into the region. Syntectonic monzonitic, syenitic, and gabbroic sills and stocks were emplaced during the thrusting.

Subsidence in the Late Triassic to Early Jurassic marine basinal province of northwest Nevada continued into Middle Jurassic time, and was controlled at least locally by extension. The northwest-trending half grabens that began to form in Early Jurassic time continued to be active; craton-derived quartz sands of the Dunlap Formation and the Boyer Ranch Formation were deposited in these half grabens (Oldow and Bartel, 1987). At ~170 Ma, a large-volume basaltic magma series, the Humboldt lopolith and associated igneous rocks, was emplaced into the Boyer Ranch Formation and locally vented submarine lava flows (8, 9 in Fig. 5) (Speed, 1963). Local peperitic contacts, as well as hydrothermal exchange reactions, indicate that the Boyer Ranch Formation was in a soft-sediment state during intrusion and eruption. The Humboldt lopolith may represent the final phases of extension in the northwest Nevada back-arc region prior to the late Middle(?) or Late Jurassic establishment of a major foreland fold and thrust regime. The sparse stratigraphic record of the northwest Nevada region probably stems from uplift and erosion during the onset of regional back-arc thrusting. It is not clear whether much of the region saw little or no deposition during Middle Jurassic time, or whether the Middle Jurassic sections were preferentially removed by uplift and erosion during subsequent thrusting.

Faulting and folding patterns and deflections in broad structural trends in the northwest Nevada region record dextral transpression distributed into the proximal back-arc region. This tectonic pattern is recorded by the initial Middle Jurassic activity along the polyphase Pine Nut fault system (Oldow and Gelber, 1987). Dextral shear related to the Pine Nut system may have continued southward and broadened to the east of the continental arc where the miogeocline edge and the Antler and Sonoma belts show broad dextral deflections (Speed, 1978; Saleeby and others, 1986). It is not clear as to how much of the deflection was inherited from the original geometry of the miogeocline (Geissmann and others, 1984) and how much was superposed by early Mesozoic bending (Speed, 1978; Saleeby and others, 1986). Possible relations between bending and early Mesozoic back-arc extension are unexplored.

In the Yerington district, voluminous volcanism and plutonism occurred in the late-Middle Jurassic following a lull in arc magmatism (3 in Fig. 5). The Yerington batholith was emplaced between 170 and 168 Ma in conjunction with the development of west-northwest–trending folds and associated cleavage. The orientations of these folds and cleavages are consistent with possible dextral transpression, but their possible relations to early deformation along the Pine Nut fault is unexplored. Shortly after folding and batholith emplacement, voluminous felsic volcanic (and hypabyssal) rocks were erupted at 165 Ma (Dilles and Wright, 1988). This major silicic volcanic pulse corresponds very closely in time to that observed in the eastern Sierra.

East-central California region—There was an apparent lull in eastern Sierra arc plutonism between ~200 and ~170 Ma (Stern and others, 1981; Chen and Moore, 1982). A similar lull (190 to 175 Ma) is reflected in the distribution of zircon ages from silicic tuffs of the Saddlebag Lake–Ritter Range section of the east-central Sierra (12 in Fig. 3), and the youngest ignimbrite in the Mineral King section of the southern Sierra is ~190 Ma (9 in Fig. 4). At ~170 Ma, major ignimbrite activity and silicic plutonism resumed in the eastern Sierra region. In particular, at ~165 Ma large volumes of silicic ash-flow tuff were erupted (Tobish and others, 1986; Saleeby and others, 1990). This volcanism may represent a distinct phase of intra-arc extension corresponding in time with the generation of the ophiolitic basin to the west. Along the east flank of the ignimbrite field, a number of alkaline plutons were emplaced at ~165 Ma (Miller, 1978).

In the Benton Range area to the east of the Ritter Range, a distinct phase of late-Early to early-Middle Jurassic east-vergent compressive ductile deformation separates the emplacement of different silicic plutonic suites (Renne and Turrin, 1987). This deformation event may correspond in time to the lull in arc activity along the eastern Sierra region and in the Yerington district, or may be an expression of synmagmatic deformation that was dispersed along the back edge of the Middle Jurassic Sierran arc.

In the southern Inyo Mountains (2 in Fig. 4), subaerial arc volcanism may have continued through Middle Jurassic time; however, direct age control is lacking. Arc plutonism was wide-

spread along the White-Inyo Mountains and in ranges to the south, and was probably contiguous with plutonism of the eastern Sierra. East-directed thrust faults and high-angle reverse faults of the eastern Sierra thrust system formed mainly along the back margin of the arc plutonic belt in Middle and Late Jurassic time (Dunne and others, 1978; Dunne, 1986). Large folds including the White-Inyo and Panamint anticlinoria formed in conjunction with the eastern Sierra system (Dunne and others, 1978; Wright and others, 1981). Exposures of deep-seated rocks along the eastern margin of the system in ranges of the Death Valley area reveal regional dynamothermal metamorphism of Middle to Late Jurassic age, related to heating and east-directed thrusting along the back edge of the arc (Labotka and Albee, 1988). Contractile structures along the back edge of the arc in this region appear to record low magnitudes of strain and displacement. In the eastern Sierra inset of Plate 5D, these structures are depicted as rooting westward into a voluminous arc plutonic infrastructure, above which there was massive upper crustal extension and ponding of ignimbrites. The contractile strain zone may have localized along a zone of major ductility contrast arising from magma and heat influx to the west. The relatively cool wall rocks along the back side of the arc were partially underthrust along the back edge of the spreading arc, possibly as a result of return flow dynamically linked to voluminous magma ascent (after Saleeby, 1990a).

Mojave Desert region—Voluminous ignimbrite activity and normal faulting occurred in late Middle Jurassic time in the Mojave Desert region (Busby-Spera and others, 1991a, 1991b). The normal faults formed largely in response to regional-scale extension and/or transtension. To the south, in the southeastern California–southwestern Arizona to southern Arizona region, the extensional arc began to be exploited, while still active, by a rift arm of the Gulf of Mexico (Busby-Spera and others, 1989). Meanwhile, to the north, transpressional (Siskiyou) deformation was occurring in the southern and western Sierra, along the Pine Nut fault system in northwest Nevada, and in the Klamath Mountains. The Mojave Desert occupied a transitional tectonic position along the plate margin; regional contractile deformation dominated in the Klamath region, a combination of contractile and extensional deformation operated along the Sierra region to the north, and regional extension dominated in the Mojave region to the south.

Craton-derived eolian quartz arenites of Middle Jurassic age in the Mojave region, indistinguishable in appearance from Early Jurassic Navajo-equivalent eolianites, continued to be deposited within the arc graben depression into the late-Middle Jurassic. Isotopic age data indicate that these eolian sands accumulated contemporaneously with flooding of the back-arc basin by the Carmel seaway in the plate interior (Busby-Spera and others, 1991). Bentonites from the Carmel Formation range in age from 170 to 162 Ma (B. Kowallis, personal commun., 1988), and volcanic rocks interstratified with eolianitic quartz sands within the arc in the Cowhole Mountains (3 in Fig. 4) range in age from 173 ± 4 Ma to 169 ± 2 Ma (Mattinson and Busby-Spera, unpublished data). This correlation is supported by the presence of eolianites in the Carmel Formation (Peterson and Pipiringos, 1979; Kocurek and Dott, 1983). It is not known whether intra-arc eolian quartz arenites in the Rodman Mountains (5 in Fig. 4) are age equivalent to the Lower Jurassic Glenn Canyon Group or the Middle Jurassic San Rafael Group (2 in Fig. 3).

The Middle Jurassic volcanic section in the Cowhole Mountains of the eastern Mojave region is dominated by outflow ignimbrites (3 in Fig. 4), whereas the coeval lower Sidewinder volcanic series of the central Mojave region is dominated by intracaldera ignimbrites (5 in Fig. 4). Compositions in both areas range from dacite and rhyolite to latite and quartz latite. The extracaldera ignimbrites in the Cowhole Mountains are each less than 200 m thick, and include both welded and nonwelded units. In the Central Mojave region, four welded ignimbrites, 480 to 1400 m thick, erupted from calderas located near Sidewinder Mountain, and a fifth welded ignimbrite, >2 km thick, probably marks the site of a caldera centered over Ord Mountain (Schermer and Busby-Spera, 1990).

Syndepositional normal faults in the Cowhole Mountains volcanic section are not associated with plutonic rocks or thick welded ignimbrites, suggesting that they reflect regional, rather than local, extension (Busby-Spera and oth ers, 1991a). Evidence for syndepositional faulting includes stripping of strata from upthrown blocks and deposition of landslide megabreccias on downthrown blocks, as well as overlapping of faults by ignimbrites or conglomerate-breccia units.

Regional subsidence during accumulation of the lower Sidewinder volcanic series may be indicated by a ~1 km thickness of outflow ignimbrites and post-caldera units, although an additional 3.6-km-thick section of ignimbrites was ponded in calderas (Schermer and Busby-Spera, 1990). Normal faults with several hundred meters of offset juxtapose the 170 Ma lower Sidewinder volcanic series against older rocks. These were active during emplacement of the Middle Jurassic plutonic suite, so it is not clear whether they represent a regionally developed system or local extension above plutons. Plutonic rocks of the Middle Jurassic suite, as well as 148 Ma dikes of the Late Jurassic Independence swarm, contain a locally well-developed foliation with the same orientation as that developed in the lower Sidewinder volcanics and the Fairview Valley Formation (5 in Fig. 4; Schermer and Busby-Spera, 1989, 1990). This suggests that no shortening occurred until after, or very near the end of, the Jurassic, although some workers have inferred a Middle Jurassic synplutonic contractional event in this region (Miller, 1978; Karish and others, 1987). This region was also unaffected by Late Jurassic intra-arc shortening and metamorphism documented in the Cronese Hills and possibly in the Shadow Mountains of the central Mojave by Walker and others (1989, 1990a, 1990b).

Southeastern California–Southwestern Arizona—Voluminous ignimbrite activity is recorded in this region by the Dome Rock sequence (6 in Fig. 4), which began its eruptive history by ~170 Ma and was active into the early-Late Jurassic (Busby-Spera and others, 1991b). The Dome Rock sequence interfingers with the underlying eolian quartz arenites, suggesting eolian

deposition in the region well into Middle Jurassic time. Plutonism and synchronous regional metamorphism also began in the area in the Middle Jurassic and extended into Late Jurassic time (Tosdal and others, 1989; Reynolds and others, 1989). The tectonic setting and precise timing of this event are not well understood because the rocks are strongly overprinted by Cretaceous plutonism and metamorphism. The event may have occurred in an intra-arc extensional environment (Tosdal and Haxel, 1987), or it may be analogous to the arc-related contractile deformation observed to the north. Plutonic rocks of the Kitt Peak–Trigo Peak superunit (6 in Fig. 4) and the Cargo Muchacho superunit of the Chocolate Mountains to the west consist of hornblende diorite, quartz monzodiorite, granodiorite, and granite with ages ranging from 173 to 160 Ma (Dillon, 1976; Tosdal and others, 1989). Middle and Upper Jurassic metamorphic rocks have great structural relief; there are downward increases in textural and metamorphic grade from argillitic and phyllitic rocks, hornfels near intrusions, through amphibolite schists with ductile shear zones, to deeper-level synkinematic granite gneisses, and associated sheets and pods of pegmatite (Reynolds and others, 1989).

Southern Arizona–northern Sonora—The same early-Middle Jurassic lull in volcanism that occurred along more northerly segments of the continental arc probably occurred in the southern Arizona–northern Sonora region. The lull in this region was also followed by late-Middle Jurassic ignimbrite activity and trapping of craton-derived eolian sand in intra-arc basins. Upper Middle to Upper Jurassic rocks of this region appear to record a gradual transition through time, from continental intra-arc extension to continental rifting associated with the opening of the Gulf of Mexico. Coarse clastic deposits of this age, referred to as Glance Conglomerate, are commonly interpreted as synrift deposits (Dickinson and others, 1986, 1987), although some can be attributed to localized caldera subsidence (Riggs and Busby-Spera, 1990).

The 175 Ma Cobre Ridge assemblage (7 in Fig. 4) records the collapse, filling, and resurgent doming of a very large (60 × 25 km) caldera centered over the present-day international border west of Nogales (Riggs and Busby-Spera, 1990). This caldera formed during the eruption of the Pajarito Tuff, one of the largest-volume ignimbrites in the geologic record (estimated volume between 1,000 and 3,000 km^3). The intracaldera ignimbrite is as much as 3 km thick in the southeast half of the caldera. The northwest half of the caldera subsided to greater depths than the southeast half, providing a depocenter for more than a 1.5-km-thick section of post-collapse volcanic and sedimentary strata. The post-collapse strata record the early stages of resurgent doming and attendant moat filling, and were intruded during the later stages of resurgence. Resurgent magmatism was concentrated in the half of the caldera that had undergone greater subsidence during caldera collapse. Eolian quartz arenite occurs as two widespread sheets within the uppermost part of the intra-caldera Pajarito Tuff, and as lenses in the post-collapse section. The post-collapse section includes accidentally ponded ignimbrites (thick but areally restricted), rhyolite dome talus cones, water-laid tuffs, and a thick, localized conglomerate-breccia deposit that represents material eroded from the caldera walls (Riggs and Busby-Spera, 1990, 1991).

The Canelo Hills volcanics (7 in Fig. 4), present in four southern Arizona ranges between Nogales and Bisbee, underlie and interfinger with conglomerates and breccias interpreted as synrift deposits (Glance), and are inferred to represent Middle to Late Jurassic rift magmatism (Dickinson and others, 1987). Trace-element geochemistry may support a rift interpretation for some of these rocks (Krebs and Ruiz, 1987), although alkaline volcanic rocks also occur in arcs characterized by deep crustal faulting or rifting (i.e., extensional and/or transtensional arcs; see general discussion in Luhr and Carmichael, 1981). The Canelo Hills volcanics may have closer stratigraphic and structural affinities to the extensional arc or, more likely, record a transition between extensional arc and continental rift volcanism and/or sedimentation. The Canelo Hills volcanics consist dominantly of silicic ignimbrites (Kluth, 1982, 1983; Vedder, 1984). Intracaldera facies occur locally, and at least some of the "Glance Conglomerate" may represent caldera-wall debris, rather than synrift deposits (Dickinson and others, 1987).

Plate interior

Early Jurassic eolian sand-sea development had ended by Bajocian time, when a shallow sea transgressed from the north and west across the region and persisted into early Callovian time (Peterson and Pipiringos, 1979; Imlay, 1980; 1, 2, 3, 5, 6 in Fig. 3). The regional erosional unconformity that caps Navajo-age eolianites from Wyoming to Arizona may have resulted from net deflationary conditions in the erg as the Carmel seaway advanced, cutting off the source of sands from north. Eolian conditions persisted along the margins of the seaway, particularly in the south (Kocurek and Dott, 1983). Carbonates deposited in the western part of the seaway interfinger with terrigenous marine rocks deposited in the eastern part (Kocurek and Dott, 1983) (2, 3, 6 in Fig. 3). Sedimentary rocks of this age are lacking throughout much of western Utah, but several lines of evidence suggest that the zero edge lay 50 to 75 km west of the present erosional edge (Blakey and others, 1983). West of that, crustal shortening may have begun to produce highlands in the eastern Great Basin, although this event may have been restricted in time to the Late Jurassic (Miller, 1990). A broad coastal plain, as much as 1,000 km wide, bounded the seaway along its eastern margin (Blakey and others, 1983).

The Middle Jurassic Carmel Formation and Page Sandstone were deposited in two major transgressive-regressive cycles of the western interior seaway, described by Blakey and others (1983). At maximum transgression, carbonate deposition peaked, and offshore open-shelf biomicrites interfingered eastward with oolitic shoals and, finally, shoreline and supratidal sabkhas. At peak clastic deposition, subtidal sand waves gave way eastward to beach and erg facies. It has been suggested that evaporites of this period, present in the Gypsum Springs, Carmel, and Arapien

formations and also the Preuss Formation of Callovian age, were associated with broad regional extension (Moulton, 1975) (2, 3, 6, 7, Fig. 3), but significant preferential structural thickening of the units by later deformation precludes evaluation of the stratigraphic facies and thickness changes pertinent to this proposition (Standlee, 1982). Alternatively, the broad spatial distribution of evaporites during this interval suggests that their deposition was a regional climatic effect (Kocurek and Dott, 1983).

LATE JURASSIC (162–144 Ma)

The major tectonic and paleogeographic features for Late Jurassic time are shown in Plates 5E and 5F. Considerable attention was focused on the Callovian-Oxfordian map (5E) in the Middle Jurassic discussion because of the close interrelations between Middle Jurassic arc magmatism, deformation, and ophiolite formation, which was most vigorous at the end of Callovian time. Likewise, our discussion of the Late Jurassic will extend back into the Callovian because of the close relations between ophiolite formation and subsequent Late Jurassic tectonism.

The Late Jurassic saw the final stages of fringing terrane accretion with the emplacement of the Late Jurassic ophiolites and the fringing arc sequences that were built along the outer edge of the ophiolitic basin system. During this event, large volumes of turbiditic sands derived from previously accreted terranes were deposited across the ophiolitic basin and fringing arc system. This turbidite ("flysch") deposition and synchronous deformation constitute the Nevadan orogeny (Blackwelder, 1914; Bateman and Clark, 1974; Schweickert and Cowan, 1975; Saleeby, 1981, 1983; Harper and Wright, 1984; Day and others, 1985). Upper Jurassic turbiditic sands were also deposited across many of the fragmented terranes that constitute the southern end of the Insular suture belt, including fragments of the Late Jurassic ophiolitic basin system that lie within the suture belt. This suggests that the ophiolitic basin system extended into the precollapsed framework of the Insular suture belt. Subduction-zone metamorphism continued through Nevadan time in the Condrey Mountain terrane of the central Klamaths, and began in the Pickett Peak terrane of the Franciscan complex and the Shuksan metamorphic suite of the Insular suture belt.

In contrast to much of earlier Mesozoic time, Late Jurassic arc magmatism appears not to have formed a regionally coherent belt along the plate edge. The disruption of the magmatic trends, as well as the Nevadan deformation patterns, are considered a result of a distinct change in plate-motion patterns. Magmatism occurred in clusters of relatively small plutons, zones of extension and transtension, scattered ignimbrite fields, and in fringing arc environments along the outer edge of the Late Jurassic ophiolite basin. Disruption of the arc magmatic patterns that typified much of earlier Mesozoic time was probably due to a combination of the rapid growth of the Late Jurassic ophiolitic basin system and a major reversal in tangential relative-motion patterns along the plate edge.

In Late Jurassic time, sinistral strike-slip faulting was important along much of the plate edge. Sinistral transpressive deformation in the north was closely related to the classical Nevadan orogeny, as well as major contractile deformation in the back-arc region of northwest Nevada. Sinistral strike-slip faults appear to have cut the back-arc, arc, and fore-arc regions. Sinistral transtension in the south resulted in displacements of perhaps several hundred kilometers along strike-slip faults in Mexico, and emplacement of dike swarms from the Sonora to eastern Sierra Nevada regions. Sinistral faulting may have also operated during the rift phase in the evolution of the Bisbee basin in southern Arizona, which formed in response to the opening of the Gulf of Mexico (Dickinson and others, 1986). This suggests a broad relation between sinistral faulting in the Cordillera and Atlantic margin tectonics. In plate 5-F, we extend the sinistral transtensional zone of the Mojave-Sonoran region northwestward into the northern California–Nevadan transpressive zone. The common tangential displacement pattern of these different tectonic regimes resulted from North America's large northwest component in absolute motion (May and others, 1989). This motion also resulted in the driving of the northern Sierran–Klamath belt and Blue Mountains arc northward into the southern end of the Insular suture belt, which may have been undergoing a major phase of postcollisional extension.

In the plate interior, there is the first evidence for the development of a foreland fold and thrust belt and related basin system, particularly in the north. A major orogenic sediment source lay to the west of the basin, and isopach maps indicate a trough adjacent to the highlands, typical of such basins. During the early part of this time interval, retreat of the Carmel seaway and associated spreading of eolian environments to the north and northwest was followed by a mid-Callovian to mid-Oxfordian marine incursion. Throughout this time interval, nonmarine deposition and many areas of nondeposition resulted from uplift of western source regions, although lower subsidence rates and possibly lowered eustatic sea level may have been contributing factors (Brenner, 1983).

Within the deeper levels of the Nevada–western Utah–southern Idaho orogenic highlands, syntectonic intrusions and zones of dynamothermal metamorphism developed in conjunction with foreland thrusting. These intrusions and related metamorphic zones are thought to be the southern extension of the Omineca crystalline belt of British Columbia. Thrust fault and ductile deformation geometries changed abruptly from central to western Nevada, where major relative transport patterns were to the south-southeast. Structural geometries changed westward into the waning arc of east-central California across the Pine Nut fault, which reversed its sense of tangential displacement to sinistral during Late Jurassic time.

Major plate kinematic patterns changed substantially in Oxfordian to Kimmeridgian time. North America's absolute motion accelerated markedly to rates of ~200 km/m.y. to the north-northwest (May and others, 1989), resulting in a substantial sinistral tangential component of relative motion along the plate edge. This relative plate motion pattern is consistent with the sinistral

displacements related to the Nevadan event, the south-southeast–directed nappe transport pattern of northwest Nevada, and the sinistral sense of shear in the southern transtensional belt. The major displacement patterns depicted in Plate 5F continued into earliest Cretaceous time (Plate 5G) and resulted in a general map pattern similar to that of the Figure 2 palinspastic base.

Plate edge

The North American plate edge in Late Jurassic time is poorly understood. Neither a coherent magmatic arc nor a throughgoing orogenic zone is clearly distinguished. Subduction metamorphic rocks are represented by the Condrey Mountain terrane and the Pickett Peak terrane. In addition, a major fore-arc basin was established along the California Great Valley by Late Jurassic ophiolite formation, although the Upper Jurassic strata constitute synorogenic-type deposits (flysch) rather than arc-derived volcaniclastic strata. One of the outstanding problems concerns the Late Jurassic ophiolites. They may, in general, represent inter-arc basin crust, which decoupled the North American sial from a fringing island-arc system and its subduction zone(s) for ~10 m.y. Alternatively, the ophiolites may have formed solely in a fore-arc setting as a loosely coupled domain within the North American plate edge. Related to this problem is the origin and possible extent of the fringing arc system. It could be represented primarily by the Rogue sequence (2 in Fig. 6) and parts of the southern Coast Range ophiolite (16 in Fig. 6). Moreover, part of Wrangellia may have occupied a fringing arc position outboard of at least the northern part of the ophiolitic basin. Such a relation is strongly suggested along the southern Insular suture belt, where Late Jurassic ophiolite fragments are imbricated inboard of Wrangellia.

Relations along the plate edge are further complicated by the synchronicity of both regional extensional and regional contractile deformations. Major extension characterized both the southern reaches of the plate edge and the Insular suture belt during Nevadan contractile deformation in the central Sierra Nevada to Blue Mountains region. Furthermore, phases of extension are clearly recorded within areas, and within the time interval, that would be considered as classical Nevadan. We interpret these complexities as resulting from the strong tangential components of relative motion along the plate edge in Late Jurassic time.

The Nevadan orogeny. The Nevadan orogeny has long been recognized as a major deformation event of regional scale (Blackwelder, 1914). Modern usage of the term in the California region generally follows Bateman and Clark's (1974) definition as a distinct, short-lived, Late Jurassic deformation event. Note that Bateman and Clark's (1974) definition predated the first published U-Pb zircon ages on any of the California fringing terranes. The view of the Nevadan orogeny as a distinct, short-lived, and dominant event persists (Schweickert and Cowan, 1975; Schweickert and others, 1984; Day and others, 1985), and stands as the basis of the collided exotic-arc model for the western Sierra–Klamath belt. Recent geochronological and structural studies in the southern Sierra Foothills (Tobisch and others, 1989) and in the western Klamath Mountains (Harper and others, 1991) demonstrate that Nevadan deformation extended into Early Cretaceous time. The highly restricted time interval assigned to the Nevada event is further questioned because of relations in the northern Sierra and western Klamaths where, in many instances, clear distinctions cannot be made between Siskiyou and Nevadan deformations. The clearest structural expression of the Nevadan orogeny is tight folding and slaty cleavage development in the upper Oxfordian–lower Kimmeridgian Mariposa and Galice formations of the western Sierra Nevada–Klamath belt. Major Nevadan structures include the Foothills fault system of the western Sierra Nevada (Clark, 1964; Schweickert and Bogen, 1983; Schweickert and others, 1984; Sharp, 1988) and major east-dipping thrust faults that bound the Josephine ophiolite in the western Klamaths (Snoke, 1977; Harper, 1980, 1984).

Nevadan temporal and kinematic relations are well determined for the Josephine ophiolite in the western Klamath Mountains (western Klamath inset, Plate 5F) (Saleeby and others, 1982; Harper and others, 1989, 1991; Pessagno and Blome, 1990; Saleeby, 1990b). Following north-northwest–directed seafloor spreading at ~162 Ma, the ophiolite was transported northward at rates in excess of 100 km/m.y. During this rapid transport, early to middle Oxfordian pelagic sedimentation was abruptly replaced by late Oxfordian–early Kimmeridgian flysch sedimentation of the Galice Formation. During or immediately following flysch deposition, the basin floor was obliquely underthrust to the north-northeast, beneath the Rattlesnake Creek terrane along the Orleans fault. The underthrusting appears to represent the final phases of eastward subduction of mainly Late Jurassic ophiolitic basin floor, older domains of the floor having been subcreted as the Condrey Mountain terrane. Geometry and overprinting patterns in underthrust structures of the Galice indicate that as the Josephine domain of the basin floor converged, its relative motion rotated into east-directed underthrusting (Harper and others, 1991). At this point in time, the outer edge of the basin detached at upper mantle levels and was obducted in a north-northwest direction over the Rogue fringing arc along the Madstone thrust (Harper and others, 1990; Saleeby, this volume). The upper mantle was still hot during its detachment and obduction, as shown by a basal amphibolite facies dynamothermal aureole. The aureole developed by 150 ± 1 Ma, indicating complete destruction of the basin within ~10 m.y. after the ophiolite formed. The change in underthrusting kinematics recorded in the Galice structures appears also to be recorded in the structural geometry and overprint patterns of the Condrey Mountain terrane and within its bounding upper thrust plate (Hill, 1985; Helper, 1986).

The rapid northward transport pattern of the Josephine basin floor, as well as its north-northeast underthrusting, follow directly from the plate kinematic models of Plate 5D and 5E. We consider the late Oxfordian–early Kimmeridgian changes from

oblique north-northwest to normal east-directed underthrusting to be a response to the northward acceleration of the absolute motion of North America (May and others, 1989). The north-northwest obduction of the outer basin floor over its fringing arc is consistent with the general kinematic pattern observed in the analogous fringing arc system to the north, where major sinistral displacements disrupted the Jurassic arc patterns of Wrangellia (Plafker and others, 1989). It is unknown whether the sinistral tangential patterns along the fringing arc system were a response to North America's motion or to tangential components in a converging oceanic plate, conceivably the Farallon plate (Engebretson and others, 1985). The existence and subduction of Late Jurassic ocean floor between the fringing arc and North America permits a wide range in the degree of coupling across the system.

Nevadan deformation resulted in the first structures shared commonly by the Wallowa terrane and the other terranes of the Blue Mountains (Oldow and others, 1984; Avé Lallemant, 1988). North to north-northwest–trending folds and related cleavages affect strata as young as early Oxfordian and occur in all terranes of the region. In conjunction with this deformation, Baker terrane rocks were thrust eastward over the Olds Ferry terrane along the Corner Creek fault and westward over the Wallowa terrane along the Rapid River thrust (Brooks and Vallier, 1978). Major longitudinal sinistral shear zones also affected rocks of the Baker terrane (Avé Lallemant, 1988). Nevadan structures within the Blue Mountains region are cut by a group of mainly Lower Cretaceous postdeformational plutons. The oldest of these is the composite Bald Mountain batholith, in which the oldest phase is 144 Ma (Walker, 1991). The structural relations outlined above indicate that the Wallowa terrane was accreted to the Blue Mountains arc prior to, or during, Nevadan time. The Wallowa terrane may have been introduced into the Blue Mountains region as a fragment of Stikine long before Nevadan deformation, or by sinistral faulting during Nevadan time. Alternatively, the Wallowa terrane could have been introduced into the region as an exotic arc fragment during the Middle Jurassic collision of the Insular superterrane.

The western Sierra Foothills show a complex pattern in Nevadan age shortening, extension, and sinistral strike-slip faulting. East-directed thrusting of the Smartville complex (9 in Fig. 6) occurred in early Nevadan time. The Smartville inset of Plate 5E shows the basal thrust for the complex as cut by the 158 ± 2 Ma Yuba Rivers pluton, both of which are in turn cut by a steep late-Nevadan, west-vergent, reverse-fault zone (Day and others, 1985; Edelman and others, 1989; Saleeby and others, 1989a). These relations constrain Nevadan thrusting of the Smartville complex to between 162 and 158 Ma, which could be considered late Siskiyou or early Nevadan. Additional age and structural relations pertinent to this problem are preserved in the lower American River area, where arc-related peridotitic to dioritic plutons and sheeted dikes of 164 ± 1 Ma age were deformed in the core of a large "Nevadan" anticline while still hot and locally magmatic (Springer, 1980; Saleeby, 1982). Wall rocks for the deformed intrusives consist of Paleozoic ophiolitic melange (Behrman, 1978; Saleeby, 1982) which, in the vicinity of the intrusives, have a dynamothermal contact aureole of ~160 Ma age. The protoclastic deformation and dynamothermal metamorphic fabrics represent deep-level deformation during late Siskiyou or early Nevadan folding. The plutonic-metamorphic core of the fold is flanked by steep sections of Callovian to Oxfordian volcanic rocks overlain by upper Oxfordian–lower Kimmeridgian flysch (10, 11 in Fig. 6). The west limb of the fold may have extended northward as the Smartville upper volcanic unit (9 in Fig. 6). The tectonic relations displayed in the Smartville and lower American River complexes indicate a close temporal sequence between the Callovian extension and "Nevadan" shortening.

East-verging Nevadan age structures also followed extension very closely in time in the northern Sierra terrane. Middle Jurassic volcanic and sedimentary rocks of the Tuttle Lake Formation accumulated in a deep extensional trough (Fisher, 1990). Shortly thereafter, the latest Callovian(?) to Late Jurassic east-vergent Taylorsville thrust system imbricated the northern Sierra terrane during the Nevadan orogeny (McMath, 1966; Harvard, 1988).

In the southern Sierra Foothills, a distinct phase of extension occurred between different phases of Nevadan contractile deformation. Tabular intrusions and sheeted-dike swarms of ~150 Ma age intrude along upright axial-plane cleavages of large Nevadan folds, but are then sequentially deformed by late Nevadan shortening strains (Tobisch and others, 1989; Wolf and Saleeby, 1990). Late-stage ductile shear along the Bear Mountains fault zone in this area may include a sinistral transpressive component of strain as well (Newton, 1986).

Nevadan structures adjacent to the Melones fault zone exhibit a complex pattern in normal shortening strains and sinistral displacement. Tangential components of Nevadan strain are well displayed along, and east of, the Melones fault zone. In the northern Sierra, fragments of lower Mesozoic McCloud arc rocks are offset in a sinistral sense from the eastern Klamath Mountains (Jayko, 1990). The western edge of the offset Klamath rocks show strong sinistral and steep reverse shear indicators adjacent to and within the Melones fault zone (Gefell and others, 1989). Farther south along the Melones fault zone, a major east-west–trending dike swarm of ~157 Ma age is present to the east of the fault within the Calaveras and Shoo Fly complexes (Sonora dike swarm of Sharp, 1980, 1988; Schweickert and Bogen, 1983). The orientation of the dike swarm relative to the fault is consistent with a sinistral transtensive strain pattern. The western margin of the dike swarm and its host rocks were deformed and truncated in Kimmeridgian time by west-vergent reverse faults of the Melones and Sonora fault systems (Schweickert and Bogen, 1983; Sharp, 1988; Sharp and Herzig, 1991). The dikes and their host also contain a spaced cleavage that is broadly parallel to the bounding faults and of probable Nevadan age (Schweickert and others, 1984).

The normal component of Nevadan compressive deformation appears to have changed in character between the western Klamaths and the Sierran Foothills. In the western Klamaths, the

Josephine basin was underthrust eastward, whereas in the Sierran Foothills, a family of local east-vergent and dominantly high-angle west-vergent thrust faults and generally upright regional folds formed. Basement terranes and superposed Middle Jurassic arc rocks that underwent this distributed shortening continue northward into the Klamath Mountains, where they underwent little upper-plate deformation during Nevadan underthrusting of the Josephine basin. Furthermore, the Great Valley fore-arc basin floor survived Nevadan deformation as a distinct bathymetric domain, while the Josephine basin was destroyed by eastward subduction and north-northwest obduction (western Klamath inset, Plate 5F). Late Jurassic ophiolitic basin floor may have been obducted eastward along the axial Great Valley during Nevadan time as well (Suppe, 1979; Saleeby and others, 1986; Blake and others, 1988). Obduction is interpreted in Plate 5F as having nucleated along a basin-floor transform (Saleeby, this volume), along which the northern Coast Range ophiolite was detached and transported over lower Kimmeridgian flysch of Galice or Mariposa affinity (Jayko and Blake, 1986). The Great Valley inset of plate N-F depicts the eastward obduction of the basin floor as having occurred above eastward subduction of more western basin-floor domains, as represented by the Pickett Peak metamorphic terrane.

Eastward obduction of Great Valley basin floor calls for even greater structural contrast between the Sierras and Klamaths. The relations outlined above imply the existence of a sinistral transverse tear system between the northern Sierras and Great Valley and the southwest Klamath Mountains (Harper and others, 1985). This tear clearly preceded, and perhaps controlled, the location of the mid-Cretaceous sinistral fault that dissected the western Sierra–Klamath belt. The transverse tear system is suggested in Plate 5F to have merged southeastward with the northern Foothills fault system, and, in doing so, transported fragments of eastern Klamath rocks into the northern Sierra (after Jakyo, 1990).

Nevadan age shortening may also be recorded along the plate edge in the southern California region. The lower Kimmeridgian Santa Monica Formation contains a slaty cleavage and is typical in structure to, as well as being a protolith of, correlative Nevadan flysch units to the north (Imlay, 1963; Jones and others, 1976). The Santa Monica Formation may have been offset southward from the southern Foothills region (Plate 5F). The Bedford Canyon Formation (18 in Fig. 6) is also cleaved, and, in conjunction with this deformation, it was overturned toward the east and eroded prior to Tithonian time (Buckley, 1975; Todd and others, 1988). Fringing terrane rocks that may have been in a position appropriate for Nevadan deformation are not exposed between the southern California region and the southern Sierra Foothills. Likely representatives of such rocks include the Rand-Pelona-Orocopia schist terrane. Polyphase deformation and metamorphism of Late Cretaceous–early Tertiary age have obscured the possible earlier structural histories of these rocks.

Nevadan magmatism and sedimentation. Arc magmatism was widespread along the plate edge during Late Jurassic time, but did not form a regionally coherent belt as in earlier Mesozoic time. Island-arc sequences formed along the outer fringes of the Late Jurassic ophiolitic basin, while continental magmatism was scattered over a broad region to the east of the basin.

Volcanic rocks of the Rogue fringing arc sequence are as young as 157 ± 1 Ma (Saleeby, 1984) and consist of arc tholeiitic pillow flows and submarine pryroclastic and volcaniclastic rocks of mafic to felsic composition, with a total thickness of >3.5 km (2 in Fig. 6; Garcia, 1982; Riley, 1987). Eruption of upper-level pyroclastic rocks ceased with overlap by the lower Kimmeridgian Galice flysch. The Galice lapped across the original boundary between the Rogue arc and the Josephine ophiolite, presumably a boundary transform or rift edge. A lower interval of volcaniclastic sedimentary rocks in the Galice, above the Josephine ophiolite, probably reflects distal Rogue volcanism. Both the Rogue and Josephine sequences were cut by arc-related dacitic to andesitic dikes as late as ~150 Ma (Dick, 1973; Saleeby and others, 1982; Saleeby, 1984; Harper and others, 1991). Furthermore, upper-plate rocks of the Rattlesnake Creek terrane immediately to the east of the Josephine ophiolite were intruded by a 153 to 149 Ma arc-related peridotitic to dioritic intrusive complex (Snoke and others, 1981; Saleeby and others, 1982). Additional plutons crosscut Nevadan structures within, and bounding, the Josephine ophiolite between ~150 and ~145 Ma (western Klamath inset, Plate 5F) (Harper and others, 1991). These plutons contain evidence of late Nevadan deformation but are not tectonically transported. The products of arc magmatism were thus intruded into the actively collapsing Josephine basin, probably reflecting an east-dipping subduction zone that lay outboard of the Rogue fringing arc during and after the Nevadan orogeny.

Basement relations varied considerably along the fringing arc system. The Rogue Formation was built on a structurally complex substrate that resembles a number of the lower Mesozoic fringing terranes to the east and south (Roure and DeWever, 1983; Saleeby, 1990b) (4, 6, 7, 10, 14 in Fig. 6). The Rogue basement is also cut by Middle and Late Jurassic arc plutonic complexes (Dick, 1973; Garcia, 1982; Saleeby, 1984, 1990b). In contrast, upper-level (Oxfordian to lower Tithonian) arc submarine pyroclastic and volcaniclastic strata of the Diablo Range segment of the Coast Range ophiolite appear to have been constructed on juvenile ophiolitic crust and mantle created during Callovian-Oxfordian spreading (Bailey and others, 1970; Evarts, 1977; Hopson and others, 1981; Sharp and Evarts, 1982). Also, as noted above, Upper Jurassic arc intrusives are widespread, although not voluminous, as late magmatic elements throughout much of the Coast Range ophiolite. These presumably reflect magmatism related to east-dipping subduction beneath the Late Jurassic basin. Within the Santa Barbara Channel Islands, western Transverse Ranges, and southern California borderland region, the relative volume of such igneous rocks predominates over more typical ophiolitic components (Sorenson, 1985, 1988), suggesting the growth of a substantial fringing arc segment in that region as well.

Volcanic-poor lithic turbidites of the Nevadan "flysch" were deposited across many of the fringing terranes in Late Jurassic time. Regions typified by Early and Middle Jurassic arc activity in the Klamath Mountains and northern Sierra Nevada appear to have been uplifted and largely nonvolcanic at that time. McCloud arc rocks and previously accreted early Mesozoic subduction complexes and some fringing terranes supplied detritus for the Nevadan flysch. The flysch units consist of the Santa Monica Formation in the western Transverse Ranges, the Mariposa and related slate units of the western Sierra Foothills (9, 10, 11, 12, 14, Fig. 6), the Galice slate of the western Klamaths (2, 3 in Fig. 6), and the basal Upper Jurassic (Knoxville) intervals of the Great Valley sequence (1, 15, 16, 17, Fig. 6). These flysch units and their contrasts with underlying strata record disruption of arc volcanic patterns and ophiolite basin formation, and the rise of a compositionally diverse orogenic source area. Stratigraphic thicknesses of the various units varied greatly, probably as a function of position on the basin floor relative to bathymetric barriers, as well as proximity to source areas and major dispersal systems. Stratigraphic and/or structural thicknesses of ~2 km are typical of Mariposa and Galice exposures, whereas ~6 km of Knoxville lies above the northern Coast Range ophiolite. In contrast, 50 to 100 m Knoxville sections are typical of the southern Oregon and southern California Coast Range ophiolite sections. The Nevadan flysch is characterized by an abundance of chert and argillite detritus with various admixtures of metamorphic and mainly reworked volcanic material (Dickinson and Rich, 1972; Behrman and Parkison, 1978; Saleeby, 1979; Seiders and others, 1979; Harper, 1983; Ingersol, 1983; Bogen, 1984; Miller and Saleeby, 1987). Metamorphic lithic material includes detrital glaucophane, ultramafic material, and metaquartzite. Detrital zircon populations throughout the Galice and from a few locations within the Mariposa slate units are dominated by Middle to Early Proterozoic species with a subordinate early Mesozoic magmatic-arc component (Miller and Saleeby, 1987; Tobisch and others, 1989). The cratonal component is probably reworked from metaclastic units of the McCloud arc basement, earlier Mesozoic fringing terranes, and perhaps the Kings sequence. Radiolaria from chert pebbles extracted from the Knoxville and Galice are Middle to Late Triassic and locally Jurassic in age (Sieders and others, 1979; Harper, 1983). Scattered limestone cobbles were derived from both McCloud and "Tethyan" late Paleozoic source rocks. Thus most of the major components of the western Sierra–Klamath belt are represented as detritus in the Nevadan flysch. Deposition in the western Sierras and Klamaths occurred in late Oxfordian–early Kimmeridgian time and lapped conformably across volcanic-arc and inter-arc basin-floor environments (2, 3, 9, 11, Fig. 6), and unconformably across early Mesozoic fringing terranes (10, 12, 14, Fig. 6). Locally in the northwestern Sierra, the flysch extends back into Callovian time (7 in Fig. 6; Imlay, 1961). Knoxville flysch deposition began in the latest Kimmeridgian across the northern Coast Range ophiolite and in the Tithonian to the south (Jones, 1975; Hopson and others, 1981). Knoxville deposition continued into Early Cretaceous time, when it was replaced by volcaniclastic and plutoniclastic turbidites of the Great Valley fore-arc basin (Dickinson and Rich, 1972; Ingersol, 1983).

Paleocurrent data indicate sediment dispersal from the north, both in the Great Valley (Ingersol, 1983) and in the western Sierra (Bogen, 1984), and from the east in the western Klamaths (Harper, 1984). These relations are consistent with the Klamath source emerging and moving westward, while much of the western Sierra moved relatively eastward and remained basinal. The Smartville complex may have formed a local bathymetric high during its eastward overthrusting, as suggested by a relatively thin, and possibly shallower-water, facies (Creely, 1965; Day and others, 1985). Basal ages of Tithonian, along with the relative thinness of the flysch units above the southern Coast Range ophiolite and the paleocurrent data, suggest that the source area(s) was primarily in the northern California and adjoining areas. There is some evidence to suggest that the Nevadan flysch basin(s) extended northward from the western Klamath region into the precollapse framework of the Insular suture belt, and perhaps southward into the northern Peninsular Ranges.

In the northern Peninsular ranges, synorogenic sedimentation is recorded for Callovian(?) to Kimmeridgian time in the Bedford Canyon Formation (18 in Fig. 6). The Bedford Canyon Formation consists of a ~5 km structural thickness of mainly turbiditic slaty sandstone, argillite, and pebble mudstone, with minor quartzite and conglomerate and lenses of metaigneous rock with fault slivers of serpentinite (Larson, 1948; Buckley, 1975; Todd and others, 1988). Petrographic and isotopic studies indicate a mixed Middle Proterozoic basement and early Mesozoic arc provenance with subordinate metachert and argillite components (Criscione and others, 1978; Gastil and others, 1988; Todd and others, 1988). Olistostromal units contain shallow-water limestone blocks of late Paleozoic(?), Bojocian, and Callovian age (Imlay, 1963, 1964; Todd and others, 1988). Imlay (1964) noted that the Bojocian fauna was a tropical form distinct from similar age fauna in terrranes to the north, suggesting a net northward latitudinal translation. The depositional and tectonic setting and provenance of the Bedford Canyon Formation indicate that it formed in a deep basin adjacent to the southwest margin of Proterozoic sialic crust, which had probably hosted early Mesozoic continental-arc activity. The basin may have formed above or adjacent to lower Mesozoic strata of the Julian and French Valley schist units (Todd and others, 1988), and may also have lapped across juxtaposed oceanic crust. Deposition and deformation may have occurred along a boundary transform along the inner margin of the Late Jurassic ophiolitic basin system (Plate 5E, 5F).

The Bedford Canyon Formation is overlain in apparent unconformity by the Tithonian Santiago Peak volcanic sequence (18 in Fig. 6; Larsen, 1948; Schoellhamer and others, 1981; Balch and others, 1984; Todd and others, 1988). The Santiago Peak volcanics are typically ~1 km in stratigraphic thickness and extend for considerable distance from the Bedford Canyon section southward into Baja California as the main western wall-rock

assemblage of the Cretaceous Peninsular Ranges batholith. The volcanic rocks were deposited primarily in a submarine environment and consist of andesitic to rhyolitic with minor basaltic flows, tuffs, and agglomerates. Volcaniclastic strata of similar composition accumulated in small submarine fans, and locally contain continental epiclastic components. The Santiago Peak volcanics are suggested to have formed as part of a fringing island-arc system (Gastil and others, 1981). As such, they could represent a southerly continuation of the fringing arc system represented to the north by parts of the Coast Range ophiolite. The apparent unconformable relations with the Bedford Canyon Formation, however, suggest that the Santiago Peak arc rocks were erupted in a position more proximal to sialic North America, inboard of Late Jurassic ophiolite basin, unless the Bedford Canyon Formation was deposited in the basin.

Nevadan-age continental magmatism in regions east of the Late Jurassic ophiolitic and flysch basin system has three major manifestations. The first is the Omineca crystalline belt, which appears to have extended southward into the United States from central British Columbia to the central Nevada region. Associated volcanism was sparse; the only known example in the west-central Nevada region consists of intermediate flows and pyroclastic and volcaniclastic rocks of the Telephone Canyon assemblage (6 in Fig. 5). Omineca belt magmatism is discussed below as an important part of foreland tectonics. Second, from the central California region southeastward, the continental arc was disrupted in conjunction with the development of a major transtensional belt, which can be related to the opening of the Gulf of Mexico (Busby-Spera, 1989). The transtensional belt magmatism is also discussed below. Third, the central Sierra region remained magmatically active through Late Jurassic time. The products of this magmatism were similar to the extensional-arc activity that was widespread during earlier Mesozoic time, and attests to at least local continuation of Sierran arc activity through Nevadan time.

Voluminous silicic volcanism is recorded for Late Jurassic time in the east-central Sierra Nevada and White Mountains region (1 in Fig. 4; Fiske and Tobisch, 1978; Tobisch and others, 1986; Hanson and others, 1987). Volcanism occurred primarily under shallow-water conditions without any expression of the Nevadan orogeny. The volume of material erupted may not have been comparable to that erupted during Callovian time; however, the preservation of kilometer-scale thickness sections in several areas suggests copious volcanism. The silicic volcanic sections occur near the northern end of the regional transtensional belts defined by the Independence dike swarm. Accordingly, in Plate 5F the transtensional belt is terminated northward into an arc-related graben depression that ponded the Upper Jurassic ignimbrites.

The southern end of the western Foothills belt was also magmatically active during Nevadan time. At ~150 Ma basaltic and trondhjemitic sheeted-dike swarms and tabular gabbro-diorite-granophyre plutons intruded along early Nevadan structures (Best, 1963; Tobish and others, 1989; Wolf and Saleeby, 1990). These intrusive rocks strongly resemble the Smartville intrusive series to the north, but they postdate the Smartville rocks by ~10 m.y. The rocks described above for the southern Foothills and the east-central Sierra region indicate that magmatism, typical of the Late Triassic to Middle Jurassic continental arc and fore-arc, locally extended into and through Nevadan time.

The Insular suture belt. The Insular suture belt consists of a small array of terranes inboard of the Insular superterrane and within a major thrust system that cuts the inner margin of the superterrane. The major west-vergent thrust system is cut by longitudinal strike-slip faults and syn- to post-tectonic plutons of the Coast batholithic belt (Brandon and others, 1988; Rubin and others, 1990b; Cowan and Bruhn, this volume). This structural configuration was attained in mid-Cretaceous to early Tertiary time, and resulted from the collapse of an elongate basin that formed along the initial suture between the Insular superterrane and the Intermontane superterrane (Plate 5, D through G). The basement framework of this basin included terranes with strong affinities to many of the fringing terranes, subduction complexes, and McCloud arc rocks that are widespread in the southwest Cordillera, as well as the facing margins of the two superterranes. Structural and stratigraphic relations in southeast Alaska indicate that the basin was in a phase of postcollisional extension during Late Jurassic time, and arc volcanism of the Gravina sequence was active within and along the margins of the basin (Saleeby and Rubin, 1990; Rubin and Saleeby, 1991).

The diversity of terranes within the Insular suture belt is displayed in the San Juan Islands–Cascades thrust system, where McCloud arc rocks and Cache Creek–affinity rocks are imbricated with younger assemblages. McCloud arc rocks are represented by the Turtleback terrane within the San Juan Islands, and the Chilliwack terrane of the northwest Cascades (Danner, 1977; Miller, 1987; Brandon and others, 1988). The Turtleback terrane consists only of Paleozoic arc-related rocks, whereas lower Mesozoic arc-related strata occur in association with the Paleozoic arc rocks of the Chilliwack terrane (4 in Fig. 7), and are apparently linked to the terrane by provenance (Brandon and others, 1988). The Mesozoic strata consist of the Upper Triassic to Lower Jurassic Cultus Formation and the Lower to Middle Jurassic Wells Creek and Harrison Lake volcanic strata (4 in Fig. 7). The Cultus Formation consists of basinal argillites with the upward appearance of siltstone and fine-grained sandstone. The volcanic units are predominantly andesitic to dacitic submarine lava and pyroclastic flows (Misch, 1966, 1977). Together, the Paleozoic and lower Mesozoic strata of the Chilliwack terrane appear to constitute a relatively complete, yet dismembered and metamorphosed, fragment of the Paleozoic through Middle Jurassic McCloud arc. Additionally, a thrust sheet of Upper Triassic–Lower Jurassic arc volcanic rocks occurs as the Hardo Formation in the San Juan Islands. This could represent a fragment of either McCloud arc rocks or an exotic arc system that arrived as part of the Insular superterrane.

The uncertain relations among early Mesozoic arc sequences embedded within the Insular suture belt are brought to focus by

considering the assertion that the Wallowa terrane of the Blue Mountains is a fragment of Wrangellia that was emplaced during Insular superterrane accretion (Davis and others, 1978; Wernicke and Klepacki, 1988). A series of stratigraphic and petrochemical arguments have been mounted to both support and refute this model (Sarewitz, 1983; Mortimer, 1986; McGroder and Umhoefer, 1989; Klepacki and Wernicke, 1989). We have included the Wallowa terrane in our Insular suture belt assemblage because of its probable emplacement into the Blue Mountains region during the initial collision of the Insular superterrane. We have raised the possibility, however, that the Wallowa terrane could represent a fragment of Stikine or some other arc system, as opposed to Wrangellia in the strict sense. The presence of the Cadwallader terrane lying within the suture belt to the north raises the possibility of a Wallowa terrane–age arc system having been situated inboard of the Insular superterrane prior to its collision (Rushmore, 1987). The arc could have constituted the inner edge of the superterrane, having been destroyed for the most part by the numerous superposed deformations that characterized the suture belt.

Cache Creek–affinity rocks of the Insular suture belt occur in the Deadman Bay terrane of the San Juan Islands (2 in Fig. 7) and the Trafton Group in the western Cascade Foothills (3 in Fig. 7; Danner, 1977; Brandon and others, 1988). As in the type Cache Creek terrane of British Columbia, these smaller fragments record ocean-island volcanism and buildups of "Tethyan" fauna carbonate reefs in the late Paleozoic, followed by Triassic to Early Jurassic chert-argillite sedimentation and local submarine volcanism. The Lower Jurassic strata in these Cache Creek affinity terranes of northern Washington also contain a turbiditic epiclastic-volcaniclastic component, in addition to chert argillite. Like the type Cache Creek terrane, these small terranes appear to have been adjacent to, or within, an arc system during early Mesozoic time.

Rocks of the Bridge River terrane occur in the Hozomeen Mountain–Ross Lake area of northern Washington (6 in Fig. 7; Misch, 1966; Tennyson and others, 1982). Triassic through Lower and Middle Jurassic strata of the Bridge River terrane are characterized by chert-argillite, submarine mafic lavas, and epiclastic-volcaniclastic turbidites, with blocks of Permian through Lower Jurassic limestone (Potter, 1983). Davis and others (1978) contrasted this assemblage with the slightly older and "Tethyan" fauna–bearing Cache Creek terrane, and also suggested that such rocks may correlate with younger elements of the North Fork and eastern Hayfork terranes of the Klamath Mountains. The more recent discovery of Early and possibly Middle Jurassic radiolaria chert in the southern Cache Creek terrane (Cordey and others, 1987) has blurred the distinction between Cache Creek and Bridge River rocks (Rushmore and others, 1988). Specifically, the Bridge River terrane may be directly equivalent to the younger western unit of the Cache Creek terrane, both of which record proximity to arc volcanism during deposition, which is in contrast to the older eastern units of the Cache Creek terrane.

Critical elements of the southern Insular suture belt also include fragments of the Late Jurassic ophiolitic basin, like those that are widespread in regions to the south (Whetten and others, 1978, 1980; Miller, 1985; Garver, 1988). The main fragments occur in the Mount Stuart area of the central Cascades (Ingalls complex; 5 in Fig. 7), in the San Juan Islands (Fidalgo complex; 2 in Fig. 7), and in the western Cascade Foothills (Haystack Mountain complex; 2 in Fig. 7). The Middle to Upper Jurassic Ingalls complex represents the most complete ophiolite sequence in northern Washington (Miller, 1985). The section is disrupted by zones of serpentinite melange and high-temperature shear zones believed to have developed on the ocean floor. Ophiolitic debris occurs within the pillow-lava section and these, along with chert and argillite, are crosscut by diabasic dikes. The Ingalls complex resembles the northern segment of the California Coast Range ophiolite (15 in Fig. 6), and has been interpreted as an oceanic transform-fault assemblage (Miller, 1985). The Fidalgo and Haystack Mountain complexes are more similar to the southern California and Oregon Coast Ranges ophiolite (1, 16 in Fig. 6). They consist of polygenetic assemblages created by sea-floor spreading and closely related island-arc volcanism and plutonism during Middle to early-Late Jurassic time (Gusey, 1978; Vance and others, 1980; Whetten and others, 1978, 1980; Brown and others, 1979). These polygenetic complexes are also overlain depositionally by Tithonian flysch, which closely resembles the Knoxville formation of the southern Coast Range ophiolite.

An additional fragment of a Middle(?) to Upper Jurassic oceanic basement complex is exposed as an inlier at Rimrock Lake in south-central Washington (Mattinson, 1972; Miller, 1989). The plutonic rocks consist of disrupted and ductilely-deformed gabbro, diorite, tonalite, and trondhjemite having arc tholeiitic geochemical affinities. The plutonic rocks are tectonically interleaved with an argillite-matrix melange. Sandstone lenses represent fragments of a submarine-fan complex derived primarily from a plagioclase-rich plutonic and chert-bearing source area. Lenses of chert and greenstone are also abundant in the melange. Radiolaria as old as Tithonian have been recovered from chert lenses and geochemical data on the greenstones have tholeiitic affinities, but are distinct from the plutonic assemblage. The Rimrock Lake complex is interpreted here as a transform zone that developed within fringing arc rocks of the Late Jurassic ophiolitic basin.

The Late Jurassic depositional history of northwest Washington is marked by a major change in most of the constituent terranes of the Insular suture belt (Brandon and others, 1988). Most of the major terranes were covered by flysch-type strata that broadly constitute a basinal overlap assemblage. The source for these strata was diverse and included dissected arc(s) and orogenic assemblages. Such strata include the Kyugot Group of southern Wrangellia, the Lummi and Sultan units above the Fidalgo and Haystack Mountain complexes, the Constitution Formation above the Deadman Bay terrane, the Nooksack Group of the Chilliwack terrane, and a number of thrust sheets

that have been detached from their substrates (1, 2, 3, 4, Fig. 7) (Whetten and others, 1978; Brandon and others, 1988; Garver, 1988; Cowan and Bruhn, this volume). Analogous basinal conditions are recorded in the Methow Valley region, to the east, by the Newby Group (7 in Fig. 7). The lower Newby Group probably represents a fore-arc basin sequence with primarily volcaniclastic sedimentary rocks inferred to be Middle to Upper Jurassic (Tennyson and Cole, 1978). The upper Newby Group grades into compositionally more diverse turbiditic strata of mainly Early Cretaceous age.

The Methow Valley sequence constitutes the southern end of an elongate Late Jurassic–Early Cretaceous basinal system which intervened between the Insular and Intermontane superterranes prior to the systems mid-Cretaceous collapse (Monger and others, 1982; Monger, 1984; Gehrels and Saleeby, 1985). Relations in southern British Columbia indicate that the Bridge River terrane and Upper Triassic arc strata of the Cadwallader terrane were overlapped by Callovian-Oxfordian strata of the basin (Rushmore and others, 1988). By inference, the Hozomeen Group of northwestern Washington also was within the basement complex of the basin. In southeastern Alaska, the basin system is represented by the Gravina sequence, which laps across the original suture between the two superterranes (Rubin and Saleeby, 1991). The western facies of the basin in southern British Columbia and northwestern Washington is interpreted as the Upper Jurassic flysch overlap that is widespread in the Cascades, San Juan Islands, and on Wrangellia (1, 2, 3, 4, Fig. 7; Gehrels and Saleeby, 1985). In Plate 5, E through G, the elongate basin is considered to have continued southward as the Late Jurassic ophiolitic basin. The basinal system to the north was at least locally extensional, while the southern ophiolitic segment was undergoing partial collapse during the Nevadan orogeny (Saleeby and Rubin, 1990).

The principal structure of the Insular suture belt in northern Washington consists of the San Juan Islands and western Cascades thrust systems, as well as thrusts within the Methow Valley area (Brandon and others, 1988; McGroder, 1988; Rubin and others, 1990b; Cowan and Bruhn, this volume). Major thrusting is well dated as mid-Cretaceous. Except for the outer and lowermost plate position of Wrangellia and the more inner position of the Methow Valley rocks, the structural stacking arrangement appears chaotic with respect to the regional paleogeographic patterns discussed above for the Triassic and most of the Jurassic. This is evident in the stacking relations between major tectonic and paleogeographic elements such as Cache Creek, McCloud arc, and Late Jurassic ophiolitic terranes (Brandon and others, 1988; Cowan and Bruhn, this volume). We believe this to have resulted from the complex history of longitudinal strike-slip displacements and related terrane-edge fragmentations having worked in conjunction with the main thrusting events. The early displacements have been obscured by thrusting, batholith emplacement, and Late Cretaceous to early Tertiary dextral-slip faulting. Distinct prethrusting strike-slip displacements along the suture belt may have been dextral for Middle Jurassic time and worked in conjunction with major thrust tectonics resulting from the northward oblique collision of the Insular superterrane. In Late Jurassic time, the suture belt became at least locally extensional, and a regional intra-arc or inter-arc basin system developed (Van der Heyden, 1989; Saleeby and Rubin, 1990; Rubin and Saleeby, 1991). The basinal system may have undergone sinistral faulting during this period and continuing into earliest Cretaceous time (Plates 5-F and G). In mid-Cretaceous time, the basin collapsed by west-vergent thick-skinned thrusting that was contemporaneous with voluminous calc-alkaline plutonism (Rubin and others, 1990b).

A hypothetical kinematic pattern for the mid-Cretaceous collapse of the Insular suture belt is presented in Plate 5G. As the Insular superterrane reimpinged on the Intermontane superterrane, the dextral tangential component of collapse was dissipated primarily within and along the back edge of the Intermontane superterrane. Stikine was driven northward within the collapse zone (modified after Wernicke and Klepacki, 1988), which resulted in episodes of sinistral relative motion along the trace of the suture belt (M. G. Miller, 1988; Greig, 1989; Hurlow, 1989; Saleeby and Rubin, 1990), in addition to the dominant dextral displacement pattern. Collapse of the suture belt was virtually completed by the time early Tertiary plutonism and renewed dextral strike-slip faulting began. The mid-Cretaceous collapse of the Insular suture belt occurred in conjunction with major phases of batholith emplacement and related dextral-sense distortion along the continental arc to the south. The displacement pattern depicted in the mid-Cretaceous frame of Plate 5G results in a map pattern that is intermediate between Figure 1 and the Figure 2 palinspastic base. Collapse of the Insular suture belt roughly corresponds in time with major subduction-zone thrusting along the Franciscan complex to the south (Cowan and Bruhn, this volume). The striking contrast between the south and the north is the absence of the Insular superterrane and Stikine, and the regionally coherent presence of the Late Jurassic ophiolitic basin system, which was trapped as fore-arc basement and also partly subducted.

Initial deposits of the Franciscan complex. The Franciscan complex is generally interpreted as a Cretaceous accretionary prism that was disrupted by major strike-slip faults during and after its formation (McLaughlin and others, 1988; Cowan and Bruhn, this volume). However, a number of Franciscan terranes began their depositional histories in the Jurassic. The Pickett Peak terrane and correlative Shuksan metamorphic suite of northwest Washington consist of metamorphosed basalt, argillite, and clastic rocks. Overall protolith characteristics and meager primary age data suggest formation of the protolith assemblage as part of the regional Late Jurassic ophiolitic basin system. The metaclastic and argillitic rocks could be Galice or Knoxville equivalents. Subduction-zone metamorphism of the Pickett Peak began by late Nevadan time ($\geqslant$145 Ma) and culminated at ~125 Ma (Lanphere and others, 1978; Suppe and Armstrong, 1972; McDowell and others, 1984). The map geometry of the Pickett Peak terrane suggests that it was subducted beneath the northern

Coast Range ophiolite (Great Valley inset, Plate 5F) and perhaps partly beneath the Josephine ophiolite and Rogue fringing arc. The structural setting of the Shuksan suite reflects later reimbrication of the metamorphic terrane within the southern Insular suture belt.

The Franciscan Yolla Bolly terrane was subducted beneath the Pickett Peak terrane during mid-Cretaceous time, although deposition of chert over a mafic volcanic substrate began as early as the Tithonian (Blake and others, 1988). Subsequent argillite and graywacke deposition and melange disruption are ascribed by these workers to evolution within a transtensional-transpressional marginal-basin system prior to subduction. This basinal system may have been inherited from the Late Jurassic ophiolitic basin system as well, and is hypothesized as such in Plate 5F and G.

Argillite matrix materials for melange units of the Franciscan central belt resemble Knoxville strata, particularly where the Knoxville is sheared into underlying Coast Range ophiolite or where it is interbedded and mixed with serpentinite protrusions and debris flows. The argillitic matrix materials are as old as Tithonian, like much of the basal Knoxville, and melange blocks of cherty tuff are Tithonian, like similar deposits that are above the southern Coast Range ophiolite (Hopson and others, 1981; Blake and others, 1988). Melange blocks of basaltic greenstone and serpentinite may also represent Coast Range ophiolite fragments, and ~162 Ma high P/T metamorphic blocks are considered to have been closely linked to Coast Range ophiolite tectonics (Blake and others, 1988; Saleeby, this volume). The protolith depositional basin for the Franciscan central belt is also hypothesized here to have been inherited from the Late Jurassic ophiolitic basin system (Plate 5, F and G).

The Franciscan complex of the southern Oregon coast (Fig. 1) contains a small fragment of an Upper Jurassic island-arc sequence (Blake and others, 1985). The Tithonian Otter Point complex consists of disrupted volcaniclastic sandstone and conglomerate, and basaltic to andesitic lava flows and submarine pyroclastic rocks. The depositional site for the Otter Point complex is unknown, although it may have been the outer fringes of the Late Jurassic ophiolitic basin system. Its position outboard of the southern Oregon Coast Range ophiolite and associated displaced Pickett Peak terrane rocks (Colebrook schist) suggest that it has undergone substantial northward displacement from its depositional site (Figs. 1, 2).

The synthesis offered above for many of the older elements of the Franciscan complex proposes that their initial petrogenetic and depositional framework was within and/or inherited from the Late Jurassic ophiolitic basin system. This follows from the broadly extant fragments of the basinal system, as well as from the age, protolith, and spatial relations of the proposed derivative Franciscan rocks. The plate tectonic model of Plate 5, E through G depicts the ophiolitic basin as having been quite large and capable of rendering vast tracts of subcreted metamorphic terranes, trapped fore-arc basin floor(s), trench-slope basins, suprasubduction-zone transform zones, and fringing arc segments. This basinal system is interpreted here as parautochthonous to the Cordilleran plate edge and, therefore, generally above the prevailing east-dipping subduction zone.

Jurassic strata of the Franciscan complex that clearly accumulated in the proto-Pacific basin, remote from the Cordilleran plate edge, are preserved in the more outer belt positions. These consist of Jurassic radiolarian chert sequences as old as the Early Jurassic, which are commonly interleaved with fragments of ocean-floor basalt and serpentinite (Murchy and Jones, 1984; Blome, 1987; Rymer, 1989). Most of these sequences are highly disrupted, although an excellent stratigraphic succession is preserved in the Marin headlands (Fig. 1), where Pliensbachian metalliferous radiolarian chert lies depositionally above ocean-floor pillow basalt (Plate 5G) (Karl, 1984; Murchey, 1984). A relatively complete pelagic depositional history is recorded for late-Early Jurassic through Early Cretaceous time above the basal deposits; an overall northward migration pattern is evident from initial equatorial paleolatitudes and the first influx of continental detritus recorded in the Cenomanian beds. The Marin Headlands terrane may represent a rare remnant of the Farallon plate, where purely pelagic depositional conditions persisted through the major Siskiyou and Nevadan Cordilleran events.

Transtensional disruption of the southern continental arc. During Late Jurassic time, the segment of the continental arc in the southern Sierra to Mojave region occupied a transitional tectonic position between the Nevadan transpressional regime to the north and a regional transtensional regime to the south (Plate 5-F). Elements of the transtensional regime include: (1) the Bisbee and McCoy basins, where rift sedimentation gradually replaced arc volcanism; (2) possible sinistral faults in Sonora, Mexico, and the Sawmill Canyon and Silver Bell–Bisbee shear zones in southern Arizona; and (3) the Independence dike swarm and associated faults in eastern California, as well as dike swarms in southwest Arizona and in Sonora. Domains and/or episodes of contractile deformation were interspersed with transtension during Late Jurassic time along part of the eastern Sierra thrust system, (Dunne and others, 1978; Dunne, 1986), and possibly along the Cronese Hills shear zone in the central Mojave region (Walker and others, 1990b). As in Middle Jurassic time, these continent-verging deformation zones underwent low-magnitude strain and displacement. The distinct southeast vergence of the Cronese Hills shear zone (Walker and others, 1990b) may reflect the sinistral tangential displacement pattern that was prevalent along the plate edge; the displacement worked in close conjunction with the transtensional zone that intersected the region of the shear zone.

The Bisbee basin is the northwest extension of the Chihuahua trough, a structural arm of the Gulf of Mexico depression (Bilodeau and Lindberg, 1983; Dickinson and others, 1986, 1987). The coarse basal units of the 3,000-m-thick Bisbee Group are interpreted as Late Jurassic to Early Cretaceous synrift deposits (Glance Conglomerate, discussed below). The finer-grained upper parts of the Bisbee Group probably accumulated during

Early Cretaceous passive thermotectonic subsidence of the region that had been affected by crustal thinning (Dickinson and others, 1986).

The Bisbee basin is an asymmetrical west-northwest–trending depositional trough with a steep northern edge and a more gentle southern edge (Bilodeau and Lindberg, 1983). Deposition of the 20–2,000-m-thick Glance Conglomerate occurred on piedmont fans during displacements along northwest-southeast- to north-northwest–south-southeast–trending high-angle faults within grabens or half grabens and calderas or other volcanotectonic depressions (Dickinson and others, 1987; Bilodeau, 1979). Clast types in the Glance Conglomerate are restricted at any one locality, reflecting extremely local derivation from Precambrian schist and granite, Paleozoic sedimentary rocks, or Jurassic igneous rocks. Abundant granitic lithic and arkosic detritus indicates unroofing of the Jurassic arc during formation of the basin.

The sparsity of Jurassic supracrustal rocks between southern Arizona and the southeastern California–southwestern Arizona region makes correlations tenuous, but the McCoy basin (Harding and Coney, 1985) has been interpreted as part of the Bisbee basin system (Robinson, 1980; Dickinson and others, 1986, 1987). Published fossil and isotopic ages show that the upper unit of the McCoy Mountains Formation is Late Cretaceous in age, and thus is not Bisbee equivalent (Stone and others, 1987; Tosdal and others, 1989), but new isotopic ages on volcanic rocks that interfinger with the lower unit of the McCoy Mountains Formation support correlation with the Bisbee basin fill (Busby-Spera and others, 1991a, 1991b). Ignimbrites of the Dome Rock sequence in southeastern California conformably underlie the McCoy Mountains Formation in gradational contact, and range in age from 171 ± 4 to 158 ± 2 Ma (6 in Fig. 4; Busby-Spera and others, 1991a). Similarly, ignimbrites of the Canelo Hills volcanic rocks that underlie and are interstratified with Glance Conglomerate range from 175 to 151 Ma (7 in Fig. 4). Other probable Upper Jurassic volcanic rocks interstratified with Glance Conglomerate include the Temporal and Bathtub formations in the Santa Rita Mountains (Bilodeau, 1979; Asmeron, 1988). In both the Bisbee and McCoy basins, the Upper Jurassic records a waning of arc volcanism and the initial accumulation of thick clastic sequences.

In contrast to the coeval Bisbee Group, rugged relief and major unroofing of the Jurassic arc are not indicated by the stratigraphy and petrology of the lower McCoy Mountains Formation. The source of the lower McCoy Mountains Formation was predominantly coeval volcanic rocks and subordinate quartz-rich sedimentary rocks (Harding and Coney, 1985), and in place of the coarse basal deposits of the Bisbee Group, the lower McCoy Mountains Formation is dominated by fine-grained sandstone and siltstone. Clearly, subsidence was required during deposition of the lower McCoy Mountains Formation, in order to accumulate up to 3,000 m of fluvial and lacustrine strata. Perhaps the Bisbee Group records relatively active rifting (McKenzie, 1978) attended by thermal doming, while the lower McCoy Mountains Formation records more passive rifting (Steckler and Brink, 1986), with a greater strike-slip component.

Sinistral displacements of up to 800 km have been suggested for the eastern Mojave to northwestern Sonora region during Late Jurassic time (Sonora-Mojave megashear of Silver and Anderson, 1974; Anderson and Schmidt, 1983). Large Mesozoic age displacements through the Mojave region appear to be precluded, however, by broad stratigraphic relations (Stewart and Poole, 1975; Walker and Wardlaw, 1989). For this reason, sinistral offsets of lower Paleozoic miogeoclinal strata and Proterozoic crystalline terranes in the Mojave region probably occurred during the late Paleozoic continental truncation event (Saleeby, 1975; Stone and Stevens, 1988). Nevertheless, a zone of possible sinistral shear is depicted in Plate 5F as part of the Late Jurassic transtensional system in the Sonora Desert region (after Silver and Anderson, 1974; Anderson and Schmidt, 1983). The megashear is suggested to tie into a fore-arc transform system that displaced Santa Monica slate from equivalent strata in the southern Sierra Foothills.

Northwest- to west-northwest–trending high-angle faults that controlled deposition of the Glance Conglomerate in southern Arizona (Plate 5F) were largely reactivated early Mesozoic (to possibly Proterozoic) structures that were again reactivated in Late Cretaceous time, making their movement history difficult to document (Drewes, 1971; Titley, 1976; Bilodeau, 1979; Bilodeau and Lindberg, 1983). Kluth (1983) interpreted conjugate patterns for some of these faults in the Canelo Hills to be the result of Late Jurassic sinistral shear broadly related to the Sonora-Mojave megashear. Bilodeau (1979), however, found no evidence for large-scale strike-slip displacement along a number of other normal faults in southern Arizona that controlled deposition of the Glance Conglomerate.

The Late Jurassic transtensional belt extends into eastern California as the 148 Ma Independence dike swarm, a northwest-trending set of dikes up to 20 km wide that extends for 500 km from the east-central Sierra Nevada southward to the eastern Transverse Ranges (Moore and Hopson, 1961; Chen and Moore, 1982; James, 1989). The dikes range in composition from lamprophyre to granodiorite and occur as single dikes up to 10 m wide or as composite sets. Geochemical data on these dikes in the central Mojave region suggest a rift-related calc-alkaline province (Karish and others, 1987). Many of the Independence dikes show sinistral shear along their margins and some were emplaced into preexisting sinistral normal faults (Moore and Hopson, 1961; Dunne and others, 1978; Karish and others, 1987). Thus, the zone represents northeast-southwest extension and northwest-southeast sinistral shear. Dikes that may be age correlative with the Independence dike swarm are present as far south as Sonora, Mexico (Silver and Anderson, 1983), although a similar mafic-felsic dike swarm in the Trigo Mountains of southwest Arizona is 10 m.y. older (Tosdal and others, 1989), suggesting that the phenomenon is not tightly restricted in age on a plate-margin scale.

The upper Sidewinder Volcanic series in the Ord, Rodman,

and Fry mountains of the central Mojave Desert (5 in Fig. 4) represents the only demonstrated eruptive equivalents of the Independence dike swarm (Karish and others, 1986; Schermer and Busby-Spera, 1989). These include mafic and silicic lava flows, domes, tuffs, and vent breccia, and hypabyssal intrusive complexes that both crosscut and are crosscut by Independence dikes (Dibblee, 1964a, 1964b; Schermer and Busby-Spera, unpublished data). Distinctive poikiotitic hornblende gabbros in several other central Mojave ranges that have a U-Pb zircon age of 148 Ma are the only known plutonic equivalents of the Independence dike swarm (Walker and others, 1990a, 1990b), except perhaps a hornblende rich gabbro-diorite complex in the Kennedy Meadows area of the southern Sierra Nevada (Dunne and others, 1991).

The northernmost extent of the Independence dike swarm corresponds roughly to the region where abundant silicic ignimbrites of Late Jurassic to Early Cretaceous age continued to accumulate in the last vestiges of the east-central Sierra early Mesozoic arc graben system (1 in Fig. 4; Tobisch and others, 1986; Hanson and others, 1987). Accordingly, the transtensional belt is terminated to the northwest in an intra-arc depression (Plate 5F). To the north and west, Nevadan compressional deformation prevailed. In addition, the Late Jurassic sinistral Pine Nut fault extends to the north of the east-central Sierra, but cannot be traced to the south; it is conceivable that it tied into the transtensional zone.

A number of structural and stratigraphic relations along the southeastern Sierra Nevada and Mojave regions indicate that Late Jurassic contractile deformation was not as intense or widespread as has commonly been assumed (Dunne, 1986; Schermer and Busby-Spera, 1989; Stone and Pelka, 1989; Busby-Spera and others, 1991b). In contrast, the region appears to have been dominated by regional sinistral transtension with local contractile deformation during Late Jurassic time. A number of the contractile deformation features interpreted in the past as Late Jurassic or even early Jurassic in age are now known to be Cretaceous.

Plate interior

Foreland deformation. Late Jurassic time marked the initiation of a foreland deformation belt from central Nevada latitudes to the Canadian Cordillera. A continuous thrust belt and associated foredeep cannot be delineated for times prior to the mid-Cretaceous, but a number of scattered deformation belts and disruption of the early Mesozoic back-arc basin and craton-edge depositional patterns in the Late Jurassic are taken as the initial rise of the foreland fold and thrust belt. In contrast, there is little to suggest a Late Jurassic initiation for foreland contractile deformation from southern Nevada latitudes southward. In general the foreland deformation belt represents a regional compressional regime that involved large-scale decollements with ramp and flat geometries typical of thin-skinned fold and thrust belts throughout the world. Documented magnitudes of displacement in the foreland thrusts are much greater than those of the earlier Mesozoic ephemeral back-arc thrusts. An additional important aspect of the Late Jurassic foreland deformation belt was the presence of relatively thick-skinned uplifts and zones of deep-seated distributed shear, associated with a broad zone of thermal-magmatic welts. Widespread volcanism apparently did not occur in association with the welts, and the broad distribution of the welts and common association of peraluminous granites suggest anatexis of thickened crust versus a purely magmatic arc origin (Speed and others, 1988).

Three distinct kinematic regimes existed in the Late Jurassic foreland belt. The Luning-Fencemaker regime reflects the response of the early Mesozoic marine basinal province of northwest Nevada to the northward acceleration of North America against the Intermontane-Insular superterrane collisional belt. The Sevier belt represents concentrated normal shortening strain in the plate interior, localized along the hinge line of the Paleozoic miogeocline (Allmendinger and Jordon, 1981; Speed and others, 1988). Between these belts in eastern and central Nevada, a complex array of reverse faults, shear zones, and uplifts records crustal shortening; however, the array formed in conjunction with magmatism and broad heating of the crust (Miller and others, 1988; Speed and others, 1988). This zone corresponds to the hinterland of Armstrong (1968, 1972) and we retain this usage, although Oldow (1984) and Speed and others (1988) asserted that a foreland-wide decollement existed beneath the hinterland, and conceivably partially coupled the Luning-Fencemaker and Sevier belts.

The Luning-Fencemaker thrust system consists of a series of south-southeast–vergent fold and thrust sheets with a southerly displacement of several hundred kilometers (Oldow, 1984). The allochthons within the thrust belt consist primarily of lower Mesozoic marine basinal strata that were transported over shelf strata. Molasse sedimentation of the King Lear Formation continued in a piggyback basin setting during transport of the Black Rock desert rocks (11 in Fig. 5; Maher, 1989). Major displacement along the Luning-Fencemaker system lasted well into the Early Cretaceous, as shown by well-determined paleolatitudinal anomalies in the Upper Jurassic–lowermost Cretaceous King Lear strata and in the underlying Happy Creek volcanic complex (Maher and Kirschvink, 1988). Callovian-age west-vergent thrusts in the Black Rock desert rocks may have been partly coincident with early Luning-Fencemaker thrusting, but King Lear depositional overlap relations indicate only limited temporal overlap, if any.

The Luning-Fencemaker thrust system was decoupled from the region of the continental arc by the Pine Nut fault, which, during the Late Jurassic–Early Cretaceous time, acted as a sinistral transpression zone (Oldow, 1984; Oldow and others, 1984). The magnitude of displacement along the fault is poorly constrained, and is likely to have varied along strike as different structural levels in hanging-wall nappes, and as the footwall rocks of the Luning-Fencemaker system are traversed along strike. The southward extension of the Pine Nut fault may, in part, correspond to sinistral shears that controlled the emplacement locus of

the 148 Ma Independence dike swarm (Dunne and others, 1978). The main kinematic role for the fault during the Late Jurassic appears to have been to decouple the Luning-Fencemaker thrust system from more-normal shortening strain patterns in the Sierra Nevada and Klamath Mountains regions.

East of the Luning-Fencemaker thrust system is another major allochthon (Willow Creek) that developed off the west flank of the Toiyabe plutonic-metamorphic welt. This is representative of hinterland deformation. The Toiyabe welt consists of Paleozoic miogeoclinal strata that underwent upper greenschist facies dynamothermal metamorphism, developed in conjunction with the emplacement of Upper Jurassic syntectonic granitic plutons (Speed and others, 1988). To the east, another major zone of dynamothermal metamorphism and plutonism of ~160 Ma age occurs in the Ruby Range and Snake Range areas. Here, penetrative east-dipping cleavage and schistosity surfaces with a west-directed sense of shear developed in conjunction with plutons and their amphibolite facies dynamothermal aureoles in lower Paleozoic strata (Miller and others, 1988).

A number of other studies suggest that significant plutonic activity and related metamorphism occurred over the hinterland between ~155 and 145 Ma; there are intrusions that yield such ages in the House Range, Newfoundland Mountains, Crater Island, Toano Range, and Whitehorse Mountains (Armstrong, 1972; Armstrong and Suppe, 1973; Slack, 1974; Allmendinger and Jordan, 1981; Moore and McKee, 1983; Allmendinger and others, 1984). Upper Jurassic plutons in the Newfoundland Mountains at Gold Hill, northwestern Utah, crosscut major east-directed thrusts. Thus, the minimum age of thrusting in northwestern Utah is Jurassic (Nolan, 1935; Allmendinger and others, 1984). A Late Jurassic dynamothermal metamorphic even in the Albian Range of southern Idaho corresponds in time with low-angle ductile shear and possibly the initiation of thrusting in that region (Armstrong, 1976; Miller and others, 1983; Allmendinger and others, 1984; Snoke and Miller, 1988). Plutonism and dynamothermal metamorphism in the hinterland correspond in time with a general absence of supracrustal rocks: this suggests the regional-scale rise of an orogenic highland with a semicontinuous plutonic-metamorphic infrastructure. In timing, nature, and overall tectonic position, the Late Jurassic hinterland of central to northeast Nevada, and adjacent Utah and Idaho, strongly resembles the Omineca crystalline belt of British Columbia. We hypothesize original continuity between the two in Plate 5F.

The precise timing for the initiation of Sevier thrusting on a regional scale is poorly known, although evidence from scattered locations along the north segment of the belt indicates zones of at least local activity during Late Jurassic time (Allmendinger, this volume). Upper Jurassic foreland-basin strata have not been identified with certainty in west-central Utah, leading to the conclusion that all thrusting in the Utah region is Cretaceous and younger in age (Armstrong, 1968). It should be noted, however, that even though the Paris-Willard thrust is generally considered part of the Idaho-Wyoming thrust belt, it is much closer in age and structural position to the pre-latest Jurassic thrusts of northwestern Utah than it is to the younger and more easterly thrusts of the Idaho-Wyoming belt, which are dated at ~90 Ma and younger (Allmendinger, this volume). Therefore, we speculate (Plate 5F) that the Paris-Willard thrust was active during Late Jurassic time. Along the southern segment of the Sevier thrust belt there is very little direct structural data to suggest Late Jurassic activity, although stratigraphic relations to the east of the belt would permit this.

Foreland basin deposits. The Upper Jurassic sedimentary record of the plate interior records the first evidence of a broad western orogenic highland flanked to the east by a regional foreland basin. Fluvial strata of the Morrison Formation represent the first foreland-basin deposits of regional extent (2, 3 in Fig. 3; Plate 5F). The uppermost Jurassic may record the first evidence of coarse detritus shed eastward from east-directed thrust sheets of the Sevier belt, although direct age data to support this are lacking. Upper Jurassic conglomeratic wedges tapering to the northeast within the Morrison Formation of northern Arizona and northwestern New Mexico may record early Sevier thrusting in the south (Silver, 1948). Correlation of the wedges with foreland thrust structures, however, has not been demonstrated.

The contrast in terrestrial depositional phenomena of the plate interior between the Early and Late Jurassic probably reflects North America's absolute motion pattern, as well as a change from back-arc to foreland basinal settings. Jurassic northward migration of the North American plate resulted in a latitudinal path where the dry trade-wind belt was traversed during the Early and early-Middle Jurassic deposition of the Navajo-Aztec-Nugget dune fields; the more humid, cooler westerly belt was then encountered during the deposition of the Upper Jurassic Morrison Formation (Steiner and Helsley, 1972, 1975; Steiner, 1975, 1978).

EVOLUTION OF THE SOUTHWEST CORDILLERAN MARGIN AND THE MOVEMENT OF LITHOSPHERIC PLATES

The major deformations, accretionary events, and terrane migration patterns of the Cordillera undoubtedly reflect relative plate-motion patterns along the plate edge. Direct data on such motions for the Triassic and Jurassic do not exist. Approximations of North American absolute motions are shown for each time interval in Plate 5, but comparable data on adjacent oceanic plates are lacking for the pre-Cretaceous and are, at best, highly speculative for Cretaceous time; therefore, gross relative-motion patterns for our overall time interval must be reconstructed from geologic, paleomagnetic, and paleobiogeographic data considered together.

The general pattern of eastward subduction of proto-Pacific and probable marginal-basin ocean floor beneath North America is clearly indicated by the persistence of a continental arc throughout most of Triassic and Jurassic time. Resolution of tangential components of motion is more complex. The literature is particularly confusing on such components; numerous argu-

ments have been advanced for both dextral and sinistral regimes (Silver and Anderson, 1974; Davis and others, 1978; Saleeby and others, 1978, 1986; Speed, 1979; Saleeby, 1981, 1983, 1989; Oldow and others, 1984, 1989; Harper and others, 1985; Avé Lallemant and Oldow, 1988).

This synthesis offers a straightforward explanation for the conflicting data sets and models for tangential plate-motion components. From Late Triassic through Cretaceous time, the general pattern of allochthonous terrane migration has been northward along the east Pacific rim. This movement pattern may reflect the dominant pattern of proto-Pacific ocean-floor movement and/or a dominant kinematic pattern within an east Pacific rim marginal-basin–fringing island-arc system. During early Mesozoic time North America also had a substantial northward component in its absolute motion. The resultant tangential component along the North American plate edge is therefore predicted to have changed from dextral to sinistral, and vice versa, as a function of the relative northward components for the oceanic plates versus the North American plate. The sensitivity for change in tangential components along the plate edge was amplified by the northwest trend of the margin, which was subparallel to North America's absolute-motion vector. The sensitivity is clearly exhibited in the distinct phase of sinistral tangential relative motion during Late Jurassic time, when North America appears to have accelerated abruptly northwestward (May and others, 1989).

The Paleozoic suture zones of the southwest Cordillera appear to have played an important role in the manifestation of the tangential components, as well as terrane-accretion patterns, throughout the Mesozoic. The suture between the northeast-trending miogeocline edge and the Antler-Sonoma thrust belts was the buttress for the southeast-directed collapse of the northwestern Nevada marine province during the rapid northwest movement of North America against the confines of the Intermontane-Insular superterrane collisional zone. The Sierran Foothills suture and the related miogeoclinal truncation zone to the east were instrumental in the propagation of tangential displacements along the Mesozoic plate edge. The suture and truncation zone originated as a boundary transform system during late Paleozoic time. Transform tectonics of uncertain displacement patterns are recorded in complex rock assemblages of Triassic to Early Jurassic age within fringing terranes that developed along the older transform locus. Dextral components of motion are suggested in structures of Early and Middle Jurassic age, and an episode of sinistral motion is recorded for Late Jurassic and possibly earliest Cretaceous time. Mid-Cretaceous time was characterized by dextral displacements and, together, the superposed tangential displacements have resulted in the polyphased truncation zone delineated in Figure 1. The locus of east-dipping subduction beneath the southwestern United States was also controlled by the late Paleozoic boundary transform system. Accordingly, subduction accretion and the geometry of plate-edge growth throughout the Mesozoic were controlled by this zone. The unique aspects of the southwestern U.S. plate edge, which stand out when viewing the *Tectonic Map of North America* (King, 1966; Muehlberger and others, in preparation), owe their origin to the late Paleozoic transform system and the subsequent propagation of tangential plate-motion components along the system and through its relatively thin veneer of accreted terranes. In contrast, regions to the north were characterized by the accretion of larger terranes whereupon tangential displacements worked much more intimately with major continental growth episodes.

Manuscript Accepted by the Society March 8, 1991

ACKNOWLEDGMENTS

Thoughtful reviews of this manuscript were provided by Richard A. Schweickert and Robert C. Speed. Saleeby acknowledges partial support from National Science Foundation (NSF) Grants EAR-8904063 and EAR-8803834; Busby-Spera acknowledges NSF Grants EAR-851924 and EAR-8803769. Final drafting of the columnar sections was done by Elizabeth Jones; Figures 1 and 2, and Plate 5 were done by Cherilyn Sunridge-Saleeby.

Printed in U.S.A.

The Geology of North America
Vol. G-3, The Cordilleran Orogen: Conterminous U.S.
The Geological Society of America, 1992

Chapter 5

Late Jurassic to early Late Cretaceous geology of the U.S. Cordillera

Darrel S. Cowan
Department of Geological Sciences, University of Washington, Seattle, Washington 98195
Ronald L. Bruhn
Department of Geology and Geophysics, University of Utah, Salt Lake City, Utah 84112

INTRODUCTION

During the period lasting from about 150 to 80 Ma, two contemporaneous but geographically distinct tectonic regimes dominated the Cordillera (Fig. 1). In the eastern part, the largely thin-skinned Cordilleran fold and thrust belt and a coeval foreland basin developed as Proterozoic, Paleozoic, and lower Mesozoic strata were contracted and emplaced eastward onto the platformal cover of the North American craton. In the western Cordillera, subduction of oceanic crust along the continental margin gave rise to the magmatic arc, forearc-basin deposits, and subduction complex that are preserved respectively in the Sierran batholithic belt, Great Valley Group, and Franciscan Complex. Upper Jurassic, Lower Cretaceous, and lower Upper Cretaceous rocks elsewhere in the western Cordillera, including Washington and southwestern California, reside in large composite terranes that may have been displaced northward with respect to continental North America on the order of 1,000 km or more since about 80 Ma. This hypothesis, supported chiefly by paleomagnetic data, is definitely controversial.

The inception of the subduction-related tectonic elements in the western Cordillera in the Late Jurassic, at about 150 Ma, is a natural time to begin this chapter of Cordilleran history. The earlier chapter closed with the Nevadan orogeny, when Upper Jurassic and older arc-related igneous rocks, ophiolitic complexes, and suites of oceanic basalts and pelagic sedimentary rocks were deformed and accreted along the continental margin, at least in the sector between about 35° and 43°N. 80 Ma also marked a time of change, when magmatism in the same sector suddenly switched off in the Sierra Nevada batholith and jumped to more restricted centers in the Rocky Mountains. The mid–Late Cretaceous also witnessed the waning of thin-skinned deformation in the Cordilleran fold and thrust belt and the inception of Laramide deformation involving contraction and uplift of the Precambrian sialic crust.

We chose to divide this review geographically into sections on the eastern (and central) and western Cordillera, rather than temporally into shorter intervals within the period 150 to 80 Ma. The reason is simply that the broadly defined tectonic regimes within each geographic domain are known to have been active from the Late Jurassic into the Late Cretaceous, and it is not yet possible to date many events more precisely. Within the western Cordillera, we further distinguish the subduction-related Sierran-Franciscan System, which formed essentially in place on the margin of North America, from terranes that include Upper Jurassic–Lower Cretaceous rocks or structures and that might have been significantly displaced. As for authorship, Bruhn wrote the section on the Cordilleran fold and thrust belt; Cowan wrote the sections dealing with the western Cordillera and was responsible for fashioning the final product. The references cited in the text and listed at the end of the chapter are purposely selective rather than exhaustive. They include papers that we used heavily in writing our own review or that we consider to be important, up-to-date contributions containing extensive lists of primary references.

EASTERN AND CENTRAL CORDILLERA

Cordilleran fold and thrust belt

A broad zone of Late Jurassic and Cretaceous contractional deformation, herein referred to as the Cordilleran fold and thrust belt, extends the length of the U.S. Cordillera (Fig. 1; Armstrong, 1968; Burchfiel and Davis, 1975). The northern part of the zone, which is contiguous with the Rocky Mountain fold and thrust belt in Canada, is locally named the Montana thrust belt or disturbed belt. The Idaho-Wyoming belt (Fig. 2) to the south bifurcates near 43°N. The eastern branch, extending through Utah and southern Nevada into southeastern California, is known as the Sevier belt. The western branch in Nevada, comprising the

Cowan, D. S., and Bruhn, R. L., 1992, Late Jurassic to early Late Cretaceous geology of the U.S. Cordillera, *in* Burchfiel, B. C., Lipman, P. W., and Zoback, M. L., eds., The Cordilleran Orogen: Conterminous U.S.: Boulder, Colorado, Geological Society of America, The Geology of North America, v. G-3.

eastern Eureka and western Winnemucca belts, is less certainly dated than the Sevier belt but probably in part contemporaneous (Oldow, 1984; Speed and others, 1988). An extensive foreland basin (Fig. 1) developed east of the Sevier, Idaho-Wyoming, and Montana belts as the craton subsided in response to the load imposed by the eastward-advancing wedge of thrust sheets (Dickinson, 1976). Still unresolved are questions about how the major thrusts root at depth to the west and how thrusting is related to Late Jurassic and Cretaceous plutonism and metamorphism in the hinterland.

Geometry and Displacements. In the Sevier, Idaho-Wyoming, and Montana belts, thrusting and folding chiefly affected the westward-thickening Proterozoic and Paleozoic miogeocline that accumulated along the subsiding, rifted edge of cratonic North America. The geometry of the thrusts and thrust sheets is best illustrated by regional, across-strike sections (Fig. 3), although these are locally difficult to construct with accuracy due to substantial Tertiary extensional faulting. In the eastern part of the belt (Fig. 3A), regional cross sections indicate that deformation was largely east vergent and thin skinned; the basal thrust lay above the crystalline basement. The eastern limit of thrusting from Nevada to Montana roughly corresponds to the "Wasatch" or "hinge" line, along which the thin platformal section to the east begins to thicken dramatically as it enters the miogeocline to the west (Cook and Bally, 1975). Restorable cross sections from the Idaho-Wyoming-Utah overthrust belt indicate that miogeoclinal and shelf strata were shortened so that the present width of the belt is about one-half the original width (Royse and others, 1975).

The basal décollement generally stepped down section toward the west, from Cretaceous rocks in the east to Paleozoic strata and finally to the base of the Precambrian sedimentary section. In several areas, the involvement of crystalline basement in the interior of the orogen has been established. In Washington and western Montana, several major thrust faults are inferred to root deeply into Precambrian crystalline basement in the western part of the thrust belt, in a region of synorogenic plutonic activity (Harrison and others, 1980). Farther south, in the Wasatch Mountains of northern Utah (Fig. 3A), a large culmination cored by Archean and lower Proterozoic gneiss and plutonic rock began to develop by middle Cretaceous and continued to form until at least Maastrichtian time (Royse and others, 1975; Bruhn and Beck, 1981). In southeastern California and southernmost Nevada, where the thrust belt changes from a southwest to southeast trend, thin sheets of Precambrian crystalline rocks were thrust eastward over Paleozoic platform strata (Burchfiel and Davis, 1975).

In the Winnemucca belt, the east- to southeast-vergent Fencemaker and Luning allochthons are made up mostly of Triassic and Jurassic sedimentary rocks that accumulated in a "back-arc" setting, east of the magmatic arc newly initiated along the edge of North America in the Triassic. Major faults beneath the allochthons, like the Fencemaker thrust, probably rooted to the west, within the magmatic arc (Oldow, 1984).

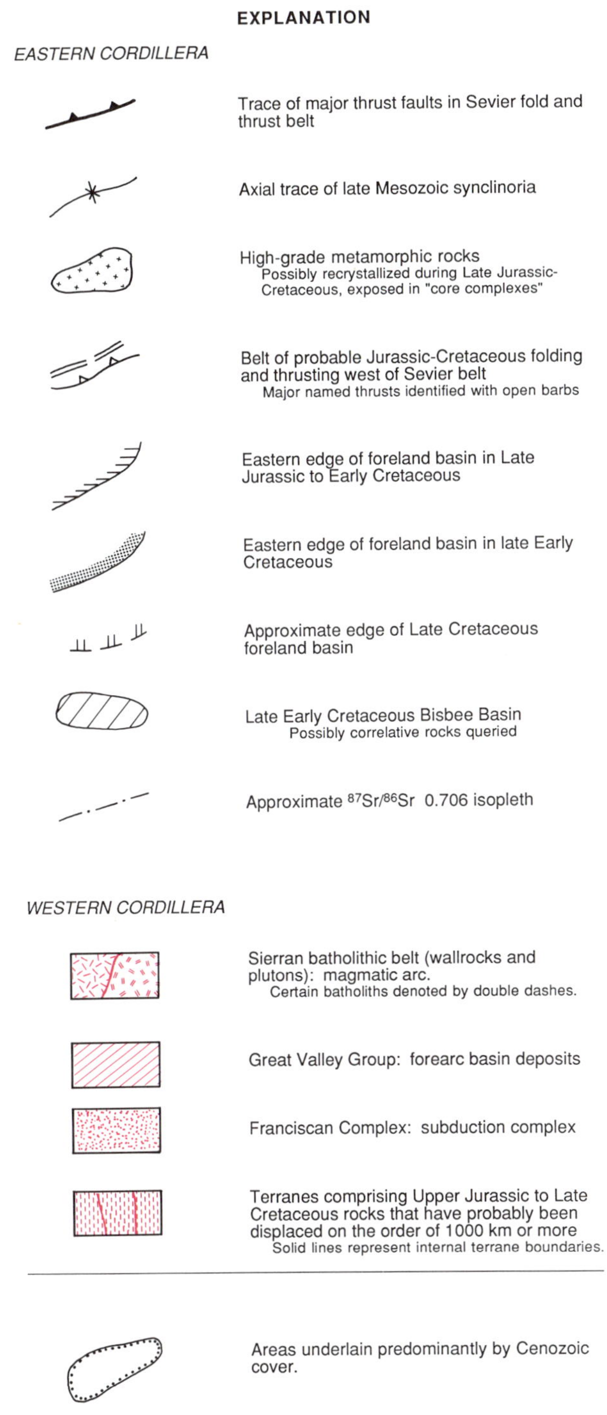

Figure 1. Major Late Jurassic to mid–Late Cretaceous geological features discussed in this chapter. Boundaries of geologic units (e.g., Sierran batholithic belt) are approximate and include large areas of continuous outcrop as well as more isolated or widely separated exposures. Limits of basin deposits are also approximate. Localities and features referred to in text: CRS = Confusion Range synclinorium; F = Fencemaker thrust; GH = Grasshopper thrust; Gu = Gualala; H = Hansel thrust; IB-A = Atlanta lobe of Idaho batholith; IB-B = Bitterroot lobe; L = Luning thrust; ML = Medicine Lodge thrust; SNB = Sierra Nevada batholith; SS = Sublett synclinorium. A-A′ = location of cross section in Figure 3A. B-B′ = location of section in Figure 3B.

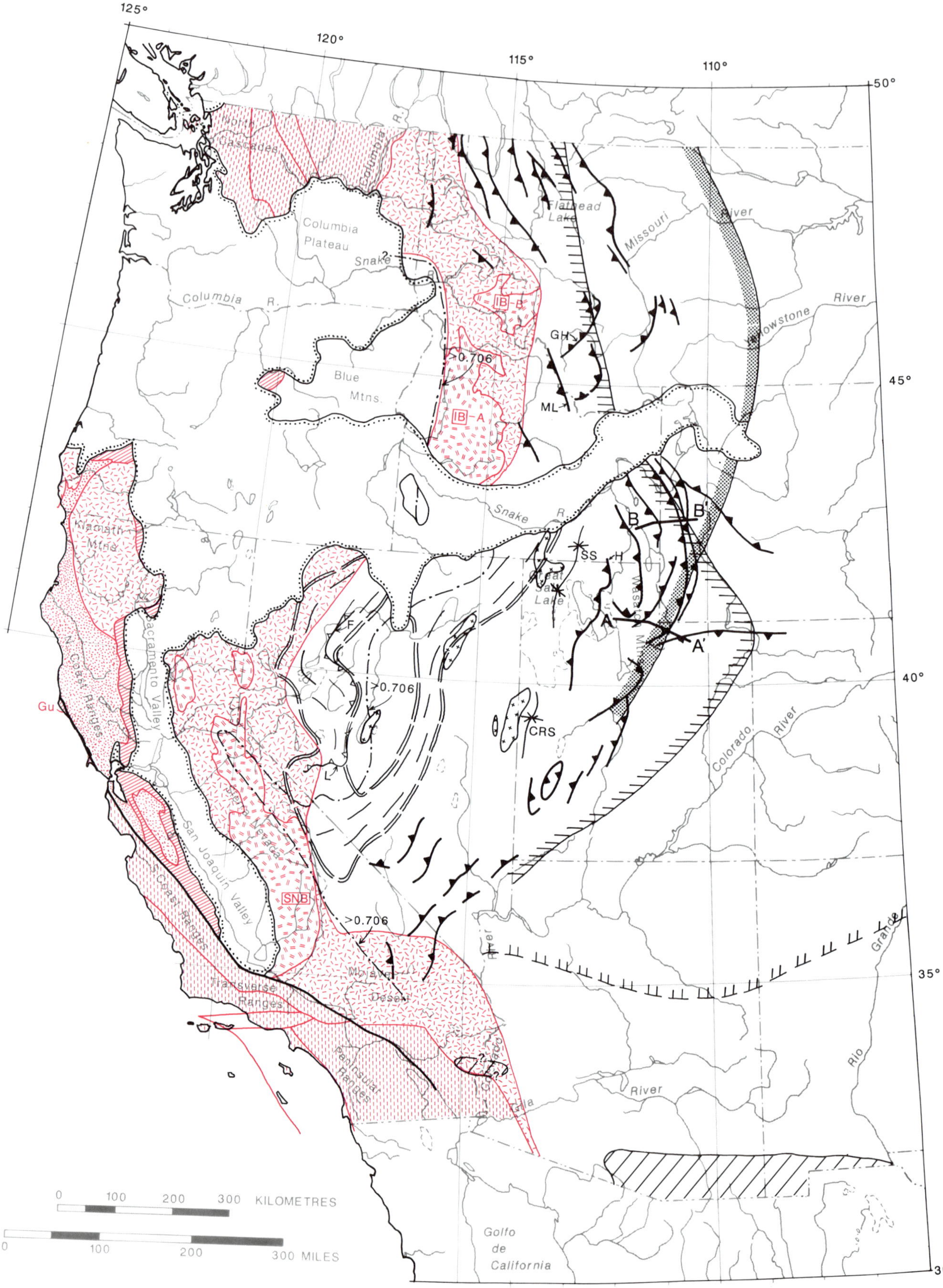

125°
120°
115°
110°
50°
45°
40°
35°
30°
Columbia Plateau
Snake R.
Columbia R.
Blue Mtns.
IB-B
IB-A
>0.706
Flathead Lake
Missouri
River
Yellowstone
River
GH
ML
B
B'
SS
H
Great Salt Lake
A
A'
F
>0.706
CRS
L
Klamath Mtns
Sacramento Valley
Coast Ranges
Gu
San Joaquin Valley
SNB
>0.706
Coast Ranges
Transverse Ranges
Mojave Desert
Peninsular Ranges
Colorado River
River
Rio Grande
Gila
River
Golfo de California
0 100 200 300 KILOMETRES
0 100 200 300 MILES

Timing of thrust movements. The chronology of thrust development in eastern Washington and Montana has been tentatively bracketed between pre–100 Ma to 50 Ma; the older thrusts occur in the western part of the belt. The Grasshopper and Medicine Lodge thrust systems in southwest Montana and east-central Idaho developed prior to 75 Ma, based on K-Ar ages from the Pioneer Batholith, which intrudes the Grasshopper thrust sheet and is interpreted as being post-thrusting. The synorogenic conglomerate and sandstone of the Beaverhead Formation were deposited in Albian time and presumably reflect uplift and erosion in the adjacent Grasshopper and/or Medicine Lodge thrust sheets to the west of the site of deposition. Thrusting of these two major sheets began no later than 100 Ma and was completed by 70 to 75 Ma. The Medicine Lodge plate is situated structurally above the Grasshopper plate and is thought to be the older of the two. Both of these thrust sheets were subsequently carried eastward on structurally lower décollements associated with formation of the younger, frontal fold and thrust zone along the eastern margin of the orogenic belt in Montana and easternmost Idaho (Ruppel and Lopez, 1984).

The chronology of thrust-system development in the Utah-Wyoming part of the orogen is better constrained than in the regions north of the Snake River Plain. The Hansel thrust system, located in the eastern Great Basin of northern Utah and south-central Idaho, developed by Middle Jurassic (Allmendinger and others, 1985). In the major thrust systems in the foreland fold and thrust belt east of the Great Basin, deformation apparently proceeded from the west toward the east, in a classic "stacked-shingle" geometry: the structurally highest and oldest fault systems were developed in the west and were carried in piggy-back fashion on sequentially younger and lower thrusts as a basal décollement propagated eastward. The Paris thrust (140 to 92 Ma), Crawford-Meade thrust system (90 to 47 Ma), and Absaroka thrust (90 to 62 Ma) (Fig. 2) represent the major thrust systems that evolved in this region between Late Jurassic and early Late Cretaceous (Wiltschko and Dorr, 1983). Deformation continued in this part of the thrust belt until the Eocene as the basal décollement continued to propagate east of the Absaroka thrust system (Armstrong and Oriel, 1965; Royse and others, 1975; Dixon, 1982).

Faulting and associated deformation in the thrust belt of southeastern California and southern Nevada occurred over an extended period of time during the Mesozoic, from pre–200 Ma to about 85 Ma (Burchfiel and Davis, 1975). Deformation on the Keystone thrust, the youngest recognized thrust system in the region, occurred prior to intrusion of magmatic rocks at 84 to 95 Ma. The older Contact or Red Springs thrust system was dated as Late Jurassic to Early Cretaceous on the basis of the age of synorogenic conglomerates. The Lavina Wash member in this conglomeratic sequence was deposited at the same time as a volcanic tuff dated at 150 Ma. The Luning and Fencemaker thrust systems in central and western Nevada were probably active from Middle or Late Jurassic through Early Cretaceous, although the notable scarcity of Upper Jurassic and Lower Cretaceous rocks in the Great Basin makes it difficult to date this deformation. In western Nevada, perhaps some of the youngest structures were developed in Albian (103 Ma) volcanic rocks and subsequently intruded by 90-Ma plutons.

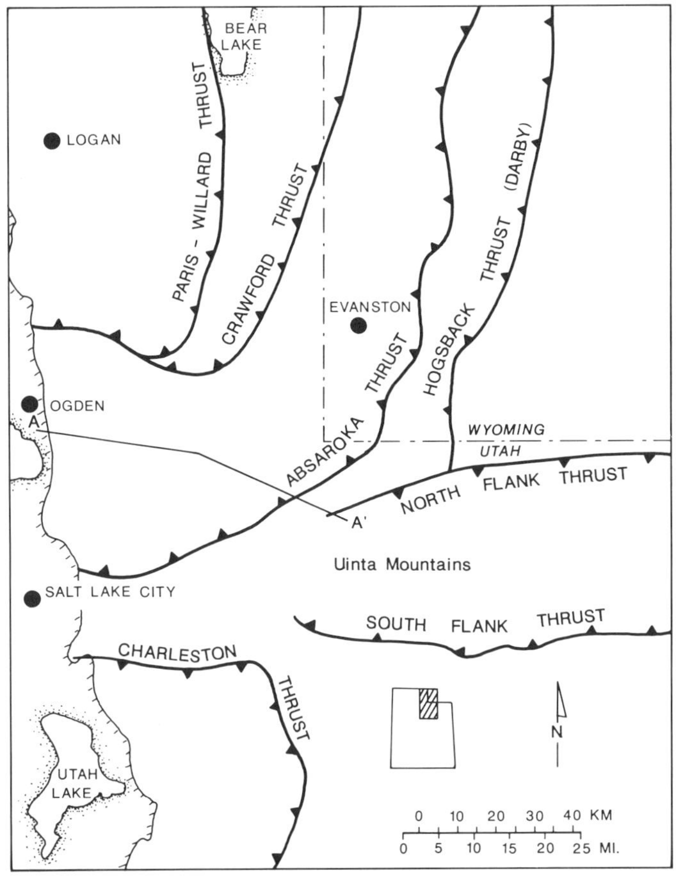

Figure 2. Major late Mesozoic to early Tertiary thrust faults in part of the Idaho-Wyoming and Sevier thrust belts. Shows location of cross section A-A′ in Figure 3A.

Foreland basin. Crustal shortening in the Cordillera during Late Jurassic and Cretaceous time resulted in the development of a large foreland basin (Fig. 1) in which sedimentary detritus eroded from orogenic highlands was deposited between the orogen and the continental interior. The basin began to evolve by Late Jurassic time; it culminated during the Late Cretaceous when it extended from southern Nevada northward into Canada and spread eastward into the western part of the present Great Plains. In cross section the foreland basin was asymmetric, with the deepest, "foredeep" part located adjacent to the up-thrusted mountains of the Sevier orogenic system. Up to 10 km of sedimentary detritus accumulated in parts of the foredeep.

The history of subsidence and deposition in the foreland basin provides critical evidence concerning the timing of deformation in the adjacent orogen and the thermal-mechanical state

of the continental lithosphere. The asymmetric cross-sectional shape of the foreland basin developed in response to elastic flexuring of the lithosphere caused by tectonic and sedimentary loading (Jordan, 1981; Heller and others, 1986). Tectonic loads were caused by crustal shortening and thickening during thrust faulting in the orogen to the west of the basin, while sedimentary loading arose from deposition of detritus in the basin. Mechanical models of basin evolution in Wyoming indicate that the lithosphere had substantial elastic rigidity, which caused subsidence to be distributed over a broad region, far to the east of the loci of crustal thrusting and sedimentary accumulation. Such broadly distributed flexuring is a common feature of foreland basins, where the lithosphere has a substantial elastic thickness because of low to moderate thermal gradients.

The initiation of regional thrusting in the Cordillera is perhaps best recorded in the sedimentary deposits of the foreland basin. Keep in mind that large-scale loading of the lithosphere by thrust faulting will cause flexure and subsidence of the lithosphere over a large area, extending several hundred kilometers away from the orogenic front. Thrust faulting in western Nevada and southeastern California was underway during much of the Jurassic; evidence includes relations between thrust faults and plutonic rocks as well as the ages of sedimentary deposits that appear to be either synorogenic or that on-lap previous thrust faults and folds. However, these events did not strongly influence depositional patterns in the epeiric seaway to the east, where shallow marine and subareal deposits, derived largely from the continental interior rather than orogenic regions to the west, accumulated.

The depositional history of the epeiric seaway changed drastically in the latest Jurassic, as nonmarine shale, sandstone, and local bodies of conglomerate spread eastward, marking the initiation of a vast, clastic wedge that spread outward from orogenic highlands in southeastern California and Nevada. The epeiric sea retreated northward into Canada and southward into the Gulf of Mexico as river and swamp deposits spread eastward, forming a large alluvial plain that marked the beginning of a regional foreland basin that evolved spasmodically throughout the Cretaceous. Nonmarine deposits interfingered with marine deposits of the Cretaceous seaway, resulting in a complex intertonguing of sedimentary facies within the foreland basin (Fouch and others, 1983). The eastward advance of thrust faulting during the Cretaceous resulted in destruction of the western edge of the foreland basin as basin deposits were caught up in thrust sheets and transported farther eastward, toward the interior of the basin. The depositional axis of the basin shifted eastward as these thrust sheets encroached from the west. The long, trough-like shape of the basin was disrupted in later Cretaceous time as isolated mountain ranges rose immediately to the east of the Sevier thrust system, within the Rocky Mountain Province. Development of these foreland uplifts continued from Late Cretaceous into Eocene time and resulted in a complex mosaic of alternating ranges and basins within the original foreland trough.

Events in the hinterland. The geometry of Late Jurassic through early Late Cretaceous thrust systems in the hinterland of the thin-skinned fold and thrust belt has proven difficult to establish, partly due to the extreme effects of Tertiary extensional deformation and the thermal effects of Tertiary plutonism and metamorphism (Armstrong, 1982; DeWitt and others, 1984; Miller and others, 1988). There is general agreement that the thrust systems rooted in the west, but the location and structural characteristics of these root zones remain controversial. Another problem concerns the relation of the evolving thrust systems to possibly coeval plutonism and metamorphism in the hinterland. Some of the medium- to high-grade metamorphic rocks, exposed in "core complexes" west of the Sevier, Idaho-Wyoming, and Montana belts (Fig. 1), are suspected to have been under metamorphic conditions during the Late Jurassic and Early Cretaceous, even though recrystallization may have begun in the Middle Jurassic and proceeded intermittently until the rocks were denuded by low-angle Tertiary extensional faulting. Available isotopic ages span this long interval. Although the bulk of the Mesozoic magmatic arc lay to the west of the Cordilleran thrust system—which was, by definition, in a "back-arc" setting—90 to 100 Ma plutons are coextensive with the southwest end of the Sevier belt in southeastern California (Fig. 1; D. Miller and others, this volume). Some isolated plutons in the eastern Great Basin are dated at 110 to 75 Ma; older, Lower Cretaceous granitoids may be present as well (Armstrong and Suppe, 1973; Miller and others, 1988). Rooting of thin-skinned thrust systems into crystalline basement should have resulted in thickening of the crust due to imbricate stacking of basement rocks and thickening by quasi-plastic flow at temperatures above the frictional/quasi-plastic transition. Recent attempts to reconstruct the crustal thickness in the hinterland at the end of thrusting have been based on assumptions concerning the mode and distribution of crustal stretching during the Cenozoic (Coney and Harms, 1984). Best estimates for pre-stretching crustal thicknesses indicate a north-trending crustal welt between 50 and 60 km thick, considerably thicker than the approximately 40-km-thick crust beneath the Rocky Mountains and Colorado Plateau to the east. These thicknesses are comparable with the present crustal thickness in the Sierra Nevada batholith. A major question remains: How did this crustal welt develop?

Early workers in the metamorphic terranes of western Utah and Nevada suggested that the Mesozoic thrust belt was marked by a regional, vertical transition in deformational styles. In this hypothesis the cooler, upper few kilometers of the crust (the suprastructure) shortened by flexural folding and translation on discrete thrust faults, while deeper levels (the infrastructure) contemporaneously thickened and shortened by large-scale quasi-plastic flow in which deformation penetrated the crust (Armstrong, 1968). Two extreme interpretations of foreland-hinterland structural relations were put forward based on the suprastructure-infrastructure hypothesis. In one, the thermal processes and penetrative crustal shortening in the infrastructure were invoked to yield vertical uplift in the hinterland, with concomitant gravity sliding of the suprastructure toward the east, into the region of the foreland thrust system. This hypothesis has now lost favor, as

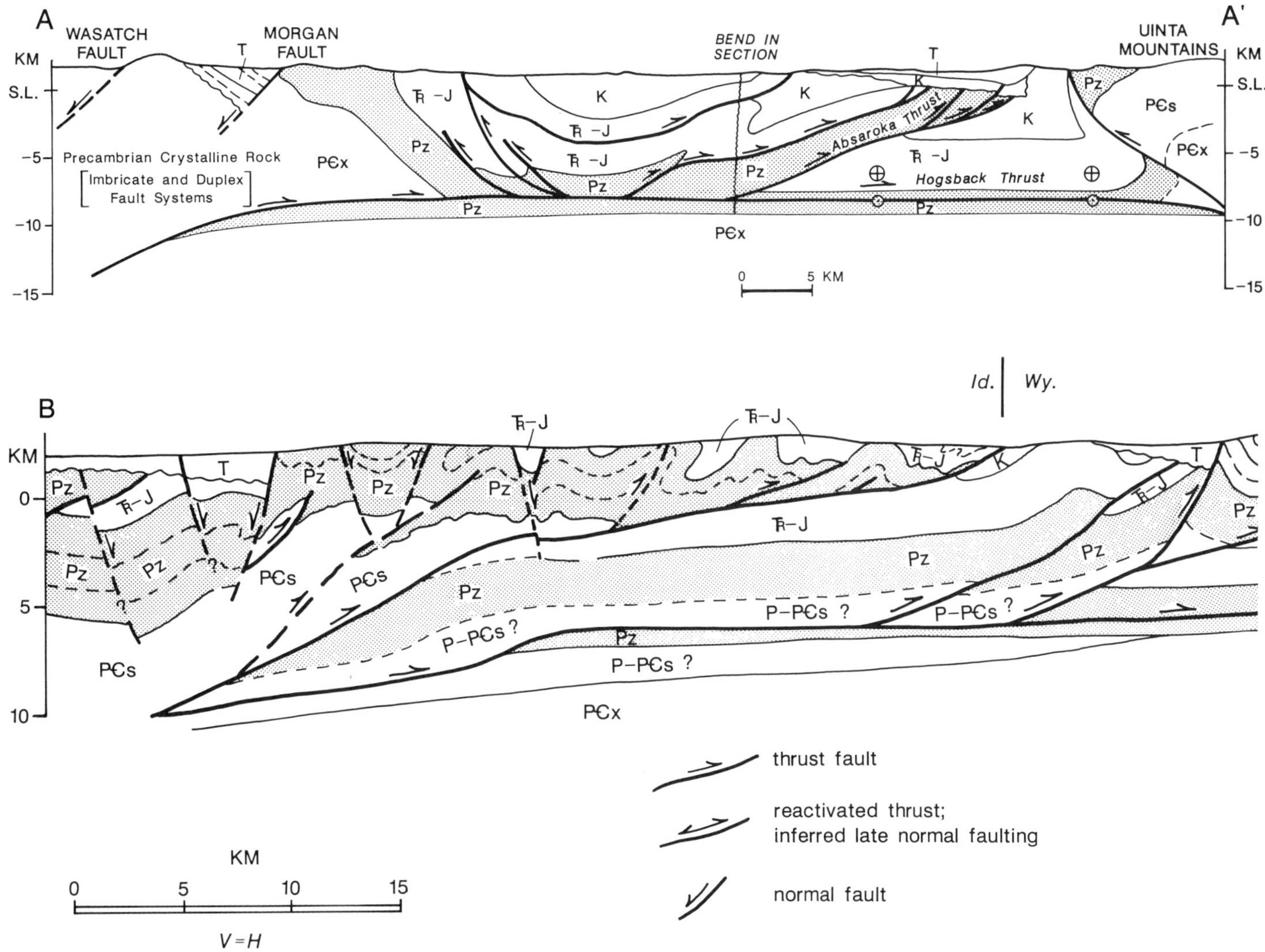

work in other compressional orogens has indicated the relatively limited role of gravitational sliding of thrust sheets. Moreover, there are apparently no large-scale Cretaceous extensional structures in the hinterland of the Sevier Belt.

Another hypothesis is that the metamorphic core complexes (Fig. 1) provide evidence for the locations of the root zones of thrust systems. The key lies in interpreting the regional pattern of metamorphic and deformational events recorded in these complexes (Armstrong, 1982; Miller and others, 1988). In general, a pulse of plutonism, thermal metamorphism, and penetrative flow during the Jurassic was followed by continuing deformation at lower temperatures during the Cretaceous, even as plutonism continued within the metamorphic terranes. This history may reflect an evolutionary record of (1) high temperature flow in the root zones of foreland thrusts followed by continued eastward translation and uplift, and (2) cooling of the metamorphic rocks as they moved eastward and were uplifted on large-scale thrust-fault ramps that cut down-section to the west.

If this latter hypothesis is valid, then one would expect to see large synclinoria of miogeoclinal strata preserved along the eastern margins of the metamorphic core complexes; these synclinoria would have formed as regionally extensive ramp-synclines in front of the uplifted zones of deeper crustal rocks. Indeed, remnants of large synclinoria, with strike lengths in excess of 100 km, are observed immediately east of the Snake Range–Kern Mountain–Deep Creek metamorphic terranes in Nevada and Utah (Confusion Range synclinorium) and Grouse Creek–Raft River–Albion complexes in northern Utah and southeastern Idaho (Sublett synclinorium) (Fig. 1). However, the exact age of these regional folds and their temporal relationships with Mesozoic deformation in the adjacent metamorphic terranes are poorly documented. Farther north, in Idaho and Montana, the foreland thrusts clearly rooted westward in the high-grade metamorphic and plutonic terrane along the eastern edge of the Idaho batholith and metamorphic complexes in eastern Washington (Harrison and others, 1980).

The root zone of the Luning and Fencemaker thrust systems in Nevada (Fig. 1) must have been located beneath the eastern

Figure 3. Cross sections through parts of the Cordilleran fold and thrust belt. 3A (above): Section A-A′ (located on Figs. 1 and 2) extends from the Wasatch Mountains to the Uinta Mountains and illustrates crystalline basement involved in thrusting. 3B (below and facing page): Section B-B′ (see Fig. 1) extends through the Idaho-Wyoming belt; after Dixon (1982).

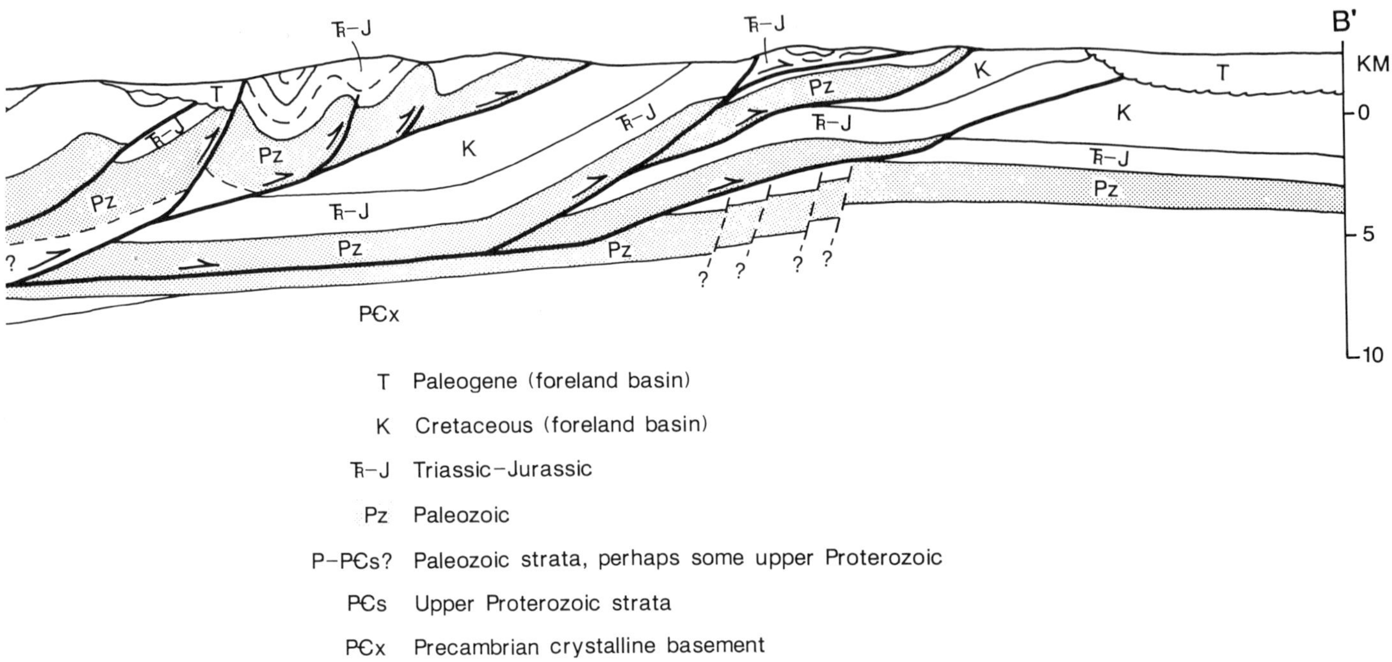

edge of the Sierra Nevada batholith. These thrusts affect Paleozoic and lower Mesozoic marine strata of the western Great Basin and partly disrupt the thrust and fold structures of the Permo-Triassic Sonoma and middle Paleozoic Antler orogenic belts. Studies of the Luning and Fencemaker allochthons indicate that (1) the rocks were deformed by folding and imbricate thrusting above a regional décollement in Paleozoic sedimentary strata, (2) multiple phases of folding occurred under low grade to nonmetamorphic conditions, and (3) thrusting overlapped in space and time with plutonism. The position and geometry of the "root zone" where the Luning and Fencemaker thrust systems penetrated into crystalline basement is unresolved. One possibility is that a simple westward-dipping décollement stepped down-section into the basement. An alternative hypothesis is that the thrusts may have rooted into a steeply dipping sinistral shear zone that penetrated into basement along the eastern edge of the Sierra Nevada batholith during the Late Jurassic and Early Cretaceous (Oldow, 1984).

Bisbee Basin and McCoy Mountains Formation

Upper Jurassic to lower Upper Cretaceous strata of the 3-km-thick Bisbee Group (Dickinson and others, 1986) crop out in southern Arizona and southwestern New Mexico (Fig. 1). Conglomeratic alluvial-fan deposits at the base are interbedded with 151 Ma (Rb-Sr) volcanics. Overlying diverse nonmarine and shallow-marine facies include an Aptian-Albian marine limestone and, locally, shallow marine strata of Albian-Cenomanian age. Rifting and subsidence in the elongate Bisbee Basin and its southeastern extension in Mexico, the Chihuahua trough, were related to the Mesozoic opening of the Gulf of Mexico. The basin can be visualized as an arm extending northwest from the gulf practically into the realm of the Sevier fold and thrust belt. The 7-km-thick, deformed and metamorphosed McCoy Mountains Formation northwest of the Bisbee Group (Fig. 1) includes an undated, lower quartzose to volcaniclastic section overlain by Upper Cretaceous arkoses (D. Miller and others, this volume).

Still unresolved is whether the formation is partly correlative with the Bisbee Group.

WESTERN CORDILLERA: SUBDUCTION-RELATED BELTS ALONG THE CONTINENTAL MARGIN

Most Upper Jurassic to Upper Cretaceous rocks in California, southwestern Oregon, and western Nevada fall into three major lithotectonic assemblages: the Sierra Nevada batholith and related plutonic rocks in the Klamath Mountains and western Nevada; the Great Valley Group; and the Franciscan Complex (Figs. 1 and 4). Soon after the kinematic tenets of plate tectonics were established in the 1960s, geologists interpreted this trinity as the on-land record of a late Mesozoic "Franciscan regime" of plate convergence along the western margin of North America. Hamilton (1969), Ernst (1970), and Dickinson (1970) summarized this model. The Sierran plutonic rocks were viewed as the roots of a magmatic arc; the Great Valley Group as the sedimentary fill of a forearc basin; and the Franciscan as a subduction complex comprising deformed and metamorphosed ocean-floor, trench, and perhaps continental-slope materials. The analogy between these largely coeval, parallel belts and their counterparts at active margins is compelling. The original model of late Mesozoic California as a classic convergent plate boundary endures, but it requires some modifications in light of additional geologic and geophysical data from California and from active convergent plate boundaries worldwide.

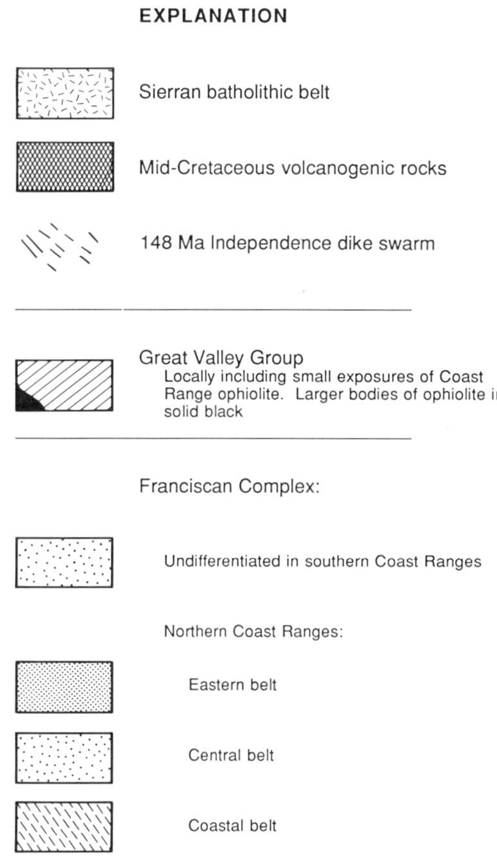

Figure 4. Three major tectonostratigraphic elements in California, SW Oregon, and W Nevada that were related to Late Jurassic to Late Cretaceous subduction along the western margin of North America. The Sierran batholithic belt includes the chiefly Cretaceous Sierra Nevada batholith (SNB) proper and isolated plutons in the northwestern Sierra Nevada, Klamath Mountains, and western Nevada. The Franciscan Complex in the northern Coast Ranges (north of San Francisco Bay) is divided into three belts. Rocks that resemble these three elements but that occur west of the San Andreas fault are depicted in Figure 11 and discussed in the section of the chapter entitled "Probable Displaced Terranes." Other specific localities and features mentioned in the text: C = Calera Limestone (largest body); CFF = Cold Fork fault zone; DR = Diablo Range; ECF = Elder Creek fault zone; H = Hornbrook Formation; MH = Marin Headlands; RR = Ritter Range roof pendant.

The Sierran batholithic belt

A belt of Upper Jurassic to Cretaceous granitoid rocks extends from southern California to northern Idaho (Fig. 1). The "Sierran batholithic belt" as defined here includes only autochthonous plutons that clearly intruded the margin of North America after the Late Jurassic Nevadan orogeny and before about 80 Ma. It comprises the great Sierra Nevada batholith, more widely dispersed plutons in the Mojave Desert, Klamath Mountains, and northwestern Nevada, part of the Idaho batholith, and mid-Cretaceous plutons in northeastern Washington and northern Idaho. Isolated Cretaceous granitoid stocks are also scattered through northern Nevada to the Utah state line, more than 400 km east of the margin of the batholithic belt as depicted on Figure 1. These stocks are probable offshoots of the Sierran belt, but their separation from it has been exaggerated by the Tertiary extension in the Basin and Range province.

Other Cretaceous granitoid rocks in the North Cascades of Washington and the Salinian block and Peninsular Ranges of California occur in composite terranes that probably have undergone postintrusion, latitudinal displacements of a thousand or more kilometers with respect to North America; these rocks are discussed under "Displaced Terranes." Plutonic rocks in the Blue Mountains of east-central Oregon yield Late Jurassic and Early Cretaceous K-Ar ages but are probably pre-Nevadan (155 to 160 Ma) in age.

The Sierra Nevada batholith (Bateman, 1981, 1983; Saleeby, 1986) (Figs. 1, 4, and 5) is an amalgam of individual plutons, comagmatic plutonic suites, and dikes; within the outline of the batholith on Figure 1, granitoid rock greatly predominates over wallrock. Plutons range widely in outcrop area and outcrop shape, although many in the southern half are elongated in a northwest direction parallel to the overall trend of the batholith and structures in the wallrocks. The batholith has probably been displaced en bloc about 200 km westward in response to Cenozoic extension in the Basin and Range province to the east. Paleomagnetic data also permit, but do not require, a few degrees

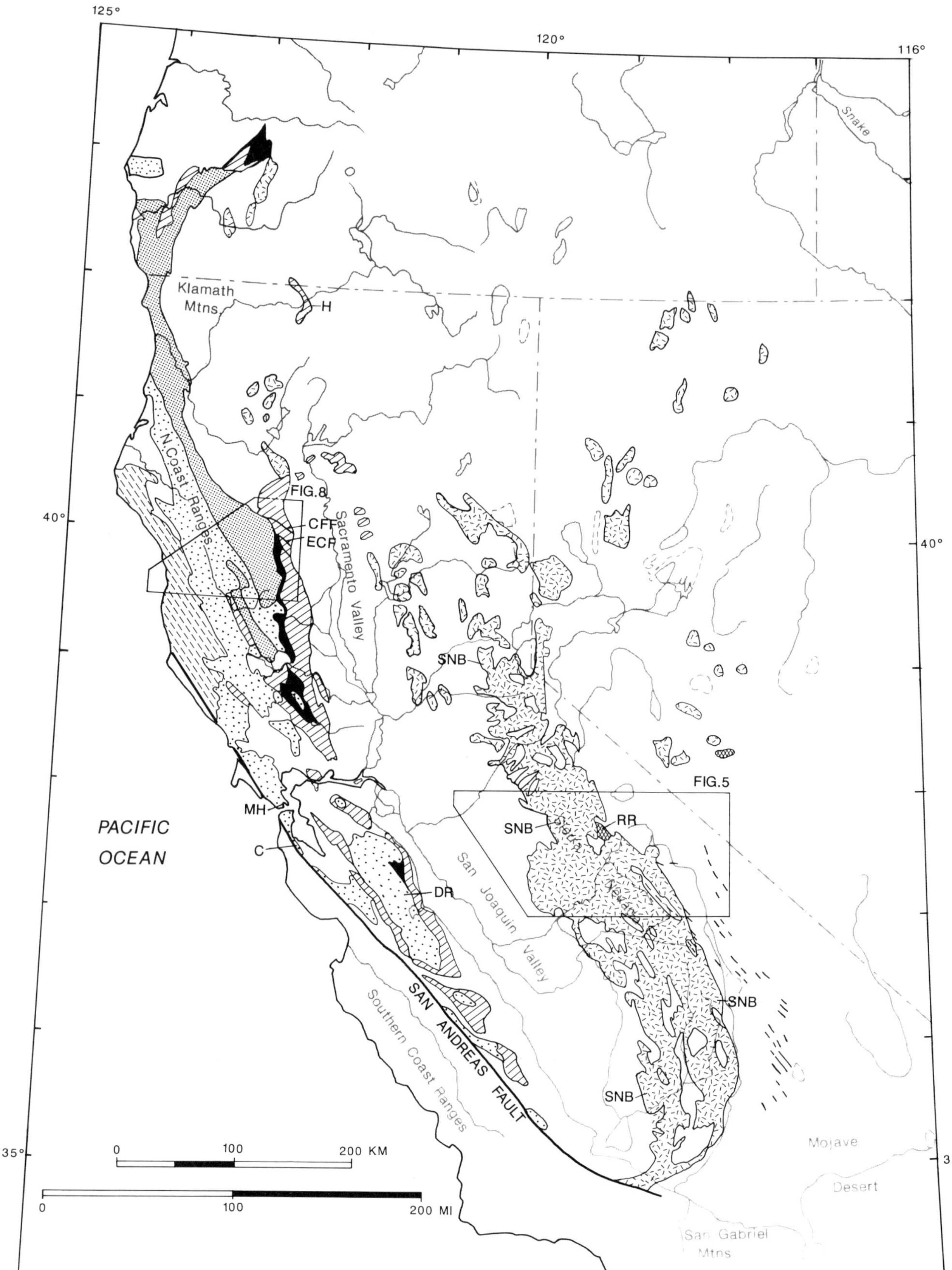
125°
120°
116°
40°
35°
Snake
Klamath
Mtns.
H
N. Coast Ranges
FIG.8
CFF
ECF
Sacramento Valley
SNB
FIG.5
RR
PACIFIC
OCEAN
MH
C
DR
San Joaquin Valley
SAN ANDREAS FAULT
Southern Coast Ranges
0 100 200 KM
0 100 200 MI
Mojave
Desert
San Gabriel
Mtns

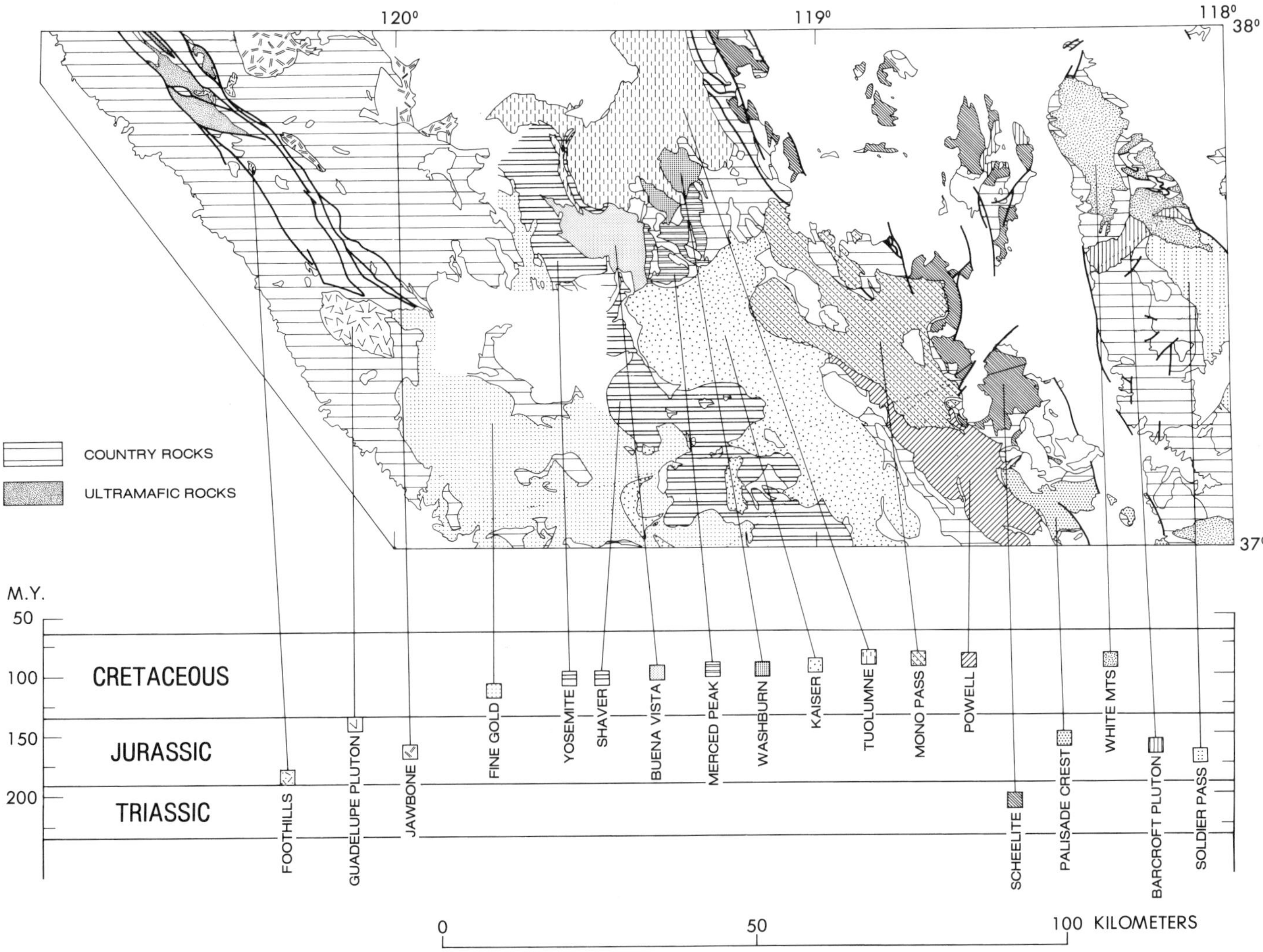

Figure 5. East-west strip across the central Sierra Nevada batholith (location shown in Fig. 4) illustrating comagmatic suites, each consisting of one or more individual plutons. Cretaceous suites are progressively younger toward the east; ages are based on U-Pb and K-Ar data. Older Jurassic and Triassic suites constitute part of the wallrocks of the main Cretaceous part of the batholith. (Slightly modified from Fig. 3 originally published in Bateman, 1983; used by permission of the Geological Society of America and the author).

of northward displacement and clockwise rotation. Cretaceous plutons exposed in isolated ranges in the Mojave Desert (Fig. 1) represent the southeastward continuation of the batholithic belt. Plutons in the Klamath Mountains (Irwin, 1985) are typically equidimensional and smaller in outcrop area than those in the Sierra Nevada; most cover about 100 km^2 or less, whereas the largest composite plutons in the central Sierra exceed 1,000 km^2. In northeastern Washington (Miller and Clark, 1975) and western Nevada (Smith and others, 1971), plutons have diverse shapes; many of those in Nevada and in the Mojave Desert are partly covered by younger deposits or truncated by Cenozoic faults. The 40,000 km^2 Idaho batholith (Hyndman, 1983; Vallier and Brooks, 1987) comprises the Atlanta lobe and Bitterroot lobe (Fig. 1), which are separated by crystalline Precambrian rocks in the Salmon River arch. Each incompletely mapped lobe has been divided into a few large component plutons.

Prebatholithic wallrocks. The Sierra Nevada batholith itself and satellitic plutons to the east in the White-Inyo Mountains intruded a variety of terranes. Lithotectonic units and structures in wallrocks strike generally northwestward. Most of the southeastern part of the batholith was emplaced in Precambrian and Paleozoic strata of the Cordilleran miogeosyncline. The northwestern part intruded a complex amalgam of Paleozoic arc-related terranes and subduction complexes. The batholith

cross-cuts a northwest-trending belt of plutonic and volcanogenic rocks that are remnants of a Triassic-Jurassic continental arc. Upper Jurassic and Lower Cretaceous plutons in the western Sierra foothills cross-cut rock units that were deformed and accreted to the continental margin during the Nevadan orogeny. The westernmost plutons are concealed beneath the Cretaceous and Cenozoic sediments of the Great Valley. Plutons in the Mojave Desert intruded Precambrian crystalline basement and its miogeosynclinal and platformal cover. In the Klamath Mountains of California and Oregon, two belts of Lower Cretaceous plutons occur in a stack of nappes comprising Paleozoic and Mesozoic volcanic, sedimentary, and ultramafic rocks. In northwestern Nevada, Cretaceous granitoid magmas intruded Paleozoic arc-related terranes that were accreted to North America in the early Mesozoic and covered with Triassic-Jurassic volcanic rocks. In central and eastern Nevada, stocks were emplaced into the Upper Proterozoic and Paleozoic shelf section. Most of the Idaho batholith, and plutons in northeastern Washington and northern Idaho, were emplaced into sialic North American basement and its cover of Proterozoic sedimentary rocks. The westernmost part of the Idaho batholith intruded Paleozoic and Mesozoic arc-related terranes lying immediately west of these Precambrian rocks.

Age. Geochronologic data are most abundant from the Sierra Nevada batholith. U-Pb ages, which most closely indicate the age of emplacement, have been obtained mainly from the central part of the batholith. Numerous K-Ar ages from throughout the batholith and from elsewhere in the batholith belt are minimum (cooling) ages, but many Cretaceous K-Ar ages on hornblende approximate the age of crystallization to within 5 m.y. The following summary of age relations is based on U-Pb and K-Ar ages reported by Evernden and Kistler (1970), Smith and others (1971), Armstrong and Suppe (1973), Miller and Clark (1975), Saleeby and Sharp (1980), Stern and others (1981), Chen and Moore (1982), Hyndman (1983), Irwin (1985), and Lund and Snee (1988).

In the Sierra Nevada, there is an age zonation defined by the eastward younging of plutons along any transect normal to the N20°W trend of the batholith (Fig. 5). The 148-Ma northwest-trending, mafic Independence dike swarm occurs mainly east of the southern Sierra Nevada (Fig. 4). Upper Jurassic and Lower Cretaceous plutons ranging in age from about 152 to 154 Ma (the oldest that cross-cut Nevadan structures) to about 130 Ma occur in the western half of the northern Sierra Nevada. A belt of mafic, Lower Cretaceous plutons about 130 to 115 Ma in age lies in the western foothills of the southern Sierra Nevada. The remainder of the batholith consists of large, predominantly granodioritic plutons ranging in age from about 110 to 80 Ma. There is a progressive eastward younging within this northwest-trending belt of plutons, which constitutes the bulk of the batholith. When dates from the entire post-Nevadan part of the batholith are taken into account, no clear episodic or periodic pattern emerges, even though 15 to 30-m.y. gaps in plutonic activity are apparent along certain transects. In addition, there may have been an overall diminution of plutonism between 130 and 150 Ma. There are, however, significant differences in the petrology of plutons in the belts outlined above.

Most K-Ar ages from late Mesozoic granitoid rocks in the Mojave Desert are either reset or are cooling dates, but a Cretaceous suite of plutons was apparently emplaced between about 100 and 75 Ma. Most of the plutons in northwestern Nevada are about 80 to 105 Ma; they represent the sporadically exposed, northern continuation of the predominantly Upper Cretaceous eastern part of the Sierra Nevada batholith. Available isotopic data suggest that Cretaceous plutons in east-central Nevada range from 110 to 75 Ma (Miller and others, 1988), although some may be a few million years older or younger. Lower Cretaceous and possibly uppermost Jurassic plutons in the Klamath Mountains are about 130 to 150 Ma; they resemble plutons of similar age in the northwestern Sierra Nevada. The oldest, tonalitic components of the Idaho batholith were emplaced about 90 to 95 Ma; the bulk of the Atlanta and Bitterroot lobes probably formed about 75 to 85 Ma. On the basis of K-Ar ages, Cretaceous plutons in northeastern Washington and northern Idaho are about 93 to 101 Ma. This part of the mid-Cretaceous batholithic belt continues northward into British Columbia.

Petrology and composition. In the Sierra Nevada and Klamath Mountains, the ages of granitoid rocks generally correspond with their petrology: Lower Cretaceous plutons are on average somewhat more mafic than those of Late Cretaceous age. 130-Ma and older plutons in the northwestern Sierras and the Klamaths are predominantly tonalite; many have foliated margins and are concentrically zoned to cores of either trondhjemite or granodiorite. In the southwestern Sierra Nevada foothills, Lower Cretaceous gabbro, quartz diorite, and tonalite were succeeded by 100-Ma and younger granodiorite. The voluminous Upper Cretaceous granitoid rocks in the Sierra Nevada and northwestern Nevada are principally granodiorite and lesser quartz monzonite and granite. Large, zoned composite plutons in this part of the batholith consist of comagmatic sequences of more mafic rocks at the margins grading to more felsic units toward the cores. Plutons of granodiorite and granite in the Mojave Desert, although less definitively dated, probably become more quartz-rich with decreasing age; some of the youngest Cretaceous(?) plutons that rose through older, sialic crust are two-mica, garnet-bearing granite. A similar pattern is inferred for central and eastern Nevada, where Lower Cretaceous tonalite and granodiorite may have been superceded by Upper Cretaceous peraluminous biotite-muscovite granite. The most distinctive compositional trends recognized to date in the Sierra Nevada batholith are the eastward increase in SiO_2, reflected in the petrology of the granitoid rocks, and an eastward increase in K_2O. An eastward increase in initial $^{87}Sr/^{86}Sr$ ratios from <0.704 in the southwestern foothills to >0.706 in the eastern Sierras reflects a change in pre-batholith wallrocks from volcanic and ophiolitic rocks accreted during the Nevadan orogeny to North American sialic basement and its miogeoclinal cover.

The younger components of the Idaho batholith are also more felsic than its oldest constituents, the 90 to 95-Ma foliated

tonalites along the western margin and a few small, approximately coeval tonalitic stocks elsewhere in the batholith. The tonalites were succeeded by biotite-bearing granodiorite and finally by the muscovite-biotite granite constituting the core of at least the Atlanta lobe. Cretaceous plutons in northeastern Washington and northern Idaho are mostly hornblende-biotite granodiorite and quartz monzonite, but some consist of the foliated, two-mica granitoid rocks that are unusual in the batholithic belt as a whole.

Syn-batholithic volcanism and deformation. The common occurrence of andalusite in contact aureoles in the central and northern Sierra Nevada and northwestern Nevada indicates that most of the Cretaceous plutons presently exposed there were emplaced at depths of less than 10 km. Originally mid-crustal levels of the batholithic complex are probably represented by relatively minor Cretaceous high-grade mafic gneisses in the southernmost Sierra Nevada and mafic Lower Cretaceous plutons in the southwestern foothills. In spite of the shallow-level nature of the batholith, there are few remnants of its original volcanic cover; most was eroded by Late Cretaceous or Eocene time. Roof pendants in the east-central Sierra Nevada and western Nevada (Fig. 4) preserve largely subaerial Lower Cretaceous lava and pyroclastic rocks, including caldera deposits as young as 99 Ma in the Ritter Range in the Sierras (Tobisch and others, 1986). These volcanogenic rocks were penetratively flattened and cleaved and locally folded during a mid-Cretaceous deformational event. East-dipping reverse faults in the western foothills of the Sierra Nevada cut Late Jurassic Nevadan structures and were locally filled with the Early Cretaceous gold-bearing quartz veins of the Mother Lode.

Evidence briefly summarized earlier, in the section on the Cordilleran fold and thrust belt, suggests that mid-Cretaceous deformation and metamorphism affected the hinterland, which had been thermally softened by contemporaneous plutonism. Another important but incompletely understood zone of mid-Cretaceous deformation and syn-kinematic metamorphism lies along the western edge of the Atlanta lobe of the Idaho batholith in west-central Idaho. This zone, comprising both steep and low-angle faults and foliations, is the boundary between Paleozoic and lower Mesozoic arc-related volcanic rocks to the west and Proterozoic metasedimentary rocks to the east that were probably deposited on North American sialic basement. Deformation and metamorphism lasted from about 118 Ma until about 90 Ma, when the early tonalitic phases of the Idaho batholith were intruded and simultaneously deformed along the zone (Lund and Snee, 1988). Initial $^{87}Sr/^{86}Sr$ ratios jump abruptly across the zone from <0.704 to >0.706. The identity of crustal elements that might have been consumed or displaced laterally along the zone and the magnitudes of the displacements themselves are unresolved.

Origin of the batholithic belt. Abundant evidence supports the interpretation that the Sierran belt, as defined in this chapter, formed in a late Mesozoic magmatic arc at a convergent plate boundary. The belt is comparable in its dimensions, form, and continuity only to the huge fields of calc-alkaline volcanic rocks that accumulate on continental margins above descending plates of oceanic lithosphere. The late Cenozoic batholith that must underlie the central Andean volcanic arc is an analog (Hamilton, 1969). In spite of the Cenozoic deformation that has disrupted the original continuity of the elongate, curvilinear Sierran belt, it is an important record of the configuration and extent of the late Mesozoic convergent margin of continental North America. Taken at face value, the dates from the belt show that a subduction zone lay west of at least the Klamath and Sierran plutons from Late Jurassic to late Early Cretaceous time. About 100 to 105 Ma, magmatism expanded northward through western Nevada into northeastern Washington and northern Idaho; apparently, the entire U.S. margin was convergent between about 105 and 90 Ma. At about 80 Ma, magmatism ceased in the Sierran batholith but continued to the north in the Idaho batholith. In the belt as a whole, the Sierra Nevada batholith in California preserves a nearly continuous record of magmatism and hence subduction from about 150 to 80 Ma. During this period, isotopic ages show that the locus of magmatism shifted progressively eastward. In several Cenozoic subduction-related arcs, the volcanic front similarly has migrated away from the trench (i.e., "landward") with time.

The Great Valley Group

From Late Jurassic to Paleogene time, sediments of the Great Valley Group (or Great Valley sequence) accumulated in an elongate basin west of and parallel to the Sierra Nevada batholith in California (Figs. 1 and 4). Although most of the unit is buried beneath the Great Valley, a representative section is exposed along the east sides of both the Sacramento Valley and the San Joaquin Valley, and Upper Cretaceous strata crop out in the northeastern Sacramento Valley. In this chapter, the term Great Valley Group is restricted to terrigenous sandstone, mudstone, and subordinate conglomerate of Tithonian and younger age that depositionally or structurally overlie diverse basement terranes flanking and underlying the Great Valley. Albian to Maastrichtian strata of the Hornbrook Formation along the eastern edge of the Klamath Mountains, middle Cretaceous strata in central Oregon, and disrupted Upper Jurassic and Lower Cretaceous clastic rocks in southwestern Oregon may originally have been contiguous with the Great Valley Group and are included in this section. Other Great Valley–like sedimentary rocks in northwestern Washington and southwestern California are discussed under "Probable displaced terranes."

Basement of the Great Valley Group. The Great Valley Group was deposited on three types of basement. Evidence from wells and from scattered outcrops in the Sierra Nevada foothills shows that beneath the eastern Great Valley an eastward-thinning wedge of Upper Cretaceous Great Valley Group lies unconformably on a basement consisting primarily of the same units exposed in the foothills: mafic, intermediate, and minor felsic plutonic rocks and low-grade metasedimentary and metavolcanic wall-

rocks. At the northern end of the Sacramento Valley, Lower Cretaceous and locally Upper Cretaceous strata rest depositionally on Paleozoic and Mesozoic basement of the Klamath Mountains, as do Albian to Coniacian beds of the Hornbrook Formation farther north. Cretaceous strata in central Oregon unconformably overlie Klamath-like basement. In the Coast Ranges flanking the west side of the valley and in southwestern Oregon, the Great Valley Group is in depositional or structural contact with the Middle and Upper Jurassic Coast Range ophiolite and its cover of sedimentary and volcanogenic strata. The affinity of basement beneath the central Great Valley is unresolved. Many wells drilled over the pronounced positive gravity and magnetic anomalies running the length of the valley have sampled gabbro, so the basement is probably either oceanic crust of the Coast Range ophiolite or the Upper Jurassic–Lower Cretaceous suite of mafic and intermediate granitoid plutons of the western Sierra Nevada batholith.

The relationship of the Great Valley Group and its basement in the Coast Ranges (Bailey and others, 1970; Hopson and others, 1981) is especially significant because it illustrates the profound change in the tectonic setting of California that was precipitated by the Nevadan orogeny. The oldest terrigenous sedimentary rocks of the Great Valley Group are Tithonian at all localities in the Coast Ranges except one, where macrofossils indicate a Kimmeridgian age. In the northern Coast Ranges, the basement of the Tithonian and younger Great Valley Group consists of variably disrupted mafic plutonic and ultramafic rocks of the Coast Range ophiolite, which formed in the Middle Jurassic about 160 Ma (based on isotopic ages; equivalent to Oxfordian). Locally, the Great Valley Group rests on a breccia derived exclusively from the underlying ophiolite. In the Diablo Range (Fig. 4; part of the Southern Coast Ranges), west of the San Joaquin Valley, the crystalline rocks of the ophiolite are overlain by a section of interbedded radiolarian chert, siliceous mudstone, tuffaceous chert and mudstone, tuff, andesite breccia, and local sills of keratophyre. Radiolaria indicate that this volcanogenic section beneath the terrigenous Great Valley Group ranges in age from Kimmeridgian to Tithonian (Late Jurassic).

After the Coast Range ophiolite formed, possibly in settings as diverse as a spreading oceanic ridge, oceanic islands, and intra-arc rift basins (Saleeby and others, this volume), it was disrupted in the north by Kimmeridgian(?) faulting that generated local scarp-related talus breccias. To the south, ophiolitic rocks were overlain by marine tuffaceous sediments and laharic breccias from a nearby volcanic arc active from Kimmeridgian to Tithonian time. Both the disruption and uplift of the ophiolite, and the cessation of proximal volcanic activity, are probably related to the Nevadan orogeny. Viewed in this broad context, the Nevadan event records more than just the accretion of Upper Jurassic ophiolites (Coast Range, Josephine, Smartville) to North America. It also marks the extinction of the Late Jurassic magmatic arc closely related to the ophiolites and the inception of a new forearc basin and magmatic arc represented by the Great Valley Group and Sierra Nevada batholith respectively.

Overall Structure and Thickness. Most of the Great Valley Group appears to lie in a huge northwest-trending asymmetric syncline broadly coincident with the Great Valley itself. The hinge of the syncline is near the west side of the valley; its west limb is a steeply dipping homocline (Fig. 6). In the northern Coast Ranges, a 15-km-thick Upper Jurassic to Upper Cretaceous section dips homoclinally eastward and apparently lacks significant stratigraphic hiatuses or unconformities. In the Diablo Range, a predominantly Upper Cretaceous section is typically about 8 to 10 km thick; locally, it overlies a few hundred meters or less of Tithonian, Valanginian, or Albian strata along a major unconformity. In the eastern limb of the syncline, the top of the Sierran basement dips 5° westward. An exclusively Upper Cretaceous section above the basement thickens westward from 0 to at least 4.5 km, which is the maximum penetrated by wells in the center of the Great Valley.

The great disparity in both thickness and age of the Great Valley Group on the west and east sides of the Sacramento Valley means that our picture of a grand syncline involving all of the Great Valley Group is too simple. Although Upper Cretaceous strata are likely continuous across the fold, and in fact define it (Fig. 6), the thick, east-dipping Upper Jurassic and Lower Cretaceous section in the northern Coast Ranges has no counterpart along the eastern flank of the Great Valley. The structural or stratigraphic complications beneath the Great Valley that eliminate part of the section are unresolved. One possible explanation (Dickinson and Seely, 1979) is that pre–Upper Cretaceous sediments were eroded off of Sierran basement when it was uplifted along an inferred major fault system on the eastern side of the Great Valley in mid-Cretaceous time (Fig. 6A). Another hypothesis, suggested by recent seismic reflection profiles from the San Joaquin Valley (Wentworth and others, 1984), is that the western homoclinal section does not flatten eastward and wrap around a hinge but rather is truncated by gently dipping or subhorizontal faults beneath the valley (Fig. 6C).

The southeast-striking, high-angle Cold Fork and Elder Creek fault zones (Fig. 4) at the northwestern end of the Sacramento Valley provide more direct evidence for mid-Cretaceous deformation of the Great Valley Group. Detailed sedimentological and paleontological studies (Jones and others, 1969; Suchecki, 1984) indicate left-lateral separation of Upper Jurassic and Lower Cretaceous strata across these faults. Although the kinematic history of the faults is incompletely resolved, it is probable that Valanginian and older Great Valley rocks, as well as pre-Cretaceous terranes of the Klamath Mountains, were displaced westward by tens of kilometers relative to rocks to the south. It is not known whether these faults are manifestations of the hypothetical faults beneath the Great Valley mentioned in the preceding paragraph.

Sedimentology and sedimentary petrology. Ingersoll (1978, 1979) and Suchecki (1984) described the sedimentology of the Great Valley Group. Upper Jurassic to lower Upper Cretaceous strata exposed on the west side of the Great Valley accumulated in deep-water (below the carbonate compensation

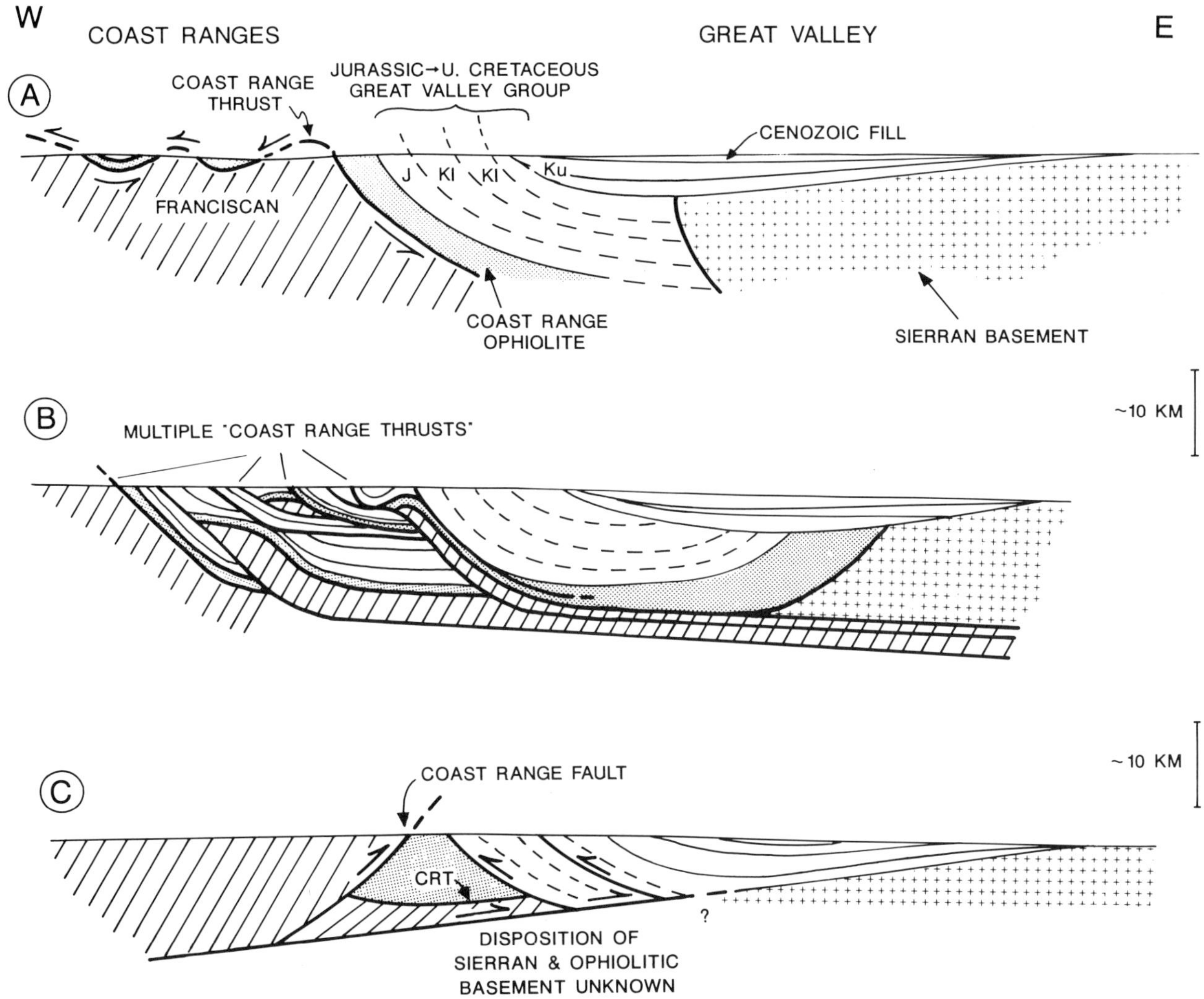

Figure 6. Schematic east-west cross sections extending from the northern Coast Ranges across the Sacramento Valley and into the western foothills of the Sierra Nevada. Each section is simplified from a different published interpretation of how the Franciscan Complex, Great Valley Group, and Sierran basement (batholith and wall rocks) are related at depth. All sections accord with certain evidence at the surface, such as the steep dip of the Coast Range fault; the east-dipping east-facing homocline of Upper Jurassic to Upper Cretaceous Great Valley Group exposed on the west side of the valley, the broadly synclinal form of the Great Valley Group, and the onlap of Upper Cretaceous strata onto Sierran basement. All recognize as well that the Upper Jurassic and Lower Cretaceous section on the west side of the syncline somehow must terminate beneath the valley. A: Modified from Ingersoll (1978) and Dickinson and Seely (1979). Depicts the widely published "conventional" interpretation of a Franciscan subduction complex carried along the Coast Range thrust beneath a forearc basin floored by a remnant of oceanic crust. In addition, mid-Cretaceous uplift of Sierran basement beneath the east side of the valley accounted for the truncation of older Great Valley Group. B: Modified from Suppe (1978). Interprets the southeastern part of the northern Coast Ranges as a thin-skinned west-vergent fold and thrust belt that formed when a late Mesozoic succession of Jurassic-Cretaceous Franciscan and structurally overlying Coast Range ophiolite and Great Valley Group was imbricated during continued Cenozoic subduction. Note the multiple (due to imbrication), locally steepened (due to folding) "Coast Range thrusts." A slab of Coast Range ophiolite, obducted onto Sierran basement, is inferred from a central magnetic high extending nearly the length of the Great Valley. C: Modified from Wentworth and others (1984). Recent reflection seismic data and magnetic data suggest that Sierran basement extends westward beneath the homocline on the west side of the valley and beneath the surface exposure of the Coast Range fault. The Coast Range ophiolite is modeled as part of an eastward-driven tectonic wedge that delaminated the Great Valley Group from its basement. The Franciscan Complex exposed in the Coast Ranges has been thrust eastward on a Coast Range fault that bounds the west side of a "triangle zone." The present geometry shown in C may not preclude an original late Mesozoic situation similar to that depicted in the western half of A; hence, a preserved Coast Range thrust (CRT) at depth.

depth) submarine fans that prograded from the east into the axis of an elongate, north-northwest trending basin. Upper Cretaceous sediments on the east side of the valley were deposited on the shelf. Paleocurrent data indicate that the overall slope of the center of the basin was southward. Nonmarine to deep-marine sediments in the Hornbrook Formation (Nilsen, 1984) were shed eastward from emergent pre-Cretaceous bedrock in the Klamath Mountains.

Upper Jurassic Great Valley strata are dominantly mudstone, locally interbedded with thin sandstones and cut by conglomerate-filled channels. Much of the mudstone-rich Tithonian section consists of slope deposits, but it also includes basin-plain and fan sediments. Most detritus was derived from the east and was deposited by west- and south-flowing currents. Some paleocurrent indicators and slump-folds, however, document eastward paleoslopes that descended from a bathymetric ridge on the western flank of the basin in the Sacramento Valley. This ridge apparently consisted of the ophiolitic basement of the Great Valley Group, for it supplied sporadic serpentine-rich debris-flow deposits and sandstones, containing abundant clasts of basalt and gabbro, that are interbedded with Tithonian mudstone.

Lower Cretaceous sand and mud were deposited primarily in basin-plain and outer-fan environments. The preponderance of south-directed paleocurrent data, together with local detrital serpentinites in the section, suggests that a bathymetric ridge formed the western flank of the basin in the Early Cretaceous as well as the Late Jurassic. By Late Cretaceous (Cenomanian to Santonian) time, the Great Valley basin had widened and submarine-fan sediment-disperal systems had correspondingly enlarged. Diverse fan-related facies associations predominate in the outcrops along the western margin of both the Sacramento and San Joaquin valleys, but strata in the latter setting were deposited nearer the Late Cretaceous shoreline. The Hornbrook Formation and the Cretaceous sections in central Oregon comprise basal shallow-marine conglomerates overlain by deep-marine clastic rocks including submarine-fan turbidites.

The petrology of sandstones in the Great Valley Group is an invaluable guide to its provenance. Many workers (for example, Ingersoll, 1983) have documented consistent, stratigraphically controlled variations in the composition of sand-sized detritus ("petrofacies;" Fig. 7). In general, the proportion of lithic detritus (aphanitic metamorphic, volcanic, and sedimentary grains) decreases stratigraphically upward while the proportion of feldspar increases. The ratio of potassium feldspar to plagioclase increases stratigraphically upward. Although some Upper Jurassic sandstones west of the Sacramento Valley are rich in debris (principally basaltic) derived locally from the underlying Coast Range ophiolite, a more general characteristic of Upper Jurassic and Lower Cretaceous sandstone and conglomerate is their consistently high percentage of chert and argillite clasts compared to stratigraphically higher strata.

These petrologic data record a temporally changing provenance. In Late Jurassic and Early Cretaceous time, deformed and locally metamorphosed supracrustal rocks in the Nevadan orogen north and northeast of the Great Valley basin were an important source of detritus. Uplifted ophiolitic rocks along the western flank of the basin also contributed sediment and local debris flows during this period. In Cretaceous time, a volcanic arc to the east in the Sierra Nevada supplied voluminous volcaniclastic detritus. The increasing proportion of feldspar and monocrystalline quartz in progressively younger Cretaceous sandstones reflects the erosional dissection and unroofing of the plutonic roots of the magmatic arc.

Paleogeography and tectonic setting of the Great Valley Group. In California, south of the Klamath Mountains, the Great Valley Group accumulated in a forearc or outer-arc basin related to the coeval, parallel Sierra Nevada magmatic arc to the east. The basin, which was morphologically and dimensionally similar to the present-day Sunda-Java forearc basin, was floored by Jurassic oceanic crust and flanked on the west by an outer-arc high in at least Late Jurassic to Early Cretaceous time. Although some sediment was supplied from this high and from the basement of the arc, the majority came from the volcanic arc itself and, in the Late Cretaceous, from its plutonic roots. Both the arc and forearc basin formed in response to post-Nevadan (Tithonian and younger) subduction along the western margin of North America. Sedimentation in the Great Valley basin continued into the Paleogene and so outlasted magmatism in the Sierra Nevada proper, which died out about 80 Ma.

At the northern end of the Late Jurassic and earliest Cretaceous Great Valley basin, the basement rocks of the Klamath Mountains formed an emergent promontory and source terrane along the continental margin. Valanginian and younger Cretaceous marine strata lying depositionally on Klamath basement probably record the enlargement and northward extension of the Great Valley basin. Due to intervening Cenozoic cover, it is not known whether middle Cretaceous strata in central Oregon were originally contiguous.

The Franciscan Complex

The Franciscan Complex is a lithologically heterogeneous assemblage of upper Mesozoic and lower Tertiary clastic sedimentary rocks, basalt, chert, and metamorphic rocks that is widely exposed in the Coast Ranges east of the San Andreas fault and west of the Great Valley and Klamath Mountains (Figs. 1 and 4). Fossils and isotopic ages indicate that most of the Franciscan is Late Jurassic (Tithonian) to Paleogene in age and therefore coeval with the Great Valley Group. Another similarity is that marine, terrigenous (land-derived volcaniclastic or arkosic) sedimentary rocks predominate in both the Great Valley Group and the Franciscan. In most other respects, these two "superunits," which are everywhere separated by faults, are utterly different. For example, in the Great Valley Group, volcanic rocks and chert occur only at its base and are exclusively Upper Jurassic; Lower, Middle, and Upper Jurassic and Lower Cretaceous cherts, and probably basalts, occur in the Franciscan. The Franciscan is famous for its blueschist and eclogite; in contrast, only the most deeply buried Great Valley strata contain the zeolite-facies min-

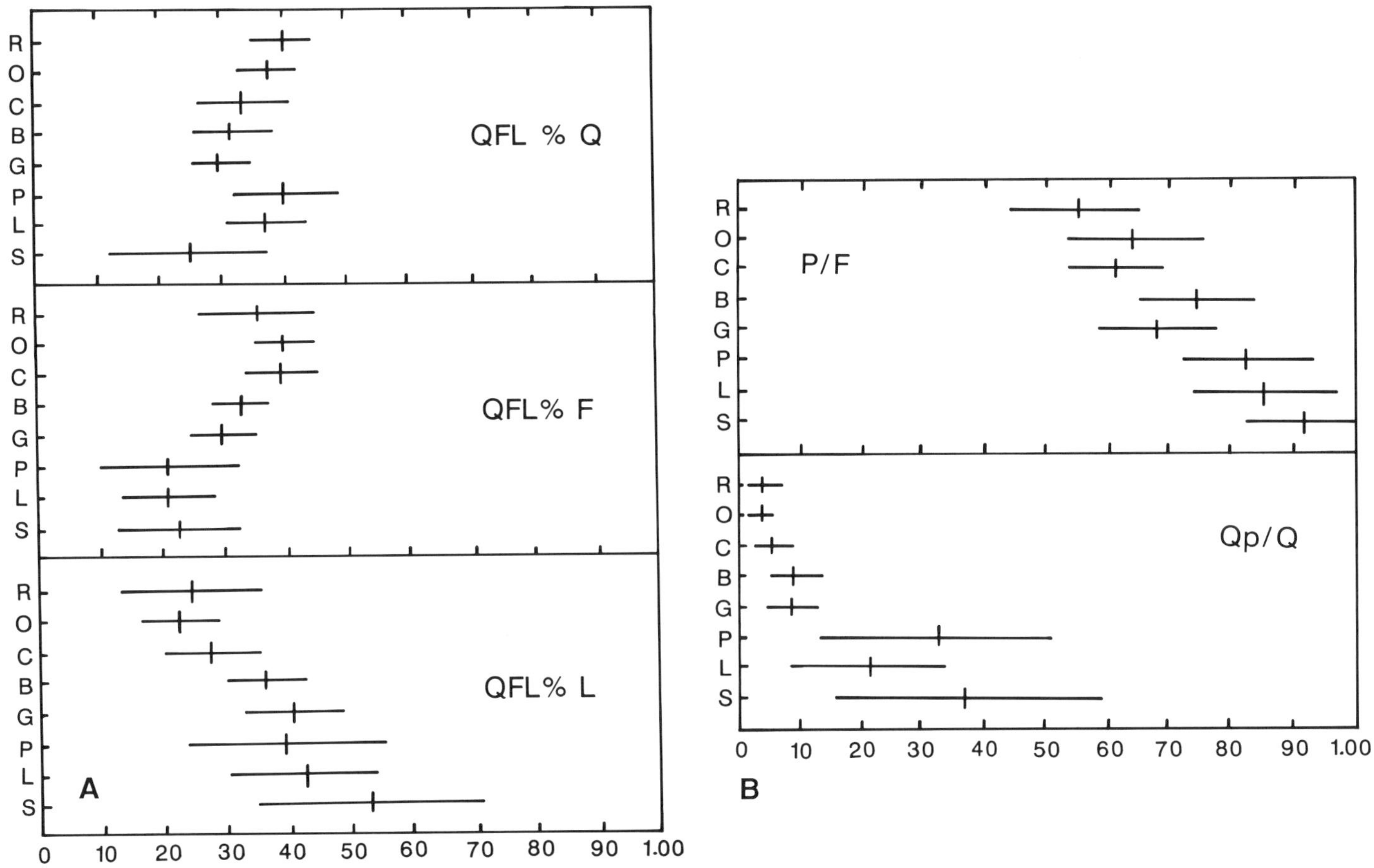

Figure 7. Petrologic data from the Great Valley Group flanking the Sacramento and San Joaquin Valleys, determined by point-counting detrital modes in Upper Jurassic to Upper Cretaceous sandstones. A: Depicts vertical stratigraphic variations in the percentages of lithic clasts (L), total feldspar grains (F), and total quartz (monocrystalline and polycrystalline) grains (Q). Vertical axis of each diagram is divided into "petrofacies," which are stratigraphic intervals defined by sandstone composition. Petrofacies are arranged largely in time-stratigraphic sequences: S is Upper Jurassic and Lower Cretaceous; L and P are Lower Cretaceous; G, B, C, and O are Cenomanian to Santonian; R is Santonian, Campanian, and Maastrichtian. Horizontal bars indicate means and standard deviations. Sandstones become less lithic and more quartzfeldspathic with decreasing age. B: Stratigraphic variations in the ratio of plagioclase (P) to total feldspar (F) and of polycrystalline quartz, including chert (Qp), to total quartz (Q). Upward increase (decreasing age) in potassium-feldspar reflects unroofing of Sierran magmatic arc. Chert clasts in lower Great Valley Group were derived from deformed wallrocks of the batholith. Same petrofacies along vertical axis as in 7A. (Modified from Ingersoll, 1983.)

eral laumontite. The most dramatic contrast is in deformational style. The well-stratified Great Valley Group is regionally tilted and locally folded on a macroscopic scale, and sedimentary structures are well preserved; much of the Franciscan, even where unmetamorphosed, is characterized by pervasive brittle faulting, stratal disruption, or a chaotic "block-in-matrix" fabric on a mesoscopic scale. Low- and high-angle faults are abundant.

This section contains an overview of the Franciscan Complex and a summary of its setting with respect to the Great Valley Group. Franciscan rocks that occur southwest of the San Andreas fault are described under "Probable displaced terranes."

Overall petrology, structure, and age. As the term "complex" implies, the Franciscan includes many locally mapped stratigraphic or lithodemic units. When these are grouped into three major northwest-trending composite packages—the Eastern Belt, Central Belt, and Coastal Belt (Figs. 4 and 8)—some important east-to-west variations in the Franciscan as a whole are evident. In general, the Franciscan youngs westward: Tithonian to Valanginian protoliths predominate in the Eastern Belt, whereas the Coastal Belt is chiefly Late Cretaceous and Paleogene in age. Both the grade and age of regional metamorphism generally decrease westward: Characteristic metamorphic minerals in gray-

wackes from the Eastern, Central, and Coastal belts are, respectively, jadeitic pyroxene, pumpellyite ± lawsonite, and laumontite. As for structure, foliated metamorphic tectonites disposed in gently dipping nappes predominate in the Eastern Belt, whereas brittle mesoscopic structures and more steeply dipping faults prevail to the west.

Eastern Belt. This package is characterized by internally coherent, east-dipping thrust sheets that were regionally metamorphosed in the blueschist facies (Suppe and Foland, 1978; Maxwell, 1981; Worrall, 1981; Blake, Jayko, and McLaughlin, 1985). In the northern Coast Ranges and the Diablo Range, the belt is juxtaposed with either the Great Valley Group or the Coast Range ophiolite along the Coast Range fault (or "thrust"); in the Klamath Mountains, the belt lies structurally beneath Jurassic rocks of the western Klamath terrane (along the South Fork fault in California). The Eastern Belt has been subdivided into several regionally extensive units that dip generally eastward (Fig. 8). In northern California the structurally highest nappe, comprising the South Fork Mountain Schist and Chinquapin Metabasalt, extends 200 km south-southeast from the Oregon-California border.

Protoliths in the Eastern Belt were dominantly sandstone and mudstone locally interbedded with radiolarian ribbon chert and minor basaltic flows and tuff. Fossils range from Tithonian to Cenomanian; some units are undated. Sandstone and mudstone were texturally reconstituted during deformation and metamorphism into multiply folded, segregated quartz-mica schists or variably foliated (typically semischistose) metasedimentary rocks. In metasedimentary rocks of the Northern Coast Ranges, lawsonite is a widespread blueschist-facies mineral. In the Diablo Range, metagraywacke in a 30-km-long thrust sheet, the structurally highest in the range, contains the high-pressure assemblage (ca. 7kb) jadeitic pyroxene + quartz. Rock units in the Eastern Belt have yielded a range of K-Ar ages, which reflect a protracted history of metamorphism and subsequent argon loss (McDowell and others, 1984). Most K-Ar ages for the South Fork Mountain Schist and the correlative Colebrooke Schist in southwestern Oregon cluster near 125 Ma, but recrystallization may have begun by at least 140 Ma. Other tectonic slices and sheets of coarse-grained metabasaltic blueschist yield K-Ar ages (white mica) as old as 152 Ma. Regional metamorphism of structurally lower units (Yolla Bolly terrane) probably occurred about 90 Ma. Displacements along certain thrust faults elevated higher-grade, originally more deeply buried rocks and emplaced them structurally above lower-grade rocks. Some thrusting postdated the 90-Ma metamorphism of the Yolla Bolly terrane.

Central Belt. The composite Central Belt (Maxwell, 1981; Blake, Jayko, and McLaughlin, 1985) can be further subdivided into three kinds of components: (1) rocks interpreted to be tectonic "outliers" of the Eastern Belt, Coast Range ophiolite, or Great Valley Group (Fig. 8); (2) fault-bounded units of diverse ages, consisting largely of graywacke and mudstone, or radiolarian ribbon chert and pillow basalt, that are characterized by relatively coherent bedding and an intact internal stratigraphy; and (3) mélange, which in this chapter is a generic term used for mudstone-rich rock bodies characterized on a mesoscopic scale by stratal disruption, penetrative anastomosing faults, or a block-in-matrix fabric. Mappable units in the Central Belt are structurally interleaved along gently to moderately east-dipping faults. Younger northwest-striking high-angle faults that probably have a significant component of dextral strike-slip are typical map-scale features of the Central Belt. Faults of this system truncate an earlier (but post–90 Ma), low-angle fault separating the Central Belt from the structurally overlying Eastern Belt.

So-called outliers of the Eastern Belt or Coast Range ophiolite (Fig. 8) are probably klippen that were originally contiguous with the main belts cropping out to the east. In many places the outliers are bounded by northwest-striking high-angle faults. Some of the klippen may have been translated northwestward tens or hundreds of kilometers by Late Cretaceous or early Tertiary dextral strike-slip along these fault zones. Some of the rocks resembling the Great Valley Group depositionally overlie the ophiolite in these outliers. Other sequences, including unmetamorphosed Tithonian to Valanginian or Lower Cretaceous to Paleocene beds, have obscure, faulted contacts and have been interpreted as Great Valley–like sediments deposited on the evolving Franciscan Complex.

Structural units in category (2) above differ widely with respect to age and stratigraphy. A glimpse of their variety is provided by some that occur within 100 km of San Francisco (Blake, 1984). The Marin Headlands terrane (Fig. 4), for example, consists of pillow basalt, overlain by an 82-m condensed sequence of Pliensbachian to upper Albian or lower Cenomanian ribbon chert, in turn overlain by graywacke. The entire section, which contains the oldest rocks identified so far in the Franciscan, is repeated along imbricate thrusts. The Novato Quarry terrane consists of Campanian sandstone, mudstone, and conglomerate. The Permanente terrane includes the Aptian to Turonian pelagic Calera Limestone (Fig. 4). Paleomagnetic and paleontological data from the Calera suggest that it accumulated much farther south at a mid-Cretaceous paleolatitude of about 20–25°N.

The mélanges in the Central Belt (discussed in Page, 1981, for example) can be visualized as a matrix underlying, separating, or enveloping the relatively more coherent outliers and structural units described above. Some mélange consists largely of disrupted sequences of black mudstone and graywacke of Tithonian to Valanginian age that was originally interbedded with green basaltic tuff, pillow basalt, and chert. Some is lithologically more heterogeneous and comprises equant blocks and elongate lenses or slabs, ranging from a few centimeters to several hundred meters in size, that are scattered through a matrix of dark mudstone. Some of the matrix is foliated and best described as *argille scagliose* or scaly clay. Typically, fragments of sandstone are most abundant; less common are massive or pillowed basalt, ribbon chert, conglomerate, variably serpentinized ultramafic rock, and "exotic" blocks. Sandstone is either devoid of metamorphic minerals or contains pumpellyite ± lawsonite ± aragonite. Most basalt is greenstone containing albite, chlorite, and generally

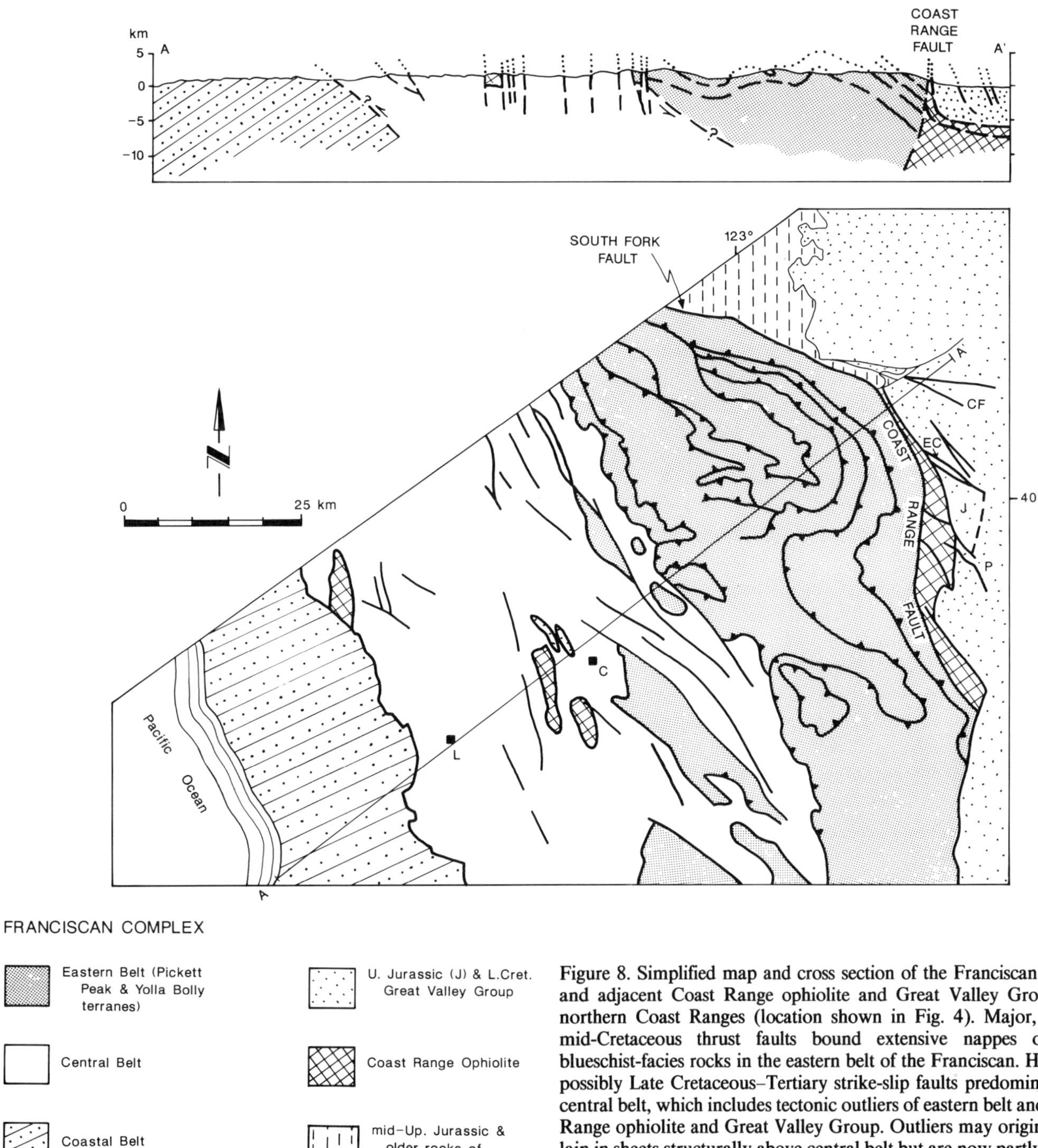

Figure 8. Simplified map and cross section of the Franciscan Complex and adjacent Coast Range ophiolite and Great Valley Group in the northern Coast Ranges (location shown in Fig. 4). Major, probably mid-Cretaceous thrust faults bound extensive nappes of chiefly blueschist-facies rocks in the eastern belt of the Franciscan. High-angle, possibly Late Cretaceous–Tertiary strike-slip faults predominate in the central belt, which includes tectonic outliers of eastern belt and of Coast Range ophiolite and Great Valley Group. Outliers may originally have lain in sheets structurally above central belt but are now partly bounded by high-angle strike-slip faults. Coast Range ophiolite includes undifferentiated serpentinite melange, gabbro, basalt, and chert. Great Valley Group locally lies depositionally on ophiolite; most of the contact is a fault. Coast Range fault juxtaposing Franciscan and ophiolite is high angle and probably Neogene. The South Fork fault juxtaposes Franciscan with older rocks of the Klamath Mountains. CF = Cold Fork fault zone. EC = Elder Creek fault zone. P = Paskenta fault zone. C = Covelo. L = Laytonville. (Modified from Blake, Jayko, and McLaughlin, 1985.)

pumpellyite. Fossils indicate that fragments and matrix range from Late Jurassic to Late Cretaceous in age.

Two kinds of "exotic" blocks are especially significant. Many mélanges in the Central Belt contain a small percentage (probably <1 percent) of coarse-grained high-grade metamorphic rocks, including blueschist, amphibolite, and eclogite. These rocks recrystallized at higher pressures (7 to 9 kb; >10 kb for eclogite) and temperatures (300 to 500°C) than did their decidedly lower-grade companions. Moreover, K-Ar and U-Pb ages point to an age of primary recrystallization of these rocks of at least 150 Ma and more likely 160 Ma. These high-grade metabasites are therefore exotic in the sense that they were metamorphosed before most of the remainder of the Franciscan had even been deposited, at depths (20 to 30 km) greater than most of the Franciscan has ever been buried. They are apparently fragments of Jurassic or older oceanic crust that was tectonically buried and metamorphosed in the Middle or early Late Jurassic. Another variety of exotic block is mid-Cretaceous pelagic limestone; an example is the Albian to Turonian pelagic limestones near Laytonville (Fig. 8). Paleomagnetic data suggest deposition at paleolatitudes of 17°S $\pm$ 7°. These limestones are exotic not only because they are far-traveled but also because they are distinctly pelagic and therefore contrast with the more voluminous associated sandstone and mudstone that clearly accumulated near a continental margin.

The origin of exotic-block mélanges in the Central Belt is still controversial and probably will be for a long time, partly because extensive landsliding has obscured critical contacts. Processes responsible for mixing fragments of such diverse ages and histories, and for the final emplacement of these mixtures, may include hydraulically driven flow of mud deep within accretionary prisms, invasive diapiric injections, muddy debris flows, and remobilization by Cenozoic faulting. The great diversity of rock units within the Central Belt makes it difficult to generalize about the history of the package as a whole, but two events are consequential for Franciscan tectonics. The assembly of the major components of the belt (excluding the outliers) took place between Cenomanian and Eocene time. Neither the duration nor the kinematics of this event is completely resolved. In addition, the Central Belt was juxtaposed with the stacked nappes of the Eastern Belt sometime during this same interval. Faulting definitely postdated the 90-Ma (Cenomanian-Turonian) high-pressure metamorphism recorded in the Eastern Belt and possibly postdated the incorporation of younger (Coniacian, Campanian) elements into the Central Belt.

Coastal Belt. Campanian to Eocene rocks in the Coastal Belt are overwhelmingly sandstone-mudstone turbidites that were structurally thickened by imbrication on east-dipping thrusts. Bedding is variably disrupted and locally folded. Rare blocks of Campanian to Maastrichtian pelagic limestone and pillow basalt occur in mélanges with a matrix of Eocene mudstone. Paleomagnetic and paleontologic data suggest that these limestones accumulated at paleolatitudes of 20–28°N. Rocks in the eastern Coastal Belt were thrust beneath the Central Belt in Eocene or Oligocene time.

The relationship of the Franciscan Complex and Great Valley Group: The Coast Range Thrust. During the 1960s, it became evident that all of the contacts between the Franciscan Complex and the main belt of Great Valley Group and Coast Range ophiolite are structural. Coincidentally, it was confirmed that the bulk of the Franciscan is coeval with the Great Valley Group, even though we now know that the Franciscan contains a few pre-Tithonian, Jurassic elements such as the Marin Headlands terrane and some high-grade blueschists. These relationships were in accord with the model that the Franciscan is a subduction complex and the Great Valley Group the fill of an adjacent forearc basin. The Franciscan was pictured as having been partly shoved beneath the Great Valley Group and its ophiolitic basement during Late Jurassic and Cretaceous subduction along a conceptual fault called the Coast Range thrust (Fig. 6A; Bailey and others, 1970). The visible tectonic contacts between the Franciscan and Great Valley Group were thought to be remnants of this major, long-lived regional thrust.

New evidence concerning the geometry and age of the faults separating the Franciscan and Great Valley Group has accumulated during the past 15 years. For example, some of the faults west of the main outcrop belt of the Great Valley Group are Neogene, high-angle dip-slip structures, locally renamed the "Coast Range fault" (Figs. 6C and 8). High-angle Tertiary faults also bound rafts of Great Valley Group in the Central Belt. Nevertheless, there is clear field evidence from the Coast Ranges as a whole that a family of moderately to gently dipping, locally imbricate faults separates the Great Valley Group and its ophiolitic basement from structurally underlying rocks of the Eastern or Central belt (for example, Suppe, 1978; Fig. 6B). Displacements on these faults postdated the 90-Ma (or younger) blueschist-facies metamorphism recorded in the Eastern Belt, and some fault blocks of Great Valley Group near Clear Lake include strata as young as Paleogene. Therefore, the age of this family of faults is Late Cretaceous or early Tertiary, not Late Jurassic or Early Cretaceous as formerly believed.

The overall strain resulting from displacements on this fault system is currently controversial. The emplacement of essentially unmetamorphosed Great Valley Group and ophiolite against originally more deeply buried, high-grade Franciscan units may be evidence for low-angle normal faulting and, hence, large-scale extension (Platt, 1986). *Some* contraction, however, was required to carry the Franciscan protoliths to the great depths demanded by the high-pressure metamorphic mineral assemblages and to emplace high-pressure rocks structurally above lower-grade rocks. There is additional evidence for contraction in the Klamath Mountains, where the Franciscan was thrust beneath the Western Jurassic Belt. In southwestern Oregon, the Franciscan was imbricated with rocks of the Western Jurassic Belt, Great Valley Group, and Coast Range ophiolite, along thrusts of probable Late Cretaceous age (Blake, Engebretson, Jayko, and Jones, 1985).

In summary, the "Coast Range Thrust" is actually a composite structure, and parts are surprisingly young. Structural contacts between the Franciscan and Great Valley Group comprise a

number of Late Cretaceous or early Tertiary faults that have been modified or replaced locally by younger high-angle dip-slip or strike-slip faults (Figs. 6B and C). On a larger scale, seismic data from the western flank of the San Joaquin Valley suggest that Franciscan rocks extend eastward in the subsurface beneath exposed Great Valley Group, but the timing and geometry of the relevant structures are not yet fully resolved.

Tectonic Interpretation of the Franciscan Complex. Although there is overwhelming agreement that the Franciscan is a subduction complex in the broadest sense of the term, a debate has arisen about its origin and evolution. The evidence linking the Franciscan to late Mesozoic and early Tertiary subduction of a "Pacific" oceanic plate beneath the American continental margin includes: (1) high-pressure metamorphic mineral assemblages that resulted from deep tectonic burial in a region of depressed isotherms, and (2) structural mixtures (stacked nappes and mélanges) of ocean-plate basalt, pelagic chert and limestone, and continent- and arc-derived terrigenous sedimentary rocks. The debate concerns whether the Franciscan formed essentially at its present latitude or whether it consists largely of materials that originated farther south.

One model, formulated in the period from 1968 to 1970 (for example, Hamilton, 1969; Dickinson, 1970; Ernst, 1970), but still in vogue, describes the Franciscan as a huge accretionary complex comprising materials that were scraped off of a descending plate, tectonically mixed, and in some cases carried to great depths and metamorphosed. The complex is envisioned to have developed from Late Jurassic to early Tertiary time immediately west of the coeval, autochthonous Great Valley Group in California and southwest Oregon. If the Great Valley Group and Sierra Nevada batholith are indeed forearc basin deposits and a magmatic arc, respectively, then the Franciscan occurs just where one would expect to find a parallel, contemporaneous *subduction complex.* Nevertheless, it is certain that the Franciscan as we see it today (Figs. 6 and 8) is not comparable to the simple steady-state *accretionary prisms* that have formed along many active convergent margins. Seismic reflection profiling and deep-sea drilling suggest that these prisms are thin (15 km or less), tapering wedges of imbricated sediments, with the internal structure of on-land, thin-skinned thrust belts. To our knowledge, steady-state subduction alone does not result in packages even remotely as heterogeneous and structurally complex as the Franciscan. Especially difficult to explain in any steady-state hypothesis is how sediments and ocean-floor basalts are carried to depths of at least 25 to 30 km—depths greater than the thicknesses of most modern prisms—and then subsequently elevated and juxtaposed, in extensive nappes, with less deeply buried rocks.

More recent models (for example, Page, 1981; Blake, 1984) accord with paleomagnetic data or analyses of relative plate-motions, collectively indicating that the Pacific family of plates converged obliquely with North America during the late Mesozoic. There is compelling paleomagnetic and paleontologic evidence, for example, that pelagic limestones like the Calera and Laytonville originated much farther south. It is possible that these and similar exotic elements moved northward on oceanic plates until they were accreted to North America at about their present latitude. Such a model recognizes that certain components of the Franciscan are far traveled but accepts that the great bulk of terrigenous sediments in the Franciscan may not have been displaced and was instead deposited opposite the Great Valley Group to the east. More extreme models hold that the terrigenous sediments themselves were transported northward before accretion, or that the Franciscan includes displaced fragments of accretionary prisms and even volcanic arcs that originally lay hundreds or thousands of kilometers farther south. Models of this genre have yet to be corroborated by a body of high-quality paleomagnetic data from clastic sedimentary rocks.

To sum up, current questions about the Franciscan are: How much of it was accreted in situ, and how much has been transported northward? What processes beyond steady-state subduction (collision, telescoping of an entire forearc) were important? Did metamorphism, deformation, and accretion of Franciscan rocks occur continuously from the Late Jurassic to early Tertiary, or was this time period punctuated by a few key metamorphic and deformational events? In formulating answers, three events will figure prominently. First, the most widespread high-pressure metamorphism recorded in the Franciscan appears to be post-Cenomanian (post–ca. 90 Ma) in age; older metamorphic rocks occur sporadically but are less abundant. Second, Franciscan rocks in the Eastern and Central belts were imbricated and emplaced beneath the Great Valley Group after the 90-Ma or younger metamorphism. This Late Cretaceous (or early Tertiary?) deformation is widespread and significant but so far enigmatic and underappreciated. Perhaps it represents the massive contraction of a wide Franciscan subduction complex due to a collision or change in plate velocities or the arrival of a large part of the complex from the south and its accretion to California. Alternatively it could record in situ extension accomplished by slip on a system of low-angle normal faults. Finally, northwest-striking high-angle faults and fault zones prevalent in the Central Belt of the Northern Coast Ranges probably record early Tertiary distributed dextral transcurrent shear in response to continued oblique subduction or the migration of a triple junction offshore.

Summary

The Sierran batholithic belt, stretching from southern California northward into eastern Washington and northern Idaho, is the most complete petrotectonic record of late Mesozoic, post-Nevadan plate convergence along the western margin of the U.S. Cordillera. Even so, activity in the magmatic arc was sporadic in space and time. In California it lasted from about 150 to 80 Ma, although the volume of preserved plutons suggests that overall magmatic activity was diminished from about 150 to 125 Ma. In northern Idaho and Washington, plutonism spanned a shorter interval in the middle Cretaceous (105 to 90 Ma) and then waned when the Idaho batholith was emplaced beginning about 90 to 95 Ma. An in-situ filled forearc basin and adjacent subduction com-

plex occur opposite the batholithic belt only in California and southwestern Oregon, where they are represented by the Great Valley Group and Franciscan Complex, respectively. Albian and Cenomanian(?) strata in central Oregon may be the northernmost outlier of the Great Valley Group. Late Jurassic and Cretaceous rocks in Washington that have been interpreted as in-situ subduction complexes or fore-arc basin deposits correlative with the packages in California may instead constitute part of a large, composite displaced terrane. If so, any counterparts to the Franciscan and Great Valley Group that originally lay west of the batholithic belt in the Pacific Northwest have been removed tectonically from the continental margin.

WESTERN CORDILLERA: PROBABLE DISPLACED TERRANES

Since the mid-1970s, many geologists—but certainly not all—have argued that parts of the North American Cordillera are out of place. They view the western Cordillera as an assemblage of tectonostratigraphic terranes, which originally were separated from one another by at least tens of kilometers and subsequently were juxtaposed along their bounding faults. Moreover, paleomagnetic data suggest that some terranes originated hundreds or thousands of kilometers south of their present latitudes; hence, they appear to have been "displaced" northward by either coast-parallel slip along major transcurrent faults, passage on a converging oceanic plate, or both.

Major paleogeographic reinterpretations like these naturally provoke controversies about which elements have been displaced and by how much, how they were accreted to the North American margin, and where to draw the boundary at any given time between displaced terranes and in-situ North America. In this chapter, the Sierran batholithic belt or magmatic arc, together with its accompanying forearc basin in California, is accepted as indisputably having been constructed on or at the western edge of North America during the period 150 to 80 Ma. Part of the Franciscan subduction complex was almost certainly assembled in California then as well. It is acknowledged here, however, that many of the rock units and tectonic elements lying to the west and north of the Sierran–Great Valley–Franciscan trinity (Fig. 1) may have been transported northward and finally emplaced at about their present latitudes in the latest Cretaceous (post–80 Ma) or early Tertiary (Umhoefer, 1987). Even though some workers have published evidence supporting an in-situ origin, the weight of paleomagnetic data and other evidence at present tips the scales toward the hypothesis that terranes in the following areas are displaced: the San Juan Islands, North Cascades, and Okanogan region of Washington; coastal southwestern Oregon; and the Salinian terrane, Sur-Obispo terrane, and Peninsular Ranges terrane in California (see Fig. 1). For this reason, the Upper Jurassic to lower Upper Cretaceous rock units and structures in these "probably displaced" terranes are discussed separately in the following sections.

San Juan Islands, North Cascades, and Okanogan region

Upper Mesozoic rocks in northwestern Washington can be grouped into four major tectonic elements (Fig. 9). Westernmost is the Late Cretaceous San Juan–Northwest Cascades thrust and nappe system, comprising Upper Jurassic to mid-Cretaceous supracrustal rocks that were imbricated with older units and metamorphosed to prehnite-pumpellyite or blueschist facies. Lying to the east is the Skagit metamorphic terrane, composed chiefly of medium- to high-grade supracrustal rocks and Upper Cretaceous to lower Tertiary synmetamorphic granitoid intrusions. Farther east are well-bedded, Jurassic to mid-Cretaceous volcaniclastic and arkosic strata of the Methow-Pasayten belt. The easternmost element is incompletely dated but includes upper Mesozoic granitoid plutons in the Okanogan Highlands and Okanogan dome. Northwest Washington also encompasses at least two prominent northwest-striking high-angle structures, the Ross Lake fault zone and the Pasayten fault, that may have been active in Early Cretaceous to early Late Cretaceous time. An overall cross section through the elements encompassed by Figure 9 is presented in Cowan and Potter (1986).

San Juan–Northwest Cascades thrust system. A stack of generally east-dipping thrust sheets (Fig. 10) in this major west-vergent Late Cretaceous thrust and nappe system underlies the San Juan Islands (Brandon and others, 1988) and the northwest Cascades west of the Straight Creek fault (Misch, 1966, 1977; Brown, 1987). Poorly exposed but broadly similar rocks in the Cascade foothills south of the Skagit River are probably also part of the thrust system, which continues at least 100 km north of the U.S.–Canada boundary to Harrison Lake. The nappes include rocks ranging in age from Precambrian to latest Early Cretaceous (Fig. 10), but only the Upper Jurassic and Cretaceous units are summarized here.

The structurally lowest upper Mesozoic unit is the Constitution Formation in the San Juan Islands. It consists principally of massive volcaniclastic sandstone, thinly interbedded sandstone and mudstone, and minor radiolarian ribbon chert and MORB-type pillow basalts. The Constitution probably accumulated on a continental margin consisting largely of arc volcanics. Radiolaria from chert indicate a Late Jurassic or Early Cretaceous age. The structurally higher Decatur terrane consists of the Middle to Upper Jurassic Fidalgo Igneous Complex, comprising an ophiolite and volcanic-arc complex, and the overlying Tithonian and younger Lummi Formation. The Lummi Formation is composed entirely of sandstone-mudstone turbidites and conglomerates and is locally rich in clasts of volcanic rock or chert. Overall, the Decatur terrane is stratigraphically similar to the Coast Range ophiolite and overlying Great Valley Group in California. The 3-km-thick Lopez Structural Complex, separating the Decatur terrane and Constitution formation, is an imbricate fault zone that includes a slice of upper Albian pillow basalt.

The three major thrust plates in the northwest Cascades are probably higher in the structural stack than the nappes exposed in the San Juan Islands. The lowest plate in the Cascades includes

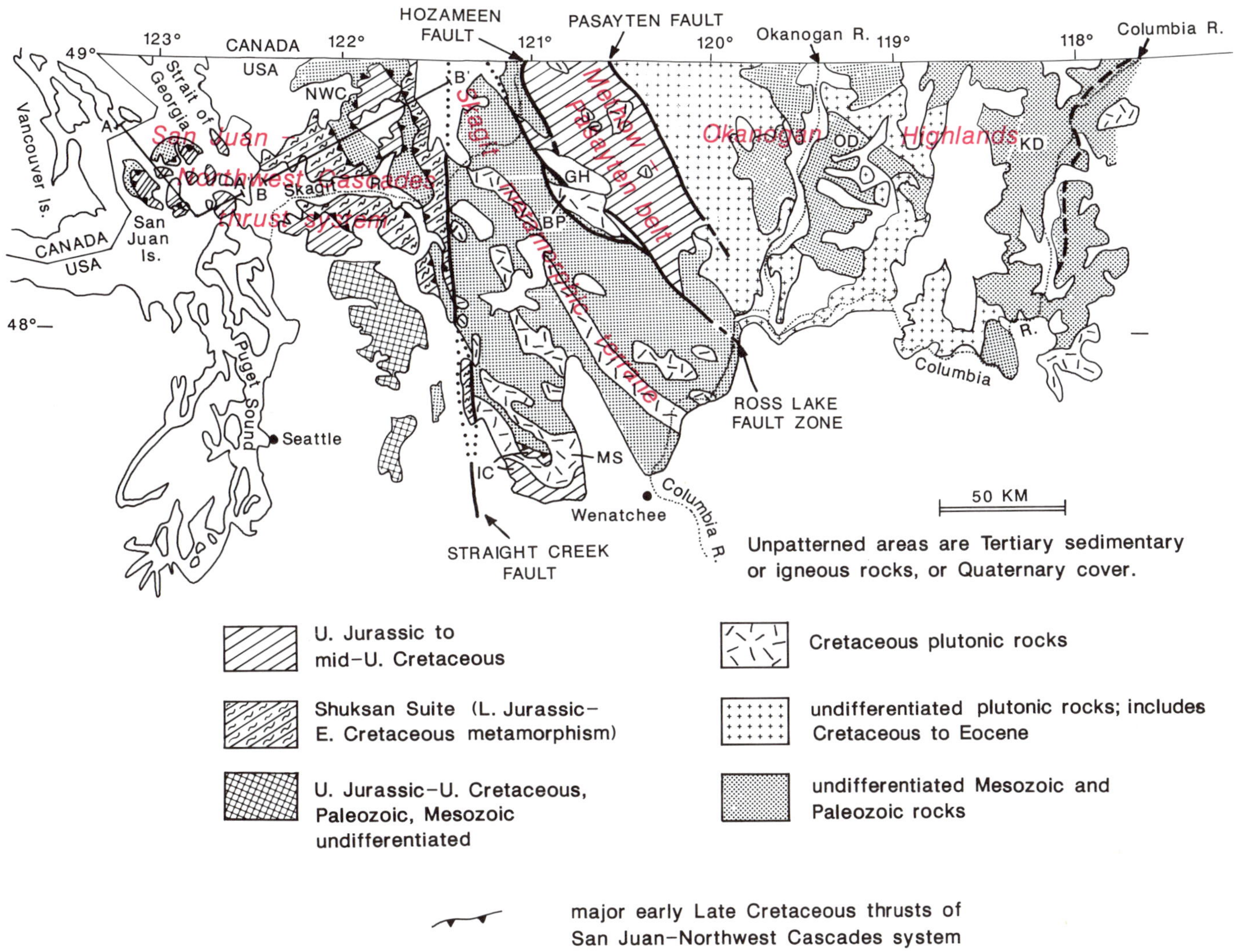

Figure 9. Simplified pre-Cenozoic geology of part of Washington. Four composite terranes, which may have been displaced from more southerly latitudes, are separated by high-angle faults. West of the early Tertiary Straight Creek fault, which continues into Canada as the Fraser River fault, the San Juan–northwest Cascades thrust system includes Upper Jurassic and Lower Cretaceous rocks in nappes juxtaposed along early Late Cretaceous imbricate thrusts. Most of the rocks between the Straight Creek fault and Ross Lake fault zone are part of the Skagit metamorphic terrane; its southern edge is structurally overlain by rocks that are related to the San Juan nappes. Upper Jurassic to lower Upper Cretaceous sedimentary and volcanic rocks constitute the Methow-Pasayten sequence between the Ross Lake and Pasayten faults. The Okanogan Highlands, extending east from the Pasayten fault, are underlain partly by Cretaceous granitoid rocks. The location of the western edge of North America at this latitude in the late Mesozoic is unresolved. It could have coincided with the as yet poorly defined fault along the Columbia River, lain somewhere within the highlands, or been marked by the Pasayten fault or a precursor. A-A′ and B-B′ are sections shown in Figure 10. Other localities and features mentioned in text: BP = Black Peak batholith; GH = Golden Horn batholith; IC = Ingalls Complex; KD = Kettle dome; MS = Mount Stuart batholith; NWC = Northwest Cascades; OD = Okanogan dome.

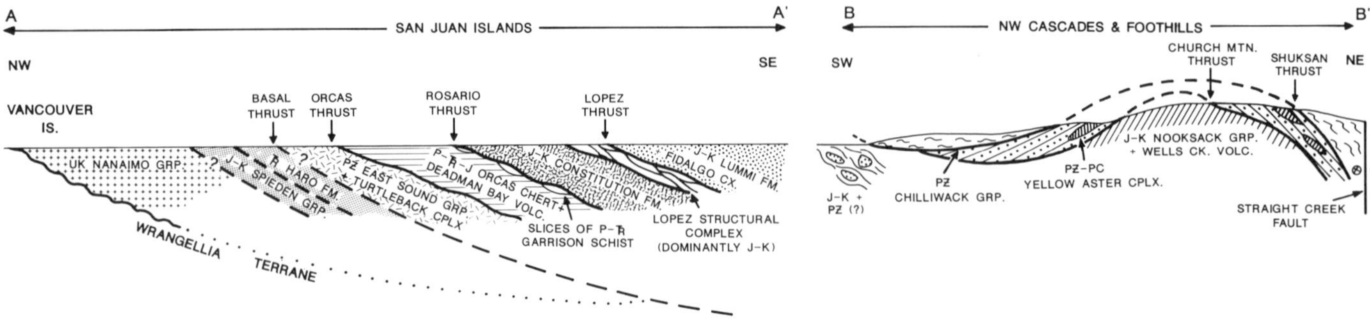

Figure 10. Diagramatic cross section (location shown on Fig. 9) illustrating the stacking order of thrust sheets in the San Juan-northwest Cascades thrust system. Nappes in the northwest Cascades are folded into a large antiform, and although the Wells Creek volcanics are the structurally lowest rocks exposed, they are undoubtedly underlain by other nappes. Rock units and thrust sheets of the San Juan Islands are inferred to be structurally lower than those in the Cascades. The actual boundary between the two sets is covered. Unit J-K + PZ(?) is largely a serpentinite melange interpreted here to consist chiefly of deformed Fidalgo Complex and Lummi Formation. The entire stack of thrust sheets is inferred to lie above a basal thrust, beneath which is a footwall consisting of the Wrangellia terrane. All geometrical relations portrayed are schematic.

Upper Jurassic and Lower Cretaceous volcaniclastic graywacke and interbedded mudstone and conglomerate of the Nooksack Group, which lies depositionally on the arc-derived Wells Creek Volcanics of probable Middle Jurassic age. A forearc or backarc setting is likely for the Nooksack. Next above is the Church Mountain plate, consisting chiefly of the Devonian to Permian Chilliwack Group but including an imbricate zone comprising slices of Chilliwack Group, Precambrian and lower Paleozoic crystalline rocks of the Yellow Aster Complex, metaperidotite, and upper Paleozoic amphibolite and blueschist. Structurally highest, above the Shuksan thrust, is the Shuksan Metamorphic Suite, consisting of the Darrington Phyllite and overlying Shuksan Greenschist. The Jurassic(?) protoliths of these units were arkosic sediments and tholeiitic basalts, respectively, but the tectonic setting in which they accumulated is uncertain. Synkinematic high-pressure metamorphism was broadly Late Jurassic to Early Cretaceous; Rb-Sr and K-Ar ages range from 105 to 164 Ma. Sedimentary rocks were converted to graphitic phyllite and lawsonite-bearing metasandstone. Basalts recrystallized to foliated actinolite-epidote greenschist interlayered with epidote-bearing blueschist. The Shuksan Suite is a structurally and stratigraphically coherent thrust sheet that presently extends southward about 110 km from the U.S.–Canada boundary. In many respects—rock types, age and grade of metamorphism, and tectonite fabric—the suite resembles the blueschist-facies South Fork Mountain Schist and interbedded metabasalts in the Eastern Belt of the Franciscan Complex.

Evidence from the San Juan Islands constrains the age of thrusting there as early Late Cretaceous: Upper Albian (about 100 Ma) rocks were not only imbricated during thrusting but also affected by a syn-thrusting, high-pressure/low-temperature metamorphism that is recorded throughout the thrust system. Lawsonite, prehnite, and aragonite-bearing metamorphic mineral assemblages are typical in the San Juan Islands; prehnite-pumpellyite facies assemblages are widespread in the North Cascades. Moreover, Upper Santonian (ca. 84 Ma) strata in the Upper Cretaceous (Santonian to Maastrichtian) Nanaimo Group, a 2.5-km-thick sequence of sandstone, mudstone, and conglomerate more widely exposed to the northwest in British Columbia, contain clasts derived from the San Juan Island nappes. Apparently, rock units in the thrust system were imbricated, tectonically buried to depths on the order of 15 km, metamorphosed, and then uplifted and exposed to erosion, in the interval from about 100 to 84 Ma.

Two important questions about the San Juan–Northwest Cascades orogen are: What is its overall geometry, and how did it develop? In a model favored here, the orogen is considered to be a geometric analog of conventional, intracontinental thrust and nappe systems, constituting an internally imbricated wedge separated from an underlying, rigid footwall block by a basal thrust (Fig. 10). The footwall block is exposed 15 km west of the San Juan Islands on Vancouver Island, where it consists of the Wrangellia terrane. The character of the San Juan–Northwest Cascades event becomes clearer if one tries conceptually to restore all

of the rock units in the orogen to their pre-thrusting configuration. Obviously, they do not fit back into a simple stratigraphic layer cake, as do nappes in many classic fold and thrust belts. Instead, the thrust system includes diverse tectonostratigraphic elements or terranes (Fig. 10), many of which had disparate geologic histories until the Late Jurassic, when they were inundated by or supplied detritus to thick clastic sequences like the Nooksack Group, Constitution Formation, and Lummi Formation. The petrology and great volume of these sediments suggest that most of the terranes, including Wrangellia, were adjacent to a magmatic arc on a large land mass, probably continental North America, in Late Jurassic and Early Cretaceous time. The Shuksan terrane, on the other hand, was deeply buried and recrystallizing under blueschist-facies conditions during this time. In the early Late Cretaceous, the terranes were telescoped, probably by at least several hundred kilometers, and nappes were driven westward onto the Wrangellia terrane. The cause of this massive shortening along the continental margin is a matter of speculation. The terranes and their sedimentary cover may have been displaced coastwise along transcurrent faults until they collided with a re-entrant in the continental margin. Alternatively, the thrusting could have been the result of a change in plate geometry or velocity, or the collision of an oceanic plateau or rise. Although the exact palinspastic configuration of the terranes is still unresolved, paleomagnetic data discussed in the following section suggest that the entire San Juan–Northwest Cascade thrust system developed at least 2,000 km south of its present latitude.

Skagit Metamorphic Terrane. In this chapter, the term Skagit terrane applies to the variably deformed and metamorphosed crystalline rocks in the North Cascades east of the early Tertiary, high-angle Straight Creek fault (Misch, 1966, 1977; Fig. 9). The Skagit is a composite terrane made up of many individual map units; migmatitic gneiss, orthogneiss, granitoid intrusions, amphibolite, schist, and pegmatite are the predominant rock types. Although isotopic ages suggest that the protoliths of at least some units are late Proterozoic or Triassic, Late Cretaceous magmatism and metamorphism were instrumental in molding the terrane that we see today.

The Skagit terrane is permeated with Upper Cretaceous granitoid intrusions, the largest of which are shown in Figure 9. Tonalite and granodiorite predominate. U-Pb and K-Ar ages (Engels and others, 1976) indicate that the onset of magmatism was about 90 Ma; most of the large, discrete plutons apparently were emplaced between 90 and 75 Ma. Minor intrusive activity, however, persisted until about 45 Ma. A widespread greenschist-facies (chlorite zone) to amphibolite-facies (kyanite or sillimanite zone) regional Barrovian metamorphism was roughly synchronous with Skagit magmatism. During this protracted Late Cretaceous to Eocene event, granitoid bodies were locally recrystallized and converted into foliated orthogneiss, and supracrustal country rock was pervasively migmatized.

The Upper Cretaceous plutons in the Skagit terrane are regionally significant. First, they represent the southeast continuation of the much more widely exposed Coast Plutonic Belt in the Coast Mountains of British Columbia. This kinship is clear if one restores the 90 km of early Tertiary dextral slip on the Straight Creek fault. Second, two of the plutons, the Mount Stuart batholith and the Black Peak batholith (Fig. 9), intrude faults that juxtapose the Skagit composite terrane with tectonic elements lying to the west and east, respectively. For example, the 90-Ma Mount Stuart batholith intruded the Windy Pass thrust separating the pelitic Chiwaukum Schist, which is a Triassic component of the Skagit terrane, from the structurally overlying Ingalls Complex. The Ingalls Complex (Fig. 9), comprising Middle and Upper Jurassic ophiolitic rocks, basalt, chert, and volcaniclastic rocks, is probably part of the San Juan–Northwest Cascades thrust system. By 90 Ma, then, the Skagit terrane had been amalgamated with the San Juan–Northwest Cascades system and with the Wrangellia terrane in British Columbia. This conclusion is significant, because paleomagnetic data from the Mt. Stuart body and from 90 to 110 Ma plutons on strike to the northwest in the Coast Plutonic Belt indicate that the plutons were emplaced far to the south—perhaps as much as 2,500 km—of their present-day latitude (Irving and others, 1985). Between about 80 and 55 Ma, the plutons and the extensive composite crustal block into which they were emplaced moved northward along the coast of North America and were rotated clockwise 40 to 70 degrees. The structures that accommodated these huge displacements have yet to be identified with certainty in either the United States or Canada.

Methow-Pasayten sequence and its bounding faults. The Methow-Pasayten sequence of Jurassic and Cretaceous sedimentary and subordinate volcanic rocks (Barksdale, 1975; Trexler and Bourgeois, 1985) lies between the Ross Lake and Hozameen faults on the west and the Pasayten fault on the east (Fig. 9). The sequence, which together with its bounding faults continues northwestward over 100 km into British Columbia, has an aggregate thickness in excess of 20 km. In Washington, fossils are sparse, so parts of the section are undated. Probable Jurassic thin-bedded volcaniclastic turbidites of the Twisp Formation are overlain by volcaniclastic rocks, tuffs, and andesitic flows of the Buck Mountain Formation, which is at least partly Hauterivian in age. Arkosic sandstone, conglomerate, and siltstone of the Aptian and Albian Harts Pass group were derived largely from the east. Albian and perhaps Cenomanian chert-clast conglomerates in the Virginian Ridge Formation are the oldest sedimentary rocks known to have been derived from a western source; they interfinger with the arkosic, east-derived Winthrop Formation. The youngest rocks are Albian or Cenomanian red beds and andesitic tuff, breccia, and flows of the Midnight Peak Formation. Parts of the section are homoclinal; elsewhere, the sequence was affected by mid-Cretaceous thrusts and kilometer-scale, locally overturned folds. Granitoid stocks and dikes about 88 Ma in age appear to cut these structures.

Although the tectonic setting of the Methow-Pasayten basin throughout its long history is still being investigated, the preponderance of volcanogenic material deposited in it suggests that it developed near a volcano-plutonic arc. Many workers favor a forearc setting related to an arc in the Okanogan Highlands to the

east. All interpretations are diminished because neither the basement of the section in Washington, nor what lay to the west of the basin prior to Albian time, is known. Clasts of volcanic rock and chert in the Virginian Ridge Formation clearly record the Albian uplift, and perhaps the transpressive accretion, of the Permian to Jurassic Bridge River–Hozameen terrane to the west.

Another complication is the poorly understood late Mesozoic history of displacements on the bounding faults. Some components of the high-angle Ross Lake fault zone are cut by the 90-Ma Black Peak batholith; pre–Late Cretaceous strike-slip on these structures is possible but has not yet been confirmed. Other structures in the Ross Lake zone record Late Cretaceous or early Tertiary strike-slip or dip-slip. The Hozameen fault, which forms the actual western boundary of the Methow-Pasayten sequence, truncates mid-Cretaceous units and structures within the sequence. Its probable southeast continuation is plugged by the 47-Ma Golden Horn batholith (Fig. 9). The Late Cretaceous and early Tertiary kinematics of this structure are unknown. The Pasayten fault, truncating the sequence on the east, is another enigma. The youngest movement on the fault was early Tertiary dip-slip (down to the west), which had ceased by about 45 Ma. The sense and timing of Cretaceous displacements, however, are unresolved. The juxtaposition of contrasting pre-Cenomanian rocks and structures along the entire 225-km trace of the fault in Washington and British Columbia could have resulted from transcurrent slip of several hundreds of kilometers or more. A more conservative interpretation, also permitted by available evidence, involves major dip-slip displacements between a long-lived Jurassic and Early Cretaceous forearc basin and an adjacent magmatic arc. Until the kinematic history of the Pasayten fault is resolved, the possibility remains that the Methow-Pasayten sequence is allochthonous with respect to both the Okanogan Highlands and North American rocks farther east.

Okanogan Highlands. The Okanogan Highlands, which here are considered to extend from the Pasayten fault to the Columbia River (Fig. 9), are underlain in part by mid-Cretaceous granitoid rocks (Armstrong and others, 1977). Isotopic ages from the plutons and their country rocks are sparse. A 30-km-wide strip adjacent to the Pasayten fault comprises northwest-trending, elongate plutons of trondhjemite, tonalite, and quartz monzonite that have yielded K-Ar cooling ages on hornblende ranging from 81 to 106 Ma. The country rocks include undated gneiss, migmatite, and granitoid bodies. Plutons of granodiorite and quartz monzonite west of the Okanogan River have yielded ages ranging from 73 to 93 Ma. They intruded upper Paleozoic and Triassic metasedimentary and metavolcanic rocks that are probably part of the Quesnellia terrane defined in southern British Columbia.

Between the Okanogan River and Columbia River, crystalline rocks are exposed in two Tertiary structural culminations, the Okanogan and Kettle domes (Cheney, 1980; Fig. 9). Isotopic ages are highly discordant as a result of an early Tertiary thermal and deformational event (Miller and Engels, 1975), so the ages of most of the crystalline units in Washington are unknown. Isotopic data from both Washington and southern British Columbia indicate that at least some of the amphibolite-facies paragneiss and orthogneiss in the domes is Precambrian. Intrusive granitoid rocks range from Jurassic to Eocene; some foliated plutons and sills are mid-Cretaceous. The intense Paleogene thermal and deformational overprint, which modified the late Mesozoic geology in this region, was imparted by magmatic activity and contemporaneous crustal extension.

The Okanogan Highlands clearly occupy a critical position between bona fide Proterozoic and younger stratified rocks and the underlying Proterozoic crystalline basement of continental North America east of the Columbia River, on the one hand, and the diverse, possibly displaced, and largely ensimatic elements to the west of the Pasayten fault, on the other. The evidence that the highlands themselves were displaced latitudinally during the Late Cretaceous consists of paleomagnetic data from lower Mesozoic plutons in the Quesnellia terrane of British Columbia that are interpreted to have been remagnetized in the Cretaceous. Alternatively, the Precambrian metamorphic rocks in Kettle and Okanogan domes may represent the westernmost extension of sialic North American crust at this latitude (for example, Cowan and Potter, 1986). If so, then Cretaceous plutonic rocks in the highlands may be an offshoot of the middle Cretaceous Sierran batholithic belt that was emplaced into North American basement in northeasternmost Washington and northern Idaho.

Summary. There are important differences in the Late Jurassic to early Late Cretaceous geologic histories of the four major tectonic elements in northern Washington. These differences are significant regardless of whether the western elements have been displaced latitudinally with respect to late Mesozoic continental North America. The effects of a major mid-Cretaceous crustal contraction are manifest in the San Juan–Northwest Cascades thrust and nappe system. This system represents the southern end of the zone of collision between the composite Insular and Intermontane superterranes, which are both much more widely exposed in the western Canadian Cordillera. The middle Cretaceous metamorphism of the Skagit terrane and Albian to Cenomanian events recorded in the Methow-Pasayten belt probably also reflect contraction, crustal thickening, and uplift associated with this event. In contrast, the middle Cretaceous geologic record in the Okanogan Highlands appears to consist largely of plutonic rocks, but whether they formed part of a wide magmatic arc standing adjacent to a Methow-Pasayten forearc basin is still a matter of conjecture.

Coastal southwest Oregon

The Gold Beach terrane, underlying a 5-km-wide coastal strip (Fig. 1), consists of a distinctive sequence of Upper Jurassic and Upper Cretaceous rocks (Blake, Engebretson, and Jayko, 1985; Bourgeois and Dott, 1985). The Tithonian Otter Point Formation is a disrupted unit of stratigraphically discontinuous volcaniclastic sandstone and conglomerate, mudstone, and volcanic rocks including flows, pyroclastic breccias, and tuffs. It is

interpreted to have formed within or adjacent to a volcanic arc. Depositionally overlying the Otter Point are at least 1,000 m of well-bedded Campanian (to possibly Maastrichtian) sandstone, mudstone, and conglomerate deposited in a tectonic setting characterized by active vertical and transcurrent faulting. The provenance for some of the coarse detritus in the Campanian section is unknown, but it apparently did not include pre–Upper Cretaceous rock units presently exposed in the Klamath Mountains to the east. The cryptic source and the wrench-tectonic setting inferred for the Campanian section, together with preliminary paleomagnetic data from the Otter Point Formation, are compatible with the overall hypothesis that the Gold Beach terrane originated hundreds of kilometers or more farther south along the continental margin of North America, whence it moved northward along transcurrent faults in the latest Cretaceous or early Tertiary.

Western and southwestern California

Late Mesozoic rocks west of the San Andreas fault can be grouped into a few distinctive lithotectonic "belts," "blocks," or "terranes" that are bounded by Cenozoic faults (Fig. 11). Most of the terrane boundaries are high-angle strike-slip faults; some are known or inferred to be thrusts. There is general agreement that these terranes, most of which strongly resemble "autochthonous," late Mesozoic lithotectonic elements east of the San Andreas fault, have been displaced relative to one another by tens to, in some cases, hundreds of kilometers of transcurrent slip. Paleomagnetic data from rocks as young as Paleocene, however, suggest that some terranes have been displaced northward even greater distances, by as much as 2,500 km. Because all of these postulated large-magnitude displacements are post-Campanian, the emphasis in this chapter is on the Late Jurassic to Campanian geology of the major terranes themselves. Regardless of how far the terranes have moved, the basic geologic distinctions among them are well established. The Salinian terrane and the Peninsular Ranges batholith and its country rocks comprise abundant Cretaceous granitoid plutons; Franciscan-like rocks underlie part of the Sur-Obispo terrane; and Middle Jurassic ophiolitic rocks and overlying strata very similar to the Great Valley Group constitute the remainder of the Sur-Obispo terrane.

Salinian terrane. The Salinian terrane (Salinian block or Salinia in some literature) is a fault-bounded slice of metamorphic country rock, Cretaceous granitoid plutons, and overlying Campanian and younger sedimentary rocks that extends at least 500 km from southern California to offshore northern California (Ross, 1978, 1984; Page, 1981). It is the proverbial sore thumb of California geology, conspicuously flanked on both the southwest and northeast by partly coeval Franciscan rocks that record an utterly different late Mesozoic geologic history (Fig. 11). The northeastern boundary of the Salinian terrane is the San Andreas fault; the southwestern boundary includes the Sur-Nacimiento fault zone and the San Gregorio–Hosgri fault. The composite Salinian terrane consists of several northwest-trending, elongate, fault-bounded blocks; some differ largely with respect to their

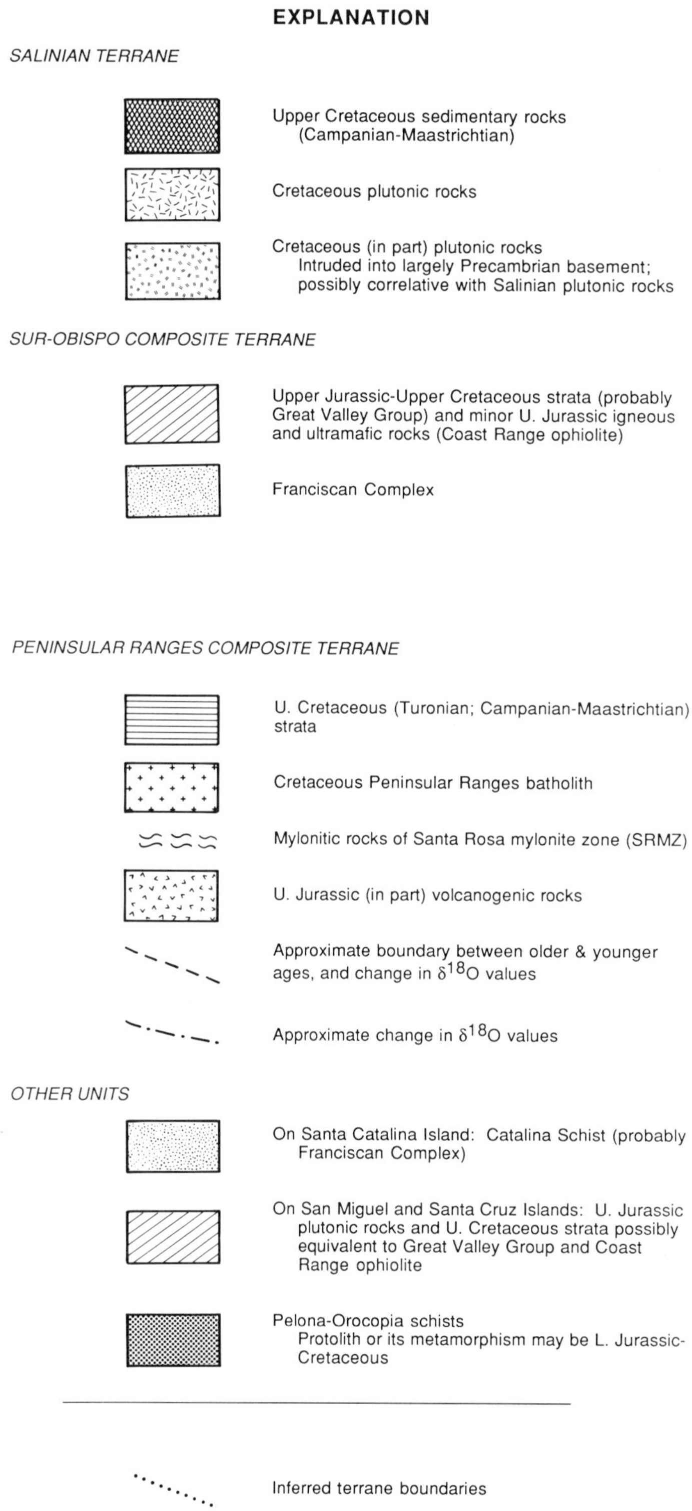

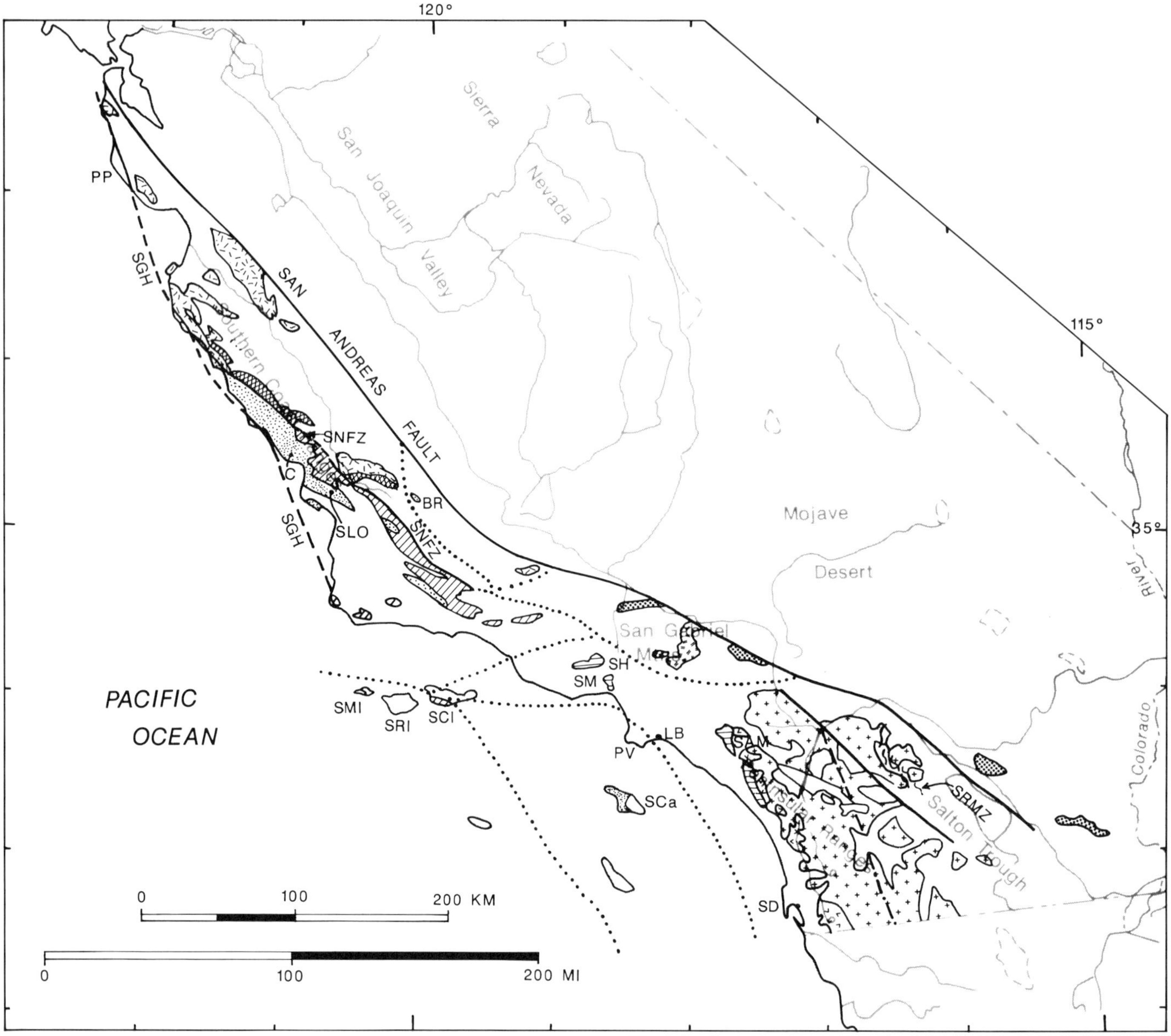

Figure 11. Upper Jurassic–Upper Cretaceous rock units in southwestern California that may have been displaced >1,000 km. In some cases, the boundary between terranes is an exposed fault; for example, the Sur-Nacimiento fault zone (SNFZ) separates the Salinian composite terrane on the east from the Sur-Obispo composite terrane. Other boundaries (for example, between the northern Peninsular Ranges composite terrane and the Sur-Obispo terrane) are inferred. Their location is adapted from Champion and Howell (1986). The "Barrett Ridge slice" of the Salinian terrane is delimited by the San Andreas fault and inferred terrane boundaries and includes exposures at Barrett Ridge (BR). The "Tujunga terrane" includes the exposures of Pelona-Orocopia schists in the San Gabriel Mtns. southwest of the San Andreas fault and east of the fault in the Salton trough. Other localities and features referred to in the text: C = Cambria; LB = Long Beach; PP = Pigeon Point; PV = Palos Verdes Peninsula; SAM = Santa Ana Mountains; SCa = Santa Catalina Island; SCI = Santa Cruz Island; SD = San Diego; SGH = San Gregorio–Hosgri fault; SH = Simi Hills; SLO = San Luis Obispo; SM = Santa Monica Mountains; SMI = San Miguel Island; SRI = Santa Rosa Island. (Explanation on facing page.)

pre-plutonic basement. One of these components, the "Barrett Ridge slice" at the southeast end of the terrane (Fig. 11), contains gneissic rocks that may originally have been contiguous with Precambrian rocks in the San Gabriel Mountains and in diverse ranges northeast of the San Andreas fault and the Salton Trough (the "Tujunga terrane" of Champion and others, 1984, for example). Pre-plutonic basement in the main part of the Salinian terrane in the Coast Ranges consists largely of amphibolite-facies metasedimentary rocks, quartzo-feldspathic gneiss, and subordinate granulite-facies metaplutonic rocks. Gneiss at Barrett Ridge yielded a U-Pb age of 1.7 Ga. All other protoliths are pre-Cretaceous, but their age is otherwise unknown.

Granitoid plutons are chiefly granodiorite, granite (quartz monzonite in older literature), and tonalite. U-Pb and Rb-Sr isotopic ages (Mattinson and James, 1985) indicate that most plutons in the northern part of the terrane were emplaced between 110 and 95 Ma; some dates are as old as 120 Ma. Dates from the southern part fall mostly between 90 and 80 Ma. K-Ar ages ranging from about 79 to 90 Ma suggest a variable cooling history. Some zircons in the plutonic rocks were inherited from a Precambrian source during either assimilation or anatexis. Initial $^{87}Sr/^{86}Sr$ ratios from Mesozoic plutonic rocks are >0.706. Values of 0.708 to 0.709 characterize the eastern part of the terrane, and granitoid rocks in the Barrett Ridge slice yielded a ratio of 0.7095.

Although the Tujunga terrane proper in the San Gabriel Mountains and southeastern California consists largely of Precambrian rocks, sparse isotopic (U-Pb; Rb-Sr) ages show that some Tujungan granitoid intrusions are late Early Cretaceous and Late Cretaceous (Campanian and older); other as yet undated plutons are suspected to be Cretaceous (Ehlig, 1981). Certain Cretaceous granitoid rocks in the Tujunga and Salinian terranes that strongly resemble one another lithologically may be evidence that the two terranes were already linked by mid-Cretaceous time. The Tujunga terrane, including its Cretaceous plutons, is floored structurally by the post–80 Ma Vincent-Orocopia-Chocolate Mountains thrust system. Beneath the thrust system are the Pelona, Rand, and Orocopia schists (Fig. 11), comprising graywacke, mafic volcanic rocks, and Mn-rich cherts that were deformed and metamorphosed under high P/T conditions (Jacobsen, 1983). Part of the protolith for the schists is pre–Late Jurassic, but some workers have pointed out the similarity of the schists to Franciscan rocks, and the possibility remains that they include Late Jurassic or Early Cretaceous protoliths. Some metamorphic recrystallization may have accompanied thrusting, but there is emerging evidence for pre–80 Ma metamorphism as well. In southeastern California, the extent and even the identity of the Tujunga terrane, and the location and nature of its boundary with North American rocks, are currently in dispute (compare Vedder and others, 1983, with D. Miller and others, this volume). On Figure 1, the boundary of the "probably displaced" Tujunga terrane is arbitrarily drawn northeast of the presently exposed Orocopia schist.

The crystalline basement of the Salinian terrane, including the Barrett Ridge slice, is locally overlain unconformably by Upper Cretaceous arkosic sandstone, conglomerate, and subordinate mudstone (Howell and others, 1977; Vedder and others, 1983). The strata are approximately Campanian (or late Campanian) to Maastrichtian and are overlain by Paleogene beds. At some localities, such as Pigeon Point (west of the San Gregorio–Hosgri fault; Fig. 11), the base of the section is not exposed and the basement is inferred to be Salinian. Another anomalous section at Gualala (Fig. 1) is included in the Salinian terrane because it lies west of the San Andreas fault, but strata there are in fault contact with altered basalts of unknown age or affinity.

Late Cretaceous detritus came predominantly from subjacent Salinian basement, but it typically includes a few percent of well-rounded porphyritic, intermediate to silicic volcanic rocks. The Upper Cretaceous section as a whole includes nonmarine as well as diverse submarine-fan facies and is interpreted to have accumulated in a tectonically active setting dominated by wrench faulting. A correlative Upper Cretaceous and Paleocene section resting on Precambrian basement in the western San Gabriel Mountains may represent an additional, late Mesozoic link between the Salinian and Tujunga terranes.

The volume and nature of Cretaceous granitoid plutonic rocks in the Salinian terrane suggest to nearly everyone that it is a displaced fragment of the great batholithic belt that formed along the western margin of North America in response to late Mesozoic subduction. Not everyone, however, currently agrees on which part of the belt the terrane came from and on how far it has moved since Late Cretaceous time. Two simple but radically different alternatives have been proposed. One hypothesis (for example, Dickinson, 1983; Ross, 1984) holds that the Salinian terrane is the southern end of the Sierra Nevada batholith proper and that it has been displaced by several hundred kilometers of latest Cretaceous and Cenozoic right-slip along the San Andreas fault and a subparallel precursor, the "proto–San Andreas." Proponents can cite isotopic data and lithologic correlations compatible with this model. The other hypothesis (for example, Page, 1982; Champion and others, 1984) is based on paleomagnetic data from Upper Cretaceous turbidites at Pigeon Point and Paleocene strata unconformably overlying Salinian granodiorite 45 km to the north. These data suggest that the Salinian terrane was derived from a magmatic belt near the continental margin at least 2,500 km to the south and was subsequently transported parallel to the coast along major transcurrent faults.

Sur-Obispo composite terrane. The Sur-Obispo terrane (Nacimiento block in older literature) includes two major upper Mesozoic tectonostratigraphic assemblages: (1) the Franciscan Complex, and (2) rocks that strongly resemble the Great Valley Group and underlying Coast Range ophiolite exposed east of the San Andreas fault (Page, 1981). The Sur-Obispo terrane is separated from the Salinian terrane to the east by the Sur-Nacimiento fault zone (SNFZ) (Fig. 11). Latest Cretaceous and Tertiary displacements along the predominantly steep faults constituting this zone must add up to at least hundreds of kilometers, because Cretaceous rocks that formed in markedly different environments

(subduction zone and magmatic arc) have been juxtaposed along it. Little else about the magnitude, timing, or sense of displacements on the SNFZ is known with certainty. Although one can view the Sur-Obispo terrane as a western belt of the "in-place" Franciscan Complex and Great Valley Group, it is clearly an allochthonous one, having been displaced by not only several hundred kilometers of Neogene right-slip on the San Andreas fault but also unresolved earlier slip on the SNFZ.

Franciscan rocks in the Sur-Obispo terrane are very similar to those in the Central Belt of the Franciscan Complex (see the relevant section above). Mélanges, consisting of blocks of various sizes dispersed in a scaly mudstone matrix, predominate. Blocks include diverse sandstones, mafic volcanic rocks, chert, conglomerate, and subordinate blueschist; high-grade eclogite and amphibolite, however, are absent. Structural trends, defined by faults, bedding attitudes, and scaly foliation in mélange, are northwesterly. Sandstones show an overall increase in metamorphic grade from southwest to northeast. Metagraywackes near the coast locally contain laumontite; those in the central part of the terrane contain pumpellyite; those in the northeastern part typically contain lawsonite (± pumpellyite, aragonite, jadeitic pyroxene). There is, however, a distinctive component of the Franciscan in the Sur-Obispo terrane that is rare in the Central Belt. Generally well-bedded but locally folded and disrupted Upper Cretaceous (probably Campanian in part) feldspathic sandstone and minor siltstone and conglomerate occur in northwest-trending "slabs" along the coast (Howell and others, 1977). The largest, near Cambria (Fig. 11), is 20 km long. These strata have discordant contacts with underlying or adjacent mélange, and some contain traces of Franciscan-derived detritus. They are interpreted as deep-marine turbidites deposited in trench-slope basins. The remainder of the Franciscan is poorly dated; some sedimentary rocks contain Early Cretaceous microfossils, and a few whole-rock K-Ar ages from metasandstones range from 70 to 88 Ma.

In a few localities, the Franciscan is structurally overlain by an igneous complex correlative with, and perhaps originally part of, the Coast Range ophiolite. These internally faulted remnants include serpentinized ultramafic rocks, gabbro, diorite and diabase dikes, and pillowed and massive basalts. Zircons from fractionated silicic rocks yielded U-Pb ages of 152, 160, and 161 Ma (Late Jurassic). The igneous rocks are depositionally overlain by up to 130 m of Upper Jurassic (Kimmeridgian and late Tithonian) tuffaceous radiolarian chert, which is in turn conformably overlain by Tithonian and, near San Luis Obispo, by Valanginian marine mudstone and thin volcaniclastic turbidites. These beds are correlated with the basal part of the Great Valley Group.

Additional Upper Jurassic, Lower Cretaceous, and Upper Cretaceous clastic sedimentary rocks, with a maximum preserved thickness of about 8 km, are exposed in the eastern and southern Sur-Obispo terrane (Vedder and others, 1983). These strata are inferred to lie on a basement of Coast Range ophiolite and Franciscan Complex, but in most places the basal contact is unexposed. With regard to age, depositional environment, and sedimentary petrology, the Upper Jurassic to Upper Cretaceous strata resemble the Great Valley Group in its main outcrop belt east of the San Andreas fault and especially along the west side of the San Joaquin Valley. Upper Jurassic and Lower Cretaceous rocks consist mostly of Tithonian and Valanginian mudstone and subordinate thin sandstone beds that accumulated in basin-plain and outer-fan environments. The section includes minor Hauterivian or Barremian and Albian rocks. Upper Cretaceous strata are chiefly Campanian and Maastrichtian submarine-fan turbidites, consisting of medium to thick-bedded sandstone and subordinate mudstone and conglomerate. Locally the Upper Cretaceous section includes Cenomanian and Turonian beds. Campanian and Maastrichtian sandstone is very similar petrologically to coeval sandstones in the Salinian terrane and in the Great Valley Group east of the San Andreas fault. All are distinctly feldspathic compared to older sandstones, and they typically contain >10 percent K-feldspar derived from felsic plutonic rocks.

An important 1.5-km-thick section of Campanian and Maastrichtian shallow-marine and nonmarine sandstone, conglomerate, and siltstone crops out next to the SNFZ for 35 km along the eastern edge of the Sur-Obispo terrane (Vedder and others, 1983). These rocks contain granitoid detritus supplied from the east; individual boulders have yielded U-Pb ages (from zircons) ranging from 102 to 126 Ma. If, as seems likely, the provenance for these strata was the Salinian terrane, they record the juxtaposition of the Salinian terrane and the Sur-Obispo terrane along the SNFZ in Campanian time, about 75 to 80 Ma.

Since the Campanian amalgamation of the Salinian and Sur-Obispo terranes, the two have been displaced northward, more or less as a unit. As noted in the preceding section, the total post–latest Cretaceous displacement of the Salinian terrane, and hence the Sur-Obispo terrane, is a matter of debate; there are competing hypotheses specifying either several hundred or at least 2,500 kilometers of transcurrent slip. The history of the Sur-Obispo terrane prior to its Late Cretaceous amalgamation with the Salinian terrane is also unresolved and controversial. The widely acknowledged similarity of Upper Jurassic and Cretaceous rock units in the Sur-Obispo terrane to the Franciscan Complex, Coast Range ophiolite, and Great Valley Group argues that the terrane contains displaced fragments of these lithotectonic belts that developed along the continental margin in response to late Mesozoic subduction. According to one model (Champion and Howell, 1986), based on preliminary paleomagnetic data from Upper Jurassic ophiolitic rocks, the terrane originated even farther south than did Salinia. In another hypothesis (Dickinson, 1983) that is compatible with geological, but not extant paleomagnetic, evidence, rock units in the Sur-Obispo terrane represent *northward* continuations of the Franciscan Complex and Great Valley Group; the whole lot was emplaced against the Salinian terrane as a result of a few hundred kilometers of Late Cretaceous *sinistral* slip on SNFZ.

Other rocks resembling the Franciscan Complex and Great Valley Group. On Santa Cruz Island in the northern Continental Borderland (Howell and others, 1985; Fig. 11), the

Willows Plutonic Complex includes 162 to 163 Ma (U-Pb ages) hornblende diorite and hornblende gabbro cut by diabase dikes; ultramafic rocks and cumulus gabbro occur in an adjacent fault zone. Major- and trace-element chemistry suggests an origin in an island-arc setting. Turonian(?) and Campanian to Maastrichtian submarine-fan turbidites underlie parts of San Miguel, Santa Rosa, and Santa Cruz islands. Although the base of the section is not exposed, these strata are typically interpreted as part of the Great Valley Group overlying a remnant of the Coast Range ophiolite (Willows complex). Similar upper Mesozoic strata apparently underlie part of the submerged Borderland to the southeast. Paleomagnetic data from Cretaceous and Tertiary rocks on the islands point to Cenozoic clockwise rotations and northward translations. The entire terrane is separated by Cenozoic faults from the Great Valley Group on the mainland.

A fault-bounded terrane of the Franciscan-like Catalina Schist (Platt, 1975) underlies part of the eastern Continental Borderland and crops out on Santa Catalina Island and the Palos Verdes Peninsula on the mainland (Fig. 11). On Catalina Island, the schist comprises three tectonic units separated by low-angle faults. Lowest are high P/T blueschist-facies metasandstone, metabasaltic rocks, and metachert. Overlying well-foliated greenschist was derived chiefly from mafic volcanic rocks; blue amphibole in some mafic schist records high P/T metamorphic conditions in this unit as well. Structurally highest is a nappe of amphibolite-facies mafic schist overlain by serpentinized ultramafic rock containing tectonic blocks of garnet amphibolite and eclogite. The low-angle, post-metamorphic faults apparently disrupted an inverted sequence of metamorphic zones that had developed beneath a hot hanging wall of ultramafic rocks. K-Ar ages of 98 to 112 Ma and a U-Pb age of 112 Ma record the minimum age of metamorphism. The original paleogeographic relationship of the Catalina Schist to the Franciscan Complex in the Coast Ranges is unresolved. The terrane has almost certainly been translated and probably rotated in response to Neogene transcurrent slip along faults in the Borderland.

Peninsular Ranges composite terrane. In this section, the "Peninsular Ranges composite terrane" is an informal name for a northwest-trending, elongate crustal block in southwestern California and the Mexican state of Baja California that is underlain largely by the Cretaceous Peninsular Ranges, or Southern California, batholith (Fig. 11). The terrane includes diverse prebatholithic wallrocks that were welded together by the plutons. Northwest-trending Cenozoic transcurrent faults slice through the Peninsular Ranges terrane. In California, its northeastern boundary is the San Andreas fault. The southwestern boundary with the Catalina Schist terrane is marked by Cenozoic strike-slip faults, including the Newport-Inglewood zone near Long Beach. The terrane is truncated at its northwestern end by east-west Cenozoic faults in the Transverse Ranges, although there is probably a sliver in this province as well. Like its cousin the Salinian terrane to the north, the Peninsular Ranges batholith is interpreted as a large, displaced fragment of the great Cretaceous magmatic arc constructed near the late Mesozoic edge of North America. Both terranes also are subjects of the current controversy precipitated by paleomagnetic data indicating post-Paleogene, northward displacements of more than 1,000 km, well in excess of those that can be restored along the San Andreas fault system.

The batholith (Silver and others, 1979; Todd and Shaw, 1979) is made up of hundreds of individual plutons and innumerable dikes. Overall, hornblende-biotite-bearing tonalite and granodiorite are most abundant, but these are part of a wide spectrum of rock types, including rare ultramafic cumulate rock, gabbro, quartz norite, and granite. Where the batholith has been studied in the most detail—east of San Diego—it can be divided into an older roughly western half and a younger eastern half that differ in several respects (Fig. 11). U-Pb ages (zircon) show that compositionally diverse plutons were emplaced throughout the western half from 120 to 105 Ma. These older plutons, which are typically elongate, northwest striking, and internally well foliated, were emplaced and deformed during a synkinematic amphibolite-facies metamorphism that affected plutons and wallrocks alike. Younger plutons in the eastern belt, dated (U-Pb) at 105 to 90 Ma, are chiefly post-tectonic tonalite and granodiorite; felsic granitoid rocks are more abundant than to the west, but large gabbro bodies are absent. The ages of plutonic rocks decrease progressively eastward.

The boundary defined by U-Pb isotopic ages approximately coincides with notable changes in the patterns of isopleths of $\delta^{18}O$ and initial $^{87}Sr/^{86}Sr$. Both sets of isopleths trend northwest, and values increase progressively from west to east across the batholiths. Near the boundary between the older and younger batholithic belts, $\delta^{18}O$ steps up abruptly from about 8.5 to 9.0 and increases eastward to nearly 13; $^{87}Sr/^{86}Sr$ increases from 0.7045 near the boundary to 0.707 along the eastern edge of the younger belt. These isotopic data alone argue for a fundamental change in the nature of the source for granitoid rocks in each of the two magmatic belts. In fact, pre-batholithic wall rocks preserved in pendants and northwest-trending screens are mostly metavolcanic west of the boundary; to the east, metamorphic rocks derived from sedimentary protoliths, including argillaceous sandstone, mudstone, quartzose sandstone, and minor limestone and dolomite, predominate.

The age of the metamorphosed screens and pendants east of San Diego is unknown. Lower-grade wallrocks intermittently exposed in a 100-km-strip along the western margins of the batholith from the Santa Ana Mountains to northeast of San Diego (Gastil and others, 1981; Schoellhamer and others, 1981) include the Santiago Peak Volcanics, which are in part Late Jurassic (Tithonian). The volcanics unconformably overlie the slaty Middle Jurassic Bedford Canyon Formation. Andesitic flows and breccias predominate; dacitic flows, tuff, volcaniclastics sandstone, and argillite are interbedded locally. In the Santa Ana Mountains, the unit is intruded by minor, probably synvolcanic, hypabyssal igneous rocks, including fine-grained diorite, tonalite, and dacite porphyry. In Baja California, along strike to the southeast of the volcanics, the western wallrocks of the batholith include the Aptian-Albian Alisitos Group, which accumulated in

a marine volcanic arc. Part of the Santiago Peak Volcanics might be Early Cretaceous as well.

Cretaceous plutons in the northeastern part of the batholith in the San Jacinto Mountains northwest of the Salton Trough intruded chiefly metasedimentary rocks. Plutons and wallrocks alike were deformed in the Santa Rosa mylonite zone (Fig. 11; Erskine and Wenk, 1985), part of the more extensive Eastern Peninsular Ranges mylonite zone developed along the eastern margin of the batholith. Mylonitic foliations in the Santa Rosa zone dip eastward; fabrics indicate that structurally high rocks to the east were thrust toward the west. The age of the mylonitic deformation is not well established; some likely occurred between 97 Ma, the age of plutonic rocks in the San Jacinto Mountains, and 62 Ma, an age of uplift and cooling determined by fission-track analyses.

Lower Cretaceous plutonic rocks that are sparsely exposed in the Transverse Ranges and the northern Continental Borderland may also be part of the Peninsular Ranges batholith, but their wallrocks differ from those in the Peninsular Ranges themselves. In the Santa Monica Mountains, quartz diorite intruded the Upper Jurassic (Oxfordian to Kimmeridgian) Santa Monica Slate. On Santa Cruz Island, metavolcanic greenschist is intruded by a 140-Ma leucocratic tonalite.

All of the elements included here within the Peninsular Ranges composite terrane are linked not only by the Peninsular Ranges batholith but also by distinctive post-batholithic Upper Cretaceous strata (Bottjer and Link, 1984). Those deposited on plutonic rocks in and near San Diego constitute probable Turonian alluvial debris overlain by upper Campanian to lower Maastrichtian shallow- and deep-marine clastic rocks. Time-correlative strata on San Miguel Island 300 km northwest (mentioned in the preceding section) were probably deposited in the same basin, but farther south, and subsequently transported northward along Cenozoic strike-slip faults. In the northern Santa Ana Mountains, Turonian alluvial-fan red beds overlying the Santiago Peak Volcanics are themselves overlain by a Turonian, Coniacian(?), Santonian(?), and Maastrichtian section that includes shallow-marine, slope, and submarine-fan facies. In the Santa Monica Mountains, correlative red beds resting on the Santa Monica Formation are overlain by Turonian and Campanian submarine-fan deposits. Over 1,800 m of similar middle Campanian and lower Maastrichtian strata also crop out in the Chatsworth Formation in the Simi Hills 20 km to the northwest, but their lower contact is unexposed. The Upper Cretaceous sections in the Santa Monica Mountains and Simi Hills were originally contiguous with the section in the Santa Ana Mountains; they have been displaced northward and subsequently rotated by Cenozoic slip on high-angle faults (Link and others, 1984). The source for the Upper Cretaceous arkosic sandstones in these three sections consisted largely of granitoid rocks in the Peninsular Ranges batholith. Paleocurrents flowed generally westward. The nonmarine Turonian sedimentary rocks and the underlying unconformity together record a regionally extensive uplift of the terrane, but its tectonic cause is unknown.

Many workers have noted that the Upper Cretaceous sedimentary rocks blanketing the Peninsular Ranges composite terrane resemble the Upper Cretaceous part of the Great Valley Group, especially with regard to petrology and provenance. Like the Great Valley Group, these strata are thought to have been deposited in a forearc basin that lay parallel to and west of the magmatic arc represented by the Peninsular Ranges batholith. A simple restoration of about 300 km of Cenozoic right-lateral strike-slip on the San Andreas and related faults closes the Gulf of California and aligns the Peninsular Ranges batholith and its related forearc basin with the coeval Sierra Nevada batholith and Great Valley Group. Once again, however, an impressive set of paleomagnetic data from Cretaceous plutonic rocks and overlying Cretaceous and Eocene sedimentary rocks in California and Baja California directly conflicts with such a restoration (Fry and others, 1985; Hagstrum and others, 1985; Champion and Howell, 1986). These data indicate that the Peninsular Ranges composite terrane was originally much farther south and attached to the continental margin in central Mexico until at least the end of the Cretaceous and possibly the end of the Eocene. The entire terrane subsequently moved northward 900 to 1,600 km prior to its post-Miocene displacement on the San Andreas fault.

EPILOGUE

We have highlighted some outstanding problems concerning the Late Jurassic to early Late Cretaceous geology of the U.S. Cordillera. For example, how are thrusts in the Cordilleran belt rooted at depth to the west? Is late Mesozoic synkinematic metamorphism in the hinterland related to deep, crustal-scale décollements? How will the paleomagnetic data suggesting large displacements be reconciled with geologic evidence requiring only modest slip along Cenozoic faults? Of the three tectonic elements in the western Cordillera related to late Mesozoic subduction, who do only the mid-Cretaceous and younger parts of the Sierran batholithic belt apparently continue north of about 43°?

Another fundamental question is: What actually drove the shortening in the Cordilleran fold and thrust belt? One hypothesis ascribes it to the active west-directed underthrusting, or subduction, of the relatively rigid North American craton beneath the western part of the continent, which had been heated and softened by arc magmatism. Such intraplate contraction is compatible with the inference that the craton was moving westward in an absolute reference frame in concert with the opening of the Atlantic Ocean. A different hypothesis calls for repeated collisions along the western convergent margin of North America to explain contractional events in the thrust belt. Clearly, the uncertainties in the late Mesozoic paleogeography of the western Cordillera need to be resolved before we can properly evaluate such a model.

REFERENCES CITED

Allmendinger, R. W., Miller, D. M., and Jordan, T. E., 1985, Known and inferred Mesozoic deformation in the hinterland of the Sevier belt, northwest Utah, *in* Kerns, J. G., ed., Geology of northwest Utah, southern Idaho, and northeast Nevada: Utah Geological Association Publication 13, p. 21–34.

Armstrong, F. C., and Oriel, S. S., 1965, Tectonic development of Idaho-Wyoming thrust belt: American Association of Petroleum Geologists Bulletin, v. 49, p. 1847–1866.

Armstrong, R. L., 1968, Sevier orogenic belt in Nevada and Utah: Geological Society of America Bulletin, v. 79, p. 439–458.

——, 1982, Cordilleran metamorphic core complexes; From Arizona to southern Canada: Annual Reviews of Earth and Planetary Science, v. 10, p. 129–154.

Armstrong, R. L., and Suppe, J., 1973, Potassium-argon geochronometry of Mesozoic igneous rocks in Nevada, Utah, and southern California: Geological Society of America Bulletin, v. 84, p. 1375–1392.

Armstrong, R. L., Taubeneck, W. H., and Hales, P. O., 1977, Rb-Sr and K-Ar geochronometry of Mesozoic granitic rocks and their Sr isotopic composition, Oregon, Washington, and Idaho: Geological Society of America Bulletin, v. 88, p. 397–411.

Bailey, E. B., Blake, M. C., Jr., and Jones, D. L., 1970, On-land Mesozoic oceanic crust in California Coast Ranges: U.S. Geological Survey Professional Paper 700–C, p. C70–C81.

Barksdale, J. D., 1975, Geology of the Methow Valley, Okanogan County, Washington: Washington Division of Geology and Earth Resources Bulletin 68, 72 p.

Bateman, P. C., 1981, Geologic and geophysical constraints on models for the origin of the Sierra Nevada batholith, California, *in* Ernst, W. G., ed., The geotectonic development of California: Englewood Cliffs, New Jersey, Prentice-Hall, p. 71–86.

——, 1983, A summary of critical relations in the central part of the Sierra Nevada batholith, California, U.S.A., *in* Roddick, J. A., ed., Circum-Pacific plutonic terranes: Geological Society of America Memoir 159, p. 241–254.

Blake, M. C., Jr., ed., 1984, Franciscan geology of northern California: Society of Economic Paleontologists and Mineralogists Pacific Section, 254 p.

Blake, M. C., Jr., Engebretson, D. C., Jayko, A. S., and Jones, D. L., 1985, Tectonostratigraphic terranes in southwest Oregon, *in* Howell, D. G., ed., Tectonostratigraphic terranes of the Circum-Pacific region: Houston, Circum-Pacific Council for Energy and Mineral Resources, p. 147–157.

Blake, M. C., Jr., Jayko, A. S., and McLaughlin, R. J., 1985, Tectonostratigraphic terranes of the northern Coast Ranges, California, *in* Howell, D. G., ed., Tectonostratigraphic terranes of the Circum-Pacific region: Houston, Circum-Pacific Council for Energy and Mineral Resources, p. 159–171.

Bottjer, D. J., and Link, M. H., 1984, A synthesis of Late Cretaceous southern California and northern Baja California paleogeography, *in* Crouch, J. K., and Bachman, S. B., eds., Tectonics and sedimentation along the California margin: Society of Economic Paleontologists and Mineralogists Pacific Section, p. 171–188.

Bourgeois, J., and Dott, R. H., Jr., 1985, Stratigraphy and sedimentology of Upper Cretaceous rocks in coastal Southwest Oregon; Evidence for wrench-fault tectonics in a postulated accretionary terrane: Geological Society of America Bulletin, v. 96, p. 1007–1019.

Brandon, M. T., Cowan, D. S., and Vance, J. A., 1988, The Late Cretaceous San Juan thrust system, San Juan Islands, Washington: Geological Society of America Special Paper 221, 82 p.

Brown, E. H., 1987, Structural geology and accretionary history of the Northwest Cascades system, Washington and British Columbia: Geological Society of America Bulletin, v. 99, p. 201–214.

Bruhn, R. L., and Beck, S. L., 1981, Mechanics of thrust faulting in crystalline basement, Sevier orogenic belt, Utah: Geology, v. 9, p. 200–204.

Burchfiel, B. C., and Davis, G. A., 1975, Nature and controls of Cordilleran orogenesis, western United States; Extensions of an earlier synthesis: American Journal of Science, v. 275-A, p. 363–396.

Champion, D. E., and Howell, D. G., 1986, Paleomagnetism of Cretaceous and Eocene strata, San Miguel Island, California, borderland and the northward translation of Baja California: Journal of Geophysical Research, v. 91, p. 11557–11570.

Champion, D. E., Howell, D. G., and Gromme, C. S., 1984, Paleomagnetic and geologic data indicating 2,500 km of northward displacement for the Salinian and related terranes, California: Journal of Geophysical Research, v. 89, p. 7736–7752.

Chen, J. H., and Moore, J. G., 1982, Uranium-lead isotopic ages from the Sierra Nevada batholith, California: Journal of Geophysical Research, v. 87, p. 4761–4784.

Cheney, E. S., 1980, Kettle dome and related structures of northeastern Washington, *in* Crittenden, M. D., Jr., and others, eds., Cordilleran Metamorphic Core Complexes: Geological Society of America Memoir 153, p. 463–483.

Coney, P. J., and Harms, T., 1984, Cordilleran metamorphic core complexes; Cenozoic extensional relics of Mesozoic compression: Geology, v. 12, p. 550–554.

Cook, T. D., and Bally, A. W., eds., 1975, Stratigraphic Atlas of North and Central America: Princeton, New Jersey, Princeton University Press, 272 p.

Cowan, D. S., and Potter, C. J., 1986, B-3 Juan de Fuca spreading ridge to Montana thrust belt: Boulder, Colorado, Geological Society of America, Centennial Continent/Ocean Transect no. 9, 3 sheets with text, scale 1:500,000.

DeWitt, E., Armstrong, R. L., Sutter, J. F., and Zartman, R. E., 1984, U-Th-Pb, Rb-Sr, and Ar-Ar mineral and whole-rock isotopic systematics in a metamorphosed granitic terrane, southeastern California: Geological Society of America Bulletin, v. 95, p. 723–739.

Dickinson, W. R., 1970, Relations of andesites, granites, and derivative sandstones to arc-trench tectonics: Reviews of Geophysics and Space Physics, v. 8, p. 813–860.

——, 1976, Sedimentary basins developed during evolution of Mesozoic-Cenozoic arc-trench system in western North America: Canadian Journal of Earth Sciences, v. 13, p. 1268–1287.

——, 1983, Cretaceous sinistral slip along Nacimiento fault in coastal California: American Association of Petroleum Geologists Bulletin, v. 67, p. 624–645.

Dickinson, W. R., and Seely, D. R., 1979, Structure and stratigraphy of forearc regions: American Association of Petroleum Geologists Bulletin, v. 63, p. 2–31.

Dickinson, W. R., Klute, M. A., and Swift, P. N., 1986, The Bisbee Basin and its bearing on late Mesozoic paleogeographic and paleotectonic relations between the Cordilleran and Caribbean regions, *in* Abbott, P. L., ed., Cretaceous stratigraphy, western North America: Society of Economic Paleontologists and Mineralogists Pacific Section, p. 51–62.

Dixon, J. S., 1982, Regional structural synthesis, Wyoming salient of western overthrust belt: American Association of Petroleum Geologists Bulletin, v. 66, p. 1560–1580.

Ehlig, P. L., 1981, Origin and tectonic history of the basement terrane of the San Gabriel Mountains, central Transverse Ranges, *in* Ernst, W. G., ed., The geotectonic development of California: Englewood Cliffs, New Jersey, Prentice-Hall, p. 253–283.

Engels, J. C., Tabor, R. W., Miller, F. K., and Obradovich, J. D., 1976, Summary of K-Ar, Rb-Sr, U-Pb, Pbα, and fission-track ages of rocks from Washington State prior to 1975 (exclusive of Columbia Plateau basalts): U.S. Geological Survey Miscellaneous Field Studies Map MF–710, scale 1:1,000,000.

Ernst, W. G., 1970, Tectonic contact between the Franciscan mélange and the Great Valley sequence; Crustal expression of a late Mesozoic Benioff zone: Journal of Geophysical Research, v. 75, p. 886–901.

Erskine, B. G., and Wenk, H.-R., 1985, Evidence for Late Cretaceous crustal thinning in the Santa Rosa mylonite zone, Southern California: Geology, v. 13, p. 274–277.

Evernden, J. F., and Kistler, R. W., 1970, Chronology of emplacement of Mesozoic batholithic complexes in California and western Nevada: U.S.

Geological Survey Professional Paper 623, 42 p.

Fouch, T. D., Lawton, T. F., Nichols, D. J., Cashion, W. B., and Cobban, W. A., 1983, Patterns and timing of synorogenic sedimentation in Upper Cretaceous rocks of central and northeast Utah, *in* Reynolds, M. W., and Dolly, E. D., eds., Mesozoic paleogeography of west-central United States: Society of Economic Paleontologists and Mineralogists Rocky Mountain Section, p. 305–336.

Fry, J. G., Bottjer, D. J., and Lund, S. P., 1985, Magnetostratigraphy of displaced Upper Cretaceous strata in Southern California: Geology, v. 13, p. 648–651.

Gastil, G., Morgan, G., and Krummenacher, D., 1981, The tectonic history of peninsular California and adjacent Mexico, *in* Ernst, W. G., ed., The geotectonic development of California: Englewood Cliffs, New Jersey, Prentice-Hall, p. 284–306.

Hagstrum, J. T., McWilliams, M., Howell, D. G., and Gromme, S., 1985, Mesozoic paleomagnetism and northward translation of the Baja California Peninsula: Geological Society of America Bulletin, v. 96, p. 1077–1090.

Hamilton, W. B., 1969, Mesozoic California and the underflow of Pacific mantle: Geological Society of America Bulletin, v. 80, p. 2409–2430.

Harrison, J. E., Kleinkopf, M. D., and Wells, J. D., 1980, Phanerozoic thrusting in Proterozoic Belt rocks, northwestern United States: Geology, v. 8, p. 407–411.

Heller, P. L., Bowdler, S. S., Chambers, H. P., Coogan, J. C., Hagen, E. S., Schuster, M. W., Winslow, N. S., and Lawton, T. F., 1986, Time of initial thrusting in the Sevier orogenic belt, Idaho-Wyoming and Utah: Geology, v. 14, p. 388–391.

Hopson, C. A., Mattinson, J. M., and Pessagno, E. A., Jr., 1981, Coast Range ophiolite, western California, *in* Ernst, W. G., ed., The geotectonic development of California: Englewood Cliffs, New Jersey, Prentice-Hall, p. 418–510.

Howell, D. G., Vedder, J. G., McLean, H., Joyce, J. M., Clarke, S. H., Jr., and Smith, G., 1977, Review of Cretaceous geology, Salinian and Nacimiento blocks, Coast Ranges of central California, *in* Howell, D. G., Vedder, J. G., and McDougall, K., eds., Cretaceous geology of the California coast ranges, west of the San Andreas fault: Society of Economic Paleontologists and Mineralogists Pacific Section, p. 1–46.

Howell, D. G., Gibson, J. D., Fuis, G. S., Knapp, J. H., Haxel, G. B., Keller, B. R., Silver, L. T., and Vedder, J. G., 1985, C–3 Pacific abyssal plain to the Rio Grande rift: Boulder, Colorado, Geological Society of America, Centennial Continent/Ocean Transect no. 5, 3 sheets with text, scale 1:500,000.

Hyndman, D. W., 1983, The Idaho batholith and associated plutons, Idaho and western Montana, *in* Roddick, J. A., ed., Circum-Pacific plutonic terranes: Geological Society of America Memoir 159, p. 213–240.

Ingersoll, R. V., 1978, Paleogeography and paleotectonics of the late Mesozoic forearc basin of northern and central California, *in* Howell, D. G., and McDougall, K. A., eds., Mesozoic paleogeography of the western United States: Society of Economic Paleontologists and Mineralogists Pacific Section, p. 471–482.

—— , 1979, Evolution of the Late Cretaceous forearc basin, northern and central California: Geological Society of America Bulletin, Part I, v. 90, p. 813–826.

—— , 1983, Petrofacies and provenance of late Mesozoic forearc basin, northern and central California: American Association of Petroleum Geologists Bulletin, v. 67, p. 1125–1142.

Irving, E., Woodsworth, G. J., Wynne, P. J., and Morrison, A., 1985, Paleomagnetic evidence for displacement from the south of the Coast Plutonic Complex, British Columbia: Canadian Journal of Earth Sciences, v. 22, p. 584–598.

Irwin, W. P., 1985, Age and tectonics of plutonic belts in accreted terranes of the Klamath Mountains, California and Oregon, *in* Howell, D. G., ed., Tectonostratigraphic terranes of the Circum-Pacific region: Houston, Circum-Pacific Council for Energy and Mineral Resources, p. 187–199.

Jacobson, C. E., 1983, Structural geology of the Pelona Schist and Vincent thrust, San Gabriel Mountains, California: Geological Society of America Bulletin, v. 94, p. 753–767.

Jones, D. L., Bailey, E. H., and Imlay, R. W., 1969, Structural and stratigraphic significance of the *Buchia* zones in the Colyear Springs–Paskenta area, California: U.S. Geological Survey Professional Paper 647–A, 24 p.

Jordan, T. E., 1981, Thrust loads and foreland basin evolution, Cretaceous, western United States: American Association of Petroleum Geologists Bulletin, v. 65, p. 2506–2520.

Link, M. H., Squires, R. L., and Colburn, I. P., 1984, Slope and deep-sea fan facies and paleogeography of the Upper Cretaceous Chatsworth Formation, Simi Hills, California: American Association of Petroleum Geologists Bulletin, v. 68, p. 850–873.

Lund, K., and Snee, L. W. 1988, Metamorphism, structural development, and age of the continent–island arc juncture in west-central Idaho, *in* Ernst, W. G., ed., Metamorphism and crustal evolution of the western United States, Rubey Vol. 7: Englewood Cliffs, Prentice-Hall, p. 296–331.

McDowell, F. W., Lehman, D. H., Gucwa, P. R., Fritz, D., and Maxwell, J. C., 1984, Glaucophane schists and ophiolites of the northern California Coast Ranges; Isotopic ages and their tectonic implications: Geological Society of America Bulletin, v. 95, p. 1373–1382.

Mattinson, J. M., and James, E. W., 1985, Salinian Block U/Pb age and isotopic variations; Implications for origin and emplacement of the Salinian terrane, *in* Howell, D. G., ed., Tectonostratigraphic terranes of the Circum-Pacific region: Houston, Circum-Pacific Council for Energy and Mineral Resources, p. 215–226.

Maxwell, J. C., compiler, 1981, Geologic cross sections, northern California coast ranges to northern Sierra Nevada, and Lake Pillsbury area to southern Klamath Mountains: Geological Society of America Map and Chart Series MC-28N, scale 1:250,000.

Miller, E. L., Gans, P. B., Wright, J. E., and Sutter, J. F., 1988, Metamorphic history of the east-central Basin and Range province; Tectonic setting and relationship to magmatism, *in* Ernst, W. G., ed., Metamorphism and crustal evolution of the western United States; Rubey Vol. 7: Englewood Cliffs, Prentice-Hall, p. 649–682.

Miller, F. K., and Clark, L. D., 1975, Geology of the Chewelah–Loon Lake area, Stevens and Spokane Counties, Washington: U.S. Geological Survey Professional Paper 806, 74 p.

Miller, F. K., and Engels, J. C., 1975, Distribution and trends of discordant ages of the plutonic rocks of northeastern Washington and northern Idaho: Geological Society of America Bulletin, v. 86, p. 517–528.

Misch, P.,1966, Tectonic evolution of the Northern Cascades of Washington State; a west-Cordilleran case history: Canadian Institute of Mining and Metallurgy Special Volume 8, p. 101–148.

—— , 1977, Bedrock geology of the North Cascades, *in* Brown, E. H., and Ellis, R. C., eds., Geological excursions in the Pacific Northwest: Bellingham, Western Washington University, p. 1–62.

Nilsen, T. H., ed., 1984, Geology of the Upper Cretaceous Hornbrook Formation, Oregon and California: Society of Economic Paleontologists and Mineralogists Pacific Section, 257 p.

Oldow, J. S., 1984, Evolution of a late Mesozoic back-arc fold and thrust belt, northwestern Great Basin, U.S.A.: Tectonophysics, v. 102, p. 245–274.

Page, B. M., 1981, The southern Coast Ranges, *in* Ernst, W. G., ed., The geotectonic development of California: Englewood Cliffs, New Jersey, Prentice-Hall, p. 329–417.

—— , 1982, Migration of Salinian composite block, California, and disappearance of fragments: American Journal of Science, v. 282, p. 1694–1734.

Platt, J. P., 1975, Metamorphic and deformational processes in the Franciscan Complex, California; Some insights from the Catalina Schist terrane: Geological Society of America Bulletin, v. 86, p. 1337–1347.

—— , 1986, Dynamics of orogenic wedges and the uplift of high-pressure metamorphic rocks: Geological Society of America Bulletin, v. 97, p. 1037–1053.

Ross, D. C., 1978, The Salinian block; A Mesozoic granitic orphan in the California coast ranges, *in* Howell, D. G., and McDougall, K. A., eds., Mesozoic paleogeography of the western United States: Society of Economic Paleontologists and Mineralogists Pacific Section, p. 509–522.

—— , 1984, Possible correlations of basement rocks across the San Andreas, San Gregorio–Hosgri, and Rinconada-Reliz–King City faults, California: U.S.

Geological Survey Professional Paper 1317, 37 p.
Royse, F., Jr., Warner, M. A., and Reese, D. L., 1975, Thrust belt structural geometry and related stratigraphic problems, Wyoming-Idaho–northern Utah, *in* Bolyard, D. W., ed., Deep drilling frontiers of the central Rocky Mountains: Rocky Mountain Association of Geologists 1975 Symposium, p. 41–54.
Ruppel, E. T., and Lopez, D. A., 1984, The thrust belt in southwest Montana and east-central Idaho: U.S. Geological Survey Professional Paper 1278, 41 p.
Saleeby, J. B., 1986, C-2 Central California offshore to Colorado Plateau: Boulder, Colorado, Geological Society of America, Centennial Continent/Ocean Transect no. 10, 2 sheets with text, scale 1:500,000.
Saleeby, J. B., and Sharp, W. D., 1980, Chronology of the structural and petrologic development of the southwest Sierra Nevada foothills, California: Geological Society of America Bulletin, Part I, v. 91, p. 317–320.
Schoellhamer, J. E., Vedder, J. G., Yerkes, R. F., and Kinney, D. M., 1981, Geology of the northern Santa Ana Mountains, California: U.S. Geological Survey Professional Paper 420-D, 109 p.
Silver, L. T., Taylor, H. P., Jr., and Chappell, B., 1979, Some petrological, geochemical, and geochronological observations of the Peninsular Ranges batholith near the international border of the U.S.A. and Mexico, *in* Abbott, P. L., and Todd, V. R., eds., Mesozoic crystalline rocks; Peninsular Ranges batholith and pegmatites, Point Sal ophiolite: San Diego, California, San Diego State University, p. 83–110.
Smith, J. G., McKee, E. H., Tatlock, D. B., and Marvin, R. F., 1971, Mesozoic granitic rocks in northwestern Nevada; A link between the Sierra Nevada and Idaho batholiths: Geological Society of America Bulletin, v. 82, p. 2933–2944.
Speed, R., Elison, M. W., and Heck, F. R., 1988, Phanerozoic tectonic evolution of the Great Basin, *in* Ernst, W. G., ed., Metamorphism and crustal evolution of the western United States; Rubey Vol. 7: Englewood Cliffs, Prentice-Hall, p. 572–605.
Stern, T. W., Bateman, P. C., Morgan, B. A., Newell, M. F., and Peck, D. L., 1981, Isotopic U-Pb ages of zircon from the granitoids of the central Sierra Nevada, California: U.S. Geological Survey Professional Paper 1185, 17 p.
Suchecki, R. K., 1984, Facies history of the Upper Jurassic–Lower Cretaceous Great Valley sequence; Response to structural development of an outer-arc basin: Journal of Sedimentary Petrology, v. 54, p. 170–191.
Suppe, J., 1978, Cross section of southern part of Northern Coast Ranges and Sacramento Valley, California: Geological Society of America Map and Chart Series MC-28B, scale 1:250,000.
Suppe, J., and Foland, K. A., 1978, The Goat Mountain schists and Pacific Ridge Complex; A redeformed but still intact late Mesozoic Franciscan schuppen complex, *in* Howell, D. G., and McDougall, K. A., eds., Mesozoic paleogeography of the western United States: Society of Economic Paleontologists and Mineralogists Pacific Section, p. 431–451.
Tobisch, O. T., Saleeby, J. B., and Fiske, R. S., 1986, Structural history of continental volcanic arc rocks, eastern Sierra Nevada, California; A case for extensional tectonics: Tectonics, v. 5, p. 65–94.
Todd, V. R., and Shaw, S. E., 1979, Structural, metamorphic, and intrusive framework of the Peninsular Ranges batholith in southern San Diego County, California, *in* Abbott, P. L., and Todd, V. R., eds., Mesozoic crystalline rocks; Peninsular Ranges batholith and pegmatites, Point Sal ophiolite: San Diego, California, San Diego State University, p. 83–110.
Trexler, J. H., Jr., and Bourgeois, J., 1985, Evidence for mid-Cretaceous wrench-faulting in the Methow Basin, Washington; Tectonostratigraphic setting of the Virginian Ridge Formation: Tectonics, v. 4, p. 379–394.
Umhoefer, P. J., 1987, Northward translation of "Baja British Columbia" along the Late Cretaceous to Paleocene margin of western North America: Tectonics, v. 6, p. 377–394.
Vallier, T. L., and Brooks, H. C., eds., 1987, Geology of the Blue Mountains region of Oregon, Idaho, and Washington; The Idaho batholith and its border zone: U.S. Geological Survey Professional Paper 1436, 196 p.
Vedder, J. G., Howell, D. G., and McLean, H., 1983, Stratigraphy, sedimentation, and tectonic accretion of exotic terranes, Southern Coast Ranges, California, *in* Watkins, J. S., and Drake, C. L., eds., Studies in continental margin geology: American Association of Petroleum Geologists Memoir 34, p. 471–496.
Wentworth, C. M., Blake, M. C., Jr., Jones, D. L., Walter, A. W., and Zoback, M. D., 1984, Tectonic wedging associated with emplacement of the Franciscan assemblage, California Coast Ranges, *in* Blake, M. C., Jr., ed., Franciscan geology of northern California: Society of Economic Paleontologists and Mineralogists Pacific Section, p. 163–173.
Wiltschko, D. V., and Dorr, J. A., Jr., 1983, Timing of deformation in overthrust belt and foreland of Idaho, Wyoming, and Utah: American Association of Petroleum Geologists Bulletin, v. 67, p. 1304–1322.
Worrall, D. M., 1981, Imbricate low-angle faulting in uppermost Franciscan rocks, South Yolla Bolly area, northern California: Geological Society of America Bulletin, Part I, p. 703–729.

Manuscript Accepted by the Society April 12, 1988

NOTES ADDED IN PROOF

Since we submitted the final version of this manuscript in March 1988, some published information has either amplified certain points in our chapter, introduced new findings that we would have included had they been available, or in two cases amended the age of rock units. The following notes are arranged by subheadings in the same sequence as they appear in the text. In some instances, we just list a reference or two that we judge to be a valuable summary or source of data for the topic at hand.

Cordilleran Fold and Thrust Belt. *Timing of Thrust Movements.* Fleck and Carr (1990) reported an $^{40}Ar/^{39}Ar$ age of 99 Ma for the Lavinia Wash unit, and so dated the final displacement on both the Keystone and Contact thrusts as no older than late Early Cretaceous.

Sierran Batholithic Belt. Preliminary U-Pb isotopic ages in Walker (1989) indicate that the plutonic rocks in east-central Oregon (cited as "probably pre-Nevadan" in text) were emplaced in the Early Cretaceous and so represent the probable northward continuation of the Early Cretaceous Sierran-Klamath belt. *Age:* Saleeby and others (1989) reported new U-Pb ages for granitoid plutons in the northwestern Sierra Nevada batholith. *Syn-batholithic Volcanism and Deformation.* More evidence for deformation within and adjacent to the Sierra Nevada batholith and in the extension of the batholithic belt in the southeastern Mojave Desert has accumulated. On the basis of displaced rock units in pendants in the east-central Sierra Nevada batholith at 38°N, Lahren and Schweickert (1990) hypothesized 400-500 km of Early Cretaceous dextral strike-slip along the north-northwest–striking intrabatholithic Mojave–Snow Lake fault. Paleogeographically, their hypothesis implies that pre-Late Cretaceous elements in the Sierra Nevada foothills, the Great Valley Group, and the Franciscan Complex lying west of the batholith also lay at least 400 km farther south prior to about 100 or 110 Ma. Also, Tobisch and others (1989) presented geologic and isotopic evidence from the western foothills for Early Cretaceous ductile deformation within and adjacent to an east-dipping reverse fault zone. In southeastern California and adjacent western Arizona at latitude 34°N, where the Cordilleran fold and thrust belt not only includes slices of Precambrian basement but is also coextensive with mid-Cretaceous intrusions, isotopic data (for example, Knapp and Heizler, 1990) show that thrusting, synkinematic plutonism, and regional metamorphism proceeded contemporaneously from ~80-90 Ma but probably continued during the Late Cretaceous.

Great Valley Group. Robertson (1990) comprehensively reviewed evidence for the fore-arc setting of the pre-upper Tithonian igneous and sedimentary basement of the Great Valley Group.

San Juan Islands, North Cascades, and Okanogan Region. McGroder (1991) summarized evidence for the interpretation that west-vergent thrusting in the San Juans and northwest Cascades, burial and regional metamorphism of the Skagit Metamorphic Terrane, and shortening in the Methow-Pasayten sequence were contemporaneous events (from about 100 to 90 Ma), recording east-west crustal shortening on the order of 400-500 km due to collision of the Insular

Superterrane (Wrangellia terrane at these latitudes). According to Rubin and others (1990), this zone of mid-Cretaceous, collision-related shortening extends for at least 1,200 km, from Washington into southeastern Alaska. Walker and Brown (1991) and Haugerud and others (1991) provided complete isotopic data for early Late Cretaceous ages of plutonism and metamorphism in the Skagit Metamorphic Terrane. Regarding the discordant paleomagnetic data from mid-Cretaceous plutons in northwestern Washington and southwestern British Columbia, Butler and others (1989) presented geologic arguments that they believe favor tilting, rather than latitudinal displacements. The issue remains unresolved.

Western and Southwestern California. *Salinian Terrane, Sur-Obispo Composite Terrane.* Lund and others (1991), citing new data from Upper Cretaceous strata on the central granitic block of the Salinian Terrane, favor the hypothesis that it has been displaced northward only a few hundred kilometers. These results imply that a separate, westernmost slice of the Salinian Terrane, or perhaps another terrane altogether, underlies the paleomagnetically discordant Upper Cretaceous strata at Pigeon Point (Fig. 11), which have apparently been displaced >2,000 km. The putative stratigraphic link between the Salinian and Sur-Obispo terranes, based on "Salinian" granitic clasts in an Upper Cretaceous conglomerate on Sur-Obispo, needs to be re-evaluated.

Peninsular Ranges Composite Terrane. Silver and Chappell (1988) is a more readily accessible reference on the petrology and geochemistry of the Peninsular Ranges batholith than Silver and others (1979).

ADDITIONAL REFERENCES

Butler, R. F., Gehrels, G. E., McClelland, W. C., May, S. R., and Klepacki, D., 1989, Discordant paleomagnetic poles from the Canadian Coast Plutonic Complex: Regional tilt rather than large-scale displacement?: Geology, v. 17, p. 691–694.

Fleck, R. J., and Carr, M. D., 1990, The age of the Keystone thrust: Laser-fusion $^{40}Ar/^{39}Ar$ dating of foreland basin deposits, southern Spring Mountains, Nevada: Tectonics, v. 9, p. 467–476.

Haugerud, R. A., Van der Heyden, P., Tabor, R. W., Stacey, J. S., and Zartman, R. E., 1991, Late Cretaceous and early Tertiary plutonism and deformation in the Skagit gneiss Complex, North Cascade Range, Washington and British Columbia: Geological Society of America Bulletin, v. 103, p. 1297–1307.

Knapp, J. H., and Heizler, M. T., 1990, Thermal history of crystalline nappes of the Maria fold and thrust belt, west central Arizona: Journal of Geophysical Research, v. 95, p. 20049–20073.

Lahren, M. M., and Schweickert, R. A., 1990, Evidence of uppermost Proterozoic to Lower Cambrian miogeoclinal rocks and the Mojave–Snow Lake fault; Snow Lake pendant, central Sierra Nevada, California: Tectonics, v. 9, p. 1585–1608.

Lund, S. P., Bottjer, D. J., Whidden, K. J., Powers, J. E., and Steele, M. C., 1991, Paleomagnetic evidence for Paleogene terrane displacements and accretion in southern California, *in* Abbott, P. L., and May, J. A., eds., Eocene Geologic History, San Diego Region: Society of Economic Paleontologists and Mineralogists, Pacific Section, v. 68, p. 99–106.

McGroder, M. F., 1991, Reconciliation of two-sided thrusting, burial metamorphism, and diachronous uplift in the Cascades of Washington and British Columbia: Geological Society of America Bulletin, v. 103, p. 189–209.

Robertson, A.H.F., 1990, Sedimentology and tectonic implications of ophiolite-derived clastics overlying the Jurassic Coast Range Ophiolite, northern California: American Journal of Science, v. 290, p. 109–163.

Rubin, C. M., Saleeby, J. B., Cowan, D. S., Brandon, M. T., and McGroder, M. F., 1990, Regionally extensive mid-Cretaceous west-vergent thrust system in the northwestern Cordillera; Implications for continent-margin tectonism: Geology, v. 18, p. 276–280.

Saleeby, J. B., Geary, E. E., Paterson, S. R., and Tobisch, O. T., 1989, Isotopic systematics of Pb/U (zircon) and $^{40}Ar/^{39}Ar$ (biotite-hornblende) from rocks of the central Foothills terrane, Sierra Nevada, California: Geological Society of America Bulletin, v. 101, p. 1481–1492.

Silver, L. T., and Chappell, B. W., 1988, The Peninsular Ranges batholith: An insight into the evolution of the Cordilleran batholiths of southwestern North America: Transactions of the Royal Society of Edinburgh-Earth Sciences, v. 79, p. 105–121.

Tobisch, O. T., Paterson, S. R., Saleeby, J. B., and Geary, E. E., 1989, Nature and timing of deformation in the Foothills terrane, central Sierra Nevada, California; Its bearing on orogenesis: Geological Society of America Bulletin, v. 101, p. 401–413.

Walker, N. W., 1989, Early Cretaceous initiation of post-tectonic plutonism and the age of the Connor Creek fault, northeastern Oregon: Geological Society of America Abstracts with Programs, v. 21, p. 155.

Walker, N. W., and Brown, E. H., 1991, Is the southeast Coast Plutonic Complex the consequence of accretion of the Insular superterrane? Evidence from U-Pb zircon geochronometry in the northern Washington Cascades: Geology, v. 19, p. 714, 717.

Printed in U.S.A.

The Geology of North America
Vol. G-3, The Cordilleran Orogen: Conterminous U.S.
The Geological Society of America, 1992

Chapter 6

Late Cretaceous to early Eocene geologic evolution of the U.S. Cordillera

David M. Miller
U.S. Geological Survey, 345 Middlefield Road, Menlo Park, California 94025
Tor H. Nilsen
Applied Earth Technology, Inc., 595 Price Avenue, Suite 1, Redwood City, California 94063
William L. Bilodeau
Department of Geology, California Lutheran University, 60 W. Olsen Rd., Thousand Oaks, California 91360

INTRODUCTION

D. M. Miller

Late Cretaceous, Paleocene, and early Eocene time (ca. 80 to 50 Ma) was marked by major changes in the loci of magmatism, tectonism, metamorphism, and sedimentation in the western United States. Prior to this period, the Cordillera was segmented into an eastern fold and thrust belt with adjacent foreland basin, a central magmatic arc with flanking fore-arc and back-arc basins, and accretionary complexes and shifting terranes along the coast (Cowan and Bruhn, this volume).

From latest Cretaceous to early Eocene time, major "Laramide" thrust-block uplifts in the craton created new patterns of localized sedimentation and thick basin deposits, destroying the previous region-wide depositional systems of the foreland basin (Fig. 1). Farther west, foreland thin-skinned thrusting waned during the latest Cretaceous and Paleocene in parts of the belt, but continued into the Eocene in other parts. Broad regions west of the thrust belt cooled from previous metamorphic conditions. However, metamorphism continued to the end of the Cretaceous in the North Cascades and, during most of the Late Cretaceous and Paleocene, the southern Cordillera underwent widespread plutonism, metamorphism, and ductile thrusting. Magmatic arcs near the continental margin became inactive over much of the central U.S. Cordillera, undergoing rapid uplift and erosion. However, magmatism continued as eastward-migrating belts across southern California and Arizona (Fig. 1) and in eastern Idaho and Montana, contemporaneous with widely distributed volcanism in the craton of Colorado far east of its former locus. Along the Pacific convergent plate margin, fore-arc basins, subduction complexes, strike-slip sedimentary basins, and accretionary tectonic slices characterized various parts of the margin in a shifting mosaic of tectonic elements.

Following the "Laramide" time period, in the middle Eocene to Oligocene, extensional processes progressively overtook much of the U.S. Cordillera. Older structures and rocks were overprinted by normal faults, thick piles of calc-alkalic volcanic rocks, and localized basin sedimentation (Christiansen and Yeats, this volume).

The changes in tectonics, sedimentation, and magmatism during the Late Cretaceous and again in the Eocene reflect major changes in fundamental continental geologic processes in the Cordillera. These changes probably resulted from reconfigurations of the Kula and Farallon oceanic plates and their interaction with the North American plate. During the Late Cretaceous, the subduction rate increased, changing from steep to shallow subduction under much of the U.S. Cordillera (Lipman and others, 1971; Coney, 1972; Engebretson and others, 1984, 1985). Along the continental margin, lateral slip of impinging seamounts, oceanic plateaus, and continental fragments apparently was caused by oblique plate convergence from the Late Cretaceous to the Paleocene. During the middle Eocene, the mode of subduction changed again; convergence rate decreased, and there was inferred steepening of subduction or dismemberment of the subducting plate.

Modern analogs for the U.S. Cordillera provide additional perspective for inferring plate-tectonic setting and intraplate stress distribution. The Andes of Argentina, with both steeply and shallowly dipping Benioff zones, appear to provide a striking analog. The Sierras Pampeanas are composed of basement uplifts with intervening deep basins, all interior from an inactive magmatic arc and above a shallowly subducting oceanic plate (Jordan and Allmendinger, 1986; *Jordan and Alonso, 1987*). In contrast, the Puna-Altiplano region is a broad, arid, high-elevation plateau

Miller, D. M., Nilsen, T. H., and Bilodeau, W. L., 1992, Late Cretaceous to early Eocene geologic evolution of the U.S. Cordillera, *in* Burchfiel, B. C., Lipman, P. W., and Zoback, M. L., eds., The Cordilleran Orogen: Conterminous U.S.: Boulder, Colorado, Geological Society of America, The Geology of North America, v. G-3.

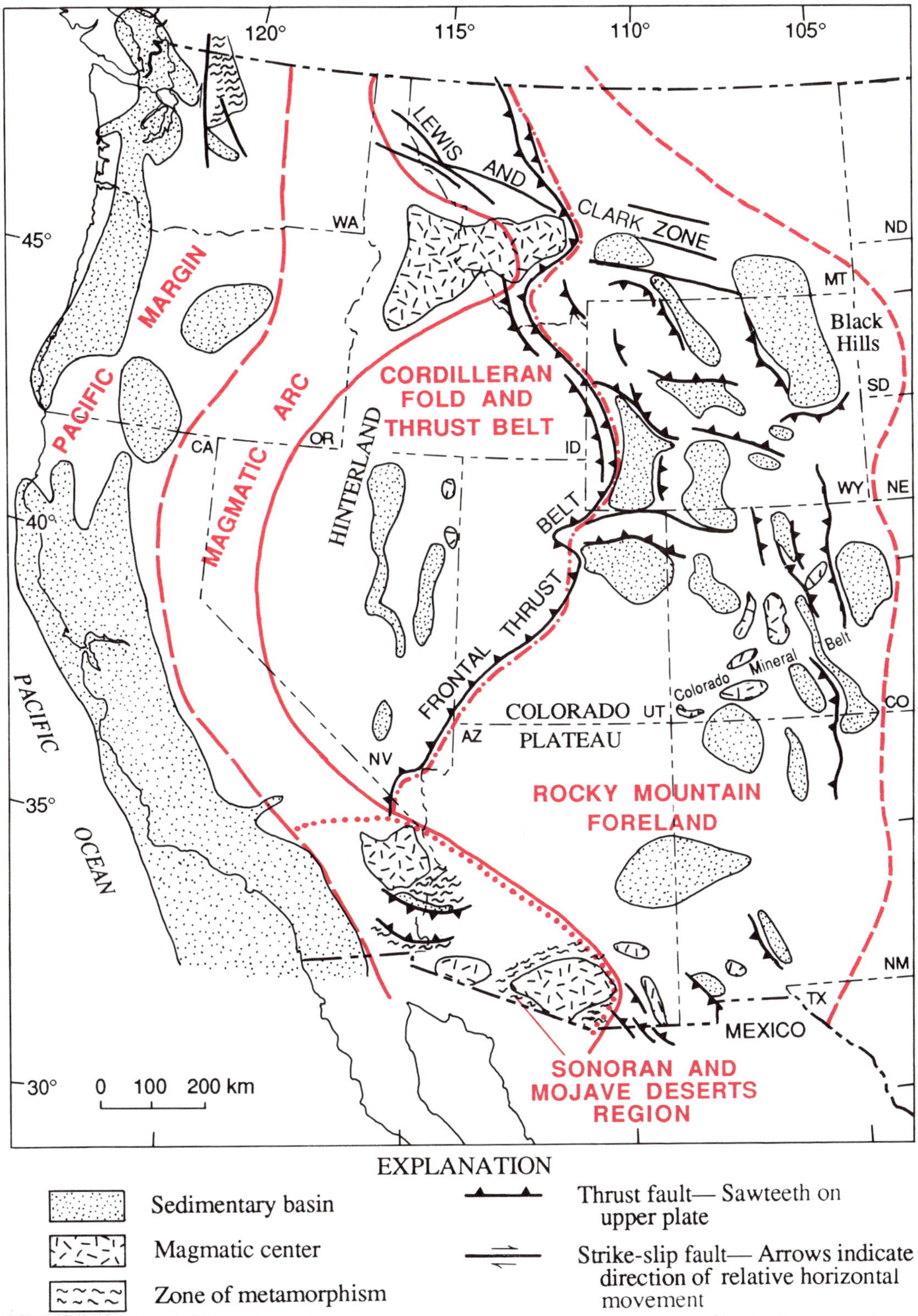

Figure 1. Map of the western U.S. Cordillera showing distribution of the major tectonic elements and locations of Late Cretaceous to early Eocene sedimentary, igneous, and metamorphic rocks. Those regions described in this chapter are outlined. Deposits east of the Rocky Mountain foreland region are described by Baars and others (1988).

between a thin-skinned thrust belt and an active magmatic arc, all above a moderate-angle subduction zone (Allmendinger, 1986). The Sierras Pampeanas may be analogous to the latest Cretaceous and Paleocene U.S. Cordillera, whereas the Puna may be analogous to the earlier Cretaceous U.S. Cordillera.

In this chapter, the U.S. Cordillera is divided into five regions, each of which is marked by distinctive tectonic, sedimentary, and magmatic attributes (Fig. 1). In the east, the Rocky Mountain foreland was characterized by deep-seated thrust faults, basin sedimentation, and localized magmatism. Westward, the Cordilleran fold and thrust belt was typified by a frontal belt of thin-skinned thrusts and cooling of a western metamorphic belt. The Sonoran and Mojave Desert region was characterized by plutonism, metamorphism, and ductile thrusting of crystalline rocks. In the magmatic arc, a large central sector became inactive while magmatism shifted eastward with time into central Idaho and southern Arizona. The Pacific margin was typified by Andean-type fore-arc basins and accretionary complexes, as well as accreting terranes and strike-slip faults. The major tectonic, metamorphic, plutonic, and depositional events in the Cordillera for latest Cretaceous to early Eocene time are synthesized on Plate 6 in map and tabular format.

Tectonism from ca. 80 to 50 Ma has been referred to classically as the Laramide orogeny, based on the ages of basement-cored uplifts and basins in the foreland. The Laramide orogeny was originally thought to be younger than the thrust-belt deformation of the late Mesozoic Sevier orogeny, which was in turn considered to be broadly contemporaneous with the Nevadan orogeny in the California batholiths. However, this terminology is less useful for our current understanding of Mesozoic events. The Nevadan orogeny in California is now restricted to a small part of Late Jurassic time, and the Sevier orogeny of Utah and Nevada now spans Cretaceous to Eocene time (Allmendinger, this volume). As currently understood, much of the deformation and intrusion in the batholiths and adjacent regions cannot be assigned to an orogeny, and the Laramide and Sevier orogenies overlap temporally and spatially.

The Cordilleran orogeny of *Drewes (1978)* is a more comprehensive term (although we do not employ his subdivisions of the orogeny or his tectonic applications) and here is expanded to encompass Mesozoic to early Tertiary orogeny spanning the width of the Cordillera. By this expanded definition, the orogeny was long lived (perhaps episodic) and multifaceted because of (1) changing plate interactions, (2) varying influences of differently constituted parts of the continental crust on tectonism, and (3) progressive thermal and mechanical modification of the continental crust. Accordingly, provinces discussed in this chapter are defined on the basis of structural and depositional style; previous orogenic nomenclature is minimized. Within the Rocky Mountain structural province, we use "Laramide structure" to describe the classical basement-uplift structures, and "Laramide time" to describe the Late Cretaceous to early Eocene interval during which Laramide structures formed.

ROCKY MOUNTAIN FORELAND

W. L. Bilodeau

INTRODUCTION

The Late Cretaceous to Eocene Rocky Mountain foreland province evolved by formation of thrust-fault–bounded, basement-cored mountain ranges and associated intermontane structural basins. The basins contain thick syntectonic sediments that were eroded from the adjacent uplifts and deposited in alluvial fan, fluvial, and lacustrine depositional settings. Many of these basins have yielded petroleum from structural and stratigraphic traps. The "thick-skinned" deformation, which affected a wide area of the western craton, disrupted a large part of the Rocky Mountain foreland basin (Jordan, 1981) that had been forming east of the Cordilleran thrust belt since Early Cretaceous time (Cross, 1986; Heller and others, 1986). This retro-arc foreland basin (Dickinson, 1976), the Western Interior Cretaceous basin of *Kauffman (1977),* Jordan (1981), and *Weimer (1986),* contained 1,500 to 6,000 m of more or less undisturbed Cretaceous and older Mesozoic sedimentary strata prior to the onset of contractional deformation (Baars and others, 1988). Thick-skinned, basement-involved deformation also affected parts of Arizona south of the Cretaceous foreland basin; this region is included within the Rocky Mountain foreland province on the basis of structural style. This structural style is referred to as Laramide style to distinguish it from the thin-skinned, decollement style of the Cordilleran thrust belt.

The distinction between the Laramide and Sevier "orogenies" is fundamentally one of the degree of basement involvement during compressional tectonism in areas of different crustal constitution (*Brown, 1984a*; Allmendinger, this volume). The Cordilleran frontal thrust belt deformation is predominantly decollement-style (thin-skinned) thrusting; associated folding is restricted to the thick miogeoclinal sedimentary wedge and bordering cratonal strata that overlie Precambrian continental crust. The thick-skinned, Laramide-style foreland uplifts and basins are restricted to the Rocky Mountain foreland basin east of the miogeoclinal wedge and hingeline, reflecting deformation in an area of relatively thin sediment cover over a prefractured, anisotropic cratonic basement. Thickness and structural fabric of the Precambrian basement beneath the foreland province vary from one region to another and they controlled aspects of the Late Cretaceous to Eocene foreland deformation (Tweto, 1975; Allmendinger, this volume). Much of the deformation was localized along preexisting zones of crustal weakness, reactivating structural trends with long histories of recurrent movement. Some studies suggest that Rocky Mountain foreland deformation can be modeled as two main phases, one during the Late Cretaceous and Paleocene and one during the Eocene (Chapin and Cather, 1981, 1983; Gries, 1983a; Cross, 1986). The direction of compression for these two phases may differ, changing from an east-northeast

to a more northeast direction (Chapin and Cather, 1981, 1983). Gries (1983a) proposed an additional latest Laramide north-south direction of compression.

The Rocky Mountain foreland province extends from southern New Mexico and southeastern Arizona to southern Montana and from the Black Hills of western South Dakota to the western margin of the Colorado Plateau in Utah (Fig. 1). Not all geologists consider the Colorado Plateau and southern parts of New Mexico and Arizona to be part of the Rocky Mountain foreland province, because they are not part of the present Rocky Mountain physiographic province. However, similarities in the timing and structural style of deformation in southeastern Arizona (Davis, 1979) and southwestern New Mexico (*Seager, 1983*; Seager and Mack, 1986) suggest that this southern region should be included in the province. The Colorado Plateau is situated between the thin-skinned Cordilleran thrust belt of Utah and ductile thrusts of Arizona on the west and southwest, and the thick-skinned basement-cored uplifts on the east. The basement-involved, monoclinal style of Late Cretaceous to Eocene deformation on the plateau (Kelley, 1955; Davis, 1978) is similar to the deformation in the Rocky Mountain foreland.

The northern boundary of the province is placed where significant Laramide-style deformation of the foreland terminates in central Montana. This boundary corresponds with the east-southeast–trending Lewis and Clark zone (Smith, 1965; Sales, 1968) of linear structural elements that extend from west-central Montana and Idaho to north of the Black Hills in South Dakota (Fig. 1). Included in this zone are the Nye-Bowler, Lake Basin, Willow Creek, and Cat Creek lineaments (Smith, 1965; Sales, 1968; *Stone, 1969*) (Fig. 2). Smith (1965) and Sales (1968) considered the Lewis and Clark zone to consist of fundamental basement fractures that reflect deep-seated, left-lateral movements on a shear zone 100 to 150 km wide and 800 km long. North of this zone, the Rocky Mountain foreland basin remained essentially undisturbed, although smaller-scale compressional and wrench structures of Late Cretaceous to Eocene age are known.

The southern boundary of the province is arbitrarily placed in northern Mexico because of a lack of precise information on where Laramide-style deformation ends and Sevier-style deformation begins. The western boundary of the province south of the Colorado Plateau is not well established because westward in southern Arizona, the exposed thrust faults of Late Cretaceous to Eocene age increasingly involve mid-crustal rocks (Haxel and others, 1984). The distinction between interpretations of these faults as the lower reaches of Laramide-style thrust faults (Davis, 1979) or as Sevier-style thrusts that cut crystalline basement rocks (*Drewes, 1978, 1981*) becomes largely one of semantics in this area of plutonism and ductile deformation.

North of the Colorado Plateau, the western boundary of the Rocky Mountain foreland province is a zone of interaction between northwest- and west-trending basement block uplifts and the youngest, easternmost decollement-style thrusts of the frontal thrust belt. The structural complexity of this zone is great, but is being deciphered by detailed mapping and stratigraphic studies (Dorr and others, 1977; *Schmidt and Hendrix, 1981; Dunn, 1983;* Lamerson, 1982; Bruhn and others, 1986; *Woodward, 1986*).

The eastern boundary of the foreland province is loosely defined by eastern margins of Laramide basins in New Mexico, Colorado, and Wyoming, and by the Black Hills uplift.

Structural and tectonic framework

The Rocky Mountain foreland province, characterized by basement uplifts and adjacent basins, encompasses the region ascribed by *Wilmarth (1938)* to the Laramide revolution. Subsequently accepted as the Laramide orogeny, it was classically interpreted as the Late Cretaceous to early Tertiary tectonic event that initially produced the synorogenic Laramie Formation in Wyoming and Colorado. Armstrong (1968) suggested use of the Sevier orogeny to distinguish the thin-skinned thrust belt of Utah, which lies west of the Laramide uplifts and includes structures older than the classic Laramide orogeny, from the basement uplifts. Since these early definitions of orogenies in the Rocky Mountain region, increasingly well-defined ages for deformational events demonstrated broad temporal overlap of the Sevier and Laramide orogenies (Allmendinger, this volume). We therefore use Laramide as a description of thick-skinned tectonic style (and associated sedimentary basins) distinct from the thin-skinned thrusting to the west.

History of interpretations of Laramide structures

One of the most heated controversies in the history of Cordilleran geology has been the debate between advocates for horizontal compression (e.g., *Berg, 1962; Blackstone, 1963,* 1980; Sales, 1968; *Lowell, 1974;* Coney, 1976) and advocates for dif-

Figure 2. Sketch map showing the Rocky Mountain foreland province with major structures and sedimentary basins. Abbreviations for basins and structures: **Sedimentary Basins:** BB, Baca; BHB, Bighorn; CJB, Carthage–La Joya; CMB, Crazy Mountains; CSB, Cutter Sag–Love Ranch; DB, Denver; EPB, Echo Park; GB, Galisteo; GRB, Green River; HB, Hanna; HPB, Huerfano Park; LB, Laramie; NP-MPB, North Park–Middle Park; PB, Piceance; PRB, Powder River; RB, Raton; RIB, Ringbone; SB, Shirley; SBB, Sierra Blanca; SJB, San Juan; SLB, San Luis; SPB, South Park; SWB, Sand Wash; UB, Uinta; WB, Washakie; WRB, Wind River. **Uplifts (U) and Arches (A):** BHU, Black Hills; BiU, Bighorn; BSU, Blacktail-Snowcrest; BTU, Beartooth; BU, Burro; CA, Casper; CCU, Circle Cliffs; DCA, Douglas Creek; DU, Defiance; ECU, Echo Cliffs; FRU, Front Range; GMU, Granite Mountains; GU, Gunnison; HiU, Hidalgo; HU, Hartville; KU, Kaibab; LU, Laramie; MBU, Medicine Bow; MGU, Madison-Gravelly and Madison-Gallatin; MU, Monument; NU, Nacimiento; OCU, Owl Creek; PU, Park; RGU, Rio Grande; RSU, Rock Springs; SAU, southern Arizona; SCU, Sangre de Cristo; SJU, San Juan; SLU, San Luis; SMU, Sierra Madre; SRU, San Rafael; SU, Sawatch; TTU, Teton-Targhee; UnU, Uncompahgre; UU, Uinta; WHU, White River; WMU, Wet Mountains; WRU, Wind River; and ZU, Zuni. **Lineaments:** CCL, Cat Creek; GFTZ, Great Falls tectonic zone; LBL, Lake Basin; NBL, Nye-Bowler; WCL, Willow Creek.

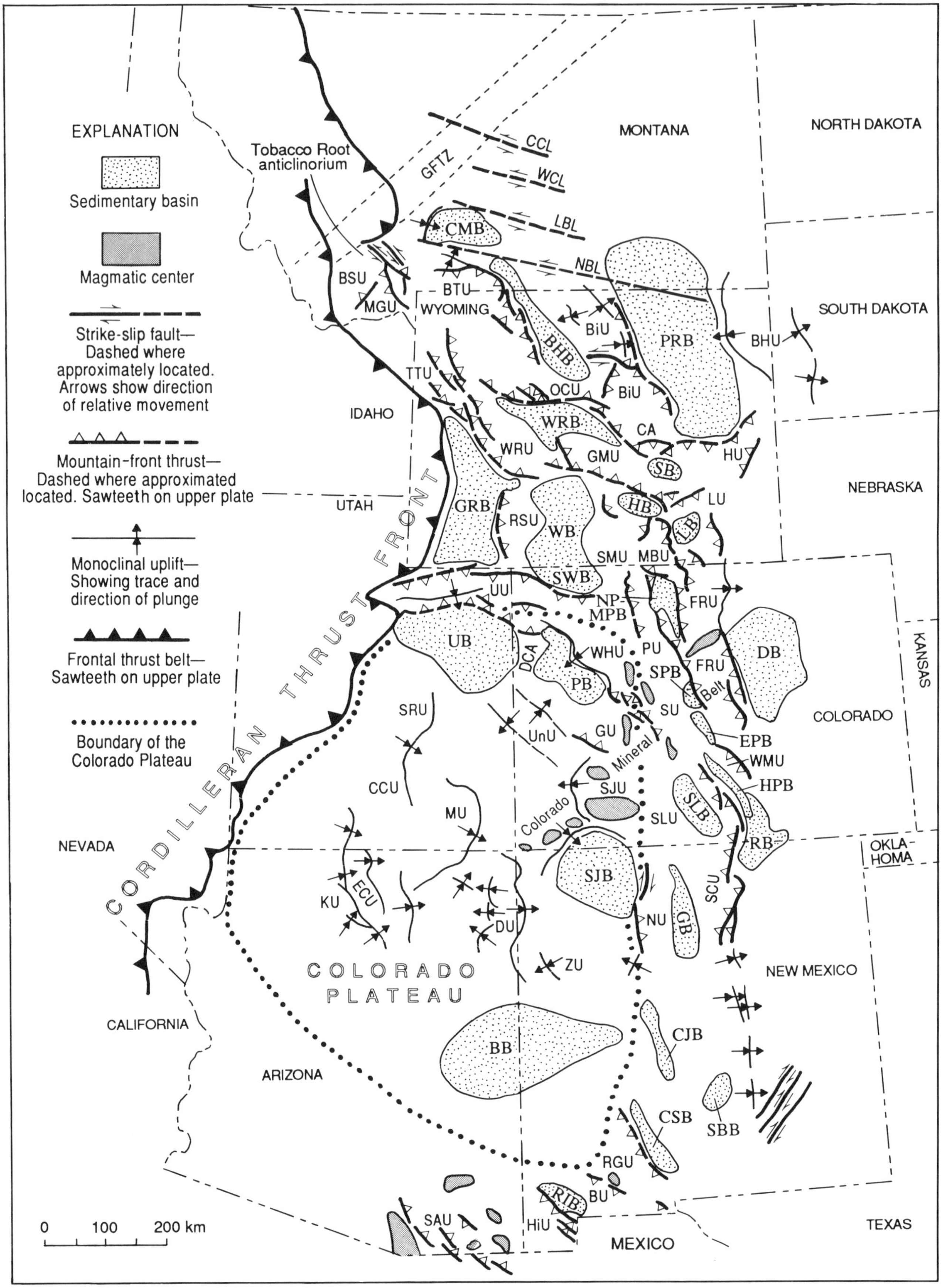
EXPLANATION
Sedimentary basin
Magmatic center
Strike-slip fault—Dashed where approximately located. Arrows show direction of relative movement
Mountain-front thrust—Dashed where approximated located. Sawteeth on upper plate
Monoclinal uplift—Showing trace and direction of plunge
Frontal thrust belt—Sawteeth on upper plate
Boundary of the Colorado Plateau
0 100 200 km
Tobacco Root anticlinorium
MONTANA
NORTH DAKOTA
SOUTH DAKOTA
WYOMING
IDAHO
UTAH
NEBRASKA
COLORADO
KANSAS
NEVADA
CALIFORNIA
ARIZONA
NEW MEXICO
OKLA-HOMA
TEXAS
MEXICO
CORDILLERAN THRUST FRONT
COLORADO PLATEAU
GFTZ
CCL
WCL
LBL
NBL
CMB
BSU
MGU
BTU
BHB
BiU
PRB
BHU
TTU
OCU
WRB
CA
WRU
GMU
SB
HU
GRB
RSU
WB
HB
LB
LU
SMU
MBU
SWB
UU
NP
MPB
FRU
UB
DCA
WHU
PU
PB
SPB
Belt
DB
SU
SRU
UnU
GU
Mineral
EPB
WMU
HPB
CCU
SJU
SLB
MU
Colorado
SLU
RB
SJB
SCU
KU
ECU
DU
NU
GB
ZU
BB
CJB
SBB
CSB
RGU
RIB
BU
SAU
HiU

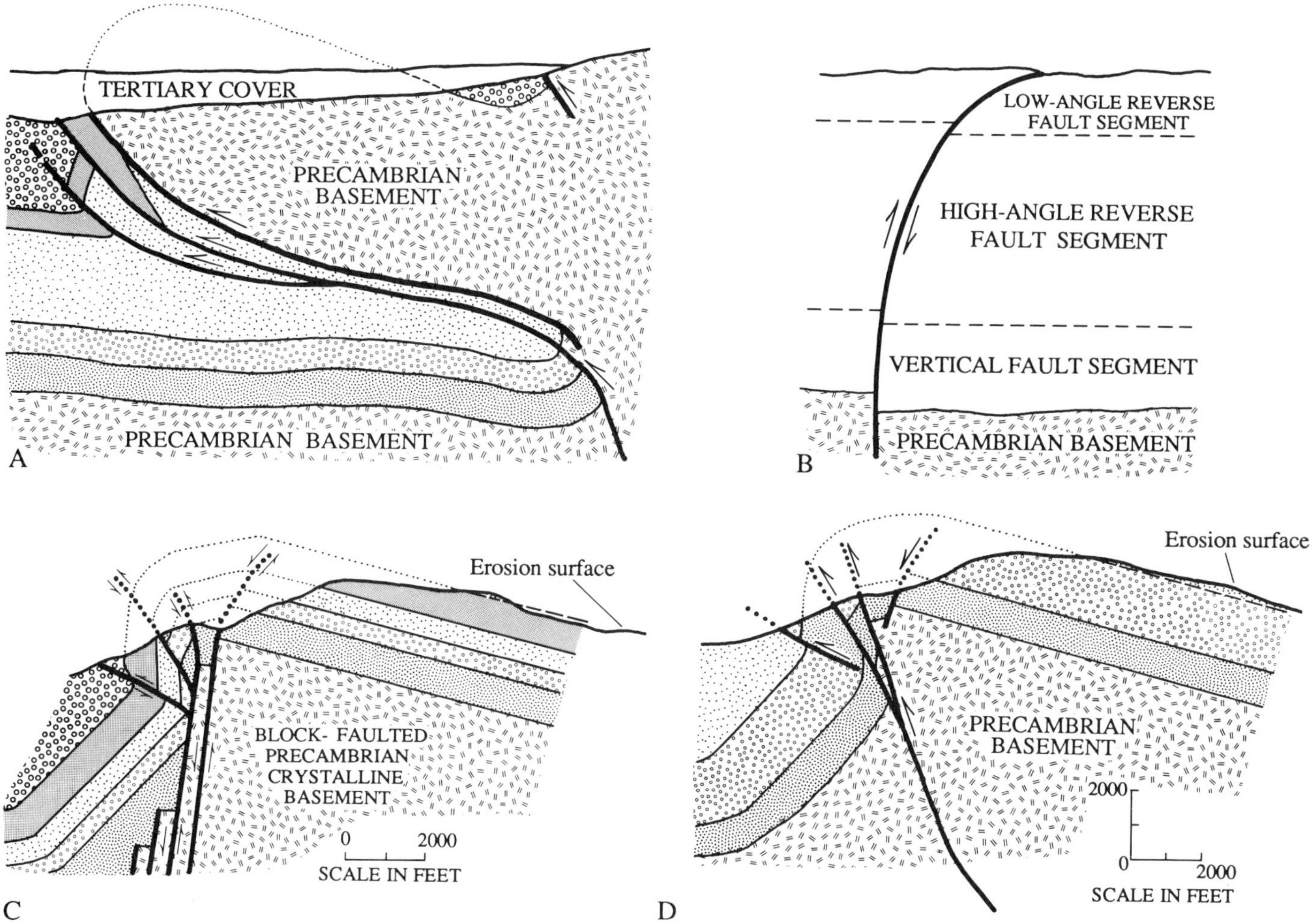

Figure 3. Previously proposed structural models for Rocky Mountain basement uplifts. (A) After Berg (1962). (B) After Prucha and others (1965). (C) After Stearns (1971). (D) After Brown (1984a).

ferential vertical tectonism (e.g., *Osterwald, 1961; Eardley, 1963; Prucha and others, 1965; Harms, 1965; Stearns, 1971, 1978*) to explain the Rocky Mountain foreland uplifts and basins. This debate centers on fundamental aspects of the kinematics and dynamics of foreland deformation. (1) Were the major structures produced strictly by horizontal compression of the basement or by differential vertical movements with no horizontal compression? (2) Was the basement folded within the cores of foreland folds or did the basement behave as a brittle material? (3) Is the subsurface geometry of the bounding faults such that the faults are high angle and steepen with depth, or are they low-angle thrusts that flatten with depth? (4) Were the folds in the sedimentary cover formed by passive draping over high-angle faults or compressional buckling over thrust faults?

Berg (1962) suggested that the major ranges and associated basins are large-scale overturned and thrust-faulted anticline-syncline pairs, supporting a general fold-thrust model for foreland structural development (Fig. 3a). *Prucha and others (1965)* countered with evidence for an initial vertical fault in the basement that curved upward through the sedimentary cover to become a low-angle reverse fault at the surface—termed an upthrust fault (Fig. 3b)—a model identical to the block-uplift model considered and discarded by *Berg (1962)*. The "verticalists" were supported by the work of *Stearns (1971, 1975, 1978)* and associates (see Matthews, 1978). The structure at Rattlesnake Mountain in northwestern Wyoming was modeled by *Stearns (1971, 1978)* as a passive drape fold over a normal fault (Fig. 3c), requiring no horizontal compression in the region. This modeling, however, considered the basement to be a homogeneous and isotropic brittle material and used laboratory models with precut or prefaulted vertical basement structures, an unrealistic assumption for the Rocky Mountain foreland (*Brown, 1984a*). An alternative model for the Rattlesnake Mountain structure (*Brown, 1984b*) consists of a listric thrust fault (Fig. 3d), and invokes shortening in the basement and rotational tilt of the mountain block. A basic flaw with all drape fold or vertical tectonic models is that the shortening evident in the folded sedimentary strata is not compensated by the basement fault geometry. Thus, balancing the shortening in

the strata with that which must occur in the basement suggests listric thrust faults that cut deep into the basement (Blackstone, 1980; *Brown, 1984a, 1984b; Stone, 1984;* Erslev, 1986).

During this horizontal-versus-vertical debate, a structural model of regional wrench or strike-slip tectonics was added by Sales (1968) and *Thomas (1971).* They explained the complexity of trends and the variety of interpreted structural geometries as being caused by regional shear wrenching along foreland strike-slip faults. *Stone (1969)* suggested that pure shear tectonics produced conjugate wrench structures throughout the Rocky Mountain foreland. In an attempt to integrate the diverse Rocky Mountain structures into one theory with a single regional cause, *Lowell (1974)* combined horizontal compression, strike-slip faulting, and vertical crustal movements with the plate tectonic theory as understood at that time. In contrast, *Woodward (1976)* recognized the diversity of structures, but stated that the principal tectonic stresses are vertical.

A Consortium for Continental Reflection Profiling (COCORP) deep reflection seismic profile across the Wind River Range in southwest Wyoming (Smithson and others, 1978, 1979; Brewer and others, 1980) demonstrated that the southwestern flank of the range is bounded by a planar thrust fault that dips northeast beneath the mountain range to more than 20 km depth, strongly supporting horizontal shortening models. Significant thrusting of Precambrian crystalline basement over adjacent sedimentary basins along mountain-front thrust faults also has been convincingly demonstrated by drilling and seismic reflection studies conducted by the petroleum industry (e.g., *Berg, 1962; Gries, 1981,* 1983b; *Jacob and Albertus, 1985*).

The unequivocal evidence for crustal shortening in the foreland province indicates that horizontal compression is the main cause of foreland deformation (Berg, 1981; Kanter and others, 1981; *Gries, 1981,* 1983b; Blackstone, 1980, 1983). Even the monoclines of the Colorado Plateau developed above high-angle reverse faults and demonstrate crustal shortening consistent with regional horizontal compressive stresses (*Reches, 1978; Huntoon, 1981;* Anderson and Barnhard, 1986). However, the 5 to 9 km of vertical structural relief between the foreland uplifts and adjoining basins, in conjunction with the large amounts of absolute subsidence in basins, have been used as an argument for mantle-driven differential vertical movements (*Woodward, 1976; Stearns, 1978*). In contrast, results of flexural modeling of the northern Bighorn and Green River basins (Hagen and others, 1985) suggest that tectonic loading from basin-margin thrust-fault blocks and sedimentary loading from basin sedimentary deposits adequately explain the total amount of subsidence in the basins and the resultant structural relief. Thus, crustal thrust faulting adequately accounts for all features of Laramide structures.

Regional tectonic setting

By Late Cretaceous time, the western Cordillera of North America had developed into an Andean-type continental margin magmatic-arc system (Coney, 1976, 1978; Dickinson, 1981). This plate-tectonic setting included three primary elements: (1) The western margin of the North American plate moved northwestward to westward to west-southwestward in an absolute sense (Coney, 1972, 1976, 1978; Engebretson and others, 1984, 1985; *Jurdy and Gordon, 1984*). (2) Arc-related batholithic magmatism ceased at about 80 Ma along much of the coastal volcanic arc (approx. 34° to 40°N) outboard of the widest zone of foreland deformation, while arc-related magmatism spread eastward into three limited areas of the foreland: southern Arizona–New Mexico, a narrow northeast-trending band in Colorado (the Colorado mineral belt), and the southwestern Montana region (Lipman and others, 1971; Armstrong, 1974; Burchfiel and Davis, 1975; Snyder and others, 1976; Coney and Reynolds, 1977; *Cross and Pilger, 1978a, 1978b;* Keith, 1978; Dickinson and Snyder, 1978). (3) There was continuous subduction of the Farallon plate. The relative rates of convergence between the Farallon and North American plates increased sharply between 75 and 70 Ma (Late Cretaceous), continued at a rapid rate from 65 to 55 or 60 Ma, and declined between ca. 55 and 40 Ma (Eocene) (Engebretson and others, 1984, 1985; Stock and Molnar, 1988).

The Late Cretaceous to Eocene Rocky Mountain foreland province has been compared with the modern tectonics of the Sierras Pampeanas of Argentina (Jordan and Allmendinger, 1986; *Jordan and Alonso, 1987*). As described in the introduction, this analog consists of an inactive volcanic arc, a fold and thrust belt, and a basement uplift province, all of which are above a zone of low-dip subduction (*Stauder, 1975; Barazangi and Isacks, 1976; Cross and Pilger, 1982;* Jordan and Allmendinger, 1986). On the basis of the magmatic pattern and using the Andean analogy, Dickinson and Snyder (1978) and *Cross and Pilger (1978a)* suggested that the arc "magmatic null" and the basement-involved Laramide-style deformation of the foreland were caused by low-angle subduction of the Farallon plate beneath that part of western North America. Lipman and others (1971), *Lowell (1974),* Burchfiel and Davis (1975), Coney (1972, 1978), and *Cross and Pilger (1978b)* also related Rocky Mountain foreland deformation, sedimentation, and/or magmatic patterns to shallow subduction.

Low-angle subduction can be related to (1) high rates of relative convergence, (2) subduction of anomalously buoyant lithosphere, such as aseismic ridges, seamount chains, or oceanic plateaus, (3) subduction of very young oceanic lithosphere and/or (4) increased rates of absolute motion of the overriding plate toward the trench (*Cross and Pilger, 1982*). All of these factors have been suggested as the cause for the low-angle subduction beneath the Rocky Mountain foreland inferred for Late Cretaceous to Eocene time. The aseismic ridge-oceanic plateau subduction theory for Laramide tectonic and magmatic patterns has been proposed and developed by *Cross and Pilger (1978b, 1982),* Livaccari and others (1981), Henderson and others (1984), and Cross (1986). This model best explains why low-angle subduction caused basement uplifts only along part of the convergent margin; the other factors listed above should have

affected the entire length of the convergent plate margin between the Farallon and North American plates.

Several mechanisms have been proposed to explain the style and timing of foreland deformation. *Lowell (1974)* stressed that the buoyancy of the subhorizontal slab would impart vertical forces to the foreland along with tangential horizontal compression and strike slip, whereas Dickinson and Snyder (1978) proposed that the horizontal shear coupling between the subducted plate and the continental foreland caused the horizontal compression and basement thrusting within the foreland. Modeling by Bird (1984) suggests that the horizontal drag forces localized at the base of the North American lithosphere by low-angle subduction of the Farallon plate caused viscous flow in the lower crust. This flow, in turn, caused crustal thickening beneath the foreland and isostatic uplift of the province, and provided the horizontal and vertical stresses necessary to create the upper crustal brittle-ductile deformation. Low-angle subduction also would provide a driving force for the northward and northeastward clockwise rotation of the Colorado Plateau suggested by Hamilton (1981, 1988) to explain the systematic change in trends of foreland uplifts around the margin of the Plateau.

Many of the models assume a single direction of horizontal compression throughout the evolution of the Rocky Mountain foreland, a simplistic approach commonly based on the misconception of orthogonal convergence at subduction zones. Not only did the absolute motion of North America undergo a major change in direction during Late Cretaceous to Eocene time, but the relative convergence direction between the North American and Farallon plates also changed (Coney, 1976, 1978; Engebretson and others, 1984, 1985). Chapin and Cather (1981, 1983) proposed that the foreland deformation was related to horizontal compression marked by motion of the Colorado Plateau, and that the direction of compression changed in the latest Paleocene to earliest Eocene from N70°E to N45°E, giving the plateau a more northerly component of motion relative to the North American craton in late Laramide time. Gries (1983a) came to similar conclusions but proposed a three-stage change of compression direction, from (1) east-west, to (2) northeast, to (3) north-south, the changes caused by changing North American plate vectors as opening of the North Atlantic and Arctic oceans progressed. These postulated changes in compression direction caused contractional or strike-slip deformation on properly oriented foreland structures at different times. Evidence for this tectonic history is found in the strata filling certain basins and the syndepositional deformation produced by the mountain flank structures (Chapin and Cather, 1981, 1983; Gries, 1983a). Two phases of Laramide tectonism are also supported by a study of structural style. Cross (1986) described evidence for a first tectonic phase lasting from about 74 to 65 Ma that was characterized by regionally extensive ductile deformation. The second phase lasted from about 60 to 45 Ma, was characterized by more brittle deformation, and was limited to within a 200-km-wide zone marginal to the Colorado Plateau. He also found no conclusive evidence for change in compression direction between the two phases.

Additional evidence for changes in stress orientation may be found in the timing and pattern of igneous activity in the central foreland. Laramide plutons were emplaced in the northeast-trending Colorado mineral belt (Fig. 1) during the Late Cretaceous and Paleocene, but not during early Eocene time (Snyder and others, 1976; Lipman, 1980). The direction of maximum compression was subparallel to the mineral belt during pluton intrusion and at a high angle to the mineral belt during igneous quiescence. Thus the change in the direction of maximum horizontal compressive stresses could have effectively closed the igneous conduits, halting plutonism in the central foreland area.

REGIONAL GEOLOGY

The Rocky Mountain foreland province can be divided into five regions on the basis of (1) structural geometry, (2) timing of deformation and magmatism, (3) characteristics and fabric of the crystalline basement, and (4) magnitude of deformation. The regions are: (1) southwest Montana, where the foreland structures have been overprinted by coeval thin-skinned thrusts and late Cenozoic extensional tectonism; (2) the Wyoming province, where most foreland uplifts and basins trend northwest; (3) Colorado–New Mexico, where most of the foreland structures trend northerly, their locations inherited from late Paleozoic structures associated with the Ancestral Rocky Mountain orogeny; (4) the Colorado Plateau, more monoclinal uplifts represent a distinctly smaller magnitude of foreland deformation; and (5) southeastern Arizona–southwestern New Mexico, where magmatism was synchronous with foreland deformation, but many relations are obscured by Neogene extensional tectonism (Fig. 2). The first two subregions have floors of Archean crust, whereas the last three have floors of Early Proterozoic crust.

Southwest Montana

The southwest Montana region exhibits a structural pattern that reflects recurrent movement of varying kinds along basement fault zones and complex tectonic overprinting by partly coeval Cordilleran thin-skinned thrusts and late Cenozoic normal faults. Late Cenozoic Basin and Range tectonism and sedimentation (*Reynolds, 1979; Ruppel, 1982*) formed the present physiography of the region, in some places burying Laramide structures and strata, and in other places uplifting and exposing them. The Late Cretaceous to early Tertiary basement-involved foreland uplifts are a series of thrust-bounded northeast- and northwest-trending anticlines (from west to east, the Blacktail-Snowcrest, Madison-Gravelly, and Madison-Gallatin uplifts) and intervening synclinoria that converge northward into the Tobacco Root anticlinorium (Fig. 2) (*Scholten, 1967;* Ryder and Scholten, 1973, *1986*; *Perry and others, 1983;* Schmidt and Garihan, 1983). The northeast-trending Blacktail-Snowcrest foreland uplift acted as a buttress to the impinging thrusts and folds of the Cordilleran belt that generally curve around it (*Perry and others, 1983;* Ruppel and Lopez, 1984). The Blacktail-Snowcrest uplift is structurally asymmetri-

cal with a low-angle basement thrust fault dipping northwestward beneath its southeastern flank (the sub-Snowcrest Range thrust of *Perry and others, 1983*). Late Cretaceous (Coniacian to Maastrichtian) syntectonic conglomerates of the Beaverhead Formation (Ryder and Scholten, 1973, *1986*) or Lima Conglomerate of the Beaverhead Group (Nichols and others, 1985) and the Sphinx Conglomerate (*DeCelles and others, 1987*) date movement on these structures. These predominantly Upper Cretaceous rocks are deformed by thin-skinned thrusts. If the Coniacian age is correct, the northeast-trending Blacktail-Snowcrest uplift may be the oldest basement-involved uplift in the Rocky Mountain foreland province (Nichols and others, 1985).

A northeast-trending zone of structures bounds the northwestern edge of the foreland province. Named the Great Falls tectonic zone by *O'Neill and Lopez (1985),* it is composed of high-angle faults and shear zones extending from the Idaho batholith across west-central Montana to southern Saskatchewan, Canada. This zone has shown recurrent structural movement from the Proterozoic to the present, controlled the intrusion and orientation of Late Cretaceous to early Tertiary dike swarms, and possibly the Pioneer and Boulder batholiths, and may mark the northwestern boundary of the Archean Wyoming Province (*O'Neill and Lopez, 1985*).

The Precambrian crystalline basement of the Montana part of the foreland is composed primarily of well-foliated and well-bedded Archean metasedimentary rocks. This basement terrane contains both northwest- and northeast-striking faults that provide a distinct structural fabric (*Maughan, 1983; Maughan and Perry, 1986*). The northwest-striking faults dip steeply to the northeast and are pervasive throughout the Archean exposures in the area. During the Late Cretaceous to early Tertiary, the main response of the foreland to regional horizontal compression was reactivation of the northwest-trending basement structures, which had formed during development of the Proterozoic Belt basin (*Schmidt and Garihan,* 1983, *1986*). Wherever northwest-striking faults cut Paleozoic or Mesozoic strata, there are associated large, asymmetrical, southwest-verging hanging-wall anticlines and footwall synclines (Schmidt and Garihan, 1983). In cores of some of the fault-related folds, the foliation in the Archean basement is folded around the same hinge as the overlying sedimentary strata. The faults underwent approximately equal left-lateral and reverse slip (Schmidt and Garihan, 1983). The easternmost basement-involved foreland uplift, the Madison-Gallatin uplift, is bounded on the southwest by a northwest-striking, steeply northeast-dipping fault system (*Garihan and others, 1983*) that is one of the longer, more continuous northwest-striking sets of faults.

The east and northeast boundary of the south-central Montana region is composed of the northwest-striking wrench faults that make up the Lewis and Clark zone. The Crazy Mountains structural basin, located between Bozeman and Billings, was formed by primarily left-lateral movement on the Lake Basin fault zone to the north-northeast and the Nye-Bowler lineament to the south-southwest (*Garrett, 1972*). The structural geometry of the basin suggests that it is a transpressional synclinal downwarp bounded to the northwest by fold-thrust structures of the Cordilleran thrust belt.

Timing of movement on structures in the Lewis and Clark zone can be constrained in two areas. At the northwestern boundary of this region, where structures of the Cordilleran thrust belt interact with foreland structures, the northwest-striking faults cut plutonic rocks of the Boulder batholith and are, in turn, deformed by Cordilleran thrusts. This brackets the movement on the westernmost faults to 75 to 70 Ma, that is, latest Campanian to early Maastrichian in age (*Schmidt and Hendrix, 1981;* Schmidt and Garihan, 1983). At the eastern edge of the southwest Montana foreland, northwest-striking faults cut rocks of the Upper Cretaceous and Paleocene Fort Union Formation but are overlain by lower Eocene strata, limiting deformation to the late Paleocene (*Roberts, 1972;* Schmidt and Garihan, 1983). Thus, deformation on northwest-striking faults youngs eastward, from Late Cretaceous to late Paleocene in age.

The foreland of southwest Montana also contains basement-involved thrust faults that strike north and dip west (Schmidt and Garihan, 1983). Uplift associated with one of these, the Scarface thrust (part of the Hilgard fault system of Tysdal and others, 1986) along the west side of the Madison-Gravelly uplift, provided the source for the syntectonic Sphinx Conglomerate (*DeCelles and others, 1987*). Middle Campanian to Maastrichtian movement on these north-striking, east-directed thrusts (Tysdal and others, 1986; *DeCelles and others, 1987*) is apparently partly coeval with movement on the northeast- and northwest-trending structures and creates a complex kinematic picture of southwest Montana foreland deformation that needs further study.

Early Cretaceous basement-involved subtle foreland structures that predate classical Laramide structures are suggested by isopach, lithofacies, mineralogic, and paleodispersal data from the Lower Cretaceous Kootenai and Blackleaf Formations north and east of Dillon, Montana (Suttner and others, 1981; Schwartz, 1982; DeCelles, 1986). These structures are mildly uplifted intra-foreland welts that trend north-northeast and northwest; they include the eventual site of emplacement of the Boulder batholith (Boulder batholith high), the Tobacco Root–Ruby Archean block (includes the Blacktail-Snowcrest and Madison-Gravelly uplifts), and the Beartooth Archean block (DeCelles, 1986).

Wyoming

The present physiography of the Wyoming part of the Rocky Mountain foreland more closely reflects its early Tertiary paleogeography than any other part of the province. Late Cenozoic regional epeirogenic uplift has exhumed older structures so that the present northwest-trending mountain ranges and basins coincide with the late Laramide foreland features. Well-exposed fault-bounded, basement-cored mountain ranges and large intermontane structural basins in Wyoming have come to characterize the structural style of the foreland province and have been the center of much of the controversy that surrounds their interpretation.

Major basement-cored uplifts of the Wyoming foreland are the Teton-Targhee, Beartooth, Bighorn, Black Hills, Owl Creek, Wind River, Granite, Laramie, Hartville, Medicine Bow, and Uinta (Fig. 2). Some of these structures are mostly outside of Wyoming (the Beartooth, Black Hills, and Uinta uplifts), but have similar styles and affect sedimentation in the Wyoming part of the foreland. Structural basins adjacent to or between these uplifts contain large volumes of nonmarine alluvial and lacustrine sedimentary deposits of Late Cretaceous to early Tertiary age. These basins include the Bighorn, Powder River, Wind River, Green River, Washakie, Sand Wash, Shirley, Hanna, and Laramie basins (Fig. 2). The crystalline basement rocks of this part of the foreland, as in southwest Montana, are Archean in age. In general, the structural trends of Wyoming foreland uplifts were not inherited from older basement structures, which mainly trend northeast. An exception is the Uinta uplift; its east-west trend was controlled by an elongate, normal-fault–bounded trough filled with Proterozoic sedimentary rocks of the Uinta Mountain Group (*Sears and others, 1982*).

Two foreland uplifts, the Uinta and Teton-Targhee, have interacted with the Cordilleran thrust belt, pinning the north and south ends of a major arcuate segment of the belt (Dorr and others, 1977; *Love, 1982; Dunn, 1983;* Bruhn and others, 1986; *Woodward, 1986*). Sedimentation in the greater Green River basin, which is between these two uplifts as well as part of the Wind River uplift and includes the Hoback, Green River, Washakie, Great Divide, and Sand Wash subbasins, also has been strongly affected by the Cordilleran thrust belt, which is its western boundary. Uplift in the thrust belt provided a source for Green River basin sediments, and thrusting in the belt eventually deformed the western margin of the basin.

The foreland uplifts are bounded by basement-involved thrust faults on one or more flanks (*Gries, 1981,* 1983a, 1983b), faults that typically emplace wedge-shaped Precambrian basement blocks many kilometers over basin-margin sedimentary deposits. Along the southern flank of the Wind River uplift, the COCORP deep-crustal seismic-reflection line (Smithson and others, 1978, 1979; Lynn and others, 1983) shows that the low-angle Wind River thrust carries Precambrian crystalline rocks 21 km southwestward over sedimentary strata of the Green River basin. Petroleum exploration drill holes through this and other mountain-front thrust faults in the Wyoming province have substantiated these structural relations (*Gries, 1981,* 1983a, 1983b). The basement-involved thrust zones commonly include several discrete thrust faults that bound narrow, intensely deformed slivers of overturned Paleozoic and Mesozoic strata.

Movement histories of major mountain-front thrust systems (shown in Fig. 2) have been dated by analyzing thickness and depositional facies trends in the synorogenic basinal sedimentary deposits, and by crosscutting relations. In many basins, subsidence is controlled primarily by crustal thickening due to thrust loading along the active margins of the basin (Hagen and others, 1985). Thus, the basin sedimentary deposits thicken and coarsen toward the active margin due to higher rates of subsidence and closer proximity to the uplifted source terrane. Sedimentation in the Wyoming basins was continuous throughout Laramide time, beginning in the Late Cretaceous but ending at different times in different places (Dickinson and others, 1988). Initial uplift and deformation in the foreland was usually marked by unconformities and erosion rather than deposition and preservation of sediments. As crustal shortening progressed and foreland uplifts grew, tectonic loading and subsidence gradually created topographic lows adjacent to the mountain fronts and trapped fluvial sediments. However, for many of the structural basins, early subsidence relative to the upthrust block was characterized by little or no deposition. Later, net subsidence was great enough for sediments to accumulate, forming a sedimentary basin. The western basins (Green River, Uinta, Bighorn, and Wind River) began to fill with syntectonic sediment in the latest Cretaceous (early Maastriachtian), younging eastward to the Powder River basin, which contains latest Maastrichtian to Paleocene syntectonic clastic deposits (Gries, 1983a). Deformation and subsidence was episodic, with significant pulses in the latest Maastrichtian to early Paleocene and the early Eocene. These pulses may be related to changes in the direction of horizontal compression within the foreland province (Chapin and Cather, 1981, 1983; Gries, 1983a).

Late Cretaceous to Paleocene sedimentation in these basins was primarily controlled by throughgoing rivers weaving eastward around the growing uplifts. By the early Eocene, this drainage pattern was disrupted by accelerated uplift of older mountain blocks, and formation of new ones, cutting off basin outlets and forming several major internal-drainage basins. Lakes developed in local basins during the Late Cretaceous and Paleogene, but not to the extent or size that existed during early to middle Eocene time. Although some of these Eocene lakes drained eastward, others possibly drained westward into the "Tyee" paleoriver of the Pacific Northwest (Dickinson and others, 1988).

Large Eocene lakes developed in the greater Green River basin (Lake Gosuite) and in the Uinta-Piceance basins (Lake Uinta), and were the loci of deposition for the Green River Formation. The Paleocene and Eocene Green River Formation is widely known for its large quantities of oil shale and trona (Surdam and Stanley, 1980). The environment in which organic- and carbonate-rich oil shale accumulated in lakes Gosuite and Uinta has been the center of much controversy. Two models have been proposed: (1) deep, chemically and thermally stratified lakes that promoted anoxic bottom conditions and preservation of organic-rich and carbonate-rich sediment, and (2) shallow playa lakes surrounded by algal mudflats that produced the organic-rich laminae (reviewed in *Picard, 1985;* Johnson, 1985). Algal stromatolite shoreline deposits, present throughout the Green River Formation, demonstrate that Lake Gosuite fluctuated between a broad, deep standing lake and a small, shallow playa salt pan (Surdam and Stanley, 1980). Fluctuations between humid and arid climatic conditions, in conjunction with tectonic effects, caused variations in the input of fresh water, which modified the size and salinity of the lakes (*Sullivan, 1985*).

Fluvial depositional systems predominated in the other Wyoming basins throughout latest Cretaceous to Eocene time. Lacustrine facies in the Powder River basin, which contains notable coal and oil shale, and part of the Bighorn basin have been interpreted as small floodbasin lakes and floodplain backswamps of a north-flowing longitudinal fluvial system (*Ethridge and others, 1981;* Flores and Ethridge, 1985). Ayers (1986) suggested that lacustrine rocks are parts of a large widespread lake that drained to the south. However, thick fluvial-floodplain and backswamp coal deposits of the Powder River and other Wyoming basins attest to the presence of long-lived meandering or anastomosing rivers.

Syntectonic, nonmarine sedimentary deposits in the Wyoming basins are 1,000 to 6,000 m thick, but are only a small fraction of the detritus eroded from the uplifts during Laramide time (Tweto, 1975). The basins were imperfect sediment traps; most of the eroded material was transported by major trunk rivers to the Gulf of Mexico, where voluminous lower Tertiary clastic sedimentary rocks are present in the subsurface (*Winker, 1982*).

Colorado–New Mexico

Foreland deformation in Colorado and New Mexico began in the latest Cretaceous (Maastrichtian) and ended by late Eocene time (Tweto, 1975). The principal basement-involved uplifts of this central foreland area are the Front Range, Park–Sierra Madre, Sawatch, Wet Mountains, San Luis, Sangre de Cristo, and Nacimiento uplifts (Fig. 2). Laramide uplifts in western Colorado are monoclinal and are part of the Colorado Plateau region (Chapin and Cather, 1981, 1983). Basement-involved uplifts and adjacent Laramide basins form a north-south band through central Colorado and New Mexico, separating the Colorado Plateau from the undisturbed craton of the mid-continent. The principal trend of uplift-bounding faults is north-northwest to northwest, partly utilizing preexisting structures of Precambrian and late Paleozoic age (Tweto, 1975). There is no complete correlation between Laramide and Ancestral Rocky Mountain uplifts and basins, but the structures correspond best in Colorado. The Precambrian basement in Colorado and New Mexico is mostly Proterozoic (1.86 to 1.5 Ga), significantly younger than the Archean basement beneath much of Montana and Wyoming.

There are two basic types of basins in this region: (1) large, bowl-shaped, structurally asymmetric basins that include the Denver, Raton, and San Juan basins, and (2) smaller, elongate or somewhat narrow, intermontane basins that form a right-stepping en echelon pattern and include, from south to north, the Galisteo, San Luis, Huerfano Park, Echo Park, South Park, and North Park–Middle Park basins (Fig. 3). The Denver, Raton, San Juan, South Park, and North Park–Middle Park basins began forming in latest Cretaceous to Paleocene time, whereas the other basins probably began forming in the early Eocene. The early Eocene origin of these latter basins, and possible modification of the South Park and North Park–Middle Park basins by a regional right-lateral wrench system, may have been caused by the rotation and northward movement of the Colorado Plateau relative to the North American craton indicated by paleomagnetic data (Chapin and Cather, 1981, 1983; Hamilton, 1981). These basins are aligned with the crest of the highest Laramide uplifts and the eastern margin of the Colorado Plateau, a zone later occupied by the Neogene Rio Grande rift; this alignment may include the Carthage–La Joya and Cutter Sag–Love Ranch basins of central to southern New Mexico (Chapin and Cather, 1981, 1983).

The Front Range uplift began to rise in the Late Cretaceous; it is bounded on both the east and west by major thrust faults (*Jacob, 1983*). Southward from Denver, on the east side of the range, the en echelon, north-northwest–trending thrust faults include the Golden, Rampart Range, and Ute Pass faults (*Harms, 1965*). These faults dip to the west beneath the Front Range, with up to 5 km of structural overhang on the Rampart Range thrust (*Jacob and Albertus, 1985*). Along the mountain front north of Denver, the exposed structures change from thrust faults to fault-controlled, en echelon, southeast-plunging asymmetric folds. Near the Wyoming border, the structural style changes again to variously trending monoclines of sedimentary strata folded along the edges of fractured basement blocks (*Matthews and Work, 1978*). Farther north, along the eastern flank of the Laramie Range in Wyoming (a continuation of the Front Range structure), west-dipping thrust faults are again encountered (Brewer and others, 1982). This segmented mountain-front structure has yet to be fully explained. Sedimentary rocks in the Denver basin are preserved only adjacent to the southernmost thrust-fault–bounded segment of the mountain front, where the Upper Cretaceous to Eocene Arapahoe Formation, and the Upper Cretaceous and Paleocene Denver Formation and Dawson Arkose attain an aggregate thickness of 820 m (*Crifasi, 1988*). These formations contain intraformational unconformities that are increasingly angular closer to the mountain front, attesting to the syntectonic nature of the basin fill (*Harms, 1965; Jacob and Albertus, 1985*).

The west side of the Front Range is also partly bounded by thrust faults: the east-dipping Elkhorn thrust in South Park and the Williams Range and Never Summer thrusts in North Park–Middle Park. The Independence Mountain thrust bounds the northern edge of North Park and has thrust Proterozoic rocks southward over rocks of the Coalmont Formation (Hail, 1965; *Park, 1977*). The Paleocene and Eocene Coalmont Formation in North Park also contains unconformities and is compressionally deformed (Hail, 1965, 1968). A major unconformity in the Coalmont separates Paleocene from Eocene strata (Hail, 1968). The Coalmont Formation is up to 3,300 m thick (Tweto, 1975), contains significant coal deposits, and crops out over most of the current basin floor (Hail, 1965, 1968). The orientation of the fluvial sandstone units in the upper part of the Coalmont changes across this boundary and signifies a possible change in basin shape and tectonism. This fluvial reorganization may be related to a change in regional compression direction and an increase in

strike-slip movement along basinal faults (Chapin and Cather, 1981, 1983).

Subsidence of South Park basin during the Late Cretaceous and Paleocene preserved 3000 m of conglomeratic South Park Formation, which contains many andesitic flows, breccias, tuffs, and tuffaceous strata (*Bryant and others, 1981*). An intraformational unconformity separates the lower two-thirds of the unit (western sediment source) from the upper third, which has an eastern source (*Bryant and others, 1981;* Chapin and Cather, 1983). Subsidence was probably due to thrust loading induced by westward movement on the Elkhorn thrust, although the principal deformation of the strata in South Park basin occurred after deposition of the South Park Formation, because it and the underlying Mesozoic sedimentary rocks are folded similarly (*Bryant and others, 1981*). This later deformation probably was caused by Eocene wrenching of the basin (Chapin and Cather, 1981, 1983).

In southern Colorado, well-exposed Laramide thrusts along the eastern flank of the Sangre de Cristo uplift (Tweto, 1975; *Lindsey and others,* 1983, *1986*) form the western edge of the asymmetrical Raton basin. Upper Cretaceous to Eocene sedimentary rocks in the basin (the Raton, Poison Canyon, Cuchara, and Huerfano Formations) are thickest to the west and thin eastward away from the mountain front (Applegate and Rose, 1985). A major angular unconformity separates the primarily Paleocene Raton and Poison Canyon Formations, which are 1,350 m in combined thickness, from the lower Eocene Cuchara Formation, which is up to 1,500 m thick, and the middle Eocene Huerfano Formation, which is about 600 m thick (Tweto, 1975). These strata appear to be several times thicker than the equivalent-age strata in the Denver basin, although similar to thicknesses in North Park and South Park. However, the Raton and Poison Canyon Formations may be facies equivalents (Smith and others, 1985), and thus their true thickness in the basin may be less than their combined stratigraphic thickness. Little is known about the west side of the Sangre de Cristo uplift because Neogene normal faulting, subsidence, and sedimentation within the San Luis basin have buried this flank. The San Luis basin apparently formed by strike-slip deformation of the western margin of the Sangre de Cristo uplift during Eocene time (Chapin and Cather, 1981, 1983). Eocene strata have been penetrated by drilling in the eastern San Luis basin, and seismic lines suggest that there are thick Eocene (and possibly older) strata deep in the basin (Gries, 1985).

Paleogene basins of New Mexico contain strata deposited in fluvial, lacustrine, and eolian environments that were shed from monoclinal and fault-bounded, basement-cored uplifts (Fig. 2). The strata can be divided into three main basin-filling events that probably reflect basin-margin tectonism: Paleocene (San Juan, Raton, and Cutter Sag–Love Ranch basins), early Eocene (San Juan, Galisteo, Raton, and Cutter Sag–Love Ranch basins), and middle to late Eocene (Galisteo–El Rito, Baca, Carthage–La Joya, Sierra Blanca, and Love Ranch basins) (Chapin and Cather, 1981, 1983; Smith and others, 1985). The San Juan and Baca basins are located within the Colorado Plateau province, and their development is generally tied to monoclinal uplifts. The other basins are part of an en echelon series along the eastern margin of the Colorado Plateau (Chapin and Cather, 1981, 1983).

Colorado Plateau

The Colorado Plateau region is characterized by monoclines in cratonal sedimentary strata that overlie Early Proterozoic basement similar to that in the Colorado–New Mexico region. Many monoclines are hundreds of kilometers long and have hundreds of meters of structural relief (Fig. 2). They are multidirectional, sinuous, and branching, although north-northwest and northeast trends predominate (Kelley, 1955; Davis, 1978). These monoclines formed in strata above high-angle basement faults that commonly had long prior histories. The basement fracture zones form a rectilinear pattern and outline a mosaic of crustal blocks (Davis, 1978), many of which coincide with major plateau uplifts and basins. The uplifts, which include the Kaibab, Echo Cliffs, Circle Cliffs, Monument, San Rafael, White River, San Juan, Uncompahgre, Defiance, and Zuni uplifts (Fig. 2; Chapin and Cather, 1981, 1983; *Huntoon, 1981*), differ from the typical Laramide foreland basement-cored uplifts primarily in magnitude and level of erosion. The amount of Laramide offset on the bounding faults is much less on the Colorado Plateau than in the typical foreland structures of Wyoming and Colorado, but the size and overall geometry of the structures probably is similar to the classic basement-cored mountain blocks early in their development.

The basement faults were reactivated in Late Cretaceous to early Tertiary time, probably by northeast-southwest regional compression (Davis, 1978; *Huntoon, 1981*). Studies of faults and fractures associated with the monoclines indicate horizontal compression during formation of the monoclines, ruling out passive drape folding during vertical block uplift (*Reches, 1978;* Anderson and Barnhard, 1986). The Uncompahgre uplift in western Colorado is a good example of a two-sided monocline-bounded plateau uplift. *Heyman and others (1986)* interpreted it to be a keystone-shaped block, bounded by inward-dipping northwest-striking reverse faults, that was uplifted in a regional northeast-oriented horizontal compressive stress field.

Sedimentary basins are confined to the north and east margins of the Colorado Plateau province. The Uinta, Piceance, San Juan, and Baca basins are all partly bounded by monoclinal uplifts, although the first three larger basins contain much sedimentary material eroded from major basement-cored uplifts adjacent to the Colorado Plateau proper (Fig. 2). Ages of the Colorado Plateau monoclines and uplifts are difficult to determine, but are best deduced from study of these sedimentary basins. Onlapping sedimentary strata indicate that the San Rafael swell in north-central Utah may have begun to form in latest Campanian time, although other sedimentary relations suggest that most of its development was Paleocene (*Lawton, 1986*). The

Uinta and Piceance basins also formed during Paleocene time and continued to subside throughout the Eocene. The San Juan basin has a similar history, whereas the Baca basin contains only Eocene strata (Chapin and Cather, 1981, 1983; Smith and others, 1985).

Chapin and Cather (1981, 1983) and Hamilton (1981) suggested that during the Eocene the Colorado Plateau moved 60 to 120 km north-northeastward relative to the rest of the foreland and craton to the east. This motion was accommodated by en echelon, right-slip faults along the eastern side of the plateau and by thrust faulting at the southern margin of the Uinta uplift (Fig. 2). The Colorado Plateau thus acted as a coherent crustal block decoupled from the craton. Independent behavior of the plateau region underscores and helps explain the lesser shortening and compressional deformation of this part of the foreland province.

Southeastern Arizona–southwestern New Mexico

The southernmost region of the foreland province is more obscured by Neogene tectonic and volcanic events than other parts of the foreland. Laramide deformation has been recognized for many years in this region, but details have proven elusive and controversial. Although it is agreed that deformation resulted from southwest-oriented horizontal compression, there are two principal interpretations: (1) thin-skinned regional thrusting of Precambrian crystalline and Phanerozoic sedimentary rocks toward the northeast (*Drewes, 1978, 1981*), and (2) thick-skinned basement-cored uplifts of Rocky Mountain foreland style (Davis, 1979; *Seager, 1983;* Seager and Mack, 1986).

The basement-cored uplift hypothesis better explains the available data. In this model, the Late Cretaceous to Eocene history is divided into three phases: (1) an early phase of compressional basement-cored block uplift, clastic sedimentation, widespread volcanism, and local plutonism, (2) an intermediate phase of major plutonism, regional uplift, and erosion, and (3) a terminal phsae of compressional tectonism, plutonism, and volcanism (*Keith and Wilt, 1985*). The boundaries between these phases overlap in time, and their distinction may be arbitrary; for example, Lipman and Sawyer (1985) showed that the plutons of the intermediate phase represent solidified magma chambers from which associated volcanic rocks (assigned by others to the first phase) were erupted.

The first phase was Late Cretaceous to Paleocene in age and was marked by developing basement-cored, fault-bounded uplifts and adjacent tectonically loaded basins. Sedimentation in these basins included the fluvial and lacustrine strata of the Fort Crittenden Formation in Arizona (*Hayes and Drewes, 1978*), the Ringbone, McRae, and Love Ranch Formations in New Mexico (Seager and Mack, 1986), and correlative rocks in the Cabullona area of northern Sonora, Mexico (*Taliaferro, 1933*). Widely overlying and interbedded with these clastic strata are andesitic to rhyolitic lavas, breccias, and tuffs of the Upper Cretaceous Salero Formation and Upper Cretaceous to early Tertiary Hidalgo Volcanics (*Hayes and Drewes, 1978; Drewes, 1981*).

Four northwest-trending, fault-bounded uplifts have been defined (Fig. 2): the Rio Grande, Burro, and Hidalgo uplifts in southwestern New Mexico (Seager and Mack, 1986), and a large unnamed uplift in southeast Arizona (Davis, 1979). Most of the bounding faults dip southwestward, providing an asymmetric form that is similar to the Wind River type. An exception is the northeast-dipping fault zone in the Huachuca Mountains, where thrust faults and folds record southwestward transport. Similar structures extend about 200 km southeastward into northern Mexico (*Rangin, 1977*). Reactivated northwest-trending basement fractures predominated in this region, as in much of the Rocky Mountain foreland province. The fault zones contain complex compressional structures, including folds of many types and reverse, thrust, and strike-slip faults (Davis, 1979; *Drewes, 1981;* Seager and Mack, 1986). The times of movement on the various structures are poorly defined and are broadly considered to be Late Cretaceous to Eocene in age.

The second phase is marked by region-wide emplacement of Paleocene granites, which are associated with porphyry copper deposits (*Marvin and others, 1978;* Dickinson, 1981; Titley, 1982). This plutonism overlaps the preceding and following phases, but is considered separate by *Keith and Wilt (1985)* because there was a pause in tectonism during the Paleocene culmination of igneous activity. The calc-alkalic, metaluminous magmas produced epizonal plutons with associated large, zoned, porphyry copper–molybdenum systems (Titley, 1982) and large related ash-flow calderas (Lipman and Sawyer, 1985). Few, if any, sedimentary rocks of this age are preserved in this region.

The third (Paleocene to Eocene) phase resulted in reactivated foreland structures by wrench tectonics and intrusion of peraluminous, two-mica granitoids (*Keith and Wilt, 1985*). Ductile deformation associated with emplacement of these granitoids in south-central Arizona is more closely allied with the ductile thrusting and plutonism occurring further west, and is described in the section on southern Arizona and southeastern California. Eocene strata have not been identified in southeastern Arizona but are present in central Arizona (Rim gravels and Baca Formation) and southern New Mexico (Baca, upper part of Love Ranch and McRae Formations) (Chapin and Cather, 1981; 1983; Smith and others, 1985). As in the rest of the Rocky Mountain foreland, these tectonic variations are interpreted to result from changes in the direction of regional horizontal compression.

IGNEOUS RECORD

Magmatic activity in the Rocky Mountain foreland was generally restricted to three areas: (1) southwest Montana–northwest Wyoming, (2) the Colorado mineral belt, and (3) southeastern Arizona–southwestern New Mexico. This igneous activity has been interpreted as eastward migration of arc magmatism from areas near the western margin of the North American plate (Lipman and others, 1971; Snyder and others, 1976; Coney and Reynolds, 1977; Coney, 1978; Lipman, 1980). Whereas magmatism in Montana and Arizona was connected to

extensive igneous terranes, the igneous activity in Colorado was regionally restricted to a narrow northeast-trending belt.

In southwestern Montana and northwestern Wyoming, igneous activity extended from earlier Mesozoic arc-related magmatism to more restricted Late Cretaceous and Paleocene magmatism. The mid-Campanian to middle Paleocene (82 to 62 Ma) igneous event produced plutonic and volcanic rocks along the northeast-trending Idaho-Montana porphyry belt (*Chadwick, 1981*), also known as the Great Falls tectonic zone (*O'Neill and Lopez, 1985*). The belt includes the Pioneer (83 to 67 Ma) batholiths and the Elkhorn (78 to 73 Ma) and Adel Mountains (Cretaceous to Paleocene) volcanic fields. Following a magmatic null, widespread early to middle Eocene (54 to 45 Ma) volcanism and laccolith intrusion was associated with the onset of extension.

Igneous rocks of the Colorado mineral belt (Pl. 6) comprise a suite of hypabyssal intrusive rocks, predominantly monzonite, granodiorite, and syenite, that were emplaced ca. 70 to 50 Ma (Tweto, 1975). The volcanism associated with these intrusions is generally preserved as andesitic volcanic and sedimentary rocks within the lower parts of the clastic fill of the Laramide basins. In the Front Range, these volcanic rocks fall into two groups: an older one between 70 and 60 Ma, and a rare younger one between 54 and 44 Ma (Gable, 1984). The younger group is of similar age to the Eocene resurgence of volcanism farther north typified by the Challis and Absaroka Volcanics. In general, there was little magmatic activity in the mineral belt from 60 to 40 Ma, suggesting a waning of Laramide magmatism (Lipman, 1980). This lessened activity possibly is related to the proposed change in the direction of regional horizontal compression from northeast to nearly north, and the closing of northeast-trending igneous conduits in the Colorado mineral belt, or could be caused by changes in subduction angle or rate.

In southeastern Arizona and southwestern New Mexico, Laramide igneous activity was discontinuous from Late Cretaceous through Eocene time and activity crudely migrated from southwest to northeast with time. This migration has been interpreted as northeastward migration of a calc-alkalic magmatic arc (Coney and Reynolds, 1977; *Keith and Wilt, 1985*) and is described more fully in the magmatic-arc section.

CORDILLERAN FOLD AND THRUST BELT

D. M. Miller

INTRODUCTION

The Cordilleran fold and thrust belt province, as used herein, includes major parts of Idaho, Utah, and Nevada (Fig. 1). The generally sharp eastern boundary with the Rocky Mountain foreland province is defined where belts of thin-skinned thrust sheets of Paleozoic, Mesozoic, and early Tertiary strata on the west (thrust belt) are bounded by thick-skinned basement-cored thrust blocks (foreland) on the east. The western boundary of the thrust-belt province is placed approximately along the eastern margins of the Sierra Nevada and Idaho batholiths. Northward, the province is continuous with the Canadian Omineca crystalline belt and provinces eastward to, and including, the Rocky Mountain thrust belt (*Davis and others, 1978*). The southern boundary of the province lies near the southern tip of Nevada where ductile-style thrusts that increasingly involve basement rocks merge with a province of thin crystalline thrusts and coeval magmatism in southwestern Arizona and southeastern California (Fig. 1).

The Late Cretaceous to early Eocene evolution of the Cordilleran thrust belt province was strongly controlled by preexisting tectonic, magmatic, and sedimentary domains. The Early Cretaceous tectonic features of the region consisted of two sinuous belts extending the length of the Cordillera, the thin-skinned thrust belt on the east, and the hinterland on the west. Thrusting on the east was localized by a hingeline between the miogeocline and craton; the thicker miogeoclinal strata were first thrust eastward over transitional to cratonal strata, and then ensuing thrusts broke cratonal strata beneath and east of the early thrust plates. The thrust faults in the hinterland were located deeper in the thick miogeoclinal strata and probably in basement rocks. In this zone of deeper thrusting, scattered Mesozoic plutons and metamorphic rocks define a broad metamorphic belt representing a thermal culmination in eastern Nevada (Armstrong, 1982); similar remnants suggest the metamorphic belt extended from eastern California to Washington and as the Omineca crystalline belt in Canada. Deep-seated bedding-plane faults in the hinterland both thickened and thinned the miogeoclinal sequence. In the western hinterland, thrusting affected Paleozoic eugeoclinal strata (*Ketner, 1984*) and lower Mesozoic back-arc strata (*Oldow, 1984*).

From Late Cretaceous to early Eocene time the intensity of tectonism and magmatism decreased over most of the thrust-belt province. Thin-skinned thrusting continued throughout the Cretaceous and declined in the Paleocene, but in some salients (eastward bulges of the belt) continued until early Eocene time. Magmatic arcs ceased activity, and hinterland metamorphic terranes were cooled during the Late Cretaceous over a segment from southernmost Nevada to central Idaho, whereas magmatism and metamorphism continued north and south of that segment. This tectonic segmentation and diachronous magmatism during Late Cretaceous and Paleocene time may have been caused by shallower subduction under the central segment (Coney, 1972; Burchfiel and Davis, 1975; Dickinson and Snyder, 1978). The shallow subduction in turn may have been caused by changing subduction regimes, such as the subduction of an aseismic ridge (Henderson and others, 1984). By Paleocene and early Eocene time, broad nonmarine basins developed in the metamorphic belt.

Beginning in the early Eocene, deformation in the northern Cordillera was characterized by widespread volcanism and developing extensional basins. Thick Eocene volcanic sequences such as the Challis Volcanics and Sanpoil Volcanics blanketed

much of northern Washington and Idaho. Eocene to Oligocene extension was greatest in the Nevada area of the Great Basin, where it distorted patterns of Cretaceous and early Tertiary tectonic belts. Total extension, comprising tectonism during early and late Tertiary time, across this area was probably 50 to 100 percent. Known details of the extension do not permit accurate palinspastic reconstructions, so data presented herein are on a modern geographic base.

Ensuing descriptions of the thrust belt are organized into three sections representing three well-studied parts of the orogen: (1) southwestern Wyoming to northwestern Nevada; (2) southern Utah to southeastern California; and (3) western Montana to northeastern Washington. The first of these regions represents the thrust belt at its widest and is, in many respects, the best studied, and thus is described in some detail. Other regions are compared and contrasted with the first to illustrate changes along strike.

WYOMING TO NORTHERN NEVADA

Frontal thrust belt

The thrust belt in southwestern Wyoming and adjacent southeastern Idaho and northern Utah (Fig. 4) forms a broad salient that contains the most complete record of synorogenic basin development in the North American Cordilleran thrust belt. Studies of the geometric and chronologic evolution of this part of the thrust belt are best summarized by Armstrong and Oriel (1965), *Oriel and Armstrong (1966),* Royse and others (1975), and *Wiltschko and Dorr (1983).* Armstrong (1968) recognized the southward extension of the belt in Utah, which he termed the Sevier orogenic belt. The overall distribution of geologic elements, geometry of structures, and many aspects of the chronology of thrusting closely resemble those of the southern Canadian thrust belt, and syntheses and descriptions of the Canadian counterpart by Bally and others (1966), *Dahlstrom (1970), Price and Mountjoy (1970),* Price (1981), and *Brown and Read (1983),* among others, have influenced interpretations in the Wyoming area.

Several characteristics throughout the Wyoming area of the thrust belt allow confident extrapolation of surface features to depth. (1) Thrust faults, with few exceptions, affected only sedimentary rocks and bottomed into a décollement near the top of crystalline basement, resulting in an eastward-thinning wedge-shaped thrusted mass (cross section in Fig. 4). (2) Deformation outside of fault zones was minor. (3) Mylonites are absent, and rocks are unmetamorphosed. (4) Thrust faults preferentially developed in weak layers composed of such materials as shale and evaporite. (5) Thrust flats are parallel to bedding, and thrust ramps crossing relatively competent rocks are steeper. (6) Tear faults and steep lateral ramps give the azimuth of thrust movement. (7) Folds are concentric and commonly are kink-like, and they are mainly generated by flexure above ramps. (8) Thrust faults cut upsection in the direction of tectonic transport. (9) Major thrust faults young to the east.

Thrust faulting in the southwest Wyoming area began in the Early Cretaceous (Armstrong and Oriel, 1965; Heller and others, 1986) and continued, perhaps episodically, to the early Eocene. During the earliest faulting, miogeoclinal strata were emplaced eastward over thinner sequences transitional to cratonal strata, but subsequent faults mainly ramped through cratonal strata and the Mesozoic and Cenozoic synorogenic sedimentary rocks within eastward-migrating foreland basins. The majority of thrust offset occurred during and after the latest Early Cretaceous (Jordan, 1981), resulting in at least 50 percent shortening of strata during this period. Many oil and gas reservoirs are controlled by structures created during thrusting and folding; migration of the hydrocarbons may have been controlled by the sequence of thrusting (Armstrong and Oriel, 1965; Royse and others, 1975).

The thrust faulting beginning in Campanian time led to sequential development of th Absaroka, Hogsback, Prospect, and La Barge thrust faults (Fig. 4). The Absaroka fault appears to have moved twice. The major movement is late Campanian in age because the fault cuts the lower Campanian Adaville Formation, but is overlapped by the upper Campanian to Maastrichtian basal Evanston Formation (Armstrong and Oriel, 1965). However, Royse and others (1975), Lamerson (1982), and DeCelles (1988) inferred a Santonian age for the early Absaroka thrust movement on the basis of unconformities. The upper Campanian and lower Maastriachtian (*Jacobson and Nichols, 1982*) Hams Fork Conglomerate Member of the Evanston Formation was interpreted as a syntectonic deposit that roughly dates late Absaroka thrusting (Armstrong and Oriel, 1965; Royse and others, 1975). The majority of oil discoveries in the Wyoming salient of the thrust belt are sited in the hanging wall of the Absaroka thrust and other parts of the belt where similar source rocks developed in similar structurally controlled reservoirs. The Hogsback fault places Cambrian strata over the lower Campanian Adaville Formation, and locally over the upper Paleocene to lower Eocene Chappo Member of the Wasatch Formation, and the fault is overlapped by upper Paleocene sandstone. Thus, all of the Hogsback fault movement postdated the Late Cretaceous, and latest movement was after the upper Paleocene. The Darby fault occupies a similar tectonic position, and presumably is about the same age. The La Barge fault, a blind thrust east of the Hogsback, is dated by an unconformity with lower Eocene upon Paleocene strata, requiring a middle early Eocene age for thrust faulting. The Prospect, Jackson, and Cliff Creek faults farther north have similar structural relations, suggesting that they also are early Eocene in age.

Thrust faults in western Wyoming extend southward into Utah and are correlated (where not traced by mapping and subsurface data) on the basis of stratigraphy in hanging walls and footwalls: the Willard thrust corresponds to the Paris, and the Mount Raymond thrust corresponds to the Absaroka and Hogsback. The latter correlation is poorly constrained because strata and structures associated with the Mount Raymond fault are exposed in a small area. The faults indicated in Figure 4 may not be continuous; they are probably en echelon, like better-exposed

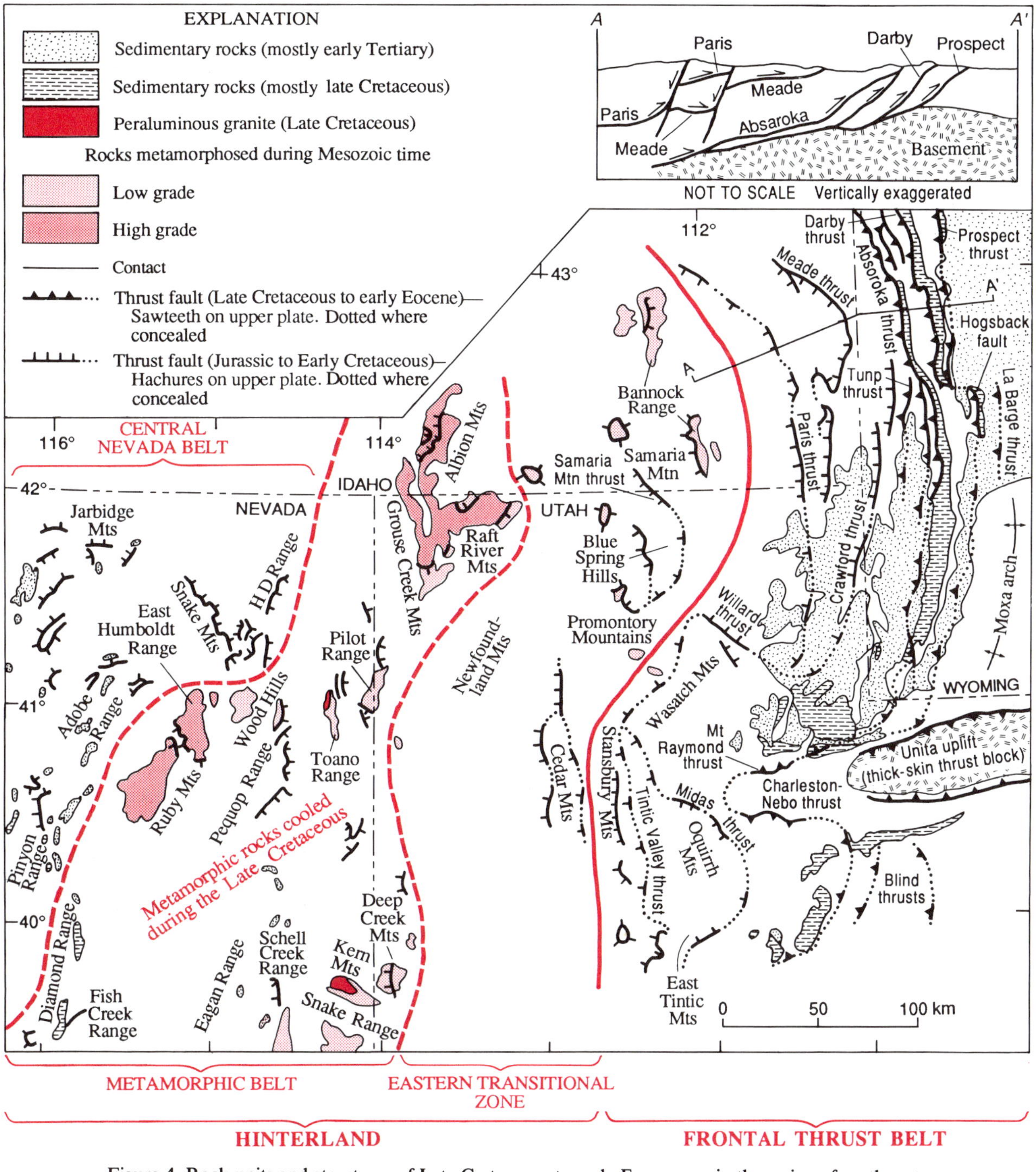

Figure 4. Rock units and structures of Late Cretaceous to early Eocene age in the region of southwest Wyoming to northeast Nevada, showing subdivision of Cordilleran thrust belt. Some low-angle faults shown in hinterland may be partly or wholly Cenozoic in age. Cross section after Royse and others (1975). Formations included in map units: Largely late Paleocene to early Eocene age: Sheep Pass Formation, Wasatch Formation, Fort Union Formation, and Hoback Formation; largely Late Cretaceous age: Evanston Formation, Adaville Formation, Echo Canyon Conglomerate, Hilliard Shale, Frontier Formation, and Newark Canyon Formation.

faults in Wyoming. Although the Willard thrust was considered by *Crittenden (1976)* to be equivalent to the Charleston-Nebo thrust farther south in the Wasatch Mountains, on the basis of hanging-wall stratigraphy it is more likely that the Charleston-Nebo and Mount Raymond thrusts are correlative. The latter two thrusts have Mesozoic strata in their hanging walls, in contrast to the Willard. The Willard more likely correlates with the Tintic Valley and Sheeprock thrusts on the basis of thick Precambrian hanging-wall strata. Timing of movement along some faults in the Wasatch Mountains is well constrained. The Mount Raymond thrust cuts the lower Upper Cretaceous Frontier Formation and is overlapped by conglomerate of the Campanian and Maastrichtian Wanship Formation, constraining movement to the early Late Cretaceous (*Crittenden, 1976*). The Charleston-Nebo thrust cuts the lower Upper Cretaceous MesaVerde Group and is overlapped by Price River or North Horn Formations, and thus is late Late Cretaceous or early Paleocene in age. The Willard thrust is older than overlapping conglomerate containing latest Late Cretaceous pollen.

South of Great Salt Lake, thrust faults west of and above the Tintic Valley thrust were suggested by *Tooker (1983)* to be younger than the Charleston-Nebo thrust and therefore out of sequence, but no definitive criteria are available. Alternatively, the faults may correlate with the Samaria Mountains and/or Blue Springs thrusts north of Great Salt Lake; *Allmendinger and others (1984)* thought that these faults were Late Cretaceous or older.

Sediments shed from the emerging mountains created by thrust faults were deposited in coalescing alluvial fans east of the mountains. Eastward, these fans merged with coastal plains and shallow epicontinental seas that spread over much of the continental interior. In Late Cretaceous time, more than 5,500 m of sediments accumulated in a foreland basin that was localized a few tens of kilometers east of the active thrust faults. With the development of basement uplifts of the Rocky Mountain foreland province during Late Cretaceous time, localized nonmarine basins developed east of the thrust belt, segmenting the older foreland basin.

Hinterland

The eastern margin of the hinterland of the Cordilleran thrust belt is here defined for the northern Great Basin by one or more of the following changes westward from the thrust belt: (1) low-angle faults place younger rocks on older, thinning stratigraphic section; (2) low-angle faults widely involve metamorphosed and/or crystalline rocks; and (3) plutonism was synchronous with thrust faulting. Thus, the hinterland low-angle faults cut to deeper tectonic levels than those in the thrust belt and do not consistently duplicate strata. The westward extent of the hinterland has not been clearly defined. Armstrong (1972) described the hinterland as far west as the front of the Roberts Mountains thrust (E. L. Miller and others, this volume, Pl. 4), and it is now known that mid-Mesozoic folding, faulting, and metamorphism extended westward to the Sierra Nevada batholith (*Coats and Riva, 1983; Ketner, 1984; Oldow, 1984*). As herein defined, the hinterland spans the distance from the foreland thrust belt to batholith. This distance, as depicted in Plate 6, has been exaggerated by subsequent Tertiary extension.

Late Cretaceous to early Eocene time in the hinterland of the thrust belt was relatively uneventful compared to earlier Mesozoic thrust faulting, folding, plutonism, and metamorphism (Cowan and others, this volume). Late Jurassic and Early Cretaceous plutons were widely emplaced, and the latest widespread plutons, 90 to 80 Ma, crosscut many folds and faults. Well-dated pre–Late Cretaceous low-angle faults (Armstrong, 1972; *Oldow, 1984; Allmendinger and others, 1984;* Miller and others, 1987) may have been partly synchronous with early thrust faults in the frontal thrust belt, but in places also predate frontal thrusts. Metamorphism and ductile deformation is dated as Jurassic and Early Cretaceous in some areas (Armstrong, 1982) and Late Cretaceous in others (Miller and others, 1988). In contrast, few faults of proven Late Cretaceous to early Eocene age are known. Metamorphism waned during the Late Cretaceous over a wide area (Miller and others, 1987; Snoke and Miller, 1988), and only a few plutons younger than 80 Ma are known in northern Nevada. Large intermontane basins developed in Late Cretaceous to early Eocene time in at least part of the hinterland.

Four broad belts within the northern Great Basin hinterland are distinguished on the basis of rock units and the inherited structural and metamorphic characteristics. From east to west they are (Fig. 4): (1) an eastern transitional zone between foreland thrust belt and hinterland, characterized by poorly dated low-angle faults in miogeoclinal strata, placing both older rocks over younger and younger over older, and by sparse plutonism; (2) the Mesozoic metamorphic belt, characterized by mainly younger over older faulting of miogeoclinal strata, strata showing varying degrees of metamorphism, and common plutons; (3) the central Nevada belt, characterized by Mesozoic low-angle faults that cut both miogeoclinal and eugeoclinal strata (formerly juxtaposed by Paleozoic thrusts), and common plutons; and (4) an early Mesozoic back-arc basin, characterized by east- and west-directed thrust faults cutting eugeoclinal strata and thick lower Mesozoic basin strata, and abundant plutons. These four belts are only briefly described because they apparently were mostly tectonically inactive during the Late Cretaceous to early Eocene.

Thrust faults of the hinterland may have created suitable traps for petroleum accumulation, but in most parts of the hinterland all but the uppermost Paleozoic strata are too mature to contain petroleum. Mississippian flysch and Paleogene nonmarine deposits are potential source rocks (*Poole and Claypool, 1984*), as are some Lower Permian to Lower Triassic strata.

Eastern transitional zone. Armstrong (1968) demonstrated by constructing a pre-Tertiary paleogeologic map that the region west of the frontal thrust belt underwent only broad folding or low-magnitude faulting at shallow levels. The structures that formed at deeper levels in the crust are not well documented, largely because of the difficulty of dating structures. In general, the rocks in the transition zone are only slightly metamorphosed,

even at stratigraphic depths approaching 12 km. Sparse Late Jurassic plutons created broad metamorphic halos, suggesting deep epizonal levels of emplacement. In the Newfoundland Mountains, Jurassic low-angle faults both attenuate and duplicate strata (*Allmendinger and Jordan, 1984*). Low-angle faults that place younger strata on older abound, but they are in part middle or late Tertiary in age, and cannot be traced laterally to bona fide thrust faults that cut upsection to the east. One discontinuity seen in several ranges, the Manning Canyon detachment, cuts Mississippian shaly rocks and in some ranges can be shown to be Cretaceous or older (Allmendinger and Jordan, 1981). *Allmendinger and others (1984)* hypothesized that ramping thrust faults in the Samaria Mountains and Blue Spring Hills represent the eastern breakout of the Manning Canyon detachment.

Metamorphic belt. West of the transitional zone, metamorphic rocks display penetrative deformational fabrics, folds, and ductile low-angle faults (Fig. 4). Although most rocks in the metamorphic belt were not deeply eroded prior to deposition of middle Tertiary strata, suggesting that only small thrusts and broad folds affected the upper part of the crust prior to middle Tertiary time (Armstrong, 1972), rocks from deeper in the crust were complexly deformed during the Mesozoic in large nappes that faced various directions (Snoke and Miller, 1988). Metamorphic minerals include sillimanite, kyanite, staurolite, diopside, and wollastonite. Although the highest grade rocks in the Ruby, Grouse Creek, Raft River, and Albion mountains in places include Tertiary granite and must be partly or entirely a product of middle Tertiary metamorphism, amphibolite-facies rocks in the Pilot Range and Ruby and Albion Mountains formed during the Mesozoic (Armstrong, 1982; *Allmendinger and others, 1984;* Miller and others, 1987; Snoke and Miller, 1988). Tectonism in the unmetamorphosed, higher level rocks within the metamorphic belt consisted of low-angle faults both attenuating and duplicating section. In some places, faulting is demonstrated to be Jurassic or older (*Riva, 1970;* Armstrong, 1972; *Snoke and Lush, 1984;* Miller and others, 1987), and in others it is known or inferred to be Early Cretaceous in age (*Coats and Riva, 1983;* Armstrong, 1982; *Allmendinger and others, 1984; Ketner, 1984*), but no structures have been shown to be Late Cretaceous or Paleocene. Although Mesozoic plutons are common in the metamorphic belt, most are older than Late Cretaceous. The few Late Cretaceous plutons are strongly peraluminous and contain muscovite. Examples include an ~75-Ma granite pluton in the Kern Mountains (*Best and others, 1974;* Miller and others, 1988), two plutons in the southern Snake Range (Miller and others, 1988), and possible Late Cretaceous plutons in the Toano Range (*Miller, 1984*) and Ruby Mountains (Snoke and Miller, 1988).

The pre–Late Cretaceous metamorphic belt, first noted as a distinct structural province by *Misch* (*1960;* the "Northeastern Nevada structural province"), was interpreted as a broad thermal culmination by Armstrong (1972, 1982). *Armstrong and Hansen (1966)* applied a model for the metamorphic belt consisting of an infrastructure (deep-seated, ductilely deformed metamorphic terrane) and a suprastructure (brittle upper crust) that were separated by a shearing zone of high thermal gradient. Alternatively, the crust may have deformed into a large-scale duplex that produced complexly folded and faulted metamorphic rocks structurally overlain by broadly warped strata (Snoke and Miller, 1988). Deep seismic reflection events have been interpreted as Mesozoic thrust duplexes (*Allmendinger and others,* 1983, *1987*).

Whatever the origin and significance of the metamorphism, it apparently terminated quickly and virtually synchronously over at least 400 km of strike during the early Late Cretaceous (*Miller, 1980; Snoke and Lush, 1984;* D. M. Miller and others, 1987; E. L. Miller and others, 1988). This interpretation of the metamorphic history is based on radiometric ages of metamorphic minerals that cluster at 88 to 62 Ma in several mountain ranges (Pl. 6B). One interpretation of the cessation of metamorphism and the juxtaposition of deep and shallow crustal levels across the eastern boundary of the metamorphic belt is that a major thrust fault was rooted in the lower crust beneath the metamorphic culmination and thrust the metamorphic rocks eastward over cold rocks during the early Late Cretaceous (Armstrong, 1982). The timing for this cooling event coincides with major movement on thrust faults in the foreland thrust belt, along which the majority of the offset in that belt occurred, suggesting a physical link between the hinterland and thrust-belt structures. Alternatively, upper crustal extensional denudation may have rapidly cooled this thickened crust of the metamorphic belt. A third alternative is regional cooling as a result of eroding the hot asthenosphere beneath the hinterland and replacing it with a much cooler subducting oceanic lithospheric plate.

Central Nevada belt. Westward, the metamorphic belt grades into a province distinguished by a general lack of metamorphism, widespread Jurassic and Early Cretaceous plutons (*Stewart, 1980*), and thrust faults placing eugeoclinal over miogeoclinal strata. Although the precise ages of many faults are unknown, several are Mesozoic (*Ketner and Smith, 1982; Coats and Riva, 1983; Ketner, 1984*). Faults formed during the Antler or Sonoma orogenies may have been reactivated in the Mesozoic. Although many faults are known to be of Late Jurassic or Early Cretaceous age (*Coats and Riva, 1983; Ketner, 1984*), only one is proven to be Late Cretaceous in age. West of Eureka and in the Fish Creek Range, *Hose (1983)* described Devonian strata that were thrust over the Cretaceous Newark Canyon Formation. The thrust is overlapped by Paleocene or Eocene deposits, thus bracketing the age of last faulting as Early Cretaceous to Eocene. The fault represents late movement in the Mesozoic orogenic belt and may point to a Late Cretaceous episode of faulting that has not been recognized elsewhere in the hinterland. Mesozoic plutons that intrude deformed rocks of the Central Nevada belt are mainly Jurassic and Early Cretaceous in age. Magmatism in the Central Nevada belt probably had ceased by Late Cretaceous time.

The nearly 1-km-thick Newark Canyon Formation of Early and Late Cretaceous (and possibly Paleocene) age (*Fouch and others, 1979*) was deposited in parts of north-central Nevada, including the Carlin-Pinyon Range (Smith and Ketner, 1976)

and the Fish Creek Range (Nolan and others, 1956). Near the Pinyon Range, thrust faults and folds formed before the deposition of the nonmarine conglomerate, siltstone, and limestone. The Upper Cretaceous nonmarine deposits represent alluvial-fan to lacustrine depositional environments within broad basins that had mountainous margins.

Broad, nonmarine basins, primarily of Paleocene and early Eocene age, are documented near the Adobe Range (Solomon and others, 1979). In several areas to the south and southeast of Elko (Fouch, 1979) more than 1.2 km of similar strata, generally referred to as the Sheep Pass Formation, were deposited in alluvial fans and broad lakes. Some Eocene sequences include coal and oil shale of economic significance (Smith and Ketner, 1976). Although the basins were not geometrically related to present basins, they had local margins of high relief from which gravity-slide breccias and blocks were derived, similar to many modern extensional basins. By late middle Eocene time (43 Ma), felsic volcanic rocks were an important component of the rock record. Data are too sparse to establish the origin and significance of the volcanic rocks, but they have been interpreted to belong to southward sweep of Paleogene volcanism from Idaho to Nevada (Dickinson and Snyder, 1978).

Early Mesozoic back-arc basin. *Speed (1978)* and *Oldow (1984)* summarized the Lower Triassic to Lower Cretaceous basin, volcanic-arc, and continental-shelf rocks in western Nevada. The rocks were deposited in a simatic back-arc basin that was the site of Middle Jurassic to Early Cretaceous folding and thrust faulting characterized by both east- and west-directed transport. Rocks derived from the volcanic arc nearby to the west are locally common in strata as young as Early Cretaceous. Upper Cretaceous strata are absent, and there is no record of late Late Cretaceous tectonism. The region may have been a high stable plateau draining to the west during the Late Cretaceous and Paleocene and not the site of deposition. Alternatively, Upper Cretaceous deposits could have been stripped during Tertiary time. By early Eocene time, drainage to the west across the batholith had been established and clasts from the eroding back-arc basin were being deposited in eastern California (Nilsen and McKee, 1979).

Synthesis

During late Late Cretaceous to early Eocene time, the Wyoming to Nevada area was mountainous, with shallow epicontinental seas in the foreland of the thrust belt; this foreland broke into smaller basins with the formation of Laramide basement uplifts. Few direct paleogeographic data are available for the hinterland. Only 1 to 2 km of stratigraphic section were eroded from the eastern hinterland during 150 m.y. (Early Jurassic to Late Eocene), suggesting relatively low rates of erosion and either little uplift or uplift as a broad plateau. Intermontane basins in the Late Cretaceous and early Cenozoic were apparently broad and fault-bounded, and some contained shallow lakes. Analogies with the Altiplano-Puna plateau of the Andes (Allmendinger, 1986) and the Sierras Pampeanas (Jordan and Allmendinger, 1986) suggest elevations of 6 to 7 km for the Nevada hinterland.

Beginning in the Campanian, tectonic and magmatic activity in much of the area waned, although frontal thrusting was widespread and continued until Paleocene and, locally, early Eocene time. Cessation of magmatic activity in much of the volcanic arc, which was caused by changing subduction configurations, probably caused the changes in tectonic and thermal patterns eastward to the thrust belt (Armstrong, 1974; Burchfiel and Davis, 1975). To scrutinize models accounting for this change, we must examine those for Late Jurassic and Early Cretaceous time.

The broad pattern of magmatism, metamorphism, and tectonism in the northern Great Basin is similar to that of the southern Canadian Cordillera, for which several tectonic models have been proposed. *Price and Mountjoy (1970)* proposed that gravity spreading in the metamorphic culmination of the Omineca crystalline belt drove thrust faults to the east. As dating of structural events was refined, it emerged that, as in the Great Basin, the metamorphic culmination largely or wholly preceded thrusting in the foreland. Recent models incorporating the new data suggest that the metamorphic culmination served as a localizing feature for subsequent intracontinental subduction (westward underthrusting of the craton), ramping and wedging the thermally weakened lower crust (Price, 1981, 1986; Bally, 1981; Armstrong, 1982; *Brown and Read, 1983; Brown and others, 1986*).

The intracontinental subduction model could apply in the northern Great Basin. In this model, a major thrust ramped up in the metamorphic belt, causing Late Cretaceous cooling of the hanging wall by uplift and eastward translation over cold rocks. This major thrust probably connected to the Late Cretaceous décollements in the frontal thrust belt. Alternatively, thrusting may have been generated by gravity spreading from the structural culmination coinciding with the metamorphic belt, thus explaining sparse west-directed thrust faults west of the metamorphic belt (Armstrong, 1982). Latest Cretaceous and Paleocene tectonic activity in the thrust belt may have had no analog in the hinterland, suggesting that those thrust belt faults ramped steeply westward into the middle crust (Royse and others, 1975) at locations near the hingeline.

Other models for thrusting lack support. Gravity gliding seems untenable because there is a lack of appropriately high mountains and large rift structures in the hinterland (Armstrong, 1972). Hamilton's (1978, 1988) model for all thrust-belt structures being driven by gravity spreading of batholiths must be modified because new data show thrusts in Utah to be 10 to 20 m.y. younger than plutonism.

After the cooling of metamorphic rocks and cessation of the volcanic arc, scattered intermontane basins developed. Sparsely distributed calc-alkalic volcanism developed across northern Nevada during the middle Eocene, starting a pattern of volcanism and continental rift-basin sedimentation in an extensional tectonic setting that was to persist throughout the remainder of the Cenozoic.

CENTRAL UTAH TO EASTERN CALIFORNIA

Southward from northern Utah and Nevada, the late Mesozoic tectonic belts are more or less continuous to the northern Mojave Desert, but major gaps in exposures of pre-Tertiary rocks and a lack of detailed studies in places make the belts locally difficult to trace. The northern Mojave Desert area and adjacent southern Nevada have been studied by B. C. Burchfiel, G. A. Davis, and their co-workers, who described the manner in which these belts appear to terminate or grade into the region of metamorphism and plutonism in the Mojave Desert. Data for the transect from southwestern Utah to eastern California (Fig. 5) are largely drawn from Burchfiel and Davis (1975, 1981) and from Armstrong (1968).

Summary of pre–Late Cretaceous tectonics

The Mesozoic magmatic arc cut across Paleozoic depositional trends southward from northwestern Nevada to the Mojave Desert, cutting successively from back-arc basin to eugeoclinal strata and Antler belt, to miogeoclinal strata, to the craton. As a result, the southern Nevada area (Fig. 5) was composed chiefly of the thrust belt (which followed the craton-miogeocline hinge southward), a poorly defined transition to a

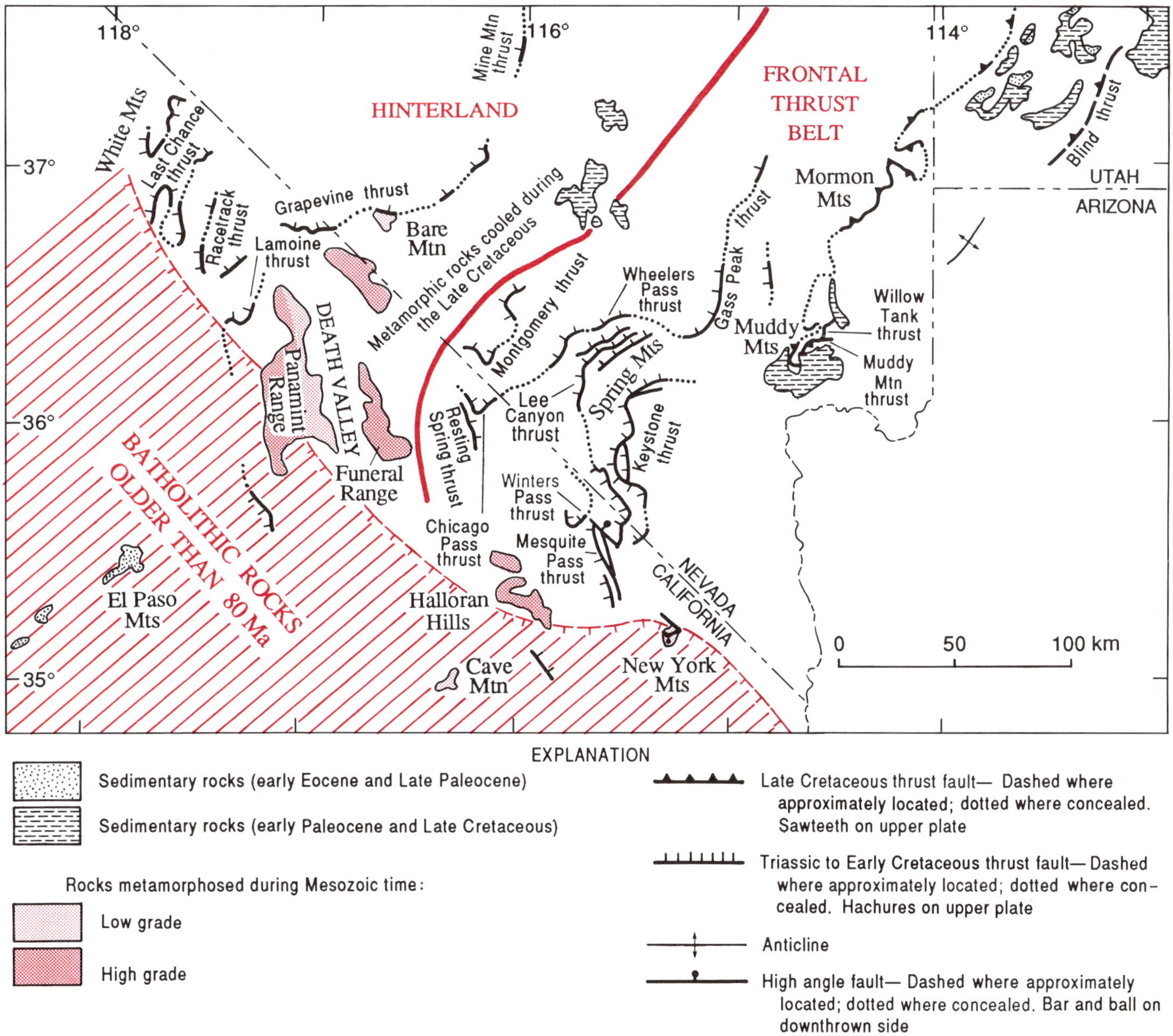

Figure 5. Rock units and structures of Late Cretaceous and Paleocene age in the Cordilleran thrust belt of southwestern Utah, southern Nevada, and southeastern California.

narrow hinterland, and the magmatic arc. Rocks of the Antler belt and back-arc basin, which are parts of the hinterland in northern Nevada, lie largely within the batholith at the southern Nevada latitude.

Frontal thrust belt. The thrust belt is nearly continuous southward from the Wasatch Mountains (Armstrong, 1968). Most easily defined is a western Early Cretaceous or older fault (Villien and Kligfield, 1986) that places miogeoclinal strata, including thick Late Proterozoic sequences, over cratonal strata; this fault occurs as several distinct, discontinuous segments (Fig. 5, Gass Peak and Wheeler Pass thrusts; possibly the Lemoine and Winters Pass thrusts; Wah Wah, Frisco, and Canyon Range thrusts). East of this thrust system is another pre–Late Cretaceous thrust system in which the main faults are the Muddy Mountain, Keystone, and Mesquite Pass thrusts. In the Muddy, Spring, and Clark Mountains, the Keystone and equivalent faults may have overridden older thrust faults and folds that appear to be at least 150 to 138 m.y. old (*Carr, 1983; Bohannon, 1983*), but these overridden faults have also been interpreted as coeval parts of a duplex structure (*Matthews, 1988*).

Departures from the general characteristics of the Cordilleran thrust belt occur at its southern terminus in the northern Mojave Desert: (1) the thrust belt cuts into the Paleozoic craton, (2) the thrusts cut crystalline basement rocks, (3) plutonism was roughly coeval and spatially coincident with thrusting, (4) mylonitic rocks and other ductile deformation features are associated with the faults, and (5) high-angle north- to northwest-striking faults were active during the same span of time as thrusting, deforming the thrusts and influencing such structures as tear faults and ramps in the thrust plates. These variations have been explained in a model by Burchfiel and Davis (1975, 1981) of southward increase in plutonism causing thermal failure of thick crystalline continental crust and overlying cratonal strata in a narrow zone and the resulting lack of well-developed décollement tectonics.

Hinterland. In southern Nevada and eastern California, several of the criteria for recognizing hinterland structures are ambiguous, because the Jurassic and Cretaceous magmatic arcs cut across middle Mesozoic tectonic belts, resulting in plutonism and metamorphism within even the frontal belt. However, ductile deformation, metamorphism, and the older tectonic events in the region of Death Valley and the White Mountains suggest that this region may be viewed as the extension of the hinterland of northeastern Nevada (Fig. 5). Exceptions are thrusts similar in style and age to those in the foreland thrust belt that occur in the Death Valley region.

The hinterland of southern Nevada and eastern California was affected by several episodes of plutonism, metamorphism, folding, and faulting (*Burchfiel and others, 1970;* Burchfiel and Davis, 1981). Folds and thrust faults of Middle Triassic to latest Jurassic age are present in the region shown in Figure 5; many thrusts were separated by only a few kilometers and were confined to zones of contemporaneous plutonism and large-scale folding. However, the Last Chance thrust is a thin-skinned thrust with large displacement and is therefore more akin to frontal thrust-belt structures. Late Jurassic to Early Cretaceous metamorphism as high as amphibolite facies affected wide regions in east-central Nevada (Miller and others, 1988) and the Death Valley area (Fig. 5; *Labotka and others, 1980; DeWitt and others, 1984*).

Late Cretaceous to Eocene

As in Wyoming, Utah, and northern Nevada, tectonic inactivity prevailed over much of southern Nevada west of the thrust belt during Late Cretaceous to early Eocene time. Minor thrusting disrupted sedimentary rocks in a foreland basin developed east of the thrust belt in central Utah. Fragmental Paleocene and Eocene deposits in the "hinterland" area attest to widespread erosion, possibly in a region typified by intermontane basins.

Frontal thrust belt. Major thrusting had ceased by early Late Cretaceous time, as indicated by crosscutting plutons with 94 to 84 Ma K-Ar ages (Burchfiel and Davis, 1981) and by Upper Cretaceous strata overlapping major thrusts (Armstrong, 1968). However, the Muddy Mountains–Glendale thrust (Fig. 5) system overrode lowermost Upper Cretaceous strata (*Bohannon, 1983*). In addition, the last movement on some thrusts and newly developed minor thrusts broke the Upper Cretaceous strata in central Utah and contributed coarse-clastic sediments to Late Cretaceous and Paleocene basins to the east (Villien and Kligfield, 1986; *Van Kooten, 1988*). In general, this southern part of the thrust belt was most active prior to the middle Late Cretaceous and underwent only minor movement during ensuing periods, when the thrust belt in southern Wyoming was undergoing considerable stratal shortening.

Foreland basin sedimentary deposits in central Utah (Pl. 6), representing alluvial systems merging eastward with shallow-marine seas, were deposited nearly continuously from Early Cretaceous to Eocene time. The deposits thicken and coarsen westward nearly to the thrust belt, where they abruptly thin. Tongues of conglomerate shed from mountains west of the foreland basins suggest three periods of tectonism in Late Cretaceous time and one in late Paleocene time (*Lawton,* 1982, *1986;* Fouch and others, 1983; Villien and Kligfield, 1986). Blind thrusts east of the Charleston-Nebo and Pavant thrusts (Pl. 6A) are required by subsurface data and unconformities (Villien and Kligfield, 1986). Foreland-basin deposits are absent in southwest Utah, southern Nevada, and northwestern Arizona, probably due to subsequent erosion. During the early Tertiary, thick-skinned thrust blocks of the Colorado Plateau were uplifted and shed gravel northward into southern Utah (Young and Hartman, 1984).

Hinterland. Some evidence suggests that the hinterland region of eastern California and southern Nevada had a history of uplift and erosion during Late Cretaceous to Eocene time much like the northern Nevada region. The metamorphic terranes apparently cooled during the late Early to earliest Late Cretaceous, as suggested by K-Ar ages (*DeWitt and others, 1984*). This cool-

ing of metamorphic rocks broadly coincided with the termination of the magmatic arc and the last major movement on thrusts to the east.

Upper Cretaceous strata are absent in the region, but the Paleocene and lower Eocene Goler Formation in the El Paso Mountains and its probable correlative, the Witnet Formation to the southwest (Fig. 5), represent fluvial and alluvial deposition in a mountainous terrain. The lower part of the 2,700- to 4,250-m-thick Goler Formation is predominantly upward-fining conglomerate representing fluvial channel and floodplain deposits, and alluvial plains and fans against mountain fronts (Cox, 1987). The upper part of the Goler is composed of westward-fining conglomerate, sandstone, and siltstone that represent channel and floodplain deposits of a westward-flowing braided river (Cox, 1987). Within this upper part are tongues of marine deposits. The thickness of the coarse deposits in the Goler Formation implies deposition within a developing basin, and the southwesterly to westerly paleocurrent measurements coupled with the marine deposits indicate that the basin was externally drained. Clasts within the Goler Formation indicate varied sources including granitoid, eugeoclinal metasedimentary, metavolcanic, and Late Proterozoic or Lower Paleozoic miogeoclinal rocks. This variation in source terranes for the clasts suggests that Mesozoic tectonic activity exposed a diverse assemblage of rocks. The Goler Basin may be analogous to Paleocene basins in the central Nevada hinterland.

WESTERN MONTANA TO NORTHEASTERN WASHINGTON

Northward from southwestern Wyoming and southern Idaho, elements of the Cordilleran thrust belt extend continuously to the Snake River Plain (Pl. 6). North of the plain, two regions of different tectonic behavior are separated by the Lewis and Clark zone (Pl. 6). In southwestern Montana and east-central Idaho, structural and stratigraphic relations are similar to those of western Wyoming, although north of southwestern Montana, plutons intruded throughout Late Cretaceous time. Farther north, geologic elements change across the long-lived zone of strike-slip faults termed the Lewis and Clark zone (Fig. 6), north of which the thrust belt is continuous with, and similar to, the Canadian thrust belt. This northern sector consists of a western metamorphosed and thrusted zone, a central basin of the Belt Supergroup that is broadly folded and cut by sparse thrust faults, and an eastern tightly folded and thrust-faulted zone.

The basic system of thrust belt impinging on foreland structures in the east, deeper and older thrusts and metamorphic rocks involved westward, and batholiths farther west is generally true for the Montana-Idaho region. However, several unique elements strongly influenced Cretaceous and Paleogene geologic development of the region: (1) as much as 20 km of strata of the Middle Proterozoic Belt Supergroup blanketed much of northwestern Montana; (2) the eugeoclinal province in western Idaho was apparently truncated and replaced by the Seven Devils terrane; (3) miogeoclinal strata in central Idaho occupy a narrower area than their Canadian and Nevada counterparts; (4) the Idaho and satellite batholiths were active until Paleocene time, and they intruded miogeocline and thrust belts; and (5) the Lewis and Clark zone and southwest Montana transverse zone controlled several tectonic and magmatic patterns. Consequences of these elements unique to Montana and Idaho include the following. (1) Pronounced salients and recesses in the frontal thrusts of the thrust belt were influenced by thick Belt strata and foreland thrust blocks. (2) Structural styles in the thrust belt were closely related to structure, metamorphism, and thickness of strata in the Belt basin. (3) Plutonism was spatially and temporally related to thrusting. (4) Major boundaries in magmatic and tectonic provinces coincide with the Lewis and Clark zone.

Primary contributions to the tectonics of the area south of the Lewis and Clark zone include Robinson and others (1968), Skipp and Hait (1977), Schmidt and O'Neill (1982), Scholten (1982), *Ruppel (1978), Ruppel and others (1981),* and Ruppel and Lopez (1984). North of the zone, primary contributions include *Harrison and others (1974, 1980),* Mudge and Earhart (1980), and *Mudge (1982).* These works form the basis for ensuing descriptions. Packages of thrust faults characterized by similar strata were grouped as "stacks of allochthons" by Skipp and Hait (1977) and as "thrust plates" by Ruppel and co-workers; the term "thrust system" is used herein.

Summary of pre-Late Cretaceous tectonics

A number of structures affected strata of the Belt basin, including probable Proterozoic faults that in places bounded the basin during its formation. Proterozoic and Paleozoic faults cut the Belt strata, and early thrusts of the Mesozoic thrust belt cut the western part of the Belt basin prior to Late Cretaceous time. The Lewis and Clark zone, a continental fault system, was intermittently active during much of Proterozoic and Phanerozoic time. Late Jurassic and Early Cretaceous tectonism in central Idaho is indicated by synorogenic deposits in western Montana (Suttner and others, 1981; Schwarz, 1982); in many locations, thrust faults in the western part of the region (Fig. 6) predate Late Cretaceous plutons. Many thrust faults in the frontal thrust belt were probably in motion at about 80 Ma, as described below. In northern and central Idaho, thrust faults moved before 100 Ma.

Late Triassic(?), Jurassic, and Early Cretaceous plutons were emplaced in a wide area of western and northern Idaho and northeastern Washington (Miller and Engels, 1975; Armstrong and others, 1977; *Fox and others, 1977*). The plutonism accompanied and in part postdated regional metamorphism that waned in the Late Cretaceous (*Fox and others, 1977; Archibald and others, 1984*); the plutonism and metamorphism chiefly affected regions underlain by eugeoclinal and miogeoclinal strata of northeast Washington, including but not restricted to, the southern part of the Omineca crystalline belt (Kootenay Arc mobile belt) in Canada. Juxtaposition of island-arc rocks of the Seven Devils terrane of eastern Oregon with eugeoclinal strata of

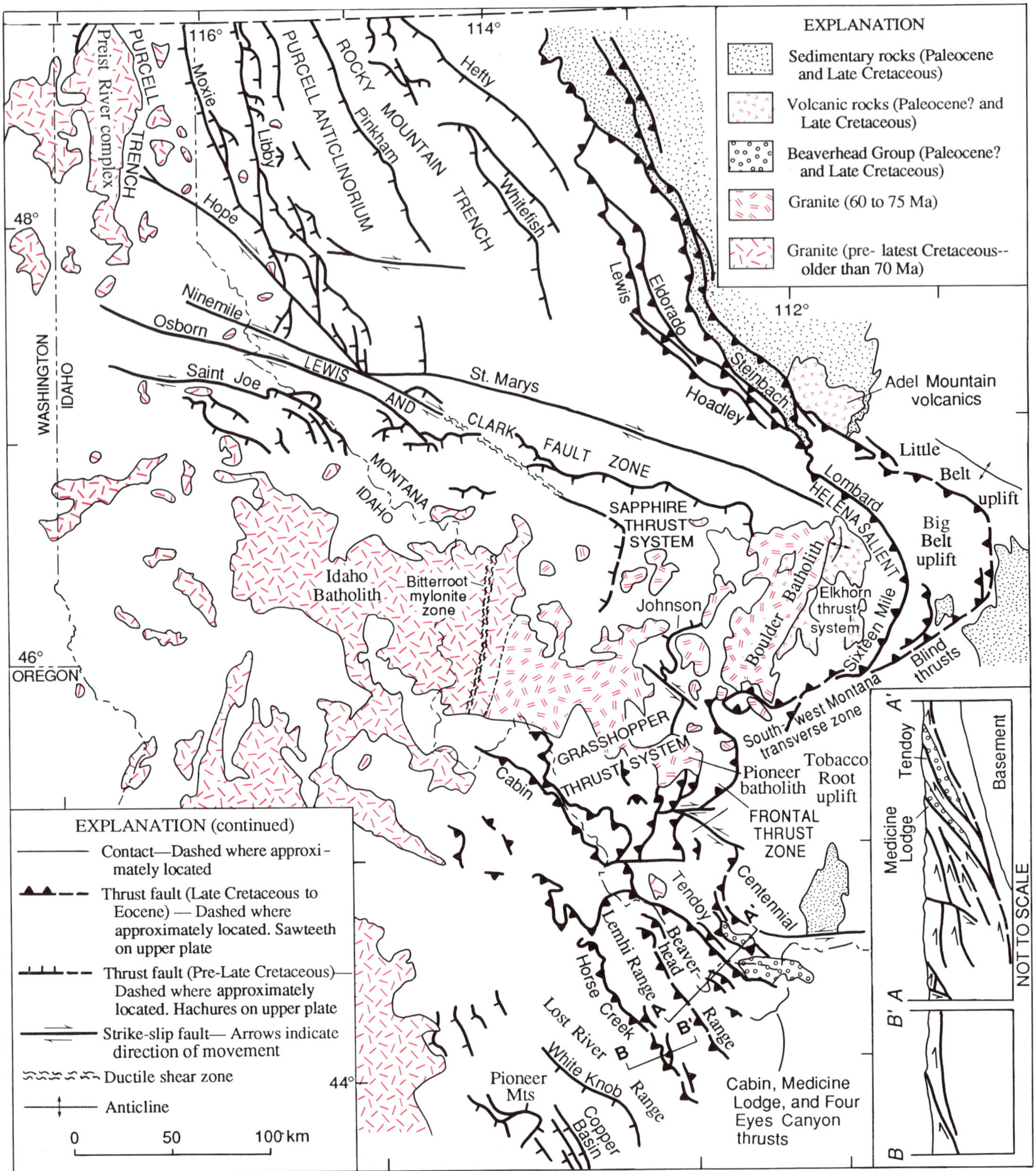

Figure 6. Rock units and structures of Late Cretaceous to early Eocene age in the Cordilleran thrust belt of western Montana, Idaho, and northeastern Washington. Cross section modified from Skipp and Hait (1977).

western Idaho was complete by earliest Late Cretaceous time when tonalite, dated at 90 to 95 Ma by Lund and Snee (1988), intruded the fault zone. Although isotopic ages for plutons in the Idaho batholith are generally Late Cretaceous and younger (Armstrong and others, 1977; *Bickford and others, 1981;* Criss and Fleck, 1987), Upper Jurassic and Lower Cretaceous volcanic-ash–bearing strata occur in western Montana, suggesting that older magmatism in the region was copious.

The paleogeography of the Idaho region at the beginning of Late Cretaceous time was characterized by thin cratonal Paleozoic strata overlying Precambrian crystalline basement on the east, wedging westward into thicker Paleozoic miogeoclinal strata of Belt strata. The thick sequences of miogeoclinal strata and Belt strata were being stacked eastward and upward along thrust faults, forming mountains that shed debris into foreland basins near the hingeline between craton and miogeocline. Farther west, active volcanoes reflected deep-seated plutonism and regional metamorphism within eugeoclinal and miogeoclinal strata.

Late Cretaceous and Paleocene

During Late Cretaceous and Paleocene time, thrusting progressed eastward and deformed progressively thinner sections of Proterozoic strata, then Paleozoic cratonal strata, and finally Mesozoic strata. Styles of thrusting changed accordingly from west to east and with time: in general, widely spaced faults and broad folds, developed in thick sections in the west, gave way to tight folds and closely spaced faults with smaller separations in the thin cratonal sequences to the east. Eastern thrust faults impinged on emerging foreland basement uplifts. Thrust faults died out north and south into the Lewis and Clark zone of oroclinal folding, later to be broken by transcurrent faults. Extensive plutonism in the Idaho and outlying batholiths during the interval 90 to 60 Ma led to volcanic edifices that shed detritus to foreland basins. In northeastern Washington, plutonism and metamorphism in the upper crust ended during Late Cretaceous time, although plutonism and volcanism renewed in the Eocene.

Frontal thrust belt. In southwest Montana and adjacent Idaho, thrust faults cut eastward upsection and young eastward. Skipp and Hait (1977) demonstrated that the clasts shed from the western Copper Basin thrust system (Fig. 6) were deposited in strata of the Upper Cretaceous Beaverhead Group and, therefore, that thrust system was active prior to Late Cretaceous time. Younger faults east of the Copper Basin are imprecisely dated except for the youngest and easternmost group (Perry and Sando, 1982), which consist of the Cabin, Medicine Lodge, Four Eyes Canyon, and Tendoy thrusts (Fig. 6). These four thrust faults represent an eastward-younging sequence that successively contributed detritus to, then overrode, synorogenic deposits represented by the Beaverhead Group. The upper part of the Beaverhead, the part that represents synorogenic debris from the thrust belt, is Campanian to Maastrichtian and possibly Paleocene in age (Nichols and others, 1985). The frontal fold and thrust zone east of and under the Medicine Lodge and related thrusts is cross-cut by a stock older than 70 Ma. Thus, the youngest thrust faults in southwest Montana are Late Cretaceous or possibly Paleocene in age. All thrust faults in the area are overlapped by Eocene volcanic rocks.

From west to east, rocks in the thrust plates in southwestern Montana change from possibly Archean rocks overlain by Paleozoic miogeoclinal strata (Cabin and Medicine Lodge plates), to Archean(?) and Early Proterozoic gneiss overlain by cratonal Paleozoic and Meszoic strata (frontal thrust zone). The changes in stratigraphy are located at boundaries between major thrust sequences. Northwest of exposures of the Medicine Lodge thrust plate and just west of the Idaho-Montana border are slightly folded Early and Middle Proterozoic Yellowjacket Formation and Archean rocks that were inferred by Ruppel and Lopez (1984) to represent the autochthon but, as convincingly demonstrated by *Skipp (1985),* are part of the Cabin thrust system.

Structures within the thrust systems of southwestern Montana differ from those in the Wyoming fold and thrust belt. East of the Tendoy thrust, the frontal fold and thrust zone is tightly folded, and small imbricate thrust faults break the fold hinges. Larger folds are probably cored by blind thrusts. Eastward-bulging salients in this zone closely correspond to small precursor sedimentary basins, and the salients are bounded north and south by tear faults such as the Centennial fault (Fig. 6). Rocks in the Cabin thrust system are tightly folded into eastward-overturned, imbricately thrust faulted sequences. Northward in Idaho, the thrust faults are discontinuous but appear to be truncated by the Idaho batholith.

In the central part of western Montana, thrust faults buttressed against the basement Tobacco Root uplift in the south continue eastward to the Helena salient of the frontal thrust zone (Fig. 6). Along the zone north of the Tobacco Root block, the southwest Montana transverse zone of Schmidt and O'Neill (1982), eastward-moving thrust plates were ramped laterally and they sliced off the tops of basement-cored folds. Eastward along the transverse zone, dextral displacement decreases as thrust faults north of the zone have less cumulative offset. Northeast of the transverse zone the Helena salient is composed of many thrust faults. There is an eastward younging of thrusts that cut upsection in the direction of transport.

Within the Helena salient, a wide zone of small-slip thrusts and closely spaced folds fan as far east as 111° longitude, where Mesozoic to middle Paleocene strata are folded. Thrust traces and folds are overlapped by Eocene volcanic rocks. The youngest thrust faults of this salient show middle to late Paleocene (*Harlan and others, 1988*) tectonism not present in southwestern Montana, where thrust faults in the frontal thrust belt are mainly older than 70 Ma. Last thrusting predated 52- to 48-Ma intrusions (*Harlan and others, 1988*), and therefore was early Eocene or older.

West of the fold and thrust zone in the Helena salient is the Elkhorn thrust system. The 74- to 76-Ma Elkhorn Mountains Volcanics in this system are folded and the lower strata are cut by thrust faults (Schmidt and O'Neill, 1982), but most folds, and

presumably thrust faults, are cut by plutons in the Boulder batholith that have K-Ar ages of 74 to 80 Ma. Apparently, thrust faults in this system were active over some 10 m.y. and were contemporaneous with plutonism and volcanism. Rocks cut by the Elkhorn system of thrust faults include the middle part of the Belt Supergroup (Ravalli Group) and overlying cratonic Paleozoic and Mesozoic strata.

West of the Elkhorn thrust system, the Sapphire thrust system and the related Grasshopper thrust system cut predominantly Belt strata and late Mesozoic plutons. Some of these faults are cut by plutons that have K-Ar ages from 82 to 70 Ma, but the details of correlating faults within the tectonic block and assigning ages to individual structures have yet to be developed. The leading thrust of the Sapphire system is overlapped by the Upper Cretaceous Elkhorn Mountains Volcanics. Rocks in the Grasshopper system are gently folded and are cut by multiple imbricate thrust faults. The western extent of the Sapphire system is unknown, but the Grasshopper system is overlapped on the west by the Cabin thrust system.

The Lewis and Clark zone bounds the northern part of the Helena salient and extends west-northwest into Idaho. Dextral faults in the zone are Eocene or younger and displace thrust faults. During thrusting (Late Jurassic or Cretaceous to Paleocene) the zone probably behaved as a giant lateral transfer zone or tear fault that accommodated different thrust geometries on either side of the zone.

North of the Lewis and Clark zone in western Montana, Late Cretaceous and Paleocene thrusting is documented. Thrust faults in the eastern part of the belt cut Paleozoic cratonal strata and then Mesozoic strata, similar to the contiguous Canadian thrust belt. Thrust faults in the western part of the belt may be older than 100 Ma. They presumably young eastward through the Belt basin, in analogy with similar structures in Canada, but faults west of the Lewis thrust are undated. The Lewis and Hoadley thrusts place Belt strata on Proterozoic to Mesozoic strata. Bentonite in Lower Cretaceous strata overridden by the Lewis thrust records K-Ar ages of 72 to 56 Ma (Mudge and Earhart, 1980), which approximately date their burial by the Lewis plate. To the east, a sill cutting the Steinbach thrust is dated at 58.3 Ma, and tightly spaced folds and small-slip faults east of the Steinbach in the Mesozoic strata deform rocks as young as the Upper Cretaceous and Paleocene Willow Creek Formation and the Paleocene(?) Adel Mountain Volcanics. Net eastward shortening on thrust faults in northern Idaho and northwestern Montana may be as much as 180 km. Large, but imprecisely estimated, offsets on the Whitefish, Hefty, Lewis, Hoadley, and Eldorado thrusts (Fig. 6) are required by abrupt facies changes across these faults.

Hinterland. In the area of the Idaho batholith, a hinterland for the thrust belt of Montana is difficult to define on the basis of criteria valid in Utah and Nevada because younger-over-older faults are apparently lacking, plutonism was coeval with thrusting, and metamorphism shifted in space and time in association with plutonism. High-grade metamorphism affected rocks south, west, and north of the Idaho batholith (*Hietanen, 1963, 1984; Dover 1969; Bickford and others, 1981*) both before and during intrusion of batholithic rocks, but details of the times of metamorphism are not available. Vestiges of pre-batholithic metamorphism may represent a metamorphic belt corresponding to the belt in Nevada, but metamorphism southeast and north of the batholith probably represents the thermal halo of the Late Cretaceous intrusions. Scholten (1982) suggested that thrust faults in southwest Montana root into crystalline basement and do not pass at depth westward to a Nevada-type hinterland.

North of the Idaho batholith, thrust faults pass to deeper levels within the thick Belt section and possibly cut basement rocks, a relation suggesting behavior transitional between thrust belt and hinterland. West of the Purcell trench are high-grade metamorphic rocks of the Priest River complex that are spatially and temporally associated with widespread plutons (Armstrong and others, 1987; Rehrig and others, 1987). Studies of these rocks (*Atwater and Rinehart, 1984*) have indicated that probably four periods of plutonism affected the region, metamorphism being associated with one or more of the first three (Triassic, Jurassic, and late Early Cretaceous) periods. Over wide areas, K-Ar ages are discordant and may be partly or wholly reset to the age of youngest plutons (Eocene), which also is the age of regional uplift and cooling.

Although details of the Mesozoic metamorphic, structural, and plutonic history of "hinterland" rocks in northeastern Washington and northern Idaho remain elusive, the general pattern of events there is similar to those described in northern continuations of the metamorphic belt, termed the Omineca crystalline belt in Canada. *Archibald and others (1984)* outlined in Canada a period of late kinematic plutonism about 165 to 170 Ma. A later group of plutons at about 100 Ma is postkinematic at shallow crustal levels, but their intrusion accompanied deeper seated metamorphism and strain. *Archibald and others (1984)* interpreted widespread discordance in K-Ar systems in the 100-Ma plutons as reflecting slow cooling during the Late Cretaceous and more rapid cooling in the Paleocene, all caused by eastward thrusting and uplift on a ramp. However, well-documented Eocene plutonism and extension in the region (Parrish and others, 1988) provide a more attractive means for "resetting" or suddenly cooling K-Ar systems in the Cretaceous plutons. The model of Cretaceous thrusting proposed by Archibald and others remains unproven.

SUMMARY

Geometric and chronologic similarities of the frontal thrust belt and hinterland extend nearly the entire length of the U.S. Cordilleran orogenic belt, suggesting that all parts of the belt were produced by similar large-scale processes. Adjacent to the Rocky Mountain foreland province, closely spaced folds and thrust faults were localized by rapid thickness changes within deformed cratonal Paleozoic and Mesozoic strata. Westward, thicker sections of the miogeocline were displaced by successively older thrust faults. West of the frontal thrust belt, deeper hinterland faults,

metamorphism, and magmatism were inconsistently developed in different areas, but evidence for deeper tectonism and accompanying metamorphism are unifying features.

Superposed upon the many laterally continuous characteristics of the Cordilleran orogenic belt are several changes along strike. Some changes were of regional scale: (1) basement involvement in the thrust belt of southern Nevada, brought about by the superposition of the batholith across the strike of Paleozoic depositional belts; and (2) diachronous plutonism and metamorphism across east-trending boundaries in central Idaho and southern California (Armstrong, 1974) that segmented the orogen. These regional-scale changes along strike probably are best explained by changes in plate tectonic processes, such as truncation of the continental margin and varying angles of subduction. Other changes were on the scale of tens to hundreds of kilometers, such as salients and recesses in the frontal thrust belt that were controlled by a combination of preexisting sedimentary basins, faulted boundaries, and uplifting tectonic blocks in the Rocky Mountain foreland. Another change, diachronous terminal thrusting along the eastern margin of the frontal belt, is somewhat enigmatic. Faults younger than Paleocene apparently are restricted to basins containing the thicker Mesozoic deposits. The temporal discordance of the youngest thrust faults may result from a combination of factors: (1) young structures are most easily and precisely dated where sedimentary deposits of similar age are preserved in salients, and thus young movement in areas lacking these deposits may go unrecognized; (2) young structures probably show exaggerated development in thicker Mesozoic sequences and retarded development elsewhere; (3) rising Rocky Mountain foreland blocks may have pinned thrust faults in some recesses of the belt (Ruppel and Lopez, 1984).

The extent and behavior of foreland thrusts as they ramped deeper to the west remains unknown. Proposals for hinterland faults range from décollement into ductilely flowing metamorphic zones (*Price and Mountjoy, 1970*), or décollement into steeply ramped thrust faults (Price, 1981; Armstrong, 1982; *Brown and Read, 1983*), or ramping into the basement essentially at the miogeocline-craton hingeline as intracontinental subduction (Scholten, 1982); all may be applicable in parts of the Cordillera. A time progression of basement ramps breaking successively eastward in response to eastward-moving thermal softening of the crust (Burchfiel and Davis, 1975) may have encompassed these models in sequence from Jurassic to Paleocene time.

SONORAN AND MOJAVE DESERTS REGION

D. M. Miller, with contributions by G. B. Haxel and R. M. Tosdal

INTRODUCTION

During Late Cretaceous to early Eocene time, the region that includes most of the Sonoran and Mojave Deserts in southern Arizona and southeastern California (Fig. 1) underwent widespread magmatism, metamorphism, and ductile thrusting of crystalline rocks, distinguishing it from the brittle deformation of the thick-skinned Cordilleran foreland and thin-skinnedfold and thrust belt. As previously described, boundaries to the east and north with the foreland uplift and fold and thrust belt provinces are gradational. The western limit of ductile thrusting is approximately in the vicinity of the southern part of the San Andreas fault system.

Tectonic framework

The Sonoran and Mojave Deserts region of southern Arizona and southeastern California is underlain by Early Proterozoic basement ranging from gneiss and granite to slightly metamorphosed sedimentary rocks. Middle Proterozoic to Mesozoic strata, 1.5 to 6 km thick, cover the crystalline basement. The crust in this region thus had many similarities with that of the cratonal foreland province, but it was pervasively intruded by scattered Triassic plutons (*Miller, 1977*) and by broad belts of Jurassic and Cretaceous plutons (*John, 1981;* Haxel and others, 1984; *Tosdal and others, 1988*). The plutonism and approximately coeval metamorphism apparently weakened the crust, because ductile deformation is documented in scattered locations for much of Mesozoic time (Keith and others, 1980; Burchfiel and Davis, 1981; *Hamilton, 1982;* Reynolds and others, 1986). Thick Mesozoic synorogenic sedimentary sequences accumulated locally (*Dickinson and others, 1986*).

During latest Cretaceous to early Eocene time, the Sonoran and Mojave Deserts region experienced widespread tectonism, magmatism, and local sedimentation. Ductile deformation belts continued south of the thin-skinned frontal thrust belt in southern Nevada, and eastward and southeastward across southern Arizona (Fig. 7). Deformation in these ductile belts was typified by ductile thrusts and nappes which were intruded by metaluminous and peraluminous plutons during the waning stages of deformation and metamorphism. Eastward, the structures are geometrically similar to the foreland uplifts of Colorado and New Mexico, suggesting a transition to structures of the foreland. In much of southeastern California, uplift rapidly followed mid-crustal deformation and plutonism, quenching isotopic systems and resulting in the deposition of middle Tertiary strata on rocks once at 10- to 20-km depths. In the western part of the area, a metamorphosed oceanic terrane represented by the Orocopia Schist and related schist was juxtaposed with epicontinental terranes along a Late Cretaceous thrust fault.

Following the magmatism and contractional tectonics of the latest Cretaceous and early Tertiary, widespread middle Tertiary volcanism and extensional faulting affected southern Arizona and southeastern California. This new orogenic realm obscured Cretaceous and Paleogene rocks and structures in many places, but also exhumed deep parts of the continental crust, providing exposures of the deep part of the older orogen.

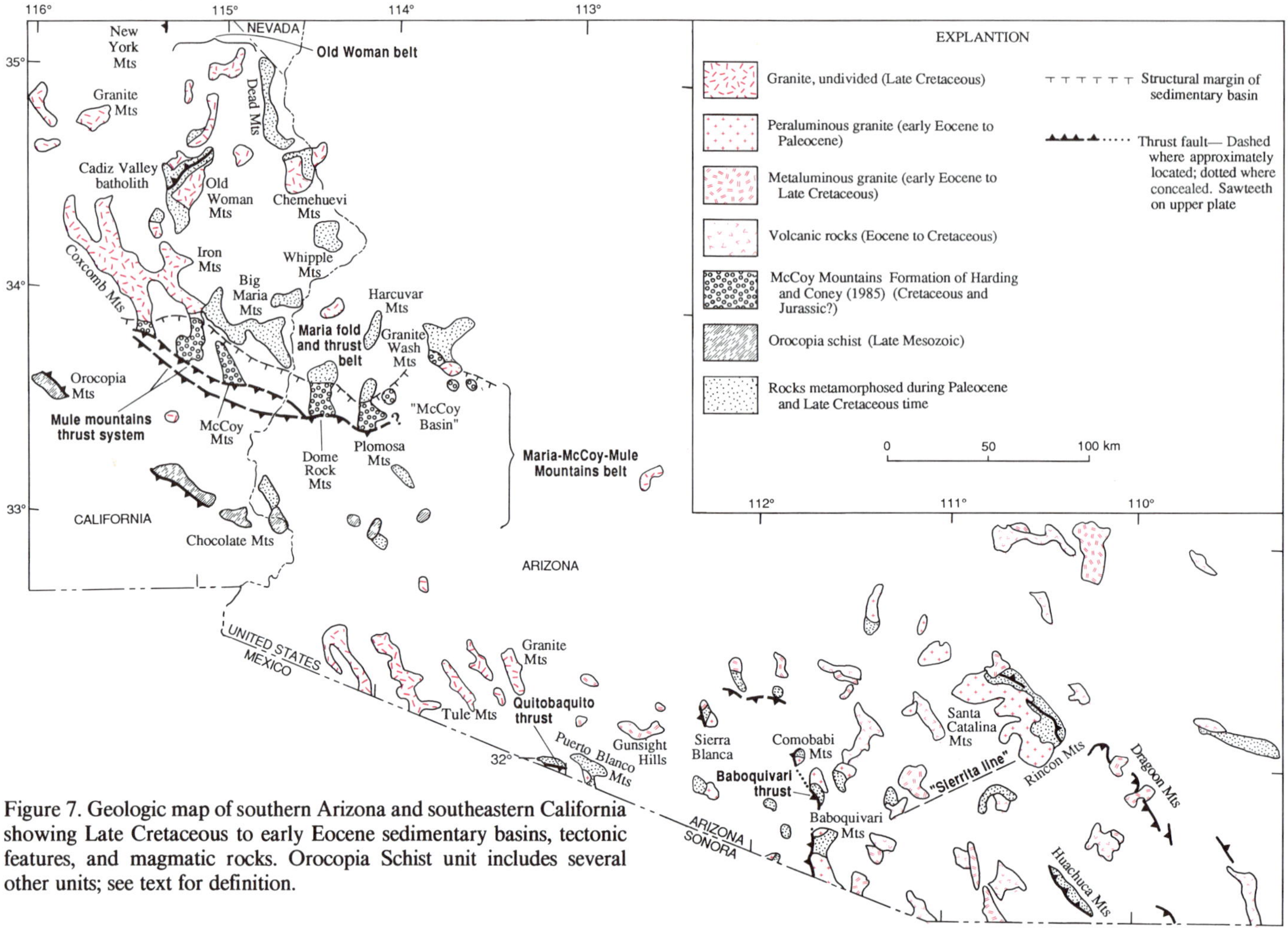

Figure 7. Geologic map of southern Arizona and southeastern California showing Late Cretaceous to early Eocene sedimentary basins, tectonic features, and magmatic rocks. Orocopia Schist unit includes several other units; see text for definition.

Regional subdivision

Within southern Arizona and southeastern California, terranes affected by Late Cretaceous and early Tertiary deformation, regional metamorphism, and granitoid plutonism represent a wide range of structural styles and crustal depths. Three regions are distinguished (Fig. 7): (1) the southeastern Arizona part of the foreland basement uplift province, which grades west into an area that underwent thrust faulting, regional metamorphism, and granitoid plutonism; (2) south-central Arizona, where thrust faulting, relatively low-grade regional metamorphism, and peraluminous granitoid plutonism occurred at intermediate crustal depths; and (3) western Arizona and southeastern California, characterized by large-scale, deep-seated ductile deformation and local synintrusive mylonitization of Late Cretaceous plutons. The Orocopia Schist of southern California and southwestern Arizona (Pl. 6) was metamorphosed in Late Cretaceous time; this schist is separated from other terranes of the southern Cordillera by a major thrust-fault system, and is described separately.

SOUTHEASTERN ARIZONA

Much of southeastern Arizona is characterized by brittle, relatively shallow deformation, including faulting and large-scale folding of supracrustal strata that produced basement uplifts as described previously in the foreland province section. Late Cretaceous to early Tertiary regional metamorphism and granitoid rocks are sparse, but the largely crystalline Santa Catalina and Rincon Mountains display ductile deformation related to granitoid plutonism and regional metamorphism. This tract evidently represents deformation at intermediate crustal levels, more akin to the region to the west. The "Sierrita line" of *Drewes (1981)* roughly separates two tectonic domains of contrasting Late Cretaceous to early Tertiary tectonic style and crustal level of exposure (Fig. 7). Southeast of this line, deformation was generally brittle, nonmetamorphic, and associated with upper crustal volcanism and plutonism. Northwest of the line, deformation was typically ductile, regional metamophism is widespread, and syntectonic plutonism is common.

Exposed in the Santa Catalina, Rincon, and Tortolita Mountains, and small ranges to the northwest, are diverse rocks of the informally designated Santa Catalina–Rincon–Tortolita complex of Keith and others (1980). Many of these rocks were deformed and metamorphosed during Late Cretaceous to early Tertiary time. Rocks in the complex include Middle Proterozoic gneissic granite, Middle and Late Proterozoic, Paleozoic, and Mesozoic strata, and late Mesozoic and Tertiary granitoids. Mesozoic and older rocks were variably deformed and metamorphosed at upper greenschist to amphibolite facies, approximately during intrusion of three granitoid suites, two from 80 to 50 Ma and one that was Oligocene in age. Calc-alkalic, epidote-biotite granodiorite ("Leatherwood suite" of Keith and others, 1980) was intruded about 75 to 60 Ma in the northern Santa Catalina Mountains. Voluminous peraluminous granite bodies, collectively designated the "Wilderness suite" by Keith and others (1980), intruded at about 50 Ma. These bodies typically are muscovite-biotite granites with common garnet, and they show variable deformation.

Much of the crystalline complex displays middle Tertiary mylonitic fabrics and low-angle normal faults (detachment faults). In and structurally below this middle Tertiary fabric, Late Cretaceous to early Tertiary fabrics, structures, and plutons are widely preserved (Keith and others, 1980, *Drewes, 1981; Keith and Wilt, 1985; Bykerk-Kauffman, 1986;* Bykerk-Kauffman and Janecke, 1987). The deformation was associated with east-directed thrusting of Proterozoic rocks over Paleozoic to Early Cretaceous rocks (*Bykerk-Kauffman, 1986*).

Several Late Cretaceous to early Tertiary low-angle faults cut through variably metamorphosed Proterozoic, Paleozoic, and Cretaceous strata and Proterozoic crystalline rocks (Fig. 7) in the northeastern Santa Catalina Mountains. The structurally highest fault has the greatest stratigraphic separation and places older rocks over younger. Other low-angle faults place younger rocks over older, and are interpreted as thrust faults that cut through a tilted stratigraphic section (Bykerk-Kauffman and Janecke, 1987). Deformation along the thrust faults and in the metamorphic rocks ranges from brittle to ductile. Here, as well as in the Rincon Mountains, kinematic indicators consistently indicate eastward overshear.

Metamorphic grade locally increases toward late metamorphic Wilderness suite granites. Reynolds and others (1988) suggested that some variations in metamorphic grade represent thrust-faulted blocks metamorphosed at different crustal levels. In addition, they noted a steep fabric associated with the Eocene Wilderness suite granites. Although fission-track and K-Ar dates for most rocks give Oligocene cooling ages, K-Ar muscovite and U-Th-Pb monazite ages are about 45 Ma (Keith and others, 1980), suggesting a period of cooling (and uplift?) following emplacement of the Wilderness suite. These relations further suggest that Late Cretaceous to early Tertiary metamorphism was more pronounced than the predominantly Neogene K-Ar ages indicate. Granite emplacement and metamorphism at about 50 Ma were at about 4 to 5 kbar (14 to 17 km depth) (Anderson, 1988).

SOUTH-CENTRAL ARIZONA

In the area from the Quitobaquito Hills and Puerto Blanco Mountains east to the Sierrita line (Fig. 7), a system of Late Cretaceous and early Tertiary ductile thrust faults places Proterozoic and Mesozoic plutonic rocks northeastward or southwestward several tens of kilometers over metamorphosed Mesozoic supracrustal rocks (Haxel and others, 1984). Thrust faults or ductile shear zones crop out in seven ranges, and concealed thrust faults are inferred adjacent to three other ranges. Exposed segments of these thrust faults are less than 5 km long, but similarities in geometry, structural style, and age relations indicate that these segments are parts of one or two regional thrust-fault systems. The thrust faults and broadly coeval regional metamorphism and plutonism probably all are genetically related.

Structural style

There are three categories of thrust faults: (1) thrusts with a single, homogeneous lower plate, (2) thrusts underlain by imbricate tectonic slides (*usage of Hutton, 1979*), and (3) ductile shear zones. The thrust faults with single lower plates are exemplified by the Window Mountain Well thrust in the Sierra Blanca (Fig. 7). The upper plate is composed of Proterozoic(?) orthogneiss and Jurassic granite, which are converted to mylonitic gneiss and blastomylonite within a zone 5 to 10 m thick just above the thrust. The lower plate is composed of schist and metaconglomerate derived from Jurassic sedimentary rocks. The upper and lower plates are intruded by Tertiary garnet-muscovite granite. Several other ranges or parts of ranges in south-central Arizona similarly are erosional and/or tectonic windows through thrust faults, with lower-plate rocks composing the core of the range and upper-plate rocks exposed on the margins. These windows vary in size from a few kilometers across to the Baboquavari Mountain chain, 70 km long by 10 km wide (Fig. 7).

Structural style in several ranges is more complex; in each of these areas, the major thrust fault is underlain by a thick stack of imbricate structural sheets separated by mylonitic tectonic slides. These tectonic slides are ductile faults or shear zones defined by gradational intensification of the blastomylonitic to crystalloblastic regional metamorphic fabric that pervades the lower plate. The Quibobaquito thrust places Proterozoic(?) gneiss and Cretaceous granite over metamorphosed Jurassic volcanic, hypabyssal, and sedimentary rocks. This thrust juxtaposes dissimilar and unrelated rocks, whereas the underlying tectonic slides juxtapose rocks that are similar, were originally related, and are locally in nontectonic contact. Displacement on the Quitobaquito thrust is probably a few tens of kilometers; movement along the tectonic slides need be no more than several kilometers.

In most of these areas of exposed thrust faults, textural metamorphic grade in the lower-plate rocks increases structurally upward toward the overlying thrust fault. In some ranges, textural metamorphic gradients increase structurally upward, but the cause of this pronounced metamorphic upgrading is hidden by

alluvium. The regional metamorphic character of higher-grade rocks and direct comparison with the areas where thrust faults are exposed, however, strongly suggest that the metamorphic rocks are overlain by a concealed thrust fault.

The third type of fault, a ductile shear zone, is displayed in the Gunsight Hills (Fig. 7). A zone of blastomylonitic to crystalloblastic orthogneiss, ca. 4 km thick, composes part of an incompletely exposed ductile shear zone within a single rock type, probably formed where a thrust fault intersected a pluton. The shear zone in the Gunsight Hills is evidently postplutonic.

The orientation of fabric elements in Late Cretaceous and early Tertiary regional metamorphic rocks in south-central Arizona is consistent in widely separate locations. Foliation typically dips moderately westward, and lineation generally plunges gently or moderately to the southwest or, less commonly, north-northeast. The direction of thrust transport in south-central Arizona, by analogy with other orogenic belts, is inferred to be parallel to the northeastward regional orientation of elongation lineation. The sense or senses and magnitude of displacement are unknown, but lithologic contrasts and similarities in juxtaposed Jurassic and Cretaceous rocks suggest that movement was several tens of kilometers. The most likely possibility is that the upper plate of the Baboquivari thrust system moved northeastward from an original position somewhere in north-central Sonora.

Interrelations of thrust faulting, metamorphism, and plutonism

These three aspects of the Late Cretaceous and early Tertiary tectonic evolution of south-central Arizona are broadly coextensive and contemporaneous, and are inferred to be genetically related. The plutons related to thrust faulting and regional metamorphism are high-silica, peraluminous granites with much associated pegmatite. Field relations and geochronologic data suggest that the typical sequence of events was thrusting, metamorphism, and then plutonism, involving considerable temporal overlap. The evidence for these generalizations is illustrated by field relations in the Sierra Blanca, where the connections between thrust faulting, regional metamorphism, and granitic plutonism, considered two at a time, are particularly instructive. Foliation and lineation are parallel in mylonitic rocks of the Window Mountain Well thrust and in the crystalloblastic lower-plate schist, suggesting that the thrust is synmetamorphic. In most outcrops, garnet-muscovite granite and related pegmatite sharply crosscut metamorphic foliation, but some dikes possess weak foliation and lineation parallel to those in the enclosing schist; the granite is thus late metamorphic to post-metamorphic. The granite is grossly post-thrusting, because it intrudes both plates of the thrust and the thrust zone. Very locally, however, pegmatitic sills are mylonitized within the thrust zone.

Further evidence for the syntectonic character of most of the peraluminous granites comes from internal fabric relations. In several ranges, an older intrusive phase bears a regional metamorphic fabric that is crosscut by a younger phase (*Wright and Haxel, 1982*). In addition, some peraluminous granite plutons are postmetamorphic where they intruded at slightly higher crustal levels than petrologically identical synmetamorphic granites.

Deformation and metamorphism in south-central Arizona are demonstrably younger than Late Cretaceous plutons dated at 85 to 69 Ma, and older than fissure veins intruded at about 53 Ma. K-Ar ages of 71 to 58 Ma probably approximate metamorphic ages (Haxel and others, 1984). The late- to post-metamorphic peraluminous granites, approximately 58 Ma, evidently represent the culmination of the orogenic episode.

Tectonic interpretation

The close spatial and temporal association of thrust faulting, regional metamorphism, and granitoid plutonism in south-central Arizona indicates that these three phenomena are related aspects of Late Cretaceous to early Tertiary orogeny. Haxel and others (1984) proposed that orogenesis was due to the conjunction of two independent causes: crustal compression and thickening, and anomalous conductive and magmatic heat flux into the crust from the underlying mantle. Emplacement of crystalline thrust sheets created a thermal blanket that, in the presence of high heat flux and a steep geothermal gradient, regionally metamorphosed the supracrustal rocks that were tectonically buried. Local partial melting in the middle crust produced silicic granitoid magmas (*Farmer and DePaolo, 1984*), which rose to intrude the regional metamorphic rocks in the upper crust. Although the thermal energy for this melting came largely from the mantle, anatexis was localized beneath areas of thermal blanketing by upper-crustal overthrusting. A flattening dip of subduction (Coney and Reynolds, 1977) may have created the conditions for both mantle heat flux and crustal overthrusting.

WESTERN ARIZONA AND SOUTHEASTERN CALIFORNIA

Late Cretaceous and early Tertiary thrust faulting, folding, and metamorphism in southeastern California and adjacent southwestern Arizona are divisible into two belts (Fig. 7). The southern belt (Maria-McCoy-Mule Mountains belt) is a west-northwest–trending zone of deformation, metamorphism, and synorogenic sedimentation that is strikingly oblique to Cretaceous magmatic and orogenic belts to the north. The northern belt (Old Woman belt) comprises ductile thrusts, nappes, and mylonitic stacks approximately colinear with the southern end of the Mesozoic Cordilleran fold and thrust belt. The relations between the belts are not well known, and the connection with the south-central Arizona zone of orogenesis is obscured by late Cenozoic cover.

Maria-McCoy-Mule Mountains belt

The Maria-McCoy-Mule Mountains belt strikes westerly from the Granite Wash and Harcuvar Mountains on the east to the Coxcomb Mountains on the west (Fig. 7). This belt of defor-

mation, sedimentation, and metamorphism contains three subparallel structural and geologic terranes from north to south: the Maria fold and thrust belt of Reynolds and others (1986), the "McCoy basin" of Harding and Coney (1985), and the Mule Mountains thrust system of Tosdal (1986). Although these terranes are described separately here, their evolutions are intertwined.

Maria fold and thrust belt. Proterozoic, Paleozoic, and Mesozoic rocks are metamorphosed and complexly deformed in the Maria fold and thrust belt (Fig. 7). South-facing folds and southward-directed thrust faults predominate (Reynolds and others, 1986; *Hamilton, 1982; Ellis, 1982; Emerson, 1982*). Superposed on these south-facing folds in the northern Dome Rock Mountains are ductile shear zones with northeast-directed sense of shear. Deformation was accompanied by regional metamorphism at conditions ranging from sub-greenschist to middle amphibolite facies at depths of less than 12 km, and also was accompanied locally by extensive fluid flow at high fluid/rock ratios (Reynolds and others, 1988; Hoisch and others, 1988). An inverted metamorphic gradient characterized by relatively high-T/low-P metamorphic mineral assemblages is present beneath some of the thrust faults in the Little Harquahala and Granite Wash Mountains (Reynolds and others, 1988). In other ranges, the highest grade metamorphic rocks are spatially related to Cretaceous(?) pegmatite swarms (e.g., Big Maria Mountains; Hoisch and others, 1988) or to early Tertiary(?) garnet–two mica granites (e.g., Harquahala Mountains; *Richard and others, 1987*). In areas far from intrusive rocks in the Big Maria Mountains, P-T data suggest elevated geothermal gradients between 35 and 50 °C/km during regional metamorphism (Hoisch and others, 1988).

Much of the deformation in this terrane is considered Cretaceous in age, although in places deformation may have been synchronous with emplacement of Middle to Late Jurassic plutons (*Hamilton, 1982; Reynolds and others,* 1986, *1987*). Folding and thrusting in much of the Maria fold and thrust belt is apparently older than granitoids and pegmatites for which K-Ar ages range from 60 to 80 Ma (*Hamilton, 1982; Martin and others, 1982;* Rehrig and Reynolds, 1980; *Reynolds, 1982; Reynolds and others, 1983,* 1986; *Keith and Wilt, 1985*); two $^{40}Ar/^{39}Ar$ plateau ages (75 Ma) from the Big Maria Mountains are interpreted to indicate that the peak metamorphic temperatures were attained in the Late Cretaceous (Hoisch and others, 1988). In the Harcuvar Mountains, the Late Cretaceous granitoids were locally deformed and metamorphosed during Late Cretaceous time (Rehrig and Reynolds, 1980). Regardless of the exact timing, deformation and metamorphism had ceased and the region had cooled between about 78 and about 60 to 55 Ma (*Shafiqullah and others, 1980; Martin and others, 1982;* Hoisch and others, 1988). In some ranges, postkinematic, latest Late Cretaceous granites intruded the regional metamorphic rocks (*Shafiqullah and others, 1980*).

"McCoy basin." Outcrops of thick sections of the Jurassic(?) and Cretaceous McCoy Mountains Formation of Harding and Coney (1985) define the limits of the "McCoy basin" of Harding and Coney (1985) (Reynolds and others, 1986; *Richard and others, 1987;* Fig. 7). Thinner sections of upper Mesozoic clastic rocks within the Maria fold and thrust belt are thought to be correlative with some part of the McCoy Mountains Formation within the "McCoy basin," although exact lithologic matches are not clear and are subject to several interpretations (Harding and Coney, 1985; Reynolds and others, 1986; *Richard and others, 1987*). The McCoy Mountains Formation is a southward-dipping homoclinal sequence of moderately metamorphosed synorogenic, continental clastic rocks, as much as 7 km thick, that records the unroofing of a magmatic arc to the north. Quartz-rich rocks at the base are succeeded upward by a section of predominantly volcanic detritus that is in turn overlain by a 3- to 5-km-thick section of coarse-grained arkosic rocks (*Pelka, 1973;* Harding and Coney, 1985). Paleocurrent data indicate source areas to the north. In the Dome Rock Mountains and vicinity, an angular unconformity between the upper arkosic part and the lower volcanic and quartzose parts of the formation implies intraformational tectonism and sudden rejuvenation of the source region to the north, but these relations do not persist in other exposures (Stone and others, 1987).

The arkosic part of the McCoy Mountains Formation contains fossil wood no older than middle Early Cretaceous in age (Stone and others, 1987) and is probably Late Cretaceous on the basis of a 78 ± 2 Ma U-Pb age of a tuff bed (Tosdal, 1986; *Tosdal and others, 1988*). The volcanic and quartzose parts, which are much more areally widespread, may be as old as Jurassic (*Haxel and others, 1985; Tosdal and others, 1988*). Although the formation has been correlated with the Upper Jurassic and Lower Cretaceous Bisbee Group of southeastern Arizona by *Dickinson and others (1986),* the age for the arkosic part makes this correlation unlikely.

The restriction of the upper thick arkosic sections of the McCoy Mountains Formation to the southern margin of the Maria fold and thrust belt suggests that the two may be related. The Maria belt and contiguous areas were being deformed, metamorphosed, and uplifted during Late Cretaceous time. These events may have preceded deposition of the arkosic part of the McCoy Mountains Formation in Arizona, but were synchronous with deposition in California (Hoisch and others, 1988). The arkosic rocks may be molasse-like depoists for early tectonic events in the Maria belt.

Deformation of the McCoy Mountains Formation and possibly correlative rocks was complex, and the tectonic environment is unclear (Harding and Coney, 1985; Reynolds and others, 1986; Tosdal, 1986). The thinner, possibly correlative, upper Mesozoic clastic rocks north of the "McCoy basin" in the Maria fold and thrust belt were deformed prior to emplacement of a Late Cretaceous pluton (*Reynolds and others, 1983,* 1986). The entire McCoy Mountains Formation within the "McCoy basin" was complexly deformed and metamorphosed during latest Cretaceous time. South-facing folds predominate, and in some areas were accompanied by northeast-directed shear of upright limbs

and formation of mesoscopic northeast-vergent folds, followed by southward-vergent shortening and tightening of preexisting folds (*Pelka, 1973;* Tosdal, 1986). The youngest structures of this deformation in the western parts of the "McCoy basin" are probably synchronous with some deformation and metamorphism in the Maria fold and thrust belt. Deformation and metamorphism in the "McCoy basin" ceased prior to intrusion of 70-Ma granitoids (*Armstrong and Suppe, 1973; Martin and others, 1982; Shafiqullah and others, 1980; Calzia and others, 1986*). At some point following metamorphism, a large tectonic block encompassing the entire McCoy Mountains Formation was tilted to the south (Stone and others, 1987).

Mule Mountains thrust system. The ductile Late Cretaceous Mule Mountains thrust system forms part of the southern boundary of the Maria-McCoy-Mule Mountains belt. It is probably the youngest structure in the belt; displacement occurred in late Late Cretaceous time, after the regional metamorphism and deformation of the underlying McCoy Mountains Formation and prior to the intrusion of 70-Ma granitic rocks and the time at which K-Ar minimum ages of ultramylonitic fault rocks were set (Tosdal, 1986; *Calzia and others, 1986*). A maximum of a few tens of kilometers of northward transport can be accommodated by the structures. The thrust system contains roof and floor thrusts separated by a zone of imbricate thrusts and folded rocks. Proposals that the roof thrust of this system represents a major terrane boundary (*Dillon, 1976; Vedder and others, 1983;* Harding and Coney, 1985) are not supported by the distribution of certain lithostratigraphic units across the fault, estimates of displacement along the fault, and the identical Pb isotopic signature of igneous rocks on either side of the thrust system (Tosdal, 1986; *Wooden and others, 1988*). The thrust system is between the recently metamorphosed and deformed "McCoy basin" terrane on the north, and a relatively cold crustal block on the south. This boundary may have been subparallel to old zones of crustal weakness inherited from the late Paleozoic to early Mesozoic truncation of the continental margin (Burchfiel and Davis, 1981; *Hamilton, 1982; Stewart and others, 1986*).

Old Woman belt

The frontal thrust belt extends discontinuously southwestward through the Great Basin region of Nevada and Utah to thrusts cutting moderately metamorphosed rocks in the New York Mountains. Southward, ductile nappes in the Old Woman Mountains (Fig. 7) may be part of the thrust belt, but the ductile styles and somewhat younger times of deformation distinguish them from the frontal belt (Howard and others, 1987a). In the Old Woman belt, the deformation is characterized by ductile, basement-involved structures that are closely related to granite emplacement and high-grade metamorphism. In the Old Woman Mountains area, metamorphosed Proterozoic and Paleozoic rocks are deformed beneath a giant crystalline nappe of Proterozoic rocks into a southeast-vergent(?), thrusted syncline. At about 78 to 75 Ma, Paleozoic rocks were metamorphosed at greenschist to upper amphibolite facies at depths of 9 to greater than 12 km (Howard and others, 1987a; Hoisch and others, 1988). The deformation and metamorphism were closely followed by intrusion of crustal-derived, Late Cretaceous metaluminous and peraluminous granitoids (Miller and others, 1982; Howard and others, 1987a; Hoisch and others, 1988), some of which were penetratively deformed during intrusion (*Howard and others, 1982b,* 1987a). In the nearby Iron Mountains, related granite also was deformed during intrusion, forming a subhorizontal zone of mylonite greater than 1 km thick (*Miller and Howard, 1985*) that appears to roof the ~72-Ma Cadiz Valley batholith and shows northeastward overshear. Conditions of mylonitization and pluton emplacement in the Iron Mountains were much shallower than for the Old Woman Mountains, at about 6 km (Anderson, 1988). Regardless of the depth of deformation, large parts of this terrane were subsequently uplifted rapidly, as evidenced by nearly concordant Late Cretaceous K-Ar and $^{40}Ar/^{39}Ar$ plateau ages on hornblende, muscovite, biotite, and orthoclase between about 78 and 66 Ma (*Miller and Howard, 1985;* Hoisch and others, 1988).

Late Cretaceous granitoids east of the Old Woman Mountains exhibit relations similar to those in the Old Woman Mountains, but ages of intrusion are generally less accurately determined. In the Chemehuevi and Dead Mountains area (Fig. 7), Late Cretaceous peraluminous to metaluminous granitoids show coeval mylonitic fabrics. In these ranges, however, subsequent Neogene core complex evolution created fabric overprints. In the Whipple Mountains, a nearly 4-km thickness of mylonitic gneiss is principally, if not exclusively, Tertiary in age (Davis and others, 1982). Earlier fabrics in 89 ± 3 Ma granitoids are not seen in 73 ± 3 Ma plutons, suggesting Late Cretaceous deformation (Anderson, 1988). Crystallization conditions for the 73-Ma plutons were about 7.5 kbar (Anderson, 1988). The ages for intrusion of peraluminous and metaluminous granite and granodiorite in the Chemehuevi, Sacramento, and Dead Mountains are not fully known, but are probably latest Cretaceous (Howard and others, 1987a; John and others, 1990). These plutons are late kinematic (Howard and others, 1987a; John and others, 1990) to synkinematic(?) (Anderson, 1988).

The Old Woman belt represents deformation and metamorphism, apparently widespread, that locally affected culminating plutons about 70 Ma. Large tracts of granite, such as the Cadiz Valley batholith, and highly metamorphosed wall rocks were deformed in several-kilometer-thick nappes and ductile shear zones. Many details of the origin and evolution of the Old Woman belt remain unknown.

OROCOPIA SCHIST AND CHOCOLATE MOUNTAINS THRUST

Distribution and structural setting

The Orocopia and Pelona Schists constitute a late Mesozoic oceanic supracrustal terrane (Fig. 8) that is structurally beneath continental and epicontinental terranes of inland southwestern

North America (Haxel and Dillon, 1978; Ehlig, 1981; Jacobson and others, 1988; *Haxel and Tosdal, 1986*). The name "Orocopia Schist" as used herein refers to both the Orocopia Schist, exposed in southeastern California and southwestern Arizona, and the Pelona Schist, offset along the Neogene San Andreas fault to the Transverse Ranges (Fig. 8). Because the Orocopia Schist is more nearly in place with reference to cratonic North America, its tectonic relations are more informative (*Dillon and others, 1987*). Three other late Mesozoic schist units in California—the Rand Schist, Catalina Schist, and schist of Sierra de Salinas—probably are related to the Orocopia Schist, but are not necessarily contiguous with it.

The structural thickness of the Orocopia Schist is 5 to 10 km, but its base is not exposed. The schist is overlain by Proterozoic to Late Cretaceous continental terranes along a regionally extensive thrust-fault system. These faults, which are individually called the Vincent, Orocopia, and Chocolate Mountains thrusts, here collectively are referred to by the last name. Latest Cretaceous and middle Tertiary low-angle normal faulting, middle to late Tertiary folding, and erosion have resulted in exposure of the Orocopia Schist in the cores of anticlinoria (*Crowell, 1962; Powell, 1981; Haxel and others, 1985*). The Orocopia Schist is believed to underlie much of the southeastern corner of California, southwesternmost Arizona, and the central and eastern Transverse Ranges, California.

Lithology and protolith

The Orocopia Schist consists predominantly of quartzofeldspathic schist derived from graywacke. Flysch-like relict bedding and gray to black porphyroblasts of graphitic albite are characteristic. Interlayered with this metagraywacke are subordinate to rare metabasalt, ferromanganiferous metachert, siliceous marble, and small scattered bodies of ultramafic rocks. The proportion of these subordinate rocks types decreases southeastward from the Transverse Ranges, where they constitute as much as 25 percent of the section, to southeasternmost California and southwestern Arizona, where they constitute less than 1 percent of the section.

Field relations suggest that the several supracrustal protoliths of the Orocopia Schist were deposited together, not juxtaposed tectonically. Major- and trace-element geochemistry (*Haxel and others, 1987;* Jacobson and others, 1988) indicates that the metagraywacke is derived from continental crust, including a continental magmatic-arc terrane. Most Orocopia metabasalts are geochemically similar to mid-ocean ridge basalt (MORB). The metacherts and marbles evidently represent marine biogenic deposits, with subordinate or minor detrital and hydrothermal components. The inferred depositional setting of the protolith of the Orocopia Schist is a continental-margin tectonic basin (*Haxel and others, 1987*).

Concordant U-Pb zircon ages for a metamorphosed dioritic dike that intruded the protolith of the Orocopia Schist indicate that deposition of at least part of the protolith was prior to 163 Ma. On the basis of this age and field relations, *Tosdal and others (1988)* and *Haxel and Tosdal (1986)* argued that the Orocopia protolith is Middle and/or Late Jurassic in age.

Deformation and metamorphism

Several lines of evidence indicate that deformation and metamorphism of the Orocopia Schist were closely related to the

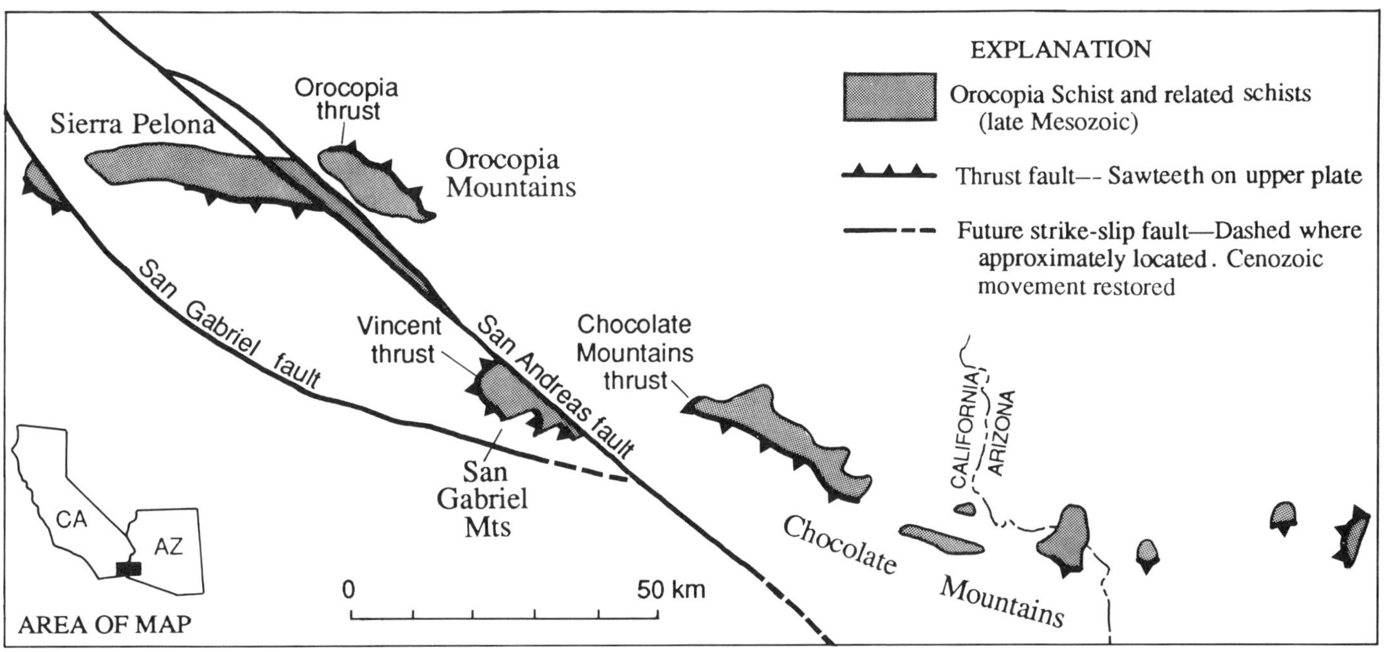

Figure 8. Distribution of the Orocopia Schist and related schists and latest Cretaceous thrust faults in southern California and adjacent Arizona. Neogene slip on the San Andreas fault system restored according to Ehlig (1981).

overlying Chocolate Mountains thrust (*Ehlig, 1958,* 1981). A pervasive foliation, synmetamorphic folds, and subordinate elongation lineations formed during a single protracted deformational and metamorphic episode (*Jacobson* (1983a, *1983b; 1984*). Foliation and lineation in the mylonitic rocks along the thrust are parallel to foliation and lineation, respectively, in the underlying schist. Metamorphic grade in the schist increases upward toward the thrust (*Graham and England, 1976; Jacobson,* 1983a, *b*), and mylonitic rocks above the thrust generally have been retrograded to the same metamorphic grade as the schist beneath the thrust. These relations have been interpreted as evidence for metamorphism of the schist during thrusting, metamorphism of the schist as a consequence of tectonic burial beneath the hanging wall, and incipient subduction of the schist.

Metamorphic mineral assemblages in the Orocopia Schist indicate upper greenschist to lower amphibolite facies, and amphiboles in metabasalts generally have compositions indicative of relatively high-pressure metamorphism (Jacobson and others, 1988). In the Sierra Pelona, *Graham and Powell (1984)* estimated metamorphic pressures of 8 to 10 kbar. In southeastern California, several mineralogic features of the schist qualitatively suggest high pressure.

Asymmetric microstructures and the collective asymmetry of thrust-zone folds indicate northeastward overthrusting along the Chocolate Mountains thrust (*Haxel and Tosdal, 1986; Dillon and others, 1987*). Paleomagnetic data in the region (*Costello, 1985*) indicate Neogene vertical-axis clockwise rotation related to the San Andreas fault system, suggesting that the original direction of overthrusting may have been northward.

The Chocolate Mountains thrust developed and the Orocopia Schist was metamorphosed after plutons with U-Pb ages of about 85 Ma were emplaced, and thrusting and metamorphism had ceased before K-Ar whole-rock and Rb-Sr mineral-isochron minimum ages of 52 to 59 Ma were set (Ehlig, 1981; *Walker and May, 1986*). Regional patterns of K-Ar and fission-track ages suggest that thrusting and metamorphism occurred during Late Cretaceous time, between ca. 85 and 74 Ma (*Mahaffie and Dokka, 1986; Haxel and others, 1985*).

Tectonic significance

Data summarized above indicate Middle(?) Jurassic deposition of the Orocopia Schist protolith in a continental-margin basin, followed by deformation and metamorphism during latest Cretaceous time. There are two broad categories of tectonic models to explain the deformation and metamorphism: subduction, and collision or accretion. The relatively high-pressure metamorphism suggests that the schist was metamorphosed as it was subducted northerly or northeasterly beneath southwestern North America (Yeats, 1968; *Crowell, 1975; Powell, 1981,* Jacobson, 1983a; Burchfiel and Davis, 1981; *Keith and Wilt, 1985; May, 1986*). Underthrusting (subduction) of the Orocopia Schist would explain its presence as far inland as southwestern Arizona, and is consistent with evidence for rapid, low-angle subduction during Late Cretaceous time (*Page and Engebretson, 1984*). However, the evidence for northeast-directed, continentward overshear on the Chocolate Mountains thrust allows scenarios in which the protolith of the schist was trapped and overridden in a convergence zone between the southwestern North American mainland and a smaller continental block now represented by the upper plate of the thrust (*Haxel and Tosdal, 1986*). If such a scenario is valid, regional correlations of basement rocks indicate that the outboard block was of North American origin (*Haxel and others, 1987*) and not a far-traveled exotic terrane as proposed by *Vedder and others (1983).* Although the predicted suture has not been identified, continentward movement along the Chocolate Mountains thrust may have represented postsubduction uplift rather than the original subduction episode (Jacobson and others, 1988). Some structures previously mapped as the Chocolate Mountains "thrust" may be faults accommodating uplift rather than tectonic burial faults. However, no direct evidence of southwestward overthrusting has yet been found, and low-angle normal faults that apparently accommodated postmetamorphic uplift of the schist are distinct from the Chocolate Mountains thrust (*Haxel and others, 1985*). The tectonic significance of the Orocopia Schist and overlying thrust fault system thus remains unclear.

MAGMATIC ARC

D. M. Miller, with contributions by G. B. Haxel

INTRODUCTION

During the interval 80 to 55 Ma, the magmatic arc that was established during Jurassic to Early Cretaceous time by the Andean-type subduction system along the entire Cordilleran Pacific margin (Fig. 1) was greatly modified. The central sector of the arc, between southern California and central Idaho, ceased activity. However, to the east, sparse and widely distributed plutons and volcanic edifices were emplaced in eastern Nevada and within the basement-uplift foreland province. The northern sector, in central Idaho and western Montana, continued magmatism until about 65 Ma in the form of eastward-migrating intrusive centers that produced small batholiths. Whether post–80 Ma magmatism continued at the locus of earlier magmatism in northeast Washington is unknown, but a magmatic-arc setting is indicated for the North Cascades and the contiguous Coast Plutonic Complex in Canada (*Monger and others, 1982*). In the southern sector—southeastern California and southern Arizona and New Mexico—magmatism continued until 60 Ma or even later as an eastward-migrating sweep of plutonism and volcanism within the continental craton (Coney and Reynolds, 1977).

The segmented magmatic arc of 80 to 50 ma was superseded by a new magmatic system in the Eocene. Beginning about 55 Ma in northern Washington, Idaho, Montana, and Wyoming, and progressing southwestward with time during the remainder

of the Eocene and Oligocene, hypabyssal intrusions and voluminous volcanic materials of such regionally extensive units as the Sanpoil, Challis, and Absaroka volcanics were associated with widespread crustal extension.

CENTRAL SECTOR

Magmatism in the Sierra Nevada batholith, and presumably its volcanic edifice, ended at about 80 Ma (*Chen and Moore, 1982*), following voluminous magmatic activity in the Early Cretaceous (Cowan and Bruhn, this volume). Similar cessation is inferred for the magmatic belt across northwest Nevada that in Early Cretaceous time connected the Sierra Nevada batholith with the Idaho batholith, as well as for the Klamath Mountains in northern California and the Blue Mountains in central Oregon. The thickened crust containing the magmatic arc probably was uplifted rapidly during the Late Cretaceous, presumably due to isostatic compensation as erosion removed the volcanic edifice. By latest Cretaceous and Paleocene time, clasts of granite from the Sierra Nevada batholith were being shed westward into the fore-arc continental borderland. In the southern Sierra Nevada, Late Cretaceous uplift must have been greater because the Paleocene(?) Witnet Formation strata are on deep-seated granulite facies granitoids (*Ross, 1983;* Cox, 1987).

Well east of the Sierra Nevada batholith and across most of Nevada, widely distributed Cretaceous plutons possibly also represent subduction-zone magmatism. Although most of the Cretaceous plutons are older than 80 Ma, as is the magmatic arc, a few plutons are as young as 75 Ma (Miller and others, 1988). These younger plutons consist of strongly peraluminous, muscovite-phenocrystic, leucocratic granite (Lee and others, 1981). They probably represent middle crustal melts, perhaps generated by Cretaceous crustal thickening.

Farther to the east, in the Rocky Mountains, intrusive-extrusive centers formed the Colorado mineral belt (Fig. 2). This magmatism was intermediate in composition: andesitic volcanic rocks and monzonite, granodiorite, and syenite hypabyssal intrusions are characteristic. The magmatism was chiefly 70 to 60 Ma (Lipman, 1980), and presumably was caused by flat subduction beneath the Rocky Mountain region. The hypabyssal complexes are widely mineralized.

NORTHERN SECTOR

The Idaho batholith and widespread plutons in northeast Washington and western Montana record Cretaceous magmatism approximately the same age as that in the Sierra Nevada batholith, but plutonism in the Idaho batholith and farther east continued after the Sierra Nevada batholith ceased activity. The plutons of the Idaho batholith and smaller batholiths to the east are host to epithermal, mesothermal, and disseminated deposits of base and precious minerals.

The earliest abundant plutons of the Idaho batholith and northeast Washington region are Jurassic in age (*McDowell and Kulp, 1969;* Miller and Engels, 1975); these are apparently restricted in the Idaho batholith to the western part and form a generally gneissic, more mafic zone that dips north under metasedimentary rocks (*Hietanen, 1984*). The generally less deformed center part of the batholith contains felsic plutons that are 62 to 100 m.y. old (*McDowell and Kulp, 1969;* Armstrong and others, 1977; Lund and Snee, 1988). Criss and Fleck (1987) summarized K-Ar data from the northern part of the batholith that indicate intrusive episodes at 82 to 75 Ma, 66 to 62 Ma, and 51 to 46 Ma. Many of the latest Cretaceous to Paleocene plutons are muscovite-biotite granite. Plutons in the western part of the batholith are associated with metamorphic assemblages, migmatites, and petrologic features that point to emplacement levels as deep as 25 km (*Hyndman and others, 1988;* Lund and Snee, 1988). To the east, metamorphic assemblages and textural features suggest progressively shallower emplacement (*Hyndman and others, 1988*), and the Pioneer and Boulder batholiths crystallized under their own volcanic cover (*Hamilton and Meyers, 1967*). However, part or all of this eastward-shallowing character may be caused by younger plutons, which are more common in the east, being emplaced at shallower levels (Lund and Snee, 1988). East of the Idaho batholith are the slightly younger Boulder and Pioneer batholiths. Plutons in the Pioneer batholith may be as young as 65 Ma, on the basis of K-Ar ages (*Marvin and others, 1983*). Plutons generally similar in age to those in the Idaho batholith (Jurassic and Cretaceous) have been identified in northeast Washington, northern Idaho, and western Montana (Miller and Engels, 1975; Armstrong and others, 1977, 1987; *Atwater and Rinehart, 1984*), but plutons younger than about 80 Ma are rare. *Asmerom and others (1988)* described muscovite-biotite granite in northeastern Washington that is about 75 Ma. In western Washington, the North Cascades contain calc-alkalic plutons mostly 90 Ma or older, but scattered plutons are as young as 75 to 65 Ma (*Tabor and others, 1987*).

The Late Cretaceous plutons in the Idaho batholith and east of the batholith range from hornblende-biotite granodiorite to biotite and biotite-muscovite granite (*Taubeneck, 1971; Marvin and others, 1983*). The muscovite-bearing granite bodies are generally the youngest and were emplaced at much shallower levels than deep-seated Early Cretaceous epidote-hornblende-biotite quartz diorite and tonalite along the western side of the batholith (Lund and Snee, 1988). Thus, not only is uplift of at least the western Idaho batholith between about 90 and 75 Ma indicated, but crustal sources for the plutons changed dramatically during that time. The Late Cretaceous granites of the northeastern part of the batholith have been characterized by Shuster and Bickford (1985) as calc-alkalic and moderately peraluminous. They argued that several features indicate a granulite facies, Proterozoic, lower-crustal source for the melts: (1) high Sr and Ba point to an aluminous, possibly pelitic source; (2) highly discordant zircons contain inherited cores giving Proterozoic ages; (3) high and heterogeneous initial Sr values indicate a crustal source; and (4) low Rb and U indicate a depleted, probably highly metamorphosed, source. In contrast, *Asmerom and others*

(1988) described muscovite-biotite granite in northeastern Washington that has enriched U and Rb as compared to granites from the Idaho batholith. They considered the granite to represent partial melting of a granodiorite source on the basis of trace-element modeling.

Many hypabyssal plutons and related volcanic sequences have been shown to significantly postdate batholith formation; these are Eocene in age and are related to the onset of widespread volcanism that heralded the new extensional tectonic regime of the middle and late Cenozoic. This northwest-trending Challis arc formed between 55 and 43 Ma across northern Washington, northeastern Oregon, and Idaho (Armstrong, 1978). Volcanic activity shifted southwestward, until about 40 Ma, to its eventual locus in the Cascade Range (*Wells and others, 1984*) and northern Nevada. The extensional tectonic manifestation included (1) deep-seated ductile shear zones such as those bordering metamorphic core complexes in northeastern Washington and along the east margin of the Bitterroot lobe of the Idaho batholith, and (2) shallower, fault-bounded grabens receiving volcanic and sedimentary rocks.

SOUTHERN SECTOR

Late Cretaceous and early Tertiary igneous rocks are widespread and locally abundant in the Sonoran-Mojave Desert region where they describe an eastward-younging pattern, but are subordinate in volume to the older rocks in the continental-margin batholiths farther west. In many places the plutons were closely associated with ductile deformation and regional metamorphism. In southern Arizona, two principal suites of granitoids—metaluminous and peraluminous—apparently are only indirectly related. By contrast, in southern California and westernmost Arizona, metaluminous and peraluminous granitoids are nearly coeval, locally forming zoned plutons or intrusive complexes.

Southern Arizona

Two contrasting granitoid suites of Late Cretaceous and early Tertiary age are known. The older and more voluminous suite consists of calc-alkalic granitoids, some hornblende bearing, that form a compositionally expanded (*Pitcher, 1979*) series from dioritic rocks through granodiorite to granite. The younger suite consists of compositionally restricted peraluminous granites that commonly contain garnet and/or primary muscovite. The primary difference in tectonic context between the two suites is crustal level: the peraluminous suite is closely associated in space and time with mid-crustal regional metamorphism, whereas the calc-alkalic suite is associated with extensive volcanic fields and hydrothermal alteration.

The calc-alkalic granitoid and volcanic rocks typically form compositionally heterogeneous or composite plutons exhibiting a variety of igneous textures, particularly porphyry textures. The associated hydrothermal disseminated copper deposits are consistently centered on porphyry intrusions (Titley, 1982) and range widely throughout southern Arizona and New Mexico. Hornblende-biotite granodiorite and monzogranite are most voluminous, and in most complexes are accompanied by subordinate early dioritic phases and late leucogranites, aplites, and minor pegmatite. In some places, late-stage leucogranite, aplite, and pegmatite contain primary muscovite or garnet, in addition to biotite. The plutons generally have Peacock indices of 57 to 60 percent SiO_2 and are metaluminous to weakly peraluminous. Geochemical differences between mineralized and unmineralized plutons are subtle (*Creasey, 1984; Farmer and DePaolo, 1984*). Widespread and thick volcanic sequences indicate caldera settings for many igneous centers (Lipman and Sawyer, 1985).

Published U-Pb zircon ages for the calcalkaline granitoids are between 74 and 64 Ma, but many K-Ar ages interpreted as crystallization ages range from 74 to 54 Ma. Some of the K-Ar ages are probably affected by loss of radiogenic argon, although several detailed K-Ar geochronologic studies (for example, Creasey, 1980) suggest that the calc-alkalic plutons may be as young as 60 Ma. Such studies also indicate that hydrothermal alteration and primary mineralization (Titley, 1982) occurred within a few million years of the intrusion and crystallization of the host pluton.

The peraluminous granites form a large composite pluton in the Santa Catalina and Rincon Mountains (Fig. 7) (Keith and others, 1980; *Banks, 1980*), and smaller plutons farther west (Haxel and others, 1984) and south (*Anderson and others, 1980*). The peraluminous rocks are largely monzogranite containing various combinations of biotite, muscovite, and garnet. Textural relations indicate that muscovite is primary (*Wright and Haxel, 1982*). Many of the peraluminous plutons consist of two phases: an older phase in which garnet is sparse, and a younger highly leucocratic and more strongly peraluminous phase, in which garnet is common. U-Pb zircon isotopic ages of 57 or 58 Ma have been determined for three peraluminous granites in south-central Arizona and northern Sonora. U-Pb, Rb-Sr, and K-Ar geochronologic data suggest that the peraluminous granites in the Santa Catalina Mountains are about 50 Ma (Keith and others, 1980).

Part or most of six ranges from the Granite Mountains west to the southwestern corner of Arizona (Fig. 7) are underlain by homogeneous leucocratic biotite and muscovite-biotite granite of probable Late Cretaceous age (*Shafiqullah and others, 1980*). This granite differs from the peraluminous granites farther east in that it apparently is not associated with regional metamorphism, and also in several petrologic details. If the exposed bodies of this distinctive granite connect beneath the intervening valleys, they represent one of the largest plutons or batholiths in the Sonoran-Mojave Desert region.

A few plutonic rock units of restricted size or distribution do not fit into the categories described above. Late Cretaceous dioritic rocks in the Tule Mountains and vicinity of southwestern Arizona (Fig. 7), and possibly similar rocks in adjacent northwestern Sonora (*Anderson and Roldan-Quintana, 1979*), may have closer affinities to the eastern Peninsular Ranges batholith than to plutons elsewhere in Arizona.

The calc-alkalic and peraluminous granitoid suites generally form separate plutons, and thus are not simply comagmatic, although the tectonic setting suggests that they are indirectly related. Differences in age, composition, and isotopic and trace-element systematics (Haxel and others, 1984; *Farmer and DePaolo, 1984; Wooden and others, 1988*) further indicate the genetic distinctness of the two granitoid suites. The Nd and Sr isotopic characteristics of the calc-alkalic granitoids are interpreted as the result of melting and assimilation of felsic continental lower crust by mantle-derived magma (*Farmer and DePaolo, 1984*). In contrast, anatexis of middle crust is indicated for the peraluminous granites (*Wright and Haxel, 1982*). Nd and Sr initial isotopic compositions for peraluminous granites coincide with those for felsic Proterozoic continental crust in the southwestern United States, suggesting that the granites contain no significant mantle component (*Farmer and DePaolo, 1984*).

The two granitoid suites are explained by the tectonic model for Late Cretaceous and early Tertiary orogenesis in southern Arizona. The calc-alkalic granitoids are a manifestation of anomalous heat flux from the mantle, and the peraluminous granites were generated by middle(?) crustal melting consequent to tectonic thickening. The time delay between upper-crustal emplacement of the calc-alkalic granitoids and the peraluminous granites represents the upper-crustal thermal relaxation time.

Southeastern California

In the eastern Mojave Desert area, compositionally heterogeneous plutons and batholiths, approximately 76 to 70 Ma (*Calzia and others, 1986;* Howard and others, 1987a), make up about 50 percent of exposed Mesozoic granitoids. These plutons and plutonic suites, such as the Cadiz Valley batholith (*John, 1981; Howard and John, 1983*) and several plutons in the Granite Mountains (*Howard and others, 1987b*), Old Woman Mountains (Miller and others, 1982; Howard and others, 1987a), and Chemehuevi Mountains (*John, 1982;* John and others, 1990), range modally from granodiorite to monzogranite and only rarely contain mafic enclaves or dikes. The plutons are calc-alkalic and compositionally expanded, including metaluminous hornblende-biotite granodiorite and peraluminous muscovite-biotite monzogranite that in many cases contains garnet. Although a general trend toward more leucocratic and peraluminous younger phases has been described (Miller and others, 1982; John and others, 1990), reversals are noted (*Miller and Howard, 1985*). A 73-Ma hornblende-biotite quartz diorite in the Whipple Mountains (*Anderson and Rowley, 1981;* Davis and others, 1982) is the most mafic pluton of this age group.

Chemical and isotopic data suggest that the metaluminous and peraluminous magmas were derived from crustal sources as separate but cogenetic intrusions that underwent some fractional crystallization. Abundance of major elements is proportional to that of SiO_2, suggesting fractional crystallization (John and others, 1989). Pb isotopic variations are independent of whole-rock composition, and indicate inheritance from Early Proterozoic sources (*Wooden and others, 1988;* John and others, 1990), in agreement with Nd data for the region (*Farmer and DePaolo, 1984; Bennett and DePaolo, 1987*). Initial Sr data also indicate an isotopically heterogeneous source, or crustal contamination (Davis and others, 1982; *Howard and others, 1982a;* Miller and others, 1982; John and others, 1990).

The Mojave Desert plutons intruded regionally metamorphosed and deformed rocks, and many early plutons were deformed (*Anderson and Rowley, 1981; Howard and others, 1982b,* 1987a; Miller and others, 1982; *Howard and John, 1983; Miller and Howard, 1985;* John and others, 1990). Relations are best established in the Old Woman Mountains, where rocks metamorphosed and deformed about 78 to 75 Ma (Ar studies of metamorphic hornblende) were intruded by ca. 72-Ma granodiorite and granite. The plutons largely postdate deformation, but local early phases were deformed (Miller and others, 1982; Howard and others, 1987a; Hoisch and others, 1988). Similar relations may hold in the Chemehuevi Mountains (*Howard and others, 1982b;* John and others, 1990), but require more accurate dating.

A Late Cretaceous thermal event causing anatexis at several levels of the crust was inferred by Miller and others (1982). Metaluminous magmas generated in the lower crust were emplaced at upper middle crustal depths (John and others, 1990; Hoisch and others, 1988) synchronous with the last deformation. As heat was transferred to shallower crustal levels, more silicic crust was melted to generate slightly younger peraluminous magmas that were emplaced with the broadly cogenetic metaluminous plutons.

Regional relations

In both Arizona and California, calc-alkalic metaluminous magmas were followed by peraluminous magmas. Models for both areas call for an upper mantle heat flux that melted lower crust to produce the calc-alkalic magmas. These magmas, in combination with crustal thickening, apparently provided a heat source that produced more silicic peraluminous melts higher in the crust.

Plutons in southern Arizona and southeastern California in Late Cretaceous time intruded crust that had been repeatedly intruded from Triassic time onward. Although the density of Mesozoic plutons in this region does not rival that of the Sierra Nevada or Peninsular Ranges batholiths, the general overlap in time and the position adjacent to these batholiths suggests a magmatic arc–related origin for eastern California and Arizona plutons. Much of this belt was included by Miller and Bradfish (1980) within their inner Cordilleran belt of peraluminous magmatism, lying east of the batholiths; they noted that the belt coincided with the margin of Proterozoic crust and broadly corresponded with areas of regional metamorphism. Coney and Reynolds (1977) suggested that the latest Cretaceous plutons of northern Sonora, southern Arizona, and southern New Mexico record an eastward sweep of magmatism resulting from flattening the subduction angle, but the progression is not clear everywhere.

The calc-alkalic granitoids may be satellites to the batholiths, but the wide and continuous range of Pb isotopes is not typical of magmas generated by plate-margin subduction (*Wooden and others, 1988*). Coastal batholiths ceased activity by about 80 Ma, as the plutonism to the east, described in this chapter, began over a broad east-southeast–trending belt. It appears that a magmatic event triggered by thermal flux from the mantle synchronously affected this region of California and Arizona at about 75 to 70 Ma. Whereas the magmatic event was quickly terminated in California, it continued episodically until about 55 Ma in Arizona.

PACIFIC MARGIN

T. H. Nilsen

INTRODUCTION

The geologic evolution, paleogeography, and tectonic history of California and the Pacific Northwest during the Late Cretaceous and early Tertiary are complex. The geologic framework of the central and southern parts of the Pacific margin during the Cretaceous was that of a convergent Andean-type margin, including from west to east: (1) a subduction complex represented by the Franciscan assemblage of California and related rocks in Oregon; (2) a system of fore-arc basins represented by the Great Valley, Hornbrook, Ochoco, and related basins; and (3) a magmatic arc represented by the batholiths of the Sierra Nevada, Mojave Desert, and Idaho. In the northern Pacific Northwest and extending northward, strike-slip faulting and pull-apart basins formed in regions within and west of the magmatic arc of northeastern Washington and possibly the North Cascades. Major disruption and reorganization of the Andean-type margin took place during latest Cretaceous and early Tertiary time, partly as a result of (1) periods of strike-slip faulting caused by oblique convergence, (2) uplift and unroofing of the accretionary wedge by detachment faults, and (3) accretion of tectonostratigraphic terranes. Our understanding of this reorganization is still preliminary.

Many concepts for the plate-tectonic evolution of continental margins have been inspired in part by studies of California geology, including the concepts of (1) paired fore-arc basins and accretionary prisms, (2) melanges as a component of accretionary prisms, and (3) continental-margin transform faulting. These concepts are still controversial, making interpretation of the Late Cretaceous and early Tertiary history of California and the Pacific Northwest difficult. Reconstructions presented herein generally are cautious ones that ignore more speculative scenarios. Data for sedimentary basins that are somewhat younger than those of other parts of the Cordillera are included in this section because major changes in tectonic framework and sedimentation patterns in many parts of the Pacific margin took place in the middle Eocene, as well as in the late Paleocene and early Eocene.

General paleogeographic framework

The Mesozoic magmatic arc represented by the western Idaho batholith, Sierra Nevada batholith, Mojave Desert area, and adjacent regions remained as a mountainous region from the latest Cretaceous into the earliest Tertiary. The mountainous region separated large fluvial deposystems, widely distributed lakes, and scattered mountain ranges on the east from complex coastal deposystems, continental borderlands, and rapidly shifting source areas and basins on the west (Fig. 1). This western region, the focus of this section, contained less extensive and possibly topographically lower uplands, smaller fluvial systems, and restricted subaerial volcanic centers, some of which were adjacent to the continental margin. The coastal parts of this western area also were characterized by locally extensive swamplands, narrow shelves, and marine slopes that shifted position through time and were cut by submarine canyons. Deep-sea trenches, fore-arc basins, and continental borderlands similar to the modern southern California Continental Borderland were also present, as well as a great variety of turbidite depositional systems. The mountains underlain by the magmatic arcs were the sources for most of the sediment deposited in these Late Cretaceous and early Tertiary basins.

Previous regional studies

The geologic history of the Pacific Northwest has been reconstructed by many workers, including Dott (1966), Dickinson (1976), Hamilton (1978), *Davis and others (1978),* Ewing (1980), *Wells and others (1984),* Heller and others (1987), *Umhoefer (1987),* and *Oldow and others (1989).* The major feature of the period from 80 to 55 Ma is the focusing of tectonism, magmatism, and sedimentation in specific areas, as compared to more widespread activity in earlier Cretaceous time; some metamorphism and magmatism continued to the end of the Cretaceous in part of the North Cascades, and rapid sedimentation continued in several small basins.

The history of basin evolution in California was first systematically described by *Reed (1933)* and *Reed and Hollister (1936).* Subsequently, the San Andreas fault was recognized by *Hill and Dibblee (1953)* as having undergone at least several hundred kilometers of Cenozoic right slip, reorganizing older principal tectonic elements. The three main rock units of (primarily) Mesozoic age in California, the Franciscan assemblage, the Great Valley sequence, and the granitic and related intrusive rocks of the Sierra Nevada, were shown by Bailey and others (1964) to be broadly coeval. The Great Valley sequence and Franciscan assemblage were interpreted as remnants of a Mesozoic fore-arc basin and subduction complex, respectively. These rocks were thought by many workers to represent an Andean-type margin, which was characterized from west to east by (1) eastward-subducting oceanic crust, (2) a submarine trench at the continental margin, (3) an uplifted and off-scraped subduction complex,

the Franciscan assemblage, (4) a fore-arc basin in which the Great Valley sequence accumulated, and (5) the magmatic arc (Hamilton and Myers, 1966; *Hsu, 1968;* Bailey and Blake, 1969; Hamilton, 1969; *Page, 1970;* Ernst, 1970; *Dickinson, 1971*). The subsequent breakup of these subduction-related belts in California during latest Cretaceous and early Tertiary time was summarized by Nilsen and Clarke (1975), Clarke and others (1975), *Nilsen* (1977, *1978, 1986,* 1987), and Nilsen and McKee (1979), who presented paleogeographic maps of the state and adjacent regions that depicted the development of numerous small early Tertiary basins.

TRANSLATED AND ROTATED TERRANES

Paleotectonic and paleogeographic reconstructions for the Pacific margin require restoration of (1) Late Cretaceous and Tertiary translation along strike-slip faults; (2) Tertiary rotation of some terranes; (3) post–middle Eocene extension, particularly in the Basin and Range province; and (4) large-scale motion of terranes that are exotic to the Pacific margin and that accreted during or after the Late Cretaceous and early Tertiary (*Umhoefer, 1987; Stone and McWilliams, 1988; Oldow and others, 1989*).

Pacific Northwest

Although major northwest-trending right-slip faults have been recognized in the Pacific Northwest, ages of movement are difficult to establish and there is little agreement on the amounts of offset. The Straight Creek fault of northern Washington (Pl. 6), the most prominent, is thought to have had 80 to 190 km of right slip, probably 110 km during the Late Cretaceous and early Tertiary (Kleinspehn, 1985). Other faults include the Ross Lake fault, with 150 km of Late Cretaceous or Tertiary right slip (Kleinspehn, 1985), and the Pasayten fault. The Entiat fault and an unnamed fault proposed by Johnson (1984a) that displaced southern Vancouver Island from northern Washington are middle Eocene or younger. The total displacement during the Late Cretaceous and Eocene between the craton and the Coast Mountains (excluding the fault proposed by Johnson) is about 450 km (Ewing, 1980). Many of these faults appear to have moved during the Late Cretaceous and early Tertiary and to have truncated the continental margin and formed strike-slip–related basins (*Johnson, 1984b; Trexler and Bourgeois, 1985*). The zone of Late Cretaceous and early Tertiary strike-slip faulting extended northward into coastal and inland Canada (*Monger and Irving, 1980; Long, 1981; Eisbacher, 1981; Yorath and Choe, 1981;* Cabrielse, 1985; Umhoefer, 1987; *Oldow and others, 1989*). In the south, the strike-slip faults are covered by Cenozoic volcanic rocks.

Paleomagnetic studies of the Pacific Northwest indicate that large clockwise tectonic rotations are typical of the convergent margin (*Beck, 1976, 1980*). In the Blue Mountains of east-central Oregon, clockwise rotation of Jurassic plutons of 60° ± 29° (*Wilson and Cox, 1980*) and of Eocene volcanic rocks of about 15° (*Gromme and others, 1986*) have been reported. In the Cascade Range, clockwise rotations were 14° ± 9° in northern California (Beck and others, 1986), 27° ± 20° in Oregon (*Magill and Cox, 1981*), and 33° ± 14° in southern Washington (*Bates and others, 1981*). Also reported are clockwise rotations of the Coast Range of Oregon of as much as 75° since the early Eocene, and of the Coast Range of Washington a lesser amount, an average of generally 20° to 52° within small blocks (*Simpson and Cox, 1977; Magill and others, 1981; Wells and Coe, 1985*). Clockwise rotations of 48° of Upper Cretaceous strata in the Methow basin and of 25° of Eocene volcanic rocks in the Okanogan highlands indicate that tectonic rotations extended into north-central and northeastern Washington (*Fox and Beck, 1985; Granirer and others, 1986*). In contrast, paleomagnetic studies of Early Cretaceous batholiths in the Klamath Mountains indicate clockwise rotations of about 25° (*Mankinen and others, 1988*) and Cretaceous and Tertiary strata on the flanks of the Klamath Mountains (*Mankinen and others, 1984*) show small amounts of clockwise rotation (12° to 14°). Western Oregon and Washington may have been rotated by dextral shear along the convergent margin driven by oblique subduction in combination with late Eocene and younger extension in the Basin and Range province (*Wells and Heller, 1988*).

Approximately east-west Cenozoic extension in the northern Basin and Range has translated the Klamath Mountains and Sierra Nevada region westward with reference to the craton. Additional extension has probably taken place under the volcanic plateaus of eastern Oregon and the Columbia Plateau of Washington, but the amounts cannot be determined. Early to middle Eocene extension in northern Washington and Idaho was associated with strike-slip fault zones (Ewing, 1980; Rehrig and others, 1987; *Orr and Cheney, 1987; Hansen and Goodge, 1988*).

Parts of the Pacific Northwest were translated northward during or after the Late Cretaceous to early Tertiary. Exotic blocks that made up the Olympic Mountains, Coast Range, and areas west of the Klamath Mountains were accreted to the Pacific Northwest margin during the Late Cretaceous to Neogene. The core of the Olympic Mountains has been interpreted by *Tabor and Cady (1978)* to be an accretionary wedge of Tertiary marine turbidite sandstone and siltstone, with lenses of basalt and gabbro. The wedge accreted to the northern end of the Oregon and Washington Coast Range, which represents an oceanic seamount chain that previously had accreted in the early Eocene. The seamount chain consists chiefly of tholeiitic and alkalic pillow basalt that may have formed either adjacent to a hot spot along a spreading ridge (*Duncan, 1982*) or as a result of accretion of thick oceanic crust and seamounts generated by reorganization of the Farallon-Kula spreading ridge (*Wells and others, 1984*). Alternatively, *Wells and others (1984)* suggested that the oceanic Coast Range basement may represent a collapsed marginal basin related to earliest Tertiary oblique rifting of the continental margin. The area of southwest Oregon west of the Klamath Mountains consists of several terranes similar to those in the Franciscan assemblage and Great Valley sequence that are thought to have

accreted during the Paleogene (Blake, 1984; *Blake and others, 1985b;* Bourgeois and Dott, 1985).

California

About 315 km of right slip took place along the San Andreas fault in Neogene time (*Clarke and Nilsen, 1973; Kanter, 1988; Graham and others, 1989*) (see Christiansen and Yeats, this volume). A cumulative total of perhaps 1,000 km of right slip took place along associated northwest-striking faults, including the Newport-Inglewood, Rinconada, San Gregorio–Hosgri, Hayward, Calaveras, San Gabriel, San Jacinto, and Elsinore faults (*Howell, 1975; Dibblee, 1976; Graham and Dickinson, 1978; Kelsey and Hagans, 1982; Fox and others, 1985*). Tens of kilometers of left slip accumulated along the east-striking Big Pine, Garlock, Santa Ynez, Malibu Coast, and White Wolf faults (*Smith, 1962; Dibblee, 1976, 1978; Carter, 1982*). The major strike-slip fault–related reconstruction required for California is the southeastward translation of areas west of the San Andreas fault.

Paleomagnetic studies and geologic reconstructions suggest that major parts of California rotated significantly in Neogene time (*Beck, 1986*). The entire Transverse Ranges province, including the Santa Barbara Channel and Channel Islands, may have rotated clockwise at least 75° (*Kamerling and Luyendyk, 1979; Luyendyk and Hornafius, 1987*). The central Sierra Nevada appears tohave rotated no more than 2° clockwise in the Tertiary (*Frie and others, 1984; Frei, 1986*), whereas the southernmost Sierra Nevada batholith rotated clockwise 45° ± 14°, probably between 80 and 20 Ma (*Kanter and McWilliams, 1982*). Baja California and the Peninsular Ranges of southwestern California appear to have rotated 32° clockwise with respect to North America since the Cretaceous (*Hagstrum and others, 1985*).

East-west extension in the Basin and Range, Modoc Plateau, and Salton Trough provinces after about 43 Ma has displaced the Sierra Nevada, Klamath Mountains, and other areas westward. Extension in the Basin and Range has been estimated to be 60–100 km (*Stewart, 1971*) or more (Hamilton and Myers, 1966; *Stewart, 1978; Zoback and others, 1981; Wernicke and others, 1982, 1988; Frei, 1986*).

Paleomagnetic, paleontologic, and stratigraphic data suggest that several large and many small exotic blocks or terranes accreted to California in Late Cretaceous to late Cenozoic time (*Coney and others, 1980; Jones and others, 1983; Vedder and others, 1983*). Many of these terranes include slivers of continental crust, pieces of oceanic crust, fragments of volcanic arcs, and other types of crustal fragments, and they appear to have migrated northward from their places of origin (*Debiche and others, 1987; Umhoefer, 1987; Stone and McWilliams, 1988*). Various fragments of oceanic crust and oceanic sedimentary rocks in the Franciscan assemblage of the central and northern California Coast Ranges probably were transported northward many thousands of kilometers (*Alvarez and others, 1980; Blake and others, 1985a; Sliter and others, 1986; Tarduno and others, 1986; Jayko and Blake, 1987*), some not accreting until post-Miocene time (*McLaughlin and others, 1982*). The Salinian block and the terranes west of it may have moved northward 2,500 km since the Cretaceous (Champion and others, 1984; *Kanter and Debiche, 1985*). The Peninsular Ranges and adjacent Baja California may have moved 11° northward (*Hagstrum and others, 1985*), apparently within the past 15 m.y., accreting to the southern California margin about 5 Ma (*Morris and others, 1986*).

Applications

Incomplete knowledge of the mechanics of terrane formation, movement, and accretion add much uncertainty to tectonic reconstructions for the Pacific margin (Engebretson and others, 1984, 1985; Debiche and others, 1987). The margin is complex as a result of rotation, translation, and extension of blocks, all of which must be restored. Terranes added to and moved along the margin in middle and late Cenozoic time must be removed or restored to their original locations, as must terranes presently located elsewhere but suspected to have originated in the Pacific Northwest and California. Delineating tectonostratigraphic terranes along the Pacific margin (*Coney and others, 1980; Blake and others, 1982; Silberling and others, 1987; Wheeler and McFeely, 1987; Wheeler and others, 1988*) provides a framework for more critical examination of the relations between them and their likely paleogeographic and paleotectonic reconstructions. The relations between Cenozoic tectonic rotations and the processes of accretion, extension, and dextral shear have been examined for the Pacific Northwest by *Wells and Heller (1988)*. They concluded that geologically constrained paleomagnetic data eliminate the process of microplate accretion as a mechanism of rotation, and suggested that dextral shear and Basin and Range extension are subequal contributors to the rotation. However, they noted that considerable freedom still exists in paleotectonic reconstructions of the continental margin because the paleomagnetic data alone do not uniquely define the relations between mobile blocks.

UPPER CRETACEOUS AND LOWER TERTIARY STRATA

The complexly distributed Upper Cretaceous and lower Tertiary strata of the Pacific margin lie within 17 areas characterized by different geologic histories. These areas will be described in order from north to south, focusing on depositional systems and tectonic events that in many cases were related to development of sedimentary basins.

Puget Sound area

Upper Cretaceous strata of the Nanaimo Group crop out east of Vancouver Island and in the Straits of Georgia (Fig. 9) (Muller and Jeletzky, 1970; *Johnson, 1978;* Pacht, 1984). The

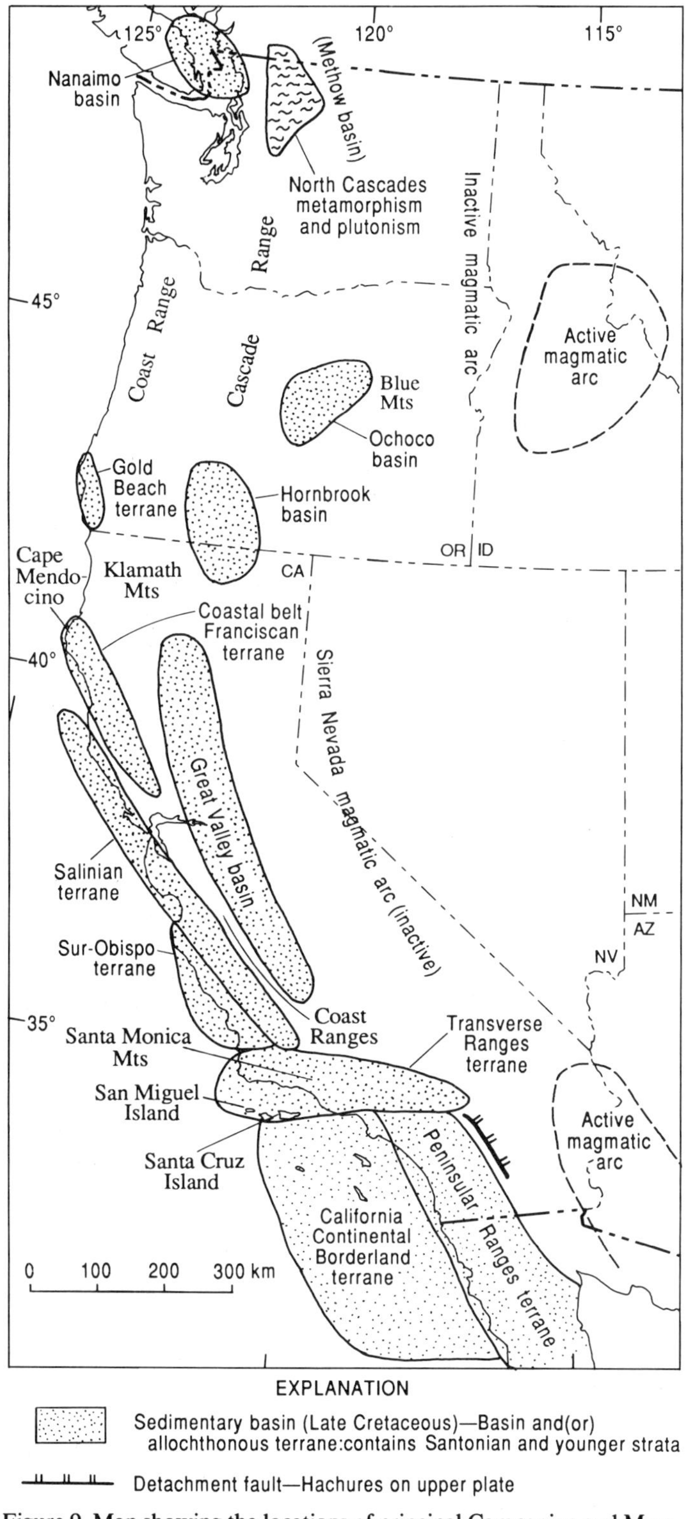

Figure 9. Map showing the locations of principal Campanian and Maastrichtian basins and allochthonous terranes of the Pacific margin.

Nanaimo Group is Turonian to Maastrichtian, consists of fluvial, shallow-marine, and deep-marine facies, and was deposited in a northwest-trending basin surrounded on all sides by source areas consisting of previously accreted terranes. The basin contained local horsts and grabens; these features and the overall setting of the basin suggest that it initially formed as a fore-arc basin under the influence of right slip (*Johnson, 1984a*). Northward-directed thrusting along the southern margin of the Nanaimo basin between 105 and 75 Ma caused the southern part of the basin to subside in response to thrust loading (*Johnson and others, 1986; Brandon and others, 1988*).

No Paleocene strata are present in the Puget Sound area. In the middle Eocene, fluvio-deltaic detritus of the lower Puget Group were transported westward and deposited into a basin in the southern part of the area (Buckovic, 1979); the basin apparently was influenced by strike-slip faulting (*Johnson, 1984a*) (see Christiansen and Yeats, this volume).

Olympic Mountains

The Neogene accretionary-wedge rocks of the core of the Olympic Mountains were added to Washington after the Coast Range was accreted in the middle Eocene (*Tabor and Cady, 1978; Heller and Ryberg, 1983*). As a result, the core rocks were not part of the Late Cretaceous and early Tertiary paleogeographic framework of the Pacific Northwest, and will not be considered further.

North Cascade Range

The North Cascades form the southern end of the Coast Plutonic complex of British Columbia, an orogenic welt consisting of allochthonous terranes that were amalgamated, accreted, tectonically thickened, and intruded by granitoid plutons during the Cretaceous and early Tertiary (*Monger and others, 1982*). Although great tectonic burial, indicated by metamorphic and igneous mineral assemblages in the complex, suggests a collisional origin, voluminous plutons suggest that a magmatic arc formed above the nearby subduction zone; numerous strike-slip faults also suggest a transpressional setting. In the North Cascades, most thrusting, metamorphism, and deformation had ended by the late Late Cretaceous, as indicated by ca. 90-Ma synkinematic plutons with K-Ar cooling ages of 90 to 80 Ma (*Tabor and others, 1987;* Cowan and Bruhn, this volume). However, K-Ar discordance increases to the northeast in rocks east of the Straight Creek fault (Fig. 9), as does the depth of pluton emplacement. In the northeastern parts of the North Cascades (Chelan Mountains terrane and Skagit Gneiss), U-Pb studies indicate metamorphism as young as 60 Ma and plutons as young as 75 to 65 Ma (*Tabor and others, 1987,* 1989). These latest Cretaceous crystalline rocks were uplifted about 25 km and cooled by middle Eocene time (*Tabor and others, 1987*), during an enigmatic Eocene event accompanied by large-scale strike-slip faulting and graben development.

Two deep, fault-controlled basins are present in the North Cascade Range (Fig. 10). Along the western margin, more than 6,000 m of the nonmarine lower to middle Eocene Chuckanut Formation of *Johnson (1984b)* were deposited by southwest-flowing streams in a small but tectonically complex basin considered by *Johnson (1984b)* to be of strike-slip origin. To the southeast, 5,000 m of nonmarine lower Eocene strata of the Swauk Formation were deposited in a similar basin that may be offset from the Chuckanut along the Straight Creek fault (*Tabor and others, 1984*). Middle and upper Eocene deposits are widespread in the North Cascade region (Christiansen and Yeats, this volume).

In the north-central North Cascade Range, the Methow basin contains Jurassic and Cretaceous strata, the youngest of which are probably Cenomanian or Turonian. This basin probably formed in response to right-lateral strike-slip faulting in a transpressive setting (*Trexler and Bourgeois, 1985*). The youngest unit in the basin is a folded 90-Ma volcanic sequence. A 700-m-thick alluvial-fan sequence unconformably resting on the Cretaceous strata has been considered Paleocene or Eocene, and probably is middle Eocene (Fig. 10).

Blue Mountains and Ochoco basin

Albian to Turonian nonmarine, shallow-marine, and deep-marine strata as thick as 1,750 m crop out within and adjacent to the Blue Mountains of central Oregon (Kleinhans and others, 1984). In the Mitchell area of the Blue Mountains, these strata were deposited chiefly as submarine-fan deposits transported to the southwest from a magmatic-arc source area to the east. In the adjacent subsurface Ochoco basin (Fig. 9), the Cretaceous strata are as thick as 4,300 m and as young as Campanian (*Thompson and others, 1984*). These strata appear to define a mid- to Late Cretaceous fore-arc basin in central Oregon that developed over an early Mesozoic fore-arc basin.

Cretaceous strata in central Oregon are overlain by Eocene to early Oligocene volcanic-arc rocks of the Clarno Formation, which is chiefly 47 to 35 Ma, although locally, ages as old as 54 Ma have been reported (*Wells and others, 1984*). North of the Blue Mountains (Fig. 10), unnamed Paleocene nonmarine strata as thick as 170 m rest unconformably on Paleozoic and Mesozoic crystalline basement rocks (*Gordon, 1985*). Fossil plants from these strata indicate a cooler climate than for the Eocene.

Cascade Range and Willamette Valley areas

No Upper Cretaceous or Paleocene strata are exposed in these regions. Cascade Range volcanism began about 42 Ma (*Lux, 1982; Verplanck and Duncan, 1987*) and the Willamette Valley west of the Cascade Range is underlain by strata of Neogene age.

Coast Range and offshore Washington and Oregon

The Coast Range is underlain by mostly submarine tholeiitic basalts that are 62 to 49 Ma in age (*Duncan, 1982*). The basalts,

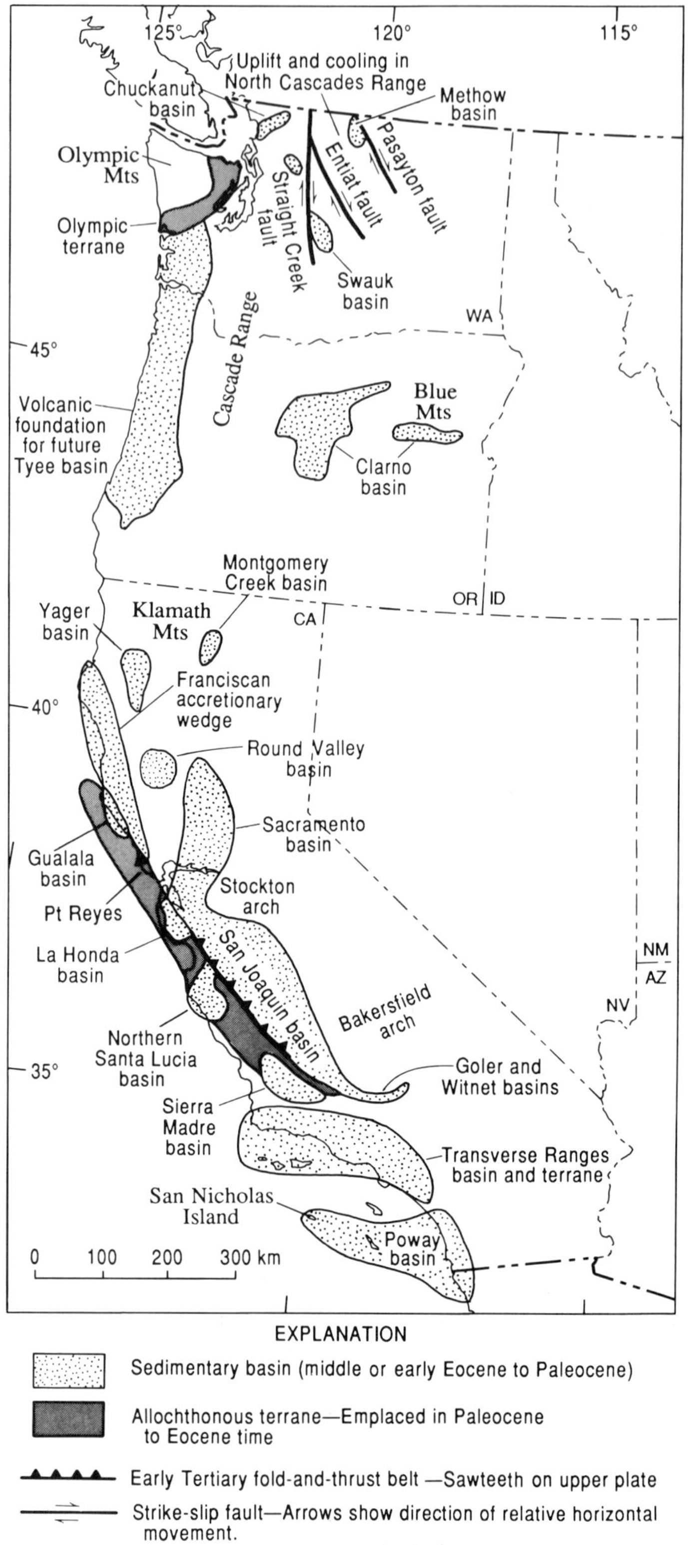

Figure 10. Map showing the locations of principal Paleocene and early Eocene basins and allochthonous terranes of the Pacific margin.

which include units in the eastern Olympic Mountains, are youngest in the central part of the range near the Columbia River and progressively older to the north and south. At the northern and southern ends of the Coast Range, continent-derived conglomerate interbedded with the volcanic rocks suggests eruption close to a continental margin. The basalts have been interpreted by most workers to be seamounts extruded on oceanic crust that were tectonically accreted to the continental margin of the Pacific Northwest in the Eocene (*Simpson and Cox, 1977; Duncan, 1982; Wells and others, 1984*). The tectonic boundary between the Coast Range and Klamath Mountains was overlapped by lower middle Eocene turbidites of the Tyee Formation, indicating that the Coast Range had accreted by about 50 Ma (*Magill and others, 1981; Heller and Ryberg, 1983;* Heller and others, 1987).

Western flank of the Klamath Mountains

Scattered outcrops of Albian(?) to Maastrichtian(?) strata of the Gold Beach terrane in coastal southwestern Oregon (Blake, 1984; Bourgeois and Dott, 1985) are separated from the Klamath Mountains by strike-slip faults (Fig. 9). Conglomerate clasts, paleocurrents, and paleomagnetic data suggest that the Klamath Mountains could not be the source terrane of these strata. These strata more closely resemble rocks of the Great Valley sequence or Cretaceous strata in the Salinian block of central California. The Upper Cretaceous strata are as thick as 1,000 m, consist of shallow-marine and deep-marine deposits, and rest unconformably on a Late Jurassic oceanic volcanic-arc complex. Paleomagnetic data suggest about 1,200 km of northward movement and probable arrival of the Gold Beach terrane in southwest Oregan in the early Tertiary or later (Blake, 1984; *Blake and others, 1985b*). Stratigraphic relations suggest deposition in a pull-apart basin associated with strike-slip faulting (Bourgeois and Dott, 1985). No Paleocene or Eocene strata are present in the Gold Beach terrane.

Eastern and southern flanks of Klamath Mountains

Cretaceous strata of the Hornbrook Formation rest unconformably on Paleozoic, Triassic, and Jurassic igneous and metamorphic rocks along the eastern flank of the Klamath Mountains (Figs. 9, 10) (Nilsen, 1984b). The Hornbrook Formation is about 1,200 m thick, ranges in age from Albian to Maastrichtian, and grades upward from fluvial strata in its lower part to deep-marine strata in its upper part. Conglomerate clasts, northeast-directed paleocurrents, and stratigraphic relations suggest that the Klamath Mountains were the source area. Paleomagnetic studies of the Hornbrook Formation indicate that it has not moved latitudinally and that it has not undergone any significant rotation (*Mankinen and others, 1984*). The Hornbrook Formation is unconformably overlain by coarse-grained fluvial deposits of the upper Eocene Payne Cliffs Formation of *McKnight (1984),* which were derived from the Klamath Mountains.

Lower and Upper Cretaceous sedimentary rocks rest unconformably on the southern flank of the Klamath Mountains and are locally preserved in downfaulted blocks within the mountains. These strata are as young as Coniacian (*Murphy and others, 1969*); younger strata that may have once been deposited in the area have been removed by erosion.

Franciscan assemblage and associated strata of the northern Coast Ranges

The Coastal belt, the northwestern part of the Franciscan assemblage (Fig. 9), is divided into the Coastal, Yager, and King Range terranes in northern California. The Coastal terrane consists of disrupted deep-marine turbidites, oceanic pillow basalt, and Paleocene to late Eocene mudstone; highly disrupted parts contain scattered blocks of chert, pelagic foraminiferal limestone, medium-grade blueschist, and volcanic rocks (Bailey and others, 1964; *Bachman, 1982; McLaughlin and others, 1982; Blake and others, 1989*). In contrast with other parts of the Franciscan assemblage, the Coastal terrane contains more potassium feldspar in sandstone beds, fewer melange units, and lesser amounts of chert, greenstone, serpentinite, and high-pressure metamorphic rocks. Limestone blocks are Campanian to Maastrichtian, and interbedded turbidites and mudstones are Paleocene to late Eocene (*McLaughlin and others, 1982; Sliter and others, 1986*). Abundant syndepositional deformation structures and pebbly mudstones suggest deposition of clastic units in a trench and on trench slopes in and adjacent to an active subduction zone (*Bachman, 1982; Underwood and Bachman, 1986*). Limestone generally overlies or is intercalated with oceanic basalt; these relations, along with paleontologic data, suggest that they were deposited on mid-ocean seamounts or plateaus (*Harbert and others, 1984; Sliter and others, 1986; Blake and others, 1989*).

Along the eastern margin of the Coastal belt, east of Cape Mendocino, are less deformed lower Tertiary turbidites of the Yager terrane (Fig. 10). The turbidites consist of mudstone, arkosic sandstone, and conglomerate; they contain Paleocene to late Eocene microfossils (*Blake and others,* 1988, *1989*). Sandstone compositions indicate a magmatic-arc source, probably a mixture of Sierran, Klamath, and Idaho provenances (*Underwood and Bachman, 1986*). The terrane appears to represent trench-slope-basin and trench-slope deposition on previously deformed Coastal belt rocks along the early Tertiary continental margin (*Underwood,* 1983, *1985*).

The King Range terrane, the youngest and westernmost part of the Coastal belt, consists of two subterranes. The Point Delgada subterrane consists of oceanic pillow basalt and overlying Coniacian or Campanian argillite. The King Peak subterrane consists of complexly broken and folded argillite and sandstone with melange, pillow basalt, and blocks of limestone and chert that yield ages from Paleocene to middle Miocene. The King Range terrane apparently accreted to the North American margin by obduction in post–middle Miocene time (*McLaughlin and others, 1982; Blake and others, 1989*).

The Franciscan assemblage has been folded, sheared, and

metamorphosed to low blueschist facies (laumontite ± calcite ± quartz assemblages) in an accretionary prism (Blake and others, 1988), followed by tectonic unroofing and strike-slip faulting. Following Late Jurassic and Cretaceous east-directed underthrusting, low-angle detachment faults apparently formed in the latest Cretaceous or earliest Tertiary, resulting in uplift of the deeply buried and highly metamorphosed parts of the Franciscan assemblage (*Platt, 1986; Jayko and others, 1987*). Following tectonic unroofing, early Tertiary right-lateral faults fragmented the Franciscan assemblage, apparently as a result of oblique subduction (*McLaughlin and others, 1988; Blake and others, 1989*). These pre–52 Ma strike-slip faults dispersed parts of the Coast Range ophiolite and overlying Great Valley sequence, as well as older and previously accreted terranes of the Franciscan assemblage.

Great Valley fore-arc basin

The Upper Jurassic to Upper Cretaceous Great Valley sequence is locally as thick as 20,000 m and was deposited in a broad fore-arc basin (Fig. 9) west of the Sierra Nevada magmatic arc (Bailey and others, 1964). On the east side of the Great Valley, the sequence generally consists of gently west-dipping, thin, shallow-marine and deltaic deposits of Turonian to Maastrichtian age that rest nonconformably on igneous and metamorphic rocks of the Sierra Nevada (Haggart and Ward, 1984; *Haggart, 1986; Russell and others, 1986*). On the west side of the Great Valley, the sequence generally consists of steeply east-dipping, thick, deep-marine deposits of Late Jurassic to latest Cretaceous age that rest on Late Jurassic ocean floor, the Coast Range ophiolite (*Bailey and others, 1970*). West of the Great Valley are many outliers of the Great Valley sequence that probably were dispersed along northwest-striking right-lateral faults (*Blake and others, 1989*). The more deeply buried parts of the Great Valley sequence have been metamorphosed to the prehnite-pumpellyite facies (*Dickinson and others, 1969*) and the upper part of the sequence contains prolific gas reservoirs (*Edmondson, 1962; Hackel, 1966;* Garcia, 1981; Cherven, 1984).

The Great Valley sequence was derived principally from the Klamath Mountains and the Sierra Nevada. Large deep-sea fan complexes were deposited by generally southward-flowing turbidity currents in the axis of the basin. The various fan complexes can be recognized and correlated by their distinctive sandstone petrofacies, which indicate progressively deeper dissection of the Sierra Nevada arc (*Ingersoll,* 1978, 1979, *1983*). Intervals of subsidence and progradation of depositional sequences were defined by *Moxon and Graham (1987)* and *Moxon (1988).*

Lower Tertiary strata locally rest conformably on the Great Valley sequence and represent a continuation of fore-arc sedimentation. However, in the latest Cretaceous and Tertiary the basin was divided by the growth of the west-trending Stockton arch (Fig. 10), forming the Sacramento basin to the north and the San Joaquin basin to the south (Nilsen and Clarke, 1975). The Sacramento basin shoaled rapidly in the early Tertiary; a series of large submarine canyons formed in the basin and funneled sediment south and west, toward the present San Francisco Bay region (*Almgren, 1978;* Dickinson and others, 1979; *Almgren and Hacker, 1984; Fischer and Cherven, 1988*). The San Joaquin basin remained deeper and during Paleocene time was characterized by deposition of basin-plain shale and several deep-sea fans fed by west-trending submarine canyons funneling sediment from the Sierra Nevada (*Graham and Berry, 1979*).

By Eocene time, the depositional framework of the San Joaquin basin changed dramatically as a result of uplift of parts of the central Coast Ranges, which caused building of several large deep-sea fans out from this western source area and the formation of major angular unconformities (Nilsen and Clarke, 1975; *Nilsen, 1981,* 1987). Recent evidence indicates that an early Tertiary fold and thrust belt formed along the western margin of the Great Valley basin, perhaps during accretion of terranes to the west. In the southern San Joaquin basin, south of the Neogene west-trending Bakersfield arch, Cretaceous and Paleocene strata are absent and lower Eocene deposits rest directly on basement rocks (*Nilsen and others, 1973*).

Sierra Nevada

Stream-channel deposits filled with Eocene auriferous gravels were deposited on Cretaceous granite that is as young as 80 Ma in the Sierra Nevada. The gold derived from the Mother Lode in the Sierra Nevada formed widespread placer deposits in gravel in the western foothills of the Sierra Nevada. These deposits grade westward into deltaic and shallow-marine deposits that crop out locally in the Sierran foothills (Yeend, 1974). Other fluvial deposits that may be as old as Paleocene crop out at the southeast end of the San Joaquin Valley and in several basins adjacent to the Neogene trace of the Garlock fault (*Nilsen and others, 1973;* Nilsen and Clarke, 1975; Cox, 1987), including the Goler basin within the eroded magmatic arc.

Salinian terrane

Upper Cretaceous strata in the Salinian terrane generally consist of deep-marine deposits that apparently were deposited in local, deep, restricted basins (*Nilsen,* 1977, *1986; Howell and others, 1977b*). Underlying metamorphic and granitoid rocks were last deformed in the Early to Late Cretaceous during emplacement of plutons that are ca. 108 to 76 Ma in age (*James and Mattinson, 1988*). The Cretaceous strata are generally arkosic, apparently were derived from adjacent granitoid and metamorphic uplands within the Salinian terrane, and are of Campanian and Maastrichtian age. Four Cretaceous turbidite basins are present within the Salinian terrane from north to south: (1) Gualala—1,500 m of west-transported plagioclase-rich sandstone and east-transported potassium-feldspar–rich sandstone; (2) Pigeon Point—2,300 m of south- and west-transported deep-sea fan to shelf deposits; (3) northern Santa Lucia—a westward-deepening basin filled with about 500 m of turbidites, shelf, and fluvial

deposits (*Grove and Graham, 1976; Grove, 1989*); and (4) Sierra Madre—south- and west-transported nonmarine to deep-marine sandstone.

Lower Tertiary strata in the Salinian terrane also consist chiefly of arkosic deep-sea fan deposits that appear to define a continental borderland setting of crystalline bedrock islands separated by restricted deep-marine basins (Nilsen and Clarke, 1975; Graham, 1978). From north to south, nine early Tertiary basins or fragments of basins can be recognized: (1) Gualala—possibly as much as 6,000 m of Paleocene and Eocene northwest-transported coarse-grained deep-sea fan deposits; (2) Point Reyes—210 m of west- to northwest-transported Paleocene deep-marine conglomerate and sandstone resting unconformably on granodioritic basement rocks; (3) Point San Pedro—at least 600 m of northwest-transported Paleocene deep-sea fan and slope deposits; (4) La Honda—at least 3,000 m of Paleocene and Eocene shallow- to deep-marine deposits that include a northeast-transported Eocene deep-sea fan deposit that is truncated by the San Andreas fault; (5) San Juan Bautista—about 400 m of west-transported Eocene deep-marine shale and siltstone unconformably overlying granitic and mafic crystalline basement rocks; (6) Point Lobos submarine-canyon deposits—250 m of west-transported Paleocene conglomerate and sandstone (*Clifton, 1984; Fitzgerald and Gorsline, 1989*); (7) northern Santa Lucia—about 1,000 m of Paleocene and Eocene shallow- to deep-marine sedimentary rocks that were deposited by northward-flowing turbidity currents and that rest nonconformably on crystalline rocks; (8) Sierra Madre—an irregularly shaped trough filled with 5,000 (?) m of Paleocene and Eocene turbidites fed by source areas to the north, east, and south; and (9) minor basins at the southeasternmost parts of the Salinian terrane in the Mount Pinos and Caliente Range areas containing turbidites derived from granitoids.

Central Coast Ranges west of the Salinian terrane

Upper Cretaceous turbidites are widespread in this region (Sur-Obispo terrane, Fig. 9) and generally resemble the Great Valley sequence (Howell and others, 1977a). The strata are as young as Campanian and Maastrichtian, overlie ophiolitic basement rocks, and appear to have been thrust over highly deformed rocks of the Franciscan assemblage. Paleocurrents generally suggest sediment transport toward the west. Lower Tertiary strata have not been recognized in inland areas west of the Salinian terrane, although they may be present offshore.

Transverse Ranges

This region has been complexly fragmented by Neogene faults. Turonian to Maastrichtian turbidite sequences as thick as several thousand meters crop out in scattered locations, including the Santa Ynez Mountains, San Miguel Island, in the subsurface of Santa Cruz Island and possibly Santa Rosa Island, the Simi Hills, and the Santa Monica Mountains. Deposits in the Santa Ynez Mountains apparently were transported westward down an elongate west-trending basin (with restoration of clockwise rotation, this basin becomes north trending) called the Santa Ynez basin. Deposits in the Santa Monica Mountains and Simi Hills, which produce oil, were transported northward to form a sand-rich deep-sea fan in a restricted-basin setting (*Link and others, 1981*). Cretaceous turbidites on San Miguel Island were transported predominantly toward the west and north (Bartling and Abbott, 1984).

Lower Tertiary deposits of the Transverse Ranges are as thick as 4,000 m and variable in character. A sequence of Paleocene and Eocene nonmarine, shallow-marine, and turbidite units in the Santa Ynez basin of the western Transverse Ranges shows transport westward down the basin axis, fed by source areas to the east and north (Nilsen, 1987; *Thompson, 1988*). Other shallow- to deep-marine strata in the central and eastern Transverse Ranges apparently form a group of smaller, restricted basins that were fed by granitoid source areas located to the north and east, probably in the area east of the San Andreas fault between the Mojave Desert and Sonora, Mexico (*Parker, 1983; Colburn and others, 1988; Colburn and Novak, 1989*).

Western Mojave Desert

Cretaceous and lower Tertiary strata crop out only in a few areas in the western Mojave Desert. Westward-transported gravel, sand, and mud were deposited in a Cretaceous deep-sea fan in the Cajon Pass area (*Sage, 1973; Kooser, 1982*). Lower Eocene marine deposits as thick as 1,035 m in the Orocopia Mountains record bathyal westward-deepening coarse-grained submarine-canyon fill (*Crowell, 1975; Advocdate and others, 1988*). Lower Tertiary nonmarine and marine deposits accumulated in the Goler and Witnet basins along the northern margins of the province (*Dibblee, 1967;* Cox, 1987).

Peninsular Ranges

Upper Cretaceous and lower Tertiary nonmarine and marine strata crop out at the northern margin of the Peninsular Ranges in the Santa Ana Mountains and along the western margin of the Peninsular Ranges near the present coastline (*Bottjer and Link, 1984*) (Fig. 9); in addition, poorly-preserved fluvial deposits are present in the central Peninsular Ranges. In the northern Santa Ana Mountains, Turonian or Cenomanian nonmarine deposits are overlain by 520 m of Turonian to Campanian shallow- to deep-marine sandstone and shale probably derived from the northern Peninsular Ranges (*Bottjer and others, 1982*).

Along the western margin of the Peninsular Ranges, gently tilted, locally oil-bearing Campanian and Maastrichtian shallow- to deep-marine strata rest unconformably on older Cretaceous fluvial strata and granitoid and metamorphic rocks. The marine strata are as thick as 1,200 m and were deposited in a westward-deepening basin. Clasts are derived chiefly from the Peninsular

Ranges batholith, except for some quartzites that suggest a more distant source (Nilsen and Abbott, 1981). In northern Baja California, the Upper Cretaceous strata were deposited in a similar westward-deepening basin (*Yeo, 1984; Cunningham and Abbott, 1986; Morris and others, 1989*).

The Peninsular Ranges batholith was being tectonically and erosionally unroofed during Late Cretaceous and Paleocene time. The last plutons were emplaced by 80 Ma (Todd and others, 1988). A ductile shear zone (Santa Rosa zone) accommodated westward overthrusting of deep parts of the batholith after pluton emplacement. The ductile overthrusting was followed by brittle faulting that led to tectonic unroofing at 62 Ma, as recorded by concordant fission-track ages for several minerals (Todd and others, 1988).

Evidence for the tectonic unroofing is also found in the sedimentary rocks of the area. Cretaceous deposits along the western flank of the Peninsular Ranges of Alta and Baja California are overlain unconformably by shallow- to deep-marine upper Paleocene and Eocene deposits (Poway basin, Fig. 10), which to the east rest directly on basement rocks (*Abbott,* 1979, *1985; May and others, 1984; Miller and Abbott, 1988*). The basal Eocene conglomerate in the San Diego area is marked by distinctive, well-rounded, siliceous, red-weathering porphyritic metarhyolitic clasts (Howell and Link, 1979; *Kies and Abbott, 1983*). These "Poway clasts" were derived by long-distance fluvial transport from a probable source terrane in northwestern Sonora, Mexico, across an erosional surface developed on the batholithic terrane of the Peninsular Ranges following regional uplift (*Steer and Abbott, 1984*).

California Continental Borderland

Upper Cretaceous rocks that appear to be chiefly of deep-marine origin are widely distributed in the California Continental Borderland, but because they are exposed only on San Miguel Island at the northern margin of the borderland, paleogeographic reconstructions are difficult (*Howell and Vedder, 1981*). Paleocene strata also are present in the subsurface within the borderland and appear to be mostly of shallow-marine origin (*Howell and Vedder, 1981*). Eocene deep-sea fan deposits exposed on San Nicholas Island (*Cole, 1975*) appear to form part of a larger deep-sea fan, named the Ferrelo fan by *Howell and Vedder (1985),* the apex of which was located to the north in the Channel Islands (*Erickson, 1975*), and which has been displaced by strike-slip faulting and rotation from its source area near San Diego.

PALEOTECTONIC AND PALEOGEOGRAPHIC SYNTHESIS

Late Cretaceous

The Campanian and Maastrichtian tectonic framework of the Pacific margin consisted of two major provinces (Pl. 6, Fig. 11). In northern Oregon, Washington, and southern British Columbia,

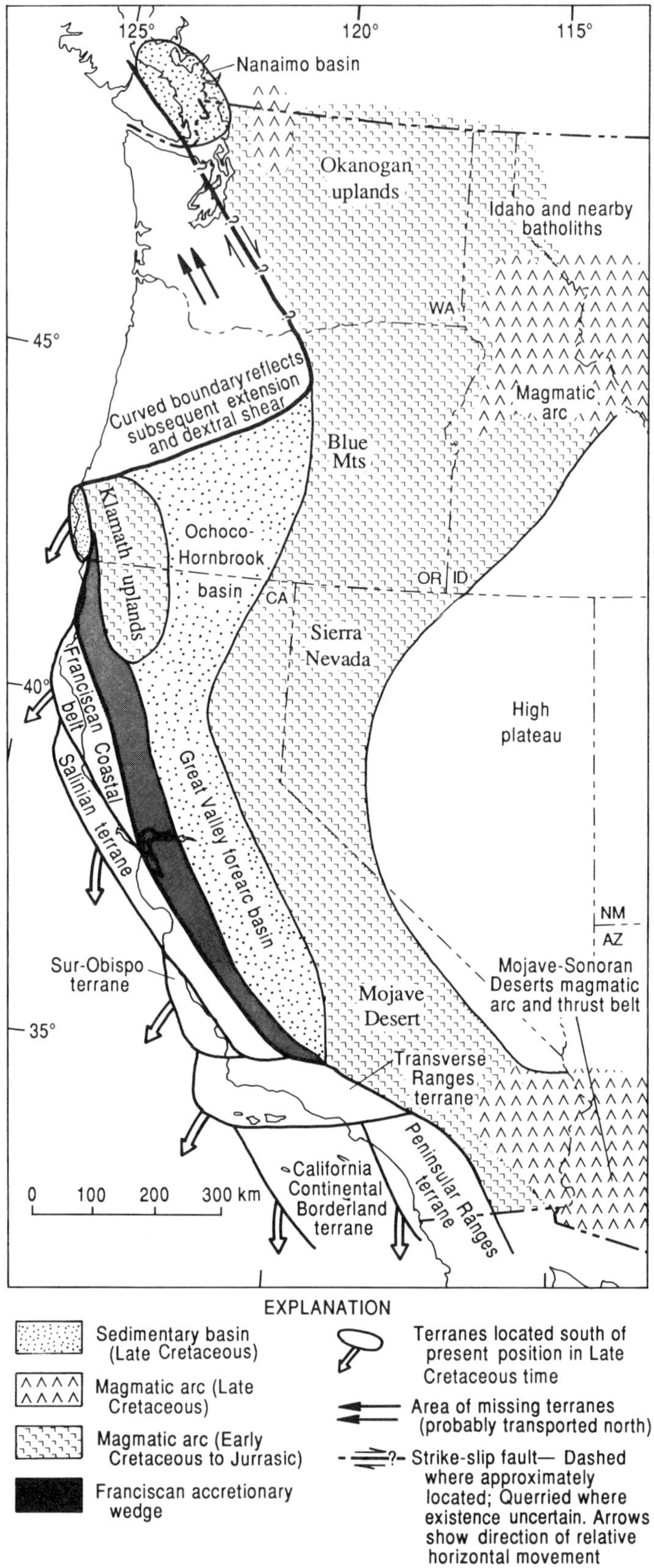

Figure 11. Paleogeographic map of the Pacific margin for Campanian and Maastrichtian time.

a skimpy rock record suggests that dextral shear influenced both tectonic unroofing of the North Cascades (Babcock and Misch, 1988) and the evolution of the fore-arc Nanaimo basin. In central and southern Oregon, California, and Baja California, fore-arc sedimentation indicates the presence of Andean-type subduction. These two provinces resulted from different plate-tectonic margins; a probable transform margin to the north between the Kula and North American plates, and a convergent margin to the south between the Farallon and North American plates (Engebretson and others, 1984, 1985; *Jurdy, 1984; Lonsdale, 1989;* Stock and Molnar, 1988).

During the Late Cretaceous, the northern province undoubtedly contained other tectonostratigraphic terranes that since have been detached and transported northward. These terranes probably included fragments of older or coeval magmatic arcs, fore-arc basins, accretionary wedges, and accreted terranes. Many terranes currently located in southern and southeastern Alaska (*Cowan, 1982; Nilsen and Zuffa, 1982; Oldow and others, 1989*) may represent fragments of the original continental margin of the northern province.

Right-slip displacement was coeval with deposition along the margins of the Nanaimo basin, in which thick, complex sequences of fluvial to deep-marine strata were deposited. Latest Cretaceous strata are missing or covered in other parts of Washington. Much of the North Cascades was probably undergoing uplift and cooling during this time, because Late Cretaceous plutons emplaced as deep as 25 km were unroofed by 55 Ma (Babcock and Misch, 1988); however, details of the unroofing processes are not known. Local parts of the North Cascades record continued metamorphism, ductile deformation, and granitoid plutonism to about 60 Ma (*Tabor and others, 1987*). These events have been attributed to compression inboard of the subduction zone (*Monger and others, 1982*) and to transpressive tectonics (Tabor and others, 1989).

In Oregon, Upper Cretaceous deposits that appear to form a fore-arc basin extend from the Hornbrook area northward to the Ochoco basin (Nilsen, 1984b; Kleinhans and others, 1984). The northern extension of the basin appears to have been truncated, and the locally thick sequence in the Ochoco basin suggests possible disruption and extension related to strike-slip faulting.

The southern province was characterized by a fore-arc basin, subduction complex, and largely inactive magmatic arc. During the Campanian and Maastrichtian, the arc was active only in the eastern Mojave Desert region. The Klamath Mountains and Blue Mountains, consisting of various types of Paleozoic, Triassic, and Jurassic terranes that had been accreted and were intruded by Jurassic plutons, represent separate blocks located west of the Cretaceous batholiths. The Blue Mountains appear to have rotated clockwise away from the Idaho batholith in Early Cretaceous time, and the Klamath Mountains appear to have translated westward away from the main batholith trend in the Early Cretaceous. The Blue Mountains and Klamath Mountains thus formed large source areas outboard, or west of, the main magmatic-arc complex during the Late Cretaceous.

Following the reorganization of the magmatic arc in the Early Cretaceous, the Great Valley fore-arc basin extended northward to the east of the Klamath Mountains as the Ochoco basin. It also may have extended farther north beneath the Columbia Plateau, and possibly to the Nanaimo basin in the Puget Sound–Vancouver Island area. The southward continuation of the Late Cretaceous Great Valley fore-arc basin is unclear because of erosion and disruption by Tertiary faults; however, similar types and ages of strata west of the Salinian terrane and along the western margin of the Peninsular Ranges and Baja California suggest that it, or a similar but possibly unconnected fore-arc basin, extended considerably farther to the south.

A subduction complex composed of the eastern and central belts of the Franciscan assemblage and possibly parts of the Coastal belt formed west of the Late Cretaceous Great Valley fore-arc basin. The Franciscan assemblage includes fragments of oceanic sea-floor, island-arc, and oceanic plateau terranes that were accreted to the continental margin after migrating northward and eastward; the subduction complex probably continued to be displaced and deformed by right shear after accretion. The Late Cretaceous subduction complex is well preserved in the northern California Coast Ranges, but probably has been removed from central and southern California, and possibly from the west side of the Klamath Mountains, by Neogene strike-slip faulting. Alternatively, the record of Late Cretaceous subduction could have been removed by complete subduction or by Late Cretaceous and early Tertiary strike-slip faulting.

Some present-day tectonic elements of California that were not present during the Late Cretaceous must be palinspastically restored to more southerly latitudes. These elements include the Salinian terrane and the areas to its west, the Transverse Ranges, the Peninsular Ranges, and the California Continental Borderland. These provinces were possibly located hundreds to thousands of kilometers south of California during the Late Cretaceous.

Other tectonic elements may have been located west of the Franciscan subduction complex in California during the Late Cretaceous. It is possible that one or more volcanic arcs were present, and the most likely candidate is represented by the Cretaceous rocks currently located west of the Klamath Mountains. However, the lack of sufficient supporting data prevents accurate reconstruction.

Early Tertiary

The Paleocene and early Eocene tectonic framework of the Pacific margin was more complex than that of the Late Cretaceous because of the disruption of the Cretaceous fore-arc basins and accretion of new terranes (Fig. 12). As in the upper Upper Cretaceous, Paleocene strata are virtually absent in Washington and Oregon, making paleogeographic reconstructions of these areas speculative at best. Paleocene nonmarine strata in north-central Oregon suggest regional uplift of the continental margin in this area and perhaps much of the Pacific Northwest. Paleocene

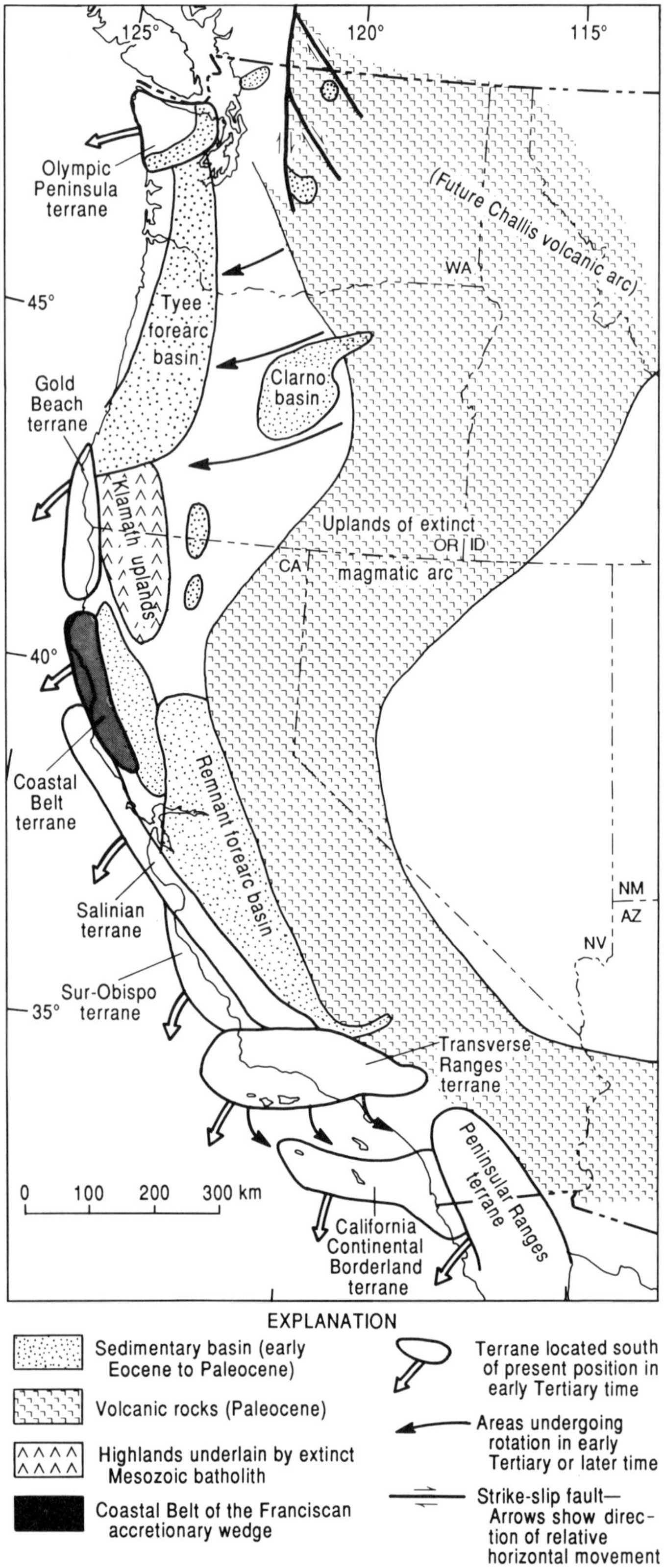

Figure 12. Paleogeographic map of the Pacific margin for Paleocene and early Eocene time.

volcanic rocks of the Oregon and Washington Coast Range formed offshore as oceanic seamounts and did not accrete until the early Eocene. Dextral shear may have taken place along this part of the Pacific margin, possibly resulting in detachment and northward translation of terranes previously located in the area. The temporal gap between completion of magmatism in the Idaho batholith (62 Ma) and initiation of volcanism in the Challis arc (55 Ma) may coincide with the interval of dextral shear.

The formation of early Eocene nonmarine basins such as the Chuckanut and Swauk in northwestern Washington (Fig. 10), which appear to be related to right-slip faulting, also suggests dextral shear. By the middle Eocene, a fore-arc basin appears to have developed along the west side of the North Cascades, including the Puget Lowland area, as the volcanic arc began to shift westward to its present locus in the Cascade Range, where volcanism began at 42 Ma.

In western Oregon and Washington, the Coast Range accreted to the continental margin in the early Eocene. This complex of oceanic seamounts subsequently rotated clockwise to its present position, and the continent-margin trench, possibly originally located west of the Blue Mountains, may have moved westward to its present position west of the accreted seamount chain. Late early Eocene and middle Eocene deep-sea fans of the Tyee Formation and related clastic sedimentary units were deposited over the subsiding seamount chain as a new fore-arc basin developed in the Coast Range, west of the older Mesozoic fore-arc basin of central Oregon.

The Klamath Mountains, Ochoco basin, and northern part of the Great Valley fore-arc basin underwent regional uplift during the Paleocene to early Eocene, as indicated by Cretaceous marine rocks that are generally overlain unconformably by Eocene nonmarine strata. The uplift may have resulted from collision and suturing of terranes such as the Gold Beach terrane (Fig. 12) outboard of the Klamath Mountains (Bourgeois and Dott, 1985), or the accretion of the oceanic seamounts of the Coast Range of Oregon and Washington.

Sedimentation and tectonics in California during the early Tertiary reflected (1) continued east-directed subduction of oceanic crust along the continental margin, (2) uplift of the Klamath Mountains and northern part of the Great Valley fore-arc basin, (3) truncation of the continental margin in southern California by apparent strike-slip faulting, (4) strike-slip translation of the Salinian and related terranes, producing late Paleocene deformation, and (5) the formation of a continental borderland off central and southern California composed of the Salinian block and related terranes. The resulting paleogeography of the continental margin was far more complex than the relatively simple paleogeography of the Late Cretaceous. Although subduction continued along the California margin during the early Tertiary, the active magmatic arc shifted far eastward, leaving California generally devoid of magmatic and volcanic activity, except for oceanic basalts that were incorporated into the subduction zone to the west.

Much of the southern extension of the Great Valley fore-arc basin was probably detached by dextral faults during the Paleo-

cene. Remnants of the fore-arc basin and associated subduction complex in southern Alaska may be the Chugach terrane and related terranes (*Nilsen and Zuffa, 1982*) and in southwestern Oregon may be the Gold Beach terrane (Bourgeois and Dott, 1985).

Two models, whose differences are yet to be resolved, explain how the Salinian and related terranes to the west were juxtaposed with southern and central California. Paleomagnetic and stratigraphic evidence suggest that the Salinian and Sur-Obispo terranes were amalgamated by the Late Cretaceous and that they migrated northward as much as 1,500 km as a single larger block (*McWilliams and Howell, 1982;* Champion and others, 1984). However, similarities between Cretaceous and older rocks in the Salinian block and those in the southern Sierra Nevada–eastern Transverse Ranges area (*James and Mattinson, 1988*) support a model calling for at most a few hundred kilometers displacement of the Salinian block, which can be accounted for by Neogene right slip along the San Andreas fault.

Movement of the Salinian terrane produced deformation throughout central and southern California. In the central Coast Ranges, uplift of the Diablo Range exposed Franciscan basement rocks and the structurally overlying Great Valley sequence, which provided local source areas for adjacent sedimentary basins in the early Tertiary. Angular unconformities in lower Tertiary strata in the central Coast Ranges may also be associated with Salinian terrane accretion (*Nilsen, 1981*). The Salinian terrane also became a major source area in the early Tertiary; several deep-sea fans of Eocene age were built eastward into the area of the Coast Ranges and western San Joaquin Valley (Tres Pinos Sandstone described by *Nilsen, 1981,* and the Point of Rocks Sandstone described by Nilsen and Clarke, 1975). East-dipping thrust faults and associated ramps and duplexes that developed in the southern Sacramento basin and along the Stockton arch during the early Tertiary probably resulted from accretion of the Salinian terrane. In the area of the southern San Joaquin Valley, Cretaceous and Paleocene strata are missing, possibly from erosional stripping caused by uplift related to the terrane accretion.

Small nonmarine early Tertiary basins formed adjacent to the modern trace of the Garlock fault (Nilsen and Clarke, 1975), possibly as pull-aparts in response to a "proto-Garlock fault," much as the collision of India with Asia in the Cenozoic produced many large strike-slip faults north of the Himalayas. Regardless of the tectonic control, these lower Tertiary deposits represent intermontane basins formed within a high plateau (Cox, 1987). Part of the clockwise rotation of the southern Sierra Nevada, Tehachapi, and San Emigdio Mountains, which *Kanter and McWilliams (1982)* concluded occurred 20 to 80 Ma, also may have resulted from accreting the Salinian block.

Farther south, in the Transverse Ranges, Mojave Desert region, southeastern California, and southwestern Arizona, the Orocopia Schist was juxtaposed with continental basement along the Chocolate Mountain thrust system. Lower Eocene transgressive marine strata that overlap the allochthon in the Orocopia Mountains indicate that thrusting was completed by that time.

Arrival of the Salinian and Sur-Obispo terranes, and probable simultaneous northward movement of the Transverse Ranges to a position opposite northwestern Mexico, formed a broad, elongate Eocene continental borderland underlain by continental and Franciscan-type crust west of central and southern California (Nilsen and Clarke, 1975). The borderland consisted of a series of small, deep, restricted basins that were separated by islands; sediment was derived from both island and continental-margin source areas (Clarke and others, 1975). The borderland persisted through the early Oligocene (*Nilsen, 1984a*), before becoming fragmented by northward transport during the Neogene along the San Andreas fault system.

SUMMARY

The Late Cretaceous and early Tertiary Pacific margin evolved rapidly as at least three major plates—the oceanic Kula and Farallon plates and the continental North American plate—underwent major changes in direction and rate at about 85 and 56 Ma (Engebretson and others, 1985). Most plate reconstructions of the Pacific margin for Late Cretaceous to early Tertiary time show two subduction realms: the Farallon plate subducting beneath North America to the south, and northward translation of the Kula plate with reference to North America (Atwater, 1970; Engebretson and others, 1985; *Jurdy and Gordon, 1984*). However, the location of the Kula-Farallon-North American triple junction during this time interval is not well constrained, permitting either right slip or convergence along the continental margin of Washington, Oregon, and California.

Most reconstructions for California indicate rapid convergence and subduction of the Farallon plate from 80 to 40 Ma. Most of the continental plates, including North America, moved faster during the early Tertiary than at present (*Jurdy and Gordon, 1984*). Episodes of oblique Farallon–North American plate subduction (*Page and Engebretson, 1984; Kanter and Debiche, 1985*) provide a northward component of relative movement during the Late Cretaceous and early Tertiary, explaining northward motion of the Salinian block and other tectonostratigraphic terranes. Dextral shear during the Late Cretaceous to early Eocene in Washington and areas farther north probably resulted from Kula–North American plate interactions (*Lonsdale, 1989*).

Convergence and right-slip along the U.S. continental margin in the Late Cretaceous and early Tertiary resulted in several large-scale tectonic displacements: (1) movement of various terranes from the Pacific Northwest and possibly outboard of the Great Valley fore-arc basin and Klamath Mountains block to regions farther north, such as British Columbia and southern Alaska; (2) probable shearing off of the southern continuation of the Great Valley fore-arc basin and associated subduction complex and translation of remnants of this complex northward; (3) movement of the Salinian composite terrane in late Paleocene time, causing deformation throughout a broad region of California; (4) accretion of oceanic seamounts of the Coast Range of Oregon and Washington in the early Eocene and their subse-

quent clockwise rotation; (5) northward movement of Baja California, the Peninsular Ranges, and the California Continental Borderland to a position opposite northwestern Mexico; and (6) northward movement of the Transverse Ranges, probably as a separate block, to a position between the southern end of the Salinian block and the northern end of the Peninsular Ranges. Tectonic unroofing of the eastern Peninsular Ranges batholith possibly accompanied the right-slip displacements of the tectonic blocks of southern California.

Late Cretaceous and early Tertiary basins were of several types (*Howell and others, 1980; Nilsen, 1986, 1987*): (1) those generated by strike-slip faulting, such as the Chuckanut, Puget, and Swauk in northern Washington; (2) those previously generated as successor or fore-arc basins and subsequently deformed by strike-slip faulting, such as the Nanaimo basin in northwestern Washington and adjacent British Columbia; (3) small Eocene fore-arc basins in western Washington (Puget basin) and western Oregon (Tyee basin); (4) the large Great Valley fore-arc basin, which may have extended northward and east of the Klamath Mountains into the Hornbrook-Ochoco basin of central Oregon; (5) trench-slope basins and trench deposits of the Yager Formation, the Coastal belt of the Franciscan assemblage, the Sur-Obispo terrane, and possibly some outliers of Upper Cretaceous and lower Tertiary deposits in the northern Coast Ranges; (6) pull-apart basins in the Salinian terrane produced during northward movement and internal slicing of this terrane prior to its accretion in the late Paleocene; (7) similar basins that developed in the Transverse Ranges; (8) marine basins in the central Coast Ranges and San Joaquin Valley east of the San Andreas fault, caused by deformation related to emplacement of the Salinian terrane; (9) tectonically controlled, largely nonmarine intermontane basins near the present Garlock fault; and (10) a remnant Late Cretaceous fore-arc basin along the western margin of the Peninsular Ranges.

The tectonic and depositional characteristics of the Pacific margin changed following the early Eocene. In the Pacific Northwest, the Cascade volcanic arc, fore-arc basins, and accretionary complexes developed in response to subduction of the Farallon plate. In contrast, Neogene tectonism in California was characterized by dextral transform faulting in a zone between the Pacific and North American plates, creating the San Andreas fault system and resulting in northward migration and local rotation of the Salinian composite terrane, the Transverse Ranges, the Peninsular Ranges, and the California Continental Borderland. New basins such as the Los Angeles, Ventura, Santa Maria, and Salton Trough (Blake and others, 1978) developed in response to the revamped Neogene tectonic setting.

CONCLUSIONS

D. M. Miller, T. H. Nilsen, W. L. Bilodeau

Concepts of the geologic evolution of the western United States have changed dramatically with increased studies of the U.S. Cordillera and with use of new earth science paradigms. Geologic studies in the nineteenth and early twentieth centuries featured detailed examination of mining districts and paleontological sites within the reconnaissance framework established by Federal surveys. The major periods of Mesozoic and Cenozoic tectonism, magmatism, and sedimentation were gradually identified and explained using the extant geosynclinal theories of orogeny. The inception of plate-tectonic models led to explanations of the Mesozoic and Cenozoic geologic evolution of the Cordillera that linked orogenies to plate-margin processes; these interpretations proved to be far more successful at explaining many geologic relations. The past two decades have seen further refinement in our understanding of the response of continental crust to the motion of oceanic plates. The concepts of tectonostratigraphic terranes and large-scale lateral slip of slices of continental crust along the continental margin are now applied, providing new insight into Cordilleran orogenies. In the most recent overview of the U.S. Cordillera, Hamilton (1988) attributed most of the Late Cretaceous to Eocene orogeny to an increased subduction rate and accompanying shallowing of subduction angle. We use Hamilton's (1988) summary as a staging point for discussing some of the remaining perplexities for Late Cretaceous to early Eocene Cordilleran orogeny.

In explaining the change from the mid-Mesozoic "Andean" tectonics to the complex Laramide tectonics, Hamilton (1988) followed many previous workers by ascribing the change to increased rate, and hence shallower angle, of subduction and consequent tectonic erosion of the base of the continental lithosphere. Hamilton interpreted the Chocolate Mountains and related thrusts, Santa Rosa mylonite zone, and thrusts southwest of the Salinian terrane as fragments of this shallow subduction zone where it had tectonically eroded some 70 km of basal continental lithosphere. By Hamilton's account, lessened activity in the magmatic arcs and the thin-skinned thrust belts also resulted from shallow-mode subduction. According to him, the frontal thrust belt was driven by gravitational spreading from overthickened crust where it was intruded by batholiths of the magmatic arc. He also explained the Laramide basement uplifts as the products of crustal shortening resulting from lithospheric drag by the subducting plate, accomplished primarily by clockwise rotation of the Colorado Plateau "plate" with reference to the craton, about an Euler pole in central New Mexico. This grand-scale scenario for Laramide tectonics provides a comprehensive framework with which to examine some unusual features of this time period and some alternative hypotheses.

The Cordilleran orogenic belt displays a striking segmentation of tectonic regimes along its length that are difficult to explain. Although a shallow angle of subduction apparently resulted from rapid plate convergence throughout the U.S. Cordillera, the central sector of Nevada, Utah, and Colorado displays features, such as cessation of the magmatic arc and basement thrusting within the craton, that are not displayed to the north and south. However, the north and south segments also underwent rapid, shallow subduction, as recorded by eastward-migrating magmatic

belts and the sea-floor record. One cause for this tectonic segmentation could be a lateral change in physical properties of the subducted crust, such as a more buoyant crust (oceanic plateau) subducted under the central segment, creating tectonic effects extending farther eastward into the continental interior. Alternatively, or additionally, thermal-mechanical changes in the continental crust prior to Laramide time may have created a segmented response to shallow subduction. Jurassic and Cretaceous plutonism was sparse but widespread within the Nevada-Utah area, some 200 to 250 km inboard of the magmatic arc (after restoration of Tertiary extension), and may have sufficiently modified the crust so that its response to shallow subduction was unique. However, the tectonic response of the Rocky Mountain province was less strikingly controlled by its crust. Archean and a variety of Proterozoic belts display the same general style of shortening, and that style apparently was not influenced by the presence or absence of late Paleozoic tectonic uplifts and basins, even though the older structures were in many places reactivated. Therefore, tectonic segmentation can be explained only partly by inherited features of the continental crust, and features of the subduction zone and/or subducted crust have exerted the primary controls on segmentation.

A huge region in and east of the Early Cretaceous magmatic arc underwent marked denudation, presumably caused by uplift, during Laramide time; the region is notable mainly for its lack of sedimentary and structural record. Much of the northern Washington, western Idaho batholith, Great Basin hinterland, Sierra Nevada, eastern Mojave Desert, and western Arizona areas were stripped of several kilometers of upper crust by tectonic and fluvial erosion at various times in the latest Cretaceous, Paleocene, and early Eocene. The presumed uplift was probably caused in some way by shallow subduction and subcrustal tectonic erosion. An enormous volume of sediment was shed from these uplifted regions, much of which presumably was transported westward to Pacific-margin basins or southeastward to the Gulf of Mexico. Only a few sedimentary basins of this age remain within the uplifted area, and the Rocky Mountain province probably formed a drainage divide limiting sediment transport to the east. This uplifted region was a boundary between largely nonmarine sedimentation to the east and largely marine sedimentation to the west.

Sedimentary basins along the margin of the Cordillera reflect the changing plate-tectonic framework of the continental margin during Laramide time. Trench, trench-slope, and fore-arc basins characterized the convergent margin south of the Kula-Farallon-North American plate triple junction. Fore-arc basins, such as the Great Valley fore-arc basin of central and northern California, were long-lived basins that originally developed in the Late Jurassic; these basins are rich in oil and gas. In the Pacific Northwest, a series of prominent strike-slip faults produced pull-apart and tipped-wedge basins associated with right-lateral displacements, all driven by the Kula–North American plate oblique subduction. Extensional basins characterized by thick accumulations of submarine-fan deposits in small, restricted basins developed in migrating terranes such as the Salinian and Gold Beach terranes. The Salinian and related terranes accreted during the early Tertiary, establishing a continental borderland along the western margin of California that shed sediment eastward into the remnant fore-arc basins. Following terrane accretion, intermontane basins possibly caused by strike-slip and (or) extensional tectonics within and east of the eroded magmatic arc, and Tertiary nonmarine intermontane basins characterized by lacustrine sedimentation formed in parts of the hinterland.

Although the Cordilleran frontal thrust belt formed almost everywhere at the west margin of the craton with its thin sedimentary cover, the driving forces may have varied with time and location. Alternatives to Hamilton's proposal of gravitational spreading from the magmatic arc to cause thrusting include (1) gravitational spreading from a thickened metamorphic belt and (2) westward intracontinental subduction of the North American craton beneath the miogeocline. Prior to the end of Sierran magmatism at 80 Ma, a gravitational wedge operating in an overall noncontractional setting may have contributed to frontal thrusting, as Hamilton (1988) proposed. However, after the Sierran arc shut down and a contractional plate-margin setting was firmly established, the central Utah thrusts continued movement until the late Paleocene, as recorded by synorogenic clastic wedges. This late thrusting is unlikely to represent a 20-m.y. gravitational decay period; it more likely indicates that a gravitational wedge was driven from a site still being tectonically thickened or that intracontinental subduction was driven by westward absolute motion of the craton. It is more plausible that younger thrusts in Wyoming, Idaho, and Montana may be tied to the magmatism in the Idaho batholith area, but the link is not compelling; the youngest thrusts in this region are 10 m.y. younger than magmatism.

One model for thrust-belt movement is that of gravitational wedges driven by an overthickened metamorphic belt east of the magmatic arc. The metamorphic belt in Nevada appears to have continued lateral shortening and vertical thickening in the Late Cretaceous, presumably because of lower crustal duplication and magma emplacement driven by subduction. This continued crustal thickening after the Sierran arc ceased activity provides a plausible source for a gravitational wedge. The gravitational wedge model is attractive because (1) body forces within the wedge drive the thrusting, eliminating the need for unreasonable mechanical properties for the thrust sheets, and (2) the direction of thrusting is directly related to an orogenic feature parallel to the continental margin, rather than parallel to oblique plate-convergence vectors.

Alternatively, intracontinental subduction may have caused frontal thrusting during the time of basement thrusting in the Rocky Mountains. Although the Cordilleran thrust belt and foreland basement structures were caused by horizontal compression, the foreland uplifts do not parallel the main orogen or the convergent plate boundary. Instead, the pattern is one of peripheral deformation around the Colorado Plateau, suggesting indentation tectonics, as proposed by Hamilton (1988). The Colorado Pla-

teau can be viewed in classical plate-tectonic terms as a microplate that moved independently from both the North American craton and the hinterland–thrust-belt block. In this view, the tectonic movement of the Colorado Plateau, induced in some fashion by the obliquely (north-northeast) subducting plate, caused the indentation tectonics. However, plateau and craton together were moving westward with reference to the thrust-belt–hinterland block, producing intracontinental subduction of the cratonal plates. This model must explain the nearly perpendicular directions of frontal thrusting in Utah and movement of the Colorado Plateau, which requires more knowledge of the interactions of the several continental and oceanic plates. Characteristics such as thrust direction and timing of the youngest thin-skinned thrusts in Utah require careful study.

The latest Cretaceous and Paleocene magmatic arcs processed continental-crust materials for the first time in Idaho and parts of easternmost California and southern Arizona, resulting in abundant silicic peraluminous granites. Sources of these leucocratic melts in the lower to middle crust are proposed to be granulite complexes and quartzo-feldspathic rocks. Associations with metaluminous granites and less common mafic rocks, along with their isotopic records, suggest a genesis involving: (1) melting near the subduction zone, which drove hot, mafic melts into the lower to middle continental crust to produce (2) rising crustal-derived granodioritic melts, which (3) melted more crust to create highly leucocratic, peraluminous granitoids. Although thickened continental crust may have contributed to the production of silicic melts, in many cases such thickening cannot be directly demonstrated.

A number of Laramide problems remain unresolved. Perhaps key among them is the paleogeography of the region from the magmatic arc and hinterland to the thrust belt. What were the times and causes of the uplift and erosion of this vast area? Was the unroofing tectonic as well as erosional? These questions are also important for resolving details in the conversion from contractional to extensional tectonic regimes during the Eocene and Oligocene. The processes involved in and timing of this tectonic conversion in different parts of the Cordillera must be better known in order to understand its plate-tectonic origin.

Within the Rocky Mountain foreland, the relative timing for uplift of the various Laramide mountain blocks is only roughly known because this knowledge depends on establishing the relations between uplift and the sedimentation in adjacent basins and dating these nonmarine strata. We are just beginning to understand the basin subsidence mechanisms and how to interpret the stratigraphic sequences of the basin fill in terms of tectonic uplift and subsidence events. Were there large interconnected fluvial systems between basins? Were they primarily internally drained, or did the tectonically controlled paleogeography change so that the main depositional systems changed through time? What characteristics of the crust controlled the heterogeneous distribution of the foreland structures and basins? In the overlap areas between subprovinces, where the structural style changes, how is this structural transformation accomplished? How much time is represented by the basal unconformities of Laramide basin-fill sequences; can it be assumed that basin subsidence was initiated at the same time as the initiation of uplift of the adjacent block? Additional detailed dating and sedimentology can help to resolve many of these problems.

Only fragments of the latest Cretaceous to early Eocene accretionary complexes and related basins remain. Were the accretionary complexes subducted with greater efficiency during rapid, shallow subduction or did probable obliquity of convergence cause lateral transport of most of the material? Were the dispersed fragments replaced by an equal volume of fragments derived from the south? Resolving these questions requires fully restoring allochthonous terranes by matching geologic histories and paleomagnetic data.

The chemical and physical changes to the lithosphere caused by shallow subduction must be explored. Do the Chocolate Mountains thrust and related thrusts represent this subduction zone? Can Laramide igneous rocks tell us about changes in the upper mantle and lower crust accompanying low-angle subduction? How did the shallowly subducting oceanic crust interact with continental crust to create the Laramide uplifts? Future geological, geophysical, and isotopic studies should be designed to infer deep-crustal properties and processes, which in turn can be integrated to provide a regional framework for Laramide orogenic processes.

REFERENCES CITED

Text references in *italics* will be found only on accompanying microfiche.

Abbott, P. L., ed., 1979, Eocene depositional systems, San Diego, California: Pacific Section, Society of Economic Paleontologists and Mineralogists, v. 4, 126 p.

Allmendinger, R. W., 1986, Tectonic development, southwestern border of the Puna Plateau, northwestern Argentine Andes: Geological Society of America Bulletin, v. 97, p. 1070–1082.

Allmendinger, R. W., and Jordan, T. E., 1981, Mesozoic evolution, hinterland of the Sevier orogenic belt: Geology, v. 9, p. 308–313.

Allmendinger, R. W., Sharp, J. W., Von Tish, D., Serpa, L., Kaufman, S., Oliver, J., and Smith, R. B., 1983, Cenozoic and Mesozoic structure of the eastern Basin and Range Province, Utah, from COCORP seismic reflection data: Geology, v. 11, p. 532–536.

Anderson, J. L., 1988, Core complexes of the Mojave-Sonoran Desert; Conditions of plutonism, mylonitization, and decompression, *in* Ernst, W. G., ed., Metamorphism and crustal evolution of the western United States: Englewood Cliffs, New Jersey, Prentice-Hall Inc., p. 502–525.

Anderson, R. E., and Barnhard, T. P., 1986, Genetic relationship between faults and folds and determination of Laramide and neotectonic paleostress, western Colorado Plateau–transition zone, central Utah: Tectonics, v. 5, p. 335–357.

Applegate, J. K., and Rose, P. R., 1985, Structure of the Raton basin from a regional seismic line, *in* Gries, R. R., and Dyer, R. C., eds., Seismic exploration of the Rocky Mountain region: Rocky Mountain Association of Geologists and Denver Geophysical Society, p. 259–265.

Armstrong, F. C., and Oriel, S. S., 1965, Tectonic development of Idaho-Wyoming thrust belt: American Association of Petroleum Geologists Bul-

letin, v. 49, p. 1847–1866.
Armstrong, R. L., 1968, Sevier orogenic belt in Nevada and Utah: Geological Society of America Bulletin, v. 79, p. 429–458.
——, 1972, Low-angle (denudation) faults, hinterland of the Sevier orogenic belt, eastern Nevada and western Utah: Geological Society of America Bulletin, v. 83, p. 1729–1754.
——, 1974, Magmatism, orogenic timing, and orogenic diachronism in the Cordillera from Mexico to Canada: Nature, v. 247, p. 348–351.
——, 1978, Cenozoic igneous history of the U.S. Cordilera from lat. 42° to 49° N.: Geological Society of America Memoir 152, p. 263–282.
——, 1982, Cordilleran metamorphic core complexes—From Arizona to southern Canada: Annual Reviews Earth and Planetary Science, v. 10, p. 129–154.
Armstrong, R. L., Taubeneck, W. H., and Hales, P. O., 1977, Rb-Sr and K-Ar geochronometry of Mesozoic granitic rocks and their Sr isotopic composition, Oregon, Washington, and Idaho: Geological Society of America Bulletin, v. 88, p. 397–411.
Armstrong, R. L., Parrish, R. R., van der Hayden, P., Reynolds, S. J., and Rehrig, W. A., 1987, Rb-Sr and U-Pb geochronometry of the Priest River metamorphic complex; Precambrian X basement and its Mesozoic-Cenozoic plutonic-metamorphic overprint, northeastern Washington and northern Idaho, *in* Schuster, J. E., ed., Selected papers on the geology of Washington: Washington Division of Geology and Earth Resources, Bulletin 77, p. 15–40.
Atwater, T., 1970, Implications of plate tectonics for the Cenozoic tectonic evolution of western North America: Geological Society of America Bulletin, v. 81, p. 3513–3536.
Ayers, W. B., Jr., 1986, Lacustrine and fluvial-deltaic depositional systems, Fort Union Formation (Paleocene), Powder River basin, Wyoming and Montana: American Association of Petroleum Geologists Bulletin, v. 70, p. 1651–1673.
Baars, D. L., and others, 1988, Basins of the Rocky Mountain region, *in* Sloss, L. L., eds., Sedimentary cover—North American Craton, U.S.: Boulder, Colorado, Geological Society of America, The Geology of North America, v. D-2, p. 109–220.
Babcock, R. S., and Misch, P., 1988, Evolution of the crystalline core of the North Cascades Range, *in* Ernst, W. G., ed., Metamorphism and crustal evolution of the western United States: Englewood Cliffs, New Jersey, Prentice-Hall, p. 214–232.
Bailey, E. H., and Blake, M. C., Jr., 1969, Late Mesozoic tectonic development of western California: Geotectonics, v. 3, p. 148–154.
Bailey, E. H., Irwin, W. P., and Jones, D. L., 1964, Franciscan and related rocks, and their significance in the geology of western California: California Division of Mines and Geology Bulletin 183, 177 p.
Bally, A. W., 1981, Thoughts on the tectonics of folded belts, *in* McClay, K. R., and Price, N. J., eds., Thrust and nappe tectonics: London, Geological Society of London, p. 13–32.
Bally, A. W., Gordy, P. L., and Stewart, G. A., 1966, Structure, seismic data, and orogenic evolution of the southern Canadian Rocky Mountains: Bulletin of Canadian Petroleum Geologists, v. 14, p. 337–381.
Bartling, W. A., and Abbott, P. L., 1984, Upper Cretaceous sedimentation and tectonics with reference to the Eocene San Miguel Island, and San Diego area, California, *in* Abbott, P. L., ed., Upper Cretaceous depositional systems, southern California–northern Baja California: Pacific Section, Society of Economic Paleontologists and Mineralogists, v. 36, p. 85–102.
Beck, M. E., Jr., Burmester, R. F., Craig, D. E., Gromme, C. S., and Wells, R. E., 1986, Paleomagnetism of middle Tertiary volcanic rocks from the Western Cascade series, northern California; Timing and scale of rotation in the southern Cascades and Klamath Mountains: Journal of Geophysical Research, v. 91, p. 8219–8230.
Berg, R. R., 1981, Review of thrusting in the Wyoming foreland, *in* Boyd, D. W., and Lillegraven, J. A., eds., Rocky Mountain foreland basement tectonics: University of Wyoming Contributions to Geology, v. 19, p. 93–104.
Bird, P., 1984, Laramide crustal thickening event in the Rocky Mountain foreland and Great Plains: Tectonics, v. 3, p. 741–758.
Blackstone, D. L., Jr., 1980, Foreland deformation; Compression as a cause: University of Wyoming Contributions to Geology, v. 18, p. 83–100.
——, 1983, Laramide compressional tectonics, southeastern Wyoming: University of Wyoming Contributions to Geology, v. 22, p. 1–38.
Blake, M. C., Jr., 1984, Tectonostratigraphic terranes in southwestern Oregon, *in* Nilsen, T. H., ed., Geology of the Upper Cretaceous Hornbrook Formation, Oregon and California: Pacific Section, Society of Economic Paleontologists and Mineralogists, v. 42, p. 159–165.
Blake, M. C., Jr., and 7 others, 1978, Neogene basin formation in relation to plate-tectonic evolution of San Andreas fault system, California: American Association of Petroleum Geologists Bulletin, v. 62, p. 344–372.
Blake, M. C., Jayko, A. S., McLaughlin, R. J., and Underwood, M. B., 1988, Metamorphic and tectonic evolution of the Franciscan Complex, northern California, *in* Ernst, W. G., ed., Metamorphism and crustal evolution of the western United States: Englewood Cliffs, New Jersey, Prentice-Hall, p. 1035–1060.
Bourgeois, J., and Dott, R. H., Jr., 1985, Stratigraphy and sedimentology of Upper Cretaceous rocks in coastal southwest Oregon; Evidence for wrench-fault tectonics in a postulated accretionary terrane: Geological Society of America Bulletin, v. 96, p. 1007–1019.
Brewer, J. A., Smithson, S. B., Oliver, J. E., Kaufman, S., and Brown, L. D., 1980, The Laramide orogeny; Evidence from COCORP deep crustal seismic profiles in the Wind River Mountains, Wyoming: Tectonophysics, v. 62, p. 165–189.
Brewer, J. A., Allmendinger, R. W., Brown, L. D., Oliver, J. E., and Kaufman, S., 1982, COCORP profiling across the Rocky Mountain front in southern Wyoming; Part 1, Laramide structure: Geological Society of America Bulletin, v. 93, p. 1242–1252.
Bruhn, R. L., Picard, M. D., and Isby, J. S., 1986, Tectonics and sedimentology of Uinta arch, western Uinta Mountains, and Uinta basin, *in* Peterson, J. A., ed., Paleotectonics and sedimentation in the Rocky Mountains, United States: American Association of Petroleum Geologists Memoir 41, p. 333–352.
Buckovic, W. A., 1979, The Eocene deltaic system of west-central Washington, *in* Armentrout, J. M., Cole, J. R., and Ter Best, H., Jr., eds., Cenozoic paleogeography of the western United States: Pacific Section, Society of Economic Paleontologists and Mineralogists Pacific Coast Paleogeography Symposium 3, p. 147–164.
Burchfiel, B. C., and Davis, G. A., 1975, Nature and controls of Cordilleran orogenesis, western United States; Extension of an earlier synthesis: American Journal of Science, v. 275-A, p. 363–396.
——, 1981, Mojave Desert and environs, *in* Ernst, W. G., ed., The geotectonic development of California: Englewood Cliffs, New Jersey, Prentice-Hall, p. 217–252.
Bykerk-Kaufman, A., and Janecke, S. U., 1987, Late Cretaceous to early Tertiary ductile deformation; Catalina-Rincon metamorphic complex, southeastern Arizona: Geology, v. 15, p. 462–465.
Champion, D. E., Howell, D. G., and Gromme, C. S., 1984, Paleomagnetic and geologic data indicating 2500 km of northward displacement for the Salinian and related terranes, California: Journal of Geophysical Research, v. 89, p. 7736–7752.
Chapin, C. E., and Cather, S. M., 1981, Eocene tectonics and sedimentation in the Colorado Plateau–Rocky Mountain area, *in* Dickinson, W. R., and Payne, W. D., eds., Relations of tectonics to ore deposits in the southern Cordillera: Arizona Geological Society Digest, v. 14, p. 173–198.
——, 1983, Eocene tectonics and sedimentation in the Colorado Plateau–Rocky Mountain area, *in* Lowell, J. D., ed., Rocky Mountain foreland basins and uplifts: Rocky Mountain Association of Geologists Guidebook, p. 33–56.
Cherven, V. B., 1984, A delta-slope-submarine fan model for the Maestrichtian part of the Great Valley sequence, southern Sacramento and northern San Joaquin Basins, California: American Association of Petroleum Geologists Bulletin, v. 67, p. 772–816.
Clarke, S. H., Jr., Howell, D. G., and Nilsen, T. H., 1975, Paleogene geography of

California, *in* Weaver, D. W., Hornaday, G. R., and Tipton, A., eds., Paleogene symposium and selected technical papers: Pacific Section, American Association of Petroleum Geologists, Society of Economic Paleontologists and Mineralogists, and Society of Exploration Geophysicists, p. 121–154.

Coney, P. J., 1972, Cordilleran tectonics and North America plate motions: American Journal of Science, v. 272, p. 603–628.

——, 1976, Plate tectonics and the Laramide orogeny, *in* Woodward, L. A., and Northrop, S. A., eds., Tectonics and mineral resources of southwestern North America: New Mexico Geological Society Special Publication 6, p. 5–10.

——, 1978, Mesozoic-Cenozoic Cordilleran plate tectonics, *in* Smith, R. B., and Eaton, G. P., eds., Cenozoic tectonics and regional geophysics of the western Cordillera: Geological Society of America Memoir 152, p. 33–50.

Coney, P. J., and Reynolds, S. J., 1977, Cordilleran Benioff zones: Nature, v. 270, p. 403–406.

Cox, B. F., 1987, Stratigraphy, depositional environments, and paleotectonics of the Paleocene and Eocene Goler Formation, El Paso Mountains, California, *in* Cox, B. F., ed., Basin analysis and paleontology of the Paleocene and Eocene Goler Formation, El Paso Mountains, California: Los Angeles, California, Pacific Section, Society of Economic Paleontologists and Mineralogists, p. 1–29.

Creasey, S. C., 1980, Chronology of intrusion and deposition of porphyry copper ores, Globe-Miami district, Arizona: Economic Geology, v. 75, p. 830–844.

Criss, R. E., and Fleck, R. J., 1987, Petrogenesis, geochronology, and hydrothermal systems of the northern Idaho batholith and adjacent areas based on $^{18}O/^{16}O$, D/H, $^{87}Sr/^{86}Sr$, K-Ar, and $^{40}Ar/^{39}Ar$ studies, *in* Vallier, T. L., and Brooks, H. C., eds., Geology of the Blue Mountains region of Oregon, Idaho, and Washington; The Idaho batholith and its border zone: U.S. Geological Survey Professional Paper 1436, p. 95–137.

Cross, T. A., 1986, Tectonic controls of foreland basin subsidence and Laramide style deformation, western United States, *in* Allen, P. A., and Homewood, P., eds., Foreland basins: International Association of Sedimentologists Special Publication 8, p. 15–39.

Davis, G. A., Anderson, J. L., Martin, D. L., Krummenacher, D., Frost, E. G., and Armstrong, R. L., 1982, Geologic and geochronologic relations in the lower plate of the Whipple detachment fault, Whipple Mountains, southeastern California; A progress report, *in* Frost, E. G., and Martin, D. L., eds., Mesozoic-Cenozoic tectonic evolution of the Colorado River region, California, Arizona, and Nevada: San Diego, California, Cordilleran Publishers, p. 409–432.

Davis, G. H., 1978, Monocline fold pattern of the Colorado Plateau, *in* Matthews, V., III, ed., Laramide folding associated with basement block faulting in the western United States: Geological Society of America Memoir 151, p. 215–233.

——, 1979, Laramide folding and faulting in southeastern Arizona: American Journal of Science, v. 279, p. 543–569.

DeCelles, P. G., 1986, Sedimentation in a tectonically partitioned, nonmarine foreland basin; The Lower Cretaceous Kootenai Formation, southwestern Montana: Geological Society of America Bulletin, v. 97, p. 911–931.

——, 1988, Lithologic provenance modeling applied to the Late Cretaceous synorogenic Echo Canyon Conglomerate, Utah; A case of multiple source areas: Geology, v. 16, p. 1039–1043.

Dickinson, W. R., 1976, Sedimentary basins developed during evolution of Mesozoic-Cenozoic arc-trench system in western North America: Canadian Journal of Earth Sciences, v. 13, p. 1268–1287.

——, 1981, Plate tectonics of the southern Cordilera, *in* Dickinson, W. R., and Payne, W. D., eds., Relations of tectonics to ore deposits in the southern Cordillera: Arizona Geological Society Digest, v. 14, p. 113–135.

Dickinson, W. R., and Snyder, W. S., 1978, Plate tectonics of the Laramide orogeny, *in* Matthews, V., ed., Laramide folding associated with basement block faulting in the western United States: Geological Society of America Memoir 151, p. 355–366.

Dickinson, W. R., Ingersoll, R. V., and Graham, S. A., 1979, Paleogene sediment dispersal and paleotectonics in northern California: Geological Society America Bulletin, v. 90, part 2, p. 1458–1528.

Dickinson, W. R., and 6 others, 1988, Paleogeographic and paleotectonic setting of Laramide sedimentary basins in the central Rocky Mountain region: Geological Society of America Bulletin, v. 100, p. 1023–1039.

Dorr, J. A., Spearing, D. R., and Steidtmann, J. R., 1977, Deformation and deposition between a foreland uplift and an impinging thrust belt; Hoback basin, Wyoming: Geological Society of America Special Paper 177, 82 p.

Dott, R. H., Jr., 1966, Eocene deltaic sedimentation at Coos Bay, Oregon: Journal of Geology, v. 74, no. 4, p. 373–420.

Ehlig, P. L., 1981, Origin and tectonic history of the basement terrane of the San Gabriel Mountains, central Transverse Ranges, *in* Ernst, W. G., ed., The geotectonic development of California: Englewood Cliffs, New Jersey, Prentice-Hall, Inc., p. 253–283.

Engebretson, D. C., Cox, A., and Thompson, G. A., 1984, Correlation of plate motions with continental tectonics; Laramide to Basin-Range: Tectonics, v. 3, p. 115–119.

Engebretson, D. C., Cox, A., and Gordon, R. G., 1985, Relative motions between oceanic and continental plates in the Pacific basin: Geological Society of America Special Paper 206, 59 p.

Ernst, W. G., 1970, Tectonic contact between the Franciscan melange and the Great Valley sequence—Crustal expression of a late Mesozoic Benioff zone: Journal of Geophysical Research, v. 75, p. 886–901.

Erslev, E. A., 1986, Basement balancing of Rocky Mountain foreland uplifts: Geology, v. 14, p. 259–262.

Ewing, T. E., 1980, Paleogene tectonic evolution of the Pacific Northwest: Journal of Geology, v. 88, p. 619–638.

Flores, R. M., and Ethridge, F. G., 1985, Evolution of intermontane fluvial systems of Tertiary Powder River basin, Montana and Wyoming, *in* Flores, R. M., and Kaplan, S. S., eds., Cenozoic paleogeography of the west-central United States: Rocky Mountain Section, Society of Economic Paleontologists and Mineralogists Rocky Mountain Paleogeography Symposium 3, p. 107–126.

Fouch, T. D., 1979, Character and paleogeographic distribution of Upper Cretaceous(?) and Paleogene nonmarine sedimentary rocks in east-central Nevada, *in* Armentrout, J. M., Cole, M. R., and Ter Best, H., Jr., eds., Cenozoic paleogeography of the western United States: Society of Economic Paleontologists and Mineralogists Pacific Coast Paleogeography Symposium 3, p. 97–111.

Fouch, T. D., Lawton, T. F., Nichols, D. J., Cashion, W. B., and Cobban, W. A., 1983, Patterns and timing of synorogenic sedimentation in Upper Cretaceous rocks of central and northwest Utah, *in* Reynolds, M. W., and Dolly, E. D., eds., Mesozoic paleogeography of the west-central United States: Society of Economic Paleontologists and Mineralogists Rocky Mountain Paleogeography Symposium 2, p. 305–336.

Gable, D. J., 1984, Geologic setting and petrochemistry of the Late Cretaceous–early Tertiary intrusives in the northern Front Range mineral belt, Colorado: U.S. Geological Survey Professional Paper 1280, 33 p.

Gabrielse, H., 1985, Major dextral transcurrent displacements along the Northern Rocky Mountain Trench and related lineaments in north-central British Columbia: Geological Society of America Bulletin, v. 96, p. 1–14.

Garcia, R., 1981, Depositional systems and their relation to gas accumulation in Sacremento Valley, California: American Association of Petroleum Geologists Bulletin, v. 65, p. 653–673.

Graham, S. A., 1978, Role of the Salinian block in the evolution of the San Andreas fault system: American Association of Petroleum Geologists Bulletin, v. 62, p. 2214–2231.

Gries, R. R., 1983a, North-south compression of Rocky Mountain foreland structures, *in* Lowell, J. D., ed., Rocky Mountain foreland basins and uplifts: Denver, Colorado, Rocky Mountain Association of Geologists Guidebook, p. 9–32.

——, 1983b, Oil and gas prospecting beneath the Precambrian of foreland thrust plates in the Rocky Mountains: American Association of Petroleum Geologists Bulletin, v. 67, p. 1–26.

——, 1985, Seismic lines in the San Luis Valley, south-central Colorado, *in* Gries, R. R., and Dyer, R. C., eds., Seismic exploration of the Rocky Moun-

tain region: Rocky Mountain Association of Geologists and Denver Geophysical Society, p. 267–274.

Hagen, E. S., Shuster, M. W., and Furlong, K. P., 1985, Tectonic loading and subsidence of intermontane basins; Wyoming foreland province: Geology, v. 13, p. 585–588.

Haggart, J. W., and Ward, P. D., 1984, Late Cretaceous (Santonian-Campanian) stratigraphy of the northern Sacramento Valley, California: Geological Society of America Bulletin, v. 95, p. 618–627.

Hail, W. J., Jr., 1965, Geology of northwestern North Park, Colorado: U.S. Geological Survey Bulletin 1188, 119 p.

—— , 1968, Geology of southwestern North Park and vicinity, Colorado: U.S. Geological Survey Bulletin 1257, 133 p.

Hamilton, W., 1969, Mesozoic California and the underflow of Pacific mantle: Geological Society of America Bulletin, v. 80, p. 2409–2430.

—— , 1978, Mesozoic terranes of the western United States, *in* Howell, D. G., and McDougall, K. A., eds., Mesozoic paleogeography of the western United States: Society of Economic Paleontologists and Mineralogists Pacific Coast Paleogeography Symposium 2, p. 33–70.

—— , 1981, Plate-tectonic mechanism of Laramide deformation, *in* Boyd, D. W., and Lillegraven, J. A., eds., Rocky Mountain foreland basement tectonics: University of Wyoming Contributions to Geology, p. 87–92.

—— , 1988, Tectonic setting and variations with depth of some Cretaceous and Cenozoic structural and magmatic systems of the western United States, *in* Ernst, W. G., ed., Metamorphism and crustal evolution of the western United States: Englewood Cliffs, New Jersey, Prentice-Hall, p. 1–40.

Hamilton, W., and Myers, W. B., 1966, Cenozoic tectonics of the western United States: Reviews of Geophysics, v. 4, p. 509–549.

Harding, L. E., and Coney, P. J., 1985, The geology of the McCoy Mountains Formation, southeastern California and southwestern Arizona: Geological Society of America Bulletin, v. 96, p. 755–769.

Haxel, G. B., and Dillon, J., 1978, The Pelona-Orocopia Schist and Vincent-Chocolate Mountains thrust system, southern California, *in* Howell, D. G., and McDougall, K. A., eds., Mesozoic paleogeography of the western United States: Pacific Section, Society of Economic Paleontologists and Mineralogists Pacific Coast Paleogeography Symposium 2, p. 453–469.

Haxel, G. B., Tosdal, R. M., May, D. J., and Wright, J. E., 1984, Latest Cretaceous and early Tertiary orogenesis in south-central Arizona; Thrust faulting, regional metamorphism, and granitic plutonism: Geological Society of America Bulletin, v. 95, p. 631–653.

Heller, P. L., and 7 others, 1986, Time of initial thrusting in the Sevier orogenic belt, Idaho-Wyoming and Utah: Geology, v. 14, p. 388–391.

Heller, P. L., Tabor, R. W., and Suczek, C. A., 1987, Paleogeographic evolution of the United States Pacific Northwest during Paleogene time: Canadian Journal of Earth Sciences, v. 24, p. 1652–1667.

Henderson, L. J., Gordon, R. G., and Engebretson, D. C., 1984, Mesozoic aseismic ridges on the Farallon plate and southward migration of shallow subduction during the Laramide orogeny: Tectonics, v. 3, p. 121–132.

Hoisch, T. D., Miller, C. F., Heizler, M. T., Harrison, T. M., and Stoddard, E. F., 1988, Late Cretaceous regional metamorphism in southeastern California, *in* Ernst, W. G., ed., Metamorphism and crustal evolution of the western United States: Englewood Cliffs, New Jersey, Prentice-Hall, p. 538–571.

Howard, K. A., John, B. E., and Miller, C. F., 1987a, Metamorphic core complexes, Mesozoic ductile thrusts, and Cenozoic detachments; Old Woman Mountains–Chemehuevi Mountains transect, California and Arizona, *in* Davis, G. H., and VandenDolder, E. M., eds., Geologic diversity of Arizona and its margins; Excursions to choice areas: Arizona Bureau of Geology and Mineral Technology Special Paper 5, p. 365–382.

Howell, D. G., and Link, M. H., 1979, Eocene conglomerate sedimentology and basin analysis, San Diego and the Southern California Borderland: Journal of Sedimentary Petrology, v. 49, p. 517–540.

Howell, D. G., Vedder, J. G., and McDougall, K., editors, 1977a, Cretaceous geology of the California Coast Ranges, west of the San Andreas fault: Pacific Section, Society of Economic Paleontologists and Mineralogists Pacific Coast Paleogeography Field Guide 2, 109 p.

Ingersoll, R. V., 1978, Submarine fan facies of the Upper Cretaceous Great Valley sequence, northern and central California: Sedimentary Geology, v. 21, p. 205–230.

—— , 1979, Evolution of the Late Cretaceous fore-arc basin, northern and central California: Geological Society of America Bulletin, v. 90, p. 813–826.

Jacobson, C. E., 1983a, Structural geology of the Pelona Schist and Vincent thrust, San Gabriel Mountains, California: Geological Society of America Bulletin, v. 94, p. 753–767.

Jacobson, C. E., Dawson, M. R., and Postlethwaite, C. E., 1988, Structure, metamorphism, and tectonic significance of the Pelona, Orocopia, and Rand Schists, southern California, *in* Ernst, W. G., ed., Metamorphism and crustal evolution of the western United States: Englewood Cliffs, New Jersey, Prentice-Hall, p. 976–997.

John, B. E., and Wooden, J. L., 1990, Petrology and geochemistry of the metaluminous to peraluminous Chemehuevi Mountains Plutonic Suite, southeastern California, *in* Anderson, J. L., ed., The nature and origin of Cordilleran magmatism: Geological Society of America Memoir 174, p. 71–98.

Johnson, R. C., 1985, Early Cenozoic history of the Uinta and Piceance Creek basins, Utah and Colorado, with special reference to the development of Eocene Lake Uinta, *in* Flores, R. M., and Kaplan, S. S., eds., Cenozoic paleogeography of the west-central United States: Rocky Mountain Section, Society of Economic Paleontologists and Mineralogists Rocky Mountain Paleogeography Symposium 3, p. 247–276.

Jordan, T. E., 1981, Thrust loads and foreland basin evolution, Cretaceous, western United States: American Association of Petroleum Geologists Bulletin, v. 65, p. 2506–2520.

Jordan, T. E., and Allmendinger, R. W., 1986, The Sierras Pampeanas of Argentina; A modern analogue of Rocky Mountain foreland deformation: American Journal of Science, v. 2;86, p. 737–764.

Kanter, L. R., Dyer, R., and Dohmen, T. E., 1981, Laramide crustal shortening in the northern Wyoming province, *in* Boyd, D. W., and Lillegraven, J. A., eds., Rocky Mountain foreland basement tectonics: University of Wyoming Contributions to Geology, v. 19, p. 135–142.

Keith, S. B., 1978, Paleosubduction geometries inferred from Cretaceous and Tertiary magmatic patterns in southwestern North America: Geology, v. 6, p. 516–521.

Keith, S. B., and 5 others, 1980, Evidence for multiple intrusion and deformation within the Santa Catalina-Rincon-Tortolita crystalline complex, *in* Crittenden, M. D., Jr., Coney, P. J., and Davis, G. H., eds., Cordilleran metamorphic core complexes: Geological Society of America Memoir 153, p. 217–267.

Kelley, V. C., 1955, Monoclines of the Colorado Plateau: Geological Society of America Bulletin, v. 66, p. 789–804.

Kleinhans, L. C., Balcells-Baldwin, E. A., and Jones, R. E., 1984, A paleogeographic reinterpretation of some middle Cretaceous units, north-central Oregon; Evidence for a submarine turbidite system, *in* Nilsen, T. H., ed., Geology of the Upper Cretaceous Hornbrook Formation, Oregon and California: Pacific Section, Society of Economic Paleontologists and Mineralogists, v. 42, p. 239–257.

Kleinspehn, K. L., 1985, Cretaceous sedimentation and tectonics, Tyaughton-Methow basin, southwestern British Columbia: Canadian Journal of Earth Sciences, v. 22, p. 154–174.

Lamerson, P. R., 1982, The Fossil basin and its relationship to the Absaroka thrust system, Wyoming and Utah, *in* Powers, R. B., ed., Geologic studies of the Cordilleran thrust belt: Rocky Mountain Association of Geologists, p. 279–340.

Lawton, T. F., 1982, Lithofacies correlations within the Upper Cretaceous Indianola Group, central Utah, *in* Nielson, D. L., ed., Overthrust belt of Utah: Salt Lake City, Utah Geological Association Publication 10, p. 199–214.

Lee, D. E., Kistler, R. W., Friedman, I., and Van Loenen, R. E., 1981, Two-mica granites of northeastern Nevada: Journal of Geophysical Research, v. 86, p. 10607–10616.

Lindsey, D. A., Johnson, B. R., and Andriessen, P.A.M., 1983, Laramide and Neogene structure of the northern Sangre de Christo Range, south-central Colorado, *in* Lowell, J. D., ed., Rocky Mountain foreland basins and uplifts: Rocky Mountain Association of Geologists Guidebook, p. 219–228.

Lipman, P. W., 1980, Cenozoic volcanism in the western United States; Implications for continental tectonics, *in* Continental tectonics: Washington, D.C., National Academy of Sciences, p. 161–174.

Lipman, P. W., and Sawyer, D. A., 1985, Mesozoic ash-flow caldera fragments in southeastern Arizona and their relation to porphyry copper deposits: Geology, v. 13, p. 652–656.

Lipman, P. W., Prostka, H. J., and Christiansen, R. L., 1971, Evolving subduction zones in the western United States, as interpreted from igneous rocks: Science, v. 174, p. 821–825.

Livaccari, R. F., Burke, K., and Sengor, A.M.C., 1981, Was the Laramide orogeny related to subduction of an oceanic plateau?: Nature, v. 289, p. 276–278.

Lund, K., and Snee, L. W., 1988, Metamorphism, structural development, and age of the continent-island arc juncture in west-central Idaho, *in* Ernst, W. G., ed., Metamorphism and crustal evolution of the western United States: Englewood Cliffs, New Jersey, Prentice-Hall, p. 296–331.

Lynn, H. B., Quam, S., and Thompson, G. A., 1983, Depth migration and interpretation of the COCORP Wind River, Wyoming, seismic reflection data: Geology, v. 11, p. 462–469.

Matthews, V., III, 1978, ed., Laramide folding associated with basement block faulting in the western United States: Geological Society of America Memoir 151, 370 p.

Miller, C. F., and Bradfish, L. J., 1980, An inner Cordilleran belt of muscovite-bearing plutons: Geology, v. 8, p. 412–416.

Miller, C. F., Howard, K. A., and Hoisch, T. D., 1982, Mesozoic thrusting, metamorphism, and plutonism, Old Woman–Piute Range southeastern California, *in* Frost, E. G., and Martin, D. L., eds., Mesozoic–Cenozoic tectonic evolution of the Colorado River region, California, Arizona, and Nevada: San Diego, California, Cordilleran Publishers, p. 561–581.

Miller, D. M., Hillhouse, W. C., Zartman, R. E., and Lanphere, M. A., 1987, Geochronology of intrusive and metamorphic rocks in the Pilot Range, Utah and Nevada, and comparison with regional patterns: Geological Society of America Bulletin, v. 99, p. 866–879.

Miller, E. L., Gans, P. B., Wright, J. E., and Sutter, J. F., 1988, Metamorphic history of the east-central Basin and Range province; Tectonic setting and relationship to magmatism, *in* Ernst, W. G., ed., Metamorphism and crustal evolution of the western United States: Englewood Cliffs, New Jersey, Prentice-Hall, p. 649–682.

Miller, F. K., and Engels, J. C., 1975, Distribution and trends of discordant ages of the pluton rocks of northeastern Washington and northern Idaho: Geological Society of America Bulletin, v. 86, p. 517–528.

Mudge, M. R., and Earhart, R. L., 1980, The Lewis thrust fault and related structures in the disturbed belt, northwestern Montana: U.S. Geological Survey Professional Paper 1174, 18 p.

Muller, J. E., and Jeletzky, J. A., 1970, Geology of the Upper Cretaceous Nanaimo Group, Vancouver Island and Gulf Islands, British Columbia: Geological Survey of Canada Paper 69-25, 77 p.

Nichols, D. J., Perry, W. J., Jr., and Haley, J. C., 1985, Reinterpretation of the palynology and age of Laramide syntectonic deposits, southwestern Montana, and revision of the Beaverhead Group: Geology, v. 13, p. 149–153.

Nilsen, T. H., 1977, Early Tertiary tectonics and sedimentation in California, *in* Nilsen, T. H., ed., Late Mesozoic and Cenozoic sedimentation and tectonics in California: Bakersfield, California, San Joaquin Geological Society Short Course, p. 86–98.

——, 1984b, Stratigraphy, sedimentology, and tectonic framework of the Upper Cretaceous Hornbrook Formation, Oregon and California, *in* Nilsen, T. H., ed., Geology of the Upper Cretaceous Hornbrook Formation, Oregon and California: Pacific Section, Society of Economic Paleontologists and Mineralogists, v. 42, p. 51–88.

——, 1987, Paleogene tectonics and sedimentation of coastal California, *in* Ernst, W. G., and Ingersoll, R. V., eds., Tertiary Basin development of coastal California: Englewood Cliffs, New Jersey, Prentice-Hall, p. 81–123.

Nilsen, T. H., and Abbott, P. L., 1981, Paleogeography and sedimentology of Upper Cretaceous turbidites, San Diego, California: American Association of Petroleum Geologists Bulletin, v. 65, p. 1256–1284.

Nilsen, T. H., and Clarke, S. H., Jr., 1975, Sedimentation and tectonics in the early Tertiary continental borderland of central California: U.S. Geological Survey Professional Paper 925, 64 p.

Nilsen, T. H., and McKee, E. H., 1979, Paleogene paleogeography of the western United States, *in* Armentrout, J. M., Cole, M. R., and Ter Best, H., Jr., eds., Cenozoic paleogeography of the western United States: Pacific Section, Society of Economic Paleotologists and Mineralogists Pacific Coast Paleogeography Symposium 3, p. 257–276.

Nolan, T. B., Merriam, C. W., and Williams, J. S., 1956, The stratigraphic section in the vicinity of Eureka, Nevada: U.S. Geological Survey Professional Paper 276, 77 p.

Pacht, J. A., 1984, Petrologic evolution and paleogeography of the Late Cretaceous Nanaimo Basin, Washington and British Columbia: Implications for Cretaceous tectonics: Geological Society of America Bulletin, v. 95, p. 766–788.

Parrish, R. R., Carr, S. D., and Parkinson, D. L., 1988, Eocene extensional tectonics and geochronology of the southern Omineca belt, British Columbia Washington: Tectonics, v. 7, p. 181–212.

Perry, W. J., Jr., and Sando, W. J., 1982, Sequence of deformation of Cordilleran thrust belt in Lima, Montana region, *in* Powers, R. B., ed., Geologic studies of the Cordilleran thrust belt: Rocky Mountain Association of Geologists, p. 137–144.

Price, R. A., 1981, The Cordilleran thrust and fold belt in the southern Canadian Rocky Mountains, *in* Price, N. J., and McClay, K. R., eds., Thrust and nappe tectonics: London, Geological Society of London, p. 427–448.

——, 1986, The southeastern Canadian Cordillera—Thrust faulting, tectonic wedging, and delamination of the lithosphere: Journal of Structural Geology, v. 8, p. 239–254.

Rehrig, W. A., and Reynolds, S. J., 1980, Geologic and geochronologic reconnaissance of a northwest-trending zone of metamorphic complexes in southern and western Arizona, *in* Crittenden, M. D., Jr., Coney, P. J., and Davis, G. H., eds., Cordilleran metamorphic core complexes: Geological Society of America Memoir 153, p. 131–157.

Rehrig, W. A., Reynolds, S. J., and Armstrong, R. L., 1987, A tectonic and geochronologic overview of the Priest River crystalline complex, northeastern Washington and northern Idaho, *in* Schuster, J. E., ed., Selected papers on the geology of Washington: Washington Division of Geology and Earth Resources Bulletin 77, p. 1–15.

Reynolds, S. J., Spencer, J. E., Richard, S. M., and Laubach, S. M., 1986, Mesozoic structures in west-central Arizona, *in* Beatty, B., and Wilkinson, P.A.K., eds., Frontiers in geology and ore deposits of Arizona and the Southwest: Arizona Geological Society Digest 16, p. 35–51.

Reynolds, S. J., Richard, S. M., Haxel, G. B., Tosdal, R. M., and Laubach, S. E., 1988, Geologic setting of Mesozoic and Cenozoic metamorphism in Arizona, *in* Ernst, W. G., ed., Metamorphism and crustal evolution of the western United States: Englewood Cliffs, New Jersey, Prentice-Hall, p. 466–501.

Robinson, G. D. Klepper, M. R., and Obradovich, J. D., 1968, Overlapping plutonism, volcanism, and tectonism in the Boulder batholith region, western Montana, *in* Coats, R. R., Hay, R. L., and Anderson, C. A., eds., Studies in volcanology: Geological Society of America Memoir 116, p. 557–576.

Royse, F., Jr., Warner, M. A., and Reese, D. L., 1975, Thrust belt structural geometry and related stratigraphic problems, Wyoming-Idaho-northern Utah: Rocky Mountain Association of Geologists 1975 Symposium, p. 41–54.

Ruppel, E. T., and Lopez, D. A., 1984, The thrust belt in southwest Montana and east-central Idaho: U.S. Geological Survey Professional Paper 1278, 41 p.

Ryder, R. T., and Scholten, R., 1973, Syntectonic conglomerates in southwestern Montana; Their nature, origin, and tectonics significance: Geological Society of America Bulletin, v. 84, p. 773–796.

Sales, J. K., 1968, Crustal mechanics of Cordilleran foreland deformation; A regional and scale-model approach: American Association of Petroleum Geologists Bulletin, v. 52, p. 2016–2044.

Schmidt, C. J., and Garihan, J. M., 1983, Laramide tectonic development of the Rocky Mountain foreland of southwestern Montana, *in* Lowell, J. D., ed., Rocky Mountain foreland basins and uplifts: Denver, Colorado, Rocky Mountain Association of Geologists Guidebook, p. 271–294.

Schmidt, C. J., and O'Neill, J. M., 1982, Structural evolution of the southwest Montana transverse zone, *in* Powers, R. B., ed., Geologic studies of the Cordilleran thrust belt: Denver, Colorado, Rocky Mountain Association of Geologists, p. 193–218.

Scholten, R., 1982, Continental subduction in the northern Rockies; A model for backarc thrusting in the western Cordillera, *in* Powers, R. B., ed., Geologic studies of the Cordilleran thrust belt: Denver, Colorado, Rocky Mountain Association of Geologists, p. 123–136.

Schwartz, R. K., 1982, Broken Early Cretaceous foreland basin in southwestern Montana; Sedimentation related to tectonism, *in* Powers, R. B., ed., Geologic studies of the Cordilleran thrust belt: Rocky Mountain Association of Geologists, p. 159–183.

Seager, W. R., and Mack, G. H., 1986, Laramide paleotectonics of southern New Mexico, *in* Peterson, J. A., ed., Paleotectonics and sedimentation in the Rocky Mountain region, United States: American Association of Petroleum Geologists Memoir 41, p. 669–685.

Shuster, R. D., and Bickford, M. E., 1985, Chemical and isotopic evidence for petrogenesis of the northeastern Idaho batholith: Journal of Geology, v. 93, p. 727–742.

Skipp, B.A.L., and Hait, M. H., Jr., 1977, Allochthons along the northeast margin of the Snake River Plain, Idaho: Wyoming Geological Association Guidebook 29, p. 499–515.

Smith, J. F., Jr., and Ketner, K. B., 1976, Stratigraphy of post-Paleozoic rocks and summary of resources in the Carlin–Pinon Range area: U.S. Geological Survey Professional Paper 867–B, 48 p.

Smith, J. G., 1965, Fundamental transcurrent faulting in the northern Rocky Mountains: American Association of Petroleum Geologists Bulletin, v. 49, p. 1398–1409.

Smith, L. N., Lucas, S. G., and Elston, W. E., 1985, Paleogene stratigraphy, sedimentation, and volcanism of New Mexico, *in* Flores, R. M., and Kaplan, S. S., eds., Cenozoic paleogeography of the west-central United States: Rocky Mountain Section, Society of Economic Paleontologists and Mineralogists Rocky Mountain Paleogeography Symposium 3, p. 293–315.

Smithson, S., Brewer, J. A., Kaufman, S., Oliver, J. E., and Hurich, C. A., 1978, Nature of the Wind River thrust, Wyoming, from COCORP deep-reflection data and from gravity data: Geology, v. 6, p. 648–652.

—— , 1979, Structure of the Laramide Wind River uplift, Wyoming, from COCORP deep reflection data and gravity data: Journal of Geophysical Research, v. 84, p. 5955–5972.

Snoke, A. W., and Miller, D. M., 1988, Metamorphic and tectonic history of the northeastern Great Basin, *in* Ernst, W. G., ed., Metamorphic and crustal evolution of the western United States: Englewood Cliffs, New Jersey, Prentice-Hall, p. 606–648.

Snyder, W. S., Dickinson, W. R., and Silberman, M. L., 1976, Tectonic implications of space-time patterns of Cenozoic magmatism in the western United States: Earth and Planetary Science Letters, v. 32, p. 91–106.

Solomon, B. J., McKee, E. H., and Andersen, D. W., 1979, Stratigraphy and depositional environments of Paleogene rocks near Elko, Nevada, *in* Armentrout, J. M., Cole, M. R., and Ter Best, H., Jr., eds., Cenozoic paleogeography of the western United States: Society of Economic Paleontologists and Mineralogists Pacific Coast Paleogeography Symposium 3, p. 75–88.

Stock, J., and Molnar, P., 1988, Uncertainties and implications of the Late Cretaceous and Tertiary position of North America relative to the Farallon, Kula, and Pacific plates: Tectonics, v. 7, p. 1339–1384.

Stone, P., Page, V. M., Hamilton, W., and Howard, K. A., 1987, Cretaceous age of the upper part of the McCoy Mountains Formation, southeastern California and southwestern Arizona and its tectonic significance; Reconciliation of paleobotanical and paleomagnetic evidence: Geology, v. 15, p. 561–564.

Surdam, R. C., and Stanley, K. O., 1980, The stratigraphic and sedimentologic framework of the Green River Formation, Wyoming, *in* Hollis, S., ed., Stratigraphy of Wyoming: Wyoming Geological Association, Annual Field Conference Guidebook, p. 205–221.

Suttner, L. J., Schwartz, R. K., and James, W. C., 1981, Late Mesozoic to early Cenozoic foreland sedimentation in southwest Montana, *in* Tucker, T. E., ed., Southwest Montana: Montana Geological Society Field Conference Guidebook, p. 93–103.

Tabor, R. W., Haugerud, R. A., Miller, R. B., Brown, E. H., and Babcock, R. S., 1989, Accreted terranes of the North Cascades Range, Washington: International Geological Congress, Field Trip T307.

Titley, S. R., 1982, Geologic setting of porphyry copper deposits, *in* Titley, S. R., ed., Advances in geology of the porphyry copper deposits southwestern North America: Tucson, University of Arizona Press, p. 37–58.

Todd, V. R., Erskine, B. G., and Morton, D. M., 1988, Metamorphic and tectonic evolution of the northern Peninsular Ranges batholith, southern California, *in* Ernst, W. G., ed., Metamorphism and crustal evolution of the western United States: Englewood Cliffs, New Jersey, Prentice-Hall, p. 894–937.

Tosdal, R. M., 1986, Mesozoic ductile deformations in the southern Dome Rock Mountains, northern Trigo Mountains, Trigo Peaks and Livingston Hills, southwestern Arizona, and Mule Mountains, southeastern California, *in* Beatty, B., and Wilkinson, P.A.K., eds., Frontiers in geology and ore deposits of Arizona and the Southwest: Arizona Geological Society Digest, v. 16, p. 62–71.

Tweto, O., 1975, Laramide (Late Cretaceous–early Tertiary) orogeny in the southern Rocky Mountains, *in* Curtis, B. F., ed., Cenozoic history of the southern Rocky Mountains: Geological Society of America Memoir 144, p. 1–44.

Tysdal, R. G., Marvin, R. F., and DeWitt, E., 1986, Late Cretaceous stratigraphy, deformation, and intrusion in the Madison Range of southwestern Montana: Geological Society of America Bulletin, v. 97, p. 859–868.

Underwood, M. B., 1983, Depositional setting of the Paleogene Yager Formation, northern Coast Ranges of California, *in* Larue, D. K., and Steel, R. J., eds., Cenozoic marine sedimentation, Pacific margin, U.S.A.: Pacific Section, Society of Economic Paleontologists and Mineralogists, v. 28, p. 81–101.

Villien, A., and Kligfield, R. M., 1986, Thrusting and synorogenic sedimentation in central Utah, *in* Peterson, J. A., ed., Paleotectonics and sedimentation in the Rocky Mountain region, United States: American Association of Petroleum Geologists Memoir 41, p. 281–307.

Yeats, R. S., 1968, Southern California structure, sea-floor spreading, and history of the Pacific Basin: Geological Society of America Bulletin, v. 79, p. 1693–1702.

Yeend, W. E., 1974, Gold-bearing gravels of the ancestral Yuba River, Sierra Nevada, California: U.S. Geological Survey Professional Paper 772, 44 p.

Young, R. A., and Hartman, J. H., 1984, Early Eocene fluviolacustrine sediments near Grand Canyon, Arizona; Evidence for Laramide drainage across northern Arizona into southern Utah: Geological Society of America Abstracts with Programs, v. 16, p. 703.

Manuscript Accepted by the Society October 12, 1989

ACKNOWLEDGMENTS

We are indebted to many colleagues who contributed data, discussed concepts, and otherwise aided in the formulation of this chapter. In particular, Richard Tosdal and Gordon Haxel assisted as contributors and reviewers. Thoughtful reviews of parts of this chapter were provided by: R. W. Allmendinger, M. D. Carr, D. S. Cowan, B. F. Cox, J. E. Harrison, K. A. Howard, F. K. Miller, S. S. Oriel, W. J. Perry, Jr., J. C. Reed, Jr., E. T. Ruppel, B.A.L. Skipp, Paul Stone, and R. W. Tabor. We especially appreciate reviews of the entire manuscript by D. L. Blackstone, Jr., R. A. Haugerud, P. W. Lipman, A. R. Palmer, S. J. Reynolds, and R. E. Wells.

Printed in U.S.A.

The Geology of North America
Vol. G-3, The Cordilleran Orogen: Conterminous U.S.
The Geological Society of America, 1992

Chapter 7

Post-Laramide geology of the U.S. Cordilleran region

Robert L. Christiansen
U.S. Geological Survey, 345 Middlefield Road, Menlo Park, California 94025
Robert S. Yeats
Department of Geosciences, Oregon State University, Corvallis, Oregon 97331

With contributions by
Stephan A. Graham
Department of Applied Earth Sciences, Stanford University, Stanford, California 94305
Wendy A. Niem and Alan R. Niem
Department of Geosciences, Oregon State University, Corvallis, Oregon 97331
Parke D. Snavely, Jr.
U.S. Geological Survey, Menlo Park, California 94025

INTRODUCTION

Before the time of the Laramide orogeny, an active orogenic and magmatic system was more or less continuous along the continental margin of western North America and had long dominated the Cordileran geologic framework; Laramide events reflected a major break in that continuity and were unusual in several respects. Contractional Laramide orogenesis affected a very wide zone, with deformation and foreland uplift extending nearly to the middle of the continent. Laramide magmatism, too, although discontinuous along strike of the orogenic region, extended locally far eastward. These events may have occurred in response to rapid westward drift of the North American Plate and extreme flattening of the Farallon subduction zone (*Dickinson and Snyder, 1978*).

Through most of the Cordilleran region, the main tectonic transition of post-Laramide time has been to an extensional or strike-slip regime, but that transition has occurred at different times and to different degrees in different parts of the region. In some measure, the transition was related to the end of plate convergence along a sector of the continental margin and formation of the San Andreas transform system, but significant extension in the continental interior began long before establishment of the present transform system.

In many, though not all parts of the region, a major phase of the long tectonic transition occurred in the middle part of the Miocene (although not necessarily during the time specifically designated as middle Miocene—16.6 to 11.2 Ma—in the time scale adopted for this series of publications). We use the term "mid-Miocene" in order to discuss the events of this phase without frequent reference to specific times that vary from place to place; the term is deliberately unspecific and depends on local context. Similarly, we use the term "mid-Tertiary" to refer to the span of time between the beginning of significant post-Laramide events—which occurred in various areas as early as middle Eocene to as late as early Miocene—and the mid-Miocene.

The principal post-Laramide tectonic elements of the Cordilleran region have been (1) an active continental margin, driven by a subduction zone whose character, extent, and continuity have changed markedly with time; (2) a near-coastal region of transform tectonics that, with time, has replaced portions of the subduction system; (3) an early post-Laramide Cordilleran interior characterized by regional magmatic activity and by relatively narrow zones of significant tectonic extension; (4) a wide interior region of distributed tectonic extension and magmatism; and (5) tectonically stable regions that had been relatively undeformed but subsiding forelands east of the east-facing late Mesozoic and earliest Tertiary orogenic belts. Transitions between the corresponding tectonic regimes did not occur synchronously throughout the region as a whole, and spatial boundaries between the tectonic elements have changed with time as well. Nevertheless, these principal elements form the framework for discussion in this chapter.

In using this framework, the presentation, except for a final summary, is not strictly chronological but, instead, focuses on the coherent evolution of major geologic systems during defined increments of time. The previous chapter (D. M. Miller and others, this volume) deals with the Laramide orogeny and associated

Christiansen, R. L., and Yeats, R. S., 1992, Post-Laramide geology of the U.S. Cordilleran region, *in* Burchfiel, B. C., Lipman, P. W., and Zoback, M. L., eds., The Cordilleran Orogen: Conterminous U.S.: Boulder, Colorado, Geological Society of America, The Geology of North America, v. G-3.

events that continued into the early Tertiary; thus, we treat only briefly here certain aspects of Paleocene and early Eocene geology that bear on both Laramide and post-Laramide geologic development. Similarly, because other volumes (*Ruddiman and Wright, 1987*; *Morrison, 1991*) discuss thoroughly the surficial geologic features of Quaternary time, they are treated here only incidentally in relation to other topics. In this chapter, for each major time period we generally discuss coastal regions first, proceeding from north to south, then discuss areas more toward the continental interior. We also generally discuss stratigraphy and sedimentary geology first and then turn toward magmatic systems. Our primary emphasis is on description and regional comparisons; where an interpretive emphasis is explicit, it focuses on general concepts that can help in visualizing and integrating the complex geologic evolution of the Cordillera.

We generally avoid much discussion of igneous and metamorphic petrogenesis. For descriptions, the chemical compositions of igneous rocks are designated by the names of common volcanic rocks of similar composition. Where any confusion might exist about the most appropriate name for a given composition, most names are taken from the compositional fields designated by *Cox and others* (*1979*) on a plot of total alkalies versus SiO_2; in some instances, names are modified according to the ratio of K_2O to Na_2O, following the same source. (The additional name rhyodacite is also used for compositions overlapping parts of the dacite and rhyolite fields of Cox and others.) This usage should be understood as an abbreviated method of comparing compositions, not as an attempt to rename or reclassify the rocks themselves, as described by the authors of the works cited.

SECTION AUTHORSHIP AND ACKNOWLEDGMENTS

Most of this chapter was written by the two primary authors. In general, Yeats drafted the sections on geology of the coastal region (except for those written by separately identified contributors), and Christiansen drafted those on geology of the Cordilleran continental interior and on magmatic evolution.

The entire chapter was reviewed by Wes Hildreth, Allison R. Palmer, Randall R. Parrish, John H. Stewart, and Mary Lou Zoback, to whom we are deeply grateful. We also acknowledge helpful comments from John T. Hagstrum, David G. Howell, Benjamin M. Page, James G. Smith, Peter Ward, and Ray E. Wells. Although we benefitted greatly from the help of each of these individuals, they are not to blame for errors or faulty judgments in the chapter.

EARLY AND MID-TERTIARY OCEANIC REALM AND CONTINENTAL MARGIN

A profound change occurred in middle Eocene time along the entire continental margin of the United States Cordilleran region, a change marked in each area by an angular unconformity or a depositional hiatus. The depositional sequences that followed typically were more provincial than the preceding Laramide sequences, occurring in separate depositional basins or continental-margin prisms. Exceptions are the core rocks of the Olympic Mountains and the Franciscan assemblage of the coastal region of northern California, where subduction-related tectonism continued into mid-Tertiary time. In Oregon, a forearc sequence built out over deformed Paleocene and early Eocene strata on the northern margin of the Klamath Mountains. In coastal California, several local basins developed in a continental borderland setting similar to the present-day borderland off southern California. On the western edge of the Cascades, Sierra Nevada, and Peninsular Ranges, Eocene fluvial and deltaic sequences graded westward into deep-sea fans.

LARAMIDE COLLISIONS AND SETTING OF THE CONTINENTAL MARGIN

During the middle Cretaceous, the successive accretion of exotic terranes in the Pacific Northwest was accompanied by contractional deformation in the San Juan Islands, North Cascade Range, Klamath Mountains, and Blue Mountains (Fig. 1). That deformation continued into the early Tertiary, concluding around 52 to 50 Ma (*Ryberg, 1984*). In the Klamath Mountains, Paleocene oceanic basalt is overlain by submarine fan- and slope-facies strata, and shallow-marine to nonmarine equivalents of these strata rest directly on basement rocks. Between the Klamath Mountains and the Canadian border, there is a complete absence of any sedimentary rocks between the Cretaceous and the Eocene; no depositional contact is exposed between the oceanic basalts of Paleocene and Eocene age in the Coast Range and older rocks. These oceanic basalts, forming the basement of an allochthonous terrane (Figs. 1, 2) appear to be seamounts that are youngest at the latitude of the Columbia River (48 Ma) and systematically older south to the Klamath Mountains and north to Vancouver Island (62 Ma; *Duncan, 1982*); the youngest flows are locally subaerial. These basalts accreted to the continent prior to deposition of the Eocene Tyee Formation of Oregon and correlative strata in Washington. During the time of this accretion of oceanic basalts, displacement was active on several strike-slip faults; the best known is the north-south Straight Creek fault of the North Cascade Range of Washington, on which strike-slip was largely completed by middle Eocene time.

In California, the Great Valley forearc basin began to evolve in the latest Jurassic; throughout the rest of the Mesozoic the basin filled as the axis of Sierran arc magmatism migrated eastward and the Franciscan subduction complex migrated westward, widening the basin with time (*Ingersoll, 1982*). By the end of the Cretaceous, the basin was nearly filled, and Sierra-derived sediment traversed it to be incorporated in the subduction complex to the west. In the Paleogene, the Great Valley region continued as a forearc shelf showing a diminished influence of magmatism, which had migrated still farther east (Dickinson and others, 1979); contributions from Sierran and Klamath sources were distributed in part by submarine canyons. The coastal belt of the Franciscan assemblage, like the Great Valley basin, repre-

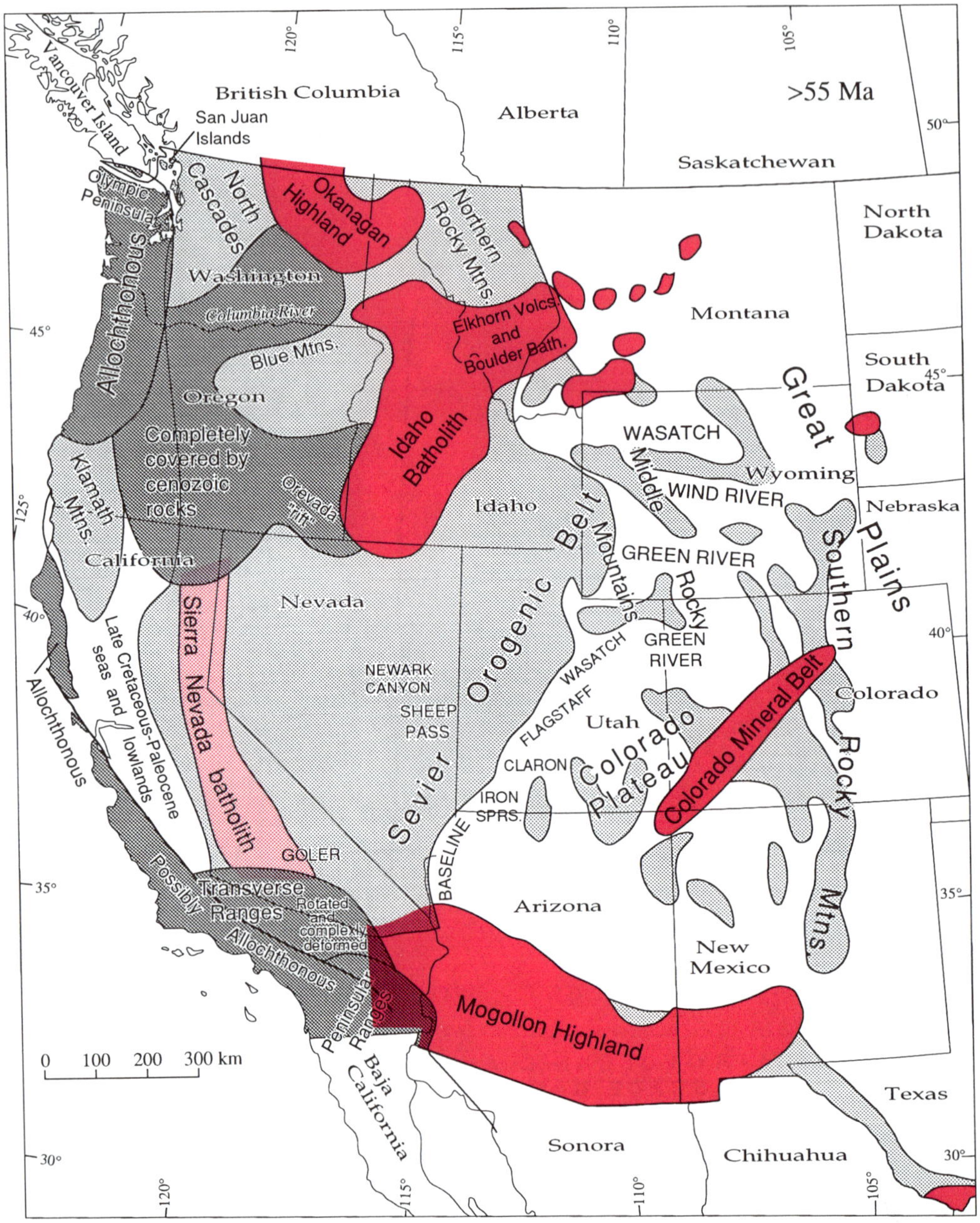

Figure 1. Index map of the United States Cordilleran region showing paleotectonic framework of the continental interior at the end of Laramide deformation (about 55 Ma). No features are shown in gray areas near the continental margin labelled "allochthonous" (accreted during or after Laramide time), in the area of southern California that was variably rotated and complexly deformed in post-Laramide time, and in areas of the Pacific Northwest so completely covered by Cenozoic rocks that geology of Laramide age cannot be interpreted. Light red, Cretaceous (>70 Ma) batholithic terrane, in which more than 50 percent of the basement lithologies are plutonic rocks; red pattern, Laramide (70 to 55 Ma) metamorphic, plutonic, and volcanic terranes; stipple, Late Cretaceous to early Eocene mountain belts and uplifts. Names in bold capitals indicate general locations of selected depositional basins. Later Cenozoic rotations and displacements not restored.

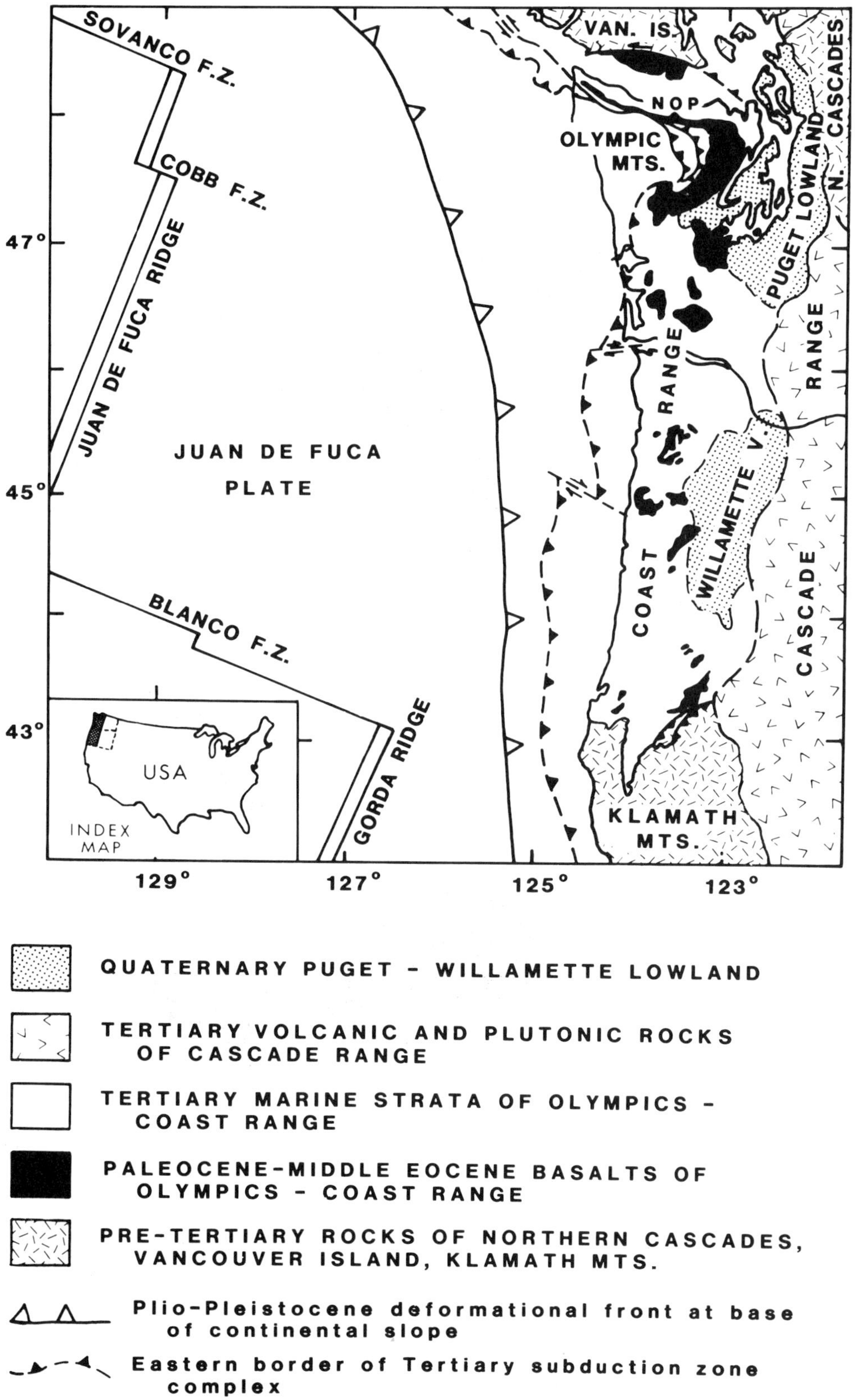

Figure 2. Generalized tectonic and geologic map of western Washington, Oregon, and the adjacent offshore area. Spreading ridges and fracture zones shown in their present positions. NOP, Northern Olympic Peninsula. Modified from *Duncan (1982)* and Snavely (1987).

sents a continuation of subduction-related processes into the Cenozoic (*Bachman and others, 1984*) with an increasing amount of accompanying strike slip (*Blake and others, 1985b*). In the central Coast Range west of the Salinian block, Upper Cretaceous strata, older Franciscan rocks, and the Coast Range ophiolite are involved in major thrusting, the age of which is uncertain because no lower Tertiary strata are known. Within the Salinian block and in the Transverse Ranges (Fig. 1), Upper Cretaceous and Paleocene rocks are found in local basins overlying crystalline bedrock unconformably. On the western margin of the Peninsular Ranges, a coastal Upper Cretaceous sequence derived from the Peninsular Ranges was deposited on a high-relief surface and deepened westward. The southwestern Transverse Ranges represent the northwestern continuation of the Peninsular Ranges, offset and rotated clockwise in the Miocene and containing evidence of thin-skinned thrusting in Paleocene time (Yeats, 1987) coeval with emplacement in the central Transverse Ranges of high-grade crystalline basement over the low-grade Pelona Schist along the Vincent thrust (see Fig. 14; *Ehlig, 1981*).

UPPER EOCENE TO LOWER MIOCENE CONTINENTAL-MARGIN SEQUENCES

The submarine basalts of the Coast Range of Washington and Oregon had already accreted to the continent when the Tyee Formation of Oregon and correlative units in Washington were deposited, and the Salinian block was already emplaced against North America, either along a proto–San Andreas fault or as a far-traveled allochthon, when the local borderland basins of a central Coast Ranges of California were formed. The following sections describe these basins up to the time of major reorganization accompanying contact of the Pacific Plate with the North America Plate in the Miocene. To discuss these basins, it is first necessary to recognize later horizontal displacements and clockwise rotations imposed by the San Andreas and related fault systems, extension of the Basin and Range Province, and possible additional terrane accretion as recently as Miocene time.

Western Washington–Oregon coastal sequence

By Wendy A. Niem, Alan R. Niem, and Parke D. Snavely, Jr.

In middle Eocene time, rapid subsidence of the recently accreted Paleocene to middle Eocene seamount and oceanic basalt terrane (the Metchosin Volcanics of Vancouver Island, Crescent Formation of the Olympic Mountains, and Siletz River Volcanics of coastal Oregon) created a 640-km-long forearc or marginal basin (Heller and Ryberg, 1983; Snavely, 1987). This linear depression extended from the pre-Tertiary Klamath Mountains northward to the North Cascades and Vancouver Island (Fig. 2); it included the present Coast Range of Washington and Oregon, Willamette Valley–Puget Lowland, and northern Olympic Peninsula. To the east, the marine strata of the forearc interfinger with and are overlain by nonmarine Tertiary and Quaternary calc-alkalic volcanic and volcaniclastic rocks of a magmatic arc. The western margin of the middle Eocene basin originally may have been partly controlled by an accretionary ridge of Paleocene to lower Eocene strata, later modified by dextral movement on the Fulmar fault (Fig. 3; *Snavely, 1984,* 1987). The depression is now bounded on the west by the middle Eocene to late Miocene subduction complex in the core of the Olympic Mountains and on the outer continental shelf and upper slope.

Micaceous lithic-feldspathic sands and silts were supplied at a high rate to the southern and southeastern part of the forearc basin from the uplifted Klamath Mountains and from a more distant continental-plutonic source, probably the Idaho batholith (Fig. 1; Chan and Dott, 1983; Heller and others, 1985). This large influx of mica-rich sediment produced the 2,000-m-thick Tyee delta (Fig. 3) in the southern Oregon Coast Range (including the Flournoy Formation of E. M. Baldwin, 1974, 1975). Turbidites cascading off the front of the Tyee delta dispersed sediment down the now north-trending axis of the basin (Snavely and others, 1964; Lovell, 1969) to form an oblong sand-rich submarine fan (Chan and Dott, 1983, 1986) or ramp (Heller and Dickinson, 1985). Tyee fan turbidites grade northward into bathyal hemipelagic siltstones and basalt breccias, lapping onto and diverted around Siletz River volcanic highs in northwestern Oregon (Snavely and others, 1964). Relatively undeformed Tyee strata overlap the imbricate thrust boundary between the Klamath pre-Tertiary terrain and the accreted Eocene oceanic basalts and lithic sandstones of the Coast Range, indicating that accretion was completed at the southern end of the Oregon Coast Range by about 50 Ma (Heller and Ryberg, 1983; Wells and others, 1984; *Snavely, 1984,* 1987; *Niem and Niem, 1990*). Paleomagnetic studies show that these postaccretion middle Eocene strata and the underlying accreted basaltic basement have been rotated as much as 72° clockwise since the middle Eocene through a combination of dextral shear between North America and north-moving oceanic plates and of basin-range extension (Simpson and Cox, 1977; Wells, 1984; Wells and Heller, 1988). The facies relations and paleomagnetic data, together with isotopic and petrographic characteristics (*Peterman and others, 1981*; Heller and Ryberg, 1983; Heller and others, 1985), show that, prior to rotation, the Oregon part of the marginal basin during the middle Eocene trended northwest and was closer to the Blue Mountains of eastern Oregon, the Idaho batholith, and the Challis volcanic belt of northern Washington, Idaho, and western Montana (*Dickinson, 1979*; Wells, 1984). To maintain geographic familiarity, however, the basin is illustrated in its post-rotation position in this chapter (Fig. 3).

The Washington part of the marginal basin formed in late middle to late Eocene time during subsidence of the early and middle Eocene oceanic basaltic crust, at least partly synchronous with and after its accretion to the pre-Tertiary continental crust of Vancouver Island and western Washington (*Muller, 1983*; Wells and others, 1984; Wells and Coe, 1985; Snavely, 1987). The northern boundary of the basin on southern Vancouver Island

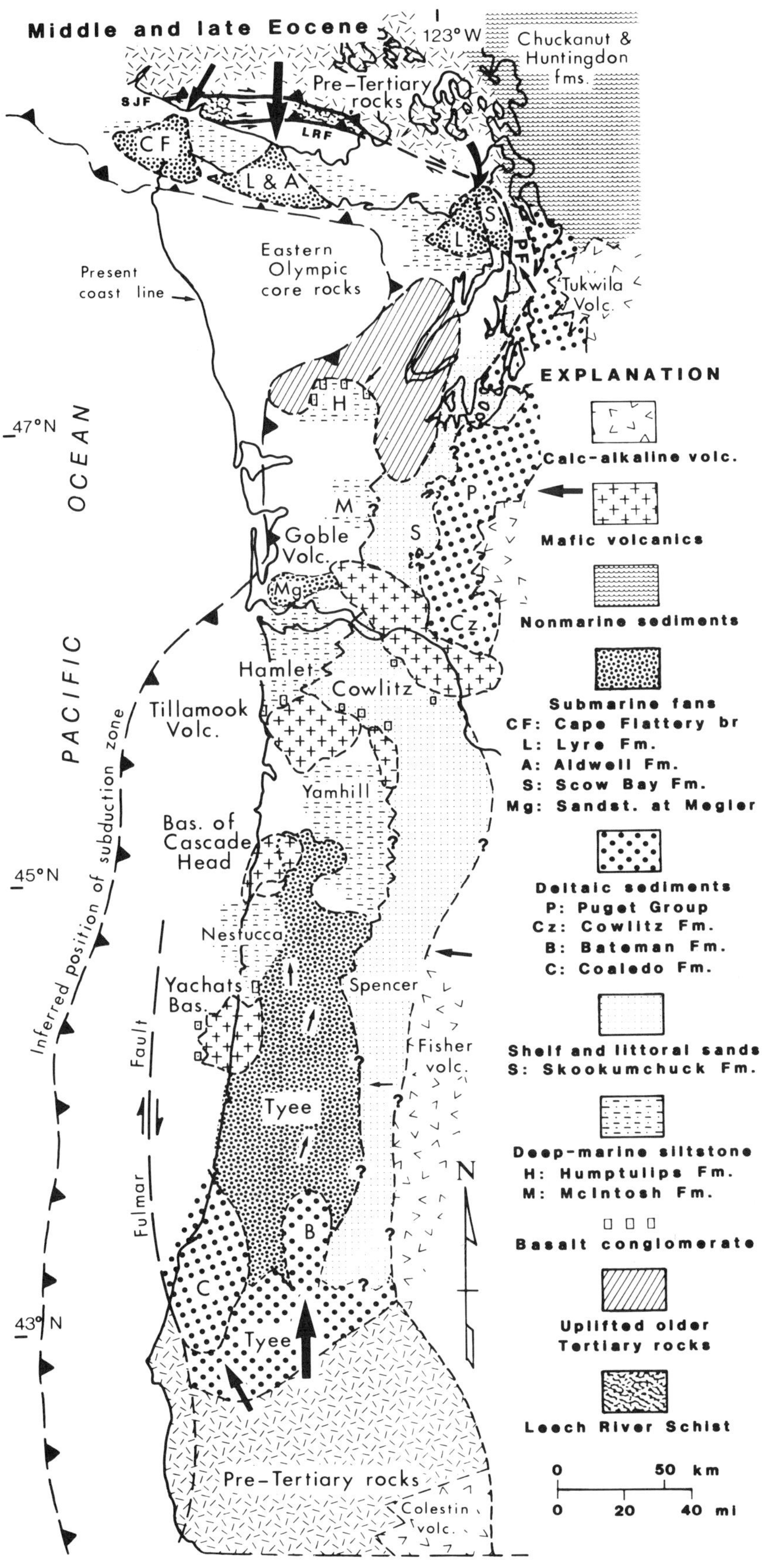

Figure 3. Paleogeographic map of western Washington and Oregon in middle and late Eocene time (52 to 39 Ma). The area labeled "Chuckanut and Huntingdon formations" in northern Washington refers to sediments deposited in nonmarine fluvial environments. Post-Eocene rotations not restored. Arrows indicate sediment dispersal directions. The Hamlet formation was informally named by Niem and Niem (1985). LRF, Leech River fault; NOP, northern Olympic Peninsula; PF, "Puget fault" of S. Y. Johnson (1985b); SJF, San Juan fault.

and the adjacent North Cascades is tectonic and may have been a rift or dextral transform margin on the San Juan–Puget fault system (SJF anad PF, Fig. 3; Johnson, 1985b). *Johnson* (*1982*, 1984b, 1985a) suggests that thick nonmarine arkosic sandstone and conglomerate of the Eocene Chuckanut Formation on the northwest flank of the pre-Tertiary North Cascades (Fig. 4) were deposited on North American continental crust in pull-apart basins created by movement along the San Juan–Puget and adjacent dextral faults during oblique plate convergence. Slope turbidites, mudstones, and lithic outer fans of the middle Eocene Aldwell Formation (Brown and others, 1960; *Marcott, 1984*), Scow Bay Formation (*Allison, 1959*; *Armentrout and Berta, 1977*; *Melim, 1984*), and sandstone at Megler (Wells, 1989; Wells and Coe, 1985) were deposited in the developing western Washington marginal basin (Figs. 3, 4). Suturing of the oceanic basalts to the continental crust certainly was completed prior to 44 Ma; isotopically dated felsic rocks of that age, interpreted by Snavely (1987) to be related to development of the continental-margin arc, intrude the early Eocene basalt of the Crescent Formation west of Bremerton, Washington, and at Striped Peak on the Strait of Juan de Fuca. Clasts of these felsic igneous rocks and of underlying Crescent basalt in the Aldwell Formation may reflect continued folding, thrusting, and local uplift within the marginal basin (Snavely, 1983). Micaceous arkosic sandstone and siltstone of the middle Eocene lower part of the Puget Group near Seattle (*Vine, 1962*; *Gard, 1968*; Buckovic, 1979) and the McIntosh Formation of southwestern Washington (Snavely and others, 1958) were derived from the North Cascades and Rocky Mountains. These sediments (Figs. 3, 4) were deposited on a broad coastal plain and in deltaic and shallow-marine conditions in the northeastern part of the marginal basin.

Several periods of igneous activity in the forearc basin followed accretion of the oceanic-basalt basement. Middle to late Eocene tholeiitic to alkalic basalt, mafic andesite, and minor dacite in the Oregon Coast Range comprise subaerial and submarine lavas, breccias, lapilli tuffs, and associated intrusions from several eruptive centers (Figs. 3, 4) and interfinger with minor volcanic conglomerate and sandstone (Snavely and MacLeod, 1974; Wells, 1981; Wells and others, 1983; *Snavely and Wells, 1984*; Niem and Niem, 1985). Oligocene igneous bodies throughout the central Oregon Coast Range include 33-Ma nepheline syenite sills and stocks, flows and intrusions of camptonite, and thick 30-Ma ferrogabbro sills (Snavely and Wagner, 1961; MacLeod and Snavely, 1973; *Loeschke, 1979*).

Throughout the forearc basin, except adjacent to the active volcanic centers, regional late middle to late Eocene downwarp-

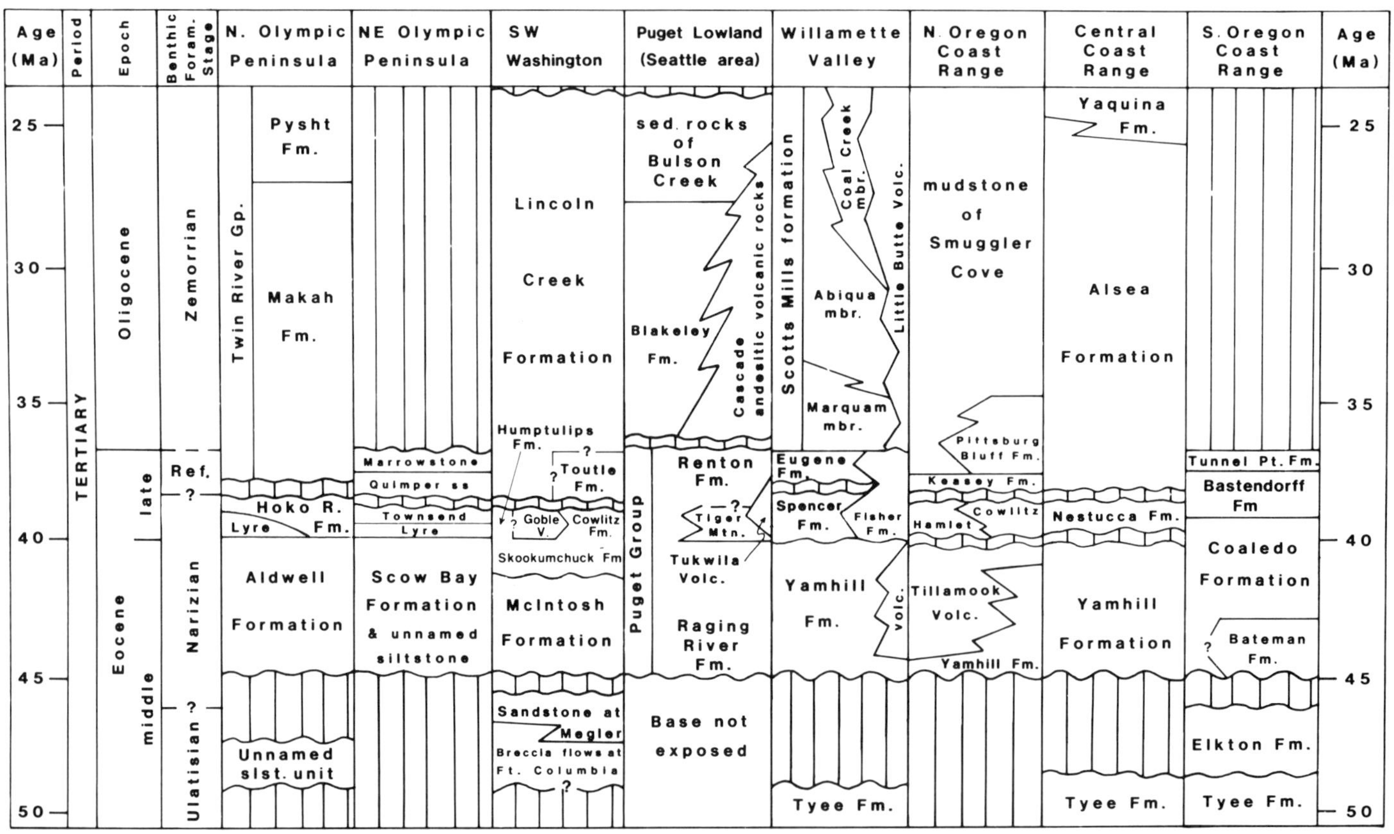

Figure 4. Correlation chart of principal middle Eocene to Oligocene formations in western Washington and Oregon. Slightly modified from Snavely (1987).

ing resulted in increased water depths. Thin-bedded bathyal siltstone and mudstone deposited in the deeper parts of the basin in the northeastern Olympic Peninsula, southwestern Washington, and northern Oregon (Fig. 3) include the Humptulips, upper part of the McIntosh, and Yamhill Formations (Snavely and others, 1951; *Baldwin and others, 1955*; Wells, 1989; Spencer, 1983; Rau, 1984, 1986).

Coal-bearing sediments along the eastern and southern margins of the forearc basin formed a broad low-lying coastal plain in a near-tropical climate (Fig. 3; Snavely and Wagner, 1963; Dott, 1966; *Wolfe and Hopkins, 1967*; Axelrod, 1968; Baldwin, 1974; Dott and Bird, 1979). Along the southern margin of the basin, arkosic and volcanic sediments were dispersed north-northwest from a Klamath Mountains source on the south and the developing volcanic-arc source on the southeast (Dott, 1966; *Ryberg, 1978*; Dott and Bird, 1979; Heller and Ryberg, 1983). This sequence of strata (Fig. 4) records progressive middle to late Eocene shoaling from upper bathyal fine-grained turbidites and sea gullies of the Elkton Formation and Lorane Siltstone (*Gandera, 1977*; Dott and Bird, 1979) to the shallow-marine and deltaic coal-bearing Bateman and Coaledo Formations (Dott, 1966; E. M. Baldwin, 1974; *Rooth, 1974*; *Ryberg, 1978*; Chan and Dott, 1986).

In late Eocene time, continued underthrusting and related transform movement along the sinistral Leech River fault (LRF, Fig. 3) on southern Vancouver Island uplifted the pre-Tertiary terranes of the northern and northeastern margins of the forearc basin (Snavely, 1983) and juxtaposed the Leech River Schist complex (Fairchild and Cowan, 1982). The schists were metamorphosed about 40 Ma, prior to emplacement at their present structural position (*Fairchild, 1979*). Large proximal-fan cones and channels (Figs. 3, 4)—the Lyre Formation of *Brown and others (1956)* and Ansfield (1972) and the Cape Flattery Breccia and Hoko River Formation (*Snavely and others, 1980b*; *DeChant, 1989*)—contain abundant detritus from uplifted Leech River Schist and older metamorphic and igneous rocks of Vancouver Island. They spread southward into the deep marginal basin on the flank of the northern Olympic Peninsula, possibly along a major submarine fault escarpment. That basin extends northwestward offshore between the Prometheus gravity high (associated with basalts of the Crescent Formation) and Vancouver Island (*MacLeod and others, 1977*; *Snavely and others, 1980b*; Snavely, 1987).

In Washington, 3,500 m of arkosic-micaceous sandstone of the upper part of the Puget Group, including the Renton Formation of late Eocene age (Figs. 3, 4), was deposited as an extensive deltaic system by streams draining granitic areas to the east (Buckovic, 1979; *Frizzell, 1979*). Two calc-alkalic volcanic centers within the upper part of the Puget Group, the Tukwila and Northcraft Formations, erupted upon this coastal plain (*Beikman and others, 1967*; Buckovic, 1979; *Hagen, 1987*). On the southern margin of the broad deltaic system, micaceous arkosic sandstone and coal beds (Skookumchuck and Cowlitz Formations, Fig. 3) accumulated in nearshore to delta-plain environments and intertongue with marine sandstone and siltstone of the McIntosh Formation to the west (Henriksen, 1956; Snavely and others, 1958; *Livingston, 1966*).

The coal-bearing Cowlitz Formation of southwestern Washington grades laterally into micaceous arkosic nearshore to wave-dominated shelf sandstone and overlying deep-water mudstone in northern Oregon (*Van Atta, 1971*; Armentrout and others, 1983; *Bruer and others, 1984*; *Farr, 1989*; *Berkman, 1990*). Natural gas is produced from Cowlitz sandstone at the Mist field in northwest Oregon, the only commercial field in the Pacific Northwest (*Newton, 1979*; Armentrout and Suek, 1985). The sandstone pinches out into deep-marine siltstone to the west (Figs. 3, 4), including the Nestucca Formation of Wells and others (1983), Hamlet mudstone of Niem and Niem (1985) and *Rarey (1986)*, and other units (Wells, 1981). A lateral- and age-equivalent micaceous arkosic shelf sandstone (Spencer Formation) extends south through the Willamette Valley (*Schlicker, 1962*; *Al-Azzaby, 1980*; *Baker, 1988*). South of Eugene, nearshore to intertidal sandstone, coal, and carbonaceous siltstone of the Spencer intertongue with and are overlain by the volcanic-rich Fisher Formation, which crops out mainly in the western Cascades (Vokes and others, 1951; Snavely and Wagner, 1963; *Gandera, 1977*).

Widespread readjustments to the underthrusting throughout the forearc basin at approximately 39 Ma created a regional angular unconformity at the base of uppermost Eocene strata (Fig. 4) throughout the basin (Snavely, 1987; *Snavely and others, 1980c*). This tectonism segmented the forearc basin into several shallow basins that deepened westward (*Snavely and others, 1980c*). Locally derived basaltic conglomerate and sandstone from uplifted parts of the Coast Range and eastern Olympic Mountains outline the developing basins, such as Grays Harbor basin (Fig. 5). The offshore continuation of these basins is reviewed by Snavely (1987).

Volcanism became more widespread in the latest Eocene and Oligocene, erupting andesitic, dacitic, and less abundant basaltic lavas and pyroclastics in the Western Cascades (Fig. 5) and bringing a marked change in the composition of sediment deposited in the forearc basin (Snavely and Wagner, 1963; *Kadri and others, 1983*; *Niem and Niem, 1984,* 1985). In the Grays Harbor basin, this change in composition is pronounced between the massive, tuffaceous shallow-marine to deep-marine siltstone of the upper Eocene to Oligocene Lincoln Creek Formation and the underlying nontuffaceous arkosic and micaceous older Eocene Cowlitz and Skookumchuck Formations and the upper part of the Puget Group (*Beikman and others, 1967*; *Armentrout, 1973*; *Armentrout and others, 1980*; *Karachewski, 1983*). Along the eastern strandline of the basin (Fig. 5), nonmarine to shallow-marine volcaniclastics of the Toutle Formation interfinger with water-laid lapilli tuff and basaltic flows from the Western Cascades (Roberts, 1958). The abundant volcanic materials from the adjacent volcanic arc either buried the source terrane of the older arkosic sediments or diverted the streams that had transported clean arkosic micaceous sands of the Cowlitz Formation and

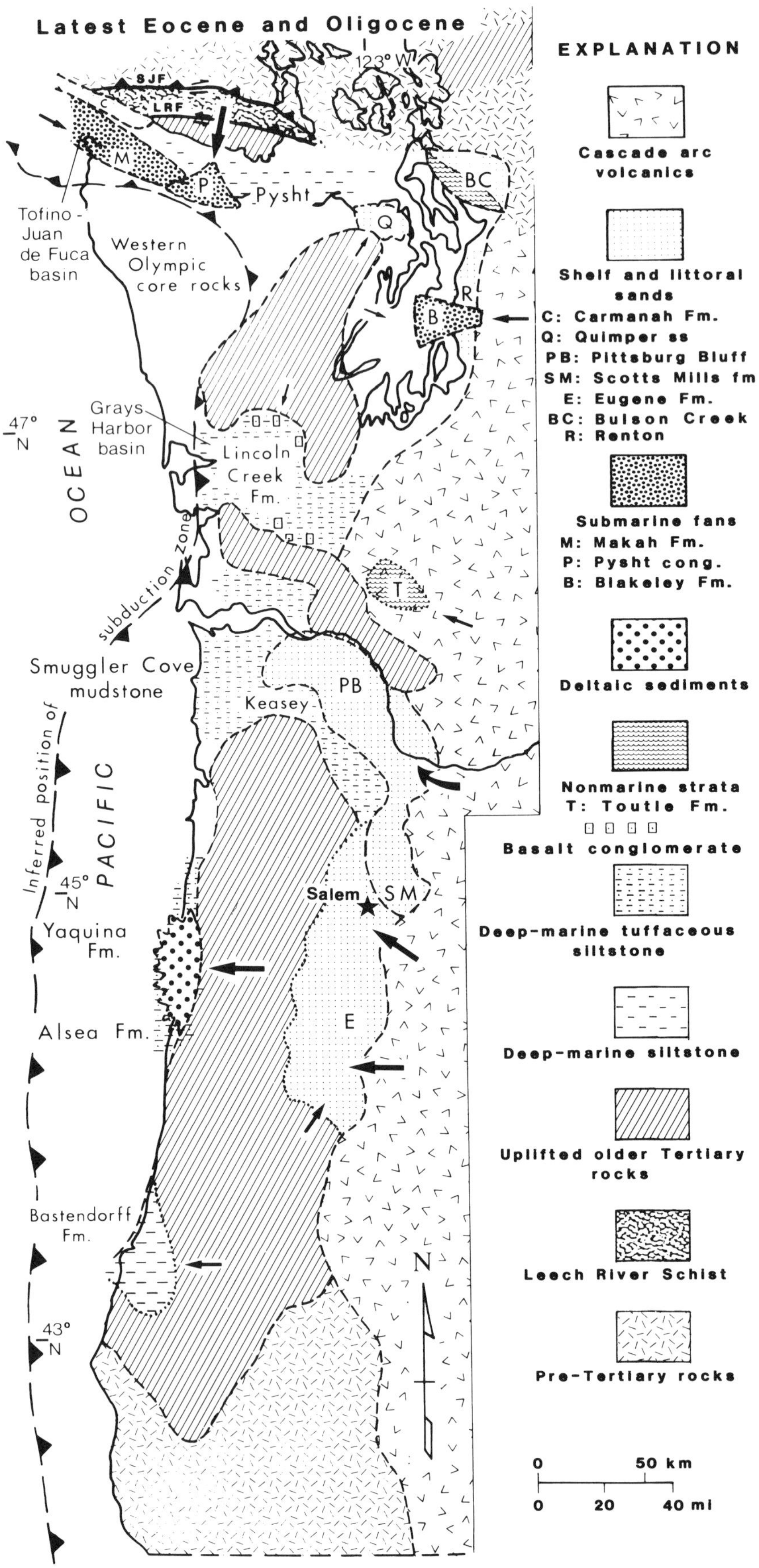

Figure 5. Paleogeographic map of western Washington and Oregon in latest Eocene and Oligocene time (39 to 24 Ma). Post-Eocene rotations not restored. Arrows indicate sediment dispersal directions.

Puget Group to the basin margin (Buckovic, 1979). The Renton Formation of the upper part of the Puget Group represents the last of the arkosic deltaic facies in the Seattle area in the latest Eocene (Fig. 4). Northwest of Seattle (Figs. 4, 5), littoral to inner-shelf sandstone and conglomerate of the Quimper Sandstone grade upward into the deep-marine Marrowstone Shale on the Quimper Peninsula, reflecting a gradual transgression in the latest Eocene (*Allison, 1959*; *Thoms, 1959, Armentrout and Berta, 1977*).

By Oligocene time, continued rapid subsidence in the Puget Sound area inundated the Renton coastal delta-plain environment in which the Puget Group had formed and created an isolated deep marginal basin on the shelf. On the north end of this isolated basin (Figs. 4, 5), 1,200 m of polymictic conglomerate (*Lovseth, 1975*) derived from uplifted North Cascade pre-Tertiary rocks was deposited in alluvial-fan channels and is overlain by nearshore sandstones (*Danner, 1966*; *Marcus, 1981*). Within the basin, volcaniclastic turbidite sandstones and debris-flow conglomerates of the Blakeley Formation were deposited as a submarine fan (Fig. 5). Paleocurrent indicators in the Blakeley strata show westward dispersal from an active volcanic terrane in the Western Cascades (*McLean, 1968, 1977*).

Subsidence continued farther northwestward during latest Eocene to Oligocene time in the deep marginal Tofino–Juan de Fuca basin along the northern Olympic Peninsula (*Snavely and others, 1980b*). Pre-Tertiary rocks on Vancouver Island provided detritus for the shallow-marine sandstones of the Carmanah Formation (Fig. 5), overlying the inactive Leech River fault on southern Vancouver Island, and for more than 2,500 m of the middle and upper parts of the Twin River Group on the northern Olympic Peninsula (Fig. 4), including siltstone and lithic-feldspathic turbidite sandstone of the Makah Formation and bathyal mudstone and local deep-marine channel conglomerate of the Pysht Formation (*Gower, 1960*; Snavely and others, 1978; *Drummond, 1979*; *Fairchild, 1979*; *Snavely and others, 1980b*). The northern Olympic units accumulated as depositional lobes and channels on submarine fans (Figs. 4, 5) that filled the deep axial part of a west-trending marginal basin (*Gower, 1960*; *Snavely and others, 1980b*).

In Oregon, shallow-marine tuffaceous sandstone and siltstone of the uppermost Eocene Eugene Formation interfinger with nonmarine volcanic materials (Figs. 4, 5) of the Little Butte Volcanic Series and Fisher Formation of the western Cascades (Vokes and others, 1951; *Beaulieu and others, 1974*; *Yeats and others, 1992*). Rapid sedimentation of volcaniclastic and arkosic sandstones, together with regional uplift of the eastern part of the forearc basin, led to regression of marine waters from the shallow Willamette embayment toward the end of the Paleogene (Snavely and Wagner, 1963). The Scotts Mills Formation (Figs. 4, 5), which consists of barnacle-plate coquina and shallow-marine to nonmarine feldspathic and tuffaceous sandstone and thin beds of coal in the foothills of the Western Cascades east of Salem, records the position of the rocky storm-dominated Oligocene strandline (*Miller and Orr, 1984, 1988*). Farther offshore, deep-marine tuffaceous mudstone of the upper Eocene Bastendorff Formation was deposited on the subsided feldspathic-lithic Coaledo deltaic system (Figs. 3, 4, 5; *Allen and Baldwin, 1944*; *Rooth, 1974*; Baldwin, 1981, p. 31). An influx of coarser tuffaceous and volcanic detritus, derived from active volcanoes of the Western Cascades arc, produced the shallow-marine, locally glauconitic Tunnel Point Formation (Fig. 4) that overlies the Bastendorff (Baldwin and Beaulieu, 1973). Fine tuffaceous detritus is also preserved in the thick upper Eocene to Oligocene Alsea Formation on the central Oregon coast (Figs. 4, 5; *Snavely and others, 1975*), recording a general shoaling as this part of the forearc basin filled with tuffaceous clastics derived from the erupting Western Cascades. Increased pyroclastic activity in the arc during the late Oligocene resulted in sedimentation rates that exceeded the rate of subsidence, producing the high-energy wave-dominated volcaniclastic delta of the Yaquina Formation (fig. 5), which prograded over the Alsea Formation (*Goodwin, 1973*; *Snavely and others, 1975*). In northern Oregon, increased supply of fine volcaniclastics to the basin is reflected in thick outer-shelf and slope tuffaceous siltstone and mudstone of the upper Eocene Keasey Formation and the informally named upper Eocene to Oligocene Smuggler Cove formation (Figs. 4, 5; *Van Atta, 1971*; *McDougall, 1975*; Niem and Niem, 1985). As the eastern part of the basin shoaled, tuffaceous to arkosic deltaic and inner to middle shelf sandstones of the upper Eocene to Oligocene Pittsburg Bluff Formation prograded westward over the deeper marine Keasey (*Van Atta, 1971*; *Moore, 1976*; *Olbinski, 1983*; *Nelson, 1985*).

Great Valley basin sequence

By S. A. Graham

Voluminous Upper Cretaceous submarine fan sequences of the Great Valley sequence in the forearc basin of central California (Ingersoll, 1979; *1983*) were succeeded by Campanian time in the eastern portion of the basin by deltaic complexes prograding from the ancestral Sierra Nevada (*Cherven, 1983*), thus evolving from a ponded terrace forearc in the late Mesozoic to a broad ridged forearc by early Tertiary time (*Dickinson and Seely, 1979*). This shoaling in the forearc was broadly synchronous with initiation of the Laramide orogeny in the Rocky Mountain region and cessation of magmatism in the Sierra Nevada continental-margin arc (*Moxon and Graham, 1987*), a change often attributed to low-angle subduction (*Dickinson and Snyder, 1978*), and with accretion of the oceanward proto-Salinian block during the Paleocene, perhaps by oblique subduction (Dickinson and others, 1979). Besides possibly contributing to uplift of the forearc basin, obliquely contractional (transpressional) emplacement of the Salinian block may also have produced structural segmentation of the longitudinally continuous forearc basin. The northern, leading edge of the Salinian block lay south of the Stockton arch (Fig. 6), a transverse structural feature that appeared in the Paleocene and divided the forearc basin into the distinct San Joaquin and Sacramento basins (Dickinson and others, 1979).

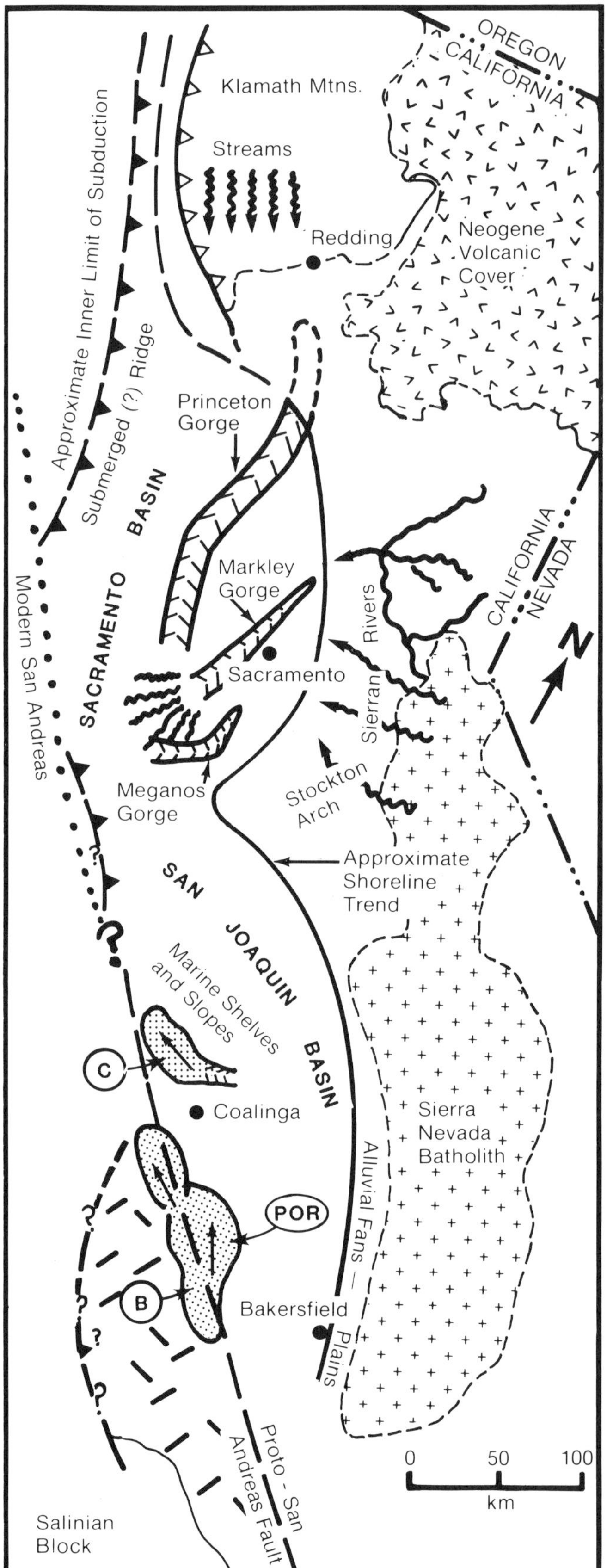

Deposition in the Paleogene Sacramento forearc basin, mainly in neritic to shallow bathyal water depths (Fig. 7), was dominated by four major filled submarine valleys or "gorges" that trended obliquely across the Paleogene shelf of the Sacramento basin (Fig. 6). These gorges, largely confined to the subsurface but well-known through hydrocarbon exploration, range between 70 and 275 km in length, 750 and 1,000 m in depth, 5 and 35 km in width, and from Paleocene to Oligocene in age (Dickinson and others, 1979). The gorges are largely nonoverlapping in map view. Stratigraphically they are separated by intervening bathyal shales and shelf sandstones, imparting a cyclicity to the basin (Fig. 7). The origin of the gorges remains somewhat uncertain. Almgren (1984) inferred a tectonic trigger for gorge erosion although elements of the Sacramento basin fill such as the deletaic sand of the Domengine Formation (and associated basal unconformity) and overlying bathyal Nortonville and Markley Formations (Fig. 7) so closely match global sea-level fluctuations postulated by *Vail and others* (*1977*) that eustatic controls on sediment packaging seem likely, as well. The fill of the gorges is largely fine-grained and bathyal, indicating that the gorges were filled, and probably for the most part incised, in a submarine environment (Dickinson and others, 1979). The gorges apparently served as sediment conduits to funnel detritus from the Sierra Nevada into a deep-sea trench in the vicinity of the present Coast Ranges (Fig. 6). Paleogene sandstones are largely arkosic to quartzose, reflecting inactivity of arc magmatism and deep dissection as well as intense subaerial weathering during a period of warming in the early Eocene (Dickinson and others, 1979). By the Oligocene, volcanism became active in the Sierra-Wasatch region, and long drainages produced volcanic-dominated sandstone and conglomerate in the subaerial valleys of the Sierra Nevada and in the fill of the youngest (Markley) gorge.

In contrast, the San Joaquin segment of the Paleogene Great Valley basin is best characterized as an element of the Paleogene Salinian borderland system, discussed elsewhere in this chapter. The basin persisted as a ponded, largely deep-marine basin, bounded on its western margin by the granitic terrane of the Salinian block (Fig. 6). Submarine canyons are less prominent in the Paleogene of the San Joaquin basin; instead, submarine fans derived from the basin-flanking Salinian and Sierran granitic terranes form important elements of the basin fill. Fan deposits (Fig.

Figure 6. Generalized paleogeographic map of the Great Valley and Salinian block in Paleogene time. Salinian block restored to position prior to Neogene right-slip on faults of the San Andreas system. Proto–San Andreas fault along which Salinian block may have been emplaced during the Paleocene shown by heavy dashed line. Selected Paleogene submarine fan deposits of the Salinian borderland: C, Cantua Sandstone Member (of Lodo Formation); B, Butano Sandstone; POR, Point of Rocks Sandstone Member (of Kreyenhagen Formation). Martinez gorge, which occurs at the southern end of Markley gorge and is truncated by it, is omitted for simplicity. Locations of selected cities and California state boundaries shown for reference. After *Graham* (*1981*) and Dickinson and others (1979) with permission of the Geological Society of America.

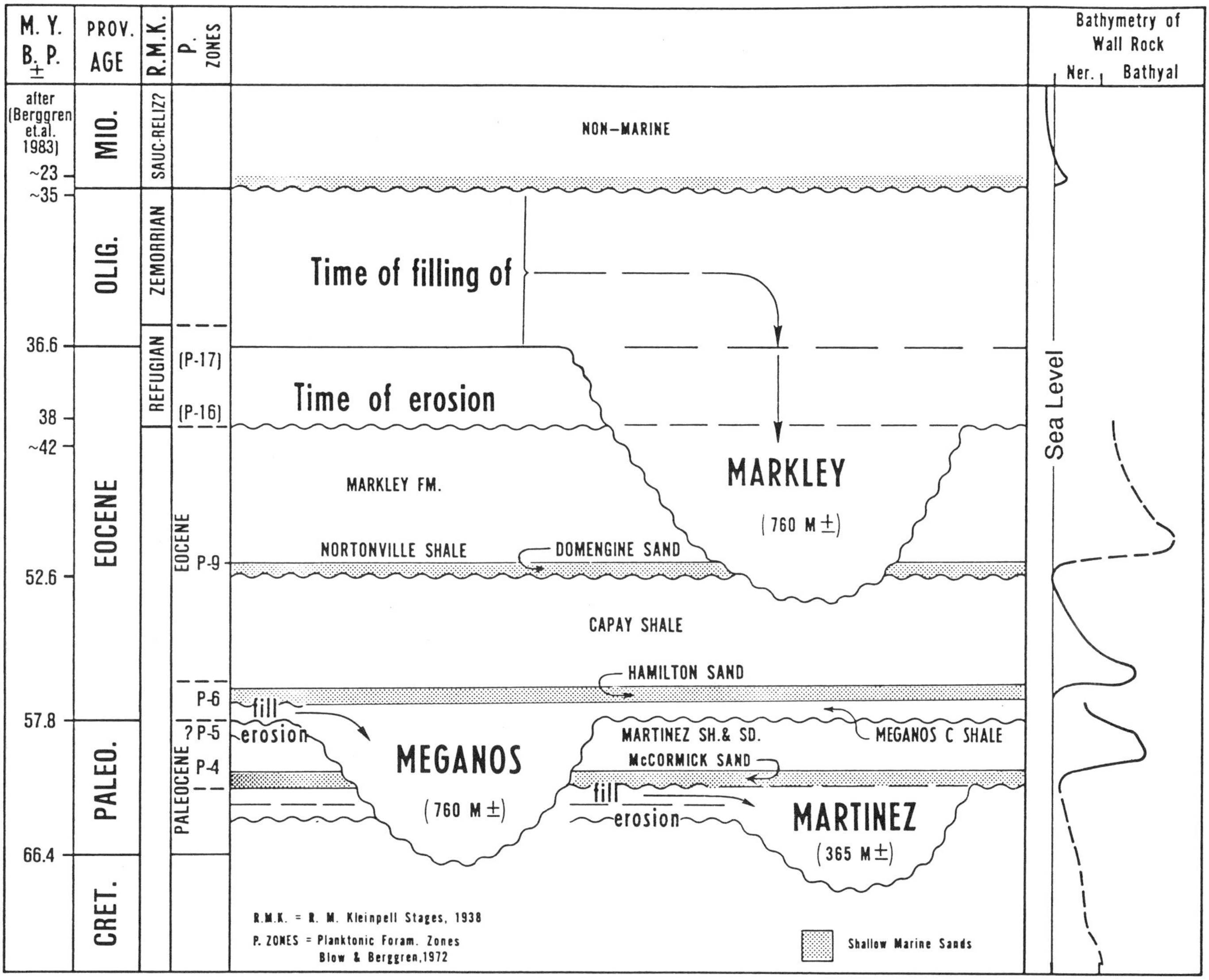

Figure 7. Diagram of Paleogene formations in the Great Valley at the latitude of Sacramento showing generalized relationships of Paleogene submarine canyons of the Sacramento basin to periods of erosion and sedimentation. "P-zones" are biostratigraphic zones of *Blow and Berggren (1972)*. After Almgren (1984) with permission of the Pacific Section, American Association of Petroleum Geologists.

6) like the Salinian-derived Point of Rocks Sandstone and the Sierran-derived Cantua Sandstone Member (of the Lodo Formation) are notable for their sand-rich character and lack of proximal-to-distal facies differentiation (*Nilsen and others, 1974*; Nilsen and Clarke, 1975; *Graham and Berry, 1979*). In addition to the submarine fan facies, best developed in upper Paleocene to middle Eocene strata, the fill of the Paleogene San Joaquin basin consists largely of shelf and slope facies (Nilsen and Link, 1975; *DeCelles, 1988*). As in the Sacramento basin, sandstone compositions strongly reflect the absence of a volcanic arc and erosional stripping in a tropical humid climate in the early Eocene (e.g., *Graham and Berry, 1979*; Graham, 1987), followed by volcanism in the sediment source region during the Oligocene (Bent, 1985).

Borderland sequence of the central Coast Ranges

Discussion of Eocene to lower Miocene sequences of the Coast and Transverse Ranges (Fig. 8) must consider the effects of later deformation. In Figure 9 we assume for discussion purposes 315 km of right-lateral offset along the San Andreas fault, 100 km of right-lateral offset along the San Gregorio–Hosgri fault, relatively small right-lateral offset on other faults of the central Coast Ranges (including the Russell and the Rinconada faults)

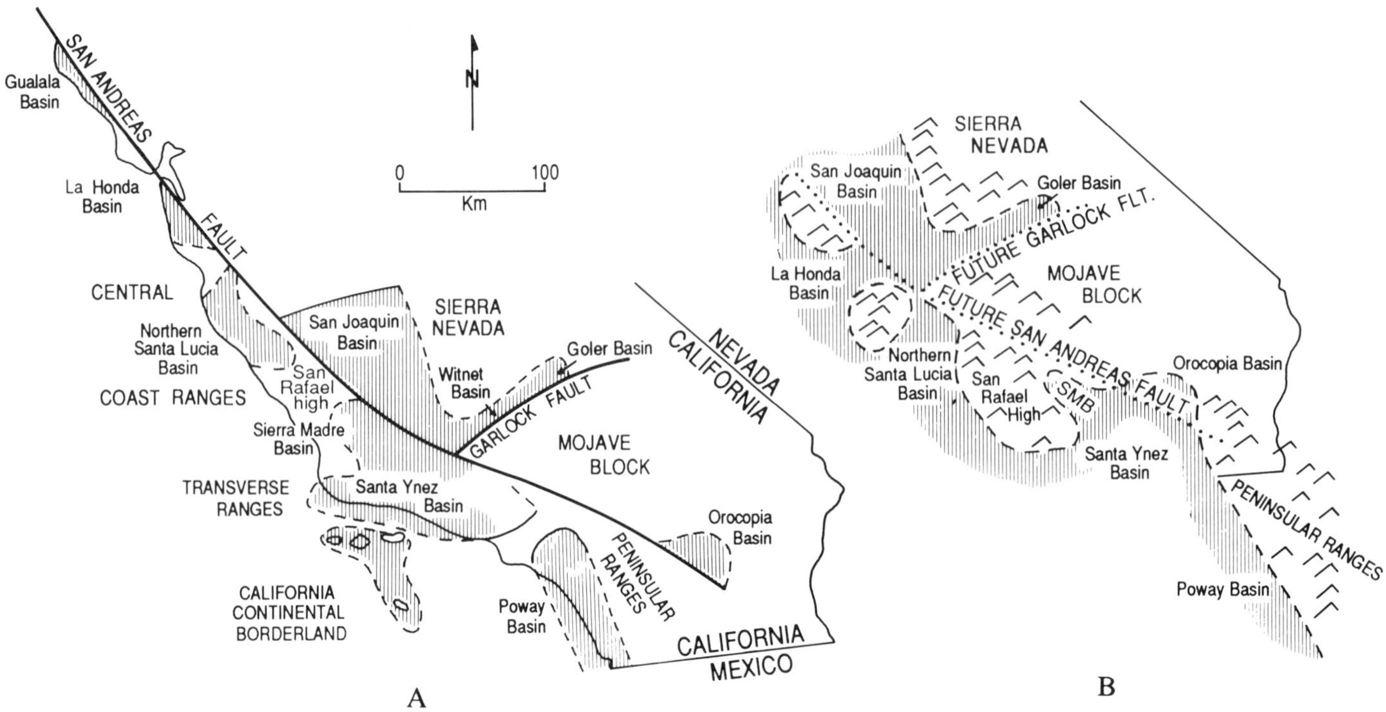

Figure 8. Eocene basins of southern California. A, basins in their present positions; B, basins in positions restored after removal of strike slip on the San Andreas, Garlock, and other faults. SMB, Sierra Madre basin. Modified from *Nilsen (1987)*.

and Peninsular Ranges (including the San Jacinto, Elsinore, and Newport-Inglewood faults), 60 km left-lateral offset on an east-trending fault system of the southern margin of the Transverse Ranges (including the Malibu Coast fault), and the clockwise rotation of much of the Transverse Ranges by 37 to 100°, summarized by Luyendyk and others (1985). The strike-slip faulting of this region is discussed in greater detail by Stewart and Crowell (this volume).

Late Cretaceous and Paleocene deposition in the Great Valley forearc basin was disrupted by the accretion to North America of the Salinian block (Dickinson and others, 1979), the northern end of the composite allochthonous Santa Lucia–Orocopia terrane (see Figs. 6, 14; *Champion and others, 1984*; Howell and others, 1987). Evidence of Paleocene to early Eocene folding related to this accretion is found in the Santa Cruz Mountains south of San Francisco (Clark, 1968) and in the Diablo Range east of the San Andreas fault (Harding, 1976; Graham, 1978). Following accretion of the Salinian block, Eocene and Oligocene strata were deposited in local basins, possibly bounded by normal or strike-slip faults and flanked by highlands (Nilsen and Clarke, 1975; Graham, 1978), a tectonic setting similar to the modern continental borderland of southern California. Faults bounding the early Tertiary borderland basins may have trended northeast, at a high angle to the younger San Andreas fault (Nilsen and Clarke, 1975).

These borderland basins are now dismembered by Neogene right slip on the San Andreas fault (Fig. 8). In the La Honda basin of the Santa Cruz Mountains, the lower and middle Eocene Butano Sandstone rests unconformably on the Paleocene Locatelli Formation and is at least 1,500 m and possibly 3,000 m thick (Clark, 1968; Nilsen and Clarke, 1975). The Butano is overlain conformably by the hemipelagic, deep-marine Twobar Shale Member of the San Lorenzo Formation with late middle Eocene microfossils. In the southwestern San Joaquin Valley, the lower and middle Eocene Point of Rocks Sandstone unconformably overlies the Paleocene and lower Eocene Lodo Formation and is conformably overlain by the Kreyenhagen Formation of late middle Eocene age (*Blueford, 1984*). Removal of about 300 km of right slip on the San Andreas fault juxtaposes these two sequences (Figs. 6, 8). Paleocurrent patterns, heavy mineral composition, and other sedimentary features suggest that they were parts of a single depositional basin, with the Butano and Point of Rocks comprising a deep-sea fan derived from a Salinian source terrane (Nilsen and Clarke, 1975).

In the northern Santa Lucia Range, as described by Nilsen and Link (1975), Eocene strata rest on a surface of moderate relief underlain by crystalline rocks and Cretaceous and Paleocene strata. The Eocene sequence of the northern Santa Lucia basin (Fig. 8) consists of the basal lower Eocene shallow-marine transgressive Junipero Sandstone, the Lucia Mudstone, a deep-water deposit of early and middle Eocene age, and the middle Eocene The Rocks Sandstone, a submarine fan deposit. This is

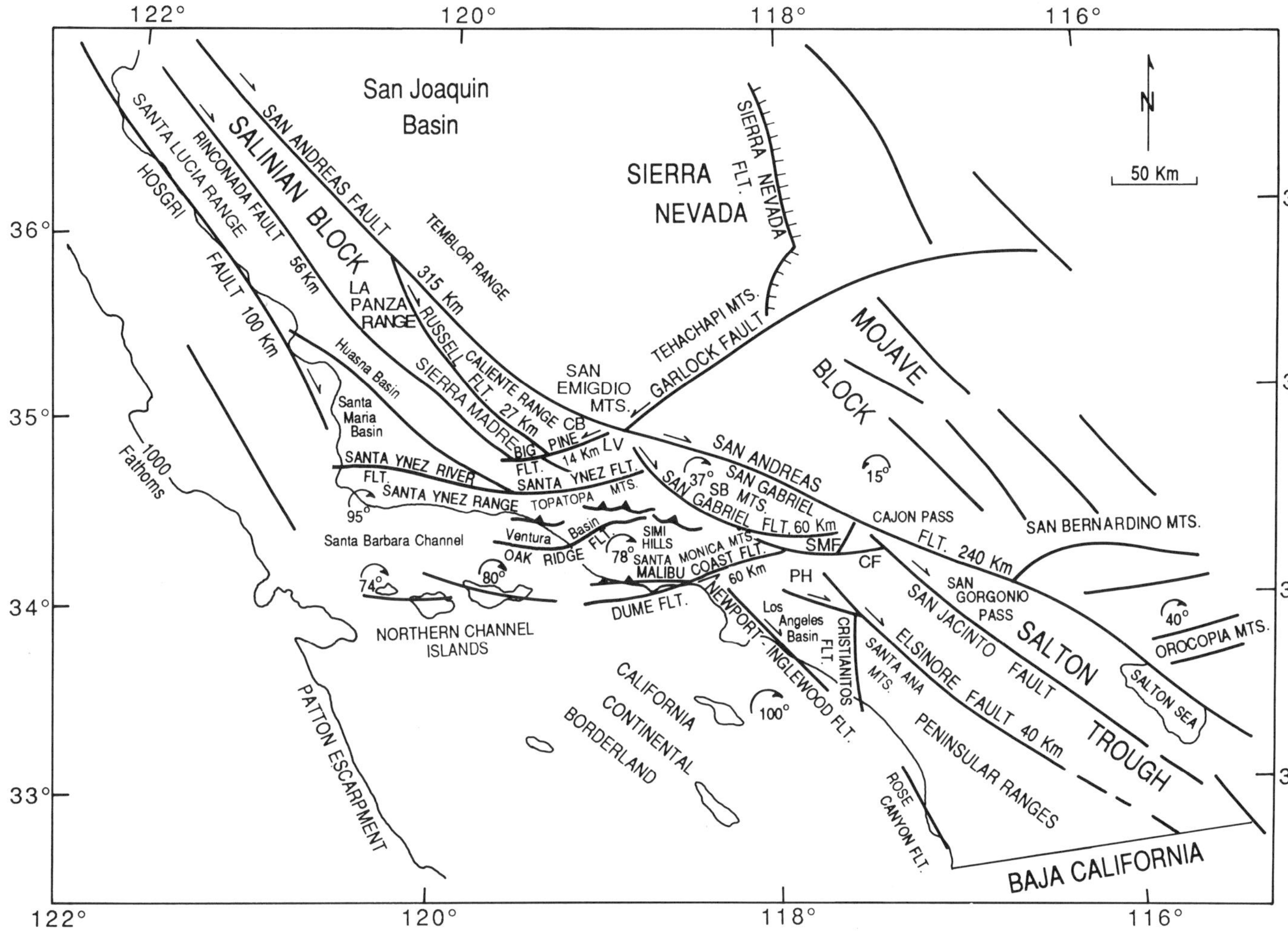

Figure 9. Fault map of southern California showing place names mentioned in text and post–early Miocene rotations and fault displacements assumed for discussions of pre-tectonic paleogeography. CB, Cuyama badlands; CF, Cucamonga fault; LV, Lockwood Valley; PH, Puente Hills; SB, Soledad basin. Modified after *Hornafius and others* (*1986*), which gives references for displacements and rotations.

overlain by the Church Creek Formation of middle to late Eocene age and by the nonmarine Oligocene(?) Berry Formation. Across the San Andreas fault at the southern margin of the San Joaquin Valley, the San Emigdio Mountains and western Tehachapi Mountains (Fig. 9) contain a Paleogene sequence similar to that in the northern Santa Lucia Range. Basement rocks are overlain by the Tejon Formation, which consists of the basal Uvas Conglomerate Member overlain by the deep-water Liveoak Shale Member of middle Eocene age, the Metralla Sandstone Member, and the Reed Canyon Siltstone Member. The Tejon Formation is overlapped by the non-marine Tecuya Formation that changes facies westward to the middle and upper Eocene shallow-marine and deep-marine San Emigdio Formation. Restoration of San Andreas offset (Fig. 8) juxtaposes these sequences to reveal a shelf and continental source area to the east and a deep-marine basin to the west (Nilsen and Link, 1975), the two being separated by a basement ridge (Graham, 1978).

A continuous sequence of Late Cretaceous to early middle(?) Eocene age up to 7,500 m thick is found in the southern Santa Lucia Range and La Panza Range (Fig. 9), and farther south, in the Sierra Madre basin (Fig. 8), a lower to middle Eocene sequence over 6,000 m thick is found (Chipping, 1972; Nilsen and Clarke, 1975). Chipping (1972) determined that the source area for most of these largely deep-water deposits was to the north and east, close to the present-day trace of the San Andreas fault. In Paleocene and early Eocene time, the Sierra Madre basin of Chipping (1972) was isolated from the Santa Ynez basin of the western Transverse Ranges by an emergent area in the vicinity of the San Rafael Mountains. In middle and late Eocene time, the San Rafael emergent area subsided enough

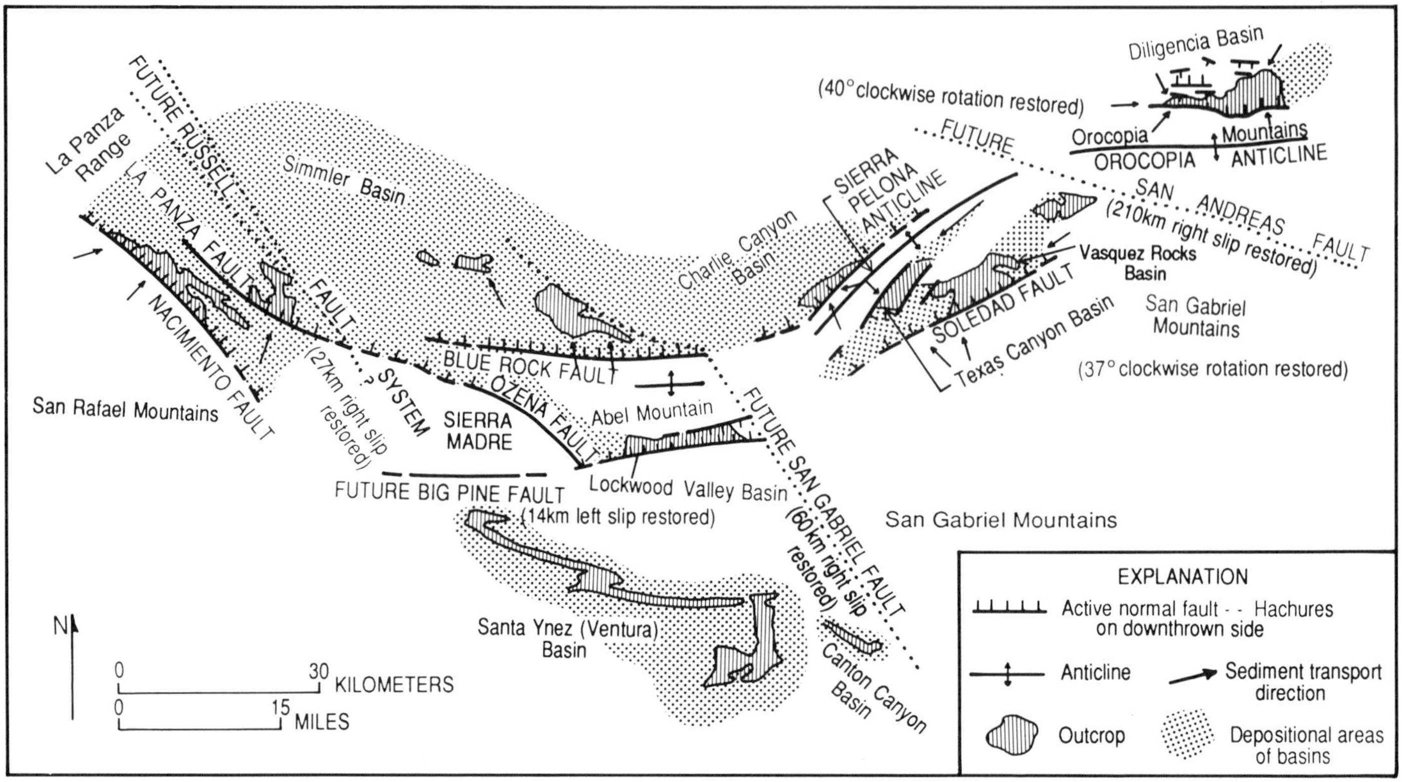

Figure 10. Mid-Tertiary nonmarine basins of the Transverse Ranges after restoration of strike slip on the San Andreas, San Gabriel, Russell, and Big Pine faults and restoration of clockwise rotation of blocks east of the San Gabriel fault. Clockwise rotation of western Transverse Ranges not restored. Modified after Bohannon (1975) and Yeats and others (1989).

to receive algal deposits (Sierra Blanca Limestone), but Eocene deposits there are thin compared to the basins flanking it on the north and south. Marine Eocene deposits are traced southeastward to the Big Pine fault (Fig. 9; Chipping, 1972). Farther east, they are known from the Cuyama Badlands, west of the San Joaquin Valley (*Dibblee, 1982b*), and in Lockwood Valley in the northern Transverse Ranges (*Carman, 1964*), where they may rest on basement rocks.

Borderland sequence of the Transverse Ranges

The Paleogene strata of the Transverse Ranges (the Santa Ynez basin of Fig. 8) can be divided into a northern and a southern succession which can only be correlated with each other in a very general way. These successions are separated by the Neogene Ventura basin and its offshore continuation in the Santa Barbara Channel (Fig. 9), the Neogene fill of which is in contact with the Paleogene to the north and south along large-displacement reverse faults that probably penetrate to the middle crust. The northern succession is found in the Santa Ynez Range, its eastern continuation in the Topatopa Mountains, and in the high country north to the Big Pine fault, along which Paleogene rocks are in fault contact with Eocene strata of the Sierra Madre in the Coast Ranges. This succession was originally continuous with strata northeast of the Salton Sea in the Orocopia Mountains, now displaced from the remainder of the terrane by the San Andreas fault (Figs. 8, 9). The southern succession is found in the area between the Oak Ridge and Malibu Coast faults and the northern Channel Islands (Fig. 9). This succession can be correlated to strata of the Peninsular Ranges and California Continental Borderland.

Dickinson (*1969*) subdivided the Paleocene and Eocene rocks of the northern succession into two units, an older Paleocene and Eocene sequence including the shallow-water transgressive Sierra Blanca Limestone, the deep-water Anita Formation, and the Juncal Formation, including sandstone deposited by turbidity currents, and a younger Eocene sequence including the Matilija Sandstone, the Cozy Dell Shale, and the regressive Coldwater Formation. These two sequences reflect two episodes of basin deepening and shallowing, with the regression between the two sequences occurring in the Matilija Sandstone (Link, 1975) at about 50 Ma. The younger sequence is considerably better sorted than the older one although both reflect the presently east-trending Santa Ynez depositional basin (Fig. 8; *Reed and Hollister, 1936*). In the eastern part of the area, the younger sequence continues upward into the nonmarine Sespe

Formation (Nilsen, 1984), deposition of which ended with the new transgression of the Vaqueros Formation in the Oligocene, about 26 to 27 Ma. The lateral equivalent of the Coldwater-Sespe regressive sequence is entirely marine in the western Santa Ynez Range, where it includes the Sacate, Gaviota, and Alegria Formations (Dibblee, 1950). North toward the Big Pine fault and east toward the San Gabriel fault, both sequences are thinner, coarser grained, and interspersed with unconformities (*Dibblee, 1982b*). The Eocene rocks are overlapped by Miocene strata west of the San Gabriel fault; the marine Eocene Maniobra Formation (*Crowell and Susuki, 1959*) of the Orocopia Mountains probably connects with Eocene strata in Lockwood Valley near the east end of the Big Pine fault rather than with the Eocene of the Topatopa Mountains.

South of the Neogene Ventura basin, the Paleogene succession is best known in the Simi Hills (Fig. 9; Squires and Filewicz, 1983), where it can be subdivided into two transgressive-regressive sequences as in the Santa Ynez Range. The papers in Squires and Filewicz (1983) show Cretaceous turbidites to be overlain disconformably by the nonmarine Simi Conglomerate and shallow-marine Las Virgenes Sandstone, which is overlain by the deep-water Paleocene–early Eocene Santa Susana Formation, including proximal turbidite fans. A thin-skinned thrust underwent most of its displacement during deposition of the Santa Susana Formation (Yeats, 1987). The Santa Susana Formation and the overlying transgressive Llajas Formation are separated by a hiatus of 4 m.y. ending at 53 Ma (*Filewicz and Hill, 1983*). The basal conglomerate of the Llajas Formation (*Squires, 1981*) is overlain by shale and locally by proximal turbidite fans, then by regressive shallow-marine strata, and finally by the nonmarine Sespe Formation. This same stratigraphy is recognized in the Santa Monica Mountains to the south and in the subsurface westward to the coast (*Seedorf, 1983*). The sequences can be correlated in a general way to the northern Channel Islands (Weaver, 1969) and to the Santa Ana Mountains of the Peninsular Ranges (Yeats and others, 1974).

Paleomagnetic evidence (*Luyendyk and others, 1980,* 1985; Hornafius, 1985) shows that the western Transverse Ranges were rotated clockwise, probably as numerous separate blocks, during and after the Miocene (Fig. 9). The Paleogene basin of the Santa Ynez Range and areas to the north and east originally trended north (Fig. 8B) and were later rotated to their present east trend (Fig. 8A), indicating that source terranes originally lay west of the basin, in addition to north and east (Hornafius, 1985).

Nonmarine faulted basins of southern California

An east-west set of half-grabens formed and was filled with nonmarine red beds in late Oligocene and early Miocene time along the northern margin of the Transverse Ranges (Fig. 10; Woodburne, 1975). These strata rest on crystalline rocks as old as Precambrian in the Soledad basin and with angular unconformity on indurated strata as young as Eocene north of the Ozena and La Panza faults (Fig. 10). A typical depression contains sediments of alluvial fan, fluvial, and lacustrine origin, coarsening upcurrent to locally derived conglomerate and sedimentary breccia adjacent to a coeval fault that bounds the basin. Local volcanic interbeds provide age control. Except for the Diligencia and Vasquez Rocks basins, these basins have not undergone tectonic rotation (*Luyendyk and others, 1980*; Luyendyk and Hornafius, 1987). Even though they are syntectonic and were formed after the Pacific and North American Plates had made contact, they predate all strike slip on the San Andreas, San Gabriel, and Russell faults (Bohannon, 1975; Yeats and others, 1989; Tennyson, 1989).

In the Soledad basin, the Vasquez Formation, with 24- to 26-Ma lavas near its base, occurs in three fault-bounded basins separated by basement ridges, including the Sierra Pelona anticline (Fig. 10; *Muehlberger, 1958*; Bohannon, 1975). The Sierra Pelona now exposes the Pelona Schist at the surface, but during the time of deposition of the Vasquez, only granitic rocks, structurally higher above the Vincent thrust (see Fig. 14), were exposed there. The Sierra Pelona anticline may be correlated across the San Andreas fault with the Orocopia anticline, also underlain by the low-grade Orocopia Schist and flanked on the north by the Diligencia basin, filled by the Diligencia Formation (*Crowell, 1975a*). Vertebrate remains 365 m above the base of the Diligencia are of early Miocene age (*Woodburne and Whistler, 1973*), and basalt flows are dated as 22 and 20 Ma (*Crowell, 1973*). Volcanic rocks dated as 32 to 25 Ma occur farther east in the Chocolate and Whipple Mountains in fault-bounded basins trending north and northwest (see Nilsen, 1984, for summary). By removing 60 km of right slip from the San Gabriel fault (Fig. 10), the southernmost half-graben in the Soledad basin is juxtaposed against the Plush Ranch Formation of *Carman (1964)* in the Lockwood Valley basin, with basalts dated as 20 to 18 Ma (*Crowell, 1973*; Bohannon, 1975), and the northernmost half-graben, Charlie Canyon basin, lines up with the Simmler Formation north of the Blue Rock fault. Each sequence is bounded on the south by a coeval normal fault—the Big Pine fault adjacent to Lockwood Valley (Crowell, 1968) and the Blue Rock fault in the eastern Cuyama basin (Bohannon, 1975).

Farther west, on the northern edge of the Sierra Madre (Fig. 10), the Caliente Formation (*Vedder, 1968*) underlies the marine Vaqueros Formation and is the lithologic equivalent of the Simmler Formation mapped elsewhere. The Caliente Formation coarsens to sedimentary breccia adjacent to the Ozena fault (Fig. 10), now a south-dipping reverse fault but in Caliente time probably a north-dipping normal fault (Yeats and others, 1989). If right slip of 27 km is removed from the Russell fault (Yeats and others, 1989), the Ozena fault lines up with the La Panza fault, which was active, north-side-down, during deposition of the Simmler Formation (Ballance and others, 1983). A separate nonmarine basin fill of the Simmler Formation is found south of the La Panza fault against the Nacimiento fault, also north-side-down during Simmler deposition (Ballance and others, 1983).

In the Santa Maria and Huasna basins along the coast (Fig. 9), the oldest widespread unmetamorphosed sedimentary unit is

the nonmarine Lospe Formation which, in its lower part, includes an unstratified rubble consisting primarily of local basement (Woodring and Bramlette, 1950). Higher parts of the section contain variegated sandstone, mudstone, gypsiferous siltstone, and tuff. In the Huasna basin, the Lospe overlies dacite and contains clasts of dacite lithologically similar to those dated as 22 to 26 Ma (*Turner, 1968*; Ernst and Hall, 1974; Hall, 1981a), and it underlies and locally interfingers with Oligocene or lower Miocene marine strata (*C. A. Hall and Corbato, 1967*; Ernst and Hall, 1974). In contrast, the Lospe at its type locality near Point Sal in the Santa Maria basin is conformable with overlying middle Miocene marine strata (Woodring and Bramlette, 1950) and, thus, postdates the half-grabens described above.

As shown in Figure 10, the mid-Tertiary half-grabens occur in an east-west belt at least 180 km long (restoring right-slip offset on faults of the San Andreas system). If the Lospe occurrences in the Huasna basin west of the Nacimiento fault are included, the belt is at least 250 km long. Luyendyk and Hornafius (1987) show that the basins east of the San Gabriel fault have undergone clockwise tectonic rotation, about 37° for the Soledad basin and 40 to 105° for basins east of the San Andreas fault. When these rotations are restored along with the strike-slip fault offsets (Fig. 10), the entire belt is arcuate, convex southward (Tennyson, 1989). Farther south, the Santa Ynez basin received a more continuous sequence from Eocene to early Miocene time (Nilsen, 1984) with contemporaneous faulting less apparent. The east-west belt of faulted basins formed at the same time as extended basins in eastern California and western Arizona and appears to be a westward continuation of that extended region (Tennyson, 1989). Tennyson (1989) also identified another belt of extensional basins farther north, from the Mojave Desert, across the San Joaquin basin, to the La Honda basin in the Santa Cruz Mountains.

California Continental Borderland and Peninsular Ranges

The Late Cretaceous and early to middle Tertiary stratigraphy of the Peninsular Ranges can be correlated northwest to the Santa Monica Mountains and Simi Hils and west to the California Continental Borderland (Fig. 9). Like the Sierra Nevada, the region of the Peninsular Ranges sloped westward in Late Cretaceous and early Tertiary time, and sediments accumulated along the western margin. Unlike the Great Valley, the western margin was subsequently disrupted by strike-slip faulting and extension so that basin analysis requires restoration of the displaced remnants to their early Tertiary positions with respect to the Peninsular Ranges (Fig. 11).

Cretaceous strata were derived largely from rocks of the Peninsular Ranges and occur in a west-deepening and west-thickening sequence including proximal turbidite fans. The Cretaceous rocks are overlain locally by Eocene strata, but elsewhere Paleocene strata intervene (*Popenoe and Woodring, 1945*; *Schoellhamer and others, 1981*), including nonmarine conglomerate and clay overlain by calcareous sandstone and shale in the Santa Ana Mountains (Silverado Formation) and Santa Cruz Island and deep-water strata on San Miguel Island (*Sage, 1973*).

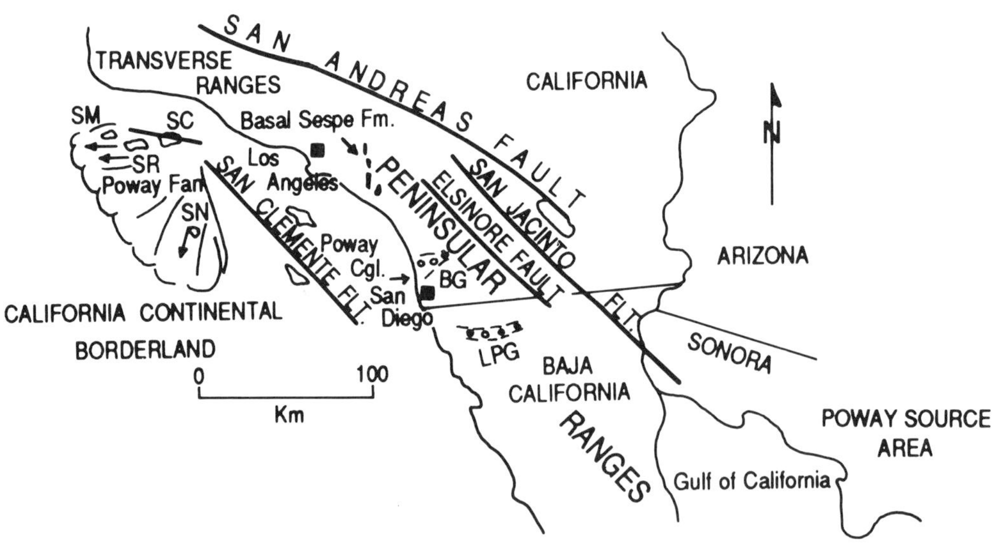

Figure 11. Relations of the Poway Conglomerate of the San Diego area. Poway Conglomerate is traced eastward to fluvial Ballena Gravel (BG), which is itself offset by oblique-slip opening of the Gulf of California from a probable source terrane in northern Mexico. The correlative Poway deep-sea fan in the California Continental Borderland is offset by strike-slip faults from the Poway exposures. Also shown are other fluvial depositional systems to the north (basal part of Sespe Formation of the Santa Ana Mountains) and south (Las Palmas Gravel). SC, Santa Cruz Island; SM, San Miguel Island; SN, San Nicolas Island; SR, Santa Rosa Island. Modified after *Link and others* (*1979*).

Unlike the Cretaceous, the Paleocene and Eocene conglomerates are characterized by exotic clasts, probably derived from northwest Mexico or southern Arizona (Fig. 11). The Eocene rocks were deposited by several major west-flowing streams, each with its own distinctive conglomerate clast population (Minch, 1979); these include the Ballena Gravel with its clast assemblage distinctive of the Poway Group, and farther south, the Las Palmas Gravel of northernmost Baja California. Farther north, the Paleogene of the Santa Ana Mountains is characterized by a still different clast assemblage (Abbott and others, 1983). At San Diego, the gravel grades westward into shallow-marine sand and turbidite fans. Offshore, correlative Eocene deep-water sedimentary deposits on San icolas Island and the northern Channel Islands contain a Poway clast assemblage, indicaitng oblique-slip offset from an original position west of San Diego (Fig. 11; *Kies and Abbott, 1983*; Abbott and others, 1983). Offset and accompanying clockwise rotation of the northern Channel Islands (Fig. 9) is predominantly Miocene or younger in age, but *Bartling and Abbott (1983)* found evidence that some offset between the mainland and San Miguel Island occurred prior to Eocene deposition. Oligocene rocks are found only on Santa Rosa Island as the nonmarine Sespe Formation.

COLLISION AND EMPLACEMENT OF ACCRETIONARY COASTAL TERRANES

In previous chapters, the western Cordillera has been shown to have grown westward by the accretion of far-derived tectonostratigraphic terranes, largely during the Mesozoic (Silberling and others, 1987). It has been further suggested that several terranes have been accreted to North America during the Cenozoic: both the Paleocene to lower Eocene basalts of the Oregon and Washington Coast Ranges (Fig. 2) and the Santa Lucia–Orocopia allochthon of California (see Fig. 14) at about 55 Ma; the Gold Beach terrane of the southwest Oregon coast in post-Eocene time; the coastal belt and King Range terrane of the Franciscan Complex of northern California at about 25 Ma; and later, and the Baja-Borderland allochthon of southern California and Baja California possibly as recently as 16 to 17 Ma. In addition, the structurally lowest formations in the core of the Olympic Peninsula are highly deformed Tertiary rocks that are in tectonic contact with both the accreted Crescent Formation and younger strata of the Coast Range of Washington (Tabor and Cady, 1978a).

Olympic Mountains core rocks

By Alan R. Niem, Parke D. Snavely, Jr., and Wendy A. Niem

The core rocks of the Olympic Mountains comprise a complexly deformed accretionary terrane of lower Eocene to middle Miocene broken formations and melanges (Tabor and Cady, 1978a, 1978b). The general structure is one of highly folded and domed curvilinear packets of turbidite fan sequences separated by imbricate west-vergent thrust faults with wide shear zones, younger to the west (Fig. 12). Surrounding the core rocks is a horseshoe-shaped outcrop belt (peripheral rocks) of lower to middle Eocene oceanic basalt of the Crescent Formation, which is overlain by 6,000 m of younger, much less deformed strata deposited in a marginal basin (*Snavely and others, 1980b*; Snavely, 1983), discussed earlier.

The core rocks can be subdivided into an eastern assemblage of generally older, lower Eocene to Oligocene rocks and a western assemblage of generally younger, middle Eocene to middle Miocene rocks (Tabor and Cady, 1978a). The eastern core rocks consist of broken formations of sparsely fossiliferous turbidite sandstone, conglomerate, semischist, and slate or phyllite that have undergone pumpellyite-prehnite and greenschist metamorphism and display a distinctive pencil cleavage (Tabor and Cady, 1978b; *Tabor, 1983*). The younger western core rocks are largely broken formation and mélange of accreted clastic submarine fans and slope deposits that have undergone calcification and laumontite metamorphism (Stewart, 1974; *Rau, 1979*; *Snavely and others, 1986*). All of these turbidite fan sequences were deposited in trench-slope settings at an undetermined distance to the west and have been added to the continent during the Tertiary as accretionary prisms which were largely thrust beneath the Crescent basaltic crust and peripheral sedimentary rocks (*MacLeod and others, 1977*; Tabor and Cady, 1978a, 1978b; *Snavely and Wagner, 1981*; *Snavely and others, 1986*; *Clowes and others, 1987*).

With the exception of an allochthonous block of Jurassic (144-Ma) hornblende-quartz diorite and gabbro and Lower Cretaceous pillow basalt, graywacke, and minor chert at Point of Arches on the Pacific coast (Fig. 12; *Snavely and others, 1986*), the oldest rocks of the Olympic Mountains are tholeiitic oceanic basalts of the lower and middle Eocene Crescent Formation that were accreted to the North American continent in middle Eocene time (Snavely, 1983). The exotic pre-Tertiary outlier at Point of Arches may represent an olistostromal block of Mesozoic Vancouver Island continental crust that was incorporated into the rock sequence of the Olympic Mountains in the middle Miocene (Snavely, 1987).

Because of the complex deformation, low-grade metamorphism, and sparse fossil control, the exact timing and nature of tectonic and sedimentological events for the Olympic core rocks are still unclear. Several events, however, are tentatively recognized.

Following oblique plate convergence in the late early Eocene, the elevated pre-Tertiary terrane of Vancouver Island shed thick deposits of lithic quartzose turbidite sandstone, siltstone, and mudflow conglomerate in middle Eocene time (Snavely, 1987). Some of this material is preserved in the northwesternmost Olympic core rocks—the Ozette mélange of *Snavely and Wagner (1981)* and the Sooes terrane of *Snavely and others (1986)*.

In the middle late Eocene, thousands of meters of lower Eocene to middle upper Eocene trench and fan sandstone, slope

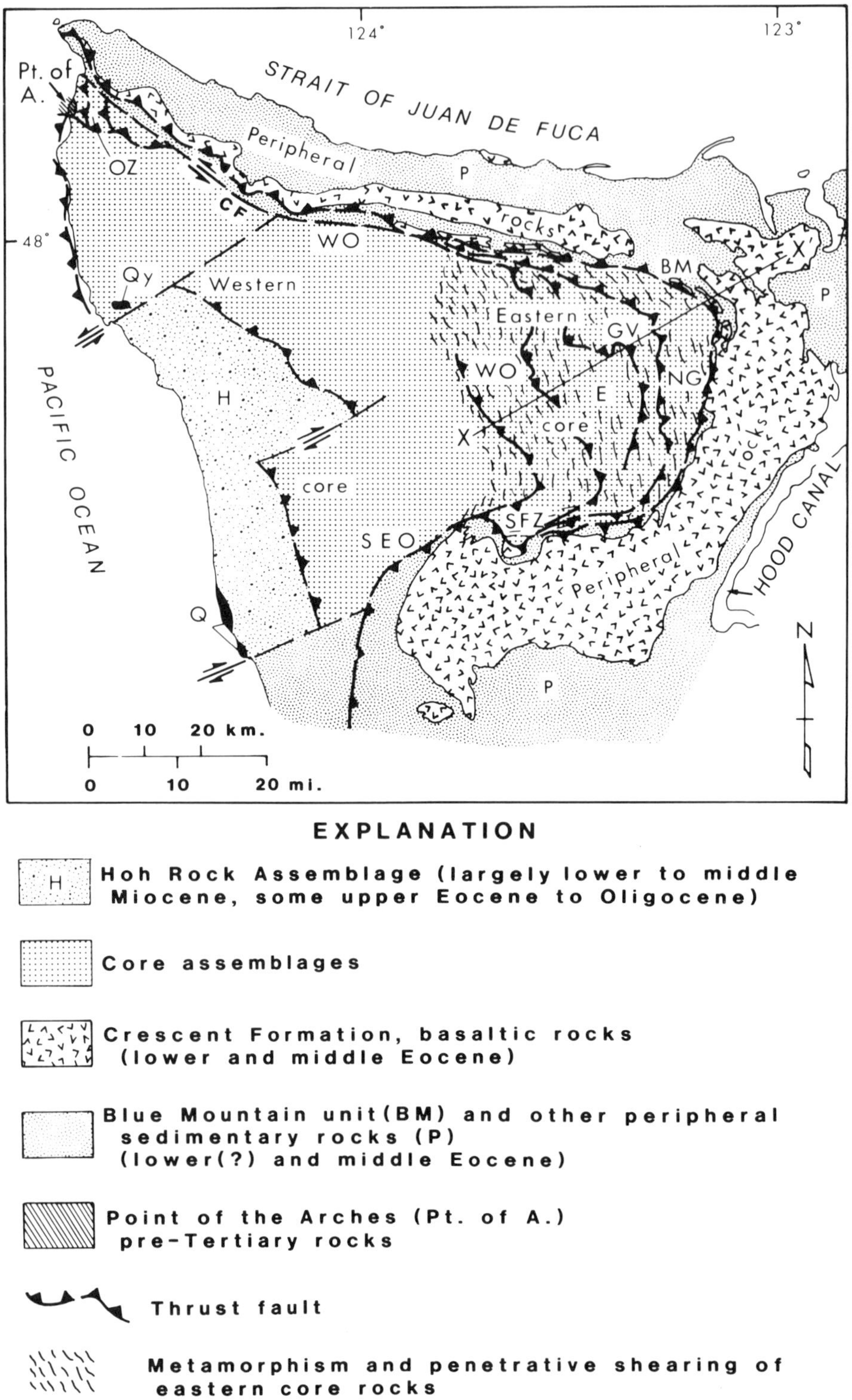

Figure 12. Major geologic terranes of the Olympic Mountains, northwestern Washington. Olympic core lithic assemblages: E, Elwha (lower to middle Eocene); GV, Grand Valley (middle? to upper? Eocene); NG, Needles–Gray Wolf (upper Eocene); OZ, Ozette (middle Eocene to Oligocene); SEO, undifferentiated assemblage of the SE core (upper Eocene and Oligocene); WO, western (upper Eocene and Oligocene). Formations along Olympic coast: Q, Quinault Formation (upper Miocene and Pliocene); Qy, Quillayute Formation (upper Miocene). CF, Calawah fault; SFZ, southern fault zone. X–X′ shows general location of schematic sections in Figure 13. Modified from Tabor and Cady (1978a, b), *Tabor (1983)*, and Snavely (1987).

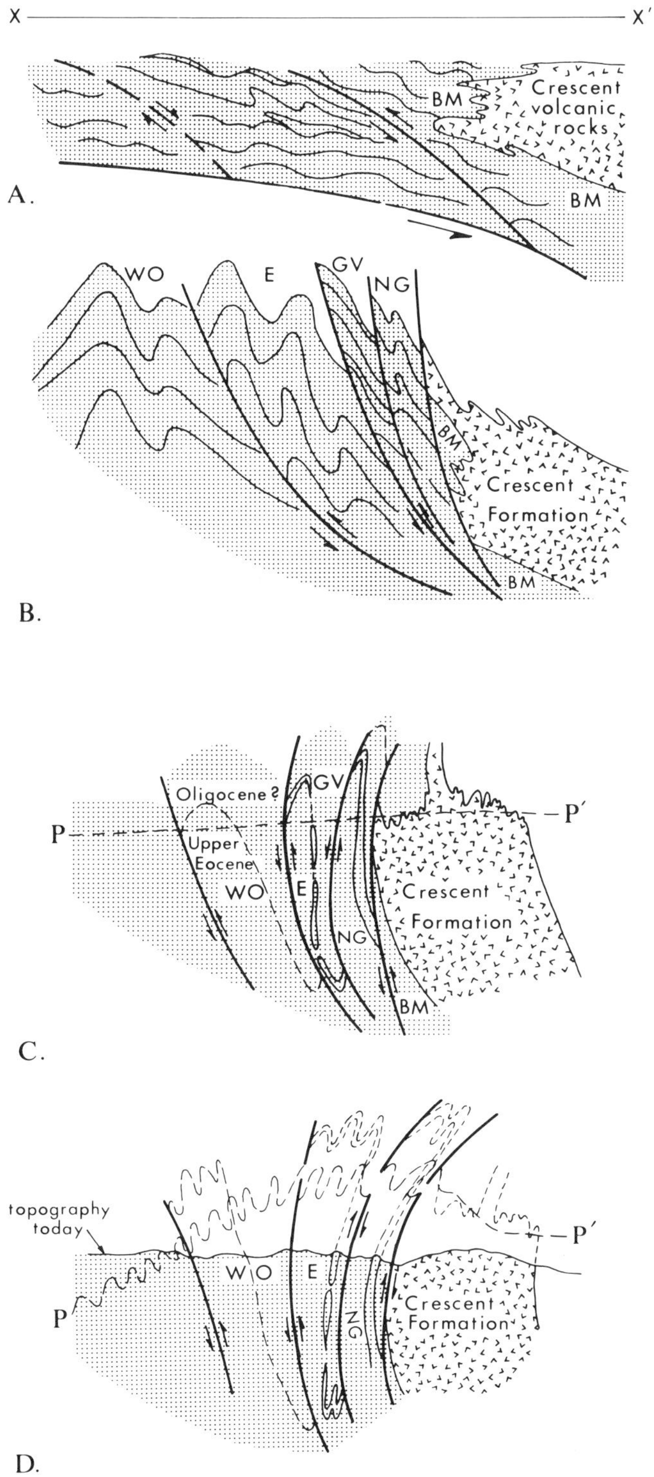

Figure 13. Schematic diagram showing development of structure in the eastern Olympic core rocks. Location shown in Figure 12. Letter symbols as in Figure 12. A, plate motion begins to imbricate sedimentary deposits; B, early imbrication and folding as sedimentary and volcanic rocks of core are uplifted and underthrust; C, after early folding and faulting, major core units established (P–P′ is form line to show later deformation in D); D, after shear folding, doming, and erosion (deformation of P–P′ exaggerated for illustration). Modified slightly from *Tabor (1983)*.

mudstone, and minor conglomerate—the Elwha, Needles–Gray Wolf, Grand Valley, and western Olympic lithic assemblages of the eastern core, and Ozette assemblage of the northwestern Olympic core (Figs. 12, 13)—were uplifted, folded, and underthrust eastward and northward in imbricate slices beneath Crescent volcanic rocks and associated turbidites (Figs. 12, 13A, 13B; Tabor and Cady, 1978b; *Tabor, 1983*). This event may have occurred while uppermost Eocene strata of the western core were being deposited in a trench and slope setting farther to the west (Tabor and Cady, 1978b). Suturing of the lower to middle Eocene eastern core rocks to the North American Plate occurred prior to intrusion of 41-Ma hornblende-quartz diorite dikes reported by *Tabor and others (1972)*.

Lower to middle Oligocene strata are generally missing within the western core rocks. It is unclear whether these rocks were subducted, were never deposited, or were deposited but later uplifted and eroded off the Olympic core rocks.

A dynamic metamorphic event of about 29 Ma in the eastern core area (Tabor, 1972) may have reflected continued eastward plate subduction and imbricate underthrusting of the turbidite strata (Tabor and Cady, 1978b) or cooling associated with initial uplift of the eastern core strata. This event produced recrystallization, pumpellyite-prehnite metamorphism, and penetrative shearing of the older Eocene to Oligocene turbidite fan strata of the eastern Olympic core (Fig. 12), now semischist, mica phyllite, and metagraywacke and conglomerate (*Tabor, 1975, 1983*).

During the early to middle Miocene, thick turbidite lithic-micaceous sandstone and conglomerate and deep-marine organic-rich siltstone of the Hoh Rock assemblage (*Rau, 1973, 1975, 1979,* 1986) were deposited in a trench or continental slope west of the Olympic Peninsula. Following Tabor and Cady (1978a), we restrict the Hoh Rock assemblage to upper Oligocene to upper middle Miocene strata that form the bulk of the western Olympic core. The western Olympic core rocks, including the Hoh Rock assemblage, western Olympic lithic assemblage, and an undifferentiated lithic assemblage of the southwest core (Fig. 12), were accreted to the older eastern core rocks and northwestern core rocks during the late middle Miocene, tilting the older late Eocene thrust faults to vertical (Fig. 13C, D). Highly deformed and disrupted upper Oligocene and Miocene strata now occur in the western Olympic Mountains as broken formation separated by mélange or by wide shear zones of east-dipping imbricate thrust slices (*Stewart, 1970*; *Rau, 1973, 1975, 1979*; Tabor and Cady, 1978a, 1978b; *Snavely and others, 1986*). Some thick-bedded Oligocene turbidite strata, as at Ozette Island, are thrust eastward over middle and upper Eocene strata (*Snavely and others, 1986*) as well as the lower middle Miocene Hoh Rock assemblage along the western coastline of the Olympic Peninsula. The middle Miocene convergence also formed tectonic phyllite breccia in the older eastern core rocks and 17-Ma quartz veins in associated shear zones (Tabor, 1972).

During deformation in the late middle Miocene, the lower to middle Miocene bathyal mudstones of the Hoh Rock assem-

blage were overturned, isoclinally folded, and faulted (*Rau, 1975, 1979,* 1986) and were thermally matured by subduction and imbricate thrusting. They may be the source of numerous oil and gas seeps along the western coast of the Olympic Peninsula (*Rau, 1973*; Snavely and others, 1977; *Rau and McFarland, 1982*; Snavely, 1987; *Snavely and Kvenvolden, 1989*); more than 12,000 barrels of oil were produced from the Hoh Rock assemblage by a well west of Grays Harbor (*McFarland, 1981*). Offshore seismic studies and onshore well data show that the Hoh Rock assemblage extends as a subduction complex (see Fig. 27) beneath the southwest Washington coast and shelf and may be widespread on the outer continental shelf of northern Oregon (Snavely and others, 1977; *Rau and McFarland, 1982*; Snavely, 1987; Duncan and Kulm, 1989).

A regional unconformity between the upper Oligocene to middle Miocene mélange and broken formation of the Hoh Rock assemblage and the overlying relatively undeformed upper Miocene to Pliocene strata in the western core of the Olympic Mountains and offshore suggests that oblique plate convergence and associated deformation waned by the late middle Miocene (*Dickinson, 1979*; *Nilsen and McKee, 1979*; *Snavely and others, 1980b*).

Coastal Klamath and Franciscan terranes

On the southwest Oregon Coast from Cape Blanco south, *Blake and others* (*1985a*) have defined the Gold Beach terrane (Fig. 14), comprising mainly Upper Jurassic and Upper Cretaceous sedimentary rocks in fault contact with previously accreted terranes of the Klamath Mountains. The provenance for the Upper Cretaceous beds does not appear to be rocks now exposed in the Klamath Mountains or northern Sierra Nevada; a southern California source is suggested (*Bourgeois and Dott, 1985*). The Gold Beach terrane may not have been emplaced until after deposition of the Eocene overlap assemblage on the northern margin of the Klamath Mountains (*Blake and others, 1985a*).

The coastal belt of the Franciscan Complex of northern California (Fig. 14) is the youngest on-land part of the accretionary prism, containing rocks of latest Cretaceous to middle Tertiary age (Bachman, 1982; *Bachman and others, 1984*). As in the Olympic Mountains, stratigraphic tops tend to face landward, and paleontological ages tend to be younger seaward, suggesting imbricate thrusting. Structural effects of subduction are intermixed with those of right-slip faulting, produced as the Mendocino triple junction moved northward and encroached on this region (*Bachman and others, 1984*). The coastal belt *sensu lato* is divided into three terranes (*Blake and others, 1985b*); from east to west, these are the Yager structural complex, the coastal belt *sensu stricto,* and the King Range terrane (Fig. 14). The informally named Yager complex of S. B. Bachman and Crouch (1987) consists of Paleogene shale, silty turbidites, and limy mudstone spanning a time period of 25 m.y.; these strata are folded but lack the penetrative stratal disruption that characterizes the older central belt of the Franciscan to the east and the coastal belt of the Franciscan to the west (*Underwood, 1983*; 1985). The coastal belt *sensu stricto* ranges in age from Late Cretaceous to Oligocene. Tectonic juxtaposition of the coastal belt, Yager complex, and central belt occurred prior to deposition of Miocene and younger strata of the Wildcat Group in the Eel River basin (*Bachman and others, 1984*) at about 25 Ma (Sliter and others, 1986). The King Range terrane consists of a "basement" of Upper Cretaceous deep-sea basalts and arkosic turbidites (Point Delgada subterrane) overlain by deep-water argillite and turbidite sandstone with fossils ranging in age from Eocene to middle Miocene or possibly younger (King Peak subterrane). These rocks are in part coeval with the Wildcat Group, indicating accretion after 15 Ma, following the beginning of sedimentation in the Eel River basin (McLaughlin and others, 1982).

Salinian block and southern California terranes

Rocks in much of coastal California, including the Salinian block and the Coast Ranges west of it, most of the Transverse Ranges, part of the Mojave block, and all of the Peninsular Ranges and California Continental Borderland (see Fig. 9), are proposed to have been emplaced into their mid-Tertiary positions relative to stable North America as two composite allochthons, one docking in the Paleocene or early Eocene, and the other probably at a similar time but at least by the mid-Miocene (Fig. 14; *Howell and others, 1985,* 1987). Both were later involved in displacement of about 300 km farther northward on the San Andreas and related faults. The history of these allochthons, however, remains in dispute (Butler and others, 1991).

The Santa Lucia–Orocopia allochthon (Fig. 14; Vedder and others, 1983) is a composite terrane that includes the Salinian block. The Salinian block (Fig. 6) was already at least largely in place at the time of deposition of Eocene borderland basins, because some of the Eocene strata have a western, Salinian source (Nilsen and Clarke, 1975). *Dickinson and others* (*1972*) explained the greater offset of pre-Eocene piercing points by a proto–San Andreas fault, but Ross (1978) stated that rocks of the Salinian block lacked characteristics that would match them with any known counterparts in the Sierra Nevada, Tehachapi Mountains, or San Emigdio Range (see Fig. 9). Other right-lateral faults, however, may produce much of the apparent discrepancy accounted for by a proto–San Andreas hypothesis (Graham, 1978). By the time of final emplacement of the allochthon, the Salinian block had been joined to the Stanley Mountain and San Simeon terranes in the Coast Ranges (Fig. 14; *Howell and others, 1977*; Vedder and others, 1983). Paleomagnetic vectors of Upper Cretaceous and Paleocene strata of the Salinian block show inclinations that are anomalously low compared to cratonic North America, suggesting that the composite Santa Lucia–Orocopia allochthon (comprising the Salinia, Stanley Mountain, and San Simeon terranes in the Coast Ranges and the Tujunga terrane of southern California) was displaced northward about 2,500 km after Cretaceous time but prior to the middle Eocene (*Champion and others, 1984*). Emplacement of the composite allochthon, in this view, is recorded in the juxtaposition of the Tujunga terrane

above the northeast-verging Vincent thrust (*Ehlig, 1981*) upon continental-margin deposits metamorphosed at high pressure (the Baldy terrane, Fig. 14, comprising the Pelona, Rand, and Orocopia Schists). *Powell* (*1981,* 1982), however, regards the Tujunga terrane as having been thrust over rocks interpreted as being part of autochthonous North America prior to the emplacement of Jurassic plutons. Both Ross (1984), in later work revising the conclusion of his 1978 paper, and *James and Mattinson* (*1988*) have found similarities between basement-rock features in the central Salinian block and the southern Sierra Nevada to cast doubt on a far-travelled source for the Santa Lucia–Orocopia allochthon.

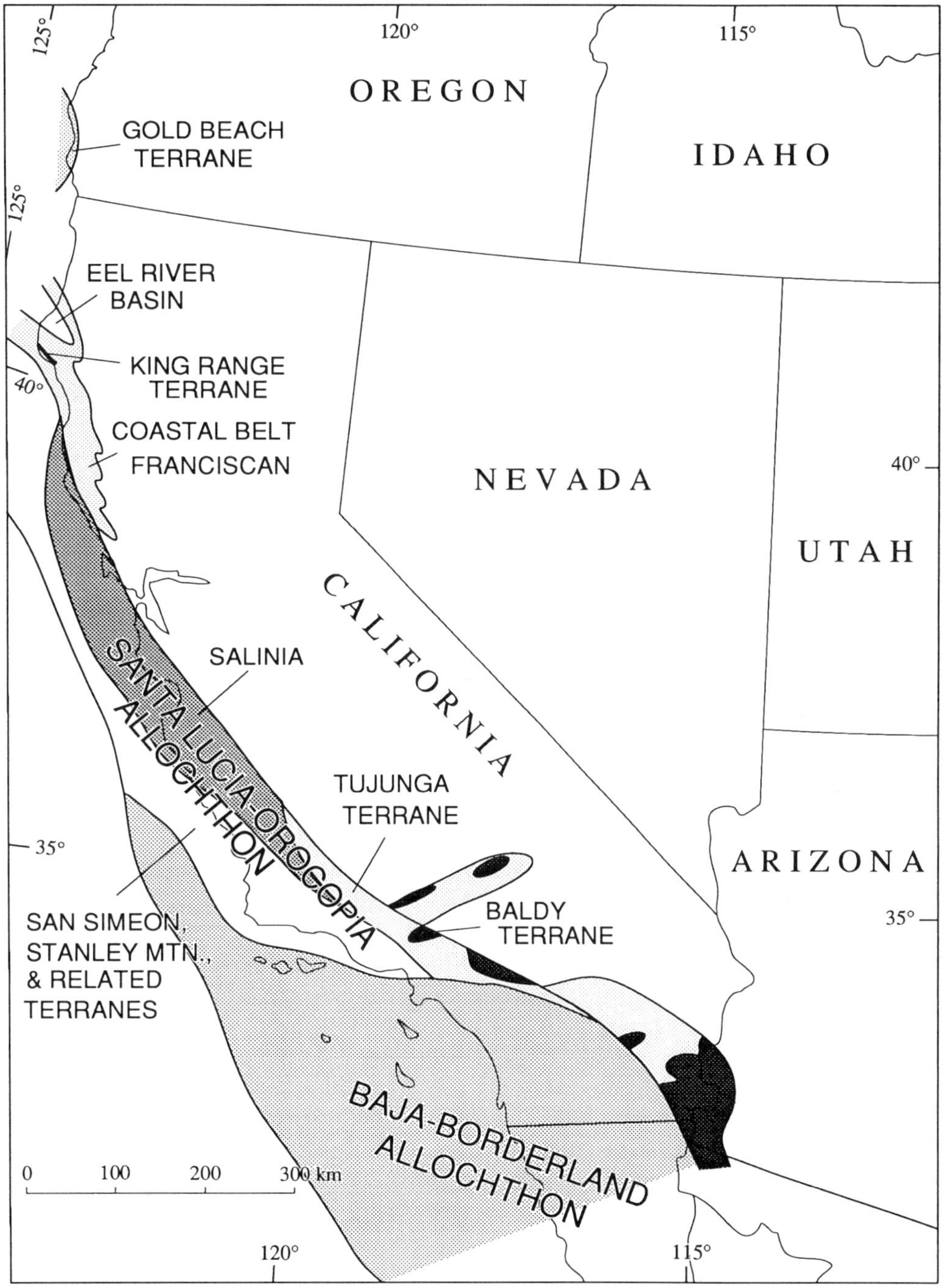

Figure 14. Terranes possibly accreted to California and Oregon in post-Paleocene time. Contact between Tujunga and Baldy terranes is the Vincent thrust. Modified from Silberling and others (1987) and Howell and others (1987).

It has already been noted that the Paleogene strata of the northern Transverse Ranges cannot be correlated easily with strata of similar age in the Channel Islands, Santa Monica Mountains, and Peninsular Ranges. In addition, the basement rocks of the two regions are very different and comprise separate terranes (Silberling and others, 1987). As in central California, rocks older than Miocene commonly have low magnetic inclinations compared to stable North America, as documented for the Southern California batholith (*Teissere and Beck, 1973*), Cretaceous to Eocene strata (*Champion and others, 1981*; *Patterson, 1984*; *Filmer and Kirschvink, 1989*), and a variety of igneous and sedimentary rocks of Mesozoic age (*Hagstrum and others, 1985*). These southern California terranes, amalgamated in the Paleocene or Eocene as the Baja-Borderland allochthon, are therefore interpreted by *Hagstrum and others* (*1985*), *Howell and others* (*1985*), *Filmer and Kirschvink* (*1989*), and *Hagstrum and Filmer* (*1990*) as having been transported 900 to 1,700 km northward during the Late Cretaceous and early to mid-Tertiary. Upper Miocene strata overlap both the central and southern California rock assemblages, and paleopoles for mid-Miocene rocks of Baja California are concordant with those for North America (Hagstrum and others, 1987). In contrast to these ideas, *Gastil and Miller* (*1984*) correlated Paleozoic and Mesozoic facies from Baja California to autochthonous North America. Biostratigraphic and paleomagnetic data for Eocene strata from Baja California imply that the allochthon was part of North America by the early Eocene (*Flynn and others, 1989*), consistent with the correlation of the Eocene Poway clast assemblage with source rocks in northwestern Sonora (Abbott and Smith, 1989). Thus, northward transport of 900 km or more would have to have occurred during the time from the Late Cretaceous to the late Paleocene (*Hagstrum and Filmer, 1990*). Butler and others (1991) have suggested that a uniform southwestward tilt of the batholithic rocks and compaction-shallowing of paleomagnetic inclinations in the overlying sedimentary rocks also could explain the discordant Cretaceous paleopoles for the Baja-Borderland allochthon. Still, the large number of anomalously flat inclinations in a variety of geologic settings and the consistency in the amounts of paleolatitudinal discordance remain difficult to explain without significant northward transport of the outboard terranes.

Early and Mid-Tertiary Continental Erosion, Sedimentation, and Extension

A pattern of relatively localized tectonic extension began to develop in parts of the Cordilleran continental interior soon after the relaxation of Laramide contraction; elsewhere, however, early post-Laramide time was marked by relative quiescence and widespread postorogenic erosion that erased part of the record of earlier events. The following sections, after a brief resumé of the Laramide setting, describe these relatively early stages of post-Laramide geologic evolution, up to the times of distributed regional extension that began in much of the Cordilleran interior in mid-Miocene time.

LARAMIDE SETTING OF THE CORDILLERAN INTERIOR

Tectonic events of the latest Cretaceous and earliest Tertiary greatly changed the geologic framework of the continental Cordillera. Notably, significant crustal shortening marked by a mosaic of thrust-faulted uplifts and basins—the "Laramide orogeny" in the strictest sense—was imprinted across a large part of what had previously been a relatively stable cratonic foreland region (*V. Matthews, 1978*; *Hamilton, 1988*).

Major latest Cretaceous to early Tertiary tectonic elements in the eastern Cordillera (Fig. 1) included reactivated Jurassic to Cretaceous thrust belts extending northward through the Northern Rocky Mountains into Canada and southward from the Mexican border region into Mexico; between, in the former foreland of the Sevier orogenic belt, lay the Laramide orogenic zone of the Middle and Southern Rocky Mountains and its newly emergent borderlands underlain by Upper Cretaceous and Paleocene marine and lowland deposits in the Great Plains and Colorado Plateau. By the end of Laramide deformation, large intermontane basins accumulated sediments; the upper parts of these orogenic sediment fills have been designated widely, especially in the earlier literature, as the Wasatch Formation. Some of the sediments accumulated in extensive lakes, such as the Green River Formation of southwestern Wyoming and the northern Colorado Plateau (*Bradley, 1964*; *Ryder and others, 1976*) and the Flagstaff Limestone (*Fouch, 1976*) and parts of the Claron (Anderson and Rowley, 1975), Sheep Pass (Winfrey, 1960; Fouch, 1979), and Newark Canyon (Smith and Ketner, 1976) Formations of the western Colorado Plateau and eastern Great Basin. Farther south, in southwestern Utah and southeastern Nevada, the Cretaceous Baseline Formation and the probably Upper Cretaceous and, possibly, early Tertiary Iron Springs Formation are orogenic basin deposits (Bohannon, 1983). Orogenic basins of sedimentation are also recognized in the southern Basin and Range region, as far west as the Paleocene and lower Eocene Goler Formation of the Mojave Desert (Fig. 8; *Cox, 1982*; *Cox and Diggles, 1986*).

Continental-margin magmatism, though probably episodic, had remained important in later Mesozoic through Laramide time from Idaho and western Montana northwestward through British Columbia and from southern California, Arizona, and New Mexico southward through the Sierra Madre Occidental (*Armstrong, 1974b*). However, through most of the region between—the latitudinal band of Laramide shortening in the former orogenic foreland—little magmatism persisted after about 70 Ma. The only notable exceptions (Fig. 1) were local centers on and near the Northern Great Plains and Black Hills (*McDowell, 1971*; *Marvin and others, 1973a, 1980*) and in a northeast-trending band of volcanoes and subvolcanic plutons—the Colorado mineral belt—across the Southern Rockies (*Tweto, 1975*).

Some areas of Laramide mountain building experienced

gradually declining orogenic deformation as they were modified by vigorous erosion; sedimentation in major basins adjacent to the principal uplifts was generally more or less continuous. Thus, typically no specific marker of the end of Laramide orogeny is obvious, and definition of the Laramide is sometimes ambiguous. In all but a few areas, however, major Laramide deformation was over by about 55 to 50 Ma. The last stages of Late Cretaceous to early Tertiary crustal shortening occurred late in early Eocene time (e.g., *Keefer, 1965*; *Tweto, 1975*; *Dorr and others, 1977*). In most interior areas of the Pacific Northwest and the Northern Rockies, the last Laramide deformation predated volcanic rocks of about 55 to 53 Ma (*Armstrong, 1974a,* 1978). In the Middle and Southern Rockies the end of the Laramide has commonly been taken to be as young as about 40 Ma (*Coney, 1976a,* 1978), but specific evidence of contractional deformation that young is generally lacking. Epis and Chapin (1975) note that a mature Eocene erosion surface is post-Laramide in the Southern Rocky Mountains; strata of the Southern Rockies and Colorado Plateau of middle Eocene or younger age generally represent quieter depositional environments than older Eocene and Paleocene sediments (Chapin and Cather, 1981; Lucas and Ingersoll, 1981). In central southern Arizona, a regional episode of thrust faulting, metamorphism, and plutonism ended at 58 to 60 Ma (*Wright and Haxel, 1982*; *Haxel and others, 1984*); magmatism certainly ended no later than 50 Ma (Shafiqullah and others, 1980). In this chapter we consider most events younger than about 55 to 50 Ma (i.e., post-"Wasatch") in a post-Laramide context.

POST-OROGENIC EROSION AND DEPOSITION

During early post-Laramide time, before tectonic extension became an important factor in much of the region, both widespread erosion and sedimentation in areally separated basins were active agents of landscape change in the continental interior (Fig. 15). Thus, much of early post-Laramide geologic history is recorded in dispersed and fragmentary evidence of drainage evolution and in sometimes discrete sediment accumulations in postorogenic basins.

Plains of the volcanic plateau region

In early post-Laramide time, eastern Oregon and adjacent parts of Idaho, Nevada, and perhaps southeastern Washington (Fig. 15)—the area that later would become volcanic plateaus of the Columbia Intermontane region and northwestern Great Basin—formed a more or less continuous lowland. Eocene deposits in much of that region are commonly included in the Clarno Formation, exposed principally in central and northeastern Oregon (Fig. 15) but with scattered correlatives elsewhere, including volcanic rocks interbedded with the marine Yamhill Formation in the southern Willamette Valley (Fig. 4; *Yeats and others, 1992*). Beneath the Clarno are Paleocene(?) and Eocene fluvial and deltaic nonvolcanic arkose and coal that have been included in the formation by some investigators but are separated from it by Walker and Robinson (1990a) on the basis of lithology. The Clarno Formation consists mainly of volcanic and volcanic-derived sedimentary rocks (as discussed later in connection with magmatic evolution). Nevertheless, it and the underlying sediments record a large piedmont basin between a northern volcanic highland along the axis of the Blue Mountains (*Swanson, 1969*) and the eastern postorogenic uplands of the Northern Rockies; a comparable basin north of the Blue Mountains may have extended northwestward toward the northern Cascades, but, if so, it would now be entirely concealed by younger basalts of the Columbia Plateau. In central Oregon, a high proportion of the Clarno consists of sandy to clayey sediments—mainly reworked tuffs and mudflow deposits (*Waters and others, 1951*; Walker and Robinson, 1990a). Volcanogenic materials predominate, but erosional debris from the bordering postorogenic highlands is present; interlayered andesitic lavas are volumetrically subordinate. Tectonic disturbances are recorded locally by unconformities within and above the Clarno, but nowhere are they great and none has been correlated widely with any certainty. Upper Eocene clastic sedimentary rocks in the Bull Run and Copper basins of northeastern Nevada, also intercalated with contemporaneous volcanic rocks (Fig. 15; *Axelrod, 1966a, b*; *Ehman, 1985*), probably represent the southeastern margin of the same broad depositional plain although any earlier connection to the Clarno would now be buried by the younger volcanic plateaus or, perhaps, severed by younger Cenozoic extension (Hamilton and Myers, 1966). The eastern margin of this sedimentary plain was a broad erosional upland at general elevations of about 600 to 1,200 m extending from northeastern Washington through central Idaho to northeastern Nevada (Axelrod, 1968).

The Clarno was succeeded in central Oregon by the at least mainly lower Oligocene to lower Miocene John Day Formation (*Peck, 1964*; Robinson, 1975; Robinson and others, 1990). The John Day contains some locally erupted lavas and pyroclastic flows but consists largely of reworked pyroclastic material derived from the Cascades volcanic arc to the west (P. T. Robinson and others, 1984).

Paleomagnetic studies (Beck, 1980; Wells and Heller, 1988; Grommé and others, 1986) indicate that older terranes within and west of the volcanic plateau region underwent significant vertical-axis tectonic rotation in early to mid-Tertiary time. The Clarno basin and adjoining crustal blocks rotated as much as 15 to 30° clockwise during this time while the Oregon Coast Range rotated 50 to 70°. Thus, although much of the rotation is superposed on the basin, it is possible that, to some extent, the basinlike region of the Clarno and John Day Formations formed initially in Eocene time by clockwise rotation and seaward migration of blocks that formerly lay closer to the late Laramide continental margin of the Pacific Northwest (Heller and Ryberg, 1983).

Drainages and basins in the montane regions

In parts of the postorogenic uplands of the Rocky Mountains and eastern Basin and Range Province, early post-Laramide sediments accumulated in basins inherited from the Laramide orogeny. Successively younger deposits typically tended to merge into more extensive aggradational plains. The sediments com-

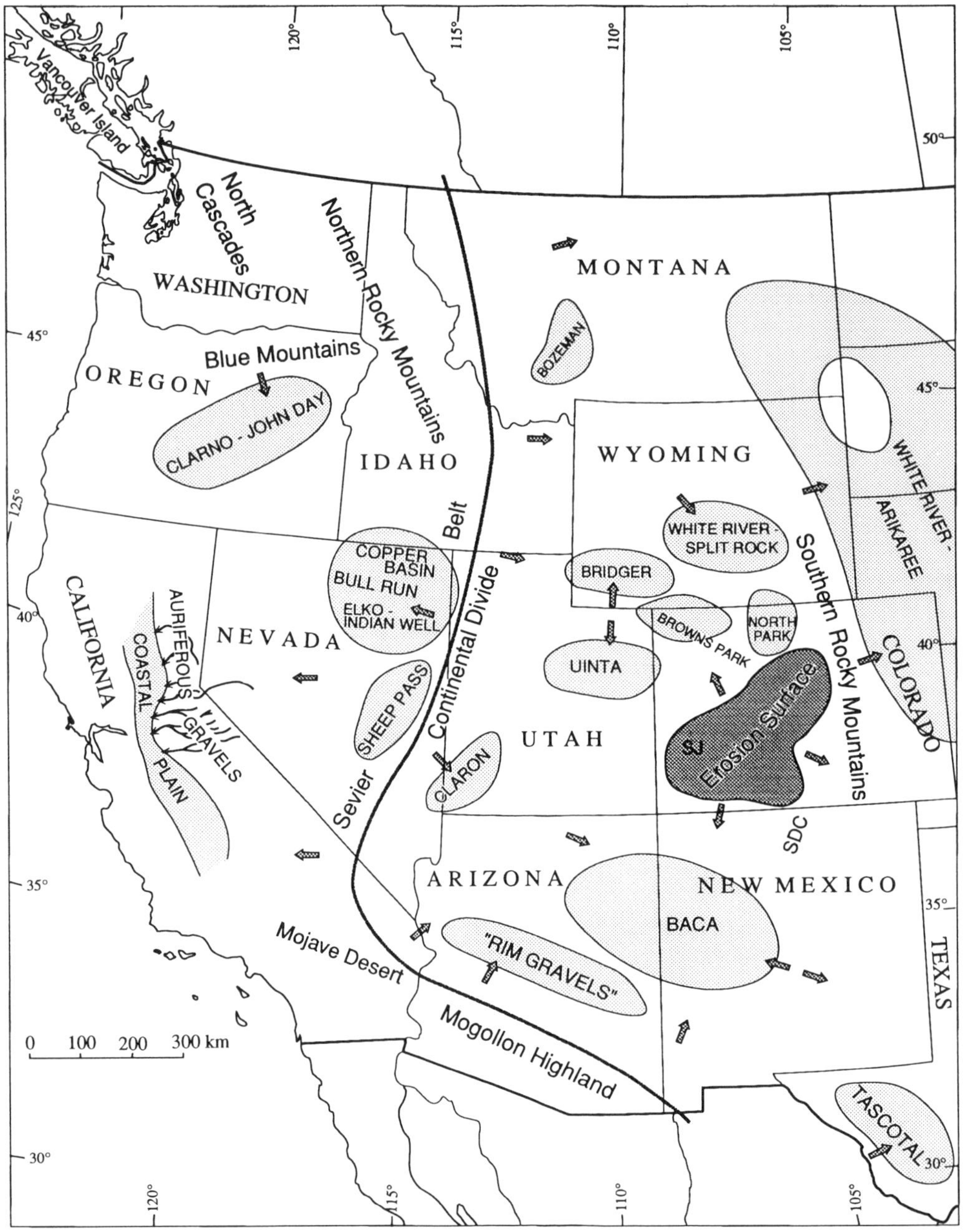

Figure 15. Map showing major depositional basins, erosional features, and stream systems of early post-Laramide time in the United States Cordilleran interior. Dark band, probable general location of continental divide. Names and shading indicate generalized depocenters of selected basins. Arrows indicate probable directions of sediment transport. SJ, San Juan Mountains; SDC, Sangre de Cristo Mountains. Later Cenozoic rotations and displacements not restored.

monly include abundant volcanic components as well as materials derived from erosion of the Laramide mountain blocks.

Early post-Laramide strata in the Northern Rocky Mountains of southern Canada, Idaho, and Montana occur both in certain zones of major localized extension, discussed later, and in remnants of Laramide basins within and east of the Late Cretaceous to Paleocene orogenic belt. In western Montana (Fig. 15), basin deposits of the Bozeman Group and associated units typically contain a lower section with a basal conglomerate and abundant tuffaceous as well as locally derived clastic materials, ranging in age from late Eocene to middle Oligocene (Pardee, 1950; *Robinson, 1961*; *Kuenzi and Fields, 1971*; Reynolds, 1979). These strata generally were deposited in broad shallow basins that span the areas of present-day basins and ranges and

are separated by a regional unconformity from middle or upper Miocene and younger deposits associated with extension on faults more or less like those of the present.

In the Middle Rockies, Wyoming basins, and northern Colorado Plateau, corresponding early post-Laramide sedimentary rocks include the middle and upper Eocene Bridger and Uinta Formations and various correlatives, the widespread Oligocene White River Group, and the Oligocene and Miocene Arikaree Formation and its several correlatives. These deposits, including abundant volcanic materials (*Denson and Chisholm, 1971*) from distant sources to the west and northwest, formed extensive intermontane lowlands that gradually expanded and merged as erosion accelerated in the uplands, perhaps with increasing aridity (Clark, 1975). Important uranium deposits in lower Eocene basin fills probably were leached and redeposited from the volcanic-rich sediments of the White River Formation (*Harshman, 1972*; *Zielinski and Naeser, 1977*). Local drainages (Fig. 15) were from the eroding Laramide and volcanic highlands into nearby basins (*Bown, 1982*); regional drainage was from the zone of major Laramide uplifts in Wyoming eastward onto the Great Plains and also eastward from the Cretaceous to Paleocene thrust belt of western Wyoming and central Utah. The uplifted Laramide ranges of the former foreland were gradually worn down by pedimentation and their flanks partly buried beneath the aggradational mantle to produce a mature regional topography (Love, 1960).

The Southern Rocky Mountains were strongly eroded in early post-Laramide time, and a regional erosion surface truncated most of the extensive upland by late Eocene time (Epis and Chapin, 1975). This surface (Fig. 15), like the aggradational surface of the Wyoming basin, records the integration and extension of a major eastward drainage system off the eroding Laramide highlands. Much of this surface in the Southern Rockies was buried by extensive volcanic deposits in Oligocene and early Miocene time, as described in a later section. Lensoid, generally conglomeratic fluvial deposits occur in places on this surface but are generally datable only as post-Laramide and older than regional Oligocene volcanism. Such deposits include the Blanco Basin Formation and Telluride Conglomerate of the San Juan Mountains (Larsen and Cross, 1956), the Echo Park Alluvium (*Epis and Chapin, 1974*; Scott and Taylor, 1975) and the Vallejo Formation (*Upson, 1941*) of southern Colorado, and similar units in the Sangre De Cristo Mountains of northern New Mexico (*Lipman, 1983*).

Laramide orogenic sedimentation, following a major late pulse of crustal shortening about 55 Ma (Chapin and Cather, 1981), had occurred in numerous basins across the southern part of the Southern Rockies, Colorado Plateau, and High Plains (Fig. 15; Lucas and Ingersoll, 1981); subsequent sedimentation gradually spread across these basins. Upper Eocene and lower Oligocene deposits of the Baca basin (*Schiebout and Schrodt, 1981*), in the Arizona–New Mexico border area, reflect mainly postorogenic northeast drainage from the Mogollon Highland of central and southern Arizona. Regional drainage farther west was northeastward onto the Colorado Plateau, and local basins accumulated sediments (Young and McKee, 1978; Young, 1979; *Nilsen and McKee, 1979*). "Rim gravels" of 54 to 28 Ma along the southern margin of the plateau (Peirce and others, 1979; Nations and others, 1982) probably represent a piedmont accumulation adjacent to the highland.

The present Basin and Range region of southern Arizona and New Mexico during the early Tertiary seems to have been an area of moderate elevation and topographic relief. The mountainous Mogollon Highland (Figs. 1, 15) was a persistent remnant of Laramide and older topography (Chapin and Cather, 1981; Reynolds and Spencer, 1985). Continued uplift of this highland into the early Tertiary was accompanied in parts of southern Arizona by an episode of epithermal fissure-vein mineralization, but no local igneous activity appears to have been related to that episode (*Tosdal, 1981*). By late Eocene and early Oligocene time, this highland was largely bevelled by erosion (Shafiqullah and others, 1980). Early Tertiary aggradation by both sedimentary and volcanic units in southwestern New Mexico, including voluminous welded ash flows, formed a broad region of low relief by about 32 Ma. Continued Oligocene and early Miocene sedimentation reflected shifting drainage patterns as local volcanic fields accumulated on this constructional surface (*Wrucke, 1961*; *Kautz and others, 1981*; *Gawne, 1981*). In part, some of the clastic accumulations in basins of southern Arizona such as the Pantano, Whitetail, San Manuel, Higley, and others (see Fig. 16; Nations and others, 1982), preserve portions of the montane drainage system that was wearing down the Mogollon Highland;

Figure 16. Map of early and mid-Tertiary extensional zones and metamorphic "core complexes" in the United States Cordilleran interior. Heavy dashed lines delimit extensional belts in the eastern Great Basin and southern Basin and Range regions. Red, "metamorphic core complexes": A, Albion Mountains; B, Buckskin Mountains; BE, Buckeye Mountains; BH, Bullfrog Hills; BM, Big Maria Mountains; BRD, Bitterroot dome; CH, Chemuevi Mountains; C-R, Catalina-Rincon Mountains; GC, Grouse Creek Mountains; Hq, Harcuvar Mountains; Hr, Harquahala Mountains; KD, Kettle dome; MR, Mineral Ridge; OD, Okanagan dome; P, Picacho; PD, Pioneer dome; PM, Pinaleno Mountains; PR, Priest Rapids; RM, Ruby Mountains; RR, Raft River Mountains; SD, Spokane dome; SM, South Mountain; SR, Snake Range; ST, Santa Teresa Mountains; T, Tortolita Mountains; TH, Trappman Hills; W, Whipple Mountains; WT, White Tank Mountains. Other designated features: BBT, Barstow-Bristol trough; BHR, Beaverhead range; CG, Chiwaukum graben; CH, Candelaria Hills; CM, Chocolate Mountains; DV, Death Valley; E-LF, Entiat-Leavenworth fault; FHD, Flathead detachment; GR, Grant Range; HMD, Heart Mountain detachment; JL, Jemez lineament; LCL, Lewis and Clark line; NDF, Newport detachment fault; NM, Newberry Mountains; PF, Puget fault; PT, Purcell Trench; RG, Republic graben; RMT, Rocky Mountain trench; SB, Saphire block; SCF, Straight Creek fault; SDC, Sangre de Cristo Mountains; SP, Spanish Peaks; TG, Toroda Creek graben; WL, Walker Lane. Hachured lines, selected extensional faults discussed in text. Arrows show trend and vergence of kinematic indicators in selected "core complexes" after Wust (1986). Selected depositional basins are named in boxes. Later Cenozoic rotations and displacements not restored.

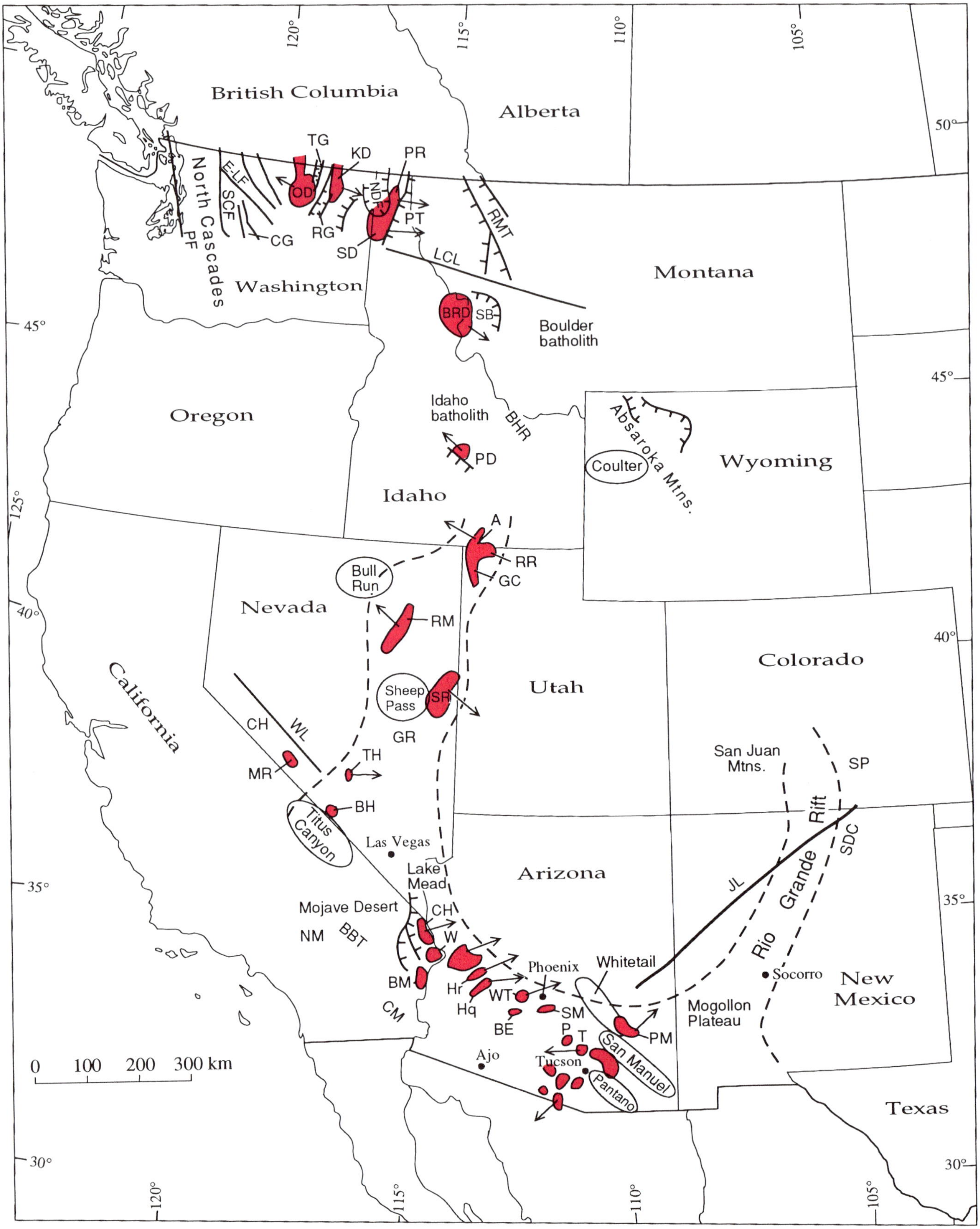
British Columbia
Alberta
Montana
Washington
Oregon
Idaho
Wyoming
Nevada
Utah
Colorado
California
Arizona
New Mexico
Texas
North Cascades
Boulder batholith
Idaho batholith
Absaroka Mtns.
Coulter
Bull Run
Sheep Pass
San Juan Mtns.
Rio Grande Rift
Titus Canyon
Las Vegas
Lake Mead
Mojave Desert
Phoenix
Whitetail
Socorro
Mogollon Plateau
Ajo
Tucson
San Manuel
Pantano
0 100 200 300 km

in greater part, however, they reflect topography formed by mid-Tertiary tectonic extension, discussed later.

The early Tertiary highland area continued westward into the Mojave Desert of southeastern California. Biotite $^{40}Ar/^{39}Ar$ data from the Mojave region (*Miller and Sutter, 1982*) indicate that not until about 45 Ma had uplift and erosion brought plutonic terranes, which had remained hot through Late Cretaceous and earlier Tertiary time, to crustal levels where they cooled to at least 280 ± 40°C. After that time, the pre-Miocene central Mojave Desert region remained as a highland undergoing regional uplift and vigorous erosion (*Hewett, 1955*) to expose a Mesozoic plutonic complex with scattered prebatholithic and synbatholithic inliers. Dokka and Glazner (1982) estimate 3 km of erosion during the early Tertiary; no Tertiary deposits older than about 25 Ma have been recognized on this regional erosion surface (Armstrong and Higgins, 1973; Woodburne and others, 1982). Drainage from the eastern part of the highland was northeast onto the Colorado Plateau (Fig. 15); faulting along the Plateau–Basin and Range boundary did not disrupt that eastward drainage until sometime after emplacement of the 18-Ma Peach Springs Tuff (Young and Brennan, 1974; Young and McKee, 1978).

During early post-Laramide time most of the present eastern Great Basin region (Fig. 15), from southern Nevada to central Idaho, was an eroding upland that shed sediments eastward toward the Colorado Plateau (Stewart, 1983b) and sustained long streams westward to the Pacific Ocean (Bateman and Wahrhaftig, 1966). By Oligocene time, volcanism became predominant and ultimately built a plateaulike surface across the region, but prevolcanic sediments locally demonstrate that there were large basins of post-Laramide deposition similar to those of the Rockies and the Colorado Plateau. Basins of Laramide age in eastern Nevada continued to receive sediment in early post-Laramide time as widespread erosion and successive reintegration of drainage occurred, just as they did farther east. In northeastern Nevada (Smith and Ketner, 1976; Solomon and others, 1979) the Elko Formation of Eocene age consists mainly of lacustrine limestone and shale and interbedded tuffs of distal origin, overlying a surface that bevels earlier Tertiary and latest Cretaceous orogenic-basin deposits. Succeeding the Elko, the Indian Well Formation of Eocene and Oligocene age contains more abundant volcanic materials but indicates persistence of the basin as a site of sediment accumulation. In east-central Nevada, prevolcanic fluvial and lacustrine strata in the lower part of the Sheep Pass Formation (Fouch, 1979; Fouch and others, 1979) probably include both Laramide orogenic conglomerates and post-Laramide, lower and middle Eocene basin fills; above, middle Eocene to lower Oligocene lacustrine sediments contain abundant volcanic detritus. The postorogenic Claron Formation is widespread in southwestern Utah across the Great Basin–Colorado Plateau transition and appears to represent deposition in a broad shallow basin nearly continuously from Laramide time until the beginning of local volcanism in the mid-Oligocene (Anderson and Rowley, 1975). In numerous areas of the northern Basin and Range Province, channel conglomerates lie beneath the oldest local volcanic deposits and preserve remnants of the streams that drained the postorogenic highland region, both to the east in western Utah (*Morris and Lovering, 1961*) and to the west in western Nevada (*Proffett and Proffett, 1976*). As elsewhere in the Cordillera, aprons of volcanogenic sediments accumulated around the margins of rapidly eroding mid-Tertiary volcanic highlands (*J. J. Anderson, 1971*).

The Sierra Nevada underwent erosion and deep weathering in the Eocene and, before the inception of major Oligocene volcanism in the headwaters region farther east, carried a major west-flowing drainage system to the Pacific (Fig. 15; *Lindgren, 1911*; Bateman and Wahrhaftig, 1966). Runoff from the initially rugged topography eroded Mesozoic gold-bearing veins in the foothills of the Sierra Nevada and recycled the gold into important placer deposits. Drainage integration reduced the Sierran slope to a subdued landscape with broad flood plains by middle or late Eocene time; with the onset of volcanism to the east, perhaps as early as 38 Ma, the streams became charged with pyroclastic material and many of the drainage courses were rearranged (*Yeend, 1974*).

To summarize, this section describes a prevalent pattern of early post-Laramide Cordilleran landscape evolution with vigorous erosion in postorogenic highlands that shed their debris first into separated successors of the Laramide orogenic basins and later into coalescing aggradational mantles. Volcanism was also important though it shifted greatly with time. By mid-Miocene time, regional drainage was highly integrated, and much of the sediment load bypassed the montane and intermontane region.

TECTONIC EXTENSION IN THE CORDILLERAN INTERIOR

The regional pattern of erosion and sedimentation before mid-Miocene time was sharply punctuated by belts of tectonic extension. Zones of extension formed in different areas at different times—generally during the Eocene in the Northern Rockies, the Oligocene in the Great Basin region, and Oligocene to middle Miocene in the southern Basin and Range Province and Southern Rocky Mountains—but in each area they accompanied significant magmatism. At any one time, this extension was confined to zones no more than a few hundred kilometers in width within the late Mesozoic and Laramide orogenic belts from the Northern Rockies through the present eastern Great Basin region to the southern Basin and Range Province and the present Rio Grande rift zone, which extends northward into the Southern Rockies (Fig. 16). Middle Eocene to middle Miocene extension in these areas, now expressed prominently by detachments and other low-angle faults, predates the distributed extension of the wider Basin and Range Province.

Among the important expressions of this mid-Tertiary extension are terranes sometimes denoted as "metamorphic core complexes" (Crittenden and others, 1980), characterized by exposures of mid-crustal metamorphic and plutonic rocks, sometimes with a gneissic or mylonitic overprint, in sheared contact with nonmetamorphosed stratified cover rocks that are attenu-

ated by normal faults, commonly with low dips. These complexes occur mainly in a zone about 100 to 200 km wide (Fig. 16). The eastern margin of this zone generally lies 100 to 200 km west of the thrust-belt front in the Northern Rocky Mountains and 200 to 300 km west of the Sevier fold and thrust belt in the eastern Great Basin. Southwest and south of the Colorado Plateau, the belt of "metamorphic core complexes" steps eastward through the Laramide Mogollon Highland; the belt appears to end eastward in southeastern Arizona and northern Sonora, 300 km west of the Laramide front in south-central New Mexico and west Texas (see Fig. 1). Some views of the significance of these complexes emphasize their spatial relation to Mesozoic compressional deformation and regard them as exposing a metamorphic infrastructure to that deformation (*Armstrong and Hanson, 1966*; *Brown and Read, 1983*). Other workers emphasize the large Cenozoic tectonic extension that juxtaposes rocks of greatly contrasted deformational histories (Davis and Coney, 1979; Coney, 1979).

Controversy remains over the ages and styles of deformation associated with these complexes. At least most of them are associated, at least in part, with closely spaced Tertiary normal faulting, large stratal rotations in hanging-wall blocks, and fault detachment of younger (including Tertiary) over older rocks along areally extensive surfaces, indicating a major role for Tertiary extension in at least their later evolution. Relations near some of the detachments indicate large relative extensions and the formation of tectonite fabrics in the lower plates at mid-crustal depths (*Davis, 1983*). Differences of opinion exist as to whether the detachment surfaces have large lateral displacements and juxtapose rocks from greatly different crustal levels (Wernicke, 1981; Bartley and Wernicke, 1984; Reynolds and Spencer, 1985; Wernicke, this volume) or are uplifted, essentially in situ exposures of a former mid-crustal transition layer between zones of brittle and ductile extension (Eaton, 1982; Miller and others, 1983; Gans and others, 1985). Furthermore, evidence in some of the complexes for Mesozoic metamorphism and high-strain deformation (Armstrong, 1982; Coney and Harms, 1984) as well as Tertiary extension, complicates the interpretation of their tectonic significance.

Zoback and others (1981) demonstrated that the state of stress throughout the region of mid-Tertiary extension was relatively uniform; axes of least principal stress (σ_3) by about 20 Ma, were horizontal and east-northeast throughout most of the region of extension. Wust (1986) showed that, regardless of age within the mid-Tertiary, extension directions of varied features over large areas are consistently perpendicular to the trend of the extensional belt and verge both ways from its central axis (Fig. 16).

Northern and Middle Rocky Mountains region

Although much of the continental interior was undergoing tectonic extension, the overall pattern of Eocene and Oligocene deformation in the northern Cascades appears to define a regime of crustal shortening by folding associated with oblique, or transpressional strike-slip faulting (Ewing, 1980; Tabor and others, 1984; Johnson, 1985b; Wernicke and others, 1987). Major strike-slip faults, including the Straight Creek, Entiat-Leavenworth, and Puget faults (Fig. 16), have strike lengths greater than 100 km, trend generally north to northwest, and bound deep basins of nonmarine deposits. Johnson (1985b) notes that rapid subsidence and uplift, abrupt thickness and facies changes, frequent drainage readjustments, and association with volcanism characterized sedimentation in these basins. A swarm of north-south dikes associated with 47-Ma volcanism (G. A. Davis, 1980; *Tabor and others, 1982*) indicates local extension. Also, the Chiwaukum graben in the eastern part of the area (Fig. 16) formed as an extensional structure in middle Eocene time during deposition of the arkosic Chumstick Formation about 45 Ma (*Gresens, 1980*). Nevertheless, the area was shortened soon afterwards, as basement-cored folds of the Chumstick are beveled beneath an unconformity at the 34-Ma base of the Wenatchee Formation. Following the erosional episode, later deformation of the Wenatchee was by more open folding and thrusting, contemporaneous with continued folding in the Cascades. Johnson (1985b) concludes that extensional features of this region developed locally and intermittently in relation to pull-apart strains between long strike-slip faults in a zone of oblique plate convergence.

Farther east, from the Okanogan Highlands eastward, extension clearly dominated Eocene tectonics (Fig. 16). A group of structural domes with "metamorphic core complexes" in northeastern Washington lies within a United States extension of the Omineca crystalline belt of southern British Columbia; in Canada the metamorphic complexes of this belt are collectively called the Shuswap terrane. The westernmost of these in the United States, the Okanogan dome, is cored by a gneissic Cretaceous pluton, increasingly foliated towards its margins and mantled by gneiss that yields K-Ar and fission-track ages of 66 to 46 Ma (Fox and others, 1977). Farther east, the Okanogan dome is separated from the Kettle dome by a complex faulted synformal structure—the Republic and Toroda Creek grabens (*Muessig, 1967*; Pearson and Obradovich, 1977)—with Eocene sedimentary and volcanic rocks deformed along the same trend as the dome axes (Cheney, 1980). Similar structures in Eocene strata but also including mylonites separate the Kettle dome from the uplifted miogeoclinal region farther east, and a thick sedimentary cover is detached from uplifted basement north of Spokane along the spoon-shaped mylonitized shear zone of the Newport detachment fault (*F. K. Miller, 1971*; Cheney, 1980; *Rhodes and Cheney, 1981*). Thus, structural relief of the metamorphic-cored uplifts is Eocene or younger in age. Many of the principal faults in the Eocene sequences have low dips. Younger Eocene strata were confined to or greatly thickened in the structural depressions and show growth relations to the faults. Parrish and others (1988) showed Eocene extension of about 30 percent on ductile and brittle normal faults across the international boundary.

Farther eastward, Eocene extension is recorded in structures related to the Purcell and Rocky Mountain trench systems that

extend into Idaho and Montana from Canada. Nonmetamorphic rocks east of the Purcell trench (Fig. 16; *Harrison and others, 1972*) are separated from the crystalline complex of Priest River west of the trench by an east-dipping normal fault (*Rhodes and Hyndman, 1984*) that is interpreted by Parrish and others (1988) as the root zone of the spoon-shaped Newport detachment fault, which overlies the Priest River Complex to the west. In southernmost Canada and Montana, the southern Rocky Mountain trench and the parallel Flathead Lake depression, which cut Laramide structures, are half grabens with downward-flattening normal faults along their east margins and Eocene sedimentary fills and graben floors that dip eastward into the bounding faults (*Leech, 1966*; *Clague, 1974*).

The Bitterroot dome lies in the northeastern part of the Idaho batholith (Figs. 1, 16) and is separated by an extensive mylonitized detachment surface from mainly nonmetamorphosed cover rocks of the Sapphire block that moved eastward across the area of the dome (Hyndman, 1980). The mylonitic zone along the detachment surface is younger than sheared Late Cretaceous and Paleocene batholithic rocks and is intruded and overlain by Eocene igneous rocks, about 50 to 40 Ma. The volcanics are brecciated, suggesting significant deformation during and after their accumulation. Rb-Sr systems (*Chase and others, 1978*), apatite fission-track ages, biotite K-Ar ages (*Armstrong, 1974a*; *Ferguson, 1975*), and O-isotopic systematics (*Kerrich and Hyndman, 1986*) in the northeastern lobe of the Idaho batholith were reset by thermal events associated with uplift and magmatism between about 40 and 45 Ma. Farther south along the east margin of the Idaho batholith, a roof of lower-grade metamorphic rocks and 50- to 44-Ma volcanics is tectonically detached from a crystalline complex in the Pioneer Mountains (Fig. 16; *Dover, 1982*; *O'Neill and Pavlis, 1988*). K-Ar ages in the core are generally 46 to 43 Ma, including a nonfoliated granodiorite that yields 46 Ma. To the east, in the Beaverhead Range, a pre-Miocene crystalline-cored domal uplift and peripheral displacement of younger strata over older on flat faults have been interpreted as Laramide contractional structures (*Scholten and Ramspott, 1968*), but the age of deformation has not been documented by specific evidence. The structural style resembles Eocene extensional structures elsewhere in the Northern Rockies.

Not only above the crystalline "core complexes" but elsewhere, Eocene sedimentary and volcanic rocks of the Northern Rockies accumulated during tectonic extension on more-or-less north-south faults. Volcanics of 51 to 40 Ma in the central Idaho batholith region (McIntyre and others, 1982) and the Boulder batholith area of Montana (Fig. 16; Smedes, 1962; Smedes and Thomas, 1965) are interpreted from stratigraphic relations to extensional faults to have erupted in association with displacement on the faults.

An unusual group of large-displacement extensional structures was associated with the growth of an early and middle Eocene volcanic field in the northeastern Absaroka Range of northwestern Wyoming and adjacent Montana (Fig. 16). The major structure was early referred to as the Heart Mountain thrust, but Pierce (1957) demonstrated its detachment origin. The detachment broke away from a headwall region in the Absaroka Range and was bounded by a basal bedding-plane fault that cut upsection to meet the ground surface. Faults in the upper plate terminate against its base. The Heart Mountain is only one of several similar, apparently related detachment structures in the Absaroka–Big Horn area. Their origin has long been controversial, but Pierce's studies (1957; *1963*, 1973) showed that they are extensional. He interpreted wide separation of recognized allochthonous blocks and the filling of intervening terranes by volcanic rocks believed to be autochthonous as having formed in a brief, even catastrophic event, the allochthon separating and dispersing into numerous independent blocks. On the other hand, Hauge (1985) suggested that many volcanic rocks between the obvious allochthonous blocks are themselves allochthonous and concluded that the Heart Mountain allochthon was continuous and was propelled by slow gravitational spreading of the growing volcanic highland.

In the Rocky Mountains of southwestern Montana and adjacent Idaho (Fig. 16), other extensional faulting is younger than the Eocene extension just reviewed. This younger extension is represented by tilted linear fault-block ranges and basins, resembling the modern Great Basin region in both structural pattern and timing (Pardee, 1950; Reynolds, 1979). In addition to these younger faults, however, broader basins of this region formed and were filled by coeval upper Eocene to middle Oligocene sediments that were bevelled by erosion before mid-Miocene time. Although bounding faults are not recognized along these early basins, tectonic extension may have affected the area in late Eocene to mid-Miocene time as well as later.

Thus, in summary, post-Laramide extension in the Northern Rockies began late in early Eocene time and is revealed in low-angle faulting and other structural features. At upper-crustal levels, broad basin formation and widespread volcanism accompanied extension. In a belt through northeastern Washington and northern Idaho, about 100 to 200 km west of the eastern limit of extension, mid-crustal rocks are juxtaposed with detached upper-crustal sections.

In the westernmost Middle Rockies, Cenozoic extension began no earlier than the time of deposition of the lower to middle Miocene Colter Formation (Fig. 16) in a deep, rather restricted basin nearJackson Hole, accompanied by mafic to intermediate volcanism from small local centers (Love, 1956; *J. D. Love and others, 1973, 1976*). For the most part, however, relief in the Middle Rockies and Wyoming Basins continued to be evened out by aggradation and pedimentation until mid-Miocene time.

Sierra-Wasatch region

A major zone of latest Eocene to middle Miocene extension spans the eastern Great Basin along a belt 200 to 300 km west of the frontal thrusts of the older Sevier orogenic belt (Figs. 15, 16). This belt of mid-Tertiary extension is characterized by low-angle

normal faults, steeply dipping strata (>40°) of early Miocene and greater age, "metamorphic core complexes," and peraluminous granitic intrusive rocks (Armstrong, 1972a; *Best and others, 1974*; Miller and Bradfish, 1980; Stewart, 1983b). The extension is distinct in both time and location from the regionally distributed extension of later Cenozoic time.

In the Nevada-Utah-Idaho border area, metamorphic complexes in the Albion, Raft River, and Grouse Creek Ranges (Fig. 16) may have evolved in Mesozoic and Tertiary time (Armstrong, 1968; *D. M. Miller, 1980*), but major features appear to date from mid-Tertiary extension. Relatively undeformed 38- to 25-Ma mesozonal stocks intrude the gneissic core rocks (*Armstrong, 1976*; Compton and others, 1977). Folding and ductile deformation is related to east-west extension that began near the end of the Eocene and continued until 25 to 20 Ma (Compton and others, 1977; *Todd, 1980*; *Compton, 1980, 1983*). A short distance to the east, late Mesozoic thrust faults may have been reactivated as low-angle extensional faults in mid-Tertiary time (*Allmendinger and Platt, 1983*).

Evidence for mid-Tertiary extension is abundant in parts of the eastern Great Basin farther south (Fig. 16). Isotopic ages of mid-Mesozoic metamorphic and plutonic rocks in the Ruby Mountains of northeastern Nevada were reset to around 36 to 21 Ma. These rocks grade upward from migmatitic gneiss through a high-strain zone of ductile deformation and are juxtaposed by low-angle normal faults against a nonmetamorphosed allochthonous cover (Snoke, 1980); the low-angle faults cut Tertiary rocks as young as 17 to 15 Ma. Unconformities above steeply dipping older Tertiary rocks (commonly greater than 35 to 45°) and north- to northeast-trending faults in the Pinon Range and Elko areas of north-central Nevada are bracketed by ages of 37 and 35 Ma, synchronous with earliest Cenozoic local magmatism (Smith and Ketner, 1977; Solomon and others, 1979). Smith and Ketner attributed the deformation to a late compressional event on the basis of a fold beneath the unconformity, but the steep dips—and even the folding—are similar to patterns elsewhere in the region of mid-Tertiary extension. Just north, in the Bull Run basin, upper Eocene sedimentary and volcanic rocks and upper Paleozoic rocks are juxtaposed against lower Paleozoic miogeoclinal sedimentary rocks along low- and high-angle normal faults (*Ehman, 1985*). Although much of the Upper Cretaceous to Oligocene Sheep Pass Formation of east-central Nevada was deposited in Laramide-age basins (Fouch, 1979; Fouch and others, 1979), normal growth faults, large incorporated slide blocks, and the distribution of coarse clastics suggest that parts of it accumulated during extensional faulting (*Newman, 1979*). In the Snake Range and vicinity, a 120-km-wide belt exposes metamorphic rocks, juxtaposed along a low-angle detachment (*Misch, 1960*) with nonmetamorphosed strata that are rotated on normal faults and emplaced against the detachment surface (Armstrong, 1972a; *Coney, 1974*). K-Ar mica ages on Precambrian to Jurassic crystalline rocks below the detachment were reset to values as young as 17 to 20 Ma (*Lee and others, 1970*). Within the Snake and Schell Creek Ranges, the detachment zone accommodated an average of 95 km of extension (250 percent), largely during the Oligocene, during local volcanism (Miller and others, 1983; Gans and others, 1985). Good stratigraphic control indicates that the rocks were subhorizontal before emplacement of the Oligocene units. There is, however, disagreement as to whether the extension was more or less in situ, virtually denuding an uplifted mid-crustal section that originally lay at a depth of about 7 km (Miller and others, 1983; Gans and others, 1985) or was part of a regional crustal shear zone extending to greater depth (Bartley and Wernicke, 1984). In various parts of the region, extensional faulting both predated and accompanied Oligocene volcanism (*W. J. Taylor and others, 1989*). Locally impressive folding of detached Tertiary strata in the adjoining region (*Moores and others, 1968*) probably reflects gravitational deformation associated with extension and uplift.

In south-central Nevada, outcrops of metamorphic rocks (the Trappman Hills, Bullfrog Hills, and Mineral Ridge, Fig. 16), separated from overlying steeply dipping volcanic rocks by low-angle faults, have generally been considered to be locally upfaulted basement blocks, but associated large stratal rotations and K-Ar data suggest that they are "metamorphic core complexes" like those farther northeast (Stewart, 1983b). The Mineral Ridge structure is considerably younger and is discussed later. Near Death Valley, in the southwestern part of the region, the Titus Canyon Formation (Fig. 16) is a thick lower Oligocene (Stock and Bode, 1935) and younger fluvial and lacustrine sequence that largely predates local volcanism. The Titus Canyon includes coarse conglomerates, sedimentary breccias, and megabreccia lenses that suggest syndepositional faulting although regional relief remained low, as shown by conformably overlying units (*Reynolds, 1969*). Other lower Oligocene to lower Miocene units in the Death Valley area also appear to be associated with synchronous faulting (*Cemen and others, 1982*) that might reflect early tectonic extension. As described later, however, the most conspicuous detachment faulting near Death Valley is of late Cenozoic age.

Although mid-Tertiary extension in the present Great Basin region was considerable—and locally extreme—it was largely confined to a narrow belt in the hinterland of the late Mesozoic Sevier orogenic belt. Nevertheless, somewhat ambiguous evidence suggests that at least some mid-Tertiary extension may have been more widely distributed. Wherever local evidence of such extension has been described (Fig. 16), it appears to have narrow limits, both spatial and temporal. Zoback and others (1981) note that high dips of fault-rotated middle Tertiary volcanic rocks also occur in the center of the Great Basin and that these volcanics must be cogenetic with large intrusive masses; if those masses added new material to the crust, local extension related to batholithic emplacement is implied. In west-central Nevada, a northwest-southeast elongate trough that accumulated extensive ash-flow tuffs of 27 to 22 Ma is interpreted by Ekren and others (1980) as a preexisting rift zone along the course of the younger right-lateral Walker Lane. A localized east-west extensional fault trough of late Oligocene to early Miocene age has

been interpreted on the basis of stratigraphic and gravity data in the Candelaria Hills of central-western Nevada (*Speed and Cogbill, 1979b*). Robinson and Stewart (1984), however, argue that it is of volcanic-collapse origin. In south-central Nevada, Ekren and others (1968), and *Anderson* (*1978*, p. 423) showed that an early generation of extensional faulting began shortly after eruption of the oldest volcanics of the area, about 27 Ma.

In contrast to these widespread but local extensional relationships, later discussion will show that extensive welded ash-flow sheets of 34 to 18 Ma were emplaced across virtually the entire region on a surface of low relief and progressively built a plateau-like constructional volcanic surface that was broken only locally—mainly by structures related to volcanic processes in the source areas of the ash flows themselves. Only in the proportion that magma in the chambers related to this volcanism originated in the mantle would significant crustal extension be implied directly by the magmatism itself. J. H. Stewart (1980) has suggested that east-west alignments of volcanic centers and volcano-related fault-bounded troughs in central Nevada may represent a middle Tertiary regime of north-south extension. Specific evidence for the corresponding stress orientation, however, is lacking; the suggestion seems to be contradicted by east-northeast directions of least principal stress interpreted from stratal rotations in those same terranes by Zoback and others (1981).

Rowley and others (*1978a*) interpreted uppermost Oligocene and younger ash-flow tuffs that are extensive in the eastern Great Basin but pinch out rapidly near the margin of the Colorado Plateau to indicate structural differentiation between these provinces to have begun as early as 26 Ma. Anderson and Mehnert (1979), however, argue that these distributions reflect changes in the source areas of the ash flows and showed convincingly that west-derived debris flows crossed the boundary after 21 Ma. Furthermore, extensive ash-flow sheets of the Great Basin show that there was no widespread development of parallel linear fault blocks before about 21 Ma.

Thus, major mid-Tertiary extension during regional volcanism in the present Great Basin, including the juxtaposition of mid-crustal crystalline rocks with brittlely deformed upper-crustal rocks, occurred in a belt west of the Mesozoic orogenic front; brief episodes of localized extension, perhaps tectonic, may have been more widespread. Any structural differentiation of the present Basin and Range Province from a higher Colorado Plateau beginning late in this period of pre–basin-range extension was minor.

Mojave-Sonoran region

In the Basin and Range Province south of the latitude of Las Vegas, early to mid-Miocene normal and detachment faulting, locally associated with crystalline "core complexes," attenuated the Tertiary section (Fig. 16) before younger high-angle normal faults blocked out at least some of the basins and ranges of present-day topography. Extreme extension related to strike-slip faulting north of Lake Mead and to low-angle faulting in the region immediately south of Las Vegas began about 14 Ma (R. E. Anderson, 1971; Anderson and others, 1972; Bohannon, 1984), later than in areas farther south and after the time period discussed in this section. Fault-controlled basins formed near Lake Mead between about 17 and 14 Ma, however, and localized sedimentary basins formed even earlier (as early as 20 Ma; Bohannon, 1983, 1984) suggesting that initial extension began before the low-angle faulting described by Anderson. Glazner and Bartley (1984) propose that, between southern Arizona and Las Vegas, major episodes of mid-Tertiary extension moved progressively northward with time.

The Mojave Desert region of eastern California was long thought to have been a persistently stable block through middle and late Tertiary time (*Hewitt, 1955*; Garfunkel, 1974), but recent work demonstrates major extension and significant stratal rotations during the early Miocene (*Miller and others, 1982*; Dokka, 1986; *Dokka and Woodburne, 1986*; *Glazner and others, 1989*). An extensional belt of thick Tertiary volcanic and sedimentary rocks, the Barstow-Bristol trough, trends northwest across the central Mojave region (Fig. 16). Early Miocene extension in and near the Newberry Mountains was along basal detachments. Displacements were northeast-southwest on an undulating surface against which a section of mainly Miocene volcanic rocks dips generally 40 to 50°; both low- and high-angle normal faults cut the detached blocks. Locally the upper-plate rocks are folded along axes parallel to the extension direction, perhaps as an effect of movement across the undulating detachment surface. The basal section of 23 to 24 Ma age that accumulated and became detached on this extending terrane is mainly volcanic but includes fluviatile and lacustrine sediments deposited across many, now separate ranges and basins of the area (Armstrong and Higgins, 1973; *Burke and others, 1982*; Woodburne and others, 1982). Younger rocks in the Barstow-Bristol trough are 22- to 18-Ma volcanics and contemporaneous lacustrine sediments, bracketing major extension between about 24 and 20 Ma.

Northeast-southwest extension in a 200-km-wide corridor along the lower Colorado River between California and Arizona (Fig. 16) was about contemporaneous with that of the central Mojave Desert. Metamorphic "core complexes" are separated by regionally extensive detachment surfaces from fault-attenuated blocks that include crystalline basement rocks, pre-Tertiary sedimentary or metasedimentary rocks, and Tertiary strata (Rehrig and Reynolds, 1980; Davis and others, 1980, *1982*, 1986; *Spencer, 1985*; Howard and John, 1987; John, 1987; Davis, 1988; *Davis and Lister, 1988*). Stratal rotations of the detached blocks, growth-fault relations, and contemporaneous volcanism and clastic sedimentation in structural basins date the detachment faulting as between 26 and 15 Ma, especially between about 20 and 17 Ma. Tilts are generally to the southwest, and the sense of separation of the crystalline basal plate is southwest relative to the detached upper plates as far west as a probable breakaway zone west of the Colorado River in southeastern California (Fig. 16). Detachment of upper-crustal rocks postdated mylonitization of

the pervasively northeast-lineated metamorphic basal plate but partly utilized the mylonite zone; chloritic breccias typically cut through the mylonites at the base of the detachment. Similar relations are documented in the crystalline complexes in the eastern part of the Colorado River corridor, where K-Ar ages in the lower plates were reset to values as young as 15 Ma. Lateral displacement on these detachments is very large; above metamorphic complexes in the Harcuvar and Harquahala Mountains of western Arizona, pre–late Miocene displacement was about 50 km (Reynolds and Spencer, 1985). Present outcrops of lower-plate rocks reflect concurrent evolution of synforms and antiforms both parallel and perpendicular to the direction of extension (*Frost, 1982*). Spencer (1984) modelled these features as consequences of isostatic unloading with tectonic denudation above the detachments. In at least one area, nearly east-west dikes in the exposed lower plate probably were contemporaneous with northeast-southwest extension in the detached plate; thus, stresses in the lower plate locally varied from those in the attenuated upper plate, probably as a result of synchronous lower-plate compression and divergence associated with warping and uplift. Some degree of early Miocene extension probably characterized all of the area from the central Mojave Desert eastward to the Colorado Plateau (*Hileman and others, 1990*).

In the Chocolate Mountains area of southeasternmost California (Fig. 16), structural evidence of tilting is no older than about 21 Ma (*Murray, 1982*), but Crowe (1978) suggested that a pre-32-Ma northwest-trending depositional basin might reflect even earlier extension. This possible earlier extension may indicate the overlap of two extensional provinces—the southern end of the lower Colorado River corridor and the western end of a southern Arizona corridor, discussed next. Farther east, however, near Ajo (Fig. 16), initial extension appears to be bracketed by ages of 27 and 23 Ma on steeply and moderately tilted strata, respectively (*Gray and others, 1985*).

Extension somewhat older than that of the Mojave–lower Colorado River area dominated mid-Tertiary tectonic evolution of central-southern Arizona (Fig. 16). Beginning mainly about 35 to 30 Ma (but locally earlier), several basins accumulated thick sections of mid-Tertiary coarse clastic sediments and intercalated volcanic rocks. Although such basins, including the Pantano, Whitetail, San Manuel, and others (Fig. 16; Nations and others, 1982) probably gathered drainage from continued degradation of the Mesozoic Mogollon Highland, their generally northwest trends and local fault relations suggest structural origins and reflect initial mid-Tertiary tectonic extension, mainly by low-angle faulting and the rotation of upper-crustal blocks (Eberly and Stanley, 1978; Scarborough and Peirce, 1978; Shafiqullah and others, 1978, 1980). Some of the low-angle faults of southeastern Arizona, regarded by some investigators as erosional remnants of regional Laramide thrust sheets (Drewes, 1981), have been shown to be low-angle normal faults related to mid-Tertiary regional extension (Davis and Hardy, 1981; Dickinson, 1984). Near the transition between the Colorado Plateau and Basin and Range Province in central Arizona, a Precambrian and Paleozoic sequence cut by Laramide thrusts was steeply east-tilted and cut by low-angle faults, partly before but mainly during and after deposition of the Oligocene and lower Miocene San Manuel Formation (*Krieger, 1974*). The San Manuel shows a growth relation to tilting and contains megabreccia lenses of rocks exposed by the tilting.

The lower part of the extensional tectonic system of southern Arizona is exposed by several metamorphic "core complexes" (Fig. 16; *G. H. Davis, 1975,* 1980; Rehrig and Reynolds, 1980; S. J. Reynolds and Rehrig, 1980; Banks, 1980; *Keith and others, 1980*). Near Phoenix, K-Ar ages of mylonitized crystalline rocks with overprinted chloritic breccias beneath detachment faults are 25 to 20 Ma. North and east of Tucson, a northwest-trending uplifted block spans the Rincon, Catalina, and Tortolita Ranges and areas to the northwest, exposing a crystalline complex that has K-Ar ages largely reset to 30 to 20 Ma (*Mauger and others, 1968*; *Banks and others, 1978*; Banks, 1980). The crystalline core is separated from overlying folded stratified rocks by low-angle faults (*Davis, 1975*). Some plutons cutting the metamorphic complex appear to give primary middle Tertiary ages (Creasey and others, 1977) and were cataclastically deformed along with older rocks (Banks, 1980; *Keith and others, 1980*). Mid-Tertiary plutons are also present in the core block of the farthest east "metamorphic core complex" recognized, in the Pinaleno and Santa Teresa Mountains (Rehrig and Reynolds, 1980); dikes as young as 25 Ma are truncated at a fault that displaces southwest-tilted 28- to 22-Ma volcanic rocks and overlying volcanic conglomerates down onto the detachment surface (Davis and Hardy, 1981).

It should be noted that not all authors accept the primary significance of mid-Tertiary extension in the origin of the "metamorphic core complexes" of southern Arizona (*Thorman, 1977*; Drewes, 1981). Features indicative of regionally pervasive metamorphism and northeast-trending deformation are interpreted by these authors to reflect a regional Laramide thrust system, the mid-Tertiary influence having been mainly to uplift the metamorphosed and deformed rocks, gravitationally denuding them and resetting their K-Ar systems. Most recent authors, however, favor alternative explanations for the parallelism of Tertiary extension and "core complex" fabrics, interpreting the cores and detachments as elements of megaboudins in an extending crust (Davis and Coney, 1979), as an uplifted mid-crustal zone of brittle-to-ductile transition (Rehrig and Reynolds, 1980), or as crustal shearing on a gently dipping zone, with the mylonites formed at deep structural levels before the basal plate was drawn out southwestward and upward into a zone of brittle deformation (Davis and others, 1986).

Detachment faults localized significant base- and precious-metal syntectonic mineralization in the Mojave–Colorado River–Sonoran region (*Spencer and Welty, 1986*; *Roddy and others, 1988*). In addition, manganese deposits and local barite veins are associated widely with the detachments and with fractures in the rotated upper plates (*Lasky and Webber, 1949*; *Berg and others, 1982*; Glazner, 1988).

Mexican Highland–Southern Rocky Mountains region

In the southeastern Basin and Range Province and Rio Grande rift (Fig. 16), the beginning of tectonic extension is recognized by a regional change from constructional volcanic plateaus, including regionally extensive ash flows, to broad fault-bounded basins of alluvial sediments and local central volcanoes. Mid-Tertiary tectonic extension of the upper crust by low-angle and listric faulting associated with rotation of the surficial plates (Morgan and others, 1986) was separated northward from the unextended Colorado Plateau by the northeast-trending Jemez lineament (Fig. 16), a structural zone of Laramide and older ancestry (Aldrich and Laughlin, 1984). In the volcanic Mogollon Plateau of southwestern New Mexico, adjacent to the Colorado Plateau, basin-range faults and sedimentary basins related to the modern topography are as old as 21 Ma. Elston and others (1976) recognize no features of regional tectonic extension older than those faults and basins although the youngest areally extensive ash flows beneath these sediments are about 26 Ma. In the southern Rio Grande Rift zone in central and southern New Mexico, thick fanglomerates with interlayered flows from local central volcanoes are as old as 26 Ma and postdate widespread ash flows of 35 Ma, apparently bracketing initial extension (Chapin and Seager, 1975). In the central part of the Rio Grande rift near Socorro, Chapin and Seager (1975) show that initial extension occurred on numerous normal faults after eruption of a 31-Ma ash-flow tuff but before emplacement of several 28-Ma epizonal intrusions. The earliest recognized alluvial basin sediments in this area are 3 to 4 m.y. younger than initial faulting; before that time erosion of new relief was accommodated by existing through-flowing drainages. In the Sandia and Sangre de Cristo Mountains in central New Mexico, *Kelley and Duncan* (*1986*) interpret fission-track data on basement rocks to indicate block uplift beginning 30 to 35 Ma. *Lipman* (*1983*) and Lipman and others (1986) interpret a distended 27- to 25-Ma caldera complex in the northern Sangre de Cristo Mountains to have evolved during early extension in the rift. Where the northern Rio Grande rift now breaks across the Southern Rocky Mountains of south-central Colorado, early extension is bracketed by tilted, areally extensive volcanic rocks of the San Juan Mountains as young as 27 Ma that are eroded and overlapped unconformably by basalts as old as 26 Ma; succeeding alluvial deposits exposed at the margins of the rift basin are as old as 20 Ma (Lipman and Mehnert, 1975). Henry and Price (1986) indicate that extension in west Texas, at the edge of the extended region, began by 28 Ma and was expressed as high-angle normal faults by 23 Ma, lasting until about 17 Ma.

Thus, in southwestern New Mexico and the present Rio Grande rift as in southern Arizona, extension apparently began as early as 31 to 28 Ma and affected drainage and volcanism widely by 26 Ma, though perhaps not in the Mogollon Plateau until as late as 21 Ma. The present Rio Grande rift, expressed by its thick sediment fill, is mainly late Miocene and younger in age and is discussed later.

Rehrig and Heidrick (*1976*) showed that mid-Tertiary extension recorded major changes in a regional stress field. Relatively weak extensional joints and veins associated with Laramide deformation indicate north-south to northwest-southeast least principal stress, consistent with regional east-northeast shortening; regional Oligocene to Miocene extension occurred with a northeast-southwest least principal stress. The stress field had the same orientation over a broad region, onto the foreland of the Laramide orogenic belt, although local timing of the change in stress field varied. *Price and Henry* (*1984*) suggest that least principal stress in Trans-Pecos Texas before about 32 Ma was northwest-southeast, unchanged from its Laramide direction, afterwards changing to northeast-southwest, as normal faulting became active. *Muller* (*1986*) concluded from dike orientations and an elastic model that the Spanish Peaks area, at the margin of the Great Plains area, at the margin of the Great Plains in southern Colorado, shifted about 22 Ma from nearly north-south least principal stress during emplacement of earlier dikes to nearly east-west during emplacement of later dikes. Although this was contemporaneous with rifting about 50 km to the west, no extensional faulting of this age has been recognized at Spanish Peaks. Aldrich and others (1986) suggest that extension within the Rio Grande rift has been east-west since its inception, even during the early stages when least principal stress in bordering areas was more nearly northeast-southwest. In peripheral areas during much of the transition time, the horizontal stress field may have been nearly uniform, as suggested by predominantly radial dike swarms in the San Juan volcanic field of southwestern Colorado (*Lipman, 1981*).

EARLY AND MID-TERTIARY MAGMATIC ARCS AND INTERIOR MAGMATIC SYSTEMS

The following sections describe post-Laramide magmatic systems of the Cordilleran region up to a regional transition in magmatic compositions and eruptive styles that occurred diachronously in mid-Miocene time. Most regions that had undergone Laramide and older late Mesozsoic crustal shortening subsequently underwent extension and associated regional magmatism, but magmatism was active at significantly different times in various regions. Some of the magmatic systems were related to continental-margin subduction, but others occur far inland in settings that suggest other factors as dominant influences. Because the intent of this chapter is mainly descriptive and because there is considerable diagreement about magmotectonic relations, several volcanic "belts" are discussed here without primary reference to their origins as arcs, rifts, or other magmotectonic elements. Relations of some zones, however, do seem sufficiently well established to justify their characterizations here. Many of the magmatic episodes described are composite, with typical arc segments forming narrow zones near the continental margin but adjoining, more or less linear magmatic belts with distinc-

tive compositions being associated with Cordilleran interior extension.

EOCENE TO EARLY MIOCENE MAGMATISM OF THE NORTHERN UNITED STATES CORDILLERA

The distribution of post-Laramide magmatism from the Northern Rockies to the central Great Basin changed progressively through early and middle Tertiary time. Arc magmatism had been more or less continuous in the western montane region of Canada and the North cascades through Laramide and, particularly, Eocene time (Souther, 1977; Ewing, 1980; *Armstrong 1988*). Magmatism migrated from this region generally southward, into and across the northwestern United States, not as a uniform sweep but as a succession of distinct belts. Each new magmatic belt within the Cordilleran interior was established rather abruptly along a more southward front. As it did so, activity in the older belts did not end as abruptly but, rather, declined and changed in character as major activity shifted to the newer one farther south. During the major activity of each interior belt, typical arc magmatism remained continuous near the continental margin while, toward the continental interior, magmatism—though remaining broadly calc-alkalic—tended to be more silicic and more potassic and to be associated with significant tectonic extension. Not only did the interior magmatic belts step southward, but the southern terminus of the magmatic arc extended southward as well. For convenience in discussion, the segments of the near-coastal arc within the United States are discussed along with the interior magmatic belts for each of the first two such belts.

The continental-margin arc and Challis magmatic belt

The oldest major belt of post-Laramide volcanism in the northern United States Cordillera was active mainly in the middle Eocene. This magmatism waned late in middle Eocene time, beginning about 43 Ma, as the magmatic pattern shifted in the first of several major steps. This first magmatic belt across the Laramide orogenic region extends from the northern Cascades to the Cordilleran front in Montana and Wyoming. The continuity of the belt was noted by Lipman and others (1972) but was demonstrated more explicitly by *Armstrong* (*1974a*, 1978), who designated its activity as the Challis volcanic episode. We restrict the term "Challis magmatic belt" somewhat from Armstrong's usage, including only that part of the Cordilleran region in which Eocene magmatism was active before 43 Ma east of the Cascades and west of the Rocky Mountain front (Fig. 17), thus excluding volcanic rocks of the North Cascades, the Northern Great Plains, and those rocks of the Clarno Formation of central Oregon that postdate 43 Ma. Within that restricted definition, the principal features of the Challis belt remain essentially as Armstrong noted them. Challis-belt magmatism was most active during the 5 m.y. from about 50 to 45 Ma but began in the late early Eocene, at least 54 Ma. Relatively minor volcanism continued widely until about 40 to 37 Ma and locally persisted to about 20 Ma but changed in character with the regional magmatic shifts.

Vance (*1982*) referred to a Challis "arc," but as specific relations between the Challis belt as defined here and contemporaneous subduction are arguable, the term "magmatic belt" is preferred here. Volcanic rocks of the Challis belt are largely calc-alkalic (but potassic) andesite to dacite but also include abundant rhyolite, especially in the central part; basalt is generally rare. In addition, a belt of small alkalic intrusions occurs in southern British Columbia and northeastern Washington (*Fox, 1973*); highly potassic shoshonites and related rocks occur along with only moderately potassic calc-alkalic andesites in the eastern part of the belt (*Nicholls and Carmichael, 1969*; *Prostka, 1973*; *Kudo and Broxton, 1985*).

Predominantly mafic to intermediate early to middle Eocene eruptive rocks in the North Cascades (Fig. 17) formed a typical magmatic arc that was in part contemporaneous with contractional folding (Tabor and others, 1984). Exposed volcanic rocks older than about 44 Ma are confined to blocks east of the preexisting Straight Creek fault. The oldest volcanics, interbedded with tightly folded fluvial sandstones of the Swauk Formation, are 52-Ma dacites of its Silver Pass Volcanic Member and probable correlatives. Unconformably above is a gently dipping sequence of calc-alkalic basalts, andesites, and some dacites, and rhyolites (the Teanaway Formation of about 47 Ma) and other correlative basalts. Plutons in the North Cascades include a linear northwest-southeast group of small 48- to 43-Ma quartz-dioritic to granodioritic stocks west of the Straight Creek fault (*Yeats and Engels, 1971*). *Tabor and others* (*1968*) dated stocks in the eastern North Cascades between 48 and 50 Ma and suggested that one of them is a satellite of the much larger 47- or 48-Ma Golden Horn granite batholith. Some plutons east of Glacier Peak are between 46 and 43 Ma (*Cater and Crowder, 1967*; Cater, 1982).

The older part of the Clarno Formation of Oregon, which has yielded K-Ar ages greater than 43 Ma, is here included as part of the arc contemporaneous with the Challis magmatic belt (Fig. 17). The only part of the Clarno that has yielded ages between 54 and 43 Ma is within the broadly antiformal uplift around the type area of the formation in north-central Oregon (*Fiebelkorn and others, 1983*; Walker and Robinson, 1990a). These rocks, mainly andesitic lavas, are overlain by 43- to 38-Ma predominantly andesitic volcaniclastics and tuffaceous sedimentary rocks, also included in the Clarno Formation but associated with a regional distribution different from that of the basal andesites and, thus, not part of the arc contemporaneous with the Challis magmatic belt.

The continental-margin magmatic arc just described in and near the central to northern Cascades is continuous into western British Columbia. However, the magmatic belt farther east, discussed following, is compositionally different and is associated with major Eocene extension in and near the northern United States Rockies.

Challis-belt magmatism in northeastern Washington (Fig.

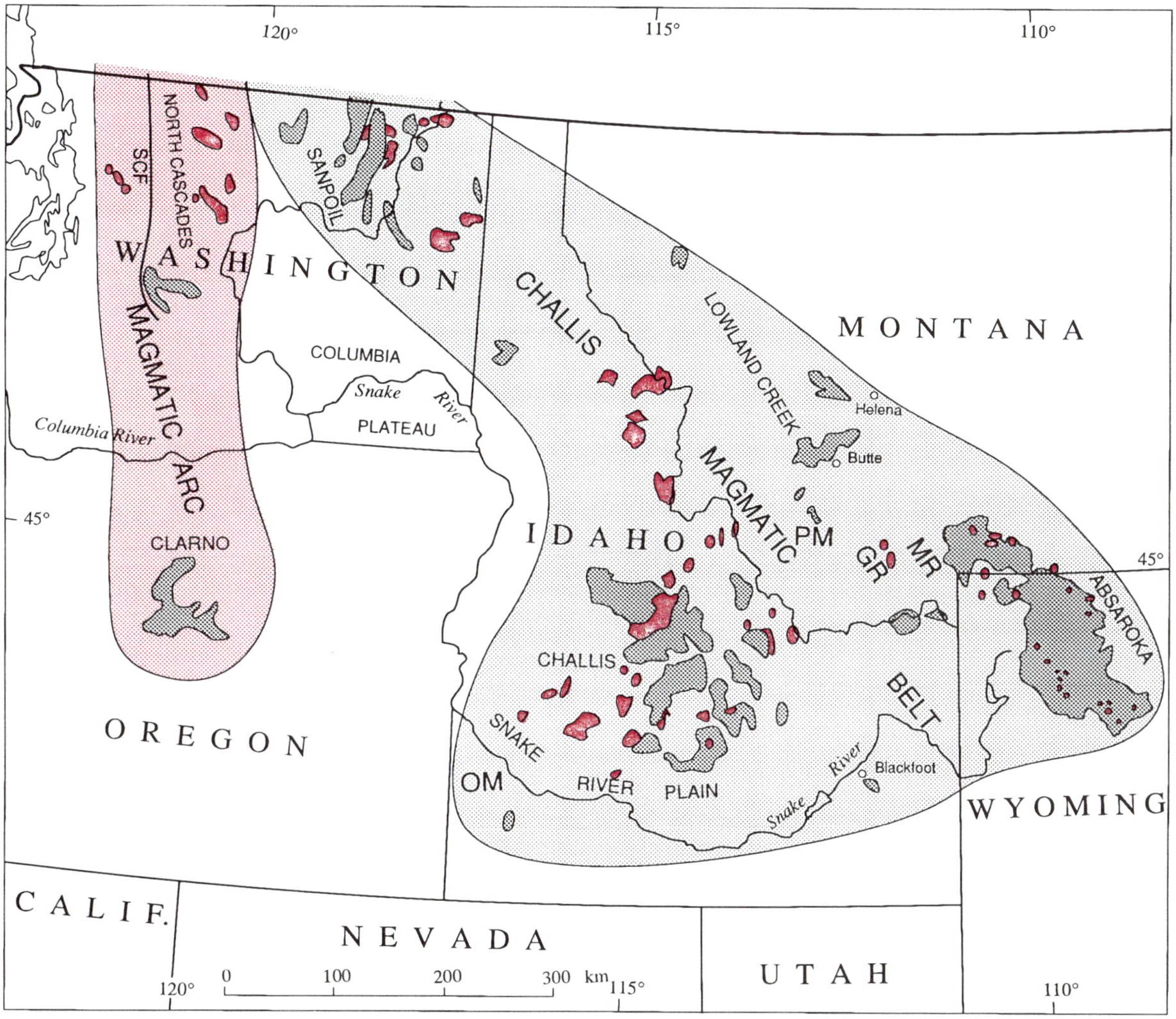

Figure 17. Map of the Challis magmatic belt (53 to 43 Ma) and contemporaneous continental-margin arc. Dark-patterned areas indicate approximate outcrop distributions; pink, probable volcanic arc; gray, andesite-rhyolite suite; red, intrusions. GR, Gravelly Range; MR, Madison Range; OM, Owyhee Mountains; PM, Pioneer Mountains; SCF, Straight Creek fault. Later Cenozoic rotations and displacements not restored.

17) was contemporaneous with the widespread resetting of older plutonic mica K-Ar and apatite fission-track ages (Miller and Engels, 1975; Fox and others, 1977). The Eocene Sanpoil Volcanics and overlying Klondike Mountain Formation (potassic andesite to rhyodacite) and associated intrusives dated as 53 and 46 Ma overlie a sequence of lower Eocene tuffaceous fluvial and lacustrine sediments in extensional basins (*Muessig, 1967*; Pearson and Obradovich, 1977). Some of the associated plutons are two-mica granites, interpreted to represent significant crustal melting (Miller and Engels, 1975; Miller and Bradfish, 1980).

Corresponding widespread volcanic rocks in the Idaho batholith region of central Idaho mapped as the Challis Volcanics (Fig. 17) give the magmatic belt its name. They range in age mainly from 51 to 43 Ma and in composition from basalt to rhyolite, but potassic andesite to rhyodacite predominates (*Armstrong, 1974a*; McIntyre and others, 1982; *Ekren, 1985*; *Bennett and Knowles, 1985*). Initial activity, probably from about 51 to 48 Ma, erupted mainly intermediate lavas from numerous centers onto a mountainous terrain. The most voluminous activity produced rhyodacitic to rhyolitic ash-flow sheets between about 48 and 45 Ma, associated with large collapse calderas. The batholithic 47-Ma Casto (*Cater and others, 1973*; McIntyre and others, 1982), which intrudes some of the caldera complexes, established an enormous hydrothermal system in a 70-km ring zone centered on the pluton (*Criss and others, 1984*). Numerous small plutons within and near the Idaho batholith are of similar ages (Bennett, 1980) and probably represent other sources of Challis volcanism. Significant activity related to caldera complexes continued until

about 43 Ma; minor activity synchronous with the younger Tuscarora magmatic belt farther southwest continued in parts of the area until about 37 Ma, as discussed later.

Igneous rocks similar in age and composition to the Challis Volcanics occur in several areas now separated from the Idaho batholith by younger basalts of the Columbia Plateau and Snake River Plain (Fig. 17). Andesites to rhyodacites and related intrusions are surrounded by younger rocks in northern Idaho near Moscow (*Jones, 1982*), in the Owyhee Mountains area of southwestern Idaho (dated at 49 to 46 Ma; *Armstrong and others, 1980*; Ekren and others, 1982), and in the Blackfoot, Idaho, area southeast of the Snake River Plain (dated at 47 to 40 Ma; *Armstrong, 1972b*; *Trimble, 1982*). These rocks probably represent outlying elements of the Challis volcanic field, later isolated by intervening basaltic lavas and by subsidence of the Snake River Plain (Axelrod, 1968).

Andesite to rhyodacite and rhyolite north and east of the Idaho batholith, generally 50 to 45 Ma, include the Lowland Creek Volcanics of the Butte-Helena area (Smedes, 1962; Smedes and Thomas, 1965; Chadwick, 1972) and the Pioneer Mountains (Zen, 1975; *Marvin and others, 1982*), and correlative lavas in west-central Montana (*Williams and others, 1976*). Small dacitic to rhyolitic intrusions in and near the northern Gravelly Range are 45 to 50 Ma (*Marvin and others, 1974a*). The easternmost end of the Challis belt, in northwestern Wyoming and southern Montana, is the Absaroka volcanic field (*Chadwick, 1970,* 1972; Smedes and Prostka, 1972; *L. L. Love and others, 1976*), mainly between 49 and 43 Ma. Igneous centers in the eastern part of this field define two parallel northwest-trending belts (*Chadwick, 1970*), the eastern of which is distinctly more potassic than the western (*L. L. Love and others, 1976*). Although highly potassic compositions are mainly in the eastern belt, some leucitic rocks also occur well to the west, in the southern Madison Range (*Hamilton, 1964*).

Several Challis-age plutons produced significant mineralization, particularly in and near the Idaho batholith, where they host porphyry-type ore bodies or produced chalcophile vein mineralization (Bennett, 1980). A 1,000-km-long belt of both Laramide and Eocene porphyry-type mineralization trends northeast from the Idaho batholith to the Great Plains of central Montana (*Rostad, 1978*; *Armstrong and others, 1978, 1982*). Similar small porphyry deposits occur in the contemporaneous arc in the North Cascades (*Armstrong and others, 1982*).

In summary, numerous separately named 53 to 43 Ma, predominantly intermediate and generally calc-alkalic plutons, volcanic fields, and volcanogenic formations extending more than 1,000 km from the edge of a continental-margin arc in the Washington Cascades as far eastward as northwestern Wyoming are erosional remnants of the once nearly continuous Challis magmatic belt. This andesite-rhyolite belt, most active about 50 to 45 Ma, accompanied extension and capped a mountainous highland (Axelrod, 1968) northeast of the broad basinlike region that would later be buried by volcanic plateaus of the Columbia Intermontane region. Although some igneous activity continued locally in the Challis belt after 43 Ma, it was generally minor and different in character, as discussed later.

Eocene foreland magmatism of the Northern Great Plains

Magmatism simultaneous with that of the Challis belt occurred at widely dispersed centers on the Cordilleran foreland of Montana, Wyoming, and South Dakota (Fig. 18). This activity was considerably different in composition and character from the generally calc-alkalic Challis magmatism. The foreland magmas were mainly alkalic, including phonolites, trachytes, and latites. This magmatism has been recognized in Montana as part of a central Montana alkalic province (*Larsen, 1940*; Chadwick, 1972), but similar magmatic centers also occur to the southeast in central Wyoming and in the Black Hills. Some compositions are sodic, but those north of the Missouri River in Montana are highly potassic (*Larsen and Buie, 1938*); also noteworthy are carbonatites in the Bearpaw Mountains (*Pecora, 1962*) and kimberlitic compositions in diatremes of the Missouri Breaks (*Hearn, 1968, 1979*). Most of the igneous centers probably erupted small fields of central volcanoes, now preserved mainly as subvolcanic stocks, sills, and laccoliths (*Hurlbut and Griggs, 1939*). Only locally are extensive lavas preserved, especially in the Bearpaw Mountains (*Hearn, 1976*) and Highwood Mountains (*E. S. Larsen and others, 1941*) of north-central Montana.

Although the Eocene alkalic centers (Fig. 18) are postorogenic, some of them coincide with Cretaceous and Paleocene igneous centers, and several of them correspond to basement uplifts in the Laramide Cordilleran foreland. Other Laramide alkalic igneous centers also occur in the province but are not discussed here.

On the Great Plains of Montana, the Eocene centers include rocks of 54 to 47 Ma, mainly 53 to 50 Ma (*McDowell, 1971*; *Marvin and others, 1973a, 1980*). Each of seven major centers (Fig. 18) appears to have been active for only a short time in the Eocene (*Marvin and others, 1980*; *Diehl and others, 1983*) although a single isolated phonolitic intrusion gave a much younger age of 27 Ma. Farther south, a group of about 40 volcanic vents and small intrusive bodies of phonolite, trachyte, and rhyolite in the Rattlesnake Hills in central Wyoming of about 44 Ma (Fig. 18; *Pekarek and others, 1974*) represents a short igneous episode (*Shive and others, 1977*). Trachytes, phonolites, and related rocks also form an east-west belt of both Laramide and post-Laramide shallow intrusive bodies along the north side of the Black Hills and their western flank in western South Dakota and eastern Wyoming (Fig. 18). K-Ar ages generally range from 59 to 49 Ma, but two rocks yielded ages of 39 to 38 Ma (*McDowell, 1971*; *Staatz, 1983*; Redden and others, 1983).

Base-metal mineralization of mid-Tertiary age occurs in the Little Belt Mountains (*Witkind, 1973*; *Armstrong and others,*

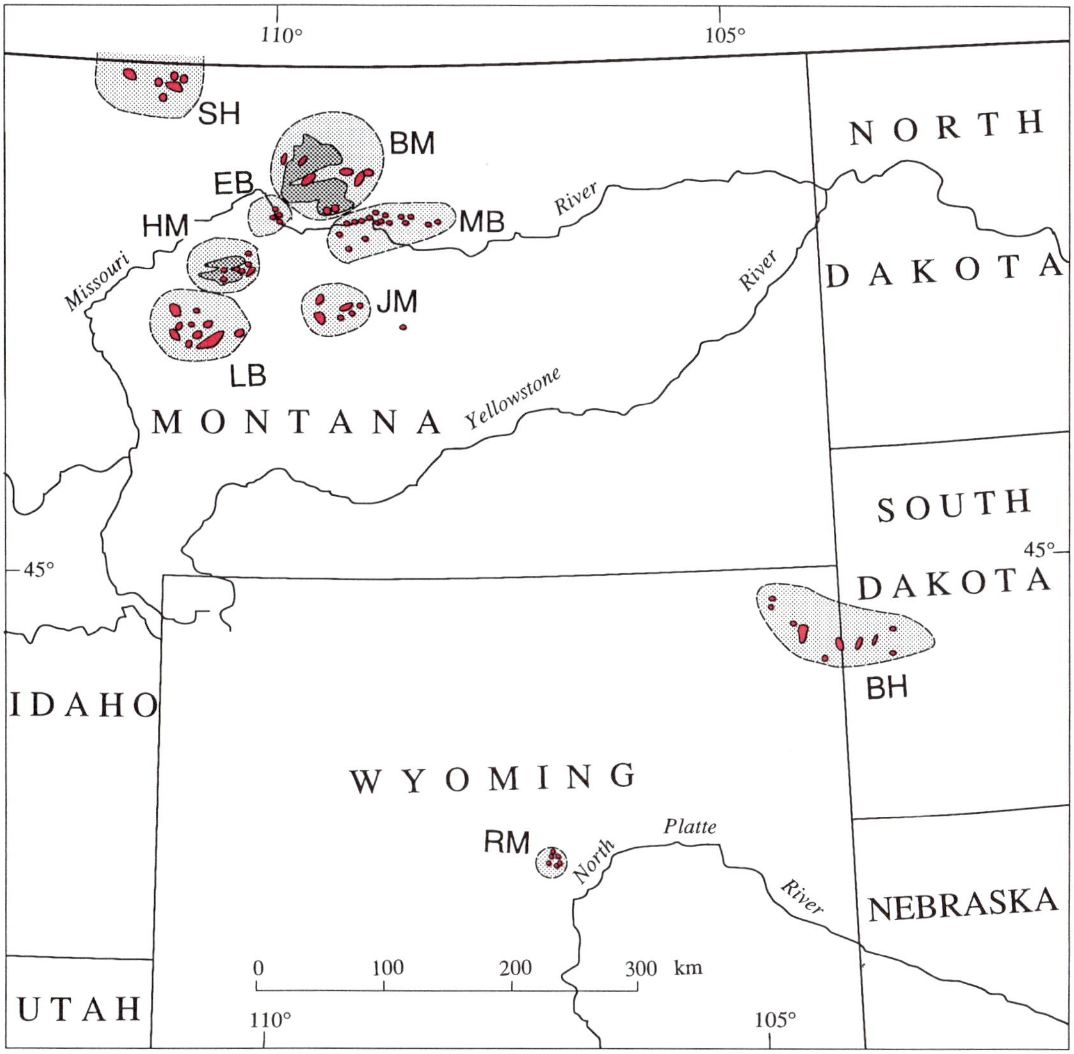

Figure 18. Map of early Tertiary magmatic centers of the Northern Great Plains region (55 to 38 Ma). Dark-patterned areas indicate approximate outcrop distributions; gray, volcanic rocks; red, intrusions. Dashed lines and light shading outline igneous fields: BH, Black Hills; BM, Bearpaw Mountains; EB, Eagle Buttes; HM, Highwood Mountains; JM, Judith Mountains; LB, Little Belt Mountains; MB, Missouri Breaks; RH, Rattlesnake Hills; SH, Sweetgrass Hills.

1978, 1982) and the Sweetgrass Hills and Bearpaw Mountains (*Marvin and others, 1980*); kimberlitic diatremes of the Missouri breaks have been prospected for diamonds (*Hearn and McGee, 1983*). Disseminated and vein deposits of thorium and rare earths are associated with alkalic centers of the Black Hills (*Staatz, 1983*). Small oil and gas fields on the flanks of intrusive centers of the Sweetgrass Hills and Bearpaw Mountains are in structural traps, partly of Laramide origin but also related to the Eocene intrusions (*Norwood, 1972*).

In summary, alkalic igneous centers on the northern Cordilleran foreland define two spatially overlapping age groups, one coinciding with Laramide deformation and the other with the time of most active volcanism in the post-Laramide Challis belt. They suggest analogies to late orogenic and postorogenic mafic and alkalic magmatism at dispersed centers associated with uplifts in the foreland belts of other orogenic systems, notably the central European foreland of the Alps.

The continental-margin arc and Tuscarora magmatic belt

Post-Laramide magmatism shifted southward and southwestward at about 43 Ma from an arc that ended in central Oregon and an interior Challis belt to an arc extending to northeastern California and an interior belt extending east-southeast into northeastern Nevada and adjacent Utah (Fig. 19; Stewart and Carlson, 1976). The new interior magmatic belt had no

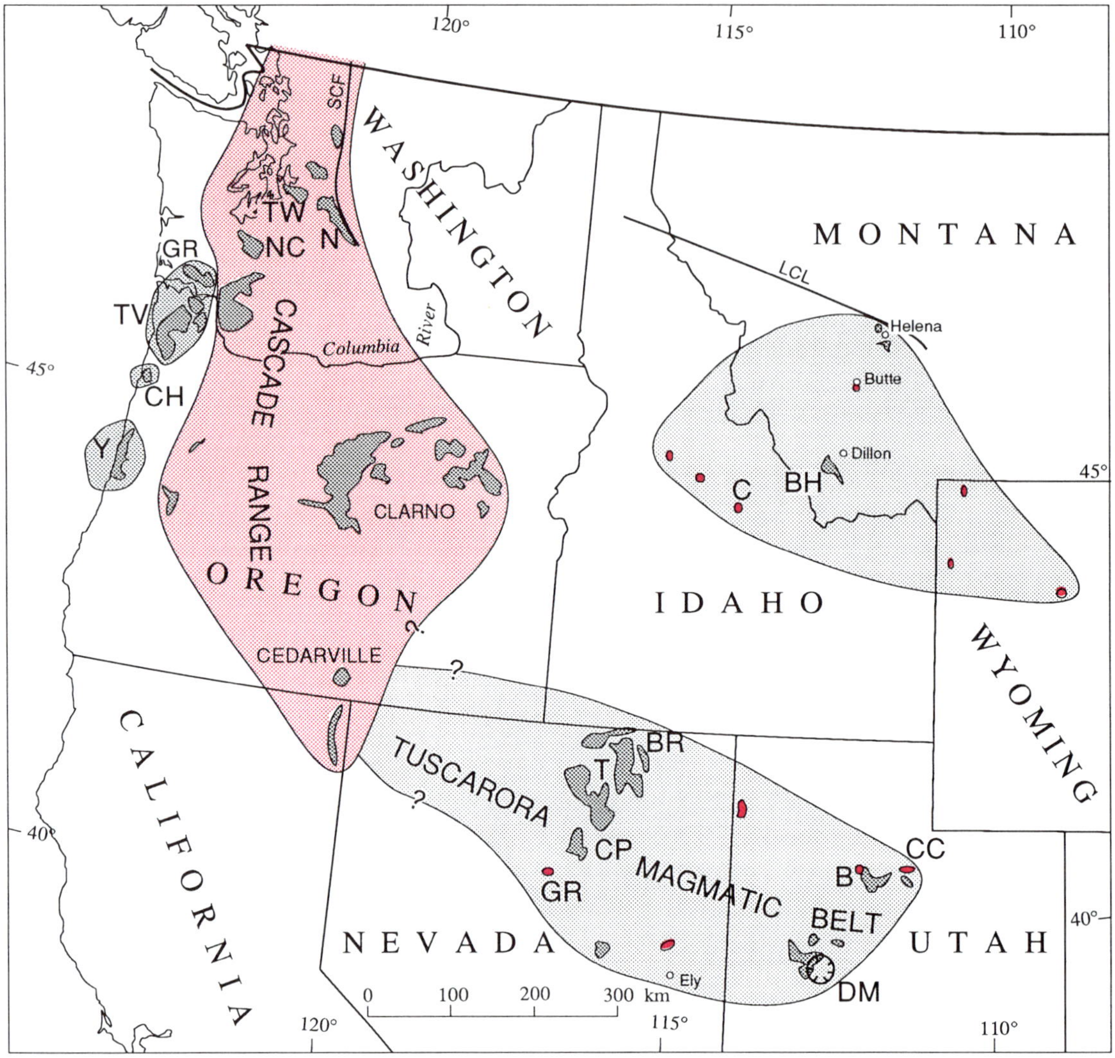

Figure 19. Map of the Tuscarora magmatic belt and contemporaneous magmatic suites (43 to 37 Ma). Dark-patterned areas indicate approximate outcrop distributions; pink, probable volcanic arc; light gray, andesite-rhyolite suite; red, intrusions; dark gray, coastal predominantly basaltic suite. B, Bingham; BH, Beaverhead area; BR, Bull Run; C, Challis; CC, Cottonwood Canyon; CH, basalt of Cascade Head; CP, Carlin–Pinon Range area; DM, Drum Mountains caldera; G, Goble Volcanics; GR, Grays River Volcanics; H, Helena; LCL, Lewis and Clark line; N, Naches Formation; NC, Northcraft Formation; SR, Shoshone Range; T, Tuscarora; TV, Tillamook Volcanics; TW, Tukwila Formation; Y, Yachats Basalt. Later Cenozoic rotations and displacements not restored.

overlap with the older Challis belt. The name "Tuscarora magmatic belt" for this new interior belt is taken here from Eocene andesites and rhyolites exposed widely in and near the Tuscarora Mountains of northern Nevada (*Axelrod, 1966b*; *Hope and Coats, 1976*; Stewart and Carlson, 1976). The Tuscarora belt became a distinct geographic entity as magmatism in the Challis belt began to decline, continuing until about 37 Ma to erupt andesitic lavas, more potassic eastward; dacite to rhyolite was prominent in the later phases and in the eastern parts of the belt. Basalt is notably rare.

In Washington, Oregon, and northeastern California, volcanism was arclike and is regarded by some as representing establishment of the Western Cascades arc (Wells and others, 1984). Volcanism of 43 to 37 Ma in and near the Western Cascades, however, extended southward only to central Oregon (*Lux, 1982*); the rest of the arc lay east of the present Cascades (Fig. 19).

In Washington, along the western foothills of the Cascades, arc activity of both Challis and Tuscarora ages is represented by interlayered volcanic rocks and ash beds in the Eocene Puget

Group as young as 40 Ma (*Turner and others, 1983*; *Frizzell and others, 1984*). Hornblende quartz-diorite dikes in the eastern Olympic Peninsula are about 43 to 42 Ma (*Tabor and others, 1972*; R. E. Wells and others, 1984). Notable within the Puget Group are the andesitic Tukwila Formation, 41 to 42 Ma (Fig. 19; *Frizzell, 1979*), and the correlative Northcraft Formation (Fig. 19; *Gard, 1968*; Wells and Coe, 1985). In the Washington Cascades, volcanics younger than about 44 or 43 Ma occur west of the Straight Creek fault (Tabor and others, 1984), including the Naches Formation (Fig. 19) with mainly rhyolite near the base but predominantly basalt and andesite above. Lower volcanics of the Naches (Tabor and others, 1984) and correlative andesite (*Frizzell and others, 1984*) erupted in the range 44 to 40 Ma. Small to moderate diorite and gabbro bodies have similar ages (*Frizzell and others, 1984*).

In northwestern Oregon, arc volcanism of 43 to 37 Ma is represented by andesitic tuffs, basaltic to andesitic debris, and some lavas in upper Eocene strata of the Coast Range, including the Keasey, Eugene, Fisher, and Alsea Formations (Armentrout and others, 1983). Volcanic units of the Western Cascades in Oregon relate mainly to a north-south magmatic arc younger than the Tuscarora belt, but the basal Little Butte Volcanics (Peck and others, 1964) interfinger in the Eocene Eugene Formation (Wolfe, 1981) and yield a few K-Ar ages between 42 and 38 Ma (Fig. 19; *Lux, 1982*).

A group of varied but generally mafic subalkalic volcanic centers lies along the continental margin of Washington and Oregon in front of the main group of arc volcanics (Fig. 19; MacLeod and Snavely, 1973; *Phillips and others, 1989*). Toward the Washington coast from the Tukwila-Northcraft zone, as far as the lower Columbia River area, tholeiitic basalt and femic andesite are interlayered with upper Eocene sedimentary rocks on the continental margin (*Livingston, 1966*; Wells and Coe, 1985; *Walsh and others, 1987*). A little farther southwest in Oregon, the Tillamook Volcanics of 44- to 40-Ma age includes compositions resembling oceanic tholeiite along with femic andesite and relatively sparse dacite and rhyolite (Magill and others, 1981; Wells and others, 1983; Niem and Niem, 1985). Other forearc centers of similar age on the continental margin include the basalt of Cascade Head (*Snavely and Wells, 1984*) and the Yachats Basalt (Snavely and MacLeod, 1974), both about 38 Ma.

Arc volcanism in Oregon contemporaneous with the Tuscarora belt is recorded mainly east of the Cascades in rocks of the Clarno Formation (*Rogers and Ragland, 1980*; Noblett, 1981). Volcanic rocks of the Clarno remain inadequately studied. The volcanism that produced the Clarno Formation began during Challis time, but more widespread volcanic rocks mapped as Clarno range between 43 and 38 Ma (*Fiebelkorn and others, 1983*; Walker and Robinson, 1990a). A section 2 km thick described by *Waters and others (1951)* consists principally of andesitic mudflow deposits and fine-grained volcanic sedimentary rocks. Elsewhere are andesitic lavas and small diorite intrusions (*Oles and Enlows, 1971*; *Enlows and Parker, 1972*; Noblett, 1981), minor basalt (Rogers and Novitsky-Evans, 1977), and rhyolite (Robinson, 1975). The predominantly andesitic Clarno appears to be entirely older than a 36-Ma welded tuff at the base of the overlying, mainly rhyolitic John Day Formation (*Swanson and Robinson, 1968*). Some andesitic rocks in north-central Oregon assigned to the Clarno Formation by *Enlows and Parker (1972)* have yielded younger K-Ar dates that probably are erroneous, based on stratigraphic relations (*Taylor, 1981b*). In Oregon south of the mapped Clarno, lithologically similar pyroxene and hornblende andesite occurs above arkosic sedimentary rocks; an olivine andesite near the top yielded an age of 40 Ma (*Wells, 1979*). In the Warner Mountains of northeastern California, the base of the Tertiary section is the andesitic "lower Cedarville series" of Russell (1928), where *Axelrod (1966b)* reports an age of 40 Ma.

Igneous activity between 43 and 37 Ma persisted locally in parts of the older, Challis belt south of the Lewis and Clark line (Fig. 19). (Even younger bimodal rhyolites and basalts in southwestern Montana—up to 20 Ma—are considered separately later.) The areally restricted post-43-Ma magmatism of the Challis region, less voluminous than earlier magmatism there, was mainly rhyolitic and produced few if any mafic magmas. A few small intrusive masses of rhyodacite (McIntyre and others, 1982), perhaps some latitic dikes, and a rhyolitic stock (*Leonard and Marvin, 1982*) were emplaced in the Idaho batholith. Farther east, in the Beaverhead area of Montana, 39-Ma plagioclase rhyolites (*Chadwick, 1978*) may be overlain by others that interfinger with Oligocene basin deposits of the Dillon area (*Myers, 1952*; *Lowell, 1965*) and farther southeast (*Scholten and others, 1955*). *Meyer and others (1968)* reported 40-Ma rhyolite dikes near Butte, and rhyolitic lavas, dikes, and sills near Helena were dated between 40 and 36 Ma (*Chadwick, 1978*). Two subvolcanic dacites near the south margin of the Absaroka volcanic field in northwestern Wyoming are 40 to 39 Ma in age (*L. L. Love and others, 1976*).

A belt of generally calc-alkalic volcanic and intrusive rocks, here designated the Tuscarora magmatic belt (Fig. 19), extended east-southeast across northern and northeastern Nevada and west-central Utah (Stewart and Carlson, 1976, *1978*; Stewart, 1980). The volcanism was contemporaneous with extension near the Utah-Nevada border. The igneous rocks are predominantly intermediate lavas—generally somewhat potassic—and subordinately rhyolite (Stewart and McKee, 1977; *Moore and McKee, 1983*). In the Carlin-Pinon Range area of north-central Nevada (J. F. Smith and Ketner, 1976), intermediate volcanism associated with the Elko Formation began about 41 Ma and was predominant until widespread 37-Ma ash flows of the Indian Well Formation erupted. Plutons of 41- to 38-Ma age occur near Tuscarora, the northern Shoshone Range, and Ely (Armstrong, 1970; McKee and Silberman, 1970; *McDowell, 1971*; *Coats and McKee, 1972*; *Theodore and others, 1973*). Initial mid-Tertiary, predominantly intermediate volcanic and intrusive rocks in and near the Drum Mountains of western Utah (Staatz and Carr, 1964; Lindsey, 1982) are 42 to 39 Ma in age. Similar activity in the Bingham mining district of central-western Utah produced

monzonite and latite stocks, dikes, and volcanic rocks, 40 to 37 Ma, with SiO_2 generally increasing through time (*Moore and others, 1968*; W. J. Moore, 1973; *Warnaars and others, 1978*). In the Cottonwood and Park City areas of the central Wasatch Range (*Crittenden and others, 1973*) initial mid-Tertiary magmatism, about 41- to 37-Ma in age, is recorded by quartz-diorite to quartz-monzonite stocks.

Voluminous ash flows older than 37 Ma occur in only two areas of this belt (Fig. 19). In northern Nevada, welded tuff in the Frost Creek Volcanics of *Axelrod* (*1966b*) in the Bull Run area is 43 Ma, the Dead Horse Tuff near Jarbidge (Coats, 1964, *1971*) is 40 Ma (*Axelrod, 1966b*), and other voluminous welded ash-flow tuff interlayered with andesite to dacite ranges from 44 to 38 Ma (Coats, 1964, *1971*; *McKee and others, 1976*; J. H. Stewart and Carlson, 1976; J. H. Stewart, 1980). In the Drum Mountains area of western Utah, eruption of the 39-Ma rhyodacitic Mount Laird Tuff and the 38-Ma rhyolitic Joy Tuff produced large calderas (*Shawe, 1972*; Lindsey, 1982).

Hydrothermal chalcophile mineral deposits are associated widely with centers of the Tuscarora magmatic belt. The largest by far is the Bingham porphyry-copper deposit of 39- to 36-Ma age in western Utah (*Warnaars and others, 1978*).

In summary, the volcanic arc in western Washington and Oregon, central and southern Oregon, and northeastern California was probably adjoined by a continuous magmatic belt that extended across a large continental-interior zone in northern to east-central Nevada and west-central Utah. A gap in known exposures in southeastern Oregon and northwestern Nevada may mainly reflect burial by extensive younger volcanic rocks. It is clear, however, that the Tuscrora belt did not overlap the earlier Challis belt and did not continue southward beyond the present-day northern Great Basin. The Tuscarora belt evolved during a period of major tectonic extension in at least part of the belt.

Mid-Tertiary magmatism in the Western Cascades arc

A second regional magmatic shift occurred in the Pacific Northwest about 38 to 37 Ma, and the new distribution persisted until about 17 Ma. The eastern part of the arc, which had extended eastward as the Clarno Formation along the Blue Mountains, declined. Whereas earlier arc volcanism in the Cascades region did not extend south of central Oregon but spread eastward through the Clarno region, later intermediate volcanism became more or less concentrated in and near the Cascades, initiating what is called here the early Western Cascades arc (Fig. 20). Predominantly andesitic magmatism in the Pacific Northwest has remained in a near-coastal zone since the early Oligocene. Volcanism continued in the Western Cascades arc well into the Miocene, but present discussion is limited to activity before 17 to 18 Ma (the early Western Cascade episode of Priest and others, 1983). Western Cascades magmas were mostly andesitic, especially earlier in the period considered here, but rhyolite erupted too—some as ash flows—especially later in the period. In some areas, there is little or no stratigraphic gap between early and late Western Cascades rocks (or even to the younger High Cascades), but elsewhere the early sequence was extensively eroded. Activity that lasted until mid-Miocene time formed a temporally and spatially continuous, broadly curving belt that is crossed by the straighter north-south trend of the latest Miocene to Quaternary High Cascades (Hammond, 1979). In southernmost British Columbia, the Pemberton belt, continuous with the early Western Cascades arc, includes ages of 29 Ma and younger on volcanic and intrusive rocks and lies east of the late Cenozoic High Cascades belt (Souther, 1977; *Berman and Armstrong, 1980*).

In Washington, the early Western Cascades arc (Fig. 20) overlapped the arc suites west of the older Challis and Tuscarora belts. Although the name "Western Cascades" is used here (following Oregon terminology), the arc in Washington more or less followed the present Cascade Crest, and some centers, such as the 25- to 30-Ma complex at Horse Lake Mountain near Wenatchee (*Tabor and others, 1982*), are located to the east. In much of the early Washington literature, Western Cascades lavas were included in the general term Keechelus Andesite, but that term lost its stratigraphic utility through generations of extension and redefinition (*Waters, 1955b*; *1961a*). In northern Washington, Western Cascades volcanic rocks are sparse. Tabor and others (1984) and *Frizzell and others* (*1984*) note upper Oligocene volcanics of 36- to 20-Ma age unconformable on the Naches; similar rocks 30 to 25 Ma in age extend nearly as far east as Wenatchee (*Tabor and others, 1982*). Because of uplift and erosion, however, early Western Cascades magmatism in the Northern Cascades is now represented largely by intrusive rocks, including numerous shallow batholiths of 34- to 22-Ma age (*Cater, 1969*; Tabor and Crowder, 1969; *Yeats and Engels, 1971*; Cater, 1982). Particularly important are the Snoqualmie batholith, 25 and 20 to 17 Ma in age (*Frizzell and others, 1984*), and the Tatoosh pluton, 26 to 22 and 18 to 14 Ma in age (Fiske and others, 1963; Mattinson, 1977). These composite intrusions represent the emplacement and evolution of andesitic magmas in large chambers as shallow as 1 to 2 km (*Erikson, 1969*) over extended periods of time (Mattinson, 1977). Intrusive and eruptive activity were contemporaneous and consanguineous. Near Mount Rainier, which now overlies the Tatoosh pluton (Fiske and others, 1963; *Frizzell and others, 1984*), thick calc-alkalic volcanic detritus with some lava (the 36- to 28-Ma Ohanapecosh Formation) is overlain unconformably by silicic pyroclastic materials, including ash-flow tuff (the 27- to 20-Ma Stevens Ridge Formation), and interfingering and overlying basaltic to andesitic lava and associated volcanic sediment (the 24- to 17-Ma Fifes Peak Formation). The folded, faulted, and generally altered Ohanapecosh is unconformable on the Puget Group, reflecting deformation concurrent with establishment of the early Western Cascades arc at about 37 Ma.

In the Cascades of western Oregon (Fig. 20), as in Washington, Oligocene calc-alkalic volcanic rocks typically overlie Eocene rocks unconformably (McBirney and White, 1982). Andesite (typically as lava) and dacite and rhyolite (commonly as

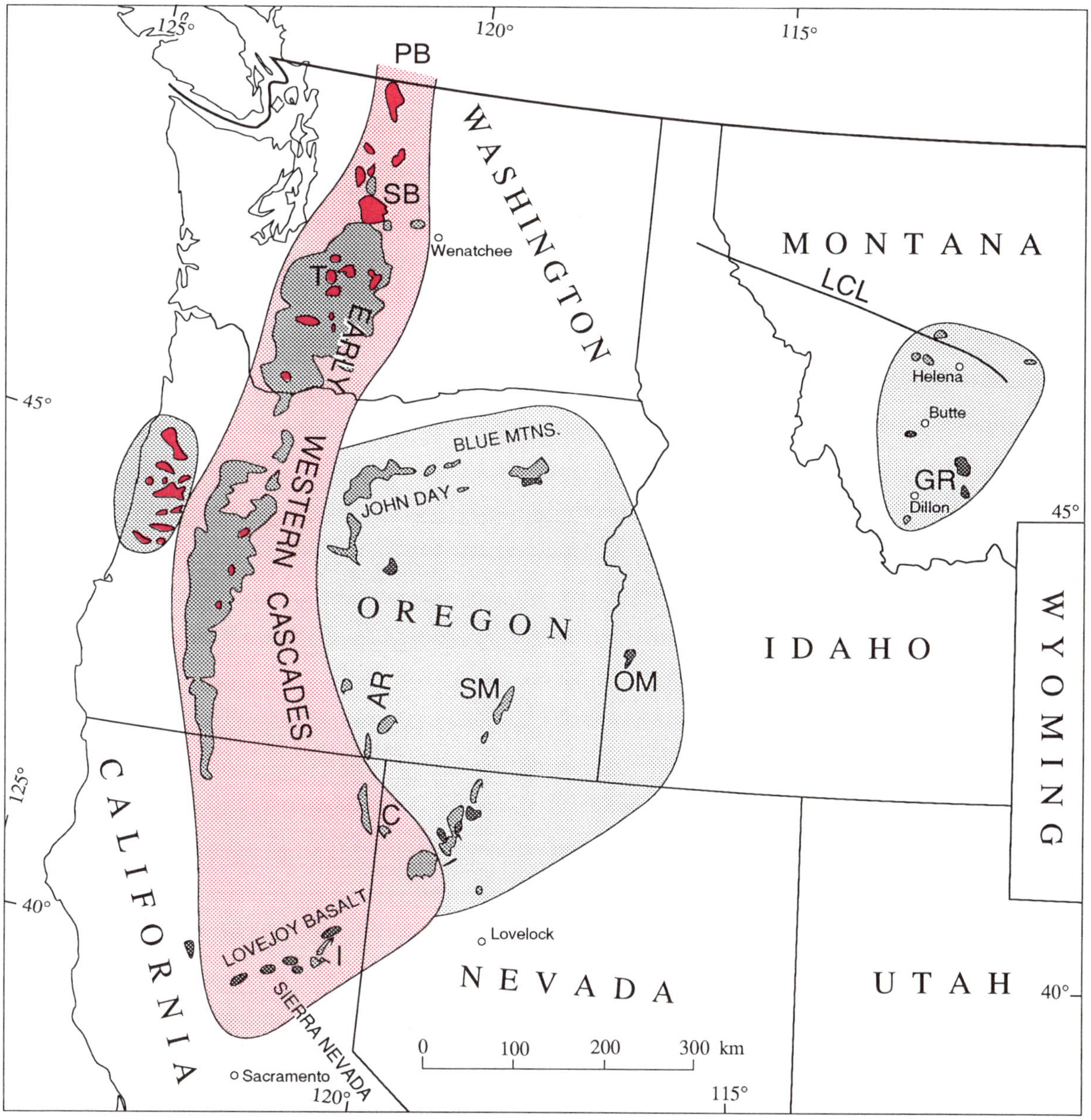

Figure 20. Map of the early Western Cascades arc and the John Day magmatic region (37 to 17 Ma). Dark-patterned areas indicate approximate outcrop distributions; pink, probable volcanic arc; red, intrusions; light gray, bimodal volcanic suite; dark gray, coastal ferrogabbro suite. AR, Abert Rim; C, Cedarville; GR, Gravelly Range; I, Ingals Formation; LCL, Lewis and Clark line; OM, Owyhee Mountains; PB, Pemberton belt; SB, Snoqualmie batholith; SM, Steens Mountain; T, Tatoosh pluton. Later Cenozoic rotations and displacements not restored.

pyroclastic rocks) were erupted from numerous early Western Cascades centers now marked in part by shallow subvolcanic intrusive bodies and by the Little Butte Volcanics (Peck and others, 1964). Although the basal part of the Little Butte Volcanics near the southern Willamette Valley is as old as 42 Ma, most of the unit is 37 to 17 Ma in age (*Sutter, 1978*; *Lux, 1982*). The Oligocene to mid-Miocene volcanic belt continues southward in the Oregon Cascades with similar lithologies and complex stratigraphy. Numerous small volcanoes built a broad low range rather than a sharp-crested volcanic chain (McBirney and White, 1982). In contrast to central Oregon, where younger rocks continue the record of arc volcanism through the Miocene, no stratigraphically continuous rocks in the southern Oregon Cascades are younger than about 17 Ma (*Smith and others, 1982*; *Fiebelkorn and others, 1983*).

As they had earlier, local mafic centers developed on the

Oregon continental margin in front of the early Western Cascades arc. Some of these centers were notably alkalic and are now represented by 33- to 30-Ma stocks, sills, and flows of nepheline syenite, camptonite, and ferrogabbro (MacLeod and Snavely, 1973; *Loeschke, 1979*). The 36- to 34-Ma calc-alkalic Goble Volcanics, northwest of Portland, comprises mainly basalt whose composition and tectonic setting suggested to *Beck and Burr (1979)* that it is a subalkalic arc-front series coeval with complex oblique subduction (Wells and Coe, 1985).

Early Western Cascades basaltic to rhyodacitic volcanism extended into northern California between 31 and 20 Ma (*Williams, 1949*; Macdonald, 1966; *Hammond, 1983*; *Vance, 1984*) although most of the older arc there is buried by High Cascades lavas. Farther south, the possibly Oligocene Reeds Creek Andesite (*Clark and Anderson, 1938*; Durrell, 1966) north of Sacramento and the Ingalls Formation (*Durrell, 1959a*), a 24- to 22-Ma (Dalrymple, 1964a) sequence of andesitic mudflow breccias in the northern Sierra Nevada, indicate continuation of the early Western Cascades arc beneath younger volcanics into the northeastern Sierra Nevada (Fig. 20).

The older arc appears wider than the present Cascades, perhaps in part because of later tectonic extension of the eastern part. In the Warner Range, 100 km east of the Cascades, andesitic debris flows and lavas and rhyolitic ash-flow tuffs of the "upper Cedarville series" of Russell (1928) range between 32 and 17 Ma (Fig. 20; Duffield and McKee, 1986). Andesite and dacite extended into northwestern Nevada, where Bonham (1969) reported a date of 31 Ma. Farther south and southeast, this calc-alkalic belt was replaced by a wide zone of voluminous, mainly rhyolitic volcanism (Stewart and Carlson, 1976), discussed in a later section. Thus, before mid-Miocene time, the early Western Cascades arc ended southeastward as a distinct entity in west-central Nevada.

Hammond (1979) denoted a significant tectonic contrast between the northern and southern sectors of the Western Cascades. As noted earlier, major uplift and folding mark the North Cascades (and their extension into British Columbia), but more or less continuous downwarping was recorded by successive eastward tilting during accumulation of the corresponding volcanic rocks from central Oregon to northern California. In northern California, *Vance (1984)* inferred that part of this tilting reflects contractional warping during a post-20-Ma volcanic hiatus. In some measure, the contrast between the northern and southern Cascades marks a transition in tectonic environments from contractional to extensional, concomitant with clockwise rotation of the Cascades (Wells, 1990).

The early Western Cascades arc produced, for the most part, a classic calc-alkalic suite (*White and McBirney, 1978*; McBirney and White, 1982), although some of the older lavas in the northwest have been considered tholeiitic by Priest and others (1983). Early in its evolution, the arc was predominantly andesitic, especially in its western part; it tended to produce more silicic lavas with time and with distance eastward. To the east, the John Day Formation of central and eastern Oregon largely preserves a downwind stratigraphic record of early Western Cascades volcanism (P. T. Robinson and others, 1984), thus tending to sample most effectively the more explosively erupted, more silicic compositions.

The John Day magmatic region

Although much of the John Day Formation of 36 to about 18 Ma age in central and eastern Oregon represents downwind fallout of early Western Cascades pyroclastic eruptions, it also records local volcanism of quite different character (Fig. 20). Significant volumes of high-silica alkalic rhyolite, including ash-flow tuffs, vented from numerous centers east of the arc, and other centers of alkali basalt to trachyandesite occur locally (*Peck, 1964*; *Fisher, 1966*; Robinson, 1975; Robinson and others, 1984). At least a few centers of predominantly intermediate volcanism, similar lithologically to the Clarno Formation, may have continued in John Day time until about 27 Ma (*Oles and Enlows, 1971*; Walker and Robinson, 1990a, b). Uplift of the Blue Mountains by northwest-southeast compression formed the north edge of the basin during John Day volcanism (*Robyn and Hoover, 1982*).

Rhyolitic to bimodal rhyolite-trachyandesite volcanism of comparable ages, east of predominantly andesitic arc rocks, continued south of the John Day Formation and represent a continuation of a similar magmatic province (Fig. 20). The volcanic rocks include rhyolites and tuffs in the Summer Lake–Goose Lake–Abert Rim area of southern Oregon (*Walker, 1960,* 1977; *Wells, 1979*) and the Pike Creek Formation near the base of Steens Mountain (Fuller, 1931; *Evernden and others, 1964*). In northwestern Nevada 26- to 24-Ma rhyolites, including the 24-Ma Ashdown Tuff, occur together with basalts and trachyandesites (Noble and others, 1970, 1973; *Smith, 1973*); the informally named Majuba igneous complex of *MacKenzie and Bookstrom (1976)* north of Lovelock is a complex shallow rhyolitic intrusion of similar age (*Silberman and Dockter, 1977*). Highly fluid valley-filling alkali-basaltic lavas formed the lower Miocene Lovejoy Basalt (*Durrell, 1959b,* 1966; Dalrymple, 1964a), filling a valley system through the western Cascades arc in the northern Sierra Nevada from an unknown source to the east, possibly within the John Day magmatic region (Fig. 20). The mainly trachyandesitic Steens Mountain Volcanics (Fuller, 1931; *Minor and others, 1987*) in south-central Oregon and volcanic series of Orevada View of *Rytuba and Conrad (1981)* in north-central Nevada are thick, rapidly erupted lava sequences, 23 to 20 Ma in age, that filled a northwest-trending graben during this volcanic episode. In the Owyhee Mountains of southwestern Idaho (Fig. 20), alkali basalts and trachyandesites of 31-Ma age lie between outliers of calc-alkalic Challis lavas and a section of mid-Miocene rhyolites (Ekren and others, 1982).

Magmas similar to those erupted in the John Day basin and surrounding areas—bimodal rhyodacites to rhyolites and basalts to trachyandesites—also erupted in Oligocene and early Miocene time in southwestern Montana (Fig. 20; Chadwick, 1972, *1978*).

These include rhyodacite of 27-Ma age southwest of Butte (*Marvin and others, 1982*), rhyolites and basalts near Helena of 37- to 20-Ma age (*Chadwick, 1978*), and basalts in the Gravelly Range–Virginia City area between 34 and 23 Ma in age (*Marvin and others, 1974a*). Basalts and rhyolites in the Lima-Dillon area interfinger with fossiliferous Oligocene basin deposits (*Myers, 1952*; *Scholten and others, 1955*; *Lowell, 1965*); Reynolds (1979) noted that this volcanism coincided with tectonic extension restricted to the same zone. Much of the abundant tuffaceous material in the Oligocene basin deposits of western Montana may represent this relatively local volcanism (*Freeman and others, 1958*) although some probably was derived from more distant voluminous Oligocene volcanism.

Sierra-Wasatch magmatic belts

Voluminous volcanism in mid-Tertiary time buried much of a large region of the present Great Basin, between the modern Sierra Nevada and Wasatch Range, to form a constructional volcanic plain. Before 43 Ma, during the time of main activity in the Challis belt, there had been no volcanism in the Sierra-Wasatch region south of a few outliers in southern Idaho (Fig. 17). Initial mid-Tertiary activity had been part of the 43- to 37-Ma Tuscarora magmatic belt (Fig. 19). During the time after about 37 Ma that major activity in the Pacific Northwest built the calc-alkalic early Western Cascades arc and scattered bimodal rhyolites and basalts to trachyandesites erupted in the John Day region, volcanism in the Sierra-Wasatch region was dominated by voluminous silicic pyroclastic eruptions up to the time of a brief mid-Miocene magmatic lull (McKee and others, 1970). Magmatism migrated southward through the region in a succession of east-west belts (Fig. 21).

As a new belt was established farther south between about 37 and 34 Ma, activity continued but declined in parts of the Tuscarora belt (Fig. 21A). The pattern shifted again at about 34 Ma (Fig. 21B), after which (until mid-Miocene time—about 17 Ma) some volcanism remained active in northern and central Nevada, but major activity continued to shift southward and was characterized by voluminous rhyolitic pyroclastic eruptions (Stewart and Carlson, 1976, *1978*; J. H. Stewart and others, 1977). At any particular time, the most voluminous activity was in a single east-west belt about 80 to 150 km wide, extending

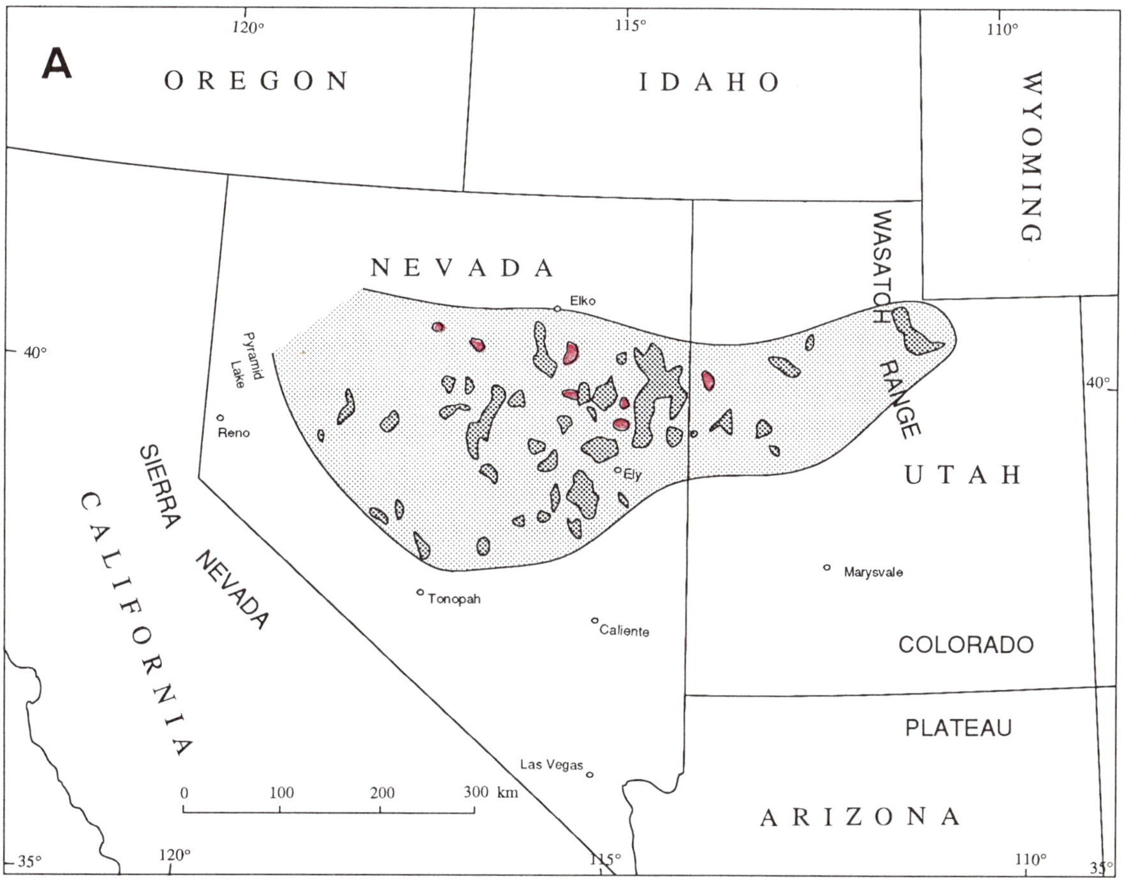

Figure 21. Maps of mid-Tertiary magmatic belts of the Sierra-Wasatch region. Dark-patterned areas indicate approximate outcrop distributions; gray, andesite-rhyolite suite; red, intrusions. Later Cenozoic displacements not restored. A, 37 to 34 Ma; B, 34 to 21 Ma. Hachured circles and ellipses indicate selected calderas. No volcanic rocks older than 28 Ma erupted south of the line marked <28.

virtually continuously from the present Sierran front to the Wasatch Range or the western margin of the Colorado Plateau. Igneous activity within this belt was generally concentrated toward its southern front, but volcanism, mainly rhyolitic, also continued in previously active areas farther north.

In each successive belt, activity evolved similarly. Initial andesites and related magmas were emplaced as hypabyssal intrusions, erupted from central volcanoes, and were distributed widely as debris flows and volcanic sediments (Armstrong, 1970; McKee and Silberman, 1970; *Wells and others, 1971*; *Blake and others, 1975*). The rocks are broadly calc-alkalic but more potassic than typical arc compositions (*Hausel and Nash, 1977*; *Wender and Nash, 1979*); potassium increases systematically eastward (*Merriam and Anderson, 1942*; Lipman and others, 1972). Similar rocks of this initial suite, with silica contents from about 55 to 62 percent, and averaging near 60 percent, are variably described as andesite (Solomon and others, 1979), high-potassium andesite (*Blake and others, 1975*), latite (W. J. Moore, 1973), or rhyodacite (Staatz and Carr, 1964). These predominantly intermediate lavas were generally succeeded by voluminous more silicic caldera-forming ash-flow eruptions; after about 34 Ma, rhyolitic ash flows accounted for the bulk of regional activity (J. H. Stewart, 1980). Intermediate lavas continued to erupt locally, but basalt is exceptionally rare.

In north-central Nevada (Fig. 21A), early Oligocene intermediate plutonism and volcanism, about 38 to 37 Ma in age, partly overprinted the Tuscarora magmatic belt but extended farther southwest and south (Armstrong, 1970; *McDowell, 1971*; *Ekren and others, 1974a*). Early andesite in the Elko area (Solomon and others, 1979) overlies volcanics emplaced from pyroclastic eruptions farther north. Younger regional lavas were generally more rhyolitic and are interlayered with locally erupted ash-flow tuffs (*Blake and others, 1975*). In the Drum Mountains area of western Utah, where significant ash-flow eruptions had occurred as part of the Tuscarora belt, voluminous rhyolitic eruptions continued from 38 to 32 Ma (*Lindsey and others, 1975*).

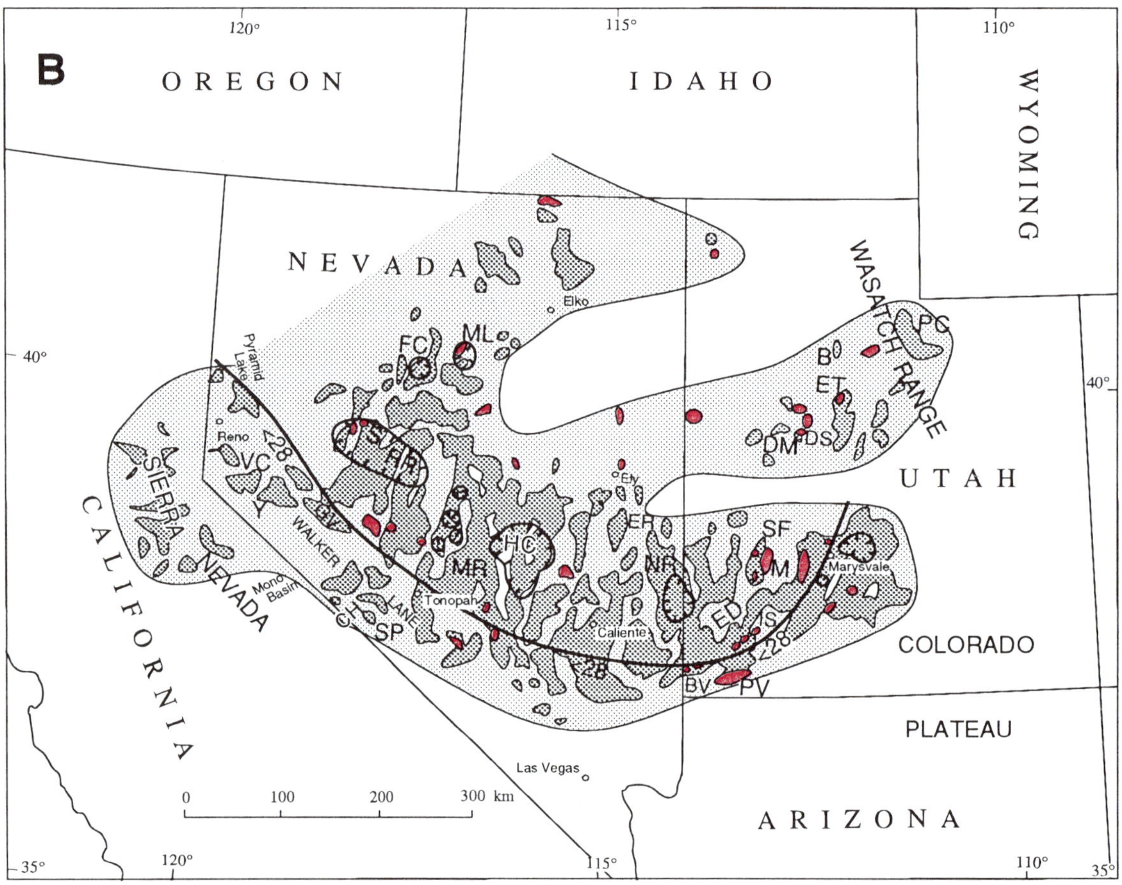

B, Bingham; BV, Bull Valley; CH, Candelaria Hills; DM, Drum Mountains; DS, Desert Mountain; ED, Escalante Desert; ER, Egan Range; ET, East Tintic Range; FC, Fish Creek Mountains; GV, Gabbs Valley Range; HC, Hot Creek Range; IS, Iron Springs; M, Mineral Mountains; ML, Mount Lewis; MR, Monitor Range; NR, Needles Range; PC, Park City; PV, Pine Valley Mountains; SP, Silver Peak Range; S-RR, Stillwater–Reese River area; SF, San Francisco Mountains; TQ, Toquima Range; VC, Virginia City; Y, Yerington.

Between 34 and 21 Ma, an east-west belt across Nevada and Utah, both overlapping earlier volcanism and extending southward beyond it, erupted voluminous ash flows (Fig. 21B). In the north-central part of this belt, *Masursky* (*1960*) and *Burke and McKee* (*1979*) considered some of the eruptions, particularly the 32-Ma Caetano Tuff, to have been from an east-west volcanic trough. *Wrucke and Silberman* (*1975, 1977*), however, proposed (despite a protest by *Gilluly, 1977*) that the Caetano erupted from the Mount Lewis area in the northern Shoshone Range to form a caldera 16 km in diameter, now represented by ring faults, intrusions, and erosional volcanic remnants. Nearby, the 25-km-diameter Fish Creek Mountains caldera formed during eruption of a 24-Ma ash-flow sheet (*McKee, 1970*). The section is capped by widespread 24- to 22-Ma highly differentiated rhyolitic ash flows, the Bates Mountain Tuff and tuff of Clipper Gap (*Sargent and McKee, 1969*; Grommé and others, 1972). Other 31- to 21-Ma ash flows (Fig. 21B), a northwest-southeast volcano-tectonic trough, and several calderas occur in the Stillwater–Reese River area (*McKee and Stewart, 1971*; *Riehle and others, 1972*; *Burke and McKee, 1979*). In the Toquima Range are several 32- to 22-Ma ash-flow sheets and calderas, again capped by the Bates Mountain Tuff and tuff of Clipper Gap (*McKee, 1974, 1976*; *Boden, 1986*). Voluminous ash flows as old as 34 Ma with sources from the Monitor Range to the Egan Range are also capped by the Bates Mountain and Clipper Gap units (Cook, 1965; *Scott, 1966*; *Quinlivan and others, 1974*; *Ekren and others, 1974b*).

East of central Nevada, the band of 34- to 21-Ma volcanics splits into two east-west belts (Fig. 21B; J. H. Stewart and others, 1977). In the northern branch, in central western Utah, a granitic stock at Desert Mountain, possibly within a resurgent caldera (*Shawe, 1972*), is about 29 Ma (Armstrong, 1970; *Lindsey and others, 1975*). In the East Tintic Mountains (*Morris, 1975*; Morris and Lovering, 1979), voluminous 33-Ma pyroclastic quartz latite and slightly younger monzonite and latite define a 15-km caldera. At Bingham, older latite and monzonite are overlain by 33- to 31-Ma rhyolite and latite (*Moore and others, 1968*; W. J. Moore, 1973). In the Cottonwood–Park City area, the younger stocks and volcanic rocks are mainly about 35- to 32-Ma and perhaps as young as 24 Ma in age (*Crittenden and others, 1973*).

In the southern branch of the 34- to 21-Ma belt (Fig. 21B), distinctive 31- to 28-Ma crystal-rich dacitic ash-flows, including the voluminous Needles Range Group, are major units (Mackin, 1960, *1963*). To the north and west, these units are generally high in a thick volcanic section, but to the south and east, they mark the base. At least four Needles Range ash-flow sheets erupted from calderas near the Utah-Nevada border, spreading to the margin of the Colorado Plateau and far westward into Nevada (Best and others, 1973; Best and Keith, 1983). Underlying andesite and rhyolite are 34 to 32 Ma in age (*Lemmon and others, 1973*; Best and Keith, 1983; Best and others, 1989). To the northeast, younger andesite to rhyolite is associated with 28- to 21-Ma granitic stocks near the San Francisco Mountains (*Erickson, 1973*; *Lemmon and others, 1973*) and with late Oligocene intrusive phases in the Mineral Mountains (Rowley and others, 1979; *Nielson and others, 1986*). The largest coherent mainly post–Needles Range volcanic field is centered on the Colorado Plateau margin near Marysvale (Anderson and Rowley, 1975; Rowley and others, 1979). Local lavas as old as 31 Ma underlie the Needles Range rocks, but the field was dominated from about 30 to 22 Ma by andesite. At least three rhyolitic ash-flow eruptions formed calderas (Steven and others, 1984a).

In an arcuate band west and south of the 34- to 21-Ma volcanic fields, initial magmatism did not occur before about 28 Ma but continued thereafter in concert with younger phases of the fields to the north (Fig. 21B). In southwestern Utah, andesitic to rhyodacitic volcanic and intrusive centers of 22 to 20 Ma (Armstrong, 1970; *Hausel and Nash, 1977*) include the Pine Valley Mountains (*Cook, 1957*), the Bull Valley District (*Blank, 1959*), and the Iron Springs district (Mackin, 1947, 1960), some with associated contact-metasomatic iron deposits. Distinctive K-rich mafic lavas erupted between 25 and 21 Ma (Best and others, 1980). Straddling the Utah-Nevada border, andesitic to dacitic lavas and tuffs older than about 26 Ma span more than 100 km around the younger Caliente cauldron complex (*Noble and McKee, 1972*; Ekren and others, 1977). Above are widespread ash-flow sheets, including the 25-Ma Isom Formation, the 24- to 21-Ma Quichipa Group, and younger units (Mackin, 1960; Cook, 1965; Armstrong, 1970; *Noble and McKee, 1972*; *Fleck and others, 1975*); most of the units older than 21-Ma are centered north of Caliente (Ekren and others, 1977). Farther west, the oldest Tertiary volcanics include four regionally extensive 27- to 24-Ma ash-flow tuffs (Ekren and others, 1971). North of Tonopah, in west-central Nevada, intermediate to rhyolitic lavas erupted between 26 and 24 Ma (Bonham and Garside, 1979). Regionally extensive ash flows between 28 and 22 Ma extend along the northwest-trending, right-lateral Walker Lane in west-central Nevada and adjacent California, from the Silver Peak Range, Candelaria Hills, and Mono Basin (C. M. Gilbert and others, 1968; *Speed and Cogbill, 1979a*; Robinson and Stewart, 1984), through the Gabbs Valley Range (Ekren and others, 1980), the Yerington district (*Proffett and Proffett, 1976*), and Virginia City and Carson City (*Bingler, 1978*), to the Reno–Pyramid Lake area (Bonham, 1969; *Bingler, 1978*).

The oldest widespread post-Eocene rocks of the Sierra Nevada are rhyolites contemporaneous with those of western Nevada. The 23- to 21-Ma Valley Springs Formation of the western Sierra foothills and the Delleker Formation farther north are rhyolitic lavas and sediments (*Durrell, 1959a*, 1966; Dalrymple, 1964a). Higher in the central Sierra, welded tuff is mainly 29 to 21 Ma in age although some may be as old as 34 Ma (Dalrymple, 1963, 1964a; *Evernden and James, 1964*). These units occupy narrow outcrop bands along west-flowing Sierran drainages that headed in the Walker Lane during the voluminous 28- to 22-Ma ash-flow eruptions (Slemmons, 1966; *Proffett and Proffett, 1976*). The existence in this area of possibly older tuffs suggest that the headwaters might have been even farther into central Nevada, where rhyolite as old as 34 Ma vented.

Numerous hydrothermal chalcophile mineral deposits were associated with mid-Tertiary magmatism throughout the Sierra-Wasatch region (*Albers and Kleinhampl, 1970*; Silberman and others, 1976). Most are associated with local centers rather than with the large caldera complexes (E. H. McKee, 1979). Some of the larger rhyolitic centers and associated sediments, however, may host uranium mineralization (*Steven and others, 1981*). In eastern Nevada and western Utah, ore deposits occur in the same two belts as the volcanism, associated with broad aeromagnetic highs (J. H. Stewart and others, 1977) and topographic and structural lineaments (*Rowley and others, 1978b*). Farther west the belts overlap, but *Ekren and others (1974a, 1976)* propose that east-west lineaments through central Nevada are their structural expressions.

In summary, this section has described a sequence of overlapping, south-stepping magmatic belts in the Sierra-Wasatch region from about 37 Ma up to a mid-Miocene magmatic transition between about 21 and 17 Ma, discussed later. The magmas, generally calc-alkalic but more potassic and considerably more silicic than those of typical magmatic arcs, included almost no basalts. The most voluminous volcanism occupied bands that were successively farther south during the periods 37 to 34 Ma, 34 to 28 Ma, and 28 to 21 Ma. Some of the same trends continued through the later magmatic transition, but a sharper distinction between an arc-related suite in the west and a predominantly basaltic or bimodal suite in the east was to characterize later evolution.

EOCENE TO EARLY MIOCENE MAGMATISM OF THE SOUTHERN UNITED STATES CORDILLERA

An "amagmatic corridor" (*Anderson, 1981*; Wernicke and others, 1987) across southernmost Nevada and adjacent California and Arizona at the latitude of Las Vegas marks a fundamental break and a line of approximate temporal symmetry in magmatic evolution (compare Figs. 21B, 22). Therefore, the description of mid-Tertiary magmatism now skips eastward to southern New Mexico to reestablish temporal continuity with Sierra-Wasatch events.

Activity in shifting magmatic belts, similar in some ways to that of the Sierra-Wasatch region, occurred in the southern Basin and Range Province of Arizona and New Mexico and in the Southern Rocky Mountains during the latest Eocene and Oligocene (Fig. 22). Magmatism with similarities to the early and middle Eocene activity of the Northern Great Plains occurred in southern Cordilleran foreland areas during late Eocene to early Miocene time (see Fig. 24).

Mid-Tertiary Sierra Madre–Rio Grande–Southern Rocky Mountains magmatism

While there was little or no post-Laramide magmatism through much of southernmost Nevada, eastern California, and western Arizona between about 55 and 25 Ma, magmatism in the Sierra Madre Occidental of Mexico remained nearly continuous from Laramide through mid-Tertiary time (*Alba and Chavez, 1974*; *McDowell and Clabaugh, 1979*; Clark and others, 1979, 1982; *Keller and others, 1982*). This continuing early Tertiary magmatism in Mexico occurred in a zone parallel to the continental margin and was, thus, analogous to that of western Canada. At about 43 Ma, as andesitic magmatism in the Pacific Northwest began to shift southwestward from the Challis to the Tuscarora belt, the long-lived Sierra Madre volcanism began to extend north and west into the Eocene magmatic gap. By the time of the regionally voluminous Sierra-Wasatch ash flows that began about 34 Ma, large areas of New Mexico and Colorado were blanketed by volcanic rocks, and activity extended to southeastern California (Fig. 22). Major caldera-forming ash-flow eruptions continued into early Miocene time, about 26 to 21 Ma; early Miocene magmatism also extended northwestward through the Mojave Desert region.

Mid-Tertiary potassic andesite-rhyolite magmatism in west Texas close to the Rio Grande (Fig. 22), distinct from coeval but more mafic and alkalic magmatism farther east in west Texas (see Fig. 24; Barker, 1977, 1979), is part of a continuous belt in northern Mexico and southwestern New Mexico. Activity in this belt may have begun as early as 49 Ma but did not culminate until caldera-forming eruptions of 38 to 28 Ma (*McDowell, 1979*; Henry and Price, 1984). Mid-Tertiary potassic calc-alkalic volcanism began in southwestern New Mexico about 43 Ma (Fig. 22) and was widespread by 37 Ma (Elston, 1984a). Pyroxene andesite to rhyodacite from numerous centers formed the Mogollon-Datil volcanic field from east of the southern Rio Grande (*Seager, 1973, 1975,* 1981) to the Arizona–New Mexico border area (Deal and others, 1978; *Marvin and others, 1978*) and northward to the volcanic Mogollon Plateau (Elston and Northrup, 1976; Chapin and Elston, 1978; Ratté and others, 1984). Initial, predominantly andesitic magmatism was succeeded by predominantly silicic caldera-forming ash-flow eruptions, but andesite to dacite continued to erupt locally. Basalt remained sparse until at least 31 Ma. The earliest large calderas in southwesternmost New Mexico formed between 39 and 35 Ma (Deal and others, 1978; Elston, 1984a); caldera-forming eruptions began in the Mogollon Plateau at about 34 Ma (Ratté and others, 1984; Elston, 1984a). Major ash flows continued to erupt throughout southwestern New Mexico until 28 Ma. After about 31 Ma, however, highly differentiated rhyolite was associated with olivine-bearing basalt to andesite, which outlasted the ash flows and became more widespread and voluminous after about 26 Ma (Chapin and Seager, 1975; *Elston, 1984b*), continuing until about 21 Ma. The voluminous 34- to 28-Ma rhyolite of the 125-km Mogollon Plateau (Fig. 22) marked the emplacement of a composite upper-crustal batholith (Elston and others, 1976; *Rhodes, 1976*; *Krohn, 1976*). Andesite, andesitic sediments, and distal ash flows of the Mogollon-Datil field overlapped into eastern Arizona along the margin of the Colorado Plateau (*Wrucke, 1961*; *Ratté and others, 1969,* 1984), and younger (26- to 20-Ma) predominantly mafic lavas were associated with local rhyolitic centers (*Richter and others, 1983*).

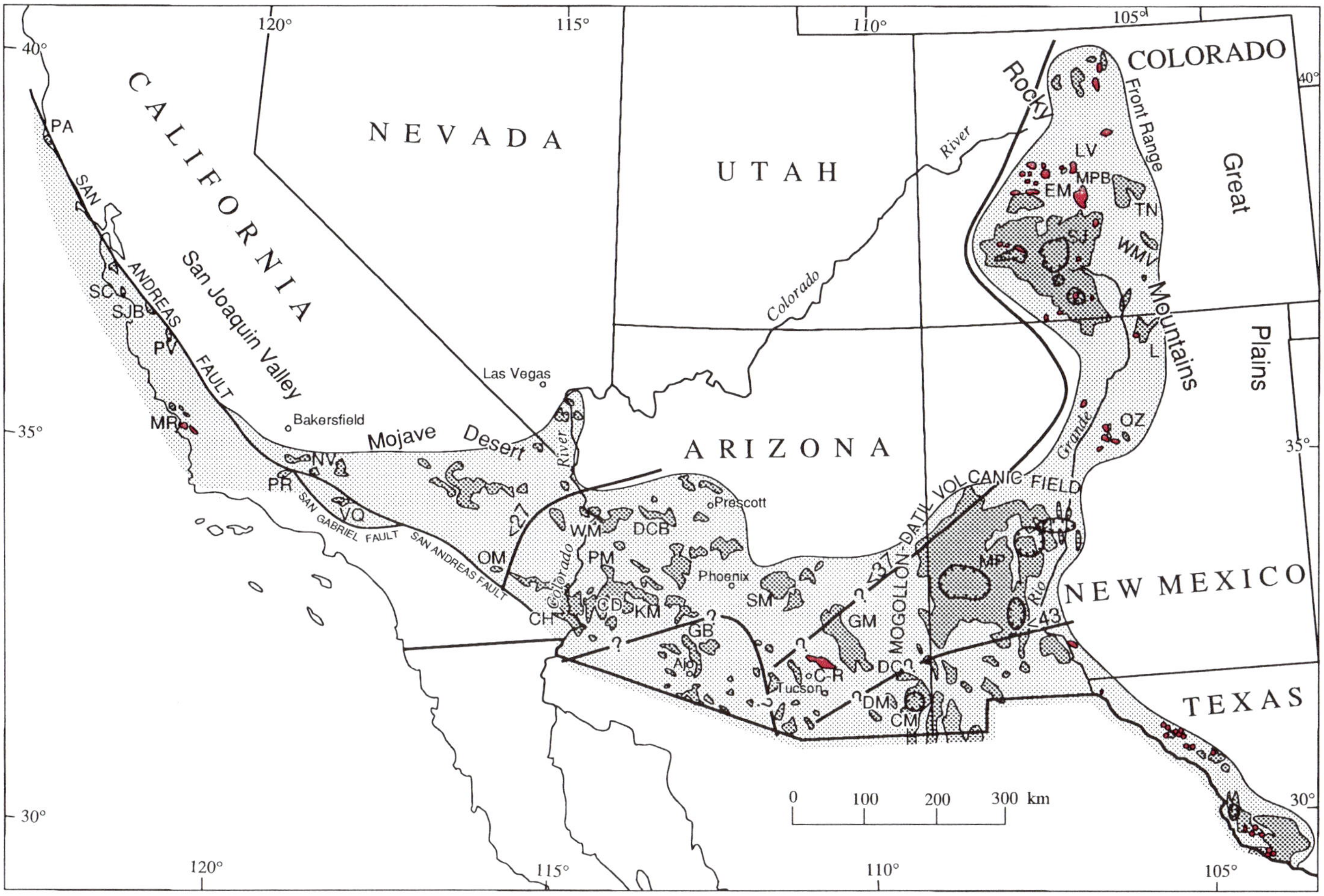

Figure 22. Map of magmatic belts of the Mojave-Sonoran, Rio Grande, and Southern Rocky Mountains regions and part of coastal California (43 to 21 Ma). Dark-patterned areas indicate approximate outcrop distributions; gray, andesite-rhyolite suite; red, intrusions. Hachured circles and ellipses indicate selected calderas. No volcanic rocks older than the ages indicated (Ma) occur northwest of the lines marked <43, <37, and <27; ages of oldest post-Laramide volcanics enclosed by the queried line in southwestern Arizona are uncertain but none >26 Ma are known. C-R, Catalina-Rincon Mountains; CD, Castle Dome Mountains; CH, Chocolate Mountains; CM, Chiricahua Mountains; DC, Dos Cabezas Mountains; DCB, Date Creek basin; DM, Dragoon Mountains; EM, Elk Mountains; GM, Galiuro Mountains; KM, Kofa Mountains; L, Latir volcanic field; LV, Leadville; MP, Mogollon Plateau; MPB, Mount Princeton batholith; MR, Moro Rock; NV, Neenach Volcanics; OM, Orocopia Mountains; OZ, Ortiz Mountains; PA, Point Arena; PM, Plomosa Mountains; PR, Plush Ranch Formation; PV, Pinnacles Volcanics; SC, Santa Cruz Mountains; SJ, San Juan Mountains; SJB, San Juan Bautista; SM, Superstition Mountains; TN, Thirtynine Mile volcanic field; VQ, Vasquez Formation; WM, Whipple Mountains; WMV, Wet Mountain Valley. Later Cenozoic extension, rotations, and right-lateral displacement (>300 km across the San Andreas fault) not restored.

In the Southern Rockies (Fig. 22), magmatism was delayed by no more than 1 to 2 m.y. from that of the Mogollon-Datil region. Potassic calc-alkalic volcanism, predominantly andesitic, began in Colorado no later than 41 Ma (*Bryant and others, 1981*) and by 36 to 35 Ma blanketed the Eocene erosion surface with coalesced lavas and breccias (Steven, 1975). In north-central New Mexico (G. O. Bachman and Mehnert, 1978; *Kautz and others, 1981*; *Hagstrum and others, 1982*; *Lipman, 1983*; Lipman and others, 1986), volcanism began with clustered 37- to 27-Ma (mainly 29- to 27-Ma) stratovolcanoes and evolved to 27-Ma caldera-forming rhyolitic ash flows. Postcollapse lavas of 27- to 26-Ma age coincided with early tectonic extension; smaller intrusions produced molybdenum mineralization about 23 to 21 Ma. In the San Juan Mountains of southwestern Colorado, early andesitic to rhyodacitic lavas and breccias began at 38 Ma but were mainly 35 to 30 Ma in age (E. S. Larsen and Cross, 1956; Lipman and others, 1970; *Varga and Smith, 1984*). Nearby to the north, andesitic stocks, sills, laccoliths, and dikes (*Godwin and Gaskil, 1964*) are flanked by volcanic breccias. Andesites in central (Epis and Chapin, 1968, *1974*; *Sanders and others, 1976*)

and northern Colorado (Izett, 1966, *1968*; *Corbett, 1968*) are mainly 36 to 27 Ma in age.

Superposed on these Oligocene intermediate lavas in Colorado are two large ash-flow fields, the surface expressions of shallow composite batholiths, the larger in the San Juan Mountains and the other overlapping the Laramide Colorado Mineral Belt around Leadville (Fig. 22). Between about 31 and 27 Ma, the San Juans produced voluminous rhyodacitic to rhyolitic ash-flow sheets and a complex of source-area calderas (Steven and Lipman, 1976; *Lipman and others, 1978a*) while intermediate lavas erupted episodically. In the Colorado Mineral Belt, high-level magmatism controlled by NE-trending Precambrian structures (*Tweto and Sims, 1968*; *Tweto, 1975*), reflects a composite upper-crustal batholith of both Laramide and mid-Tertiary ages (Steven, 1975). Epizonal 41- to 35-Ma quartz diorite to quartz monzonite, some with molybdenum mineralization, and local volcanic remnants crop out from the central Front Range to the Leadville-Climax area (Taylor and others, 1968; Steven, 1975; *Bryant and others, 1981*). Epizonal granodiorites of 34- to 29-Ma age a little farther west (*Obradovich and others, 1969*; *Lipman and others, 1969*; *Bryant, 1979*) have quench textures and fluid inclusions, that indicate at least some surface venting (*Cunningham, 1976*). Voluminous caldera-forming ash-flow tuffs include the 36-Ma Wall Mountain Tuff (*Lipman and others, 1969*; Steven, 1975) that flowed through drainages across the Front Range to the Great Plains (Epis and Chapin, 1968; *1974*; *Izett and others, 1969*; *Chapin and Lowell, 1979*), and a 34-Ma tuff with a concentrically zoned composite stock in its resurgent-cauldron core (*Fridrich and Mahood, 1984*).

In the Mogollon Highland of Arizona (see Fig. 15), early magmatism may have coincided with 43- to 35-Ma activity in New Mexico and Colorado (Fig. 22), but the record is obscure. Only single K-Ar dates in a few localities suggest volcanism between 48 and 39 Ma as far westward as the vicinity of Tucson (Fig. 22; Gilluly, 1956; *Bikerman and Damon, 1966*; *Marvin and others, 1978*), and these dates could be spurious. The oldest widespread post-Laramide activity in southeastern Arizona produced andesite to rhyodacite flows and granodiorite stocks and dikes between 37 and 25 Ma (*Bikerman and Damon, 1966*; *Bikerman, 1967*; *Percious, 1968*; Drewes, 1972; *Hoggatt and others, 1977*; *Marvin and others, 1978*; Shafiqullah and others, 1978; *Banks and others, 1978*; Erickson, 1981; *Thorman, 1981*). The principal intermediate rocks are potassic andesite although they have been designated variably in the Arizona literature as basalt, basaltic andesite, andesite, potassic andesite, latite, and doreite. Conspicuously porphyritic rocks have been called "turkey track porphyry" (*Cooper, 1961*; *Mielke, 1965*). High initial $^{87}Sr/^{86}Sr$ suggests pervasive interactions in the lower crust (Damon, 1971; Shafiqullah and others, 1978). Rocks reported as "ultrapotassic" trachyte and rhyolite (*Shafiqullah and others, 1976*, 1980; *Rehrig and others, 1980*) do not represent magmatic compositions (cf. *Stewart, 1979*) but reflect regional metasomatism associated with detachment faulting (Brooks, 1986; Glazner, 1988; *Roddy and others, 1988*).

A belt of voluminous rhyolite extends northwest through southeastern Arizona from the Chiricahua Mountains to the Superstition Mountains (Fig. 22). In the Chiricahua Mountains (*Enlows, 1955*; *Sabins, 1957*; *Fernandez and Enlows, 1966*; Shafiqullah and others, 1978; *Drewes, 1982*), basal 32- to 29-Ma andesites are succeeded by diverse 28- to 23-Ma calc-alkalic lavas, including abundant rhyolite, continuous northward into the Pinaleno Mountains (*Thorman, 1981*). Climactic 25-Ma rhyolitic ash flows erupted to form a 20-km caldera (*Marjaneimi, 1968, 1969*). In and near the Galiuro Mountains (Cooper and Silver, 1964; *Simons, 1964*; Creasey and Krieger, 1978; Krieger, 1979; *Creasey and others, 1981*), basal 29- to 28-Ma andesite to rhyodacite is overlain by 26- to 23-Ma rhyodacite and rhyolite, including voluminous ash-flow tuffs from unknown sources. Intrusive rocks with comparable ages in the Catalina-Rincon block to the west (Creasey and others, 1977) may be displaced with respect to the Galiuro sequence. An overlapping volcanic field occurs in the Superstition Mountains–Superior area east of Phoenix (*Peterson, 1968*; *Sheridan and others, 1970*; Stuckless and Sheridan, 1971; Sheridan, 1978; *Peterson, 1979*), where 29- to 25-Ma basalt and potassic andesite are overlain by voluminous 24- to 20-Ma rhyodacitic to rhyolitic ash flows, interlayered lavas, and bedded tuffs.

Many of the best-known mineral deposits of the Sierra Madre–Rio Grande region are Laramide; mid-Tertiary deposits are generally smaller. Elston (1978) pointed out that the older, less differentiated, andesitic to rhyolitic mid-Tertiary centers host chalcophile base-metal deposits whereas lithophile metals such as lithium, beryllium, and tin occur with the younger rhyolitic centers. Among the significant mid-Tertiary deposits of the region are 30- to 23-Ma molybdenum porphyries of the Southern Rocky Mountains from northern New Mexico to central Colorado (*Lipman, 1981*).

In summary, magmatism encroached northward about 43 Ma from the Sierra Madre Occidental of Mexico into a post-Laramide magmatic gap. Some activity had reached the Rocky Mountains of central Colorado by at least 41 Ma and, perhaps, central-southern Arizona by 40 Ma. Early activity was diverse but produced mainly intermediate potassic calc-alkalic lavas and intrusives. Later activity was more silicic, locally producing batholithic magma chambers and voluminous caldera-forming ash-flow eruptions. After about 26 Ma the predominant activity became more femic in many areas, accompanied locally by silicic eruptions.

Mid-Tertiary Sonoran-Mojave–Coast Ranges magmatism

The regional time-space pattern of mid-Tertiary volcanism in southwestern Arizona and adjacent California (Fig. 22), known only through broad reconnaissance and a few detailed studies, remains unclear. Magmatism began in a belt through this region as early as 35 to 27 Ma. Potassic calc-alkalic lavas in the Prescott-Jerome area are 27 to 23 Ma in age (*Krieger and others, 1971*). Early mafic to silicic volcanics in and near the Gila Bend

Mountains are 30 to 27 Ma in age (Eberly and Stanley, 1978; Shafiqullah and others, 1980), coinciding with what Eberly and Stanley interpret as an Oligocene to early Miocene structural trough. Calc-alkalic magmas intruded and erupted near the Chocolate Mountains, on both sides of the lower Colorado River, between 35 and 22 Ma (*Olmsted and others, 1973*; Crowe, 1978; Crowe and others, 1979; Shafiqullah and others, 1980). Magmatism there at first was compositionally diverse but became predominantly silicic between 26 and 22 Ma, including large plutons, thick ash-flow tuffs, and caldera complexes (*Gutmann, 1982*; *Grubensky and Bagby, 1990*). Lavas younger than about 22 to 20 Ma are mainly basaltic to andesitic but include some dacite to rhyolite. Similar lavas of 23- to 20-Ma age also occur in the lower Miocene Diligencia Formation (Spittler and Arthur, 1982) in the Orocopia Mountains, just northwest. Volcanic rocks older than 30 Ma occur northward as far as the Whipple Mountains (Davis and others, 1980, *1982*), where 32- to 26-Ma andesite and basalt that predate detachment faulting are overlain by 19- to 17-Ma synextensional andesite. Thus, a potassic calc-alkalic magmatic belt extended between 35 and 17 Ma from southern New Mexico through southern Arizona to the lower Colorado River area. Within that belt, however, are many areas in which dates older than about 25 Ma are unknown, such as the Castle Dome (*Gutmann, 1982*), Kofa (*Dahm and Hankins, 1982*), and Plomosa Mountains (*Miller, 1970*; *Miller and McKee, 1971*). Plutonic and volcanic rocks older than 23 Ma, where preserved, commonly are greatly tilted and extended.

South of the Gila Bend area, it is not clear from published data whether any magmatism was as old as 35 to 30 Ma. No published ages in this more southerly zone, which encompasses and overlaps a distinct province of Laramide thrusting, metamorphism, and plutonism (*Haxel and others, 1984*), are greater than 25 Ma (Fig. 22). In the ranges around Ajo (Gilluly, 1946; *May and others, 1981*; *Gray and Miller, 1984*; *Gray and others, 1985*; *Tosdal and others, 1986*), a 25- to 16-Ma volcanic sequence is dominated by latite to potassic andesite but includes basalt and rhyolite. Nevertheless, altered and highly extended but largely undated post-Laramide igneous rocks underlie these lavas unconformably. Locally, an earlier generation of post-Laramide lamprophyric dikes is cut by 24-Ma rhyolitic dikes (*Wright and Haxel, 1982*).

North of a Prescott–Whipple Mountains band (Fig. 22), initial post-Laramide magmatism in the lower Colorado River extensional corridor occurred no earlier than about 25 to 26 Ma (Eberly and Stanley, 1978; Shafiqullah and others, 1980; Rehrig and Reynolds, 1980), about the time of earliest dated local extension. In westernmost Arizona, steeply tilted and altered calc-alkalic basalt to rhyolite includes abundant intermediate compositions but commonly yields poor K-Ar dates; vertebrate fossils and some K-Ar ages indicate that these rocks are at least as old as 24 to 25 Ma (*Rehrig and others, 1980*; Otton, 1982; Suneson and Lucchitta, 1983; *Brooks, 1985*). Similar volcanic sections extend westward to the central Mojave region, where the pre-Miocene highland began to accumulate a sedimentary and volcanic cover at about 25 to 22 Ma (Dokka and Glazner, 1982; Woodburne and others, 1982; *Moseley and others, 1982*; *Burke and others, 1982*; *Nielson and Glazner, 1986*). Basal 24- to 20-Ma basalt to rhyolite is generally dominated by andesite. Above are 20- to 17-Ma basaltic lavas and silicic bedded tuffs from local dacitic to rhyolitic centers. The Peach Springs Tuff (Young and Brennan, 1974; *Buesch, 1986*; Glazner and others, 1986; *Wells and Hillhouse, 1989*; *Nielson and others, 1990*), a distinctive and voluminous 18-Ma rhyolitic ash-flow sheet, is a regional marker from the western Colorado Plateau to the central Mojave Desert. Relations apparently similar to those of the central Mojave are relatively poorly known through the western Mojave Desert, where undated erosional remnants of rhyolitic lavas and tuffs are locally associated with andesite and basalt (Dibblee, 1967). At the southwestern margin of the area, rhyolite with minor andesite and dacite of the Neenach Volcanics, resting on arkosic conglomerates and sandstones above granitic basement (Dibblee, 1967), is 24 to 19 Ma in age (see Fox and others, 1985). Near the southern end of the San Joaquin Valley, about 50 km farther northwest, 25- to 22-Ma calc-alkalic basalt to dacite erupted across an early Miocene marine shoreline.

The lower Colorado River extensional corridor terminates in Nevada south of Las Vegas (Fig. 22), where two compositionally diverse calc-alkalic assemblages are distinguished by their relations to detachment faulting (Longwell, 1963; R. E. Anderson, 1971, *1978*; Anderson and others, 1972). The older Patsy Mine Volcanics, similar to sections in the Mojave region, is probably 23 to 15 Ma. In contrast with post-16-Ma basaltic volcanism of areas farther south, however, younger magmatism continued to be calc-alkalic and predominantly intermediate, as discussed later.

The southwestern margins of several 27- to 20-Ma volcanic fields are offset more than 300 km northwest along the San Andreas fault system (Fig. 22). Andesite and some dacite of the Diligencia Formation in the Orocopia Mountains are offset to the 26- to 22-Ma Vasquez Formation *Corwell, 1973*; Weigand, 1982) and to lavas in the Plush Ranch Formation (*Crowell, 1973*; Weigand, 1982) that yield 18- to 20-Ma ages. Despite the differences in measured K-Ar ages, identical offsets in basement lithologies and overlying mid-Miocene sedimentary rocks confirm that these three areas were initially parts of a continuous volcanic field (see Fig. 10). Similarly, the Neenach Volcanics of the western Mojave Desert is offset by the San Andreas fault to the Pinnacles area in the central Coast Ranges (Turner, 1970; *Matthews, 1973*; 1976; Weigand, 1982), and volcanics of the southern San Joaquin Valley that spanned a shoreline are offset to near San Juan Bautista (Turner, 1970; Huffman, 1972). Calc-alkalic volcanism of 27- to 20-Ma age continued westward in a 700-km coastal block from the San Gabriel Mountains to Point Arena (Fig. 22), including the calc-alkalic, mainly dacitic plugs, necks, and domes of the Morro Rock–Islay Hill chain and related eruptive dacite and rhyolite of the Cambria Felsite, 27 to 22 Ma in age (Ernst and Hall, 1974), 23- to 20-Ma submarine basalt in the Santa Cruz Mountains (Turner, 1970), and the 24-Ma Iversen

Basalt (*Weaver, 1944*; Turner, 1970) near Point Arena. Restoring the mid-Miocene and younger right slip on the San Andreas and related faults places these volcanic rocks between about the latitudes of Las Vegas and Tucson during their eruption, suggesting that they were a continuous part of the mid-Tertiary calc-alkalic magmatism of the Mojave-Sonoran region (Fig. 22; Plate 7D).

In summary, post-Laramide magmatism older than 35 Ma was mainly restricted to New Mexico and Colorado but perhaps reached westward to Tucson. Between 35 and 32 Ma, magmatism began across a zone from southeastern Arizona to southeastern California but probably did not begin in the northern Sonoran Desert, the Mojave Desert, or the Coast Ranges (subsequently offset by the San Andreas fault system) until 27 to 25 Ma. As to the east, early activity was mainly intermediate and calc-alkalic but somewhat potassic, later becoming more silicic and locally producing voluminous ash flows. After 22 to 21 Ma, magmatism in the eastern part of the region declined, and lavas farther west became predominantly femic. Thus, as in the Sierra-Wasatch region, mid-Tertiary magmatism from the Rio Grande and Southern Rockies to the Mojave Desert (and its offset extensions in the Coast Ranges) was active in successive belts that extended across much of the active region, with some magmatism continuing in older belts as newer ones were established. In the southern region, each belt extended generally farther west and northwest than its predecessor. This view (Fig. 23A) contrasts somewhat with that of Coney and Reynolds (1977; Fig. 23B), who proposed that a wave of post-Laramide magmatism swept rapidly westward across the region, and with that of Lipman and others (1986), who proposed a southward sweep through the Rockies.

Mid-Tertiary southern foreland magmatism

Alkalic magmatism was active in late Eocene to early Miocene time near the Laramide front and its foreland in the southern Cordillera—a mainly younger analogue of the Eocene foreland magmatism of the Northern Great Plains. During the Eocene, the southern foreland belt was confined to Mexico, west Texas, and southernmost New Mexico; during the Oligocene, similar activity spread northward along the old Laramide front.

Early Tertiary alkalic centers of the Trans-Pecos magmatic province, east of the potassic calc-alkalic belt (Fig. 22; Barker, 1977), extend several hundred kilometers into Mexico but end northward in southernmost New Mexico near El Paso (Fig. 24; Maxwell and others, 1967; Barker, 1977, 1979; *McDowell, 1979*; D. F. Parker and McDowell, 1979; Henry and Price, 1984). Magmatism may have begun as early as 49 Ma but culminated between 38 and 32 Ma. The magmas intruded their sedimentary cover with stocks, sills, and laccoliths and erupted to form large shield volcanoes. Alkalic basalt is present, but the most abundant rocks define two contemporaneous series—one of hawaiite, mugearite, nepheline trachyte, and phonolite; and the other of trachyandesite, quartz trachyte, and peralkalic rhyolite. These compositional suites resemble those of continental rift zones or oceanic islands more than they do the contemporaneous magmatic suites that lie immediately adjacent to them and extend 1,000 km westward. Sr-isotopic compositions in the alkalic foreland belt vary widely and show some effects of crustal contamination, but initial $^{87}Sr/^{86}Sr$ as low as 0.7031 (*Barker and others, 1977*) suggests relatively little modification of mantle values; the invariably higher Sr-isotopic ratios in igneous rocks farther west suggest ubiquitous crustal magmatic components. After about 32 Ma, Trans-Pecos alkalic foreland magmatism waned in Texas. In the south, some rhyolite but more abundant basalt and trachyte continued to erupt (*Parker and others, 1988*); in the north, only mafic compositions erupted after 32 Ma. The youngest dated basalt, near the northern end, is 16 Ma.

The Eocene Trans-Pecos province ended abruptly a few kilometers north of the New Mexico border (Barker, 1979), but similar volcanism during the Oligocene continued northward (Fig. 24). Swarms of lamprophyric and other mafic and intermediate dikes were emplaced to the east, on the Great Plains of southeastern New Mexico, between 32 and 35 Ma (*Calzia and Hiss, 1978*; *Brookins, 1980*). In central New Mexico, 150 km north of the Texas Trans-Pecos province, the Ruidoso-Carrizozo area includes 35- to 25-Ma syenite to granitic stocks, sills, and laccoliths and trachyandesitic to trachytic lavas, breccias, and ash-flow tuffs in an area of 75 by 150 km (Kelley and Thompson, 1964; *Thompson, 1972*), locally hosting molybdenite deposits (*Giles and Thompson, 1972*). Another group of 37- to 26-Ma alkalic stocks, sills, dikes, and flows in the Laughlin Peak area, 300 km farther northeast, includes mugearite, trachyte, and phonolite; thorium- and rare-earth-bearing veins are probably related to the phonolite (*Staatz, 1985*). In south-central Colorado, the Spanish Peaks intrusions comprise two stocks and both radial and parallel dike swarms, the roots of a volcanic center that was active about 26 to 20 Ma (*Johnson, 1968*; Stormer, 1972a; *R. P. Smith, 1978, 1979*). The rocks are barely silica-saturated to slightly oversaturated, mainly augite-bearing trachybasaltic to trachyandesitic porphypry with minor trachyte, rhyolite, and lamprophyre. The Spanish Peaks center was emplaced at the time of early extension in the Rio Grande rift and has been interpreted by *R. P. Smith* (1979) as early rift magmatism but lies entirely outside the area of extensional faulting. Smaller but similar 32- to 26-Ma centers in the Wet Mountain Valley, northwest of Spanish Peaks, show some mineralization (*Siems, 1968*; Scott and Taylor, 1975; *Kleinkopf and others, 1979*). The northernmost alkalic center, in the southern Front Range, includes the Cripple Creek gold district within a deeply eroded remnant of a 34-Ma trachyte-phonolite volcano (*Koschmann, 1949*; *McDowell, 1971*; Steven, 1975).

Mid-Tertiary magmatism of the Colorado Plateau

By this point, the distribution of post-Laramide, pre–mid-Miocene magmatism has been shown to reflect regional late Mesozoic and earliest Tertiary orogenic patterns. Most regions of

earlier Cenozoic and Mesozoic crustal shortening subsequently underwent extension and regional magmatism; regions unaffected by those earlier orogenies were little affected by mid-Tertiary magmatism except for widely spaced alkalic centers at the Laramide front or its immediate foreland. The Colorado Plateau defined the foreland of the Late Cretaceous Sevier thrust belt in the Wasatch–eastern Great Basin region but was ultimately separated from the stable continental interior by large-scale Laramide uplift and shortening of the Middle and Southern Rocky Mountains. The plateau itself was only lightly affected by Laramide

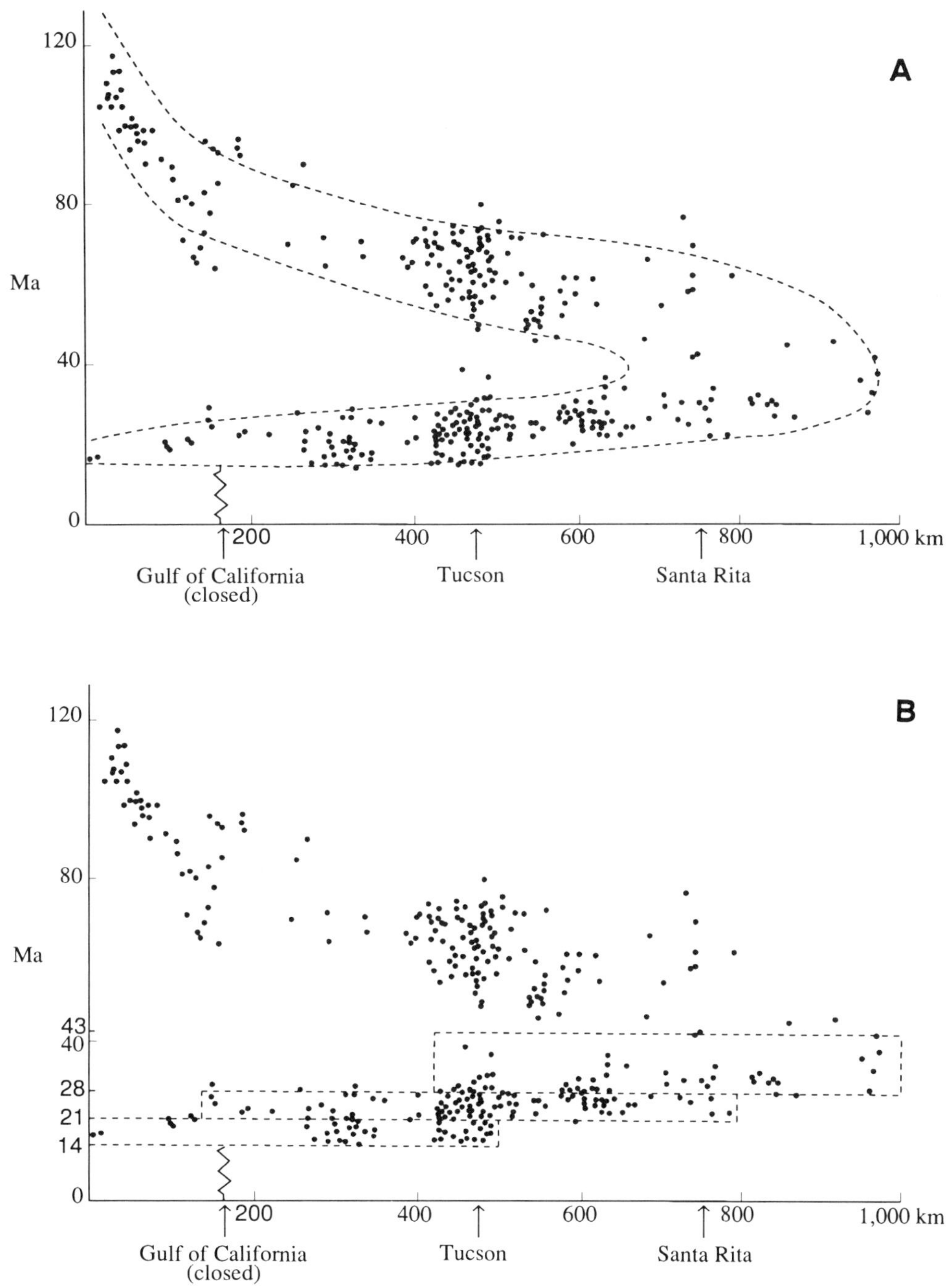

Figure 23. Distribution of radiometric ages in the southern United States and northern Mexican Cordillera comparing sequential establishment of magmatic belts as proposed in this paper to westward sweep of magmatism as proposed by Coney and Reynolds (1977). Data shown in both A and B are those of Coney and Reynolds. A, Interpretation of Coney and Reynolds (1977), with smooth eastward Laramide sweep of activity followed by westward post-Laramide sweep. B, Boxes indicate regional post-Laramide magmatic belts recognized here.

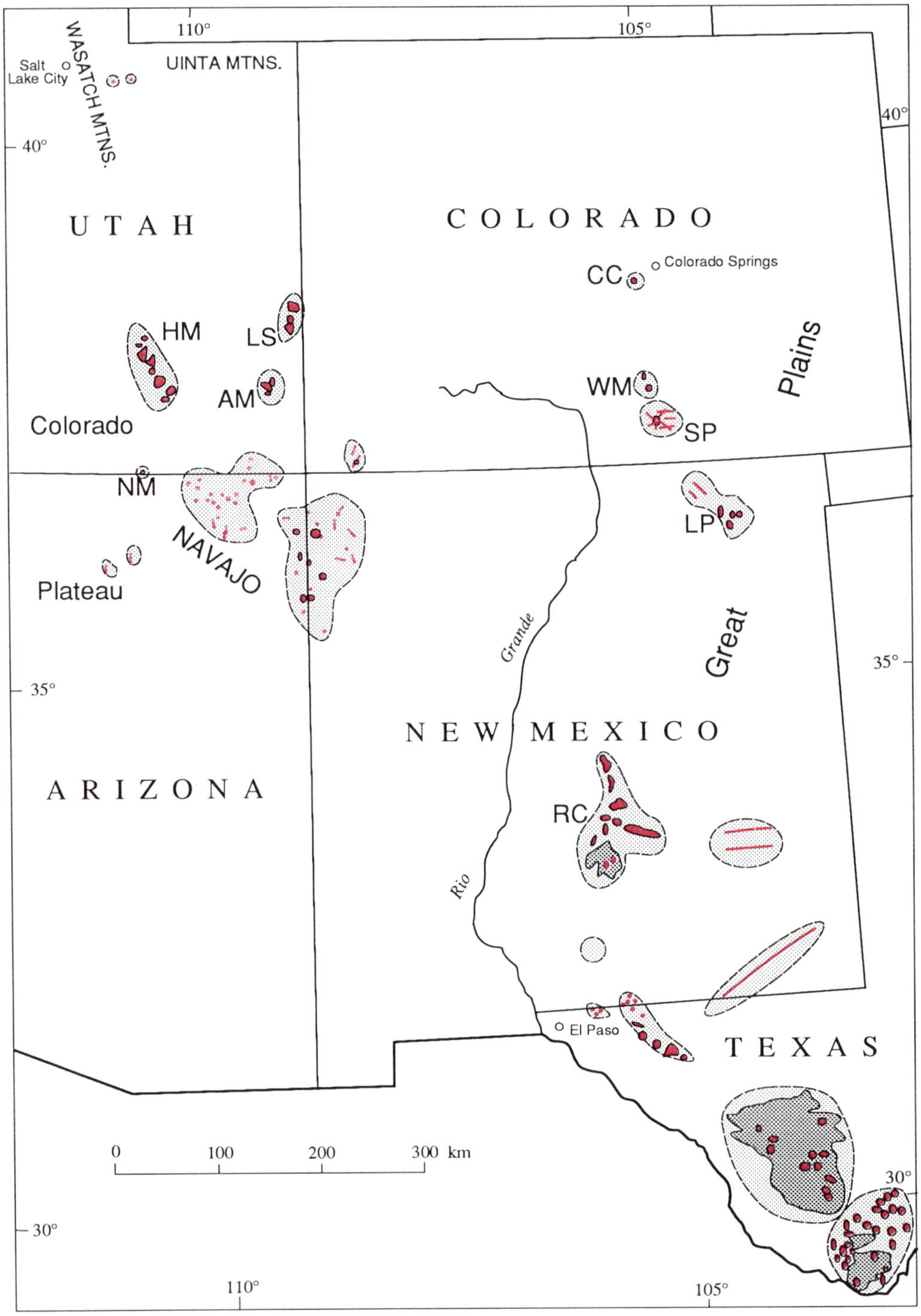

Figure 24. Maps of early to mid-Tertiary magmatic centers of the southern Cordilleran foreland in the Trans-Pecos and southern Great Plains regions, 47 to 17 Ma, and the Colorado Plateau, 48 to 24 Ma. Dark-patterned areas indicate approximate outcrop distributions; gray, volcanic rocks; red, intrusions. Dashed lines and light shading outline igneous fields: AM, Abajo Mountains; CC, Cripple Creek; HM, Henry Mountains; LP, Laughlin Peak; LS, La Sal Mountains; NM, Navajo Mountain; RC, Ruidoso-Carizozo area; SP, Spanish Peaks; WM, Wet Mountain Valley area.

deformation, and Laramide magmatism was confined to the linear Colorado Mineral Belt, which extended onto the plateau near the Four Corners area (see Fig. 1; Armstrong, 1969; *Tweto, 1975*). Thus, although surrounded by orogenic belts, the Colorado Plateau remained a foreland, and the mid-Tertiary subvolcanic complexes of the plateau (Fig. 24) are broadly analogous in composition and tectonic setting to the Eocene Montana-Wyoming centers and the Eocene to early Miocene Texas-Colorado centers.

The Navajo volcanic field (*Williams, 1936*) lies mainly in northwestern New Mexico and northeastern Arizona but laps into adjacent Utah and Colorado (Fig. 24). The rocks are biotite- and leucite-bearing potassic trachybasalt and minette; some are similar to kimberlite although lacking phlogopite (*Watson, 1967*). They occur mainly as volcanic and wall-rock breccias that filled deeply eroded volcanic feeders, now preserved as diatremes, dikes, and plugs. These centers were active about 30 Ma (Armstrong, 1969; *Naeser, 1971*) and probably fed maar volcanoes at the surface (Shoemaker and others, 1962; *McGetchin and Silver, 1972*).

The mid-Tertiary stock and laccolith complexes of the Abajo, La Sal, and Henry Mountains near the center of the Colorado Plateau (Fig. 24) are the eroded roots of volcanic complexes (Hunt, 1956; Armstrong, 1969), consisting predominantly of hornblende diorite porphyry of sodic trachyandesite composition, with abundant hornblende-rich inclusions. The structural dome of Navajo Mountain is also intrusive and has hydrothermally altered breccias (Hunt, 1956), but a single small outcrop of syenite porphyry is the only exposed igneous rock (*Condie, 1964*). Of stocks, sills, and laccoliths in the Abajo Mountains (*Witkind, 1964a*), one is dated as 27 to 28 Ma; those of the La Sal Mountains (Hunt, 1958) include one body dated at 24 Ma. The dominant hornblende diorite of the La Sal Mountains differentiated to monzonite, syenite, and sodalite syenite; hydrothermal alteration accompanied intrusion of the differentiated magmas, and pipelike breccias associated with aegerine rhyolite porphyry attest to late explosive activity. In the stocks and laccolith complexes of the Henry Mountains, where the predominant hornblende-diorite porphyry is associated with augite-bearing monzonite and minor rhyolite and alkali basalt (Hunt and others, 1953; *Engel, 1959*), Armstrong (1969) dated materials from one intrusion at 48 and 44 Ma; in view of the discordance, however, this Eocene age is unconfirmed.

A small mid-Tertiary igneous center lies between the Wasatch and Uinta Mountains, at the northwest margin of the Colorado Plateau (*Best and others, 1968*). Several bodies of mica peridotite and lamproite—a group of highly potassic, magnesian, and silica-poor igneous rocks—yielded late Eocene K-Ar ages of 40 to 37 Ma, but one gave a middle Miocene age of 13 Ma.

LATE CENOZOIC CONTINENTAL MARGIN OF THE PACIFIC NORTHWEST

During Neogene time, the coastal region from Washington to northern California was uplifted and deformed while zones farther inland subsided as forearc basins. Consequently, sedimentation in basins became successively more provincial with time. The sources of sediments to these basins were uplifted Mesozoic orogenic terranes and a volcanic arc that extended the length of the region.

SEDIMENTARY EMBAYMENTS OF THE WASHINGTON-OREGON COAST

By Wendy A. Niem, Alan R. Niem, and Parke D. Snavely, Jr.

The Neogene was a time of emergence and structural deformation of onshore western Washington and Oregon. Uplift of the Olympic Mountains and Coast Range was accompanied by sagging of the Puget Lowland and Willamette Valley to form a modern forearc basin (see Fig. 27). In the early Miocene (about 21 Ma), regional uplift of the older forearc basin restricted marine sedimentation to the Grays Harbor, Astoria, Tillamook, Newport, and Coos Bay embayments on the western flank of the Washington and Oregon Coast Range (Fig. 25; Snavely and Wagner, 1963; *Snavely and others, 1969*). Shallow-marine and nonmarine sedimentation dominated the Tofino–Juan de Fuca basin, along the northern Olympic Peninsula. The pre-Tertiary Klamath Mountain, Vancouver Island, and North Cascades terranes continued to contribute much lithic sediment to the southern and northern ends of the deformed forearc basin, and volcaniclastic sediment was derived from the Cascades arc. An ancestral Columbia River transported detritus to the Neogene shoreline from the Cascades and from extrabasinal granitic and metamorphic sources in eastern Oregon, Idaho, Montana, and northeastern Washington. Uplifted Coast Range and Olympic rocks, including the Tillamook Volcanics and Crescent Formation, provided intrabasinal sediment sources.

Along the Pacific coast of the Olympic Peninsula, thick-bedded lithic-feldspathic sandstone and thin-bedded siltstone and turbidite sandstone of late Oligocene to early middle Miocene age were laid down in bathyal environments unconformably over Eocene mélange and broken formation. The Hoh Rock assemblage is a thick late Eocene to middle Miocene accretionary wedge of mélange and broken formation (Figs. 25, 26; *Rau, 1975, 1979*; Snavely, 1987). Along the northern Olympic Peninsula, lower Miocene coal-bearing feldspathic sandstone and lithic conglomerate of the Clallam Formation (Fig. 25), derived from Vancouver Island and the North Cascades, accumulated in a nearshore delta front (*Gower, 1960*; *Addicott, 1976*; *Anderson, 1985*). In a small nonmarine basin west of Seattle, more than 1,000 m of lignitic claystone and fluvial basaltic conglomerate, the Blakely Harbor Formation (*Fulmer, 1975*), was derived from Crescent basalts of the Olympic Mountains on the west and from the Cascades volcanic arc on the east (Figs. 25, 26).

In the Grays Harbor embayment, the lower to middle Miocene Astoria(?) Formation is predominantly deep-marine carbonaceous siltstone with minor fine-grained micaceous feldspathic sandstone in the western part of the embayment (Fig. 25;

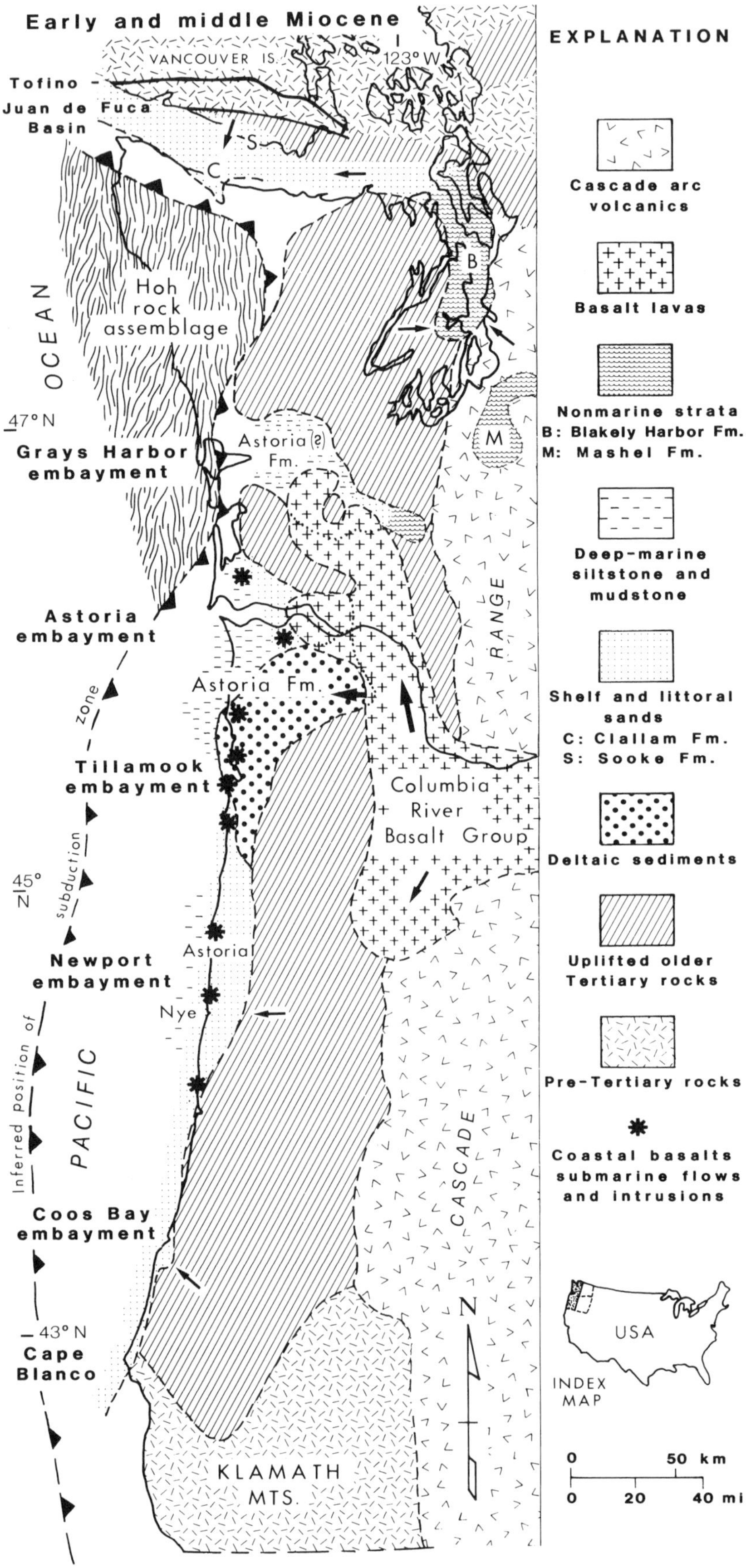

Figure 25. Paleogeographic map of western Washington and Oregon during early and middle Miocene time (24 to 11 Ma). Post-Eocene rotations not restored. Arrows indicate sediment dispersal directions.

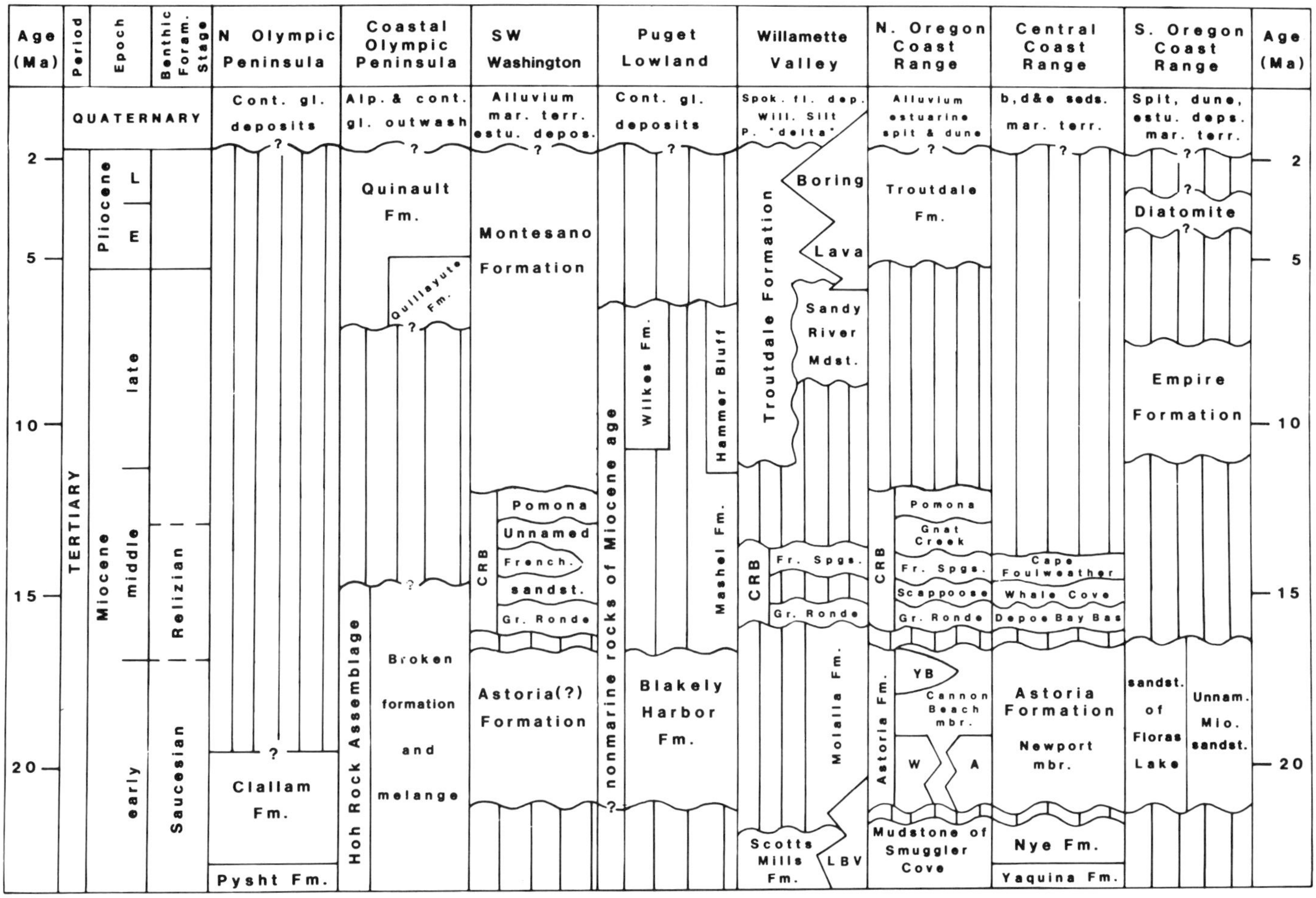

Figure 26. Correlation chart of principal Miocene to Quaternary formations in western Washington and Oregon. CRB: Columbia River Basalt Group. In southwestern Washington column, "unnamed sandstone" interbedded with the Columbia River Basalt Group is called Astoria Formation by *Pease and Hoover (1957)* and Wells (1981). In Willamette Valley column: LBV, Little Butte Volcanics. In northern Oregon Coast Range column: Astoria Formation of Niem and Niem (1985)—informal members: YB, Youngs Bay member; W, Wickiup Mountain member; A, Angora Peak member. Modified from Snavely (1987).

Rau, 1967). The unit coarsens and becomes more tuffaceous eastward, where it was deposited in shallow-marine to continental conditions (*Pease and Hoover, 1957*; Snavely and others, 1958). An unconformity in southwestern Washington suggests local uplift and gentle folding of the Lincoln Creek Formation prior to deposition of the Astoria(?) (Snavely and others, 1958; *Armentrout, 1973*; *1987*; *Armentrout and others, 1980*).

In the Astoria embayment, the lower to middle Miocene Astoria Formation, informally subdivided into members by Niem and Niem (1985), formed in nearshore to deep-marine environments at and near the mouth of the ancestral Columbia River (Figs. 25, 26). The basal coal-bearing high-energy deltaic sandstone (Angora Peak member) and laterally equivalent inner to middle shelf sandstone (Wickiup Mountain member) overlie deep-marine upper Eocene to lower Miocene mudstone (Smuggler Cover Formation) with slight angular unconformity (*Cressy, 1974*; *Cooper, 1981*; Niem and Niem, 1985; *Parker, 1990*). The Angora Peak member, which was deposited as a large high-energy wave-dominated delta, interfingers with and is overlain by deeper-water turbidites and mudstone of the Cannon Beach member (*Cooper, 1981*). Upward fining of the feldspathic sandstone of the inner-shelf Wickiup Mountain member into bathyal mudstone of the Cannon Beach member reflects a gradual middle Miocene transgression. Channeling into and interfingering with the Cannon Beach delta-slope turbidites and mudstone are thick-bedded carbonaceous and micaceous arkosic sandstones filling the head of a submarine canyon (the informally named Youngs Bay member of *Coryell, 1978*; and *Nelson, 1978*). Provenance areas for the 600-m-thick Astoria Formation include Eocene basalts of the Oregon Coast Range, the coeval calc-alkalic Western

Cascades arc, and distant extrabasinal source areas farther east, which contributed sedimentary quartzite, metamorphic, and granitic rock fragments as well as abundant mica via an ancestral Columbia River drainage system (*Cressy, 1974*; *Smith, 1975*; *Cooper, 1981*).

Between about 16 and 12 Ma, flood basalts of the Columbia River Basalt Group that erupted from fissures far to the east followed the valley of the Columbia River westward through a 100-km wide gap in the Cascade Range (Fig. 25; *Beeson and Moran, 1979*; *Tolan and others, 1984*) and advanced into the Willamette Valley and marine embayments farther west (Snavely and others, 1958; Snavely and Wagner, 1963; *Gower and Pease, 1965*; Wells, 1981; *Beeson and others, 1985*). Contemporaneously, pillow lavas, submarine breccias, flows, and associated intrusions were emplaced onto the sea floor in the Grays Harbor, Astoria, Tillamook, and Newport embayments (asterisks on Fig. 25; Snavely and others, 1973), extending onto the inner shelf (*Snavely and others, 1980a*; *Snavely and Wells, 1984*; *Niem and others, 1990, 1991*). The coastal basalts, which are equivalent in age, petrography, chemisty, and stratigraphy to units of the Columbia River Basalt Group, are invasive extensions of the flows erupted on and adjacent to the Columbia Plateau (Beeson and others, 1979; Niem and Niem, 1985; *Pfaff and Beeson, 1989*; *Wells and others, 1989*). Interaction of the flood-basalt lavas with sea water at the Miocene shoreline formed thick lava deltas of palagonitized breccias and pillow flows capped by subaerial flows (*Murphy, 1981*; Niem and Niem, 1985). Lateritic and pisolitic paleosols developed on some of the flows in western Oregon, indicating a warm, humid, and temperate climate (*Niem and Van Atta, 1973*; *Hoffman, 1981*). Rapid changes in water depth at the basin margin are recorded by sedimentary interbeds as thick as 200 m between lava flows. The sediments accumulated in fluvial-marsh, shelf-slope, and submarine-channel environments (*Murphy and Niem, 1980*; *Murphy, 1981*; Niem and Niem, 1985).

Rapid subsidence in the Newport embayment on the central coast of Oregon brought a gradual end to progradation of the Oligocene to earliest Miocene Yaquina volcaniclastic delta (Fig. 5; *Goodwin, 1973*). The overlying Nye Mudstone, 1,300 m of massive dark organic-rich mudstone, was deposited at bathyal depths in anoxic, cold, open-marine water (*Snavely and others, 1969*; *McKeel and Lipps, 1972*). Early Miocene uplift and shoaling of the embayment resulted in an unconformity between the Nye and the overlying bioturbated Newport sandstone unit (*Cooper, 1981*) of the Astoria Formation (Fig. 26), which accumulated at inner to middle shelf depths (*Snavely and others, 1969*; *Cooper, 1981*).

In the Coos Bay basin, a late middle Miocene unconformity punctuates the Neogene stratigraphic record of nearly continuous shoaling with regional uplift (Fig. 26). Unnamed lower to middle Miocene feldspathic sandstones that floor part of the modern bay were deposited in low-energy, warm-temperate waters on the inner to middle shelf (*E. J. Moore, 1963*; *Armentrout, 1980*). The youngest unit, silty feldspathic sandstone and conglomerate of the inner sublittoral upper Miocene Empire Formation, unconformably overlies the older beds (Figs. 26, 27; *Armentrout, 1967, 1980*).

South of Coos Bay, in the Cape Blanco area, the Neogene sequence records rapid responses to abrupt tectonic events (*Bourgeois and Leithold, 1983*). The oldest unit in this sequence is cross-bedded polymictic conglomerate and arkosic sandstone derived from the Klamath Mountains and deposited in a wave-dominated nearshore environment (Fig. 25; *Addicott, 1980*; *Armentrout, 1980*; *Leithold and Bourgeois, 1983, 1984*). Volcanic detritus and tuff beds reflect calc-alkalic volcanism in the Cascades arc. A late middle Miocene regional unconformity, possibly related to a change in the subduction regime of the Juan de Fuca Plate, separates the shallow-marine lower and middle Miocene sandstone from overlying upper Miocene sublittoral sandstone of the Empire Formation (Figs. 26, 27; Snavely and others, 1977, *1980a*; *Armentrout, 1980*). Locally, a lower Pliocene diatomaceous siltstone that is more widespread on the adjacent continental shelf caps the sequence (*Fowler and others, 1971*; *Orr and others, 1971*).

By late Miocene to Pliocene time, continued underthrusting of the oceanic plate beneath the North American plate resulted in the emergence of most of western Washington and Oregon above sea level (Fig. 27; Snavely and Wagner, 1963; Snavely and others, 1977). A corresponding late Miocene regional unconformity is widely recognized seismically in the Oregon and Washington offshore area (*Kulm and Fowler, 1974*; Snavely and others, 1977, *1980a*; Snavely, 1987; Duncan and Kulm, 1989; *Niem and others, 1990, 1991*). As uplift continued in the late Miocene and Pliocene, lithified eastern core rocks of the Olympic Mountains were further compressed against the buttress of Crescent volcanics (Figs. 13C, 13D), yielding upward and outward by shear folding to form a mushroom-like dome (Tabor and Cady, 1978a, b; *Tabor, 1983*; *Brandon and Calderwood, 1990*). Pervasive pencil cleavage developed in the domed and refolded eastern-core semischists and slates (Tabor and Cady, 1978b). The earlier formed east-dipping thrusts were backfolded (Fig. 13D), and movement continued as eastward reverse faulting (*Tabor, 1983*).

Locally along the northern Olympic coast during late Miocene and Pliocene time, gently dipping lithofeldspathic strata accumulated over highly deformed Eocene rocks and the upper Oligocene to upper middle Miocene Hoh Rock assemblage in small shelf basins (Figs. 26, 27). These strata include shallow-marine sandstone, siltstone, and conglomerate lenses of the upper Miocene Quillayute Formation (*Rau, 1979*) and the littoral to deltaic upper Miocene and Pliocene Quinault Formation (*Horn, 1969*; *Rau, 1970*). Sediment sources for the Quinault and Quillayute include uplifted Crescent volcanics and deformed turbidites in the western core of the Olympics. Overpressured shale diapirs of the Hoh Rock assemblage locally intrude the Quinault, both onshore (*Rau and Grocock, 1974*) and offshore (Snavely and others, 1977; Snavely, 1987). In the Grays Harbor embayment the Astoria(?) Formation and underlying Lincoln Creek Formation were broadly deformed in the late middle Miocene

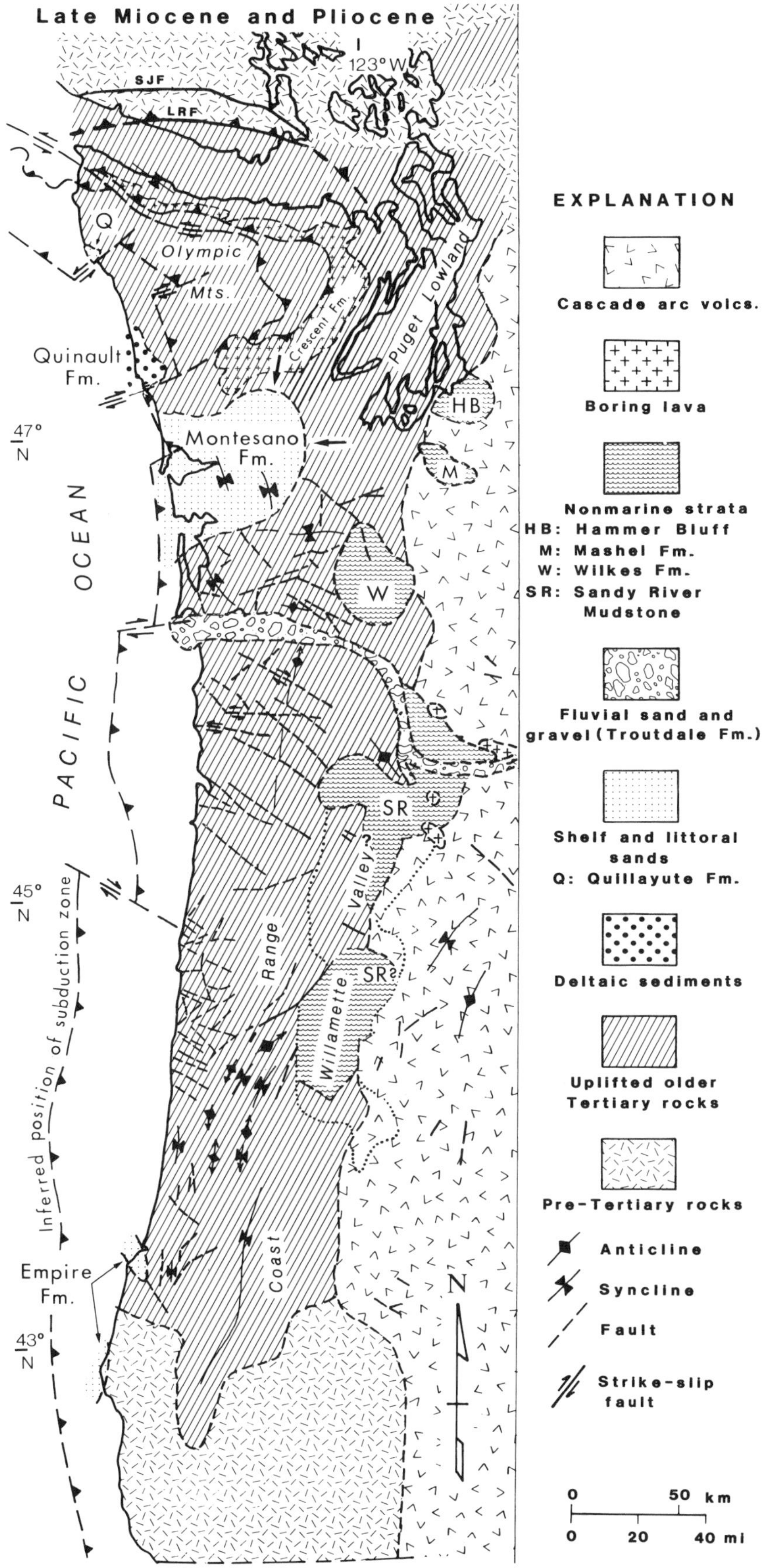

Figure 27. Paleogeographic map of western Washington and Oregon during late Miocene and Pliocene time (11 to 2 Ma). Arrows indicate sediment dispersal directions.

into anticlines and synclines cut by strike-slip faults (*Rau, 1967,* 1986). Renewed subsidence in Grays Harbor basin, however, increased water depths from littoral to more than 900 m, as recorded by planktonic and benthonic foraminifers, mollusks, and diatoms in the upper Miocene Montesano Formation (*Fowler, 1965*). Sandstones in the Montesano were derived from the uplifted Olympic Mountains and the Cascades arc (Fig. 27). In latest Miocene and Pliocene time, marine waters had withdrawn from the uplifted Grays Harbor basin, and the Montesano was gently folded (*Gower and Pease, 1965*; *Rau, 1967,* 1986).

Paleomagnetic studies of Columbia River basaltic flows show as much as 16° of clockwise rotation of a rigid Western Cascades–Oregon Coast Range–Klamath Mountains block (the Willamette Plate of Magill and others, 1982) about an apparent pivot in southwestern Washington since the middle Miocene (R. E. Wells, 1984). That rotation may have resulted from both oblique subduction of the Juan de Fuca Plate and extensional opening of the basin-range region (*Magill and Cox, 1980*; Magill and others, 1982; Wells and Heller, 1988).

The major structures of the Oregon Coast Range are a broad syncline in the south and a northward-plunging anticline in the north (Fig. 27). In addition, numerous closely spaced high-angle northwest-trending and subordinate conjugate northeast-trending oblique-slip faults truncate middle Miocene and older folded rocks in the northern and central Oregon Coast Range and southwestern Washington (Snavely and others, 1976a, b; Wells, 1981, 1989; *Olbinski, 1983*; Wells and others, 1983; *Nelson, 1985*; Niem and Niem, 1985). Small blocks between the transcurrent faults, as shown by recent paleomagnetic studies (*Nelson, 1985*; Wells and Coe, 1985), have been rotated clockwise. The folding and oblique shearing may have been driven by oblique convergence of the Juan de Fuca Plate beneath the North American Plate (Magill and others, 1982; Wells and Coe, 1985). No significant post–middle Miocene rotations have been detected in the Olympic Mountains except for 45° clockwise rotations in marginal-basin rocks on the northwest coast of the Olympic Peninsula (*Moyer and others, 1985*), possibly related to late Miocene underthrusting and uplift recorded in the western Olympic core rocks.

During the late Miocene and early Pliocene, nonmarine sediments accumulated in low-lying structural downwarps (Snavely and others, 1958) within the Puget Lowland–Willamette Valley trough south of Seattle and in the Portland area. Stream gravels and tuffaceous lacustrine sediments of the informally named Mashel, Wilkes, and Hammer Bluff formations (Fig. 26), derived from the Cascades, were deposited at the base of the range (Fig. 27; Roberts, 1958; Snavely and others, 1958; *Walters, 1965*; *Noble and Wallace, 1966*; *Gard, 1968*; *Livingston, 1971*; *Hammond, 1980*). Nonmarine strata in the Portland basin are 500 m thick and include lacustrine muds, silts, and sands of the Sandy River Mudstone (Trimble, 1963; *Mundorff, 1964*), tuffaceous sediments, and fluvial volcaniclastic deposits derived from Western Cascades volcanoes. Farther south, a proto–Willamette River sequence includes channel gravels lying generally east of the modern Willamette River and its overbank clays to the east and west (*Yeats and others, 1992*).

Middle Miocene to Pliocene fluvial conglomerates and micaceous sandstones of the Troutdale Formation (Figs. 26, 27) were deposited by the Columbia River; these fluvial deposits both interfinger with and unconformably overlie the Sandy River Mudstone in the Portland basin (Trimble, 1963), which, therefore, is middle Miocene in age. The conglomerates have up to 30 percent of quartzite pebbles derived from Mesozoic, Paleozoic, and older rocks in the upper reaches of the drainage basin. The Troutdale is traceable upriver from Portland into the Columbia River gorge (*Tolan and others, 1984*) and downstream to the mouth of the river, where age-equivalent quartzite conglomerate caps ridges both north and south of the river (*Schlicker and others, 1972*; Niem and Niem, 1985). The present elevation of the quartzite-bearing fluvial terraces at the mouth of the Columbia indicates 100 to 150 m of uplift since the Pliocene. Numerous small shield volcanoes in the Portland basin erupted the Boring Lava during and subsequent to Troutdale deposition (*Treasher, 1942*; *Allen, 1975*).

Pliocene and Pleistocene isostatic uplift and doming of the Olympic Mountains (Snavely and others, 1977; Tabor and Cady, 1978b; *Tabor, 1983*) was greatest in the central eastern core rocks, which now form the high Olympics. Pleistocene glaciation, left-lateral slip on the Calawah fault (see Fig. 12), landslides, and historic seismicity record continued evolution of the Olympic terrane (Snavely, 1987).

The Quaternary history of western Oregon and Washington is complex, reflecting repeated advances and retreats of alpine and continental glaciers in northwestern Washington, reversal of major stream drainages, invasion by the sea of the Puget Lowland, and catastrophic flooding in western Washington and Oregon due to collapse of glacial ice dams on the upper reaches of the Columbia River drainage system. The headlands, sea stacks, drowned estuaries, spits, and dune fields of the scenic coastline formed during the Holocene rise in sea level.

BASINS OF THE MENDOCINO REGION AND GREAT VALLEY

Geologic conditions similar to those of the Washington-Oregon shelf and coastal region during Neogene time extended southwards into northern California, both offshore and onshore, but are generally less well known. The difference in tectonic settings reflects formation of the San Andreas transform-fault system farther south and gradual northward migration into northern California of the Mendocino triple junction.

Eel River basin

Most of the Eel River basin lies off the northern California coast (*Hoskins and Griffiths, 1971*), but the southern corner is onshore near Eureka (Fig. 28) and has yielded limited hydrocarbon production. After Oligocene deformation of the informally named Paleogene Yager complex of Bachman and Crouch

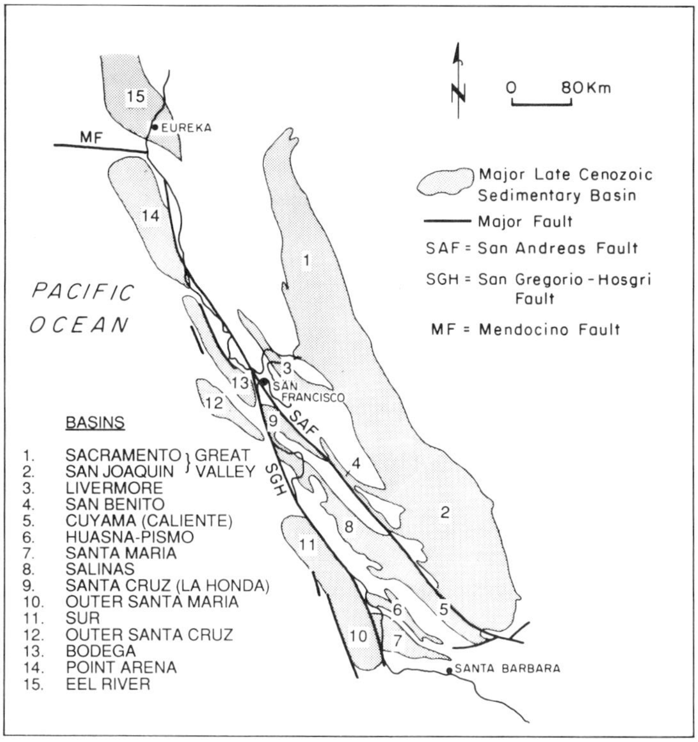

Figure 28. Distribution of late Cenozoic sedimentary basins in northern and central California. Modified from Blake and others (1978).

(1987) and the Franciscan coastal belt *sensu stricto,* strata of the onshore part of the basin were deposited across both of these belts, forming a forearc basin. Basal rocks of the Eel River basin include submarine debris flows and range in age from late Oligocene to middle Miocene or early late Miocene (*Bachman and others, 1984*; Bachman and Crouch, 1987). These beds are overlain by the upper Miocene to Pleistocene Wildcat Group (Ogle, 1953), which makes up most of the basin fill and is 1,900 m thick in beach exposures but may be as much as 3,670 m thick elsewhere (Ogle, 1953). The Wildcat Group consists of a basal deep-water sandstone overlain by siliceous shale grading eastward to shallower-water facies. These beds are overlain by facies representing progressively shallower deposition, including mudstone and a sandstone of deep-sea fan facies (Eel River Formation), shelf sediments including fan-channel sands (Rio Dell Formation), shallow-marine sandstone (Scotia Bluffs Sandstone), and, at the top, the nonmarine Carlotta Formation. The younger part of the sequence has been correlated to the open-ocean time scale using microfauna, ash beds, and magnetic stratigraphy (*Dodd and others, 1977*; *Keller and Ingle, 1981*).

Deposition in the Eel River basin overlapped in part the deposition, accretion, and thrusting of the King Range terrane in the Miocene (McLaughlin and others, 1982; Fig. 14). The Wildcat Group is folded and faulted, and reverse faults including the Mad River fault zone show evidence of late Quaternary activity (Kesley and Carver, 1988).

Sacramento forearc basin

By S. A. Graham

The ridged forearc configuration of Paleogene north-central California described earlier persisted through the late Cenozoic. The Sacramento forearc basin (Fig. 28) was an interior lowland or shallow-marine embayment separated from the Pacific Ocean by the uplifted Mesozoic subduction complex that underlies the northern Coast Ranges. Although this broad tectonic setting is well established, stratigraphic details of the upper Cenozoic strata of the Sacramento basin are poorly known, and stratigraphic nomenclature is inconsistent. The strata are poorly dated, laterally variable, mostly nonmarine rocks, largely confined to the shallow subsurface. Names of parts of the Miocene section beneath the east side of the Sacramento basin, such as the Mehrten and Valley Springs Formations, are often applied to Neogene volcanic units of the Sierra Nevada, consistent with the Sierran volcanic provenance of much of the section (Hackel, 1966). On the outcropping west margin of the basin, the Neroly, Tehama, and Red Bluff Formations generally denote Miocene, Pliocene, and Pleistocene units (Hackel, 1966).

Depositional environments of the Cenozoic strata are only broadly delimited. Marine conditions are documented for the middle and late Miocene by the shallow-marine Neroly Formation and associated strata in the region now occupied by the Sacramento River Delta and the Coast Ranges near San Francisco Bay. The distribution of these rocks suggests that the southern Sacramento basin resembled an enlarged version of San Francisco Bay (Hackel, 1966). Erosional remnants of Neogene shallow-marine and nonmarine strata crop out in the Coast Ranges north of San Francisco and west of the Sacramento basin. These may have been contiguous parts of a southern extension of the Eel River basin, which Nilsen and Clarke (1989) considered to be a Cenozoic forearc basin outboard of and separate from the Sacramento basin. By Pliocene time, the Sacramento Valley was permanently established as a terrestrial lowland. Then, as now, the nonmarine basin was characterized by the longitudinal Sacramento River with transverse tributaries draining the bordering Sierra Nevada, Coast Ranges, and Klamath Mountains.

For most of the late Cenozoic, the Sacramento basin was a forearc basin. However, with northward migration of the Mendocino triple junction and propagation of the San Andreas transform-fault system, Sierran arc volcanism ceased, its southern termination migrating progressively northward (Christiansen and Lipman, 1972; *Snyder and others, 1976*). In the process, the Sacramento basin became a strike-slip successor basin in a manner similar to that of the San Joaquin basin somewhat earlier in the Tertiary (discussed later). This change is recorded in the fill of the basin by a shift from volcaniclastic sediments to sediments dominated by clastics derived from the rising Coast Ranges as well as the dissected Sierran region. Such relations are best documented along the western flank of the Sacramento basin near San Francisco Bay, where the volcaniclastic arc-derived Miocene

Neroly Formation was succeeded by nonmarine clastics derived from the Mesozoic accretionary wedge that was progressively uplifted to the west in the evolving San Andreas continental-margin transform system (Graham and others, 1984).

LATE CENOZOIC TRANSFORM TECTONIC SYSTEM

The San Andreas transform system came into being with the first contact of the Pacific Plate with the North American Plate in the late Oligocene, about 29 Ma (Atwater, 1970, *1989*; Stock and Molnar, 1988). Contact of the Pacific and North American Plates separated the subducting Farallon and Vancouver Plates into a northern segment (Juan de Fuca Plate) with the Mendocino transform and Juan de Fuca Ridge system as its boundaries with the Pacific Plate, and a southern segment (Rivera Plate) with the East Pacific Rise as its boundary with the Pacific Plate. The Mendocino and Rivera triple junctions, controlled by these boundaries, moved respectively northwest and southeast relative to the point of initial contact, thereby extending the transform system to its present length. An implication of this lengthening process is that subduction of the Vancouver Plate beneath North America continued to take place northwest of the Mendocino triple junction at the same time that strike slip was occurring along the transform, as documented in coastal northern California (*Bachman and others, 1984*). Passage of these triple junctions coincided with nearby volcanism, rifting, and pulses of initial basin subsidence (Dickinson and Snyder, 1979a, b).

TECTONICS OF THE SAN ANDREAS TRANSFORM SYSTEM

Because the San Andreas fault and other strike-slip faults of California are described in detail elsewhere by Stewart and Crowell (this volume), this section emphasizes their effects on Neogene basin tectonics and on volcanism. The San Andreas fault trends generally N41°W, a few degrees counterclockwise from the Pacific–North American Plate slip vector at this latitude (*Minster and Jordan, 1987*; *DeMets and others, 1987*), with the largest departure from the average trend (the "big bend") in the Transverse Ranges, where left-stepping bends in the fault produce reverse faults, folding, uplift, and basin subsidence in adjacent areas (Fig. 9).

Neogene basins north of the Transverse Ranges are elongate parallel to the San Andreas fault or more westerly than the fault (Fig. 28); these include the Point Arena, Bodega, Outer Santa Cruz, Sur, Outer Santa Maria, Santa Cruz (La Honda), Salinas, Huasna-Pismo, Livermore, and San Benito basins. To the south, the Ventura basin is embedded within the western Transverse Ranges and trends east-west (see Fig. 39). Basins at the margins of the Transverse Ranges having one side with an east-west (Transverse Ranges) trend and the other with a northwest (Coast Ranges or Peninsular Ranges) trend (see Fig. 36) include the Los Angeles, Santa Monica, Santa Maria, Cuyama, and South San Joaquin basins. The South San Joaquin basin is part of the physiographic Great Valley, but the Neogene basin has a tectonic history similar to others in the Transverse Ranges (Crowell, 1987). Thicknesses of Neogene sediments in Transverse Ranges basins are related to positions with respect to the "big bend"; maximum thicknesses of strata of 4 Ma and younger are greatest in the Ventura, Los Angeles, and South San Joaquin basins, close to the "big bend," suggesting that it had already formed by that time (*Yeats, 1978*).

The Salton trough is a pull-apart basin bounded by the San Andreas and San Jacinto faults (see Fig. 9). Unlike other basins to the north, it contains an active spreading center beneath the southern end of the Salton Sea (*Elders and others, 1972*; Fuis and Kohler, 1984). The tectonic style of the Salton basin is related to that of the Gulf of California, where rhombic basins are bounded by right-slip faults and passive continental margins, including the Peninsular Ranges, move obliquely away from North America. The only significant difference is that, whereas oceanic crust occurs in the centers of the basins of the central and southern Gulf of California, the Salton basin consists of attenuated continental crust overlain by lacustrine sediments and by deltaic sediments of the Colorado River.

The earliest right slip on any member of the San Andreas system in central California occurred on the Russell fault (Fig. 9), about 22 Ma (Yeats and others, 1989) although the Pacific and North American Plates came into contact about 7 m.y. earlier. The first strike-slip offsets probably occurred at the foot of the continental escarpment (Crowell, 1987), seaward of the Salinian block and central Coast Ranges. The influence of the transform system broadened and reached inland to form the east-trending normal-faulted basins filled with nonmarine sediments and mafic volcanic rocks. The date of 22 Ma for the earliest documented offset in the central Coast Ranges is taken from the rapid formation and strike-slip displacements of new deep-marine troughs during deposition of the Vaqueros Formation. At this time or shortly thereafter, strike slip began to occur on the San Andreas fault itself (Huffman, 1972). Isopachs of strata in the southwestern San Joaquin Valley indicate the beginning of en echelon folding late in the early Miocene, before 15 Ma (Harding, 1976). Gradually, strike slip became concentrated along the San Andreas fault; the Russell and San Gabriel faults (and perhaps others as well) had little or no strike slip after about 4 to 5 Ma.

Throughout coastal California, the Miocene was marked by volcanism and extension, in contrast to the Paleogene and the post-Miocene (Pilger and Henyey, 1979; Weigand, 1982). During the middle Miocene, the western Santa Monica Mountains moved westward with respect to the Peninsular Ranges (Yeats, 1968), accumulating about 60 km (*Sage, 1973*) to 90 km (Campbell and Yerkes, 1976) of left slip on a precursor of the Malibu Coast–Hollywood fault system. Also largely during this time, the entire Transverse Ranges, together with local blocks in the Mojave Desert, central Coast Ranges, and California borderland, underwent as much as 90° of clockwise rotation (Luyendyk and Hornafius, 1987), accompanied by strike-slip displacement between adjacent blocks within the Coast Ranges, Peninsular

Ranges, and Transverse Ranges (Fig. 9). The strike-slip displacements may have contributed to the formation of rhomboid basins at the margins of the Transverse Ranges (*Luyendyk and others, 1980*), the precursors of basins formed in the last 5 m.y. Another major change took place at about 5 Ma with a change in Pacific Plate motion to more northerly (*Cox and Engebretson, 1985*). At that time or shortly thereafter, the Gulf of California opened (Larson and others, 1968; Moore and Buffington, 1968), and Neogene basins in southern California developed their modern configuration and began to subside at an accelerated rate (*Yeats, 1978*). From the Transverse Ranges northward, Pacific–North American plate motion was increasingly concentrated on the modern San Andreas fault. Total cumulative strike slip on the San Andreas fault system (including related older faults in southern California) is about 330 km (Stewart and Crowell, this volume).

LATE CENOZOIC BASINS OF CENTRAL CALIFORNIA

By S. A. Graham

The initiation and subsequent northwestward migration of the Mendocino triple junction heralded a change from convergent-margin to transform-margin setting and a consequent change in patterns of sedimentation along the central California coast. In some instances, Cretaceous to Paleogene depocenters persisted into the Neogene with minor modifications; in others, older basins were uplifted and eroded; elsewhere, entirely new Neogene basins appeared. Principal late Cenozoic depocenters of coastal California include some 14 more-or-less distinct sedimentary basins onshore and offshore (Fig. 28; Blake and others, 1978), all of which evolved under the influence of the San Andreas transform fault system.

Salinas basin

The late Cenozoic Salinas basin (Figs. 28, 29) represents the reshaping of a Paleogene depocenter (Graham, 1978, 1987). The Paleogene Salinas basin was one of several structurally controlled deep-marine basins that characterized the continental borderland developed with emplacement of the Salinian block along the California margin in Paleocene time (Nilsen and Clarke, 1975). Eocene sand-rich submarine fans and associated deep-marine strata (Fig. 30) of the Salinian borderland system were extensively eroded prior to the Oligocene and are preserved only locally in outcrops in the Santa Lucia and La Panza Ranges along the northwestern and southern basin margins (Fig. 29).

In simplest terms, middle and upper Cenozoic strata form a northwest-southeast elongated sediment prism comprising a transgressive-regressive sequence (Fig. 30). At its thickest, the Oligocene to Holocene section may be 4 km, but the northern, eastern, and southern basin-margin sections thin markedly and lap onto Salinian basement (Graham, 1978). The western limit of the basin is arbitrarily defined by the present limit of outcrops in the Santa Lucia Range; original facies limits thus are poorly known, and the basin may have been relatively open to the Pacific on the west.

The oldest elements of the upper Cenozoic Salinas basin fill are Oligocene gravel-rich braided-stream deposits with minor associated fine-grained red beds. These are locally preserved (Fig. 29) as the Oligocene(?) Berry Formation of the Santa Lucia Range (Fig. 30) and the Simmler Formation of the La Panza Range southeast of Paso Robles. Overlying conformably are transgressive, shallow-marine sandstones generally included in the Vaqueros Formation. The thickness of this nonmarine/shallow-marine couplet varies from a few meters to hundreds of meters, with the nonmarine unit often absent. Similarly, the age of the sequence ranges widely, from Oligocene to middle Miocene, depending on position relative to basin margins (Graham, 1978).

Overlying the basal clastic strata abruptly but conformably are fine-grained basinal facies assigned to the Monterey Formation (Fig. 30). These range in age from earliest Miocene to latest Miocene, depending on stratigraphic and spatial position in the basin, and are dominated by terrigenous shales in lower Miocene sections, calcareous nannoplankton- and foraminifer-rich mudstones in the middle Miocene, and siliceous shales, porcelanites, and diatomites in upper Miocene sections (Durham, 1974; Graham, 1978). All represent bathyal marine deposition of fine-grained, largely biogenic sediments under periodically oxygen-poor bottom-water conditions (*Mertz, 1984*). These rocks constitute the principal oil-source rock in the petroliferous Salinas basin (Graham, 1987).

Basinal environments yielded to late Miocene shallow-marine (Pancho Rico Formation) and Pliocene and Pleistocene nonmarine (Paso Robles Formation) environments (Durham, 1974) in response to regionally extensive deformation and uplift. Paleogene to Miocene deep-marine depocenters of the western part of the Salinas basin were elevated to form the Santa Lucia Range, as the depocenter of coarse nonmarine detritus shifted eastward to the vicinity of the present Salinas River. In Pliocene time, the Paso Robles Formation likely was deposited in two distinct alluvial depocenters in the northern and southern portions of the Salinas Valley region (*Dohrenwend, 1979a*), but with headward capture, the northern system eventually was established as the northwest-flowing Salinas River.

In sum, transform-margin tectonics strongly influenced patterns of late Cenozoic sedimentation in the Salinas basin. The Oligocene to Holocene section, partitioned stratigraphically from the underlying Paleogene section, records an early phase of rapid subsidence, attributable to extension near the northwest-migrating Mendocino triple junction. This extension might reflect either the geometry of the plates interacting near the triple junction or thermal contrasts associated with plates of different ages adjacent to it (Graham, 1978). Subsequently, basinal Monterey strata were deposited in structurally controlled settings, with marked facies differences in fine-grained facies evident when examined in detail (*Graham, 1976,* 1978; *Mertz, 1984*). Marine sedimentation

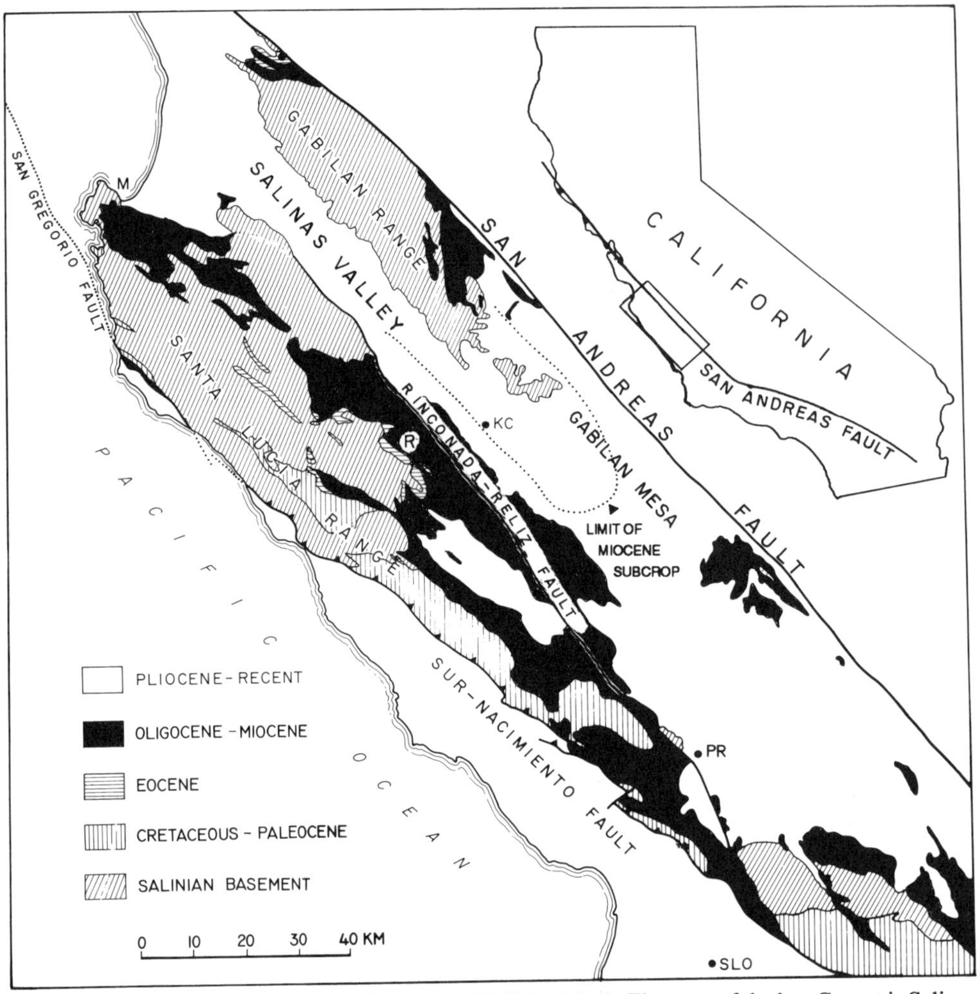

Figure 29. Generalized geologic map of the central Salinian block. Elements of the late Cenozoic Salinas basin include Oligocene to Holocene strata that crop out in the Santa Lucia Range as well as the subsurface section of the Salinas Valley. KC, King City; M, Monterey; PR, Paso Robles; R, Reliz Canyon; SLO, San Luis Obispo. After Graham (1978) with permission of the American Association of Petroleum Geologists.

ceased during the Pliocene and Pleistocene transpressive structural event that elevated the central California Coast Ranges. Nonmarine sedimentation has continued from the Pleistocene through the Holocene in the areally reduced Salinas basin, with fluvial drainage strongly influenced by the interplay of gradients on the surfaces of tilted basement blocks and bounding northwest-trending faults of the San Andreas system (*Tinsley, 1975*; *Dohrenwend, 1979b*).

Santa Cruz basin

The Santa Cruz and Salinas basins (Fig. 28) offer fine examples of structural inversion of topography. The long-lived Paleogene deep-marine depocenters of these two basins now form the high elevations of the Santa Cruz Mountains and the Santa Lucia Range, respectively. Conversely, the ridge of Salinian crystalline basement that separated the two basins and supplied much coarse clastic detritus to each during most of the Tertiary (Graham, 1978) has subsided since the latest Neogene and currently underlies the lowlands around Monterey Bay.

With the Monterey Bay basement block forming its southern margin, the late Cenozoic Santa Cruz basin represents continued sedimentation in a borderland-style basin established in the Paleogene with emplacement of the Salinian block (Nilsen and Clarke, 1975). The basin is notable for its small size, great stratigraphic variability and thickness (up to perhaps 11 km total Tertiary fill, including lower and middle Tertiary strata), and tendency toward fault-block partitioning of stratigraphy (Clark and others, 1984; *Stanley, 1984*). The latter reflects two causes: (1) synsedimentary faulting in cases like the southern basin-

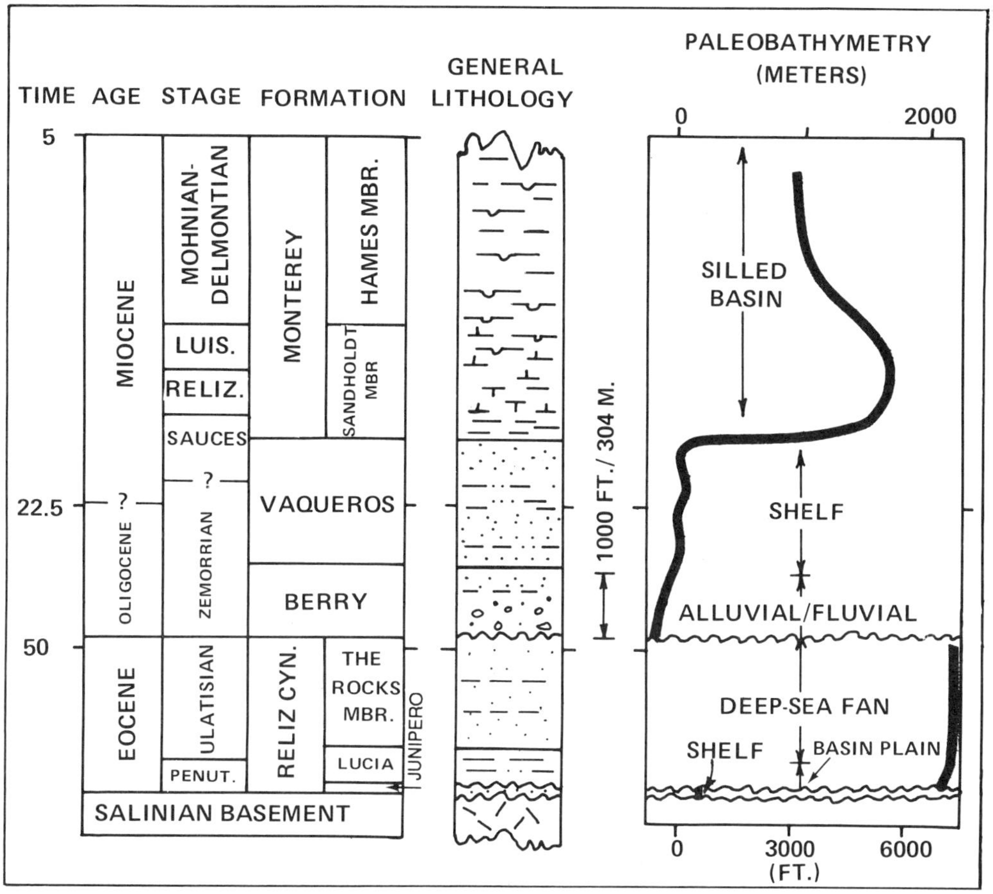

Figure 30. Stratigraphy and paleobathymetry of the Reliz Canyon section of the Salinas basin. This section was selected as a fairly complete yet typical Tertiary section of the basin; location of Reliz Canyon shown in Figure 29. Provincial benthic foraminiferal stages are those of *Kleinpell (1938)*. After Graham (1978) with permission of the American Association of Petroleum Geologists.

margin Zayante fault (*Clark and Rietman, 1973*), where faulting is related to regional extension in late Oligocene time (Tennyson, 1989); and (2) juxtaposition of unrelated terranes by major right-lateral strike slip in cases like the San Gregorio–Hosgri fault (Graham and Dickinson, 1978; Clark and others, 1984).

Salinian granitic basement and associated metamorphic wall rocks crop out in the northern and southern Santa Cruz Mountains. These basement rocks bound the Tertiary basin (*Stanley, 1984*) and are capped by thin middle or upper Miocene basin-margin onlap sequences (*Clark, 1966*). In contrast, basement is not exposed in the intervening central Santa Cruz Mountains, where a nearly complete, largely deep-marine Tertiary section crops out. Post-Eocene strata (Fig. 31) include deep-marine mudstones and submarine fan sandstones of late Eocene to early Miocene age, mapped variously as the San Lorenzo Formation, Vaqueros Formation, and Lambert Shale. Locally interstratified with these units are great thicknesses (up to 1,200 m) of the submarine-erupted Mindego Basalt. The overlying lower and middle Miocene section consists of a couplet of basal transgressive bioclastic sandstone (Lompico Sandstone) and bathyal siliceous mudstone (Monterey Formation). An unconformity, characterized by extensive stripping of pre–middle Miocene strata, is especially evident beneath the Lompico Sandstone along the basin margins (*Clark, 1966*). Similarly, this sequence is bevelled beneath the basal unconformity of the upper Miocene to Pliocene sediment package, which includes the spectacularly cross-bedded, tidally dominated Santa Margarita Sandstone (*Phillips, 1983*) and overlying thick (up to 2,700 m) bathyal siliceous mudstones of the Santa Cruz Mudstone. The Santa Cruz Mudstone is conformably overlain by shelfal to basinal siltstones, sandstones, and mudstones of the Miocene and Pliocene Purisima Formation (*Gavigan, 1984*).

Subsidence history analysis (Fig. 31) suggests that sediment accumulation in the Tertiary Santa Cruz basin was strongly controlled by tectonics. Lower Tertiary deep-marine sequences, discussed in an earlier section, were deposited in a tectonic regime related to the structural emplacement of Salinia. The cause of marked uplift and basaltic volcanism in the Oligocene and early

Miocene remains unresolved. Some workers have related these to migration of the Mendocino triple junction and initial displacement along the San Andreas fault (Graham, 1978; Dickinson and others, 1987) whereas Tennyson (1989) related these phenomena to interarc spreading that predated transform faulting. Middle Miocene tectonic pulses and wholesale basin uplift in the Pliocene and Pleistocene likely reflect discrete structural episodes, perhaps transpressional in nature, during the strike-slip history of the basin (Dickinson and others, 1987; Graham, 1987).

Bodega and Point Arena basins

Portions of largely offshore late Cenozoic basins crop out in limited exposures onshore on the Point Reyes Peninsula and near Point Arena in the narrow strip of land west of the San Andreas fault along the northern California coast (Fig. 28). Sections of both basins display similarities to each other and to other Neogene basins of the Salinian block, as well as important differences.

The extension of the Bodega basin exposed onshore on the Point Reyes Peninsula rests directly on Salinian granodiorite or on a very locally preserved erosional remnant of Paleocene deep-marine strata and consists of distinctive stratigraphic packages of Neogene sedimentary rocks (Fig. 32; Clark and others, 1984). The oldest package is a basal middle Miocene arkosic sandstone (Laird Sandstone), about 50 m thick, overlain by a thick sequence (about 1,000 m) of basinal middle and upper Miocene porcelanite assigned to the Monterey Formation (Galloway, 1977; Clark and others, 1984). The overlying package begins with the basal transgressive glauconitic Santa Margarita Sandstone (about 20 m), followed by a thick (up to 2,000 m) section of upper Miocene and Pliocene bathyal siltstone and siliceous mudstone (Santa Cruz Mudstone and Purisima Formation). Uplift characterized the region after the Pliocene, and thin terrace deposits form the only record of geologically recent deposition.

In contrast, limited outcrops of the Point Arena basin just south of Point Arena, about 100 km along the coast to the north, represent generally older and deeper-water facies than those of Point Reyes. At Point Arena (Fig. 33), the 23-Ma Iversen Basalt of *Weaver* (*1944*), vesicular and probably submarine, unconformably overlies Paleogene submarine fan deposits and conformably underlies sand-rich submarine fan facies of the lower Miocene Skooner Gulch Formation (*Phillips and others, 1976*; *Welton, 1980*). In turn, the latter is overlain by shaley distal fan facies of the Gallaway Formation of *Weaver* (*1944*). Above the Gallaway, the Point Arena Formation of *Weaver* (*1944*) is a lower Miocene shale and porcelanite sequence lithologically identical to rocks included in the Monterey Formation elsewhere in central California. All of these lower Miocene strata apparently were deposited in extremely deep-marine waters (*Phillips and others, 1976*).

Tectonic context is extremely important in interpreting the rocks of the onshore portions of the Bodega and Point Arena basins. Both display stratigraphic packaging and lithologies reminiscent of basins like the Santa Cruz basin that are clearly related to San Andreas transform tectonics. However, palinspastic restoration of about 300 km of post-Oligocene right-lateral strike slip on the San Andreas fault leaves the Bodega and Point Arena basins north of the predicted position of the Mendocino triple junction (hence, the northern end of the propagating San Andreas fault) during the early and middle Miocene (*Snyder and others, 1976*). Although still speculative, this apparent contradiction may be resolved by recognition of major right-slip along the San Gregorio–Hosgri fault (Fig. 28). Restoration of the 70 to 150 km of right-slip proposed for that fault juxtaposes the Neogene section of Point Reyes with similar sections in the Santa Cruz Mountains and at Monterey Bay (Graham and Dickinson, 1978; Clark and others, 1984).

The Point Reyes sequence rests depositionally on Salinian crystalline basement. In contrast, the basement of the Point Arena sequence may consist of spilitic basalt (*Wentworth, 1968*). Paleomagnetic data (*Kanter, 1984*) suggest that Cretaceous to Eocene strata of the Point Arena area may be allochthonous with respect to North America. However, the overlying Iversen Basalt (*Weaver, 1944*) is paleomagnetically *in situ* within the uncertainty of known Neogene right slip on the San Andreas system. Thus, the Iversen Basalt may simply represent a former continuation of the Mindego Basalt eruptive center of the Santa Cruz basin, dislocated by Neogene right-slip on the San Gregorio–Hosgri fault system (*Graham and Peabody, 1981*).

San Joaquin basin

The San Joaquin basin (Fig. 28) figures prominently in the late Cenozoic history of central California because it is the largest sedimentary basin in the region and because it persisted as a depocenter during the change from a convergent margin to a transform margin. Plate-tectonic reconstructions (Fig. 34) indicate northwestward migration of the Mendocino triple junction and propagation of the San Andreas fault along the western margin of the San Joaquin basin during the early and middle Miocene. In the process, the San Joaquin basin evolved from a forearc basin to a strike-slip successor basin (Graham and others, 1984). Geologic manifestations of this transformation include a change from sandstones dominated by volcanic detritus to sandstones dominated by granitic detritus and recycled sedimentary and metasedimentary materials (Bent, 1985) and an episode of uplift and subsidence that migrated northwestward up the basin more or less synchronously with passage of the Mendocino triple junction (*Glazner and Loomis, 1984*; Graham and Williams, 1985).

The Oligocene to Holocene stratigraphy varies considerably across the length and breadth of this large basin (*Dibblee, 1973b*; *Bartow and McDougall, 1984*; Bartow, 1991). However, generalized east-west cross sections (Fig. 35) across the central and southern portions of the basin display salient structural and stratigraphic relations. The basin is elongated northwest-southeast, with a marked cross-sectional asymmetry. The gently dipping east flank of the basin features mostly shelfal and nonmarine

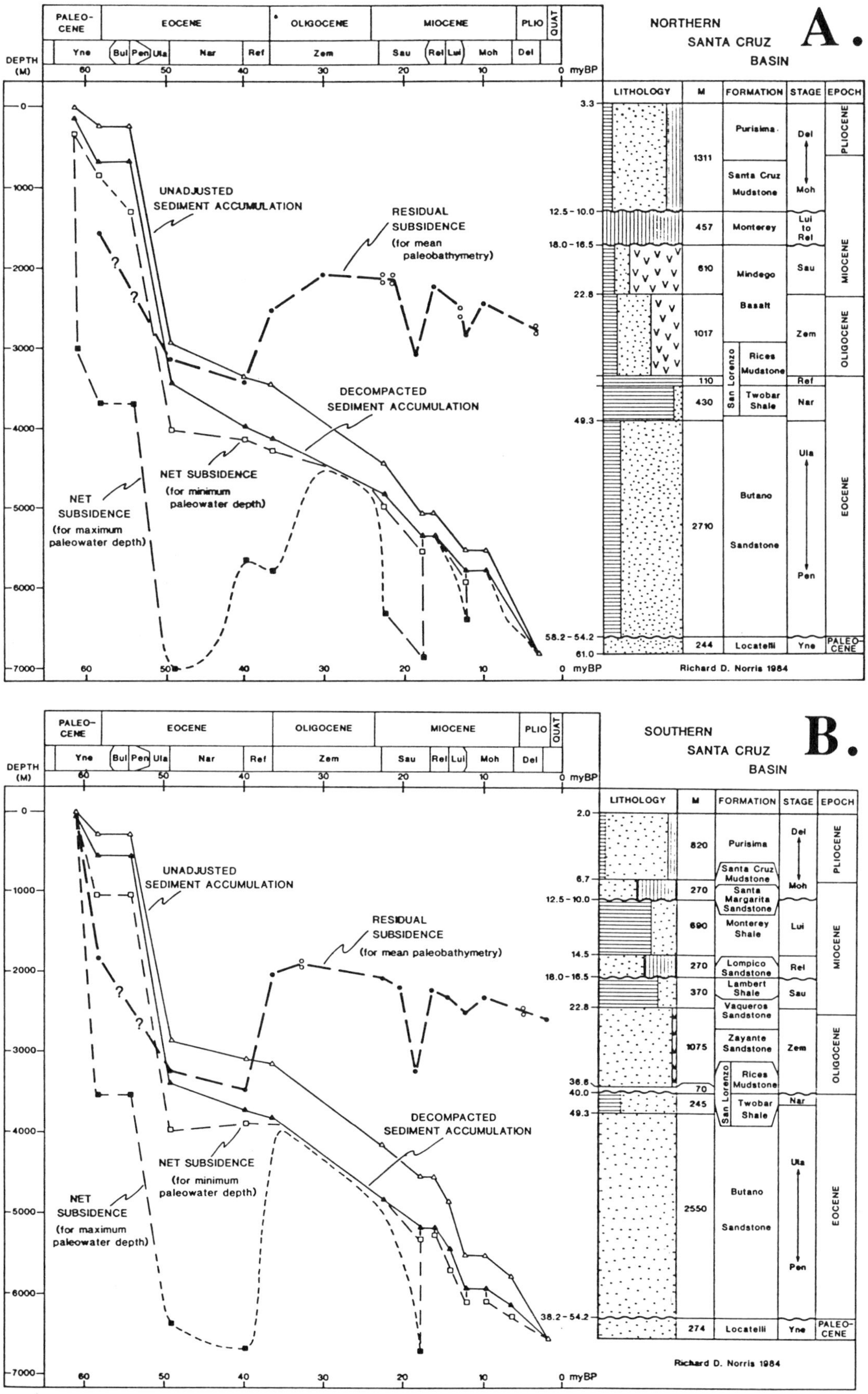
A.
NORTHERN SANTA CRUZ BASIN
PALEOCENE
EOCENE
OLIGOCENE
MIOCENE
PLIO
QUAT
Yne Bul Pen Ula Nar Ref Zem Sau Rel Lui Moh Del
DEPTH (M)
myBP
UNADJUSTED SEDIMENT ACCUMULATION
RESIDUAL SUBSIDENCE (for mean paleobathymetry)
DECOMPACTED SEDIMENT ACCUMULATION
NET SUBSIDENCE (for minimum paleowater depth)
NET SUBSIDENCE (for maximum paleowater depth)
LITHOLOGY
M
FORMATION
STAGE
EPOCH
Purisima
Santa Cruz Mudstone
Monterey
Mindego
Basalt
San Lorenzo
Rices Mudstone
Twobar Shale
Butano Sandstone
Locatelli
Lui to Rel
PLIOCENE
MIOCENE
OLIGOCENE
EOCENE
PALEOCENE
1311
457
610
1017
110
430
2710
244
Richard D. Norris 1984
B.
SOUTHERN SANTA CRUZ BASIN
UNADJUSTED SEDIMENT ACCUMULATION
RESIDUAL SUBSIDENCE (for mean paleobathymetry)
DECOMPACTED SEDIMENT ACCUMULATION
NET SUBSIDENCE (for minimum paleowater depth)
NET SUBSIDENCE (for maximum paleowater depth)
Purisima
Santa Cruz Mudstone
Santa Margarita Sandstone
Monterey Shale
Lompico Sandstone
Lambert Shale
Vaqueros Sandstone
Zayante Sandstone
Rices Mudstone
Twobar Shale
Butano Sandstone
Locatelli
820
270
690
270
370
1075
70
245
2550
274
Richard D. Norris 1984

facies resting on Sierran basement. The eastern basin section is complicated by numerous basin-margin unconformities and facies changes, likely arising from eustatic as well as tectonic events (Graham and Williams, 1985; Bartow, 1991). In contrast, the western flank of the basin features upper Cenozoic strata involved in complicated structures associated with the San Andreas fault. Stratigraphic packaging is also strongly influenced by San Andreas wrench and thrust tectonics, with numerous unconformities, stratal wedging, and facies changes associated with the growth of specific structures (Harding, 1976; Davis and Lagoe, 1988). The geology of the southern basin margin was also complicated by early Miocene extensional tectonism, shared with the formerly contiguous Santa Cruz basin across the San Andreas fault, as reflected in synsedimentary normal faulting and basaltic volcanism (Hirst, 1988).

The center of the basin preserves the most complete record of sedimentation (Fig. 35). Continuous, deep-marine sedimentation apparently persisted from late Eocene through late Miocene shale (Fig. 34; Bandy and Arnal, 1969). The Temblor Formation is a sandstone and shale sequence, ranging from Oligocene to middle Miocene in age and from shallow-marine along the basin margin to deep-marine in the basin center (*Carter, 1985*). Temblor clastics compositionally track the conversion to a strike-slip setting (Bent, 1985). Deep-marine shales, siliceous shales, and porcelanites of the overlying Monterey Formation, the most important petroleum-source rocks of the region, more clearly reflect the basinwide impact of the change to a transform setting (Graham, 1987). Basin reorganization, best monitored through subsidence-history analysis, permitted accumulation of fine-grained Monterey lithologies in periodically oxygen-poor deep-marine environments (Graham and Williams, 1985). By the end of the Miocene, the San Joaquin basin filled to relatively shallow water depths.

During its entire Oligocene through Miocene history as a marine depocenter, the San Joaquin basin was in communication with the Pacific Ocean through passages across the Salinian block along the western margin of the basin (Fig. 34). Middle Tertiary depocenters can be correlated across the fault from the western San Joaquin basin to the Santa Cruz basin (*Stanley, 1984*; Bent, 1985; Graham and others, 1989). Nevertheless, the Salinian block was a sill that confined deep-water sediment-gravity flows to the bottom of the San Joaquin basin. This structural control on bathymetry and sediment dispersal is best illustrated by the late Miocene Stevens submarine fan system, an important petroleum reservoir that is encased in basinal siliceous shales of the Monterey Formation (Fig. 35B) and was largely fed from eastern deltaic sources represented by the Santa Margarita Formation of the Bakersfield area (*MacPherson, 1978*; Webb, 1981).

The San Joaquin basin persisted into the Pliocene as a large shallow-marine embayment in the California coastline, bounded on the west by the Coast Ranges, which were rising along the San Emigdio and Temblor Range thrust belts adjacent to the San Andreas fault (Davis and Lagoe, 1988). Thick accumulations of highly fossiliferous, shallow-marine clastics of the Etchegoin and Jacalitos Formations (Fig. 35; *Stanton and Dodd, 1976*) are overlain by a thick complex of poorly documented alluvial-fluvial-lacustrine sequences of Pliocene to Holocene age that record eventual isolation of the San Joaquin basin from the Pacific Ocean. In general, relatively coarse-grained nonmarine facies are assigned to the Kern River Formation along the east basin margin, whereas facies of the western basin margin—strongly influenced by San Andreas structures but otherwise comparably coarse-grained and nonmarine—are included in the Tulare Formation (*Bartow and Pittman, 1983*). The basin has persisted through the Quaternary as a structural and topographic lowland, characterized mainly by longitudinal drainage to the northwest but also with local areas of internal drainage, and was heavily influenced by Pleistocene glaciation and deglaciation of the adjacent Sierra Nevada (Lettis, 1988).

LATE CENOZOIC BASINS OF SOUTHERN CALIFORNIA

With initiation and evolution of the San Andreas transform system and, especially, with formation of its "big bend" in the Transverse Ranges region, new fault-bounded uplifts and deep sedimentary basins were blocked out (Fig. 36). Consequently, Neogene depositional patterns in most of southern California differed considerably from earlier Tertiary patterns and have changed considerably with time.

Santa Maria, Pismo, and Huasna basins

The Santa Maria basin is shaped like a wedge opening westward (Fig. 37), with its northeastern margin along the uplifted Pismo and Huasna basins—part of the southern Coast Ranges—and its southern margin along the Santa Ynez Mountains—part of the Transverse Ranges. The basement is mainly the Franciscan Complex with lesser amounts of Coast Range ophiolite and overlying Great Valley sequence. This is overlain directly by the nonmarine Lospe Formation, conformably overlain in turn by the marine lower Miocene Point Sal Formation (Woodring and Bramlette, 1950). In contrast, the Santa Ynez Mountains to the south contain a sequence of Upper Cretaceous and Paleogene strata that thin northward (Dibblee, 1950) and are cut off abruptly by the Lompoc-Solvang fault of Hall (1981a, b), also called the Santa Ynez River fault (Fig. 37;

Figure 31. Geohistory diagrams for the Santa Cruz basin. Display differs slightly from typical geohistory diagrams (e.g., *Van Hinte, 1978*) in using the top of the sediment column rather than sea level as the datum. Paleobathymetry and uncertainties in its estimation are reflected in net subsidence curves. Most important for current discussion are residual subsidence curves, which are generally interpreted as reflecting tectonic components of subsidence and uplift. Thickness and depth (M) in meters. Provincial benthic foraminiferal stages are those of *Kleinpell (1938)*. Reprinted from Dickinson and others (1987) with permission of Prentice-Hall, Inc.

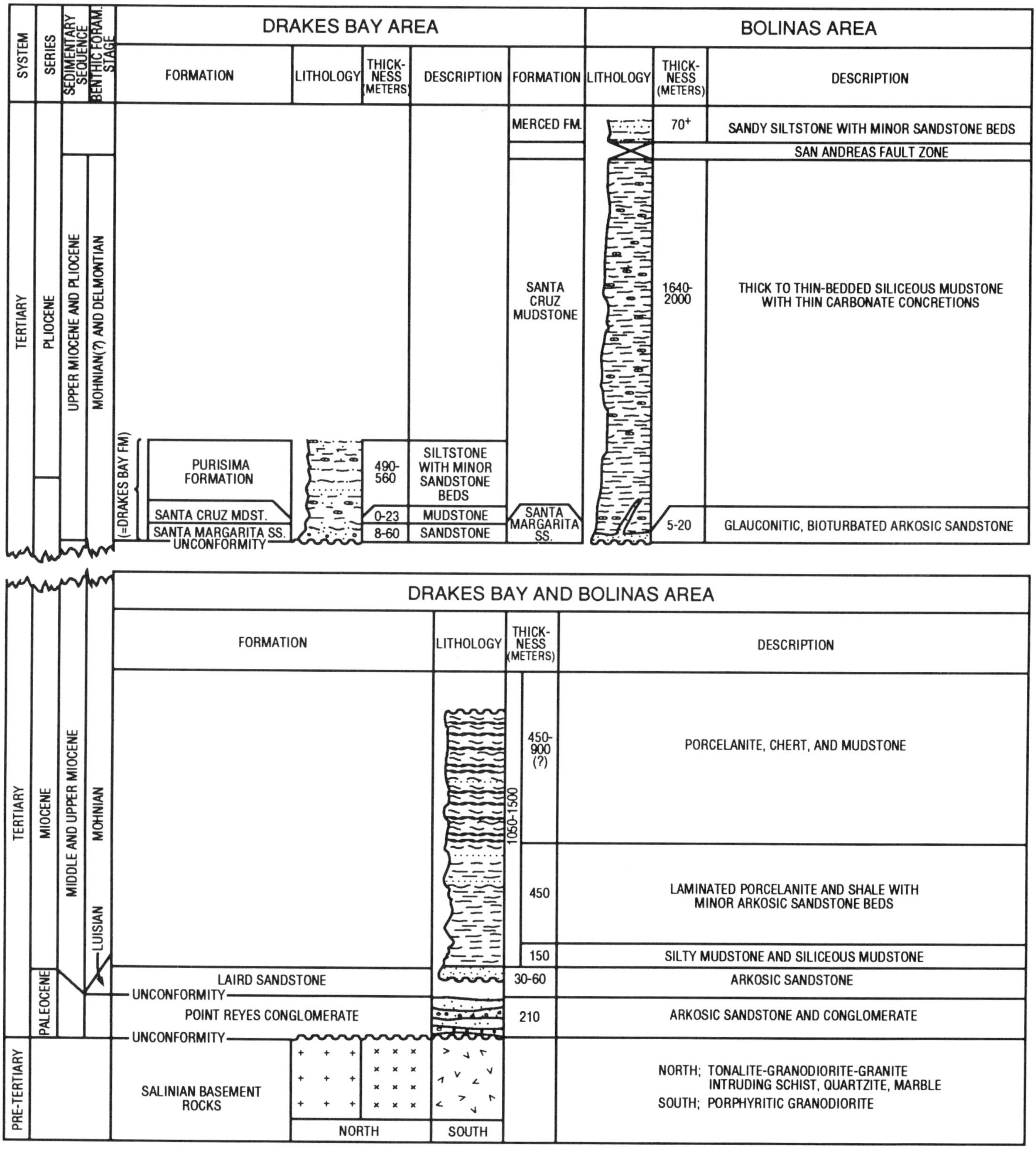

Figure 32. Composite stratigraphic sections of the Bodega basin, exposed onshore on the Point Reyes Peninsula. Provincial benthic foraminiferal stages are those of *Kleinpell* (*1938*). Simplified from J. C. Clark and others (1984) with permission of the Pacific Section, Society of Economic Paleontologists and Mineralogists.

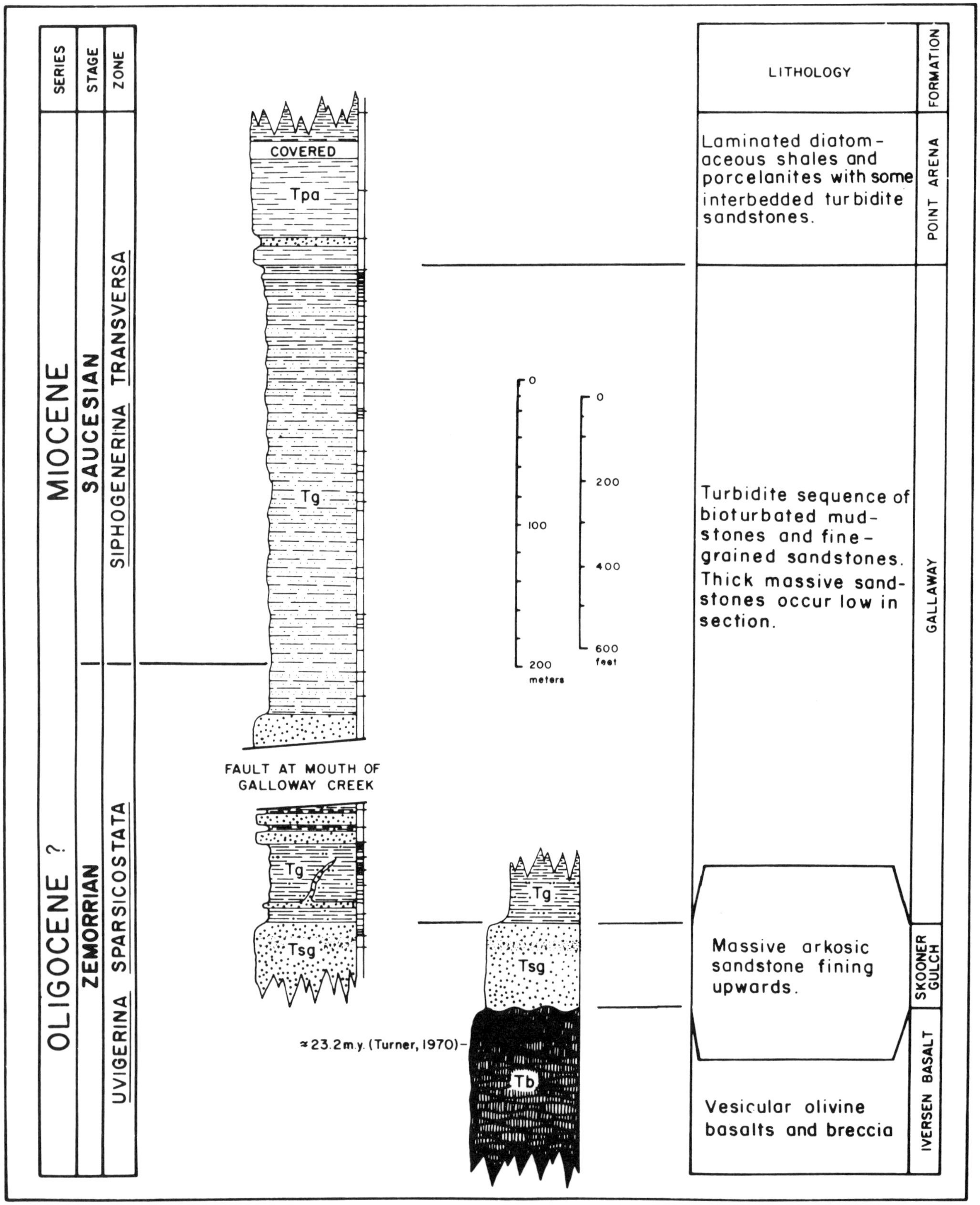

Figure 33. Stratigraphic sections of the Point Arena basin, exposed along the northern California coast south of Point Arena. Provincial benthic foraminiferal stages are those of *Kleinpell* (*1938*). After *Phillips and others* (*1976*) with permission of the Pacific Section, Society of Economic Paleontologists and Mineralogists.

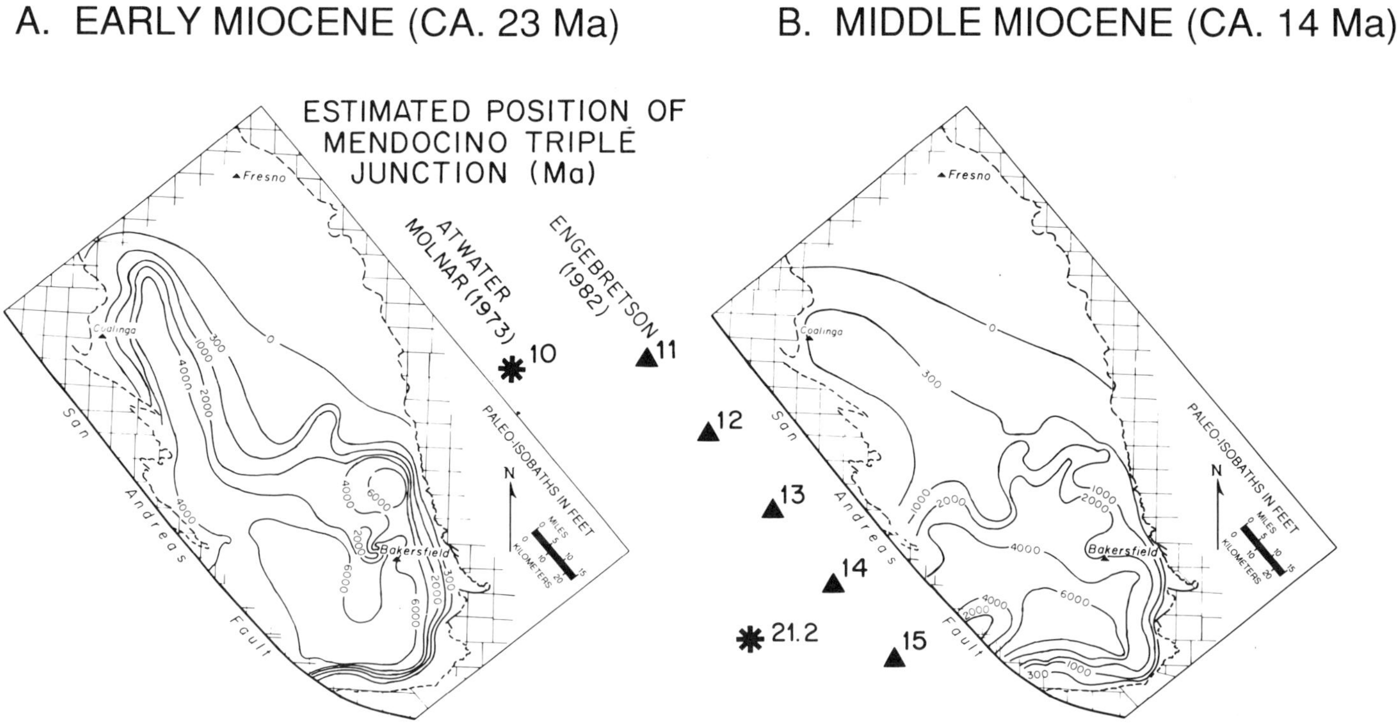

Figure 34. Paleobathymetry (Bandy and Arnal, 1969) of the San Joaquin basin compared with two reconstructions of the latitudinal shift of the Mendocino triple junction with time. A, earliest Miocene time; B, middle Miocene time. After Graham and Williams (1985) with permission of the American Association of Petroleum Geologists.

Sylvester and Darrow, 1979). The Santa Ynez Mountains also contain volcanic rocks dated as 17 to 20 Ma (Turner, 1970; Hall, 1981a). On the northeast, the basin is bounded by the Santa Maria River fault (Hall, 1981a, b). The ranges to the north comprise the Pismo and Huasna basins, which have the same basement as the Santa Maria basin but, unlike the Santa Maria basin, contain Oligocene igneous and nonmarine sedimentary rocks, the lower Miocene Vaqueros Sandstone and Rincon Shale, and overlying volcanic rocks of the Obispo Formation dated as 17 to 15 Ma (Turner, 1970; Hall, 1981a). The Obispo Formation is overlain by the Monterey Formation of early middle Miocene age (Turner, 1970).

Hall (1981a, b) suggested that the volcanic rocks in the Huasna and Pismo basins delineated a feature he called the San Luis Obispo transform fault. This "leaky transform" produced a slivering of fault-bounded blocks as they underwent right-lateral shear and clockwise rotation during and after deposition of the Obispo Formation (*Surdam and Hall, 1984*). The Obispo Formation grades laterally eastward to the Point Sal Formation, predominantly mudstone and siltstone with local sandstone (*Crawford, 1971*; *Surdam and Hall, 1984*); in the Santa Maria basin, only the Point Sal Formation is present, separated from the Obispo Formation by a basement ridge. The Point Sal and Obispo are both overlain by diatomite, porcelanite, dolomite, phosphatic shale, and siliciclastic siltstone of the middle and upper Miocene Monterey Formation (Surdam and Stanley, 1984; see also *Bramlette, 1946*; *Regan and Hughes, 1949*; *Redwine, 1981*), deposited in the pull-apart Santa Maria, Pismo, and Huasna basins as they formed. The Monterey is overlain in the Pismo basin by mudstone, diatomaceous mudstone, diatomite, and sandstone of the upper Miocene and Pliocene Pismo Formation (Stanley and Surdam, 1984) and in the Santa Maria basin by the much thicker but coeval Sisquoc Formation (mudstone and diatomite), Foxen Mudstone, and Careaga Sandstone (marine sandstone and conglomerate). The Sisquoc grades northeastward to shallow-marine sandstone in the northeastern Santa Maria basin (*Crawford, 1971*) and is overlain with angular unconformity by the Pleistocene nonmarine Paso Robles Formation (Woodring and Bramlette, 1950), which is itself deformed. The pull-apart basins continued to evolve during the late Miocene and Pliocene, with deposition continuing longest in the more southwesterly basins, particularly the Santa Maria basin.

Deformation in the Miocene and Pliocene was controlled by the Hosgri and San Andreas right-slip faults on the west and east, respectively, and by the Lompoc-Solvang (or Santa Ynez River) right-slip fault on the south (Hall, 1981a, b), producing slivering along northwest-elongated blocks and developing local basins that became more restricted in the Pliocene. The slivering was

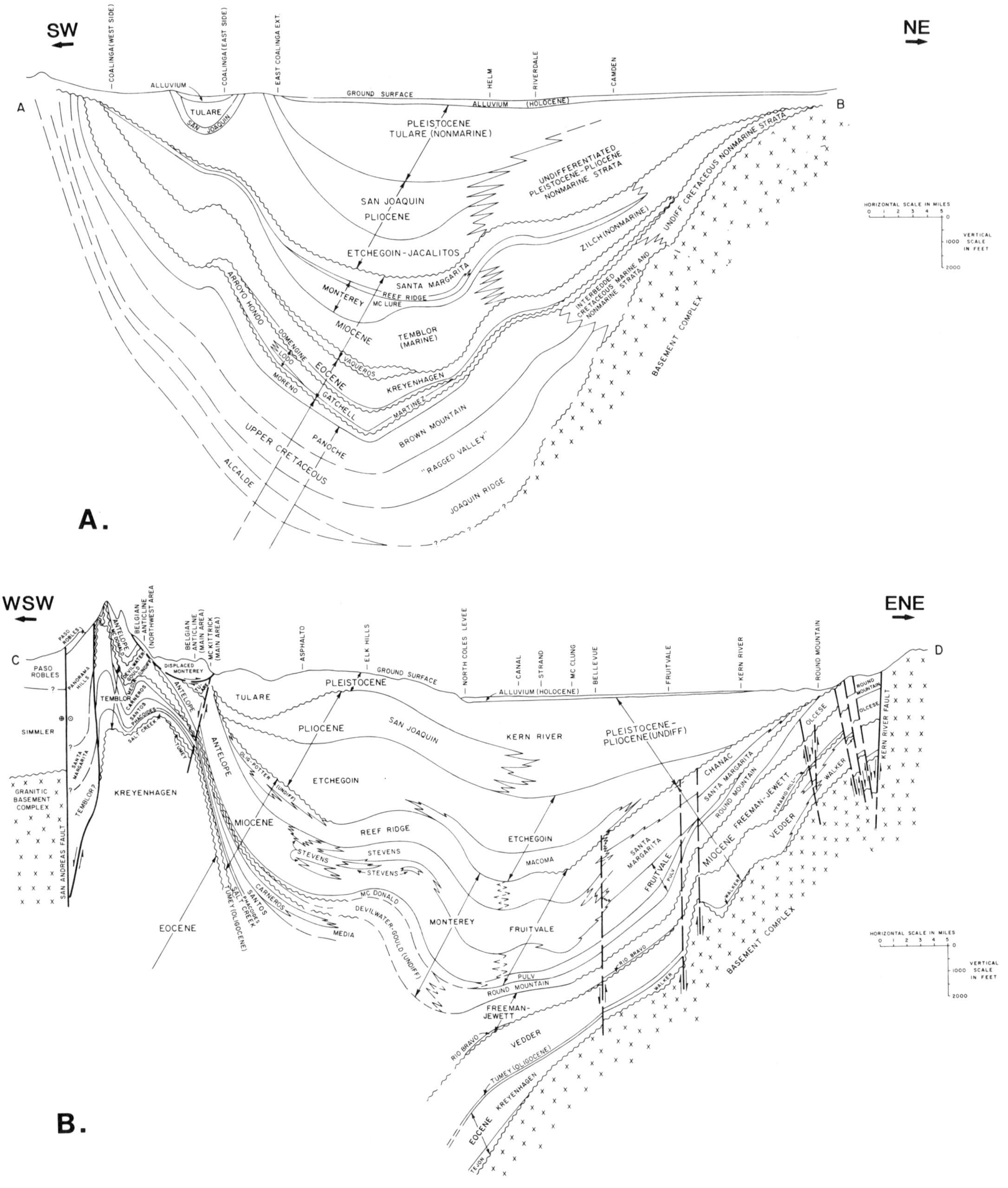

Figure 35. Generalized cross sections across the San Joaquin basin showing key structural and stratigraphic relations. After *California Division of Oil and Gas* (*1973*). A, NE-SW section across the central San Joaquin basin through Riverdale and Coalinga; B, ENE-WSW section across the southern San Joaquin basin through Bakersfield (Fruitvale) and Elk Hills.

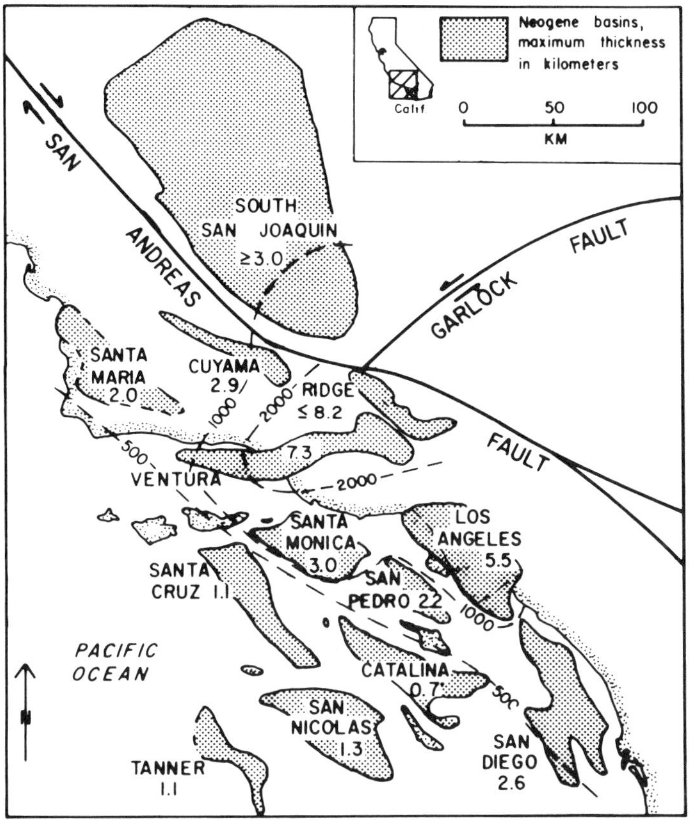

Figure 36. Neogene basins of southern California showing maximum thickness (in km) of post-Miocene sedimentary rocks. Contours in dashed lines indicate net subsidence rates in m/m.y. for thicker parts of each basin. Modified after *Yeats (1978)*.

accompanied by clockwise rotation of the Santa Ynez Mountains from an original northerly orientation (Hornafius, 1985). The Santa Maria basin itself comprises, from north to south, the Santa Maria Valley syncline, Los Alamos syncline, and Santa Rita syncline (Fig. 37; *Krammes and Curran, 1959*), of which at least the northern two reached depths and temperatures sufficient to generate petroleum, which migrated to structures in three producing trends. From north to south, these are the Guadalupe–Santa Maria Valley–Cat Canyon (north of the Santa Maria Valley syncline), the Casmalia-Orcutt (south of that syncline), and the Lompoc trends. Oil was generated in the Pismo basin as well (*Heasler and Surdam, 1984*), but later deformation and erosion brought reservoirs to the surface and degraded much of the oil. Oil is found in sandstone in the Pismo basin and the northeastern Santa Maria basin, but the most important occurrences are in fractured rocks of the Monterey Formation (*Regan and Hughes, 1949*), with dolomite being an important reservoir (*Redwine, 1981*). The fractured reservoirs of the Santa Maria basin are an onshore analog to oil fields in the western Santa Barbara Channel and the offshore part of the Santa Maria basin (Ogle and others, 1987).

Cuyama basin

Like the Santa Maria basin, the Cuyama basin is part of the southern Coast Ranges, but its southern boundary, the Big Pine fault, is part of the Transverse Ranges (Fig. 37; *Dibblee, 1973a, b, 1982b*). Mesozoic and older crystalline rocks are overlain unconformably by Upper Cretaceous and lower Tertiary sedimentary rocks, and all were strongly deformed and deeply eroded prior to deposition of Oligocene red beds (Simmler Formation) in normal-faulted basins (Fig. 38; *Bartow, 1974*; Ballance and others, 1983). The evolution of the Cuyama basin itself is recorded in the members of the Vaqueros Formation. The basin began with deposition of the transgressive Quail Canyon Sandstone Member, of early Miocene age followed by the deep-water Soda Lake Shale Member (Hill and others, 1958; *Cross, 1962*; Dickinson and others, 1987); depositional patterns reflect coeval right slip on the Russell fault beginning at 22 to 23 Ma (Yeats and others, 1989). These units are followed by the shallow-marine Painted Rock Sandstone Member of early Miocene age (Hill and others, 1958). Where local subsidence did not occur, the Soda Lake Shale Member is not found, and the Quail Canyon and Painted Rock sandstones are mapped together as the Vaqueros Formation (*Dibblee, 1973b*). Renewed deepening of the basin was reflected in deposition of the Monterey Shale (*Lagoe, 1984, 1985*) followed by the Santa Margarita Formation of late Miocene age. The Miocene sequence shallows eastward toward the San Andreas fault (Clifton, 1981) to the shallow-marine Branch Canyon Sandstone and the nonmarine Caliente Formation (Fig. 38), and the interfingering facies permit correlation of vertebrate, molluscan, and microfaunal assemblages (*Repenning and Vedder, 1961*; Vedder, 1973). The Pliocene and Pleistocene strata are completely nonmarine, including the Quatal, Morales, and Paso Robles Formations (Hill and others, 1958; *Dibblee, 1973b, 1982b*).

Miocene depositional patterns were strongly influenced by coeval northwest-trending faults (Fig. 37), the Rinconada fault (Dibblee, 1976), Russell fault (and its northern extension, the Chimeneas and San Juan fault; Yeats and others, 1989), and possibly the San Andreas fault (Huffman, 1972). These patterns are not expressed in the modern topography; lower and middle Miocene sandstone, claystone, and siliceous shale are found in the Hurricane Deck area of the San Rafael Mountains (*Vedder and Brown, 1968*), and middle Miocene marine strata occupy much of the higher parts of the Sierra Madre (Fig. 37; *Fritsche, 1969*; *Dibblee, 1973a*). In the Pliocene and Pleistocene, the modern topography developed along with north- and south-dipping reverse faults. The oil fields of the Cuyama basin occur mainly in Oligocene to lower Miocene marine sandstone in traps controlled by the Russell fault (Schwade and others, 1958; *Cross, 1962*). Part of the Cuyama Valley is a broad lowland with a Transverse

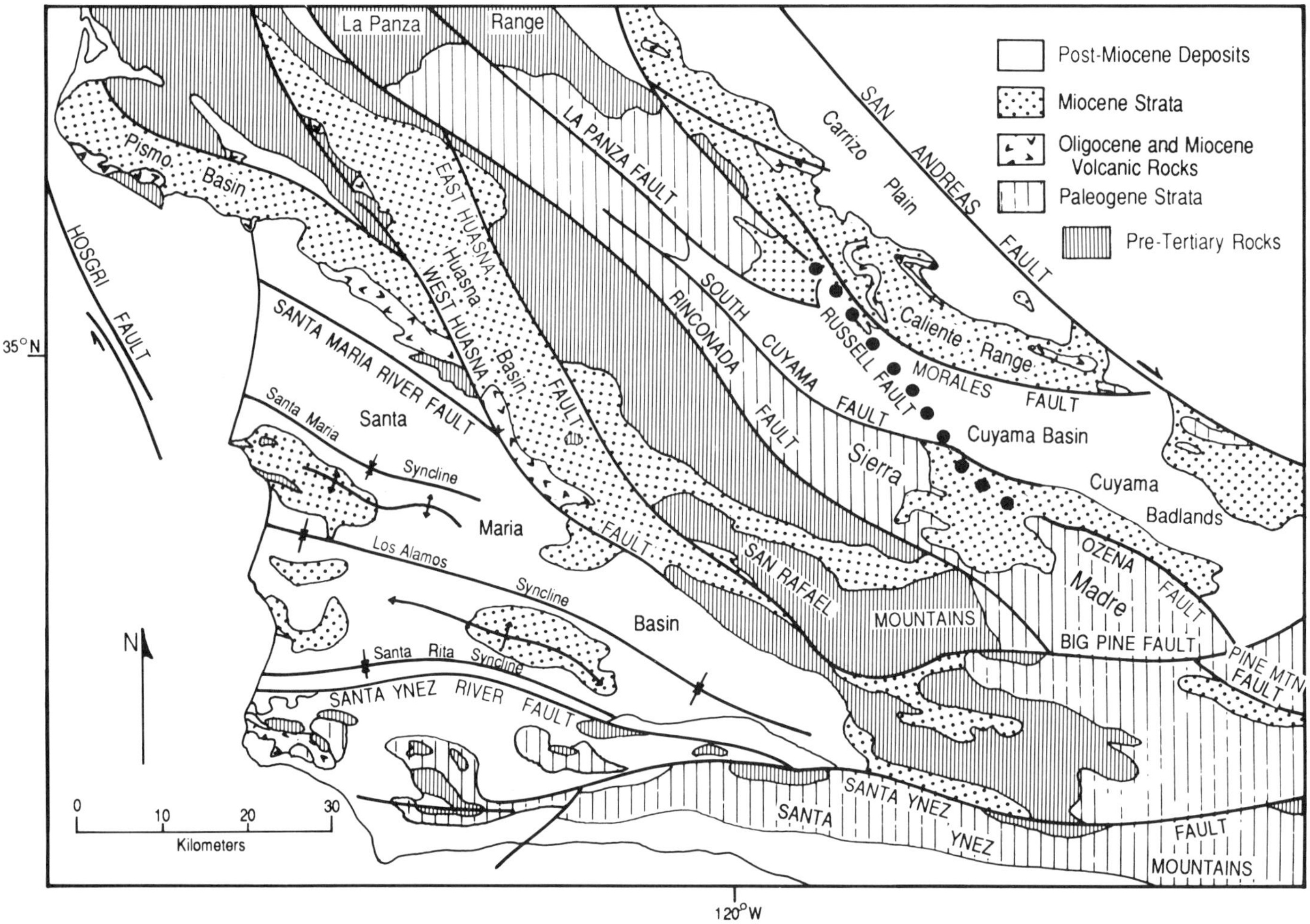

Figure 37. Tectonic map of part of southern Coast Ranges showing Santa Maria, Pismo, Huasna, and Cuyama basins. The Russell fault, known from the subsurface, is dotted. Geology east of the San Andreas fault not shown.

Range trend between the Caliente Range, thrust southward on the Whiterock and Morales faults, and the Sierra Madre, thrust northward on the South Cuyama and Ozena faults. The Carrizo Plain is another broad lowland between the Caliente Range on the southwest and the San Andreas fault and Temblor Range on the northeast.

Ventura basin

West of the San Gabriel fault, the western Transverse Ranges consist of the Santa Ynez Mountains and Topatopa Mountains on the north and the northern Channel Islands and Santa Monica Mountains on the south separated by the Ventura basin, which formed mainly in the Pliocene and Pleistocene (Fig. 39; Dibblee, 1982a). The western part of the Ventura basin below sea level is the Santa Barbara Channel (Vedder and Brown, 1969); the eastern part comprises the lower Santa Clara Valley, Oxnard Plain, and interbasin highlands including Oak Ridge, the Simi Hills, and the Santa Susana Mountains (*Nagle and Parker, 1971*). Farther east, but still west of the San Andreas fault, the San Gabriel Mountains (Fig. 9) and highlands to the northwest along the "big bend" of the San Andreas fault are the dominant features (*Ehlig, 1981*; Dibblee, 1982a). Intermontane basins in this area include the East Ventura and Ridge basins (successors to the middle Tertiary Soledad basin) and the San Fernando and San Gabriel Valleys. Across the San Andreas fault lie the eastern Transverse Ranges dominated by the San Bernardino Mountains, part of the Mojave block, and the Techachapi and San Emigdio Mountains—the southernmost tail of the Sierra Nevada rotated into a Transverse Ranges trend. To the south are

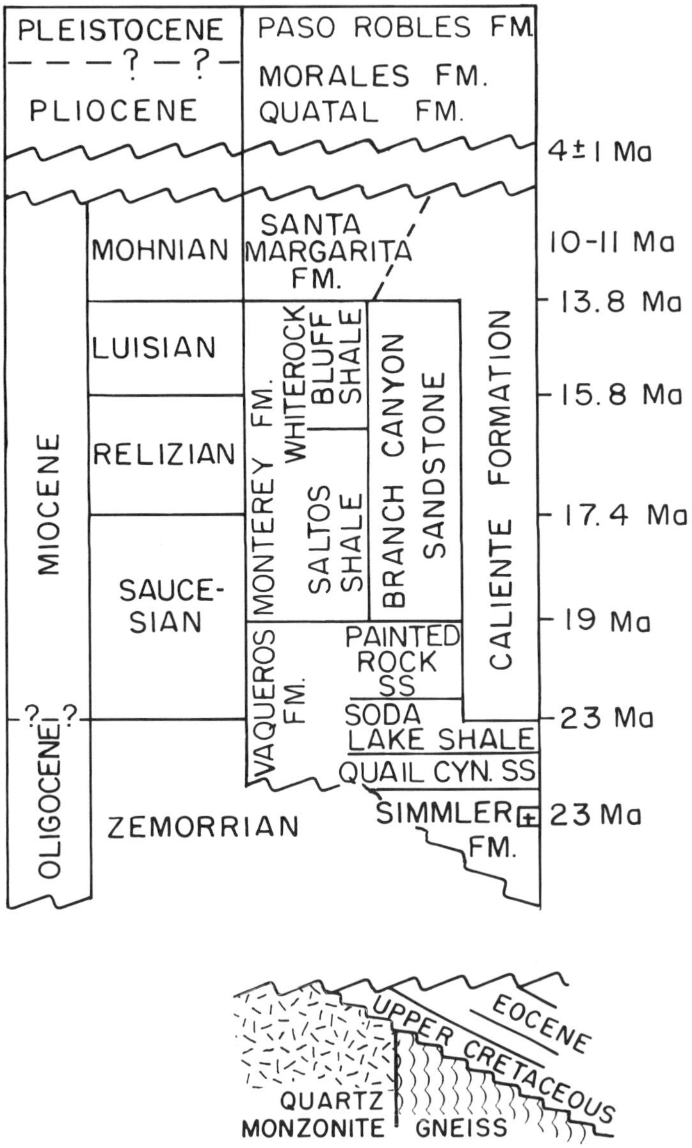

Figure 38. Stratigraphy of Cuyama basin. Mesozoic basement rocks and Upper Cretaceous to Eocene strata were tilted prior to deposition of the nonmarine Simmler Formation, dated by an interbedded basalt flow. Miocene strata grade from marine on the west to nonmarine on the east. Provincial benthic foraminiferal stages are those of *Kleinpell (1938)*. From Yeats and others (1989).

the northwest trends of the California Continental Borderland, Los Angeles basin, and Peninsular Ranges (Fig. 9).

The Oligocene regression marked by the nonmarine Sespe Formation in the Ventura basin was followed by transgressive deposition of the shallow-marine Vaqueros Formation of Oligocene age followed, north of the Ventura basin, by the lower Miocene Rincon Shale (*Edwards, 1971*) and the middle and upper Miocene Monterey Formation (Dibblee, 1950, 1966). A bentonite bed marking the Rincon-Monterey contact may be equivalent to volcanic rocks of the Obispo Formation of the Pismo basin to the north. East of Ventura, the Rincon Shale contains interbeds of shallow-marine sandstone. East of the Ojai Valley, as far as the San Gabriel fault, fine-grained siliceous and organic shale lithologically equivalent to the Monterey Formation is interbedded with deep-water sandstone, and the assemblage is called the Modelo Formation (Winterer and Durham, 1962; *Yeats and others, 1985*; Stitt, 1986). North of the Santa Ynez fault, an overturned syncline of Miocene strata occurs on the downthrown side of the Pine Mountain fault (Fig. 37). In this syncline (*Dickinson and Lower, 1966*), the Sespe is overlain by the Vaqueros Formation (mudstone, limestone, and sandstone), Monterey Formation (siliceous shale, calcareous mudstone, and cross-bedded sandstone), and Santa Margarita Formation (white sandstone, phosphorite, and gypsum). Locally, the Rincon Formation overlies the Vaqueros (*Reid, 1978*).

On the south side of the Ventura basin (Fig. 39), the lower Miocene section commonly consists of two sandstone units and an intervening mudstone unit (*Edwards, 1971*) in the western Santa Monica Mountains, the area between the Simi Valley and Oak Ridge (Yeats, 1976), and in the northern Channel Islands (Weaver, 1969). In the eastern Santa Monica Mountains, the lower Miocene section is nonmarine (Yerkes and Campbell, 1979), but in the westernmost exposures in the northern Channel Islands, it is nearly all marine mudstone (Weaver, 1969). In the Santa Monica Mountains, the lower Miocene section is overlain by the thick middle Miocene Topanga Group, including shallow-marine and deep-marine sandstone overlain by the Conejo Volcanics, itself overlain by marine turbidites of late-middle Miocene age (Yerkes and Campbell, 1979). To the north and west, the Conejo Volcanics (17 to 14 Ma), spread outward from a caldera complex in the Santa Monica Mountains and ponded against an ancestral Oak Ridge (Fig. 40; Yeats, 1987). North of the Santa Cruz Island fault, the Santa Cruz Island Volcanics, of about the same age as the Conejo Volcanics (Turner, 1970; *Crowe and others, 1976*) similarly spread out northward into the Santa Barbara Channel. These rocks are overlain with angular unconformity by the Modelo Formation—a basal sandstone and overlying siliceous shale. In the Santa Monica Mountains, the basal part of the Modelo is late Miocene in age, but elsewhere the basal part is middle Miocene. In contrast to coeval Modelo and Monterey north of the Oak Ridge fault, the Modelo south of the fault is relatively thin, deposited on a structural shelf. The hinge line between the Modelo shelf and turbidite trough follows the Santa Susana fault, with the now-upthrown Santa Susana Mountains side downdropped with respect to the structural shelf during Modelo deposition (Yeats, 1979).

The deep central trough of the Ventura basin itself (Fig. 39) began to form at the end of the Miocene, beginning with deposition of the Sisquoc Formation and continuing with the Pliocene Repetto Formation (of former usage) and the Pliocene and Pleistocene Pico Formation, predominantly turbidites, the Pleistocene Santa Barbara Formation, mainly mudstone, and the Saugus Formation, predominantly coarse-grained shallow-marine and

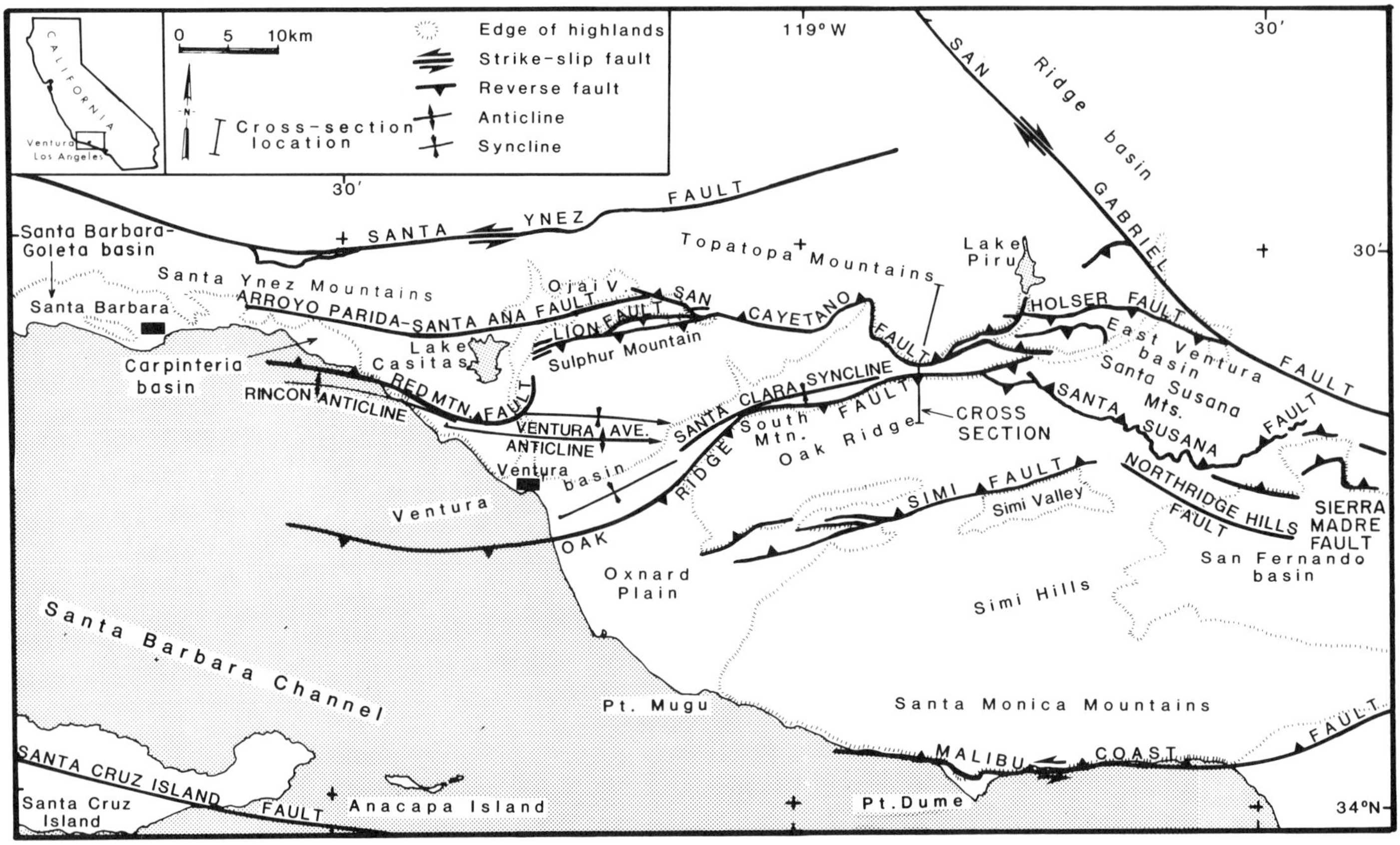

Figure 39. Map of the central Transverse Ranges showing location of the Ventura basin. Reverse faults indicated with teeth on hanging-wall side; strike-slip faults indicated by arrows. From Yeats (1979).

nonmarine conglomerate and sandstone (Yeats, 1976). Tephrochronology, magnetostratigraphy, and amino-acid stereochemistry show that San Pedro deposition ended about 0.2 to 0.4 Ma (*Lajoie and others, 1982*; Yeats, 1983, *1988*). Farther east, the basin is shallower, and facies boundaries migrate downward. The main turbidites are in the Miocene and Pliocene Towsley Formation, and the Pico Formation comprises fine-grained mudstone and shallow-marine sandstone (Winterer and Durham, 1962). The Pico is overlain by the coarse-grained, largely nonmarine Saugus Formation, which is as young as 0.4 Ma, based on magnetostratigraphy and tephrochronology (*Levi and others, 1986*).

Thick Pliocene and Pleistocene deposits, including the world's thickest Pleistocene marine sequence, are found in the Santa Clara Valley and the eastern part of the Santa Barbara Channel, bounded on the south by the Oak Ridge reverse fault and on the north by the Red Mountain and San Cayetano reverse faults (Figs. 39, 41). South of the Oak Ridge fault, the Pliocene to Pleistocene sequence onlaps the Miocene section over a submarine slope (*Yeats, 1965*), with the original depositional edge north of the Santa Monica Mountains and Channel Islands. North of the Red Mountain fault, shallow-marine Santa Barbara Formation and nonmarine Casitas Formation fill the Carpinteria and Santa Barbara–Goleta intermontane basins (Fig. 39) bounded on the south by reverse faults (Dibblee, 1966; Jackson and Yeats, 1982). North-south convergence in late Quaternary time produced the rootless, overpressured Ventura Avenue and Rincon anticlines (Yeats, 1983; *Yeats and others, 1988*), large displacement on the Red Mountain, San Cayetano, Santa Susana, and Oak Ridge faults, and a fold belt in the east Ventura basin (*Yeats and others, 1985*).

The San Gabriel fault is the only major right-slip fault crossing the western Transverse Ranges. It acquired most of its displacement in the Miocene when it was the main strand of the San Andreas fault (Crowell, 1954, 1975b). As it formed, part of the crust immediately to the east downdropped to form the Ridge pull-apart basin (Fig. 42; Crowell and Link, 1982). Grabens filled with the predominantly Oligocene Vasquez Formation are overlain unconformably by nonmarine Miocene strata, including the Mint Canyon Formation, which responded to early displacement on the San Gabriel fault and may be the offset continuation of nonmarine strata in Lockwood Valley (Ehlig and others, 1975). The upper Miocene marine Castaic Formation (Crowell, 1954) unconformably overlies the Mint Canyon and is the basal part of the sediment fill of the Ridge basin (Fig. 42; *Stanton, 1960*). The Castaic is overlain by the nonmarine Ridge Basin Group (*Axelrod, 1950*), including alluvial fans entering the basin from the

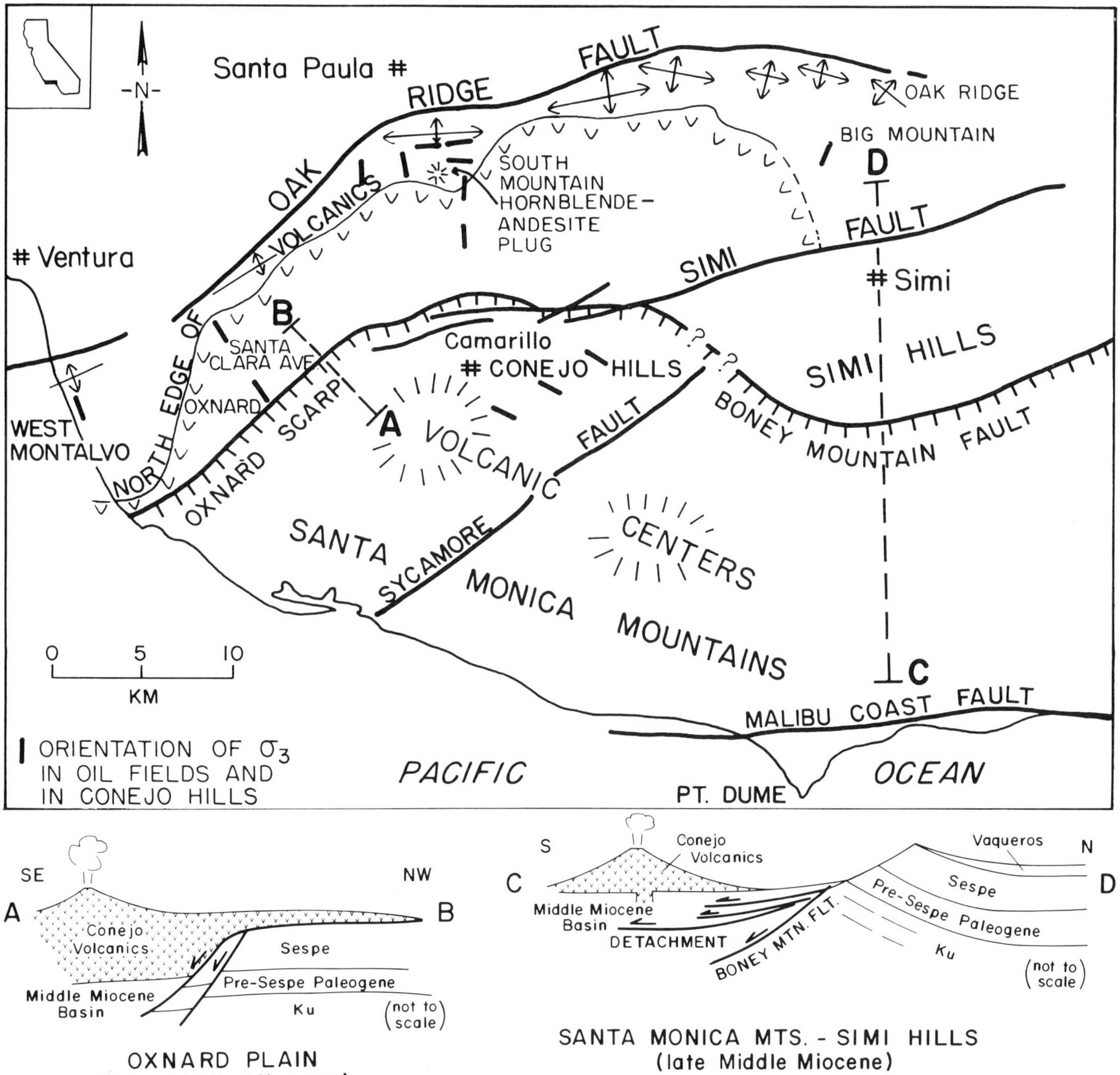

Figure 40. Middle Miocene tectonics of the southern Ventura basin. Structures south of the Oak Ridge fault follow an annular pattern around Miocene volcanic centers of the western Santa Monica Mountains. South-facing scarp of middle Miocene age (Oxnard scarp and Boney Mountain fault of *Campbell, 1973*) is S-shaped in plan, convex north near Camarillo and convex south at Simi Hills. Scarp follows Simi fault for about 5 km; otherwise it is north or south of it. Short bars show orientation of σ_3 based on normal-fault pattern in oil fields or orientation of dikes (Conejo Hills, after *Williams, 1983*). σ_3 tends to be normal to volcanic centers. Edge of Conejo volcanics is controlled by anticlines south of the Oak Ridge fault. The Oak Ridge fault itself, although younger, probably was originally controlled by the middle Miocene annular structure. Insets are cross sections restored to late middle Miocene time for the Oxnard area (A–B), and the Simi Hills–Santa Monica Mountains area (C–D). Reprinted from Yeats (1987) with permission of Prentice-Hall, Inc.

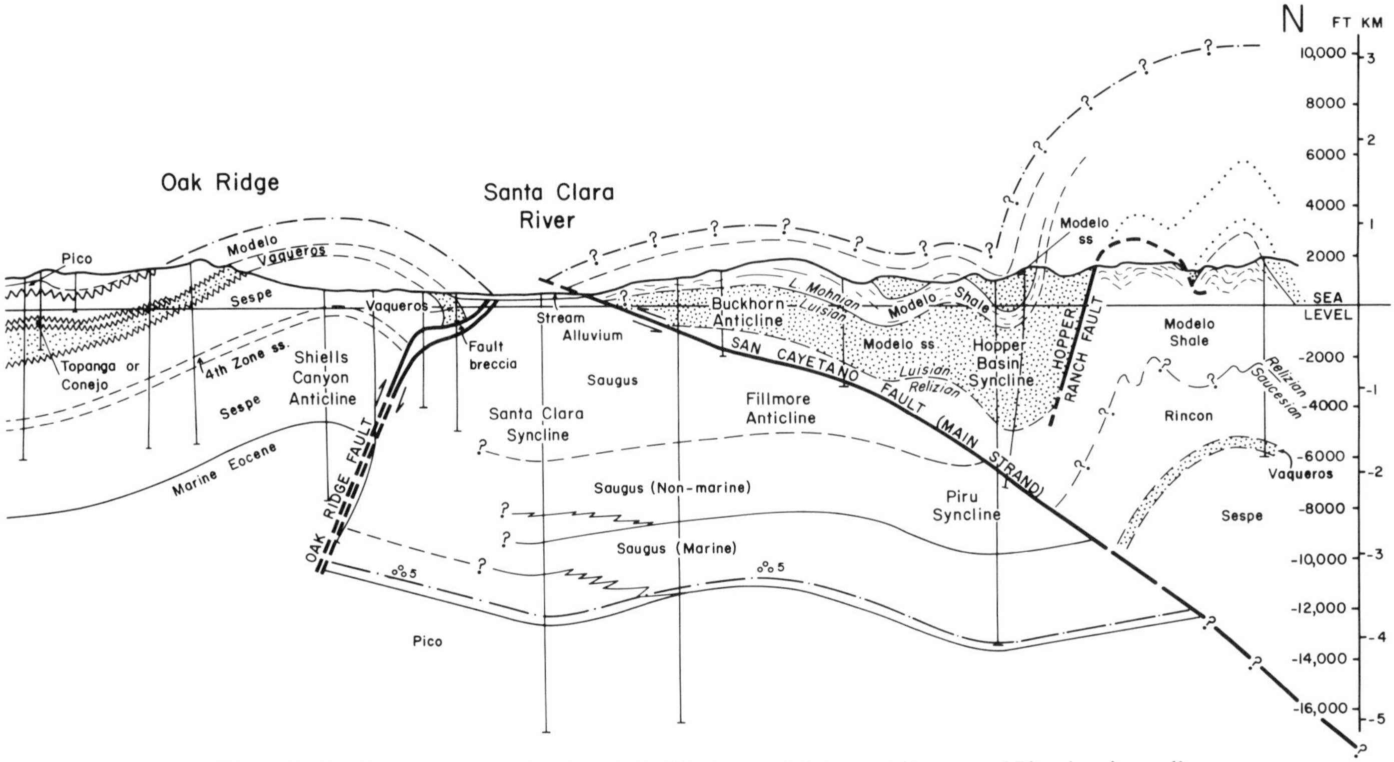

Figure 41. North-south cross section through the Ventura basin between Fillmore and Piru showing well control. Open circles with "5" and dot-dash pattern mark a unit dated as 1 Ma in age near Ventura and correlated east to this region; projection of this horizon north of the San Cayetano fault is highly speculative. Provincial benthic foraminiferal stages are those of *Kleinpell (1938)*. Reprinted from Yeats (1983) with permission of American Geophysical Union.

east, lacustrine deposits in the center, and the Violin Breccia (Crowell, 1954), deposited in both the Castaic Formation and the Ridge Basin Group from the scarp of the coeval San Gabriel fault (Crowell, 1975b; Crowell and Link, 1982). Above the Ridge Basin Group is the Hungry Valley Formation (*Crowell, 1950*), the upper part of which overlaps and postdates the San Gabriel fault. In the East Ventura basin, the Pliocene Pico and Pleistocene Saugus Formations are found on both sides of the San Gabriel fault, showing that most of the strike slip on the San Gabriel fault accumulated prior to Pico deposition (Stitt, 1986).

San Fernando Valley

The San Fernando Valley is representative of several late Cenozoic basins that formed on the south side of the San Gabriel Mountains, presumably related to the Sierra Madre and Cucamonga reverse faults that have thrust the San Gabriel Mountains southward (Figs. 9, 39). The San Fernando Valley is underlain by basement rocks correlated with those of the Santa Monica Mountains to the south but unlike the basement rocks of the San Gabriel Mountains. Modelo turbidite fans of late Miocene age extended southward into a lowland now occupied by the Santa Monica Mountains, responding to local topographic ridges in the area of the present Santa Monica Mountains (*Sullwold, 1960*) and Simi Hills. The northwest part of the basin contains the marine Pliocene and nonmarine coarse-clastic Pleistocene Saugus Formation, with the thickest accumulation in the northern corner of the San Fernando basin, adjacent to the San Gabriel Mountains (*Oakeshott, 1958*; Wentworth and others, 1971; *Yerkes, 1973*). In contrast, the present ground-water basin drains southward to the southern margin of the San Fernando basin (*Johnson and Duke, 1973*). To the east, the San Gabriel Valley is separated from the San Fernando Valley by the Verdugo Hills and is bounded on the north by the Sierra Madre fault (Fig. 43). Still farther east, the Ontario Valley is bounded on the north by the Cucamonga fault and on the northeast by the San Jacinto fault (Fig. 9).

Los Angeles basin

The northwest grain of the Los Angeles basin is parallel to that of the Peninsular Ranges to the east and southeast and the California Continental Borderland to the west, but its northern border with the Santa Monica Mountains is a fold belt with a Transverse Range trend (Fig. 9; *McCulloh, 1960*; Yerkes and others, 1965). The central basin, containing a great thickness of

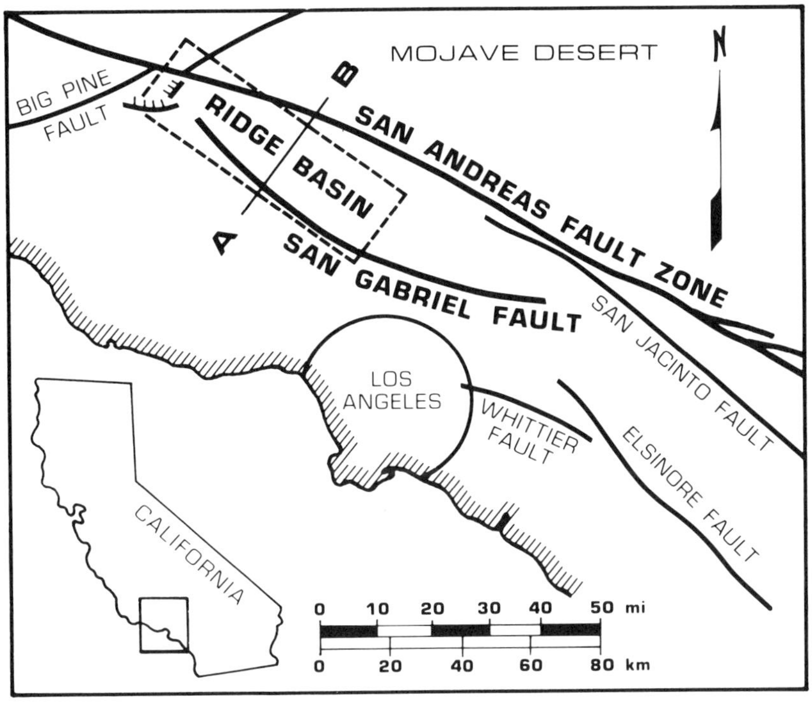

A

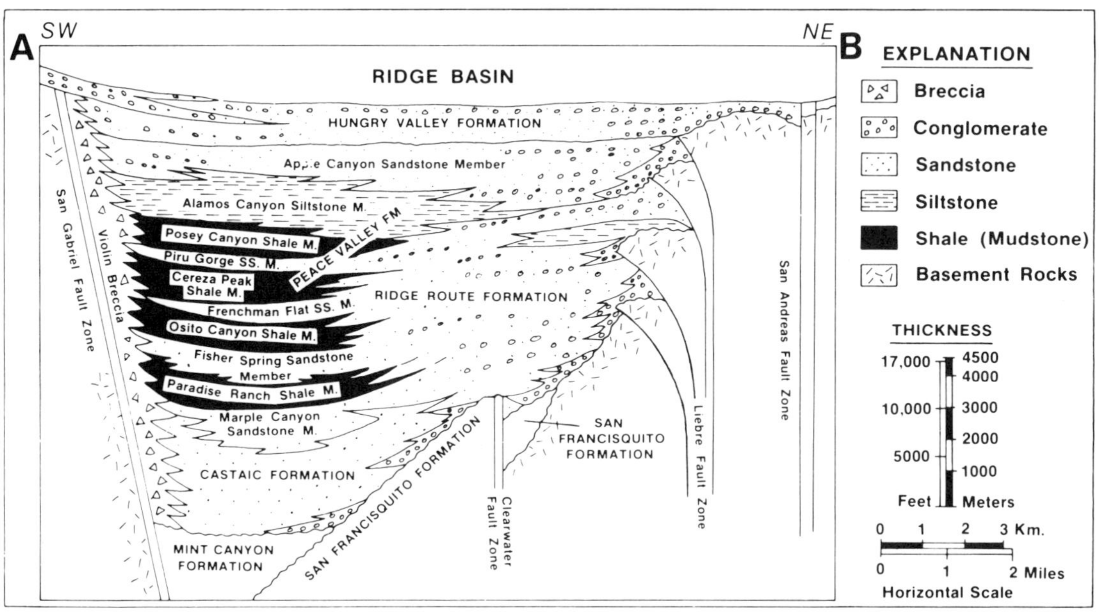

B

Figure 42. Map and section of Ridge basin, southern California, showing facies of the Castaic Formation and the Ridge Basin Group. From Crowell and Link (1982). A, Location of Ridge basin showing line of cross section, A–B; B, Diagramamatic cross section showing major stratigraphic and structural relations.

post-Miocene strata, is flanked on the northeast by the Whittier fault and on the southwest by the Newport-Inglewood fault zone (Figs. 43, 44). To the southeast, the basin floor rises gently to the Anaheim Nose and the Santa Ana Mountains, part of the Peninsular Ranges, underlain by the Southern California batholith and associated eugeoclinal wallrocks and by a west-thickening sequence of Late Cretaceous to early Miocene age (*Schoellhamer and others, 1981*). A similar sequence underlies the San Joaquin Hills (*Vedder and others, 1957*), Puente Hills (Durham and Yerkes, 1964; Yerkes, 1972), and the eastern Santa Monica Mountains (Fig. 43; *Hoots, 1931*). In contrast, the middle Miocene in and west of the Newport-Inglewood fault zone is underlain by blueschist-facies rocks of the Catalina Schist and associated gabbroic rocks (*Schoellhamer and Woodford, 1951*; *Woodford, 1960*; Yeats, 1973), an assemblage similar to that of the inner California Continental Borderland.

The nonmarine Sespe Formation of Oligocene age and the overlying and partly interbedded Vaqueros Formation show no tendency to thicken toward the Los Angeles basin. The overlying middle Miocene Topanga Formation is largely shallow-marine and is interbedded with volcanic rocks and cut by dikes and sills, also of middle Miocene age (Woodring and others, 1946; *Vedder and others, 1957*; Yerkes and others, 1965). This sequence also shows no tendency to thicken toward the basin, and it is locally very thick in adjacent areas that are now highlands, indicating that the Los Angeles basin formed later. The Peninsular Ranges and the Catalina Schist must have been juxtaposed by late early Miocene time because the San Onofre Breccia, bearing clasts of Catalina Schist, overlies Vaqueros and older strata along the western edge of the Peninsular Ranges (Woodford, 1925; *Stuart, 1976,* 1979). The middle Miocene was characterized by crustal extension (Campbell and Yerkes, 1976), and during this time, the

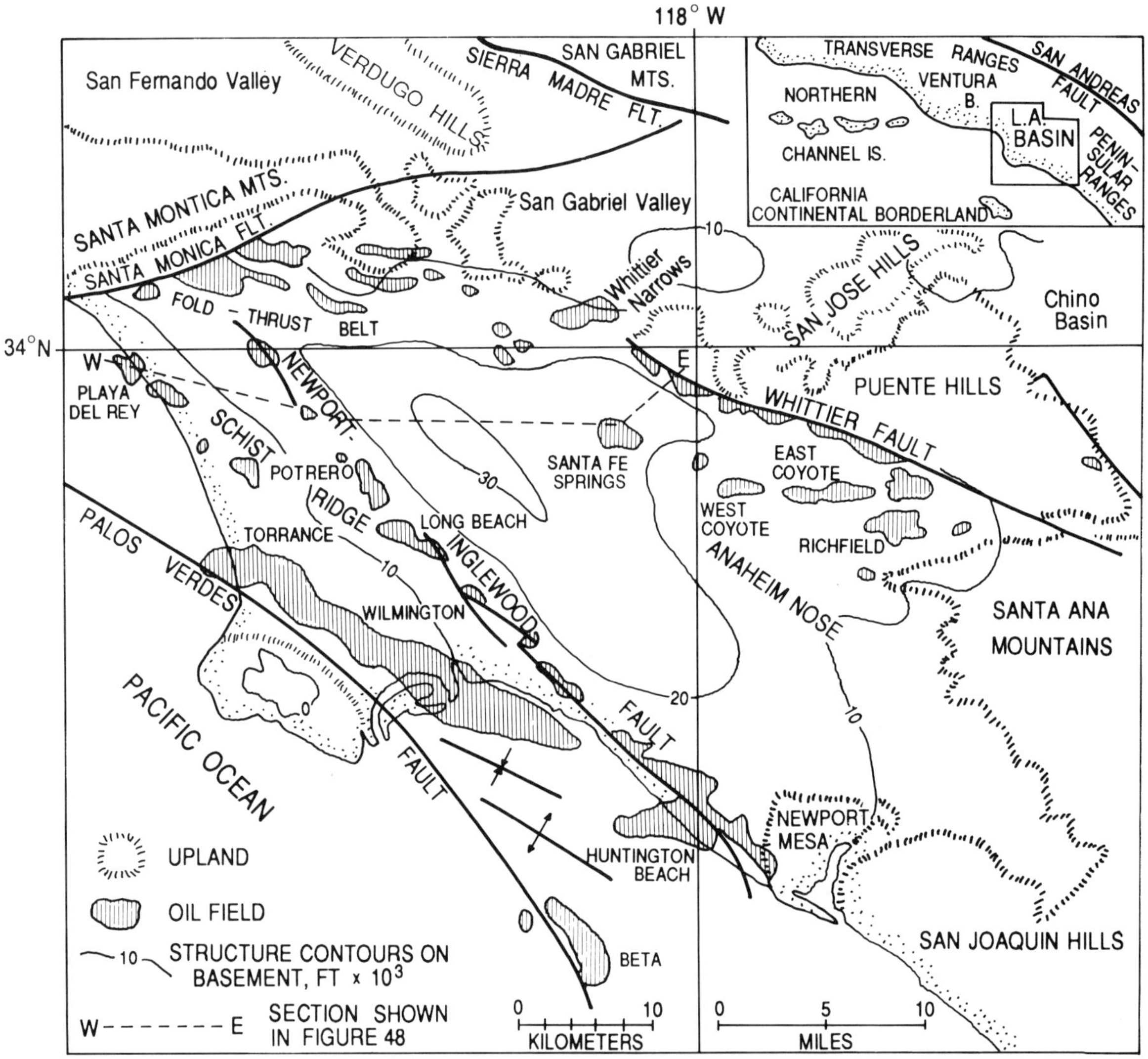

Figure 43. Structure contours on the basement of the Los Angeles basin (contour interval, 10,000 ft) showing locations of cross section of Figure 44. Map shows major faults (heavy lines) and oil fields (shaded pattern). Modified after Yeats and Beall (1991).

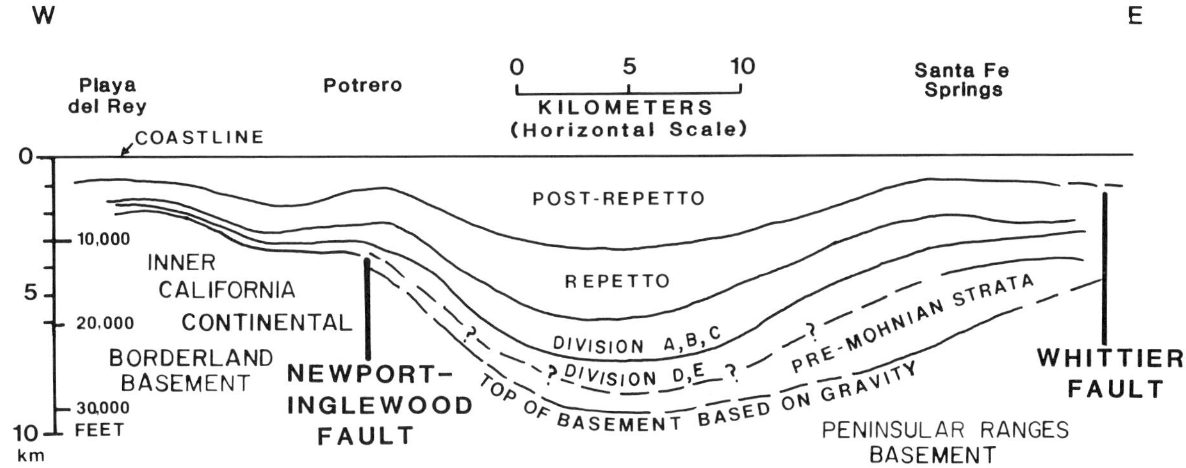

Figure 44. Restored cross section of the Los Angeles basin prior to late Pleistocene deformation. Base of Pliocene is the contact between "Repetto" and "Division A,B,C." Base of "Division E" is middle Miocene. Modified after Yeats and Beall (1991).

Santa Monica Mountains may have moved as much as 90 km westward with respect to the Santa Ana Mountains along a left lateral fault zone at the northern margin of the Los Angeles basin (Yeats, 1968; Campbell and Yerkes, 1976).

West of the Newport-Inglewood fault, the oldest strata to be preserved atop schist basement consist of a middle Miocene basal conglomerate deposited on a surface of moderate relief. In the Palos Verdes Hills (Woodring and others, 1946) and along the coast to the north, the Miocene consists of cherty, phosphatic, and diatomaceous shale and mudstone lithologically similar to the Monterey Shale elsewhere, but throughout most of the basin, fine-grained strata are interbedded with sandstone turbidites and are called the Puente Formation. The Puente Formation is thicker in the Puente Hills than in known subsurface sequences in the Los Angeles basin, suggesting that the basin had not yet achieved its present form (Yeats and Beall, 1991).

The Pliocene strata are conformable with the Puente Formation except on basin margins, but in contrast to the Puente Formation, these strata thicken toward the center of the basin (Fig. 44; *Conrey, 1967*). Microfossils are assigned to several foraminiferal stages of the late Miocene and Pliocene by *Natland and Rothwell* (*1954*), and in most of the basin they are similar to assemblages found today in deep water. Shallow-water equivalents are found on the margins of the basin. The Pleistocene, best known around the flanks of the Palos Verdes Hills, includes the San Pedro Formation, composed mainly of shallow-marine sand and gravel but including its Timms Point Silt and Lomita Marl Members. These units also thicken toward the center of the basin (Yerkes and others, 1965; *Johnson and Duke, 1973*).

Many of the oil fields of the Los Angeles basin occur on anticlines and upturned stratigraphic traps aligned over northwest-trending basement faults with a strong component of right slip (Fig. 43; *Harding, 1974*). The Whittier fault is the northwestern extension of the right-slip Elsinore fault, and it also has a reverse-slip component with the Puente Hills upthrown with respect to the Coyote and Richfield oil fields to the south (Durham and Yerkes, 1964; Yerkes, 1972; *Harding, 1974*; *Truex, 1975*). The Whittier fault also marks a hinge line between thick strata of the Puente Formation to the north in the Puente Hills and thin Puente strata to the south in the Coyote and Richfield oil fields. Farther west, the Newport-Inglewood fault zone is expressed at the surface mainly as aligned anticlines housing oil fields (*Harding, 1973*; Yeats, 1973; Barrows, 1974); total post-Miocene right slip is small. Farther west, the Wilmington and Torrance oil fields occur on broad, northwest-trending folds that began to form in the late Miocene (*Truex, 1974*). Oilfields along the northern margin of the Los Angeles basin (*Lamar, 1970*) are related to the fold-and-thrust belt formed due to convergence between the Santa Monica Mountains and the main Los Angeles basin (Davis and others, 1989).

The Los Angeles basin is complex in detail but simple overall, and this has led to models of origin based on thermal isostatic subsidence (Turcotte and McAdoo, 1979) and of lithospheric thinning (Mayer, 1987). The basin is one of the most prolific oil and gas producers in the world for several reasons (Yeats and Beall, 1991): (1) it is underlain by highly organic strata in the Monterey and Puente Formations that were buried to depths where temperatures were sufficient to generate hydrocarbons, (2) organic strata are interbedded with layers of turbidite sandstone that provide avenues of migration from the generating source region, (3) strike-slip faults in basement rocks produce anticlines of sufficient structural relief to trap hydrocarbons but not so much structural relief that the traps were breached by erosion, and (4) turbidite sand edges formed stratigraphic traps at positions far enough downdip that they were not subsequently breached by erosion. In some cases, these stratigraphic traps were

further concentrated by anticlinal folding over wrench structures. West of the Los Angeles basin in the California Continental Borderland, there is less subsidence, hence less thermal generation of petroleum, and less sand to form reservoirs. East of the Los Angeles basin, the organic source beds are not present.

Western margin of Peninsular Ranges

Along the coast south of the Los Angeles basin, Miocene and younger strata rest unconformably on lower Miocene, Eocene, and Upper Cretaceous strata and on basement. The San Onofre Breccia of late middle Miocene age rests on progressively older rocks from northwest (Topanga in the San Joaquin Hills, Fig. 43) to southeast (Eocene at Camp Pendleton, east of the Christianitos fault, Fig. 9; *Stuart, 1976,* 1979) and appears in northernmost Baja California underlying middle Miocene sedimentary rocks and basalts of the Rosarito Beach Formation (Minch, 1967; *Minch and others, 1970*). In San Diego east of the La Nación fault, Eocene strata are overlain by sandstone, grit, and mudstone correlated to the Rosarito Beach Formation (*Artim and Pinckney, 1973*). These are overlain by the upper Pliocene and lower Pleistocene shallow-marine San Diego Formation (Kennedy, 1975; Kennedy and Peterson, 1975; *Demere, 1982*), which is itself overlain by Pleistocene marine and nonmarine terrace deposits. The coastal plain is cut by north-northwest–trending faults that are in part right slip, including the Rose Canyon fault that may be the southern extension of the Newport-Inglewood fault of the Los Angeles basin (Fig. 9; Kennedy, 1975). Several of these faults cut upper Quaternary sediments (*Artim and Pinckney, 1973*; Kennedy, 1975).

TECTONICS OF THE NORTHWESTERN GULF OF CALIFORNIA REGION

The Salton trough forms the northern end of the Gulf of California province, a series of oblique spreading centers across which the Peninsular Ranges—part of the Pacific Plate—have been moving northwestward from the rest of North America for the past 4 m.y. (Moore and Buffington, 1968; Larson and others, 1968; *Elders and others, 1972*). The area is bounded on the northeast by the San Andreas fault, beyond which is a series of relatively low ranges, including the Little San Bernardino, Orocopia, and Chocolate Mountains (Fig. 45). On the southwest are the Peninsular Ranges, cut by the San Jacinto and Elsinore right-lateral faults, which enter the basin obliquely. On the north, the San Andreas fault turns west and terminates the basin on the north at San Gorgonio Pass (Figs. 9, 45). On the southeast, the basin is separated topographically from the Gulf of California only by the delta of the Colorado River (Crowell and Sylvester, 1979). The geology of areas flanking the basin is summarized by Dibblee (1954, *1984*), Sharp (1967), *Bartholomew* (*1970*), Babcock (1974), and Sylvester and Smith (1976), who list references to more detailed studies.

Strata as young as the marine Eocene Maniobra Formation (*Crowell and Susuki, 1959*) and nonmarine Oligocene Diligencia Formation (*Crowell, 1975a*), deposited prior to the initiation of the Gulf of California province and the San Andreas fault, have been described in an earlier section. Younger strata comprise an older sequence of Miocene and Pliocene age, now found in the low ranges flanking the Salton trough, and the basin fill itself, largely of Quaternary age (Crowell and Baca, 1979). Strata on the southwest include locally derived nonmarine conglomerate and sandstone (Anza and Split Mountain Formations; *Merriam and Bandy, 1965*) interbedded with mafic volcanic rocks dated as 18 to 13 Ma (Gastil and others, 1979; Crowell and Sylvester, 1979). These grade northeastward toward the basin to marine sandstone, shale, mudstone, and limestone of the Pliocene Imperial Formation (*Woodard, 1974*; Crowell and Baca, 1979; *Kerr, 1984*). Overlying Pleistocene strata show the same facies change from the Canebrake and Ocotillo Conglomerates (Dibblee, 1954) northeastward to the Palm Spring Formation, bearing local marine intercalations, to lake beds of the Borrego and Brawley Formations (Dibblee, 1954). On the north and northeast sides of the basin, the San Gorgonio Pass area, Indio Hills, and Mecca Hills contain a similar stratigraphy, with coarse-grained locally-derived sediments grading basinward to finer-grained strata (Dibblee, 1954). Marine strata mapped as the Imperial Formation in the Salton trough extend northward up the Colorado River, where they are mapped as the Bouse Formation (see inset, Fig. 45; Metzger, 1968; Smith, 1970; Eberly and Stanley, 1978; *Buising, 1990*). Marine deposits may extend still farther north near Lake Mead as the upper Miocene Hualapai Limestone Member of the Muddy Creek Formation (Blair, 1978), although Lucchitta (1979) challenges that view with evidence that the Hualapai accumulated in a series of closed interior basins.

Sediments within the basin are difficult to correlate with those of its flanks because of metamorphism, but they are probably entirely Pliocene and Quaternary in age, deposited in the newly formed rift as it widened (*Muffler and White, 1969*). Quaternary volcanic domes at the southern end of the Salton Sea (Salton Buttes, Fig. 45) are dated as 16,000 years old (*Muffler and White, 1969*; Robinson and others, 1976). Holocene sediments of the Salton basin include sediments of Lake Cahuilla, deltaic deposits of the Colorado River, fan gravels from adjacent ranges, and sand dunes (*Van de Kamp, 1973*).

The tectonic history of the Salton trough may be divided into two stages, comparable to the history of the Gulf of California: a Miocene proto-Gulf stage (*Karig and Jensky, 1972*; D. G. Moore, 1973) that appears to be part of the basin-range extension discussed later, and northwest-southeast rifting and transform faulting accompanying sea-floor spreading in the Gulf of California for the past 4 m.y. (*Elders and others, 1972*; Sharp, 1982; Lachenbruch and others, 1985). Whereas the spreading centers in the Gulf of California expose young oceanic crust, the Salton Buttes and Cerro Prieto spreading centers in the Salton trough only attenuate the continental crust. Densities in the lower crust, however, are higher beneath the trough than beneath adjacent ranges (*Fuis and others, 1982*). The San Andreas fault cannot be

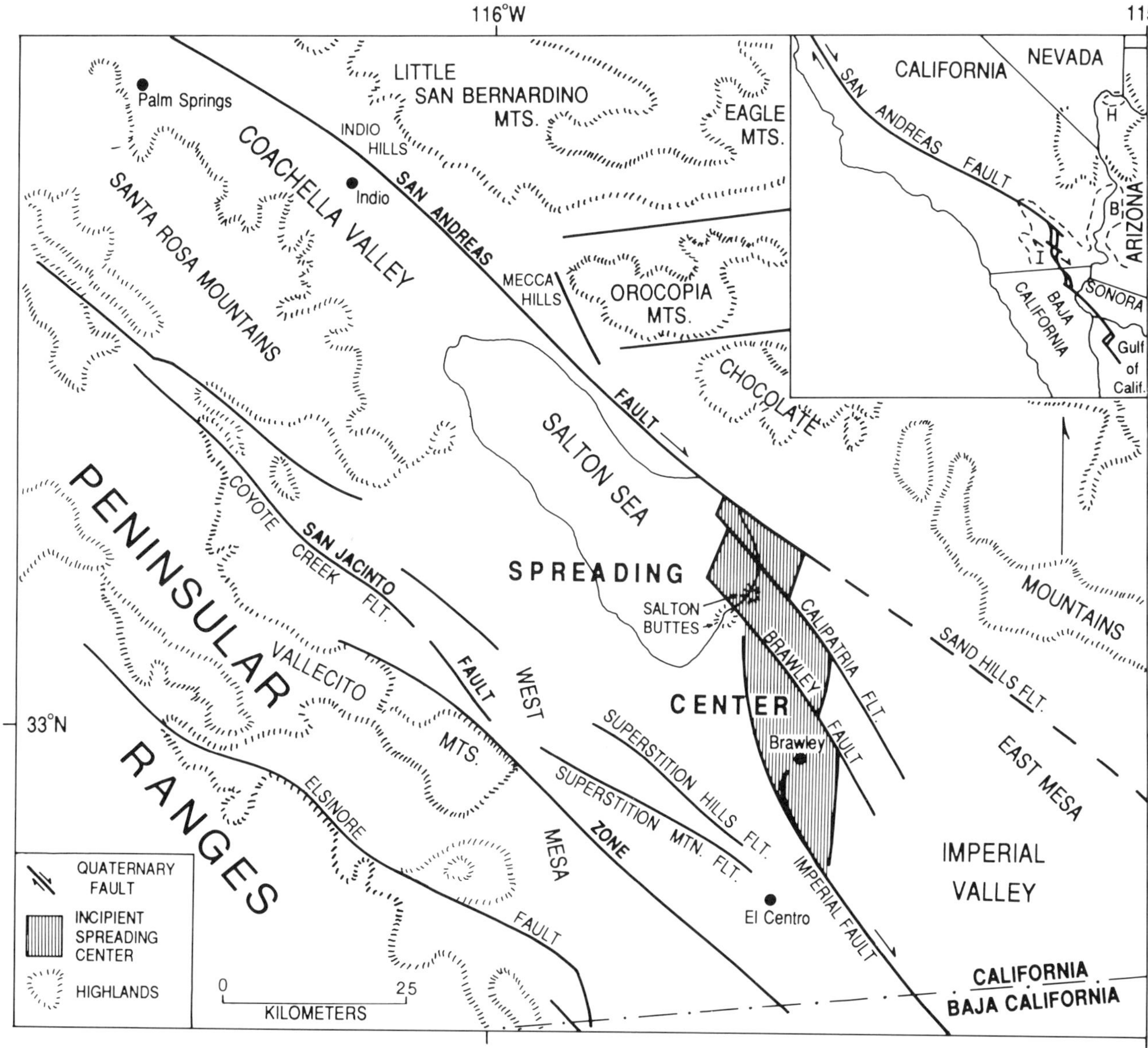

Figure 45. Tectonic map of the Salton basin. Incipient spreading center (shaded pattern) is marked by high heat flow, high seismicity, and transition between oceanic and continental crust. Inset shows Pacific–North American plate boundary extending southeastward into the Gulf of California. Dashed lines in inset show areas of exposure of the Imperial Formation (I), Bouse Formation (B), and Hualapai Limestone (H).

clearly mapped southeast of the Salton Sea, and the San Jacinto fault zone is succeeded to the southeast by several right-lateral faults within the Salton trough, including the Imperial fault, which produced major earthquakes in 1940 and 1979 (Sharp, 1982). Clockwise rotation accompanied strike-slip faulting in the hills west of the basin (*Johnson and others, 1983*). The heat flow accompanying spreading is high enough to produce hydrothermal circulation and high-temperature geothermal reservoirs, and Pleistocene sediments are being subjected to medium-grade metamorphism (*Muffler and White, 1969*).

MAGMATISM OF LATE CENOZOIC ARCS AND THE COASTAL TRANSFORM SYSTEM

The tectonic transition from a largely stable early Tertiary Cordilleran interior with extension primarily localized in distinct belts to an unstable late Cenozoic region of widely distributed extension was accompanied by regional magmatic changes. Volcanism in the early Western Cascades arc had been active no farther south than northwestern Nevada. In the central Sierra-Wasatch region (the present-day Great Basin region), where ear-

lier magmatic belts had propagated southward, magmatic activity declined in mid-Miocene time, between about 20 and 17 Ma (McKee and others, 1970). During that time, a predominantly intermediate calc-alkalic belt parallel to the continental margin became nearly continuous through the western part of the Cordilleran region as far south as southern Nevada and adjacent California; less abundant, generally more mafic and alkalic olivine-bearing lavas or bimodal mafic and silicic suites erupted locally farther inland. Simultaneously, a calc-alkalic magmatic belt, now west of the San Andreas fault, became active farther south in northern Mexico, originally separated from the Nevada-California belt by a gap of about 400 km. These two belts define a continental-margin magmatic arc with a distinct gap between its northern and southern segments. With time the gap widened episodically, but the arc remained active on both sides of the widening gap. Later magmatism farther west, near the northern end of the continental-margin transform-fault system, produced discrete isolated volcanic fields in a north-propagating pattern.

THE MID- TO LATE MIOCENE MAGMATIC ARC

In mid-Miocene time as before, continental-margin magmatism was active in and near the Cascade Range (Fig. 46). A mid-Miocene magmatic belt was continuous northward into Canada, where it is preserved east of the modern arc as igneous rocks of the Pemberton belt, mainly between 22 and 8 Ma (*Berman and Armstrong, 1980*). In Washington, mid-Miocene andesitic magmatism occurred near the modern Cascades axis and as far as 30 km east (*Tabor and others, 1980*). Large composite granodioritic plutons are notable, particularly rocks of 18- to 14-Ma age in the Tatoosh pluton beneath Mount Rainier and of 20- to 17-Ma age in the Snoqualmie batholith to the north. Mid- to later Miocene volcanism in the Washington Cascades is recorded in dacitic to andesitic volcaniclastic rocks of the Ellensburg Formation that accumulated episodically on the Columbia Plateau between eruptions of 17- to 14-Ma flood basalts farther east (*Waters, 1955a*; Swanson and others, 1979a; *Tabor and others, 1982*) and continued until at least 10 Ma (*Evernden and James, 1964*). Correlative similar continental sediments occur in Washington west of the Cascades (Mullineaux and others, 1959; *Gard, 1968*).

Magmatism in northern to central Oregon has been more or less continuous from initiation of the early Western Cascades arc through the youngest High Cascades activity. Traditional stratigraphic divisions in the Western Cascades break at the lowest flows of the Columbia River Basalt Group, about 17 Ma, although those lavas do not extend the length of the range. Western Cascades volcanic rocks coeval with or younger than the Columbia River Basalt Group are included by Peck and others (1964) in the Sardine Formation and associated intrusive rocks, but mapped units also include the Rhododendron (e.g., Trimble, 1963), Outerson (Thayer, 1937), and Deschutes Formations (*Taylor, 1973*) and others. Lavas and related volcanic sedimentary rocks were voluminous about 18 to 8 Ma in an episode that Priest and others (1983) designate as the late Western Cascades. The rocks are mainly calc-alkalic pyroxene andesites and basalts. Activity in this sector of the Cascades declined after about 13 to 14 Ma but continued until at least about 8 to 10 Ma (*McBirney and others, 1974*; McBirney and White, 1982).

In southern Oregon and northern California, magmatism ended in the Western Cascades about 17 Ma (*Smith and others, 1982*; *Fiebelkorn and others, 1983*). The mid-Miocene arc probably lay farther east, where it is now largely buried by younger rocks, represented in part by rhyodacitic to andesitic tuffs and sediments in the Abert Lake–Goose Lake area (Walker, 1977). The arc did not extend as far east as the Warner Range, however (Fig. 46), as bimodal basalts and rhyolites erupted there between 17 and 18 Ma (Duffield and McKee, 1986).

Predominantly andesitic to dacitic calc-alkalic magmas erupted in a linear band through east-central California and northwestern Nevada in mid-Miocene time (Fig. 46). In northern California this episode produced the andesitic Bonta and Penman Formations of the northern Sierra Nevada (*Durrell, 1959a,* 1966). Farther south in the Sierra, middle and upper Miocene andesites and related rocks are commonly breccias of both eruptive and debris-flow origin in the Mehrten Formation (Curtis, 1954). Slemmons (1966) demonstrated that the source volcanoes, most of them older but some younger than 9-Ma latites, erupted mainly near the present range crest and farther east, but a few vents are preserved as dacitic to rhyodacitic domes in the western foothills (*Rose, 1959*). Volcanic debris flows repeatedly filled the Sierran drainages and mantled the surface before the range was uplifted and separated from the Great Basin (Bateman and Wahrhaftig, 1966). In central western Nevada, the 17- to 12-Ma Alta and Kate Peak Formations and correlative units such as the Chloropagus Formation are mainly lavas and breccias of pyroxene andesite and hornblende-pyroxene andesite, minor basalt, and dacite from a chain of stratovolcanoes (Thompson and White, 1964; *Rose, 1969*; Bonham, 1969; *Silberman and McKee, 1972*).

Similar predominantly andesitic rocks extend farther southeastward in western Nevada and adjacent California to about latitude 38°N (Fig. 46; Stewart and Carlson, 1976; Stewart, 1980). Andesites and basalts between Yerington and Tonopah are 21 to 15 Ma (Gilbert and Reynolds, 1973; *Marvin and Mehnert, 1977*; *Kleinhampl and Ziony, 1985*). Near Mono Lake, andesites postdating 22-Ma rhyolitic ash flows were most voluminous near 13 to 8 Ma (Gilbert and others, 1968; *Chesterman, 1968*; *Silberman and others, 1972*; *Kleinhampl and others, 1975*). Andesites to dacites in the Silver Peak Range and vicinity are probably between about 16 and 11 Ma (Robinson and others, 1968; *Crowder and others, 1972*; *Albers and Stewart, 1972*). Calc-alkalic andesites to rhyodacites formed stratovolcanoes in the mining districts of Tonopah and Goldfield about 20 Ma (Bonham and Garside, 1979) and eastward to south-central Nevada between about 21 and 15 Ma (Anderson and Ekren, 1968; Ekren and others, 1971). A single 13-Ma andesite-dacite center to the south (*Poole and others, 1966*; *Ponce, 1984*) was

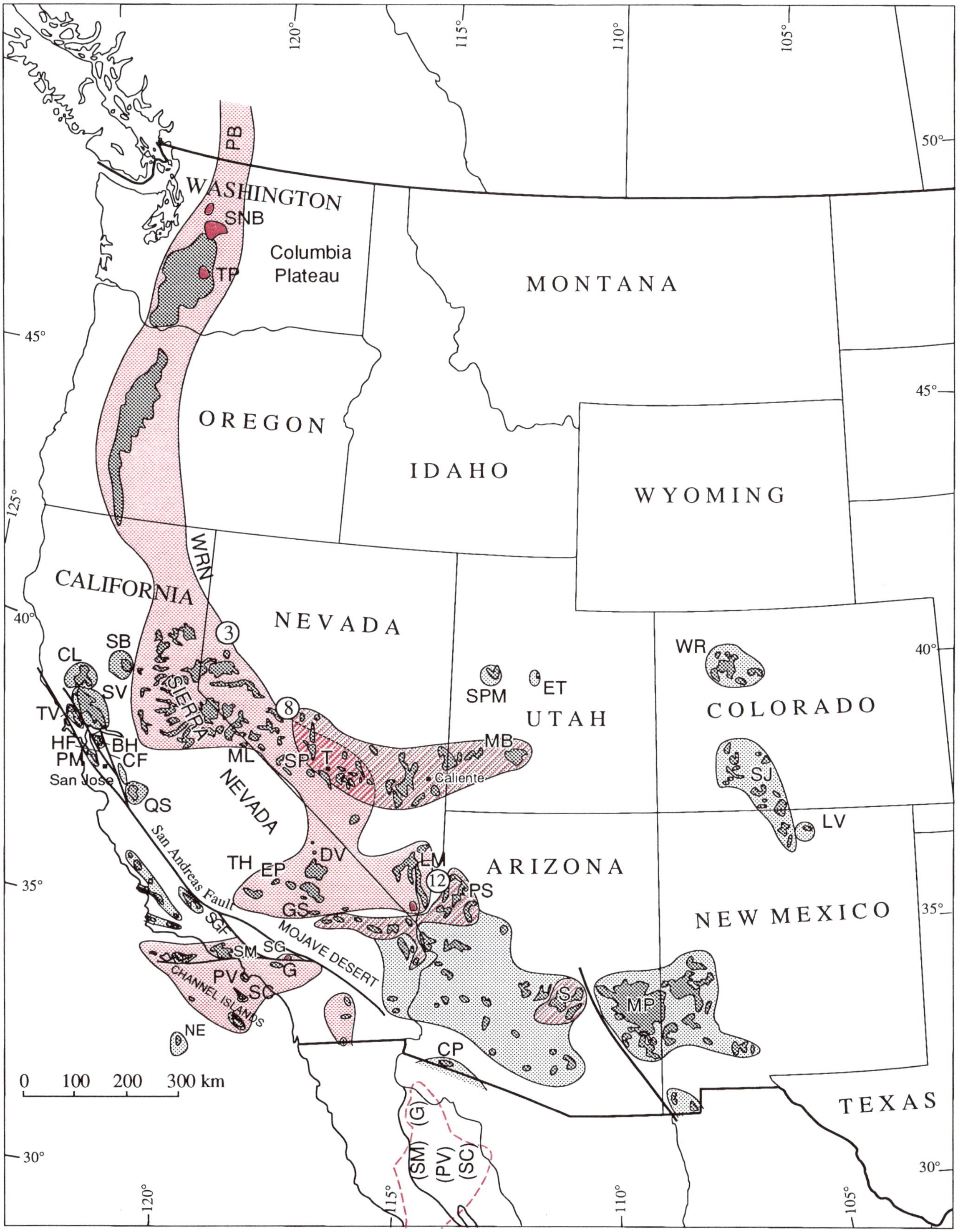
WASHINGTON
OREGON
IDAHO
MONTANA
WYOMING
CALIFORNIA
NEVADA
UTAH
COLORADO
ARIZONA
NEW MEXICO
TEXAS
Columbia Plateau
PB
SNB
TP
WRN
SIERRA NEVADA
San Andreas Fault
MOJAVE DESERT
CHANNEL ISLANDS
CL
SB
SV
TV
HF
BH
PM
CF
San Jose
QS
ML
SP
T
MB
Caliente
SPM
ET
WR
SJ
LV
TH
EP
DV
LM
PS
GS
SGF
SM
SG
G
PV
SC
NE
CP
S
MP
(SM)
(G)
(PV)
(SC)
3
8
12
0 100 200 300 km
120°
115°
110°
105°
125°
50°
45°
40°
35°
30°

contemporaneous with nearby, more voluminous bimodal rhyolite and basalt.

Extensive areas of calc-alkalic andesite and dacite in eastern California, probably mainly middle Miocene in age, have been little studied but are prominent from Death Valley through the central Mojave Desert (Fig. 46; Luedke and Smith, 1984). They include 14- to 12-Ma hypabyssal granitic stocks west of Death Valley (*McDowell, 1964*; *Stern and others, 1966*), 14- to 12-Ma volcanic rocks in and near the Owlshead Mountains (*Wagner, 1988*), and undated Miocene volcanic rocks in the Goldstone Lake area (*Jennings and others, 1962*). In the Barstow basin, andesite and dacite of about 18 Ma age underlie 17- to 14-Ma basalt and associated volcanics to the north and west (Woodburne and others, 1982; *Burke and others, 1982*). In the Lava Mountains, just south of the Garlock fault, andesite and dacite both underlie and overlie fossiliferous sedimentary rocks (*Smith, 1964*) now regarded as late Miocene. In the El Paso Mountains, at the northwestern margin of the Mojave block, the basal Ricardo Formation contains interlayered basalt, andesite, and dacite of 18- to 17-Ma age beneath 17- to 15-Ma basalts (*Dibblee, 1952*; *Cox and Diggles, 1986*). A diverse but mainly andesitic sequence interfingers with middle Miocene sedimentary rocks in the southern Sierra Nevada (Dibblee, 1967).

From near Lake Mead to the southernmost point of Nevada and adjacent areas, calc-alkalic predominantly intermediate magmatism continued from early through mid-Miocene time (Anderson, 1973, *1978*; Anderson and others, 1972), contrasting with areas farther south, where volcanism shifted in the mid-Miocene to mainly basaltic or bimodal compositions. The older Patsy Mine Volcanics are overlain by a voluminous 15-Ma rhyodacitic welded tuff; the overlying 15- to 12-Ma volcanics and associated plutons, emplaced during extreme local extension on low-angle faults (R. E. Anderson, 1971), include basalt to rhyodacite but are mainly intermediate. The 13-Ma Hamblin-Cleopatra stratovolcano north of Lake Mead, is andesitic (Anderson, 1973; Bohannon, 1979). Basaltic magmatism associated with high-angle block faulting near Lake Mead is younger than 10 to 12 Ma (R. E. Anderson, 1971; Bohannon, 1984).

Farther south in the lower Colorado River trough, the earlier calc-alkalic volcanism was succeeded by mid-Miocene, predominantly basaltic to bimodal rhyolite-basalt magmatism associated with high-angle basin-range faulting (Fig. 46; Crowe, 1978; Suneson and Lucchitta, 1983). A gap in the mid-Miocene andesitic arc extends from the central Mojave Desert and southern Nevada to the San Andreas fault; to restore this gap to its mid-Miocene context, the distributions of volcanic sections west of the San Andreas fault must be considered in relation to their original positions before late Cenozoic offset (Fig. 46; Plates 7E, 7F).

Calc-alkalic, predominantly intermediate igneous rocks of about 20 to 12 Ma were widespread in coastal southern California from the western Transverse Ranges, through the California Continental Borderland, into Mexico (Fig. 46). Contrasting mid-Miocene bimodal basalt and rhyolite of the southern Coast Ranges and northwesternmost Transverse Ranges, described later, define the northern limit of this band of calc-alkalic, predominantly intermediate magmatism. In the western and central Santa Monica Mountains, the 16- to 13-Ma Conejo Volcanics (Figs. 40, 46), basalt to rhyodacite as much as 4 km thick, erupted during tectonic extension (Yerkes and Campbell, 1979; Weigand, 1982). The 15-Ma Zuma Volcanics, south of the Malibu Coast fault (Yerkes and Campbell, 1979), is a more silicic facies, equivalent to the Conejo (Weigand, 1982). Lithologically similar volcanic rocks extend westward beyond the coast to the northern Channel Islands (*Crowe and others, 1976*; *Hurst, 1982*). The low-K_2O rocks have been described as tholeiitic (Hurst, 1979, *1982*; Hurst and others, 1982) but show little or no iron-enrichment and may best be regarded as a low-K_2O calc-alkalic series (Weigand, 1982; *Johnson and O'Neil, 1984*). Farther south in the Continental Borderland, subaerial andesite to rhyodacite on San Clemente are 16 to 13 Ma (Turner, 1970; *Merifield and others, 1971*). On Santa Catalina, a quartz diorite stock and related dikes are 19 Ma in age, and subaerial to shallow-marine 15- to 13-Ma basalt to rhyodacite (but predominantly andesite) erupted during subsidence (*Vedder and others, 1979*; Hurst and others, 1982). In the Palos Verdes Hills, at the west end of the Los Angeles basin, 15-Ma andesite and basalt are invasive into middle Miocene marine sedimentary rocks (Woodring and others, 1946; Turner, 1970; *Schleicher, 1974*).

As to the west, abundant mid-Miocene calc-alkalic, predominantly andesitic volcanic rocks in the central and eastern Los

Figure 46. Map of mid-Miocene magmatic belts of the western Cordilleran region (21 to 17 Ma in most of the region; 26 to 17 Ma east of heavy line in southeastern Arizona; arc suite includes some rocks as young as 12 Ma). Dark-patterned areas indicate approximate outcrop distributions; pink, probable volcanic arc; dark gray, andesite-rhyolite suite; red, intrusions; light gray, basaltic or mainly basaltic bimodal volcanic suite; gray with red ruling, mainly rhyolitic bimodal volcanic suite. Dashed red outline in northwestern Baja California indicates approximate location of mid-Miocene arc terranes before later Cenozoic rotation and strike-slip displacement into southern California; location symbols in parentheses in this area refer to present locations indicated in Southern California. Other later Cenozoic displacements and rotations not restored. Circled numbers in western Nevada and Arizona indicate approximate time of the end of arc volcanism in the vicinity (in Ma). B, Barstow; BH, Berkeley Hills; CF, Calaveras fault; CL, Clear Lake volcanic field; CP, Cabeza Prieta; CR, Caliente Range; DV, Death Valley; EP, El Paso Mountains; ET, East Tintic Mountains; G, Glendora Volcanics; GS, Goldstone Lake; HF, Hayward fault; L, Lava Mountains; LM, Lake Mead; LV, Latir volcanic field; MB, Mount Belknap Volcanics; MCF, Malibu Coast fault; MP, Mogollon Plateau; NE, Northeast Bank; PB, Pemberton volcanic belt; PM, Page Mill Basalt; PS, Peach Spring Tuff; PV, Palos Verdes Hills; QS, Quien Sabe Volcanics; S, Superstition-Superior volcanic field; SB, Sutter Buttes; SC, Santa Catalina; SG, San Gabriel Mountains; SGF, San Gabriel fault; SJ, San Juan Mountains; SM, Santa Monica Mountains; SNB, Snoqualmie batholith; SP, Silver Peak Range; SPM, Spor Mountain; SV, Sonoma Volcanics; T, Tonopah; TH, Tehachapi; TP, Tatoosh pluton; TV, Tolay Volcanics; WR, White River.

Angeles basin accumulated during subsidence (*Eaton, 1958*; *Higgins, 1976*). In the east, the undated Glendora Volcanics (Fig. 46) are mainly andesite interlayered in coarse-clastic shallow-water sedimentary rocks of middle Miocene age (*Shelton, 1955*). The probably correlative but more femic El Modeno Volcanics to the south (*Yerkes, 1957*; Weigand, 1982) are dated as 14 Ma (Turner, 1970). In the southeastern San Gabriel Mountains, shallow-intrusive granodiorite and dacite are 26 to 17 Ma in age (Hsü and others, 1963).

Basalt and andesite of 20- to 15-Ma age near and beyond the Mexican border (Hawkins, 1970; Gastil and others, 1979) are calc-alkalic but more K_2O-rich than rocks of the Transverse Ranges, Continental Borderland, and Los Angeles basin, probably reflecting their position on crystalline basement and more extensive interactions with continental crust. Weigand (1982) noted both subalkalic and alkalic compositions. Gastil and others (1979) describe 22- to 8-Ma calc-alkalic basalts to rhyolites on both sides of the Gulf of California, among which andesites are predominant; a chain of 24- to 12-Ma andesitic stratovolcanoes was more or less continuous along the east coast of Baja California (*Sawlan and Smith, 1984*).

Since mid-Miocene time, the northern Channel Islands and Santa Monica Mountains have been rotated 70 to 130° clockwise, laterally disrupted, and displaced northward along the San Andreas and related faults (Fig. 47; Plates 7E, 7F), implying formation as a crustal block parallel to the coast opposite the Los Angeles basin or even farther southeast (*Howell and others, 1974*; Kamerling and Luyendyk, 1979, *1985*; Luyendyk and others, 1985). When its position is restored, the mid-Miocene coastal igneous province is seen to be part of a linear calc-alkalic, predominantly intermediate belt that extended southward from the present latitude of northern Mexico across an extending and subsiding oceanic or transitional crust onto continental crust. Thus, a gap in a nearly continuous continental-margin magmatic arc spanned about 300 to 400 km southward from the central Mojave Desert and southern Nevada to northwestern Mexico; tectonic extension accompanied calc-alkalic magmatism at both ends of the gap.

In summary, a major change in magmatic patterns occurred in mid-Miocene time. Before about 21 Ma the mid-Tertiary Western Cascades arc had ended in northwestern Nevada; the Cordilleran interior had been characterized by zones of voluminous, andesite-rhyolite magmatism and by distinct belts of crustal extension. Starting about 21 Ma and at least by 17 Ma, however, a band about 100 km wide of predominantly andesitic calc-alkalic centers spanned the western Cordillera nearly continuously from north of the Canadian border to the latitude of southernmost Nevada; following a gap of 300 to 400 km, the belt continued southward in Mexico. East of this continental-margin arc, basalts and bimodal rhyolites and basalts erupted during widespread extension; extension appears to have overlapped 21- to 12-Ma calc-alkalic, predominantly intermediate centers in places, notably near the ends of the magmatic gap—in the Lake Mead region and coastal southern California.

Some interpretive syntheses (Lipman and others, 1972; Lipman, 1980) consider pre–mid-Miocene volcanism in the Cordilleran interior as a wide arc; other interpretations relate that volcanism to crustal spreading following Laramide contraction and crustal thickening (Coney and Harms, 1984; Wernicke and others, 1987; *Sonder and others, 1987*). Thus, the mid-Miocene andesitic belt south of northwestern Nevada represents either the establishment or rapid reorganization of a linear continental-margin arc. A magmatic gap in the mid-Miocene arc occupies a zone in which subduction had ceased after the Pacific-Farallon spreading center came in contact with the North American Plate between about 28 and 25 Ma (Atwater, 1970, *1989*; Stock and Molnar, 1988). Triple junctions at the ends of the subduction gap delimit the evolving San Andreas transform system (Dickinson and Snyder, 1979a). The magmatic gap has widened with time, its northern boundary roughly tracking northward propagation of the Mendocino triple junction (Christiansen and Lipman, 1972; *Snyder and others, 1976*) although probably in discrete jumps rather than as a smooth northward retreat (Fig. 46).

THE LATEST MIOCENE TO QUATERNARY HIGH CASCADES MAGMATIC ARC

Continental-margin magmatism is now represented by active systems in the Cascade Range, extending from northern California into southern British Columbia. Where older magmatic arcs of the Cascades are geographically and geomorphically distinct from this later activity, the latest Miocene to Quaternary volcanic zone is referred to as the High Cascades (Fig. 47). The distributions of volcanic vents in the Cascades since the end of the Miocene indicate that the arc is segmented above parts of the subducting plate that have different dips and varied seismicity (Guffanti and Weaver, 1988). At least some of the segments are controlled by major crustal structures (*Blakely and Jachens, 1990*).

In northern and central Washington, where magmatism lapsed between about 9 to 10 Ma and 4 to 6 Ma, the northern part of the High Cascades arc comprises individual andesitic to dacitic volcanoes and associated intrusive systems, spaced more or less regularly. The northernmost volcanic systems, from Mount Garabaldi in southern British Columbia (*Matthews, 1958*) to Glacier Peak, form a northwest-southeast line, perpendicular to present subduction of the Juan de Fuca Plate beneath North America (Fig. 47; *Riddihough, 1984*). From Glacier Peak to Mount St. Helens the trend changes at nearly right angles toward the southwest. Although the direction of subduction remains about the same as to the north, regional crustal strain in central Washington is more complex, including a zone of right-slip seismicity northwestward through Mount St. Helens (*Weaver and Smith, 1983*).

Virtually all upper Miocene and younger volcanic rocks in northern Washington are within a few long-lived systems, centered near the modern volcanoes and having diameters of 30 to 40 km. Mount Baker, a mainly late Pleistocene andesitic strato-

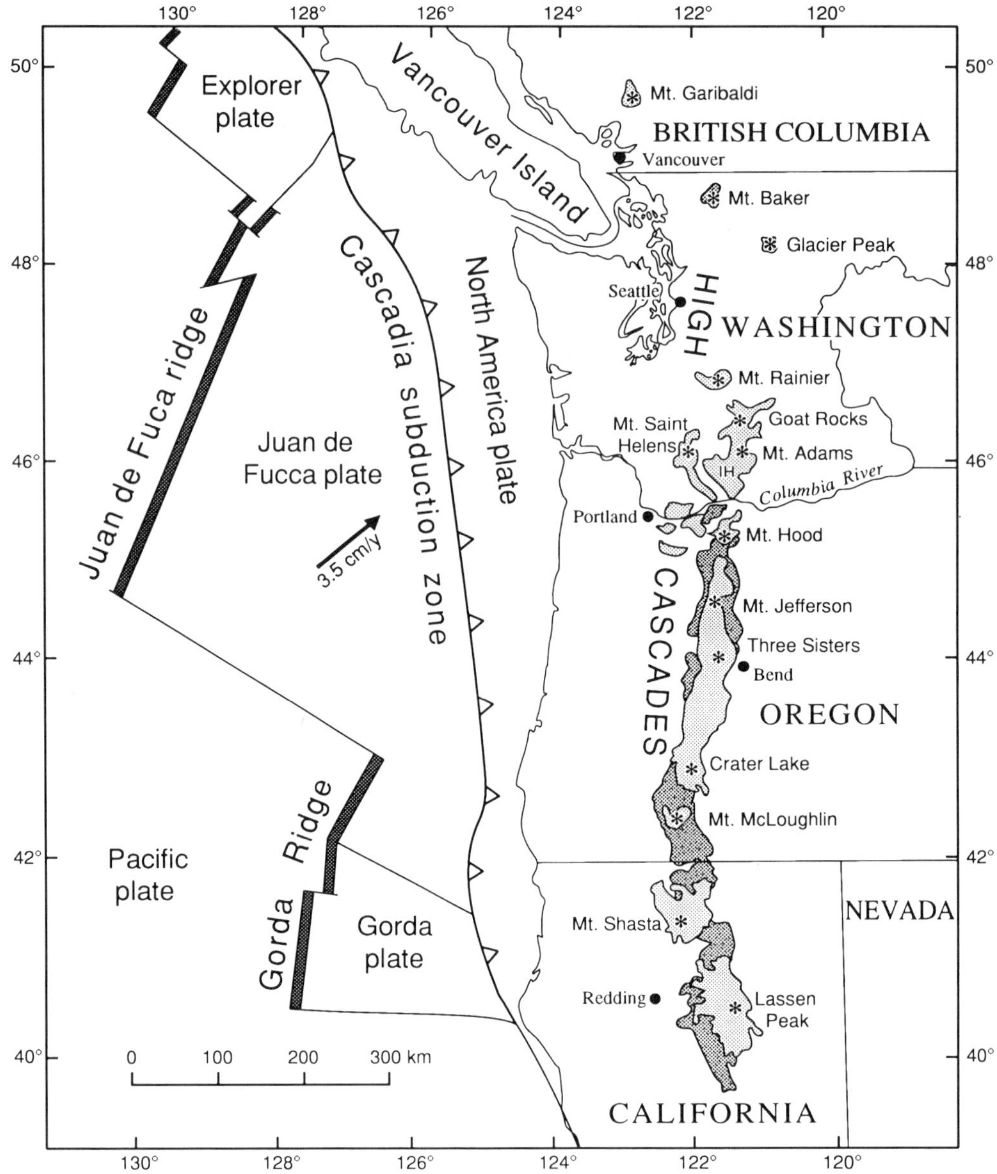

Figure 47. Map of late Miocene and younger magmatic systems of the High Cascades and modern plate-tectonic elements of the adjacent Pacific basin. Arrow is vector of relative convergence between Juan de Fuca and North America Plates. Light stipple, High Cascades volcanic rocks of 7- to 2-Ma age; dark stipple, High Cascades volcanic rocks of 2- to 0-Ma age. B, Boring Lava; IH, Indian Heaven.

cone on the eroded remnants of a 400,000-year-old cone, has produced minor Holocene eruptions—as recently as the mid-1800s—and has an active solfataric and hydrothermal system (*Coombs, 1939*; *Frank and others, 1977*; *Hyde and Crandell, 1978*). Within about 30 km to the northeast are 4- to 3-Ma dacitic shallow intrusives and volcanics (*Engels and others, 1976*; *Luedke and others, 1983*). Glacier Peak, 90 km southeast, is mainly a complex of dacitic domes, flows, and pyroclastic deposits but with eroded remnants of an andesitic cone and numerous surrounding dikes of basalt to dacite (Tabor and Crowder, 1969; *Beget, 1982*; *Luedke and others, 1983*). About 5 km^3 of pumiceous ash erupted as a pair of widespread layers within 10 to 25 years of each other about 11,250 years ago (Powers and Wilcox, 1964; *Mehringer and others, 1977*). Mount Rainiier, the highest but not the most voluminous Cascades volcano, is a mainly Pleistocene pyroxene-andesite stratocone (Fiske and others, 1963) that has erupted relatively little pumice (*Mullineaux, 1974*) but has produced numerous Holocene debris flows, some reaching the urbanized Puget lowland (Crandell, 1971). Rainier last erupted between 1820 and 1854 (*Mullineaux and others, 1969*). Mount St. Helens is only about 40,000 years old (Mullineaux and Crandell, 1981) and is historically the most active volcano in the

Cascades (*Crandell and Mullineaux, 1978*). Mainly dacites erupted before about 2,500 years ago; since then pyroxene andesites, a few basalts, and several dacitic domes have built the present stratovolcano. Mount St. Helens has produced many pyroclastic eruptions and erupted violently in May, 1980. Following the two-month intrusion of a dacitic cryptodome, the magma column was unroofed by a catastrophic landslide that removed much of the volcano, triggering a hydrothermal blast, mudflows, and a plinian eruption (R. L. Christiansen and Peterson, 1981). Since then, a composite dacitic dome has grown intermittently within the large crater left by the events of May, 1980 (*Swanson and others, 1983*).

In southern Washington, the character of the volcanic High Cascades changes. Weaver and Michaelson (1985) and *Michaelson and Weaver (1986)* note that seismicity of the Juan de Fuca Plate is segmented, with a transition about where this change begins. To the south, subduction is steeper, oblique to the volcanic front (Fig. 47), and more strongly coupled to the continental plate. In addition to regularly spaced long-lived volcanic systems like those farther north, numerous short-lived vents extend nearly north-south but are gently convex westward from southern Washington to northern California (*Luedke and others, 1983*). These vents erupted mainly mafic andesite ("basaltic andesite" in much of the Cascades literature) and basalt, forming a band about 50 km wide of overlapping, mainly 6- to 3-Ma mafic shield volcanoes (*McBirney and others, 1974*; McBirney and White, 1982; *Sherrod and Smith, 1990*). Each vent was active only briefly; probably even the shields remained active for no more than a few thousand years. It is these mafic lavas that are sometimes referred to as a High Cascades "platform" (*Taylor, 1978,* 1981a), but some now eroded andesitic to dacitic volcanoes probably were coeval with them and similar mafic lavas have erupted among the younger stratocones. Thus, the "platform" and the more felsic stratocones probably are contemporaneous elements of the volcanic chain. In the north, the mafic lavas fill a graben that offsets older rocks (Taylor, 1981a, 1990; *D. L. Williams and others, 1982*; Hughes, 1990). No such structure has been mapped in the apparently downwarped southern Cascades (*Williams, 1949*; Hammond, 1979), although regional aeromagnetic and gravity anomalies suggest that it may continue southward to Mount Shasta (Blakely and others, 1985). A crustal root beneath this zone thickens from about 40 km in the north to about 48 km in the south (*Leaver and others, 1984*). The southward change in volcanic and petrologic character of the Cascades corresponds to a southward change from uplift and contraction to extension within and adjacent to the arc, associated with a block rotation of the Cascades (Wells, 1990).

At intervals of 60 to 130 km within the chain of mafic shield volcanoes from southern Washington to northern California, longer-lived systems form large andesitic stratocones (Fig. 47). To *Hughes and others (1980)* their alignments and offsets suggest segmentation of the subduction zone, just as the other young volcanic vents indicate (Guffanti and Weaver, 1988). Some of these volcanoes have wide compositional ranges; others are more restricted (*McBirney, 1968*). All, however, erupted voluminous andesite more silicic than that of the mafic shields; most also erupted at least some dacite. Farthest north is the now largely eroded White Pass–Goat Rocks center (*Ellingson, 1972*), roughly midway between Mount Rainier and Mount Adams. Mount Adams has erupted mainly pryoxene andesites but also some basalts to dacites from numerous peripheral vents as well as the main summit vent, including at least seven Holocene eruptions (*Hildreth and Fierstein, 1983*). About 90 km south, Mount Hood is mainly a late Pleistocene stratovolcano but has erupted several sequences of Holocene dacitic domes and related pyroclastic flows (*Wise, 1968, 1969a*; White, 1980; Crandell, 1980), the youngest about 200 to 300 years ago; minor pyroclastic eruptions occurred in 1859 and 1865. Mount Jefferson, 80 km south of Mount Hood, is a late Pleistocene pyroxene andesite stratovolcano (*Greene, 1968*; *Condie and Swenson, 1973*). About 70 km farther south a long-lived system centered near the southern of the Three Sisters has erupted diverse upper Pleistocene and Holocene basalts, andesites, dacites, and rhyodacites (*Williams, 1944*; *Taylor, 1978,* 1981a; *Luedke and others, 1983*). Several andesitic to rhyodacitic ash-flow tuffs occur near Bend—two of them dated at 4 to 3 Ma (*Fiebelkorn and others, 1983*)—and at least some had sources in this system although most of the specific sources have not been identified (Taylor, 1981a; *Mimura, 1984*). A gap of 130 km separates the Three Sisters from the large magmatic system of ancient Mount Mazama, a cluster of andesitic to dacitic stratocones that evolved over at least 400,000 years in late Pleistocene time (Williams, 1942; Bacon, 1983). About 7,000 years ago, the system erupted rhyodacitic domes, flows, and associated pumice from vents north of the complex summit area, followed 6,800 years ago by the climactic pyroclastic summit eruption of 50 to 60 km^3 of compositionally zoned magma, distributed as pyroclastic flows and as fallout ash over much of the Pacific Northwest (Powers and Wilcox, 1964; *Kittleman, 1973*; *Mehringer and others, 1977*). The roof of the partially emptied magma chamber collapsed to form Crater Lake, the largest recognized caldera of the Cascades. About 60 km south of Crater Lake is the eroded late Pleistocene stratocone of Mount McLoughlin. Another 120 km beyond is Mount Shasta, the largest andesite-dacite stratocone of the Cascades, which in the past 450,000 years has evolved through several episodes of rapid cone building separated by longer periods of erosion and few eruptions (Christiansen, 1982a); Holocene eruptions, have included large pyroclastic flows (*Miller, 1978, 1980*). An ancestral cone was largely destroyed by a catastrophic landslide 300,000 years ago (*Crandell, 1989*).

South of Mount Shasta, the chain of mafic shields and basaltic lavas, 4 to 6 Ma in age and younger, ends (Fig. 47); following a gap of about 50 km, however, Pliocene and Quaternary volcanism resumed for another 150 km. The oldest High Cascades eruptions in this southern zone formed the Tuscan Formation, probably no older than 3.5 Ma (*Harwood and others, 1981*). The

Tuscan comprises mainly andesitic mudflow breccias (*Anderson, 1933*) but includes andesitic, dacitic, and rhyolitic lava flows in areas closer to the source volcanoes (Lydon, 1968) as well as one or more extensive rhyolitic ash flows—notably the Nomlaki Tuff Member that once covered most of the Sacramento Valley (*Anderson and Russell, 1939*; *Sarna-Wojcicki, 1976*). The southernmost major High Cascades volcano, near the southern margin of the subducting Gorda Plate (*Jachens and Griscom, 1983*), is Lassen Peak, one of a cluster of late Quaternary dacitic domes on the eroded flank of a 500,000-year-old stratovolcano (Williams, 1932; *Crandell and others, 1974*; *Heiken and Eichelberger, 1980*; *Clynne, 1990*). Small to moderate steam blasts in 1914 to 1917 from a summit crater on Lassen Peak climaxed in May, 1915, with eruption of a lava flow, devastating avalanches and pyroclastic eruptions on the east flank, and mudflows and floods more than 30 km downstream (*Day and Allen, 1925*; *Loomis, 1926*). Seismicity (*Klein, 1979*) and hydrothermal discharge (*Day and Allen, 1925*; *Ingebritson and Sorey, 1985*) are notably active.

Major Quaternary basaltic fields are interspersed among the predominantly andesitic High Cascades (Fig. 47), including Indian Heaven near the north end, between Mount St. Helens and Mount Adams (*Hammond and others, 1976*), the Boring Lava of the Portland area (Trimble, 1963), flows in central Oregon north of the Three Sisters (Taylor, 1968, 1990; *Hughes, 1990*), and others near Mount Shasta (*Williams, 1949*; Hart and others, 1984) and Lassen Peak (Williams, 1932; *Anderson, 1940*). Some of the basalts are high-alumina olivine tholeiites (*Anderson and Gottfried, 1971*; Hart and others, 1984), typically having low K_2O and incompatible trace-element contents and relatively unfractionated rare-earth elements (REE). Other subalkalic basalts, perhaps best characterized as calc-alkalic, typically have higher K_2O, higher incompatible trace-elements, and are enriched in light but depleted in heavy REE.

In summary, arc magmatism has continued in the Pacific Northwest with northward retreat of its southern terminus as a subduction gap has widened since mid-Miocene time. In the past 8-9 Ma, the arc has formed the volcanic High Cascades, still virtually continuous from near Lassen Peak to north of the Canadian border. The chain varies along its length with changes in its geometric relation to subduction and to regional tectonics of the Cordilleran interior. In the north, where subduction is perpendicular to the arc, magmatism is focused in regularly spaced long-lived silicic-andesite to dacite systems with diameters of no more than 30 to 40 km. Farther south, where subduction is more oblique to the chain and where tectonic extension characterizes late Cenozoic evolution of the adjacent region, magmatism is distributed in numerous shorter-lived systems and includes more abundant basalts and mafic andesites. Nevertheless, regularly spaced long-lived and more evolved magmatic systems continue along the chain; a few of them have erupted voluminous ash flows, and others may be capable of such eruptions in the future (Christiansen, 1982a).

MID-MIOCENE AND LATER MAGMATISM OF THE CALIFORNIA COAST RANGES

In late Cenozoic time, with establishment of the coastal transform system and consequent cessation of subduction, widening of the subduction gap, and episodic northward retreat of the southern termination of the volcanic-arc front, localized magmatic systems were active in the California Coast Ranges. At least some of these systems appear to have propagated northward with time (Dickinson and Snyder, 1979a). The way in which this propagation is understood, however, depends on which magmatic systems are presumed to be related to it. In this chapter, we have already discussed some coastal-region magmatic systems in connection with two other regional patterns. First, Oligocene to early Miocene compositionally diverse but potassic and commonly silicic magmatism associated with tectonic extension in the continental interior continued through the western Mojave Desert and lower Colorado River regions; parts of the westernmost volcanic fields were later offset northward along the San Andreas and related faults. Second, mid-Miocene predominantly intermediate calc-alkalic magmatic arc affected the Transverse Ranges and the California Continental Borderland. In this section, only mid-Miocene and younger systems of the Coast Ranges that appear to be distinct from these other two regional patterns are discussed. Distinctions made here between these magmatic suites of the coastal region are similar to those of Weigand (1982) except that Oligocene and lower Miocene volcanics of the Coast Ranges are related here to rocks of similar ages farther south. Furthermore, two types of magmatic systems can be distinguished among the mid-Miocene and younger volcanic fields associated uniquely with the Coast Ranges. One is basaltic or bimodal and seems to have been extensive along the transform fault system early in its evolution. The other is compositionally diverse, characterized by relatively voluminous silicic and intermediate magmas, and appears to have propagated northwestward with time, staying a little behind the northwestward-migrating Mendocino triple junction.

Now-disrupted volcanic sequences in the southern Coast Ranges and westernmost Transverse ranges (Turner, 1970; Hall, 1981a; Weigand, 1982) were discussed earlier in their stratigraphic and structural context; they were once parts of a single system of basalt and rhyolitic tuffs and flows related to eruption along a mid-Miocene part of the San Andreas fault system between 20 and 15 Ma. About 60 km east, nearly adjacent to the San Andreas fault, basalt interbedded with mid-Miocene marine sedimentary rocks of the Caliente Range (Figs. 37, 46) is dated at 16 to 14 Ma (Turner, 1970; Vedder, 1975); on the San Francisco Peninsula, 380 km to the northwest but just across the San Andreas fault, the 14-Ma Page Mill Basalt is also interlayered with marine sedimentary rocks (Turner, 1970). As the separation between these latter two localities is opposite to the sense of fault offset (Fig. 46), early mid-Miocene submarine basaltic magmatism occurred along parts of a coastal transform system that prob-

ably was at least 700 km long. Small volumes of younger basalts also occur along the continental-margin transform system (Fig. 46), including 4-Ma basalts near the San Andreas fault south of San Francisco Bay (Fox and others, 1985) and 600 km to the south at Northeast Bank, near the margin of the California Continental Borderland (*Hawkins and others, 1971*).

Volcanic fields that appear to define a north-propagating system, in or near the Diablo Range and in structurally continuous blocks north of San Francisco Bay, are entirely east of the San Andreas fault (Fig. 46), although they are displaced by smaller right-lateral faults that branch from it (Fox and others, 1985). Southernmost of these volcanic fields is the andesitic to rhyolitic 15- to 9-Ma Quien Sabe Volcanics of the central Diablo Range (*Taliaferro, 1948*; *Leith, 1949*; Fox and others, 1985). Basalts to rhyolites of 12 to 8 Ma age in the Berkeley Hills (*Creely and others, 1982*; Fox and others, 1985) are about 130 km northwest of the Quien Sabe Volcanics, but about 30 km of that distance represents later displacement on the Calaveras fault. The volcanic rocks of the Berkeley Hills were shown by Fox and others (1985) to be offset about 40 km farther northwestward on the Hayward–Rogers Creek fault system to the Tolay Volcanics of *Morse and Bailey* (*1935*). About 40 km north of the Berkeley Hills but separated from the Tolay Volcanics only by a right-lateral fault of uncertain displacement, the Sonoma Volcanics are compositionally diverse and range between 8 and 3 Ma in age (*Mankinen, 1972*; Fox, 1983; Fox and others, 1985). The lower part of the section consists of basalts to rhyolites with andesites predominant; the upper part is predominantly rhyolitic, including ash-flow tuffs. Some ash beds associated with this volcanism are widespread markers in diverse late Tertiary sedimentary environments of the northern Coast Ranges (*Sarna-Wojcicki, 1976*). The youngest of the propagating Coast Range volcanic fields, overlapping the northern end of the Sonoma Volcanics, comprises the basaltic to rhyolitic Clear Lake Volcanics, which range in age from 2 Ma to 10,000 years (*Anderson, 1936*; Hearn and others, 1981; Donnelly-Nolan and others, 1981). After early, mainly basaltic eruptions, later volcanism was mainly dacitic and rhyolitic, reflecting complex fractional crystallization, variable melting, and assimilation in crustal magma chambers. The volumes of young silicic volcanic rocks, gravity and aeromagnetic anomalies (*Isherwood, 1976*), and seismic-wave delays (*Iyer and others, 1981*) suggest a still active large silicic magma chamber beneath the central part of the field. The Geysers, an exceptionally large geothermal field associated with this system, produces geothermal steam for power generation. Hydrothermal mercury deposits in the volcanic field include the late Pleistocene Sulphur Bank mine, associated with still-active hot springs (*White and Roberson, 1962*; *Sims and White, 1981*). The McLaughlin mine, just east near the boundary of the Coast Ranges and Sacramento Valley, is an important gold deposit associated with hot springs of this system (*Lehrman, 1986*).

The Sutter Buttes (Fig. 46) rise abruptly from the center of the Sacramento Valley and are cored by 2.5- to 1.5-Ma, mainly rhyolitic intrusions, with subordinate dacites and andesites, that arched and uplifted a thick sedimentary section by more than 750 m, breaching the surface as a cluster of volcanic domes and associated ejecta (*Williams and Curtis, 1977*). Contemporaneous rhyolitic intrusives nearby are now buried more than 1,700 m beneath the Sacramento Valley. The lavas lie west of the High Cascades volcanic front in the structural depression of the Sacramento Valley, are considerably more potassic than those of the Cascades, and were coeval with a wave of contractional deformation followed by extension that swept northward through the Valley, possibly related (Harwood and Helley, 1987) to passage of the Mendocino triple junction. Thus, the Sutter Buttes magmatism seems related to the transform-dominated Coast Ranges and not to the subduction-related High Cascades.

Fox and others (1985) concluded that magmatism migrated northward through the Coast Ranges (with adjustments for later disruption by right-lateral faulting) at nearly 4 cm/yr from 25 to 12 Ma, dropping to a little more than 1 cm/yr from 12 Ma to the present (Fig. 48). That conclusion is not substantially contradicted here by considering only those volcanic fields not already regarded as parts of other regional systems and not as part of the north-propagating system. Northward propagation of this system has lagged consistently about 80 to 100 km behind the north-propagating Mendocino triple junction, possibly related to local extensional adjustments to passage of the triple junction. Consistently high $^{87}Sr/^{86}Sr$ ratios, high ^{18}O, and relatively high incompatible trace elements (*Futa and others, 1981*; *Johnson and O'Neil, 1984*) as well as high K_2O and disequilibrium textures indicate significant roles for crustal melting and magmatic mixing in the origins of the magmatic suite.

LATE CENOZOIC UPLIFT AND REGIONAL EXTENSION OF THE CORDILLERAN INTERIOR

Although tectonic extension had followed the end of Laramide contraction in different parts of the Cordilleran interior by as little as 10 m.y.—or possibly less—and by as much as 50 m.y. (Wernicke and others, 1987), the pattern of continental extension changed throughout the entire region in mid-Miocene time, generally between about 18 and 10 Ma. Belts of earlier extension had been superposed on rapidly degrading Laramide highlands and ultimately became parts of widespread depositional or volcanic plains, but later extension is associated with regional uplift of wide areal extent and is characterized by regionally distributed linear block faulting. Present elevations (Fig. 49A) and the evolution of regional drainage systems provide important records of uplift, and distributed extension is manifest in the physiography of the Basin and Range Province, parts of the Rocky Mountains, and adjoining regions. A broad, more or less circular region 1,600 km across and centered near the High Plateaus, at the western margin of the Colorado Plateau, has been uplifted epeirogenically

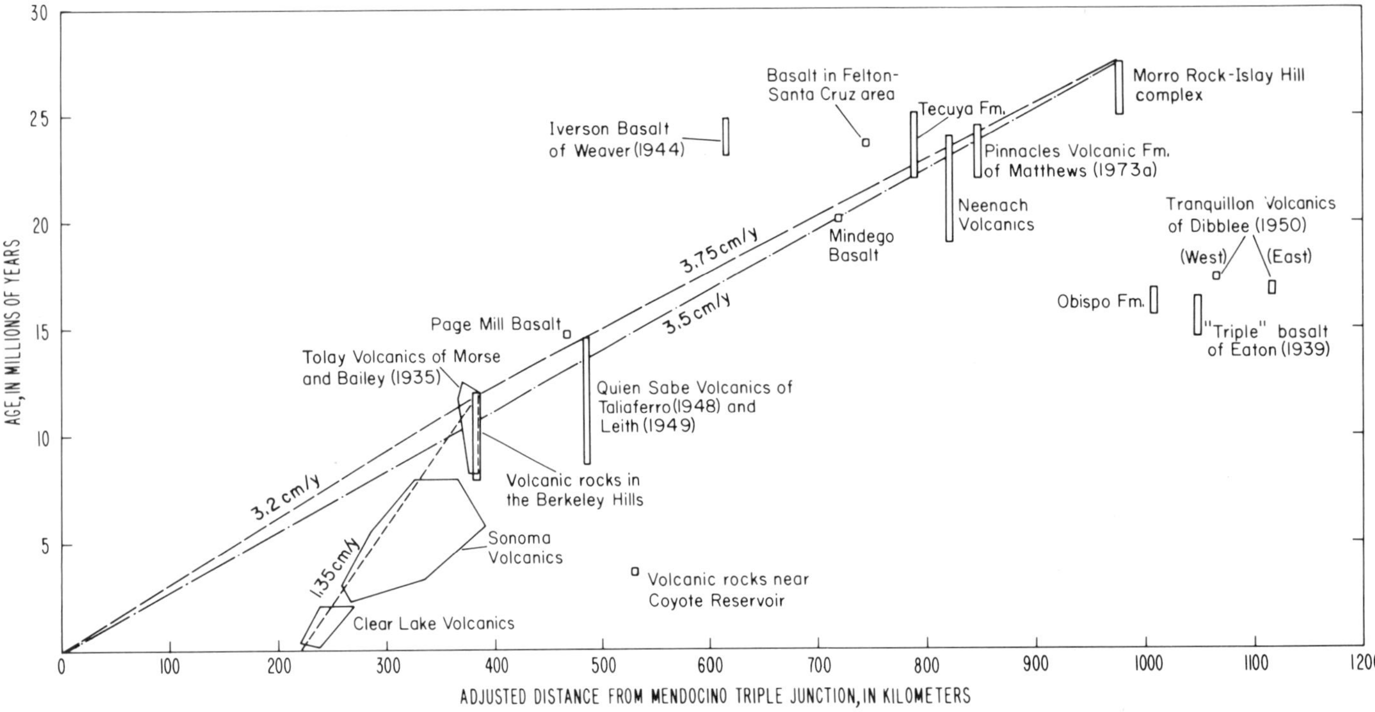

Figure 48. Ages of volcanic rocks in the California Coast Ranges as a function of adjusted distance from the Mendocino triple junction. From Fox and others (1985).

during and since the late Miocene, concurrently with widespread normal faulting in the Basin and Range Province and elsewhere.

UPLIFT OF THE BASIN-RANGE REGION AND COLORADO PLATEAU

Some definitions may help to clarify discussion of late Cenozoic tectonics in this section. The term "Basin and Range" is applied almost universally to a physiographic province, the boundaries and subprovinces of which are generally well defined (*Fenneman, 1931*). Significant differences in history and physiography divide the northern and southern parts of the uplifted and extended region near the latitude of Las Vegas. The northern Basin and Range Province is currently more active than the southern, but uplift and extension reach beyond the province boundaries. The term "basin-range" is used here as a structural or tectonic term in the region of late Cenozoic extensional faulting, not confined to the Basin and Range Physiographic Province. In some contexts, it is convenient to use the term "Great Basin region" for the region of basin-range structure from the Sierra Nevada to the Middle Rocky Mountains; the term includes areas outside the hydrographic limits of the Great Basin, however, as should be clear from context. Some confusion in the literature results from use of the term "basin-range" by some investigators for virtually all Cenozoic extensional structures, including mid-Tertiary, mid-Miocene, and later Cenozoic features of contrasting styles (*Loring, 1976*; Crowe, 1978). Here we follow Scarborough and Peirce (1978), Zoback and others (1981), and Damon and others (1984) in restricting "basin-range" to features expressed in the modern regional topography.

The Basin and Range Province as a region of gently tilted normal-fault blocks, marked by linear ranges and intervening alluvium-floored valleys, is familiar to most geologists. Not always fully appreciated, however, is the high average elevation of the northern Basin and Range region. By contrast, high regional elevation of the Colorado Plateau is frequently remarked upon. Nevertheless, almost the entire Great Basin, except for the Bonneville and Lahonton basins at its eastern and western margins respectively, stands at average elevations above 1.5 km (Fig. 49A). Average elevations in both the Great Basin and the Colorado Plateau are about 1.7 km (*Diment and Urban, 1981*); substantial parts of both provinces are above 2 km. Only the rim of the High Plateaus, which rises between them, is higher by about 500 m in average elevation than the centers of the two provinces.

As already described, early Tertiary eastward drainages from the Laramide and older Sevier orogenic highland along the eastern margin of the present Great Basin region emptied into successors of Laramide basins in the Rocky Mountains and Colorado Plateau and ultimately through the cratonic interior (see Fig. 15). The broad lakes of the Colorado Plateau stood at relatively low elevations, probably connected without precipitous drops, and drained through Wyoming (*Bradley, 1964*) and ultimately

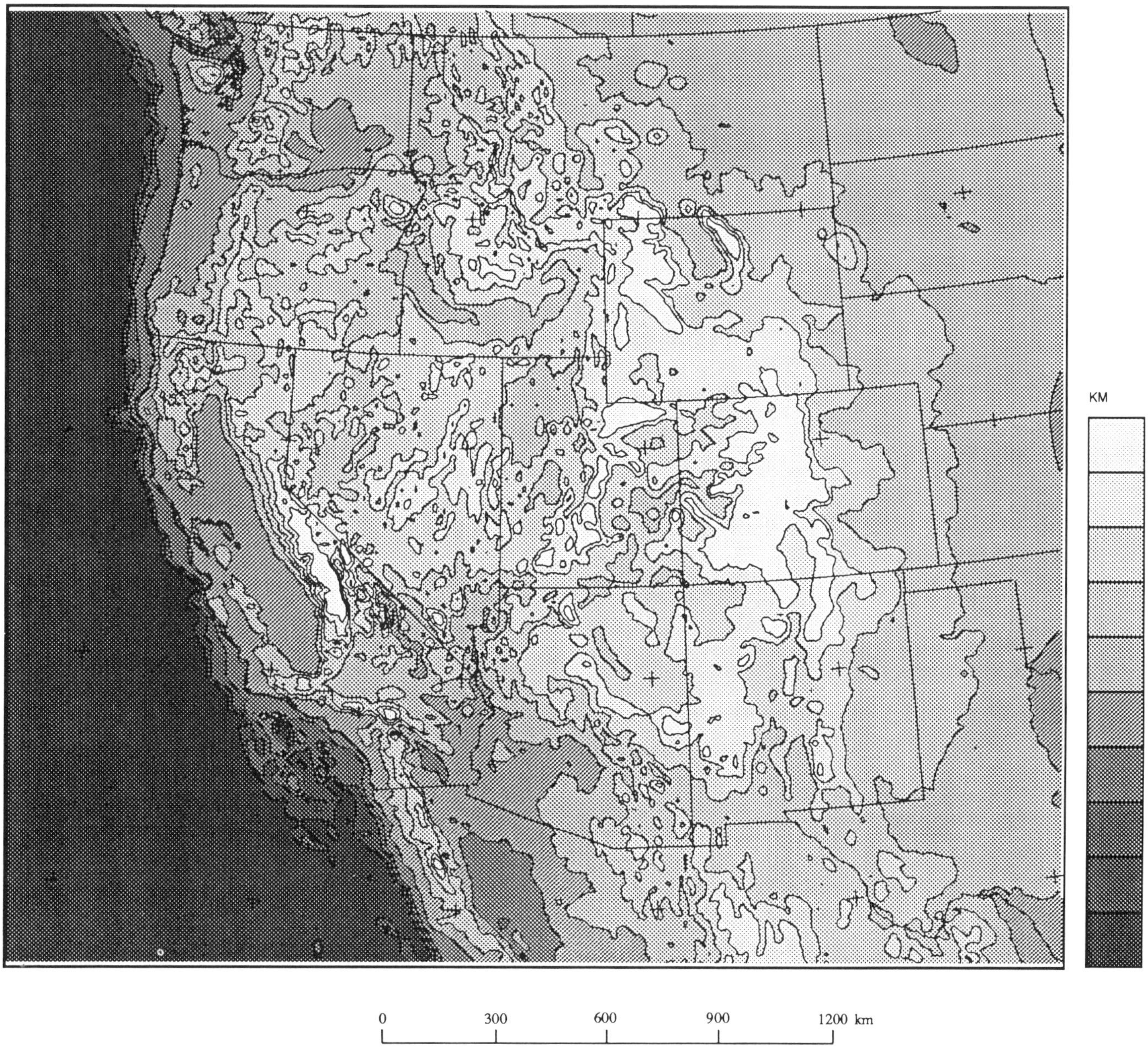
KM
2.5
2.0
1.5
1.0
0.5
0.0
-0.5
-1.0
-1.5
0
300
600
900
1200 km

A

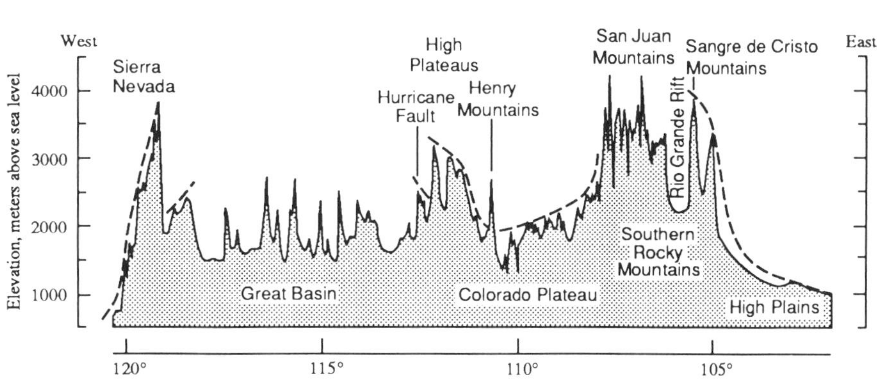
West
East
Elevation, meters above sea level
4000
3000
2000
1000
Sierra Nevada
High Plateaus
Hurricane Fault
Henry Mountains
San Juan Mountains
Sangre de Cristo Mountains
Rio Grande Rift
Great Basin
Colorado Plateau
Southern Rocky Mountains
High Plains
120°
115°
110°
105°

B

through the ancestral North Platte River system to the Gulf of Mexico. Westward drainages down the slope of the post-Laramide highland extended across the present Sierra Nevada to coastal basins in California (Fig. 15). Thus, a record of late Cenozoic regional uplift is found in the structures and elevations of rocks on both sides of the old drainage divide and in the late Cenozoic drainage evolution of the Sierran block, the Snake River region, and the Colorado Plateau.

By late in the mid-Tertiary, the early Tertiary regional headwaters near the center of the Mesozoic thrust-belt highland were buried by a constructional plain of silicic ash flows. Distribution patterns of the extensive mid-Tertiary ash flows show no linear grain although they are now broken by more or less north-south normal faults, active mainly since about 14 to 10 Ma (Christiansen and McKee, 1978; Zoback and others, 1981; Stewart, 1983b). The distributions of younger ash-flow sheets, by contrast, show clear effects of basin-range physiography.

Extensional breakup of the Sierra-Wasatch region and uplift of its western bounding block are recorded in drainage evolution of the long western slope of the Sierra Nevada (Fig. 49B). Early to middle Tertiary streams deposited rock debris from eastern sources in channels across the Sierran block and in more westerly lowland and marine sedimentary sequences, beginning with the Eocene Ione Formation on the east side of the Great Valley (*Allen, 1929*; *Palmer and Merrill, 1982*). The old stream channels, many of them important sources of placer gold, show the late Cenozoic northern Sierran block to be nearly intact, uplifted along its eastern margin, and tilted westward (*Lindgren, 1911*); the southern Sierra is more complexly uplifted and internally warped and faulted. Christensen (1966) showed that major uplift of the Sierra Nevada began after initial mid-Miocene andesitic volcanism of the Mehrten Formation but before the 9-Ma eruption of latites into drainages cut within those andesites; most uplift postdates 9 Ma but predates 3 Ma. The great eastern scarp of the Sierra Nevada may be largely Pleistocene in age. Thus, uplift was concurrent with normal faulting in the Great Basin, much of which probably was uplifted along with the Sierra Nevada; both regions were broken by extensional faulting concurrent with uplift. Huber (1981) analyzed the evolution of the San Joaquin River (see Fig. 51) in the central Sierra Nevada. The data are not definitive before about 10 Ma but suggest that, although uplift might have been underway by 25 Ma, it was less than 0.03 mm/y. Uplift of the San Joaquin headwaters varies from about 2,000 to more than 3,400 m and since 10 Ma has accelerated to about 0.3 mm/y and may still be increasing. Volcanic rocks and gravels document an ancestral course of the Tuolumne River across the Sierra Nevada and demonstrate that a highland of moderate relief near the present crest of the Sierra Nevada did predate 10 Ma (*Huber, 1990*). Coarse clastic deposits in the southeastern San Joaquin Valley indicate initial mid-Miocene uplift of the southern Sierra (*Bartow, 1984*).

The Snake River drainage provides a different perspective on regional uplift and deformation (see Fig. 51). The drainage originated with formation of the Snake River Plain by downwarping and graben faulting in mid-Miocene and later time, concurrent with uplift along the margins of the plain (*Kirkham, 1931*; Malde and Powers, 1962; Axelrod, 1968). From hydrologic gradients, barbed tributaries, and a regional unconformity in lacustrine and fluvial deposits that would now be regarded as about 9 Ma in age, *Wheeler and Cook* (*1954*) proposed that the present lower course of the Snake was established by spillover and capture in the latest Pliocene (Malde, 1991). Before then, the eastern Blue Mountains Province formed a divide, with north-flowing streams tributary to the Salmon River and southward streams flowing into the Snake. Late Miocene lakes impounded along the Snake River remained fresh, indicating an outlet. Biogeographic data suggest that the lower course of the Snake may have been through southeastern Oregon, northern Nevada, and northeastern California into an ancestral Pit River-Klamath River system (*Wheeler and Cook, 1954*; *Malde, 1965*; 1991). Thus, regional late Miocene uplift was accompanied by subsidence of the Snake River Plain and establishment of drainage from it across the northern Great Basin region; with continued regional uplift and block-faulting to the south, that course was blocked to form a persistent sedimentary basin that was ultimately captured by the Salmon and Columbia River systems.

The central Colorado Plateau and northern Basin and Range Province of similar average elevations are separated by the transition zone of the High Plateaus (Fig. 49A, B). The higher average elevations in this transition accompany north-south normal faulting and intermittent late Cenozoic basaltic volcanism. Although *Rowley and others* (*1978a*) argued that the boundary between the High Plateaus and the Great Basin reflects elevation differences as old as 26 Ma, Anderson and Mehnert (1979) showed convincingly that westward-derived units as young as 21 Ma crossed the boundary. That boundary was not marked by faults or basaltic volcanism until the much younger basin-range episode, no earlier than 14 and mainly after 10 Ma (Best and Hamblin, 1978; Anderson and Mehnert, 1979).

The southern Basin and Range Province, in contrast to the northern, is consistently lower than the Colorado Plateau, separated from it by a series of escarpments and a mountainous transition zone (Fig. 49A). Incised eastward drainage from a Laramide highland in the present Basin and Range Province entered the southwestern Colorado Plateau near sea level (Young and McKee, 1978; Young, 1979; Lucchitta, 1979). That drainage was partly filled by the 18-Ma Peach Springs Tuff from a source to the west and was subsequently disrupted by normal faulting, volcanism, and uplift of the plateau relative to the adjacent structural basins (Young and Brennan, 1974). The westward course of

Figure 49. Regional uplift of the United States Cordilleran interior. A, Topographic map of the United States Cordilleran region. Contour interval 500 m. B, Topographic profile at latitude 38°N. Vertical exaggeration is about 150:1; dashed line indicates accordance of nonvolcanic summits with broad features of elevation profile. After Cordell (1978).

the Colorado River from the plateau through the southern Basin and Range Province resulted from successive captures in Pliocene and later time. Uplift in the past 5.5 m.y., about half by broad regional upwarping and half across province-bounding normal faults, has been about 900 m at a cumulative rate of about 0.16 mm/y (Lucchitta, 1979).

Gravels and associated basalts on the south-central margin of the plateau in central Arizona have been interpreted as reflecting a drainage reversal from northward to southward between 10 and 5 Ma (*McKee and McKee, 1972*). However, Peirce and others (1979) point to topographically low Oligocene(?) and lower Miocene sediments on an erosion surface in the transition zone as recording a much earlier lowering of base levels to the south and the formation of southward drainages from the plateau rim during mid-Tertiary tectonic extension in southern Arizona. Eocene and Oligocene "rim gravels" had been deposited by strong northeast-flowing streams from the Mogollon highland farther south; these streams were disrupted by a new drainage divide with 600 m of relief along the present plateau margin, forming limestone cliffs as a permanent drainage barrier. Basin-range faulting superposed on this escarpment is younger than 13 Ma, and the present boundary between the provinces is largely erosional rather than structural.

The southeastern Colorado Plateau stands above the Rio Grande rift (see Fig. 50), where extension began at 28 to 26 Ma (Chapin and Seager, 1975; Lipman and Mehnert, 1975; Lipman and others, 1986). Gravels in the Mogollon-Datil volcanic sequence indicate elevation of the Colorado Plateau relative to the southern Basin and Range Province in New Mexico in earliest Miocene time, about 24 Ma. Late Cenozoic regional uplift and crustal dilation have produced minor, perhaps still active rejuvenation of Precambrian fault zones in the southern Colorado Plateau basement, generally along northeast trends (*Shoemaker and others, 1978*).

The northern Basin and Range Province, although high in average elevation, has a thin crust—about 30 km near the center and thinner toward the margins (R. B. Smith, 1978; *Prodehl, 1979*). The Sierra Nevada is underlain by a Mesozoic crustal root of more than 50 km (*Eaton, 1966*), but the range did not rise until the mid-Miocene. These seeming paradoxes point out that present elevations of both the Sierra and the Great Basin reflect not crustal thickness but anomalously low densities in the upper mantle (*Crough and Thompson, 1977*), consistent with compressional-wave velocities in the uppermost Basin and Range mantle of about 7.8 to 7.9 km/s (R. B. Smith, 1978; *Prodehl, 1979*), compared to about 8.2 km/s under the stable continental crust of the High Plains, about 50 km thick (*Stewart and Pakiser, 1962*).

Geophysical properties of the Colorado Plateau crust and upper mantle are generally intermediate between those of the basin-range regions and the stable crust of the High Plains (*Thompson and Zoback, 1979*). The crustal thickness of about 40 km is insufficient to account for apparent isostatic equilibrium of the plateau, indicating the importance of low-density upper mantle. Thompson and Zoback conclude that the Colorado Plateau lithosphere is about 80 km thick, compared to about 60 km beneath the Basin and Range and Rio Grande rift and about 120 km beneath the High Plains. Heat-flow data support this picture, with values of about 60 mW/m^2 in the interior of the Colorado Plateau, a transition zone with 80 to 90 mW/m^2, and values greater than 100 mW/m^2 in the Basin and Range Province (Lachenbruch and Sass, 1978; *Bodell and Chapman, 1982*), suggesting Tertiary lithospheric thinning and a transient thermal response, uplift lagging behind initial thinning by 15 to 20 Ma. Lateral warming and weakening of the Colorado Plateau probably began about 20 Ma, early in the regional mid-Miocene tectonic transition.

Thus, although part of the elevation of the Colorado Plateau relative to its western, southwestern, southern, and southeastern borders occurred during adjacent mid-Tertiary tectonic extension and basin formation, the entire Colorado Plateau–Basin and Range region has been broadly uplifted and the province boundary sharpened during and since the mid-Miocene. Regional differences between the northern and southern extended regions reflect greater continued uplift of the northern basin-range region and Colorado Plateau during Pliocene and Quaternary time.

UPLIFT OF THE ROCKY MOUNTAINS AND GREAT PLAINS

In late Mesozoic and early Tertiary time, a continuous orogenic highland through the Northern Rockies, eastern Great Basin region, and Mogollon Highland probably formed the continental divide (see Fig. 15). With late Cenozoic uplift of the Rocky Mountains, the continental divide has moved to the eastern periphery of the Cordilleran region, and the Middle and Southern Rocky Mountains have the highest average elevations in the United States (Fig. 49A). Late Cenozoic uplift of the Rockies has been broader and more regional than earlier uplifts associated with Laramide crustal shortening and thickening.

Much of the aggradational mantle that had buried basins and range flanks of the Middle and Southern Rockies by mid-Miocene time has been erosionally dissected or removed. Much of the earlier part of that mantle was eroded and redeposited into the upper Miocene Ogallala Formation and its correlatives (*Frye and Leonard, 1959*), but even most of that upper Miocene alluvial cover has been stripped with continued regional uplift of the High Plains as well as the Rockies (Love, 1960; Izett, 1975; *Scott, 1982*). Virtually all of a once thick middle Eocene and younger sedimentary fill has been removed from the Bighorn Basin since the Miocene (Fig. 49A; *McKenna and Love, 1972*); detached masses of Eocene volcanic rocks moved downward and across parts of the resulting stripped surface in and near the Absaroka Range (*Bown, 1982*). The Ogallala Formation on the Great Plains of eastern Colorado buries and preserves paleostream courses that were less well integrated than the modern streams (*Borman and Meredith, 1983*; *Borman and others, 1984*), documenting contemporaneous rise of the Southern Rockies (*Scott, 1982*). With these changes, the dominant assemblages of clastic heavy minerals changed from reflecting abundant volcanic source

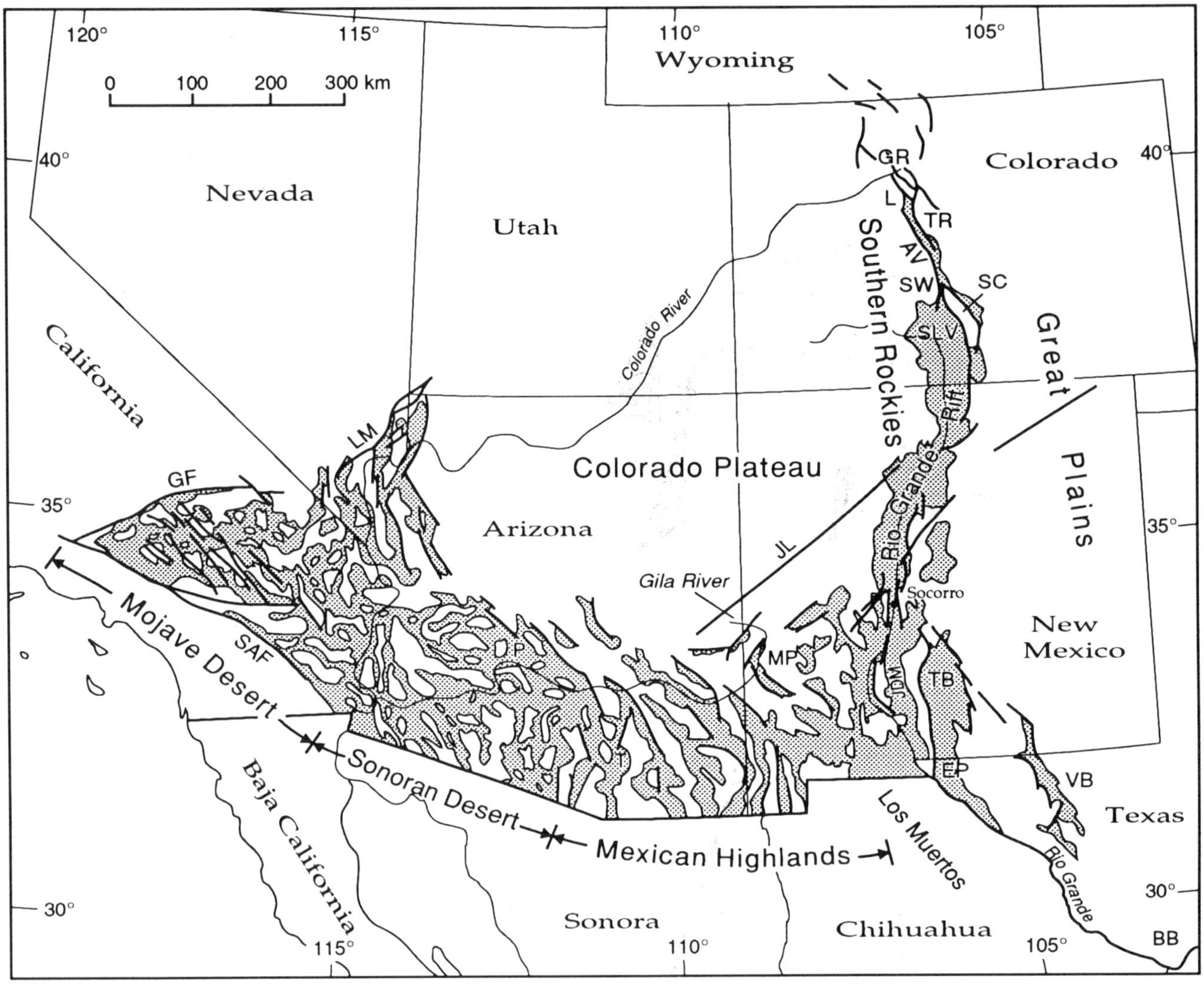

Figure 50. Map showing major late Cenozoic extensional features of the Southern Rocky Mountains, Rio Grande rift system, and southern Basin and Range region. Stipple, late Cenozoic basins. Hachured lines, normal faults. AV, Arkansas Valley; BB, Big Bend; EP, El Paso; GF, Garlock fault; GR, Gore Range; JDM, Jornada del Muerto; L, Leadville; LM, Lake Mead fault zone; MP, Mogollon Plateau; P, Phoenix; SAF, San Andreas fault; SC, Sangre de Cristo Mountains; SLV, San Luis Valley; SW, Sawatch Range; T, Tucson; TB, Tularosa Basin; TR, Tenmile Range; VB, Valentine basin.

materials in the early Miocene to predominantly plutonic and metamorphic sources in the late Miocene, indicating regional uplift and deep erosion of the source areas (*Denson, 1969*). Resulting drainage changes in central Wyoming caused the leaching of uranium from volcanic-rich Oligocene and Miocene deposits and its deposition and economic concentration in buried Eocene strata (*Harshman, 1972*). Early ideas of a late Cenozoic peneplain cycle in the Southern Rockies (*Atwood and Mather, 1932*; *Atwood and Atwood, 1938*) have been shown to be incorrect (*Steven, 1968*; Scott, 1975); in the San Juan Mountains, many of the surfaces once correlated with such a peneplain have been shown to be either stripped surfaces or constructional mid-Tertiary volcanic surfaces, virtually unmodified until the widespread erosion and canyon cutting that was associated with mid-Miocene and later epeirogenic uplift.

The United States Northern Rocky Mountains comprise two distinct physiographic regions. Linear belts north of the Lewis and Clark line (see Fig. 51), continuous with the Canadian Rockies and intermontane orogenic belts, have undergone little late Cenozoic uplift. To the south, mid-Miocene and younger basin-range faulting are widespread (Pardee, 1950; Reynolds, 1979). Average elevations over a region north of the Snake River Plain nearly 600 km across are greater than 1.5 km and are more than 2 km in most of the Idaho batholith whereas none of the mountains farther north reach 2 km and only a narrow band along the continental divide, less than 100 km across, averages

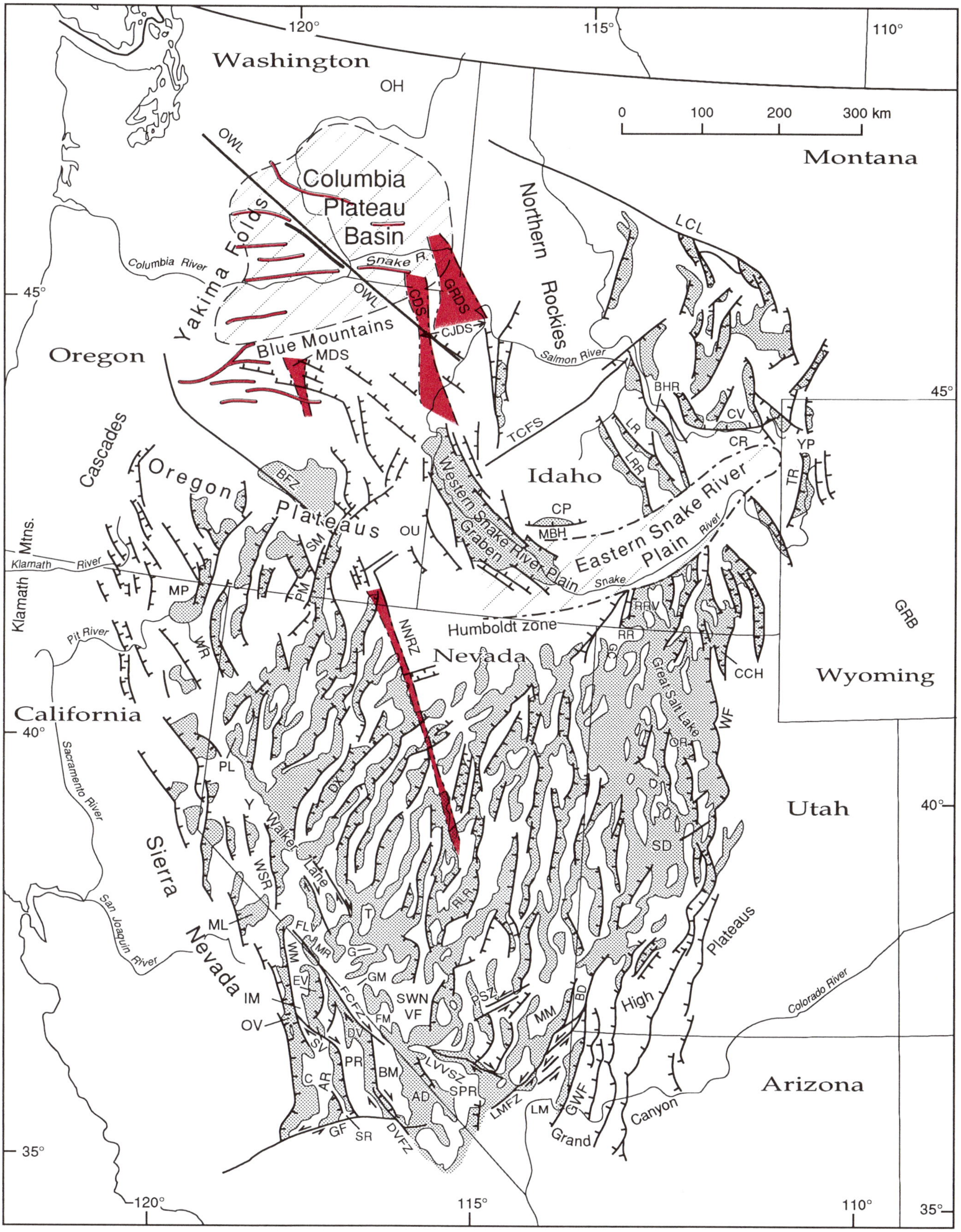
Washington
Oregon
California
Idaho
Nevada
Montana
Wyoming
Utah
Arizona
Columbia Plateau Basin
Yakima Folds
Blue Mountains
Oregon Plateaus
Cascades
Klamath Mtns.
Sierra Nevada
Northern Rockies
Western Snake River Plain Graben
Eastern Snake River Plain
Humboldt zone
Great Salt Lake
High Plateaus
Grand Canyon
Walker Lane
Columbia River
Snake R.
Salmon River
Klamath River
Pit River
Sacramento River
San Joaquin River
Colorado River
0 100 200 300 km
OWL
OH
CDS
GRDS
CJDS
MDS
TCFS
LCL
BHR
CV
CR
YP
TR
LR
LRR
CP
MBH
BFZ
SM
PM
OU
MP
WR
NNRZ
RR
RRV
GC
CCH
GRB
WF
OR
SD
PL
Y
DX
WSR
RLR
T
ML
FLV
MR
G
GM
WM
EV
CFZ
FCFZ
SWN VF
FM
PSZ
MM
BD
IM
OV
SV
DV
PR
BM
LVVSZ
SPR
AD
C
AR
GF
SR
DVFZ
LMFZ
LM
GWF
120°
115°
110°
45°
40°
35°

more than 1.5 km (Fig. 49A). Post-Miocene uplift south of the Lewis and Clark line has caused erosional removal of much mid-Tertiary basin fill, pedimentation across mid-Tertiary and older structures, and exhumation of parts of the underlying erosion surfaces (*Wolfe, 1964*); earlier concepts of these elements as a late Cenozoic peneplain probably are as invalid here as in the Southern Rockies. Reynolds (1979) demonstrated that present elevations of mid-Tertiary rocks in southwestern Montana define a broad arch, nearly 250 km across and more than 600 m high. In northwestern Wyoming, 9-Ma lacustrine deposits in Jackson Hole preserve a palynological record of uplift and increasing continentality of climate (*Barnosky, 1984*). Uplift has been regional into the Quaternary even in the Bighorn basin of Wyoming and Montana, an area commonly considered to be tectonically quiescent (*Reheis, 1985*).

Regional uplift of the Southern Rocky Mountains too was accompanied by normal faulting (*Colman, 1985*). Alluvial channel deposits, graben fills, and piedmont deposits record initial uplift in the early Miocene and subsequent acceleration of uplift (Izett, 1975; Taylor, 1975; Scott, 1975). Miocene paleovalleys that drained the mid-Tertiary volcanic fields now stand high. Now separated elements of the widespread Eocene erosion surface and associated deposits demonstrate 1.5 to 3 km of Miocene and later uplift (Epis and Chapin, 1975). Thermal modeling based on fission-track ages of apatites from uplifted crystalline rocks in central Colorado indicate mainly mid-Miocene and younger erosional removal of 1.5 km from terrain now 4.3 km above sea level (*Bryant and Naeser, 1980*). Basaltic lavas and associated fine-grained basin sediments in northwestern Colorado as old as 24 Ma show that early to mid-Miocene relief was low, the climate semiarid, and deposition quiet (Larson and others, 1975). Between 10 and 8 Ma, uplift produced increased precipitation and canyon cutting, establishing the upper Colorado River. Pliocene acceleration of regional uplift produced enhanced canyon cutting, continuing into the Quaternary (Scott, 1975), matched on the Great Plains by the end of widespread alluviation and by dissection of the Ogallala Formation (*Scott, 1982*).

Thus, the highest average elevations in the Cordillera are in the Southern Rockies, where late Cenozoic uplift coincides with both Laramide uplift and the late Paleozoic ancestral Rockies. Late Cenozoic uplift there may have been enhanced by the low-density crustal roots of earlier uplifts, but its primary cause must lie deeper. Little is known of Middle and Southern Rocky Mountains crustal structure, but the data available suggest moderate thickness—perhaps about 45 km (Prodehl and Lipman, 1989). Crustal thickness in the uplifted Northern Rockies is generally 45 to 50 km, somewhat less than that beneath the Montana Great Plains, but compressional-wave velocities in the uppermost mantle are 7.9-8.0 km/s beneath the Northern Rockies compared to 8.2 to 8.4 km/s beneath the adjacent plains (*McCamy and Meyer, 1964*; *Smith and others, 1982*). Crust beneath the Canadian Rockies is only about 37 km thick *Hales and Nation, 1973*).

In summary, the linear Cordilleran belt as preserved in western Canada and Mexico is broken in the western United States by a broad regional uplift, 1,600 km across and centered near the western margin of the Colorado Plateau. This epeirogenic uplift involves several physiographic and tectonic provinces with differing late Cenozoic geologic histories and is largely a result of changes in the upper mantle. A low upper-mantle density is required by isostasy in each of the uplifted tectonic provinces. Upper-mantle seismic velocities are lower than in the stable continental interior. Scattered evidence suggests that the epeirogenic process started slowly near the beginning of the Miocene, significant regional uplift began in the mid-Miocene, but greatest uplift has been in the last 10 m.y.; the process remains active.

Figure 51. Map showing major late Cenozoic extensional features of the Great Basin, Columbia Intermontane, and Northern Rocky Mountain regions. Gray stipple, late Cenozoic basins; diagonal ruling outlined by dashed lines, basalt plateaus; red lines, anticlinal axes of Yakima fold system; red stipple, mid-Miocene basaltic dike swarms: CJDS, Chief Joseph dike swarm, comprising GRDS (Grande Ronde dike swarm) and CDS (Cornucopia dike swarm); MDS, Monument dike swarm; NNRZ, northern Nevada rift zone. Hachured lines, normal faults. AD, Amargosa Desert; AR, Argus Range; BD, Beaverdam Mountains; BFZ, Brothers fault zone; BHR, Beaverhead Range; BM, Black Mountains; C, Coso Range; CCH, Cache Valley; CP, Camas Prairie; CR, Centennial Range; CV, Centennial Valley; DV, Death Valley; DVFZ, Death Valley fault zone; DX, Dixie Valley; EV, Eureka Valley; FCFZ, Furnace Creek fault zone; FLV, Fish Lake Valley; FM, Funeral Mountains; G, Goldfield; GC, Grouse Creek Mountains; GF, Garlock fault; GM, Grapevine Mountains; GRB, Green River basin; GWF, Grand Wash fault; IM, Inyo Mountains; LCL, Lewis and Clark line; LM, Lake Mead; LMFZ, Lake Mead fault zone; LR, Lemhi Range; LRR, Lost River Range; LVVSZ, Las Vegas Valley shear zone; MBH, Mount Bennett Hills; ML, Mono Lake; MM, Mormon Mountains; MP, Modoc Plateau; MR, Mineral Ridge; OH, Okanagan Highlands; OR, Oquirrh Range; OU, Owyhee Uplands; OV, Owens Valley; OWL, Olympic-Wallowa lineament; PM, Pueblo Mountains; PR, Panamint Range; PSZ, Pahranagat shear zone; RLR, Railroad Valley; RR, Raft River Range; RRV, Raft River Valley; SD, Sevier Desert; SM, Steens Mountain; SPR, Spring Mountains; SR, Slate Range; SV, Saline Valley; SWNVF, southwestern Nevada volcanic field; T, Tonopah; TCFS, Trans-Challis fault system; TR, Teton Range; WF, Wasatch fault; WM, White Mountains; WR, Warner Range; WSR, Wassuk Range; YP, Yellowstone Plateau.

RIO GRANDE RIFT SYSTEM

Mid-Miocene and later uplift of the Southern Rocky Mountains region has been accompanied by extension along the Rio Grande "depression" or "rift," a linear system of faulted basins and uplifts associated with predominantly basaltic volcanism (*Bryan, 1938*; *Kelley, 1952*). The characteristic feature of this rift system, a chain of physiographic and structural basins and bordering uplifts, is most prominent from central Colorado to central New Mexico, near the axis of late Cenozoic uplift of the Southern Rockies and separating the relatively unfaulted Colorado Plateau and Great Plains (Fig. 50). Farther north, a weaker system is marked mainly by distributed normal faults; south of central New Mexico, the rift widens, branches, and overlaps with the Basin and Range Province from west Texas through southwestern New

Mexico and beyond. Some authors (e.g., Seager and Morgan, 1979) consider the Rio Grande rift continuous as a distinct feature through southern New Mexico east of the Basin and Range Province into west Texas and northern Mexico; others (e.g., Cordell, 1978) implicitly consider the rift to merge with the broader Basin and Range Province and to lose its distinct identity south of central New Mexico. Here, the connected system of linear fault-bounded rift basins and bordering uplifts is discussed as a component of the broader regional system of basin-range extension; certain geophysical properties characteristically associated with the rift are regionally varying quantities rather than unique identifying features. This view, thus, contrasts with other published definitions that consider the Rio Grande rift to comprise either all extensional features east of the Colorado Plateau and Mexican Highlands or all features associated with particular lithospheric properties.

The present rift throughout its length has been established since mid-Miocene time, with concurrent deposition of the upper part of the Santa Fe Group largely confined to the modern basins (*Bryan, 1938*; *Galusha and Blick, 1971*; *Baltz, 1978*; *Manley, 1979,* 1984; *May, 1984*), and thus is contemporaneous with regional uplift. Before mid-Miocene time, the broad sedimentary basins of the region did not define a continuous system of linear fault-bounded troughs; the younger connected basins of the Rio Grande rift have opened across the older basins, mainly since about 13 to 10 Ma (Morgan and others, 1986), and are nearly filled by deposits of the upper part of the Santa Fe Group and its correlatives. In the centers of the older basins, sediments of the lower part of the Santa Fe Group nearly 2 km thick range from Oligocene to late Miocene in age and are unbroken by significant unconformities (Chapin and Seager, 1975; *Seager and others, 1984*). These older parts of the Santa Fe extend beyond the later basin limits and appear to record extension between about 30 to 26 Ma and about 18 Ma, coeval with mid-Tertiary extension through a broad Mojave–Rio Grande region by broad downwarping and, at least locally, by low-angle normal faulting and large structural rotations (Chapin and Seager, 1975; *Lipman, 1983*; Morgan and others, 1986). In contrast, significant stratal rotations do not accompany the younger faulting (Morgan and others, 1986; *Hagstrum and Lipman, 1986*). Although seismic-reflection profiles suggest that extension has been on listric faults during both the earlier and the later extension (Cape and others, 1983), respective dips of the earlier and later strata exposed within and adjacent to the basins contrast markedly. Some normal faults that bound or break up the rift basins are reactivated older faults (*Slack, 1975*; Ramberg and others, 1978; *Muehlberger and others, 1978*; Tweto, 1979), but major regional structures transect the Laramide and older structural grain (Chapin, 1979). Integration of the Rio Grande drainage along the chain of rift basins dates only from the Pliocene (*Bryan, 1938*; Bachman and Mehnert, 1978). Although rift faulting in many places has continued into the Quaternary (*Scott, 1970*; *Muehlberger and others, 1978*; Bachman and Mehnert, 1978; *Manley, 1979*; *Golombek, 1983*; Personius and Machette, 1984), major faulting was mainly before about 3 Ma (Chapin and Seager, 1975; *Baltz, 1978*). Seismicity is generally low except for local microearthquake swarms, especially near Socorro (Sanford and others, 1979), and no strain accumulation is measurable across the rift (*Savage and others, 1985*).

The northern part of the extensional zone (Fig. 50), included in the rift system by Tweto (1979), is marked by distributed normal faults and associated Miocene and younger sediments from southernmost Wyoming to central Colorado (Izett, 1975; *Colman, 1985*). No connected axial depression, however, extends much north of Leadville, Colorado. Farther south, the rift axis is a graben along the Arkansas and San Luis Valleys, widening from a taper near Leadville to about 40 km in the northern San Luis Valley (Taylor, 1975; Scott and Taylor, 1975; Tweto, 1979). The rift system also includes bordering late Miocene and younger horst blocks, including the Gore and Tenmile Ranges north of Leadville, the Sawatch Range west of the upper Arkansas Valley, and the Sangre de Cristo Mountains east of the San Luis Valley; the broader extensional zone reaches to the Rocky Mountain front. Santa Fe fluvial deposits and their correlatives also occur locally in fault-bounded basins beyond the axial graben, including the Wet Mountain Valley east of the Sangre de Cristo Mountains. Although early subsidence in south-central Colorado began shortly before 26 Ma (Lipman and Mehnert, 1975), horst-and-graben faulting is mainly late Miocene and Pliocene in age (Taylor, 1975).

From the San Luis Valley south nearly to Socorro (Fig. 50), the main feature of the rift is a series of right-stepping en echelon asymmetrical grabens or half-grabens with large fault displacements (commonly greater than 6 km) against an uplifted block along one margin, a complexly warped and faulted opposite margin, and sagged or tilted basin fill (*Kelley, 1952,* 1979; *Baltz, 1978*; Cordell, 1978). The basins are structurally and physiographically connected through constrictions. The en echelon offsets reflect transverse northeast-trending structures, across which the sense of basin asymmetry commonly reverses and along which crustal strain is concentrated (Chapin, 1979).

From about Socorro to Texas and Chihuahua, the rift widens to a complex belt at least 150 km across. The major basins branch into several strands, one extending southward along the Rio Grande south of El Paso, perhaps nearly as far as Big Bend, and another to the Los Muertos basin in northern Chihuahua (Fig. 50; *Muehlberger and others, 1978*; Seager and Morgan, 1979). Though mid-Tertiary extensional basins began to form about 30 to 29 Ma (Chapin and Seager, 1975; Chapin, 1979; Seager and Morgan, 1979), the superposed normal faulting is 23 to 21 Ma and younger in age (*Dasch and others, 1969*; Elston and others, 1976; *McDowell, 1979*; Henry and Price, 1986), most of it associated with 14- to 13-Ma and younger basalts. Faulting continued into the Quaternary although it culminated between 8 and 3 Ma (Seager and Morgan, 1979).

The Rio Grande rift is geometrically and kinematically a zone of enhanced crustal extension (*Woodward and DuChene, 1975*; Ramberg and others, 1978; Golombek and others, 1983)

in a larger region of extension. Ideas of the rift as a collapsed vault, with early regional doming as the cause of rift subsidence (Cordell, 1978), are invalidated by the presence of thick pre- and early extensional fills in the rift basins (Chapin, 1979). Geologic estimates suggest that, although total extension north of central New Mexico is no more than a few kilometers, it is significantly more than the vertical displacement (*Woodward, 1977*; *Golombek, 1983*). As the rift zone widens southward and lies within the Basin and Range Province, it accounts for tens of kilometers of extension, and the ratio of extension to uplift increases, implying clockwise rotation of the Colorado Plateau relative to the Great Plains (Hamilton and Myers, 1966; Golombek and others, 1983). Reprocessed seismic-reflection profiles show moderately dipping smaller faults within the rift north of Socorro, either planar or slightly listric, truncated by a deeper listric master fault (de Voogd and others, 1986a). Orientations of dikes and normal faults indicate a change in the regional stress field from northeast-southwest mid-Tertiary extension to a dominant nearly east-west least principal stress in the late Cenozoic (Zoback and others, 1981; Aldrich and Laughlin, 1984; Aldrich and others, 1986; P. Morgan and others, 1986).

The southern Rio Grande rift (Ramberg and others, 1978; *Sinno and others, 1986*; *Daggett and others, 1986*) has an anomalously low density upper mantle, with a P-wave velocity of 7.7 km/s, and a thin crust, about 26 km, thickening to about 33 km in central and northern New Mexico (*Olsen and others, 1979*). This compares to a 30-km crust and a 7.8- to 8.0-km/s upper mantle beneath the Basin and Range Province to the west (*Gish and others, 1981*). In sharper contrast to both, the crust beneath the Southern Great Plains is about 50 km thick, and upper-mantle velocity is about 8.2 km/s (*Stewart and Pakiser, 1962*). Heat flow in the rift is high, averaging about 80 mW/m^2 in the central part and about 95 mW/m^2 in the southern part (*Reiter and others, 1975, 1986*), higher than in the Basin and Range Province (*Decker and Smithson, 1975*; *Decker and others, 1984*). A mid-crustal magma body may exist beneath the Socorro area (*Sanford and others, 1977*; *Rinehart and others, 1979*; *Larsen and others, 1986*; *Keshet and Hermance, 1986*). Together, these features indicate thinning of the lithosphere beneath the rift, the asthenosphere perhaps even reaching the base of the crust (*Bridwell, 1976*; Seager and Morgan, 1979; *Sinno and others, 1986*). Teleseismic delays indicate anomalous mantle to depths of 70 to 100 km (*Davis and others, 1984*). Although Seager and Morgan (1979) and Sinno and others (*1986*) suggest that these properties distinguish the southern Rio Grande rift from the Basin and Range Province, the Rio Grande rift is only one of several zones of enhanced extension and correspondingly accentuated physical properties within the larger basin-range region.

A comprehensive view of the Rio Grande rift as having a two-stage history has been formulated in recent decades; this chapter adopts a generally consistent view but one that, in the context of broader regional Cordilleran evolution, differs in emphasis. Early rift evolution is commonly considered to have begun in the late Oligocene (about 30 Ma in the south to perhaps 26 Ma in the north) along a zone of Laramide and earlier Rocky Mountain uplifts, reflecting extension by basin formation and low-angle faulting. Later rift extension—mainly late Miocene and Pliocene in age—by normal faulting along a system of connected grabens or half grabens and associated uplifts was coeval with regional uplift. The emphasis adopted here considers the earlier activity as part of a broader regional pattern of mid-Tertiary extension reaching westward from the edge of the High Plains to the Mojave Desert and the California coastal region, and northward to southern Colorado and the Lake Mead area. That mid-Tertiary extension resembled coeval extension in the present Great Basin region and may have propagated northward as extension in the Sierra-Wasatch region propagated southward. During this mid-Tertiary extension, the Rio Grande–Mojave region subsided relative to the Colorado Plateau. The modern Rio Grande rift with its connected graben faulting and high bordering uplifts evolved later, during regional uplift of the central Cordillera, as a highly integrated zone within a broader region of more widely distributed extension.

GILA-SONORAN-MOJAVE REGION

Mid-Miocene and younger extension in the southern Basin and Range Province—the Mogollon Plateau and Mexican Highlands of southwestern New Mexico and southeastern Arizona and the Sonoran and Mojave Desert regions of southwestern Arizona, southeastern California, and adjacent Nevada—has occurred in two stages. Both of these stages were superposed on belts of earlier, mid-Tertiary extension. In the mid-Miocene stage, extension was widespread and was characterized by broad basin formation and mafic to intermediate calc-alkalic magmatism. This episode was succeeded in parts of the region by a magmatic and perhaps a tectonic lull beginning about 18 Ma (Chapin, 1979), then by a later stage of regional, basin-range extension expressed at exposed levels by high-angle normal-fault blocks and predominantly basaltic volcanism. In southern Arizona, Damon and others (1984) characterize the mid-Miocene extension and the succeeding lull as a 12 m.y. transition period between the mid-Tertiary (36 to 24 Ma) and basin-range (12 to 0 Ma) extensional periods.

The earlier stage of late Cenozoic extension in the Mogollon Plateau and Mexican Highlands region (Fig. 50) began in southwestern New Mexico at about 21 Ma (Elston and others, 1976). Sometime later the Mogollon Plateau, underlain by a mid-Tertiary granitic batholith, acquired its present form as a structural block bordered by marginal uplifts and grabens (*Coney, 1976b*). This early extension was not demonstrably associated with basin-range faulting but is recognized from clastic sedimentary rocks commonly mapped as the Gila Conglomerate and interlayered volcanics of 21- to 19-Ma age, from structural control of volcanic vents, and from broad basinward tilts in sedimentary units and associated mafic andesites (*Rhodes and Smith, 1972*; *Coney, 1976b*; Deal and others, 1978). Mid-Miocene extension is also suggested near the northwestern edge of the Mexi-

can Highlands, where mid-Tertiary low-angle faults are cut by high-angle faults partly contemporaneous with mid-Miocene basin sediments and interlayered and overlying 17- to 14-Ma ash flows (*Krieger and others, 1971*; *Krieger, 1974*). Mid-Miocene drainage across an early Tertiary erosion surface in the central Arizona transition zone deposited clastic sediments of the Hickey Formation in stream channels and broad basins, interlayered with basalts dated at 15 to 10 Ma (Anderson and Creasey, 1958; *Lehner, 1958*; Krieger, 1965; McKee and Anderson, 1971; *McKee and McKee, 1972*). Normal faults and differential warping produced displacements of 300 to 1,000 m.

The later, basin-range stage of extension in the Mogollon Plateau and Mexican Highlands began or accelerated at about 13 to 10 Ma. In contrast to the Gila mapped in the Mogollon Plateau, the Gila Conglomerate in southeastern Arizona is younger, filled the precursors of modern basins, and was cut by throughflowing streams in Pliocene and Quaternary time (*Richter and others, 1983*). This fill began to accumulate between 15 and 10 Ma, generally after 13 to 12 Ma (*Krieger and others, 1974*; Scarborough and Peirce, 1978) and corresponded with a change from predominantly intermediate to predominantly basaltic volcanism (Shafiqullah and others, 1978). The typical basins are grabens or half grabens; thick evaporites drilled in some suggest that the lowest basins were closed but partly interconnected to provide sediment traps (*Peirce, 1976*). Mapped late Miocene and younger normal faults trend predominantly northwest and subordinately north-south, dip steeply, and have typical structural relief of 2 to 4 km and a maximum of more than 6 km (Scarborough and Peirce, 1978; Peirce, 1984). Basin-range faulting ended as the dominant influence on landscape evolution in the Mexican Highlands about at 6 Ma (*Menges and McFadden, 1981*; *Elston, 1984b*) although minor faulting continued through Pliocene and Quaternary time (*Morrison and others, 1981*) and minor seismicity remains active (R. B. Smith, 1978). Average elevation remains above 1 km (Fig. 49A). Since 6 to 3 Ma, pediments have cut into many of the ranges and buried the range-bounding faults; more distal source areas are represented in some of the conglomerates (*Richter and others, 1983*). During the late Cenozoic, just as in the mid-Tertiary episode, extension was bounded on the north by the Jemez lineament (Fig. 50), a northeast-trending zone of Precambrian ancestry that had previously influenced Laramide tectonism (Aldrich and Laughlin, 1984). Differential extension across the Jemez zone is taken up by both normal and strike-slip faulting.

The Sonoran Desert of southwestern Arizona is now a tectonically quiescent and topographically low part of the southern Basin and Range Province. Average elevations are between 100 and 1,000 m (Fig. 49A). Broad sedimentary basins began to form across tilted mid-Tertiary units sometime after 19 Ma, mainly after 15 Ma but were broken by northwest-trending normal faults between 13 and 8 Ma (Eberly and Stanley, 1978; Shafiqullah and others, 1980). The mid-Tertiary rocks commonly dip steeply to moderately northeast or southwest, whereas units younger than 14 to 15 Ma dip gently or are undeformed (Reynolds, 1980; Otton, 1982). Some mid-Tertiary or older northeast-trending topographic elements were reactivated or exhumed but do not indicate the late Cenozoic stress field (Eberly and Stanley, 1978; *Reynolds, 1982*). Upper Miocene and younger sedimentary rocks in the grabens of this system include thick evaporites, and associated volcanic rocks are predominantly basalt. Younger pediments are graded toward throughgoing drainages that evolved between 11 and 6 Ma. In contrast to both the Mexican Highlands and the margins of the Colorado Plateau, virtually no Quaternary faulting is recognized in southwestern Arizona (*Morrison and others, 1981*). Without tectonic renewal, the fault-block landscape is subdued, and range-bounding scarps are obliterated beneath the wide pediments. Since about 5 Ma the Gila River has become fully integrated through these basins and flowed to the Gulf of California (Safiqullah and others, 1980).

The lowest part of the Basin and Range Province occurs along the lower Colorado River and adjacent Mojave Desert. (Excluded from this discussion is the Salton Trough, a tectonic continuation of the modern spreading system of the Gulf of California, treated earlier.) In the southern part of the area, exposed mid-Tertiary detachments and steeply tilted upper-plate blocks older than 16 to 14 Ma are cut by high-angle normal and strike-slip faults; immediately succeeding strata dip gently beneath little-deformed uppermost Miocene and younger units, including the marine and brackish-water Bouse Formation and basalts of about 7 Ma and younger age (*Miller and McKee, 1971*; Davis and others, 1980, *1982*; *Hamilton, 1982*; Sunneson and Lucchita, 1983; *Buising, 1990*). To the north, in the Lake Mead area, there was no mid-Miocene tectonic or magmatic lull; extreme extension with large stratal rotations above low-dipping faults occurred between 15 and 13 Ma (Anderson, 1971; Anderson and others, 1972). Younger high-angle normal faults and the present ranges and basins are associated with sedimentary rocks, landslide breccias, and basalts of the Muddy Creek Formation and immediately underlying units (Longwell, 1963; Anderson and others, 1972; Bohannon, 1984); most graben formation and gentle block tilting occurred between 12 and 10 Ma but continued as the basins were largely filled by Muddy Creek deposits until about 5 Ma. Although these late Miocene faults indicate tectonic extension, both total extension and extensional strain rates were lower than in the earlier extension. Present topography near the lower Colorado River reflects more the domal uplifts associated with earlier detachment faulting than any younger modification by high-angle faults (G. A. Davis, 1980). The northern boundary of this tectonic region is the northeast-trending Lake Mead fault system (Fig. 50), along which both extreme mid-Miocene extension and later high-angle normal faulting were transformed into left lateral displacements (Anderson, 1973; Bohannon, 1984).

The central and western Mojave Desert is continuous with the eastern Mojave, lower Colorado River, and Sonoran regions,

but some of its distinctive features reflect its proximity to the San Andreas fault system. Only a few late Miocene and younger northwest-trending normal faults with minor associated stratal tilts resemble those to the east, and displacements generally are small (Dokka, 1986). Since about 20 to 17 Ma, the tectonics of the western and central Mojave region has been dominated by parallel northwest-trending right-lateral faults (Garfunkel, 1974) of individually small displacements; cumulative displacement is about 25 to 40 km (Dokka, 1983). Mid-Miocene fluvial and lacustrine deposits of the 17- to 12-Ma Barstow Formation, unconformable on strongly tilted mid-Tertiary units near the center of the Mojave Desert, were deposited in a broad basin with an east-west northern margin, now broken by the northwest-trending faults (Woodburne and others, 1982; *Woodburne and Tedford, 1982*). The Barstow Formation is folded into an east-west syncline, perhaps reflecting stresses within blocks between stepped strike-slip faults. The northern boundary of the Mojave block is the left-lateral Garlock fault, a transform structure between a northern region of large late Cenozoic extension and a southern region of oblique late Cenozoic deformation and relatively little extension (Davis and Burchfiel, 1973; Wernicke and others, 1982). Domains of short east-west left-lateral faults occur in the northeastern part of the central Mojave Desert (Garfunkel, 1974) and in the Transverse Ranges at its southern margin. Strike-slip seismicity of the central and western Mojave region ends abruptly in the eastern Mojave (Hill, 1978).

Crustal thickness in the southern Basin and Range Province decreases from about 32 km beneath southwestern New Mexico to 28 km beneath southeastern Arizona and 24 km beneath southwestern Arizona, consistent with regional elevation changes; upper-mantle seismic velocities are low, 7.6 to 7.8 km/s (*Sinno and others, 1981*; *Gish and others, 1981*). Heat flow is generally high, about 90 mW/m^2 (*Reiter and others, 1975, 1986*; *Lachenbruch and Sass, 1981*). Farther west, the late Cenozoic tectonic transition from normal faulting in the Sonoran region to strike slip in the Mojave is paralleled by an increase in crustal thickness to about 30 km (*Fuis, 1981*) and a decrease in heat flow to about 70 mW/m^2 (*Lachenbruch and Sass, 1981*), with rather sharp boundaries at the break in active seismicity. Least principal stress throughout the southern basin-range region before about 10 Ma was consistently horizontal and northeast-southwest (Zoback and others, 1981). Younger dikes and faults record a rotation of regional least principal stress to west-northwest (*Rehrig and Heidrick, 1976*; Zoback and others, 1981).

Economic deposits of evaporite minerals and clays have accumulated in some of the mid-Miocene and younger closed, sediment-starved basins, including halite (*Peirce, 1976*), lithium minerals (*Bohannon, 1976*), and borates—notably the Kramer borate deposit of the western Mojave Desert (*Barnard and Kistler, 1966*). Low-temperature hydrothermal convection systems of possible interest for geothermal energy are related to the relatively high heat flow eastward from the lower Colorado River area (*Mariner and others, 1983*), but higher-temperature systems (>90°C) occur only in the belt of still weakly active seismicity in the transition zone and Mexican Highlands and in the Rio Grande rift (*Brook and others, 1979*).

SIERRA-WASATCH REGION

Although belts of tectonic extension began to form in the northern Basin and Range Province as early as the late Eocene and were important during Oligocene and early Miocene time, the present Great Basin and associated modern tectonic and physiographic features date mainly since about 10 Ma (Zoback and others, 1981). As in the southern Basin and Range Province, the mid-Miocene (about 21 to 14 Ma) was a period of transition. Widespread welded ash-flow sheets of Oligocene and early Miocene age were broken by late Miocene and younger faults across the entire region. Distributed extension began with broad mid-Miocene subsiding basins and, at least in places, with low-angle faults and associated large stratal rotations. This mid-Miocene extension was accompanied locally by mafic and intermediate volcanism but regionally by a magmatic lull (McKee and others, 1970). Parallel linear ranges and basins broke across mid-Miocene and older features and by some time between 14 and 10 Ma dominated the region, which remains tectonically active, especially near its margins (Christiansen and McKee, 1978; Stewart, 1978). Estimates of total post-mid-Miocene extension vary greatly. Earlier views generally proposed several tens of kilometers (e.g., Thompson and Burke, 1974), but much greater extension had been advocated by Hamilton and Myers (1966). Most present workers now agree that the region has extended by a large percentage of its pre-late Eocene width, but the amounts, timing, and style of extension have varied with locale.

Mid-Miocene basins and extensional faulting

The mid-Miocene formation of broad depositional basins possibly reflects crustal thinning across areas much larger than the present individual basins or ranges, but the early basins were bounded and at least locally broken by normal faults. Remnant features of this early broad-basin phase of regional extension are found across much of Nevada.

In southern Nevada, lacustrine and fluvial deposits of the Horse Spring Formation of 21 to 12 Ma (Anderson and others, 1972; Bohannon, 1984) accumulated during extension and rotation on listric or low-angle faults (Guth, 1981; Bohannon, 1984). The Horse Spring extends from the southeastern margin of the Great Basin region near Lake Mead (Fig. 51), into the ranges north of Las Vegas Valley (*Longwell and others, 1965*; Anderson and others, 1972; Guth, 1981).

About 100 km northwest of Las Vegas (Fig. 51), a mid-Miocene depositional basin extended across several present ranges near the margin of the younger southwestern Nevada vol-

canic field (Christiansen and others, 1977). Deposits of 29 Ma in this area (*Marvin and others, 1970*) were correlated by *Hinrichs* (*1968*), *Sargent and Stewart* (*1971*), and *Barnes and others* (*1982*) with the Horse Spring Formation but are considerably older and correlate with the Oligocene part of the Titus Canyon Formation near northern Death Valley (C. Stock and Bode, 1935; *Reynolds, 1969,* 1976). Stratigraphically above both the Titus Canyon (*Cornwall and Kleinhampl, 1964*; *Reynolds, 1969*) and the correlative deposits northwest of Las Vegas Valley, the tuffaceous rocks of Pavits Spring filled a broad basin (*Poole and others, 1965*; *Hinrichs, 1968*; *Sargent and Stewart, 1971*; *Barnes and others, 1982*). Fossil fragments suggest a Miocene age (*Poole and others, 1965*); conglomerate clasts include the 18-Ma Hiko Tuff from a source nearly 100 km to the east (Armstrong, 1970; *Noble and McKee, 1972*; *Barnes and others, 1982*). The overlying Crater Flat Tuff is 14 Ma in age (*Marvin and others, 1970*). Lithologically and stratigraphically similar sandstone and claystone in the southern Amargosa Desert (*Denny and Drewes, 1965*), possibly correlative with the rocks of Pavits Spring, include fanglomerate and megabreccia lenses deposited in a fault-block basin.

Lower Miocene alluvial-fan deposits succeeded by lacustrine beds east of central Death Valley define a 25- to 20-Ma basin sequence unconformably overlain by a mid-Miocene alluvial-fan sequence; the earliest associated evidence of normal block faulting, about 20 Ma, is high in the lower sequence (*Cemen and others, 1982*). The upper basin sequence onlaps a source area between the present Funeral and Black Mountains (Fig. 51); both sequences are steeply east-tilted beneath 14-Ma deposits in a trough related to the present topography of both ranges and to the right-lateral Furnace Creek fault zone between them. Thus, extension appears to date from about 20 Ma but present physiography only from about 14 Ma.

At the western edge of the region, just north of the Garlock fault, the Ricardo Formation overlies lower Tertiary strata that had been tilted 10 to 30° northeastward; the Ricardo basin accumulated coarse basal alluvial fans, overlain by an 18- to 15-Ma volcanic sequence (*Cox and Diggles, 1986*). Thus, mid-Miocene basin deposits, associated with local extensional faulting but covering large areas broken by later faults, characterized the entire width of the southern Great Basin region.

Farther north, in southern central Nevada, widespread clastic sedimentary rocks and conformably overlying ash-flow tuff (mapped as the Fraction Tuff but probably including several units between 21 and 16 Ma in age) covered an erosional surface of considerable relief (Ekren and others, 1971; Bonham and Garside, 1979). Near Goldfield and Tonopah, Nevada (Fig. 51), tuffaceous sedimentary and volcanic rocks mapped as the Siebert Tuff overlie the Fraction Tuff and older rocks and in turn are tilted beneath younger Miocene strata and intruded by 16-Ma dacites (*Albers and Stewart, 1972*). Thus, widespread extension began in the region from Goldfield and Tonopah eastward into central Nevada shortly before about 18 Ma, accompanied at least locally by basin sedimentation and broadly by the filling of a low region by ash flows. Farther east in south-central Nevada, mid-Miocene extension was compartmented by strike-slip faults into smaller domains. For example, G. A. Davis (1980) suggests that left-lateral faults of the northeast-trending Pahranagat shear zone (Fig. 51) separate areas of nonaligned normal faulting to the north and south.

In northeastern Nevada, a wide range of sedimentary rocks was once included in the Humboldt Formation. Because the older strata were related to a different tectonic setting (*Van Houten, 1956*)—discussed in an earlier section—the formation has been restricted and redefined to include only Miocene strata (Smith and Ketner, 1976). Thus restricted, the Humboldt consists mainly of fine-grained tuffaceous fluvial and lacustrine deposits with some coarser clastic rocks and limestones. Conglomerates record the progressive denudation of Tertiary volcanic rocks, miogeoclinal strata, and crystalline rocks from the Ruby Mountains. The Humboldt Formation, unconformable on Oligocene and older units, partly underlies and partly overlies a 15-Ma rhyolite; zircon from a tuffaceous unit is 10 Ma. Smith and Ketner note that normal faulting preceded and accompanied deposition of the Humboldt, but at least much of its deposition predates present topographic and structural relief (*Van Houten, 1956*).

Northern Nevada rift zone

The broad structural basins and their sedimentary fill in the northern Great Basin region were broken along a northwest-trending rift zone late in the mid-Miocene extensional episode (Zoback and Thompson, 1978). In Nevada, this rift (Fig. 51) is marked by a linear swarm of dikes and associated lavas of 17 to 14 Ma age, by a regional aeromagnetic anomaly (the source of which may extend to a depth of 15 km), and by a fault-bounded trough (*Robinson, 1970*; Christiansen and McKee, 1978; Zoback and Thompson, 1978). *Stewart and others* (*1975*) suggested that this zone in Nevada might be an arcuate continuation of the Brothers fault zone trending west-northwest across central Oregon. Greater similarities of age and tectonic function, however, are suggested (Christiansen and McKee, 1978; Zoback and Thompson, 1978) with rift structures and dikes of 17- to 14-Ma age along the western Snake River Plain (*Mabey, 1976*; Mabey and others, 1978) and the Chief Joseph dike swarm (*Taubeneck, 1970*) in western Idaho, northeastern Oregon, and southeastern Washington (Fig. 51)—the principal source of basaltic eruptions of the Columbia Plateau.

Interpretation of the northern Nevada rift, western Snake River Plain, and Chief Joseph dike swarm as tectonically related has several important implications. Zoback and Thompson (1978) showed that the trend of the rift zone is consistent with west-southwest regional mid-Miocene least principal stress before it shifted clockwise about 10 Ma to its present northwest orientation. The apparent right offset of the Nevada part of the zone relative to the western Snake River Plain by 70 to 100 km might indicate either a mid-Miocene transform relation or subsequent right offset of an originally continuous 700-km rift zone. In the

TOPICAL SYNTHESES

Plates
(in accompanying slipcase)

Contents

first interpretation, southwestward extrapolation along the Humboldt zone of the N55°E eastern Snake River Plain trend, suggested from regional aeromagnetic anomalies by Mabey and others (1978), would be an appropriate transform orientation. In the second interpretation, 100 km of offset of a once-continuous feature would imply about 100 km of extension on basin-range structures across the north-central Great Basin and very little extension north of the Humboldt zone. The N58°W Brothers fault zone, part of the northern transform boundary of younger basin-range extension (*Lawrence, 1976*; Christiansen and McKee, 1978), would be an appropriate orientation for right-lateral offset. Extensive younger volcanic rocks of the Owyhee uplands (compare Figs. 51, 53) cover and obscure the critical relationships that might help distinguish between these interpretations. Both factors may account for parts of the apparent offset of the northern Nevada rift zone.

The late Miocene to Quaternary Great Basin region

The Great Basin region typifies modern basin-range structure and physiography, with distributed extension on roughly parallel normal faults. The elongate and subparallel fault blocks generally trend between north and northeast, are 100 to 200 km long, and have crest-to-crest spacing of 25 to 35 km. Stewart (1978, 1980) noted that large domains are characterized by similar tilts, separated from areas of opposite tilt by longitudinal "antiformal" or "synformal" axes or by linear transverse offsets.

The tectonic northern and southern boundaries of the region are late Cenozoic strike-slip systems that accommodate differential extension in a manner analoguous to oceanic transform faults (Fig. 51). Faults in these bounding systems vary in trend and history; preexisting structures guided at least some of them. Important left-lateral features of the southern boundary are the Lake Mead fault zone in Nevada (Anderson, 1973; Bohannon, 1979, 1983) and the Garlock fault in California. The central Garlock fault has displacement of about 65 km, best shown by an offset Jurassic dike swarm (*Smith, 1962*), transforming differential extension between the distended region to the north and the relatively stable Mojave block (Davis and Burchfiel, 1973; Wernicke and others, 1982). *Kistler and Peterman* (*1978*) conclude from Sr-isotopic compositions of granitic rocks that part of the Garlock fault follows the old rifted margin of Precambrian continental crust. East-west seismicity through southern Nevada delimits the higher, more active Great Basin region from the contrasting Mojave-Sonoran region (Smith and Sbar, 1974; Eaton and others, 1978; Best and Hamblin, 1978). The northern boundary is broader and more diffuse but similarly transforms differential extension. In the northeast, the principal boundary is the eastern Snake River Plain, probably following Precambrian structures (*Eaton and others, 1975*; Christiansen and McKee, 1978); southwest extrapolation of the eastern Snake River Plain along the Humboldt zone may trace a mid-Miocene transform boundary. Lesser extension north of the Snake River Plain is accommodated along the rejuvenated Trans-Challis fault zone (Bennett, 1986) and the Lewis and Clark line (Reynolds, 1979). Right-lateral faults in Oregon accommodate differential extension farther west (*Lawrence, 1976*). For convenience, the present discussion is restricted to areas south of the eastern Snake River Plain, the Humboldt zone, and its southwestward extrapolation; tectonic extension farther north is discussed in the following section of this chapter.

The east margin of the Great Basin extends northward from the Grand Wash Cliffs, near the western Grand Canyon (Fig. 51). Late Miocene and younger north- to northeast-trending normal faults postdate the mid-Miocene Horse Spring Formation, and sediments of the Muddy Creek Formation and related units were deposited in the resulting basins after 12 to 10 Ma (Bohannon, 1984). A wide transition zone lies mainly within the High Plateaus (Fig. 51; Best and Hamblin, 1978; Anderson and Mehnert, 1979). The High Plateaus were not differentiated from areas farther west before at least 21 Ma (Anderson and Mehnert, 1979). Faulting has coincided with accelerated regional uplift after 14 to 12 Ma (*Callaghan, 1939*; Best and Hamblin, 1978; *Rowley and others, 1978b*). Rowley and others (1979) speculate that clastic accumulations in the adjacent Basin and Range Province might be as old as 20 Ma. In a large structural domain west of the High Plateaus, consistently eastward tilt initiated with low-angle faults from west to east during deposition of the mid-Miocene Horse Spring Formation (Guth, 1981) but evolved fully after 14 Ma (Wernicke and others, 1985) with formation of the Las Vegas Valley shear zone (Fig. 51), having as much as 65 km of right-lateral faulting and crustal bending (Longwell, 1960, 1974; Ekren and others, 1968; Bohannon, 1984). Increasing tilt from east to west suggests continued rotation of older blocks above younger faults that initiated farther east and below them; total extension may be nearly 100 percent. In the Mormon Mountains (Fig. 51), Wernicke and others (1985) interpret at least 8 km of extension, with westward transport above a detachment that formed at an initial westward dip of 20 to 25° across older thrust faults, with flexure of the higher blocks by rotation on structurally lower detachments. Northwest of the High Plateaus, the Sevier Desert region (Fig. 51; Zoback, 1983) is topographically and structurally low and had late Miocene to Quaternary volcanism, but seismicity is low. A listric fault, the Sevier Desert detachment, passes at least 15 km through the crust from a normal-fault zone at the surface (Allmendinger and others, 1983; Smith and others, 1989). Planar and listric normal faults, including some of late Quaternary age (*Bucknam and Anderson, 1979*; Crone and Harding, 1984), are truncated by the detachment below 3 to 4 km. Other faults in the area show evidence of consistent strike slip (Zoback, 1989).

Farther north, the eastern boundary of the Great Basin is the recurrently active, west-dipping Wasatch fault (Fig. 51; *Gilbert, 1928*). Measured dips average 60°, both on the Wasatch fault (Zoback, 1983) and on the west-bounding fault of the Oquirrh Range (*Gilluly, 1928*). Seismicity indicates average fault dips of about 40 to 60° but correlates poorly with mapped faults. The faulting is largely younger than 12-Ma basalts (*Moore and*

McKee, 1983). In northern Utah and adjacent Nevada and Idaho, the upper Miocene and Pliocene Salt Lake Formation and correlative strata record basin-range extension beginning at about 10 Ma (*Mansfield, 1927*; *Morris and Lovering, 1961*; *Heylmunn, 1965*). East of the Grouse Creek–Raft River "core complex," conglomerate, megabreccia, and vitric tuff are unconformable on a tectonically rooted unmetamorphosed section (Todd, 1983), but farther west they occur only in a composite extensional allochthon of Tertiary and unmetamorphosed Triassic and Permian rocks separated from the core rocks by low-angle faults (Armstrong, 1968; *D. M. Miller, 1980*; *Compton, 1983*). Emplacement of this composite allochthon postdates included basalts and rhyolites of 14- to 12-Ma age. Brecciated sheets of pre-Tertiary rocks east of the core ranges, some carrying 11-Ma basin sediments, form a Miocene denudation sequence that preserves an inverted "core complex" stratigraphy (Todd, 1983). North of the Raft River Mountains, the exposed Salt Lake Formation includes 11- to 7-Ma basalts and rhyolites; the adjacent Raft River basin has 1,700 m of faulted and hydrothermally altered fill with no interlayered volcanics but is cut by east-dipping listric normal faults that flatten within the fill, indicating basin opening between 10 and 5 Ma (Williams and others, 1982; Covington, 1983). Similar stratigraphic relations continue eastward south of the Snake River Plain (*Mapel and Hail, 1956*; *Williams, 1964*; Trimble, 1976; Trimble and Carr, 1976). Normal faults in eastern Idaho displace a widespread 6-Ma ash-flow tuff on an erosion surface of low relief by more than 600 m (*Allmendinger, 1982*); above, the Salt Lake Formation includes basin deposits related to uplift of adjacent ranges between 6 and 5 Ma. Although *Allmendinger* (*1982*) regards this as the initial basin-range faulting of the area, regional onlaps of the ash flows show that the ranges and basins were blocked out earlier. Lacustrine and fluvial sedimentary rocks on the flanks of the Teton Range in eastern Idaho and western Wyoming were deposited in basins about 9 Ma (J. D. Love, 1956; *Love and others, 1973*). Late Miocene and Pliocene normal faulting, tilting, and basin formation farther south merge with pedimentation and aggradation north of the Green River basin (*Dorr and others, 1977*).

The western margin of the Great Basin shows both similarities and differences relative to the eastern margin. North of the Garlock fault near the southernmost Sierra Nevada (Fig. 51), the mid-Miocene basal part of the Ricardo Formation is overlain unconformably by 14- to 8-Ma deposits of the middle part of the Ricardo, eroded from the high-standing Mojave block, and post-8-Ma alluvial fans from the rising Sierra Nevada (*Cox and Diggles, 1986*). These alluvial strata, correlative with the Kern River Formation west of the Sierra, indicate late Miocene uplift of the range.

Near Death Valley (Fig. 51), faulting and tilting have been active from the mid-Miocene through the Holocene (Hunt and Mabey, 1966; *Hooke, 1972*) although the area is seismically quiet. In the southern Death Valley region, sheets of brecciated Precambrian, Paleozoic, and Tertiary rocks mapped as the "Amargosa chaos" by Noble (1941) lie against striated fault surfaces of basement rocks known as "turtlebacks." East-dipping basin fill and 14- to 5-Ma eruptive rocks are faulted down onto the turtlebacks and other west-dipping faults (*Drewes, 1963*; Hunt and Mabey, 1966; Hill and Troxel, 1966; Fleck, 1970; *Cemen and others, 1982*). Coherent upper Tertiary cover sections were deposited in adjacent basins and tilted against listric faults as the basins widened (Wright and others, 1974; *Otton, 1976*). Similar features continue at least 100 km southeast (Hewitt, 1956; *Kupfer, 1960*) and southwest (*Wagner, 1988*). Central Death Valley is a pull-apart basin between the Black and Panamint Mountains (Burchfiel and Stewart, 1966) and the right-lateral Furnace Creek and Death Valley fault zones (Hill and Troxel, 1966). The Furnace Creek fault zone has had 50 to 80 km of right slip since the mid-Miocene (Stewart, 1967; McKee, 1968; *Cemen and others, 1982*). The Death Valley fault zone, with right slip of only a few kilometers, ends near the Garlock fault (Wright and Troxel, 1967; Davis and Burchfiel, 1973; *Davis, 1977*; *Spencer, 1990*). *Stewart* (*1983a*) and *McKenna and Hodges* (*1990*) suggest that the Panamint Mountains (Fig. 51) once structurally overlay the Black Mountains, moving tens of kilometers on the Amargosa detachment system. Folded Miocene detachments within the Panamint Mountains overlie a metamorphosed lower plate, probably above a buried active detachment (Wernicke and others, 1986; *Hodges and others, 1990*). North-trending normal faults in the Grapevine Mountains, northeast of Death Valley, flatten downwards into a west-dipping detachment with large recumbent west-verging folds in the highly extended upper plate (Reynolds, 1976). *Snow and White* (*1990*) suggest that the Cottonwood Mountains west of northern Death Valley, analogous to the Panamint Mountains to the south, were detached from a position above the Grapevine Mountains east of Death Valley.

A larger regional system of latest Cenozoic oblique extension reaches beyond Death Valley in the western Great Basin. To the east, ranges tilt eastward toward a west-dipping normal fault at the west base of the Spring Mountains with 3.5 km of displacement (Fig. 51; Burchfiel and others, 1974, 1983). Low-dipping normal faults within the ranges are inactive; the range-bounding faults may merge into detachments at depth, suggesting a continuum of late Cenozoic extension. To the north, sedimentary and volcanic deposits of 13 to 10 Ma include the Esmeralda Formation and other strata in fault-bounded basins (Robinson and others, 1968; Stewart and Diamond, 1990) superposed on the broad mid-Miocene basin of 21- to 16-Ma welded tuffs. The Esmeralda and underlying rocks have been detached along an extensive low-angle fault from the metamorphic rocks of Mineral Ridge below and later broken up into the ranges and basins of the present topography (Stewart and Diamond, 1990). Much of southwestern Nevada and eastern California is characterized by right-lateral faulting and bending on northwest trends. Aggregate right-lateral displacement of nearly 200 km (*Albers, 1967*; Stewart and others, 1968) records the imprint of San Andreas–type shear on basin-range extension (Atwater, 1970; Zoback and others, 1981). The eastern boundary of this zone is the Walker

Lane, a topographically low, discontinuous alignment of faults and abrupt changes in range and basin trends (*Locke and others, 1940*); the entire right-lateral zone is designated the "Walker Lane belt" by Stewart (1988). In southern Nevada, drag on the Las Vegas Valley shear zone is younger than 14 Ma (Ekren and others, 1968). Preexisting structures, however, controlled many late Cenozoic structures of the Walker Lane belt (*Speed and Cogbill, 1979c*), and Mesozoic displacements may account for significant offset (Stewart, 1988). Zoback (1989) interpreted the coexistence of strike-slip, normal, and oblique-slip faulting in this region as indicating near equality of vertical principal stress and maximum horizontal principal stress, the north-south principal stress varying with time.

Central western Nevada, from the Wassuk Range to Yerington (Fig. 51), actively eroded after burial by arc volcanics between about 21 and 15 Ma, became an integrated depositional basin after 13 Ma; up to 2.5 km of sediments accumulated during normal faulting on northwest trends (Gilbert and Reynolds, 1973). Although local normal faulting was as old as 18 to 17 Ma (Proffett, 1977), it was most intense between 13 and 10 Ma (Gilbert and Reynolds, 1973). Active east-dipping listric faults extended to depths of at least 13 to 16 km; overlying strata and inactive faults rotated westward as new faults continued to initiate, producing extension of more than 100 percent and crustal thinning by half (Proffett, 1977). By 8 Ma, the once integrated basin was broken into numerous smaller basins and ranges; general quiescence followed until the more widely spaced modern basins and ranges formed about 4 to 3 Ma (Gilbert and Reynolds, 1973). Northward-retreating mid-Miocene arc volcanism continued to be succeeded by upper Miocene basins farther north, such as that of the Truckee Formation (Thompson and White, 1964; Bonham, 1969), and tilted as much as 30°.

The youngest initiation of extension in the Great Basin region has been in the southwestern corner, from west of Death Valley to the Sierra Nevada (Christensen, 1966; Bachman, 1978; Duffield and others, 1980). Before latest Miocene and Pliocene time, a continuous gentle erosion surface bevelled the area of the present east-tilted Panamint, Slate, and Argus Ranges and the west-tilted Coso and Inyo Ranges, White Mountains, and Sierra Nevada (Fig. 51). Upper Miocene to Pleistocene volcanic and alluvial-fan deposits west of the Panamint Mountains dip eastward into a listric fault and were separated from similar deposits in the Panamints less than 4 m.y. ago (*Hall, 1971*; *Coleman and Walker, 1990*). Extension from the Coso Range to the southern Owens Valley began by 7 to 8 Ma (*Bacon and others, 1982*; *Schweig, 1989*). The earliest recognized faulting was 3 Ma and now includes right-slip, normal, and oblique-slip faults (*Weaver and Hill, 1978*; Duffield and others, 1980; Zoback, 1989). Early lacustrine deposits of 2.3 Ma in Owens Valley are now offset by as much as 2.3 km (Bachman, 1978). Major seismicity passes northward through the White Mountains, Owens Valley, and Sierran blocks into west-central Nevada (Hill, 1978; R. E. Wallace, 1984b; *Hill and others, 1985*). A chain of young deep basins east of the seismic belt—Panamint Valley, Saline Valley, and Eureka Valley—and faulted Pliocene volcanic rocks cap a pull-apart basin nearly 2 km deep (*Blakely and McKee, 1985*). The Pliocene Mono Lake basin is a southwest-plunging extensional warp surrounded and partly filled by volcanic and sedimentary rocks, 4 to 2 Ma in age, with 1.8 km of vertical displacement at the southwest end (Gilbert and others, 1968; Christensen and others, 1969).

Although regional physiography demonstrates widely distributed normal and strike-slip faulting in the Great Basin region through late Tertiary and Quaternary time, Holocene faulting and seismicity are concentrated near the eastern, southern, and western margins and in a belt extending north-northeast into central Nevada (Smith and Sbar, 1974; R. B. Smith, 1978; *Howard and others, 1978*; R. E. Wallace, 1984b; *Hill and others, 1985*; Zoback, 1989). Since mid-Miocene time, regional least principal stress (exept in some areas near the margins) has rotated from west-southwest to west-northwest (*Zoback and Zoback, 1989*; Zoback and others, 1981; Zoback, 1983; *Vetter and Ryall, 1983*), but old structures continue to take up much of the displacement by oblique slip or selective reactivation of favorable segments (*Thompson and Burke, 1973*; *Wallace, 1984a*). Transitions to strike-slip deformation along the western and southern margins suggest an imprint of the relative motion of the Pacific and North American Plates (*Zoback and Zoback, 1989*; Zoback and others, 1981; Sbar, 1982).

No comprehensive picture of basin-range extension and kinematics is wholly satisfactory. Scarcely more than 25 years ago, agreement on the extensional origin of basin-range structures was still being consolidated (Gilluly, 1963; Hamilton and Myers, 1966). Wernicke (1985; Wernicke and others, 1985) suggests that deeply penetrating listric normal faults become master detachments in a system of uniform simple shear through much or all of the lithosphere. Control of the regular spacing and size of basin-range blocks, however, is not explicit in the model (Zoback, 1983). *Hamilton* (*1980, 1982*) favors shallow brittle fracturing and block rotation above listric faults that merge in a mid-crustal zone of partly brittle, partly ductile behavior to produce giant boudins. Eventual exposure of detachment zones in crustal arches may reflect tectonic denudation and isostatic unloading (Spencer, 1984) or reverse-drag flexure of the upper plates (Wernicke and others, 1985). Various hypotheses consider strike-slip faulting to be minor (*Scholz and others, 1971*), or that both normal and strike-slip deformation accommodate oblique regional extension (Hamilton and Myers, 1966; Atwater, 1970; Christiansen and McKee, 1978), or that normal faulting is secondary to dextral shear (*Livaccari, 1979*). Strike-slip zones also transform differential extension between crustal domains (Davis and Burchfiel, 1973; G. A. Davis, 1980; *Greensfelder and others, 1980*; Wernicke and others, 1982). The view of extension as a collapsed uplifted arch (Best and Hamblin, 1978) is no longer tenable as horizontal extension greatly exceeds vertical uplift. Middle- and lower-crustal extension might be accommdated by igneous intrusion (Anderson, 1971; Wright and Troxel, 1973; Thompson and Burke, 1974), ductile flow (Compton and others,

1977; Proffett, 1977; *Smith and Bruhn, 1984*), or lithospheric shearing displaced laterally from associated surficial faulting (Wernicke, 1985; Davis and others, 1986). Each of these mechanisms may have been involved to different degrees in different parts of the basin-range region (*Anderson and others, 1983*).

Estimates of overall extension range widely with assumptions. Assuming deeply penetrating faults with average dips of 60° (Stewart, 1971) or extrapolating observed extension in the 1954 Dixie Valley earthquake and the offset of Pleistocene shorelines (*Thompson and Burke, 1973*) yielded estimates of 10 to 15 percent. Hamilton (1978) estimated nearly 70 percent extension by thinning a crust like that of the Colorado Plateau. Davis and Burchfiel (1973) recognized 100 percent extension north of Garlock fault, and Wernicke and others (1982) summed late Miocene and younger transform displacements along the entire southern margin of the Great Bsain to at least 140 km, an increase of 65 percent. Later, more complete reconstructions demonstrate nearly 250 km of net west-northwest extension between the southern Sierra Nevada and the Colorado Plateau during the past 15 m.y. (*Wernicke and others, 1988*). Calculations based on seismic-monment relase indicate extension across the Great Basin region at about 10 mm/y (Smith and others, 1989), consistent with geologically (*Wernicke and others, 1988*) and geodetically (*Minster and Jordan, 1987*) determined rates.

Crustal properties of the Great Basin region are distinctive, with average elevation and bouguer gravity regionally symmetrical about a north-south axis (Eaton and others, 1978; *Mabey and others, 1983*). The region is in approximate isostatic equilibrium, but individual basins and ranges are supported by the strength of the lithosphere. The lack of magnetic character or seismic expression of the western edge of Precambrian crystalline basement (*Kistler, 1983*), suggests thermal homogenization (J. H. Stewart and others, 1977; Mabey and others, 1978). The crust is 20 to 35 km thick (R. B. Smith, 1978; *Prodehl, 1979*), with a transition to the Colorado Plateau about 100 km wide (*Shuey and others, 1973*; *Keller and others, 1975*; *Smith and others, 1975,* 1989) and minimum thickness in the northwest, beneath the Lahontan basin (R. B. Smith, 1978; *Stauber, 1983*). Upper-mantle seismic velocities are low, averaging 7.5 to 7.7 km/s beneath the margins and 7.7 to 7.9 km/s beneath the center (R. B. Smith, 1978). The thin lithosphere may be nearly equivalent to the crust itself or only slightly thicker (*Priestly and others, 1980*; *Olsen and others, 1980*; *Patton and Taylor, 1984*). *Thompson and others* (*1989*) document a model for the western Great Basin in which a crust-mantle transition zone of varied thickness has its base at a relatively constant depth of about 35 km. Conductive heat flow throughout the region is high, generally between 60 and 100 but as high as 140 mW/m^2, and can be accounted for by lithospheric extension and associated basaltic intrusion without anomalous conductive flux from the asthenosphere (Lachenbruch and Sass, 1978).

Mineral and energy resources in the Great Basin region reflect basin-range faulting, high heat flow, and basin sedimentation (in addition to magmatic systems discussed later). Upper crustal fracturing and high regional heat flow promote deep circulation, heating, and convection of meteoric water (*Yeamans, 1983*) and a significant geothermal resource potential (*Mariner and others, 1983*), particularly in the zones of active seismicity near the eastern and western margins or in the Battle Mountain heat-flow high along the Humboldt zone (*Brook and others, 1979*). Only Railroad Valley in central Nevada has had significant petroleum production (*Duey, 1983*), but other basins have unexploited hydrocarbon potential (*Bortz, 1983*). Oil is generated from various Paleozoic to lower Cenozoic source rocks and matured by the high regional heat flow (*Poole and others, 1983*). The basins also hold significant saline mineral resources. Most of the known borate deposits of the western United States are in mid-Miocene and younger basins of the Mojave Desert and Great Basin (*Smith, 1985*). Searles Lake, north of the Garlock fault, has produced more than $1 billion worth of borates, bromine, lithium, and phosphates (G. I. Smith, 1979).

In summary, late Cenozoic extension in the Great Basin region contrasts with mid-Tertiary extension, which was locally intense but not regionally distributed. Mid-Miocene extension began with broad shallow basins bounded by normal faults and evolved by continued extension and successive breakup into complexes of basins and ranges with low-angle faults and prominent detachments. Regional least principal stress was west-southwest. Beginning about 17 Ma, an axial rift opened along a north-northwest trend from the central Great Basin through the western Snake River Plain to the eastern Columbia Plateau. By 14 Ma, extension was distributed across most of the region except the southwestern sector. After about 10 Ma, the regional stress field rotated 45° clockwise, perhaps reflecting a superposed influence of coastal transform tectonics; renewed extension produced the present physiography. In latest Miocene to Quaternary time extension has encroached southwestward to the Sierra Nevada and overprinted slightly older structures in the western Great Basin.

COLUMBIA INTERMONTANE REGION AND NORTHERN ROCKY MOUNTAINS

Oregon east of the Cascades and south of the Blue Mountains Province, and adjacent parts of Idaho, Nevada, and California were designated the Oregon Plateaus by Waters (1962) in order to distinguish them clearly from the Columbia Plateau, with which they are often included as a physiographic province (*Fenneman, 1931*) but from which they differ in tectonic and magmatic evolution. For the larger region, including the Oregon Plateaus, Columbia Plateau, and the High Lava Plains of Oregon and Idaho, we adapt a term from *Thornbury* (*1965*) and designate a large, predominantly volcanic-plateau region as the "Columbia Intermontane region," following Christiansen and McKee (1978) in enlarging Thornbury's province to include all of the Oregon Plateaus.

Early in mid-Miocene time the Oregon Plateaus (Fig. 51) were a great tectonic basin, bounded on the west and north by the constructional volcanic Cascades and the uplifted Blue Mountains region. To the northeast, now across the graben of the western Snake River Plain, was the uplifted Idaho batholith. The south was open to the central Great Basin. Since then, the Oregon Plateaus basin has accumulated thick volcanic strata and sedimentary rocks of volcanic derivation (*Shotwell and others, 1963*; Kittleman and others, 1965; Walker, 1977, *1979*), broken by normal faults into gently tilted blocks (*Donath, 1962*; Walker, 1977). Numerous short northwest- and north-trending faults near the southwestern margin form a zone 20 to 50 km wide east of the High Cascades (Fig. 51). The Modoc Plateau of northeastern California and south-central Oregon has a surface of upper Miocene and lower Pliocene basaltic lavas broken by relatively small northwest-trending faults (Macdonald, 1966; *Peterson and McIntyre, 1970*; McKee and others, 1983). Normal faults farther east form a reticulated set of ranges trending generally north to north-northeast and smaller northwest-trending blocks. The faults displace mid-Miocene and younger volcanic rocks (Walker, 1977, *1979*); many of the scarps are only slightly eroded. Some faults are nearly vertical to a few hundred meters depth (*Donath, 1962*; *Donath and Kuo, 1962*), but range-bounding faults dip 50 to 60° (*Gilbert, 1928*; *Fuller and Waters, 1929*). Block faulting and tilting in the Warner Range, southwesternmost of the major ranges, initiated sometime between 14 and 7 Ma (Duffield and McKee, 1986). Steens Mountain, the Pueblo Mountains, and parallel valleys in southern Oregon mainly postdate a 9-Ma regional ash-flow sheet (*Walker, 1979*; Rytuba and McKee, 1984). Several northwest-trending, mainly right-lateral fault zones across eastern Oregon accommodate differential extension that decreases incrementally northward (*Lawrence, 1976*; *Lillie and Couch, 1979*; Malde, 1991). Most prominent of these is the Brothers fault zone, a 240-km-long zone of short en echelon shears younger than 6 Ma (*Stewart and others, 1975*; Walker, 1977; *Walker and Nolf, 1981). Rytuba and Conrad (1981*) describe an "Orevada rift" (see Fig. 1), parallel to the Brothers fault zone near the Nevada-Oregon border; pre-Tertiary rocks are exposed widely to the south but are not exposed to the north until they rise again on the flank of the Blue Mountains. The Owyhee Upland of the eastern Oregon Plateaus is a larger counterpart of the little-deformed Modoc Plateau on the west (*Kittleman and others, 1967; Ekren and others, 1981*). The earliest upper Cenozoic volcanic and sedimentary rocks were deformed late in mid-Miocene time, and the resulting basins were largely filled in the late Miocene and Pliocene.

The progressive northward decrease in crustal extension through the Oregon Plateaus imparts vertical-axis rotations to their western and northern margins (Magill and others, 1982; Grommé and others, 1986; Wells and Heller, 1988). Paleomagnetic declinations show that Eocene volcanic rocks of the Clarno Formation north of the plateaus, Oligocene to lower Miocene rocks of the Cascades, and Upper Cretaceous to Oligocene sedimentary rocks of the Klamath block have rotated 15 to 30° clockwise and, thus, are not part of a rigid block with the Oregon Coast Range, which has 50 to 70° of clockwise post-Eocene rotation (Simpson and Cox, 1977; *Beck and Plumley, 1980*; Magill and others, 1982). This decoupling of blocks might reflect a superposed distributed right shear (Wells and Coe, 1985; Wells and Heller, 1988), but it can be interpreted as resulting from the differential extension of crustal domains (Grommé and others, 1986). Reconstruction of the rotations suggests 225 to 400 km of post-Eocene extension at the latitude of the southern Oregon border (Grommé and others, 1986; Wells and Heller, 1988).

The Blue Mountains region, along the northern margin of the Oregon Plateaus, was initially uparched in the early Oligocene, continuing through early Miocene deposition of the John Day Formation (*Swanson and Robinson, 1968*; *Robyn and Hoover, 1982*). The mid-Miocene and younger volcanic section of the Oregon Plateaus onlaps the John Day and older formations on the south flank of the Blue Mountains (Walker, 1990). Initial volcanism of the Columbia River Basalt Group at 18 to 17 Ma accompanied initial west-southwest extension, marked by dike swarms (Fig. 51) near the western and eastern ends of the Blue Mountains and by small north-trending left-oblique normal faults (*Thayer, 1957*; *Robyn and Hoover, 1982*). Later east-west folds and reverse faults between 10 and 6 Ma (*Rogers, 1966*; *Robyn and Hoover, 1982*) accompanied basin-range extension to the south and rotation of the Blue Mountains (Grommé and others, 1986).

The Snake River Plain, a topographic low that spans 600 km across southern Idaho (Fig. 51), exposes only upper Cenozoic volcanic and sedimentary rocks. The western and eastern segments of the plain contrast in trend and origin although their surfaces merge smoothly. The western segment is a northwest-trending graben about 50 km across and 250 km long at an average elevation of about 900 m (*Malde, 1959*; Mabey, 1982). A thick sedimentary and basaltic fill, the Idaho Group, is overlain by the mainly basaltic Snake River Group (Malde and Powers, 1962). Near its northern end, normal faults of the western plain overlap a southeastern extension of the Columbia Plateau (*Fitzgerald, 1982*). The graben formed in the mid-Miocene, possibly continuous with the northern Nevada rift zone (Zoback and Thompson, 1978), and has continued to grow southeastward in later Miocene and Pliocene time to intersect the eastern segment (Malde, 1991), which averages 100 km wide and rises northeastward for 350 km from less than 1 km to 2 km elevation where it adjoins the volcanic Yellowstone Plateau (Christiansen, 1982b). The eastern plain is a linear downwarp between uptilted margins (*Kirkham, 1931*; Hamilton and Myers, 1966; Mabey, 1982). Local marginal faults (Carr and Trimble, 1963) entirely postdate volcanic rocks erupted from adjacent parts of the plain, suggesting that the faults are local adjustments to subsidence rather than rift structures (*Allmendinger, 1982*). The eastern plain transects basin-range faults almost perpendicularly on both sides and may reflect a buried regional northeast-trending Precambrian structure continuous with the Humboldt zone of the northern Great Basin (*Eaton and others, 1975*; Mabey and others, 1978). In two areas, adjacent east-west fault blocks dip toward the north margin of the eastern plain, the Mount Bennett Hills and Camas

Prairie near the western end (*Walton, 1962*; *Rember and Bennett, 1979*), and the Centennial Range and Centennial Valley near the eastern end (*Myers and Hamilton, 1964*; *Witkind, 1975*). The eastern Snake River Plain marks a zone of accommodation between regions of great extension to the south and lesser extension to the north (R. L. Christiansen and McKee, 1978). Upper crustal seismicity and young faulting define a "bow wave" pattern symmetrical about the eastern plain and converging toward the Yellowstone Plateau (Christiansen and McKee, 1978; *Scott and others, 1985*; Pierce and Morgan, 1992). Least principal crustal stress rotates counterclockwise from west-northwest in the Great Basin region south of the Yellowstone Plateau, to northeast in the Yellowstone region, and north-south north of the eastern Snake River Plain.

The Columbia Plateau is a structural basin that deepened as its thick fill of flood basalts accumulated, mainly from about 17 to 14 Ma but continuing less actively until 6 Ma (Swanson and others, 1979a; *Baksi, 1989*; Tolan and others, 1989). Fluvial and marginal-marine sediments accumulated locally (*Van Atta and Kelty, 1985*; *Smith and others, 1989*). On the south, the basin abuts the uplifted Blue Mountains along a buried but nearly straight boundary that extrapolates beneath the Cascade Range to the north margin of the Klamath Mountains, suggesting a faulted continental margin (*Hamilton, 1969*; *Riddihough and others, 1986*). The northern and eastern margins of the plateau are the Okanagan Highlands and part of the northern Rocky Mountains. The extensional setting of the plateau basin is shown by coeval dike swarms (Fig. 51). The Chief Joseph dike swarm in the eastern part of the area, comprises two linear subswarms in a north-northwest zone of 265 by 80 km (*Taubeneck, 1970*; Swanson and others, 1975). The smaller Monument dike swarm intersects the Blue Mountains (*Waters, 1961b*; *Fruchter and Baldwin, 1975*). Although G. A. Davis (1980) suggested that the Chief Joseph dike swarm marks the weak boundary between old North American crust and accreted oceanic terrane fragments, it lies entirely west of that boundary (*Armstrong and others, 1977*; Hamilton, 1978). A functional relation to the northern Nevada rift zone reflects late Cenozoic extensional stress across a heterogeneous lithosphere (R. L. Christiansen and McKee, 1978; Zoback and Thompson, 1978). Syneruptive subsidence of the Columbia Plateau (*Mackin, 1961*) had little relation to the eruptive withdrawal of magma but, rather, reflects isostatic loading by thick basaltic fill (Swanson and others, 1975; *Hooper and Camp, 1981*).

The Yakima belt of east-west to northwest-trending folds and reverse faults (Fig. 51) on the Columbia Plateau near and southwest of the Olympic-Wallowa lineament—which divides the plateau basin northwest-southeast—grew with the basin (*Hooper and Camp, 1981*; *N. P. Campbell and Bentley, 1981*; *Campbell, 1989*) as basin-range areas to the south extended and the Blue Mountains rotated clockwise against the plateau margin. The eastern Columbia Plateau has not rotated significantly since the mid-Miocene (Grommé and others, 1986), but rotation increases progressively from the central plateau to the northern Oregon coast (Magill and others, 1982; Sheriff, 1984; *Reidel and others, 1984*).

Mid-Miocene and younger tectonics of parts of the Northern Rocky Mountains link them to basin-range extension. Displacement of the Lost River, Lemhi, and Beaverhead Ranges and their intervening valleys (Fig. 51) has been since sometime between 7 and 4 Ma (*Baldwin, 1951*; *Ruppel, 1982*; *Scott and others, 1985*). The faults terminate against the northeast-trending Trans-Challis fault system of Eocene ancestry (Bennett, 1986), reactivated as a local northern boundary of basin-range extension. The magnitude-7.3 Borah Peak earthquake of 1983 on the Lost River fault produced left-oblique displacement of more than a meter on a fault that dips about 49° to as deep as 14 km (*Crone and Machette, 1984*; *Barrientos and others, 1987*). High seismicity extends eastward from these faults toward the Yellowstone Plateau (Smith and Sbar, 1974; Smith and others, 1977; R. B. Smith, 1978; Smith and Christiansen, 1980), where the magnitude-7.3 Hebgen Lake earthquake of 1959 produced surface displacements as great as 4 to 7 m on two subparallel faults that dip 45 to 50° to at least 10 to 15 km (*Witkind, 1964b*; *Barrientos and others, 1987*). These faults form part of an east-west zone that includes the Centennial Range north of the Snake River Plain (*Myers and Hamilton, 1964*). If modern extension is mainly on low-angle detachments, the two historic events suggest that detachment must be by mid-crustal creep rather than by seismic faulting (*Barrientos and others, 1987*). Basin-range faulting continues northward in western Montana (Pardee, 1950; Hamilton and Myers, 1966; Reynolds, 1979), but Quaternary displacements become fewer and seismicity lower. Reynolds (1979) demonstrated late Cenozoic extension of 10 to 15 percent in western Montana south of the Lewis and Clark line, which acts as a transform northern boundary. Least principal stress is about east-west, nearly parallel to that in the Great Basin region (*Zoback and Zoback, 1989*).

Geophysical properties reflect the tectonic similarities and differences between the Northern Rockies–Columbia Intermontane region and the Basin and Range Province. The crust beneath the central Columbia Plateau is about 25 km thick, comparable to the Great Basin, but both crustal and upper-mantle seismic velocities are higher; the velocity just below the Moho is about 8.1 km/s (*Hill, 1972,* 1978), compatible with both the mid-Miocene subsidence and subsequent relative stability of the plateau and with the emplacement and subsequent crystallization of basaltic magma deep in the crust. The crust thickens southward, and beneath the Oregon Plateaus is 35 to 40 km. Seismic velocities are similar to Sierran crust and upper mantle (Hill, 1978; *Zucca and others, 1986*; *Fuis and others, 1987*), in apparent contradiction of the proposal (Hamilton and Myers, 1966; *Hamilton, 1980*) that the plateaus are a volcanic rift in a drastically thinned or ruptured crust. A thick, dense lower crust underlies the western Snake River Plain (Hill, 1978). The crust in the Northern Rocky Mountains, like that of the Oregon Plateaus, is about 40 km thick, consistent with lesser extension of both regions than of the Great Basin. High heat flow, however, is comparable to that of the Great Basin (*Blackwell, 1969,* 1978). In the eastern Snake River Plain, although surface heat flow is dominated by its great volcanic aquifer, crustal heat flow beneath the aquifer is high and increases progressively eastward toward the magmatically active Yellowstone Plateau (*Brott and others, 1978, 1981*). Heat flow beneath the Columbia Plateau, however, appears to be low, consistent with its relative stability.

As in the Great Basin, late Cenozoic extensional faulting and high heat flow in parts of the Columbia Intermontane and Northern Rocky Mountains regions favor resources related to deep hydrothermal circulation. High-temperature geothermal systems exist in the Snake River Plain and the Cretaceous Idaho and Boulder batholiths (*Brook and others, 1979*), but they are generally smaller and fewer than in the more active Great Basin region. There are many lower temperature hot-spring systems (*Mariner and others, 1983*), however, some of which are utilized for space heating, particularly in Klamath Falls, Oregon (*Lund and others, 1976*), and Boise, Idaho (*Kunze and others, 1976*).

In summary, this section reinforces the point emphasized in earlier syntheses (Gilluly, 1963; Hamilton and Myers, 1966; *Blackwell, 1969*) that mid-Miocene and younger tectonic features of the Columbia Intermontane and the Northern Rocky Mountain regions are extensional and define a pattern generally consistent with that of the Great Basin region. The amount of extension, however, is significantly less; in a sense, these northern regions form a somewhat diffuse zone of transition between major extension to the south and relative late Cenozoic stability farther north (Christiansen and McKee, 1978). Late in the mid-Miocene tectonic transition, the northern Nevada rift zone and its continuation broke northward through a heterogeneous lithosphere to control early evolution of the western Snake River Plain and Columbia Plateau. Later extension became more widely distributed in Oregon, Idaho, Montana, and the Yellowstone Plateau, producing normal faults in subregions bounded northward by strike-slip transform structures, some of which were controlled by preexisting crustal features. The eastern Snake River Plain and the Brothers fault zone function as accommodation structures between the highly extended Great Basin region and the only slightly extended Northern Rocky Mountains and Columbia Plateau.

MAGMATISM ASSOCIATED WITH LATE CENOZOIC EXTENSION AND UPLIFT

Magmatism of generally simpler character than that of earlier extensional episodes has accompanied mid-Miocene and younger regional uplift from the Sierra Nevada to the Rocky Mountains and basin-range extension of the Basin and Range Province, Columbia Intermontane region, and Northern Rocky Mountains. Over most of this large region, magmatism has been predominantly basaltic and typically has produced dispersed fields of mainly small, simple volcanoes. In addition, however, more extensive, longer-lived volcanic fields have produced more evolved magmas and more complex volcanoes; some compositionally bimodal volcanic fields have erupted relatively abundant basalts and rhyolites but few or no intermediate lavas. In the Southern Rocky Mountains and southeastern Basin and Range Province, predominantly basaltic and bimodal magmatism replaced calc-alkalic andesite-rhyolite magmatism locally as early as about 26 Ma and widely by 21 Ma; in parts of the Great Basin region too, similar magmatism dates from as early as 21 Ma. Most of the activity discussed here, however, dates mainly from about 18 to 17 Ma, following widespread magmatic lulls. A further change in magmatic patterns accompanied tectonic changes between about 14 and 10 Ma in some regions.

SOUTHERN ROCKY MOUNTAINS AND RIO GRANDE–MOJAVE REGION

Magmatism has been intermittently active in the region of the Southern Rocky Mountains during late Cenozoic uplift, both in and near the linear zone of extension and subsidence near an axis of uplift—the Rio Grande rift—and, along the western slopes of the Rockies (Fig. 52). Much of this activity was Miocene in age, but it has continued into Quaternary time.

Volcanic and intrusive activity along the western slopes has been mainly from small scattered centers (Figs. 46, 52). Clusters of northwest-trending basaltic and related dikes and flows and small sanidine-phyric dacitic to rhyolitic intrusions emplaced between about 12 and 8 Ma in north-central Colorado and adjacent Wyoming represent contaminated or mixed rhyolitic and more mafic magmas (*Segerstrom and Kirby, 1969*; *Segerstom and Young, 1972*; *Marvin and others, 1974b*; *Snyder, 1978*; *Luedke and Smith, 1978b*). Scattered dikes and flows of alkali basalt and related differentiates were emplaced widely through central Colorado, mainly between 24 and 8 Ma, but a few are as young as Holocene (*Yeend, 1969*; *Marvin and others, 1974b*; Larson and others, 1975). A 12-Ma biotite granite is exposed in the Elk Mountains, near the southern end of this central Colorado group (*Obradovich and others, 1969*), and other granites and felsites are present in the southern and eastern parts of the area (Tweto, 1979). Basaltic to trachyandesitic and bimodal volcanism in and near the San Juan Mountains (*Lipman, 1969*; Lipman and others, 1970, *1978a*; Lipman and Mehnert, 1975) produced calc-alkalic lavas that are somewhat high in alkalies, particularly in K_2O, and are generally more femic than earlier calc-alkalic rocks of the region. The oldest basalts are about 26 Ma in age; local activity continued until about 5 Ma and even into the Pleistocene in northern New Mexico. Besides the predominant mafic rocks, which carry abundant evidence of crustal contamination (*Lipman, 1969*; *Doe and others, 1969*), small bodies of rhyolite or mixed rhyolite and basalt represent crustal melts. Furthermore, the youngest voluminous rhyolitic caldera complex of the San Juan Mountains erupted near Lake City about 23 Ma, 4 to 5 m.y. after the other large San Juan calderas (*Lipman, 1976*). The 26- to 23-Ma Latir volcanic field in northern New Mexico (Fig. 22) overlapped the magmatic transition during extreme local extension (*Lipman, 1983*; Lipman and others, 1986).

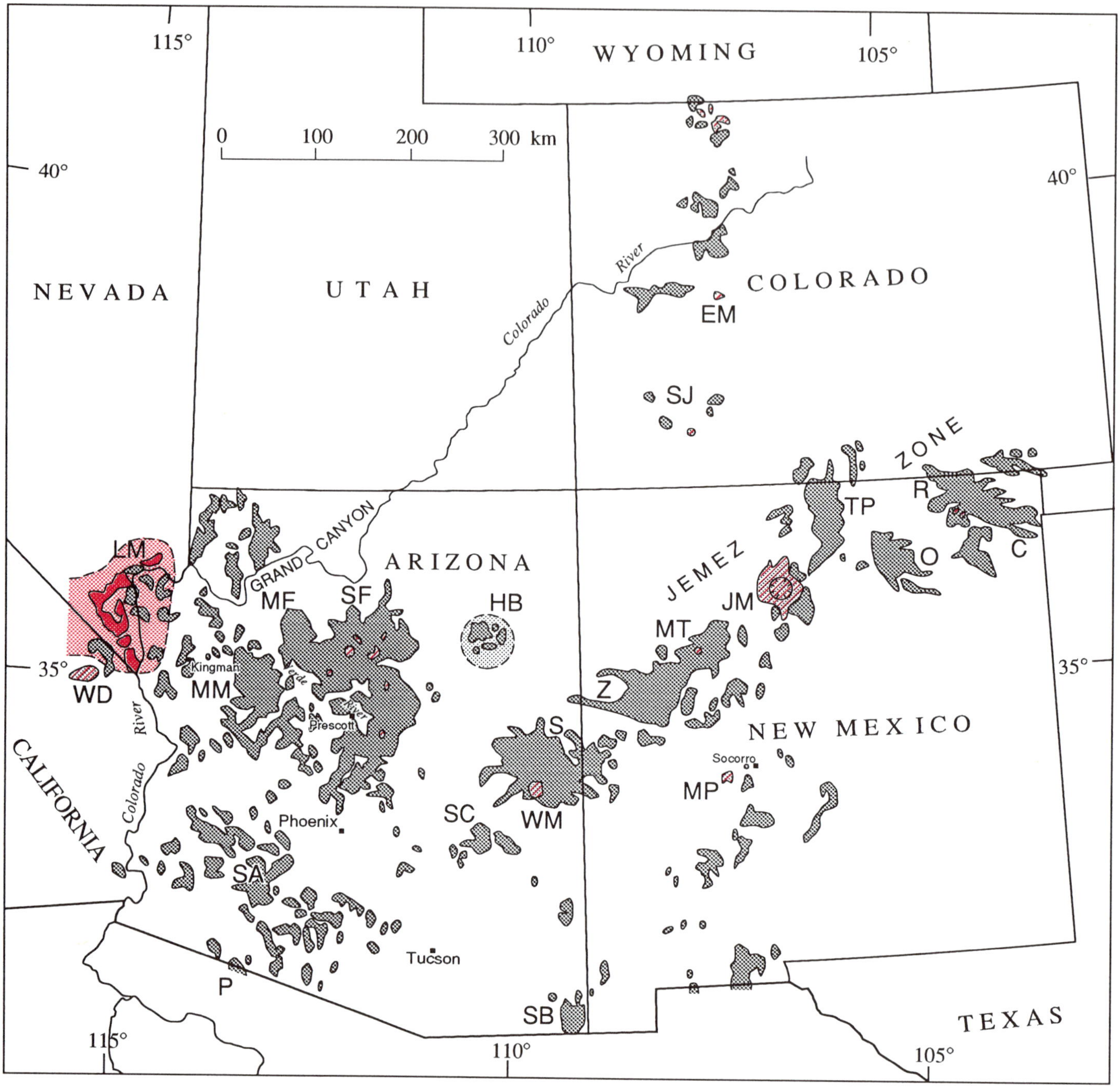

Figure 52. Map showing distribution of igneous rocks of the Southern Rocky Mountains and Rio Grande–Mojave Region younger than 17 Ma. Dark-patterned areas indicate approximate outcrop distributions; pink, probable volcanic arc; gray, basaltic or mainly basaltic bimodal volcanic suite; gray with red ruling, mainly rhyolitic bimodal volcanic suite or local intermediate centers. Hachured circle indicates caldera. C, Clayton; EM, Elk Mountains; HB, Hopi Buttes (dashed line and light shading enclose volcanic field); JM, Jemez Mountains; LM, Lake Mead; MF, Mount Floyd; MM, Mahon Mountains; MP, Magdalena Peak; MT, Mount Taylor; O, Ocate volcanic field; R, Raton; S, Springerville; SA, Sentinel Plain–Arlington volcanic field; SB, San Bernardino (also known as Geronimo) volcanic field; SC, San Carlos volcanic field; SF, San Francisco volcanic field; SJ, San Juan Mountains; TP, Taos Plateau; WD, Woods Mountains; WM, White Mountains; Z, Zuni.

Basaltic and related magmatism in the northern Rio Grande rift (Fig. 52) dates to the early Miocene (Chapin and Seager, 1975), but most is younger. On the Taos Plateau of northern New Mexico and Colorado, untilted olivine tholeiite of 5- to 2-Ma age and small intermediate to rhyolitic central volcanoes (*Ozima and others, 1967*; *Lipman, 1969*; Lipman and Mehnert, 1975; 1979a) nearly cover the rift and are discordant on tilted Miocene calc-alkalic basalt and trachyandesite (*Thompson and others, 1986*; *McMillan and Dungan, 1986*). Rift volcanism south of the Jemez zone is less voluminous than on the Taos Plateau. Early activity produced abundant calc-alkalic "basaltic andesites" (Chapin and Seager, 1975), interlayered with basin-filling gravel, between 26 and 20 Ma. Magmatism was sparse between 20 and 14 Ma, but beginning at 14 to 12 Ma and mainly during the Pliocene and Pleistocene, numerous small single-vent basaltic volcanoes erupted (Chapin and Seager, 1975; Bachman and Mehnert, 1978). In central New Mexico, the basalt is both alkalic and tholeiitic, with less abundant lavas related to fractional crystallization and slight crustal assimilation (*Kudo and others, 1971*; Chapin and Seager, 1975; *Kelley and Kudo, 1978*; *Baldridge, 1979*). Bimodal rhyolite and basalt erupted locally, as at the 14-Ma Magdalena Peak center near Socorro (*Weber and Bassett, 1963*; Chapin and Seager, 1975). A mid-crustal magma body is interpreted in the rift near Socorro from seismic reflections and phase delays, active uplift, and an electrical conductivity anomaly (*Sanford and others, 1977*; *Rinehart and others, 1979*; *Brocher, 1981*). Since about 10 Ma, mainly since about 5 Ma accompanying increased faulting, the southern rift (broadly transitional with the Basin and Range Province, Fig. 50) has erupted alkali basalt, basanite, and related lavas (*Renault, 1970*; *Hoffer, 1971, 1976*; *Aoki and Kudo, 1976*; *Seager and others, 1984*). Pleistocene basalt near the edge of the southern Great Plains (*Allen, 1951*; *Weber, 1964*; *Renault, 1970*) resembles the calc-alkalic older basalt marginal to the rift in southern Colorado and northern New Mexico, possibly reflecting thermal and chemical interactions with thicker, cooler crust marginal to the rift.

In the Mexican Highlands region of the Basin and Range Province in southwestern New Mexico and southeastern Arizona (Figs. 46, 52), Miocene and younger basaltic and related magmatism was generally similar to that of the Rio Grande rift. Early Miocene mafic calc-alkalic lavas succeeded a mid-Tertiary andesite-rhyolite suite between about 24 and 19 Ma. Activity was minor between about 20 and 14 Ma, but in latest Miocene to Quaternary time, small basaltic eruptions again become widespread (Elston and others, 1976; Deal and others, 1978). The earlier Miocene suite comprises mainly basalt and "basaltic andesite" but includes minor dacite to rhyolite (Elston and others, 1976; *Richter and others, 1981, 1983*). The calc-alkalic basalt to trachyandesite probably reflect combined fractional crystallization and crustal assimilation (*Fodor, 1975*). Early to mid-Miocene basaltic activity in southeastern Arizona overlapped extensively with voluminous silicic magmatism. Basalt and "basaltic andesite" of about 24- to 14-Ma age erupted as far west as the Tucson region (*Bikerman, 1967*; Shafiqullah and others, 1978), and voluminous rhyolitic ash flows continued to erupt from the Chiricahua Mountains to the Superstition-Superior area (Fig. 22) between 25 and 20 Ma (*Peterson, 1968*; Sheridan, 1978; Creasey and Krieger, 1978; Krieger, 1979), where later alkali basalt and basanite, together with rhyolitic to rhyodacitic tuffs and lavas (Fig. 46) are 18 to 15 Ma (*Sheridan and others, 1970*; Stuckless and Sheridan, 1971; Sheridan, 1978). Only small single-vent alkali-basalt eruptions have accompanied basin-range faulting since 12 Ma (Damon, 1971; Shafiqullah and others, 1978). In the San Bernardino (or Geronimo) field near the southeast corner of Arizona (*Lynch, 1978*), generally post-7 Ma, alkalic, and variously fractionated inclusion-bearing basalt (*Fodor, 1975*; *Lynch, 1978*; *Evans and Nash, 1979*) represents magma generated in a somewhat depleted upper mantle and relatively little crustal assimilation (*Menzies and others, 1983*).

The Jemez zone is a northeast-trending zone that divides the southern Basin and Range Province from bordering regions, from the High Plains to the White Mountains region. Although important tectonically in Precambrian, Laramide, and mid-Tertiary episodes, the Jemez zone has been a major volcanic locus only since about 14 to 12 Ma (Fig. 52; *Luedke and Smith, 1978a*; Smith and Luedke, 1984), especially in Pliocene and Quaternary time. The zone has been simultaneously active along its entire length; 12-Ma basalts in the Springerville field near its southwest end and at least 8-Ma activity in the Raton volcanic field at its northeast end (Lipman, 1980) show that it is not a propagating "hot-spot" as suggested by *Suppe and others* (*1975*). Mafic and alkalic centers that range in age from 8 Ma to 4,500 years on the High Plains near Raton and Clayton, New Mexico, overprint a mid-Tertiary foreland alkalic field with alkali basalt, basanite, nephelinite, and dacite (*Baldwin and Muehlberger, 1959*; Stormer, 1972a; *Staatz, 1985*) as well as quartz-xenocryst calc-alkalic basalt (Stormer, 1972a, *1972b*). The Ocate volcanic field, closer to the Rio Grande rift, erupted both alkali basalt and silicic xenocrystic basalt, trachyandesite, and dacite from 8 to 0.8 Ma (*Nielson and Dungan, 1985*). West of the Rio Grande rift, three other predominantly basaltic fields on the Colorado Plateau margin mark the Jemez zone. In the Mount Taylor field, a 4.3- to 1.5-Ma basalt-trachyte field surrounds the andesite-rhyolite stratovolcano of Mount Taylor, dating from 3.0 to 2.5 Ma (*Baker and Ridley, 1970*; *Lipman and Moench, 1972*; *Lipman and others, 1979*; *Lipman and Mehnert, 1979b*; *Crumpler, 1980*). Almost immediately adjacent on the southwest is the Pliocene and Quaternary Zuni-Bandera lava field (*Renault, 1970*; *Laughlin and others, 1971*; *1972a*), whose youngest flow is dated archaeologically at about 700 A.D. (*Nichols, 1946*) and varies from olivine tholeiite to quartz tholeiite, perhaps erupting during a crustal assimilation event (Leeman, 1970; *Laughlin and others, 1972b*; *Carden and Laughlin, 1974*). The White Mountains–Springerville field in eastern Arizona is more voluminous and longer lived than the two fields to the northeast. The glaciated andesite-dacite Mount Baldy stratovolcano of about 9-Ma age lies within a basaltic platform of 12- to 6-Ma age (*McKee and McKee, 1972*; *Merrill and Péwé, 1977*; *Laughlin and others, 1979, 1980*; *Condit and others, 1989*). On the north side, the younger Springerville basaltic lavas date between 3 Ma and 0.8 Ma and include both alkali basalt and olivine tholeiite (*Laughlin and others, 1979, 1980*).

At the intersection of the Jemez zone with the west margin of the Rio Grande rift (*Aldrich, 1986*) is the voluminous Jemez Mountains volcanic field (Fig. 52; R. L. Smith and others, 1970). Although some alkali basalts erupted as long ago as 17 Ma, the more or less continuous history of the field defines three sequences dating from about 14 Ma (*Gardner and others, 1986*). The first sequence began with basalt in the southern part of the field, evolving to 13- to 10-Ma bimodal basalt and rhyolite, then a mafic to silicic series between 10 and 7 Ma (*Bailey and others, 1969*; Smith and others, 1970; *Guilbeau and Kudo, 1985*; *Gardner and others, 1986*). A second sequence centered in the northern part of the field mainly between 7 and 3 Ma produced voluminous basalt, andesite, dacite, and rhyolite (*Bailey and others, 1969*; R. L. Smith and others, 1970; *Luedke and Smith, 1978a*), probably a magma-mixing series (*Gardner and others, 1986*). The third sequence reflects a voluminous chemically and thermally zoned rhyolitic magma chamber in the north-central part of the field, erupting climactically to produce two major ash-flow sheets of the Bandelier Tuff at 1.4 and 1.1 Ma (Smith and Bailey, 1966; *Doell and others, 1968*) and a compound resurgent caldera with postcollapse rhyolitic ring-fracture lavas between 0.9 and 0.1 Ma (*R. L. Smith and others, 1961*; Smith and Bailey, 1968; *Doell and others, 1968*).

Basaltic volcanism continued southwest across the Basin and Range Province along an extrapolation of the Jemez zone (Fig. 52). At the foot of the Colorado Plateau southwest of the White Mountains, the San Carlos volcanic field of Pliocene and Quaternary age (*Bromfield and Shride, 1956*) erupted alkali basalt and basanite (*Holloway and Cross, 1978*), some of it explosively through lacustrine sediments (*Wohletz, 1978*) to carry large inclusions of both cognate and mantle origins (*Frey and Prinz, 1978*). Farther southwest, little-studied upper Miocene basalt from small but widespread centers occurs across the Sonoran Desert in central-southern Arizona (*Luedke and Smith, 1978a*). A few dates (Eberly and Stanley, 1978; Shafiqullah and others, 1980; *Tosdal and others, 1986*) are as old as 17 to 16 Ma—the time of waning mid-Tertiary deformation—but most are 13 Ma and younger—coeval with basin-range faulting. Basalt erupted intermittently in the 3.3- to 1-Ma Sentinel Plain–Arlington volcanic field southwest of Phoenix (Shafiqullah and others, 1980). The Pleistocene and Holocene Pinacate field of alkali basalt and hawaiite includes numerous cinder cones and explosive hydromagmatic cones and craters, mainly in Sonora but overlapping the Arizona border (*Wood, 1974*; *Gutmann, 1976*; *Gutman and Sheridan, 1978*).

In and near the lower Colorado River area, a transition from calc-alkalic intermediate to predominantly basaltic volcanism accompanied extension between 21 and 16 Ma; between 13 and 5 Ma, volcanism coeval with basin-range faulting became largely basaltic but included local rhyolite (*Olmsted and others, 1973*; Crowe, 1978; Crowe and others, 1979; Shafiqullah and others, 1980; *Murray, 1982*; *Dahm and Hankins, 1982*; Otton, 1982; Suneson and Lucchitta, 1983; *J. E. Nielson and Glazner, 1986*; *Moyer and Nealey, 1989*). During the mid-Miocene basaltic activity in the south, a calc-alkalic arc was active to the north, near Lake Mead. Northward retreat of the arc terminus after about 13 to 12 Ma was followed by basaltic volcanism near Lake Mead between 11 and 4 Ma (R. E. Anderson and others, 1972; Shafiqullah and others, 1980). One large center west of the Colorado River trough in the eastern Mojave Desert produced voluminous 16- to 15-Ma rhyolite, including two ash-flow sheets and a caldera overlain by mafic lavas of 10-Ma age (*McCurry, 1988*). For 200 km westward across the Mojave Desert, little or no volcanism occurred between about 21 and 10 Ma. Even farther west, however, sparse 17- to 14-Ma basalt and trachyandesite is associated with widespread basin deposits (*Burke and others, 1982*; *Cox and Diggles, 1986*). Post–10-Ma basaltic volcanism in the central Mojave Desert is related to that of the western Great Basin, discussed later.

In a wide transition zone between the Basin and Range Province and the southwestern Colorado Plateau, from north of the Colorado River to the southwest end of the Jemez zone (Fig. 52), basaltic and related volcanism has migrated northeastward (Smith and Luedke, 1984). North of the Grand Canyon, basalt in the upper Miocene to middle Pliocene Muddy Creek Formation was partly coeval with and partly succeeded by hawaiite and alkali basalt volcanism of 7- to 3-Ma age along faults on the plateau margin. That activity migrated northeastward at about 1 cm/y, partly filling old drainages (*McKee and others, 1968*; *Hamblin, 1970, 1974*; Best and Brimhall, 1974; Best and others, 1980). Synchronous basalt to trachyandesite farther south included a poorly known differentiated complex at Mahon Mountain, southeast of Kingman, and basalt of 10- to 9-Ma age west of Prescott (Shafiqullah and others, 1980). Basalt of 14- to 17-Ma age (*McKee and McKee, 1972*) surrounds a pair of rhyolitic centers near Mount Floyd on the Colorado Plateau (*Luedke and Smith, 1978a*). In the Prescott–Verde Valley region, hypersthene-normative basalt and related trachyandesite, including the Hickey Formation, are between 15 and 8 Ma in age (Anderson and Creasey, 1958; Krieger, 1965; *Krieger and others, 1971*; E. H. McKee and Anderson, 1971; *McKee and Elston, 1980*). Dacitic tuffs and flows erupted late in this episode (*McKee and Elston, 1980*). Basaltic lavas that erupted on the margins of the Colorado Plateau flowed down drainages into the Verde Valley between about 6 and 3 Ma. Basaltic eruptions began in the San Francisco volcanic field as early as 9 Ma but have been fairly continuous since about 6 Ma (*Damon and others, 1974*; Smith and Luedke, 1984). Basalt and related lavas in that field erupted from hundreds of vents between about 2.5 Ma and 900 years ago (*Robinson, 1913*; *Smiley, 1958*; *Damon and others, 1974*; Moore and others, 1976; Wolfe, 1984; *Tanaka and others, 1986*). A few longer-lived centers of trachyandesite and andesite to rhyolite evolved together with the basalt. The largest of these, San Francisco Mountain in the eastern part of the field, is a mainly andesitic stratocone built on rhyolite and rhyodacite domes as old as

2.8 Ma (*Damon and others, 1974*; Wolfe, 1984). A small rhyolite dome and associated pyroclastics (*Sheridan and Updike, 1975*) and a large rhyodacitic center have erupted east of San Francisco Mountain within the past 250,000 years. Basalt of the San Francisco field is mainly alkalic; some of the more evolved lavas define an alkali basalt-hawaiite-mugearite series but the more voluminous intermediate to silicic centers are calc-alkalic (Moore and others, 1976; *Wenrich-Verbeek, 1979*).

SALTON TROUGH

Volcanism of small volume (Robinson and others, 1976) is associated with the Salton trough (see Figs. 45, 53), superposed across a late Miocene basin-range extensional terrane in southernmost California and northwestern Mexico (*Dokka and Merriam, 1982*) and associated with opening of the Gulf of California since about 4 to 5 Ma (Larson and others, 1968; *Elders and others, 1972*). Along with very high heat flow, this activity indicates significant interaction between oceanic spreading-center magmatism of the gulf and the thinning continental crust (Lachenbruch and others, 1985). Five small rhyolite domes are of late Pleistocene age. In addition, however, tholeiite like that of mid-ocean ridges occurs as inclusions in the rhyolite and as sills, dikes, and subsurface flows. Additional inclusions of granitic basement rocks, xenocrysts, and high Sr-isotopic ratios of the rhyolite indicate a significant crustal component. The tholeiitic magmatism is probably the direct result of lithospheric spreading. The resulting heat input partially melts the thinning continental crust, components of which are further assimilated during emplacement. Cerro Prieto, just south of the border in Mexico, is a late Pleistocene composite dacite dome and associated pyroclastics of about 1 km^3, interpreted as crustal partial melts related to basaltic intrusions (*Reed, 1984*). The Salton Trough magmatism sustains several high-temperature geothermal systems, of which the highly saline Salton Sea system is one of the hottest known (*Kasameyer and others, 1984*) and is actively metamorphosing the sediment fill of the Salton trough (*Muffler and White, 1969*).

SIERRA-WASATCH—HIGH PLATEAUS REGION

Magmatism associated with mid-Miocene and younger extension in the Great Basin region has evolved through several distinct patterns, some of them relatively short-lived but others persisting through regional tectonic changes. During the mid-Miocene, between about 21 and 14 Ma, a predominantly andesitic arc occupied the western Great Basin region. Farther east there was relatively little volcanism in the Great Basin for a few million years around 17 Ma (McKee and others, 1970) except for a persistent belt of voluminous silicic volcanism across southern Nevada and southwestern Utah (Fig. 46; Noble, 1972; *Best, 1986*). As the regional magmatic pace picked up again after about 17 Ma, basaltic and related volcanism was notable first in the northern Nevada rift zone, but after about 14 Ma tended to concentrate near the margins of the region and in a single northeast–trending zone through south-central Nevada (Stewart and Carlson, 1976; Stewart and others, 1977).

Other than the southern Nevada belt, described following, there was little volcanism during the mid-Miocene broad-basin phase of extension in the central and eastern Great Basin. A few exceptions in western Utah (Fig. 46) produced highly evolved rhyolite of 21 Ma age, significant for beryllium mineralization, at Spor Mountain in the southern Thomas Range (Staatz and Carr, 1964; Lindsey, 1982; *Christiansen and others, 1984*) and some 18-Ma potassic rhyodacite in the East Tintic mining district (*Laughlin and others, 1969*; Morris and Lovering, 1979). The northern Nevada rift zone, which broke along a line reaching north-northwest from the center of the Great Basin between about 17 and 14 Ma, produced basaltic lavas along its entire length, but they are best considered together with mid-Miocene basaltic and bimodal lavas of the Columbia Intermontane region in the next section of this chapter.

Mid-Miocene magmatism of the southern Nevada volcanic belt (Figs. 46, 53) was generally K-rich; its axis lay just south of the 28- to 21-Ma axis of mid-Tertiary rhyolitic ash flows (Fig. 21B). In the west, near Tonopah and Goldfield, rhyolite and K-rich mafic lavas between 21 and 18 Ma in age overlap and intertongue with the mid-Miocene arc (*Armstrong and others, 1972*; *Silberman and McKee, 1972*; Bonham and Garside, 1979). Several cooling units of the voluminous 21- to 18-Ma Fraction Tuff erupted near Tonopah, and trachyte to rhyolite were common until 15 Ma (Bonham and Garside, 1979; *Kleinhampl and Ziony, 1985*). Farther east, rhyolite of 19 Ma age is preserved as dikes (*Ekren and others, 1973*). Caldera-forming 18- to 16-Ma ash-flows—correlated by Ekren and others (1971) with the Fraction Tuff but shown by Bonham and Garside (1979) to be younger—and other rhyolites erupted in the Kawich Range (*Kistler, 1968*; Ekren and others, 1971). Widespread 21- to 18-Ma ash flows in eastern Nevada, many of them distinctive phenocryst-rich rhyodacite, probably erupted from a cauldron complex south of Caliente (*Noble and McKee, 1972*; Ekren and others, 1977). Rhyolite and K-rich trachandesite to latite erupted between 22 and 18 Ma in southwestern Utah through active northeast-trending faults (*Rowley and others, 1978b*; *Mehnert and others, 1978*; Best and others, 1980; Best and Keith, 1983; *Best, 1986*). At the east end of the belt, rhyolite of the 21- to 17-Ma Mount Belknap Volcanics and basalt as old as 22 Ma record the last major episode of the mainly mid-Tertiary Marysvale volcanic field (Rowley and others, 1979; *Cunningham and Steven, 1979*; Steven and others, 1979, 1984a). A large granitic intrusion in the Mineral Mountains may also represent part of that episode (*Nielson and others, 1986*).

After the mid-Miocene (Fig. 53), volcanism near the center of the southern Nevada volcanic belt, formed the complex southwestern Nevada volcanic field (Christiansen and others, 1977). The 15- to 13-Ma Silent Canyon center of subalkalic and peralkalic trachyte to rhyolite (Noble and others, 1968; *Orkild and others, 1968*) overlapped the longer lived Timber Mountain–Oasis Valley caldera complex, which emplaced nine subalkalic

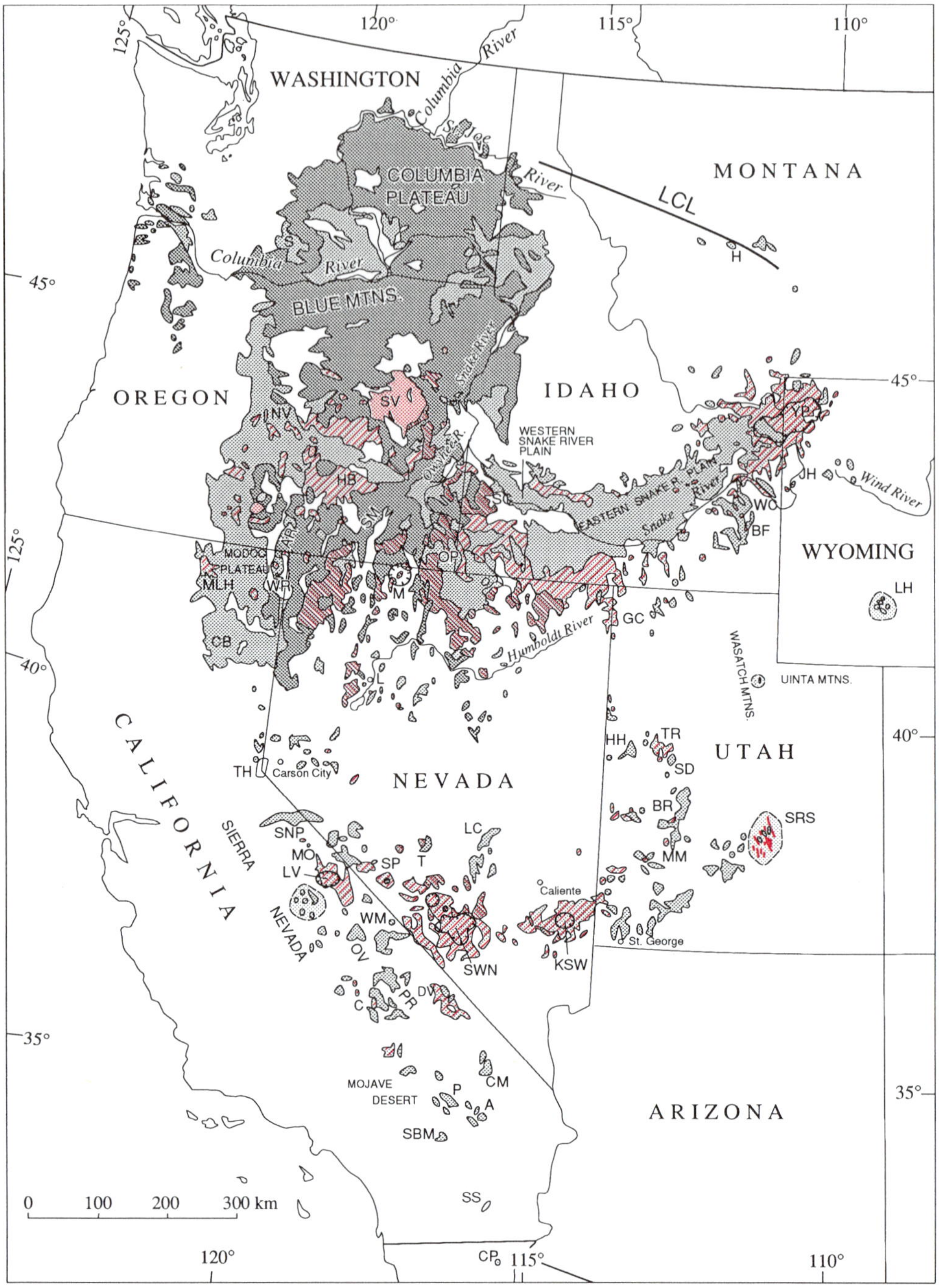

Figure 53. Map showing distribution of igneous rocks of the Great Basin, Columbia Intermontane, and Northern Rocky Mountains Regions younger than 17 Ma. Pink, calc-alkalic volcanic suite (17 to 14 Ma); gray, basaltic or mainly basaltic bimodal volcanic suite (dark, 17 to 14 Ma); gray with red ruling, mainly rhyolitic bimodal volcanic suite (dark, 17 to 14 Ma). Hachured circles and ellipses indicate calderas; dashed lines and light shading enclose fields of small alkalic bodies. A, Amboy Crater; AR, Abert Rim; BF, Blackfoot lava field; BR, Black Rock Desert; C, Coso Range; CB, Caribou volcanic field; CM, Cima volcanic field; CP, Cerro Prieto; DV, Death Valley; GC, Grouse Creek Mountains; GV, Gem Valley lava field; H, Helena; HB, Harney Basin; HH, Honeycomb Hills; JH, Jackson Hole; KSW, Kane Springs Wash; L, Lovelock; LC, Lunar Crater; LCL, Lewis and Clark line; LH, Leucite Hills; LV, Long Valley caldera; M, McDermitt caldera; ML, Medicine Lake Highland; MM, Mineral Mountains; MO, Mono Lake; NV, Newberry volcano; OP, Owyhee Plateau; OV, Owens Valley; P, Pisgah Crater; PR, Panamint Range; S, Simcoe volcanic field; SBM, San Bernardino Mountains; SC, Silver City; SD, Sevier Desert; SM, Steens Mountain; SNP, Sonora Pass; SP, Silver Peak Range; SRS, San Rafael Swell; SS, Salton Sea; SV, Strawberry Volcanics; SWN, southwestern Nevada volcanic field; T, Tonopah; TH, Lake Tahoe; TR, Thomas Range; WC, Willow Creek lava field; WM, White Mountains; WR, Warner Range; YP, Yellowstone Plateau.

ash-flow sheets over 11,000 km^2 between 16 and 11 Ma (*Byers and others, 1976a,* 1976b; Christiansen and others, 1977; *Christiansen, 1979*). Basalts erupted around the field from time to time. Farther east, trachytic to rhyolitic peralkalic ash flows erupted between 16 and 12 Ma (*Noble, 1968*; *Noble and McKee, 1972*; Novak, 1984), including the 14- to 13-Ma Kane Springs Wash center. Alkali basalt and high-silica topaz-bearing rhyolite in southwestern Utah are 13 to 10 Ma in age (*Mehnert and others, 1978*; Best and others, 1980; Best and Keith, 1983; *Best, 1986*). The youngest activity of the southern Nevada volcanic belt includes the 8- to 7-Ma Black Mountain and 6-Ma Stonewall Mountain caldera complexes of trachytic to rhyolitic lavas and pyroclastics, both peralkalic and subalkalic, in the southwestern Nevada volcanic field (Noble and Christiansen, 1974; *Christiansen, 1979*; *Noble and others, 1984*). Farther west, the Silver Peak center erupted K-rich trachyandesite and rhyolite of 6- to 5-Ma age (Robinson and others, 1968; *Robinson, 1972*).

In the Mojave Desert, at the south edge of the western Great Basin (Fig. 53), widely but sparsely distributed basalt to basanite (*Wise, 1966, 1969b*; *Smith and Carmichael, 1969*) of 10 Ma and younger age penetrates a crust that extended during the mid-Miocene but has been relatively stable since then. Alkali basalt of 10- to 6-Ma age overlies a late Tertiary erosion surface in the eastern San Bernardino Mountains and adjacent Mojave Desert (*Oberlander, 1972*; *Neville and Chambers, 1982*; *Neville and others, 1985*). The Cima volcanic field of the northeastern Mojave Desert had three periods of activity between about 8 Ma and Holocene time (*Katz and Boettcher, 1980*; Dohrenwend and others, 1984). Other mainly Quaternary centers include Pisgah Crater (*Wise, 1966, 1969b*), Dish Hill (*Wilshire and others, 1980*), Amboy Crater (*Parker, 1959, 1963*), Malapai Hill (*Stull and McMillan, 1973*), and others, many of them with ultramafic inclusions of mantle origin (*Wise, 1966, 1969b*; *Stull and McMillan, 1973*; *Wilshire and others, 1980*; *Neville and others, 1985*). A zone of hawaiite and alkali basalt of 8 Ma to Holocene age extends from the edge of the Mojave Desert discontinuously north-northeast across the middle of the southern Nevada volcanic belt nearly to the center of Nevada (Fig. 53; *Trask, 1969*; *Scott and Trask, 1971*; *Marvin and others, 1973b*; *Ekren and others, 1973*; *Vaniman and others, 1982*; Crowe and others, 1983). Relative to other Basin and Range basalts (Leeman, 1970), these lavas have high Sr (averaging about 1,200 ppm), high initial $^{87}Sr/^{86}Sr$, and low $^{143}Nd/^{144}Nd$, suggesting generation in part of a heterogeneous upper mantle enriched in incompatible elements (*Hedge and Noble, 1971*; Leeman, 1982a; *Menzies and others, 1983*).

Miocene and younger volcanism was also characteristic of the southwestern Great Basin, between central Death Valley and the southern Sierra Nevada (Fig. 53). In the Black Mountains and Greenwater Range east of Death Valley 14- to 6-Ma rhyolite and basalt to trachyandesite are overlain by rhyolite and basalt of 5 Ma and younger age with interlayered fanglomerate and other basin deposits (*Drewes, 1963*; Fleck, 1970; *McAllister, 1970, 1973*; *Haefner, 1976*; *Cemen and others, 1982*). Basaltic volcanism near Death Valley continues locally into the Quaternary; *de Voogd and others (1986b)* interpret a "bright spot" seismic reflector as a magma chamber 15 km deep. Volcanism has been intermittent in and near the Coso Range since 6 Ma (*Lanphere and others, 1975*; *Duffield and Smith, 1978*; Duffield and others, 1980; Bacon and others, 1980, *1981, 1982*; Novak and Bacon, 1986). Latest Miocene and Pliocene activity progressed from early monogenetic basalt vents, through later polygenetic mixed-magma andesite and dacite, to rhyodacite and rhyolite from a longer-lived center. Pleistocene volcanism, mainly younger than 0.5 Ma, produced bimodal rhyolite and basalt in eruptions that are time-predictable from the volume of the preceding eruption (*Bacon, 1982*). Small basalts and rhyolites extruded onto the uplifted Sierran block, immediately west, between 3.6 and 0.2 Ma (Dalrymple, 1963; Moore and Dodge, 1980; *Bacon and Duffield, 1981*). Post-6-Ma alkali basalt to hawaiite, mugearite, and trachyandesite, as well as rhyolite, are also abundant north of the Coso Range and west of northern Death Valley (Fig. 53; McKee, 1968; *Ross, 1970*; *Hall, 1971*; *Blakely and McKee, 1985*). The youngest are Holocene maar volcanoes at the north end of the Panamint Range (*Crowe and Fisher, 1973*). A Pleistocene, mainly basaltic field in Owens Valley near Big Pine includes a single small rhyolite dome (*J. G. Moore, 1963*; *Bateman, 1965*; *Luedke and Smith, 1981*). Small potassic alkali basalt and trachyandesite flows of 11 to 9 Ma span the northern Owens Valley in both the White Mountains and the Sierra Nevada (Dalrymple, 1963; Moore and Dodge, 1980). Additional small eruptions of alkali basalt, potassic alkali basalt, and leucite basanite, tephrite, and phonolite within the Sierran block were as old as 4.5 Ma and sent flows down the drainages of the western slope as recently as 120 ka (*Van Kooten, 1980*; Moore and Dodge, 1980; *Dodge and Moore, 1981*; *Gillespie and others, 1984*).

A large, long-lived magmatic system indents the Sierran front between Long Valley and Mono Lake (Fig. 53; Bailey and others, 1976; Bailey, 1989). Volcanism between about 4 and 2 Ma produced potassic alkali basalt, trachyandesite, and minor leucitic basanite, tephrite, and scattered rhyodacite and rhyolite (Dalrymple, 1963, 1964a; *Huber and Rinehart, 1967*; Gilbert and others, 1968; Bailey and others, 1976). A major rhyolitic system developed near Long Valley about 2 Ma and by 1.1 Ma had formed a large integrated magma chamber (*Metz and Mahood, 1985*), which was disrupted by catastrophic eruption of the Bishop Tuff at 0.73 Ma (*Dalrymple and others, 1965*; Hildreth, 1979; *Hildreth and Mahood, 1986*). Ash was deposited over much of the western United States, at least as far east as Nebraska (Izett and others, 1970), and at least 700 km^3 of magma erupted as ash flows while the chamber roof collapsed into the 30-km Long Valley caldera (Bailey and others, 1976; *Hildreth and Mahood, 1986*). Postcollapse activity includes intracaldera volcanism and uplift (Bailey and others, 1976; *Mankinen and others, 1986*), and magma remains at high levels (*Sanders, 1984*; *Savage and Cockerham, 1984*; *Hill and others, 1985*). Since 0.2 Ma, mafic lavas have erupted to the west and northwest along the Sierran front (*Putnam, 1949*; *Huber and Rinehart,*

1967), past the north side of Mono Lake (*Christensen and Gilbert, 1964*). Rhyolitic vents younger than 30 ka, the Inyo and Mono domes and craters, overlap the Long Valley caldera and extend 50 km north (*Dalrymple, 1967*; Bailey and others, 1976; *Wood, 1977*). Their most recent eruptions were about 600 years ago from aligned vents above dikes (Miller, 1985; *Sieh and Bursik, 1986*; *Vogel and others, 1989*).

Volcanism younger than mid-Miocene from small scattered centers north of the Long Valley area (Fig. 53) was varied. Latitic to basaltic lavas and ash flows, 11 to 9 Ma in age, are interlayered in the upper Miocene andesitic arc sequence (Dalrymple, 1963, 1964a; Slemmons, 1966; Gilbert and others, 1968; *Noble and others, 1974*), including a latitic caldera center on the Little Walker River considered by *Noble and others* (1976) as a possible extension of the southern Nevada volcanic belt. Upper Miocene to upper Pleistocene basalts and rhyolites as old as 7 Ma occur between Sonora Pass and the Lake Tahoe area (Dalrymple, 1964a, *1964b*). Remnants of the mid-Miocene volcanic arc from the Bodie Hills, northeast of Mono Lake, nearly to Carson City are overlain by basalt and rhyolite from small central eruptions between 8 and 3 Ma in age (*Chesterman, 1968*; Gilbert and others, 1968; *Silberman and Chesterman, 1972*; *Silberman and others, 1972*; Gilbert and Reynolds, 1973; *Speed and Cogbill, 1979a*). Basalt erupted in northwestern Nevada east and southeast of Lovelock between about 13 and 11 Ma (*Morton and others, 1977*). A suite of basalt, trachyandesite, and rhyolite east of Lake Tahoe is 11 to 7 Ma in age (Thompson and White, 1964; Bonham, 1969; *Silberman and others, 1979*; *Luedke and Smith, 1981*), and another is 2.5 to 1 Ma (*Birkeland, 1963*; Dalrymple, 1964a; *Silberman and others, 1979*), including a center near the active Steamboat Springs hydrothermal system (White and others, 1964). Farther north in the Sierra Nevada, small areas of basalt unconformably overlie arc andesite as young as 11 Ma but are themselves undated (*Durrell, 1959a*, 1966; *Luedke and Smith, 1981*).

Basaltic and bimodal volcanism has accompanied basin-range faulting since about 10 Ma along the eastern, just as along the western margin of the Great Basin (Fig. 53). Basalt and sparse rhyolite of 9 to 5 Ma age in northeast-trending belts (Best and others, 1980; *Rowley and others, 1981*) span the region from the basalt-hawaiite fields of the Colorado Plateau margin in northwestern Arizona to the east end of the southern Nevada volcanic belt. Younger alkali basalt to hawaiite and quartz-bearing basalt to trachyandesite (3 Ma and less) also form a northeast-trending belt—the St. George zone of Smith and Luedke (1984)—across the north-south normal faults of the western Colorado Plateau margin (*Lowder, 1973*; Best and Brimhall, 1974; *Hausel and Nash, 1977*), commonly but not consistently younger toward the northeast (*Fleck and others, 1975*; Best and Hamblin, 1978; Best and others, 1980). In and near the Mineral Mountains are a few 9- to 8-Ma rhyolite domes (*Lipman and others, 1978b*; *Evans and Steven, 1982*) and a 12-km-long cluster of Pleistocene high-silica rhyolite lavas and tuffs (*Lipman and others, 1978b*; *Nielson and others, 1986*). Tholeiitic basalt of about 2.5 Ma to Holocene age is abundant in the Black Rock and Sevier Deserts (*Hogg, 1972*; *Condie and Barsky, 1972*; *Pushkar and Condie, 1973*; *Hoover, 1974*; Best and others, 1980). Associated 3- to 2-Ma rhyolitic domes and flows near Cove Fort may represent small leaks from a chemically zoned magma chamber (Armstrong, 1970; *Mehnert and others, 1978*; *Crecraft and others, 1981*; *Carrier and Chapman, 1981*). In an east-west belt of mafic and silicic lavas extending 100 km eastward from the Nevada border Quaternary basalt overlaps the east end of a large field of highly evolved 10- to 6-Ma rhyolite around Keg Mountain and the Thomas Range (Staatz and Carr, 1964; *Shawe, 1972*; *Lindsey and others, 1975*; *Mehnert and others, 1978*; Lindsey, 1982; *Christiansen and others, 1984*, 1986). On the west, the Honeycomb Hills contain 5-Ma K-rich trachybasalt, latite, and topaz-bearing rhyolite (*Hogg, 1972*; *Turley and Nash, 1980*; Christiansen and others, 1986).

The eastern marginal volcanic zone of the Great Basin widens northward. Middle and late Miocene bimodal volcanism, mainly 16-8 Ma in age and dominated by rhyolite, extended northward along the Utah-Nevada border from the Honeycomb Hills nearly to Idaho (Armstrong, 1970; Armstrong and others, 1975, *1976*; *Compton, 1983*). Small isolated centers occur to the east in the Bonneville basin (*Armstrong and others, 1976*; *Luedke and Smith, 1978b*), at least some of them quite potassic (*Hogg, 1972*). Pleistocene olivine tholeiite and clustered high-silica rhyolite domes of 1.5-Ma and 100- to 50-ka age just south of the eastern Snake River Plain form the Gem Valley, Blackfoot, and Willow Creek lava fields (*Mansfield, 1927*; Armstrong and others, 1975; *Fiesinger and others, 1982*; *Pierce and others, 1982*; E. H. Christiansen and others, 1986). Small rhyolite flows and tuff of 9 to 5 Ma extend eastward to Jackson Hole (Love, 1956; *Love and others, 1973*; *Naeser and others, 1980*), and 4-Ma basalt occurs in the upper Wind River drainage (*Blackstone, 1966*). Basalt of 8 to 4 Ma occurs in Montana north of the eastern Snake River Plain and Yellowstone Plateau to the Lewis and Clark line (*Chadwick, 1969*, 1972, *1978*; Reynolds, 1979).

Both mineral deposits and high-temperature geothermal systems are related to the non-arc magmatism of 22 Ma and younger age in the Great Basin region. Concentrations of light metals such as Be, Li, and Sn, as well as fluorite deposits, are associated with the widely distributed high-silica, topaz-bearing rhyolites (Staatz and Carr, 1964; Christiansen and others, 1986). Uranium occurs in tuffaceous sedimentary rocks in some of the eastern rhyolitic centers (Lindsey, 1982; *Cunningham and others, 1982*; *Steven and others, 1984b*), and intrusive bodies related to some of these fields host W and Mo (Best and Keith, 1983). Christiansen and others (1986) propose that fields of topaz-bearing rhyolites, as in the Thomas Range, might overlie substantial porphyry Mo deposits. High-temperature geothermal systems are sustained by several young volcanic centers, and others may remain to be discovered, especially in systems with young rhyolites (*Smith and Shaw, 1975, 1978*; *Brook and others, 1979*). Particularly significant are geothermal systems in the Coso Range, Long Valley caldera, and Steamboat Springs in the western Great Basin and Roosevelt Hot

Springs near the Mineral Mountains of western Utah. Locally, epithermal Au-Ag vein deposits are related to the active geothermal systems or to similar, now extinct systems associated with late Cenozoic volcanism (D. E. White and Heropoulos, 1983).

In summary, there were two major episodes of late Cenozoic basaltic or bimodal rhyolite-basalt magmatism in the Great Basin region east of the mid-Miocene arc and south of the Oregon Plateaus and Snake River Plain. The earlier, notably K-rich, occurred during extensional broad basin formation between 22 and 17 Ma, mainly in the southern Nevada volcanic belt and especially its eastern sector. This belt continued the succession of south-migrating east-west volcanic belts that had prevailed in the region since the beginning of mid-Tertiary tectonic extension and magmatism. A regional magmatic lull at about 17 Ma was only gradually replaced by widespread renewed activity coincident with the establishment of basin-range faulting. Since about 14 Ma, and especially since about 10 Ma, this magmatism has concentrated near the margins of the region and in a single north-northeast-trending belt of dispersed basaltic centers reaching from the southwest margin into central Nevada. Magmatism marginal to the extending region has migrated slowly onto the southwestern Colorado Plateau and has overprinted a large area of Pliocene and younger extension in the southwestern Great Basin, from Death Valley to the Sierran front. Activity continued in the southern Nevada belt, from the Marysvale field to the eastern Sierra Nevada. The youngest major magmatic system is the Long Valley caldera, which climaxed in the early Pleistocene but remains active. Much of the region's rhyolite is highly evolved, representing relatively long-lived, compositionally zoned magma chambers in which crustal heating and concentrations of light metals may be of economic significance. Basaltic and bimodal magmatism has replaced Miocene arc magmatism successively northward with time in the western Great Basin.

COLUMBIA INTERMONTANE REGION AND NORTHERN ROCKY MOUNTAINS

In a remarkable series of eruptive events, referred to by Watkins and Baksi (1974) as the Columbia volcanic episode, great volumes of both basalt and rhyolite poured out between about 17 and 14 Ma, marking a major regional tectonic and magmatic change between the northern Great Basin and the Columbia Intermontane region. Most of the earliest voluminous volcanism was basaltic; in some areas, voluminous rhyolite overlapped or followed the basalt. Basaltic and rhyolitic volcanism have continued in later Cenozoic time but with different spatial distributions.

The mid-Miocene Columbia volcanic episode

Magmatic activity of 17 to 14 Ma in the northern Nevada rift zone increased progressively northward (Christiansen and McKee, 1978; Zoback and Thompson, 1978). Near its south end in central Nevada, the rift is marked by a dike swarm up to 15 km wide and by erosional remnants of subalkalic basalt to trachyandesite (*Gilluly and Masursky, 1965*; Armstrong, 1970; McKee and Silberman, 1970; *Larson and others, 1971*; *Wells and others, 1971*; Christiansen and McKee, 1978) with scattered younger rhyolitic flows and domes (*Wells and others, 1971*; *Stewart and others, 1975*; *Erickson and others, 1978*; Christiansen and others, 1986). To the north-northwest, 17- to 14-Ma lavas are more widely preserved to both the east and the west (*Luedke and Smith, 1981*); by northernmost Nevada they cover nearly the full width of the state (Fig. 53). Rhyolite becomes more abundant northward and upsection; less abundant andesite and dacite bridge the predominant compositions (*LeMasurier, 1968*). Some of the intermediate rocks are tholeiitic Fe-rich differentiates (*Wallace and others, 1980*). Others are calc-alkalic and lie on straight-line trends with the basalts and rhyolites. A second volcanic zone west of Lovelock extends northward and converges toward the northern Nevada rift zone near the Nevada-Oregon border (*McKee and Marvin, 1974*; *Luedke and Smith, 1981*), where the 20,000-km^2 McDermitt volcanic field of 16- to 15-Ma age (*Rytuba and Conrad, 1981*; Rytuba and Mckee, 1984) produced seven large ash-flow sheets, aggregating 1,700 km^3 around a caldera complex. These voluminous peralkalic rhyolites preserve magmatic compositional zonations that existed in crustal magma chambers, with roofward concentrations of incompatible elements (*Conrad, 1984*).

Voluminous basaltic and bimodal volcanism of 17- to 14-Ma age was widespread in the central Oregon Plateaus (Fig. 53). At Steens Mountain in south-central Oregon, the voluminous 16-Ma subalkalic Steens Basalt commonly has large plagioclase phenocrysts and notably high Al_2O_3 (*Baksi and others, 1967*; *Gunn and Watkins, 1970*; Watkins and Baksi, 1974; *Hart and Mertzman, 1982*; Rytuba and McKee, 1984). Dikes at Steens Mountain trend north-northeast along the mountain front (*Walker and Repenning, 1965*). Extensive sheet-like flows of 17- to 14-Ma age lithologically similar to the Steens Basalt extend westward through the Oregon Plateaus (*Walker, 1970*) to northwestern Nevada (Bonham, 1969; *Smith, 1973*; *Noble and others, 1973*), the Warner Range of northeastern California (Duffield and McKee, 1986), the Abert Rim area of Oregon (Walker, 1977; *Wells, 1979*; *Hart and Mertzman, 1982*), and northward across Harney Basin (Walker, 1977, *1979*). The Steens and related basalts rest on a regional erosion surface of low relief cut in rocks of John Day age (about 36 to 18 Ma) and older (*Avent, 1969*). Widespread, locally voluminous rhyolitic welded tuffs and lavas of 16- to 14- Ma age generally overlie but are also interlayered with these basalts (Bonham, 1969; Noble and others, 1970, *1973*; *Korringa, 1973*; *Greene, 1984*; Duffield and McKee, 1986). The most voluminous rhyolites are peralkalic. In south-central and central Oregon, silicic centers of 15- to 14-Ma age are smaller but widely distributed (MacLeod and others, 1976; Walker, 1977, *1979*; *Wells, 1979*). In the Blue Mountains region of north-central and northeastern Oregon, rhyolites of 16- to 14-Ma age include the Dinner Creek Welded Tuff as well as rhyolitic flows and breccias (*Brooks and others, 1976*; Walker, 1977; *Fiebelkorn and others, 1983*).

Similar mid-Miocene stratigraphy continues east of Steens Mountain through the Owyhee uplands to the Snake River Plain (Fig. 53). Although these volcanic rocks are sometimes described as part of a Snake River Plain province (*McKee and Mark, 1971*; *Hart and Aronson, 1983*), they are spatially and temporally part of the Oregon Plateaus. Voluminous sheetlike flows of 17- to 14-Ma age subalkalic basalt to trachyandesite in eastern Oregon include the Owyhee Basalt and other units (Kittleman and others, 1965; Watkins and Baksi, 1974; *Bottomly and York, 1976*; *Brown and Petros, 1985*). Similar lavas of about 17 Ma age in the Owyhee Mountains of western Idaho (*McIntyre, 1972*; Pansze, 1975; Armstrong and others, 1975; *Ekren and others, 1981*, 1982) may have vented beneath the western Snake River Plain. Others occur as far south as northeastern Nevada (Coats, 1964; *Armstrong and others, 1976*; *Coats and others, 1977*). These basalts are overlain widely by 17- to 14-Ma rhyolites. In and near northeastern Nevada, they include the Jarbidge Rhyolite (Coats, 1964; *Coats and others, 1977*; Christiansen and others, 1986) and other rhyolites and dacites (*McKee and others, 1976*). In southwestern Idaho and adjacent Oregon (Pansze, 1975; Armstrong and others, 1975; *Ekren and others, 1981*, 1982, *1984*; *Hart and Aronson, 1983*), voluminous rhyolites of 17- to 14-Ma age erupted at high temperatures and have the form of viscous lava flows although Ekren and others interpret them as rheomorphic ash flows. Similar rhyolite in eastern Oregon is interlayered with basin sediments deposited on the Owyhee and other basalt and trachyandesite (Kittleman and others, 1965; *Ekren and others, 1984*).

The Columbia River Basalt Group spans from 17 to 6 Ma, but probably 99 percent of the volume (Fig. 53) erupted between 17 and 14 Ma (Watkins and Baksi, 1974; *McKee and others, 1981*; Tolan and others, 1989; *Baksi, 1989*). From the first detailed modern studies, it was evident that single flows of great volume and extent have distinctive compositions (*Mackin, 1961*; *Waters, 1961b*). Flows of 17- to 14-Ma age now included in the Imnaha, Picture Gorge, Grande Ronde, and Wanapum Basalts of the Columbia River Basalt Group are marked by distinctive chemical types (Wright and others, 1973; Swanson and others, 1979a). The lavas vary stratigraphically upward toward more evolved compositions (*Osawa and Goles, 1970*; Wright and others, 1973; *Nathan and Fruchter, 1974*; *McDougall, 1976*; *Reidel, 1983*; *Nelson, 1983*; *Hooper, 1984*; *Carlson, 1984*; *Church, 1985*). Although the rocks are almost always described as basalt, many are actually mafic andesite. The voluminous flows erupted at high rates from linear fissure swarms near the southeastern and southern margins of the plateau (Fig. 51; *Taubeneck, 1970*; Swanson and others, 1975; *Fruchter and Baldwin, 1975*; *Hooper, 1982*; Mangan and others, 1986) and accumulated in a subsiding basin (Swanson and others, 1979b). The flows backed up to the flank of the Rocky Mountains (*Camp and others, 1982*; *Reidel, 1982, 1983*), blocking drainages, locally forming pillow-palagonite complexes, interfingering with sediments such as the Latah Formation (*Griggs, 1976*; *Camp, 1981*), and quenching to form zones of hackly radiating joints where deranged drainage flooded the basin (*Long and Wood, 1986*). Such flooding also may have produced local steam explosions, now recorded in small circular ridges and fractures (*Hodges, 1978*). Basalt interfingers on the west with volcanic sediments of the Ellensburg Formation derived from the Cascades (*Schmincke, 1967b*; Swanson and others, 1979a; *Smith and others, 1989*) and, on the south, with the calc-alkalic Strawberry Volcanics and the sedimentary Mascall Formation derived from it (*Thayer, 1957*; *Thayer and Brown, 1966*; Robyn, 1979).

Basalts of the Columbia Plateau spilled through an ancestral Columbia River Gorge across the Cascade Range (Waters, 1973; *Beeson and Moran, 1979*; *Tolan and Beeson, 1984*). A few flows spread into the basin between the Cascades and Coast Ranges and into the sea. Stratigraphically and chemically identical lavas occur in middle Miocene marine sediments of the Coast Range as both flows and intrusive breccias (Snavely and others, 1958, *1973*; *Niem and Cressy, 1973*; *Nathan and Fruchter, 1974*; Wells, 1981; *Van Atta and Kelty, 1985*; *Beeson and others, 1985*, 1989; *Wells and others, 1989*). The intrusives were interpreted by Snavely and others (1958) as locally fed but were later interpreted as invasive into preexisting sediments (Beeson and others, 1979, 1989).

Roughly coeval with the major 17- to 14-Ma Columbia Plateau outpourings, the Strawberry Volcanics of the Blue Mountains area comprise varied 15- to 12-Ma basalt to rhyolite, including abundant pyroxene andesite, mainly calc-alkalic but with some Fe-enrichment sequences (*Thayer, 1957*; *Thayer and Brown, 1966*; Robyn, 1979; Walker, 1990). The Strawberry Volcanics is by far the largest part of a northeast-trending group of predominantly intermediate calc-alkalic centers through central Oregon, probably along a zone of thickened crust (Robyn, 1979; *Robyn and Hoover, 1982*; *Bunker and others, 1982*; Walker and Robinson, 1990b).

Later Miocene to Quaternary volcanism of the Columbia and Oregon plateaus

Younger lavas of the Columbia River Basalt Group were less voluminous but continued to 6 Ma with mainly canyon-filling flows erupted at increasingly long intervals, included stratigraphically in the Saddle Mountains Basalt (Swanson and others, 1979a, b; Tolan and others, 1989). These lavas are considerably more diverse and, in some instances, more highly evolved than the older Columbia Plateau lavas. (References for these data are the same as for the older lavas.) Most of the lavas erupted from linear fissures in the same swarms as the earlier eruptions, and at least one flow—the Pomona Member—reached the Pacific Coast (Snavely and others, 1973; Beeson and others, 1979; *Anderson, 1980*; Magill and others, 1982). Sediments commonly divide the flows on the Columbia Plateau, and some lavas were invasive into wet sediments (*Schmincke, 1967a*). Unlike most of the Columbia Plateau flows, at least one of the youngest, at about 9 Ma, erupted near the center of the plateau (*Helz and others, 1976*; Swanson and others, 1979a, b). Some others might have erupted from now-buried vents within the basin. Relatively minor basalt erupted in the St. Joe River area of northern Idaho, east of the plateau, at about 3 Ma (*Dort, 1967*; *Gray and Kittleman, 1967*; *Bishop, 1969*).

A transition in the Oregon Plateaus, as in the Great Basin region, from broad basin formation to distributed block faulting between about 14 and 10 Ma was accompanied by changes in bimodal volcanism. Basaltic volcanism declined, and voluminous rhyolite erupted mainly in the Owyhee Plateau. On the margins of the Snake River Plain near the junction of the western and eastern plains (Figs. 53, 54), 12- to 8-Ma rhyolite includes the Idavada Volcanics (Malde and Powers, 1962; *Malde and others, 1963*; Armstrong and others, 1975; Bonnichsen, 1982a). Like many 17- to 14-Ma rhyolites, some of the 12- to 10-Ma rhyolites and rhyodacites of the Owyhee Plateau erupted at high temperatures, have viscous-lava emplacement structures, and are interpreted by Ekren and others (*1981*, 1982, *1984*) as rheomorphic ash flows. Although much of the younger rhyolite erupted east of the older centers, now covered by younger basalts of the western Snake River Plain, similar lavas extend westward into southeastern Oregon (Kittleman and others, 1965, *1967*; Walker, 1977) and northern Nevada (*Coats, 1968*). The Cougar Point Tuff of 12- to 11-Ma age comprises at least nine voluminous densely welded ash-flow sheets in northern Nevada and southwestern Idaho (Coats, 1964, *1971*; Bonnichsen and Citron, 1982). They define three cycles of high-temperature pigeonite-bearing rhyolite from more salic to more femic stratigraphically upwards. They erupted from the subsided 55- by 95-km Bruneau-Jarbidge eruptive center in Idaho, now buried by younger volcanic rocks, as did rhyolitic lavas as young as 8 Ma (Bonnichsen, 1982a, *1982b*; *Hart and Aronson, 1983*). Physiographically, the Bruneau-Jarbidge center might properly be regarded as the southwest end of the eastern Snake River Plain (Bonnichsen, 1982a).

Following the volcanic transition of 14 to 10 Ma, basaltic volcanism again became widespread in the Oregon Plateaus (Fig. 53). East of the northern Nevada rift zone, basalts and associated gravels extending south to the Humboldt River and northeast to the western Snake River Plain, mainly 11 to 8 Ma in age, are commonly mapped as the Banbury Basalt (Malde and Powers, 1962; *Mark and others, 1975*; Stewart, 1980; *Ekren and others, 1981*), although differing in age and composition from the type Banbury (Hart and others, 1984). From northwestern Nevada through central and eastern Oregon, basalt—mainly of 10- to 9-Ma and Pleistocene age—caps sedimentary units that overlie a middle Miocene unconformity (Bonham, 1969; *Morton and others, 1977*; *Walker, 1979*; *Hart and Mertzman, 1982*; Hart and others, 1984; *Carlson, 1984*; *Greene, 1984*). The most widespread type is high-Al olivine tholeiite with low K and Rb contents, relatively high light rare-earth elements and Ba, high $^{87}Sr/^{86}Sr$ (0.7052 to 0.7076), and low $^{143}Nd/^{144}Nd$ (to 0.51237); these characteristics may reflect magmas generated from a chemically evolved upper mantle and little crustal interaction during rapid ascent through a thinned and extended crust (*Carlson, 1984*).

A zone just east of the Cascade Range, marking a transition between the High Cascades and the volcanic plateaus, erupted voluminous basalt younger than 10 Ma (Fig. 53). Both high-Al, low-K olivine tholeiite, notable in northeastern California and south-central Oregon (Macdonald, 1966; McKee and others, 1983; Hart and others, 1984; Luedke and Smith, 1984; Duffield and McKee, 1986), and calc-alkalic basalt are present. Several large systems produced voluminous basalts and rhyolites but much less intermediate lava. The southernmost is the Quaternary Caribou basaltic field east of Lassen Peak (Williams, 1932; Macdonald, 1966). To the north, east of Mount Shasta, the bimodal Medicine Lake Highland is a predominantly basaltic Quaternary shield with a partly filled caldera (*Anderson, 1941*; Donnelly-Nolan, 1988). Holocene basalt is extensive on its flanks (Donnelly-Nolan and others, 1990). Upper Holocene rhyolitic lavas and pumice erupted high on the shield (*Heiken, 1978*; Fink, 1983; Donnelly-Nolan and others, 1990), and older rhyolite crops out near the base (*Mertzman, 1983*). Some dacite represents mixed basaltic and rhyolitic magmas (*Eichelberger, 1975*), but sparsely porphyritic andesite is a complex product of magma mixing, fractional crystallization, and crustal assimilation (*Gerlach and Grove, 1982*; *Grove and Donnelly-Nolan, 1986*). The Quaternary Newberry volcano, east of the central Oregon Cascades, is similar to the Medicine Lake Highland and probably overlies a small magma body (*Williams, 1935*; *Higgins, 1973*; *Fitterman, 1988*; MacLeod and Sherrod, 1988). Between Medicine Lake and Newberry is a faulted but similar 5-Ma shield (MacLeod and others, 1976). Upper Miocene basalt continues northward east of the Cascades nearly to the Columbia River (*Farooqui and others, 1981*; *Bunker and others, 1982*). The Quaternary Simcoe basaltic field overlies the Columbia Plateau at the margin of the Cascades east of Mount Adams (Fig. 53; *Sheppard, 1967*). The predominantly basaltic Medicine Lake, Newberry, and Simcoe systems are commonly included in discussions of High Cascades volcanism but are petrologically and tectonically more akin to the Oregon Plateaus.

Propagating volcanism of the High Lava Plains

Volcanism since the mid-Miocene has evolved in a distinctly different pattern in two zones sometimes called collectively the High Lava Plains. One of these zones extends northwest across eastern Oregon and the other is the northeast-trending eastern Snake River Plain. Voluminous rhyolitic volcanism has propagated linearly outward from the center of the region to the distal ends of each of the two zones.

Rhyolitic magmatism has propagated northeastward along the eastern Snake River Plain (Fig. 54) at 3 to 4 cm/y, from the 12- to 8-Ma Bruneau-Jarbidge center to the still active Yellowstone Plateau field (Christiansen and Lipman, 1972; Armstrong and others, 1975; Christiansen and McKee, 1978). Basaltic lavas form a constructional surface on the downwarped eastern plain, which is, thus, distinct from the incised filled graben of the western plain. Voluminous rhyolitic ash flows that mark the north-

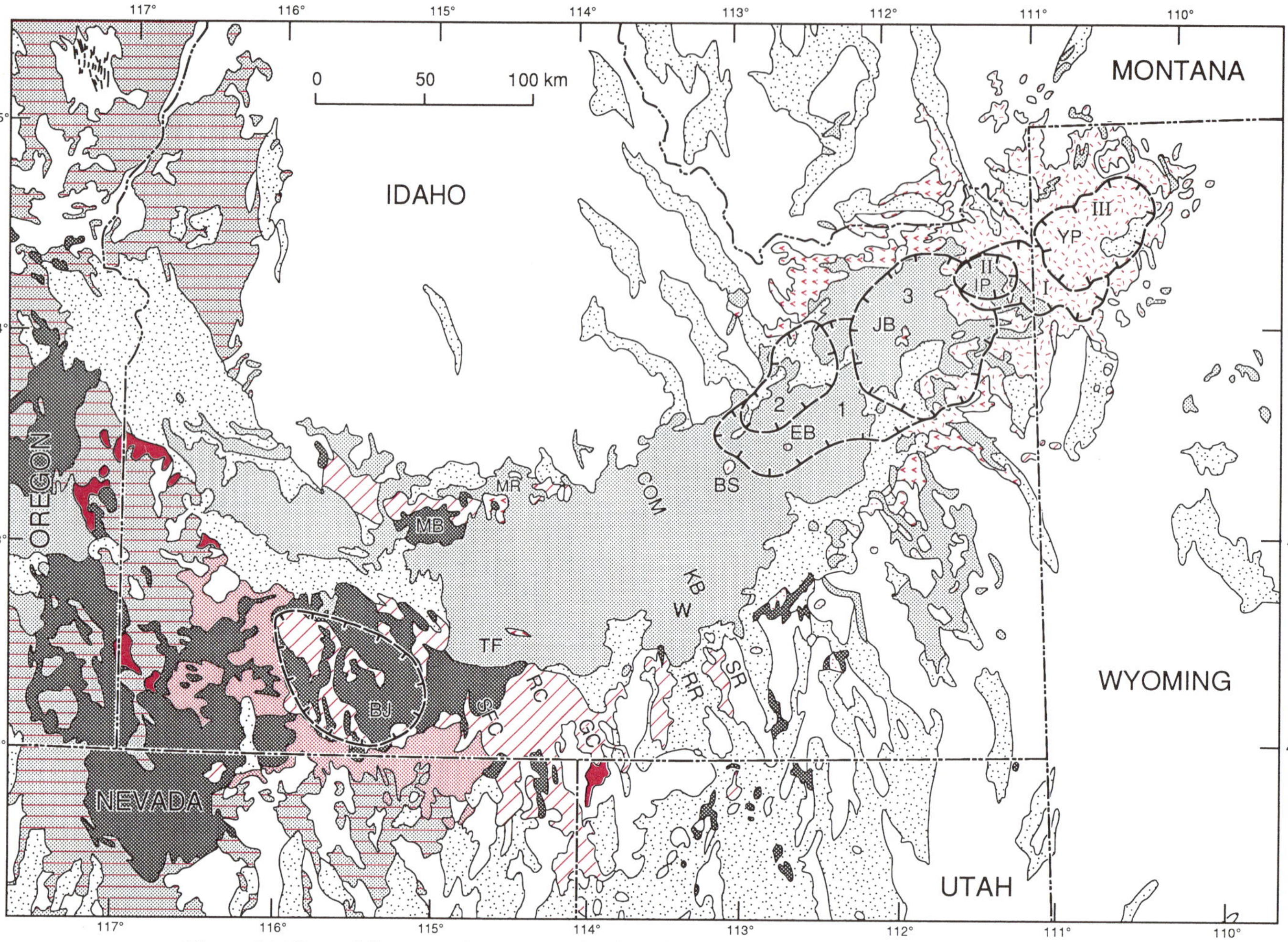

Figure 54. Upper Miocene and younger volcanic units of the eastern Snake River Plain and vicinity showing eastward progression of voluminous rhyolitic magmatism and associated basalts and subsequent basaltic and minor rhyolitic volcanism near the subsided axis of the plain. Gray with red ruling, basalts and rhyolites (17 to 14 Ma); dark red, rhyolites (14 to 12 Ma); pink, Cougar Point Tuff and Little Jacks Rhyolite (12 to 11 Ma); red cross hatch, rhyolites (11 to 8 Ma); red stipple, rhyolites of Starlight Formation (8 to 7 Ma); red V pattern, rhyolites of Heise volcanic field (7 to 4 Ma); red hash marks, rhyolites of Yellowstone Plateau volcanic field (<3 Ma); dark gray, basalts (11 to 6 Ma); light gray, basalts (<6 Ma); sand pattern, basin sediments. Hachured dashed lines, calderas: BJ, Bruneau-Jarbidge volcanic center; 1, 2 and 3, calderas of Heise volcanic field (order of decreasing age); I, II, and III, calderas of Yellowstone Plateau volcanic field (order of decreasing age); AF, American Falls; BS, Big Southern Butte; COM, Craters of the Moon; EB, East Butte; GC, Goose Creek; IP, Island Park; JB, Juniper Buttes; KB, Kings Bowl; MB, Mount Bennett Hills; MR, Magic Reservoir; P, Pocatello; RR, Raft River; SFC, Salmon Falls Creek; SR, Sublette Range; RC, Rock Creek; TF, Twin Falls; W, Wapi lava field; YP, Yellowstone Plateau.

eastward progression are exposed only at the margins, interlayered with both rhyolitic and basaltic lavas and sedimentary basin-fill. After the 12- to 11-Ma Cougar Point Tuff spread widely from the Bruneau-Jarbidge center, similar welded tuffs and lavas of 11- to 8-Ma age, the Idavada Volcanics, erupted from now buried sources farther east that may have been parts of the same long-lived, composite center. The Idavada extends along the south margin of the plain as far eastward as the Sublette Range (*Mapel and Hail, 1956*; Armstrong and others, 1975, *1980*; *Hart and Aronson, 1983*) and along the north margin in the Mount Bennett Hills (*Armstrong and others, 1980*; *Ekren and others, 1984*; *Honjo and others, 1986*). Thick rhyolite of 11- to 9-Ma age also was penetrated in a 3-km drill hole north of Big Southern Butte, 150 km northeast of Twin Falls (Morgan and others, 1984). Rhyolite and basalt of 9- to 8-Ma age just west of the Raft River may relate to some of these sources. Younger 8- to 7-Ma welded tuff along the south margin occurs in tuffaceous sediments, rhyolite, and basalt of the Starlight Formation (stratigraphically and lithologically equivalent to the Salt Lake Formation) near American Falls and Pocatello (Carr and Trimble, 1963; Armstrong and others, 1975; Trimble and Carr, 1976; Trimble, 1976). Farther northeast along both margins, the Heise volcanic field, dominated by three 7- to 4-Ma rhyolitic ash-flow sheets (Fig. 54; *McBroome and others, 1981*; *Embree and others, 1982*; Morgan and others, 1984), also erupted minor rhyolite, basalt, and trachyandesite (*Prostka and Hackman, 1974*; *Armstrong and others, 1980*; Morgan and others, 1984).

The active Yellowstone Plateau volcanic field (Fig. 54; *Christiansen and Blank, 1972*; *Christiansen, 1979, 1982b,* 1984) marks the northeast end of the propagating eastern Snake River Plain system, astride the eastern margin of basin-range extension (Fig. 51). This strongly bimodal field has erupted only basalts and rhyolites since 2.5 Ma, including small mixed rhyolite-basalt complexes (*Wilcox, 1944*). Three major caldera-forming ash-flow eruptions produced the Yellowstone Group with ages of 2.0, 1.2, and 0.6 Ma, ranging in volume from 300 to 2,500 km^3. Each climaxed a cycle of precaldera and postcaldera rhyolitic eruptions. The nested and overlapping calderas extend 110 km from Island Park in easternmost Idaho to the east margin of the Yellowstone Plateau, filled with more than 1,000 km^3 of rhyolitic lavas as young as 70 ka and hosting the world's largest active hydrothermal system. Coeval basalts are present around the margins but are absent within the youngest caldera, suggesting that rhyolitic magma still prevents the rise of denser basaltic magma. Geophysical evidence supports the existence of a shallow, still partly molten batholith (*Eaton and others, 1975*; Smith and Christiansen, 1980; *Smith and Braile, 1984*). Petrologic relations show that the rhyolitic magmas formed by partial melting of lower-crustal rocks during long-lived basaltic magmatism from the mantle and extensive modification in the shallow crust (*Doe and others, 1982*; Christiansen, 1984; *Hildreth and others, 1984,* 1991).

Volcanism continues on the Snake River Plain (Figs. 53, 54), largely basaltic but also including small rhyolites. Basalts, mainly Pliocene and younger, mantle the plain (Malde and Powers, 1962; Armstrong and others, 1975, *1980*; *Amini and others, 1984*). On the western plain, they are interstratified with basin fill and other sediments (Malde and Powers, 1962; *Malde and others, 1963*). On the eastern plain, the Quaternary Snake River Group forms a nearly continuous, largely uneroded surface, including several Holocene lava fields (Kuntz and others, 1982, *1986*). The basalt is generally olivine tholeiite with high Na_2O+K_2O, TiO_2, P_2O_5, and $^{87}Sr/^{86}Sr$ and low $^{143}Nd/^{144}Nd$, interpreted by *Leeman and Manton* (*1971*), Leeman (1982a), and *Menzies and others* (*1984*) as derived from an Archaean lithospheric mantle. Local older basalts associated with the voluminous rhyolites typically are more primitive (e.g., having lower Fe/Mg) but nevertheless more strongly modified within the crust (Leeman, 1982a). Less common more evolved mafic lavas, as at Craters of the Moon, may reflect advanced fractional crystallization in the lower crust and significant upper-crustal assimilation (*Leeman, 1976, 1982b*; *Menzies and others, 1984*). The largest associated rhyolite bodies, near Magic Reservoir, date from 6 to 3 Ma (*Malde and others, 1963*; *Leeman, 1982c*; *Struhsacker and others, 1982*; *Honjo and others, 1986*). Rhyolite of 6-Ma age also is exposed beneath the surficial basalts at the head of the Snake River Canyon near Twin Falls (*Malde and others, 1963*; Armstrong and others, 1975). Farther east, between Craters of the Moon and Idaho Falls, are three rhyolite domes of less than 1-Ma age, Big Southern Butte (*Spear and King, 1982*) and Middle and East Buttes. Juniper Buttes, between Idaho Falls and Island Park, is an uplifted rhyolite-basalt complex (*Kuntz, 1979*).

In the High Lava Plains that span the northern Oregon Plateaus (Fig. 55), rhyolitic volcanism since about 11 Ma shows significant parallels to that of the eastern Snake River Plain. Whereas rhyolite had been associated widely with 17- to 14-Ma basalt in the Oregon Plateaus, subsequent rhyolitic volcanism propagated northwestward across eastern Oregon (*Walker, 1974*; MacLeod and others, 1976; *Walker and Nolf, 1981*; Smith and Luedke, 1984; Duffield and McKee, 1986). In contrast to the eastern Snake River Plain, where the largely buried rhyolitic centers are inferred mainly from distributions of ash flows (Fig. 54), the dome fields of eastern Oregon clearly record rhyolitic centers in a band nearly along the Brothers fault zone (Figs. 51, 55). The numerous domes of each center are similar in age and composition over areas as great as 3,000 km^2, forming two parallel belts. Between 10 and 5 Ma, propagation was at about 3 cm/y, comparable to the average rate along the eastern Snake River Plain; since 5 Ma the Oregon system has turned more northward and slowed to about 1 cm/y. Near the east end of the zone, three voluminous caldera-related units, the Devine Canyon, Prater Creek, and Rattlesnake Ash-flow Tuffs, erupted between 9 and 7 Ma near Harney Basin, fitting the contoured age progression of the domes (*Greene, 1973*; MacLeod and others, 1976; *Enlows, 1976*; *Walker, 1979*; *Walker and Nolf, 1981*). Several fields of Pliocene to Holocene basalt also erupted along the trace of the northern rhyolitic belt (MacLeod and others, 1976; *Walker, 1979*; *Walker and Nolf, 1981*), ranging from low-K high-Al olivine tholeiite to alkali basalt (Leeman and Rogers, 1970). The Newberry volcano, a large bimodal shield just east of the Cascade front, is the northwesternmost end of the propagating zone. These systematic relations suggest either that we observe the propagating southeast Oregon and Snake River–Yellowstone systems at unique times in their history, just as they arrive coincidentally at the Cascade front and the margins of active extension or that their activity is inherently tied to the contemporary state of regional tectonism.

Mineral resources associated with bimodal volcanism of the Columbia Intermontane region resemble those of the Basin and Range Province. Epigenetic Au, Hg, Sb, Li, nd U deposits are associated with some of the young rhyolite. U is concentrated in sedimentary rocks associated with highly evolved silicic centers, particularly (but not solely) with peralkalic magmatism. Most notable are deposits in the McDermitt caldera (*Rytuba and Glanzman, 1979*; Rytuba and McKee, 1984) and west of the Abert Rim near Lakeview, Oregon (*Castor and Berry, 1981*).

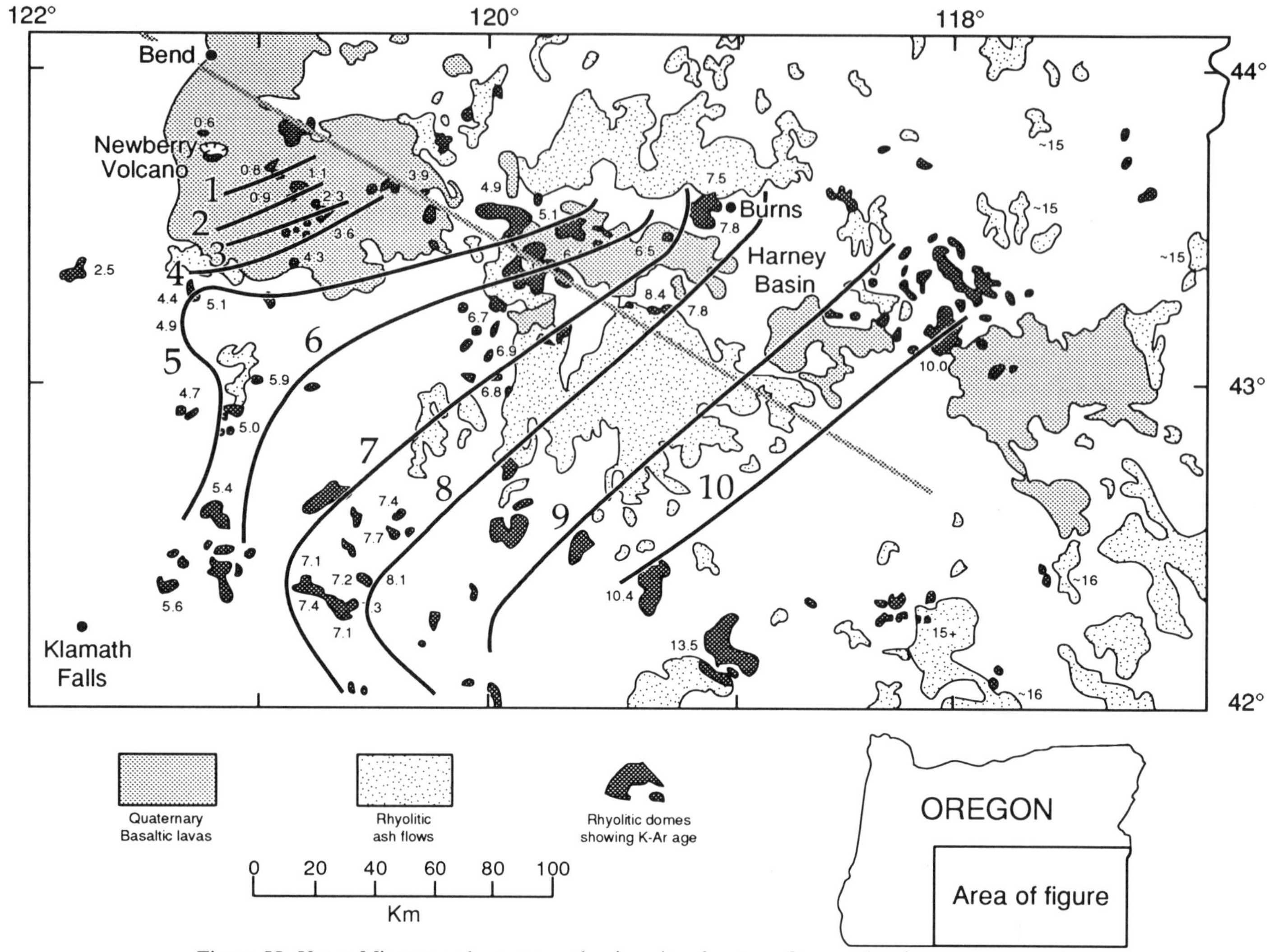

Figure 55. Upper Miocene and younger volcanic units of eastern Oregon showing northwestward progression of rhyolitic dome fields and continued subsequent basaltic volcanism. Volcanic belts obliquely cross the Brothers fault zone (heavy northwest-trending line; see Fig. 51). Approximate age contours of domes labelled in increments of 1 Ma. After MacLeod and others (1976).

Young volcanism associated with tectonic extension also yields the potential for both high- and low-temperature geothermal resources (*Brook and others, 1979*; *Mariner and others, 1983*).

Reviewing the magmatic history of the entire Columbia Intermontane region and adjacent Northern Rocky Mountains, basaltic and bimodal rhyolite-basalt magmatism has characterized the region since mid-Miocene time. Early activity, between about 17 and 14 Ma, accompanied extension in the northern Nevada rift zone, western Snake River Plain, and southeastern Columbia Plateau fissure swarm. Basaltic eruptions dominated early activity, but rhyolite was voluminous and widespread in the Oregon Plateaus, especially later in the episode. Volcanism in the Oregon Plateaus declined with the long transition from 17- to 14-Ma rifting to more distributed extension after 11 to 10 Ma through most of the region except the Owyhee Plateau. After about 11 Ma, basalts erupted widely and episodically, notably adjacent to the Cascade front. Voluminous rhyolitic magmatism was concentrated in two zones more or less symmetrical about the 17- to 14-Ma rift axis, propagating both northeastward and northwestward. Although the eastern Snake River Plain produced greater volumes than the southeast Oregon zone, large fields of rhyolitic domes in southeast Oregon suggest significant magmatic volumes. At present, active systems at the ends of each propagating volcanic arm are at unique tectonic positions marginal to basin-range extension. The eastern Snake River Plain–Yellowstone system is cited frequently as a continental "hot spot", and its late Miocene and subsequent N55°E propagation relative to the North American Plate at 3 to 4 cm/y is close to that calculated by summing relative-motion vectors for the North American, Pacific, and Juan de Fuca Plates and assuming the Hawaiian hot spot to be fixed (*Morgan, 1972*; Smith and Sbar, 1974; Armstrong and others, 1975). Its late mid-Miocene de novo origin, its symmetrical opposite in Oregon, its temporal relation to regional basin-range extension, and the unique tectonic positions of the two active systems, however, raise doubts about simple models of active subasthenospheric origin (Christiansen and McKee, 1978). These same factors argue against its continuity with an Eocene Coast Range "hot spot," as suggested by *Duncan* (*1982*).

UPLIFTED REGIONS OF MINOR EXTENSION

Minor volumes of post–mid-Miocene highly alkalic volcanics occur in several uplifted but little-extended areas bordering the basin-range region. Their tectonic settings are, in some respects, comparable to the more widespread early and mid-Tertiary alkalic foreland volcanism.

Late Cenozoic mafic alkalic dikes and pipes occur in three areas of the Colorado Plateau. The strongly alkalic Hopi volcanic field erupted intermittently in the southern plateau between 7 and 2 Ma (Fig. 52), about 60 km across in a late Miocene lacustrine basin (*Williams, 1936*; Hunt, 1956; Shoemaker and others, 1962; *Naeser, 1971*; *Sutton, 1974*; *Luedke and Smith, 1978a*). Nearly 200 small dikes and diatremes fed explosive eruptions of alkali basalt, analcite basalt, limburgite, and monchiquite—a sodic series that contrasts with the potassic mid-Tertiary Navajo Buttes to the north. Uranium mineralization occurs in associated carbonates (Shoemaker and others, 1962). Farther north, in the San Rafael Swell of central Utah (Fig. 53), 6- to 4-Ma alkali-basaltic and trachytic dikes, sills, and intrusive breccias strike north-northwest in a northeast-trending zone (*Gilluly, 1927*; *Delaney and others, 1986*). As mentioned earlier, a single dike of phlogopite-olivine lamproite in the narrow neck of the northwestern Colorado Plateau between the Wasatch and Uinta Mountains (Fig. 53) dates at 13 Ma (*Best and others, 1968*).

The Leucite Hills of south-central Wyoming (Fig. 53) contain 1-Ma intrusive bodies and flow remnants of highly potassic and magnesian compositions (*Bradley, 1964*; *Carmichael, 1967*; *McDowell, 1971*). Characteristic minerals include leucite, phlogopite, olivine, diopside, and—in some rocks—sanidine. Compositions, mineralogy, and melting temperatures indicate that the suite is of mantle origin, possibly reflecting partial melting of phlogopite-bearing rocks.

Leucitic lavas are also part of an ultrapotassic suite of 4- to 3-Ma age in the Sierra Nevada, as noted earlier in describing young basaltic suites west of the Great Basin. The leucitic rocks occur at the western edges of large dispersed volcanic fields that include more voluminous basalt but lie almost entirely within the little-faulted Sierran block (*Van Kooten, 1980*; Moore and Dodge, 1980; *Dodge and Moore, 1981*). Basanite, potassic alkali basalt, and phonolite are associated with the leucitic rocks. This suite too probably reflects small degrees of partial melting in a phologopite-rich mantle, possibly enriched in alkalies and other incompatible elements by a fluid-rich phase during crustal uplift (*Van Kooten, 1980*).

Similar petrogenetic interpretations of the highly alkalic rocks in several areas marginal to basin-range extension suggest that the upper mantle has been widely altered by the separation of fluid-rich phases during regional uplift. Near the margins of Cordilleran uplift, the small amount of extension indicates limits on the degree of partial melting to produce these ultra-alkalic suites, and special structural factors localize their ascent.

SUMMARY: POST-LARAMIDE HISTORY OF THE CORDILLERAN REGION

Because the major descriptive parts of this chapter follow a topical rather than a strictly chronological or geographic order, we conclude with a brief chronological summary of the post-Laramide geologic history of the United States Cordillera. Time intervals are selected that represent widespread changes, but, of course, other events spanned these chosen intervals. Most sections of the summary describe events from the continental margin, to the intermontane region, to the Rocky Mountains and High Plains; within each of these regions, discussion is generally from north to south.

LARAMIDE PALEOGEOGRAPHY AND PHYSIOGRAPHY

The Laramide orogeny and contemporaneous events showed a major break from previously continuous Cordilleran orogenic and magmatic systems parallel to the continental margin and produced a framework that controlled much of the post-Laramide regional evolution (Fig. 1). The continental divide extended down the main axis of late Mesozoic thrusting from the Canadian Rockies, through the United States Northern Rockies, the Sevier orogenic belt, and the Mogollon Highland, to the western cordillera of Mexico. Subduction-related magmatic arcs ended southward in Idaho and Montana and northward in southern Arizona and New Mexico, leaving a wide magmatic gap in much of the western United States after about 70 Ma. Only locally in the Colorado Rockies was there any igneous activity between these latitudes.

Along the continental margin, an assemblage of basalt-floored oceanic terranes was accreted to the Pacific Northwest by about 52 to 50 Ma; the granitic and metamorphic Salinian block arrived at its pre–San Andreas position along the California coast at about 55 Ma, producing uplift and structural segmentation of the adjacent continental margin. Thus, a marine and lowland region of varied width, perhaps in part allochthonous, bordered the rapidly eroding mountain belt west of the continental divide.

East of the divide, the Laramide Middle and Southern Rocky Mountains formed as thrust-bounded uplifts between deep basins that filled with orogenic sediments, major deformation ending by 55 to 50 Ma.

POST-LARAMIDE EOCENE TO MID-MIOCENE (55 TO 21 Ma)

Major post-Laramide events summarized by time slices are illustrated synoptically in the nine panels of Plate 7, corresponding to the following text sections. In order for the map reconstructions to properly illustrate the text, the base maps for Plate 7 show displacements on the San Andreas fault system restored for mid-Miocene and earlier times and the western Transverse Ranges rotated back by 90° before the mid-Miocene. No other displacements or rotations are restored on the base maps.

55 to 43 Ma (Plate 7A)

Throughout the coastal region, a profound change in middle Eocene time produced an angular unconformity or a depositional hiatus. In the Pacific Northwest, the newly accreted terranes subsided and were buried in middle and upper Eocene submarine-fan and continental-slope deposits that grade eastward into shallow-marine, deltaic, and fluvial equivalents. Before later clockwise rotation and translation, the marine units trended more northwesterly and were closer to the orogenic uplands of northern Washington, eastern Oregon, and Idaho. Thick nonmarine arkosic sandstone and conglomerate on the northwest flank of the North Cascades filled pull-apart basins created by oblique subduction. A typical arc suite of mafic and intermediate igneous rocks at least as old as 48 Ma was associated with contractional tectonism from British Columbia into the North Cascades. The arc extended southward into part of the Clarno region of west-central Oregon.

In California, continued deformation folded the submarine continental margin, further segmenting it into local basins. In northern California, oblique submarine gorges across the continental shelf funnelled sediment from major drainages of the continental interior into a deep-sea trench. In central California, submarine fans built outward to the accreted Salinian block and graded into outer shelf and slope facies. In southern California, a cycle of basin deepening and shallowing began at about 50 Ma.

While the crust of the North Cascades was being shortened by folding and strike-slip faulting, areas inland were dominated by coeval extension and volcanism. Over a wide region of the Northern Rocky Mountains from southern British Columbia through the Idaho batholith region, detached upper-crustal sections were juxtaposed with mid-crustal rocks, now exposed discontinuously as "metamorphic core complexes" in structural domes. Between the domes, graben-like basins were extended by tens of kilometers on generally low-dipping normal faults and filled with thick sediments and contemporaneous volcanics. Igneous activity ranging from basalt to rhyolite but dominated by potassic andesite to rhyodacite formed the Challis magmatic belt from the North Cascades to the Cordilleran front in Montana and Wyoming. The belt was active mainly between about 50 and 45 Ma but began at least as early as 54 Ma and waned after about 43 Ma.

During this time, virtually no tectonism or magmatism was recorded in the broad depositional lowlands farther south, now mainly in the Basin and Range Province. Magmatism, however, remained nearly continuous from Laramide through mid-Tertiary time in the Sierra Madre Occidental of Mexico, and potassic andesite-rhyolite magmatism may have overlapped into the Rio Grande Valley in west Texas as early as 49 Ma.

East of the continental divide, intermontane basins of the postorogenic upland accumulated sediments, some of them in extensive lakes. Regional drainage in lowlands adjacent to the Northern and Middle Rockies was eastward onto the Great Plains and southward onto the Colorado Plateau as the Laramide foreland ranges were worn down by pedimentation. The Southern Rockies underwent regional erosion. Strongly alkalic magmatism, both sodic and potassic and coeval with the Challis belt, occurred at scattered foreland centers in Montana, Wyoming, and South Dakota. Foreland alkalic magmatism was also active in Mexico, west Texas, and southernmost New Mexico, adjacent to the southern Cordilleran potassic-andesite and rhyolite belt.

43 to 37 Ma (Plate 7B)

The tectonic patterns of middle Eocene time in the coastal region continued with little change until about 39 Ma, when uplifts and segmentation of the forearc produced shallow basins that deepened westward. Clockwise tectonic rotation of the coastal region progressed at a rate of more than 1° per million years. Local centers erupted basalt and minor intermediate lavas in the subsiding forearc on the continental margin of Washington and Oregon. A more or less continuous magmatic arc of 43- to 37-Ma age extended from western British Columbia to northwestern Oregon and eastward into the younger part of the Clarno basin of central Oregon and to northeastern California. Marine sedimentation in California became generally shallower with time as the basins gradually filled.

Beginning in latest Eocene time, major extension and related magmatism, similar to that earlier in the Northern Rockies, became active farther south. By about 43 Ma, extension was unimportant in the Northern Rockies, and the Challis belt waned to small local centers south of central western Montana. Major extension became active in northeastern Nevada and adjacent Utah and Idaho, now represented mainly by a belt of detachment faults and rocks of juxtaposed crustal levels 200 to 300 km west of the Mesozoic Sevier frontal thrusts. Coarse clastic deposits accumulated in extensional basins although regional relief remained low; the Tuscarora belt of potassic andesite to rhyolite was established from the continental-margin arc of the Pacific Northwest southeastward into northern Nevada and northwestern Utah.

During this same period, potassic andesite to rhyodacite volcanism of the Sierra Madre Occidental in Mexico began to extend into southwestern New Mexico and farther north; by at least 41 Ma it had reached the Southern Rockies of central Colorado. Numerous centers formed the Mogollon-Datil volcanic field of southwestern New Mexico, and voluminous caldera-forming ash-flow eruptions occurred as early as 39 Ma. Magmatism may have extended westward into Arizona but left only an obscure record.

Along and east of the continental divide, intermontane basins of the postorogenic upland continued to accumulate sediments. In the Northern and Middle Rockies and adjacent lowlands, the deposits merged with time into aggradational mantles that expanded as erosion accelerated in the uplands. In the Southern Rockies, a regional erosion surface truncated most of the extensive upland by the late Eocene. Centers of alkalic magmatism continued locally in the northern foreland, especially in the Black Hills, and in the southern Cordilleran foreland of west Texas and southern New Mexico. Very minor alkalic volcanism also occurred in the Colorado Plateau, the foreland of the Late Cretaceous Sevier thrust belt.

37 to 28 Ma (Plate 7C)

By Oligocene time, subsidence in the Puget Sound area had inundated the coastal plain and deltas and created a deep marginal basin on the shelf; coarse alluvial fans at the north end graded into a submarine fan within the basin. Farther south, rapid sedimentation and uplift caused marine regression in western Oregon. Arc volcanism that previously had extended east of the present Cascades and had adjoined the Challis and Tuscarora magmatic belts became concentrated in and near the Cascades, initiating the early Western Cascades arc of 37 to 17 Ma. Tectonic rotation of the forearc region continued, probably by combined dextral shear along the coast and differential extension within and behind the arc. After about 31 Ma, the arc extended southward through western Oregon to the northern Sierra Nevada and eastward into northwestern Nevada.

In central and southern California, a marine regression was gradually replaced by nonmarine deposition that lasted through this time period except in the westernmost parts of the basins. In the southern coastal region, a transform-fault system originated with contact of the Pacific and North American Plates at about 29 Ma, separating the northern subducting plate (the Vancouver Plate, which became the Juan de Fuca Plate) from the southern plate (the Farallon Plate, which became the Rivera Plate). The transform system joined these plates at the Mendocino and Rivera triple junctions. This Oligocene transform at the base of the continental slope entirely predated establishment of the San Andreas fault system.

East of the early Western Cascades arc, the John Day basin formed by uparching of the Blue Mountains province at its northern margin. Downwind ash from the Cascades accumulated in the John Day basin as it subsided between 36 and 18 Ma. Rhyolite and alkali basalt to trachyandesite also erupted locally in the basin as well as in southern Oregon and nearby Nevada and Idaho and eastward in southwestern Montana.

In the present Great Basin region, the extensional belt continued to evolve southward through eastern Nevada, and voluminous mid-Tertiary volcanism formed successively south-stepping belts from the present Sierra Nevada to the Wasatch Range, creating a broad plain. With decline of the Tuscarora belt, a southward shift formed another andesite-rhyolite belt between about 37 and 34 Ma. After 34 Ma, major activity again shifted southward and became dominated by voluminous rhyolitic pyroclastic eruptions. East of central Nevada, the active volcanic belt split into two branches across central western and southwestern Utah.

Tectonic extension moved progressively northward and westward from the Mexican border region. From the present Rio Grande into southeastern Arizona, extension and associated local magmatism began between 35 and 30 Ma and affected drainage widely by 26 Ma. Magmatism was widespread from about 37 Ma. Voluminous caldera-forming ash-flow eruptions occurred as far north as the Mogollon Plateau by 34 Ma and continued throughout southwestern New Mexico until 28 Ma; after 31 Ma, highly differentiated rhyolite and olivine-bearing basalt to andesite dominated. Volcanism blanketed the Southern Rockies by 36 to 35 Ma; composite batholithic intrusions produced voluminous ash flows and calderas in central Colorado between 36 and 34 Ma and in southwestern Colorado between 31 and 27 Ma. Sometime before 28 Ma, a northwest-trending basin spanned southwestern Arizona, across the lower Colorado River, and northward nearly to the Colorado Plateau. Magmatism may have begun in that basin as early as 35 to 30 Ma, but south of it, dated mid-Tertiary magmatism was no older than 25 Ma.

Drainage on the lowlands east of the continental divide became more integrated. Oligocene stream-laid sediments that blanket the northern High Plains contain abundant ash from voluminous rhyolitic fields far to the west, and upper Eocene and Oligocene deposits accumulated on the piedmont north and northeast of the degrading Mogollon Highland from western Arizona to western New Mexico. Alkalic magmatism culminated in west Texas between 38 and 32 Ma and reached north in the High Plains and Rocky Mountain front of New Mexico and Colorado between about 37 and 20 Ma. Alkalic magmatic centers were also active on the central Colorado Plateau.

28 to 21 Ma (Plate 7D)

The Gold Beach terrane arrived in southwest Oregon sometime after the Eocene, but its time of accretion to the North America Plate is uncertain. Three oceanic terranes in northern California were emplaced by imbricate thrusting about 25 Ma and later. Submarine debris flows of late Oligocene to middle or early late Miocene age in the Eel River basin of northwestern California partly overlapped accretion and thrusting of the easternmost of them. The early Western Cascades arc remained active from British Columbia into northeastern California and northwestern Nevada.

The transform-fault system at the base of the continental slope lengthened with time as its limiting triple junctions migrated apart. Passage of the Mendocino triple junction coincided with local volcanism, rifting, and basin subsidence along the present central California Coast Ranges. At about 22 Ma, the transform system began to broaden and extend inland. With a change in the Pacific–North American slip vector from more northerly to more westerly, offset formed new deep-marine troughs, and strike slip began on the San Andreas fault itself. In southern California two zones of extension were continuous from the Coast Ranges into the interior, both now offset by the San Andreas fault; one spanned across the southern San Joaquin Valley to the Mojave Desert and the other through the Transverse Ranges. Widespread nonmarine deposition in these pre–San Andreas extensional basins of the coastal region came to an end with a latest Oligocene marine transgression at 27 or 26 Ma.

East of the early Western Cascades arc, downwind ash and tuffaceous sediments continued to accumulate in the John Day basin. Broad extensional basins also formed by tectonic extension across parts of southwestern Montana.

To the south, extension continued to migrate southward in a belt along the east side of the present Great Basin region in Nevada and east-central California, and major volcanism again stepped southward. Rhyolitic volcanism was voluminous in west-central Nevada and adjacent California, erupting voluminous ash flows that extended down the drainages of the present Sierra Nevada.

The earlier northwest-trending extensional basin of southeastern California and western Arizona was overlapped by major early Miocene extension along the lower Colorado River, beginning as early as 26 Ma but not prominent until after about 20 Ma. Lavas of about 27- to 20-Ma age, mainly basalt to andesite but with locally voluminous dacite to rhyolite, erupted in extensional zones reaching westward from the lower Colorado River region, across the central Mojave Desert and eastern Transverse Ranges, subsequently offset by the San Andreas fault into the extensional zone of the western Transverse Ranges and Coast Ranges.

The southern Basin and Range Province subsided relative to the Colorado Plateau in the mid-Tertiary, probably in the early Miocene. Sediments accumulated in broad extensional basins, extending well beyond the present basin limits. Magmatism between about 27 and 22 Ma, dominated by latite to potassic andesite but including basalt to rhyolite, was active eastward to the Tucson region until 25 Ma. A northwest-trending belt through southeastern Arizona produced successive caldera-forming rhyolitic eruptions from about 25 Ma in the southeast to 24 to 20 Ma at the Colorado Plateau transition zone. In southern New Mexico and the Southern Rocky Mountains, predominantly basaltic and bimodal magmatism replaced andesite-rhyolite magmatism locally as early as about 26 Ma and widely by 21 Ma. Between 26 and 23 Ma, the bimodal suite included voluminous rhyolitic caldera complexes in northern New Mexico and southwestern Colorado.

On the High Plains, alkalic foreland magmatism waned in West Texas through early Miocene time but remained active farther north near the Rocky Mountain front until about 20 Ma. The Colorado Plateau experienced alkalic magmatism from dispersed centers much like some of those of the High Plains, mainly between about 28 and 24 Ma.

MID-MIOCENE (21 TO 14 Ma)

21 to 17 Ma (Plate 7E)

Accretion of oceanic terranes to the continental margin continued in the core of the Olympic Peninsula into the middle Miocene. The Olympic core comprises broken formations and mélanges derived from turbidite fan sequences deposited farther west in the trench and slope. Both a lower Eocene to Oligocene eastern assemblage, yielding K-Ar ages of 29 Ma, and a middle Eocene to middle Miocene western assemblage were thrust beneath the older accreted terranes.

In northern California, the Mendocino triple junction migrated gradually northward. The Eel River basin overprinted parts of the accreted Coast Range oceanic terranes, grading upward and easward into shallower-water and nonmarine facies. The San Joaquin basin, the largest in the region, remained in communication with the Pacific Ocean across the Salinian block but mainly received deep-water sediment-gravity flows. Strata of its gently dipping east flank rest on Sierran basement while the west flank dips more steeply and reflects San Andreas tectonics by unconformities, stratal wedging, and abrupt facies changes. Basaltic or bimodal rhyolite-basalt magmatism was active locally along the transform system, especially early in its evolution. The southernmost Coast Ranges were dominated by slivering and clockwise rotation along the San Luis Obispo "leaky transform" fault and other complexly interfering right-slip faults. The western Santa Monica Mountains moved westward with respect to the Peninsular Ranges along a precursor of the Malibu Coast–Hollywood fault system. The San Gabriel fault was the main strand of the San Andreas system across the western Transverse Ranges.

As before, the early Western Cascades magmatic arc was more or less continuous from Canada through the Cascade Range to northern California. In addition, however, the arc lengthened rapidly southward along the west margin of the Great Basin; although the early Miocene arc had ended in northwestern Nevada, by about 20 Ma it was continuous to the latitude of Las Vegas. South of Las Vegas, mid-Miocene magmatism was predominantly basaltic to bimodal rhyolite and basalt associated with normal faulting. Arc magmatism resumed, however, after a gap of about 300 to 400 km although the corresponding volcanic rocks were subsequently translated and rotated into the Transverse Ranges, California continental borderland, and Los Angeles basin.

By about 21 to 20 Ma, extensive mid-Tertiary welded ash flows had blanketed the Great Basin region with a volcanic plateau unbroken by significant faults. There was little or no structural differentiation of the present Basin and Range Province from the Colorado Plateau until the subsequent mid-Miocene transition to widely distributed basin-range extension. Regional uplift of a 1,600-km region encompassing both the northern Basin and Range Province and the Colorado Plateau, might have begun as early as 25 Ma but probably was not significant before about 17 Ma. Broad basins, bounded and at least locally broken by normal faults, formed across areas much larger than the present individual basins or ranges, probably reflecting crustal thinning. In the center of the region, where earlier magmatic belts had propagated southward, magmatism declined between about 20 and 17 Ma although a belt across the southern part of the region produced bimodal rhyolite and basalt.

The southern Basin and Range Province resembled the northern. An early to mid-Miocene corridor of major extension along the lower Colorado River, about 200 km wide, was active between 26 and 15 Ma, especially between about 20 and 17 Ma. Near the north end, localized basins began to form at 20 Ma, extension and predominantly basaltic and bimodal volcanism began at about 21 Ma, and a voluminous ash flow tuff of 18 Ma was emplaced across the corridor from the western Colorado Plateau to the central Mojave Desert. Both to the west, across the Mojave Desert, and eastward across the entire southern Basin and Range Province, tectonic and magmatic activity declined during this time.

17 to 14 Ma (Plate 7F)

An unconformity suggests that oblique plate convergence and associated deformation waned in the Olympic Peninsula area by the late middle Miocene. Uplift of the Olympic Mountains and the Coast Range of Washington and Oregon began, with marine sedimentation increasingly restricted to a series of embayments. The Puget Lowland and Willamette Valley subsided to receive west-flowing lavas of the Columbia River Basalt Group, which passed as broad intracanyon flows through a low ancestral Coast Range to reach the sea. A basal mid-Miocene unconformity is overlain by sediments eroded from pre-Tertiary rocks of uplands to the east, volcanic materials of the early Western Cascades arc, and detritus from the continental interior through the ancestral Columbia River.

From late in the mid-Miocene, the Western Cascades–Oregon Coast Range–Klamath Mountains block rotated clockwise by as much as 16°, possibly showing the combined influences of oblique subduction of the Juan de Fuca Plate on the west and extensional opening of the basin-range region to the southeast. In addition, small blocks between oblique-transcurrent faults rotated individually.

The ridged basin configuration of north-central California persisted. The King Range terrane of the northern Coast Ranges contains deep-water turbidite as young as middle Miocene or possibly even younger, indicating continued accretion after 15 Ma. The Sacramento forearc basin was an interior lowland to shallow-marine embayment separated from the Pacific Ocean by the uplifted subduction complex of the coast ranges. Mid-Miocene and older volcanic units of the Sierra Nevada provided detritus to shallow-marine sediments of the Sacramento basin. Deep-marine sedimentation characterized borderland basins in central and southern California. Volcanic complexes of 17- to 14-Ma age are widespread in the southwestern Transverse ranges and southward. Basaltic or bimodal rhyolite-basalt volcanism continued locally along the coastal transform system until about 14 Ma. The terranes of the Baja-Borderland allochthon of southern California and regions surrounding the Gulf of California were stabilized in their present relative configurations by about 16 to 17 Ma.

The late Western Cascades arc produced abundant lavas and volcanic sedimentary rocks between 18 and 8 Ma in Washington and northern and central Oregon. In the present Cascade Range of southern Oregon and northern California, magmatism virtually ended at about 17 Ma, but the mid-Miocene arc was probably continuous farther east. It was particularly well developed between 17 and 12 Ma from northeasternmost California through western and southwestern Nevada. The Columbia River gorge occupied a gap through the Cascades between Washington and Oregon. Volcanic debris flows repeatedly filled the Sierran drainages before uplift and separation of the range from the Great Basin. As earlier, the northern segment of the arc continued south to the central Mojave Desert, the vicinity of Las Vegas, and Lake Mead but was separated by a gap from a southern segment at the present latitude of Mexico. Small bimodal rhyolite-basalt fields erupted within that gap.

The Columbia Plateau subsided as flood basalts erupted and accumulated, mainly between about 17 and 14 Ma; fluvial to marginal-marine sediments also accumulated in parts of the basin. Although eruption of the Columbia River Basalt Group accompanied crustal extension, folds and reverse faults reflect syneruptive basin subsidence, greater extension of the Oregon Plateaus to the south, and clockwise rotation of the Blue Mountains against the plateau margin. The Oregon Plateaus were also a great extensional basin with widespread volcanism. In addition to varied basalts, rhyolite was erupted voluminously in some areas, especially in the Owyhee Plateau on the east, and pyroxene andesite erupted in a northeast-southwest zone through central Oregon.

The tectonics of the Great Basin was linked to that of the Columbia Intermontane region by a major regional rift system. The 17- to 14-Ma principal dike swarm that fed Columbia Plateau volcanism trended directly into the graben of the Western Snake River Plain, which filled with coeval sediments and basalt. Following a similar trend but probably offset at least some distance to the west along a transform zone, the broad basins of the northern Great Basin region were broken by the northern Nevada rift zone, marked by a linear dike swarm, associated lavas and a fault-bounded trough. The bimodal rhyolite-basalt magmatism of the Great Basin region is mostly younger than 17 Ma, following the widespread magmatic lull. By about 17 Ma significant uplift had begun to be a dominant factor in the 1,600-km region that encompassed the Great Basin region, the Columbia Intermontane region, and surrounding areas.

The southern Basin and Range continued to evolve much like the northern. In the western and central Mojave Desert, northwest-trending right-lateral faults of moderate to small displacement became the dominant structures. Near Lake Mead, although localized subsidence had begun as early as 20 Ma, fault-controlled basins were prominent between about 17 and 14 Ma, and extension was extreme at about 14 Ma. Mid-Miocene and younger extension eastward to the Rio Grande rift, distributed regionally on high-angle normal-faults accompanied by local, predominantly basaltic volcanism, followed the magmatic and perhaps tectonic lull and was superposed on earlier mid-Tertiary extension. Alkali basalt and basanite and moderately voluminous rhyolite to rhyodacite erupted in part of the southern Colorado Plateau transition zone around 18 to 15 Ma. Minor basaltic volcanism was local in the Colorado Rockies.

MID-MIOCENE TO QUATERNARY (14 TO 0 Ma)

14 to 10 Ma (Plate 7G)

Uplift and clockwise rotation continued in the Coast Range of the Pacific Northwest, and ridged basins received sediments in parts of north-central California. As the San Andreas fault system lengtheded into the Coast Ranges, andesite-rhyolite volcanism just east of the San Andreas fault propagated northward within about 100 km of the north-migrating Mendocino triple junction, starting at about 15 Ma in the central Diablo Range.

Although the late Western Cascades arc remained active, erupted volumes declined after about 13 to 14 Ma. Arc magmatism in northern and central Washington lapsed about 9 to 10 Ma. Arc activity remained vigorous in the western part of the present Great Basin until about 12 Ma to the latitude of southern Nevada; after that, the southern terminus of the volcanic arc began to retreat incrementally northward. The previous southern segment of the arc at the latitude of northern Mexico became slivered by strike-slip faults and began to rotate piecemeal with displacement northward into California. The entire Transverse Ranges and local blocks elsewhere in southern California underwent clockwise rotation, locally as much as 90°, yet the western blocks remained at bathyal depths, blanketed by biogenous sediments.

Regional tectonic extension in the continental interior changed to widely distributed normal faulting between about 14 and 10 Ma with accelerated uplift of the region from the Sierra Nevada to the High Plains. As the Oregon Plateaus became extensively faulted after about 14 Ma, basaltic volcanism declined and rhyolite erupted only in parts of the region, particularly between 12 and 10 Ma in the Owyhee region, the starting area for progressive northeastward migration of rhyolitic magmatism along the eastern Snake River Plain. By 14 to 10 Ma, extension and local mafic and intermediate volcanism were scattered widely over much of the present Great Basin region. Faulting differentiated the Colorado Plateau and eastern Great Basin more strongly after about 12 Ma. Uplift and basin-range faulting beheaded streams draining the Sierran slope, forming the hydrographic Great Basin, redirected the drainage of the Snake River into the Columbia, and moved the continental divide from the old Sevier highland to the Middle and Southern Rockies. Volcanism of the Great Basin tended to concentrate with time near its margins.

Extreme extension with large stratal rotations above low-dipping faults in the Lake Mead area between 15 and 13 Ma were succeeded by younger high-angle normal faults. The present ranges and basins are associated with sedimentary rocks, landslide breccias, and basalts as old as 11 Ma. The southern Basin and Range Province had subsided relative to the Colorado Plateau in the mid-Tertiary. Fill began to accumulate in the precursors of modern basins in southeastern Arizona between 15 and 10 Ma, and predominantly basaltic volcanism became prominent about 13 to 12 Ma. Superposed basin-range faulting is mainly younger than 13 Ma. The continuity of the Rio Grande rift through central and northern New Mexico and central Colorado dates mainly since about 13 to 10 Ma. The voluminous Jemez Mountains volcanic field at the intersection of the rift with the Jemez zone has erupted more or less continuously since about 14 Ma. In central Colorado, normal faulting and small-scale basaltic volcanism were wisespread, and small granitic bodies intruded to high structural levels between about 12 and 8 Ma.

10 to 5 Ma (Plate 7H)

By late Miocene time, most of western Washington and Oregon had emerged above sea level. The Coast Range was uplifted above the Puget lowland and Willamette Valley, and a corresponding regional unconformity extends offshore. With continued late Miocene uplift, the lithified eastern Olympic core rocks were deformed against the buttress of accreted Eocene basaltic basement, bulged into a mushroom-like dome, and thrust eastward on backfolded older faults. Locally along the northern Olympic coast, gently dipping shallow-marine, littoral, and deltaic strata were deposited unconformably over highly deformed older rocks around the perimeter of the Olympic uplift. Upper Miocene and lower Pliocene nonmarine sediments accumulated locally in basins in the Puget Lowland and Willamette Valley and along the Columbia River. Farther south, local basins formed near the strandline in late Miocene and Pliocene time, but most of the Coast Range was uplifted, gently folded, and cut by numerous high-angle faults of dominant northwest and subordinate northeast trends.

With northward migration of the Mendocino triple junction, the Sacramento basin became a strike-slip successor basin as the San Joaquin basin had somewhat earlier. The latter basin, as well as other outboard basins of southern California, shoaled from bathyal to neritic or nonmarine near the end of the Miocene. The southern terminus of the continental-margin arc migrated episodically northward east of the present Sierra crest with the Mendocino triple junction as the San Andreas transform system continued to replace part of the continental-margin subduction zone.

Within the continental interior, subsidence and volcanism decreased in the Columbia Plateau to 6 Ma, with canyon-filling flows erupted at increasingly longer intervals. Most of the normal faulting and tilting related to present topography in the Oregon Plateaus postdates about 10 Ma, ending northward against several long northwest-trending fault zones across Oregon but gradational southward into the Great Basin region. East-west folding and reverse faulting in and adjacent to the Blue Mountains between 10 and 6 Ma accompanied basin-range extension to the south. After 10 Ma, basaltic volcanism, characterized by low-K olivine tholeiite, again became widespread in the Oregon Plateaus and adjoining areas, continuing eastward with associated gravel to the western Snake River Plain and southeastward to the Humboldt River. Linear extensional fault blocks active east of the Idaho batholith terminated northward against rejuvenated strike-slip fault systems. Rhyolitic magmatism continued its northeastward propagation along the eastern Snake River Plain. Rhyolitic magmatism of smaller volume also began to propagate at a similar rate northwestward across the Oregon Plateaus after about 11 Ma.

In the Great Basin, as in the Oregon Plateaus, most present topography reflects normal faulting about 10 Ma and younger in age, which was active across virtually the entire width of the region by the late Miocene. Basalt of 11- to 10-Ma age occurs in several Nevada basins west of the northern Nevada rift zone, but especially since about 10 Ma, magmatism has concentrated near the margins of the Great Basin and in a single belt of dispersed centers that reaches from the southwest margin toward central Nevada. After 10 Ma, the southern Nevada volcanic belt was less active than before but did produce local basalts and at least two centers of voluminous peralkalic rhyolites.

At the north end of the Colorado River trough, significant extension (though less than in the mid-Miocene) was as young as about 5 Ma, and basaltic volcanism was active between 11 and 4 Ma. Farther south and southwest, mid-Tertiary detachments and steeply tilted upper-plate blocks older than 16 to 14 Ma were cut by high-angle normal and strike-slip faults. Extension along the lower Colorado River area, accompanied after 7 Ma by local basaltic and bimodal volcanism, formed a proto–Gulf of California in which marine and brackish-water sediments accumulated (although the gulf itself did not form until later). In the southern Basin and Range Province, upper Miocene and younger sediments and volcanics—predominantly alkali basalt—accumulated in closed basins. Younger pediments graded toward throughgoing drainages that evolved between 11 and 6 Ma. Faults related to modern physiography began or accelerated in the Mexican Highlands and Mogollon Plateau at about 10 Ma and slowed or ended around 6 to 3 Ma. The northeast-trending Jemez zone between the southeastern Colorado Plaetau and the Basin and Range Province became a significant volcanic locus after 12 Ma. In the southern Rio Grande rift, transitional with the Basin and Range Province, alkali basalt and related lavas were relatively widespread after about 10 Ma. Tholeiite and alkali basalt both erupted in the central rift, and olivine tholeiite dominated in the northern rift. Voluminous mafic and silicic volcanism occurred in several cycles in the Jemez Mountains but was minor in the Colorado Rockies and vicinity.

Late Miocene mafic alkalic magmas formed lavas and shallow intrusions in three areas of the Colorado Plateau: the Hopi volcanic field in the south, the San Rafael Swell of central Utah, and a small center between the Wasatch and Uinta Mountains.

The Northern Rockies in Washington, northern Idaho, and northwestern Montana have been relatively stable in late Cenozoic time. To the south, however, an area centered on the Idaho batholith has been uplifted more than 1.5 km since about 10 Ma. Uplift increased generally southward through the Rockies and caused erosional removal of mid-Tertiary basin fill, pedimentation of older structures, and exhumation of underlying erosion surfaces. Much of the earlier aggradational mantle was removed, being first redeposited into the upper Miocene Ogallala Formation and correlatives, and later stripped from both the High Plains and Rocky Mountains, which have undergone 1.5 to 2.5 km of uplift since 10 to 8 Ma.

5 to 0 Ma (Plate 7I)

Deformation and sedimentation in the coastal region in Pliocene and Quaternary time have mainly continued the trends of the late Miocene. Uplift continued in the ranges of western Washington and Oregon as nonmarine sediments accumulated in the Puget Lowland and Willamette Valley and in local basins farther south. Latest Cenozoic activity in the northern High Cascades arc is restricted to regularly spaced, long-lived individual centers in a line perpendicular to plate convergence. Southeast of Mount Rainier, subduction is steeper and oblique to the volcanic front, and numerous shorter-lived vents have erupted mainly mafic andesite and basalt among the longer-lived andesitic centers. In northern to central Oregon, lavas fill a graben, but farther south to the end of the arc in northern California, the fundamental structure may be a downwarp.

Pliocene to Pleistocene transpression uplifted the central California Coast Ranges. Basins that had persisted as bathyal depocenters throughout the Tertiary were tectonically inverted, and Quaternary sedimentation is nonmarine. By Pliocene time, the Sacramento and San Joaquin Valleys were a nearly completely nonmarine basin dominated by the longitudinal Sacramento and San Joaquin Rivers, with tributaries from the surrounding nonvolcanic ranges. Andesite-rhyolite volcanism continued to propagate northward opposite the Mendocino triple junction north of San Francisco Bay and in the Sutter Buttes near the center of the Sacramento Valley. Thicknesses of sediments in the Transverse Ranges relate systematically to their positions with respect to the "big bend" of the San Andreas fault; maximum thicknesses of 4-Ma and somewhat older strata are greatest close to the bend, indicating that it had formed by that time.

In the continental interior, mid-Miocene basins of the Oregon Plateaus that largely filled in late Miocene and Pliocene time, now stand at 1,500 to 2,000 m above sea level. Both olivine tholeiite and calc-alkalic basalt similar to that of the Cascades are notable in the transition zone between the western Oregon Plateaus and the High Cascades, where several complexly evolved volcanic systems formed long-lived volcanic shields of basalt and rhyolite with smaller volumes of intermediate lavas. The linear fault blocks east of the Idaho batholith grew rapidly between 7 and 4 Ma. Rhyolitic magmatism has continued to propagate northeastward at 3 to 4 cm/y along the eastern Snake River Plain and is now focused in the very large caldera system of the Yellowstone Plateau at the eastern margin of basin-range extension; basaltic lavas bury the downwarped eastern Snake River Plain. Rhyolitic volcanism has also continued to propagate northwestward across the Oregon Plateaus and is now active on the Newberry Volcano at the Cascades front.

The youngest initiation of extension in the Great Basin region is westward from Death Valley to the Sierra Nevada front, where a Pliocene erosion surface is now broken by five major ranges. Major extension, active seismicity, and minor basaltic and bimodal volcanism continue northward along the western margin of the Great Basin and in a north-northeast-trending belt into central Nevada. Within the zone at the Sierran front, a long-lived volcanic field climaxed in the Pleistocene with voluminous rhyolitic ash flows to form the Long Valley caldera. Quaternary faulting, seismicity, and basaltic to bimodal volcanism are similarly concentrated near the east margin of the Great Basin region and have migrated onto the margin of the Colorado Plateau.

A major change in Pacific Plate motion at about 6 to 5 Ma opened the Gulf of California and accelerated subsidence of the modern basins of southern California. The Salton basin, pulled apart between the San Andreas and San Jacinto faults during the past 4 m.y., is a continuation of the Gulf of California oceanic spreading center, where small volumes of basalt and rhyolite, high heat flow, and active metamorphism demonstrate interaction between the spreading center and a thinning continental crust.

Although basin-range faulting ended as the dominant influence on landscape evolution in the southern Basin and Range Province with opening of the Gulf of California, minor faulting continues in the Mexican Highlands and maintains a discernible rejuvenating influence. Alkali basalt and related lavas are moderately abundant in the southern Basin and Range, and volcanism has migrated northeastward across the transition zone onto the southwestern margin of the Colorado Plateau at accelerating rates of 1 to 3 cm/y; lavas on the faulted plateaus north of the Grand Canyon show that the Colorado River had cut to nearly its present depth before 1 Ma. Although most of the southern Basin and Range Province is tectonically quiet, the Rio Grande rift is relatively active. Magma production in the central rift has increased since about 5 Ma along with increased faulting. Volcanism in the Jemez Mountains culminated with major caldera-forming ash-flows in the early Pleistocene.

Small highly alkalic magmatic bodies have penetrated the crust around the margins of the broad uplifted region centered on the Great Basin and Colorado Plateau, including the Sierran block, Mojave Desert, High Plains, and Wyoming Basin.

REFERENCES CITED

Text references in *italics* will be found only on accompanying microfiche.

Abbott, P. L., and Smith, T. E., 1989, Sonora, Mexico, source for the Eocene Poway Conglomerate of southern California: Geology, v. 17, p. 329–332.

Abbott, P. L., Kies, R. P., Bachman, W. R., and Natenstedt, C. J., 1983, A tectonic slice of Eocene strata, northern part of California Continental Borderland, *in* Larue, D. K., and Steel, R. J., eds., Cenozoic marine sedimentation Pacific margin, U.S.A.: Society of Economic Paleontologists and Mineralogists, Pacific Section, p. 151–168.

Aldrich, M. J., Jr., and Laughlin, A. W., 1984, A model for the tectonic development of the southeastern Colorado Plateau boundary: Journal of Geophysical Research, v. 89, p. 10207–10218.

Aldrich, M. J., Jr., Chapin, C. E., and Laughlin, A. W., 1986, Stress history and tectonic development of the Rio Grande rift, New Mexico: Journal of Geophysical Research, v. 91, p. 6199–6211.

Allmendinger, R. W., Sharp, J. W., Von Tish, D., Serpa, L., Brown, L., Kaufman, S., Oliver, J., and Smith, R. B., 1983, Cenozoic and Mesozoic structure of the eastern Basin and Range Province, Utah, from COCORP seismic-reflection data: Geology, v. 11, p. 532–536.

Almgren, A. A., 1984, Timing of Tertiary submarine canyons and marine cycles of deposition in the southern Sacramento Valley, California, *in* Almgren, A. A., and Hacker, P. D., eds., Paleogene submarine canyons of the Sacramento Valley, California: American Association of Petroleum Geologists Symposium, Pacific Section, v. 1, p. 1–16.

Anderson, C. A., and Creasey, S. C., 1958, Geology and ore deposits of the Jerome area, Yavapai County, Arizona: U.S. Geological Survey Professional Paper 308, 185 p.

Anderson, J. J., and Rowley, P. D., 1975, Cenozoic stratigraphy of southwestern High Plateaus of Utah, *in* Anderson, J. J., Rowley, P. D., Fleck, F. J., and Nairn, A. E. M., eds., Cenozoic geology of southwestern High Plateaus of Utah: Geological Society of America Special Paper 160, p. 1–51.

Anderson, R. E., 1971, Thin-skinned distension in Tertiary rocks of southeastern Nevada: Geological Society of America Bulletin, v. 82, p. 43–58.

——, 1973, Large-magnitude late Tertiary strike-slip faulting north of Lake Mead, Nevada: U.S. Geological Survey Professional Paper 794, 18 p.

Anderson, R. E., and Ekren, E. B., 1968, Widespread Miocene igenous rocks of intermediate composition, southern Nye county, Nevada, *in* Eckel, E. B., ed., Nevada Test Site: Geological Society of America Memoir 110, p. 57–63.

Anderson, R. E., and Mehnert, H. H., 1979, Reinterpretation of the history of the Hurricane fault in Utah, *in* Newman, G. W., and Goode, H. D., eds., Basin and Range Symposium and Great Basin Field Conference: Rocky Mountain Association of Geologists, p. 145–165.

Anderson, R. E., Longwell, C. R., Armstrong, R. L., and Marvin, R. F., 1972, Significance of K-Ar ages of Tertiary rocks from the Lake Mead region, Nevada-Arizona: Geological Society of America Bulletin, v. 83, p. 273–288.

Ansfield, V. J., 1972, The stratigraphy and sedimentology of the Lyre Formation, northwestern Olympic Peninsula, Washington [M.S. thesis]: Seattle, University of Washington, 77 p.

Armentrout, J. M., and Suek, D. H., 1985, Hydrocarbon exploration in western Oregon and Washington: American Association of Petroleum Geologists Bulletin, v. 69, p. 627–643.

Armentrout, J. M., Hull, D. A., Beaulieu, J. D., and Rau, W. W., 1983, Correlation of Cenozoic stratigraphic units of western Oregon and Washington: Oregon Department of Geology and Mineral Industries Oil and Gas Investigation 7, 90 p.

Armstrong, R. L., 1968, Mantled gneiss domes in the Albion Range, southern Idaho: Geological Society of America Bulletin, v. 79, p. 1295–1314.

—— , 1969, K-Ar dating of laccolithic centers of the Colorado Plateau and vicinity: Geological Society of America Bulletin, v. 80, p. 2081–2086.

—— , 1970, Geochronology of Tertiary igneous rocks, eastern Basin and Range Province, western Utah, eastern Nevada, and vicinity, U.S.A.: Geochimica et Cosmochimica Acta, v. 34, p. 203–232.

—— , 1972a, Low-angle (denudation) faults, hinterland of the Sevier orogenic belt, eastern Nevada and western Utah: Geological Society of America Bulletin, v. 79, p. 429–458.

—— , 1978, Cenozoic igneous history of the U.S. Cordillera from lat 42° to 48°N, *in* Smith, R. B., and Eaton, G. P., eds., Cenozoic tectonics and regional geophysics of the Western Cordillera: Geological Society of America Memoir 152, p. 265–282.

—— , 1982, Cordilleran metamorphic core complexes; From Arizona to southern Canada: Annual Review of Earth and Planetary Sciences, v. 10, p. 129–154.

Armstrong, R. L., and Higgins, R. E., 1973, K-Ar dating of the beginning of Tertiary volcanism in the Mojave Desert, California: Geological Society of America Bulletin, v. 84, p. 1095–1100.

Armstrong, R. L., Leeman, W. P., and Malde, H. E., 1975, K-Ar dating, Quaternary and Neogene volcanic rocks of the Snake River Plain, Idaho: American Journal of Science, v. 275, p. 225–251.

Atwater, T., 1970, Implications of plate tectonics for the Cenozoic tectonic evolution of western North America: Geological Society of America Bulletin, v. 81, p. 3513–3536.

Axelrod, D. I., 1968, Tertiary floras and topographic history of the Snake River basin, Idaho: Geological Society of America Bulletin, v. 79, p. 713–734.

Babcock, E. A., 1974, Geology of the northeast margin of the Salton trough, Salton Sea, California: Geological Society of America Bulletin, v. 85, p. 321–332.

Bachman, G. O., and Mehnert, H. H., 1978, New K-Ar dates and the late Pliocene to Holocene geomorphic history of the central Rio Grande region, New Mexico: Geological Society of America Bulletin, v. 89, p. 283–292.

Bachman, S. B., 1978, Pliocene-Pleistocene break-up of the Sierra Nevada–White–Inyo Mountains block and formations of Owens Valley: Geology, v. 6, p. 461–463.

—— , 1982, The Coastal Belt of the Franciscan; Youngest phase of northern California subduction, *in* Trench-forearc geology; Sedimentation and tectonics on modern and ancient active plate margins: Geological Society of London Special Publication 10, p. 401–417.

Bachman, S. B., and Crouch, J. K., 1987, Geology and Cenozoic history of the northern California margin; Point Arena to Eel River, *in* Ingersoll, R. V., and Ernst, W. G., eds., Cenozoic basin development of coastal California; Rubey Vol. VI: Englewood Cliffs, New Jersey, Prentice-Hall, p. 124–145.

Bacon, C. R., 1983, Eruptive history of Mount Mazama and Crater Lake caldera, Cascade Range, U.S.A.: Journal of Volcanology and Geothermal Research, v. 18, p. 57–115.

Bacon, C. R., Duffield, W. A., and Nakamura, K., 1980, Distribution of Quaternary rhyolite domes of the Coso Range, California; Implications for the extent of the geothermal anomaly: Journal of Geophysical Research, v. 85, p. 2425–2433.

Bailey, R. A., 1989, Geologic map of the Long Valley caldera, Mono-Inyo craters volcanic chain, and vicinity, eastern California: U.S. Geological Survey Miscellaneous Investigations Map I-1933, scale 1:62,500.

Bailey, R. A., Dalrymple, G. B., and Lanphere, M. A., 1976, Volcanism, structure, and geochronology of Long Valley caldera, Mono County, California: Journal of Geophysical Research, v. 81, p. 725–744.

Baldwin, E. M., 1974, Eocene stratigraphy of southwestern Oregon: Oregon Department of Geology and Mineral Industries Bulletin 83, 40 p.

—— , 1975, Revision of the Eocene stratigraphy of southwestern Oregon, *in* Weaver, D. W., Hornaday, G. R., and Tipton, A., eds., Paleogene Symposium and Selected Papers; Conference on Future Energy Horizons of the Pacific Coast: American Association of Petroleum Geologists, Society of Economic Paleontologists and Mineralogists, and Society of Exploration Geophysicists, Pacific Sections, p. 49–64.

—— , 1981, Geology of Oregon (third edition): Dubuque, Iowa, Kendall/Hunt, 170 p.

Baldwin, E. M., and Beaulieu, J. D., 1973, Geology and mineral resources of Coos County, Oregon: Oregon Department of Geology and Mineral Industries Bulletin 80, 82 p.

Ballance, P. F., Howell, D. G., and Ort, K., 1983, Late Cenozoic wrench tectonics along the Nacimiento, South Cuyama, and La Panza faults, California, indicated by depositional history of the Simmler Formation, *in* Anderson, D. W., and Rymer, M. J., eds., Tectonics and sedimentation along faults of the San Andreas system: Society of Economic Paleontologists and Mineralogists, Pacific Section, p. 1–9.

Bandy, O. L., and Arnal, R. E., 1969, Middle Tertiary basin development, San Joaquin Valley, California: Geological Society of America Bulletin, v. 80, p. 783–820.

Banks, N. G., 1980, Geology of a zone of metamorphic core complexes in southeastern Arizona, *in* Crittenden, M. D., Coney, P. J., and Davis, G. H., eds., Cordilleran metamorphic core complexes: Geological Society of America Memoir 153, p. 177–215.

Barker, D. S., 1977, Northern Trans-Pecos magmatic province; Introduction and comparison with the Kenya rift: Geological Society of America Bulletin, v. 88, p. 1421–1427.

—— , 1979, Cenozoic magmatism in the Trans-Pecos province; Relation to the Rio Grande rift, *in* Riecker, R. E., ed., Rio Grande rift; Tectonics and magmatism: Washington, American Geophysical Union, p. 382–392.

Barrows, A. G., 1974, A review of the geology and earthquake history of the Newport-Inglewood structural zone, southern California: California Division of Mines and Geology Special Report 114, 115 p.

Bartley, J. M., and Wernicke, B. P., 1984, The Snake Range décollement interpreted as a major extensional shear zone: Tectonics, v. 3, p. 647–657.

Bartow, J. A., 1991, The Cenozoic evolution of the San Joaquin Valley, California: U.S. Geological Survey Professional Paper 1501, 40 p.

Bateman, P. C., and Wahrhaftig, C., 1966, Geology of the Sierra Nevada, *in* Bailey, E. H., ed., Geology of northern California: California Division of Mines and Geology Bulletin 190, p. 107–172.

Beck, M. E., Jr., 1980, Paleomagnetic record of plate-margin tectonic processes along the western edge of North America: Journal of Geophysical Research, v. 85, p. 7115–7131.

Beeson, M. H., Perttu, R., and Perttu, J., 1979, The origin of the Miocene basalts of coastal Oregon and Washington: Oregon Geology, v. 41, p. 159–166.

Beeson, M. H., Tolan, T. L., and Anderson, J. L., 1989, The Columbia River Basalt Group in western Oregon; Geologic structures and others factors that controlled flow emplacement patterns, *in* Reidel, S. P., and Hooper, P. R., eds., Volcanism and tectonism in the Columbia River flood-basalt province: Geological Society of America Special Paper 239, p. 223–246.

Bennett, E. H., 1980, Granitic rocks of Tertiary age in the Idaho batholith and their relation to mineralization: Economic Geology, v. 75, p. 278–288.

—— , 1986, Relationship of the trans-Challis fault system in central Idaho to Eocene and Basin and Range extensions: Geology, v. 14, p. 481–484.

Bent, J. V., 1985, Provenance of upper Oligocene–middle Miocene sandstones of the San Joaquin basin, California, *in* Graham, S. A., ed., Geology of the Temblor Formation, western San Joaquin basin, California: Society of Economic Paleontologists and Mineralogists, Pacific Section, v. 44, p. 97–120.

Best, M. G., and Brimhall, W. H., 1974, Late Cenozoic alkalic basaltic magmas in the western Colorado Plateaus and the Basin and Range transition zone, U.S.A., and their bearing on mantle dynamics: Geological Society of America Bulletin, v. 85, p. 1677–1690.

Best, M. G., and Hamblin, W. K., 1978, Origin of the northern Basin and Range Province; Implications from the geology of its eastern boundary, *in* Smith, R. B., and Eaton, G. P., eds., Cenozoic tectonics and regional geophysics of the Western Cordillera: Geological Society of America Memoir 152, p. 313–340.

Best, M. G., and Keith, J. D., 1983, Mid-Tertiary history of the central Pioche-Marysvale igneous belt, southwestern Utah, *in* Gurgel, K. D., ed., Geologic excursions in volcanology; Eastern Snake River Plain (Idaho) and southwestern Utah: Utah Geological and Mineralogical Survey Special Studies 61, p. 35–55.

Best, M. G., Shuey, R. T., Caskey, C. F., and Grant, S. K., 1973, Stratigraphic relations of members of the Needles Range Formation at type localities in southwestern Utah: Geological Society of America Bulletin, v. 84, p. 3269–3278.

Best, M. G., McKee, E. H., and Damon, P. E., 1980, Space-time-composition patterns of late Cenozoic mafic volcanism, southwestern Utah and adjoining areas: American Journal of Science, v. 280, p. 1035–1050.

Best, M. G., Christiansen, E. H., and Blank, H. R., Jr., 1989, Oligocene caldera complex and calc-alkaline tuffs and lavas of the Indian Peak volcanic field, Nevada and Utah: Geological Society of America Bulletin, v. 101, p. 1076–1090.

Blackwell, D. D., 1978, Heat flow and energy loss in the western United States, *in* Smith, R. B., and Eaton, G. P., eds., Cenozoic tectonics and regional geophysics of the Western Cordillera: Geological Society of America Memoir 152, p. 175–208.

Blair, W. N., 1978, Gulf of California in the Lake Mead area of Arizona and Nevada during late Miocene time: American Association of Petroleum Geologists Bulletin, v. 62, p. 1159–1170.

Blake, M. C., Jr., and 7 others, 1978, Neogene basin formation in relation to plate-tectonic evolution of San Andreas fault system, California: American Association of Petroleum Geologists Bulletin, v. 62, p. 344–372.

Blakely, R. J., Jachens, R. C., Simpson, R. W., and Couch, R. W., 1985, Tectonic setting of the southern Cascade Range as interpreted from its magnetic and gravity fields: Geological Society of America Bulletin, v. 96, p. 43–48.

Bohannon, R. G., 1975, Mid-Tertiary conglomerates and their bearing on Transverse Range tectonics, southern California: California Division of Mines and Geology Special Report 118, p. 75–82.

——, 1979, Strike-slip faults of the Lake Mead region of southern Nevada, *in* Armentrout, J. M., Cole, M. R., and TerBest, H., Jr., eds., Cenozoic paleogeography of the western United States: Society of Economic Paleontologists and Mineralogists, Pacific Section, Pacific Coast Paleogeography Symposium 3, p. 129–139.

——, 1983, Mesozoic and Cenozoic tectonic development of the Muddy, North Muddy, and northern Black Mountains, Clark Colunty, Nevada, *in* Miller, D. M., Todd, V. R., and Howard, K. A., eds., Tectonic and stratigraphic studies in the eastern Great Basin: Geological Society of America Memoir 157, p. 125–148.

——, 1984, Nonmarine rocks of Tertiary age in the Lake Mead region, southeastern Nevada and northwestern Arizona: U.S. Geological Survey Professional Paper 1259, 72 p.

Bonham, H. F., 1969, Geology and mineral deposits of Washoe and Storey Counties, Nevada: Nevada Bureau of Mines Bulletin 70, 140 p.

Bonham, H. F., Jr., and Garside, L. J., 1979, Geology of the Tonopah, Lone Mountain, Klondike, and northern Mud Lake Quadrangles, Nevada: Nevada Bureau of Mines and Geology Bulletin 92, 142 p.

Bonnichsen, B., 1982a, The Bruneau-Jarbidge eruptive center, southwestern Idaho, *in* Bonnichsen, B., and Breckenridge, R. M., eds., Cenozoic geology of Idaho: Idaho Bureau of Mines and Geology Bulletin 26, p. 237–254.

Bonnichsen, B., and Citron, G. P., 1982, The Cougar Point Tuff, southwestern Idaho and vicinity, *in* Bonnichsen, B., and Breckenridge, R. M., eds., Cenozoic geology of Idaho: Idaho Bureau of Mines and Geology Bulletin 26, p. 255–281.

Brooks, W. E., 1986, Distribution of anomalously high K_2O volcanic rocks in Arizona: Metasomatism at the Picacho Peak detachment fault: Geology, v. 14, p. 339–342.

Brown, R. D., Jr., Gower, H. D., and Snavely, P. D., Jr., 1960, Geology of the Lake Crescent–Port Angeles area, Washington: U.S. Geological Survey Oil and Gas Investigations Map OM-203, scale 1:62,500.

Buckovic, W. A., 1979, The Eocene delta system of west-central Washington, *in* Armentrout, J. M., Cole, M. R., and Terbest, H., Jr., eds., Cenozoic Paleogeography of the western United States; Pacific Coast Paleogeography Symposium 4: Society of Economic Paleontologists and Mineralogists, Pacific Section, p. 147–163.

Burchfiel, B. C., and Stewart, J. H., 1966, "Pull-apart" origin of the central segment of Death Valley, California: Geological Society of America Bulletin, v. 77, p. 439–442.

Burchfiel, B. C., Fleck, R. J., Secor, D. T., Vincelette, R. R., and Davis, G. A., 1974, Geology of the Spring Mountains, Nevada: Geological Society of America Bulletin, v. 85, p. 1013–1022.

Burchfiel, B. C., Hamill, G. S., IV, and Wilhelms, D. E., 1983, Structural geology of the Montgomery Mountains and the northern half of the Nopah and Resting Spring Ranges, Nevada and California: Geological Society of America Bulletin, v. 94, p. 1359–1376.

Butler, R. F., Dickinson, W. R., and Gehrels, G. E., 1991, Paleomagnetism of coastal California and Baja California; Alternatives to large-scale northward transport: Tectonics, v. 10, p. 561–576.

Byers, F. M., Jr., Carr, W. J., Orkild, P. P., Quinlivan, W. D., and Sargent, K. A., 1976b, Volcanic suites and related cauldrons of Timber Mountain–Oasis Valley caldera complex, southern Nevada: U.S. Geological Survey Professional Paper 919, 70 p.

Campbell, R. H., and Yerkes, R. F., 1976, Cenozoic evolution of the Los Angeles basin area; Relation to plate tectonics, *in* Howell, D. G., ed., Aspects of the geologic history of the California Continental Borderland: Los Angeles, American Association of Petroleum Geologists, Pacific Section, Miscellaneous Publication 24, p. 541–558.

Cape, C. D., McGeary, S., and Thompson, G. A., 1983, Cenozoic normal faulting and the shallow structure of the Rio Grande rift near Socorro, New Mexico: Geological Society of America Bulletin, v. 94, p. 3–14.

Carr, W. J., and Trimble, D. E., 1963, Geology of the American Falls Quadrangle, Idaho: U.S. Geological Survey Bulletin 1121-G, 44 p.

Cater, F. W., 1982, Intrusive rocks of the Holden and Lucerne Quadrangles, Washington; The relation of depth zones, composition, textures, and emplacement of plutons: U.S. Geological Survey Professional Paper 1220, 108 p.

Chadwick, R. A., 1972, Volcanism in Montana: Northwest Geology, v. 1, p. 1–20.

Chan, M. A., and Dott, R. H., Jr., 1983, Shelf and deep-sea sedimentation in Eocene forearc basin, western Oregon; fan or non-fan?: American Association of Petroleum Geologists Bulletin, v. 67, no. 11, p. 2100–2116.

——, 1986, Depositional facies and progradational sequences in Eocene wave-dominated deltaic complexes, southwestern Oregon: American Association of Petroleum Geologists Bulletin, v. 70, no. 4, p. 415–429.

Chapin, C. E., 1979, Evolution of the Rio Grande rift; a summary, *in* Riecker, R. E., ed., Rio Grande rift; Tectonics and magmatism: Washington, American Geophysical Union, p. 1–5.

Chapin, C. E., and Cather, S. M., 1981, Eocene tectonics and sedimentation in the Colorado Plateau–Rocky Mountain area: Arizona Geological Society Digest, v. 14, p. 173–198.

Chapin, C. E., and Elston, W. E., eds., 1978, Field guide to selected cauldrons and mining districts of the Datil-Mogollon volcanic field, New Mexico: New Mexico Geological Society Publication 7, 149 p.

Chapin, C. E., and Seager, W. R., 1975, Evolution of the Rio Grande rift in the Socorro and Las Cruces areas: New Mexico Geological Society Guidebook, 26th Field Conference, p. 297–321.

Cheney, E. S., 1980, Kettle dome and related structures of northeastern Washington, *in* Crittenden, M. D., Coney, P. J., and Davis, G. H., eds., Cordilleran metamorphic core complexes: Geological Society of America Memoir 153,

p. 463–483.
Chipping, D. H., 1972, Early Tertiary paleogeography of central California: American Association of Petroleum Geologists Bulletin, v. 56, p. 480–493.
Christensen, M. N., 1966, Late Cenozoic crustal movements in the Sierra Nevada of California: Geological Society of America Bulletin, v. 77, p. 163–182.
Christensen, M. N., Gilbert, C. M., Lajoie, K. R., and Al-Rawi, Y., 1969, Geological-geophysical interpretation of Mono basin, California-Nevada: Journal of Geophysical Research, v. 74, p. 5221–5239.
Christiansen, E. H., Sheridan, M. F., and Burt, D. M., 1986, The geology and geochemistry of Cenozoic topaz rhyolites from the western United States: Geological Society of America Special Paper 205, 82 p.
Christiansen, R. L., 1982a, Volcanic hazard potential in the California Cascades, *in* Martin, R. C., and Davis, J. F., eds., Status of volcanic prediction and emergency response capabilities in volcanic hazard zones of California: California Division of Mines and Geology Special Publication 63, p. 41–59.
——, 1984, Yellowstone magmatic evolution; Its bearing on understanding large-volume explosive volcanism, *in* Explosive volcanism; Inception, evolution, and hazards: Washington, National Academy of Sciences, p. 84–95.
Christiansen, R. L., and Lipman, P. W., 1972, Cenozoic volcanism and plate tectonic evolution of the western United States; Part II, Late Cenozoic: Royal Society of London Philosophical Transactions, ser. A, v. 271, p. 249–284.
Christiansen, R. L., and McKee, E. H., 1978, Late Cenozoic volcanic and tectonic evolution of the Great Basin and Columbia Intermontane regions, *in* Smith, R. B., and Eaton, G. P., eds., Cenozoic tectonics and regional geophysics of the Western Cordillera: Geological Society of America Memoir 152, p. 283–311.
Christiansen, R. L., and Peterson, D. W., 1981, Chronology of the 1980 eruptive activity, *in* Lipman, P. W., and Mullineaux, D. R., eds., The 1980 eruptions of Mount St. Helens, Washington: U.S. Geological Survey Professional Paper 1250, p. 17–30.
Christiansen, R. L., Lipman, P. W., Carr, W. J., Byers, F. M., Jr., Orkild, P. P., and Sargent, K. A., 1977, Timber Mountain–Oasis Valley caldera complex of southern Nevada: Geological Society of America Bulletin, v. 88, p. 943–959.
Clark, J., 1975, Controls of sedimentation and provenance of sediments in the Oligocene of the central Rocky Mountains, *in* Curtis, B. F., ed., Cenozoic history of the Southern Rocky Mountains: Geological Society of America Memoir 144, p. 95–117.
Clark, J. C., 1968, Correlation of the Santa Cruz Mountains Tertiary; implications for San Andreas history, *in* Dickinson, W. R. and Grantz, A., eds., Proceedings of conference on geologic problems of San Andreas fault system: Stanford University Publications Geological Sciences, v. 11, p. 166–180.
Clark, J. C., Brabb, E. E., Greene, H. G., and Ross, D. C., 1984, Geology of Point Reyes peninsula and implications for San Gregorio fault history, *in* Crouch, J. K., and Bachman, S. B., eds., Tectonics and sedimentation along the California margin: Society of Economic Paleontologists and Mineralogists, Pacific Section, v. 38, p. 67–85.
Clark, K. F., Foster, C. T., and Damon, P. E., 1982, Cenozoic mineral deposits and subduction-related magmatic arcs in Mexico: Geological Society of America Bulletin, v. 93, p. 533–544.
Clifton, H. E., 1981, Progradational sequence in Miocene shoreline deposits, southeastern Caliente Range, California: Journal of Sedimentary Petrology, v. 51, p. 165–184.
Coats, R. R., 1964, Geology of the Jarbidge Quadrangle, Nevada-Idaho: U.S. Geological Survey Bulletin 1141-M, 24 p.
Compton, R. R., Todd, V. R., Zartman, R. F., and Naeser, C. W., 1977, Oligocene and Miocene metamorphism, folding, and low-angle faulting in northwestern Utah: Geological Society of America Bulletin, v. 88, p. 1237–1250.
Coney, P. J., 1978, Mesozoic-Cenozoic Cordilleran plate tectonics, *in* Smith, R. B., and Eaton, G. P., eds., Cenozoic tectonics and regional geophysics of the Western Cordillera: Geological Society of America Memoir 152, p. 33–50.
——, 1979, Tertiary evolution of metamorphic core complexes, *in* Armentrout, J. M., Cole, M. R., and TerBest, H., Jr., eds., Cenozoic paleogeography of the western United States: Society of Economic Paleontologists and Mineralogists, Pacific Section, Pacific Coast Paleogeography Symposium 3, p. 15–28.
Coney, P. J., and Harms, T. A., 1984, Cordilleran metamorphic core complexes: Cenozoic extensional relics of Mesozoic compression: Geology, v. 12, p. 550–554.
Coney, P. J., and Reynolds, S. J., 1977, Cordilleran Benioff zones: Nature, v. 270, p. 403–406.
Cook, E. F., 1965, Stratigraphy of Tertiary volcanic rocks in eastern Nevada: Nevada Bureau of Mines Report 11, 61 p.
Cooper, J. R., and Silver, L. T., 1964, Geology and ore deposits of the Dragoon Quadrangle, Cochise County, Arizona: U.S. Geological Survey Professional Paper 416, 196 p.
Cordell, L., 1978, Regional geophysical setting of the Rio Grande rift: Geological Society of America Bulletin, v. 89, p. 1073–1090.
Covington, H. R., 1983, Structural evolution of the Raft River basin, Idaho, *in* Miller, D. M., Todd, V. R., and Howard, K. A., eds., Tectonic and stratigraphic studies in the eastern Great Basin: Geological Society of America Memoir 157, p. 229–237.
Crandell, D. R., 1971, Postglacial lahars from Mount Rainier Volcano, Washington: U.S. Geological Survey Professional Paper 677, 75 p.
Crandell, D. R., 1980, Recent eruptive history of Mount Hood, Oregon, and potential hazards from future eruptions: U.S. Geological Survey Bulletin 1492, 81 p.
Creasey, S. C., and Krieger, M. H., 1978, Galiuro Volcanics, Pinal, Graham, and Cochise Counties, Arizona: U.S. Geological Survey Journal of Research, v. 6, p. 115–131.
Creasey, S. C., Banks, N. G., Ashley, R. P., and Theodore, T. G., 1977, Middle Tertiary plutonism in the Santa Catalina and Tortolita Mountains, Arizona: U.S. Geological Survey Journal of Research, v. 5, p. 705–717.
Crittenden, M. D., Coney, P. J., and Davis, G. H., eds., 1980, Cordilleran metamorphic core complexes: Geological Society of America Memoir 153, 490 p.
Crone, A. J., and Harding, S. T., 1984, Relationship of late Quaternary fault scarps to subjacent faults, eastern Great Basin, Utah: Geology, v. 12, p. 292–295.
Crowe, B. M., 1978, Cenozoic volcanic geology and probable age of inception of basin-range faulting in the southeasternmost Chocolate Mountains, California: Geological Society of America Bulletin, v. 89, p. 251–264.
Crowe, B. M., Crowell, J. C., and Krummenacher, D., 1979, Regional stratigraphy, K-Ar ages, and tectonic implications of Cenozoic volcanic rocks, southeastern California: American Journal of Science, v. 279, p. 186–216.
Crowe, B., Self, S., Vaniman, D., Amos, R., and Perry, F., 1983, Aspects of potential magmatic disruption of a high-level radioactive waste repository in southern Nevada: Journal of Geology, v. 91, p. 259–276.
Crowell, J. C., 1954, Strike-slip displacement of the San Gabriel fault, southern California: California Division of Mines Bulletin 170, chapter 4, p. 49–52.
——, 1968, Movement histories of faults in the Transverse Ranges and speculations on the tectonic history of California, *in* Dickinson, W. R., and Grantz, A., eds., Proceedings, Conference on Geologic Problems of San Andreas Fault System: Stanford University Publications Geological Sciences, v. 11, p. 323–341.
——, 1975b, The San Gabriel fault and Ridge basin, *in* Crowell, J. C., ed., San Andreas fault in southern California: California Division of Mines and Geology Special Report 118, p. 208–233.
——, 1987, Late Cenozoic basins of onshore southern California; Complexity is the hallmark of their tectonic history, *in* Ingersoll, R. V., and Ernst, W. G., eds., Cenozoic basin development of coastal California; Rubey Vol. VI: Englewood Cliffs, New Jersey, Prentice-Hall, p. 207–241.
Crowell, J. C., and Baca, B., 1979, Sedimentation history of the Salton trough, *in* Crowell, J. C., and Sylvester, A. C., eds., Tectonics of the juncture between the San Andreas fault system and the Salton trough, southeastern California: Santa Barbara, University of California, Department of Geological Sciences,

p. 101–123.
Crowell, J. C. and Link, M. H., eds., 1982, Geologic history of Ridge basin, southern California: Society of Economic Paleontologists and Mineralogists, Pacific Section, 304 p.
Crowell, J. C., and Sylvester, A. G., eds., 1979, Tectonics of the juncture between the San Andreas fault system and the Salton trough, southeastern California: Santa Barbara, University of California, Department of Geological Sciences, 193 p.
Curtis, G. H., 1954, Mode of origin of pyroclastic debris in the Mehrten Formation of the Sierra Nevada: University of California Publications in Geological Sciences, v. 29, p. 453–502.
Dalrymple, G. B., 1963, Potassium-argon dates of some Cenozoic volcanic rocks of the Sierra Nevada, California: Geological Society of America Bulletin, v. 74, p. 379–390.
—— , 1964a, Cenozoic chronology of the Sierra Nevada, California: University of California Publications in the Geological Sciences, v. 47, 41 p.
Damon, P. E., 1971, The relationship between late Cenozoic volcanism and tectonism and orogenic-epierogenic periodicity, *in* Turekian, K. K., ed., Conference on the late Cenozoic glacial ages: New York, Wiley, p. 15–35.
Damon, P. E., Lynch, D. J., and Shaffiqullah, M., 1984, Cenozoic landscape development in the Basin and Range Province of Arizona, *in* Smiley, T. L., Nations, J. D., Pewe, T. L., and Schafer, J. P., eds., Landscapes of Arizona: Lanham, Maryland, University Press of America, p. 175–206.
Davis, G. A., 1980, Problems of intraplate extensional tectonics, western United States, *in* Continental tectonics: Washington, National Academy of Sciences, p. 84–95.
—— , 1988, Rapid upward transport of mid-crustal mylonitic gneisses in the footwall of a Miocene detachment fault, Whipple Mountains, southeastern California: Geologische Rundschau, v. 77, p. 191–209.
Davis, G. A., and Burchfiel, B. C., 1973, Garlock fault: An intracontinental transform structure, southern California: Geological Society of America Bulletin, v. 84, p. 1407–1422.
Davis, G. A., Anderson, J. L., Frost, E. G., and Shackelford, T. J., 1980, Mylonitization and detachment faulting in the Whipple-Buckskin-Rawhide Mountains terrane, southeastern California and western Arizona, *in* Crittenden, M. D., Coney, P. J., and Davis, G. H., eds., Cordilleran metamorphic core complexes: Geological Society of America Memoir 153, p. 79–183.
Davis, G. A., Lister, G. S., and Reynolds, S. J., 1986, Structural evolution of the Whipple and South Mountains shear zones, southwestern United States: Geology, v. 14, p. 7–10.
Davis, G. H., 1980, Structural characteristics of metamorphic core complexes, southern Arizona, *in* Crittenden, M. D., Coney, P. J., and Davis, G. H., eds., Cordilleran metamorphic core complexes: Geological Society of America Memoir 153, p. 36–78.
Davis, G. H., and Coney, P. J., 1979, Geologic development of the Cordilleran metamorphic core complexes: Geology, v. 7, p. 120–124.
Davis, G. H., and Hardy, J. J., Jr., 1981, The Eagle Pass detachment, southeastern Arizona: Product of mid-Miocene listric(?) normal faulting in the southern Basin and Range: Geological Society of America Bulletin, v. 92, pt. 1, p. 749–762.
Davis, T. L., and Lagoe, M. B., 1988, A structural interpretation of major tectonic events affecting the western and southern margins of the San Joaquin Valley, California, *in* Graham, S. A., ed., Studies of the geology of the San Joaquin basin, California: Society of Economic Paleontologists and Mineralogists, Pacific Section, v. 60, p. 65–88.
Davis, T. L., Namson, J., and Yerkes, R. F., 1989, A cross section of the Los Angeles area; Seismically active fold and thrust belt, the Whittier Narrows earthquake, and earthquake hazard: Journal of Geophysical Research, v. 94, p. 9644–9664.
Deal, E. G., Elston, W. E., Erb, E. E., Peterson, S. L., and Reiter, D. E., 1978, Cenozoic volcanic geology of the Basin and Range Province in Hidalgo County, southwestern New Mexico: New Mexico Geological Society Guidebook, 29th Field Conference, p. 219–229.
de Voogd, B., Brown, L. D., and Merey, C., 1986a, Nature of the eastern boundary of the Rio Grande rift from COCORP surveys in the Albuquerque basin, New Mexico: Journal of Geophysical Research, v. 91, p. 6305–6320.
Dibblee, T. W., Jr., 1950, Geology of southwestern Santa Barbara County, California: California Division of Mines Bulletin 150, 95 p.
—— , 1954, Geology of the Imperial Valley region: California Division of Mines Bulletin 170, chapter 2, p. 21–28.
—— , 1966, Geology of the central Ynez Mountains, Santa Barbara County, California: California Division of Mines and Geology Bulletin 186, 99 p.
—— , 1967, Areal geology of the western Mojave Desert, California: U.S. Geological Survey Professional Paper 522, 153 p.
—— , 1976, The Rinconada and related faults in the southern Coast Ranges, California, and their tectonic significance: U.S. Geological Survey Professional Paper 981, 55 p.
—— , 1982a, Regional geology of the Transverse Ranges province of southern California, *in* Fife, D. L., and Minch, J. A., eds., Geology and mineral wealth of the California Transverse Ranges: South Coast Geological Society, p. 7–26.
Dickinson, W. R., 1984, Reinterpretation of Lime Peak thrust as a low-angle normal fault; Implications for the tectonics of southeastern Arizona: Geology, v. 12, p. 610–613.
Dickinson, W. R., and Snyder, W. S., 1979a, Geometry of triple junctions related to San Andreas transform: Journal of Geophysical Research, v. 84, p. 561–572.
Dickinson, W. R., and Snyder, W. S., 1979b, Geometry of subducted slabs related to San Andreas transform: Journal of Geology, v. 87, p. 609–627.
Dickinson, W. R., Ingersoll, R. V., and Graham, S. A., 1979, Paleogene sediment dispersal and paleotectonics in northern California: Geological Society of America Bulletin, v. 90, pt. 1, p. 897–898, pt. 2, p. 1458–1528.
Dickinson, W. R., Armin, R. A., Beckvar, N., Goodlin, T. C., Janecke, S. U., Mark, R. A., Norris, R. D., Radel, G., and Wortman, A. A., 1987, Geohistory analysis of rates of sediment accumulation and subsidence for selected California basins, *in* Ingersoll, R. V., and Ernst, W. G., eds., Cenozoic basin development in coastal California; Rubey Vol. VI: Englewood Cliffs, New Jersey, Prentice-Hall, p. 1–23.
Dohrenwend, J. C., McFadden, L. D., Turrin, B. D., and Wells, S. G., 1984, K-Ar dating of the Cima volcanic field, eastern Mojave Desert, California; Late Cenozoic volcanic history and landscape evolution: Geology, v. 12, p. 163–167.
Dokka, R. K., 1983, Displacements on late Cenozoic strike-slip faults of the central Mojave Desert, California: Geology, v. 11, p. 305–308.
—— , 1986, Patterns of early Miocene crustal extension, central Mojave Desert California, *in* Mayer, L., ed., Extensional tectonics of the southwestern United States; A perspective on proceses and kinematics: Geological Society of America Special Paper 208, p. 75–95.
Dokka, R. K., and Glazner, A. F., 1982, Aspects of early Miocene extension of the central Mojave Desert, *in* Cooper, J. D., ed., Geologic excursions in the California desert: Anaheim, California, Geological Society of America, Cordilleran Section 78th Annual Meeting, p. 31–45.
Donnelly-Nolan, J. M., 1988, A magmatic model of Medicine Lake volcano, California: Journal of Geophysical Research, v. 93, p. 4412–4420.
Donnelly-Nolan, J. M., Hearn, B. C., Jr., Curtis, G. H., and Drake, R. E., 1981, Geochronology and evolution of the Clear Lake Volcanics, *in* McLaughlin, R. J., and Donnelly-Nolan, J. M., eds., Research in the Geysers–Clear Lake geothermal area, northern California: U.S. Geological Survey Professional Paper 1141, p. 47–60.
Donnelly-Nolan, J. M., Champion, D. E., Miller, C. D., Grove, T. L., and Trimble, D. A., 1990, Post-11,000 year volcanism of Medicine Lake volcano, Cascade Range, northern California: Journal of Geophysical Research, v. 95, p. 19693–19704.
Dott, R. H., Jr., 1966, Eocene deltaic sedimentation at Coos Bay, Oregon: Journal of Geology, v. 74, p. 373–420.
Dott, R. H., Jr., and Bird, K. J., 1979, Sand transport through channels across an Eocene shelf and slope in southwestern Oregon, U.S.A., *in* Doyle, L. J., and

Pilkey, O. H., eds., Geology of continental slopes: Society of Economic Paleontologists and Mineralogists Special Publication 27, p. 327–342.

Drewes, H., 1972, Cenozoic rocks of the Santa Rita Mountains, southeast of Tucson, Arizona: U.S. Geological Survey Professional Paper 746, 66 p.

—— , 1981, Tectonics of southeastern Arizona: U.S. Geological Survey Professional Paper 1144, 96 p.

Duffield, W. A., and McKee, E. H., 1986, Geochronology, structure, and basin-range tectonism of the Warner Range, northeastern California: Geological Society of America Bulletin, v. 97, p. 142–146.

Duffield, W. A., Bacon, C. R., and Dalrymple, G. B., 1980, Cenozoic volcanism, geochronology, and structure of the Coso Range, Inyo County, California: Journal of Geophysical Research, v. 85, p. 2381–2404.

Duncan, R. A., and Kulm, L. D., 1989, Plate tectonic evolution of the Cascades arc-subduction complex, *in* Winterer, E. L., Hussong, D. D., and Decker, R. W., eds., The eastern Pacific Ocean and Hawaii: Boulder, Colorado, Geological Society of America, The Geology of North America, v. N, p. 413–438.

Durham, D. L., 1974, Geology of the southern Salinas Valley: U.S. Geological Survey Professional Paper 819, 111 p.

Durham, D. L., and Yerkes, R. F., 1964, Geology and oil resources of the eastern Puente Hills area, southern California: U.S. Geological Survey Professional Paper 420-B, 62 p.

Durrell, C., 1966, Tertiary and Quaternary geology of the northern Sierra Nevada, *in* Bailey, E. H., ed., Geology of northern California: California Division of Mines and Geology Bulletin 190, p. 185–197.

Eaton, G. P., 1982, The Basin and Range Province; Origin and tectonic significance: Annual Review of Earth and Planetary Sciences, v. 10, p. 409–440.

Eaton, G. P., Wahl, R. R., Prostka, H. J., Mabey, D. R., and Kleinkopf, M. D., 1978, Regional gravity and tectonic patterns; Their relation to late Cenozoic epeirogeny and lateral spreading in the western Cordillera, *in* Smith, R. B., and Eaton, G. P., eds., Cenozoic tectonics and regional geophysics of the Western Cordillera: Geological Society of America Memoir 152, p. 51–91.

Eberly, L. D., and Stanley, T. B., Jr., 1978, Cenozoic stratigraphy and geologic history of southwestern Arizona: Geological Society of America Bulletin, v. 89, p. 921–940.

Ehlig, P. L., Ehlert, K. W., and Crowe, B. M., 1975, Offset of the upper Miocene Caliente and Mint Canyon Formations along the San Gabriel and San Andreas faults, *in* Crowell, J. C., ed., San Andreas fault in southern California: California Division of Mines and Geology Special Report 118, p. 83–92.

Ekren, E. B., Rogers, C. L., Anderson, R. E., and Orkild, P. P., 1968, Age of Basin and Range normal faults in Nevada Test Site and Nellis Air Force Range, Nevada, *in* Eckel, E. B., ed., Nevada Test Site: Geological Society of America Memoir 110, p. 247–250.

Ekren, E. B., Anderson, R. E., Rogers, C. L., and Noble, D. C., 1971, Geology of northern Nellis Air Force Base Bombing and Gunnery Range, Nye County, Nevada: U.S. Geological Survey Professional Paper 651, 91 p.

Ekren, E. B., Orkild, P. P., Sargent, K. A., and Dixon, G. L., 1977, Geologic map of Tertiary rocks, Lincoln County, Nevada: U.S. Geological Survey Miscellaneous Investigations Series I-1041, scale 1:250,000.

Ekren, E. B., Byers, F. M., Jr., Hardyman, R. F., Marvin, R. F., and Silberman, M. L., 1980, Stratigraphy, preliminary petrology, and some structural features of Tertiary volcanic rocks in the Gabbs Valley and Gillis Ranges, Mineral County, Nevada: U.S. Geological Survey Bulletin 1464, 54 p.

Ekren, E. B., McIntyre, D. H., Bennett, E. H., and Marvin, R. F., 1982, Cenozoic stratigraphy of western Owyhee County, Idaho, *in* Bonnichsen, B., and Breckenridge, R. M., eds., Cenozoic geology of Idaho: Idaho Bureau of Mines and Geology Bulletin 26, p. 215–235.

Elston, W. E., 1978, Mid-Tertiary cauldrons and their relationship to mineral resources, southwestern New Mexico; A brief review, *in* Chapin, C. E., and Elston, W. E., eds., Field guide to selected cauldrons and mining districts of the Datil-Mogollon volcanic field, New Mexico: New Mexico Geological Society Special Publication, p. 107–113.

—— , 1984a, Mid-Tertiary ash flow tuff cauldrons, southwestern New Mexico: Journal of Geophysical Research, v. 89, p. 8733–8750.

Elston, W. E., and Northrop, S. A., eds, 1976, Cenozoic volcanism in southwestern New Mexico: New Mexico Geological Society Special Publication 5, 151 p.

Elston, W. E., Rhodes, R. C., Coney, P. J., and Deal, E. G., 1976, Progress report on the Mogollon Plateau volcanic field, southwestern New Mexico, no. 3; Surface expression of a pluton, *in* Elston, W. E., and Northrup, S. A., eds., Cenozoic volcanism in southwestern New Mexico: New Mexico Geological Society Special Publication 5, p. 3–28.

Epis, R. C., and Chapin, C. E., 1968, Geologic history of the Thirtynine Mile volcanic field, central Colorado, *in* Epis, R. C., ed., Cenozoic volcanism in the Southern Rocky Mountains: Colorado School of Mines Quarterly, v. 63, no. 3, p. 51–85.

—— , 1975, Geomorphic and tectonic implications of the post-Laramide, late Eocene erosion surface in the Southern Rocky Mountains, *in* Curtis, B. F., ed., Cenozoic history of the Southern Rocky Mountains: Geological Society of America Memoir 144, p. 45–74.

Ernst, W. G., and Hall, C. A., Jr., 1974, Geology and petrology of the Cambria Felsite, a new Oligocene formation, west-central California Coast Ranges: Geological Society of America Bulletin, v. 85, p. 523–532.

Ewing, T. E., 1980, Paleogene tectonic evolution of the Pacific Northwest: Journal of Geology, v. 88, p. 619–638.

Fairchild, L. H., and Cowan, D. S., 1982, Structure, petrology and tectonic history of the Leech River complex northwest of Victoria, Vancouver Island: Canadian Journal of Earth Sciences, v. 19, p. 1817–1835.

Fiske, R. S., Hopson, C. A., and Waters, A. C., 1963, Geology of Mount Rainier National Park, Washington: U.S. Geological Survey Professional Paper 444, 93 p.

Fleck, R. J., 1970, Age and tectonic significance of volcanic rocks, Death Valley, California: Geological Society of America Bulletin, v. 81, p. 2807–2816.

Fouch, T. D., 1979, Character and paleogeographic distribution of Upper Cretaceous(?) and Paleogene nonmarine sedimentary rocks in east-central Nevada, *in* Armentrout, J. M., Cole, M. R., and TerBest, H., Jr., eds., Cenozoic paleogeography of the western United States: Society of Economic Paleontologists and Mineralogists, Pacific Section, Pacific Coast Paleogeography Symposium 3, p. 97–111.

Fouch, T. D., Hanley, J. H., and Forester, R. M., 1979, Preliminary correlation of Cretaceous and Paleogene lacustrine and related nonmarine sedimentary and volcanic rocks of the eastern Great Basin of Nevada and Utah, *in* Newman, G. W., and Goode, H. D., eds., Basin and Range Symposium and Great Basin Field Conference: Rocky Mountain Association of Geologists, p. 305–312.

Fox, K. F., Jr., 1983, Tectonic setting of late Miocene, Pliocene, and Pleistocene rocks in part of the Coast Ranges north of San Francisco, California: U.S. Geological Survey Professional Paper 1239, 33 p.

Fox, K. F., Rinehart, C. D., and Engels, J. C., 1977, Plutonism and orogeny in north-central Washington; Timing and regional context: U.S. Geological Survey Professional Paper 989, 27 p.

Fox, K. F., Jr., Fleck, R. J., Curtis, G. H., and Meyer, C. E., 1985, Implications of the northwestwardly younger age of the volcanic rocks of west-central California: Geological Society of America Bulletin, v. 96, p. 647–654.

Fuis, G. S., and Kohler, W. M., 1984, Crustal structure and tectonics of the Imperial Valley region, California, *in* Rigsby, C. A., ed., The Imperial Basin; Tectonics, sedimentation, and thermal aspects: Society of Economic Paleontologists and Mineralogists, Pacific Section, p. 1–13.

Fuller, R. E., 1931, The geomorphology and volcanic sequence of Steens Mountain in southeastern Oregon: University of Washington Geologic Publications, v. 3, p. 1–130.

Galloway, A. J., 1977, Geology of the Point Reyes peninsula, Marin County, California: California Division of Mines and Geology Bulletin 202, 72 p.

Gans, P. B., Miller, E. L., McCarthy, J., and Ouldcott, M. L., 1985, Tertiary extensional faulting and evolving ductile-brittle transition zones in the northern Snake Range and vicinity; New insights from seismic data: Geology, v. 13, p. 189–193.

Garfunkel, Z., 1974, Model of the late Cenozoic tectonic history of the Mojave Desert, California, and for its relation to adjacent areas: Geological Society of America Bulletin, v. 85, p. 1931–1944.

Gastil, G., Krummenacher, D., and Minch, J., 1979, The record of Cenozoic volcanism around the Gulf of California: Geological Society of America Bulletin, Part I, v. 90, p. 839–857.

Gilbert, C. M., and Reynolds, M. W., 1973, Character and chronology of basin development, western margin of the Basin and Range Province: Geological Society of America Bulletin, v. 84, p. 2489–2510.

Gilbert, C. M., Christensen, M. N., Al-Rawi, Y., and Lajoie, K. R., 1968, Structural and volcanic history of Mono basin, California-Nevada, *in* Coats, R. R., Hay, R. L., and Anderson, C. A., eds., Studies in volcanology; A memoir in honor of Howel Williams: Geological Society of America Memoir 116, p. 275–329.

Gilluly, J., 1946, The Ajo mining district, Arizona: U.S. Geological Survey Professional Paper 209, 112 p.

——, 1956, General geology of central Cochise County, Arizona: U.S. Geological Survey Professional Paper 287, 169 p.

——, 1963, The tectonic evolution of the western United States: Geological Society of London Quarterly Journal, v. 119, p. 133–174.

Glazner, A. F., 1988, Stratigraphy, structure, and potassic alterations of Miocene volcanic rocks in the Sleeping Beauty area, central Mojave Desert, California: Geological Society of America Bulletin, v. 100, p. 424–435.

Glazner, A. F., and Bartley, J. M., 1984, Timing and tectonic setting of Tertiary low-angle normal faulting and associated magmatism in the southwestern United States: Tectonics, v. 3, p. 385–396.

Glazner, A. F., Nielsen, J. E., Howard, K. A., and Miller, D. M., 1986, Correlation of the Peach Springs Tuff, a large-volume Miocene ignimbrite sheet in California and Arizona: Geology, v. 14, p. 840–843.

Golombeck, M. P., McGill, G. E., and Brown, L. L., 1983, Tectonic and geologic evolution of the Espanola basin, Rio Grande rift; Structure, rate of extension, and relation to the state of stress in the western United States: Tectonophysics, v. 94, p. 483–507.

Graham, S. A., 1978, Role of Salinian block in evolution of San Andreas fault system, California: American Association of Petroleum Geologists Bulletin, v. 62, p. 2214–2231.

——, 1987, Tectonic controls on petroleum occurrence in central California, *in* Ingersoll, R. V., and Ernst, W. G., eds., Cenozoic basin development of central California; Rubey Vol. VI: Englewood Cliffs, New Jersey, Prentice-Hall, p. 47–63.

Graham, S. A., and Dickinson, W. R., 1978, Evidence for 115 kilometers of right-slip on the San Gregorio–Hosgri fault trend: Science, v. 199, p. 178–181.

Graham, S. A., and Williams, L. A., 1985, Tectonic, depositional, and diagenetic history of Monterey Formation (Miocene), central San Joaquin basin, California: American Association of Petroleum Geologists Bulletin, v. 69, p. 385–411.

Graham, S. A., McCloy, C., Hitzman, M., Ward, R., and Turner, R., 1984, Basin evolution during change from convergent to transform continental margin in central California: American Association of Petroleum Geologists Bulletin, v. 68, p. 233–249.

Graham, S. A., Stanley, R. G., Bent, J. V., and Carter, J. B., 1989, Oligocene and Miocene paleogeography of central California and displacement along the San Andreas fault: Geological Society of America Bulletin, v. 101, p. 711–730.

Grommé, C. S., McKee, E. H., and Blake, M. C., Jr., 1972, Paleomagnetic correlations and potassium-argon dating of middle Tertiary ash-flow sheets in the eastern Great Basin, Nevada and Utah: Geological Society of America Bulletin, v. 83, p. 1619–1638.

Grommé, C. S., Beck, M. E., Jr., Wells, R. E., and Engebretson, D. C., 1986, Paleomagnetism of the Tertiary Clarno Formation of central Oregon and its significance for the tectonic history of the Pacific Northwest: Journal of Geophysical Research, v. 91, p. 14089–14103.

Guffanti, M., and Weaver, C. S., 1988, Distribution of late Cenozoic volcanic rocks in the Cascade Range; Volcanic arc segmentation and regional tectonic considerations: Journal of Geophysical Research, v. 93, p. 6513–6529.

Guth, P. L., 1981, Tertiary extension north of the Las Vegas Valley shear zone, Sheep and Desert Ranges, Clark County, Nevada: Geological Society of America Bulletin, v. 92, pt. 1, p. 763–771.

Hackel, O., 1966, Summary of the geology of the Great Valley, *in* Bailey, E. H., ed., Geology of northern California: California Division of Mines and Geology Bulletin 190, p. 217–238.

Hagstrum, J. T., Sawlan, M. G., Hausback, B. P., Smith, J. G., and Grommé, C. S., 1987, Miocene paleomagnetism and tectonic setting of the Baja California Peninsula, Mexico: Journal of Geophysical Research, v. 92, p. 2627–2639.

Hall, C. A., Jr., 1981a, San Luis Obispo transform fault and middle Miocene rotation of the western Transverse Ranges, California: Journal of Geophysical Research, v. 86, p. 1015–1031.

——, 1981b, Evolution of the western Transverse Ranges microplate; Late Cenozoic faulting and basinal development, *in* Ernst, W. G., ed., Rubey Volume I: The geotectonic development of California: Englewood Cliffs, New Jersey, Prentice-Hall, p. 559–582.

Hamilton, W., 1978, Mesozoic tectonics of the western United States, *in* Howell, D. G., and McDougall, K. A., eds., Mesozoic paleogeography of the western United States: Society of Economic Paleontologists and Mineralogists, Pacific Section, Pacific Coast Paleogeography Symposium 2, p. 33–70.

Hamilton, W., and Myers, W. B., 1966, Cenozoic tectonics of the western United States: Reeviews of Geophysics, v. 4, p. 509–549.

Hammond, P. E., 1979, A tectonic model for evolution of the Cascade Range, *in* Armentrout, J. M., Cole, M. R., and TerBest, H., Jr., eds., Cenozoic paleogeography of the western United States: Society of Economic Paleontologists and Mineralogists, Pacific Section, Pacific Coast Paleogeography Symposium 3, p. 219–237.

Harding, T. P., 1976, Tectonic significance and hydrocarbon trapping consequences of sequential folding synchronous with San Andreas faulting, San Joaquin Valley, California: American Association of Petroleum Geologists Bulletin, v. 60, p. 356–378.

Hart, W. K., Aronson, J. L., and Mertzman, S. A., 1984, Areal distribution and age of low-K, high-alumina olivine tholeiite magmatism in the northwestern Great Basin: Geological Society of America Bulletin, v. 95, p. 186–195.

Harwood, D. S., and Helley, E. J., 1987, Late Cenozoic tectonism of the Sacramento Valley, California: U.S. Geological Survey Professional Paper 1359, 46 p.

Hauge, T. A., 1985, Gravity-spreading origin of the Heart Mountain allocthon, northwestern Wyoming: Geological Society of America Bulletin, v. 96, p. 1440–1456.

Hawkins, J. W., 1970, Petrology and possible tectonic significance of late Cenozoic volcanic rocks, southern California and Baja California: Geological Society of America Bulletin, v. 81, p. 3323–3338.

Hearn, B. C., Jr., Donnelly-Nolan, J. M., and Goff, F. E., 1981, The Clear Lake Volcanics; Tectonic setting and magma sources, *in* McLaughlin, R. J., and Donnelly-Nolan, J. M., eds., Research in the Geysers–Clear Lake geothermal area, northern California: U.S. Geological Survey Professional Paper 1141, p. 25–46.

Heller, P. L., and Dickinson, W. R., 1985, Submarine ramp facies model for delta-fed, sand-rich turbidite systems: American Association of Petroleum Geologists Bulletin, v. 69, no. 6, p. 960–976.

Heller, P. L., and Ryberg, P. T., 1983, Sedimentary record of subduction to forearc transition in the rotated Eocene basin of western Oregon: Geology, v. 11, p. 380–383.

Heller, P. L., Peterman, Z. E., O'Neil, J. R., and Shafiqullah, M., 1985, Isotopic provenance of sandstones from the Eocene Tyee Formation, Oregon Coast Range: Geological Society of America Bulletin, v. 96, no. 6, p. 770–780.

Henriksen, D. A., 1956, Eocene stratigraphy of the lower Cowlitz River–eastern Willapa Hills area, southwestern Washington: Washington Division of Mines and Geology, Bulletin 43, 122 p.

Henry, C. D., and Price, J. G., 1984, Variations in caldera development in the Tertiary volcanic field of Trans-Pecos Texas: Journal of Geophysical Research, v. 89, p. 8765–8786.

——, 1986, Early Basin and Range development in Trans-Pecos Texas and adjacent Chihuahua; Magmatism and orientation, timing, and style of deformation: Journal of Geophysical Research, v. 91, p. 6213–6224.

Hewett, D. F., 1956, Geology and mineral resources of the Ivanpah Quadrangle, California and Nevada: U.S. Geological Survey Professional Paper 275, 172 p.

Hildreth, W., 1979, The Bishop Tuff; Evidence for the origin of compositional zonation in silicic magma chambers, *in* Chapin, C. E., and Elston, W. E., eds., Ash-flow tuffs: Geological Society of America Special Paper 180, p. 43–75.

Hildreth, W., Halliday, A. N., and Christiansen, R. L., 1991, Isotopic and chemical evidence concerning the genesis and contamination of basaltic and rhyolitic magma beneath the Yellowstone plateau volcanic field: Journal of Petrology, v. 32, p. 63–138.

Hill, D. P., 1978, Seismic evidence for the structure and Cenozoic tectonics of the Pacific Coast States, *in* Smith, R. B., and Eaton, G. P., eds., Cenozoic tectonics and regional geophysics of the Western Cordillera: Geological Society of America Memoir 152, p. 145–174.

Hill, M. L., and Troxel, B. W., 1966, Tectonics of Death Valley region, California: Geological Society of America Bulletin, v. 77, p. 435–438.

Hill, M. L., Carlson, S. A., and Dibblee, T. W., Jr., 1958, Stratigraphy of Cuyama Valley–Caliente Range area, California: American Association of Petroleum Geologists Bulletin, v. 42, p. 2973–3000.

Hirst, B., 1988, Early Miocene tectonism and associated turbidite deposystems of the Tejon Area, Kern County, California, *in* Graham, S. A., ed., Studies of the geology of the San Joaquin basin, California: Society of Economic Paleontologists and Mineralogists, Pacific Section, v. 60, p. 207–222.

Hornafius, J. S., 1985, Neogene tectonic rotation of the Santa Ynez Range, western Transverse Ranges, California, suggested by paleomagnetic investigation of the Monterey Formation: Journal of Geophysical Research, v. 90, p. 12503–12522.

Howard, K. A., and John, B. E., 1987, Crustal extension along a rooted system of imbricate low-angle faults; Colorado River extensional corridor, California and Arizona, *in* Coward, M. P., Dewey, J. F., and Hancock, P. L., eds., Continental extensional tectonics: Geological Society [London] Special Publication 28, p. 299–311.

Howell, D. G., Champion, D. E., and Vedder, J. G., 1987, Terrane accretion, crustal kinematics and basin evolution, southern California, *in* Ernst, W. G., and Ingersoll, R. V., eds., Cenozoic basin development of coastal California; Rubey Vol. VI: Englewood Cliffs, New Jersey, Prentice-Hall, p. 242–258.

Huber, N. K., 1981, Amount and timing of late Cenozoic uplift and tilt of the Central Sierra Nevada, California; Evidence from the upper San Joaquin River basin: U.S. Geological Survey Professional Paper 1197, 28 p.

Huffman, O. F., 1972, Lateral displacement of upper Miocene rocks and the Neogene history of offset along the San Andreas fault in central California: Geological Society of America Bulletin, v. 83, p. 2913–2946.

Hunt, C. B., 1956, Cenozoic geology of the Colorado Plateau: U.S. Geological Survey Professional Paper 279, 99 p.

——, 1958, Structural and igneous geology of the La Sal Mountains, Utah: U.S. Geological Survey Professional Paper 294-I, 364 p.

Hunt, C. B., and Mabey, D. R., 1966, Stratigraphy and structure, Death Valley, California: U.S. Geological Survey Professional Paper 494-A, 162 p.

Hunt, C. B., Averitt, P., and Miller, R. L., 1953, Geology and geography of the Henry Mountains region, Utah: U.S. Geological Survey Professional Paper 28, 234 p.

Hurst, R. W., 1979, Geologic and geochemical investigations of Miocene volcanics in the Transverse Ranges, southern California, *in* Armentrout, J. M., Cole, M. R., and TerBest, H., Jr., eds., Cenozoic paleogeography of the western United States: Society of Economic Paleontologists and Mineralogists, Pacific Section, Pacific Coast Paleogeography Symposium 3, p. 257–276.

Hurst, R. W., Wood, W. R., and Hume, M., 1982, The petrologic and tectonic evolution of volcanic rocks in the Southern California Borderland; A transitional tectonic environment, *in* Frost, E. G., and Martin, D. L., eds., Mesozoic-Cenozoic tectonic evolution of the Colorado River region, California, Arizona, and Nevada: San Diego, Cordilleran Publishers, p. 287–297.

Hyndman, D. W., 1980, Bitterroot dome–Sapphire tectonic block, an example of a plutonic-core gneiss complex with its detached suprastructure, *in* Crittenden, M. D., Coney, P. J., and Davis, G. H., eds., Cordilleran metamorphic core complexes: Geological Society of America Memoir 153, p. 427–443.

Ingersoll, R. V., 1979, Evolution of the Late Cretaceous forearc basin, northern and central California: Geological Society of America Bulletin, v. 90, p. 813–826.

Izett, G. A., 1966, Tertiary extrusive volcanic rocks in Middle Park, Grand County, Colorado: U.S. Geological Survey Professional Paper 550-B, p. B42–B46.

——, 1975, Late Cenozoic sedimentation and deformation in northern Colorado and adjoining areas, *in* Curtis, B. F., ed., Cenozoic history of the Southern Rocky Mountains: Geological Society of America Memoir 144, p. 179–209.

Izett, G. A. Wilcox, R. E., Powers, H. A., and Desborough, G. A., 1970, The Bishop ash bed, a Pleistocene marker bed in the western United States: Quaternary Research, v. 1, p. 121–132.

Jackson, P. A., and Yeats, R. S., 1982, Structural evolution of Carpinteria basin, western Transverse Ranges, California: American Association of Petroleum Geologists Bulletin, v. 66, p. 805–829.

John, B. E., 1987, Geometry and evolution of a mid-crustal extensional fault system, Chemuevi Mountains southeastern California, *in* Coward, M. P., Dewey, J. F., and Hancock, P. L., eds., Continental extensional tectonics: Geological Society [London] Special Publication 28, p. 313–335.

Johnson, S. Y., 1984a, Evidence for a margin-truncating transcurrent fault (pre-late Eocene) in western Washington: Geology, v. 12, p. 538–541.

——, 1984b, Stratigraphy, age, and paleogeography of the Eocene Chuckanut Formation, northwest Washington: Canadian Journal of Earth Sciences, v. 21, p. 92–106.

——, 1985a, Stratigraphy and sedimentology of the Eocene Chuckanut Formation on Bellingham Bay, Washington, *in* Templeman-Kluit, D., ed., Field guides to geology and mineral deposits in the southern Canadian Cordillera: Geological Association of Canada, Field Guide, p. 6-1 to 6-21.

——, 1985b, Eocene strike-slip faulting and nonmarine basin formation in Washington, *in* Biddle, K. T., and Christie-Blick, N., eds., Strike-slip deformation, basin formation, and sedimentation: Society of Economic Paleontologists and Mineralogists Special Publication 37, p. 283–302.

Kamerling, M. J., and Luyendyk, B. P., 1979, Tectonic rotations of the Santa Monica Mountains region, western Transverse Ranges, California, suggested by paleomagnetic vectors: Geological Society of America Bulletin, v. 90, pt. I, p. 331–337.

Kelley, V. C., 1979, Tectonics, middle Rio Grande rift, New Mexico, *in* Riecker, R. E., ed., Rio Grande rift; Tectonics and magmatism: Washington, American Geophysical Union, p. 57–70.

Kelley, V. C., and Thompson, T. B., 1964, Tectonics and general geology of the Ruidoso-Carrizozo region, central New Mexico: New Mexico Geological Society 15th Field Conference Guidebook, p. 110–121.

Kelsey, H. M., and Carver, G. A., 1988, Late Neogene and Quaternary tectonics associated with northward growth of the San Andreas transform fault, northern California: Journal of Geophysical Research, v. 93, p. 4797–4819.

Kennedy, M. P., 1975, Geology of the San Diego metropolitan area, California; Section A, Western San Diego metropolitan area: California Division of Mines and Geology Bulletin 200, p. 9–39.

Kennedy, M. P., and Peterson, G. L., 1975, Geology of the San Diego metropolitan area, California; Section B, Eastern San Diego metropolitan area: California Division of Mines and Geology Bulletin 200, p. 43–56.

Kittleman, L. R., Green, A. R., Hagood, A. R., Johnson, A. M., McMurray, J. M., Russell, R. G., and Weeden, D. A., 1965, Cenozoic stratigraphy of the Owyhee region, southeastern Oregon: University of Oregon Museum of Natural History Bulletin 1, 45 p.

Krieger, M. H., 1965, Geology of the Prescott and Paulden Quadrangles, Arizona:

U.S. Geological Survey Professional Paper 467, 127 p.
——, 1979, Ash-flow tuffs of the Galiuro Volcanics in the northern Galiuro Mountains, Pinal County, Arizona: U.S. Geological Survey Professional Paper 1104, 32 p.

Kuntz, M. A., Champion, D. E., Spiker, E. C., Lefebvre, R. H., and McBroome, L. A., 1982, The Great Rift and the evolution of the Craters of the Moon lava field, Idaho, *in* Bonnichsen, B., and Breckenridge, R. M., eds., Cenozoic geology of Idaho: Idaho Bureau of Mines and Geology Bulletin 26, p. 423–437.

Lachenbruch, A. H., and Sass, J. H., 1978, Models of an extending lithosphere and heat flow in the Basin and Range Province, *in* Smith, R. B., and Eaton, G. P., eds., Cenozoic tectonics and regional geophysics in the western Cordillera: Geological Society of America Memoir 152, p. 209–250.

Lachenbruch, A. H., Sass, J. H., and Galanis, S. P., Jr., 1985, Heat flow in southernmost California and the origin of the Salton Trough: Journal of Geophysical Research, v. 90, p. 6709–6736.

Larsen, E. S., Jr., and Cross, W., 1956, Geology and petrology of the San Juan region, southwestern Colorado: U.S. Geological Survey Professional Paper 258, 303 p.

Larson, E. E., Ozima, M., and Bradley, W. C., 1975, Late Cenozoic basic volcanism in northwestern Colorado and its implications concerning tectonism and the origin of the Colorado River system, *in* Curtis, B. F., ed., Cenozoic history of the Southern Rocky Mountains: Geological Society of America Memoir 144, p. 155–178.

Larson, R. L., Menard, H. W., and Smith, S. M., 1968, Gulf of California; A result of ocean-floor spreading and transform faulting: Science, v. 161, p. 781–784.

Leeman, W. P., 1970, The isotopic composition of strontium in late-Cenozoic basalts from the Basin-Range Province, western United States: Geochimica et Cosmochimica Acta, v. 34, p. 857–872.

——, 1982a, Olivine tholeiitic basalts of the Snake River Plain, Idaho, *in* Bonnichsen, B., and Breckenridge, R. M., eds., Cenozoic geology of Idaho: Idaho Bureau of Mines and Geology Bulletin 26, p. 181–191.

Leeman, W. P., and Rogers, J.J.W., 1970, Late Cenozoic alkali-olivine basalts of the Basin-Range Province, USA: Contributions to Mineralogy and Petrology, v. 25, p. 1–24.

Lettis, W. R., 1988, Quaternary geology of the northern San Joaquin Valley, *in* Graham, S. A., ed., Studies of the geology of the San Joaquin basin, California: Society of Economic Paleontologists and Mineralogists, Pacific Section, v. 60, p. 333–350.

Lindsey, D. A., 1982, Tertiary volcanic rocks and uranium in the Thomas Range and northern Drum Mountains, Juab County, Utah: U.S. Geological Survey Professional Paper 1221, 71 p.

Link, M. H., 1975, Matilija Sandstone; A transition from deep-water turbidite to shallow-marine deposition in the Eocene of California: Journal of Sedimentary Petrology, v. 45, p. 63–78.

Lipman, P. W., 1980, Cenozoic volcanism in the western United States; Implications for continental tectonics, *in* Studies in geophysics; Continental tectonics: Washington, D.C., National Academy of Sciences, p. 161–174.

Lipman, P. W., and Mehnert, H. H., 1975, Late Cenozoic basaltic volcanism and development of the Rio Grande depression in the southern Rocky Mountains, *in* Curtis, B. F., ed., Cenozoic history of the Southern Rocky Mountains: Geological Society of America Memoir 144, p. 119–154.

——, 1979a, The Taos Plateau volcanic field, northern Rio Grande rift, New Mexico, *in* Riecker, R. E., ed., Rio Grande rift; Tectonics and magmatism: Washington, American Geophysical Union, p. 289–311.

Lipman, P. W., Steven, T. A., and Mehnert, H. H., 1970, Volcanic history of the San Juan Mountains, Colorado, as indicated by potassium-argon dating: Geological Society of America Bulletin, v. 81, p. 2329–2352.

Lipman, P. W., Prostka, H. J., and Christiansen, R. L., 1972, Cenozoic volcanism and plate tectonic evolution of the Western United States; I, Early and middle Cenozoic: Royal Society of London Philosophical Transactions, ser. A., v. 271, p. 217–248.

Lipman, P. W., Mehnert, H. H., and Naeser, C. W., 1986, Evolution of the Latir volcanic field, northern New Mexico, and its relation to the Rio Grande rift, as indicated by potassium-argon and fission track dating: Journal of Geophysical Research, v. 91, p. 6329–6345.

Longwell, C. R., 1960, Possible explanation of diverse structural patterns in southern Nevada: American Journal of Science, v. 258-A, p. 192–203.

——, 1963, Reconnaissance geology between Lake Mead and Davis Dam, Arizona-Nevada: U.S. Geological Survey Professional Paper 374-E, 51 p.

——, 1974, Measure and date of movement on Las Vegas Valley shear zone, Clark County, Nevada: Geological Society of America Bulletin, v. 85, p. 985–990.

Love, J. D., 1956, Cretaceous and Tertiary stratigraphy of the Jackson Hole area, northwestern Wyoming: Wyoming Geological Association Guidebook, 11th Annual Field Conference, p. 75–94.

——, 1960, Cenozoic sedimentation and crustal movement in Wyoming: American Journal of Science, v. 258A, p. 204–214.

Lovell, J.P.B., 1969, Tyee Formation; Undeformed turbidites and their lateral equivalents; Mineralogy and paleogeography: Geological Society of America Bulletin, v. 80, no. 1, p. 9–22.

Lucas, S. G., and Ingersoll, R. V., 1981, Cenozoic continental deposits of New Mexico; An overview: Geological Society of America Bulletin, v. 92, p. 917–932.

Lucchitta, I., 1979, Late Cenozoic uplift of the southwestern Colorado Plateau and adjacent lower Colorado River region: Tectonophysics, v. 61, p. 63–95.

Luedke, R. G., and Smith, R. L., 1984, Map showing distribution, composition, and age of late Cenozoic volcanic centers in the western conterminous United States: U.S. Geological Survey Miscellaneous Investigations Map I-1523, scale 1:2,500,000.

Luyendyk, B. P., and Hornafius, J. S., 1987, Neogene crustal rotations, fault slip, and basin development in southern California, *in* Ingersoll, R. V., and Ernst, W. G., eds., Cenozoic basin development of coastal California; Rubey Vol. VI: Englewood Cliffs, New Jersey, Prentice-Hall, p. 259–283.

Luyendyk, B. P., Kamerling, M. J., Terres, R. R., and Hornafius, J. S., 1985, Simple shear of southern California during Neogene time suggested by paleomagnetic declinations: Journal of Geophysical Reserach, v. 90, p. 12454–12466.

Lydon, P. A., 1968, Geology and lahars of the Tuscan Formation, California, *in* Coats, R. R., Hay, R. L., and Anderson, C. A., eds., Studies in volcanology; A memoir in honor of Howel Williams: Geological Society of America Memoir 116, p. 441–475.

Mabey, D. R,. 1982, Geophysics and tectonics of the Snake River Plain, Idaho, *in* Bonnichsen, B., and Breckenridge, R. M., eds., Cenozoic geology of Idaho: Idaho Bureau of Mines and Geology Bulletin 26, p. 139–153.

Mabey, D. R., Zietz, I., Eaton, G. P., and Kleinkopf, M. D., 1978, Regional magnetic patterns in part of the Cordillera in the western United States, *in* Smith, R. B., and Eaton, G. P., eds., Cenozoic tectonics and regional geophysics of the Western Cordillera: Geological Society of America Memoir 152, p. 93–106.

Macdonald, G. A., 1966, Geology of the Cascade Range and Modoc Plateau, *in* Bailey, E. H., ed., Geology of northern California: California Division of Mines and Geology Bulletin 190, p. 65–96.

Mackin, J. H., 1947, Some structural features of the intrusions in the Iron Springs District: Utah Geological Society Guidebook to the Geology of Utah, no. 2, 62 p.

——, 1960, Structural significance of Tertiary volcanic rocks in southwestern Utah: American Journal of Science, v. 258, p. 81–131.

MacLeod, N. S., and Sherrod, D. R., 1988, Geologic evidence for a magma chamber beneath Newberry volcano, Oregon: Journal of Geophysical Research, v. 93, p. 10067–10079.

MacLeod, N. S., and Snavely, P. D., Jr., 1973, Volcanic and intrusive rocks of the central part of the Oregon Coast Range, *in* Beaulieu, J. D., ed., Geologic field trips in northern Oregon and Southern Washington: Oregon Department of Geology and Mineral Industries Bulletin 77, p. 47–74.

MacLeod, N. S., Walker, G. W., and McKee, E. H., 1976, Geothermal significance of eastward increase in age of upper Cenozoic rhyolitic domes in southeastern Oregon, *in* Second United Nations Symposium on the Development and Use of Geothermal Resources: Washington, D.C., U.S. Government Printing Office, v. 1, p. 465–474.

Magill, J., Cox, A., and Duncan, R., 1981, Tillamook Volcanic Series; Further evidence for tectonic rotation of the Oregon Coast Range: Journal of Geophysical Research, v. 86, p. 2953–2970.

Magill, J. R., Wells, R. E., Simpson, R. W., and Cox, A., 1982, Post-12 m.y. rotation of southwestern Washington: Journal of Geophysical Research, v. 87, p. 3761–3776.

Malde, H. E., 1991, Quaternary geology and structural history of the Snake River Plain, Idaho and Oregon, *in* Morrison, R. B., ed., Quaternary nonglacial geology; Conterminous U.S.: Boulder, Colorado, Geological Society of America, The Geology of North America, v. K-2, p. 251–282.

Malde, H. E., and Powers, H. A., 1962, Upper Cenozoic stratigraphy of western Snake River Plain: Geological Society of America Bulletin, v. 73, p. 1197–1220.

Mangan, M. T., Wright, T. L., and Swanson, D. A., 1986, Regional correlation of Grande Ronde Basalt flows, Columbia River Basalt Group, Washington, Oregon, and Idaho: Geological Society of America Bulletin, v. 97, p. 1300–1318.

Manley, K., 1984, Brief summary of the Tertiary geologic history of the Rio Grande rift in northern New Mexico, *in* Baldridge, W. S., Dickerson, P. W., Riecker, R. E., and Zidek, J., eds., Rio Grande rift; Northern New Mexico: New Mexico Geological Society Guidebook, 35th Field Conference, p. 63–66.

Matthews, V., III, 1976, Correlation of the Pinnacles and Neenach Volcanic Formations and their bearing on the San Andreas fault problem: American Association of Petroleum Geologists Bulletin, v. 60, p. 2128–2141.

Mattinson, J. M., 1977, Emplacement history of the Tatoosh volcanic-plutonic complex, Washington; Ages of zircons: Geological Society of America Bulletin, v. 88, p. 1509–1514.

Maxwell, R. A., Lonsdale, J. T., Hazzard, R. T., and Wilson, J. A., 1967, Geology of Big Bend National Park, Texas: Texas Bureau of Economic Geology Publication 6711, 320 p.

Mayer, L., 1987, Subsidence analysis of the Los Angeles basin, *in* Ingersoll, R. V., and Ernst, W. G., eds., Cenozoic basins of southern California; Rubey Vol. VI: Englewood Cliffs, New Jersey, Prentice Hall, p. 299–320.

McBirney, A. R., and White, C. M., 1982, The Cascade province, *in* Thorpe, R. S., ed., Andesites; Orogenic andesites and related rocks: New York, Wiley, p. 115–135.

McIntyre, D. H., Ekren, E. B., and Hardyman, R. F. 1982, Stratigraphic and structural framework of the Challis Volcanics in the eastern half of the Challis 1°x2° Quadrangle, Idaho, *in* Bonnichsen, B., and Breckenridge, R. M., eds., Cenozoic geology of Idaho: Idaho Bureau of Mines and Geology Bulletin 26, p. 3–22.

McKee, E. H., 1968, Age and rate of movement of the northern part of the Death Valley–Furnace Creek fault zone, California: Geological Society of America Bulletin, v. 79, p. 509–512.

——, 1979, Ash-flow sheets and calderas; Their genetic relationship to ore deposits in Nevada, *in* Chapin, C. E., and Elston, W. E., eds., Ash-flow tuffs: Geological Society of America Special Paper 180, p. 205–211.

McKee, E. H., and Anderson, C. A., 1971, Age and chemistry of Tertiary volcanic rocks in north-central Arizona and relation of the rocks to the Colorado Plateau: Geological Society of America Bulletin, v. 82, p. 2767–2782.

McKee, E. H., and Silberman, M. L., 1970, Geochronology of Tertiary igneous rocks in central Nevada: Geological Society of America Bulletin, v. 81, p. 2317–2328.

McKee, E. H., Noble, D. C., and Silberman, M. L., 1970, Middle Miocene hiatus in volcanic activity in the Great Basin area of the western United States: Earth and Planetary Science Letters, v. 8, p. 93–96.

McKee, E. H., Duffield, W. A., and Stern, R. J., 1983, Late Miocene and early Pliocene basaltic rocks and their implications for crustal structure, northeastern California and south-central Oregon: Geological Society of America Bulletin, v. 94, p. 292–304.

McLaughlin, R. J., Kling, S. A., Poore, R. Z., McDougall, K., and Beutner, E. C., 1982, Post–middle Miocene accretion of Franciscan rocks, northwestern California: Geological Society of America Bulletin, v. 93, p. 595–605.

Metzger, D. G., 1968, The Bouse Formation (Pliocene) of the Parker-Blythe-Cibola area, Arizona and California: U.S. Geological Survey Professional Paper 600-D, p. D126–D136.

Miller, C. D., 1985, Holocene eruptions at the Inyo volcanic chain, California; Implications for possible eruptions in Long Valley caldera: Geology, v. 13, p. 14–17.

Miller, C. F., and Bradfish, L. J., 1980, An inner Cordilleran belt of muscovite-bearing plutons: Geology, v. 8, p. 412–416.

Miller, E. L., Gans, P. B., and Garing, J. D., 1983, The Snake Range décollement: An exhumed mid-Tertiary ductile-brittle transition: Tectonics, v. 2, p. 239–263.

Miller, F. K., and Engels, J. C., 1975, Distribution and trends of discordant ages of the plutonic rocks of northeastern Washington and northern Idaho: Geological Society of America Bulletin, v. 86, p. 517–528.

Minch, J. A., 1967, Stratigraphy and structure of the Tijuana–Rosarito Beach area, northwestern Baja California, Mexico: Geological Society of America Bulletin, v. 78, p. 1155–1178.

——, 1979, The late Mesozoic-early Tertiary framework of continental sedimentation, northern Peninsular Ranges, Baja California, Mexico, *in* Abbott, P. L., ed., Eocene depositional systems San Diego, California: Society of Economic Paleontologists and Mineralogists, Pacific Section, p. 43–67.

Moore, D. G., 1973, Plate edge deformation and crustal growth, Gulf of California structural province: Geological Society of America Bulletin, v. 84, p. 1883–1906.

Moore, D. G., and Buffington, E. C., 1968, Transform faulting and growth of the Gulf of California since the late Pliocene: Science, v. 161, p. 1238–1241.

Moore, J. G., and Dodge, F.C.W., 1980, Late Cenozoic volcanic rocks of the southern Sierra Nevada, California; I, Geology and petrology; Summary: Geological Society of America Bulletin, pt. I, v. 91, p. 515–518.

Moore, R. B., Wolfe, E. W., and Ulrich, G. E., 1976, Volcanic rocks of the eastern and northern parts of the San Francisco volcanic field, Arizona: U.S. Geological Survey Journal of Research, v. 5, p. 549–560.

Moore, W. J., 1973, Igneous rocks in the Bingham mining district: U.S. Geological Survey Professional Paper 629-B, 42 p.

Morgan, L. A., Doherty, D. J., and Leeman, W. P., 1984, Ignimbrites of the eastern Snake River Plain; Evidence for major caldera-forming eruptions: Journal of Geophysical Research, v. 89, p. 8665–8678.

Morgan, P., Seager, W. R., and Golombek, M. P., 1986, Cenozoic thermal, mechanical and tectonic evolution of the Rio Grande rift: Journal of Geophysical Research, v. 91, p. 6263–6276.

Morris, H. T., and Lovering, T. S., 1979, General geology and mines of the East Tintic Mining district, Utah and Juab Counties, Utah: U.S. Geological Survey Professional Paper 1024, 203 p.

Mullineaux, D. R., and Crandell, D. R., 1981, The eruptive history of Mount St. Helens, *in* Lipman, P. W., and Mullineaux, D. R., eds., The 1980 eruptions of Mount St. Helens, Washington: U.S. Geological Survey Professional Paper 1250, p. 3–15.

Mullineaux, D. R., Gard, L. M., and Crandell, D. R., 1959, Continental sediments of Miocene age in the Puget Sound lowland, Washington: American Association of Petroleum Geologists Bulletin, v. 43, p. 688–696.

Nations, J. D., Landye, J. J., and Hevly, R. H., 1982, Location and chronology of Tertiary sedimentary deposits in Arizona; A review, *in* Ingersoll, R. V., and Woodburne, M. O., eds., Cenozoic nonmarine deposits of California and Arizona: Los Angeles, Society of Economic Paleontologists and Mineralogists, Pacific Section, p. 107–122.

Niem, A. R., and Niem, W. A., 1985, Oil and gas investigation of the Astoria basin, Clatsop and northernmost Tillamook counties, northwest Oregon: Oregon Department of Geology and Mineral Industries Oil and Gas Investigation 14, 8 p.

Nilsen, T. H., 1984, Oligocene tectonics and sedimentation, California: Sedimentary Geology, v. 38, p. 305–336.

Nilsen, T. H., and Clarke, S. H., Jr., 1975, Sedimentation and tectonics in the early Tertiary continental borderland of central California: U.S. Geological Survey Professional Paper 925, 64 p.

——, 1989, Late Cenozoic basins of northern California: Tectonics, v. 8, p. 1137–1158.

Nilsen, T. H., and Link, M. H., 1975, Stratigraphy, sedimentology and offset along the San Andreas fault of Eocene to lower Miocene strata of the northern Santa Lucia Range and the San Emigdio Mountains, Coast Ranges, central California, *in* Weaver, D. W., Hornaday, G. R., and Tipton, A., eds., Paleogene symposium and selected technical papers: American Association of Petroleum Geologists, Society of Economic Paleontologists and Mineralogists and Society of Exploration Geophysicists, Pacific Sections, p. 367–400.

Noble, D. C., 1972, Some observations on the Cenozoic volcano-tectonic evolution of the Great Basin, western United States: Earth and Planetary Science Letters, v. 17, p. 142–150.

Noble, D. C., and Christiansen, R. L., 1974, Black Mountain volcanic center, *in* Guidebook to the geology of four Tertiary volcanic centers in Nevada: Nevada Bureau of Mines and Geology Report 19, p. 27–34.

Noble, D. C., Sargent, K. A., Mehnert, H. H., Ekren, E. B., and Byers, F. M., Jr., 1968, Silent Canyon volcanic center, Nye County, Nevada, *in* Eckel, E. B., ed., Nevada Test Site: Geological Society of America Memoir 110, p. 65–75.

Noble, D. C., McKee, E. H., Smith, J. G., and Korringa, M. K., 1970, Stratigraphy and geochronology of Miocene volcanic rocks in northwestern Nevada: U.S. Geological Survey Professional Paper 700-D, p. D23–D32.

Noble, L. F., 1941, Structural features of the Virgin Spring area, Death Valley, California: Geological Society of America Bulletin, v. 52, p. 941–1000.

Noblett, J. B., 1981, Subduction-related origin of the volcanic rocks of the Eocene Clarno Formation near Cherry Creek, Oregon: Oregon Geology, v. 43, p. 91–99.

Novak, S. W., 1984, Eruptive history of the rhyolitic Kane Springs Wash volcanic center, Nevada: Journal of Geophysical Research, v. 89, p. 8603–8615. [Correction *in* 1985, v. 90, p. 2070.]

Novak, S. W., and Bacon, C. R., 1986, Pliocene volcanic rocks of the Coso Range, Inyo County, California: U.S. Geological Survey Professional Paper 1383, 44 p.

Ogle, B. A., 1953, Geology of Eel River Valley area, Humboldt County, California: California Division of Mines Bulletin 164, 128 p.

Ogle, B. A., Wallis, W. S., Heck, R. G., and Edwards, E. B., 1987, Petroleum geology of the Monterey Formation in the offshore Santa Maria/Santa Barbara areas, *in* Ingersoll, R. V., and Ernst, W. G., eds., Cenozoic basin development of coastal California; Rubey Vol. VI: Englewood Cliffs, New Jersey, Prentice-Hall, p. 382–406.

Otton, J. K., 1982, Tertiary extensional tectonics and associated volcanism in west-central Arizona, *in* Frost, E. G., and Martin, D. L., eds., Mesozoic-Cenozoic tectonic evolution of the Colorado River region, California, Arizona, and Nevada: San Diego, Cordilleran Publishers, p. 143–157.

Pansze, A. J., Jr., 1975, Geology and ore deposits of the Silver City–DeLamar–Flint region, Owyhee County, Idaho: Idaho Bureau of Mines and Geology Pamphlet 161, 79 p.

Pardee, J. T., 1950, Late Cenozoic block faulting in western Montana: Geological Society of America Bulletin, v. 61, p. 359–406.

Parker, D. F., and McDowell, F. W., 1979, K-Ar geochronology of Oligocene volcanic rocks, Davis and Barrilla Mountains, Texas: Geological Society of America Bulletin, v. 90, pt. 1, p. 1100–1110.

Parrish, R. R., Carr, S. D., and Parkinson, D. L., 1988, Eocene extensional tectonics and geochronology of the southern Omineca belt, British Columbia and Washington: Tectonics, v. 7, p. 181–212.

Pearson, R. C., and Obradovich, J. D., 1977, Eocene rocks in northeast Washington; Radiometric ages and correlation: U.S. Geological Survey Bulletin 1433, 41 p.

Peck, D. L., Griggs, A. B., Schlicker, H. G., Wells, F. G., and Dole, H. M., 1964, Geology of the central and northern parts of the Western Cascade Range in Oregon: U.S. Geological Survey Professional Paper 449, 56 p.

Peirce, H. W., 1984, Some late Cenozoic basins and basin deposits of southern and western Arizona, *in* Smiley, T. L., Nations, J. D., Pewe, T. L., and Schafer, J. P., eds., Landscapes of Arizona: Lanham, Maryland, University Press of America, p. 207–227.

Peirce, H. W., Damon, P. E., and Shafiqullah, M., 1979, An Oligocene Colorado Plateau edge in Arizona: Tectonophysics, v. 61, p. 1–14.

Personius, S. F., and Machette, M. N., 1984, Quaternary and Pliocene faulting in the Taos Plateau region, northern New Mexico, *in* Baldridge, W. S., Dickerson, P. W., Riecker, R. E., and Zidek, J., eds., Rio Grande rift; Northern New Mexico: New Mexico Geological Society Guidebook, 35th Field Conference, p. 83–90.

Pierce, K. L., and Morgan, L. A., 1992, The track of the Yellowstone hotspot: Volcanism, faulting, and uplift, *in* Link, P. K., Kuntz, M. A., and Platt, L. W., eds., Regional geology of eastern Idaho and western Wyoming: Geological Society of America Memoir (in press).

Pierce, W. G., 1957, Heart Mountain and South Fork detachment thrusts of Wyoming: American Association of Petroleum Geologists Bulletin, v. 41, p. 591–626.

——, 1973, Principal features of the Heart Mountain fault and the mechanism problem, *in* DeJong, K. A., and Scholten, R., eds., Gravity and tectonics: New York, Wiley & Sons, p. 457–471.

Pilger, R. H., and Henyey, T. L., 1979, Pacific-North American plate interaction and Neogene volcanism in coastal California: Tectonophysics, v. 57, p. 189–209.

Powell, R. E., 1982, Crystalline basement terranes in the southern eastern Transverse Ranges, California, *in* Cooper, J. D., ed., Geologic excursion in the Transverse Ranges, southern California: 78th annual meeting of the Cordilleran Section of the Geological Society of America, Anaheim, California, April 19–21, 1982, p. 109–151.

Powers, H. A., and Wilcox, R. E., 1964, Volcanic ash from Mount Mazama (Crater Lake) and from Glacier Peak: Science, v. 144, p. 1334–1336.

Priest, G. R., Woller, N. M., Black, G. L., and Evans, S. H., 1983, Overview of the geology of the central Oregon Cascade Region, *in* Priest, G. R., and Vogt, B. F., eds., Geology and geothermal resources of the central Oregon Cascade Range: Oregon Department of Geology and Mineral Industries Special Paper 15, p. 3–28.

Prodehl, C., and Lipman, P. W., 1989, Crustal structure of the Rocky Mountain region, *in* Pakiser, L. C., and Mooney, W. D., Geophysical framework of the continental United States: Boulder, Colorado, Geological Society of America Memoir 172, p. 249–284.

Proffett, J. M., Jr., 1977, Cenozoic geology of the Yerington district, Nevada, and implications for the nature and origin of Basin and Range faulting: Geological Society of America Bulletin, v. 88, p. 247–266.

Ramberg, I. B., Cook, F. A., and Smithson, S. B., 1978, Structure of the Rio Grande rift in southern New Mexico and west Texas based on gravity interpretation: Geological Society of America Bulletin, v. 89, p. 107–123.

Ratté, J. C., Marvin, R. F., Naeser, C. W., and Bikerman, M., 1984, Calderas and ash-flow tuffs of the Mogollon Mountains, southwestern New Mexico: Journal of Geophysical Research, v. 89, p. 8713–8732.

Rau, W. W., 1984, The Humptulips Formation; A new Eocene formation of southwest Washington: Washington Geologic Newsletter, v. 12, no. 4, p. 1–5.

——, 1986, Geologic map of the Humptulips Quadrangle and adjacent areas, Grays Harbor County, Washington: Washington Department of Natural Resources Geologic Map GM-33, scale 1:62,500.

Rehrig, W. A., and Reynolds, S. J., 1980, Geologic and geochronologic reconnaissance of a northwest–trending zone of metamorphic core complexes in southern and western Arizona, *in* Crittenden, M. D., Coney, P. J., and Davis, G. H., eds., Cordilleran metamorphic core complexes: Geological Society of America Memoir 153, p. 131–157.

Reynolds, M. W., 1976, Geology of the Grapevine Mountains, Death Valley, California; A summary, *in* Troxel, B. W., and Wright, L. A., eds., Geologic Features of Death Valley, California: California Division of Mines and Geology Special Report 106, p. 19–25.

——, 1979, Character and extent of Basin-Range faulting, western Montana and east-central Idaho, *in* Newman, G. W., and Goode, H. D., eds., Basin and

Range symposium and Great Basin field conference: Rocky Mountain Association of Geologists, p. 305–312.

Reynolds, S. J., 1980, Geologic framework of west-central Arizona: Arizona Geological Society Digest, no. 10, p. 1–16.

Reynolds, S. J., and Rehrig, W. A., 1980, Mid-Tertiary plutonism and mylonitization, South Mountains, central Arizona, *in* Crittenden, M. D., Coney, P. J., and Davis, G. H., eds., Cordilleran metamorphic core complexes: Geological Society of America Memoir 153, p. 159–176.

Reynolds, S. J., and Spencer, J. E., 1985, Evidence for large-scale transport on the Bullard detachment fault, west-central Arizona: Geology, v. 13, p. 353–356.

Roberts, A. E., 1958, Geology and coal resources of the Toledo–Castle Rock district, Cowlitz and Lewis Counties, Washington: U.S. Geological Survey Bulletin 1062, 71 p.

Robinson, P. T., 1975, Reconnaissance geologic map of the John Day Formation in the southwestern part of the Blue Mountains and adjacent areas, north-central Oregon: U.S. Geological Survey Miscellaneous Geologic Investigations Map I-872, scale 1:125,000.

Robinson, P. T., and Stewart, J. H., 1984, Uppermost Oligocene and lowermost Miocene ash-flow tuffs of western Nevada: U.S. Geological Survey Bulletin 1557, 53 p.

Robinson, P. T., McKee, E. H., and Moiola, R. J., 1968, Cenozoic volcanism and sedimentation, Silver Peak region, western Nevada and adjacent California, *in* Coats, R. R., Hay, R. L., and Anderson, C. A., eds., Studies in volcanology, a memoir in honor of Howel Williams: Geological Society of America Memoir 116, p. 577–611.

Robinson, P. T., Elders, W. A., and Muffler, L.J.P., 1976, Quaternary volcanism in the Salton Sea geothermal field, Imperial Valley, California: Geological Society of America Bulletin, v. 87, p. 347–360.

Robinson, P. T., Brem, G. F., and McKee, E. H., 1984, John Day Formation of Oregon; A distal record of early Cascade volcanism: Geology, v. 12, p. 229–232.

Robinson, P. T., Walker, G. W., and McKee, E. H., 1990, Eocene(?), Oligocene, and lower Miocene rocks of the Blue Mountains region, *in* Walker, G. W., ed., Geology of the Blue Mountains region of Oregon, Idaho, and Washington; Cenozoic Geology of the Blue Mountains Region: U.S. Geological Survey Professional Paper 1437, p. 29–62.

Robyn, T. L., 1979, Miocene volcanism in eastern Oregon; An example of calc-alkaline volcanism unrelated to subduction: Journal of Volcanology and Geothermal Research, v. 5, p. 149–161.

Rogers, J.J.W., and Novitsky-Evans, J. M., 1977, The Clarno Formation of central Oregon, U.S.A.; Volcanism on a thin continental margin: Earth and Planetary Science Letters, v. 34, p. 56–66.

Ross, D. C., 1978, The Salinian block; A Mesozoic granitic orphan in the California Coast Ranges, *in* Howell, D. G., and McDougall, K. A., eds., Mesozoic Paleogeography of the Western United States; Pacific Coast Paleogeography Symposium 2: Society of Economic Paleontologists and Mineralogists, Pacific Section, p. 509–522.

——, 1984, Possible correlations of basement rocks across the San Andreas, San Gregorio–Hosgri, and Rinconada-Reliz–King City faults, California: U.S. Geological Survey Professional Paper 1317, 37 p.

Rowley, P. D., Steven, T. A., Anderson, J. J., and Cunningham, C. G., 1979, Cenozoic stratigraphic and structural framework of southwestern Utah: U.S. Geological Survey Professional Paper 1149, 22 p.

Russell, R. J., 1928, Basin and Range structure and stratigraphy of the Warner Range, northeastern California: University of California Department of Geological Sciences Bulletin, v. 17, p. 387–496.

Rytuba, J. J., and McKee, E. H., 1984, Peralkaline ash flow tuffs and calderas of the McDermitt volcanic field, southeast Oregon and north central Nevada: Journal of Geophysical Research, v. 89, p. 8616–8628.

Sanford, A. R., Olsen, K. H., and Jaksha, L. H., 1979, Seismicity of the Rio Grande rift, *in* Riecker, R. E., ed., Rio Grande rift; Tectonics and magmatism: Washington, American Geophysical Union, p. 145–168.

Sbar, M. L., 1982, Delineation and interpretation of seismotectonic domains in western North America: Journal of Geophysical Research, v. 87, p. 3919–3928.

Scarborough, R. B., and Peirce, H. W., 1978, Late Cenozoic basins of Arizona: New Mexico Geological Society Guidebook, 29th Field Conference, p. 253–260.

Schwade, I. T., Carlson, S. A., and O'Flynn, J. B., 1958, Geologic environment of Cuyama Valley oil fields, California, *in* Weeks, L. G., ed., Habitat of oil: American Association of Petroleum Geologists, p. 78–98.

Scott, G. R., 1975, Cenozoic surfaces and deposits in the Southern Rocky Mountains, *in* Curtis, B. F., ed., Cenozoic history of the Southern Rocky Mountains: Geological Society of America Memoir 144, p. 227–248.

Scott, G. R., and Taylor, R. B., 1975, Post-Paleocene Tertiary rocks and Quaternary volcanic ash of the Wet Mountain Valley, Colorado: U.S. Geological Survey Professional Paper 868, 15 p.

Seager, W. R., 1981, Geology of the Organ Mountains and southern San Andres Mountains, New Mexico: New Mexico Bureau of Mines and Mineral Resources Memoir 36, 97 p.

Seager, W. R., and Morgan, P., 1979, Rio Grande rift in southern New Mexico, west Texas, and Chihuahua, *in* Riecker, R. E., ed., Rio Grande rift: Tectonics and magmatism: Washington, D.C., American Geophysical Union, p. 87–106.

Shafiqullah, M., Damon, P. E. Lynch, D. J., Kuck, P. H., and Rehrig, W. A., 1978, Mid-Tertiary magmatism in southeastern Arizona: New Mexico Geological Society Guidebook, 29th Field Conference, p. 231–241.

Shafiqullah, M., Damon, P. E., Lynch, D. J., Reynolds, S. J., Rehrig, W. A., and Raymond, R. H., 1980, K-Ar geochronology and geologic history of southwestern Arizona and adjacent areas: Arizona Geological Society Digest, v. 12, p. 201–260.

Sharp, R. V., 1967, San Jacinto fault zone in the Peninsular Ranges of southern California: Geological Society of America Bulletin, v. 78, p. 706–730.

——, 1982, Tectonic setting of the Imperial Valley region: U.S. Geological Survey Professional Paper 1254, p. 5–14.

Sheridan, M. F., 1978, The Superstition cauldron complex, *in* Guidebook to the geology of central Arizona: Arizona Bureau of Geology and Mineral Technology Special Paper 2, p. 85–96.

Sheriff, S. D., 1984, Paleomagnetic evidence for spatially distributed post-Miocene rotation of western Washington and Oregon: Tectonics, v. 3, p. 397–408.

Shoemaker, E. M., Roach, C. H., and Byers, F. M., Jr., 1962, Diatremes and uranium deposits in the Hopi Buttes, Arizona, *in* Engel, A.E.J., James, H. L., and Leonard, B. F., eds., Petrologic studies: Geological Society of America Buddington Volume, p. 327–355.

Silberling, N. L., Jones, D. L., Blake, M. C., Jr., and Howell, D. G., 1987, Lithotectonic terrane map of the western conterminous United States: U.S. Geological Survey Miscellaneous Field Studies Map MF-1874-C, scale 1:2,500,000.

Silberman, M. L., Stewart, J. H., and McKee, E. H., 1976, Igneous activity, tectonics, and hydrothermal precious-metal mineralization in the Great Basin during Cenozoic time: Society of Mining Engineers Transactions, v. 260, p. 253–263.

Simpson, R. W., and Cox, A., 1977, Paleomagnetic evidence for tectonic rotation of the Oregon Coast Range: Geology, v. 5, p. 585–589.

Slemmons, D. B., 1966, Cenozoic volcanism of the central Sierra Nevada, California, *in* Bailey, E. H., ed., Geology of northern California: California Division of Mines and Geology Bulletin 190, p. 199–208.

Sliter, W. V., McLaughlin, R. J., Keller, G., and Evitt, W. R., 1986, Paleogene accretion of Upper Cretaceous oceanic limestone in northern California: Geology, v. 14, p. 350–353.

Smedes, H. W., 1962, Lowland Creek Volcanics, an upper Oligocene formation near Butte, Montana: Journal of Geology, v. 70, p. 255–266.

Smedes, H. W., and Prostka, H. J., 1972, Stratigraphic framework of the Absaroka Volcanic Supergroup in the Yellowstone National Park: U.S. Geological Survey Professional Paper 729-C, 33 p.

Smedes, H. W., and Thomas, H. H., 1965, Reassignment of the Lowland Creek Volcanics to Eocene age: Journal of Geology, v. 73, p. 508–510.

Smith, G. I., 1979, Subsurface stratigraphy and geochemistry of late Quaternary evaporites, Searles Lake, California: U.S. Geological Survey Professional Paper 1043, 130 p.

Smith, J. F., Jr., and Ketner, K. B., 1976, Stratigraphy of post-Paleozoic rocks and summary of resources in the Carlin–Pinon Range area, Nevada: U.S. Geological Survey Professional Paper 867-B, 48 p.

—— , 1977, Tectonic events since early Paleozoic in the Carlin–Pinon Range area, Nevada: U.S. Geological Survey Professional Paper 867-C, 19 p.

Smith, P. B., 1970, New evidence for Pliocene marine embayment along the lower Colorado River area, California and Arizona: Geological Society of America Bulletin, v. 81, p. 1411–1420.

Smith, R. B., 1978, Seismicity, crustal structure, and intraplate tectonics of the interior of the western Cordillera, *in* Smith, R. B., and Eaton, G. P., eds., Cenozoic tectonics and regional geophysics of the western Cordillera: Geological Society of America Memoir 152, p. 111–144.

Smith, R. B., and Christiansen, R. L., 1980, Yellowstone Park as a window on the Earth's interior: Scientific American, v. 242, no. 2, p. 84–95.

Smith, R. B., and Sbar, M. L., 1974, Contemporary tectonics and seismicity of the western United States with emphasis on the intermountain seismic belt: Geological Society of America Bulletin, v. 85, p. 1205–1218.

Smith, R. B., Shuey, R. T., Pelton, J. R., and Bailey, J. P., 1977, Yellowstone hot spot; Contemporary tectonics and crustal properties from earthquake and aeromagnetic data: Journal of Geophysical Research, v. 82, p. 3665–3676.

Smith, R. B., Nagy, W. C., Julander, K. A., Viveiros, J. J., Barker, C. A., and Gants, D. G., 1989, Geophysics and tectonic framework of the eastern Basin and Range–Colorado Plateau–Rocky Mountain transition, *in* Pakiser, L. C., and Mooney, W. D., eds., Geophysical framework of the continental United States: Geological Society of America Memoir 172, p. 205–233.

Smith, R. L., and Bailey, R. A., 1966, The Bandelier Tuff; A study of ash-flow eruption cycles from zoned magma chambers: Bulletin Volcanologique, v. 34, p. 83–104.

—— , 1968, Resurgent cauldrons, *in* Coats, R. R., Hay, R. L., and Anderson, C. A., eds., Studies in volcanology; A memoir in honor of Howel Williams: Geological Society of America Memoir 116, p. 613–662.

Smith, R. L., and Luedke, R. G., 1984, Potentially active volcanic lineaments and loci in western conterminous United States, *in* Explosive volcanism; Inception, evolution, and hazards: Washington, D.C., National Academy of Sciences, p. 47–66.

Smith, R. L., Bailey, R. A., and Ross, C. S., 1970, Geologic map of the Jemez Mountains, New Mexico: U.S. Geological Survey Miscellaneous Geologic Investigations Map I-571, scale 1:125,000.

Snavely, P. D., Jr., 1983, Peripheral rocks; Tertiary geology of the northwestern part of the Olympic Peninsula, Washington, *in* Muller, J. E., Snavely, P. D., Jr., and Tabor, R. W., eds., The Tertiary Olympic terrane, southwest Vancouver Island and northwestern Washington: Geological Association of Canada, Mineralogical Association of Canada and Canadian Geophysical Union, Joint Annual Meeting, Field Trip no. 12, p. 6–31.

—— , 1987, Tertiary geologic framework, neotectonics, and petroleum potential of the Oregon-Washington continental margin, *in* Scholl, D. W., Grantz, A., and Vedder, J. G., eds., Geology and resource potential of the continental margin of western North America and adjacent ocean basins; Beaufort Sea to Baja California: Circum-Pacific Council for Energy and Mineral Resources Earth Science Series, no. 6, p. 305–335.

Snavely, P. D., Jr., and MacLeod, N. S., 1974, Yachats Basalt; An upper Eocene differentiated volcanic sequence in the Oregon Coast Range: U.S. Geological Survey Journal of Research, v. 2, p. 395–403.

Snavely, P. D., Jr., and Wagner, H. C., 1961, Differentiated gabbroic sills and associated alkalic rocks in the central part of the Oregon Coast Range, Oregon: U.S. Geological Survey Professional Paper 424-D, p. D156–D161.

—— , 1963, Tertiary geologic history of western Oregon and Washington: Washington Division of Mines and Geology Report of Investigations 22, 25 p.

Snavely, P. D., Jr., Rau, W. W., Hoover, L., and Roberts, A. E., 1951, McIntosh Formation, Centralia-Chehalis coal district, Washington: American Association of Petroleum Geologists Bulletin, v. 35, p. 1052–1061.

Snavely, P. D., Jr., Brown, R. D., Jr., Roberts, A. E., and Rau, W. W., 1958, Geology and coal resources of the Centralia-Chehalis district, Washington: U.S. Geological Survey Bulletin 1053, 159 p.

Snavely, P. D., Jr., Wagner, H. C., and MacLeod, N. S., 1964, Rhythmic-bedded eugeosynclinal deposits of the Tyee Formation, Oregon Coast Range: Kansas Geological Survey Bulletin 169, p. 461–480.

Snavely, P. D., Jr., MacLeod, N. S., and Wagner, H. C., 1973, Miocene tholeiitic basalts of coastal Oregon and Washington and their relations to coeval basalts of the Columbia Plateau: Geological Society of America Bulletin, v. 84, p. 387–424.

Snavely, P. D., Jr., MacLeod, N. S., Wagner, H. C., and Rau, W. W., 1976a, Geologic map of the Yaquina and Toledo Quadrangles, Lincoln County, Oregon: U.S. Geological Survey Miscellaneous Investigations Series, Map I-867, scale 1:62,500.

—— , 1976b, Geologic map of the Cape Foulweather and Euchre Mountain Quadrangles, Lincoln County, Oregon: U.S. Geological Survey Miscellaneous Investigations Series, Map I-868, scale 1:62,500.

Snavely, P. D., Jr., Pearl, J. E., and Lander, D. L., 1977, Interim report on petroleum resources potential and geologic hazards in the outer continental shelf; Oregon and Washington Tertiary province: U.S. Geological Survey Open-File Report 77–282, 64 p.

Snavely, P. D., Jr., Niem, A. R., and Pearl, J. E., 1978, Twin River Group (upper Eocene to lower Miocene); defined to include the Hoko River, Makah, and Pysht Formations, Clallam County, Washington: U.S. Geological Survey Bulletin 1457-A, p. 111–120.

Snoke, A. W., 1980, The transition from infrastructure to suprastructure in the northern Ruby Mountains, Nevada, *in* Crittenden, M. D., Jr., Coney, P. J., and Davis, G. H., eds., Cordilleran metamorphic core complexes: Geological Society of America Memoir 153, p. 287–334.

Solomon, B. J., McKee, E. H., and Andersen, D. W., 1979, Stratigraphy and depositional environments of Paleogene rocks near Elko, Nevada, *in* Armentrout, J. M., Cole, M. R., and TerBest, H., Jr., eds., Cenozoic paleogeography of the western United States: Society of Economic Paleontologists and Mineralogists, Pacific Section, Pacific Coast Paleogeography Symposium 3, p. 75–88.

Souther, J. G., 1977, Volcanism and tectonic environments in the Canadian Cordillera; a second look, *in* Baragar, W.R.A., Coleman, L. C., and Hall, J. M., eds., Volcanic regimes in Canada: Geological Association of Canada Special Paper 16, p. 3–24.

Spencer, J. E., 1984, Role of tectonic denudation in warping and uplift of low-angle faults: Geology, v. 12, p. 95–98.

Spencer, P. K., 1983, Age and paleoecology of marine siltstones exposed on the Bolton Peninsula near Quilcene, Jefferson County, Washington, *in* LaRue, D. K., and Steel, R. J., eds., Cenozoic marine sedimentation, Pacific margin, U.S.A.: Society of Economic Paleontologists and Mineralogists, Pacific Section, p. 197–203.

Spittler, T. E., and Arthur, M. A., 1982, The lower Miocene Diligencia Formation of the Orocopia Mountains, southern California; Stratigraphy, petrology, sedimentology and structure, *in* Ingersoll, R. V., and Woodburne, M. O., eds., Cenozoic nonmarine deposits of California and Arizona: Los Angeles, Society of Economic Paleontologists and Mineralogists, Pacific Section, p. 83–99.

Squires, R. L., and Filewicz, M. V., eds., 1983, Cenozoic geology of the Simi Valley area, southern California: Society of Economic Paleontologists and Mineralogists, Pacific Section, 266 p.

Staatz, M. H., and Carr, W. J., 1964, Geology and mineral deposits of the Thomas and Dugway Ranges, Juab and Tooele Counties, Utah: U.S. Geological Survey Professional Paper 415, 188 p.

Stanley, K. O., and Surdam, R. C., 1984, The role of wrench fault tectonics and relative changes of sea level on deposition of upper Miocene–Pliocene Pismo Formation, Pismo syncline, California, *in* Surdam, R. C., ed., A guidebook to the stratigraphic, tectonic, thermal, and diagenetic histories of the Monterey Formation, Pismo and Huasna basin, California: Society of Economic Paleontologists and Mineralogists Guidebook No. 2, August 13–16, 1984, San Jose, California, p. 21–37.

Steven, T. A., 1975, Middle Tertiary volcanic field in the Southern Rocky Mountains, *in* Curtis, B. F., ed., Cenozoic history of the Southern Rocky Mountains: Geological Society of America Memoir 144, p. 75–94.

Steven, T. A., and Lipman, P. W., 1976, Calderas of the San Juan volcanic field, southwestern Colorado: U.S. Geological Survey Professional Paper 958, 35 p.

Steven, T. A., Cunningham, C. G., Naeser, C. W., and Mehnert, H. H., 1979, Revised stratigraphy and radiometric ages of volcanic rocks and mineral deposits in the Marysvale area, west-central Utah: U.S. Geological Survey Bulletin 1469, 40 p.

Steven, T. A., Rowley, P. D., and Cunningham, C. G., 1984a, Calderas of the Marysvale volcanic field, west-central Utah: Journal of Geophysical Research, v. 89, p. 8751–8764.

Stewart, J. H., 1967, Possible large right-lateral displacement along fault and shear zones in the Death Valley–Las Vegas area, California and Nevada: Geological Society of America Bulletin, v. 78, p. 131–142.

——, 1971, Basin and range structure; A system of horsts and grabens produced by deep-seated extension: Geological Society of America Bulletin, v. 82, p. 1019–1044.

——, 1978, Basin-range structure in western North America; A review, *in* Smith, R. B., and Eaton, G. P., eds., Cenozoic tectonics and regional geophysics of the Western Cordillera: Geological Society of America Memoir 152, p. 1–31.

——, 1980, Geology of Nevada: Nevada Bureau of Mines and Geology Special Publication 4, 136 p.

——, 1983b, Cenozoic structure and tectonics of the northern Basin and Range Province, California, Nevada, and Utah, *in* The role of heat in the development of energy and mineral resources in the northern Basin and Range Province: Geothermal Resources Council Special Report No. 13, p. 25–40.

——, 1988, Tectonics of the Walker Lane belt, western Great Basin; Mesozoic and Cenozoic deformation in a zone of shear, *in* Ernst, W. G., ed., Metamorphism and crustal evolution of the western United States; Rubey Volume VII: Englewood Cliffs, New Jersey, Prentice-Hall, p. 683–713.

Stewart, J. H., and Carlson, J. E., 1976, Cenozoic rocks of Nevada; Four maps and a brief description of distribution, lithology, age, and centers of volcanism: Nevada Bureau of Mines and Geology Map 52, scale 1:1,000,000.

Stewart, J. H., and Diamond, D. S., 1990, Changing patterns of extensional tectonics; Overprinting of the basin of the middle and upper Miocene Esmeralda Formation in western Nevada by younger structural basins, *in* Wernicke, B. P., ed., Basin and Range extensional tectonics near the latitude of Las Vegas, Nevada: Geological Society of America Memoir 176, p. 447–475.

Stewart, J. H., and McKee, E. H., 1977, Geology; Pt. I, *in* Geology and mineral deposits of Lander County, Nevada: Nevada Bureau of Mines and Geology Bulletin 88, p. 1–59.

Stewart, J. H., Albers, J. P., and Poole, F. G., 1968, Summary of regional evidence for right-lateral displacement in the western Great Basin: Geological society of America Bulletin, v. 79, p. 1407–1414.

Stewart, J. H., Moore, W. J., and Zietz, I., 1977, East-west patterns of Cenozoic igneous rocks, aeromagnetics anomalies, and mineral deposits, Nevada and Utah: Geological Society of America Bulletin, v. 88, p. 67–77.

Stewart, R. J., 1974, Zeolite facies metamorphism of sandstone in the western Olympic Peninsula, Washington: Geological Society of America Bulletin, v. 85, p. 1139–1142.

Stitt, L. T., 1986, Structural history of the San Gabriel fault and other Neogene structures of the central Transverse Ranges, California, *in* Ehlig, P. L., ed., Neotectonics and faulting in southern California: Guidebook and volumes prepared for the 82nd Annual Meeting of the Cordilleran Section of the Geological Society of America, Los Angeles, California, March 25–28, 1986, p. 43–102.

Stock, C., and Bode, F. D., 1935, Occurrence of lower Oligocene mammal-bearing beds near Death Valley, California: National Academy of Sciences Proceedings, v. 21, p. 571–579.

Stock, J., and Molnar, P., 1988, Uncertainties and implications of the Late Cretaceous and Tertiary position of North America relative to the Farallon, Kula, and Pacific Plates: Tectonics, v. 7, p. 1339–1384.

Stormer, J. C., Jr., 1972a, Ages and nature of volcanic activity on the southern High Plains, New Mexico and Colorado: Geological Society of America Bulletin, v. 83, p. 2443–2448.

Stuart, C. J., 1979, Lithofacies and origin of the San Onofre Breccia, coastal southern California, *in* Stuart, C. J., ed., A guidebook to Miocene lithofacies and depositional environments, coastal southern California and northwestern Baja California: Society of Economic Paleontologists and Mineralogists, Pacific Section, p. 25–42.

Stuckless, J. S., and Sheridan, M. F., 1971, Tertiary volcanic stratigraphy in the Goldfield and Superstition Mountains, Arizona: Geological Society of America Bulletin, v. 82, p. 3235–3240.

Suneson, N. H., and Lucchitta, I., 1983, Origin of bimodal volcanism, southern Basin and Range Province, west-central Arizona: Geological Society of America Bulletin, v. 94, p. 1005–1019.

Surdam, R. C., and Stanley, K. O., 1984, Stratigraphic and sedimentologic framework of the Monterey Formation, Pismo syncline, California, *in* Surdam, R. C., ed., A guidebook to the stratigraphic, tectonic, thermal, and diagenetic histories of the Monterey Formation, Pismo and Huasna basin, California: Society of Economic Paleontologists and Mineralogists Guidebook No. 2, August 13–16, 1984, San Jose, California, p. 75–83.

Swanson, D. A., Wright, T. L., and Helz, R. T., 1975, Linear vent systems and estimated rates of magma production and eruption for the Yakima Basalt on the Columbia Plateau: American Journal of Science, v. 275, p. 877–905.

Swanson, D. A., Wright, T. L., Hooper, P. R., and Bentley, R. D., 1979a, Revisions in stratigraphic nomenclature of the Columbia River Basalt Group: U.S. Geological Survey Bulletin 1457-G, 59 p.

Swanson, D. A., and 7 others, 1979b, Reconnaissance geologic map of the Columbia River Basalt Group, Pullman and Walla Walla Quadrangles, southeast Washington and adjacent Idaho: U.S. Geological Survey Miscellaneous Geologic Investigations Map I-1139, scale 1:250,000.

Sylvester, A. G., and Smith, R. R., 1976, Tectonic transpression and basement-controlled deformation in San Andreas fault zone, Salton Trough, California: American Association of Petroleum Geologists Bulletin, v. 60, p. 2081–2102.

Tabor, R. W., 1972, Age of the Olympic metamorphism, Washington; K-Ar dating of low-grade metamorphic rocks: Geological Society of America Bulletin, v. 83, p. 1805–1816.

Tabor, R. W., and Cady, W. M., 1978a, The structure of the Olympic Mountains, Washington; Analysis of a subduction zone: U.S. Geological Survey Professional Paper 1033, 38 p.

——, 1978b, Geologic map of the Olympic Peninsula, Washington: U.S. Geological Survey Miscellaneous Investigations Series Map I-994, scale 1:125,000.

Tabor, R. W., and Crowder, D. F., 1969, On batholiths and volcanoes; Intrusion and eruption of late Cenozoic magmas in the Glacier Peak area, North Cascades, Washington: U.S. Geological Survey Professional Paper 604, 67 p.

Tabor, R. W., Frizzell, V. A., Jr., Vance, J. A., and Naeser, C. W., 1984, Ages and stratigraphy of lower and middle Tertiary sedimentary and volcanic rocks of the central Cascades, Washington; Applications to the tectonic history of the Straight Creek fault: Geological Society of America Bulletin, v. 95, p. 26–44.

Taylor, E. M., 1968, Roadside Geology, Santiam and McKenzie Pass highways, Oregon, *in* Dole, H. M., ed., Andesite Conference guidebook: Oregon Department of Geology and Mineral Industries Bulletin 62, p. 3–33.

——, 1981a, Central High Cascade roadside geology Bend, Sisters, McKenzie Pass, and Santiam Pass, Oregon, *in* Johnston, D. A., and Donnelly-Nolan, J., eds., Guides to some volcanic terranes in Washington, Idaho, Oregon, and northern California: U.S. Geological Survey Circular 838, p. 55–83.

——, 1990, Volcanic history and tectonic development of the central High Cascade Range, Oregon: Journal of Geophysical Research, v. 95, p. 19611–19622.

Taylor, R. B., 1975, Neogene tectonism in south-central Colorado, *in* Curtis, B. F., ed., Cenozoic history of the Southern Rocky Mountains: Geological Society of America Memoir 144, p. 211–226.

Taylor, R. B., Theobald, P. K., and Izett, G. A., 1968, Mid-Tertiary volcanism in the central Front Range, Colorado, *in* Epis, R. C., ed., Cenozoic volcanism in the Southern Rocky Mountains: Colorado School of Mines Quarterly, v. 63, no. 3, p. 39–50.

Tennyson, M. E., 1989, Pre-transform early Miocene extension in western California: Geology, v. 17, p. 792–796.

Thayer, T. P., 1937, Petrology of later Tertiary and Quaternary rocks of the north-central Cascade Mountains in Oregon: Geological Society of America Bulletin, v. 48, p. 1611–1652.

Thompson, G. A., and Burke, D. B., 1974, Regional geophysics of the Basin and Range Province: Annual Reviews of Earth and Planetary Sciences, v. 2, p. 213–238.

Thompson, G. A., and White, D. E., 1964, Regional geology of the Steamboat Springs area, Washoe County, Nevada: U.S. Geological Survey Professional Paper 458-A, 52 p.

Todd, V. R., 1983, Late Miocene displacement of pre-Tertiary and Tertiary rocks in the Matlin Mountains, northwestern Utah, *in* Miller, D. M., Todd, V. R., and Howard, K. A., eds., Tectonic and stratigraphic studies in the eastern Great Basin: Geological Society of America Memoir 157, p. 239–270.

Tolan, T. L., Beeson, M. H., Anderson, J. L., Fecht, K. R., and Swanson, D. A., 1989, Revisions to the estimates of the areal extent and volume of the Columbia River Basalt Group, *in* Reidel, S. P., and Hooper, P. R., eds., Volcanism and tectonism in the Columbia River flood-basalt Province: Geological Society of America Special Paper 239, p. 1–20.

Trimble, D. E., 1963, Geology of Portland, Oregon, and adjacent areas: U.S. Geological Survey Bulletin 1119, 119 p.

—— , 1976, Geology of the Michaud and Pocatello Quadrangles, Bannock and Power Counties, Idaho: U.S. Geological Survey Bulletin 1400, 88 p.

Trimble, D. E., and Carr, W. J., 1976, Geology of the Rockland and Arbon Quadrangles, Power County, Idaho: U.S. Geological Survey Bulletin 1399, 115 p.

Turcotte, D. L., and McAdoo, D. C., 1979, Thermal subsidence and petroleum generation in the southwestern block of the Los Angeles basin: Journal of Geophysical Research, v. 84, p. 3460–3464.

Turner, D. L., 1970, Potassium-argon dating of Pacific Coast Miocene foraminiferal stages, *in* Bandy, O. L., ed., Radiometric dating and paleontologic zonation: Geological Society of America Special Paper 124, p. 91–129.

Tweto, O., 1979, The Rio Grande rift system in Colorado, *in* Riecker, R. E., ed., Rio Grande rift: Tectonics and magmatism: Washington, American Geophysical Union, p. 33–56.

Underwood, M. B., 1985, Sedimentology and hydrocarbon potential of Yager structural complex; Possible Paleogene source rocks in Eel River basin, northern California: American Association of Petroleum Geologists Bulletin, v. 69, p. 1088–1100.

Vedder, J. G., 1973, Geologic framework and correlation of Miocene rocks in the Caliente Range, *in* Fischer, P. J., ed., Sedimentary facies changes in Tertiary rocks; California Transverse and southern Coast Ranges: Society of Economic Paleontologists and Mineralogists, Pacific Section, Field Trip 2, 1973 Annual Meeting, p. 42–53.

—— , 1975, Juxtaposed Tertiary strata along the San Andreas fault in the Temblor and Caliente Ranges, California, *in* Crowell, J. C., ed., San Andreas fault in southern California: California Division of Mines and Geology Special Report 118, p. 234–240.

Vedder, J. G., and Brown, R. D., Jr., 1969, Geologic framework of the Santa Barbara Channel region: U.S. Geological Survey Professional Paper 679, p. 1–11.

Vedder, J. G., Howell, D. G., and McLean, H., 1983, Stratigraphy, sedimentation, and tectonic accretion of exotic terranes, southern Coast Ranges, California: American Association of Petroleum Geologists Memoir 34, p. 471–496.

Vokes, H. E., Snavely, P. D., Jr., and Myers, D. A., 1951, Geology of the southern and southwestern border areas, Willamette Valley, Oregon: U.S. Geological Survey Oil and Gas Investigations Map OM-110, scale 1:62,500.

Walker, G. W., 1977, Geologic map of Oregon east of the 121st meridian: U.S. Geological Survey Miscellaneous Geologic Investigations Map I-902, scale 1:500,000.

—— , 1990, Miocene and younger rocks of the Blue Mountains region, exclusive of the Columbia River Basalt Group and associated mafic flows, *in* Walker, G. W., ed., Geology of the Blue Mountains region of Oregon, Idaho, and Washington; Cenozoic geology of the Blue Mountains region: U.S. Geological Survey Professional Paper 1437, p. 101–118.

Walker, G. W., and Robinson, P. T., 1990a, Paleocene(?), Eocene, and Oligocene rocks of the Blue Mountains region, *in* Walker, G. W., ed., Geology of the Blue Mountains region of Oregon, Idaho, and Washington; Cenozoic geology of the Blue Mountains region: U.S. Geological Survey Professional Paper 1437, p. 13–27.

—— , 1990b, Cenozoic tectonism and volcanism of the Blue Mountains region, *in* Walker, G. W., ed., Geology of the Blue Mountains region of Oregon, Idaho, and Washington; Cenozoic geology of the Blue Mountains region: U.S. Geological Survey Professional Paper 1437, p. 119–135.

Wallace, R. E., 1984b, Patterns and timing of late Quaternary faulting in the Great Basin province and relation to some regional tectonic features: Journal of Geophysical Research, v. 89, p. 5763–5769.

Waters, A. C., 1962, Basalt magma types and their tectonic associations; Pacific northwest of the United States, *in* Macdonald, G. A., and Kuno, H., eds., The crust of the Pacific basin: American Geophysical Union Monograph 6, p. 158–170.

—— , 1973, The Columbia River Gorge; Basalt stratigraphy, ancient lava dams, and landslide dams, *in* Bealieu, J. D., ed., Geologic field trips in northern Oregon and southern Washington: Oregon Department of Geology and Mineral Industries Bulletin 77, p. 133–162.

Watkins, N. D., and Baksi, A. K., 1974, Magnetostratigraphy and oroclinal folding of the Columbia River, Steens and Owyhee Basalts in Oregon, Washington, and Idaho: American Journal of Science, v. 274, p. 148–189.

Weaver, C. S., and Michaelson, C. A., 1985, Seismicity and volcanism in the Pacific Northwest; Evidence for the segmentation of the Juan de Fuca Plate: Geophysical Research Letters, v. 12, p. 215–218.

Weaver, D. W., 1969, Geology of the northern Channel Islands: American Association of Petroleum Geologists and Society of Economic Paleontologists and Mineralogists, Pacific Sections, Special Publication, 200 p.

Webb, G. W., 1981, Stevens and earlier Miocene turbidite sandstones, southern San Joaquin Valley, California: American Association of Petroleum Geologists Bulletin, v. 65, p. 438–465.

Weigand, P. W., 1982, Middle Cenozoic volcanism of the western Transverse Ranges, *in* Fife, D. L., and Minch, J. A., eds., Geology and mineral wealth of the California Transverse Ranges (Mason Hill volume): Santa Ana, South Coast Geological Society, p. 170–188.

Wells, R. E., 1981, Geologic map of the eastern Willapa Hills, Cowlitz, Lewis, Pacific, and Wahkiakum Counties, Washington: U.S. Geological Survey Open-File Map 81-674, scale 1:62,500.

—— , 1984, Paleomagnetic constraints on the interpretation of early Cenozoic Pacific Northwest paleogeography, *in* Nilsen, T. H., ed., Geology of the Upper Cretaceous Hornbrook Formation, Oregon and California: Society of Economic Paleontologists and Mineralogists, Pacific Section, v. 42, p. 231–237.

—— , 1989, Geologic map of the Cape Disappointment–Naselle River area, Pacific and Wahkiakum Counties, Washington: U.S. Geological Survey Miscellaneous Investigations Series Map I-1832, scale 1:62,500.

—— , 1990, Paleomagnetic rotations and the Cenozoic tectonics of the Cascade arc, Washington, Oregon, and California: Journal of Geophysical Research, v. 95, p. 19409–19417.

Wells, R. E., and Coe, R. S., 1985, Paleomagnetism and geology of Eocene volcanic rocks of southwest Washington, implications for mechanisms of tectonic rotation: Journal of Geophysical Research, v. 90, p. 1925–1947.

Wells, R. E., and Heller, P. L., 1988, The relative contribution of accretion, shear, and extension to Cenozoic tectonic rotation in the Pacific Northwest: Geological Society of America Bulletin, v. 100, p. 325–338.

Wells, R. E., Niem, A. R., MacLeod, N. S., Snavely, P. D., Jr., and Niem, W. A., 1983, Preliminary geologic map of the west half of the Vancouver (Wa.-Ore.) 1 degree × 2 degree Quadrangle, Oregon: U.S. Geological Survey Open-File Report 83-591, scale 1:250,000.

Wells, R. E., Engebretson, D. C., Snavely, P. D., Jr.,and Coe, R. S., 1984, Cenozoic plate motions and the volcano-tectonic evolution of western Oregon and Washington: Tectonics, v. 3, p. 275–294.

Wentworth, C. M., Yerkes, R. F., and Allen, C. R., 1971, Geologic setting and activity of faults in the San Fernando area, California: U.S. Geological Survey Profesional Paper 733, p. 6–16.

Wernicke, B., 1981, Low-angle normal faults in the Basin and Range Province; nappe tectonics in an extending orogen: Nature, v. 291, p. 645–648.

——, 1985, Uniform-sense normal simple shear of the continental lithosphere: Canadian Journal of Earth Sciences, v. 22, p. 108–125.

Wernicke, B., Spencer, J. E., Burchfiel, B. C., and Guth, P. L., 1982, Magnitude of crustal extension in the southern Great Basin: Geology, v. 10, p. 499–502.

Wernicke, B., Walker, D. J., and Beaufait, M. S., 1985, Structural discordance between Neogene detachments and frontal Sevier thrusts, central Mormon Mountains, southern Nevada: Tectonics, v. 4, p. 213–246.

Wernicke, B. P., Hodges, K. V., and Walker, J. D., 1986, Geological setting of the Tucki Mountain area, Death Valley National Monument, California, *in* Dunne, G. C., ed., Mesozoic and Cenozoic structural evolution of selected areas, east-central California, guidebook and volume, trips 2 and 14: Los Angeles, Cordilleran Section, Geological Society of America, p. 67–80.

Wernicke, B. P., Christiansen, R. L., England, P. C., and Sonder, L. J., 1987, Tectonomagmatic evolution of Cenozoic extension in the North American Cordillera, *in* Coward, M. P., Dewey, J. F., and Hancock, P. L., eds., Continental extensional tectonics: Geological Society [London] Special Publication 28, p. 203–221.

White, C. M., 1980, Geology and geochemistry of Mt. Hood volcano: Oregon Department of Geology and Mineral Industries Special Paper 8, 26 p.

White, D. E., and Heropoulos, C., 1983, Active and fossil hydrothermal-convection systems of the Great Basin, *in* The role of heat in the development of energy and mineral resources in the northern Basin and Range Province: Geothermal Resources Council Special Report No. 13, p. 41–53.

White, D. E., Thompson, G. A., and Sandberg, C. H., 1964, Rocks, structure, and geologic history of Steamboat Springs thermal area, Washoe County, Nevada: U.S. Geological Survey Professional Paper 458-B, 63 p.

Williams, H., 1932, Geology of the Lassen Volcanic National Park, California: University of California Department of Geological Sciences Bulletin, v. 21, p. 195–385.

——, 1942, The geology of Crater Lake National Park, Oregon, with a reconnaissance of the Cascade Range south to Mount Shasta: Carnegie Institution of Washington Publication 540, 162 p.

Williams, P. L., Covington, H. R., and Pierce, K. L., 1982, Cenozoic stratigraphy and tectonic evolution of the Raft River basin, Idaho, *in* Bonnichsen, B., and Breckenridge, R. M., eds., Cenozoic geology of Idaho: Idaho Bureau of Mines and Geology Bulletin 26, p. 491–504.

Winfrey, W. M., Jr., 1960, Stratigraphy, correlation, and oil potential of the Sheep Pass Formation, east-central Nevada: Intermountain Association of Petroleum Geologists 11th Annual Field Conference, p. 126–133.

Winterer, E. L., and Durham, D. L., 1962, Geology of the southeastern Ventura basin, Los Angeles County, California: U.S. Geological Survey Professional Paper 334-H, p. 275–366.

Wolfe, E. W., 1984, The volcanic landscape of the San Francisco volcanic field, *in* Smiley, T. L., and others, eds., Development of the Arizona landscape; The geologic story: Lanham, Maryland, University Press of America, p. 111–136.

Wolfe, J. A., 1981, A chronologic framework for Cenozoic megafossil floras of northwestern North America and its relation to marine geochronology: Geological Society of America Special Paper 184, p. 38–47.

Woodburne, M. O., 1975, Cenozoic stratigraphy of the Transverse Ranges and adjacent areas, southern California: Geological Society of America Special Paper 162, 91 p.

Woodburne, M. O., Miller, S. T., and Tedford, R. H., 1982, Stratigraphy and geochronology of Miocene strata in the central Mojave Desert, California, *in* Cooper, J. D., ed., Geologic excursions in the California desert: Geological Society of America, Cordilleran Section, 78th annual meeting, Anaheim, California, p. 47–64.

Woodford, A. O., 1925, The San Onofre Breccia: California University Publications, Bulletin of the Department of Geological Sciences, v. 15, p. 159–280.

Woodring, W. P.,and Bramlette, M. N., 1950, Geology and paleontology of the Santa Maria district, California: U.S. Geological Survey Professional Paper 222, 185 p.

Woodring, W. P., Bramlette, M. N., and Kew, W. S. W., 1946, Geology and paleontology of the Palos Verdes Hills, California: U.S. Geological Survey Professional Paper 207, 145 p.

Wright, L. A., and Troxel, B. W., 1967, Limitations on right-lateral, strike-slip displacement, Death Valley and Furnace Creek fault zones, California: Geological Society of America Bulletin, v. 78, p. 933–950.

——, 1973, Shallow-fault interpretation of Basin and Range structure, southwestern Great Basin, *in* de Jong, K. A., and Scholten, R., eds., Gravity and tectonics: New York, Wiley, p. 397–407.

Wright, L. A., Otton, J. K., and Troxel, B. W., 1974, Turtleback surfaces of Death Valley viewed as phenomena of extensional tectonics: Geology, v. 2, p. 53–54.

Wright, T. L., Grolier, M. J., and Swanson, D. A., 1973, Chemical variation related to the stratigraphy of the Columbia River Basalt: Geological Society of America Bulletin, v. 85, p. 371–386.

Wust, S. L., 1986, Regional correlation of extension directions in Cordilleran metamorphic core complexes: Geology, v. 14, p. 828–830.

Yeats, R. S., 1968, Rifting and rafting in the southern California borderland, *in* Dickinson, W. R., and Grantz, A., ed., Proceedings, Conference on Geologic Problems of San Andreas Fault System: Stanford University Publications Geological Sciences, v. 11, p. 307–322.

——, 1973, Newport-Inglewood fault zone, Los Angeles basin, California: American Association of Petroleum Geologists Bulletin, v. 57, p. 117–135.

——, 1976, Neogene tectonics of the central Ventura basin, California, *in* Fritsche, A. E., Ter Best, H., Jr., and Wornardt, W. W., eds., The Neogene Symposium: Society of Economic Paleontologists and Mineralogists, Pacific Section, p. 19–32.

——, 1979, Stratigraphy and paleogeography of the Santa Susana fault zone, Transverse Ranges, California, *in* Armentrout, J. M., Cole, M. R., and Ter Best, H., eds., Cenozoic paleogeography of the western United States: Society of Economic Paleontologists and Mineralogists, Pacific Section, Pacific Coast Paleogeography Symposium 3, p. 191–204.

——, 1983, Large-scale Quaternary detachments in Ventura basin, southern California: Journal of Geophysical Research, v. 88, p. 569–583.

——, 1987, Changing tectonic styles in Cenozoic basins of southern California, *in* Ingersoll, R. V., and Ernst, W. G., eds., Cenozoic basin development of coastal California; Rubey Vol. VI: Englewood Cliffs, New Jersey, Prentice-Hall, p. 284–298.

Yeats, R. S., and Beall, J. M., 1991, Stratigraphic controls of oil fields in the Los Angeles basin, a guide to migration history, *in* Biddle, K. T., ed., Active margin basins: American Association of Petroleum Geologists Memoir 52, p. 221–237.

Yeats, R. S., Cole, M. R., Merschat, W. R., and Parsley, R. M., 1974, Poway fan and submarine cone and rifting of the inner southern California borderland: Geological Society of America Bulletin, v. 85, p. 293–302.

Yeats, R. S., Calhoun, J. A., Nevins, B. B., Schwing, H. F., and Spitz, H. M., 1989, The Russell fault; An early strike-slip fault of the California Coast Ranges: American Association of Petroleum Geologists Bulletin, v. 73, p. 1089–1102.

Yerkes, R. F., 1972, Geology and oil resources of the western Puente Hills area, southern California: U.S. Geological Survey Professional Paper 420-C, 63 p.

Yerkes, R. F., and Campbell, R. H., 1979, Stratigraphic nomenclature of the central Santa Monica Mountains, Los Angeles County, California: U.S. Geological Survey Bulletin 1457-E, 31 p.

Yerkes, R. F., McCulloh, T. H., Schoellhamer, J. E. and Vedder, J. G., 1965, Geology of the Los Angeles basin; an introduction: U.S. Geological Survey Professional Paper 420-A, 57 p.

Young, R. A., 1979, Laramide deformation, erosion and plutonism along the southwestern margin of the Colorado Plateau: Tectonophysics, v. 61, p. 25–47.

Young, R. A., and Brennan, W. J., 1974, The Peach Springs Tuff; Its bearing on structural evolution of the Colorado Plateau and development of Cenozoic drainage in Mohave County, Arizona: Geological Society of America Bulletin, v. 85, p. 83–90.

Young, R. A., and McKee, E. H., 1978, Early and middle Cenozoic drainage and erosion in west-central Arizona: Geological Society of America Bulletin, v. 89, p. 1745–1750.

Zen, E., 1975, Eocene volcanic rocks in the Pioneer Mountains, Montana: U.S. Geological Survey Professional Paper 975, 48 p.

Zoback, M. L., 1983, Structure and Cenozoic tectonism along the Wasatch fault zone, Utah, *in* Miller, D. M., Todd, V. R., and Howard, K. A., eds., Tectonic and stratigraphic studies in the eastern Great Basin: Geological Society of America Memoir 157, p. 3–27.

—— , 1989, State of stress and modern deformation of the northern Basin and Range Province: Journal of Geophysical Research, v. 94, p. 7105–7128.

Zoback, M. L., and Thompson, G. A., 1978, Basin and Range rifting in northern Nevada; Clues from a mid-Miocene rift and its subsequent offsets: Geology, v. 6, p. 111–116.

Zoback, M. L., Anderson, R. E., and Thompson, G. A., 1981, Cainozoic evolution of the state of stress and style of tectonism of the Basin and Range Province of the western United States: Philosophical Transactions of the Royal Society of London, pt. A, v. 300, p. 407–434.

MANUSCRIPT ACCEPTED BY THE SOCIETY NOVEMBER 27, 1991

Printed in U.S.A.

The Geology of North America
Vol. G-3, The Cordilleran Orogen: Conterminous U.S.
The Geological Society of America, 1992

Chapter 8

Tectonic overview of the Cordilleran orogen in the western United States

B. C. Burchfiel
Department of Earth, Atmospheric, and Planetary Sciences, Massachusetts Institute of Technology, Cambridge, Massachusetts 02139
Darrel S. Cowan
Department of Geological Sciences, University of Washington, Seattle, Washington 98195
Gregory A. Davis
Department of Geological Sciences, University of Southern California, Los Angeles, California 90089-0740

INTRODUCTION

The Cordilleran orogen, with an evolution that has spanned the entire Phanerozoic, is one of the longest lived orogenic belts on the planet. A reason for its long history is that since at least Cambrian time there has been a large region underlain by oceanic lithosphere to the west of continental North America, and such continent-ocean lithospheric boundaries are tectonically unstable. Interaction between these two types of lithosphere resulted in intermittent tectonic activity during Paleozoic time, and nearly continuous tectonic activity during the Mesozoic and Cenozoic. Because Pacific oceanic plates are still present west of North America, the Cordilleran orogen continues to evolve. Plate-boundary interactions along the western boundary of North America have included diverse types of convergent, divergent, and transform activity as well as combinations of these plate-boundary interactions. Fortunately, the preservation of a large tract of Mesozoic and Cenozoic oceanic lithosphere west of North America enables the evolution of Pacific Basin plates to be reconstructed for Cenozoic time and, with less certainty, back to middle Mesozoic time (Engebretson and others, 1985; Stock and Molnar, 1988). Such reconstructions permit correlation between continental geology and lithospheric plate interactions to a degree uncommon in most post-Paleozoic orogens, and provide a testing ground for relating oceanic plate tectonics to continental deformation and evolution.

The Cordilleran orogen consists of western and eastern regions. The former is underlain to a large extent by accreted terranes of mainly oceanic affinity that are separated by important, but often cryptic tectonic boundaries from the eastern region, the basement of which has been part of North America since Precambrian time (Fig. 1). A few of the western accreted terranes are "native" (Gray, 1986) and can be tied directly to North America, whereas most have more uncertain origins. Some, although we believe not many, appear to be far traveled and exotic to North America. All of these terranes form a composite accretionary belt that makes up from 20 to 50 percent of the width of the Cordillera at different latitudes. Accretion of tectonic elements to North America began in middle Paleozoic time and has continued to the present. Older fragments of lithosphere accreted to western North America during ancient plate boundary activity were modified during younger accretion of lithospheric fragments and reworked in nonaccretionary plate-boundary systems.

In this chapter we summarize and interpret key elements of the tectonic history of this complex orogen within the western United States. Our temporal coverage of orogenic development is not even handed. We emphasize the Mesozoic and older history of the mountain system, primarily because of the extensive coverage of its younger history in other chapters in this volume. Nevertheless, we do treat selected aspects of Cenozoic tectonics, most importantly pre–Basin and Range extensional tectonics throughout the orogen and accretionary tectonics in coastal areas.

PRECAMBRIAN EVOLUTION

Middle Proterozoic events and units

Archean to Middle Proterozoic rocks forming the cyrstalline basement of the North American craton underlie the eastern part of the Cordilleran orogen. These rocks belong to tectonic elements that evolved and were accreted together during several episodes, the last of which culminated in deformation and metamorphism at about 1,700 Ma (Fig. 2; Hoffman, 1988). The structural grain in western portions of the Precambrian North American craton generally trends east or northeast at a high-angle to the younger Cordillera. The pre-Cordilleran trends can be

Burchfiel, B. C., Cowan, D. S., and Davis, G. A., 1992, Tectonic overview of the Cordilleran orogen in the western United States, *in* Burchfiel, B. C., Lipman, P. W., and Zoback, M. L., eds., The Cordilleran Orogen: Conterminous U.S.: Boulder, Colorado, Geological Society of America, The Geology of North America, v. G-3.

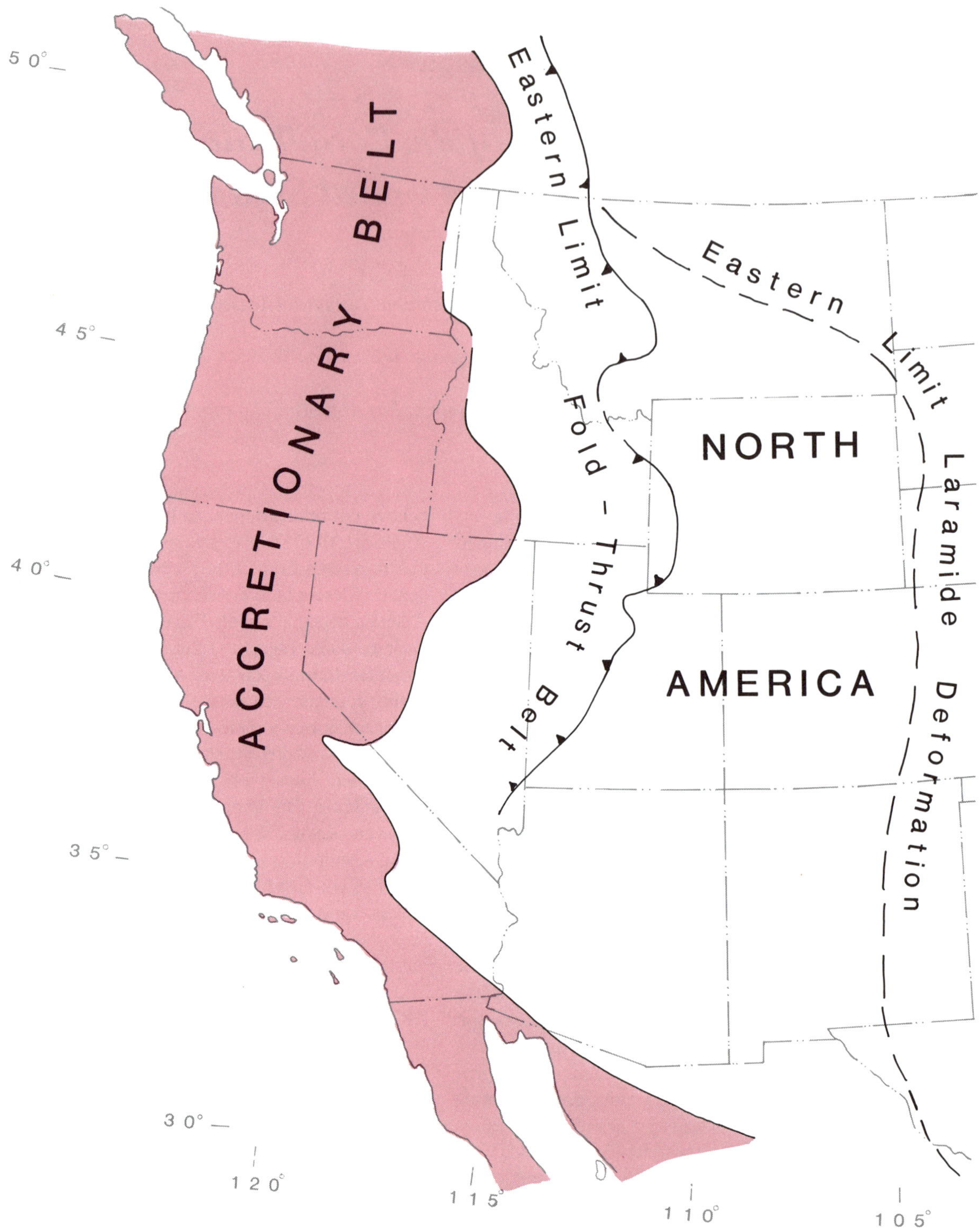

Figure 1. First-order subdivisions of the Cordilleran orogen. An accretionary belt (light red) consisting of native and exotic fragments of oceanic, island-arc, and continental elements that have been added to the North American craton forms the western part of the orogen. The eastern limits of North American rocks deformed in Sevier foreland fold and thrust belt and in the Laramide foreland belt are shown for reference and illustrate the eastern extent of Cordilleran deformation of the craton.

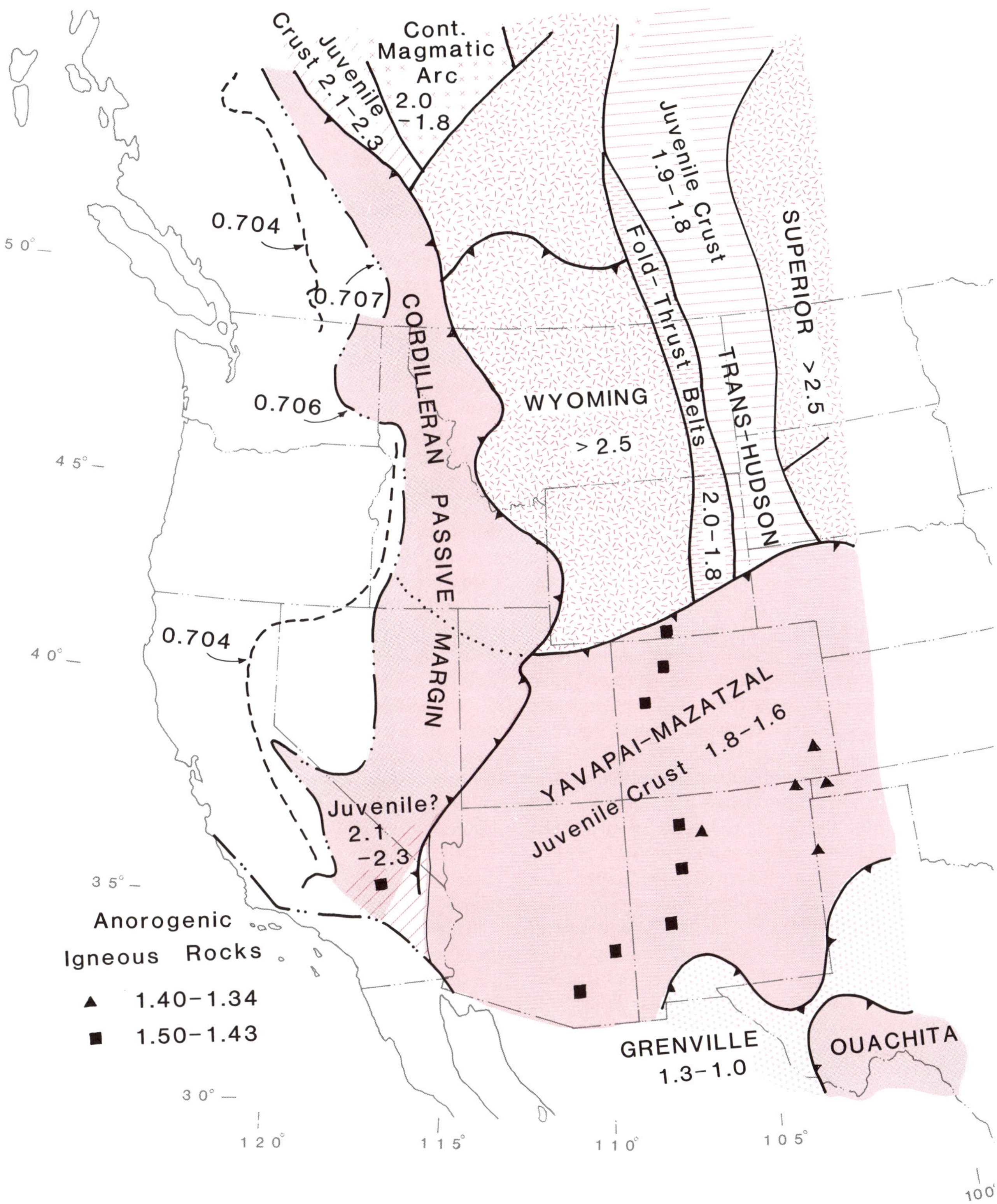

Figure 2. Ages of pre-Cordilleran Precambrian orogenic belts that form the western part of the North American craton and underlie eastern parts of the passive margin (gray) of the Cordilleran orogen (modified from Hoffman, 1988). The western extent of North American crystalline basement is generally taken to be near the Sri = 0.706. Sri = 0.704 and 0.707 (Canada) are also shown. The eastern limit of the Sevier thrust belt, which generally follows the eastern boundary of the Paleozoic passive margin, is shown for reference.

followed on gravity and magnetic maps well into the eastern part of the Cordillera in southern Canada, and are directly observable in isolated inliers of older basement rocks within the eastern Cordillera of the United States (Fig. 3). In contrast, early Cordilleran continental margin elements trend generally north (all directions are given in present-day coordinates), an indication that pre-Cordilleran orogenic belts were first truncated during rifting along the west side of North America and then transported away (cf., Sears and Price, 1978).

Late Middle Proterozoic sedimentary rocks (P_1, P_2)[1] were deposited on the cratonal crystalline basement rocks in several areas now within the Cordilleran orogen, but the imprecise dating of all the Middle Proterozoic rocks precludes both their detailed correlations and Cordilleran-wide interpretations of their tectonic settings. These rocks were probably deposited during at least three intracontinental extensional events. The oldest and most extreme extensional event took place in Belt-Purcell time (ca., 1,500 to 1,300 Ma), and was followed by lesser extensional events in Pahrump, Unkar, and Apache-Troy time (ca., 1,150 to 1,070 Ma), and again in Late Proterozoic time (ca., 900 to 800 Ma). Convergent deformation occurred in the western Belt basin at about 1,300 Ma. A fourth extensional event in latest Proterozoic time led to the initiation of the throughgoing Cordilleran miogeocline and, in our opinion, to the inception of the Cordillera orogen. Deformational events that occurred prior to the formation of the miogeocline are hence regarded as pre-Cordilleran. The extensional pre-Cordilleran events were accompanied by the deposition of sequences of Middle Proterozoic sedimentary and igneous rocks of variable thickness (Fig. 3). These sequences commonly influenced the development of the younger Cordilleran orogen.

Structures and rocks formed during the pre-Cordilleran events are present mostly in the eastern part of the orogen, several hundred kilometers from the present continental margin. Their relations to the throughgoing continental margin of latest Precambrian age are characteristically obscured by the superposition of numerous younger orogenic and depositional events. Rocks deposited during these events comprise the Belt (and its equivalent in Canada, the Purcell), Uinta Mountains, Pahrump, Grand Canyon, and Apache Supergroups or Groups (Fig. 3; P_1). Their tectonic settings and relations to Middle and Late Proterozoic continental margin development and plate-boundary interaction remain among the critical unresolved problems of North American evolution.

Belt/Purcell Supergroup. Rocks of the Belt/Purcell Supergroup (Fig. 3; P_1) form a thick sedimentary wedge tapering from an inferred erosional pinchout against cratonal basement rocks in the northeast to sections more than 15 to 20 km thick in the west. Belt (United States)/Purcell (Canada) deposition took place during active faulting along the southeast side of the basin on east- (*McMannis, 1963*) and/or northwest-striking faults (*Schmidt and Garihan, 1986*), and along several east- to northeast-striking faults in southern Canada (McMechan and Price, 1982) and the United States (*Reynolds, 1984*). Sediment sources were the craton to the northeast, highlands on the southwest, and along active faults within the basin (Harrison, 1972; *Savage, 1973*). The original boundaries of the Belt basin are poorly known. Reynolds (*1984*) argues that most of the present eastern and southern boundaries of Belt rocks result from pre-Cambrian erosion and that the original basin was much larger. The basin has an apparent eastward salient, the Central Montana Trough or Helena Embayment, that was a site of more rapid sedimentation during Belt time as well as several times during the Phanerozoic. *Reynolds (1984)* suggests that the salient is not a Belt age trough, but a broad syncline of post-Belt and pre-Cambrian age. Nevertheless, the preserved distribution and facies of Belt strata suggest that the southeastern part of the basin formed a prominent salient into the craton. Its southwestern and western boundaries are virtually unknown because they have been modified or truncated by younger tectonic, igneous, and metamorphic activity.

The age of Belt Supergroup strata remains controversial. Belt strata rest unconformably on older metamorphic rocks that in places yield metamorphic ages of about 1,700 Ma (Young, 1984). Pre-Belt augen gneiss near present western limits of Belt exposure in the Priest River area, northeastern Washington, has yielded an age of 1,576 ± 13 Ma (*Evans and Fischer, 1986*). Mafic sills (Moyie) intrude the lower part of the Belt/Purcell Supergroup and have been dated at about 1,433 ± 10 Ma (U/Pb) by Zartman and others (*1982*). Younger ages have been reported for the middle and upper part of the Belt sequence. Mafic lava flows (Purcell lavas) and sills that occur in the middle part of the Belt have yielded ages of 1,075 to 1,200 Ma, but these ages may not be very reliable. *Harrison (1984)* reports unpublished ages that range from 1.237 to 0.818 Ga for these rocks. Rb/Sr ages of 1,165 ± 20 and 1,170 ± 20 Ma on glauconite from the upper part of the Belt are also reported in *Harrison (1984).* Diabase dikes and sills that are younger than Belt strata are about 750 to 830 Ma (*Harrison, 1977*). Thus, some workers have argued that the age of the Belt Supergroup ranges from about 1.5 Ga to 900 Ma (cf., *Obradovich and others, 1984*). Other workers have proposed that a deformational episode, the East Kootenay orogeny, ended deposition of the supergroup at about 1,200 to 1,300 Ma (McMechan and Price, 1982). In the central and western part of the Belt basin, strata metamorphosed and deformed during the East Kooteny orogeny were intruded by the post-tectonic Hellroaring Creek stock in southern Canada dated at 1,305 ± 52 Ma (Rb/Sr whole-rock isochron; *Ryan and Blenkinsop, 1971*). Elsewhere, metamorphosed Belt rocks and synkinematic granitic gneisses have yielded a wide variety of ages, but the most reliable range from 1,330 ± 45 Ma (*Obradovich and Peterman, 1968*) to about 1,377 to 1,368 Ma (*Evans and Fischer, 1986*). Paleomagnetic data for a Proterozoic polar wander path presented by Elston and McKee (1982) also suggest that the Belt Supergroup is entirely older than about 1,250 Ma. Thus, the best available data

[1]Designations following rock or tectonic units refer to the Tectonostratigraphic Map, Plate 1.

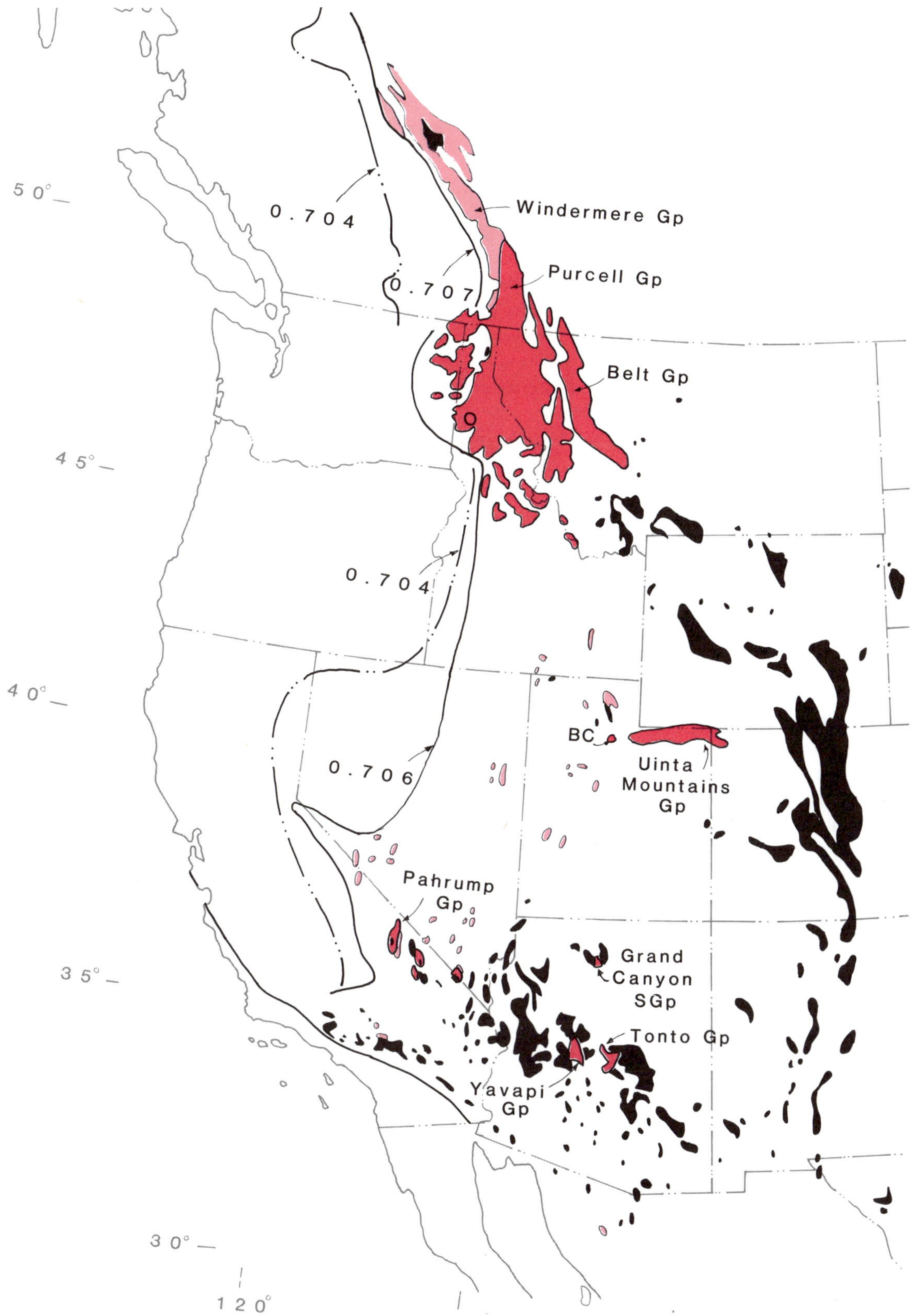

Figure 3. Precambrian crystalline (black) and sedimentary rocks present within the Cordilleran orogen. The locations of the various Middle (red) and Late (pink) Proterozoic sedimentary rock sequences mentioned in the text are indicated and the positions of the Sri = 0.706 and 0.704 lines are shown. BC, Big Cottonwood Mountains; O = Orofino.

suggest that the age of the Belt Supergroup ranges between about 1,600 and 1,300 Ma; younger ages reported for the middle and upper Belt strata are considered suspect and need to be reevaluated.

The tectonic setting of the depositional environment of the Belt Supergroup, like its age, is also unclear. The position of the Belt basin relative to older metamorphic rocks of the North American craton and the occurrence within it of synsedimentary faults with large vertical throw and multiple episodes of intrabasin mafic magmatism all suggest that the basin was formed by extension within continental crust. There is no direct evidence that extension was accompanied by major separation of the ancient craton and the formation of oceanic lithosphere. However, sedimentary sequences as thick as that of the Belt Supergroup must be deposited on either very attenuated continental crust or on oceanic crust.

Some workers (e.g., Winston and others, 1984; *Grotzinger, 1986*) have presented evidence that there was a continental source of sediments to the west or southwest. Two tectonic scenarios can be proposed for the setting of such a continent-bordered Belt basin. The first (favored by one of the authors, GAD) is that it represents a Middle Proterozoic aulacogen, a rifted reentrant into continental North America that connected northwestward with a more northerly trending rifted, passive margin (Gabrielse, 1972; Harrison and others, 1974; Burchfiel and Davis, 1975; Gabrielse and Yorath, 1989). If it is an aulacogen, it formed 800 to 900 m.y. prior to development of the more extensive Late Proterozoic passive margin of western North America that initiated the Cordilleran orogenic system. A second scenario (favored by another of the authors, BCB) is that the Belt basin was essentially ensialic and had no or little direct connection to a proto-Pacific Ocean basin to the west (cf., Hoffman, 1989). If so, modern analogs for the basin might be the Black Sea (*Neprochnov and others, 1974*) or the Caspian Sea (*Berbarian, 1983*).

An aulacogenic origin for the Belt basin is bolstered by the contention of Gabrielse (1972), *Delaney (1981),* and Gabrielse and Yorath (1989) that the Belt-Purcell and Wernecke Supergroup assemblages of sedimentary rocks, the latter in the northern Canadian Cordillera, are of miogeoclinal character. Burchfiel and Davis (1975) and *Dickinson (1977)* proposed that two miogeoclines or continental terrace wedges, Belt-Purcell-Wernecke and Cordilleran, would presumably require two periods of continental fragmentation and drift. Both Belt and Wernecke sedimentary assemblages were affected by a Middle Proterozoic tectonic event, called the East Kootenay orogeny (ca., 1,300 Ma) in southern areas (McMechan and Price, 1982) and the Hayhook orogeny in northern areas. Rocks in the western and northwestern parts of the Belt-Purcell basin were deformed, metamorphosed, and intruded by granite stocks. The fact that Belt rocks appear to have stratigraphic continuity from east to west across the basin suggests that no major suture lies within it. Hence, the East Kootenay/Hayhook orogeny is most easily explained as the result of convergent plate boundary activity near, but to the west, of the Belt and Wernecke depocenters. If these two basinal assemblages represent aulacogenic reentrants into a passive margin, their proximity to subsequent convergent deformation along that margin is readily explained. The fact that no correlative tectonic events are known in the United States south of the Belt region would argue that a Middle Proterozoic rifted continental margin did not extend much to the south of the present United States-Canadian border (*Dickinson, 1977*).

As mentioned above, possible analogs for a completely ensialic, land-locked Belt basin are the Black and Caspian Seas. Both of these landlocked basins contain about 15 km of sedimentary rocks that rest on thinned continental crust and, locally, on oceanic crust. Hoffman (1988) states that sedimentological and stable isotope evidence suggests that the Belt water column was stagnant and stratified and the basin increasingly isolated during time; strata-bound sulfide deposits in lower Belt turbidites may indicate a euxinic sedimentologic environment comparable to that of the Black Sea, although similar deposits are present within rifted troughs such as the Red Sea. An important characteristic of the Black and Caspian Sea basins is that they have been filled rapidly within the last 100 m.y. and probably were formed by several rifting events. By analogy, the Belt basin may either have been filled in an equally short period of time or have formed during multiple rifting events as suggested by the existence of several periods of mafic magmatism within the basin. The cooling of oceanic lithosphere beneath basins such as the Black Sea creates a lithosphere that consists mostly of mantle and may, therefore, be stronger than continental lithosphere. Thus, the Belt basin may have been somewhat immune to subsequent deformation until its basement was softened by heating, probably associated with plutonism. This might explain why part of the Belt-Purcell basin functioned as a relatively stable block for much of Phanerozoic time and was only deformed when heated by Mesozoic magmatism. A problem with an ensialic origin for the Belt basin is that tectonic models for its presence are difficult to envision. No intra-orogenic setting comparable to that of the Black and Caspian Seas has been postulated for the regions in which Belt and Wernecke Supergroup sediments were deposited, and obvious reasons for isolated, intracontinental rifting are not forthcoming. Likewise, the subsequent deformation, metamorphism , and intrusion of Belt strata during the East Kootenay orogeny is difficult to explain if the Belt basin did not develop in proximity to an ocean-continent boundary.

At the time the Belt basin was forming, that part of the North American continent that was later to become the southern Cordilleran region in the United States was the site of numerous large intrusions of anorogenic rapakivi granite, mangerite, and anorthosite (Fig. 2). These intrusions generally range in age from 1,300 to 1,500 Ma (Hoffman, 1989) and are part of a northeast-trending belt of similar intrusions that extends across the continent to Labrador in southeastern Canada (Anderson, 1983). There appears to be no relationship between this plutonism and development of the Belt basin much farther to the north. The age, as young as 1,300 Ma, and northeast trend of this plutonic belt indicate that Cordilleran events had not yet been initiated in the southwestern United States.

Pahrump, Grand Canyon, and Apache-Troy Groups. In the southwestern part of the Cordillera the Middle Proterozoic Pahrump, Grand Canyon, and Apache-Troy Groups (P_1) lie unconformably on a basement that was intruded, deformed, and metamorphosed at ca. 1,700 Ma, and intruded by 1,300 to 1,500 Ma anorogenic granites. The lower parts of these sedimentary sequences are intruded by diabase sills dated in Arizona at ca. 1,070 to 1,150 Ma. The Pahrump Group of the Death Valley region, California, constitutes a 3- to 4-km-thick, west- to northwest-trending sequence of marine and nonmarine sedimentary rocks (*Hewitt, 1940*; Wright and Troxel, 1966; Wright and others, 1974). Pahrump strata rest unconformably on a basement that was deformed and metamorphosed about 1,600 to 1,700 Ma and intruded by anorogenic granites at about 1,400 Ma (*Lanphere and others, 1964; Wasserburg and others, 1959*). This cover-to-basement relationship indicates that Pahrump rocks are younger than the lower part of the Belt (which was intruded by the Moyie sills at 1,433 Ma). The Pahrump sequence has traditionally been divided into three units, a lower terrigenous sequence with some carbonate rocks (Crystal Spring Formation), a middle stromatolitic dolomite (Beck Spring Dolomite), and an upper terrigenous sequence characterized by thick conglomerate and diamictite (Kingston Peak Formation) with subordinate carbonate rocks. Crystalline basement rocks and the Crystal Spring Formation are intruded by regionally extensive diabase sills and dikes. Hammond (1986) reports that upper members of the Crystal Springs Formation were still wet and unconsolidated at the time of intrusion. Many unsuccessful attempts have been made to date the Pahrump diabase intrusions, and most workers suggest that they are about 1,070 to 1,150 Ma based on correlation with similar rocks in Arizona (e.g., *Hammond, 1983*). We accept this correlation and age assignment and believe that the diabase and the lower two Pahrump units are distinctly older than all or part of the uppermost Pahrump unit, the Kingston Peak Formation (see below). Some faulting has been reported to be contemporaneous with deposition of the two older units (*Wrucke, 1966*). The present distribution of these rocks in a belt 150 km long and 60 km wide has led to the interpretation that the Pahrump Group was deposited in an east-trending aulacogen (Wright and others, 1974). However, Crystal Spring and Beck Spring units do not contain strata suggesting deposition near active faults, and we believe it more likely that their present outcrop patterns are the consequence of younger faulting and erosion known to be contemporaneous with the deposition of the uppermost "Pahrump" unit.

Rocks of the lower part of the Pahrump Group are similar to marine and nonmarine sedimentary rocks of the lower part of the Grand Canyon Supergroup (Elston and McKee, 1982) and the Apache Group–Troy Quartzite (*Shride, 1967*) to the east in Arizona (Fig. 3). Both units rest unconformably on crystalline basement rocks that have been intruded at other locales by 1,400-Ma anorogenic granitic rocks. The Unkar Group, lowest unit of the Grand Canyon Supergroup, was intruded by diabase dikes and sills and is capped by mafic lavas that date (Rb/Sr) 1,070 ± 30 Ma and 1,070 ± 70 Ma, respectively (Elston and McKee, 1982). These rocks are paraconformably overlain by sedimentary rocks of the Chuar Group that contains no igneous intrusive or extrusive rocks. Each group is about 2 km thick. The Apache Group, which crops out farther to the southeast in Arizona (Fig. 3), consists of only a few hundred meters of shale, quartzite, and limestone capped by basalt flows. It is overlain by 400 m of Troy Quartzite. The Apache Group and Troy Quartzite are shelf deposits of uniform lithology that crop out over more than 40,000 km^2. Both units were faulted, and intruded by diabase dikes and sills that in central Arizona have yielded U/Pb ages of about 1,150 Ma (Anderson, 1989; Conway and Sliver, 1989).

The lower Pahrump, Unkar, and Apache-Troy units have such a similar post-1,400-Ma depositional history that we suggest that they are broadly correlative. Original outcrop patterns of the lower Pahrump units and the Unkar Group have been influenced by subsequent faulting and erosion, but the thinner Apache Group–Troy Quartzite crops out over a broad area suggesting deposition in a stable shelf environment. All three units were probably deposited over wide areas. Deposition was associated with mafic magmatism that where dated is ca. 1,070 to 1,150 Ma. Mafic magmatism was at least locally accompanied by faulting and we suggest that all three rock units were deposited within one or more broad intracratonic basins of extensional origin on the North American craton. The relatively thin stratigraphic sections of these basins require only minor extension and subsidence of the craton. We can find no evidence that deposition of these rocks relates to a major rifting event that created a continental margin in what is now the southern part of the United States Cordillera. Dates from the mafic igenous rocks associated with these sequences suggest a temporal correlation with similar rocks (Purcell) in the middle part of the Belt Group. The faulting and mafic magmatism associated with deposition of all these units may be related to widespread extensional events that affected the North American craton at about this time, such as the Keweenawan rift of the north-central United States in which sedimentary deposits of the rift contain thick mafic intrusive and extrusive rocks that date between 1,145 and 1,115 ± 15 Ma (*Silver and Green, 1963, 1972*).

Uinta Mountains Group. The Uinta Mountains Group (P_1), exposed in the Uinta Mountains of north-central Utah (Fig. 3; Hansen, 1965), consists of up to 8 km of coarse alluvial fan deposits that grade southward into finer grained sediments, some of which are possibly marine. The distribution of facies within the Uinta Mountains Group suggests that it was bounded on the north by an east-west–striking basin margin fault (Sears and others, 1982; *Crittenden and Wallace, 1973*). This fault is parallel to and may have been controlled by the suture between the Archean

Wyoming Province in the north and a younger (ca., 1,800 to 1,600 Ma) Proterozoic arc province to the south (Fig. 2). Wells drilled on the Colorado Plateau about 150 km south of the Uinta Mountains indicate that the group is absent between crystalline basement rocks and an overlying Paleozoic cover and may have been confined to a pre-Paleozoic east-trending trough to the north.

Unlike similar sequences of sedimentary rocks in this broad time interval, the Uinta Mountains Group contains no igneous rocks; thus precise dating of its deposition is impossible. The group rests on Archean and Early Proterozoic metamorphic rocks and is overlain unconformably by Middle Cambrian quartzite. In the northern part of the Uinta Mountains, strata of the group were deposited unconformably on Early Proterozoic quartzite and its Archean basement rocks within an Early Proterozoic belt of deformation (the Cheyenne belt) that fringes the Archean Wyoming Province (Sears and others, 1982; Karlstrom and Houston, 1984). The Big Cottonwood Formation, which we regard as correlative to the Uinta Mountains Group, crops out about 50 km west of the Uinta Mountains (Fig. 3). It rests on a metamorphic basement that may also be Archean in age, and is unconformably overlain by the Late Proterozoic diamictite-bearing Mineral Fork Formation. These relations place deposition of the Uinta Mountains Group between about 1,700 and 800 Ma. *Crittenden and Peterman (1975)* have reported a whole-rock Rb/Sr age of 950 to 925 Ma from shales in the uppermost formation (Red Pine Shale) of the group. The same shales contain the problematic metazoan fossil *Chuaria* whose maximum age range is given by *Hoffman (1977)* as 1,150 to 570 Ma. This suggests that at least the upper part of the Uinta Mountains Group is yougner than the Unkar, Apache-Troy, and lower Pahrump units, but the age of its lower part remains uncertain.

The east-west–trending trough in which the Uinta Mountains Group was deposited has been referred to, probably incorrectly, as an aulacogen related to Cordilleran margin formation by rifting to the west (e.g., Burchfiel and Davis, 1975; Sears and others, 1982). The age of the trough is too old to be related to development of the Cordilleran margin, and there is no independent evidence that Middle Proterozoic rifting ever extended much to the south of the United States–Canadian border. Geological data support a newer interpretation that deposition of the Uinta Mountains Group took place in a half-graben with a fault on its north side, but the location and nature of its southern boundary remains unknown. Uinta Mountains sediments may have been deposited in a successor basin along or across the older Cheyenne belt suture, with reactivation of the suture playing an important role in delineating the northern edge of the basin.

Chuar Group–Sixtymile Formation. In the Grand Canyon area of Arizona the Chuar Group, the upper part of the Grand Canyon Supergroup (P_1; Fig. 3), unconformably overlies the Unkar Group and is overlain unconformably by the Sixtymile Formation (P_2?). The unconformities between these units may record a significant period of time, but the lower units were not folded prior to erosion. The Chuar Group consists of 2 km of marine sandstone, shale, and limestone (Elston and McKee, 1982). Their lack of mafic igneous rocks suggests they are younger than the Apache Group–Troy Quartzite. Paleomagnetic data presented by Elston and McKee (1982) suggest that the Chuar Group may be as old as 1,050 Ma, and its sediments also contain *Chuaria.* Both of these tenuous pieces of data would place these rocks in the upper Middle to Late Proterozoic and possibly correlative with upper parts of the Uinta Mountains and Belt (?) sequences. The unconformably overlying clastic rocks of the Sixtymile Formation were deposited during active faulting and contain breccia interpreted to have formed as landslide deposits from fault scarps. These faults bound tilted blocks that consist of homoclinally dipping strata of the Grand Canyon Supergroup and Sixtymile Formation that were transgressed by the subhorizontal Middle Cambrian Tonto Group. Some of the faults have displacements up to 3.2 km. Elston and McKee (1982) interpret K/Ar cooling ages of about 823 Ma in the 1,070-Ma-old lavas of the underlying Unkar Group to represent the time of faulting.

Correlation and tectonic setting for Middle Proterozoic rocks. The imprecise dating of Middle Proterozoic rocks in the eastern part of the Cordillera preclude their detailed correlation and the development of a well-documented interpretation of their tectonic setting. The relations discussed above suggest that all these sequences of sedimentary rocks were deposited during extensional events. The most extreme extensional event was the oldest, from about 1,500 to 1,300 Ma, that was responsible for formation of the early Belt-Purcell basin, and interpretations of its tectonic setting have been presented above.

We regard the Middle Proterozoic Pahrump (except its uppermost unit; see below), Unkar, and Apache-Troy stratigraphic sequences as having been deposited during at least one extensional event at about 1,150 to 1,070 Ma that was associated with regional mafic magmatism. This event may be broadly related to widespread extension within the craton that formed the clearly defined Keweenawan rift system far to the northeast. The limited thickness (~ 2 km) of sediments in these southwestern U.S. sequences requires only modest extension of the continental crust and did not lead to the formation of oceanic crust. Rocks of the Apache Group and Troy Quartzite of eastern Arizona are thin and widespread, suggesting that they were deposited in a shelf environment marginal to more extended areas. There is no evidence that these extensional basins opened westward or southwestward to an adjacent continental margin and thus they are not regarded by us to be aulacogens.

All the Middle Proterozoic rock units were affected by at least one and possibly two younger extensional events. The Chuar Group, Sixtymile Formation, and upper Belt Supergroup are devoid of mafic magmatism. They were cut by normal faults that are thought to have been active about 800 to 900 Ma (Elston and McKee, 1982; McMechan and Price, 1982). In northern Canada, mafic sills with ages of 766 and 769 Ma (*Armstrong and others, 1982*) were intruded into faulted and tilted rocks that lie unconformably below latest Proterozoic (Windermere) strata. They may have accompanied a poorly dated period of extension about 800 Ma. Based mainly on reset K/Ar ages from mafic igneous rocks in Washington and Arizona, Elston and McKee (1982) and McMechan and Price (1982) suggested that a significant rifting event affected much of the western part of North America about 800 to 900 Ma. The event was called the Grand Canyon–Mackenzie Mountains disturbance by Elston and McKee, and the Goat River orogeny for the deformation in southern Canada by McMechan and Price. Most of the dating of this event is based on interpreting reset K/Ar data and cannot be considered very reliable, but there is enough evidence to suggest that an extensional event did occur at about this time.

A still younger extensional event, associated with deposition of Late Proterozoic diamictite, that led to the formation of the Paleozoic Cordilleran passive margin is also present in the western Cordillera (see below). Because the diamictite is not present in the Grand Canyon area, faulting that affected the Grand Canyon Supergroup could belong to either of the younger two extensional events. The uppermost unit of the Pahrump Group (Kingston Peak Formation) was deposited contemporaneously with faulting and mafic magmatism younger than the 1.1? Ga diabase sills and dikes in underlying units (Miller, 1985), but these younger mafic rocks are undated. We tentatively interpret faults in the Kingston Peak Formation to be related to latest Proterozoic rifting, not the earlier 900- to 800-Ma extensional event (see below).

In summary, the tectonic history of Middle Proterozoic time (ca., 1,600 to 900 Ma) in what is now western North America was dominated by at least three important extensional events: the first and most important between 1,500 and 1,300 Ma, the second about 1,150 to 1,070 Ma, and a third about 800 to 900 Ma. None of these extensional events appears to be related to the formation of the Cordilleran continental margin in latest Proterozoic time (see below). Only one Middle Proterozoic convergent event, the East Kootenay orogeny, has been recognized within areas now within the U.S. Cordilleran system. It affected rocks in the western part of the Belt basin at about 1,300 ± 50 Ma and has no recognized counterparts in areas farther to the south.

Late Proterozoic rift events and inception of the Cordilleran passive margin

Geological evidence suggests widespread rifting, mafic igneous activity and sedimentation along the entire western margin of North America during latest Proterozoic time (Stewart, 1972; McMechan and Price, 1982; *Devlin and others, 1985; Devlin and Bond, 1988*). These events led either directly or intermittently to continental separation, ocean-floor spreading, and the formation of a passive, Atlantic-type margin at about 600 Ma. Within the eastern part of the Cordillera a sequence of marine and nonmarine terrigeneous clastic rocks interbedded locally with carbonate rocks forms a westward thickening wedge referred to as the Windermere sequence (P_2; Fig. 4). The distribution of the latest Proterozoic to earliest Paleozoic Windermere sequence is generally parallel to the trend of the Cordilleran orogen and oblique to trends in locally underlying Middle Proterozoic rocks. Although most Windermere rocks are believed to have been deposited along a passive continental margin as the basal sediments of the Cordilleran miogeocline (Stewart, 1972; *Dickinson, 1977;* Bond and Kominz, 1984), some lowermost Windermere strata may record localized deposition in rift basins that predated complete continental separation. For example, basal mafic igneous rocks that probably formed in a rift setting (*Devlin and others, 1985, Devlin and Bond, 1988*) are associated with sedimentary deposits that commonly contain diamictites, some of which can be shown to be of glacial origin (e.g., *Link, 1983; Christie-Blick, 1983;* Miller, 1985). Geochemical studies have supported a rift setting for the Huckleberry volcanic rocks (*Devlin and others, 1985*) of Windermere affiliation in southern British Columbia and northeastern Washington, as have similar studies on Late Proterozoic volcanic rocks in southern Idaho (Harper and Link, 1986). More generally, however, Late Proterozoic diamictite-bearing Windermere sequences lie conformably below fossiliferous Lower Cambrian strata of the Cordilleran miogeocline (e.g., *Crittenden and others, 1971, 1983;* Stewart and Suczek, 1977).

Reliable dates for the base of the Windermere sequence come from Canada. *Armstrong and others (1982)* report Rb/Sr ages of 766 and 769 Ma from mafic sills that intrude pre-Windermere strata in the McKenzie Mountains of northwest Canada, and Evenchick and others (1984) obtained Pb/U ages of 728 Ma on gneissic granites that underlie Windermere rocks in the Deserters Range 500 km farther south. These data suggest that the base of the Windermere is younger than 728 to 769 Ma. Bond and Kominz (1984) and Armin and Mayer (1983) analyzed the subsidence history of Windermere and younger Paleozoic miogeoclinal rocks and suggested an age of 550 to 600 Ma for continental separation following rifting and the onset of thermal subsidence. This age is significantly younger than that proposed by earlier workers, such as Stewart (1972), Burchfiel and Davis (1975), and McMechan and Price (1982), who hypothesized that rifting began between 800 and 900 Ma. These older ages for the initiation of rifting came from interpretations of reset K/Ar ages on older igneous rocks (McMechan and Price, 1982) and from K/Ar ages on the Huckleberry volcanic rocks (preferred ages of 827 and 918 Ma; Fig. 4) in northeastern Washington (*Miller and others, 1973*). *Devlin and others (1985)* attempted to redate the Huckleberry volcanics by both Rb/Sr and Sm/Nd methods and concluded that the isotopic systems were so disturbed that no meaningful age could be determined.

The Death Valley region, California, is one of the few areas in the U.S. Cordillera where Windermere diamictites and mafic volcanic rocks lie unconformably below younger Windermere strata at the base of the Cordilleran miogeocline. In this region, the Late Proterozoic Kingston Peak Formation contains abundant diamictite locally associated with mafic magmatism (Wright and Troxel, 1966; Wright and others, 1974; Miller, 1985). Conglomerate in the Kingston Peak contains clasts derived from underlying Pahrump units (Crystal Spring, Beck Spring Formations)

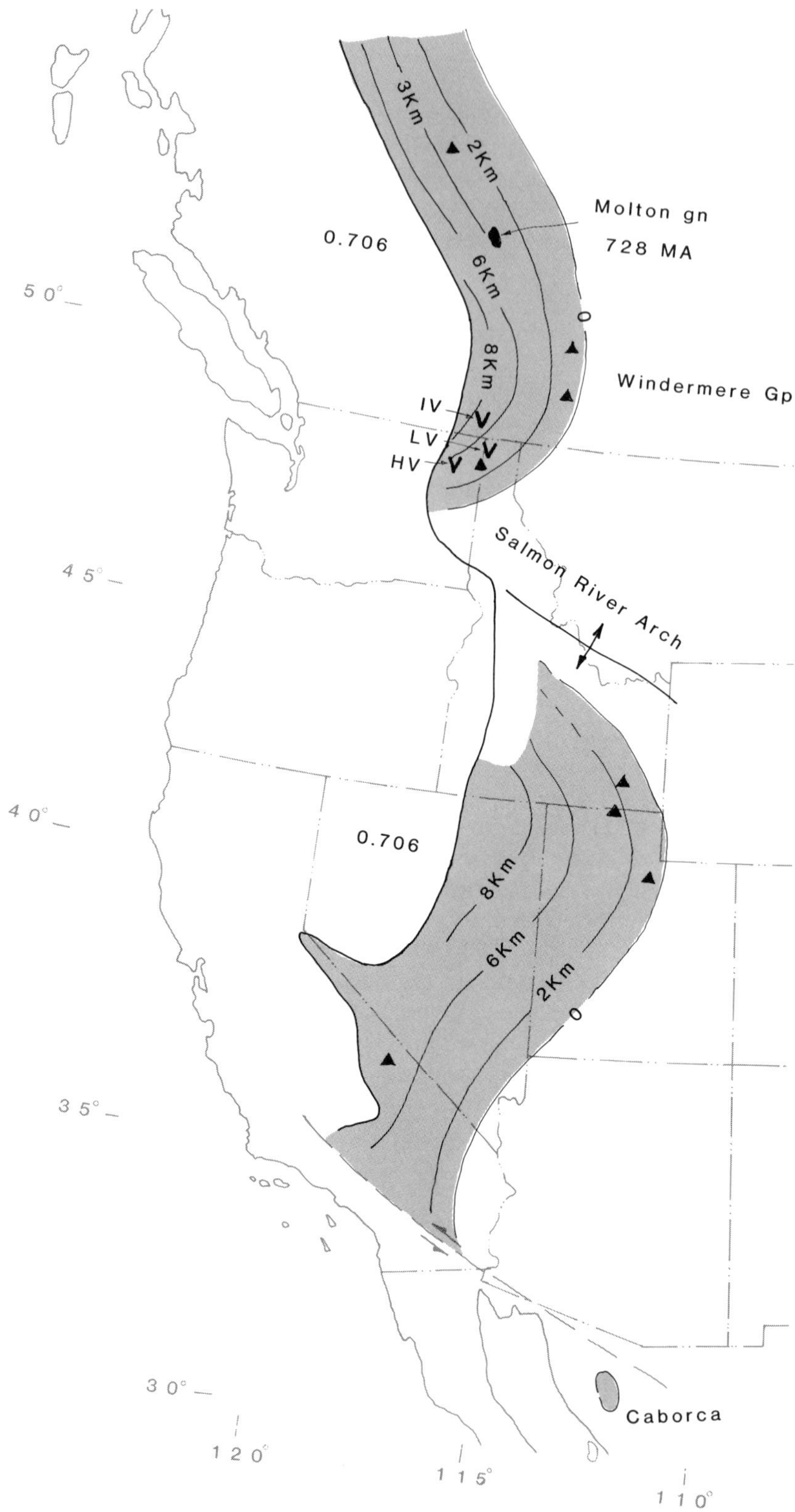

Figure 4. Generalized map showing the distribution and thicknesses of latest Proterozoic and Early Cambrian terrigenous rocks (light gray) that indicate the formation of the Cordilleran passive margin (compare their distribution to the rock sequences in Fig. 3). Localities containing tillites are indicated by black triangles. Prominent truncation of facies belt that formed in late Paleozoic time is also shown in the southern Cordillera. Truncation of the facies belt west of the Salmon River arch occurred in late Mesozoic time. V's indicate the location of Windermere Irene (IV), Leola (LV), and Huckleberry (HV) volcanic rocks.

and from Pahrump-age diabase sills. The mafic dikes and volcanic rocks in the Kingston Peak Formation are chemically distinct from mafic rocks intruding the older units (*Hammond, 1983*). Mapping in the Death Valley area (Miller, 1985) and at the type area of the Kingston Peak Formation (Walker and others, 1986) clearly demonstrates that faulting was associated with deposition of the formation and was responsible for rapid lateral changes in facies and thickness. In the latter (i.e., the type) area, Kingston Peak strata are more than 3 km thick, and some syndepositional fault displacements must have throws greater than that thickness. Some of the major faults cutting the Kingston Peak Formation are overlapped by an internally conformable sequence of Noonday Dolomite and younger (latest Proterozoic to Early Cambrian) clastic rocks at the base of the miogeocline; in the same area, a somewhat younger normal fault cuts upper Kingston Peak and lower Noonday rocks, but is overlapped unconformably by higher Noonday strata (Walker and others, 1986).

We regard crustal extension associated with Kingston Peak and Noonday sedimentation as an expression of the latest Proterozoic rift phase that led to development of the Cordilleran continental margin. If so, the southern Death Valley/Kingston Peak region is one of the few areas in the U.S. Cordillera where rift phase sediments and their associated normal faults are recognized. In reality, this simple statement regarding synchronous rifting and sedimentation is misleading. The geology of the Kingston Peak area is complicated by recognition of an intraformational unconformity within rocks traditionally assigned to the Kingston Peak Formation in this, its type area (Walker and others, 1986). The unconformity developed after Late Proterozoic folding and cleavage development in lower Kingston Peak rocks and underlying Pahrump units, and it was offset by normal faults prior to deposition of the upper part of the formation; both upper and lower Kingston Peak sequences contain diamictites. These compressional structures and the unconformity above them indicate either a significant hiatus within the period of "Kingston Peak" sedimentation, or that extensional faulting and synchronous Kingston Peak sedimentation were locally associated with transpressive(?) deformation.

Major unresolved problems concerning the formation of the U.S. Cordilleran passive margin have to do with (1) the general lack of recognition of faults that are part of the rift phase of margin development, and (2) the absence in the preserved geologic record of continental slope and rise sediments of Windermere age. There is little structural evidence for major extension during deposition of the latest Proterozoic to early Paleozoic part of the sequence, except in the Death Valley area mentioned above. Presumably, most syn-rift faulting occurred in areas farther west where shelf-edge, slope, and rise facies of Windermere sediments must have been deposited. The western extent of the crystalline basement rocks of North America is interpreted to follow the sinuous trace of the Sri = 0.706 line (Figs. 2 and 3; cf., Poole and others, this volume; Farmer, 1988), a line that also approximates the western limits of shallow-water miogeoclinal strata (= continental shelf deposits). Although deeper water slope and rise sediments of Cambrian and younger age have been preserved (e.g., in the Roberts Mountains allochthon; see below), the absence in the geologic record of deep-water sediments of Windermere age is perplexing.

LATEST PROTEROZOIC TO DEVONIAN EVOLUTION

Passive margin development

From latest Proterozoic through Late Devonian time the miogeocline of the Cordillera formed along the western side of the present North American craton (Figs. 4 and 5). Although the sequence was deposited without significant stratigraphic breaks, its stratigraphy, especially from Middle Cambrian time on, was influenced by major sealevel fluctuations (*Vail and others, 1977*). The Windermere sequence (P_2), as described above, forms a thick basal miogeoclinal sequence dominated by interbedded marine and nonmarine clastic rocks derived from the craton to the east and containing minor interbedded carbonates. These rocks form a wedge that thickens from a stratigraphic pinchout and hingeline in the east to more than 6 km in the west (Fig. 4). Beginning in Middle Cambrian time this clastic wedge was overlain by a thick succession of Paleozoic carbonate strata (€Jcs) that was interrupted only rarely prior to Mississippian time by the influx of terrigeneous material (Fig. 5). All of the pre-Mississippian detritus had a source from the craton to the east. Facies belts within this wedge trend obliquely to the trends of underlying metamorphic basement rocks and Middle Proterozoic sedimentary deposits. Deposition of the lower part of the miogeocline has a subsidence history that fits a thermal subsidence curve for a continental passive margin initiated by rifting at ca. 550 to 600 Ma (Bond and Kominz, 1984; Armin and Mayer, 1983).

The miogeocline (€Jcs) has a sinuous trace with a major eastward reentrant in the central part of the western United States and two salients, one in the area of the Belt basin and the other in southern Nevada and southeastern California (Figs. 4 and 5: Stewart, 1972; Poole and others, this volume). It appears from this pattern of sedimentation that the Belt basin functioned as a stable part of the North American craton during most of Paleozoic time. The southern part of the northern salient was not only a sediment source for the Proterozoic Belt basin, but was a positive area, called the Lemhi or Salmon River arch, at several times during the Paleozoic (Armstrong, 1975; Ruppel, 1986). There are two prominent discontinuities in the miogeocline at the two salients that are the result of younger tectonic modification. These discontinuities are expressed by the truncation of isopachs (Fig. 4) and facies trends (Fig. 5) in miogeoclinal strata and are discussed below under the appropriate time intervals for their formation.

The evolution of the miogeocline was not entirely a response to passive subsidence. The eastern part of the miogeocline and its cratonal foreland were areas of low-amplitude deformation at various times during the early Paleozoic (Poole and others, this

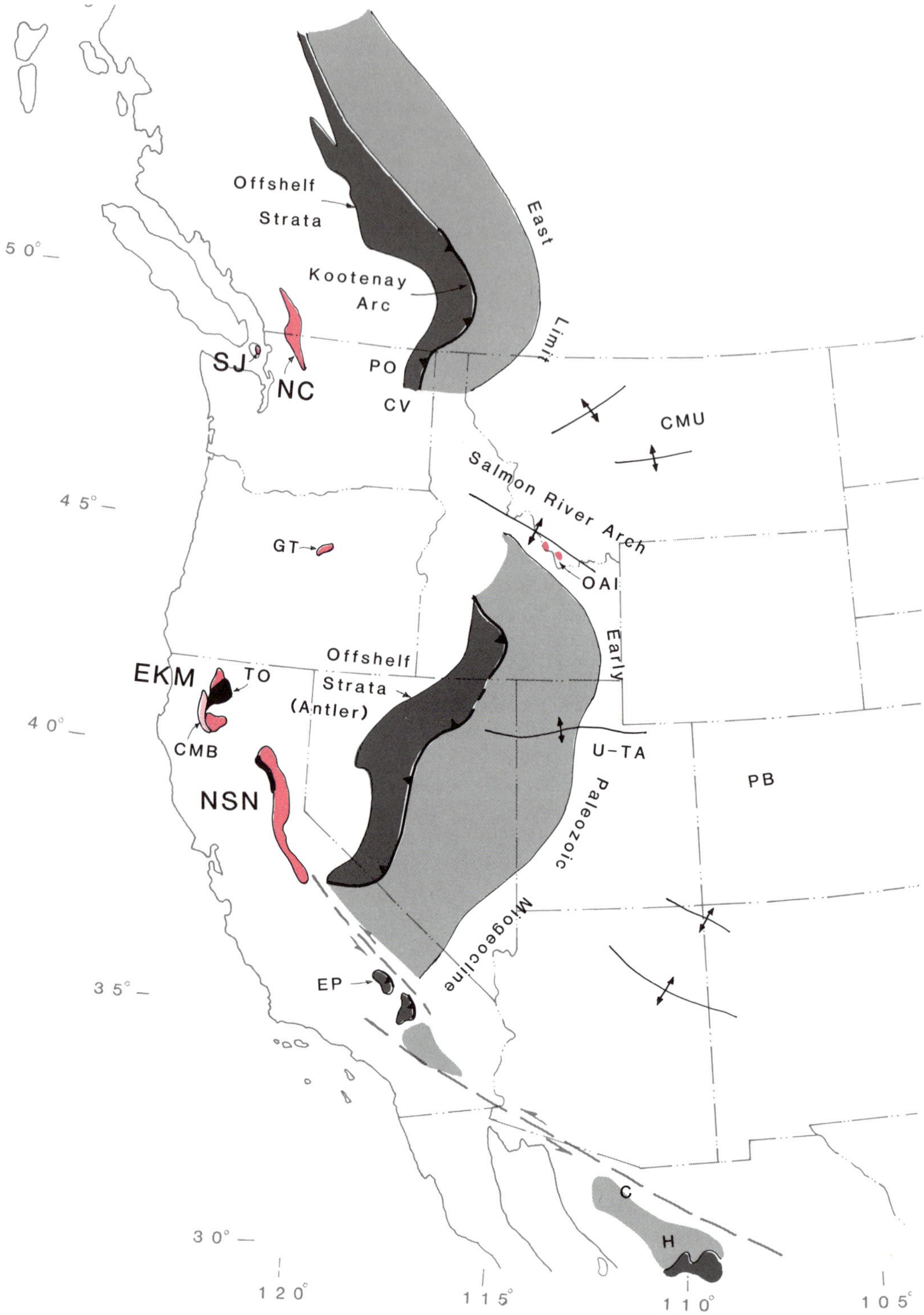

Figure 5. Paleogeographic and tectonic map of the Cambrian to Late Devonian Cordillera. Patterned units include Cambrian to Late Devonian allochthonous offshelf units (gray) emplaced onto shelf rocks of the passive margin (light gray) in latest Devonian to Middle Mississippian time (see Fig. 6), volcanic arc sequences (red), and the Devonian Central Metamorphic Belt of the Klamath Mountains (pink). C, Caborca; CMB, Central Metamorphic belt; CMU, Central Montana uplift; CV and PO, Corvada and Pend d'Oreille Groups; EP, El Paso Mountains; EKM, eastern Klamath Mountains; GT, Grindstone terrane; H, Hermosillo; NC, northern Cascades; NSN, northern Sierra Nevada; PB, Paradox basin; SJ, San Juan Islands; TO - Trinity ophiolite (black); U-TA, Uinta-Toole arch; OAI, Ordovician alkali intrusions.

volume). Uplift took place intermittently from Cambrian through Upper Devonian time along the axis of the Proterozoic Uinta Mountains trough and its western continuation into central Nevada, the Uinta-Tooele arch (Fig. 5). Uplift along the Lemhi or Salmon River arch took place at several times; it was probably the source for arkosic debris that was shed into the miogeocline during the Cambrian (e.g., *Suzeck, 1977;* Stewart and Suzeck, 1977). Near the axis of the arch, alkalic igneous rocks (Oi) were intruded during Ordovician time (*Olsen and others, 1977; Loring and Armstrong, 1980;* Evans and Zartman, 1988). Faulting of Cambrian age in the Paradox basin (*Baars, 1966*), and Upper Devonian uplift along the Central Montana uplift are further evidence of intracontinental deformation probably associated with continental margin activity (Fig. 5). Rocks of the continental slope and rise also contain evidence of low-amplitude(?) recurrent extension or possibly other types of tectonic activity including mafic igneous activity (E. Miller and others, this volume; see below). Thus it seems clear that the early Paleozoic "passive" continental margin was the site of ill-defined deformation(s) that may have extended well into the craton.

Western offshore elements

There is no evidence for the formation of oceanic lithosphere west of the North American continent until the end of Proterozoic time. Mafic plutonic blocks in an eastern Klamath Mountains mélange (see below) have an age compatible with ophiolite generation at the time of initial continental rifting of the western North American margin. In addition, the blocks contain xenocrystic Precambrian zircons from an unknown continental source, a relationship that suggests proximity of a possible spreading center to a continental margin (*Walling, 1989a*).

Lying in widespread thrust contact above western parts of the Cordilleran miogeocline are sequences of Cambrian to Upper Devonian sedimentary rocks of deep-water facies (Fig. 5). These rocks, called the Antler assemblage in Nevada and the Pend d'Oreille and Covada Groups (€Dos) in Idaho and Washington, consist of chert, hemipelagic sediments, turbidites, quartzite, and greenstone. The sequences contain stratigraphic units that probably tie them to North America. For example, in Nevada they contain Middle Ordovician and Upper Devonian quartzite of the same age as widespread quartzites deposited on the miogeoclinal continental shelf (Poole and others, this volume), and in Washington they share stratigraphic characteristics with miogeoclinal strata (Gehrels and Smith, 1987).

Cambrian and younger rocks of the Antler assemblage (€dos) were deposited in deep water, and have traditionally been interpreted as having been deposited in a continental slope and rise setting west of the continental shelf (= miogeocline; Roberts and others, 1958; *Kay and Crawford, 1964;* Stewart, 1972; Burchfiel and Davis, 1972, 1975; Stewart and Poole, 1974; Poole and others, this volume). However, more recent stratigraphic analysis suggests that the basement for Antler strata may have been tectonically thinned continental crust (Turner and others, 1989; Poole and others, and E. Miller and others, this volume). A few fragments of serpentinite are present along the Roberts Mountains thrust at the base of the Antler assemblage, but no ophiolites are present in the allochthon. Mafic extrusion during deposition of Antler strata occurred from Middle Cambrian to Late Devonian time, apparently during episodes of continued intraplate extension (Poole and others, this volume). The geochemistry of some of the mafic volcanic rocks present in the Antler assemblage indicated to *Rogers and others (1974)* that they are oceanic tholeiites, but more detailed studies by *Madrid (1987)* support a predominant alkalic lineage and suggest an intracontinental plate setting such as a rifted outer continental margin. Watkins and Brown (1989) have recently proposed that Middle Ordovician basaltic seamounts formed in deep marine settings in the vicinity of the continental rise. In Early Mississippian time, Antler basin sedimentary and volcanic strata were thrust eastward across the continental shelf deposits of the miogeocline during the Antler orogeny. A transitional facies between the two environments is locally present, but it is in fault contact with both groups of rocks (Poole and others, this volume).

Within the northern Sierra Nevada, Klamath Mountains, central Oregon, San Juan Islands, and northern Cascades are early Paleozoic rock assemblages that were formed in depositional environments within or adjacent to islanid arcs (Poole and others, this volume; E. Miller and others, this volume). Most field and geochemical data, some of which are reviewed below suggest that the arcs were developed on oceanic crust, but in a setting or settings not far removed from a Precambrian continent(s). Arc-volcanic rocks are associated with limestone, interpreted as fringing reefs or shallow-water sediments deposited during volcanic quiescence; terrigeneous sedimentary rocks, some of which are quartz and K-feldspar–rich; and peridotite, gabbro, quartz diorite, and plagiogranite that can be interpreted as representing ophiolitic or oceanic island-arc roots. Roure and others (1989) have recently presented geochemical evidence (including Sm/Nd, REE, O) that the mid-Paleozoic volcanics of the northern Sierra Nevada represent a mature island arc constructed on a continental fragment.

The most extensively studied early Paleozoic ophiolitic assemblage is the Trinity ophiolite of the eastern Klamath Mountains (Saleeby, this volume). Mafic rocks from the ophiolite have Ordovician ages in the range of 440 to 475 Ma. Boudier and others (1989) have recently interpreted the ophiolite as the consequence of slow sea-floor spreading, perhaps in a marginal sea near the North American continent. Tonalite blocks lithologically similar to some in the Trinity ophiolite occur within a blueschist block-bearing mélange (Skookum Gulch; OSos) that either overlies (Wallen and others, 1988) or structurally underlies (*Cotkin, 1989*) the ophiolite. The tonalite contains primary latest Proterozoic (565 to 570 Ma) zircons, as well as older xenocrystic Proterozoic zircons of unknown origin (Wallin and others, 1988; *Wallin, 1989a*). The mélange itself is pre–Late Silurian and yields block and matrix metamorphic ages ranging from 439 to 451 Ma (*Haessig, 1989;* Wallin and others, 1988). Thrust-faulted sedimentary rocks (Yreka-Callahan assemblage; OSos) that lie structurally above the mélange range in age from Ordovician to Middle Devonian (*Potter and others, 1977*). Some units contain zircons from a Precambrian continental crystalline provenance (e.g., Duzel Phyllite, Antelope Mountain Quartzite), whereas others have a zircon population (432 to 455 Ma) derived from a Late Ordovician–Early Silurian(?) felsic volcanoplutonic terrane (*Walling, 1989b*). Taken collectively, the Trinity ophiolite, the Skookum Gulch blueschist-bearing mélange, and craton- and arc-derived detritus in some units of the Yreka-Callahan assemblage attest to Ordovician-Silurian oceanic arc and marginal(?) basin activity not far from a Precambrian continental margin or an offshore microcontinental block (*Cotkin, 1989; Wallin, 1989a*).

A similar geologic history can be inferred from the geology of the pre–Late Devonian Sierra City mélange and the Shoo Fly Complex (both OSos) in the northern Sierra Nevada (e.g., *Girty and Pardini, 1987*). Zircons of Early Proterozoic age (2.09 Ga) are present in some siliclastic sediments of the Shoo Fly Complex, as are Cambrian zircons (in more feldspathic sediments) from an unknown magmatic provenance (506 Ma; Girty and Wardlaw, 1985). Severe deformation of these rocks occurred prior to emplacement of a major Devonian granitic pluton (Di) within them (375 Ma). We tentatively propose that early Paleozoic clastic sedimentation in eastern Klamath and western Sierran regions occurred west of a microcontinental block rifted from the North American margin, since clastic sediments in these areas appear coarser grained than coeval terrigeneous sediments in the Antler assemblage to the east of the inferred microcontinent. This paleogeographic scenario is supported by the recent studies of Rouer and others (1989), who believe that mid-Paleozoic volcanic rocks (e.g., Sierra Buttes and Taylor Formations) that rest unconformably on deformed Shoo Fly units have geochemical characteristics of a mature island arc with a continental basement. Additional evidence for an offshore microcontinental block or sliver comes from sedimentologic studies of Antler assemblage (€Dos) rocks in Nevada. *Madrid (1987, in* Poole and others, this volume) offers evidence for a subareal western "highland" for some arkosic detritus in Early Ordovician to Late Devonian Antler strata of the Antler basin. Rocks exposed in that western provenance presumably included arkosic, continent-derived strata of the Cambrian Harmony Formation or similar lithologic units; there is no evidence for volcanic components.

The early Paleozoic Sierran-Klamath arc and a nearby(?) back- or intra-arc oceanic basin (Trinity ophiolite) may have been geographically far removed from the Antler basin and not, therefore, North American in their affinities. Alternatively, the arc may have been constructed near or on a microcontinental sliver(?) derived from the North American rifted margin—the alternative that we favor. The polarity of pre–Late Devonian subduction associated with the hypothesized offshore arc is inferred to have been eastward (with reference to present geography) because of (1) the stacking sequence of thrust-bounded early Paleozoic units in the eastern Klamaths, (2) a reported westward direction of thrusting for some low-angle bounding faults in the Yreka-Callahan assemblage (*Cotkin, 1989*), and (3) the west-directed vergence for pre–Sierra Buttes deformation in the Shoo Fly Complex of the western Sierra Nevada (*Girty and Guthrie, 1989*). It must be stressed that latitudinal controls on the position of early Paleozoic rock assemblages in the accretionary belt of the U.S. Cordillera are completely lacking. However, Late Ordovician and Early Devonian faunas within the Klamath-Sierran arc, including both conodonts and megafossils, are reported to have North American affinities (*Boucot and Potter, 1977; Hannah and Moores, 1986*).

Unlike the rather cryptic record of offshore subduction and arc-related magmatic activity in Ordovician and Silurian time reviewed above, the Klamath-Sierran offshore arc exhibits much clearer evidence of east-dipping subduction and related diachronous magmatic activity from Early Devonian (Klamaths) to Late Devonian (Sierras) time on. On the western side of the Klamath-Sierran Devonian arc are rock units and structures that strongly suggest eastward subduction. In the central Klamath Mountains, the Central Metamorphic Belt (Dm) consists of amphibolite and overlying metasediments (= Duzel of Yreka-Callahan assemblage) that were metamorphosed in Early Devonian time (Fig. 6; 370 to 399 Ma; *Lanphere and others, 1968; Hotz, 1977; Cashman, 1980*). The amphibolite is interpreted by Davis (1968) and Peacock (1987) to be oceanic crust that was metamorphosed as it was thrust (subducted) eastward beneath the hot Trinity ophiolite and its overlying stack of thrust-bounded units. Subduction appears to have been accompanied by development of an immature eastern Klamath island arc represented by Early Devonian Copley and Balaklala volcanic rocks (DPa; Lapierre and others, 1985) and the coeval Mule Mountain trondhjemitic pluton (400 Ma; Miller, 1987; Fig. 6). Likewise, in the northwestern Sierra Nevada deformed Ordovician and Silurian sedimentary rocks, tuffs, and mélange within west-directed, thrust-bounded packages (Shoo Fly Complex; OSos) are unconformably overlain by Upper Devonian arc rocks (Sierra Buttes Formation) and intruded by their plutonic equivalents (Fig. 6; 378 +5/−10 Ma; Saleeby and others, 1987; Hanson and others, 1988).

Paleozoic sections overlying Devonian volcanic rocks in the eastern Klamaths and northern Sierras (both DPa) contain similar, but not identical units and are generally regarded as parts of the same, stratigraphically complex offshore arc (Davis, 1969; *Hannah and Moores, 1986;* Miller, 1987; Harwood, 1988; E. Miller and others, this volume). In addition, Upper Devonian and younger Paleozoic volcanic and basinal rocks in the northern Sierras have lithologic and paleontologic counterparts in the Antler (€Dos) and Havallah (DPos) basins to the east (Figs. 5 and 7; cf., Miller and Harwood, 1989). Stratigraphic reasons for tying the Antler and equivalent basins to the north to a position outboard of the early Paleozoic Cordilleran miogeocline have already been alluded to and are referenced more thoroughly in Poole and others (this volume) and E. Miller and others (this volume).

Pre–Late Devonian arc-related plutonic and volcanogenic rocks are present in the San Juan Islands and northwest Cascades

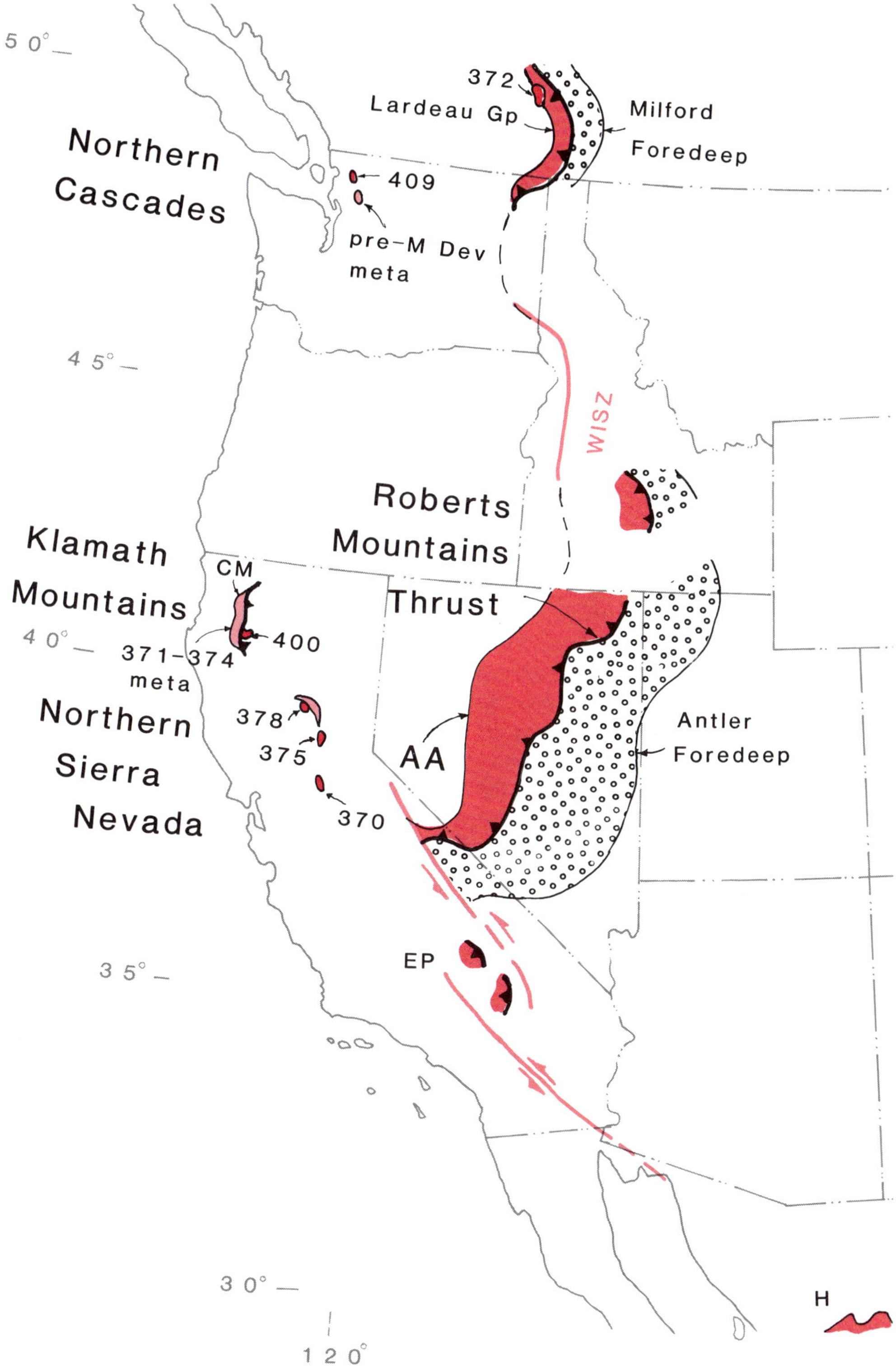

Figure 6. Tectonic elements of the Mississippian Antler orogeny and Devonian metamorphic (pink) and plutonic (red) events in offshore arcs. Deformation in the offshelf sequences southeast of Hermosillo, Mexico, is Late Mississippian and younger than that in the Antler belt of Nevada (Stewart and others, 1990). Numbers refer to ages on plutons (red) except where noted as ages of metamorphism. Shown in light gray are faults of younger age that disrupt Antler elements. AA, Antler allochthon (gray); CM, Central Metamorphic Belt (pink); H, Hermosillo; EP, El Paso Mountains; WISZ, West Idaho suture zone.

(Fig. 5); they include the Yellow Aster and Turtleback igneous complexes and the East Sound Group (DPa; Brandon and others, 1988), the latter being an arc-volcanic sequence of Early Devonian to Permian age with minor interbedded limestone. Discordant U-Pb data from zircon in clinopyroxene gneiss in the Yellow Aster suggest an original age of at least 1,452 Ma; younger tonalites are older than 425 Ma, and probably about 460 Ma (Ordovician; *Mattinson, 1972*). Gabbro and voluminetrically more important tonalite in the Turtleback Complex yield Cambrian K-Ar and U-Pb dates (507 to 554 Ma; Brandon and others, 1988). Arc-related strata of the East Sound Group lie unconformably on Turtleback rocks, just as correlative(?) strata of the Cascade Chilliwack Group were presumably deposited on a Yellow Aster basement. All of these early to late Paleozoic units were imbricated with younger elements to form the mid-Cretaceous San Juan–Northwest Cascades thrust system. They may represent fragments or slices of the early Paleozoic western offshore arc in the Klamaths and northern Sierras that were displaced to the northwest prior to thrusting.

MIDDLE PALEOZOIC ANTLER OROGENY AND ASSOCIATED DEFORMATION

In latest Devonian and Early Mississippian time, rocks of the Antler assemblage (€Dos) were thrust eastward about 200 km onto miogeoclinal rocks of the North American continental slope and shelf (Roberts and others, 1958; Poole and others, this volume). The basal thrust for this complex allochthon, the Roberts Mountains thrust, is probably only preserved in a few places, because the base of the allochthon at most locales has been modified by younger faults (Fig. 6). The structure of the allochthon consists of numerous thrust slices that have been interpreted by some workers (e.g., Speed, 1977; Oldow, 1984b) as developing in an accretionary prism related to west-dipping subduction. Loading of the Antler allochthon onto the North American margin produced a flexural downwarp of the continental lithosphere and a consequent foredeep (Mfd) that was filled by several kilometers of Mississippian and Pennsylvanian clastic debris derived from the Antler allochthon (Speed and Sleep, 1982). The age of west-derived Antler debris in the foredeep and the overlap of Mississippian foredeep deposits onto the allochthon in Middle Mississippian time date its final thrust emplacement as Early Mississippian (*Little, 1987*).

The southwestward continuation of the Antler allochthon can be traced into the eastern part of the Sierra Nevada (Schweickert and Lahren, 1987) where it appears to be sharply terminated. Rocks farther west belong to the Klamath-Sierran arc discussed above. However, rocks correlated with the Antler allochthon are present to the south in the El Paso Mountains north of the Garlock fault and in several ranges in the northern Mojave Desert (Fig. 6; Burchfiel and Davis, 1981; *Carr and others, 1980, 1984;* €Dos). These rocks are interpreted to lie within a sliver displaced southward by late Paleozoic left-slip truncation (see below) of the southwestern part of the North American craton (Davis and others, 1978; Burchfiel and Davis, 1981; Walker, 1988). The trunction event removed rocks in the southern continuation of both the miogeocline and the Antler belt and shifted them into northern Mexico where similar rocks occur southeast of Hermosillo (Fig. 6; *Poole, 1981, 1983*). It should be noted that the deformation of Antler-correlative(?) strata in Mexico appears to be slightly younger than classic Antler deformation in Nevada (Stewart and others, 1990).

The Antler allochthon can also be traced northward from central Nevada into southern Idaho where some rocks are easily assigned to the allochthon. Other rocks, however, such as the Salmon River assemblage of Upper Cambrian to Upper Devonian oceanic rocks (€dos; *Nilsen, 1977;* Poole and others, this volume) are more difficult to assign to a specific paleogeographic element. We regard all these rocks as parts of a geometrically and stratigraphically complex Antler allochthon and suggest that the incompatibility of sediment source direction in the Salmon River assemblage to the rest of the allochthon, as reported by *Nilsen (1977),* might be explained by rotation of thrusted units during emplacement and the effects of younger structural overprinting. The northward continuation of the allochthon is obscured by the Idaho batholith, but it appears to trend into the region of western Idaho where Paleozoic facies belts have been truncated on the west by Mesozoic structures (see below).

The boundary between early Paleozoic miogeoclinal and continental rise/slope facies reappears in northeastern Washington. Here, the two rock sequences are in contact along a fault zone of variable dip direction that is probably post–middle Paleozoic in age and has modified original relationships between them. This contact can be followed into the Kootenay arc region of southern Canada, where recent work by Gehrels and Smith (1987) has suggested the presence of a major thrust fault contact within the early Paleozoic rocks of the Lardeau Group (Fig. 6). Overlap by the Upper Mississippian Milford Group of the two deformed and metamorphosed rock assemblages separated by this thrust within the Lardeau Group suggests the possibility of an Antler age event in the northernmost United States and adjacent southern Canada. The age of the deformation in this area is not as well constrained as in Nevada, but did occur between Devonian and Late Mississippian time (Klepacki and Wheeler, 1985). Deformational style was almost certainly different from that of the orogeny in Nevada. Some clasts, up to boulder size, in post-orogenic Milford conglomerates are of metamorphic and granitic (Ordovician?) rocks, unlike debris in the Nevada Antler foredeep that consists only of rocks derived from the Roberts Mountains allochthon. Early Paleozoic plutons (Cambro-Ordovician, Ordovician, and Devonian) and at least one pre-Milford phase of regional deformation and metamorphism have been documented in the Shuswap Complex west of the Kootenay arc (Okulitch, 1985). Their occurrence is reminiscent of little understood events of comparable age in the offshore Klamath-Sierran terrane to the south, but their original spatial and genetic relationships to the Canadian continental margin are uncertain.

Early Devonian deformation in the Klamath Mountains,

pre–Late Devonian deformation in the northern Sierra Nevada, and pre-Devonian deformation in the San Juan Islands might suggest an association with the Antler orogeny of Late Devonian–Early Mississippian age. Some workers have argued that the Antler orogeny was caused by a collision between the continent and an east-facing offshore arc (Burchfiel and Davis, 1972, 1975; Speed, 1979; Oldow, 1984b; Speed and others, 1988) and that deformation in the arc terrane was a precursor to the Antler orogeny. In this scenario tectonically imbricated strata in the Roberts Mountains allochthon were deformed in a forearc setting prior to and during collision. However, several lines of evidence do not support this hypothesis. The arc may have been west-facing as discussed above, and Devonian deformation in the northern Sierran arc is more than 15 m.y. older than the beginning of the Antler orogeny in Nevada. No arc-volcanic debris is present within the Antler foredeep deposits (Harbaugh and Dickinson, 1981), and no volcanic rocks of arc affinity that would represent a collided arc lie either above or immediately west of the Antler allochthon. Some workers have proposed that the colliding arc would have subsided soon after collision (Speed and Sleep, 1982), or was rifted from the margin shortly after collision. However, rocks of the late Paleozoic Havallah assemblage (DPos; see below) that are widely interpreted to have been deposited in deep water west of the Antler allochthon have been shown to range from Late Devonian to Permian in age (Fig. 7; Miller and others, 1984). Thus, deposition of the Havallah assemblage occurred prior to, contemporaneous with, and following emplacement of the Antler allochthon. The Havallah strata originated in a western position where a pre-Antler colliding arc should have been, if arc-continent collision took place. If there was a pre-Antler offshore arc it appears not to have collided with the continental margin at the time of emplacement of the Roberts Mountains allochthon.

Other workers (e.g., Burchfiel and Davis, 1972, 1975; Miller and others, 1984; E. Miller and others, this volume; Oldow and others, 1989) have hypothesized that emplacement of the Roberts Mountains allochthon was not the result of an arc collision, but was caused by partial closure of a marginal basin in which allochthonous Antler rocks came from its eastern part. Closure was incomplete and part of the marginal sea (Antler basin) remained open to receive sediments of the late Paleozoic Havallah basin. This interpretation is supported by Paleozoic sections in the northern Sierras, where sediments apparently correlative with those of the Havallah basin rest stratigraphically above Late Devonian phosphatic sediments comparable to those in both the Roberts Mountains allochthon and the Cordilleran miogeosyncline (Miller and Harwood, 1989).

Recent studies in parts of the Alpine system of Europe suggest an analog for the Antler orogeny (*Moretti and Royden, 1988; Royden, 1988a, b;* and Royden and Burchfiel, 1989). Several thrust belts within the Alpine system, for example, the eastern Carpathians and the Apennines, have been formed by retrograde movement of a subducting slab. In these belts the subduction zone migrated toward its foreland, probably in part due to the negative buoyancy of the subducted slab, and created extension in the overriding plate. Thus, formation of the thrust belts within the subduction zone was contemporaneous with extension 50 to 100 km away within the upper or overriding plate. In the Carpathians, extension caused thinning of the continental crust and formation of the Pannonian basin, whereas in the Apennines upper plate extension formed the oceanic crust of the Tyrrhenian Sea (*Kastens and Mascle, 1990*). Both thrust belts are associated with arc volcanism in the overriding plate, but because areas of extension migrated laterally with the subduction zone boundary, the arcs themselves were affected by extension and subsidence during or shortly after their formation. Thus, the upper plate extended areas became regions of sedimentation that developed contemporaneously with thrusting. When the thick continental crust enters the subduction zone it may be subducted a short distance depending upon the strength of the slab pull, but eventually subduction is terminated when the buoyancy of the continental lithosphere becomes greater than the vertical load on the downgoing slab. Mountain belts of this type characteristically have arcuate shapes toward the downgoing slab, have thrust belts that consist mainly of flysch-type sediments, and lack high mountain ranges that expose crystalline basement and belts of plutons or high-grade metamorphic cores.

The Antler orogenic belt has many of the characteristics of these types of Alpine thrust belts, and a similar interpretation can be proposed for it. Middle Devonian deformation in the west-facing Sierra-Klamath arc, or its along-strike equivalent, might have led to the initiation of east-facing subduction in the oceanic or marginal sea behind the arc. A thrust belt (e.g., the accretionary prism of the east-facing subduction zone) developed and retrograde motion within the subduction zone caused the thrust belt to migrate eastward toward the continental margin. An enlarging extensional region formed to the west (early Havallah basin) and farther west, the west-facing Sierran-Klamath arc remained active but was decoupled from east-facing Antler subduction. The thrust belt (accretionary prism) was emplaced onto the miogeocline as the North American continental margin (slope, shelf) was drawn down into the subduction zone. Thrust faulting and emplacement of the Antler allochthon eventually ceased when the negative buoyance of the subducting slab (westernmost North America) was lost. During early emplacement of the allochthon, Late Devonian sedimentation began in the extended region to the west (Havallah basin). Extension in the upper plate of these convergent systems occurs synchronously with convergence and is not an unrelated aftereffect. Such thrust belts do not require collision of an arc against continent, nor is it necessary to postulate that a collided arc must be immediately rifted away or subsided thermally to explain its nonpresence. Although most earlier workers have concluded that western portions of the original Antler orogenic belt are somehow missing, it is likely that the entire Antler orogenic belt is preserved in the western United States. The Antler orogen, as we view it, consisted of only a narrow upper plate thrust belt (accretionary prism) of oceanic rise and continental slope sedimentary rocks and a contemporaneous

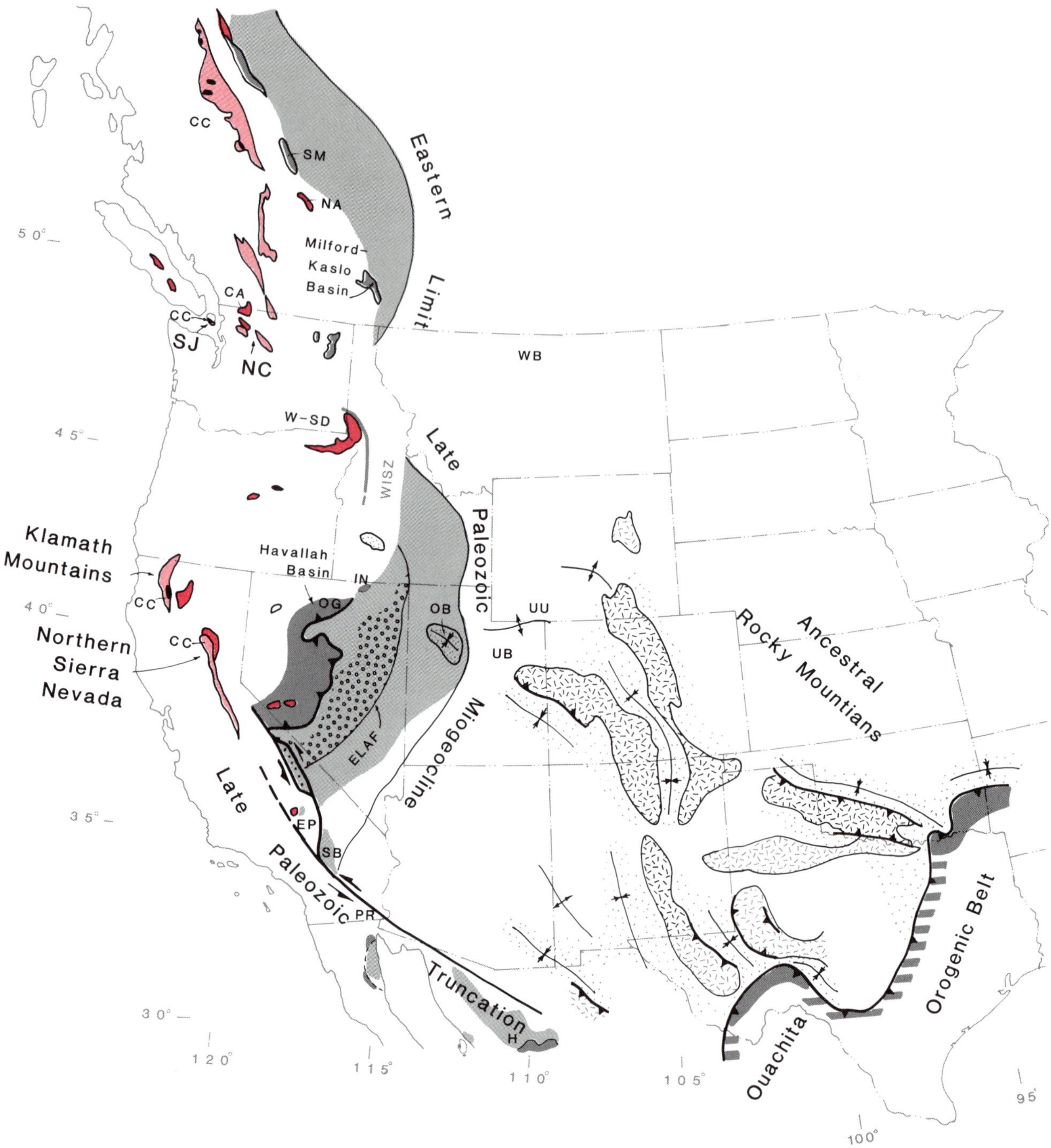

Figure 7. Paleogeographic and tectonic map of late Paleozoic time showing Cordilleran elements as well as intracratonic Ancestral Rocky Mountains (dark patterned where uplift was relatively large, and gray patterned where uplift was relatively moderate) and Ouachita orogenic belt (dashed where present in subsurface). Light dot pattern for Ancestral Rocky basins adjacent to uplifts. Ophiolites are black. Truncation of the southern margin of the Cordilleran belt took place during Pennsylvanian and Permian times. Distribution of rocks in Mexico from Stewart and others (1990). Other rock units: ophiolites (black); miogeoclinal rocks (light gray); offshelf rocks (gray); arc-volcanic rocks (red). CA, Chilliwack arc (red); CC, Cache Creek assemblage (pink); ELAF, eastern limit of Antler foredeep; EP, El Paso Mountains; H, Hermosillo; IN, Independence Mountains; NA, Nicola arc (red); NC, northern Cascade Mountains; OB, Oquirrh basin; OG, Osgood Mountains; PR, Peninsular Ranges; SB, San Bernardino Mountains; SJ, San Juan Islands; SM, Slide Mountain; UU, Uinta uplift; UB, Uinta basin; WB, Williston basin; WISZ, future location of West Idaho suture zone (gray); W-SD, Wallowa-Seven Devils arc red.

basin of sedimentation immediately to its west. The most important piece of evidence to support this interpretation is the contemporaneity of the basal sedimentary rocks of the Havallah sequence with the Antler orogeny to the east (Miller and others, 1984; E. Miller and others, this volume).

Other features associated with the Apennine/Carpathian type of thrust belts are their limited lateral extent, the filling of reentrants in continental margins, diachronous thrusting along the belt, and associated strike-slip faults. It may be significant that the Antler allochthon is present in the southern reentrant of the miogeocline. Complex structues at the ends of the reentrant, for example at its southern end in central Nevada (Oldow, 1984b) and at its northern end in Idaho (*Nilsen, 1977*), may be associated with strike-slip movement or oblique subduction. Depending upon the original geometry of the reentrant in the miogeocline, the structural style of the Antler belt in central Nevada may not have extended far beyond its present outcrops. It probably did extend some distance to the southwest, as fragments of the belt are present in the Mojave Desert (Burchfiel and Davis, 1981) and perhaps in Sonora, Mexico (*Poole, 1981, 1983*), where they were displaced by left slip during Paleozoic time (see below).

LATE PALEOZOIC EVOLUTION

The late Paleozoic evolution of the Cordillera is dominated by three events. (1) Extensive deformation within the North American craton for the first time since the Middle and Late Proterozoic formed uplifts and basins within the crystalline basement rocks of the craton from the Ouachitas in the southeast to the western Cordillera on the northwest. (2) The truncation of structures and facies belts along a transform boundary in the southwestern part of the Cordilleran belt developed a new northwest-trending margin that controlled Mesozoic evolution in this part of the orogen. (3) The closure of a marginal basin and the accreton of an island-arc terrane in the western Cordillera during the Permo-Triassic Sonoma orogeny caused a significant westward shift of the continental margin from a position it had maintained in central Nevada since Cambrian time.

North American craton and the Ancestral Rocky Mountains

During Mississippian time low-amplitude deformation on the craton produced east-west–trending troughs in the Williston and Uinta basins east of the central part of the miogeocline (*Rose, 1976*), and uplifts along the Montana arch and the Oquirrh-Uinta axis (Fig. 7). These mild deformations within the craton gave way in Early Pennsylvanian time to extensive uplifts cored by Precambrian crystalline rocks, called the Ancestral Rocky Mountains, that were flanked by deep sedimentary basins (Fig. 7). These uplifts lasted into Early Permian time and locally into Late Permian time when the Front Range and Uncompahgre uplifts remained sediment sources (*McKee and others, 1967, 1975; Bachman, 1975*). The formation of uplifts and basins was not uniform, and deformation was diachronous and most intense during Pennsylvanian time. In many places, differential uplift exceeded 6 km. The uplifts and basins extended northwestward from southern Oklahoma to northeastern Nevada and perhaps to southern Idaho, and spanned a north-south breadth from west Texas to central Colorado. Many of the uplifts are bounded by thrust faults, some of which followed older faults of Early Proterozoic (*Muehlberger and Dickerson, 1989*) to Cambrian age. These uplifts are called the Ancestral Rocky Mountains, because some of them were rejuvenated during Late Cretaceous and early Tertiary time to form the Laramide Rocky Mountains.

The origin of the Ancestral Rockies is uncertain and remains an important tectonic problem of North American geology. The dynamic system that produced these structures produced faults that cut almost across the entire width of the craton. Burchfiel (1979), Kluth and Coney (1981), and *Kluth (1986)* recognized that the formation of the Ancestral Rockies occurred at the same time as the formation of the Ouachita Mountains along the southern boundary of the North American craton. They hypothesized that this deformation occurred in response to diachronous continent-continent collision along this boundary. Recent recognition of important late Paleozoic left-slip truncation of the continental margin in the southwestern part of the Cordillera (see below) might suggest that it played a role in the formation of the Ancestral Rockies, although the contractional character, geometry, and areal distribution of these structures suggest that this plate margin may have had a complex late Paleozoic history.

Cordilleran Miogeocline

The late Paleozoic tectonostratigraphic elements of the U.S. Cordillera are best and most completely developed along an east-west transect at about the latitude of central Nevada. Within the eastern part of the miogeocline late Paleozoic carbonate sedimentation (€Jcs) continued unbroken from early Paleozoic time with only minor influxes of fine clastic material from the Antler foredeep (Mfd). In this part of the miogeocline there is essentially no indication that the Antler orogeny had taken place to the west. Pennsylvanian and Permian deformation from the Ancestral Rockies extended west into the miogeocline and caused significnt west-trending uplifts in northwest Utah, adjacent eastern Nevada, and perhaps southern Idaho. The Uinta-Oquirrh uplift, which partly follows its early Paleozoic predecessor the Uinta-Tooele arch, and the Deep Creek–Tintic uplift to the south, flanked the subsiding Uinta basin during Pennsylvanian time (*Jordan and Douglass, 1980;* Fig. 7). The Oquirrh basin in north-central Utah, which may have partly connected with the Uinta basin, received more than 6 km of Pennsylvanian and Early Permian carbonates and clastics. The remainder of the eastern part of the miogeocline continued to receive sediment in areas of broader and more subdued subsidence that were controlled at least locally by uplifts of low amplitude. Along the northwest part of the miogeoclinal salient in southern Idaho, the Salmon River/Lemhi arch (Fig. 4) ceased to be a positive area in late Paleozoic time and carbonate deposition spread across it to the northeast.

Thick Antler foredeep deposits within the western part of the miogeocline accumulated above early Paleozoic shelf carbonates from Late Mississippian time into Permian time. Locally, sediments of Mississippian (Merrimac and Chester) age were deposited unconformably on the Antler allochthon (*Poole and Sandberg, 1977;* Speed, 1977; Miller and others, 1984; *Little, 1987*). Pennsylvanian and Permian sediments (PPcs) transgressed across eroded Antler highlands (= the Antler allochthon) and had covered them by Late Pennsylvanian and Early Permian time (Saller and Dickinson, 1982). In places such as the Independence and Osgood Mountains and near Carlin (*Dott, 1955*), Nevada, faulting accompanied late Paleozoic sedimentation (Fig. 7). In the Independence Mountains, east-west–striking faults were active during clastic sedimentation and the eruption of basalts. In the Osgood Mountains, Pennsylvanian sediments overlapped not only the Antler allochthon, but locally were deposited on low-grade metamorphic rocks of the autochthon/parautochthon (*Hotz and Wilden, 1964*). Some of these sequences show progressive deepening of water depths during sedimentation. Stratigraphic relations suggest that significant extensional faulting occurred along the western part of the continental margin during late Paleozoic time (E. Miller and others, this volume). Some faults active during this time may be westernmost expressions of Ancestral Rocky Mountains structures, but there is considerable evidence that most of the widespread deformation in the western part of the miogeocline was related to tectonic activity along the post-Antler continental margin farther west.

Western offshore elements

In western Nevada, the Havallah assemblage (DPos), a thick sequence of fine- and coarse-grained clastic sediments, chert, and greenstone, makes up the Golconda allochthon that was thrust eastward onto the North American continental margin in Late Permian to Early Triassic time during the Sonoma orogeny (Fig. 8). Rocks of the Havallah assemblage are mainly of Late Mississippian to Permian (locally Guadalupian) age, but in a few places they are as old as Late Devonian (Famennian; Schoonover Formation, Miller and others, 1984). Volcanic detritus of Early Mississippian (Kinderhookian) age in the Schoonover Formation of the Havllah assemblage is inferred to have come from a western island-arc source, but detritus in somewhat younger Kinderhookian clastic rocks is clearly derived from the Antler allochthon on the continental shelf to the east. The upper part of the Havallah assemblage contains an increasing amount of westward-derived arc-volcanic material indicating that a volcanic arc clearly lay to the west by Permian time. The repeated occurrence throughout Havallah stratigraphic sections of basalts (e.g., Late Devonian, Mississippian, Permian) has been interpreted by E. Miller and others (1984; this volume) to indicate recurrent extension within the Havallah basin and along the continental slope and rise to the east. With a volcanic arc to the west and the North American continent to the east, the Havallah assemblage is believed to have been deposited in a marginal basin that experienced extension (or transtension) during its evolution (E. Miller and others, this volume).

Late Paleozoic rocks exposed in the northern Sierra Nevada and eastern Klamath Mountains (Fig. 7) comprise broadly correlative portions of a volcanic-arc complex (DPa). Structural and stratigraphic relations between coeval rocks in the Havallah basin to the east (DPos) and the arc complex are difficult to establish because the intervening area is mostly covered by younger rocks or intruded by Mesozoic plutons. The upper Paleozoic arc-related rocks lie depositionally above middle Paleozoic arc sequences (Klamaths: Copley and Balaklala units; Sierras: Sierra Butte and Taylor Formations) and thus are younger parts of an evolving arc complex. From late Middle Devonian time (Saleeby and others, 1987) arc volcanism waxed and waned throughout the late Paleozoic. Although volcanic rocks or volcaniclastic sedimentary rocks are present throughout the upper Paleozoic succession, much of the sequence is poorly dated. Major periods of intense volcanic activity occurred in Late Devonian, Pennsylvanian, and Permian (Leonard-Guadalupe) times (Miller, 1987; Harwood, 1988). Granitic rocks within the arc are rare, but recent work has shown the presence of Middle to Late Devonian (Di; 364 to 385 Ma, Saleeby and others, 1987; Hanson and others, 1988) and Late Permian (Pi; 259 Ma, Wright quoted in E. Miller, this volume) plutons. Stratigraphic correlations between northern Sierran and eastern Klamath sections are complicated by pronounced facies changes within the volcanic archipelago (e.g., *Hannah and Moores, 1986;* Harwood, 1983; Watkins, 1985; *Miller, 1986*). Deep- and shallow-water sedimentary rocks are variably present within the arc and were deposited both during times of little or no volcanic activity and/or at distances from active volcanic centers. The Permian McCloud Limestone of the eastern Klamath Mountains contains a late Paleozoic fusulinid and coral fauna that is distinct from either coeval North American or Tethyan faunas now found in area to the east and west, respectively. However, elements of both faunas are found in McCloud rocks and they are considered to have a transitional fauna (cf., Miller, 1987). The distinctive McCloud fossil assemblage is, however, found within other late Paleozoic arc sequences present within the Cordillera from the northern Sierra Nevada to northern Canada (e.g., Chilliwack, Quesnellia), a relation that indicates regional paleogeographic and tectonic ties between them (Miller, 1987; see below).

West of the Klamath-Sierran arc is a belt of highly disrupted chert- and shale-rich strata that contain rocks that range in age from Devonian(?) to Early Jurassic (PJap). This belt, called informally the Cache Creek assemblage from its type locality in southern Canada, has been interpreted as a mélange terrane (i.e., an accretionary prism) formed in late Paleozoic to early Mesozoic time (Davis and others, 1978). Within the chert and shale mélange are blocks of sandstone, limestone, and mafic volcanic rocks and other ophiolite sequence lithologies. One characteristic of this mélange is the presence of a late Paleozoic Tethyan fauna in some of its limestone units. Blueschists, both exotic blocks and intact metamorphic terranes, are locally present within it and

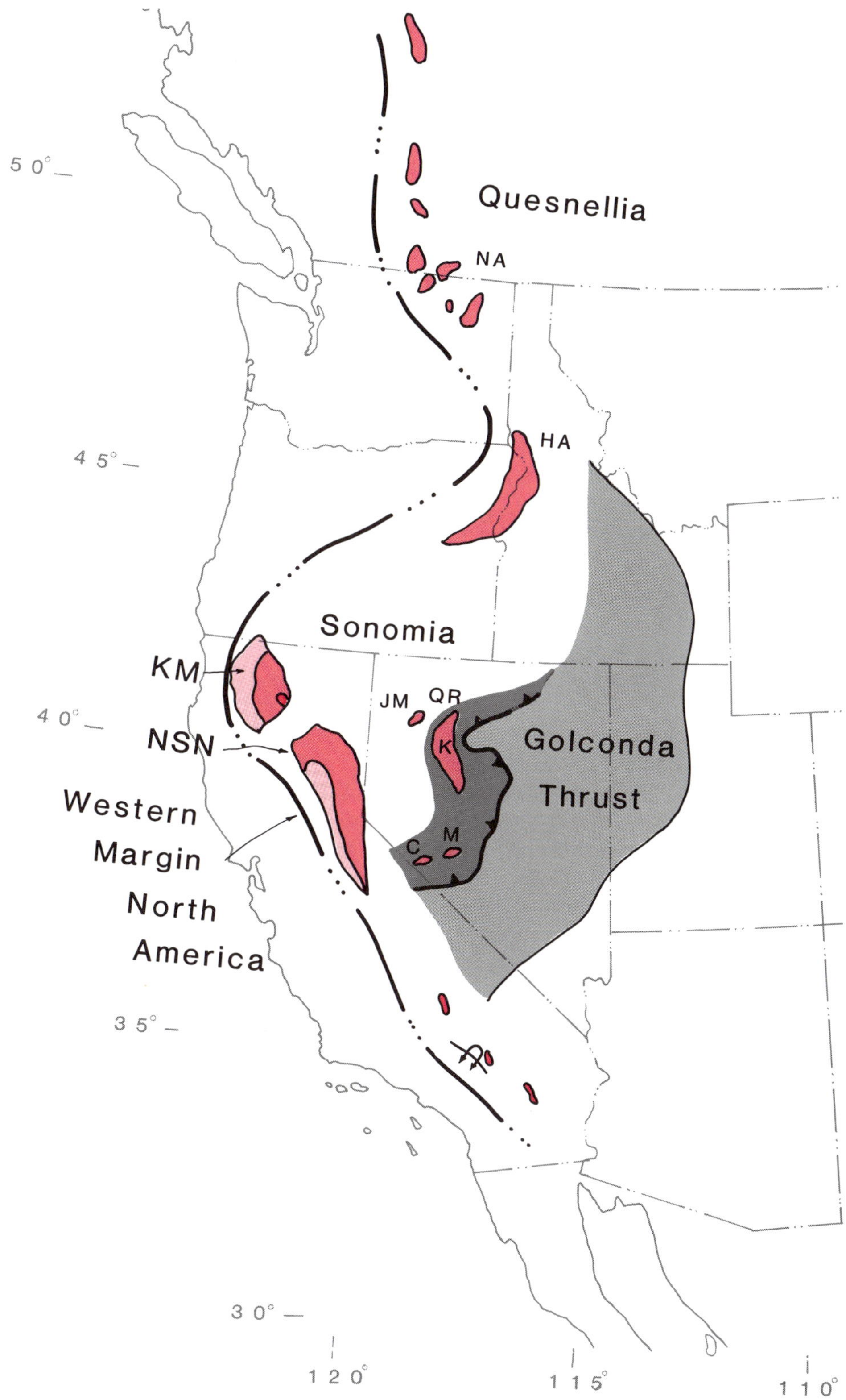

Figure 8. Tectonic elements of the Sonoma orogeny (sensu lato). Patterned units include: late Paleozoic miogeocline and Antler overlap assemblages (light gray); offshelf and marginal basin sediments of Havallah basin (gray); Permian plutons (deep red); volcanic rocks (red); western mélange (pink). C, Candelaria; HA, Huntington arc (red); JM, Jackson Mountains; K, Koipato volcanic rocks (red); KM, Klamath Mountains; M, Mina; NA, Nicola arc (red); NSN, northern Sierra Nevada; QR, Quinn River.

generally yield Late Triassic ages in the range of ca. 225 to 210 Ma, although Permian high-pressure schists are present in the same belt in western Washington (see below).

Collectively, the late Paleozoic (to earliest Mesozoic) tectonostratigraphic elements along the east-west transect reviewed above (Fig. 7) can be interpreted as indicating that a marginal basin, fringing island arc, and subduction-related accretionary prism (mélange belt) lay west of the North American continental margin. The location of the Permian(?)-Triassic mélange belt west of the late Paleozoic Sierran-Klamath arc strongly suggests that the arc was west facing. The incorporation of oceanic rocks and carbonate blocks containing Tethyan fossils into the mélange indicates that a broad expanse of oceanic lithosphere was subducted eastward beneath the arc. The breadth of the Havallah (marginal?) basin between arc and continent is a matter of conjecture. Extensional deformation within the basin and along the North American continental margin may be partly a result of back-arc extension. However, the basin might have been wide enough that deformation within it was related to transtension caused by oblique motions within a plate-boundary system located between the arc and western North America.

Truncation of the southwestern Cordilleran margin and establishment of a new continental margin

Tracing these tectonostratigraphic elements southward from the east-west transect at the latitude of central Nevada indicates that they experienced considerable tectonic modification and paleogeographic change during late Paleozoic and later time. The miogeocline (€Jcs) strikes southwestard across southern Nevada and southeastern California where it trends into Antler rocks (€Dos) in the northern Mojave Desert (Fig. 7). In the San Bernardino Mountains and eastern Peninsular Range of southwestern California, rocks belonging to the eastern part of the miogeocline or the craton (€jc) are presently juxtaposed against rocks to the west that are in part of oceanic character and probably mostly Mesozoic in age (Jos). The contact between these different rock units is obscured by Mesozoic plutons. However, there are no representatives of the late Paleozoic western miogeocline, Havallah basin, arc, or mélange belts. This area represents the southern of the two major discontinuities in the Paleozoic paleogeographic elements of the Cordillera. We favor the interpretation that the southwest part of the Cordillera was truncated by left-slip displacement, and that the continuation of some of the Cordilleran elements were displaced southward into Mexico (Fig. 7) where they are present at Caborca (*Cameron, 1982*), in northern Baja California, and southeast of Hermosillo (Stewart and others, 1990). Antler rocks, both the allochthon and its foredeep and overlap assemblages, present in the northwest Mojave Desert and El Paso Mountains (*Carr and others, 1984*) are interpreted to be a sliver of these rocks translated southward along the left-slip boundary, but left behind as the main truncated fragment was displaced into northern Mexico (Davis and others, 1978; Burchfiel and Davis, 1981).

Recent studies by Walker (*1985,* 1988) and Stone and Stevens (1988) have indicated that the left-slip displacement took place during Early or Middle Pennsylvanian to late Early Permian time and that it was complete by Early Triassic or Late Permian time. Northwest-trending turbidite basins in the western Death Valley region developed across miogeoclinal facies trends in Pennsylvanian and Permian times. These basins are interpreted as pull-apart basins developed along a left-slip transform fault zone that was mostly transtensional, but at times transpressional (Stevens and Stone, 1988). The inferred left-slip faults have been deformed by younger tectonic events, intruded by abundant Mesozoic plutons, and are not exposed. Granitic plutons of Late(?) Permian or Early Triassic age (252 to 241 Ma) form a poorly defined, northwest-trending belt that crosses the eastern boundary of the displaced sliver (Burchfiel and Davis, 1981; Walker, 1988). Early Triassic sedimentary rocks (Ŧ s) are present on both sides of the fault (Walker, 1988). These relations indicate that displacement on the fault had been completed by the Early Triassic. The plutons suggest that the truncated boundary may have already been the locus of eastward subduction by Late (?) Permian time. This northwest-trending continental margin established in late Paleozoic time controlled the trend of subsequent Mesozoic and Cenozoic tectonics in this part of the Cordillera.

In southern California Late(?) Permian–Early Triassic plutons (Pi) intrude deformed and metamorphosed cratonal, miogeoclinal, and Antler rocks (Figs. 8 and 9), and Early Triassic rocks (Ŧ s) overlap the plutons and the deformed rocks. These relations indicate that truncation was completed and an Andean tectonic setting had developed along the truncated margin by this time. This Late(?) Permian–Early Triassic Andean-type deformation contrasts with the Sonoma orogeny of probably equivalent age farther north that was characterized by the emplacement of Havallah assemblage strata (Golconda allochthon) eastward onto the North American continental shelf. A continuation of the Andean belt to the southeast is at present unknown, but it is permissible given uncertainties in radiometric age data that its continuation to the north is the Sierran-Klamath arc (see below).

Relations between late Paleozoic tectonostratigraphic elements in the southwestern and central parts of the Cordillera remain uncertain because of superposition of younger magmatic and structural events. In the El Paso Mountains just north of the Garlock fault (Figs. 6, 7) Antler rocks are overlain unconformably by a thick succession of late Paleozoic carbonate rocks and fine-grained clastic rocks that are intercalated with a thick section of andesitic volcanic rocks of Permian (Leonard-Guadalupe) age (ℙPcs) and are intruded by a 252-Ma pluton. Some Pennsylvanian sedimentary rocks in this sequence are similar to Havallah assemblage rocks at the southern end of the Golconda allochthon. Arc debris within the Havallah assemblage has been dated at 256 to 267 Ma (*Tomlinson and Wright, 1986*). Volcanic rocks (Pa) tectonically incorporated in the southern part of the Golconda allochthon are interpreted by Speed (1977) as part of the arc that lay west of the Havallah basin and have been dated as 253 ± 3 Ma (Black Dike Formation) and 255 ± 8 Ma (Mina

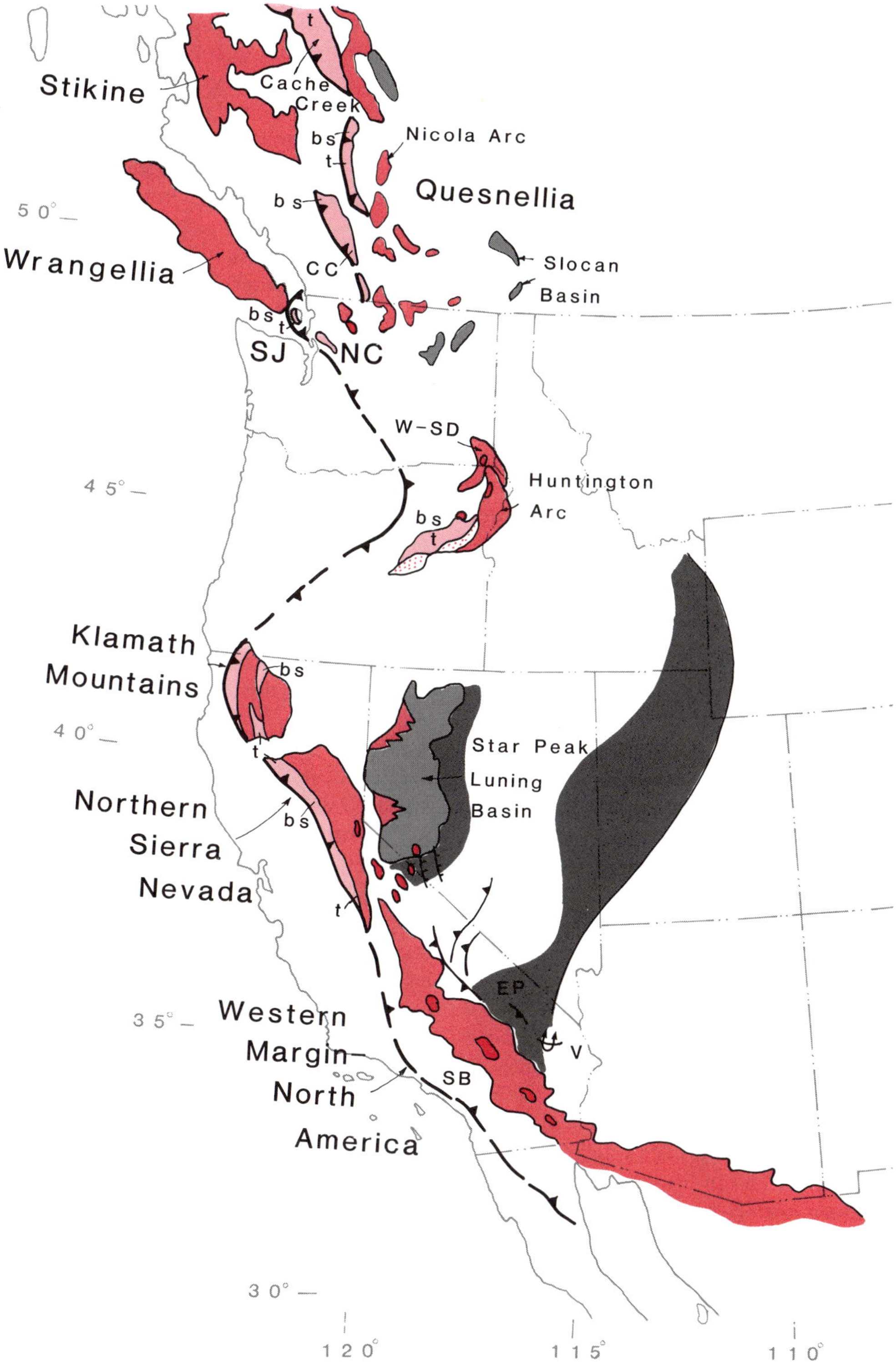

Figure 9. Paleogeographic and tectonic map of Triassic to Middle Jurassic time showing the Huntington and Nicola arcs. Patterned units include: Triassic plutons (deep red) and volcanic rocks (red); fore-arc basins (red dots); marginal basins (Havallah, Slocan, etc.) developed on oceanic crust and/or extended continental crust shown (gray); western Cache Creek mélange belt (pink). Light gray shows present distribution of thick sedimentary rocks in the eastern part of the Cordillera. Localities of blueschist-facies metamorphism (bs) and those containing Tethyan faunas (t) are indicated. CC, Cache Creek assemblage (pink); EP, El Paso Mountains; NC, northern Cascades; SB, San Bernardino Mountains; SJ, San Juan Islands; V, Victorville; W-SD, Wallowa–Seven Devils arc.

Formation; Speed, 1977; E. Miller and others, this volume). In the northern Sierra Nevada a major pulse of volcanism also occurs in Leonard time (Harwood, 1988), and the Pit River stock in the eastern Klamath Mountains has an age of 260 Ma (*Irwin, 1985*). All these volcanic and plutonic rocks are permissibly the same age. Based on this admittedly sparse database we interpret that all these rocks are expressions of the same Permian arc. Paleogeography during Late Permian time consisted of an offshore island arc in the north that crossed the continental margin to become an Andean arc in southern California (E. Miller and others, this volume). The back-arc Havallah basin narrowed and ended to the south, and the Permian arc was superposed on a complex older paleogeography in southern California.

Recognition that the left-slip truncation of the southwestern Cordilleran margin could be as old as Pennsylvanian–Early Permian (Walker, *1985,* 1988; Stone and Stevens, 1988) leads to uncertainties concerning the southern continuation of the Klamath-Sierran arc and its interrelations with the late Paleozoic left-slip fault. Left-slip motion could very well have been contemporaneous with Pennsylvanian–Early Permian volcanic arc activity to the north. Two interpretations can be envisioned that represent end members of a spectrum of possible plate tectonic geometries. First, the left-slip fault zone was a transform boundary that joined an eastward-dipping subduction zone beneath the more northerly arc. This geometry would not require any of the Pennsylvanian–Early Permian tectonostratigraphic elements of the central transect to have originally continued southwestward past the transform fault, although such a continuation is not precluded. Second, the left-slip boundary is a fault zone that lies inboard of an obliquely convergent east-dipping subduction zone present along the entire western margin of the Cordillera. In Pennsylvanian/Permian time sinistrally oblique convergence became dominant and produced an upper plate, trench-parallel transcurrent fault that cut across all tectonic elements, moving those elements west of the fault southward. This geometry would require that at least the early late Paleozoic tectonic elements would have originally continued to the south. At present there is not enough data to indicate which interpretation, or some combination of them, might be correct. The evidence necessary to support these interpretations lies in discovering what tectonostratigraphic elements were present west of the southern Cordilleran margin in early late Paleozoic time. In either interpretation, by Late Permian time east-dipping subduction would have been continuous along the entire margin and would have initiated an Andean belt in the south that changed northward to an island-arc setting. The south-to-north paleogeography of the southwestern United States would then have been similar to the present east-to-west geometry of southern Alaska and the Aleutians (Oldow and others, 1989; E. Miller and others, this volume).

Younger strike-slip displacements have further complicated attempts to reconstruct late Paleozoic paleogeography in California. The few known or postulated Paleozoic and Mesozoic strike-slip faults that have disrupted western Cordilleran tectonic elements have been intruded by the Mesozoic Sierra Nevada batholith and are difficult to recognize. Schweickert and Lahren (1987) have made a strong case that western Mojave-area miogeoclinal rocks and structures have been displaced by several hundred kilometers of right slip into the eastern part of the Sierra Nevada, probably in late Mesozoic time (see below). Reversing 500 km of dip along the Snow Lake fault responsible for this displacement places the northern Sierran arc west of the El Paso Mountains (Fig. 7) in early Mesozoic time. The likely existence of this fault indicates that the late Paleozoic Sierra-Klamath island arc has moved parallel to the Cordilleran margin since its formation, and probably has been in its present position with respect to coeval tectonostratigraphic elements to the east since only late Mesozoic time. Thus, the arc that lay west of the Havallah basin during much of late Paleozoic time may have been northern parts of a greater northern Sierra–eastern Klamath–Chilliwack arc system (DPa; Miller, 1987), not the part that is presently there.

Continuity of late Paleozoic elements in the northern part of the Cordillera

Most of the late Paleozoic tectonostratigraphic elements present at the latitude of central Nevada are missing in western Idaho and eastern Oregon; this absence defines the northern of the two major discontinuities in Cordilleran tectonic elements (Figs. 6 and 7). Parts of a late(?) Paleozoic/early Mesozoic arc and related rocks (TR a, PJa), in part correlative with Klamath-Sierran tectonostratigraphic terranes (Miller, 1987), are juxtaposed across a faulted and intruded north-trending tectonic zone against various members of the Cretaceous Idaho batholith and metamorphosed pendants within it of Belt strata and probable miogeoclinal rocks of latest Proterozoic and/or Paleozoic age (Lund and Snee, 1988). This major and enigmatic tectonic boundary, here called the West Idaho suture zone, is also called the Salmon River suture by Lund and Snee (1988). All Paleozoic paleogeographic and tectonic elements in the east-west transect through central Nevada between the miogeocline and the eastern Klamath/northern Sierra arc are missing along the trace of the suture zone. The presence of paleogeographic elements similar to those of the central Nevada transect along the Canadian–U.S. border suggests that these elements were also originally present at the latitude of eastern Oregon and western Idaho in late Paleozoic and early Mesozoic time, but have somehow been removed (as discussed below). In southern Canada early Paleozoic miogeoclinal shelf, slope, and rise sequences (€Dos), as well as parts of an Antler-age foredeep (Milford Group, Mfd) are preserved. Late Paleozoic rocks equivalent to the Havallah assemblage and parts of a western arc assemblage are also present (Milford and Kaslo Groups; *Klepacki, 1985;* Mount Roberts–Anarchist Group, Read and Okulich, 1977; MTR os), and record poorly dated deformations of mid- to Late Permian (*Klepacki, 1985*) and pre-Late Triassic time (Read and Okulich, 1977) that might be approximately equivalent to the Sonoma orogeny (see below). Lying unconformably upon these units is the upper Triassic, west-facing (Mortimer, 1986) Nicola arc (TR Ja), west of which is the type-area of the Cache Creek mélange belt (PJap) with its carbonate blocks containing Permian Tethyan fauna.

Possible counterparts of some of these northeastern Washington and adjacent British Columbia units reappear to the west in the northern Cascades and the San Juan Islands of northwestern Washington. For example, the western equivalent of the Cache Creek assemblage is the Deadman Bay terrane (PTR oc), including the Trafton (PJap) and Elbow Lake (PTR oc) units in the northwestern Cascades (Brandon and others, 1988). Associated with these units are Permian to Early Triassic high-pressure rocks of the Vedder Complex (range ca. 271 to 238 Ma; Armstrong and others, 1983) and Garrison Schist (286 to 242 Ma; Brandon and others, 1988) that include greenschists, albite-epidote amphibolite, and blueschist. In the Chelan Mountains terrane in the northern Cascades (Tabor and others, 1987), Upper Triassic plutonic rocks of the Marblemount belt and associated metavolcanic rocks may represent a western repetition of the Triassic Nicola arc (TR Ja).

Davis and others (1978) suggested that late Paleozoic tectonic elements once continuous along the northwestern margin of the United States from the northern Sierra into southern Canada were removed from the area of eastern Oregon during late Mesozoic time by right-slip faulting (see below). These elements were displaced northwestward into their present position in the northern Cascades, thus producing a doubling-up of the tectonic elements in western Washington and adjacent southwestern British Columbia (see below). The timing and possible geometry of this inferred displacement is discussed below.

LATE PALEOZOIC–EARLY MESOZOIC SONOMA OROGENY

Late Paleozoic and early Mesozoic deformation affected rocks along most of the western edge of the North American Cordillera (Gabrielse and others, 1983). Deformational events were clearly diachronous and of variable structural style, but we regard these events as part of a Sonoma orogeny sensu lato. "Sonoma" orogenic events in southern California were different in style from those of the type orogeny in central Nevada. In southern California plutonism and deformation occurred in a continental margin (i.e., Andean setting), as discussed earlier. The difference in tectonic setting of the two areas is related to truncation of the continental margin in the southern California region, followed by eastward subduction—events that did not occur in central Nevada. Somewhat cryptic Permian and/or pre–Late Triassic deformation(s) are also known within the volcanic arc regions of southern Canada (*Klepacki, 1985;* Gabrielse and others, 1983; Read and Okulitch, 1977), but cannot be proven within the northern part of the adjacent U.S. Cordillera.

The Sonoma orogeny was first described from central Nevada where rocks of the Havallah assemblage (DPos) were thrust eastward on the Golconda thrust above the western part of the Antler allochthon (€Dos) and its late Paleozoic cover (PPcs; Fig. 8). The structural style and lithologies of the allochthon are very similar to those of the middle Paleozoic Antler allochthon (cf., Brueckner and Snyder, 1985). Eastward translation of the allochthon was at least 50 km without removing the effects of Cenozoic extension. The age and tectonic setting for the Sonoma orogeny in its type area remain problems. Rocks of the Golconda allochthon (Havallah assemblage) are as young as Late Permian. In north-central Nevada upper Lower Permian rocks of the allochthon are unconformably overlain by Lower Triassic volcanic rocks (TR a, Koipato Formation; Silberling, 1973), but the volcanic rocks do not rest on the autochthon as well and could have been carried piggyback atop the allochthon during its final emplacement. In the Mina-Candelaria region of west-central Nevada, Speed (1977) demonstrates that the Golconda allochthon lies structurally above Early Triassic rocks of the Candelaria Formation (PTR s). This formation consists of hemipelagic sediments overlain by prograding fan deposits. An upward increase in volcanic debris in the fan deposits is interpreted by Speed to indicate the advance of the Golconda allochthon and an associated island arc. The allochthon and its internal structure are interpreted by Speed (1977) and Brueckner and Snyder (1985) as part of an accretionary prism formed in a west-dipping subduction zone. Final implacement of the allochthon terminated Candelaria sedimentation on the continental margin during arc-/basin-continent collision.

Oldow (1984b), however, has suggested that the thrust relationship in the Mina-Candelaria area may be late Mesozoic in age and does not provide direct dating of the Sonoma orogeny. In northwest Nevada the allochthon is overlain by thick (4 to 5 km) Upper and Middle Triassic (Oldow and others, 1989, have indicated Lower Triassic as well) fine-grained sedimentary rocks (Fig. 9; TR Jsb) that are believed to be lateral equivalents of thin shelf deposits (TR s) that rest on the autochthon to the east (*Lupe and Silberling, 1985*). Taken together all these relations would indicate that thrusting was completed by Early(?) or Middle Triassic time.

Several problems remain unsettled with current interpretations of the Sonoma orogeny. Speed (1977) suggested that the arc which collided with the continental margin included all late Paleozoic rocks in the northern Sierra–eastern Klamath arc and scattered outcrops of coeval sedimentary and volcanic rocks in northwest Nevada. He called this large (composite?) arc complex Sonomia and regarded it as having been accreted to North America during the Sonoma orogeny. Others, however, have suggested the Golconda allochthon may have been emplaced in Mesozoic time (*cf., Ketner, 1984*) and that stratigraphic sequences are present in some areas within Sonomia that indicate no deformation during Permo-Triassic time. For example, a section in the Quinn River Crossing area of northwest Nevada (Fig. 8) preserves a record of continuous Late Permian and Early Triassic shallow-water sedimentation (Ketner and Wardlaw, 1981; *Jones and others, 1988*), and a structurally separate section in the same area contains an Early Permian fauna of McCloud affinity (Miller, 1987) characteristic of the Klamath-Sierran arc. A sequence of rocks in the Jackson Mountains (Fig. 8) also contains a succession of volcanic rocks (Pa) and limestone (DPos) that is continuous across the Permo-Triassic boundary (McGill Canyon Formation; *Russell, 1981*). We regard the evidence for latest Permian to Early Triassic emplacement of the Golconda allochthon as convincing, however, and suggest that the stratigraphic data for continuity of Permian and Triassic rocks in northwestern Nevada can be incorporated into the tectonic setting for the Sonoma orogeny.

An Apennine tectonic setting has been proposed above for the earlier Antler orogeny, and such a setting could also explain some of the unresolved geologic relations related to the Sonoma orogeny. The Sonoma orogeny, like the Antler orogeny, juxtaposes a continental slope and rise and/or marginal basin sequence atop the continental shelf, and the similarity between the two orogenies suggests a similar tectonic setting (Burchfiel and Davis, 1972). Final closing of the Havallah basin can be interpreted to have occurred along a west-dipping retrograde subduction zone of Apennine-type that carried an arc in its upper plate. Volcanism migrated eastward as the thrust front propagated eastward, and Late Permian (Black Dike and Mina Formations) and possibly even Early Triassic (Koipato) volcanic rocks were erupted during basin closure. In the latter case (Early Triassic volcanism), the volcanic rocks were deposited on already deformed Havallah rocks, but could still have been syntectonic with respect to development of structures in the easternmost part of the Golconda accretionary prism. The upper plate rocks were extended during thrusting, and local basins within the arc terrane received sediments (Quinn River and Jackson Mountains) that were deposited contemporaneously with thrusting farther east.

It is conceivable that the Lower(?), Middle, and Upper Triassic basin (Star Peak/Luning, see below) that overlies the Golconda allochthon in northwest Nevada was caused by similar extension, but available data appear to support an earlier termination of west-dipping subduction. Final closure of the Havallah basin took place when the thick continental lithosphere of North America entered the subduction zone, thus cancelling the negative buoyancy of the downgoing slab. Continental lithosphere was not pulled into the subduction zone very far before termination of subduction because overlap of the Golconda allochthon onto the continental margin may have been as little as 50 km (E. Miller and others, this volume; see also Speed and others, 1988). The arc itself did not ride over the thick continental lithosphere but was simply accreted passively when subduction ceased. Shortening within the arc is not necessary in such a setting as the arc was mainly in extension during its accretion to North America. Such a scenario would explain the lack of high-grade metamorphism and plutonism normally associated with "collisional" events. The lack of a foredeep could result from the part of the Golconda allochthon overlying the continental lithosphere being so thin that it did not produce a flexural depression in front of it. It remains uncertain whether the west-dipping retrograde subduction system that closed the Havallah basin was contemporaneous with east-dipping subduction on the west side of an eastern Klamath–northern Sierra arc terrane, although we consider it more likely than not (see below).

TRIASSIC TO LATE JURASSIC EVOLUTION

The accretionary belt

Numerous geologists have proposed over the years that the late Paleozoic and earliest Mesozoic "fringing" arc terranes of the northern Sierra Nevada and eastern Klamath region (= Sonomia) have geologic counterparts in central Oregon (Sonomia?), northeastern Oregon, northern Washington, and British Columbia (e.g., Monger, 1975, 1977; Davis and others, 1978; Saleeby, 1983; Miller, 1987). As we have seen, the more southerly terranes comprising Sonomia were apparently accreted to North America in latest Permian or Early Triassic time during the Sonoma orogeny, and it is possible that similar and broadly correlative events took place in southern Canada. In early Mesozoic time a series(?) of back-arc basins were superposed on rocks affected by "Sonoman" events from southern Canada to southern Nevada. These basins formed east of a geographically diverse fringing volcanic archipelago until Middle to Late Jurassic time. In that interval of time, foreshortening of arc, basin, and continental margin assemblages occurred when rocks of back-arc basins floored by oceanic and/or extended continental crust were thrust eastward. In southernmost Canada, an offshore amalgamated superterrane (1) with arc (Quesnellia, Stikinia) and ocean-basin components (Kaslo/Slocan/Slide Mountain, Cache Creek) was sutured against the continental margin in the Middle Jurassic, ca. 180 to 160 Ma (Monger and others, 1982; Armstrong, 1988; Gabrielse and Yorath, 1989). In contrast, in northern California a westward-migrating Middle Jurassic arc was constructed across the Triassic–Early Jurassic suture between rocks of Sonomia and the accreted Cache Creek oceanic assemblage, and thus occupied a distal continental margin position (see below); an epicontinental marine basin lay to the east. This arc experienced phases of Middle Jurassic through Late Jurassic internal telescoping with predominantly west-vergent thrust imbrication; basinal rocks to the east were displaced predominantly eastward and southeastward. South of southern Nevada, an Andean type margin developed during early Mesozoic time, but epicontinental Triassic marine sedimentation east of the arc had given way to continental sedimentation by Late Triassic time.

Our understanding of the geology and geometry of this fringing late Paleozoic–early Mesozoic arc system and its western margin is woefully inadequate because of the disruptive effects of younger deformation and the erosion, burial, or intrusion of critical exposures. The Klamath Mountains of northwestern California, and nearby areas of the northern Sierra Nevada, afford the best opportunities for reconstructing the paleogeography and geology of this fringing archipelago. Accordingly, this discussion of Mesozoic tectonics in the accretionary belt of the U.S. Cordillera emphasizes the two northern Californian regions. Both contain several parallel, early Mesozoic tectonic elements—an eastern arc, a western accretionary complex, and an intervening Late Triassic blueschist belt—that once extended from central California to cental British Columbia (Fig. 9; Davis and others,

1978; Saleeby and others, this volume). Although these regions have not escaped severe Mesozoic deformation and widespread regional metamorphism, their late Paleozoic–early Mesozoic geologic record remains more intact than in any other parts of the fringing arc system within the conterminus United States. In the Klamaths, for example, Cretaceous and Cenozoic cover across the bedrock terranes is scanty, Mesozoic plutonism is widespread, but not pervasive, and the results of paleomagnetic studies indicate, surprisingly, no evidence for demonstrable latitudinal translation of Klamath terranes with respect to North America (e.g., Fagin and Gose, 1983; Mankinen and others, 1989; Beck, this volume).

In contrast, terranes of late Paleozoic and early Mesozoic age now comprising the Cordilleran collage in Oregon and Washington are generally allochthonous to varying degrees. Geologic relations within them are treated here and by Saleeby and others (this volume), but times and modes of their accretion outboard of the early Mesozoic blueschist belt are poorly understood.

Early Triassic continental margin, California

In Early Triassic time we assume, although it cannot be demonstrated unequivocally, that the western edge of North America lay along the outboard margin of newly accreted Sonomia (Fig. 9). If so, the Paleozoic terranes in eastern Klamath region (in eastward ascending structural order: Central Metamorphic, Dm; Trinity ophiolite; Yreka-Callahan, OSos; eastern Klamath, DPa) all lay within western areas of the newly enlarged continent. The polarity of the Sonomia arc is controversial, but as discussed above we favor east-directed subduction beneath it both prior to and following the emplacement of the Golconda allochthon in its back-arc region. Permian volcanism (DPa) in the eastern Klamaths was pronounced, and continued, albeit at reduced levels of activity from Late Permian through Middle Jurassic time (TR Ja; Miller, 1987; Saleeby and others, this volume). Permian volcanism also occurred in the northern Sierran region, but its continuation into the Triassic cannot be demonstrated because Upper Triassic carbonate rocks rest unconformably on Permian and older Paleozoic units. To the south and east, Saleeby and others (this volume) document widespread, but volumetrically sparse Early Triassic igneous activity in central Nevada (Koipato, TR a), the central California–Nevada border area (TR Ja, TR Jac), and the eastern Mojave (TR Ja)—all, with the possible exception of the silicic Koipato volcanics (see above), presumably related to eastward subduction. Oblique (sinistral?) subduction, or even local transform tectonism along the Early Triassic Sierran-Klamath suture belt (Saleeby and others, this volume), is neither precluded nor confirmed by existing data.

Triassic through Middle Jurassic evolution of the Klamath Mountains

Volcanic and volcaniclastic rocks throughout the Permian to Middle Jurassic section of the eastern Klamath province (DPa, TR Ja) argue for long-term convergence between oceanic and continental realms. The paleogeography of the largely submerged continental margin was that of a volcanic archipelago with both shallow-marine platforms, some characterized by carbonate deposition (e.g., Hosselkus Limestone), and deeper marine basins with turbidite and hemipelagic sedimentation (Pit Formation; Miller, 1987). Farther east, much of Sonomia lay beneath a moderately deep, epicontinental basin (Star Peak/Luning basin, TR Jsb) of Triassic age (Fig. 9) that was bordered on the east and south by shallow-marine (shelf) environments, deltaic systems, and the terrestrial continental realm (PTR s, TR s; Speed, 1978; *Lupe and Silberling, 1985;* Saleeby and others, this volume). Unfortunately no stratigraphic record is preserved of forearc sedimentary deposits that presumably once lay between the Triassic trench and the eastern Klamath arc to the east.

The paleogeography of the Pacific realm to the west is much less definite. In the Klamath Mountains, pre–Middle Jurassic rocks within the western Paleozoic and "Triassic" of *Irwin (1966)* lie west of and structurally below Sonomia as defined above and form a tectonically imbricated assemblage of rocks of oceanic affinity (Fig. 10). They range in age from the Pennsylvanian or Lower Permian North Fork ophiolite (Poc; 256 to 310 Ma; *Ando and others, 1983*) to Lower Jurassic (Pliensbachian) radiolarian cherts in the structurally overlying North Fork terrane (*Blome and Irwin, 1983*). Rock sequences in this western assemblage represent ophiolites, ophiolitic mélanges, nondisrupted chert-argillite sections, metasedimentary mélanges, basaltic seamounts, and carbonate atolls (PJap; PTR ap). Details of the evolution and disruption of these sequences are discussed below and have been treated in detail by Davis and others (1978), Wright (1982), Mortimer (1985), Gray (1986), Wright and Fahan (1988), and Saleeby and others (this volume). Collectively, in age, lithology, and sparse indigenous Tethyan faunal elements, they closely resemble the Cache Creek assemblage of western Canada (Fig. 9; Monger, 1975; Davis and others, 1978; *Cordey and others, 1987*). Calaveras rocks in the northwestern Sierra Nevada (IPJap) represent a southward continuation of the Klamath assemblage, although an indigenous(?) Tethyan fauna has been reported for only the westernmost terrane (Tuolumne River, see below).

The western Klamath and Sierran rocks of Cache Creek affinity are generally regarded as comprising an accretionary complex accreted against and beneath the eastern Klamath Paleozoic terranes along the Sierran-Klamath suture. This view is supported by the occurrence of a tectonic slice of Late Triassic blueschists (TR m, ca., 214 to 222 Ma; Fort Jones/Stuart Fork terrane) separating rocks of the Central Metamorphic terrane (Dm) from the North Fork terrane (PTR ap). Blueschists of similar age occur in close spatial association with Cache Creek or Cache Creek-like rocks in southern and central British Columbia, northern Washington (PTR oc; PJap), central Oregon (PJap), and the northern Sierras (Fig. 9). They provide one of the strongest lines of evidence for latitudinally extensive, east-directed subduction of the Pacific realm beneath one or more Late Triassic volcanic arcs (Davis and others, 1978; Miller, 1987). To date, the youngest

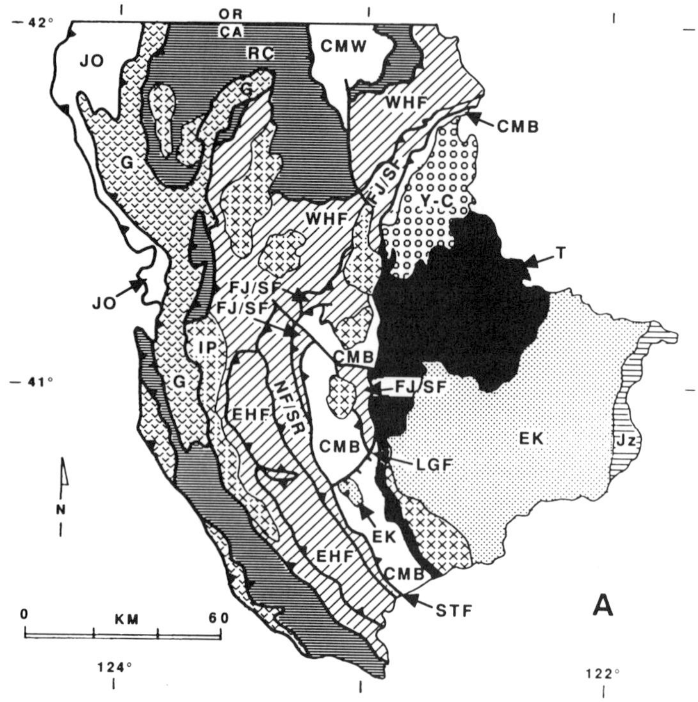

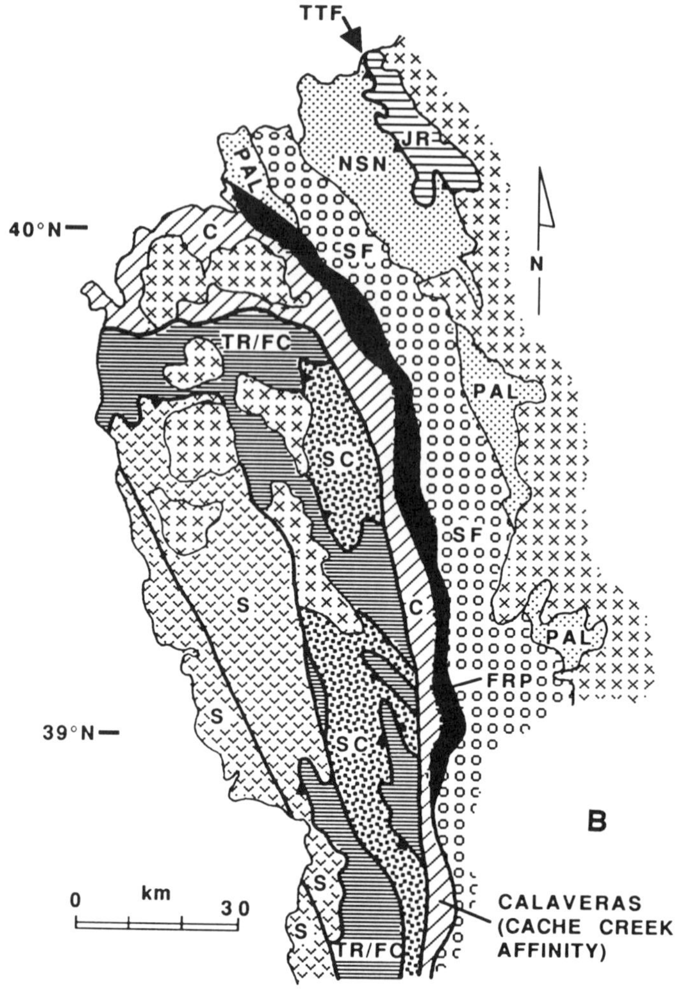

Figure 10. Generalized terrane/tectonic maps of Klamath Mountains (A) and northern Sierra Nevada (B), California. Similar patterns in (A), largely after Wright and Fahan (1988) and (B), from Dilek (1989), represent broadly correlative terranes (t.) or units as described in text. Klamath Mountain terranes/units (A) are from west to east: JO, Josephine ophiolite; G, Galice intra-arc basin; RC, Rattlesnake Creek t.; IP, Ironside Mountain Pluton; WHF, western Hayfork t.; EHF, eastern Hayfork t.; NF/SR, North Fork/Salmon River t.; FJ/SF, Fort Jones/Stuart Fork t. (blueschist); CMW, Condrey Mountain window; STF, Siskioyou thrust fault; CMB, Central Metamorphic Belt; LGF, La Grange fault; T, Trinity ophiolite; Y-C, Yreka-Callahan assemblage; EK, Eastern Klamath belt (Devonian through Triassic strata); Jz, Jurassic strata. Granitic plutons (x). Sierra Nevada terranes/units (B) are from west to east: S, Smartville assemblage (undifferentiated) of Late Jurassic arc and intra-arc basin units; TR/FC, Tuolumne River/Fiddle Creek t.; SC, Slate Creek t. (may be all or in part equivalent with TR/FC and basement units in S); C, Calaveras t.; FRP, Feather River peridotite (contains undifferentiated Red Ant blueschists in northern area); SF, Shoo Fly complex; NSN, northern Sierra Nevada belt (Devonian through Permo-Triassic strata); PAL, undifferentiated Paleozoic strata; TTF, Taylorsville thrust; JR, Jurassic strata. Granitic plutons (x).

sedimentary rocks dated in the 1,000-km-long Cache Creek belt are Early Jurassic siliceous tuff and chert (Pliensbachian) in the Klamath Mountains (*Blome and Irwin, 1983*) and Early to Middle Jurassic siliceous argillite (Pliensbachian to Bajocian) in southern British Columbia (*Cordey and others, 1987*). In both areas, deformed Cache Creek rocks and rocks of arc affinity are intruded by Jurassic granitic plutons that indicate final closure of Cache Creek "seas" by Middle Jurassic time.

The mode and time(s) of assembly of the various terranes of Cache Creek affinity west of the Klamath Triassic–Early Jurassic suture have been controversial topics since Hamilton (1969) first proposed that much of the western belt had largely formed by accretion of diverse stratotectonic elements. The former composite western Paleozoic and "Triassic" belt of *Irwin (1966),* now known to contain widespread Jurassic strata, constitutes a structurally stacked sequence of six terranes—from lowest to highest and west to east (Fig. 10; Plate 8): Rattlesnake Creek/Marble Mountains (Ŧ Ja, Jac); western Hayfork (Ja); eastern Hayfork (PŦ ap); North Fork ophiolite/Salmon River (Poc); North Fork (PŦ ap); and Fort Jones/Stuart Fork (Ŧ m). Boundaries between the terranes are known to be shallow-dipping, west-vergent thrust faults that root eastward.

The predominantly chert-argillite sequences of the eastern Hayfork and North Fork terranes may be correlative (Davis and others, 1980; Wright, 1982), although the eastern Hayfork terrane (Permian to Late Triassic) is chaotically disrupted (a mélange), whereas the structurally higher North Fork terrane (Permian to Early Jurassic) is stratigraphically intact. The North Fork ophiolite or Salmon River terrane (Mortimer, 1985) is best interpreted as the dismembered late Paleozoic oceanic basement of these two terranes. It occupies the core of a regional, west-overturned antiform that physically separates eastern Hayfork (western) and North Fork (eastern) sequences (*Ando and others, 1983*). Greenschist metasedimentary and metavolcanic rocks of the North Fork terrane are lithologically similar and in part coeval with structurally overlying Fort Jones/Stuart Fork rocks of high-P, low-T type. Collectively, the eastern Hayfork, North Fork, and Fort Jones ocean basin assemblages can be interpreted as an evolving subduction-related accretionary complex coupled with North America between Late Triassic and early Middle Jurassic time (prior to North American Jurassic plutonism, ca., 170 to 160 Ma, that stitches the terranes together, see below).

The quest for preserved sutures in the Cordillera, that is, for tectonic zones or contacts along which oceanic lithosphere was lost by subduction (*Dewey, 1977*) is a frustrating one (Davis and others, 1978). Most original plate boundaries between convergent tectonic elements in the Cordilleran collage have been disrupted or reactivated by younger, and not necessarily kinematically compatible fault zones, or obliterated by the effects of plutonism and/or metamorphism, or covered by younger units. However, the original Sierran-Klamath suture(s) between Cache Creek assemblage rocks and the western edge of North America may locally be preserved in both the Klamath Mountains and Sierra Nevada.

The Siskiyou thrust fault of the central Klamath Mountains (Fig. 10) separates Devonian metamorphic rocks of the Central Metamorphic terrane from Late Triassic high-pressure–low-temperature metasedimentary and metavolcanic rocks of the Ft. Jones/Stuart Fork Terrane (Davis, 1968). Davis (*in* Davis and others, 1978) erroneously considered Stuart Fork metacherts in the central Klamath Mountains (Cecilville Quadrangle) to be derived from structurally lower Early Jurassic cherts of the North Fork terrane, and concluded that the ductile Siskiyou thrust had to be younger than the Pliensbachian cherts. Goodge (1989), however, has mapped a major thrust fault (Salmon River thrust) between the top of the North Fork terrane and overlying Stuart Fork metacherts, a discovery that is permissive of the still higher Siskiyou thrust being synkinematic with Late Triassic blueschist metamorphism (Stuart Fork/Fort Jones) in its lower plate. South of the Cecilville Quadrangle Central Metamorphic belt rocks (Dm) are in fault contact with North Fork sedimentary rocks of low greenschist grade (PJap). The "Siskiyou" fault here must be of early Middle Jurassic age (post-Pliensbachian, pre-161-Ma East Fork Pluton) and could be the Salmon River fault mapped by Goodge between Stuart Fork and North Fork rocks farther north. It could represent (1) the ocean-continent plate boundary active during final accretion of the North Fork and related terranes to the prograding continental margin (Central Metamorphic plus younger Fort Jones/Stuart Fork terrane), (2) a reactivated late Early to early Middle Jurassic suture, or (3) an out-of-sequence thrust across a preassembled accretionary complex.

A Jurassic ocean-continent suture may also be preserved in the central Sierra Nevada, where Schweikert and others (1988) have mapped a ductile, west-directed thrust fault between upper plate Shoo Fly rocks (OSos) and a lower plate chert-argillite assemblage with Calaveras/Cache Creek affinities (ℙŦ ap). *Sharp (1988)* presents evidence that deformation and metamorphism accompanying thrust faulting is Middle Jurassic (ca. 174 Ma); the possible suture (or out-of-sequence thrust fault) is intruded by the Middle Jurassic Standard Pluton (167 Ma, Ji).

The westernmost early Mesozoic Klamath terrane, the Rattlesnake Creek/Marble Mountains terrane (Ŧ Ja), and its Sierra Nevada equivalent—the Tuolumne River terrane (or Fiddle Creek Complex; Fig. 10)—are structurally and stratigraphically complex. They consist of a Middle Triassic–Early Jurassic metasedimentary and metavolcanic cover sequence deposited on an oceanic basement of serpentinitic mélange. The supracrustal units are intruded by Early Jurassic plutons of mafic composition (212 to 192 Ma in the Klamath terrane, Saleeby and others, this volume). The paleogeographic setting of the Rattlesnake Creek/Tuolumne River (Fiddle Creek) cover sequence is controversial. Saleeby and others (this volume) favor the interpretation that these sequences developed during active arc-related volcanism in the eastern Klamath region, but place them in the tectonic context of a forearc igneous belt superposed atop a Triassic accretionary prism (Plate 5). The controlling subduction zone dipped eastward beneath North America, but oblique rifting and basin formation characterized the forearc region above it. Edelman and others (1989b) believe that the Fiddle Creek terrane may have been an intraoceanic arc that was not accreted to the more easterly Calaveras/Cache Creek terrane until the early Middle Jurassic. If they are correct, then its northern counterpart, the Rattlesnake Creek terrane, is the only candidate for an exotic(?) early Mesozoic oceanic arc in the Klamath Mountains.

We prefer the interpretation that the Rattlesnake Creek/Tuolumne River/Fiddle Creek terrane is more native than exotic. The uppermost stratigraphic units in, or on, the Fiddle Creek Complex are quartzose clastic sediments (Colfax Sequence? of probable Callovian-Kimmeridgian age) possibly derived from siliciclastic rocks of the North American Shoo Fly Complex (Edelman and others, 1989b). More important in terms of its provenance, fossils collected from carbonate blocks in this Klamath-Sierran terrane indicate its proximity to the eastern Klamaths during Middle to Late Triassic through Early Jurassic sedimentation. A Devonian(?) coral fauna of Eastern Klamath affinity has been reported from the Klamath Rattlesnake Creek terrane (Gray, 1986). Blocks of limestone up to 100 m in size are found in a Permian siliceous diamictite that Edelman and Sharp (1989) assign to the Sierran Tuolumne River terrane. The blocks contain a distinctive Early Permian fusulinid and tetracoral fauna known only from the McCloud Limestone of the eastern Klamaths (Watkins and others, 1987). Watkins and others interpret the boulder-bearing unit as a Permian submarine landslide deposit derived from the escarpment of a McCloud carbonate bank somewhere to the east. Blocks of Permian Tethyan limestone have been found at two localities within Middle Triassic to Early Jurassic strata of the Tuolumne River terrane (Edelman and Sharp, 1989); their source is not known, but may have been Calaveras/Cache Creek units previously accreted against Sonomia. The tectonic significance of Tethyan versus eastern Klamath faunas has recently been blurred by the discovery of a Late Permian Tethyan coral indigenous to both the eastern Hayfork terrane (Cache Creek assemblage) and the McCloud belt of the eastern Klamaths (Miller and Wright, 1987). Newton (1988) has postulated a possible pantrophic distribution of "Tethyan" faunas and pointed out that there remain great uncertainties in the use of paleobiogeographic data to resolve longitudinal tectonic displacements.

Another exotic terrane candidate in the northwestern Sierra Nevada is the Slate Creek terrane (Ŧ Ja), a pseudostratigraphic assemblage of plutonic (range 209 to 203 Ma) and volcanic rocks also believed by Edelman and others (1989a, b) to represent an immature intraoceanic arc. It is likely that Slate Creek rocks constitute the basement for the Late Jurassic Smartville Complex (Dilek, 1989). Rocks of the Slate Creek Complex are reported to have been thrust eastward prior to 165 Ma over rocks of the Fiddle Creek Complex along a west-rooting thrust fault (Slate Creek thrust) that has no known Klamath Mountain equivalents. It is not clear to us that an unequivocal distinction has been made between Smartville basement, Slate Creek, and Fiddle Creek lithologic units (all with plutonic components dated at ca 210 to 200 Ma); Dilek (1989) appears to have reached a similar conclusion. We believe that neither the existence of the Slate Creek thrust, juxtaposing similar(?) assemblages of Slate Creek (upper plate) and Fiddle Creek (lower plate) rocks, nor its unusual direction of displacement (eastward) have been adequately demonstrated. We do not dispute the existence of the east-directed Taylorsville thrust in the northeastern Sierra Nevada (Fig. 9) with its footwall rocks of late Middle Jurassic (Callovian) age. If the Slate Creek thrust is a valid structure it may have formed during the same, rather cryptic deformation responsible for the Taylorsville fault.

Middle Jurassic magmatism and intra-arc deformation, Klamaths/northern Sierra Nevada

Charlton (1979) and Wright (1982) believe that the western Hayfork terrane, of Middle Jurassic age, was a marine volcanic arc deposited atop a tectonically disrupted eastern Hayfork accretionary prism (Fig. 10). *Charlton (1979), Donato (1987),* and Gray (1986) favor the hypothesis that the western Hayfork arc was also deposited upon older deformed rocks in the Rattlesnake Creek/Marble Mountain terrane to the west, although thrust faults now separate the two. Andesitic volcanism (177 to 168 Ma) in the western Hayfork arc terrane began shortly after the mid-Bajocian termination (ca. 180 Ma) of arc volcanism in the eastern Klamath Mountains. But Hayfork arc-related plutonic activity (174 to 159 Ma) affected all six of the western terranes plus the Central Metamorphic, Trinity (Harper and Wright, 1984; Mortimer, 1985; Wright and Fahan, 1988), and eastern Klamath (*Renne and Curtis, 1986*) terranes within Sonomia. Because intrusion of Middle Jurassic igneous rocks related to the Hayfork arc occurred as far east as the easternmost Klamaths, volcanic equivalents of this Middle Jurassic arc presumably once lay across the Triassic/Early Jurassic Sierran-Klamath suture(s). Proximity of the Hayfork and eastern Klamath arc sequences to each other in Middle Jurassic time is also suggested by the occurrence in western Hayfork conglomerates of carbonate cobbles bearing a distinctive Permian McCloud Limestone fauna.

Development of the Middle Jurassic arc in the Klamath Mountains (177 to 159 Ma), and in the Sierra Nevada region to the south (Edelman and Sharp, 1989; Saleeby and others, this volume), was accompanied by the profound internal telescoping of the arc along east-rooting thrust faults (Davis and others, 1978; Harper and Wright, 1984; Wright and Fahan, 1988). The event appears to be unusual because most active magmatic arcs worldwide are characterized by internal extensional stresses (e.g., Hamilton, 1988a). However, the tectonic contacts between all six terranes of the Klamath western Paleozoic and "Triassic" subprovince are thrust faults that telescoped the previously formed Cache Creek–like accretionary complex. Because these thrusts displace western Hayfork volcanic rocks (177 to 166 Ma), Early Jurassic (Pliensbachian) cherts of the North Fork terrane, and 170-Ma-old plutons, but were intruded by the Wooley Creek (161 Ma), Vesa Bluffs (164 Ma), and East Fork (164 Ma) Plutons, they are clearly of Middle Jurassic age and fall within the bracketing ages for arc magmatism, ca., 177 to 159 Ma (Wright and Fahan, 1988). Mortimer (1985) believes that regional metamorphism occurred across the terrane boundaries after their tectonic assembly, but prior to intrusion of the Vesa Bluffs Pluton across five of these boundaries. Wright and Fahan (1988) and Saleeby and others (this volume) believe that a more complicated temporal interplay occurred between magmatism, metamorphism, and thrust faulting in areas south of that studied by Mortimer. They present evidence that several igneous plutons (171 to 168 Ma) were emplaced during synchronous regional deformation and metamorphism, and that the thrust fault now separating western Hayfork and Rattlesnake Creek terranes developed contemporaneously with regional metamorphism at ca. 168 to 169 Ma.

Collectively, the relationships described above support the contention of Davis and others (1978), Burchfiel and Davis (1981), and Wright and Fahan (1988) that (1) a complex Middle Jurassic arc (Ja, Jac) developed across the previously sutured plate boundary between Triassic and Early Jurassic North America and an accretionary complex of Cache Creek affinity (ℙJap) to the west; and (2) that this arc experienced internal telescoping along intraplate, not interplate, faults. Eastward subduction must have occurred beneath this continental margin arc because no ocean basin could have lain to the east of it. This view has been refined and greatly augmented by Wright (1982), Harper and Wright (1984), Mortimer (1985), Gray (1986), and Wright and Fahan (1988), and extended by Edelman and others (1989), Edelman and Sharp (1989b) and Saleeby and others (this volume) into the Sierran region to the south. However, the Middle Jurassic deformational history of the northwestern Sierra Nevada may include an added tectonic phase not present, or at least not preserved, in the Klamath Mountains. West-directed Middle Jurassic thrusts separating Red Ant (blueschist), Calaveras, and Fiddle Creek/Tuolumne River terranes are comparable to, and possibly equivalent with those of the Klamath Mountains, but they may be truncated upwards by the east-directed, Middle Jurassic Slate Creek thrust. Edelman and others (1989a) report that the thrust is intruded by the Middle Jurassic Scales Pluton (165 Ma). As described above, we are skeptical about the existence of this structure.

Late Jurassic arc development, intra-arc ophiolites, and the Nevadan Orogeny

Late Callovian–Early Oxfordian rifting of the Middle Jurassic Hayfork arc at ca. 164 to 162 Ma accompanied a westward shift in the locus of Jurassic Klamath magmatism to the Rogue-Chetco arc (Ja, Ji) and led directly to formation of the rifted Galice basin (Jac) with its Josephine oceanic lithosphere (Fig. 10; Harper and Wright, 1984; Wright and Fahan, 1988). Saleeby and others (this volume) report evidence for even earlier (172 to 170 Ma) rifting and Galice-like sedimentation within the Middle Jurassic arc, but it is not known whether this rifting is related to the formation of the Galice basin or to a geographically separate area. Geologic evidence summarized by Harper and Wright (1984), Harper and others (1985), and Saleeby and others (this volume) indicates that the Galice rift developed across the Hayfork continental margin arc, that sea-floor spreading occurred parallel to the continental margin, that the eastern and western margins of the Galice basin were transform faults, presumably sinistral, and that its tectonic setting may have resembled the present Andaman Sea. This setting would have been one of oblique subduction beneath the Rogue arc, and upper plate, intra-arc Galice rifting more-or-less parallel to the strike of the subduction zone. Basement rocks beneath the Rogue arc in southwestern Oregon resemble in age and lithology those of the Rattlesnake Creek terrane to the east of the Galice basin, further tying this Late Jurassic island arc to Middle Jurassic North America. Both the arc and the basin were short-lived.

Between ca. 154 to 150 Ma strong compression assigned to the controversial Nevadan orogeny, and reminiscent of the Middle Jurassic thrusting events only 10 to 20 m.y. before, led to a closing of the Galice basin. The Galice basin was subducted beneath the Hayfork remnant arc and its older basement (PR ap + Ja), just as the Rogue arc was emplaced beneath the back-arc Galice basin and its oceanic lithosphere. Harper and others (1985) propose that the locations of these major thrust faults were controlled by preexisting, basin-bounding transform faults. The location of the Condrey Mountain window to the east (Fig. 10) indicates that a Galice basin-like sequence of rocks (Jm), albeit somewhat older, was displaced beneath the older arc terrane for at least 75 km and metamorphosed under transitional greenschist-blueschist facies conditions. Despite profound telescoping of the Galice basin, igneous activity presumably related to continued eastward subduction beneath the region did not cease. Post-tectonic dikes and plutons that cut foliated Galice rocks are as young as 144 Ma (Harper and Wright, 1984; Wright and Fahan, 1988).

Late–Middle Jurassic to earliest Late Jurassic (Callovian-Oxfordian, ca., 160 ± 5 Ma) ophiolites were not restricted to the Galice basin (Figs. 10, 11). They formed widely along the North American Cordilleran margin, mostly in arc-proximal settings that are reminiscent of, but not identical to that in the western Klamath Mountains (Saleeby and others, this volume). One Sierran ophiolite-like complex, the Smartville (164 to 162 Ma), formed during intra-arc rifting across the older (ca., 200 Ma) Slate Creek Complex, but rifting did not apparently lead to the formation of new oceanic lithosphere since depleted ultramafic tectonites are not a component of the ophiolitic sequence (*Beard and Day, 1987*). Its geologic and geographic setting, like that of the Josephine ophiolite, establishes it as a "native" California complex.

Examples of other Late Jurassic ophiolites include those of Cedros Island off Baja California, the Great Valley (= Coast Range) ophiolite of the California Coast Ranges, the Snow Camp Mountain and Wild Rogue Complexes of southwestern Oregon, and the Rimrock, Ingalls, and Fidalgo ophiolites of Washington (Fig. 11). It should be emphasized that present locations are not likely to have been the initial sites of generation for most of these ophiolites (e.g., McLaughlin and others, 1988, on the Coast Range ophiolite), given geologic and paleomagnetic evidence for latitudinal transport along the continental margin at different times and with different senses of displacement relative to North America (see below). Evidence for ocean-spreading at high angles to the continental margin is presented for several ophiolite complexes (Josephine, Wild Rogue, southern Coast Range). Harper and others (1985) and Saleeby and others (this volume) discuss these complexes at great length; their findings and interpretations cannot be treated in detail here.

Until recently, the prevailing tectonic concept in the northern Sierras has been that the Late Jurassic Nevadan orogeny was the consequence of the collision of a western island arc with an eastern, continental margin arc along a suture that is now the Melones fault zone (e.g., Schweickert and Cowan, 1975; *Schweickert and others, 1984;* Moores and Day, 1984; *Day and others, 1985;* Ingersoll and Schweickert, 1986). Initially, there were only a few dissenters to this oceanic arc–continental arc collision hypothesis (e.g., Davis and others, 1978; Burchfiel and Davis, 1981; Saleeby, 1981, 1982), but recent studies in the region indicate (1) that the Nevadan orogeny is clearly diminishing in orogenic stature, at least in its type Sierra Nevada area, and (2) that collisional models for Nevadan deformation are increasingly difficult to accept. Northern Sierran workers (e.g., Edelman and others, 1989a, b; Saleeby and others, this volume) are now recognizing that some (perhaps most) structures previously assigned to the Nevadan orogeny (e.g., west-directed thrust faults between northwestern Sierran terranes and the east-directed(?) Slate Creek thrust) formed prior to Middle Jurassic plutonism (ca., 165 Ma) and subsequent Smartville intra-arc rifting (164 to 162 Ma).

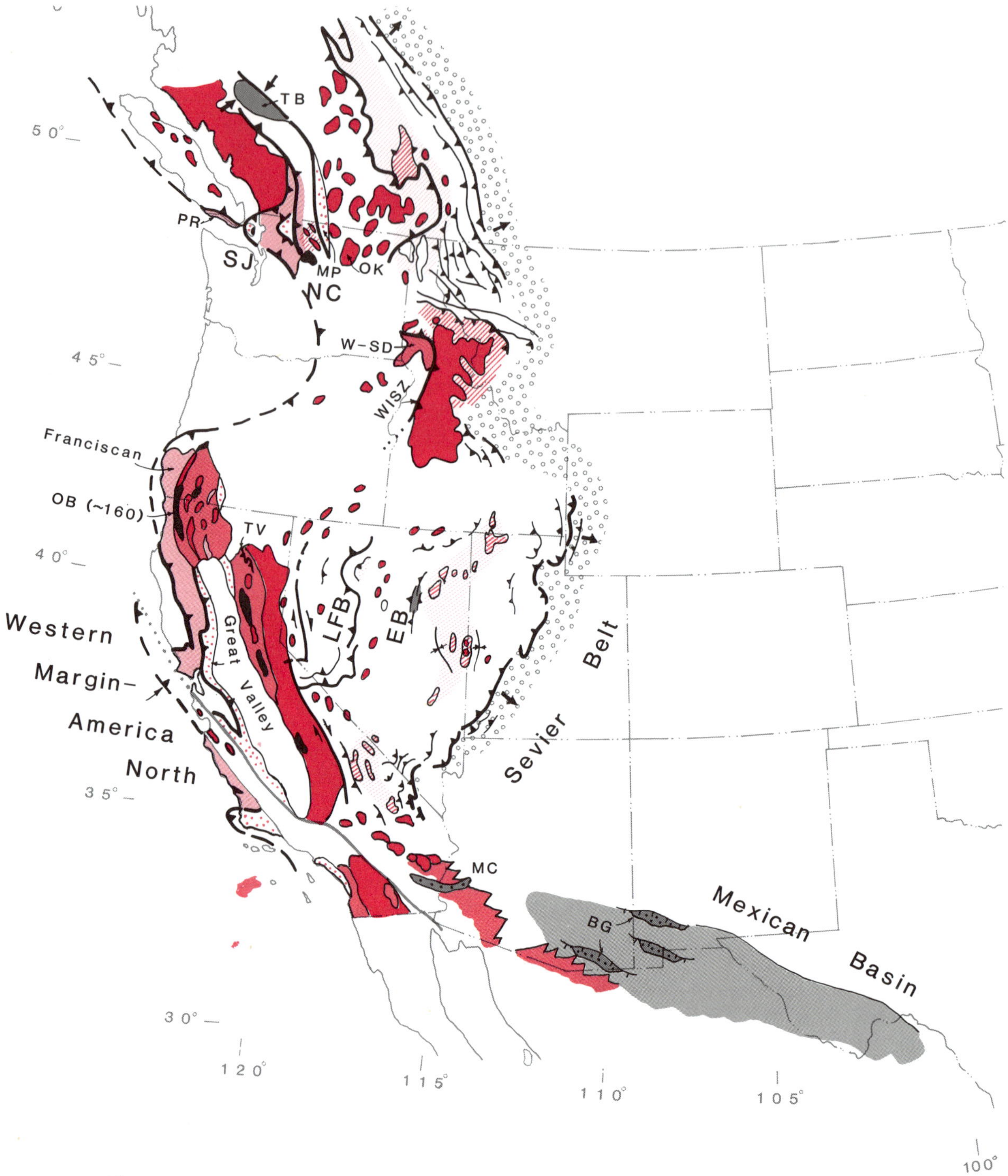

Figure 11. Paleogeographic and tectonic map of Middle Jurassic to Late Cretaceous time. Cordilleran foredeep shown in gray circles and directions of sediment transport shown with arrows. Local basins: Tyaughton (Canada), Newark Canyon (central Nevada), McCoy Mountains (southern California) and Bisbee/Glance (Arizona) shown in gray. Other patterned units: volcanic rocks (red); plutons (deep red); mélange belts (pink); fore-arc basins (red dots). Metamorphic rocks are shown with red oblique lines (high grade) and with pink oblique lines (low grade). Future location of San Andreas fault shown in gray. Shallow-marine rocks of Mexican basin are light gray. Black is ~165 to 160 Ma ophiolite belt. BG, Bisbee/Glance extensional basins (gray with dots); EB, Eureka belt; LFB, Luning/Fencemaker belt; MC, McCoy Mountains; MP, Methow-Pasayten foredeep (red dots); NC, northern Cascades; OK, Okanogan Mountains; PR, Pacific Rim sequence; SJ, San Juan Islands; TB, Tyaughton successor basin (gray); TV, Taylorsville thrust; WISZ, West Idaho suture zone; W-SD, Wallowa–Seven Devils arc.

Opinions still differ, however, on the extent of Nevadan deformation in the northern Sierra. For example, Saleeby and others (this volume) believe that the Smartville intra-arc basin and its sedimentary cover, the Mariposa Formation, experienced compressional "Nevadan" deformation between 162 and 158 Ma that led to east-directed thrust faulting. In contrast, Edelman and others (1989b) regard such thrust faulting as being limited to emplacement of the older Slate Creek thrust. They consider Nevadan deformation to be intraplate and unaccompanied by pervasive regional metamorphism. To the south, workers in the central Sierra Nevada have also downgraded the importance of the Nevadan orogeny. Tobisch and others (1989) propose that some ductile "Nevadan" structures (e.g. the regionally important Bear Mountains fault zone), are partly or wholly Cretaceous in age because they postdate intrusion of 147-Ma pluton (Guadelupe). Tobisch and others (1987) conclude that Nevadan-age structures are not developed as regional penetrative structures east of the Melones fault zone and in general tend to increase in intensity westward away from it.

What is most damaging to island arc–continental arc collisional models for the Nevadan orogeny, however, is a growing awareness that the Middle Jurassic magmatic episode described for the Klamath Mountains is present in the Sierra Nevada as well. Here too, plutons of 170 to 160 Ma have stitched together rock sequences formerly assigned to Late Jurassic island and continental arcs or their basements, for example, the Standard Pluton across the Calaveras–Shoo Fly Triassic–Early Jurassic thrust (*Sharp, 1988*). Furthermore, this Jurassic plutonic belt continues southward across Sierra Nevada rocks east of the early Mesozoic suture in the Klamaths and Sierras and into Mojave and Sonoran regions of southeastern California and southwestern Arizona (Saleeby, 1983; Busby-Spera, 1988; Wright and Fahan, 1988). There is simply no longer any evidence for the existence of an open ocean basin between western and eastern Sierran regions (or western and eastern Klamath regions) after Middle Jurassic time. The focus on Nevadan deformation has thus shifted to regions and plate interactions lying outboard of the Middle Jurassic arc and the younger intra-arc Smartville and Josephine ophiolites superposed on it.

The analysis of Late Jurassic relative Farallon–North American Plate motions by Engebretson and others (1985) indicates east-west convergence at the latitude of the Klamath Mountains (see below) and requires that such convergence, acting on a continental margin assumed to trend northwest, would produce both normal and tangential (sinistral) strain components. Saleeby (1981; Saleeby, 1982) may have been the first geologist to infer that the close spatial association of the Callovian-Oxfordian ophiolites with contemporaneously active volcanic arcs could best be explained by a tectonic setting of transtension along the continental margin. Saleeby (1981) proposed that intra-arc rifting within the Andaman Sea, in response to strongly oblique subduction along the Indian-Asian Plate boundary, provides a good tectonic analogy for Josephine intra-arc rifting. Although presumably one would not have to look beyond Farallon–North American Plate interactions for postulating sinistral transtensional strain in the leading edge of the North American Plate, another source for such strain has been proposed.

Interrelationships have been hypothesized between two seemingly unrelated sea-floor spreading events, the rifting of the Gulf of Mexico at about 165 to 160 Ma and the formation of Cordilleran ophiolites at the same time (Harper and others, 1985; Saleeby and others, this volume). The two regions may be kinematically linked by the Sonoran(-Mojave) megashear of Silver and Anderson (1974, *1983*), a controversial sinistral transform fault believed to have accommodated rhombochasmic opening of the continent-bound Gulf of Mexico. The fault, with up to 800 km of left slip, is postulated to have extended across northern Mexico from a position south of the gulf rifting axis and to have merged at a low angle with the North America–Pacific Plate boundary in California (e.g., Dickinson, 1981; Harper and others, 1985). Gulf rifting and the feeding of rift-related motion into the California continental margin may have interrupted prevailing patterns of eastward oblique(?) subduction and modified North American–Pacific Plate interactions, thus leading to transtensional rifting of the continental margin arc along northwest-striking transform faults (Saleeby and others, this volume). Sea-floor spreading directions in the resultant basins should largely have been parallel to the margin; dikes in the intra-arc Smartville Complex with north-northwest strikes (*Xenophontos and Bond, 1978*) are difficult to reconcile with the megashear hypothesis.

The geometric relations between Klamath, Sierran, and Great Valley ophiolitic basins are extremely conjectural. Harper and others (1985) believe that separate spreading centers may have been linked by left-stepping transform faults (broadly comparable to a mirror image pattern of fault-linked spreading basins in the modern Gulf of California), and that some ophiolites may have formed along Cayman Trough–like leaky transform zones. McLaughlin and others (1988) have postulated that the lower part of the Coast Range ophiolite (169 to 160 Ma) formed in an arc-proximal position at equatorial latitudes, and was then translated northward (156 to 146 Ma) and accreted along the California coast (146 to 145 Ma) before deposition of Great Valley sequence sediments across it.

Tectonic scenarios for Jurassic compressional deformation in Sierran-Klamath regions

Reasons for Middle Jurassic intra-arc thrust faulting and the Late Jurassic subduction of intra-arc ophiolitic sequences within an evolving Klamath magmatic arc are not readily apparent. Two general tectonic scenarios exist. The first, which we favor, relates compressional strain in the continental margin arc to noncollisional subduction-related processes that may have affected the degree of coupling between upper and lower plates—processes, for example, related to changes in convergence rate, or dip of the subducting slab, or age of subducted lithosphere. The second scenario attributes the Middle Jurassic and Nevadan orogenies to collision(s) between the continental margin of North America and more westerly arc(s), microcontinent(s), or superterranes. If synplutonic thrusting within the Sierran-Klamath magmatic arc resulted from its collision with an arc or a microcontinent in the trench region farther west, the colliding entity is no longer preserved in the region.

The Slate Creek and Rattlesnake Creek/Tuolumne River/Fiddle Creek arc(?) terranes cannot be responsible for Middle Jurassic deformation through collision with North America, even if one or both were nonnative terranes. Both terranes lay at the time of hypothetical Middle Jurassic collision (= thrust faulting) above east-dipping subduction zones and within the telescoping Klamath/Sierran magmatic arc. The case for Middle Jurassic arc-continent or superterrane-continent collision is also weakened by a continuous record of western Klamath magmatism before, during, and after the hypothesized collision—syntectonic plutonism (ca. 171 to 168 Ma), and ensuing development (164 to 150 Ma) across western parts of the thrust complex of the west-facing Rogue-Chetco island arc and its rifted Josephine ophiolite/Galice Formation back-arc basin (Saleeby, 1982; Harper and Wright, 1984). The overall continuity of western Klamath magmatic activity between ca. 177 an 144 Ma appears to refute any scheme of Middle Jurassic clogging of an evolving, east-dipping subduction zone by collision with a phantom arc or migrating superterrane.

Conditions responsible for tectonic collapse of the continental margin arc were repeated again during the Late Jurassic "Nevadan" orogeny of the Klamath Mountains (154 to 150 Ma) with the telescoping of the Rogue arc, the Galice back-arc basin, and the Hayfork remnant arc and continental margin. The cause(s) of this deformational event are as cryptic as those of the Middle Jurassic orogeny that preceded it by 10 to 20 m.y. Harper and Wright (1984), Harper and others (1985), Wright and Fahan (1988), and Saleeby and others (this volume) consider the two events to be part of a protracted period of compressional tectonism within the continental margin, separated only by the brief phase (ca., 164 to 155 Ma) of intra-arc extension (transtension?) and ophiolite generation. The Klamath, Sierran, and southern Coast Ranges areas experienced Nevadan compression differently, although Saleeby and others (this volume) regard the subduction of various small, arc-proximal ocean basins to be the unifying hallmark of the orogeny. East(?)-directed subduction of the Galice basin and the formation of transitional greenschist-blueschist lower plate rocks have already been described. Compression of the intra-arc Smartville basin (162 to 158 Ma; Saleeby and others, this volume) may have had the opposite polarity and may have been a consequence of westward subduction of the Great Valley ophiolite beneath the neighboring Diablo arc (Dilek, 1989; Saleeby and others, this volume), with subsequent collision of the Sierra and Diablo arcs or arc remnants. Such subduction could explain the origin of Late Jurassic blueschists and eclogitic rocks that are now found as widespread exotic components within younger Franciscan mélanges, although an alternative interpretation is present below. If Diablo arc–Sierra arc collision was related to westward subduction, post-collisional Jura-Cretaceous convergence apparently continued by a flipping to east-directed subduction (Franciscan) of oceanic lithosphere along the western edge of the Diablo arc. In Klamath areas to the north, eastward subduction appears to have been continuous through the Middle Jurassic to latest Jurassic interval treated here. Such geometric and tectonic complexities in this relatively small area are not improbable. Comparable complexities in the subductive behavior of small inter-arc basins can be observed today in the southern Philippines area, where the South China, Sula, Celebes, and Molucca Seas are experiencing multivergent closure (cf., Hamilton, 1978).

We conclude, somewhat tentatively, that major episodes of Middle and Late Jurassic compressional deformation in Sierran and Klamath regions were not driven by collision of the western continental margin with either an offshore microcontinent or a non-"native" volcanic arc(s). Collisional models for Middle and Late Jurassic orogenesis require a collidee and we are unable to verify the existence of either an offshore microcontinent or a far-traveled, exotic arc or arc. The impressive continuity of Jurassic magmatism between ca. 177 and 144 Ma in the Klamath Mountains argues most strongly against offshore collisions at times of Middle and Late Jurassic telescoping of the evolving Klamath arc.

We prefer the interpretation that tectonic imbrication of the Middle Jurassic arc by thrust-faulting of the same age was the consequence of strong interplate coupling along the east-dipping subduction zone. Eastward subduction of Farallon lithosphere is inferred for both the Sierras and Klamaths, but the two areas might have been separated by a trench-trench transform fault (Harper and others, 1985; Dilek, 1989) that formed after uniform west-directed Middle Jurassic thrusting in both regions and just prior to east-directed thrusting in the northwestern Sierras (e.g., Taylorsville thrust, Slate Creek thrust?). Callovian-Oxfordian rifting of the continental margin arc could have been transtensional (e.g., Josephine-Galice basin?) and related to an extra-Cordilleran event, the rifting of the Gulf of Mexico. Alternatively, rifting of the margin may have been the consequence of changes (steepening?) in the geometry of the subducting oceanic plate, changes that led to intra-arc extension of the upper plate at high angles (e.g., Smartville?) to the continental margin. Intra-arc spreading was short-lived. Closure of the basins by thrust-faulting (Smartville?) and subduction (Galice) characterizes the Nevadan orogeny in northern California (Saleeby and others, this volume). Renewed or continued operation of a hypothetical east-west–striking Sierran-Klamath transform during Late Jurassic deformation may have enabled Klamath and Coast Range/Sierran regions to deform independently and with opposite vergences of thrusting/subduction during basin closure.

The period from 203 to 150 Ma between J1 and J2 cusps in the apparent polar wander path of North America was one of constant angular plate velocity related to opening of the central Atlantic Ocean (May and Butler, 1986). At the J2 cusp, which approximates, in time the Nevadan orogeny, the mean angular velocity of North America increased five-fold and was accompanied by a considerable poleward change in absolute plate motion. Any tectonic explanation for the Nevadan orogeny attached to this observation (e.g., see May and others, 1989), however, is put into question by the existence of a major pre-Nevadan compressional deformation 10 to 20 m.y. earlier along the western margin of the continent.

Back-arc region

From late Permian to Late Cretaceous time, the southern and northern segments of the back-arc region of the Cordillera evolved differently. The boundary between the two segments shifted temporally, but generally it is located at the latitude of southern Nevada. During Late Permian, Triassic, and Early Jurassic time deformation began within the back-arc region in the transitional area between the northern and southern segments (Fig. 9). In southeastern California, North American miogeoclinal and cratonal rocks were intruded by granitic plutons and locally overlain by volcanic rocks of Late Permian (Pi), Triassic, and Early Jurassic age (Ŧ Ja). The northwest trend of these igneous rocks crosses, from south to north, the Paleozoic boundaries between cratonal, miogeoclinal, and offshore arc facies belts and indicates eastward subduction along the previously truncated North American margin (compare Figs. 9 and 11 to Figs. 5, 6, and 7; Davis and others, 1978; Stone and Stevens, 1988; Walker, 1988). However, rocks that belonged to the western parts of the upper plate of the subduction system (the forearc region) are no longer present because they have been moved by younger displacements. Major overprinting by younger igneous and tectonic events has made Permian and early Mesozoic arc rocks and structures difficult to recognize.

Permian deformation of small magnitude and of probable limited extent had been documented in the southwestern U.S. Cordillera by *Stone (1984).* This deformation may be responsible for the Late Permian or earliest Triassic folds known in the El Paso Mountains and Victorville areas of California (*Carr and others, 1984; Miller and Cameron, 1982; Miller and Sutter, 1982*). Triassic deformation is reported from the eastern Sierra Nevada (pre-219 Ma) by Schweikert and Lahren (1987), from east of the southern Sierra Nevada by *Dunne and others (1978),* and by Burchfiel and Davis (1981; these structures are pre–184 Ma and could be Early Jurassic). Some structures in the east-central Mojave Desert (*Novitski-Evans, 1978*) and in the San Bernardino Mountains (*Cameron, 1982*) may be either Triassic or Early Jurassic in age. All of these structures occur within the eastern part of the magmatic belt, or lie just east of it, and all have a general eastward vergence. Their relation to the magmatic arc suggests that during part of its history the latest Paleozoic/early Mesozoic arc was under compression, and that foreshortening of the orogen occurred along its eastern margin (Fig. 9).

Within the southern segment of the Cordillera Triassic igneous rocks are rare, although volcanic detritus in Upper Triassic rocks of northern Arizona suggest a southern, but unidentified, volcanic source (*Stewart and others, 1972*). Major magmatic activity began in Early Jurassic time and continued throughut most of the rest of Mesozoic time. Evidence presented by Tosdal and others (1989) and Riggs and Haxel (1990) indicates that Early Jurassic magmatism was accompanied by caldera subsidence, but evidence for regional subsidence related to regional extension (cf., Busby-Spera, 1988) remains uncertain. With present data, there is little evidence for important regional shortening or extensional tectonism within the southern segment of the Cordillera during early Mesozoic time.

In contrast, in the northern segment of the Cordillera beginning in Middle Triassic time a deep marine basin, here referred to as the Star Peak/Luning basin (Ŧ Jsb), formed in west-central Nevada above the collided Sonoman arc (Fig. 9). The Star Peak/Luning basin was in a back-arc position to the Sierran-Klamath–southern Oregon volcanic arc (Ŧ Ja, Ŧ a) and was flanked by a shallow-marine shelf to the east (Ŧ s) and south (PŦ s), and a volcanic arc, since displaced by younger events, on the west. Middle and Late Triassic rocks within the basin are characterized by several kilometers of fine-grained sediments (Nichols and Silberling, 1977; Speed, 1978). Subsidence and marine deposition in the Star Peak/Luning basin persisted into Early and locally Middle Jurassic time when deposition was terminated by deformation (*Oldow, 1981,* 1984a). The basin may have had an extensional orgin. Early Jurassic normal faults were present during deposition of the Dunlap Formation at its southern terminus (Oldow and Bartel, 1987) and the large mafic Humbolt lopolith (Ji) was intruded into its eastern shelf. However, demonstrable extension is much younger than the Middle Triassic inception of the basin. Facies changes are rapid from basin to shelf sediments, and may mark the location of normal faults, although subsequent thrusting has obscured the original basin margins. Thus, while back-arc extension can be postulated, the origin of the Star Peak/Luning basin remains uncertain. The northern continuation of this basin is covered by extensive Cenozoic sedimentary and volcanic rocks in southern Oregon. The Slocan basin in southern Canada (Fig. 9) has a similar age, tectonic setting, and basin fill, and it is intriguing to postulate that these two basins were once continuous (Wernicke and Klepacki, 1988). However, rocks equivalent to the Star Peak/Luning basin are not exposed in western Idaho and eastern Oregon and may have been removed in late Mesozoic time (see below).

During Early and Middle Jurassic time the magmatic arc became well developed and can be traced the entire length of the Cordillera. East of the arc and back-arc basin, in both the southern and northern segments, was a broad area of shallow-marine and nonmarine deposition, and areas of probable low relief that were the source areas for terrigenous material. Deformation east of the arc appears to have been minor even though evidence for plate convergence and intra-arc deformation is widespread within the westernmost part of the Cordillera. As in Triassic time, most of the convergent structures are present in the transition region between the northern and southern segments of the Cordillera within the eastern part or just east of the Andean-type arc in southeastern California. East-vergent thrusts and folds have been documented in the western Death Valley region (east Sierran or Swansea-Coso thrust sytem of Middle Jurassic age; Last Chance and related thrusts [which, however, could be older]; *Dunne and others, 1978*), the central Mojave Desert (169 to 154 Ma; *Walker and others, 1990;* Burchfiel and Davis, 1981; *Dewitt and others, 1984*); and possibly western Arizona (Tosdal, 1986). Recently, Busby-Spera (1988) has postulated that the arc region from western Nevada southward was the site of extension from Late Triassic to early Middle Jurassic time. Compressional structures of this age are known, however, particularly in the transitional area between the two segments of the Cordillera, and the temporal and spatial relations between extension and compresssion within the arc remain to be established. The eastern part of the Cordillera and adjacent craton remained the site of either shallow-marine or nonmarine deposition or persisted as areas of low positive relief subject to erosion.

LATE JURASSIC TO LATE CRETACEOUS EVOLUTION

Accretionary belt: Cretaceous subduction and contraction in the western Cordillera

Two major belts of Cretaceous deformation lie along the western margin of the U.S. Cordillera: the Franciscan subduction complex (JKap, uKITap) in California and southwest Oregon, and the Cascade orogen (Ŧ Jap, JKfa) in northwestern Washington. Both belts preserve a clear record of mid- to Late Cretaceous contraction. The Franciscan consists dominantly of Upper Jurassic and Lower Cretaceous sediments and oceanic crust that were deformed, metamorphosed, and accreted to the continental margin in a subduction zone setting. In contrast, contraction in the Cascade orogen largely affected Paleozoic and Mesozoic tectonic elements that had already been incorporated into the continental margin of North America. An important question is whether these belts were contiguous in the Late Cretaceous, or, stated another way, whether the Cascade orogen represents the northern continuation of the Franciscan belt. In this section, we explore some possible answers to this question, which has obvious implications for the Late Cretaceous paleogeography of the western United States.

Late Mesozoic California: The Franciscan subduction system

Three Cretaceous tectonic elements stand out in California: the Franciscan Complex (JKap); the Great Valley Group (or sequence; JKfa); and the Sierra Nevada batholith (Ji, lKi, uKi; Fig. 11). The genetic link among these coeval late Mesozoic elements was established in the late 1960s, when they were first interpreted as "type examples" of rock assemblages formed at an ancient convergent plate margin (e.g., Hamilton, 1969; Dickinson, 1970; Ernst, 1970). In this still widely accepted model, the Sierra Nevada batholith represents the roots of a Late Jurassic and Cretaceous magmatic arc. The Great Valley Group consists of detritus, eroded largely from the magmatic arc and its wall rocks, that accumulated in a marine fore-arc basin across a western basement of the Middle and Upper Jurassic Coast Range ophiolite and an eastern basement of Sierran crystalline rocks. The Franciscan Complex includes scraps of oceanic lithosphere and overlying pelagic sediments, including radiolarian cherts, plus terrigenous clastic sediments of trench and inner slope-basin facies, that were swept together to form a subduction complex adjacent to and beneath the Great Valley Group. Some Franciscan carbonate rocks (Laytonville, Calera) were originally deposited in mid-ocean positions far to the southwest of California and North America (Tarduno and others, 1986), but the exotic nature of these off-scraped accretionary complex rocks and their ophiolitic or seamount substrates does not require that the trench sediments that now enclose them were also far traveled. The plate convergence responsible for this "Franciscan subduction system" and its distinctive products lasted from the close of the Nevadan orogeny (Tithonian; ca. 150 Ma) to at least 80 Ma, when magmatism ceased in the Sierra Nevada. More information on the Sierran batholithic belt, Great Valley Group, and Franciscan Complex is in Cowan and Bruhn (this volume). Here, we focus on the Cretaceous deformational history of the Franciscan.

There are several aspects of the Franciscan that strongly corroborate the interpretation that it originated as a subduction complex. Most of the complex consists of volcaniclastic and subordinate arkosic sediments (graywacke turbidites) that were deposited on an oceanic substrate. Overall, the Franciscan is more highly deformed than the Great Valley Group; it is riddled with contractional low-angle faults and, locally, high-angle transcurrent faults. Chaotic mélanges, comprising diverse blocks in a matrix of scaly mudstone, formed where partly dewatered sediments were caught up in shear zones and diapirs (Cowan, 1985). More important, much of the Franciscan was metamorphosed at moderate to high pressures (3 to 7 kb) and relatively low temperatures (150 to 300°C). Rocks containing higher pressure metamorphic mineral assemblages occur either as thrust sheets intercalated with lower pressure assemblages (e.g., *Suppe and Foland, 1978*), or as isolated blocks in mudstone- or serpentinite-matrix mélanges. Some of the higher grade exotic blocks have Kimmeridgian ages (ca., 150 to 155 Ma; *Coleman and Lanphere, 1971; Suppe and Foland, 1978*) that require the existence of a pre-Franciscan subduction-related metamorphic event (see above). Although the overall composition, structure, and metamorphism of the Franciscan are well explained by a general model of late Mesozoic convergence along the California margin, the detailed history of subduction-related deformation and metamorphism within the Franciscan itself is incompletely resolved. For example, were deformation and metamorphism essentially continuous from the Late Jurassic (ca., 145 Ma) to the early Late Cretaceous (ca. 80 Ma), or alternatively, were there instead discrete events that might have resulted from changes in convergence velocity or plate configurations?

The available isotopic data from the Franciscan Complex and granitoid rocks in the Sierra Nevada and Klamath Mountains seem to support the latter alternative. In the Franciscan, high-grade blocks of blueschist and eclogite, most of which originated as amphibolite and garnet amphibolite (Blake and others, 1988), have yielded isotopic dates as old as 164 Ma (data reviewed in Mattinson, 1988). Their metamorphism was probably due to the subduction responsible for development of the Middle and Late Jurassic arc and closely associated ophiolite complexes in northern California (reviewed above), but the geometry of the subduction zone or zones is controversial (Saleeby and others, this volume; Saleeby, this volume). In the eastern Franciscan belt in northern California, metabasalt and schist of the Pickett Peak terrane (Klm) have yielded isotopic dates that cluster between 110 and 145 Ma (data summarized in Jayko and Blake, 1989). We interpret the initial high-pressure metamorphism to be at least 145 Ma (~Tithonian) and the younger dates to have resulted from protracted cooling. Blake and others (1988) describe another event, at approximately 90 ma, when terranes in the eastern Franciscan belt were metamorphosed under blueschist-facies conditions and imbricated to form a stack of west-vergent thrust sheets.

In the Sierra Nevada batholith and its outliers in the Klamath Mountains, the magmatic history is still being clarified by U-Pb dates on zircons, which provide more reliable controls on age of intrusion than do K-Ar dates on hornblende and biotite. Not only the isotopic data but also the volumes of granitoid rocks suggest that the tempo of subduction-related arc magmatism varied. Pre-150-Ma plutonic and hypabyssal granitoid rocks related to the late Middle Jurassic to early Late Jurassic magmatic arcs were locally deformed by the Nevadan orogeny. An episode of post-Nevadan plutonism occurred between 150 and 140 Ma (Harper and Wright, 1984); plutons of this age are scattered through the Klamath Mountains and northwestern Sierra Nevada. A widely acknowledged lull in magmatic activity, recorded by volumetrically insignificant plutonic and volcanic rocks, lasted from about 140 to 120 Ma in both the U.S. and Canadian Cordillera (e.g., Armstrong, 1988). The bulk of the Sierra Nevada batholith and its outliers to the northeast was emplaced between 120 and 80 Ma (lKi, uKi).

Considering the Franciscan Complex and Sierra Nevada batholith and its outliers collectively, we propose the following Late Jurassic and Early Cretaceous history of subduction in California. At the close of the Late Jurassic Nevadan event described above, an east-dipping Franciscan subduction system was established along the western edge of the Late Jurassic system of arcs and rifted intra-arc basins. We tentatively conclude that the ca. 145-Ma high-pressure metamorphism of the Pickett Peak terrane affected rift-generated oceanic crust and its sedimentary cover that descended eastward during the inception of Franciscan subduction. To us, such a geometry is more consistent with the structural position of the eastern Franciscan belt, beneath the western part of the Great Valley Group, than is the alternative geometry involving west-dipping subduction (Saleeby, this volume). The 150- to 140-Ma plutonic rocks in the Sierras and Klamath Mountains represent the latest Jurassic magmatic arc formed during this early phase of Franciscan subduction. The Early Cretaceous magmatic lull reflects a substantial diminution, 20 to 25 m.y. long, in the velocity of convergence and in the volume of oceanic crust and overlying sediments that descended beneath the margin. We would predict that the volume of material added to the Franciscan Complex by frontal accretion or underplating was correspondingly small during this period. Possible kinematic explanations for diminished subduction and the resulting magmatic lull include: (1) an apparent polar wander path for the continent (May and others, 1989) that suggests rapid northward absolute motion of North America (and consequently, more highly oblique convergence along its margin) between about 150 and 135 Ma; or (2) hot-spot (relative motion) models (Page and Engebretson, 1984) that suggest a diminished normal component of convergence between the Farallon and North American Plates between 145 and 135 Ma. The lull, however, does not coincide exactly with either of the changes based on plate-kinematic models. Rapid, predominantly normal, rather than highly oblique, convergence probably gave rise to the major magmatic pulse lasting from 120 to 80 Ma. The dominantly Tithonian and Valanginian protoliths of the extensive Yolla Bolly terrane in the eastern Franciscan belt were subducted (underplated), metamorphosed to blueschist facies assemblages at depths of about 18 to 27 km, and imbricated with the Pickett Peak terrane during this period of renewed convergence.

Late Mesozoic tectonostratigraphic elements in the Pacific Northwest

The following summary of late Mesozoic geology is partly based on a review by Cowan and Bruhn (this volume). Here, we emphasize Early to early Late Cretaceous tectonics in an east-to-west transect centered on 49°N, from northeastern Washington to Vancouver Island. In this transect, late Mesozoic rock units and structures fall naturally into five fault-bounded tectonic elements: the Okanogan highlands; the Methow-Pasayten belt; the Skagit metamorphic terrane; the San Juan–northwest Cascades thrust system; and the Wrangellia terrane (Fig. 12). In the next section, we consider possible tectonostratigraphic correlations of these elements with the Franciscan Complex, Great Valley Group, and Sierra Nevada batholith in California.

The Okanogan highlands (Figs. 11, 12), extending roughly 150 km east of the Pasayten fault, are underlain by upper Paleozoic and lower Mesozoic, largely volcanogenic rocks of the Quesnellia terrane (MTR os), and structurally separated amphibolite-facies metasedimentary rocks and orthogneiss (CMm). Some of the gneiss is Early Proterozoic and it may represent the western margin of cratonic North America that was overridden by Quesnellia in the early Jurassic (*Armstrong and van der Heyden, 1989*). Post–Middle Jurassic granitoid plutons are abundant, but very few have been dated in the United States using U-Pb isotopes in zircons. East of the Pasayten fault in

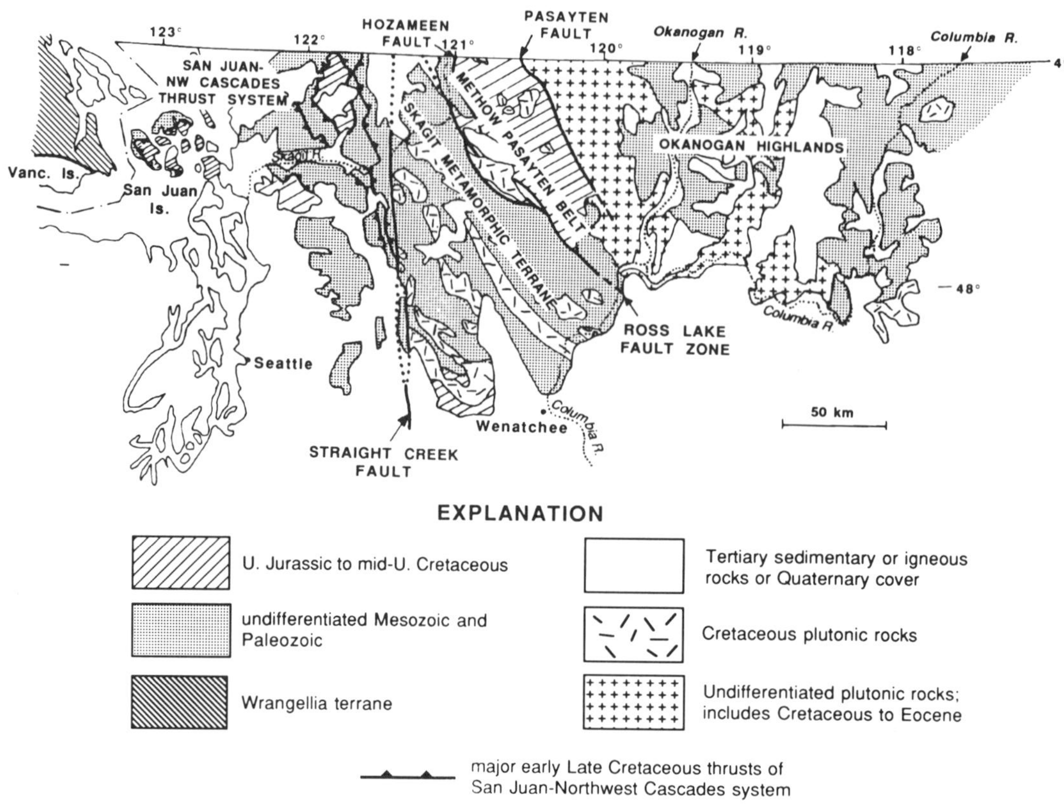

Figure 12. Simplified map of pre-Cenozoic geology of Washington and southeasternmost Vancouver Island, modified from Figure 9 in Cowan and Bruhn (this volume). The five major elements that participated in late Mesozoic tectonic events (see text) are: Wrangellia terrane; San Juan–Northwest Cascades thrust system; Skagit metamorphic terrane; Methow-Pasayten belt; and Okanogan Highlands.

Washington, K-Ar dates cluster between 106 and 85 Ma. The Kaniksu batholith in southernmost British Columbia and northern Idaho yielded a U-Pb date of 94 Ma. Considered collectively, the plutonic rocks extending from the Pasayten fault to the Kaniksu are part of a belt of 90- to 115-Ma (K-Ar) plutons extending north into eastern British Columbia (Fig. 13; *Archibald and others, 1984*; Armstrong, 1988). These plutons as a group straddle the Middle Jurassic tectonic boundary between the Belt Supergroup and its crystalline basement (both North America), and the Kootenay and Quesnellia terranes.

West of the Pasayten fault, the Methow-Pasayten belt (Fig. 12) includes well-stratified volcaniclastic and arkosic sedimentary rocks and minor andesitic volcanic rocks ranging in age from Early Jurassic to Albian and probably Cenomanian. Dickinson (1976), *Tennyson and Cole (1978),* and Davis and others (1978) proposed that these strata accumulated in a fore-arc basin from Jurassic to Early Cretaceous time. More recent work has shown that the Methow-Pasayten belt, and the originally contiguous Tyaughton basin to the northwest in British Columbia, evolved through diverse tectonic settings from Middle Jurassic to Middle Cretaceous time. *Umhoefer (1989),* for example, hypothesizes a fore-arc setting in the Middle Jurassic and a transform-rift setting from Oxfordian to Valanginian time. Hauterivian volcanogenic rocks in the Buck Mountain Formation are overlain by east-derived Aptian-Albian arkosic sandstone, which Tennyson and Cole (*1978*) suggested was derived from the unroofing of a magmatic arc in the Okanogan Highlands to the east. In the Albian, sedimentation in the Methow basin changed dramatically. Sediments supplied from an uplifted Bridge River–Hozameen terrane (PJap) to the west accumulated in foreland or piggy-back basins in the midst of a mid-Cretaceous fold and thrust belt involving older Methow strata. McGroder (1989) has established that the primarily east-vergent structures that developed in the Methow basin during this short-lived event, which ended by about 88 Ma, accommodated 50 km or more of shortening.

We believe that the sedimentological evidence from the Methow-Pasayten basin supports a fore-arc setting from Hauterivian time until at least the onset of contractional deformation in the Albian. Moreover, if we interpret the belt of late Early Cretaceous plutons extending from the Okanogan Highlands through northeastern Washington into northern Idaho and adjacent British Columbia as a magmatic arc active from about 115 to 94 Ma (Aptian to Cenomanian), then the Methow basin remained in a fore-arc setting with respect to this arc even during its contraction. The subduction zone responsible for the arc lay somewhere west of the Methow basin and the bounding Ross Lake fault system; we discuss its probable location and character in the following section.

The two tectonic elements lying west of the Methow-Pasayten belt and the Ross Lake fault zone along its western boundary are the Skagit metamorphic terrane (the crystalline core of the northern Cascade Range, CMm), and the San Juan–Northwest Cascades (SJ-NWC) thrust system (e.g., Misch, 1988; Brandon and others, 1988). In many respects, these two elements are different, but they both experienced deformation, at approximately 90 to 100 Ma, and metamorphism that, in our view, establishes a link between them. The Skagit is a composite terrane. It comprises diverse pre–Middle Cretaceous supracrustal schists and gneisses, and variably deformed granitoid plutonic rocks ranging in age from about 90 to 45 Ma. Part of the Skagit experienced deep burial (28 to 34 km; Whitney and McGroder, 1989) and medium- to high-grade Barrovian synkinematic metamorphism about 90 to 100 Ma. The SJ-NWC system consists of generally west-vergent thrust sheets that were assembled into an imbricate stack between about 85 and 100 Ma. The nappes include a wide variety of lithotectonic units of diverse ages. Most of these units record a high-pressure, low-temperature metamorphism that was probably caused by rapid tectonic burial accompanying thrusting. Although we believe that the evidence strongly supports a net horizontal shortening on the order of several hundred kilometers, Brown (1987) alternatively interprets the SJ-NWC system as a dominantly transcurrent orogen.

The critical geologic relations between the SJ-NWC system and the Skagit terrane are still unresolved. In the United States, the two are now largely separated by the high-angle Straight

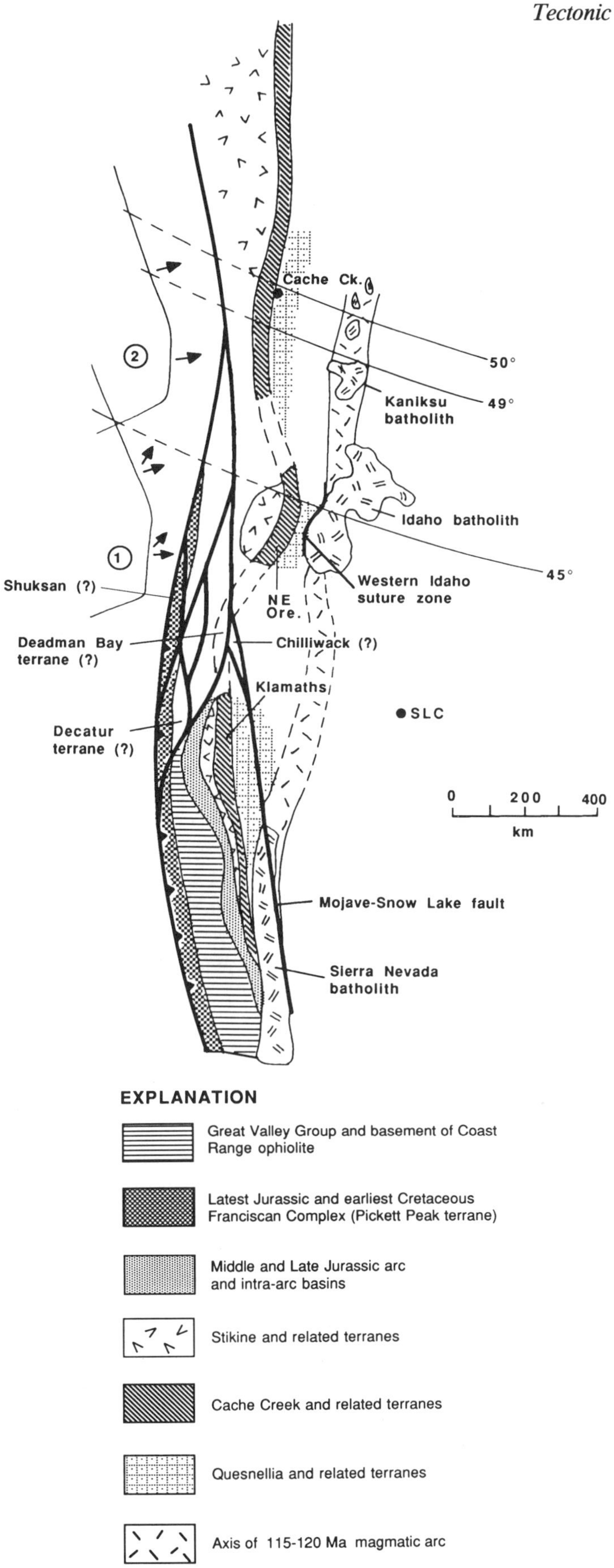

Figure 13. Paleogeographic map of the western Cordillera and continental margin in late Early Cretaceous time (ca. 120 to 115 Ma). Palinspastic base constructed/modified: (a) from Oldow and others (1989, their Plate 5B), but incorporating a restoration of ca. 325 km of Cenozoic basin-and-range extension; (b) restoring ca. 500 km of post-120-Ma dextral slip along Mojave–Snow Lake fault (Lahren and Schweickert, 1989); (c) restoring 60° of post-120-Ma clockwise rotation of Pz/Mz terranes in northeast Oregon. Kaniksu and Idaho batholiths are in northern extension of 120 Ma and younger Sierran magmatic arc (single dashes where inferred). Kaniksu and Idaho batholiths, SLC (Salt Lake City), and solid lines of latitude are in the same position as at present with respect to cratonic North America. A continental borderland developed in the gap between Klamath Mountains and northeast Oregon or in a displaced position farther north. The borderland was bounded on the west by the obliquely convergent Franciscan subduction zone. Fault-bounded slices (hypothetical) within the borderland labeled "Shuksan?," Deadman Bay terrane?," "Decatur terrane?," and "Chilliwack terrane?" illustrate hypothetical former positions for these terranes now present in the San Juan–Northwest Cascades thrust system. (1) and (2) show hypothetical positions of the Wrangellian (Insular) superterrane. In (1), it collided with the margin in earliest stages of borderland development and subsequently moved northward. In (2), it collided with a borderland that had itself been displaced northward.

Creek fault (Fig. 12), which accumulated 90 to 180 km of dextral transcurrent slip in the Paleogene. In southwest British Columbia, however, the crystalline core lies west of the Fraser–Straight Creek fault and structurally overlies the SJ-NWC on an east-dipping thrust (Journeay and Csontos, 1989). *Brandon and Cowan (1985)* and McGroder (1989) use the contemporaneity of the short-lived deformational events at 90 to 100 Ma in both elements to model the SJ-NWC system and the Skagit terrane as part of a single Alpine-type contractional orogen of mid-Cretaceous age. They hypothesize that the SJ-NWC nappes were driven westward over part (McGroder) or all (*Brandon and Cowan*) of the Skagit terrane and in fact accounted for its rapid burial to midcrustal depths at 90 to 100 Ma. McGroder envisions the root zone for the nappes to be within the crystalline core; Brandon and Cowan derive the nappes from a mid-Cretaceous root zone in the vicinity of what is now the Ross Lake fault zone (Fig. 12); this early Tertiary fault zone separates the core from the Bridge River–Hozameen terrane and Methow-Pasayten basin to the east. The timing of contraction in the adjacent Methow-Pasayten belt (90 to 100 Ma) is too similar to be fortuitous; it represents synchronous, thin-skinned, east-vergent back-thrusting that accommodated an order of magnitude less shortening than did the west-vergent SJ-NWC system. All mid-Cretaceous deformation in the Cascade orogen was presumably driven by the collision of the Wrangellia terrane (e.g., Davis and others, 1978), which constitutes the foreland to the SJ-NWC thrust and nappe system and inferentially lies below it.

When viewed as a whole, the SJ-NWC thrust system includes a remarkable diversity of units. In the San Juan Islands, the structurally lowest thrust sheet comprises the lower Paleozoic tonalite, gabbro, and minor ultramafic rocks of the Turtleback Complex, and probably related Devonian to Permian volcanogenic rocks of the East Sound Group (DPa). Next above is a thin internally imbricated sheet of Permian-Triassic seamount basalt and originally overlying Triassic-Jurassic radiolarian chert. Together these units constitute the Deadman Bay terrane (PT̄ oc), which is further distinguished because it includes in situ (not tectonically mixed) blocks of limestone containing "Tethyan" Permian fusulinids. The Deadman Bay terrane is separated from the overlying Jurassic-Cretaceous Constitution Formation (JKfa) by the Rosario thrust zone, which contains tectonic lenses of Permian-Triassic amphibolite (Garrison Schist). Volcaniclastic graywacke, radiolarian chert, and pillow basalt of the Constitution Formation are structurally overlain by the Decatur terrane, consisting of the Middle Jurassic ophiolite and superposed arc of the Fidalgo Complex and its Upper Jurassic clastic cover, the Lummi Group. The Lopez Complex, a shear zone separating the Constitution Formation and Decature terrane (Fidalgo and Lummi), contains a slice of uppermost Albian pillow lava. These are the youngest rocks found so far in the SJ-NWC system.

In the northern Cascades, the structurally lowest element in the SJ-NWC system consists of the Middle(?) Jurassic Wells Creek Volcanics (Ja) and overlying Upper Jurassic and Lower Cretaceous volcaniclastic Nooksack Group (JKfa). Next above is an imbricate structural complex, principally comprising the upper Paleozoic arc-related Chilliwack Group (DPa), and structurally overlying all are the greenschist, blueschist, and phyllite of the Shuksan Metamorphic Suite (Joc). The probable Jurassic protoliths of the Shuksan were metamorphosed in the Late Jurassic and Early Cretaceous, prior to their inclusion in the SJ-NWC system. The imbricate complex below the Shuksan contains tectonic slices of lower Paleozoic and Precambrian(?) metaplutonic rocks of the Yellow Aster Complex (similar in part to the Turtleback Complex), Permian-Triassic amphibolite and blueschist of the Vedder Complex (closely resembling the Garrison Schist), and chert and basalts of the Elbow Lake unit (similar to Deadman Bay terrane). Although the stacking order of nappes both in the northwest Cascades (Misch, 1988) and in the San Juan Islands (Brandon and others, 1988) is well established, the geometrical relation between the two sequences of nappes is still uncertain. *Cowan and Potter (1986)* and Misch (1988) considered the San Juan nappes to extend eastward beneath the stack in the northern Cascades, but Brandon (1989) postulated that the entire San Juan sequence is the equivalent of the Church Mountain Plate and its imbricate zone.

Following *Cowan and Potter (1986)* and Brandon (1989), we envision the SJ-NWC system as a westward-tapering wedge of nappes that was emplaced generally westward (present coordinates) onto Wrangellia during the mid-Cretaceous Cascade orogeny. Fundamentally, this orogeny records the collision of the Wrangellia terrane (Davis and others, 1978). As such, the SJ-NWC system comprises tectonic fragments of whatever Wrangellia collided with. Davis and others (1978) first noted that many of the elements in the SJ-NWC system and Skagit core appear to be equivalent to elements in the Methow basin and Okanogan highland region. They hypothesized that the Early Cretaceous margin had been doubled-up by dextral transcurrent faulting and that the western displaced sliver was subsequently telescoped by the collision of Wrangellia.

We now modify this hypothesis but preserve its central premise that the SJ-NWC system consists of fragments of tectonostratigraphic elements that were already part of the North American continental margin when they were telescoped and imbricated during the mid-Cretaceous. We suggest, as did Brandon and others (1988), that many of the SJ-NWC elements closely resemble units in the Klamath-Sierran orogen in southwestern Oregon and northern California. For example, the Deadman Bay terrane is equivalent to diverse terranes of Cache Creek affinity in the Klamath Mountains that lie west of and below the Triassic–Early Jurassic suture against Sonomia. The Vedder Complex and Garrison schist resemble the Fort Jones–Stuart Fork schists, and the Chilliwack Group is lithologically and faunally (for at least Permian rocks) equivalent to the Devonian-Permian eastern Klamath arc (Miller, 1987). Moreover, these Klamath-Sierran elements have been tectonically mixed with probable fragments of the Franciscan Complex (Shuksan; Brown and Blake, 1987) and the Great Valley Group–Coast Range ophiolite (Decatur terrane; Garver, 1988).

In the following section, we propose a specific model for how these Klamath-Sierran-Franciscan elements were transported northward.

The Cascade orogenic event terminated during the interval from about 93 to 83 Ma. In the San Juan Islands, Santonian beds of the Nanaimo Group (uKITfa) contain clasts derived from nappes in the SJ-NWC system. The Windy Pass thrust in the south-central Cascades is cut by the approximately 90-Ma Mt. Stuart batholith, and contractional structures in the Methow basin are cut by approximately 88-Ma granitoid intrusive rocks. These plutonic rocks are part of an extensive, voluminous suite, 90 Ma and younger in age, which represents the southeastern extension of the Late Cretaceous Coast Plutonic Complex of British Columbia. They represent a magmatic arc constructed in and across the mid-Cretaceous Cascade collisional orogen.

Reconciling late Mesozoic tectonics in California and the Pacific Northwest: The "borderland" hypothesis

In our view, the first-order differences between late Mesozoic California and the Pacific Northwest can be succinctly summarized as follows. First, there is an incomplete record in the northern region (e.g., lat 49°N) of the Tithonian subduction system so well represented in California by the post-Nevadan, latest Jurassic, and earliest Cretaceous elements of the Franciscan (subduction complex), Great Valley Group (forearc basin deposits), and Sierra Nevada batholith and its outliers (magmatic arc). Only fragments of the Tithonian Franciscan (= Shuksan) and Great Valley Group (= Lummi Formation) are found at 49°N, where they occur in a mid-Cretaceous thrust and nappe system tectonically interleaved with fragments of Paleozoic and Mesozoic rocks that strongly resemble the pre-Nevadan basement of the Klamath Mountains and Sierra Nevada. Second, no clear equivalent of the ca. 150- to 140-Ma, post-Nevadan Sierra batholith has yet been found in the 49°N transect, nor, for that matter is a clear record preserved of the Middle and Late Jurassic rifting, ophiolite genesis, and Nevadan telescoping of more southerly regions (except for the fragments of Coast Range ophiolite involved in the SJ-NWC thrust system). The earliest that the post-Nevadan Sierran magmatic arc seems to have extended northward was in the Hauterivian, about 120 Ma. At 49°, the arc is represented by the belt of ca. 115- to 94-Ma plutons in northern Idaho, northeastern Washington, and adjacent British Columbia, and by Hauterivian volcanogenic rocks in the Methow basin.

We propose a model, involving both transcurrent displacements and contraction, to explain these differences. To account for the tectonic interleaving of Franciscan, Great Valley, and Klamath-Sierran elements in the San Juan–northwest Cascades thrust system, we suggest that fragments of the continental margin in northernmost California and Oregon were transported northward several hundred kilometers and assembled into a tectonic borderland. These events occurred at some time after the development of Tithonian units in the Franciscan and Great Valley sequences, but prior to the collision of Wrangellia with the borderland in the vicinity of 49°N at about 100 Ma. We further hypothesize that, prior to about 120 Ma, when convergence and a related magmatic arc were reestablished along the entire continental margin under consideration here, the margin north of about 45° was dominated by coast-parallel, transcurrent displacements. Consequently, Nevadan telescoping and Tithonian subduction were largely restricted to California.

Our hypothesis is built on several important postulates or assumptions. Our first postulate is that by the beginning of Late Jurassic time (about 160 Ma) all of the diverse pre–Late Jurassic terranes from northern California into British Columbia—except those of the San Juan Islands–northwestern Cascades region, components of the Wrangellian superterrane (= Superterrane II) and, perhaps, the diverse terranes of northeastern Oregon—had been incorporated into the margin of North America at essentially their present latitudes with respect to the continent (Fig. 13). In British Columbia, the Late Jurassic margin lay west of Stikinia and, consquently, west of the Cache Creek terrane and Quesnellia (Fig. 9). Our second postulate is that these terranes have essentially remained at these positions since the Late Jurassic. The greatest unknown in reconstruction of this region, as discussed below, is the location and correlation of the amalgamated (by about 150 Ma) terranes of northeastern Oregon and westernmost Idaho (Huntington/Olds Ferry, TR a, and Wallowa–Seven Devils arcs, ℙJa, Baker [= Cache Creek] oceanic assemblage, ℙJap). The corollaries that follow from our assumptions are as follows. Any major zones of post–Middle Jurassic collisional suturing must lie west of the Klamath Mountains and the Intermontane superterrane (= Superterrane I), which would be west of Stikinia, and possibly west of west-central Idaho if the western Idaho suture represents the closing of a Late Jurassic–Early Cretaceous ocean (implicitly favored by Lund and Snee, 1988, but not our view; see below). Any post–Middle Jurassic transcurrent displacements of large displacement must also lie west of the Intermontane superterrane at the latitude of 49°N, but could pass southward into the West Idaho suture zone, between the Klamath and Blue Mountains (northeastern Oregon), west of the Klamath Mountains, or along all three generalized traces.

Our first and second postulates essentially follow Davis and others (1978), Monger and others (1982), Mortimer (1986), and Miller (1987). They contrast sharply with ideas expressed by Umhoefer (1987), Avé Lallement and Oldow (1988), Oldow and others (1989), and other syntheses that are based on the "Baja British Columbia" model suggested by Irving and others (1985) and on paleomagnetic data in general. We know of no data that require that Stikinia, Canadian portions of the Cache Creek terrane, or Quesnellia were ever far south (i.e., >1,000 km) of their present latitudes in the Early Cretaceous. We have indicated above our preference for the paleomagnetic analysis of May and Butler (1986) in which Superterrane I originated at essentially its present latitude, and we prefer the recent interpretation of Butler and others (1989) that accounts for the apparent discordant paleomagnetic data (see below; cf., Beck, this volume).

The history of displacement and accretion of the pre-Cretaceous terranes in northeastern Oregon, and regional correlation of the Wallowa–Seven Devils terrane, are controversial topics. Wernicke and Klepacki (1988) and Lund and Snee (1988), among others, have hypothesized major transcurrent slip (dextral?) along the fault system juxtaposing the terranes with North American rocks across the West Idaho suture zone (also called the Salmon River suture). In spite of many field studies, however, kinematic indicators from the "suture" record exclusively dip-slip displacements (e.g., *Strayer and others, 1987; Strayer, 1988*). Moreover, the paleomagnetic data of Wilson and Cox (1980) indicate negligible latitudinal displacements of these pluton-stitched western terranes since Late Jurassic–Early Cretaceous time (Beck, this volume). We favor a dominantly contractional history for Cretaceous tectonic events (ca. 118 to 88 Ma; Lund and Snee, 1988) along the "suture," but cannot rule our precontractional transcurrent displacement between the terranes now juxtaposed along it. Pre-Cretaceous terranes now west of it may have originally been sutured against North America in Middle or early Late Jurassic time, prior to the amalgamation of those terranes by Late Jurassic plutons (ca. 160 Ma). One of the components of this amalgamated terrane, the Baker terrane, has Cache Creek affinities and seems to lie within the belt of Cache Creek assemblage rocks in northern California and southern British Columbia that was accreted to North America by the Middle Jurassic (Fig. 13; cf., Davis and others, 1978, their Fig. 4). As for the Wallowa–Seven Devils terrane, we can report only that some workers (e.g., Wernicke and Klepacki, 1988) favor the original suggestion of Jones and others (1977) that it represents part of Wrangellia, whereas others (e.g., Mortimer, 1986) deny this correlation and support its pre–Late Jurassic contiguity with a fringing arc such as Stikinia.

If our assumptions are correct, they have important implications for the northwest extension of a dextral transcurrent fault recently hypothesized to lie in the eastern Sierra Nevada. Lahren and Schweickert (1989) present compelling evidence for approximately 500 km of dextral slip in the interval of time between 148 and 110 Ma on their proposed northwest-trending Mojave–Snow Lake fault (Fig. 13); for reasons presented below, we favor post–120-Ma dispalcement. The northwest trace of this fault has not yet been defined. We propose that it probably passed east of the pre-Tertiary rocks in the northern Sierras and Klamath Mountains and west of pre-Tertiary rocks in the Blue Mountains (Fig. 13), but its location along the subsequent site of the Cretaceous West Idaho suture zone is an alternative possibility.

Many tectonic syntheses, even those using palinspastic base maps (e.g., Hamilton, 1978; *Dickinson and Thayer, 1978*; Wells and Heller, 1988), depict or infer that the Klamath pre-Tertiary terranes were originally contiguous with those in northeastern Oregon. We suggest that a restoration of 500 km of dextral slip on the northwest projection of the Mojave–Snow Lake fault (Fig. 13) leaves a major gap between terranes presently in the Klamaths and northeastern Oregon or, possibly but less likely, along the West Idaho suture zone. This gap (preferred alternative) was presumably largely filled by terranes originally adjacent to those of the Klamaths and Coast Ranges to the south and those of the Blue Mountains (Olds Ferry, Baker, and Wallowa) to the north. In short, the pre–Late Jurassic terranes that constituted the continental margin in British Columbia extended southward as a continuous belt through eastern Oregon into the Klamaths and northern Sierras, which were then (i.e., KM and NS) 500 km farther south.

The Franciscan–Great Valley–Sierra Nevada system was newly established along this margin but extended only to about 45°N; there is no evidence of a Tithonian arc north of 45°. The position or even presence of the Wrangellian superterrane is unconstrained at this time. During the magmatic lull (ca. 140 to 120 Ma), the kinematics of continental margin tectonics based on geologic evidence are, in our opinion, uncertain. Nevertheless, plate-motion reconstructions (Engebretson and others, 1985) and paleomagnetic studies (e.g., May and others, 1989) propose left-oblique convergence between North America and the Farallon Plate prior to either 120 Ma (Engebretson and others) or between 150 to 135 Ma (May and others). If the case for sinistral plate convergence is valid, dextral displacement on the Mojave–Snow Lake fault system could probably not have occurred until sometime after either 135 or 120 Ma. We will assume, to avoid unresolvable debate, that the Mojave–Snow Lake fault did not form until after the younger age. Upon formation, this important fault extended northwestward from the east-central Sierra Nevada, possibly to connect with transcurrent (transform) faults north of 45°. It crossed the pre–Late Jurassic accreted terranes and the northern end of the Tithonian Franciscan system and fore-arc basin. Distributed dextral slip created a topographically and structurally complex southern California–type borderland, wherein fragments of the pre–Late Jurassic terranes were juxtaposed with Tithonian elements along dominantly transcurrent faults; contractional and extensional faults may have locally been active as well. Rocks in the borderland were transported at least to 49°N. Calling upon the plutonic record, convergence was reestablished ca. 120 Ma along the entire margin from British Columbia to California. Wrangellia approached and finally collided with the borderland about 90 to 100 Ma. Fragments of the borderland were then driven westward onto and across Wrangellia to form the San Juan–northwest Cascades thrust system.

Back-arc region: Middle Jurassic to Late Cretaceous

Areas east of the narrow Triassic to Middle Jurassic magmatic arc that developed along the western edge of the continental United States had been subject to little or no early Mesozoic tectonism. However, beginning in Middle Jurassic time these areas began to experience deformation, metamorphism, and plutonism that eventually extended more than 1,000 km eastward through the Paleozoic miogeocline and well into the North American craton. There are significant differences in the Middle Jurassic to Late Cretaceous evolution of northern and southern segments of the Cordillera. A poorly defined transition between these segments lies in the vicinity of the California-Nevada border southwest of Las Vegas. In the northern segment major shortening occurred, whereas in the southern segment shortening appears to have been less and extensional deformation more important, particularly during latest Jurassic and Early Cretaceous time (Fig. 11).

Tectonism in the northern segment of the U.S. Cordillera. The eastward spread of deformation and magmatism began in Middle Jurassic time, but is treated in this time slice, because it begins a major episode of eastward migration of deformation and magmatism that continues into the early Cenozoic. Events occurring during this period are best and most completely developed in a transect across central Nevada. They are preserved in three deformational belts called, from west to east, the Luning-Fencemaker, Eureka, and Sevier belts (Fig. 11).

Luning-Fencemaker belt. The Luning-Fencemaker belt forms a generally north-trending, but locally arcuate, belt of thrust faults and folds in western Nevada (Oldow, 1983, 1984a; Fig. 11). The structures record the closure of the Star Peak/Luning basin (Ꞧ Jsb) and deformations of its eastern shelf (Ꞧ s; Speed and others, 1988; *Smith and others, 1989*). The most important structures contributing to the closure of the basin are thrust faults of the Lunning/Fencemaker thrust system (Oldow, 1984a; Speed and others, 1988). This system moved basinal (Fencemaker thrust) and shelf (Luning thrust) sediments east and southeast perhaps several hundred kilometers over adjacent shelf deposits (Oldow, 1983, 1984a; Oldow and others, 1984; Speed, 1978; Speed and others, 1988). The youngest basinal rocks involved in the thrusting are of early Middle Jurassic age, but the timing of thrusting is poorly constrained. Speed and others (1988) suggest that deformation may have begun in Early or Middle Jurassic time and continued into the Cretaceous. Thrusting could have begun in late Middle Jurassic time, although there is little direct evidence for tectonism until the Late Jurassic or Early Cretaceous (Oldow, 1984a). It is important to establish the onset of deformation in the Star Peak/Luning basin because it heralds the beginning of back-arc compressional tectonics in the northern segment of the U.S. Cordillera.

Within the Luning-Fencemaker belt, deformation of probable Late Jurassic and Early Cretaceous age has been reported. Along the west side of the Star Peak/Lunning basin in the Jackson Mountains, plutons (~160 Ma) invaded the basin rocks and were followed by formation of the northwest-vergent Deer Creek thrust system and deposition of syntectonic coarse clastic rocks of the Upper Jurassic–Lower Cretaceous King Lear Formation (Kb; *Maher and Saleeby, 1988*; Speed and others, 1988). Synthrusting sills have yielded ages of ca. 115 Ma (*Maher and Saleeby, 1988*). In the region of the Triassic shelf east of the Luning/Fencemaker thrust system the west-vergent Willow Creek thrust lies along the west side of a narrow north-trending metamorphic-plutonic culmination that has yielded 160 to 150 Ma K/Ar hornblende ages from syntectonic plutons (Speed and others, 1988).

The Luning/Fencemaker thrust system can be traced from north-central Nevada, where it emerges from below Cenozoic cover, southwestward into westernmost Nevada where it has no obvious continuation to the west. Oldow (1984a; personal communication, 1989) has postulated that the thrust system ends against the unexposed northwest-striking Pine Nut fault, which he interprets as an intra-arc or back-arc transpressional fault system (sinistral) that bounded the evolving allochthon on its western side. We do not understand clearly the nature or displacement history of the Pine Nut fault, or its geologic relationships with the Luning/Fencemaker thrust system. Oldow and others (1989) state: (1) that mid-Jurassic mylonites in an ancestral Pine Nut fault system have fabrics that suggest pre-170-Ma dextral transpression; (2) that the fault had sinistral, transpressional displacements of 500 to 700 km in the Jurassic and Early Cretaceous; and (3) that a western strand of the Pine Nut fault system experienced approximately 250 km of right-lateral motion after emplacement of the mid-Cretaceous Sierra Nevada batholith (Oldow and others, 1989, Fig. 9). It seems to us, that there is little hard evidence for the latter two kinematic/displacement determinations. Jurassic–Early Cretaceous movements of the fault system are described by Oldow and others (1989) and Oldow (personal communication, 1989) as permitting decoupling of faulted and folded rocks in the contemporaneous Luning-Fencemaker thrust system from the more stable arc to the west; Oldow and others (1989). state the Luning-Fencemaker thrust system merges southwestward with the Pine Nut fault. It seems to us that these geometries would produce dextral, not sinistral displacements between the foreshortened basin on the east and the stable arc to the west, a conclusion apparently shared by Speed and others (1988, their Fig. 22-7). Relationships between the sinistral(?) Pine Nut fault of Oldow (1984a) and the dextral, possibly coeval Mojave–Snow Lake fault of Lahren and Schweickert (1989; 500 km of Late Jurassic–Early Cretaceous dextral slip; see above) are unresolved at the time of this writing. Lahren and Schweickert have proposed that these two faults may be offset equivalents of each other separated by dextral east-striking faults in west-central Nevada (Stewart, 1985), and that the displacement history of both is dextral.

Eureka belt. The north-trending Eureka belt (Speed, 1983; Speed and others, 1988) of east-central Nevada (Fig. 11) consists of folds and thrust faults with general eastward vergence. Within the belt are nonmarine sedimentary rocks of the Newark Canyon Group (Kb), which range from Aptian to Maastrichtian in age (*Schmitt, 1989*). *Schmitt (1989)* interpreted these rocks to have formed within an active fold thrust belt. Recently, *Bartley* (*1990*) has suggested that the belt contains thrusts that are both east- and west-vergent and that the Newark Canyon rocks overlie some of these structures unconformably. Thus the age of the deformation in the Eureka belt remains unclear. The amount of shortening within the belt appears to be minor compared to the belts flanking it on both sides.

Sevier belt and its hinterland. In Late Jurassic time, deformation and plutonism became active in northeast Nevada, east-central Nevada, northwestern Utah, and southern Idaho (Saleeby and others, this volume; Allmendinger, this volume; Miller and others, this volume; and *Thorman and others, 1990*). In this region plutons of early Late Jurassic age (165 to 150 Ma) cut Mesozoic deformational fabrics, folds, and thrust faults. Ductile deformation developed complex structures with variable- (Snoke and Miller, 1988) or west-vergence (*Miller and others, 1988*). In the northern part of this area, folds and thrust faults are both southeast- and, locally, northwest-vergent (*Thorman and others, 1990*). These structures involve both Antler (€Dos) and Havallah (DPos) assemblage rocks emplaced in Paleozoic times, and miogeoclinal rocks (€Jcs, ℙPcs). Sediments eroded from these Mesozoic structures may have reached eastward into the craton, now part of the Colorado Plateau, by late Middle Jurassic time (*Peterson, 1990*). Although these deformed rocks lie in the hinterland of the Sevier foreland fold and thrust belt to the east, there is no regional or local evidence for thrust faults of this age in the Sevier belt. In other areas, such as eastern Nevada and west-central Utah, where Late Jurassic metamorphism and plutonism have been documented, Cenozoic rocks rest with little angular discordance on Paleozoic strata indicating that Mesozoic deformation lacked significant surface manifestation (Armstrong, 1968).

The Sevier belt of Late Jurassic (?) to late Late Cretaceous age (Armstrong, 1968) forms a major throughgoing structural element of the eastern Cordillera from southeastern California to northern Canada (Fig. 11). This east-vergent foreland fold and thrust belt lies east of the Late Jurassic plutons and synmetamorphically deformed rocks, and may be entirely younger. The time of inception of the Sevier Cordilleran foreland fold and thrust belt is, unfortunately, not clear. Armstrong and Oriel (1965), and most subsequent workers, considered that the east-vergent Willard/Paris allochthon, which carried sedimentary rocks of the Late Proterozoic–Paleozoic miogeocline in its hanging wall and shed initial coarse clastics eastward into the embryonic Cordilleran foredeep (Kfd), developed in latest Jurassic to Early Cretaceous time. However, Heller and others (1986) have presented evidence to suggest that thrusting did not begin until about late Early Cretaceous time. Clearly by this time a continuous belt of east-vergent thrusting occurred from Canada to southeastern California.

During the formation of the Sevier thrust belt a foredeep (Kfd) developed along its east margin, and received up to 6 km of debris eroded from the rising mountains. Dating of the coarse clastic wedges within the foredeep sequence provides the main evidence for dating events within the thrust belt (Armstrong and Oriel, 1965; Jordan, 1981), although the relationship between the development of coarse clastic wedges and episodes of thrust faulting can be interpreted differently (cf., Heller and others, 1986). The thrust faults and the axis of the foredeep migrated eastward, with older and more westerly parts of the foredeep becoming involved in the thrusting. The foredeep was asymmetric, being deeper in the west and shallower toward the east where the nonmarine alluvial apron from the thrust belt graded eastward into the marine sediments of the Cretaceous intracontinental seaway. Formation of the foredeep was probably the result of flexural loading of the foreland by the advancing thrust belt (*Price, 1973*; Jordan, 1981).

Continuity of back-arc tectonic elements along the Cordillera. *Luning-Fencemaker belt.* The southern continuation of the Luning-Fencemaker belt has been discussed briefly above. We suggest, following Lahren and Schweickert (1989), that several hundred kilometers of Late Jurassic to Early Cretaceous right slip (we favor post-120-Ma, displacements) along a Mojave-Snow Lake–Pine Nut(?) fault would have offset southwestern portions of a somewhat older Luning/Fencemaker thrust system. Reversing displacement on this combined fault might well place the Luning/Fencemaker thrust system in a position adjacent to east-vergent structures (e.g., Taylorsville thrust) in the northern Sierra Nevada. Such a reconstruction would tie this thrust system directly into tectonic activity much nearer the plate margin. This suggestion is controversial. Oldow (personal communications, 1989) doubts the existence of east-west–striking faults that might connect offset portions of a Pine Nut–Mojave–Snow Lake fault, and believes that the kinematics and pre-Callovian age of the Taylorsville thrust preclude its association with the younger(?) Luning-Fencemaker system. An alternative explanation for the Mojave–Snow Lake versus Pine Nut fault problem is possible. That is, the two faults are not equivalent and are not coeval. Sinistral displacement along the Pine Nut fault would have occurred prior to dextral displacement after 135 or 120 Ma (see earlier discussion) along the somewhat more western(?) Mojave–Snow fault zone.

Even though the timing of closure of the Star Peak/Luning basin is poorly dated, Middle to Late Jurassic convergent deformation may have been widespread along the Cordillera, albeit somewhat diachronous. Major accretion of offshore (within plate?) tectonic elements took place in southern Canada during Middle Jurassic time. Rocks of the Slocan basin and an early Mesozoic arc and associated rocks that comprise Quesnellia (all part of the previously amalgamated Superterrane I) were thrust eastward onto the western margin of North America. This Middle Jurassic deformation was accompanied by metamorphism and intrusion of large plutons beginning the Mesozoic development of the Omineca crystalline belt. The oldest post-tectonic plutons are 173 to 182 Ma (Armstrong, 1988). Accretion of oceanic rocks, largely of mélange character, continued into late Middle Jurassic time to the west in the Bridge River area and perhaps into Late Jurassic time suggesting that eastward subduction continued beneath North America following emplacement of the Intermontane superterrane. Depending upon the age of deformation in the Star Peak/Luning basin, its closure could be about the same age or, more likely, slightly younger than the closure of the Slocan basin farther north. The closure of these two basins records tectonically similar events: the eastward telescoping across the North America Plate of back-arc basins, and their associated western arc elements. Thus, tectonic elements and events of a similar character are present both in southern Canada and northwestern Nevada, but whether they had continuity through the area of the West Idaho suture zone remains uncertain. All that can be said is that these tectonic elements and events have not yet been identified within rocks on either side of the suture.

Eureka belt. Continuity of Eureka belt structures into areas north and south of central Nevada is poorly documented. To the north they may trend more to the northeast and merge with structures within the western part of the Sevier belt in southern Idaho (Fig. 11). In northeastern Nevada, they may form part of a poorly understood belt of east-striking thrust faults called the Elko County belt by Speed and others (1988). Some of these structures are dated as Late Jurassic by *Thorman and others (1990).* These east-striking thrust faults may also mark the eastern fringes of the Luning-Fencemaker belt. The southern continuation of the Eureka belt is virtually unknown because of extensive Cenozoic cover. In addition, the timing of Mesozoic events in south-central Nevada is very poorly constrained because rocks of appropriate ages to date the deformation(s) are rare.

Sevier belt. The Sevier foreland fold and thrust belt of Early Cretaceous to late Late Cretaceous age is one of the most continuous tectonic elements in the Cordillera (Figs. 11, 14; Allmendinger, this volume). Typically, the thrust faults of the Sevier belt have thin-skinned geometries, eastward directions of transport relative to underlying rocks, and become progressively younger from west to east. There are, however, significant changes in their style and paleogeographic control along strike. In Montana, more westerly thrust faults are largely controlled by the bedding anisotropy of sedimentary strata in the Middle Proterozoic Belt Supergroup (P_1). Along the eastern margin of the thrust belt, thrusts are controlled by younger cratonal Paleozoic and Mesozoic rocks (€Jc), but many may be of younger Laramide age (Fig. 14). In southern Idaho some of the thrusts carry crystalline basement rocks (p€m) where they cross a paleogeographic high (Salmon or Lemhi arch) that marked the southern uplands for the Belt basin and was the site of intermittent cratonal Paleozoic deposition. Thrust faults in the western part of the Idaho segment of the belt cut through Paleozoic off-shelf sedimentary rocks west of the miogeocline (e.g., *Skipp and Hait, 1977*; Skipp, 1987). From south of the Snake River Plain to southern Nevada, folds and thrust faults are developed within the eastern part of the Late Proterozoic and Paleozoic miogeocline (Allmendinger, this volume). Each more westerly thrust fault carries a thicker miogeoclinal section in its hanging wall and has a ramp-flat geometry that cuts upsection toward the east (Armstrong and Oriel, 1965; Royse and others, 1975). Along the eastern edge of the thrust belt the miogeoclinal-cratonal depositional hinge zone is involved in the thrusting. Most thrust faults carry only sedimentary rocks in their hanging walls, although the hanging wall of the Willard thrust of central Utah contains a thick section of crystalline basement rocks, indicating that locally the thrust faults ramped laterally into basement. In southeastern California, near the mapped southern terminus of the Sevier belt, the thrust belt turns southward away from the miogeoclinal hinge and crosses into cratonal strata and their underlying Precambrian crystalline basement (*Burchfiel and Davis, 1976,* 1981, 1988).

Although nearly all the folds and thrust faults of the Sevier belt are of décollement style in their exposed eastern parts, we interpret them ultimately to involve crystalline basement rocks, largely contemporaneous Mesozoic plutons, and regionally metamorphised rocks to the west in the hinterland of the thrust belt. In the western hinterland, deformation is associated with metamorphism and plutonism and represents deeper level involvement of the crust in the thrust belt (*Allmendinger and others, 1984*; Allmendinger and Jordan, 1984; Allmendinger, this volume), but outcrops of these rocks are only locally present in the Albion, Ruby, and Snake Ranges or can be inferred on seismic lines such as in eastern Idaho (Royse and others, 1975). Rocks in these ranges record a complex Mesozoic history of ductile deformation, intrusion, and metamorphism. Recently Miller and Gans (1989) have made a strong case that the inception of thrust faulting in the Sevier belt may have predated peak metamorphism and intrusion of granites in the hinterland by 10 to 15 Ma, but thrusting continued during and ended simultaneously with the end of metamorphism at about 70 Ma. These exposures of more ductilely deformed parts of thrust belt initially may have been brought closer to the surface as thrust-related hanging wall anticlines (Armstrong, 1972) or on crustal duplexes (Snoke and Miller, 1988), but owe part of their exposure to Cenozoic extensional unroofing.

The location where Sevier thrust faults ramp into deeper crustal crystalline rocks is controlled by preexisting paleogeography, the temporal and spatial position of plutonism and ductile crust, and the evolution of an orogenically thickened crust. The first two controls can be partly evaluated from exposed rocks, but the third can be only inferred. Detachment beneath the Belt basin may have been at depths of 15 to 20 km near the base of the Belt Supergroup, at much shallower depths across the ancient Salmon River arch to the south, and then perhaps at greater depths of 5 to 12 km beneath the miogeocline farther south. The Mesozoic magmatic arc was also superimposed across a contrasting paleogeography creating a region of hot ductile crust at shallow, but variable depth. The configuration of the ductile crust relative to the eastern margin of the thrust belt was highly variable. Present exposures of rocks ductilely deformed during times of thrusting only partly reflect their original distribution; present exposures more likely reflect the consequences of younger deformations—both contractional and extensional—that have carried such rocks to higher structural levels. Shortening occurred east of the magmatic arc by relative westward underthrusting of colder continental lithosphere into and beneath the ductile crust of the arc. Within the region of ductile deformation crustal thickening accommodated the brittle shortening experienced within higher structural levels east of the arc (e.g., *Burchfiel and Davis, 1976*).

Regional setting of back-arc convergence. We interpret the spread of Late Jurassic and Cretaceous plutonism, metamorphism, and deformation into eastern areas of the Cordillera that were relatively undeformed since Precambrian time to be directly related to events at the western plate margin and within the accretionary belt. Throughout Triassic to early Middle Jurassic

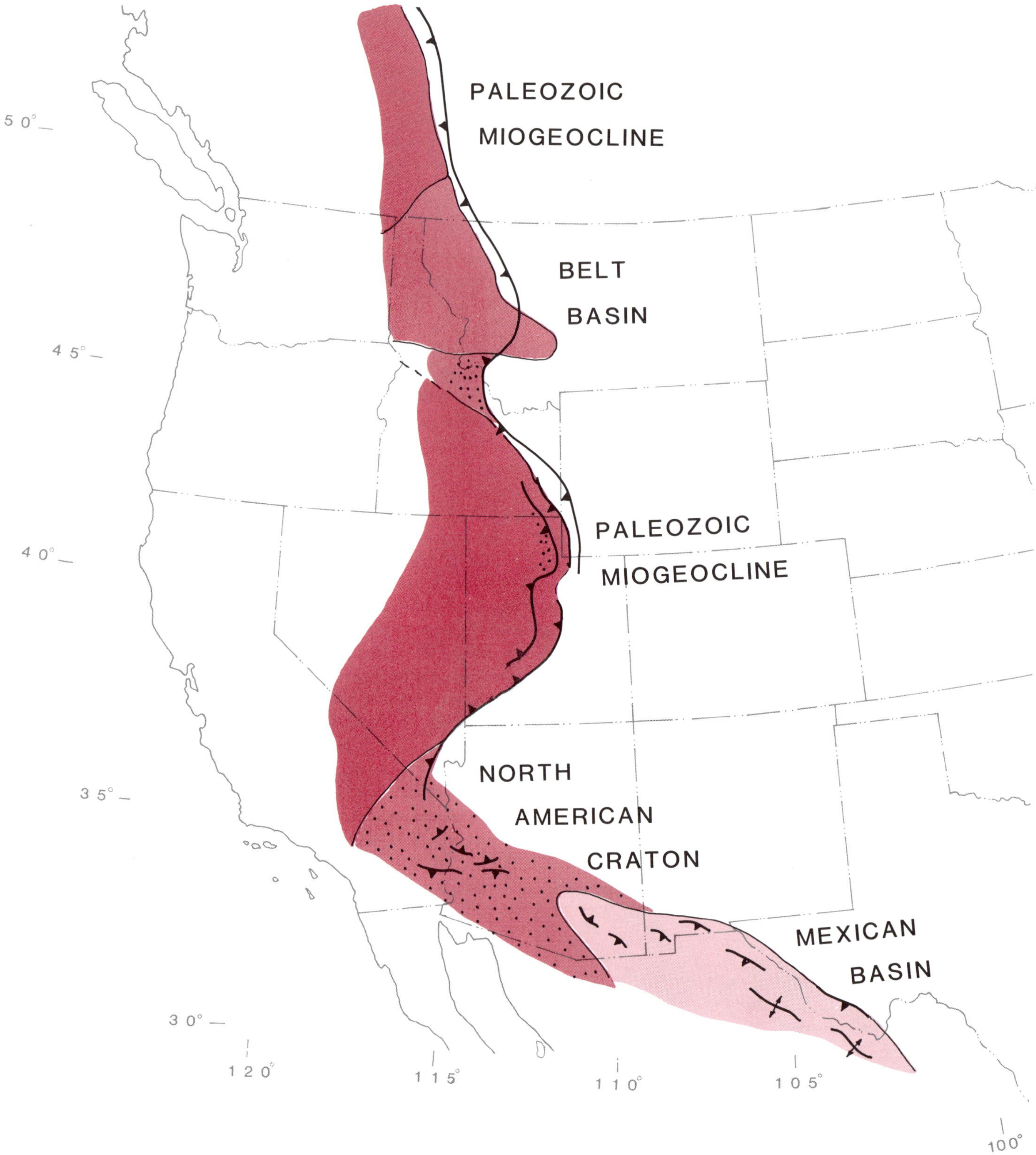

Figure 14. Generalized sketch map to show relation of the foreland fold and thrust belt of the Cordilera to major paleogeographic elements of different ages. Dotted areas are where Precambrian crystalline rocks are involved in thrust-belt deformation.

time the western margin of the North American continent was marked by a narrow continental-margin arc, forming a largely submerged volcanic archipelago. The late Middle to Late Jurassic archipelago included a western region of small, locally ocean-floored, intra-arc basins that acted as a tectonic buffer zone between Pacific oceanic plates to the west and the North American continent to the east, on which marine and nonmarine sedimentary rocks were being deposited (Saleeby and others, this volume). During Callovian/Oxfordian collapse of the offshore arc region, coupling between the subducting oceanic plate(s) and western North America is believed to have increased as the intra-arc buffer zone was eliminated. Such a region of strong coupling is believed to have occurred from the central Sierra Nevada northward through the Blue Mountains of central Oregon and into Canada.

This major change in tectonic style, from weak coupling in early Mesozoic time to strong coupling in late Mesozoic time, can be related to changes in global plate geometry (Coney, 1972), and, most likely, to resultant changes along the Cordilleran continental margin in subduction zone parameters (e.g., convergence rate, absolute velocities of upper and lower plates, angle of slab descent). During most of Mesozoic time, arc magmatism along the length of the Cordillera indicates subduction of oceanic lithosphere beneath North America. With the onset of opening of the North Atlantic in Middle Jurassic time (ca. 175 Ma; see *Klitgord and Schouten, 1986,* for a recent review), convergence rates along the western continental margin increased and remained relatively high for most of late Mesozoic time (Engebretson and others, 1985). Although there is no quantitative evidence, it could be argued that the ill-defined tectonic style of the early Mesozoic Cordillera was the result of low ocean-continent convergent rates. Weak coupling between subducting oceanic plates and the North American Plate led to little or no back-arc deformation. During late Mesozoic time, increased convergence rates produced increased coupling between oceanic plates and the North American Plate, possibly due to the active overriding of the offshore subducting plates by the west-moving continental plate. Increased interplate coupling led to a period of broad and significant intraplate convergence in Cordilleran intra-arc and back-arc regions. Events such as the development of Nevadan and pre-Nevadan west-directed thrusts in the Klamath Mountains, the closure of small intra-arc ocean basins (e.g., Galice), east-directed thrusting in the northern Sierra (Taylorsville thrust), and the beginning of closure of the Star Peak/Luning basin in late Middle Jurassic to early Late Jurassic time are viewed as expressions of the onset of strong interplate coupling along the western margin of the Cordillera.

Late Mesozoic back-arc intraplate deformation in Nevada and Utah was accompanied by an east-west broadening of arc magmatism that created a region of ductile crust from Jurassic through most of the Cretaceous time. Crustal shortening associated with high-level deformation in the Luning/Fencemaker, Eureka, and Sevier belts was transferred to ductile shortening at depth (i.e., thickening of the deeper crust). Thermal reequilibration during crustal thickening led to crustal melting and advection of heat to shallower crustal levels, further enhancing ductility. Farther south in southeastern California and Arizona coupling between plates appears to have been weaker and/or more strongly oblique so that Triassic, Jurassic, and Early Cretaceous deformation was not important even though arc magmatism continued (Tosdal and others, 1989).

Timing of deformation is critical in resolving the role of crustal thickening in the eastward spread of plutonism (e.g., see Speed and others, 1988), and the temporal partitioning of shortening within the Cordillera. Recent dating of structures in the back-arc region by *Smith and others (1989)* suggests a general eastward progression in initiation of deformation, followed by concentration of shortening in the Sevier belt with diffuse shortening across the entire Cordillera. Middle and Late Jurassic plutonism and deformation do lie in a region west of the Sevier belt, but the temporal and spatial relations of deformation are not well known enough for Middle Jurassic to Early Cretaceous times to be certain of an eastward progression. Deformation within the Sevier belt generally progressed eastward in Early and Late Cretaceous time (Allmendinger, this volume) and appears to be broadly coeval with major deformation across the arc region to the west (as reviewed above). Localization of deformation within the Luning-Fencemaker belt, apparently along the east side of the Middle Jurassic arc, may have been controlled by the region of probable thin, weak crust beneath the Star Peak/Luning basin in western Nevada (Speed and others, 1988). Deformation appears to have begun in the Luning/Fencemaker and possibly also part of the Eureka belt before folding and thrusting in the Sevier belt. Thrusting within the Sevier belt was largely or entirely younger than that of the Luning-Fencemaker belt if the interpretation of Heller and others (1986) is correct. Such timing relations can be interpreted (Speed and others, 1988) to suggest that deformation began in the eastern Sevier belt only after deformation in the Luning-Fencemaker belt had sufficiently thickened the crust beneath the Star Peak/Luning basin to permit shortening stresses to be transmitted farther east—even though the eastern limit of magmatism had already shifted into the eastern region. The region between the two belts was also affected by plutonism and metamorphism, but deformation there appears to have been relatively minor (Toiyabe uplift zone and Eureka belt). This discussion has focused on a Nevada-Wyoming/Utah transect, because it contains the most complete cross section of the central and eastern Cordillera in the conterminous United States.

It has been postulated (e.g., Nur and Ben-Avraham, 1982) that contractional Cordilleran tectonics may have been largely driven by collisions with exotic components of the accretionary belt. We prefer increased coupling between the leading edge of the continent and east-subducting oceanic lithosphere as the driving force for episodes of intraplate shortening. The foreland fold and thrust belt formed a continuous belt from southeastern California to Alaska, although the accretion of terranes was more restricted both spatially and temporally. In California only the suturing of the Rogue and Diablan arcs to the continental margin can be inferred during Late Jurassic time, and no accretion of large exotic blocks or fragments during Cretaceous time can be identified within the Franciscan assemblage. In northwestern Washington, mid-Cretaceous deformation in the northern Cascades and San Juan Islands did occur in response to the collision of the composite Wrangellian superterrane against a preassembled continental borderland. But although collision, dramatic shortening, dextral strike-slip displacements, and a westward shift in the locus of eastward subduction and associated plutonism all occurred within this part of the Cordillera, there are no distinct changes within the foreland fold and thrust belt that can be correlated with these events. Perhaps with better data on the absolute timing of events within the foreland belt and the subduction system, such relations might emerge. We disagree with the interpretation of Saleeby and others (this volume), that collision of the Stikine terrane, now in the Canadian Cordillera, was responsible for Late Jurassic intraplate deformation in the U.S. Cordillera. We find no evidence for collision along the U.S. Cordilleran margin at this time, and Late Jurassic intraplate deformational styles can be matched in the modern noncollisional part of the Andes (e.g., *Allmendinger and others, 1983*).

Estimates of a 60- to 70-km-thick crust developed at the end of thrusting have been made by Hamilton and Myers (1966) and by Coney and Harms (1984). Along the eastern topographic slope above the thickened crust, the thrust belt probably developed a critical taper and evolved as a deforming thrust wedge (*Dahlen and others, 1984; Dahlen and Barr, 1989*). Recent studies have shown that regions of thick continental crust may develop extensional structures contemporaneous with crustal shortening and thickening (*Molnar and Tapponnier, 1975; Burchfiel and Royden, 1985; Burchfiel and others, 1987, 1988*). Such structures have not been positively identified in the Cordillera, but the Manning Canyon fault in northern Utah (Allmendinger and Jordan, 1984) and structures described in the Raft River and Black Pine Mountains, northwestern Utah and southern Idaho, by Wells and others (*1990*) might be possible examples.

Tectonism in the southern segment of the Cordillera. During Late Jurassic to late Late Cretaceous time, evolution of the southern segment of the Cordillera was different from the northern segment. The region of southeastern California, southern Arizona, and southwestern New Mexico was affected primarily by extension and, possibly, by strike-slip deformation. Thrust fault-related structures of this age are uncommon and those that can be recognized appear to be short lived and do not have displacements as large as those in the northern segment. Burchfiel and Davis (*1971*) report a basement-involved, east-vergent recumbent fold in southeastern California that is cored by a syntectonic pluton dated at 145 Ma (Walker, personal communication, 1988). Metamorphism and deformation, locally west-vergent (*Wernicke and others, 1989*), in the Death Valley area and Halloran Hills to the south are permissibly Late Jurassic but could be older (*Dewitt and others, 1984; Dunne and others, 1978*). Recent work by *Walker and others (1990)* has demonstrated late Middle or early Late Jurassic deformation (169 to 154 Ma) in the Chronese Hills area of the north-central Mojave Desert. All these structures appear to involve only small magnitudes of shortening. Early Cretaceous thrust faults in the Clark and New York Mountains areas of southeastern California (*Burchfiel and Davis, 1971, 1976,* 1988) may have several tens of kilometers of eastward shortening. These thrusts, some of which carry crystalline basement rocks, represent the southernmost known extent of Sevier belt deformation.

Much of the southern segment of the Cordillera appears to be characterized by Late Jurassic to early Late Cretaceous extensional deformation. The northwest-trending, 20-km-wide Independence dike swarm (*Moore and Hopson, 1961*) can be traced through southeastern California into the central Mojave Desert (James, 1989; Karish and others, 1987) and possibly as far as southwestern Arizona (Tosdal and others, 1989). These mafic dikes have yielded ages of 148 Ma (*Chen and Moore, 1982*). Their orientation suggests northeast-southwest extension with evidence for some left-slip shear (Saleeby and others, this volume). Generally, sedimentary and volcanic deposits in the southern segment suggest deposition in rifted basins controlled by northwest-striking faults (*Bilodeau, 1982*), and some faults of this trend in southeastern California can be shown to have strike-slip displacement (Burchfiel and Davis, 1981). In southeastern Arizona the deposition of coarse-grained sediments began in latest Jurassic (Glance Conglomerate) and continued into Cretaceous time (Bisbee Group) in northwest-trending basins interpreted to be of extensional origin (Kb; Fig. 11; *Bilodeau, 1982; Bilodeau and Lindberg, 1983*). A belt of Upper Jurassic arc volcanic rocks lies to the south of these basins in Arizona (Ja; *Kluth and others, 1982; Kluth, 1983*). These basins may be part of an extensional system that propagated from the Gulf of Mexico northwest into Arizona (Dickinson, 1989; Tosdal and others, 1989) and, perhaps, into southeastern California where deposition of the poorly dated lower part of the McCoy Mountains sequence (Fig. 11) may have continued from Middle Jurassic into Late Jurassic time. The upper part of the McCoy sequence is Cretaceous (Stone and others, 1987), but how much of the sequence was deposited in Cretaceous time remains uncertain.

The general tectonic setting of the southern segment of the Cordillera was probably Andean, but with locally significant subsidence in the arc region (Busby-Spera, 1988). What tectonic elements were present at this time along the plate boundary in southern California remain unclear because reconstructions of displacements on younger strike-slip faults, such as the San Andreas (Crowell and Stewart, this volume), and the late Mesozoic fault recognized by Lahren and Schweickert (1989), remain uncertain. However, what data are present would suggest eastward subduction of oceanic rocks beneath North America forming a northwest-trending belt of Upper Jurassic and Lower to lower

Upper Cretaceous arc rocks. The magmatic belt remained narrow and did not spread eastward as it did farther north. All deformation was intra-arc and probably of small magnitude.

LATEST CRETACEOUS TO EARLY TERTIARY EVOLUTION

The accretionary belt

Paleomagnetic data from 90 to 100 Ma plutons in the Pacific Northwest have been interpreted by some workers (e.g., Beck, this volume; *Beck and others, 1981*; Irving and others, 1985) to mean that rocks of the San Juan Islands, northern Cascades, and regions farther east were part of a larger terrane, Baja British Columbia (Umhoefer, 1987), that originated 2,000 to 2,500 km farther south and was still migrating northward along the continental margin during the interval from 75 to 60 Ma. This interpretation is contrary to established pre–mid-Cretaceous links between western parts of the Cordillera and North America (Price and Carmichael, 1986; Armstrong, 1988). Recent data presented by Butler and others (1989) indicate that the apparently discordant paleomagnetic data result from tilting (~30° northeast-side-up) of parts of the Coast Plutonic Complex and the northern Cascade Range. Their data account for the discordant paleomagnetic directions without large-scale displacement of these rocks relative to North America.

During the interval from about 75 to 40 Ma, the tectonics of northern Washington and adjacent British Columbia were dominated by strike-slip faulting (Fig. 15), driven by the strongly oblique convergence of the Kula(?) Plate against North America (Engebretson and others, 1985). Arc-related magmatism was sporadic in space and time. Contractional events were even more localized and short-lived, even though both thin-skinned and thick-skinned contraction (Fig. 15) was prevalent elsewhere in the Cordillera. East-west extension became a dominant style in northeastern Washington and parts of southern British Columbia beginning about 58 Ma (Fig. 16). Although the kinematic history of high-angle faults in the region is still being investigated, we suggest that such familiar Paleogene events as the accumulation of the Coast Range basalts (lToi; Metchosin, Crescent, Siletz), deposition of thick nonmarine sediments, and perhaps even low-angle extensional faulting, were fundamentally related to transcurrent deformation along the continental margin. It is also likely that strike-slip faulting episodically removed large slices from this segment of the continental margin and transported them to Alaska.

We recognize and briefly review below three intervals within the period from 75 to 40 Ma that were characterized by different tectonic styles: 75 to 60 Ma, 60 to 51 Ma, and 51 to 40 Ma (all ages are about ±2 m.y.). For each, we highlight the role of strike-slip faulting and its possible relation to selected geologic features and events. A more comprehensive review of Late Cretaceous and Paleogene geology of the U.S. Pacific Northwest is presented in D. Miller and others (this volume) and Christiansen and Yeats (this volume).

75 to 60 Ma: Late Cretaceous and Paleocene. Tectonic events during the 75- to 60-Ma interval are only dimly perceived. In the crystalline core of the northern Cascades, granitoid plutons have yielded U-Pb dates of 73 to 75 Ma (Ki) and 68 to 59 Ma (lTi; Miller and others, 1989). In general, the younger suite lies northeast of the older and may record the same general Late Cretaceous–Paleocene northeastward migration of magmatisms noted in the Coast Plutonic Complex of British Columbia (Armstrong, 1988). The plutons range from tonalite to granodiorite and probably represent the roots or underpinnings of a Late Cretaceous magmatic arc. Some of the Paleocene plutons are now lineated and foliated orthogneiss; these were intruded during amphibolite facies metamorphism and deformation, or were emplaced in or adjacent to active dextral fault zones (Miller and others, 1989). Synintrusive deformation in the Cascade core is characterized in part by northwest-trending, gently plunging ductile stretching lineations. Localized dextral transcurrent slip (probably on the order of tens of kilometers) along high-angle faults during this period has so far been documented only along the Ross Lake fault zone in Washington (Fig. 15; *Miller, 1989*), and the Yalakom and Castle Peak fault zones in British Columbia (*Umhoefer, 1989*).

In summary, the tectonic history of the greater Cascade region for the interval from 75 to 60 Ma encompasses presumably arc-related magmatism, synplutonic deformation and metamorphism, and dextral displacements recorded both in ductilely stretched metamorphic rocks and along northwest-striking transcurrent faults. These events are compatible with the dextral-oblique convergence of the Kula Plate that has been hypothesized on the basis of hot-spot reconstructions (Engebretson and others, 1985). Because the position of the Kula-Farallon Plate boundary is unknown, however, Engebretson and others (1985) noted that the less obliquely convergent Farallon Plate, instead of the Kula Plate, may have been consumed during this interval.

60 to 51 Ma: Paleocene to early middle Eocene. In contrast to the preceding time interval, magmatism in the crystalline core of the northern Cascades was greatly subdued or absent from 60 to 50 Ma. East of the core in southern British Columbia and northern Washington, however, granitoid plutonic rocks (62 to 52 Ma) were intruded prior to, during, and after an important episode of crustal extension that was accommodated primarily along east-dipping, low-angle normal faults (Fig. 16). Southwest of the Cascade core, basalts (lToi; 62 to 49 Ma) accumulated in coastal Oregon (Roseburg and Siletz River volcanics), Washington (Crescent Formation), and southern Vancouver Island (Metchosin Formation) to form the basement of the Coast Ranges. Localized basins related to strike-slip faults also opened in the early Eocene and began receiving thick sequences of nonmarine sediments (lTr).

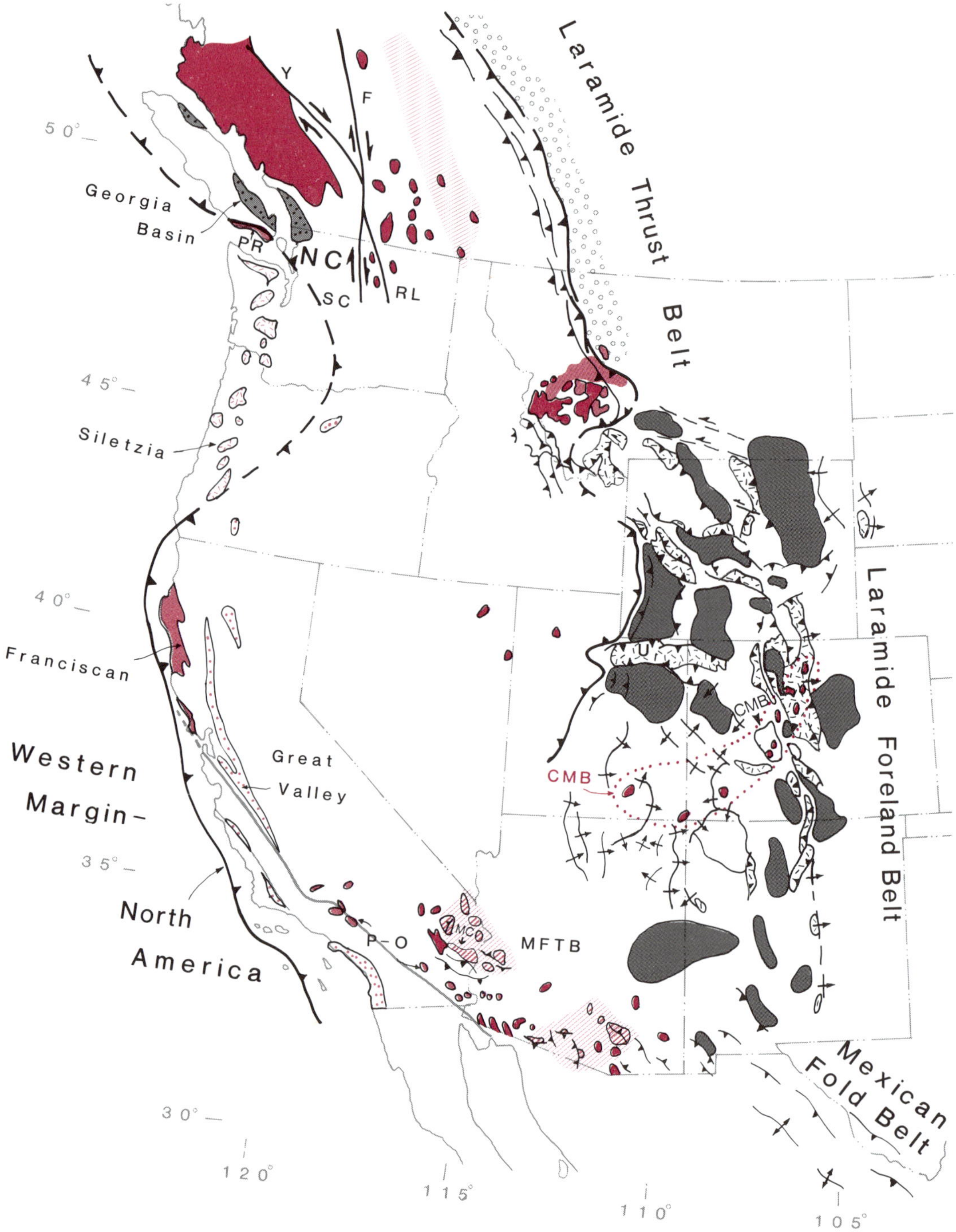

Figure 15. Paleogeographic and tectonic map of latest Cretaceous to middle Eocene time. Laramide thrust blocks (patterned where Precambrian basement exposed, except in the Uinta Mountains where only Precambrian sedimentary rocks are involved), monoclines, and adjacent basins (light gray) are detailed. Laramide foredeep deposits are shown with gray circles. Other patterned units: Georgia successor basin, Canada (gray with dots); plutonic rocks (deep red, except for Colorado Mineral Belt plutons, CMB, outlined with red dots); volcanic rocks of Idaho and Montana (red); fore-arc basins (red dots); Siletz volcanic rocks (red pattern); mélange belt (pink). High-grade metamorphic rocks are shown with oblique red lines; low-grade rocks with oblique pink lines. Future location of San Andreas fault shown in gray. CMB, Colorado Mineral Belt (red); F, Frazer fault zone; MC, McCoy Mountains; MFTB, Maria fold and thrust belt; NC, northern Cascades; P-O, Pelona/Orocopia Schist; PR, Pacific Rim Complex (pink); RL, Ross Lake fault zone; SC, Straight Creek fault; U, Uinta Mountains; Y, Yalakom fault zone.

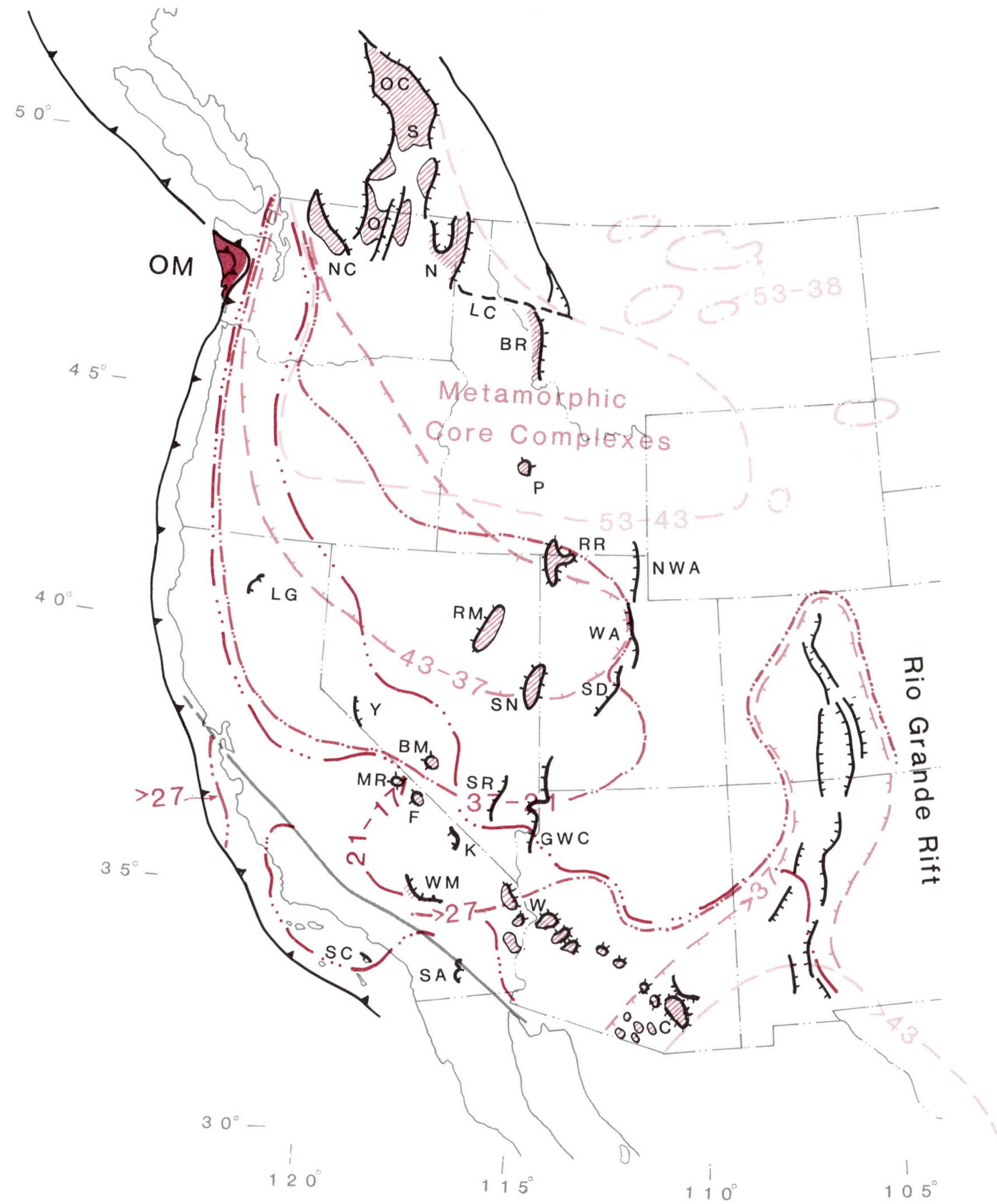

Figure 16. Map showing the distribution of major extensional fault zones and arc-type volcanic rocks extruded from 53 to 17 Ma and subdivided into the indicated time slices. Volcanic rocks compiled and modified from Christiansen and Yates (this volume). Alkaline rocks of 53 to 38 Ma are illustrated in the northern part of the Cordillera to show the eastern limits of igneous activity. Location of the Rio Grande rift and "metamorphic core complexes" (medium red oblique lines) are also shown. Ticks on faults are on hanging wall side. Future location of San Andreas fault zone shown in gray. BM, Bare Mountain; BR, Bitterroot mylonite zone; C, Catalina Mountains; F, Funeral Mountains; GWC, Grand Wash Cliffs fault zone; K, Kingston Range; LC, Lewis and Clark line; LG, La Grange fault (Klamath Mountains); MR, Mineral Ridge; N, Newport fault; NC, northern Cascades; NWA, northern Wasatch fault; O, Okanogan Mountains; OC, Omenica crystalline zone; OM, Olympic Mountains (red; shows material accreted during this time interval); P, Pioneer Mountains; RM, Ruby Mountains; RR, Raft River/Albion Ranges; S, Shuswap; SA, Santa Rosa fault zone; SC, Santa Catalina; SD, Sevier Desert fault zone; SN, Snake Range; SR, Sheep Range; W, Whipple Mountains; WA, Wasatch fault; WM, Western Mojave; Y, Yerrington.

The origin of the Coast Range basalts, including both tholeiitic and alkalic varieties, is controversial. One school of thought (Duncan, 1982) holds that they represent a chain of seamounts generated near a spreading ridge that was near the continental margin. According to an alternative hypothesis (Wells and others, 1984; Snavely, 1987) favored here, the basalts accumulated in one or more rift basins that formed when part of the North American continental margin was laterally transported from this latitude in response to strongly coupled, dextral-oblique subduction of the Kula Plate. In support of the "rifting" hypothesis is some indirect evidence suggesting that several transcurrent faults might have been contemporaneously active onshore. The nonmarine Swauk and Chuckanut basins began rapidly subsiding about 55 Ma (lTr; Johnson, 1985). The Swauk basin lies between the Leavenworth and Straight Creek faults; the Chuckanut basin is bounded by the Straight Creek fault on the east and the inferred "Puget fault" beneath the Puget Lowlands on the west (Johnson, 1985). Thick (up to 6 km), predominantly alluvial sequences in both basins preserve abundant sedimentological and structural features indicating deposition in fault-controlled settings. The Ross Lake fault zone was active during this period (*Miller, 1989*), and there might have been contemporaneous dextral slip on other northwest-striking high-angle faults (West Coast, Yalakom, Foggy Dew, and Harrison). Tight geologic and isotopic controls in timing are still lacking.

51 to 40 Ma: Middle to late Eocene. Some of the tectonic features established in the earlier interval, such as nonmarine fault-related basins, and low-angle extension faults in northern Washington and south-central British Columbia, persisted through the middle Eocene (Fig. 16). A notable flare-up (phase?) of calc-alkaline (oceanward) to alkaline (continentward) magmatism at 51 Ma gave rise to the widespread Challis volcanic field or arc (lTa; Fig. 16). Several northwest-striking dextral transcurrent faults, including the Straight Creek–Fraser River system (Fig. 15), were probably active as well. Basaltic volcanism in the Coast Ranges ceased at about the same time, and much of this volcanic basement subsequently subsided and was inundated by continentally derived marine sands and muds.

The Challis flare-up represents a significant change from the relative magmatic quiescence of the early Eocene. Volcanic rocks are widespread in a southeast-trending swath from south-central British Columbia through northeastern Washington into Idaho and northwestern Wyoming (Fig. 16; e.g., Ewing, 1980). These volcanic rocks are locally interbedded in northwestern Washington with the nonmarine sedimentary rocks of the fault-related Chumstick, Swauk, and Puget-Naches basins (lTr). In southern British Columbia and adjacent northeastern Washington, volcanic and sedimentary rocks accumulated while the crust was being extended along predominantly west-dipping low-angle normal faults (in contrast to the somewhat older east-dipping faults; Parrish and others, 1988). Abundant isotopic dates from this region show that metamorphic rocks presently exposed in structural culminations as extensional core complexes were tectonically elevated and denuded about 45 to 50 Ma. These complexes resemble the younger metamorphic core complexes of the central and southern U.S. Cordillera, but may differ from them in geometric style and tectonic setting (see below).

The Challis magmatic realm is commonly interpreted as a subduction-related, but entirely subaerial, arc. Part of the realm experienced synmagmatic extension and transcurrent faulting. The Intermontane middle and late Eocene tectonic regime in the Pacific Northwest was "transtensional," a term that emphasizes the relationship between dextral slip on north- and northwest-striking transcurrent faults (particularly the Tintina–northern Rocky Mountain trench, and Fraser River–Straight Creek system) and west-northwest–directed extension on normal faults (e.g., Ewing, 1980; Price and Carmichael, 1986). Another expression of dextral strain occurs in the ductile crystalline core of the northern Cascades, where 45-Ma granitoid rocks contain subhorizontal to northwest-plunging stretching lineations adjacent to the Ross Lake fault zone (*Miller, 1989*).

In summary, the middle and late Eocene magmatic arc, and contemporaneous dextral transcurrent faulting onshore are compatible with dextral-oblique convergence of the Kula Plate offshore—a regime that apparently originated earlier, during the Late Cretaceous (Engebretsen and others, 1985). The reasons why magmatism episodically subsided and flared up, or why both local and regional extension were significant from about 56 to 45 Ma, are not known. These changes do not correspond with obvious changes in the velocities of North America–Kula convergence, although there is a clear correlation of the global ca. 43-Ma plate reorganization with the 42- to 37-Ma extinction of the Challis arc, cessation of dextral strike-slip faulting (especially on the Straight Creek–Fraser River system), and inception of the north-south–trending Cascade arc. Cumulative dextral displacement in northern British Columbia across faults of late Mesozoic to early Cenozoic age have been estimated by Gabrielse (1985) to be in excess of 1,000 km, although such large cumulative displacement cannot yet be reconciled by geologic relationships across more southerly faults in the British Columbia–Washington border area. This observation has been used (e.g., Price and Carmichael, 1986) to explain Eocene crustal extension in the Omineca crystalline belt on both sides of the border as an accommodation for greater and lesser amounts of nonaligned dextral shear in the northern and southern regions, respectively. We are not convinced, however, that profound extension in the metamorphic core complexes of the Omineca belt is kinematically linked to dextral faults. Accumulating evidence (Parrish and others, 1988) suggests the bulk of the extension was east-west, not northwest-southeast as predicted by the "linkage" models (e.g., Price and Carmichael, 1986).

Although basaltic volcanism in the Coast Ranges ceased about 50 Ma, there is a difference of opinion about when the Siletzia terrane was accreted or structurally juxtaposed with pre-Tertiary rocks along the continental margin in Washington and southwestern British Columbia (Fig. 15). In southern Oregon, the timing is well established as about 50 Ma, when sediments of the Tyee Formation (lTfa) overlapped the contact between volcanic rocks of Siletzia and older basement rocks of the Klamath Mountains (*Heller and Ryberg, 1983*). Many workers (e.g., Snavely, 1987) claim a similar age for accretion to the north. The only visible fault, however, the Leech River fault on southern Vancouver Island, is late Eocene to early Oligocene (ca. 36 Ma; Rusmore and Cowan, 1985). So even though the Metchosin and Crescent lavas (lToi) were erupted close enough to land to be interbedded with continentally derived detritus, we believe that the evidence indicates that the basaltic basement was first thrust beneath (subducted?) the continental margin in late Eocene time (Clowes and others, 1987). A widespread disconformity in the upper Eocene sedimentary section in the Coast Ranges (Christiansen and Yeats, this volume) may record this evnt.

Back-arc region of the Cordillera

Beginning in latest Cretaceous time major changes in magmatic and deformational patterns occurred within the U.S. Cordillera (e.g., Armstrong, 1974; Burchfiel and Davis, 1975) that were related to a marked increase in the rate of convergence (Engebretson and others, 1985) along the western margin of the North American Plate and to an inferred change in the geometry of the subduction system. Deformational events of latest Cretaceous to early Cenozoic time, the Laramide orogeny, separated the Cordillera into three segments: a central segment between the latitude of southern California and central Montana, and northern and southern segments that extend into Canada and Mexico, respectively (Fig. 15).

Central segment. The greatest change took place within the central segment where magmatic activity largely ceased by 85 to 75 Ma within the Sierra Nevada arc and in Nevada and western Utah, but locally spread eastward in southern Montana and far to the east in Colorado—regions that had not been affected by igneous activity since Proterozoic time (Lipman, this volume). Concurrently, major contractional deformation was initiated in former cratonal areas east of the foreland fold and thrust belt (Figs. 1, 15). Deformation within the thrust belt declined and appears to have ceased along much of this segment before early Cenozoic time; in the southern part of the segment, thrust faulting had ceased entirely before latest Cretaceous time. To the east in the former craton, large irregular blocks cored by Precambrian crystalline basement were thrust in various directions along a generally convex northeast arcuate trend within the foreland of the Cordillera (Fig. 15; *Tweto, 1975*). The thrusted blocks disrupted the widespread foredeep deposits (Kfd) associated with the foreland fold and thrust belt, and were flanked by local deep basins (lTc; see Dickinson and others, 1988; Dickinson, this volume). Basins between uplifted block express crustal subsidence that in some cases exceeded uplift of the blocks by up to two or three times. Less intense deformation formed large monoclines within the Colorado Plateau to the south (Fig. 15; *Davis, 1978*). This deformation generally followed preexisting weaknesses in southern regions (formed during late Paleozoic Ancestral Rocky and Proterozoic deformations), and northwestern regions (Precambrian deformations), although older structural controls appear to be lacking throughout much of Wyoming. Generally, thrust faults record east-west shortening in the south, and northeast-southwest shortening in the north, and vergence of the structures is variable. North- and south-vergent thrust structures are present along the flanks of the Uinta Mountains, which are cored by Middle(?) to Late Proterozoic sedimentary rocks; the Laramide structures are probably controlled by Proterozoic faults.

Laramide foreland structures interfered with the fold and thrust belt in southern Montana and northern Utah, forming three distinct salients within the eastern part of the fold and thrust belt (cf., Schmidt and Perry, 1988). In southern Montana, structures of the two tectonic regimes are contemporaneous (Schmidt and Garihan, 1983), but in northern Utah, deformation of the foreland Uinta Mountains appears to be largely younger than thrust faulting to the west. In Montana, Laramide foreland structures strike obliquely to the thrust belt and are associated with northwest-striking left-slip faults (*Lorenz, 1984*). Far to the south, in New Mexico and west Texas, Laramide structures merge with structures in the southern segment (Fig. 15). The timing of foreland deformation is diachronous. It may have begun as early as Coniacian time in southern Montana (Perry and others, 1983; Ruppel and Lopez, 1984; D. Miller and others, this volume), although some workers regard the beginning of deformation as Maastrichtian in age and synchronous throughout most of the foreland (e.g., see Dickinson and others, 1988). Deformation ceased in the northern part of the foreland by early Eocene time, but lasted until latest Eocene time in the southern part (Dickinson and others, 1988).

Northern segment. Within the northern segment of the Cordillera during Laramide time, magmatism and east-vergent thrusting within the fold and thrust belt continued from Early Cretaceous into early Cenozoic time, but it is unclear whether deformation was continuous or episodic. Thin-skinned thrusting migrated eastward involving Proterozoic strata (P_1; Belt Supergroup), very eastern parts of the miogeocline (€Jcs), western parts of the cratonal Paleozoic sequence (€Jc), and early parts of the foredeep sequence (Kfd). Deformation ended in the northern segment by middle Eocene time. Extensional tectonism began west of the thrust belt in Eocene time, but it is uncertain whether the beginning of extension was contemporaneous with the final phases of thrusting or occurred very soon after its cessation.

Southern segment. Within the southern segment, magmatism continued from Early Cretaceous into early Cenozoic time, but during latest Cretaceous to early Cenozoic time it became more voluminous and migrated to the northeast (Coney and Reynolds, 1977; Keith and Wilt, 1986). Significant shortening is marked by a belt of generally west-northwest–striking thrust faults and folds. Deformation is represented in the California-Arizona border region by the Maria fold-and-thrust belt (Fig. 15; *Reynolds and Spencer, 1990*; Reynolds and others, 1988) with a predominant southward vergence (southwest, south, southeast), and by the north-vergent Mule Mountain thrust farther south; in some areas the vergence is unclear (such as in southern Arizona; cf., Haxel and others, 1984). In the western part of this belt, deformation is largely ductile and associated with metamorphism, locally of amphibolite grade (Hamilton, 1987; Hoish and others, 1988; Tosdal, 1986; Haxel and others, 1984; Reynolds and others, 1988). Thrust faulting and folding involved Precambrian crystalline rocks (P€m), Mesozoic igneous rocks, the cratonal cover of Paleozoic and Mesozoic sedimentary rocks (€jc), and locally thick sequences of Mesozoic sedimentary rocks, such as those deposited in the McCoy Mountains (JKb) and Bisbee Group (Kb) basins. Deformation in the western part of the belt occurred within the late Mesozoic magmatic arc. Ductile thrust faults in the Maria fold and thrust belt are cross-cut by post-tectonic plutons with ages ca. 70 to 75 Ma.

In the eastern part of the belt the structures also involved Precambrian and younger cratonal rocks, but they were associated with more brittle deformation (Reynolds and others, 1988) and appear to merge with the Laramide foreland structures of the central segment in southern New Mexico. Where these structures pass south into Mexico and west Texas they involved thick marine Mesozoic deposits (lKs) of the Mexican basin and continental interior seaway and developed a thin-skinned fold and thrust belt style (Figs. 11, 15). The shortening that replaced earlier extension in the southern segment probably began in early Late Cretaceous time, although its inception is not well established, and ends by 53 Ma (D. Miller and others, this volume). The age and trends of these Laramide structures suggest that there is no direct connection between the deformation in the southern Laramide segment and the older Sevier foreland fold and thrust belt of southeastern Nevada and adjacent southeastern California.

Tectonic setting of Laramide deformation

The changes in the tectonic patterns during Late Cretaceous to early Cenozoic time (compare Figs. 11 and 15) are interpreted as resulting from changes in the geometry of the subduction system. Evidence from Franciscan and related rocks in the westernmost Cordillera indicates that eastward subduction continued through Laramide time (Dickinson, this volume; Cowen and Bruhn, this volume; D. Miller and others, this volume), but convergence rates increased markedly. The termination of arc magmatism in central Cordilleran regions (Armstrong, 1974) is inferred to be the result of flattening of the subducting slab beneath this segment of the Cordillera, whereas slab-dip remained moderate beneath both northern and southern segments. In the southern segment, subduction may have shallowed somewhat during late Laramide time causing magmatism to migrate eastward (Lipman and others, 1971; Armstrong, 1974; Burchfiel and Davis, 1975; Coney and Reynolds, 1977; Keith and Wilt, 1986). In the central segment flat-slab subduction caused an eastward shift of sparse magmatism into the foreland. Deformation there largely reactivated older structures within the craton. The different directions of shortening within the Laramide province are regarded by some workers as resulting from three periods of deformation (Gries, 1983), or clockwise rotation of a block that was to become the Colorado Plateau (Hamilton, 1981, 1988b), or changes in the direction of shortening accompanied by dextral shear along the eastern part of the Laramide structures (*Chapin and Cather, 1981*). We interpret the Laramide structures to be detached near the Moho from the underlying mantle lithosphere, and the deformation and rotation to be due to a partitioning of northeast-southwest shortening on different structures during deformation within an inhomogeneous crust. Shortening within the foreland was part of a regional convergent system and was related to shear stress transmitted across the shallow-dipping boundary between the two lithospheric plates (e.g., see Bird, 1988). Coupling between the two plates was incomplete; the irregularity of the deformation in the crust was not a direct reflection of displacements within the underlying mantle lithosphere or subducted slab.

Geological relations in southern California provide evidence for a period of flat subduction during Laramide time. In many areas, high-pressure/low-temperature metasedimentary and metaigneous rocks (Pelona, Orocopia, and Rand Schists; KTm; Fig. 15) tectonically underlie Precambrian, Paleozoic, and Mesozoic crystalline rocks of the North American Plate (for a recent review see Jacobson and others, 1988). Although the emplacement of the schists remains controversial, we agree with the interpretation of Yeats (1968) that the schists were emplaced from a Pacific source, most likely during the latest Cretaceous to earliest Cenozoic phase of flat-slab subduction. At this time, lithosphere of the lower crust and mantle beneath southern California was displaced eastward and replaced by Pelona and correlative schists that represent Middle Jurassic and younger sedimentary and volcanic oceanic rocks formed west of the continent (Burchfiel and Davis, 1981). Recent work on the timing of this event suggests emplacement ages between 79 and 60 Ma for the schists (Jacobson and others, 1988). We disagree with Tosdal and others (1989) and Saleeby and others (this volume) that the protolith of the schist was deposited in a small ocean-floored pull-apart basin within continental crust related to the Middle Jurassic (ca. 164 Ma) Mojave-Sonora megashear system. We can find no evidence for marginal facies or extensional structures that would be related to formation of such a basin, nor is there evidence for a suture that would be related to its closure. The present exposure of the HP/LT schists can be shown in many places to be the result of unroofing by Cenozoic low-angle extensional faulting (see below).

We believe that there is little fundamental difference in the basic cause for shortening within the three segments of the Laramide Cordillera. The main difference in structural styles within the segments lies in the level of detachment within the crust, a level controlled by such factors as the angle of subduction beneath the Cordillera, by the continental thermal regime, and by preexisting geological relationships. In the northern segment, the level of crustal detachment in the eastern part of the thrust belt lies within the Belt and younger sedimentary sequences, but ramps downward to the west into a ductilely shortened middle crust heated by magmatic intrusions; the regions of midcrustal ductile shortening and shallow brittle shortening are separated by east-climbing thrust faults that follow bedding anisotropy. In the middle segment, the level of detachment is within an inherently weak lower crust (*Brace and Kohlstead, 1980*). The detachment must underlie the Laramide foreland structures, most of the eastern Cordillera, and extend westward into a deeper region of shortening within the crust or mantle lithosphere. In principal we agree with Oldow and others (1989) that the tectonic elements of the Cordilleran orogen are decoupled from deeper crustal and mantle lithosphere structure. However, in contrast to their orogenwide detachment or basal decoupling system beneath deformed Cordilleran areas (their "orogenic float"), we believe that such detachments must become considerably more diffuse at depth beneath Cordilleran magmatic belts and may not be continuous across the ductile lithosphere present there. In the southern segment, shortening occurs essentially within the arc. Only a narrow, presumably steeper detachment is present between the ductile crust of the arc and the brittle, shallow crust to the northeast where thrust faulting occurs within thin cratonal sequences (Fig. 15). The presence of cratonal rocks within the southern segment is a paleogeographic inheritance from the late Paleozoic truncation of the southwestern North American margin. The geometry and processes of contraction within the Cordillera during Laramide time have modern analogs in the Andes (*Burchfiel and Davis, 1976*), including Laramide-style foreland deformation in northwestern Argentina (e.g., Jordan and Allmendinger, 1986). Sevier thin-skinned style counterparts are also present in the Andean chain where, for example, back-arc shortening of more than 200 km occurs in Bolivia along a shallow detachment controlled by sedimentary stratal anisotropy (*Sheffels, 1988*).

Tectonic evolution of the northern U.S. Cordillera and formation of the Columbia embayment

The position of the Sri = 0.706 line (Fig. 3) is generally taken to mark the western edge of Precambrian cratonal rocks of North America, but it is a composite line with a location throughout the Cordillera controlled by different events of different ages. From the Mexican border to the central Sierra Nevada its location has been controlled by a series of strike-slip fault-related events: first, by left-slip faults in late Paleozoic time; second, by right-slip faults in mid- to late Mesozoic time; and third, by right-slip faults in Cenozoic time. In central Nevada its location may be close to the original site of latest Precambrian rifting, albeit modified by the effects of Paleozoic and Mesozoic shortening and Cenozoic extension. In the northern part of the Cordillera its location is very sharply defined in west-central Idaho by the Mesozoic West Idaho suture zone (Armstrong and others, 1977; Fleck and Criss, 1985; Lund and Snee, 1988). Farther north the Sri = 0.706 in east-central Washington trends west, then north.

The trace of the Sri = 0.706 line in western Idaho and southeastern Washington approximately defines the eastern and northeastern margins of the Columbia embayment (Fig. 12), a concave westward domain in which few exposures of pre-Tertiary rocks are found. The present configuration of the Columbia embayment, filled largely by Miocene Columbia River Basalt (Fig. 17) and bounded by pre-Tertiary terranes on the north (Okanogan, northern Cascades) and south (Blue Mountains), may have developed during multiple tectonic events: one in mid-Cretaceous time; a second in early Cenozoic time; and a third at the time of Basin-Range extension to the south. As suggested by Davis and others (1978) and Brandon and others (1988), among others, the locus of west-vergent mid-Cretaceous thrusting at the latitude of 49°N lay along the eastern boundary of the colliding Wrangellian superterrane. At the latitude of central Idaho, the generally north-striking West Idaho suture zone (Fig. 11) also records mid-Cretaceous contraction (e.g., Lund and Snee, 1988; Hyndman and others, 1988) localized along the boundary between pre–Late Jurassic accreted terranes and continental North America. Mylonitic granitoid rocks contain kinematic criteria for west-vergent thrusting (*Strayer and others, 1987; Strayer, 1988*). Available isotopic data suggest that contractional deformation occurred between ca. 118 and 88 Ma (Lund and Snee, 1988; C. Manduca, personal communication, 1988) and may in part have been contemporaneous with San Juan–northwest Cascades thrusting in areas to the northwest.

We speculate that the Cretaceous contraction in western Idaho is an intraplate expression of east-directed subduction located farther west, but that no major accretion (collision) of a large tectonic element, such as Wrangellia, took place along the subduction zone at the latitude of central Oregon. The similarity of Paleozoic and Mesozoic paleogeographic and tectonic elements between regions north and south of the West Idaho suture zone (e.g., the Star Peak/Luning–Slocan basin; see previous discussions and Wernicke and Klepacki, 1988) suggest that at least some of these elements were once continuous throughout the region (see above). The fact that they are now missing has at least two candidate explanations or, perhaps, some combination of the two. One, the missing elements may have been removed by transcurrent, presumably dextral, faulting during Late Jurassic to mid-Cretaceous time and prior to the intraplate shortening now recorded along the suture (cf., Davis and others, 1978; Hyndman and others, 1988). A second possibility is that the missing terrains may have been dramatically foreshortened in a purely contractional deformation, that they once occupied high-level divergent nappe systems along the suture, and that they have since been removed by erosion (cf., Lund and Snee, 1988, their Fig. 11-8). In support of this latter interpretation, a late syntectonic and/or post-tectonic Cretaceous tonalite now exposed in the suture region (age ⩽93 Ma) may have been emplaced at depths of 25 km or deeper (Lund and Snee, 1978).

It is interesting to note that the Idaho batholith, related satellite plutons, and their metamorphic aureoles lie almost 300 km to the east of correlative rocks in northern Idaho and northeastern Washington (Fig. 11). Farther east, the foreland fold and thrust belt shows a comparable, but much less dramatic westward deflection going from south to north (Fig. 11). The reasons for the apparent offset of the batholithic axis are not completely clear, but seem in part related to faulting along a diffuse and controversial west-northwest–trending zone of deformation known as the Lewis and Clark line. Folds north of the Idaho batholith appear to be deflected left laterally into the zone, and a northeast-vergent, northwest-trending belt of thrust faults involves rocks of its contact aureole (Skipp, 1987; *Sears and others, 1989*). Some of these thrust faults curve to trend southward at their eastern ends and are intruded by plutons that date about 70 to 80 Ma. (These are not the easternmost thrusts in the foreland belt, which are younger.) This pattern suggests that during Late Cretaceous time the region around the batholith was the site of northeast-southwest shortening accompanied by a broad zone of left-lateral shear north of the batholith—perhaps with left slip along some faults within and parallel to the Lewis and Clark line and/or some clockwise rotation of the batholith and its adjacent aureole.

The left-lateral shear does not appear to be large enough to explain all of the apparent 300 km offset of the Cretaceous batholithic axis for two reasons. First, the foreland fold and thrust belt is not sinistrally deflected a comparable distance across eastward projections of the shear zone. Second, the magnitude of thrusting east of the batholith (and south of the zone of left-lateral shear) does not appear to be larger than that farther north, a relationship that would be expected if all of the apparent offset of the batholithic belt had occurred by tear or transform faulting confined to the thrust belt. These relationships suggest that a large part of the offset of the batholithic belt dates from its inception; that is, the "offset" may reflect either a preexisting eastward stepping to the south of the continental margin subduction zone (along the northern margin of the Columbia embayment) or, less likely, lateral changes in the dip of the subducting slab (steeper to the north). Speculatively, the east-west offset in the continental margin might have been related to operation of a mid-Cretaceous sinistral transform fault that connected the subduction zone in the SJ-NWC system to the intraplate contractional West Idaho suture zone. The position of this hypothetical fault, buried beneath Miocene Columbia Plateau basalt, may coincide with the segment of the Sri = 0.704 and 0.706 isopleths that extends westward from near Orofino, Idaho (Fig. 3).

Thus, it is possible that following mid-Cretaceous events part of the northern boundary of the embayment was formed. The east-west northern boundary of the embayment was probably further developed, or modified, during Eocene time when differential extension took place within the northwest part of the Cordillera (Fig. 16). Large-magnitude east-west extension took place from the Purcell Trench in northern Idaho westward across northern Washington (Parrish and others, 1988). The southern limit of this extended terrane is hidden by the Columbia River basalts, but the normal faults responsible for the deformation are not present in areas west of the West Idaho suture. East and northeast of the suture, Eocene extension along the Bitterroot mylonite zone (*Garmezy and Sutter, 1983*) appears to terminate northward along strands of the Lewis and Clark fault zone, and some of this extension could have been transferred along the zone to extended areas farther north and west, such as in the Priest River area (Fig. 16; *Rehrig and Reynolds, 1981*). Such a transfer of motion would cause right slip on the Lewis and Clark fault zone. There is likely a causal relation between the location of Eocene extension and magmatism and regions and earlier centers of Late Cretaceous plutonism where the crust had been thickened, heated, and softened. Because these regions were already offset from each other in Cretaceous time, the northern and southern domains of Eocene extension probably also developed with a primary offset. Extension along still younger north-striking faults, such as those in Flathead Valley and numerous valleys in southwestern Montana and eastern Idaho, also appears to end against strands of the Lewis and Clark fault zone (Fig. 18). These faults, with mid- to late Cenozoic displacement, are more numerous south of the fault zone; their presence suggests that throughout the Tertiary transfer movement on the Lewis and Clark fault zone has been right slip (*Reynolds, 1979*).

CENOZOIC EVOLUTION

The Cenozoic is a time of severe tectonic compartmentalization of the U.S. Cordillera. It is a time when structural domains with variable internal states of stress and strain response exist side by side in ever-changing patterns related to evolving plate interactions and consequent magmatic activity. Throughgoing tectonic elements, such as the Mesozoic foreland fold and thrust belt in the east and the accretionary belt of the western margin, gave way in the Paleogene to geographically more restricted domains of compressional, extensional, and strike-slip deformation. The most profound compartmentalization is evident after initiation of the San Andreas transform fault system in early Miocene time (Stewart and Crowell, this volume). Because of the great geographic and temporal diversity of Cenozoic tectonic styles, and because of the exhaustive treatment of this era in Yates and others (this volume), we depart from our more systematic discussion of earlier times and events and treat only selected aspects of Cenozoic tectonics—specifically, extensional tectonics prior to the inception of Basin-and-Range structure, and accretionary tectonics along the continental margin.

Extensional tectonics

Metamorphic core complexes. Our discussion of Cenozoic extensional tectonics will not attempt to duplicate the broader, more detailed coverage of Wernicke (this volume), but will instead concentrate on several fundamental matters of continuing controversy. Progress since 1980 in understanding Cordilleran extensional tectonics has primarily focused in shallow-dipping normal faults (detachment faults) of regional extent that characteristically separate footwall rock assemblages of significantly deeper crustal origin than the rocks in their hanging walls (cf., Davis and Lister, 1988). Many examples of such faults are exposed on the flanks of individual mountain ranges that possess domal or antiformal geometries (Davis and Coney, 1979). Lower

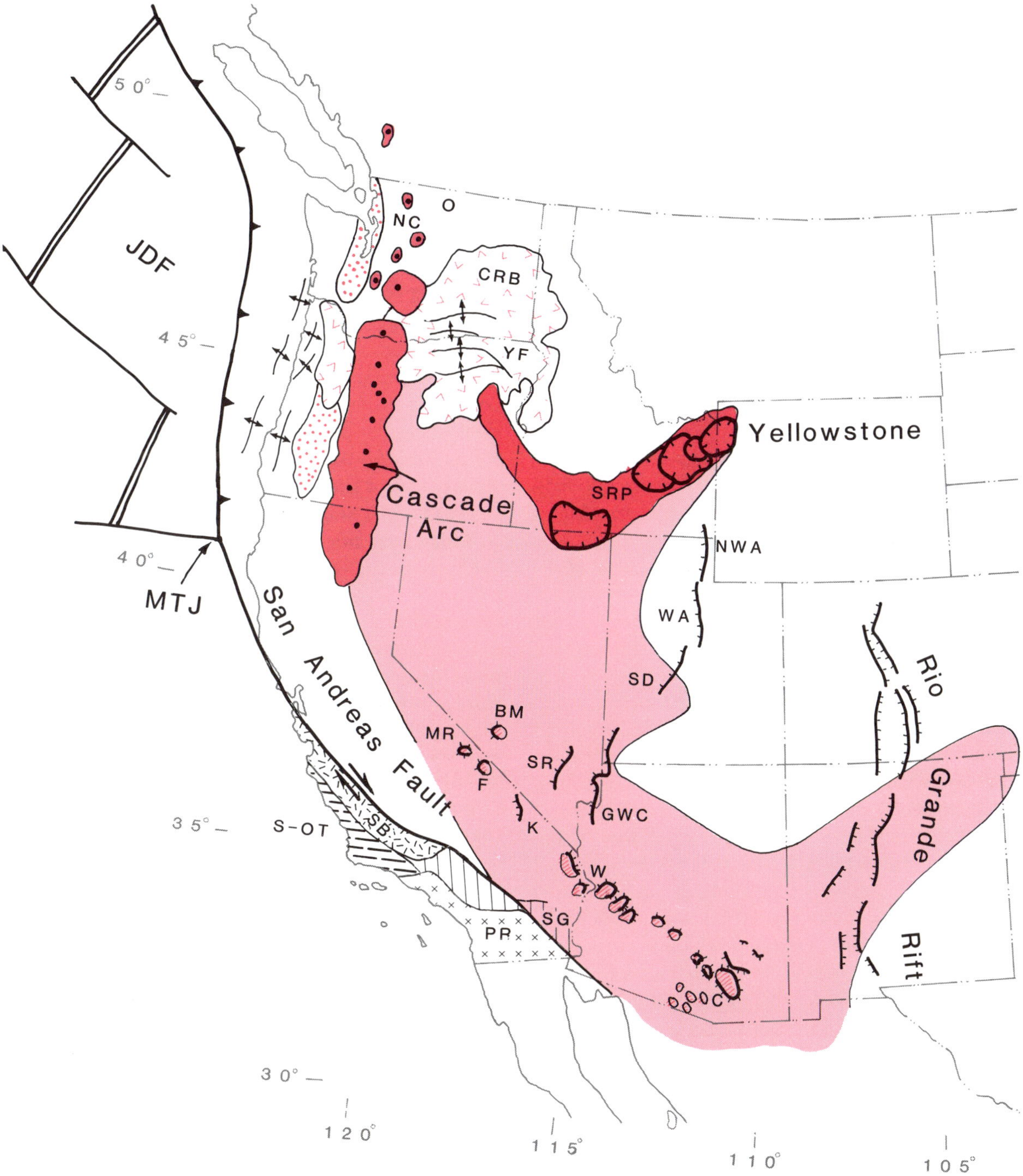

Figure 17. Major tectonic elements of late Cenozoic age and volcanic rocks extruded from 17 Ma to the present. Major normal faults shown by heavy lines with ticks on their hanging walls. Patterned units include: Cascade volcanic arc (red); Snake River and Newberry volcanic tracks (deep red); fore-arc rocks (red dots). Also shown are the general area of bimodal volcanism in the Basin and Range Province (pink, see Fig. 18), Columbia River basalt (CRB, medium red v's), and calderas (circular ticked lines) along the Snake River Plain (SRP). Metamorphic core complexes in the southern Cordillera are patterned with medium red oblique lines. BM, Bare Mountain; C, Catalina Mountains; CRB, Columbia River basalt; F, Funeral Mountains; JDF, Juan de Fuca Plate; K, Kingston Range; MR, Mineral Ridge; MTJ, Mendocino triple junction; NC, northern Cascade volcanoes (red); NWA, northern Wasatch fault zone; O, Okanogan Mountains; PR, Peninsular Ranges terrane (x-pattern); S-OT, Sur-Obispo composite terranes (line pattern); SB, main Salinian block (patterned); SD, Sevier Desert fault zone; SG, San Gabriel terrane (vertical lines); SR, Sheep Range; SRP, Snake River Plain; W, Whipple Mountains; WA, Wasatch fault zone; YF, Yakama folds.

plate rocks in the cores of these ranges are typically more highly metamorphosed and ductilely deformed than rocks in the upper plate—hence the term "metamorphic core complexes." The term is confusing to some geologists in that "metamorphic core" does not follow traditional practice in referring to a metamorphic axis or culmination along the length of the orogen, but is applied on the local scale of individual mountain ranges. In most complexes, lower plate crystalline rocks have experienced pre-Cenozoic (pre-extensional) metamorphic and igneous events. In the majority of these, ductile deformation and metamorphic recrystallization also accompanied Cenozoic crustal extension (e.g., Shuswap, Snake, Whipple, Santa Catalina; Fig. 16). However, in a smaller number of complexes, all lower plate deformation during crustal extension is observed to be brittle at present levels of exposure (e.g., Santa Rosa Mountains–Borrego Springs area of Peninsular Ranges, Funeral Range [Death Valley], Kingston Range, Klamath Mountains—all in California). Nevertheless, ductile deformation synchronous with extension can be postulated to have occurred at deeper, unexposed levels of the lower plates of these complexes (cf., Lister and Davis, 1989).

Detachment faulting versus basin-and-range faulting. Basin-and-range normal faults (Fig. 18), which had preoccupied geologic thinking about crustal extension in the western United States prior to 1980, are now widely recognized as younger, steeper faults commonly superimposed across earlier detachment fault complexes. We do not believe, as do some workers (e.g., Eaton, 1980, 1982; Zoback and others, 1981; Anderson and others, 1983; Gans and others, 1985, 1989, their Fig. 20), that the two structural styles are simply shallow (basin-and-range) and deep (detachment faults) crustal expressions of a single extensional faulting/tectonics phenomenon. According to these publications, detachment faults of initial low dip may have formed only at depth while separating deeper domains of ductile flow from overlying block-faulted domains of basin-and-range style. The 30- to 40-km-wide spacing of range- and basin-front faults in the Great Basin has no geometric counterpart in the metamorphic core/detachment fault complexes where distended upper plates may be broken by steep normal faults less than 2 or 3 km apart. The steep dips (>50°) of some basin-and-range faults can be traced by seismic reflection surveys to depths of 8 to 10? km (Anderson and others, 1983; Okaya and Thompson, 1985; Effimoff and Pinezich, 1986), whereas some active detachment faults have almost certainly breached the earth's surface at low angle (Miller and John, 1988; Davis and Lister, 1988) and others preserve evidence for initially shallow dips (<20 to 25°) within a few kilometers of the earth's surface (Davis, 1980; Spencer, 1985; Burchfiel and others, 1987; Hodges and others, 1989; Burchfiel and Davis, 1988; Davis and Lister, 1988).

Another general distinction between basin-and-range and core/detachment complex faulting concerns their tectonic environments. Metamorphic core/detachment fault complexes in the eastern Cordillera vary in age along the length of the Cordillera: Eocene (southern British Columbia and northern Washington and Idaho); Oligocene (eastern Great Basin); Oligo-Miocene and Miocene (southeastern California, southern Nevada, and southwestern Arizona); and late Miocene(?), Pliocene to Recent (Death Valley–Panamint Range area). All but the latter are associated closely in time and space with magmatism of arc affinity that accompanied eastward subduction of oceanic lithosphere beneath the continent (Fig. 16; cf., Gans and others, 1989). They are intra-arc phenomena. In contrast, faulting in the Great Basin Province was and continues to be generally associated with bimodal volcanism (basalts + rhyolites) and a variable tectonic environment—from a back-arc setting prior to northward passage of the Mendocino triple junction to a transform tectonics zone afterwards, at least in the western part of the province Figs. 17, 18).

Detachment faults: Geometric considerations. The question of the shallow versus steep initial dip of Cordilleran detachment faults is a highly controversial one. Field geologists concerned with the mapping of such faults over areas of thousands of square kilometers generally opted for an initial, shallow-dipping (<30°) configuration, regardless of differences of opinion regarding their origin (e.g., Davis and others, 1980; Wernicke, 1981; Miller and others, 1983). But more recently, the inability of rock methanicists to develop stress field models appropriate for the initiation and operation of such low-angle normal faults (for an important exception, cf., *An, 1989*) has encouraged alternative views involving initially steeper normal faults (e.g., Buck, 1988). Some geologists have begun to question or revise their initial field-based interpretations favoring shallow original dips for some detachment faults (e.g., Wernicke and Axen, 1988). Still others believe that all detachment faults initially had steeper dips prior to rotational flattening (e.g., Davis, 1983) or flexural rotation (Buck, 1988) during crustal stretching. Concerns about the existence of active, shallow-dipping normal faults have also come from seismologists (e.g., *Jackson, 1987*; Jackson and White, 1989), who find little or no evidence in worldwide areas of ongoing crustal extension for earthquakes along normal faults dipping less than 30°.

We believe that these negative seismological findings are not too surprising, because none of the seismically active areas that have been compared with the Tertiary Cordillera (e.g., Tibet, East Africa, Aegean Sea, Baikal, the present Great Basin) are characterized by convergent-parallel crustal extension within the magmatic arcs of active subduction systems. But, we remain convinced that some Cordilleran detachment faults propagated to or very near the earth's surface at shallow angles of dip. Arguments from southwestern Cordilleran examples include: (1) observable footwall and hanging-wall geometric constraints on late Miocene through Recent detachment faults (northwestern Panamint Range; Burchfiel and others, 1987; Hodges and others, 1989); (2) close parallelism in dip of some detachment faults with nearby shallow-dipping Mesozoic thrust faults (Mormon Mountains, Kingston Range, Tucson area); (3) the upward, shallow-angle climb (≤20°) of some detachment faults out of crystalline basement into overlying subhorizontal strata (Whipple Mountains, South Mountain?); (4) the truncation from below of upper

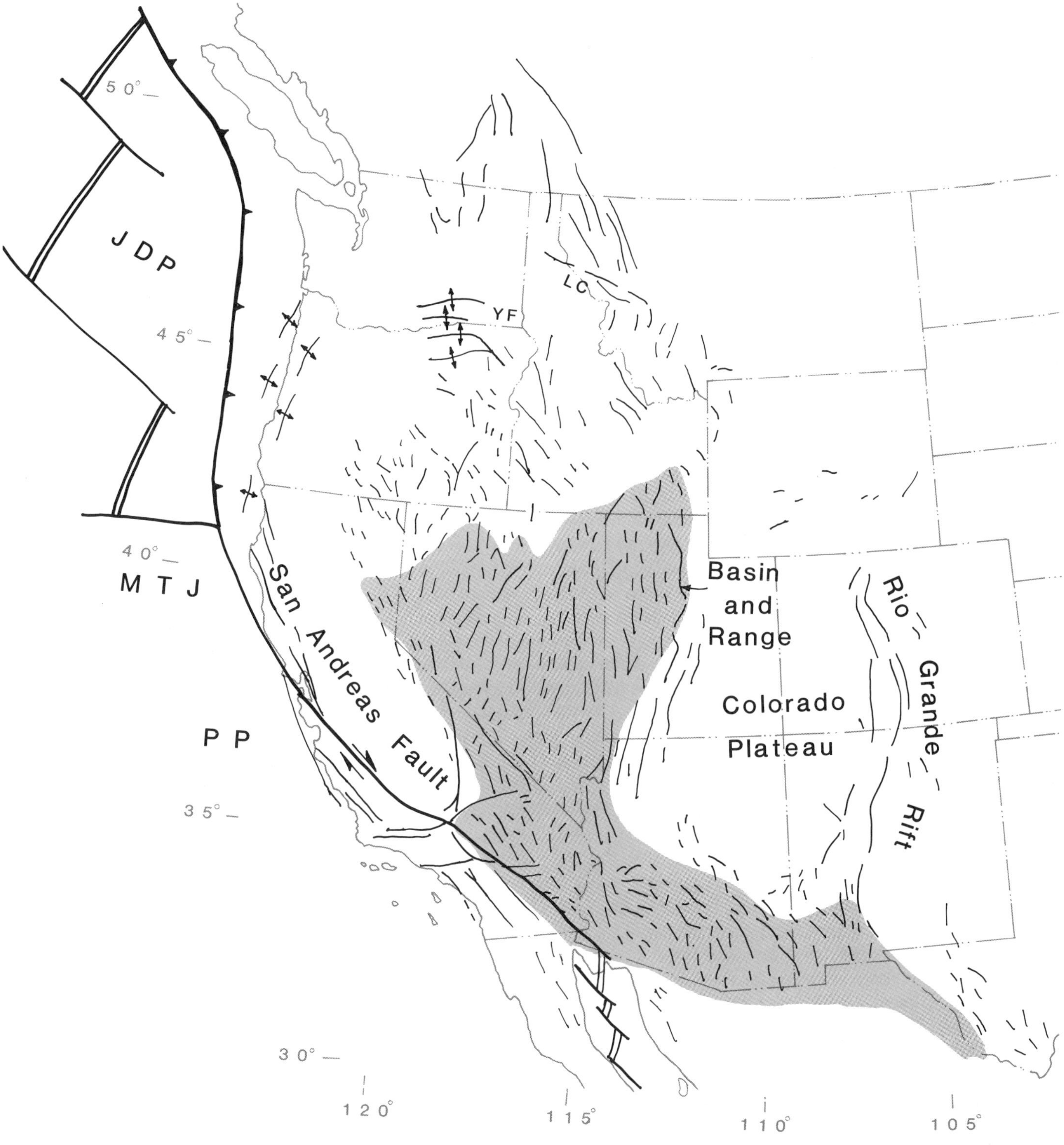

Figure 18. Present distribution of tectonic elements of the Cordillera. Trends of faults in the Basin and Range and Rio Grande rift area are shown, as well as the Yakima folds (YF) and tectonic elements along the plate boundaries. JDP, Juan de Fuca Plate; LC, Lewis and Clark line; MTJ, Mendocino triple junction; PP, Pacific Plate.

plate horst and graben structure (symmetrical about a vertical sigma 1 axis) by the Whipple detachment fault; and (5) the truncation from above of a lower plate, fan-shaped dike swarm (symmetrical about a vertical sigma 1 axis), also by the Whipple fault. Subsequent, additional flattening of detachment faults, by rotation of their dips during crustal stretching, isostatically induced warping of the differentially thinned crust (Spencer, 1984; *Howard and others, 1982*), or lower crustal ductile flow (Gans, 1987; Block and Royden, 1990) is not in dispute.

The configuration at greater depth of Cordilleran detachment faults is also imperfectly known. Sheetlike midcrustal plutons that lie in the lower plate of the Whipple fault and are parallel to it (*Anderson and Rowley, 1981*; Davis and Lister, 1988; see also John, 1988, for nearby Chemehuevi Mountains) suggest that the fault must have flattened to a subhorizontal orientation in the upper middle crust just above the plutons. Similarly, the regionally persistent metamorphic grade of footwall rocks (upper greenschist to middle amphibolite facies) for tens of kilometers downdip within the lower plate argues that the fault that has carried them to the surface sampled only a midcrustal horizon of subhorizontal orientation. Given that some Cordilleran detachment faults have displacements of tens of kilometers (Reynolds and Spencer, 1985; *John, 1987*; Davis and Lister, 1988), faults that initially dipped steeply through the entire crust should display progressively higher grade metamorphic rocks in the deeper exposures of their footwalls. That this is not seen in southwestern U.S. core complexes is the strongest argument that the detachment faults of this region originally flattened into upper midcrustal levels governed by the continental stress guide (Lister and Davis, 1989). This argument is bolstered by CALCRUST and COCORP seismic reflection surveys that suggest linkage in extended central and southern Cordilleran regions between surface exposures of Tertiary mylonitic gneisses and subhorizontal, midcrustal reflections of ductilely sheared rocks near and below the brittle-ductile transition (Allmendinger and others, 1987; *Okaya and Frost, 1986*; Davis, 1988).

Seismic reflection records of the LITHOPROBE program across early Tertiary metamorphic core complexes in the Omineca terrane of the southern Canadian Rocky Mountains (Fig. 16) present a contrasting picture to that described above for the more southerly core complexes described above (Cook and others, 1988; Parrish and others, 1988). Major, moderately dipping (30°) Eocene normal faults such as the Slocan Lake fault zone appear to extend to depths in excess of 20 to 25 km, and possibly as deep as 30 to 33 km—well below the upper midcrustal level of detachment (≤16 ±5 km; Anderson and others, 1988) for the Whipple and similar faults in southeastern California and Arizona. Although the Canadian faults are termed "detachment" faults by some workers, we wonder if some of these faults might not be fundamentally different from the younger, mid- to Late Tertiary detachment faults of Utah, Nevada, California, and Arizona. The deeply rooted normal faults of the Canadian Rocky Mountains have generally steeper and more planar dips than those we either infer for U.S. faults south of the Snake River Plain (Fig. 17) or see imaged in seismic reflection surveys. More important, perhaps, the Canadian faults have rather tightly constrained early Cenozoic ages (very latest Paleocene through middle Eocene, 58 to 45 Ma), which suggest that the oldest of them might have been active concurrently with waning stages of Laramide thrust faulting in the foreland fold and thrust belt farther east (see Sevigney and others, 1990, for possible confirmation of this relationship). Paleocene strata are widely involved in thrusting in the eastern foothills, but early and middle Eocene strata are lacking there and the end of compressive deformation is thus not well constrained (Price and Mountjoy, 1970). In Wyoming to the south, however, the Jackson-Prospect, Darby, and Cache Creek thrust faults were active into early Eocene time based on relationships with early Cenozoic strata.

No such possible contemporaneity between crustal extension and thrust faulting is seen in Cordilleran areas south of Wyoming, where thrust faulting ended in mid- to Late Cretaceous time (Sevier deformation) and was followed by an eastward shifting of Laramide compressive deformation into the Rocky Mountain craton. The onset of detachment-style crustal extension did not occur until Oligocene time, ca. 36 Ma ago, in east-central Nevada and west-central Utah (Gans and others, 1989). We suggest that extensional faults of the Canadian Rockies formed in a stress regime that was different from that of more southerly Cordilleran regions. If true, not all Cordilleran "metamorphic core complexes" and "detachment" faults have comparable origins. Arguments against initially shallow-dipping geometries for detachment faults using as evidence the moderately dipping, crust-transecting "detachment" faults of the Canadian Rockies may not be valid.

Some workers (Armstrong, 1972; Coney and Harms, 1984; Wernicke and others, 1987) have suggested that crustal (lithospheric) thickening by thrust faulting is a necessary precursor for subsequent crustal relaxation and attenuation along low-angle normal faults. In U.S. regions south of the Snake River Plain, triggering of continental extension in previously thrust fault-thickened regions apparently occurred during and after post-Laramide resteepening of the subducting, shallow-dipping Farallon Plate. Such resteepening, as reflected in a Cenozoic reestablishment of subduction-related Cordilleran magmatism following the 70 to 40 Ma magmatic lull (Christensen and Yeats, this volume; Lipman, this volume), may have (1) led to thermal weakening of the thickened lithosphere, either due to plutonism within it or the rise of asthenosphere beneath it; and/or (2) induced an extensional stress state in that gravitationally unstable crust. The latter possibility is supported by patterns of intraplate compressional and extensional stresses along the length of the modern South American Andes that reflect the shallow versus steep configuration, respectively, of the actively subducting Nazca Plate (e.g., *Burchfiel and Davis, 1976*; Cross and Pilger, 1982; *Jordan and others, 1983*). Thus a change from compressional to extensional intraplate stress in the Cenozoic western United States may have accompanied a change in configuration of the subducting Farallon Plate from shallow (Laramide) to steep (post-Laramide) dips. No such dramatic changes in subducting plate configuration or tectonic history are evident from the magmatic and deformational record of eastern Cordilleran regions north of the Wyoming Province (Armstrong, 1974; Burchfiel and Davis, 1975). Here, subduction-related magmatism and east-directed thrust faulting persisted between Late Jurassic and early Tertiary time. It is tempting to draw analogies between synthrusting, early Cenozoic extensional tectonics in the Canadian Rocky Mountains and synconvergent extension within the Miocene Himalayas (Royden and Burchfiel, 1987).

Wust (1986) plotted trends and "vergence" of extension directions in more than a score of Cordilleran metamorphic core complexes—"vergence" being defined by the direction of upper plate motion relative to a presumably fixed lower plate. West- and east-"vergent" complexes, the detachment faults of which root westward and eastward, respectively, are generally separated by a geographic axis that more or less parallels the trace of the foreland fold and thrust belt (to the east) and closely coincides with what Coney and Harms (1984) regard as the axis (reconstructed) of a tectonically thickened late Mesozoic–early Cenozoic crustal welt. No such axis can be defined south of the latitude of Las Vegas, southern Nevada. In lower Colorado River areas the east-rooting Whipple detachment system lies along strike with the more northerly, west-rooting Eldorado-McCullough Mountains system (Davis and Lister, 1988; Howard and John, 1987), and no west-rooting detachment systems are known in western areas of the Mojave Desert.

Wust (1986, p. 828), erroneously we believe, regards the upper plates of all Cordilleran detachment faults as the active elements, stating that these plates "tended to move outward from an inferred tectonic axis." With arrows of supposed extensional vergence pointing outward both eastward and westward from the axis of the crustal welt described above, Wust and subsequent workers (e.g., Best, 1988; Schweickert and Irwin, 1989) have postulated lateral spreading (gravitational collapse) of a structurally thickened and, by inference, topographically high crustal welt. While Cenozoic tectonic thinning of the crust has obviously occurred, the kinematic scenario described above is misleading in its emphasis on active upper plates moving away from the axis of an overthickened crust. Such a picture may give readers the mistaken impression of symmetrical, gravitationally induced spreading (rooted into the crust) or sliding (nonrooted) of the flanks of the welt away from its axis. Core complexes in southeastern California, for example, are considered by Wust to have had active upper plates that moved eastward from an inferred tectonic axis. Yet, it seems clear from the geometry of detachment faulting in this area that the footwalls (lower plates) of the east-rooting detachment systems were the active structural elements. Specifically, while midcrustal rocks ($\geqslant$10 km deep) were being transported to the surface in the footwalls of Miocene detachment faults, coeval strata (ca. 2 km thick) deposited on the hanging wall (upper plate) during faulting were never displaced into deep basins or graben that received significant post-tectonic sedimentation. Such hanging wall sections are still largely preserved today and lie tectonically above midcrustal mylonitic gneisses. Thus we believe that crustal extension in the lower Colorado River region along detachment faults such as the Whipple fault occurred by the active westward displacement of lower plate rocks upwards and out from beneath the southwestern margin of the Colorado Plateau (Reynolds and Spencer, 1985; Davis and Lister, 1988). If this movement mode is also appropriate for east-rooting detachment systems in the Great Basin to the north, then the true vergence of crustal extension along these faults was westward (= active lower plates), not eastward (= active upper plates). A corollary of this conclusion is that the axis of the hypothesized crustal welt must also have shifted westward with respect to the continent, a geometric necessity recognized by Coney and Harms (1984). Thus, the overthickened Cordilleran crust was thinned in Tertiary time by asymmetric lateral displacement of the thickened crust away from the stable continent, not by symmetric, gravitationally induced spreading away from the axis of prior thickening. This history accounts for the nearly 2:1 predominance of east-rooting detachment systems, that is, those lying between the inferred axis of thickening and the continent, over west-rooting systems (Wust, 1986).

Although data are limited, the rates of extension along some of these east-rooting faults can be shown or inferred to be surprisingly fast; some begin to approach plate-motion rates despite their intraplate setting. Although detachment faulting may typically occur over a 5- to 10-m.y.-long interval, mineral thermal histories from fission-track and $^{40}Ar/^{39}Ar$ studies suggest that most extension is apt to occur within 1- to 3-m.y.-long episodes of rapid fault displacements. For example, geologically conservative estimates for horizontal and vertical components of slip along the Whipple fault over the time interval from 20 to 16 Ma are 7.2 mm/yr and 3 mm/yr, respectively (Davis, 1988); even higher rates of slip are deemed likely for the 20- to 18-Ma period when geologic evidence indicates that most uplift and rapid cooling of footwall mylonitic gneisses occurred. Rapid cooling of lower plate rocks over a 2- to 3-m.y.-period has also been documented by geochronologic studies in the Ruby Mountains, Nevada (Dokka and others, 1986; Dallmeyer and others, 1986), and central Mojave Desert (Dokka, 1989). Such cooling is best attributed to rapid uplift and tectonic denudation.

Low-angle normal faults in western portions of the U.S. Cordillera. Until relatively recently (i.e., the late 1980s), recognition of low-angle normal faults had essentially been confined to areas of the central and eastern Cordillera (cf., Davis and Lister, 1988, Fig. 1). Most examples lie within either southern or northern portions of the Basin and Range Province, but the occurrence of detachment faults in such disparate geologic areas as the western Intermontane Belt of Canada (*Friedman and Armstrong, 1988*), the Omineca Belt of Canada, Washington and Idaho, the central Mojave Desert (Dokka, 1989), and the Peninsular Ranges of southwestern California indicates the independence of this structural style from areas of basin-and-range extension. Miocene detachment faults of both the central Mojave system and the lower Colorado River system of Arizona and California appear to have been offset northwestward several hundred kilometers by the younger San Andreas fault (cf., Dokka, 1989; *Frost and others, 1986*), thus placing detachment fault systems very near to the continental margin.

Cenozoic low-angle normal faults were slow to be recognized in the Cordillera because geologists were conditioned by traditional fault mechanics against accepting shallow-dipping faults as having an extensional origin. Many were interpreted as strange, highly planar unconformities. When shallow-dipping younger-over-older Cenozoic faults were observed, they were either considered to be thrust faults, rotated steep normal faults, or gravity slide surfaces. Recognition of low-angle normal faults probably occurred initially in the eastern Cordillera because of their great abundance there and because of the excellent exposures throughout this arid to semiarid region. It is not surprising that the discovery of regional or subregional low-angle normal faults in more poorly exposed and stratigraphically less continuous areas of the western Cordillera has taken longer—nor is it surprising that geologists the world over are quickly abandoning stereotypical views about the absence of extensional faults in contractional terranes.

Tectonically thinned ophiolite and stratigraphic sequences in the California Coast Ranges are now being interpreted as the consequence of extensional tectonics active during evolution of the subduction-related accretionary prism of that region (e.g., Jayco and others, 1987; *Platt, 1986*). The Orocopia-Chocolate Mountains "thrust" fault east of the Salton Sea, with its high-P, low-T footwall assemblage of Orocopia Schist, is now increasingly recognized as a subduction-related thrust fault, but one that has been extensively overprinted, reactivated, and excised by Tertiary extensional faulting of both Paleogene and Miocene age (e.g., *Frost and Martin, 1983*). The low-angle Sortan normal fault of the Chocolate Mountains juxtaposes only incipiently recrystallized Mesozoic sedimentary rocks atop Orocopia Schists, a telescoping of metamorphic grades from crustal levels once in excess of 10 km apart (Jacobsen and others, 1988). Intrusion of the normal fault by a 60-Ma pluton (minimum K/Ar age) demonstrates that uplift of Orocopia Schists to relatively shallow crustal levels followed soon after subduction-related metamorphism at ca. 74 Ma (minimum age). Rapid cooling from ca. 500 to 100° C of Pelona (= Orocopia) Schists between 70 and 57 Ma (fission-track ages) corroborates such a history and has been postulated as the result of rapid uplift (*Mahaffie and Dokka, 1986*).

The stacked sequence of Mesozoic thrust sheets and the overlying Devonian Trinity ultramafic allochthon in the Klamath Mountains has also not escaped major extension. A major southeast-rooting low-angle normal fault, the La Grange fault (Figs. 10A, 16) of the central California Klamaths, separates a footwall of central metamorphic belt units and the overlying Trinity allochthon from a hanging-wall assemblage containing units as young as the Oligocene-Miocene Weaverville Formation (Schweickert and Irwin, 1989). At one locality, a laminated, lineated mafic "mylonite" or ultraclasite up to 20 cm thick lies directly below the fault. It and underlying chloritized, sheared Salmon hornblende schists are reminiscent of lower plate rock assemblages beneath detachment faults in eastern Cordilleran regions (Davis and Lister, 1988), although footwall rocks, with the possible exception of the "mylonite," have not been ductilely deformed. The La Grange detachment fault dips 20 to 22° at the La Grange Mine, omits several kilometers of structural thickness of rock units at that locality, and has a broadly domed lower plate that is analogous to the antiformal domes seen in metamorphic core complexes to the east.

To the north, on Vancouver Island, two major low-angle faults have recently been described as lying within a younger-over-older stacked sequence of Triassic Karmutsen metavolcanics, Cretaceous (Nanaimo) sedimentary rocks, and Tertiary volcanics and quartz diorite. The poorly exposed faults dip as shallowly as 5 to 8°, and the lowest is underlain by thin, dark, flinty cataclasite and brecciated, chloritized metabasalt up to 300 m thick. *Muller (1988)* considers these faults to be extensional detachment faults and predicts that they or similar structures will be recognized in the future as widespread and important features of the island's geology. His prediction can probably be safely extended to other areas within the accretionary belt.

The origins of low-angle faults in the western Cordillera are apt to be quite diverse—ranging from (1) synsubduction extension within high levels of accretionary prisms in order to achieve critical wedge taper (California Coast Ranges?); to (2) rapid, isostatically induced(?) uplift of recently underplated oceanic metasedimentary and metavolcanic rocks following the cessation of subduction (Pelona/Orocopia Schist?); to (3) possible lateral spreading of previously thrust-fault thickened crust (Klamath Mountains?) analogous to eastern Cordilleran metamorphic core complexes. Undoubtedly, still other causes for low-angle faulting in this and other orogenic belts await discovery (cf., *Dewey, 1988; Burchfiel and Royden, 1989*).

Accretionary and transcurrent tectonics

Pacific Northwest. In contrast to the Canadian Cordillera where Cenozoic accretion produced little in the way of new subareal terrane, the U.S. continental margin was the site of significant accretion in all Pacific margin states. Accretion of the early Tertiary basalts of Siletzia from a position proximal to the continental margin, and without significant latitudinal displacement, probably occurred in late Eocene time. Some outboard units of northern Siletzia appear to have been subsequently sliced away and transcurrently displaced northward (e.g., Yakatat terrane of southern Alaska; *Davis and Plafker, 1986*), but Coast Range basalts left behind were later underthrust by an assemblage of now-metamorphosed oceanic volcanic rocks and associated sediments as old as Eocene and Oligocene. Continued Cenozoic accretion of sediments along the Cascadia subduction zone between the Juan de Fuca and North American Plates has constructed one of the widest accretionary prisms (ca. 150 to 200 km) on the planet, both by offscraping and underplating of subducted materials along the continental margin (Christiansen and Yeats, this volume; Cook and others, 1988). A part of this accretionary complex is exposed in the Olympic Mountains of western Washington (lTap, uTap; Fig. 16).

California. The Cenozoic accretionary history of coastal California is discussed by Christensen and Yeats (this volume). While there seems little doubt that some major elements of the central California Coast and Peninsular (= southern Coast) Ranges are nonnative to Californian North America, controversy exists among various geological and paleomagnetic studies regarding the degree of allochthoneity of some terranes and the timing of terrane emplacement. Our discussion will treat briefly only the two major candidates for accreted terranes of Cenozoic age in California—the Santa Lucia allochthon (= Salinian block of Christensen and Yeats, this volume) and the Peninsular Ranges terrane (= Baja-Borderland allochthon, of Christensen and Yeats, this volume); it draws heavily upon an excellent recent overview of the history of both terranes by Lund and Bottjer (1991).

The Santa Lucia allochthon is an amalgamated entity consisting principally of the coastal Sur-Obispo composite terrane of "Franciscan," "Great Valley," and "Coast Range" ophiolite counterparts, and the continental, largely granitic, Salinian block (Fig. 17; *Vedder and others, 1983*). Lund and Bottjer, summarizing and analyzing all available paleomagnetic data from this amalgamated terrane conclude that it has had a three-fold history: (1) internal amalgamation at equatorial paleolatitudes in the Cretaceous, (2) northward displacement (ca. 2°/Ma) with respect to North America between 70 and 50 Ma, and (3) accretion to coastal California in the Eocene. The paleomagnetic data supporting northward transport of some 40° latitude are described by these authors as "remarkably coherent." However, controversy exists as to whether the Salinian block, which now lies east of the Sur-Obispo terrane and is separated from it by the Cenozoic Nacimiento fault, was part of the Santa Lucia amalgamation and ever lay as far south as did the Sur-Obispo component. Some workers favor this interpretation (*Vedder and others, 1983*; Champion and others, 1984; Lund and Bottjer, 1990), but others (e.g., Hamilton, 1969; Burchfiel and Davis, 1981; *Powell, 1981*) have argued from geologic grounds that the Salinian block is a native terrane, offset northward only several hundreds of kilometers from an original position in the western Mohave Desert and at the southwestern end of the Sierra Nevada.

The conflict between paleomagnetic and geologic data may be resolved (S. Lund, personal communication, 1989) if "Salinian" sedimentary rocks yielding Late Cretaceous and Paleocene/Eocene equatorial paleolatitudes (Champion and others, 1984; *Kanter and Debiche, 1985*) were not deposited on granitic basement rocks of the main Salinian block sensu stricto (= Central block of Ross, 1978), but are instead the cover for plutonic assemblages that lay to the south of Salinia in Cretaceous and early Paleogene time. The present fault-bounded distribution of these sedimentary rock sites and their occurrence north of, and outboard from the main (central) Salinian block permits this view.

We postulate a Cretaceous position for Salinia southwest of the southwestern Sierra Nevada (Tehachapi Mountains), west of the present western Mojave Desert, and west and north(?) of the San Gabriel Mountain block. Geologic evidence supports an original Salinia–southern Sierra Nevada/Mojave Desert juxtaposition based on similarities between plutonic and metamorphic rocks of the two regions (cf., Burchfiel and Davis, 1981, their Fig. 9-9; *Mattinson and James, 1985*; correlative lithologies include Rand (Mojave) and Sierra de Salinas schists (Salinia) and gabbroic rocks at Logan (Salinia) and in the San Emigdio Mountains (southwesternmost Sierra Nevada). By Late Jurassic time (148 Ma), dikes of the Independence dike swarm had stitched together the southern Sierra Nevada, Mojave Desert, and terranes (e.g., eastern San Gabriel) of the eastern Transverse Ranges (James, 1989). Structural linkage of the eastern San Gabriel terrane to North America prior to intrusion of distinctive Triassic (Mt. Lowe, 218 Ma) and Middle Jurassic plutons across the tectonic boundary between them is a strong possibility (*Barth and others, 1988; Barth, 1989*).

Mesozoic geologic ties between the western San Gabriel terrane and the main Salinian block are not as strong since the two areas were not along orogenic strike. Salinia with its voluminous granitic plutons of central and eastern Sierran character presumably lay west of the San Gabriels, which contain only scattered plutons of Cretaceous age. The western portion of the original Salinian batholith probably resembled the western half of the Sierran batholith, but it and associated early and middle Mesozoic accreted terranes were tectonically removed in some manner prior to the emplacement of the Sur-Obispo terrane. The former geographic juxtaposition of the main Salinian and San Gabriel terranes is supported by the young age of Cretaceous plutons (80 to 90 Ma U/Pb) in both the southern half of Salinia and in the western San Gabriel Mountains; plutons in the northern half of the main Salinian block are distinctly older (95 to 120 Ma). In addition, *Ehlig and Joseph (1977)* document similarities in age and composition between plutonic rocks in the southern Salinian and San Gabriel terranes. Most impressive is the essentially identical age (72.0 and 72.4 Ma, Rb/Sr) and initial strontium ratios (Sri = 0.7128 and 0.7136) of a highly unusual "polka dot granite" found as dikes in the La Panza Range of Salinia and in the San Gabriel Mountains terrane west of the San Andreas.

Given geologic relationships briefly described above, we strongly favor excluding the main (central) Salinian block from the Santa Lucia allochthon, leaving within that allochthon only (1) the Sur-Obispo composite terrane with its Franciscan- and Great Valley–like stratigraphic assemblages and basements, and (2) northern "Salinian"-like terranes (northern block of Salinia; Ross, 1978). Paleomagnetic data supporting large (40°) northward transport of a smaller Santa Lucia allochthon seem firm, and accord well with independent data that the Coast Range ophiolite east of the San Andreas formed at similarly low latitudes before being transported northward at a time prior to Great Valley sequence sedimentation across it (cf., McLaughlin and others, 1988).

The second major terrane of southwestern California considered as allochthonous by some paleomagnetists is the Peninsular Ranges terrane (or Baja-Borderland allochthon, Christensen and Yeats, this volume). This composite terrane (Fig. 17), consisting mostly of the Guerrero terrane (cf., Coney, 1989) of the Peninsular Ranges and Baja California, is somewhat less controversial than the more northerly Santa Lucia allochthon, perhaps because paleomagnetic measurements indicate that it is less far traveled. Such data are best interpreted as indicating a North American derivation for the arc and fore-arc assemblages within the Peninsular Ranges terrane, but from a position along the Mexican continental margin some 1,000 km (ca. 15°) south of present latitudes. Northward transport of the terrane, presumably by margin-parallel transcurrent faulting occurred some time between 40 and 20 Ma (Lund and Bottjer, 1990). Similarities in this timing and displacement to the history of transcurrent disruption of Paleogene units in the Pacific Northwest, as reviewed above, suggest the possibility of a throughgoing, late Paleogene, dextral transcurrent fault system along the western margin of the continent. The term "transcurrent" is preferred to "transform" because most of the Vancouver Plate (≃northern Farallon)/North American Plate boundary during this 40 to 20 Ma time interval was one of dextrally oblique subduction (Engebretson and others, 1985; Stock and Molnar, 1988). Northward displacements of the Peninsular Ranges terrane, Leech River–like schists, and Crescent/Metchosin/Yakatat basalts (Cowan, 1982) probably occurred within one or more fore-arc slivers (cf., Jarrard, 1986). An earlier episode of fore-arc slivering of the continental margin (Cretaceous, ca. 120 to 100 Ma) has been postulated above to explain the northward dispersal of Californian Franciscan and Great Valley sequences into the Pacific Northwest prior to its collision with Superterrane II.

Estimates for the time of accretion of the Peninsular Ranges terrane to North America and the previously accreted Santa Lucia allochthon are not well constrained and fall within a time interval between 40 and 20 Ma. The suture between North American terranes and the Peninsular Ranges allochthon is delineated today (but not necessarily in Tertiary time) by the Oak Ridge fault in the western Transverse Range, but lies hidden beneath alluviated areas along the southern flank of the San Gabriel Mountains west of the San Andreas fault (Fig. 17). Although the Peninsular Ranges allochthon now lies south of the Santa Lucia allochthon (Sur-Obispo composite terrane), paleomagnetic data indicate that the reverse was true in Cretaceous time. Thus, northward transport of the Santa Lucia allochthon must have occurred outboard of the still, at that time, autochthonous Peninsular Ranges terrane. No direct physical evidence for this transport scenario is known. Younger northward displacement of the Peninsular Ranges, ca. 1,000 km with respect to North America, was probably not entirely transcurrent. A major component of convergence between the two continental terranes is indicated by the occurrence of a deep and steeply north-dipping lithospheric root beneath the San Gabriel, San Bernardino, and western Transverse Ranges. The root, which extends to depths as great as 250 km below the San Bernardino Mountains, is interpreted by Bird and Rosenstock (1984) as mantle lithosphere. It likely records an impressive, but largely cryptic phase of continental subduction of northern portions of the Peninsular Ranges lithosphere beneath North American rocks of the Transverse Ranges. Subsequent Neogene tectonic rotations in this region are widespread and pronounced, and have severely complicated reconstructions of pre–middle Miocene paleogeography (e.g., Luyendyk and others, 1985; Christensen and Yeats, this volume) across the now nearly east-west suture. For example, rocks in the western Transverse Ranges and San Gabriel Mountains block have experienced net clockwise Neogene rotations of 68, 131, and 35°, respectively (Luyendyk and others, 1985).

TECTONIC SUMMARY

The Cordilleran orogenic belt of the western United States can be considered to have had its inception with the formation by rifting of a latest Proterozoic/Early Cambrian passive margin underlain by Precambrian crystalline rocks and irregularly preserved sequences of Middle and Late Proterozoic sedimentary rocks. Since the inception of this continent/ocean interface the Cordilleran orogen has evolved and continues to evolve by the activity of numerous and varied types of plate boundaries. The Cordilleran orogen can be divided into two parts: (1) an eastern part formed on older North American Precambrian craton; and (2) a western accretionary belt consisting of fragments from Phanerozoic oceanic, island-arc, and continental environments accreted to the North American craton from mid-Paleozoic to Cenozoic time. Essentially autochthonous sedimentary basins and accretionary prisms above subduction zones are additional major components of the accretionary belt. Tectonic and sedimentary accretion along the western boundary of the orogen has led to a westward shifting of the continental margin, but truncation and lateral removal of rocks from that margin have also been important. Rock units in both the accretionary and eastern belts of the Cordillera have been reworked by tectonic activity and variably metamorphosed and intruded by igneous rocks of both mantle and crustal origin.

Our view of the role of accretionary tectonics (Coney and others, 1980) in the construction of the western U.S. Cordillera is somewhat conservative. While accepting the non–North American origin of some oceanic terranes, for example, the late Paleozoic Cache Creek assemblage (largely on the basis of its Tethyan faunas) and some Mesozoic oceanic rock assemblages of the Franciscan Complex, we find that most tectonostratigraphic terranes of the western United States have various ties (or hints of ties) to North America that suggest that they are not exotic components of the orogen. This conclusion does not clash with our recognition that episodes of transcurrent displacements along the evolving plate margin have truncated and dispersed, usually northward, North American terranes such as the Middle to Late Jurassic fore-arc region of the western Klamath Mountains (which is now completely missing), outboard portions of the younger Great Valley sequence and its ophiolitic substrate, and the Mexican(?) Peninsular Ranges composite terrane.

The Cordilleran passive margin developed across the North American craton, which consists of a northern region characterized by the aggregation of several Archean fragments sutured together during a short period of Early Proterozoic time, and a southern region characterized by the progressive southward accretion of younger, largely primitive tectonic elements added from about 2.3 to 1.0 Ga. The southern part of the craton was affected by anorogenic granite/rhyolite magmatism from 1.5 to 1.34 Ga. Mafic dikes and sills invaded the craton at about 1.4, 1.0 to 1.2, and 0.7 to 0.8 Ga during periods of widespread extension.

The passive, Atlantic-style margin with its miogeoclinal sedimentary cover evolved through Devonian time. Early Paleozoic island arcs are now present farther west within the accretionary belt, but their position in early Paleozoic time relative to North America and the Antler sedimentary basin that lay west of the continent remains uncertain. The arcs contain important early Paleozoic, principally Devonian, magmatism and local metamorphism, but rocks of the passive margin, which contain local mafic magmatism and very local alkaline plutonism, show no evidence of convergent activity until Late Devonian to Mississippian time during the Antler and related orogenic events. During the Antler orogeny continental slope and rise rocks were emplaced eastward onto miogeoclinal continental shelf strata with little associated magmatism or metamorphism. This convergent deformation was unaccompanied by an arc-continent collision and may have resembled in some respects the modern, noncollisional plate convergent zone of the Apennine Mountains in Italy. Emplacement of allochthonous strata atop the continental margin created an orogenic highlands that shed debris eastward into a well-developed foredeep.

During late Paleozoic time the Cordilleran passive margin was reestablished as the Antler highlands was worn away. Arc magmatism continued within the offshore arcs, probably separated from North America only by marinal seas such as the Havallah basin. The continental interior and the eastern part of the Cordillera were affected by the amagmatic foreland-type Ancestral Rocky deformation during Pennsylvanian-Permian time (probably not a Cordilleran event) at the same time that the southwestern part of the continent was truncated along a probable left-slip transform boundary. This truncation established a new margin in southwestern California along which Mesozoic tectonic and magmatic elements trend almost at right angles across all older elements; farther north, early and late Paleozoic tectonic and magmatic elements are subparallel to those of the Mesozoic. Granitic plutonism and associated deformation as old as Permian affected North American rocks along part of the newly truncated continental margin in the southernmost part of the Cordillera. During the Permo-Triassic Sonoma orogeny, an offshore arc (Sonomia)—perhaps continuous southward into the Permian magmatic belt just described—and associated marginal basin rocks (Havallah assemblage) were accreted to the central part of the U.S. continental margin. The style of this deformational event may have closely paralleled that of the earlier Antler orogeny. With the accretion of Sonomia to North America, the western margin of the continent had shifted markedly westward.

Early Mesozoic time saw the establishment of an Andean-type margin along the southern part of the Cordillera, whereas to the north the new continental margin (west of Sonomia) was characterized by a partly submerged volcanic archipelago and an inboard, back-arc Star Peak/Luning basin. A Late Triassic to Early Jurassic convergent plate boundary separated eastern arc rocks from a complex accretionary prism of late Paleozoic–early Mesozoic oceanic realm rocks (Cache Creek assemblage) along the length of the Cordillera. The volcanic archipelago had a complicated early and middle Mesozoic history with episodes of both severe contraction and intra-arc telescoping of older units, and extension or transtension opening small ocean-floored basins. Middle Jurassic time saw the beginning of the closure of the Star Peak/Luning basin, profound internal foreshortening of the magmatic arc to the west, and widespread Cordilleran mountain building from at least central California north in Canada. Back-arc deformation spread eastward, thickening the crust and eventually forming the thin-skinned, east-vergent Sevier foreland fold and thrust belt within Precambrian through Mesozoic strata along most of the central and northern Cordillera. In contrast, the southernmost, cratonal part of the Cordillera was characterized by intra-arc extension and localized compression during most of Mesozoic time. Thick and broad zones of subduction mélange, blueschist, and ophiolitic rocks were accreted along the evolving Mesozoic Cordilleran margin west of the magmatic arc at times ranging from the latest Paleozic or earliest Triassic through the mid-Cretaceous.

Although transcurrent faults may have been active during early Mesozoic time, it is only during late Mesozoic time (i.e., after the Middle Jurassic), that evidence for such structures is clear. Transcurrent faults, mainly right lateral in our view, disrupted older accretionary zone terranes, magmatic belts, and North American cratonal and miogeoclinal elements. These faults doubled up terranes in areas such as the San Juan Islands and northern Cascades, and truncated them elsewhere, such as along the northern California margin and, perhaps, the West Idaho suture zone. Accretion of the Wrangellian superterrane in mid-Cretaceous time caused major west-vergent thrusting in the northern Cascades and San Juan Islands, but not farther south.

During latest Cretaceous and early Cenozoic time a gap developed in the Andean magmatic arc within the central U.S. part of the Cordillera. East of this gap and the older (but locally synchronous) foreland fold and thrust belt, the North American craton experienced strong contractional deformation with the uplift of large, thrust-bounded basement-cored blocks and the subsidence of deep, intervening basins; magmatic activity in these, the Laramide Rocky Mountains, was sparse. Some Laramide structures appear to be reactivated structures inherited from the Ancestral Rocky Mountains deformation. Andean-style magmatic and tectonic activity, including a back-arc fold and thrust belt of Sevier-type, continued in regions both south and north of the

Laramide-disturbed craton. These relations suggest that the areas of the magmatic gap and the Laramide Rocky Mountains formed above a very shallow-dipping segment of the subducting Farallon Plate. Steep subduction angles, characteristic of the pre-latest Cretaceous subduction beneath all segments of the U.S. Cordillera, persisted in areas to the north and south. Plate reconstructions indicate a rapid increase in the convergence rate across the Farallon–North American Plate boundary at this time.

In Eocene time extension began within parts of the northwestern U.S. and adjacent Canadian Cordillera. Even though arc magmatism was widespread north and south of the magmatic gap, dextral transcurrent movements along and within the western parts of the Cordilleran margin continued to be significant, particularly in the Pacific Northwest and along the southern California margin. Tertiary strike-slip fault displacements of more southerly North and Central(?) American–derived terranes continued the process of tectonic slivering and shuffling initiated in Mesozoic, and perhaps earlier, times. During Oligocene and Miocene time extension affected much of the Cordilleran region and was responsible for formation of numerous detachment fault-related "metamorphic core complexes." The magmatic arc began to sweep westward and the magmatic gap was completely closed by about 20 Ma. Progressive subduction of the Pacific-Farallon oceanic spreading ridge beginning between 30 and 25 Ma caused concomitant lengthening of a transform Pacific–North American Plate boundary along the central part of the Cordilleran margin, gradual shortening of surviving subduction boundaries along the margin, and a related cessation of inboard arc volcanism. In the Pacific Northwest to the north, Miocene back-arc extension was accompanied by voluminous basaltic volcanism in the northwestern part of the United States forming the broad Columbia River Plateau. Basalt/rhyolite volcanism and broadly synchronous basin-and-range normal faulting became widespread within the ever-lengthening region of ongoing extension east of the Pacific–North American transform boundary. During the past 15 m.y. basaltic and bimodal volcanic rocks have been extruded progressively eastward along the Snake River Plain, currently reaching the active calderas at Yellowstone. The present high elevation of the Colorado Plateau and most of the Cordilleran region occurred in late Cenozoic time. Even though the Basin and Range is underlain by extended and thin crust, it stands at an average elevation of 1.5 to 2.0 km because the underlying mantle lithosphere is also thin and the asthenosphere is significantly elevated.

REFERENCES CITED

Text references in *italics* will be found only on accompanying microfiche.

Allmendinger, R. W., and Jordan, T. E., 1984, Mesozoic structure of the Newfoundland Mountains, Utah; Horizontal shortening and subsequent extension in the hinterland of the Sevier belt: Geological Society of America Bulletin, v. 95, p. 1280–1292.

Allmendinger, R. W., and 7 others, 1987, Overview of the COCORP 40°N transect, western United States; The fabric of an orogenic belt: Geological Society of America Bulletin, v. 98, p. 364–372.

Anderson, E. R., Zoback, M. L., and Thompson, G. A., 1983, Implications of selected subsurface data on the structural form and evolution of some basins in the northern Basin and Range Province, Nevada and Utah: Geological Society of America Bulletin, v. 94, p. 1055–1072.

Anderson, J. L., 1983, Proterozoic anorogenic granite plutonism of North America, *in* Medaris, L. H., Jr., Byers, C. W., Mickelson, D. M., and Shanks, W. C., eds., Proterozoic geology: Geological Society of America Memoir 161, p. 133–154.

Anderson, J. L., Barth, A. P., and Young, E. D., 1988, Mid-crustal Cretaceous roots of Cordilleran metamorphic core complexes: Geology, v. 161, p. 366–369.

Anderson, P., 1989, Proterozoic plate tectonic evolution of Arizona, *in* Jennings, J., and Reynolds, S., eds., Geological evolution of Arizona: Arizona Geological Society Digest 17, p. 17–55.

Armin, R. A., and Mayer, L., 1983, Subsidence analysis of the Cordilleran miogeocline; Implications for timing of Late Proterozoic rifting and amount of extension: Geology, v. 11, p. 702–705.

Armstrong, G. C., and Oriel, S. S., 1965, Tectonic development of Idaho-Wyoming thrust belt: American Association of Petroleum Geologists Bulletin, v. 49, p. 1847–1866.

Armstrong, R. L., 1968, Sevier orogenic belt in Nevada and Utah: Geological Society of America Bulletin, v. 79, p. 429–458.

—— , 1972, Low angle (denudation) faults, hinterland of the Sevier orogenic belt, eastern Nevada and western Utah: Geological Society of America Bulletin, v. 83, p. 1729–1754.

—— , 1974, Magmatism, orogenic timing, and orogenic diachronism in the Cordillera from Mexico to Canada: Nature, v. 247, p. 348–351.

—— , 1975, Precambrian (1,500 million year old) rocks of central Idaho; The Salmon River arch and its role in Cordilleran sedimentation and tectonics: American Journal of Science, v. 275-A, p. 437–467.

—— , 1988, Mesozoic and early Cenozoic magmatic evolution of the Canadian Cordillera, *in* Clark, S. P., Jr., Burchfiel, B. C., and Suppe, J., eds., Processes in continental lithospheric deformation: Geological Society of America Special Paper 218, p. 55–91.

Armstrong, R. L., Taubeneck, W. H., and Hales, P. O., 1977, Rb-Sr and K-Ar geochronometry of Mesozoic granitic rocks and their SR isotopic composition, Oregon, Washington, and Idaho: Geological Society of America Bulletin, v. 88, p. 397–441.

Armstrong, R. L., Bernardi, M. L., and Rady, P. M., 1983, Late Paleozoic high-pressure metamorphic rocks in northwestern Washington and southwestern British Columbia; The Vedder Complex: Geological Society of America Bulletin, v. 94, p. 451–458.

Avé Lallemant, H. G., and Oldow, J. S., 1988, Early Mesozoic southward migration of Cordilleran transpressional terranes: Tectonics, v. 7, p. 1057–1075.

Best, M. G., 1988, Early Miocene change in direction of least principal stress, southwestern United States; Conflicting inferences from dikes and metamorphic core-detachment fault terranes: Tectonics, v. 7, p. 249–260.

Bird, P., 1988, Formation of the Rocky Mountains, western United States; A continuum computer model: Science, v. 239, p. 1501–1507.

Bird, P., and Rosenstock, R. W., 1984, Kinematics of present crust and mantle flow in southern California: Geological Society of America Bulletin, v. 95, p. 946–957.

Blake, M. C., Jr., Jayko, A. S., McLaughlin, R. J., and Underwood, M. B., 1988, Metamorphic and tectonic evolution of the Franciscan Complex, northern California, *in* Ernst, W. G., ed., Metamorphism and crustal evolution of the western United States; Rubey Volume 7: Englewood Cliffs, New Jersey, Prentice-Hall, p. 1035–1060.

Block, L., and Royden, L. H., 1990, Core complex geometries and regional scale flow in the lower crust: Tectonics, v. 9, p. 557–567.

Bond, G. D., and Kominz, M. A., 1984, Construction of tectonic subsidence curves for the early Paleozoic miogeocline, southern Canadian Rocky Mountains; Implications for subsidence mechanisms, age of breakup, and crustal thinning: Geological Society of America Bulletin, v. 95, p. 155–173.

Boudier, F., Le Sueur, E., and Nicholas, A., 1989, Structure of an atypical ophiolite; The Trinity complex, eastern Klamath Mountains, California: Geological Society of America Bulletin, v. 101, p. 820–833.

Brandon, M. T., 1989, Geology of the San Juan–Cascade nappes, northwestern Cascade Range and San Juan Islands, *in* Joseph, N. L., and 13 others, eds., Geologic guidebook for Washington and adjacent areas: Washington Division of Geology and Earth Resources Information Circular 86, p. 137–162.

Brandon, M. T., Cowen, D. S., and Vance, J. A., 1988, The Late Cretaceous San Juan thrust system, San Juan Islands, Washington: Geological Society of America Special Paper 221, 81 p.

Brown, E. H., 1987, Structural geology and accretionary history of the Northwest Cascades System, Washington and British Columbia: Geological Society of America Bulletin, v. 99, p. 201–214.

Brown, E. H., and Blake, M. C., Jr., 1987, Correlation of Early Cretaceous blueschists in Washington, Oregon, and northern California: Tectonics, v. 6, p. 795–806.

Brueckner, H. K., and Snyder, W. S., 1985, Structure of the Havallah sequence, Golconda allochthon, Nevada; Evidence for prolonged evolution in an accretionary prism: Geological Society of America Bulletin, v. 96, p. 1113–1130.

Buck, W. R., 1988, Flexural rotation of normal faults: Tectonics, v. 7, p. 959–973.

Burchfiel, B. C., 1979, Geological evolution of the west-central United States; International Association for the Genesis of Ore Deposits Symposium: Nevada Bureau of Mines and Geology, Report 33, p. 1–11.

Burchfiel, B. C., and Davis, G. A., 1972, Structural framework and evolution of the southern part of the Cordilleran orogen, western United States: American Journal of Science, v. 272, p. 97–118.

——, 1975, Nature and controls of Cordilleran orogenesis, western United States; Extensions of an earlier synthesis: American Journal of Science, v. 275-A, p. 363–396.

——, 1981, Mojave Desert and environs, *in* Ernst, W. G., ed., The geotectonic development of California; Rubey Volume 1: Englewood Cliffs, New Jersey, Prentice-Hall, p. 217–252.

——, 1988, Mesozoic thrust faults and Cenozoic low-angle normal faults, eastern Spring Mountains, and Nevada, and Clark Mountains thrust complex, California, *in* Wide, D. L., and Faber, M. L., eds., This extended land; Geological Journeys in the southern Basin and Range; Geological Society of America Cordilleran Section Field Trip Guidebook: Las Vegas, University of Nevada Department of Geoscience, p. 87–106.

Burchfiel, B. C., Hodges, K. V., and Royden, L. H., 1987, Geology of Panamint Valley–Saline Valley pull-apart system, California; Palinspastic evidence for low-angle geometry of a Neogene range-bounding fault: Journal of Geophysical Research, v. 92, p. 10422–10426.

Busby-Spera, C. J., 1988, Speculative tectonic model for the early Mesozoic arc of the southwest Cordilleran United States: Geology, v. 12, p. 1121–1125.

Butler, R. F., Gehrels, G. E., McClelland, W. C., May, S. R., and Klepacki, D., 1989, Discordant paleomagnetic poles from the Canadian coast plutonic complex; Regional tilt rather than large-scale displacement?: Geology, v. 17, p. 691–694.

Champion, D. E., Howell, D. G., and Gromme, C. S., 1984, Paleomagnetic and geologic data indicating 2,500 km of northward displacement for the Salinian and related terranes, California: Journal of Geophysical Research, v. 89, p. 7736–7752.

Clowes, R. M., and others, 1987, LITHOPROBE—southern Vancouver Island; Cenozoic subduction complex imaged by deep seismic reflections: Canadian Journal of Earth Sciences, v. 24, p. 31–51.

Coney, P., 1972, Cordilleran tectonics and North America Plate motion: American Journal of Science, v. 272, p. 603–628.

Coney, P. J., 1989, Structural aspects of suspect terranes and accretionary tectonics in western North America: Journal of Structural Geology, v. 11, p. 107–126.

Coney, P. J., and Harms, T. A., 1984, Cordilleran metamorphic core complexes; Cenozoic extensional relics of Mesozoic compression: Geology, v. 12, p. 550–554.

Coney, P. J., and Reynolds, S. J., 1977, Cordilleran Benioff zones: Nature, v. 270, p. 403–406.

Coney, P. J., Jones, D. L., and Monger, J.W.H., 1980, Cordilleran suspect terranes: Nature, v. 288, p. 329–333.

Conway, C. M., and Silver, L. T., 1989, Early Proterozoic rocks of 1710–1610 Ma age in central to southwestern Arizona, *in* Jennings, J., and Reynolds, S., eds., Geological evolution of Arizona: Arizona Geological Society Digest 17, p. 165–186.

Cook, F. A., and 9 others, 1988, Lithoprobe seismic reflection structure of the southeastern Canadian Cordillera; Initial results: Tectonics, v. 7, p. 157–180.

Cordey, F., Mortimer, N., DeWever, P., and Monger, J.W.H., 1987, Significance of Jurassic radiolarians from the Cache Creek terrane, British Columbia: Geology, v. 15, p. 1151–1154.

Cowan, D. S., 1982, Geological evidence for post 40-m.y. B.P. large-scale northwestward displacement of part of southeastern Alaska: Geology, v. 10, p. 309–313.

——, 1985, Structural styles in Mesozoic and Cenozoic mélanges in the western Cordillera of North America: Geological Society of America Bulletin, v. 96, p. 451–462.

Cross, T. A., and Pilger, R. H., Jr, 1982, Controls on subduction geometry, location of magmatic arcs, and tectonics of arc and back-arc regions: Geological Society of America Bulletin, v. 93, p. 545–562.

Dallmeyer, R. D., Snoke, A. W., and McKee, E. H., 1986, The Mesozoic-Cenozoic tectothermal evolution of the Ruby Mountains, East Humboldt Range, Nevada; A Cordilleran metamorphic core complex: Tectonics, v. 5, p. 931–954.

Davis, G. A., 1968, Westward thrust faulting in the south-central Klamath Mountains, California: Geological Society of America Bulletin, v. 79, p. 911–934.

——, 1969, Tectonic correlations, Klamath Mountains and western Sierra Nevada: Geological Society of America Bulletin, v. 80, p. 1095–1108.

——, 1980, Problems of intraplate extensional tectonics, western United States, *in* Continental tectonics: National Academy of Science National Research Council, p. 84–95.

——, 1988, Rapid upward transport of mid-crustal mylonitic gneisses in the footwall of a Miocene detachment fault, Whipple Mountains, southeastern California: Geologische Rundschau, v. 77, p. 191–209.

Davis, G. A., and Lister, G. S., 1988, Detachment faulting in continental extension; Perspectives from the southwestern U.S. Cordillera, *in* Clark, S. P., Burchfiel, B. C., and Suppe, J., eds., Processes in continental lithospheric deformation: Geological Society of America Special Paper 218, p. 133–159.

Davis, G. A., Monger, J.W.H., and Burchfiel, B. C., 1978, Mesozoic construction of the Cordilleran "collage," central British Columbia to central California, *in* Howell, D. G., and McDougall, K. A., eds., Mesozoic paleogeography of the western United States: Los Angeles, California, Pacific Section Society of Economic Paleontologists and Mineralogists Pacific Coast Paleogeography Symposium 2, p. 1–32.

Davis, G. A., Anderson, J. C., Frost, E. G., and Shackelford, T. C., 1980, Mylonitization and detachment faulting in the Whipple-Buckskin–Rawhide Mountains terrane, southeastern California and western Arizona, *in* Crittenden, M. D., Jr., Coney, P. J., and Davis, G. H., eds., Cordilleran metamorphic core complexes: Geological Society of America Memoir 153, p. 79–129.

Davis, G. H., 1983, Shear-zone model for the origin of metamorphic core complexes: Geology, v. 11, p. 342–347.

Davis, G. H., and Coney, P. J., 1979, Geologic development of the Cordilleran metamorphic core complexes: Geology, v. 7, p. 120–124.

Dickinson, W. R., 1970, Relations of andesites, granites, and derivative sandstones to arc-trench tectonics: Reviews of Geophysics and Space Physics, v. 8, p. 813–860.

—— , 1976, Sedimentary basins developed during evolution of Mesozoic-Cenozoic arc-trench system in western North America: Canadian Journal of Earth Sciences, v. 13, p. 1268–1287.

—— , 1981, Plate tectonic evolution of the southern Cordillera, *in* Dickinson, W. R., and Payne, W. D., eds., Relation of tectonics to ore deposits in the southern Cordillera: Tucson, Arizona Geological Society Digest, v. 14, p. 113–135.

—— , 1989, Tectonic setting of Arizona through geologic time, *in* Jenney, J. P., and Reynolds, S. J., eds., Geological evolution of Arizona: Tucson, Arizona Geological Society Digest, v. 17, p. 1–16.

Dickinson, W. R., and 6 others, 1988, Paleogeographic and paleotectonic setting of Laramide sedimentary basins in the central Rocky Mountain region: Geological Society of America Bulletin, v. 100, p. 1023–1039.

Dilek, Y., 1989, Structure and tectonics of an early Mesozoic oceanic basement in the northern Sierra Nevada metamorphic belt, California; Evidence for transform faulting and ensialic arc evolution: Tectonics, v. 8, p. 999–1014.

Dokka, R. K., 1989, The Mojave extensional belt of southern California: Tectonics, v. 8, p. 363–390.

Dokka, R. K., Mahaffie, M. J., and Snoke, A. W., 1986, Thermochronologic evidence of major tectonic denudation associated with detachment faulting, northern Ruby Mountains–East Humboldt Range, Nevada: Tectonics, v. 5, p. 995–1006.

Duncan, R. A., 1982, A captured island chain in the Coast Range, Oregon and Washington: Journal of Geophysical Research, v. 87, p. 10,827–10,837.

Eaton, G. P., 1980, Geophysical and geological characteristics of the crust of the Basin and Range Province, *in* Continental tectonics: National Academy of Sciences National Research Council, p. 84–95.

—— , 1982, The Basin and Range Province; Origin and tectonic significance: Annual Reviews of Earth and Planetary Sciences, v. 8, p. 409–440.

Edelman, S. H., and Sharp, W. D., 1989, Terranes, early faults, and pre-Late Jurassic amalgamation of the western Sierra Nevada metamorphic belt, California: Geological Society of America Bulletin, v. 101, p. 1420–1433.

Edelman, S. H., Day, H. W., and Bickford, M. E., 1989a, Implications of U-Pb zircon ages for the tectonic settings of the Smartville and Slate Creek complexes, northern Sierra Nevada, California: Geology, v. 17, p. 1032–1035.

Edelman, S. H., and 5 others, 1989b, Structure across a Mesozoic ocean-continent suture zone in the northern Sierra Nevada, California: Geological Society of America Special Paper 224, 56 p.

Effimoff, I., and Pinezich, A. R., 1986, Tertiary structural development of selected basins in Basin and Range Province, northeastern Nevada, *in* Mayher, L., ed., Extensional tectonics of the southwestern United States; A perspective on processes and kinematics: Geological Society of America Special Paper 208, p. 31–42.

Elston, D. P., and McKee, E. H., 1982, Age and correlation of the Late Proterozoic Grand Canyon disturbance, northern Arizona: Geological Society of America Bulletin, v. 93, p. 681–699.

Engebretson, D. C., Cox, A., and Gordon, R. G., 1985, Relative motions between oceanic and continental plates in the Pacific Basin: Geological Society of America Special Paper 206, 59 p.

Ernst, W. G., 1970, Tectonic contact between the Franciscan mélange and the Great Valley sequence; Crustal expression of a late Mesozoic Benioff zone: Journal of Geophysical Research, v. 75, p. 886–901.

Evans, K. V., and Zartman, R. E., 1988, Early Paleozoic alkalic plutonism in east-central Idaho: Geological Society of America Bulletin, v. 100, p. 1981–1987.

Evenchick, C. A., Parrish, R. P., and Gabrielse, H., 1984, Precambrian gneiss and Late Proterozoic sedimentation in north-central British Columbia: Geology, v. 12, p. 233–237.

Ewing, T. E., 1980, Paleogene tectonic evolution of the Pacific Northwest: Journal of Geology, v. 88, p. 619–638.

Fagin, S. W., and Gose, A. W., 1983, Paleomagnetic data from the Redding section of the eastern Klamath belt, northern California: Geology, v. 11, p. 505–508.

Farmer, G. L., 1988, Isotope geochemistry of Mesozoic and Tertiary igneous rocks in the western United States and implications for the structure and composition of the deep continental lithosphere, *in* Ernst, W. G., ed., Metamorphism and crustal evolution of the western United States; Rubey Volume 7: Englewood Cliffs, New Jersey, Prentice-Hall, p. 87–109.

Fleck, R. J., and Criss, R. C., 1985, Strontium and oxygen isotope variations in Mesozoic and Tertiary plutons of central Idaho: Contributions of Mineralogy and Petrology, v. 90, p. 291–308.

Gabrielse, H., 1972, Younger Precambrian of the Canadian Cordillera: American Journal of Science, v. 272, p. 521–536.

—— , 1985, Major dextral transcurrent displacements along the Northern Rocky Mountain Trench and related lineaments in north-central British Columbia: Geological Society of America Bulletin, v. 96, p. 1–14.

Gabrielse, H., and Yorath, C. J., 1989, The Cordilleran orogen in Canada: Geoscience Canada, v. 16, p. 67–83.

Gabrielse, H., Snyder, W. S., and Stewart, J. H., 1983, Penrose Conference report; Sonoma orogeny and Permian to Triassic tectonism in western North America: Geology, v. 11, p. 484–486.

Gans, P. B., 1987, An open-system, two layer crustal stretching model for the eastern Great Basin: Tectonics, v. 6, p. 1–12.

Gans, P. B., Miller, E. L., McCarthy, J., and Ouldcott, M. L., 1985, Tertiary extensional faulting and evolving ductile-brittle transition zones in the northern Snake Ranges and vicinity; New insights from seismic data: Geology, v. 13, p. 189–193.

Gans, P. B., Mahood, G. A., and Schermer, E., 1989, Synextensional magmatism in the Basin and Range Province; A case study from the eastern Great Basin: Geological Society of America Special Paper 233, 53 p.

Garver, J. I., 1988, Fragment of the Coast Range ophiolite and the Great Valley sequence in the San Juan Islands, Washington: Geology, v. 16, p. 948–951.

Gehrels, G. E., and Smith, M. T., 1987, "Antler" allochthon in the Kootenay arc?: Geology, v. 15, p. 769–770.

Girty, G. H., and Wardlaw, M. S., 1985, Petrology and provenance of pre-Late Devonian sandstones, Shoo Fly Complex, Sierra Nevada, California: Geological Society of America Bulletin, v. 96, p. 516–521.

Goodge, J. W., 1989, Evolving early Mesozoic convergent margin deformation, central Klamath Mountains, northern California: Tectonics, v. 8, p. 845–864.

Gray, G. C., 1986, Native terranes of the central Klamath Mountains, California: Tectonics, v. 7, p. 1043–1054.

Gries, R., 1983, North-south compression of Rocky Mountain foreland structures, *in* Lowell, J. D., ed., Rocky Mountain foreland basins and uplifts: Rocky Mountain Association of Geologists Guidebook, p. 9–32.

Hamilton, W. B., 1969, Mesozoic California and the underflow of Pacific mantle: Geological Society of America Bulletin, v. 80, p. 2409–2430.

—— , 1978, Mesozoic tectonics of the western United States, *in* Howell, D. G., and McDougall, K. A., eds., Mesozoic paleogeography of the western United States: Pacific Section, Society of Economic Paleontologists and Mineralogists, p. 33–70.

—— , 1981, Plate-tectonic mechanism of Laramide deformation: University of Wyoming Contributions to Geology, v. 19, p. 87–92.

—— , 1987, Mesozoic geology and tectonics of the Big Maria Mountains region, southeastern California; Mesozoic rocks of southern Arizona and adjacent areas: Arizona Geologic Society Digest, v. 18, p. 33–48.

—— , 1988a, Plate tectonics and island arcs: Geological Society of America Bulletin, v. 100, p. 1594–1615.

—— , 1988b, Tectonic setting and variations with depth of some Cretaceous and Cenozoic structural and magmatic systems of the western United States, *in* Ernst, W. G., ed., Metamorphism and crustal evolution of the western United States; Rubey Volume 7: Englewood Cliffs, New Jersey, Prentice-

Hall, p. 1–40.
Hamilton, W., and Myers, W. B., 1966, Cenozoic tectonics of the western United States: Reviews of Geophysics, v. 4, p. 509–549.
Hammond, J. G., 1986, Geochemistry and petrogenesis of Proterozoic diabase intrusions in the southern Death Valley region of California: Contributions to Mineralogy and Petrology, v. 93, p. 312–321.
Hanson, R. E., Saleeby, J. B., and Schweickert, R. A., 1988, Composite Devonian island-arc batholith in the northern Sierra Nevada, California: Geological Society of America Bulletin, v. 100, p. 446–457.
Harbaugh, D. W., and Dickinson, W. R., 1981, Depositional facies of Mississippian clastics, Antler foreland basin, central Diamond Mountains, Nevada: Journal of Sedimentary Petrology, v. 51, p. 1223–1234.
Harper, G. D., and Link, P. K., 1986, Geochemistry of Upper Proterozoic rift-related volcanics, northern Utah and southeastern Idaho: Geology, v. 14, p. 864–867.
Harper, G. D., and Wright, J. E., 1984, Middle to Late Jurassic tectonic evolution of the Klamath Mountains, California-Oregon: Tectonics, v. 3, p. 759–772.
Harper, G. D., Saleeby, J. B., and Norman, E.A.S., 1985, Geometry and tectonic setting of seafloor spreading for the Josephine ophiolite, and implications for Jurassic accretionary events along the California margin, *in* Howell, D. G., ed., Tectonostratigraphic terranes of the Circum-Pacific region: Houston, Texas, Circum-Pacific Council for Energy and Mineral Resources, Earth Sciences Series 1, p. 239–258.
Harrison, J. E., 1972, Precambrian Belt basin of northwestern United States; Its geometry, sedimentation, and copper occurrences: Geological Society of America Bulletin, v. 83, p. 1215–1240.
Harrison, J. E., Groggs, A. B., and Wells, K. D., 1974, Tectonic features of the Precambrian Belt basin and their influence in post-Belt structures: U.S. Geological Survey Professional Paper 866, 15 p.
Harwood, D. S., 1983, Stratigraphy of upper Paleozoic volcanic rocks and regional unconformities in part of the northern Sierra terrane, California: Geological Society of America Bulletin, v. 94, p. 413–422.
——, 1988, Tectonism and metamorphism in the northern Sierra terrane, northern California, *in* Ernst, W. G., ed., Metamorphism and crustal evolution of the western United States; Rubey Volume 7: Englewood Cliffs, New Jersey, Prentice-Hall, p. 764–788.
Haxel, G., Tosdal, R. M., May, D. J., and Wright, J. E., 1984, Latest Cretaceous and Early Tertiary orogenesis in south-central Arizona; Thrust faulting, regional metamorphism, and granitic plutonism: Geological Society of America Bulletin, v. 95, p. 631–653.
Heller, P. L., and 7 others, 1986, Time of initial thrusting in the Sevier orogenic belt, Idaho-Wyoming and Utah: Geology, v. 14, p. 388–391.
Hodges, K. V., and 8 others, 1989, Evolution of extensional basins and Basin and Range topography west of Death Valley, California: Tectonics, v. 8, p. 453–468.
Hoffman, P. F., 1988, Belt basin; A landlocked remnant oceanic basin? (Analogous to the south Caspian and Black Seas): Geological Society of America Abstracts with Programs, v. 7, p. A50.
——, 1989, Precambrian geology and tectonic history of North America, *in* Bally, A. W., and Palmer, A. R., eds., The Geology of North America; An overview: Boulder, Colorado, Geological Society of America, The Geology of North America, v. A, p. 447–512.
Hoisch, T. D., Heizler, M. T., Harrison, T. M., Miller, C. F., and Stoddard, E. F., 1988, Late Cretaceous regional metamorphism in southeastern California, *in* Ernst, W. G., ed., Metamorphism and crustal evolution of the western United States; Rubey Volume 7: Englewood Cliffs, New Jersey, Prentice-Hall, p. 538–571.
Howard, K. A., and John, B. E., 1987, Crustal extension along a rooted system of low-angle faults; Colorado River extensional corridor, California and Arizona, *in* Coward, M., Dewey, J., and Hancock, P., eds., Continental extension tectonics: Geological Society of London Special Paper 28, p. 299–311.
Hyndman, D. W., Alt, D., and Sears, J. W., 1988, Post-Archean metamorphic and tectonic evolution of western Montana and northern Idaho, *in* Ernst, W. G., ed., Metamorphism and crustal evolution of the western United States; Rueby Volume 7: Englewood Cliffs, New Jersey, Prentice-Hall, p. 332–361.
Ingersoll, R. V., and Schweickert, R. A., 1986, A plate-tectonic model for Late Jurassic ophiolite genesis, Nevadan orogeny, and forearc initiation, northern California: Tectonics, v. 5, p. 901–912.
Irving, E., Woodsworth, G. J., Synne, P. J., and Morrison, A., 1985, Paleomagnetic evidence for displacement from the south of the Coast Plutonic Complex, British Columbia: Canadian Journal of Earth Sciences, v. 22, p. 584–598.
Jackson, J. A., and White, N. J., 1989, Normal faulting in the upper continental crust; Observations from regions of active extension: Journal of Structural Geology, v. 11, p. 15–36.
Jacobson, C. E., Dawson, M. R., and Postlethwaite, C. E., 1988, Structure, metamorphism, and tectonic significance of the Pelona, Orocopia, and Rand schists, southern California, *in* Ernst, W. G., ed., Metamorphism and crustal evolution of the western United States; Rubey Volume 7: Englewood Cliffs, New Jersey, Prentice-Hall, p. 976–997.
James, E. W., 1989, Southern extension of the Independence dike swarm of eastern California: Geology, v. 17, p. 587–590.
Jarrard, R. D., 1986, Terrane motion by strike-slip faulting of forearc slivers: Geology, v. 14, p. 780–783.
Jayko, A. S., and Blake, M. C., Jr., 1989, Deformation of the eastern Franciscan Belt, northern California: Journal of Structural Geology, v. 11, p. 375–390.
Jayko, A. S., Blake, M. C., Jr., and Harms, T., 1987, Attenuation of the Coast Range ophiolite by extensional faulting, and nature of the Coast Range "thrust," California: Tectonics, v. 6, p. 475–488.
John, B. E., 1988, Structural reconstruction and zonation of a tilted mid-crustal magma chamber; The felsic Chemehuevi Mountains plutonic suite: Geology, v. 16, p. 613–617.
Johnson, S. Y., 1985, Eocene strike-slip faulting and nonmarine basin formation in Washington, *in* Biddle, K. T., and Christie-Blick, N., eds., Strike-slip deformation, basin formation, and sedimentation: Society of Economic Paleontologists and Mineralogists Special Publication 37, p. 283–302.
Jones, D. L., Silberling, N. J., and Hillhouse, J., 1977, Wrangellia; A displaced terrane in northwestern North America: Canadian Journal of Earth Sciences, v. 14, p. 2565–2577.
Jordan, T. E., 1981, Thrust loads and foreland basin evolution, Cretaceous, western United States: American Association of Petroleum Geologists Bulletin, v. 65, p. 2509–2520.
Jordan, T. E., and Allmendinger, R. W., 1986, The Sierras Pampeanas of Argentina; A modern analogue of Laramide deformation: American Journal of Science, v. 286, p. 737–764.
Journeay, J. M., and Cosontos, L., 1989, Preliminary report on the structural setting along the southeast flank of the Coast Belt, British Columbia: Geological Survey of Canada Paper 89-1E, p. 172–189.
Karish, C. R., Miller, E. L., and Sutter, J. F., 1987, Mesozoic tectonic and magmatic history of the central Mojave Desert; Mesozoic rocks of southern Arizona and adjacent areas: Arizona Geological Society Digest, v. 18, p. 15–32.
Karlstrom, K. E., and Houston, R. S., 1984, The Cheyenne Belt; Analysis of Proterozoic suture in southern Wyoming: Precambrian Research, v. 25, p. 415–446.
Keith, S. B., and Wilt, J. C., 1986, Laramide orogeny in Arizona and adjacent regions; A stratotectonic synthesis, *in* Beatty, B., and Wilkinson, P.A.K., eds., Frontiers in geology and ore deposits of Arizona and the Southwest: Arizona Geological Society Digest, v. 16, p. 502–554.
Ketner, K. B., and Wardlaw, B. R., 1981, Permian and Triassic rocks near Quinn River Crossing, Humboldt County, Nevada: Geology, v. 9, p. 123–126.
Kluth, C. F., and Coney, P. J., 1981, Plate tectonics of the Ancestral Rocky Mountains: Geology, v. 9, p. 10–15.
Lahren, M. M., and Schweickert, R. A., 1989, Proterozoic and Lower Cambrian miogeoclinal rocks of Snow Lake pendant, Yosemite-Emigrant Wilderness, Sierra Nevada, California; Evidence for major Early Cretaceous dextral

translation: Geology, v. 17, p. 156–160.

Lapierre, H., Albarede, F., Albers, J., Coulon, C., and Cabanis, B., 1985, Early Devonian volcanism in the eastern Klamath Mountains, California; Evidence for an immature island arc: Canadian Journal of Earth Sciences, v. 22, p. 214–227.

Lipman, P. W., Christiansen, R. L., and Prostka, H. J., 1971, Evolving subduction zones in the western United States, as interpreted from igneous rocks: Science, v. 148, p. 821–825.

Lister, G. S., and Davis, G. A., 1989, The origin of metamorphic core complexes and detachment faults formed during Tertiary continental extension in the northern Colorado River region, U.S.A.: Journal of Structural Geology, v. 11, p. 65–94.

Lund, S. P., and Bottjer, D. J., 1991, Paleomagnetic evidence for microplate tectonic development of southern and Baja California: American Association of Petroleum Geologists Memoir 47 (in press).

Lund, S. P., and Snee, L. W., 1988, Metamorphism, structural development, and age of the continent; Island arc juncture in west-central Idaho, *in* Ernst, W. G., ed., Metamorphism and crustal evolution of the western United States; Rubey Volume 7: Englewood Cliffs, New Jersey, Prentice-Hall, p. 396–331.

Luyendyk, P. B., Kamerling, M. J., Terres, R. R., and Hornafius, J. S., 1985, 1985, Simple shear of southern California during Neogene time suggested by paleomagnetic declinations: Journal of Geophysical Research, v. 90, p. 12454–12466.

Mankinen, E. A., Irwin, W. P., and Gromme, C. S., 1989, Paleomagnetic study of the Eastern Klamath terrane, California, and implications for the tectonic history of the Klamath Mountains province: Journal of Geophysical Research, v. 94, p. 10444–10472.

Mattinson, J. M., 1988, Constraints on the timing of Franciscan metamorphism; Geochronological approaches and their limitations, *in* Ernst, W. G., ed., Metamorphism and crustal evolution of the western United States; Rubey Volume 7: Englewood Cliffs, New Jersey, Prentice-Hall, p. 1023–1034.

May, S. R., and Butler, R. F., 1986, North American Jurassic apparent polar wander; Implications for plate motion, paleogeography, and Cordilleran tectonics: Journal of Geophysical Research, v. 91, p. 11519–11544.

May, S. R., Beck, M. E., Jr., and Butler, R. F., 1989, North American apparent polar wander, plate motion, and left-oblique convergence; Late Jurassic–Early Cretaceous orogenic consequences: Tectonics, v. 8, p. 443–452.

McGroder, M. F., 1989, Structural geometry and kinematic evolution of the eastern Cascades fold belt, Washington and British Columbia: Canadian Journal of Earth Sciences, v. 26, p. 1585–1602.

McLaughlin, R. J., Blake, M. C., Jr., and Griscom, A., 1988, Tectonics of formation, translation, and dispersal of the Coast Range ophiolite of California: Tectonics, v. 7, p. 1033–1056.

McMechan, M. E., and Price, R. A., 1982, Superimposed low-grade metamorphism in the Mount Fisher area, southeastern British Columbia; Implications for the East Kootenay orogeny: Canadian Journal of Earth Sciences, v. 129, p. 476–489.

Miller, E. L., and Gans, P. B., 1989, Cretaceous crustal structure and metamorphism in the hinterland of the Sevier thrust belt, western U.S. Cordillera: Geology, v. 17, p. 59–62.

Miller, E. L., Gans, P. B., and Garing, J., 1983, The Snake Range décollement; An exhumed mid-Tertiary ductile-brittle transition: Tectonics, v. 2, p. 239–263.

Miller, E. L., Holdsworth, B. K., Whiteford, W. B., and Rogers, D., 1984, Stratigraphy and structure of the Schoonover sequence, northern Nevada; Implications for Paleozoic plate margin tectonics: Geological Society of America Bulletin, v. 95, p. 1063–1076.

Miller, J.M.G., 1985, Glacial and syntectonic sedimentation; The Upper Proterozoic Kingston Peak Formation, southern Panamint Range, eastern California: Geological Society of America Bulletin, v. 96, p. 1537–1553.

Miller, J.M.G., and John, B. E., 1988, Detached strata in a Tertiary low-angle normal fault terrane, southeastern California; A sedimentary record of unroofing, breaching, and continued slip: Geology, v. 16, p. 645–648.

Miller, M. M., 1987, Dispersed remnants of a northeast Pacific fringing arc; Upper Paleozoic terranes of Permian McCloud faunal affinity, western U.S.: Tectonics, v. 6, p. 807–830.

Miller, M. M., and Harwood, D. S., 1989, Penrose Conference Report; Paleozoic and early Mesozoic paleogeographic relations between the Klamath Mountains, northern Sierra Nevada, and western North America: Geology, v. 17, p. 369–372.

Miller, M. M., and Wright, J. E., 1987, Paleogeographic implications of Permian Tethyan corals from the Klamath Mountains, California: Geology, v. 15, p. 266–269.

Miller, R. B., Bowring, S. A., and Hoppe, W. J., 1989, Paleocene plutonism and its tectonic implications, North Cascades, Washington: Geology, v. 17, p. 846–849.

Misch, P., 1988, Tectonic and metamorphic evolution of the North Cascades; An overview, *in* Ernst, W. G., ed., Metamorphism and crustal evolution of the western United States; Rubey Volume 7: Englewood Cliffs, New Jersey, Prentice-Hall, p. 179–195.

Monger, J.W.H., 1975, Correlation of eugeosynclinal tectonostratigraphic belts in the North American Cordillera: Geoscience Canada, v. 2, p. 4–9.

——, 1977, Upper Paleozoic rocks of the western Canadian Cordillera and their bearing on Cordilleran evolution: Canadian Journal of Earth Sciences, v. 14, p. 1832–1859.

Monger, J.W.H., Price, R. A., and Templeman-Kluit, D. J., 1982, Tectonic accretion and the origin of the two major metamorphic and plutonic welts in the Canadian Cordillera: Geology, v. 10, p. 70–75.

Moores, E. M., and Day, H. W., 1984, Overthrust model for the Sierra Nevada: Geology, v. 12, p. 416–419.

Mortimer, N., 1985, Structural and metamorphic aspects of Middle Jurassic terrane juxtaposition, northeastern Klamath Mountains, California, *in* Howell, D. G., ed., Tectonostratigraphic terranes of the Circum-Pacific region: Houston, Texas, Circum-Pacific Council of Energy and Mineral Resources, Earth Sciences Series 1, p. 201–214.

——, 1986, Late Triassic, arc-related, potassic igneous rocks in the North American Cordillera: Geology, v. 14, p. 1035–1038.

Newton, C. R., 1988, Significance of "Tethyan" fossils in the American Cordillera: Science, v. 242, p. 385–391.

Nichols, K. M., and Silberling, N. J., 1977, Stratigraphy and depositional history of the Star Peak Group (Triassic), northwestern Nevada: Geological Society of America Special Paper 178, 73 p.

Nur, A., and Ben-Avrahim, A., 1982, Oceanic plateaus, the fragmentation of continents, and mountain building: Journal of Geophysical Research, v. 87, p. 3644–3661.

Okaya, D. A., and Thompson, G. A., 1985, Geometry of Cenozoic extensional faulting; Dixie Valley, Nevada: Tectonics, v. 4, p. 107–125.

Okulitch, A. V., 1985, Paleozoic plutonism in southeastern British Columbia: Canadian Journal of Earth Sciences, v. 22, p. 1409–1424.

Oldow, J. S., 1983, Tectonic implications of a late Mesozoic fold and thrust belt in northwestern Nevada: Geology, v. 11, p. 542–546.

——, 1984a, Evolution of a late Mesozoic back-arc fold and thrust belt, northwestern Great Basin, USA: Tectonophysics, v. 102, p. 245–274.

——, 1984b, Spatial variability in the structure of the Roberts Mountains allochthon, western Nevada: Geological Society of America Bulletin, v. 95, p. 174–185.

Oldow, J. S., and Bartel, R. L., 1987, Early to Middle(?) Jurassic extensional tectonism in the western Great Basin; Growth faulting and synorogenic deposition of the Dunlop Formation: Geology, v. 15, p. 740–743.

Oldow, J. S., Avé Lallemant, H. G., and Schmidt, W. J., 1984, Kinematics of plate convergence deduced from Mesozoic structures in the western Cordillera: Tectonics, v. 3, p. 201–227.

Oldow, J. S., Bally, A. W., Avé Lallemant, H. G., and Leeman, W. P., 1989, Phanerozoic evolution of the North American Cordillera; United States and Canada, *in* Bally, A. W., and Palmer, A. R., eds., The Geology of North America; An overview: Boulder, Colorado, Geological Society of America, The Geology of North America, v. A, p. 139–232.

Page, B. M., and Engebretson, D. C., 1984, Correlation between the geologic record and computed plate motions for central California: Tectonics, v. 3, p. 133–155.

Parrish, R. R., Carr, S. D., and Parkinson, D. L., 1988, Eocene extensional tectonics and geochronology of the southern Omineca belt, British Columbia and Washington: Tectonics, v. 7, p. 181–213.

Peacock, S. M., 1987, Serpentinization and infiltration metasomatism in the Trinity Peridotite, Klamath province, northern California; Implications for subduction zones: Contributions to Mineralogy and Petrology, v. 95, p. 55–77.

Perry, W. J., Jr., Wardlaw, B. R., Bostick, N. H., and Maughan, E. K., 1983, Structure, burial history, and petroleum potential of frontal thrust belt and adjacent foreland, southwest Montana: American Association of Petroleum Geologists Bulletin, v. 67, p. 725–743.

Poole, F.G., and Sandberg, C. A., 1977, Mississippian paleogeography and tectonics of the western United States, *in* Stewart, J. H., Stevens, C. H., and Fritsche, A. E., eds., Paleozoic paleogeography of the western United States: Los Angeles, California, Pacific Section, Society of Economic Paleontologists and Mineralogists Pacific Coast Paleogeography Symposium 1, p. 67–85.

Price, R. A., and Carmichael, D. M., 1986, Geometric test for Late Cretaceous–Paleogene intracontinental transform faulting in the Canadian Cordillera: Geology, v. 14, p. 468–471.

Price, R. A., and Mountjoy, E. W., 1970, Geological structure of the Canadian Rocky Mountains between Bow and Athabasca Rivers; Progress report: Geological Association of Canada Special Paper 6, p. 7–25.

Read, P. B., and Okulitch, A. V., 1977, The Triassic unconformity of south-central British Columbia: Canadian Journal of Earth Sciences, v. 14, p. 606–638.

Reynolds, S. J., and Spencer, J. E., 1985, Evidence for large-scale transport on the Bullard detachment fault, west-central Arizona: Geology, v. 13, p. 353–356.

Reynolds, S. J., Richard, S. M., Haxel, G. B., Tosdal, R. M., and Laubach, S. E., 1988, Geologic setting of Mesozoic and Cenozoic metamorphism in Arizona, *in* Ernst, W. G., ed., Metamorphism and crustal evolution of the western United States; Rubey Volume 7: Englewood Cliffs, New Jersey, Prentice-Hall, p. 466–501.

Riggs, N. R., and Haxel, G. B., 1990, The Early to Middle Jurassic magmatic arc in southern Arizona; Plutons to sand dunes, *in* Gehrels, G. E., and Spencer, J. E., eds., Geologic excursions through the Sonoran Desert region; Arizona and Sonora: Arizona Geological Survey Special Paper 7, p. 90–103.

Roberts, R. J., Hotz, P. E., Gilluly, J., and Ferguson, H. G., 1958, Paleozoic rocks of north-central Nevada: American Association of Petroleum Geologists Bulletin, v. 42, no. 12, p. 2813–1857.

Ross, D. C., 1978, Salinia in the California Coast Ranges, *in* Howell, D. G., and McDougall, K. A., eds., Mesozoic paleogeography of the western United States: Los Angeles, California, Pacific Section, Society of Economic Paleontologists and Mineralogists Pacific Coast Paleogeography Symposium 2, p. 509–522.

Roure, O., Lapierre, H., Coulon, C., and Michard, A., 1989, New petrological and geochemical data on mid-Paleozoic island-arc volcanics of northern Sierra Nevada, California; Evidence for a continent-based island arc: Canadian Journal of Earth Sciences, v. 26, p. 2465–1478.

Royden, L. H., and Burchfiel, B. C., 1987, Thin-skinned north-south extension within the convergent Himalayan region; Gravitational collapse of a Miocene topographic front: Geological Society of London Special Publication 28, p. 611–619.

——, 1989, Are systematic variations in thrust belt styles related to plate boundary processes?; The western Alps versus the Carpathians: Tectonics, v. 8, p. 51–61.

Royse, F., Warner, M. A., and Reese, D. L., 1975, Thrust belt structural geometry and related stratigraphic problems, Wyoming-Idaho-northern Utah, *in* Bolyard, D. W., ed., Deep drilling frontiers in the central Rocky Mountains; Rocky Mountain Association of Geologists, p. 41–54.

Ruppel, E. T., 1986, The Lemhi Arch; A Late Proterozoic and early Paleozoic landmass in central Idaho, *in* Peterson, J. A., ed., Paleotectonics and sedimentation in the Rocky Mountain region, United States: American Association of Petroleum Geologists Memoir 41, p. 119–130.

Ruppell, E. T., and Lopez, D. A., 1984, The thrust belt in southwestern Montana and east-central Idaho: U.S. Geological Survey Professional Paper 1278, 41 p.

Rusmore, M. E., and Cowan, D. S., 1985, Jurassic-Cretaceous rock units along the southern edge of the Wrangellia terrane on Vancouver Island: Canadian Journal of Earth Sciences, v. 22, p. 1223–1232.

Saleeby, J., 1981, Ocean floor accretion and volcano-plutonic arc evolution of the Mesozoic Sierra Nevada, California, *in* Ernst, W. G., ed., Geotectonic development of California; Rubey Volume 1: Englewood Cliffs, New Jersey, Prentice-Hall, p. 132–181.

——, 1982, Polygenetic ophiolite belt of the California Sierra Nevada; Geochronological and tectonostratigraphic development: Journal of Geophysical Research, v. 87, p. 1803–1824.

——, 1983, Accretionary tectonics of the North American Cordillera: Annual Review of Earth and Planetary Science, v. 15, p. 45–73.

Saleeby, J., Hannah, J. L., and Varga, R. J., 1987, Isotopic constraints on middle Paleozoic deformation in the northern Sierra Nevada, California: Geology, v. 15, p. 757–760.

Saller, A. H., and Dickinson, W. R., 1982, Alluvial to marine facies transition in the Antler overlap sequence, Pennsylvanian and Permian of north-central Nevada: Journal of Sedimentary Petrology, v. 52, p. 925–940.

Schmidt, C. J., and Garihan, J. M., 1983, Laramide tectonic development of the Rocky Mountain foreland of southwestern Montana, *in* Lowell, J. D., ed., Rocky Mountain foreland basins and uplifts: Rocky Mountain Association of Geologists, p. 271–294.

Schmidt, C. J., and Perry, W. J., eds., 1988, Interaction of the Rocky Mountain foreland and the Cordilleran thrust belt: Geological Society of America Memoir 171, 582 p.

Schweickert, R. A., and Cowan, D. S., 1975, Early Mesozoic tectonic evolution of the western Sierra Nevada, California: Geological Society of America Bulletin, v. 86, p. 1329–1336.

Schweickert, R. A., and Irwin, W. P., 1989, Extensional faulting in southern Klamath Mountains, California: Tectonics, v. 8, p. 135–150.

Schweickert, R. A., and Lahren, M. M., 1987, Continuation of the Antler and Sonoma orogenic belts to the eastern Sierra Nevada, California, and Late Triassic thrusting in a compressional arc: Geology, v. 15, p. 270–273.

Schweickert, R. A., Bogen, N. L., and Merguerian, C., 1988, Deformational and metamorphic history of Paleozoic and Mesozoic basement terranes in the western Sierra Nevada metamorphic belt, *in* Ernst, W. G., ed., Metamorphism and crustal evolution of the western United States; Rubey Volume 7: Englewood Cliffs, New Jersey, Prentice-Hall, p. 789–822.

Sears, J. W., and Price, R. A., 1978, The Siberian connection; A case for Precambrian separation of the North American and Siberian cratons: Geology, v. 6, p. 267–270.

Sears, J. W., Graff, P. J., and Holden, G. S., 1982, Tectonic evolution of Lower Proterozoic rocks, Uinta Mountains, Utah and Colorado: Geological Society of America Bulletin, v. 93, p. 990–997.

Sevigney, J. H., Parrish, R. R., Donelick, R. A., and Ghent, E. D., 1990, Northern Monashee Mountains, Omineca crystalline belt, British Columbia; Timing of metamorphism, anatexis, and tectonic denudation: Geology, v. 18, p. 97–192.

Silberling, N. J., 1973, Geologic events during Permian-Triassic time along the Pacific margin of the United States, *in* Logan, A., and Hill, L. V., eds., The Permian and Triassic systems and their mutual boundary: Canadian Society of Petroleum Geologists Memoir 2, p. 345–362.

Silver, L. T., and Anderson, T. H., 1974, Possible left-lateral early to middle Mesozoic disruption of the southwestern North American craton margin: Geological Society of America Abstracts with Programs, v. 6, p. 955.

Skipp, B., 1987, Basement thrust sheets in the Clearwater orogenic zone, central Idaho and western Montana: Geology, v. 15, p. 220–224.

Snavely, P. D., Jr., 1987, Tertiary geologic framework, neotectonics, and petroleum potential of the Oregon-Washington continental margin, *in* Scholl,

D. W., Grantz, A., and Veddera, J. G., eds., Geology and resource potential of the continental margin of western North America and adjacent ocean basins; Beaufort Sea to Baja California: Houston, Texas, Circum-Pacific Council for Energy and Mineral Resources, Earth Science Series 6, p. 305–335.

Snoke, A. W., and Miller, D. M., 1988, Metamorphic and tectonic history of the northeastern Great Basin, *in* Ernst, W. G., ed., Metamorphic and tectonic history of the western United States; Rubey Volume 7: Englewood Cliffs, New Jersey, Prentice-Hall, p. 606–648.

Speed, R. C., 1977, Island-arc and other paleogeographic terranes of late Paleozoic age in the western Great Basin, *in* Stewart, J. H., Stevens, C. H., and Fritsche, A. E., eds., Paleozoic paleogeography of the western United States: Los Angeles, California, Pacific Section, Society of Economic Paleontologists and Mineralogists Pacific Coast Paleogeography Symposium 1, p. 349–362.

—— , 1978, Paleogeographic and plate tectonic evolution of the early Mesozoic marine province of the western Great Basin, *in* Howell, D. G., and McDougall, K. A., eds., Mesozoic paleogeography of the western United States: Los Angeles, California, Pacific Section, Society of Economic Paleontologists and Mineralogists Pacific Coast Paleogeography Symposium 2, p. 253–270.

—— , 1979, Collided Paleozoic microplate in the western United States: Journal of Geology, v. 87, p. 279–292.

—— , 1983, Evolution of the sialic margin in the central-western United States, *in* Watkins, J. S., and Drake, C. L., eds., Studies in continental margin geology: American Association of Petroleum Geologists Memoir 34, p. 457–468.

Speed, R. C., and Sleep, N. H., 1982, Antler orogeny and foreland basin; A model: Geological Society of America Bulletin, v. 93, p. 815–828.

Speed, R. C., Elison, M. W., and Heck, F. R., 1988, Phanerozoic tectonic evolution of the Great Basin, *in* Ernst, W. G., ed., Metamorphism and crustal evolution of the western United States; Rubey Volume 7: Englewood Cliffs, New Jersey, Prentice-Hall, p. 572–605.

Spencer, J. E., 1984, Role of tectonic denudation in uplift and warping of low-angle normal faults: Geology, v. 12, p. 95–98.

—— , 1985, Miocene low-angle normal faulting and dike emplacement at Homer Mountain and surrounding areas, southeastern California and southernmost Nevada: Geological Society of America Bulletin, v. 96, p. 1140–1155.

Stewart, J. H., 1972, Initial deposits in the Cordilleran geosyncline; Evidence of a late Precambrian (<850 m.y.) continental separation: Geological Society of America Bulletin, v. 83, p. 1345–1360.

—— , 1985, East-trending dextral faults in the western Great Basin; An explanation for anomalous trends of pre-Cenozoic strata and Cenozoic faults: Tectonics, v. 4, p. 547–564.

Stewart, J. H., and Suczek, C. A., 1977, Cambrian and latest Precambrian paleogeography and tectonics in the western United States, *in* Stewart, J. H., Stevens, C. H., and Fritsche, A. E., eds., Paleozoic paleogeography of the western United States: Los Angeles, California, Pacific Section, Society of Economic Paleontologists and Mineralogists Pacific Coast Paleogeography Symposium 1, p. 1–17.

Stewart, J. H., and 5 others, 1990, Tectonics and stratigraphy of the Paleozoic and Triassic southern margin of North America, Sonora, Mexico, *in* Gehrels, G. E., and Spencer, J. E., eds., Geological excursions through the Sonoran Desert region, Arizona and Sonora: Arizona Geological Survey Special Paper 7, p. 183–202.

Stock, J., and Molnar, P., 1988, Uncertainties and implications of the Late Cretaceous and Tertiary position of North America relative to the Farallon, Kula, and Pacific Plates: Tectonics, v. 7, no. 6, p. 1339–1384.

Stone, P., and Stevens, C. H., 1988, Pennsylvanian and Early Permian paleogeography of east-central California; Implications for the shape of the continental margin and the timing of continental truncation: Geology, v. 16, p. 330–333.

Stone, P., Page, V. M., Hamilton, W., and Howard, K. A., 1987, Cretaceous age of the McCoy Mountains Formation, southeastern California and southwestern Arizona, and its tectonic significance; Reconciliation of paleobotanical and paleomagnetic evidence: Geology, v. 15, p. 561–564.

Tabor, R. W., Zartman, R. E., and Frizzell, V. A., Jr., 1987, Possible tectonostratigraphic terranes in the North Cascades crystalline core, Washington, *in* Schuster, E. J., ed., Selected papers on the geology of Washington: Washington Division of Geology and Earth Resources Bulletin 77, p. 107–128.

Tarduno, J. A., McWilliams, M., and others, 1986, Southern Hemisphere origin of the Cretaceous Laytonville Limestone of California: Science, v. 231, p. 1425–1428.

Tobisch, O. T., Paterson, S. R., Longiaru, S., and Battacharyya, T., 1987, Extent of the Nevadan orogeny, central Sierra Nevada, California: Geology, v. 15, p. 132–135.

Tobisch, O. T., Paterson, S. R., Saleeby, J. B., and Geary, E. E., 1989, Nature and timing of deformation in the Foothills terrane, central Sierra Nevada, California; Its bearing on orogenesis: Geological Society of America Bulletin, v. 101, p. 401–413.

Tosdal, R. M., 1986, Mesozoic ductile deformations in the southern Dome Rock Mountains, northern Trigo Mountains, Trigo Peaks and Livingston Hills, southwestern Arizona, and Mule Mountains, southeastern California, *in* Beatty, B., and Wilkinson, P.A.K., eds., Frontiers in geology and ore deposits of Arizona and the Southwest: Arizona Geological Society Digest, v. 16, p. 62–71.

Tosdal, R. M., Haxel, G. B., and Wright, J. E., 1989, Jurassic geology of the Sonoran Desert region, southern Arizona, southeastern California, and northernmost Sonora; Construction of a continental-margin magmatic arc, *in* Jenney, J. P., and Reynolds, S. J., eds., Geological Evolution of Arizona: Arizona Geological Society Digest, v. 17, p. 397–434.

Turner, R.J.W., Madrid, R. J., and Miller, E. L., 1989, Roberts Mountains allochthon; Stratigraphic comparison with lower Paleozoic outer continental margin strata of the northern Canadian Cordillera: Geology, v. 17, p. 341–344.

Umhoefer, P. J., 1987, Northward translation of "Baja British Columbia" along the Late Cretaceous to Paleocene margin of western North America: Tectonics, v. 6, p. 377–394.

Walker, J. D., 1988, Permian and Triassic rocks of the Mojave Desert and their implications for the timing and mechanisms of continental truncation: Tectonics, v. 7, p. 685–709.

Walker, J. D., Klepacki, D. W., and Burchfiel, B. C., 1986, Late Precambrian tectonism in the Kingston Range, southern California: Geology, v. 14, p. 15–16.

Wallin, E. T., Matinson, J. M., and Potter, A. W., 1988, Early Paleozoic magmatic events in the eastern Klamath Mountains, northern California: Geology, v. 16, p. 144–148.

Watkins, R., 1985, Volcaniclastic and carbonate sedimentation in late Paleozoic island-arc deposits, eastern Klamath Mountains, California: Geology, v. 13, p. 709–713.

Watkins, R., and Brown, Q. J., 1989, An Ordovician continental-margin sequence of turbidite and seamount deposits in the Roberts Mountains allochthon, Independence Range, Nevada: Geological Society of America Bulletin, v. 101, p. 731–741.

Watkins, R., Reinheimer, C. E., Wallace, J. W., and Anestell, M. K., 1987, Paleogeographic significance of a Permian sedimentary megamictite in the central belt of the northern Sierra Nevada: Geological Society of America Bulletin, v. 99, p. 771–778.

Wells, R. E., and Heller, P. L., 1988, The relative contribution of accretion, shear, and extension to Cenozoic tectonic rotation in the Pacific Northwest: Geological Society of America Bulletin, v. 100, p. 325–338.

Wells, R. E., Engebretson, D. C., Snavely, P. D., Jr., and Coe, R. S., 1984, Cenozoic plate motions and the volcano-tectonic evolution of western Oregon and Washington: Tectonics, v. 3, p. 275–294.

Wernicke, B., 1981, Low-angle normal faults in the Basin and Range Province; Nappe tectonics in an extending orogen: Nature, v. 291, p. 645–648.

Wernicke, B., and Axen, G. J., 1988, On the role of isostasy in the evolution of normal fault systems: Geology, v. 16, p. 848–851.

Wernicke, B., and Klepacki, D. W., 1988, Escape hypothesis for the Stikine

block: Geology, v. 16, p. 461–464.
Wernicke, B., Christiansen, R. L., England, P. C., and Sonder, L. J., 1987, Tectonomagmatic evolution of Cenozoic extension in the North American Cordillera, *in* Coward, M. P., Dewey, J. F., and Hancock, P. L., eds., Continental extensional tectonics: Geological Society of London Special Publication 28, p. 203–221.
Whitney, D. L., and McGroder, M. F., 1989, Cretaceous crustal section through the proposed Insular-Intermontane suture, North Cascades, Washington: Geology, v. 17, p. 555–558.
Wilson, D., and Cox, A., 1980, Paleomagnetic evidence for tectonic rotation of Jurassic plutons in Blue Mountains, eastern Oregon: Journal of Geophysical Research, v. 85, p. 3681–2689.
Winston, D., Woods, M. O., and Byer, G. B., 1984, The case for an intracratonic Belt-Purcell basin, *in* McBane, J. D., and Garrison, P. B., eds., Northwest Montana and adjacent Canada: Billings, Montana, Montana Geological Society 1984 Field Conference and Symposium Guidebook, p. 103–118.
Wright, J. E., 1982, Permo-Triassic accretionary subduction complex, southwestern Klamath Mountains, northern California: Journal of Geophysical Research, v. 87, no. B5, p. 3805–3818.
Wright, J. E., and Fahan, M. R., 1988, An expanded view of Jurassic orogenesis in the western United States Cordillera; Middle Jurassic (pre-Nevadan) regional metamorphism and thrust faulting within an active arc environment, Klamath Mountains, California: Geological Society of America Bulletin, v. 100, p. 859–876.
Wright, L. A., and Troxel, B. W., 1966, Strata of late Precambrian–Cambrian age, Death Valley region, California-Nevada: American Association of Petroleum Geologists Bulletin, v. 50, p. 846–857.
Wright, L. A., Troxel, B. W., Williams, E. G., Roberts, M. T., and Diehl, P. E., 1974, Precambrian sedimentary environments of the Death Valley region, eastern California, *in* Geological Society of America Cordilleran Section Meeting Guidebook, Field Trip 1: Shoshone, California, Death Valley Publishing Co., p. 27–35.
Wust, S. L., 1986, Regional correlation of extension directions in Cordilleran metamorphic core complexes: Geology, v. 14, p. 828–830.
Yeats, R. S., 1968, Southern California structure, seafloor spreading, and history of the Pacific Basin: Geological Society of America Bulletin, v. 79, p. 1693–1702.
Young, G. M., 1984, Proterozoic plate tectonics in Canada with emphasis on evidence for a Late Proterozoic rifting event: Precambrian Research, v. 25, p. 233–256.
Zoback, M. L., Anderson, R. E., and Thompson, G. A., 1981, Cainozoic evolution of the state of stress and style and tectonism of the Basin and Range Province of the western United States: Royal Society of London Philosophical Transactions, series A, v. 300, p. 407–434.

MANUSCRIPT ACCEPTED BY THE SOCIETY OCTOBER 29, 1990

ACKNOWLEDGMENTS

We thank J.W.H. Monger and John S. Oldow for thorough and constructive reviews of an early draft of this manuscript. Appreciation is extended to A. R. Palmer, Geological Society of America Centennial Science Program Coordinator, for his patience and persistence in overseeing the preparation of this chapter. The writers' interests in Cordilleran tectonics have long been supported by research grants from the Earth Sciences Division of the National Science Foundation, to which we express our thanks.

Printed in U.S.A.

The Geology of North America
Vol. G-3, The Cordilleran Orogen: Conterminous U.S.
The Geological Society of America, 1992

Chapter 9

Magmatism in the Cordilleran United States; Progress and problems

Peter W. Lipman
U.S. Geological Survey, MS 910, 345 Middlefield Road, Menlo Park, California 94025

INTRODUCTION

The widespread igneous rocks in the western United States have now been studied for more than a century; they provide primary geologic constraints on many types of ore deposits, key information on regional geologic history, and the basis for development of broad concepts applicable to global magmatism. Studies have been especially intense in the past few decades, fueled by societal needs such as volcanic hazard evaluations and the search for geothermal and mineral resources. Also important have been an explosion of new techniques in geochronology and isotope geochemistry, increased availability and precision of mineral and rock chemical analyses, and detailed field structural studies of batholiths and stratigraphic studies of volcanic fields. Improved understanding of eruptive mechanisms has resulted from combined studies of historic volcanism and prehistoric deposits, and the rich interactions between studies of igneous activity and innovative plate tectonic concepts have led to new ideas on the origins of continental-margin magmatism.

Earlier chapters in this volume have emphasized geologic history and areal comparisons for key time intervals across the Cordilleran region. Here, the focus is on process and product: styles of volcanism and plutonism, magma genesis, differentiation mechanisms, relation to tectonic setting. Such topics are hardly limited to the Cordilleran United States, but the attempt here is to summarize the bearing of recent work in this region on critical problems of igneous geology. Emphasis is on Phanerozoic rocks of the U.S. Cordillera, as Precambrian rocks are the topic of a separate volume. In addition, selective hemispheric comparisons with igneous rocks along the broadly plate-convergent continental margins to the north (Canadian Cordillera) and south (Mexico to the Andes) provide additional insights into the relevant general magmatic processes.

Though research during the past 25 years is emphasized, such studies have built on late 19th- and early 20th-century work in the western U.S. that has contributed importantly to modern concepts of igneous processes. Examples include the recognition of laccolithic structures in the Henry Mountains of Utah by *Gilbert (1877;* see also *Jackson and Pollard, 1988*); documentation of complex geometric relations between batholithic plutons and their wall rocks in the Sierra Nevada that began with the folio maps by W. Lindgren and H. W. Turner late in the 19th century and was continued by *Calkins (1930), Cloos (1936),* and *Mayo (1941)*; early recognition of the association between porphyritic intrusions and disseminated Cu-Mo ore bodies (e.g., *Ransome, 1919*); important progress in the understanding of pyroclastic deposits in the Yellowstone Park area by *Iddings* (*1899;* also see *Boyd, 1961*); and concepts of caldera formation developed by *Diller and Patton (1902)* and Williams (1942) at Crater Lake, Oregon.

Also notable was early recognition that silicic volcanic units now known to be welded ash-flow sheets spread regionally as thin tabular stratigraphic marker units in the San Juan Mountains of Colorado (*Cross and Larsen, 1935,* and earlier maps and reports), despite arguments from experimental petrologists that such silicic magmas must be too viscous to flow widely. Petrographic and mineral chemistry studies of volcanic rocks of the San Juan field by *Larsen and others (1938)* documented the importance of magma mixing long before such processes were widely recognized. The significance of mixing between rhyolitic and basaltic magmas in the Yellowstone area was also demonstrated by Wilcox (1944).

A notable step toward the understanding of relations between regional tectonism and magmatic activity was Lindgren's (1915) recognition that igneous activity migrated and became more alkalic eastward across the U.S. Cordillera in early Cenozoic time, a progression that many now consider related to increased rates of plate convergence and flattening of the associated subducted slab beneath the Cordilleran region (Lipman and others, 1971; Coney and Reynolds, 1977). Similarly, *Lindgren and others (1910)* and also *Callahan (1951)* recognized early that the Basin and Range Province of the western U.S. was regionally characterized by Tertiary volcanic sequences in which voluminous rhyolites were widely underlain by andesitic rocks

Lipman, P. W., 1992, Magmatism in the Cordilleran United States; Progress and problems, *in* Burchfiel, B. C., Lipman, P. W., and Zoback, M. L., eds., The Cordilleran Orogen: Conterminous U.S.: Boulder, Colorado, Geological Society of America, The Geology of North America, v. G-3.

Figure 1. Mount St. Helens in eruption, May 18, 1980. Aerial view looking northeast shows cauliflower-like structure of Plinian eruption column rising from the central crater (*Christiansen and Peterson, 1981*). Photography by J. G. Rosenbaum, U.S. Geological Survey.

and overlain by basalt flows. This three-fold succession is now interpreted by many as recording the sequential replacement of (1) early to mid-Tertiary andesitic arc volcanism, by (2) more silicic igneous assemblages related to inter- and back-arc extension in the middle Tertiary, and then by (3) later Cenozoic fundamentally basaltic volcanism as the plate-convergence regime was replaced by oblique extension east of the San Andreas transform (Christiansen and Lipman, 1972; *Elston and Bornhorst, 1979*).

IGNEOUS PROCESSES AND MAGMATIC TYPES

Although the presence of diverse magma types in the western U.S. was apparent early, and resulting observations were providing key contributions to developing broad concepts of igneous processes (e.g., *Daly, 1914; Lindgren, 1915*), the complex interrelations between magmatism and the broader geologic evolution of the Cordilleran region has been an especially rich topic for research in the past quarter century. Important Phanerozoic igneous types and provinces include: Mesozoic to recent volcanic arcs and associated batholithic plutons; evolved silicic ash-flow tuffs associated with large calderas; broadly basaltic and bimodal mafic-silicic magmas in extensional regions; and lower crustal and mantle igneous rocks brought to the surface by eruptive and tectonic processes.

Continental-margin arc volcanism

Many of the Mesozoic-Cenozoic igneous rocks of intermediate to silicic compositions in the western United States have been interpreted broadly as products of continental-margin arc volcanism resulting from complex subduction-related processes along the western margin of the American plate.

Cascade Range. Subduction-related volcanism is especially evident for the Quaternary Cascade volcanoes, which continue to be active adjacent to the convergent boundary between the American and Juan de Fuca plates (Fig. 1). The Quaternary volcanoes overlie and truncate an earlier eroded Cascade arc, largely Miocene in age, the crest of which strikes obliquely eastward from the Quaternary arc (*McBirney and White, 1982; Everts and others, 1987; Swanson and others, 1989*).

The young Cascade volcanoes have permitted detailed studies of volcanic and petrologic processes. Mount St. Helens, built mostly during the last few thousand years on a basement of eroded volcanic and plutonic rocks of the Tertiary western Cascade arc, has been the most active volcano in the Cascades during the past few thousand years (*Crandell and Mullineaux, 1978*). The 1980 eruptions of Mount St. Helens provided the first well-documented example of eruptive processes accompanying massive sector collapse and associated debris slides. A similar, but larger, prehistoric deposit lies adjacent to the north slopes of Mount Shasta (*Crandell and others, 1984; Crandell, 1989*). At Mount St. Helens in 1980, the debris slides depressurized a developing magma body at shallow depth beneath the volcano, leading to massive pyroclastic explosions that had overlapping characteristics of pyroclastic flow and surge processes (*Hobblit and others, 1981; Moore and Sisson, 1981; Waitt, 1981*). Subsequent incremental growth of a dacitic lava dome within the 1980 crater has greatly improved understanding of emplacement mechanisms of these historically infrequent volcanic events (Swanson and others, 1983). Some aspects of the 1980 Mount St. Helens eruptions appear to have been anticipated, in smaller scale, by the eruptive activity at Mount Lassen in 1915 (*Eppler and Malin, 1990*). Combined seismic and petrologic studies at Mount St. Helens have also led to an exceptional understanding of the geometry, depth, and physical conditions of the subsurface magma chamber (*Scandone and Malone, 1985;* Rutherford and others, 1985).

Other Cascade volcanoes have also been targets of recent study (*Muffler and others, 1989*). At Mount Mazama/Crater Lake, stratigraphic, petrologic, and geochronologic data demonstrate that the 6,850 B.P. climactic caldera-forming eruption developed from a complex magmatic system in which at least six separate vents intermittently erupted andesitic to dacitic lavas during the preceding several hundred thousand years (Bacon, 1983). Caldera-related eruptions at Crater Lake shifted from a central-vent geometry during the initial Plinian phase to ring-fault vents during later ash-flow eruptions; the eruptions drained a compositionally zoned magma chamber, in which several distinct magma batches were sequentially emplaced and fractionated (*Bacon and Druitt, 1988*).

Like Mazama and St. Helens, volcanoes such as Newberry, Medicine Lake, Hood, and Shasta have ranged in eruptive compositions from primitive basalt to lavas and tuffs of evolved dacite and rhyolite. Such volcanoes have the greatest potential for future explosive eruptions. In contrast, Rainier and Baker volcanoes have erupted dominantly basaltic andesite and andesite lavas, with only minor volumes of more silicic magma (*McBirney and White, 1982*), and large pyroclastic eruptions are unlikely in the near future. In addition to the central volcanoes, large Quaternary basalt fields have developed in the Cascades, especially in Oregon and northern California.

Studies of the Tertiary Western Cascade arc have progressed more slowly because of heavy vegetative cover and poor exposures over large areas, as well as limited economic incentives for detailed studies. Particularly interesting in this region are subvolcanic intrusions that were broadly cogenetic with eruptive activity; these offer exceptional opportunities to study the consolidated crustal conduits and magma chambers of arc volcanoes (*Fiske and others, 1963; Cater, 1969; Tabor and Crowder, 1969; Hopson and others, 1987*).

All the Cascade volcanoes are largely basaltic andesite to dacite in composition, and they are low in incompatible elements in comparison with igneous rocks emplaced on the craton farther east in the Cordillera (*Peterman and others, 1970; Church and Tilton, 1973*). Such data indicate that any crustal component of the Cascade magmas is overwhelmingly Phanerozoic in age, but controversy continues regarding the relative roles of mantle versus crustal contributions and fractionation mechanisms in generating the observed compositions. Much petrologic work has focused on distinguishing components in arc lavas derived from melting of the subducting slab versus magma generated in the overlying mantle wedge (*Gill, 1981*). Many petrologists favor magma generation in the mantle wedge where melting is fluxed by devolatilization of subducted oceanic crust, but the presence of the short-lived cosmogenic radionuclide ^{10}Be in Cascade and other lavas provides evidence of a component from subducted oceanic sediment (*Terra and others, 1986*). The diverse compositions among Cascade lavas have been attributed to mixing of mantle-derived basalt with dacitic magma (Eichelberger, 1975), but most petrologists favor more complex models for evolution of such arc magmas. For example, combined experimental studies and elemental and isotopic variations indicate that calc-alkaline lavas, such as at Medicine Lake in northern California, have resulted from combined crystal fractionation, crustal assimilation, and magma mixing (Grove and others, 1982). The diverse petrologic assemblages possible from such processes in the Cascade arc are documented strikingly by the close proximity of calc-alkaline, shoshonitic, and primitive tholeiitic lavas in mongenetic volcanoes near Crater Lake, Oregon (*Bacon, 1990*).

Mesozoic batholiths and associated volcanic suites. The great Mesozoic batholiths of the western U.S., including the Peninsular, Sierra Nevada, and Idaho segments, are now widely recognized as the plutonic underpinnings of Andean-type continental-margin arc magmatism on a vast scale (Hamilton and Myers, 1967; *Saleeby, 1981;* Hyndman, 1983; *Silver and Chappell, 1988*). These batholiths include a broad range of compositional and structural features that reflect variable environments of magma generation, processes of magmatic fractionation, and depths of solidification.

Upper-crustal emplacement levels, as exposed in the northern and central Sierra and much of the Peninsular batholith, are marked by roof rocks consisting largely of broadly contemporaneous volcanic strata and by the presence of multiple discrete plutons with diverse structural and petrologic features. In addition to abundant intermediate-composition lavas of volcanic-arc type, roof pendants in the central Sierra include ash-flow caldera remnants of Cretaceous and Jurassic age (Schweickert, 1976; *Fiske and Tobisch, 1978; Busby-Spera, 1984*). Mesozoic volcanic-arc assemblages are even better preserved in southern Arizona, where the axial zones of Triassic-Jurassic arc magmatism lie east of the volumetrically overwhelming Cretaceous

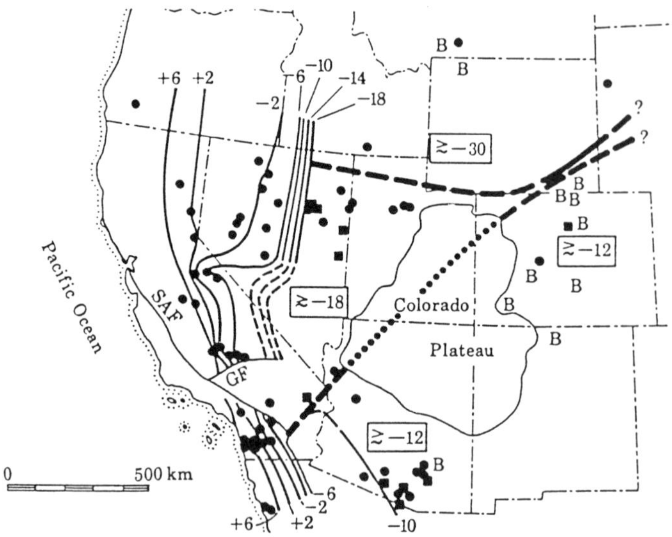

Figure 2. Summary of regional variations in initial ϵ_{Nd} values of Mesozoic and Tertiary granitic rocks (DePaolo and Farmer, 1984). Abrupt gradient to large negative values marks margin of the Precambrian craton, as reflected by a substantial crustal component in the granitic magmas. Heavy dashed lines demark Precambrian age provinces: from north to south, 2.6 to 2.9 Ga, ca 1.8+ Ga (?), and 1.65 to 1.8 Ga. CAF, San Andreas fault; GF, Garlock fault; B, location of analyzed basement sample; solid circles, Mesozoic or Tertiary biotite, or biotite and hornblende-bearing (metaluminous) granitic rock; solid squares, Mesozoic or Tertiary muscovite-bearing (peraluminous) granitic rock.

batholithic rocks, and the area occupied by granitic plutons is accordingly smaller. There, great dacitic to rhyolitic ash-flow sheets are locally well preserved, and remnants of at least 10 to 15 caldera sources have been identified based on thick intracaldera tuff accumulations and interleaved caldera-collapse megabreccias (*Lipman and Sawyer, 1985*).

Compositional and structural features of multiple discrete plutons are well exposed at relatively shallow crustal levels in the Peninsular and Sierra Nevada batholiths. Most such bodies are characterized by sharp and steep external contacts, relatively narrow thermal metamorphic gradients in country rocks, well-developed internal flow structures that are typically steep, internal compositional zones becoming gradationally or discontinuously more silicic inward, and the widespread presence of rounded mafic inclusions. Structurally similar discrete plutons in the Klamath Mountains of California, the Blue Mountains of Oregon, and the Northern Cascades in Washington, are characterized by many of the same features, although interpretation of their genesis and original tectonic setting is complicated by terrane-accretion events roughly concurrent with the magmatism.

At deeper crustal levels, the Cordilleran batholiths consist of more heterogeneous gneissic granitic rocks of less silicic average compositions, characterized by poorly defined boundaries between discrete plutons, irregular compositional variations, common gradational and migmatitic contact zones, and dynamothermal metamorphic gradients in adjacent country rocks. Such features typify the Mesozoic rocks of the Idaho batholith (much complicated by widespread Eocene intrusions) and the southernmost Sierra Nevada (Hyndman, 1983; Ross, 1985). A migmatitic basal zone, exposed in the central Idaho batholith, consists of gneisses and amphibolites with mineralogy indicating formation at 12- to 20-km depths (Hyndman, 1983), and mafic dikes provide direct evidence for the important role of subcrustal basaltic magma in inducing crustal anatexis (*Hyndman and Foster, 1988*). Although largely interpreted as representing crystallization from magma, some features within the granitic bodies at these levels appear due to ultrametamorphism and partial melting of metasedimentary wall rocks (Saleeby and others, 1987).

Transverse chemical and isotopic gradients, which help constrain sources of the magmas and mechanisms of differentiation, are well defined in the Mesozoic batholithic rocks, especially in California. Across the Peninsular and Sierran batholiths, elements such as K, Rb, U, and Th increase, while Ca, Sr, and related elements decrease eastward at constant differentiation indices (*Bateman and Dodge, 1970;* Silver and Chappell, 1988; *Ague and Brimhall, 1988*). Along with these chemical changes, the granitic rocks show eastward increases in radiogenic Sr and Pb, and in ^{18}O (Kistler and Peterman, 1973; DePaolo and Farmer, 1984; *Gromet and Silver, 1987*). These compositional changes, though in places broadly gradational, define steps in the Peninsular batholith that are interpreted as reflecting abrupt eastward changes in composition and age of basement source rocks. In the central Sierra Nevada, such eastward increases in incompatible elements, defined by granitic plutons ranging in age from Triassic to Late Cretaceous, were interpreted as reflecting variable depth to a fossil Benioff zone and subducted slab by *Dickinson (1970),* but a more complex origin for the variations, involving dominant compositional influence by crustal sources, is required by the volume and dominance of silicic magma, and by gradients in isotopic composition that map out the truncated margin of the Precambrian craton. Broadly, the eastward compositional progressions in the granitic rocks are paralleled by eastward transitions in the country rocks and inferred basement of the batholith from accreted, immature, volcanic arc and associated sediments to miogeoclinal sediments overlying Precambrian craton (*Ague and Brimhall, 1987*). DePaolo and Farmer (1984) use Nd and Sr isotopic compositions of such Mesozoic and Cenozoic granitic rocks to infer that the magmas were derived largely from the crust (Fig. 2), with regional variations due mainly to (1) variations in the flux of mantle magma entering the crust, (2) crustal thickness, and (3) crustal age.

Latest Cretaceous–Tertiary rocks of the eastern Cordillera. At the end of the Mesozoic, igneous and tectonic activity migrated rapidly eastward and became generally more alkalic, spreading into regions where magmatism had been largely absent earlier in the Phanerozoic. The eastward Late Cretaceous–early Tertiary (Laramide) spread of magmatism was broadly systematic (Fig. 3), though discontinuous in space and somewhat episodic in time (Lipman and others, 1971; Armstrong, 1974; Snyder and others, 1976; Christiansen and Yeats, this volume).

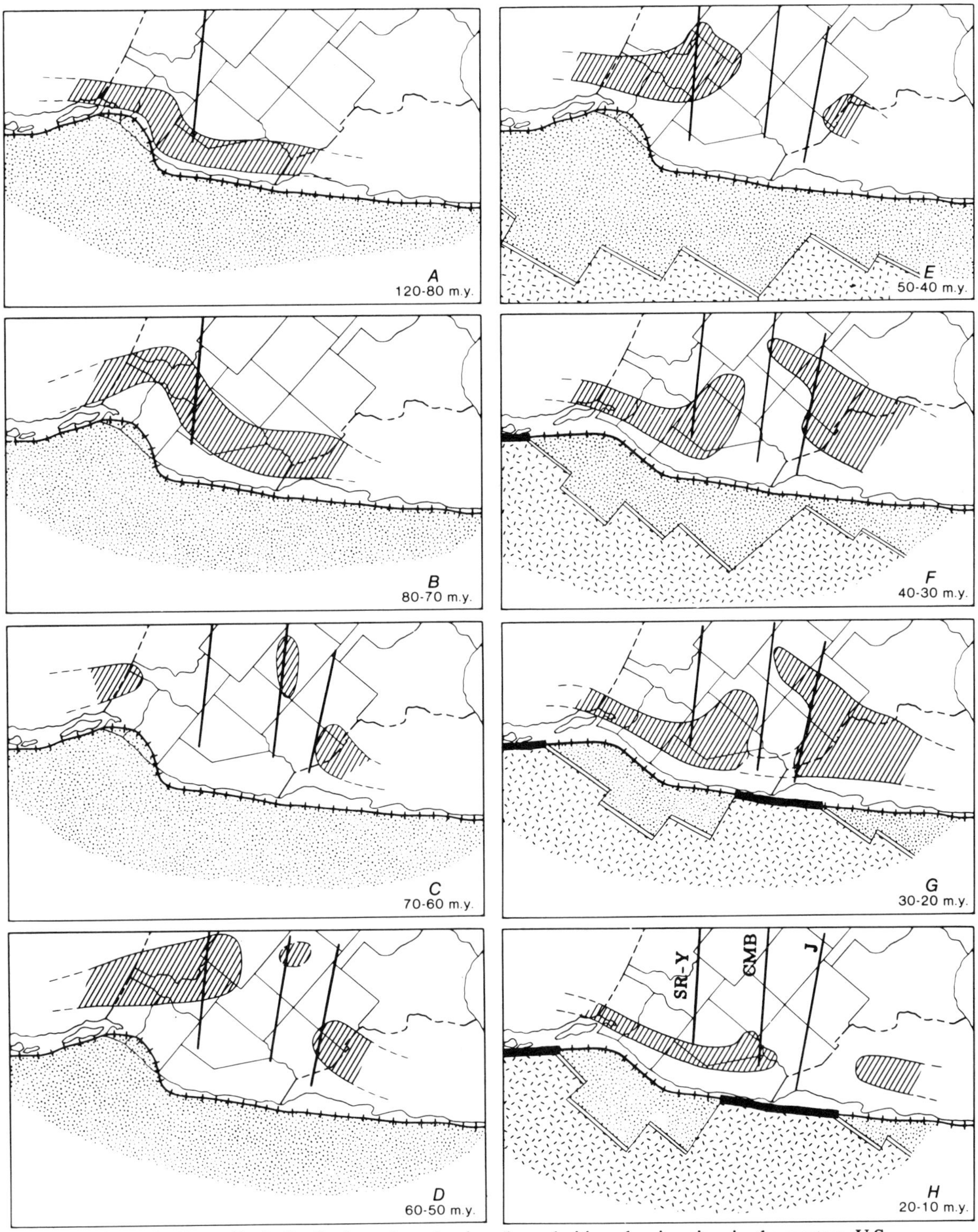

Figure 3. Generalized distribution of predominantly andesitic volcanic suites in the western U.S., inferred to be related to subduction (Lipman, 1980). No attempt has been made to remove the effects of late Cenozoic extensional and rotational deformation. Lined patterns, andesitic rocks; fine stipple, Farallon plate; coarse stipple, Pacific plate; unpatterned, American plate. Double parallel line, divergent plate boundary; fine line, transform offsets in divergent boundary; railway-track line, convergent plate boundary; heavy line, oblique transform plate boundary. Northeast-trending lines mark approximate traces of the Snake River–Yellowstone zone, the Colorado mineral belt, and the Jemez zone.

The origin of these patterns is related by many to increased rates of plate convergence (Coney, 1976; Dickinson and Snyder, 1978; *Hamilton, 1981*), although processes remain controversial. Whatever the ultimate cause, the eastward spread of igneous activity and associated tectonism at the end of the Mesozoic blocked out for the first time the full width of the present Cordilleran region. Notable features include the complex and nonsynchronous space-time patterns of activity, the pronounced influence on distribution and composition of igneous bodies by the Precambrian craton, and striking petrologic similarities with continental-margin arc magmatism elsewhere despite the broad region affected.

Voluminous magmatism first reached the eastern margin of the present Cordilleran region in northern Montana, at 80 to 75 Ma, as marked by the Elkhorn Mountain volcanic field (*Smedes, 1966*), associated intrusion of the Boulder batholith (Robinson and others, 1968), and scattered outlying igneous centers; some activity may have occurred as early as 85 Ma in southwestern Montana (*Meen and Eggler, 1987*). These rocks dominantly constitute a high-K calc-alkaline suite (Tilling, 1974), but more alkalic rocks were emplaced locally, especially during late-stage activity and in outlying and eastern areas (*Meen and Eggler, 1987*). Magmatism also spread eastward in the southwest, into southern Arizona at about 75 to 70 Ma, but did not reach its full eastward extent in New Mexico and adjacent Trans-Pecos Texas until 60 to 55 Ma (Coney and Reynolds, 1977; Fig. 3). In the southwest, high-K calc-alkaline suites also dominated, becoming more alkalic eastward and during the waning of magmatism (*Keith, 1978*). In the east-central Cordillera, such early Tertiary igneous activity was sparsely developed, mainly as scattered small high-level intrusions in eastern Nevada, Utah, and the northeast-trending magmatic loci of the Colorado mineral belt (Fig. 3). The easternmost activity, along the Colorado mineral belt, peaked between about 70 and 60 Ma, thus intermediate in time between the easternmost magmatism in Montana and that in New Mexico and Texas. The Colorado intrusions are compositionally broadly similar to those to the north and south—predominantly high-K calc-alkaline, but including volumetrically minor alkalic bodies generally emplaced late in the local intrusive sequence (*Simmons and Hedge, 1978*). This Laramide magmatism (70 to 60 Ma) was also associated with major mineralization, including the great copper porphyry systems of southern Arizona and adjacent New Mexico, the Butte copper deposits in Montana, and scattered precious-metal veins along the Colorado mineral belt and elsewhere.

In each of the three eastern Cordilleran regions (north, central, and southern), after an early Tertiary magmatic lull, volcanism began anew at slightly differing times in the middle Tertiary—about 55 Ma in the north, 45 to 40 Ma in the central, and 40 to 35 Ma in the south (Fig. 3; Christiansen and Yeats, this volume). As in the earlier pulse, the middle Tertiary activity was largely high-K calc-alkaline in the Rocky Mountains, but highly alkalic magmas (trachytic and commenditic volcanics; syenitic intrusions) appeared in increasing volume along and just east of the Rocky Mountain front from Montana to Texas (*Larsen, 1940; Barker, 1987*). Within the Rocky Mountains, the major volcanic fields include Challis, Absaroka, San Juan, and Mogollon-Datil. These all began with predominantly intermediate-composition lavas and associated volcaniclastic rocks erupted from central volcanoes; in all but the Absaroka field, the central-volcano eruptions were followed by widespread outpouring of more silicic ash-flow sheets from caldera centers (*Elston, 1984a;* Steven and Lipman, 1976; *McIntyre and others, 1984*). In the interior Cordillera, similar volcanism spread southward across Nevada and Utah in the Eocene and Oligocene, and also renewed in Arizona (Stewart and others, 1977; Steven and others, 1984; *Damon and others, 1981*).

Petrologic and isotopic data show that many intermediate to silicic Cenozoic volcanic suites in the eastern Cordillera contain as major constituents variable melts of lower crust, energized by mantle-generated basaltic magmas (Lipman and others, 1978; Hildreth, 1981; DePaolo and Farmer, 1984; *Johnson and Lipman, 1988*). Intermediate-composition volcanic rocks of the Absaroka field define secondary lead isotope isochrons with a Cenozoic lower intercept and an upper intercept of about 2.6 Ga, the age of the underlying Archean craton (Peterman and others, 1970). Analogously in the southern Rocky Mountains, predominantly intermediate-composition lavas and associated more silicic ash-flow tuffs of Oligocene age show isotopic evidence for a major crustal component derived from material similar in age to the dominant exposed cratonic rocks (1.5 to 1.7 Ga), in which incompatible elements such as U and Rb were depleted at the time of craton formation, presumably during granulite-facies metamorphism of lower crust (Lipman and others, 1978; Johnson and others, 1990).

Despite the differences in isotopic systematics and ages of basement rocks, the Laramide and middle Tertiary magmatism in the eastern Cordillera are strikingly similar in eruptive styles, petrography, and elemental compositions. Even though these rocks are as much as 800 to 1,000 km from the plate margin (even when corrected for later Basin and Range extension), they closely resemble those of interior portions of other continental-margin volcanic arcs such as the Peruvian and North Chilean Andes. Notable are (1) the predominance of high-K calc-alkaline types—becoming more alkalic and less voluminous toward the craton interior, (2) the broad compositional correlation with crustal thickness—becoming more silicic and alkalic on thicker and older cratonic crust, (3) the voluminous silicic ash-flow eruptions and associated calderas, and (4) the geochemical evidence for major lower crustal components in the magmas. Accordingly, these early and middle Tertiary igneous suites of the central and eastern Cordillera have been considered by many to be inner parts of a broad continental-margin arc (Lipman and others, 1971; Snyder and others, 1976; Coney and Reynolds, 1977; Cross and Pilger, 1978; *Keith, 1978*).

During the times of early and middle Tertiary magmatic activity in the central and eastern Cordillera, the great Mesozoic batholithic belt near the plate margin from southern California to southwestern Idaho was magmatically quiescent. The Tertiary volcanic fields, however, are continuous with igneous regions of similar age to the south and north: the great Sierra Madre Occidental in Mexico (*Cameron and others, 1980; McDowell and Clabaugh, 1979*), and Eocene activity preserved largely as intrusions in British Columbia and forming a virtually continuous igneous belt as far north as southeastern Alaska (*Souther, 1977; Ewing, 1980; Armstrong, 1988*). The U.S. sector of these rocks thus defines a broad, diffuse eastern belt of plate-margin magmatism on a continental scale, at a time when several "eastern Pacific" plates appear to have been undergoing prolonged subduction along the western margin of the American plate.

Silicic magmatism: Ash-flow tuffs and calderas

Relations between voluminous ash-flow deposits and associated caldera sources have been a major focus of U.S. volcanic studies during the past quarter century. The largest known ash-flow sheets exceed 10^3 km^3 in volume, and associated calderas are as much as 60 km in diameter. Such deposits and structures represent volcanic windows into the upper parts of shallow granitic batholiths (Buddington, 1959; Smith, 1960a; Lipman, 1984). In addition to posing potential problems of volcanic hazards, such silicic volcanic activity provides important controls on many of the Earth's ore deposits and geothermal energy sources, as well as offering key insights into the generation and evolution of large-volume silicic magma bodies (Smith, 1979; Hildreth, 1981). Especially influential has been the conceptual feedback between detailed studies of several well-preserved, large Quaternary caldera systems (Valles, Long Valley, Yellowstone, Crater Lake) and concurrent work on eroded Tertiary ash-flow fields, where deeper parts of the magmatic system are exposed (San Juan, Mogollon-Datil, Marysvale, Trans-Pecos, and smaller fields along the Rio Grande rift and in the Basin and Range Province).

Building on studies by Williams (1942) at Crater Lake, Robert L. Smith and others (Smith, 1960a, *1960b; Ross and Smith, 1961; Smith and Bailey, 1966*) developed a conceptual basis of depositional, welding, crystallization, and compositional zones in ash-flow sheets, based on detailed studies of the Jemez volcanic field and associated Valles caldera in New Mexico, which inspired an explosion of work in the western U.S. and elsewhere in the world. The Valles studies also led to recognition of the importance of resurgent processes in caldera evolution (Smith and Bailey, 1968). Recent work in the Jemez volcanic field has utilized textural studies of the pyroclastic deposits and drill-hole data resulting from geothermal exploration to constrain locations of pyroclastic vents and associated multiple caldera subsidences (*Nielson and Hulen, 1984;* Self and others, 1986; Heiken and others, 1990).

Work on the bimodal rhyolite-basalt volcanic assemblages in the eastern Snake River Plain and Yellowstone Park region (Armstrong and others, 1975; Christiansen, *1982,* 1984; *Morgan and others, 1984*) has documented multiple Pleistocene caldera cycles, in which eruption of voluminous ash-flow sheets of silicic rhyolite resulted from protracted crustal-magmatic evolution driven by energy transfer from the mantle by mafic magmas. The Snake River–Yellowstone magmatic system is strongly bimodal, has migrated northeastward at rates of several centimeters per year since inception at about 15 Ma in southwestern Idaho, and appears to be at least partly localized by an ancient structural zone within the Precambrian craton (Eaton and others, 1975). The youngest caldera Yellowstone, formed at 0.6 Ma, is among the world's largest, with dimensions of 40 × 65 km; it is underlain by a shallow and still partially molten batholith, and hosts the world's largest active geothermal system (Christiansen, 1984; Smith and Braille, 1984).

At the Pleistocene Long Valley caldera (Bailey and others, 1976, *1989*), detailed petrologic studies of compositional gradients in the outflow ash-flow sheet (Bishop Tuff) have permitted evaluation of alternative mechanisms for the evolution of compositional gradients in silicic magma chambers (Hildreth, 1979; *Halliday and others, 1984;* Cameron, 1984). In addition, geophysical studies in response to seismic unrest and geothermal exploration drilling have provided exceptional insights into the three-dimensional caldera structure and the active magmatic and tectonic processes at Long Valley (Hill and others, 1985).

Studies on these major Pleistocene ash-flow systems have been complemented by work on Tertiary calderas where deeper erosional levels and structural disruption permit a more complete sampling of the eruptive products, although obtaining suitable material for detailed petrologic study can be more difficult in the older volcanic systems. Results include studies of pyroclastic eruption and emplacement mechanisms, determination of structural features leading to improved models of caldera formation (Lipman, 1984), and evidence for fractionation mechanisms in ash-flow magmas (Hildreth, 1981). Correlations between large rhyolitic ash-fall beds and their source calderas document that as much as one third of the total eruptive volume accumulates over vast regions as down-wind airfall deposits (*Izett, 1981*).

Volcanic fields along the crest of the Southern Rocky Mountains and adjacent flanks of the Rio Grande rift zone in Colorado and New Mexico have yielded especially interesting results. In the San Juan field, 15 to 20 major ash-flow eruptions (each 100 to 3,000 km^3) are associated with subsidence of successive clustered calderas over an area 75 by 150 km that is interpreted to mark composite cupolas along the roof of a shallow subvolcanic batholith (Fig. 4; Steven and Lipman, 1976). For one large-volume Oligocene ash-flow sheet (Fish Canyon Tuff) in the San Juan field, combined geothermometry and geobarometry based on microprobe mineral chemistry suggested to Stormer and Whitney (1985) that the phenocryst assemblage of this caldera-related tuff crystallized at depths of about 25 km. These results have led to some controversial interpretations of relations between magma chambers and caldera evolution (*Grunder and Boden, 1987;* Johnson and Rutherford, 1989). At the 34-Ma Grizzly Peak caldera in central Colorado, compositional gradients in the intra-caldera tuff fill (more than 2.7 km thick) define step functions that are interpreted to require the presence of multiple compositional layers in the source magma chamber (*Fridrich and Ma-*

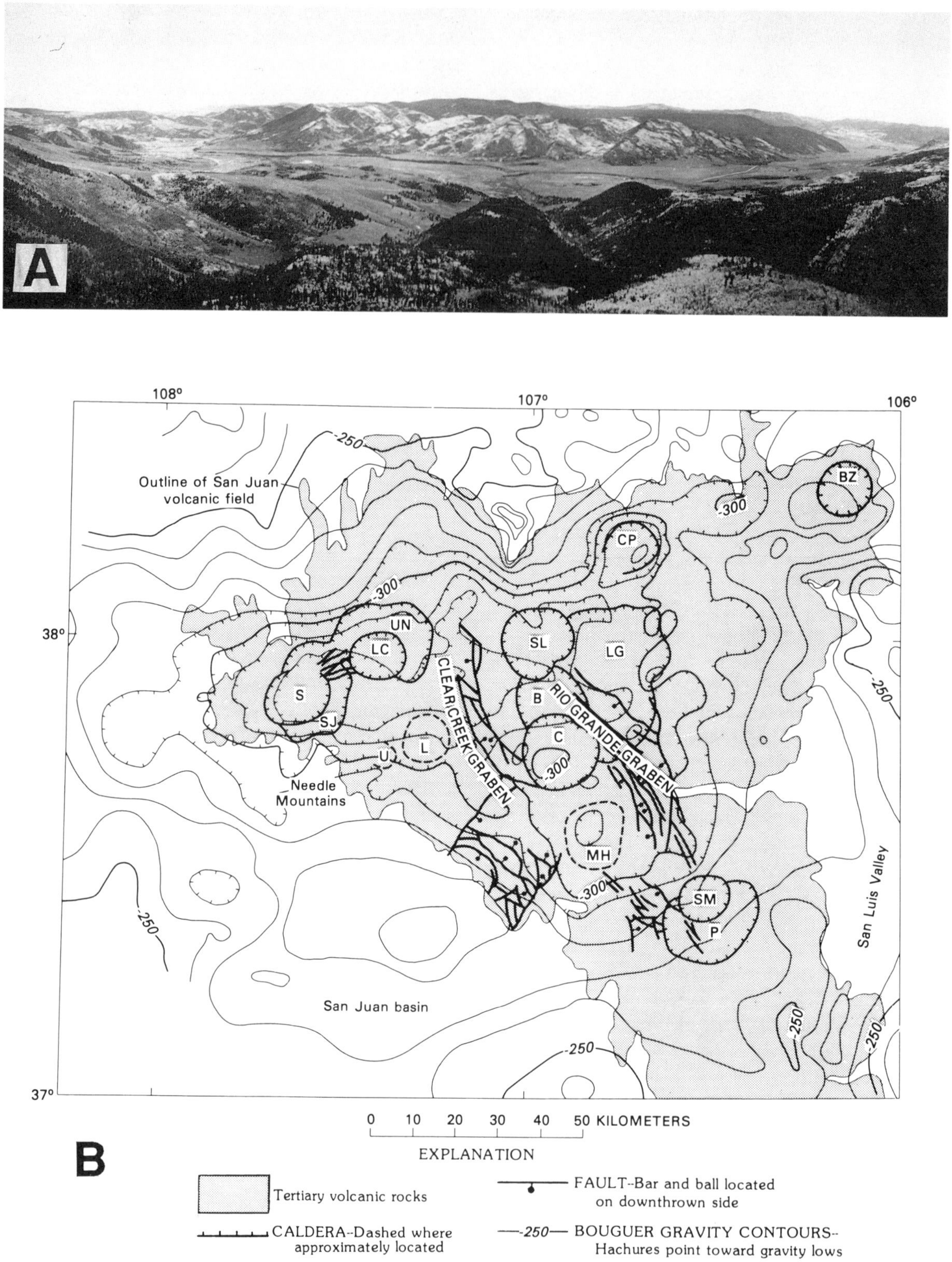

Figure 4. Calderas in the San Juan volcanic field. A. Creede caldera (Oligocene) from the north, showing central resurgent dome, outer topographic wall, and structural moat. Soft lacustrine caldera sediments of Oligocene age that filled the moat have been largely excavated by the Rio Grande. From *Ratté and Steven* (*1967,* color frontispiece). B. Calderas in relation to Bouguer gravity field (Steven and Lipman, 1976).

hood, 1987; Johnson and Fridrich, 1990). An oblique erosional section through the nearby Mount Aetna caldera complex exposes a transition from caldera fill to ring-fracture structures in underlying caldera-floor rocks, displaying mixed brittle and ductile shear structures (*Shannon, 1989*). Subvolcanic batholiths, interpreted as constituting the solidified magma chambers for ash-flow eruptions, are exposed in the Latir and Organ Mountains, New Mexico (Lipman, *1983,* 1988; Seager and McCurry, 1988). At shallower levels in the caldera environment, voluminous megabreccias in the thick fills of several deeply eroded San Juan calderas were derived by landsliding from the caldera walls and record times of catastrophic collapse (*Lipman, 1976*). As documented by *Chapin and Lowell (1979),* an outflow tuff sheet erupted from a caldera in central Colorado and deposited in paleovalleys preserves distinctive structures that document a complex interplay between primary and secondary flow structures and permit interpretation of emplacement processes. General features of other major Tertiary ash-flow fields and calderas in the Cordilleran region are summarized by Elston (*1984a*), *Ratté and others (1984),* Steven and Lipman (1976), Steven and others (1984), *Henry and Price (1984),* Christiansen and others (1977), *Byers and others (1989),* and *Best and others (1989).*

Much recent research has also focused on physical conditions and processes that lead to nonexplosive eruption and emplacement of silicic magma as lava flows or domes. Such flows and domes have widely preceded and followed major caldera-forming pyroclastic eruptions in the Cordillera, and some ash-flow deposits have undergone sufficient secondary flow to resemble lavas. As a result, distinctions between pyroclastic and lava eruptions have become somewhat blurred, even including distinguishing features of some resulting deposits. For example, large-volume Miocene rhyolites (10 to 200 km^3) in southwestern Idaho, which had been interpreted as densely welded remobilized ash-flow deposits (Ekren and others, 1984), are alternatively thought to be lavas erupted at sufficiently high temperatures and high effusion rates to have formed far-flowing sheets rather than steep-sided domes (Bonnichsen and Kauffman, 1987). Features of lava domes near Long Valley, California, are interpreted by Eichelberger and others (1986) to indicate that silicic magma can erupt quiescently as vapor-poor lavas if magma ascends sufficiently slowly to permit subsurface gas release from a permeable foam that collapses during extrusion. Conditions of shallow ascent, rather than pre-eruption volatile concentrations, are thought to control eruptive behavior. Mapping and petrographic study of glassy and pumiceous textures in such rhyolite flows and domes indicate that development of the textural stratigraphy within the flows is controlled by a combination of cooling rates, microfracturing, and migration of gases released by crystallization (*Fink and Manley, 1987*).

Basaltic magmatism and continental extension

Volcanism that can be regarded as fundamentally basaltic (Christiansen and Lipman, 1972; Christiansen and Yeats, this volume) gradually superseded the middle Tertiary volcanic association of intermediate-composition lavas and more silicic ash-flow tuffs, beginning at about 27 to 26 Ma along the incipient Rio Grande rift in the southern Rocky Mountains, but later (23 to 20 Ma) in the Great Basin. Such basaltic volcanism includes basalt plateaus, bimodal assemblages of basaltic and silicic rocks, and fractionated basaltic sequences in which basalt is volumetrically dominant. Although all magmatism can be broadly described as fundamentally basaltic in the sense of Hildreth (1981), including subduction-related arc magmas (however silicic) that were energized by mantle-derived basaltic melts, it is useful to distinguish between the basaltic and arc-related Cenozoic igneous suites of the Cordillera. Fundamentally basaltic volcanism in the western U.S. became more widespread westward and northward during the Miocene, and now characterizes the entire Cordilleran region except for the Cascade Range. In addition to providing tracers of volcano-tectonic evolution and examples of diverse eruptive processes, such basalts are geochemical probes for the evolution of the subcontinental lithospheric and asthenospheric mantle and for fractionation processes in a continental environment.

Discontinuously distributed late Cenozoic basalts in the southeastern Cordillera, along the Rio Grande rift and adjacent areas, show especially large compositional and isotopic variations in space and time. Tholeiitic basalts are most abundant along the rift axis and along transverse crustal flaws such as the Jemez zone; more alkalic basalt types and bimodal assemblages occur mainly along the rift flanks. Such variations are interpreted to record variable components of crust, "enriched" lithospheric mantle, and "depleted" asthenospheric mantle; the component of asthenospheric mantle correlates with intensity of lithospheric extension and thinning (Fig. 5; *Baldridge, 1979;* Perry and others, 1987). Intersections between the rift and transverse structures such as the Jemez zone have been loci of the most voluminous and compositionally diverse activity, further demonstrating lithospheric structural influences on the basaltic volcanism. Tholeiitic basalt flows within the Rio Grande rift also show evidence of complex open-system magmatic evolution: basaltic magmas were both variably modified by crystal fractionation and mixed with andesitic and dacitic magmas that themselves had previously assimilated granulite-facies Proterozoic lower crust (Dungan and others, 1986; *McMillan and Dungan, 1986*). Miocene-Pliocene rocks of the San Juan field on the west flank of the Rio Grande rift are a bimodal assemblage of basalt and rhyolite, in which the rhyolites are interpreted as partial melts of lower crust, with the thermal energy for melting provided by concurrent basaltic volcanism (Lipman and others, 1978). On the High Plains, east of the Rio Grande rift, petrologically diverse basalt flows, including highly mafic alkalic types, indicate variable depths and amounts of melting in mantle sources (*Stormer, 1972*). In the San Francisco volcanic field, on the southwest margin of the Colorado Plateau in Arizona, combined petrologic, age, and paleomagnetic data document that volcanic loci have migrated northeastward during Quaternary time and that variations in migration rate correlate with magma production and eruptive frequency; these features

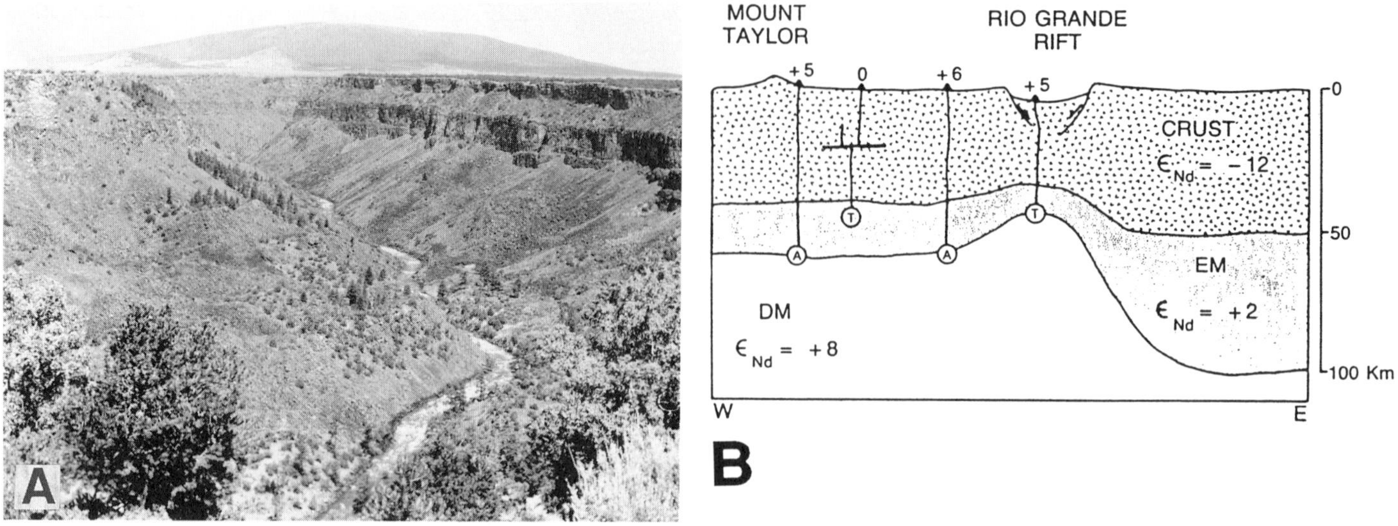

Figure 5. Basaltic volcanism along the Rio Grande rift. A. Rio Grande gorge along axis of Rio Grande rift, cut into flat-lying tholeiitic basalt flows of Servilleta Basalt (Pliocene) and interlayered rift-fill sediments. In background is Cerro de la Olla, a prominent olivine-andesite shield volcano of the Taos Plateau volcanic field (4 Ma). Photo by P. W. Lipman, U.S. Geological Survey. B. Schematic cross section showing mantle structure and its relation to basalt genesis and neodymium isotopic compositions beneath the Colorado Plateau–Basin and Range transition zone and the central Rio Grande rift. The E_{Nd} of enriched lithospheric mantle (EM) and depleted asthenopsheric mantle (DM) represent values of derived melts and imply nothing about mantle homogeneity. Numbers at the surface are E_{Nd} values of the erupted basalts. Alkalic basalts (A) equilibrate at the DM/EM boundary and ascend without crustal interaction to erupt with E_{Nd} values that reflect the relative proportions of DM and EM components incorporated from the mantle source. Tholeiites (T) from the Mount Taylor area equilibrate within EM but, because of a longer crustal residence time, are contaminated by minor amounts of middle to upper crustal rocks and are shifted slightly toward crustal E_{Nd} values. Tholeiites from the central Rio Grande rift are derived from the elevated DM/EM boundary and ascend through the crust with less interaction, perhaps due to the more pronounced extension along the rift (Perry and others, 1987).

are interpreted by *Tanaka and others (1986)* as related to absolute western motion of the North American plate.

Late Cenozoic basaltic rocks are widely but discontinously present in the Great Basin and adjacent parts of the Basin and Range Province, where they commonly are interbedded with basin-fill sediments and overlie steeply tilted and strongly extended mid-Tertiary volcanic assemblages containing abundant intermediate to silicic lavas and tuffs (*Anderson, 1971; Proffett, 1977*). The mid-Tertiary volcanic rocks have attributes of both plate-convergence and extensional petrologic suites, and associated extension locally reaches several hundred percent along low-angle detachment faults in association with tectonically unroofed middle-crustal rocks exposed in core complexes. The broad tectonic implications and relation to plate-tectonic mechanisms of the extended mid-Tertiary suite—alternatively intra-arc extension during waning of plate convergence, or intraplate gravitational spreading driven by thermal relaxation and upwelling of mantle basalt—has generated considerable controversy that remains incompletely resolved (Elston, 1984b; Glazner and Bartley, 1984; Wernicke and others, 1987; Gans and others, 1989).

The severe extension that has tectonically exposed middle crustal levels (10 to 15 km depths) in core complexes in the eastern Basin and Range Province projects along a belt from southern Canada to northern Mexico and is associated in time and space with peak middle Tertiary arc volcanism; extension typically began a few million to as much as 10 m.y. later (*Armstrong, 1982; Coney, 1987;* Gans and others, 1989; Best and Christiansen, 1991). Most reasonably, emplacement of arc magmas into the lower and middle crust raised temperatures and reduced crustal viscosity and strength, subsequently permitting extensional faulting in the more brittle upper crust (Thompson and Burke, 1974; Lachenbruch and Sass, 1978). The time correlation between magmatic intrusion and extensional formation of core complexes is clear locally, as at South Mountains, Arizona (*Reynolds and others, 1986*). Additional evidence for magmatic control of extension in the upper crust is provided by extreme local attenuation and stratal rotation on low-angle faults, which geometrically mimic upper-plate detachment faulting of core complexes, in roof rocks above shallow, caldera-related magma chambers along the Rio Grande rift, at Questa and probably Socorro, New Mexico (*Lipman, 1983,* 1988; *Chamberlin, 1983*). At Questa, faults bounding the rotated blocks sole into an exposed, 26 to 23 Ma, upper crustal batholith emplaced to within a few kilometers of the regional surface, and the brittle-ductile transition is telescoped within the upper crust. Here, crustal magmatism is the unambiguous cause of intense extension. Not all major loci of middle Tertiary arc volcanism were succeeded by extreme extension: just those located in suitable stress environments and at crustal sites weakened by prior magmatism and tectonism. Notable middle Tertiary volcanic loci that escaped subsequent extensional dismemberment include the Absaroka field within the Wyoming foreland (*Smedes and Prostka, 1972*), much of the San Juan, Mogollon-Datil, and eastern Utah fields near the margins of the Colorado Plateau (Steven and Lipman, 1976; *Elston and others, 1984a;* Steven and others, 1984; Best and Christiansen, 1991), and the great Sierra Madre Occidentale field in northern Mexico (*McDowell and Clagbaugh, 1979*).

Farther west, relatively small, discontinuous volcanic fields developed along the growing transform fault system in California, and decrease in age northward, tracking plate evolution of the Mendocino triple junction (Dickinson and Snyder, 1979; *Pilger and Henyey, 1979; Johnson and O'Neil, 1984*).

Tertiary basalts are especially voluminous in the northwestern parts of the Cordillera, where they form great plateaus of widespread lava flows: the Columbia River basalt field, Oregon plateaus, and Snake River Plain. Voluminous basaltic volcanism during the past 20 m.y., as preserved on the Columbia River basalt field and in the adjacent Oregon plateaus, has been interpreted as broadly due to back-arc spreading in a continental-margin environment; areal variations in isotopic compositions of the lavas document contrasts in age and chemical composition of the crust and mantle source regions, as well as generally decreased crustal contamination as volcanism progressed (*Carlson and Hart, 1987*). Peak basaltic activity, between 17 and 14 Ma, was concurrent with maximum eruption rates in the Cascade arc to the west (*McBirney, 1978*) and inception of the Yellowstone "hot spot." Magma production rates for the Columbia River flows from linear dike swarms appear to have been similar to those at major oceanic spreading centers and the largest intraplate (Hawaiian) volcanoes (Swanson and others, 1975). The large volumes and areal extents of individual Columbia River flows may be due to sustained but relatively infrequent eruptions from large, deep magma reservoirs. Physical and chemical constraints require that Columbia River basalt flows were fed from large (>700 km^3) magma reservoirs near the crust-mantle boundary and prohibit significant crystal fractionation or crustal assimilation between reservoir and surface (*Hooper, 1984*). In general structural setting, basaltic volcanism on the Columbia and Oregon Plateaus appears related to crustal extension behind the active Cascade volcanic arc, in an area where continental crust was absent or weakly developed. Inception of basalt eruption prior to sizable extension in the Columbia Plateau may indicate the primary role of an upwelling mantle plume (*Hooper, 1990*), or alternatively, that physical conditions in the upper mantle were so close to melting that only minor initial extension could trigger basalt magmatism.

Textural and structural features of the widespread Columbia River basalts, associated with the large volumes and widespread distribution, have also attracted attention. Textural features can provide indications of flow directions and interactions with wet sediments or ponded water (*Waters, 1961; Schmincke, 1967*). Joint structures and groundmass petrographic textures of Columbia River basalt flows are not in accord with conductive cooling models; rather, they indicate quenching by infiltrated water, interpreted as due to flooding from drainages deranged by eruption of the large-volume lavas (*Long and Wood, 1986*).

Basalts of the Snake River Plain in Idaho are interstratified with basin-fill and other sediments, along with earlier rhyolitic tuffs and lavas (Armstrong and others, 1975). These basalts are mainly olivine tholeiites interpreted by *Leeman (1982)* to result from partial melting of Archean ($\sim$2.5 Ga) subcontinental lithosphere, without major contamination by crustal material or contribution from asthenospheric or deeper mantle sources such as those of oceanic basalts. Less common evolved mafic lavas may reflect fractional crystallization in the lower crust and some upper-crustal assimilation. The youngest volcanics on the Snake River Plain, along the Great Rift, are compositionally diverse, basaltic lavas and associated, volumetrically minor, silicic types that are interpreted to have resulted from significant crustal contamination. These lavas have erupted at nearly constant output rates for the past 15 kyr; recurrence intervals have been a few hundred to a few thousand years (Kuntz and others, 1986).

The eastern Snake River Plain–Yellowstone system is cited frequently as a continental "hot spot," and its northeasterly propagation direction is close to that predicted for the American plate (Armstrong and others, 1975). Problems with any simple model of an active asthenospheric plume origin for this magmatic system include: (1) the association with a symmetrically opposing magmatic trend in Oregon (Fig. 6; *MacLeod and others, 1975*), (2) geometric and temporal association with development of regional basin and range extension, (3) relatively recent inception—in middle Miocene time, and (4) coincidence with preexisting northeasterly structural trends in the craton (Eaton and others, 1975; Christiansen and McKee, 1978). More likely, the Yellowstone trend has resulted from interaction between crustal structures and mantle dynamics during late Cenozoic regional extension induced by plate-boundary interactions.

A provocative alternative is that many flood basalt provinces, including the mid-Miocene magmatism over a large region of the Pacific Northwest (Columbia and Oregon Plateaus, west Snake River Plain, and the adjacent northern Basin-Range region), were initiated by large planetary impacts (*Alt and others, 1988*). Phanerozoic collisions of planetary bodies with Earth have been frequent (*Wetherill and Shoemaker, 1982; Grieve, 1982*), and large impacts may be capable of generating pressure-relief magmatism, as early proposed for Sudbury, Canada (*Dietz, 1964*) and Bushveldt-Vredefort, South Africa (*Hamilton, 1970*). Eleven episodes of flood basalt eruptions associated with continental separations in the last 250 m.y. have been quasi-periodic, with a mean cycle time of 32 ± 1 m.y. that is similar to a 26- to 29-m.y. periodicity inferred for mass extinctions (*Raup and Sepkoski, 1986; Sepkoski, 1988*), suggesting close interrelations between impacts, flood basalts, and faunal extinctions (*Rampino and Stothers, 1988*). The complex space-time sequence of Mio-

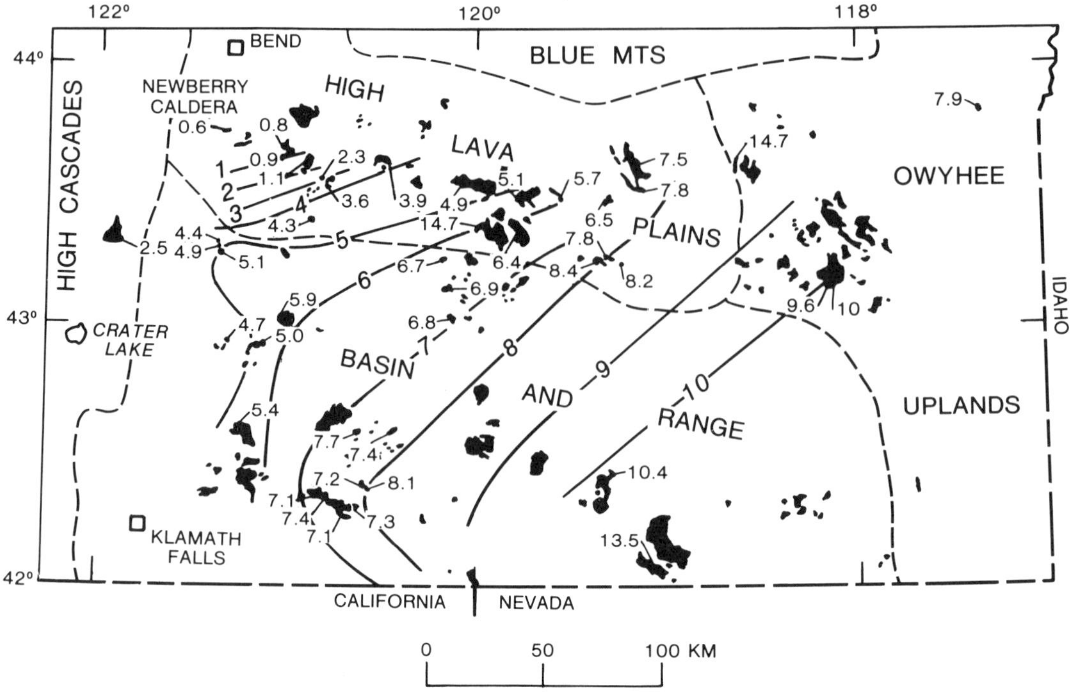

Figure 6. Age pattern of silicic lava domes (black) in eastern Oregon (*MacLeod and others, 1975*). Domes, associated with rhyolitic ash-flow tuffs and basalt fields, are distributed over several major geologic provinces, yet define a northwestward propagation trend toward Newberry caldera that mirrors the northeastward age trend of silicic volcanism along the eastern Snake River Plain toward Yellowstone. Ages of individual domes (tie lines) are the basis for generalized contours, both in Ma.

cene magmatic events in the Pacific Northwest make this a plausible place to pose the question of relation to a planetary impact, but specific evidence supporting a large impact event—such as ejecta deposits, platinum-group metal anomalies, concurrent fossil extinctions, or other environmental trauma—is thus far lacking. To the contrary, several features suggest that an impact cause is unlike: (1) inception of mid-Miocene magmatism in the Pacific Northwest correlates closely with regional tectonic evolution related to the northward-spreading influence of the Mendocino triple junction, as summarized above, and would require "good aim" for any inferred triggering impact; (2) inception of volcanism in the Columbia Plateau and Pacific Northwest seems to have been gradual, and peaking of regional magmatism nonsynchronous (17 to 15 Ma); and (3) the Columbia Plateau flows were fed from fissure vents (Chief Joseph dike swarm) far north of the impact site proposed by *Alt and others (1988),* rather than simply having flowed farther north than inferentially associated rhyolitic lavas.

Lower crustal and mantle samples

Interpretation of processes involved in magma generation, as well as many geophysical models, are dependent on evaluation of lower crustal and mantle compositions and physical properties. Accordingly, possible samples of such materials are of special petrologic importance. Thus, much recent study has been devoted to basaltic maars and diatremes in the western U.S., both to interpret the petrologic and structural implications of entrained xenoliths from deep crustal and continental mantle sources, and also to model phreatomagmatic processes in explosive volcanism. Another major result of studies during the past quarter century has been recognition that most ultramafic bodies in the western Cordillera are tectonically disrupted fragments of ophiolitic suites, representing former oceanic lithosphere, that have been tectonically accreted to the North American plate. Such ophiolites provide additional key samples of mantle materials.

Diatremes, phreatomagmatic explosions, and lithospheric xenoliths. Maars and tuff rings, produced by eruptions (commonly phreatomagmatic) from diatremes and basaltic breccia pipes, are petrologically important far beyond their small eruptive volumes and sporadic distribution because some of the deposits carry lower crustal and mantle xenoliths. These provide the only direct samples of such materials from deep within the North American plate.

Breccia pipes (diatremes) are exceptionally exposed in the Eocene alkalic province of central Montana (Hearn, 1968) and in the Hopi Buttes region of the Colorado Plateau (Shoemaker and others, 1962). These areas provide critical structural evidence that, in addition to transporting xenoliths from the lower crust and upper mantle, wall-rock material can also be carried downward, especially near the walls of the breccia pipe (Fig. 7). Where subsidence of wall-rock materials dominates in such

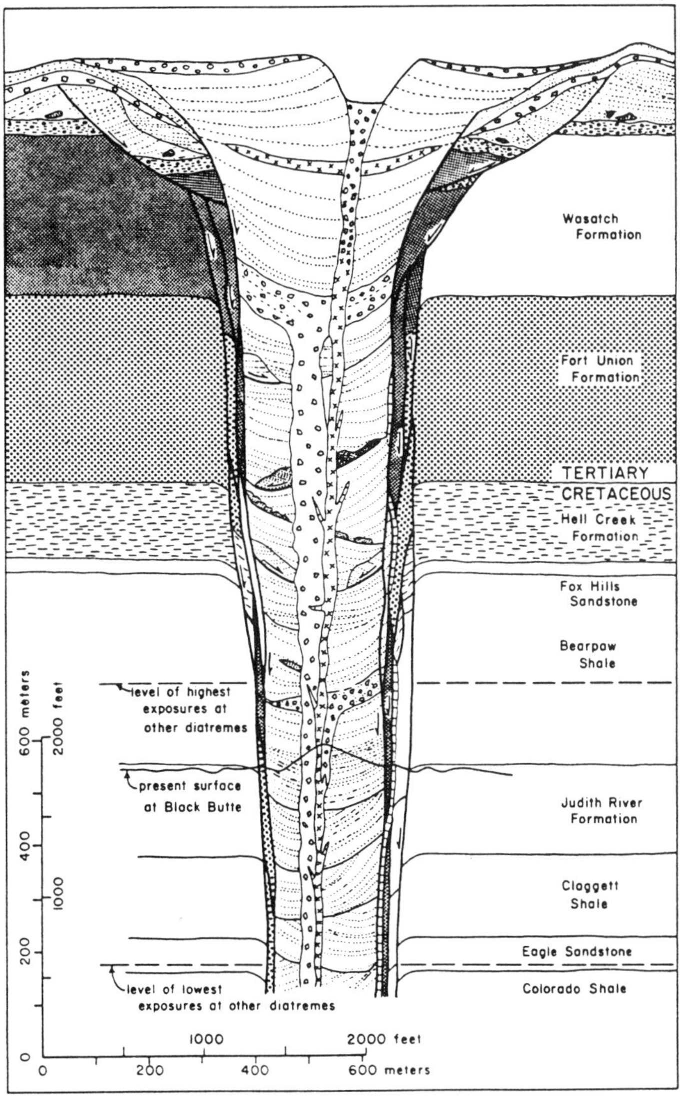

Figure 7. Generalized vertical section of diatremes in north-central Montana (Hearn, 1968).

funnel-shaped vents, diatremes merge in structural geometry and process of formation with small caldera subsidences (Lipman, 1984).

The volatile components of many alkalic diatreme-forming magmas are dominated by mantle-derived CO_2 (*Eggler, 1976*), but textural studies of pyroclastic ejecta have shown that abundant external water may be incorporated at shallow levels, leading to violent phreatomagmatic eruptions and formation of maars (*Wohletz and Sheridan, 1979*). Since base surges were first observed during the 1965 eruption of Taal volcano in the Philippines (*Moore, 1967*), cross-bedded deposits at prehistoric maar volcanoes, such as the Ubehebe Craters in California, have been increasingly recognized to be of phreatomagmatic base-surge origin. Identification of such surge deposits, versus airfall and fluvially reworked tuffs (*Fisher and Waters, 1969; Crowe and Fisher, 1973*), has major implications for the frequency of explosive eruptions at mafic volcanoes and attendant hazards. Such deposits also have provided materials for much recent work on the relative roles of magmatic volatiles versus externally derived meteoric water in generating the textural diversity of phreatomagmatic deposits (*Wohletz, 1986*).

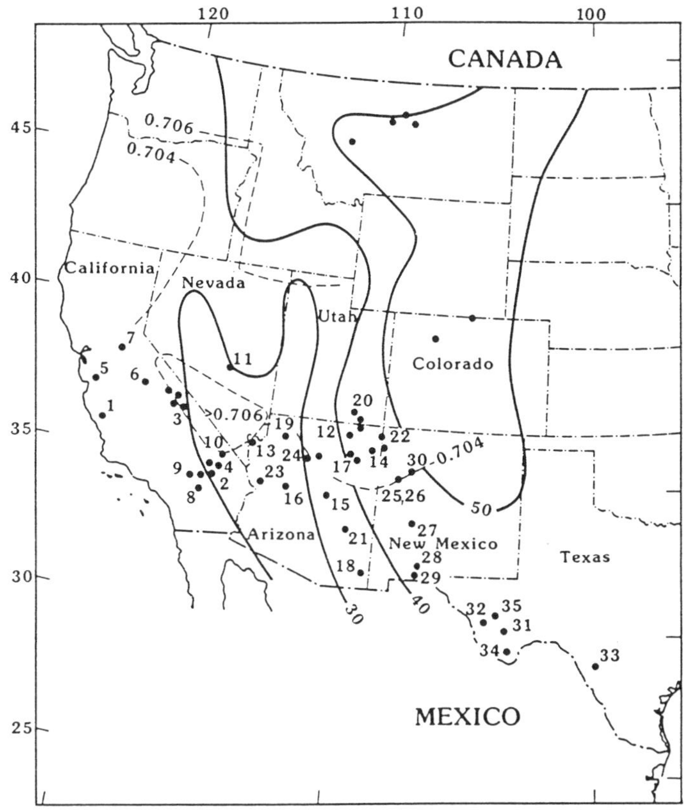

Figure 8. Xenolith localities in the southwestern U.S. Changes in crustal thickness (30, 40, 50 km) and isotopic domains of Cenozoic volcanic rocks ($^{87/86}$Sr <0.704, 0.704 to 0.706, >0.706) are also shown (*Menzies and others, 1987*).

Xenoliths and high-pressure megacrysts brought to the surface by such basaltic eruptions in the western U.S. constitute the only direct samples of the deep crust and subcontinental mantle of the region (Fig. 8). Accordingly, such samples have become the targets of intense study, leading to interpretation of vertical and lateral compositional variations within the continental lithosphere and implications for magma genesis (Wilshire and others, 1988). The size, abundance, and petrologic character of xenoliths in an intrusive feeder for a kimberlitic diatreme on the Colorado Plateau were used by McGetchin and Silver (1972) to develop a model for the crust and upper mantle of the region; they found that the lower crust consists mainly of mafic metavolcanic and metaplutonic rocks, and the mantle is compositionally and structurally complex.

Subsequent detailed petrologic studies of xenoliths from this and other localities in the western U.S. have documented complex compositional variations in the lower crustal and upper mantle sections sampled by the xenolith populations (*Menzies and others, 1987*). For example, the late Cenozoic Geronimo volcanic field in southeastern Arizona (*Kempton and others, 1984*) provides evidence for an exceptional diversity of mantle host rocks (type I), interpreted to represent variably modified, mid-ocean ridge basalt (MORB)–type mantle beneath the southwestern U.S., as well as xenoliths thought to result from in-place crystallization of the host alkalic basalt magma at mantle depths (type II). Sparse granulite xenoliths from the Geronimo field indicate that the lower crust in this region is heterogeneous in composition and age, and depleted in U and Th as would have been expected from Pb isotopic systematics of Tertiary silicic magmas in the area (Menzies and others, 1985). The Geronimo granulites also suggest mixing with Cenozoic mafic magmas inferred to be related to a recent lithospheric underplating event.

A broadly similar suite of crustal and mantle xenoliths from Kilborne Hole, along the Rio Grande rift in southern New Mexico, documents crustal growth at about 1.6 Ga, without further major geochemical modification (*Padovani and Reid, 1989*). Unlike the Geronimo suite, lower crustal xenoliths from Kilbourne Hole show little evidence of subsequent depletion events that might be associated with partial melt extraction. Small-scale modification of phase equilibria in the Kilbourn Hole xenoliths are inferred to reflect mainly recent reheating of the lower crust related to late Cenozoic extension along the Rio Grande rift.

In contrast, some xenoliths from Colorado Plateau diatremes are petrologically unusual eclogites, containing low-temperature but high-pressure minerals characteristic of blueschist terranes, which are somewhat controversially interpreted as derived from an oceanic crustal plate subducted beneath the Colorado Plateau (Helmstaedt and Doig, 1975). Related hydration of mantle materials above a dehydrating subducted slab has also been inferred from xenoliths from some of these same Colorado Plateau localities by *D. Smith (1979)* and *Helmstaedt and Schulze (1979)*. Seismic tomographic evidence also suggests the continuing presence of remnants of the Farallon plate beneath the eastern Cordillera and adjacent High Plains (Spence and Gross, 1990).

Such xenolith suites also provide critical evidence for chemical modification (metasomatism) of mantle materials, resulting from intertwined effects of passage of mafic magmas and more dilute, alkali-bearing aqueous or CO_2-rich fluids through the lithospheric mantle, perhaps derived by dehydration of a subducted slab at asthenospheric depth (e.g., *Wilshire, 1984*). Such metasomatism of mantle source regions has also been inferred to be important for the generation of mafic alkalic magmas in general (*Eggler, 1976; Menzies and others, 1987*).

Ophiolitic suites. Structurally complex ultramafic bodies in the Coast Ranges of California, the western Sierra Nevada, the Klamath Mountains, and the Northern Cascades have long been widely recognized as tectonically emplaced fragments of mantle material. Following recognition of the tabular geometry of several large ultramafic bodies that are much disrupted tectonically (*Irwin and Lipman, 1964; Bailey and others, 1970*), most are now interpreted as ophiolitic fragments of oceanic lithosphere emplaced along thrust and transform faults in plate-convergent tectonic settings (*Coleman, 1971; Snoke, 1977; Harper, 1984*). Many are Triassic-Jurassic in age and were emplaced or accreted to North America later in the Mesozoic, but the larger Trinity ophiolite includes lower Paleozoic oceanic crustal rocks (*Lindsley-Griffin, 1977*), as does much of the Sierran foothills belt (Saleeby, 1982). Structural mechanisms by which the ophiolites have been emplaced and preserved in the Cordilleran region remain controversial (*Coleman, 1971; Moores, 1982*).

Most well-studied Cordileran ophiolites are now interpreted as oceanic crust generated in back-arc basins and having transitional affinities to mafic components of island-arc volcanics, rather than having been generated along MORB-type spreading centers in open oceans (*Quick, 1981;* Saleeby, 1982; *Snoke and others, 1982; Harper, 1984*). A partial exception may be the large Middle to Late Jurassic Coast Range ophiolite (*Bailey and others, 1970;* Hopson and others, 1981), where ocean-island, intra-arc basin, and possibly ocean-ridge components appear to be present (*Shervais and Kimbrough, 1985*). Little of the vast MORB-type oceanic lithosphere, which must have formed on the east flanks of Pacific spreading centers in the Paleozoic and Mesozoic, is preserved in Cordilleran ophiolites; virtually all has been completely subducted.

Additional complexities result from tectonic juxtaposition of several discrete ophiolite assemblages, as in the western Sierra belt where ages of about 300, 200, and 160 Ma are documented by detailed geochronologic data and contrasting metamorphic and deformational features (Saleeby, 1982). The structural complexities of the Sierran ophiolites are further interpreted by Saleeby (this volume) to indicate emplacement during terrane accretion and oblique plate convergence. The mantle peridotites of some ophiolites also retain evidence of complex high-pressure histories, such as the apparent intrusion in the Trinity ophiolite of partial melts from depths greater than 40 to 60 km into depleted mantle material at depths of about 30 km (*Quick, 1981*). In contrast, Jurassic plagiogranite associated with the Josephine ophiolite locally contains xenocrystic Precambrian zircons, suggesting incorporation of terrigenous supracrustal strata (*Wright and Wylid, 1986*). Despite much recent detailed work on Cordilleran ophiolites, some features of the structurally complex, western U.S. bodies can best be interpreted only by comparison with more completely preserved and better exposed examples elsewhere, such as in Cyprus and Oman. Such problems are considered more fully in another chapter (Saleeby, this volume).

MAGMATIC EVOLUTION

The relative contributions to magmatic evolution of alternative processes—mafic melts from the mantle, partial melting of lower crust, crystal fractionation in magma chambers, open-system magma replenishment and mixing, and upper-crustal assimilation—are long-standing problems that require study in a global context. Recent work in the U.S. Cordillera has provided significant new data and conceptual models bearing on such problems. A common theme has been the trend toward more complex models, involving multistage processes at differing crustal levels (e.g., Open-Magmatic Systems conference: special issue of Journal of Geophysical Research, May 1986). A corollary is the need to study magmatic systems of similar petrologic type over a range of depth levels in the crust.

Relations between volcanism and associated plutonism

Large-volume volcanic eruptions in the western U.S., especially those involving ash flows and associated caldera subsidence (Fig. 9), are increasingly recognized as high-level manifestations of emplacement and consolidation of subvolcanic granitic plutons of batholithic dimensions, as reviewed by Buddington (1959), Hamilton and Myers (1967), Smith and Bailey (1968), and Lipman (1984). The eruptive history, structural features, and petrologic evolution of ash-flow calderas provide data on early stages of the evolution of the associated subvolcanic magmatic system. Broadly cogenetic, erosionally unroofed, granitic plutons record mainly the late stages of emplacement and crystallization. Combined studies of volcanic fields and associated subvolcanic plutons, necessary to decipher the overall evolution of a magmatic system, are currently a major focus for research in the western U.S.

An informative example is provided by the compositionally diverse plutons of the upper Cretaceous Boulder batholith of Montana, which is interpreted by Hamilton and Myers (*1974*) to have formed as a thin and shallow mass that spread over a floor of premagmatic rocks and that was roofed mostly by its own volcanic ejecta. Similarly, in northern New Mexico, exceptional structural and topographic relief provides a cross section through the Oligocene Latir volcanic field, associated Questa caldera, and cogenetic composite batholithic rocks that represent the source magma body for the volcanic rocks and into which the Questa caldera collapsed (*Lipman, 1983,* 1988; *Johnson and others, 1989*). Several middle Tertiary caldera fragments and cogenetic intrusions are also exposed in the rugged Sawatch Range of Colorado (*Fridrich and Mahood, 1984, 1987; Shannon, 1989*). Remnants of caldera structures, in proximity to apparently cogenetic granitic intrusions, are also widely preserved among the Mesozoic arc rocks of southeastern Arizona (*Lipman and Sawyer, 1985*) and locally in roof pendants in the Sierra Nevada (*Fiske and Tobisch, 1978; Busby-Spera, 1984*). In the younger environment of the Pleistocene Yellowstone and Long Valley calderas, combined geological and geophysical data indicate the underlying presence of silicic batholiths that are, in part, still molten (Eaton and others, 1975; Iyer and others, 1981; Sanders, 1984).

Geologic and geophysical evidence indicates that such composite granitic batholiths beneath volcanic fields must be broadly tabular (Hamilton and Myers, 1967). Gravity data, combined with surface geology, show that the batholithic magma chambers beneath Rocky Mountain volcanic fields such as the San Juan, Mogollon-Datil, and Latir rose to within a few kilometers of the surface, at least beneath calders clusters (Lipman, 1988). More broadly, the largest exposed Cretaceous-Tertiary batholiths in the Rocky Mountains (Boulder in Montana; Mount Princeton in Colorado) and those inferred beneath the Tertiary volcanic fields, which are 50 km or more in map dimensions, must be roughly laccolithic in order to remain within the middle and upper crust.

The tabular shapes of shallow batholiths have important implications for emplacement mechanisms and the much debated question of how space for intrusion of granite is accommodated. Chemical, isotopic, and field evidence require that silicic magmatism in the U.S. Cordillera involve large-scale melting and assimilation of the lower crust as a result of the rise of mantle-derived basaltic melts (Lipman and others, 1978; Hildreth, 1981; DePaolo and Farmer, 1984). Melting of lower crust to form granitic magma and transfer of the magma to a near-surface environment provide space, both by the removal of material at depth and by isostatic subsidence of the middle crust as a result of near-surface loading. Thus, the Boulder batholith, which is a plausible analog for the Tertiary batholiths inferred to underlie Tertiary volcanic fields such as the San Juan or Latir fields (Lipman, 1984), occupies a broad synclinal sag defined by wall rocks (*Hamilton and Myers, 1974*). In addition to subsidence of the floor below a batholith, its roof will be broadly lifted and spread as the magma body approximates a position of neutral buoyancy. These roof-zone processes are demonstrated by the uplifts and extensional structures observed over crests of the Tertiary batholiths at Questa and in the San Juan Mountains (Steven and Lipman, 1976; Lipman, 1988). Additional space is generated by block stoping, as evidenced by net-veined roof rocks, although no large-scale subterranean "cauldron-subsidence" has been recognized in shallow batholiths of the U.S. Cordillera, comparable to that speculatively inferred to be common in the Peruvian Andes (*Pitcher, 1978*). The disruption and rotational tilting of large roof pendants at the Questa caldera (*Lipman, 1983,* 1988) suggest a further mechanism by which large slabs of the batholith roof may founder and settle rapidly in an extensional environment above the magma chamber. Steeply dipping roof pendants, which are common within and between plutons in the Sierra Nevada batholith, may reflect such rotational spreading and sinking of roof rocks at deeper crustal levels (Tobisch and others, 1986). Possibly, some tilted roof blocks sink more or less intact to deep batholithic levels, eventually to recrystallize, partially melt, and constitute parts of mid-crustal migmatitic floors of batholiths.

Mid-crustal depth levels, exposed in tectonically unroofed "core complexes" along a north-south belt the length of the central Cordillera, include characteristic peraluminous two-mica granites, commonly containing garnet (*Miller and Bradfish, 1980; Kistler and others, 1981;* Miller and Barton, 1990). Such Cretaceous and Tertiary two-mica granites are in striking contrast to the metaluminous granitic rocks, commonly containing biotite and hornblende as dominant mafic minerals, that are volumetrically predominant in upper crustal granitic plutons and batholiths. Peraluminous phenocryst mineralogy is also absent from

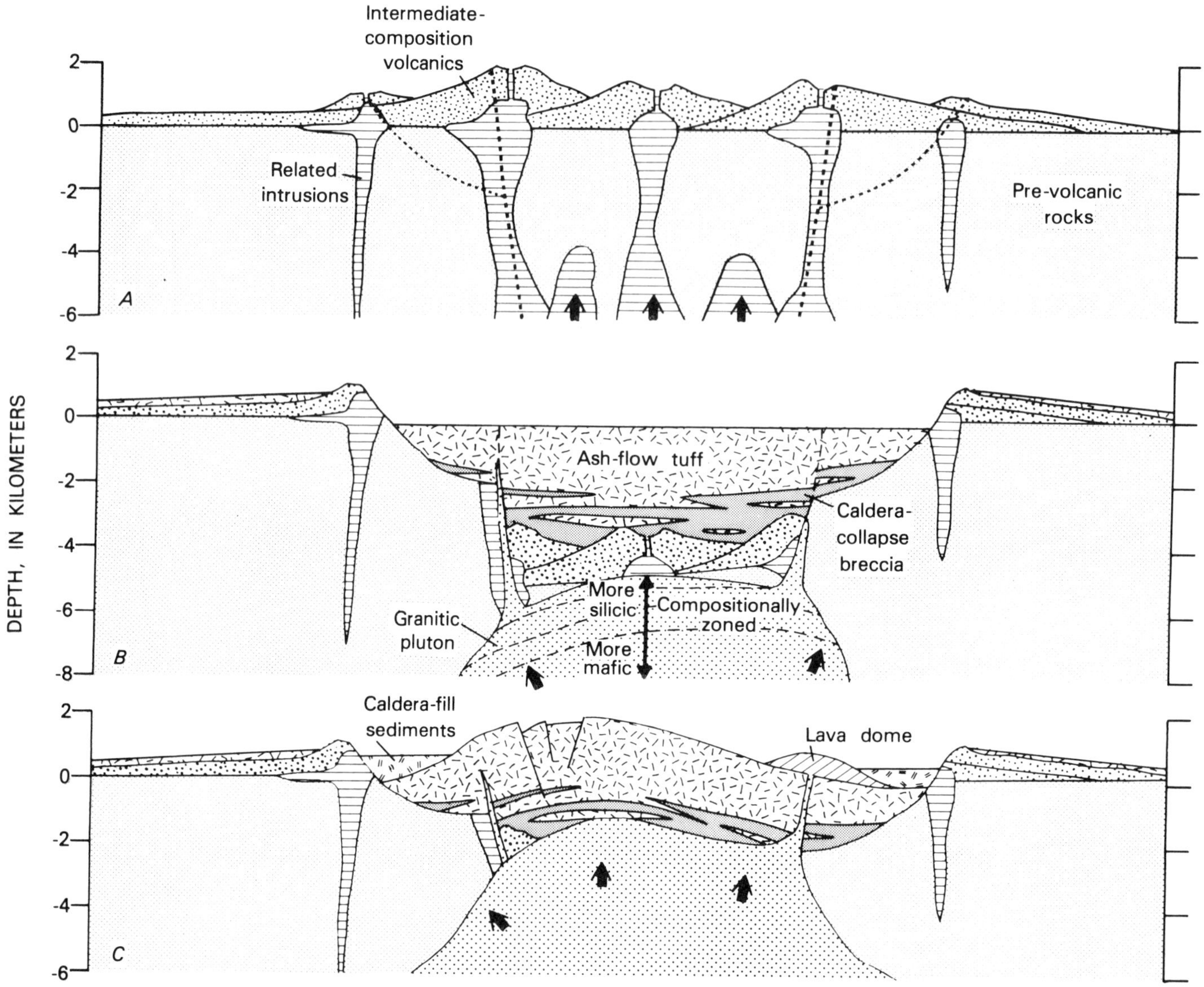

Figure 9. Model for ash-flow caldera cycle (Lipman, 1984; modified from Smith and Bailey, 1968). These diagrams incorporate many common elements of ash-flow calderas; complex alternative scenarios are possible. A. Precollapse volcanism. Clustered intermediate-composition stratovolcanoes grow over isolated small, high-level plutons that mark beginning of accumulation of batholithic-size silicic magma body that will feed ash-flow eruptions. Uplift related to emplacement of plutons leads to development of arcuate ring fractures; site of subsequent caldera collapse indicated by dotted lines. Heavy arrows indicate upward movement of magma. B. Caldera geometry just after ash-flow eruptions and concurrent caldera collapse. Central area of clustered earlier volcanoes caves into collapsed area. Intracaldera tuff ponds during subsidence and is an order of magnitude thicker than cogenetic outflow ash-flow sheet. Initial collapse along ring faults is followed by slumping of oversteepened caldera walls and accumulation of voluminous collapse breccias that interfinger with ash-flow tuff in the caldera-fill sequence. Caldera floor subsides asymmetrically and is tilted to the left side of diagram. Main magma body underlies entire caldera area and is compositionally zoned (or was prior to eruptions), becoming more mafic downward. C. Resurgence and postcaldera deposition. Resurgence is asymmetric, with greatest uplift in area of greatest prior collapse. Extensional graben faults form over crest of the dome. Some resurgent uplift is accommodated by movement along the ring faults in a sense opposite that during caldera subsidence. Magma body has risen into volcanic pile and intrudes cogenetic intracaldera welded tuff. Original caldera floor has been almost entirely obliterated by rise of the magma chamber to near the level of prevolcanic land surface. Caldera moat is partly filled by lava domes and volcaniclastic sediments. Hydrothermal activity and mineralization become dominant late in cycle.

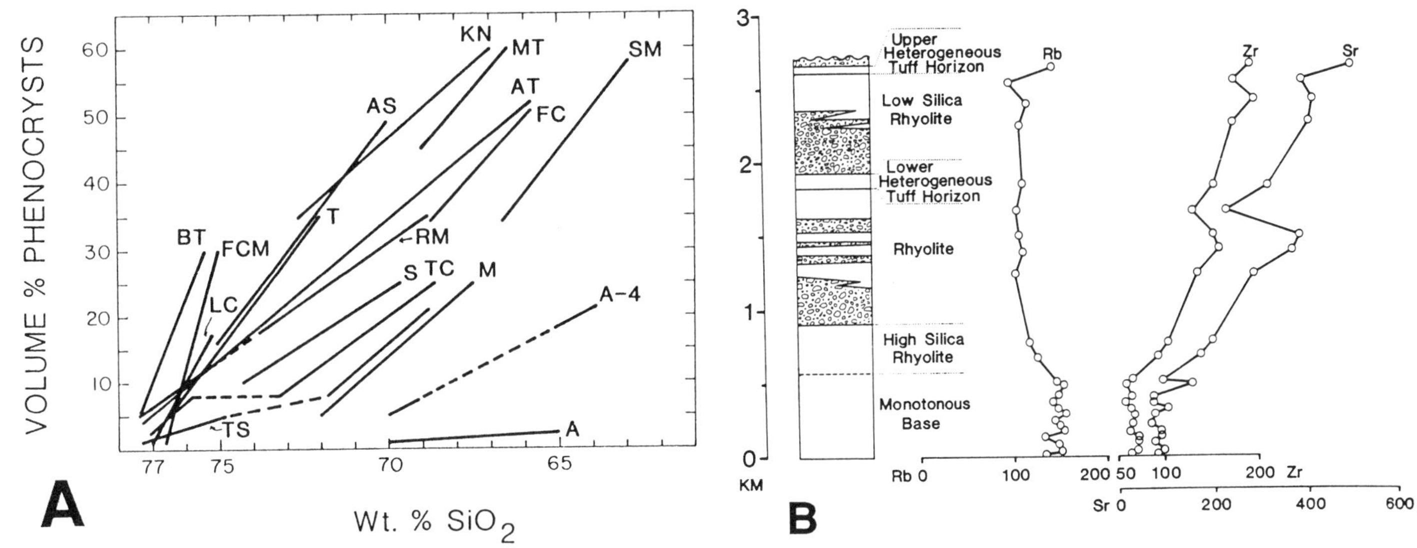

Figure 10. Compositional gradients in ash-flow tuffs. A. Generalized variations of phenocryst contents with bulk-rock SiO_2 for selected eruptions. Western U.S. units: AS, Apache Spring Tuff, Arizona; AM Ammonia Tanks tuff from Southwest Nevada field; BT, Bishop Tuff from Long Valley caldera; FC, Fish Canyon Tuff from San Juan field; FCM, Fish Creek Mountains Tuff, Nevada; KN, Kneeling Nun Tuff, New Mexico; LC, Lava Creek Tuff from Yellowstone caldera; M, Mazama eruption from Crater Lake; MT, Monotony Tuff, Nevada; RM, Rainier Mesa tuff from Southwest Nevada field; SM, Snowshoe Mountain Tuff from San Juan field; T, Tshirege tuff from Valles caldera; TC, Tiva Canyon tuff from Southwest Nevada field; TS, Topopah Spring tuff from Southwest Nevada field (Hildreth, 1981). B. Thick section of intracaldera Grizzly Peak Tuff, Colorado, showing variations in trace-element abundances (ppm) in welded tuff. Conglomerate pattern indicates caldera-collapse breccia zones (*Fridrich and Mahood, 1987*).

volcanic suites of the western U.S., and is exceedingly rare elsewhere in the world (exceptions have been documented in Peru and Bolivia: *Noble and others, 1984; Koeppen and others, 1987*), indicating that such magmas rarely rise into or through the upper crust. The micaceous mineralogy and peraluminous normative compositions indicate derivation from hydrous aluminous source materials; high volatile contents are also indicated by widespread pegmatite development. The high water content of rising peraluminous magma would lead to saturation and separation of a volatile phase at relatively great crustal depth, raising the granite solidus. Rise of such magmas would also lead to crystallization, simply by crossing the solidus (*Hyndman, 1981*). Rapid crystallization would thus result at mid-crustal levels, inhibiting the rise of magma to subvolcanic environments. Scarcity of such peraluminous compositions, or magmatic epidote that can be stable in high-pressure calc-alkaline magmas (*Zen and Hammarstrom, 1984*), in Cordilleran volcanic deposits also casts doubt on petrogenetic models that invoke voluminous eruptions of silicic magmas rapidly from middle or lower crustal depths (Whitney and Stormer, 1985; Stormer and Whitney, 1985).

Compositional gradients in magma chambers

Compositional variations in magmas are ephemeral and difficult to isolate for study, being readily disrupted by eruptive processes and overprinted by later crystallization in plutonic environments. Particularly attractive for detailed petrologic study are individual volcanic deposits that preserve quenched records of compositional gradients in their source magma chamber (Fig. 10).

Many ash-flow deposits contain vertical compositional gradients, typically becoming more mafic and phenocryst-rich upward, and implying compositional zoning or layering that became more mafic downward in the source magma chamber (Williams, 1942; *Smith and Bailey, 1966;* Lipman and others, 1966; Hildreth, 1981). Interpreted equilibrium conditions, based on the phenocryst and chemical gradients, also indicate downward increases in crytallization temperatures and decreases in volatile contents. Well-documented examples from the western U.S. include the culminating 6,850-B.P. eruption from Crater Lake (Williams, 1942; *Bacon and Druitt, 1988*); the 0.7-m.y.-old Bishop Tuff from Long Valley caldera (Hildreth, 1979; *Halliday and others, 1984*); several Miocene tuff sheets from the southwestern Nevada volcanic field (*Lipman, 1971;* Byers and others, 1976, *1989; Broxton and others, 1989; Schuraytz and others, 1989; Vogel and others, 1989*); several tuff sheets from the McDermitt caldera complex of Nevada and Oregon (*Conrad, 1984*); and tuffs from the Grizzly Peak, Lake City, and Bachelor calderas in Colorado (*Fridrich and Mahood, 1987; Hon and Lipman, 1989;* Whitney and others, 1988). Rare examples of "reverse" gradients, in which the erupted material becomes more silicic during the eruptive sequence (*Ekren and others, 1980*), probably result mainly from atypical geometry of the vent and conduit, initially drawing from lower levels along the sides of a zoned magma chamber. Alternatives involving mafic magma overlying silicic magma in the source chamber would be gravitationally unstable.

The origin of such compositional gradients remains controversial, but increasing evidence from mineralogically diverse deposits favors a dominant role for early crystal fractionation. The observed phenocrysts in many zoned tuff units appear to have crystallized largely in place, after the compositional gradients were established. Favored fractionation models invoke crystallization along walls or bottom of the magma body, accompanied by buoyant rise of more-evolved interstitial melt (*McBirney, 1980; Bacon and Druitt, 1988; Sawka and others, 1990*). Dynamic modelling and petrologic observations suggest that such crystallizing magma chambers can develop multiple doubly diffusing convecting cells in their density-stratified upper parts, leading to stable, compositionally uniform layers or gradations during continued crystallization (Sparks and others, 1984; *Fridrich and Mahood, 1987*).

Fractionation by crystal settling within upper parts of the magma chamber that were subsequently erupted can be shown to be improbable, both from the lack of appropriate crystals in late-erupted tuffs, which could have settled from higher levels in the source magma reservoir (Lipman and others, 1966; Hildreth, 1979), as well as by rheological limitations on extensive settling in silicic melts (Sparks and others, 1984; *McBirney and others, 1985; Marsh, 1988*). Some concerns that the chemical gradients cannot be explained by crystal fractionation, based on modelling using the observed phenocrysts of the erupted rocks, have led to appeals to poorly understood processes of liquid fractionation (Hildreth, 1979, 1981). Such difficulties now seem largely explained by: (1) the probability that much of the crystal fractionation occurred at depths below the sampled upper part of the system, involving somewhat different phenocryst phases and compositions than those contained in the eventually erupted differentiate; and (2) major roles for accessory minerals such as sphene, apatite, allanite, and monazite in accounting for trace-element contents of the magma (Cameron, 1984; *Johnson and Lipman, 1988*). Such accessories have highly variable trace-element contents and distribution coefficients that appear to vary greatly with bulk compositions, compositional gradients, pressure, temperature, and volatile species (*Mahood and Hildreth, 1983*). Even transitions from calc-alkaline to peralkaline compositions have been modelled by such fractionation processes (*Conrad, 1984; Johnson and others, 1989*). Some role for volatile-driven fractionation, perhaps involving halogen complexing, also seems possible, especially in alkaline and peralkaline systems (*Novak and Mahood, 1987; Johnson and others, 1989*).

A further complexity in interpreting such compositional gradients is evidence for the existence of a second discrete magma type in several systems, typically as a volumetrically subordinate mafic component, that cannot be cogenetic in any simple fractionation process. Examples include the concurrent low- and high-Sr magmas at Crater Lake (*Bacon and Druitt, 1988*) and several examples from the Oligocene-Miocene San Juan field: low- and high-Zr magmas at the Platoro caldera complex; calc-alkaline rhyolite and mafic alkalic high-Ba magmas at the Bachelor caldera; and simultaneously active, dominant alkaline, and subordinate calc-alkaline magma systems at Lake City caldera (*Dungan and others, 1989; Hon and Lipman, 1989;* Whitney and others, 1988). In each system, emplacement of a volumetrically subordinate mafic component accompanied, and perhaps triggered, the ash-flow eruptions and caldera subsidence. These relations, based on detailed petrologic study of discrete pumice and scoria populations, suggest that large upper crustal magmatic systems typically evolve rapidly by complex processes of differentiation and magma replenishment prior to voluminous pyroclastic eruptions, for which much of the evidence is obscured by convective mingling in the magma chamber and mechanical mixing during eruption and emplacement.

In contrast to such geometries inferred for subvolcanic magma bodies, vertical compositional changes are rare in solidified granitic plutons, especially at mesozonal depths. Such vertical gradients have been documented along roof zones of a few bodies (Barnes, 1983; *John, 1988; Johnson and others, 1989*); more typically, mesozonal granitic plutons vary laterally from mafic margins to silicic interiors (Bateman and Chappell, 1979; Noyes and others, 1983). Such lateral compositional transitions are locally abrupt, but commonly are at least partly gradational. In carefully studied examples such as the Cretaceous Tuolomne Intrusive Suite in the central Sierra Nevada, crystal-fractionation models can account for much of the variation, but isotopic variations indicate that additional magma batches were also involved (*Kistler and others, 1986; Hill and Silver, 1988*). Thus, such variations in mesozonal magma bodies appear to result from processes similar to those inferred from volcanically quenched deposits, involving combined mixing of discrete magma batches and progressive crystallization along walls and floors of the magma reservoir, associated flow of residual melt to the interior of the pluton, and continued upward movement of lower-density residual liquid in the core (*McBirney, 1980; Sawka and others, 1990*).

Such features in mesozonal plutons probably largely record somewhat later or deeper stages in the fractionation and crystallization of a magma body than the volcanic deposits, as evidenced by the virtually ubiquitous development of mineral foliation during flow. High-level subvolcanic granitic plutons typically lack a mineral foliation, indicating that upward movement was largely complete prior to extensive crystallization. In addition, rapid crystallization in upper parts of such plutons may have resulted from eruption-related volatile release, and upper parts of many large plutons probably had brief crystallization histories in comparison with deeper levels of the same bodies. Granitic plutons also tend to be less silicic than associated volcanic rocks, indicating that the more evolved magmas are preferentially erupted, before the residuum solidifies to form the granitic body (Lipman, 1984; *Mahood, 1987*).

A few subvolcanic plutons show "reverse" compositional gradients, in which the pluton becomes more mafic toward the interior (*Fridrich and Mahood, 1984; Nabelek and others, 1986; Johnson and others, 1989*). These are interpreted as recording late-stage upward movement and displacement of previously more nearly horizontal compositional gradients in the magma chamber, rather than any equilibrium geometry during crystallization and differentiation.

Magma chamber dynamics

The complex behavior of large bodies of silicate melt, in which crystal and volatile contents, thermal gradients, and other physical properties vary vertically and laterally, has been much studied theoretically, modelled in laboratories, and applied to observed features of igneous bodies. In the western U.S., building on studies by Sparks and others (1984, among many papers) and *McBirney and others (1985),* much emphasis has been placed on the role of doubly diffusive convection in upper parts of silicic magma bodies, accompanied by magma replenishment and sidewall crystallization in lower parts. Such mechanisms help account for: (1) discrepancies between the minerals inferred to fractionate to produce the observed chemical gradients versus the phenocryst assemblages actually present; (2) compositional gaps documented within some zoned ash-flow sheets; (3) the sustained productivity of some upper crustal magmatic systems; and (4) the contrasts between zoned ash-flow sheets, which appear to have tapped vertical compositional gradients in subvolcanic magma bodies, and zoned plutons in which compositional gradients are dominantly lateral. Most of these features have already been mentioned; a few deserve further comment.

Some compositional gaps in volcanic units and lateral composition boundaries within shallow plutons may also record varying draw-down rates during the eruption, accompanied by disruption of the initially subhorizontal gradients in magma reservoirs by differential draining during eruptions (*Spera and others, 1986; Blake and Ivy, 1986*). The sequence of compositions erupted is influenced by the ratio of magma-chamber diameter to conduit diameter, which affects the amount of magma mixing, and the amount of evolved viscous magma left in the chamber. Dynamic aspects of magma rheology also may account for recovery processes in large magma chambers that have been disturbed by large explosive eruptions, leading to caldera resurgence (*Marsh, 1984; Gebhart, 1987*).

Many petrologists working on Cordilleran magmatic systems have ascribed a significant role for magma mixing (*Larsen and others, 1938;* Eichelberger, 1975; *Anderson, 1976*), based on evidence from both volcanic and plutonic systems (Fig. 11). In silicic volcanic rocks, ovoid inclusions of basalt or andesite, characterized by embayed outlines and fine-grained margins, have been interpreted as resulting from quenching of mafic magma injected into a dominant volume of more silicic melt characterized by a lower liquidus temperature (Bacon, 1986; *Barnes and others, 1986; Sparks and Marshall, 1986*). Variably resorbed

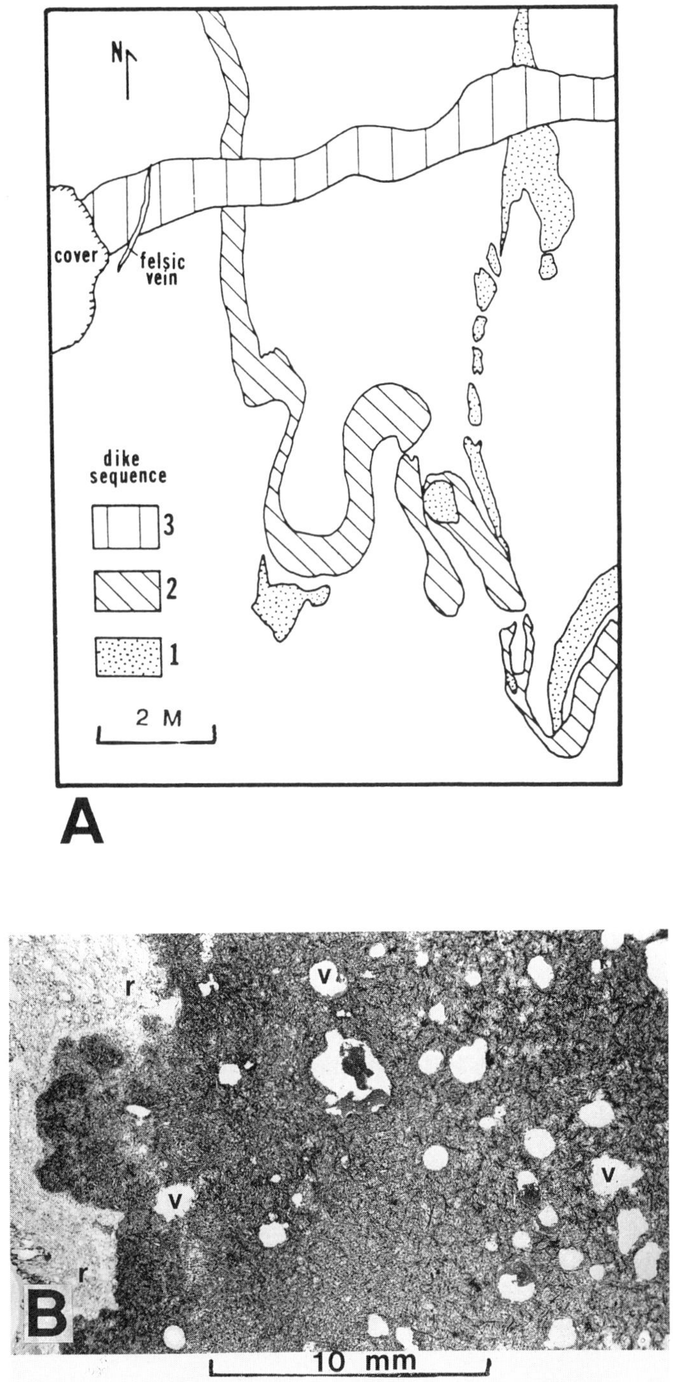

Figure 11. Textural evidence for mixing of mafic and silicic magmas. A. Sequence of intrusions and deformation of mafic dikes in the mixed zone of the Wooley Creek batholith, northern California (*Barnes and others, 1986*). Such synplutonic dike emplacement is a probable common process in generating discrete mafic inclusions. Earliest dike shows greatest amount of disruption and was locally engulfed by second dike. Dike 3 averages 1 m in width; it is cut by a felsic vein. B. Photomicrograph of rounded embayed basaltic inclusion in glassy rhyolite host (r) showing fine-grained quenched margin and vesicular (v), more coarsely crystallized interior (Bacon, 1986). Such textures are interpreted to indicate that liquid basalt quenched against rhyolitic melt with a lower solidus temperature.

disequilibrium phenocrysts in andesite lavas, such as quartz, alkali feldspars, and biotite, have similarly been interpreted as due to mixing and nearly complete disaggregation of silicic melt into a more mafic dominant composition (*Larsen and others, 1938;* Eichelberger, 1975; *Johnson and Lipman, 1988*). In Cordilleran granitic terranes, disruption of synmagmatic mafic dikes, leading to formation of rounded mafic inclusions (pillows), has been repeatedly interpreted as evidence of injection of still-liquid granitic magma by more mafic magma (*Roddick and Armstrong, 1959; Lipman, 1963; Furman and Spera, 1985; Barnes and others, 1986; Frost and Mahood, 1987; Hyndman and Foster, 1988*).

Seemingly identical mafic inclusions have been cited by Australian petrologists in support of the presence of large volumes of restite (incompletely melted remnants of the source terrane) in granitic and volcanic magmas (*Wyborn and Chappell, 1986; Chappell and others, 1987*). Some component of restite in Phanerozoic silicic magmas is demonstrated directly by the presence of zircon crystals containing resorbed cores of Precambrian age in Tertiary granitic plutons in New Mexico (*Johnson, 1989*), and probably widely elsewhere in the Cordilleran region. The key problem is the relative importance of such restite material in silicic magmas versus magma-mixing processes.

Little evidence exists for any voluminous restite component in Cordilleran magmas, especially as represented by volcanic phenocrysts. Chappell and associates interpret the phenocrysts of Paleozoic crystal-rich (40 to 60 percent) dacitic tuffs in Australia as largely restite in origin, based on bulk-rock compositional similarities with associated granites. In broadly similar tuffs in the U.S. Cordillera, including the voluminous Fish Canyon Tuff (3,000 km^3) of the San Juan field, which is cited by the Australian group as a likely example of restite "phenocrysts" (*Wyborn and Chappell, 1986*), all the phenocryst phases are characterized by extensive solid solution of non-ideal components (high Ca and Ba contents of alkali feldspar, high K in plagioclase, high Mn in magnetite and ilmenite, etc.). Such solid solution indicates crytallization at high temperature and relatively low pressure, in comparison with the mineral chemistry of high-grade metamorphic rocks that could provide a restite component during partial melt in the lower crust. In the case of the Fish Canyon Tuff, candidates for a possible restite component among the major phenocryst constituents seem limited to rare resorbed, calcic cores of plagioclase phenocrysts that show less solid solution than rim compositions and define a discrete trend in the ternary feldspar system (Whitney and Stormer, 1985, Fig. 2). Such plagioclase cores, which constitute only a fraction of a percent of the total feldspar population, provide an upper limit to any possible restite component; alternatively, such cores may be relict from magma mixing events.

In upwelling, relatively dry partial melts of lower crustal material, most initially entrained restite (other than highly refractory minerals such as zircon) is probably completely melted as a consequence of adiabatic depressurization during slow rise of the melt diapir. Only rarely, where the magma rises exceptionally rapidly and is quenched by volcanic eruptions or in a shallow plutonic environment, will a sizeable restite component survive. Nevertheless, more rigorous evaluation of possible restite components is needed, especially for accessory phases such as sphene and allanite, which can be complexly resorbed and strongly zoned (*Czamanske and Dillet, 1988*). Such accessories are primary repositories of rare earth and other trace elements that provide key constraints for fractionation modelling in silicic magmas; any significant restite component would invalidate some basic assumptions of the fractionation modelling. Additional insights into such problems of crystallization/resorption equilibria may be obtainable from quantitative studies of phenocryst size and shape populations, similar to those used to trace the crystallization history of Mount St. Helens dacite (*Cashman, 1988*). Such studies provide information on equilibrium rates, interruptions and discontinuities in crystallization, degree of undercooling, and eruptive probability (*Marsh, 1981*).

Magma sources

The Cordilleran region provides a rich assemblage of alternative sources of varying ages and compositions for the generation of magmas, including asthenosphere, lithospheric mantle, lower crust, and upper crust. Combined isotopic and petrologic approaches to this potential geochemical nightmare now indicate that most Cordilleran igneous rocks, both basaltic and silicic, owe their dominant compositions largely to lower crustal and mantle components ofthe continental lithosphere. Changing ages and compositions of the lithosphere are reflected by regional isotopic patterns in strontium (Leeman, 1970; *Armstrong and others, 1977*), lead (Zartman, 1974), and neodymium (DePaolo and Farmer, 1984; *Farmer and DePaolo, 1984*).

Both basalts and rhyolites of the late Cenozoic Snake River Plain–Yellowstone region lie on a single Pb-isotope trend, which is commensurate with derivation of the lead from continental lithosphere stabilized at the time of formation of the exposed Archean upper crust (Peterman and others, 1970; Doe and others, 1982). The basalts, including tholeiites approaching ocean-ridge types in composition, are thought to be largely melts of the mantle keel of the continental lithosphere without significant asthenospheric components (*Leeman, 1982*), whereas the associated rhyolites are dominantly derived from lower crustal materials.

Similarly, the isotopic systematics of middle Tertiary volcanic rocks of the southern Rocky Mountains indicate predominantly lithospheric sources, but of Proterozoic age commensurate with the known age of exposed cratonic basement in the region (Lipman and others, 1978; Dungan and others, 1986; Johnson and others, 1990). Though these volcanic rocks are predominantly of intermediate to silicic compositions, they are widely thought to result from mantle processes (Fig. 12) involving ponding of basaltic melt near the base of the crust, thermal interchange, and attendant partial melting of lower crust (Hildreth, 1981). A complicating factor, particularly for Sr isotopic compositions, is the conspicuous effect of small amounts of assimilated

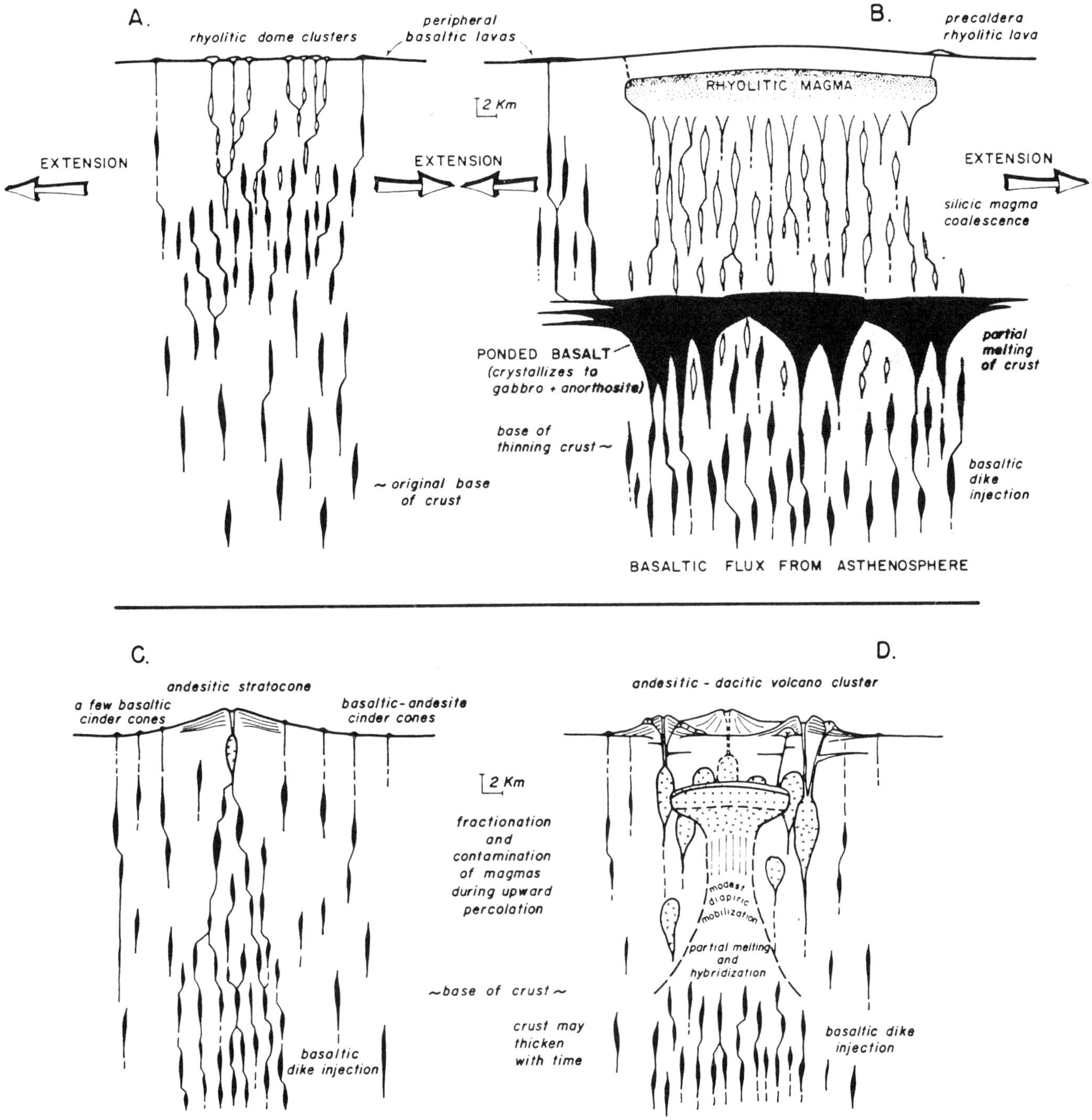

Figure 12. Contrasting styles of lithosphere magmatism (Hildreth, 1981). Top pair of cartoons depicts rhyolitic–basaltic magmatism under stress conditions favoring marked crustal extension: (A) modest power input; (B) large power input. Shaded regions indicate partial melting of crustal rocks sufficient to permit separation of rhyolitic magmas. Lower pair of cartoons depicts two possible stages in development of volcanic systems where tectonic extension, if any, is subordinate and shallow: (C) early stage; (D) intermediate stage. This model applies to island-arc, continental-margin arc, and continental-interior systems that produce abundant intermediate-composition magmas.

upper crust on evolved silicic magmas containing lower Sr concentrations (*Noble and Hedge, 1969; Halliday and others, 1984; Tegtmeyer and Farmer, 1990*). Controversial suggestions that the apparent lower crustal component in Cordilleran magmas is melted from subducted sediments, and that crustal melting is minor (*Cameron and Cameron, 1985; Cameron and others, 1989:* compare *Ruiz and others, 1988*), seem regionally improbable as documented by isotopic studies in the Rocky Mountains, where the chemical signature of thick, old cratonic crust is strong (Lipman and others, 1978; Doe and others, 1982; Johnson and others, 1990) and also in the South American Andes (Hildreth and Moorbath, 1988). Large apparent mantle components in some large-volume silicic systems in the U.S. Cordillera (*Halliday and others, 1984; Musselwhite and others, 1989; Tegtmeyer and Farmer, 1990*), and also in the Peruvian Andes (*Boily and others, 1989*), likely reflect hybridization of the lower crust by earlier mantle-derived basalt magmas that ponded in or near the base of the crust, rather than the absence of significant lower crustal melting (Johnson and others, 1990).

Phanerozoic granitic rocks of the Cordillera record a generally similar story of dominant, lower crustal sources, inferred to have been mobilized by input of mantle basalt (DePaolo and Farmer, 1984; *Farmer and DePaolo, 1984; Ague and Brimhall, 1987, 1988*). Some geochemical contrasts between such granitic rocks and their volcanic counterparts appear to be due to fractionation under differing volatile contents and degrees of crystallization. Subvolcanic granitic magmas appear typically to preserve mainly the late stages of evolution of the magmatic system, at times when crystal content was relatively high and volatile content low (*Johnson and others, 1989*). The result is a reduced component of upper-crustal assimilation.

Only along the plate margin and in cratonic areas of extreme extension do voluminous basaltic rocks contain a distinct asthenospheric signature. Examples include some late Cenozoic basalts from the southern Rio Grande rift, characterized by Sr and Nd isotopic values that are less evolved than middle Tertiary igneous rocks of the same region (*Lipman and Mehnert, 1975;* Perry and others, 1987). Similarly, the voluminous and compositionally diverse basalt fields of the Pacific Northwest have been interpreted to reflect variable source contributions from enriched (asthenospheric) and depleted (cratonic) mantle components, and further modified by modest crustal assimilation in a continental back-arc tectonic setting (*Carlson and Hart, 1987*). Less voluminous alkalic basalts, erupted peripherally to the Rio Grande rift and other extensional areas, may represent more variable mixtures of asthenospheric and lithospheric sources, as indicated by varied elemental and isotopic compositions (Leat and others, 1988; Roden and others, 1990).

Magmatism in relation to tectonic setting

Early 20th century work emphasized broad correlations between magmatic types and regional tectonic environment, especially in relation to concepts of the geosynclinal cycle (*Barrell, 1917; Knopf, 1948; Waters, 1955*). Ideas about relations between Phanerozoic volcanism and regional tectonism in the western U.S. have changed greatly in the past two decades under the influence of developing plate tectonic models for this complex region, but many uncertainties and controversies remain.

The western margin of the Phanerozoic Cordilleran craton was established during the late Proterozoic and early Paleozoic by a series of rifting events, with which are associated volumetrically minor tholeiitic basalt (*Stewart, 1972; Larson and others, 1985*). Most of these rocks are variably metamorphosed and little studied petrologically, but they appear broadly similar to rift-related continental tholeiites elsewhere.

During much of the Paleozoic, the Cordilleran region was a passive plate margin lacking associated magmatism (Poole, this volume), and the character of local and volumetrically minor igneous activity associated with sporadic Paleozoic tectonism remains incompletely deciphered (*Evans and Zartman, 1988*). Little magmatism appears to be associated with the broadly concurrent foreland compression the late Paleozoic Antler orogenic belt. Most Paleozoic igneous rocks in the Klamath Mountains and Sierran foothills of northern California, including andesitic arc volcanics, associated granitic plutons, and ultramafic ophiolitic suites, constitute accreted exotic terranes that were not generated within the American plate. Various reconstructions would have these marginal igneous belts initially developed as small offshore ocean basins and fringing volcanic arcs, or alternatively, as far-travelled exotic terranes (Beck, this volume).

Mesozoic time marked the initiation of sustained continental-margin arc volcanism and associated batholithic plutonism along the western margin of the American plate. Early Triassic Sonomian deformation was accompanied by a discontinuous volcanic arc along much of the western Cordilleran margin (*Dilles and Wright, 1988;* Saleeby and others, this volume). Continental-arc magmatism appears to have been nearly continuous along the plate edge throughout Late Triassic and Jurassic time, and magmatic composition became more alkalic eastward (*Miller, 1978*). In the central Sierra Nevada, where especially detailed geochronologic data are available, the earliest plutons—at 215 Ma—were restricted to the east side of the eventual batholith; in contrast, Jurassic plutons were emplaced discontinuously across the full width of the batholith belt (*Chen and Moore, 1982*). Later, the volumetrically dominant Cretaceous pltonism appears to have migrated steadily eastward between 120 and 80 Ma. A remaining complexity is the degree to which this shifting pattern of magmatism is related to varying rates of plate convergence, as opposed to recording intermittent docking of accreted terrains.

At the end of the Mesozoic, the distribution and compositions of igneous rocks in the Cordilleran region underwent major changes. Understanding of the complex age-composition patterns has blossomed in recent years as abundant radiometric dates and chemical compsitions became available. Lipman and others (1971) interpreted the changing distribution and compositions of Cenozoic volcanic deposits (Fig. 3) in relation to the evolving

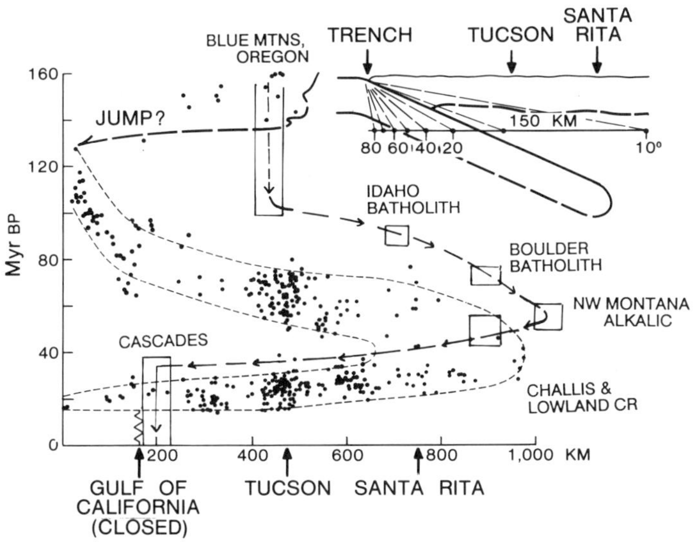

Figure 13. Ages of igneous activity versus distance from western margin of American plate. Dots are data plotted by Coney and Reynolds (1977); boxes are approximate ages of major igneous features in the northern U.S. Cordillera, compiled from multiple sources. In both southern and northern regions, the early Tertiary eastward sweep of magmatism has been interpreted as related to flattening of the oceanic crustal slab being subducted beneath North America; the middle Tertiary magmatic flare-up has been variably interpreted as due to retrograde slab steepening (Coney and Reynolds 1977; *Keith, 1978*), intra-arc extension (*Elston and Bornhorst, 1979;* Lipman, 1980), or intra-craton extension unrelated to plate-margin processes (Wernicke and others, 1987). Though the patterns are similar, the timing of eastward migration, followed by westward step is asynchronous between the two regions.

geometry of interactions between the American and various Pacific oceanic plates (Atwater, 1970). The early to middle Tertiary subduction geometry appears to have involved an abrupt hinge between steeply and gently dipping subduction segments (Lipman, 1980), rather than an "imbricate" double system as originally inferred. Age and composition data on late Mesozoic–early Tertiary igneous rocks in southwestern North America suggest that the magmatic arc that was initially near the continental margin swept about 1,000 km northeast as the Benioff zone flattened to <15° (Coney and Reynolds, 1977). A roughly similar pattern of volcanic migration occurred about 20 m.y. earlier in the northern Rocky Mountains (Fig. 13). The role of changing rates of plate convergence on patterns of volcanism, influenced by the gravitational instability of oceanic lithosphere, was further explored by Cross and Pilger (1978) and *Glazner and Supplee (1982)* for the Cordilleran. Tectonic accretion continued locally along the plate margin in the Tertiary, adding the Eocene oceanic basalt terranes to coastal Oregon and Washington (*Snavely and others, 1968*).

The late Cenozoic volcanic evolution of the Great Basin and adjacent regions has been interpreted in terms of basalt and bimodal rhyolite-basalt suites in an extensional tectonic environment developed in back-arc and transform settings (Christiansen and Lipman, 1972; Christiansen and McKee, 1978; Dickinson and Snyder, 1979). Faulting and volcanism were increasingly concentrated near margins of the extended region and along linearly propagating volcanic systems in Idaho and Oregon, which mark transverse accommodation boundaries for the extension.

A striking feature of the Cenozoic volcano-tectonic evolution was the development of a unified stress field across the entire width of the Cordillera beginning during Laramide deformation, as indicated by orientation of dikes, faults, and other stress indicators (Rehrig and Heidrick, 1976; Zoback and Zoback, 1980; Aldrich and others, 1986). This continuity indicates the dominance of broad regional causes, rather than local anomalies, for the complex magmatic and tectonic patterns of the Cenozoic. For example, a change in stress geometry from southwestward- to westward-directed extension accompanied Miocene changes in magmatic and tectonic styles in the northern and southern Basin-Range (*Zoback and Thompson, 1978; Shafiqullah and others, 1980*) and in the Rio Grande rift (*Lipman, 1981;* Morgan and others, 1986).

The above studies and others have mostly related the volcano-tectonic patterns to changing plate velocities and plate-boundary interactions, although significant effects from preexisting zones of weakness and previously deformed and thermally conditioned regions within the continental crust were recognized to concentrate recurrent volcanic activity along discrete magmatic loci (Snyder and others, 1976). In contrast, *Coney and Harms (1984)* and Wernicke and others (1987) proposed that gravitational stresses within the North American plate, remaining from late Mesozoic compression and crustal thickening, have dominated over plate-interaction forces as controls of the Cenozoic tectonomagmatic evolution. Others infer generation of eastern Cordilleran Laramide to mid-Tertiary magmas as a consequence of localized mantle upwelling, independently of plate-margin subduction (*Mutschler and others, 1988*). Such processes, while probably contributing to the overall complex evolution in the western Cordillera, are inconsistent with many features of Cordilleran geology, such as the coherent magmato-tectonic patterns across the full width of the Cordillera, the timing of some volcanotectonic events, petrologic similarities between eastern Cordilleran igneous rocks and those on cratonic sides of other continental-margin arcs such as the Andes, and petrologic and seismic tomographic evidence for presence of the subducted Farallon plate under the eastern Cordilleran region.

Particularly problematic for interpretations that attribute complexities in the distribution of Cenozoic magmatism and extension to earlier compression and crustal thickening (Wernicke and others, 1987) are inconsistencies in the time-space relations between crustal thickening, magmatism, and extension along the Cordillera. For example, the major regional uplift and crustal thickening in the eastern Cordillera and adjacent High Plains (*Axelrod and Bailey, 1976; Trimble, 1980;* Eaton, 1987) has occurred mainly since late Miocene time, well after (rather than before) the peak of mid-Tertiary magmatism and inception of extension. In the northern Rockies, extension began within the

Eocene magmatic arc immediately following crustal shortening and inferred thickening, but in southern Arizona, inception of extension within the region of the Laramide arc followed the termination of compression by at least 20 to 30 m.y. Extension rates also peaked at significantly differing times within some nearby areas: for example, 35 to 30 Ma in east-central Nevada, versus 17 to 19 Ma at Yerington, Nevada (Gans and others, 1989; *Proffett, 1977*).

These alternative interpretations, and the problems resulting from any one in isolation, suggest that the complex patterns of middle Tertiary magmato-tectonic evolution record the interplay between multiple processes, including: continuing subduction of the Farallon plate; influence of preexisting cratonic structures, including lithospheric flaws of Precambrian ancestry as well as more recent thermotectonic instabilities inherited from Laramide events; and effects of emplacing voluminous mantle-derived basalt into an increasingly inhomogeneously weakened Cordilleran lithosphere as extension became dominant over compressional plate processes. In the broadest sense, the complexities seem plausibly related fundamentally to the dying subduction system of a diminishing Farallon plate beneath North America, although detailed evaluation of such interrelated effects on the complex space-time variations in magmatic processes is likely to provide the basis for many years of continuing research and controversy.

A related concern has been apparent discrepancies between seemingly steady-state plate motions and episodic magmatism in the western U.S. (*Gilluly, 1973; McBirney and others, 1975*). Some of the apparent discontinuties result from failure to examine the igneous record for the full width of the Cordillera; others may result from the apparent paucity of magmatism resulting from low-angle subduction, as under present-day Peru (Pilger, 1984). Still, some fluctuations in magma production with time seemingly do not correlate with known changes in plate motions. Others have evaluated timing between magmatic episodes from different tectonic plates as tests of global plate models (*Erlich and Melekestsev, 1971; Kennett and others, 1977*); some regularities have emerged, but available data seem too incomplete and inconsistent among regions to permit meaningful generalizations without further analysis.

Much remains to be deciphered concerning relations between volcanism and plate tectonics in the western United States. Such attempts are handicapped because probable geometric and structural complexities within the Farallon and other plates east of the East Pacific Rise (*Jurdy, 1984;* Engebretson and others, 1985) have now been subducted. For example, the southward migration of peak magmatism along the eastern Cordillera and Rocky Mountain front in Eocene and Oligocene time (Figs. 3 and 13), which is seemingly inconsistent with Tertiary motions between the American and Pacific plates, possibly is related to oblique subduction of an oceanic plateau for which a partial record survives on the Pacific plate (*Henderson and others, 1984*).

Crustal evolution

The timing and location of Cenozoic magmatism in the eastern Cordilleran region appear closely associated with modifications in thickness and composition of the Precambrian cratonic crust. At present, crustal thickness varies only modestly (<5 km) across boundaries between the Rocky Mountains and the High Plains, despite large changes in mean elevation (*Prodehl and Lipman, 1989;* Smith, this volume). Regional seismic-refraction data indicate that isostatic balance is maintained mainly by lower average densities (from seismic velocities) in the crust and upper mantle beneath the Rocky Mountains in comparison with the adjacent High Plains. Shallow Mesozoic seas once spread across this entire region, and the present elevations and changed crsutal sections must be due largely to Cenozoic events. Some crustal thickening must have resulted from regional compression in the early Tertiary Laramide orogeny (*Bird, 1984*), but botanical and sedimentation data indicate that mean elevations were only about half the present values in the middle Tertiary (*Epis and Chapin, 1975; Axelrod and Bailey, 1976; Trimble, 1980*).

Local and regional uplift accompanied the middle Tertiary magmatism, as indicated by structural highs over the Tertiary batholiths, but most of the regional uplift is associated with extension along the Rio Grande rift and related structures. Uplift associated with rifting has continued to the present, as indicated by Holocene fault scarps, geomorphic features, and geochronologic evidence for uplift during the past few million years (*Scott, 1975; Taylor, 1975; Kelley and Duncan, 1986*). This late Cenozoic uplift of 1 to 1.5 km must be partly due to thermal expansion resulting from emplacement of magma as discussed above; a regional decrease in lithospheric thickness is probably also important (Eaton, 1987). The regional topography of the Rocky Mountains and adjacent flanks is geometrically comparable to an oceanic spreading ridge (*Cordell, 1978;* Eaton, 1987). Though the formative mechanisms must differ, a comparable thermal structure extending to the base of the lithosphere or deeper seems required.

The density of the lower crust and upper mantle may be reduced by ponding and underplating by mantle-generated basaltic magma at the base of the crust (Fig. 12). Widespread partial melting of lower crust due to heating by emplaced basalt, necessary to account for the voluminous middle Tertiary silicic magmatism, should be accompanied by crystallization and fractionation of the basalt (Lipman, 1988; Johnson and others, 1990). Such fractionation would contribute additional low-density material, either to reside in the lower crust or to augment the rising silicic magma bodies. Ultramafic cumulates from basalt fractionation may reside at the evolving crust-mantle petrologic boundary, effectively "overplating" the uppermost mantle, or possibly descend gravitationally deeper into the mantle (*Arndt and Goldstein, 1989*). Either process would effectively reduce the mean density of the lower crust and upper mantle as required by seismic-refraction data. Such open-system crust-mantle transfers also seem required to explain the high present-day heat flow along the Rio Grande rift and southern Rocky Mountains (*Decker and others, 1988*). Late Cenozoic continental alkalic basalts that reached the surface as lava must have interacted less with the lower crust than the middle Tertiary mantle melts that ponded near the base of the crust (*Cameron and others, 1989*). The state of crustal stress is probably the main control on whether mantle-derived mafic magma ponds and interacts low within the crust (plate convergence) or rises through the crust without major compositional modification (plate extension).

Inferences that significant volumes of basaltic magma have ponded buoyantly as tabular bodies in the middle crust of the Cordilleran region, based on seismic data (*Sanford and others, 1977; Brown and others, 1980*) and density considerations (*Glazner and Ussler, 1989*), remain unsubstantiated by geological observation. Large-volume sill-like bodies of basaltic composition are seemingly unknown in mid-crustal sections tectonically exposed in core complexes (*Crittenden and others, 1980*), while such bodies are common in stratified upper-crustal sections at depths down to a few kilometers (*Mudge, 1968*). Tabular gabbros that represent crystallized basalt ponded low in the crust are as much as 10 km thick in rare exposed lower-crustal sections, such as in the Ivrea zone of Italy and Kohistan section of northern Pakistan (*Zingg, 1983; Pin and Sills, 1986;* Hamilton, 1989). Alternatively, seismic-reflection bright spots, such as observed at mid-crustal levels in the Socorro, New Mexico, and Death Valley, California areas (*Brown and others, 1980; de Voogd and others, 1988*), may represent hydrothermal fluids concentrated along the brittle-ductile transition zone or at solidification interfaces in tabular granitic intrusions, especially in regions undergoing detachment-style crustal extension. Geologically exposed middle-and upper-crustal intrusions are predominantly intermediate to silicic in composition, and large negative Bouguer gravity anomalies asosciated with such plutons indicate that neutral buoyancy is incompletely achieved in magmatic processes.

Oceanic crust, subducted at low angle during early- and mid-Tertiary time, may also have been significant in initially generating melts as the subducted crust became heated and thermally equilibrated with adjacent asthenosphere. These melts could then have interacted with overlying lithospheric mantle and lower crust, even though little remains in the magmas that reached the surface that is geochemically identifiable as a slab contribution. Any plausible plate-tectonic reconstruction requires that thousands of kilometers of ocenic lithosphere were recycled into the mantle beneath the U.S. Cordillera during the middle Tertiary, but the role of the subducted slab remains obscure. Subduction of young, hot, near-ridge oceanic crust in the middle Tertiary, progressive termination of spreading along segments of the plate margin, and generation of a "slab window" (Dickinson and Snyder, 1979; *Damon, 1979*) provided additional contributions to the developing complexity of the asthenospheric thermal regime beneath the Cordilleran region. The mid-Tertiary high-K calc-alkaline igneous rocks of the Rocky Mountains are petrologically similar to products of other plate-convergence volcanism where thick crust overlies a subduction system, as in the central Andes. Inferred paleodepths to the subducted slab, based on compositional analogs with active volcanic arcs (Lipman and others, 1971; *Keith, 1978*), are likely to be only qualitatively meaningful, though, because of chemical interactions between mantle melts and lower crust. The regional compositional asymmetries of Cenozoic magmatism in the Cordillera thus appear largely related to the thickness and composition of the lithospheric mantle and crust traversed by the rising magmas, where their dominant chemical identities were established (Moore, 1962; Lipman, 1980; Elston, 1984b; Hildreth and Moorbath, 1988).

MAGMATISM AND SOCIETAL CONCERNS

Recent studies of igneous rocks in the U.S. Cordillera have been stimulated by national needs to evaluate volcanic hazards, to appraise thermal energy resources, and to improve understanding of metallic ore deposits. Emerging additional concerns involve handling and storing toxic wastes, for which several promising sites exist in igneous terranes.

Active volcanism and volcanic hazards

Prior to 1980, geophysical monitoring of volcanic activity in the United States, leading to hazard evaluations and forecasts, was largely confined to the active volcanoes of the Hawaiian Islands. Only a few volcanic events had occurred in the western U.S. within recorded historic times, that is, since about 1800. Most notable were several eruptions of Mount St. Helens in the middle 19th century and at Mount Lassen in 1915.

A program of volcanic hazard studies, initiated in the 1960s on potentially active volcanoes in the Cascade Range, used stratigraphic and geochronologic data to determine the frequency and style of past eruptions and to assess future hazards (Crandell and others, 1979). One early result was recognition that Mount St. Helens was the youngest and most active of the Cascade volcanoes, and was likely to erupt again "perhaps before the end of this century" (*Crandell and Mullineaux, 1978*).

This prediction, of course, came true with the catastrophic eruption of May 18, 1980 (*Christiansen and Peterson, 1981*). Immediate lessons of the May 1980 events included a new appreciation of the significance and hazards associated with explosive flank eruptions involving sector collapse, related debris avalanches, and lateral blasts. Evidence of similar prehistoric volcanic collapse events has since been recognized widely from deposits elsewhere, though not everywhere involving a major magmatic blast component (*Ui, 1983; Crandell and others, 1984; Siebert, 1984*). Also informative at Mount St. Helens was the documented complexity of deposits emplaced within minutes to hours on May 18, for which the causative events would be impossible to document reliably from the geologic record. Continuing lower-level activity at Mount St. Helens since 1980 has made this volcano a laboratory and case study for methods in volcano monitoring, hazards assessments, and interactions with responsible agencies and officials (Fig. 14; Miller and others, 1981). Swanson and others (1983) document successful forecasts of 13 eruptions at Mount St. Helens between June 1980 and December 1982, based on precursory seismicity, deformation, and gas emissions. The studies at Mount St. Helens since 1980 are far too numerous to present adequately here; major collections of reports include USGS *Professional Paper 1250,* collected papers in *Science* (Sept. 1983), and a special issue of the *Journal of Geophysical Research* (Sept. 1987).

Further volcanic hazard concerns were triggered in 1980 by several magnitude 6 earthquakes localized near the south margin of the 0.7-Ma Long Valley caldera in eastern California; subsequent detailed studies documented continued uplift on the intracaldera resurgent dome and suggested the presence of magma in cupolas as shallow as 4 to 5 km (Hill and others, 1985; Sanders, 1984). Because eruptions have occurred in the Long Valley area at intervals of about 500 yr during the past few thousand years, especially along the Inyo/Mono Craters (*Miller, 1985*), this region will continue to be a focus for volcano hazard and monitoring studies.

Studies of inactive and extinct volcanic areas also are providing key information about eruptive probabilities; an example is the time-predictable relation of repose periods between eruptions and volume of the preceding eruption documented at the Coso volcanic field in southeastern California and on the Snake River Plain in Idaho (*Bacon, 1982;* Kuntz and others, 1986). Combined seismic, gravity, heat flow, and geodetic data have been used by Smith and Braile (1984) to constrain the state of the magma body that presently underlies the 0.6-Ma Yellowstone caldera; anomalously low P-wave velocities, increased seismic attenuation, a negative gravity anomaly, high heat flow, and active uplift indicate the continuing presence of silicic magma at shallow depth and the probability of future volcanic activity (*Dzurisin and Yamashita, 1987*).

Along with other recent catastrophic volcanic activity elsewhere in the Cordilleran belt, especially the 1982 eruption of El Chichon in Mexico and the 1985 activity of Nevado del Ruiz in Columbia, future studies will necessarily focus on improving techniques for hazard assessment based on past volcanic history and present morphology and eruption forecasts, largely from geophysical monitoring. Notable current trends include improved geochronologic methods, research into the physics of eruption processes, and computer-assisted modelling of eruption effects. The range of atmospheric effects due to volcanic eruptions is another emerging topic, especially with regard to aviation problems caused by volcanic particulates (e.g., during the 1989–1990 eruptions of Mount Redoubt: *Alaskan Volcano Observatory Staff, 1990*), effects of sulfur aerosols and other volcanic volatiles on global climate (*Kennett and Thunell, 1977;* Sigurdsson, 1990), and possible volcanic contributions to geochemical anomalies associated with global extinction events that are more commonly interpreted to be caused by planetary impacts (*Alvarez and others, 1980; Officer and Drake, 1985*).

Geothermal resources

Most of the proved and probable geothermal resources in the western U.S. are spatially associated with large, young, silicic magmatic systems (Smith and Shaw, 1975).

The largest geothermal development in the world is at The Geysers, within the basaltic to rhyolitic Clear Lake volcanic field in northern California (*Hearn and others, 1981*). Geothermal power development at The Geysers exceeds 1,250 megawatts (in 1983) and provides most of the electrical needs of the San Francisco metropolitan area. The geothermal field is located along a faulted anticline in prevolcanic rocks, above the west margin of an inferred large, shallow magma body that is still partially molten, as demarked by a pronounced gravity low and teleseismic P-wave delays. As well as providing geothermal heat, this intrusion is thought to have been the source of the Clear Lake Volcanics, erupted between 2.1 Ma and 10 ka. The Clear Lake area is a locus of hydrothermal mercury deposits and large, low-grade disseminated epithermal gold resources associated with still-active hot springs.

Geothermal power generated at Roosevelt Hot Springs in the Mineral Mountains, Utah, is related to a Pleistocene silicic magmatic system (*Nielson and others, 1986*). The Mineral Mountains intrusive complex records the interaction between magmatic, structural, and hydrothermal processes since Oligocene time. The Roosevelt Hot Spring geothermal system is structurally controlled by range-front faults and is driven by a pluton emplaced during the Pleistocene. Associated rhyolitic flows, domes, and pyroclastic deposits were emplaced along the range crest at 0.8 to 0.5 Ma.

Geothermal studies at the Valles caldera have included both commercial exploration focused within the caldera and a program of research drilling to evaluate possible energy recovery from hot dry rock (*Heiken and Goff, 1983*). Exploration drilling on the resurgent dome has encountered an economically marginal wet-steam system, with maximum temperatures of 341°C at a depth of 3.2 km, and has also documented previously unknown great thickness and stratigraphic and structural complexities of the caldera fill (*Nielson and Hulen, 1984*). Drillholes into granitic basement a few kilometers west of the caldera have also encountered high geothermal gradients and constitute the first phase of heat-exchange experiments involving transferring fluids along artificially propagated fractures between holes (*Laughlin and others, 1983*). The present distribution of isotherms has been modeled by cooling of a crystallizing silicic pluton emplaced under the caldera at 1.1 Ma, the time of last major caldera collapse (*Kolstad and McGetchin, 1978*), but perturbations of the thermal gradients indicate magmatic or hydrothermal events within the last few tens of thousands of years (*Harrison and others, 1986*).

At the Long Valley caldera (Hill and others, 1985), drilling results indicate that a shallow convective hydrothermal system, dominated by meteoric water and circulating within the caldera-fill deposits (Fig. 15), masks conductive heat flux from deeper

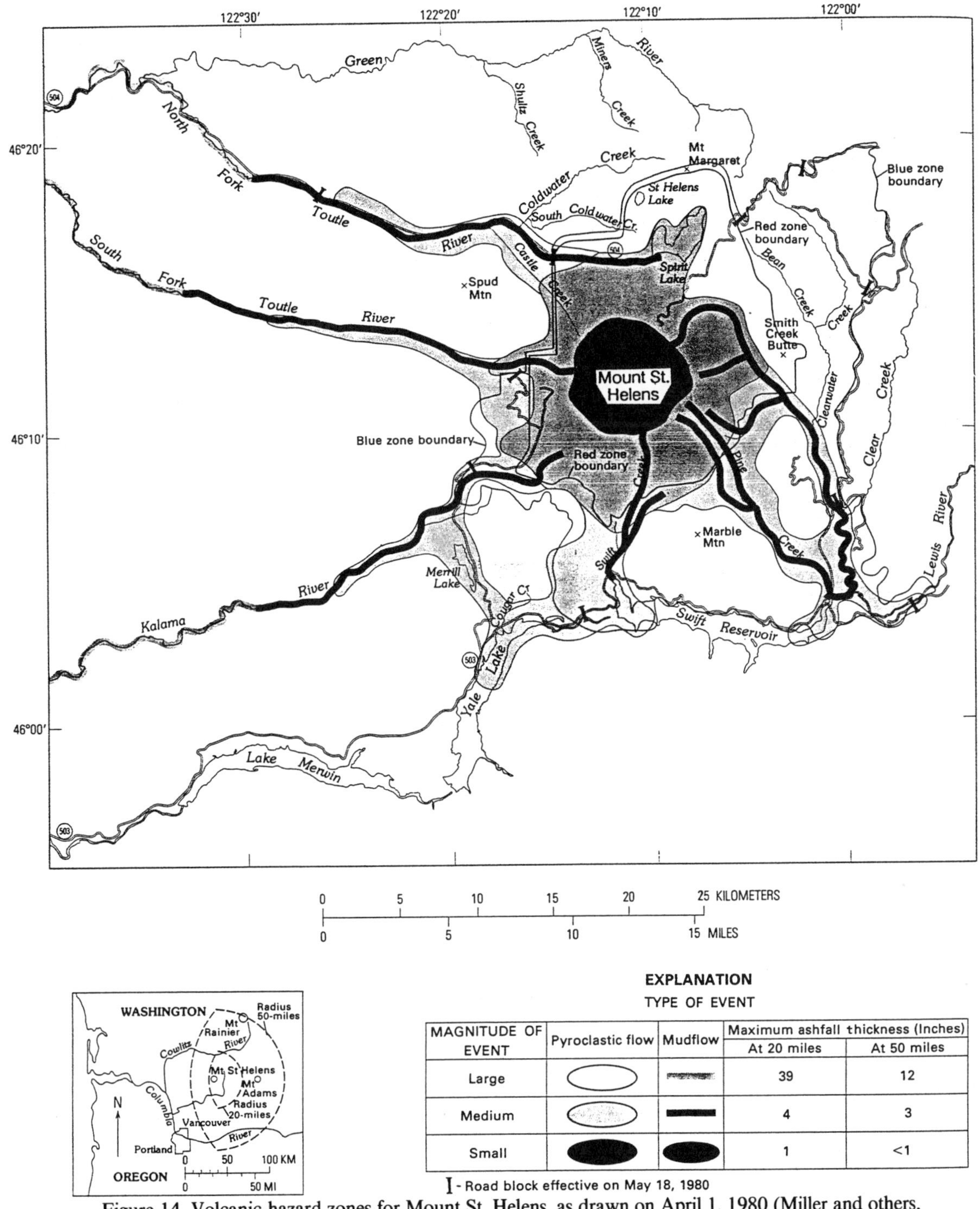

Figure 14. Volcanic-hazard zones for Mount St. Helens, as drawn on April 1, 1980 (Miller and others, 1981). This map shows kinds and extents of hazards for eruptions of three different magnitudes and was used for hazards briefings during April and early May, 1980. Inset shows boundaries of two ash-fall-hazard zones that reach 3 and 80 km north, east, and south of the volcano. Also shown are Washington State closure zones and roadblocks around the mountain. The red zone was closed to all except scientists and law-enforcement and administrative personnel; the blue zone was open during daytime to landowners with permits.

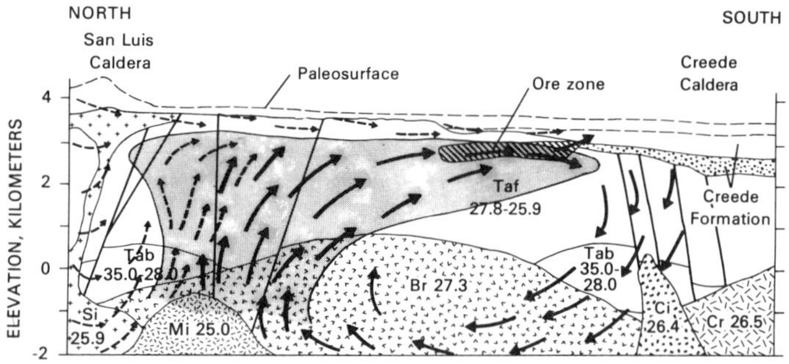

Figure 15. Geothermal model for Long Valley caldera (*Sorey, 1985*). Solid arrows, inferred flow directions in present-day hydrothermal system; open arrows, deeper circulation beneath the west moat during previous activity. Temperature profiles are from wells along line of section in caldera. Dashed contacts for Bishop Tuff are based on seismic and gravity data.

magmatic sources (*Blackwell, 1985; Sorey, 1985*). Combined geophysical and geologic evidence favors a long-lived deep heat source beneath the western part of the caldera, but additional deeper drilling is needed to penetrate the convective hydrothermal system and sample the conductive gradients above the magma chamber. Thermal effects from magma cupolas as shallow as 4.5 to 5 km, as determined seismically (Sanders, 1984), have not yet perturbed the thermal regime of the upper 2 km of caldera fill, suggesting that this magma has probably been in place for less than 60,000 years. Availability of magma in the western part of the caldera is also confirmed by eruptions along the Inyo Craters chain about every 500 yr for the last few thousand years (*Miller, 1985*).

Federally supported geothermal investigations by DOE, USGS, and academic institutions have also evaluated young igneous systems in the Cascade Range and elsewhere in the western U.S. Commercially exploitable geothermal resources have proved elusive in the Cascade Range, despite the youth of many of the volcanoes. Deeply circulating cool meteoric water, probably related to the heavy rainfall of the region, conceals volcano-related thermal anomalies, and surface hot springs are rare and probably not closely related to magmatic sources at depth. Research drilling in Newberry Volcano reached water temperatures of 265°C at a depth of 930 m, with a nearly linear gradient of 505°C/km below 700 m, indicating that blind thermal resources may exist more widely than suggested by shallow indicators (*MacLeod and Sammel, 1982*).

Several major igneous-related geothermal areas, notably those within the 0.6-Ma Yellowstone caldera, are in national parks and are therefore inappropriate for commercial exploration.

Thermal engines and metal sources for ore deposits

Many economically important ore deposits in the Cordilleran region, including all porphyry copper and molybdenum deposits, most epithermal veins, and many tungsten skarns and replacement iron deposits, are closely tied to post-Paleozoic igneous activity (Hutchinson and Albers, this volume). The association between such ore-deposit types and Cordilleran magmatic belts has long been recognized (e.g., Lindgren, 1915), yet improved understanding of the structural and petrologic evolution of such volcanoes and associated intrusions remains a valid goal for successful mineral exploration and resource appraisal. Major igneous-related ore-deposit types are noted briefly here, but emphasis is on applying concepts of magmatic activity to improved understanding of ore-forming processes.

Major ore-deposit types associated with Mesozoic and Cenozoic intermediate-composition to silicic intrusions in the Cordilleran magmatic belt include copper-molybdenum porphyry deposits, precious metal veins, tin-tungsten skarns, and some replacement iron ores. Especially notable are the great porphyry copper deposits associated with upper crustal intrusions in southern Arizona, and roughly similar copper deposits at Butte, Montana, and Bingham, Utah. All these are associated with intrusions emplaced during the eastward sweep of magmatism in Late Cretaceous and early Tertiary time (Titley, 1982). Gold-silver veins associated with the great Mesozoic subduction-related batholiths, such as the California Mother Lode belt, were particularly significant historically in providing early incentives for exploring and settling the Cordilleran region. Several tungsten skarn deposits have been economically important along the eastern Sierra Nevada and Idaho batholiths, and Mesozoic-Cenozoic replacement iron deposits are locally associated with granitic intrusions, as in the Iron Springs district of Utah.

Many economically important base and precious metal deposits in the western U.S. are associated with the flare-up of middle Tertiary volcanism. Notable examples include the bonanza precious metal deposits of Goldfield, Tonopah, and similar volcanogenic deposits in the Great Basin; and the base- and precious-metal veins of the San Juan field in Colorado (Fig. 16). These veins are largely localized by structures developed during volcanic activity. Calderas have been demonstrated to be major loci for mineralization, typically not during the pyroclastic phase that culminates in caldera subsidence, but rather by providing strong structural controls for later intrusion and mineralization (*Steven and others, 1974; McKee, 1979; Rytuba, 1981*). Structural control by caldera volcanism seems also important for many of the major porphyry copper deposits of Late Cretaceous to early Tertiary age in southern Arizona (*Lipman and Sawyer, 1985*).

Cordilleran ore deposits also relate directly to changes in tectonic setting. When continuous plate-margin subduction terminated in the middle Tertiary and widespread extension began in the Cordilleran region, contrasting metal assemblages were deposited. These include economically important Mo, U, and Be deposits, mostly associated with highly differentiated silicic rhyolites and small plutons of compositionally comparable evolved granite (*Christiansen and others, 1983*), which constitute the silicic components of developing regional bimodal magmatism. Ex-

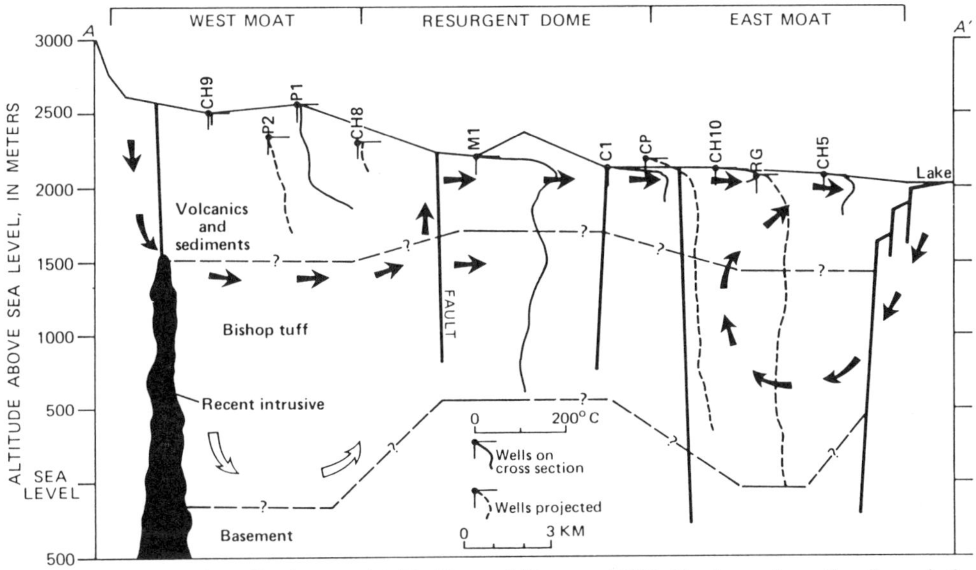

Figure 16. Creede mineralization model (*Bethke and Lipman, 1987*). North–south section through the mining district shows inferred geometry of hydrothermal system in relation to caldera structure and mineralized zones. Arrows, fluid circulation and mixing paths; heavy solid arrows, playa lake brines derived from Creede Formation; heavy dashed arrows, deeply circulating meteoric water that equilibrates with igneous rocks; thin dashed arrows, circulating shallower waters retaining meteoric compositions. Volcanic units: Taf, ash-flow caldera fill (numbers indicate radiometric ages); Tab, intermediate-composition lavas of caldera floor. Inferred intrusions: Br, resurgent intrusion of Bachelor caldera; Cr and Ci, resurgent and ring intrusions of Creede caldera; Si, ring intrusion of San Luis caldera; Mi, mineralization-related intrusion.

amples include the important Climax-type molybdenum porphyry deposits of the southern Rocky Mountains (White and others, 1981); beryllium deposits at Spor Mountain, Utah; and volcanic-related uranium-mercury-lithium ores in such volcanic fields as McDermitt, Nevada-Oregon. Much additional uranium ore, deposited syngenetically in Mesozoic sediments of the Colorado Plateau, is also volcanic related in the sense that the uranium was leached from silicic ash deposited downwind from volcanic fields above the Mesozoic batholiths to the west.

A long-standing question concerns the sources of such metals in igneous-related mineral deposits. The relative roles of transport from great depths by magmas, versus concentration of ores from upper crustal country rocks, has been controversial since the 19th century. Oxygen isotope studies have shown that most igneous-related ores were deposited from hydrothermal fluids, in which deeply circulating meteoric water was a major or dominant component (*Taylor, 1974*), and the role of the associated igneous bodies may have been mainly as a thermal engine to drive a convecting hydrothermal system (*Norton, 1982*). For example, lead in epithermal veins in the San Juan volcanic field is more radiogenic than in associated igneous rocks and must have been largely leached by hydrothermal fluids from Precambrian rocks that underlie the volcanic field (Doe and others, 1982). Some would argue that most Phanerozoic ore deposits represent such recycling of metals originally concentrated in the Precambrian, especially as massive sulfide deposits generated during processes of crustal formation (*Hutchinson, 1980*). Other metals, such as the Mo, U, and Be deposits emplaced during regional extension, are associated with distinctively evolved silicic magmas and are thought to have been concentrated as incompatible components during magmatic differentiation (*Christiansen and others, 1983; Stein and Hannah, 1985*). In a more general sense, though, even these metals, along with their host magmas, are probably largely recycled from lower-crustal material of the craton that has been variably melted and mobilized by mantle-generated mafic magmas.

Toxic waste disposal

In the past decade, feasibility evaluations for the containment of radioactive and other toxic wastes have supported regional studies and detailed work in several volcanic areas. These areas include Columbia River basalt flows in the vicinity of the Hanford nuclear processing site in Washington, basalt flows of the Snake River Plain near the National Reactor Testing Station in Idaho, and the silicic ash-flow plateau of the southwestern Nevada volcanic field in the vicinity of the Nevada Test Site (*Roseboom, 1983*).

On the Columbia Plateau, thick, widespread basalt flows are interleaved with fine-grained sediments and silicic tuff beds deposited in shallow basins (Swanson and others, 1975). In the Snake River Plain region, combined geological and geophysical studies have documented the presence of silicic volcanic centers buried beneath the basalt plain, including several large ash-flow calderas that become successively younger northeastward (Armstrong and others, 1975; *Morgan and others, 1984*). Complex stratigraphic interfingering between the widespread basalt flows in these regions, which are highly porous due to gas cavities and cooling joints, and the sediments and silicic tuffs, which are both relatively impermeable, presents substantial hydrogeologic problems for evaluating possible fluid flow in the vicinity of the proposed storage sites. At the same time, the site studies, including drilling, have greatly enhanced knowledge of the regional volcanic stratigraphy and provided special opportunities for detailed petrologic investigations.

Extended stratigraphic and petrologic studies of volcanic rocks of the southwestern Nevada field, initiated in the late 1950s in connection with underground nuclear weapons testing and more recently focused on possible development of a nuclear waste repository at Yucca Mountain, have provided an exceptional three-dimensional understanding of a structurally complex and petrologically diverse volcanic field. Between about 14 and 7 Ma, peralkaline and calc-alkaline magmatic systems were active intermittently from several adjacent centers, concurrently with basin-range extensional faulting (Christiansen and others, 1977; *Noble and others, 1984; Sawyer and Sargent, 1989*). At several of the centers, sequential large-volume eruptions of highly evolved silicic tuffs that sampled compositionally zoned or layered magma bodies document rapid fractionation at the tops of the associated subvolcanic magma chambers (*Lipman, 1971;* Byers and others, 1976; *Vogel and others, 1989*). The intermittent eruption of basaltic lavas within the same centers demonstrates the important role of mantle-derived mafic melts in generating such dominantly silicic volcanism. Detailed evaluation for a waste repository site has now focused on the Yucca Mountain area, on the south side of the caldera complex, where a thick sequence of outflow tuff sheets may provide suitable conditions in the unsaturated zone above the water table (*Carr and Yount, 1988*).

PERSPECTIVES

Cordilleran igneous rocks have been foci for a diversity of productive studies in the past, as briefly reviewed here, but what lies ahead? Several directions are obvious. Despite more than a century of intensive work, much remains to be learned from detailed field-based studies. Many areas have yet to be mapped in detail, and others have not been studied by scientists knowledgeable about active igneous processes. For example, many mineralized igneous areas of economic importance have been studied primarily by exploration geologists who have limited experience with active volcanoes, while specialists in such volcanic processes have tended to avoid the complexities inherent in altered and structurally intricate mineralized areas.

Petrologists have now convincingly demonstrated that many igneous provinces result from interplay of multiple differentiation processes, including interaction of a basaltic mantle melt with various levels of the crustal section, magma mixing, crystal fractionation, and probably also volatile-driven transport of some elements. The relative importance of contributions from these diverse processes remains controversial, and to some degree, undoubtedly varies with each magmatic system. Currently, much effort in the study of Cordilleran igneous systems is spent on such problems as: (1) evaluating the reliability of applying experimentally determined geothermometers and geobarometers, based on simplified chemical systems, to more compositionally complex natural minerals; (2) developing criteria for evaluating equilibrium mineral assemblages versus those that may represent juxtaposition from differing P-T conditions of crystallization by processes of magma mixing; (3) evaluating the role of accessory minerals in accounting for minor-element fractionation; and (4) improving understanding of melt-crystal distribution coefficients in relation to magma composition, especially for evolved silicic systems. Undoubtedly, others could make this list much longer and/or quite different.

Much progress can be expected from developing techniques in geochemistry and geochronology, both of which have already contributed much to recent progress. Despite the great advances in isotopic geochronology, the complex thermal histories of many parts of the Cordillera and the ever-present desire for more precise resolution of events will continue to provide future challenges and breakthroughs. $^{40/39}$Ar incremental-heating methods, which are now being widely applied, promise to increase resolution and to provide tests for complexities in the thermal history of a sample. Similarly, laser-fusion $^{40/39}$Ar methods, which allow determinations on single phenocrysts, are permitting study of previously intractable problems where sample homogeneity is at issue. Pb-isotopic determinations on single grains of igneous zircon, both by high-resolution conventional mass spectrometry and by ion-probe techniques, are yielding increasingly precise crystallization ages and information on source regions for the magmas. U-series disequilibrium studies hold promise for dating times of growth of individual phenocryst groups in young volcanics. Application of multiple geochronologic methods, especially involving fission-track techniques, also holds promise of providing supplemental information on times of thermal closure and thus cooling history and uplift rates.

Additional important constraints have been coming from seismic, electrical, and other geophysical studies that place limits on the third dimension of igneous systems (below exposed levels), and also by research drilling. Magma appears to have been detected beneath the Yellowstone and Long Valley calderas (Iyer and others, 1981; Sanders, 1984), and the character of the shallowest body at Long Valley, at an apparent depth of 4 to 6 km, may be testable by deep drilling. Electrical geophysical methods are proving especially valuable in delimiting regions of fluid flow and hydrothermal alteration (*Stanley and others, 1976; Hermance and others, 1984; Fitterman and others, 1988*). Seismic reflection methods, as used by COCORP and others, have yet to be applied rigorously to exploring the shapes of upper crustal igneous bodies, but such methods might help determine the geometry and depth of large upper crustal batholiths such as that inferred to occupy a 50 by 75-km area beneath the San Juan volcanic field (*Plouff and Pakiser, 1972;* Steven and Lipman, 1976). Scientific drilling, such as that proposed to explore the roots of the Creede epithermal vein system within the San Juan field (*Bethke and Lipman, 1987*), may also yield information on the depth to this batholith as well as a complete section through a major ash-flow caldera. Drilling for geothermal exploration and other research goals has already provided a vastly improved understanding of the geometry of the Valles caldera in New Mexico (*Heiken and Goff, 1983; Nielson and Hulen, 1984; Goff and Gardner, 1988*).

Theoretical modelling and experimental simulation of deep-seated igneous processes such as melt segregation and rise, magma chamber processes, crystallization kinetics, and eruption dynamics are currently rapidly developing fields that provide important complements to observations made on actual igneous rocks. These and related topics, such as experimental petrology, mineral physics, and magma rheology are all broad research fields that, though not notably site specific to the Cordilleran region, can be expected to yield rich research rewards when applied to the diverse igneous features preserved in the western U.S.

Increasingly, magma generation concepts applied to the Cordilleran region are emphasizing a major role for the continental lithosphere, both crust and mantle keel. Both these materials and the processes that affect them remain perhaps the least understood aspect of continental magmatism. Little in the way of lower crust is available for study in the Cordilleran region, other than crustal xenoliths carried up by alkalic basalts and kimberlite diatremes. But are these representative samples, or potentially anomalous, along with the unusual volcanic processes that bring them to the surface? How does the process of exchange between continental lithosphere and asthenosphere, either by physical emplacement of asthenospheric magma in the lithosphere or by thermal reconfiguration of one into the other, affect magmas that reach the upper crust? Much remains to be learned, including many questions for which we do not even know effective methods of approach!

REFERENCES CITED

Text references in *italics* will be found only on accompanying microfiche.

Aldrich, M. J., Jr., Chapin, C. E., and Laughlin, A. W., 1986, Stress history and tectonic development of the Rio Grande rift, New Mexico: Journal of Geophysical Research, v. 91, p. 6199–6211.

Armstrong, R. L., 1974, Magmatism, orogenic timing, and orogenic diachronism in the Cordillera from Mexico to Canada: Nature, v. 247, p. 348–351.

Armstrong, R. L., Leeman, W. P., and Malde, H. E., 1975, K-Ar dating, Quaternary and Neogene volcanic rocks of the Snake River Plain, Idaho: American Journal of Science, v. 275, p. 225–251.

Atwater, T., 1970, Implications of plate tectonics for the Cenozoic tectonic evolution of western North America: Geological Society of America Bulletin, v. 81, p. 3513–3536.

Bacon, C. R., 1983, Eruptive history of Mount Mazama, U.S.A.: Journal of Geophysical Research, v. 12, p. 57–115.

——, 1986, Magmatic inclusions in silicic and intermediate volcanic rocks: Journal of Geophysical Research, v. 91, p. 6091–6112.

Bailey, R. A., Dalrymple, G. B., and Lanphere, M. A., 1976, Volcanism, structure, and geochronology of Long Valley caldera, Mono County, California: Journal of Geophysical Research, v. 81, p. 725–744.

Barnes, C. G., 1983, Petrology and upward zonation of the Wolley Creek batholith, Klamath Mountains, California: Journal of Petrology, v. 24, p. 495–537.

Bateman, P. C., and Chappell, B. W., 1979, Crystallization, fractionation, and solidification of the Tuolumne intrusive series, Yosemite National Park, California: Geological Society of America Bulletin, v. 90, p. 465–482.

Best, M. J., and Christiansen, E. H., 1991, Limited extension during peak Tertiary volcanism, Great Basin of Nevada and Utah: Journal of Geophysical Research (in press).

Bonnichsen, B., and Kauffman, D. F., 1987, Physical features of rhyolite lava flows in the Snake River Plain volcanic province, southwestern Idaho, *in* Fink, J. H., ed., The emplacement of silicic domes and lava flows: Geological Society of America Special Paper 212, p. 119–145.

Buddington, A. F., 1959, Granite emplacement with special reference to North America: Geological Society of America Bulletin, v. 70, p. 671–748.

Byers, F. M., Jr., Carr, W. J., Orkild, P. P., Quinlivan, W. D., and Sargent, K. A., 1976, Volcanic suites and related cauldrons of the Timber Mountain–Oasis Valley caldera complex, southern Nevada: U.S. Geological Survey Professional Paper 919, p. 1–70.

Cameron, K. L., 1984, The Bishop Tuff revisited; New rare earth element data consistent with crystal fractionation: Science, v. 224, p. 1338–1340.

Christiansen, R. L., 1984, Yellowstone magmatic evolution; Its bearing on understanding large-volume explosive volcanism, *in* Explosive volcanism, inception, evolution, and hazards: Washington, D.C., National Academy Press, p. 84–95.

Christiansen, R. L., and Lipman, P. W., 1972, Cenozoic volcanism and plate-tectonic evolution of the western United States; 2, Late Cenozoic: Philosophical Transactions, Royal Society of London, Ser. A, v. 271, p. 249–284.

Christiansen, R. L., and McKee, E. H., 1978, Late Cenozoic volcanic and tectonic evolution of the Great Basin and Columbia intermontane regions, *in* Smith, R. B., and Eaton, G. P., eds., Cenozoic tectonics and regional geophysics of the western Cordillera: Geological Society of America Memoir 152, p. 283–311.

Christiansen, R. L., and 5 others, 1977, Timber Mountain–Oasis Valley caldera complex of southern Nevada: Geological Society of America Bulletin, v. 88, p. 943–959.

Coney, P. J., 1976, Plate tectonics and the Laramide orogeny: New Mexico Geological Society Special Publication 6, p. 5–10.

Coney, P. J., and Reynolds, S. J., 1977, Cordilleran Benioff zones: Nature, v. 270, p. 403–406.

Crandell, D. R., Mullineaux, D. R., and Miller, C. D., 1979, Volcanic-hazards studies in the Cascade Range of the western United States, *in* Sheets, P. D., and Grayson, D. K., eds., Volcanic activity and human ecology: London, Academic Press, p. 195–219.

Cross, T. A., and Pilger, R. H., 1978, Constraints on absolute motion and plate interactions inferred from Cenozoic igneous activity in the western United States: American Journal of Science, v. 278, p. 865–902.

DePaolo, D. J., and Farmer, G. L., 1984, Isotopic data bearing on the origin of Mesozoic and Tertiary granitic rocks in the western United States: Philosophical Transactions, Royal Society of London, v. A310, p. 743–754.

Dickinson, W. R., and Snyder, W. S., 1978, Plate tectonics of the Laramide orogeny, *in* Matthews, V., III, ed., Laramide folding associated with basement block faulting in the western United States: Geological Society of America Memoir 151, p. 355–366.

—— , 1979, Geometry of subducted slabs related to the San Andreas transform: Journal of Geology, v. 87, p. 609–627.

Doe, B. R., Leeman, W. P., Christiansen, R. L., and Hedge, C. E., 1982, Lead and strontium isotopes and related trace elements as genetic traces in the Upper Cenozoic rhyolite basalt association of the Yellowstone Plateau volcanic field: Journal of Geophysical Research, v. 87, p. 4785–4806.

Dungan, M. A., and 5 others, 1986, Open-system evolution of the Taos Plateau volcanic field, northern New Mexico; 1, The petrology and geochemistry of the Servilleta Basalt: Journal of Geophysical Research, v. 91, p. 5999–6028.

Eaton, G. P., 1987, Topography and origin of the southern Rocky Mountains and the Alvarado Ridge: Geological Society of London Special Paper 28, p. 355–370.

Eaton, G. P., and 7 others, 1975, Magma beneath Yellowstone National Park: Science, v. 188, p. 787–796.

Eichelberger, J. C., 1975, Origin of andesite and dacite; Evidence of mixing at Class Mountain in California and at other circum-Pacific volcanoes: Geological Society of America Bulletin, v. 86, p. 1381–1391.

Eichelberger, J. C., Carrigan, C. R., Westrich, H. R., and Price, R. H., 1986, Non-explosive silicic volcanism: Nature, v. 323, p. 598–602.

Ekren, E. B., McIntyre, D. H., and Bennett, E. H., 1984, High-temperature large-volume lavalike ash-flow tuffs without calderas in southwestern Idaho: U.S. Geological Survey Professional Paper 1272, 76 p.

Elston, W. E., 1984b, Subduction of young oceanic lithosphere and extensional orogeny in southwestern North America during mid-Tertiary time: Tectonics, v. 3, p. 229–250.

Engebretson, D. C., Cox, A., and Gordon, R. G., 1985, Relative motions between oceanic and continental plates in the Pacific Basin: Geological Society of America Special Paper 206, 59 p.

Gans, P. B., Mahood, G. A., and Schermer, E., 1989, Synextensional magmatism in the Basin and Range province; A case study from the eastern Great Basin: Geological Society of America Special Paper 233, 53 p.

Glazner, A. F., and Bartley, J. M., 1984, Timing and tectonic setting of Tertiary low-angle normal faulting and associated magmatism in the southwestern United States: Tectonics, v. 3, p. 385–396.

Grove, T. L., Gerlach, D. C., and Sando, T. W., 1982, Origin of calc-alkaline series lavas at Medicine Lake Volcano by fractionation, assimilation, and mixing: Contributions to Mineralogy and Petrology, v. 80, p. 160–182.

Hamilton, W. B., 1989, Crustal geologic processes of the United States, *in* Pakiser, L. C., and Mooney, W. D., eds., Geophysical framework of the continental United States: Geological Society of America Memoir 172, p. 743–781.

Hamilton, W., and Myers, W. B., 1967, The nature of batholiths: U.S. Geological Society Professional Paper 544-C, 30 p.

Hearn, B. C., Jr., 1968, Diatremes with kimberlitic affinities in north-central Montana: Science, v. 159, p. 622–625.

Heiken, G., and 6 others, 1990, The Valles/Toledo caldera complex, Jemez volcanic field, New Mexico: Annual Reviews of Earth and Planetary Sciences, v. 18, p. 27–53.

Helmstaedt, H., and Doig, R., 1975, Eclogite nodules from kimberlite pipes of the Colorado Plateau; Samples of subducted Franciscan-type oceanic lithosphere: Physics and Chemistry of the Earth, v. 9, p. 95–111.

Hildreth, W., 1979, The Bishop Tuff; Evidence for the origin of compositional zoning in silicic magma chambers, *in* Chapin, C. E., and Elston, W. E., eds., Ash-flow tuffs: Geological Society of America Special Paper 180, p. 43–75.

—— , 1981, Gradients in silicic magma chambers; Implications for lithospheric magmatism: Journal of Geophysical Research, v. 86, p. 10153–10192.

Hildreth, W., and Moorbath, S., 1988, Crustal contributions to arc magmatism in the Andes of central Chile: Contributions to Mineralogy and Petrology, v. 98, p. 455–589.

Hill, D. P., Bailey, R. A., and Ryall, A. S., 1985, Active tectonic and magmatic processes beneath Long Valley caldera, eastern California; An overview: Journal of Geophysical Research, v. 90, p. 11111–11120.

Hopson, C. A., Mattinson, J. M., and Pessagno, E. A., 1981, Coast Range ophiolite, western California, *in* Ernst, W. G., ed., The geotectonic development of California: Englewood Cliffs, New Jersey, Prentice-Hall, p. 419–510.

Hyndman, D. W., 1983, The Idaho batholith and associated plutons, Idaho and western Montana, *in* Roddick, J. A., Circum-Pacific plutonic terranes: Geological Society of America Memoir 159, p. 213–240.

Iyer, H. M., and 5 others, 1981, A deep low-velocity body under the Yellowstone caldera, Wyoming; Delineation using teleseismic P-wave residuals and tectonic interpretation; Summary: Geological Society of America Bulletin, Part 1, v. 92, p. 792–798.

Johnson, C. M., Lipman, P. W., and Czamanske, G. K., 1990, H, O, Sr, Nd, and Pb isotope geochemistry of the Latir volcanic field and cogenetic intrusions, New Mexico, and relations between evolution of a continental magmatic center and modification of the lithosphere: Contributions to Mineralogy and Petrology, v. 104, p. 99–124.

Johnson, M. C., and Rutherford, M. J., 1989, Experimentally determined conditions in the Fish Canyon, Colorado, magma chamber: Journal of Petrology, v. 30, p. 711–738.

Kistler, R. W., and Peterman, Z. E., 1973, Variations in Sr, Rb, K, Na, and initial $^{87}Sr/^{86}Sr$ in Mesozoic granitic rocks and intruded wall rocks in central California: Geological Society of America Bulletin, v. 84, p. 3489–3512.

Kuntz, M. A., Champion, D. E., Spiker, E. C., and Lefebvre, R. H., 1986, Contrasting magma types and steady-state, volume-predictable basaltic volcanism along the Great Rift, Idaho: Geological Society of America Bulletin, v. 97, p. 579–594.

Lachenbruch, A. H., and Sass, J. H., 1978, Models of an extending lithosphere and heat flow in the Basin and Range province, *in* Smith, R. B., and Eaton, G. P., eds., Cenozoic tectonics and regional geophysics of the western Cordillera: Geological Society of America Memoir 152, p. 209–250.

Leat, P. T., Thompson, R. N., Morrison, M. A., Hendry, G. A., and Dickin, A. P., 1988, Compositionally diverse Miocene–Recent rift related magmatism in northwest Colorado; Partial melting and mixing of mafic magmas from 3 different asthenospheric and lithospheric mantle sources, *in* Menzies, M. A., and Cox, K., eds., Oceanic and continental lithosphere: Journal of Petrology Special Issue, p. 351–377.

Leeman, W. P., 1970, The isotopic composition of strontium in late Cenozoic basalts from the Basin-Range province, western United States: Geochimica et Cosmochimica Acta, v. 34, p. 857–872.

Lindgren, W., 1915, The igneous geology of the Cordilleras and its problems, *in* Problems of American geology: New Haven, Connecticut, Yale University Press, p. 234–286.

Lipman, P. W., 1980, Cenozoic volcanism in the western United States; Implications for continental tectonics, *in* Burchfiel, B. C., Oliver, J. E., and Silver, L. T., eds., Studies in geophysics, continental tectonics: Washington, D.C., National Academy of Sciences, p. 161–174.

—— , 1984, The roots of ash-flow calderas in western North America; Windows into the tops of granitic batholiths: Journal of Geophysical Research, v. 89, p. 8801–8841.

—— , 1988, Evolution of silicic magma in the upper crust; The mid-Tertiary Latir volcanic field and its cogenetic granitic batholith, northern New Mexico, USA: Transactions of the Royal Society of Edinburgh, v. 79, p. 265–288.

Lipman, P. W., Christiansen, R. L., and O'Connor, J. J., 1966, A compositionally zoned ash-flow sheet in southwestern Nevada: U.S. Geological Survey Professional Paper 524-F, 47 p.

Lipman, P. W., Christiansen, R. L., and Prostka, H. J., 1971, Evolving subduction zones in the western United States, as interpreted from igneous rocks: Science, v. 148, p. 821–825.

Lipman, P. W., Doe, B. R., Hedge, C. E., and Steven, T. A., 1978, Petrologic evolution of the San Juan volcanic field, southwestern Colorado; Pb and Sr isotope evidence: Geological Society of America Bulletin, v. 89, p. 59–82.

McGetchin, T. R., and Silver, L. T., 1972, A crustal-upper mantle model for the Colorado Plateau based on observation of crystalline rock fragments in the Moses Rock dike: Journal of Geophysical Research, v. 77, p. 7022–7037.

Menzies, M. A., Kempton, P. D., and Dungan, M. A., 1985, Interaction of continental lithosphere and asthenospheric melts below the Geronimo volcanic field, Arizona: Journal of Petrology, v. 26, p. 663–693.

Miller, C. D., Mullineaux, D. R., and Crandell, D. R., 1981, Hazards assessments at Mount St. Helens: U.S. Geological Survey Professional Paper 1250, p. 789–802.

Miller, C. F., and Barton, M. D., 1990, Phanerozoic plutonism in the Cordilleran Interior, U.S.A., *in* Kay, S. M., and Rapela, C. W., eds., Plutonism from Antarctica to Alaska: Geological Society of America Special Paper 241, p. 213–232.

Moore, J. G., 1962, K/Na ratios of Cenozoic igneous rocks in the western United States: Geochimica et Cosmochimica Acta, v. 26, p. 101–130.

Morgan, P., Seager, W. R., and Golombeck, M. P., 1986, Cenozoic thermal, mechanical, and tectonic evolution of the Rio Grande rift: Journal of Geophysical Research, v. 91, p. 6263–6276.

Noyes, H. J., Frey, F. A., and Wones, D. R., 1983, A tale of two plutons; Geochemical evidence bearing on the origin and differentiation of the Red Lake and Eagle Peak plutons, central Sierra Nevada, California: Journal of Geology, v. 91, p. 487–509.

Perry, F. V., Baldridge, W. S., and DePaolo, D. J., 1987, Role of asthenosphere and lithosphere in the genesis of late Cenozoic basaltic rocks from the Rio Grande rift and adjacent regions of the southwestern United States: Journal of Geophysical Research, v. 92, p. 9193–9214.

Peterman, Z. E., Doe, B. R., and Prostka, H. J., 1970, Lead and strontium isotopes in rocks of the Absaroka volcanic field, Wyoming: Contributions to Mineralogy and Petrology, v. 27, p. 121–130.

Pilger, R. H., Jr., 1984, Cenozoic plate kinematics, subduction, and magmatism, South American Andes: Journal of the Geological Society of London, v. 141, p. 793–802.

Rehrig, W. A., and Heidrick, T. L., 1976, Regional tectonic stress during the Laramide and late Tertiary intrusive periods, Basin and Range province, Arizona: Arizona Geological Society Digest, v. 10, p. 205–229.

Robinson, G. D., Klepper, M. R., and Obradovich, J. D., 1968, Overlapping plutonism, volcanism, and tectonism in the Boulder batholith region, western Montana, *in* Coates, R. R., Hay, R. L., and Anderson, C. A., eds., Studies in volcanology: Geological Society of America Memoir 116, p. 557–576.

Roden, M. F., Smith, D., and Murthy, V. R., 1990, Chemical constraints on lithosphere composition and evolution beneath the Colorado Plateau: Journal of Geophysical Research, v. 95, p. 2811–2831.

Roseboom, E. H., Jr., 1983, Disposal of high-level nuclear waste above the water table in arid regions: U.S. Geological Survey Circular 903, 21 p.

Ross, D. C., 1985, Mafic gneissic complex (batholithic root?) in the southernmost Sierra Nevada, California: Geology, v. 13, p. 288–291.

Rutherford, M. J., Sigurdsson, H., Carey, S., and Davis, A., 1985, The May 18, 1980, eruption of Mt. St. Helens; 1, Melt composition and experimental phase equilibria: Journal of Geophysical Research, v. 90, p. 2929–2947.

Saleeby, J. B., 1982, Polygenetic ophiolite belt of the California Sierra Nevada; Geochronologic and tectonostratigraphic development: Journal of Geophysical Research, v. 87, p. 1803–1824.

Saleeby, J. B., Sams, D. B., and Kistler, R. W., 1987, U/Pb zircon, strontium, and oxygen isotopic and geochronological study of the southernmost Sierra Nevada batholith, California: Journal of Geophysical Research, v. 92, p. 10443–10466.

Sanders, C. O., 1984, Location and configuration of magma bodies beneath Long Valley, California, determined from anomalous earthquake signals: Journal of Geophysical Research, v. 89, p. 8287–8302.

Schweickert, R. A., 1976, Shallow-level plutonic complexes in the eastern Sierra Nevada, California, and their tectonic implications: Geological Society of America Special Paper 176, 58 p.

Seager, W. R., and McCurry, M., 1988, The cogenetic Organ cauldron and batholith, south central New Mexico; Evolution of a large-volume ash flow cauldron and its source magma chamber: Journal of Geophysical Research, v. 93, p. 4421–4434.

Self, S., Goff, F., Gardner, J. N., Wright, J. V., and Kite, W. M., 1986, Explosive rhyolitic volcanism in the Jemez Mountains; Vent locations, caldera development, and relation to regional structure: Journal of Geophysical Research, v. 91, p. 1779–1798.

Shoemaker, E. M., Roach, C. H., and Byers, F. M., Jr., 1962, Diatremes and uranium deposits in the Hopi Buttes, Arizona: Geological Society of America Buddington Volume, p. 327–355.

Sigurdsson, H., 1990, Assessment of the atmospheric impact of volcanic eruptions, *in* Global catastrophes in Earth history: Impacts, volcanism, and mass mortality: Geological Society of America Special Paper 247, in press.

Silver, L. T., and Chappell, B. W., 1988, The Peninsular Ranges batholith; An insight into the evolution of the Cordilleran batholiths of southwestern North America: Transactions of the Royal Society of Edinburgh, v. 79, p. 105–22.

Smith, R. B., and Braile, L. W., 1984, Crustal structure and evolution of an explosive silicic volcanic system at Yellowstone National Park, *in* Explosive volcanism, inception, evolution, and hazards: Washington, D.C., National Academy Press, p. 96–109.

Smith, R. L., 1960a, Ash flows: Geological Society of America Bulletin, v. 71, p. 795–842.

——, 1979, Ash-flow magmatism, *in* Chapin, C. E., and Elston, W. E., eds., Ash-flow tuffs: Geological Society of America Special Paper 180, p. 5–27.

Smith, R. L., and Bailey, R. A., 1968, Resurgent cauldrons, *in* Coates, R. R., Hay, R. L., and Anderson, C. A., eds., Studies in volcanology: Geological Society of America Memoir 116, p. 613–662.

Smith, R. L., and Shaw, H. R., 1975, Igneous related geothermal systems: U.S. Geological Survey Circular 726, p. 58–83.

Snyder, W. S., Dickinson, W. R., and Silberman, M. L., 1976, Tectonic implications of space-time patterns of Cenozoic magmatism in the western United States: Earth and Planetary Science Letters, v. 32, p. 91–106.

Sparks, R.S.J., Huppert, H. E., and Turner, J. S., 1984, The fluid dynamics of evolving magma chambers: Philosophical Transactions of the Royal Society of London, v. 310, p. 511–534.

Spence, W., and Gross, R. S., 1990, A tomographic glimpse of the upper mantle source of magmas of the Jemez lineament, New Mexico: Journal of Geophysical Research, v. 95, p. 10,829–10,850.

Steven, T. A., and Lipman, P. W., 1976, Calderas of the San Juan volcanic field, southwestern Colorado: U.S. Geological Survey Professional Paper 958, 35 p.

Steven, T. A., Rowley, P. D., and Cunningham, C. G., 1984, Calderas of the Marysvale volcanic field, west central Utah: Journal of Geophysical Research, v. 89, p. 8751–8764.

Stewart, J. H., Moore, W. J., and Zietz, I., 1977, East-west patterns of Cenozoic igneous rocks, aeromagnetic anomalies, and mineral deposits, Nevada and Utah: Geological Society of America Bulletin, v. 88, p. 67–77.

Stormer, J. C., Jr., and Whitney, J. A., 1985, Two-feldspar and iron-titanium oxide equilibria in silicic magma and depth of origin of large volume ash-flow tuffs: American Mineralogist, v. 70, p. 52–64.

Swanson, D. A., Wright, T. L., and Heltz, R. T., 1975, Linear vent systems and estimate rates of magma production and eruption for the Yakima Basalt on the Columbia Plateau: American Journal of Science, v. 275, p. 877–905.

Swanson, D. A., and 5 others, 1983, Predicting eruptions at Mount St. Helens, June 1980 through December 1982: Science, v. 221, p. 1369–1375.

Thompson, G. A., and Burke, G. B., 1974, Regional geophysics of the Basin and Range province: Annual Reviews of Earth and Planetary Sciences, v. 2, p. 213–233.

Tilling, R. I., 1974, Composition and time relations of plutonic and associated volcanic rocks, Boulder batholith region, Montana: Geological Society of America Bulletin, v. 85, p. 1925–1930.

Titley, S. R., 1982, Geologic setting of porphyry copper deposits, southeastern Arizona, *in* Titley, S. R., ed., Advances in geology of the porphyry copper deposits: Tucson, University of Arizona Press, p. 37–58.

Tobisch, O. T., Saleeby, J. B., and Fiske, R. S., 1986, Structural history of continental volcanic arc rocks, eastern Sierra Nevada; A case for extensional tectonics: Tectonics, v. 5, p. 65–94.

Wernicke, B. P., Christiansen, R. L., England, P. C., and Sonder, L. J., 1987, Tectonomagmatic evolution of Cenozoic extension in the North American

Cordillera: Geological Society of London Special Publication 28, p. 203–221.

White, W. H., and 6 others, 1981, Character and origin of Climax-type molybdenum deposits: Economic Geology 75th Anniversary Volume, p. 270–316.

Whitney, J. A., and Stormer, J. C., Jr., 1985, Mineralogy, petrology, and magmatic conditions from the Fish Canyon Tuff, central San Juan volcanic field, Colorado: Journal of Petrology, v. 26, p. 726–762.

Whitney, J. A., Dorais, M. J., Stormer, J. C., Jr., Kline, S. W., and Matty, D. J., 1988, Magmatic conditions and development of chemical zonation in the Carpenter Ridge Tuff, central San Juan volcanic field, Colorado: American Journal of Science, v. 288A, p. 16–44.

Wilcox, R. E., 1944, Rhyolite-basalt complex of Gardiner River, Yellowstone Park, Wyoming: Geological Society of America Bulletin, v. 55, p. 1047–1080.

Williams, H., 1942, The geology of Crater Lake National Park, Oregon, with Wohletz, K. H., 1986, Explosive magma-water interactions; Thermodynamics, explosion mechanism, and field studies: Bulletin of Volcanology, v. 48, p. 245–264.

Wilshire, H. G., and 6 others, 1988, Mafic and ultramafic xenoliths from volcanic rocks of the western United States: U.S. Geological Survey Professional Paper 1443, 179 p.

Zartman, R. E., 1974, Lead isotopic provinces in the Cordillera of the western United States and their geological significance: Economic Geology, v. 69, p. 792–805.

Zoback, M. L., and Zoback, M., 1980, State of stress in the conterminous United States: Journal of Geophysical Research, v. 85, p. 6113–6156.

Manuscript Accepted by the Society July 26, 1990

ACKNOWLEDGMENTS

Any review of this sort reflects the author's passions and prejudices, while overlooking some favorites of others. I hope at least to have been thought provoking. My apologies to any who feel shorted. Also, my thanks to some long-term associates and colleagues who have tried to encourage some sense of balance in my thinking over the years, especially Bill Dickinson, Bob Christiansen, Tom Steven, Warren Hamilton, Patrick Muffler, Jack Lockwood, Bob Tilling, Gordon Eaton, Mike Dungan, Phil Bethke, John Pallister, and David Sawyer. Obviously, they cannot be faulted for omissions or other failures here. Thanks also to Myron Best, Don Hyndman, Wes Hildreth, and John Pallister for thoughtful comments on this paper.

Printed in U.S.A.

The Geology of North America
Vol. G-3, The Cordilleran Orogen: Conterminous U.S.
The Geological Society of America, 1992

Chapter 10

Metamorphism of the western Cordillera and its relationship to tectonics

W. G. Ernst
School of Earth Sciences, Stanford University, Stanford, California 94305-2210

INTRODUCTION

Phanerozoic lithologic assemblages associated with specific plate-tectonic regimes are reasonably well understood, especially the siting of igneous and sedimentary rock suites. Less appreciated is the relationship between differential plate motions and subsolidus phase equilibria. Metamorphism is a complex but quantifiable function of the conditions attending recrystallization; in turn, the thermal-depth parameters of contrasting segments of the crust are related to individual plate-tectonic histories.

In examining metamorphic complexes of the western Cordillera, conterminous United States, an attempt will be made to place metamorphism within a regional framework. Extrapolating backward through time, the record of early events in ancient rocks has been partly or completely obliterated by more recent episodes of deformation and recrystallization. Subsequent rifting, drifting, and amalgamation of exotic terranes have dispersed formerly contiguous units, and juxtaposed far-traveled, unrelated lithotectonic entities, thereby greatly complicating the task of inferring the origin and petrotectonic evolution of particular regions. Systematic relationships are partially preserved, however, and will be emphasized in this summary.

PHYSICAL CONDITIONS OF METAMORPHISM AND PLATE-TECTONIC REGIMES

Recrystallization of rocks takes place in response to changes in the physical and/or chemical environment. At elevated temperatures and pressures, and in the presence of catalysts such as hydrothermal solutions, reaction rates are enhanced but are nevertheless sufficiently sluggish that metamorphic pressure-temperture (P-T) conditions must be sustained for many thousands or even millions of years to produce a recognizable effect. Such mineralogic changes typify orogenic belts, deeper portions of the continental crust, and the uppermost mantle.

Experimental phase equilibrium relations have been determined for synthetic and natural rock and mineral systems. These data, taken together with $^{18}O/^{16}O$ and deuterium/hydrogen isotopic geothermometers, as well as diverse mineralogic thermobarometers, allow the erection of a petrogenetic grid. Metamorphic facies illustrated schematically in Figure 1 are appropriate for rocks of roughly basaltic composition, at oxygen fugacities near values defined by the univariant condensed assemblage fayalite + magnetite + quartz and—except for granulitic lithologies where low f_{H_2O} prevails—at a relatively high fugacity of H_2O. Also projected onto this diagram are phase relations for Al_2SiO_5 polymorphs that would be stable in spatially associated aluminous schists.

The thermal structures of divergent and convergent plate boundaries have been modeled by numerous workers (e.g., *Oxburgh and Turcotte, 1971*; *Goto and others, 1987*; *van den Beukel and Wortel, 1987*). Employing computed P-T arrangements, the geologic disposition of metamorphic mineral assemblages presented in the facies grid of Figure 1 is shown for simplified, idealized constructive and destructive lithospheric plate junctions in Figures 2 and 3 respectively. The very high heat-flow regime characterized by an oceanic spreading center accounts for the observed telescoping of the metamorphic zonation, and relatively shallow-level development of high-rank hornfels beneath the ridge (Fig. 2). In contrast to this bilaterally symmetrical pattern, a convergent plate-tectonic setting is characterized by a broad, relatively high heat-flow environment in the volcanic/plutonic arc, reflected by a general upwarp of the metamorphic facies assemblages plus anatexis of basal portions of thick continental crust, and a spectacular downward projection of relatively low-temperature, high-pressure phase compatibilities in the vicinity of the descending lithospheric slab (Fig. 3).

Regions of crustal attenuation—such as the Basin and Range, the Rio Grande Rift, and the Salton Trough—are typified by abnormally high heat flow, and hence, elevated geothermal gradients. Unusually thick sialic crust, produced for instance, by continental collision and overthrusting or by crust-lithosphere delamination (*Bird, 1978*) such as the Colorado Plateau, is characterized by low heat flow, and thus possesses a lower than

Ernst, W. G., 1992, Metamorphism of the western Cordillera, and its relationship to tectonics, *in* Burchfiel, B. C., Lipman, P. W., and Zoback, M. L., eds., The Cordilleran Orogen: Conterminous U.S.: Boulder, Colorado, Geological Society of America, The Geology of North America, v. G-3.

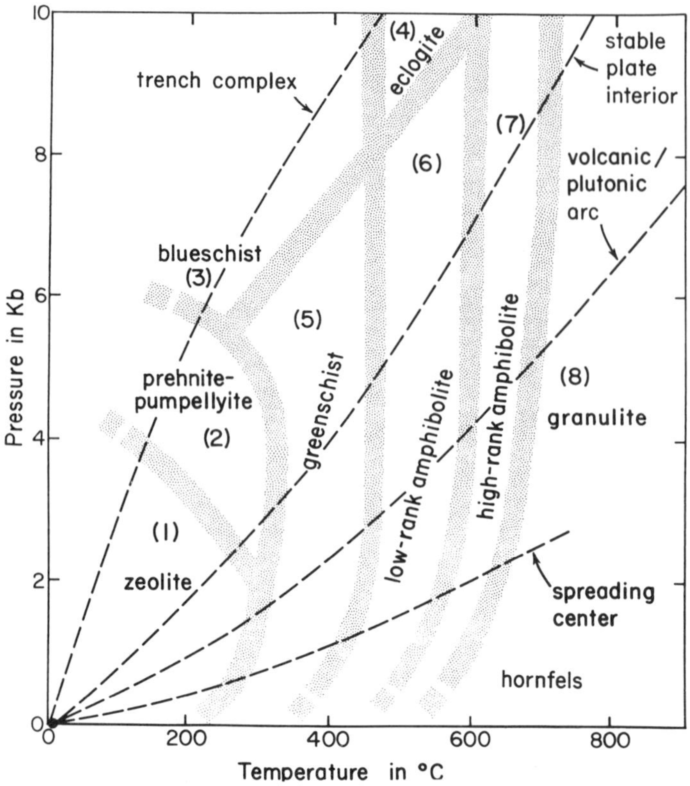

Figure 1. Diagrammatic petrogenetic grid for basaltic rock types, at intermediate oxidation states with P-T assignments of metamorphic facies, after *Ernst (1976)*. An aqueous fluid is assumed to be present except for high-rank metamorphic environments where fluid may be absent or, if present, presumably contains only low proportions of H_2O. In addition, kyanite, sillimanite, and andalusite fields of stability (*Holdaway, 1971*) in aluminous pelites are indicated. Typical examples of metamorphic P-T trajectories or field gradients (*Richardson, 1970*; *Spear and others, 1984*) are also illustrated.

average continental geotherm. Due to gradual, long-term thermal relaxation, stable cratons also are typically regions of low heat flow and low metamorphic P-T gradients. Oblique subduction boundaries and strike-slip zones display heterogeneous P-T conditions, depending on the degree of lithospheric closure or spreading due to a subduction or rifting component, and the rate of advective heat transport as a consequence of magma ascent and aqueous fluid flow. In all cases, physical conditions—hence attendant metamorphism—reflect the overall plate dynamics.

METAMORPHISM AND EXOTIC TERRANES

The previous discussion has treated segments of the continental crust implicitly as consisting of autochthonous or parautochthonous units during deformation and recrystallization. However, the disjuncture, transport, and suturing of disparate lithotectonic entities result in the amalgamation of terranes that, in the general case, may have been metamorphosed at different times and places under contrasting P-T conditions (Coney and others, 1980; Jones and others, 1983). The apparently fragmentary and accreted nature of continental assemblies complicates the task of unraveling the geotectonic history. Furthermore, the processes of rifting and suturing (plate divergence and oblique or head-on convergence) are accompanied by evolutionary changes in physical and chemical conditions; hence, accretionary collages are characterized by polymetamorphism.

Even if the western Cordillera were perfectly exposed, its post-Paleozoic metamorphic and crustal development would be difficult to interpret, but incomplete preservation of the evidence insures that aspects of the evolution may never be known. These uncertainties are compounded for more ancient lithotectonic assemblages. Recognizing the severe limitations of the data base, we now turn to brief summaries for specific areas. Because of more concentrated study, the emphasis is on petrotectonic accretion and metamorphism in California.

REGIONAL METAMORPHIC SUMMARIES FOR THE WESTERN CORDILLERA

Overview

Magmas represent partial fusion samples of deep-crustal and upper-mantle protoliths, providing geochemical constraints on the nature of the basement source. Mesozoic and Cenozoic igneous rocks of the western United States reflect a continental crust-capped lithosphere typified by discrete lead isotopic provinces (Zartman, 1974; Wooden and others, 1988b), which exhibit variable degrees of light rare-earth element enrichment; the underlying asthenosphere appears to be isotopically similar in some ways to depleted suboceanic mantle. The Cordilleran sialic crust in general becomes more oceanic in lithologic and geochemical character proceeding away from the continental nucleus. The model age of separation of the basement from a uniform(?) mantle reservoir monotonically decreases from the Wyoming province (>2.6 Ga) through the central Great Basin–western Mojave (2.0 to 2.3 Ga) to the Sonoran Desert (~1.7 Ga). Within the Mesozoic calc-alkaline belts, a systematic westward decrease in the influence of cratonal rocks, as expressed by $^{87}Sr/^{86}Sr$ initial ratios and ϵ_{Nd} signatures of volcanics and plutonics, is well documented (*Armstrong and others, 1977*; Kistler and Peterman, 1978; *Farmer and DePaolo, 1984*; Bennett and DePaolo, 1987).

The metamorphic development of the western United States during Late Jurassic, Cretaceous, and Tertiary time is related to dominantly early crustal thickening (convergence) followed by later thinning (perhaps analogous to back-arc spreading). This history appears to reflect Late Cretaceous–Eocene near-horizontal subduction (Dickinson and Snyder, 1978), and mantle delamination plus crustal thickening and mid-Cenozoic uplift plus extension (Bird, 1988). Expansive tracts are characterized by relatively high-temperature, low- and moderate-pressure recrystallization associated with calc-alkaline pluton emplacement (Hamilton, 1988). Local zonations, culminating in andalusite-bearing hornfels, are recognized as contact aureoles encircling

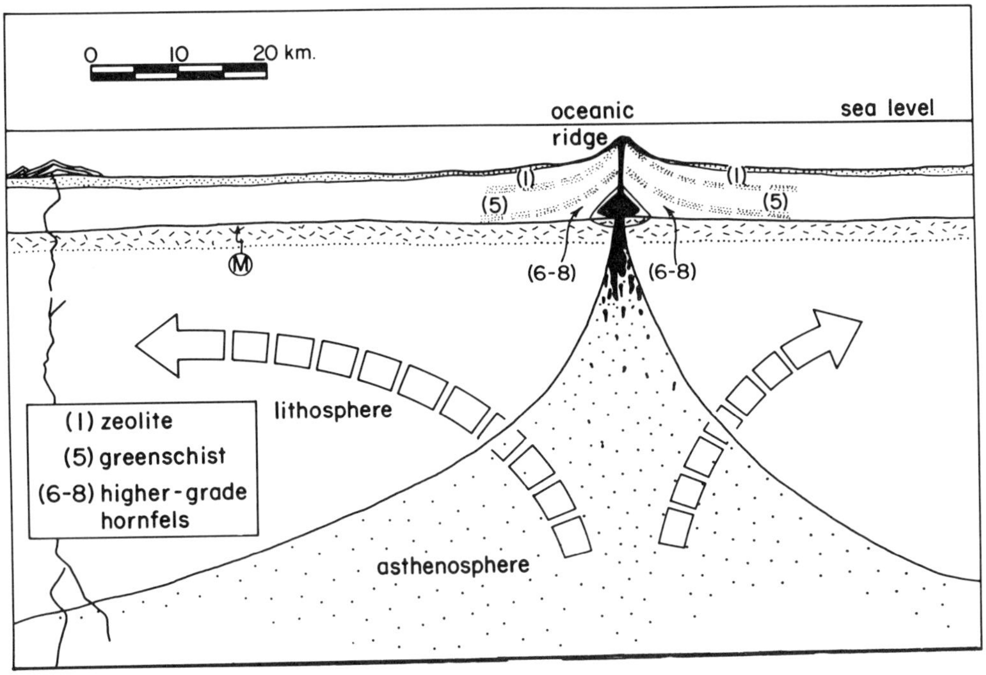

Figure 2. Schematic distribution of metamorphic facies in the high heat-flow regime of a simplified divergent plate boundary, after *Ernst (1976).* Individual facies are numbered as in Figure 1.

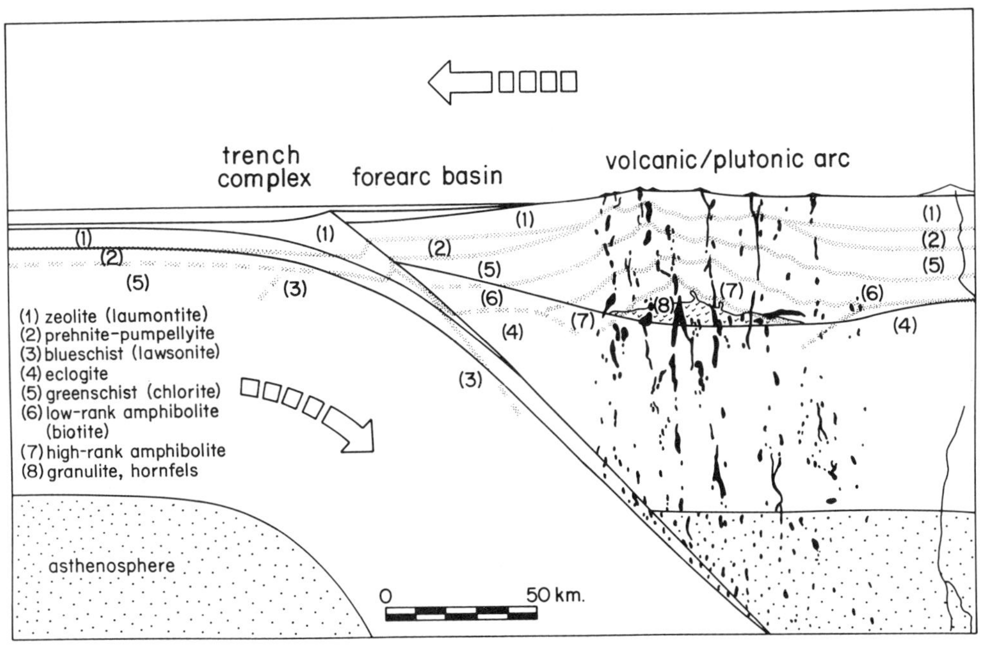

Figure 3. Schematic distribution of metamorphic facies in the low heat-flow trench complex environment, and high heat-flow volcanic/plutonic arc environment appropriate to a simplified convergent plate boundary, after *Ernst (1976).* Individual facies are numbered as in Figure 1. The broad expanse of high-T assemblages in the arc, and the remarkable downward projection of high-P assemblages in the subducting slab are a consequence of magmatic heat transport and lithospheric plate descent, respectively.

isolated stocks within the brittle upper crust (5 to 15 km depths); more extensive, sequential invasion of ductile middle crustal levels (~20 km depths) by abundant batholiths has resulted in the repeated generation and overlapping of thermally induced parageneses and in the wholesale, pervasive regional upgrading of metamorphic assemblages to ranks as high as sillimanite zone (*Barton, 1987*). In the northwest Cordillera, between Idaho and southeastern Alaska, more profound depths of intrusion of magmatic epidote-bearing granitoids into thick (45 to 60 km), kyanite-bearing crust—both suspected exotic and native North American terranes—have been exposed by subsequent erosion (Zen, 1988). If plutons rise nearly adiabatically, and impose local or regional thermal overprints on various crustal levels, the apparent P/T ratio of a metamorphic regime may be more a reflection of penecontemporaneous magma ascent and the extent of postrecrystallization uplift and erosion than a measure of overall crustal geothermal gradient (Barton and others, 1988).

In many cases, outboard from the high heat-flow regimes are narrow, coeval discontinuous mélange belts characterized by lenses of high-P, low-T rocks belonging to the blueschist and eclogite facies; these suites represent assemblages produced during subduction-zone metamorphism (Ernst, 1984). Surviving, intensely tectonized blueschist terranes are rich in disrupted ophiolitic and serpentinitic materials, but are devoid of contemporaneous calc-alkaline igneous rocks. Such oceanic protolithic assemblages are typical of the Pacific margin in rocks of Phanerozoic age, but are absent within the continental interior in rocks of any age. Paired metamorphic belts and an obvious relationship to plate-tectonic features are lacking, for the most part, in Precambrian complexes.

The areal distribution of metamorphic rocks of the western United States is presented in highly generalized form in Plate 9. Only the most widely distributed, main facies assemblages are indicated in this provisional summary; relict, overprinted assemblages are not figured. Serpentinites and peridotites are shown in black. Some, but by no means all, important structural boundaries are also provided. Most complexes are extensively covered by virtually unrecrystallized younger rocks, alluvium, or sea water; the metamorphic record is therefore incomplete. Compared to Figure 1, note that blueschist and eclogite facies have been combined, as have low- and high-rank amphibolites (the latter have been split into low- and high-pressure groups, however, based on aluminosilicate polymorphs in associated metapelitic schists). Plate 9 also indicates very approximate ages of principal metamorphic recrystallization events characteristic of the dominant metamorphic assemblages. Ages of relict assemblages are not illustrated.

Pacific Northwest

The Oregon Coast Ranges (not shown on Plate 9) and the Olympic Peninsula of northwest Washington are the sites of chiefly Paleogene terrigenous debris and both calc-alkaline and tholeiitic lavas. Situated on the Pacific side of the currently active volcanic arc, more deeply buried portions of these young supracrustal units have been subjected to incipient zeolite facies recrystallization. Within the core of the Olympic Peninsula, chiefly lower Tertiary rocks pass eastward into deformed pumpellyite-bearing and, possibly, low-grade greenschist assemblages (Tabor and Cady, 1978). East of Puget Sound lies the structurally complex and higher grade metamorphic basement of the northern Cascade Ranges.

The north-south–trending Straight Creek fault juxtaposes structurally mixed, relatively high-P and lower temperature northwest Cascades system terranes against more easterly Skagit high-T tonalitic gneisses and amphibolites (Misch, 1988). On the west, imbricated terranes of the northwest Cascades system include the Vedder blueschist complex of oceanic affinities and Permian metamorphic age, the lower Paleozoic or older Yellow Aster complex of continental para- and orthogneisses, the 130- to 150-Ma Shuksan suite of blueschists, and upper Paleozoic Chilliwack Group strata, now recrystallized to prehnite-pumpellyite facies (Brown, 1986; Galagher and others, 1988). Movement along the northwest Cascades thrust system took place during Late Cretaceous time. East of the Straight Creek fault, northwest-trending metamorphosed supracrustals and quartz dioritic to granodioritic orthogneiss belts exhibit emplacement and recrystallization ages of 220, 90 to 110, and 45 Ma, involving the repeated development of Barrovian-type parageneses (Babcock and Misch, 1988). East of the Ross Lake mylonite zone (*Kriens, 1987*), feebly recrystallized but intensely folded prehnite-pumpellite facies assemblages developed in Upper Jurassic and Cretaceous clastic strata of the Methow-Pasayten belt are juxtaposed against the more westerly Skagit gneiss core; within the fault zone itself, Permo-Triassic Hozameen Group volcanic and sedimentary rocks also have been recrystallized to prehnite-pumpellyite facies assemblages.

Permo-Triassic subduction was followed in Late Jurassic–Early Cretaceous time by renewed convergence, but a large component of strike slip insured that relics of the earlier orogeny were disaggregated and dispersed within the northwest Cascades system. Telescoping by relatively northwestward overriding of oceanic subduction zone units by the sialic arc was succeeded by Late Cretaceous northward post-recrystallization drift, and final docking at the southern terminus of Wrangellia on Vancouver Island (Brown, 1988). The continental margin or island arc represented by the Skagit Crystalline Core has been displaced an uncertain amount by dextral shear along the Ross Lake fault system, and achieved its present location prior to middle Eocene time.

Northern Rocky Mountains

Proceeding inland from the northern Cascades, metamorphic basement is exposed in a series of isolated or regionally contiguous uplifts—the Okanogan, Kettle-Lincoln, and Priest River domes. These features are representative of an extensive zone of metamorphic core complexes sited almost exclusively west of the Laramide fold-and-thrust belt, and predominantly within the Cordilleran miogeocline; their regional disposition is illustrated in Figure 4. Such metamorphic/structural highs represent segments of midcrustal levels that display the effects of high-rank metamorphism and ductile deformation; subsequent brittle extension and feeble recrystallization typify superjacent rocks lying above detachment faults that bound the complexes. This region is characterized by structures reflecting late Mesozoic crustal thickening and Cenozoic extension (*Potter and others, 1986*). In general, older, deeper levels of early stage metamorphism are exposed in the northern part of the belt; progressively younger, less deep-seated core complexes are to the south.

The Okanogan complex (Hansen and Goodge, 1988) bears evidence of three metamorphic episodes: (1) early, upper amphibolite–facies regional recrystallization; (2) greenschist-facies metamorphism associated with mylonitization; and (3) late-stage zeolitic and chloritic alteration accompanying brecciation during extensional deformation. Similar parageneses typify other core complexes of the eastern Great Basin and northern Rockies (Armstrong, 1982). The early, sillimanite-producing event is inferred to reflect long-sustained plate subduction during Jura-Cretaceous time: at mid-crustal levels (15 to 25 km depth), advective heat transfer reflecting pluton emplacement ultimately caused thermal weakening and buoyant rise of the basement, accompanied *en route* by recrystallization-induced grain-size reduction and retrograde metamorphism. A stretching regime was superimposed in early Tertiary time, possibly due to asthenospheric upwelling plus divergence and renewed volcanism; brittle detachment resulted at shallow crustal levels.

Greenschist-facies units constitute Quesnellia, a parautochthonous, westward-rooting eugeoclinal terrane, which rests structurally on the core complexes. Late Triassic–Early Jurassic paleomagnetic data yield concordant inclinations—hence this terrane apparently was at the same magnetic latitude as the North American craton during its construction (May and Butler, 1986). To the east, Quesnellia is in tectonic contact with the Kootenay Arc, an in situ but folded miogeoclinal section, recrystallized from the lower greenschist (~west) to upper amphibolite facies (~east). The Rocky Mountain fold-and-thrust belt lies even farther to the east; it consists of greenschist facies to practically unmetamorphosed, well-ordered Proterozoic and Paleozoic North American strata. Metamorphism in the broad region of Belt Supergroup outcrop is shown diagrammatically as prehnite-pumpellyite facies in Figure 4, but reconnaissance study of illite recrystallization (Maxwell and Hower, 1967) demonstrated that western portions in northern Idaho are as high as biotite zone of the greenschist facies, with eastward gradation to diagenetically altered (anchimetamorphic) strata in western Montana (Harrison, 1972). After the approximately mid-Jurassic suturing of Quesnellia to the continental margin through inferred westward underflow of the North American crust-capped plate, lithospheric descent evidently reversed polarity and stepped outboard of this accreted native terrane; eastward subduction then resulted in the construction of a new calc-alkaline magmatic arc, and in granitoid invasion of the Kootenay Arc and Quesnellia during the period 150 to 100 Ma (Rhodes and Hyndman, 1988).

The east-west–trending Baker and associated terranes of eastern Oregon appear to provide a link between lithotectonic assemblages of oceanic affinities in the northwest Cascades system and the western Triassic and Paleozoic belt of the Klamath Mountains (Kays and others, 1988; *Avé Lallemant (1989)*. The Baker complex, sutured against in situ calc-alkaline arc rocks of the Seven Devils–Wallowa terrane (*Follo, 1987*) on the east and north, represents a preserved segment of the Upper Triassic accretionary prism. On the north, argillite+chert+basalt protoliths dominate the lower Mesozoic greenschist belt. An ophiolitic serpentinite-matrix mélange is characteristic of the southern belt; oceanic rocks are Permo-Triassic in age of formation, with the time of localized blueschist metamorphism closely following at about 225 Ma (Hotz and others, 1977). The southern margin of the mélange belt is marked by the Canyon Mountain peridotite, which lies along a thrust fault structurally above the high-P rock-bearing chaotic zone. The region was later subjected to oroclinal bending and Cenozoic development of the Cascades and Columbia River volcanic fields across the preexisting basement.

In west-central Idaho, a north-south–trending, nearly vertical, Early Cretaceous plate boundary places outboard, amalgamated, early Mesozoic island-arc rocks of the native Seven Devils–Wallowa terrane of subgreenschist facies, and yet more westerly, oceanic, in part higher pressure Baker complex against autochthonous, multicycle, shallow-water Proterozoic and Paleozoic strata of the North American continent on the east (*Brooks and Vallier, 1979*; *Selverstone and others, 1987*; Lund and Snee, 1988). The suture probably underwent dextral shear during dynamothermal recrystallization at 93 to 118 Ma. Metamorphic grade culminated in amphibolite-facies assemblages (Hamilton, 1963) along the juncture where the deepest seated rocks are exposed, and declined toward greenschists both eastward and westward. A final dynamothermal episode accompanied emplacement of tonalitic to granitic plutons of the Idaho batholith 70 to 90 m.y. ago (Lund and Snee, 1988).

The Idaho batholith (a more southerly Atlanta, and a less extensive, northerly Bitterroot lobe) is itself a reflection of renewed plate convergence at the western margin of North America during Late Cretaceous time (Hyndman and others, 1988). A broad expanse of regionally metamorphosed rocks surrounds the crustally contaminated igneous complex, grading from a zone containing sillimanite ± other Al_2SiO_5 polymorphs near the contact to a chlorite zone in Belt Supergroup strata more than 50 km away (Hietanen, 1984; *Grover and others, 1987*); higher grade assemblages were produced during the latest Cretaceous anatectic event as an overprint on Proterozoic units feebly recrystallized but not deformed at approximately 1.2 Ga. Deep-seated S-type intrusions were emplaced at depths of at least 25 km in the northwest where kyanite was produced in the country rocks; exposed crustal depths are somewhat less to the southeast where andalusite is common in metapelites, and discrete aureoles surround hypabyssal plutons of the Idaho batholith.

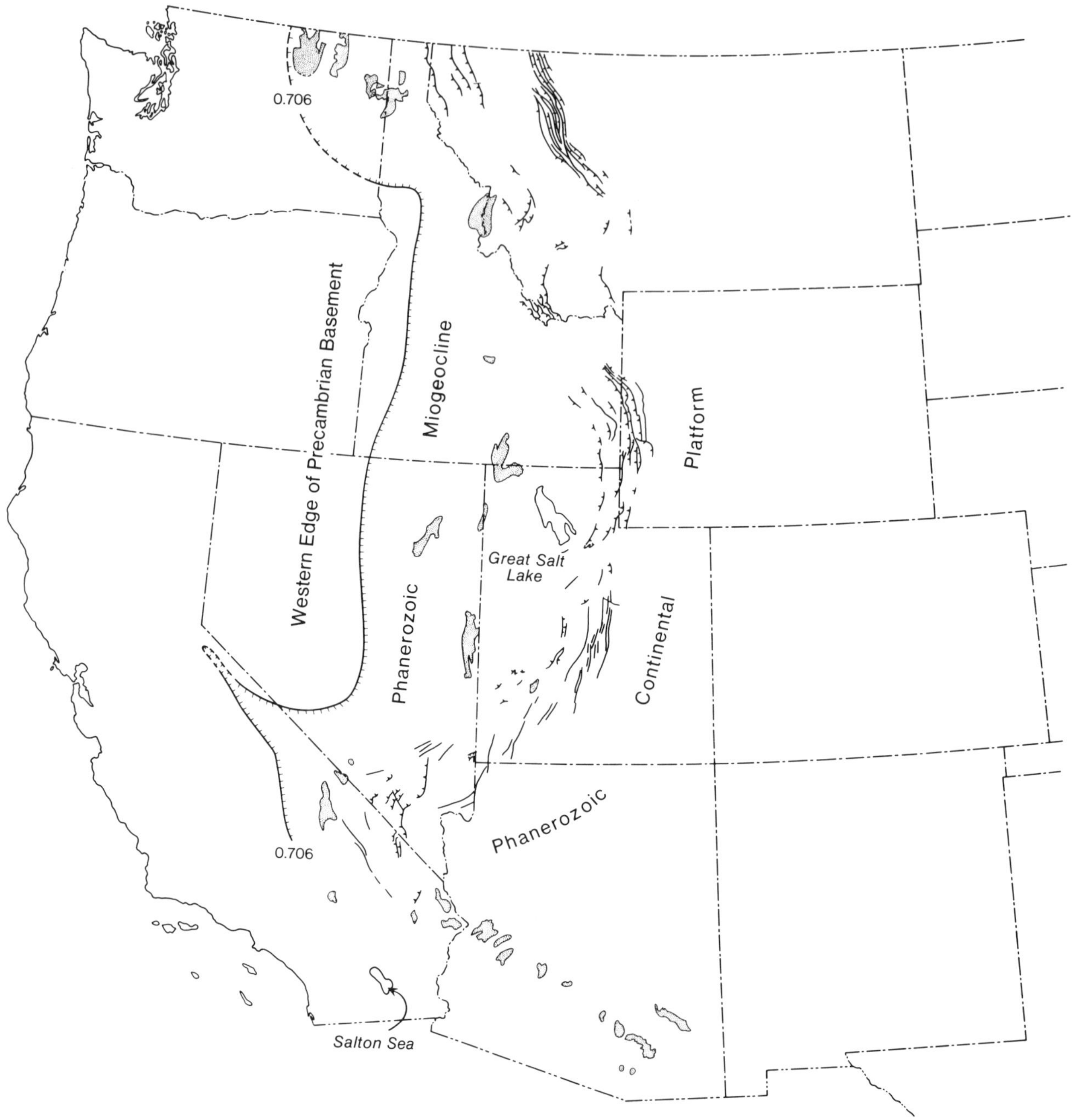

Figure 4. Metamorphic core complexes in the western conterminous United States, after Coney (1980), Armstrong (1982), and Anderson (1988). The eastern limit of the Cordilleran fold-and-thrust belt, nearly identical with the hinge line between the Phanerozoic miogeocline on the west, and the platform on the east, is illustrated. Also shown is the latest Precambrian(?) continental margin as indicated by the $^{87}Sr/^{86}Sr$ = 0.706 line in late Phanerozoic igneous rocks (*Armstrong and others, 1977*; Kistler and Peterman, 1978). To the west, initial strontium isotopic ratios for Mesozoic and Cenozoic silicic igneous units are less than 0.706, indicating an important mantle contribution to the melts, whereas to the east, initial ratios exceed 0.706, suggesting substantial involvement of old sialic crust.

Middle Rocky Mountains

The northwest Beartooth Mountains apparently contain a major tectonic boundary between contrasting Archean lithotectonic assemblages of the Wyoming province (Condie, 1976; Mogk and Henry, 1988). The region to the east—including most of the Beartooth complex, the Bighorn Mountains, and sporadic samples of basement from eastern Wyoming and Montana drill cores—is intensely invaded by granitoids; the area to the north and west—consisting of the North Snowy block of the Beartooths, and the Gallatin, Madison, Ruby, and Tobacco Root Mountains—is more sparsely intruded by calc-alkaline plutons, and seems to be a collage of supracrustal terranes dominated by protoliths consisting of shallow-water sediments.

In the more easterly lithotectonic belt, ancient platform sediments were transformed to polymetamorphic granulite-grade rocks at about 3.3 Ga (Wooden and others, 1988a). These units are associated with 2.8-Ga andesites, and occur as xenoliths in voluminous 2.75-Ga granitoids; the latter invaded the superjacent section contemporaneous with amphibolite-facies regional overprinting. The more westerly lithotectonic belts consist of high-grade paragneissic sequences, which display evidence of intensive crustal thickening by tectonic imbrication and isoclinal folding, and carry relict granulite-facies orthogneisses as old as 3.4 Ga (*Spencer and Kozak, 1975*; *Vitaliano and others, 1979*; *Erslev, 1983*). The juxtaposition of these eastern continental and western marginal basin ensialic terranes in late Archean time is regarded as a manifestation of subduction, Andean magmatic arc construction, and continental collision (*Mueller and Wooden, 1987*). Mineralogic thermobarometers indicate physical conditions of the earlier, multistage granulitic metamorphism on the order of 6 to 8 kb and 750° ± 50°C. The imbricated sialic crust in the Beartooth Mountains and adjacent areas at 3.4 Ga evidently was at least 25 to 30 km thick (Mogk and Henry, 1988). Central portions of the Wyoming province show similar early Archean ages of formation of the sialic crust (*Grisham and others, 1987*; *Aleinikoff and others, 1987*).

Throughout this ancient Precambrian nucleus, several episodes of granulite-facies metamorphism have been overprinted by amphibolite-facies assemblages, but Plate 9 shows only the dominant phase configurations. In the west-central part in the Wind River Range, relict greenstone belts are preserved in slightly lower grade rocks.

The southern margin of the Archean craton is marked by the Cheyenne belt of sheared units, juxtaposing accretionary fore-arc basin and island-arc–type Proterozoic supracrustals against the more ancient, northerly shield and lower Proterozoic miogeoclinal cover series (*Hills and Houston, 1979*; Duebendorfer and Houston, 1986, 1987); the suture trends east-west across northern Utah and southeastern Wyoming. Metasedimentary and mafic metavolcanic rocks in the Medicine Bow and Sierra Madre Mountains have been metamorphosed under greenschist to low-rank amphibolite-facies conditions and invaded by late Archean granites (*Karlstrom and others, 1983*); grade increases along a narrow zone where this Archean basement is juxtaposed against mylonitized Proterozoic miogeoclinal strata plus mafic dikes and sills of amphibolite grade in the Cheyenne belt. Archean and early Proterozoic quartzofeldspathic gneisses and overlying strata are in contact along this trend in the northeast margin of the Uintas, where metamorphic grade reached P-T conditions appropriate to the aluminosilicate triple point (Sears and others, 1982). Recrystallization along and south of this shear zone probably reflects suturing of offshore arcs against the Wyoming craton due to marginal basin collapse.

The Black Hills contain small gneissic remnants and younger granitoids of the Archean shield, but are dominated by Proterozoic supracrustal and plutonic units (Gosselin and others, 1988). Mid-crustal levels are indicated by the presence of sillimanite±kyanite.

In the Wasatch Mountains (Bryant, 1988), sedimentary units containing terrigenous debris as old as 3.6 Ga were deposited on oceanic crust, then intruded by tholeiitic dikes and sills at about 3.4 Ga. The section was intensely metamorphosed to low granulite grade, producing relict hypersthene-bearing assemblages, at approximately 2.6 Ga. On the west, sialic material was added to the Archean nucleus in early Proterozoic time. Subsequently, the entire complex was migmatized during widespread amphibolite-facies recrystallization at 1.8 Ga; country rocks were transformed to felsic, micaceous sillimanite-bearing paraschists+ gneisses, and mafic orthoamphibolites, invaded by anatectic, synkinematic quartz monzonite gneiss. Contemporaneously, new continental crust formed south of the east-west cratonal boundary using early Proterozoic time, as described in the next section.

Southern Rocky Mountains

The basement in Colorado (*Barker, 1969*; Condie, 1986; Bickford, 1988a) consists of lower Proterozoic bimodal metamorphosed equivalents of tholeiites and dacites-rhyolites, andesitic metagraywackes, and nonvolcanic, chemically more mature metasediments; these superjacent units have been invaded by several generations of middle Proterozoic calc-alkaline igneous rocks. Better stratified pelitic and quartzofeldspathic units occur in the Front Range on the east, with more arc-like supracrustals situated farther west in central Colorado.

Except for practically unrecrystallized middle and upper Proterozoic strata in the Uinta Mountains of northwest Colorado, and largely covered basement rocks of late Precambrian age in easternmost Colorado, differential uplifts reveal the regional development of low- to high-rank amphibolite grade metamorphism and locally, lower granulite facies in the southernmost Front Range (Reed and others, 1987). Recrystallization was apparently contemporaneous with widespread intrusion of granitoids at approximately 1.7 to 1.8 Ga, accompanying at least two periods of isoclinal folding and migmatization. Nonfoliated, postkinematic granitic plutons were emplaced throughout the southern Rocky Mountains about 1.4 Ga and again at ~1.1 Ga (*Bickford and others, 1986, 1987*).

The recrystallization of sedimentary units plus bimodal volcanics, and invasion by granitoids, attest to progressive accretionary events taking place at the southern margin of the Wyoming craton during early and middle Proterozoic time. Accretion of mafic volcanic-dominated arcs was followed by telescoping and sinistral(?) shear along the east-west Cheyenne belt, a transpression boundary between adjacent basement terranes (Karlstrom and Houston, 1984; *Houston and Duebendorfer, 1987*; *Giles, 1987*). Subsequently, local reheating spatially associated with the Midcontinent rift system occurred during late Proterozoic time. The late Precambrian midcrustal levels exposed by subsequent Phanerozoic uplift reflect moderate continental geothermal gradients, and neither unusually thick nor thin crust.

Rio Grande Rift area

Principally Neogene extension in the southwestern United States has resulted in the marked attenuation of continental crust, especially in the Basin and Range Province and in the Rio Grande Rift. Abnormally high heat flow plus Miocene and younger volcanism reflect the relatively shallow approach of hot asthenosphere; partial fusion of basal portions of the continental crust, and high-T, low-P, regional metamorphism undoubtedly attend the inferred (ongoing) lithospheric thinning, but the petrotectonic effects of these deep-seated processes are still hidden from direct examination.

Bordering the Rio Grande Rift, scattered basement uplifts have, however, revealed the widespread extent of Proterozoic metamorphism over broad tracts of northern and central New Mexico (*Grambling, 1986*). Protoliths include mafic plus felsic volcanics; minor banded iron formations; and shaly, arkosic, and quartzose sedimentary units. These supracrustal units date from approximately 1.7 Ga, with formational ages of native sialic crust and allochthonous oceanic crust in general decreasing southward (*Condie, 1987*). Six contrasting lithotectonic terranes have been recognized by Grambling and others (1988). Similar to geologic relationships in Colorado, synkinematic calc-alkaline plutons are of middle Proterozoic age, with a few younger anorogenic granitoids dating from 1.4 to 1.5 Ga (*Silver and others, 1977*).

Superjacent rocks were metamorphosed under physical conditions near 4 kb, corresponding to a depth of about 14 km, and a temperature maximum of 520°C, close to the Al_2SiO_5 triple point (Grambling, 1981, 1988). Accordingly, Plate 9 shows a simplified mixture of high- and low-P amphibolite facies complexes in this region. At least three stages of penecontemporaneous deformation resulted in a 20 percent thickening of the recrystallized Proterozoic crust in the vicinity of the present-day Rio Grande Rift. Postkinematic cooling and retrograde metamorphism apparently took place virtually isobarically, prior to uplift and erosion. Remarkably, isograds and isobars are subparallel to a mid-Paleozoic unconformity, which itself has remained in a near-horizontal to gently west-dipping attitude to the present time. Thus, cratonization of this part of the North American shield was completed in pre-Mississippian, probably in late Precambrian time.

Colorado Plateau

The Colorado Plateau consists of a flat-lying sequence of thin-bedded, unmetamorphosed Paleozoic and Mesozoic strata resting on an, in part, intensely deformed and recrystallized Proterozoic basement. The latter is well exposed in northern Arizona in the Grand Canyon, but is covered elsewhere in the province (*Breed and Roat, 1976*).

The oldest rocks in the Grand Canyon belong to the tightly folded Vishnu Schist, protoliths of which were laid down in early Proterozoic time. The metamorphic age of this unit, consisting predominantly of micaceous and quartzofeldspathic schists, with minor paragneisses, migmatites, and amphibolites, is 1.7 Ga. It has been invaded by synkinematic and postkinematic granitic to dioritic plutons (Babcock and others, 1979). Metamorphic grade of the Vishnu Schist is preponderantly amphibolite facies, but thermal overprintings of higher rank assemblages are present adjacent to intrusive igneous rocks (Brown and others, 1979; Clark, 1979). The complex is unconformably overlain by virtually unmetamorphosed upper Proterozoic miogeoclinal strata, perhaps equivalent in age to the Belt and/or Uinta groups. A second major angular unconformity represents latest Precambrian tilting, uplift, and erosion, followed by epeiric sea incursion and deposition of the capping Cambrian and younger platform strata.

Great Basin

The northern and central Basin and Range Province, including virtually all of Nevada, easternmost California, northwestern Utah, and southeastern Idaho, will be discussed in this section. Portions of southern Arizona, southeastern California, and southwestern New Mexico constitute the southern extension of this geologic province, but will be treated in the section dealing with the Mojave-Sonoran Desert. The Basin and Range Province represents a widespread crustal response to chiefly Neogene attenuation of the underlying lithosphere (Zoback and others, 1981; *Wernicke and others, 1982*). It is also the main site of numerous basement uplifts, including metamorphic core complexes (see Fig. 4). The eastern part of the Great Basin is characterized by latest Archean (*Lush and others, 1988*) and Proterozoic basement (Allmendinger and others, 1987), the western part by Precambrian, and accreted Paleozoic and Mesozoic terranes.

An evolving Atlantic-type continental margin existed in late Precambrian and Paleozoic time between the southern Sierra Nevada and central Utah (*Stewart, 1972*; Speed and others, 1988). Miogeoclinal strata are chiefly fine-grained clastics in lower portions of the column, limestones and dolomites in the upper, Paleozoic part of the section. Passive-margin sedimentation was interrupted twice by the arrival from the west of oceanic allochthons (chert/argillite + mafic volcanic rocks) that overrode the outer shelf in Devono-Mississippian and Permo-Triassic times; west-rooting décollements are the Roberts Mountains and Golconda thrusts, respectively. Intense metamorphism is not characteristic of either of these events, but greenschist-facies overprinting is present locally; the orogenies are interpreted as reflecting westward descent of marginal basins leading the American Plate into collision with offshore island arcs (*Poole and Sandberg, 1977*). After a Pennsylvanian to earliest Triassic rifting event, which apparently truncated the south-southwest extension of the North American continent, a northwest-trending Pacific-type convergent margin was established in Middle or Late Triassic time (Burchfiel and Davis, 1972). Calc-alkaline volcanism, plutonism, and the deposition of volcanogenic detrital sedimentary units took place semicontinuously over the next 150+ m.y., reflecting the presence of eastward-descending paleopacific plate(s) beneath the encroaching North American lithosphere. Volumetric maxima in peraluminous, anatectic, granitoid pluton emplacement (and metamorphism) occurred at about 160 and 75 Ma, according to Armstrong and Suppe (1973), especially in southern Idaho and the Mojave Desert. A third, as yet buried, plutonic culmination probably is associated with Neogene extension. Although shown in Plate 9 as of Late Jurassic–Tertiary age, some of the preserved phase assemblages in the north-central Basin and Range may reflect somewhat earlier, mid-Paleozoic recrystallization, but control is poor on the geochronologic relationships.

Accompanying this crustal growth, imbricate, east-vergent thrusting folded the section like an accordian intermittently during the Late Jurassic to Paleocene. Amphibolite to greenschist facies recrystallization at intermediate crustal levels took place in mid- and late Mesozoic time (Snoke and Miller, 1988; *Miller and others, 1987*; Miller and others, 1988). Some pelitic schists retain relict Jurassic and Cretaceous phase assemblages, including sillimanite±kyanite or andalusite, whereas Tertiary parageneses more typically carry andalusite±sillimanite. Metamorphic core complexes occur along and west of the Cordilleran fold-and-thrust belt and the spatially coincident late Precambrian cratonal hinge line. Mylonitization and buoyant rise of such basement highs reflect Late Oligocene and, especially, Miocene normal faulting and thinning of the thermally softened continental crust; stretching presumably was related to upward asthenospheric flow and heat transfer. Polymetamorphism and multiple deformations of contrasting styles (compressional versus extensional) are thus characteristic of these complex geologic areas. Metamorphism in the Basin and Range is scattered and ill constrained, due to poor exposures, intensive post-recrystallization faulting, and the plentiful occurrence of carbonate units bearing undiagnostic mineral assemblages, especially in the east. Accordingly, some of the ranges shown in Plate 9 as greenschist facies may actually be of lower grade, whereas others presumed to be unmetamorphosed may represent fairly intensely recrystallized carbonate-rich sections. Generally, the magnitude of uplift is reflected in the observed metamorphic grade. Core complexes typically exhibit amphibolite facies assemblages (shown on Plate 9 as sillimanite-bearing, whereas some actually contain kyanite), with adjacent ranges slightly to substantially lower in grade.

The Walker Lane belt, a complex northwest-trending Mesozoic-Cenozoic dextral shear zone approximately 700 km long and 100 to 300 km wide (Stewart, 1988), defines the western margin of the Great Basin. It marks the juxtaposition of predominantly Paleozoic basement (Precambrian present but covered) and relatively thin crust on the east, and the exposed Proterozoic section of easternmost California within Walker Lane on the west, which in turn gives way to an apparently thicker sialic section in the vicinity of the Sierra Nevada (Knuepfer and others, 1987). In this southeast portion of the batholith, pendants metamorphosed at high temperatures contain protoliths correlated with the miogeoclinal section of the westernmost Basin and Range.

In the Death Valley region, middle Proterozoic gneisses are overlain by well-ordered platform carbonates and quartzites of late Proterozoic and Paleozoic age (Labotka and Albee, 1988). Similar strata are widely exposed on the west in the White-Inyo Range (Dunne and others, 1978; *Ernst and Hall, 1987*; Nelson and others, 1991). To the south, pelitic schists in the Panamints were metamorphosed to andalusite or sillimanite ± cordierite-bearing assemblages, whereas north across the Furnace Creek fault, the northern Funeral Mountains contain kyanite and staurolite. The 3.0 ± 1 kb confining pressures attending approximately Late Jurassic recrystallization in the Panamint Mountains and the White-Inyo Range may be ascribed to the overlying stratigraphic column, whereas in the Funerals, a lithostatic pressure of 6.5 ± 1 kb during Early Cretaceous metamorphism indicates the local emplacement of the east-vergent Inyo Range nappe complex (Labotka and Albee, 1988). Moderately high temperatures of prograde metamorphism (maximum = 650 to 700°C) evidently reflect the intrusion of calc-alkaline plutons episodically over the interval 140 to 170 Ma. Retrograde production of greenschist-facies assemblages occurred throughout the region in Late Cretaceous time.

Mojave-Sonoran Desert

The Precambrian gneissic basement of this province formed about 1.6 to 1.9+ Ga, was deformed repeatedly, and was pervasively metamorphosed prior to final stabilization in middle Proterozoic time (Karlstrom and others, 1987). Granulite-facies sillimanite (locally kyanite)+cordierite+garnet–bearing pelitic gneisses and migmatites, and opx+cpx–bearing cummingtonitic amphibolites in southernmost Nevada/northwesternmost Arizona pass southward along the California-Arizona border to high-rank muscovite-bearing amphibolitic and quartzofeldspathic gneisses (Theodore and others, 1987; Thomas and others, 1988); the former possess crystallization temperatures of ~700°C, the latter ~650°C, both at low lithostatic pressures of about 3 ± 1 kb. The age of metamorphism was approximately 1.75 Ga. These high-grade Proterozoic basement rocks give way in southern and southeastern Arizona to 1.65-Ga greenschists. The nature of this contact is not known but is probably tectonic.

In general, Sonoran Desert Proterozoic anorogenic granites tend to be biotite ± muscovite-bearing and peraluminous, whereas the Mojave Desert and Colorado River areas are typified by metaluminous biotite ± hornblende–bearing anorogenic granites (Anderson, 1988; Thomas and others, 1988). The more aluminous nature of the granitoids on the east suggests a greater involvement of metasedimentary material in the derivation of such magmas.

A thin basement cover of Paleozoic platform strata is overlain by variable but in general much thicker successions of Mesozoic and Cenozoic miogeoclinal and calc-alkaline volcanogenic units. Phanerozoic metamorphism took place at various times from Permo-Triassic to early Tertiary time in the Mojave Desert, southern Arizona, and southwestern New Mexico in conjunction with ductile nappe emplacemnt, thrusting plus mylonitization, granitic intrusion, and hydrothermal fluid circulation (DeWitt and others, 1984; Haxel and others, 1984; *Hoisch, 1987*; Hoisch and others, 1988; Reynolds and others, 1988). Pre-Neogene metamorphism as now exposed ranges from greenschist to amphibolite facies, depending in some cases on the amount of uplift and overburden removed and proximity to thrust surfaces or granitoid contacts. Many reported phase assemblages have prograde maxima near the Al_2SiO_5 triple point. Ductile deformation is indicated by the ubiquity of a penetrative schistosity, extreme attenuation, and ductile thrusts and nappes.

Retrograde metamorphism, most commonly of greenschist to lower amphibolite facies, accompanied middle Tertiary (15 to 35 Ma) mylonitization, including production of subhorizontal foliation and lineations, overprinted by chloritic brecciation and detachment faulting (*Reynolds and others, 1986*; *Bykerk-Kauffman and Janecke, 1987*). This deformation is characteristic of core complexes, as in the Whipple and Sacramento Mountains of southeasternmost California (Anderson, 1988). Here, calc-alkaline plutons of several ages (e.g., 89, 73, 26, and 19 Ma) invaded the intensely mylonitized, amphibolitized, 1.7-Ga gneissic basement complex (*Orrell and Anderson, 1987*)—the early Miocene body intruding contemporaneously with mylonitic deformation. The Late Cretaceous peraluminous and metaluminous plutons contain phase assemblages that formed at 760° ± 25°C and at intermediate to deep crustal levels of 20 to 30 km; intrusion has resulted in regional overprinting and upgrading of preexisting amphibolite facies assemblages in the host gneisses. Mylonitization and Neogene intrusion of metaluminous plutons occurred at progressively shallower levels, as decompression recrystallization and more brittle deformation attending extension were taking place in the retrograded country rocks.

Klamath Mountains

The Klamath Mountains consist of five composite terranes (Irwin, *1960,* 1972; Burchfiel and Davis, 1981): the eastern Klamath belt; the central metamorphic belt; the Stuart Fork Formation; the western Paleozoic and Triassic belt (WTrPz); and the western Jurassic belt. These terranes, rich in associated mafic + ultramafic masses, have been juxtaposed along east-rooting thrust faults. The region has been invaded by Middle and Late Jurassic calc-alkaline plutons (*Lanphere and others, 1968*; Irwin, 1985). Sedimentary units in general are right side up, and dip predominantly eastward. The effects of at least five areally distinct metamorphic episodes, spanning the interval from Middle Devonian to Late Jurassic time are contained in rocks of this province (Coleman and others, 1988).

The eastern Klamath terrane consists of Paleozoic to mid-Mesozoic metasedimentary strata and intercalated metavolcanic units overlying the Trinity peridotite along a contact of uncertain nature. This gigantic ultramafic massif consists dominantly of harzburgite, dunite, plagioclase±spinel lherzolite, and serpentinized equivalents, cut by gabbros, norites, and diorites (*Quick, 1981*); the complex may have originated as a diapir within the uppermost mantle. Radiometric ages of the mafic intrusives are in the ranges 565 to 570 Ma and 450 to 480 Ma (*Mattinson and Hopson, 1972*; *Wallin and others, 1988*), compatible with Early Cambrian and Middle Ordovician igneous events and recrystallization of the ultramafic host (*Jacobsen and others, 1984*). Pillow basalts, interpreted by *Lindsley-Griffin (1977)* as overlying the plutonic sequence, occur locally in the northwestern part of the complex.

Two areally distinct packets of island-arc strata are present above the Trinity peridotite: the Yreka section north of the massif, and the Redding section to the southeast. The former consists of fossiliferous, feebly recrystallized, chloritic metashales and metagraywackes of Ordovician to Early Devonian age (*Potter and others, 1977*; *Hotz, 1977*); more intensely metamorphosed semischists and phyllites, recrystallized at about 370 Ma, are assigned to the central metamorphic belt by *Cashman (1980)*. The more extensive Redding section is lithologically diverse (Irwin, 1981; Miller, 1987): lower units include a thick series of Devonian greenstones and metarhyolites that pass stratigraphically upward and eastward into upper Paleozoic–Jurassic marbles, slates, and calc-alkaline metavolcanics. Lower Permian limestones contain a temperate-water North American fauna (*Skinner and Wilde, 1965*; *Miller and Wright, 1987*), and remnant magnetic signature (Renne and others, 1988), indicating that major latitudinal drift could not have taken place after Early Permian time. The eastern Klamath section has been incipiently recrystallized to lower greenschist facies assemblages, most recently during the Middle Jurassic. The enigmatic Skookum Gulch blueschists appear to have been metamorphosed outboard of the arc in Late Ordovician time (*Cotkin and Armstrong, 1987*), prior to tectonic insertion into the section.

The central metamorphic belt, lying beneath the Trinity thrust, is made up of two major units: calcareous/quartzitic schists, and hornblende schists (Davis and Lipman, 1962); the pair has been interpreted by Burchfiel and Davis (1972) as the sedimentary cover and underlying oceanic crust of a Devonian or older lithic assemblage. The eastward progression of biotite-bearing high-rank greenschist, to epidote amphibolite and garnet amphibolite assemblages adjacent to the Trinity massif, indicates a somewhat more intense recrystallization for this terrane than for other Klamath belts (Davis and others, 1965). The age of metamorphism is 380 to 399 Ma (*Lanphere and others, 1968*; *Hotz, 1977*). Foliation in the schists coincides with the penetrative deformation exhibited by serpentinized western portions of the structurally overlying Trinity peridotite; metamorphic grade increases upward (eastward) into the ultramafic massif. Fluids evolved during prograde recrystallization of the central metamorphic belt appear to have migrated upward, causing extensive serpentinization and metasomatism of the overlying Trinity peridotite (*Peacock, 1987*). Thus, the two terranes are regarded as having been juxtaposed at considerable depth along the east-dipping Trinity thrust fault during recrystallization in mid-Devonian time (*Lipman, 1964*; *Davis and others, 1980*; Irwin, 1981). This date probably represents the time of subduction and accretion of an oceanward terrane to the continental margin or island arc (Hamilton, 1969), and precedes or perhaps heralds the onset of Roberts Mountains thrusting in eastern Nevada.

The east-dipping Siskiyou thrust juxtaposes the western Paleozoic and Triassic composite belt and tectonically overlying Stuart Fork beneath the structurally higher central metamorphic belt. The Stuart Fork Formation is a series of interleaved metapelitic and metamafic units that bound the northwestern portion of the central metamorphic belt on the north, and crop out in a window along the east side of the central metamorphic belt farther south. Lawsonite- and glaucophane-bearing assemblages locally characterize the northern part of the Stuart Fork (Hotz, 1973; *Borns, 1984*); to the south, these units give way to epidote+crossite–bearing metamorphic rocks (*Goodge, 1985*). Geochronologic data indicate that these blueschists were produced about 220 Ma (Hotz and others, 1977), the time of formation of high-pressure rocks in the Baker terrane to the northeast in eastern Oregon (Kays and others, 1988). In turn, the Stuart Fork apparently has been thrust over the WTrPz (*Goodge, 1986*). According to *Lanphere and others (1968),* much of the Stuart Fork has been overprinted by a 158-Ma greenschist facies recrystallization, perhaps reflecting later invasion by calc-alkaline plutons of Nevadan age (Irwin, 1985).

Three geographically distinct, north-south–trending lithotectonic segments of the WRrPz have been recognized (Irwin, 1972). These are, from east to west: the North Fork terrane; the Hayfork terrane, which itself has been recognized as a faulted amalgamation of eastern and western belts by Wright (1982); and the Rattlesnake Creek terrane. All three contain metamorphosed chert, dismembered ophiolitic fragments, and volcanogenic sediments, variably mélanged. The North Fork and Rattlesnake Creek belts are particularly rich in serpentinized peridotites and metagabbroic and metabasaltic rocks, whereas the Hayfork carries abundant andesitic detritus. Late Paleozoic and Triassic fossils have been described in exotic limestone blocks from all three WTrPz terranes; in addition, Early Jurassic radiolaria have been reported in siliceous sediments from both North Fork and Rattlesnake Creek terranes (Irwin, 1981). Although most of the cherts and pillow basalts are probably far-traveled, terrigenous debris in some WTrPz units appears to have been derived chiefly from the central metamorphic belt and the eastern Klamath terrane (Wright, 1982; Gray, 1986). The three WTrPz terranes of Irwin are not clearly recognized northward in the Klamath Mountains; the Marble Mountains terrane (Mortimer, 1985) has age and lithologic characteristics similar to the Rattlesnake Creek terrane. On the Oregon side of the state line, parts of the Applegate Group of *Wells and others (1949)* appear to be analogous to the Hayfork assemblage. In the northernmost Klamaths, the May Creek Schist is lithologically and metamorphically comparable to the deepest parts of the Marble Mountains terrane (Kays and others, 1988).

Most of the WTrPz belt has been metamorphosed to greenschist facies assemblages, with biotite widespread in metapelitic rocks on the north (Hill, 1985; Mortimer, 1985). Grade imperceptibly rises northward to amphibolite facies in structurally deepest portions framing the Condrey Mountain window. Scattered blueschist blocks have been reported in the central North Fork terrane (*Davis, 1968*), but like mélanges of the southern Baker terrane in eastern Oregon, tectonic significance is obscure. Perhaps they represent olistoliths derived from blueschists of the Stuart Fork Formation (Ando and others, 1983), implying that the North Fork section was assembled more or less in place relative to the more easterly Stuart Fork.

Meta-andesite of the western Hayfork possesses a Middle Jurassic minimum radiometric age, and pre-Nevadan calc-alkaline stocks intrusive into the western Paleozoic and Triassic belt range up to 171 Ma (*Lanphere and others, 1968*; Wright, 1982; Snoke and others, 1982; *Cotkin and others, 1985*); accordingly, this belt must have accreted and been metamorphosed by Middle Jurassic time at the latest.

The western Jurassic belt lies to the west, tectonically beneath the WTrPz. Its western, basal unit is the Josephine ophiolite and Chetco arc, an enormous mafic-ultramafic sheet exposed in southwestern Oregon and northernmost California. Southern segments are chiefly serpentinized harzburgite, a sheeted-dike complex, and pillow basalts, whereas gabbroic and amphibolitic units crop out in Oregon (*Garcia, 1979*). To the east lies the native Preston Peak mafic-ultramafic body, which evidently was constructed on disrupted basement of the WTrPz (*Snoke, 1977*). The age of formation for these ophiolites is about 160 Ma (Saleeby and others, 1982). The Josephine complex was emplaced and immediately overlain by rocks of the Rogue and Galice Formations, a metamorphosed intra-arc assemblage of mafic igneous rocks, basal hemipelagic and dominantly flyschoid slates, graywackes, shales, andesitic flows, and pyroclastics. The sedimentary age of this thick, volcanogenic clastic deposit is Late Jurassic, based on the contained fossils (Irwin, 1981). The source of the detritus was apparently the more easterly Klamath provinces (*Snoke, 1977*; Saleeby and others, 1982). Metamorphic effects are pervasive but weak and, eastward, do not rise above the chlorite zone of the greenschist facies. Recrystallization occurred about 150+ m.y. ago (Lanphere and others, 1978).

East of the main outcrop zone of the western Jurassic belt, the Condrey Mountain Schist is exposed in the core of a dome, as units beneath a window through the overlying WTrPz (*Donato and others, 1980*; Helper, 1986a). The Condrey Mountain Schist contains epidote and crossite, and represents blueschist P-T conditions transitional toward those of the greenschist facies. It is similar petrologically and perhaps in recrystallization age (*Helper, 1986b*) to the Colebrook and South Fork Mountain Schists of the Franciscan complex, and the Shuksan suite of the northwest Cascades (*Brown and Blake, 1987*). It seems clear that the WTrPz—including the May Creek Schist in southwestern Oregon—constitutes a thin flap of rocks resting on the underthrust Galice and Condrey Mountain units. If the latter is tentatively correlated with the western Jurassic belt, rather than the Franciscan, an eastward metamorphic intensity progression from prehnite-pumpellyite and chloritic greenschist zones on the west toward an eastern high-pressure blueschist zone would seem to be indicated.

Cretaceous Great Valley strata resting on the eastern Klamaths and Sierra Nevada have remnant magnetism nearly coincident with the paleomagnetic field attributable to these localities at that time (*Mankinen and Irwin, 1982*). Hypothesized latitudinal drifting of basement must have been completed prior to the deposition of these sediments. Partially covered in central Oregon by younger rocks, Klamath-like lithotectonic units crop out in the northeasternmost part of the state as the previously described Baker complex (Kays and others, 1988).

Sierra Nevada

The Sierra Nevada appears to represent the southeastern continuation of the Klamath province (Davis, 1969; Hietanen, 1981; Day and others, 1988). Although terrane boundaries are nearly vertical, most authors compare the tectonic imbrication to that of the Klamaths. However, pre-Cretaceous vergence was east-directed in the northwest Sierra, according to Day and others (1985). Most of the lithotectonic units can be correlated across the northern end of the Sacramento Valley; however, the Klamath central metamorphic belt apparently has no exact counterpart in the Sierran foothills. Moreover, the Klamath Mountains consist chiefly of metamorphosed country rocks intruded by discrete, pre-Cretaceous calc-alkaline plutons, whereas the Sierra Nevada consists principally of coalescing, Jura-Cretaceous batholithic units, incompletely separated from one another by thin metamorphic septa. The northwest foothills belt is the only major segment of Sierran country rocks preserved adjacent to the plutonic series, although isolated roof pendants occur scattered throughout the range. The Sierra Nevada may represent a deeper level of crustal exposure, especially at its southern extremity where high-rank amphibolitic to low-rank granulitic meta-igneous rocks beneath the batholith have been exhumed (Sams and Saleeby, 1988).

Similar to the Klamaths, suture zones bounding individual Sierran belts are marked by serpentinized peridotites, but the volumetric abundance of ultramafic material is reduced in the southeastern extension of the wallrock complex. Different terranes (Sharp, 1988) include, from east to west: the early to mid-Paleozoic Shoo Fly terrane, the Feather River peridotite and Melones fault zone blueschists, the late Paleozoic–early Mesozoic Calaveras-Melones complex, and the western Jurassic belt. These lithotectonic units display a steeply east-dipping imbrication; isoclinally folded sections have been recognized, but where primary flow tops and sedimentary laminations are preserved, sections typically face east (Clark, *1964,* 1976). In general, age of formation, metamorphic grade (at least, in the south), and structural complexity increase eastward (Schweickert and others, 1988).

The Shoo Fly assemblage, apparently of local provenance (*Hannah and Moores, 1986*), constitutes some of the most ancient rocks exposed in the Sierra Nevada, and comprises several imbricate thrust sheets. It includes metamorphosed fine-grained siliceous sediments, quartzites, and intercalated phyllites suggestive of continental shelf, slope, and rise sedimentation; cherts, a greenstone+limestone sequence, and a serpentinite-bearing mélange are also present as oceanic components. The complex, at least partly Silurian in age, is overlain unconformably by Upper Devonian–Permian arc rocks, which are succeeded eastward by stratified Mesozoic calc-alkaline volcanogenic units as young as Jurassic (Schweickert, 1981; *Varga and Moores, 1981*). The nature and age range of the stratigraphic section and its eastward facing suggest correlation with the eastern Klamath belt. Similar to the latter, the Shoo Fly is tectonically underlain on the west by a major, steeply dipping ultramafic slice, the Feather River peridotite, approximately 315 to 387 Ma (*Ehrenberg, 1975*; *Standlee, 1978*; *Saleeby and Moores, 1979*). The northern portion of the intervening fault zone encloses the peridotite, and juxtaposes Shoo Fly and Feather River mafic-ultramafic units against the upper Paleozoic–lower Mesozoic Calveras mélange complex on the west. Farther south, the Calaveras belt lies between the Melones fault on the west and the Shoo Fly terrane on the east, from which it is separated by the foothills suture, an east-dipping thrust fault (Schweickert, 1981; *Saleeby, 1982*).

Greenschist and prehnite-pumpellyite facies assemblages characterize much of the northern part of the Shoo Fly (Day and others, 1988; Harwood, 1988), but rocks reach amphibolite or higher metamorphic grade east of the foothills suture in the southern portion of this terrane and in roof pendants distributed along the length of the batholith (e.g., *Ernst, 1974*). The times of recrystallization are not well constrained, and probably occurred before, during, and after Permo-Triassic suturing of the outboard Calaveras-Melones complex against this terrane (Schweickert, 1981); accretion was completed prior to emplacement of the 170- to 172-Ma Standard pluton (Snoke and others, 1982), which crosscuts the Shoo Fly–Calaveras boundary in the southern Sierra Nevada foothills.

Just south of the Feather River peridotite, blueschist-facies rocks occur in slices within the Melones fault zone (Hietanen, 1981). The tectonic position between Shoo Fly on the east, and western Paleozoic and Triassic lithotectonic units on the west suggests that these blueschists may be relics of a belt analogous to the Klamath Stuart Fork Formation. Because of the likelihood of argon loss, the K-Ar radiometric age of 174 Ma (*Schweickert and others, 1980*) provides a minimum date of metamorphism for these Feather River blueschists.

The somewhat chaotic Calaveras-Melones complex comprises several different lithotectonic entities of contrasting origins, brought together along the anastomosing shear system of the western Sierra Nevada through oblique convergence. Each segment is rich in metachert, meta-argillite, phyllite, fine-grained marble, and metavolcanics. Ophiolitic lenses and carbonate blocks in mélange, some of which contain cosmopolitan late Paleozoic Tethyan faunas (Douglass, 1967), are evidently exotic, in contrast to associated terrigenous materials. In aggregate, this collection of units is correlated with the WTrPz belt of the Klamath Mountains (Hietanen, 1981).

The main Calaveras belt lies east of the southern part of the Melones fault zone, and west of the foothills suture. It occupies the proper structural position to represent a continuation of the Klamath North Fork terrane, but is not correspondingly rich in serpentinized peridotites; mafic metavolcanics are abundant, however. In the northern Sierran foothills, a much less extensive belt of apparent Pennsylvanian depositional age occurs directly west of the Feather River peridotite; it, too, is a possible equivalent of the North Fork terrane. To the west, it is juxtaposed against the Franklin Canyon Formation (Hietanen, 1981). The upper Paleozoic–Triassic Franklin Canyon unit consists principally of metamorphosed basalt, andesite, dacite, and rhyodacite, and minor lenses of phyllite, somewhat analogous to the Klamath Hayfork terrane. The westernmost Calaveras-Melones belt in the northern foothills is the Horseshoe Bend Formation, an imbricate mélange (Hietanen, 1981); its western boundary is the Big Bend fault. A conjectural along-strike continuation of the Rattlesnake Creek terrane of the Klamath Mountains, it consists of calc-alkaline metavolcanics, metasedimentary rocks, and mafics + ultramafics of Permo-Triassic age. To the south, the Bear Mountains and Big Bend faults merge, constituting the western boundary of the terrane. In this southern region, the Peñon Blanco volcanics and mafic + ultramafic plutonic complex constitute a major portion of the complex. Tectonic blocks of various metamorphic grades, but including crossite-bearing metabasaltic materials, are associated with this ophiolitic slice (*Morgan, 1976*; *Saleeby, 1982*); the 196- to 200-Ma minimum radiometric apparent age of the latter suggests the possibility of a similar relationship to the Feather River blueschists as the North Fork blocks bear to Stuart Fork blueschists.

Other than this isolated occurrence of high-pressure tectonic blocks, the Calaveras-Melones complex displays widespread development of biotite (locally garnet) in pelitic units, and both epidote amphibolite and garnet amphibolite metamorphic facies in mafic schists, especially in the northern part of the belt.

Farther to the southeast, both Melones and Bear Mountains fault systems are obliterated by plutonic rocks of the Sierra Nevada batholith. The 200-Ma Kings metaophilite complex, and the approximately 300+ Ma Bear Mountains and Kaweah meta-ophiolite mélanges, remnants of oceanic crust + uppermost mantle, crop out along the extension of these shear zones (*Behrman, 1978*; *Saleeby, 1982*). They are probably correlative with the Peñon Blanco mafic + ultramafic body and the Feather River peridotite, respectively. Judging from the equatorial fauna contained in the associated Permian limestone blocks, these southernmost pre-Mesozoic mafic-ultramafic complexes evidently represent far-traveled units that were swept against the accreting North American continent in Triassic time (Saleeby, 1981). Amphibolite-facies metamorphism, widespread in this zone, may reflect docking of these oceanic complexes along the developing margin of California. However, in spite of abundant ophiolitic material, much of the lithotectonic belt lying west of the Melones fault zone is calc-alkaline arc debris, which probably accumulated nearly in place (*Behrman, 1978*).

The outboard Jurassic belt, probably the southeastern continuation of the western Jurassic belt of the Klamaths, in general is bounded on the east by the Big Bend–Bear Mountains fault system, and on the west by the Great Valley. Along the fault zone, a narrow belt of Jurassic slates apparently rests with angular unconformity on Calaveras-Melones basement. Two lithotectonic entities make up much of the western Jurassic belt: the Mariposa Slate, including interlayered calc-alkaline volcanogenic metasedimentary and metavolcanic rocks, correlated with the Klamath Galice Formation; and the apparently native Smartville ophiolite (an analogue of the Josephine and Preston Peak massifs of the Klamath Mountains), a mafic (+ ultramafic) complex overlain by weakly metamorphosed basalts, andesites, and dacites (Xenophontos and Bond, 1978; Beard and Day, 1987). The smaller Pine Hill pluton and dike swarm occurs farther south in a somewhat similar geologic setting. Sedimentary protoliths of the Mariposa were laid down during late Jurassic time but are no younger than Oxfordian-Kimmeridgian; the ophiolite complexes, associated with older basement, have crystallization ages of about 160 Ma (*Saleeby, 1982*). The entire belt has been pervasively but weakly chloritized, with metamorphic grade confined to the prehnite-pumpellyite facies, transitional to lower greenschist facies. It was assembled by accumulation of debris from nearby calc-alkaline volcanism and preexisting units of the more easterly Calaveras and northern Sierra Nevada–southern Klamath provinces, according to *Behrman and Parkison (1978).*

Northern Franciscan–Great Valley

Franciscan and Great Valley units are areally associated from southern Oregon to at least Cedros Island and Vizcaino Peninsula, west-central Baja California. Both belts consist chiefly of clastic, recrystallized sedimentary rocks, and each has a depositional range from latest Jurassic to Paleogene time. The great ensimatic, tectonically imbricated prism of partly chaotic (Hsü, *1968,* 1971), chiefly east-dipping and east-facing Franciscan strata lies closest to the Pacific Ocean; it is separated from the roughly contemporaneous Klamath/Sierran arc by the well-ordered Great Valley sequence. These two terrigenous belts have been regarded as a classic example of trench and fore-arc basin sedimentary deposits, respectively (Ernst, 1970; *Dickinson, 1972, 1976*). Current vector, petrofacies, and conglomerate studies suggest a common, principally calc-alkaline volcanic + plutonic

Klamath and/or Sierran source for Franciscan and Great Valley detritus (*Jacobson, 1978*; Dickinson and others, 1982; Seiders and Blome, 1984). This late Mesozoic-Paleogene pair of tectono-sedimentary belts appears to be analogous to the modern Indonesian, Andean, and Aleutian convergent plate regimes, where deposition is occurring in parallel trench and fore-arc basin + arc environments (*Hamilton, 1979*), chiefly by axial filling.

However, paleomagnetic and fossil data suggesting the far-traveled nature of terranes accreted to the western margin of North America in post-Paleozoic time (Coney and others, 1980; Beck, 1986) have evoked hypotheses involving juxtaposition of exotic, unrelated Franciscan and Great Valley terranes (e.g., Blake and Jones, *1978*, 1981). The present intimate spatial association of the two assemblages reflects relative westward thrusting of Great Valley strata along an ancestral Coast Range thrust over the higher pressure but coeval Franciscan complex (Bailey and others, 1964, 1970). New seismic reflection and refraction profiles suggest that the Franciscan complex may have behaved like a tectonic indentor, wedging eastward between overlying Great Valley strata and the Sierran basement (Wentworth and others, 1984). Locally, Upper Cretaceous Great Valley rocks rest with angular unconformity on the underlying Franciscan accretionary wedge, representing ponded terrigenous debris shed across a rising mass of decoupled trench mélange (e.g., Maxwell, 1974; *Smith and others, 1979*). Four major lithotectonic belts (divided into more numerous tectonostratigraphic terranes) crop out from east to west in the northern Coast Ranges, and on a smaller scale, in the southern Coast Ranges (Blake and others, *1984, 1985*, 1988): the Great Valley sequence, the eastern Franciscan belt, the central Franciscan mélange belt, and the coastal belt of Franciscan.

The well-bedded Great Valley clastic prism floors both Sacramento and San Joaquin Valleys, constituting an asymmetric synclinorium with a near-vertical western limb and gently west-dipping eastern limb (Hackel, 1966). The assemblage is transgressive onto the North American continent such that, although uppermost Jurassic distal turbidites on the west lie on oceanic basement of the Coast Range ophiolite, Upper Cretaceous mid-fan and/or proximal clastic rocks near the axis and on the east side of the Great Valley overlie the granitoid-bearing western Sierran foothills. Somewhere beneath the Great Valley, sialic crust passes westward to oceanic crust; this westward transition to thinner, more mafic crust is also displayed by the Sierra Nevada (e.g., see Bateman, 1981).

The Coast Range ophiolite, exposed along the west side of both Sacramento and San Joaquin Valleys, includes variably serpentinized peridotite tectonite, overlying mafic + ultramafic plutonics, minor plagiogranite, sheeted mafic dikes and sills, and a surmounting group of massive and pillowed tholeiitic basalts + breccias (Bailey and others, 1970; *Bailey and Blake, 1974*). It was generated at approximately 165 Ma at a spreading center (Hopson and others, 1981), and is locally associated with slightly younger (~153 Ma) calc-alkaline arc rocks (*Evarts, 1977*); however, most segments of the ophiolite, particularly in the southern Coast Ranges, are capped by Upper Jurassic deep-sea radiolarian cherts, which pass upward to rhythmically layered, thin-bedded Great Valley distal mudstones. Some occurrences of the ophiolite are interlayered with arc rocks and may have been generated nearly in situ (*Sharp and Evarts, 1982*), whereas other segments having remnant paleomagnetic inclinations compatible with near-equatorial formation (*Luyendyk, 1982*) probably are exotic to the Californian continental margin.

The intensity of sporadically developed ridge metamorphism decreases vertically in the ophiolite. Actinolite and/or pyroxene hornfels of the plutonic series pass upward in the extrusive units to hydrothermally altered and zeolitic assemblages. Overprinted in these earlier, oceanic-phase associations are the effects of laumontite zone load metamorphism, which extend into the deeply buried lower portions of the superjacent Great Valley sequence (*Dickinson and others, 1969*; *Bailey and Jones, 1973*).

The eastern Franciscan belt and its southern extension in the Diablo Range consist of a relatively coherent series of tectonically imbricated phyllitic schists, quartzofeldspathic metagraywackes, dark metashales, minor lenses of greenstone+serpentinite, and widespread but volumetrically insignificant chert layers. The complex is bounded on the east by the Coast Range thrust or related high-angle fault and the structurally overlying ophiolite + Great Valley sequence, and on the west by a fault juxtaposing the tectonically lower central Franciscan mélange belt. Imbricate structures within the eastern belt, as well as bounding thrusts, dip gently to steeply eastward (Suppe, 1973). The depositional ages of turbiditic rocks of this belt range from latest Jurassic to mid-Cretaceous.

Regional metamorphism has imprinted several distinctly different aspects to portions of the lawsonitic eastern belt: (a) Mafic tectonic blocks of high-grade garnet-bearing blueschist, amphibolite, and eclogite occur as sparsely scattered fragments. Actinolite+chlorite or talc rinds indicate that the blocks were not in chemical equilibrium with the enclosing olistostrome, tectonic mélange, or serpentinite. Recrystallization ages of these high-grade blocks typically are on the order of 150 to 165 Ma (Coleman and Lanphere, 1971; Mattinson, 1988)—older than the matrix in which they are found. (b) Interlayered schistose metapelitic and metabasaltic blueschists occur in the northeastern part of the Franciscan complex directly beneath the Coast Range thrust (Blake and others, 1967, *1969*; *Irwin and others, 1974*; *Brown and Ghent, 1983*). This unit, the South Fork Mountain Schist, represents the highest structural entity in a stack of imbricate, east-dipping sheets (*Worrall, 1981*), Slabs of contrasting metamorphic grades are tectonically interlayered (*Suppe, 1972*). The minimum metamorphic age of these schists is about 120 Ma (Lanphere and others, 1978). (c) Relatively undeformed metagraywacke+metashale sequences, especially extensive along the east side of the Diablo Range, contain neoblastic aragonite and jadeitic clinopyroxene+quartz (*McKee, 1962a, b; Ernst, 1971a*). The time of this very high-pressure, low-temperature recrystallization seems to have been about 90 to 120 Ma (*Ernst, 1971b*; Suppe and Armstrong, 1972; *Bostick, 1974*; *Mattinson and Echeverria, 1980*).

The central Franciscan belt consists predominantly of chaotic mélange, with minor proportions of more coherent strata. It also contains the most abundantly dispersed high-grade mafic tectonic blocks (eclogite, garnet amphibolite, blueschist) and serpentinized peridotite lenses within the Franciscan complex. The sedimentary age of the shaly mélange matrix appears to be chiefly mid- and Late Cretaceous, although included blocks and slabs range in age from Late Jurassic to Late Cretaceous, and are of both cognate and foreign lithologies (Blake and Jones, 1981). Fractured and attenuated blocks may be regarded as boudins, whereas with lesser degrees of disruption, tectonic mélange passes gradually into semistratified broken formation (*Hsü, 1968, 1974*). The nature of laminar flow within the Franciscan subduction zone has been modeled by *Cowan and Silling (1978),* and by Cloos (1982). Exotic blocks evidently have been spalled off from different loci along the margins of the subduction-induced mélange circulation system; this phenomenon, coupled with the different settling velocities of blocks during flow, accounts for both the lithologic variation and size distribution of the tectonic fragments (Cloos, 1982).

Metamorphism of the central Franciscan belt produced pumpellyite and lawsonite, but not jadeitic pyroxene (Blake and others, 1967; Suppe, 1973; Cloos, 1983). Moderate pressures and low temperatures are indicated. The recrystallization age of the matrix has not been determined, but is probably Late Cretaceous, judging by the contained fossils.

Exotic pelagic limestone blocks occur in tectonized mélange near Laytonville, California, along the western margin of the central belt (Bailey and others, 1964; *Gucwa, 1975*). Employing remnant magnetic and abundant microfossil data, Alvarez and others (1980) and Tarduno and others (1986) demonstrated that this unit formed about 88 to 101 Ma at 14° to 17°S latitude. The origin in mid-Cretaceous time was possibly as deep-water carbonate fringing a seamount. These, as well as other Upper Cretaceous limestone blocks evidently have been supplied to the present lithotectonic environment by northeastward spreading, which brought oceanic crust and bathyal sediments to the Franciscan trough of deposition (*Sliter and others, 1986*). The allochthonous nature of such pelagic limestone provides few constraints regarding provenance of the surrounding Franciscan clastic wedge.

The coastal belt of Franciscan is bounded on the east by the apparently east-dipping coastal belt thrust, which separates it from the structurally higher central Franciscan mélange belt. On the west, the coastal belt underlies the continental margin and slope of North America. Stratal continuity is characteristic of this rather arkosic lithotectonic unit. The sediments were deposited during Late Cretaceous to Miocene time (McLaughlin and others, 1982). Premetamorphic rock types include deep-water pelagic strata on the west, apparently associated with an altered basaltic substrate and chert; to the east, this section passes by degrees to mid-fan turbiditic, andesitic graywacke, and voluminous quartzofeldspathic graywacke (*Bachman, 1978*; Blake and Jones, 1981). Greenstone and serpentinite lenses are scarce, and high-grade tectonic blocks rare in this belt (Blake and others, 1988).

Metamorphism in the coastal belt of Franciscan is very weak, and has not been investigated in detail. Laumontite apparently is widely distributed as a vein mineral. A few metasedimentary parageneses include prehnite and/or pumpellyite, but in situ blueschists have not been reported (Bailey and others, 1964; McLaughlin and others, 1982). Assemblages appear to be similar to those described from the most deeply buried portions of the Great Valley sequence. Evidently, as currently exposed, this seaward belt of Franciscan was not subjected to profound underflow prior to decoupling from the subducting plate.

Salinian-Nacimiento block

The southern California Coast Ranges west of the San Andreas fault are underlain by a pair of bedrock series—the Pacific margin Franciscan mélange of the Nacimiento block, and the more landward granitic + metamorphic terrane of the Salinian block—in many ways a small-scale equivalent to that which occurs east of the modern Pacific–North American plate junction (Page, 1966; *Ross, 1978*). These oceanic and continental entities were amalgamated along the Sur-Nacimiento fault system, across which lowermost Miocene sediments have been deposited with little apparent offset (Page, *1970a, b,* 1981). Only the silicic eastern(?) portion of the Salinian plutonic arc is preserved. A hypothesized western, more mafic magmatic + country rock sequence, similar to the Sierran foothills belt, evidently was removed or left behind (Page, 1982; *Silver, 1982*) during transport prior to amalgamation; most of the sedimentary strata that would have occupied the fore-arc basin intervening between arc and trench are also missing. Proximal Great Valley–type arkosic sandstones of Late Cretaceous age lie with angular unconformity on Salinian granites, whereas west of the suture, fragments of the uppermost Jurassic section of fine-grained Great Valley distal turbidites resting on ribbon chert + underlying ophiolite (*Page, 1972*) occur in fault-bounded slices juxtaposed against Franciscan mélange of the Nacimiento block. Because this late Mesozoic continental margin has been duplicated during Cenozoic strike slip, both Nacimiento and Salinian complexes are clearly far travelled relative to analogous units east of the San Andreas fault. Similar to relationships previously described for northern California, the Nacimiento block Franciscan rocks may be parautochthonous with respect to the fragmented Salinian arc.

Metamorphism within the Salinian block ranges from amphibolite facies chiefly in the southeast, to rare garnet granulite facies near the Sur-Nacimiento fault zone (*Compton, 1966*; *Ross, 1976a, b, 1977*; *John, 1981*). Country rocks include well-ordered platform carbonates, graywackes, and mafic igneous protoliths; some metaclastics of the central Salinian block probably had a lithic precursor similar to Pelona-type schists of the Transverse Ranges (*Ross, 1976b*). For the various metamorphosed superjacent rocks, U/Pb data reflect the involvement of 1.7-Ga crust (James and Mattinson, 1988). Metamorphism probably preceded and/or accompanied synkinematic to postorogenic emplacement of the granitoids at 78 to 130 Ma (*Ross, 1972*; Page, 1981; Mattinson and James, 1985); recrystallization terminated before deposition of the Upper Cretaceous Great Valley arkoses.

The Franciscan paragenesis in the Nacimiento block reflects, on a much smaller scale, the broad eastward intensification of high-pressure metamorphism exhibited by the northern Coast Ranges. Westernmost quartzose metasedimentary sections contain very rare occurrences of laumontite, medial portions of the belt are characterized by pumpellyite, and eastern tracts carry lawsonite ± jadeitic pyroxene (*Ernst, 1980*). Scanty bulk-rock K-Ar radiometric data indicate an apparent Late Cretaceous time of high-pressure, low-temperature recrystallization (Suppe and Armstrong, 1972).

Transverse Ranges

The Transverse Ranges in southern California are disposed along, and transected by, the San Andreas fault system. During the Neogene, dextral slip aggregated between 300 and 500 km; this process has dispersed and rotated formerly contiguous structural blocks, and juxtaposed geologically unrelated terranes (Luyendyk and others, 1985; Carter and others, 1987). Earlier movement on a proto–San Andreas transform system seems to have been occasioned by early Miocene overriding of the East Pacific Rise by the North American lithospheric plate, but the manner and extent of offsets are not yet completely understood (*Atwater, 1970*; Engebretson and others, 1985).

Because the San Andreas fault cuts obliquely across the late Mesozoic–Paleogene continental margin, right-lateral strike slip has duplicated the volcanic/plutonic arc–fore-arc basin–trench complex in the central California Coast Ranges. Moreover, although the western Transverse Range appear to be underlain by outboard Franciscan and inboard Salinian basement, farther to the southeast, Proterozoic blocks of the North American craton have been strung out along the transform zone in and adjacent to the San Gabriel Mountains. Also scattered about this major Cenozoic plate boundary to the southeast, Mesozoic terrigenous strata and cherts deposited on oceanic protoliths have been converted to relatively high-P, low-T schists (*Graham and England, 1976*; Peacock, 1988).

In the central Transverse Ranges, upper-plate 1.7-Ga amphibolites, quartzofeldspathic granulitic gneisses, and synkinematic granitoids were intruded by an anorthosite+gabbro+syenite complex at 1.2 Ga (Ehlig, 1981); after subsequent deposition of Phanerozoic sediments, the complex was partly recrystallized during deep crustal intrusion by 220-Ma and younger granodiorite plutons (*Barth, 1987*). Similar Mesozoic granitic rocks and migmatized miogeoclinal Paleozoic(?) strata occur in the San Bernardino Mountains on the east. Much of the high-grade metamorphism of the ancient basement rocks, therefore, could reflect a Mesozoic overprint.

The allochthonous San Gabriel section has been thrust over lower-plate, relatively high-P, low-T, Pelona Schists of apparent Late Cretaceous metamorphic age (Ehlig, 1981). In addition to occurrences in the central Transverse Ranges, Pelona-type schists occur as enigmatic isolated fragments along the Garlock fault in the northwestern Mojave Desert, and are disposed to the southeast of the San Andreas in southern California and southwestern Arizona (*Haxel and Dillon, 1978*; Jacobson and others, 1988). All are situated below low-angle thrusts that bring higher grade rocks of continental affinities over oceanic metabasalts and related metasediments. Lower plate units possess mineral assemblages transitional between blue- and greenschist metamorphic facies, with calculated temperatures of recrystallization increasing toward the overlying thrust contacts. Preservation of the inverted thermal gradient reflected in mineral parageneses in the Pelona Schists, as well as in the Catalina terrane and the Klamath central metamorphic belt, appears to require progressive refrigeration of the subduction complexes due to continued underflow (Peacock, 1988).

Peninsular Ranges

This geologic province is characterized by three major prebatholithic lithotectonic belts (Gastil and others, 1981; Gastil, 1985). Upper Jurassic and Lower Cretaceous volcanogenic detrital sediments and calc-alkaline lavas + pyroclastics dominate the western zone along the coast of southern California and Baja California; older Mesozoic volcanic and plutonic rocks constitute the basement of this belt. In aggregate, these sparsely fossiliferous layered rocks and subjacent units reflect the construction of a nearby island arc. Eastward across an east-dipping suture, Mesozoic turbidites represent a submarine-fan complex, which developed at the western margin of North America prior to Early Cretaceous amalgamation with the outboard-arc terrane. The marginal flysch basin gives way on the east to a westward-younging wedge of Phanerozoic miogeoclinal and platform strata. Subsequent invasion by Early Cretaceous gabbros and tonalites commenced in the western, accreted arc; gradually, magmatism shifted eastward as plutons of more granodioritic and granitic compositions were emplaced during later Cretaceous (and in Arizona, Paleogene) time.

The Pacificward calc-alkaline arc was metamorphosed under greenschist facies conditions in the west, and under low-pressure andalusite ± sillimanite-bearing amphibolite facies in the east (Todd and others, 1988). Moving into the continent from the island-arc/flysch basin suture, the apparent grade of recrystallization gradually increases to sillimanite zone assemblages, with anatectic migmatites largely confined to eastern, miogeoclinal wall rocks. Regional dynamothermal metamorphism throughout was roughly contemporaneous with pluton emplacement at 80 to 140 Ma and with thermal softening of midcrustal levels. Plutons situated on the west tend to be more oceanic in bulk-rock chemistry, whereas toward the east, granitoids exhibit increasing contributions from continental-crust progenitors. Late Cretaceous and Paleogene mylonitization of the eastern Peninsular Ranges signals an evolutionary change in deformational style, from a ductile response attending convergence, crustal thickening, and compression, to granulation and shearing accompanying differential—chiefly dip—slip, localized along the former Paleozoic cratonal boundary, and finally to Tertiary extension and crustal attenuation.

Southern California Borderland

Two contrasting lithotectonic entities make up the basement of the southern California Borderland (Howell and Vedder, 1981): Franciscan and related high-pressure, low-temperature metamorphics and mélanged units; and ophiolitic oceanic + calkalkaline arc rocks and associated volcanogenic sediments, and their low-P, low- to high-T recrystallized equivalents. As in the California Coast Ranges, these belts were formed and metamorphosed during Late Jurassic to Late Cretaceous time. Four distinct terranes are juxtaposed in the borderland, reflecting transform faulting and chiefly Neogene northward drift, obscuring the original configuration of the late Mesozoic convergent plate margin (Howell and Vedder, 1981).

The originally outboard subduction complex consists of blueschists, relatively high-pressure greenschists, and amphibolitized eclogites, derived from precursor graywackes, shales, cherts, oceanic tholeiites, and ultramafics (*Platt, 1975*; *Sorensen, 1986*). Metamorphism took place at about 91 to 112 Ma (Suppe and Armstrong, 1972; *Mattinson, 1986*). The initially landward arc complex consists of feebly metamorphosed slates, phyllites, schists, and contact-zone hornfels derived from volcanogenic sediments and igneous protoliths (Sorensen, *1985*, 1988). Metamorphism is at least in part as old as 141 ± 3 Ma (*Mattinson and Hill, 1976*), but is undoubtedly younger for aureoles surrounding Cretaceous hypabyssal plutons.

PETROTECTONIC SUMMARY OF THE WESTERN CORDILLERA

Metamorphic overview

The geologic map schematically summarizing metamorphism in the western Cordillera, presented as Plate 9, is much simplified but nevertheless demonstrates the complicated, dislocated nature of crustal recrystallization in this region. Although the Mesozoic history of the continental margin is moderately well known, younger, relatively intact metamorphic tracts are still buried. In contrast, more ancient pre-Mesozoic belts have been exhumed and eroded, or overprinted at depth by later thermotectonic processes, and thus, at best, are preserved in only fragmentary fashion. Moving progressively into the North American continent, we lose both (1) the direct relationship between lithotectonic assemblages and inferred plate-tectonic motions, and (2) the comparatively high degree of completeness of the rock record. What, then, can be tentatively concluded regarding metamorphism and tectonism in the U.S. Cordillera?

As is clear from Plate 9, Archean metamorphism is confined to Wyoming and adjacent parts of Montana and South Dakota; it is chiefly of high-T, low- to moderate-P amphibolite and granulite facies. Cratonization was completed by late Archean time (i.e., ~2.7 Ga), and the section was not significantly affected by later dynamothermal metamorphism. Mineral assemblages that developed in upper Proterozoic shelf and slope sediments of the Belt Supergroup in northwestern Montana are of very low grade. Lower Proterozoic rocks of the accretionary continental crust lying to the south of the Archean nucleus in Colorado, Utah, and New Mexico are dominantly of intermediate-P amphibolite type, whereas a nonsynchronous regional trend from greenschists in southeastern Arizona to low-P amphibolites and low-rank granulites in western Arizona is evident. Yet farther to the west, in the Mojave Desert and eastern Transverse Ranges, relict high-rank, low-P amphibolites of Proterozoic age are scattered along the broad, modern dextral shear system. In California, east-central Oregon, and northwestern Washington, fragmentary lithologic remains of Ordovician through Jurassic recrystallization events are preserved as segments of Pacific rim convergent plate junctions; the sutures are marked by tectonized meta-ophiolites. Pre-Cretaceous, oceanward subduction zone, high-P metamorphic belts are juxtaposed against landward, high-T, calc-alkaline continental margin complexes.

Westward from the North American latest Precambrian cratonal hinge line, sections have been partly to thoroughly overprinted by Late Jurassic–Cretaceous and early Cenozoic intracontinental polymetamorphism accompanying widespread igneous activity. Products include inboard, relatively high-rank metamorphic rocks best displayed in core complexes and other pronounced uplifts (see Fig. 4), and broad-scale development of greenschist and low-P amphibolite facies assemblages in the Basin and Range, Mojave-Sonoran Desert, and Sierra Nevada–Peninsular Ranges; greater overall depths of emplacement, followed by uplift and erosion, are recorded by the area surrounding the northern part of the Idaho batholith, where high-P amphibolites of Late Cretaceous age are exposed. Outboard from the Late Jurassic North American margin in western California and northwestern Washington, accretionary prisms containing abundant oceanic as well as terrigenous materials have been subjected to blueschist-facies metamorphism, probably during Cretaceous and younger suturing of these belts against the continent.

A distinct younging trend exists for metamorphic terranes moving away from the Wyoming continental nucleus toward the Pacific Basin and the Gulf of Mexico. Moreover, where detailed geologic relationships are available within any one belt (e.g., Cheyenne belt, Colorado–New Mexico, Klamaths, Sierra Nevada, California Coast Ranges), both formational ages of lithotectonic assemblages and times of recrystallization decrease progressively seaward. The same relationship may also hold for the northern Cascades, where the margin has been doubled through strike-slip. Within broad expanses of Cordilleran terranes invaded by late Mesozoic–Paleogene granitoids, the more northerly Idaho batholith region represents greater depths of intrusion and later exhumation compared to coeval calc-alkaline belts in Nevada and California. The Rocky Mountains of Utah, Wyoming, Colorado, and New Mexico display shallow-level structural evidence of Laramide orogeny, but neither widespread emplacement of calc-alkaline magmas nor extensive dynamothermal metamorphism of Phanerozoic age (except in southwestern Montana). The deep-seated post-Paleozoic orogeny was sited far to the west–or at least has been more thoroughly uplifted, eroded, and exposed to view.

Only bordering the Pacific Ocean is there a preserved duality of metamorphic belts. Inland, the older metamorphic terranes exhibit lithologic assemblages produced mainly in continental-margin and island-arc–type settings. Conspicuous by their absence are blueschist belts and extensive tracts of ophiolite; either such associations were not abundantly formed in ancient plate-tectonic environments, or they have been selectively obliterated by subsequent dynamothermal events. For instance, later crustal heating/softening would result in the overprinting and destruction of high-P mineral assemblages and in the downward relative movement of dense peridotitic and ophiolitic massifs.

Metamorphic P-T regimes

Metamorphism in the western Cordillera and elsewhere can be related to two principal continental-margin petrotectonic realms—the subduction zone and the volcanic/plutonic arc. Contrasts in P-T-time paths are a function of the departure of the prograde metamorphic trajectories from a steady-state intraplate geotherm, and reflect the low thermal conductivities of rocks. The subduction-zone environment is characterized by protoliths of oceanic affinity; by low-angle thrusts that dip beneath the stable, nonsubducted arc or continental margin; by oceanward vergence of nappes; and by blueschist parageneses. These constitute the outboard, relatively narrow Phanerozoic belts shown in Plate 9. A metamorphic P-T regime typified by a clockwise path involving high-P, low-T progradation and more or less isothermal decompression reflecting adiabatic rise toward the surface is illustrated in Figure 5; such a topology is characteristic of environments involving tectonic thickening. Where high-pressure metamorphic assemblages slowly move back up the plate junction during continued convergence and refrigeration (not shown, but the probable case for the Franciscan), retrograde reactions would closely mimic the prograde P-T trajectory. In contrast, the arc is characterized by calc-alkaline igneous rocks and associated volcanogenic sedimentary assemblages, open folding and buoyant rise of subjacent, gregarious batholiths; and by predominantly low- and intermediate-pressure regional and contact metamorphic assemblages. Broad regions—including the Peninsular Ranges, Sierra Nevada and Idaho batholiths, and portions of the Basin and Range—constitute this inboard petrotectonic province. A metamorphic P-T trajectory involving a counterclockwise path, typified by high-T, low-P progradation and nearly isobaric cooling, is shown in Figure 5. Such a path requires advective heat transport into the sialic crust as a consequence of igneous thickening. Some granulitic complexes appear to have been subjected to nearly isobaric thermal relaxation (*Bohlen, 1987*). If tectonic thickening precedes and outweighs magma and hydrothermal fluid injection, mid- and upper-crustal levels would be characterized by a clockwise-type P-T-time path (not shown, but probably appropriate for pre-Cenozoic metamorphic episodes in the Basin and Range). Of course, broad tracts of the middle and lower continental crust, neither in the vicinity of arcs nor of trenches, have experienced either: (1) compression, deep levels of recrystallization and P-T trajectories deviating from intraplate geotherms (*Clark and Ringwood, 1964*) toward the subduction-zone type of metamorphism; or (2) attenuation, mid-crustal levels of recrystallization, and P-T paths transitional toward those attending volcanic/plutonic arc metamorphism. Ductile deformation ac-

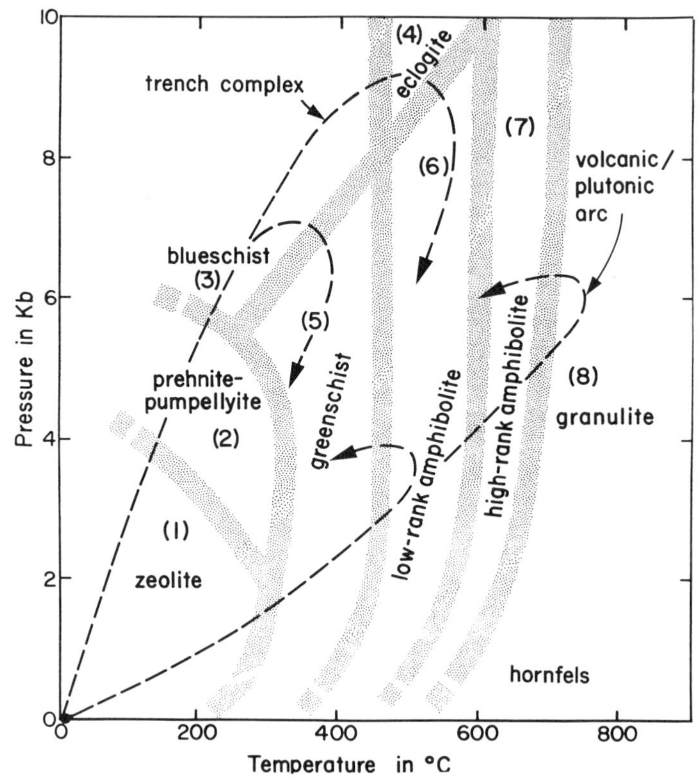

Figure 5. Diagrammatic prograde and retrograde metamorphic P-T trajectories for subduction zone (e.g., Stuart Fork) and volcanic/plutonic arc (e.g., Sierra Nevada) environments superimposed on the facies grid of Figure 1. The contrast between tectonically and magmatically thickened sections is evident. A steady-state intraplate geotherm would lie between the two end-member prograde paths. Arrows show the direction of geothermal gradient evolution with time. The extent of departure of the retrograde trajectory from prograde trajectory depends on the rapidity of decompression, rock conductivities, tectonism, and heat transfer through aqueous fluid flow. Retrograde P-T paths are shown as approaching the steady-state intraplate geotherm. However, if subduction refrigeration continues as rock packets ascend slowly towards the surface, the retrograde trajectory may be displaced to only slightly lower pressures than the prograde path. Similarly, if magmatic heat continues to be supplied to an uplifting arc section, the retrograde path may approximate the prograde trajectory; or, if tectonic thickening preceeds melt emplacement, an intermediate but clockwise P-T path is expected.

companies both of these tectonometamorphic environments except at shallow depths. Coeval calc-alkaline and basaltic magma emplacement in general is proportional to the inferred geothermal gradient, and is virtually confined to those metamorphic belts characterized by high T/P ratios.

The effects of each of the various sorts of subsolidus environment described above, as well as burial metamorphism of superjacent strata and ridge + sea-floor recrystallization of oceanic crust, are preserved imperfectly in the western United States (Plate 9). However, the volumetrically dominant metamorphic type represented in Mesozoic belts, and apparently also in the Precambrian basement, is that associated with calc-alkaline volcanic/plutonic arcs. Where produced at shallow levels characterized by brittle fracture, discrete aureoles surround hypabyssal plutons; in deeper zones of ductile deformation, high-temperature, moderate-pressure metamorphism is developed on a regional scale with a less obvious relationship to intrusive contacts. Clearly, the elevated thermal regime attending metamorphism in the western Cordillera is typified by the presence of granitoids, but whether the plutons were a cause or a consequence of the high temperatures has yet to be established (Barton and others, 1988).

Because of advective heat contributed to the thermal regime of a magma-invaded lithotectonic complex, high-T metamorphism tends to transgress terrane boundaries. In contrast, subducted packets of rock are tectonically and thermally isolated from neighboring terranes due to the rapidity of juxtaposition; hence, high-P metamorphic parageneses are restricted to the subduction complex itself.

Cordilleran subduction and transform motion in western California

A portion of the record of Middle Jurassic and younger sea-floor spreading is contained in the oceanic crust and overlying hemipelagic sediments of the Pacific Basin (*Pitman and others, 1974*; Engebretson and others, 1985). The western limbs of paleo-Pacific plates are still extant. In contrast, eastern limbs of these oceanic spreading systems have been overridden by the advancing North American plate; hence, over the past 165 m.y., at least 10,000 to 11,500 kms of subduction component must have been accommodated (Ernst, 1984; *Engebretson and others, 1988*). Thus, in spite of extensive northward drift of oceanic crust-capped lithosphere, underflow was the dominant mechanism responsible for production of paired metamorphic belts and voluminous calc-alkaline igneous activity reflecting Cordilleran continental growth over the late Mesozoic–Cenozoic time interval.

Based on currently observed high heat flow in the California Coast Ranges, *Cloos and Dumitru (1987)* have demonstrated that rapid thermal obliteration of high-pressure metamorphic mineral assemblages is apparently occurring today at depths of 5 to 10 km and greater. Thus, because of the widespread existence of such rocks at the surface, the present dominantly strike-slip regime of western California must be a recently imposed change on a chiefly convergent-type margin. The process of subduction evidently has refrigerated blueschist lithologies nearly continuously since formation of the original high-pressure rocks in late Mesozoic time, thereby allowing their preservation.

Both underflow of great expanses of post–Middle Jurassic paleo-Pacific oceanic crust, and recent change in thermal structure of the western continental margin resulting in the ongoing rapid destruction of long-preserved high-pressure mineral assemblages argue for sustained subduction as the chief plate-tectonic process operating in the western Cordilleran regime during mid-Mesozoic to mid-Cenozoic time.

Geochemical provinces and crustal evolution of the western Cordillera

Study of metamorphic belts suggests a progressive westward growth of the western conterminous United States during Phanerozoic time. But what about the Precambrian? The western interior consists of a number of geochemical provinces, as indicated by contrasting radiometric and isotopic signatures (Zartman, 1974; Kistler and Peterman, 1978; Condie, *1981,* 1986; *Farmer and DePaolo, 1984*; Farmer, 1988; Bickford, 1988a, b; Bennett and DePaolo, 1987; Wooden and others, 1988a, b). These terranes are arranged as belts bordering the Archean cratonal nucleus in Wyoming, and exhibit overall Proterozoic and Phanerozoic accretion toward the modern continental margins. Preserved lithotectonic belts apparently developed asymmetrically and episodically rather than concentrically and continuously, indicating a sequential growth in certain sectors rather than a gradual circumferential enlargement. Major truncations of these terranes also suggest occasional rifting and removal of previously accreted segments of North American continental crust.

Data are few but allow the provisional recognition of systematic geochronologic and geochemical trends in Precambrian crustal evolution within the several lithotectonic provinces; nevertheless, it is difficult to decide whether this portion of the continent grew chiefly by the parautochthonous reworking of preexisting sialic materials and the development of native continental crust surmounting convergent plate junctions, or by the random accretion of allochthonous oceanic terranes of unrelated geochemistry. However, the preponderantly sialic nature of the crust appears to favor the parautochthonous growth mechanism.

Because of the observed monotonic decrease in ages and isotopic affinities of the belts proceeding outward from the >2.6-Ga Wyoming craton, it is evident that addition of exotic microcontinental terranes containing sialic basement distinctly older than the in situ developing continental margin did not occur in the western conterminous United States. An exception to this generalization appears to be the Salinian continental margin salient west of the San Andreas fault, but how far northward this sialic fragment has drifted since rifting is still uncertain (*Hill and Dibblee, 1953*; Page, 1982). The westward step in the $^{87}Sr/^{86}Sr$ initial ratio east of the Sierra Nevada (Kistler and Peterman, 1978) may represent a major fault displacing the Late Precambrian continental margin—another possible rifting event. Finally, the Yellow Aster complex of the northwest Cascades (Misch, 1988) may be an exotic microcontinental scrap, but it is very small indeed.

True continental growth by the separation of sialic material from the mantle, in contrast to rearrangement of pieces of already extant crust, requires an important component of convergent plate motion (Ernst, 1984). Judging by the Phanerozoic rock record (especially the Mesozoic), the gradual oceanward decreases in initial $^{87}Sr/^{86}Sr$ ratios, and increases in ϵ_{Nd} toward positive values in sialic crust of the western United States (DePaolo, *1980,* 1981) are compatible with the hypothesis that, although possibly far-traveled ophiolitic debris is abundant, and exotic microcontinental slabs are present but rare, much of the crustal growth occurred nearly in situ. Accretion involved the sweeping back into the continent of terrigenous debris of local or regional provenance, and of extensive lateral rearrangement of parautochthonous slices of the margin plus offshore arcs, as well as the episodic removal of continental margin/island-arc sections and docking at distant sites. However, oblique and head-on subduction evidently were responsible for most of the real growth documented for the conterminous United States sector of the western Cordillera (e.g., see *Ernst, 1983*).

REFERENCES CITED

Text references in *italics* will be found only on accompanying microfiche.

Allmendinger, R. W., and 7 others, 1987, Overview of the COCORP 40°N transect, western United States—the fabric of an orogenic belt: Geological Society of America Bulletin, v. 98, p. 308–319.

Alvarez, W., Kent, D. V., Silva, I. P., Schweickert, R. A., and Larson, R. A., 1980, Franciscan Complex limestone deposited at 17° South paleolatitude: Geological Society of America Bulletin, v. 91, p. 476–484.

Anderson, J. L., 1988, Core complexes of the Mojave-Sonoran Desert—conditions of plutonism, mylonitization, and decompression, *in* Ernst, W. G., ed., Metamorphism and Crustal Evolution of the Western United States: Englewood Cliffs, New Jersey, Prentice-Hall, Inc., p. 502–525.

Ando, C. J., Irwin, W. P., Jones, D. L., and Saleeby, J. B., 1983, The ophiolitic North Fork terrane in the Salmon River region, central Klamath Mountains, California: Geological Society of America Bulletin, v. 94, p. 236–252.

Armstrong, R. L., 1982, Cordilleran metamorphic core complexes—from Arizona to southern Canada: Annual Review of Earth and Planetary Science, v. 10, p. 129–154.

Armstrong, R. L., and Suppe, J., 1973, Potassium-argon geochemistry of Mesozoic igneous rocks in Nevada, Utah, and southern California: Geological Society of America Bulletin, v. 84, p. 1375–1392.

Babcock, R. S., and Misch, P., 1988, Evolution of the crystalline cores of the North Cascades Range, *in* Ernst, W. G., ed., Metamorphism and Crustal Evolution of the Western United States: Englewood Cliffs, New Jersey, Prentice-Hall, Inc., p. 214–232.

Babcock, R. S., Brown, E. H., Clark, M. D., and Livingston, D. E., 1979, Geology of the older Precambrian rocks of the Grand Canyon. Part II. The Zoroaster plutonic complex and related rocks: Precambrian Research, v. 8, p. 243–275.

Bailey, E. H., Irwin, W. P., and Jones, D. L., 1964, Franciscan and related rocks, and their significance in the geology of western California: California Division of Mines and Geology Bulletin 183, 171 p.

Bailey, E. H., Blake, M. C., Jr., and Jones, D. L., 1970, On-land Mesozoic oceanic crust in California Coast Ranges: U.S. Geological Survey Professional Paper 700-C, p. 70–81.

Barton, M. D., and 8 others, 1988, Mesozoic contact metamorphism in the western United States, *in* Ernst, W. G., ed., Metamorphism and Crustal Evolution of the Western United States: Englewood Cliffs, New Jersey, Prentice-Hall, Inc., p. 110–178.

Bateman, P., 1981, Geologic and geophysical constraints on models for the origin of the Sierra Nevada batholith, California, *in* Ernst, W. G., ed., The Geotectonic Development of California: Englewood Cliffs, New Jersey, Prentice-Hall, Inc., p. 71–86.

Beard, J. S., and Day, H. W., 1987, The Smartville intrusive complex, Sierra Nevada, California—the core of a rifted volcanic arc: Geological Society of America Bulletin, v. 99, p. 779–791.

Beck, M. E., 1986, Model for late Mesozoic-early Tertiary tectonics of coastal California and western Mexico and speculation on the origin of San Andreas fault: Tectonics, v. 5, p. 49–64.

Bennett, V. C., and DePaolo, D. J., 1987, Proterozoic crustal history of the western United States as determined by Neodymium isotopic mapping: Geological Society of America Bulletin, v. 99, p. 674–685.

Bickford, M. E., 1988a, The accretion of Proterozoic crust in Colorado—igneous, sedimentary, deformational, and metamorphic history, *in* Ernst, W. G., ed., Metamorphism and Crustal Evolution of the Western United States: Englewood Cliffs, New Jersey, Prentice-Hall, Inc., p. 411–430.

——, 1988b, The formation of continental crust—Part 1. A review of some principles; Part 2. An application to the Proterozoic evolution of southern North America: Geological Society of America Bulletin, v. 100, p. 1375–1391.

Bird, P., 1988, Formation of the Rocky Mountains, western United States; A continuum computer model: Science, v. 239, p. 1501–1507.

Blake, M. C., Jr., and Jones, D. L., 1981, The Franciscan assemblage and related rocks in northern California—a reinterpretation, *in* Ernst, W. G., ed., The Geotectonic Development of California: Englewood Cliffs, New Jersey, Prentice-Hall, Inc., p. 307–328.

Blake, M. C., Jr., Irwin, W. P., and Coleman, R. G., 1967, Upside-down metamorphic zonation, blueschist facies, along a regional thrust in California and Oregon: U.S. Geological Survey Professional Paper 575-C, p. 1–9.

Blake, M. C., Jr., Jayko, A. S., McLaughlin, R. J., and Underwood, M. B., 1988, Metamorphic and tectonic evolution of the Franciscan Complex, northern California, *in* Ernst, W. G., ed., Metamorphism and Crustal Evolution of the Western United States: Englewood Cliffs, New Jersey, Prentice-Hall, Inc., p. 1035–1060.

Brown, E. H., 1986, Geology of the Shuksan Suite, North Cascades, Washington, U.S.A., *in* Evans, B. W., and Brown, E. H., eds., Blueschists and Eclogites: Geological Society of America Memoir 164, p. 143–154.

——, 1988, Metamorphic and structural history of the Northwest Cascades, Washington and British Columbia, *in* Ernst, W. G., ed., Metamorphism and Crustal Evolution of the Western United States: Englewood Cliffs, New Jersey, Prentice-Hall, Inc., p. 196–213.

Brown, E. H., Babcock, R. S., Clark, M. D., and Livingston, D. E., 1979, Geology of the older Precambrian rocks of the Grand Canyon. Part I. Petrology and structure of the Vishnu complex: Precambrian Research, v. 8, p. 219–241.

Bryant, B., 1988, Evolution and early Proterozoic history of the margin of the Archean continent in Utah, *in* Ernst, W. G., ed., Metamorphism and Crustal Evolution of the Western United States: Englewood Cliffs, New Jersey, Prentice-Hall, Inc., p. 431–445.

Burchfiel, B. C., and Davis, G. A., 1972, Structural framework and evolution of the southern part of the Cordilleran orogen, western United States: American Journal of Science, v. 272, p. 97–118.

——, 1981, Triassic and Jurassic tectonic evolution of the Klamath Mountains–Sierra Nevada geologic terrane, *in* Ernst, W. G., ed., The Geotectonic Development of California: Englewood Cliffs, New Jersey, Prentice-Hall, Inc., p. 50–70.

Carter, J. N., Luyendyk, B. P., and Terres, R. T., 1987, Neogene clockwise tectonic rotation of the eastern Transverse Ranges, California, suggested by paleomagnetic vectors: Geological Society of America Bulletin, v. 98, p. 199–206.

Clark, L. D., 1976, Stratigraphy of the north half of the western Sierra Nevada metamorphic belt, California: U.S. Geological Survey Professional Paper 923, 25 p.

——, 1979, Geology of the older Precambrian rocks of the Grand Canyon. Part III. Petrology of mafic schists and amphibolites: Precambrian Research, v. 8, p. 277–302.

Cloos, M., 1982, Flow mélanges—numerical modeling of geological constraints on their origin in the Franciscan subduction complex, California: Geological Society of America Bulletin, v. 93, p. 330–345.

——, 1983, Comparative study of mélange matrix and metashales from the Franciscan subduction complex with the basal Great Valley sequence, California: Journal of Geology, v. 91, p. 291–306.

Coleman, R. G., and Lanphere, M. A., 1971, Distribution and age of high-grade blueschists, associated eclogites, and amphibolites from Oregon and California: Geological Society of America Bulletin, v. 82, p. 2397–2412.

Coleman, R. G., Mortimer, N., Donato, M. M., Manning, C. E., and Hill, L. B., 1988, Tectonic and regional metamorphic framework of the Klamath Mountains and adjacent Coast Ranges, California and Oregon, *in* Ernst, W. G., ed., Metamorphism and Crustal Evolution of the Western United States: Englewood Cliffs, New Jersey, Prentice-Hall, Inc., p. 1061–1097.

Condie, K. C., 1976, The Wyoming Archean Province in the western United States, *in* Windley, B. F., ed., The Early History of the Earth: London, Wiley, p. 499–511.

——, 1986, Geochemistry and tectonic setting of early Proterozoic supracrustal rocks in the southwestern United States: Journal of Geology, v. 94, p. 845–864.

Coney, P. J., 1980, Cordilleran metamorphic core complexes—an overview, *in* Crittenden, M. D., Jr., Coney, P. J., and Davis, G. H., eds., Cordilleran metamorphic core complexes: Geological Society of America Memoir 153, p. 7–31.

Coney, P. J., Jones, D. L., and Monger, J.W.H., 1980, Cordilleran suspect terranes: Nature, v. 288, p. 329–333.

Davis, G. A., 1969, Tectonic correlations, Klamath Mountains and western Sierra Nevada, California: Geological Society of America Bulletin, v. 80, p. 1095–1108.

Davis, G. A., and Lipman, P. W., 1962, Revised structural sequence of pre-Cretaceous metamorphic rocks in the southern Klamath Mountains, California: Geological Society of America Bulletin, v. 73, p. 1547–1552.

Davis, G. A., Holdaway, M. J., Lipman, P. W., and Romey, W. D., 1965, Structure, metamorphism, and plutonism in the south-central Klamath Mountains, California: Geological Society of America Bulletin, v. 76, p. 933–966.

Day, H. W., Moores, E. M., and Tuminas, A. C., 1985, Structure of the northern Sierra Nevada: Geological Society of America Bulletin, v. 96, p. 436–450.

Day, H. W., Schiffman, P., and Moores, E. M., 1988, Metamorphism and tectonics of the northern Sierra Nevada, *in* Ernst, W. G., ed., Metamorphism and Crustal Evolution of the Western United States: Englewood Cliffs, New Jersey, Prentice-Hall, Inc., p. 737–763.

DePaolo, D. J., 1981, A neodymium and strontium isotopic study of the Mesozoic calc-alkaline granite batholiths of the Sierra Nevada and Peninsular Ranges, California: Journal of Geophysical Research, v. 86, p. 10470–10488.

DeWitt, E., Armstrong, R. L., Sutter, J. F., and Zartman, R. E., U-Th-Pb, Rb-Sr, and Ar-Ar mineral and whole-rock isotopic systematics in a metamorphosed granitic terrane, southeastern California: Geological Society of America Bulletin, v. 95, p. 723–739.

Dickinson, W. R., and Snyder, W. S., 1978, Plate tectonics of the Laramide orogeny: Geological Society of America Memoir 151, p. 355–365.

Dickinson, W. R., Ingersoll, R. V., Cowan, D. S., Helmold, K. P., and Suczek, C. A., 1982, Provenance of Franciscan graywackes in coastal California: Geological Society of America Bulletin, v. 93, p. 95–107.

Douglass, R. C., 1967, Permian Tethyan fusilinids from California: U.S. Geological Survey Professional Paper 583-A, p. 7–43.

Duebendorfer, E. M., and Houston, R. S., 1986, Kinematic history of the Cheyenne belt, Medicine Bow Mountains, southeastern Wyoming: Geology, v. 14, p. 171–174.

——, 1987, Proterozoic accretionary tectonics at the southern margin of the Archean Wyoming craton: Geological Society of America Bulletin, v. 98, p. 554–568.

Dunne, G. C., Gulliver, R. M., and Sylvester, A. G., 1978, Mesozoic evolution of rocks of the White, Inyo, Angua, and Slate ranges, eastern California, *in* Howell, D. G., and McDougall, K., eds., Mesozoic Paleogeography of the Western United States: Society of Economic Paleontologists and Mineralogists, p. 189–207.

Ehlig, P. L., 1981, Origin and tectonic history of the basement terrane of the San Gabriel Mountains, central Transverse Ranges, *in* Ernst, W. G., ed., The Geotectonic Development of California: Englewood Cliffs, New Jersey, Prentice-Hall, Inc., p. 253–283.

Engebretson, D. C., Cox, A., and Gordon, R. G., 1985, Relative motions between oceanic and continental plates in the Pacific Basin: Geological Society of America Special Paper 206, 59 p.

Ernst, W. G., 1970, Tectonic contact between the Franciscan mélange and the Great Valley sequence, crustal expression of a Late Mesozoic Benioff zone: Journal of Geophysical Research, v. 75, p. 886–901.

——, 1984, California blueschists, subduction, and the significance of tectonostratigraphic terranes: Geology, v. 12, p. 436–440.

Farmer, G. L., 1988, Isotope geochemistry of Mesozoic and Tertiary igneous rocks in the western U.S. and implications for the structure and composition of the deep continental lithosphere, *in* Ernst, W. G., ed., Metamorphism and Crustal Evolution of the Western United States: Englewood Cliffs, New Jersey, Prentice-Hall, Inc., p. 87–109.

Gallagher, M. P., Brown, E. H., and Walker, N. W., 1988, A new structural and tectonic interpretation of the western part of the Shuksan blueschist terrane, northwestern Washington: Geological Society of America Bulletin, v. 100, p. 1415–1422.

Gastil, G., 1985, Terranes of Peninsular California and adjacent Sonora, *in* Howell, D. G., ed., Tectonostratigraphic Terranes of the Circum-Pacific Region: Circum-Pacific Council Energy Mineral Resources Earth Science Series 1, p. 273–283.

Gastil, G., Morgan, G., and Krummenacher, D., 1981, The tectonic history of Peninsular California and adjacent Mexico, *in* Ernst, W. G., ed., The Geotectonic Development of California: Englewood Cliffs, New Jersey, Prentice-Hall, Inc., p. 284–306.

Gosselin, D. C., Papike, J. J., Zartman, R. E., Peterman, Z. E., and Laul, J. C., 1988, Archean rocks of the Black Hills, South Dakota—reworked basement from the southern extension of the trans-Hudson orogen: Geological Society of America Bulletin, v. 100, p. 1244–1259.

Grambling, J. A., 1981, Kyanite, andalusite, sillimanite, and related mineral assemblages in the Truchas Peaks region, New Mexico: American Mineralogist, v. 66, p. 702–722.

——, 1988, A summary of Proterozoic metamorphism in northern and central New Mexico—the regional development of 520°C, 4 kb rocks, *in* Ernst, W. G., ed., Metamorphism and Crustal Evolution of the Western United States: Englewood Cliffs, New Jersey, Prentice-Hall, Inc., p. 446–465.

Grambling, J. A., Williams, M. L., and Mawer, C. K., 1988, Proterozoic tectonic assembly of New Mexico: Geology, v. 16, p. 724–727.

Gray, G. G., 1986, Native terranes of the central Klamath Mountains, California: Tectonics, v. 5, p. 1043–1054.

Hackel, O., 1966, Summary of the geology of the Great Valley, *in* Bailey, E. H., ed., Geology of northern California: California Division of Mines and Geology Bulletin 190, p. 217–238.

Hamilton, W., 1963, Metamorphism in the Riggins region, western Idaho: U.S. Geological Survey Professional paper 436, 95 p.

——, 1969, Mesozoic California and the underflow of the Pacific mantle: Geological Society of America Bulletin, v. 80, p. 2409–2430.

——, 1988, Tectonic setting and variations with depth of some Cretaceous and Cenozoic structural and magmatic systems of the western United States, *in* Ernst, W. G., ed., Metamorphism and Crustal Evolution of the Western United States: Englewood Cliffs, New Jersey, Prentice-Hall, Inc., p. 1–40.

Hansen, V. L., and Goodge, J. W., 1988, Metamorphism, structural petrology, and regional evolution of the Okanogan Complex, northeastern Washington, *in* Ernst, W. G., ed., Metamorphism and Crustal Evolution of the Western United States: Englewood Cliffs, New Jersey, Prentice-Hall, Inc., p. 233–270.

Harrison, J. E., 1972, Precambrian Belt basin of northwestern United States—its geometry, sedimentation, and copper occurrences: Geological Society of America Bulletin, v. 83, p. 1215–1240.

Harwood, D. S., 1988, Tectonism and metamorphism in the northern Sierra terrane, northern California, *in* Ernst, W. G., ed., Metamorphism and Crustal Evolution of the Western United States: Englewood Cliffs, New Jersey, Prentice-Hall, Inc., p. 764–788.

Haxel, G. B., Tosdal, R. M., May, D. J., and Wright, J. E., 1984, Latest Cretaceous and early Tertiary orogenesis in south-central Arizona—Thrust faulting, regional metamorphism, and granitic plutonism: Geological Society of America Bulletin, v. 95, p. 631–653.

Helper, M. A., 1986a, Deformation and high P/T metamorphism in the central part of the Condrey Mountain window, north-central Klamath Mountains, California and Oregon, *in* Brown, E. H., and Evans, B. W., eds., Blueschists and Eclogites: Geological Society of America Memoir 164, p. 125–141.

Hietanen, A., 1981, Petrologic and structural studies in the northwestern Sierra Nevada, California: U.S. Geological Survey Professional Paper 1226-C, 59 p.

——, 1984, Geology along the northwest border zone of the Idaho batholith, northern Idaho: U.S. Geological Survey Bulletin 1608, 17 p.

Hill, L. B., 1985, Metamorphic, deformational, and temporal constraints on terrane assembly, northern Klamath Mountains, California, *in* Howell, D. G., ed., Tectonostratigraphic Terranes of the Circum-Pacific Region: Circum-Pacific Council for Energy and Mineral Resources, no. 1, p. 173–186.

Hoisch, T. D., Miller, C. F., Heizler, M. T., Harrison, T. M., and Stoddard, E. F., 1988, Late Cretaceous regional metamorphism in southeastern California, *in* Ernst, W. G., ed., Metamorphism and Crustal Evolution of the Western United States: Englewood Cliffs, New Jersey, Prentice-Hall, Inc., p. 538–571.

Hopson, C. A., Mattinson, J. M., and Pessagno, E. A., Jr., 1981, Coast Range ophiolite, western California, *in* Ernst, W. G., ed., The Geotectonic Development of California: Englewood Cliffs, New Jersey, Prentice-Hall, Inc., p. 418–510.

Hotz, P. E., 1973, Blueschist metamorphism in the Yreka–Fort Jones area, Klamath Mountains, California: U.S. Geological Survey Journal of Research, v. 1, p. 53–61.

Hotz, P. E., Lanphere, M. A., and Swanson, D. A., 1977, Triassic blueschists from northern California and north-central Oregon: Geology, v. 5, p. 659–663.

Howell, D. G., and Vedder, J. G., 1981, Structural implications of stratigraphic discontinuities across the southern California Borderland, *in* Ernst, W. G., ed., The Geotectonic Development of California: Englewood Cliffs, New Jersey, Prentice-Hall, Inc., p. 535–558.

Hsü, K. J., 1971, Franciscan mélanges as a model for eugeosynclinal sedimentation and underthrusting tectonics: Journal of Geophysical Research, v. 76, p. 1162–1170.

Hyndman, D. W., Alt, D., and Sears, J. W., 1988, Post-Archean metamorphic and tectonic evolution of western Montana and northern Idaho, *in* Ernst, W. G., ed., Metamorphism and Crustal Evolution of the Western United States: Englewood Cliffs, New Jersey, Prentice-Hall, Inc., p. 332–361.

Irwin, W. P., 1972, Terranes of the western Paleozoic and Triassic belt in the southern Klamath Mountains, California: U.S. Geological Survey Professional Paper 800-C, p. 103–111.

——, 1981, Tectonic accretion of the Klamath Mountains, *in* Ernst, W. G., ed., The Geotectonic Development of California: Englewood Cliffs, New Jersey, Prentice-Hall, Inc., p. 29–49.

——, 1985, Age and tectonics of plutonic belts in accreted terranes of the Klamath Mountains, California and Oregon, *in* Howell, D. G., ed., Tectonostratigraphic Terranes of the Circum-Pacific Region: Circum-Pacific Council for Energy and Mineral Resources Earth Science Series 1, p. 187–199.

Jacobson, C. E., Dawson, M. R., and Postlethwaite, C. E., 1988, Structure, metamorphism, and tectonic significance of the Pelona, Orocopia, and Rand schists, southern California, *in* Ernst, W. G., ed., Metamorphism and Crustal Evolution of the Western United States: Englewood Cliffs, New Jersey, Prentice-Hall, Inc., p. 976–997.

James, E. W., and Mattinson, J. M., 1988, Metamorphic history of the Salinian block—an isotopic reconnaissance, *in* Ernst, W. G., ed., Metamorphism and Crustal Evolution of the Western United States: Englewood Cliffs, New Jersey, Prentice-Hall, Inc., p. 938–952.

Jones, D. L., Howell, D. G., Coney, P. J., and Monger, J.W.H., 1983, Recognition, character, and analysis of tectonostratigraphic terranes in western North America, *in* Hashimoto, M., and Uyeda, S., eds., Accretion Tectonics in the Circumpacific Region: Tokyo, Terra Scientific Publishing Co., p. 21–35.

Karlstrom, K. E., and Houston, R. S., 1984, The Cheyenne belt—analysis of a Proterozoic suture in southern Wyoming: Precambrian Research, v. 25, p. 415–446.

Karlstrom, K. E., Bowring, S. A., and Conway, C. M., 1987, Tectonic significance of an Early Proterozoic two-province boundary in central Arizona: Geological Society of America Bulletin, v. 99, p. 529–538.

Kays, M. A., Ferns, M. L., and Brooks, H. C., 1988, Metamorphism of Triassic-Paleozoic belt rocks—a guide to field and petrologic relations in the oceanic mélange, Klamath and Blue Mountains, California and Oregon, *in* Ernst, W. G., ed., Metamorphism and Crustal Evolution of the Western United States: Englewood Cliffs, New Jersey, Prentice-Hall, Inc., p. 1098–1142.

Kistler, R. W., and Peterman, Z. E., 1978, A study of regional variations of initial strontium isotopic composition of Mesozoic granitic rocks in California: U.S. Geological Survey Professional Paper 1071, 17 p.

Knuepfer, P.L.K., and 5 others, 1987, Crustal structure of the Basin and Range–Sierra Nevada transition from COCORP deep seismic-reflection profiling: Geological Society of America Bulletin, v. 98, p. 488–496.

Labotka, T. C., and Albee, A. L., 1988, Metamorphism and tectonics of the Death Valley region, California, *in* Ernst, W. G., ed., Metamorphism and Crustal Evolution of the Western United States: Englewood Cliffs, New Jersey, Prentice-Hall, Inc., p. 714–736.

Lanphere, M. A., Blake, M. C., Jr., and Irwin, W. P., 1978, Early Cretaceous metamorphic age of the South Fork Mountain schist in the northern Coast Ranges of California: American Journal of Science, v. 278, p. 798–815.

Lund, K., and Snee, L. W., 1988, Metamorphism, structural development, and age of the continent-island arc juncture in west-central Idaho, *in* Ernst, W. G., ed., Metamorphism and Crustal Evolution of the Western United States: Englewood Cliffs, New Jersey, Prentice-Hall, Inc., p. 296–331.

Luyendyk, B. P., Kamerling, M. J., Terres, R. R., and Hornafius, J. S., 1985, Simple shear of southern California during Neogene time suggested by paleomagnetic declinations: Journal of Geophysical Research, v. 90, p. 12454–12466.

Mattinson, J. M., 1988, Constraints on the timing of Franciscan metamorphism; Geochronological approaches and their limitations, *in* Ernst, W. G., ed., Metamorphism and Crustal Evolution of the Western United States: Englewood Cliffs, New Jersey, Prentice-Hall, Inc., p. 1023–1034.

Mattinson, J. M., and James, E. W., 1985, Salinian block U/Pb age and isotopic variations—implications for origin and emplacement of the Salinian terrane, *in* Howell, D. G., ed., Tectonostratigraphic Terranes of the Circum-Pacific Region: Circum-Pacific Council for Energy and Mineral Resources Earth Science Series 1, p. 215–226.

Maxwell, D. T., and Hower, J., 1967, High-grade diagenesis and low-grade metamorphism of illite in the Precambrian Belt Series: American Mineralogist, v. 52, p. 843–857.

Maxwell, J. C., 1974, Anatomy of an orogen: Geological Society of America Bulletin, v. 85, p. 1195–1204.

May, S. R., and Butler, R. F., 1986, North American Jurassic apparent polar wanderer—implications for plate motions, paleogeography, and Cordilleran tectonics: Journal of Geophysical Research, v. 91, p. 11519–11544.

McLaughlin, R. J., Kling, S. A., Poore, R. Z., McDougall, K., and Beuther, E. C., 1982, Post-middle Miocene accretion of Franciscan rocks, northwestern California: Geological Society of America Bulletin, v. 93, p. 595–605.

Miller, E. L., Gans, P. B., Wright, J. E., and Sutter, J. F., 1988, Metamorphic histories of the east-central Basin and Range province—tectonic setting and relationship to magmatism, *in* Ernst, W. G., ed., Metamorphism and Crustal Evolution of the Western United States: Englewood Cliffs, New Jersey, Prentice-Hall, INc., p. 649–682.

Miller, M. M., 1987, Dispersed remnants of a northeast Pacific fringing arc—upper Paleozoic terranes of Permian McCloud faunal affinity, western U.S.: Tectonics, v. 6, p. 807–830.

Misch, P., 1988, Tectonic and metamorphic evolution of the North Cascades—an overview, *in* Ernst, W. G., ed., Metamorphism and Crustal Evolution of the Western United States: Englewood Cliffs, New Jersey, Prentice-Hall, Inc., p. 179–195.

Mogk, D. W., and Henry, D. J., 1988, Metamorphic petrology of the northern Archean Wyoming Province, southwestern Montana—evidence for Archean collisional tectonics, *in* Ernst, W. G., ed., Metamorphism and Crustal Evolution of the Western United States: Englewood Cliffs, New Jersey, Prentice-Hall, Inc., p. 362–382.

Mortimer, N., 1985, Structural and metamorphic aspects of Middle Jurassic terrane juxtaposition, northeastern Klamath Mountains, California, *in* Howell, D. G., ed., Tectonostratigraphic Terranes of the Circum-Pacific Region: Circum-Pacific Council for Energy and Mineral Resources Earth Science Series 1, p. 201–214.

Nelson, C. A., Hall, C. A., Jr., and Ernst, W. G., 1991, Geologic history of the White–Inyo Range, *in* Hall, C. A., Jr., ed., Natural history of the White–Inyo Range, eastern California: Berkeley, University of California Press, Natural History Guides, v. 55, p. 42–74.

Page, B. M., 1966, Geology of the Coast Ranges of California, *in* Bailey, E. H., ed., Geology of Northern California: California Division of Mines and Geology Bulletin 190, p. 255–276.

——, 1981, The southern Coast Ranges, *in* Ernst, W. G., ed., The Geotectonic Development of California: Englewood Cliffs, New Jersey, Prentice-Hall, Inc., p. 329–417.

——, 1982, Migration of Salinian composite block, California, and disappearance of fragments: American Journal of Science, v. 282, p. 1694–1734.

Peacock, S. M., 1988, Inverted metamorphic gradients in the westernmost Cordillera, *in* Ernst, W. G., ed., Metamorphism and Crustal Evolution of the Western United States: Englewood Cliffs, New Jersey, Prentice-Hall, Inc., p. 953–975.

Reed, J. C., Jr., Bickford, M. E., Premo, W. R., Aleinikoff, J. N., and Pallister, J. S., 1987, Evolution of the Early Proterozoic Colorado province—Constraints from U-Pb geochronology: Geology, v. 15, p. 861–865.

Renne, P. R., Scott, G. R., and Bazard, D. R., 1988, Multicomponent paleomagnetic data from the Nosoni Formation, eastern Klamath Mountains, California—cratonic Permian primary directions with Jurassic overprints: Journal of Geophysical Research, v. 93, p. 3387–3400.

Reynolds, S. J., Richard, S. M., Hazel, G. B., Tosdal, R. M., and Laubach, S. E., 1988, Geologic setting of Mesozoic and Cenozoic metamorphism in Arizona, *in* Ernst, W. G., ed., Metamorphism and Crustal Evolution of the Western United States: Englewood Cliffs, New Jersey, Prentice-Hall, Inc., p. 466–501.

Rhodes, B. P., and Hyndman, D. W., 1988, Regional metamorphism, structure, and tectonics of northeastern Washington and northern Idaho, *in* Ernst, W. G., ed., Metamorphism and Crustal Evolution of the Western United States: Englewood Cliffs, New Jersey, Prentice-Hall, Inc., p. 271–295.

Ross, D. C., 1978, The Salinian block—a Mesozoic granitic orphan in the California Coast Ranges, *in* Howell, D. G., and McDougall, K. A., eds., Mesozoic Paleogeography of the western United States: Pacific Section, Society of Economic Paleontologists and Mineralogists Pacific Coast Paleogeography Symposium 2, p. 509–522.

Saleeby, J. B., 1981, Ocean floor accretion and volcanoplutonic arc evolution of the Mesozoic Sierra Nevada, *in* Ernst, W. G., ed., The Geotectonic Development of California: Englewood Cliffs, New Jersey, Prentice-Hall, Inc., p. 132–181.

Saleeby, J. B., Harper, G. D., Snoke, A. W., and Sharp, W. N., 1982, Time relations and structural-stratigraphic patterns in ophiolite accretion, west central Klamath Mountains, California: Journal of Geophysical Research, v. 87, p. 3831–3848.

Sams, D. B., and Saleeby, J. B., 1988, Geology and petrotectonic significance of crystalline rocks of the southernmost Sierra Nevada, California, *in* Ernst, W. G., ed., Metamorphism and Crustal Evolution of the Western United States: Englewood Cliffs, New Jersey, Prentice-Hall, Inc., p. 865–893.

Schweickert, R. A., 1981, Tectonic evolution of the Sierra Nevada Range, *in* Ernst, W. G., ed., The Geotectonic Development of California: Englewood Cliffs, New Jersey, Prentice-Hall, Inc., p. 87–131.

Schweickert, R. A., Merguerian, C., and Bogen, N. L., 1988, Deformational and metamorphic history of Paleozoic and Mesozoic basement terranes in the western Sierra Nevada metamorphic belt, *in* Ernst, W. G., ed., Metamorphism and Crustal Evolution of the Western United States: Englewood Cliffs, New Jersey, Prentice-Hall, Inc., p. 798–822.

Sears, J. W., Graff, P. J., and Holden, G. S., 1982, Tectonic evolution of lower Proterozoic rocks, Uinta Mountains, Utah and Colorado: Geological Society of America Bulletin, v. 93, p. 990–997.

Seiders, V. M., and Blome, C. D., 1984, Clast compositions of upper Mesozoic conglomerates of the California Coast Ranges and their tectonic significance, *in* Blake, M. C., Jr., Franciscan Geology of Northern California: Pacific Section, Society of Economic Paleontologists and Mineralogists, v. 43, p. 135–148.

Sharp, W. D., 1988, Pre-Cretaceous crustal evolution in the Sierra Nevada region, California, *in* Ernst, W. G., ed., Metamorphism and Crustal Evolution of the Western United States: Englewood Cliffs, New Jersey, Prentice-Hall, Inc., p. 823–864.

Snoke, A. W., and Miller, D. M., 1988, Metamorphic and tectonic history of the northeastern Great Basin, *in* Ernst, W. G., ed., Metamorphism and Crustal Evolution of the Western United States: Englewood Cliffs, New Jersey, Prentice-Hall, Inc., p. 606–648.

Snoke, A. W., Sharp, W. D., Wright, J. E., and Saleeby, J. B., 1982, Significance of mid-Mesozoic peridotitic to dioritic intrusive complexes, Klamath Mountains–western Sierra Nevada, California: Geology, v. 10, p. 160–166.

Sorensen, S. S., 1988, Tectonometamorphic significance of the basement rocks of the Los Angeles Basin and the Inner California Continental Borderland, *in* Ernst, W. G., ed., Metamorphism and Crustal Evolution of the Western United States: Englewood Cliffs, New Jersey, Prentice-Hall, Inc., p. 988–1022.

Speed, R., Elison, M. W., and Heck, F. R., 1988, Phanerozoic tectonic evolution of the Great Basin, *in* Ernst, W. G., ed., Metamorphism and Crustal Evolution of the Western United States: Englewood Cliffs, New Jersey, Prentice-Hall, Inc., p. 572–605.

Stewart, J. H., 1988, Tectonics of the Walker Lane belt, western Great Basin—Mesozoic and Cenozoic deformation in a zone of shear, *in* Ernst, W. G., ed., Metamorphism and Crustal Evolution of the Western United States: Englewood Cliffs, New Jersey, Prentice-Hall, Inc., p. 683–713.

Suppe, J., 1973, Geology of the Leech Lake Mountain–Ball Mountain region, California: University of California Publications in the Geological Sciences, v. 107, p. 1–82.

Suppe, J., and Armstrong, R. L., 1972, Potassium-argon dating of Franciscan metamorphic rocks: American Journal of Science, v. 272, p. 217–233.

Tabor, R. W., and Cady, W. M., 1978, Geologic map of the Olympic Peninsula, Washington: U.S. Geological Survey Map I–994, scale 1:250,000.

Tarduno, J. A., McWilliams, M., Sliter, W. V., Cook, H. E., Blake, M. C., Jr., and Premoli-Silva, I., 1986, Southern hemispheric origin of the Cretaceous Laytonville Limestone of California: Science, v. 231, p. 1425–1428.

Theodore, T. G., Blair, W. N., and Nash, J. T., 1987, Geology and gold mineralization of the Gold Basin–Lost Basin mining districts, Mojave County, Arizona: U.S. Geological Survey Professional Paper 1361, 167 p.

Thomas, W. M., Clark, H. S., Young, E. O., Orrell, S. E., and Anderson, J. L., 1988, Proterozoic high-grade metamorphism in the Colorado River region, Nevada, Arizona, and California, *in* Ernst, W. G., ed., Metamorphism and Crustal Evolution of the Western United States: Englewood Cliffs, New Jersey, Prentice-Hall, Inc., p. 526–537.

Todd, V. R., Erskine, B. G., and Morton, D. M., 1988, Metamorphic and tectonic evolution of the northern Peninsular Ranges batholith, southern California, *in* Ernst, W. G., ed., Metamorphism and Crustal Evolution of the Western United States: Englewood Cliffs, New Jersey, Prentice-Hall, Inc., p. 894–937.

Wentworth, C. M., Blake, M. C., Jr., Jones, D. L., Walter, A. W., and Zoback, M. D., 1984, Tectonic wedging associated with emplacement of the Franciscan assemblage, California Coast Ranges, *in* Blake, M. C., Jr., ed., Franciscan Geology of Northern California: Pacific Section, Society of Economic Paleontologists and Mineralogists, v. 43, p. 163–173.

Wooden, J. L., Mueller, P. A., Mogk, D. W., 1988a, A review of the geochemistry and geochronology of the Archean rocks of the northern part of the Wyoming Province, *in* Ernst, W. G., ed., Metamorphism and Crustal Evolution of the Western United States: Englewood Cliffs, New Jersey, Prentice-Hall, Inc., p. 383–410.

Wooden, J. L., Stacey, J. S., Doe, B. R., Howard, K. A., and Miller, D. M., 1988b, Pb isotopic sevidence for the formation of Proterozoic crust in the southwestern United States, *in* Ernst, W. G., ed., Metamorphism and Crustal Evolution of the Western United States: Englewood Cliffs, New Jersey, Prentice-Hall, Inc., p. 69–86.

Wright, J. E., 1982, Permo-Triassic accretionary subduction complex, southwestern Klamath Mountains, northern California: Journal of Geophysical Research, v. 87, p. 3805–3818.

Xenophontos, C., and Bond, G. C., 1978, Petrology, sedimentation, and paleogeography of the Smartville terrane (Jurassic)—Bearing on the genesis of the Smartville ophiolite, *in* Howell, D. E., and McDougall, K. A., eds., Mesozoic Paleogeography of the western United States: Pacific Section: Society of Economic Paleontologists and Mineralogists Pacific Coast Paleogeography Symposium 2, p. 291–302.

Zartman, R. E., 1974, Lead isotope provinces in the Cordillera of the western United States and their geologic significance: Economic Geology, v. 69, p. 792–805.

Zen, E-an, 1988, Tectonic significance of high-pressure plutonic rock in the western Cordillera of North America, *in* Ernst, W. G., ed., Metamorphism and Crustal Evolution of the Western United States: Englewood Cliffs, New Jersey, Prentice-Hall, Inc., p. 41–67.

Zoback, M. L., Anderson, R. E., and Thompson, G. A., 1981, Cenozoic evolution of the state of stress and style of tectonism of the Basin and Range province of western U.S.: Philosophical Transactions of the Royal Society of London, Series A., v. 300, p. 407–434.

Manuscript Accepted by the Society December 30, 1988
UCLA Institute of Geophysics and Planetary Physics Publication No. 3033

ACKNOWLEDGMENTS

The author gratefully acknowledges the scientific help of participants at a University of California, Los Angeles, Rubey Colloquium on the subject "Metamorphism and crustal evolution of the western United States," which took place in January, 1986. That symposium, and the summary volume subsequently produced (see text citations for individual contributors) provided the necessary framework for the present chapter. Field and laboratory support for this latter work was provided by UCLA and by the National Science Foundation (NSF), most recently through grant EAR83-12702. The following scientists reviewed all or part of the first-draft of this chapter: J. L. Anderson, V. C. Bennett, M. C. Blake, Jr., E. H. Brown, B. Bryant, R. G. Coleman, H. W. Day, J. W. Goodge, T. D. Hoisch, D. W. Hyndman, E. W. James, M. A. Kays, D. M. Miller, E. L. Miller, D. W. Mogk, D. C. Ross, S. M. Peacock, and V. R. Todd. UCLA, the NSF, and especially the contributors to the Rubey volume, are thanked for their help, without which the present summary could not have been written. Additional help was provided by attendees of Geological Society of America meetings at Phoenix (1987) and Las Vegas (1988) where preliminary versions of the map were presented.

Printed in U.S.A.

The Geology of North America
Vol. G-3, The Cordilleran Orogen: Conterminous U.S.
The Geological Society of America, 1992

Chapter 11

Cordilleran sedimentary assemblages

William R. Dickinson
Laboratory of Geotectonics, Department of Geosciences, University of Arizona, Tucson, Arizona 85721

INTRODUCTION

The nature of sedimentary assemblages is controlled by interactions of (a) provenance, which refers to source rocks as modified by the effects of weathering and varied relief; (b) mechanisms of erosion and sediment transport; (c) dispersal paths as a function of paleogeography; (d) depositional processes operative in different environments of sedimentation; and (e) diagenesis, which is influenced strongly by the thermal history and geochemical evolution of sedimentary basins. All these factors are dependent primarily upon tectonics and climate.

Even though the governing influences of tectonic setting and climatic regime can be subtle in detail, a general overview of Cordilleran sedimentary assemblages is best achieved with reference to evolving paleotectonic and paleogeographic patterns. Salient regional constraints must be considered against a background of changing paleolatitude and eustatics, which reflect trends in global tectonics and climate.

This chapter outlines overall correlations between different types of Cordilleran sedimentary assemblages and inferred combinations of paleotectonics, paleogeography, paleolatitude, and eustatics. As all parameters involved in the discussion are viewed more properly as spectra, rather than as quanta, the resulting treatment is inherently arbitrary to some degree. Sedimentary assemblages are here grouped primarily in terms of their bulk compositions and the dominant depositional systems responsible for their formation. Precambrian sequences deposited within the present Cordilleran region, but prior to the rifting events that delineated the Cordilleran continental margin, are not discussed.

KEY CORDILLERAN ASSEMBLAGES

No classification of Cordilleran sedimentary assemblages can be fully satisfactory, given their wide variability. To some extent, each succession is unique. Table 1 presents a provisional array of key assemblages, whose recognition is based jointly upon dominant rock types and characteristic depositional systems in the context of paleotectonic and paleogeographic setting (e.g., Dickinson, 1972). The names adopted for each type should be viewed simply as convenient labels, and not as complete descriptions. Designation of the four major assemblage types (A to D) is based upon the character of the substratum.

Ocean-basin assemblages (A) were deposited upon volcanogenic basement in regions that lay beyond the edge of the continental block. They are divided here into three main groups representative of (1) the ocean floor of open oceans or marginal seas including local seamounts and submerged plateaus, (2) the positive relief of intraoceanic island arcs, and (3) the ocean flanks where turbidites derived from continental sources build subsea fans and blankets of abyssal-plain sediment across the sea floor or within marginal trenches. By inference, the substratum depositionally beneath ocean-basin assemblages lacked any true continental basement within the crustal profile.

Continental-margin assemblages (B) were deposited within zones of transitional crust along the continent-ocean interface. They are divided here into three main groups representative of the three kinds of plate boundaries that control the morphology of continental margins: (1) miogeoclinal prisms deposited along passive continental margins where continental basement, thinned by rifting, subsides thermotectonically following drift separation of continental blocks (Bond and others, 1983; Armin and Mayer, 1983); (2) forearc basins developed between magmatic arcs and associated trenches during continuing or intermittent subduction along active continental margins (Dickinson, 1976); and (3) transform-related basins formed by deformation of the continental margin within a belt of complex strike-slip motions (Crowell, 1974a). In part, these distinctions are somewhat arbitrary, because local or subordinate transform motions can accompany either rifting or subduction, but they can be applied successfully on the basis of dominant tectonic setting.

Continental-block assemblages (C) were deposited within the interior of the continental block upon a full thickness of continental basement rocks, either cratonic or orogenic in nature. They include (1) platform cover formed in marine settings reflecting eustatic highstands, in terrestrial settings by aggradation of fluvial or eolian deposits, or in both marine and terrestrial settings during downflexure of the platform margin beneath miogeoclinal sediment loads; (2) overlap sequences that buried dormant oro-

Dickinson, W. R., Cordilleran sedimentary assemblages, 1992, *in* Burchfiel, B. C., Lipman, P. W., and Zoback, M. L., eds., The Cordilleran Orogen: Conterminous U.S.: Boulder, Colorado, Geological Society of America, The Geology of North America, v. G-3.

TABLE 1. KEY CORDILLERAN SEDIMENTARY ASSEMBLAGES

A. **Ocean-Basin Assemblages*** Extracontinental settings of varied bathymetry within oceanic basins	1. ocean floor 2. island arc 3. ocean flank
B. **Continental-Margin Assemblages*** Settings along continent-ocean interface, including deformed edge of continental block	1. miogeoclinal prism 2. forearc basin 3. transform margin
C. **Continental-Block Assemblages*** Intracontinental settings across flooded or depressed surface of continental block	1. platform cover 2. orogen overlap 3. foreland basin
D. **Continental-Rupture Assemblages**† Settings within basins whose subsidence was induced by breakage of the continental block through extensional rifting or compressive thrusting	1. Ancestral Rockies 2. Bisbee Basin 3. Laramide Rockies 4. Basin and Range 5. Salton Trough

*Divided into genetic assemblage groups, each of varied age and distribution.
†Divided into regional assemblages, each having a particular age span.

genic belts following deep erosion initiated by prior deformation and uplift; and (3) the marine or terrestrial fill of foreland basins (Beaumont, 1981) developed across the surface of the continental block by downflexure under the load of thrust sheets emplaced along the flanks of adjacent orogenic belts. Relatively thin platform-cover and orogen-overlap assemblages commonly grade laterally into thicker miogeoclinal-prism, foreland-basin, or continental-rupture assemblages.

Continental-rupture assemblages (D) accumulated within intracontinental basins formed by tectonic rupture of the continental block. Styles of deformation causing rupture have included extensional rifting, compressive thrusting, and transform torsion, or more than one of those styles acting in combination. Subsidence of thinned continental crust and downflexure of tectonically loaded continental basement are the major mechanisms of subsidence in such intracontinental basins. As details of their evolution are still poorly understood, the continental-rupture assemblages of intracontinental basins are discussed here in five subregional groupings of different age (see Table 1 for designation).

CORDILLERAN LITHOFACIES ASSOCIATIONS

In recent years, the intensive study of varied depositional systems has made it possible to distinguish a steadily expanding number of different lithofacies within stratigraphic sequences (Davis, 1983). Table 2 is an attempt to distill from this rich detail a select few overall lithofacies associations having regional significance within Cordilleran sedimentary assemblages. Again, the designations used here for the three marine (M1 to M3) and three nonmarine (N1 to N3) lithofacies associations are merely convenient labels, and not complete descriptions of each.

Only the most general classification of sedimentary realms is embodied by recognition of the six characteristic lithofacies associations. The three marine associations respectively embrace essentially all subsiding shelves (M3), all constructional bodies composed dominantly of turbidites (M2), and all deep bottoms starved for coarse clastic sediment (M1). The three nonmarine associations include progradational marine deltas (N1), aggradational terrestrial sediment cover (N2), and interior basin fills where base level is lacustrine (N3) rather than marine. Gradational or transitional sedimentary suites that involve various pairs of these broad lithofacies associations clearly occur, and are not meant to be precluded by the simplistic nature of the discussion here.

COMPOSITIONAL SEDIMENTARY SUITES

The gross compositions of different Cordilleran sedimentary assemblages reflect important aspects of their origins. For example, varying proportions of terrigenous clastics, as opposed to carbonates and evaporites, carry important information about paleogeography or paleoclimate, or both. Similarly, variations in grain size reflect changing patterns of provenance and dispersal. From a paleotectonic and paleogeographic standpoint, the most significant division of terrigenous clastic strata in terms of grain

TABLE 2. CHARACTERISTIC LITHOFACIES ASSOCIATIONS WITHIN CORDILLERAN SEDIMENTARY ASSEMBLAGES

Marine Lithofacies Associations

M1. **Starved basin**: pelagites and hemipelatites deposited on starved basin floors and adjacent submarine slopes, together with subordinate intercalated turbidite beds.

M2. **Turbidite fan**: deep-marine turbidite fan bodies, together with deposits backfilling turbidite channels and submarine canyons incised near the base of nearby submarine slopes.

M3. **Marine shelf**: strata deposited on shallow-marine bottoms and carbonate platforms, together with interfingering strandline complexes and slope sediments deposited near the shelfbreak.

Nonmarine Lithofacies Associations

N1. **Deltaic complex**: progradational fluviodeltaic or coastal-plain prisms including mixed terrestrial and marginal-marine strata, together with shaly prodelta sediments or sandy turbidite ramp deposits of the foredelta region.

N2. **Terrestrial cover**: laterally extensive terrestrial deposits of fluvial plains and eolian ergs commonly forming aggradational redbed successions transitional to strandline complexes deposited in arid to semiarid environments.

N3. **Lacustrine basin**: fluviolacustrine basin fills composed of central open-lacustrine or playa facies, together with bordering marginal-lacustrine and fluvial facies of floodplains, braidplains, or alluvial fans and bajadas.

size is into the following three gross suites: (a) shaly suites devoid of coarse clastic detritus, (b) mixed or interbedded sand-shale suites lacking a significant gravel fraction, and (c) conglomeratic suites in which a gravel fraction is prominent.

Thin-section studies of sandstones allow the sand fraction of terrigenous clastic suites to be classified compositionally in terms of petrofacies, rather than in terms of grain size or lithofacies (Dickinson, 1985). The study of sandstone petrofacies has proved to be especially useful for Cordilleran provenance studies because it allows some definition of provenance type to be made without reliable information about paleocurrents or paleogeography. Petrofacies analysis thus provides an independent data set to be used with other data sets to support paleotectonic interpretations. This additional conceptual leverage is important where the positions of various potential provenance terranes may have shifted through time in response to complex plate motions.

The major provenance types that have contributed terrigenous detritus to Cordilleran sedimentary assemblages can be discerned from petrofacies analysis of selected sandstone suites (Table 3), although sediments of mixed provenance are also common. Quartzose sands in orthoquartzites and related rocks were derived repeatedly from the weathered craton or from older sediments whose ultimate source was the craton, but locally were also derived from deeply weathered younger batholiths. Feldspatholithic sands are typically volcaniclastic sediment derived from island arcs or continental volcanic fields, and are rich in volcanic rock fragments. The quartzofeldspathic sands of arkoses were derived principally from granitoid and gneissoid rocks exposed in deeply eroded plutons of magmatic arcs or in locally uplifted blocks of continental basement. Volcanoplutonic sands (Dickinson, 1982), derived jointly from the volcanic cover and the plutonic roots of magmatic arcs, are typically lithofeldspathic sands of complex composition containing varied types of lithic fragments. Feldspar-poor quartzolithic sands (Dickinson, 1985) containing varied sedimentary and metasedimentary lithic fragments were commonly derived either from deformed strata exposed in fold-thrust belts, including subduction complexes, or from cover sequences stripped off basement uplifts. Calcareous grains in hybrid sandstones (Zuffa, 1980) include both intrabasinal clasts identical to those found in calcarenites, and detrital limeclasts recycled from older carbonate successions.

PALEOLATITUDE AND EUSTATICS

Changes in paleolatitude and eustatic fluctuations through time can be assessed independently of Cordilleran paleotectonics and paleogeography. Thus, both become independent variables for interpreting the intricate mosaic of Cordilleran sedimentary assemblages. Some important aspects of Cordilleran depositional history were influenced most strongly by paleolatitude or by eustatics, each controlled by global tectonic patterns, rather than by strictly Cordilleran paleotectonics or details of Cordilleran paleogeography.

TABLE 3. SALIENT SANDSTONE PETROFACIES WITHIN CORDILLERAN SEDIMENTARY ASSEMBLAGES*

Petrofacies Type	Framework Composition*	Provenance Interpretation
Quartzose	Dominantly Qm with minor Qp and F (K > P)	Sources in eroded and deeply weathered cratonic landmass or batholith, or recycled from quartzose sedimentary/metasedimentary sources.
Volcaniclastic	Dominantly Lv (Lv > F) and F (P > K) with low Qm content	Sources in active magmatic arcs or more extensive volcanic fields having similar character.
Arkosic	Dominantly F and Qm with variable K/P ratio and low Lt	Sources in uplifted cratonic basement or deeply eroded plutons of magmatic arcs.
Volcanoplutonic	Mixed Qt, F, and L content with P > K, Lv > Ls, and Qm > Qp	Sources in associated volcanic, plutonic, and minor metamorphic rocks of eroded magmatic arcs.
Quartzolithic	Mixed Qm, Qp, and Ls content, with minor Lv and F; variable Qt/L and Qm/Lt ratios	Sources in uplifted strata of fold-thrust belts adjacent to peripheral (collision) or retroarc foreland basins, or recycled from analogous more local uplifts.

*Grain types:
Qt, total quartzose grains (Qt = Qm + Qp)
Qm, monocrystalline quartz
Qp, polycrystalline quartz (largely chert/metachert)
F, feldspar (P, plagioclase; K, K-feldspar)
L, total unstable lithic fragments (L = Lv + Ls; Lt = L + Qp)
Lv, volcanic/metavolcanic lithic fragments
Ls, sedimentary/metasedimentary lithic fragments

Figure 1 shows the inferred paleolatitude trajectory of the Cordilleran continental margin through Phanerozoic time. Future data will doubtless improve the resolution and accuracy of the curve, but general trends are clear from present information. Throughout Paleozoic time, and into early Mesozoic time, the Cordilleran margin lay within the tropics, with its southern limit typically athwart the equator and its northern limit still well within the trade-wind belt. Adjacent seas were warm and readily capable of sustaining coralgal reefs, whereas land areas were potentially susceptible to the rain shadows of Hercynian highlands or the Pangaean landmass lying in the path of the trade winds. During mid-Mesozoic time, the continent made a rapid transit northward into the temperate zone. The exact timing of this transit is still uncertain pending general acceptance of a Jurassic paleowander path for the craton, and the degree of allochthoneity of many suspect terranes is dependent upon the way in which this question is ultimately resolved (Gordon and others, 1984; May, 1985). In any case, the Cretaceous and Tertiary setting was broadly similar to that of the present, with interior parts of the Cordilleran region subject to the rain shadows of Cordilleran highlands blocking the path of the prevailing westerlies.

Figure 2 is a general plot of global eustatic trends during Phanerozoic time as inferred by previous authors. Nearly all details of past eustatic fluctuations remain in doubt, but gross trends are well established (e.g., Wise, 1974). Lowstands of sea level near the beginning and end of Paleozoic time, as well as at the present, alternated with highstands of sea level in mid-Paleozoic and Cretaceous times. Major transgressions associated with highstands were linked to times of continental fragmentation by rifting, whereas major regressions associated with lowstands were linked to times of continental assembly through crustal collisions that terminated intervals of oceanic closure (Valentine and Moores, 1972; Heller and Angevine, 1985). Superimposed upon these overall variations were rapid glacio-eustatic fluctuations of sea level from mid-Mississippian to mid-Permian time in response to episodes of glaciation in Gondwanaland (Wanless and Shepard, 1936). The broad highstands and lowstands caused commensurate transgression and regression of marine environments across the continental platform, whereas the glacial effects were responsible for markedly cyclic patterns of sedimentation in many environments during late Paleozoic time. Within the Cordilleran region, stratigraphic cyclicity in response to glacio-eustatic fluctuations in sea level was most pronounced for shallow-marine and marginal-marine environments, in which continental-block and continental-rupture assemblages of Pennsylvanian and Early Permian age were deposited inland from the continental margin.

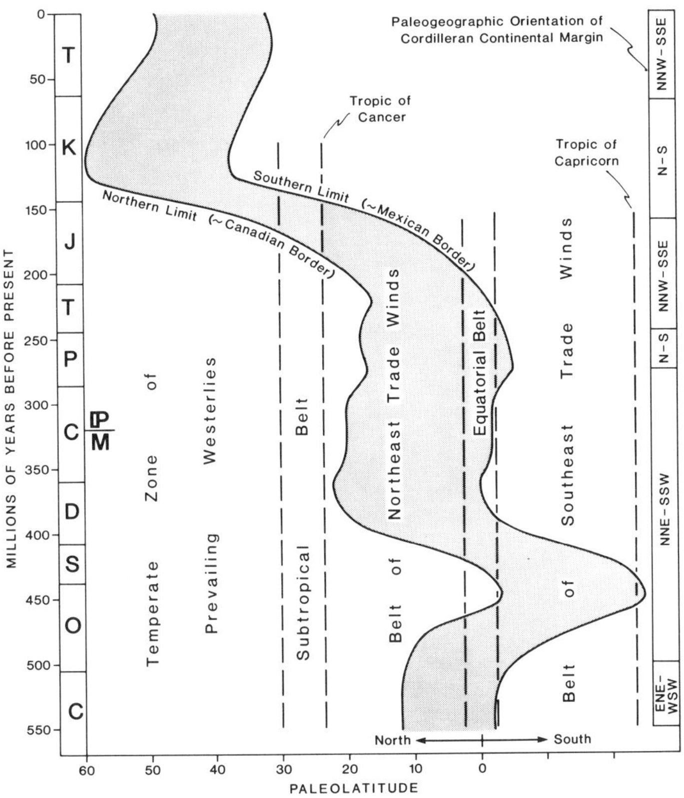

Figure 1. Paleolatitude trajectory and paleogeographic orientation of Cordilleran continental margin during Phanerozoic time, as inferred from paleopoles presented by Van Alstine and de Boer (1978), Irving (1979), Van der Voo (1981), Harrison and Lindh (1982), Steiner (1983), Gordon and others (1984), and May (1985).

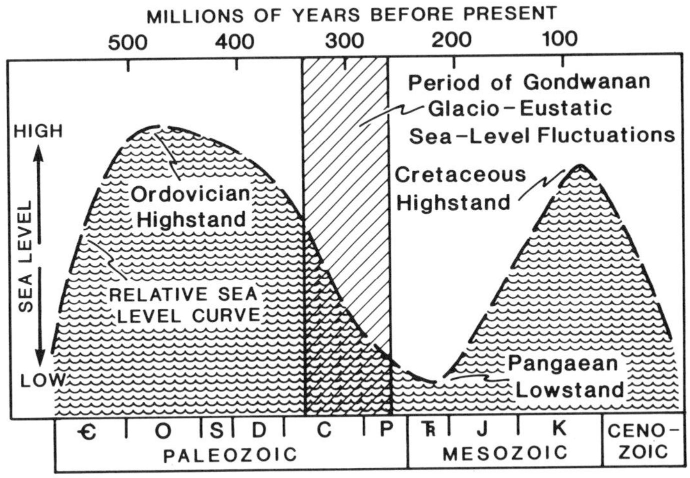

Figure 2. Overall global eustatic fluctuations during Phanerozoic time adapted from Worsley and others (1984); duration of Gondwanan glacio-eustatic fluctuations inferred from Crowell (1978) and Ross and Ross (1985).

OCEAN-BASIN ASSEMBLAGES

Ocean-basin assemblages now within the Cordilleran region (Fig. 3) form significant parts of a vast tectonic collage of suspect terranes, many of which are exotic to some degree with respect to North America (Helwig, 1974; Coney and others, 1980). Cordilleran ocean-basin assemblages of mid-Mesozoic and younger age formed within the Pacific Ocean basin, whose floor is Jurassic or younger and which has been rimmed since its inception by segments of the circum-Pacific orogenic belt. Older ocean-basin assemblages of Phanerozoic age formed within the antecedent Paleo-Pacific Ocean basin, part of Panthalassa whose pre-Jurassic floor was formed by sea-floor spreading during the assembly of Pangaea by oceanic closures elsewhere on the globe. Ocean-basin assemblages fall into three broad lithologic groups, reflecting three major kinds of oceanic settings:

(a) Ocean-floor assemblages (A1, Table 1), deposited in deep water beyond the direct influence of terrestrial sediment sources, are composed dominantly of starved-basin lithofacies (M1, Table 2), among which chert-argillite sequences are most diagnostic. The substratum is commonly pillow basalt and hyaloclastite. Subordinate occurrences of pelagic or platform limestone that accumulated on seamounts or raised oceanic plateaus are locally associated with chert-argillite terranes.

(b) Island-arc assemblages (A2, Table 1) include volcanogenic and volcaniclastic strata deposited in varying water depths within and adjacent to active volcanic arcs. Petrologic types range from basalt and basaltic andesite through andesites to dacite and rhyodacite. Sandstone petrofacies are dominantly volcaniclastic (Table 3). Depositional facies that interfinger with or pass laterally into arc volcanics include turbidites, shelf deposits, and local fluvial strata.

(c) Ocean-flank assemblages (A3, Table 1) spread into ocean basins from terrestrial sources along the margin of the continental block, or within its interior, are composed dominantly of turbidite-fan lithofacies (M2, Table 2). Quartzose petrofacies are most typical of sands transported across the passive Paleozoic continental margin. Highlands along active Mesozoic and Cenozoic continental margins typically contributed either volcanoplutonic petrofacies derived from eroded magmatic arcs or arkosic petrofacies derived from uplifted basement rocks exposed along disrupted transform margins. Thick ocean-flank turbidite successions depositionally overlie thinner ocean-floor pelagites and hemipelagites deposited either (a) before turbidite-fan systems prograded into the ocean basin, (b) before the oceanic substratum moved into a marginal trench marking a subduction zone, or (c) before the oceanic substratum was carried against the continental margin along a bounding transform.

The present structural condition and tectonic setting of ocean-basin assemblages within the Cordilleran region is highly

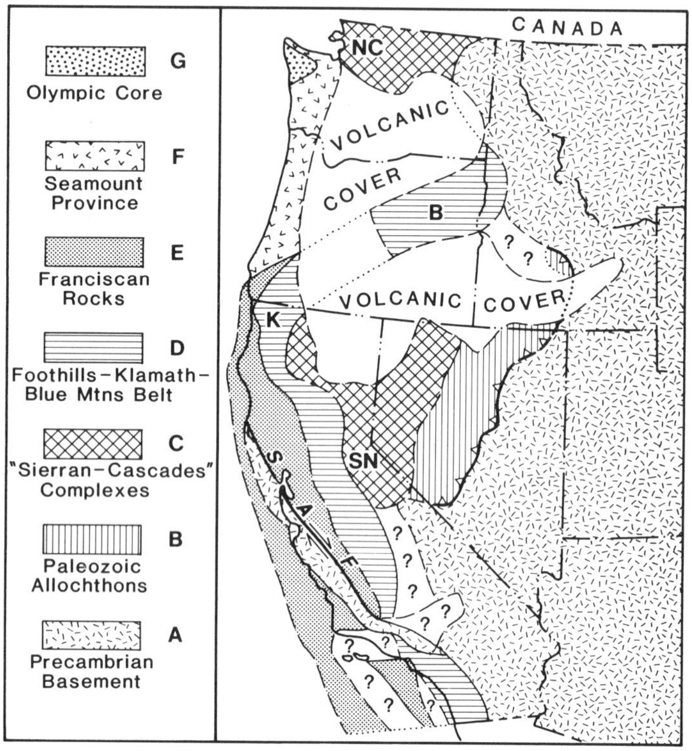

Figure 3. Major belts of Cordilleran oceanic assemblages: A, region underlain by Precambrian continental basement (but nature of Salinian block west of San Andreas fault, SAF, is uncertain); B, Roberts Mountains and Golconda allochthons of deformed Paleozoic rocks thrust over miogeoclinal prism during Antler and Sonoma orogenies; C, complex Paleozoic and Mesozoic island arcs and associated deformed strata of Sierra Nevada (SN) and North Cascades (NC) and neighboring regions; D, accreted Mesozoic island arcs and associated deformed strata of Sierra Nevada foothills and Klamath (K) and Blue (B) Mountains; E, upper Mesozoic and lower Cenozoic Franciscan subduction complex of California Coast Ranges, southwesternmost Oregon, and nearby offshore areas; F, Paleogene seamount province of Pacific Northwest coastal ranges; G, underthrust mid-Tertiary core rocks of Olympic Mountains; queried areas obscured by thick cover or extensive granitic intrusions. All the accreted oceanic assemblages shown are now incorporated into the continental block, and are masked in part by various overlap assemblages.

variable. Paleozoic and Mesozoic island-arc assemblages (A2, Table 1), although metamorphosed and deformed to variable extent, were commonly accreted intact to the edge of the continental block, and display limited internal disruption. The same is true of large seamount masses associated with ocean-floor assemblages (e.g., F of Fig. 3). Other ocean-basin assemblages were intricately folded and thrusted by intense deformation within subduction zones during the very process that led to their incorporation into the continental block. Thus, oceanic strata of contrasting or mixed ocean-floor and ocean-flank facies are commonly imbricated together in complex fashion within subduction complexes that include mélanges and thrust panels. Recycled quartzolithic petrofacies (Table 3) derived from uplifted parts of growing subduction complexes are incorporated locally.

Most subduction complexes of Paleozoic age were emplaced structurally above miogeoclinal strata inland from the edge of the continental block during the Antler and Sonoma orogenies (Fig. 3). During these orogenic events, the passive continental margin was underthrust beneath the flanks of colliding or accreting island arcs, which formed either well offshore within the Paleo-Pacific Ocean (Speed, 1979; Speed and Sleep, 1982) or as fringing arcs bounding marginal seas lying just offshore (Burchfiel and Davis, 1972, 1975; Miller and others, 1984). These obducted masses were assembled at sites removed from the Cordilleran margin, and were emplaced in their entirety as composite allochthons during final phases of thrusting (Oldow, 1984; Brueckner and Snyder, 1985). The allochthons of such collisional orogenic systems typically contain a high proportion of starved-basin (chert-argillite) marine lithofacies (M1, Table 2), deposited as ocean-floor assemblages (A1, Table 1).

By contrast, most subduction complexes of Mesozoic and Cenozoic age were assembled incrementally along the edge of the continental block. These accreted masses, whose volume grew sequentially through time, formed during continued subduction of Pacific lithosphere beneath the active continental margin (Hamilton, 1969). They are composed of successively underthrust sequences that were stacked progressively within evolving subduction zones of the circum-Pacific belt. In general, the ratio of ocean-flank (A3, Table 1) to ocean-floor facies within the Mesozoic-Cenozoic subduction complexes increased with time as the growing bulk of the associated arc-trench system provided more sediment sources along the Cordilleran continental margin. The most voluminous ocean-flank turbidites (M2, Table 2) form the bulk of the well-known Franciscan Complex (Bailey and others, 1964) in coastal California, but similar turbidites also occur in the underthrust core of the Olympic Mountains (Fig. 3).

CONTINENTAL-MARGIN ASSEMBLAGES

Cordilleran continental-margin assemblages formed in three fundamentally different kinds of plate settings: (a) along the passive Paleozoic continental margin adjacent to the Paleo-Pacific Ocean basin, (b) along the active Mesozoic and Cenozoic continental margin as part of the circum-Pacific subduction system, and (c) along the Cenozoic transform margin of California. Each of the three different types of assemblages display contrasting lithofacies and petrofacies associations.

The Cordilleran miogeoclinal prism (B1, Table 1), of latest Precambrian to mid-Paleozoic age (Fig. 4), is composed dominantly of marine-shelf lithofacies (M3, Table 2) including both clastic and carbonate strata (Stewart and Poole, 1974). Mid-Cambrian horizons mark an approximate regional transition between underlying successions dominated by interbedded sandstone and shale, and overlying successions dominated by interbedded limestone or dolostone and shale. Sandstone petrofa-

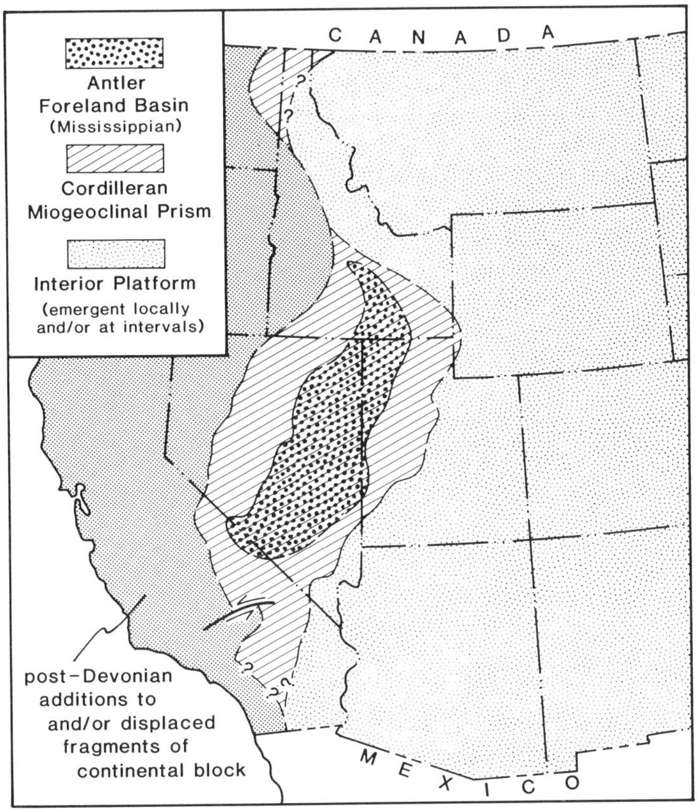

Figure 4. Pre-Pennsylvanian Phanerozoic continental assemblages.

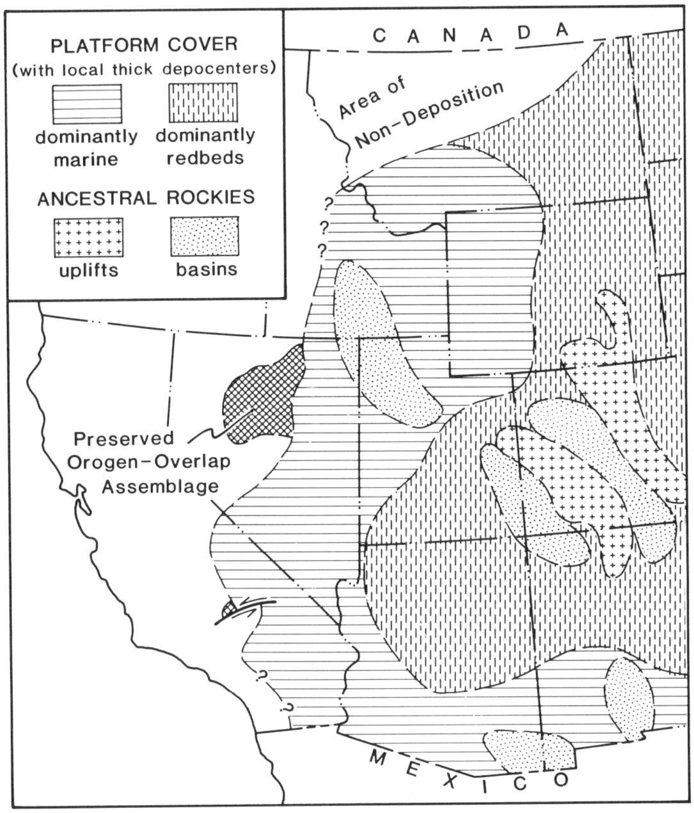

Figure 5. Pennsylvanian to mid-Triassic continental assemblages.

cies (Table 3) are dominantly quartzose, although subordinate arkosic petrofacies occur in basal horizons of the prism. The quartzose nature of the sandstones presumably reflects their derivation from a tectonically quiescent craton subject to deep tropical weathering at low paleolatitudes (Fig. 1). The prevalence of carbonate sedimentation along the passively subsiding continental margin doubtless also reflects the tropical setting, where reef-building and other carbonate-producing organisms could flourish. Widespread marine transgression across the edge of the continental block was evidently aided by the sea-level highstand (Fig. 2), which followed breakup of a Late Proterozoic supercontinent that included the North American craton (Stewart, 1976; Sears and Price, 1978).

Sequences deposited near the continental margin during late Paleozoic and early Mesozoic time were influenced by tectonism during the Antler and Sonoma orogenies, and are discussed below as orogen-overlap assemblages (Fig. 5). Terrane accretion of ocean-basin assemblages along the adjacent continental margin repeatedly modified its morphology between mid–Late Devonian and mid–Late Jurassic time (Fig. 4). Strata that accumulated within the accreting belt during that time span were typically deformed together with the accreted assemblages, and are nearly impossible to distinguish from the latter with confidence (e.g., Harper and Wright, 1984). Distinctions between upper Paleozoic and lower Mesozoic sequences transported to their present locations as parts of mobile terranes and those deposited in place following terrane accretion are beyond the scope of this discussion.

Elongate forearc basins (B2, Table 1) began to develop along the expanded Cordilleran continental margin by Late Jurassic time within a belt that lay between active magmatic arcs on the east and developing subduction complexes or accretionary terranes on the west (Figs. 6, 7). Development of the post-Pangaean Cretaceous highstand of sea level (Fig. 2) may have aided the flooding of forearc regions. Prolonged subsidence uninterrupted by significant intrabasinal tectonics was characteristic of some forearc basins, particularly in central California (Ingersoll, 1978, 1983; Suchecki, 1984), but in others, transpression related in part to terrane accretion probably modified sedimentary patterns. Forearc evolution continued along the continental margin during Laramide deformation inland (Fig. 8), but was progressively terminated in California by Cenozoic evolution of the San Andreas transform system (Atwater, 1970), and has continued to the present only in the Pacific Northwest (Fig. 9). Turbidite-fan and deltaic-complex lithofacies (M2 and N1, Table 2), separately or in combination, are dominant in most forearc-basin sequences,

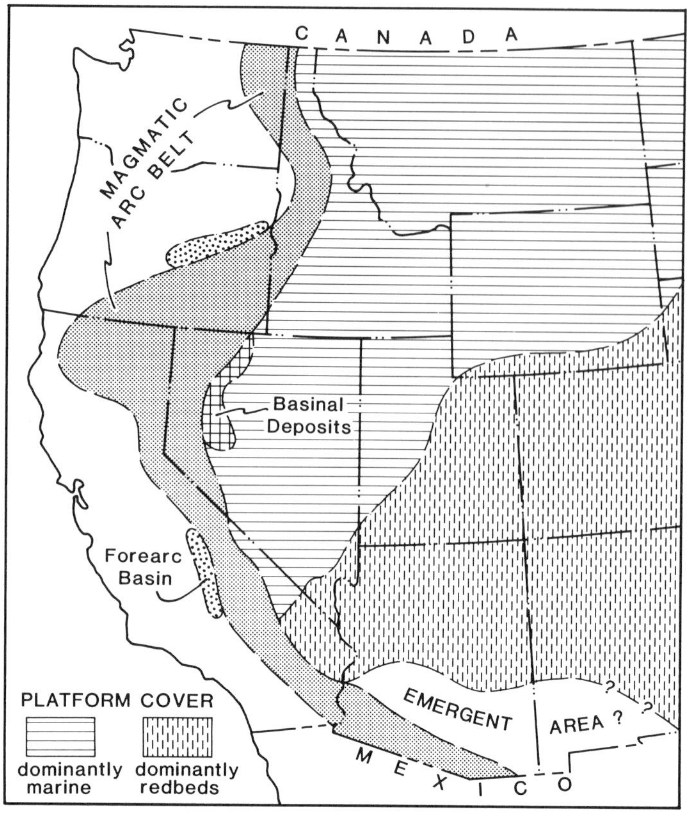

Figure 6. Mid-Triassic to mid-Upper Jurassic continental assemblages.

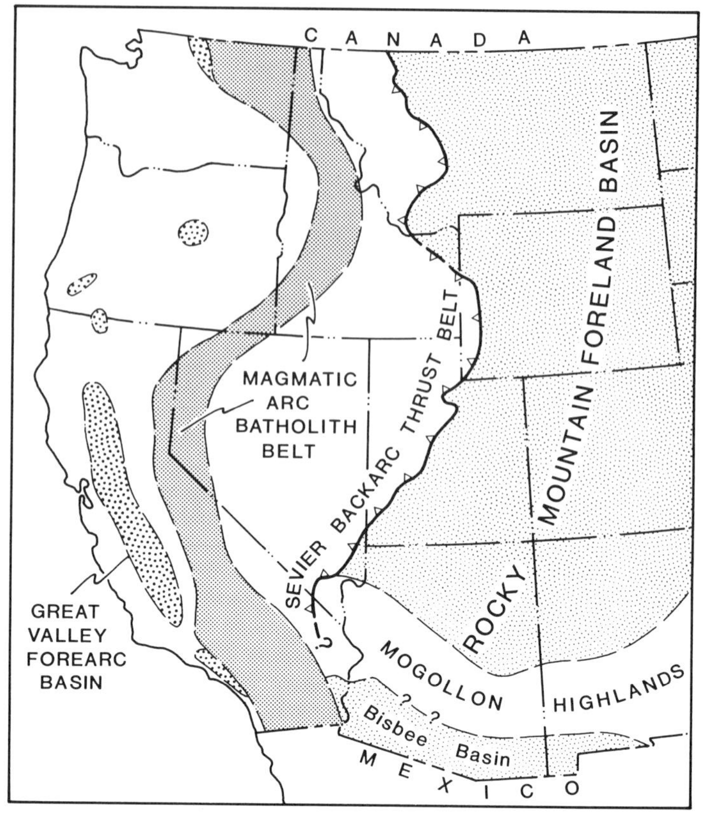

Figure 7. Mid–Upper Jurassic to Upper Cretaceous continental assemblages.

although clastic marine-shelf lithofacies (M3, Table 2) are also present in subordinate amounts (Cherven, 1983). Volcanoplutonic petrofacies of arc derivation are most characteristic of sandstone suites, although volcaniclastic and arkosic petrofacies are also prominent as end members of the volcanoplutonic array of sandstone compositions (Table 3). Locally, quartzolithic petrofacies derived from uplifted subduction complexes form parts of the forearc-basin sequences. Conglomeratic horizons attest to the proximity of bedrock sources. Some forearc successions grade laterally to interbedded volcanic and volcaniclastic rocks of adjacent magmatic arcs (Figs. 6, 7, 9), whose internal relations are not discussed here.

Transform-margin assemblages (B3, Table 1) accumulated in both onshore and offshore Cenozoic basins along the California continental margin. The most clear-cut examples of these assemblages occur within Neogene basins (Fig. 9) associated with the San Andreas system (Crowell, 1974b), but Paleogene analogues (Fig. 8) may have been related to the inferred proto–San Andreas system (Nilsen and Clarke, 1975). Rapid lateral and vertical facies changes are common within individual basins of limited extent formed by transtensional pullapart or transpressional wrench deformation. Starved-basin lithofacies (M1, Table 2) on basin floors and slopes include thick successions of organic-rich chert, porcellanite, diatomite, and siliceous shale deposited under anoxic conditions (Lagoe, 1984, 1985). Basinal strata interfinger in complex patterns with the deposits of turbidite fans, deltaic complexes, shallow shelves, and fluvial systems of various kinds including conglomeratic alluvial fans (Graham, 1976). Arkosic petrofacies (Table 3) derived from uplifted basement terranes are dominant in typical sandstone suites, but volcaniclastic rocks derived from local volcanic centers are also present.

CONTINENTAL-BLOCK ASSEMBLAGES

As designated here, continental-block assemblages are composed of laterally extensive sedimentary cover deposited across regions underlain by normal thicknesses of Precambrian continental basement, or upon younger rock masses previously incorporated into the continental block within the Cordilleran region. Such widespread accumulation of sediment masking the surface of the continental block has occurred in response to three fundamental processes: lithospheric downflexure, eustatic flooding, or nonmarine aggradation. Broad downflexure of the continental block was achieved under the sedimentary load of the

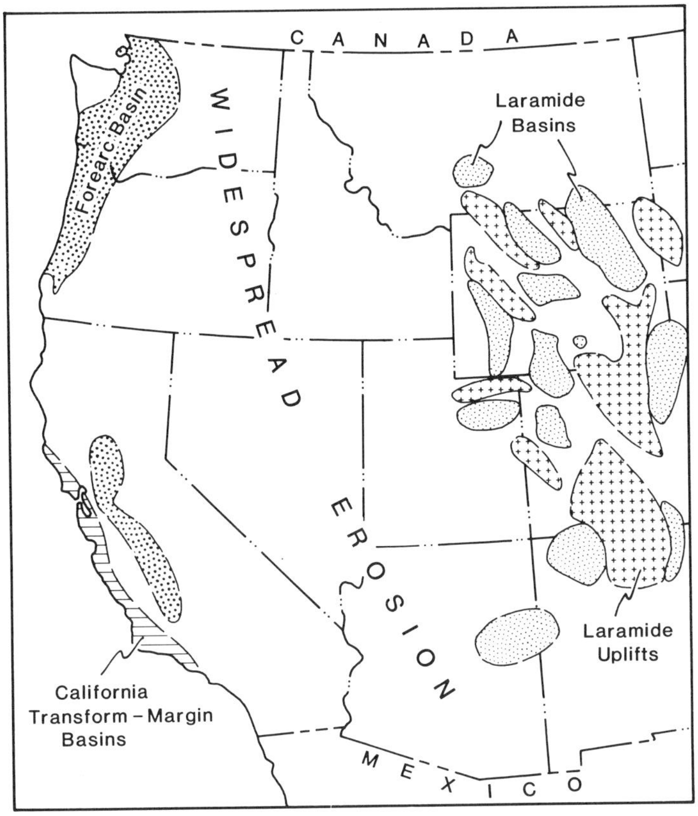

Figure 8. Uppermost Cretaceous to mid-Oligocene continental assemblages.

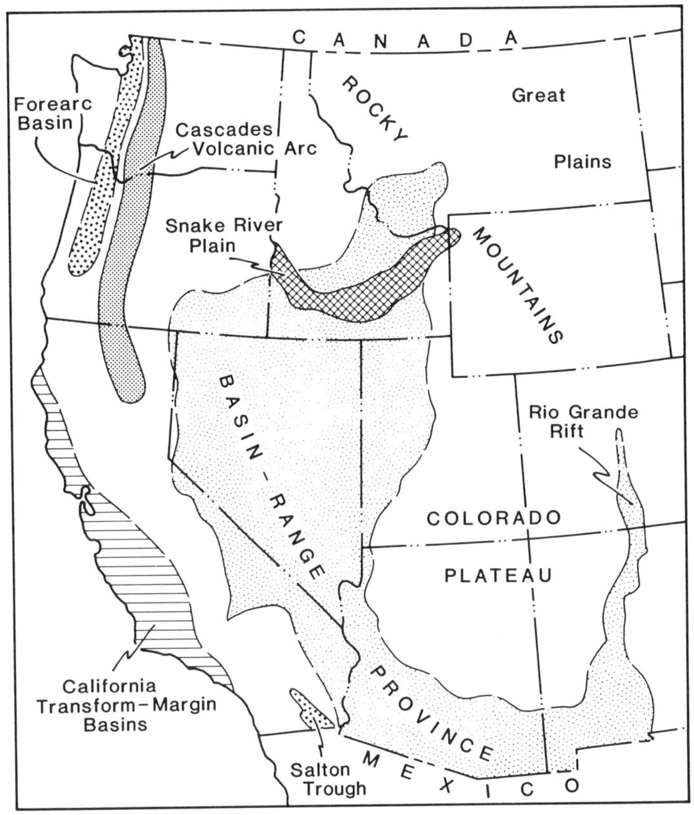

Figure 9. Mid-Oligocene to Holocene continental assemblages.

miogeoclinal prism (Watts and Ryan, 1976) during early Paleozoic time, and under the tectonic load of thrust sheets in parts of the Cordilleran foreland during late Paleozoic and late Mesozoic time (Speed and Sleep, 1982; Jordan, 1981). Prominent foreland basins developed at different sites during latest Devonian and Mississippian time in response to Antler thrusting (Fig. 4), and during latest Early Cretaceous and Late Cretaceous time in response to Sevier thrusting (Fig. 7). Conglomeratic facies are present near the thrust fronts in both foreland-basin successions.

Paleozoic platform-cover assemblages (C1, Table 1) of largely marine-shelf lithofacies (M3, Table 2) in Mississippian and older strata were deposited to varying thicknesses across inland parts of the Cordilleran region (Fig. 4), but are interrupted by multiple disconformities and paraconformities (Gutschick and Sandberg, 1983). Three factors were probably jointly or sequentially responsible for this widespread marine sedimentation: (a) downflexure of the flank of the continental block under the sediment load of the thick lower Paleozoic miogeoclinal prism farther offshore (Fig. 4), (b) the pronounced eustatic highstand of mid-Paleozoic time (Fig. 2), and (c) renewed downflexure of the inland flank of the late Paleozoic Antler foreland region (Fig. 5). The gradational boundary between the platform-cover assemblage and laterally equivalent strata of the miogeoclinal prism is taken as the structural hinge line, beyond which the miogeoclinal strata thicken progressively westward (Stewart and Poole, 1974). A prominent shelfbreak later separated the platform succession from deep water of the Antler foreland basin (Rose, 1976). Quartzose petrofacies (Table 3) in the Paleozoic platform-cover assemblage reflect derivation of clastic sediment principally from the weathered craton, and the dominantly carbonate strata in many areas reflect sedimentation in warm tropical seas (Fig. 1).

Orogen-overlap assemblages (C2, Table 1) of late Paleozoic and earliest Mesozoic age (Fig. 5) developed locally where sedimentation resumed above the eroded remnants of the Antler and Sonoma highlands (Silberling and Roberts, 1962). These strata, dominantly of marine-shelf lithofacies, rest unconformably on deformed ocean-basin assemblages of the Roberts Mountains and Golconda allochthons, and include basal nonmarine deposits that are conglomeratic locally. Quartzolithic petrofacies (Table 3) in constituent sandstones were derived mainly from deformed sequences of argillite, chert, and quartzite in the eroded orogens. Laterally equivalent platform-cover assemblages farther east (Cook and Bally, 1975) include similar marine-shelf lithofacies (M3, Table 2) on the west and north, but inland and to the south include terrestrial-cover lithofacies (N2, Table 2) of nonmarine and marginal-marine red beds (Fig. 5). The red-bed sedimentation probably records arid to semiarid coastal settings on the leeward side of the Pangaean landmass in the trade-wind belt (Fig. 1). Craton-derived quartzose petrofacies are dominant in both marine and red-bed sandstones, although widespread admixtures of feldspathic detritus reflect the influence of uplifted basement sources in the Ancestral Rockies (see below).

Mesozoic platform-cover assemblages (C1, Table 1) of mid-Triassic to mid-Cretaceous strata (Cook and Bally, 1975) were deposited as discontinuous blankets of terrestrial-cover lithofacies (N2, Table 2), in part conglomeratic, across inland parts of the Cordilleran region (Fig. 6). Sedimentation spanned the time when the Cordilleran continental margin was in rapid transit from the tropics to the temperate zone. Older intervals of the Mesozoic platform-cover assemblage consequently include the deposits of large eolian ergs that formed along the leeward flank of the continental mass while it lay within the trade-wind belt. The deposits of broad fluvial braidplains, alluviated coastal plains, and extensive tidal flats are also present in widespread red-bed sequences. Toward the north and west, the red-bed successions pass laterally into marine deposits of shelf seas (Fig. 6), and deltaic complexes occur locally above older orogen-overlap assemblages in basinal areas (Lupe and Silberling, 1985). Dominantly quartzose petrofacies in the Mesozoic platform cover reflect derivation of sediment both from the craton and by recycling of older sedimentary sequences.

Elongate foreland-basin assemblages (C3, Table 1) developed along much of the length of the Cordilleran region when thrust masses were emplaced upon the continental block from the west during the Antler and Sevier orogenies. During the Antler event (Fig. 4), the ocean-basin assemblages of a subduction complex were thrust as a complex allochthon above the pre-Carboniferous miogeoclinal prism during arc collision with the previously passive continental margin near the end of Devonian time. Deltaic-complex and turbidite-fan lithofacies (N1 and M2, Table 2) were deposited in the Mississippian peripheral foreland basin that subsided east of the Antler thrust front (Dickinson and others, 1983). The dispersal system for clastic detritus of quartzolithic petrofacies (Table 3) derived from the adjacent Antler orogen linked piedmont fan–deltas eastward to turbidite fans that accumulated in deep marine settings. Analogous emplacement of a similar allochthon, also composed of deformed ocean-basin assemblages, during the younger Sonoma event evidently did not cause sufficient subsidence in the foreland region to produce any extensive sedimentary record of the thrusting.

During the Sevier event (Fig. 7), miogeoclinal strata flanking the craton were thrust eastward toward the continental interior within an intracontinental backarc thrust system lying inland from the Mesozoic batholith belt, which was located closer to the Cordilleran continental margin (Dickinson, 1976). The resulting retroarc foreland basin that subsided to shelf depths within the Rocky Mountain region, which lay east of the Sevier thrust front, received deltaic-complex and marine-shelf lithofacies (N1 and M3, Table 2). Sediment sources within the Sevier orogenic belt contributed both quartzolithic and recycled quartzose petrofacies (Table 3) to the foreland basin (Dickinson and others, 1985). Recent work indicates that the main phases of Sevier thrusting did not begin until mid-Cretaceous time and were completed by the end of Cretaceous time (Bohannon, 1983; Lawton, 1985; Heller and others, 1986). Widespread sheetlike mid–Upper Jurassic to mid-Cretaceous platform-cover assemblages, derived from the orogenic region to the west, antedated the onset of subsidence within the Sevier foreland region, and represent nonmarine sedimentary aggradation of the surface of the continental block.

CONTINENTAL-RUPTURE ASSEMBLAGES

At intervals during the evolution of the Cordilleran region, tectonism along the continental margin was accompanied by disruption of continental basement well within the interior of the continental block. Where extension or thrusting of continental crust was sufficient to induce isostatic or flexural subsidence, local intracontinental basins were formed in which special suites of sedimentary rocks accumulated. These strata are here termed continental-rupture assemblages (D, Table 1), and are discussed in chronologic order. Their geodynamic linkages to plate interactions along the continental margin are not yet well understood, but coincidences in timing support the inferred genetic relationships outlined below.

In Pennsylvanian time, subsidence began within a number of northwest-trending intracontinental basins of the Ancestral Rockies province (Fig. 5). Adjacent basement uplifts provided local sources for coarse conglomeratic detritus of arkosic petrofacies (Table 3). Within basin interiors, the locally derived arkosic strata and more distally derived quartzose petrofacies interfinger with carbonates and evaporites in marine-shelf lithofacies (M3, Table 2). Starved-basin and turbidite-fan lithofacies (M1 and M2, Table 2) accumulated locally in the deeper water of some basins. Initiation of basin subsidence was coincident with the Ouachita orogeny along the southeastern flank of the continental block. Deformation of the continental interior to cause local subsidence in discrete basins evidently occurred in response to stresses generated by continental collision along the Ouachita-Marathon segment of the Hercynian orogenic belt (Kluth and Coney, 1981). Hercynian suturing of Laurasia to Gonwanaland was the final step in the assembly of Pangaea. To date, structural studies within the Ancestral Rockies province have not resolved the extent to which extensional, compressional, or torsional strains were involved in the intracontinental deformation that gave rise to the late Paleozoic uplifts and basins of the Rocky Mountain region.

In Jurassic time, Mesoamerican continental rifting to the southeast resulted in the opening of the oceanic Gulf of Mexico. The Chihuahua Trough in northern Mexico was an extensional arm of the Gulf of Mexico depression and extended by Late Jurassic time into the United States, where its northwestern extremity was the Bisbee Basin (Fig. 7). Subsidence within the rifted

Bisbee Basin extended from Late Jurassic to mid-Cretaceous time, as complexly intertonguing lacustrine-basin, marine-shelf, and deltaic-complex lithofacies (Table 2) were deposited within an elongate belt of crust thinned by incipient rifting (Bilodeau, 1982). Arkosic and quartzose petrofacies (Table 3), derived respectively from uplifted basement and its sedimentary cover, are dominant in sandstones of the Bisbee Group, but volcaniclastic petrofacies derived from underlying or nearby arc volcanics are also present (work in progress). Basal syntectonic horizons are conglomeratic alluvial-fan deposits. Crustal extension that produced the Bisbee Basin essentially represented a distal arm of the Atlantic spreading system, which had propagated into a backarc setting with respect to the circum-Pacific subduction system.

In latest Cretaceous and Paleogene time, Laramide deformation formed a complex pattern of basement-cored uplifts and intervening basins within the interior of the Cordilleran region (Fig. 8). Laramide deformation of the continental block was probably caused by shear between Cordilleran lithosphere and a subhorizontal subducted slab of oceanic lithosphere in the mantle beneath (Dickinson and Snyder, 1978; Bird, 1984). Partial coupling of the obliquely subducted slab to overriding lithosphere in the forearc region may have caused proto–San Andreas deformation near the continental margin (Dickinson and others, 1979). Structural loads imposed by thrust-bounded flanks of Laramide uplifts were apparently responsible for flexural subsidence within adjoining basins (Hagen and others, 1985). Lacustrine-basin or terrestrial-cover lithofacies (N3 and N2, Table 2) are characteristic of basin fills. Arkosic petrofacies (Table 3) derived from the erosion of basement uplifts are perhaps most typical of Laramide sandstones (Dickinson and others, 1985). Also present, however, are quartzose petrofacies that reflect the stripping of sedimentary cover from basement, and volcaniclastic petrofacies derived from local subduction-related volcanic fields. Sandstones of mixed petrofacies are common (Mack, 1984). Organic-rich lacustrine shales and carbonates typify the axial facies of some Laramide basins.

Beginning in mid-Oligocene time (Fig. 9), the Basin and Range province, including the Rio Grande rift, began to experience polyphase extensional deformation that has continued in one guise or another to the present (Eaton, 1982). Although the deformation evidently represents incipient continental rifting, no oceanic crust has yet been formed. Early phases of Basin-Range extension took place in intraarc and(or) backarc tectonic settings, probably as steepening of slab descent in the mantle beneath the Cordilleran region relieved Laramide compressive stresses. Later phases have been coeval with evolution of the San Andreas transform system to the west, and were probably influenced by oblique shear and unstable triple junctions along the continental margin. Early phases of extension commonly involved movement on gently dipping detachment faults and the lateral transport of rotated tiltblocks (Wernicke, 1981, 1985). Later phases commonly embody classic block faulting, with grabens and half-grabens bounded by steeply to moderately dipping normal faults. Basinal deposits are nearly all of varied lacustrine-basin lithofacies (N3, Table 1), including not only thick evaporite successions but also voluminous conglomeratic sequences deposited on alluvial fans and coarse braidplains. Markedly heterogeneous petrofacies reflect the varied nature of diverse fault-block sources.

The Salton Trough of southern California forms the northern extension of the Gulf of California depression (Fig. 9). Oblique separation of the crustal sliver of Baja California from the continental block of mainland Mexico since mid-Miocene time has formed a series of pocket basins underlain by oceanic and(or) transitional crust along the axis of the gulf. The Salton Trough is the most northerly of these basins, each of which occupies a pullapart structure within the San Andreas transform system (Lachenbruch and others, 1985). Within the Salton Trough, two mixed lithofacies associations overlie a transitional crust that is intermediate in thickness between continental and oceanic crust. Lacustrine-basin lithofacies (N3, Table 2) contain locally derived arkosic petrofacies (Table 3), whereas deltaic-complex lithofacies (N1, Table 2) contain quartzolithic petrofacies (Table 3) derived from the Colorado River drainage network (Van de Kamp, 1973).

SUMMARY

The foregoing outline of Cordilleran sedimentary assemblages is little more than a skeletal framework, but may provide an adequate basis for more searching inquiry. Few regions of the globe afford any greater variety of strata. There can be little doubt that this variety reflects a remarkable diversity of tectonic settings through Phanerozoic time. Fortuitous exposures of the diverse assemblages in readily accessible areas offer incomparable opportunities for further study.

REFERENCES CITED

Armin, R. A., and Mayer, L., 1983, Subsidence analysis of the Cordilleran miogeocline; Implications for timing of Late Proterozoic rifting and amount of extension: Geology, v. 11, p. 702–705.

Atwater, T., 1970, Implications of plate tectonics for the Cenozoic tectonic evolution of western North America: Geological Society of America Bulletin, v. 81, p. 3513–3536.

Bailey, E. H., Irwin, W. P., and Jones, D. L., 1964, Franciscan and related rocks and their significance in the geology of western California: California Division of Mines and Geology Bulletin 183, 177 p.

Beaumont, C., 1981, Foreland basins: Geophysical Journal of the Royal Astronomical Society, v. 65, p. 291–329.

Bilodeau, W. L., 1982, Tectonic models for Early Cretaceous rifting in southern Arizona: Geology, v. 10, p. 466–470.

Bird, P., 1984, Laramide crustal thickening event in the Rocky Mountain foreland and Great Plains: Tectonics, v. 3, p.741–758.

Bohannon, R. G., 1983, Mesozoic and Cenozoic tectonic development of the Muddy, North Muddy, and northern Black Mountains, Clark County, Nevada, *in* Miller, D. M., Todd, V. R., and Howard, K. A., eds., Tectonic and stratigraphic studies in the eastern Great Basin: Geological Society of America Memoir 157, p. 125–148.

Bond, G. C., Kominz, M. A., and Devlin, W. J., 1983, Thermal subsidence and eustasy in the lower Palaeozoic miogeocline of western North America: Nature, v. 306, p. 775–779.

Brueckner, H. K., and Snyder, W. S., 1985, Structure of the Havallah sequence,

Golconda allochthon, Nevada; Evidence for prolonged evolution in an accretionary prism: Geological Society of America Bulletin, v. 96, p. 1113–1130.

Burchfiel, B. C., and Davis, G. A., 1972, Structural framework and evolution of the southern part of the Cordilleran orogen, western United States: American Journal of Science, v. 272, p. 97–118.

——, 1975, Nature and controls of Cordilleran orogenesis, western United States; Extensions of an earlier hypothesis: American Journal of Science, v. 275-A, p. 363–396.

Cherven, V. B., 1983, A delta-slope-submarine fan model for Maastrichtian part of Great Valley Sequence, Sacramento and San Joaquin basins, California: American Association of Petroleum Geologists Bulletin, v. 67, p. 772–816.

Coney, P. J., Jones, D. L., and Monger, J.W.H., 1980, Cordilleran suspect terranes: Nature, v. 288, p. 329–333.

Cook, T. D., and Bally, A. W., eds., 1975, Stratigraphic atlas of North and Central America: Princeton, New Jersey, Princeton University Press, 272 p.

Crowell, J. C., 1974a, Sedimentation along the San Andreas fault, California, *in* Dott, R. H., Jr.,and Shaver, R. H., eds., Modern and ancient geosynclinal sedimentation: Society of Economic Paleontologists and Mineralogists Special Publication 19, p. 292–303.

——, 1974b, Origin of late Cenozoic basins in southern California, *in* Dickinson, W. R., ed., Tectonics and sedimentation: Society of Economic Paleontologists and Mineralogists Special Publication 22, p. 190–204.

——, 1978, Gondwanan glaciation, cyclothems, continental positioning, and climatic change: American Journal of Science, v. 278, p. 1345–1372.

Davis, R. A., Jr., 1983, Depositional systems: Englewood Cliffs, New Jersey, Prentice-Hall, 669 p.

Dickinson, W. R., 1972, Evidence for plate tectonic regimes in the rock record: American Journal of Science, v. 272, p. 551–576.

——, 1976, Sedimentary basins developed during evolution of Mesozoic-Cenozoic arc-trench system in western North America: Canadian Journal of Earth Science, v. 13, p. 1268–1287.

——, 1982, Compositions of sandstones in circum-Pacific subduction complexes and fore-arc basins: American Association of Petroleum Geologists Bulletin, v. 66, p. 121–137.

——, 1985, Interpreting provenance relations from detrital modes of sandstones, *in* Zuffa, G. G., ed., Provenance of arenites: Dordrecht, Holland, Reidel, p. 333–361.

Dickinson, W. R., and Snyder, W. S., 1978, Plate tectonics of the Laramide orogeny, *in* Matthews, V., III, ed., Laramide folding associated with basement block faulting in the western United States: Geological Society of America Memoir 151, p. 355–366.

Dickinson, W. R., Ingersoll, R. V., and Graham, S. G., 1979, Paleogene sediment dispersal and paleotectonics in northern California: Geological Society of America Bulletin, v. 90, pt. I, p. 897–898; pt. II, p. 1458–1528.

Dickinson, W. R., Harbaugh, D. H., Saller, A. H., Heller, P. L., and Snyder, W. S., 1983, Detrital modes of upper Paleozoic sandstones derived from Antler orogen in Nevada; Implications for nature of Antler orogeny: American Journal of Science, v. 283, p. 481–509.

Dickinson, W. R., Lawton, T. F., and Inman, K. F., 1985, Sandstone detrital modes, central Utah foreland region; Stratigraphic record of Cretaceous-Paleogene tectonic evolution: Journal of Sedimentary Petrology, v. 56, p. 276–293.

Eaton, G. P., 1982, The Basin and Range Province; Origin and tectonic significance: Annual Reviews of Earth and Planetary Science, v. 10, p. 409–440.

Gordon, R. G., Cox, A., and O'Hare, S., 1984, Paleomagnetic Euler poles and the apparent polar wander and absolute motion of North America since the Carboniferous: Tectonics, v. 3, p. 499–537.

Graham, S. A., 1976, Role of Salinian block in evolution of San Andreas fault system, California: American Association of Petroleum Geologists Bulletin, v. 62, p. 2214–2231.

Guschick, R. C., and Sandberg, C. A., 1983, Mississippian contineṭnal margins of the conterminous United States, *in* Stanley, D. J., and Moore, G. T., eds., The shelfbreak; Critical interface on continental margins: Society of Economic Paleontologists and Mineralogists Special Publication 33, p. 79–96.

Hagen, E. S., Schuster, M. W., and Furlong, K. P., 1985, Tectonic loading and subsidence of intermontane basins, Wyoming foreland province: Geology, v. 13, p. 529–592.

Hamilton, W., 1969, Mesozoic California and the underflow of Pacific mantle: Geological Society of America Bulletin, v. 80, p. 2409–2430.

Harper, G. D., and Wright, J. E., 1984, Middle to Late Jurassic tectonic evolution of the Klamath Mountains, California and Oregon: Tectonics, v. 3, p. 759–772.

Harrison, C.G.A., and Lindh, T., 1982, A polar wandering curve for North America during the Mesozoic and Cenozoic: Journal of Geophysical Research, v. 87, p. 1903–1920.

Heller, P. L., and Angevine, C. L., 1985, Sea-level cycles during the growth of Atlantic-type oceans: Earth and Planetary Science Letters, v. 75, p. 417–426.

Heller, P. L., and 7 others, 1986, Time of initial thrusting in the Sevier orogenic belt, Idaho-Wyoming and Utah: Geology, v. 14, p. 388–391.

Helwig, J., 1974, Eugeosynclinal basement and a collage concept of orogenic belts, *in* Dott, R. H., Jr., and Shaver, R. H., eds., Modern and ancient geosynclinal sedimentation: Society of Economic Paleontologists and Mineralogists Special Publication 19, p. 359–376.

Ingersoll, R. V., 1978, Submarine fan facies of the Upper Cretaceous Great Valley Sequence, northern and central California: Sedimentary Geology, v. 21, p. 205–230.

——, 1983, Petrofacies and provenance of late Mesozoic forearc basin, northern and central California: American Association of Petroleum Geologists Bulletin, v. 67, p. 1125–1142.

Irving, E., 1979, Paleopoles and paleolatitudes of North America and speculations about displaced terrains: Canadian Journal of Earth Science, v. 16, p. 669–694.

Jordan, T. E., 1981, Thrust loads and foreland basin evolution, Cretaceous, western United States: American Association of Petroleum Geologists Bulletin, v. 65, p. 2506–2520.

King, P. B., 1969, Tectonic map of North America: United States Geological Survey, scale 1:5,000,000.

Kluth, C. F., and Coney, P. J., 1981, Plate tectonics of the Ancestral Rocky Mountains: Geology, v. 9, p. 10–15.

Lachenbruch, A. H., Sass, J. H., and Galanis, J. P., Jr., 1985, Heat flow in southernmost California and the origin of the Salton Trough: Journal of Geophysical Research, v. 90, p. 6709–6736.

Lagoe, M. B., 1984, Paleogeography of Monterey Formation, Cuyama Basin, California: American Association of Petroleum Geologists Bulletin, v. 68, p. 610–627.

——, 1985, Depositional environments in the Monterey Formation, Cuyama Basin, California: Geological Society of America Bulletin, v. 96, p. 1296–1312.

Lawton, T. F., 1985, Style and timing of frontal structures, thrust belt, central Utah: American Association of Petroleum Geologists Bulletin, v. 69, p. 1145–1159.

Lupe, R., and Silberling, N. J., 1985, Genetic relationship between lower Mesozoic continental strata of the Colorado Plateau and marine strata of the western Great Basin; Significance for accretionary history of Cordilleran lithotectonic terranes, *in* Howell, D. G., ed., Tectonostratigraphic terranes of the circum-Pacific region: Circum-Pacific Council for Energy and Mineral Resources Earth Science Series, v. 1, p. 263–271.

Mack, G. H., 1984, Exceptions to the relationship between plate tectonics and sandstone composition: Journal of Sedimentary Petrology, v. 51, p. 1247–1258.

May, S. R., 1985, Paleomagnetism of Jurassic volcanic rocks in southeastern Arizona and North American Jurassic apparent polar wander [Ph.D. thesis]: Tucson, University of Arizona, 227 p.

Miller, E. L., Holdsworth, B. K., Whiteford, W. B., and Rodgers, D., 1984, Stratigraphy and structure of the Schoonover sequence, northeastern Nevada; Implications for Paleozoic plate-margin tectonics: Geological Society of America Bulletin, v. 95, p. 1063–1076.

Nilsen, T. H., and Clarke, S. H., Jr., 1975, Sedimentation and tectonics in the early Tertiary continental borderland of central California: U.S. Geological Survey Professional Paper 925, 64 p.

Oldow, J. S., 1984, Spatial variability in the structure of the Roberts Mountains allochthon, western Nevada: Geological Society of America Bulletin, v. 95, p. 174–185.

Rose, P. R., 1976, Mississippian shelf margins, western United States: U.S. Geological Survey Journal of Research, v. 4, p. 449–466.

Ross, C. A., and Ross, J.R.P., 1985, Late Paleozoic depositional sequences are synchronous and worldwide: Geology, v. 13, p. 194–197.

Sears, J. W., and Price, R. A., 1978, The Siberian connection; A case for Precambrian separation of the North American and Siberian cratons: Geology, v. 6, p. 267–270.

Silberling, N. J., and Roberts, R. J., 1962, Pre-Tertiary stratigraphy and structure of northwestern Nevada: Geological Society of America Special Paper 72, 58 p.

Speed, R. C., 1979, Collided Paleozoic microplate in the western United States: Journal of Geology, v. 87, p. 279–292.

Speed, R. C., and Sleep, N. H., 1982, Antler orogeny; A model: Geological Society of America Bulletin, v. 93, p. 815–828.

Steiner, M. B., 1983, Mesozoic apparent polar wander and plate motions of North America, *in* Reynolds, M. W., and Dolly, E. D., eds., Mesozoic paleogeography of west-central United States: Society of Economic Paleontologists and Mineralogists, Rocky Mountain Section, Rocky Mountain Paleogeography Symposium 2, p. 1–11.

Stewart, J. H., 1976, Late Precambrian evolution of North America; Plate tectonics implications: Geology, v. 4, p. 11–15.

Stewart, J. H., and Poole, F. G., 1974, Lower Paleozoic and uppermost Precambrian Cordilleran miogeocline, Great Basin, western United States, *in* Dickinson, W. R., ed., Tectonics and sedimentation: Society of Economic Paleontologists and Mineralogists Special Publication 22, p. 27–58.

Suchecki, R. K., 1984, Facies history of the Upper Jurassic–Lower Cretaceous Great Valley Sequence; Response to structural development of an outer-arc basin: Journal of Sedimentary Petrology, v. 54, p. 170–191.

Valentine, J. W., and Moores, E. M., 1972, Global tectonics and the fossil record: Journal of Geology, v. 80, p. 167–184.

Van Alstine, D. R., and de Boer, J., 1978, A new technique for constructing apparent polar wander paths and the revised Phanerozoic path for North America: Geology, v. 6, p. 137–139.

Van de Kamp, P. C., 1973, Holocene continental sedimentation in the Salton Basin, California; A reconnaissance: Geological Society of America Bulletin, v. 84, p. 827–848.

Van der Voo, R., 1981, Paleomagnetism of North America; A brief review, *in* McElhinny, M. W., and Valencio, D. A., eds., Paleoreconstruction of the continents: American Geophysical Union Geodynamics Series, v. 2, p. 159–176.

Wanless, H. R., and Shepard, F. P., 1936, Sea level and climatic change related to late Paleozoic cycles: Geological Society of America Bulletin, v. 47, p. 1177–1206.

Watts, A. B., and Ryan, W.B.F., 1976, Flexure of the lithosphere and continental margin basins: Tectonophysics, v. 36, p. 275–287.

Wernicke, B., 1981, Low-angle normal faults in the Basin and Range Province; Nappe tectonics in an extending orogen: Nature, v. 291, p. 645–648.

—— , 1985, Uniform-sense normal simple shear of the continental lithosphere: Canadian Journal of Earth Sciences, v. 22, p. 108–125.

Wise, D. U., 1974, Continental margins, freeboard, and the volumes of continents and oceans through time, *in* Burk, C. A., and Drake, C. L., eds., The geology of continental margins: New York, Springer-Verlag, p. 45–58.

Worsley, T. R., Nance, D., and Moody, J. B., 1984, Global tectonics and eustasy for the past 2 billion years: Marine Geology, v. 58, p. 373–400.

Zuffa, G. G., 1980, Hybrid sandstones: Journal of Sedimentary Petrology, v. 50, p. 21–29.

Manuscript Accepted by the Society June 25, 1987

ACKNOWLEDGMENTS

This paper is based on insights gained from discussions with numerous students and colleagues during the past three decades. Its preparation was supported by the Laboratory of Geotectonics in the Department of Geosciences at the University of Arizona. Sketch maps were compiled from numerous sources using the magnificent work of King (1969) as a drafting base.

NOTES ADDED IN PROOF

1. The Jurassic paleowander path for North America was clarified by May, S. R., and Butler, R. F., 1986, North American Jurassic polar wander: Implications for plate motion, paleogeography, and Cordilleran tectonics: Journal of Geophysical Research, v. 91, p. 11,519–11,544 (which reaches conclusions compatible with the relations depicted on Figure 1).

2. The duration of Gondwanan glaciations was clarified by Veevers, J. J., and Powell, C. McA., 1987, Late Paleozoic episodes in Gondwanaland reflected in transgressive-regressive depositional sequences in Euramerica: Geological Society of America Bulletin, v. 98, p. 475–487 (which reaches conclusions compatible with the relations depicted on Figure 2).

3. Although the initial positions of accreted upper Paleozoic and lower Mesozoic island arc assemblages (A2) of Figure 3 remain uncertain, recent analyses (Miller, M. M., 1987, Dispersed remnants of a northeast Pacific fringing arc: Upper Paleozoic terranes of Permian McCloud faunal affinity, western U.S.: Tectonics, v. 6, p. 807–830; Newton, C. R., 1987, Biogeographic complexity in Triassic bivalves of the Wallowa terrane, northwestern United States: Oceanic islands, not continents, provide the best analogues: Geology, v. 15, p. 1126–1129) suggest that the parent island arcs formed a fringing arc system standing near the eastern edge of the Paleo-Pacific Ocean at some presently indeterminate but modest distance from the continental margin.

4. Prior to the Laramide basement deformation depicted by Figure 8, insertion of a dense slab of subhorizontally subducted oceanic lithosphere into the mantle beneath the Rocky Mountain region apparently induced widespread subsidence, as shown by Figure 7, extending far to the east of the narrower belt affected directly by supracrustal thrust loading along the Sevier orogen (Cross, T. A., and Pilger, R. H., Jr., 1978, Tectonic controls of late Cretaceous sedimentation, western interior, USA: Nature, v. 274, p. 653–657).

Printed in U.S.A.

The Geology of North America
Vol. G-3, The Cordilleran Orogen: Conterminous U.S.
The Geological Society of America, 1992

Chapter 12

Cenozoic extensional tectonics of the U.S. Cordillera

Brian Wernicke
Department of Earth and Planetary Sciences, Harvard University, Cambridge, Massachusetts 02138

INTRODUCTION

The U.S. Cordillera is typical of orogenic belts in its preservation of multiple episodes of extensional, strike-slip, and compressional deformation. Widespread, latest Proterozoic extension established an early Paleozoic passive margin (*Stewart, 1972*; *Bond and Kominz, 1984*). Other events, probably of lesser overall magnitude and extent, include mid-Proterozoic rifting in the Pacific Northwest, resulting in the accumulation of the Belt Supergroup, late Paleozoic rifting along the continental margin arc (*Miller and others, 1984*) and perhaps also within the craton and miogeocline (*Kluth, 1986*), and Mesozoic extensional tectonism in the "hinterland" of the Mesozoic fold and thrust belt (*Allmendinger and Jordan, 1984*).

The most clearly expressed extensional event occurred during Cenozoic time, creating one of the Earth's few well-exposed regions of large-magnitude intracontinental extension. While the crust is thinned as often as it is thickened, major extension and thinning ultimately result in subsidence and burial. Most of the planet's extensional tectonic record lies beneath thick syn- and post-rift sediments, or has been overprinted by compressional events. The Cenozoic extended region in the Cordillera remains above sea level, despite large extensional strains recorded in upper crustal rocks. This situation results from thin lithosphere in the region and the fact that most of the extended terrain is still underlain by continental crust 25 to 35 km thick (*Thompson and Zoback, 1979*; *Braile and others, 1989*), which may have been twice as thick prior to extension. The well-exposed, unmodified, and widely developed Cenozoic extensional tectonism in the Cordillera allows study of extensional processes that are difficult to observe in outcrop elsewhere, such as on passive continental margins.

During the 1980s, there was a major rethinking of Cordilleran-extension concepts. In the mid-1970s, after nearly a century of doctrine attributing Basin and Range physiography to block faulting (*Gilbert, 1874*; *Davis, 1903*), Cordilleran extension was viewed as being modest (10 to 15 percent widening), accommodated in the upper crust mainly on widely spaced, steeply dipping fault planes (Stewart, 1971; Thompson and Burke, 1974). Based on concepts independent of this doctrine, large-magnitude Cenozoic extension (100 percent widening), mainly predating the development of basin-range fault-block structure, had been proposed as early as the mid-1960s (Hamilton and Myers, 1966). The 1980s witnessed a shift of consensus toward this latter view. With this shift came renewed skepticism that all Basin and Range physiography is controlled by block faulting (as it is described in Stewart, 1971, for example), reminiscent of dissenting views expressed around the turn of the century (e.g., *Spurr, 1901*). The principal catalyst of the new consensus was the discovery of large, shallowly dipping normal faults (detachments) that often juxtapose upper crustal rocks on metamorphic tectonites from the middle crust (a structural association known as a Cordilleran metamorphic core complex; Crittenden and others, 1980). The impact of this consensus has been felt worldwide, prompting reappraisal of long-studied field relations not only in the Cordillera, but in orogens such as the Alps and Caledonides (e.g., *Selverstone, 1988*; *Serrane and Siguret, 1987*).

This chapter synthesizes the results of this upheaval for the Cordillera. The discussion first covers observations of extensional tectonism in the upper crust, constrained primarily by geologic data, and summarizes geophysical observations relevant to understanding the deeper structure of the lithosphere. These observations are synthesized to present an interpretive cross-sectional model of extension of the Cordilleran lithosphere in the central part of the Basin and Range province, which emphasizes fluid behavior of the middle part of the crust during the early stages of extension. Finally, the distribution of extensional tectonism in space and time is reviewed to provide a basis for discussion of the causes of extension. For readers with limited time, the figures and summary section constitute a self-contained overview of the chapter.

UPPER CRUSTAL OBSERVATIONS

Pre-Cenozoic framework and extensional provinces

The pre-Cenozoic Cordillera is divisible into three main paleogeographic elements (Fig. 1): cratonic North America, the Paleozoic Cordilleran miogeocline, and rocks accreted to North

Wernicke, Brian, 1992, Cenozoic extensional tectonics of the U.S. Cordillera, *in* Burchfiel, B. C., Lipman, P. W., and Zoback, M. L., eds., The Cordilleran Orogen: Conterminous U.S.: Boulder, Colorado, Geological Society of America, The Geology of North America, v. G-3.

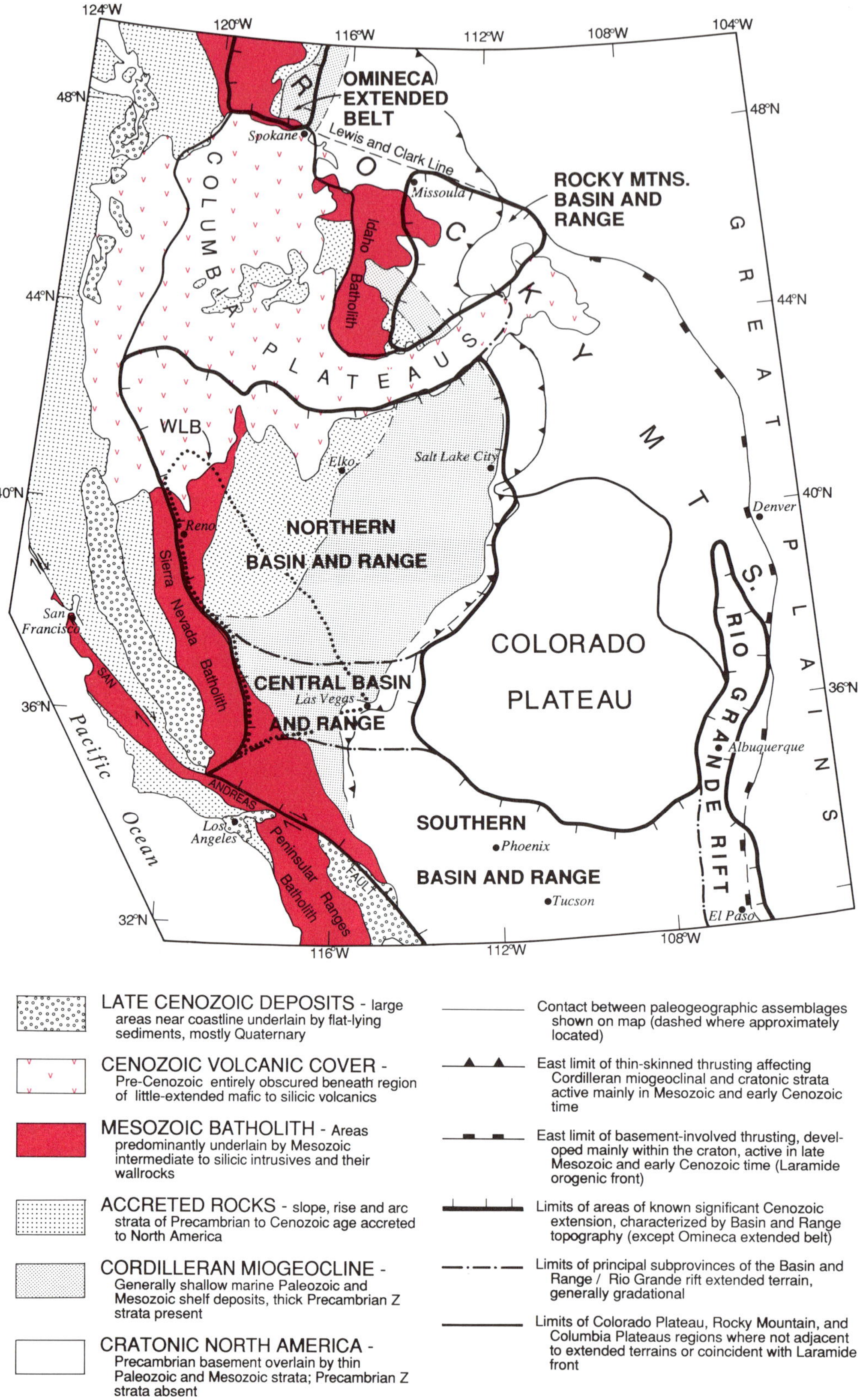

Figure 1. Map showing distribution of extensional provinces relative to major pre-extensional tectonic elements. See text for discussion.

America during various Paleozoic and Mesozoic orogenies. Following most of the accretion, during Cretaceous and early Tertiary time, large areas of cratonic North America and previously undeformed portions of the miogeocline experienced crustal shortening (*Allmendinger, this volume*), broadly synchronous with the emplacement of granitic batholiths along the western side of the orogen (Fig. 1). Where developed within the miogeocline, this shortening was accommodated by east-vergent thin-skinned thrusting of the miogeoclinal wedge and adjacent cratonic cover, forming the Cordilleran thrust and fold belt. In cratonic areas, shortening was accommodated by basement-involved thrusting of variable trend and vergence in latest Cretaceous and early Tertiary time (Laramide orogeny; Fig. 1).

Upper crustal extensional tectonism in the Cordillera is best developed within areas of pre-Cenozoic tectonism. While Cenozoic normal faults occur throughout the Cordillera, four subregions or extensional provinces have been identified that have experienced substantial extensional strain, generally in excess of 10 to 15 percent elongation over original width. These include the Basin and Range province, Rio Grande rift, Rocky Mountains Basin and Range, and Omineca extended belt (Fig. 1). The latter two names are introduced here, although extension in these areas has long been recognized. Within the Basin and Range province, which has traditionally been subdivided into a northern and southern part (the northern being mostly coincident with the Great Basin, a large region of interior drainage), a threefold division into a northern, central, and southern Basin and Range now seems appropriate based on contrasting histories and structural styles, elaborated below (Fig. 1). Another commonly distinguished subprovince in the Basin and Range includes a broad zone in the western parts of the central and northern Basin and Range known as the Walker Lane belt (e.g., Stewart, 1988; Fig. 1, WLB), wherein the trends of basin ranges are highly irregular, in contrast to the more-uniform northerly trends of ranges to the east.

Regions outside of these provinces include: (1) those that have not experienced major extension; and (2) those in which major extension might have occurred, but is either obscured by younger deposits or not yet recognized owing to structural complexity. The Colorado Plateau and most of the Rocky Mountains have been locally mildly extended, but are mainly intact. The Columbia Plateaus comprise a large region of the Pacific Northwest underlain by little-extended Neogene volcanic strata, beneath which substantial extension may have taken place (Hamilton and Myers, 1966) but is not yet documented (Fig. 1). Areas west of the Mesozoic batholiths (Cascades, Coast Ranges, and Continental Borderland off southern California) are all areas where significant extensional tectonism has been proposed (e.g., *Platt, 1986*; *Jayko and others, 1987*), but there is no consensus as to the age, distribution, or magnitude of extension in these areas.

A factor in the pre-Cenozoic framework of noncratonic regions that appears to predispose the upper crust to remain stable during Cenozoic time is the presence of intense Cretaceous plutonism, as in the Peninsular Ranges, Sierra Nevada, and Idaho batholiths (Fig. 1). The western boundaries of the extensional provinces lie within or near the eastern margins of the batholiths. Thus, the Mesozoic magmatism has apparently strengthened the upper crust (but not necessarily the whole lithosphere) in areas where it was most intense, creating a ribbon of firm ground that was later pulled away from cratonic North America, leaving the extensional provinces in their wake (Hamilton and Myers, 1966). The eastern boundaries of the extensional provinces follow specific pre-extensional boundaries locally, but do not have a simple overall pattern. Thus, although upper crustal extension generally does not affect cratonic areas that were not shortened in Cretaceous or early Tertiary time, the eastern limit of extension may lie anywhere from internal portions of the miogeoclinal thrust belt (Omineca extended belt) to the outermost areas of Laramide deformation bordering the Great Plains (Rio Grande rift).

Domainal character and asymmetry of large-scale extensional tectonism

Within the extensional provinces, the magnitude of extension and upper crustal thinning is strongly heterogeneous, partitioned between areas of low and high strain (e.g., *Davis and Burchfiel, 1973*; Proffett, 1977; Guth, 1981; Wernicke and others, *1982,* 1988; Chamberlin, 1983; Miller and others, 1983), especially in the early stages of extension. Plate 8 subdivides the extensional provinces of the Cordillera into two main tectonic elements: strongly extended domains and stable blocks (or simply extended domains and blocks). Cordilleran metamorphic core complexes, described in more detail in the next section, lie within the extended domains, but not all of the extended domains contain core complexes.

The partitioning of strain within the upper crust is documented in many areas, although portions of Plate 8 are interpretive. The distribution of extended domains and stable blocks in some areas is highly uncertain, particularly in the western part of the Basin and Range. Strongly extended domains in the southern Basin and Range are best understood in the northeast, including the Pinaleño, Catalina, South Mountains, Whipple, and Mojave extended domains (Pl. 8), active mainly in Oligocene and early Miocene time. Areas to the southwest are also strongly extended, but the details of their distribution, particularly near the Mexican border and southward, are still emerging. The distribution of extended domains in the central Basin and Range is well understood across the entire province, including the Sheep Range and Death Valley domains to the west and the Lake Mead domain to the east (Pl. 8), active principally in mid-Miocene and later time. As in the southern Basin and Range, the distribution of extended domains is better understood in the eastern part of the northern Basin and Range, especially in a transect across the Sevier Desert, Snake Range, and Ruby Mountains domains (Pl. 8). Activity in these areas ranges in age from Oligocene to Recent. Farther north in the Rocky Mountains Basin and Range, the Pioneer Mountains and Bitterroot extended domains form a narrower belt of strong extension than in the northern Basin and Range (Pl. 8), active principally in Eocene and Oligocene time. A wider belt of strong Eocene extension exists farther north in the Omineca extended belt, including the Okanagan and Newport extended domains (Pl. 8).

Significant extension is by no means restricted to the strongly extended domains. The magnitude of extension within the stable blocks varies, but is generally less than a few tens of percent increase over original width. Within the extended domains, elongations are at least an order of magnitude greater, and in many cases are so large that the entire upper crust, except for a few volumetrically minor tectonic fragments, has been entirely removed from the domain. Thus, many of the extended domains shown on Plate 8 are nearly as wide as the magnitude of extension within them. Plate 8 suggests a scale of partitioning such that stable blocks and extended domains alternate with a wavelength of 100 to 200 km, or a distance as wide as 3 to 5 typical basin-range pairs.

A second major property of the extensional provinces is that, within a given extended domain, the normal faults tend to dip in the same direction. The asymmetry suggests a distinction of their across-strike boundaries according to whether the normal faults dip away from or toward the adjacent stable block (Pl. 8). Boundaries where the faults dip away from the stable block will be referred to here as proximal, and those where they dip toward them distal (Wernicke, 1985), in reference to the position of the adjacent stable block with respect to the structurally lowest normal fault in the system. Proximal boundaries, where the structurally lowest fault breaches the Earth's surface, are also referred to as breakaway zones (e.g., Howard and John, 1987). The patterns of asymmetric faulting correspond approximately to the tilt domains outlined by Stewart (1979) on the basis of Tertiary-strata dip direction in the Basin and Range. The tilt domains include both extended domains and stable blocks shown on Plate 8. Stewart (1979) identified across-strike tilt boundaries as synformal (strata dip toward the boundary) and antiformal (strata dip away from the boundary), which generally lie within stable blocks whose boundaries are both either proximal or distal, respectively.

In map view, the strongly extended domains typically are elongate perpendicular to their extension direction, and tend not to continue with constant net extension for large distances along strike, often terminating quite abruptly (Pl. 8). Domains with opposite asymmetry of normal faulting may closely interact with one another, both along and across strike of normal faults within the domains. Three principal end-member classes of termination and interaction of extended domains along strike appear to be most common (Fig. 2). One is a transform- or transfer-type boundary (*Burchfiel and Stewart, 1966*; *Davis and Burchfiel, 1973*), in which the extension direction is parallel to strike-slip faulting. Another is a mixture of transform and pure wrench faulting, in which crudely hexagonal blocks separate along strike-slip faults (Hill, 1982), such that the overall horizontal extension direction does not parallel the strike-slip faults. With this mechanism, the strain field for the crust as a whole is constrictional, with local, upper crustal shortening possible with an axis perpendicular to the extension direction. The third principal class of interaction, which may or may not exhibit strike-slip faulting, is referred to as an accommodation zone (Fig. 2), consisting of two main types: high relief and low relief, depending on the opposed-asymmetry geometry of interacting domains (Fig. 2; terminology after Rosendahl, 1987).

It is possible that these mechanisms may act in concert near the ends of domains, although there appear to be good examples

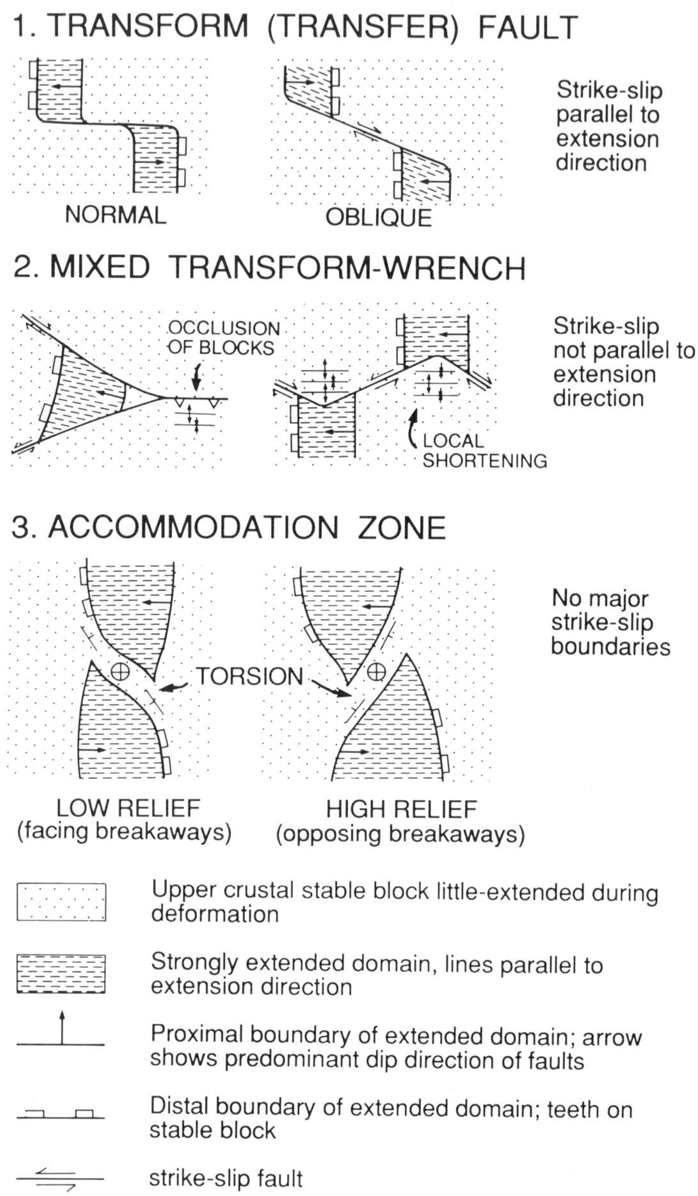

Figure 2. Models of along-strike termination of strongly extended domains and interactions with stable blocks. Examples of oblique transfer and mixed transform-wrench faulting appear to be common in the central Basin and Range and Walker Lane belt, while more-diffuse accommodation-zone terminations may typify the remainder of the extended Cordillera (Pl. 8).

of all of the types shown in Figure 2 at various places in the Cordillera, with the exception of the normal transfer boundary (Fig. 2; Pl. 8). The best examples of transform and mixed transform-wrench boundaries, at a variety of scales, are found in the central Basin and Range and a portion of the northern Basin and Range that includes the Walker Lane belt (e.g., *Anderson, 1973*; Stewart, 1988). Strike-slip faulting in association with extension is pervasive there, and is one of the criteria that distinguishes the central Basin and Range and much of the Walker Lane belt from the other subprovinces designated on Figure 1 (e.g., *Wright, 1976*). The strike-slip faulting appears to accommodate regional constriction of the upper crust in the central Basin and Range (e.g., Wernicke and others, 1988), although examples of transfer-type faults are found in that area as well (e.g., Burchfiel and others, 1987). Outside of the central Basin and Range/Walker Lane belt subprovinces, the occurrence of discrete strike-slip faults along the boundaries of strongly extended domains is limited, with most of the interactions dominated by accommodation-zone-style tectonism (Fig. 2; e.g., *Faulds and others, 1990*). One exception is the Lewis and Clark line, interpreted to be in part an oblique transfer structure (e.g., *Rehrig and Reynolds, 1981*; Pl. 8). A number of transfer structures have been described from mildly extended areas, such as in the mildly extended volcanic cover in the northwestern corner of the Basin and Range (*Lawrence, 1976*).

A potentially important deformation style within the extended Cordillera involves rotation of crustal blocks about vertical axes. Such rotations are suggested by the accommodation zone geometry shown in Figure 2, where the ribbon of crust separating domains is not only bent by torsion about a horizontal axis, but is also rotated about a vertical axis. An end-member case would be a domain wherein strain was accommodated by domino-style rotation of blocks and bounding strike-slip faults about vertical axes with no vertical strain (*Freund, 1974*). This mechanism and derivatives have been suggested to explain mapped fault patterns and paleomagnetic declination anomalies in the southwestern U.S. Cordillera, especially in the western Transverse Ranges area near Los Angeles (*Luyendyk and Hornafius, 1987*). Deformation of this type has been proposed for the central Basin and Range and Walker Lane belts on a more local scale (*Ron and others, 1986*; *Nelson and Jones, 1987*; Stewart, 1988), however the relative importance of this mechanism in the domains outlined on Plate 8 is highly uncertain.

Cross-sectional kinematics of extended upper crust

Generalities. The classical model of extensional tectonics, derived from observations of faults bounding a number of the present-day ranges in the Basin and Range province (e.g., *Gilbert, 1874*; *Davis, 1903*), holds that the upper crust extends via brittle fracturing on moderately to steeply dipping (typically 60°, mostly >40°), widely spaced, major, normal faults, whose net displacements are a small fraction of spacing between range blocks. This model implies crustal extension to be about 10 to 15 percent increase over original width (*Thompson, 1960*; Hamilton and Myers, 1966; Stewart, 1971; Thompson and Burke, 1974). Increasing documentation of shallowly inclined Cenozoic normal faults during the 1970s and 1980s (many of which had already been described and attributed to either compression or mass-wasting phenomena during the first 70 years of the century) has required significant revision of this model as a complete description of Cenozoic extensional tectonism in the Cordillera (e.g., Hamilton and Myers, 1966; Anderson and others, 1983; *Coney, 1987*).

The field observation most significant to the new conceptual framework is the widespread occurrence of large subhorizontal faults of Cenozoic age across which stratigraphic or structural succession is omitted, and which serve as basal dislocations for mosaics of imbricate normal faults, first documented by *Ransome and others (1910)* in the central Basin and Range. A number of studies in the early 1970s (presaged by *Gilbert, 1928*) identified shallowly inclined normal faults of limited extent (less than 10 km of exposed width parallel to slip direction for a single fault) as an important mode of accommodation of large extensional strain in the crust, at least in some areas (e.g., Anderson, 1971; Armstrong, 1972; Wright and Troxel, 1973; Proffett, 1977). By 1980, consensus held that in large areas of the Basin and Range and Pacific Northwest extension was accommodated by a class of faults that are typically exposed as single subhorizontal fault zones across distances of a few tens to as much as 50 to 60 km parallel to their transport directions. These faults emplace strongly extended upper crustal fault blocks similar to those mapped by Anderson (1971) above deeper crustal rocks.

These faults, known as detachments or detachment faults (e.g., Davis and others, *1980*, 1986), reveal in outcrop how upper crustal fault blocks interact with deeper structural levels (Fig. 3). The fundamental observation is that the upper crustal faults either merge with, or arc truncated by, the detachments. Figure 3 illustrates a typical example of these faults, and some of the complexities typically encountered where steep faults are truncated by shallow ones.

Footwall metamorphic tectonite. The footwalls of detachments are often composed of metamorphic tectonites that yield Cenozoic cooling ages for thermochronometers with closure temperatures in the range of 100° to 500 °C (e.g., *Armstrong and Hansen, 1966*; *Keith and others, 1981*; *Dokka and others, 1986*; *Miller and others, 1988*), and contain metamorphic assemblages that suggest depths of Mesozoic and Cenozoic metamorphism in excess of 10 km, up to as much as 30 km (e.g., *Labotka, 1980*; *Davis and others, 1980*; *Hamilton, 1982*; *Wernicke, 1982*; Bartley and Wernicke, 1984; *Labotka and Albee, 1988*; *Anderson, 1988*; *Hodges, 1988*; *Hodges and Walker, 1990*). The idea that the tectonites may to some extent be genetically related to extension was first suggested by *Armstrong (1963)* in his discussion of the Raft River, Ruby Mountains, and Snake Range areas of

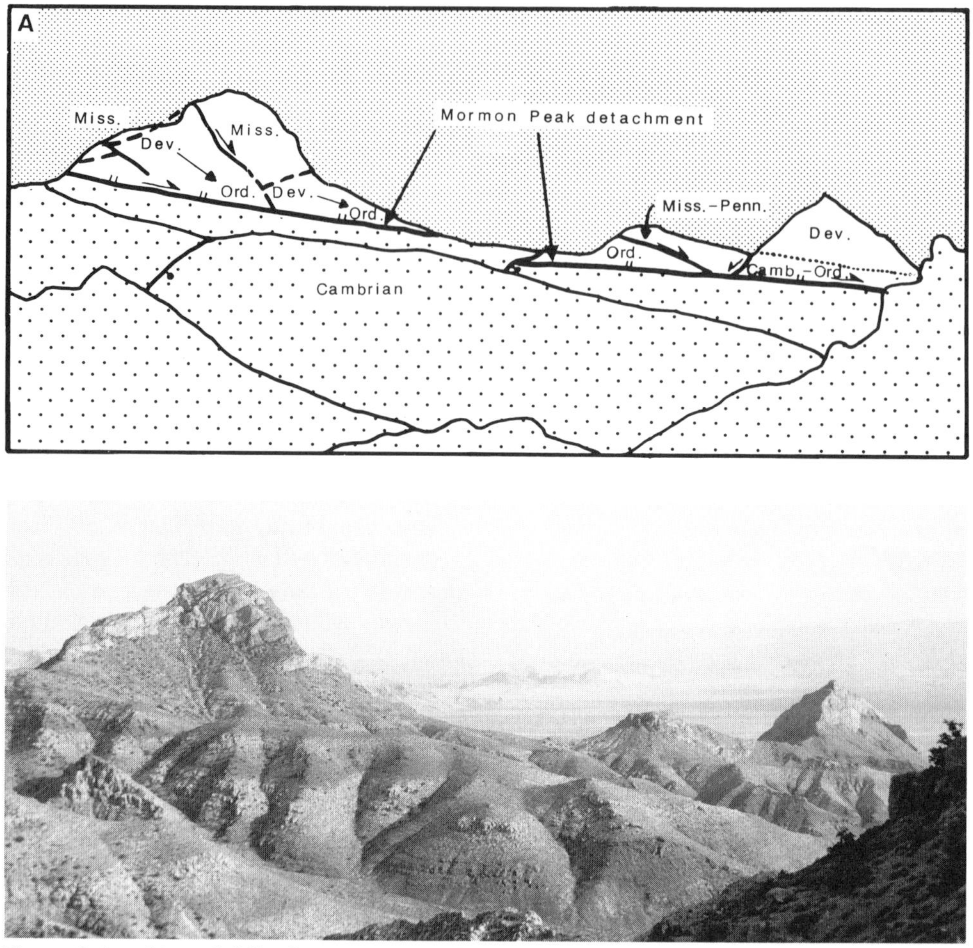

Figure 3 (on this and following two pages). Examples of detachment systems, central Basin and Range. (A) View south of Mormon Peak detachment (see Fig. 4a, lowest frame). Tilted hanging wall blocks are cut by normal faults that do not cut the detachment. Detachment cuts gently (about 10°) downsection to the west through unmetamorphosed footwall strata. Vertical relief is about 800 m. (B) View north of Mormon Peak detachment (Fig. 4a, lowest frame), showing timing relations among associated faults. Fault 1 places highest Devonian (D3) and lowest Mississippian (M1) strata against lower Devonian units (D1, D2) and is truncated by fault 3, which merges with the detachment at very low angle. The truncation of high-angle faults by low-angle faults creates a situation where hanging wall imbricate faults sharply abut the detachment, making them difficult to restore palinspastically (e.g., Davis and Lister, 1988). Vertical relief is about 100 m. (C) View north of Tucki Wash fault (arrows show trace), Death Valley region. Hanging wall has moved to the left, relative to the footwall. The fault cuts steeply downsection through hanging wall Cambrian strata on the right, curving into parallelism toward the left. Eocambrian strata in the footwall are subparallel to the fault along both steep and shallow portions, apparently monoclinally warped as they were dragged out from beneath the hanging wall (analogous to point C in second frame from the top of Figure 5b). Near upper arrow, a younger subhorizontal fault cuts the Tucki Wash fault, which may have been flexurally rotated once in the footwall of younger, structurally higher faults (e.g., as in Fig. 5b). Curvature of the fault plane lay at about 6 km below surface while fault was active.

east-central Nevada. He referred to these ranges as "arch ranges," stating (p. 133):

> The three best examples of arch ranges are composed of rocks which were probably deep below the surface before normal faulting. Because of this, the rocks of which they are composed would have been more plastic than the shallow superstructure. As a consequence, these ranges could have formed during uplift, perhaps developing beneath horsts formed at shallower tectonic levels.

The tectonites are usually lineated and foliated mylonites whose foliation is broadly parallel to overlying detachments, and whose stretching directions are observed to be roughly parallel to both upper plate extension direction and the slip direction on the detachment (Davis and Coney, 1979; G. H. Davis, 1980; *Rehrig and Reynolds, 1980*). Further, S-C fabrics (*Simpson and Schmid, 1983*) in the tectonites in some instances suggest that they devel-

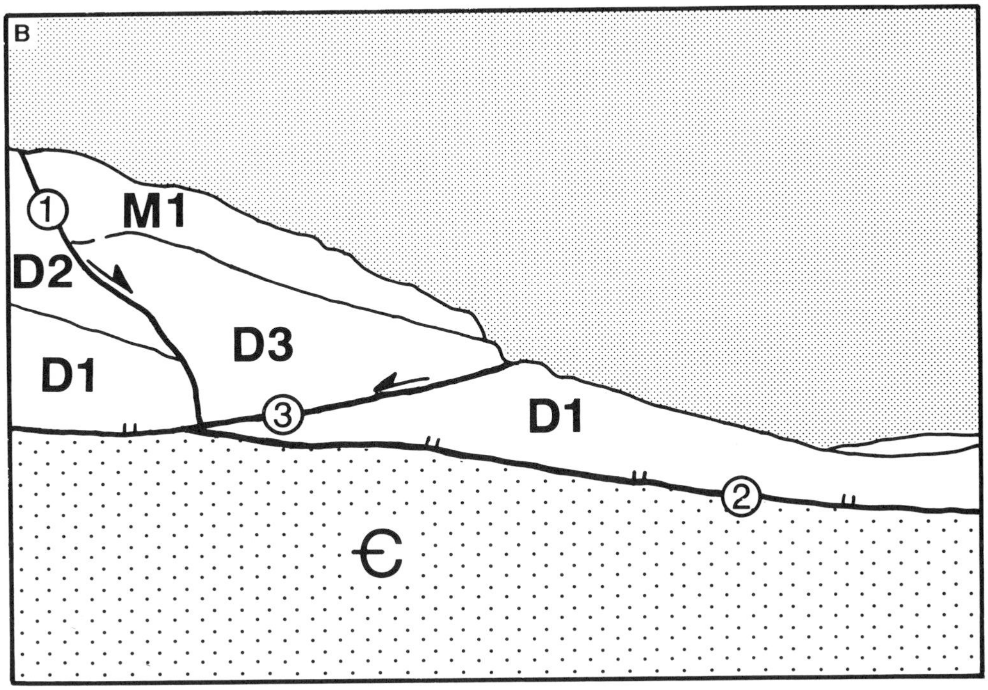

oped in part via simple shear in the same sense as displacement on the overlying detachments (e.g., *Snoke and Lush, 1984*; Davis and others, 1986; *Malavieille, 1987*; *Gaudemer and Tapponnier, 1987*; Lee and others, 1987; Parrish and others, 1988).

Areas containing the threefold association of detachment, hanging wall imbricate normal fault blocks, and footwall metamorphic tectonite are referred to as Cordilleran metamorphic core complexes, or simply core complexes (Davis and Coney, 1979; articles in Crittenden and others, 1980; Armstrong, 1982). Their origin is widely believed to result from tectonic denudation of deep structural levels along normal shear zones that initially penetrated to depths of 10 to 15 km and have net displacements measured in tens of kilometers (e.g., *Howard and others, 1982*; Spencer, *1982,* 1984, *1985*; *Davis, 1983*; Allmendinger and others, 1983; Bartley and Wernicke, 1984; Reynolds, 1985; Reynolds and Spencer, 1985; *Wernicke and others, 1985*; Dokka, *1986,* 1989; Davis and others, 1986; *John, 1987*; *Hodges and others, 1987*; Davis and Lister, 1988; Snoke and Miller, 1988;

C

Hamilton, 1988; Parrish and others, 1988; *Glazner and others, 1989*). In this manner, some of the tectonites may form along the deep-seated portions of the detachment fault zones early in their history, and be progressively overprinted by brittle deformation and finally the detachment fault itself during ascent to the surface (e.g., Davis and others, 1986).

A major complication in interpreting the tectonites as products of extension is that because they are uplifted from mid-crustal levels, they often contain older metamorphic fabrics developed during Mesozoic or even Precambrian time that are difficult to date because they were at high temperatures in the Cenozoic (e.g., *Armstrong and Hills, 1967*). Thus, while the age and tectonic significance of many of these tectonites have been controversial, geochronological and structural analyses in many of the core complexes have resulted in substantial progress in testing kinematic models for their origin (e.g., *Reynolds and Rehrig, 1980*; Reynolds, 1985; *Dokka and others, 1986*; *Wright and others, 1986*; *Wright and Snoke, 1986*; Lee and others, 1987; *DeWitt and others, 1986*; *Miller and others, 1988*; Parrish and others, 1988; Walker and others, 1990; *Asmerom and others, 1990*). The 1990s promise to be a decade of major progress in understanding the relationship between upper crustal and deep crustal levels as more data from the core complexes is gathered. In particular, understanding the role of pre-mylonitic, relatively high-temperature fabrics found in some core complexes that are of Tertiary age may add significant new insight into deep crustal processes active during extension (*Wright and Snoke, 1986*; *Wright and others, 1986*; Lee and others, 1987), as discussed below.

The strongly extended domains on Plate 8 are characterized by core complexes and detachment faults, but some detachments do not expose metamorphic tectonite in their footwalls (e.g., Figs. 3a and b). In addition, some highly extended domains do not contain detachments at current levels of exposure, because of insufficient erosion of hangingwall imbricate fault mosaics, inefficiency of tectonic denudation, or both. Thus the regional distribution of core complex tectonites (e.g., Coney and Harms, 1984; Wust, 1986) does not completely reveal the distribution of strongly extended crust.

Magnitude of extension. Sequentially restored regional cross sections across two extended domains, the Mormon Mountains and Snake Range domains (Fig. 4; Pl. 8), illustrate the salient characteristics of upper crustal fault kinematics encountered in the extended domains. The amount of extension in these examples is a large fraction of the exposed width of the domains (>80 percent in both cases), which is typical of domains for which reconstructions have been attempted. Despite roughly equivalent amounts of extension, the Snake Range domain contains a core complex (the Snake Range) while the Mormon Mountains area does not, reflecting the greater depths brought to the surface by the normal faults in the Snake Range domain. Only a small volume of the upper 7 to 10 km of the crust is preserved within the domains relative to their widths, with overall extension of the upper crust in excess of 300 to 400 percent increase over original width. In contrast, the stable domains are extended a few tens of percent or less, generally along steep normal faults. In other words, the two stable blocks on either side of the domain are separated along a system of faults whose pre-extension trajectories through the upper crust occupy a volume that is small relative to that of the stable blocks.

Total extension across the Basin and Range province is about 250 km based on reconstruction of the extended domains in the central Basin and Range (Wernicke and others, 1988). Assuming the large batholithic blocks on western margins of the

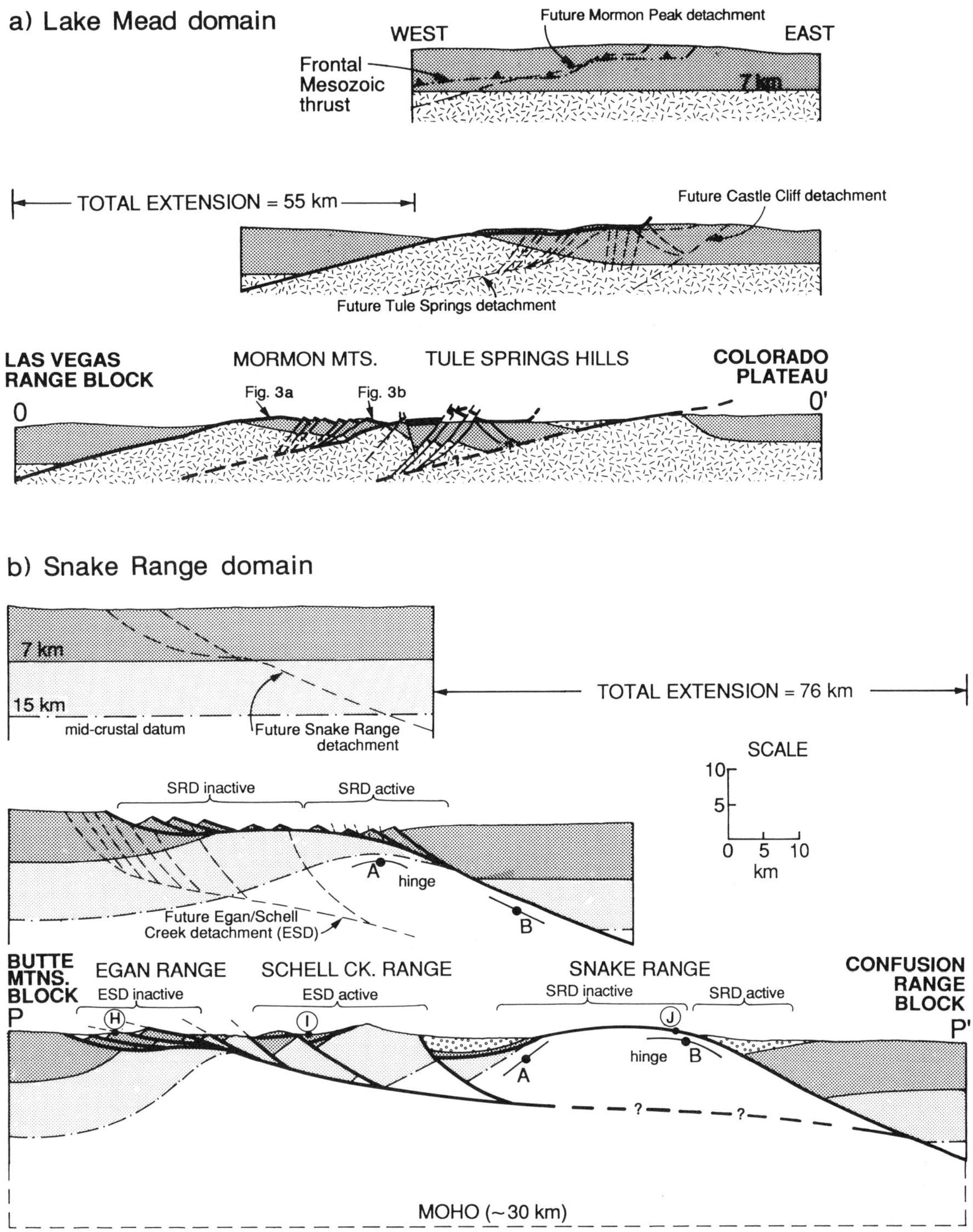

Figure 4. Stepwise-restored cross sections through two highly extended domains. See Pl. 8 for locations of sections. Scale is the same for both sections. (a) Mormon Mountains domain, showing position of frontal thrust fault of the foreland fold and thrust belt, which in the Mormon Mountains area is a decollement in Middle Cambrian strata. Pre-extension datum at 7 km corresponds approximately to the nonconformity between Proterozoic crystalline basement and Phanerozoic sediments (after Axen and others, 1990). (b) Snake Range domain. Pre-extension depths of 7 and 15 km are not intended to correspond to any particular stratigraphic horizon, although most of the faulting is developed within regionally conformable miogeoclinal sediments (after Bartley and Wernicke, 1984, with some modifications based on Hauser and others, 1987). Note migrating hinge in footwall of Snake Range decollement. Localities H, I, and J correspond to those on Figure 6. See text for discussion.

extensional provinces have not appreciably rotated with respect to the interior of North America during extension, the magnitude of extension in the northern and southern Basin and Range may be comparable (e.g., *Bogen and Schweickert, 1985*; *Frei, 1986*; Levy and Christie-Blick, 1989). This gives elongations of a factor of 3.5 for the central Basin and Range and 2.0 for the northern Basin and Range.

Initial dip of detachments. Detachment faults are typically regionally subhorizontal, but their initial trajectory through the crust has been a subject of debate. On one hand, studies of earthquake focal mechanisms from rapidly extending areas, including the Basin and Range, indicate that seismic slip on low-angle (<30°) normal faults seldom if ever occurs (Jackson, 1987). On the other hand, geologic examples of some detachments (e.g., Allmendinger and others, 1983; Bartley and Wernicke, 1984; Reynolds, 1985; *Spencer, 1985*; *Wernicke and others, 1985*; Burchfiel and others, 1987; *John, 1987*; Howard and John, 1987; Davis and Lister, 1988; Dokka, 1989; Axen and others, 1990; Foster and others, 1990) suggest that they initiated at dips of 30° or less, and were active at even shallower dips in the upper crust throughout their history of movement.

The footwalls of many normal faults contain indicators showing increasing structural depth in the direction of hanging wall transport, such as downcutting through stratified rocks, increasing metamorphic grade, and decreasing cooling ages. For example, the Mormon Peak detachment initiated with an average dip of 20 to 25° through the upper 7 to 8 km of crust (*Wernicke and others, 1985*; Axen and others, 1990), cutting gently (15 to 20°) downsection to the west across the basal decollement of the Mesozoic fold and thrust belt and into autochthonous Precambrian basement (O–O′, Pl. 8; Figs. 3a, b, and 4a). Though controversial (cf. Miller and others, 1983; Bartley and Wernicke, 1984; *Gans and others, 1988*), the Snake Range detachment probably initiated with an average dip of 30 to 40° through the upper 10 km of the crust (Fig. 4b), based on the author's interpretation of: (1) the angle its subsurface trace makes with the adjacent Confusion Range block (Allmendinger and others, 1983); (2) the average fault-bed angle in hanging wall rocks that restore palinspastically against the west side of the Confusion Range block (angles measured from Fig. 6 *in* Miller and others, 1983, and Figs. 16 and 20 *in Rogers, 1987*); and (3) metamorphic and thermochronometric gradients in its footwall (Miller and others, 1983; Lee and others, 1987). However, a large proportion of normal faults within the upper 5 to 10 km of the crust have steeper average initial dips, such as the Egan–Schell Creek detachment (Fig. 4b; Armstrong, 1972; *Wernicke, 1981*; *Gans and others, 1989*) and many of the faults in the central Basin and Range (Figs. 3c and 7; *Wernicke and others, 1989*).

While detachments may initiate with steep dips in the upper 5 to 10 km of the crust, the width of exposed footwall of most core complex detachments relative to changes in structural depth across strike seem too large to permit average initial dips through the upper 15 km of crust to exceed 30 to 40° (e.g., *Compton and others, 1977*; John, 1988; Wernicke and others, 1988; Spencer and Reynolds, 1989; Richard and others, 1990; Foster and others, 1990). If we consider detachment footwalls to show a range of between 10 and 15 km of downcutting from the breakaway zone to deepest exposed metamorphic tectonite, across a 30 to 60 km width of relatively intact footwall, then initial dips would lie in the range of 10 to 27°. If the detachments characteristically dipped 60° in at least the upper 5 km, then average initial dips would range from 5 to 21° below 5 km. Combined with seismic reflection data from the Basin and Range and passive margin settings (e.g., Allmendinger and others, 1983; Smith and Bruhn, 1984; *Cheadle and others, 1987*; *Tankard and Welsink, 1987, 1990*), the weight of evidence suggests that, in the depth range of 5 to 15 km, initial dips may vary from less than 10° up to 60°, but are commonly 30° or less.

Footwall uplift and flexure. Strongly extended domains are characterized by uplift and flexure on their margins, apparently driven by the isostatic forces that accompany footwall unloading (e.g., Spencer, *1982,* 1984, *1985*; *Howard and others, 1982*; Bartley and Wernicke, 1984; *Wernicke and Axen, 1988*; *Buck, 1988*). In cases where the detachment is inclined steeply through a large fraction of the upper crust, flexure of footwall strata accompanied by steep faulting may accommodate uplift on the proximal boundaries (e.g., the Colorado Plateau and Butte Mountains blocks, Fig. 4; *Wernicke and Axen, 1988*). In cases where the faults are initially gently inclined, the uplifts may form broad domes (such as beneath the Mormon Peak detachment, Fig. 4a, or the Snake Range detachment, Fig. 4b), which characterize core complex detachments (Davis and Coney, 1979; Spencer, 1984).

Sequential development. Strongly extended domains usually display complex overprinting relationships among multiple generations of normal faults, both within individual hanging walls of detachments, and among multiple generations of detachments (Fig. 4). Two kinematic processes are commonly suggested (Fig. 5). In one (Fig. 5a), arrays of fault blocks rotate synchronously, as in a row of toppling dominos (e.g., *Chamberlin, 1978*). When the first array reaches too low a dip for continued movement, a new one develops, rotating the earlier faults to still shallower dips, possibly even reversing their dip direction (e.g., *Morton and Black, 1975*; Proffett, 1977; Chamberlin, 1983; *Gans and Miller, 1983*). Within such a system, rotation of fault blocks occurs simultaneously across a domain, and a large fraction of the rotation of initially steep faults to shallow dips is accomplished while the faults are inactive.

In another process (Fig. 5b), normal fault blocks sequentially detach, meatslicer style, from the distal block, with rotation occurring principally as a result of isostatic rebound (Spencer, 1984; Bartley and Wernicke, 1984; *Wernicke and Axen, 1988*; *Hamilton, 1988*; *Buck, 1988*). Consequences of this process are that within a given hanging wall, cessation of normal faulting progresses from proximal to distal areas (i.e., diachronous rotation), and that nearly all of the rotation of individual faults may occur without the development of another generation of faults. An additional consequence is the migration of a monoclinal flexure through the footwall, such that the footwall hinge about which rotation of the detached slices occurs (points A and B, Fig. 4b; points A through D, Fig. 5b) propagates in wave-like fashion from proximal to distal sides of the domain. Structural analysis of a core complex in the Mojave extended domain (Pl. 8) supports the passage of such a monoclinal flexure through footwall metamorphic tectonite during extension (*Bartley and others, 1990*).

These two processes are not mutually exclusive, and likely interact at a variety of scales. For example, each sequentially detached block in Figure 5b could be divided into splays such as those shown in Figure 5a, which could then experience two periods of domino-style rotation. This is likely the case for the development of blocks in the upper plate of the Snake Range decollement in the Snake Range (Miller and others, 1983). However, within the same system, there is evidence for eastward migration of fault block rotation above the Snake Range decollement, in that the fault mosaic above the decollement in the Schell Creek Range area is unconformably overlapped by strata dated at 27 Ma (locality I, Figs. 4b and 6b), yet thermochronology on footwall rocks in the Snake Range suggests significant movement on the decollement until at least 20 to 22 Ma (locality J, Figs. 4b and 6b). A similar pattern is apparent for the structurally lower Egan Range system, as structurally lowest faults in that system are cut by 35 Ma granites (locality J) while higher splays to the east in the Schell Creek Range involve younger rocks (locality I). Correlations of these fault systems across the three ranges are, however, partly interpretive (Bartley and Wernicke, 1984; Hauser and others, 1987; cf. *Miller and others, 1988*).

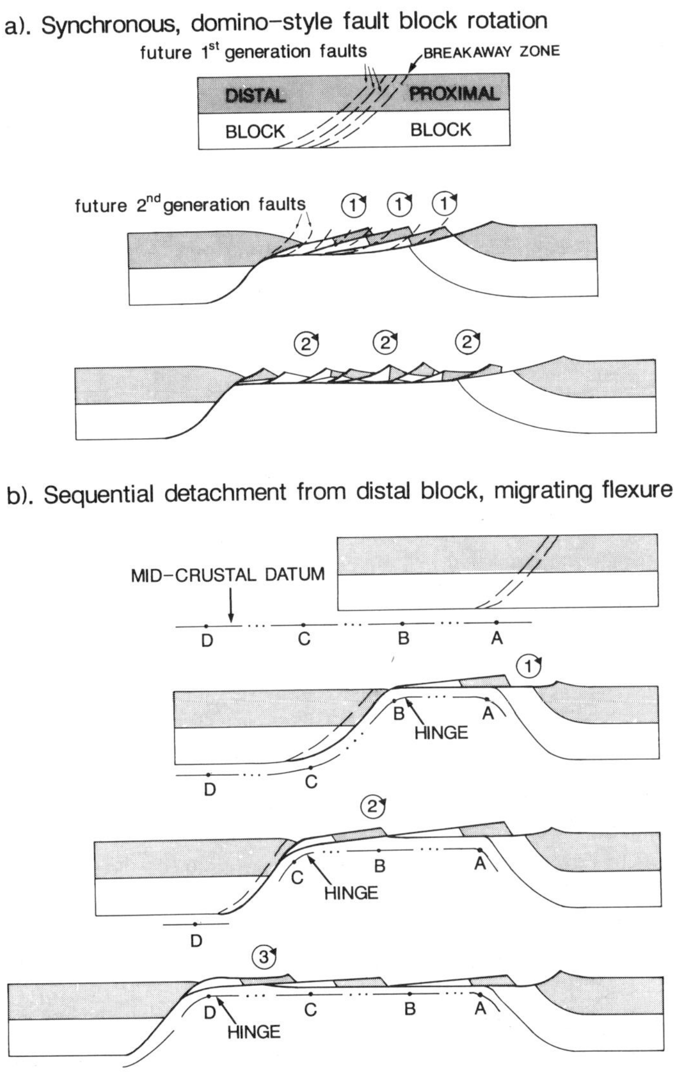

Figure 5. End-member kinematic models for the development of imbricate fault mosaics in the hanging walls of detachments. In (a), rotation of fault blocks occurs simultaneously, perhaps in successive generations. In (b), thin slivers of crust are detached sequentially, such that rotation of each block occurs at different times. Depending on the details of sub-detachment strain, a hinge may or may not migrate through the lower plate in (a). Kinematics in (b) seem to require new material to shear out from beneath the distal block for each successive sliver.

An example of regional sequential detachment is found across the central Basin and Range (Fig. 6a), where the cessation of faulting and times of major rotation of range blocks migrate eastward (*Wright and others, 1984*; *Hamilton, 1988*; Wernicke and others, 1988). Even at the scale of individual ranges, a progression of faulting within narrow fault blocks in the Panamint Range (Locality A, Fig. 6a) propagates upward, such that the footwalls of some faults rotate prior to their hanging walls (Fig. 3c; *Wernicke and others, 1989*).

At present, it is not clear whether one or the other of these mechanisms predominates within the extended domains, although the bulk of evidence suggests that the domino mechanism may be restricted to smaller subdomains and that the sequential detachment mechanism is important domain-wide. However, in both of the domains shown in Figure 4, more than one detachment system affects them. In both cases, the detachment systems propagate downward, such that the unloaded footwalls of earlier detachments become incorporated into the hanging walls of later ones, opposite to the pattern in Figure 5b. The breakaway zones of the younger systems tend to lie within a few kilometers of those of the older ones. Thus, in map view, extension does not propagate very far into the proximal block, but incorporates a large volume of the unloaded footwall of the earlier fault system into its hanging wall. This progression causes younger imbricate faults to rotate (and thereby deactivate) portions of older systems, amplifying the domes created by isostatic rebound. These kinematics create a large-scale pattern of thick fault blocks resembling dominos, but rotation of adjacent blocks is not synchronous. In both the Mormon and Snake Range areas, however, the oldest detachments may have been active in their most distal segments while the younger, structurally lower systems were active.

Rate and duration of extension. In several of the strongly extended domains, the timing and magnitude of extension are well enough constrained to permit assessment of the spreading rate. While the cooling histories and upper crustal unconformities associated with block tilting typically show that individual blocks are up-ended over an interval of 1 to 5 m.y. (e.g., *Anderson and*

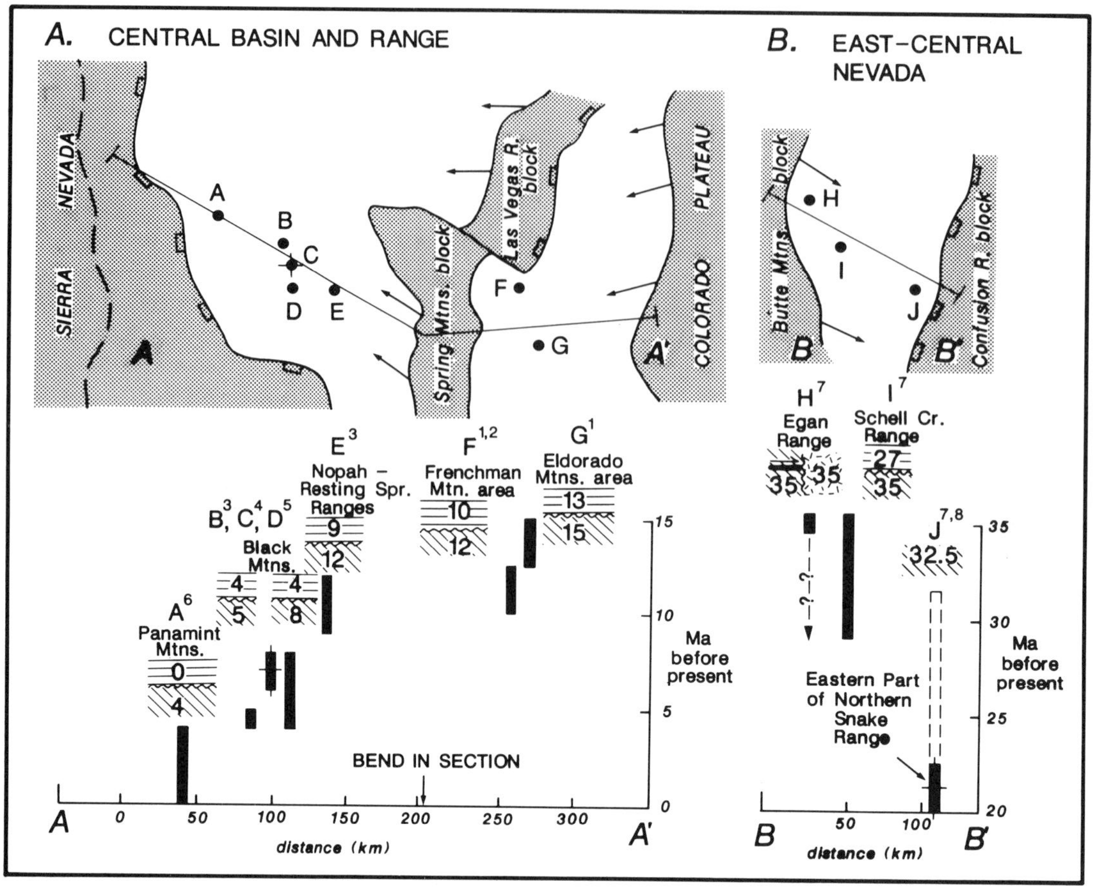

Figure 6. Timing constraints on extension for (A) central Basin and Range domains and (B) Snake Range domain in east-central Nevada (Pl. 8). Constraints are projected onto lines of section, which are not precisely coincident with lines of section shown on Plate 8. Black bars indicate intervals of rotation of fault blocks based on ages of tilted strata, faulting, and intrusion as schematically shown next to symbol. Black bars ornamented with thin tick marks show interval at which deepest portions of exposed core complexes cooled through the blocking temperatures of Ar in micas (about 250 to 350 °C). See text for discussion. Sources: 1, *Anderson and others, 1972*; 2, *Bohannon, 1984*; 3, *Wright and others, 1984*; 4, *Fleck, 1970*; 5, *Holm and others, 1989*; 6, Hodges and others, 1989; 7, *Gans and others, 1989*; 8, Lee and others, 1987.

others, 1972; Dokka and others, 1986; *Holm and others, 1989*; *Davis, 1988*), the duration of major extension across a given domain may be significantly longer, particularly if the domain contains several major blocks up-ended in sequence (Fig. 6). For example, tilting of individual fault blocks in the Lake Mead, Death Valley, and Snake Range domains occurred in as little as 1 m.y., but rapid extension appears to have taken place for 5 to 10 m.y. in the Lake Mead domain, 10 to 12 m.y. in the Death Valley domain, and 10 to 15 m.y. in the Snake Range domain, because the major block rotations are not necessarily synchronous across the domains.

The spreading of these extended domains, assuming 75 km of extension in the Snake Range domain, 100 km in the Lake Mead domain (southern part), and 150 km in the Death Valley domain, are thus 5 to 8 mm/a, 10 to 20 mm/a, and 10 to 13 mm/a, respectively. In the central Basin and Range, the timing and magnitude of extension suggest that the total rate of westward motion of the Sierra Nevada away from the Colorado Plateau was at least 20 mm/a in Miocene time (ca. 10 Ma), but has since slowed to approximately 10 mm/a in the last 5 m.y. (Wernicke and others, 1988).

Complexities in the coupling between upper crustal strain and that of deeper layers (discussed below) and in the three-dimensional kinematics of extension preclude any simple relationship between upper crustal displacement rates and the strain rates of deeper layers. Further, if strain in deeper layers is localized into relatively narrow shear zones, strain rates there would be far greater than those calculated for the lithosphere as a whole. Nonetheless, if we consider the case of the central Basin and Range, averaging the westward displacement rate of the Sierra Nevada across a uniformly extending lithosphere yields strain rates as great as 2×10^{-14} sec^{-1}, although conspiring uncertainties in timing and rate in the other direction yields values an order of magnitude lower (Wernicke and others, 1988). These rates are comparable to or greater than an earlier estimate of strain rates in the strongly extended domains of about 3 to 6 × 10^{-15} sec^{-1} (Zoback and others, 1981). For active deformation, particularly in the northern Basin and Range where strain is spread out over a broad area and is relatively slow, the average rate is in the 10^{-16} sec^{-1} range (Zoback and others, 1981).

Relation of extended domains to Basin and Range topography. In a number of areas, both stable blocks and strongly extended domains contain widely spaced, domino-style faults that control modern Basin and Range topography. For example, a number of regularly spaced ranges in the western part of the northern Basin and Range formed in late Miocene to Recent time as large, simultaneously rotating dominos, whose bounding faults penetrate to substantial depths but accommodate relatively little extension (classical Basin and Range style; e.g., Stewart, 1978; Okaya and Thompson, 1986; Allmendinger and others, 1987).

However, within most of the remainder of the province, Basin and Range topography is not as clearly an expression of deep-seated steep block faults (e.g., *Blackwelder, 1928*). Although partly due to block faulting, much of the topography may result from structures associated with large-magnitude extension, such as the eroded remnants of domiform detachment footwalls (e.g., the Snake Range and Mormon Mountains, Fig. 4), or the topography that may result from the isolation of thin, sequentially detached slices atop rebounded middle crustal rocks (Fig. 5b). This includes the eastern part of the northern Basin and Range, the central Basin and Range (especially the Death Valley region; Fig. 7), and the southern Basin and Range (Wernicke, *1981,* 1985; *Hamilton, 1982*; Anderson and others, 1983; Allmendinger and others, 1983; Burchfiel and others, 1987; Hamilton, 1987; *Dickinson and others, 1987*; Hodges and others, 1989). Thus, the widespread belief that Basin and Range topography represents a late event overprinting the strongly extended domains, and thus predominantly expresses late-stage moderate crustal extension (e.g., Zoback and others, 1981; Eaton, 1982; Coney and Harms, 1984; *Coney, 1987*), explains some of the physiography well, but such block faulting may not be the primary control on the topography in much of the province (e.g., Anderson, 1989).

Areas of large-magnitude extension need not develop Basin and Range topography. For example, the topography within the Omineca extended belt, while generally consisting of northerly trending ranges, lacks the wide alluvial valleys characteristic of the Basin and Range province.

DEEP-CRUSTAL OBSERVATIONS

The geophysical framework of extended portions of the Cordillera, in particular the Basin and Range province, was well established by the end of the 1970s (Thompson and Burke, 1974; Smith and Eaton, 1978; *Lachenbruch and Sass, 1978*; Eaton and others, 1978; *Thompson and Zoback, 1979*). Nonetheless, concomitant with the progress in understanding upper crustal fault kinematics through geologic studies, the 1980s saw major progress in imaging lower crustal and upper mantle structure, through the acquisition of deep seismic-reflection profiles (e.g., Allmendinger and others, 1983, 1987; Smith and Bruhn, 1984; Potter and others, 1986; Okaya and Thompson, 1986; Klemperer and others, 1986; Hauser and others, 1987, *1987b*; *Goodwin and Thompson, 1988*; McCarthy and Thompson, 1988; *Valasek and others, 1989*) and combined seismic-reflection and refraction profiling (*Mooney and Brocher, 1987*; *Catchings and others, 1988*). These studies provide a considerably improved basis for developing models of extensional tectonism at lithospheric scale, which became particularly numerous with the recognition of the core complexes in the late 1970s and early 1980s (e.g., Davis and Coney, 1979; *Rehrig and Reynolds, 1980*; Eaton, *1979,* 1982; *Wernicke, 1981*; *Hamilton, 1982*). A complete survey of all deep geophysical measurements is beyond the scope of this chapter, but there are several observations, in particular that bear on deep crustal structural evolution, that warrant discussion here. These include the distribution of long-wavelength variations in topography and gravity with respect to the strongly extended domains, and the interpreted position of the seismic Moho beneath the extended areas, particularly near the boundaries of strongly extended domains.

The domainal character of upper crustal extension, and the observation that the upper 7 to 15 km of crust has been partially or completely removed from the strongly extended domains, raises important questions about how such a geometry relates to strain at deeper crustal levels. For example, if normal faulting completely denuded the middle crust, then the amount of isostatic rebound of the denuded crust, and hence the magnitude of the topographic anomaly produced by such denudation, should be a function of the density of the medium upon which the lithosphere floats. Similarly, one might expect the tectonic removal of the upper one-third to one-half of the crust to lead to upwarping of the Moho, with an amplitude roughly equal to the amount of material denuded, or perhaps greater if the lower portions of the crust extend preferentially beneath the strongly extended domains. These responses make simple predictions about the density structure of the crust in and near the extended domains (e.g., *Cady, 1980* Wernicke, *1983,* 1985, *1989,* 1990; Spencer and Reynolds, *1984,* 1989; Okaya and Thompson, 1986; *Thompson and McCarthy, 1986, 1990*; Block and Royden, *1988,* 1990).

Regional gravity and topographic patterns in the Basin and Range province show variations at wavelengths of 30 km (scale of basins and ranges) and 150 to 200 km (Eaton and others, 1978; *Fletcher and Hallet, 1983*; *Froidevaux, 1986*) showing a rough correlation with the distribution of Stewart's (1979) tilt domains (Zuber and others, 1986), and thus also with the distribution of extended domains and stable blocks (Pl. 8). For example, in the Omineca extended belt, the broad region of crustal denudation is associated with depressions of 100 to 300 m (*Cady, 1980*), while in the central Basin and Range, a transect from the Colorado Plateau to the Sierra Nevada (Q–Q′, Pl. 8)

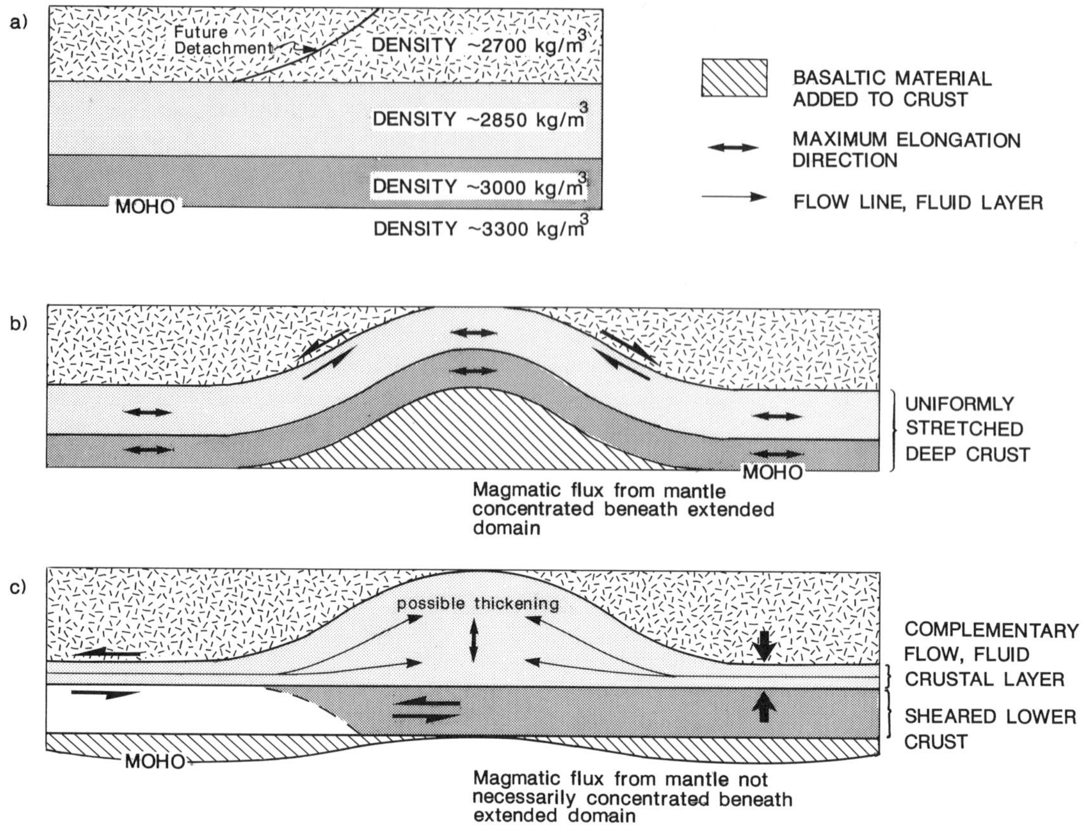

Figure 7. Contrasting modes of deep-crustal response to extension. (a) Unstrained crust. (b) Deep crust stretches uniformly, causing displacement of stretched layer out from beneath stable blocks and into extended domains. Mantle magmatic flux is concentrated beneath extended domain (possibly mixing with and remobilizing old deep crust), thereby smoothing the Moho. (c) Deep crust is divided into an upper, quartzose, fluid layer and a lower, mafic, relatively viscous layer. Fluid layer flow is the primary mechanism for keeping relief on the Moho low, and is governed in large part by flotational equilibrium of upper crustal stable blocks as they separate. Thus, fluid layer may thicken substantially beneath extended domains. Mantle magmatic flux need bear no particular relation to the position of the extended domain or stable blocks on either side, but flotational flow may serve as a mechanism to concentrate either magmatic additions or remelted crust into the extended domains relative to stable blocks via simple shear. Situation in (c) may typify early extensional history of thickened crust (e.g., northern Basin and Range), while the latest stages of extension prior to sea-floor spreading may be more similar to the situation in (b) (e.g., the Salton trough).

shows two broad depressions of 500 to 800 m centered on the strongly extended domains. Similar patterns are apparent across many of the other domains (Wernicke, *1989,* 1990; cf. *Thompson and McCarthy, 1986, 1990*). Stable blocks tend to stand higher and have broad negative Bouguer anomalies across them. However, stable blocks that have distal boundaries on either side, such as the Confusion Range block, tend to lie near topographic lows (Pl. 8; Wernicke, 1990).

Deep seismic-reflection profiling of the Basin and Range and the Omineca extended belt shows a prominent set of reflections at the Moho (Klemperer and others, 1986; Potter and others, 1986), which is locally corroborated by combined reflection-refraction imaging in the northern Basin and Range province (*Catchings and others, 1988*; *Thompson and others, 1990*). In addition, studies in the Basin and Range and other extended terrains suggest that the Moho lies at or near an abrupt downward cutoff of pervasive strong reflections characterizing much of the lower crust. This cutoff typically occurs at nearly constant depth over large regions. Throughout most of the Basin and Range, this depth is about 30 ± 3 km (Klemperer and others, 1986). The reflective lower crust contrasts with the upper crust and upper mantle, which tend to be transparent, except for discrete panels of dipping reflections. Such a pattern appears to typify the extended crust around the British Isles (*Matthews and Cheadle, 1986*; *Cheadle and others, 1987*) and the Basin and Range, except prominent upper mantle reflections have not been imaged in the Basin and Range (Allmendinger and others, 1987).

The subhorizontal Moho has been imaged across several of

the boundaries between extended domains and stable blocks, in particular on either side of the Snake Range domain (Fig. 4b; Klemperer and others, 1986), across the northwestern boundary of the Whipple domain (*Hauser and others, 1987b*), and across most of the domain boundaries in the Omineca extended belt (Potter and others, 1986). Further, in some of the domains, the strongly reflective character of the lower crust is observed in the upper crust, coincident with regions of core complex tectonites, implicating extensional tectonism as the cause of the reflectivity (e.g., *Hurich and others, 1985*; Frost and others, 1987; McCarthy and Thompson, 1988; *Valasek and others, 1989*; *Goodwin and Thompson, 1988*). In the western part of the northern Basin and Range, where strongly extended domains are apparently scarce and core complex tectonites absent, the upper crust lacks pervasive reflections (Allmendinger and others, 1987).

LITHOSPHERE-SCALE KINEMATICS AND DYNAMIC IMPLICATIONS

The observations discussed in the preceding two sections are synthesized below in the context of a large-scale model for lithospheric extension in the U.S. Cordillera. Previous large-scale models have emphasized three main effects. One is the consequences of adding magma to the crust during extension, and redistributing crustal mass via magmatic transport, since extension is often broadly coeval with magmatic activity (e.g., *Mackin, 1960*; *Thompson, 1960*). A second effect is rheological layering of the lithosphere, where it is presumed that discrete brittle faulting in the upper crust is accommodated by more uniform stretching of a ductile substratum (e.g., Stewart, 1971; Eaton, *1979,* 1982; Smith and Bruhn, 1984). A third is that since core complex detachments have large offsets, the simple shear on these faults may be relayed laterally through the lower crust and lithosphere, such that thinning of various lithospheric layers may be strongly heterogeneous in any given vertical column (e.g., *Wernicke, 1981*; Davis and others, 1986). Few if any workers regard magmatism, uniform stretching, and simple shear of the deep crust as mutually exclusive processes, as they may occur in varying proportions in the history of Cordilleran extension. However, the interdependence of these processes, if any, and their implications for physical models of extension, are topics of much current debate. In this section, the debate is illuminated in the context of a recent model, advocated below, that invokes fluid behavior of the deep crust during extension. This model, while not yet generally accepted, is intended to serve as a synthetic device and focal point for discussing the merits and weaknesses of previous models.

The fluid crustal layer

The principal problem arising from synthesis of surface and deep-crustal observations is reconciling the heterogeneity of upper crustal strain with the apparent uniformity of deep-crustal structure across domain boundaries. If the deep-crustal structure of an extended domain continues beneath an adjacent stable block, then flow of deep crust out from underneath the stable block and into the extended domain is indicated (e.g., Wernicke, *1983,* 1985; Spencer and Reynolds, *1984,* 1989; Gans, 1987; Block and Royden, *1988,* 1990). This interpretation has been used from time to time to argue for the importance of magmatism, uniform stretching, and simple shear of the deep crust in extensional models (e.g., Okaya and Thompson, 1986; Gans, 1987; and Wernicke, 1985, respectively).

Existing data permit some or all of these concepts relating extension in the upper crust to deeper levels to be invoked in varying proportions. However, the observations of (1) gentle regional topographic sags (ca. a few hundred meters) and gravity highs broadly coincident with most of the extended domains, and (2) a low-relief Moho transecting strong gradients in upper crustal thinning, independently suggest an important dynamic control on the development of extension, which may reconcile these diverse concepts and observations with considerable ease: flotation of the stable blocks within a layer of intracrustal fluid, at least during the early phases of extension. This concept in essence applies the "tilted buoyant block" model for classical basin-range structure (e.g., J. K. Sales, *in* Stewart, 1978) to the strongly extended domains (e.g., Block and Royden, *1988,* 1990; *Buck, 1988*; *Kruse and others, 1989, 1991*; Wernicke, *1989,* 1990).

In the extended domains, the removal of at least the upper 10 km of crust has resulted in sags in topography in the range of 0 to 0.1 times that amount (Pl. 8). A survey of the amount of denudation in the extended domains versus the amount of topographic depression suggests a typical proportion of about 0.05, although some domains show little or no depression (Wernicke, 1990). Barring unlikely complications addressed elsewhere (e.g., Block and Royden, 1990; Wernicke, 1990; *Kruse and others, 1991*), this implies that the density of the compensating fluid lies within approximately 150 kg/m^3 of the density of the upper crust. In areas where topographic depression is minimal or absent, the medium would be nearly equal or perhaps even slightly greater than average upper crust (e.g., in the Snake Range and Ruby Mountains domains, Pl. 8; Block and Royden, 1990). Even in areas where the topographic depression is relatively large (e.g., the Death Valley and Lake Mead domains, Pl. 8), the average density of the compensating fluid would be less than 2850 to 2900 kg/m^3, and could therefore only lie within the crust. The compensating medium would be expected to contain a relatively small proportion of quartz-poor rocks thought to characterize the lower crust (e.g., mafic diorite or gabbro).

The lack of Moho relief across domain boundaries independently supports the concept of intracrustal isostasy (e.g., Block and Royden, 1990). If the separating stable blocks float in a sea of middle crust too weak to maintain shear stresses on geologic timescales, then the development of strong, upper crustal gradients in strain would not exert differential vertical loads on layers beneath the fluid (by the definition of a fluid), and they should remain undeflected unless affected from below. The crux of the argument is that the existence of a fluid layer (not simply a weak or ductile layer) requires complementarity of vertical strain between the fluid and upper crustal layers. That is, the combined thickness of the fluid layer plus upper crust would be areally constant at any given time during extension (assuming the bottom of the fluid layer is relatively flat), no matter how complex or heterogeneous the three-dimensional strain field of the upper crust might be.

Intracrustal isostasy is compatible with, but not required by, laboratory experiments on the strengths of rocks likely to exist in the deep crust. The rheology of the continental lithosphere is thought to be characterized by weak, ductile, lower crust and relatively strong upper crust and upper mantle (e.g., *Brace and Kohlstedt, 1980*), often likened to a "jelly sandwich" following *Matthews and Cheadle (1986).* The jelly sandwich analogy is apt for a broad range of possible lithospheric strength profiles in which the lower crust is weak and ductile, but is improperly scaled with reference to the fluid layer concept, because bread does not float in jelly. If we separate two pieces of flat-lying bread resting on jelly, the horizontal pressure difference between bread-loaded and unloaded jelly is insufficient to drive flow into the void between the bread as it is pulled apart. In other words, the viscosity of the jelly is large in proportion to the imposed loads and the timescale at which even the most patient experimentalist would separate the bread, so the jelly remains far out of fluid equilibrium. A proper culinary analogy could only be drawn with substances that are impractically runny for sandwiches.

The fluid layer concept, if correct, has major implications for the roles of magmatism, ductility, and simple shear between lithospheric layers (Fig. 7). It shares the notion of rheological layering as a control of strain localization (e.g., Eaton, *1979,* 1982; *Matthews and Cheadle, 1986*), contrary to the suggestion that the lower crust may extend on inclined shear zones collinear with upper crustal detachments and upper mantle shear zones (e.g., *Wernicke, 1986*), at least during the early stages of extension. It is compatible with the concept that detachments are crustal shear zones that cut downward across the brittle-ductile transition of quartzose rocks at shallow angle (e.g., Wernicke, *1981,* 1985; Davis and others, 1986). The concept of whole-lithosphere normal simple shear (e.g., Wernicke, 1985; *Lister and others, 1986*) is also compatible with the fluid layer concept kinematically, because discrete upper mantle (e.g. *Klemperer, 1988*) and perhaps lowermost crustal shear may be transferred to the upper crust in the fluid layer. However, during the early stages of extension, discrete shear zones developed above and below the fluid layer would not be dynamically coupled in any simple way, and could develop independently of one another (e.g. *Matthews and Cheadle, 1986*; *Reston, 1990*).

Within the crust, the fluid layer model specifies (1) the relation of rheological layering to core-complex detachments, (2) a degree of weakness of the layer sufficient to cause fluid behavior on geologic timescales, (3) flow complementary to upper crustal thinning within the layer as the primary mechanism for keeping Moho relief low, and (4) deep-crustal behavior as a function of its probable strength and composition, limiting somewhat the participation of mafic rocks in the compensation process (Fig. 7b).

Earlier models stressing the importance of rheological layering and magmatism as controls on crustal strain (e.g., Eaton, *1979,* 1982; Miller and others, 1983) interpret core-complex detachments as exposures of a regional decoupling horizon between brittle deformation above and ductile deformation below, across which displacement is a small fraction of total extension, accommodating small variations in strain between the layers. For geological reasons elaborated below, it is improbable that core-complex detachments represent the interface between the fluid and nonfluid parts of the crust. Rather, they more likely represent shear zones that transect the upper crust above the fluid layer, crossing the brittle-ductile transition at low-angle (e.g., Wernicke, *1981,* 1985), although perhaps locally exhuming the fluid layer.

To the extent that mechanical thinning of a subdetachment ductile crust is uniform across domain boundaries, relief on the pre-extension Moho is equivalent to differences in upper crustal thinning between extended domains and stable blocks (Fig. 7b). Thus, if the ductile crust extends uniformly beneath strong horizontal gradients in upper crustal thinning, addition of magma from the mantle in greater proportion beneath the extended domains is required to maintain constant depth to Moho (Fig. 7b; e.g., Okaya and Thompson, 1986; *Thompson and McCarthy, 1986*; Gans, 1987). While such an end-member model is consistent with low strength, a layer that maintains constant thickness during unloading would also maintain a substantial lateral-pressure gradient and remain out of fluid equilibrium (Fig. 7c). If the ductile layer were to contain a mafic layer at its base, then the model would also require substantial relief to develop on a mafic-nonmafic interface within the uniformly stretched layer (Fig. 7b).

Thus, while broadly similar, these earlier concepts differ somewhat with the fluid-layer model in: (1) interpretation of the significance of detachments as localizing at the brittle-ductile transition; (2) the implied exclusion of fluid behavior for the deep crust; (3) the solution to the differential thinning problem primarily with a subcrustal magmatic flux beneath extended domains rather than complementary vertical strain of a fluid layer; and (4) the assumption of pure shear of the subdetachment lithosphere rather than complementary flow of a siliceous layer decoupled from relatively viscous mafic and ultramafic layers below. Such flow includes the possibility of major vertical thickening of the fluid layer in precisely those areas where upper crustal extension is greatest, quite in contrast to the kinematics of uniform thinning of deep crust (Fig. 7; Wernicke, 1990).

Role of magmatism. Redistribution of crustal mass via heterogeneous flow of a flotational medium for upper crustal blocks is a simpler mechanism than systematic concentration of the subcrustal magmatic flux in explaining the deep crustal mass budget (Fig. 7), although the two processes may act in concert. The complexities of mantle heat flux, decompression melting (in both crust and mantle), melt extraction, migration, mixing with old crust, and emplacement would all have to be incorporated into a physical model that explained the mass budget by magmatic additions alone, or by magmatic additions superimposed on a uniformly stretched deep crust.

Once derived, the prospects for a purely magmatic model (e.g., of the type proposed by *McKenzie, 1984,* and its derivatives) to keep crustal thickness uniform during heterogeneous upper crustal extension seem remote. Additions of mantle magma to the crust would as likely create relief on the Moho as smooth it, especially given the sensitivity of the amount of melt generated to parameters such as upper mantle temperature (*White and McKenzie, 1989*), which would probably not correlate in any simple way with local variations in upper crustal thinning. Because a fluid layer would eliminate differential loads on the upper mantle beneath it, its existence would preclude the possibility of decompression melting occurring solely beneath the extended domains, i.e., the mantle would have no way of knowing where overlying extended domains are. Magmatic additions are clearly an important part of the crustal mass budget during extension, and they undoubtedly profoundly modifiy the entire lithospheric column (Anderson, 1989), but collusion between the asthenosphere and upper crust to inflate the crust with precisely the amount of magma needed to simulate the simple physics of intracrustal flotation during the early stages of extension seems improbable.

Syn-rift magma centers were emplaced over broad regions of the Cordillera at approximately the same time as extension, indiscriminately across stable blocks and extended domains (Anderson, 1989). This also suggests that the mantle magmatic flux is not concentrated preferentially beneath extended domains, and thus it is unlikely to control their localization, as advocated in some models (e.g., *Gans and others, 1989*). The degree to which mantle-derived magma underlies strongly extended domains in preference to stable blocks may be governed by flotationally driven, lateral transport of magma entrained in the fluid layer, rather than by variations in subcrustal flux (Fig. 7). The fluid layer may be rich in synrift magma, whose rheology and latent heat may substantially contribute to weakening the layer over broad regions. In this sense, magmatism may be a primary control on its development. Determining whether a magmatic flux is required to create a fluid layer, or whether a fluid layer might generally be present in continental lithosphere in the absence of a magmatic flux from the mantle, is a relevant and challenging problem in lithospheric physics.

Application. The fluid-layer concept, and some of its consequences, seem best explored by application to the central Basin and Range because of the large amounts of extension in the strongly extended domains and their abrupt boundaries with stable blocks (section Q–Q′, Pl. 8 and Fig. 8; Wernicke and others, 1988), although it should apply equally well to most of the other domains. The model combines knowledge of upper crustal fault kinematics (Wernicke and others, 1988; *Snow and Wernicke, 1989*) with the fluid-layer concept. In addition, it is assumed that the lowermost crust, if present (mafic, density 2,900 to 3,100 kg/m^3), does not participate to a major extent in the intracrustal flow, but instead accommodates shear between the upper and fluid crustal layers and the mantle. New material is shown to be added to the crust in layers near the Moho, with maximum thickness beneath volcanic regions known from the surface geology. The strain pattern in the upper mantle is shown as being breached along a localized shear zone, in a fashion suggested by the pattern of discrete inclined reflections imaged in the upper mantle beneath the extended terrain around the British Isles (e.g., the Flannan reflections; *Matthews and Cheadle, 1986*; *Reston, 1990*), although such mantle reflections have not been recorded anywhere in the Basin and Range province. The upper crustal blocks are shown sliding westward off of a rigid upper mantle breached beneath the Sierra Nevada, based on geophysical and geochemical studies in the region (Jones, 1987; *Farmer and others, 1989*), although a variety of other kinematic models involving penetrative stretching, detachment, and sinking of mantle lithosphere are tenable.

The presence of a middle crustal layer weak enough to behave as a fluid on geologic timescales is enhanced by a thickened crust and a high geotherm, as may have existed in most of the Cordillera immediately prior to major extension. Based on experimental data on the strengths of rocks, wet quartzose crust would be substantially weaker than a relatively dry, quartz-free, basal crustal layer (Fig. 8a), perhaps inhibiting the participation of the lowermost crust in regional flow relative to the middle part of the crust (summaries of Smith and Bruhn, 1984; *Kirby and Kronenberg, 1987*; and *Kuznir and Park, 1987*).

Core complex tectonites and the fluid layer. The fluid-layer concept raises the question of whether the tectonites of the metamorphic core complexes belong to the layer, and if so, whether they may be used to test the kinematics shown in Figure 8. As shown in Figure 8b, the fluid layer is not exposed along the line of section, always buried beneath the sequentially detached blocks or alluvial fill accumulated between them. Following Zuber and others (1986), the boundary between a fluid layer and overlying crust would probably form below the brittle-ductile transition within a ductile zone where strength decreases downward (Fig. 9; e.g., *Brace and Kohlstedt, 1980*). It would thus neither represent the brittle-ductile transition nor a detachment fault, possibly lying at substantially greater depth than either of these two features. Most of the core complexes may be variably rotated fragments that originally lay above the fluid layer, such that the tectonites would mainly record simple shear related to denudation (Fig. 9). This accounts for the observation that in some core complexes, penetrative ductile strain appears to diminish structurally downward away from the detachment faults (*Compton and others, 1977*; *Reynolds and Rehrig, 1980*; G. H. Davis, 1980; *Malavieille, 1987*). In this case, core-complex footwalls would represent a layer of relatively strong mid-crustal rock, rotated and flexed above unexposed fluid beneath it.

One of the best-studied core complexes both geologically and geophysically, the Whipple Mountains in the Whipple extended domain (Davis and Lister, 1988; Pl. 8) exposes a good candidate for the boundary between the fluid layer and the upper crust. In the footwall of the Whipple Mountains detachment, crystalline rocks largely undeformed during extension give way structurally downward across a sharp boundary (not a detachment) to strongly deformed mylonitic gneisses. The boundary (mylonitic front of Davis and Lister, 1988) is inclined toward the proximal boundary of the extended domain, and is aligned with a band of dipping seismic reflections that flatten out at mid-crustal levels beneath the proximal block, apparently continuing beneath it at regional scale (Frost and others, 1987; Davis and Lister, 1988; *Flueh and Okaya, 1989*). The mylonitic front thus appears to represent a surface exposure of a boundary between regional ductile flow below and relatively rigid crust above (Frost and others, 1987; Gans, 1987), possibly representing the top of the fluid layer where the ductile crust strengthens rapidly upward (Fig. 9). This situation may be geometrically analogous to the west sides of the Spring Mountains and Colorado Plateau blocks (Fig. 8b), except that conditions of erosion and denudation in the Whipples were favorable to yield a glimpse of the boundary.

The majority of kinematic indicators in the Whipple mylonites show the same sense of shear as on the younger detachment that cuts them (Davis and others, 1986). This would appear to be opposite to that predicted by the fluid-layer model (Fig. 8b), in which flow out from beneath a proximal block would impart a sense of shear on the top of the layer opposite to that of the detachment. However, as shown in Figure 9, an upper boundary to the fluid layer initially situated several tens of kilometers away from the breakaway zone (beneath the flat segment of the fault at points A and B, Fig. 9) would exhibit the same sense of shear as the detachment, because the fluid layer initially flows toward the breakaway zone (Fig. 9a). These kinematics should continue until the wave of uplift reaches the to-be-exhumed mylonite front, whereupon the sense of shear might briefly reverse. Given that a small percentage of the shear indicators in the Whipple mylonites are opposite to that on the detachment (Davis and Lister, 1988), a test of the model shown in Figure 9 would be to determine if the oppositely directed indicators are younger than those sympathetic with the detachment.

The pre-mylonitic fabrics present in some of the core complexes mentioned earlier (Ruby Mountains domain, *Wright and Snoke, 1986*; Snake Range domain, Lee and others, 1987) might be an expression of the fluid layer. As discussed below, the kinematics of the fluid layer might in general contrast sharply with those of uplift and exposure via tectonic denudation.

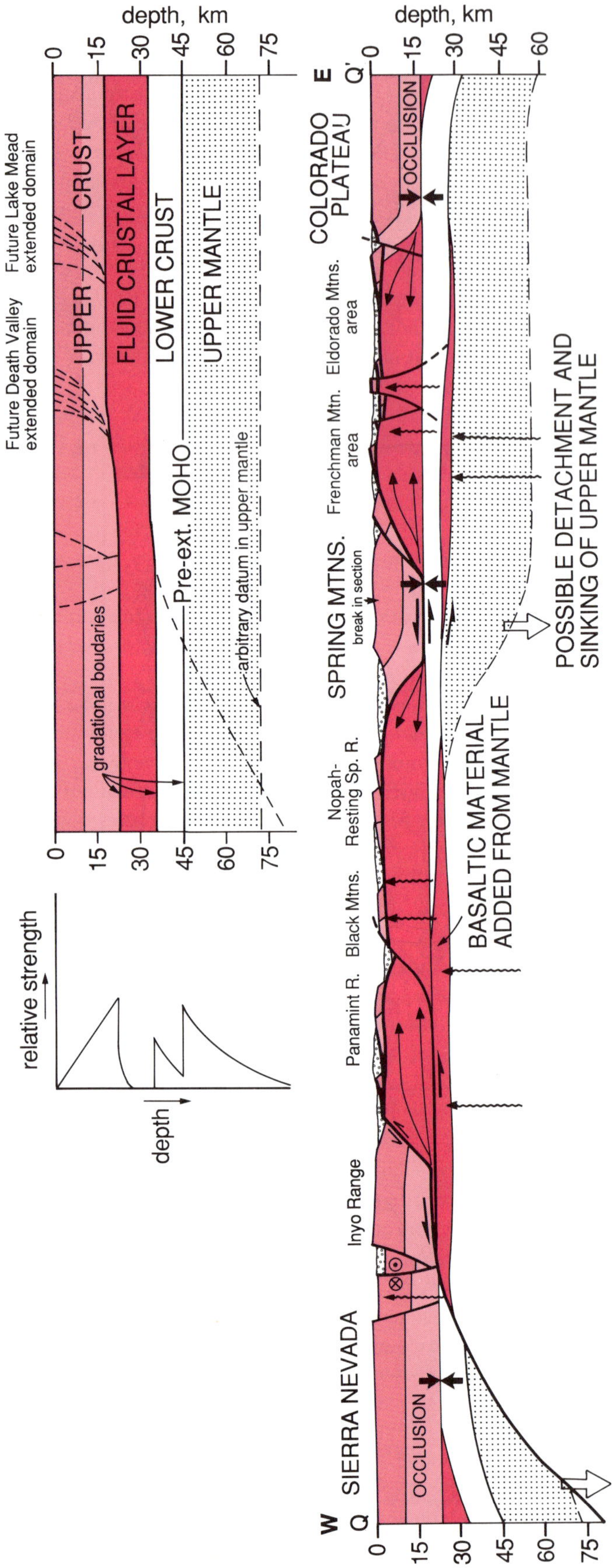

Figure 8. Application of fluid layer concept to the central Basin and Range, with upper crustal kinematics after Wernicke and others (1988). Line of section shown on Plate 8. Datum in upper mantle is not intended to be an isotherm or mechanical boundary. Circle with "X" denotes motion of block away from viewer; circle with dot denotes motion toward the viewer. Complementary flow within fluid layer allows upper crustal blocks to separate with minimal internal fragmentation, keeping thickness of upper crustal plus fluid layers nearly constant. Occlusion of stable blocks with lower crust promotes the development of high-angle faults within both extended domains and stable blocks, and the eruption of mafic lavas previously restricted to zones of mixing in the lower crust and fluid layer. See text for discussion. Lithosphere shown to fail by large-scale simple shear, but may be accompanied by detachment and sinking of the mantle part of the lithosphere.

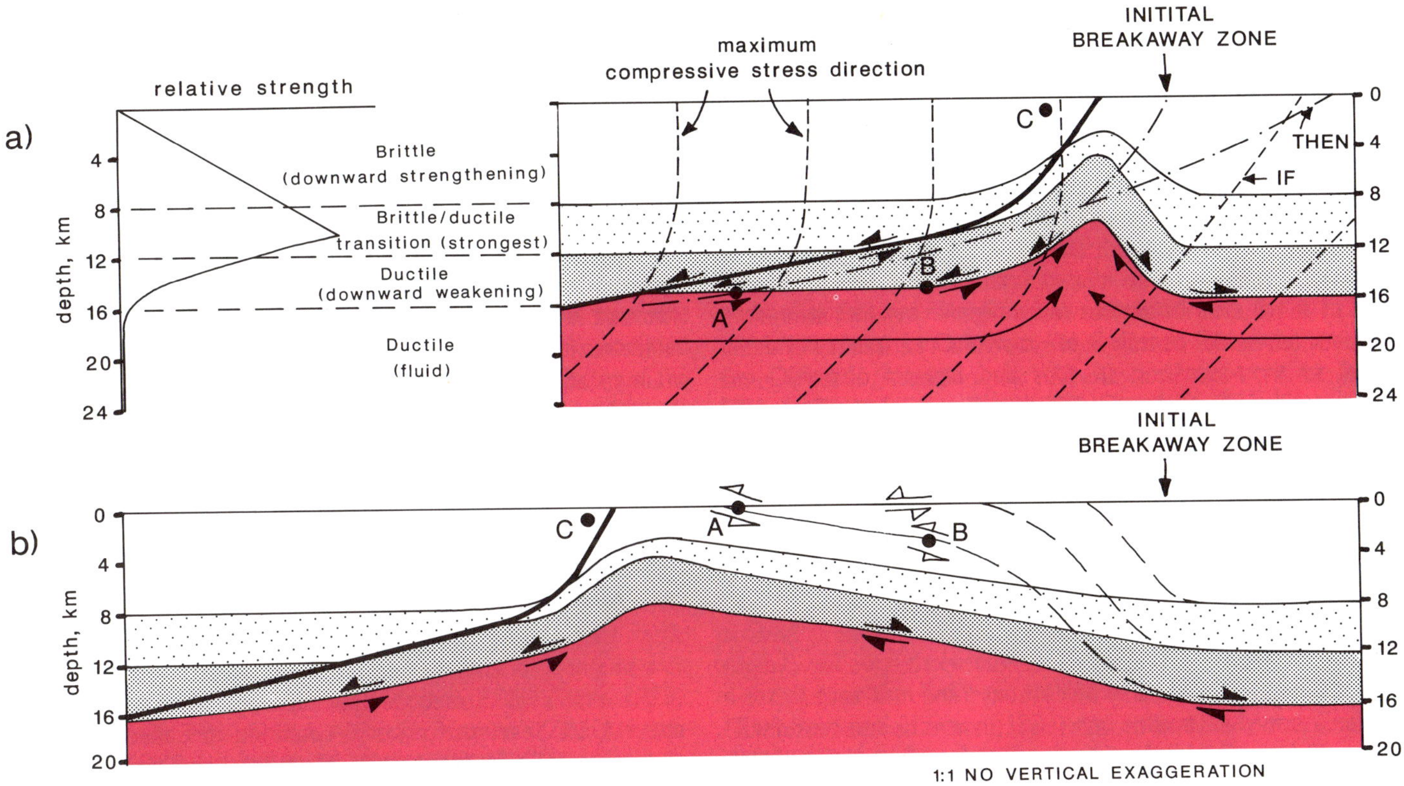

Figure 9. Cross section showing possible relations between detachments and various rheological layers. (a) Initial position of detachment shown as dot-dash line; active position shown as heavy line. Detachment flattens across strongest portion of crust, such that its initial dip through the ductile crust is low, due to rotated stress trajectories (dashed lines; e.g., *Yin, 1989*; Melosh, 1990). In some instances, inclined stress trajectories can be present in the brittle layer (marked IF), resulting in low initiation angles at shallow levels (marked THEN). Initial movement on the detachment forces fluid flow toward the breakaway zone (solid arrows at top of fluid layer), resulting in top-to-the-left shear at points A and B. Detachment is not the brittle-ductile transition, but a zone of shear that cuts downward across it at low angle (e.g., *Wernicke, 1981*). Neither is the top of the fluid layer the brittle-ductile transition, but rather the lower limit of ductile crust strong enough to remain out of fluid equilibrium. Perturbation of fluid layer, which can be magma rich at the onset of extension, might trigger volcanism soon after extension begins. (b) With continued movement, translating hanging wall and reference point C about 40 km to the left, uplifted ductile and fluid layers cool, forming new brittle crust. Long dashed lines show positions of rheologic boundaries at the onset of extension. Model predicts fossil flow direction across ductile/fluid boundary (arrows at points A and B) in the same sense as shear on the exhumed detachment. Drawing intended to schematically reproduce observations in the Whipple domain, such that point A is equivalent to the surface trace of the Whipple mylonite front (see text for discussion).

Scale and direction of flow. An interesting problem posed by the fluid layer concept is the scale at which flow might occur. At what lengthscale would the fluid layer be influenced by the separation of two stable blocks? Figure 8 is drawn to suggest that middle crust under the Sierra Nevada and Colorado Plateau is involved in flow to a limited extent near their edges, but this is unconstrained. The crust beneath the central Colorado Plateau is 10 to 20 km thicker than in the Basin and Range (e.g., *Hauser and Lundy, 1989*), perhaps indicating that the scale is limited to a range of 100 to 200 km.

However, the reflective crust apparently connected to the Whipple mylonite tectonites appears to be developed province-wide in the southern Basin and Range (Frost and others, 1987), with a tendency to rise toward the surface near core complexes (McCarthy and Thompson, 1988). In addition, the western part of the northern Basin and Range does not seem to be greatly extended at the surface, yet has crustal thickness equal to the eastern side of the province, raising the possibility that most of the crustal thinning in the western part of the province occurred by massive flow (100s of km) of the middle crust laterally out of the region. The west side may have functioned as a large stable block, which developed classical Basin and Range structure only once its fluid layer had been hydraulically expelled into the strongly extended domains to the east. Seismic-reflection profiles across the northern Basin and Range show pervasive subhorizontal reflections in the deep crust over a broad region of basin-range blocks on the western side of the province, with a relatively transparent upper crust (Allmendinger and others, 1987). These reflections appear to rise eastward into the extended domains, perhaps supportive of this concept. Alternatively, the crust may have originally been thinner on the west side (Coney and Harms, 1984).

The pattern of fluid flow need not be two-dimensional, especially given the complicated map-view distribution of extension (Pl. 8). Flow in a direction perpendicular to the regional extension direction may result across transverse boundaries between strongly extended domains and stable blocks, or could be directed radially toward extended domains if they are equant. Thus, although Figure 8 roughly balances the area of fluid in cross section, in regions such as the central Basin and Range where extended domains end abruptly along strike, cross-sectional area need not be conserved.

Implications for necking instability models. Physical models have been developed to account for the two wavelengths of variation in topography and gravity (about 30 and 175 km) discussed above. These models relate them to necking instabilities in stretched viscous or plastic layers of the lithosphere (*Fletcher and Hallett, 1983*; *Froidevaux, 1986*; Zuber and others, 1986; *Ricard and Froidevaux, 1986*). Extension of strong, upper crustal and upper mantle layers separated by a weak lower crust may result in two wavelengths of boudinage, a short one corresponding to the upper crust and a long one corresponding to the upper mantle, which in turn may correspond to the scale of individual ranges and tilt domains, respectively (Zuber and others, 1986). The origin of the spacing of basin-range blocks likely corresponds to these physics. However, the model shown in Figure 8 ascribes the long-wavelength anomalies to lateral density contrasts within the upper crust and fluid layer, maintaining a flat lower boundary on the fluid layer. The necking instability models simulate only a small amount of initial extensional strain, and, as such, an initial long-wavelength disruption of the upper mantle would somehow have to influence the development of upper crustal strain.

Occlusion of the fluid layer. The model in Figure 8 predicts that at some point, stable blocks in the upper crust and mafic lower crust will occlude, eliminating the fluid layer. The fluid layer would then occupy only the volume between the separating stable blocks, having been squeezed out from beneath them. The fluid layer probably cools during extension (Block and Royden, 1990), assuming upward heat transport due to magmatism is at some point overcome by the effect of advecting the fluid layer toward the surface (Block and Royden, 1990). Thus, its history as a fluid, particularly in its upper portions near the breakaway zone, may be relatively brief once extension begins (Fig. 9). Occlusion and freezing modify the overall rheology of the crust, and thus might be expected to result in a change in structural style once the fluid layer is eliminated.

The most obvious potential effect of elimination of the layer, alluded to above for the origin of the western part of the northern Basin and Range, is that once a stable block is beached atop the relatively strong lower crust and upper mantle, it may tend to fragment internally into large fault blocks. Indeed, the domainal character of extension and the narrow zone of initial breakage between the stable blocks (Fig. 8a) may be due to the smooth sailing (minimal basal traction) enjoyed by the stable blocks early in extension, relative to the increased basal traction on them once the fluid layer is occluded. This effect provides a simple explanation for the development of late high-angle faults observed in many stable blocks (akin to the downward-stepping brittle-ductile transition concept of Eaton, 1982), and is perhaps a control on the downstepping of major detachments observed in some of the domains (Figs. 4 and 7). In the case of the Death Valley region, this would include the late-stage fragmentation of the Inyo Range area (which lies in a stable block) and the active faulting in Death Valley proper (which lies in an extended domain; Fig. 8b).

Post-occlusion tectonics. Once the stage shown in Figure 8b is reached, and depending on whether sufficient driving force is available to continue extension, the entire lithosphere may fail in a manner analogous to the upper crust, perhaps along localized shear zones developed prior to occlusion that link together in some way to begin shearing the asthenosphere toward the surface. In this mode, isostatic compensation would occur in the upper mantle, such that deep topographic depressions may form with relatively little additional strain. Such a style of extension, accompanied by magmatic additions to the crust from the mantle concentrated along the rift zone, may typify the final stages of rifting prior to sea-floor spreading, as in the Salton trough and Gulf of California across the western extreme of the southern Basin and Range (*Fuis and others, 1984*; Lachenbruch and others, 1985).

The pre-existence of a fluid layer may be a criterion that distinguishes wide rifts that retain a substantial thickness of continental crust despite large separations of their bounding blocks (e.g., the Basin and Range) from narrow rifts that evolve quickly into ocean basins (e.g., the Red Sea). The fluid layer spreads extensional strain over a broad area, thereby inhibiting localized upwelling of asthenosphere to shallow levels.

Composition of magmas. Another observation bearing on the fluid-layer concept is the progression from predominantly intermediate to silicic magmatism characteristic of the early phases of extensional magmatism to the basaltic or bimodal eruptive sequences typical of the later phases (e.g., Anderson, 1971; Wernicke and others, 1987; *Gans and others, 1989*). An overall cooling of the crust during extension should enhance brittle deformation, perhaps increasing the likelihood that mafic magmas ponded in the lower crust may be erupted at the surface (e.g., *de Voogd and others, 1988*; *Gans and others, 1989*). In other words, the existence of the fluid layer may inhibit the ascent of mafic magma in the upper crust, perhaps because it is more difficult to breach with the brittle fissures presumably necessary to allow the ascent of mafic magma.

LARGE-SCALE TEMPORAL EVOLUTION OF CENOZOIC EXTENSION

Patterns in timing of Cenozoic extension are complex within strongly extended domains and areas that are only moderately extended, and are highly diachronous on the scale of the western U.S. Cordillera as a whole. The timing of extension in the strongly extended domains broadly follows the pattern of major intermediate to silicic magmatism that swept through the Cordillera in Cenozoic time (Coney, 1980; Lipman, 1980), but many of the details of the timing of maximum extension within a given area and its relation to volcanic activity are unresolved. In many of the domains, strong extension was accompanied by little or no magmatic activity, while conversely many of the most important magmatic centers lie in areas that are well removed from synvolcanic extension. Nonetheless, a broad correlation between magmatism and extension is apparent.

The overall pattern of Cenozoic extension, including its timing and vergence (direction of movement of detachment hanging walls), has been considerably refined over the last decade (e.g., *Reynolds, 1979*; *Chapin, 1979*; *G. A. Davis, 1980*; Coney, 1980; *Anderson, 1981*; Zoback and others, 1981; Chamberlin, 1983; Glazner and Bartley, 1984; Wust, 1986; Wernicke and others, 1987, 1988; Reynolds and others, 1988; *Baars and others, 1988*; Dokka, 1989; Taylor and others, 1989; *Janecke, 1989*; Anderson, 1989; *Gans and others, 1989*; Keller and others, 1990). The discussion below is derived principally from these studies and references therein.

Extension is strongly diachronous at large scale (Fig. 10), spanning most of Cenozoic time. A number of the extended domains appear to follow a four-stage evolutionary history, despite the diachronism of extension on a Cordillera-wide scale (Wernicke and others, 1987). These stages include: (1) the formation of early intermontane basins, probably signaling the onset of at least limited extension; (2) the eruption of predominantly intermediate to silicic volcanic rocks; (3) large-magnitude crustal extension, occurring during or immediately after second-stage magmatism; and (4) basaltic or bimodal volcanism, accompanied by lesser amounts of extension. The period of major extension within a given strongly extended domain may last as much as 10 to 15 m.y. or more (Fig. 6), although because extension has a tendency to migrate within a given domain, the period of tilting of an individual, steeply tilted block is often brief, on the order of 1 to 3 m.y. or less (Fig. 6).

The earliest extension began in the Pacific Northwest in the Omineca extended belt and along the western margin of the Rocky Mountains Basin and Range about 50 to 55 Ma, and was complete by about 35 to 40 Ma (Fig. 10a). There is currently no discernible migratory pattern in either magmatism or extension. Extension verges away from a central stable block in the Omineca extended belt, apparently transformed southeastward into the Rocky Mountains Basin and Range along the Lewis and Clark line. South of the line, vergence flip-flops along strike of the belt, going from southeastward in the north to northwestward or west-southwestward in the south (Fig. 10a).

South of the Columbia Plateaus, major extension apparently did not begin until after 40 Ma (Fig. 10b). Magmatic belts in both the northern and southern Basin and Range were active at their northern and southern extremes, respectively, by early Oligocene time, and migrated with crude symmetry toward one another along the long axis of the province, arriving at the central Basin and Range in middle Miocene time. Major extension in the northern Basin and Range was restricted to an elongate north-trending belt in the middle of the subprovince during this interval, while magmatism migrated southward in an east-west–trending wave that spanned nearly the entire width of the province. The strongly extended domains, as in the Omineca belt, show vergence directed away from a central block, with the most intense strain apparently alternating systematically from one side of the central stable block to the other (Fig. 10b), perhaps a large-strain analog to flip-flopping rift segments separated by overlapping high-relief accommodation zones documented in East Africa (Pl. 8 and Fig. 2; Rosendahl, 1987). There is no clear pattern of propagation of extension from north to south, although future studies might reveal one, as understanding of the magnitude and timing of extension improves. By comparison, the southern Basin and Range appears to show a pattern of northward propagation of extension, but has a more complicated pattern of vergence than the northern Basin and Range (Pl. 8). The extension directions within each of the two regions are generally parallel during this interval, but are northwesterly in the northern Basin and Range and northeasterly in the southern Basin and Range, forming a radial pattern about the convex-west margin of the Colorado Plateau. Modest extension affected the entire Rio Grande rift in late Oligocene time (ca. 28 Ma), and was locally severe.

Once extension and magmatism began to affect the central Basin and Range in middle Miocene time (ca. 17 Ma), profound changes in extensional architecture took place. Extension directions and least-principal-stress orientations were both southwest-northeast throughout the province by this time. Afterwards, while modest extension continued within the older belts, the locus of strongly extended domains apparently shifted to a position along the frontal portion of the Cordilleran fold and thrust belt (Figs. 1 and 10c). In addition, probable strong extension affected the Rio Grande rift and Yerington domains, with variable vergence. While major extension in the Socorro region of the Rio Grande rift occurred prior to 16 Ma (especially between about 28 and 20 Ma), a substantial amount of extension there and elsewhere in the rift is younger. Despite similarities in their pre-middle Miocene evolution, the northern Basin and Range developed into a broad region of large-scale tilted fault blocks accommodating moderate amounts of extension, while the southern Basin and Range, although somewhat faulted, was far less active. The Rocky Mountains Basin and Range became active (although perhaps in part prior to 16 Ma), accommodating moderate extension.

During late Miocene time in the Basin and Range, the over-

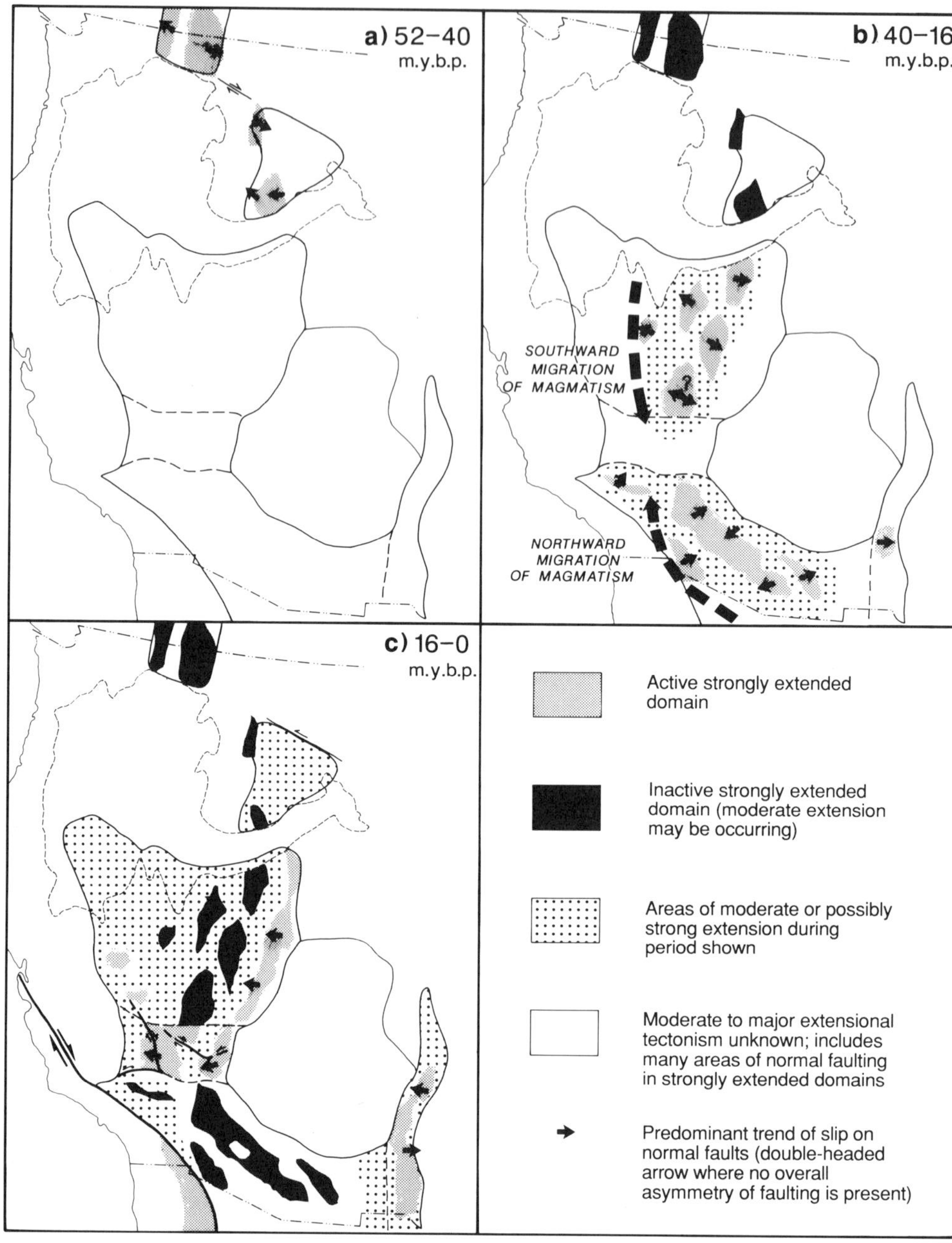

Figure 10. Evolution of Cordilleran extension in time and space. Extensional province boundaries from Figure 1 are included for reference. Heavy arrows in (b) show migration direction of mid-Tertiary magmatism. Note outward movement of major extension between (b) and (c). See text for discussion.

all extension and least-principal-stress directions appear to have rotated clockwise, changing from northeast to northwest. For example, vergence in the older eastern part of the central Basin and Range is predominantly southwestward, while in the younger western part it is mainly northwestward (Fig. 10c). This shift is also apparent in the moderately extended portions of the northern Basin and Range. The shift coincides with, and may be genetically related to the growth of the Pacific–North America transform plate boundary, including the development of the San Andreas fault (Figs. 1 and 10; Pl. 8; Atwater, 1970).

Overall, the locus of extensional strain appears to have migrated from the center of the extended provinces to their periphery as extension proceeded, albeit crudely (Fig. 10c). Most of the extension in the southern Basin and Range was transferred to the opening of the Salton trough–Gulf of California rift in the west, and to the Rio Grande rift in the east. In the north, while broadly distributed, the locus of major extension shifted from the center of the province to the margins, forming a U-shaped belt around the older domains (Fig. 10c). Around the Columbia Plateaus and northward, the development of the Neogene Rocky Mountains Basin and Range and the extended portions of the Columbia Plateaus (which could include areas due west of the Idaho batholith in the central region of the Plateaus) suggests a similar pattern of outward migration of the locus of extension. The eruption of basaltic and bimodal volcanic suites in the Columbia Plateaus and northern Basin and Range appear to have followed a similar overall pattern in the Neogene, tending to concentrate on the edges of the provinces over the last few million years (Christiansen and McKee, 1978). The scarcity of significant post-Eocene normal faulting in the Omineca belt is noteworthy, perhaps a reflection of its position far to the north of the San Andreas fault.

The central Basin and Range is one area where extension did not migrate outward, but migrated westward across the entire width of the Basin and Range in post-middle Miocene time. It seems to have represented a stronghold between westward-migrating batholithic blocks and the interior of the North American plate, failing only once the blocks had already begun to drift at other latitudes. Once failure occurred, however, it developed into the most strongly extended subprovince of the Basin and Range, accommodating about 250 km of extension across a current width of 350 km.

DISCUSSION

Origin of forces causing extension

Forces responsible for spreading may include three, broad, nonexclusive classes: (1) body forces within the lithosphere that cause it to spread under its own weight (*Molnar and Chen, 1983*; Coney and Harms, 1984; *Coney, 1987*; Sonder and others, 1987; Wernicke and others, 1987); (2) tractions applied to the edges of the North American plate, for example as a result of changing relative plate motions (Atwater, 1970; Coney, 1978, *1987*; *Engebretson and others, 1984*); and (3) tractions applied to the base of the lithosphere by flow in the underlying asthenosphere (e.g., *Scholz and others, 1971*; *Elston, 1984*; *Humphreys, 1987*). There is no consensus among Cordilleran geologists regarding the causes of extension, but certain aspects of distribution and direction of spreading with time appear to bear on the importance of each of these classes of force at different times.

Forces originating within the North American plate. The importance of body foces in the early history of extension is suggested by the concentration of extension in central areas where post-compressional crustal thickness would be expected to be maximum (Coney and Harms, 1984) and by the variability of extension directions, which tend to be perpendicular to the boundaries of shortened areas (Pl. 8 and Fig. 10). Further support for the importance of gravitational effects, which also explains the diachronism of the onset of extension at various latitudes, is an apparent correspondence between the temperature of the upper mantle immediately following compression (implied by the degree of plutonism) and the time lag between compression and extension. At lithosphere scale, the geotherm should decrease upon thickening because material advection is faster than thermal reequilibration (*England and Richardson, 1977*). This effect, combined with the increased thickness of relatively weak crustal rocks, may cause the strength of the lithosphere to decrease markedly with time as thermal reequilibration occurs (*Glazner and Bartley, 1985*). Thus, in areas where maximum thickening coincided most closely with the Cretaceous batholiths (Omineca extended belt and the Rocky Mountains Basin and Range), extensional collapse occurred soon after compression because the lithosphere was initially weak. In areas where thickening lay east of the batholith belt and where the crust was only mildly intruded (northern and southern Basin and Range), the delay was longer because more time was needed to warm and weaken the lithosphere, either by conductive relaxation or upwelling of arc magmas. Areas where thickening was accompanied by relatively little plutonism would require the most time to weaken sufficiently (central Basin and Range). Heat carried by upwelling asthenosphere, perhaps related to either subduction of plates offshore or detachment and sinking of the mantle part of the North American plate, might also influence the strength and gravitational potential of the lithosphere, contributing to the effect of body forces (Sonder and others, 1987; Anderson, 1989).

Another group of observations possibly bearing on the importance of body forces is the position of the constrictional strain field in the central Basin and Range with respect to variations in regional topography and width of the province (Wernicke and others, 1988). This region lies where the northern Basin and Range narrows from about 800 km to less than 400 km wide at the latitude of Las Vegas (Fig. 1). The batholithic western margin of the Basin and Range was not rotated about a vertical axis to any significant degree during extension (*Frei, 1986*). Therefore, the percent extensional strain (and therefore net decrease in gravitational potential of the lithosphere; Sonder and others, 1987) strongly increases southward, assuming crustal thickness was approximately the same prior to extension. The abrupt south-facing monoclinal step in topography observed along the northern boundary of the central Basin and Range separates the two subprovinces, whose average elevations differ by about 1,000 m (Eaton and others, 1978; Fig. 1 and Pl. 8). The step is a direct indication of a contrast in potential. The strong component of constrictional strain in the central Basin and Range (and perhaps portions of the Walker Lane belt) may be explained by the tendency of material in the north to flow down the potential gradient, thereby causing north-south shortening during extension.

Forces applied to the edge of the North American plate. The principal plate boundary effects during Paleogene time that are likely to have influenced the development of extension include: (1) a slowing of relative plate convergence, (2) decreasing age of the subducted plate (e.g., *Engebretson and others, 1984*; Stock and Molnar, 1988), and (3) increasing dip of a nearly horizontal Benioff zone inherited from Laramide time (Coney and Reynolds, 1977).

Observations of modern subduction zones show a strong correlation between the stress regime of the overriding plate and the age of the subducted oceanic lithosphere (Molnar and Atwater, 1978), such that extension predominates where old lithosphere is consumed (e.g., western Pacific) and compression where young lithosphere is consumed (e.g., eastern Pacific). The age of subducted lithosphere also roughly correlates with slab dip, which tends to be steeper in regions of old downgoing lithosphere, presumably because of its greater negative buoyancy. Convergence rate, while possibly an important factor, may be comparably fast or slow in either tectonic regime. Thus, the three observations discussed above conflict somewhat. The steepening Benioff zone and perhaps decreased convergence rates would favor the development of extension, but the decreasing age of the subducting plate would hinder it. It is paradoxical that the downgoing slab would steepen as the age of the subducted lithosphere decreased. A slowing of westward drift of North America relative to the Atlantic hotspots at this time may have had an influence in some way (Coney, 1978, *1987*).

The mid-Miocene and younger history of the Cordillera is characterized by a regional clockwise rotation in spreading direction and least-principal-stress direction within extended areas, from east-northeast to west-northwest (*Zoback and Thompson, 1978*; Zoback and others, 1981; *Angelier and others, 1985*), concurrent with the growth of the Pacific–North America transform plate boundary. The change in plate boundary configuration may have changed the orientation of traction exerted on the edge of the North American plate from being approximately normal to the continental margin to being parallel to it (e.g., Anderson, 1989). Thus, as the transform boundary grew, the orientation of diffuse spreading within the continent changed such that spreading could absorb some component of right-lateral shear (Atwater, 1970).

Another effect proposed for the evolution of plate boundary stresses points to the Mendocino triple junction (*Ingersoll, 1982*), which is unstable and requires westward spreading to occur within the North American plate as the triple junction migrates. A related concept is the effect of the disparity in age of the downgoing plate across the Mendocino transform prior to the formation of the triple junction (Glazner and Bartley, 1984). The greater age and lesser buoyancy of the downgoing plate to the north of the transform might tend to induce extensional deformation, as the pattern of extension and volcanism in the southern Basin and Range correlates reasonably well with the calculated positions of the fracture zone through mid-Tertiary time. However, the apparently symmetrical pattern of southward migration of magmatism in the northern Basin and Range at this time is not explained by this effect.

Forces exerted on the base of the North American plate. The magmatism associated with early extension is probably related to subduction (e.g., Lipman, 1980; Coney, 1980), because subduction occurred continuously offshore during Paleogene time, and there is no other expression of arc magmatism toward the trench. This arc was unusual in comparison with most modern arcs, however, in that it consisted of migrating belts of magma about 600 to 800 km long, oriented at high angle to the trench, with localized centers more than 1,000 km inland (Lipman, 1980).

These observations suggest that forces exerted on the crust from below, driven by convective flow related to subduction, may be an important factor in the development of extension. In addition, detachment and sinking of the mantle lithosphere destabilized by compression (a physical concept potentially related to the flat-slab hypothesis for Laramide time) may play an important role in generating a flux of magma from the mantle (*Houseman and others, 1981*). Some workers have stressed the importance of a flux of mantle-derived magma (e.g., *Dickinson and Snyder, 1979*), perhaps localized over the highly extended domains (*Gans and others, 1989*), as a driving mechanism for extension. The width of mid-Tertiary magmatic belts and their widespread occurrence outside areas of strong extension indicate that the thermal disturbance of the lithosphere also occurred over broad regions, seemingly independent of any control of previous tectonic features of the crust (Anderson, 1989). The localized distribution of early extension relative to these magmatic belts casts doubt on a simple relationship between forces that might be associated with mantle magmatic flux and the localization of upper crustal extension.

The observation that magmatism in the extended domains occurs either synchronously with or slightly prior to severe (but not all) extension in a given region (Wernicke and others, 1987) suggests that a partially molten deep crust existed when major extension began (Sonder and others, 1987). Because the generation of these early magmas implies a flux of mafic material from the mantle prior to that time (Anderson, 1989), such a flux is difficult to attribute to decompression melting associated with extension, as in the model of *McKenzie (1984)* and derivatives. Complexities in patterns of sedimentation, volcanism, and extension (e.g., Bartley and others, 1988; Taylor and others, 1989), and the fact that syn-rift sedimentary basins typically predate eruptive activity (Wernicke and others, 1987) indicate caution in generalizing along these lines. A magma-rich fluid layer developed prior to extension raises the possibility that, at the onset of extension, the initial upwelling of the fluid layer near the breakaway zone triggers eruptive activity within the extended domains (Fig. 9a). By this mechanism, volcanism would be expected after a modest degree of extension but before or during the most intense phases, thereby accounting for the common occurrence of pre-eruptive sedimentary basins. Syn-extensional eruptive centers may thus be viewed as a passive response to perturbations of the upper crust. This would explain why moderate-volume eruptive centers localize early in the history of some extensional systems near proximal boundaries (areas summarized in *Gans and others, 1989*), but would allow the most voluminous eruptive centers to lie near zones of major asthenospheric upwelling, independent of the details of the timing and distribution of major extensional strain (e.g., Bartley and others, 1988; Taylor and others, 1989). From a theoretical point of view, assessment of the mechanical influence of arc magmatism on extension, particularly in regard to how mantle flow might exert tractions on the base of the lithosphere, is a difficult problem (*Coney, 1987*).

Variations in the age of the downgoing plate may affect the distribution of forces applied to the base of the North American plate. As the age of the downgoing plate beneath North America progressively decreases through the Tertiary, the time that the slab remains cold after subduction decreases (Severinghaus and Atwater, 1990). Thus, the sinking of an old plate may strongly influence the distribution of flow in the upper mantle (and hence also the forces applied to the base of the lithosphere) because it remains relatively rigid as it sinks, while the subduction of very young oceanic lithosphere might exert less influence on upper mantle flow because the plate is short-lived.

Frontiers

One of the principal problems in studying the extended lithosphere in the Cordillera and elsewhere is the existence of low-angle normal faults indicated from field and seismic-reflection data, and their absence from records of seismicity in extending regions, including the Basin and Range (Jackson, 1987; Smith and others, 1989). The paradox may in part be resolved with the sequential uplift model shown in Figure 5. Here, a regional, shallowly inclined or subhorizontal shear zone may acquire its mylonitic and brittle overprint as it moves transiently up the migrating flexure, such that all activity in the brittle crust occurs along the relatively steep (generally >30°) flexure (*Wernicke and Axen, 1988*; *Buck, 1988*; *Hamilton, 1988*).

However, the evidence for shallow initial dips along the sides of a number of stable blocks cited above is not reconciled by this argument. The observation that some well-studied Basin and Range faults (such as the Borah Peak fault in the Rocky Mountains Basin and Range) project steeply into the mid-crust (e.g., Smith and others, 1989), sometimes cited as evidence against the existence of low-angle normal faults, sheds no light on the problem, as they are simply examples of large steep normal faults, not in conflict with any known concept of Cordilleran extension.

The mechanical unlikelihood of low-angle normal faults may not be a problem when viewed in light of a low-viscosity lower crust. For a variety of assumptions, flow within a viscous layer at the base of the brittle crust is capable of causing rotations of the stress field with depth, such that the initiation of low-angle normal faults, assuming coulombic behavior in the brittle crust, may be expected (e.g., *Yin, 1989*; Melosh, 1990; Fig. 9a). The strong part of the crust may serve as a "stress guide," whose differential motion with respect to lower layers creates subhorizontal shear tractions that rotate the principal stress planes (*Lister and Davis, 1989*). Figure 9a shows that a 45° rotation of the stress field with depth is sufficient to cause initiation of subhorizontal fractures. Faults with low initial dip in the uppermost crust may simply reflect a shallower depth of upward steepening of the maximum compressive stress (Fig. 9a), although special circumstances such as preexisting anisotropies may be necessary to allow the stress field to remain rotated at depths of 0 to 5 km.

Assuming the geological data indicating initiation and active slip on low-angle normal faults is correct, the problem of reconciling slip on these faults with seismicity data on normal faults is still unresolved. The earthquakes analyzed by Jackson (1987), who concluded that seismic slip on normal fault planes inclined at less than 30° is "almost unknown," involves a sampling of 40 earthquakes over the past two decades whose fault plane solutions are well constrained and unambiguously indicate primarily normal slip. Even with the ambiguity of which nodal plane represents the fault (it is resolved for 15 of the events), it is clear that few if any dip less than 30°. Although the data are only certain to within 10 to 15°, the vast majority of the events are clearly steeper, mostly in the 40 to 60° range.

Several possibilities exist to reconcile Jackson's (1987) compilation with the existence of active low-angle normal faults. One is low recurrence intervals on shallow faults, such that the small sample of events did not record them in equal frequency to events on steeper planes. Since they transect the brittle crust at low angle, they have large surface areas and are capable of absorbing proportionately more elastic strain than steep faults. Thus, probable active low-angle normal faults such as the Sevier Desert detachment, which dips at 10° to 15° across the upper 12 to 15 km of crust (Allmendinger and others, 1983), may have 4 to 5 times the surface area of a steep basin-range fault, and have an accordingly lower recurrence interval. Thus, Jackson's (1987) sampling may not have included enough events to document an earthquake on a low-angle fault. Another possibility is that low-angle normal faults are for some reason aseismic (Jackson, 1987). It may be that steep faults are more common in the upper 7 to 10 km than shallow ones, tending to decrease in dip downward across the brittle-ductile transition zone and remain shallow across the ductile crust above the fluid layer (Fig. 9). Such a geometry would favor creep along the shallowly dipping portions at 7 to 15 km depth (Jackson, 1987). Whatever the reason, this problem is sure to remain a lively topic of debate in geodynamics.

Progress in assessing the importance of the numerous proposed causes of extension is clearly a frontier area. Improved knowledge of the timing, direction and rate of extension through time, the structure of the upper mantle, and careful continuum modeling, should lead not only to major advances in evaluating the importance of each of the effects described above, but also to the discovery of processes not yet considered.

Promising along these lines are efforts to use isotope geochemistry to constrain the source regions of magmas (e.g., Farmer, 1988; *Farmer and others, 1989*) and high-resolution tomographic studies of the structure of the upper mantle (e.g., *Humphreys, 1987*). In combination with knowledge of the upper crustal displacement field determined from geologic studies and shallow-reflection profiling, these techniques may ultimately constrain the three-dimensional strain field of the entire lithosphere during extension, shedding new light on long-debated questions such as the relationship between strain in the upper and lower parts of the lithosphere.

SUMMARY

The Cordillera has experienced a number of episodes of extension throughout its long history, most notably in Cenozoic time. Cenozoic extensional provinces are developed mainly within areas that were shortened during Mesozoic and early Cenozoic time (Fig. 1), although not all shortened areas have been significantly extended. Extension is best expressed east of the great Mesozoic batholiths, which were largely resistant to upper crustal extension (Figs. 1 and 10; Pl. 8), although important Cenozoic extension exists to the west of them as well. The magnitude of extension, variably oriented but roughly east-west (Pl. 8; Fig. 10), lies in the range of 200 to 300 km in the central and northern Basin and Range, possibly somewhat less in the southern Basin and Range and areas in the Pacific Northwest.

The mode of extension appears to vary with time in any given area. The early history of extension tends to be domainal and asymmetrical, characterized by the separation of stable upper crustal blocks along narrow zones of breakage (initially as little as 10 km wide or less) within which faults are inclined in a single direction (Figs. 4 and 8), broadly accompanied by intermediate to silicic magmatism. These blocks were able to separate from tens of kilometers up to as much as 150 km from one another (Death Valley domain), in effect completely removing the upper 7 to 15 km of crust from the strongly extended domains between them (Figs. 4 and 8). The fault zones (detachment systems) were initially inclined at average dips ranging from less than 10° to as much as 60° through the upper crust, probably on average shallowing downward. Hanging walls experienced multiple generations of domino-style normal faults (perhaps mostly at the scale of individual ranges; Fig. 5a), and sequential detachment and flexural rotation of thin fault blocks above a migrating wave of mid-crustal uplift (especially at large scale; Figs. 3c, 5b, and 6). Superimposed on this pattern, major detachments may sequentially step downward, such that higher detachments may be deformed by movement along imbricate splays of deeper ones in the same extended domain (Figs. 4 and 8). In many of the strongly extended domains, denudation has been sufficient to exhume metamorphic tectonites developed during extension along single systems of detachment and ductile simple shear, now exposed across tens of kilometers parallel to their transport directions (metamorphic core complexes). Not all detachments, however, display metamorphic tectonite in their footwalls (Figs. 3a and b).

The stable blocks are typically situated on broad topographic highs and Bouguer anomaly lows. Stable blocks and extended domains alternate with a wavelength of 150 to 200 km, about 3 to 5 times the width of a typical basin-range pair. The topographic difference between the extended areas and stable blocks suggests that the blocks are afloat on a fluid layer in the crust whose average density is probably between 2,800 and 2,900 kg/m^3 (Fig. 7). The existence of a fluid layer is independently supported by an apparent lack of deflection of the Moho across boundaries between strongly extended domains and stable blocks (Fig. 8). Detachment faults do not represent the boundary between the fluid and upper crustal layers, which probably occurs well within the ductile crust below the brittle-ductile transition of quartz (Fig. 9). Rather, they appear to be shear zones developed above the fluid layer that cut gently downward across the brittle-ductile transition and through a substantial thickness of crust below it, perhaps locally exhuming rocks initially within the fluid layer (Fig. 9).

The broad synchronism of early extension and magmatic activity suggests that the development and maintenance of the fluid layer may depend on magmatic heat. The occurrence of magmatic centers outside of strongly extended domains, including some of the most voluminous, suggests that mantle magmatic flux does not occur preferentially beneath strongly extended domains, and thus it is improbable that such a flux either localizes or provides the driving forces for their development in any simple way. Rather, synextensional magmatism within the extended domains may be a passive response to extension, triggered by the perturbation of a magma-rich fluid layer as the stable blocks initially separate.

The later stages of extension are characterized by breakage of both extended domains and stable blocks by broadly distributed, high-angle faults (classical Basin and Range block faulting) that accommodate relatively little extensional strain, and are often accompanied by mafic or bimodal magmatism. This structural style is best developed in the western part of the northern Basin and Range province, where the earlier mode of extension seems to be least well developed (Fig. 10c). Elsewhere in the extended Cordillera, block faulting may be less important than early large-magnitude extension in the development of the topography (e.g., the southern Basin and Range). In the Omineca extended belt, classical Basin and Range topography did not develop. Speculatively, the development of deep-seated block faulting in the Basin and Range may have resulted from the elimination of the fluid layer, either by expulsion from beneath the stable blocks (thereby increasing shear tractions on their undersides) or "freezing" of the layer squeezed upward into the extended domains, as predicted from the kinematics of the fluid-layer model (Figs. 7, 8, and 9). Physical models of lithosphere-scale necking of viscous or plastic layers of variable strength may explain the two wavelengths of observed deformational and geophysical variations.

The timing of initiation of major extension varies from about 50 to 55 Ma in the Pacific Northwest to as recently as about 16 Ma in the central Basin and Range. Following magmatic patterns, initial extension may have propagated along the axis of the Cordillera in two waves moving toward one another through the northern and southern subprovinces of the Basin and Range during Oligocene and early Miocene time, ultimately converging on the central Basin and Range in middle Miocene time. By this time, the extension direction and least-principal-stress direction throughout the Basin and Range province were northeast-southwest, in contrast to Oligo-Miocene and earlier times when extension directions may have been more variable (Fig. 10). The middle Miocene was apparently a time of major reorganization of the extending system. As extension was concentrated along a central axis of the Cordillera in pre-mid-Miocene time, it expanded to include the margins afterwards (Fig. 10). By late Miocene time, the extension direction and least-principal-stress direction rotated clockwise from northeast-southwest to their current northwest-southeast orientations. There is no one time at which a Cordillera-wide shift from domainal large-magnitude extension to classical Basin and Range block faulting occurred, because strong extension occurred at different times in different places throughout the Cenozoic (Fig. 10), including the development of metamorphic core complexes.

Proposals for the causes of extension are varied, complexly interrelated, difficult to model physically, and as a rule not mutually exclusive. An origin of early extension, in part related to buoyancy forces within the lithosphere inherited from Cretaceous and early Tertiary crustal thickening, explains a number of aspects of the development of extension. Similarly, the Miocene clockwise rotation in extension direction and stress field seems best explained as a consequence of the growth of the Pacific–North America transform plate boundary. Changes in relative plate motion in the early Tertiary and the effects of tractions exerted on the base of the lithosphere by mantle flow probably influenced the development of extension, but it remains difficult to assess their importance quantitatively.

The notion that normal faults that initiate or slip at low angles are mechanically unlikely may be erroneous, given the likelihood of strongly variable viscosity of deep-crustal layers. However, observations of normal faults that were active at low dips in the brittle crust apparently conflict with studies indicating that seismic slip on normal faults dipping less than 30° is rare. Resolution of key issues will require improved coordination between geochemistry (particularly isotope tracer work), seismic imaging (especially of the upper mantle), determination of the displacement field of the upper crust, and physical modeling.

REFERENCES CITED

Text references in *italics* will be found only on accompanying microfiche.

Allmendinger, R. W., and 7 others, 1983, Cenozoic and Mesozoic structure of the eastern Basin and Range from seismic reflection data: Geology, v. 11, p. 532–536.

Allmendinger, R. W., Hauge, T. A., Hauger, E. C., Potter, C. J., and Oliver, J., 1987, Tectonic heredity and the layered lower crust in the Basin and Range province, western United States, *in* Coward, M. P., Dewey, J. F., and Hancock, P. L., eds., Continental extensional tectonics: Geological Society of London Special Publication 28, p. 223–246.

Anderson, R. E., 1971, Thin-skin distension in Tertiary rocks of southwestern Nevada: Geological Society of America Bulletin, v. 82, p. 43–58.

——, 1989, Tectonic evolution of the intermontane system; Basin and Range, Colorado Plateau, and High Lava Plains, *in* Pakiser, L. C., and Mooney, W. D., Geophysical framework of the continental United States: Geological Society of America Memoir 172, p. 163–177.

Anderson, R. E., Zoback, M. L., and Thompson, G. A., 1983, Implications of selected subsurface data on the structural form and evolution of some basins in the northern Basin and Range province, Nevada and Utah: Geological Society of America Bulletin, v. 95, p. 1055–1072.

Armstrong, R. L., 1972, Low-angle (denudation) faults, hinterland of the Sevier orogenic belt, eastern Nevada and western Utah: Geological Society of America Bulletin, v. 83, p. 1729–1754.

——, 1982, Cordilleran metamorphic core complexes—from Arizona to southern California: Annual Reviews of Earth and Planetary Sciences, v. 10, p. 129–154.

Atwater, T., 1970, Implications of plate tectonics for the Cenozoic tectonic evolution of western North America: Geological Society of America Bulletin, v. 81, p. 3513–3536.

Axen, G. J., Wernicke, B. P., Skelly, M. J., and Taylor, W. J., 1990, Mesozoic and Cenozoic tectonics of the Sevier thrust belt in the Virgin River Valley area, southern Nevada, *in* Wernicke, B. P., ed., Basin and Range extensional tectonics near the latitude of Las Vegas, Nevada: Geological Society of America Memoir 176, p. 123–154.

Bartley, J. M., and Wernicke, B. P., 1984, The Snake Range decollement interpreted as a major extensional shear zone: Tectonics, v. 3, p. 647–659.

Bartley, J. M., Axen, G. J., Taylor, W. J., and Fryxell, J. E., 1988, Cenozoic tectonics of a transect through eastern Nevada near 38°N. latitude, *in* Weide, D. L. and Faber, M. L., eds., This extended land; Geological journeys in the southern Basin and Range: Geological Society of America Field Trip Guidebook, Cordilleran Section, p. 1–20.

Block, L. and Royden, L., 1990, Core complex geometries and regional scale flow in the lower crust: Tectonics, v. 9, p. 557–567.

Braile, L. W., Hinze, W. J., von Frese, R.R.B., and Keller, G. R., 1989, Seismic properties of the crust and uppermost mantle of the conterminous United States and adjacent Canada, *in* Pakiser, L. C., and Mooney, W. D., eds., Geophysical framework of the continental United States: Geological Society of America Memoir 172, p. 655–680.

Burchfiel, B. C., Hodges, K. V., and Royden, L. H., 1987, Geology of Panamint Valley–Saline Valley pull-apart system, California; Palinspastic evidence for low-angle geometry of a Neogene range-bounding fault: Journal of Geophysical Research, v. 92, p. 10422–10426.

Chamberlin, R. M., 1983, Cenozoic domino-style crustal extension in the Lemitar Mountains, New Mexico; A summary, *in* Chapin, C. E., and Callender, J. F., eds., Soccorro Region II: Soccorro, New Mexico Geological Society, Guidebook 34, p. 111–118.

Christiansen, R. L., and McKee, E. H., 1978, Late Cenozoic volcanic and tectonic evolution of the Great Basin and Columbia intermontane regions, *in* Smith, R. B., and Eaton, G. P., eds., Cenozoic tectonics and regional geophysics of the western Cordillera: Geological Society of America Memoir 172, p. 283–312.

Coney, P. J., 1978, Mesozoic-Cenozoic Cordilleran plate tectonics, *in* Smith, R. B., and Eaton, G. P., eds., Cenozoic tectonics and regional geophysics of the western Cordillera: Geological Society of America Memoir 152, p. 33–50.

——, 1980, Cordilleran metamorphic core complexes—an overview, *in* Crittenden, M. D., Coney, P. J., and Davis, G. H., eds., Cordilleran metamorphic core complexes: Geological Society of America Memoir 153, p. 4–34.

Coney, P. J., and Harms, T. A., 1984, Cordilleran metamorphic core complexes; Cenozoic extensional relics of Mesozoic compression: Geology, v. 12, p. 550–554.

Coney, P. J., and Reynolds, S. J., 1977, Cordilleran Benioff zones: Nature, v. 270, p. 403–406.

Crittenden, M. D., Jr., Coney, P. J., and Davis, G. H., eds., 1980, Cordilleran metamorphic core complexes: Geological Society of America Memoir 153, 490 p.

Davis, G. A., and Lister, G. S., 1988, Detachment faulting in continental extension; Perspectives from the southwestern U.S. Cordillera, *in* Clark, S. P., Jr., Burchfiel, B. C., and Suppe, J., eds., Processes in continental lithospheric deformation: Geological Society of America Special Paper 218, p. 133–160.

Davis, G. A., Lister, G. S., and Reynolds, S. J., 1986, Structural evolution of the Whipple and South Mountains shear zones, southwestern United States: Geology, v. 14, p. 7–10.

Davis, G. H., 1980, Structural characteristics of metamorphic core complexes, southern Arizona, *in* Crittenden, M. D., Coney, P. J., and Davis, G. H., eds., Cordilleran metamorphic core complexes: Geological Society of America Memoir 153, p. 131–158.

Davis, G. H., and Coney, P. J., 1979, Geologic development of the Cordilleran metamorphic core complexes: Geology, v. 7, p. 120–124.

Dokka, R. K., 1989, The Mojave extensional belt of southern California: Tectonics, v. 8, p. 363–390.

Eaton, G. P., 1982, The Basin and Range province; Origin and tectonic significance: Annual Reviews of Earth and Planetary Sciences, v. 8, p. 409–440.

Eaton, G. P., Wahl, R. R., Prostka, H. J., Mabey, D. R., and Kleinkopf, M. D., 1978, Regional gravity and tectonic patterns; Their relation to Late Cenozoic epeirogeny and lateral spreading in the western Cordillera, *in* Smith, R. B., and Eaton, G. P., eds., Cenozoic tectonics and regional geophysics of the western Cordillera: Geological Society of America Memoir 152, p. 51–92.

Farmer, G. L., 1988, Isotope geochemistry of Mesozoic and Tertiary igneous rocks in the western United States and implications for the structure and composition of the deep continental lithosphere, *in* Ernst, W. G., ed., Metamorphism and crustal evolution of the western United States, Rubey Volume VII: Englewood Cliffs, New Jersey, Prentice Hall, p. 87–109.

Foster, D. A., Harrison, T. M., Miller, C. F., and Howard, K. A., 1990, The $^{40}Ar/^{39}Ar$ thermochronology of the eastern Mojave Desert, California, and adjacent western Arizona with implications for the evolution of metamorphic core complexes: Journal of Geophysical Research, v. 95, p. 20002–20024.

Frost, E. G., and 9 others, 1987, Crustal transect; Colorado Plateau–detachment terrane–Salton Trough, *in* Davis, G. H., and VandenDolder, E. M., eds., Geologic diversity of Arizona and its margins; Excursions to choice areas: Arizona Bureau of Geology and Mineral Technology Special Paper 5, p. 398–422.

Gans, P. B., 1987, An open-system, two-layer crustal stretching model for the eastern Great Basin: Tectonics, v. 6, p. 1–12.

Glazner, A. F., and Bartley, J. M., 1984, Timing and tectonic setting of Tertiary low-angle normal faulting and associated magmatism in the southwestern United States: Tectonics, v. 3, p. 385–396.

Guth, P. L., 1981, Tertiary extension north of the Las Vegas Valley shear zone, Sheep and Desert Ranges, Clark County, Nevada: Geological Society of America Bulletin, v. 92, p. 763–771.

Hamilton, W. B., 1987, Crustal extension in the Basin and Range province, southwestern United States, *in* Coward, M. P., Dewey, J. F., and Hancock, P. L., eds., Continental extensional tectonics: Geological Society of London Special Publication 28, p. 155–176.

Hamilton, W., and Myers, W. B., 1966, Cenozoic tectonics of the western United States: Reviews of Geophysics, v. 4, p. 509–549.

Hauser, E. C., and 9 others, 1987, Crustal structure of eastern Nevada from CO-CORP deep seismic reflection data: Geological Society of America Bulletin, v. 99, 833–844.

Hill, D. P., 1982, Contemporary block tectonics; California and Nevada: Journal of Geophysical Research, v. 87, p. 5433–5450.

Hodges, K. V., and 8 others, 1989, Evolution of extensional basins and Basin and Range topography west of Death Valley, California: Tectonics, v. 8, p. 453–469.

Howard, K. A., and John, B. E., 1987, Crustal extension along a rooted system of low-angle normal faults: Colorado River extensional corridor, California and Arizona, *in* Coward, M. P., Dewey, J. F., and Hancock, P. L., eds., Continental extensional tectonics: Geological Society of London Special Publication 28, p. 299–312.

Jackson, J. A., 1987, Active normal faulting and crustal extension, *in* Coward, M. P., Dewey, J. F., and Hancock, P. L., eds., Continental extensional tectonics: Geological Society of London Special Publication 28, p. 3–18.

John, B. E., 1988, Structural reconstruction and zonation of a tilted mid-crustal magma chamber; The felsic Chemehuevi Mountains plutonic suite: Geology, v. 16, p. 613–617.

Jones, C. H., 1987, Is extension in Death Valley accommodated by thinning of the mantle lithosphere beneath the Sierra Nevada, California?: Tectonics, v. 6, p. 449–475.

Keller, G. R., Morgan, P., and Seager, W. R., 1990, Crustal structure, gravity anomalies and heat flow in the southern Rio Grande rift and their relationship to extensional tectonics: Tectonophysics, v. 174, p. 21–37.

Klemperer, S. L., Hauge, T. L., Hauser, E. C., Oliver, J. E., and Potter, C. J., 1986, The Moho in the northern Basin and Range province, Nevada, along the COCORP 40°N seismic-reflection transect: Geological Society of America Bulletin, v. 97, p. 603–618.

Lachenbruch, A. H., Sass, J. H., and Galanis, S. P., 1985, Heat flow in southernmost California and the origin of the Salton trough: Journal of Geophysical Research, v. 90, p. 6709–6736.

Lee, J., Miller, E. L., and Sutter, J. F., 1987, Ductile strain and metamorphism in an extensional tectonic setting; A case study from the northern Snake Range, Nevada, USA, *in* Coward, M. P., Dewey, J. F., and Hancock, P. L., eds., Continental extensional tectonics: Geological Society of London Special Publication 28, p. 267–298.

Levy, M., and Christie-Blick, N., 1989, Pre-Mesozoic palinspastic reconstruction of the eastern Great Basin (western United States): Science, v. 245, p. 1454–1462.

Lipman, P. W., 1980, Cenozoic volcanism in the western United States; Implications for continental tectonics, *in* Burchfiel, B. C., Oliver, J. E., and Silver, L. T., Continental Tectonics: Washington, D.C., National Research Council, National Academy of Sciences, Studies in Geophysics, p. 161–175.

McCarthy, J., and Thompson, G. A., 1988, Seismic imaging of extended crust with emphasis on the western United States: Geological Society of America Bulletin, v. 100, p. 1361–1374.

Melosh, H. J., 1990, Mechanical basis for low-angle normal faulting in the Basin and Range province: Nature, v. 343, p. 331–335.

Miller, E. L., Gans, P. B., and Garing, J., 1983, The Snake Range decollement; An exhumed mid-Tertiary ductile-brittle transition: Tectonics, v. 2, p. 239–263.

Molnar, P., and Atwater, T., 1978, Interarc spreading and Cordilleran tectonics as alternates related to the age of subducted oceanic lithosphere: Earth and Planetary Science Letters, v. 41, p. 330–340.

Okaya, D. A., and Thompson, G. A., 1986, Involvement of deep crust in extension of Basin and Range province, *in* Mayer, L., ed., Extensional tectonics of the southwestern United States; A perspective on processes and kinematics: Geological Society of America Special Paper 208, p. 15–22.

Parrish, R. R., Carr, S. D., and Parkinson, D. L., 1988, Eocene extensional tectonics and geochronology of the southern Omineca belt, British Columbia and Washington: Tectonics, v. 7, p. 181–212.

Potter, C., and 7 others, 1986, COCORP deep seismic reflection traverse of the interior of the North American Cordillera, Washington and Idaho: Tectonics, v. 5, p. 1007–1027.

Proffett, J. M., Jr., 1977, Cenozoic geology of the Yerington district, Nevada, and implications for the nature and origin of Basin and Range faulting: Geological Society of America Bulletin, v. 88, p. 247–266.

Reynolds, S. J., 1985, Geology of the South Mountains, central Arizona: Arizona Bureau of Geology and Mineral Technology Bulletin 195, 61 p.

Reynolds, S. J., and Spencer, J. E., 1985, Evidence for large-scale transport on the Bullard detachment fault, west-central Arizona: Geology, v. 13, p. 353–356.

Reynolds, S. J., Richard, S. M., Haxel, G. B., Tosdal, R. M., and Laubach, S. E., 1988, Geologic setting of Mesozoic and Cenozoic metamorphism in Arizona, *in* Ernst, W. G., ed., Metamorphism and crustal evolution of the western United States, Rubey Volume VII: Englewood Cliffs, New Jersey, Prentice-Hall, p. 466–501.

Richard, S. M., Fryxell, J. E., and Sutter, J. F., 1990, Tertiary structure and thermal history of the Harquahala and Buckskin Mountains, west-central Arizona; Implications for denudation by a major detachment fault system: Journal of Geophysical Research, v. 95, p. 19973–19989.

Rosendahl, B. R., 1987, Architecture of continental rifts with special reference to east Africa: Annual Reviews of Earth and Planetary Sciences, v. 15, p. 445–503.

Severinghaus, J., and Atwater, T., 1990, Cenozoic geometry and thermal state of the subducting slabs beneath western North America, *in* Wernicke, B. P., ed., Basin and Range extensional tectonics near the latitude of Las Vegas, Nevada: Geological Society of America Memoir 176, p. 1–22.

Smith, R. B., Nagy, W. C., Julander, K. A., Viveiros, J. J., Barker, C. A., and Gants, D. G., 1989, Geophysical and tectonic framework of the eastern Basin and Range–Colorado Plateau–Rocky Mountain transition, *in* Pakiser, L. C., and Mooney, W. D., eds., Geophysical framework of the continental United States: Boulder, Colorado, Geological Society of America Memoir 172, p. 205–233.

Smith, R. B., and Bruhn, R. L., 1984, Intraplate extensional tectonics of the eastern Basin and Range; Inferences on the structural style from seismic reflection data, regional tectonics, and thermal-mechanical models of brittle-ductile deformation: Journal of Geophysical Research, v. 89, p. 5733–5762.

Smith, R. B., and Eaton, G. P., 1978, Cenozoic tectonics and regional geophysics of the western Cordillera: Geological Society of America Memoir 152, 388 p.

Snoke, A. W., and Miller, D. M., 1988, Metamorphic and tectonic history of the northeastern Great Basin, *in* Ernst, W. G., ed., Metamorphism and crustal evolution of the western United States, Rubey Vol. VII: Englewood Cliffs, New Jersey, Prentice Hall, p. 606–648.

Sonder, L. J., England, P. C., Wernicke, B. P., Christiansen, R. L., 1987, A physical model for Cenozoic extension of western North America, *in* Coward, M. P., Dewey, J. F., and Hancock, P. L., eds., Continental extensional tectonics: Geological Society of London Special Publication 28, p. 187–202.

Spencer, J. E., 1984, Role of tectonic denudation in warping and uplift of low-angle normal faults: Geology, v. 12, p. 95–98.

Spencer, J. E., and Reynolds, S. J., 1989, Middle Tertiary tectonics of Arizona and adjacent areas, *in* Jenny, J. P., and Reynolds, S. J., eds., Geologic evolution of Arizona: Tucson, Arizona Geological Society Digest, v. 17, p. 539–574.

Stewart, J. H., 1971, Basin and Range structure; A system of horsts and grabens produced by deep-seated extension: Geological Society of America Bulletin, v. 82, p. 1019–1044.

——, 1978, Basin-range structure in western North America; A review, *in* Smith, R. B., and Eaton, G. P., eds., Cenozoic tectonics and regional geophysics of the western Cordillera: Geological Society of America Memoir 152, p. 1–32.

——, 1979, Regional tilt patterns of late Cenozoic basin-range fault blocks, western United States: Geological Society of America Bulletin, v. 91, pt. 1, p. 460–464.

——, 1988, Tectonics of the Walker Lane belt, western Great Basin; Mesozoic and Cenozoic deformation in a zone of shear, *in* Ernst, W. G., ed., Metamorphism and crustal evolution of the western United States, Rubey Vol. VII: Englewood Cliffs, New Jersey, Prentice Hall, p. 683–713.

Stock, J., and Molnar, P., 1988, Uncertainties and implications of the Late Cretaceous and Tertiary positions of North America relative to the Farallon, Kula, and Pacific plates: Tectonics, v. 7, p. 1339–1384.

Taylor, W. J., Bartley, J. M., Lux, D. R., and Axen, G. J., 1989, Timing of Tertiary extension in the Railroad Valley–Pioche transect, Nevada: Journal of Geophysical Research, v. 94, p. 7757–7774.

Thompson, G. A., and Burke, D. B., 1974, Regional geophysics of the Basin and Range province: Annual Reviews of Earth and Planetary Sciences, v. 2, p. 213–238.

Walker, J. D., Bartley, J. M., and Glazner, A. F., 1990, Large-magnitude Miocene extension in the central Mojave Desert; Implications for Paleozoic to Tertiary paleogeography and tectonics: Journal of Geophysical Research, v. 95, p. 557–570.

Wernicke, B., 1985, Uniform-sense normal simple shear of the continental lithosphere: Canadian Journal of Earth Sciences, v. 22, p. 108–125.

——, 1990, The fluid crustal layer and its implications for continental dynamics, *in* Salisbury, M., and Fountain, D. M., eds., Exposed cross sections of the continental crust, NATO Advanced Studies Institute: Dordrecht, Holland, Kluwer Academic Publishers, D. Riedel, p. 509–544.

Wernicke, B., Christiansen, R. L., England, P. C., and Sonder, L. J., 1987, Tectonomagmatic evolution of Cenozoic extension in the North American Cordillera, *in* Coward, M. P., Dewey, J. F., and Hancock, P. L., eds., Continental extensional tectonics: Geological Society of London Special Publication 28, p. 203–221.

Wernicke, B., Axen, G. J., and Snow, J. K., 1988, Basin and Range extensional tectonics at the latitude of Las Vegas, Nevada: Geological Society of America Bulletin, v. 100, p. 1738–1757.

Wright, L. A., and Troxel, B. W., 1973, Shallow-fault interpretation of Basin and Range structure, southwestern Great Basin, *in* DeJong, K. A., and Scholten, R., eds., Gravity and tectonics: New York, John Wiley and Sons, p. 397–407.

Wust, S. L., 1986, Regional correlation of extension directions in Cordilleran metamorphic core complexes: Geology, v. 14, p. 828–830.

Zoback, M. L., Anderson, R. E., and Thompson, G. A., 1981, Cainozoic evolution of the state of stress and style of tectonism in the Basin and Range province of the western United States: Philosophical Transactions of the Royal Society of London, v. A300, p. 407–434.

Zuber, M. T., Parmentier, E. M., and Fletcher, R. C., 1986, Extension of continental lithosphere; A model for two scales of Basin and Range extension: Journal of Geophysical Research, v. 91, p. 4826–4838.

MANUSCRIPT ACCEPTED BY THE SOCIETY APRIL 24, 1991

ACKNOWLEDGMENTS

This chapter was improved by the constructive reviews of R. E. Anderson, G. A. Davis, W. B. Hamilton, J. McCarthy, E. L. Miller, S. J. Reynolds, G. A. Thompson, M. L. Zoback, and an anonymous reviewer. Conversations with R. L. Armstrong, G. J. Axen, J. M. Bartley, D. K. Holm, S. Kruse, M. K. McNutt, R. J. O'Connell, B. M. Sheffels, J. K. Snow, and G. A. Thompson substantially influenced views on the tectonics of the middle and lower crust expressed in the manuscript. This work was supported by NSF grants EAR-84-51181 and EAR-86-17869, and grants from Texaco Incorporated and Exxon Production Research Company. Special thanks to Carolyn White and Lilo Gallagher for their tireless efforts in the difficult tasks of producing multi-colored figures and compiling more than 200 references, and to Mary Lou Zoback for editorial rigor beyond expectation.

Printed in U.S.A.

The Geology of North America
Vol. G-3, The Cordilleran Orogen: Conterminous U.S.
The Geological Society of America, 1992

Chapter 13

Fold and thrust tectonics of the western United States exclusive of the accreted terranes

Richard W. Allmendinger
Department of Geological Sciences, Cornell University, Ithaca, New York 14853

INTRODUCTION

Compressional tectonism within the Cordilleran orogen of the western United States has occurred sporadically, both temporally and spatially, since at least the end of the Devonian Period and has continued until the Present. The focus of this chapter, however, will be on the Mesozoic and early Cenozoic deformation, when the Cordillera experienced the widest and most diverse development of contractional structures. These structures are relatively completely described and illustrate a full range of structural styles. The provinces treated here are: the basement uplifts of the Rocky Mountain foreland, and the Cordilleran thrust belt and hinterland (as far west as the Sierra Nevada and Idaho Batholiths), including the southern Cordillera of Arizona and southeastern California (Fig. 1).

This chapter describes the structural geometry and kinematics of the various types of contractional structures. Specifically addressed are "thick-skinned" versus "thin-skinned" deformation, basement involvement in compressional deformation, magnitudes of shortening in different parts of the region, and the relations between preexisting paleotectonic features and contractional structural geometry. Although other chapters in this volume deal more directly with events during particular time periods, some attention to timing must be given here, particularly in relation to structural progression and shortening rates.

The following fold and thrust provinces, mostly the accreted terranes, have been omitted completely due to space limitations and their coverage in other parts of this volume: the Franciscan of the California Coast Ranges, the Klamath and Sierran Foothills, the Olympic Mountains and Northern Cascades, and thrusts associated with the Antler and Sonoma orogenies. Much of our understanding of these provinces comes from stratigraphic data; see the chapters in this volume by Saleeby and others, E. Miller and others, and D. Cowan and others. Also omitted are contractional structures associated with the San Andreas and related strike-slip fault system (Crowell and Stewart, this volume[1]).

OROGENIC NOMENCLATURE

Much of the deformation described here occurred during the broad span of time between the mid-Jurassic and the Eocene (>120 m.y.) and has variously been assigned to the Nevadan, Sevier, and Laramide orogenies. The Nevadan orogeny, defined by Wilmarth (1938) as Late Jurassic and earliest Cretaceous diastrophism, is associated with folds and faults in the Sierra Nevada Mountains that are now known to be Oxfordian and Kimmeridgian in age (163 to 152 Ma; Nokleberg and Kistler, 1980; *Schweickert* in Ernst, 1981). Armstrong (1968) defined the Sevier orogeny as occurring approximately between the beginning of the Cretaceous and the Campanian (~145 to 75 Ma), although he recognized the uncertainty of its upper and lower bounds. The Laramide orogeny is generally recognized to have occurred between the Maastrichtian and the middle Eocene (Wilmarth, 1938; *Gries* in Lowell, 1983).

This orogenic nomenclature has become increasingly cumbersome in recent years for several reasons. The Nevadan has become so tightly defined spatially and temporally that it is difficult to apply it in a strict sense to any region outside of the Sierra Nevada, although many parts of the Cordillera have deformation that occurred generally between the mid-Jurassic and the Early Cretaceous. The principal contribution of Armstrong's (1968) classic paper on the Sevier belt was to emphasize the distinction between the structural styles of the Cordilleran thrust belt and the basement deformation of the Rocky Mountain foreland. However, those structural styles have been equated with the orogenic names "Sevier" and "Laramide," with the further implication that there was little significant overlap between the two. In fact, foreland basement and thrust-belt deformation had similar timing.

[1]The extensive number of references and unavoidable space limitations have necessitated the use of a two-tier system of literature citations. Those references appearing in normal type may be found at the back of this chapter. Those in italics are listed in a supplemental reference list on microfiche in the back of the book. Articles in books are listed with the author name in italics and the editor and year in normal type (for example: *Smith* in Jones, 1982); the book, listed under its editor's name, will be found in the main reference list with the complete citation listed with the supplemental list. Other articles in the supplemental list include those superceded by newer publications, abstracts, theses, articles that address primarily extensional deformation, and articles on the Cordillera outside of the United States.

Allmendinger, R. W., 1992, Fold and thrust tectonics of the western United States exclusive of the accreted terranes, *in* Burchfiel, B. C., Lipman, P. W., and Zoback, M. L., eds., The Cordilleran Orogen: Conterminous U.S.: Boulder, Colorado, Geological Society of America, The Geology of North America, v. G-3.

More than half of the total shortening (55 out of a total of 104 km in section X–X′ of Royse and others, 1975) in the Idaho–Wyoming salient of the Sevier orogenic belt occurred at the same time as the formation of the Laramide basement uplifts of the Rocky Mountain foreland (times as defined by Armstrong, 1968). Furthermore, new age data and reevaluation of existing data have called into question whether any of the deformation in the original Sevier orogenic belt north of southern Nevada is older than mid-Cretaceous (~Albian in age; *Lawton* in Nielson, 1982; Lawton, 1985; Heller and others, 1986; *Villien and Kligfield* in Peterson, 1986).

In general, thrusting and folding occurred throughout the Cordillera at numerous different times in different places (Fig. 2); attempts to force individual structures into the threefold division have mostly resulted in confusion. This sporadic spatial and temporal distribution should not be surprising in light of recent concepts of subduction zone segmentation and terrane accretion at the plate margin. Thus, the terms Nevadan, Sevier, and Laramide are used sparingly in this chapter.

BASEMENT UPLIFTS OF THE ROCKY MOUNTAIN FORELAND

The Rocky Mountains foreland comprises a number of broad structural blocks of Precambrian crystalline basement (Fig. 3). The structural style of the region commonly is characterized as "thick-skinned" in contrast to the "thin-skinned" thrust belt to the west. These terms have their roots in Chamberlin's (1919) description of the Colorado Rockies as "thick-shelled" and the Appalachians as "thin-shelled" mountain ranges and took their present form in Rodger's (1949) discussion of the Appalachians. The distinction is somewhat artificial (*Kulik and Schmidt* in Schmidt and Perry, 1988); some thin-skinned belts involve basement, and some faults in thick-skinned provinces have dips as low as those found in thin-skinned regions. As used here, thick-skinned refers to structural detachment at middle or deep crustal levels (if detached within the crust at all); it carries no connotation either with respect to fault dip, or with respect to basement involvement.

Much controversy has surrounded the kinematic and dynamic origin of Rocky Mountain foreland deformation, resulting in strikingly different interpretations of subsurface fault geometry (Fig. 4; compare Matthews, 1978, with Lowell, 1983). Other significant elements of the debate include the large variation in structure trends exhibited by the region and the roles of basement and sedimentary cover in the deformation (*Stearns* in Matthews, 1978; Brown, 1984). Much of the seismic reflection and industry borehole data published in the last ten years (e.g., Smithson and others, *1978,* 1979; Gries, 1983), and described below, clearly favors a kinematic interpretation of the region with horizontal contraction dominating over vertical uplift.

Three-dimensional structural geometry

Map distribution. The region of Rocky Mountain foreland deformation can be divided into four subprovinces (Fig. 3):

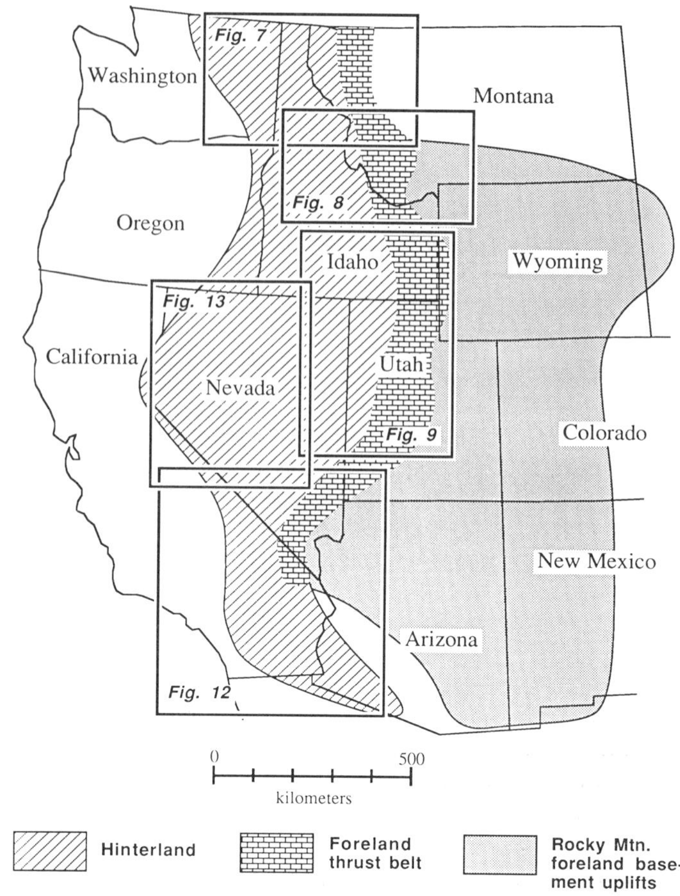

Figure 1. Regional map of the western United States showing the locations of the three major provinces discussed in the text. Also shown are the locations of Figures 7, 8, 9, 12, and 13. The area of the Rocky Mountain foreland basement uplifts is shown in detail in Figure 3.

(I) Wyoming and northeastern Utah, (II) the Rockies of Colorado and New Mexico, (III) the Colorado Plateau, and (IV) a western overlap region. The first subprovince displays the broadest variation in structural trends, with virtually every compass azimuth represented. Northwest (Wind River Range, Bighorns, Beartooths) and west-northwest (Granite and Owl Creek Mountains) trends are particularly prominent, although the east-trending Uinta Mountains, located at the southern margin of the first subprovince, are a major anomaly. The Wyoming–northern Utah subprovince is also characterized morphologically by isolated uplifts and braod, flat bains commonly one to two times wider than the adjoining ranges, although in cross section the basins are highly asymmetric with thick Tertiary depocenters near the uplifts. The Rockies of Colorado and New Mexico vary much less in their orientation, with northerly to north-northwest trends dominating. The intervening basins are narrow and deep. The southern part of the second subprovince has been overprinted by Cenozoic extensional tectonism associated with the Rio Grande Rift. Laramide faults and folds of the Colorado Plateau have much smaller structural relief, with the principal surface manifestation a series of broad flexures with monoclinal boundaries. Although a variety of structural trends are apparent, N20°W and N55°E are particularly well represented (*Davis* in Matthews, 1978; Kelley, 1955). Finally, along the western edge of the Rocky Mountains, there is an overlap between thin-skinned structures of the Cordilleran thrust belt and the thick-skinned uplifts of the foreland.

Relations to preexising tectonic features. A number of authors have suggested that older paleotectonic features controlled the map pattern of the Laramide Rocky Mountain foreland uplifts. The Rocky Mountains of Colorado and New Mexico correlate closely with the region of the Pennsylvanian Ancestral Rocky Mountains (Mallory, 1972; *Chapin and Seager, 1975*; Chapin, 1979) and also occur within the region of Proterozoic crust lying to the south of the Nash Fork–Mullen Creek shear zone of the Colorado Lineament (*Tweto and Sims, 1963*; Tweto, 1975; Warner, 1978; *Allmendinger and others,* 1982). For the monoclines of the Colorado Plateau, which also are underlain by Proterozoic crust, *Davis* (in Matthews, 1978) proposed that their orientation was controlled by old fractures within the basement. In contrast, the crust beneath the Wyoming–northern Utah subprovince is Archean and was not as strongly affected by the Ancestral Rockies event. The Uinta Mountains expose a very thick sequence of Proterozoic quartzites not found to the north or south, and is considered by some to be the site of a Precambrian aulacogen (Hoffman and Crittenden, cited in Stewart and Poole, 1974). *Bruhn and others* (*1983,* in Peterson, 1986) proposed that the Precambrian aulacogen was inverted and experienced left-lateral slip during Laramide shortening. In a somewhat similar fashion, the upper Paleozoic Snowcrest trough was also inverted during subsequent deformation to form the Blacktail-Snowcrest uplift (*Perry and others* in Schmidt and Perry, 1988). The northern limit of the Rocky Mountains, including both the Wyoming and overlap subprovinces, generally coincides with the edge of the Precambrian Belt Basin of central and northern Montana, another proposed Precambrian aulacogen (Harrison and others, 1974; *Schmidt and Garihan* in Lowell, 1983; Ruppel and Lopez, 1984). Finally, the entire province lies east of the uppermost Precambrian to mid-Devonian hingeline of the Cordilleran miogeocline (*Kulik and Schmidt* in Schmidt and Perry, 1988).

Despite these obvious and probably important spatial correlations between paleotectonic features and the map distribution of foreland basement uplifts, there are few places where individual fault reactivation has been conclusively demonstrated, and in many places, Rocky Mountain foreland structures crosscut older trends. The best cases for individual fault reactivation can be made for the Precambrian shear zones in the northern Front Ranges of Colorado and southeastern Wyoming, and for Precambrian fractures and faults in southwestern Montana and underlying Laramide monoclines in the Colorado Plateau (Hoppin and Palmquist, 1965; *Davis* in Matthews, 1978; *Reches* in Matthews, 1978; Weimer, 1980; *Schmidt and Garihan* in Lowell, 1983).

Geometry of major faults. The difference between horizontal shortening and vertical uplift schools is most acute in the interpretation of subsurface fault geometries (Fig. 4). The problem is exacerbated by the commonly poor exposure and/or limited structural relief of the mountain-bounding faults. Nonetheless, the available outcrop (e.g., Rattlesnake Mountain: Stearns, 1971; Colorado Plateau: *Davis* in Matthews, 1978; *Reches* in

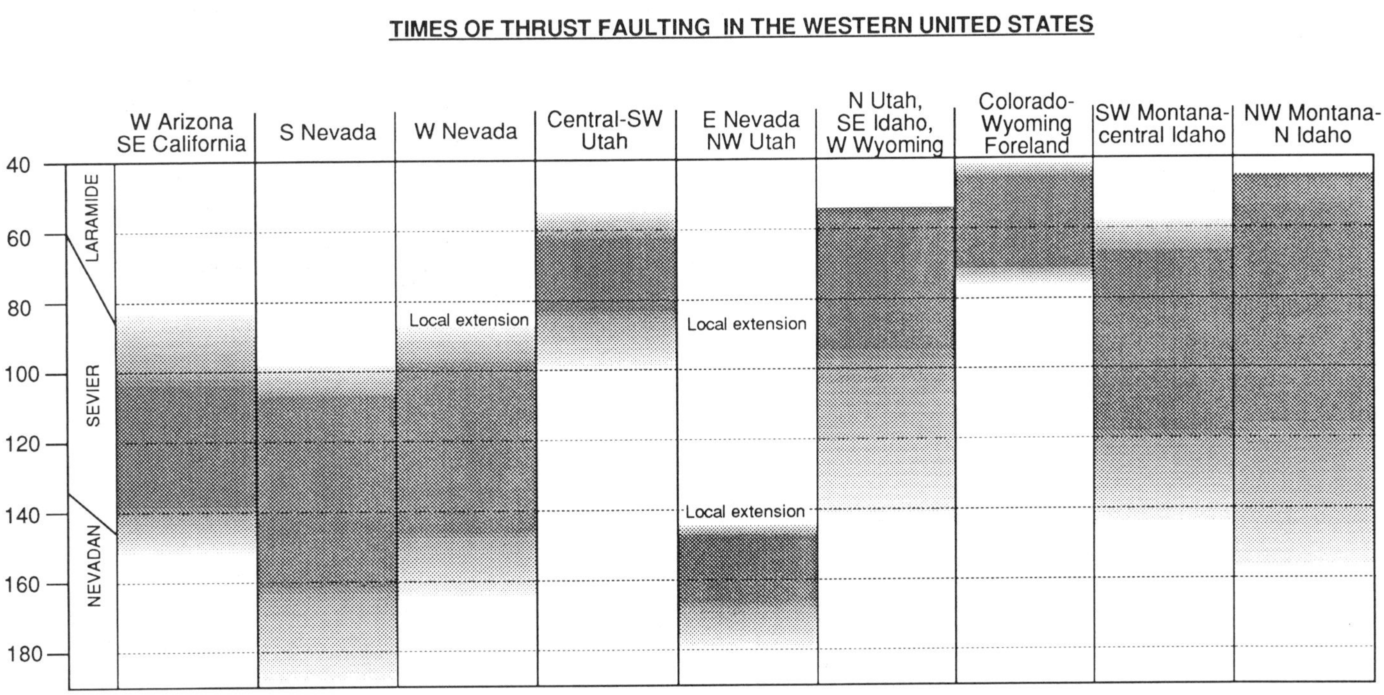

Figure 2. Summary of times of deformation in the Cordillera of the western United States. Orogenic nomenclature on left from Armstrong (1968).

Matthews, 1978; northern Wind Rivers: Mitra, 1984; Big Horn Mountains: Hoppin, 1970; *Palmquist* in Matthews, 1978) shows moderate to high-angle (30° to 90° dipping) reverse faults associated with the uplifts. Very low-angle thrust faults commonly are observed in secondary structures in the sedimentary cover (see below) and locally also are associated with the major uplifts of basement over cover (e.g., *O'Neill, 1981*).

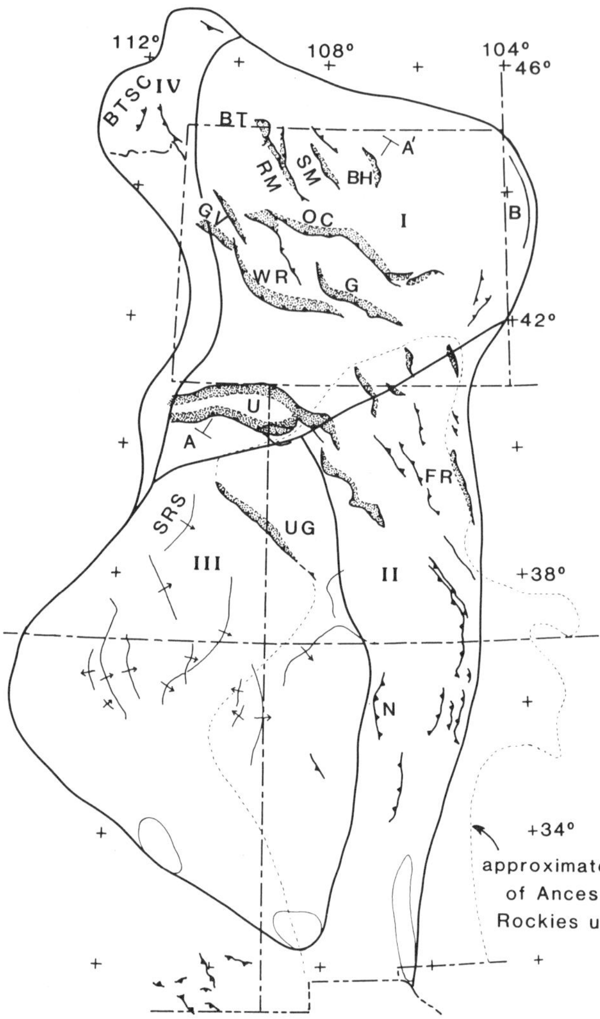

Figure 3. Generalized structure map of the Rocky Mountain foreland basement uplifts. Bold lines and Roman numerals I to IV show the locations of the four subprovinces described in the text. The boundary between subprovinces I and II corresponds to the transition between Archean crust to the north and Proterozoic crust to the south. Light dashed line shows the general extent of Ancestral Rockies uplifts (from Mallory, 1972; *Chapin and Seager, 1975*). Stippled region behind major thrust faults in Wyoming and northern Utah and Colorado shows the approximate amount of overlap of basement over Phanerozoic strata (from Gries, 1983); shortening along the line A–A′ is ~56 km or ~15 percent. Vergence of monoclines on the Colorado Plateau indicated with a single arrow. B = Black Hills; BH = Bighorn Mountains; BT = Bear Tooth Mountains; BTSC = Blacktail-Snowcrest uplift; FR = Front Ranges; G = Granite Mountains; GV = Gros Ventre Mountains; N = Nacimiento fault zone; OC = Owl Creek Mountains; RM = Rattlesnake Mountain; SM = Sheep Mountain; SRS = San Rafael Swell; U = Uinta Mountains; UG = Uncompahgre uplift; WR = Wind River Mountains.

Seismic reflection and industry borehole data provide particularly valuable insight into the nature of the bounding structures at depth. Berg (1962) marshalled evidence from drilling data to demonstrate that some of the major uplifts exhibited considerable overhang of basement over sedimentary cover. More recently, Gries (1983, in Lowell, 1983) summarized data from 16 wells that had drilled Phanerozoic strata beneath thrust plates of crystalline basement as well as numerous industry seismic data to show that many of the uplifts are bounded by gently to moderately dipping thrust faults. Using data on overhangs presented by Gries (1983, in Lowell, 1983; Fig. 3), an average northeast-southwest shortening of ~15 percent can be determined for the Wyoming subprovince. Other industry seismic reflection data that convincingly demonstrate the dominance of horizontal shortening have been published by *Wellborn* (in Veal, 1977), Clement (1983), *Frahme and Vaughn* (in Lowell, 1983), *Ray and Berg, Basham and Martin, Stone,* and *Livesey* (all in Gries and Dyer, 1985), and *Thompson and Hill* (in Barazangi and Brown, 1986). In contrast, several interpretations of industry seismic data (e.g., *Beggs* and *Davis and Young* both in Veal, 1977; *Sacrison* in Matthews, 1978) have emphasized vertical uplift over horizontal shortening, but many of those profiles either do not extend sufficiently far into the basement uplift or the fault trajectory is not defined in the basement beneath hanging wall and footwall sedimentary sections.

The COCORP Wind River profile provides some of the most important data bearing on the nature of the bounding faults (Smithson and others, *1978,* 1979; *Brewer and others, 1980*). These deep reflection data (Fig. 5) show virtually unequivocal

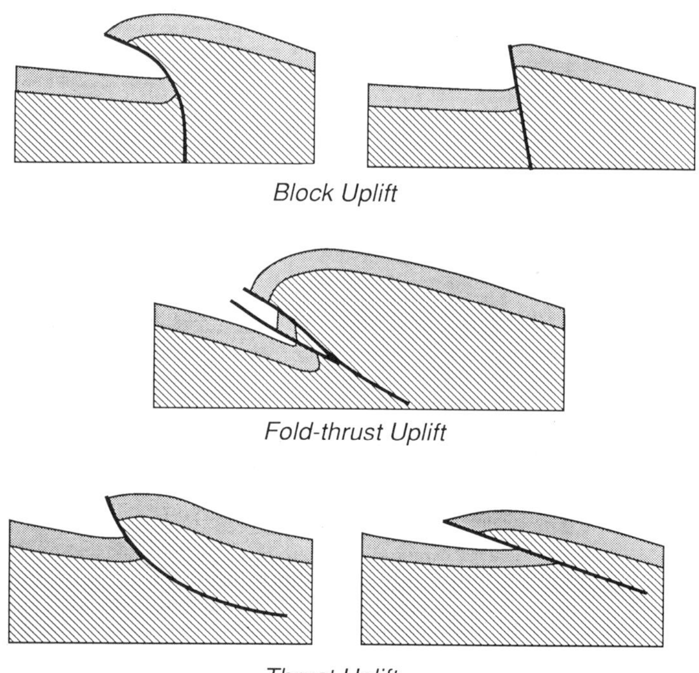

Figure 4. Alternative generalized cross sections of Rocky Mountain foreland basement uplifts (slightly modified from *Smithson and others, 1978*).

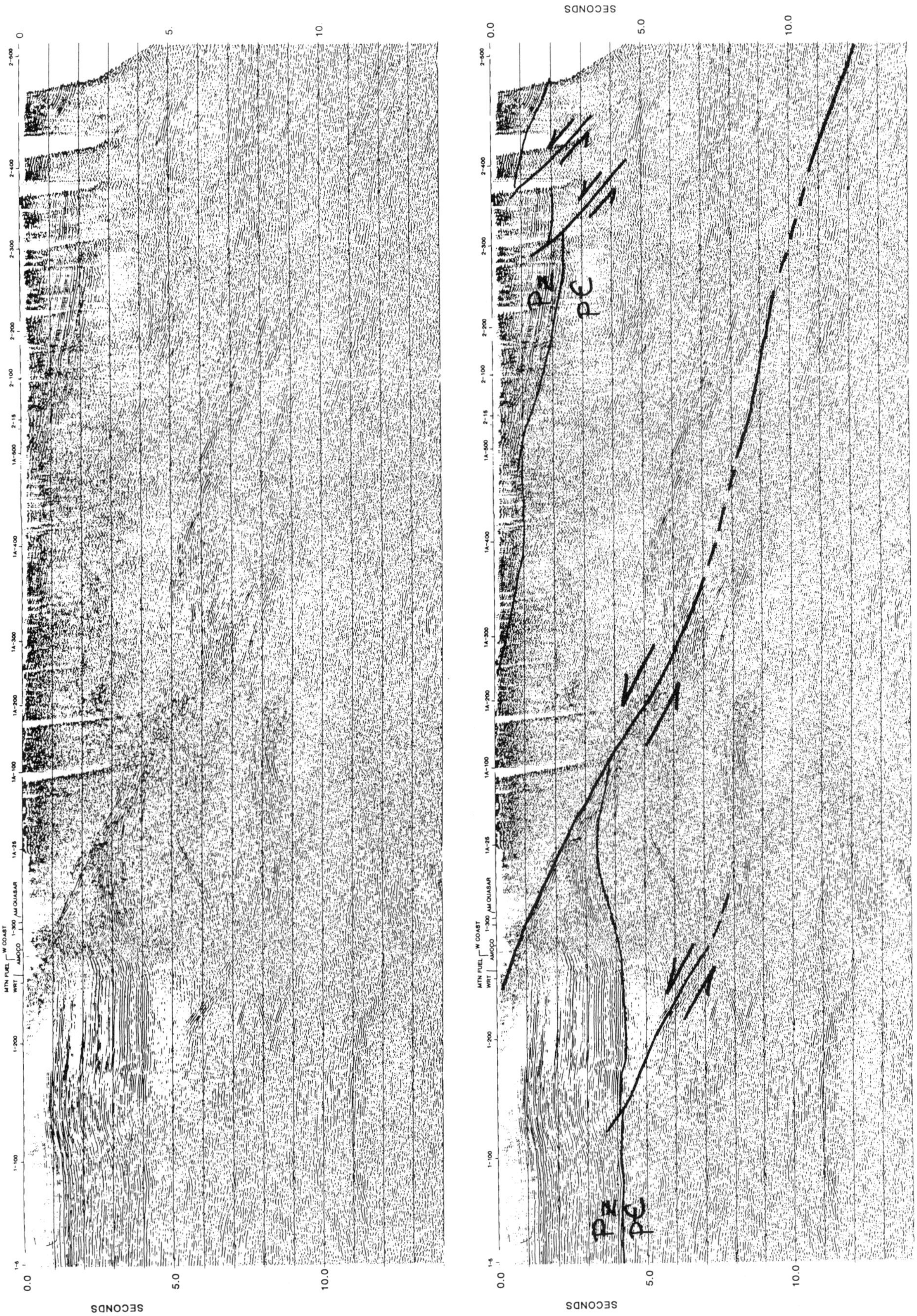

Figure 5. COCORP Wind River lines 1, 1a, and 2. Time sections (1 s ≈ 3 km) from *Sharry and others* (in Barazangi and Brown, 1986). Top is uninterpreted, bottom interpreted. For alternative interpretations see Smithson and others (*1978*, 1979) and *Sharry and others* (in Barazangi and Brown, 1986). Illustrates the thick-skinned nature of Rocky Mountain foreland basement uplifts.

evidence that the Wind River thrust extends to 24 to 36 km depth with an average dip of 36°. Recent depth migrations of those data indicate that the thrust may actually shallow at depth and break into numerous flat splays (*Lynn and others, 1983; Sharry and others* in Barazangi and Brown, 1986) between 22 and 26 km depth. COCORP data from the east side of the Laramie Range in southeastern Wyoming have been interpreted to show either a shallow or a steep bounding reverse fault (Brewer and others, 1982; *Johnson and Smithson, 1985*).

The orientations of faults beneath the Colorado Plateau monoclines are much less certain. Limited field evidence favors relatively steep to near-vertical faults in the Precambrian basement (*Davis* and *Reches* both in Matthews, 1978), but the exposed structural relief is very small. A major basement fault beneath the San Rafael Swell was not imaged by either of two COCORP profiles across it (*Allmendinger and others* in Barazangi and Brown, 1986; Keach, personal communication, 1986); this could be due to a near-vertical fault geometry or to fault displacement too small to produce adequate acoustic impedance. Alternatively, *Allmendinger and others* (in Barazangi and Brown, 1986) argued that the swell could be a large-scale arch or flexure decoupled above a prominent flat reflecting horizon in the mid-crust at approximately 27 to 29 km depth.

Folds and secondary structures. Most folds in the Rocky Mountain foreland have a parallel geometry and are cored at depth by a reverse fault in the basement. Some of the folds formed near the terminations of the major structural blocks have a conical geometry due to decreasing shortening along strike. The prevalence of bedding plane slip (e.g., Hennier and Spang, 1983) indicates that, kinematically, many folds are of the flexural slip style.

In a genetic sense, folds in the Rocky Mountain foreland have been variously referred to as "passive" or "drape" folds to indicate that the sedimentary cover was not deformed by horizontal compression but was passively draped over faulted and uplifted blocks of basement rocks, which were mechanically unable to fold (*Stearns* in Matthews, 1978). This concept has been forcefully challenged by numerous authors, including Brown (1984, in Lowell, 1983) who demonstrated that drape folds require pronounced stratal thinning or décollements at the basement-sediment interface to balance with vertical faulting in the basement. In fact, many foreland folds have thickened strata in their steep limbs (Brown, 1984), including some specifically interpreted as drape folds (e.g., *Cook and Stearns, 1975*). Although the monoclines have been interpreted as "passive drape folds" (*Cook and Stearns, 1975; Stearns and Jamison* in Veal, 1977), *Reches* (in Matthews, 1978; *Reches and Johnson* in Matthews, 1978) presented an elegant analysis demonstrating their origin in a regime of horizontal maximum compression.

Borehole data show that many of the uplifts have overturned panels of Paleozoic strata as well as an overturned sedimentary-basement unconformity associated with their faulted margins (Fig. 6; Berg, 1962; Gries, 1983, in Lowell, 1983). These important observations demonstrate the active participation of basement in the folding and support the fold-thrust model of Berg (1962; Fig. 4, 6).

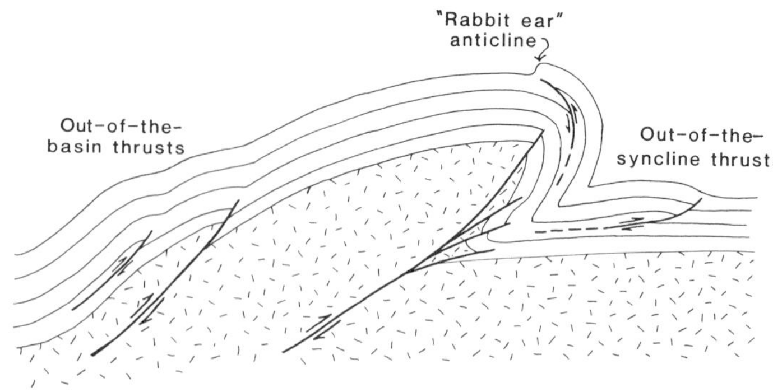

Figure 6. Schematic cross section showing characteristic structures associated with a Rocky Mountain foreland basement uplift. Note overturned basement-sedimentary unconformity in fault zone.

Not all folds in the Rocky Mountain foreland are cored by basement; many are detached above low-angle thrust faults in the sedimentary section (*Petersen* in Lowell, 1983). These thrust faults are commonly associated with tight synclines in the footwall of the major uplifts (hence the name "out-of-the-syncline thrust," see Fig. 6; Brown, 1984). They are interpreted as kinematic adjustments to space problems within the cores of the parallel folds (*Petersen* in Lowell, 1983; Brown, 1984). A possibly analogous type of structure is the "out-of-the-basin" thrust (Fig. 6), a fault that involves basement on the flank of a basin and dips toward the basinal axis (Brown, 1984). Examples of basement-cored, out-of-the-basin structures include (Fig. 3) many faults and folds on the east side of the northern Colorado Front Ranges between Boulder and Fort Collins (Boos and Boos, 1957; LeMasurier, 1970), the Sheep Mountain anticline on the east side of the Bighorn Basin (Hennier and Spang, 1983; *Brown* in Lowell, 1983), the Emigrant Trail thrust in the Wind River Basin, and the Woodside anticline on the northeast flank of the San Rafael Swell (*Allmendinger and others* in Barazangi and Brown, 1986). All of these structures appear to be antithetic to the major uplifts (although in several cases the major faults are on the opposite side of the uplift, as in Fig. 6) and could be related to synclinal flexure of the adjoining basins.

Strike-slip or oblique-slip faulting has figured prominently in a number of tectonic syntheses of the Rocky Mountain foreland province (Sales, 1968; Stone, 1969; *Chapin and Cather* in Lowell, 1983; *Bruhn and others, 1983,* in Peterson, 1986), but to date it has mostly been supported by indirect evidence based on regional structural trends and large-scale basin geometries. The best field structural evidence demonstrating probably important strike-slip displacements comes from the Nacimiento fault zone east of the San Juan Basin (Baltz, 1967; *Chapin and Cather* in Lowell, 1983).

Normal faults associated with Laramide thrusting or with younger Cenozoic extension also occur throughout much of the Rocky Mountain foreland province. These faults commonly occur in, and appear to be confined to, the hanging-wall blocks of the older basement uplifts, causing a "collapse" of the upper plate. This geometry is particularly obvious in the Granite Mountains of central Wyoming, but also can be seen in the Gros Ventre–Teton area, the Blacktail-Snowcrest uplift, and the Owl Creek Mountains (Wise, 1963; Love, 1970, 1978; *Sales* in Lowell, 1983; *Perry and others* in Schmidt and Perry, 1988).

Tectonic and mechanical models of foreland basement deformation

The diversity of structural trends in the Rocky Mountain foreland, particularly when compared to the thin-skinned fold-thrust belt, has commonly been cited as evidence of dominant vertical uplift (Prucha and others, 1965; *Stearns* in Matthews, 1978). Sales (1968) and Stone (1969), based on simple shear and pure shear models (respectively), proposed that the diversity was due to a combination of thrust faulting and wrench faulting in the province. A significant element of most recent models is the inference of northward or northeastward translation or rotation of the Colorado Plateau perhaps by as much as 60 to 120 km (Hamilton, 1981; *Chapin and Cather, Chapin,* and *Gries* all in Lowell, 1983). *Gries* (in Lowell, 1983), from an examination of basin isopachs and ages, suggested that the major structural trends developed at different times, with north-south trends formed during the Campanian to Eocene, northwest-trending structures formed during the Paleocene, and east-west structures formed during the Eocene. Establishment of timing relations in the Rocky Mountain foreland has been notoriously difficult, although several important studies by Keefer (1965), Love (1970, 1978), Tweto (1975), Dorr and others (1977), *Chapin and Cather* (in Lowell, 1983), *Lawton* (in Lowell, 1983; in Peterson, 1986), *Bruhn and others* (in Peterson, 1986), and Nichols and others (1985) indicate much promise for future work.

Fletcher (1984) presented an analytical model, based on a very viscous layer over a much less viscous half space, which predicts that the wavelengths of first-order flexures in the Rocky Mountain foreland should be 4 to 6 times greater than the thickness of the more viscous layer. Based on this model, the base of the highly viscous layer beneath the San Rafael Swell should be 27.5 to 18 km deep, in good agreement with conclusions based on the COCORP seismic data from the same area, discussed above (*Allmendinger and others* in Barazangi and Brown, 1986). Fletcher's model has been extended to the entire province by *Schmidt and others* (*1985*) and *Kulik and Schmidt* (in Schmidt and Perry, 1988) to infer that the depth to the "brittle-ductile transition" in the crust during Laramide deformation of the foreland was 11 to 16 km beneath southwest Montana and 25 to 38 km beneath Wyoming. However, it should be noted that neither of the COCORP profiles in Wyoming have identified anything in the crust that might be interpreted as a subhorizontal rheological boundary. In fact, the Wind River profile indicates that a localized shear zone extends nearly all the way to the base of the crust.

On a grander scale, Rocky Mountain foreland deformation correlates with a north-south gap in latest Cretaceous to Eocene age magmatism, which has been interpreted to indicate that the subduction zone beneath this segment of North America was nearly horizontal during the Laramide (Armstrong, 1974; Burchfiel and Davis, 1975; Coney and Reynolds, 1977; *Dickinson and Snyder* in Matthews, 1978). A segmented subduction zone would explain why the Rocky Mountain foreland basement uplifts are not present in Canada. This model is most convincing in light of the strikingly similar structural style of the amagmatic Sierras Pampeanas of western Argentina, which are today underlain by a nearly flat segment of the subducted Nazca plate, and that of the Rocky Mountain foreland (Jordan and others, 1983; *Jordan and Allmendinger, 1986*). An observation particularly pertinent to the question of crustal rheology during Laramide deformation is that crustal seismicity in the Sierras Pampeanas occurs as deep as 40 km, indicating that the entire crust is behaving in a brittle manner, at least at rapid strain rates.

Using the flat subduction model as a starting point for a quantitative model, Bird (1984) suggested that viscous drag on the thinned base of the continental lithosphere was the driving mechanism for the Rocky Mountain foreland uplifts. This drag produced generally eastward flow in the ductile lower crust, producing a significant thickening of the crust beneath Colorado and Wyoming, which, in turn, was responsible for the mid-late Cenozoic uplift of the modern Rockies. The evidence from deep seismic reflection data argues against uniform ductile flow of the lower crust in the Rocky Mountain foreland of Wyoming, although it might have occurred farther south beneath Utah and Colorado. Also unlike Colorado where crustal thicknesses exceed 50 km, there is no evidence for a crust much thicker than 40 km beneath most of Wyoming (Smith, 1978; *Prodehl, 1979; Allmendinger and others, 1982*).

THE CORDILLERAN THRUST BELT AND HINTERLAND

The Cordilleran thrust belt is one of the largest and most continuous geologic provinces in western North America (Fig. 1). At the broadest scale, the belt displays structural geometries typical of thin-skinned fold-thrust belts throughout the world (*Dahlstrom, 1970; Boyer and Elliott, 1982*): ramp and flat geometry of the thrust faults that cut across the sedimentary strata and sole into a basal décollement, and horizontal shortening of the section on the order of 50 percent. In general, the deformation in the Cordilleran thrust belt occurred between the mid-Jurassic and the Eocene (Fig. 2) and progressed from west to east (Armstrong and Oriel, 1965; *Oriel and Armstrong, 1966*; Burchfiel and others, 1974; *Price* in McClay and Price, 1981).

These general characteristics, however, belie the significant differences that exist along strike in the belt. Five major segments, demarcated by regional salients and recesses, form the focus of this discussion (Fig. 1): northern Montana, southwest Montana–Idaho, Idaho-Wyoming–northern Utah, west-central Utah, and southern Nevada–southeastern California. Three of these

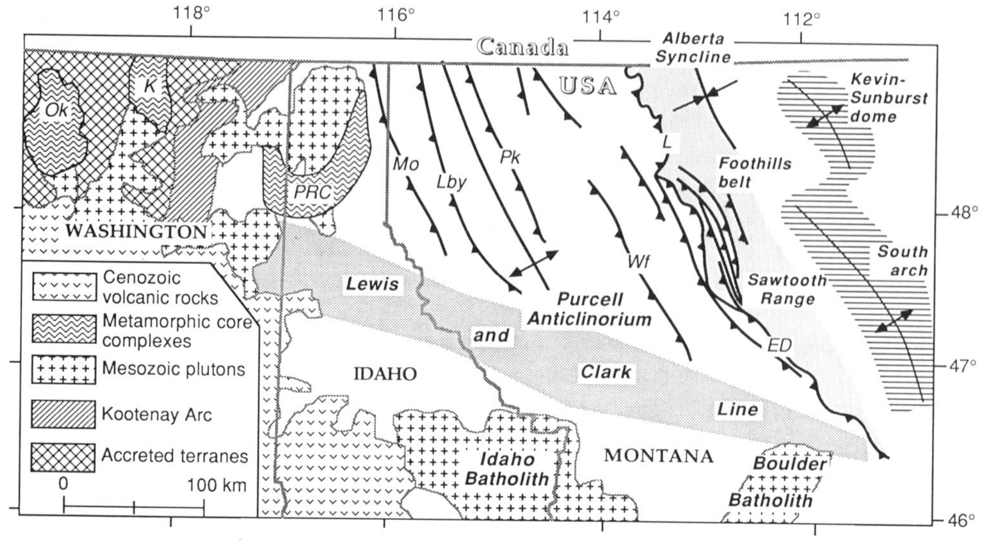

Figure 7. Generalized tectonic map of northern Montana, Idaho, and northeast Washington showing important contractional structures. *ED* = El Dorado; *K* = Kettle Dome; *L* = Lewis thrut; *Lby* = Libby thrust; *Mo* = Moiye thrust; *Ok* = Okanagan Dome; *Pk* = Pinkham thrust; *PRC* = Priest River Complex; *Wf* = Whitefish thrust. The Kevin-Sunburst dome and the South arch combined comprise the Sweetgrass arch.

segments are illustrated with balanced cross sections (Plate 10). The following discussion is divided into two parts: the foreland thrust belt and the hinterland. The boundary between these two is defined here as the zone west of which the basal décollement cut into crystalline basement; as discussed below, the foreland-hinterland boundary must have migrated with time, progressively eastward to its present position. In addition to basement involvement, the hinterland is characterized by low-grade metamorphism in the sedimentary section, plastic deformation at the base of thrust plates, and the presence of Mesozoic and/or early Cenozoic plutons.

Structural geometry and shortening in the foreland thrust belt

Northern Montana. The northern Montana part of the thrust belt (Fig. 7), referred to by some as the "Montana disturbed belt" (*Mudge, 1970; Woodward, 1981*), can be divided into two segments, one north of 48°20′ latitude and the other to the south. The northern segment displays, from east to west: (1) the Kevin-Sunburst dome, (2) the Alberta syncline and the Foothills belt, and (3) a broad western belt of allochthonous strata belonging to the Precambrian Belt Supergroup lying west of the Lewis thrut fault. From east to west across the southern segment are: (1) the South arch, (2) the Foothills belt (minus the Alberta syncline), (3) the Sawtooth Range, and (4) allochthonous Belt Supergroup west of the El Dorado thrust. The northwest-trending, *en echelon* Kevin-Sunburst dome and South Arch together compose the north-trending Sweetgrass basement arch (*Perry, 1928; Lorenz* in Powers, 1982).

Available age data for this segment of the thrust belt indicate that at least the first two regions were deformed between 72 and 46 Ma (Hoffman and others, 1976; Schmidt, 1978; *Mudge* in Powers, 1982). The Lewis thrust can only be constrained as post–early Campanian and pre–late Eocene (*Price* in McClay and Price, 1981).

The Foothills are widest at the United States–Canada border and narrow southward as the Sweetgrass arch impinges on the northwest-trending thrust belt. They are characterized by folds and closely spaced imbricate thrust faults, which deform Upper Cretaceous strata and sole above the Mississippian carbonate section, in a manner probably similar to, though less well known than, the Alberta Foothills (Ross, 1959; *Mudge* in Powers, 1982). The Alberta syncline extends 40 km south of the international border and marks the eastern edge of the foothills belt; as in Canada, it probably forms the eastern edge of a triangle zone created by duplexes and blind thrusts at depth beneath the uppermost Cretaceous section (*Price* in McClay and Price, 1981; *Jones* in Powers, 1982). Little evidence indicates its presence farther south, perhaps due to the closer proximity of the thrust belt and the Sweetgrass arch.

The Sawtooth Range contains exposures of Cambrian to Upper Cretaceous strata in closely spaced imbricate thrust slices and duplexes (Mudge, 1972, in Powers 1982). Its structure is similar to that of the Foothills but appears to have a décollement in the Cambrian rather than Mississippian section. Industry seismic data and balanced cross sections show that basement beneath the Foothills and Sawtooth Range is probably not involved in the thrust faulting; it lies 1,525 to 2,750 m below sea level in eastern and western parts of the region, respectively (*Mudge* in Powers, 1982).

The northern Montana segment of the thrust belt is dominated by the Lewis thrust fault (Fig. 7), which, in the region of maximum displacement near Glacier National Park, places Precambrian Belt Supergroup over Upper Cretaceous strata and can be traced along strike for more than 450 km from southern Alberta, Canada, to Augusta, Montana (Willis, 1902; Ross, 1959; Mudge and Earhart, 1980; *Mudge* in Powers, 1982). The Lewis, along with the El Dorado and Hoadley thrusts in the southern part of the segment, forms the eastern structural boundary of the allochthonous Belt Supergroup. The maximum displacement on the Lewis thrust at the latitude of Glacier National Park is probably somewhere between 50 and 80 km (Mudge and Earhart, 1980; *Mudge* in Powers, 1982; *Price* in McClay and Price, 1981; Bally, 1984; *Yoos, 1988*). The upper estimate is based on interpretations of industry seismic reflection data that suggest that the autochthonous margin of the Belt basin lies west of, or beneath, the Rocky Mountain trench (Bally and others, 1966; *Yoos, 1988*). The structure in the allochthonous Belt rocks is characterized by broad open folds and widely spaced thrust faults, in marked contrast to the Foothills and the Sawtooth Range.

The structure and stratigraphy of the Lewis thrust plate was clarified considerably by recent industry seismic and borehole data (Harrison and others, 1985). The Arco-Marathon No. 1 Gibbs well was spudded in the lower part of the exposed Belt sequence and reached a total depth of 5,417 m (17,774 ft.). There is little evidence for the presence of significant thrust faults in the hole; Harrison and others (1985) interpreted the Pinkham thrust at about 2,377 m (7,800 ft), but a repeated stratigraphic section was not encountered. However, industry and COCORP seismic data do show clear evidence of deeper seated thrust repetition of the Belt sequence (*Yoos, 1988*).

The total amount of shortening in the northern Montana segment was estimated by Bally (1984) at a minimum of 165 km between the front of the belt (in southern Canada) and Moyie thrust along the Idaho-Montana border (Plate 10A). Because of the thick Belt Supergroup strata, the location of easternmost basement involvement in thrusting is much less certain. It is clearly not involved, at least as far west as the Rocky Mountain trench, as shown by industry seismic reflection data and inferred from regional gravity and magnetic data and balanced cross sections in the United States and in southern Canada (Bally and others, 1966; *Price* in McClay and Price, 1981; Bally, 1984; *Yoos, 1988*). Based on a 30 to 40 mgal residual Bouguer gravity relative high, *Wynn and others* (*1977*) and Harrison and others (1980) concluded that basement was anomalously high beneath the Purcell anticlinorium and thus was involved in thrusting west of the Rocky Mountain trench. *Price* (in McClay and Price, 1981) concluded that basement was not involved beneath the Purcell anticlinorium; instead, he suggested that the anticlinorium formed as a hanging wall anticline over a major ramp formed by the old continental shelf-slope break of North America. Available seismic reflection data are not conclusive on this point; recent COCORP and industry results indicate that the anticlinorium is cored by a duplex at depth and that basal décollement may be as deep as 22 km at the Montana-Idaho state line (*Yoos, 1988*).

The Lewis and Clark lineament and Helena Salient occur at the southern limit of this segment of the thrust belt (Figs. 7 and 8). The former is a 20 to 30 km broad zone of west-northwest–trending faults that includes the St. Mary's, Ninemile, St. Joe, and Osburn faults, and has probably been active from late Precambrian until recent (Harrison and others, 1974). Displacement along the lineament has occurred with both right- and left-lateral senses at different times (Harrison and others, 1974; *Reynolds, 1979; Reid and others, 1981; Bennett, 1986*). The Helena salient of the thrust belt, located just south of the Lewis and Clark line, coincides with the Helena embayment of the Belt basin, and is, apparently, an example of paleotectonic control of the thrust belt geometry (Harrison and others, 1974; *Woodward, 1981*).

Southwestern Montana–Idaho. This segment of the thrust belt is complicated by the pronounced impingement of Rocky Mountain foreland structure represented by outcrops of Archean basement (Fig. 8), which may have acted as a buttress to the thrust belt (*Beutner* in Heisey and others, 1977; *Skipp* in Schmidt and Perry, 1988). Faults at the northern end of this segment trend nearly east-west and are inferred to have right-lateral displacement (*Woodward, 1981; Schmidt and O'Neill* in Powers, 1982; Ruppel and Lopez, 1984). Farther south, the nature of the allochthons and sequence of faulting has been most intensively studied in the region west of Lima, although controversy exists over the names and correlations of thrust faults (Scholten and others, 1955; *Skipp and Hait* in Heisey and others, 1977; *Ruppel, 1978; Perry and Sando* in Powers, 1982; Perry and others, 1983; *Perry and others* in Schmidt and Perry, 1988; Ruppel and Lopez, 1984; Skipp, 1984, 1987).

The three easternmost thrusts in the area are, from east to west (Fig. 8), the Tendoy, Four Eyes Canyon, and Medicine Lodge (*Skipp and Hait* in Heisey and others, 1977; *Perry and Sando* in Powers, 1982; Perry and others, 1983). *Perry and Sando* (in Powers, 1982) documented a west-to-east age progression in these thrusts and suggested that all were surface-breaking faults that overrode their own debris. There is no evidence yet for thin-skinned thrusting younger than latest Cretaceous in age, even though Paleocene and early Eocene thrusting is documented in segments both north and south. The name "Medicine Lodge thrust" has been used in both a general (Scholten and others, 1955; *Ruppel, 1978*; Ruppel and Lopez, 1984) and in a restricted sense (*Skipp and Hait* in Heisey and others, 1977; *Perry and Sando* in Powers, 1982; Skipp, 1984, 1987); the latter usage is followed here.

West of the Medicine Lodge thrust, *Skipp and Hait* (in Heisey and others, 1977) and Skipp (1987) have identified seven major thrusts (from east to west, Fig. 8): the Cabin, Fritz Creek, Hawley Creek, Lost River–Arco Hills, White Knob, Copper Basin, and Wood River allochthons. The Cabin and Medicine Lodge thrusts are probably part of the same system (Skipp, 1987). Movement of the Medicine Lodge, Cabin, Fritz Creek, and Hawley Creek thrusts (formerly assigned to the "Lemhi stack" by *Skipp and Hait* in Heisey and others, 1977) was probably during the Late Cretaceous, whereas movement of the west-

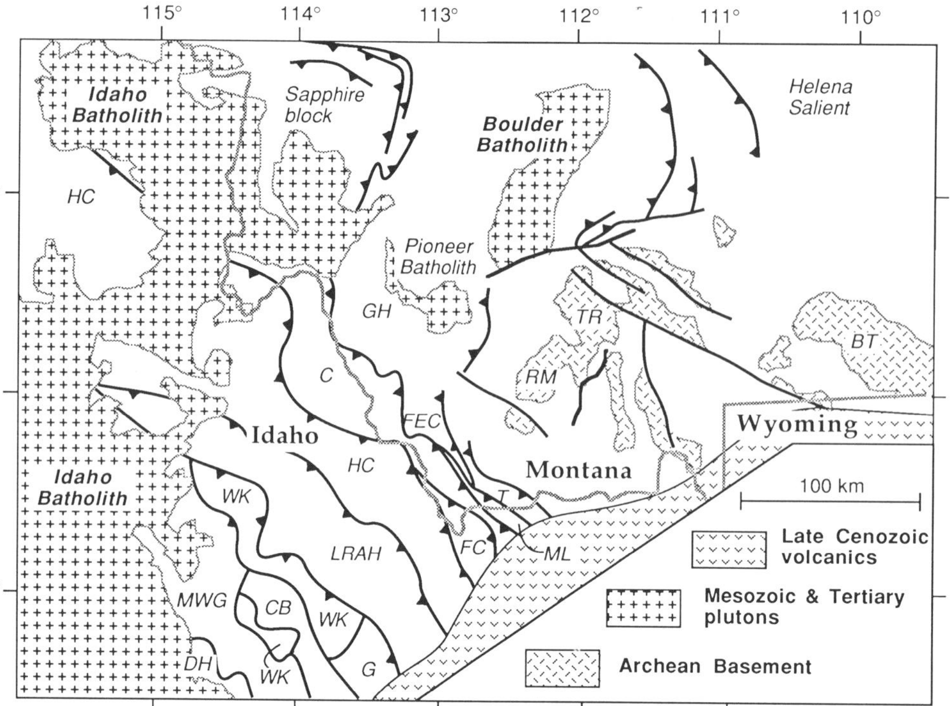

Figure 8. Generalized tectonic map of southwestern Montana and central Idaho, showing important contractional structures. *BT* = Beartooth Mountains; *C* = Cabin thrust plate; *CB* = Copper Basin plate; *DH* = Dollarhide thrust complex; *FC* = Fritz Creek plate; *FEC* = Four Eyes Canyon plate; *G* = Grouse plate; *GH* = Grasshopper plate; *HC* = Hawley Creek plate; *LRAH* = Lost River–Arco Hills plate; *ML* = Medicine Lodge plate; *MWG* = Milligen–Wood River–Grand Prize thrust complex; *RM* = Ruby Mountains; *T* = Tendoy plate; *TR* = Tobacco Root Mountains; *WK* = White Knob plate. *RM* and *TR* comprise the Blacktail-Snowcrest uplift. Modified from *Schmidt and Garihan* (in Lowell, 1983) and Skipp (1987). Note that faults shown in Idaho are mostly concealed by younger deposits.

ern thrusts plates (formerly assigned to the "Lemhi and Pioneer stacks" by *Skipp and Hait* in Heisey and others, 1977) was probably pre–Late Cretaceous in age.

The pre-Belt, Dillon Granite Gneiss and metasedimentary gneisses and schists are incorporated in the upper plate of the Cabin–Medicine Lodge thrust system and probably also in the Tendoy thrust (*Dubois* in Powers, 1982; *Scholten* in Powers, 1982; Skipp, 1984, 1987, *1988*), although not in the Tendoy as restricted to the Lima-Dell, Montana region by *Perry and others* (in Schmidt and Perry, 1988). The basement of the Tendoy (*sensu lato*) has cratonic affinities whereas that in the Cabin thrust has a thick Proterozoic cover (B. Skipp, personal communication, 1986). Thus, basement involvement occurs in a much more easterly position, relative to the leading edge of the thrust belt, than anywhere else in the belt.

The total shortening on the three frontal thrusts of the segment is poorly constrained, but may be about 40 km on the Cabin–Medicine Lodge thrust system and less on the Tendoy and related thrust faults (*Skipp and Hait* in Heisey and others, 1977). Displacement of the thrust plates west of the White Knob Mountains has been estimated at between 50 and 150 km, the larger estimate being derived from the difference between the easternmost continental edge strata and the location of the Paleozoic continental margin inferred from isotopic data (*Skipp and Hall, 1975; Skipp and Hait* in Heisey and others, 1977). Thus, the total amount of shortening across the southwest Montana–Idaho segment of the thrust belt remains poorly known.

Idaho-Wyoming–northern Utah. This segment of the Cordilleran thrust belt (Fig. 9) is the best known of all those in the United States and rivals any in the world in terms of documented ages of thrust faulting. It makes up the northernmost part of Armstrong's (1968) "Sevier orogenic belt" and has been described in classic papers by Veatch (1907), Mansfield (1927, *1952*), Rubey and Hubbert (1959), Armstrong and Cressman (1963), Armstrong and Oriel (1965), *Oriel and Armstrong* (*1966*), and Royse and others (1975). Today, it continues to be the focus of detailed structural, stratigraphic, and mechanical analyses (Dorr and others, 1977; *Dorr and Gingerich, 1980*; Jordan, 1981; *Lamerson* in Powers, 1982; Dixon, 1982; Allmendinger, 1982; Wiltschko and Dorr, 1983; Woodward, 1986) because of its demonstrated hydrocarbon potential, excellent regional maps, relatively simple setting, and well-known timing relations.

Four major thrust systems, defined by major ramp zones, compose the 100 to 150 km wide Idaho-Wyoming–northern Utah thrust belt (generally, from east to west): the Darby system (including the Prospect, LaBarge, Hogsback, and Darby thrusts), the Absaroka system (early and late Absaroka, Commissary

Ridge thrusts), Crawford-Meade system, and Paris-Willard system. They form an arcuate salient ~350 km long and bounded to the north and south by Rocky Mountain foreland uplifts. The curvature at the north end of the segment appears to be due to deformation and rotation of the Prospect–Granite Creek sheet against the Gros Ventre Range buttress, as demonstrated by geometric relations and paleomagnetic studies (*Grubbs and Van der Voo, 1976; Beutner* in Heisey and others, 1977; Schwartz and Van der Voo, 1984). The major faults formed generally from west to east (Figs. 9 and 10; Armstrong and Oriel, 1965; *Oriel and Armstrong, 1966*), but important imbrications were commonly "backward-breaking" thrusts formed in east-to-west order, such as the Prospect, Bear, and Cliff Creek thrusts and parts of the Meade system (Allmendinger, 1981; *Lamerson* in Powers, 1982; *Royse, 1985*). Some faults probably moved more than once (e.g., the Paris thrust, *Oriel and Armstrong, 1966*) or have had different splays move at different times (the early and late Absaroka movements, Royse and others, 1975; *Lamerson* in Powers, 1982).

Crosscutting and overlapping relations unambiguously define the ages of major thrusts on the external side of thrust belts. However, as in all mountain belts, dating becomes increasingly less precise and more indirect in the internal parts. Following the approach in Spieker's (1946) classic work, synorogenic conglomerates have been used to determine the ages of the older thrusts in the Idaho-Wyoming–northern Utah segment (Armstrong and Cressman, 1963; Armstrong and Oriel, 1965; *Oriel and Armstrong, 1966*; Royse and others, 1975). Poor preservation and difficulty in dating the conglomerates, as well as recent studies in young mountain belts indicating that the appearance of conglomerates may postdate thrusting by many millions of years (*Jordan and others, 1988*), make the older ages uncertain. This problem is most severe for the Paris thrust system; its presumed latest Jurassic–earliest Cretaceous age has recently been challenged (Heller and others, 1986).

In contrast to the thrust belt of southern Alberta, the Idaho-Wyoming–northern Utah thrust belt lacks a well-defined foothills belt of deformed and imbricated Mesozoic syntectonic strata, and a simple Alberta-style triangle zone is also missing or is poorly developed. Wyoming has two major, ramp-forming competent units, the Paleozoic carbonates and Jurassic Nugget Sandstone (F. Royse, personal communication, 1986). All major thrusts except the Prospect have exposures of the Cambrian-Devonian or older strata in their upper plates, and the Prospect is inferred to cut down to the Cambrian at depth (Rubey and Hubbert,1959; Armstrong and Oriel, 1965; Royse and others, 1975). A ramp and flat geometry is particularly well displayed by a seismic section across the Darby thrust at the eastern margin of the thrust belt (Fig. 11).

Structural studies at outcrop scale and microscopic scale show that most of the strain within thrust plates, manifested as brittle fractures, intracrystalline plastic strain, and locally, pressure solution cleavage, occurred as early layer-parallel shortening with some subsequent strain at the late stages of the folding

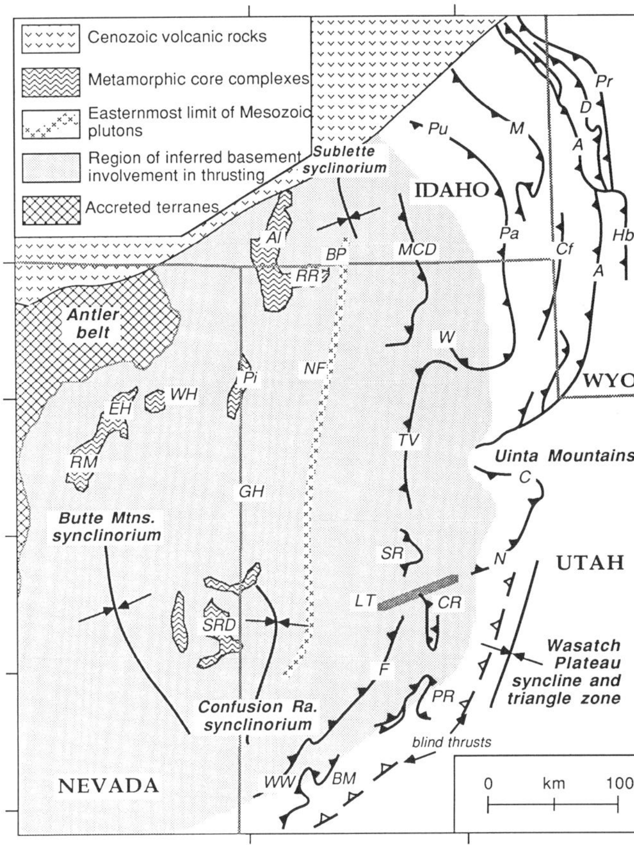

Figure 9. Generalized tectonic map of southwestern Wyoming, southern Idaho, western Utah, and eastern Nevada showing important contractional structures. *A* = Absaroka thrust; *Al* = Albion Range core complex; *BM* = Blue Mountains thrust; *BP* = Black Pine Mountains; *C* = Charleston thrust; *Cf* = Crawford thrust; *CR* = Canyon Range thrust; *D* = Darby thrust; *EH* = East Humboldt Range; *F* = Frisco thrust; *GH* = Gold Hill; *Hb* = Hogsback thrust; *LT* = Leamington "transcurrent fault"; *M* = Meade thrust; *MCD* = Manning Canyon detachment; *N* = Nebo thrust; *NF* = Newfoundland Range; *Pa* = Paris thrust; *Pi* = Pilot Range; *Pr* = Prospect thrust; *PR* = Pavant Range thrust; *Pu* = Putnam thrust; *RM* = Ruby Mountains core complex; *RR* = Raft River Range core complex; *SR* = Sheeprock thrust; *SRD* = Snake Range core complex; *TV* = Tintic Valley thrust; *W* = Willard thrust; *WH* = Wood Hills; *WW* = Wah Wah thrust.

process (Allmendinger, 1982; *Craddock and Wiltschko, 1983*; Mitra and others, 1984). Deformation along thrusts is difficult to characterize because the faults are seldom exposed, but it appears to be entirely low-temperature brittle cataclastic deformation west to, and including, the Absaroka thrust. Plastic deformation of limestones and, to a very limited extent, in sandstones has been observed along the Meade thrust, and a narrow zone of plastic deformation of quartzites is present at the base of the Willard thrust (Crittenden, 1972; Bruhn and Beck, 1981; Allmendinger, 1982; *Bruhn and others, 1983*).

Based on regional industry seismic reflection data and magnetic data, Royse and others (1975) and Dixon (1982) have

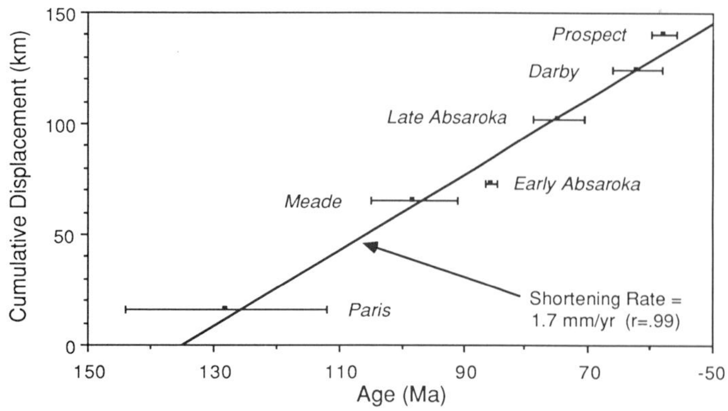

Figure 10. Plot of age versus cumulative displacement for the Idaho–Wyoming–northern Utah segment of the thrust belt. The linear regression line gives a bulk average shortening rate (see discussion in text). Note increase in scale of error beds with increasing age. Based on data from Armstrong and Oriel (1965), *Oriel and Armstrong* (*1966*); Royse and others (1975); Jordan (1981); Wiltschko and Dorr (1983).

shown that crystalline basement is 3,600 to 6,100 m below sea level beneath the east edge of the thrust belt (deeper in the south, shallower in the north) and deepens to 9,200 m below sea level beneath the Bear River Range. However, the Archean Farmington Complex (*Hedge and others* in Miller and others, 1983) is exposed in the northern Wasatch Range just west of the Bear River Range and is cut by the Ogden and Taylor thrust faults just east of Ogden, Utah (*Sorensen and Crittenden, 1972;* Bruhn and Beck, 1981).

Royse and others (1975) proposed that all of the exposed basement is in the upper plate of the Absaroka thrust. The uplift of the basement during thrust faulting is confirmed by studies of fission tracks in apatite from the top of the northern Wasatch, which were last annealed during the Late Cretaceous (*Naeser and others* in Miller and others, 1983). The eastward tilt of rocks and structures in the northern Wasatch Range (including the Willard thrust) was only partly a function of rotation during mid–late Cenozoic extension; the attitudes of strata unconformably underlying Paleocene and Eocene units (*Mullens and Laraway, 1964; Sorensen and Crittenden, 1979*) indicate that the rocks of the range dipped 10° to 70°E prior to the early Tertiary. The Farmington Complex is in the core of a large hanging-wall anticline formed over the ramp of the Absaroka out of the basement, and the presently east-dipping Willard thrust has been folded in the upper plate (*Allmendinger and Jordan, 1981; Bruhn and others, 1983*). Thus, the eastern boundary of the hinterland coincides with the northern Wasatch, Bannock, and Portneuf Ranges in southern Idaho and northern Utah.

Shortening on individual thrust faults varies in a north-south direction, and different faults have different magnitudes of displacement (Fig. 10). Overall, as much as 140 km of east-west shortening can be documented in the region (Royse and others, 1975; Jordan, 1981).

West-central Utah. South of the Uinta Mountains, the thrust belt has been strongly overprinted by mid–late Cenozoic extension (Fig. 9). Nonetheless, many of the major thrusts directly south of the Uinta axis can be correlated tentatively with the better known thrusts to the north. For example, based on the presence or absence of the thick Proterozoic quartzite sequence, the Willard probably correlates with the Tintic Valley, Sheeprock, and Canyon Range thrusts. Similarly, the Absaroka may correlate with the Charleston-Nebo thrusts because both are overlapped by Maastrichtian strata (although some disagreement exists; see Crittenden, 1972; *Allmendinger and Jordan, 1981; Christie-Blick* and *Tooker* both in Miller and others, 1983). As with the Absaroka thrust, crystalline basement rocks crop out in the upper plate of the Charleston-Nebo thrust plate just east of the Wasatch normal fault near the town of Santaquin, Utah.

South of the Leamington "transcurrent fault" near Nephi, Utah (Fig. 9; *Morris* in Miller and others, 1983), which Royse (personal communication, 1986) relates to a lateral ramp rather than a tear fault, the thrust belt changes orientation to a northeasterly strike, and only two major thrusts are exposed. The structurally higher Canyon Range thrust carried the thick Proterozoic quartzite sequence over Paleozoic carbonates, and the lower Pavant Range thrust, which now dips east or southeast, juxtaposed Paleozoic carbonates and Cambrian Tintic quartzite over Mesozoic strata (Christiansen, 1952; *Burchfiel and Hickcox, 1972; Hintze, 1980; Baer and others* in Nielson, 1982). Despite discontinuous exposures of Mesozoic–early Tertiary structures in southwestern Utah, stratigraphic relations and relative structural position support a correlation of the Canyon Range thrust with the Frisco and Wah Wah thrusts, and the Pavant Range with the Beaver Lake–Blue Mountain thrust (Armstrong, 1968; however, see *Morris* in Miller and others, 1983).

To the east of these two thrust faults lies a zone of folded Mesozoic and lower Tertiary strata, 40 to 50 km wide, including the highly deformed Jurassic Arapien shale and evaporite sequence. The eastern front of this deformation can be identified by the eastward decrease in angularity of the unconformity beneath the lower Tertiary strata of the Wasatch Plateau monocline (Spieker, 1946), although a minor blind thrust may continue 65 km farther east to the Huntington anticline on the west flank of the San Rafael Swell. Recent industry data and COCORP studies confirm that the deformed Jurassic rocks (Arapien Formation) are underlain by one or more blind thrusts (*Standlee* in Powers, 1982; *Allmendinger and others* in Barazangi and Brown, 1986; *Villien and Kligfield* in Peterson, 1986). A regional syncline in Jurassic and Cretaceous strata unconformably beneath the Eocene strata in the Wasatch monocline is probably analogous to the Alberta syncline, and east-dipping faults on the easternmost edge of the Gunnison Plateau are interpreted as a normal-fault reactivated part of a triangle zone (*Standlee* in Powers, 1982; *Allmendinger and others* in Barazangi and Brown, 1986).

Evidence for possible basement involvement comes from the Placid No. 1-2 WXC-USA well, located approximately 15 km south of Scipio Utah on the west side of the Valley Mountains (Plate 10C). That well encountered Proterozoic crystalline base-

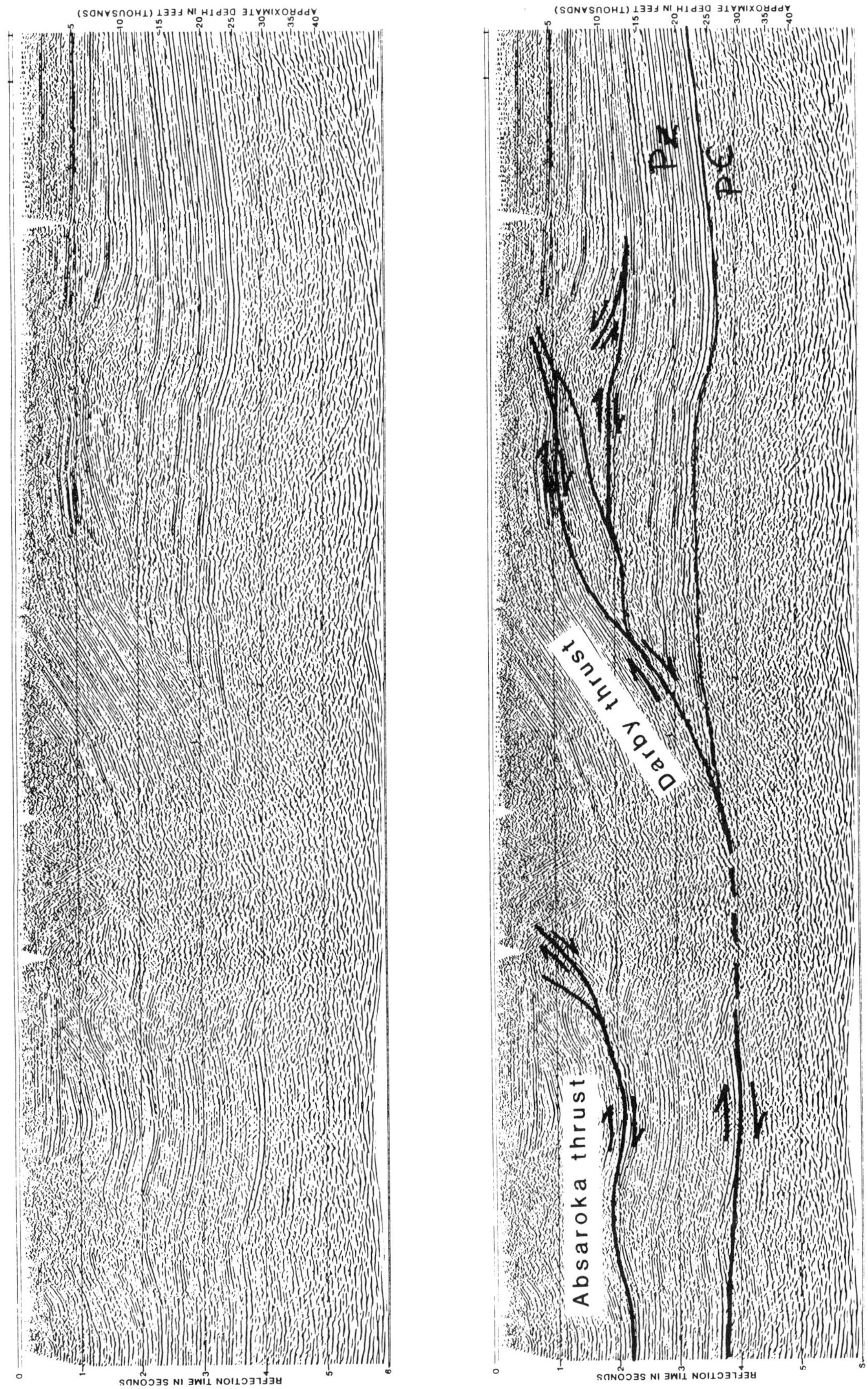

Figure 11. Migrated seismic reflection profile of the front of the Idaho–Wyoming–northern Utah segment of the thrust belt in southwestern Wyoming, from *Williams and Dixon* (in Gries and Dyer, 1985). The two thrust systems are the Darby to the east and the Absaroka to the west. Arch in basement in eastern half of line is partly due to a velocity pull-up due to thrusting of high velocity rocks in the Darby thrust and the underlying duplex. Top section uninterpreted, bottom interpreted.

ment (K-Ar isotopic ages on biotite of 1154 Ma and 1101 Ma) at 5,540 m depth (D. Sprinkle, personal communication, 1985; F. Royse, personal communication, 1986; J. Baer, personal communication, 1986). Basement almost certainly is involved in the Pavant Range thrust plate west of and beneath the House Range, where regional geology between the Confusion and House Ranges, and dipping reflectors on COCORP Utah line 1, indicate an 8 to 10 km high ramp out of basement (Plate 10C; Armstrong, 1982; Allmendinger and others, 1983, in Barazangi and Brown, 1986).

Mesozoic–early Tertiary shortening in west-central Utah is difficult to determine because of blind thrusts and the flow of Jurassic salt in the east and ~70 percent late Cenozoic cover and superimposed extensional deformation in the west. There are no exposed hanging-wall cutoffs for either the Canyon or Pavant Range thrusts. First-order estimates of shortening based on balanced cross sections constrained by seismic and bore-hole data indicate between 100 and 120 km (Plate 10D; *Allmendinger and others* in Barazangi and Brown, 1986).

Southern Nevada–southeastern California. This segment of the thrust belt (Fig. 12) displays some of the broadest departures from "typical" thrust belt geometry. The Las Vegas shear zone, largely or completely a buried late Cenozoic fault with right-lateral displacement (Longwell, *1960,* 1974; *Guth, 1981; Wernicke and others, 1982;* however, see also *Royse, 1983*), divides the segment into two parts. Both north and south of the shear zone, three major correlative thrusts make up the thrust belt (*Longwell, 1922, 1926, 1928;* Longwell and others, 1965; Burchfiel and others, 1974; Burchfiel and Davis, *1972,* 1975, in Ernst, 1981; *Bohannon* in Miller and others, 1983; Carr, 1983; Wernicke and others, 1985), which are from east to west: (1) the Mormon thrust north of the shear zone and the Red Spring and Contact thrusts to the south; (2) the Glendale and Muddy Mountain thrusts to the north and the Keystone thrust to the south of the shear zone; and (3) Gass Peak thrust north of the shear zone and Wheeler Pass thrust to the south. As discussed below, these correlations are not universally accepted. Movement on all of these thrusts is inferred to have occurred between the mid-Jurassic and the early Late Cretaceous (see review in *Burchfiel and Davis* in Ernst, 1981; *Carr and others, 1986*).

Each of the two frontal fault systems have décollement levels in the Cambrian Bonanza King Dolomite, rather than, as one might expect, in the underlying Bright Angel Shale (Longwell and others, 1965; Burchfiel and others, 1974; *Bohannon* in Miller and others, 1983; Wernicke and others, 1985). Burchfiel and others (1982) have pointed out that this structural and stratigraphic association is apparently at variance with the dogma that thrust décollements should form in "incompetent" shale sequences. However, these field observations may best be understood in light of recent experimental work by *Weeks and Tullis (1984)* showing that wet dolomite has a very low coefficient of friction (0.08 to 0.54). Parts of the Muddy Mountain, Keystone, Red Spring, and Contact thrust plates overrode their own erosional detritus, which comprised fluvial sandstone and conglomerate probably of latest Jurassic to Late Cretaceous age

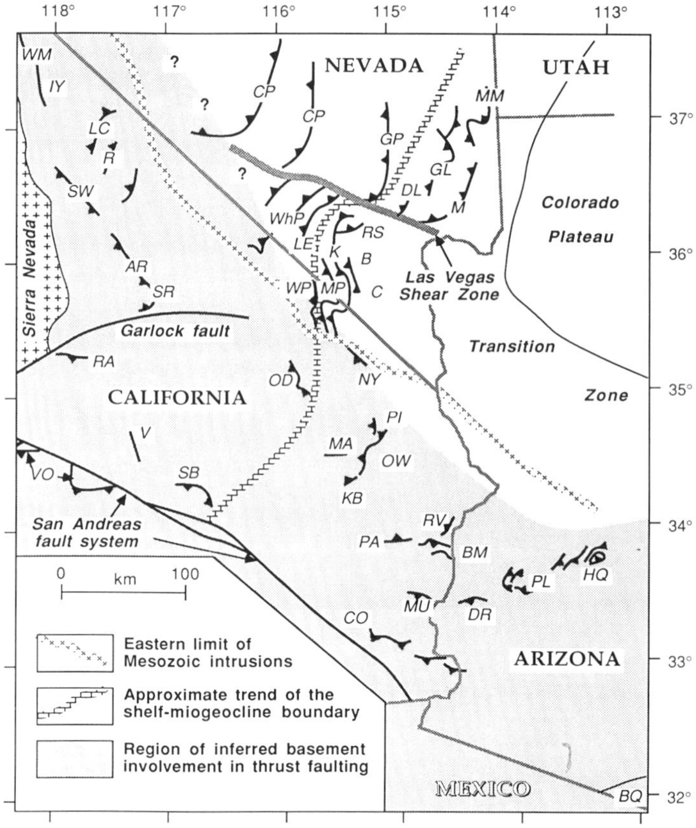

Figure 12. Generalized tectonic map of the southern part of the Cordilleran thrust belt in southern Nevada, southeastern California, and southwestern Arizona. Modified from *Burchfiel and Davis* (in Ernst, 1981) and *Martin and others* (in Frost and Martin, 1982). *AR* = Argus and Layton Well thrusts; *B* = Bird Spring thrust; *BM* = Big Maria Mountains; *BQ* = Baboquivari and related thrusts; *C* = Contact thrust; *CO* = Chocolate Mountains thrust; *CP* = CP thrust; *DL* = Dry Lake thrust; *DR* = Dome Rock Mountains; *GL* = Glendale thrust; *GP* = Gass Peak thrust; *HQ* = Harquahala Mountains; *IY* = Inyo Mountains; *K* = Keystone thrust; *KB* = Kilbeck Hills; *LC* = Last Chance thrust; *LE* = Lee Canyon thrust; *M* = Muddy Mountains thrust; *MA* = Marble Mountains; *MM* = Mormon Mountains thrust; *MP* = Mesquite Pass thrust; *MU* = Mule Mountains; *NY* = New York Mountains; *OD* = Old Dad Mountains; *OW* = Old Woman Mountains; *NY* = New York Mountains; *PA* = Palen Mountains; *PI* = Piute Mountains; *PL* = Plomosa thrust; *R* = Racetrack thrust; *RA* = Rand Mountains; *RV* = Riverside Mountains; *RS* = Red Spring thrust; *SB* = San Bernadino Mountains; *SW* = Swansea-Coso thrusts; *V* = Victorville; *VO* = Vincent-Orocopia thrust; *WhP* = Wheeler Pass thrust; *WM* = White Mountains; *WP* = Winters Pass thrust.

(*Longwell, 1926;* Burchfiel and others, 1974; Brock and Engelder, 1977; Carr, *1980,* 1983). In the southern Spring Mountains, the easternmost Contact and Red Spring thrust plates moved before and were subsequently overridden by the more westerly Keystone plate, a reverse stacking order atypical of the Cordilleran thrust belt (Burchfiel and others, 1974; Carr, *1980,* 1983). The reverse sequence of thrusting, interpreted by some as backward-breaking, hanging-wall imbrication (F. Royse, personal communication, 1986), has resulted in truncated folds and thrusts that cut both up and down section in the direction of transport.

The Gas Peak and Wheeler Pass thrust plates contain exposures of the thick Proterozoic clastic sequence and thus are analogous to the Canyon Range, Sheeprock, Tintic Valley, Willard, Paris, and Putnam thrust plates to the north. Preexisting structures in the lower plates of the Gass Peak and Wheeler Pass thrusts are truncated by those thrusts (Burchfiel and others, 1974; *Burchfiel and Davis,* in Ernst, 1981).

Due to the pronounced overprinting of Cenozoic extensional structures and the limited amount of industry exploration in the region, the extent of basement involvement and the amount of shortening on Mesozoic–early Cenozoic thrust faults in southern Nevada is not well known. Wernicke and others (1985) suggest a minimum of 30 km for the Mormon thrust system. Farther south, Burchfiel and others (1974) estimated total minimum shortening across the four major thrusts exposed in the Spring Mountains to be between 36 and 75 km.

The Cordilleran thrust belt, as commonly accepted, can be traced only as far south as the Clark and New York Mountains in southeastern California (Fig. 12). In the former, three thrusts are recognized (from east to west): the Keystone, Mesquite Pass, and Winters Pass faults. Structures in the Keystone plate at the south end of the Clark Mountains are a result of early Mesozoic deformation combined with late Mesozoic movement on the Keystone. The Mesquite Pass and Winters Pass thrusts contain a Proterozoic clastic section as well as underlying crystalline basement in their upper plates (Burchfiel and Davis, 1977; *Nelson and Burchfiel* in Newman and Goode, 1979). The Winters Pass thrust zone is characterized by locally extensive mylonite zones in both upper and lower plates. In the New York Mountains, the easternmost thrusts in the belt are thought to correlate with the Keystone thrust of the southern Clark Mountains. However, this easternmost thrust carried crystalline basement as well as greenschist facies Paleozoic and Mesozoic strata in its upper plate, suggesting that the Cordilleran thrust belt progressively cut into the craton southward before dying out or radically changing form (Burchfiel and Davis, 1977, in Ernst, 1981). In both the Clark and New York Mountains, Late Cretaceous plutons intrude the easternmost thrust (Fig. 12).

Arizona–southeasternmost California. The continuity of the Cordilleran thrust belt between the regions described above and northern Chihuahua in Mexico has long been a subject of intense debate. The presence of fold and thrust structures in Arizona and southeasternmost California is undeniable (some will be described in the "Hinterland" section), but their tectonic significance is poorly known and their geometric and mechanical similarity to the thrust belt to the north is doubtful. The problem has been compounded by the misidentification of low-angle normal faults as thrust faults.

The concept of a continuous, thin-skinned thrust belt in Arizona (Drewes, *1978,* 1981; *Hansen and others, 1980*) was tested in spectacular fashion by the 5,490-m-deep Phillips' Tombstone State A-1 well, which demonstrated that the proposed thrusts were probably mid–late Cenozoic low-angle normal faults (Reif and Robinson, 1981; *Robinson* in Powers, 1982). Similarly, parts of Drewes' (*1978,* 1981) Hidalgo overthrust, the proposed regional décollement for a thin-skinned thrust belt, in the field have been shown to be low-angle normal faults (*Dickinson, 1984*). The idea is not completely without merit; the Maria fold and thrust belt of west-central Arizona (described below) was active between the Jurassic and Late Cretaceous (Reynolds and others, 1986, in Ernst, 1988). Thrust faults also are present in southeastern Arizona and southwestern New Mexico, but their geometric style is much more closely akin to laterally discontinuous Rocky Mountain foreland basement uplifts (Davis, 1979; *Brown and Clemons, 1983;* Seager, 1983), although some have interpreted them as laterally continuous overthrusts (*Corbitt and Woodward, 1973;* Woodward and DuChene, 1981, in Powers, 1982).

The simplest explanation for the absence of typical Cordilleran thrust belt structures in Arizona is the lack of a thick miogeoclinal wedge of sedimentary strata overlying the basement. The southern termination of the thrust belt in southeastern California coincides closely with the region where the Mesozoic magmatic arc cuts across the miogeocline and into the craton (Fig. 12; Burchfiel and Davis, *1972,* 1975). North of this region, the Mesozoic Cordilleran orogen essentially parallels the early-mid Paleozoic continental margin of western North America, but to the south it follows the magmatic arc, cutting obliquely across the old continental trends and leaving the passive margin farther to the west. In fact, the atypical geometries and structural sequences in the thrust belt south of Las Vegas may also be due to the belt diverging from the old miogeoclinal trends and cutting into the craton (Burchfiel and Davis, *1972,* 1975, in Ernst, 1981). In general, thrusting and magmatism south of Las Vegas were temporally and spatially coincident, exerting a significant control on structural style (Armstrong, 1974).

The Cordilleran Hinterland

The structural geometry and tectonics of the Cordilleran Hinterland are more complex and laterally variable than the thrust belt, and thus are less well known. Furthermore, the hinterland has everywhere been the site of subsequent and, locally at least, coeval extension. At present, it is virtually impossible to gauge the amount of Mesozoic–early Cenozoic shortening (or extension) in any segment of the hinterland, and the detailed relations between structures, metamorphism, and magmatism are poorly known. For these reasons, the hinterland represents the frontier area for future structural and tectonic studies of the U.S. Cordillera. It is in the hinterland that the tests of mechanical models of thin-skinned thrusting lie, where the relations between terrane accretion and foreland deformation may be defined, and where the influence of metamorphism and magmatism on fault-zone rheology may be observed.

Although metamorphism and intrusion clearly play key roles, they are beyond the scope of this chapter (for reviews, see

Ernst, this volume; and Ernst, 1988). A discussion of hinterland tectonics cannot avoid use of the phrase "metamorphic core complex" (Davis and Coney, 1979; Crittenden and others, 1980) because these regions contain exposures of the deepest structural levels in the Cordillera. There is much confusion concerning this term, in part due to the polygenetic nature of all of those core complexes north of Arizona (Armstrong, 1982). I regard a "metamorphic core complex" as fundamentally a zone of Cenozoic detachment faulting and metamorphism, which has exposed, only partly coincidentally, mid-crustal levels of the Mesozoic–early Tertiary compressional orogen. Three distinctly different segments of the hinterland will be reviewed: (1) eastern and central Washington, (2) eastern Nevada–western Utah, and (3) southeastern California–southwestern Arizona.

Eastern and central Washington. West of the Purcell anticlinorium and the easternmost Cretaceous batholiths lies the Priest River Complex and the Spokane Dome (Fig. 7). The former is a terrane of poorly dated, amphibolite-grade metamorphic rocks, and the latter refers to a 4-km-thick, arched zone of mylonitic gneisses within the complex that straddle the Washington-Idaho border (Rhodes and Hyndman, 1984). Although their age is poorly known, mylonites on both sides of the dome show top-to-the-east sense of shear, which is in the same direction as the Cordilleran thrust belt to the east (Rhodes and Hyndman, 1984).

The Kootenay Arc swings southwestward out of Canada and crosses into the United States near the Idaho-Washington state line (Fig. 7). The arc comprises Proterozoic and lower Paleozoic sedimentary rocks deposited in a slope or ocean basin that was proximal to North America, as well as the easternmost accreted rocks ("Intermontane terrane") in this segment of the Cordillera. The rocks are tightly folded and metamorphosed, in contrast to the open folds and low-grade metamorphism of the Purcell anticlinorium to the east. In Canada, cleavage fans, back folds or refolded folds, and folded thrust faults characterize the Kootenay Arc and its partial northern equivalent Selkirk Fan (*Brown* in McClay and Price, 1981; *Brown and others, 1986; Price, 1986,* in McClay and Price, 1981); similar, though less well exposed structures are present in northeastern Washington (Miller and Clark, 1975). The Kootenay Arc probably marks the westernmost edge of the autochthonous North American craton (*Price* in McClay and Price, 1981).

Farther west lie two metamorphic core complexes referred to as the Kettle and Okanogan domes (or "gneiss domes"; Fig. 7), which are separated by the Cenozoic Republic graben (Fox and others, 1977; *Cheney* in Crittenden and others, 1980). Medium- to high-grade metamorphic rocks in the cores of both domes have been correlated with the Monashee Group, which forms the core and autochthonous cover of the Shuswap complex in Canada (Fox and others, 1976, 1977; *Brown* in McClay and Price, 1981). $^{87}Sr/^{86}Sr$ in Mesozoic igneous rocks indicate that the region is underlain by continental crust (Armstrong and others, 1977), but Paleozoic and lower Mesozoic "eugeoclinal" strata are part of the easternmost of the Cordilleran collage of accreted terranes (Coney and others, 1980). Both domes have widespread mylonites and cataclastic rocks, but the ages and kinematic significance of these fault zones are poorly known (Fox, 1976, 1977; *Cheney* in Crittenden and others, 1980; *Hurich and others, 1985*). The east-dipping Kettle River fault that bounds the east side of the Kettle dome is characterized by both a mylonitic foliation and a superposed chloritic breccia zone; the age of the mylonitization is uncertain, but the chloritic breccia zone is probably related to Tertiary extension (Rhodes and Cheney, 1981).

A COCORP deep seismic reflection transect across northern Washington indicates a complex deep crustal structure (Potter and others, 1986). Thick packages of west-dipping reflectors can be mapped discontinuously from a near-surface position beneath the Kootenay Arc to near the base of the crust beneath the Okanogan dome. These reflectors are interpreted as thrust faults in the crystalline basement that form the roots of the Cordilleran thrust belt, and have subsequently been offset by Tertiary low-angle normal faults (Potter and others, 1986). Their geometry is generally consistent with *Brown and others'* (1986) basement duplex model for the deep structure of the Shuswap complex.

Eastern Nevada–western Utah. A structural belt analogous to the Kootenay Arc is missing or remains buried in this segment of the hinterland. Instead, the region between the thrust belt and the metamorphic core complexes is characterized by homoclinal sections and generally broad, open folds in little-metamorphosed Paleozoic to lower Mesozoic strata (Allmendinger and others, 1984). Close-spaced imbricate thrust faults with exposed ramp zones are rare; thrust faults probably parallel bedding for long distances perpendicular to strike. In northern Utah and southern Idaho, bedding-parallel detachment in the Carboniferous Manning Shale is common (*Allmendinger and Jordan, 1981;* Allmendinger and others, 1984). In west-central Utah and east-central Nevada, no major thrust faults crop out over a distance of more than 100 km west of the Canyon Range (Fig. 9), and seismic reflection data show that for much of that distance, the Canyon Range thrust probably parallels bedding (Armstrong, 1968; *Stewart and Carlson, 1978;* Stewart, 1980; *Hintze, 1980;* Allmendinger and others, 1983; *Miller and others, 1983; Smith and Bruhn, 1984*).

The large-scale structural geometry thus suggests little horizontal shortening at structural levels presently exposed. However, at least in northern Utah and southern Idaho, poorly dated penetrative strain within the allochthons (locally 50 to 200 percent extension) is much greater than in the thrust belt to the east. In the Newfoundland Mountains of the northern Utah hinterland, layer parallel shortening and thrusting predated latest Jurassic and thus was older than any demonstrable deformation in the thrust belt (Allmendinger and others, 1984; *Allmendinger and Jordan, 1984a*). Pre-latest Jurassic thrust faults with generally top-to-the-east displacement have also been documented in the Pilot Range (Miller and others, 1987) and in the Gold Hill area south of Wendover (Nolan, 1935; Allmendinger and others, 1984).

Crystalline basement is exposed only in the Raft River–Albion Range and the Ruby Mountains core complexes (Fig. 9;

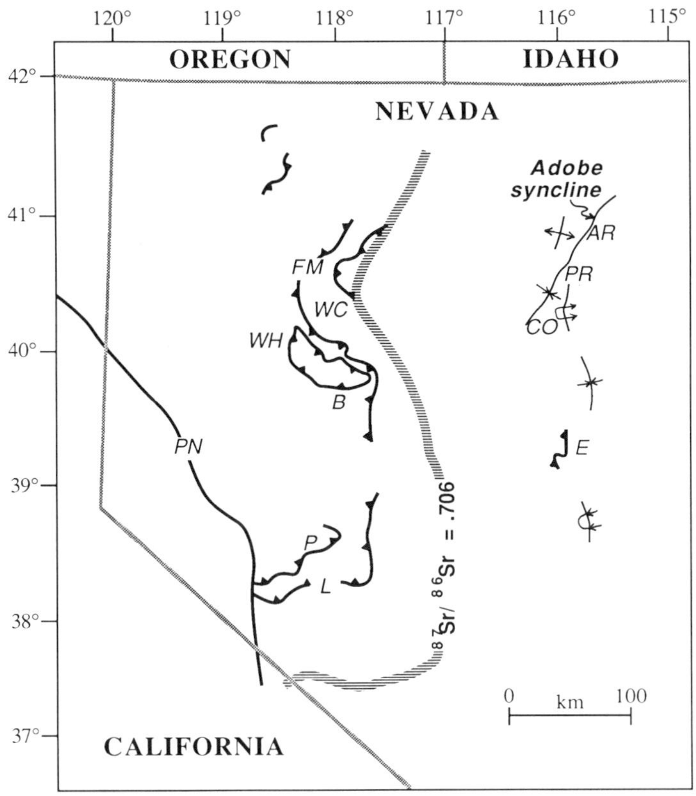

Figure 13. Generalized tectonic map of Mesozoic thrust faults in central and western Nevada. Modified from Stewart (1980) and *Speed and others* (in Ernst, 1988). *AR* = Adobe Range; *B* = Boyer thrust; *CO* = Cortez Range; *E* = Eureka; *FM* = Fencemaker thrust; *L* = Luning thrust; *P* = Pamlico thrust; *PN* = Pin Nut fault; *PR* = Piñon Range; *WC* = Willow Creek thrust; *WH* = Wild Horse thrust.

Compton and others, 1977; Snoke and Lush, 1984). In the Raft River Range, the little-deformed Archean basement lies just below a Tertiary low-angle normal fault, but inferences based on regional data indicate that it lies above the westward projection of the Absaroka thrust (*Allmendinger and Jordan, 1981;* see also *Crittenden, 1979; Compton and Todd, 1979*) and may coincide with the basement ramp of the Willard thrust (*Beck and Bruhn, 1983*). Isotopic dating of syntectonic mineral growths related to first and second directions in the Albion Range suggest Jurassic and Cretaceous ages (*Snoke and Miller* in Ernst, 1988). Fold axes related to these deformations trend northeasterly and northwesterly, respectively, with generally easterly vergences (Compton and others, 1977; *Miller* in Crittenden and others, 1980).

In the East Humboldt Range (Fig. 9), remobilized Archean basement occurs in the core of a recumbent nappe, which was emplaced on an upright lower Paleozoic sequence (Snoke and Lush, 1984; *Snoke and Miller* in Ernst, 1988). Similar Mesozoic structures characterize the core of the Ruby Range, but it is not yet known whether basement there is Precambrian. Facing directions and sense of shear in Mesozoic structures is variable in the Ruby–East Humboldt terrane; there and in the Wood Hills, at least some of the folds have westerly vergence (*Snoke* and *Howard* both in Crittenden and others, 1980; Snoke and Howard, 1984; *Snoke and Miller* in Ernst, 1988).

Indications of basement involvement also come from the positions of regionally developed synclinoria next to the core complexes and from deep seismic reflection data. Armstrong (1982) interpreted the Confusion Range and Sublette Range synclinoria (Fig. 9), which he referred to as "dejected zones," to have formed above major basement ramps. COCORP Utah line 1 supports this contention and, in combination with field relations, suggests that the basement ramp must have had a vertical relief of ~10 km (Plate 10D; Allmendinger and others, 1983, in Barazangi and Brown, 1986). One implication of Armstrong's (1982) model is that some of the uplift and arching of the Albion and Snake Range core complexes must have occurred during shortening in the thrust belt to the east. Some support for this notion comes from K-Ar cooling ages of middle-late Cretaceous on micas from the hinterland region (Miller and others, 1987; *Snoke and Miller* in Ernst, 1988).

Central and western Nevada. West of the core complexes in central Nevada, surface indications of Mesozoic shortening are meager. Minor thrusts and folds are present near Eureka and to the north in the Cortez, Piñon, and Adobe Ranges where they have westerly vergence (*Ketner and Smith, 1974;* Smith and Ketner, 1977; Stewart, 1980; Ketner, 1984); this belt has been named the Eureka thrust belt by Speed (1982, *and others* in Ernst, 1988). Several ranges in northeastern Nevada record a complex history of Triassic through Cretaceous thrusting with both easterly and southerly vergences (Riva, 1970; Thorman, 1970; *Coats and Riva* in Miller and others, 1983). Ketner (Ketner and Smith, 1982; Ketner, 1984) has emphasized the possibility that many thrust faults in this region that have been assigned to the Paleozoic Antler and Sonoma orogenies (not discussed here) could be Mesozoic in age.

In western Nevada, substantial Mesozoic shortening has been documented in Triassic and Jurassic marine strata (Silberling and Roberts, 1962; *Speed* in Howell and McDougall, 1978). The Lunning, Pamlico, and Fencemaker thrust systems (Fig. 13) were formed in four episodes between the Late Jurassic and Late Cretaceous or early Tertiary and resulted in the thrusting of Mesozoic basinal facies over coeval shelf facies (Speed, 1982, in Howell and McDougall, 1978; *Oldow, 1981a, 1981b, 1983;* Oldow and others, 1984). The formation of the marine depositional basin and subsequent localization of the thrusts appears to have been controlled by the buried edge of North American Precambrian basement and the suture of the late Paleozoic–early Mesozoic Sonomia terrane (Speed, 1982, in Howell and McDougall, 1978). Fold axes related to these episodes trend dominantly northeast and northwest, with minor north and east sets (Oldow and others, 1984). Shortening in the region may be as much as 100 km (*Oldow, 1981a, b*).

Southeastern California–southwestern Arizona. In southeastern California, west of the southern Nevada thrust belt and north of the Garlock fault, lies a broad terrane of Mesozoic metamorphism, plutonism, and contractional deformation that has been severely overprinted by younger extension. In the White, Inyo, Argus, and Slate Ranges (Fig. 12), Mesozoic deformation occurred in three pulses: (1) Middle Triassic–Early Jurassic; (2) late–Middle to early–Late Jurassic; and (3) Late Cretaceous (*Burchfiel and others, 1970; Dunne and others* in Howell and McDougall, 1978). During the first pulse, northeast- and northwest-trending folds, both sets with easterly vergence, and the Last Chance thrust system were formed. Left-slip fractures and the Swansea-Coso-Argus thrust system were formed during the second episode, and the White Mountain fault zone formed during the third (Fig. 12). In the Death Valley area, amphibolite-grade metamorphism and folding occurred between the Late Jurassic and the Late Cretaceous (Labotka and others, 1980; *Burchfiel and Davis* in Ernst, 1981).

The final termination of the Cordilleran fold and thrust belt occurs in the Old Dad, Old Woman, Piute, and Marble Mountains and the Kilbeck Hills (Fig. 12; Dunne, 1977; *Burchfiel and Davis* in Ernst, 1981; *Howard and others, 1981; Miller and others* in Frost and Martin, 1982), where the structural geometries and basement involvement classify the zone as "hinterland" according to the definition used here. Remnants of fold and thrust structures of post-Jurassic and pre–Late Cretaceous age are scattered across the eastern Mojave Desert and western Arizona in the Big Maria, Riverside, Palen, Mule, Dome Rock, Plomosa, and Harquahala Mountains and surrounding ranges (Fig. 12; Reynolds and others, *1980,* 1986, in Ernst, 1988; *Burchfiel and Davis* in Ernst, 1981; and articles by *Baltz, Ellis, Hamilton, Lyle, Martin and others, Reynolds, Tosdal,* all in Frost and Martin, 1982). These structures make up the "Maria fold and thrust belt" in which thrusts with southwest, south, north, and northeast vergences truncate south to southeast-verging folds (Reynolds and others, 1986, in Ernst, 1988). In most of these areas, Paleozoic cratonic strata and the Mesozoic McCoy Mountains Formation, metamorphosed to upper greenschist facies or higher, are commonly faulted and folded together with crystalline Precambrian basement rocks. In the Vincent and Chocolate Mountains and nearby ranges (Fig. 12), however, Precambrian crystalline basement and Mesozoic granitoids are thrust over the Pelona-Orocopia Schist, a metamorphosed oceanic suite of rocks comprising graywacke, chert, limestone, and mafic and ultramafic rocks (*Haxel and Dillon* in Howell and McDougall, 1978).

In south-central Arizona, Haxel and others (1984) have identified a number of Late Cretaceous–early Tertiary thrust faults that place Precambrian crystalline basement and Mesozoic plutons over metamorphosed Jurassic and Cretaceous igneous and sedimentary rocks ("BQ" on Fig. 12). The fault zones have a well-developed mylonitic foliation with a northeast-southwest elongation lineation. The translation direction is poorly known, although a few folds suggest a southwesterly direction (Haxel and others, 1984).

SUMMARY OF TOPICAL ISSUES

Thrust-belt mechanics and the Cordilleran orogen

A thorough discussion of mechanical models of thrust faulting is beyond the scope of this chapter; what follows are comments on various models as they have been or might apply to the Cordillera. Those models can be divided into three basic groups: gravity sliding, gravity spreading, and horizontal tectonic compression.

The gravity-sliding model has been applied to the northern part of the U.S. Cordillera by Rubey and Hubbert (1959), *Mudge (1970), Scholten* (in de Jong and Scholten, 1973), and *Hose and Danes* (in de Jong and Scholten, 1973), following the elegant analysis of *Hubbert and Rubey (1959).* This model, however, has now been abandoned by virtually all workers in the region because of the overwhelming evidence that the basal décollement of the thrust belt has always dipped westward toward the hinterland (e.g., Bally and others, 1966; Price, 1971).

The gravity spreading model, in which fold-thrust belts are likened to spreading continental ice caps, was applied to the Cordillera by Price (1971, in de Jong and Scholten, 1973) and *Elliott (1976)* based on the model by *Bucher (1956).* The key aspect of the model is an upwelling and spreading infrastructure in the core of the orogen, which drives the fringes of the mountain belt over the craton. *Hamilton* (in Howell and McDougall, 1978) proposed that the spreading of the magmatic arc was the driving mechanism. The model has lost favor in recent years for several reasons (see Bally and others, 1966; *Cambell, 1973*): (1) lack of evidence for regions of hinterland extension coeval with, and of the same magnitude as, foreland shortening; (2) recognition that deformation in the core of the Canadian part of the orogen completely predates Rocky Mountain foreland shortening; and (3) the recognition of the importance of accreted terranes in orogenesis.

The horizontal tectonic compression models of *Chapple (1978)* and Davis and others (1983) incorporate gravity, particularly as an important factor in the cratonward-propagation of a thrust belt. However, gravity becomes important only after horizontal shortening, vertical thickening, and topographic development have occurred; thus tectonic compression is the key driving mechanism. The Davis and others (1983) model requires continued shortening and thickening in the back end of the deforming wedge to maintain its self-similar shape as erosion reduces the wedge taper (which is dependent on friction along the base of the wedge). Presently available data from the best dated part of the thrust belt in Idaho, Wyoming, and northern Utah shows only limited evidence for continued shortening in the western part of the belt as more easterly thrusts became active (see also Woodward, 1987). A provocative aspect of their model, however, is that topography should flatten where the basal décollement crossed the brittle-plastic transition. This would suggest that the hinterland of the Cordillera, beneath which the basal décollement was probably everywhere in the plastic regime, was at a high elevation but had relatively flat relief, much as the modern Andes and the Altiplano Plateau appear today.

Finally, as explicitly proposed by *Price* (in de Jong and Scholten, 1973), structural thickening in the thrust belt should load the edge of the craton and induce flexural subsidence. This scenario provides an excellent explanation of the highly asymmetric foreland basins or foredeeps observed in the Cretaceous and lower Tertiary of the Cordillera (e.g., McGookey and others, 1972) and has been treated analytically by Jordan (1981) and *Beaumont (1981)*. It has also been proposed that the Moxa and Sweetgrass arches, which occur mostly east of the thrust belt, are forebulges of the flexures, perhaps localized by preexisting basement features and modified by subsequent deformation (Jordan, 1981; *Lorenz* in Powers, 1982). The general lack of a pronounced, asymmetric foreland basin south of southwestern Utah (e.g., McGookey and others, 1972) could be related to either (1) decreasing shortening and hence loading southward; (2) the thrust belt cutting progressively into thicker and stiffer craton southward; or (3) the prevalence of out-of-sequence thrusting in the southern segment.

Basement involvement in contractional deformation

A through-going theme of this chapter has been the nature and location of basement involvement in fold and thrust deformation. From this review, it seems that there are two fundamental controls: the thermal regime of the crust and the location of the Paleozoic hingeline west of which the basement was significantly thinned during the latest Proterozoic rifting along the western margin. Burchfiel and Davis (1975) first pointed out the correlation between the locations of the easternmost Mesozoic plutons and the position of basement involvement. The extreme southern end of the thrust belt in the Old Woman Mountains lies within the Mesozoic magmatic arc; basement is involved in all of the thrusts that occur within the magmatic arc (Fig. 12). Burchfiel and Davis (1975) attributed this to thermal softening of the basement, effectively reducing the rheological contrast between it and the overlying sedimentary section and allowing it to be involved in thin-skinned thrusting.

Farther north in Utah, however, basement involvement in the thrust belt occurred >100 km (in present distances) east of the easternmost Mesozoic igneous rocks. There, basement involvement generally coincides with the location of the Paleozoic hingeline (Armstrong and Oriel, 1965; Royse and others, 1975; *Allmendinger and others* in Barazangi and Brown, 1986), and it seems likely that a combination of deep burial (and hence warmer thermal regime) of the basement and its previously stretched and deformed nature allow it to become involved in thrusting. In northwestern and west-central Montana, basement was involved only within or west of the Belt basin.

Basement involvement in the Rocky Mountain foreland uplifts appears to be, in a general way, a function of both thermal state of the crust and paleogeography. The nearly complete cessation of magmatism west of the province during that deformation must have cooled the crust, allowing stresses to be transmitted farther eastward to the locus of thick-skinned deformation. The observations that the Wind River thrust can be traced to depths of ~25 km and that earthquakes occur to 40 km depth in an analogous modern province (*Jordan and Allmendinger, 1986*), indicate that the crust during deformation was cool enough to prohibit large-scale ductile flow, except perhaps near its base. However, thermal regime alone does not explain why thin- and thick-skinned deformation occurred simultaneously, and the answer must lie in the paleogeography and preexisting deformation such as the Ancestral Rockies. The main difference between the regions of thin- and thick-skinned deformation is that the former occurs largely in the miogeocline with its thinned Precambrian crust, whereas the latter formed in the thicker cratonal crust with its thin Phanerozoic cover.

Rates of foreland shortening and comparison to plate convergence

Because of the relatively well-known ages of individual faults, rates of shortening can be calculated for the Idaho-Wyoming–northern Utah segment of the thrust belt. A bulk rate of shortening of 1.7 mm/yr can be determined from a linear regression fit to a plot of cumulative shortening versus thrust fault age (Fig. 10). In general, shortening rates appear faster during the Late Cretaceous and early Tertiary (~3 mm/yr) and slower during the Early to mid-Cretaceous (1.4 to 1.7 mm/yr), although these figures are subject to the considerable uncertainty in the age of the Paris-Willard thrust system (e.g., Heller and others, 1986).

To compare foreland shortening directly with plate convergence, the displacement rates of individual thrusts are of greater interest than the bulk rate (i.e., whether the regression line in Fig. 10 should instead be a step function). Such information is available for only a few faults (Wiltschko and Dorr, 1983; Warner and Royse, 1987; Royse, personal communication, 1986): the Prospect thrust north of Snider Basin moved about 12 km in 2 m.y. (= 6 mm/yr), and the younger Absaroka thrust moved about 16 km in 4 m.y. (= 4 mm/yr). Overall, the Absaroka thrust moved 26 km in 21 m.y. (= 1.24 mm/yr). These rates are comparable to those observed in the Subandean belt and the Precordillera of the modern Andes (3 to 6 mm/yr over the last 9 m.y.; *Allmendinger and others, 1983; Jordan and Gardeweg, 1988*).

These western United States foreland shortening rates were one to two orders of magnitude slower than plate convergence rates—50 to 150 mm/yr—during the same interval (Engebretson and others, 1984). The Andean rates are, likewise, 15 to 30 times slower than current convergence at the Nazca–South American plate boundary. These data suggest that there is no direct coupling between plate-margin processes (terrane accretion) and foreland deformation.

Hinterland topography and extension during foreland shortening

The involvement of basement in the shortening requires significant thickening of the continental crust beneath the hinter-

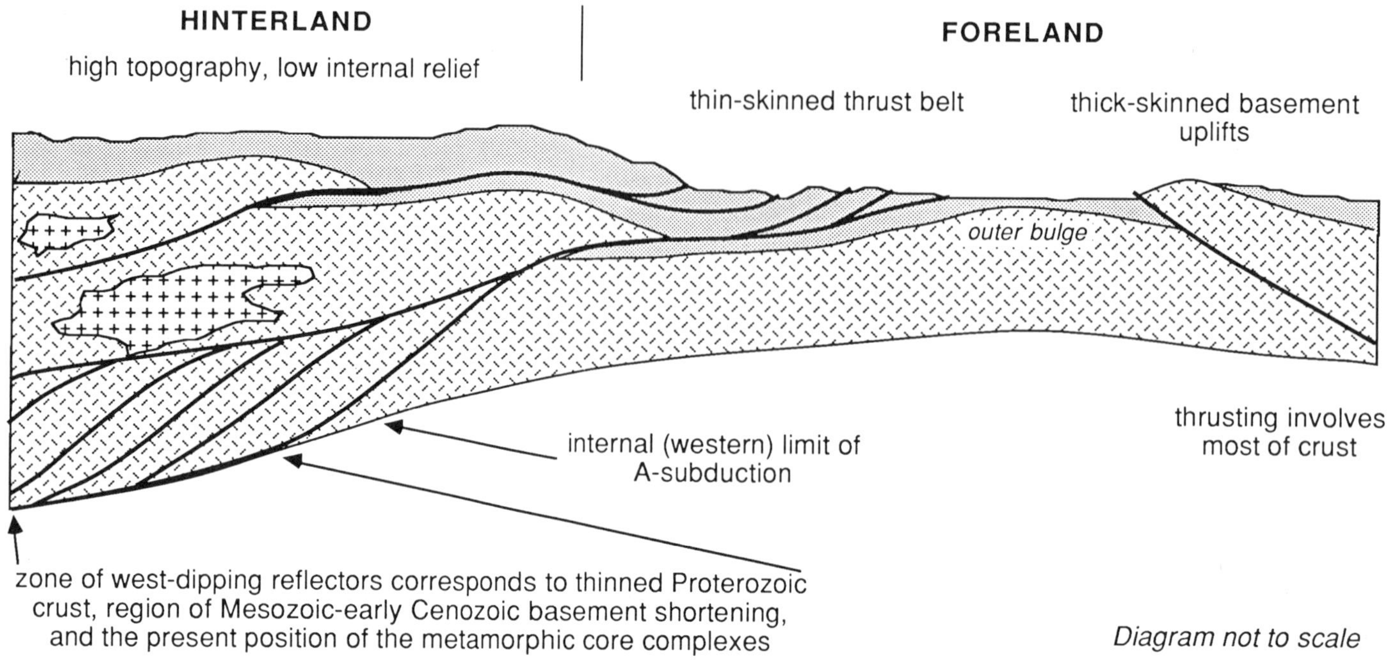

Figure 14. Schematic cross section of the Mesozoic–early Cenozoic compressional orogen of the western United States. Not to scale and not balanced.

land. Crustal thickening should have had two related effects: (1) isostatically driven topographic uplift of the hinterland, and (2) the development of gravitational instability that could result in horizontal extension (*Allmendinger and Jordan, 1984b;* Coney and Harms, 1984). Evidence for the former remains controversial. Armstrong (1972) pointed out that the ubiquitous presence of upper Paleozoic and lower Mesozoic strata unconformably beneath widespread mid-Tertiary deposits in western Utah and eastern Nevada, as well as lack of detritus in the foreland derived from an eroded hinterland, were evidence for little significant erosion and, by inference, little topographic relief in the hinterland. However, Miller and others (1987) recently reviewed regional geochronologic evidence showing a Late Cretaceous episode of cooling, perhaps initiated by uplift over a deep-seated thrust or extension. Also, the modern Altiplano-Puna Plateau of the Andes, analogous in many ways to the hinterland, is internally drained, and the local relief is relatively small (Jordan and others, 1983; *Allmendinger and Jordan, 1984b; Jordan and Alonso, 1987*). The little-studied Cretaceous and Lower Tertiary clastic rocks of Nevada (*Fouch and others* in Newman and Goode, 1979; Stewart, 1980) are thus a prime target for research.

Extension related to compressional orogenesis arises from two distinct mechanisms: (1) superficial extension due to topography, as modeled by McTigue and Mei (1981) and *Burchfiel and Royden (1985);* and (2) crustal-scale extension related to the buoyancy of the crustal root and the topography (Froidevaux and Isacks, 1984). The post-thrusting, mid-Cenozoic extension in the Cordillera is widely recognized and may well be related to the second mechanism (e.g., Coney and Harms, 1984). Less well known but equally important is the evidence for extension *during* shortening in the thrust belt, just as there is extension in the hinterland of the Andes and the Himalaya, coeval with foreland shortening (*Dalmayrac and Molnar, 1981; Suárez and others, 1983; Allmendinger and Jordan, 1984b; Burchfiel and Royden, 1985; Sébrier and others, 1985; Molnar, 1986; Allmendinger and others, 1988*). Only a few localities, most notably the Newfoundland Mountains, Pilot Range, and Crater Island in northwestern Utah and a few areas in north-central Nevada, have unequivocal evidence of Mesozoic or early Tertiary extension (Berger and Taylor, 1980; *Allmendinger and Jordan, 1984a;* Allmendinger and others, 1984). However, the hinterland has many exposures of undated low-angle faults and, though the dogma of the 1950s and 1960s was to call them Mesozoic thrust faults, the dogma of the 1970s and 1980s has been to call them Tertiary extensional detachments. It may turn out that many of those low-angle faults are Mesozoic extensional structures.

The evidence of extension in the hinterlands of compressional mountain belts (*Dalmayrac and Molnar, 1981; Allmendinger and Jordan, 1984a, b;* Allmendinger and others, 1984; *Burchfiel and Royden, 1985*) might be interpreted as supporting the gravity-spreading model of thrust belts. However, in all known cases, the extension *follows* shortening. Thus, it seems more likely to be related to crustal thickening and, presumably, topographic development.

Perspectives from deep seismic data and schematic cross section of the compressional orogen

The Mesozoic–early Cenozoic Cordilleran orogen in the western United States has been crossed by three COCORP deep seismic reflection transects, one in northern Montana, Idaho, and

Washington, another across central Utah, Nevada, and California, and a third in southeastern California and western Arizona (Allmendinger and others, 1987, *1987a, 1987b;* Potter and others, 1986; Hauser and others, 1987). These data make it possible to construct a cartoon cross section of the compressional orogen (Fig. 14). Interpretations of zones of west-dipping reflectors on each of these transects suggest that thrust faults cut from a position entirely or mostly in the sedimentary section to the base of the *present* crust across a present lateral distance of 150 to 200 km (probably 75 to 150 km prior to regional extension). If the reflection Moho observed today is the same as that during the Mesozoic, then this zone marks the hindward or interior limit of underthrusting of relatively undeformed basement (A-subduction). However, if the reflection Moho is late Cenozoic in age, as seems likely (Allmendinger and others, 1987, *1987a*), the nature of deformation at the base of the crust during shortening remains unknown. In general, however, it is a zone in which upper crustal displacements are transferred to the lower crust.

This zone of crustal penetrating thrusts begins at the foreland-hinterland boundary and coincides with the region of Proterozoic transitional crust. The metamorphic core complexes also appear to be restricted to this zone of dipping reflectors, further suggesting a relation between Proterozoic rifting, Mesozoic–early Cenozoic thrusting, and mid–late Cenozoic extension.

CONCLUSIONS

The Cordilleran orogen of the western United States exhibits a broad range of contractional structural styles, from thick-skinned to thin-skinned and basement-involved to detachments within the sedimentary section. The only style for which there are relatively few well-documented exposures are contractional nappes involving basement in high-grade metamorphic terranes, and all of those occur in thrusts that formed within the magmatic arc (e.g., Old Woman Mountains, Big Maria Mountains, and Harquahala Mountains, Fig. 12; *Howard and others, 1981; Hamilton* in Frost and Martin, 1982; *Reynolds* in Frost and Martin, 1982; Reynolds and others, 1986; Reynolds and others in Ernst, 1988). Part of this lack may be more illusory than real; differentiating extensional and contractional structures in the metamorphic hinterland remains a major problem, and thus a major goal for future research.

Other topics that appear ripe for additional research include (in no particular order): (1) documentation of coeval contraction and extension and its significance in terms of mountain-building models; (2) the role of topography and erosion both as a result of and as a modifier of deformation in a compressional mountain belt; (3) the relations between magmatism and shortening; (4) the controls on the changes in structural geometry between foreland and hinterland deformation and concomitant changes in strain mechanisms; (5) the role of fluids in contractional deformation; and (6) the coupling of deformation across a magmatic arc and the relations of plate geometry, plate convergence, and sporadic events such as terrane accretion at the plate margin to fold and thrust deformation in the hinterland and foreland of the orogen. This list is necessarily a personal one and thus incomplete at best.

Increasingly, these problems require an interdisciplinary approach. Field geology, the traditional means of documenting contractional deformation, is not, as some have claimed, out of vogue, it is now and always has been an indispensable *technique* for studying the earth. However, the scientific questions of today demand that it be combined with other techniques, be they geophysical, geochemical, mechanical, or microstructural. Also, no longer can one distinguish between regional and detailed studies; as the question of process rather than description becomes more important, it is clear that the most minute observation can have vast implications for understanding huge regions.

REFERENCES CITED

Text references in *italics* will be found only on accompanying microfiche.

Allmendinger, R. W., 1981, Structural geometry of Meade thrust plate in northern Blackfoot Mountains, southeastern Idaho: American Association of Petroleum Geologists Bulletin, v. 65, p. 509–525.

—— , 1982, Analysis of microstructures in the Meade plate of the Idaho–Wyoming foreland thrust belt, U.S.A.: Tectonophysics, v. 85, p. 221–251.

Allmendinger, R. W., Sharp, J. W., Von Tish, D., Serpa, L., Kaufman, S., Oliver, J., and Smith, R. B., 1983, Cenozoic and Mesozoic structure of the eastern Basin and Range Province, Utah, from COCORP seismic reflection data: Geology, v. 11, p. 532–536.

Allmendinger, R. W., Miller, D. M., and Jordan, T. E., 1984, Known and inferred Mesozoic deformation in the hinterland of the Sevier belt, northwest Utah, *in* Kerns, G. J., and Kerns, R. L., eds., Geology of northwest Utah, southern Idaho, and northeast Nevada: Utah Geological Association Publication 13, p. 21–34.

Allmendinger, R. W., Hauge, T., Hauser, E. C., Potter, C. J., Klemperer, S. L., Nelson, D. K., Knuepfer, P., and Oliver, J., 1987, Overview of the COCORP 40°N Transect, western United States; The fabric of an orogenic belt: Geological Society of America Bulletin, v. 98, p. 308–319.

Armstrong, F. C., and Cressman, E. R., 1963, The Bannock thrust zone, southeastern Idaho: U.S. Geological Survey Professional Paper 374–J. 22 p.

Armstrong, F. C., and Oriel, S. S., 1965, Tectonic development of Idaho–Wyoming thrust belt: American Association of Petroleum Geologists Bulletin, v. 49, p. 1847–1866.

Armstrong, R. L., 1968, Sevier orogenic belt in Nevada and Utah: Geological Society of America Bulletin, v. 79, p. 429–458.

—— , 1972, Low-angle (denudation) faults, hinterland of the Sevier orogenic belt, eastern Nevada and western Utah: Geological Society of America Bulletin, v. 83, p. 1729–1754.

—— , 1974, Magmatism, orogenic timing, and orogenic diachronism in the Cordillera from Mexico to Canada: Nature, v. 247, p. 348–351.

—— , 1982, Cordilleran metamorphic core complexes; From Arizona to southern Canada: Annual Reviews of Earth and Planetary Science, v. 10, p. 129–154.

Armstrong, R. L., Taubeneck, W. H., and Hales, P. O., 1977, Rb-Sr and K-Ar geochronology of Mesozoic granitic rocks and their Sr isotopic composition, Oregon, Washington, and Idaho: Geological Society of America Bulletin, v. 88, p. 397–411.

Bally, A. W., 1984, Tectogenèse et sismique réflexion: Bulletin Societe Géologique France, v. 26, p. 279–285.

Bally, A. W., Gordy, P. L., and Stewart, G. A., 1966, Structure, seismic data, and orogenic evolution of southern Canadian Rocky Mountains: Bulletin of Canadian Petroleum Geology, v. 14, p. 337–381.

Baltz, E. H., 1967, Stratigraphy and regional tectonic implications of part of Upper Cretaceous and Tertiary rocks, east-central San Juan Basin, New Mexico: U.S. Geological Survey Professional Paper 552, 101 p.

Barazangi, M., and Brown, L. D., eds., 1986, Reflection seismology; The continental crust: American Geophysical Union, Geodynamics Series, v. 14, 339 p.

Berg, R. R., 1962, Mountain flank thrusting in Rocky Mountain foreland, Wyoming and Colorado: American Association of Petroleum Geologists Bulletin, v. 46, p. 2029–2032.

Berger, B. R., and Taylor, B. E., 1980, Pre-Cenozoic normal faulting in the Osgood Mountains, Humboldt County, Nevada: Geology, v. 8, p. 594–598.

Bird, P., 1984, Laramide crustal thickening event in the Rocky Mountain foreland and Great Plains: Tectonics, v. 3, p. 741–758.

Boos, C. M., and Boos, M. F., 1957, Tectonics of eastern flank and foothills of Front Range, Colorado: American Association of Petroleum Geologists Bulletin, v. 41, p. 2603–2676.

Brewer, J. A., Allmendinger, R. W., Brown, L. D., Oliver, J. E., and Kaufman, S., 1982, COCORP profiling across the northern Rocky Mountain Front in southern Wyoming; Part I, Laramide structure: Geological Society of America Bulletin, v. 93, p. 1242–1252.

Brock, W. G., and Engelder, T., 1977, Deformation associated with the movement of the Muddy Mountain overthrust in the Buffington window, southeastern Nevada: Geological Society of America Bulletin, v. 88, p. 1667–1677.

Brown, W. G., 1984, Basement involved tectonics foreland areas: American Association of Petroleum Geologists, Continuing Education Course Note Series, v. 26, 92 p.

Bruhn, R. L., and Beck, S. L., 1981, Mechanics of thrust faulting in crystalline basement, Sevier orogenic belt, Utah: Geology, v. 9, p. 200–204.

Burchfiel, B. C., and Davis, G. A., 1975, Nature and controls of Cordilleran orogenesis, extensions of an earlier synthesis: American Journal of Science, v. 275–a, p. 363–396.

—— , 1977, Geology of the Sagamore Canyon–Slaughterhouse Spring area, New York Mountains, California: Geological Society of America Bulletin, v. 88, p. 1623–1640.

Burchfiel, B. C., Fleck, R. J., Secor, D. T., Vincelette, R. R., and Davis, G. A., 1974, Geology of the Spring Mountains, Nevada: Geological Society of America Bulletin, v. 85, p. 1013–1022.

Burchfiel, B. C., Wernicke, B., Willemin, J. H., Axen, G. J., and Cameron, S. C., 1982, A new type of décollement thrust: Nature, v. 300, p. 513–515.

Carr, M. D., 1983, Geometry and structural history of the Mesozoic thrust belt in the Goodsprings district, southern Spring Mountains, Nevada: Geological Society of America Bulletin, v. 94, p. 1185–1198.

Chamberlin, R. T., 1919, The building of the Colorado Rockies: Journal of Geology, v. 27, p. 145–251.

Chapin, C. E., 1979, Evolution of the Rio Grande rift; A summary, *in* Riecker, R. E., ed., Rio Grande rift; Tectonics and magmatism: American Geophysical Union, p. 1–5.

Christiansen, F. W., 1952, Structure and stratigraphy of the Canyon Range, central Utah: Geological Society of America Bulletin, v. 63, p. 717–740.

Clement, J. H., 1983, North flank of the Uinta Mountains, Utah, *in* Bally, A. W., ed., Seismic expression of structural styles: American Association of Petroleum Geologists, Studies in Geology Series no. 15, v. 3, p. 3.2.2-29–3.2.2-32.

Compton, R. R., Todd, V. R., Zartman, R. E., and Naeser, C. W., 1971, Oligocene and Miocene metamorphism, folding, and low-angle faulting in northwestern Utah: Geological Society of America Bulletin, v. 88, p. 1237–1250.

Coney, P. J., and Harms, T., 1984, Cordilleran metamorphic core complexes; Cenozoic extensional relics of Mesozoic compression: Geology, v. 12, p. 550–554.

Coney, P. J., and Reynolds, S. J., 1977, Cordilleran Benioff zones: Nature, v. 270, p. 403–406.

Coney, P. J., Jones, D. L., and Monger, J.W.H., 1980, Cordilleran suspect terranes: Nature, v. 288, p. 329–333.

Crittenden, M. D., Jr., 1972, Willard thrust and Cache allochthon, Utah: Geological Society of America Bulletin, v. 83, p. 2871–2880.

Crittenden, M. D., Jr., Coney, P. J., and Davis, G. H., eds., 1980, Cordilleran metamorphic core complexes: Geological Society of America Memoir 153, 490 p.

Davis, D., Suppe, J., and Dahlen, F. A., 1983, Mechanics of fold-and-thrust belts and accretionary wedges: Journal of Geophysical Research, v. 88, p. 1153–1172.

Davis, G. H., 1979, Laramide folding and faulting in southeastern Arizona: American Journal of Science, v. 279, p. 543–569.

Davis, G. H., and Coney, P. J., 1979, Geologic development of the Cordilleran metamorphic core complexes: Geology, v. 7, p. 120–124.

de Jong, K. A., and Scholten, R., eds., 1973, Gravity and tectonics: New York, Wiley-Interscience, 502 p.

Dixon, J. S., 1982, Regional structural synthesis, Wyoming salient of western overthrust belt: American Association of Petroleum Geologists Bulletin, v. 66, p. 1560–1580.

Dorr, J. A., Jr., Spearing, D. R., and Steidtmann, J. R., 1977, Deformation and deposition between a foreland uplift and an impinging thrust belt; Hoback Basin, Wyoming: Geological Society of America Special Paper 177, 82 p.

Drewes, H., 1981, Tectonics of southeastern Arizona: U.S. Geological Survey Professional Paper 1144, 96 p.

Dunne, G. C., 1977, Geology and structural evolution of Old Dad Mountain, Mojave Desert, California: Geological Society of America Bulletin, v. 88, p. 737–748.

Engebretson, D. C., Cox, A., and Thompson, G. A., 1984, Correlation of plate motions with continental tectonics; Laramide to Basin-Range: Tectonics, v. 3, p. 115–119.

Ernst, W. G., ed., 1981, The geotectonic development of California: Englewood Cliffs, New Jersey, Prentice-Hall, 706 p.

—— , 1988, Metamorphism and crustal evolution of the western United States: Englewood Cliffs, New Jersey, Prentice-Hall, 1008 p.

Fletcher, R. C., 1984, Instability of lithosphere undergoing shortening; A model for Laramide foreland structures: Geological Society of America Abstracts with Programs, v. 16, p. 83.

Fox, K. F., Jr., Rinehart, C. D., Engels, J. C., and Stern, T. W., 1976, Age of emplacement of the Okanogan gneiss dome, north-central Washington and British Columbia: Geological Society of America Bulletin, v. 87, p. 1217–1224.

Fox, K. F., Jr., Rinehart, C. D., and Engels, J. C., 1977, Plutonism and orogeny in north-central Washington; Timing and regional context: U.S. Geological Survey Professional Paper 989, 27 p.

Froidevaux, C. and Isacks, B. L., 1984, The mechanical state of the lithosphere in the Altiplano–Puna segment of the Andes: Earth Planetary Science Letters, v. 71, p. 305–314.

Frost, E. G., and Martin, D. L., eds., 1982, Mesozoic–Cenozoic tectonic evolution of the Colorado River Region, California, Arizona, and Nevada: San Diego, Cordilleran Publishers, 608 p.

Gries, R., 1983, Oil and gas prospecting beneath Precambrian of foreland thrust plates in Rocky Mountains: American Association of Petroleum Geologists Bulletin, v. 67, p. 1–28.

Gries, R. R., and Dyer, R. C., eds., 1985, Seismic exploration of the Rocky Mountain region: Denver, Rocky Mountain Association of Geologists, 300 p.

Hamilton, W., 1981, Plate tectonic mechanism of Laramide deformation: University of Wyoming Contributions to Geology, v. 19, p. 87–92.

Harrison, J. E., Griggs, A. B., and Wells, J. D., 1974, Tectonic features of the Precambrian Belt basin and their influence on post-Belt structures: U.S. Geological Survey Professional Paper 866, 15 p.

Harrison, J. E., Kleinkopf, M. D., and Wells, J. D., 1980, Phanerozoic thrusting in Proterozoic belt rocks, northwestern United States: Geology, v. 8, p. 407–411.

Harrison, J. E., Cressman, E. R., and Kleinkopf, M. D., 1985, Regional structure, the Atlantic Richfield–Marathon Oil No. 1 Gibbs borehole, and hydrocarbon potential west of the Rocky Mountain trench in northwestern Montana: U.S. Geological Survey Open-File Report 85–249, 8 p.

Hauser, E. C., Gephart, J., Latham, T., Oliver, J., Kaufman, S., Brown, L., and Lucchitta, I., 1987, COCORP Arizona transect; Strong crustal reflections and offset Moho beneath the transition zone: Geology, v. 15, p. 1103–1106.

Haxel, G. B., Tosdal, R. M., May, D. J., and Wright, J. E., 1984, Latest Cretaceous and early Tertiary orogenesis in south-central Arizona; Thrust faulting, regional metamorphism, and granitic plutonism: Geological Society of America Bulletin, v. 95, p. 631–653.

Heisey, E. L., Lawson, D. E., Norwood, E. R., Wach, P. H., and Hale, L. A., eds., 1977, Rocky Mountain thrust belt geology and resources: Casper, Wyoming Geological Association, 787 p.

Heller, P. L., Bowdler, S. S., Chambers, H. P., Coogan, J. C., Hagen, E. S., Shuster, M. W., Winslow, N. S., and Lawton, T. F., 1986, Time of initial thrusting in the Sevier orogenic belt, Idaho–Wyoming and Utah: Geology, v. 14, p. 388–391.

Hennier, J., and Spang, J. H., 1983, Mechanisms for deformation of sedimentary strata at Sheep Mountain anticline, Bighorn Basin, Wyoming: Wyoming Geological Association Guidebook, v. 34, p. 97–111.

Hoffman, J., Hower, J., and Aronson, J. L., 1976, Radiometric dating of time of thrusting in the Disturbed belt of Montana: Geology, v. 4, p. 16–20.

Hoppin, R. A., 1970, Structural development of the Five Springs Creek area, Bighorn Mountains, Wyoming: Geological Society of America Bulletin, v. 81, p. 2403–2416.

Hoppin, R. A., and Palmquist, J. C., 1965, Basement influence on later deformation; The problem, techniques of investigation, and examples from Bighorn Mountains, Wyoming: American Association of Petroleum Geologists Bulletin, v. 49, p. 993–1003.

Howell, D. G., and McDougall, K. A., eds., 1978, Mesozoic paleogeography of the western United States: Society of Economic Paleontologists and Mineralogists, Pacific Section, Pacific Coast Paleogeography Symposium 2, Los Angeles, 573 p.

Jordan, T. E., 1981, Thrust loads and foreland basin evolution, Cretaceous, western United States: American Association of Petroleum Geologists Bulletin, v. 65, p. 2506–2520.

Jordan, T. E., Isacks, B. L., Allmendinger, R. W., Brewer, J. A., Ramos, V. A., and Ando, C. J., 1983, Andean tectonics related to geometry of subducted Nazca plate: Geological Society of America Bulletin, v. 94, p. 341–361.

Keefer, W. R., 1965, Stratigraphy and geologic history of the uppermost Cretaceous, Paleocene, and lower Eocene rocks in the Wind River Basin, Wyoming: U.S. Geological Survey Professional Paper 495–A, 77 p.

Kelley, V.C., 1955, Monoclines of the Colorado Plateau: Geological Society of America Bulletin, v. 66, p. 789–804.

Ketner, K. B., 1984, Recent studies indicate that major structures in northeastern Nevada and the Golconda thrust in north-central Nevada are of Jurassic or Cretaceous age: Geology, v. 12, p. 483–486.

Ketner, K. B., and Smith, J. F., Jr., 1982, Mid-Paleozoic age of the Roberts thrust unsettled by new data from northern Nevada: Geology, v. 10, p. 298–303.

Labotka, T. C., Albee, A. L., Lanphere, M. A., and McDowell, S. D., 1980, Stratigraphy, structure, and metamorphism in the central Panamint Mountains (Telescope Peak Quadrangle), Death Valley area, California: Geological Society of America Bulletin, v. 91, pt. 1, p. 125–129, pt. 2, p. 843–933.

Lawton, T. F., 1985, Style and timing of frontal structures, thrust belt, central Utah: American Association of Petroleum Geologists Bulletin, v. 69, p. 1145–1159.

LeMasurier, W. E., 1970, Structural study of a Laramide fold involving shallow-seated basement rock, Front Range, Colorado: Geological Society of America Bulletin, v. 81, p. 421–434.

Longwell, C. R., 1974, Measure and date on movement Las Vegas Valley shear zone, Clark County, Nevada: Geological Society of America Bulletin, v. 85, p. 985–990.

Longwell, C. R., Pampeyan, E. H., Bower, B., and Roberts, R. J., 1965, Geology and mineral deposits of Clark County, Nevada: Nevada Bureau of Mines Bulletin 62, 218 p.

Love, J. D., 1970, Cenozoic geology of the Granite Mountains area, central Wyoming: U.S. Geological Survey Professional Paper 495–C, 154 p.

—— , 1978, Cenozoic thrust and normal faulting, and tectonic history of the Badwater area, northeastern margin of Wind River Basin, Wyoming: Wyoming Geological Association Guidebook, p. 235–238.

Lowell, J. D., ed., 1983, Rocky Mountain foreland basins and uplifts: Denver, Rocky Mountain Association of Geologists, 392 p.

Mallory, W. W., comp., 1972, Regional synthesis of the Pennsylvanian System, *in* Geologic atlas of the Rocky Mountain region: Denver, Rocky Mountain Association of Geologists, p. 111–132.

Mansfield, G. R., 1927, Geography, geology, and mineral resources of part of southeastern Idaho: U.S. Geological Survey Professional Paper 152, 453 p.

Matthews, V., III, ed., 1978, Laramide folding associated with basement block faulting in the western United States: Geological Society of America Memoir 151, 370 p.

McClay, K. R., and Price, N. J., eds., 1981, Thrust and nappe tectonics: Geological Society of London Special Publication 9, 539 p.

McGookey, D. P., coord., 1972, Cretaceous System, *in* Geologic atlas of the Rocky Mountain region: Denver, Rocky Mountain Association of Geologists, p. 190–228.

McTigue, D. F., and Mei, C. C., 1981, Gravity-induced stresses near topography of small slope: Journal of Geophysical Research, v. 86, p. 9268–9278.

Miller, D. M., Todd, V. R., and Howard, K. A., eds., 1983, Tectonic and stratigraphic studies in the Eastern Great Basin: Geological Society of America Memoir 157, 327 p.

Miller, D. M., Hillhouse, W. C., Zartman, R. E., and Lanphere, M. A., 1987, Geochronology of intrusive and metamorphic rocks in the Pilot Range, Utah and Nevada, and comparison with regional patterns: Geological Society of America Bulletin, v. 99, p. 866–879.

Miller, F. K., and Clark, L. D., 1975, Geology of the Chewelah–Loon Lake area, Stevens and Spokane counties, Washington: U.S. Geological Survey Professional Paper 806, 74 p.

Mitra, G., 1984, Brittle to ductile transition due to large strains along the White Rock thrust, Wind River Mountains, Wyoming: Journal of Structural Geology, v. 6, p. 51–62.

Mitra, G., Yonkee, W. A., and Gentry, D. J., 1984, Solution cleavage and its relationship to major structures in the Idaho–Utah–Wyoming thrust belt: Geology, v. 12, p. 354–358.

Mudge, M. R., 1972, Structural geology of the Sun River Canyon and adjacent areas, northwestern Montana: U.S. Geological Survey Professional Paper 663–B, 52 p.

Mudge, M. R., and Earhart, R. L., 1980, The Lewis thrust fault and related structures in the Disturbed belt, northwestern Montana: U.S. Geological Survey Professional paper 1174, 18 p.

Newman, G. W., and Goode, H. D., eds., 1979, Basin and Range Symposium: Denver, Rocky Mountain Association of Geologists, 662 p.

Nichols, D. J., Perry, W. J., Jr., and Haley, J. C., 1985, Reinterpretation of the palynology and age of Laramide syntectonic deposits, southwestern Montana, and revision of the Beaverhead Group: Geology, v. 13, p. 149–153.

Nielson, D. L., ed., 1982, Overthrust belt of Utah: Salt Lake City, Utah Geological Association Publication 10, 335 p.

Nokleberg, W. J., and Kistler, R. W., 1980, Paleozoic and Mesozoic deformations, central Sierra Nevada, California: U.S. Geological Survey Professional Paper 1145, 24 p.

Nolan, T. B., 1935, The Gold Hill mining district, Utah: U.S. Geological Survey Professional Paper 177, 172 p.

Oldow, J. S., Avé Lallemant, H. G., and Schmidt, W. J., 1984, Kinematics of plate convergence deduced from Mesozoic structures in the western Cordillera: Tectonics, v. 3, p. 201–227.

Perry, W. J., Jr., Wardlaw, B. R., Bostick, N. H., and Maughan, E. K., 1983, Structure, burial history, and petroleum potential of the frontal thrust belt and adjacent foreland, southwest Montana: American Association of Petroleum Geologists Bulletin, v. 67, p. 725–743.

Peterson, J. A., ed., 1986, Paleotectonics and sedimentation in the Rocky Mountain region, United States: American Association of Petroleum Geologists Memoir 41, 693 p.

Potter, C. J., Sanford, W. E., Yoos, T. R., Prussen, E. I, Keach, R. W., II, Oliver, J. E., Kaufman, S., and Brown, L. D., 1986, COCORP deep seismic reflection traverse of the interior of the North American Cordillera, Washington, and Idaho; Implications for orogenic evolution: Tectonics, v. 5, p. 1007–1025.

Powers, R. B., ed., 1982, Geologic studies of the Cordilleran thrust belt: Denver, Rocky Mountain Association of Geologists, 976 p.

Prucha, J. J., Graham, J. A., and Nickelson, R. P., 1965, Basement controlled deformation in Wyoming province of Rocky Mountain foreland: American Association of Petroleum Geologists Bulletin, v. 49, p. 966–992.

Price, R. A., 1971, Gravitational sliding and the foreland thrust and fold belt of the North American Cordillera; Discussion: Geological Society of America Bulletin, v. 82, p. 1133–1138.

Reif, D. M., and Robinson, J. P., 1981, Geophysical, geochemical, and petrographic data and regional correlation from the Arizona State A-1 well, Pimal County, Arizona: Arizona Geological Society Digest, v. 13, p. 99–109.

Reynolds, S. J., Spencer, J. E., Richard, S. M., and Laubach, S. E., 1986, Mesozoic structures in west-central Arizona, *in* Beatty, B., and Wilkinson, P.A.K., eds., Frontiers in geology and ore deposits of Arizona and the southwest: Arizona Geological Society Digest, v. 16, p. 35–51.

Rhodes, B. P., and Cheney, E. S., 1981, Low-angle faulting and the origin of Kettle dome, a metamorphic core complex in northeastern Washington: Geology, v. 9, p. 366–369.

Rhodes, B. P., and Hyndman, D. W., 1984, Kinematics of mylonites in the Priest River "metamorphic core complex," northern Idaho and northeastern Washington: Canadian Journal of Earth Sciences, v. 21, p. 1161–1170.

Riva, J., 1970, Thrusted Paleozoic rocks in the northern and central HD Range, northeastern Nevada: Geological Society of America Bulletin, v. 81, p. 2689–2716.

Rodgers, J., 1949, Evolution of thought on structure of middle and southern Appalachians: American Association of Petroleum Geologists Bulletin, v. 33, p. 1643–1654.

Ross, C. P., 1959, Geology of Glacier National Park and the Flathead region, northwestern Montana: U.S. Geological Survey Professional Paper 296, 125 p.

Royse, F., Warner, M. A., and Reese, D. L., 1975, Thrust belt structural geometry and related stratigraphic problems, Wyoming–Idaho–northern Utah, *in* Bolyard, D. W., ed., Deep drilling frontiers of the central Rocky Mountains: Denver, Rocky Mountain Association of Geologists, p. 41–54.

Rubey, W. W., and Hubbert, M. K., 1959, Role of fluid pressure in mechanics of overthrust faulting; II, Overthrust belt in geosynclinal area of western Wyoming in light of fluid-pressure hypothesis: Geological Society of America Bulletin, v. 70, p. 167–206.

Ruppel, E. T., and Lopez, D. A., 1984, The thrust belt in southwest Montana and east-central Idaho: U.S. Geological Survey Professional Paper 1278, 41 p.

Sales, J. K., 1968, Crustal mechanics of Cordilleran foreland deformation; A regional and scale-model approach: American Association of Petroleum Geologists Bulletin, v. 52, p. 2016–2044.

Schmidt, C. J., and Perry, W. J., eds., 1988, Interaction of the Rocky Mountain foreland and the Cordilleran thrust belt: Geological Society of America Memoir 171, 582 p.

Schmidt, R. G., 1978, Rocks and mineral resources of the Wolf Creek area, Lewis and Clark and Cascade counties, Montana: U.S. Geological Survey Bulletin 1441, 91 p.

Scholten, R., Keemon, K. A., and Kupsch, W. O., 1955, Geology of the Lima region, southwestern Montana and adjacent Idaho: Geological Society of America Bulletin, v. 66, p. 345–404.

Schwartz, S. Y., and Van der Voo, R., 1984, Paleomagnetic study of thrust sheet rotation during foreland impingement in the Wyoming–Idaho overthrust belt: Journal of Geophysical Research, v. 89, p. 10077–10086.

Seager, W. R., 1983, Laramide wrench faults, basement-cored uplifts, and complementary basins in southern New Mexico: New Mexico Geology, v. 5, p. 69–76.

Silberling, N. J., and Roberts, R. J., 1962, Pre-Tertiary stratigraphy and structure of northwestern Nevada: Geological Society of America Special Paper 72, 58 p.

Skipp, B., 1987, Basement thrust sheets in the Clearwater orogenic zone, central Idaho and western Montana: Geology, v. 15, p. 220–224.

Smith, J. F., Jr., and Ketner, K. B., 1977, Tectonic events since early Paleozoic in the Carlin–Piñon Range area, Nevada: U.S. Geological Survey Professional Paper 867–C, p. C1–C18.

Smith, R. B., 1978, Seismicity, crustal structure, and intraplate tectonics of the interior of the western Cordillera, *in* Smith, R. B., and Eaton, G. P., eds., Cenozoic tectonics and regional geophysics of the western Cordillera: Geological Society of America Memoir 152, p. 111–144.

Smithson, S. B., Brewer, J. A., Kaufman, S., Oliver, J., and Hurich, C., 1979, Structure of the Laramide Wind River uplift, Wyoming, from COCORP deep reflection data and from gravity data: Journal of Geophysical Research, v. 84, p. 5955–5972.

Snoke, A. W., and Howard, K. A., 1984, Geology of the Ruby Mountains–East Humboldt Range, Nevada, *in* Lintz, J., Jr., ed., Western geological excursions: Geological Society of America 1984 Annual Meeting, Reno, Nevada, v. 4, p. 260–303.

Snoke, A. W., and Lush, A. P., 1984, Polyphase Mesozoic–Cenozoic deformational history of the northern Ruby Mountains–East Humboldt Range, Nevada, *in* Lintz, J., Jr., ed., Western geological excursions: Geological Society of America 1984 Annual Meeting, Reno, Nevada, v. 4, p. 232–260.

Speed, R. C., 1982, Evolution of the sialic margin in the central western United States, *in* Watkins, J., and Drake, C., eds., Geology of continental margins: American Association of Petroleum Geologists Memoir 34, p. 457–468.

Spieker, E. M., 1946, Late Mesozoic and early Tertiary history of central Utah: U.S. Geological Survey Professional Paper 205–D, p. 117–161.

Stearns, D. W., 1971, Mechanisms of drape folding in the Wyoming province: Wyoming Geological Association, 23rd Annual Field Conference, p. 125–143.

Stewart, J. H., 1980, Geology of Nevada: Nevada Bureau of Mines and Geology Special Publication 4, 136 p.

Stewart, J. H., and Poole, F. G., 1974, Lower Paleozoic and uppermost Precambrian Cordilleran miogeocline, Great Basin, western United States, *in* Dickinson, W. R., ed., Tectonics and sedimentation: Society of Economic Paleontologists and Mineralogists Special Publication 22, p. 28–57.

Stone, D. S., 1969, Wrench faulting and Rocky Mountain tectonics: Mountain Geologist, v. 6, p. 67–79.

Thorman, C. H., 1970, Metamorphosed and nonmetamorphosed Paleozoic rocks in the Wood Hills and Pequop Mountains, northeast Nevada: Geological Society of America Bulletin, v. 81, p. 2417–2448.

Tweto, O., 1975, Laramide (Late Cretaceous–early Tertiary) orogeny in the southern Rocky Mountains, *in* Curtis, B. M., ed., Cenozoic history of the southern Rocky Mountains: Geological Society of America Memoir 144, p. 1–44.

Veal, H. K., ed., 1977, Exploration frontiers of the central and southern Rockies: Rocky Mountain Association of Geologists, Denver, 498 p.

Veatch, A. C., 1907, Geography and geology of a portion of southwestern Wyoming with special reference to coal and oil: U.S. Geological Survey Professional paper 56, 178 p.

Warner, L. A., 1978, The Colorado Lineament: A middle Precambrian wrench fault system: Geological Society of America Bulletin, v. 89, p. 161–171.

Warner, M. A., and Royse, F., 1987, Thrust faulting and hydrocarbon generation: American Association of Petroleum Geologists Bulletin, v. 71, p. 882–889.

Weimer, R. J., 1980, Recurrent movement on basement faults, a tectonic style for Colorado and adjacent areas, *in* Colorado geology: Denver, Rocky Mountain Association of Geologists, p. 23–35.

Wernicke, B., Walker, D. J., and Beaufait, M. S., 1985, Structural discordance between Neogene detachments and frontal Sevier thrusts, central Mormon Mountains, southern Nevada: Tectonics, v. 4, p. 213–246.

Willis, B., 1902, Stratigraphy and structure, Lewis and Livingston ranges, Montana: Geological Society of America Bulletin, v. 13, p. 305–352.

Wilmarth, M. G., 1938, Lexicon of geologic names of the United States (including Alaska): U.S. Geological Survey Bulletin 896, 2396 p.

Wiltschko, D. V., and Dorr, J. A., Jr., 1983, Timing of deformation in overthrust belt and foreland of Idaho, Wyoming, and Utah: American Association of Petroleum Geologists Bulletin, v. 67, p. 1304–1322.

Wise, D. U., 1963, Keystone faulting and gravity sliding driven by basement uplift of the Owl Creek Mountains, Wyoming: American Association of Petroleum Geologists Bulletin, v. 47, p. 586–598.

Woodward, L. A., and DuChene, H. R., 1981, Overthrust belt of southwestern New Mexico; Comparison with Wyoming–Utah overthrust belt: American Association of Petroleum Geologists Bulletin, v. 65, p. 722–729.

Woodward, N. B., 1986, Thrust fault geometry of the Snake River Range, Idaho and Wyoming: Geological Society of America Bulletin, v. 97, p. 178–193.

—— , 1987, Geological applicability of critical-wedge thrust-belt models: Geological Society of America Bulletin, v. 99, p. 827–832.

Manuscript Accepted by the Society June 2, 1988

ACKNOWLEDGMENTS

This chapter is dedicated to Steven S. Oriel and Max D. Crittenden, Jr., two great thrust belt geologists who, had they had more time, should have been the ones to write this chapter. The following people are sincerely thanked for careful and perceptive reviews of the manuscript. A. W. Bally, G. A. Davis, M. Elison, P. Flemings, D. M. Miller, A. Palmer, W. Perry, C. J. Potter, B. Skipp, T. Yoos, and M. L. Zoback. I am particularly indebted to F. Royse who had his arm severely twisted to be a coauthor but somehow managed to refuse, while providing much useful criticism of the manuscript. My own work on thrust belts and Cordilleran-style fold and thrust deformation has been supported by National Science Foundation Grants EAR-7018758, EAR-8218617, EAR-8206178, and EAR-8519037. This is Institute for the Study of the Continents Contribution No. 866.

NOTES ADDED IN PROOF

A wealth of new data on various parts of the Cordilleran thrust belt has appeared since this chapter was accepted three and a half years ago. A review of most of this new work is beyond the scope of this brief note. However, two developments deserve special mention. They are, first, new data on the age of thrust motion in the southern Nevada segment of the thrust belt and, second, new data on hinterland extension coeval with shortening in the thrust belt.

The age of motion of the Keystone and Contact thrust plates in southern Nevada is most closely constrained by the Lavinia Wash Tuff. At the time this review was first accepted, that age was thought to be Late Jurassic-Early Cretaceous based on existing K/Ar dating. New laser fusion $^{40}Ar/^{39}Ar$ geochronology published by Fleck and Carr (1990), however, has identified problems of alteration and detrital mineral contamination in the unit and suggests that the Lavinia Wash Tuff is late Early Cretaceous. As Fleck and Carr (1990) point out, the new data establishes a maximum age for the Keystone and Contact thrusts of late Early Cretaceous and strengthens previous correlations of Contact–Red Springs and Keystone thrusts with the Cretaceous Summit–Willow Tank and Muddy Mountain–Glendale thrusts in the Muddy Mountains. Consequently, the column for southern Nevada in Figure 2 of this paper is in error and the age range of thrusting should be shifted upward in time. Nonetheless, the point made with this figure—that thrusting was *not* the same age everywhere within the Cordilleran thrust belt—remains valid.

In 1988, I hinted at the possibility of extension in the hinterland coeval with thrusting in the foreland. Two years later, some of the best evidence to date for this process was published by Wells and others (1990). Their detailed structural and geochronological study of the eastern Raft River Mountains (Fig. 9) clearly documented ~90 Ma low-angle, younger-over-older faults that are interpreted as normal faults. These faults were subsequently isoclinally folded and then cut by a second phase of younger-over-older faults that are possibly of Early Tertiary age. Wells and others interpret the early Late Cretaceous extension as related to development of hinterland topography, although the age of the extension coincides with the ages of some of the older thrusts in the Idaho-Wyoming thrust belt farther east.

In conclusion, it seems likely that the future will bring substantial revisions in the recognized ages of many of the structures associated with Cordilleran fold and thrust belt. Not only are geochronologic methods improving all the time, but new concepts of how to interpret the geometry of growth strata in thrust belts (e.g., Medwedeff, 1989; DeCelles and others, 1991) will probably spur a reexamination of many of the structures that are, at present, thought to be well-dated by synorogenic strata.

ADDITIONAL REFERENCES

DeCelles, P. G., and 6 others, 1991, History and kinematics of foreland uplift from synorogenic conglomerate, Beartooth Range, Wyoming and Montana: Geological Society of America Bulletin, v. 103, p. 1458–1475.

Fleck, R. J., and Carr, M. D., 1990, The age of the Keystone thrust: Laser-fusion $^{40}Ar/^{39}Ar$ dating of foreland basin deposits, southern Spring Mountains, Nevada: Tectonics, v. 9, p. 467–476.

Medwedeff, D. A., 1989, Growth fault-bend folding at southeast Lost Hills, San Joaquin Valley, California: American Association of Petroleum Geologists Bulletin, v. 73, p. 54–67.

Wells, M. L., Dallmeyer, R. D., and Allmendinger, R. W., 1990, Late Cretaceous extension in the hinterland of the Sevier thrust belt, northwestern Utah and southern Idaho: Geology, v. 18, p. 929–933.

Printed in U.S.A.

The Geology of North America
Vol. G-3, The Cordilleran Orogen: Conterminous U.S.
The Geological Society of America, 1992

Chapter 14

Strike-slip tectonics in the Cordilleran region, western United States

John H. Stewart
U.S. Geological Survey, 345 Middlefield Road, Menlo Park, California 94025
John C. Crowell
Institute for Crustal Studies and Department of Geological Sciences, University of California, Santa Barbara, California 93106

INTRODUCTION[1]

Strike-slip faulting is a major element in the tectonic development of the Cordilleran region. Since Mesozoic time, major strike-slip displacements have dominated the tectonic development of the Cordillera, and important Precambrian strike-slip systems may exist as well. Major displacements have been mostly parallel, or subparallel, to the western margin of North America and related to lateral movements between major crustal plates. The lateral movement is associated with either transform faulting, such as the present-day San Andreas system, or with oblique subduction that produced intra-arc or back-arc strike-slip faulting. Inland faults are related to distributed shear caused by the continental margin movements, to adjustments between regions of differential extension, or to shear within fold and thrust belts.

Strike-slip faults are but one type of major tectonic manifestations of deformation, along with thrust faults, normal faults, folds, and other structures, in the Cordillera, and are singled out for treatment because of evidence accumulated over the past few decades that displacements on some may amount to hundreds and even thousands of kilometers. As more information comes in concerning the lateral mobility of crustal units, strike-slip faults are being found to play a significant role. They are usually first recognized because of a mismatch or discontinuity in the inferred geologic history of the two sides and where the resolution and uniting of the mismatch requires lateral movements. Strike-slip faults may be difficult to recognize, and even overlooked, in slightly deformed strata because the slip direction may be parallel to the trace of stratal units on the fault surface and because of the difficulty of establishing distant correlations (Crowell, 1962). Some fault zones between tectonostratigraphic terranes may be sutures, or joins between far-traveled blocks, and some of these sutures may in turn be strike-slip zones (Howell, 1985; Crowell, 1985).

Here we treat strike-slip systems by age (Precambrian, Paleozoic, Mesozoic and early Cenozoic, and late Cenozoic), summarizing information on the distribution, amount of offset, age of movement, and geological problems of these systems. At the end of the chapter, a short section on "Tectonics" summarizes current ideas on the causes of strike-slip displacements in the Cordillera.

PRECAMBRIAN STRIKE-SLIP SYSTEMS

Strike-slip faults with clearly defined Precambrian movement have been identified only rarely in the Cordilleran orogen. One possible system of such faults, referred to as the Colorado lineament by Warner (1978), occurs in a 1,100-km-long and 160-km-wide, northeast-trending belt from Arizona to Wyoming (Fig. 1).

In the Arizona part of the Colorado lineament, possible Precambrian strike-slip faults (both right lateral and left lateral) include a group in central Arizona (Fig. 1; Shylock, SH; Chaparral, CH; Moore Gulch, MG; faults in the Tonto Basin, TB) and in northern Arizona (Fig. 1; Mesa Butte, MB; Bright Angel, BA; and Sinalaya, SI). The Shylock fault (*Anderson, 1967*) is a north-trending, high-angle fault that cuts Precambrian rocks as young as 1,750 Ma (*Anderson and others, 1971*) and is overlapped by unbroken Cambrian rocks at the north end, thus proving a Precambrian age of faulting. *Anderson (1967)* postulated a minimum of 8 km of right slip based on displacement of a quartz-diorite intrusive body. This interpretation is controversial; some geologists (see summary by *Karlstrom and Conway, 1986a, b*) have indicated primarily vertical movement, while *Darrach and others (1986)* have indicated left slip on a fault zone adjacent to the Shylock fault. The northeast-trending Chaparral fault is interpreted as right lateral on the basis of analysis of small-scale structural features (*Krieger, 1965*; see also *Karlstrom and Conway, 1986b*). The time of displacement on the Chaparral fault, al-

[1]Letters after named faults in the text are designations used on Figures 1 and 2.

Stewart, J. H., and Crowell, J. C., 1992, Strike-slip tectonics in the Cordilleran region, western United States, *in* Burchfiel, B. C., Lipman, P. W., and Zoback, M. L., eds., The Cordilleran Orogen: Conterminous U.S.: Boulder, Colorado, Geological Society of America, The Geology of North America, v. G-3.

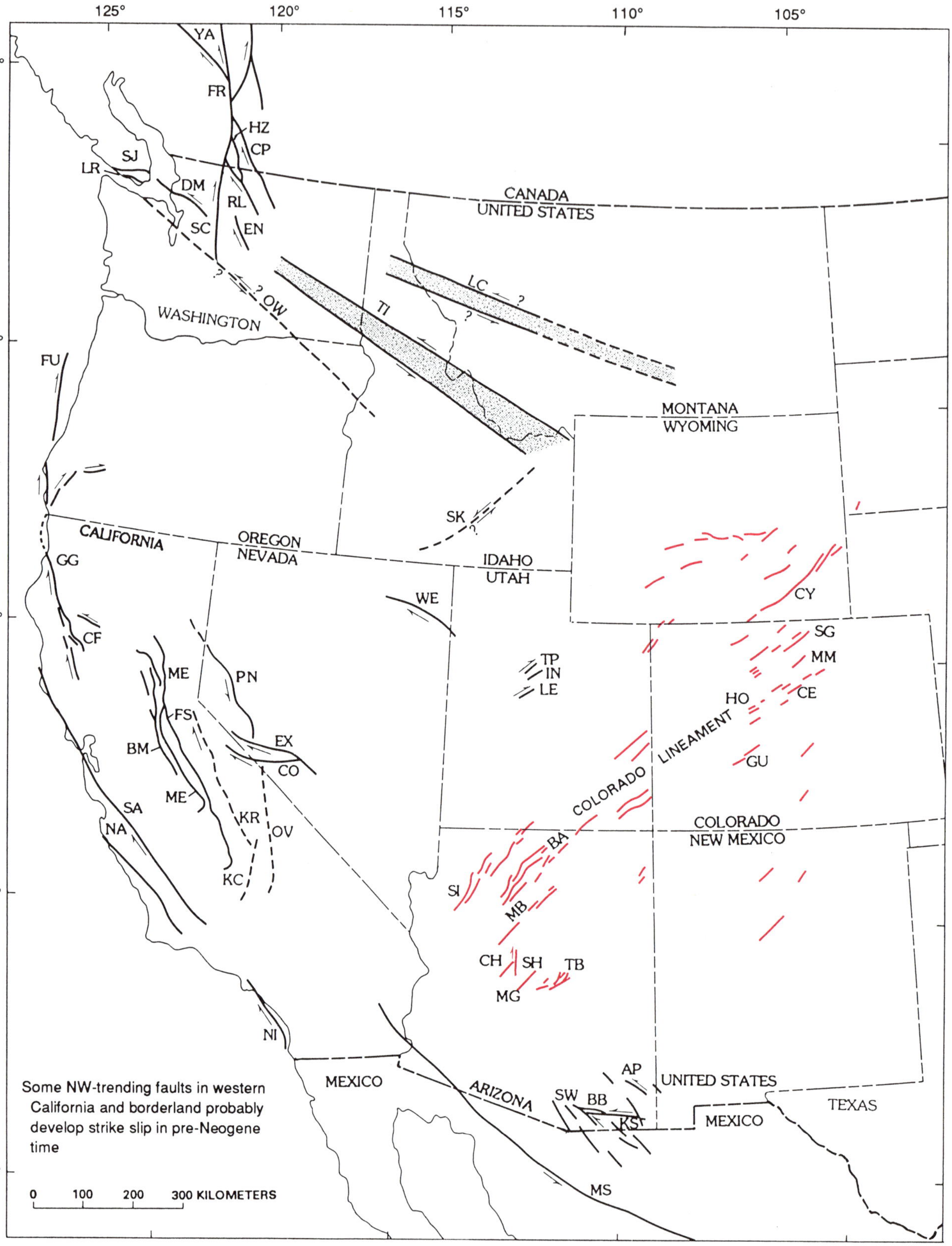

Figure 1. Precambrian, Paleozoic(?), Mesozoic, and early Cenozoic strike-slip systems in western North America. Explanation on facing page.

though not clearly established, is considered to be Precambrian (*Anderson, 1967*). The Moore Gulch fault appears to be a major boundary between terranes of different geologic history (*Karlstrom and others, 1987*) and may be a strike-slip fault, but its kinematic history is poorly known.

Faults in the Tonto Basin (TB) are northeast to north-trending, high-angle structures of possible left slip as well as major, generally east-side-down, vertical movement (*Conway, 1976*). Left slip is inferred from as much as 10 km of apparent left-lateral offset of intrusive rocks and from the apparent counter-clockwise rotation of one of the faults and an associated fold. The faults appear to postdate the folding and thrusting of the Mazatzal orogeny (1,675 to 1,625 Ma), but present evidence is uncertain about whether these faults should be considered the final phase of the Mazatzal orogeny or a completely separate Proterozoic event (*Conway and Silver, 1989*).

The northeast-trending Sinalaya (SI), Bright Angel (BA), and Mesa Butte (MB) fault systems in the Grand Canyon region of northern Arizona cut Phanerozoic rocks, and a Precambrian ancestry for some of them can be demonstrated. Here the Bright Angel fault shows at least seven and possibly as many as nine distinct episodes of displacement (*Maxson, 1961; Sears, 1973*), most of it in the Precambrian. Although *Maxson's (1961)* suggestion of a significant strike-slip component of displacement on these faults has been strongly challenged by *Sears (1973)*, Shoemaker and others (1978) consider the Bright Angel and Mesa Butte fault systems to be analogous to the Precambrian right-lateral systems that include the Shylock and Chaparral faults of central Arizona.

In southeastern Utah, the Colorado lineament, as described by Warner (1978), consists of northeast-trending Phanerozoic structures that may represent reactivation of Precambrian structures. These structures trend parallel to magnetic and gravity anomalies that are interpreted as associated with faults in Precambrian rocks. In Colorado, the Colorado lineament includes the Colorado mineral belt with its northeast-trending faults (Fig. 1; Central Front Range, CE; Gunnison, GU; Homestake, HO; Moose Mountain, MM) of Precambrian age (Tweto and Sims, 1963), although strike-slip displacement has not been demonstrated. In Wyoming, the Colorado lineament includes the Cheyenne belt (CY), a major shear zone separating Archean rocks of the Wyoming province to the north from Early Proterozoic rocks to the south. The belt has been considered to be a strike-slip zone by Warner (1978) and *Houston and others (1979)*, although more recent work indicates that the belt may be a thrust zone that has been rotated to a subvertical orientation (Karlstrom and Houston, 1984) and subsequently modified by distributed right-lateral strike slip (*Duebendorfer and Houston, 1986*).

A Precambrian ancestry has also been proposed for the Lewis and Clark line (LC), a zone of middle and late Cenozoic right-slip movement in northwest Montana and northern Idaho (*Harrison and others, 1974; Reynolds and Kleinkopf, 1977;* Reynolds, 1979). The zone is marked by a Middle Proterozoic sedimentary trough, by Late Proterozoic block faulting, and by Late Proterozoic diabase sills that thin outward, or end, at one of the major faults in the zone (*Harrison and others, 1974*). *Reynolds and Kleinkopf (1977)* suggest that the Lewis and Clark line has had recurrent movements from Late Proterozoic into Tertiary time, but Precambrian strike slip, though likely, is not documented.

Other strike-slip zones in the western Cordillera may also have had a Precambrian ancestry, but proof of such an old age is generally lacking.

EXPLANATION

Precambrian faults, mostly after Warner (1978)

Strike-slip fault, arrows indicate relative movement

Fault or fault zone indicated by letters:

BA, Bright Angel
CE, Central Front Range
CY, Cheyene belt
CH, Chaparral
GU, Gunnison
HO, Homestake
MB, Mesa Butte
MG, Moore Gulch
MM, Moose Mountain
SG, Skin Gulch
SH, Shylock
SI, Sinalaya
TB, faults in Tonto Basin

Paleozoic(?), Mesozoic, and early Cenozoic faults

--?- Strike-slip fault or fault zone, dashed where inferred or uncertain, arrows indicate relative movement, query indicates direction of slip uncertain

Fault or fault zone indicated by letters:

AP, Apache Pass
BB, Babomari
BM, Bear Mountain
CF, Cold Fork, Elder Creek, and Paskenta
CO, Coaldale
CP, Chewack-Pasayten
DM, Devils Mountain
EN, Entiat
EX, Excelsior
FS, Foothill suture
FR, Fraser River
FU, Fulmar
GG, Grogan
HZ, Hozameen
IN, Inez
KC, Kern Canyon
KS, Kino Spring
KR, Kings River
LC, Lewis and Clark line
LE, Leamington
LR, Leech River
ME, Melones
MS, Mojave-Sonora megashear
NA, Nacimiento
NI, Newport-Inglewood
OV, Owens Valley-White Mountain
OW, Olympic-Wallowa lineament
PN, Pine Nut
RL, Ross Lake
SA, proto-San Andreas
SC, Straight Creek
SJ, San Juan, Survey Mountain
SK, Snake River
SW, Sawmill Canyon
TI, Trans-Idaho discontinuity
TP, Tintic-Prince
WE, Wells
YA, Yalakom

PALEOZOIC STRIKE-SLIP SYSTEMS

Little is known about Paleozoic strike-slip systems, but those that have been proposed are mostly right-lateral, northwest-trending faults in the Rocky Mountains region and in Texas (*Muehlberger, 1965; Pindell, 1985; Stevenson and Baars, 1985*). They are most likely inland effects of the collision of North America with South America–Africa during the late Paleozoic Ouachita-Marathon orogeny (*Pindell, 1985*). In addition, *Poole and others (1977)* have proposed 200 km of Paleozoic left-lateral offset on a fault along the trend of the Snake River (SK) plain in Idaho. *Coats and Riva (1983)*, on the other hand, suggest right-lateral Mesozoic displacement along the same trend.

MESOZOIC AND EARLY CENOZOIC STRIKE-SLIP SYSTEMS

The western margin of North America changed markedly during the Mesozoic and early Cenozoic. During this time, microcontinental, island arc, and oceanic terranes were accreted to western North America (Coney and others, 1980; Jones and others, 1982; Howell, 1985). Some of these terranes appear to be exotic fragments derived from unknown sources in the Pacific region and transported thousands of kilometers before colliding with and being added to the continental margin (Jones and others, 1982). Once assembled, these fragments, as well as fragments of previously existing North America, were apparently transported for large distances laterally by strike-slip faulting (Beck, 1980). Paleomagnetic studies indicate that transport, including movement in the late Cenozoic, was generally to the north and that some terranes were transported several thousand kilometers (*Beck, 1976;* Beck, 1980; *Jones and others, 1977; Alvarez and others, 1980;* Beck and others, 1981; Champion and others, 1984; *Hagstrum and others, 1985; Tarduno and others, 1985, 1986; Umhoefer, 1987*).

Mesozoic and early Cenozoic strike-slip systems along which large-scale northward transport took place have been recognized in western Canada, Alaska, and northernmost Washington (Gabrielse, 1985), but in much of the western Cordillera of the United States, strike-slip faulting of this age is only locally recognized (Fig. 1). Strike-slip faults that accommodated the inferred northward movement of terranes are either locally hidden and perhaps unrecognized in areas of complex structures or are overprinted along zones of late Cenozoic strike-slip faulting.

Northern Washington

A well-defined belt of north-northwest–trending, strike-slip faults extends from Alaska, across western Canada, and terminates in northern Washington (Gabrielse, 1985). Total dextral offset in this belt in Canada is probably as much as 950 km on an eastern set of faults and as much as 300 km on a western set (Gabrielse, 1985).

In northern Washington and adjacent parts of Canada, five known or possible right-lateral faults (Fig. 1; Straight Creek, SC; Entiat, EN; Ross Lake, RL; Hozameen, HZ; and Chewack-Pasayten, CP) and three problematical left-lateral faults (San Juan, SJ; Leech River, LR; and Devils Mountain, DM) have been described (*Misch, 1966, 1977; Davis, 1977b;* Davis and others, 1978; *Gresens, 1982;* Tabor and others, 1984; *Whetten, 1978; Ewing, 1980; Johnson, 1984*). As described by *Davis (1977b)* and Davis and others (1978), the oldest of these is the Ross Lake fault, along which the Cascade crystalline core is believed to have moved relatively northward at least 160 km dextrally on the west side of the fault. Earliest movement, dated by deformation of sedimentary rocks adjacent to the fault, is latest Early Cretaceous or earliest Late Cretaceous. Movement may have ceased by 88 Ma when a late syntectonic pluton was emplaced across the fault (Davis and others, 1978), although dynamothermal events dated by isotopic methods indicate recurrent movements along the fault zone as late as mid-Cenozoic time (*Hoppe, 1982*).

The Entiat (EN), Hozameen (HZ), and Chewack-Pasayten (CP) faults are less clearly defined as strike-slip faults. The Entiat fault bounds the east side of a major 8-km-wide Eocene graben that may have formed either as a rhombochasm related to right-lateral displacement on the Entiat fault or by right-lateral displacement and associated rotation of the western side of the graben away from the Entiat fault (*Gresens, 1982*). The Hozameen fault has been considered a strike-slip fault by *Misch (1977),* although as discussed by *Davis (1977b),* map relations apparently preclude major displacement. It is older than a small 84-Ma pluton that intrudes it (*Okulitch and others, 1977*). The Chewack-Pasayten fault was considered by *Coats (1974)* to have a component of right-lateral displacement, although he considered it to be primarily a normal-slip fault. The possibility of strike-slip displacement was indicated by the lack of correspondence of clasts in Cretaceous conglomerates and crystalline source rocks to the east of the fault. *Lawrence (1978),* on the other hand, on the basis of well-documented microstructural and petrofabric studies, proposed that offset was left lateral with perhaps a few tens of kilometers of displacement. The Chewack-Pasayten fault is cut by the Late Cretaceous Straight Creek fault (see below) and is thus older than Late Cretaceous. Reactivation of the Chewack-Pasayten fault in post-Paleocene time is indicated by the offset of Paleocene rocks (*Barksdale, 1975*).

The Straight Creek (SC) fault cuts the Ross Lake (RL), Hozameem (HZ), and Chewack-Pasayten (CP) faults and is the best-documented strike-slip fault in northern Washington. *Misch (1977)* indicated 190 km of right-lateral offset on the Straight Creek fault based on offset crystalline rocks, whereas *Anderson (1977), Okulitch and others (1977),* Kleinspehn (1985), and *Price and others (1985)* indicate right-lateral offset of 160 km, 120 km, 110 km, and 80 km, respectively, based on various interpretations and correlations of offset rock bodies and faults. The Straight Creek fault and its southern splays curve eastward near its southern margin and merge with the Olympic-Wallowa lineament (Tabor and others, 1984). Major strike-slip offset on the Straight Creek fault in Canada probably began in latest Cretaceous time and ceased prior to intrusion of Oligocene batholithic rocks (Kleinspehn, 1985). In Washington, major strike slip on the Straight Creek fault apparently was over before deposition of early Eocene rocks (*Tabor and others, 1991*).

Major left-lateral faults with a westerly trend have been proposed in the Vancouver Island–Puget Sound region of northwestern Washington and southwestern British Columbia. On Vancouver Island, such faults include the San Juan fault (SJ), along which at least 50 km of left slip have been proposed, and the Leach River fault (LR) (*Fairchild and Cowan, 1982; Cowan, 1982; Brandon, 1985*). A recent seismic reflection profile (*Yorath and others, 1985; Clowes and others, 1986*), however, indicates that these faults are not strike-slip faults, but northeast-dipping thrusts. To the east, the left-lateral Devils Mountain (DM) fault has been proposed by *Whetten (1978)* in the San Juan Island–Cascade Range region of Washington. The fault appears to offset Mesozoic mélange sequences about 60 km left laterally. Major movement is no older than Late Cretaceous, although the fault cuts rocks as young as Eocene or Oligocene, indicating at least minor displacement on the fault zone in the Tertiary. In a different interpretation, *Johnson (1984)* has suggested that a major right-lateral transcurrent fault (not shown on Fig. 1) truncated the continental margin in Late Cretaceous and early Tertiary time in northwest Washington. Similarly, Wells and others (1984) speculated that major right-lateral slip and northward transport of outboard terranes accompanied oblique rifting of the western part of Oregon and Washington to produce the early Tertiary oceanic basement of the Coast Range.

Trans-Idaho discontinuity, Olympic Wallowa lineament, and Lewis and Clark line

A major discontinuity in trends of Paleozoic and Mesozoic rocks, recognized in the northwestern part of the conterminous United States, was named the Trans-Idaho discontinuity (TI) by Yates (1968). He noted that such features as the distribution of Late Proterozoic volcanic rocks, isopach trends of Paleozoic miogeosynclinal strata, the eastern limit of Permian and Triassic volcanic rocks, and the eastern limit of major belts of Mesozoic alpine ultramafic rocks are apparently offset about 300 to 650 km left laterally across in a northwest-trending zone. To the evidence outlined by Yates can be added the apparent offset of the Wrangellia terrane exposed on Vancouver Island and the Wallowa terrane of northeast Oregon and adjacent parts of Idaho and Washington. These two terranes contain similar sequences of late Paleozoic and early Mesozoic rocks (*Jones and others, 1977; Muller, 1977*) that are apparently displaced left laterally about 550 km along the Trans-Idaho discontinuity. The Trans-Idaho discontinuity, as delineated by Yates, is largely buried by the Columbia River basalts and engulfed by the Cretaceous and early Tertiary Idaho batholith. It is not yet clear whether it represents abrupt changes in trend along an irregular Proterozoic to Mesozoic continental margin, is an orocline, or is a transcurrent fault, but the abruptness of the discontinuity suggests a fault zone. If a fault, displacement was younger than early Mesozoic rocks in the apparently displaced Wrangellia and Wallowa terranes and older than the emplacement of the Cretaceous and early Tertiary Idaho batholith.

Yates (1968) noted that other structural features such as the Lewis and Clark line (LC) and the Olympic-Wallowa lineament (OW) (*Raisz, 1945; McKee, 1967*) are parallel to the inferred Trans-Idaho discontinuity, but not in the correct position to account for the major observed disruption of Proterozoic to Mesozoic trends. These other structural features, however, may either be subparallel left-lateral fault zones or fault zones that join with the Trans-Idaho discontinuity in an as yet undetermined manner. The Lewis and Clark line is defined primarily as a Cenozoic zone of *right-lateral* displacement, but that does not rule out the possibility of Mesozoic *left-lateral* movement related to left-lateral movement on the Trans-Idaho discontinuity. *Smith (1965),* in fact, on the basis of the pattern of oroclinal folds and of en echelon folds and faults, has proposed that *left-lateral* movements played an important role in developing the Lewis and Clark line.

The Olympic-Wallowa lineament (OW), which was originally defined entirely on the basis of present-day geomorphic features (*Raisz, 1945*), is a zone of discontinuous northwest-trending faults and folds and has been characterized diversely as an important structural zone (*McKee, 1967*) and as "a fictional structural element" (*Davis, 1977b*). Speculatively, the Olympic-Wallowa lineament may truncate the Straight Creek fault (SC; Tabor and others, 1984), and the eastward curvature of the southern end of the Straight Creek fault as it joins the Olympic-Wallowa lineament suggests drag on a system of left-lateral faults that, in this view, produced the Olympic-Wallowa lineament. Such an interpretation requires major left-lateral movement on the Olympic-Wallowa zone prior to eruption of the Miocene Columbia River Basalt Group, which is not displaced significantly along the lineament. Local, late Cenozoic northwest-trending faults along the trend of the lineament are apparently related to dextral shear (H. Laubscher, cited in *Davis, 1977b*).

Southwest Washington, western Oregon and its borderland, and northwest California

On the Oregon continental shelf, *Snavely and others (1980)* and Snavely (1987) have described the Fulmar fault (FU) along which juxtaposition of lower Eocene arkosic wacke and basalt indicates a minimum strike slip of 200 km before overlap of upper Eocene strata. Offshore and in the Coast Ranges of Washington and Oregon, faults with straight traces trending generally northwest or northeast are common and have been interpreted as strike slip by several authors (Wells and Coe, 1985; Snavely, 1987). Where studied in detail, these faults had initial movement in the early to middle Eocene and accommodated block rotation along the convergent margin. Displacement on some of these faults continued into the late Cenozoic.

In the Coast Ranges of northwest California and southwest Oregon, *Blake (1985;* written communication, 1986) and Blake and others (1985a, *b*) have described a northwest-, north-, and northeast-trending system of Late Cretaceous and/or early Cenozoic faults with 300 km or more of right slip. This system includes the Grogan fault (GG), along which 75 km of offset had previously been proposed by *Kelsey and Hagans (1982).* The faults described by *Blake (1985)* and Blake and others (1985a, *b*) cut Mesozoic accretionary terranes, and displace slabs of these northwestward. Timing of the lateral faulting is constrained to postdate the youngest fossils and metamorphism (Late Cretaceous) in the displaced rocks and to predate early Eocene overlap strata. In addition to these right-lateral faults described by Blake, the Cold Fork, Elder Creek, and Paskenta faults (CF), as well as possible tear faults in the upper plate of the Coast Range thrust

(*Jones and Irwin, 1971*) collectively offset a north-trending Early Cretaceous shoreline about 100 km left laterally.

Sierra Nevada and western Great Basin, California and Nevada

Late Paleozoic or early Mesozoic (most likely Early Triassic) strike-slip faulting (not shown on Fig. 1) or rifting has been proposed in the Sierra Nevada region to explain the apparent southwest truncation of the Cordilleran miogeoclinal belt (Hamilton, 1969; Burchfiel and Davis, 1972, 1981; Saleeby, 1981). The location of a truncated edge has not been clearly defined, and *Stone (1984)* has suggested that truncation is not required and that the Cordilleran belt may simply bend to the south.

Strike-slip displacement has also been proposed on a major system of high-angle faults (the north-northwest–trending Foothills fault system of *Clark, 1960),* which extend several hundred kilometers along the west side of the Sierra Nevada. This belt includes the Melones (ME), Bear Mountain (BM), and other faults, but the nature of offset on these faults is uncertain. Right-lateral displacement thrusting, or a combination of these mechanisms have been proposed (*Clark, 1960;* Saleeby, 1981, *1982;* Schweickert and others, 1984; *Nokleberg, 1983*). The Melones fault zone has been considered a terrane boundary (*Schweickert and Cowan, 1975*), but *Saleeby (1982)* discounts this interpretation because similar Jurassic volcanic rocks occur on both sides of the fault. Offset facies or structural trends have not been observed along these faults, but their high angle of dip and long trace suggests a strike-slip system. The system is Late Jurassic in age on the basis of the youngest rocks cut by the fault and the oldest ages of igneous rocks that intrude the fault zone (*Clark, 1960; Nokleberg, 1983*).

Elsewhere in the Sierra Nevada, Mesozoic right slip has also been proposed along the Foothill suture (FS) (Saleeby, 1981), Kings River fault (KR) (Saleeby, 1981; *Nokleberg, 1983*), and Kern Canyon fault (KC) (*Moore and du Bray, 1978;* Schweickert, 1981), and left-lateral offset on the Owens Valley–White Mountains fault zone (OV) (Schweickert, 1981; Saleeby, 1981). Mesozoic strike slip on these faults has not been clearly demonstrated. *Moore and du Bray (1978)* demonstrate 13 km of right-lateral offset on the Kern Canyon fault that can be dated no more precisely than younger than 80 Ma and older than 3.5 Ma. Post–Middle or Late Jurassic displacement on the Owens Valley–White Mountain fault zone can be no more than a few kilometers because a Middle Jurassic pluton (*Ross, 1962;* R. W. Kistler, oral communication, 1985) and Late Jurassic dikes (*Moore and Hopson, 1961; Chen and Moore, 1979*) extend across the fault zone without major offset.

In the east-central part of the Sierra Nevada and adjacent parts of the Great Basin, *Kistler and others (1980)* and R. W. Kistler in *Stewart and others (1984a)* have proposed right-lateral offset totalling 320 km on four northwest-trending zones on the basis of apparent offset of the isotopic boundary between granitic rocks with initial $^{87}Sr/^{86}Sr$ ratios greater than 0.7060 and those with ratios less than 0.7080. A similar large right-lateral displacement (500 km) has also been proposed (*Lehren and Schweickert, 1989*) to explain the anomalous position of possible uppermost Precambrian and Lower Cambrian rocks in the eastern Sierra Nevada.

In the western Great Basin, various models that involve strike-slip displacement have been proposed (*Speed, 1978; Oldow, 1983; Stewart, 1985*). The left-lateral Pine-Nut fault (PN) in western Nevada has been proposed (*Oldow, 1983*) to account for the juxtaposition of two Mesozoic terranes with different structural histories. In western Nevada and eastern California, *Stewart (1985)* proposed two major southeast- to east-trending faults (the Coaldale, CO, and Excelsior faults, EX) with a total right-lateral displacement of 105 to 135 km. The Coaldale and Excelsior faults are believed to be Mesozoic in age because Mesozoic plutons are not cut by these faults.

Southeastern California and southern Arizona

The Mojave Desert region of southeastern California and the southern part of Arizona contain a few strike-slip faults that have been considered to be Mesozoic in age. The region is characterized, however, by many northwest-trending faults and lineaments (Fig. 2) that are mainly Tertiary in age. Conceivably some of these faults were initiated as strike-slip faults in the Mesozoic.

In southeastern Arizona, Drewes (1981) has mapped several northwest-trending faults (Fig. 1; Sawmill Canyon, SW; Apache Pass, AP; Babomari, BB; and Kino Spring, KS) that he considers to have had left-lateral displacement during the Mesozoic. Strike-slip displacement is suggested by the vertical attitude and straight trace of the faults, and left-lateral offset is indicated by drag features and the pattern of splay faults. The faults may in part have had a Precambrian ancestry, but major movement is probably Mesozoic and early Tertiary (Drewes, 1981). Some of these faults or related features are considered by *Bilodeau (1982)* primarily to be high-angle, normal-slip faults bounding Early Cretaceous fault blocks. A pattern of northwest-trending Mesozoic linear tectonic features, in part related to strike-slip faulting, has been suggested to lie in southeastern Arizona by *Titley (1976).*

Mojave-Sonora megashear

A major northwest-trending left-lateral fault zone, the Mojave-Sonora megashear (MS), has been proposed by *Silver and Anderson (1975)* and *Anderson and Silver (1979)* to explain complex structural and stratigraphic patterns in northern Mexico and the southwestern United States. Two terranes of Precambrian orogenic and magmatic rocks, one ranging in age from 1.8 to 1.7 Ga and the other from 1.7 to 1.6 Ga, extend southwest across the United States, but in southeasternmost California and in northwestern Mexico the pattern is apparently disrupted and the 1.8 to 1.7 Ga terrane apparently is displaced at least 600 km left laterally along the proposed megashear. The 1.7- to 1.6-Ga terrane is not recognized southwest of the 1.8- to 1.7-Ga terrane in the offset segment, but no outcrops of Precambrian rocks exist in northern Mexico where this terrane should lie. Total offset on the megashear is estimated at 700 to 800 km on the basis of possible offsets of Late Proterozoic and Paleozoic rocks (*Silver and Anderson, 1974; Anderson and Silver, 1979;* Anderson and Schmidt, 1983; *Stewart and others, 1984b*). Considerable uncertainty exists about where the megashear (if it exists) extends across southeastern California, because similar Late Proterozoic and Paleozoic miogeoclinal rocks (*Stewart and Poole, 1975;* Burchfiel and Davis, 1981; *Stewart and others, 1984b*) occur on both sides of the megashear where it was originally drawn by *Silver and Anderson (1974)* in California. A more southerly position of the megashear and (or) later tectonic disruption has been suggested to explain this anomaly (*Stewart and others, 1984b*). The megashear, as interpreted by Anderson and Schmidt (1983), is related to the Jurassic opening of the Gulf of Mexico and the necessary rearrangement of continental fragments in the Middle America, Gulf of Mexico, and Caribbean Sea regions.

Strike-slip deformation along coastal California and the proto–San Andreas fault

The time and place of origin of the San Andreas fault system and its history through time are problematic. Tectonic overprinting along the Pacific margin of North America has apparently obscured earlier strike-slip movements that predate the birth of the late Cenozoic San Andreas system. Working backward through time, after conceptually removing late Cenozoic displacement on the San Andreas fault system, earlier episodes of major deformation, probably including strike slip, are needed to reconstruct the continental margin satisfactorily. An earlier stage of activity, referred to as occurring on the "proto–San Andreas fault" (*Suppe, 1970*), apparently took place in northern and central California near the end of the Cretaceous or early in the Tertiary (Nilsen and Clarke, 1975) but it has not been recognized in southern California (*Crowell, 1981*). Evidence for this earlier episode of right slip includes the modern position of the northern part of the batholithic Salinia block west of the present-day San Andreas. This composite block appears to be displaced about 600 km from its speculative source adjacent to the Peninsular Ranges (Hill and Dibblee, 1953; *Ross, 1978*; Page, 1982, Mattinson and James, 1985). In central California, pre-Eocene displacement—attributed perhaps to the proto–San Andreas—resulted and played a role in forming a borderland consisting of islands and intervening straits and basins that controlled Eocene facies (Nilsen and Clarke, 1975). However, even earlier strike-slip displacements may be involved in the accretion of terranes (Howell, 1985).

Western California, and especially the Salinian block, is made up of long slices now accreted to the continent. In recent years, paleomagnetic data, supplemented with sparse stratigraphic information disclosing mismatches across major faults, suggest that some of these faults may be either sutures or major strike-slip zones inasmuch as they separate discrete blocks with characteristic geologic histories: the tectonostratigraphic terranes (Howell, 1985). For example, though controversial, Cretaceous strata may be displaced right laterally as much as several thousand kilometers from unrecognized sources along the Pacific coasts of central or South America (*Howell and others, 1977; Page, 1981,* 1982; Champion and others, 1984; Page and Engebretson, 1984; Blake and others, 1985a; *Hagstrum and others, 1985; Kanter and Debiche, 1985;* Mattinson and James, 1985). *Dickinson (1983)* on the other hand, has speculated that the northwest-trending Nacimiento fault zone (NA) may have had left-lateral displacement of many hundreds of kilometers prior to movement on the proto–San Andreas system. Nevertheless, an understanding of the style and timing of deformation in this complex region awaits much more paleomagnetic, isotopic, stratigraphic, and petrographic work along the whole border of western North America.

In southern California, however, none of the late Cenozoic strands of the San Andreas system exposed on land to the east of the Newport-Inglewood zone (NI; Fig. 1), such as the Elsinore (EL; Fig. 2), San Jacinto (SN; Fig. 2), and San Andreas (SA; Fig. 2) itself, shows convincing evidence of pre-Miocene movement. Some sutures with strike-slip components may lie at depth in order to account for the juxtaposition of basement units, such as the emplacement of presumed allochthonous schist terranes (Pelona, Orocopia) against the North American craton, but this interpretation is controversial (*Vedder and others, 1983*). The active Newport-Inglewood fault zone, known to have had displacement before late Miocene time, may be a strand of the proto–San Andreas system (*Suppe, 1970*) or of some other fault or subduction system (*Hill, 1971;* Yeats, 1973; Platt and Stuart, 1974). Proto–San Andreas fault strands may lie along the eastern coast of southern Baja California, but these proposed pre-Miocene faults have not been traced into the United States (*Gastil and others, 1972; Minch and James, 1974;* but see *Gastil, 1985*). The strike-slip history of faults in the California borderland, and at the subsea edge of the continent, is unknown, although many of the linear escarpments in this region are known or inferred to have had late Cenozoic activity, and those with a northwestern trend are commonly considered to have right slip upon them.

Inland areas

Inland of the Mesozoic western margin of North America, few major Mesozoic or early Cenozoic strike-slip systems have been recognized. The most important are (1) the Wells fault (WE) of Nevada, along which 65 to 80 km of right-lateral offset has been proposed (*Thorman and Ketner, 1979*), although *Stevens (1981)* and *Coats and Riva (1983)* have challenged this proposal; and (2) right-lateral faults (Tintic Prince, TP; Inez, IN; Leamington, LE) in western Utah (*Morris and Shepard, 1964; Morris, 1983*). Elsewhere in the western United States, minor Mesozoic and early Cenozoic strike-slip systems have been documented in a variety of structural settings (*Cook, 1978; Anderson and Barnhard, 1986*), such as tears in thrust systems, breaks in folds, and conjugate sets in areas of contraction.

LATE CENOZOIC STRIKE-SLIP SYSTEMS

The plate-tectonic setting of the western conterminous United States changed in the late Cenozoic from a convergent margin to a complex margin characterized by convergence in its northern part and transform faulting along the San Andreas fault system in its southern part (Atwater, 1970; Atwater and Molnar, 1973; *Minster and Jordan, 1978;* Page and Engebretson, 1984; *Engebretson and others, 1984*). According to plate reconstructions, this transform plate boundary began to develop about 29 Ma (late Oligocene) when the Pacific lithospheric plate on the west, bordered by the East Pacific Rise, reached the North American plate on the east. The intervening Farallon plate was then segmented into two parts: the remainders today are the Rivera plate on the southeast and the Juan de Fuca on the northwest. The place of impingement was near the United States–Mexico border, relative to the North American side of the San Andreas fault system (Atwater and Molnar, 1973; Engebretson and others, 1985; Crowell, 1987). By about 22 Ma (early Miocene) the California borderland and coastal region had been fragmented and stretched as the result of an extensional component of movement between the two plates. Deep basins, such as the Los Angeles, Ventura, and southern San Joaquin, subsided as a consequence. Following these earlier displacements, at about 5 Ma, plate readjustments resulted in opening of the Gulf of California and changes in the fault pattern in California. Today the transform belt extends from the mouth of the Gulf of California on the southeast, where it joins the East Pacific Rise, to the vicinity of Cape Mendocino, northern California, where it meets both the Mendocino fracture zone extending westward across the deep floor of the Pacific Ocean and the convergent boundary between the Juan de Fuca plate and the North American plate extending on northward.

In addition to the San Andreas fault system, inland zones of strike-slip faulting across a broad region of the western United States have been related to right-lateral shear, perhaps to the same transform motion that produced the San Andreas fault (Fig. 2). Focal mechanisms of earthquakes indicate a broad zone of strike-slip displacement extending across much of California and the western Great Basin (*Smith and Lindh, 1978; Zoback and Zoback, 1980a; Rogers and others, 1983; Vetter and Ryall, 1983*), and locally elsewhere in the western United States (*Hill, 1978; Smith and Lindh, 1978; Arabasz and Julander, 1986*).

San Andreas fault

The San Andreas fault (SA) proper, as distinct from the many subparallel faults in California constituting the San Andreas system, first received widespread attention following the 1906 San Francisco earthquake. The ground was broken along it, and it was recognized as the culprit fault for the earthquake (*Lawson and others, 1908*). It had been mapped and named previously in the vicinity of San Francisco (*Lawson, 1895*), and after the earthquake was traced as a nearly continuous fault southward to the southeastern corner of the Salton Sea. The fault (Fig. 2) has several distinct segments (Allen, 1968; *Jennings,*

EXPLANATION

⇐ – –?– Strike-slip fault, dashed where inferred or uncertain, arrow indicates relative movement, query indicated direction of slip uncertain. Includes a few lineaments with inferred strike-slip movement.

.... Lineaments or straight faults that may have late Cenozoic strike-slip movement

Faults or lineament indicated by letters:

AB, Agua Blanca
AC, Agua Caliente
AP, Agai-Pah Hills
BA, Banning
BC, Blue Cut
BD, Bitter Ridge
BL, Blackwater
BM, Bristol Mountain
BP, Big Pine
BR, Brothers
BS, Benton Spring-Pilot Mountains
BT, Bartlett Springs
BU, Bullion
BV, Bitter Spring Valley
BY, Brawley
CA, Candelaria
CB, Cabin Canyon
CC, Chiriaco
CD, Concord
CH, Chino
CI, Calico
CL, Carson lineament
CO, Coaldale
CR, Camp Rock
CS, Cane Spring
CU, Cucamonga
CV, Calaveras
CW, Clemens Well
CY, Coyote Lake
DU, Dume
DV, Death Valley
ED, Eugene-Denio
EL, Elsinore
EX, Excelsior
FC, Furnace Creek
FR, Franklin
GA, Garlock
GB, Gold Butte
GG, Grogan
GH, Gumdrop Hills
GM, Granite Mountain
GN, Greenville
GR, Green Valley
GV, Grizzly Valley
HA, Harper Lake
HB, Hamblin Bay
HE, Helendale
HG, Hosgri
HL, Honey Lake
HO, Hope
HU, Huasna
HY, Hayward
IH, Indian Head
IM, Imperial
KC, Kern Canyon
KO, Kokoweef
KS, Kane Spring Valley
LC, Last Chance
LE, Lenwood
LI, Likely
LO, Lockhart
LR, Lime Ridge
LU, Ludlow
LV, Las Vegas Valley
MA, Maacama-Bennett Valley
MB, Malibu Coast
MC, McLoughlin
MD, Midland-Sweitzer
ME, Mendocino
MM, Mine Mountain
MX, Manix
NA, Nacimiento
NE, Newport-Inglewood
NI, Ninemile
OL, Olinghouse
OR, Oak Ridge
OS, Osborne
OV, Owens Valley-White Mountain
OZ, Ozena
PA, Pahranagat
PE, Palos Verde
PG, Pisgah
PH, Pahrump
PI, Pinto Mountain
PL, Pyramid Lake
PR, Point Reyes
PS, Petrified Spring-Bettles Well
PV, Panamint Valley-Hunter Mountain
RE, Reliz
RG, Rogers Creek-Healdsburg
RI, Rinconada
RO, Rodman
RS, Rose Canyon
RV, Rock Valley
SA, San Andreas
SB, Santa Lucia Bank
SC, San Juan-Chimeneas
SE, San Clemente
SG, San Gabriel
SH, Sand Hills-Algodones
SI, Santa Cruz Island
SJ, St. Joe
SL, Slaughterhouse
SM, Santa Monica
SN, San Jacinto
SO, Santa Rosa-Cortes Ridge
SR, Santa Rosa
SU, Sur
SV, Santa Ynez River
SW, St. Marys
SX, Santa Maria River
SY, Santa Ynez
SZ, San Gregorio
VA, Vale
WA, Wahmonie
WH, Whittier
WL, Wabuska Lineament
WS, Warm Springs Valley
WW, White Wolf

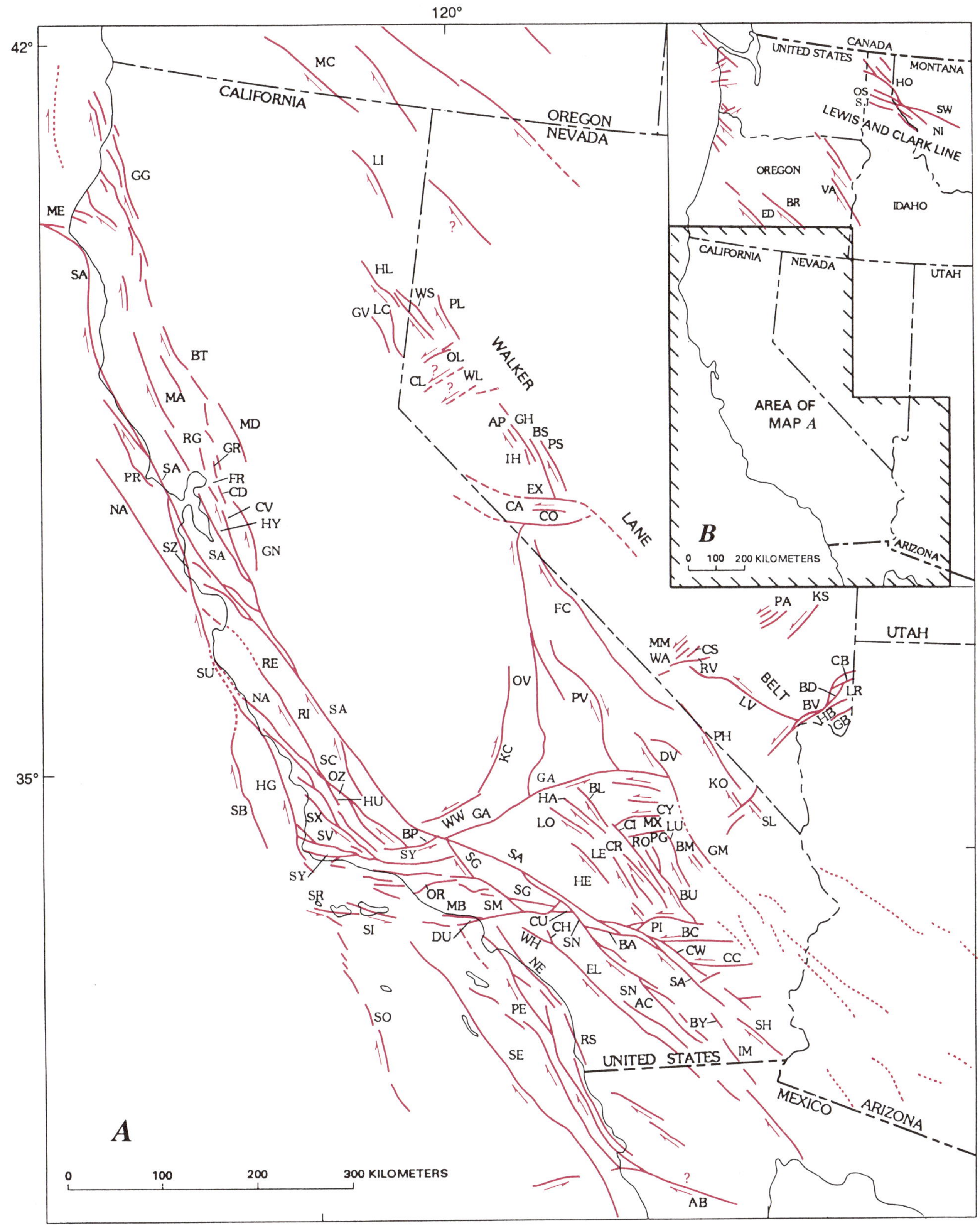

Figure 2. Late Cenozoic strike-slip faults in western United States and northwestern Mexico. A, detail of California and contiguous areas; B, western United States. Explanation on facing page.

1975). Along its northern reach, largely offshore, it displays a curved trace disclosed by subsea bathymetry. From where it comes onshore about 170 km northwest of San Francisco to the San Francisco Bay region, its course is straight and the fault zone is relatively narrow. Southeast of San Francisco Bay, however, it trends through a gentle arc to the region where the Calaveras and Hayward faults trend more northerly. From where these faults join the San Andreas proper, a straight stretch extends on southeastward to the northern flanks of the Transverse Ranges. This is the reach of the fault zone where its trace is nearly parallel to the relative movement between the Pacific and North American lithospheric plates (Page and Engebretson, 1984). It is also the region characterized by displacement distributed between creep and relatively frequent shallow earthquakes of moderate magnitudes, such as the Parkfield Earthquake of 1966 (Brown and others, 1967; Allen, 1968, 1981; *Eaton, 1985*).

The trace of the San Andreas fault takes a "big bend" (roughly 40° to the east) within the Transverse Ranges, and follows an east-southeasterly course to join with the Salton Trough and the northwestern part of the Gulf of California. This stretch of the fault is complex, and also is the site of historic earthquakes. Landforms showing right slip are characteristic (*Sharp, 1954*). Along several stretches, however, the linear zone is widened and interrupted, and consists of braided faults in an anastomosing belt (*Wallace, 1949;* Noble, 1954; *Matti and others, 1985*). At the southeastern end of the San Andreas proper within the Salton Trough, plate displacement sidesteps to several subparallel strike-slip faults, such as the Brawley and Imperial faults. These faults, and others, were the sites of earthquakes in historic time (Johnson and others, 1982). During these earthquakes, surface ruptures took place near the boundaries of a set of pull-apart basins that can be followed on southeastward along the floor of the Gulf of California (*Curray and Moore, 1984; Lonsdale, 1985*).

In western California, fault scarps, offset streams, and other landforms document late Cenozoic, including Holocene, movements along the San Andreas fault zone. Investigations of these features supplement those of earthquakes and geodetic measurements showing modern activity. At several places, alluvial and sag-pond deposits dated by ^{14}C methods display crosscutting relations, buried sand blows, and other features documenting that earthquakes have recurred along it at intervals (Allen, 1968, 1981; *Wallace, 1970; Sieh, 1978,* 1984; *Davis and Duebendorfer, 1982; Rust, 1982;* Weldon and Sieh, 1985). On the San Andreas fault north of Los Angeles, for example, 12 earthquakes have been identified, with an average recurrence interval of about 145 years (Sieh, 1984). Elsewhere along the San Andreas and other faults, similar displacement magnitudes and rates are extended back into earlier Quaternary time (Sieh and Jahns, 1984; *Wallace, 1986*).

The San Andreas fault (SA) in central California has a total right slip of about 330 km, acquired since the present main strand apparently originated in Miocene time. Eocene and Oligocene strata, both marine and nonmarine, as well as a transected volcanic edifice, about 22 m.y. old, all show the same offset (Hill and Dibblee, 1953; Addicott, 1968; Huffman, 1972; Matthews, 1976; *Howell, 1975; Crowell, 1979;* Stanley, 1987). In addition, folds believed related to early simple-shear displacements on the San Andreas fault also originated in early Miocene time as shown by subsurface data in oil fields along the western border of the San Joaquin Valley (Harding, 1976). In the late Miocene, in the same region, coarse submarine debris flows were shed eastward from a growing San Andreas scarp as land sources rose. These sedimentary tongues have since been displaced right laterally from their sources for about 240 km (*Fletcher, 1967;* Huffman, 1972). The difference of about 90 km between their displacement and the total displacement recognized therefore must have taken place during early and middle Miocene time.

In southern California, from within the Transverse Ranges and southeastward into the Salton Trough, the San Andreas fault has displaced identifiable terranes a total of about 270 km (Crowell, 1962, 1982; *Crowell and Walker, 1962;* Ehlig and others, 1975; *Powell, 1981*). The youngest strata showing this total displacement are late Miocene in age, assigned to the Caliente and Mint Canyon Formations (Ehlig and others, 1975). This segment of the fault probably originated in late Miocene time, and as much as 90 km of displacement may have taken place along the San Andreas zone before the birth of the San Gabriel fault (SG) (*Powell, 1981; Stitt, 1986*). Along the San Gabriel fault, rocks older than about 11 Ma are displaced about 60 km (Crowell, 1962, 1982; Carman, 1964; Ehlig and others, 1975; *Stitt, 1986*). From about 11 until 5 Ma this fault was active but has since been abandoned as the site of active displacements in the region except locally (Crowell, 1982). When the Gulf of California proper began to open, beginning about 5.5 Ma (*Larson and others, 1968; Moore and Buffington, 1968;* Elders and others, 1972; *Curray and Moore, 1984*), the San Andreas was reactivated, and at least 160 km has accumulated since then (*Ramirez, 1983; Frizzell and others, 1986*). About 20 km of displacement is deemed to have occurred on other strands of the system (e.g., *Smith, 1977*). The total displacement of terranes along the San Andreas–San Gabriel and closely related faults (90 + 60 + 160 + 20 = 330 km) matches the 330 km identified in central California.

Faults associated with San Andreas system in central and northern California

In central and northern California, fault strands lead from the vicinity of the Andreas fault itself, but with a more northerly trend. These include the Hayward (HY), Calaveras (CV), and related faults (Aydin and Page, 1984). These complex linear fault zones at places consist of several faults with significant components of dip slip. Long, attenuated blocks lie between these faults and are deformed internally by faults and folds that trend subparallel to the major faults, or trend into them obliquely, suggesting that simple-shear deformation has prevailed (*Wilcox and others, 1973*). Much of the right-slip faulting as well as the deformation within blocks between the major faults is Pliocene and Quaternary in age inasmuch as strata of these ages are involved in the deformation (Aydin and Page, 1984).

Extending northwestward in northern California from the Hayward, Calaveras, and related faults in the San Francisco Bay region is a broad system of faults east of the San Andreas proper. Those extending northwestward from the Calaveras fault include the Franklin (FR), Rogers Creek–Healdsburg (RG), Maacama–Bennett Valley (MA), and perhaps Concord (CD) and Green Valley faults (GR) (*Fox, 1983; Kelsey and Cashman, 1983*). Still farther to the east lie other subparallel fault zones, most of which are considered to be parts of the San Andreas system with Quaternary right slip upon them, but some also with Neogene and pre-Neogene displacements. These include the Greenville fault (GN; Aydin and Page, 1984), Bartlett Springs (BT), Midland-Sweitzer (MD), and others (*Jennings, 1977*). To the north, the system includes the Grogan fault (GG), (*Kelsey and Hagans, 1982*), with major early Cenozoic (*Blake, 1985*) and minor Pliocene and Quaternary strike slip (Kelsey and Carver, 1988). The system of strike slip-faults extends into the offshore near the Oregon border (Clarke, 1987). The inland system of strike-slip faults in northern California may indicate that the principal strand of the modern plate boundary may be moving inland from the present San Andreas fault, and that a new small plate, named the Humboldt plate (Herd, 1978), is evolving, or alternatively, the region is a soft plate margin with a wide zone of deformation (Hearn and others, 1988).

A complex system of faults related to both translational and contractional strain occurs at the northern terminus of the San Andreas fault onshore and offshore near the Mendocino triple junction. As interpreted by Kelsey and Carver (1988), coupling of the North American plate with the subducting oceanic Gorda plate north of the triple junction may produce compressional faults and folds. By this interpretation, the compressional structures are overprinted by translational strike-slip faulting as the Mendocino triple junction and the San Andreas fault migrates northward.

West of the San Andreas fault proper, in central California, the composite Salinian block consists of a northern block lying largely offshore (west of the San Gregorio fault, SZ) and several terranes further south within the central Coast Ranges. The Nacimiento fault (NA) zone lies within the central Coast Ranges and borders the Salinian block on the southwest. Displacements upon it took place largely in pre-Neogene time (Page, 1982; *Dickinson, 1983*). At the south, however, the Neogene Rinconada fault (RI) has cut obliquely across these older faults and locally constitutes the margin of the Salinian block. Right-lateral offset of Upper Cretaceous and lower Tertiary stratigraphic units across the Rinconada fault may total about 60 km, whereas offset of lower Pliocene units totals about 18 km (Dibblee, 1976; Ross, 1984). On the northwest, the Rinconada fault may merge with the Reliz fault (RE) and perhaps with others in the vicinity of Monterey Bay, 100 km south of San Francisco. These in turn probably merge with the San Gregorio fault (SZ). On the southeast, the Rinconada fault is transected and perhaps offset by the Big Pine fault (BP).

To the east and northeast of the Rinconada fault, and within the Salinian block, are other faults, including the San Juan–Chimeneas fault zone (SC). This fault zone meets the San Andreas fault on the north, and bounds a long slice of gneissic basement consisting of rocks similar to those within the central Transverse Ranges to the southeast. On the southeast, the slice enters the Transverse Ranges and is probably offset by the left-lateral Big Pine fault. The San Juan–Chimeneas fault zone does not cut upper Pliocene strata, and it may be an important strand of the San Andreas system that was active in Miocene time (*Smith, 1977*) but inactive now. Faults with minor amounts of presumed Neogene strike-slip displacements are numerous elsewhere in the Salinian block (*Compton, 1966a, b; Clark, 1981; Page, 1981; Stanley and McCaffrey, 1983;* Ross, 1984).

Major Neogene faults southwest of the Salinian block, which crosscut it to the north, include the north-northwest–trending San Gregorio (SZ) and Hosgri fault (HG) zone (*Hoskins and Griffiths, 1971;* Hall, 1975; *Silver and Normark, 1978; Graham and Dickinson, 1978; Page, 1981,* 1982; *Gibson, 1983; Nagle and Mullins, 1983; Crouch and others, 1984;* McCulloch, 1987). The fault zone is about 380 km long, lies mainly offshore, and converges with the San Andreas fault on the north near the entrance to San Francisco Bay. Its southeastern extent is problematic; whether it splays out or continues into or around the western Transverse Ranges, or terminates, is under investigation. Throughout its length it consists of a set of faults, many of which dip steeply to the northeast.

The total displacement on the San Gregorio and Hosgri fault zone is disputed. Restored mismatches of Miocene facies suggest different amounts of offset at different places ranging from 70 to 115 km (Hall, 1975; *Greene, 1977;* Graham, 1978; *Graham and Dickinson, 1978; Clark and Brabb, 1978*). Along the southern part of the fault zone (Hosgri fault), changes in magnetic anomaly patterns and grain of structural trends of Cretaceous rocks indicate an early strike-slip history (McCulloch, 1987). Since the late Miocene, however, the Hosgri fault has been the locus of thrusting (*Crouch and others, 1984*). According to this hypothesis, the Hosgri fault may be the inclined margin of a detached crustal slab or slabs, including parts of the composite Salinian block.

Offshore of northern and central California, other faults with northwesterly trends are known, namely, the offshore Nacimiento fault (NA) (offset to the north along the San Gregorio and Hosgri fault zone), the Point Reyes fault (PR) (a late Cenozoic high-angle reverse fault), and to the south, the Santa Lucia Bank fault (SB). The Santa Lucia Bank fault along the eastern edge of Santa Lucia Bank roughly parallels, and lies 45 km west of, the San Gregorio and Hosgri fault zone (McCulloch, 1987). Strike-slip displacement on the Santa Lucia fault is indicated by its linear trace, great changes of stratigraphic thickness across the fault, and change in the magnetic anomaly pattern (McCulloch, 1987).

Big Pine, Santa Ynez, and Santa Monica faults systems in southern California and its borderland

A region of generally east- to northeast-trending faults, forming most of the Transverse Ranges west of the San Andreas fault, extends from the Big Pine fault (BP) on the north to the Santa Monica system (SM) on the south. This region also includes the eastern stretch of the San Gabriel fault (SG) zone, interpreted as an older strand of the San Andreas fault (SA) (see discussion under "San Andreas fault").

The Big Pine fault (BP) extends southwestward from the San Andreas fault, and then curves westward into the southern Coast Ranges (Hill and Dibblee, 1953; Crowell, 1962; *Vedder and others, 1973*). It is doubtful that this fault is active at present, as inferred from subdued offset streams and other fault landforms. The fault has displaced upper Miocene units, and perhaps pre-Miocene older faults and folds, about 12 km left laterally, probably during the late Neocene and Quaternary. During Oligocene–early Miocene time the Big Pine fault formed a steep fault scarp, indicating an earlier episode with dip separation. At that time, alluvial fans were shed northward from the scarp. These fanglomerates have since been displaced 12 km left laterally from identifiable source areas of Precambrian gneisses south of the scarp (Crowell, 1962). At the western end, the Big Pine fault bifurcates, apparently becomes oblique slip, and merges with several smaller faults (J. G. Vedder, written communication, 1987).

The Santa Ynez fault (SY), including its western branch (known as the Santa Ynez River fault, SV), has an east-west trend through the western Transverse Ranges for about 170 km (*Dibblee, 1950, 1966; Jennings, 1977; Hall, 1981a; Sylvester and Darrow, 1979*). On the east it splays into several faults: northern branches apparently fail to meet the San Gabriel fault (SG), and another branch passes beneath a thrust fault near the San Gabriel. On the west the Santa Ynez fault zone also splits into several branches: a southern one strikes southwesterly into the offshore, a central branch (the Santa Ynez River fault, SV) continues on to the west, and several splays trend northwesterly (*Hall, 1981a, b*). These northwesterly splays may in turn be associated with the San Gregorio and Hosgri fault zone.

The Santa Ynez fault (SY) and related fault zones appear to separate two tectonic terranes with vastly different amounts of clockwise rotation: relatively little on the north, and more than 90° to the south (*Luyendyk and others, 1980,* Luyendyk and others, 1985; *Hornafius, 1985*). Whether these faults and several others of similar trend within the western Transverse Ranges and Santa Barbara Channel are throughgoing left-lateral faults or boundaries between displaced and rotated terranes is as yet unknown.

The Santa Monica fault system (SM), named by *Barbat (1958),* is a broad belt or chain of east-west–trending faults that borders the Transverse Ranges on the south and separates them from the southern California borderland, the Los Angeles basin, and the Peninsular Ranges. From west to east, the chain includes offshore faults (*Vedder and others, 1969; Yeats, 1983; Clark and others, 1985*) along the southern margin of the Santa Barbara Channel and along the submarine ridge surmounted by the Channel Islands. Associated faults include the Santa Cruz Island fault (SI), several subsea faults between the Channel Islands and the mainland, and the Malibu Coast (MB) and Dume faults (DU). From the shoreline eastward, the east-west system consists of several faults, including the Santa Monica fault (SM) proper and Cucamonga fault (CU). The Cucamonga fault is crosscut by the San Jacinto fault (SN), and disconnected fault strands within the Banning fault system (BA) to the east may be offset counterparts of these faults and the San Gabriel fault (SG) (*Matti and others, 1985*).

The Santa Monica system or belt (SM) is one of the most complex in southern California and has undergone fault displacements of different senses during the Neogene and Quaternary, and at places perhaps even earlier. Fault displacements were accompanied by stretching, block rotation, and later, shortening (*Crowell, 1976; Luyendyk and others, 1980,* Luyendyk and others, 1985; *Yeats, 1983*). Fault scarps and offset streams and ridges, underlain by displaced Quaternary deposits, are noticeable, demonstrating that some of the faults in the system are active. A complex structural knot has been mapped where the system apparently interacts with the Whittier fault zone (WH) to the south (*Lamar, 1970*). To the west of this complex region, primarily on the Santa Monica fault proper, in the subsurface, Miocene deep-sea fans have been displaced left-laterally according to unpublished reports. Farther to the west there is a mismatch in the character of volcanic rocks on either side of the Santa Cruz Island fault (SI), but the displacement sense and magnitude to produce this mismatch are unknown (*Howell and others, 1976*). Deflections of stream drainages on this fault (*Junger, 1979*), however, suggest up to 300 m of late Quaternary left-lateral displacement. Interpretation of the submarine geology west of Santa Cruz Island suggest left-lateral displacements of at least several kilometers to form a small, unnamed pull-apart basin (*Junger, 1976*).

Contrast in pre-Miocene geology across the Santa Monica system is striking. Precambrian basement rocks occur in the San Gabriel Range north of this fault system and are not known in the northern Peninsular Ranges to the south (*Ehlig, 1981; Gastil and others, 1981*). Cretaceous and Paleogene strata, on the other hand, include similar beds that have apparently been offset left-laterally about 60 km (*Sage, 1973; Campbell and Yerkes, 1976;* Link and others, 1984). According to the hypothesis of Luyendyk and others (1980, *1985*), the Santa Monica fault system is the southern boundary of rotated terranes within the Transverse Ranges, and the Santa Ynez fault zone and its fault associates, is the northern boundary.

San Jacinto, Elsinore, and other faults in southern California and its borderland

From the Imperial Valley region near the United States–Mexico border several major strike-slip faults trend northwestward from the buried divergent plate boundary at the head of the

Gulf of California into the northern Peninsular Ranges. These faults (San Jacinto, SN; Elsinore, EL; and others) extend northwest from pull-apart basins at the divergent plate boundary of the Salton Trough (*Johnson and others, 1982*). The zones are characterized by many discontinuous and subparallel faults, some of which are left-stepping en echelon. The active San Jacinto fault zone, the site of several damaging earthquakes since 1912 (*Sharp, 1967; Allen and Nordquist, 1972;* Sharp and others, 1972; Allen, 1981; *Johnson and others, 1982; Ziony and Yerkes, 1985*) extends into the San Andreas fault but apparently does not join the most active strand of this zone at the surface (*Matti and others, 1985*).

The total right slip on the San Jacinto fault zone is probably about 30 km, as shown by the map separation of high-angle basement-rock contacts (Sharp, 1967; *Bartholomew, 1970*). The timing of origin of the fault zone is uncertain, but judged to be late Miocene or younger, based on the age of the preserved strata showing possible facies changes away from the fault (Sharp, 1967; *Crowell and Ramirez, 1979*).

Between 20 and 25 km southwest of the San Jacinto belt of faults lies the Elsinore fault zone (EL), largely within the crystalline rocks of the northern Peninsular Ranges. This zone extends northwestward from spreading centers near the United States–Mexico border beneath the deep alluvium of Imperial Valley, and the Laguna Salada in Mexico. Along the eastern scarp of the Peninsular Ranges, the Elsinore fault zone is irregular, and there is controversy concerning its course and whether it is truly a right-slip fault in this region (*Lowman, 1980*). Farther northwest, however, there is a belt of subparallel faults with a width of 10 or 12 km, trending southeastward between the Elsinore zone and the San Jacinto fault zone. Perhaps this set of straight and long faults, including the Agua Caliente fault (AC), is transferring right slip from the northwestern reaches of the Elsinore zone into the Salton Trough region near the San Jacinto zone. Farther northwest, the Elsinore fault zone splits into the Chino fault (CH) on the east, trending more northerly, and the Whittier fault (WH), trending more westerly.

Parts of the Elsinore fault zone are active, as shown by landforms and low-magnitude earthquakes along it (*Gray, 1961; Langenkamp and Combs, 1974; Weber, 1977; Lamar and Rockwell, 1986; Ziony and Yerkes, 1985*). The Chino and Whittier faults also display landform evidence of Quaternary activity. The time of origin of the Elsinore belt is problematic, however; it may be only 5 m.y. old south of its juncture with the Whittier and Chino faults in view of its apparent kinematic relation to the divergent plate boundary within the Gulf of California. The Whittier fault originated at an older time, however, and played a role in the Miocene evolution of the Los Angeles basin (*Yerkes and others, 1965;* Crowell, 1987). Total right slip in the northern part of the Elsinore fault zone is probably about 40 km, as shown by offset facies-change and pinchout lines in a Paleocene sequence (*Lamar, 1970; Sage, 1973; Crowell and Ramirez, 1979*).

To the west of the Elsinore fault, the Newport-Inglewood belt (NE) of strike-slip faults and related faults and folds extends from the western part of the Los Angeles region southeastward along the coast of southern California, and into the Mexican borderland (*Ziony and Yerkes, 1985*). At the northwest, it crosses the Los Angeles basin and is marked by a set of hills and discontinuous fault scarps at the surface that overlie significant differences in basement rocks beneath Miocene strata (*Yerkes and others, 1965; Hill, 1971;* Yeats, 1973; *Barrows, 1974;* Platt and Stuart, 1974). The belt extends offshore on the southeast into a discontinuous set of escarpments (*Vedder and others, 1974; Clarke and others, 1985*). Onshore branches include the Rose Canyon fault (RS) (*Legg and Kennedy, 1979*). Folds trending more northwesterly meet the belt of discontinuous faults obliquely, indicating that they were probably formed by right simple shear above a zone of fundamental breaks in the basement (Harding, 1973; *Wilcox and others, 1973;* Platt and Stuart, 1974). The total displacement, character, and age of the old fault zone deemed to lie beneath Neogene deformed sediments are unknown (*Hill, 1971;* Yeats, 1973; Platt and Stuart, 1974).

The California borderland south of the Channel Islands contains many linear fault scarps, with a northwestern trend (*Vedder and others, 1974; Howell, 1976; Legg and Kennedy, 1979; Crouch, 1981;* Howell and Vedder, 1981; *Legg, 1985;* Vedder, 1987). Earthquake focal mechanisms and offset bathymetric features on the sea floor show that many are active with right-lateral displacements.

In this borderland region, major fault zones with northwest trends separate four terranes in the southern California borderland (Howell and Vedder, 1981; Vedder, 1987). The juxtaposition of these distinctly different terranes is probably the result of major right slip. The western terrane, east of the Patton fault, consists of rocks of the Franciscan assemblage (Lower Cretaceous and lower Tertiary), overlain by Neogene strata, including volcanic rocks. The next terrane to the east, about 80 km wide, is characterized by a thick section of Cretaceous and Paleogene strata underlying Oligocene and Neogene rocks. The third terrane, lying near the coast but oceanward from the Newport-Inglewod belt of deformation, is similar to the first. The fourth, lying largely onshore to the east of the Newport-Inglewood belt, includes rocks of the southwestern flank of the Peninsular Ranges, and has a basement largely of continental character.

Paleogene deep-sea fan deposits, now preserved on some of the offshore islands within the third terrane, contain clasts that are probably derived from source areas well to the south and across fault zones bordering the terrane on the east (*Howell and others, 1974; Minch and others, 1976; Crouch, 1979;* Howell and Vedder, 1981). Right slip of many tens of kilometers is implied. Moreover, major right slip of the terranes is suggested by their present arrangement. The first and third terranes are underlain by rocks formed within an accretionary prism during Cretaceous and early Tertiary times when subduction prevailed between the North American plate and plates coming into the continent from Pacific regions. The second terrane is viewed as consisting of fore-arc basin rocks, and originally lay inboard or on the continental side of the subduction zone and accretionary prism.

During Neogene times, major right slip upon faults trending slightly obliquely across these units may have rearranged and duplicated them. Right-slip displacements of as much as 500 km are implied by this concept and are in part supported by paleomagnetic data (*Crouch, 1979; Luyendyk and others, 1980;* Luyendyk and others, 1985).

Garlock fault, Mojave Desert, and southern Sierra Nevada

The Garlock fault (GA), bounding the Mojave Desert on the north, is a major northeast-trending fault that has displaced dike rocks, Paleozoic units, and a thrust fault left laterally 48 to 64 km (Smith, 1962; *Smith and Ketner, 1970;* Davis and Burchfiel, 1973). The fault is active, especially along its central reaches, as shown by landforms and offset of Quaternary deposits (*Clark, 1973; Carter, 1980; LaViolette and others, 1980;* Clark and others, 1984). Sag ponds and elongate Quaternary pull-apart basins filled with thick sedimentary sequences occur along it. Davis and Burchfiel (1973) view the Garlock fault as an intracontinental transform structure related to greater basin-range extension north of the fault than to the south. Seismic profiling across the Garlock fault suggests that the fault zone extends to a depth of only 8 or 10 km (Cheadle and others, 1986), implying that the terranes on either side may be allochthonous above this depth.

The main part of the Mojave Desert is segmented by many subparallel northwest-trending faults (Fig. 2, HE, LO, HA, BL, LE, CR, CI, RO, PG, LU, BM, GM, BU), and in the north-central and southern part by east-trending faults (Fig. 2, CY, MX, PI, BC, CC). Right offset on individual northwest-trending faults is generally less than a kilometer to perhaps as much as 14 km, based on offset of Paleozoic to Cenozoic rocks and Cenozoic faults and folds (*Dibblee, 1961,* 1967a, *1980; Miller and Morton, 1980; Miller and others, 1982; Dokka, 1983*). At many places the faults are marked by active scarps or ground-water barriers, and earthquakes have occurred upon some of them (*Allen and others, 1965; Hutton and others, 1980*). Cumulative displacement on these major right-lateral faults across the Mojave Desert is estimated at 27 to 38 km (*Dokka, 1983*). The left-lateral offset on the east-trending Pinto Mountain (PI) and Blue Cut faults (BC) in the southern Mojave Desert is about 16 km and 5 to 6 km, respectively (*Dibblee, 1967b; Hope, 1969*). Other east-trending faults (*Powell, 1975, 1981, 1982; Crowell and Ramirez, 1979*) in the southern and north-central Mojave Desert and the Colorado Desert to the south are also considered left lateral (*Luyendyk and others, 1980,* Luyendyk and others, 1985). Several tectonic models have been proposed for the Mojave Desert region, some that involve large internal shear with or without rotation (*Garfunkel, 1974; Luyendyk and others, 1980;* Luyendyk and others, 1985) and others that consider the Mojave Desert and adjacent regions to be a relatively coherent block undergoing rotation (Bird and Rosenstock, 1984). Available paleomagnetic data in the Mojave Desert support a model in which fault-bounded elongate domains have undergone differential clockwise and counter-clockwise rotation (*Burke and others, 1982; Hillhouse and Wells, 1986; Golombek and Brown, 1986; Carter and others, 1987*).

Two major faults, the Kern Canyon (KC) and White Wolf (WW), occur in the southern Sierra Nevada. The Kern Canyon fault (KC) (*Moore and du Bray, 1978*), mentioned previously under the discussion of Mesozoic and early Cenozoic faults, offsets 80-Ma granitic plutons as much as 13 km right-laterally, but apparently does not offset a 3.5-Ma basalt. The White Wolf fault is a high-angle fault with a large component of left slip, along which surface rupture occurred during a 7.2-magnitude earthquake in 1952 (*Buwalda and St. Amand, 1955; Stein and Thatcher, 1981; Bartow, 1984*).

Walker lane belt

A northwest-trending structural zone 700 km long and from 100 to 300 km wide containing both right-lateral and left-lateral faults separates the Sierra Nevada region to the west from areas of typical basin-range topography on the east. The term "Walker Lane belt" (Fig. 2) is used (*Carr, 1984;* Stewart, 1988) to describe this structural zone rather than the term "Walker Lane," which was originally defined (*Giannella and Callaghan, 1934; Billingsley and Locke, 1939, 1941; Locke and others, 1940;* Albers, 1967) as the *line* that separates areas of typical basin-range topography to the east from areas of diverse topography to the west. The Walker Lane, as originally defined, is the eastern margin of the Walker Lane belt as used here.

Rather than a simple, throughgoing system of right-lateral faults as was implied by *Giannella and Callaghan (1934)* and other geologists, the Walker Lane belt is a complex zone composed of structural sections (large structural blocks as well as relatively narrow zones) that have undergone either right-lateral shear, left-lateral shear, or remained relatively undeformed by shear (Stewart, 1988).

In the northwest part of the Walker Lane belt, displacement on northwest-trending faults, (Grizzly Valley, GV; Last Chance, LC; Honey Lake, HL; Warm Springs Valley, WS; and Pyramid Lake, PL) (*Bonham, 1969; Bell, E. J., and Slemmons, 1979; Bell, J. W., 1984a, b; Bell, E. J., 1984*) includes both normal dip slip and right-lateral slip. The Pyramid Lake fault is characterized by a pattern of en-echelon Riedel and conjugate Riedel shears and young geomorphic features indicative of Holocene strike-slip movement. *Bonham (1969)* suggested a cumulative dextral offset on the faults in the northern part of the Walker Lane belt of 32 km based on the apparent offset of the northern outcrop limit of a group of Cenozoic tuffs, although he indicates that this estimate is conjectural.

Next to the south are several poorly defined northeast-trending zones (Olinghouse fault zone, OL; Carson lineament, CL; Wabuska lineament, WL) of known or presumed left-lateral displacement (*Rogers, 1975; Sanders and Slemmons, 1979;* Slemmons and others, 1979; *Bell, J. W., 1984a, b*). Surface rupture occurred on the Olinghouse fault zone during an earthquake of about magnitude 7 in 1869. Maximum displacement was 3.65 m of left slip.

South of the zones of northeast-trending faults is a region of northwest-trending dextral slip faults (Agai-Pah Hills, AP; Indian Head, IH; Gumdrop Hills, GH; Benton Spring–Pilot Mountains; BS; Petrified Spring–Bettles Well, PS). Cumulative dextral displacement across this group of faults is estimated at 48 to 60 km (*Nielsen, 1965; Hardyman and others, 1975; Ekren and others, 1980; Hardyman, 1984; Ekren and Byers, 1984*), mainly on the basis of apparent offset of the boundaries of Oligocene and Miocene tuff units. Right-normal oblique-slip surface rupture occurred a short distance east of the Bettles Well fault during the 7.2- to 7.3-magnitude Cedar Mountain earthquake of 1932 (*Gianella and Callaghan, 1934; Molinari, 1984*).

A peculiar cross zone of east-west faults (Excelsior, EX; Candelaria, CA; Coaldale, CO) occurs to the south. During the Mesozoic, major right-lateral displacement is inferred to have taken place on the Excelsior and Coaldale faults (see discussion under Mesozoic and early Cenozoic strike-slip systems), but reactivation of these faults in the Cenozoic appears to have offset Oligocene and Miocene tuffs in the opposite sense as much as 20 to 25 km. A component of oblique normal left slip is also indicated on the Cenozoic Candelaria fault (*Speed and Cogbill, 1979*) lying between the Excelsior and Coaldale faults, and left slip was reported (*Callaghan and Giannella, 1935*) on surface ruptures near the Excelsior fault zone during the 6.3-magnitude Excelsior Mountains earthquake in 1934.

South of the Excelsior, Candelaria, and Coaldale system, major right-lateral faults or shear zones include the Owens Valley–White Mountains (OV), the Panamint Valley–Hunter Mountain (PV), Death Valley (DV), Furnace Creek (FC), Pahrump (PH), and Las Vegas Valley (LV) systems.

1. The Owens Valley–White Mountains fault system (OV) is a complex zone of normal and right-lateral faults along which surface rupture occurred in the 1872 earthquake that had an estimated magnitude of about 8 and right-lateral displacement as much as 6 m (*Hill, 1972; Oakeshott and others, 1972;* Beanland and Clark, 1992). Total Cenozoic right-lateral offset along the fault zone, however, appears to be small because a major system of Late Jurassic dikes (*Chen and Moore, 1979*) crosses Owens Valley with no obvious offset (Moore and Hopson, 1961). This conclusion is supported by the continuity of a 170-Ma (R. W. Kistler, oral communication, 1985) pluton across Owens Valley (*Ross, 1962*).

2. The Panamint Valley–Hunter Mountain fault zone (PV) is a complex zone characterized by right-oblique slip (*Smith, 1979*) and right slip (*Burchfiel and others, 1987*). The central section of the fault zone is interpreted by *Burchfiel and others (1987)* as a right-slip fault with 8 to 10 km of displacement connecting pull-apart basins on opposite sides of the fault zone.

3. The Death Valley fault zone (DV) has had less than 8 km of right-lateral offset based on apparently little-disrupted Precambrian stratigraphic trends (*Wright and Troxel, 1967; Davis, 1977a*), although *Brady (1984)* estimates possibly 20 km and *Butler and others (1988)* estimate about 35 km of right-lateral offset during the late Cenozoic, based on provenance studies of gravels and conglomerate.

4. Right-lateral displacement on the Furnace Creek fault zone (FC) has been estimated at 40 to 100 km on the basis of apparent offsets of (1) stratigraphic trends of upper Precambrian and Paleozoic rocks, (2) granitic plutons, and (3) structural trends (*Stewart, 1967;* Stewart, 1988; *Poole and others, 1967, 1977;* Stewart and others, 1968; *McKee, 1968; Pelton, 1966; Moore, 1976; Poole and Sandberg, 1977; Cooper and others, 1982; Snow and Wernicke, 1988;* Wernicke and others, 1988). Other workers (*Wright and Troxel, 1967, 1970; Prave and Wright, 1986; Cemen and others, 1986*) have challenged this concept of large offset, at least near the southern end of the fault. The southern part of the Furnace Creek fault zone has been considered (*Wright and Troxel, 1970; Stewart, 1983*) to mark the boundary between a region of greater extension on the southwest relative to the northeast. If so, the main displacement on the Furnace Creek fault zone is late Cenozoic in age, because late Cenozoic sedimentary and volcanic rocks are cut by high- and low-angle extensional faults, and major extension is clearly late Cenozoic in age, southwest of the fault zone. Latest movement on the fault zone is Quaternary, because the fault zone cuts Quaternary deposits along much of its length (*Brogan, 1979*).

5. Right-lateral offset on the Pahrump fault zone (PH) (*Liggett and Childs, 1973*), which is also called the Stewart Valley fault system (Stewart and others, 1968) in the northern part, has been estimated at 16 to 19 km on the basis of offset of Precambrian and Paleozoic formations and consistent indications of drag on either side of the fault zone (R. L. Christiansen, written commun., *in* Stewart and others, 1968). An unknown, but perhaps considerable, amount of strike slip is inferred for the Kokoweef fault (KO) and apparently correlative Slaughterhouse fault (SL) (*Burchfiel and Davis, 1977*) approximately online with the Pahrump fault zone in easternmost California. According to *Liggett and Childs (1973)*, the Pahrump fault zone cuts Tertiary and Quaternary basin sediments, indicating that the youngest movement on at least part of the fault zone is late Cenozoic in age.

6. The Las Vegas Valley shear zone (LV) is a complex structure with 40 to 67 km of right-lateral displacement based on offset trends of Paleozoic strata and Mesozoic structures (Stewart and others, 1968; *Longwell, 1974;* Wernicke and others, 1982). Although some of this offset is on faults covered by alluvium, offset includes drag that is evident from the curving of structural trends as they approach the deformational belt (*Longwell, 1960; Nelson and Jones, 1987*). Displacement is considered late Cenozoic in age (*Fleck, 1970; Anderson and others, 1972*), and offset in relation to large-scale extension north of the shear zone relative to that to the south (*Guth, 1981;* Wernicke and others, 1982; *Wernicke and others, 1983, 1984*). *Royse (1983)*, on the other hand, has argued that the shear zone developed as a lateral ramp during Cretaceous thrusting.

Two northeast-trending zones of left-lateral faults occur in the southeastern part of the Walker Lane belt. The northern of the two zones includes the Mine Mountain (MM), Wahmonie (WA), Cane Springs (CS), and Rock Valley faults (RV) (*Carr, 1984*). Offset is primarily late Cenozoic in age, and 2 km or less on individual faults. The southern of the two zones is the Lake Mead fault system consisting of the Bitter Spring Valley (BV), Hamblin Bay (HB), Cabin Canyon (CB), Bitter Ridge (BD), Lime Ridge (LR), and Gold Butte faults (GB). Cumulative offset on the Lake Mead fault zone, based on offset of Mesozoic and unconformably overlying Cenozoic rocks, is as much as 65 km (Anderson, 1973; *Bohannon, 1979,* Bohannon, 1984).

Basin and Range Province exclusive of Walker Lane belt and Mojave Desert region

The extent of strike-slip structures in the Basin and Range Province outside of the Walker Lane belt and Mojave Desert region is poorly understood. *Shawe (1965)* and *Slemmons (1967)* have proposed a widespread system of conjugate north- to northwest-trending right-lateral faults and northeast-trending left-lateral faults in the northern Basin and Range Province. Wright (1976) and Anderson (*1984*) indicate that most such conjugate systems are in the western, southern, and eastern parts of the Great Basin. R. E. Anderson (personal communication, 1985) indicates that in southwestern Utah, strike-slip faults related to the conjugate systems, although great in number, are generally short in length. *Anderson (1984)* relates these conjugate systems to the same stress field that has produced extensional basin-range structures and has shown that normal faulting and strike-slip faulting are intermixed and contemporaneous. These systems are related to a stress field in which the least principal stress direction is horizontal (*Anderson, 1984*) and the intermediate and maximum principal stresses change periodically from regimes in which the maximum principal stress is vertical, producing normal faulting, to regimes in which the maximum principal stress is horizontal, producing strike-slip faulting (*Donath, 1962;* Wright, 1976; *Angelier and others, 1985; Anderson and Bernhard, 1984; Anderson, 1984*; Zoback, 1989).

Other strike- or oblique-slip systems in the Basin and Range Province are related to reactivation of preexisting fault systems that are parallel or oblique to a later extensional directions (*Zoback, 1979; Zoback and Zoback, 1980b*). Still others are related to variations in strike of extensional faults, producing normal movement on segments perpendicular to a regional extension direction and oblique-slip on segments oblique to the extension direction (*Thompson and Burke, 1973*). Finally, strike-slip systems form parallel to the extension direction along the boundary between areas that differ in the amount, rate, or sense of normal faulting (*Anderson, 1984*). Such strike-slip faulting parallel to the extension direction should be expected, particularly along transverse zones (*Stewart, 1980*) separating regions with different tilt directions of basin-range blocks.

In the southern Basin and Range Province in Arizona and southeastern California, a well-defined northwest-trending grain of tectonic features is evident, although how much of this is related to late Cenozoic strike-slip faulting is unknown. In New Mexico, in the southeastern part of the Basin and Range Province, left-slip of as much as 28 km along the north-trending Rio Grande rift, based on a mismatch of features across the rift, has been described by *Kelley (1979).* Other left-slip features in New Mexico have been described by *Muehlberger (1979).*

Linear belts of structures, volcanic features, and mineral deposits in the Basin and Range Province are accredited to strike-slip displacement, but in general their origin is obscure (*Suppe and others, 1975; Ekren and others, 1976; Shawe and Stewart, 1976; Rowley and others, 1978, 1979; Rytuba and Conrad, 1981*).

Washington, Oregon, and northeast California

Several major and many minor northwest- and northeast-trending faults or fault systems in southern Washington, Oregon, and northeast California are considered to have had lateral offset (*Pease, 1969;* Lawrence, 1976; *Davis, 1977b; Walker, 1977; Hannah, 1977; Lillie and Couch, 1979; Bentley, 1979; Ross, 1979;* Wells and Coe, 1985; Snavely, 1987). The Brothers fault zone (BR), for example, consists of a series of en echelon, right-stepping faults consistent with an interpretation of Riedel shears in a right-lateral system (Lawrence, 1976). Offset may have been minor and related to relatively greater extension to the south of the fault zone than to the north. The Eugene-Denio (ED) and the McLoughlin (MC) zones are less well defined but appear to coincide with a right lateral shift in the position of the High Cascade volcanic trend by 10 to 20 km and 15 to 20 km, respectively.

Elsewhere, *Donath (1962)* implied that an early system of strike-slip faullts controlled a later rhombic pattern of normal faults in south-central Oregon. In the Coast Range of Washington and Oregon, abundant northwest- and northeast-trending faults of a few kilometers displacement accommodated right-lateral simple shear strain and accompanying block rotation (Wells and Coe, 1985; Snavely, 1987). In the Cascade Range at southwest Washington, a right-slip system 90 km long passing under Mount St. Helens has been defined on the basis of small- to moderate-scale earthquakes and local mechanisms (*Weaver and Smith, 1983*).

Lewis and Clark line

The Lewis and Clark line (Fig. 2, HO, OS, SW, SJ, NI) is a west-northwest–trending linear structural zone about 15 to 50 km wide and at least 500 km long extending across western Montana and northern Idaho (*Billingsley and Locke, 1939; Wallace and others, 1960; Smith, 1965; Weidman, 1965; Hobbs and others, 1965; Harrison and others, 1974;* Reynolds, 1979). It is characterized primarily by right-lateral faults with total displacement across the zone of about 40 to 50 km, on the basis of offset stratigraphic and structural features, and it may have had a long history, with recurrent movements since the Precambrian (*Harri-*

son and others, 1974; Reynolds, 1979). A Mesozoic or early Cenozoic episode of left-lateral displacement (*Smith, 1965,* see discussion above) followed by right-lateral displacement on the same faults in the late Cenozoic is possible, although disputed (*Harrison and others, 1974*). The Lewis and Clark line, according to Reynolds (1979), accommodates extension on basin-range structures to the south with respect to a relatively more stable block to the north. It terminates to the east by translation of strike-slip movement into dip-slip on faults that splay southward.

TECTONICS

The Mesozoic and Cenozoic strike-slip systems in western North America have been related to two general types of tectonic models. In the first model, called California-type by Beck and others (1981), blocks are detached from the North American continent and moved laterally along it by transform faulting. This style is exemplified by the present-day San Andreas transform system in California and northern Mexico. A similar kinematic model was proposed by *Carey (1958),* but he proposed oroclinal warping and block rotation as well as strike-slip faulting. In the second model, called Sunda-type by Beck and others (1981), blocks move parallel to the continental margin in response to oblique subduction (*Fitch, 1972;* Beck, 1983).

Whether either the transform-fault or oblique-subduction model can be applied to Precambrian strike-slip faults is uncertain. The belt of strike-slip faults (Colorado lineament) proposed by Warner (1978) to extend from Arizona to Wyoming could perhaps represent a continental-margin strike-slip system related to transform faulting or oblique subduction. On the other hand, Karlstrom and Houston (1984) consider part of this belt in Wyoming to be a suture zone rather than a strike-slip zone. In addition, the Colorado lineament in Wyoming is along a boundary between rocks dated generally as 2.5-Ga and rocks dated at 1.72 to 1.80 Ga, but southwest of there it cuts across a northeast-trending belt of 1.72- to 1.80-Ga rocks in Colorado, Utah, and Arizona (*Condie, 1982*). Thus, the belt of proposed strike-slip faulting cuts across Precambrian terranes and does not follow closely a Precambrian continental margin.

During the late Mesozoic and early Cenozoic, northward oblique subduction of the Kula plate (Engebretson and others, 1985) has been considered the cause of right-lateral shear in the northern part of the Cordillera (Beck and others, 1981) and along the continental margin of Oregon and Washington (Snavely, 1987). This explanation may apply to more than just the northern part of the Cordillera, if the Kula plate originally lay outboard of most of North America, as proposed in one plate-tectonics model (*Page and Engebretson, 1984,* Fig. 4). A transform-fault model (Davis and others, 1978; *Ewing, 1980;* and Wells and others, 1984), on the other hand, has also been applied to the northern Cordillera. In this model, the spreading ridge between the Kula and Farallon plates, or a spreading ridge related to some other plate geometry, is viewed as extending inland beneath the continent. In this explanation, the area to the north of the inward-extending ridge moved northward in response to spreading at the ridge.

A different plate geometry dominated the southern part of the United States and northern Mexico during the Mesozoic. Here the left-lateral Mojave-Sonora megashear, and perhaps proposed left-lateral faults in southern Arizona, are related to transform faulting that allowed for the Jurassic opening of the Gulf of Mexico (Anderson and Schmidt, 1983).

Late Cenozoic strike-slip systems are related to transform faulting that resulted from contact of the western margin of North America with the northwest-moving Pacific Plate starting about 29 Ma (Atwater, 1970; Atwater and Molnar, 1973; Engebretson and others, 1985). At this time, the spreading ridge between the Farallon and Pacific plates first impinged on the North American continent, and the margin was progressively converted from one of plate convergence to one of right-lateral transform faulting, dominated by the San Andreas system (Atwater, 1970; Atwater and Molnar, 1973; Page and Engebretson, 1984). Inland zones of strike-slip faulting across a broad region of the western United States have also been related to this dextral shear produced by the transform system (Atwater, 1970). Dextral shear is also reflected in the widespread distribution of earthquakes with strike-slip focal mechanisms in the western United States. In addition to faults produced by dextral shear in a transform system, many late Cenozoic faults in the western United States have been interpreted as part of an extensional system in which strike-slip faults are generally parallel to the extension direction and mark boundaries between regions that are extending at different rates (Davis and Burchfiel, 1973; *Liggett and Childs, 1977; Anderson, 1984*).

REFERENCES CITED

Text references in *italics* will be found only on accompanying microfiche.

Addicott, W. O., 1968, Mid-Tertiary zoogeographic and paleogeographic discontinuities across the San Andreas fault, California, *in* Dickinson, W. R., and Grantz, A., eds., Proceedings, Conference on Geologic Problems of San Andreas Fault System: Stanford, California, Stanford University Publications in the Geological Sciences, v. 11, p. 144–165.

Albers, J. P., 1967, Belt of sigmoidal bending and right-lateral faulting in the western Great Basin: Geological Society of America Bulletin, v. 78, p. 143–156.

Allen, C. R., 1968, The tectonic environments of seismically active and inactive areas along the San Andreas fault system, *in* Dickinson, W. R., and Grantz, A., eds., Proceedings, Conference on Geologic Problems of San Andreas Fault System: Stanford, California, Stanford University Publications in the Geological Sciences, v. 11, p. 70–82.

—— , 1981, The modern San Andreas fault, *in* Ernst, W. G., ed., The geotectonic development of California, Englewood Cliffs, New Jersey, Prentice-Hall, Inc., p. 511–534.

Anderson, R. E., 1973, Large-magnitude late Tertiary strike-slip faulting north of Lake Mead, Nevada: U.S. Geological Survey Professional Paper 794, 18 p.

Anderson, T. H., and Schmidt, V. A., 1983, The evolution of Middle American and the Gulf of Mexico-Caribbean Sea region during Mesozoic time: Geological Society of America Bulletin, v. 94, p. 941–966.

Atwater, T., 1970, Implications of plate tectonics for the Cenozic tectonic evolution of western North America: Geological Society of America Bulletin, v. 81, p. 3513–3536.

Atwater, T., and Molnar, P., 1973, Relative motion of the Pacific and North American plates deduced from sea-floor spreading, *in* Kovach, R. L., and Nur, A., eds., Proceedings, Conference on Tectonic Problems of the San Andreas Fault System: Stanford, California, Stanford University Publications in the Geological Sciences, v. 13, p. 136–148.

Aydin, A., and Page, B. M., 1984, Diverse Pliocene-Quaternary tectonics in a transform environment, San Francisco Bay region, California: Geological Society of America Bulletin, v. 95, p. 1303–1317.

Beanland, S. and Clark, M. M., 1992, The Owens Valley fault zone and surface rupture associated with the 1872 earthquake: U.S. Geological Survey Bulletin (in press).

Beck, M. E., Jr., 1980, Paleomagnetic record of plate-margin tectonic processes along the western edge of North America: Journal of Geophysical Research, v. 85, no. B12, p. 7115–7131.

——, 1983, On the mechanism of tectonic transport of oblique subduction: Tectonophysics, v. 93, p. 1–11.

Beck, M. E., Jr., Burmester, R. F., Engebretson, D. C., and Schoonover, R., 1981, Northward translation of Mesozoic batholiths, western North America; Paleomagnetic evidence and tectonic significance: Geofisica International, v. 20, p. 143–162.

Bird, P., and Rosenstock, R. W., 1984, Kinematics of present crust and mantle flow in southern California: Geological Society of America Bulletin, v. 95, p. 946–957.

Blake, M. C., Jr., Jayko, A. S., and McLaughlin, R. J., 1985a, Tectonostratigraphic terranes and the northern Coast Range, California, *in* Howell, D. G., ed., Tectonostratigraphic terranes of the Circum-Pacific region: Circum-Pacific Council for Energy and Mineral Resources Earth Science Series 1, p. 159–172.

Bohannon, R. G., 1984, Nonmarine sedimentary rocks of Tertiary age in the Lake Mead region, southeastern Nevada and northwestern Arizona: U.S. Geological Survey Professional Paper 1259, 72 p.

Brown, R. D., Jr., and 7 others, 1967, The Parkfield-Cholame, California, earthquakes of June-August, 1966: U.S. Geological Survey Professional Paper 579, 66 p.

Burchfiel, B. C., and Davis, G. A., 1972, Structural framework and evolution of the southern part of the Cordilleran orogen, western United States: American Journal of Science, v. 272, p. 97–118.

——, 1981, Mojave Desert and environs, *in* Ernst, W. G., ed., The geotectonic development of California: Englewood Cliffs, New Jersey, Prentice-Hall, p. 217–252.

Carman, M. F., Jr., 1964, Geology of the Lockwood Valley area, Kern and Ventura Counties, California: California Division of Mines Special Report 81, 62 p.

Champion, D. E., Howell, D. G., and Grommé, C. S., 1984, Paleomagnetic and geologic data indicating 2,500 km of northward displacement for the Salinian and related terranes, California: Journal of Geophysical Research, v. 89, no. B9, p. 7736–7752.

Cheadle, M. J., and 8 others, 1986, The deep crustal structure of the Mojave Desert, California, from COCORP seismic reflection data: Tectonics, v. 5, p. 293–320.

Clark, M. M., and 12 others, 1984, Preliminary slip-rate table and map of late Quaternary faults of California: U.S. Geological Survey Open-File Report OF 84–106, scale 1:1,000,000.

Clarke, S. H., Jr., 1987, Geology of the California continental margin north of Cape Mendocino, *in* Scholl, D. W., Grantz, A., and Vedder, J. G., eds., Geology and resource potential of the continental margin of western North America and adjacent ocean basins; Beaufort Sea to Baja California: Houston, Texas, Circum-Pacific Council for Energy and Mineral Resources Earth Science Series, v. 6, p. 337–351.

Coney, P. J., Jones, D. L., and Monger, J.W.H., 1980, Cordilleran suspect terranes: Nature, v. 288, p. 329–333.

Crowell, J. C., 1962, Displacement along the San Andreas fault, California: Geological Society of America Special Paper 71, 61 p.

——, 1982, The tectonics of Ridge Basin, southern California, *in* Crowell, J. C., and Link, M. H., eds., Geologic history of Ridge Basin, southern California: Pacific Section, Society of Economic Paleontologists and Mineralogists, p. 25–41.

——, 1985, The recognition of transform terrane dispersion within mobile belts, *in* Howell, D. G., ed., Tectonostratigraphic terranes of the Circum-Pacific Region: Circum-Pacific Council for Energy and Mineral Resources Earth Science Series, no. 1, p. 51–62.

——, 1987, Late Cenozoic basins of onshore southern California; Complexity is the hallmark of their tectonic history, *in* Ingersoll, R. V., and Ernst, W. G., eds., Cenozoic basin development of coastal California: Englewood Cliffs, New Jersey, Prentice-Hall, p. 207–241.

Davis, G. A., and Burchfiel, B. C., 1973, Garlock fault; An intracontinental transform structure, southern California: Geological Society of America Bulletin, v. 84, p. 1407–1422.

Davis, G. A., Monger, J.W.H., and Burchfiel, B. C., 1978, Mesozoic construction of the Cordilleran "collage," central British Columbia to central California, *in* Howell, D. G., and McDougall, K. A., eds., Mesozoic paleontology of the western United States: Pacific Section, Society of Economic Paleontologists and Mineralogists Pacific Coast Paleogeography Symposium 2, p. 1–32.

Dibblee, T. W., Jr., 1967a, Areal geology of the western Mojave Desert: U.S. Geological Survey Professional Paper 522, 153 p.

——, 1976, The Rinconada and selected faults in the southern Coast Ranges, California, and their tectonic significance: U.S. Geological Survey Professional Paper 981, 55 p.

Drewes, H., 1981, Tectonics of southern Arizona: U.S. Geological Survey Professional Paper 1144, 96 p.

Ehlig, P. L., Ehlert, K. W., and Crowe, B. M., 1975, Offset of the upper Miocene Caliente and Mint Canyon Formations along the San Gabriel and San Andreas faults, *in* Crowell, J. C., eds., San Andreas fault in southern California: California Division of Mines and Geology Special Report 118, p. 83–92.

Elders, W. A., Rex, R. W., Meidav, T., Robinson, P. T., and Biehler, S., 1972, Crustal spreading in southern California: Science, v. 178, p. 15–24.

Engebretson, D. C., Cox, A., and Gordon, R. G., 1985, Relative motions between oceanic and continental plates of the Pacific Basin: Geological Society of America Special Paper 206, 59 p.

Fitch, T. J., 1972, Plate convergence, transcurrent faults, and internal deformation adjacent to Southeast Asia and the Western Pacific: Journal of Geophysical Research, v. 77, p. 4432–4461.

Gabrielse, H., 1985, Major dextral transcurrent displacements along the northern Rocky Mountain Trench and related lineaments in north-central British Columbia: Geological Society of America Bulletin, v. 96, p. 1–14.

Graham, S. A., 1978, Role of Salinian block in evolution of San Andreas fault system: American Association of Petroleum Geologists Bulletin, v. 62, p. 2214–2231.

Hall, C. A., Jr., 1975, San Simeon-Hosgri fault system, coastal California; Economic and environmental implications: Science, v. 190, p. 1291–1294.

Hamilton, W., 1969, Mesozoic California and the underflow of the Pacific mantle: Geological Society of America Bulletin, v. 80, p. 2409–2430.

Harding, T. P., 1973, The Newport-Inglewood trend, California; An example of wrenching style of deformation: American Association of Petroleum Geologists Bulletin, v. 57, p. 97–116.

——, 1976, Tectonic significance and hydrocarbon trapping consequences of sequential folding synchronous with San Andreas faulting, San Joaquin Valley, California: American Association of Petroleum Geologists Bulletin, v. 60, p. 356–378.

Herd, D. G., 1978, Intracontinental plate boundary east of Cape Mendocino, California: Geology, v. 6, p. 721–725.

Hill, M. L., and Dibblee, T. W., Jr., 1953, San Andreas, Garlock, and Big Pine faults, California: Geological Society of America Bulletin, v. 64, p. 443–458.

Howell, D. G., 1985, ed., Tectonostratigraphic terranes of the Circum-Pacific region: Houston, Texas, Circum-Pacific Council for Energy and Mineral Resources Earth Sciences series, number 1, 581 p.

Howell, D. G., and Vedder, J. G., 1981, Structural implications of stratigraphic discontinuities across the southern California borderland, *in* Ernst, W. G., ed., The geotectonic development of California: Englewood Cliffs, New Jersey, Prentice-Hall, p. 535–558.

Huffman, O. F., 1972, Lateral displacement of upper Miocene rocks and the Neogene history of offset along the San Andreas fault in central California: Geological Society of America Bulletin, v. 83, p. 2913–2946.

Johnson, C. E., and many others, 1982, The Imperial Valley earthquake of October 15, 1979: U.S. Geological Survey Professional Paper 1254, 451 p.

Jones, D. L., Cox, A., Coney, P., and Beck, M., 1982, The growth of western North America: Scientific American, v. 247, p. 70–83.

Karlstrom, K. E., and Houston, R. S., 1984, The Cheyenne belt; Analysis of a Proterozoic suture in southern Wyoming: Precambrian Research, v. 25, p. 415–446.

Kelsey, H. M., and Carver, G. A., 1988, Late Neogene and Quaternary tectonics associated with northward growth of the San Andreas transform fault, northern California: Journal of Geophysical Research, v. 93, no. 85, p. 4797–4819.

Kleinspehn, K. L., 1985, Cretaceous sedimentation and tectonics, Tyaughton-Methow basin, southwestern British Columbia: Canadian Journal of Earth Sciences, v. 22, no. 2, p. 154–174.

Lawrence, R. D., 1976, Strike-slip faulting terminates the Basin and Range Province in Oregon: Geological Society of America Bulletin, v. 87, p. 846–850.

Link, M. H., Squires, R. L., and Colburn, I. P., 1984, Slope and deep-sea fan facies and paleogeography of Upper Cretaceous Chatsworth Formation, Simi Hills, California: American Association of Petroleum Geologists Bulletin, v. 68, p. 850–873.

Luyendyk, B. P., Kamerling, M. J., Terres, R. R., and Hornafius, J. S., 1985, Simple shear of southern California during Neogene time suggested by paleomagnetic declinations: Journal of Geophysical Research, v. 90, p. 12454–12466.

Matthews, V., III, 1976, Correlation of Pinnacles and Neenach volcanic formations and their bearing on San Andreas fault problem: American Association Petroleum Geologist Bulletin, v. 60, p. 2128–2141.

Mattinson, J. M., and James, E. W., 1985, Salinian block U/Pb age and isotopic variations; implications for origin and emplacement of the Salinian terrane, *in* Howell, D. G., ed., Tectonostratigraphic terranes of the Circum-Pacific region: Houston, Texas, Circum-Pacific Council for Energy and Mineral Resources Earth Science Series, no. 1, p. 215–226.

McCulloch, D. S., 1987, Regional geology and hydrocarbon potential of offshore central California, *in* Scholl, D. W., Grantz, A., and Vedder, J. G., eds., Geology and resource potential of the continental margin of western North America and adjacent ocean basins; Beaufort Sea to Baja California: Houston, Texas, Circum-Pacific Council for Energy and Mineral Resources Earth Science Series, v. 6, p. 353–401.

Nilsen, T. H., and Clarke, S. H., 1975, Sedimentation and tectonics in the early Tertiary continental borderland of central California: U.S. Geological Survey Professional Paper 925, 62 p.

Noble, L. F., 1954, The San Andreas fault zone from Soledad Pass to Cajon Pass, California: California Division of Mines Bulletin 170, chapter 4, p. 37–48.

Page, B. M., 1982, Migration of Salinian composite block, California, and disappearance of fragments: American Journal of Science, v. 282, p. 1694–1734.

Page, B. M., and Engebretson, D. C., 1984, Correlation between the geologic record and computed plate motions for central California: Tectonics, v. 3, p. 133–155.

Platt, J. P., and Stuart, C. J., 1974, Discussion *on* 'Newport-Inglewood fault zone, Los Angeles basin, California': American Association of Petroleum Geologists Bulletin, v. 58, p. 877–898.

Reynolds, M. W., 1979, Character and extent of Basin-Range faulting, western Montana and east-central Idaho, *in* Newman, G. W., and Goode, H. D., eds., Basin and Range Symposium and Great Basin Field Conference: Rocky Mountain Association of Geologists and Utah Geological Association, p. 185–193.

Ross, D. C., 1984, Possible correlations of basement rocks across the San Andreas, San Gregorio–Hosgri, and Rinconada-Reliz–King City faults, California: U.S. Geological Survey Professional Paper 1317, 37 p.

Saleeby, J., 1981, Ocean floor accretion and volcanoplutonic arc evolution of the Mesozoic Sierra Nevada, *in* Ernst, W. G., ed., The geotectonic development of California: Englewood Cliffs, New Jersey, Prentice-Hall, p. 132–181.

Schweickert, R. A., 1981, Tectonic evolution of the Sierra Nevada Range, *in* Ernst, W. G., ed., The geotectonic development of California: Englewood Cliffs, New Jersey, Prentice-Hall, p. 87–131.

Schweickert, R. A., Bogen, N. L., Girty, G. H., Hanson, R. E., and Merguerian, C., 1984, Timing and structural expression of the Nevadan orogeny, Sierra Nevada, California: Geological Society of America Bulletin, v. 95, p. 967–979.

Sharp, R. V., 1967, San Jacinto fault zone in the Peninsular Ranges of southern California: Geological Society of America Bulletin, v. 78, p. 705–730.

Sharp, R. V., and others, 1972, The Borrego Mountain earthquake of April 9, 1968: U.S. Geological Survey Professional Paper 787, 207 p.

Shoemaker, E. M., Squires, R. L., and Abrams, M. J., 1978, Bright Angel and Mesa Butte fault systems of northern Arizona, *in* Smith, R. B., and Eaton, G. P., eds., Cenozoic tectonics and regional geophysics of the western Cordillera: Geological Society of America Memoir 152, p. 341–365.

Sieh, K. E., 1984, Lateral offsets and revised dates of large pre-historic earthquakes at Pallett Creek, southern California: Journal of Geophysical Research, v. 89, p. 7641–7670.

Sieh, K. E., and Jahns, R. H., 1984, Holocene activity of the San Andreas fault at Wallace Creek, California: Geological Society of America Bulletin, v. 95, p. 883–896.

Slemmons, D. B., Van Wormer, D., Bell, E. J., and Silberman, M. L., 1979, Recent crustal movements in the Sierra Nevada–Walker Lane region of California-Nevada; Part 1, Rate and style of deformation: Tectonophysics, v. 52, p. 561–570.

Smith, G. I., 1962, Large lateral displacement on the Garlock fault, California, as measured from offset dike swarm: American Association of Petroleum Geologists Bulletin, v. 46, p. 85–104.

Snavely, P. D., Jr., 1987, Tertiary geologic framework, neotectonics, and petroleum potential of the Oregon-Washington continental margin, *in* Scholl, D. W., Grantz, A., and Vedder, J. G., Jr., eds., Geology and resource potential of the continental margin of western North America and adjacent ocean basins; Beaufort Sea to Baja California: Houston, Texas, Circum-Pacific Council for Energy and Mineral Resources Earth Science Series, v. 6, p. 305–335.

Stanley, R. G., 1987, New estimates of displacement along the San Andreas fault in central California based on paleobathymetry and paleogeography: Geology, v. 15, p. 171–174.

Stewart, J. H., 1988, Tectonics of the Walker Lane belt, western Great Basin; Mesozoic and Cenozoic deformation in a zone of shear, *in* Ernst, W. G., ed., Metamorphism and crustal evolution of the western United States, Ruby Volume 7: Englewood Cliffs, New Jersey, Prentice-Hall, p. 683–713.

Stewart, J. H., Albers, J. P., and Poole, F. G., 1968, Summary of regional evidence for right-lateral displacement in the western Great Basin: Geological Society of America Bulletin, v. 79, p. 1407–1413.

Tabor, R. W., Frizzell, V. A., Jr., Vance, J. A., and Naeser, C. W., 1984, Ages and stratigraphy of lower and middle Tertiary sedimentary and volcanic rocks of the central Cascades, Washington; Application to the tectonic history of the Straight Creek fault: Geological Society of America Bulletin, v. 95, p. 26–44.

Tweto, O., and Sims, P. K., 1963, Precambrian ancestry of the Colorado mineral belt: Geological Society of America Bulletin, v. 74, p. 991–1014.

Vedder, J. G., 1987, Regional geology and petroleum potential of the southern

California borderland, *in* Scholl, D. W., Grantz, A., and Vedder, J. G., eds., Geology and resource potential of the continental margin of western North America and adjacent ocean basins; Beaufort Sea to Baja California: Houston, Texas, Circum-Pacific Council for Energy and Mineral Resources Earth Science Series, v. 6, p. 403–447.

Warner, L. A., 1978, The Colorado lineament; A middle Precambrian wrench fault system: Geological Society of America Bulletin, v. 89, p. 161–171.

Weldon, R. J., II, and Sieh, K. E., 1985, Holocene rate of slip and tentative recurrence interval for large earthquakes on the San Andreas fault, Cajon Pass, southern California: Geological Society of America Bulletin, v. 96, p. 793–812.

Wells, R. E., and Coe, R. S., 1985, Paleomagnetism and geology of Eocene volcanic rocks of southwest Washington; Implications for mechanisms of tectonic rotation: Journal of Geophysical Research, v. 90, no. B2, p. 1925–1947.

Wells, R. E., Engebretson, D. C., Snavely, P. D., Jr., and Coe, R. S., 1984, Cenozoic plate motions and the volcano-tectonic evolution of western Oregon and Washington: Tectonics, v. 3, p. 275–294.

Wernicke, B., Spencer, J., Burchfiel, B. C., and Guth, P. L., 1982, Magnitude of crustal extension in the southern Great Basin: Geology, v. 10, p. 499–502.

Wernicke, B., Axen, G. J., and Snow, J. K., 1988, Basin and Range extension tectonics at the latitude of Las Vegas, Nevada: Geological Society of America Bulletin, v. 100, p. 1738–1757.

Wright, L. A., 1976, Late Cenozoic fault patterns and stress fields in the Great Basin and westerrn displacement of the Sierra Nevada block: Geology, v. 4, p. 489–494.

Yates, R. G., 1968, The Trans-Idaho discontinuity: 23rd International Geological Congress, v. 1, p. 117–123.

Yeats, R. S., 1973, Newport-Inglewood fault zone, Los Angeles basin, California: American Association of Petroleum Geologists Bulletin, v. 57, p. 117–135.

Zoback, M. L., 1989, State of stress and modern deformation of the northern Basin and Range province: Journal of Geophysical Research, v. 94, p. 7105–7128.

Manuscript Accepted by the Society May 17, 1989

ACKNOWLEDGMENTS

We wish to thank M. C. Blake, Jr., S. H. Clarke, Jr., D. S. McCulloch, R. J. McLaughlin, P. D. Snavely Jr., R. W. Tabor, J. G. Vedder, R. E. Wallace, R. J. Weldon, and R. E. Wells for their help.

Printed in U.S.A.

The Geology of North America
Vol. G-3, The Cordilleran Orogen: Conterminous U.S.
The Geological Society of America, 1992

Chapter 15

Metallogenic evolution of the Cordilleran region of the western United States

Richard W. Hutchinson
Department of Geology and Geological Engineering, Colorado School of Mines, Golden, Colorado 80401
John P. Albers
U.S. Geological Survey, 345 Middlefield Road, Menlo Park, California 94025

INTRODUCTION

Metallogenic evolution of a region is related to evolving stratigraphic, structural, and magmatic processes through time, and all these are controlled by the region's tectonic evolution. This chapter considers the broad evolutionary litho- and tectono-stratigraphic relations that control the distribution of major types of mineral deposits throughout the western United States. By this approach, ore deposits—in addition to their economic significance—become valuable guides to paleotectonic processes. The approach is intended to provide different insights from those based on detailed studies of individual ore deposits or types. Two maps (Figs. 1, 2) outline the distribution of deposits of widely differing ages and types within rocks belonging to ten successive time intervals, but in the discussion the latter are grouped into seven major successive and evolutionary tectonic intervals: >1.65 Ga, 1.65 Ga to 400 Ma, 400 to 208 Ma, 208 to 80 Ma, 80 to 40 Ma, 40 to 22 Ma, and 22 to 0 Ma. In combination, the two maps show the distribution of rocks of the ten time intervals; Figure 1 of the older basement rocks with the younger overlying ones of Figure 2 stripped away. The locations of major deposits or types of deposits discussed in the text are also plotted on the two figures, but the scale prevents plotting of all examples mentioned. Some deposits appear on both figures because metallogenic relations suggest that they have been formed by dual, overlapping processes of differing ages. The deposits plotted are discussed to illustrate key metallogenic relations, but the map scale and page constraints prevent plotting and discussion of all significant deposits. Metallogenic relations are considered mainly for major deposit types and on a regional scale. Possible broad relations between diverse types and ages of deposits are also considered. But complex local metallogenic relations and unresolved genetic problems cannot be treated comprehensively. Though the expanse and diversity of deposits also preclude detailed descriptions, definitive publications are referenced where these problems and descriptions are outlined.

DEPOSITS FORMED DURING ESTABLISHMENT OF THE CRATON, FROM ARCHEAN TO MID-PROTEROZOIC TIME: >1.65 Ga

In the Cordilleran region of the western United States, establishment of cratonic greenstone-granitic gneiss terranes continued from Archean to mid-Proterozoic time until about 1.65 Ga. At that time, major volcanism ceased in the southern Cordillera and was followed by widespread granitic plutonism (Condie, 1982a, b). Although Condie attributed the volcanism to plate-tectonic subduction, its regional extent throughout much of North America and in other shield areas of the world raises doubts about this interpretation. In any case, significant metalliferous deposits, including Algoma-type iron formations, volcanogenic massive base-metal sulfides, gold lodes, and magmatic cobalt-nickel-platinum ores occur in rocks of Archean and Early Proterozoic age in widely separate parts of the Cordillera. These may be separated into two time-space groups.

Late Archean deposits: >2.5 Ga

Deposits of this age lie north of the Wyoming shear zone (Condie, 1982a) in Wyoming, Montana, and South Dakota, although this boundary also extends westward across northern Utah (Bryant, 1988) where, however, no deposits of Archean age are known. The most important are gold deposits, hosted in various rocks of volcano-sedimentary or greenstone belts. They include Homestake, South Dakota (Noble, 1950; Rye and Rye, 1974), the similar but much smaller Jardine deposit, Montana (Hallager, 1983), and occurrences in the South Pass area, Wyoming (Bayley and others, 1973). Algoma-type iron formations (Gross, 1965) occur at Atlantic City, Wyoming, and in the Dillon region, Montana (Bayley and James, 1973; Fig. 1). Also important, though not yet significantly productive, are layered chromite and platinum deposits of the Stillwater Complex, Montana (Raedeke, 1983). The rare-element pegmatites in the Black Hills,

Hutchinson, R. W., and Albers, J. P., 1992, Metallogenic evolution of the Cordilleran region of the western United States, *in* Burchfiel, B. C., Lipman, P. W., and Zoback, M. L., eds., The Cordilleran Orogen: Conterminous U.S.: Boulder, Colorado, Geological Society of America, The Geology of North America, v. G-3.

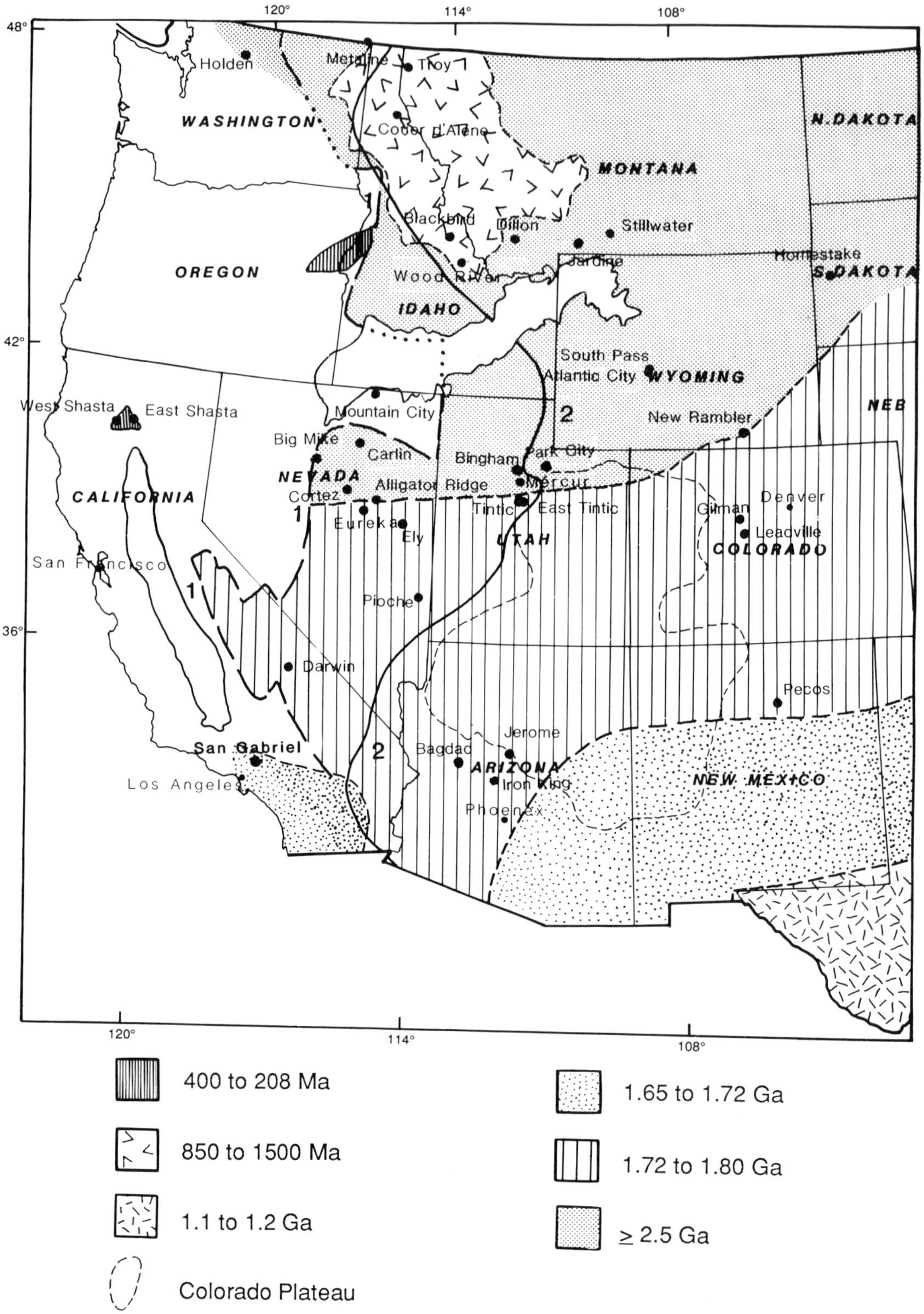

Figure 1. Ages of Archean to uppermost Triassic basement rocks in the U.S. Cordillera. In places these basement rocks are covered by younger strata. Precambrian age boundaries after Condie (1982a). Craton boundary defined by strontium 0.706 line. Dashed line (labeled 1) shows western edge of craton; solid line (labeled 2) marks hinge between Paleozoic miogeoclinal slope-facies rocks to the west and carbonate platformal rocks to the east. Deposits shown are discussed in text.

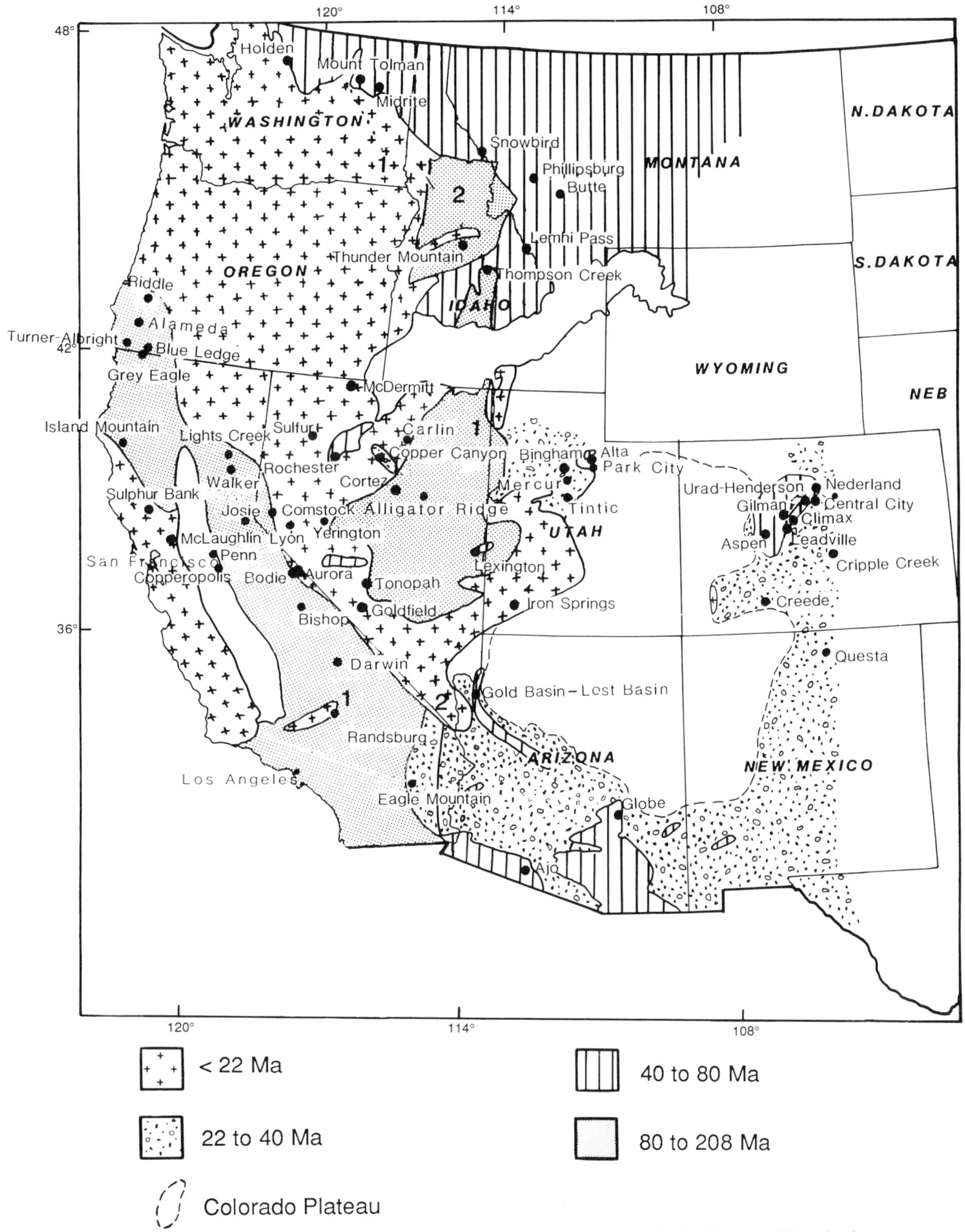

Figure 2. Ages of rocks with corresponding patterns of mineralization in the U.S. Cordillera in time slices from beginning of Jurassic (208 Ma) to present. Dashed line (labeled 1) shows western edge of craton; solid line (labeled 2) marks hinge between Paleozoic miogeoclinal slope-facies rocks to the west and platformal carbonate rocks to the east. Deposits shown are discussed in text.

South Dakota, also belong within the Wyoming Province and apparently formed during major orogeny and granitic plutonism that deformed and invaded the greenstone belts at about 2.5 Ga. Base-metal mineralization is minor (Hausel and others, 1985).

Early Proterozoic deposits: 2.5 to 1.65 Ga

Early Proterozoic deposits in southeastern Wyoming, Colorado, New Mexico, and Arizona are mostly associated with volcanic rocks dated at 1.8 to 1.72 Ga (Fig. 1), which were intruded by granite at 1.5 to 1.4 Ga (Condie, 1981, 1982b). Volcanogenic massive base-metal sulfide deposits are economically the most important. They include the United Verde and Verde Extension deposits at Jerome, Arizona (Anderson and Nash, 1972), the Iron King and others near Prescott, Arizona (Gilmour and Still, 1968), those of the Bagdad district, Arizona (Baker and Clayton, 1968), the Pecos greenstone belt, New Mexico (Riesmeyer and Robertson, 1979; Sumner, 1983), and Chaffee and Fremont Counties, Colorado (Raymond and Sheridan, 1980). All are past producers of copper and zinc with significant by-product gold, silver, and rarely, lead. Unlike the Archean terrane farther north, gold lodes (Lakes, 1906; Horlacher, 1987) and iron formations (Bayley and James, 1973) in this region are small and have not been significant producers.

Comparative metallogeny of Archean and Early Proterozoic terranes

Massive sulfide and gold ores are commonly coregional and broadly contemporaneous in Archean belts and in some Early Proterozoic and Phanerozoic sequences (Hutchinson, 1981a, b)—a repeated spatial and temporal association that suggests a common genetic link. The rarity of significant gold deposits in other lower Proterozoic greenstone belts is a metallogenic enigma. This unexplained contrast, noted in the Canadian Shield (Hutchinson, 1982; Hutchinson and Burlington, 1984), is apparently also true in the Archean and Proterozoic terranes in the western Cordillera.

Notwithstanding their differing ages and their differing abundances in the older and younger greenstone belts, the Archean and Early Proterozoic gold and massive base-metal deposits of the U.S. Cordilleran region are geologically comparable. All occur in supracrustal greenstone belts or volcanosedimentary successions within greenstone-granitic gneiss terranes. Geologically and genetically they resemble those of the better exposed and structurally less complicated late Archean and Early Proterozoic Superior, Slave, and Churchill structural provinces of the Canadian Shield (Stockwell, 1982).

Thus, the Homestake and Jardine gold deposits of the Archean Wyoming Province are like those of the Lupin (formerly Contwoyto Lake) Mine (McConnell, 1964; Kerswill, 1984) in the Archean Slave Province of Canada's Northwest Territories. The iron formations of the Wyoming Province resemble those of the Sherman and Adams Mines in the Archean Superior Province of eastern Ontario (Boyum and Hartvikson, 1970; Dubuc, 1966), and the Early Proterozoic massive base-metal sulfides in Colorado, New Mexico, and Arizona are probable metamorphosed analogues of those at Flin Flon–Snow Lake in the Early Proterozoic rocks of the Churchill Province in northern Manitoba (Franklin and Thorpe, 1982).

Unusual hydrothermal copper-palladium-platinum-nickel-cobalt deposits of the New Rambler Mine (McCallum and others, 1976) occur in metamorphosed ultramafic-mafic igneous rocks of southeastern Wyoming. Their undated host rocks are similar to those of the Archean Stillwater Complex to the northwest, but lie immediately south of the Wyoming shear zone. Although they seem therefore to be of Early Proterozoic age, they might be Archean, perhaps like Homestake to the north, but affected by the pervasive metamorphism of the more southerly region. Other similar nickel-copper-cobalt occurrences are also present in the Colorado Rockies (Goddard and Lovering, 1941).

The rare-element pegmatites of Gunnison County, Colorado, were probably formed during major orogeny and granitic plutonism that affected the rocks of this region at about 1.7 to 1.5 Ga.

DEPOSITS FORMED AT A PASSIVE RIFTED MARGIN FROM MID-PROTEROZOIC TO EARLY DEVONIAN TIME: 1.65 Ga TO 400 Ma

Following establishment of the North American craton by 1.65 Ga, important and diverse metalliferous deposits were formed in sedimentary rocks deposited thereafter until about 400 Ma, along what was apparently a passive, extensional western cratonic margin. Although covered in many areas by younger strata, intruded by younger batholiths and in places complexly deformed by later tectonism, these rocks and their deposits are widespread throughout the northwestern Cordillera of the United States where they can be divided into two time-space groups.

Mid-to-Late Proterozoic deposits: 1.65 Ga to 750± Ma

The oldest and most important deposits of the passive margin are the sediment-hosted exhalative copper-cobalt-silver-lead-zinc ores in fine-grained, clastic sedimentary rocks of the Belt Supergroup in Idaho and Montana. Based on metal content, these appear to be of two subtypes.

The first, cobalt-copper-silver–rich, is represented by the deposits of the Blackbird district, Idaho (Hughes, 1983), the Troy deposit in northwestern Montana (Fig. 1; Balla, 1983), and the Sheep Creek occurrence of central Montana (Hutchinson, 1988). These occurrences are in the southern, northwestern, and eastern margins of the Belt basin, respectively. The basin's tectonic evolution was controlled by the Lewis and Clarke line, or Coeur d'Alene trough (Harrison and others, 1974; Harrison, 1972), the Purcell Trench, and a probable southward extension of the more northerly Rocky Mountain Trench (Hutchinson, 1988; Kanasewich, 1968). These regional lineaments apparently represent extensional basin-margin or deep basinal faults that formed an aulacogenic depository and controlled the basin's stratigraphy, sedimentation, and metal distribution. The Blackbird deposits in rocks of the Yellowjacket Formation, formerly considered the lowermost unit of the Belt Supergroup, are associated with mafic tuffs, products of volcanism that accompanied earliest sedimentation in the Belt basin. Whether and how the volcanism influenced their high cobalt-copper content are not known.

Deposits of the second subtype are lead-zinc-silver–rich and mainly in the northwestern parts of the Belt basin. The Sullivan orebody in southern British Columbia (Hamilton and others, 1982) is the prime example, but stratiform lead-zinc mineralization also occurs in the Burke, Revett, and St. Regis Formations in the Coeur d'Alene district, Idaho (Harrison, 1972). The host rocks are mainly turbidites, apparently deposited in deep intracratonic troughs or trenches along the extended western margin of the craton. The ores are believed to have formed as primary, metal-rich chemical precipitates from sea-floor hydrothermal discharge that accompanied the rifting (Hughes, 1983; Hamilton and others, 1982). These deposits were generated by the same process as the earlier base-metal sulfides of the greenstone belts, but are tabular and stratiform rather than lenticular and stratabound, and sediment hosted rather than volcanic hosted. They are characteristic of mid-Proterozoic rocks of similar lithofacies throughout the world, suggesting a period of global extension and expansion (Hutchinson, 1980). Examples are the deposits of Broken Hill, Hilton, Mt. Isa, and McArthur River in Australia, Rajasthan in India, and Ammeberg in Sweden. Like their comparable, earlier Precambrian volcanic-hosted antecedents, ores of this type may be enriched local examples of more widespread but weakly metalliferous strata that were favorable source beds for subsequent metal remobilization and redistribution.

The titaniferous magnetite occurrences of the San Gabriel anorthosite in southern California and of the Laramie anorthosite in southeastern Wyoming (Fig. 1; Higgs, 1954; Hagner, 1968) are metallogenically comparable to the titaniferous iron ores of the Grenville Province in northeastern North America (Gross, 1968) and Norway (Frietsch and others, 1979). Where dated, their associated anorthosites are 1.5 to 1.4 Ga but all are younger than about 1.7 and older than about 1.0 Ga. Although controversial, a rift-, or failed rift–related, anorogenic tectonic environment is favored for the anorthositic petrogenesis (Emslie, 1978; Moore, 1982). This suggests an extensional environment for the vast belt of distinctive anorthositic intrusions and their related Fe-Ti ores, and accords with a passive extensional margin to the west.

Latest Proterozoic and early Paleozoic deposits: 750± to 400 Ma

Mid-Proterozoic deep-water clastic sedimentation in the Belt Basin was superseded in Late Proterozoic time by deposition of shallower water clastic strata of the Windermere Formation, shallow clastic-carbonate strata of the Creston Formation, and finally in early Paleozoic time, by very shallow water, miogeoclinal clastic-calcareous sedimentation. This may reflect infilling of the deep aulacogenic, mid-Proterozoic rifted basin and resulting westward migration of the late Precambrian continental margin. The miogeoclinal strata were flanked farther east by platformal carbonates (Figs. 1, 2).

The resulting Cambro-Ordovician dolostones of the eastern miogeoclinal assemblage (Roberts and others, 1958; Silberling and Roberts, 1962) contain the lead-zinc deposits of the Van Stone Mine and Metaline district in northeastern Washington, in the adjacent Salmo district of southern British Columbia, at Pioche and Eureka, Nevada, and at East Tintic, Utah (Cox, 1968; McConnell and Anderson, 1968; Fyles and Hewlett, 1959; Gemmill, 1968; Nolan and Hunt, 1968; Shepard and others, 1968). Although hosted in miogeoclinal rocks, these deposits lie close to the seaward margin of the carbonate platform (Fig. 1).

Van Stone and Metaline (Morris and others, 1973) have been compared to Mississippi Valley–type ores, but high precious-metal content at Pioche, Eureka, and East Tintic is unlike that of most Mississippi Valley–type deposits. The Metaline deposits appear mainly epigenetic, although some of the sulfides are remarkably stratabound (McConnell and Anderson, 1968), suggesting possible synsedimentary or early diagenetic origin. The Metaline-Salmo deposits have been folded but not intruded. A possibly significant difference is the presence at Pioche, Eureka, and East Tintic, but not at Metaline, Van Stone, or Salmo, of extensive Cretaceous or Tertiary extrusive, pyroclastic, and subvolcanic porphyritic intrusive rocks. The prevalent genetic interpretation is that this magmatism generated the metalliferous hydrothermal fluids that formed the ores. Considering the similarity of Van Stone, Metaline, and Salmo to Mississippi Valley–type deposits and the absence of magmatism in these, it is possible that hydrothermal activity related to this magmatism at Pioche, Eureka, and East Tintic added precious metals to, and remobilized early Paleozoic lead-zinc mineralization in the miogeoclinal Cambro-Ordovician rocks. Remobilization and hydrothermal additions (McConnell and Anderson, 1968) could explain the many epigenetic characteristics of these deposits as well as their differences, both from one another and from the essentially undeformed and unintruded Mississippi Valley–type ores in undisturbed and unintruded platformal strata of the midcontinent.

If these lead-zinc ores were indeed penecontemporaneous with their host rocks, the source of their generative hydrothermal fluids is unknown. But basinal dewatering due to subsidence and burial of older, deep-water siliciclastic Beltian sediments may have generated such fluids for the Van Stone–Metaline–Salmo deposits, although not for the more southerly and easterly deposits of Nevada and Utah. A similar process has been suggested (Beales and Jackson, 1966), for some Mississippi Valley–type ores. Such fluids, generated and heated by burial and lithification of older, weakly metalliferous strata, and carrying metals derived by fluid-rock reactions in them, may have been forced upward by compactional lithostatic pressure into the younger miogeoclinal sequences to form these ores, either syndepositionally or diagenetically. Alternatively, these metals might have been derived from weathering of the same older metalliferous Beltian rocks on flanking terrane, with subsequent reprecipitation in the miogeoclinal carbonates due to reduction. Although disconformities and local unconformities exist, no major angular unconformity is known between the Proterozoic source strata and the ore-bearing early Paleozoic carbonates to suggest an intervening period of weathering and erosion.

A common heritage is suggested for the metals in older clastic-hosted and younger veins, shear zones, or carbonate-hosted ores in the Wood River district of south-central Idaho (Hall and others, 1978). Here, lead-zinc-silver sulfides occur in shear zones near an 82-Ma granodiorite pluton. The deposits are in hornfelsed Devonian argillaceous rocks beneath a regional thrust fault. All isotopic, fluid-inclusion, and geological data are consistent with circulation of hydrothermal meteoric water through faulted and shattered rocks near the pluton. Metals and sulfur dissolved from the country rocks were redeposited in favorable beds beneath the thrust. Although these relatively small deposits are attributed to the small size of this particular system, the generally overlooked principle of metal inheritance and generation of younger deposits by reworking of older deposits or rocks may be more important in metallogenesis than is generally recognized. Similar principles and relations have recently been suggested to explain metalogenic evolution of the coregional, but very different ages of Pb-Zn-Ag deposits in the northwestern United States (Hutchinson, 1988).

The Mountain City copper deposit in northeastern Nevada (Coats and Stephens, 1968) is apparently a cupreous pyrite body formed in oceanic crust at a spreading rift, or in a back-arc basin of Ordovician age, but emplaced during obduction of the Roberts Mountain allochthon in Late Devonian or Early Mississippian time. If so, it reflects either an opening Ordovician spreading axis in the paleo-Pacific, west of the North American continental margin, or an off-shore island arc and back-arc basin over an oceanic subduction zone. In either case, as a tension-generated deposit it is confirmatory of a passive, rifted, and extensional continental margin in Ordovician time.

Stratiform, bedded barite is widespread in interbedded fine clastic and cherty strata that belong to a detrital-volcanic western assemblage (Roberts and others, 1958; Silberling and Roberts, 1962) as in the Devonian Slaven Chert and Ordovician Valmy and Vinini Formations of Nevada. Although thrust over miogeoclinal carbonate rocks of an eastern assemblage during the early Mississippian Antler orogeny, the former were originally deposited in a deeper eugeoclinal environment flanking the miogeocline to the west. These barite deposits are probably synsedimentary exhalites and, like their associated chert, may be products of sea-floor hydrothermal activity. They contain minor lead-zinc (Walker, 1985), suggesting, in view of the common association of bedded barite with sediment-hosted lead-zinc ores, a genetic relation with the lead-zinc ores in the Cambro-Ordovician carbonate rocks of the miogeoclinal assemblage. Perhaps the extension along the early Paleozoic continental margin involved rifting along the breaks in slope between the easternmost platformal carbonate and more westerly miogeoclinal sequences, and between the latter and the most westerly eugeoclinal sequences. This might have provided channelways for movement to the sea floor of barium-lead-zinc–rich brines, derived from either sub-sea-floor hydrothermal convection or dewatering of deeper, buried strata. Here the baritic strata were deposited in the deep marine shale-chert environment, and lead-zinc ores in the shallower miogeoclinal and platformal carbonate strata farther east.

DEPOSITS FORMED AT AN ACTIVE COMPRESSIONAL, LOCALLY PASSIVE MARGIN, MID-DEVONIAN TO EARLY JURASSIC TIME: 400 Ma TO 208 Ma

By mid-Devonian time, the prolonged extensional tectonism along the western continental margin gave way to a margin that was still, at times and locally, passive but at other times and elsewhere was characterized by subduction-related compression of a consuming plate boundary (Miller and others, this volume). The metallogenic results of this change are, like the tectonic environments, complex, varied both from place to place and through time, and the original distribution of the various deposits has been drastically revised by later thrusting. The deposits are considered here in a time-space sequence from oldest to youngest and west to east.

The massive sulfide deposits of West Shasta, California, occur in subaqueous quartz keratophyres of Middle Devonian age (Albers and others, 1981; Albers and Bain, 1985) that are the upper part of a compositionally varied mafic-felsic volcanic succession. These rocks represent an island arc formed on oceanic lithosphere, presumably over the subducted oceanic plate outboard of the newly activated, compressional continental margin. The West Shasta deposits are pyritic zinc-copper-gold-silver ores, distinctly poor in lead. They are a primitive type of exhalative volcanogenic massive base-metal sulfide (Hutchinson, 1973, 1980; Doe and others, 1985) similar to those in Archean rocks, in rocks generated during subduction beneath oceanic island arcs and along early Paleozoic consuming plate margins. Bedded barite in the Devonian Kennett Formation of northern California may be a distal-lateral lithofacies equivalent of the West Shasta sulfides.

Mid- to late Paleozoic eugeoclinal rocks of northern Nevada may have been deposited in a deep back-arc basin, flanking the West Shasta volcanic arc on its east along the continental margin. Key relations are obscured by younger volcanic rocks of the Cascade Range, by sedimentary strata of the Great Valley sequence, or by major Sierra Nevadan batholiths. Alternatively, some paleomagnetic and tectonic evidence suggests that the Nevada rocks were formed at far-distant paleo-Pacific locations and subsequently accreted to the continental margin. But in either case they were subsequently thrust eastward during the Antler and Sonoman orogenies in Devono-Mississippian and Permo-Triassic time, by compression along the newly evolving collisional plate boundary. This complicates the original stratigraphic relations throughout the region. These rocks include eugeoclinal chert, shale, argillite, and tholeiitic basalt of the Vinini-Valmy and Havallah-Pumpernickel sequences in northern Nevada that belong to the upper plates of the Roberts Mountains and Golconda allochthons respectively. These tectonically overlie lower-plate rocks that are partly transitional slope facies and partly shelf facies (Roberts and others, 1958), including limy argillite and impure magnesian-ferroan carbonate. Although these strata in part may have been deposited off the early Paleozoic continental margin, the presence of the mid-Devonian volcanic island arc-generated rocks and volcanogenic massive base-metal sulfide ores of West Shasta suggests that at least the upper Paleozoic strata were deposited in a deep back-arc basin between an offshore volcanic arc and the continent. The older Vinini-Valmy strata may have been deposited, before appearance of this arc, on the continental slope and flanking ocean floor between the continental edge and an Ordovician paleo-Pacific spreading axis where the older ophiolitic rocks and cupreous pyrite ores of Mountain City, Nevada, were formed.

Like the older Vinini-Valmy strata, the upper Paleozoic eugeoclinal rocks contain important mineral deposits, and although the origin of some is controversial, others and perhaps all might have formed in back-arc basinal environments. The Big Mike cupreous pyrite body in northern Nevada occurs in Pennsylvanian-Permian eugeoclinal pillowed basalt and carbonaceous chert-argillite of the Havallah sequence (Fig. 1; Rye and others, 1984). This deposit may be of either the cupreous pyrite or the Besshi types (Hutchinson, 1980), both of which may form along subductive margins in deep, local, rift-controlled back-arc basins, between off-shore island arcs and continental shorelines.

The eugeoclinal Devono-Mississippian rocks of central Nevada, like the similar Cambro-Ordovician strata, also contain bedded barite, as at East Northumberland Canyon, Nevada. These too are apparently chemical sediments from sea-floor hydrothermal activity. They occur characteristically in fine, siliciclastic, eugeoclinal shale-chert sequences that are similar to the Ordovician Vinini-Valmy rocks of the extensional, passive margin. The mid-Paleozoic barite was perhaps deposited in a more restricted extensional back-arc basin in comparison to the lower Paleozoic setting, but still represents a comparable tectonostratigraphic habitat. The mid-Paleozoic rocks have been thrust eastward to form an upper plate overlying correlative strata of an originally more easterly, shallower water-deposited miogeoclinal slope and shelf assemblage.

The miogeoclinal lower-plate rocks host the gold deposits of Alligator Ridge, Carlin, Dee, Getchell, Gold Acres, Jerritt Canyon, and Maggie Creek in north-central Nevada, and Mercur, in Utah. The Nevada deposits are all in autochthonous rocks beneath the Roberts Mountains allochthon near the craton boundary (Fig. 1; Tooker, 1985), whereas Mercur is in the innermost part of the Paleozoic miogeocline, near the platformal hinge line, and almost astride the basement boundary between Archean and Proterozoic rocks.

The deposits are characterized by micron-sized gold particles disseminated in impure carbonate and carbonaceous sedimentary rocks of Cambrian to Mississippian age. Typically stratiform, they contain 0.5 to 3.0 million oz and grade from 0.1 to 0.24 oz gold per ton. Although in a region of extensive Mesozoic-Tertiary magmatism, igneous rocks in the deposits are minor to absent. The host strata are depleted in carbon and carbonate, but enriched in clay minerals and silica.

These deposits are generally considered to be hydrothermal replacements, but a genetic tie to regional magmatism is questionable. Rye (1985) postulated a shallow meteoric water-dominated hydrothermal system activated by igneous activity, and that the ore components, all mobile in shallow hydrothermal systems, were derived from Paleozoic sedimentary host rocks. The deposits contrast markedly with the epithermal hot spring and vein gold deposits in Tertiary subaerial volcanic rocks. These relations suggest the possibility that the sediment-hosted deposits formed contemporaneously with their miogeoclinal host rocks in the Paleozoic extensional environment along the craton boundary in north-central Nevada, slightly inboard of the craton margin (Fig. 1; Tooker, 1985). Here gold, leached either from older crustal rocks by deep hydrothermal convection, or from a subcrustal source, may have been discharged into the miogeocline, to form auriferous, carbonate-rich sediments, either during deposition or diagenetically. The gold was perhaps subsequently remobilized, redistributed, and overprinted by widespread Tertiary subaerial volcanism and hydrothermal activity throughout the Great Basin, to form the younger, volcanic-hosted, epithermal ores.

Moreover, the Nevada deposits resemble early Paleozoic greywacke-hosted gold deposits as at Muruntau in the Uzbek Republic of the Soviet Union, Ballarat-Bendigo in Victoria, Australia, and Nova Scotia, Canada. Deposits of this general age and type constitute a distinct metallogenic group (Keppie and others, 1986; Haynes, 1986; Hutchinson, 1987). They resemble the older greenstone belt lodes, of which they may be evolutionary successors, but occur mainly in lower Paleozoic fine clastic sedimentary successions of slope- or trough-deposited greywacke, argillite, or slate with some chert, impure deep water carbonate, and fewer mafic volcanic rocks than their greenstone belt predecessors.

A cratonal facies of mid- to late Paleozoic carbonate rocks in eastern Utah and Colorado lies east of the deeper miogeoclinal assemblages, and extends southwestward through New Mexico, Arizona, and southeastern California. These rocks, probably shelf-deposited facies equivalents of the deeper miogeoclinal rocks, host the important lead-zinc-silver (copper-gold) ores of Leadville and Gilman, Colorado, and probably, those of Tintic and Park City, Utah, which are closer to the platform-miogeocline boundary. The nonporphyry ores of Bingham, Utah, Santa Rita, New Mexico, the Magma and Christmas mines, Arizona, and Darwin, California, are also in late Paleozoic, dominantly carbonate rocks that are either platformal or miogeoclinal but near the platform boundary. All these districts have undergone Mesozoic or Tertiary porphyritic subvolcanic intrusion or volcanism.

The origin of the Leadville ores is controversial. Thompson and others (1983) favored classic epigenetic replacement, relating all mineralization to Tertiary intrusive activity. Devoto (1983) suggested mineralization during Late Mississippian emergence, karstification, and erosion, with Tertiary plutonism remobilizing and adding precious metals to the older lead-zinc ores. Considering the time-stratigraphic restriction of both the Cambro-Ordovician and mid- to late Paleozoic carbonate-hosted ores and their high precious-metal content (where associated with later magmatism) in comparison to the otherwise similar Mississippi Valley–type ores in undeformed and unintruded platformal carbonate rocks of the same general ages in the midcontinent, the addition of precious metals to Paleozoic carbonate-hosted lead-zinc ores by Tertiary igneous-related hydrothermal activity seems plausible. One wonders to what extent older metallization has been remobilized and redistributed by the latter.

The dual-stage genetic hypothesis demands consideration of the repeated, apparently fortuitous spatial coincidence of younger, precious-metal-contributing Tertiary magmatism with older lead-zinc-bearing Paleozoic carbonates. This coincidence may not be fortuitous. Both these mineralizing environments are near common, older, major geologic-tectonic boundaries (Fig. 1). Bingham and Park City lie along the ancient Archean-Proterozoic boundary (Condie, 1982b) and also at the Salt Lake City crossroads of Billingsley and Locke (1941). And many of the carbonate-hosted deposits lie near a flexure at the boundary between platformal and miogeoclinal facies that persisted throughout Paleozoic time. If both Paleozoic and Tertiary hydrothermal-magmatic activity were localized and controlled along these, or other unrecognized tectonic boundaries, their repeated coincidence would not be fortuitous, but is explained by tectonic reactivation (Hutchinson, 1988).

Finally, continuing subduction somewhere to the west of the continental margin generated a late Paleozoic, Pennsylvanian-Permian-Triassic volcanic arc, as evidenced by the rocks and ores of East Shasta, California, the Wallowa–Seven Devils terrane in eastern Oregon, and possibly by the highly metamorphosed (thus more problematic) deposits at Walker, California, and Holden, Washington. These rocks and ores belong to suspect terranes that have been accreted to the western continental margin during continued collisional tectonism. The location of their original generation in the paleo-Pacific is controversial, and some evidence suggests that they are far travelled. But the metallogenic relationships cited below suggest that their generative arc lay continentward from the older Devonian arc, developing in the latter's back-arc basin as this closed by continuing compression. It here constituted the latter's younger, inner, evolutionary magmatic successor, extending from northern California at least to northern Washington. It has been deformed, metamorphosed, and intruded during accretion and subsequent orogenies. In northeastern Oregon its rocks and ores have been displaced eastward by continued compression along the continental margin. Some of its deposits in East Shasta and eastern Oregon are of the Kuroko or polymetallic type, higher in lead-silver but lower in copper-gold than those of the older, Devonian (West Shasta) arc, and also unlike the latter, commonly baritic. They are hosted in shallower-water–deposited volcanic-sedimentary rocks that are typical of more evolved volcanic arcs, formed closer to, or on continental margins, perhaps by later tectonic reworking of older, more primitive oceanic arcs (Hutchinson, 1973, 1980).

The deposits of this late Paleozoic arc lie consistently east of and inboard from those of the Devonian arc, not only in California, but also in places north of the Canadian border (Hutchinson, 1988). Random later accretion of many terranes from far-distant paleo-Pacific locations seems unlikely to have brought into their present consistent positions, first the rocks and ores of a Pennsylvanian-Permian arc of continental affinity, and subsequently those of a more primitive, oceanic-affiliated Devonian arc. But accretionary growth involving generation of a younger, more evolved Pennsylvanian-Permian arc in the closed back-arc basin of an older, more primitive offshore Devonian arc appears to explain them.

DEPOSITS EMPLACED ALONG AN ACCRETIONARY MARGIN, EARLY JURASSIC TO LATE CRETACEOUS TIME: 208 Ma TO 80 Ma

General comments

During much of Mesozoic time the North American plate apparently moved southwestward relative to the Pacific plate, overriding oceanic plates east of the East Pacific Rise. During the resulting collision, fragments of oceanic island arcs and crust were swept against the continental margin, forming exotic terranes of diverse, and often enigmatic, "foreign" generation from Alaska to Peru. This accretion culminated in Jurassic-Cretaceous time, and was accompanied by intense deformation and variable metamorphism in accreted terranes, closely followed by intrusion of large Cordilleran batholiths in eastern California and Idaho. These events marked the beginning of regional, calc-alkaline arc magmatism which dominated the later metallogenic evolution of the western Cordillera of the United States. The mineral deposits emplaced or generated during this portion of Mesozoic time are diverse, but may be divided into those indigenous to the accreted terranes, and those related to Mesozoic plutonism.

Deposits indigenous to the accreted terranes

Two types of massive base-metal sulfide deposits occur, mainly in Jurassic volcanic rocks in various accreted terranes; the cupreous pyrite type in slices of oceanic crust, and the polymetallic (Kuroko) type (Hutchinson, 1980) in rocks of oceanic or continental-margin volcanic arcs. Examples of the former include the Turner-Albright deposit in the western Klamath Mountains of southern Oregon, the Island Mountain and Leona Heights deposits in the California Coast Ranges, and the Copperopolis and Newton Mines in the Sierra foothills of California (Hutchinson, 1973). Their spatial distribution suggests an early accretionary event, when oceanic crust containing the more easterly deposits of Copperopolis and the western Klamaths were emplaced, and a later event, when the similar but more westerly ophiolitic deposits of the California Coast Ranges were accreted. Also characteristic in these rocks are manganiferous cherts of coastal California (Jenkins, 1943), podiform chromitite occurrences in Washington and the Klamath Mountains (Wells and others, 1946), short-fiber asbestos at Copperopolis, California (Merritt, 1962), and the nickel laterites of Riddle, Oregon (Cumberlidge and Chace, 1968).

Examples of the polymetallic type are the Penn Mine and many occurrences of the Foothills copper-zinc belt along the western Sierra Nevada (Heyl, 1948), the Blue Ledge and Grey Eagle deposits of northern California, and the Almeda deposit in southern Oregon (Koski, 1981).

The gold deposits of the California Mother Lode belt also were probably indigenous to accreted island-arc terrane. They include both epigenetic veins and, in the south, stratiform, ankeritic gray ores that are associated with serpentinites and highly sheared, Jurassic, mafic metavolcanic rocks of the Pinon Blanco Formation (165 to 170 Ma). The veins cut highly schistose rocks of widely varying lithology and age, including upper Paleozoic Calaveras Formation, metavolcanic and metapyroclastic rocks of the Jurassic Pinon Blanco Formation, metasedimentary rocks of the Jurasic Mariposa Formation, and probably Jurassic ultramafic serpentinites. At Grass Valley–Nevada City, most veins are in a stock of the Sierran batholith.

The ankeritic ores are remarkably similar to those of Larder Lake, northeastern Ontario, which are also associated with sheared serpentinite, tholeiitic basalt, and graywacke, and have been interpreted (Ridler, 1970) as of submarine exhalative-hydrothermal origin. Their host rocks make up the lower, most primitive, undifferentiated, ultramafic-mafic, "oceanic" portion of the Blake River Group, a thick Archean volcanosedimentary succession that includes stratigraphically higher calc-alkaline and felsic volcanic rocks containing volcanogenic massive base-metal sulfide deposits (Hutchinson, 1984). Perhaps the ankeritic ores and host rocks of the Mother Lode represent an analogous lower segment of primitive, ultramafic-mafic Jurassic oceanic crust on which were laid down the stratigraphically higher calc-alkaline to felsic volcanic rocks and polymetallic massive sulfide deposits of the Foothills copper-zinc belt. This may represent a Mesozoic replica of the common coregionality and contemporaneity of these two types of deposits in Archean rocks (Hutchinson, 1984).

In the Mother Lode, both the Mesozoic and late Paleozoic rocks were deformed together during Nevadan orogeny and plutonism. Gold from primary auriferous exhalites of the Jurassic oceanic crust may have been remobilized and redistributed into epigenetic veins and stockworks that cut the intrusions, the Mesozoic source rocks, and the adjacent older rocks alike.

Deposits related to Mesozoic plutonism

The tungsten skarns at Bishop in east-central California (Bateman, 1956; Gray and others, 1968) are the oldest of this group, but similar deposits occur nearby in Nevada (Schilling, 1964). The Bishop deposits occur in silicic and calcareous metasedimentary rocks of probable Paleozoic age, which are believed to be overlain by Mesozoic metavolcanic rocks. All have been deformed and metamorphosed by earliest plutons of the Sierran batholith dated at 240 Ma, which converted the Paleozoic calcareous strata to tactite. These intrusions, thought to have been derived from partial melting of the older sedimentary-volcanic rocks, are also believed to have introduced tungsten-copper-molybdenum into the tactites. The Bishop deposits belong to a regional belt of comparable deposits located just inboard from the craton boundary that extends from northern Mexico to Alaska. These deposits occur repeatedly in western North America in the lowest, often Cambrian miogeoclinal carbonates (Newberry and Einaudi, 1981). Alternatively, these deposits might be explained by earlier, syndepositional mineralization introduced along a flexure-fracture near the craton boundary, with metal redistribution by later plutonism along the same reactivated lineament.

The Eagle Mountain magnetite ores in southern California occur similarly (DuBois and Brummett, 1968) in silicic and impure calcareous metasedimentary strata of inferred Paleozoic age. These rest on an older Precambrian metasedimentary succession, but their stratigraphic relations are obscured by Mesozoic porphyritic quartz-monzonitic intrusions. The primary magnetite (pyrite) ores, which have been extensively weathered to form secondary, hematitic ores, are stratabound in two dolomitic members. They are interpreted as replacements of favorable carbonates by fluids derived from the monzonitic intrusions, which also caused hornfelsic alteration and generated diopside, epidote, and actinolite-tremolite. Comparable ores occur elsewhere in southeastern California, and at the Lyon and other prospects in western Nevada (Reeves, 1964; Reeves and others, 1958). Moreover, they are found regionally along the coastal Cordillera of both South and North America, as at Marcona, Peru, and Texada, British Colombia. Unlike the inference at Eagle Mountain, these other North and South American deposits are in Jurassic, not Paleozoic, carbonate rocks but are also intruded by late Mesozoic plutons. Although this metallogenic comparison raises doubt about the inferred Paleozoic age of the carbonates at Eagle Mountain, still other skarn iron deposits occur in the Permian McCloud limestone, which is intruded by the 260-Ma Pit River stock in the East Shasta district of California. Here late Mesozoic plutonism believed responsible for generation of the Jurassic-hosted deposits has not been recognized. Presumably, therefore, these deposits are much older than the latter, and represent an earlier but similar episode of ore formation. Whatever their age and origin, these skarn iron deposits never appear in accreted oceanic crust, but always in accreted island-arc terranes, or in the craton, associated with calc-alkaline intrusions.

The Eagle Mountain deposits are also similar to those of the Iron Springs district, southwestern Utah (Mackin, 1968). The latter, like those of the Pacific coast, are in Jurassic carbonates, although, as at Eagle Mountain, their host rocks were formerly assigned a Paleozoic (Carboniferous) age. Also as at Eagle Mountain, the Iron Springs ores are stratabound, semiconcordant within the carbonate strata, and invaded at their base by sill-like, semiconcordant mid-Tertiary intrusions dated at 28 to 29 Ma (Bullock, 1973).

All these skarn iron ores have been attributed to complete, selective replacement of favorable carbonate beds by iron-rich fluids from the ubiquitous calc-alkaline intrusions so characteristic of a consuming plate boundary with which they are associated. But their broad regional geologic and metallogenic relations and many similarities suggest, alternatively, that they may have been deposited, either synchronously with, or by diagenetic replacement of, fringing carbonates during calc-alkaline volcanic island-arc magmatism by related submarine hydrothermal activity. If so, two generative periods are indicated, one accompanying formation of the Permian arcs and a second with generation of the Early Jurassic arcs that were subsequently accreted to the continental margin. The ores were recrystallized and variably remobilized by late Paleozoic, late Mesozoic Coast Range, or other intrusions that may have invaded the older accretionary sutures, explaining the repeated spatial coincidence of later intrusions and earlier mineralized strata.

These skarn iron deposits compose a metallogenic province that extends from southern Peru northward to Alaska, with an embayment from Eagle Mountain in southeastern California northeastward through the Lyon and western Nevada prospects, to Iron Mountain in southwestern Utah (Fig. 1). Its continental extent and continuity, and its embayment in the southwestern United States, are difficult to explain by accretion of many far-distant exotic terranes.

Radiometric dating indicates that important porphyry copper deposits at Lights Creek, California; Ely and Yerington, Nevada; and Bisbee, Arizona, are related to Jurassic or Early Cretaceous calc-alkalic porphyritic intrusions. These are older than the many similar deposits associated with Late Cretaceous–Tertiary porphyritic plutons elsewhere in Mexico, Arizona, New Mexico, Utah, Montana, and Washington.

Uranium deposits

Uranium deposits, although known in accreted terranes, are most important in the cratonic platform. Most are peneconcordant tabular layers in lenticular Devonian to Pliocene sandstones (Finch, 1967; Young, 1977). The major hosts are Triassic and Jurassic, apparently deposited mainly in shallow continental depressions, intermontane basins, and coastal plains. They apparently formed by precipitation of uranium and other metals from oxygenated solutions in strongly reducing environments after sandstone deposition (Finch, 1967). Closely related to these peneconcordant deposits are discordant "rolls" and veins. These probably represent redistribution of uranium from earlier-formed deposits at times of geomorphic uplift, weathering, and erosion, by solution in oxygenated ground waters, followed by redeposition in reducing environments. Some crosscutting veins may be hydrothermal, but distinguishing criteria are unavailable. Copper, vanadium, and chromium commonly accompany uranium in deposits of the Colorado Plateau. The uranium deposits vary widely in isotopic ages, with 12 Colorado Plateau deposits ranging from about 295 to 54 Ma (Finch, 1967). Rosenberg and Hooper (1982) reported fission-track ages of about 6.7 Ma for redistributed ore near Grants, New Mexico, and ages of 11,000 to 150,000 years are reported for some unoxidized uranium minerals in Wyoming (Finch, 1967).

Gruner's genetic hypothesis (1956) of multiple migration-accretion, endorsed by Garrels (1957) and Finch (1967), perhaps best explains the widespread distribution, wide variation in age, and paucity of igneous rocks that characterize these deposits. Gruner suggested that the huge volume of Precambrian rocks eroded in post-Pennsylvanian time in the western United States could have provided at least 400 million tons of uranium. In a semiarid interior basin, this uranium, leached and transported as calcium, magnesium, or sodium bicarbonates in CO_2-rich ground waters, was deposited in reducing sites to form reduced black ores. Large concentrations could result from successive cycles of oxidation-solution-migration-deposition.

The age and lithologic similarities of uraniferous Precambrian rocks in the Canadian shield to those in the Colorado Front Range (Nash and others, 1981) suggest continuity of uranium-rich basement terranes beneath the western United States and support Gruner's thesis. Alternative uranium sources may be the abundant volcanic detritus in Mesozoic continental clastic sedimentary rocks or the widespread, weakly radioactive Tertiary rhyolites and ignimbrites of the western United States. But these may have been generated by deep partial melting of the same granitic basement source rocks.

DEPOSITS FORMED DURING RAPID SUBDUCTION OF A GENTLY DIPPING PLATE, LATE CRETACEOUS AND EARLY TERTIARY (LARAMIDE) TIME: 80 TO 40 Ma

General

The metallogenic signature of the Laramide orogeny, from Late Cretaceous through mid-Tertiary time (Coney, 1976), is visible in places throughout the eastern Cordillera (Fig. 2). Its tectonics, magmatism, and relation to plate motions have been extensively discussed (Lipman and others, 1972; Coney, 1976, 1978b; Dickinson and Snyder, 1978; Damon and others, 1981). In summary, the Cretaceous, calc-alkaline magmatic arc that had distinguished the Pacific margin deteriorated in Late Cretaceous time due to acceleration and change in direction of subduction of the Farallon plate. This resulted in a flattened dip (Lipman and others, 1972) of the subducted slab to as little as 20°, which was manifest as far inboard as 1,500 km from the trench. This caused gaps in, and migration of, magmatism over the region (Coney, 1978b). Dickinson and Snyder (1978) believed the subducted slab flattened even further, scraping beneath the Cordilleran crust, perhaps with minimal penetration of the asthenosphere. The dominantly northwest-trending compressive structures of the region are attributable to Laramide tectonism (Kelley, 1955; Gilluly, 1963; Sales, 1968; Coney, 1976), and indicate northeast-southwest–directed regional compression. This compression is apparently reflected by east-northeast–trending Laramide dikes and veins throughout the region from Idaho to Arizona, which indicate northwest-southeast extension. An area 650 by 800 km, later to become the Colorado Plateau, largely escaped Laramide deformation and magmatism; perhaps its underlying crust was thicker than in its surroundings (Gilluly, 1963; Coney, 1976).

The ages of intrusive and volcanic rocks show that magmatism and mineralization in the southern Cordillera occurred almost entirely from 80 to 55 Ma (Late Cretaceous to earliest Eocene), but in the north, although active during this time, they continued during the later Eocene Challis episode from 55 to 40 Ma. The early Cenozoic magmatism, like its Mesozoic counterpart, was largely calc-alkalic but increasingly alkalic eastward (Lipman and others, 1972).

Mineral deposits that formed during the early Laramide interval (80 to 55 Ma) are primarily the important porphyry coppers of Arizona, New Mexico, Sonora, Mexico, and Butte, Montana. Others (Fig. 2) include porphyry molybdenum in Idaho; silver-gold veins near Rochester, Nevada; Gold Basin–Lost Basin, Arizona; and Central City, Colorado; and probably the tungsten-bearing veins at Nederland, Colorado. Generally, Laramide deposits occur irregularly across a vast region from the southwestern Great Basin to the Rocky Mountain front (Fig. 2).

Southwestern United States

In the southwest, the Laramide orogeny involved compression, volcanism, plutonism, and uplift, and swept northeastward across the region (Coney, 1978a). The 34 productive porphyry copper deposits, excluding Bisbee (Titley, 1982), that are associated with this magmatism range in age from about 72 to 52 Ma. Lowell (1974) noted 88 porphyry copper occurrences in an area 650 km (northwest) by 560 km (northeast), including northern Sonora, Mexico; more than half of these are subeconomic. This broader review defines distinctive trends within the cluster more clearly than consideration of only productive deposits.

Their regional extent, distance from the continental margin, and occurrence in platformal rocks of the craton distinguish these deposits from most other porphyry coppers of the American Cordilleras, and has led to controversy over their origin (Lowell, 1974; Sillitoe, 1972, 1975). Andean deposits lie in linear belts parallel to continental margins associated with calc-alkaline magmatic arcs. These closely parallel offshore trenches marking subduction zones. Their distance from the trenches averages about 300 km, and their generation has been ascribed to partial melting of subducted wet oceanic crust at depths of 150 to 250 km (Mitchell and Garson, 1972). Their distinctive calc-alkalic porphyritic plutons were emplaced as shallow stocks beneath a cover of comagmatic volcanic rocks in the magmatic-arc environment (Sillitoe, 1975).

The southwestern United States deposits were generated (Sillitoe, 1972) above an imbricate, flat-dipping subduction zone. Citing Titley (1972), Sillitoe (1975) noted their association with calc-alkaline rocks, as in the Andean deposits. While advocating a regional compressional environment of formation, Sillitoe considered that crustal lineament intersections also influenced their distribution, as earlier advocated (Billingsley and Locke, 1941; Lowell, 1974).

Subduction-zone generation implies regional compression, whereas the contention of lineament control suggests instead a tensional environment. But local extension occurs in subduction regions, and this appears to be so in the southwestern Cordillera. One distinctive belt of deposits, too conspicuous to be dismissed as fortuitous, trends N60°E from Ajo to Globe, Arizona (Clark, 1972), and transects the main northwest-trending belt, 650 km long, of deposits extending from La Caridad, Sonora, to Mineral Park, Arizona (Fig. 3). These two nearly rectilinear belts may represent extension along a shear-couple induced by the regional subduction-compression.

Rehrig and Heidrick (1972), and Heidrick and Titley (1982), described fracture and dike systems near copper-bearing porphyries and in surrounding areas. Fracturing near some deposits is concentric and radial, interpreted (Heidrick and Titley, 1982) as due to the energetics of porphyry intrusion. Regionally, however, dikes, veins, and fractures trend east-northeast, coincident with the Ajo-Globe alignment and also with four other defined trends (Heidrick and Titley, 1982). This preferred orientation is also temporarily restricted to pre-mid-Tertiary rocks.

The northwest-trending La Caridad–Mineral Park belt of copper-bearing plutons is parallel to, but southeast of, the late Mesozoic (90 to 79 Ma; Evernden and Kistler, 1970) calc-alkaline magmatic arc of the Sierra Nevada and Peninsular Ranges batholiths. The copper porphyries may therefore represent a slightly younger, final stage of the preceding, long-lived, calc-alkaline arc magmatism that accompanied flattening of the subduction zone due to eastward acceleration. The southeastward younging of the porphyries may reflect final southeasterly waning of arc activity. If the northwest-trending arc interpretation is correct, the temporally related east-northeast extensional features are oriented essentially normal to the compressional regime, although this may reflect a shear direction rather than simple extension.

The northwest belt of early Laramide mineralization in extreme western Arizona (Fig. 2) is also indicated by 69- to 65-Ma gold veins in the Gold and Lost Basin districts near Las Vegas (Theodore and others, 1982). Gold in these veins was believed remobilized from 712- and 822-Ma Proterozoic veins that cut 1,400-Ma granite.

Northwestern United States

Porphyry and other epigenetic 80- to 55-Ma deposits are less abundant in the northwestern Cordillera than in the southwestern United States. Most deposits in the northwestern United States are probably latest Laramide (Eocene). In Idaho, and re-

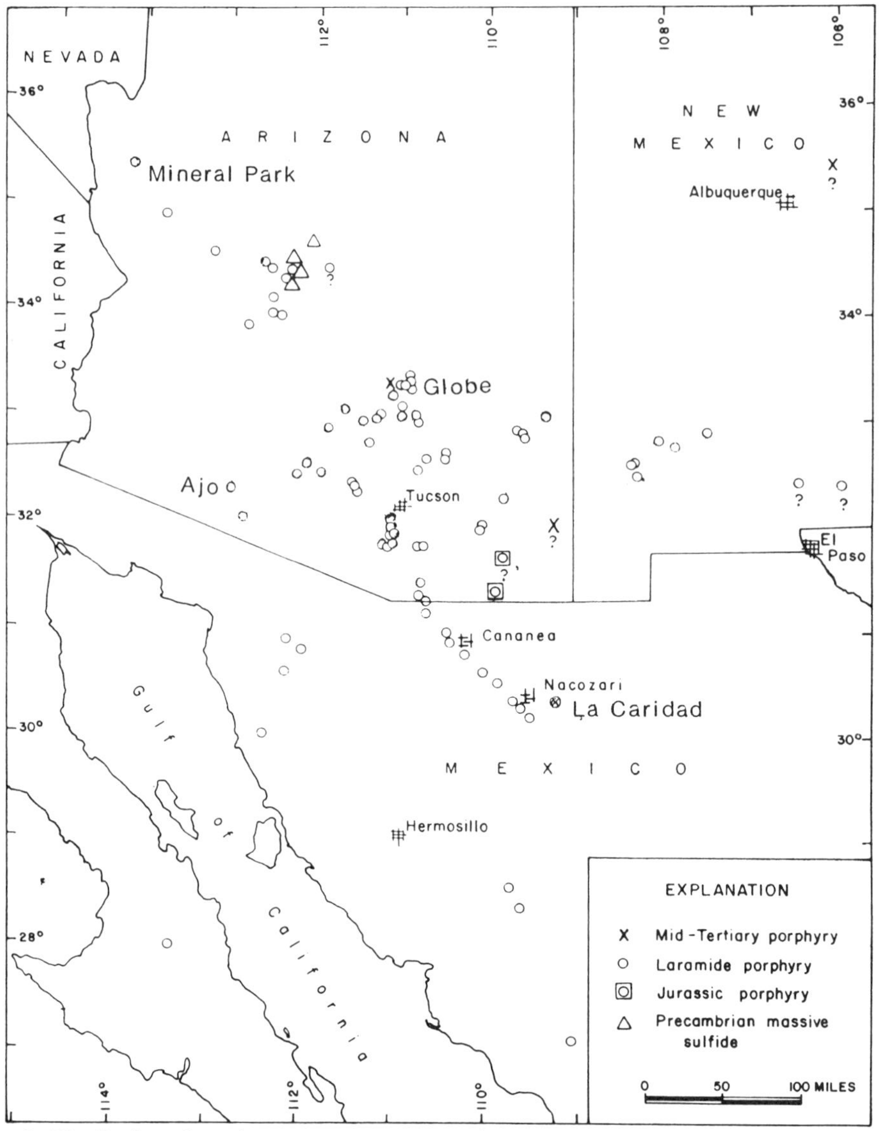

Figure 3. Distribution of porphyry copper occurrences in the southwest Laramide Province. Deposits plotted have porphyry-type alteration and contain at least 20,000,000 tons of 0.1 percent Cu. After Lowell (1974, Fig. 2).

gionally throughout the Cordillera from 42° to 54° N latitude, there is a striking uniformity in Cenozoic volcanic-plutonic chronology (Armstrong, 1974). Magmatism began at about 54 Ma and terminated at 40 Ma, with peak activity about 45 Ma in mid-Eocene time. Thus, Bennett (1980) identified 42 small Tertiary stocks or plutons near the Idaho batholith. Based on only a few age determinations, most were in the range of 50 to 40 Ma, the youngest possibly post-Laramide. Many contain molybdenum, uranium, silver, gold, or base-metal mineralization, apparently from accompanying hydrothermal systems.

A northeast-trending belt of porphyry molybdenum copper deposits, associated with many northeast-trending dikes, extends from west-central Idaho to north-central Montana. First recognized by Ross (1934), it has been named the "Idaho-Montana porphyry belt" (Rostad, 1978; Armstrong and others, 1978). It is 1,000 km long and contains eight known molybdenum deposits ranging in age from 86 to 41 Ma. Early- and late-stage mineralization at Butte, located midway in this belt, are 63 and 58 Ma, respectively (Meyer and others, 1968). Except for the 86-Ma Thompson Creek deposit, the age of mineralization in this belt is Laramide (80 to 40 Ma), but only half the dated deposits are of the early Laramide age that prevails in the southwestern porphyry copper region.

Although the deposits along the belt are not perfectly aligned, the overall trend defined by deposits at the two extremities is N50°E, similar to that of the older Ajo-Globe alignment far to the southeast. This again suggests extension or shear along a cross lineament to the main, northwest-trending front of compressional Laramide tectonism. Significantly, many hot springs occur along the western portion of this N50°E trend, perhaps reflecting the latest, waning stages of the late Eocene hydrothermal activity that formed these deposits. All the deposits lie within the North American craton as defined by the strontium 0.706 line (Fig. 1), and about half—in the center—are underlain by Belt Supergroup basement rocks.

Geologic settings of molybdenum deposits in the North American Cordillera differ widely (Bookstrom, 1981; Westra and Keith, 1981; White and others, 1981); consequently, their genesis is problematic, with no apparent common source of metal (Clark, 1972). Clark favored a mantle source and suggested subduction-related derivation of subcrustal components. This may apply in the west, but crustal metal sources have been postulated for the important younger deposits of Colorado–New Mexico. Tungsten in the deposits may also have had multiple sources.

Mineralization at Butte is in the 71-Ma Butte quartz monzonite of the Boulder batholith (Klepper, 1973), which was emplaced long before regional eastward thrusting that affected late Paleocene strata (Knopf, 1964) but ended prior to emplacement of a 58-Ma sill (Mehnert and Schmidt, 1971). The major Eldorado thrusts emerge far to the east of the batholith, implying that the batholith and the associated Butte deposit are in the upper plate, thus rootless in their present locale. The Butte region is a wedge-shaped tract between the easterly trending Lewis and Clark lineament about 78 km north of Butte, and a major east-west lineament about 40 km to its south (Meyer and others, 1968). This tract has undergone recurrent subsidence since Precambrian time and coincides with a large eastward embayment in the Belt basin; it has migrated eastward relative to the north and south blocks. Its tectonic and stratigraphic relations to the underlying Beltian and older granitic basement rocks are not known, but the Belt strata are regionally mineralized with Cu, Co, Ag, Pb, and Zn. They might constitute a deep source for Butte metals, remobilized by hydrothermal activity associated with intrusion of the 71-Ma Butte quartz monzonite, a possible example of metal inheritance and magmatic overprinting accompanying structural reactivation (Hutchinson, 1988).

Various epigenetic deposits in western Montana–northern Idaho, including Phillipsberg, define a linear trend between Butte and the Coeur d'Alene district 300 km to the northwest. Most of this trend lies within the Butte central block (Meyer and others, 1968); thus, the great vein deposits of Butte are at the intersection of major northeast- and northwest-trending lineaments that existed by Laramide time, if not earlier. A great polymetallic ore deposit like Butte, located at the intersection of major crustal lineaments, might imply a subcrustal metal source. Alternatively, Butte's proximity to surrounding and possibly underlying metalliferous sedimentary rocks of the Belt Supergroup suggest the possibility of these as a metal source.

Laramide ages of porphyry and Butte vein deposits in the northwest are supplemented by an age of 43 Ma for an epithermal gold-silver deposit in the Thunder Mountain district, Idaho (Adams, 1984), and by an age of 71 Ma (Metz and others, 1985) for the Snowbird rare-earth/fluorspar deposit on the Idaho-Montana boundary. The Silver Dyke deposit, Montana, is 47 m.y. old; the Mount Tolman porphyry copper-molybdenum deposit, Washington, is 51 m.y. old (Armstrong and others, 1982); and uranium mineralization at the Midnite Mine, Washington, dates from 51 Ma (Ludwig and others, 1981). Engels and others (1976) reported a K-Ar age of 44 Ma for gangue from the Holden Mine in Washington. Nold (1983) and the writers consider this to be a volcanogenic deposit in metamorphosed Permian silicic volcanic rocks, and the 44-Ma age to be a metamorphic, not a primary one. These ages, with the observations of Bennett (1980) and Armstrong (1974), suggest that epigenetic mineralization throughout much of the northwestern U.S. Cordillera (Fig. 2) spans nearly the full range of Laramide and earliest post-Laramide time.

Central Colorado

Laramide deposits are also prominent in central Colorado (Fig. 2), although their distribution is uncertain owing to overprinting by younger deposits and to controversy over the age of carbonate-hosted mineralization. Laramide mineralization has not been established isotopically, but was discussed in relation to Late Cretaceous–middle Tertiary dikes and plutons by Tweto (1968, p. 694), who stated that "mineralization at Leadville probably is in some part correctly referred to the 'Laramide,' or more specifically to the Paleocene or early Tertiary. It is not proved to be of this age, however, and wholly or in part, it could be appreciably younger. Available geologic evidence requires only that it be of pre-Miocene age." But Devoto (1983) argued that lead-zinc-silver mineralization in Paleozoic strata of central Colorado is Pennsylvanian and variably remobilized by Laramide to mid-Tertiary intrusive activity. The writers consider these deposits to be probably of dual generation and, hence, include them on both Figures 1 and 2.

Precious-metal and uranium mineralization at Central City in the Colorado mineral belt dates from 59 Ma (Phair, 1979; Rice and others, 1982); Laramide stocks along the mineral belt range from 65 to 55 Ma. Although base and precious metallization cannot be related to specific intrusions, it apparently also occurred within this 10-m.y. interval, and all stages of mineralization at Central City probably fall within about 2 m.y. (Rice and others, 1985).

In summary, central Colorado includes the easternmost Laramide deposits of the U.S. Cordillera, and the mineralization is related to sporadic magmatism generated by low-angle subduction. Most representative are deposits of the Colorado mineral belt, but Laramide deposits are not restricted to this belt. The northeast-trending belt is an extensional or shear-related tectonic lineament subparallel to similar prominent belts in New Mexico and southern Arizona, and also contains younger deposits.

Nevada and Utah

Some ten dated deposits in western Nevada fall predominantly in the 80- to 70-Ma early Laramide age range and are surrounded by younger or older deposits (Fig. 2). They include precious-metal veins in the Rochester district (Vikre, 1981) and some tungsten skarns. Near the Utah boundary, the Lexington silver- and tungsten-bearing deposit has been dated at 64 Ma (Schilling, 1965), and farther east, in the Alta district, a pluton spatially related to lead-silver-zinc deposits has yielded ages of 61 and 72 Ma (James and McKee, 1985).

These deposits extend from 118°W longitude eastward to the Colorado Front Range; thus, early Laramide mineralization was irregularly distributed over an area more than 1,000 km wide (Fig. 2). The oldest 75- to 80-Ma deposits lie farthest west, and sparse data suggest an eastward younging to the 60- to 55-Ma range. This supports a shallow, flat-dipping, fast-moving subducted slab beneath the central and eastern Cordillera during Laramide time. Local, perhaps sporadic partial melting of the slab could have supplied metals along extensional lineaments perpendicular to the compressive northeastward-directed regional stress.

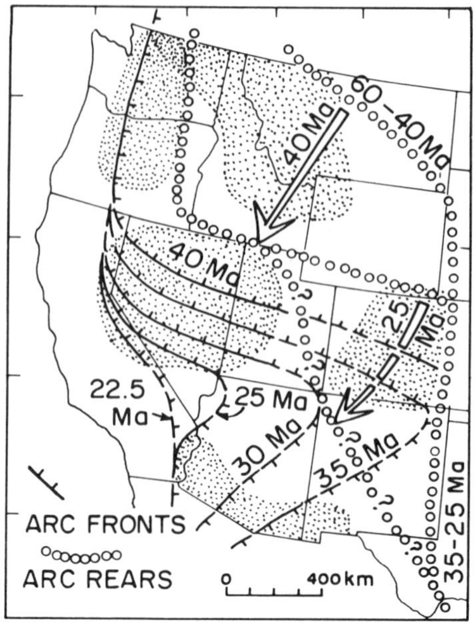

Figure 4. Diagrammatic map showing mid-Cenozoic migration of arc volcanism in the western United States. After Dickinson and Snyder (1978, Fig. 5).

DEPOSITS FORMED DURING SOUTHWESTWARD REGRESSION OF THE MAGMATIC ARC, MIDDLE TERTIARY: 40 TO 22 Ma

Late Laramide time from 55 to 40 Ma saw a lull in magmatism and accompanying mineralizing processes in the southern and eastern Cordillera, while activity continued north of the 42nd parallel. This lull terminated at about 40 Ma when the northeastward-migrating calc-alkalic magmatic arc began rapid retrogression toward the continental margin (Coney and Reynolds, 1977), apparently due to steepening subduction, generating broad extensional domains (Coney, 1978a, b; Lipman, 1981). Throughout mid-Tertiary time, renewed magmatism produced dominantly calc-alkalic rocks, as previously during Laramide time. Important mineral deposits formed during this interval include copper and molybdenum porphyries, precious-metal veins, carbonate-hosted base metal, and skarn iron ores. This interrelated tectonism, magmatism, and mineralization apparently resulted from reorganization of the Pacific–North American–Farallon plate system. Most post-Paleozoic mineralization is related to calc-alkalic magmatism generated in subduction-zone environments.

The most important deposit of this interval was the 36-Ma Bingham porphyry copper (Moore and Lanphere, 1971), one of the world's largest copper deposits (James, 1978). Bingham's peripheral vein and "replacement" deposits have also yielded abundant lead, zinc, and other metals. Whether the latter are closely related in age to the porphyry, however, is debatable. A second porphyry deposit, Copper Canyon, Nevada, about 500 km west of Bingham, is 38 m.y. old (Theodore and others, 1973). Both deposits may have formed along an easterly trending, southward-retreating arc front (Dickinson and Snyder, 1978; Fig. 4). On the other hand, Bingham lies near the hinge line between Paleozoic miogeoclinal and platformal strata, close to the boundary between Archean and Proterozoic rocks (Fig. 1; Bryant, 1988). This area is the "Salt Lake crossroads" (Billingsley and Locke, 1941), where ancient intersecting fracture systems may penetrate the crust and provide pathways for mantle-derived metals. Isotopic data suggest a deep-crustal or upper-mantle source for the Bingham stock and sulfur, but that the peripheral ore lead was remobilized from Precambrian basement (Einaudi and others, 1978).

Carbonate-hosted lead-zinc-silver deposits at Tintic, East Tintic, Park City, and the peripheral replacements at Bingham are all classically considered to be mid-Tertiary, based largely on their spatial relation to nearby intrusions of that age. However, these age and genetic relations are largely circumstantial.

Porphyry molybdenum deposits in the southern Rocky Mountains also formed late in mid-Tertiary time. The Urad-Henderson, Climax, and Questa deposits (Wallace and others, 1968, 1978; White and others, 1981) range from 30 to 23 Ma. Clark (1972) and Lipman (1981) related these to silicic subvolcanic plutonism that followed earlier calc-alkalic magmatism. Lipman (1981) also suggested that faulting associated with this silicic magmatism, as in the Questa caldera, was a local expression of widespread inter- and back-arc extension as subduction steepened and the arc migrated westward. Clark (1972) considered that the varying geologic environments, together with available isotopic data, indicated a subcrustal source for the metals. However, Wallace and others (1978) cited evidence for derivation of molybdenum at Climax and Urad-Henderson from crustal rocks.

Mid-Tertiary precious-metal deposits are widespread throughout the eastern Cordillera. Ages of deposits at Cripple Creek, Silverton, Summitville, and Creede, Colorado; and Cortez, Nevada, are consistent with retreating arc-volcanism (Dickinson and Snyder, 1978; Fig. 4), although their individual ages do not exactly coincide with the Dickinson and Snyder time lines.

The stress pattern, inferred from orientations of dikes associated with mid-Tertiary volcanic rocks of the southern Rocky Mountains, was apparently isotropic or nondirectional during the early mid-Tertiary calc-alkaline magmatism, rather than regionally directed (Lipman, 1981). This was attributed to anorogenic transition from northeasterly Laramide compression to late Tertiary extension. Ore deposits associated with the mid-Tertiary intermediate intrusions differ little from those related to similar earlier Laramide rocks, but are generally small producers (Lipman, 1981). Sources of metals in these mid-Tertiary deposits appear to be multiple. Some metals were derived from geochemically evolved magmas; others, especially lead, from Precambrian crust (Lipman, 1981).

Other mid-Tertiary deposits include the vein-like rare earths–thorium deposits at Lemhi Pass, Idaho (Staatz and others, 1972, 1973).

DEPOSITS FORMED DURING WANING ARC MAGMATISM, LATE TERTIARY AND QUATERNARY: 22 TO 0 Ma

The easterly trending, southward-retreating mid-Tertiary volcanic arc straightened into a northerly trending, continuous arc parallel to the continental margin by earliest Miocene time (Dickinson and Snyder, 1978). Meanwhile, termination of extensional tectonism along the East Pacific Rise due to development of transform motion along the continental margin about 30 Ma brought the prolonged subduction to a close. At this time, most of the Cordilleran region became influenced by a right-lateral stress regime that produced oblique extension in the Great Basin (Atwater, 1970, Fig. 14), and an accompanying transition from calc-alkaline to "fundamentally basaltic" volcanism (Christiansen and Lipman, 1972). The southern Basin-Range region underwent the transition to basaltic volcanism and extensional faulting by about 25 Ma, but predominantly andesitic volcanism persisted farther north where basin-range faulting had not yet begun. By 10 Ma this tectonovolcanic transition had progressed northeastward, restricting predominantly calc-alkalic volcanism to areas near the continental margin farther northwest.

Not surprisingly, deposits generated during this time interval coincide with the areas of calc-alkalic volcanism; in some areas, mainly peripheral to the Great Basin, mineralization accompanies rhyolitic phases of bimodal basaltic-rhyolitic volcanism (Christiansen and Lipman, 1972). But virtually no mineralization is associated with the great silicic ash flows erupted from calderas that are so prominent in the Great Basin (McKee, 1979). Where mineralization accompanies silicic volcanism in a caldera setting, as at Goldfield, Nevada, it appears related to later resurgence, but only structurally and not in the magmatic context, to earlier caldera formation. The richest precious-metal deposits of Silverton in the San Juan Mountains of Colorado were emplaced 5 to 15 m.y. after caldera formation (Lipman and others, 1976). The mineralization is associated instead with emplacement along caldera structures of many small, late quartz-porphyritic intrusions, and this time-space association is common in many districts of the Rocky Mountain region.

The most important deposits formed from 22 to 0 Ma include vein and disseminated gold, and vein silver, some with associated lead-zinc, antimony, and mercury ores. Unimportant are porphyry-type copper, vein uranium, and manganese ores. The prolific work of M. L. Silberman, E. H. McKee, and other geochronologists has provided radiometric dates for many epigenetic deposits in the Great Basin, many of which are late Tertiary to Quaternary.

The principal late Tertiary epithermal precious-metal deposits are in the north and central Great Basin and in the Walker Lane, a northwesterly trending belt lying east of the Sierra Nevada. Epithermal deposits accompany all stages of volcanic activity in the northern Great Basin (Silberman, 1985). Four dated deposits formed during early intermediate activity, and three during dominant silicic ash-flow tuff eruption. Deposits in the northern Great Basin are characterized by their well-defined time-grouping, particularly in the 16 to 12 Ma range, rather than by association with particular host rocks.

In contrast, deposits along the northwesterly trending Walker Lane in southwestern Nevada show no clear time-related grouping. Ages of 20 deposits range from 20 Ma at Goldfield to 2 Ma at Sulfur. In addition to Goldfield, the great bonanza deposits of Comstock, Tonopah, Aurora, and Bodie are within the Walker Lane, all associated with andesitic host rocks (Silberman, 1985). Five deposits in the Walker Lane date from 5 Ma or earlier, showing that mineralization continued after subduction, which evidently ended here between 10 and 5 Ma.

Late Tertiary, mainly precious-metal deposits beyond the Great Basin include those in the western San Juan Mountains of Colorado (Naesser and others, 1979), the Randsburg district in southern California (Hulin, 1925), and McLaughlin in northern California (Bonham, 1985). A base- and precious-metal sulfide vein dated at 13.8 Ma occurs along a branch of the San Andreas fault near the present location of the Mendocino triple junction (McLaughlin and others, 1985). The California occurrences are believed related to the waning stages of arc magmatism.

Studies of epithermal deposits (Silberman, 1985), particularly precious-metals systems, suggest a common duration of hydrothermal activity of 0.5 to 1.5 m.y. This duration compares closely with that for some porphyry copper deposits of 1 m.y., although some have double this duration. Hydrothermal activity and mineralization apparently follow the principal magmatism by 1 to several million years. This may reflect cooling of the main magmatic bodies by convective fluid circulation. Several large, low-grade disseminated gold-silver deposits in Nevada formed at least in part from thermal-spring activity (White, 1974), as did the similar McLaughlin deposit in California. Age ranges for most thermal-spring systems are similar to those of porphyry and vein deposits.

Deposits in the northern and central Great Basin do not occur in a specific type of host rock (Silberman, 1985). In the Walker Lane, however, deposits are clearly associated with andesite. On a broader scale, nonplacer gold-producing mines and districts in the Cordillera occur preferentially in accreted oceanic and island-arc terranes (Albers, 1981), whereas silver-rich deposits exhibit an even stronger preference for the craton. An exception is northwestern Nevada, where deposits of both metals abound. Northwestern Nevada consists mostly of accreted arc terrane, although continental crust apparently underlies its eastern part. This marked regional terrane preference of gold and silver deposits implies different sources, and that different underlying or nearby rocks could have provided differing metals for most deposits.

Many gold and silver veins appear to be lineament-controlled by regional cryptic fracture systems. Such systems are commonly concealed by younger materials, and hence are difficult to map. But linear trends of numerous deposits or intrusions, often marked by parallel vein systems in individual deposits (Landwehr, 1967), strongly indicate throughgoing fracture systems at the time of mineralization. Most lineaments, as defined principally by deposits of Laramide and younger age in the Cordillera, trend northeasterly, but some, particularly in the Great Basin, trend northwest (Roberts, 1966; Shawe, 1977; Madrid and Bagby, 1986). Some may be inherited from the underlying Precambrian basement, but most appear related to late Mesozoic to late Tertiary differential drift with internal wrenching or tension of the continental plate (Shawe, 1977).

Important late Tertiary mercury deposits occur in the California Coast Ranges, northwestern Nevada, Oregon, and Idaho. Most are Miocene and Pliocene in age (Bailey and Everhart, 1964), but at Sulphur Bank, California, mercury deposition from hot springs persists to the present. The large deposits of the McDermitt caldera on the Nevada-Oregon border are about 16 m.y. old (McKee, 1976). Although mercury deposits have doubtless formed throughout geologic time, older ones are rare, probably because most ancient shallow epithermal deposits have been eroded, or perhaps covered and so hidden. In the western U.S. Cordillera, mercury deposits are most numerous in accreted terranes. However, they are hosted by many rock types, and their typical occurrence in restricted belts appears to be governed by the volcanic and tectonic activity that generated them (Bailey and others, 1973). Their distribution coincides well with areas of late Tertiary volcanism.

Other late Tertiary deposits include subeconomic porphyry coppers from 13 to 21 Ma in age in the Cascade Range of Oregon-Washington (Powers and others, 1981), antimony occurrences in northwestern Nevada, and perhaps uranium veins in eastern California. Some mineralizing processes in the Cordillera continue to the present, as evidenced by mercury and antimony deposition in hot spring areas in California, Nevada, and Idaho (Dickson and Tunell, 1968).

DISCUSSION

Regional metallogenic studies can define broad empirical and genetic relations that are not evident from detailed investigations of single deposits or types. Evolutionary metallogeny is also a guide to paleotectonic reconstructions (Hutchinson, 1981b); examples are discussed below.

Inheritance of metal assemblages

Regional metal assemblages constituting metallogenic provinces may be largely governed by processes of early crustal evolution (Titley, 1987). Especially important were the broad tectonic environments and related hydrothermal activity that accompanied initial formation of volcanic, sedimentary, and granitic rocks. In the U.S. Cordillera, these include: first, generation of Archean and Early Proterozoic greenstone belts; second, extensive sedimentation along the passive mid-Proterozoic to mid-Paleozoic continental margin; and finally, succeeding volcano-magmatic arc tectonism. During early Precambrian time, extensive sea-floor hydrothermal activity, as in modern mid-ocean rifts (Ballard and Francheteau, 1983), accompanied the submarine volcanism and associated sedimentation that generated the greenstone belt successions (Hutchinson, 1981a, b). Chemical precipitation from the discharged hydrothermal fluids formed both base- and precious-metal deposits on the sea floors (Hutchinson, 1973; Barnett and others, 1982; Kerswill, 1984) throughout the evolving cratonic regions from northeastern to southwestern North America. With its parental volcanism, this continued throughout late Archean and Early Proterozoic time, both before and after an intervening major orogeny at about 2.5 Ga. Where diluted with admixed pyroclastic or epiclastic detritus, weakly metalliferous, mixed exhalative-pyroclastic-sedimentary strata were deposited (Hutchinson and Burlington, 1984; Hallager, 1983). Although not of ore grade, these strata, together with comparable mid-Proterozoic Beltian metalliferous sediments of the later passive margin, constituted favorable, widespread, pre-enriched sources for later metal remobilization. This was effected by hydrothermal systems associated with subsequently superimposed plutonomagmatic and metamorphic activity to form different, younger types of deposits. Oldest of these episodes was the medium- to high-rank metamorphism of the Hudsonian orogeny that affected much of the craton, including parts of the U.S. Cordillera, about 1.7 Ga (Stockwell, 1982). Later Nevadan and Laramide orogenies and widespread Tertiary magmatism were also important. Metal assemblages developed during early continental growth have been variably overprinted by younger events; although some new deposit types have resulted, broad features of the primary assemblages have survived. The result is coregional deposits of differing type but broadly similar metal content in rocks of widely differing age. Although difficult to document because of the broad-scale processes involved, regionally coherent ages and metal types within metallogenic provinces, together with isotopic data (Zartman and Stacey, 1971), provide strong evidence for inheritance of assemblages.

A good example of metal inheritance is the lead-zinc province of Idaho, Montana, and adjacent British Columbia. Similar metal assemblages characterize sediment-hosted deposits in Proterozoic rocks of the Belt Supergroup, Paleozoic carbonates in northern Washington–southern British Columbia, and Cretaceous veins in the Coeur d'Alene district. Precious-metal deposits in the Rocky Mountains of Colorado and New Mexico, including minor lodes associated with iron formations in Proterozoic greenstones and epithermal veins in Tertiary volcanic rocks, are other possible examples.

Copper-rich volcanogenic massive sulfide deposits in Early Proterozoic greenstones of Arizona and New Mexico are broadly coregional with, and may have been progenitor sources for, the widespread Tertiary porphyry copper deposits (Fig. 3). If the Carlin-type gold deposits were originally deposited as Paleozoic miogeoclinal-eugeoclinal auriferous sediments, their epigenetic characteristics may have been superimposed by Tertiary, volcanic-related hydrothermal activity. The auriferous sediments may have been progenitor metal sources for the young epithermal deposits in the same region. The mid-Proterozoic uraniferous granitic basement rocks in Wyoming, Colorado, Utah, and New Mexico may have been sources, through deep-crustal partial melting, for the detrital volcanic ash in Mesozoic continental clastic sedimentary rocks and for the weakly radioactive Tertiary rhyolites and ignimbrites of the same region. Weathering of any of these may have provided uranium for the widespread hydrogenic sandstone-hosted deposits of the region.

Metal inheritance explains, at least in part, many metallogenic provinces. It may be useful in mineral exploration because younger types of deposits may be indicators of the possible older types, and vice versa.

Structural and tectonic reactivation

Regional tectonic elements, particularly major tectonostratigraphic terranes, terrane boundaries, and lineaments have controlled the distribution of many mineral deposits through time (Woodward, 1984; Bryant and Nichols, 1988). Repeated reactivation along the boundaries, lineaments, and at their intersections was accompanied by repeated magmatism and related hydrothermal activity. These, in conjunction with metal inheritance, resulted in superposition of younger ore-forming systems on the products of older ones, greatly complicating the distinction between the older and younger deposits. Genetic hypotheses for ores have traditionally assumed single-state, single-process generation, but evidence for metal inheritance and tectonic reactivation raises doubts about such simple genetic theories, and dictates consideration of more complex, multistage and multiprocess explanations.

Massive base-metal sulfide deposits are an example. Postdepositional metamorphism of these sea-floor–deposited ores remobilizes, recrystallizes, and reconstitutes them, producing characteristics that were formerly interpreted as definitive of epigenetic replacement. The distinction between earlier-formed but metamorphosed deposits and later superimposed deposits formed by metal inheritance and tectonic reactivation becomes increasingly difficult with the intensity and variety of the superimposed processes. The definitive characteristics of these two end-members become gradational, blurred, and increasingly complex. These difficulties are much greater with broader scale, repeated and varied tectonomagmatic reactivation than with local metamorphism of a massive sulfide body.

A regional hinge line (Figs. 1 and 2) from southern California to Idaho separates miogeoclinal, platformal, and shelf carbonate strata from slope-facies rocks deposited in deeper water. Another such hinge farther west presumably separates the slope-facies rocks from eugeoclinal strata, and coincides approximately with the margin of the craton. Both these hinge lines may mark the loci of prolonged Paleozoic syndepositional faulting. Numerous lead-zinc-copper-silver deposits along the easterly hinge, in both lower and middle Paleozoic carbonates, indicate—if the ores are syndepositional or diagenetic—two periods of Paleozoic sea-floor hydrothermal discharge, both followed by Tertiary tectonomagmatic reactivation along this structure. Where these ores are rich in precious metals, as at Pioche, Eureka, Mercur, and East Tintic, the rocks are cut by Tertiary intrusions; where intrusions are absent, as at Metaline, Van Stone, and Salmo to the northwest, the ores lack precious metals. This suggests Tertiary intrusive reactivation and metal additions along the easterly hinge. The Mississippi Valley–type deposits in undeformed and unintruded rocks of the midcontinent also occur in broadly correlative lower and middle Paleozoic carbonates, but lack many of the discordant relations of deposits in deformed Cordilleran terranes, and also lack the precious metals of the districts affected by Tertiary magmatism.

The deposits at Bingham and Park City, Utah, also lie near the same easterly regional hinge line, where this would be intersected at depth by a projection of the much older Wyoming lineament. At Bingham, Pennsylvanian, carbonate-chert–hosted, polymetallic ores are coregional with the porphyry copper ores of the Tertiary Bingham stock. All have been classically explained by a single-stage, single-process genetic hypothesis invoking Tertiary ore emplacement by fluids uniquely generated from the stock. But the structural setting at Bingham and the possibilities of metal inheritance and tectonic reactivation suggest a more complex, multistage origin. The high-grade ores in upper Paleozoic rocks of these districts may have formed either synchronously or diagenetically in the Paleozoic sediments from sea-floor hydrothermal discharge at structural intersections along the easterly, carbonate platform-slope facies miogeoclinal flexure. Tectonic reactivation may have generated Tertiary magmatism at the same intersections, causing overprinting and reworking of older ores by intrusive-related hydrothermal systems.

Although complicated by eastward thrusting of the Antler and Sonoman orogenies, the slope facies–eugeoclinal hinge coinciding roughly with the craton margin farther west in Nevada (Figs. 1 and 2) was apparently also reactivated periodically throughout Paleozoic time, and may have controlled the distribution of another group of similar deposits in rocks of differing age. Bedded barite in eugeoclinal greenstone-chert of the Ordovician Valmy-Vinini Formations is coregional with similar deposits of the Devonian Slaven chert. And the many sediment-hosted gold deposits of Nevada occur just within the craton near the slope facies–eugeoclinal hinge, in lithologically similar, mixed impure carbonate and fine clastic slope-facies rocks that range from Cambrian to Mississippian in age. Among these sediment-hosted gold deposits, only Mercur in Utah lies farther east, near the easterly carbonate platform-slope facies hinge and also astride a projection of the older basement Archean-Proterozoic terrane boundary.

Paleotectonic reconstructions

Certain ore deposits are generated uniquely in specific tectonostratigraphic and magmatic environments. Examples are ophiolite-hosted cupreous pyrite deposits, volcanic island-arc–generated polymetallic massive sulfides, sediment-hosted and rift- or aulacogen-affiliated lead-zinc-silver ores, and subduction and magmatic arc-affiliated porphyry coppers. Metallogenic relations in complex orogenic belts provide additional guides to paleotectonic evolution and an independent constraint on paleotectonic reconstructions from other geological data. The lead-zinc-silver ores of the Proterozoic Belt Supergroup and the lower Paleozoic barite deposits of Nevada confirm the likelihood of a passive, rifted, western continental margin from mid-Proterozoic through Devonian time. So also do the Ordovician oceanic, or back-arc, rift-generated, cupreous pyrite ores of Mountain City, Nevada. These were obducted by compression along the continental margin into their present site during the Late Devonian Antler orogeny, and their emplacement accords with a change at that time to an active compressive margin.

Metallogenic relations of volcanogenic massive sulfides along the western continental margin raise questions about the many exotic accretionary terranes inferred from tectonic, paleomagnetic, or paleontologic evidence, about their generative and docking ages, and about their migration from far-distant paleo-Pacific locations. Deposits of a primitive, Devonian oceanic volcanic arc at West Shasta, California, lie outboard of deposits of a more evolved, Permo-Triassic continental arc at East Shasta. These metallogenic relations are best explained by generation of both arcs along a continental margin in a trench-arc–back-arc basin system over a subduction zone from mid-Devonian through Permo-Triassic time; first an older Devonian arc formed on oceanic crust and later, in the back-arc basin, a continental affiliated Permo-Triassic arc (Hutchinson, 1973). Subsequently, both arcs were welded to the continent by continuing subduction-induced compression.

Other paleomagnetic, faunal, and tectonic evidence, however, suggests that these terranes were accreted to the continental margin from far-distant paleo-Pacific locations. This conflicts with the metallogenic relations because, to produce the present space and age relations of these two arcs, accretion from Pacific locations requires both unique and coincidental timing, and tectonic mechanisms for their generation, eastward migration, and docking. Moreover, similar space-age relations between volcanic-hosted massive sulfide deposits persist northward along the Pacific margin in British Columbia and southeastern Alaska (Hutchinson, 1988). Accretion of so many exotic terranes from far-distant Pacific locations could only coincidentally produce such widespread and consistent metallogenic relationships.

Farther south in California, ophiolitic cupreous pyrite ores at Copperopolis occur in the same Foothills belt as the many small, felsic, metavolcanic-hosted polymetallic deposits (Hutchinson, 1973), indicating that slices of oceanic crust have been incorporated into the rocks of a felsic, continent-margin volcanic arc during its accretion. The temporal and spatial relations of both their generation and emplacement, compared to those of the more westerly Franciscan ophiolitic rocks and ores, are obscure and merit investigation.

Archean metallogenic relations (Hutchinson, 1984) indicate that gold deposits like those of the Mother Lode, and massive base-metal sulfides like those of the Foothills belt, are coregional in many Precambrian successions, as they are in the Jurassic rocks of California where they parallel one another for 322 km along the Sierran foothills. The gold ores occur in stratigraphically lower, ultramafic-mafic volcanic and sedimentary rocks, whereas the base-metal ores occur in stratigraphically higher, felsic, volcanic-pyroclastic rocks. Perhaps the two California belts also represent differing lithofacies and stratigraphic portions of a single, Jurassic volcanosedimentary succession. The Mother Lode stratiform gray ores are primary, auriferous, ankeritic exhalites formed in the lower, mafic part, whereas the Foothills belt and its base-metal ores are the upper felsic part. The entire succession was isoclinally accreted to the continent in Late Jurassic time, and subsequently intruded by the Jurassic-Cretaceous Sierra Nevadan batholiths. These events deformed and recrystallized the sulfide ores and rocks of the Foothills belt, and intrusion-related hydrothermal activity formed epigenetic, discordant gold veins in the similarly deformed Mesozoic and upper Paleozoic rocks of the Mother Lode belt by remobilization of vein constituents from the primary, auriferous exhalites.

Metallogeny of the last 80 m.y. was subduction-related, as evidenced by the spatial, temporal, and genetic association of mineralization with plutonic, epizonal, and extrusive rocks formed by migrating magmatic arcs that swept across the entire region, due apparently to changing subductive mechanics. Related hydrothermal systems formed the porphyry copper and molybdenite deposits and the epithermal precious-metal ores of the region. Some of the contained metals were supplied by the magmas, others by older rocks through metal inheritance. Whatever the source of metals and fluid, magmatism provided the heat that drove the generative hydrothermal systems. This metallogenic environment differs markedly from the extensional tectonics of earliest Paleozoic time, which evolved into those of an active subductive margin with island-arc magmatism in the later Paleozoic–early Mesozoic. Rapid continental drift with resulting accretion in late Mesozoic time culminated in continental-margin arc magmatism in the latest Mesozoic-Cenozoic.

Local extension accompanied the formation of such disparate deposits as Tertiary precious-metal ores and intracratonic porphyry coppers, oceanic- and island-arc–affiliated Mesozoic-Paleozoic massive base-metal sulfides, and Proterozoic sediment-hosted lead-zinc ores. Extension is apparently critical to the hydrothermal fluid circulation that forms all of these. Local extension, a predictable affiliate of the earlier, rifted passive margin, somehow developed briefly or episodically within the regionally compressive regime of the later consuming subductive margin, and these tensional episodes coincided with hydrothermal ore generation.

REFERENCES CITED

Adams, O., 1984, Geology and ore deposits of the Thunder Mountain mining district, Valley County, Idaho [abs.]: Geological Society of America Abstracts with Programs, v. 16, p. 425.

Albers, J. P., 1981, A lithologic-tectonic framework for the metallogenic provinces of California: Economic Geology, v. 76, no. 4, p. 765–790.

Albers, J. P., and Bain, J.H.C., 1985, Regional setting and new information on some critical geologic features of the West Shasta District, California: Economic Geology, v. 80, no. 8, p. 2072–2091.

Albers, J. P., Kistler, R. W., and Kwak, L., 1981, The Mule Mountain Stock, an early Middle Devonian pluton in northern California: Isochron/West, no. 31, p. 17.

Anderson, C. A., and Nash, J. T., 1972, Geology of the massive sulfide deposits at Jerome, Arizona; A reinterpretation: Economic Geology, v. 67, no. 7, p. 845–863.

Armstrong, R. L., 1974, Geochronometry of the Eocene volcanic-plutonic episode in Idaho: Northwest Geology, v. 3, p. 1–15.

Armstrong, R. L., Hollister, V. F., and Harakal, J. E., 1978, K-Ar dates for mineralization in the White Cloud–Cannivan porphyry–Mo belt of Idaho and Montana: Economic Geology, v. 73, no. 1, p. 94–95.

Armstrong, R. L., Harakal, J. E., and Hollister, V. F., 1982, Eocene mineralization at Mount Tolman (Keller), Washington, and Silver Dyke, Montana: Isochron/West, no. 33, p. 9–10.

Atwater, T., 1970, Implications of plate tectonics for the Cenozoic evolution of North America: Geological Society of America Bulletin, v. 81, p. 3513–3536.

Bailey, E. H., and Everhart, D. L., 1964, Geology and quicksilver deposits of the New Almaden district, Santa Clara County, California: U.S. Geological Survey Professional Paper 360, 206 p.

Bailey, E. H., Clark, A. L., and Smith, R. M., 1973, Mercury, *in* Brobst, D. A., and Pratt, W. P., eds., United States mineral resources: U.S. Geological Survey Professional Paper 820, p. 401–414.

Baker, A., and Clayton, R. L., 1968, Massive sulfide deposits of the Bagdad District, Yavapai County, Arizona, *in* Ridge, J. D., ed., Ore deposits of the United States 1933–1967, Graton-Sales Volume: American Institute of Mining, Metallurgical, and Petroleum Engineers, v. 2, p. 1311–1327.

Balla, J. C., 1983, Geology of the Troy deposit, northwestern Montana, *in* Symposium on the genesis of Rocky Mountain ore deposits; Changes with time and tectonics: Denver Region Exploration Geologists Society, p. 29–30.

Ballard, R. D., and Francheteau, J., 1983, Geologic processes of the mid-ocean ridge and their relation to sulfide deposition, *in* Rona, P. A., Bostrom, K., Laubier, L., Smith, K. L., Jr., eds., Hydrothermal processes at seafloor spreading centers, NATO Conference Series 4: Marine Sciences, v. 12, p. 17–25.

Barnett, E. S., Hutchinson, R. W., Adamcik, A., and Barnett, R., 1982, Geology of the Agnico-Eagle gold deposit, Quebec, *in* Hutchinson, R. W., Spence, C. J., and Franklin, J. M., eds., Precambrian sulfide deposits: Geological Association of Canada Special Paper 25, p. 403–426.

Bateman, P. C., 1956, Economic geology of the Bishop tungsten district, California: California Division of Mines Special Report 47, 87 p.

Bayley, R. W., and James, H. L., 1973, Precambrian iron formations of the United States: Economic Geology, v. 68, no. 7, p. 934–959.

Bayley, R. W., Proctor, P. D., and Condie, K. C., 1973, Geology of the South Pass area, Fremont County, Wyoming: U.S. Geological Survey Professional Paper 793, 39 p.

Beales, F. W., and Jackson, S. A., 1966, Precipitation of lead-zinc ores in carbonate reservoirs as illustrated by Pine Point ore field, Canada: Transactions of the Institution of Mining and Metallurgy, Section B, v. 75, p. 278–285.

Bennett, E. H., 1980, Granitic rocks of Tertiary age in the Idaho batholith and their relation to mineralization: Economic Geology, v. 75, no. 2, p. 278–288.

Billingsley, P. R., and Locke, A., 1941, Structure of ore districts in the continental framework: Transactions of the American Institute of Mining Engineers, v. 144, p. 9–64.

Bonham, H. F., Jr., 1985, Characteristics of bulk minable gold-silver deposits in Cordilleran and island-arc settings, *in* Tooker, E. W., ed., Geologic characteristics of sediment- and volcanic-hosted disseminated gold deposits; Search for an occurrence model: U.S. Geological Survey Bulletin 1646, p. 71–77.

Bookstrom, A. A., 1981, Tectonic setting and generation of Rocky Mountain porphyry molybdenum deposits, *in* Dickinson, W. R., and Payne, W. D., eds., Relations of tectonics to ore deposits in the southern Cordillera: Arizona Geological Society Digest, v. 14, p. 215–226.

Boyum, B. H., and Hartviksen, R. C., 1970, General geology and ore grade control at the Sherman Mine, Temagami, Ontario: Canadian Institute of Mining and Metallurgy Bulletin, v. 63, no. 701, p. 1059–1068.

Bryant, B., 1988, Evolution and early Proterozoic history of the margin of the Archean continent in Utah, *in* Ernst, W. G., ed., Metamorphism and crustal evolution of the western United States; Rubey Colloquium 7: Englewood Cliffs, New Jersey, Prentice-Hall, p. 432–445.

Bryant, B., and Nichols, D. J., 1988, Late Mesozoic and early Tertiary reactivation of an ancient crustal boundary along the Uinta Trend and its interaction with the Sevier Orogenic belts, *in* Perry, W. J., and Schmidt, C. J., eds., Interactions of the Rocky Mountain foreland and Cordilleran thrust belt: Geological Society of America Memoir 171, p. 411–430.

Bullock, K. C., 1973, Geology and iron deposits of Iron Springs district, Iron County, Utah: Provo, Utah, Brigham Young University Geology Studies, v. 20, pt. 1, p. 27–64.

Christiansen, R. L., and Lipman, P. W., 1972, Cenozoic volcanism and plate tectonic evolution of the western United States; 2, Late Cenozoic: Philosophical Transactions of the Royal Society of London, Series A, v. 271, no. 1213, p. 249–284.

Clark, K. F., 1972, Stockwork molybdenum deposits in the western Cordillera of North America: Economic Geology, v. 67, no. 6, p. 731–758.

Coats, R. R., and Stephens, E. C., 1968, Mountain City copper mine, Elko County, Nevada, *in* Ridge, J. D., ed., Ore deposits of the United States 1933–1967, Graton-Sales Volume: American Institute of Mining Metallurgical, and Petroleum Engineers, v. 2, p. 1074–1101.

Condie, K. C., 1981, Evolving parallel tectonic styles in adjacent Proterozoic crustal provinces of the southwestern United States: Geological Society of America Abstracts with Programs, v. 13, p. 50.

——, 1982a, Plate-tectonics model for Proterozoic continental accretion in the southwestern United States: Geology, v. 10, p. 37–41.

——, 1982b, Plate tectonics and crustal evolution: Oxford, Pergamon Publishing Co., 318 p.

Coney, P. J., 1976, Plate tectonics and the Laramide orogeny: New Mexico Geological Society Special Publication 6, p. 5–10.

—— , 1978a, The plate tectonic setting of southeastern Arizona, *in* Callender, J. F., and others, eds., Land of Cochise, southeastern Arizona: New Mexico Geological Society Annual Field Conference Guidebook 29, p. 285–290.

—— , 1978b, Mesozoic-Cenozoic Cordilleran plate tectonics, *in* Smith, R. B., and Eaton, G. P., eds., Cenozoic tectonics and regional geophysics of the western Cordillera: Geological Society of America Memoir 152, p. 33–50.

Coney, P. J., and Reynolds, S. J., 1977, Cordilleran Benioff zones: Nature, v. 270, p. 403–406.

Cox, M. W., 1968, Van Stone Mine area (lead-zinc), Stevens County, Washington, *in* Ridge, J. D., ed., Ore deposits of the United States 1933–1967, Graton-Sales Volume: American Institute of Mining, Metallurgical, and Petroleum Engineers, v. 2, p. 1511–1520.

Cumberlidge, J. T., and Chace, F. M., 1968, Geology of the Nickel Mountain mine, Riddle, Oregon, *in* Ridge, J. D., ed., Ore deposits of the United States, 1933–1967, Graton-Sales Volume: American Institute of Mining, Metallurgical, and Petroleum Engineers, v. 2, p. 1128–1147.

Damon, P. E., Shafiqullah, M., and Clark, K. F., 1981, Age trends of igneous activity in relation to metallogenesis in the southern Cordillera, *in* Dickinson, W. R., and Payne, W. D., eds., Relations of tectonics to ore deposits in the southern Cordillera: Arizona Geological Society Digest, v. 14, p. 137–154.

Devoto, R. H., 1983, Central Colorado karst-controlled lead-zinc-silver deposits (Leadville, Gilman, Aspen, and others), a late Paleozoic Mississippi Valley-type district, *in* Symposium on the genesis of Rocky Mountain ore deposits; Changes with time and tectonics: Denver Region Exploration Geologists Society, p. 51–70.

Dickinson, W. R., and Snyder, W. S., 1978, Plate tectonics of the Laramide orogeny, *in* Matthews, V., III, ed., Laramide folding associated with basement block faulting in the western United States: Geological Society of America Memoir 151, p. 355–366.

Dickson, F. W., and Tunell, G., 1968, Mercury and antimony deposits associated with active hot springs in the western United States, *in* Ridge, J. D., ed., Ore deposits of the United States 1933–1967, Graton-Sales Volume: American Institute of Mining, Metallurgical, and Petroleum Engineers, v. 2, p. 1673–1701.

Doe, B. R., Delevaux, M. H., and Albers, J. P., 1985, The plumbotectonics of the West Shasta Mining District, eastern Klamath Mountains, California: Economic Geology, v. 80, no. 8, p. 2136–2148.

DuBois, R. L., and Brummett, R. W., 1968, Geology of the Eagle Mountain mine area, *in* Ridge, J. D., ed., Ore deposits of the United States 1933–1967, Graton-Sales Volume: American Institute of Mining, Metallurgical, and Petroleum Engineers, v. 2, p. 1592–1606.

Dubuc, F., 1966, Geology of the Adams Mine: Canadian Institute of Mining and Metallurgy Bulletin, v. 59, no. 646, p. 176–181.

Einaudi, M. T., Moore, W. J., and Wilson, J. C., 1978, An issue devoted to the Bingham mining district; Introduction: Economic Geology, v. 73, no. 7, p. 1215–1217.

Emslie, R. F., 1978, Anorthosite massifs, Rapakivi granites, and Late Proterozoic rifting of North America: Precambrian Research, v. 7, p. 61–98.

Engels, J. C., Tabor, R. W., Miller, F. K., and Obradovich, J. D., 1976, Summary of K-Ar, Rb-Sr, U-Pb, Pb, and fission track ages of rocks from Washington State prior to 1975 (exclusive of Columbia Plateau basalts): U.S. Geological Survey Miscellaneous Field Studies Map MF–710, scale 1:1,000,000.

Evernden, J. F., and Kistler, R. W., 1970, Geochronology of emplacement of Mesozoic batholithic complexes in California and western Nevada: U.S. Geological Survey Professional Paper 623, 42 p.

Finch, W. I., 1967, Geology of epigenetic uranium deposits in sandstone in the United States: U.S. Geological Survey Professional Paper 538, 121 p.

Franklin, J. M., and Thorpe, R. I., 1982, Comparative metallogeny of the Superior, Slave, and Churchill Provinces, *in* Hutchinson, R. W., Spence, C. J., and Franklin, J. M., eds., Precambrian sulphide deposits: Geological Association of Canada Special Paper 25, p. 3–90.

Frietsch, R., Papunen, H., and Vokes, F. M., 1979, The ore deposits of Finland, Norway, and Sweden; A review: Economic Geology, v. 74, no. 5, p. 975–1001.

Fyles, J. T., and Hewlett, C. G., 1959, Stratigraphy and structure of the Salmo lead-zinc area: British Columbia Department of Mines Bulletin, v. 41, 157 p.

Garrels, R. M., 1957, Geochemistry of "sandstone-type" uranium deposits, *in* Advances in nuclear engineering, Vol. 2: Nuclear Engineering and Science Congress, Philadelphia, Proceedings, 2nd Conference, p. 288–293.

Gemmill, P., 1968, The geology and ore deposits of the Pioche District Nevada, *in* Ridge, J. D., ed., Ore deposits of the United States 1933–1967, Graton-Sales Volume: American Institute of Mining, Metallurgical, and Petroleum Engineers, v. 2, p. 1128–1147.

Gilluly, J., 1963, The tectonic evolution of the western United States: Quarterly Journal of the Geological Society of London, v. 119, no. 2, p. 133–174.

Gilmore, P., and Still, A. R., 1968, The geology of the Iron King Mine, *in* Ridge, J. D., ed., Ore deposits of the United States 1933–1967, Graton-Sales Volume: American Institute of Mining, Metallurgical, and Petroleum Engineers, v. 2, p. 1238–1257.

Goddard, E. N., and Lovering, T. S., 1941, Nickel deposit near Gold Hill, Boulder County, Colorado: U.S. Geological Survey Bulletin 931–O, p. 349–362.

Gray, R. F., Hoffman, V. J., Bagan, R. J., and McKinley, H. L., 1968, Bishop tungsten district, California, *in* Ridge, J. D., ed., Ore deposits of the United States, 1933–1967, Graton-Sales Volume: American Institute of Mining, Metallurgical, and Petroleum Engineers, v. 2, p. 1531–1554.

Gross, G. A., 1965, Principle types of iron-formation and derived ores: Canadian Institute of Mining and Metallurgy Bulletin, v. 59, no. 646, p. 150–153.

Gross, S. O., 1968, Titaniferous ores of the Sanford Lake District, New York, *in* Ridge, J. D., ed., Ore deposits of the United States 1933–1967, Graton-Sales Volume: American Institute of Mining, Metallurgical, and Petroleum Engineers, v. 1, p. 140–153.

Gruner, J. W., 1956, Concentration of uranium in sediments by multiple migration-accretion: Economic Geology, v. 51, no. 6, p. 495–520.

Hagner, A. F., 1968, The titaniferous magnetite deposit at Iron Mountain, Wyoming, *in* Ridge, J. D., ed., Ore deposits of the United States 1933–1967, Graton-Sales Volume: American Institute of Mining, Metallurgical, and Petroleum Engineers, v. 1, p. 665–680.

Hall, W. E., Rye, R. O., and Doe, B. R., 1978, Wood River, Idaho intrusion related lead-silver deposits derived from country rock source: U.S. Geological Survey Journal of Research, v. 6, no. 5, p. 579–592.

Hallager, W. S., 1983, Geological and geochemical evidence for exhalative gold deposition at Jardine, Montana, *in* Symposium on the genesis of Rocky Mountain ore deposits; Changes with time and tectonics: Denver Region Exploration Geologists Society, p. 13–18.

Hamilton, J. M., Bishop, D. T., Morris, H. C., and Owens, O. E., 1982, Geology of the Sullivan Orebody, Kimberly, B. C., Canada, *in* Hutchinson, R. W., Spence, C. J., and Franklin, J. M., eds., Precambrian sulfide deposits: Geological Association of Canada Special Paper 25, p. 597–665.

Harrison, J. E., 1972, Precambrian Belt basin of northwestern United States; Its geometry, sedimentation, and copper occurrences: Geological Society of America Bulletin, v. 83, p. 1215–1240.

Harrison, J. E., Griggs, A. B., and Wells, J. D., 1974, Tectonic features of the Precambrian Belt Basin and their influence on post-Belt structures: U.S. Geological Survey Professional Paper 866, 15 p.

Hausel, W. D., Graff, P. J., and Albert, K. G., 1985, Economic geology of the Copper Mountain supracrustal belt, Owl Creek Mountains, Fremont County, Wyoming: Wyoming Geological Survey Report of Investigations 28, 33 p.

Haynes, S. J., 1986, Geology and geochemistry of turbidite-hosted gold deposits, greenschist facies, eastern Nova Scotia, Canada, *in* Keppie, J. D., Boyle, R. W., and Haynes, S. J., eds., Turbidite-hosted gold deposits: Geological Association of Canada Special Paper 32, p. 161–177.

Heidrick, T. L., and Titley, S. R., 1982, Fracture and dike patterns in Laramide plutons and their structural and tectonic implications, *in* Titley, S. R., ed., Advances in geology of the porphyry copper deposits, southwestern North America: Tucson, The University of Arizona Press, p. 73–91.

Heyl, G. R., 1948, Foothill copper-zinc belt of the Sierra Nevada, California, *in* Copper in California: California Division of Mines Bulletin 144, p. 11–29.

Higgs, D. V., 1954, Anorthosite and related rocks of the western San Gabriel Mountains, southern California: University of California Publications in Geologic Sciences, v. 30, no. 3, p. 171–222.

Horlacher, C. F., 1987, Precambrian geology and mineralization in the vicinity of Ohio, Gunnison County, Colorado [M.S. thesis]: Golden, Colorado School of Mines, 223 p.

Hughes, G. J., Jr., 1983, Basinal setting of the Idaho cobalt belt Blackbird Mining District, Lemhi County, Idaho, *in* Symposium on the genesis of Rocky Mountain ore deposits; Changes with time and tectonics: Denver Region Exploration Geologists Society, p. 21–27.

Hulin, C. D., 1925, Geology and ore deposits of the Randsburg Quadrangle, California: California Division of Mines and Geology Bulletin 95, 152 p.

Hutchinson, R. W., 1973, Volcanogenic sulfide deposits and their metallogenic significance: Economic Geology, v. 68, no. 8, p. 1223–1246.

—— , 1980, Massive base-metal sulfide deposits as guides to tectonic evolution, *in* Strangeway, D. W., ed., The continental crust and its mineral deposits: Geological Association of Canada Special Paper 20, p. 659–684.

—— , 1981a, Metallogenic evolution and Precambrian tectonics, *in* Kroner, A., ed., Precambrian plate tectonics: New York, Elsevier Science Publishing Co., p. 733–759.

—— , 1981b, Mineral deposits as guides to supracrustal evolution, *in* O'Connell, R. J., and Fyfe, W. S., eds., Evolution of the Earth: American Geophysical Union Geodynamic Series, v. 5, p. 120–140.

—— , 1982, Syndepositional hydrothermal processes and Precambrian sulfide deposits, *in* Hutchinson, R. W., Spence, C. J., and Franklin, J. M., eds., Precambrian sulfide deposits: Geological Association of Canada Special Paper 25, p. 761–791.

—— , 1984, Archean metallogeny; A synthesis and review: Journal of Geodynamics, v. 1, p. 339–358.

—— , 1987, Metallogeny of Precambrian gold deposits; Space and time relationships: Economic Geology, v. 82, no. 8, p. 1993–2007.

—— , 1988, Regional metallogeny in the northwestern United States and its significance in ore deposits geology, *in* Kisvarsanyi, G., and Grant, S. K., eds., Proceedings, North American Conference on Tectonic Control of Ore Deposits and the Vertical and Horizontal Extent of Ore Systems: Rolla, University of Missouri, p. 5–13.

Hutchinson, R. W., and Burlington, J. L., 1984, Some broad characteristics of greenstone belt gold lodes, *in* Foster, R. P., ed., Gold '82; The geology, geochemistry, and genesis of gold deposits: Geological Society of Zimbabwe Special Paper 1, p. 339–371.

James, L. P., 1978, The Bingham copper deposits, Utah, as an exploration target; History and pre-excavation geology: Economic Geology, v. 73, no. 7, p. 1218–1227.

James, L. P., and McKee, E. H., 1985, Silver-lead-zinc ores related to possible Laramide plutonism, Salt Lake County, Utah: Economic Geology, v. 80, no. 2, p. 497–504.

Jenkins, O. P., 1943, Manganese in California: California Division of Mines Bulletin 125, 387 p.

Kanasewich, E. R., 1968, Precambrian rift; Genesis of stratabound ore deposits: Science, v. 161, p. 1002–1005.

Kelley, V. C., 1955, Regional tectonics of the Colorado Plateau and relation to the origin and distribution of uranium: University of New Mexico Publications in Geology, no. 5, p. 120.

Keppie, J. D., Boyle, R. W., and Haynes, S. J., eds., 1986, Turbidite-hosted gold deposits: Geological Association of Canada Special Paper 32, 186 p.

Kerswill, J. A., 1984, The Lupin gold deposit, Contwoyto Lake area, Northwest Territories; Styles of gold distribution and possible genetic models [abs.], *in* Geological Association of Canada-Mineralogical Association of Canada 1984 program with abstracts: Geological Association of Canada, v. 9, p. 78.

Klepper, M. R., 1973, Geology of the southern part of the Boulder batholith [abs.], *in* Society of Economic Geologists Butte Field Meeting Guidebook: Butte, Montana, Anaconda Co., p. B.

Klepper, M. R., Robinson, G. D., and Smedes, H. W., 1974, Discussion *of* 'Nature of the Boulder batholith of Montana' Geological Society of America Bulletin, v. 85, p. 1953–1958.

Knopf, A., 1964, Time required to emplace the Boulder batholith, Montana; A first approximation: American Journal of Science, v. 262, no. 10, p. 1207–1211.

Koski, R. A., 1981, Volcanogenic massive sulfide deposits in ocean crust and island arc terranes, northwestern Klamath Mountains, Oregon and California, *in* Silverman, M. L., ed., Proceedings of the Symposium on Mineral Deposits of the Pacific Northwest: U.S. Geological Survey Open-File Report 81–0355, p. 197–212.

Lakes, A., 1906, The Gunnison gold belt of Colorado: The Mining World, v. 25, p. 576.

Landwehr, W. R., 1967, Belts of major mineralization in western United States: Economic Geology, v. 62, no. 4, p. 494–501.

Lipman, P. W., 1981, Volcano-tectonic setting of Tertiary ore deposits, southern Rocky Mountains, *in* Dickinson, W. R., and Payne, W. D., eds., Relations of tectonics to ore deposits in the southern Cordillera: Arizona Geological Society Digest, v. 14, p. 199–214.

Lipman, P. W., Prostka, H. J., and Christiansen, R. L., 1972, Cenozoic volcanism and plate tectonic evolution of the western United States; 1, Early and Middle Cenozoic: Philosophical Transactions of the Royal Society of London, Series A, v. 271, no. 1213, p. 217–248.

Lipman, P. W., Fisher, F. S., Menhert, H. H., and others, 1976, Multiple ages of mid-Tertiary mineralization and alteration in the western San Juan Mountains, Colorado: Economic Geology, v. 71, no. 3, p. 571–588.

Lowell, J. D., 1974, Regional characteristics of porphyry copper deposits in the southwest: Economic Geology, v. 69, no. 5, p. 601–617.

Ludwig, K. R., Nash, J. T., and Naeser, C. W., 1981, U-Pb isotope systematics and age of uranium mineralization, Midnite mine, Washington: Economic Geology, v. 76, no. 1, p. 89–110.

Mackin, J. H., 1968, Iron ore deposits of the Iron Springs district, southwestern Utah, *in* Ridge, J. D., ed., Ore deposits of the United States 1933–1967, Graton-Sales Volume: American Institute of Mining, Metallurgical, and Petroleum Engineers, v. 2, p. 992–1019.

Madrid, R. J., and Bagby, W. C., 1986, Structural alignment of sediment hosted gold deposits in north-central Nevada; An example of inherited fabrics [abs.]: Geological Society of America Abstracts with Programs, v. 18, p. 393.

McCallum, M. E., Loucks, R. R., Carlson, R. R., Cooley, E. F., and Doerge, T. A., 1976, Platinum metals associated with hydrothermal copper ores of the New Rambler Mine, Medicine Bow Mountains, Wyoming: Economic Geology, v. 71, no. 7, p. 1429–1450.

McConnell, G. W., 1964, Notes on similarities between some Canadian gold deposits and the Homestake deposits of South Dakota: Economic Geology, v. 59, no. 4, p. 719–720.

McConnell, R. H., and Anderson, R. A., 1968, The Metaline District, Washington, *in* Ridge, J. D., ed., Ore deposits of the United States 1933–1967, Graton-Sales Volume: American Institute of Mining Metallurgical, and Petroleum Engineers, v. 2, p. 1460–1480.

McKee, E. H., 1976, Origin of the McDermitt caldera in Nevada and Oregon and its related mercury deposits [abs.]: Economic Geology, v. 71, no. 3, p. 701.

—— , 1979, Ash-flow sheets and calderas; Their genetic relationship to ore deposits in Nevada, *in* Chapin, C. E., and Elston, W. E., eds., Ash-flow tuffs: Geological Society of America Special Paper 180, p. 205–211.

McLaughlin, R. J., Sorg, D. H., Morton, J. L., Theodore, T. G., and Meyer, C. E., 1985, Paragenesis and tectonic significance of base and precious metal occurrences along the San Andreas fault at Point Delgada, California: Economic Geology, v. 80, no. 2, p. 344–359.

Mehnert, H. H., and Schmidt, R. G., 1971, Dating the Eldorado thrust: U.S. Geological Survey Professional Paper 705–A, p. A37.

Merritt, P. C., 1962, California asbestos goes to market: Mining Engineering, v. 14, no. 9, p. 57–60.

Metz, M. C., Brookins, D. G., Rosenberg, P. E., and Zartman, R. E., 1985, Geology and geochemistry of the Snowbird deposit, Mineral County, Montana: Economic Geology, v. 80, no. 2, p. 394–409.

Meyer, C., Shea, E. P., and Goddard, C. G., Jr., 1968, Ore deposits at Butte, Montana, *in* Ridge, J. D., ed., Ore deposits of the United States 1933–1967, Graton-Sales Volume: American Institute of Mining, Metallurgical, and Petroleum Engineers, v. 2, p. 1373–1416.

Mitchell, A.H.G., and Garson, M. S., 1972, Relationship of porphyry coppers and circum-Pacific tin deposits to palaeo-Benioff zones: Transactions of the Institute of Mining and Metallurgy, Section B, v. 81, no. 783, p. B10–B25.

Moore, S. A., 1982, A partisan review of Proterozoic anorthosites: American Mineralogist, v. 67, p. 1087–1100.

Moore, W. J., and Lanphere, M. F., 1971, The age of porphyry-type copper mineralization in the Bingham mining district, Utah; A refined estimate: Economic Geology, v. 66, no. 2, p. 331–332.

Morris, H. T., Heyl, A. V., and Hall, R. B., 1973, Lead, *in* Brobst, D. A., and Pratt, W. P., eds., United States mineral resources: U.S. Geological Survey Professional Paper 820, p. 313–332.

Naesser, C. W., Cunningham, G. C., Marvin, R. F., and Obradovich, J. D., 1979, Pliocene intrusion and mineralization, Rico, Colorado [abs.]: Geological Society of America Abstracts with Programs, v. 11, p. 485.

Nash, J. T., Granger, H. C., and Adams, S. S., 1981, Geology and concepts of genesis of important types of uranium deposits, *in* Skinner, B. J., ed., Economic Geology Seventy-fifth Anniversary Volume, 1905–1980: Economic Geology Publishing Co., p. 63–116.

Newberry, R. J., and Einaudi, M. T., 1981, Tectonic and geochemical setting of tungsten skarn mineralization in the Cordillera, *in* Dickinson, W. R., ed., Relation of tectonics to ore deposits in the southern Cordillera: Arizona Geological Society Digest, v. 14, p. 99–111.

Noble, J. A., 1950, Ore mineralization in the Homestake gold mine, Lead, South Dakota: Geological Society of America Bulletin, v. 61, no. 3, p. 221–252.

Nolan, T. B., and Hunt, R. N., 1968, The Eureka Mining District, Nevada, *in* Ridge, J. D., ed., Ore deposits of the United States 1933–1967, Graton-Sales Volume: American Institute of Mining, Metallurgical, and Petroleum Engineers, v. 1, p. 966–991.

Nold, J. L., 1983, The Holden mine; A metamorphosed volcanogenic deposit in the Cascade Range of Washington: Economic Geology, v. 78, no. 5, p. 944–953.

Phair, G. G., 1979, Interpretation of lead-uranium ages of pitchblende deposits in the central Front Range, Colorado: Geological Society of America Bulletin, v. 90, p. 858–870.

Power, S. G., Field, C. W., Armstrong, R. L., and Harakal, J. E., 1981, K-Ar ages of plutonism and mineralization, western Cascades, Oregon and southern Washington: Isochron/West, no. 31, p. 27–29.

Raedeke, L. D., 1983, Platinum group metals in the Stillwater Complex, *in* Symposium on the genesis of Rocky Mountain ore deposits; Changes with time and tectonics: Denver Region Exploration Geologists Society, p. 31–39.

Raymond, W. A., and Sheridan, D. M., 1980, Data on some stratabound Precambrian deposits containing zinc, copper, lead, silver, and gold in the Royal Gorge 15′ Quadrangle, Colorado: U.S. Geological Survey Open-File Report 80–833, 9 p.

Reeves, R. G., 1964, Metallic mineral resources; Iron, *in* Mineral and water resources of Nevada: U.S. 88th Congress, 2nd session, Senate Doc. 87, p. 101–112.

Reeves, R.G.L., Shawe, F. R., and Kral, V. E., 1958, Iron ore deposits of west-central Nevada, Pt. B, *of* Iron ore deposits of Nevada: Nevada Bureau of Mines Bulletin 53, p. 33–78.

Rehrig, W. A., and Heidrick, T. L., 1972, Regional fracturing in Laramide stocks of Arizona and its relationship to porphyry copper mineralization: Economic Geology, v. 67, no. 2, p. 198–213.

Rice, C. M., Lux, D. R., and Macintyre, R. M., 1982, Timing of mineralization and related intrusive activity near Central City, Colorado: Economic Geology, v. 77, no. 7, p. 1655–1666.

Rice, C. M., Harmon, R. S., and Shepherd, T. J., 1985, Central City, Colorado; The upper part of an alkaline porphyry molybdenum system: Economic Geology, v. 80, no. 7, p. 1769–1796.

Ridler, R. H., 1970, Relationship of mineralization to volcanic stratigraphy in the Kirkland–Larder Lakes area, Ontario: Geological Association of Canada Proceedings, v. 21, p. 33–42.

Riesmeyer, W. D., and Robertson, J. M., 1979, Precambrian geology and ore deposits of the Pecos Mine, San Miguel County, New Mexico, *in* Ingersoll, R. V., and others, eds., Guidebook of Santa Fe Country: New Mexico Geological Society Annual Field Conference Guidebook, v. 30, p. 175–179.

Roberts, R. J., 1966, Metallogenic provinces and mineral belts in Nevada: Nevada Bureau of Mines Report 13, p. 47–72.

Roberts, R. J., Hotz, P. E., Gilluly, J., and Ferguson, H. G., 1958, Paleozoic rocks of north-central Nevada: American Association of Petroleum Geologists Bulletin, v. 42, no. 12, p. 2813–2857.

Rosenberg, P. E., and Hooper, R. L., 1982, Fission-track dating of sandstone-type uranium deposits: Geology, v. 10, no. 9, p. 481–485.

Ross, C. P., 1934, Geology and ore deposits of the Casto Quadrangle, Idaho: U.S. Geological Survey Bulletin 854, 135 p.

Rostad, O. H., 1978, K-Ar dates for mineralization in the White Cloud–Cannivan porphyry molybdenum belt of Idaho-Montana; A discussion: Economic Geology, v. 73, no. 7, p. 1366–1367.

Rye, D. M., and Rye, R. O., 1974, Homestake Gold Mine, South Dakota; 1, Stable isotope studies: Economic Geology, v. 69, no. 3, p. 293–317.

Rye, R. O., 1985, A model for the formation of carbonate-hosted disseminated gold deposits based on geologic, fluid-inclusion, geochemical, and stable-isotope studies of the Carlin and Cortez deposits, Nevada, *in* Tooker, E. W., ed., Geologic characteristics of sediment- and volcanic-hosted disseminated gold deposits; Search for an occurrence model: U.S. Geological Survey Bulletin 1646, p. 35–42.

Rye, R. O., Roberts, R. J., Snyder, W. S., Lahusen, G. L., Motica, J. E., 1984, Textural and stable isotope studies of the Big Mike cupriferous volcanogenic massive sulfide deposit, Pershing County, Nevada: Economic Geology, v. 79, no. 1, p. 124–140.

Sales, J. K., 1968, Crustal mechanics of Cordilleran foreland deformation; A regional and scale model approach: Bulletin of the American Association of Petroleum Geologists, v. 52, p. 2016–2044.

Schilling, J. H., 1964, Tungstein: Nevada Bureau of Mines Bulletin 65, p. 155–161.

—— 1965, Isotopic age determinations of Nevada rocks: Nevada Bureau of Mines Report 10, 79 p.

Shawe, D. R., 1977, Continent-margin tectonics and ore deposits, western United States: Transactions of the American Institute of Mining, Metallurgical, and Petroleum Engineers, v. 262, p. 185–190.

Shepard, W. M., Morris, H. T., and Cook, D. R., 1968, Geology and ore deposits of the East Tintic Mining District, Utah, *in* Ridge, J. D., ed., Ore deposits of the United States 1933–1967, Graton-Sales Volume: American Institute of Mining, Metallurgical, and Petroleum Engineers, v. 1, p. 941–965.

Silberman, M. L., 1985, Geochronology of hydrothermal alteration and mineralization; Tertiary epithermal precious-metal deposits in the Great Basin, *in* Tooker, E. W., ed., Geologic characteristics of sediment- and volcanic-hosted disseminated gold deposits; Search for an occurrence model: U.S. Geological Survey Bulletin 1646, p. 55–70.

Silberling, N. J., and Roberts, R. J., 1962, Pre-Tertiary stratigraphy and structure of northwestern Nevada: Geological Society of America Special Paper 72, 58 p.

Sillitoe, R. H., 1972, Relation of metal provinces in western America to subduction of oceanic lithosphere: Geological Society of America Bulletin, v. 83, p. 813–818.

—— , 1975, Subduction and porphyry copper deposits in southwestern North America; A reply to recent objections: Economic Geology, v. 70, no. 8, p. 1474–1477.

Staatz, M. H., Shaw, V. E., and Wahlberg, J. S., 1972, Occurrence and distribution of rare earths in the Lemhi Pass thorium veins, Idaho and Montana: Economic Geology, v. 67, no. 1, p. 72–82.

Staatz, M. H., Tabor, R. W., Weis, P. L., and others, 1973, Geology and mineral resources of the northern part of the North Cascades National Park, Washington: U.S. Geological Survey Bulletin 1359, 132 p.

Stockwell, C. H., 1982, Proposals for time classification and correlation of Precambrian rocks and events in Canada and adjacent areas of the Canadian Shield; Part 1, A time classification of Precambrian rocks and events: Geological Survey of Canada Paper 80–19, 135 p.

Sumner, W., 1983, Geology and mineralization of the Jones Hill massive sulfide prospect, Santa Fe County, New Mexico, *in* Symposium on the genesis of Rocky Mountain ore deposits; Changes with time and tectonics: Denver Region Exploration Geologists Society, p. 19–20.

Theodore, T. G., Silberman, M. L., and Blake, D. W., 1973, Geochemistry and potassium-argon ages of plutonic rocks in the Battle Mountain mining district, Lander County, Nevada: U.S. Geological Survey Professional Paper 798–A, 24 p.

Theodore, T. G., McKee, E. H., Nash, J. T., and Antweiler, J. C., 1982, Genesis of Late Cretaceous gold mineralization in the Gold Basin–Lost Basin mining districts, Mojave County, Arizona: Geological Society of America Abstracts with Programs, v. 14, p. 240.

Thompson, T. B., Arehart, G. B., Johansing, R. J., Osborne, L. W., Jr., and Landis, G. P., 1983, Geology and geochemistry of the Leadville District, Colorado, *in* Symposium on the genesis of Rocky Mountain ore deposits; Changes with time and tectonics: Denver Region Exploration Geologists Society, p. 101–115.

Titley, S. R., 1972, Some geological criteria applicable to the search for southwestern North America porphyry copper deposits: Mining, Metallurgical Institute of Japan and American Institute of Mining, Metallurgical, and Petroleum Engineers, Joint Meeting, Tokyo, 1972, Papers, G–1, no. 2, 16 p.

—— , 1982, Geologic setting of porphyry copper deposits, southeastern Arizona, *in* Titley, S. R., ed., Advances in geology of porphyry copper deposits, southwestern North America: Tucson, University of Arizona Press, p. 37–58.

—— , 1987, The crustal heritage of silver and gold ratios in Arizona ores: Geological Society of America Bulletin, v. 99, no. 6, p. 814–826.

Tooker, E. W., ed., 1985, Geologic characteristics of sediment- and volcanic-hosted disseminated gold deposits; Search for an occurrence model: U.S. Geological Survey Bulletin 1646, 150 p.

Tweto, O., 1968, Leadville district, Colorado, *in* Ridge, J. D., ed., Ore deposits of the United States, 1933–1967, Graton-Sales Volume: American Institute of Mining, Metallurgical, and Petroleum Engineers, v. 1, p. 681–705.

Vikre, P. G., 1981, Silver mineralization in the Rochester district, Pershing County, Nevada: Economic Geology, v. 76, no. 3, p. 580–609.

Walker, C., 1985, Stratigraphy, sedimentation, and mineralization of the Vinini Formation, Elko County, Nevada [M.S. thesis]: Golden, Colorado School of Mines, 87 p.

Wallace, S. R., and 5 others, 1968, Multiple intrusion and mineralization at Climax, Colorado, *in* Ridge, J. D., ed., Ore deposits of the United States 1933–1967, Graton-Sales Volume: American Institute of Mining, Metallurgical, and Petroleum Engineers, v. 1, p. 605–640.

Wallace, S. R., MacKenzie, W. B., Blair, R. G., and Muncaster, N. K., 1978, Geology of the Urad and Henderson molybdenite deposits, Clear Creek County, Colorado, with a section on a comparison of these deposits with those at Climax, Colorado: Economic Geology, v. 73, no. 3, p. 325–368.

Wells, F. G., Cater, F. W., Jr., and Rynearson, G. A., 1946, Geological investigations of chromite in California; Part 1, Klamath Mountains: California Division of Mines Bulletin 134, 76 p.

Westra, G., and Keith, S. B., 1981, Classification and genesis of stockwork molybdenum deposits: Economic Geology, v. 76, no. 4, p. 844–873.

White, D. E., 1974, Diverse origins of hydrothermal ore fluids: Economic Geology, v. 69, no. 6, p. 954–973.

White, W. H., and 6 others, 1981, Character and origin of Climax-type molybdenum deposits, *in* Skinner, B. J., ed., Economic Geology Seventy-fifth anniversary volume, 1905–1980: Economic Geology Publishing Co., p. 270–316.

Woodward, L. A., 1984, Basement control of Tertiary intrusions and associated mineral deposits along Tijeras-Canoncito fault system, New Mexico: Geology, v. 12, p. 531–533.

Young, R. G., 1977, Sandstone depositional systems in uraniferous sandstones of Colorado Plateau [abs.]: Bulletin of the American Association of Petroleum Geologists, v. 61, no. 5, p. 843.

Zartman, R. E., and Stacey, E. S., 1971, Lead isotopes and mineralization ages in the Belt Supergroup rocks, northwestern Montana and northern Idaho: Economic Geology, v. 66, no. 6, p. 849–860.

Manuscript Accepted by the Society May 17, 1989

ACKNOWLEDGMENTS

Mr. Luis A. Fraticelli of the U.S. Geological Survey provided invaluable assistance with preparation of the initial drawings. Final drafting was completed by the U.S. Geological Survey in Menlo Park with assistance from Dr. E. H. McKee. Drs. J. F. Guilbert, P. Lipman, R. L. Christiansen, A. R. Palmer, and J. H. Stewart gave constructive comments for revision of the manuscript in helpful reviews. The writers appreciate and acknowledge all this assistance.

NOTES ADDED IN PROOF

Because of its long delayed publication, editors of this volume have thoughtfully offered its authors the opportunity to appropriately update the content of their chapters and add any relevant comments. My coauthor, John Albers, whose contribution to our chapter was major, died during manuscript preparation, but I believe that he would have agreed with these additional comments.

The vast region, immense span of time, and tremendous diversity of deposits encompassed in this chapter make a comprehensive metallogenic synthesis virtually impossible. In attempting to adequately outline the metallogeny of the different terranes in the region by assigning them to specific "time slices" during which they developed, we may have overlooked fundamental, broad elements of metallogeny that may link the differing terranes and their deposits through time. In other words, the forest may have been obscured by the trees. Metal inheritance and structural reactivation are two basic processes that influence metallogenic evolution in supracrustal rocks (Hutchinson, 1988) but are often ignored in detailed, site-specific studies of mineral deposits, particularly in genetic interpretations.

Current understandings of global tectonic processes and studies of sea-floor hydrothermal systems have revealed that the initial introduction of metals into supracrustal rocks takes place along both accreting and consuming plate boundaries. Here both concentrated ores and metal-enriched strata are generated. These are subsequently reworked by subduction, or obducted and accreted to continent margins by plate collisions; cratonized by orogenic processes; uplifted; and their contents redistributed by intracratonic magmatic, supergene, and metamorphic processes. Although this cycle and its processes are inferred for the modern Earth, Archean rocks and ores resemble those of early stages and Proterozoic ones those of later stages. This observation suggests that analogous tectonic processes have been active throughout Earth's history, albeit apparently on differing spatial and temporal scales (Hutchinson, 1981b).

Metallogenic evolution must be considered in this broad context. During the later stages of this cycle, what happens to the metals that were originally introduced into supracrustal rocks in the early stages? Clearly and logically, they must have been reworked, recycled and variably redistributed by the many, varied superimposed processes. The redistribution, however, is inhomogeneous because of its great variety. Some metals (elements) are further concentrated by certain processes to form new types of ores; others may be dissipated by the same processes or in turn concentrated by different ones. The scale of redistribution varies from local to district to regional. Old ores are metamorphosed in situ, mainly isochemically. New ores, with some similarities to the old ones, are formed near the latter, and totally different ones appear elsewhere in the region.

It is in this context that metal inheritance plays its important role. Are coregional but markedly differing types and ages of Pb-Zn-Cu-Ag ores in the Pacific Northwest totally unrelated (Hutchinson, 1988)? Or have their major metals been recycled through time, inherited from older strata by younger deposits that formed during superimposed tectonic activity? Similarly, are the consistently bimodal Au/Ag ratios reported by Titley (1991; 1987) in the Phanerozoic ores of the southwestern United States somehow controlled by older deposits or distributions of the precious metals in the underlying basement substrata?

Structural reactivation is equally important and works in conjunction with metal inheritance to reshape older ore deposits and generate new ones. Old, syndepositional structures such as plate margins, failed rifts, transform faults, and fault-controlled basin margins and hinges are weaknesses in the continental crust after their incorporation into the latter by cratonization and accretion. They become preferred loci for later intracontinental movement, magmatism, and related hydrothermal activity. Ores and metal-enriched strata formed along them in the early stages thus become preferred sites for renewed hydrothermal activity. This reworks and redistributes their metal content, sometimes adding its own differing suite of metals, to form differing types of ore near the old ones, but related to or in younger rocks.

Were the widespread Tertiary volcanism and its related epithermal gold deposits of Nevada somehow related to reactivation of old syndepositional basement faults that previously controlled the easterly carbonate shelf, central miogeoclinal, and westerly eugeoclinal facies of Paleozoic sediments throughout the Great Basin? If so, was Tertiary magmatism the source of gold, or could the latter have been recycled from some earlier gold-rich strata? Are the Carlin-type ores, which have some similarities to, but also some marked differences from "classic" epithermal deposits (Hutchinson, 1987), truly of Tertiary epithermal replacement origin? Or might they be older, synsedimentary turbidite-hosted Paleozoic deposits that were reworked by, and that added gold into, superimposed Tertiary hydrothermal systems? Are the Mississippian carbonate-chert–hosted polymetallic ores of the Lark and other mines at Bingham, Utah, and the argillite-hosted skarn Cu ores of the same district all epigenetic replacements related to intrusion of the Tertiary Bingham stock? Or might they be older, synsedimentary Mississippian deposits reworked by intrusion-related Tertiary hydrothermal activity? Bingham lies at the "Salt Lake Crossroads" of Billingsley and Locke (1941), a most distinctive structural location. It is the intersection of a southwestward projection of the Wyoming Lineament, a major ancient Archean/Proterozoic tectonic boundary (Condie, 1982a), with a hinge line separating Paleozoic platformal carbonate strata to the east from miogeoclinal strata to the west.

Admittedly speculative, and surely contradicted by all current understandings and innumerable publications on the Nevadan and Bingham deposits, these possibilities have not been considered or investigated. An overriding and dogmatic assumption is inherent in genetic interpretations about most ore deposits in the western United States. It is assumed ipso facto that they are primarily and genetically related to the pervasive Tertiary magmatism of the region that included the metals as integral components of the magmas. It is suggested that this assumption merits thorough testing and re-evaluation in the broader context of metallogenic evolution of the region, particularly in relation to the processes of metal inheritance and structural reactivation.

ADDITIONAL REFERENCE

Titley, S. R., 1991, Correspondence of ores of silver and gold with basement terranes in the American southwest: Mineralium Deposita, v. 26, no. 1, p. 66–71.

Printed in U.S.A.

The Geology of North America
Vol. G-3, The Cordilleran Orogen: Conterminous U.S.
The Geological Society of America, 1992

Chapter 16

Petrotectonic and paleogeographic settings of U.S. Cordilleran ophiolites

Jason B. Saleeby
Division of Geological and Planetary Sciences, California Institute of Technology, Pasadena, California 91125

INTRODUCTION

Ophiolites are "distinctive associations of mafic to ultramafic rocks . . . which when completely developed comprise from bottom to top: (1) an ultramafic complex, consisting of variable proportions of harzburgite, lherzolite, and dunite, usually with a metamorphic tectonite fabric (more or less serpentinized); (2) a gabbroic complex, ordinarily with cumulus textures commonly containing cumulus peridotites and pyroxenites and usually less deformed than the ultramafic complex; (3) a mafic sheeted dike complex; and (4) a mafic volcanic complex, commonly pillowed. Rocks commonly associated with ophiolites include: (1) an overlying sedimentary section, which typically includes ribbon chert, thin shale interbeds, and minor limestone; (2) podiform bodies of chromite generally associated with dunite; and (3) sodic felsic intrusive and extrusive rocks" (Anonymous, 1972). It is generally thought that ophiolites represent on-land fragments of oceanic crust and upper mantle (Dewey and Bird, 1971; Gass and others, 1973; Miyashiro, 1975; Coleman, 1977; Moores, 1982). The plate tectonic settings of ophiolite petrogenesis and continent edge emplacement are the subjects of major debate. Two general ophiolite types, Cordilleran and Tethyan, were differentiated in a major ophiolite review by Moores (1982). Cordilleran ophiolites were defined as complete to partial, or dismembered, sequences lacking continental-affinity structural basement; Tethyan ophiolites are more often complete crust–upper mantle sequences and commonly rest tectonically above continent edge basement. The contrast between Cordilleran and Tethyan type ophiolites is a useful starting point for this synthesis. Cordilleran ophiolites are considered here to be representative of the circum-Pacific orogenic style. The governing tectonic phenomenon in the circum-Pacific is the movement of large oceanic plates beneath and along the edges of continental plates with or without intervening island arc–marginal basin systems. Tethyan ophiolites, in contrast, formed in ocean basins separating converging continental masses and were for the most part emplaced during ensuing collisions.

The U.S. sector of the North American Cordillera offers, at present, the most extensively studied suite of Cordilleran-type ophiolites. By considering the petrotectonic and paleogeographic settings of these ophiolites, a broad understanding of the ophiolite tectonics can be gained. A fuller understanding of these ophiolites also offers a foundation for understanding the basement tectonics and the depositional basin framework for the mid-Paleozoic through Mesozoic active margin phase of the U.S. Cordillera.

The approach taken in this chapter is to discuss briefly the circum-Pacific environment of ophiolite genesis and emplacement and then focus on the U.S. Cordilleran ophiolites, using modern western Pacific examples as generalized templates for modeling the ancient Cordilleran ophiolites. This treatment is parallel to the early Mesozoic time-slice chapter by Saleeby and Busby-Spera (this volume), which interprets the early Mesozoic of the U.S. Cordillera to have been dominated by a style of tectonics that was similar to the present western Pacific regime.

CIRCUM-PACIFIC ENVIRONMENT OF OPHIOLITE GENESIS AND EMPLACEMENT

Vast expanses of ophiolitic crust and upper mantle underlie the Pacific Ocean Basin, mainly as the Pacific plate. It is not clear, however, that fragments of such crust have been, or are being, emplaced onto or adjacent to continental crust as intact ophiolite sequences. The fate of most abyssal lithosphere appears to be subduction back into the mantle. Nevertheless, on-land and marine geological studies reveal a multitude of ophiolitic rock assemblages along the modern western Pacific arc system (Davies, 1971; Ewart and Bryan, 1972; Karig, 1972; Varne and Rubenach, 1972; Prinzhofer and Nicolas, 1980; Hawkins, 1980; Bloomer and Hawkins, 1983; Sharaskin and others, 1983; Silver and others, 1983; Jakes and Miyake, 1984). Careful consideration of the tectonic settings and petrochemistry of these assemblages suggest that many circum-Pacific ophiolites formed within immature island arc, extensional fore-arc, and interarc basin environments. Intimately related to these environments are transform fracture zones that have become plate convergence zones by changes in relative plate motions; these changes make them important ophiolite generation and emplacement zones (Casey and Dewey, 1984; Saleeby, 1984a). Additionally, fragments of the Pacific plate are apparently embedded in the Philippine Sea and

Saleeby, J. B., 1992, Petrotectonic and paleogeographic settings of U.S. Cordilleran ophiolites, *in* Burchfiel, B. C., Lipman, P. W., and Zoback, M. L., eds., The Cordilleran Orogen: Conterminous U.S.: Boulder, Colorado, Geological Society of America, The Geology of North America, v. G-3.

in the Bering Sea behind the respective arcs (Cooper and others, 1977; Hilde and others, 1977; Hilde and Lee, 1983), and are likely to be widespread constituents of western Pacific oceanic arc basements.

Consideration of the modern, western Pacific, tectonic regime in conjunction with observations within the deep oceanic crustal sections offered by the U.S. Cordilleran ophiolites leads to the conclusion that island arc and related tectonic environs spawn most circum-Pacific ophiolites. The ophiolite nature of interarc basin crust was emphasized by Karig (1972) and subsequently documented more thoroughly by Hawkins (1977), Lawver and Hawkins (1978), and Hawkins and Melchior (1985). Ophiolitic rocks are also common as uplifted basement within fore-arc environments (Ewart and Bryan, 1972; Hawkins and others, 1979; Bloomer, 1981, 1983; Aubouin and others, 1982; Kobayashi, 1983; Sharaskin and others, 1983; Jakes and Miyake, 1984; Hawkins and Falvey, 1985; Ishii, 1985; Beccaluva and others, 1986). In addition to MORB-like igneous assemblages, volcanic rocks recovered from the ophiolitic fore-arcs include boniniticseries pillowed lavas, highly porphyritic tholeiitic-affinity pyroclastic deposits, and local alkalic to shoshonitic lavas and pyroclastic deposits (Dietrich and others, 1978; Meijer, 1980; Natland, 1981; Ogawa and others, 1985; Crawford and others, 1986). Boninitic lavas are typically silica-saturated high-Mg, low-Ti, high-Ni and Cr basalts, andesites and rare dacites with one or more varieties of pyroxene phenocrysts, minor olivine phenocrysts, and a paucity of plagioclase or hornblende phenocrysts (Cameron and others, 1979; Bloomer and Hawkins, 1987).

The fore-arc igneous sequences appear to represent thin mafic crust constructed on disrupted oceanic basement that was accreted to the hanging wall of the respective subduction zone. Diapiric uplift of fore-arc serpentinites and mafic igneous activity represent phases of extension within the fore-arc environment. Extreme extension leads to rifting whereby an interarc basin forms by sea-floor spreading (Hawkins and Falvey, 1985; Ishii, 1985; Ogawa and others, 1985). Subsequent to rifting, the main arc volcanic chain relocates along the front edge of the spreading basin, with the extinction of the original chain and its displacement along the rear edge of the basin as a remnant arc. The cycle is similar to that envisaged by Karig (1972) except that extension and rifting may initiate within the fore-arc region or along the frontal margin of the initial arc.

Careful analysis of the sequence of events outlined above leads to the expectation that interarc basin to fore-arc regimes can create complex polygenetic ophiolite assemblages. Primitive arc and fore-arc igneous activity produce mafic volcanic–plutonic sequences above older disrupted oceanic crust and upper mantle rocks. Interarc basin formation potentially disrupts and disperses such sequences and inserts juvenile oceanic-crust/upper-mantle sequences beneath the basins. Subsequent arc plutonic and volcanic activity may superpose stratocone constructs and/or pyroclastic aprons across any of these preexisting assemblages shortly after their genesis. It has been difficult for workers in active island arcs to gain detailed insight into the deep architecture of such complex assemblages due to the rarity of uplifts above sea level as well as burial by the proximal arc products. In this regard it is asserted that basement studies in the U.S. Cordilleran ophiolites strongly complement the dominantly marine geological studies of modern island arcs. As will be discussed below, many of the features and complexities of modern island arc, fore-arc and back-arc regimes can be recognized in the ophiolitic exposures of the U.S. Cordillera.

Detachment of circum-Pacific ophiolites from host or parent oceanic lithosphere is considered from two perspectives: (1) detachment of abyssal crust during the nucleation of a new subduction-arc system, and (2) accretion of fore-arc/arc/back-arc ophiolitic sequences to continent edges during collisions.

As noted above, major fragments of the Pacific plate appear to be embedded in the Philippine Sea and Bering Sea. Magnetic anomaly patterns within these stranded plates trend at high angles to the adjacent arcs (Cooper and others, 1977; Hilde and others, 1977; Hilde and Lee, 1983), suggesting a fracture zone control for the nucleation of intra-oceanic subduction, and the stranding of the oceanic plate fragment along the hanging wall of the newly created subduction zone. Reconstructions of west Pacific, major plate motion patterns for the late Mesozoic and Cenozoic also suggest initiation of the Tonga-Kermadec as well as Mariana-Bonin subduction-arc systems by such a mechanism (Hilde and others, 1977). Nucleation of subduction along a major fracture zone is presently in its intermediate stages along the Macquarie Ridge–Hjort Trench juncture in the extreme southwest Pacific (Varne and others, 1969; Varne and Rubenach, 1972). Consideration of the relations observed along the Macquarie Ridge–Hjort Trench juncture and the Mariana arc system, inclusive of the Philippine Sea and the Bering Sea and its adjacent Aleutian Arc, leads to the conclusion that fracture zone nucleation of intra-oceanic subduction results in the stranding of fragments of abyssal oceanic crust within island arc frameworks. The stranded fragments are basement nuclei upon and about which oceanic crust of the island arc domain grows.

Fracture zone complexes are also susceptible to en masse incorporation into accretionary prisms and, in turn, emplacement onto continent edges if collision follows. In east Taiwan, a fracture zone complex that formed in the South China Sea has been incorporated into the accretionary prism of the northern segment of the Philippine arc system (Liou and others, 1977). This segment of the arc is in a collisional state with the China mainland, resulting in the thrusting of the accretionary prism and its encased ophiolite complex over the China mainland (Suppe and others, 1981).

The island arc tectonic regime offers a myriad of possible geometries for the emplacement of ophiolites above or along the edges of oceanic plateaus, remnant arcs, active arc segments, and continents. A general sampling of such geometries is offered below by brief overviews of some case studies. Along the Papuan ophiolite belt in New Guinea, a series of remnant arcs, interarc basins, and the related active arc of the west Melanesian system are molded together at their western extremity along a collision zone with the northern passive margin of Australia, where the ophiolite belt is thrust over the continental margin (Davies, 1971; Karig, 1972; Davies and Jacques, 1984). The ophiolite belt may have originated as a fore-arc complex or, alternatively, as an interarc basin. Similar uncertainty in ophiolite heritage is exhibited along the Sulawesi arc where a major ophiolite sheet is thrust over the microcontinental fragment of the Sula Platform, where it has converged obliquely along the south-facing trench of the arc (Silver and others, 1983). The ophiolite roots into the Sulawesi fore-arc basin floor; however, it may have originated as an interarc basin with its current fore-arc setting related to an arc reversal. Thrusting of back-arc basin ophiolite over adjacent active arcs is recorded in the Yap arc-trench system and the Zambales Range of the Philippine system (Hawkins and Batiza, 1977; Hawkins, 1980; Hawkins and Evans, 1983; Schweller and others, 1983). In both instances, thrusting appears to have been related to the collision of seamount chains or non-volcanic ridges, with the respective subduction zones. The unpredictability of this mechanism, in terms of its kinematic consequences, is exhibited in the eastern Sunda arc where collision with the northwestern Australian margin has, in contrast, initiated an arc reversal with the thrusting of back-arc basin crust beneath the adjacent active arc (Silver and others, 1983). This is just a small sampling of the tectonic situations that lead to the emplacement of arc-related ophiolites.

Basic uncertainties arising in the cases noted above also pervade the interpretations of the U.S. Cordilleran ophiolites. Prime examples are the possible pre-collision short-term residence of back-arc basin crust within a fore-arc setting following an arc reversal (Papua and Sulawesi), and the sense of interarc basin thrusting relative to adjacent arc segments (Yap and Zambales Range versus eastern Sunda). In contrast to the complex geometries noted above, we must also consider the relatively simple geometry of a single subduction zone/fore-arc/arc system. Studies of the western margin of Guatemala reveal submerged fore-arc ophiolite having the characteristics of accreted, abyssal, fracture zone crust, but with a superposed fore-arc igneous assemblage (Aubouin and others, 1982; Ogawa and others, 1985). The Guatemala fore-arc ophiolite complex is quite comparable to a number of U.S. Cordilleran polygenetic ophiolite belts. In the case of the U.S. Cordilleran ophiolites, the problem arises as to whether some of the major ophiolite belts represent multiple arc/fore-arc/interarc basin systems like those noted above, or if they represent longitudinally shuffled and dispersed fore-arc ophiolites of a single west-facing system. The uncertainties and complexities outlined above for both generation and emplacement of western Pacific ophiolites help to pose problems for the U.S. Cordilleran ophiolites.

TECTONIC FRAMEWORK OF U.S. CORDILLERAN OPHIOLITES

Major tectonic belts and locations of ophiolitic assemblages of the U.S. Cordillera are shown in Figure 1. The Figure 1 inset shows similar but generalized information plotted on a palinspastic base for Early Cretaceous time (Saleeby and Busby-Spera, this volume). The palinspastic base is useful in the consideration of Cordilleran ophiolite tectonics because all petrogenetic and most emplacement phenomena occurred by the end of Jurassic time. Major crosscutting batholiths and overlap sequences are not shown (Fig. 1 inset). The U.S. Cordilleran ophiolitic assemblages belong to one of five regional tectonic regimes. These regimes reflect the interpretive basis of this synthesis. Numerous ophiolitic terranes or complexes have been delineated within each of these regimes, but the focus here is on broad petrotectonic and paleogeographic patterns, which requires a broad perspective. The regimes are: (1) the Paleozoic McCloud arc/back-arc basin system; (2) the upper Paleozoic and Triassic Central belt; (3) the early Mesozoic extensional fore-arc and possible interarc basin igneous assemblages; (4) the Callovian-Oxfordian interarc basin ophiolites; and (5) the Franciscan subduction and trench-linked transform complex.

Throughout much of the North American Cordillera, there is a repeated association of structurally complex, lower Paleozoic, continental rise strata with lower to middle Paleozoic, arc plutonic and volcanic rocks, and adjacent, well-stratified, mid-Paleozoic to lower Mesozoic, island arc sequences with uncertain basement relations (Rubin and others, 1990). The upper Paleozoic arc strata are distinguished by the presence of the McCloud fauna, which formed in carbonate reefs that covered a fractured volcanic ridge during the Permian (Miller, 1987). The continental rise–McCloud arc association is considered to be the remnant of a major, northeast Pacific, fringing arc system that developed outboard of the Cordilleran miogeocline. The eastern Klamath and northern Sierra terranes are typical of this association, as well as Quesnel and Stikine, which are major terranes of British Columbia and whose southern reaches are shown in Figure 1. Collectively, these terranes are referred to as the McCloud arc (Miller, 1987). Volcanism throughout much of the McCloud arc extended into Middle Jurassic time (Saleeby and others, this volume), although it is the Paleozoic history of the McCloud arc that is most intimately involved with ophiolitic tectonics. Within the Paleozoic arc and fore-arc basement, there are detached fragments of Panthalassa lithosphere and in situ, suprasubduction zone, ophiolitic assemblages (Saleeby, 1990). To the east of the McCloud arc lie the remnants of a major back-arc basin system that closed sequentially during the Devonian-Mississippian Antler orogeny, the Permo-Triassic Sonoma orogeny, and then finally in the Middle to Late Jurassic Siskiyou-Nevadan events (Miller and others, this volume; Saleeby and Busby-Spera, this volume). Within the Paleozoic allochthons of the Antler and

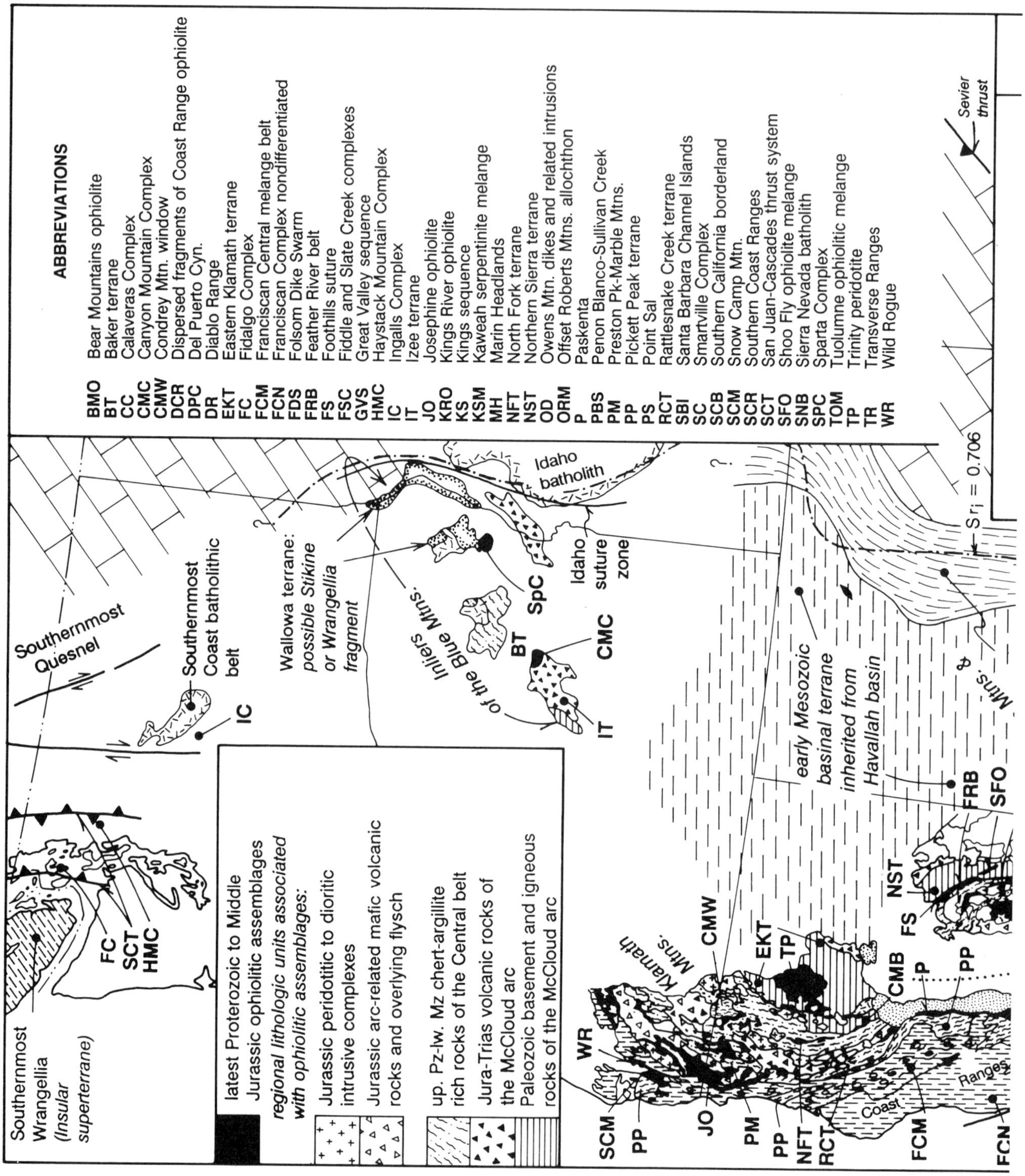
ABBREVIATIONS
BMO Bear Mountains ophiolite
BT Baker terrane
CC Calaveras Complex
CMC Canyon Mountain Complex
CMW Condrey Mtn. window
DCR Dispersed fragments of Coast Range ophiolite
DPC Del Puerto Cyn.
DR Diablo Range
EKT Eastern Klamath terrane
FC Fidalgo Complex
FCM Franciscan Central melange belt
FCN Franciscan Complex nondifferentiated
FDS Folsom Dike Swarm
FRB Feather River belt
FS Foothills suture
FSC Fiddle and Slate Creek complexes
GVS Great Valley sequence
HMC Haystack Mountain Complex
IC Ingalls Complex
IT Izee terrane
JO Josephine ophiolite
KRO Kings River ophiolite
KS Kings sequence
KSM Kaweah serpentinite melange
MH Marin Headlands
NFT North Fork terrane
NST Northern Sierra terrane
OD Owens Mtn. dikes and related intrusions
ORM Offset Roberts Mtns. allochthon
P Paskenta
PBS Penon Blanco-Sullivan Creek
PM Preston Pk-Marble Mtns.
PP Pickett Peak terrane
PS Point Sal
RCT Rattlesnake Creek terrane
SBI Santa Barbara Channel Islands
SC Smartville Complex
SCB Southern California borderland
SCM Show Camp Mtn.
SCR Southern Coast Ranges
SCT San Juan-Cascades thrust system
SFO Shoo Fly ophiolite melange
SNB Sierra Nevada batholith
SPC Sparta Complex
TOM Tuolumne ophiolitic melange
TP Trinity peridotite
TR Transverse Ranges
WR Wild Rogue
Southernmost Wrangellia (Insular superterrane)
latest Proterozoic to Middle Jurassic ophiolitic assemblages
regional lithologic units associated with ophiolitic assemblages:
Jurassic peridotitic to dioritic intrusive complexes
Jurassic arc-related mafic volcanic rocks and overlying flysch
up. Pz-lw. Mz chert-argillite rich rocks of the Central belt
Jura-Trias volcanic rocks of the McCloud arc
Paleozoic basement and igneous rocks of the McCloud arc
Southernmost Quesnel
Southernmost Coast batholithic belt
IC
FC
SCT
HMC
Wallowa terrane: possible Stikine or Wrangellia fragment
Inliers of the Blue Mtns.
BT
SpC
CMC
IT
Idaho suture zone
Idaho batholith
early Mesozoic basinal terrane inherited from Havallah basin
Sri = 0.706
Sevier thrust
Mtns. &
Klamath Mtns.
SCM
PP
WR
JO
PM
PP
NFT
RCT
CMW
EKT
TP
CMB
P
PP
FS
NST
FRB
SFO
Coast Ranges
FCM
FCN

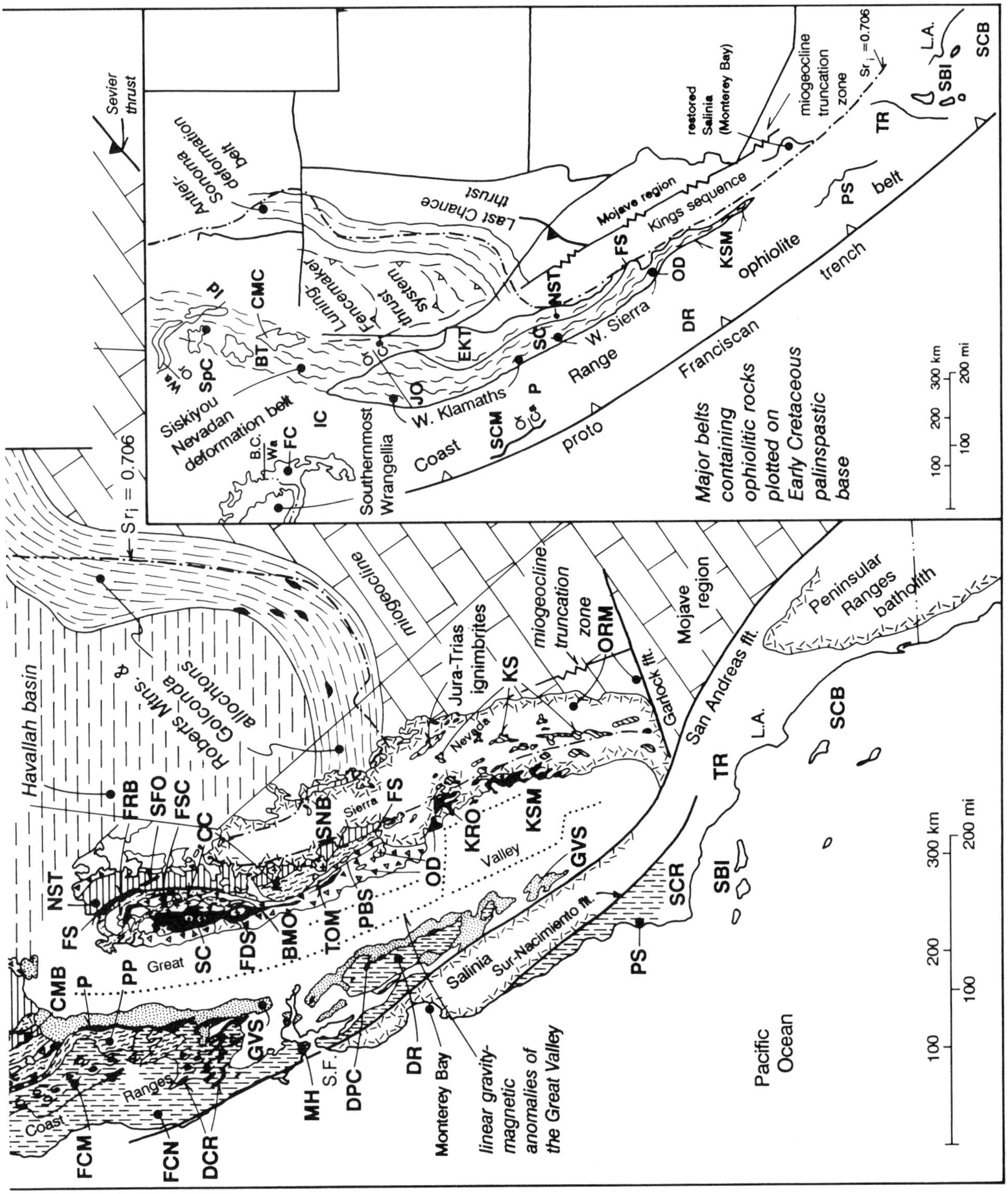

Figure 1. Generalized tectonic map of the western U.S. Cordillera emphasizing rock units and structures closely associated with ophiolitic assemblages, and showing locations of ophiolitic assemblages discussed in text. References are given in text. Inset shows major deformation belts and selected ophiolitic assemblage locations plotted on palinspastic base for Early Cretaceous time (after Saleeby and Busby-Spera, this volume).

Sonoma belts lie small detached fragments of back-arc basin oceanic lithosphere.

To the west of the McCloud arc rocks of the eastern Klamath, northern Sierra, and Quesnel terranes runs a highly disrupted belt that is characterized by upper Paleozoic ophiolitic fragments, upper Paleozoic to lower Mesozoic chert-argillite deposits, upper Paleozoic Tethyan-affinity limestone blocks, and local Triassic blueschist slabs and blocks (Davis and others, 1978). This association is referred to as the Central belt in the western Sierra Nevada, terranes of the western Paleozoic and Triassic belt in the Klamath Mountains, and the Cache Creek terrane in British Columbia. Such rocks also occur as the Baker terrane in the Blue Mountains of eastern Oregon. The regional belt formed by a combination of subduction and transform tectonics; subsequent to a major Triassic phase of subduction accretion, the belt resided as a regional Jurassic fore-arc and possibly intra-arc basin basement complex. We will refer to these rocks as the Sierra-Klamath Central belt, which is distinct from the early to mid-Paleozoic Central metamorphic belt of the eastern Klamaths and the Jurassic-Cretaceous Central mélange belt of the Franciscan complex.

Intimately interwoven with and along the western margin of the Sierra-Klamath Central belt lie submarine flow, pyroclastic, and hypabyssal complexes with overlying cherty, tuffaceous, and epiclastic strata that range from Upper Triassic to Lower and locally lower–Middle Jurassic. These volcanic-sedimentary accumulations are commonly associated with multi-stage serpentinite mélange, and are locally shown to have been built on upper Paleozoic ophiolitic basement of the Central belt. Locally, the sequences resemble relatively complete ophiolite associations, although they are shown to represent polygenetic igneous assemblages. These complex associations are interpreted as mainly extensional fore-arc, volcanic and sedimentary sequences. In a number of localities, siliceous sedimentation related to these sequences led to the accumulation of chert and tuffaceous chert, which upon metamorphism are not readily distinguished from older chert sequences of the Central belt. Along the western margin of the Sierra-Klamath extensional fore-arc rocks run a number of relatively intact, Callovian to Kimmeridgian, mafic, submarine volcanic and slaty flysch sequences. Such strata were laid down on a composite basement consisting of the fore-arc sequence and older Central belt rocks. In general, the volcanic strata represent island arc sequences, although some may also represent extensional fore-arc eruptions. Swarms of sheeted dikes and shallow-level intrusions associated with such eruptives in the northern Sierra Nevada constitute the Smartville complex, and give the overall assemblage an ophiolitic character. The rifting event recorded in the Smartville complex is considered closely related to the interarc basin formation event represented to the west in the Coast Range and Josephine ophiolite belts.

A regionally semicontinuous belt of Callovian-Oxfordian suprasubduction zone ophiolites runs along the California and southern Oregon Coast Ranges and western Klamath Mountains, and fragments of it are dispersed to central through northwestern Washington and underlie parts of the Santa Barbara Channel Islands and southern California borderland. The ophiolites are interpreted here as interarc basin, rifted fore-arc, and related fringing arc lithosphere. The western half of the California Great Valley is considered to be underlain by such basement, which surfaces as the Coast Range ophiolite to the west. The Josephine ophiolite of the western Klamath Mountains represents another excellent exposure of such lithosphere, and the southern Oregon Coast Ranges to northern Washington exposures are extensions of the Coast Range belt that were offset by dextral-slip faulting during the Cretaceous.

The Franciscan complex contains numerous tectonic slices of ultramafite and pillow lava–chert sequences that represent either tectonic outliers of the Coast Range ophiolite that have been ground into Franciscan mélanges by late-stage transcurrent faulting (McLaughlin and others, 1988), or fragments of probable Farallon plate imbricated into the Cretaceous trench.

The five tectonic regimes outlined above formed sequentially from earliest Paleozoic through Cretaceous time. The geologic record for the lower to middle Paleozoic assemblages is highly fragmented. The upper Paleozoic through Jurassic assemblages occur in regional belts; thus, a more coherent petrotectonic and paleogeographic history may be deduced for these assemblages. The Franciscan assemblages are highly fragmented, but those that were derived from the Coast Range ophiolite belt can be more fully interpreted in the context of their parent assemblage. The emphasis of this chapter is the petrotectonic interpretation of ophiolitic assemblages, and the application of the interpretation toward a fuller insight into the paleogeography and important tectonic events that shaped the western U.S. Cordillera.

SCATTERED FRAGMENTS OF EARLY PALEOZOIC OCEANIC LITHOSPHERE

Fragments of early Paleozoic oceanic lithosphere occur in three general tectonic settings in the western U.S. Cordillera. These include: (1) ophiolitic fault slices and larger massifs lying within the composite basement of the mid-Paleozoic to early Mesozoic McCloud arc; (2) ophiolitic blocks and slabs dispersed within late Paleozoic to early Mesozoic ophiolitic mélange of the Sierra-Klamath Central belt; and (3) ultramafic fault slices associated with the Roberts Mountains allochthon. Of these three associations, only the fragments associated with the Roberts Mountains allochthon can be interpreted in a regionally coherent paleogeographic framework. Ophiolitic fragments of the former two associations carry unique attributes at essentially point locations along the orogen, thereby prohibiting a regionally coherent paleogeographic interpretation. Nevertheless, some interesting local interpretations can be made.

The oldest ophiolitic rocks of the Cordillera with probable oceanic crustal affinities are within the Shoo Fly complex of the northern Sierra Nevada. They consist of uppermost Precambrian ophiolitic mélange that is imbricated into lower Paleozoic Shoo

Fly continental rise strata (Schweickert and others, 1984; Saleeby, 1990). The ophiolitic protolith probably represents oceanic lithosphere created by sea-floor spreading off the North American passive margin following rift generation of the margin, as discussed in Poole and others (this volume). The Shoo Fly complex represents a thrust-imbricated vestige of the early Paleozoic North American continental rise that suffered tectonism and arc magmatism superimposed on the rise environment by the initiation of early Paleozoic subduction (Schweickert and others, 1984; Saleeby and others, 1987). Major, lower Paleozoic, oceanic lithosphere remnants that appear to have been detached from their underpinnings and emplaced into this active plate juncture system include the Trinity peridotite of the eastern Klamath Mountains and possibly the Kings River ophiolite of the southwestern Sierra Nevada. The embryonic arc that developed along the continental rise became a major mid- to upper Paleozoic fringing arc system (McCloud arc), which consumed Panthalassa lithosphere by primarily east-dipping subduction (Miller, 1987; Miller and others, this volume). A large fragment of the basaltic crustal levels of such lithosphere is presently wedged eastward beneath the eastern Klamath McCloud arc sequence as the Central metamorphic belt, presumably reflecting the original subduction geometry (Lipman, 1964; Peacock and Norris, 1989).

The Trinity peridotite and Kings River ophiolite share unique isotopic and trace element geochemical features, which indicate they were derived from a similar, highly depleted, sub-Panthalassa mantle during mid-Ordovician time (Jacobsen and others, 1984; Shaw and others, 1987; Saleeby, 1990). Following their ocean ridge generation and initial lithosphere detachment along the Cordilleran margin, the Trinity peridotite and Kings River ophiolite diverged markedly in their tectonic evolutions. The Trinity peridotite was stripped of any Ordovician oceanic crustal rocks that it might have sat beneath, and was then intruded and covered by island arc tholeiitic to boninitic series rocks during Silurian and Devonian time (Lanphere and others, 1968; Mattinson and Hopson, 1972; Lindsley-Griffin, 1977; Quick, 1981; Jacobsen and others, 1984; Brouxel and Lapierre, 1988; Brouxel and others, 1988; Rouer and others, 1989). This suprasubduction zone magmatism represents a phase of extension within the early history of the McCloud arc. The polygenetic Trinity complex might have sat in back-arc basin, embryonic arc, or extensional fore-arc settings. Its present geometric configuration is most consistent with an extensional fore-arc (Saleeby, 1990).

In contrast to the Trinity peridotite, the Kings River ophiolite retained its complete, mid-Ordovician, spreading ridge, crust to upper mantle sequence (Saleeby, 1978). Following its mid-Ordovician genesis, the next entry into its geologic record was diking and tectonic disruption within a late Paleozoic transform system (Saleeby, 1978, 1979, 1980; Shaw and others, 1987). The gap of more than 150 m.y. in its geologic history suggests that it drifted within the Panthalassa basin for an extended period of time, or was trapped as an intact fragment of Panthalassa lithosphere within the distal reaches of a McCloud arc marginal basin or fore-arc, where it escaped arc magmatism and penetrative deformation until the late Paleozoic transform disruption of the arc and basinal system.

Rocks of the Roberts Mountain allochthon are most easily interpreted as having resided in a back-arc basin setting relative to the early to mid-Paleozoic rocks of the McCloud arc system (Poole and others, this volume; Miller and others, this volume). These back-arc basin strata formed off the outer edge of the Cordilleran miogeocline with periods of basaltic volcanism presumably recording diffuse spreading within the back-arc basin/continental rise environment. Tectonic lenses of serpentinized harzburgite and dunite occur in association with the basaltic and basinal strata of the Roberts Mountains allochthon (Poole and Desborough, 1973). It is unknown as to whether the ultramafites represent new mantle lithosphere created during back-arc basin spreading, or older mantle lithosphere created by spreading off the edge of the miogeocline prior to the onset of east-dipping subduction and McCloud arc activity.

Possible late- to post-Antler paleogeographic settings of the McCloud arc, its back-arc basin, and the adjacent miogeocline are shown diagrammatically in Figures 2a and 3a (modified from Miller and others, this volume). Figures 2a and 3a are starting points in alternative tectonic models for late Paleozoic rifting and transform faulting, which greatly modified the southwestern U.S. Cordillera, led to the development and emplacement of the Sierra-Klamath Central belt ophiolitic assemblages, and established the early Mesozoic southwestern Cordilleran plate edge.

LATE PALEOZOIC RIFTING AND TRANSFORM FAULTING

Rifting and large-magnitude transform faulting were widespread in the Cordillera during late Paleozoic time, affecting the McCloud arc, its back-arc basin, the subsided Antler orogenic belt, and parts of the miogeocline (Eisbacher, 1983; Walker, 1988; Saleeby, 1990; Smith and Miller, 1990; Miller and others, this volume). Back-arc basin rifting is well recorded for Late Mississippian to Early Permian time by submarine basaltic volcanic rocks within the chert-rich Havallah sequence (Miller and others, this volume). Spreading within the Havallah basin appears to have been diffuse without the production of complete ophiolitic crust and upper mantle sequences. Fault slices of serpentinized ultramafites are locally associated with Havallah rocks emplaced as part of the Golconda allochthon (Poole and Desborough, 1973; Speed, 1978; Schweickert and Lahren, 1987). These may represent new mantle lithosphere created during Havallah basin spreading, or alternatively, older mantle fragments reworked from the Roberts Mountains allochthon. The Havallah basin apparently sat partly above subsided McCloud arc rocks also, as recorded in the Peale chert of the northern Sierra terrane (Harwood, 1988). As a result of rifting and transform faulting, much of the McCloud arc was fractured into inert volcanic ridges that were capped with carbonate reef deposits bearing the distinct McCloud shallow-water fauna (Miller, 1987).

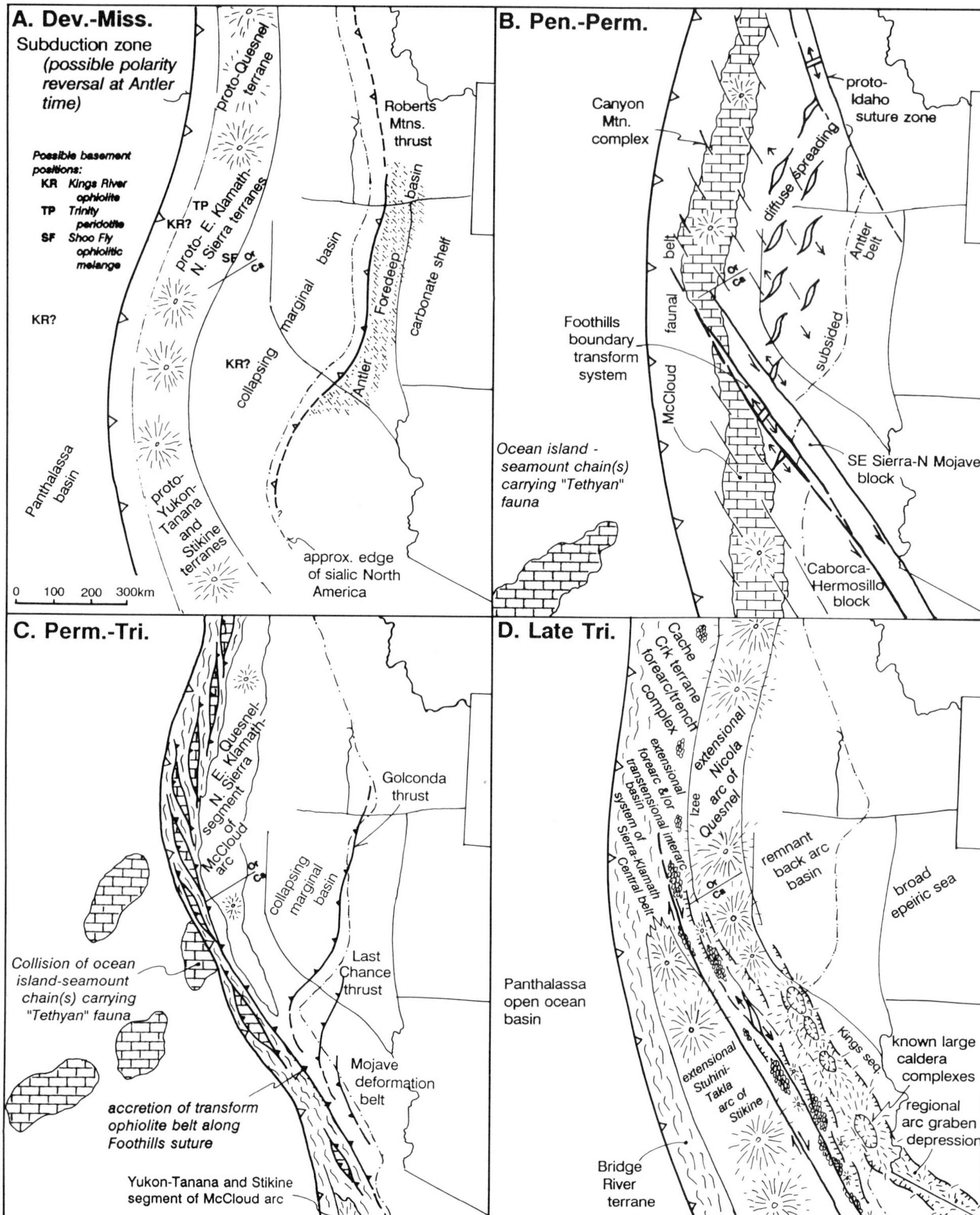

Figure 2. Plate tectonic model for relationships between late Paleozoic rifting and transform faulting along southwestern U.S. Cordillera and earlier tectonic paleogeographic features as well as superposed Sonoma orogenic deformation and the early Mesozoic continental arc. This model considers all elements of the McCloud arc to have remained in a Cordilleran suprasubduction zone environment. Alternative model is presented in Figure 3.

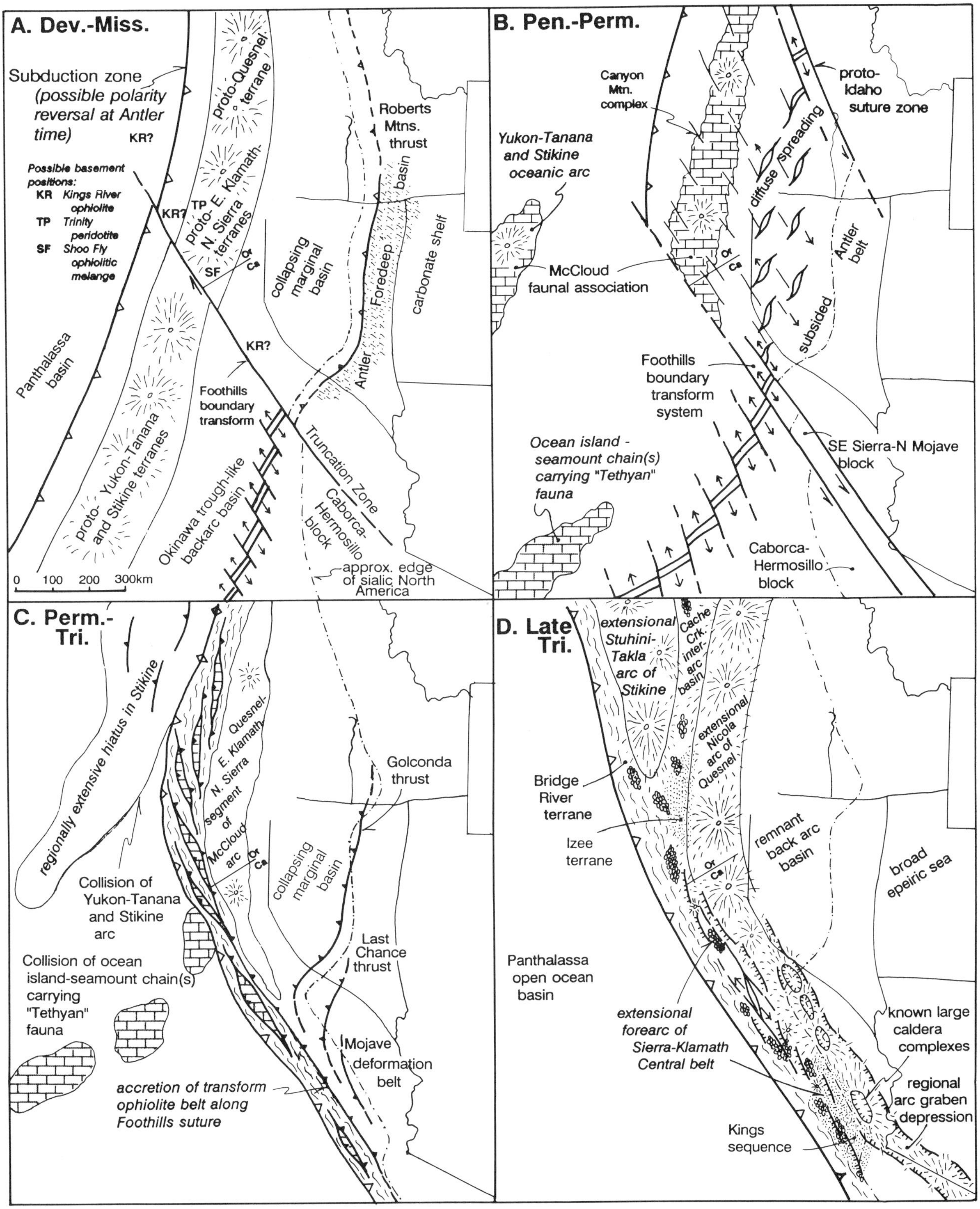

Figure 3. Plate tectonic model for relationships between Paleozoic rifting and transform faulting along southwestern U.S. Cordillera and earlier tectonic palegeographic features as well as superposed Sonoma orogenic deformation and the early Mesozoic continental arc. This model considers parts of the McCloud arc to have been dislodged from the Cordilleran subduction-arc system, and to have migrated within the greater Panthalassa ocean basin prior to accretion in the Pacific Northwest.

Rifting of the McCloud arc and its Havallah back-arc basin corresponded, in time, with a major turning point in the tectonic development of the southwestern Cordillera. This was the disruption of Paleozoic NE-SW–trending facies patterns in the miogeocline and tectonic trends along the outer shelf such as the Antler belt and the Havallah basin shelf-slope break, as well as the truncation of the southern McCloud arc. These features ran continuously from the southern Idaho to northern Nevada region through eastern California prior to their truncation to the southwest and the superpositioning of NW-SE structural trends and allochthonous tectonic belts along the truncation zone (Hamilton and Meyers, 1966; Burchfiel and Davis, 1972). The truncation zone apparently ran along the axis of the Sierra Nevada and, subsequent to the truncation event, the California segment of the Cordilleran plate edge followed this trend (Saleeby and Busby-Spera, this volume). An offset fragment of Antler belt rocks lies along the truncation locus in the southeastern Sierra to northern Mojave region (Fig. 1); its sinistral offset is constrained to Permo-Carboniferous time (Walker, 1988). The Permo-Carboniferous timing of this event is corroborated by the contemporaneous superpositioning of continental borderland strata and structures across shelf strata along the east-central California region (Stone and Stevens, 1988). Permo-Carboniferous sinistral transform truncation was also suggested for the southern Sierra Nevada region based on the tectonic position and structural-stratigraphic relations of the Kings-Kaweah ophiolite belt (Saleeby, 1975), and as a mechanism for sinistral offsets in miogeoclinal strata reported by Silver and Anderson (1974). Subsequent work suggests that this transform ophiolite belt extends the entire length of the western Sierra Nevada (Saleeby, 1982, 1990), and that it may be directly related to the continental truncation event. In this context, the Foothills ophiolite belt is interpreted to have formed along the truncation boundary transform with its subsequent thrust accretion along the zone forming the Foothills suture.

Late Paleozoic transform ophiolites

Structurally complex ophiolitic rocks of the western Sierra Nevada are regionally interleaved with belts of lower Mesozoic island-arc-affinity strata and upper Paleozoic to lower Mesozoic, chert-rich and continent-derived, clastic basinal strata (Saleeby, 1982, 1990). In some localities, the ophiolitic rocks can be shown to have constituted basement for the basinal and volcanic strata and to have been metamorphosed and highly disrupted prior to unconformable overlap. In most cases, however, the contacts between basement and superjacent strata are high-angle faults (Clark, 1964, 1976). From south to north, the major ophiolitic exposures consist of the Kings-Kaweah belt, Tuolumne River belt, Bear Mountains belt, and Feather River belt (Fig. 1). Additionally, there are numerous, thin, discontinuous serpentinite slivers that locally contain metabasite, metachert, and carbonate blocks, which are structurally encased in younger strata, and were probably derived from the ophiolitic basement. Collectively, the chert-rich and ophiolitic units constitute the western Sierra Central belt. Paleozoic rocks of the Central belt appear to constitute the basement for lower Mesozoic rocks of the western Sierra metamorphic belt. The transform fracture zone affinity of this ophiolitic basement complex is most clearly displayed in the Kings-Kaweah belt.

The Kings-Kaweah ophiolite belt is exposed for about 125 km along the southwestern Sierra Nevada foothills (Saleeby, 1977, 1978, 1979). The Lower Ordovician Kings River ophiolite occurs as several large tectonic slabs at the northern end of the belt. Smaller additional slabs and blocks of the ophiolite are dispersed southwards along the belt within the Kaweah serpentinite mélange. Narrow belts of Kaweah-type mélange, as well as ductile shear zones, bound the Kings River slabs. The ophiolite belt exhibits all of the major features that are typical of major oceanic transform fracture zones. These include: (1) the injection and solidification of tholeiitic magma into zones of high shear strain, with the production of flaser gabbros and metamorphic tectonites ranging from upper amphibolite to low greenschist (dynamically retrograded) assemblages; (2) development of widespread high-strain mylonitic fabrics superposed over high-temperature mantle flow fabrics in peridotites; (3) evidence for seawater infiltration into lower crustal–upper mantle rocks during high-temperature deformation; (4) widespread serpentinization of mantle rocks under a wide range of physical conditions yielding high- and low-temperature, as well as highly schistose to isotopic, serpentinites; (5) widespread diapiric rise of serpentinites and protrusive breachment onto the sea floor contemporaneous with basaltic volcanism and pelagic sedimentation; (6) a criss-cross pattern in dikes, faults, shear zones, and ultramafic protrusions, which is a common structural feature of spreading ridge–transform intersection zones; (7) complex accumulations and surface remobilizations of polymict sedimentary serpentinite and ophicalite; and (8) chaotic soft-sediment deformation of overlying pelagic sediments, and continued upward rise and surface breachment of serpentinites resulting in highly mixed assemblages.

Uranium/lead zircon ages on flaser diorite and plagiogranite dikes and Sm/Nd isochron ages on these dikes, as well as flaser gabbro and diabase dikes, indicate a Permo-Carboniferous age for the transform assemblage (Saleeby and Sharp, 1980; Shaw and others, 1987). In the Kings River area, the dikes, high-strain zones, and narrow serpentinite mélange belts are superposed over Ordovician ridge crest ophiolite. Detailed resolution of the relative proportions of Ordovician versus Permo-Carboniferous rocks within the Kaweah mélange to the south has not been attempted.

Fragmentation and incorporation of the Kings River ophiolite into the transform mélange seemingly requires the propagation of a major fracture zone and presumed spreading ridge segments into the early Paleozoic Panthalassa crust, in conjunction with major transform displacement. As noted above, it may have been abyssal crust, a large marginal basin behind the McCloud arc or within the McCloud fore-arc. Fragments of

McCloud arc rocks scattered elsewhere along the Central belt are consistent with the latter (Saleeby, 1990). The Kings-Kaweah segment of the transform system appears to have cut into the Cordilleran miogeocline and its Antler belt (Roberts Mountains) allochthon, as shown by the juxtaposition of the ophiolite belt along their truncation locus.

Permo-Carboniferous ophiolitic crustal rocks are widespread in disrupted mafic-ultramafic complexes of the Tuolumne,, Bear Mountains, and Feather River belts as well (Saleeby 1982, 1990; Saleeby and others, 1989a). Features that were cited above as typical of oceanic transform faults are present in each of these complexes, and tectonic lenses and diked screens of early Paleozoic rocks that are typical of the eastern Klamath McCloud arc or its structural basement are present in the Bear Mountains and Feather River belts (Sharp, 1988; Saleeby and others, 1989a; Saleeby, 1990). One might surmise that the late Paleozoic basement framework for the western Sierra Nevada was a gigantic ophiolitic mélange with fragments of eastern Klamath-like basement rocks dispersed southward within a Permo-Carboniferous, leaky transform, mafic-ultramafic matrix. The Permian to Middle Triassic depositional history along this regional basement belt further records this tectonic history.

Chert-rich olistostrome belts

Distinct belt-like exposures of chaotic argillaceous and cherty strata run the entire length of the western Sierra Central belt and extend through the Klamath Mountains (Schweickert and others, 1977; Behrman, 1978; Irwin and others, 1978; Saleeby, 1979; Blome and Irwin, 1983; Edelman and others, 1989). The main belt in the Sierra Nevada is the Calaveras complex, which runs semi-continuously along the western margin of the Feather River belt in the north and along the western margin of the southern Shoo Fly complex to the south (Fig. 1). The eastern border of the Calaveras complex is a polyphase tectonic boundary, which experienced its last major phase of ductile deformation during Middle Jurassic time (Sharp, 1988). Additional belts of Calaveras-like rocks occur to the west of the main belt interleaved with Paleozoic ophiolitic basement rocks and Jurassic arc-related strata.

The Calaveras complex consists of massive, rhythmically bedded, and diamicticeous chert; siliceous argillite; siltstone and subordinate quartz arenite with blocks and slabs of chert; metabasalt; marble and quartz arenite; and lenses of ultramafic and subordinate mafic rock (Schweickert and others, 1977; Edelman and others, 1989; Saleeby and others, 1986; Sharp, 1988). Chert inclusions in particular show abundant evidence for soft-sediment deformation occurring as pebbles in diamictites and as chaotic blocks ranging as much as tens of meters in diameter. Locally, hundred-meter-scale accumulations of bedded chert, some complexly folded, appear to be totally encased in argillite.

A thick amphibolite-marble unit occurs near the southern end of the Calaveras complex along its western margin. Field relations, and petrographic and geochemical data show that the unit represents an alkali basalt-affinity, volcanic shoal and reef sequence that was partly interbedded with Calaveras argillite and also shed slide blocks eastward into deeper-water chert-argillite olistostromes. The age of the volcanic structure is apparently Permo-Carboniferous, as suggested by horn coral remnants recovered from a marble olistolith (Schweickert and others, 1977). Fossils recovered from bedded intervals of the Calaveras range from Permo-Carboniferous to Middle Triassic in age (Hietanen, 1981; Bateman and others, 1985; Standlee and Nestell, 1985). A lower Permian carbonate bed at the northern end of the belt contains fauna that are diagnostic of the McCloud arc sequence of the eastern Klamaths (Standlee and Nestell, 1985). At the extreme northwestern end of the Sierran Foothills metamorphic belt, a large mass of chaotic olistostromes contains abundant blocks and contorted beds containing Permian McCloud fauna (Watkins and others, 1987), suggesting soft-sediment sliding from the fragmented McCloud arc directly into the Foothills transform trough.

Basement for the Calaveras complex is unknown, although the widespread lenses of ultramafic rock, typically serpentinite or metaserpentinite, and metabasite strongly suggest a disrupted Paleozoic ophiolitic substrate. The volcanic shoal sequence along the southern end of the belt has been interpreted as a seamount built on Paleozoic oceanic crust (Schweickert and others, 1977; Saleeby, 1981; Sharp, 1988). Alternatively, it could represent the volcanic and sedimentary carapace of a transform ridge constructed parallel to the olistostrome-filled trough. Chert-argillite olistostromes with quartz arenite admixtures and limestone blocks accumulated in a deep structural trough within the Kaweah mélange (Saleeby, 1979). The Kaweah olistostrome complex is very similar to the Calaveras complex and may represent the southern continuation of it. The limestone blocks in the Kaweah olistostromes carry a Permian Tethyan-affinity fauna. Similar Tethyan-affinity limestone blocks are reworked into Jurassic olistostromes between the Tuolumne and Bear Mountains ophiolitic complexes (Douglass, 1967; Behrman, 1978; Schweickert and Bogen, 1983).

Structurally complex argillaceous and cherty rocks of the North Fork terrane in the central Klamath Mountains strongly resemble rocks of the Calaveras complex (Irwin and others, 1978; Ando and others, 1983). Radiolaria dates from cherts range from Permian to Early Jurassic (Blome and Irwin, 1983). Ophiolitic basement rocks are exposed in the core of a large antiform. The ophiolite is highly disrupted, yields a Permo-Carboniferous U/Pb zircon age, and consists of both MORB and within-plate-like basaltic lavas. The North Fork terrane has strong similarities to the belt of Carboniferous ophiolitic rocks and coextensive Calaveras complex of the western Sierra Central belt. It lies to the west of, and structurally beneath, the Central Metamorphic belt and may, in part, represent the protolith of Late Triassic or older, high-P, metamorphic blocks of the Fort Jones terrane (Hotz and others, 1977). Ophiolitic basement rocks of the North Fork terrane and overlying chaotic chert-argillite are considered to be the northern continuation of the western Sierra Central belt.

Chert-argillite deposits of the Sierra-Klamath Central belt are considered to have never constituted a coherent sedimentary sequence, or to have rested on a coherent basement. In contrast, a tectonically active environment of deposition is envisaged where pelagic and continent-derived clastics accumulated in and along the flanks of linear troughs where ophiolitic basement rocks were being laterally sheared, and differentially uplifted and down-dropped by transform and subsequent subduction–fore-arc tectonics. During this complex depositional history, the transform system became the locus of a new subduction zone, resulting in its accretion to North America. This subduction accretion event probably occurred in latest Permian to Early Triassic time and corresponds in time with the Sonoma orogeny (Saleeby and Busby-Spera, this volume).

Tectonics of the Late Paleozoic rifting and transform truncation event

Alternative models for the sequence of events envisaged for the late Paleozoic truncation of the southwestern Cordillera are shown in Figures 2 and 3. The critical aspects of the models are in relating the transform development of the Sierran Foothills ophiolite belt and disruption of the McCloud arc to events and stratigraphic relations in regions to the south and east. The sequence of events shown spans the initial disruption of the McCloud arc, Antler orogenic belt and miogeocline, dispersion of derivative fragments along the transform zone, and the subsequent accretion of the Foothills ophiolite belt by subduction initiation along the transform zone. The Figure 2 model considers this tectonic regime to have developed totally within a suprasubduction zone setting along the Cordilleran plate edge, whereas Figure 3 involves a more intimate involvement of oceanic plates comprising part of the greater Panthalassa basin.

In Figure 2a the Antler orogenic belt and the outer McCloud arc and back-arc basin system are uniformly molded against the outer edge of the Cordilleran miogeocline, which along its inner edge remained a carbonate shelf through Antler time (Miller and others, this volume). The southern segment of the McCloud arc is hypothesized to have consisted of the proto-Yukon-Tanana terrane currently of central to northern British Columbia, parts of the Yukon territory, and eastern and southeastern Alaska. This interpretation is based primarily on the current tectonic position of the Yukon-Tanana terrane and its superjacent McCloud arc rocks of the Stikine terrane, outboard of the Cache Creek terrane, and on a distinct late Proterozoic detrital zircon component within Yukon-Tanana metaclastic rocks, which appears to be unique to the craton basement that is adjacent to the truncated southwestern U.S. Cordilleran margin (Mortensen, 1990). The southern McCloud arc, its back-arc basin, and the southwest reaches of the miogeocline arc shown in Figure 2b to have been disrupted by a Permo-Carboniferous boundary transform and related spreading system. Deformation, large-magnitude displacement, and spreading along the Sierran Foothills boundary transform occurred in conjunction with diffuse spreading in the Havallah basin and extensional tectonism in the miogeocline. The transform system gave rise to the Foothills ophiolite mélange belt, with the dispersion of McCloud arc basement fragments into the mélange (Saleeby, 1990), sinistral offset of a large fragment of the miogeocline edge (Caborca-Hermosillo block) to the Sonora (Mexico) region (Silver and Anderson, 1974), and an analogous offset of Antler belt rocks to the southeast Sierra–northern Mojave region (Walker, 1988). Analogous disruption of the miogeocline edge in southern British Columbia (Eisbacher, 1983) is suggested to have extended southward to the proto-Idaho suture zone. Rifting and transform faulting also disrupted volcanic patterns in the McCloud arc resulting in a series of largely nonvolcanic ridges, which in Early Permian time hosted major carbonate reef buildups of the McCloud limestone (Miller, 1987; Miller and others, this volume). Rifting and possible transform fragmentation of the McCloud arc in the Blue Mountains is further recorded by the Permo-Carboniferous construction of the arc tholeiitic Canyon Mountain ophiolite complex (Gerlach and others, 1981; N. W. Walker, personal communication, 1991), and its fragmentation into serpentinite matrix mélange.

In Figure 3a, rifting and transform faulting along and to the west of the Sierran Foothills truncation zone are considered to have begun in Antler time. The plate geometry depicted is similar to the active Okinawa Trough and its lateral transformation into the Taiwan collision zone (Herman and others, 1979; Suppe, 1984). In the Cordilleran case, the northern Sierra–eastern Klamath and Quesnel segments of the McCloud arc were outboard of the convergent orogenic zone like the northern extremity of the Luzon arc, whereas the Yukon-Tanana/Stikine segment was outboard of the spreading basin like the Ryukyu arc. Continued spreading in the hypothesized basin freed the southern segment of the McCloud arc as an ephemeral intra-oceanic arc within the northeastern Panthalassa basin (Fig. 3b).

Both models depict the encroachment of a Panthalassa, equatorial, oceanic island–seamount chain during Permian time. This chain, or chains, hosted a Tethyan-affinity shallow-water fauna that grew in carbonate fringing reefs (Monger and Ross, 1971; Nestel, 1980; Ross and Ross, 1983). The ensuing accretion of this oceanic island–seamount chain(s) along the Cordilleran plate edge gave rise to the Cache Creek terrane of British Columbia, and analogous "Tethyan" limestone-bearing mélanges of the Blue Mountains, western Klamaths, and western Sierra Foothills (Sierra-Klamath Central belt).

Figures 2c and 3c depict the plate tectonic setting of the Permo-Triassic Sonoma orogeny. This event is defined classically by the collapse of the Havallah basin and the eastward thrusting of the Golcanda allochthon across the miogeocline edge (Miller and others, this volume). Other major continent verging thrust systems related to the Sonoma orogeny include the Last Chance system and structures of the central and western Mojave Desert region (Saleeby and Busby-Spera, this volume). The Sonoma orogeny is suggested in Figures 2c and 3c to have been driven by the initiation of subduction along the Foothills boundary transform system with the resulting accretion of the Foothills ophiolite mélange belt to the hanging wall of the new subduction zone; also shown diagrammatically in Figure 4a. Consumption of Panthalassa ocean floor resulted in the collision and accretion of the equatorial oceanic island–seamount chain(s), and in the case of Figure 3c the added collision of the Yukon-Tanana/Stikine oceanic arc. The collision(s) may have been instrumental in driving the Sonoma orogeny, possibly triggering a short-lived polarity reversal in the McCloud arc.

Geologic relations along the southwestern U.S. truncation zone have been greatly complicated by superposed Mesozoic deformations and igneous activity. The truncation zone along the axial Sierra Nevada is particularly obscure due to the crosscutting Sierra batholithic belt. Metamorphic rocks to the east of the Foothills ophiolite belt can, in a broad sense, be related to regional stratigraphic patterns to the east; as such, they probably represent remnants of the truncated continent edge, marginal basin, and McCloud arc. In the northern half of the range, the Shoo Fly complex is juxtaposed against the Feather River ophiolite belt and the Calaveras complex. The Shoo Fly complex is considered a transitional simatic-sialic basement domain within the McCloud arc–Havallah basin transition zone. Stratigraphic ties between the Shoo Fly complex and the Havallah basin are demonstrated by biostratigraphic data from the Carboniferous Peale chert, which lies stratigraphically above the Shoo Fly complex, and from cherty strata of the Havallah sequence (Harwood and Murchey, 1989). It follows that the Golconda thrust, along its southwestern reaches, may root westward beneath the Shoo Fly complex, as suggested by Speed (1979).

In the southern Sierra Nevada, the boundary between the Foothills ophiolite belt and Paleozoic rocks to the east appears to have been overlapped depositionally by lower Mesozoic fore-arc strata of the Kings sequence (Saleeby and others, 1978). The Kings sequence represents the main metamorphic framework assemblage of the southern Sierra region (Bateman and Clark, 1974; Saleeby and Busby-Spera, this volume). The distribution of the main pendants is shown in Figure 1. Western turbiditic and olistostromal facies of the Kings sequence were deposited on the Kings-Kaweah ophiolite belt and Calaveras-like chert-argillite. Thick quartzo-feldspathic sands, carbonates, turbidites, and silicic volcanic rocks of the eastern facies may have been deposited on displaced miogeocline edge rocks and/or displaced Antler belt rocks. It is also possible that much or all of the Kings sequence was deposited on the Foothills ophiolite belt outboard of the main truncation (Kistler, 1990). Kings-sequence-like rocks occur to the north in thrust contact with the Calaveras complex, and in pendants to the south in Salinia and the northern Peninsular Ranges batholith (Saleeby and Busby-Spera, this volume). A number of relations in regions to the south suggest deposition on or along the edge of the truncated miogeocline. The only other location along the U.S. Cordilleran plate edge where thick terrigenous fore-arc basin strata of early Mesozoic age occur is in the "Izee terrane" of the Blue Mountains, where the basinal strata lap across Baker terrane and McCloud arc fragments (Dickinson and Thayer, 1978; Saleeby and Busby-Spera, this volume).

The subduction accretion of the Foothills ophiolite belt in Permo-Triassic time produced a large tract of disrupted and mixed oceanic crust, mantle, and sedimentary rocks, which served as fore-arc basement for the early Mesozoic southwestern U.S. continental arc. This basement complex hosted frontal arc and extensional fore-arc volcano-plutonic associations during the Late Triassic through Jurassic time (Saleeby and Busby-Spera, this volume). In the southern Sierra region, the arc–fore-arc system developed along the miogeocline–Antler belt truncation zone. In the northern Sierra to Klamath Mountains region, it developed along the truncated McCloud arc. This regional pattern is depicted in Figures 2d and 3d, which show the Cordilleran continental arc in relation to the truncation zone in Late Triassic time.

FORE-ARC AND INTERARC BASIN GENERATION OF JURASSIC OPHIOLITIC ASSEMBLAGES

The plate geometries depicted in Figures 2d and 3d are taken as possible initial conditions for treatment of the petrogenesis and tectonics of Jurassic ophiolitic assemblages of California, southern Oregon, and northwestern Washington. The original strike length of the accreted boundary transform complex along the early Mesozoic fore-arc probably exceeded that shown; however, Cretaceous and Cenozoic tectonics have obliterated the pertinent geologic history in southern Alta and Baja California, and in much of Oregon and Washington. Geologic relations along the western Sierra Nevada and Klamath Mountains indicate that the Jurassic fore-arc regions underwent two distinct phases of extension that resulted in volcanic and plutonic construction of distinct ophiolitic assemblages, at least partly within the basement framework of the regional Paleozoic ophiolitic mélange. The first phase of extension occurred during Late (probably latest) Triassic to Early Jurassic time; the second phase occurred during the late Middle Jurassic and included the production of the California Coast Range ophiolite. A plate tectonic model for the two cycles in fore-arc extension and an intervening contractile phase of tectonism is covered in depth in Saleeby and Busby-Spera (this volume). A highly diagrammatic cross-sectional model is given in Figure 4. Discussion of the Late Triassic to Early Jurassic phase of extension will focus on Figures 2d and 3d in addition to Figure 4.

Late Triassic to Early Jurassic fore-arc extension

Distinct volcanic and shallow-level plutonic sequences were produced in an apparent fore-arc setting along the entire length of the Sierra-Klamath Central belt during Late Triassic to Early Jurassic time. Some of the sequences are presently situated as nappes totally detached from their basement rocks, while some still retain vestiges of primary relations with ophiolitic mélange basement rocks. Early phase ophiolite research considered these polygenetic assemblages to be complete yet broken consanguineous ophiolite sections. Subsequent work has revealed the polygenetic development of these sequences with a clear arc affinity to

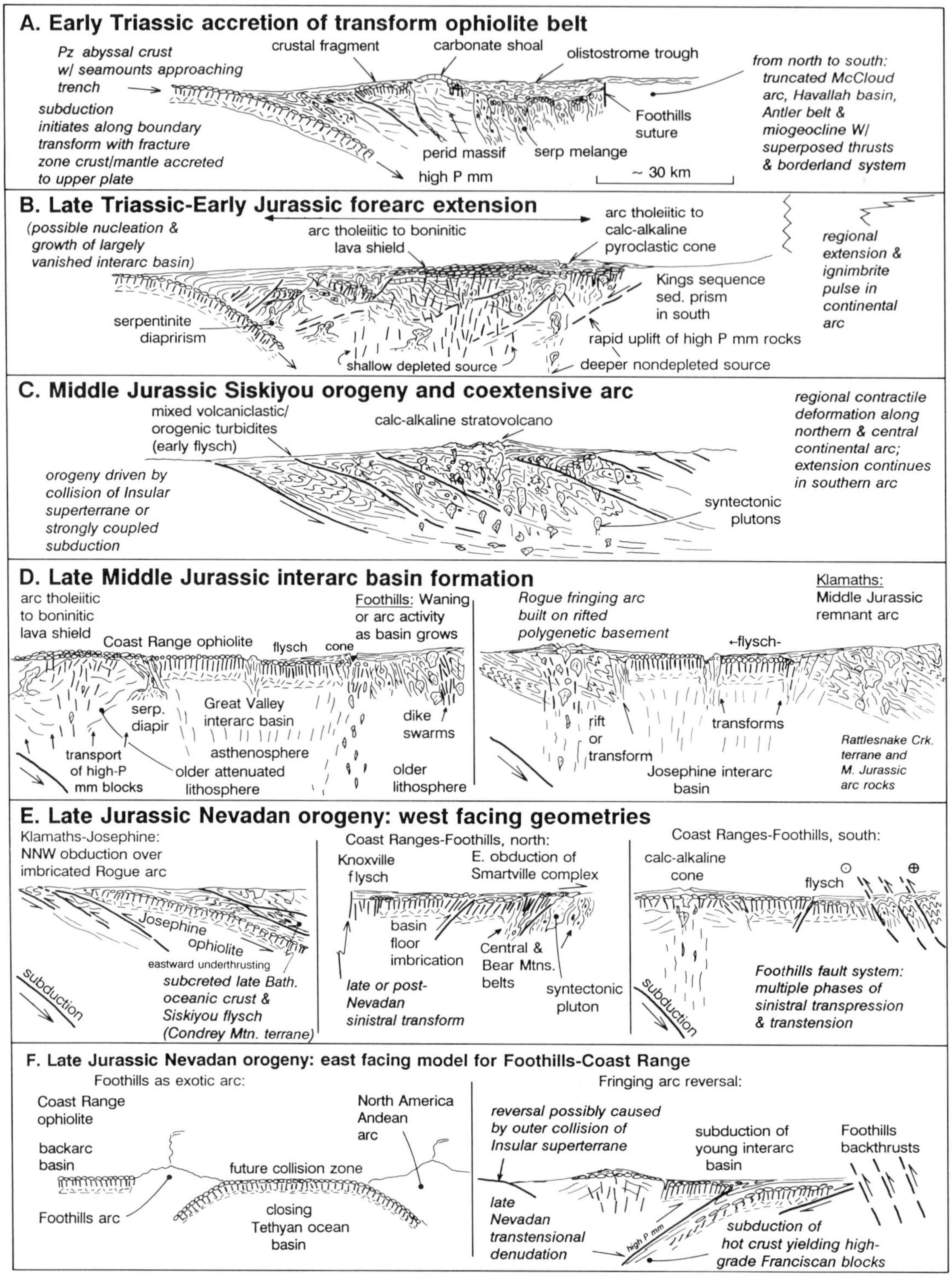

Figure 4. Series of diagrammatic cross sections depicting the Permo-Triassic accretion of the Sierran Foothills transform ophiolite belt along the trace of the Foothills suture and the subsequent forearc basement evolution of the ophiolite belt leading to the generation of late Middle Jurassic interarc basin ophiolites including the Coast Range and Josephine. E and F depict different geometries of the Late Jurassic Nevadan orogeny, which involved at least local closure of the interarc basin system (after Schweickert and Cowan, 1975; Snoke, 1977; Saleeby and others, 1982; Harper and Wright, 1984; Saleeby and others, 1986; Saleeby and Busby-Spera, this volume).

the volcanic intervals. The better-studied sequences include, from north to south, the southern Rattlesnake Creek terrane, Fiddle Creek complex, Slate Creek complex, Penon Blanco Formation, Sullivan Creek terrane, and the western facies of the Kings sequence (reviewed in Saleeby, 1990). Petrogenetic ages on most of the better-studied sequences cluster at 200 ± 10 Ma (Irwin and others, 1978; Wright, 1981; Saleeby, 1982, 1990, unpublished data; Bateman and others, 1985; Saleeby and others, 1989a). The sequences are characterized by mainly mafic submarine pillow lavas and pyroclastic rocks, and shallow-level intrusive rocks ranging mainly from peridotite to diorite in composition. Sequences of radiolarian and tuffaceous chert are associated with pillow lavas, but are rare or absent in thick pyroclastic intervals. Volcanic and dike rocks show arc tholeiitic, boninitic, and locally calc-alkaline and shoshonitic petrochemical affinities (Wright, 1981; Wright and Wyld, 1985; Dilek and others, 1990; Saleeby, 1990). Sets of sheeted dikes are closely associated with some volcanic sections, and in most instances the dikes cross-cut older polydeformed ophiolitic basement rocks. A generalized model for the extensional fore-arc is given in Figure 4b.

The tectonic framework envisioned for the early Mesozoic fore-arc of the California region is much like the mid-Tertiary of the Mariana-Bonin fore-arc, where deep-water sequences of pillowed arc-tholeiite, boninite, and marianite were erupted across disrupted oceanic crust of the fore-arc basement (Ujiie and Matsumaru, 1977; Natland, 1981; Bloomer, 1983; Umino, 1985; Dobson, 1986). Following the construction of the broad-pillow lava carapaces, highly porphyritic pyroclastic aprons were built. Hemipelagic sedimentation continued throughout the eruptive history and was periodically swamped out during major eruptive phases or sediment influxes from the main constructional arc. Abyssal or fracture zone ophiolitic basement of the fore-arc was locally uplifted by faulting and diapirism during and following volcanism, resulting in the reworking of ophiolitic debris into the fore-arc volcanic-sedimentary sequences.

Extensional tectonism within arc systems may have a variety of manifestations. Such tectonism and volcanism in the Mariana-Bonin arc led to the development of the West Mariana and Shikoku interarc basin system, marking the generation of a substantial tract of juvenile oceanic crust and upper mantle. Such is not directly recorded for the Late Triassic to Early Jurassic of California; complete consanguineous ophiolite sections of the appropriate age are not represented. Either extensional tectonism did not progress to such an extreme, or the resultant interarc basin vanished by subduction or by attrition due to later rifting and transform faulting. An extensional framework is clearly implied, however, by the remnants of sheeted dikes and widespread serpentine diapirism, in addition to the volcanic associations. Furthermore, thick tuffaceous chert-pillow lava sequences, such as those constituting the younger portions of the North Fork terrane of the Klamath Mountains (Irwin and others 1978) and parts of the northern Sierra Fiddle Creek complex (Saleeby, 1990), record the development of a substantial basinal system, possibly reflecting diffuse spreading. Local transform zones within the fore-arc are also marked by tabular flaser gneiss bodies and mylonitic overprints in older peridotitic and amphibolitic basement rocks. A broad extensional framework for the Late Triassic to Early Jurassic of the southwestern Cordillera is also suggested by widespread relations along the main constructional arc to the east of the fore-arc domain. A regional graben complex formed, and there was an associated distinct and voluminous pulse of silicic ignimbrite activity (Busby-Spera, 1988; Saleeby and Busby-Spera, this volume). The broad extensional framework for the early Mesozoic Cordilleran arc system may reflect the normal retreat component of North America's absolute motion relative to the plate edge (May and Butler, 1986; Saleeby and Busby-Spera, this volume), with the added effect of the subduction of old, cool, Panthalassa lithosphere inherited from the Pangaea regime (Busby-Spera and others, 1990).

Two alternative plate tectonic geometries for early Mesozoic extension along the Cordilleran fore-arc are depicted in Figures 2d and 3d. In Figure 2d, the western Sierra-Klamath extensional basin system served as a major transtensional dislocation zone for the outboard migration of the Yukon-Tanana/Stikine microplate. In this view, the basinal system constituted an ephemeral interarc basin during the microplates migration inasmuch as Stikine was experiencing arc magmatic activity (Monger and others, 1992). In Figure 3d, the western Sierra-Klamath basinal system constituted a broad, volcanically active fore-arc with interleaved basin floor domains and volcano-tectonic ridges analogous to the Ogasawara Ridge, Trough, and paleolands of the modern Bonin fore-arc (Yuasa and others, 1982).

In both models (Figs. 2d and 3d), the western Sierra-Klamath basinal system is considered to have been in broad geographic continuity with the Cache Creek and Bridge River basinal system of northern Washington and British Columbia (Saleeby and Busby-Spera, this volume). The Cache Creek terrane remained at least partly basinal during Late Triassic to possibly as late as Middle Jurassic time, with the deposition of tuffaceous chert and volcaniclastic sands (Cordey and others, 1987; Monger and others, 1992). The youngest parts of the Cache Creek terrane resemble the Bridge River terrane and also the younger parts of the North Fork terrane of the Klamath Mountains and parts of the northern Sierra Central belt. Upper Triassic arc tholeiitic rocks of the Sparta "ophiolite" of the Blue Mountains may represent another igneous complex within this basinal system. The distinguishing feature betwen the younger parts of the Cache Creek terrane, the Bridge River terrane, and the early Mesozoic basinal system of the western Sierra and Klamaths is the position of the Stikine arc fragment between the former two. This fragment is distinct by its construction on a large tract of Paleozoic McCloud arc rocks. Its lower Mesozoic, marine, pyroclastic sequences strongly resemble those of the Sierra-Klamath Central belt, however (Saleeby and Busby-Spera, this volume), and have at least local facies relations with younger portions of the Cache Creek terrane (Monger and others, 1992). These scattered bits of stratigraphic links lead to the conclusion that the lower Mesozoic pyroclastic sequences and their basinal facies constituted a broad suprasubduction zone overlap system spread across previously accreted, albeit loosely coupled, Cache Creek and related Central belt oceanic assemblages and McCloud arc rocks.

Early Jurassic extensional tectonism in the western Sierra-Klamath fore-arc may have led to the denudation of high-pressure metamorphic rocks. At present, only sparse data exists, but the data that does exist is thought-provoking. Polyphase serpentinite mélange zones of the Bear Mountains, Tuolumne, and Kaweah ophiolite belts contain high-P amphibolite and locally eclogitic blocks (Behrman, 1978; Saleeby and Sharp, 1980; Sharp, 1988). The most-reliable K/Ar and $^{40}Ar/^{39}Ar$ dates available cluster at ~200 Ma, corresponding to the main pulse of fore-arc extensional volcanism. Such relations suggest that the high-pressure metamorphic environment beneath the fore-arc was partly disrupted by extensional tectonism, and that the ages recorded in the tectonities reflect the extensional sampling event. Crustal-level upward transport appears to have involved significant serpentinite diapirism. Concomitant fore-arc volcanism in such a scenario could not have been long-lived, or involved the production of large magma chambers at depth, owing to the preservation of high-P metamorphic assemblages. This is consistent with the temporal distinctiveness of the eruptive phase, and with the tholeiitic and boninitic compositions, which approximate mantle-derived liquids. On a broader tectonic scale, Davis and others (1978) have delineated a nearly Cordilleran-wide belt that contains blueschist-facies metamorphic blocks having isotopic ages that cluster in the Late Triassic. The belt coincides with the Cache Creek terrane and similar rocks extending southward through the Blue Mountains and Sierra-Klamath Central belt. In the context of the above discussion, it seems quite possible that the clustering of ages in the Late Triassic could represent abrupt disruption of the high-pressure metamorphic environment and denudation related to regional extension/transtension along the entire Cordilleran arc/fore-arc system. The actual subduction metamorphic environment may have begun in Permo-Triassic time with longevity of the high-P/T conditions maintained by subduction of very old and cold sub-Panthalassa lithosphere inherited from the Pangaea regime (Busby-Spera and others, 1990).

Subsequent to the Late Triassic to Early Jurassic extensional tectonic regime discussed above, the Cordilleran arc and fore-arc from the central California to British Columbia regions underwent a major crustal shortening event during Middle Jurassic time referred to as the Siskiyou orogeny (Coleman and others, 1988; Saleeby and Busby-Spera, this volume). Alternative tectonic models for the Siskiyou orogeny include a phase of strongly coupled subduction, which drove overriding-plate thick-skinned thrusting, or the oblique collision of a major exotic terrane conglomeration referred to as the Insular superterrane (Wrangellia, the Alexander and southern Alaska Peninsular terranes). The subduction coupling model requires a north-south segmentation in subduction zone dynamics inasmuch as the southern Cordilleran arc remained strongly extensional and otherwise unaffected by Siskiyou shortening (Busby-Spera, 1988; Saleeby and Busby-Spera, this volume). The subduction coupling model suffers from the apparent north to south migration in the initiation of Siskiyou deformation, which broadly matches the paleolatitudinal shift recorded in the Insular superterrane as well as its accretion record, and from a lack of any obvious change in plate motion patterns, at least as far as North America is concerned. Finally, Siskiyou deformation was abruptly replaced by renewed regional extension in late Middle Jurassic time. Thus, an oblique collision model within a generally weakly coupled subduction regime is favored here, and in Saleeby and Busby-Spera (this volume).

Late Middle Jurassic fore-arc and interarc basin ophiolites

Deformation and metamorphism of the Middle Jurassic Siskiyou event resulted in the west-vergent imbrication of Lower Jurassic fore-arc sequences and their ophiolitic mélange basement complexes in the northern Sierra and Klamath Mountains regions (Harper and Wright, 1984; Wright and Fahan, 1988; Coleman and others, 1988; Saleeby and Busby-Spera, this volume). During and at the close of Callovian time, the fore-arc margin of the Siskiyou deformation zone began to rift apart and then a family of interarc basins rapidly formed. Rift complexes are preserved in the western Sierra Nevada and Klamath Mountains, and interarc basin and fringing arc or rifted fore-arc igneous complexes are preserved in the western Klamath Mountains and Coast Ranges. Rifting is well recorded in the northwestern Sierra Nevada between 164 and 162 Ma by major swarms of basaltic to plagiogranite sheeted dikes and parallel tabular intrusions (Saleeby, 1982; Saleeby and others, 1989a). At the northern end of the Bear Mountains ophiolitic belt, sheeted dikes and coeval gabbro and diorite lenticular stocks were injected along the preexisting ophiolitic mélange structural fabric and accompanied by dynamothermal contact aureoles (Springer, 1980; Saleeby, 1982, 1990). The Folsom dike swarm resembles a sheeted complex except for the fact that it is bounded by, and locally encases, screens of Paleozoic host rocks. The dike complex extends for more than 30 km along strike. To the south, it feathers into the Bear Mountains belt and to strands of the Bear Mountain fault zone. The Bear Mountains fault zone is commonly interpreted as a Late Jurassic (Nevadan) thrust. The relations outlined above in conjunction with common attenuation of stratigraphic units, younger over older and low metamorphic grade over high-grade rocks suggest that early phases of low-angle normal faulting may have occurred along the Bear Mountains fault zone in conjunction with late Middle Jurassic extensional magmatism. The northern end of the Folsom dike swarm merges into the upper volcaniclastic unit of the Smartville complex. This complex consists of a well-preserved pseudostratigraphic sequence of minor layered gabbro, larger gabbro-diorite and tonalite tabular plutons, massive diabasic hypabyssal rock, pillow basalt, mafic and plagiogranite sheeted dikes, and submarine pillowed and volcaniclastic rocks of basaltic to locally dacitic composition, all of Middle Jurassic age (Xenophontos and Bond, 1978; Menzies and others, 1980; Beard and Day, 1987; Saleeby and others,

1989a). Petrographic and field data along with rare earth and major element chemical data indicate a tholeiitic to calc-alkaline evolution from early to late magmas and lavas. The early (lower) mafic flows are typically pillowed and aphyric, while later (upper) basaltic to andesitic pyroclastic rocks are porphyritic. A distinctive feature of the Smartville complex is a regional-scale (70 × 15 km) sheeted dike complex that is similar in age, composition, and structure to the Folsom dike swarm.

In the west-central Klamath Mountains, the Rattlesnake Creek terrane in the area of Preston Peak and the Marble Mountains also records a phase of Middle Jurassic rifting. At Preston Peak, dikes, irregular hypabyssal bodies, near-vent breccias, and local pillow flows of an arc tholeiitic series are well preserved in a section that intrudes and unconformably overlies older serpentinite-matrix mélange (Snoke and others, 1977; Snoke and Whitney, 1979; Saleeby and others, 1982). The mélange basement contains mainly large blocks of tectonitic harzburgite and lherzolite with smaller mafic amphibolite blocks typical of the laterally continuous Rattlesnake Creek terrane to the south. Sparse isotopic and radiolarian age data suggest a Middle Jurassic age for the cover rocks (Saleeby and others, 1982; Harper and others, 1992). To the east, in the Marble Mountains area, high-grade ophiolitic mélange is considered equivalent to the Rattlesnake Creek terrane, but there it has undergone amphibolite-facies metamorphism en masse within the Siskiyou thrust belt (Hill, 1985; Donato, 1987; Coleman and others, 1988). The interlayered mafic amphibolites, metaperidoties, and siliceous schists represent metamorphosed ophiolitic mélange. The remnants of a late Middle Jurassic sheeted dike complex were recognized within one of the larger amphibolite layers (Hill, 1985; Harper and others, 1992).

From the above relations it is clear that the polygenetic ophiolitic basement of the western Sierra-Klamath belt experienced a significant phase of extension, and actually hosted major sheeted dike swarms in late Middle Jurassic time. The age of this extensional event matches very closely with ages of both the Coast Range and Josephine ophiolites to the west (Hopson and others, 1981; Whetten and others, 1978, 1980; Saleeby and others, 1982, 1984; Miller, 1985; Harper and others, 1989, 1992; Saleeby, 1990). The sequence of events idealized in Figures 4b through 4d covers the superpositioning of Siskiyou thrusting on the Late Triassic to Early Jurassic extensional fore-arc assemblages of the Sierra-Klamath Central belt, and then renewed extension in late Middle Jurassic time. In contrast to the earlier phase of extension, the later phase can be shown to have produced significant domains of new oceanic lithosphere, as recorded in the Josephine ophiolite and parts of the Coast Range ophiolite.

Coast Range ophiolite—northern Great Valley region

Geophysical and basement core data from the northern and central Great Valley suggest that its basement consists of a gently west-dipping slab of oceanic crust (Suppe, 1979; Saleeby and others, 1986; Blake and others, 1989). The mafic composition of the crust is shown by regional gravity and seismic velocity data in addition to the basement cores. A gentle west dip to the mafic slab is suggested by the main gravity-magnetic lineaments, which run down the axis of the valley. The gross west-dipping structure may be interpreted as major east-directed overthrust (Suppe, 1979; Saleeby and others, 1986), or alternatively, as a deformed rift edge (Harper and others, 1985). Regardless of the interpretation favored, the Great Valley oceanic basement is of late Middle Jurassic age and represents the vestiges of an interarc or fore-arc rift basin. The Great Valley basement surfaces as the Coast Range ophiolite along the transition between the western Great Valley and Coast Ranges.

The Coast Range ophiolite along the northern Great Valley region is represented primarily by a belt of sheared serpentinite and serpentinite-matrix mélange, which separates the Great Valley sequences from the Franciscan complex (Hopson and others, 1981). Remnants of oceanic crustal rocks are scattered along the serpentinite belt, with the most complete association in the Paskenta area (Bailey and others, 1970; Hopson and others, 1981; Blake and others, 1987a, b). Major and trace element geochemical data from mafic rocks at Paskenta and elsewhere along the northern Great Valley belt are similar to data from enriched MORB or ocean island tholeiites (Bailey and Blake, 1974; Shervais and Kimbrough, 1985; MacPherson and Phipps, 1988). The Paskenta ophiolite section was disrupted early in its history by submarine faulting (Bailey and others, 1970; Hopson and others, 1981; Blake and others, 1987a, b). A talus breccia as much as 1 km thick, consisting of ophiolitic volcanic and plutonic clasts, rests unconformably on the ophiolite at structural levels ranging as deep as ultramafic cumulates. Radiolarian chert locally lies along the base of the breccia and yields Oxfordian to Kimmeridgian ages. A rugged submarine topography is implied by rapid changes in thickness of the breccia and changes in substrate structural level in the ophiolite. The breccia is conformably overlain by Upper Kimmeridgian (~152 Ma) flysch of the Knoxville Formation (Jones, 1975). Major ophiolitic olistostromes occur in analogous stratigraphic positions along much of the northern Coast Range belt; some were deposited directly on diked serpentinite (MacPherson and Phipps, 1988).

Several features of the northern Coast Range assemblage are consistent with the interpretation that it represents interarc or extensional fore-arc basin crust. Fine tephra deposits are interlayered with radiolarian cherts (Hopson and others, 1981), which probably represent distal products of arc activity. Calc-alkaline hornblende diorite dikes of ~155 Ma age (Lanphere, 1971) crosscut the main igneous sequence and predate the unconformity at the base of the submarine talus breccia. Silicic-to-intermediate thin-bedded tuffs locally occur in the lower intervals of the overlying Knoxville Formation. Each of these signals of distal or proximal arc activity is also characteristic of the Oxfordian to Kimmeridgian history of the Josephine ophiolite, as discussed below. The intense submarine faulting and denudation recorded in the Paskenta assemblage and throughout the northern Coast

Range belt, and the initial phases in the development of the serpentinite mélange may reflect a major longitudinal transform fault system within the northern Great Valley interarc basin floor, similar to that described for the Josephine ophiolite.

Coast Range ophiolite—Diablo Range

Ophiolitic rocks are exposed as an upper plate tectonic envelope around the Franciscan core of the Diablo antiform and as fault slivers along the San Andreas system to the west (Fig. 1). These exposures represent oceanic basement for the southern Great Valley sequence (Bailey and others, 1970; Hopson and others, 1981). Well-studied sections of this segment of the Coast Range belt occur in the Del Puerto Canyon and Mount Diablo areas of the Diablo Range (Evarts, 1977; Hopson and others, 1981; Williams, 1983). These studies show great contrast between rocks of the Diablo Range and the northern Great Valley belt, indicating a fore-arc igneous and/or fringing arc site of petrogenesis for the former.

In the Del Puerto Canyon area, members of a complete ophiolite section are broken into several large fault blocks (Evarts, 1977). The volcanic members of the sequence consist of basaltic and keratophyric pillow lavas overlain by a thick, submarine, pyroclastic and volcaniclastic sequence of mainly andesitic composition (Lotta Creek tuff). The lower levels of the volcanic sequences are cut by numerous dikes and sills of mainly keratophyre and quartz keratophyre. U/Pb zircon and K/Ar hornblende age data show that the plutonic, volcanic, and volcaniclastic intervals all formed in a very short time span (Sharp and Evarts, 1982). The Lotta Creek tuff is overlain with apparent conformity by upper Tithonian mudstone of the Knoxville Formation. The volcanic section of the Del Puerto sequence shows a clear island arc affinity. The lower pillow lavas contain low-TiO_2 basaltic andesites, some with strong boninitic affinities (Bailey and Blake, 1974; Robertson, 1989; J. H. Natland, written communication, 1983). The overlying Lotta Creek tuff represents distal to medial products of calc-alkaline volcanic activity, which is well recorded in correlative strata above pillowed sequences in ophiolite remnants as much as 150 km south of Del Puerto Canyon (Hopson and others, 1981). The abundance of felsic intrusives within Del Puerto and other Diablo Range remnants is also suggestive of a strong arc component. Such arc magmatism appears to have persisted throughout Oxfordian time.

An additional important Coast Range ophiolite remnant occurs at the northern end of the Diablo Range, near Mount Diablo (Williams, 1983). Here, the upper part of the sequence is intact and well preserved, but the deep plutonic and upper mantle harzburgites are highly disrupted. A significant feature of this ophiolite remnant is that it contains the only well-developed sheeted dike complex of the entire Diablo Range segment of the Coast Range belt. The dike complex has a thickness of about 2.5 km and feeds a deep-water pillow basalt sequence of about 1.5 km thickness. The sheeted dikes and pillow lavas have tholeiitic geochemical affinities, although screens of calc-alkaline intrusive rock lie within the dike complex. Paleomagnetic studies on the Mount Diablo ophiolite remnant show substantial rotation (Williams, 1983), which upon correction leads to a transverse orientation for the sheeted dikes relative to the plate edge. These relations suggest that the Mount Diablo remnant represents part of an interarc basin floor with local screens of rifted arc rocks, which were split off tangentially along the plate edge.

Diablo Range ophiolite exposures to the south of Mount Diablo contain an abundance of arc-related material and may have been bounded to the north by interarc basin floor, represented by the Mount Diablo remnant. Stratigraphic thickness and more-proximal volcaniclastic units of the upper calc-alkaline volcanic levels increase southward along the Diablo Range belt (Hopson and others, 1981), suggesting a major arc segment once ran southward from the northern Diablo Range. If the Diablo Range and northern Great Valley segments of the ophiolite belt have not moved appreciably, relative to one another, the northern Great Valley segment may represent the remnants of this basin with the Mount Diablo remnant lying near the rift edge (Saleeby and Busby-Spera, this volume). The paleomagnetic data for the Mount Diablo remnant show little resolvable latitudinal translation (Williams, 1983), whereas data obtained on pillow lavas and a thick volcaniclastic cover from the southern end of the Diablo Range are equivocal (Beebe, 1986).

Coast Range ophiolite fragments have been translated northward from the Diablo Range by dextral faulting into the northern Franciscan complex (Blake and others, 1985a). These ophiolitic fragments have clear affinities to rocks of the Diablo Range, as opposed to those of the northern Great Valley. The affinities include a strong island arc geochemical signature to the lavas (Bailey and Blake, 1974; Shervais and Kimbrough, 1985), a relative abundance of felsic intrusive rocks and tephra in the sedimentary record, and a distinctly younger age for the basal beds of the overlying Knoxville Formation (Hopson and others, 1981).

The Diablo Range segment of the Coast Range ophiolite belt is interpreted as a fore-arc igneous and/or outer fringing arc complex. The tholeiitic and boninitic lavas might have begun accumulating in a fore-arc regime during the early phases of late Middle Jurassic rifting, or actually represent the outer reaches of the interarc basin floor. The overlying pyroclastic rocks represent fringing arc development above the older rifted fore-arc basement sequence or directly on the outer reaches of the interarc basin floor.

Southern Coast Range ophiolite

Small remnants of the Coast Range ophiolite are dispersed along the southwestern Coast Ranges west of the San Andreas and Sur-Nacimiento faults (Fig. 1). These remnants lack the substantial submarine pyroclastic and volcaniclastic strata that are so well developed in the Diablo Range segment, although a modest amount of arc-derived tephra is dispersed within radiolarian cherts lying above pillow lava sections (Hopson and others,

1981). The most-complete and best-studied section in the southern Coast Range is found along the Coast at Point Sal (Hopson and others, 1981). Geochemical data on the pillow lavas indicate basaltic andesite compositions, some having boninitic affinities (Bailey and Blake, 1974; Shervais and Kimbrough, 1985; Robertson, 1989; J. H. Natland, written communication, 1983). Lead, neodymium, and strontium isotope data also indicate an arc affinity for the ophiolite (Chen and Shaw, 1982).

The southern Coast Range ophiolite remnants could represent vestiges of interarc basin crust or fore-arc igneous complexes. They clearly remained for a substantial period of time beyond the reach of proximal arc activity prior to Knoxville overlap. The geological record in this region is too fragmented to warrant more detailed interpretations. Hopson and others (1981) suggested that dismembered metavolcanic and plutonic rocks of Santa Cruz island about 100 km southeast of Point Sal (Fig. 1) represent the southward continuation of the Coast Range belt. A more detailed petrologic study of the Santa Cruz rocks shows, however, that they represent exposures of a regional Middle to Late Jurassic island arc terrane that extends southward to the southern California borderland (Sorensen, 1985). This arc terrane is interpreted to be a fringing segment that might have been outboard of or along strike, of the southern Coast Range ophiolite (Saleeby and Busby-Spera, this volume).

The southern Coast Range ophiolite belt attests to en masse transport of an ophiolite terrane via large-magnitude strike-slip faulting. Not only does the current transport pattern of the San Andreas system display this, but paleomagnetic and radiolarian paleobiogeographic data from the Point Sal remnant indicate major early-stage northward transport during the Late Jurassic (Beebe, 1986; Pessagno and Blome, 1990). The strike-slip transport of Coast Range and related ophiolite fragments has been complex and polyphase. In addition to the tangential transport patterns suggested within the spreading basins that generated the ophiolite belt and the current San Andreas transform regime, an important Late Cretaceous to early Tertiary phase of dextral transport dispersed fragments into the Pacific Northwest region (Blake and others, 1985a; Garver, 1988).

Coast Range ophiolite—Pacific Northwest fragments

Two main clusters of Coast Range ophiolite-related fragments occur in the Pacific Northwest region. The first is found within the southern Oregon Coast Ranges, the second in northwestern Washington. Two distinct fragments of Coast Range ophiolite occur in the southern Oregon Coast Ranges. The more western Snow Camp Mountain fragment occurs as an isolated nappe above, and adjacent to, the Dothan complex, which is equivalent to the Franciscan complex (Blake and others, 1985b). The Snow Camp Mountain ophiolitic nappe is bounded by a dextral-slip fault along its eastern margin. The nappe contains all elements of a complete ophiolite sequence with tholeiitic affinities, but it is in a highly faulted state. Depositionally above the pillow lavas sit turbidites of the Upper Jurassic Myrtle Formation, which is equivalent to the Knoxville Formation of the Great Valley. The Myrtle is much thinner than the northern Great Valley Knoxville, suggesting a more distal outer basin setting for the Snow Camp Mountain sequence, or perhaps transport from a more southerly position. Northwest of the Snow Camp Mountain nappe lies another belt-like exposure of ophiolitic rocks that runs for about 50 km along strike. Dismembered harzburgite tectonite and various mafic rocks are poorly exposed along much of the belt. A well-exposed, structurally intact assemblage occurs in the Wild Rogue Wilderness area (Ramp, 1980; Gray and others, 1982; Saleeby, 1984b, 1990). Here a well-developed mafic sheeted dike complex and a mafic pillow lava and felsic submarine pyroclastic sequence are intruded and separated by distinctly younger (~155 Ma) dioritic and leuco-tonalitic intrusives. Local screens of pillow lava and aquagene tuff lie within the sheeted complex as well. The Wild Rogue assemblage represents a rifted arc complex. A Middle Jurassic submarine arc was rifted during the late Middle Jurassic, and shortly thereafter voluminous arc-related plutonic rocks invaded the complex. The sequence of events is similar to that observed in the Josephine ophiolite and its neighboring terranes. Additionally, after rotational restoration (Saleeby and Busby-Spera, this volume), the orientations of the sheeted dikes in the Wild Rogue complex are at high angles to the plate edge, like the nearby Josephine ophiolite. Original relations between the Wild Rogue assemblage and the Snow Camp nappe are unknown, as are possible affinities of these fragments with the two distinct segments of the Coast Range belt to the south. It is possible that the Wild Rogue assemblage could alternatively be a tectonic outlier of Josephine ophiolite–Rogue arc affinity rocks. The Snow Camp Mountain ophiolite has clear Coast Range affinities, as shown by its Myrtle overlap.

The main ophiolite fragments of northwestern Washington occur in the Mount Stuart area of the central Cascades (Ingalls complex) and in the San Juan Islands (Fidalgo complex; Fig. 1). The Ingalls complex represents the most-complete ophiolite assemblage in the region (Miller, 1985). The ophiolite section is disrupted by zones of serpentinite mélange and high-temperature shear zones believed to have developed on the ocean floor. Ophiolitic talus deposits occur within the pillow lava section and these, along with chert and argillite, are crosscut by later-stage diabase dikes. Geochemical data on the Ingalls complex are variable and range from MORB-like tholeiites to intra-plate alkaline basalts (Vance and others, 1980). The Ingalls complex has been interpreted to be an oceanic transform fault assemblage (Miller, 1985). It resembles parts of the northern Great Valley belt, and is likewise interpreted in Saleeby and Busby-Spera (this volume) to be a transform fault assemblage from the late Middle to Late Jurassic interarc basin system of the southwestern Cordillera.

In the San Juan Islands, Brown and others (1979) interpreted a structurally complex suite of ultramafic rocks, gabbro, intermediate and silicic intrusive rocks, and mafic to intermediate volcanic rocks of the Fidalgo complex to be an ophiolite. Mafic to felsic, volcanic rocks of the ophiolite are interbedded with upper Kimmeridgian to lower Tithonian tuffaceous cherts, argillite, and sandstones that represent the products of arc volcanism (Gusey, 1978). The older gabbroic and ultramafic rocks that host the intrusive-volcanic suite are interpreted to be remnants of Callovian oceanic crust (Whetten and others, 1978; Brandon and others, 1988). The main intrusive and volcanic parts of the Fidalgo complex represent a late Middle to Late Jurassic volcanic arc built on slightly older oceanic crust. It is overlain depositionally by the Lummi Group, which consists of Tithonian flysch at its base. Strong similarities between the Fidalgo complex and overlying Lummi and the Diablo Range segment of the Coast Range belt and its Knoxville cover have been discussed by Garver (1988), implying substantial northward offset of the Fidalgo complex. Additional fragments of the Fidalgo complex occur in the western Cascade foothills as the Haystack Mountain complex (Whetten and others, 1980).

Both the Fidalgo and Ingalls ophiolitic complexes rest as high-level nappes across a myriad of displaced and highly fragmented terranes that were imbricated during the mid-Cretaceous (Brandon and others, 1988; Cowan and Bruhn, this volume; Saleeby and Busby-Spera, this volume). Various members of the structural sequence range from Precambrian to late Early Cretaceous in age. The tectonic association is unlike anything present in the Franciscan complex beneath the Coast Range belt, except for the Shuksan metamorphic suite, which resembles the Pickett Peak high-P metamorphic terrane of the Franciscan complex (Brown and Blake, 1987).

Josephine ophiolite

Remnants of oceanic crust and upper mantle that formed in an interarc basin are well preserved as the Josephine ophiolite, which runs semi-continuously along the western edge of the Klamath Mountains for more than 200 km (Snoke, 1977; Saleeby and others, 1982; Harper, 1984). The most-complete section of the ophiolite lies immediately south of the Oregon-California border (Harper, 1984). Here, the ophiolite consists of harzburgite tectonite, cumulate ultramafites and gabbro, high-level gabbro and diorite, a 1.5 km-thick sheeted dike complex, and 0.4 km of pillow lava and breccia (Harper, 1980, 1984). Pillow lavas show geochemical features of arc tholeiites and boninites (Harper, 1988). Structural relations within the Josephine ophiolite, including upper mantle flow fabrics and sheeted dike and transform fault orientations, indicate sea-floor spreading, plate transport, and thrust emplacement kinematics with strong tangential components relative to the Cordilleran plate edge (Harper and others, 1985, 1989, 1990, 1992). Such relations suggest that rift edges to the Josephine basin originally contained significant boundary transform segments.

The Josephine ophiolite is conformably overlain by the Galice Formation, which consists of a thin siliceous pelagic sequence overlain by a thick sequence of turbidites having both volcanic arc and recycled continental sources. Relations along the base of the Galice are critical for the interpretation of the tectonic evolution of the Josephine ophiolite. Northward along the Galice outcrop belt, the substrate geology changes dramatically. In southern Oregon, the lower Galice is interstratified with pyroclastic apron deposits of the Upper Jurassic Rogue Formation, which represents the volcanic levels of a fringing arc built outboard of the Josephine basin (Garcia, 1979, 1982). The Josephine ophiolite lies structurally between the Rogue arc sequence and the Rattlesnake Creek terrane along Late Jurassic (Nevadan) thrust faults. The thrusts appear to have nucleated along boundary transforms or rift edges between the Josephine interarc basin and adjacent rifted arc segments represented by the Rogue and Rattlesnake Creek terrane. This is suggested by the conformable overlap of the Galice Formation across both the ophiolite and the Rogue sequence, and by the presence of rifted screens of Rattlesnake Creek terrane within the southern Josephine ophiolite (Harper and others, 1985, 1992; Wyld and Wright, 1985). Furthermore, fragments of Rattlesnake Creek terrane-like rocks occur within the Rogue basement, suggesting that the Josephine basin opened within the confines of older western Sierra-Klamath belt lithosphere, and then shortly thereafter the basin closed, leaving the relatively narrow ophiolite belt as its remnant (Saleeby and others, 1982; Yule and Saleeby, 1991; Harper and others, 1992).

Imbrication of the Josephine ophiolite above the Rogue arc sequence and beneath the Rattlesnake Creek terrane proper occurred during the Late Jurassic Nevadan orogeny. The age relations and structural stacking arrangement suggest west- and north-northwest–vergent collapse of a remnant arc (Rattlesnake Creek terrane and related Middle Jurassic arc rocks), interarc basin (Josephine ophiolite), and fringing arc (Rogue sequence) system. Detailed structural and geochronological studies demonstrate the oblique collapse of the interarc basin immediately following its formation (Harper and others, 1990, 1992). Integrated $^{40}Ar/^{39}Ar$ and U/Pb zircon geochronological and structural studies, along the amphibolite sole of the Josephine ophiolite, where it rests tectonically above fringing arc plutonic rocks, indicate that the ophiolite was detached at upper mantle levels within 10 m.y. of its formation, and obducted primarily northward over the arc terrane and overlapping Galice Formation (Harper and others, 1990, 1992). Such thrusting was completed by 150 Ma. Within the same time interval, 160 to 150 Ma, perhaps as much as 100 km of the Josephine basin was subducted eastward beneath the Rattlesnake Creek terrane (Klein, 1977; Helper, 1986; Harper and others, 1992). During and at the end of Josephine basin closure, numerous calc-alkaline stocks and dikes cut all major plates of the thrust system, suggesting the migration of arc magmatism to the locus of the closed interarc basin, presumably above an outer east-dipping subduction zone.

Interarc basin tectonics

Strong similarities in igneous ages and suprasubduction zone petrochemical affinities, and broad patterns in tectonic position and Late Jurassic (Nevadan) flysch overlap deposits, lead to the conclusion that the Josephine and the Coast Range ophiolites represent different segments of a major interarc basin and fringing arc system that formed mainly in Callovian to earliest Oxfordian time along the Cordilleran plate edge (Saleeby, 1981; Harper and others, 1985; Saleeby and Busby-Spera, this volume). The original strike-length of the system may have exceeded 1,500 km (Fig. 1 inset). Internal structural features of the ophiolites that were outlined above, and paleobiogeographic data on radiolaria, locally corroborated by paleomagnetic data (Hopson and others, 1986; Harper and others, 1989, 1992; Passagno and Blome, 1990), indicate a sea-floor spreading trajectory and latitudinal transport pattern that was northward parallel or sub-parallel to the Cordilleran plate edge. This implies the existence of transform structures parallel to the plate edge. In such a geometry, the outer fringes of the spreading interarc basin may or may not be fringed by a constructional chain of arc volcanos (i.e., Andaman Sea; Curray and others, 1979). Two different fringing arc relations appear to be preserved in the late Middle Jurassic ophiolites (Figs. 4d and e). In the case of the Coast Ranges and Great Valley, particularly at latitudes of the Diablo Range, constructional arc-type pyroclastic cones accumulated on either the outer fringes of new interarc basin floor, or on older and highly attenuated fore-arc mantle basement. In the case of the Josephine ophiolite, a significant fragment of preexisting arc/fore-arc basement, with its polygenetic igneous and metasedimentary assemblages, was rifted to the frontal edge of the basin and capped by the Rogue constructional arc sequence.

Reconstruction of the basin-spreading kinematics is almost prohibited by thick sedimentary overlap sequences and by superposed deformations that affected the Josephine and Coast Range ophiolites in distinctly different manners. Despite the tremendous strike-length of exposure preserved for the Coast Range belt, such reconstructions are particularly difficult due to a distinctly later phase of Franciscan underthrusting and related deformation, and then broad superposed extension that highly attenuated most Coast Range ophiolite exposures (Jayko and others, 1987). In contrast, the Josephine ophiolite was stitched into its emplacement geometry by crosscutting plutons immediately following the Nevadan emplacement event (Saleeby and others, 1982; Harper and others, 1989, 1992). Furthermore, it did not experience the same degree of late-stage extensional attenuation and, thus, the original emplacement structural relations with adjacent terranes are well preserved (Harper and others, 1989, 1990, 1992). These structural relations along with a wealth of radiometric and biostratigraphic age data facilitates some semblance of a rigorous plate kinematic model for Josephine ophiolite spreading genesis, plate transport, and tectonic emplacement. The model is considered here to be broadly applicable to the generation and plate-transport history of the Coast Range ophiolite as well. Distinctions between the Josephine and Coast Range belt arise from progressively younger Kimmeridgian flysch building out southward onto basin floor environments from the Siskiyou and subsequent Nevadan orogenic zone, and from a general lack of Nevadan deformation in the Coast Range belt.

A model that tracks the spreading kinematics, plate transport, and emplacement history of the Josephine ophiolite is developed from a much broader data base presented in Saleeby and Busby-Spera (this volume). The model is summarized in Figure 5 and considers the history of the Josephine and Coast Range ophiolites in a broad paleogeographic and tectonic context. Critical aspects that are directly related to the Josephine ophiolite are based on the work of Harper and others (1985, 1989, 1990, 1992), Pessagno and Blome (1990), and Yule and Saleeby (1991). A suprasubduction zone spreading geometry that is similar to the modern Andaman Sea (Curray and others, 1979) is hypothesized. A right-stepping system of ridge segments is shown based on the strong northward latitudinal shift of the ophiolite implied by the radiolarian paleobiogeographic data, as well as emplacement structural data. The Josephine ophiolite proper was generated at 162 ± 1 Ma based on U/Pb zircon data. The basin is suggested, however, to have begun spreading as early as ~170 Ma based on the ages of ophiolitic rocks that occur as subcreted metamorphic lenses or terranes between the Josephine ophiolite proper and its Rattlesnake Creek terrane upper plate (Klein, 1977; Hill, 1985; Saleeby and others, 1984; Harper and others, 1992). We will refer to this latest Bathonian to earliest Oxfordian basin loosely as the Josephine basin. Spreading of the Josephine basin initiated during Siskiyou contractile deformation, based on the age relations outlined above. A similar age range in igneous ages is represented in the Coast Range belt as well. One possible explanation for this apparent contradiction is that basin spreading was kinematically linked to the syn and post-collisional northward migration of the Insular superterrane. Linkage may have occurred following the oblique collision of the superterrane with the Cordilleran margin, or alternatively, the superterrane may have migrated wholly as a fore-arc microplate during Siskiyou transpressive deformation along its northeastern margin while the Josephine basin opened to the south in its wake. By ~160 Ma, part of the Josephine basin floor began to underthrust the Rattlesnake Creek terrane obliquely to the northeast. The Josephine ophiolite proper remained in a basinal setting for between 5 and 10 m.y., however, and was overlapped by upper Oxfordian to lower Kimmeridgian flysch derived from Siskiyou orogenic highlands to the northeast. Northward plate transport rates may have been as high as about 100 km/m.y. based on age and radiolarian paleobiogeographic data.

Overlap of the Galice flysch across the Rogue fringing arc sequence to the west of the Josephine basin implies that the fringing arc was coupled with the migrating basin floor. The Rattlesnake Creek-type basement rocks for the fringing arc sequence may have been rifted off of the western Sierra-Klamath belt nearly anywhere along its aggregate ~1,000-km strike length.

The plate kinematics of the Josephine basin changed dramatically during Kimmeridgian time. This change probably resulted from an abrupt northwestward acceleration in North America's absolute motion (May and Butler, 1986; May and others, 1989). Following its northeast underthrusting beneath the Rattlesnake Creek terrane the ophiolite and its flysch cover were further underthrust east-southeast beneath the upper plate terrane. At about this point in time, the outer edge of the Josephine basin was obducted north-northeast over the Rogue fringing arc. This

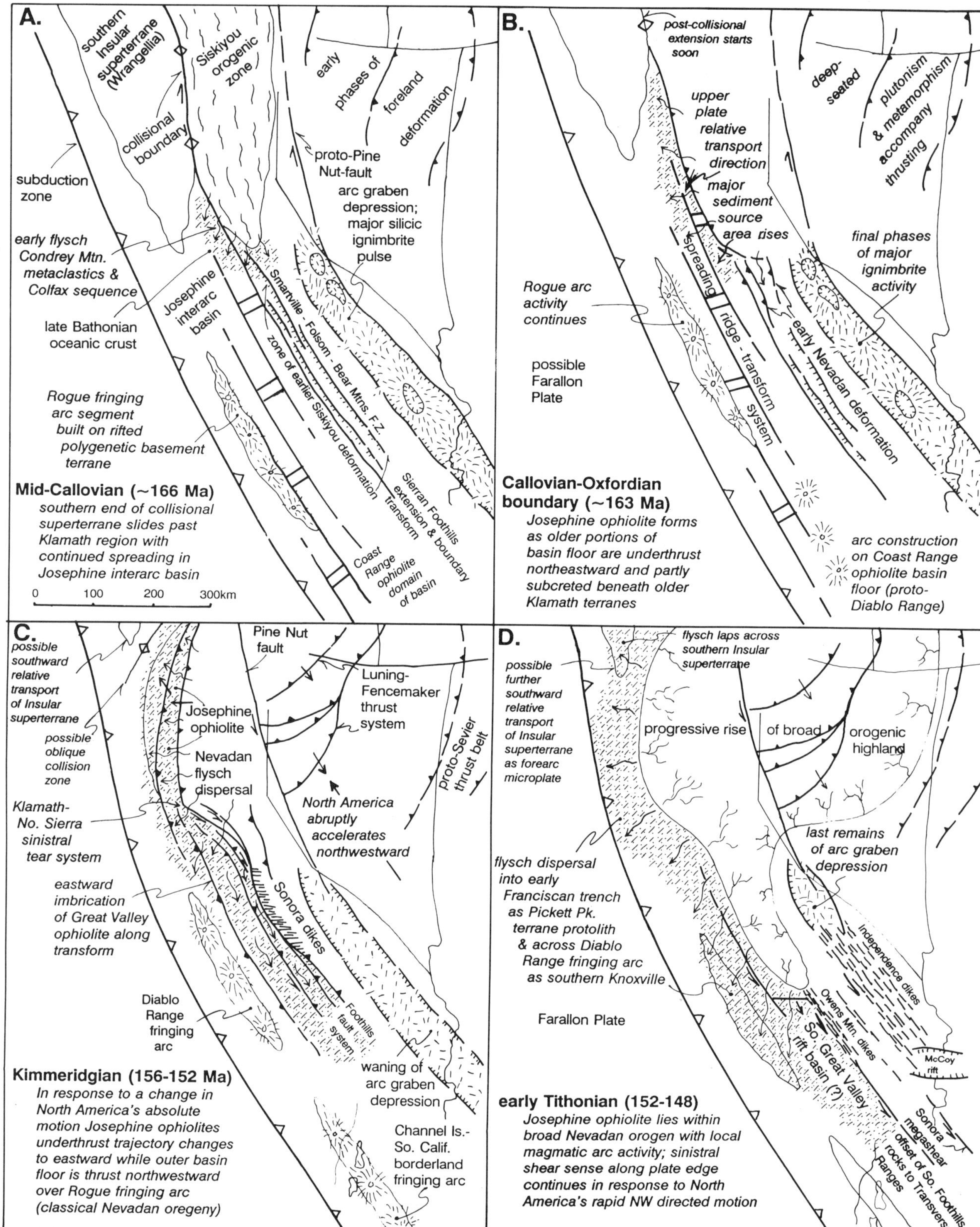

Figure 5. Plate tectonic model for evolution of southwestern Cordillera, late Middle Jurassic, interarc basin system based on broad paleogeographic and tectonic patterns discussed in Saleeby and Busby-Spera (this volume), and on sea-floor spreading, transport, and emplacement kinematics deduced for Josephine ophiolite (after Saleeby and others, 1982; Harper and others, 1989, 1990, 1992).

deformation pattern was complete by about 150 Ma and corresponds to the classical Nevadan orogeny. At 150 to 145 Ma, arc plutons, presumably related to continued east-directed subduction, cut across upper and lower thrust systems of the Josephine ophiolite.

The "Klamath-Josephine" section of Figure 4e shows the Nevadan structural stacking arrangement for the Josephine ophiolite. In contrast to the Coast Range ophiolite–Great Valley basement segments of the interarc basin system, the Josephine segment was apparently totally collapsed by Nevadan contractile deformation. The northern Great Valley and western Sierra Nevada appear to have experienced a more limited eastward-directed overthrusting, calling for a sinistral tear system separating the northern Great Valley and Klamath provinces. The Coast Range sections of Figure 4e depict the preservation of a large tract of the interarc basin system, which in Cretaceous time formed a major fore-arc basin between the Sierran magmatic arc and the Franciscan subduction complex.

The Kimmeridgian northwest acceleration of North America promoted regional sinistral tangential components to contractile and extensional deformation of Nevadan age. The north-northeast obduction of the Josephine ophiolite over the Rogue arc reflects such a pattern. Sinistral strike-slip and reverse faulting also affected the western Sierra Foothills, and the Pine Nut fault system to the east reversed its displacement pattern to sinistral as well (Saleeby and Busby-Spera, this volume). Three major mafic dike swarms were emplaced in the Sierra Nevada region between 158 and 148 Ma, and each of them suggests a sinistral sense to the regional strain; these are shown as the Sonora, Owens Mountain, and Independence swarms (Figs. 5c and d). The Owens Mountain complex strongly resembles the Smartville complex in composition and in its preferential emplacement along preexisting north-northwest–trending fabrics and structures. The Owens Mountain swarm and nearby related rocks are exposed along the extreme western Foothills in the southern Sierra (Saleeby and others, 1989; Wolf and Saleeby, 1990). Geophysical anomalies in the adjacent Great Valley, suggested by Harper and others (1985) to have resulted from sinistral transtensional rifting, and the apparent southward offset of southwestern Foothills rocks to the Transverse Ranges pose the possibility that the Owens Mountains dike swarm represents the edge of a southern Great Valley rift that opened in late Nevadan time with the opposite sense as the opening of the pre-Nevadan interarc basin (Fig. 4e, Coast Ranges-Foothills south; Fig. 5d). This rifting and tangential displacement pattern presumably constituted part of a more-regional sinistral transtensive system that penetrated the southern California to Sonora, Mexico, region, and which tied into the initial opening of the Gulf of Mexico (Saleeby and Busby-Spera, this volume).

Coast Range ophiolite and high-pressure metamorphism

The Franciscan complex contains a number of distinct high-P/T metamorphic terranes that range in age from Middle Jurassic to Late Cretaceous. The oldest is represented by high-grade blocks of eclogite, garnet amphibolite, and glaucophane schist that were derived primarily from mafic protoliths (Coleman and Lanphere, 1971). The blocks occur in scaly clay and serpentinite mélanges of the Franciscan Central belt mélange as well as in serpentinite mélanges of the Coast Range ophiolite, and protrusive and sedimentary serpentinite bodies of the lower Great Valley sequence. The serpentinite mélanges of the ophiolite belt and the protrusive serpentinites within the lower Great Valley sequence demonstrate that upwardly mobile serpentinite penetrated upper mantle, igneous crust, and sedimentary veneer levels of the Great Valley interarc/fore-arc basin and, in doing so, sampled a deep cryptic metamorphic terrane. The older matrix materials and metabasite blocks of the Franciscan Central mélange belt have broad protolith affinities to Great Valley oceanic basement and basal sediments. A highly pertinent aspect of the high-grade blocks, with respect to the Coast Range ophiolite and Great Valley basement, is that the main cluster of ophiolite igneous generation ages, discussed above, is identical to the upper range of the high-pressure metamorphic ages (Coleman and Lanphere, 1971; Suppe and Armstrong, 1972; Suppe and Foland, 1978; Ross and Sharp, 1988).

Several models are considered here to account for the regional distribution and age relations of the high-grade blocks. Model 1 is based on the model introduced above for Late Triassic to Early Jurassic high-pressure rocks of the western Sierran-Klamath belt and its continuation to the Pacific Northwest. This model asserts that distinct phases in fore-arc extension may disrupt the underlying high-pressure metamorphic environment and drive rapid denudation of the derivative metamorphic blocks. The association of arc tholeiitic and boninitic lavas, erupted during such phases of extension, suggest a lithosphere-scale dynamic pattern that should conceivably affect the subduction metamorphic environment. The key question is whether the high-pressure phases could survive the possible effects of a nearby mantle anatectic environment. The fact that tholeiitic and boninitic lavas approximate upper mantle liquids offers the best possible situation for such a mechanism to work, since magma ascent should be relatively rapid and large upper mantle–lower crustal magma chambers need not be present.

The extensional denudation model is depicted in Figure 4d for the late Middle Jurassic phase of fore-arc extension and related interarc basin formation. A high-pressure metamorphic environment was established during the Siskiyou phase of subduction (Fig. 4c). As the fore-arc evolved into an extensional tectonic regime, the deep metamorphic environment was drawn up and rapidly denudated, terminating the metamorphic conditions and thereby setting the isotopic age systems. Survival of high-pressure metamorphic blocks was restricted to zones of preexisting lithosphere that were rifted apart as sub-arc asthenosphere upwelled and created new lithosphere beneath interarc basins. The isotopic systems within the metamorphic blocks, for the most part, record the initial denudation, which coincides with the igneous production of fore-arc and interarc basin ophiolites. Subsequent serpentinite diapirism and tectonic transport patterns further dispersed the metamorphic blocks into their current structural positions.

Model 2 is depicted in Plate 5-E of Saleeby and Busby-Spera (this volume). In this model, ephemeral subduction of new interarc basin floor along lengthy transform faults followed by rapid transtensional denudation along the transforms led to rapid production and preservation of high-P metamorphic slices having protolith heritage to the interarc basin crust. Rapid changes in the polarity of the normal stress and related displacement strain along lengthy transforms are to be expected, particularly in the type of complex microplate environment envisaged for the generation of the Coast Range ophiolite (Dewey, 1975).

Model 3 is based on the plate tectonic geometry envisaged by Schweickert and Cowan (1975) and Day and others (1985) for the Nevadan orogeny in the western Sierra Nevada. This is depicted in Figure 4f and shows the western Foothills belt to be an exotic east-facing arc and the Great Valley to be a trailing back-arc basin. The Nevadan orogeny in this scenario occurred when the arc collided with North America, along the western Sierra Foothills. To account for a number of difficulties in this model, Saleeby and others (1986) suggested that the Great Valley basin formed over an east-dipping subduction zone, as envisaged in Model 1, and that the Sierran segment of the Nevadan orogeny resulted from a short-lived arc reversal event. This resulted in the westward subduction of young, hot, interarc basin crust beneath the fringing arc, as it impinged on the western Foothills. High-pressure mafic tectonites formed by near synchronisity of interarc basin spreading and subduction, like the current subduction of the actively spreading Woodlark Basin beneath the recently reversed Solomans arc (Doutch and others, 1981). At the close of the Nevadan orogeny, the deep-seated metamorphic terrane was disrupted and fragments of the terrane were dispersed to their current structural positions. Possible disruption mechanisms include eastward underthrusting of an outer collisional mass (Insular superterrane?), which drove the arc reversal like the Ontong-Java plateaus collision with the Solomans arc, and/or late to post-Nevadan transform faulting and disruption along the reversed fringing arc.

A second major belt of Franciscan high-pressure metamorphic rocks is also pertinent to Coast Range and Josephine ophiolite tectonics. This belt constitutes the Pickett Peak terrane of northern California and southern Oregon (Blake and others, 1985a and b) and the related Shuksan metamorphic suite in northwestern Washington (Brown and Blake, 1987). The Pickett Peak terrane was thrust eastward and metamorphosed beneath the northern Great Valley segment of the Coast Range belt as well as the Rogue-Josephine-Rattlesnake Creek structural stack in Latest Jurassic to Early Cretaceous time. The protolith assemblage of the Pickett Peak terrane is basaltic pillow lava with minor mafic intrusive and ultramafic rock, chert, argillite, and graywacke. Scattered igneous ages are similar to those of the Coast Range ophiolite. It follows that the Pickett Peak protolith assemblage could represent Coast Range and/or Josephine ophiolite-affinity ocenic crust and overlying Nevadan flysch (Knoxville and/or Galice). In Figure 5d, Pickett Peak metaclastic rocks are depicted as part of the late Nevadan flysch that was shed into the proto-Franciscan trench and involved in subduction tectonics and metamorphism.

OPHIOLITIC ROCKS OF THE FRANCISCAN COMPLEX

The Franciscam complex is commonly referred to as the type example of an exhumed subduction zone accretionary prism complex. In this regard, it is interesting that there are no known accreted fragments of abyssal crust and upper mantle within the complex, implying that oceanic lithosphere is for the most part subducted back into the mantle, or perhaps partly subcreted down-dip along the base of the upper plate. Fragments of the Coast Range ophiolite are scattered through the Franciscan complex, but these record a suprasubduction zone heritage like their contiguous parental ophiolite belt to the east. The Coast Range ophiolite fragments were dispersed into the Franciscan complex by Cretaceous and Cenozoic trench-linked and boundary transform faulting, respectively (McLaughlin and others, 1988), and by extensional attenuation of the original thrust geometry of Franciscan rocks beneath the Coast Range ophiolite (Jayko and others, 1987).

The Marin Headlands terrane of the Franciscan complex preserves an intact section through abyssal basaltic crust and overlying pelagic sediments of probable Farallon plate heritage. In this terrane, Pliensbachian metalliferous radiolarian chert lies depositionally above pillowed ocean floor basalt (Karl, 1984; Murchey, 1984). The pelagic sedimentary section is continuous into the Cenomanian, where continental detritus first occurs. During the nearly 100 m.y. of pelagic sedimentation, the terrane migrated northward from equatorial to northern mid-latitudes. Tectonic accretion into the Franciscan trench occurred in Late Cretaceous time during the influx of thick-bedded continent-derived turbidites. Other vestiges of possible abyssal plates occur within the Franciscan complex as radiolarian chert blocks, some as old as Early Jurassic (Murchy and Jones, 1984; Blome, 1987; Rymer and others, 1989), associated with metabasalt blocks and serpentinite slivers, and as large masses of tholeiitic to alkalic basalt interpreted as accreted seamounts (MacPherson, 1983; Shervais and Hanan, 1989). Intact or semi-intact, crust to upper mantle sections have yet to be observed, however.

CONCLUSIONS

Ophiolitic assemblages of the western U.S. Cordillera record a wealth of plate tectonic and petrogenetic phenomena spanning latest Precambrian to Late Jurassic time. Remnants of oceanic lithosphere, created as abyssal crust and as probable extensional fore-arc or interarc basin complexes, are represented by the dispersed fragments of uppermost Precambrian and lower Paleozoic assemblages. Fragments of mantle ultramafite and pillow lava sequences within the Roberts Mountains and Golconda allochthons probably record episodes of diffuse spreading during Paleozoic time behind a long-lived East Pacific arc system, which established itself on the North American continental rise off the flank of the Cordilleran miogeocline. Regionally extensive ophiolitic belts were developed in late Paleozoic through Jurassic time along the southwestern Cordillera. These may be related with some confidence to major tectonic events along the greater Cordilleran margin, the oldest of which was the late Paleozoic transform truncation of the Paleozoic miogeocline, and outboard orogenic and island arc terranes, followed by the Permo-Triassic initiation of east-dipping subduction beneath the truncated margin of southwestern North America. Accretion of a regional ophiolitic mélange belt during the transform truncation to subduction transition established the fore-arc basement regime for the Mesozoic convergent plate edge. Multiple episodes of early Mesozoic fore-arc extension, interarc basin nucleation and growth, and contractile deformation characterize the regionally extensive fore-arc basement. This style of continent-ocean plate interaction appears typical of the circum-Pacific orogen, and in this regard the Cordilleran ophiolites offer an excellent opportunity to perform on-land studies of processes that are common to the modern submarine realm.

A broad conclusion that can be drawn from the U.S. Cordilleran ophiolites, and which is consistent with ophiolitic tectonic and petrogenetic studies of the late Cenozoic West Pacific, is that the general island arc environment is the common spawning ground of ophiolites. Not only do a variety of ophiolitic assemblages develop in back-arc basin, embryonic arc, fore-arc, and arc-related transform environments, but abyssal crust–upper mantle sections can be incorporated into island arc systems as arc/fore-arc basement or trapped marginal basin floor upon subduction initiation or other major plate motion changes. Steady-state subduction and accretionary prism growth does not appear to be an important mechanism for ophiolite emplacement.

REFERENCES CITED

Ando, C. J., Irwin, W. P., Jones, D. L., and Saleeby, J. B., 1983, The North Fork ophiolitic terrane of the central Klamath Mountains, California: Geological Society of America Bulletin, v. 94, p. 236–252.

Anonymous, 1972, Penrose field conference on ophiolites: Geotimes, v. 17, p. 24–25.

Aubouin, J., and 13 others, 1982, Leg 84 of the Deep Sea Drilling Project, subduction without accretion; Middle America Trench off Guatemala: Nature, v. 297, p. 458–460.

Bailey, E. H., and Blake, M. C., 1974, Major chemical characteristics of Mesozoic Coast Range ophiolite in California: U.S. Geological Survey Journal of Research, v. 2, no. 6, p. 637–656.

Bailey, E. H., Blake, M. C., and Jones, D. L., 1970, On-land Mesozoic oceanic crust in California Coast Ranges: U.S. Geological Survey Professional Paper 700-C, p. C70–C80.

Bateman, P. C., and Clark, L. D., 1974, Stratigraphic and structural setting of the Sierra Nevada batholith, California: Pacific Geology, v. 8, p. 78–89.

Bateman, P. C., Harris, A. G., Kistler, R. W., and Krauskopf, K. B., 1985, Calaveras reversed—westward facing is indicated: Geology, v. 13, p. 338–342.

Beard, J. S., and Day, H. W., 1987, The Smartville intrusive complex, Sierra Nevada, California; The core of a rifted volcanic arc: Geological Society of America Bulletin, v. 19, p. 779–791.

Beebe, W. J., 1986, A paleomagnetic study of the southern Coast Range ophiolite, California, and tectonic implication [Ph.D. thesis]: Santa Barbara, University of California, 148 p.

Beccaluva, L., Serri, G., and Dostal, J., 1986, Geochemistry of volcanic rocks from the Mariana, Yap and Palau trenches bearing on the tectono-magmatic evolution of the Mariana trench-arc-backarc system, in Wezel, F. C., ed., the Origin of Arcs: New York, Elsevier, 148 p.

Behrman, P. G., 1978, Pre-Callovian rocks west of the Melones fault zone, central Sierra Nevada foothills, *in* Howell, D. E., and McDougal, K. A., eds., Mesozoic paleogeography of the western United States: Pacific Section, Society of Economic Paleontologists and Mineralogists, Pacific Coast Paleogeography Symposium, v. 2, p. 337–348.

Blake, M. C., Jr., Engebretson, D. C., Jayko, A. S., and Jones, D. L., 1985a, Tectonostratigraphic terranes in southwest Oregon, *in* Howell, D. G., ed., Tectonostratigraphic terranes of the Circumpacific region: Circumpacific Council for Energy and Mineral Resources Earth Science Series, No. 1, p. 147–157.

Blake, M. C., Jr., Jayko, A. S., McLaughlin, R. J., 1985b, Tectonostratigraphic terranes of the northern Coast Ranges, California, *in* Howell, D. G., ed., Tectonostratigraphic terranes of the Circumpacific region: Circumpacific Council for Energy and Mineral Resources Earth Science Series, No. 1, p. 159–171.

Blake, M. C., Jr., Jayko, A. S., Jones, D. L., and Rogers, B. W., 1987a, Unconformity between Coast Range ophiolite and part of the lower Great Valley sequence, south fork of Elder Creek, Tehama County, California, *in* Hill, M. L., Cordilleran Section: Boulder, Colorado, Geological Society of America Centennial Field Guide, v. 1, p. 279–282.

Blake, M. C., Jr., Jayko, S. S., and Murchey, B. L., 1987b, Structure, age and tectonic significance of the Coast Range ophiolite and related rocks near Paskenta, California [abs.]: Geological Society of America Abstracts with Programs, v. 19, no. 7, p. 590.

Blake, M. C., Jr., and 18 others, 1989, Mendocino triple junction to North American craton: Boulder, Colorado, Geological Society of America, Centennial Continent/Ocean Transect No. 12, 3 sheets with text, C-1 scale 1:500,000.

Blome, C. D., 1987, Paleogeographic significance of lower Mesozoic radiolarians from the Brooks Range, Alaska, *in* Taillery, I., and Weimer, P., eds., Alaskan North Slope geology: Bakersfield, California, Society of Economic Paleontologists and Mineralogists, Pacific Section, v. 1, p. 371–380.

Blome, C. D., and Irwin, W. P., 1983, Tectonic significance of late Paleozoic to Jurassic radiolarians from the North Fork terrane, Klamath Mountains, California, *in* Stevens, C. H., Pre-Jurassic rocks in western North America suspect terranes: Los Angeles, California, Society of Economic Paleontologists and Mineralogists, Pacific Section, p. 77–89.

Bloomer, S. H., 1981, Forearc ophiolite and extension: EOS Transactions of the American Geophysical Union, v. 62, p. 1086.

——, 1983, Distribution and origin of igneous rocks from the landward slopes of the Mariana Trench; Implications for its structure and evolution: Journal of Geophysical Research, v. 88, no. B9, p. 7411–7428.

Bloomer, S. H., and Hawkins, J. W., 1983, Gabbroic and ultramafic rocks from the Mariana Trench; Implications for its structure and evolution: Geophysical Monograph 27, p. 297–317.

——, 1987, Petrology and geochemistry of boninite series volcanic rocks from the Mariana Trench: Contributions to Mineral Petrology, v. 97, p. 361–377.

Bogen, N. L., 1985, Stratigraphic and sedimentologic evidence of a submarine island-arc volcano in the lower Mesozoic Penon Blanco and Jasper Point Formations, Mariposa County, California: Geological Society of America Bulletin, v. 96, p. 1322–1331.

Brandon, M. T., Cowan, D. S., and Vance, J. A., 1988, The Late Cretaceous San Juan thrust system, San Juan Islands, Washington: Geological Society of America Special Paper 221, 82 p.

Brouxel, M., and Lapierre, H., 1988, Geochemical study of an early Paleozoic island-arc–back-arc system, Part 1; The Trinity Ophiolite (northern California): Geological Society of America Bulletin, v. 100, p. 1111–1119.

Brouxel, M., Lapierre, H., Michard, A., and Albarede, F., 1988, Geochemical study of an early Paleozoic island-arc volcanic rocks (northern California): Geological Society of America Bulletin, v. 100, p. 1120–1130.

Brown, E. H., and Blake, M. C., Jr., 1987, Correlation of Early Cretaceous blueschists in Washington, Oregon and Northern California: Tectonics, v. 6, p. 795–806.

Brown, E. H., Bradshaw, J. Y., and Mustoe, G. E., 1979, Plagiogranite and keratophyre in ophiolite on Fidalgo Island, Washington: Geological Society of America Bulletin, Part I, v. 90, p. 493–507.

Burchfiel, B. C., and Davis, G. A., 1972, Structural framework and evolution of the southern part of the Cordilleran orogen, western United States: American Journal of Science, v. 272, p. 97–118.

Busby-Spera, C. J., 1988, Speculative tectonic model for the Lower Mesozoic arc of the southwest Cordilleran United States: Geology, v. 16, p. 1121–1125.

Busby-Spera, C. J., Schermer, E. R., Riggs, N. R., and Mattinson, J. M., 1990, Controls on basin subsidence within continental magmatic arcs: Nottingham England, International Association of Sedimentologists, 13th International Congress (abst.): p. 70–71.

Cameron, W. E., Nisbet, E. G., and Dietrich, V. J., 1979, Boninites, komatiites and ophiolitic basalts: Nature, v. 280, p. 550–553.

Casey, J. F., and Dewey, J. F., 1984, Initiation of subduction zones along transform and accreting plate boundaries, triple-junction evolution, and forearc spreading center implications for ophiolitic geology and obduction, *in* Gass, I., Lippard, S. J., and Shelton, A. W., eds., Ophiolites and oceanic lithosphere: Geological Society of London, p. 269–290.

Chen, J. H., and Shaw, H. F., 1982, Pb-Nd-Sr isotopic studies of California ophiolites [abs.]: Geological Society of America Abstracts with Programs, v. 14, p. 462.

Clark, L. D., 1964, Stratigraphy and structure of part of the western Sierra Nevada metamorphic belt, California: U.S. Geological Survey Professional Paper 410, 70 p.

——, 1976, Stratigraphy of the north half of the western Sierra Nevada metamorphic belt, California: U.S. Geological Survey Professional Paper 923, 26 p.

Coleman, R. G., 1977, Ophiolites, ancient oceanic lithosphere?: New York, Springer verlag, 240 p.

Coleman, R. G., and Lanphere, M. A., 1971, Distribution and age of high-grade blueschists, associated eclogites, and amphibolites from Oregon and California: Geological Society of America Bulletin, v. 82, p. 2397–2412.

Coleman, R. G., Manning, C. E., Mortimer, N., Donato, M. M., and Hill, L. B., 1988, Tectonic and regional metamorphic framework of the Klamath Mountains and adjacent Coast Ranges, California and Oregon, *in* Ernst, W. G., Metamorphism and crustal evolution of the western United States, Rubey Volume: Englewood Cliffs, New Jersey, Prentice-Hall, p. 1061–1097.

Cooper, A. K., Marlow, M. S., and Scholl, D. W., 1977, The Bering Sea—a multifarious marginal basin, *in* Talwani, M., and Pitman, W. D., III, eds., Island arcs, deep sea trenches, and backarc basins: Washington, D.C., American Geophysical Union, Maurice Ewing Series 1, p. 437–450.

Cordey, F. Mortimer, N., DeWever, P., and Monger, J.W.H., 1987, Significance of Jurassic radiolarians from the Cache Creek terrane, British Columbia: Geology, v. 15, p. 1151–1154.

Crawford, A. J., Beccaluva, L., Serri, G., and Dostal, J., 1986, Petrology, geochemistry and tectonic implications of volcanics dredged from the intersection of the Yap and Mariana trenches: Earth and Planetary Science Letters, v. 80, p. 265–280.

Curray, J. R., and 6 others, 1979, Tectonics of the Andaman Sea and Burma, *in* Watkins, J. S., and 6 others, eds., Geological and geophysical investigations of continental margins: American Association of Petroleum Geologists Memoir 29, p. 189–198.

Davies, H. L., 1971, Peridotite-gabbro-basalt complex in Eastern Papua; An overthrust plate of oceanic mantle and crust: Commonwealth of Australia Bureau of Mineral Resources, Geology and Geophysics Bulletin, v. 128, 48 p.

Davies, H. L., and Jacques, A. L., 1984, Emplacement of ophiolite in Papua New Guinea, *in* Gass, I. G., Lippard, S. J., and Shelton, A. W., eds., Ophiolites and oceanic lithosphere: Boston, Blackwell Scientific Publications, p. 341–349.

Davis, G. A., Burchfiel, B. C., and Monger, J.W.H., 1978, Mesozoic construction of the Cordilleran "collage," central British Columbia to central California, *in* Howell, D. G., and McDougall, K. A., eds., Mesozoic paleogeography of the western United States, Pacific Coast Paleogeography Symposium 2: Los Angeles, California, Pacific Section, Society of Economic Paleontologists and Mineralogists, p. 1–32.

Day, H. W., Moores, E. M., and Tuminas, A. C., 1985, Structure and tectonics of the northern Sierra Nevada: Geological Society of America Bulletin, v. 96, p. 436–450.

Dewey, J. F., 1975, Finite plate implications; Some implications for the evolution of rock masses at plate margins: American Journal of Science, v. 275-A, p. 260–284.

Dewey, J. F., and Bird, J. M., 1971, Origin and emplacement of the ophiolite suite; Appalachian ophiolites in Newfoundland: Journal of Geophysical Research, v. 76, p. 3179–3206.

Dickinson, W. R., and Thayer, T. P., 1978, Paleogeographic and paleotectonic implications of Mesozoic stratigraphy and structure in the John Day inlier of central Oregon, *in* Howell, D. G., and McDougall, K. A., eds., Mesozoic paleogeography of the western United States: Los Angeles, California, Pacific Section, Society of Economic Paleontologists and Mineralogists, p. 147–161.

Dietrich, V., Emmermann, R., Oberhansli, R., and Puchelt, H., 1978, Geochemistry of basaltic and gabbroic rocks from the West Mariana Basin and Mariana Trench: Earth and Planetary Science Letters, v. 39, p. 127–144.

Dilek, Y., Thy, P., Moores, E. M., and Grunvig, S., 1990, Late Paleozoic–early Mesozoic oceanic basement of a Jurassic arc terrane in the northwestern Sierra Nevada, California, *in* Harwood, D. S., and Miller, M. M., eds., Late Paleozoic and early Mesozoic paleogeographic relations; Klamath Mountains, Sierra Nevada and related terranes: Geological Society of America Special Paper 255, p. 351–370.

Dobson, P. F., 1986, The petrogenesis of boninite; A field, petrologic and geochemical study of the volcanic rocks of Chichi-jima, Bonin Islands, Japan [Ph.D. thesis]: Palo Alto, California, Stanford University, 163 p.

Donato, M. M., 1987, Evolution of an ophiolitic tectonic mélange, Marble Mountains, northern California Klamath Mountains: Geological Society of America Bulletin, v. 99, p. 448–464.

Douglass, R. C., 1967, Permian Tethyan fusulinids from California: U.S. Geological Survey Professional Paper 583-A, p. 7–43.

Doutch, H. F., and 8 others, 1981, Plate tectonic map of the circumpacific region, southwest quadrant: American Association of Petroleum Geologists, scale 1:10,000,000.

Edelman, S. H., and 5 others, 1989, Structure across a Mesozoic ocean-continent suture zone in the northern Sierra Nevada, California: Geological Society of America Special Paper 224, 56 p.

Eisbacher, G. H., 1983, Devonian-Mississippian sinistral transcurrent faulting along the cratonic margin of western North America; A hypothesis: Geology, v. 11, p. 7–10.

Evarts, R. C., 1977, The geology and petrology of the Del Puerto ophiolite, Diablo Range, central California Coast Ranges, *in* Coleman, R. G., and Irwin, W. P., eds., North American ophiolites: Oregon Department of Geology and Mineral Industries Bulletin 95, p. 121–139.

Ewart, A., and Bryan, W. B., 1972, Petrography and geochemistry of the igneous rocks from Eua, Tongan Islands: Geological Society of America Bulletin, v. 80, p. 3281–3298.

Garcia, M. O., 1979, Petrology of the Rogue and Galice Formation, Klamath Mountains, Oregon; Identification of a Jurassic island-arc sequence: Journal of Geology, v. 86, p. 29–41.

——, 1982, Petrology of the Rogue River island-arc complex, southwest Oregon: American Journal of Science, v. 282, p. 783–807.

Garver, J. I., 1988, Fragment of the Coast Range ophiolite and the Great Valley sequence in the San Juan Islands, Washington: Geology, v. 16, p. 948–951.

Gass, I. G., Smith, A. G., and Vine, F. J., 1973, Origin and emplacement of ophiolites: *in* U.K. Geodynamics Report: National Environmental Resources Council of London, p. 54–64.

Gerlach, D. C., Ave Lallemant, H. G., and Leeman, W. P., 1981, An island arc origin for the Canyon Mountain ophiolite complex, eastern Oregon, U.S. A.: Earth and Planetary Science Letters, v. 53, p. 255–265.

Gray, F., Ramp, L., Moring, B., Douglas, I., and Donahoe, J., 1982, Geologic map of the Wild Rogue Wilderness, Coos, Curry and Douglas Counties, Oregon: U.S. Geological Survey Map MF-1381-A, scale 1:48,000.

Gusey, D. L., 1978, The geology of southwestern Fidalgo island [M.S. thesis]: Bellingham, Western Washington University, 85 p.

Hamilton, W., and Myers, W. B., 1966, Cenozoic tectonics of the western United States: Reviews in Geophysics, v. 4, p. 509–549.

Harper, G. D., 1980, The Josephine ophiolite—remains of a Late Jurassic marginal basin in northwestern California: Geology, v. 8, p. 333–337.

—— , 1984, The Josephine ophiolite, northwestern California: Geological Society of America Bulletin, v. 95, p. 1009–1026.

—— , 1988, Episodic magma chambers and amagmatic extension in the Josephine ophiolite: Geology, v. 16, p. 831–834.

Harper, G. D., and Wright, J. E., 1984, Middle to Late Jurassic tectonic evolution of the Klamath Mountains, California-Oregon: Tectonics, v. 3, p. 759–772.

Harper, G. D., Saleeby, J. B., and Norman, E.A.S., 1985, Geometry and tectonic setting of a sea-floor spreading for the Josephine ophiolite, and implications for Jurassic accretionary events along the California margin, *in* Howell, D. G., ed., Tectonostratigraphic terranes of the Circum-Pacific region: Tulsa, Oklahoma, Circum-Pacific Council for Energy and Mineral Resources, Earth Science Series, v. 1, p. 239–258.

Harper, G. D., Saleeby, J. B., Pessagno, E. A., Jr., and Heizler, M., 1989, The Josephine ophiolite; Generation, translation, and emplacement of the Late Jurassic parautochthonous back-arc lithosphere, Klamath Mountains, California-Oregon [abs.]: Geological Society of America Abstracts with Programs, v. 21, p. A28.

Harper, G. D., Grady, K., and Wakabayashi, J., 1990, A structural study of a metamorphic sole beneath the Josephine ophiolite, western Klamath terrane, California-Oregon, *in* Harwood, D. S., and Miller, M. M., eds., Paleozoic and Early Mesozoic paleogeographic relations; Sierra Nevada, Klamath Mountains, and related terranes: Geological Society of America, Special Paper 255, p. 379–396.

Harper, G. D., Saleeby, J. B., and Heizler, M., 1992, Isotopic age of emplacement of the Josephine ophiolite and overlying flysch during the Late Jurassic Nevadan orogeny, Klamath Mountains, Oregon-California: Journal of Geophysical Research (in press).

Harwood, D. S., 1988, Tectonism and metamorphism in the northern Sierra terrane, northern California, *in* Ernst, W. G., ed., Metamorphism and crustal evolution of the western United States: Englewood, N.J., Prentice-Hall, p. 764–788.

Harwood, D. S., and Murchey, B., 1989, Upper Paleozoic correlations between the northern Sierra terrane, California, and the Havallah sequence, northern Nevada [abs.]: Geological Society of America Abstracts with Programs, v. 21, no. 5, p. 91.

Hawkins, J. W., Jr., 1977, Petrologic and geochemical characteristics of marginal basin basalts, *in* Talwani, M., and Pitman, W. D. III, eds., Island arcs, deep sea trenches, and backarc basins: Washington, D.C., American Geophysical Union, Maurice Ewing Series 1, p. 355–365.

—— , 1980, Petrology of backarc basins and island arcs; Their possible role in the origin of ophiolites: *in* Panayiotou, A., ed., Ophiolites: Nicosia, Cyprus, Proceedings of the 1979 International Ophiolite Symposium, p. 244–254.

Hawkins, J. W., Jr., and Batiza, R., 1977, Metamorphic rocks of the Yap arc-trench system: Earth and Planetary Science Letters, v. 37, p. 216–229.

Hawkins, J. W., Jr., and Evans, C. A., 1982, Geology of the Zambales Range, Luzon, Philippine Islands; Ophiolite derived from an island arc-backarc basin pair, *in* Hayes, D. E., ed., The tectonic and geologic evolutionn of southeast Asian seas and islands, part 2: Washington, D.C., American Geophysical Union Geophysical Monograph Series, v. 27, p. 95–123.

Hawkins, J. W., Jr., and Falvey, D. A., 1985, Petrology of andesitic dikes and flows from 'Eua, Tonga,*in* Scholl, D. W., and Vallier, T. L., eds., Geology and offshore resources of Pacific island arcs—Tonga region: Circumpacific Council for Energy and Mineral Resources Earth Science Series, v. 2, p. 269–279.

Hawkins, J. W., Jr., and Melchior, J. T., 1985, Petrology of Mariana Trough and Lau Basin basalts: Journal of Geophysical Research, v. 90, no. B13, p. 11431–11468.

Hawkins, J. W., Jr., Bloomer, S., Evans, C., and Melchior, J., 1979, Mariana arc-trench system; Petrology of the inner trench wall: EOS Transactions of the American Geophysical Union, v. 46, p. 968.

Helper, M. A., 1986, Deformation and high P/T metamorphism in the central part of the Condrey Mountain window, north-central Klamath Mountains, California and Oregon, *in* Evans, B. W., and Brown, E. H., Blueschists and eclogites: Geological Society of America Memoir 164, p. 125–141.

Herman, B. M., Anderson, R. N., and Truchan, M., 1979, Extensional tectonics in the Okinawa trough, *in* Watkins, J. S., Montadert, L., and Dickerson, P. W., eds., Geological and geophysical investigations of continental margins: American Association of Petroleum Geologists Memoir 29, p. 199–208.

Hietanen, A., 1973, Geology of the Pulga and Bucks Lake Quadrangles, Butte and Plumas Counties, California: U.S. Geological Survey Professional Paper 731, 66 p.

—— , 1981, Petrologic and structural studies in the northwestern Sierra Nevada, California: U.S. Geological Survey Professional Paper 1226-A-C, 59 p.

Hilde, T.W.C., Lee, C. S., 1983, Origin and evolution of the West Philippine Basin; A new interpretation: Tectonophysics, v. 102, p. 85–104.

Hilde, T.W.C., Uyeda, S., and Kroenke, L., 1977, Evolution of the Western Pacific and its margin: Tectonophysics, v. 38, p. 145–165.

Hill, L. B., 1985, Metamorphic, deformational, and temporal constraints on terrane assembly, northern Klamath Mountains, California, *in* Howell, D. G., ed., Tectonostratigraphic terranes of the Circum-Pacific region: Tulsa, Oklahoma, Circum-Pacific Council for Energy and Mineral Resources, Earth Science Series, v. 1, p. 173–186.

Hopson, C. A., Mattinson, J. M., and Pessagno, E. A., Jr., 1981, Coast Range ophiolite, western California, *in* Ernst, W. G., ed., The geotectonic development of California, Rubey Volume 1: Englewood Cliffs, New Jersey, Prentice-Hall, p. 418–510.

Hopson, C. A., Beebe, W. J., Mattinson, J. M., Pessagno, E. A., Jr., and Blome, C. D., 1986, California Coast Range ophiolite; Jurassic tectonics (abs.): EOS Transactions of the American Geophysical Union, v. 67, p. 1232.

Hotz, P. E., Lanphere, M. A., and Swanson, D. A., 1977, Triassic blueschist from northern California and north-central Oregon: Geology, v. 5, p. 659–663.

Irwin, W. P., Jones, D. L., and Kaplan, T. A., 1978, Radiolarians from pre-Nevadan rocks of the Klamath Mountains, California and Oregon, *in* Howell, D. G., and McDougall, K. A., eds., Mesozoic paleogeography of the western United States: Society of Economic Paleontologists and Mineralogists, Pacific Section, p. 303–310.

Ishii, T., 1985, Dredged samples from the Ogasawara fore-arc seamount or "Ogasawara Paleoland" - "fore-arc ophiolite", *in* Nasu, K., Kobayashi, K., Uyeda, S., Kushiro, I., and Kagami, H., eds., Formation of active ocean margins: Tokyo, Terra Scientific Publishing Co., p. 307–342.

Jacobsen, S. B., Quick, J. E., and Wasserburg, G. J., 1984, A Nd and Sr isotopic study of the Trinity peridotite; Implications for mantle evolution: Earth and Planetary Science Letters, v. 68, p. 361–378.

Jakes, P. and Miyake, Y., 1984, Magma in forearcs; Implication for ophiolite generation: Tectonophysics, v. 106, p. 349–358.

Jayko, A. S., Murchey, B. L., Wardlaw, B. R., and Stevens, C. H., 1987, Revised interpretation of the pre-Cenozoic geology of the Butt Lake area, northern Sierra Nevada, California [abs.]: Geological Society of America Abstracts with Programs, v. 19, p. 391.

Jones, D. L., 1975, Discovery of Buchia rugosa of Kimmeridgian age from the base of the Great Valley Sequence [abs.]: Geological Society of America Abstracts with Programs, v. 7, no. 3, p. 330.

Karig, D. E., 1972, Remnant arcs: Geological Society of America Bulletin, v. 93, p. 1057–1068.

Karl, S. M., 1984, Sedimentologic, diagenetic, and geochemical analysis of upper Mesozoic ribbon cherts from the Franciscan assemblage at the Marin Headlands, California, *in* Blake, M. C., Jr., ed., Franciscan geology of northern California: Society of Economic Paleontologists and Mineralogists, Pacific Section, v. 43, p. 71–88.

Kistler, R. W., 1990, Two different lithosphere types in the Sierra Nevada, California, *in* Anderson, J. L., ed., The nature and origin of Cordilleran magmatism: Geological Society of America Memoir 174, p. 271–281.

Klein, C. W., 1977, Thrust plates of the north-central Klamath Mountains near Happy Camp, California: California Division of Mines and Geology Special Publication 28, p. 23–26.

Kobayashi, K., 1983, Forearc volcanism and cycles of subduction, *in* Shimozuru, D., and Yokoyama, I., eds., Arc volcanism: Physics and Tectonics: Tokyo, Terra Scientific Publishing Co., p. 153–163.

Lanphere, M. A., 1971, Age of the Mesozoic oceanic crust in the California Coast Ranges: Geological Society of America Bulletin, v. 81, p. 3209–3212.

Lanphere, M. A., Irwin, W. P., and Hotz, P. E., 1968, Isotopic age of the Nevadan orogeny and older plutonic and metamorphic events in the Klamath Mountains, California: Geological Society of America Bulletin, v. 79, p. 1027–1052.

Lawver, L. A., and Hawkins, J. W., 1978, Diffuse magnetic anomalies in marginal basins; Their possible tectonic and petrologic significance: Tectonophysics, v. 45, p. 323–339.

Lindsley-Griffin, N., 1977, The Trinity ophiolite, Klamath Mountains, California, *in* Coleman, R. G., and Irwin, W. P., eds., North American ophiolites: Oregon Department of Geology and Mineral Industries Bulletin, v. 95, p. 107–120.

Liou, J. G., Lan, C. Y., Suppe, J., and Ernse, W. G., 1977, The east Taiwan ophiolite—its occurrence, petrology, metamorphism and tectonic setting: Taipei, Taiwan, Mining Research and Service Organization Special Report No. 1, 212 p.

Lipman, P. W., 1964, Structure and origin of an ultramafic pluton in the Klamath Mountains, California: American Journal of Science, v. 262, p. 199–222.

MacPherson, G. J., 1983, The Snow Mountain complex; An on-land seamount in the Franciscan terrane: California Journal of Geology, v. 91, p. 73–92.

MacPherson, G. J., and Phipps, S. P., 1988, Geochemistry and petrology of mafic volcanic rocks from olistostromes in the basal Great Valley Sequence, Northern California Coast Range: Geological Society of America Bullletin, v. 100, p. 1770–1779.

Mattinson, J. M., and Hopson, C. A., 1972, Paleozoic ophiolitic complexes in Washington and northern California: Carnegie Institution Year Book 71, p. 578–583.

May, S. R., and Butler, R. F., 1986, North American Jurassic apparent polar wander—implications for plate motions, paleogeography and Cordilleran tectonics: Journal of Geophysical Research, v. 91, p. 1519–1544.

May, S. R., Beck, M. E., Jr., and Butler, R. F., 1989, North American apparent polar wander, plate motion, and left-oblique convergence; Late Jurassic–Early Cretaceous orogenic consequences: Tectonics, v. 8, no. 3, p. 443–451.

McLaughlin, R. J., Blake, M. C., Jr., and Griscom, A., 1988, Tectonics of formation, translation, and dispersal of the Coast Range ophiolite of California: Tectonics, v. 7, no. 5, p. 1033–1056.

Meijer, A., 1980, Primitive arc volcanism and a boninite series; Examples from Western Pacific island arcs, *in* Hayes, D. E., ed., The tectonic and geologic evolution of Southeast Asian seas and islands, Part 1: American Geophysical Union Geophysical Monographs, v. 23, p. 269–282.

Menzies, M. D., Blanchard, D., and Xenophonotos, C., 1980, Genesis of the Smartville arc-ophiolite, Sierra Nevada foothills, California: American Journal of Science, v. 208A, p. 329–344.

Miller, M. M., 1987, Dispersed remnants of a northeast Pacific fringing arc—Upper Paleozoic island arc terranes of Permian McCloud faunal affinity, western U.S.: Tectonics, v. 6, p. 807–830.

Miller, R. B., 1985, The ophiolitic Ingalls complex, north-central Cascade Mountains, Washington: Geological Society of America Bulletin, v. 96, p. 27–42.

Miyashiro, A., 1975, Classification, characteristics and origin of ophiolites: Journal of Geology, v. 83, p. 249–281.

Monger, J.W.H., and Ross, C. A., 1971, Distribution of fusulinaceans *in* the western Canadian Cordillera: Canadian Journal of Earth Sciences, v. 8, p. 259–278.

Monger, M.W.H., and 9 others, 1992, Cordilleran terranes, *in* Gabrielse, H., and Yorath, C. J., eds., The Cordilleran Orogen: Ottawa, Canada, Geological Survey of Canada, Geology of Canada Series, v. 4 (Geological Society of America, The Geology of North America, v. G-2) ch. 8, part b, p. 281–327.

Moores, E. M., 1982, Origin and emplacement of ophiolites: Reviews of Geophysics and Space Physics, v. 20, no. 4, p. 735–760.

Mortensen, J. K., 1990, Significance of U-Pb ages for inherited and detrital zircons from Yukon-Tanana terrane, Yukon and Alaska [abs.]: Geological Society of Canada Abstracts with Programs, v. 15, p. A91.

Murchey, B. L., 1984, Biostratigraphy and lithostratigraphy of chert in the Franciscan complex, Marin Headlands, California, *in* Blake, M. C., Jr., ed., Franciscan geology of northern California: Society of Economic Paleontologists and Mineralogists, Pacific Section, v. 43, p. 51–70.

Murchey, B. L., and Jones, D. L., 1984, Age and significance of chert in the Franciscan complex in the San Francisco Bay region, *in* Blake, M. C., Jr., ed., Franciscan geology of northern California: Society of Economic Paleontologists and Mineralogists, Pacific Section, v. 43, p. 23–30.

Nestell, M. K., 1980, Permian fusilinacean provinces in the Pacific Northwest are tectonic juxtapositions of ecologically distinct faunas: Geological Society of America Abstracts with Programs, v. 12, p. 144.

Natland, J. H., 1981, Crystal morphologies and pyroxene compositions in boninites and tholeiitic basalts from Deep Sea Drilling Project Holes 458 and 459B in the Mariana forearc region: Initial Reports of the Deep Sea Drilling Project, v. LX, p. 681–707.

Ogawa, Y., Naka, J., and Taniguchi, H., 1985, Ophiolite-based forearcs; A particular type of plate boundary, *in* Nasu, N., and others, eds., Formation of active ocean margins: Tokyo, Terra Scientific Publication Co., p. 71–746.

Peacock, S. M., and Norris, P. J., 1989, Metamorphic evolution of the Central metamorphic belt, Klamath province, California; An inverted metamorphic gradient beneath the Trinity peridotite: Journal of Metamorphic Geology, v. 7, p. 191–209.

Pessagno, E. A., Jr., and Blome, C. D., 1990, Implications of new Jurassic stratigraphic, geochronometric, and paleolatitudinal data from the western Klamath terrane (Smith River and Rogue Valley subterranes): Geology, v. 18, p. 665–668.

Poole, F. G., and Desborough, G. A., 1973, Alpine-type serpentinites in Nevada and their tectonic significance [abs.]: Geological Society of America Abstracts with Programs, v. 5, p. 90.

Prinzhofer, A., and Nicolas, A., 1980, The Bogota Peninsula, New Caledonia; A possible oceanic transform fault: Journal of Geology, v. 88, p. 387–398.

Quick, J. E., 1981, Petrology and petrogenesis of the Trinity peridotite, an upper mantle diapir in the eastern Klamath Mountains, northern California: Journal of Geophysical Research, v. 87, p. 3831–3848.

Ramp, L., 1980, Sheeted dikes of the Wild Rogue Wilderness, Oregon: Oregon Geology, v. 42, no. 7, p. 119–124.

Robertson, A.H.F., 1989, Palaeoceanography and tectonic setting of the Jurassic Coast Range ophiolite, central California; Evidence from the extrusive rocks and the volcaniclastic sediment cover: Marine and Petroleum Geology, v. 6, p. 194–220.

Ross, C. A., and Ross, R. P., 1983, Late Paleozoic accreted terranes of western North America, *in* Stevens, C. H., ed., Pre-Jurassic rocks in western North America suspect terranes: Society of Economic Paleontologists and Mineralogists, Pacific Section, p. 7–22.

Ross, J. A., and Sharp, W. D., 1988, The effects of sub-blocking temperature metamorphism on the K/Ar systematics of hornblendes; $^{40}Ar/^{39}Ar$ dating of polymetamorphic garnet amphibolite from the Franciscan complex, California: Contributions to Mineral Petrology, v. 100, p. 213–221.

Rouer, O., Lapierre, H., Mascle, G., Coulon, C., and Albers, J., 1989, Geodynamic implications of Devonian silicic arc magmatism in the Sierra Nevada and Klamath Mountains, California: Geology, v. 17, p. 177–180.

Rubin, C. M., Miller, M. M., and Smith, G. M., 1990, Tectonic development of Cordilleran mid-Paleozoic volcano-plutonic complexes; Evidence for convergent margin tectonism, *in* Harwood, D. S., and Miller, M. M., eds., Paleozoic and early Mesozoic paleogeographic relations of the Sierra Nevada, Klamath Mountains, and related terranes: Geological Society of America Special Paper 255, p. 1–16.

Rymer, M. J., 1989, Lower Jurassic rocks in the Franciscan assemblage [abs.]: Geological Society of America Abstracts with Programs, Cordilleran Section, v. 21, no. 5, p. 138.

Saleeby, J. B., 1975, Structure, petrology and geochronology of the Kings-Kaweah mafic-ultramafic belt, southwestern Sierra Nevada foothills, California [Ph.D. thesis]: Santa Barbara, University of California, 289 p.

—— , 1977, Fracture zone tectonics, continental margin fragmentation, and emplacement of the Kings-Kaweah ophiolite belt, southwest Sierra Nevada, California, *in* Coleman and Irwin, eds., International Geological Correlation Program, North American ophiolite volume: Oregon State Geology Department, p. 123–140.

—— , 1978, Kings River ophiolite, southwest Sierra Nevada foothills, California: Geological Society of America Bulletin, v. 89, p. 617–636.

—— , 1979, Kaweah serpentine mélange, southwest Sierra Nevada foothills, California: Geological Society of America Bulletin, v. 90, p. 29–46.

—— , 1981, Ocean floor accretion and volcano-plutonic arc evolution of the Mesozoic Sierra Nevada, California, *in* Ernst, W. G., ed., Geotectonic development of California, Rubey Volume 1: Englewood Cliffs, New Jersey, Prentice-Hall, p. 132–181.

—— , 1982, Polygenetic ophiolite belt of the California Sierra Nevada—geochronological and tectonostratigraphic development: Journal of Geophysical Research, v. 87, p. 1803–1824.

—— , 1984a, Tectonic significance of serpentinite mobility and ophiolitic mélange, *in* Raymond, L. A., ed., Mélanges; Their nature, origin and significance: Geological Society of America Special Paper 198, p. 153–168.

—— , 1984b, Pb/U zircon ages from the Rogue River area, western Jurassic belt, Klamath Mountains, Oregon [abs.]: Geological Society of America Abstracts with Programs, v. 16, p. 331.

—— , 1987, Discordance patterns in Pb/U zircon ages of the Sierra Nevada and Klamath Mountains [abs.]: EOS Transactions of the American Geophysical Union, v. 68, p. 1514.

—— , 1990, Geochronological and tectonostratigraphic framework of Sierran-Klamath ophiolitic assemblages, *in* Harwood, D. S., and Miller, M. M., eds., Paleozoic and early Mesozoic paleogeographic relations of the Sierra Nevada, Klamath Mountains, and related terranes: Geological Society of America Special Paper 255, p. 93–114.

Saleeby, J. B., and Sharp, W. D., 1980, Chronology of the structural and petrologic development of the southwest Sierra Nevada foothills, California: Geological Society of America Bulletin, v. 91, part 2, p. 1416–1535.

Saleeby, J. B., Busby, C., Goodin, W. D., and Sharp, W. D., 1978, Early Mesozoic paleotectonic-paleogeographic reconstruction of the southern Sierra Nevada region, California, *in* Howell, D., ed., Mesozoic paleogeography of the western United States: Society of Economic Paleontologists and Mineralogists, Pacific Section, p. 311–336.

Saleeby, J. B., Harper, G. D., Sharp, W. D., and Snoke, A. W., 1982, Time relations and structural-stratigraphic patterns in ophiolite accretion, west-central Klamath Mountains, California: Journal of Geophysical Research, v. 87, p. 3831–3848.

Saleeby, J. B., Blake, M. C., and Coleman, R. G., 1984, Pb/U zircon ages on thrust plates of the west-central Klamath Mountains and Coast Ranges, northern California and southern Oregon: EOS Transactions of the American Geophysical Union, v. 65, p. 1147.

Saleeby, J. B., and 12 contributors, 1986, Central California offshore to Colorado Plateau: The Decade of North America, Centennial Continent/Ocean Transect C-2, scale 1:500,000, 87 p.

Saleeby, J. B., Hannah, J. L., and Varga, R. J., 1987, Isotopic age constraints on middle Paleozoic deformation in the northern Sierra Nevada, California: Geology, v. 15, p. 757–760.

Saleeby, J. B., Shaw, H. F., Niemeyer, S., and Moores, E. M., 1989a, U/Pb, Sm/Nd and Rb/Sr geochronological and isotopic study of northern Sierra Nevada ophiolitic assemblages, California: Contributions to Mineralogy and Petrology, v. 102, p. 205–220.

Saleeby, J. B., Geary, E. E., Paterson, S. R., and Tobisch, O. T., 1989b, Isotopic systematics of Pb/U (zircon) and $^{40}Ar/^{39}Ar$ (biotite/hornblende) from rocks of the Central Foothills terrane, Sierra Nevada, Cailfornia: Geological Society of America Bulletin, v. 101, p. 1481–1492.

Schweickert, R. A., and Bogen, N. L., 1983, Tectonic transect of Sierran Paleozoic through Jurassic accreted belts: Los Angeles, California, Society of Economic Paleontologists and Mineralogists, Pacific Section, Field Trip Guide, p. 381–394.

Schweickert, R. A., and Cowan, D. S., 1975, Early Mesozoic tectonic evolution of the western Sierra Nevada, California: Geological Society of America Bulletin, v. 86, p. 1329–1336.

Schweickert, R. A., and Lahren, M. M., 1987, Continuation of Antler and Sonoma orogenic belts to the eastern Sierra Nevada, California, and Late Triassic thrusting in a compressional arc: Geology, v. 15, p. 270–273.

Schweickert, R. A., Saleeby, J. B., Tobisch, O. T., and Wright, W. H., III, 1977, Paleotectonic and paleogeographic significance of the Calaveras complex, western Sierra Nevada, California, *in* Stewart, J. H., Stevens, C. H., and Fritsche, A. E., Paleozoic paleogeography of the western United States: Society of Economic Paleontologists and Mineralogists, Pacific Section, Paleogeography Symposium 1, p. 381–394.

Schweickert, R. A., Harwood, D. S., Girty, G. H., and Hanson, R. E., 1984, Tectonic development of the northern Sierra terrane; An accreted late Paleozoic island arc and its basement, *in* Lintz, J., Jr., ed., Western geological excursions 4: Reno, Nevada, Mackay School of Mines Department of Geological Sciences, p. 1–65.

Schweller, W. J., Karig, O. E., and Bachman, S. B., 1983, Original setting and emplacment history of the Zabales ophiolite, Luzon, Philippines, from stratigraphic evidence, *in* Hayes, D. E., ed., The tectonic and geologic evolution of southeast Asian seas and islands, part 2: Washington, D.C., American Geophysical Union Geophysical Monograph Series, v. 27, p. 124–138.

Sharaskin, A. Ya., Pustchin, I. K., Zlobin, S. K., and Kolesov, G. M., 1983, Two ophiolite sequences from the basement of the northern Tonga Arc: Ofioliti, v. 8, no. 3, p. 411–430.

Sharp, W. D., 1988, Pre-Cretaceous crustal evolution in the Sierra Nevada region, California, *in* Ernst, W. E., ed., Geotectonic development of northern California, Rubey Volume 7: Englewood Cliffs, New Jersey, Prentice-Hall, p. 823–865.

Sharp, W. D., and Evarts, R. C., 1982, New constraints on the environment of formations of the Coast Range ophiolite at Del Puerto Canyon, California [abs.]: Geological Society of America Abstracts with Programs, v. 14, no. 4, p. 233.

Shaw, H. F., Chen, J. H., Saleeby, J. B., and Wasserburg, G. J., 1987, Nd-Sr-Pb systematics and age of the Kings River ophiolite, California: Contributions to Mineral Petrology, v. 96, p. 281–290.

Shervais, J. W., and Hanan, B. B., 1989, Jurassic volcanic glass from the Stonyford volcanic complex, Franciscan assemblage, northern California Coast Ranges: Geology, v. 17, p. 510–514.

Shervais, J. W., and Kimbrough, D. L., 1985, Geochemical evidence for the tectonic setting of the Coast Range ophiolite; A composite island arc-oceanic crust terrane in western California: Geology, v. 13, p. 35–38.

Silver, E. A., McCaffrey, R., Joyodiwiryo, Y., and Stevens, S., 1983, Ophiolite emplacement by collision between the Sula platform and the Sulawesi island arc, Indonesia: Journal of Geophysical Research, v. 88, no. B11, p. 9419–9435.

Silver, L. T., and Anderson, T. H., 1974, Possible left-lateral early to middle Mesozoic disruption of the southwestern North American craton margin [abs.]: Geological Society of America Abstracts with Programs, v. 6, p. 955–956.

Smith, D. L., and Miller, E. L., 1990, Late Paleozoic extension in the Great Basin, western United States: Geology, v. 18, p. 712–715.

Snoke, A. W., 1977, A thrust plate of ophiolitic rocks in the Preston Peak area, Klamath Mountains, California: Geological Society of America Bulletin, v. 88, p. 1641–1659.

Snoke, A. W., and Whitney, S. E., 1979, Relict pyroxenes from the Preston Peak ophiolite, Klamath Mountains, California: American Mineralogist, v. 64, p. 865–873.

Snoke, A. W., Bowman, H. R., and Hebert, A. J., 1977, The Preston Peak ophiolite, Klamath Mountains, California, an ancient immature island arc; Petrochemical evidence: Short Contributions to California Geology, Special Report 129, p. 67–79.

Sorenson, S., 1985, Petrologic evidence for Jurassic, island-arc-like basement rocks in the southwestern Transverse Ranges and California Continental Borderland: Geological Society of America Bulletin, v. 96, p. 887–1006.

Speed, R. C., 1978, Basinal terrane of the early Mesozoic marine province of the western Great Basin, *in* Howell, D., ed., Mesozoic paleogeography of the western United States: Society of Economic Paleontologists and Mineralogists, Pacific Section, p. 237–252.

—— , 1979, Collided Paleozoic microplate in the western United States: Journal of Geology, v. 87, p. 279–292.

Springer, R. K., 1980, Geology of the Pine Hill intrusive complex, a layered gabbroic body in the western Sierra Nevada foothills, California; Summary 1: Geological Society of America Bulletin, v. 91, p. 381–385.

Standlee, L. A., and Nestell, M. K., 1985, Age and tectonic significance of terranes adjacent to the Melones fault zone, northern Sierra Nevada, California [abs.]: Geological Society of America Abstracts with Programs, v. 17, p. 410.

Stone, P., and Stevens, C. H., 1988, Pennsylvanian and Early Permian paleogeography of east-central California; Implications for the shape of the continental margin and timing of continental truncation: Geology, v. 16, p. 330–333.

Suppe, J., 1979, Structural interpretation of the southern part of the northern Coast Ranges and Sacramento Valley, California; Summary: Geological Society of America Bulletin, v. 90, p. 327–330.

—— , 1984, Kinematics of arc-continent collision, flipping of subduction, and back-arc spreading near Taiwan: Geological Society of China Memoir 6, p. 21–33.

Suppe, J. and Armstrong, R. L., 1972, Potassium-argon dating of Franciscan metamorphic rocks: American Journal of Science, v. 272, p. 217–233.

Suppe, J., and Foland, K. A., 1978, The Goat Mountain schists and Pacific Ridge complex; A redeformed but still intact late Mesozoic schuppen complex, *in* Howell, D. G., and McDougall, K. A., eds., Mesozoic paleogeography of the western United States: Society of Economic Paleontologists and Mineralogists, Pacific Section, Paleogeography Symposium 2, Los Angeles, California, p. 431–451.

Suppe, J., Liou, J. G., and Ernst, W. G., 1981, Paleogeographic origins of the Miocene east Taiwan ophiolite: American Journal of Science, v. 281, p. 228–246.

Ujiie, H., and Matsumaru, K., 1977, Stratigraphic outline of Haha-jima (Hillsborough Island), Bonin Islands: National Science Museum Memoir 10, p. 5–18.

Umino, S., 1985, Volcanic geology of Chichi-jima, the Bonin Islands (Ogasawara Islands): Journal of the Geological Society of Japan, v. 91, p. 505–523.

Vance, J. A., Dungan, M. A., Blanchard, D. P., and Rhodes, J. M., 1980, Tectonic setting and trace element geochemistry of Mesozoic ophiolitic rocks in western Washington: American Journal of Science, v. 280-A, p. 359–388.

Varne, R., and Rubenach, M. J., 1972, Geology of Macquarie Island and its relationship to oceanic crust, *in* Hayes, D. E., ed., Antarctic oceanology II, the Australian–New Zealand sector: American Geophysical Union, p. 251–266.

Varne, R., Gee, D. R., and Quilty, P.G.J., 1969, Macquarie Island and the cause of oceanic linear magnetic anomalies: Science, v. 166, p. 230–233.

Walker, J. D., 1988, Permian and Triassic rocks of the Mojave Desert and their implications for timing and mechanisms of continental truncation: Tectonics, v. 7, p. 685–709.

Watkins, R., Reinheimer, C. E., Wallace, J. W., and Nestell, M. K., 1987, Paleogeographic significance of a Permian sedimentary megamictite in the Central belt of the northern Sierra Nevada: Geological Society of America Bulletin, v. 99, p. 771–778.

Whetten, J. T., Jones, D. L., Cowan, D. S., and Zartman, R. E., 1978, Ages of Mesozoic terranes in the San Juan Islands, Washington, *in* Howell, D. G., and MeDougall, K. A., eds., Mesozoic paleogeography of the western United States: Society of Economic Paleontologists and Mineralogists, Pacific Section, Paleogeography Symposium 2, Los Angeles, California, p. 117–132.

Whetten, J. T., Zartman, R. E., Blakely, R. J., and Jones, D. L., 1980, Allochthonous Jurassic ophiolite in northwest Washington: Geological Society of America Bulletin, v. 91, p. 359–368.

Williams, K. M., 1983, The Mount Diablo ophiolite, Contra Costa County, California [Ph.D. thesis]: San Jose, California, San Jose State University, 156 p.

Wolf, M. B., and Saleeby, J. B., 1990, Crustal extension during the Nevadan orogeny in the southwestern Sierra Nevada Foothills terrane (FT), CA: Geological Society of America Abstracts with Programs, v. 22, p. A275.

Wright, J. E., 1981, Geology and Pb/U geochronology of the western Paleozoic and Triassic subprovince, Klamath Mountains, California [Ph.D. thesis]: Santa Barbara, University of California, 350 p.

Wright, J. E., and Fahan, M. R., 1988, An expanded view of Jurassic orogenesis in the western United States Cordillera; Middle Jurassic (pre-Nevadan) regional metamorphism and thrust faulting within an active arc environment, Klamath Mountains, California: Geological Society of America Bulletin, v. 100, p. 859–876.

Wright, J. E., and Wyld, S. J., 1985, Multi-stage serpentinite matrix mélange development; Rattlesnake Creek terrane, southwestern Klamath Mountains [abs.]: Geological Society of America Abstracts with Programs, v. 17, no. 6, p. 419.

Wyld, S. J., and Wright, J. E., 1985, The Late Jurassic Galice Basin and its ophiolitic and serpentine mélange basement, Klamath Mountains, California [abs.]: Geological Society of America Abstracts with Programs, v. 17, no. 6, p. 419.

Xenophontos, C., and Bond, G. C., 1978, Petrology, sedimentation, and paleogeography of the Smartville terrane (Jurassic)—Bearing on the genesis of the Smartville ophiolite, *in* Howell, D. E., and McDougall, K. A., eds., Mesozoic paleogeography of the western United States: Society of Economic Paleontologists and Mineralogists, Pacific Section, Paleogeography Symposium 2, Los Angeles, California, p. 291–302.

Yuasa, M., Honza, E., Tamaki, K., Tanahashi, M., and Nishimura, A., 1982, Geological map of the southern Ogasawara and northern Mariana trenches: Geological Survey of Japan, Marine Geology Map Series 18, scale 1:10,000,000.

Yule, J. D., and Saleeby, J. B., 1991, Older oceanic crust and a protracted fracture zone assemblage from beneath the Josephine ophiolite, Illinois River area, southwestern Oregon [abs.]: Geological Society of America Abstracts with Programs, v. 23, no. 2, p. 112.

Manuscript Accepted by the Society March 1, 1991

ACKNOWLEDGMENTS

Support of work on U.S. Cordillera ophiolites is from NSF Grants EAR-7708691, EAR-7925998, EAR-8206382, and EAR-8407788, and the Alfred P. Sloan Foundation is gratefully acknowledged. Lively discussions and field excursions with M. C. Blake, R. G. Coleman, P. F. Dobson, S. H., Edelman, B. R. Hacker, G. D. Harper, J. W. Hawkins, C. A. Hopson, D. L. Jones, J. M. Mattinson, E. M. Moores, A. Nicholas, R. A. Schweickert, W. D. Sharp, A. W. Snoke, J. E. Wright, and W. H. Wright III were of great assistance in my ophiolite research and in the construction of this synthesis. Helpful reviews of the initial manuscript were provided by R. G. Coleman, M. M. Miller, E. M. Moores, and F. G. Poole.

Printed in U.S.A.

The Geology of North America
Vol. G-3, The Cordilleran Orogen: Conterminous U.S.
The Geological Society of America, 1992

Chapter 17

Tectonic significance of paleomagnetic results for the western conterminous United States

Myrl E. Beck, Jr.
Department of Geology, Western Washington University, Bellingham, Washington 98225

INTRODUCTION

Tectonic models for the evolution and growth of the North American Cordillera have become unprecedentedly mobilistic in the last 15 or 20 years. This is largely the result of paleomagnetic studies, which have themselves proliferated within the Cordillera at an even more unprecedented rate. For example, my first attempt to summarize paleomagnetic results for the western Cordillera ten years ago (Beck, 1976) involved only 17 studies. Now at least 88 entries are required (Tables 2–5, this paper), even though all data from north of southern British Columbia have been relinquished to other reviewers! At the present rate of growth, by the early 1990s we should have available networks of closely spaced, high-precision paleomagnetic studies for many crucial areas within the Cordillera, as well as adequate reconnaissance coverage of the entire range. Examples of the unparalleled usefulness of paleomagnetic data for regional tectonic analysis already exist (e.g., Wells and Coe, 1985; Luyendyk and others, 1985).

The reason that paleomagnetism is such a powerful tool for tectonic analysis has been discussed many times (e.g., Irving, 1979; Beck, 1980). Its exceptional usefulness arises from the fact that paleomagnetic data are inherently quantified and objective, and that directional paleomagnetic results are ideally sensitive to many kinds of block displacements. Methods of analysis are described in the articles cited above, as well as in many other references (e.g., Demarest, 1983; McWilliams, 1984; Magill and others, 1981; Yole and Irving, 1980), and the underlying principles of rock and earth magnetism are explained in several excellent books (e.g., Irving, 1964; McElhinny, 1973; Tarling, 1983; Merrill and McElhinny, 1983). The reader is referred to these sources for questions of procedure.

The data summarized in this paper are derived mainly from the westernmost part of the conterminous United States. However, tectonic units have little respect for political boundaries, so for the sake of completeness it was necessary to include some data from Baja California and southern British Columbia. The paleomagnetism of British Columbia rocks (and some units from northern Washington) is also discussed by Irving and Wynne (1992).

DATA AND DATA PROCESSING

Paleomagnetic results for the westernmost conterminous United States (and neighboring parts of Mexico and Canada) are summarized in Tables 2 to 5, and pole positions are plotted in Figures 2 to 5. Figure 1 shows the approximate location of each of the 88 studies. Data selection follows criteria given in Beck (1980). Not all entries are of equal quality; readers should consult the originals in all cases. Studies of very young rocks (<5 Ma) and of far-travelled limestone boulders in Franciscan mélange (e.g., Tarduno and Alvarez, 1985) have not been considered. Rotations (R) and poleward translations (P) are calculated following Beck and others (1986). These statistics, with their respective 95 percent confidence limits, reflect the probability that a sampling area has been displaced relative to the stable interior of the continent. Positive (negative) vales of R signify clockwise (counterclockwise) rotation. Similarly, positive (negative) values of P indicate that the sampling area has moved northward (southward) relative to the craton, the motion being given in degrees of paleolatitude. Poleward motions are directed toward (or away from) the appropriate reference pole (Table 1), not the present pole. Thus, for instance, positive values of P for rocks of Cretaceous age indicate transport to the northwest (in present coordinates) because the Cretaceous paleomagnetic pole for North America was located near the Bering Strait.

Figure 1 has been coded to show which paleomagnetic studies are statistically "discordant," and thus may indicate significant tectonic displacements. Solid symbols show results for which P or R (often both) exceed their 95 percent confidence limits. Exceptions are poles CR, IG, SL, and GG, which are shown as concordant even though each has a value of either P or R that exceeds its confidence limit. For these poles the "discordant" R or P values are extremely small (<8°). Values of "discordance" this small tend to press the limits of the method. For the purposes of constructing Figure 1, I considered a result with both R and P <10° to be discordant only if the "discordant" statistic exceeded its confidence limit by a factor of two.

The overwhelming preponderance of solid symbols on Figure 1 dramatizes the degree to which microplate tectonic processes have modified the leading edge of North America.

Beck, M. E., Jr., 1992, Tectonic significance of paleomagnetic results for the western conterminous United States, *in* Burchfiel, B. C., Lipman, P. W., and Zoback, M. L., eds., The Cordilleran Orogen: Conterminous U.S.: Boulder, Colorado, Geological Society of America, The Geology of North America, v. G-3.

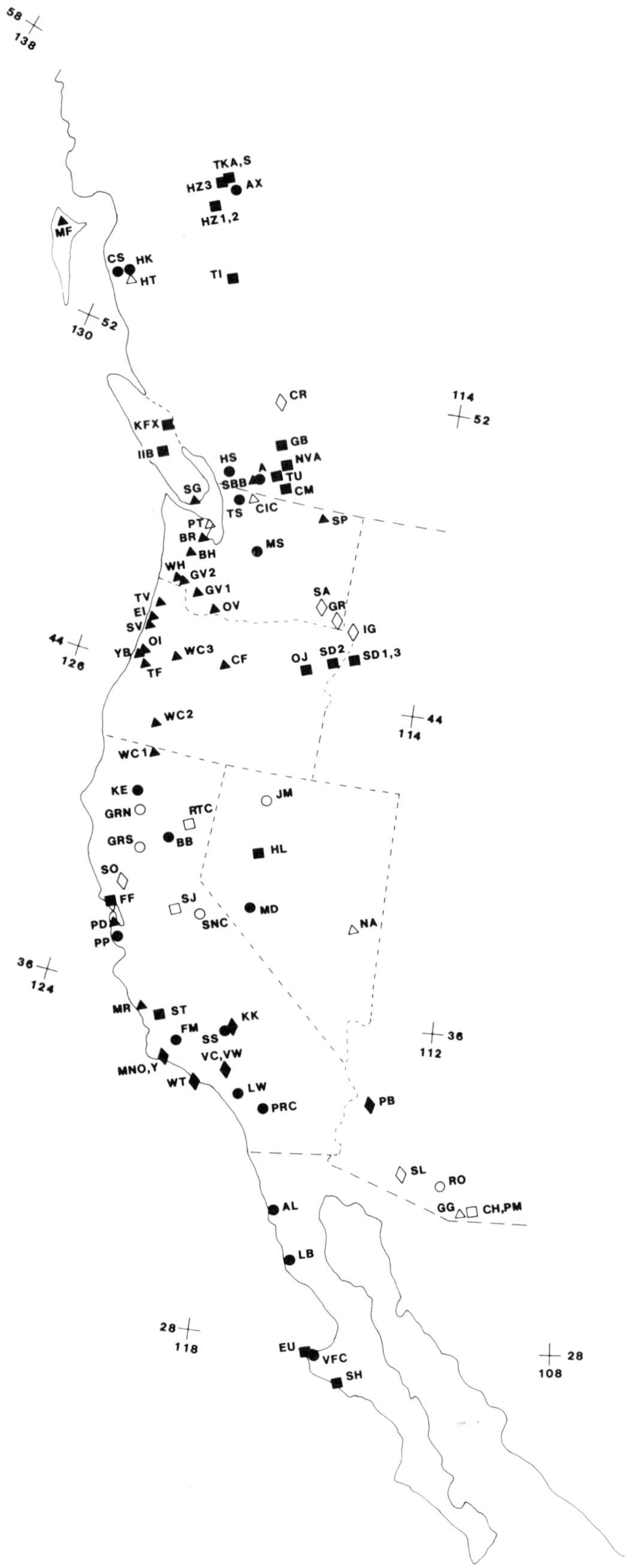

TABLE 1. REFERENCE POLES, NORTH AMERICAN CRATON*

Age	λ ,°N	ϕ,°E	A_{95}
20	83	148.0	4°
30	83	148.0	4
40	83	170.5	4
50	83	170.5	4
60	81.5	192.5	4
70	74	190	7
80	68	191	7
90	67	186	4
100	67	184	10
110	69	186	6
120	68	189	19
130	68	171	19
140	66	163	16
150	67	138	17
160	70	123	19
170	65.5	110	11
180	64	95	11
190	62.5	80.5	11
200	61	66.5	11
210	60	70.5	11
220	58.5	88	11
230	54.5	102.5	11
240	49	113.5	11
250	45	119.5	11

*λ and ϕ are latitude and longitude of reference pole, with 95% confidence interval (A_{95}). Poles 20 to 60 from Diehl and others (1983), 70 to 90 from Irving and Irving (1982), and 170 to 250 from Gordon and others (1984).

Figure 1. Sampling areas for paleomagnetic studies. Filled symbols identify discordant poles and suggest that significant block displacement has taken place. Letter designations are keyed to Tables 2 to 5. Diamonds, Neogene (Table 2); triangles, Paleogene (Table 3); circles, Cretaceous (Table 4); squares, Triassic or Jurassic (Table 5).

Reference poles

As has been much discussed recently (e.g., Gordon and others, 1984; May and Butler, 1986) the interpretation of paleomagnetic data from an orogenic belt depends as much on the apparent polar-wander (APW) path for the stable interior as it does on results from the orogen itself. Fortunately, for North America later than about 120 Ma the reference APW path seems to be reasonably well established. However, for Late Triassic through Early Cretaceous time, details of the APW path are currently being discussed. As shown below, the choice of APW path strongly influences the tectonic interpretation of paleomagnetic data for Cordilleran rocks of early and middle Mesozoic age.

Reference poles used in this compilation are given in Table 1. For the period 20 to 50 Ma the entries are taken from Diehl and others (1983). These are mean values for large sets of reliable virtual geomagnetic poles (VGP) from the craton; they are unlikely to be in error by more than a few degrees. From 60 through 160 Ma the poles in Table 1 are taken from Irving and Irving (1982). The Irving poles are 30-m.y. running averages, in which mean poles for entire formations (not VGP for individual sampling sites) are used as input data. The confidence circle about an Irving reference pole tends to be conservative (slightly too large), whereas with the Diehl method it may be slightly too small. To be consistent I have used 4° (the smallest value obtained using the Irving method) as the 95 percent confidence circle about the Diehl reference points (Table 1, last column). The values for A_{95} actually obtained by Diehl range from 3.2° to 4.1°.

For the period 170 through 250 Ma the entries in Table 1 are taken from Gordon and others (1984). The Gordon APW path is constructed in a way that is fundamentally different from other reference curves. In the Gordon method, acceptable paleomagnetic data are arranged in chronologic sequence, to which a series of small circles is fitted. The pole to each small circle (called a paleomagnetic Euler pole, PEP, by Gordon and others) is regarded as a valid estimate of the pole of absolute motion of the continent from which the APW data were obtained. A PEP thus is analogous to absolute motion poles obtained from hotspot data, and in fact the two methods give very similar (but apparently not identical) results. The reference path presented by Gordon and others (1984) differs from equivalent Irving and Irving (1982) poles by small but significant angles (up to about 12°). Differences in the results of the two methods arise to some extent from a smoothing effect of the running-average technique at points of deflection in the APW path, but important differences also arise because certain Jurassic poles included in the Irving curve were rejected by Gordon and others. Plus, the time scales used are different, which also influences the results. In deciding which APW path to use, I was influenced by the fact that May and Butler (1986) independently constructed an APW path for Late Triassic and Jurassic time and obtained a result that is not much different from the Gordon curve. Thus the 170-to-250-Ma poles given in Table 1 are taken from Gordon and others (1984). The PEP technique does not yield conventional confidence limits; the value I adopted for Table 1 (11°) is about equal to the 95 percent confidence limit on the least well determined pole used in the PEP analysis, and also is larger than all but two of the confidence circles for Irving and Irving's (1982) mean poles for 170 to 250 Ma.

The way in which reference poles for very young rocks (< 20 Ma) were calculated requires some explanation. It has been well demonstrated (e.g., Wilson, 1971) that paleomagnetic studies of very young rocks consistently yield paleomagnetic poles that are "far-sided", meaning that there is a marked tendency for such poles to lie on or near the meridian of the sampling site, but to be displaced a few degrees beyond the spin axis. Thus very young rocks typically have paleomagnetic inclinations that appear too shallow by several degrees, if compared to the present spin axis. The cause of this far-sidedness seems to be some sort of permanent or quasipermanent nondipole element in the geomagnetic field (Cox, 1975; Merrill and McElhinny, 1977). To deal with this problem for very young rocks I have used a fictitous "zero-age" reference pole for the calculation of R and P values for rocks younger than 20 Ma. The coordinates of this pole are given by (λ = 87°N; ϕ_s + 180°) where ϕ_s is the longitude of the sampling site. Thus for pole MNY (Table 2), which is roughly 10 Ma, the reference pole used was the vector sum of 83.0N, 148.0E (the 20-Ma pole, from Diehl and others, 1983) and 87.0N, 59.9E (the appropriate zero-age pole for the longitude of MNY). For MNO the calculation was similar, but because the rocks are approximately 15 Ma in age the Diehl pole was given a weight of three. The confidence interval (A_{95}) for all such reference poles is 4°.

Description of results

Neogene. Paleomagnetic results for Neogene rocks are summarized in Table 2 and Figure 2. Not surprisingly, many yield poles that fall near the reference pole. This is particularly true of rocks from southern British Columbia and the Pacific Northwest (CR, SA, GR, IG). These four have small, statistically insignificant amounts of clockwise rotation (positive R), together with equally small and insignificant negative values of P. However, it is not correct to conclude from these concordant poles that the Pacific Northwest (PNW) has experienced no tectonic block displacements since the Miocene; elegant single-flow paleomagnetic studies by Magill and others (1982) and Sheriff (1984) show that clockwise rotation continued locally in the PNW until very recently. The failure of conventional paleomagnetic studies to detect rotation may simply reflect the inherent limitations of the method. The fact that all four PNW poles have small negative P values is an interesting observation for which I do not have an explanation (see also nearby Paleogene poles SP and CPC).

In contrast to the PNW, Neogene poles from southern California (VC, VW, MNO, MNY, WT) are conspicuously discordant, with clockwise rotations of up to 90°. Several of these

TABLE 2. NEOGENE PALEOMAGNETIC DATA, WESTERN CORDILLERA*

Unit	Age	N	λ/φ	λ'/φ'	ΔR ± R	ΔP ± P	References
CR	10	65	51.7/121.6	84.2/218.2	1.7 ± 6.6	-7.0 ± 4.0	Symons, 1969a, 1969b
SA	10	11	46.4/118.0	85.6/144.2	1.3 ±14.4	-1.1 ±10.1	Choiniere and Swanson, 1979
GR	15	20	46.2/117.7	85.8/207.4	6.2 ±11.1	-4.3 ± 7.3	Choiniere and Swanson, 1979
IG	15	30	46.0/116.8	84.5/210.9	2.3 ± 8.2	-6.2 ± 5.4	Hooper and others, 1979
SO	5	14	38.6/122.5	79.4/ 83.8	5.3 ±11.3	6.6 ± 7.8	Mankinen, 1972
KK	20	12	35.2/118.3	78.3/223.2	3.1 ±10.7	-11.7 ± 7.6	Kanter and McWilliams, 1982
PB	15	11	34.0/113.9	66.3/145.9	-20.2 ± 5.4	5.4 ± 4.7	Calderone and Butler, 1984
VC	25	7	34.5/118.2	62.4/323.6	41.9 ± 6.4	-0.1 ± 5.3	Terres and Luyendyk, 1985
VW	25	10	34.5/118.4	29.0/303.9	86.1 ± 7.6	-3.8 ± 6.1	Terres and Luyendyk, 1985
MNO	15	26	34.5/120.1	19.0/308.3	91.6 ± 4.5	5.3 ± 3.8	Hornafius, 1985
MNY	10	13	34.5/ 12.1	60.6/328.4	38.8 ± 4.6	2.5 ± 3.9	Hornafius, 1985
WT	20	52	34.0/119.6	25.1/322.3	81.0 ± 4.7	13.4 ± 4.2	Kamerling and Luyendyk, 1985
SL	20	28	32.5/112.9	75.8/123.8	- 4.9± 4.6	7.1 ± 4.1	Calderone and Butler, 1984

*Age estimated to nearest 5 m.y.; paleontological ages converted using Harland and others (1982). N is number of sites; λ and φ are north latitude and west longitude of sampling site; λ' and φ' are north latitude and east longitude of paleomagnetic pole. R and P are rotation (clockwise positive) and poleward transport (north positive), calculated by method of Beck and others (1986). ΔR, ΔP are 95% confidence intervals, after Beck and others (1986). Some entries are recalculated.

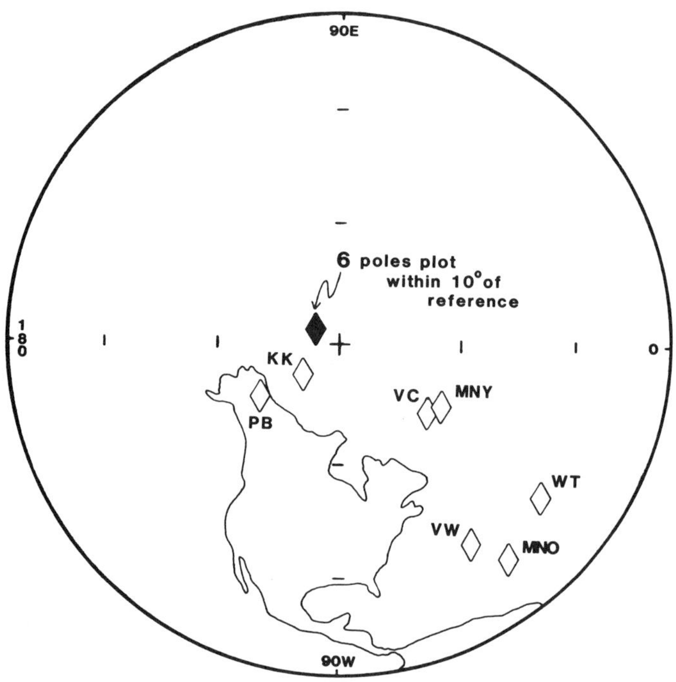

Figure 2. Equal-area projection of Neogene paleomagnetic poles, from Table 2. Solid symbol is reference pole, from Diehl and others (1983).

studies show that rotation was going on while the rocks were forming. Luyendyk and others (1985) offer a tectonic model to account for these rotations involving dextral simple shear in the continental margin driven by Pacific–North American plate interaction.

Luyendyk and others (1985) note that much of their data also show shallower-than-expected paleomagnetic inclinations (P>0). Although they recognize that northward transport of coastal California makes some contribution to these shallow inclinations, they feel that the amount of transport required is unreasonable. To assist northward transport in producing the flattened inclinations, they invoke arguments from geomagnetism (persistent nondipole sources) and structural geology (undetected initial dip). However, my calculations present no such problem; only study area WT, which may be part of a mobile borderland block, has a significantly flat inclination. The difference in part is the result of the reference pole chosen; Luyendyk and co-workers used an early Irving (1979) reference pole, whereas Table 2 was constructed as described above, using the Diehl and others (1983) 22- to 38-Ma pole, modified where needed by the appropriate "zero-age" pole.

The two remaining discordant Neogene poles (PB, KK) are less easy to account for. PB has a paleomagnetic declination that is rotated distinctly counterclockwise. Calderone and Butler (1984) speculate that it may have experienced the same sort of Neogene tectonics as has the Mojave block, for which counter-clockwise rotations have been predicted (Garfunkel, 1974). Pole KK has a statistically significant negative value of P, implying southward transport of the southern Sierra Nevada block since

TABLE 3. PALEOGENE PALEOMAGNETIC DATA, WESTERN CORDILLERA*

Unit	Age	N	λ/φ	λ'/φ'	ΔR ± R	ΔP ± P	References
MF	60	8	53.7/132.2	72 /263	37.5 ±24.4	-6.0 ± 9.2	Hicken and Irving, 1977
HT	45	11	53.3/129.0	79.8/226.0	-9.0 ±23.7	-7.2 ±10.4	Symons, 1977a
SBB	30?	4	49.3/121.2	86.5/308.8	-15.8 ±10.3	-1.7 ± 6.6	Irving and others, 1985
CIC	30	15	48.5/121.4	85.1/206.4	6.2 ± 7.8	-4.6 ± 5.0	Beck and others, 1982
SP	50	102	48.5/118.8	79.4/304.2	25.9 ± 6.4	-2.0 ± 4.0	Fox and Beck, 1985
SG	40	36	47.8/123.5	69.5/150.9	-20.0 ± 8.4	5.0 ± 5.6	Symons, 1973a
PT	50	9	47.9/122.7	74.5/187.7	-11.4 ±29.8	-6.0 ±15.2	Beck and Engebretson, 1982
BR	50	13	47.6/122.7	86.6/324.6	15.2 ±11.6	2.3 ± 7.7	Beck and Engebretson, 1982
BH	55	17	47.0/123.1	77.3/309.7	28.9 ± 9.0	1.7 ± 5.8	Globerman and others, 1982
WH	50	70	46.4/123.6	73.3/326.4	33.2 ± 7.7	4.9 ± 5.3	Wells and Coe, 1985
GV2	40	10	46.3/123.2	77.2/5.7	22.3 ±15.0	11.2 ±11.7	Wells and Coe, 1985
GV1	40	37	46.0/122.5	75.5/345.5	27.8 ± 7.5	8.3 ± 5.4	Beck and Burr, 1979
OV	35	28	45.7/122.0	70.0/317.9	38.8 ± 5.8	1.0 ± 4.1	Bates and others, 1981
TV	55	32	45.6/123.5	65.5/312.7	44.9 ± 9.4	3.5 ± 6.4	Magill and others, 1981
EI	40	10	44.9/123.7	58.4/313.2	53.9 ±12.4	4.1 ± 8.9	Beck and Plumley, 1980
SV	55	33	44.6/123.6	45 /307	72.7 ± 9.8	7.1 ± 7.1	Simpson and Cox, 1977
CF	40	46	44.6/120.4	83.9/277.6	15.3 ±10.2	-2.6 ± 6.7	Gromme and others, 1986
OI	30	15	44.5/123.7	66.8/324.8	41.1 ± 9.9	3.6 ± 7.4	Beck and Plumley, 1980
WC3	30	14	44.3/122.7	70.7/284.8	34.9 ±11.4	-11.0 ± 6.8	Magill and Cox, 1980
YB	35	8	44.2/124.0	58 /308	55.3 ±23.7	0.3 ±16.3	Simpson and Cox, 1977
TF	45	40	44.0/123.7	49 /305	67.6 ±13.2	2.4 ± 9.4	Simpson and Cox, 1977
WC2	30	10	42.7/122.7	74.1/302.2	31.1 ±16.2	-5.1 ±10.9	Magill and Cox, 1980
WC1	30	53	41.9/122.4	86.9/347.1	13.2 ± 8.5	0.8 ± 6.4	Beck and others, 1986
NA	30	14	38.5/115.0	80.5/120.5	1.2 ±12.3	4.5 ±10.2	Irving and others, 1976, Pole 1-442
PD	60	53?	37.5/122.5	———	—— ——	22.3 ± 5.7	Champion and others, 1984
MR	25	7	35.3/120.7	46.4/329.3	57.1 ± 8.9	9.4 ± 8.0	Greenhaus and Cox, 1979
GG	60	25	31.5/110.8	77.0/201.0	-3.8 ± 4.2	-3.7 ± 3.4	Barnes and Butler, 1980

*See Table 2 for explanation.

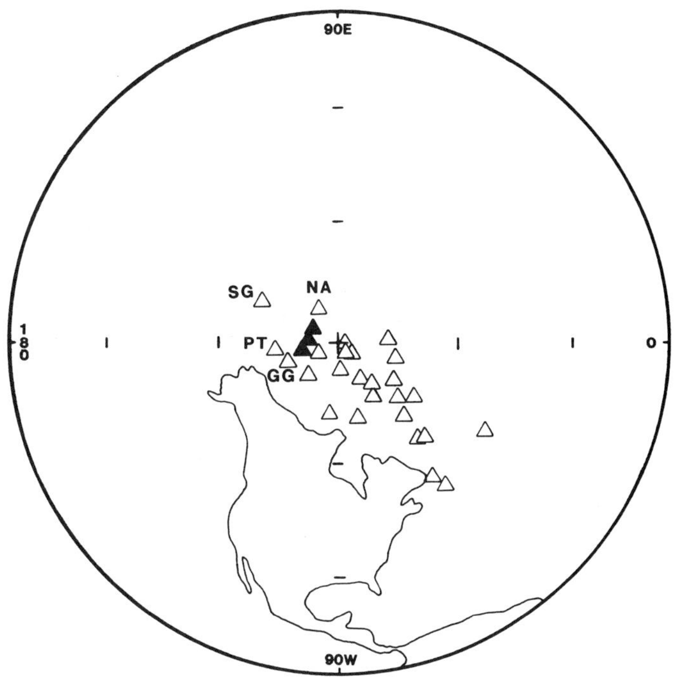

Figure 3. Equal-area projection of Paleogene paleomagnetic poles, from Table 3. Solid symbols are reference poles, from Diehl and others (1983).

about 20 Ma. Southward transport of more than a few kilometers seems entirely unreasonable. Although 12 sites are involved in study KK, they are restricted to three rather limited areas, each of which seems to have its own characteristic direction (Kanter and McWilliams, 1982, Fig. 9). Perhaps fewer than 12 independent measurements of the ancient geomagnetic field are involved, raising the possibility that secular variation has not been completely averaged.

Paleogene. Data for rocks of Paleogene age are given in Table 3 and Figure 3. Twenty-seven poles are represented, all but five of which (NA, GG, HT, CIC, PT) are discordant. The sampling areas for NA and GG are located well inboard from the continental margin, apparently far enough to have escaped the effects of plate-edge tectonics. Poles CIC and HT are from plutonic rocks in the PNW; northward movement of crustal blocks apparently had ceased in that area by about 40 Ma. Concordant pole PT (and discordant, counterclockwise rotated pole SG) may be the result of oroclinal bending accompanying creation of the Olympic Mountains (Beck and Engebretson, 1982).

The remaining 21 studies of Paleogene rocks show evidence of large-scale clockwise rotation. Some northward translation also is indicated. The pattern of poles streaked out into the area of the Atlantic shown in Figure 3 suggests displacement by dextral shear (Beck, 1976). Most of these studies show large clockwise rotations (R values up to 90°), and a few show significant poleward displacements. An exception is entry PD, which appears to have been moved relatively northward by several thousand kilometers. The rocks measured in study PD are continental-margin sedimentary rocks that may be a little older than the other rocks represented in Table 3. Their apparent northward transport fits better with Cretaceous results (see below) than with other Paleogene rocks further north.

Cretaceous. Cretaceous paleomagnetic poles form a pattern (Fig. 4) that suggests regional clockwise rotation and northward transport. Eighteen of 23 poles are discordant, and all but two of these are deflected to the right of the reference curve. The pattern is very similar to the pattern of deflection of Paleocene poles (Fig. 3), but with one important difference: most Cretaceous poles have been displaced substantially away from their sampling sites, reflecting relative northward transport of outboard terranes during the Late Cretaceous and early Tertiary (see below).

Poles from studies located near the northern end of the Sierra Nevada (JM, GRN, GRS, SNC) are not identified on Figure 4, but they form a dense cluster located near the reference poles. Several of these deal with layered rocks that appear to have been remagnetized during an episode of pervasive reheating in the Late Cretaceous. This "Sierran" cluster of concordant poles indicates that the Sierra Nevada block has not moved north or south more than a few hundred kilometers since Late Cretaceous time. Scanty and preliminary data from the Klamath block (discussed further below) indicate that it too may not have participated in Late Cretaceous–early Tertiary northward translation. Rotations seem to be common in the Klamath block, but are found rarely in the Sierra. The remaining Cretaceous studies all show evidence of substantial northward transport and clockwise rotation; some displacements may exceed 3,000 km.

Two clearly anomalous poles (TS, LW) are shown in Figure 4. TS represents the Twin Sisters dunite in the North Cascades of northwest Washington. The dunite was apparently transported to shallow depths in the crust by some sort of thrusting event, perhaps about 75 Ma. It is assumed here that it acquired its magnetization at that time. Although the Twin Sisters block is several kilometers thick, it has the geometrical aspect of a thin lens. Agreement between directions at scattered sites shows that it has not been internally deformed, but it nevertheless could have been tilted after magnetization was acquired. Presumably the peculiar location of pole TS reflects local deformation in addition to the prevailing pattern of dextral shear. Pole LW is based on results from Late Cretaceous sedimentary rocks in a structurally complex terrain in southern California. Its positive value of P fits the pattern of nearby Cretaceous studies; its anomalous location on Figure 4 is due to nearly 12° of counterclockwise rotation for which there seems to be no obvious explanation.

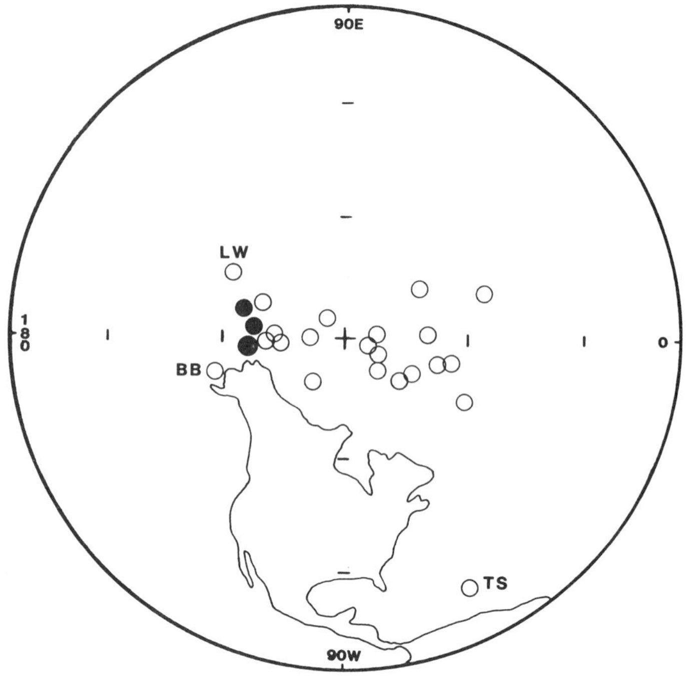

Figure 4. Equal-area projection of Cretaceous paleomagnetic poles, from Table 4. Solid symbols are reference poles, from Irving and Irving (1982). Large filled circle indicates Late Cretaceous pole, which experienced no movement that can be resolved at this scale for 40 m.y.

Jurassic and Triassic. Paleomagnetic data for rocks of this age are summarized in Table 5 and in Figure 5. Paleomagnetic results for rocks of early Mesozoic age present some important problems, the most serious of which concerns polarity. From the beginning of Mesozoic time until the mid-Cretaceous, the polarity of the geomagnetic field alternated every few million years or so. This means that directions of remanent magnetization in a rock unit of this age theoretically can correspond to a pole in either the (paleo) Northern or Southern Hemisphere. Usually, no problem arises if the study area is firmly embedded in a continental block with known APW path. However, if the study area occupies an outboard position where it may have moved relative to the continent, interpretation encounters a colossal ambiguity—the paleomagnetic data themselves afford no clue as to whether the crustal fragment originated north or south of the paleoequator. This dilemma has generated much debate where certain Cordilleran terranes (e.g., "Wrangellia"; Panuska and Stone, 1985) are concerned.

The problem can be illustrated by a hypothetical example.

TABLE 4. CRETACEOUS PALEOMAGNETIC DATA, WESTERN CORDILLERA*

Unit	Age	N	λ/φ	λ'/φ'	ΔR ± R	ΔP ± P	References
AX	100?	13	56.2/126.1	76 /327	68.3±26.0	11.8± 9.3	Monger and Irving, 1980
HK	70	15	53.3/129.0	73.5/323.2	49.9±18.3	13.9±10.2	Symons, 1977a
CS	110	23	53.2/129.5	70 / 3	53.9±13.2	26.4± 7.4	Symons, 1977b
HS	90	17	49.4/123.2	83.4/129.2	27.1±15.2	12.3± 9.9	Symons, 1973b
SBA	100	8	49.3/121.2	65.0/345.1	67.4±13.1	20.5± 9.4	Irving and others, 1985
TS	75?	17	48.7/122.0	27.5/297.5	12.0±13.2	18.5± 8.0	Beck, 1975
MS	90	17	47.5/121.0	67.9/ 34.6	44.8± 6.2	30.0± 4.2	Beck and others, 1981a
JM	80?	24	41.3/118.4	74.6/190.8	9.7±10.4	2.6± 6.4	Russell and others, 1982
KE	135?	10	40.7/122.3	56.9/332.0	70.3±21.0	13.2±15.2	Achache and others, 1982
GRN	70?	4	40.3/122.6	81.4/177.6	7.9±10.7	5.6± 7.2	Mankinen and Irwin, 1982
BB	140	9	39.9/121.3	57.6/194.8	-11.1±20.9	-15.4±14.2	Gromme and others, 1967
GRS	70?	3	39.4/122.4	68.0/156.8	-10.5±15.5	9.5±11.4	Mankinen and Irwin, 1982
MD	90	10	38.4/118.3	76.9/234.0	26.1± 8.9	-3.3± 5.6	Geissman and others, 1984
SNC	90	54	38.1/119.8	71.1/180.9	5.1± 6.4	2.9± 4.3	Gromme and Merrill, 1965 and Frei and others, 1984
PP	70	24?	37.2/121.0	—	—	25.3± 8.8	Champion and others, 1984
SS	85	8	35.2/118.4	71.9/321.8	48.9± 7.2	9.8± 5.2	Kanter and McWilliams, 1982
FM	85	10	34.7/119.8	54.4/ 18.4	48.6± 8.9	40.0± 7.6	McWilliams and Howell, 1982
LW	80	20	33.7/117.6	57.9/149.6	-11.9± 9.1	18.5± 6.8	Fry and others, 1985
PRC	85	39	33.5/117.0	84.7/346.2	31.8± 6.8	11.9± 5.1	Hagstrum and others, 1985
RO	70	9	32.1/111.5	73.6/176.0	-1.7± 6.9	3.7± 5.5	Vugteveen and others, 1981
AL	115?	4	31.0/116.5	81.3/336.1	33.8±15.4	10.9±12.1	Hagstrum and others, 1985
LB	90?	3	30.0/115.8	79.5/316.6	37.5± 7.3	7.0± 6.0	Hagstrum and others, 1985
VFC	100?	19	27.5/114.5	82.8/ 13.2	31.4±10.4	13.2± 8.5	Hagstrum and others, 1985

*See Table 2 for explanation.

Suppose investigators study a 200-Ma rock unit currently located near San Francisco. If the unit originated in its present location it should have the following direction or remanent magnetization (the expected direction): Declination (D), 356°; Inclination (I), 18°. Assume that the direction actually found (by investigators presumed to be reasonably competent!) is: D, O°; I, –30°. Clearly the unit is discordant in inclination, although its declination is essentially concordant. Two interpretations, equally valid, are possible. (1) The unit may have originated south of the paleo-equator, at a time when the geomagnetic polarity was normal. In that case it has moved northward a little over 25° of (paleo) latitude (about 2,800 km), relative to stable North America—without rotating much along the way. (2) Alternatively, the crustal fragment may have been in the Northern Hemisphere at 200 Ma, nearly 7° *north* of its current position. Scenario (2) assumes a reverse geomagnetic polarity, and it requires the fragment to have rotated nearly end-for-end (176°) during roughly 800 km of southward transport.

This example is similar to the actual situation for Wrangellian rocks on Vancouver Island (KFX, Table 5). To resolve the hemisphere problem the investigator usually chooses whichever alternative seems to do the least damage to current notions of regional tectonics or current models for plate motions in the Pacific. If the crustal fragment is large enough so that an APW path extending into the mid-Cretaceous can be constructed, the problem is sometimes simplified because the polarity of mid- and Late Cretaceous rocks is known. Note that a "minimum travel" rule often makes little sense; at a typical plate velocity (say 5 cm/yr) the hypothetical crustal block in this example could have made its long trip from the Southern Hemisphere in less than 60 m.y.

For reasons discussed below, most of the entries in Table 5 and all of the poles shown in Figure 5 assume an origin in the Northern Hemisphere. Because the case of Wrangellia is hotly debated, Table 5 entries KFX, SD1, and SD2 show both Northern Hemisphere and Southern Hemisphere options.

One other feature of Table 5 should be explained. In studies IIB and HL, the investigators were uncertain of the age of magnetization. I have calculated rotations and poleward translations for both the nominal age of the rock and the suspected age of remagnetization. Several other entries in Table 5 also have uncertain ages of magnetization.

TABLE 5. JURASSIC AND TRIASSIC PALEOMAGNETIC DATA, WESTERN CORDILLERA*

Unit	Age	N	λ/φ	λ'/φ'	ΔR ± R	ΔP ± P	References
TKA	225	6	56.7/126.4	38 /133	-34.5± 12.5	2.2± 11.0	Monger and Irving, 1980
TKS	225	8	56.6/126.5	24 /146	-53.1± 2.5	7.1± 11.2	Monger and Irving, 1980
HZ3	205	4	56.5/126.8	40 /144	-57.3± 30.4	-5.2± 25.2	Monger and Irving, 1980
HZ2	205	4	55.8/126.6	17 /185	-108.5± 23.9	-9.9± 19.4	Monger and Irving, 1980
HZ1	190	7	55.6/126.4	70 / 57	12.5± 16.5	-5.7± 13.7	Monger and Irving, 1980
TI	145	17	54.1/125.0	58.4/121.5	5.9± 24.7	19.8± 15.1	Symons, 1983a
GB	200	13	50.5/121.0	52.3/ 13.3	31.6± 11.0	1.9± 10.3	Symons, 1983b
KFX	230	28	50 /125.5	-23 /232	-149.9± 10.0	4.9± 9.3	Yole and Irving, 1980
				3 / 52	30.1± 10.0	38.9± 9.3	
NVA	200?	26	49.9/120.6	52.7/348.7	44.4± 40.5	-7.6± 9.7	Symons, 1985a
IIB	190?	10	49.6/125.5	71.0/ 35.0	20.1± 12.0	-7.4± 10.7	Symons, 1985b
	130?				42.6± 28.2	23.1± 16.4	
TU	175?	10	49.5/120.9	49.2/119.3	-18.1± 18.8	7.8± 17.1	Symons, 1974
CM	200	11	49.3/120.6	57.3/ 11.8	29.5± 9.9	-3.2± 9.3	Symons and Litalien, 1984
SD3	145?	17	45.3/116.7	60.2/295.3	78.9± 20.9	-12.1± 14.0	Hillhouse and others, 1982
SD2	230	6	45.3/117.0	46.6/130.0	-19.5± 15.0	-3.8± 14.2	Hillhouse and others, 1982
				-46.6/310	16.0± 15.0	34.4± 14.2	
SD1	230	5	45.2/116.7	7.6/170.3	-74.4± 17.0	-2.2± 16.2	Hillhouse and others, 1982
				-7.6/350.3	105.6± 17.0	32.4± 16.2	
OJ	145?	9	44.5/117.5	68.7/322.0	63.7± 20.2	-0.6± 14.3	Wilson and Cox, 1980
RTC	145?	7	40.2/120.9	71.1/159.9	4.9± 23.7	-4.2± 18.2	Hannah and Verosub, 1980
HL	200?	15	39.9/118.2	35.1/135.2	-50.7± 10.2	0.0± 10.0	Hidson and Geissman, 1984
	155?				-29.9± 17.4	19.0± 15.0	
SJ	155	10	38.1/120.8	64.5/165.2	9.7± 18.6	-11.8± 15.8	Bogen and others, 1985
FF	150?	7	38.0/122.7	26 /310	104.9± 18.3	2.4± 15.7	Gromme and Gluskoter, 1965
ST	160	8	35.2/120.5	27.1/ 13.9	60.7± 17.4	38.9± 15.8	McWilliams and Howell, 1982
CH	150	15	31.5/110.5	62.2/130.3	-2.1± 15.0	5.5± 14.0	Kluth and others, 1982
PM	170	12	31.4/110.7	61.8/116.0	-4.4± 10.3	1.4± 10.0	May and others, 1986
EU	145	8	27.6/114.6	—	—	10.4± 18.6	Hagstrum and others, 1985
SH	215	3	27.0/114.0	29.7/ 22.0	46.5± 15.4	16.6± 14.8	Hagstrum and others, 1985

*See Table 2 for explanation.

Figure 5 is different than polar plots for younger rocks presented earlier (Figs. 2 to 4) in that it shows discordant poles distributed about equally to the left and right of the reference curve. This may be an artifact of the convention discussed above, by which only poles corresponding to origin in the North Hemisphere have been plotted. If not, then older rocks from the Wrangellia (Jones and others, 1977) and Stikinia (Monger and others, 1982) have moved generally southward and rotated generally counterclockwise since they formed. This interpretation runs counter to much prevailing opinion, and if either the "Northern Hemisphere" option or the Gordon and others (1984) APW path are substantially wrong, it will have to be modified. This is discussed further below.

Triassic and Jurassic poles from further south show a different pattern. Poles OJ and SD3 from the Pacific Northwest agree well with one another. Both have been rotated clockwise through large angles, with no significant north-south movement. Note that the age of magnetization (and thus the correct reference pole) for both of these poles is uncertain.

Poles from California and Baja California (FF, ST, EU, SH) also generally indicate clockwise rotation, but unlike poles of similar age from the PNW (which if anything show southward movement), they all have positive values of P. Several of these values are significant statistically and call for relative northward transport of up to several thousand kilometers. Entry EU from Table 5 is not plotted in Figure 5; scatter in declination makes calculation of a paleomagnetic pole for this unit impossible.

Only three paleomagnetic poles from Table 5 are concordant; these (RTC, CH, PM) are not shown in Figure 5. Field areas for CH and PM probably are located far enough inland to have escaped the effects of continental-margin tectonics. RTC gives results from remagnetized rocks adjacent to the Sierra Nevada block, and the fact that it is concordant supports Cretaceous data from the same area indicating that the Sierra block has been

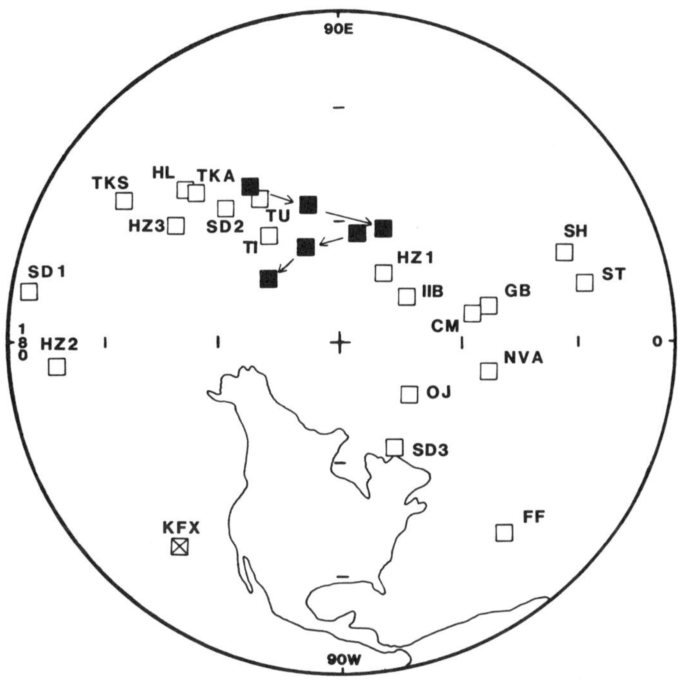

Figure 5. Equal-area projection of Jurassic and Triassic paleomagnetic poles, from Table 5. Solid symbols are reference poles, from Gordon and others (1984). Pole KFX plots in the Southern Hemisphere.

substantially free from large-block displacements since the end of the Jurassic. Finally, pole HL comes from intrusive rocks in a structurally complex part of Nevada. Its counterclockwise rotation probably represents local tectonics.

Analysis

The paleomagnetic data summarized above show, without much room for doubt, that block displacements on an impressive scale are common in the westernmost Cordillera.

As early as the mid 1970s it was possible to conclude from the accumulated paleomagnetic data that north-oblique convergence of oceanic plates with North America has played a fundamental role in Cordilleran tectonics. North-oblique convergence (and dextral transform behavior) apparently have (1) attached numerous fragments of non–North American crust to the western edge of the continent, (2) transported slivers of the leading edge of North America northward along the continental margin, and (3) rotated many crustal blocks clockwise about nearly vertical axes. This pattern of displacement and its general cause were evident in the mid-1970s, even though many problems remained (e.g., "provenance" of exotic crustal blocks, size of typical displaced crustal block, exact mechanism of displacement).

This pattern and general cause remain valid for rocks along the western edge of the continent that are younger than about 125 Ma (compare Figs. 2 to 4 with Fig. 5 of Beck, 1976). This is not surprising, in view of the record of plate interactions. For instance, Engebretson and others (1985), extending the work of Atwater (1970) and others, show mainly north-oblique convergence affecting the western United States back to about 125 Ma. However, Figure 5 suggests that the tectonic pattern may have been quite different earlier in the Mesozoic.

There appear to me to be three lines of evidence that favor southward transport of allochthonous crustal blocks (and thus the Northern Hemisphere option for their origin) for times earlier than about 125 Ma. None is particularly convincing by itself, but their sum is persuasive. The three arguments come from plate-motion studies, geology, and paleomagnetism.

1. Engebretson and others (1985) show an abrupt change in plate velocities for the Farallon and North American plates at about 125 Ma. This change is of such a nature as to favor strong right-oblique interaction after 125 Ma. Prior to that time, nearly normal or slightly left-oblique relative motion is indicated.

2. Many early Mesozoic structures in the Great Basin suggest vergence to the southeast (R. Speed, personal communication, 1984). Examples of published references include Oldow and others (1984) and Speed (1979). If local rotations have not obscured the record, then southeast vergence may indicate underthrusting of the craton from the northwest. Counterclockwise rotation of the Humboldt lopolith (entry HL, Table 5) also may be related to southeastward underthrusting (Hudson and Geissman, 1984). Left-lateral displacement on the Mojave-Sonora megashear (Silver and Anderson, 1974; Anderson and Schmidt, 1983) also implies relative southward transport in pre-Cretaceous time.

3. Panuska and Stone (1985) have constructed a plot of latitude versus time for a large, allochthonous part of southern Alaska, their Southern Alaska Superterrane, which they see as having formed by amalgamation of Wrangellia with the Alexander and Peninsula terranes far to the south of their present locations. Alaskan paleomagnetic data are beyond the limits of this paper, but the latitude versus time plot of Panuska and Stone is significant in that it shows movement of the Alexander and Wrangellia terranes southward until sometime after about 175 Ma. According to their plot, rapid northward movement must have begun prior to about 125 Ma.

Panuska and Stone (1985) also show the paleolatitude of North America (specifically, southern Alaska) as a function of time, and this may be the strongest argument of all in support of relative southward transport of outboard Cordilleran terranes. The significant feature, in my view, is the behavior of western North America during the period roughly 210 to 125 Ma, during which time (according to the diagram of Panuska and Stone) it gained about 45° in latitude. Using the pole positions given in Table 1, I have calculated north-south velocities for stable North America at two points, near Seattle and near San Diego. Both sets of calculations show a northward component of motion beginning at about 210 Ma, accelerating to a maximum of over 60 km/m.y. (6 cm/yr.) at about 150 Ma and then decreasing to zero at about 110 Ma. After 110 Ma the west coast of North

America maintained roughly its present latitude, or moved slowly southward.

It seems to me that this behavior of North America argues convincingly against northward transport of allochthonous terranes during the Jurassic. If North America were itself moving northward at what amounts to a fairly rapid plate velocity, it would require unusually rapid movement on the part of a neighboring oceanic plate to displace crustal fragments *relatively* northward. Relative southward transport during this period, on the other hand, would have been easy to accomplish. Northward transport of allochthonous material is most likely to have occurred during the period 125 Ma to present, when North America's northward component slowed to zero and then reversed.

The arguments just presented for southward-relative transport during the Jurassic are suggestive, but clearly not definitive. The direction of tectonic transport during the Jurassic remains one of the most important questions of Cordilleran tectonics.

What follows are some brief tectonic summaries based mainly on the paleomagnetic data cited above. Some of these interpretations are highly speculative.

Northwest Washington and southern British Columbia. Cretaceous batholithic rocks of the Coast Plutonic Complex and the North Cascades were translated northward along the continental margin by as much as several thousand kilometers during the Late Cretaceous and early Tertiary (HK, CS, HS, SBA, MS). Most of the transport was completed by about 40 Ma (HT, SBB, CIC). The period 100 to 40 Ma corresponds to a time of rapid north-oblique relative motion between North America and the Kula and Farallon plates, according to Engebretson and others (1985). Beck (1986) proposed a tectonic model to account for northward displacement that employs oblique convergence as a driving mechanism.

Paleomagnetic studies from Stikinia and attached crustal fragments (Superterrane I; Monger and others, 1982) present a difficult problem of interpretation. Many of these studies (TKA, TKS, HZ1–3, TI, GB, CM, NVA, TU, AX, SP) have been cited as evidence for northward transport of Stikinia prior to about 45 or 50 Ma. However, this interpretation is no longer valid if the Gordon and others' (1984) APW path is adopted. For instance, Monger and Irving (1980, Fig. 5) show the sampling area of the Hazelton Formation (poles HZ1 to 3) lying approximately 13° of latitude south of its present location in Early Jurassic time. This implies about 1,300 km of northward transport. However, if the Gordon and others' pole for 175 Ma is used, the latitude discrepancy is essentially eliminated. This occurs because the reference poles used by Monger and Irving (1980) place North America at a much higher latitude during the Triassic and Jurassic than does the Gordon path. The same is true for other Triassic and Jurassic poles from Stikinia; using the Gordon poles, values of poleward transport for TKA, TKS, HX1–3, GB, NVA, TU, and CM all are small and not significant statistically, and they are about equally positive and negative. Only the Topley Intrusions (TI) shows a statistically significant value for (north) poleward transport, and it is much younger than other Stikinia Mesozoic poles.

One might be tempted to conclude that Stikinia, and perhaps all of Superterrane I, is substantially autochthonous and has been badly miscast in the role of a displaced terrane. This would ignore other important evidence, however, particularly the paleomagnetism of the Axelgold Intrusion (AX, Table 4). The Axelgold Intrusion is a layered mafic body that intrudes sediments of the Cache Creek Group, a part of Superterrane I. Radiometric mineral ages from the interior of the body fall near 100 Ma, although it may be considerably older (Monger and Irving, 1980). Also, the Axelgold Intrusion is relatively small and may have undergone some undetected tilt. If, however, the Axelgold is 100 m.y. old and untilted, Table 4 shows that it has moved northward by about 1,000 km since it acquired its magnetization. This would agree with the Topley Intrusion result mentioned above. Alternatively, if the Axelgold's magnetization dates from the Early Cretaceous (a real possibility, according to Monger and Irving, 1980) then it is concordant in paleolatitude.

Gabrielse (1985) summarizes geological arguments for more than 900 km of dextral strike slip on the Northern Rocky Mountain Trench. Additional dextral offset is likely on faults further west. Gabrielse believes that most of this offset took place during the interval mid-Cretaceous to Eocene or Oligocene. Offset of this magnitude is a strong argument in favor of northward displacement of Stikinia.

One obvious compromise scenario suggests itself. Perhaps Triassic and Jurassic rocks of Stikinia originated at about their current position relative to North America, but were later transported southward. This would allow the Axelgold Intrusion to have been emplaced at its anomalously low paleolatitude. Northward transport after about 125 Ma then might be invoked to restore Stikinia to its original (early Mesozoic) position. This example of "yo-yo" tectonics is not impossible, but it involves too much coincidence to be very appealing.

Coast Range and Cascades of Oregon, Washington, and northern California. There are 16 entries for this province, all Paleogene. Fifteen show positive values of R that are significant at the 95 percent confidence level. Obviously, clockwise rotation has been an important feature of the tectonics of the area. Values of P (Table 3) tend to be near zero, and only three are significant at 95 percent. However, the great preponderance of positive values of P found in Coast Range entries (but not the Cascades) suggests a small amount of northward transport (Beck, 1984). Most Coast Range studies deal with basalts that are interpreted as accreted seamounts; if this is true, they probably moved northward at least several hundred kilometers before they accreted to the continental margin.

The size of the individual rotated Coast Range–Cascade blocks is a matter of debate. Extreme estimates range from blocks roughly the size of the entire province (Magill and others, 1981) to blocks a few tens of kilometers on a side (Wells and Coe, 1985). The mechanism of rotation also is not completely understood; most models invoke either rotation during accretion, simple shear in the continental margin caused by oblique subduction, or rigid-body rotation driven by extension in the Basin and Range

Province (Magill and others, 1981, 1982; Wells and Coe, 1985; Heller and Ryberg, 1983; Beck and others, 1986). Probably all three contribute. Wells (1984) and Wells and Heller (in preparation) conclude that shear rotation and rigid-body rotation driven by extension both have been major contributors to PNW rotation, whereas rotation during accretion has been relatively unimportant.

Blue Mountains, Oregon and Idaho. Paleomagnetic studies in the Blue Mountains include two on rocks of Late Jurassic/Early Cretaceous age (SD3, OJ). These agree in calling for large clockwise rotations. Both also have negative values of P that are not significant at 95 percent confidence. Paleogene pole CF also shows a clockwise rotation and a small, nonsignificant negative value of P. The simplest explanation for these magnetic results calls for (1) amalgamation of the Huntington and Seven Devils arcs and other crustal fragments and their accretion to North America; (2) intrusion of Late Jurassic/Early Cretaceous plutons; (3) clockwise rotation through an angle of about 55°, between the Early Cretaceous and the time of deposition of the Clarno Formation (CF, Table 3) in the late Eocene or early Oligocene; (4) 15° of additional clockwise rotation since Clarno time. The large (but not statistically significant) negative value of P found for entry SD3 suggests transport from the north.

Klamath/Sierra block, California. As discussed earlier, paleomagnetic results for this block show no evidence for significant northward or southward transport, at least since the end of the Jurassic. This is surprising in view of results for rocks of similar age located further west (discussed below), which appear to have been transported relatively northward by up to several thousand kilometers. Perhaps the Sierra/Klamath block was located sufficiently inboard to escape the results of shear-induced northward displacement.

Paleomagnetic studies from the Sierra and Klamath Mountains differ in one important respect; Sierran studies show little if any rotation, whereas results from the eastern Klamaths suggest clockwise rotations ranging up to 90° or more. Mankinen and others (1984) present preliminary data that indicate that large amounts of rotation took place between Permian and Jurassic time, and again between the Jurassic and Early Cretaceous. Rotation was virtually complete by Early Cretaceous time. Early Cretaceous is also the most probable time for the rifting apart of the Klamath-Sierra block, according to Blake and Jones (1977). Failure of Sierran paleomagnetic studies to find clockwise rotations similar to those measured in the Klamaths may simply be because no pre–Early Cretaceous results for the Sierra have yet been obtained.

Southern California and Baja California. Jurassic, Cretaceous, and Paleogene poles from this region support very large amounts of northward transport (ST, EU, SH, PP, FM, PRC, AL, LB, VFC, PD). Most of the rocks involved are arc or forearc deposits that probably originated along the margin of North America. Agreement between so many studies, involving many different rock types and investigators, on so startling and important a conclusion is extremely impressive. Transport northward was mostly from western Mexico, probably during the Late Cretaceous and early Tertiary. During this interval, convergence of the Farallon and Kula plates with North America was both relatively rapid and strongly north-oblique (Engebretson and others, 1985). Probably oblique subduction gave rise to a system of strike-slip faults in the leading edge of the continent along which the various crustal fragments were transported northward (Beck, 1986).

Beck and others (1981b) and Irving and others (1985) point out a curious circumstance; according to paleomagnetically determined paleolatitudes, all pre-Tertiary batholiths now located along the edge of the continent, from Baja California to British Columbia, formed south of about present latitude 40°N. Irving and others (1985), in fact, show that these Cretaceous plutons wre grouped fairly tightly near the top of (present) Baja California. The most probable interpretation is that arc and forearc rocks formed at southern latitudes during the Mesozoic and later were dispersed along the edge of the continent (Irving and others, 1985; Beck, 1980). Creation of arc and forearc rocks exclusively at southern latitudes may have been a consequence of the shape of the continent, or controlled by the location of the Farallon/Kula/North America triple junction (Beck, 1986).

Miocene results for southern California were discussed earlier. Large clockwise rotations are found in rocks as young as 10 Ma. Clearly, dextral shear in the continental margin was active during Neogene time, as it is today. The large values of R found by Luyendyk and co-workers (over 90° for MNO) suggests that individual rotating blocks are quite small, probably of the order of a few tens of kilometers in lateral dimensions. Intense sampling of Neogene rocks in southern California has made it possible to propose fairly specific models for regional deformation (Kamerling and Luyendyk, 1979).

CONCLUSIONS

Data from paleomagnetic studies show that much of the western edge of North America is composed of crustal blocks that originated elsewhere. Accretion, continental-margin translation, and in-situ block rotations all are extremely common. Information derived from conventional geological studies such as stratigraphy, structure, and geochemistry contribute to this geotectonic picture, which has been referred to as the "collage" concept (Davis and others, 1978). Comparison of stratigraphy between neighboring crustal blocks has given rise to another important concept, that of the "tectonostratigraphic terrane" (Berg and others, 1979). Through stratigraphy and the terrane concept it is possible to correlate crustal fragments now widely dispersed along the margin of the continent, and say something about their place of origin. The peculiar usefulness of paleomagnetic studies consists in their ability to extend "terrane-type" analysis to blocks of crust that have no stratigraphy (e.g., batholiths), and to detect certain kinds of displacements (e.g., in situ block rotations and continental-margin translations of more than 1,000 km) that are difficult to recognize geologically.

REFERENCES CITED

Achache, J., Cox, A., and O'Hare, S., 1982, Paleomagnetism of the Devonian Kennett limestone and the rotation of the eastern Klamath Mountains, California: Earth and Planetary Science Letters, v. 61 p. 365–380.

Anderson, T. H., and Schmidt, V. A., 1983, The evolution of Middle America and the Gulf of Mexico–Carribean Sea region during Mesozoic time: Geological Society of America Bulletin, v. 94, p. 941–966.

Atwater, T., 1970, Implications of plate tectonics for the tectonic evolution of western North America: Geological Society of America Bulletin, v. 81, p. 3513–3536.

Barnes, A. E., and Butler, R. F., 1980, A Paleocene paleomagnetic pole for the Gringo Gulch volcanics: Geophysical Research Letters, v. 7, p. 545–548.

Bates, R. G., Beck, M. E., Jr., and Burmester, R. F., 1981, Tectonic rotations in the Cascade Range of southern Washington: Geology, v. 9, p. 184–189.

Beck, M. E., Jr., 1975, Remanent magnetism of the Twin Sisters dunite intrusion and implications for the tectonics of the western Cordillera: Earth and Planetary Science Letters, v. 26, p. 263–268.

——, 1976, Discordant paleomagnetic pole positions as evidence of regional shear in the western Cordillera of North America: American Journal of Science, v. 276, p. 694–712.

——, 1980, Paleomagnetic record of plate-margin tectonic processes along the western edge of North America: Journal of Geophysical Research, v. 85, p. 7115–7131.

——, 1984, Has the Washington-Oregon Coast Range moved northward?: Geology, v. 12, p. 737–740.

——, 1986, Model for late Mesozoic-early Tertiary tectonics of coastal California and western Mexico and speculations on the origin of the San Andreas fault: Tectonics, v. 5, p. 49–64.

Beck, M. E., Jr., and Burr, C. E., 1979, Paleomagnetism and tectonic significance of the Goble volcanic series, southwestern Washington: Geology, v. 7, p. 175–179.

Beck, M. E., Jr., and Engebretson, D. C., 1982, Paleomagnetism of small basalt exposures in the west Puget Sound area, Washington, and speculations on the accretionary origin of the Olympic Mountains: Journal of Geophysical Research, v 87, p. 3755–3760.

Beck, M. E., Jr., and Plumley, P. W., 1980, Paleomagnetism of intrusive rocks in the Coast Range of Oregon; Microplate rotations in middle Tertiary time: Geology, v. 8, p. 573–577.

Beck, M. E., Jr., Burmester, R. F., and Schoonover, R., 1981a, Paleomagnetism and tectonics of the Cretaceous Mount Stuart batholith of Washington; Translation or tilt?: Earth and Planetary Science Letters, v. 56, p. 336–342.

Beck, M. E., Jr., Burmester, R. F., Engebretson, D. C., and Schoonover, R., 1981b, Northward translation of Mesozoic batholiths, western North America; Paleomagnetic evidence and tectonic significance: Geofisica Internacional, v. 20, p. 144–162.

Beck, M. E., Jr., Burmester, R. F., and Schoonover, R., 1982, Tertiary paleomagnetism of the North Cascade Range, Washington: Geophysical Research Letters, v. 9, p. 515–518.

Beck, M. E., Jr., Burmester, R. F., Craig, D. E., Gromme, C. S., and Wells, R. E., 1986, Paleomagnetism of middle Tertiary volcanic rocks from the Western Cascade Series, northern California: Journal of Geophysical Research, v. 91, p. 8219–8230.

Berg, H. C., Jones, D. L., and Coney, P. J., 1979, Pre-Cenozoic tectonostratigraphic terranes in southeastern Alaska and adjacent areas: U.S. Geological Survey Open-File Report 78–1085, 2 sheets, scale 1:1,000,000.

Blake, M.C., Jr., and Jones, D. L., 1977, Plate tectonic history of the Yolla Bolly junction, northern California: Geological Society of America Guidebook, 73rd Annual Cordilleran Section Meeting, 13 p.

Bogen, N. L., Kent, D. V., and Schweickert, R. A., 1985, Paleomagnetism of Jurassic rocks in the western Sierra Nevada metamorphic belt and its bearing on the structural evolution of the Sierra Nevada block: Journal of Geophysical Research, v. 90, p. 4627–4638.

Calderone, G., and Butler, R. F., 1984, Paleomagnetism of Miocene volcanic rocks from southwestern Arizona; Tectonic implications: Geology, v. 12, p. 627–630.

Champion, D. E., Howell, D. G., and Gromme, C. S., 1984, Paleomagnetic and geological data indicating 2,500 km of northward displacement for the Salinian and related terranes, California: Journal of Geophysical Research, v. 89, p. 7736–7752.

Choiniere, S. R., Swanson, D. A., 1979, Magnetostratigraphy and correlation of Miocene basalts of the northern Oregon coast and Columbia Plateau, southeastern Washington: American Journal of Science, v. 279, p. 755–777.

Cox, A., 1975, The frequency of geomagnetic reversals and the symmetry of the nondipole field: Reviews of Geophysics and Space Physics, v. 13, p. 35–51.

Davis, G. A., Monger, J.W.H., and Burchfiel, B. C., 1978, Mesozoic construction of the Cordilleran "collage," central British Columbia to central California, *in* Howell, D. C., and McDougall, K. A., eds., Mesozoic paleogeography of the western United States: Pacific Section, Society of Economic Paleontologists and Mineralogists, p. 1–33.

Demarest, H. H., Jr., 1983, Error analysis for the determination of tectonic rotation from paleomagnetic data: Journal of Geophysical Research, v. 88, p. 4321–4328.

Diehl, J. F., Beck, M. E., Jr., Beske-Diehl, S., Jacobson, D., and Hearn, B. C., Jr., 1983, Paleomagnetism of the Late Cretaceous and early Tertiary north-central Montana alkalic province: Journal of Geophysical Research, v. 88, p. 10593–10609.

Engebretson, D. C., Cox, A., and Gordon, R. G., 1985, Relative motions between oceanic and continental plates in the Pacific basin: Geological Society of America Special Paper 206, 59 p.

Fox, K. F., Jr., and Beck, M. E., Jr., 1985, Paleomagnetic results for Eocene volcanic rocks from northeastern Washington and the Tertiary tectonics of the Pacific Northwest: Tectonics, v. 4, p. 323–341.

Frei, L. S., Magill, J. R., and Cox, A., 1984, Paleomagnetic results from the central Sierra Nevada; Constraints on reconstructions of the western United States: Tectonics, v. 3, p. 157–177.

Fry, J. G., Bottjer, D. J., and Lund, S. P., 1985, Magnetostratigraphy of displaced Upper Cretaceous strata in southern California: Geology, v. 13, p. 648–651.

Gabrielse, H., 1985, Major dextral transcurrent displacements along the Northern Rocky Mountain Trench and related lineaments in northcentral British Columbia: Geological Society of America Bulletin, v. 96, p. 1–14.

Garfunkel, Z., 1974, Model for the late Cenozoic tectonic history of the Mohave Desert, California, and its relation to adjacent regions: Geological Society of America Bulletin, v. 85, p. 1931–1944.

Geissman, J. W., Callian, J. T., Oldow, J. S., and Humphries, S. E., 1984, Paleomagnetic assessment of oroflexural deformation in west-central Nevada and significance for emplacement of allochthonous assemblages: Tectonics, v. 3, p 179–200.

Globerman, B. H., Beck, M. E., Jr., and Duncan, R. A., 1982, Paleomagnetism and tectonic significance of Eocene basalt from the Black Hills, Washington Coast Range: Geological Society of America Bulletin, v. 93, p. 1151–1159.

Gordon, R. G., Cox, A., and O'Hare, S., 1984, Paleomagnetic Euler poles and the apparent polar wander and absolute motion of North America since the Carboniferous: Tectonics, v. 3, p. 499–537.

Greenhaus, M. R., and Cox, A., 1979, Paleomagnetism of the Morro Rock–Islay Hill complex as evidence for crustal block rotations in central coastal California: Journal of Geophysical Research, v. 84, p. 2393–2400.

Gromme, C. S., and Gluskoter, H. J., 1965, Remanent magnetization of spilite and diabase in the Franciscan Formation, western Marin County, California: Journal of Geology, v. 73, p. 74–94.

Gromme, C. S., and Merrill, R. T., 1965, Paleomagnetism of Late Cretaceous granitic plutons in the Sierra Nevada, California; Further results: Journal of Geophysical Research, v. 70, p. 3407–3420.

Gromme, C. S., Merrill, R. T., and Verhoogen, J., 1967, Paleomagnetism of Jurassic and Cretaceous plutonic rocks in the Sierra Nevada, California, and its significance for polar wandering and continental drift: Journal of Geophysical Research, v. 72, p. 5661–5684.

Gromme, C. S., Beck, M. E., Jr., Wells, R. E., and Engebretson, D. C., 1986, Paleomagnetism of the Tertiary Clarno Formation of central Oregon and its significance for the tectonic history of the Pacific Northwest: Journal of Geophysical Research (in press).

Hagstrum, J. T., McWilliams, M., Howell, D. G., and Gromme, C. S., 1985, Mesozoic paleomagnetism and northward translation of the Baja California peninsula: Geological Society of America Bulletin, v. 96, p. 1077–1090.

Hannah, J. L., and Verosub, K. L., 1980, Tectonic implications of remagnetized upper Paleozoic strata of the northern Sierra Nevada: Geology, v. 8, p. 520–524.

Harland, W. B., Cox, A. V., Llewellyn, P. G., Picton, C.A.G., Smith, A. G., and Walters, R., 1982, A geological time scale: Cambridge, Cambridge University Press, 128 p.

Heller, P. L., and Ryberg, P. F., 1983, Sedimentary record of subduction to forearc transition in the rotated Eocene basins of western Oregon: Geology, v. 11, p. 380–383.

Hicken, A., and Irving, E., 1977, Tectonic rotation in western Canada: Nature, v. 268, p. 219–220.

Hillhouse, J. W., Gromme, C. S., and Vallier, T. L., 1982, Paleomagnetism and Mesozoic tectonics of the Seven Devils volcanic arc in northeastern Oregon: Journal of Geophysical Research, v. 87, p. 3777–3794.

Hooper, P. R., Knowles, C.L.R., and Watkins, N. D., 1979, Magnetostratigraphy of the Imnaha and Grande Ronde basalts in the southeast part of the Columbia Plateau: American Journal of Science, v. 279, p. 737–754.

Hornafius, J. S., 1985, Neogene tectonic rotation of the Santa Ynez Range, Western Transverse Ranges, California, suggested by paleomagnetic investigation of the Monterey Formation: Journal of Geophysical Research, v. 90, p. 12503–12522.

Hudson, M. R., and Geissman, J. W., 1984, Preliminary paleomagnetic data from the Jurassic Humboldt lopolith, west-central Nevada; Evidence for thrust belt rotation in the Fencemaker allochthon: Geophysical Research Letters, v. 1l, p. 828–831.

Irving, E., 1964, Paleomagnetism and its application to geological and geophysical problems: New York, John Wiley and Sons, 399 p.

——, 1979, Paleopoles and paleolatitudes of North America and speculations about displaced terrains: Canadian Journal of Earth Sciences, v. 16, p. 669–694.

Irving, E., and Irving, G. A., 1982, Apparent polar wander paths Carboniferous through Cenozoic and the assembly of Gondwana: Geophysical Surveys, v. 5, p. 141–188.

Irving, E., and Wynne, P. J., 1992, Paleomagnetism; Review and tectonic implications, *in* Gabrielse, H., and Yorath, C. J., eds., The Cordilleran Orogen: Ottawa, Canada, Geological Survey of Canada, Geology of Canada series, v. 4 (Geological Society of America, The Geology of North America, v. G-2) ch. 3, p. 61–86.

Irving, E., Woodsworth, G. J., Wynne, P. J., and Morrison, A., 1985, Paleomagnetic evidence for displacement from the south of the Coast Plutonic Complex, British Columbia: Canadian Journal of Earth Sciences, v. 22, p. 584–598.

Irving, E., Tanczyk, E., and Hastie, J., 1976, Catalogue of paleomagnetic directions and pole positions; 5, Cenozoic results: Earth Physics Branch, Department of Energy, Mines, and Resources Ottawa, Canada, v. 6, 70 p.

Jones, D. L., Silberling, N. J., and Hillhouse, J., 1977, Wrangellia; A displaced terrane in northwestern North America: Canadian Journal of Earth Sciences, v. 14, p. 2565–2577.

Kamerling, M. J., and Luyendyk, B. P., 1979, Tectonic rotations of the Santa Monica Mountains region, western Transverse Ranges, California, suggested by paleomagnetic vectors: Geological Society of America Bulletin, v. 90, p. 331–337.

Kamerling, M. J., and Luyendyk, B. P., 1985, Paleomagnetism and Neogene tectonics of the northern Channel Islands, California: Journal of Geophysical Research, v. 90, p. 12485–12502.

Kanter, L. R., and McWilliams, M., 1982, Rotation of the southernmost Sierra Nevada, California: Journal of Geophysical Research, v. 87, p. 3819–3830.

Kluth, C. F., Butler, R. F., Harding, L. E., Shafiqullah, M., and Damon, P. E., 1982, Paleomagnetism of Late Jurassic rocks in the northern Canelo Hills, southeastern Arizona: Journal of Geophysical Research, v. 87, p. 7079–7086.

Luyendyk, B. P., Kamerling, M. J., Terres, R. R., and Hornafius, J. S., 1985, Simple shear of southern California during Neogene time suggested by paleomagnetic declinations: Journal of Geophysical Research, v. 90, p. 12454–12466.

Magill, J., and Cox, A., 1980, Tectonic rotation of the Oregon Western Cascades: State of Oregon, Department of Geology and Mineral Industries Special Paper 10, 67 p.

Magill, J., Cox., A., and Duncan, R., 1981, Tillamook Volcanic Series; Further evidence for tectonic rotation of the Oregon Coast Range: Journal of Geophysical Research, v. 86, p. 2953–2970.

Magill, J., Wells, R. E., Simpson, R. W., and Cox, A., 1982, Post–12 m.y. rotation of southwestern Washington: Journal of Geophysical Research, v. 87, p. 3761–3776.

Mankinen, E. A., 1972, Paleomagnetism and potassium-argon ages of the sonoma Volcanics, California: Geological Society of America Bulletin, v. 83, p. 2063–2072.

Mankinen, E. A., and Irwin, W.. P., 1982, Paleomagnetic study of some Cretaceous and Tertiary sedimentary rocks of the Klamath Mountains province, California: Geology, v. 10, p. 82–87.

Mankinen, E. A., Irwin, W. P., and Gromme, C. S., 1984, Implications of paleomagnetism for the tectonic history of the eastern Klamath and related terranes in California and Oregon, *in* Nilsen, T. H., ed., Geology of the Upper Cretaceous Hornbrook Formation, Oregon and California: Pacific Section, Society of Economic Paleontologists and Mineralogists, v. 42, p. 221–229.

May, S. R., and Butler, R. F., 1986, North American apparent polar wander; Implications for plate motions, paleogeography, and Cordilleran tectonics; Journal of Geophysical Research B., v. 91, p. 11, 519-11, 544.

May, S. R., Butler, R. F., Shafiqullah, M., and Damon, P. E., 1986, Paleomagnetism of Jurassic volcanic rocks in the Patagonia Mountains, southeastern Arizona; Implications for the North American 170 Ma reference pole: Journal of Geophysical Research B, v. 91, p. 11, 545-11, 555.

McElhinny, M. W., 1973, Paleomagnetism and plate tectonics: Cambridge, Cambridge University Press, 358 p.

McWilliams, M., 1984, Confidence limits on net tectonic rotation: Geophysical Research Letters, v. 11, p. 825–827.

McWilliams, M., and Howell, D. G., 1982, An exotic origin for terranes of western California: Nature, v. 297, p. 215–217.

Merrill, R. T., and McElhinny, M. W., 1977, Anomalies in the time-averaged paleomagnetic field and their implications for the lower mantle: Reviews of Geophysics and Space Physics, v. 15, p. 309–323.

——, 1983, The earth's magnetic field; Its history, origin, and planetary perspective: London, Academic Press, 401 p.

Monger, J.W.H., and Irving, E., 1980, Northward displacement of north-central British Columbia: Nature, v. 285, p. 289–293.

Monger, J.W.H., Price, R. A., and Tempelman-Kluit, D. J., 1982, Tectonic accretion and the origin of the two major metamorphic and plutonic welts in the Canadian Cordillera: Geology, v. 10, p. 70–75.

Oldow, J. S., Ave Lallemant, H. G., and Schmidt, W. J., 1984, Kinematics of plate convergence deduced from Mesozoic structures in the western Cordillera: Tectonics, V. 3, p. 201–228.

Panuska, B. C., and stone, D. B., 1985, Latitudinal motion of the Wrangellia and Alexander terranes and the southern Alaska superterrane, *in* Howell, D., ed., Tectonostratigraphic terranes of the Circumpacific region: American Association of Petroleum Geologists Earth Science Series 1, p. 109–120.

Russell, B. J., Beck, M. E., Burmester, R. F., and Speed, R. C., 1982, Cretaceous magnetizations in northwestern Nevada and tectonic implications: Geology, v. 10, p. 423–428.

Sheriff, S. D., 1984, Paleomagnetic evidence for spatially distributed post-Miocene rotation of western Washington and Oregon: Tectonics, v. 3,

p. 397–408.

Silver, L. T., and Anderson, T. H., 1974, Possible left-lateral early and middle Mesozoic disruption of the southwestern North American craton margin [abs.]: Geological Society of America Abstracts with Program, v. 6, p. 955–956

Simpson, R. W., and Cox, A., 1977, Paleomagnetic evidence for tectonic rotation of the Oregon Coast Range: Geology, v. 5, p. 585–589.

Speed, R. C., 1979, Collided Paleozoic microplate in the western United States: Journal of Geology, v. 87, p. 279–292.

Symons, D.T.A., 1969a, Paleomagnetism of four late Miocene gabbroic plugs in south-central British Columbia: Canadian Journal of Earth Sciences, v. 6, p. 653–662.

——, 1969b, Paleomagnetism of the late Miocene plateau basalts in the Cariboo region of British Columbia: Geological Survey of Canada Paper 69-43, 16 p.

——, 1973a, Paleomagnetic zones in the Oligocene East Sooke Gabbro, Vancouver Island, British Columbia: Journal of Geophysical Research, v. 78, p. 5100–5109.

——, 1973b, Concordant Cretaceous paleolatitudes from felsic plutons in the Canadian Cordillera: Nature-Physical Science, v. 241, p. 59–61.

——, 1974, Paleomagnetism of the Lower Jurassic Tulameen ultramafic-gabbro complex, British Columbia: Geological Survey of Canada Paper 74-1, Part B, p. 177–183.

——, 1977a, Geotectonics of Cretaceous and Eocene plutons in British Columbia; A paleomagnetic fold test: Canadian Journal of Earth Sciences, v. 14, p. 1246–1262.

——, 1977b, Paleomagnetism of Mesozoic plutons in the westernmost Coast Plutonic Complex of British Columbia: Canadian Journal of Earth Sciences, v. 14, p. 2127–2139.

——, 1983a, Further paleomagnetic results from the Jurassic Topley Intrusives in the Stikinia subterrane of British Columbia: Geophysical Research Letters, v. 10, p. 1065–1068.

——, 1983b, New paleomagnetic data for the Triassic Guichon batholith of south-central British Columbia and their bearing on Terrane I tectonics: Canadian Journal of Earth Sciences, v. 20, p. 1340–1344.

——, 1985a, Paleomagnetism of the Triassic Nicola Volcanics and geotectonics of the Quesnellia subterrane of Terrane I, British Columbia; Journal of Geodynamics, v. 2, p. 229–244.

——, 1985b, Paleomagnetism of the Westcoast Complex and the geotectonics of the Vancouver Island segment of the Wrangellian subterrane: Journal of Geodynamics, v. 2, p. 211–228.

Symons, D.T.A., and Litalien, C. R., 1984, Paleomagnetism of the Lower Jurassic Copper Mountain intrusions and the geotectonics of Terrane I, British Columbia: Geophysical Research Letters, v. 11, p. 685–688.

Tarduno, J. A., and Alvarez, W., 1985, Paleolatitudes of Franciscan limestones: Geology, v. 13, p. 741.

Tarling, D. H., 1983, Paleomagnetism: London, Chapman and Hall, 379 p.

Terres, R. R., and Luyendyk, B. P., 1985, Neogene tectonic rotation of the San Gabriel region, California, suggested by paleomagnetic vectors: Journal of Geophysical Research, v. 90, p. 12467–12484.

Vugteveen, R. W., Barnes, A. E., and Butler, R. F., 1981, Paleomagnetism of the Roskruge and Gringo Gulch volcanics, southeastern Arizona: Journal of Geophysical Research, v. 86, p. 4021–4028.

Wells, R. E., 1984, Paleomagnetic constraints on the interpretation of early Cenozoic Pacific Northwest paleogeography, *in* Nilsen, T. H., ed., Geology of the Upper Cretaceous Hornbrook Formation, Oregon and California: Pacific Section, Society of Economic Paleontologists and Mineralogists, p. 231–237.

Wells, R. E., and Coe, R. S., 1985, Paleomagnetism and geology of Eocene volcanic rocks of southwest Washington; Implications for mechanisms of tectonic rotation; Journal of Geophysical Research, v. 90, p. 1925–1947.

Wilson, D., and Cox, A., 1980, Paleomagnetic evidence for tectonic rotation of Jurassic plutons in Blue Mountains, eastern Oregon: Journal of Geophysical Research, v. 85, p. 3681–2689.

Wilson, R. L., 1971, Dipole offset; The time-averaged paleomagnetic field over the past 25 million years: Geophysical Journal of the Royal Astronomical Society, v. 22, p. 491–504.

Yole, R. W., and Irving, E., 1980, Displacement of Vancouver Island; Paleomagnetic evidence from the Karmutsen Formation: Canadian Journal of Earth Sciences, v. 17, p. 1210–1228.

Manuscript Accepted by the Society June 25, 1987.

ACKNOWLEDGMENTS

This paper owes a great deal to thoughtful reviews by E. Mankinen and R. Butler. Discussions over the years with R. Burmester and D. Engebretson also are gratefully acknowledged.

NOTES ADDED IN PROOF

In the 5.7 years since I submitted this chapter there have been many important developments in Cordilleran paleomagnetism. I have been alloted enough space for this "note" to outline perhaps two. The two I have selected are not necessarily the most important, but they may be the most interesting. For a general overview of recent developments, see Kodama (1991).

The first new development concerns the reference path. In 1986 I was confident that the paleomagnetic Euler pole (PEP) approach was correct, although I had reservations about "tracks" that extended over many tens of millions of years. I still believe that the PEP approach is valid, but difficulties continue to appear. The main difficulty at this juncture concerns the Jurassic reference pole. Van Fossen and Kent (1990) have shown that the high-latitude Middle Jurassic reference pole of Irving and Irving (1982) probably is correct, contrary to earlier conclusions by Gordon and others (1984) and May and Butler (1986). These authors both relied heavily on the Corral Canyon pole of May and others (1986), which now must be considered suspect. The principal tectonic consequence of a high-latitude Jurassic reference pole is to call into question the conclusion that most Jurassic terranes in the Cordillera are concordant; with a high-latitude reference pole, most Jurassic terranes appear to have been displaced northward, as interpreted originally (e.g., Beck, 1976). This interesting development may obviate my somewhat strained explanation (main article) that Jurassic terranes first went southward, then bounced back to the north. However, there appear to be good geological reasons to suspect that southward relative transport in the latest Jurassic and Early Cretaceous did precede northward relative transport in the Late Cretaceous (e.g., Ave Lallemant and Oldow, 1988). Elsewhere I have argued that this episode of south-vergent plate convergence may be the principal cause of the Nevadan orogeny (Beck, 1987).

A second important development concerns alternatives to northward transport, and mainly concerns the Cretaceous data set (Fig. 4). Butler and others (1989, 1991) have questioned northward transport, apparently because it is difficult for them to reconcile such transport with their current interpretation of Cordilleran geology. To explain the discordance shown in Figure 4 they invoke a combination of tilt (for batholiths and remagnetized rock units), inclination flattening (for sedimentary units), and *in situ* block rotations about the vertical. Many paleomagnetic studies are dismissed by them as poorly determined. Some of their re-analyses are ingenious and meticulously argued, but their central thesis founders on one observation; it is the *pattern* of discordance, not the individual studies, that matters. Thus, to disprove northward transport it is necessary to find an alternative explanation for the systematic way in which Cordilleran paleomagnetic poles are discordant; attacks on individual interpretations are almost irrelevant. This is discussed in Beck (1991); problems with the tilt alternative for large batholiths are analyzed in Beck (1992).

ADDITIONAL REFERENCES

Ave Lallemant, H., and Oldow, J., 1988, Early Mesozoic southward migration of Cordilleran transpressive terranes: Tectonics, v. 7, p. 1057–1075.

Beck, M., 1987, North American APW and terrane transport, Late Jurassic to Early Cretaceous: Orogenic consequences [abs.]: EOS Transactions of the American Geophysical Union, v. 68, p. 1254.

Beck, M., 1991, Case for northward transport of Baja and coastal southern California: Paleomagnetic data, analysis, and alternatives: Geology, v. 19, p. 506–509.

Beck, M., 1992, Some thermal and paleomagnetic consequences of tilting a batholith: Tectonics (in press).

Butler, R., and four others, 1989, Discordant paleomagnetic poles from the Canadian Coast Plutonic Complex: Regional tilt rather than large-scale displacement?: Geology, v. 17, p. 691–694.

Butler, R., Dickinson, W., and Gehrels, G., 1991, Paleomagnetism of coastal California and Baja California: Alternatives to large-scale northward transport: Tectonics, v. 10, p. 561–577.

Kodama, K., 1991, Contributions in geomagnetism and paleomagnetism, U.S. National Report 1987–1990, 20th General Assembly IUGG, published by the American Geophysical Union, pp. 343–442.

Van Fossen, M., and Kent, D., 1990, High-latitude paleomagnetic poles from Middle Jurassic plutons and moat volcanics in New England and the controversy regarding Jurassic APW for North America: Journal of Geophysical Research, v. 95, p. 17503–17516.

Printed in U.S.A.

Index

[Italic page numbers indicate major references]

Typeset by WESType Publishing Services, Inc., Boulder, Colorado
Printed in U.S.A. by Malloy Lithographing, Inc., Ann Arbor, Michigan

Hawaiian Islands
130°
125°
120°
115°
110°
105°
100°